BEEF CATTLE SCIENCE

(Animal Agriculture Series)

All flesh is grass!

BEEF CATTLE SCIENCE

(Animal Agriculture Series)

by

M. E. ENSMINGER, B.S., M.A., PH.D.

Formerly: Assistant Professor in Animal Science,
University of Massachusetts

Chairman, Department of Animal Science,
Washington State University

Consultant, General Electric Company
Nucleonics Department (Atomic Energy Commission)

Currently: President, Consultants-Agriservices,
Clovis, California

President, Agriservices Foundation

Collaborator, U.S. Department of Agriculture

Adjunct Professor,
California State University

Distinguished Professor,
University of Wisconsin

THE INTERSTATE
PRINTERS & PUBLISHERS, INC.
Danville, Illinois

First Edition, 1951
Second Edition, 1955
Third Edition, 1960
Fourth Edition, 1968
Fifth Edition, 1976

Fourth Edition translated into Spanish under the direction of
Dr. Mauricio B. Helman, Professor, Veterinary Sciences,
University of Argentina; and published by El Ateneo,
Florida 340/344, Buenos Aires, Argentina.

Library of Congress Catalog Card No. 67-12112

Reorder No. 1752

Printed in U.S.A.

To
Mr. J. L. Campbell
—MY VOCATIONAL AGRICULTURE TEACHER
WHO TAUGHT THAT ANY CALLING IS GREAT
WHEN GREATLY PURSUED

TO CATTLEMEN—artists and scientists. Not artists whose tools are the clay and marble of the sculptor, but artists whose materials are the "green pastures and still waters" that have inspired musicians to capture their beauty in pastoral symphonies, and painters to reproduce their splendor in landscape designs; artists whose materials are the living flesh and blood of animals, molded to perfection through heredity and environment. Not scientists who look through a microscope or shake a test tube, but scientists who, from the remote day of domestication, have given attention to the breeding, feeding, care and management, and marketing of animals.

TO CATTLEMEN—who take pride in their brands, boots, hats, and canes. To them they are much more than a trademark of the profession; they are symbols of service, pledges of integrity of the men behind them, and marks of courage, character, and wisdom. They are indicative of the quality of the cattle raised, the class of bulls used, the condition of the pasture or range, and the kind of caretakers connected with the outfit.

PREFACE TO THE FIFTH EDITION

Mankind makes constant progress and nature undergoes constant change—they never remain the same. For sheer survival, therefore, cattlemen must go on researching, discovering, creating, and advancing.

The United States beef cattle industry is characterized by five eras: (1) the Longhorn; (2) the golden purebred era; (3) the grain-fed cattle binge; (4) the coming of the exotics and more crossbreeding; and (5) the back-to-roughage movement.

● *The Longhorn*—The cattle industry of this country began with the Longhorn. Introduced by the Spanish in the 16th Century, they were long on horns, hardiness, and longevity; and they had the ability to fight and fend for themselves as they reverted to the wild. In the movie, "The Rare Breed," Jimmie Stewart described them as meatless, milkless, and murderous. Yet their tough, sinewy muscles made it possible for them to travel long distances to sparsely located water holes on the early-day western ranges, and even longer distances—often 1,000 miles or more—to the railheads.

● *The golden purebred era*—As human living standards improved and the desire for better steaks developed, the golden purebred era was born early in the 19th Century. By 1850, the "battle of the breeds," which for many years had divided English stockmen into rival camps, was transferred to the United States, where it has raged ever since. First, came the aristocratic Shorthorn from Scotland. True to their name—they shortened the murderous weapons on the head of the Longhorn. Additionally, the Shorthorn x Longhorn crosses showed marked improvement in mothering ability, milk production, early maturity, weaning weight, beef qualities, and disposition. Yet, the hybrids retained the hardiness of the Longhorn. It was the perfect example of one breed complementing another. Gradually, the Longhorn became extinct and commercial herds of this country were dominated by Shorthorns. However, with the increase in Shorthorn blood and the decrease in heterosis, there was a decrease in the ability of the animals to withstand the rigors of the western range. At this point in history, the Hereford was lofted into prominence on the beef cattle scene. Over the next 30 years, the cow herds of the beef producing areas of the nation became predominantly Herefords, and the Longhorn and Shorthorn were relegated to minor importance. Later, the Angus entered, resulting in the "black baldies."

It is noteworthy that, for a considerable period of time, crossbreeding strengthened the position of purebred breeders and made for breed growth; first with the Shorthorn which was crossed on the Longhorn, second with the Hereford which was crossed on the Shorthorn, and third with the Angus which was crossed on the Hereford.

• *The grain-fed cattle binge*—Following World War II, the time was ripe for the "chain reaction" that spawned the era of grain-fed cattle. Grain bins bulged with surpluses, and consumers were able and willing to pay for Choice, grain-fed beef, which was ably promoted and merchandized by self-service chain stores. Commercial cattle feeders, acutely attuned to consumer demands, set out to fill the need. Cattle feeding went big—and modern. In 1947, only 3.6 million head of cattle were grain-fed, representing 35 percent of the slaughter cattle. By 1972, 26.8 million head of fed cattle were marketed, representing nearly 75 percent of all the cattle slaughtered that year. Equally remarkable, a mere 2,107 feedlots, each with capacity of 1,000 head or more, marketed 62 percent of the nation's fed cattle in 1972.

• *The coming of the exotics and more crossbreeding*—Next came the exotics! Back of the rage for the exotics was the desire to use them in crossbreeding. Among commercial cattlemen, the tidal wave of exotics engendered enthusiasm and excitement such as had not been seen for some years. But among some purebred breeders of the British breeds, the exotics produced animosities reminiscent of the range wars in the days of intruding sheepmen and nesters. The "establishment" was riled because crossbreeding was being extolled, and bulls of these stark newcomers were being used. Arguments waxed hot; many oldtime purebred breeders became emotional and explosive. This was surprising because, as pointed out earlier, the practice of crossbreeding had been common in this country since the birth of the cattle business. The commercial cattleman had always exploited the benefits of heterosis—first by using Shorthorns on Longhorns, thence followed by Herefords and Angus. Moreover, at one time or another, all breeds of cattle in North America, including the Longhorn and the British breeds, were exotics—none of them were indigenous to this country. They are no more native than the white man. In the present century, the Brahman was the first exotic; next came the Charolais. Then the tidal wave broke, facilitated by the quarantine station which Canada established on Grosse Isle, in the St. Lawrence River, and opened in 1965. It was then possible to bring cattle from parts of Europe previously closed because of the disease situation. The race was on! An influx of new breeds followed. A popular textbook published in 1920 listed 16 breeds of cattle, whereas the fourth edition of this book, *Beef Cattle Science*, published in 1968, listed 21 breeds. Hence, only 5 more breeds were added in the 48-year period. In this edition, 53 breeds of cattle are covered—that's 32 new breeds, or more than double the number listed 7 years earlier in the preceding edition of this same book.

The time was ripe for the exotics and more crossbreeding. During much of the 1950s and 1960s, the cow-calf man was hurting financially. He recognized that profits could be increased by producing for less, as well as selling for more. Crossbreeding with its demonstrated potential for producing for less through complementary genes and heterosis offered new hope—hope for survival. Artificial insemination facilitated the movement.

• *The back-to-roughage era*—By 1973, 77 percent of the United States beef production came from grain-finished cattle, and only 23 percent from grass cattle. Then, suddenly there were grain shortages—and a rude awakening. The United States cattle industry was experiencing a prelude to world food shortages. The message was loud and clear. Neither beef production nor consumption is elastic at any price. They are subject to the old law of supply and demand. Moreover, more and more United States cereal grains will be used for human food, just as has been true, historically, in much of the rest of the world. In the future, beef cattle will increasingly be "roughage burners." Cattlemen will rely upon the ability of the ruminant to convert coarse forage, grass, and by-product feeds, along with a minimum of grain, into palatable and nutritious food for human consumption, thereby competing less for humanly edible grains. Increasingly, beef will be grass.

Indeed, the final chapter of the convulsive 20th Century, which is now being written, may be the most revolutionary and important of all to the cattle industry. The century that witnessed the discovery of flight, two major world wars, the atomic age, space exploration, and the energy crisis, is now being subjected to world food shortages, with beef in the most preferred category. As the ghost of hunger, foretold by parson Thomas Robert Malthus back in 1798, stalks the world, the focus is on cattle—roughage to food converters par excellence.

The world will continue to grow more populous and richer. Sociological developments will slow population growth—to some extent, and in some places. Scientific and technological developments will speed food production increases. This edition of *Beef Cattle Science* distills the maze of new developments in beef cattle and charts the future. The author will feel amply rewarded if this book makes the dreams of more people throughout the world come true, faster and more abundantly, with more beef in their future.

The author gratefully acknowledges the contributions of all those who participated in this major revision. Special appreciation is expressed to my staff, who worked long and diligently, especially to Audrey Ensminger, for her inspiration, encouragement, and assistance; to Pat Logoluso Long, who was responsible for editing, format, and typing; to Ruth Geary, who did much of the art work; and to all the rest of my staff, who contributed directly or indirectly. Also, I wish to thank most sincerely, but inadequately, all those authorities who reviewed certain portions of the manuscript; and all those who responded so liberally to my call for pictures and information, due acknowledgment of which is given at the appropriate places in the book.

M. E. Ensminger

September, 1976
Clovis, California

REFERENCES

The following books are by the same author and the same publisher as *Beef Cattle Science:*

> *Animal Science*
> *Dairy Cattle Science*
> *Sheep and Wool Science*
> *Swine Science*
> *Horses and Horsemanship*
> *Poultry Science*
> *The Stockman's Handbook*

Animal Science presents a perspective or panorama of the far-flung livestock industry; whereas each of the specific class-of-livestock books presents specialized material pertaining to a class of farm animals.

The Stockman's Handbook is a modern know-how, show-how book which contains, under one cover, the pertinent things that a stockman needs to know in the daily operation of a farm or ranch. It covers the broad field of animal agriculture, concisely and completely, and, whenever possible, in tabular and outline form.

SELECTED GENERAL REFERENCES ON BEEF CATTLE

Title of Publication	Author(s)	Publisher
Beef Cattle	R. R. Snapp A. L. Neumann	John Wiley & Sons, New York, N.Y., 1969
Beef Cattle Book	B. E. Fichte	The Progressive Farmer Company, Birmingham, Ala., 1967
Beef Cattle in Florida	L. H. Lewis T. J. Cunha G. N. Rhodes	Department of Agriculture, State of Florida, Tallahassee, Fla., 1962
Beef Cattle Production	K. A. Wagnon R. Albaugh G. H. Hart	The Macmillan Co., New York, N.Y., 1960
Beef Production in the South	S. H. Fowler	The Interstate Printers & Publishers, Danville, Ill., 1969
Beef Cattle Science Handbook	Ed. by M. E. Ensminger	Agriservices Foundation, Clovis, Calif., pub. annually since 1964
Beef Production	R. V. Diggins C. E. Bundy	Prentice-Hall, Inc., Englewood Cliffs, N. J., 1958
California Beef Production, Manual 2	H. R. Guilbert G. H. Hart	California Agricultural Experiment Station and Extension Service
Commercial Beef Cattle Production	C. C. O'Mary I. A. Dyer	Lea & Febiger, Philadelphia, Penn., 1972
Livestock Book, The	W. R. Thompson, et al.	Vulcan Service Co., Birmingham, Ala., 1952

(Continued)

Title of Publication	Author(s)	Publisher
Practical Beef Production	J. Widmer	Charles Scribner's Sons, New York, N.Y., 1946
Problems and Practices of American Cattlemen	M. E. Ensminger M. W. Galgan W. L. Slocum	Washington Agricultural Experiment Station Bulletin 562, Washington State University, Pullman, Wash., 1955
World Cattle	J. E. Rouse	University of Oklahoma Press, Norman, Okla., Vols. I-II, 1970; Vol. III, 1973

CONTENTS

PART I: GENERAL BEEF CATTLE

PART II: COW-CALF SYSTEM; STOCKERS

PART III: CATTLE FEEDLOTS; PASTURE FINISHING

APPENDIX

PART I

GENERAL BEEF CATTLE

Broadly classified there are two phases of beef cattle production: (1) the cow-calf system, and (2) cattle finishing. Sometimes, both phases are conducted on a single farm or ranch as successive steps of a continuous process—they're integrated. For example, a cow-calf man may carry his home-produced calves through the stocker stage, and he may even finish them out. More often, however, each phase is conducted to the exclusion of the other, not only on individual operations, but also in agricultural regions. Nevertheless, the fundamentals of beef cattle production are much the same in both cow-calf production and cattle finishing. That is, the general principles—such as business aspects, feeding, management, buildings and equipment, health, marketing, and meats—apply to both phases, regardless of whether they are conducted as an integrated operation or as separate phases. For this reason, organizationally, these are covered in Part I, General Beef Cattle—the first 16 chapters of this book; and they are not repeated. Instead, the practical application of these principles follows in Part II, Cow-Calf System—Stockers, and in Part III, Cattle Feedlots—Pasture Finishing.

CHAPTER 1

HISTORY AND DEVELOPMENT OF
THE BEEF CATTLE INDUSTRY[1]

Contents   **Page**

Cattle are the most important of all the animals domesticated by man, and, next to the dog, the most ancient. There are 1.18 billion cattle in the world.[2]

The word "cattle" seems to have the same origin as chattle, which means possession. This is a very natural meaning, for when Rome was in her glory a man's wealth was often computed in terms of his cattle possessions, a practice which still persists among primitive people in Africa and Asia. That the ownership of cattle implied wealth is further attested by the fact that the earliest known coins bear an ox head; and the Roman word "pecunia" for money (preserved in our adjective pecuniary) was derived from the Latin word *pecus*, meaning cattle. It is also noteworthy that the oldest known treatise on agriculture, written by the Greek poet Hesiod, referred to cattle. Apparently having had some disturbing experience with young oxen, Hesiod advised: "For draught and yoking together, nine-year-old oxen are best because, being past the mischievous and frolicsome age, they are not likely to break the pole and leave the plowing in the middle."

---

[1]In the preparation of Chapter 1, the author was especially fortunate in having the valued counsel and suggestions of Mr. Karl P. Schmidt, formerly Chief Curator of the Department of Zoology, Chicago Natural History Museum, Chicago, Ill., who so patiently and thoroughly reviewed this historical material.

[2]United Nations. (Courtesy, Dr. R. S. Temple, Food and Agriculture Organization of the United Nations, Rome, Italy)

ORIGIN AND DOMESTICATION OF CATTLE

It seems probable that cattle were first domesticated in Europe and Asia during the New Stone Age. In the opinion of most authorities, today's cattle bear the blood of either or both of two ancient ancestors—namely, *Bos taurus* and *Bos indicus*. Other species or subspecies were frequently listed in early writings, but these are seldom referred to today. Perhaps most, if not all, of these supposedly ancestral species were also descendants of *Bos taurus* or *Bos indicus* or crosses between the two.

Fig. 1-1. Ancient drawing of a bison on a rock, made by Paleolithic (Old Stone Age) man. Even prior to their domestication, man revered animals, according them a conspicuous place in the art of the day. (Courtesy, The Bettman Archive)

Bos Taurus

Bos taurus includes those domestic cattle common to the more temperate zones, and it, in turn, appears to be derived from a mixture of the descendants of the Aurochs (*Bos primigenius*) and the Celtic Shorthorn (*Bos longifrons*).

Most cattle, including the majority of the breeds found in the United States, are believed to have descended mainly from the massive Aurochs (also referred to as "Uri," "Ur," or "Urus"). This was the mighty wild ox that was hunted by our forefathers. It roamed the forests of central Europe down to historic times, finally becoming extinct about the year 1627. About the year 65 B.C., Caesar mentioned this ox in his writings, but it was domesticated long before (perhaps early in the Neolithic Age) and probably south of the Alps or in the Balkans or in Asia Minor. Caesar referred to these animals

Fig. 1-2. Artist's conception of an Aurochs (*Bos primigenius*) based on historical information. This was the mighty wild ox that was hunted by our ancestors. Most cattle are believed to have descended mainly from the Aurochs. (Drawing by R. F. Johnson)

as "approaching the elephant in size but presenting the figure of a bull." Although this is somewhat of an exaggeration as to the size of the Aurochs, it was a tremendous beast, standing 6 or 7 feet high at the withers, as is proved by complete skeletons found in bogs.

In addition to the Aurochs, another progenitor of some of our modern breeds and the earliest known domestic race of cattle was the Celtic Shorthorn or Celtic Ox. These animals, which have never been found except in a state of domestication, were the only oxen in the British Isles until 500 A.D., when the Anglo-Saxons came, bringing with them animals derived from the Aurochs of Europe. The Celtic Shorthorn was of smaller size than the Aurochs and possessed a dished face. It may have had a still different wild ancestor, or may have been an independent domestication from the Aurochs.

Bos Indicus

Bos indicus includes those humped cattle common to the tropical countries that belong to the Zebu (or Brahman) group. They are wholly domestic creatures, no wild ancestors having been found since historic times. It has been variously estimated that cattle of this type were first domesticated anywhere from 2100 to 4000 B.C. The Zebu is characterized by a hump of fleshy tissue over the withers (which sometimes weighs as much as 40 to 50 pounds), a very large dewlap, large drooping ears, and a voice that is more of a grunt than a low. These peculiar appearing animals seem to have more resistance to certain diseases and parasites and to heat than the descendants of *Bos taurus*. For this reason, they have been crossed with some of the cattle of Brazil and in the southern states of this country, especially in the region bordering the Gulf of Mexico.

Fig. 1-3. Zebu (*Bos indicus*). These wholly domestic animals were the ancestors of the humped cattle common to the tropical countries. (Drawing by R. F. Johnson)

POSITION OF OXEN IN THE ZOOLOGICAL SCHEME

Domesticated cattle belong to the family *Bovidae*, which includes ruminants with hollow horns. Members of this family possess one or more enlargements for food storage along the esophagus, and they chew their cuds. In addition to what we commonly call oxen or cattle, the family *Bovidae* (and the sub-family *Bovinae*) includes the true buffalo, the bison, musk-ox, banteng, gaur, gayal, yak, and zebu.

The following outline shows the basic position of the domesticated cow in the zoological scheme:

Kingdom *Animalia:*   Animals collectively; the animal kingdom.

Phylum *Chordata:*   One of approximately 21 phyla of the animal kingdom in which there is either a backbone (in the vertebrates) or the rudiment of a backbone, the chorda.

Class *Mammalia:*   Mammals or warm-blooded, hairy animals that produce their young alive and suckle them for a variable period on a secretion from the mammary glands.

Order *Artiodactyla:*   Even-toed, hoofed mammals.

Family *Bovidae:*   Ruminants having polycotyledonary placenta; hollow, nondeciduous, up-branched horns; and nearly universal presence of a gall bladder.

Genus *Bos:* Ruminant quadrupeds, including wild and domestic cattle, distinguished by a stout body and hollow, curved horns standing out laterally from the skull.

Species *Boss taurus* and *Bos inducus: Bos taurus* includes the ancestors of the European cattle and of the majority of the cattle found in the United States; *Bos inducus* is represented by the humped cattle (Zebu) of India and Africa and the Brahman breed of America.

USE OF CATTLE IN ANCIENT TIMES

Like other animals, cattle were first hunted and used as a source of food and other materials. As civilization advanced and man turned to tillage of the soil, it is probable that the domestication of cattle was first motivated because of their projected value for draft purposes. Large, well-muscled, powerful beasts were in demand; and any tendency to fatten excessively or to produce more milk than was needed for a calf was considered detrimental rather than desirable. Not all cattle were used for work purposes, however, in the era following their domestication. Instead of planting seeds, some races of people chose a pastoral existence—moving about with their herds as they required new pastures. These nomadic people lived mainly on the products of their herds and flocks.

As populations became more dense, feed became more abundant, and cattle became more plentiful, man became more interested in larger production of meat and milk. The pastoral people adopted a more settled life and began selecting out those animals that possessed the desired qualities— including rapid growth, fat storage, and milk production. Following this transformation, Biblical and other literature referred to milk cows, the stall-fed ox, and the fatted calf.

In contrast with the very great importance of cattle in western Asia and Europe in both ancient and modern times, it is noteworthy that cattle were never very highly valued in China, Japan, or Korea. The people of these countries have never used much beef, milk, butter, and cheese. In India, on the other hand, cattle play as important a role as in our western civilization and still retain a great religious significance.

CATTLE IN MEDIEVAL FARMING

The best of medieval farms would excite the scorn or contempt of a modern farmer. Except for plowing and carting with oxen, all labor was done by hand. Although the fields were small, several oxen were often yoked to the plow. As few farmers owned many head, it frequently was necessary for an entire village to pool its oxen and plow the fields in common.

Cattle fared badly in these early days. Pastures were overgrazed and winter feed was scarce. In the fall of the year, it was the common practice to kill and salt the carcasses of all those animals not needed for draft or breeding purposes. Prior to slaughter, aged animals and worn-out oxen were grass fattened, after a fashion. Those that were wintered over were fed largely on straw and the forage they could glean from the fields. Often by spring they were so thin that they could hardly walk.

Very little cow's milk was available, most of it being produced during the grazing season. In fact, more goat's milk than cow's milk was consumed in liquid form. Even in the 13th Century, when farming methods had improved, one writer indicated that three cows could be expected to produce only 3½ pounds of butter per week. Most cow's milk was used in cheese making.

LATE MIDDLE-AGES AGRICULTURE OF ENGLAND

During the Middle Ages (500 to 1500) in England, as elsewhere, rotation and improvement of crops and improved breeding methods were not a necessity because virgin soil was abundant and worn-out lands could be deserted for new. Increasing population and the establishment of settlements were later to make improved husbandry a dire necessity.

Examples of the open-field system could still be found in England up to the 18th Century. However, shortly before 1500, feudalism in England practically ceased to exist, and with its passing the system of enclosures and individual ownership became more prevalent.

IMPROVEMENTS IN ENGLISH CATTLE

English agrarian conditions began to improve during the reign of Elizabeth (1558-1603). No well-directed efforts toward the improvement of cattle were made, however, even in England, until late in the 18th Century. By 1700, from one-third to one-half of the arable land was still cultivated on the open field system. No individual owner could attempt to improve his herd when all the cattle of the village grazed together on the same common.

Enclosing started about 1450, but progress in this direction was slow. Animals on the common were often half starved, and it was said that 5 acres of individually owned pasture was worth more than the pasture rights over 250 acres of common.

During the 18th Century, agricultural progress in England quickened. With the coming of field cultivation of clover and seeded grasses, sometime after 1600, and the introduction and cultivation of the turnip somewhat later, a great impetus was given to agriculture and livestock breeding. Winter feed could now be had, more livestock kept, more manure produced, and better crops grown. Indeed, the progress in stock raising in the 18th Century cannot be understood apart from the progress made at the same time in general agriculture.

In cattle, size was the main criterion in selection, though power at the yoke and milking quality were not overlooked. Perhaps the ultimate in cattle size was represented by the Lincolnshire ox, standing 19 hands high and measuring 4 yards from his face to his rump, a worthy descendent of the Aurochs.

BAKEWELL'S IMPROVEMENT OF ENGLISH CATTLE

Robert Bakewell of Dishley (1726 to 1795)—an English farmer of remarkable sagacity and hard, common sense—was the first great improver of cattle in England. His objective was to breed cattle that would yield the greatest quantity of good beef rather than to obtain great size. Bakewell had the imagination to picture the future needs of a growing population in terms of meat and set about creating a low-set, blocky, quick-maturing type of beef cattle. He paid little or no attention to fancy points. Rather, he was intensely practical, and no meat animal met with his favor unless it had the ability to put meat on the back.

Fig. 1-4. Robert Bakewell of Dishley (1726-1795), noted agriculturalist and the first great improver of cattle in England. Bakewell also contributed greatly to the improvement of the Leicester breed of sheep and the Shire horse. (Courtesy, Picture Post Library, London, England)

Bakewell's efforts with cattle were directed toward the perfection of the English Longhorn, a class of cattle common to the Tees River Area. He also contributed greatly to the improvement of the Leicester breed of sheep, and the Shire horse. Success crowned his patient skill and unwearied efforts. But success in breeding was no mere happenstance in Bakewell's program. Careful analysis of his methods reveals that three factors were paramount: (1) a definite goal as evidenced by the joints that he preserved in pickle and the skeletons of the more noted animals that adorned his halls, (2) a breeding system characterized by "breeding the best to the best" regardless of relationship, rather than crossing breeds as was the common practice of the time, and (3) a system of proving sires by leasing them at fancy prices to his neighbors, rather than selling them. Because of Bakewell's methods and success, he has often been referred to as the founder of animal breeding.

Bakewell's experiments were the top news of the day, and his successes the subject of much comment, both oral and written. The American poet, Emerson, for example, said of the British farmer, "he created sheep, cows, and horses to order . . . the cow is sacrificed to her bag, the ox to his sirloin."

By the beginning of the Napoleonic Wars, Bakewell's methods were widely practiced in England, and sheep and cattle were raised more for their flesh than formerly. A new era in livestock improvement was born. As an indication of this change, it is interesting to observe the increase in weights of animals at the famous Smithfield market. In 1710, beeves had averaged 370 lb, calves 50 lb, sheep 28 lb, and lambs 18 lb; whereas in 1795 they had reached 800, 148, 80, and 50 lb, respectively. Although the effect of improved agriculture is not to be minimized, the main influence in this transformation can be attributed to Robert Bakewell, whose imagination, initiative, and courage put a firm foundation under improved methods of livestock breeding.

THE INTRODUCTION OF CATTLE TO AMERICA

Cattle are not native to the Western Hemisphere. They were first brought to the West Indies by Columbus on his second voyage in 1493. According to historians, these animals were intended as work oxen for the West Indies colonists. Cortez took cattle from Spain to Mexico in 1519. Then, beginning about 1600, other Spanish cattle were brought over for work and milk purposes in connection with the chain of Christian missions which the Spaniards established among the Indians in the New World. These missions

Fig. 1-5. Texas Longhorn steer. (Courtesy, N. H. Rose Collection, San Antonio, Tex.)

extended from the east coast of Mexico up the Rio Grande, thence across the mountains to the Pacific Coast. Here, in a land of abundant feed and water, these Longhorns multiplied at a prodigious rate. By 1833, the Spanish priests estimated that their missions owned a total of 424,000 head of cattle,[3] many of which were running in a semi-wild state. The hardy Texas Longhorn, animals of Spanish extraction, were of little commercial value except for their hides. Today, only a few of these animals remain, more as a novelty and for show purposes than for use as meat producers.

The colonists first brought cattle from England in 1609. Other English importations followed, with Governor Edward Winslow bringing a notable importation to the Plymouth Colony in 1623. The latter shipment included three heifers and a bull. Three years later, at a public court, these animals and their progeny—and perhaps some subsequent importations—were appropriated among the Plymouth settlers on the basis of one cow to six persons. It is further reported that three ships carried cattle to the Massachusetts Bay Colony in 1625. Other colonists came to the shores of New England, bringing with them their oxen from the mother country. As would be expected, the settlers brought along the kind of cattle to which they had been accustomed in the mother country. This made for considerable differences in color, size, and shape of horns, but all of these colonial-imported cattle possessed ruggedness and the ability to perform work under the yoke.

For a number of years, there were very few cattle in the United States. Moreover, those animals that the colonists did possess went without winter feed and shelter, and the young suffered the depredations of the wolves. It was difficult enough for the settlers to build houses for themselves, and they could barely raise enough corn in their fields to sustain human life.

Conditions presently changed for the better. The cattle of earlier importations multiplied, new shipments were received, and feed supplies became more abundant. Cambridge, Massachusetts enjoyed the double distinction of being the seat of Harvard College, the first institution of higher learning in what later came to be the United States, and the most prosperous cattle center in early New England. In order to provide ample grass and browse for the increased cattle population, it was necessary that the animals range some distance from the commons (the town pasture). Thus the tale that the streets of Boston were laid out along former cowpaths is not legend but fact. Usually in their travels, the cattle were under the supervision of a paid "cowkeeper" whose chief duty consisted in safely escorting the cattle to and from pasture.

In the village economy, the bull was an animal of considerable importance. Usually the town fathers selected those animals that they considered most desirable to retain as sires, and those citizens who were so fortunate as to own animals of this caliber were paid an approved service fee on a per-head basis.

---

[3] *Yearbook of Agriculture*, USDA, 1921, p. 233.

DRAFT OXEN MORE PRIZED BY COLONISTS THAN BEEF

From the very beginning, the colonists valued cattle for their work, milk, butter and hides; but little importance was attached to their value for meat. In fact, beef was considered as much a by-product as hides are today. After all, wild game was plentiful, and the colonists had learned to preserve venison, fish, and other meats by salting, smoking, and drying. So necessary were cattle for draft purposes that, in some of the early-day town meetings, ordinances were passed making it a criminal offense to slaughter a work oxen before he had passed the useful work age of seven or more years. The work requirement led to the breeding of large rugged cattle, with long legs, lean though muscular bodies, and heavy heads and necks. Patient oxen of this type were well adapted for clearing away the forest and turning the sod on the rugged New England hillsides, for hauling the harvested produce over the rough roads to the seaport markets, and for subsisting largely on forages.

Fig. 1-6. Oxen hauling logs to the saw mill. Draft oxen were more prized by the colonist than beef.

AMERICAN INTEREST IN BREEDING

Interest in obtaining well-bred cattle on this side of the Atlantic was a slow one. All through the colonial period, the American farmer let his animals shift for themselves, never providing shelter and rarely feeding them during the winter months. Eventually the lot of the colonist improved, and with it came the desire to secure blooded stock. Fortunately for the United States, the stockman could draw on the improved animals already developed in Britain and on the Continent. Thus, it is not surprising to find that the vast majority of the older breeds of United States beef cattle originated across the Atlantic.

In 1783, three Baltimore gentlemen—Messrs. Patton, Goff, and Ringold—sent to England for the best cattle obtainable. They could not have had any particular breed in mind, for at that time no distinct breed can be said to have existed in England, unless it was Bakewell's Longhorns. Hubback, the celebrated foundation Shorthorn bull, was only six years old in 1783, and the fame of the Devon and Hereford was purely local. Other importations followed. Gradually the native stock was improved.

In due time, animals of this improved breeding were to make their influence evident on the western range. The Texas Longhorns, which had thrived since the time of their importation by the Spaniards in the 16th Century, were decidedly lacking in early maturity and development in the regions of the high priced cuts. Thus, the infusion of blood of cattle of English ancestry resulted in a marked improvement in the beef qualities of the range cattle, but it must be admitted that no admixture of breeding could have improved the Texas Longhorns in hardiness and in ability to fight and to fend for themselves.

In order to supply the increased range demand for high-class bulls, a considerable number of purebred herds were established, especially in the central states. In addition to selling range bulls, these breeders furnished foundation heifers and herd bulls for other purebred breeders.

EFFECT OF THE CIVIL WAR ON THE CATTLE INDUSTRY

In 1860, just prior to the Civil War, stock raising was on the threshold of becoming one of the nation's leading industries. At that time, the aggregate value of United States livestock was more than a billion dollars, representing an increase of more than 100 percent since 1850. Texas, which had been admitted to the Union in 1845, was the leading cattle producing state, and Chicago was the foremost packing center.

With the outbreak of the Civil War, the cattle of the Southwest could no longer reach the normal markets to the north and east. Union gunboats patrolled the waterways and the Northern armies blocked land transportation. Not even the Confederate armies of the South could be used as an outlet. Prices slumped to where the best cattle in Texas could be purchased at $4.00 to $6.00 per head.

In sharp contrast to the conditions in the South and Southwest, the Civil War made for a very prosperous cattle industry in the North. The war-made industrial prosperity of the densely populated East and the food needs of the Union Army produced an abnormal demand for beef. Inflated prices followed, with the result that choice steers were selling up to $100 per head at the close of hostilities. Many cattlemen amassed modest fortunes.

At the close of the Civil War, therefore, a wide difference existed between cattle values in the North and in the Southwest. With the return of normal commerce between the states, this condition was soon rectified, only to receive another and more serious jolt with the outbreak of cattle tick fever (Texas fever) in 1868, which, a year later, was spread through cattle shipments northward to Illinois and eastward to the Atlantic coast.

THE ABILENE, KANSAS SHIPPING POINT

Until 1867, the only convenient shipping point for Texas cattle was at Sedalia, Missouri, on the Missouri Pacific Railroad. But distance was not the only hazard to early-day trailing. At that time, the Missouri Ozarks were the chief hideaway and point of operation of numerous bands of cattle thieves and robbers. Sometimes these outlaws operated under the guise of sheriffs or other local officials who pretended to be enforcing laws that prohibited the passage of Texas cattle. In any event, the end result of their handiwork was always the same: (1) stampeding the cattle and making away with a large number of them before the drovers could get them under control; (2) beating or otherwise torturing the drovers until they were glad to abandon the herd and flee for their very lives; or (3) killing those drovers who resisted. Because of these treacherous bands, it soon became necessary for the Texas drovers to travel farther to the west through eastern Kansas.

Finally, in a desperate effort to circumvent the outlaw hazards of the Missouri Ozarks, Mr. Joe G. McCoy, a prominent Illinois stockman, in 1867, conceived the idea of establishing a rail shipping point further to the west. To this end, Mr. McCoy personally inspected the Kansas Pacific Railroad route through eastern Kansas in an effort to select a site where cattle could be grazed while awaiting shipment. Abilene (now famous as the hometown of former President Eisenhower) was decided upon because (1) it was lo-

Fig. 1-7. *Trail herd—Flank Riders*, from an etching by Edward Borein. The six flank riders held the herd in line. The five riders at the tail of the herd were known as "drag" riders who kept watch on the crippled, sick, and exhausted animals. The chuck wagon is in the center background and the remuda (extra horses) in the left background. (Courtesy, Armour and Company)

Drover's Cottage, 1871   Lithographic Drawing by Birger Sandzen

THE END OF THE TEXAS CATTLE TRAIL, ABILENE, KANSAS

Fig. 1-8. Early Abilene, Kansas shipping point. This end of the Texas cattle trail in eastern Kansas on the Kansas Pacific Railroad was established in 1867, for the purpose of providing safe transportation to the East, unmolested by the Ozark outlaws. (Courtesy, Abilene Chamber of Commerce)

cated on a railroad (the Kansas Pacific Railroad, which was then being extended to Denver), (2) the surrounding country was sparsely populated, and (3) there was an abundance of grass and water upon which cattle could be held pending shipment. At the time, Abilene merely consisted of 12 small log huts, most of them with dirt floors. In less than 2 months, Joe McCoy transformed Abilene into a thriving cattle town. He built stockyards, cattle pens, a livery stable, and an 80-room hotel, called the Drover's Cottage, for trail bosses and Eastern buyers. Then the cattle came. (See Table 1-1.)

In 1871, Abilene received 600,000 head of cattle. Both shipping and grazing facilities had been seriously overtaxed. With the arrival of winter, thousands of head of cattle remained unsold, and insufficient feed supplies were available. The severe winter that followed brought heavy losses. With

TABLE 1-1

CATTLE SHIPMENTS TO ABILENE, KANSAS,
1867-1871

Year	Arrival of Cattle at Abilene
1867	35,000
1868	75,000
1869	150,000
1870	300,000
1871	600,000

the coming of spring, it was estimated that 250,000 head of cattle had starved to death within sight of Abilene. This disaster, coupled with the opening up of more plentiful and convenient shipping points, marked the rapid decline of Abilene as a shipping center.

THE OUTBREAK OF CATTLE TICK FEVER (Texas Fever)

Following the Civil War, the Texas cattle trade had received real encouragement with the establishment of the Abilene, Kansas shipping point in 1867, thus alleviating the hazards of the Missouri outlaws. In 1868, however, the Texas cattle trade received a serious setback. That summer, a group of Chicago cattlemen shipped 40,000 head of cattle from Texas to Tolono, Illinois, in Champaign County, where they were sold to the local farmers (part of this shipment was taken over into neighboring Indiana) for grazing and wintering. Soon after the arrival of the Texas shipment, the native cattle, with which they were turned to pasture, became mysteriously sick and died in great numbers; whereas the southern cattle— although apparently responsible for spreading the disease—remained in perfect health. According to reports, in some infested areas nearly every native cow died. In one township only one milk cow survived. The cause of the malady was unknown, and there was no cure. Farmers became panicky. In despair, many infected herds were shipped to eastern markets, thereby spreading the disease all the way to the Atlantic Coast.

Although the cause of the disease was unknown, the evidence pointed toward the Texas cattle as being the carriers. Strong prejudice against Texas cattle developed. Wild laws and regulations aimed at controlling the movement of Texas cattle and preventing the spread of the disease were enacted by several states. At this point, the price of Texas steers fell, and many cattle held at the Abilene, Kansas shipping point could not be sold. The Texas cattlemen, however, were not to be outdone. Some of the more ingenious among them conceived a novel advertising campaign. This consisted of (1) shipping to St. Louis and Chicago a carload of buffalos decorated with placards extolling the virtues of Texas cattle and beef; (2) pointing out that Texas cattle were more hardy than northern cattle, for none of them contracted the disease when shipped north; and (3) propagandizing the reputed superior carcasses and greater tenderness and palatability of beef from Texas cattle. Soon the pendulum swung back, with the result that Texas cattle became very popular, even commanding a premium over comparable animals native to the North or East.

THE FAR-WESTERN EXPANSION OF THE CATTLE INDUSTRY

From the very beginning, cattle raising on a large scale was primarily a frontier activity. As the population of eastern United States became more dense, the stock raising industry moved farther inland. The great westward push came in the 19th Century. By 1800, the center of the cow country was west of the Alleghenies, in Ohio and Kentucky; in 1860, it was in Illinois and Missouri; and by the 1880s, it was in the Great Plains. The ranches and cowboys of the far West were the counterpart of the New England commons and cowdrivers of the 17th and 18th Centuries.

The western range was recognized as one of the greatest cattle countries that the world had ever known. Plenty of water and unlimited grazing area were free to all comers, and the market appeared to be unlimited. Fantastic stories of the fabulous wealth to be made from cattle ranching caused a rush comparable to that of the gold diggers of 1849. All went well until the severe winter of 1886. It was the type of winter that is the bane of the cattleman's existence. Then, but all too late, it was realized that too many cattle had been kept and too little attention had been given to storing up winter feed supplies. The inevitable happened. With the melting of the snow in the spring of 1887, thousands of cattle skeletons lay weathering on the western range, a grim reminder of overstocking and inadequate feed supplies. Many ranchers went broke, and the cattle industry of the West suffered a crippling blow that plagued it for the next two decades. Out of this disaster, however, the ranchers learned the never-to-be-forgotten lessons of avoiding over-expansion and too-close grazing, and the necessity of an adequate winter feed supply.

QUESTIONS FOR STUDY AND DISCUSSION

1. How do you account for the fact that most of the cattle in the U.S. are descendants of *Bos taurus* rather than *Bos indicus*?

2. Throughout the ages, and in many sections of the world, cattle have been used
for work purposes more than horses and mules. Why has this been so?

3. Compare Robert Bakewell's breeding methods with those used in modern production testing programs. What three factors contributed most to his success as
an animal breeder?

4. Trace the use of pedigrees in (a) humans, (b) Arabian horses, and (c) the modern manner.

5. Of what significance are each of the following in the history and development
of the U.S. beef cattle industry?
   a. Cortez
   b. The Spanish missions
   c. The colonists
   d. The Civil War
   e. Abilene, Kansas
   f. Joe G. McCoy
   g. Cattle Tick Fever
   h. The winter of 1886

SELECTED REFERENCES

Title of Publication	Author(s)	Publisher
American Cattle Trails 1540-1900	C. M. Brayer H. O. Brayer	Western Range Cattle Industry, Study and American Pioneer Trails Assn., Bayside, N.Y., 1952
Animals and Men	H. Dembeck	The American Museum of Natural History, The Natural History Press, Garden City, N.Y., 1965

(Continued)

Title of Publication	Author(s)	Publisher
Cattle and Men	C. W. Towne E. N. Wentworth	University of Oklahoma Press, Norman, Okla., 1955
Cattlemen, The	M. Sandoz	Hastings House, New York, N.Y., 1958
Cowboys, The	Ed. by Time-Life Books, text by W. M. Forbis	Time-Life Books, New York, N.Y., 1973
Encyclopaedia Britannica		Encyclopaedia Britannica, Inc., Chicago, Ill.
History of Domesticated Animals, A	F. E. Zeuner	Harper & Row, New York, N.Y.
History of Livestock Raising in the United States 1607-1860	J. W. Thompson	U.S. Department of Agriculture, Agric. History Series No. 5, Washington D. C., Nov., 1942
Livestock Book, The	W. R. Thompson J. McKinney	Vulcan Service Co., Inc., Birmingham, Ala., 1952
Our Friendly Animals and Whence They Came	K. P. Schmidt	M. A. Donohue & Co., Chicago, Ill., 1938
Principles of Classification and a Classification of Mammals, The	G. G. Simpson	Bulletin of the American Museum of Natural History, Vol. 85, N.Y., 1945
Stock Raising in the Northwest 1884	H. O. Brayer G. Weis	The Branding Iron Press, Evanston, Ill., 1951
Yearbook of Agriculture 1921		U.S. Department of Agriculture, Washington, D.C., 1921, pp. 232-264

WORLD AND U.S. CATTLE AND BEEF[1]

[1]The author is very grateful to the following, each of whom provided valuable source material for this chapter: Dr. R. S. Temple, Senior Officer, Livestock Policy and Planning Unit, Animal Production and Health Division, Food and Agriculture Organization of the United Nations, Rome, Italy; Dr. James P. Hartman, Director, Livestock and Meat Products Division, USDA Foreign Agricultural Service, Washington, D.C.; and Dr. L. S. Pope, Dean, College of Agriculture and Home Economics, New Mexico State University, Las Cruces, N.M.

The final chapter of the convulsive 20th Century, which is now being written, may be the most revolutionary and important of all. The century that witnessed the discovery of flight, two major world wars, the atomic age, putting a man on the moon, and the energy crisis, is now being subjected to world food shortages,[2] with beef in the most preferred category.

Of the basic needs of man, none is more important than food. A hungry man knows no god, no country, and no boundary. Neither will he respect any treaty nor keep the peace. Indeed, each person is his brother's keeper. As the ghost of hunger, foretold by parson Thomas Robert Malthus back in 1798, stalks the world, the focus is on cattle and beef.

Never in history have there been so many people in the world; and never before have the people of the world had so much money to spend. This has had a two-pronged effect on the demand for food: (1) more food needed to feed more mouths; and (2) more people wanting high quality foods, like beef.

Increased numbers of people plus increased money (purchasing power) equals demand. This demand is being reflected in the market place for resource industries like agriculture.

Agriculture has an additional market demand factor—the protein principle. As soon as people get enough calories in their diets, and as they approach affluency, they start turning away from a starch-oriented diet to one based on animal protein, with beef at the top of the list. In the United States, we now consume almost as much beef and veal per capita as potatoes (111.1 lb vs 126 lb, respectively).

Animal protein is costly—it's resource-expensive. For example, it takes about 6 pounds of feed (grain, roughage, and protein supplement) to produce a pound of beef. Additionally, it takes resources to slaughter the animal, to process the carcass, and to keep it in refrigeration through the marketing chain. All in all, it takes about 7 times the resources to deliver a pound of beef that it takes to deliver a pound of cereals. Thus, it follows that beef is probably 7 times as resource-expensive as cereals.

Even though animal proteins are expensive, everywhere in the world that people have achieved higher income, the demand for animal protein has soared. But the demand for beef is not just people plus money plus their

---

[2]In addition to food shortages, there is the equally vital issue of the nutritional adequacy of available supplies within countries and the extent of undernutrition and malnutrition. According to the United Nations (*United Nations World Food Conference*, Rome, Nov. 5-16, 1974, p. 5), out of 97 developing countries, 61 had a deficit in food energy supplies in 1970. In the Far East and Africa, 25 to 30 percent of the population is estimated to suffer from significant undernutrition. Altogether in the developing world (excluding the Asian centrally planned economies for which insufficient information is available) malnutrition affects around 460 million people.

The same report referred to above gives (Table 18, p. 86) the per capita calories and proteins for 1970, and projections for 1985, as follows:

Year	Per Capita Calories/Day (Kilocalories)			Per Capita Proteins/Day (Grams)		
	World	Developed Countries	Developing Countries	World	Developed Countries	Developing Countries
1970	2,480	3,150	2,200	69.0	96.4	57.4
1985	2,610	3,220	2,400	72.6	100.0	63.3

desire to eat better. Rather, the demand for beef is people times income growth times the beef multiplier that requires seven times the production resources.

The world will continue to grow more populous and richer. But sociological and technological developments, which both slow population growth and speed food production increases, will make the dreams of more people throughout the world come true, faster and more abundantly, with more beef in their future.

WORLD HUMAN POPULATION

The population of the world first topped 1 billion about 1830. It took from the dawn of man until nearly 2,000 years after the birth of Christ for the number of people in the world to build up to that point. Disease had a major role in holding population down, of course.

It took another 100 years for the world to add its second billion people, which we reached about 1930. It took a mere 30 years more to reach 3 billion. Today, the total is approaching 4 billion; by 1985, we expect to have 4.9 billion[3]; and 21 years later by the year 2007 A.D.—there will be 7.5 billion people in the world. Table 2-1 shows, by regions and selected countries, where the people are, where the growth areas are, and who has the money

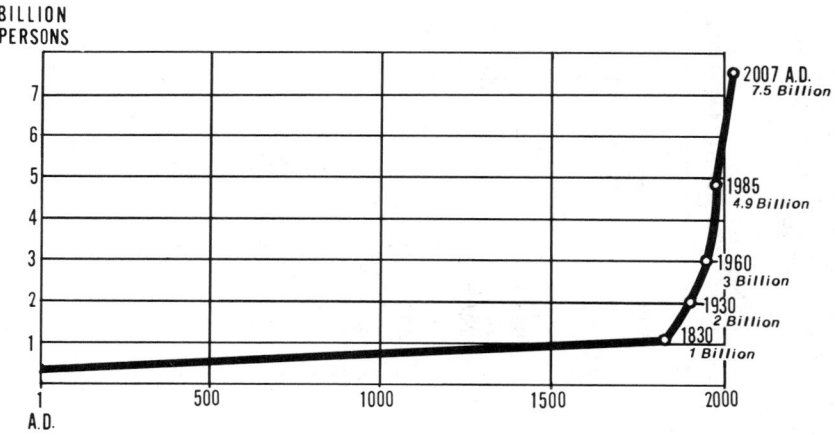

Fig. 2-1. World population, from the year 1 A.D. to 2000 A.D.

[3]The United Nations gives (*United Nations World Food Conference*, Rome, Nov. 5-16, 1974, Table 24, p. 102) the following population projections for 1985:

	People	Rate of Population Growth
	(million)	(percent/year/compound)
Developed Countries	1,227	0.9
Developing Countries	3,631	2.4
World	4,858	2.0

TABLE 2-1

WORLD HUMAN POPULATION DATA, 1972[1]

Region or Country	Mid-1972	Annual Growth Rate	Number Years to Double Population	1985 Projections	Gross National Product/Capita
	(Mil.)	(%)		(Mil.)	(US$)
WORLD[2]	3,782.0	2.0	35	4,933.0	—
AFRICA	364.0	2.6	27	530.0	—
ASIA	2,154.0	2.3	30	2,874.0	—
India	584.8	2.5	28	807.6	110
People's Republic of China	786.1	1.7	41	964.6	150[3]
Japan	106.0	1.2	58	121.3	1,430
CANADA	22.2	1.7	41	27.3	2,650
UNITED STATES	209.2	1.0	70	246.3	4,240
LATIN AMERICA	300.0	2.8	25	435.0	—
Mexico	54.3	3.3	21	84.4	580
Argentina	25.0	1.5	47	29.6	1,060
EUROPE	469.0	0.7	99	515.0	—
United Kingdom	56.6	0.5	139	61.8	1,890
U.S.S.R.	248.0	0.9	77	286.9	1,200
OCEANIA	20.0	2.0	35	27.0	—
Australia	13.0	1.9	37	17.0	2,300
New Zealand	3.0	1.7	41	3.8	2,230

[1]1972 World Population Data Sheet, Population Reference Bureau, Inc., Washington, D.C.
[2]37% of population under 15 years of age.
[3]Ensminger, M.E. and Audrey Ensminger, *China–the impossible dream*, p. 233

(gross national product, or GNP, which is the total national output of goods and services valued at market prices).

Currently, world population is compounding at the rate of two percent a year—a deceptively small number.[4] In actual numbers, the population increase amounts to two additional people per second, some 200,000 more per day, over 6 million more per month, and about 74 million per year.

The underdeveloped countries have about two-thirds of the world's population and the highest birth rates. For every birth in the developed countries, there are five in the underdeveloped countries. The only populous less-developed country which appears to have overcome the logistical and economic obstacles through providing family planning services is the People's Republic of China.(See Table 2-1.)

Do these population figures foretell the fulfillment of the "doomsday" prophecy of Thomas Malthus, made nearly two centuries ago, that world population grows faster than man's ability to increase food production? The author's answer to this question is, "not immediately." The Reverend Malthus did not foresee scientific-technological and sociological developments which would both speed food production increases and slow population growth. For example, the Green Revolution, which bought an extra decade or more of time, was outside Malthus' reckoning; nor did Malthus foresee the use of "the pill," which has reduced the population growth in Europe, Japan, Oceania, and North America to a fraction of the biological potential which he cited.

Ultimately, curbing population growth will be required to maintain the balance between production and demand for food. But supply can keep up with demand, and perhaps exceed it, from now to the year 2000 A.D. However, there will be substantial disparities between the developed and developing nations, between the rich and the poor, between the "haves" and "have nots," the satisfying of which will require large food shipments and greater international trade in farm commodities. But, from purely a production standpoint, there will be no worldwide food shortage. The crux to the problem will be who's going to pay for it, and how is it going to be transported and stored.

Farmers around the world have always demonstrated their willingness to respond to prices and profits. Remember that farmers are people, and that people do those things which are most profitable to them. Remember, too, there is a time lag, which, unfortunately, all too many consumers fail to realize. For example, a heifer cannot be bred until she is about 1½ years of age, the pregnancy period requires another 9 months, and, finally, the young are usually grown 6 to 12 months before being sold to cattle feeders, who finish them from 4 months to a year. Thus, under the most favorable conditions, this manufacturing process, which is under biological control and cannot be speeded up, requires about 4 years in which to produce a new generation of market cattle. Most consumers also fail to realize that, for various reasons, only an average of 80 (88% calf crop minus 8% calf loss) out of each

---

[4]See footnote 3.

100 cows bred in the United States wean off young; and that in addition to cattle and feed costs, there are shipping charges, interest on borrowed money, death losses, marketing charges, taxes and numerous other costs, before the steer finally reaches the packer.

WORLD CATTLE AND BEEF

It is important that cattlemen and those who counsel with them be well informed concerning worldwide beef production in order to know which countries are potential competitors. Like the price of all commodities in a free commerce, the price of beef is determined chiefly by supply and demand—that is, by the demand existing in those countries that do not produce enough to meet their domestic needs and by the supply which can be spared by those nations producing a surplus.

In 1970, the 5 most populous nations—People's Republic of China, India, U.S.S.R., United States, and Pakistan—accounted for:

1. 53% of the world's human population.
2. 48% of the world's cattle population.
3. 44% of the world's production of beef and veal.
4. 47% of the world's consumption of beef and veal.

As a group, they were beef deficit. It is not anticipated that these general relationships will change greatly by 1985.

World Cattle Numbers and Beef Production

The production of beef cattle is worldwide. Table 2-2 gives the size and density of cattle population of the 10 leading cattle-producing countries of the world, by rank. In 1972, world cattle numbers totaled nearly 1.3 billion head (up 1.6% from the previous year); about one cow for every 3 people, or 25 head per square mile.

But cattle numbers alone do not tell the whole story. It's the production of beef and veal that counts. Large numbers of cattle are kept for work and milk in India, the U.S.S.R., and the People's Republic of China. Besides, cattle are sacred to the Hindus of India (about 80% of the population). The United States leads the world, by a wide margin, in the production of beef and veal (Table 2-3), with the ranking of the other countries as shown. It is not expected that these rankings will change materially by 1985.

Efficiency of production is indicated by the annual production of beef per head of inventory, although this is not totally accurate. Values based on this are given in Fig. 2-2.

BEEF PRODUCTION IN INDIA

India, land of sacred cows and native home of the U.S. Brahman breed, is the leading cattle country in the world in numbers and density. There are 140 cattle per square mile vs 34 in the United States; that's 4 times greater density in India than in the United States. But India's cattle are of very

TABLE 2-2

SIZE AND DENSITY OF CATTLE POPULATION OF TEN LEADING
CATTLE-PRODUCING COUNTRIES OF THE WORLD IN 1972, BY RANK

Country	Cattle[1]	Human Population[2] (1972)	Cattle per Capita	Area	Cattle Per Square Mile
	(thousand head)	(millions)		(sq. mi.)	
India[3]	176,600[4]	584.8	.30	1,261,597	140
United States	121,990	209.2	.58	3,548,974	34
U.S.S.R.	104,000	248.0	.42	8,655,890	12
Brazil	97,300[4]	98.4	.99	3,286,270	30
People's Republic of China[3]	63,150[4]	786.1	.08	2,279,134	28
Argentina	52,312	25.0	2.09	1,072,700	49
Pakistan	44,200[4]	146.6	.30	365,529	121
Australia	28,975	13.0	2.23	2,971,081	10
Mexico	26,830	54.3	.49	758,259	35
Ethiopia	26,310	26.2	1.00	457,256	58
WORLD TOTAL	1,298,598	3,782.0	.34	52,403,746	25

[1]Foreign Agriculture Circular, Livestock and Meat, FLM 13-73, USDA, Foreign Agricultural Service, Washington, D.C., July 1973, p. 7, except as noted.
[2]1972 World Population Data Sheet, Population Reference Bureau, Inc., Washington, D.C.
[3]In addition to the cattle numbers listed, India has 53,550,000 buffaloes. (Source: Brief on Indian Agriculture 1971, Office of the Agricultural Attache, American Embassy, New Delhi, India, p. 14); and the People's Republic of Cina has 29,400,000 buffaloes. (Source: Footnote 1 reference gives combined figure for cattle and buffaloes; footnote 4 reference gives cattle alone—difference equals the number of buffaloes.)
[4]Statistical Yearbook 1972, United Nations, New York, 1973, p. 118.

TABLE 2-3

MAJOR BEEF AND VEAL PRODUCING NATIONS
—with projections to 1985

Country	Production Beef and Veal (1,000 metric tons)		Cattle Population (including Buffaloes) (millions)	
	1973[1]	1985[2]	1974[3]	1985[2]
United States	9,787.7	14,439	127.5	177.6
U.S.S.R.	5,487.0	6,715	106.2	115.8
Brazil	2,450.0	3,578	90.1	112.5
People's Republic of China	2,192.0[4]	3,034[4]	92.9	108.9
Argentina	2,152.0	3,946	56.5	57.9
Australia	1,493.6	1,665	31.5	23.8
France	1,454.0	2,327	23.6	24.9
Germany (West)	1,193.0	1,573	14.4	14.8
Canada	896.5	1,441	13.4	13.0
United Kingdom	875.9	1,338	14.7	14.2
Italy	710.0	811	8.5	12.2
WORLD TOTAL	40,520.0[5] (1970-72 av.)	61,008[5]	1,311.5	1,630.7[6]

[1]Foreign Agriculture Circular, Livestock and Meat, USDA, Foreign Agricultural Service, Washington, D.C., FLM8-74, June 1974, p. 10. Except China, which is FAO.
[2]1985 projections made by author, based on 1970-1980 trend projections from Agricultural Commodity Projections, 1970-1980, Volume II. Food and Agriculture Organization of the United Nations, Rome, 1971, Table 33, pp. 76-78. The author extended the projections to 1985. Except U.S. cattle numbers for 1985, which are the author's projections—see Table 17-1 of this book.
[3]Foreign Agriculture Circular, Livestock and Meat, USDA, Foreign Agricultural Service, Washington, D.C., FLM7-74, June 1974, pp. 8-9.
[4]Agricultural Commodity Projections, 1970-1980, Volume II, Food and Agriculture Organization of the United Nations, Rome, 1971, Table 33, p. 78.
[5]FAO estimate, provided with a personal communication to the author from Dr. R. S. Temple, FAO, United Nations, Rome.
The United Nations gives (United Nations World Food Conference, Rome, Nov. 5-16, 1974, Table 15, p. 80) the following world beef and veal figures: average annual consumption 1969-71, 39 million metric tons; projected demand 1985, 60 million metric tons.
To meet these 1985 projections calls for a total increase (1985/70) of 54.7%, or 3.0% per annum compounded.
[6]1985 projection made by author, based on estimated 2% yearly world increase from 1974.

BEEF PRODUCTION PER INVENTORY OF CATTLE PER YEAR

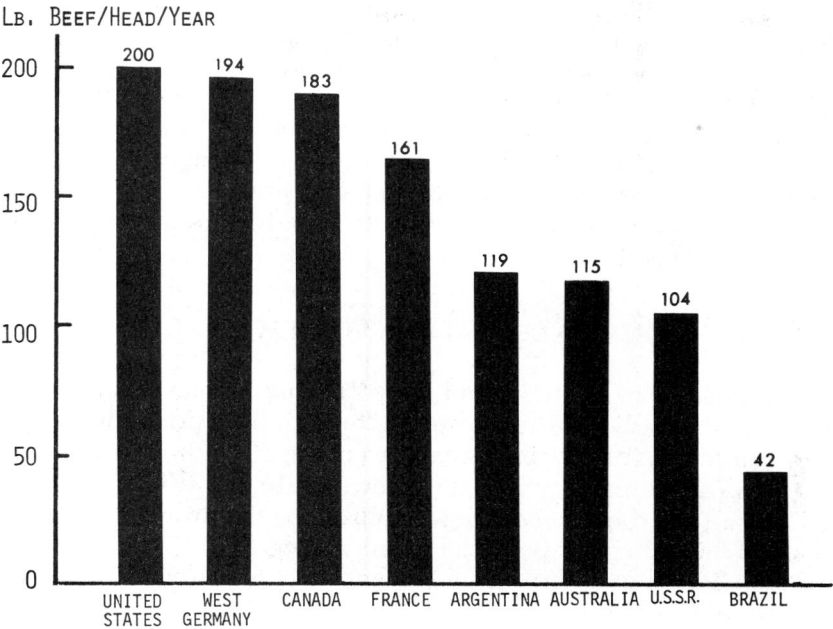

Fig. 2-2. Pounds of beef produced per inventory of cattle per year.

Fig. 2-3. Scene in India, showing cattle in a village. Usually Indian cattle are herded by the young or the old. (Photo by Dr. A. D. Weber, The Ford Foundation).

negligible importance from the standpoint of meat production, due to the large number that are either sacred or used for draft or milk purposes. The humped cattle of India pillage crops in rural areas and roam the streets of villages and cities —gentle and traffic-wise. Some are homeless, others are turned loose by owners who do not wish to pay for their keep, and still others are just AWOL. To the Hindu (approximately 80% of India's population), the cow is regarded as a mother and an object of reverence; and the eating of beef is taboo. Although India's cattle population puts a serious drain on the nation's resources, no politician dares twist a cow's tail, or even flick a hair. To do so would not be unlike an American politician campaigning against motherhood.

BEEF PRODUCTION IN AUSTRALIA AND NEW ZEALAND

Beef production in Australia and New Zealand increased sharply in the late 1960s and the early 1970s in response to high beef prices and low wool prices. Beginning in 1972, wool prices strengthened, so it is probable that the shift from cattle to sheep will slow down, although it will not likely be reversed. Of all the beef producing countries of the world, Australia and New Zealand have the best potential for increase.

Australia is a natural cattle country, and it is free from foot-and-mouth disease. Most of the cattle are grazed year-round on unfenced ranges, herded by musterers—the counterpart of the American cowboy. Slaughter animals consist of 2- to 4-year-old steers which are grass finished, although there is a growing trend to market younger animals and grain finish. The vast majority of the cattle operations in Australia are very large, ranging in size from 5,000 acres in the more developed southeastern part of Australia to over 3 million acres in the Northern Territory, and with 10,000 to 50,000 cattle per unit.

Shorthorns, Herefords, and Angus are the leading breeds. In the tropical areas of the North, Brahman and Santa Gertrudis have been introduced and are increasing in numbers. Crossbreeding is practiced widely. Many stations (we call them ranches) are inadequately fenced and watered, and there is room for improvement in their nutrition and husbandry. As evidence of the latter statement, it is noteworthy that in northwestern Queensland, Northern Territory, and parts of Western Australia, only a 45 to 55 percent calf crop is raised to branding age. Also, on some properties it is standard practice to write off a 12 to 15 percent mortality each year.

The beef industry of Australia is subjected to recurrent droughts. Until recently, the chief obstacle to further expansion of the nation's beef production was the great distance to the consumer markets of Europe and the United States. However, improved technology in processing and transporting beef is gradually overcoming this handicap. Also, a new and relatively nearby market for beef in Japan has opened up. Thus, the improved market for beef, along with government policies favorable toward the development of the cattle industry of the country, indicates a bright future for the expanding beef industry of Australia. Also, it is noteworthy that the cost of beef production in Australia is much lower than in the United States.

Fig. 2-4. A fine herd of Santa Gertrudis cattle on Eidsvold Station, Queensland, one of Australia's finest stations (ranches), owned by E. B. "Barney" Joyce, shown in front. Eidsvold is a mecca for many distinguished people, including royalty. Prince Charles, of England, is shown trailing the above herd (center, and immediately behind the cattle).

New Zealand is a small, picturesque country, about the size of Colorado. The climate is temperate, with plentiful sunshine, adequate rainfall, and no great extremes of heat or cold. Year-round grazing is available, on fenced holdings. Cattle and sheep share many areas, to the advantage of each other, with cattle utilizing the coarser vegetation and sheep the finer grasses and legumes.

In 1973, there were 9,360,000 cattle in New Zealand. Very few farmers devote themselves exclusively to beef production. In general, the raising and fattening of beef is carried on in conjunction with sheep farming. The dairy sector provides a large contribution from its cull cows and surplus calves.

BEEF PRODUCTION IN EUROPE, INCLUDING EC-9

Beef production throughout Europe is largely from dual-purpose cattle, animals bred to produce both milk and meat. This poses the problem of how to increase beef output from such herds without pushing up milk surpluses.

In the EC-6 (Belgium, France, Germany, Luxembourg, the Netherlands, and Italy), about 45% of the beef and veal production comes from cull dairy cows and milk-fed calves. But cull dairy cows and calves in the 3 new member countries (United Kingdom, Ireland, and Denmark) account for only about 20% of the total beef and veal production. By comparison, in the United States, dairy cows and calves merely account for about 15% of Federally inspected beef and veal production.

In addition to obtaining beef from cull dairy cows, throughout Europe bull calves (not steers) are fattened out and slaughtered as yearlings.

In 1973, the EC-9 accounted for about 32% of world beef imports, compared to the United States with 36%, and Japan with 7%. However, cattle numbers are increasing in the EC, especially in the United Kingdom and Ireland.

Most EC farmers would have difficulty making a transition from dairy to beef cattle because the farms are too small to produce grass-fed beef profitably, as is done in Argentina and Australia. Further, in the United Kingdom and Ireland, where there are larger farms and beef production from grass-fed animals is profitable, high EC grain prices may result in some pasture being diverted to grain or other crop production.

In much of Europe, dairy cattle are selected for their beef qualities. For example, stud bulls that are widely used in artificial insemination are often selected on the basis of rate and efficiency of gain, very much as beef bulls are selected on performance test in the United States.

The EC has had several programs to pay farmers to convert from dairy to beef cattle, but payments were not large enough to equal current income from milk sales. The only way EC farmers could substantially increase beef production at a profit would be through confined feeding, provided the price of feed grains were favorable.

Fig. 2-5. Milking ability is a "must" in beef cows in the United Kingdom. These calves give evidence that their dams possess this trait in abundance. (Courtesy, Devon Cattle Breeders' Society, of England)

Cattle numbers are not a limiting factor for beef production in the EC. For example, the EC-6 now has enough cattle to produce some 20 percent more beef if current calf slaughter were reduced to U.S. rates and the calves were fed in feedlots. However, to produce this much beef from grass and grain would require a longer turn-around period and larger cattle numbers. Also, this could result in increased milk production, which is in surplus already.

All indications are that the EC-9 will continue to be a net importer of beef until 1985, and perhaps beyond, even if local seasonal surpluses, especially of the cheaper cuts, occur from time to time.

BEEF PRODUCTION IN CANADA

Canada is still a frontier type of country with almost unlimited opportunities for expansion of the beef cattle industry. In general, Canadian cattle are noted for their size, scale, and ruggedness. This is due to the fact that in the great expanses of frontier agriculture, cattle production is on a cost-per-head rather than on a cost-per pound basis; that is, it costs little more to produce a sizable beast than to produce a small one. The main obstacles to increased beef production in Canada are: (1) the long severe winters in much of the cattle country centered primarily in the eastern and western provinces where up to seven months feeding is required; (2) the high duty and frequently closed borders for exports to the United States, the most natural potential market; and (3) the need for a permanent outlet for stocker and feeder cattle, as Canada has no finishing area comparable to the Corn Belt.

Fig. 2-6. Cattle roundup in Canada. On their way to summer pasture, this herd fords the Milk River in southern Alberta. Canada had 12.7 million head of cattle in 1973. (Courtesy, Canadian National Film Board)

In 1973, Canada had 12.7 million cattle. That year, she exported 363,500 head of cattle, largely feeders, to the United States.

The cattlemen of Canada appear to be optimistic about the future of the industry. It is predicted that more and more cattle will be finished on the small grains which are produced in great abundance.

BEEF PRODUCTION IN MEXICO

Mexico ranks ninth among the leading cattle countries of the world (Table 2-2).

Since January 1,1955, Mexico has been free of foot-and mouth disease, and the border has been open, subject to the usual quotas and duties.

Factors unfavorable to beef production in Mexico are: (1) the ravages of parasites, particularly the Texas tick; (2) lack of improved breeding, which is made difficult because of the susceptibility of newly imported cattle to diseases and parasites; (3) frequent droughts; and (4) political uncertainties and government policies unfavorable to the development of cattle units of adequate size to permit practical and economic operation in the present era.

Despite all the difficulties now existing in Mexico, the fact remains that cattle are afforded a long grazing season and labor is cheap and abundant. Cattle can be produced very cheaply. Also, in recent years, the better cattlemen of Mexico have made marked progress in improving both the quality of their cattle and the efficiency of their production.

Each year, Mexico provides several thousand head of feeder cattle for growing on the ranges of the Southwest or finishing in U.S. feedlots. We received 629,200 feeders from Mexico in 1973. However, Mexico has a growing domestic market. Already she is beginning to feel the drain of live cattle to the United States. As a result, Mexico will not be in a position to increase

Fig. 2-7. Part of a fine herd of Herefords on the ranch of Guillermo Finan, Hacienda Valle, Columbia, Muzquiz, Coahuila, Mexico. This herd would be considered outstanding anywhere—in Mexico, in the United States, or in Canada. (Courtesy, Mr. and Mrs. Guillermo Finan)

significantly beef or cattle exports in the near future. Rather, she will continue to control them through quotas established by the government.

BEEF PRODUCTION IN SOUTH AMERICA

Of the South American countries, Argentina, which ranks fifth in world cattle numbers, is recognized as the outstanding beef producer. In fact, taken as a whole, Argentine cattle probably possess better breeding and show more all-round beef excellence than do the cattle of any other country in the world. The excellence of the Argentine cattle can be attributed to two factors—their superior breeding and the lush pastures of the country. Beginning in 1850 and continuing to the present time, large numbers of purebred animals have been imported from England, Scotland, and the United States. No price has been considered too high for bulls of the right type; and, again and again, British and American breeders have been outbid by Argentine estancieros in the auction rings of Europe. These bulls and their progeny have been crossed on the native stock of Spanish extraction (Criollo cattle). Today, Herefords, Angus, and Shorthorns are the most numerous breeds of the country.

The finest cattle pastures of the Argentine are found along the La Plata River, in the region known as the Pampas, a vast, fertile plains area embracing about 250,000 square miles, which slopes ever so gently toward the sea. It's a dreamland of cattle and grass, and the "beef basket" of South America. Much of this fertile area is seeded to alfalfa upon which cattle are pastured year-round. Instead of finishing cattle largely on grains, as we do, the cattlemen in the Argentine finish their stock on alfalfa pastures. The corn of the Pampas region, which represents an acreage one-half as great as that devoted

Fig. 2-8. Well-bred cattle on lush pastures in Argentina. (Courtesy, Counselor Office Cultural Relations, Republic of Argentina, Washington, D.C.)

to alfalfa, is largely exported. Usually 2- and 3-year-old steers are finished by turning them into a lush alfalfa pasture for a period of 4 to 8 months prior to marketing. The surplus beef of Argentina is marketed as frozen or chilled beef to the European countries, especially to Great Britain. None of the frozen or chilled beef from the Argentine is admitted into the United States because of the hazard of foot-and-mouth disease; it must be canned or fully cured (i.e., corned beef).

Other South American countries of importance in beef production are Brazil, Colombia, Uruguay, and Paraguay.

Generally speaking, Brazil, which is slightly larger than the United States, produces hardy cattle of rather low quality, predominately of Zebu breeding.

Colombia is handicapped by lack of improved breeding, poor transportation facilities, and limited refrigeration, although beef production is one of the nation's principal industries.

Uruguay, which is but little larger than the state of Missouri, is noted (1) as an ideal cattle country (because of its rich pastures, abundant water supply, and temperate climate), (2) for Hereford and Shorthorn cattle of good breeding, although they are not equal in quality to the cattle in Argentina, (3) as one of the most highly specialized beef cattle countries in the world, and (4) as a beef exporting country, despite its small size (80% of the nation's exports consisting of animal products).

Paraguay, which is about 2½ times larger than Uruguay, produces cattle of similar breeding and quality to those in Brazil.

As in the Argentine, year-round grazing constitutes the basis of the beef cattle industry of the other South American countries. Virtually no grain is used in finishing animals, except for those being fitted for show. No attempt is made to finish steers until they are fully mature.

In general, the foremost obstacles, or unfavorable factors, affecting South American beef production are:

1. The ever-present foot-and-mouth disease, which, though seldom fatal, results in enormous economic losses through retarded growth and emaciation and which limits the foreign sale of both beef and cattle on foot.

2. Droughts are rather frequent in many of the cattle sections, and they are likely to be of rather long duration.

3. Parasites and certain diseases other than foot-and-mouth disease are rather prevalent in the warmer sections.

4. Prices are very much dependent upon the export trade, thus making for an uncertain market.

5. Local markets are often unsatisfactory; modern packing plants are not too plentiful; and refrigeration facilities are limited. Many of the cattle slaughtered in the more isolated areas of South America, especially in Brazil and Paraguay, are still made into jerked or salted beef.

6. Transportation facilities are few and far between.

7. Except for the cattle of Argentina and Uruguay, much improvement in breeding is needed; but the introduction of improved blood is difficult because of the heavy infestation of diseases and parasites to which the native and zebu cattle are more resistant.

Because of the glowing reports about the cattle industry of Argentina, many young men from the United States have, from time to time, been interested in establishing a cattle enterprise in South America. Without exception, experienced U.S. cattlemen who have visited in South America in person and who know whereof they speak point out the almost impossible odds of success in such a venture. In the first place, the land is in the hands of a comparatively few families who hold a monopoly on the cattle industry; and, secondly, the political unrest in these countries is usually not conducive to such private foreign investments in land or cattle.

BEEF PRODUCTION IN CENTRAL AMERICA AND PANAMA

Central America and Panama are the tropical land mass connecting North America and South America, consisting of Guatemala, Honduras, Belize (formerly British Honduras), Nicaragua, El Salvador, Costa Rica, and Panama. On an individual basis, none of these countries produces or exports sufficient beef to be much of a factor. However, as a group, in 1972 they had 10,400,000 cattle and they exported 195,513,000 pounds of beef and veal to the United States. Moreover, it is estimated that their exports will increase by 90 percent (nearly double) by 1985.

Compared with coffee, bananas, and cotton, grazing is a relatively old economy in Central America and Panama. It dates back to the early colonial period, and in terms of land utilization, quality of stock, and disposal of products, it has changed very little through the years. Primary emphasis has been on cattle of unimproved breeding, low in both yields and quality, but

Fig. 2-9. Part of a herd of 800 head of Criollo X Brahman cattle in a corral at Santa Clara Ranch, owned by Sr. Jorge Cordero, Cordero Ranches, Guatemala City, Guatemala. (Photo by Audrey Ensminger)

high in resistance to ticks and other environmental handicaps of the region. Cattle are predominantly Criollo, although efforts have been made to improve their beef cattle in recent years by importing breeding stock from the United States, primarily Brahman and Santa Gertrudis.

Livestock are concentrated chiefly along the Pacific Coast, where they fatten on grass. Pasture is abundant, except during the dry season, when it is short for about three months. Some of the more progressive ranchers are (1) ensiling grass, corn, and/or sorghum, or (2) irrigating pastures to provide forage during the dry season.

Corn is high in price, because of the demand for human consumption. Whole cottonseed, cottonseed hulls, and cane molasses are relatively cheap. Rising land values are prompting interest in supplemental pasture finishing.

Among the factors *favorable* to beef production in Central America and Panama are: abundant grass, relatively cheap land and labor, freedom from foot-and-mouth disease, and tax deferment (up to 15 years).

Among the *unfavorable* factors are: scarce and high-priced cereal grains; heavy insect infestation—especially flies and ticks; each country controlling its live cattle exports more carefully each year (for example, since 1972 Guatemala has prohibited feeder cattle exports because of the desire to increase their weight in order to take full advantage of beef exports); little culling; and low percentage calf crop.

World Meat Consumption

In general, meat consumption is highest in countries that have extensive grasslands, temperate climates, well-developed livestock industries, and that are sparsely populated. In many of the older and more densely populated regions of the world, insufficient grain is produced to support the human population when consumed directly. This lessens the possibility of keeping animals, except for consuming forages and other inedible feeds. Certainly, when it is a choice between the luxury of meat and animal by-products or starvation, people will elect to accept a lower standard of living and go on a grain diet. In addition to the available meat supply, food habits and religious restrictions affect the kind and amount of meat consumed.

Figure 2-10 shows the total per capita consumption of all red meat in certain specified countries.

From 1966 to 1972, New Zealand was the world's largest per capita consumer of red meat, followed closely by Australia. In 1972, Australia took the lead. The United States is the third largest per capita meat consumer.

Japan, with only 31 pounds per capita red meat consumption in 1972, is still only one-sixth the level of red meat consumption of the United States.

FACTORS AFFECTING WORLD BEEF CONSUMPTION

The major factors which influence the total demand for beef are (1) the increase in human population, and (2) the increase in the buying power of the population (Table 2-4). The gross domestic product (GDP), which is

PER CAPITA CONSUMPTION OF RED MEAT

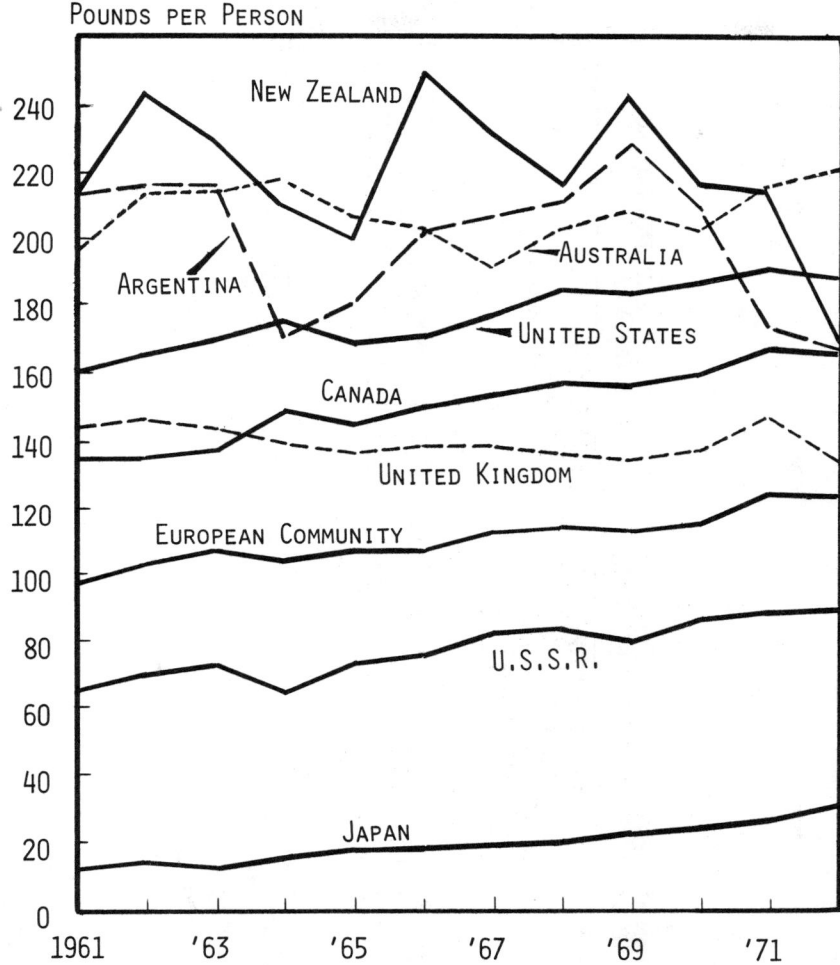

Fig. 2-10. Total per capita consumption of all red meat in certain specified countries, 1961-1972. (Source: 1961-1971 data from *Foreign Agriculture*, USDA, Foreign Agricultural Service, April 30, 1973, p. 2; 1972 data from *Foreign Agriculture Circular*, Livestock and Meat, FLM 1-74, USDA, Foreign Agricultural Service, January 1974, p. 2)

highly related to domestic buying power, is expected to rise considerably in the years ahead, especially in the developed nations. The rate of growth of the national income in the developing nations is expected to be less than that in the developed nations.

People with low incomes consume relatively low priced carbohydrates. With higher incomes, higher quality foods, notably beef, are eaten. Thus, per capita beef consumption is a good barometer of GDP and standard of living of a country. Beef is a status symbol.

TABLE 2-4

HUMAN POPULATION AND PER CAPITA GROSS DOMESTIC PRODUCT (GDP; EARNING POWER) OF FIVE NATIONS WITH GREATEST DEMAND FOR BEEF AND VEAL

Country	Human Population (millions)		Per Capita GDP (US$)		Per Capita Beef and Veal Consumption				Demand (1,000 metric tons)	
	1970[1]	1985[2]	1970[1]	1985[3]	1970[1] (lb) (kg)		1985[3] (lb) (kg)		1970[1]	1985[3]
United States	205.2	246.3	4,798	7,560	117.7	53.5	140.0	63.6	10,979	17,241
U.S.S.R.	242.6	286.9	1,295	2,617	44.7	20.3	61.7	28.1	4,915	8,851
Argentina	24.3	29.6	950	1,405	194.5	88.4	160.3	72.9	2,150	2,372
People's Republic of China	847.7	964.6	93	116	5.5	2.5	7.3	3.3	2,130	3,521
Brazil	93.6	142.6	422	662	38.7	17.6	49.2	22.4	1,650	3,508

[1]Figures for 1970 are estimated trend values, used for a base period for the projection and differ slightly, in some cases, from actual 1970 figures. (Sources: *Agricultural Commodity Projections, 1970-1980,* Volumes I and II, Food and Agriculture Organization of the United Nations, Rome, 1971; 1970 "Human Population" figures, pp. 4-7, Volume II; 1970 "Per Capita GDP" figures, pp. 17-19, Volume II; 1970 "Per Capita Beef and Veal Consumption" figures and 1970 "Demand" figures, pp. 136-137, Volume I.)

[2]1972 *World Population Data Sheet,* Population Reference Bureau, Inc., Washington, D.C.

[3]1985 projections made by author. Same sources as footnote 1 above, with their projections extended by author to 1985.

It is emphasized, however, that national demand for beef and veal is not a simple matter of merely multiplying the human population times the per capita gross domestic product. Other factors produce noteworthy changes. For example, meat prices are clearly a factor in determining variations in consumption levels. Also, such barriers to world trade as transport costs, import and export taxes, and tariffs prevent equalization of meat prices between countries.

Major meat exporters such as Uruguay, Argentina, New Zealand, and Australia have lower meat price levels and their people consume more meat in relation to income levels than the rest of the world. Also, in the South American countries exports have been restricted to ensure adequate supplies, hold domestic prices down, and keep traditional high consumption levels intact. In Fig. 2-11, these countries, along with Australia and New Zealand, are shown above the line that indicates a very clear relationship between per capita disposable income and red meat consumption. This means that they are consuming more meat than their income would indicate.

Countries such as the United States, Canada, and the United Kingdom, which have internal grain prices at world levels and generally free access to their meat markets, can be considered to have meat consumption levels in

RELATIONSHIP BETWEEN INCOME AND RED MEAT CONSUMPTION
IN SELECTED COUNTRIES, 1971

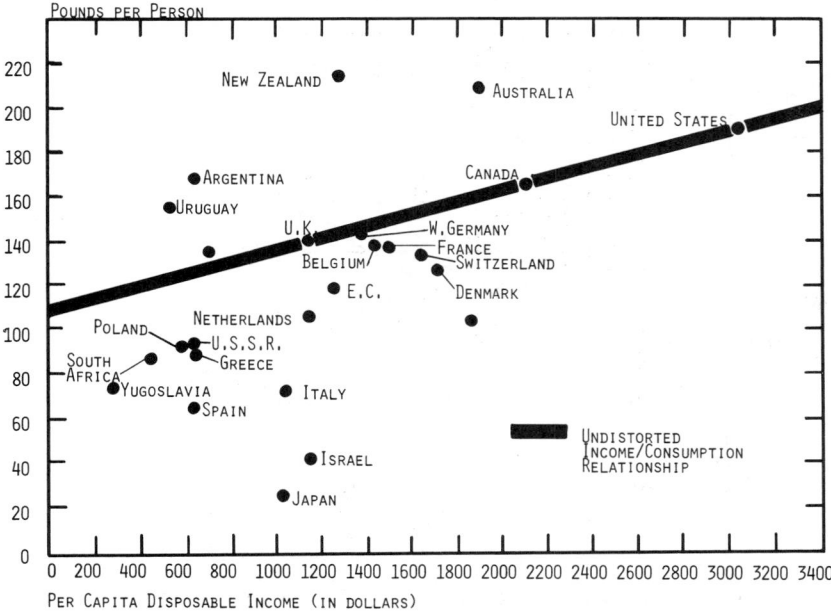

Fig. 2-11. Relationship between income and red meat consumption in selected countries, 1971. (Source: *Foreign Agriculture*, USDA, Foreign Agricultural Service, April 30, 1973, p. 3)

undistorted relation to their income levels. Here, beef and pork prices are influenced by world grain prices and meat imports from other sources can compete freely. A line connecting these countries is shown in Fig. 2-11.

Countries with sufficient protection in the grain and/or meat sector to put consumer meat prices above world levels—such as the European Community (EC) countries and Switzerland—have consumption levels below what disposable income would indicate; hence, they fall below the line in Fig. 2-11.

Some countries maintain very tight import controls, often through quotas and/or high tariffs, and the resultant very high meat prices offset higher income levels, causing per capita meat consumption to lag. Japan and Sweden are two such countries.

In addition to price and income, traditional eating habits influence meat consumption levels.

WORLD BEEF PRICES

U.S. consumers concerned over rising food prices need not feel alone. Their views have been echoed around the world, and with increased fervor as inflation continues to mount in Western Europe, Japan, and many of the developing countries. Governments have responded with stiffer price controls and, in some cases, freer import policies. But halting the food price

CLIMBING COST OF WORLD FOOD

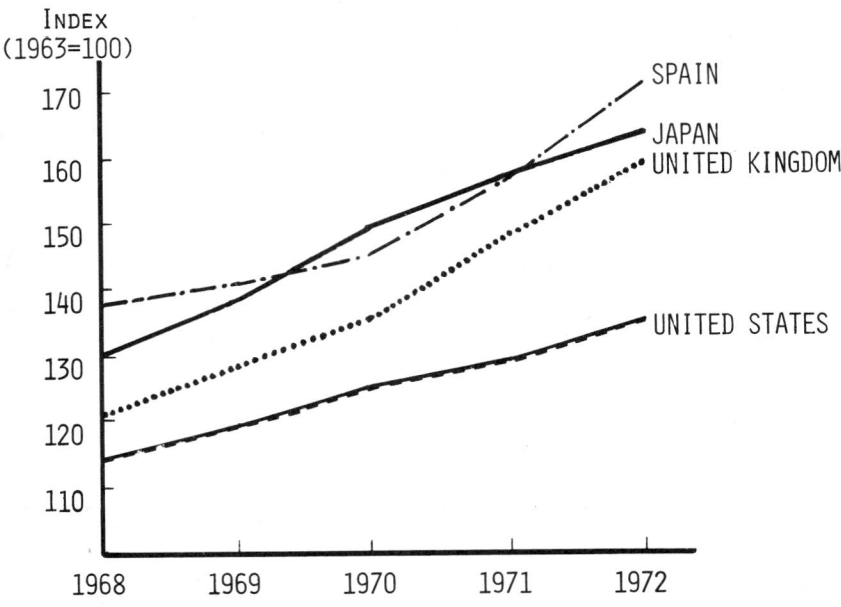

Fig. 2-12. Spiraling world food prices. (Source: *Foreign Agriculture*, USDA, Foreign Agricultural Service, April 2, 1973, p. 4)

spiral remains an elusive goal, complicated by such problems as soaring demand for meat and high quality products desired by people as they become more affluent, crop failures, and protective trade policies of the European Community (EC), Japan, and other countries and regional groups (see Fig. 2-12).

Meat prices—and beef and veal in particular, which have more than doubled since 1956 (Fig. 2-13)— account for much of the increase in food prices. Other meats have not had such large price rises, and the price of poultry meat has even declined. Of course, beef has spiraled in price because people all over the world are eating more beef, and they're able and willing to pay for it. Noteworthy, too, is the fact that beef prices in many other countries are higher than in the United States. Because of differences in cuts and quality, prices are not strictly comparable, but Fig. 2-14 gives the retail beef prices in selected cities around the world in mid-November 1973, in dollar equivalent.

Fig. 2-14 reveals that sirloin steaks cost less in the United States in 1973 than in any of the other selected capital cities of the world. Also, the following two points are pertinent when considering the comparative price of beef or other foods between countries:

1. In no country is the cost of living in terms of working time as favorable as in the United States. In time required to buy a sirloin steak, here are

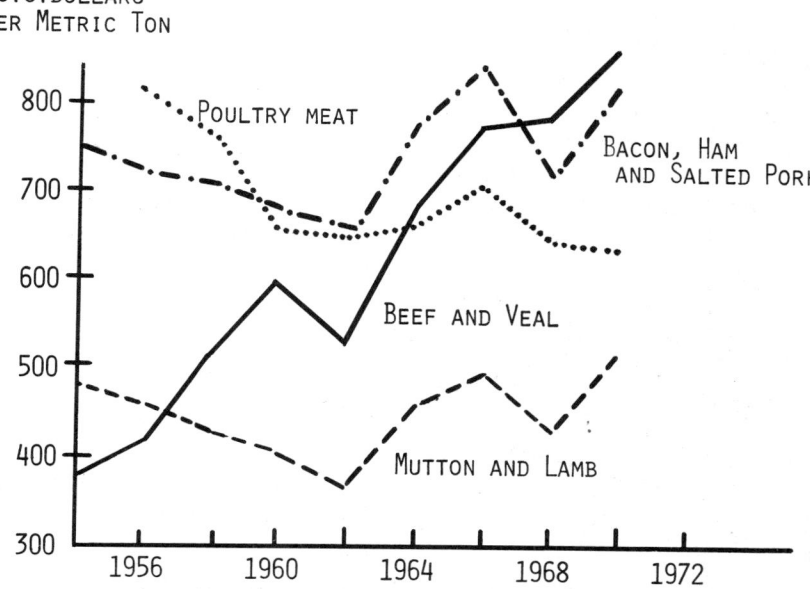

Fig. 2-13. World Market price for red meats and poultry, 1956-70. (Source: "The World Market for Beef and Other Meat" by B. E. Hill, *World Animal Review*, Food and Agriculture Organization of the United Nations, No. 4, 1972, p. 1)

THE COST OF SIRLOIN STEAK, IN U.S. $

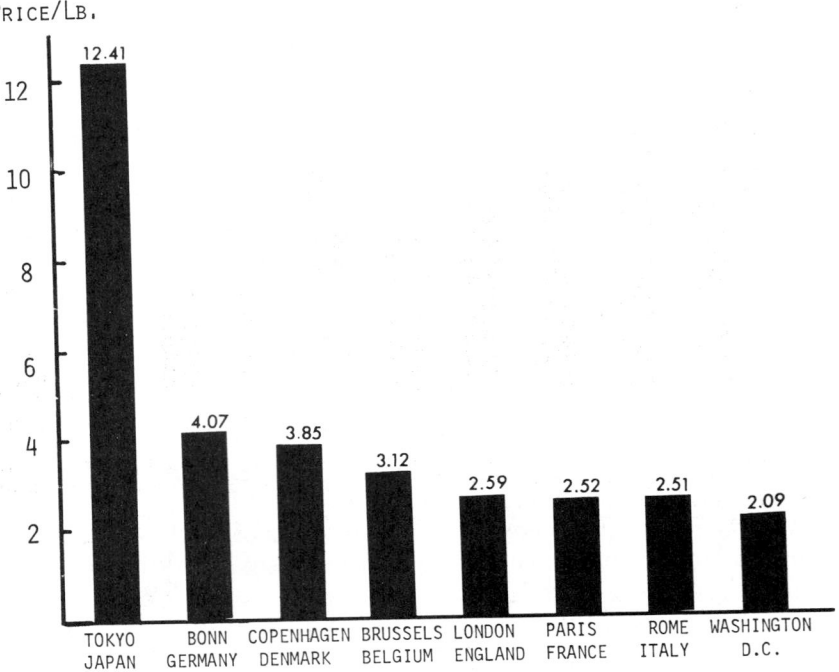

Fig. 2-14. Retail sirloin steak prices in selected cities in mid-November 1973.

the comparative figures: In the United States, 24 minutes; in Canada, 26 minutes; in Brazil, 1 hour; in France, almost 1½ hours; in the U.S.S.R., over 1½ hours; and in Japan, nearly 2 hours.

2. In no other country of the world do consumers spend such a small proportion of their income for food as in the United States. Here are some 1971 comparisons: In the United States, 16%; Australia, 17%; Canada, 21%; Japan, 27%; Western Europe, 19-34%; Eastern Europe, 35-46% (1969); and India, 55-60%.

WORLD BEEF CONSUMPTION

Fig. 2-15 shows the world's beef eaters, based on per capita consumption of beef and veal. Although Australia and New Zealand rank at the top in total meat consumption (Fig. 2-10), they drop to fifth and sixth places, respectively, on beef and veal consumption. Of course, the explanation is that Australians and New Zealanders eat enormous quantities of mutton and lamb (Fig. 2-16).

In terms of total consumption of beef and veal (Table 2-5), the major consuming nations are much the same as the major producing nations. Also, a very similar type of picture is projected for 1985. Per capita consumption, however, shows a considerably different type of ranking of nations. Argen-

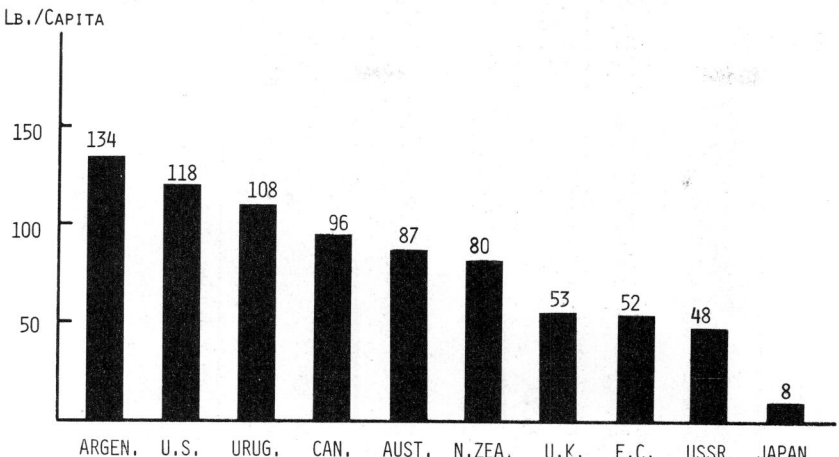

Fig. 2-15. Per capita beef and veal consumption in specified countries, 1972. (Source: *Foreign Agriculture Circular*, FLM 1-74, USDA, Foreign Agricultural Service, Washington, D.C., January 1974, p. 3)

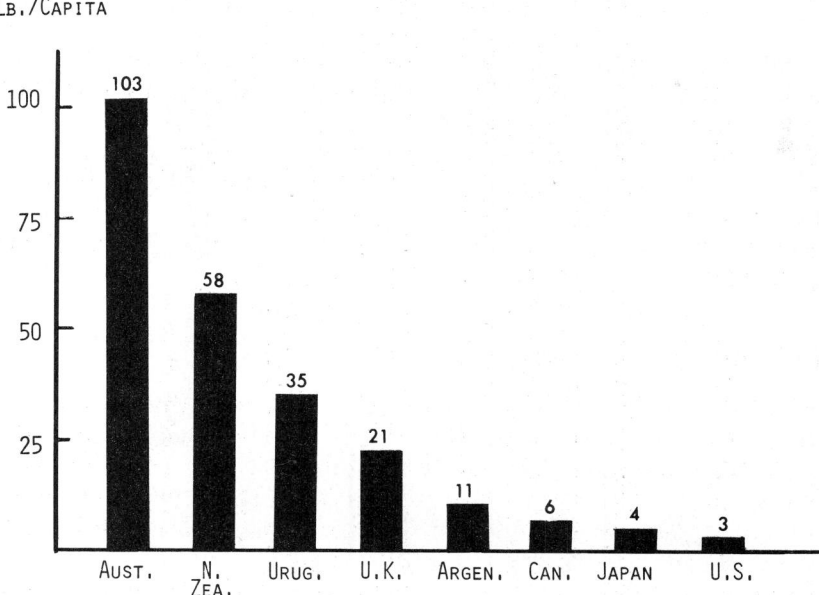

Fig. 2-16. Per capita mutton and lamb consumption in specified countries, 1972. (Source: *Foreign Agriculture Circular*, FLM 1-74, USDA, Foreign Agricultural Service, Washington, D.C., January 1974, p. 5)

TABLE 2-5

MAJOR BEEF AND VEAL CONSUMING NATIONS

Country	Demand (1,000 metric tons)		Cattle Population (millions)		Per Capita Beef and Veal Consumption			
	1970[1]	1985[2]	1974[3]	1985[4]	1970[1] (lb)	(kg)	1985[4] (lb)	(kg)
United States	10,979	17,241	127.5	177.6	117.7	53.5	140.0	63.6
U.S.S.R.	4,915	8,851	106.2	115.8	44.7	20.3	61.7	28.1
Argentina	2,150	2,372	56.5	57.9	194.5	88.4	160.3	72.9
People's Republic of China	2,130	3,521	92.9	108.9	5.5	2.5	7.3	3.3
Brazil	1,650	3,508	90.1	112.5	38.7	17.6	49.2	22.4
France	1,523	2,402	23.6	24.9	66.0	30.0	83.4	37.9
Germany (West)	1,428	2,071	14.4	14.8	51.5	23.4	66.5	30.2
United Kingdom	1,310	1,714	14.7	14.2	51.5	23.4	55.5	25.2
Italy	1,107	2,052	8.5	12.2	45.5	20.7	68.4	31.1

[1]*Agricultural Commodity Projections, 1970-1980*, Volume I, Food and Agriculture Organization of the United Nations, Rome, 1971, Table 4, pp. 136-137.

[2]1985 projections made by author. Same source as footnote 1, with their projections extended by author to 1985.

[3]*Foreign Agriculture Circular, Livestock and Meat*, USDA, Foreign Agricultural Service, Washington, D.C., FLM 7-74, June 1974, pp. 8-9.

[4]1985 projections made by author based on 1970-1980 trend figures from *Agricultural Commodity Projections, 1970-1980*, numbers for 1985, *for which see Chapter 17, Table 17-1, as to source and method of computing.*

tina ranks first with 195 lb of beef and veal consumed per person per year, with Uruguay second with 189 lb (not shown in Table 2-5), then New Zealand with 154 lb per person, followed by Australia with 134 lb. The United States ranked fifth with 118 lb per person.

World production of beef averaged 40,520 metric tons in 1970-72. (See Table 2-3) The United States produced nearly 10 billion pounds of this, or about 25 percent of world production.

Tables 2-6 and 2-7 show the meat production and per capita consumption of meats in selected countries and the favored position of beef.

World Beef Trade

Meat is produced principally for domestic markets; only about 5% of carcass meat enters world trade. In terms of value, agricultural products account for 25% of world trade, and meat and livestock about 3%. Nevertheless, meat has had a tremendous impact on international trade. Countries and people struggling to catch up, once they approach affluency, want meat. This has led to increased meat prices and trade. There has been a long-term upward trend of meat prices; they rose by approximately 34% between 1960 and 1970, much more than any other class of agricultural product. World red meat exports rose from 7 to 12.1 billion pounds from 1961 to 1971—an increase of over 72%. Out of the 1971 total, about half of the world trade was for beef and veal, about one-third was pork, and most of the remainder was lamb and mutton. Horsemeat accounted for less than 3% of the world trade.

Of the trade in meat and livestock, about 64% is accounted for by fresh, chilled or frozen meat, a further 14% by canned meat, and 22% consists of live animals. For the most part, trade in live animals is between countries which are relatively near neighbors. Most movements of cattle are among European countries, although there are fairly large movements from Canada and Mexico into the United States. But carcass meat is the major component of world meat and livestock trade.

Basically, the trade in world meat reacts to the law of supply and demand, but with several country to country exceptions such as the following: protecting their own livestock industry by quotas and levies, agreements on nonagricultural products, which may hinder the normal flow of meat, differences in currency values, problems in transportation and storage, and even political decisions.

WORLD BEEF AND VEAL EXPORTS

Beef and veal are the most important meats in world trade. In 1972, bovine meat accounted for 53 percent of world exports.

The major beef and veal exporting nations are, with the exception of EC, countries with extensive land areas and low labor costs. They are: Argentina, the EC, Australia, Ireland, and New Zealand (Fig. 2-17 and Table 2-8). In 1971, Australia became the world's largest beef and veal exporter, accounting for 18.8 percent of the world exports. Argentina ranked second and New

TABLE 2-6

MEAT PRODUCTION AND PER CAPITA CONSUMPTION IN SPECIFIED COUNTRIES, 1972

Country (Leading Countries, by Rank, of All Meats, 1972)	Total Meat Production[1]	Beef and Veal Production[1]	Per Capita Consumption of All Meats[2]	Per Capita Consumption of Beef and Veal	Percent Beef and Veal of All Meats Consumed
	(mil. lb)	(mil. lb)	(lb)	(lb)	(%)
United States	37,047.0	22,851.0	188	118	62.8
U.S.S.R.	22,255.2	11,481.7	90	48	53.3
West Germany	7,879.3	2,621.3	147	53	36.1
France	6,688.8	3,207.7	140	62	44.3
Brazil	6,060.0	4,453.3	56	40	71.4
Argentina	5,701.6	4,856.8	167	134	80.2
Australia	5,111.9	2,574.7	222	87	39.2
Poland	3,727.6	1,174.6[3]	98	29	29.6
Canada	3,413.5	1,977.0	163	96	58.9
Italy	3,208.8	1,815.3	75	44	58.7
Total 59 Countries	146,085.9	75,802.5	—	—	—

[1]Agricultural Statistics 1973, USDA, 1973, Table 504, p. 344.
[2]Foreign Agriculture Circular, Livestock and Meat, FLM 1-74, USDA, Foreign Agricultural Service, January 1974, pp. 2-3.
[3]Used 1971 figure; 1972 figure not available.

TABLE 2-7

PER CAPITA MEAT CONSUMPTION IN SPECIFIED COUNTRIES, 1972

Country (Leading Countries, by Rank, of All Meats)	All Meats (Excluding Lard)[1]	Beef and Veal	Percent Beef and Veal of All Meats
	(lb)	(lb)	(%)
Australia	222	87	39.2
United States	188	118	62.8
New Zealand	171	80	46.8
Argentina	167	134	80.2
Canada	163	96	58.9
Uruguay	157	108	68.8
Belgium-Luxembourg .	149	61	40.9
West Germany	147	53	36.1
France	140	62	44.3
Austria	138	49	35.5
Switzerland	138	58	42.0
United Kingdom	138	53	38.4

[1]Foreign Agriculture Circular, Livestock and Meat, FLM 1-74, USDA, Foreign Agricultural Service, January 1974, pp. 2-3.

BEEF AND VEAL EXPORTS, 1964-72

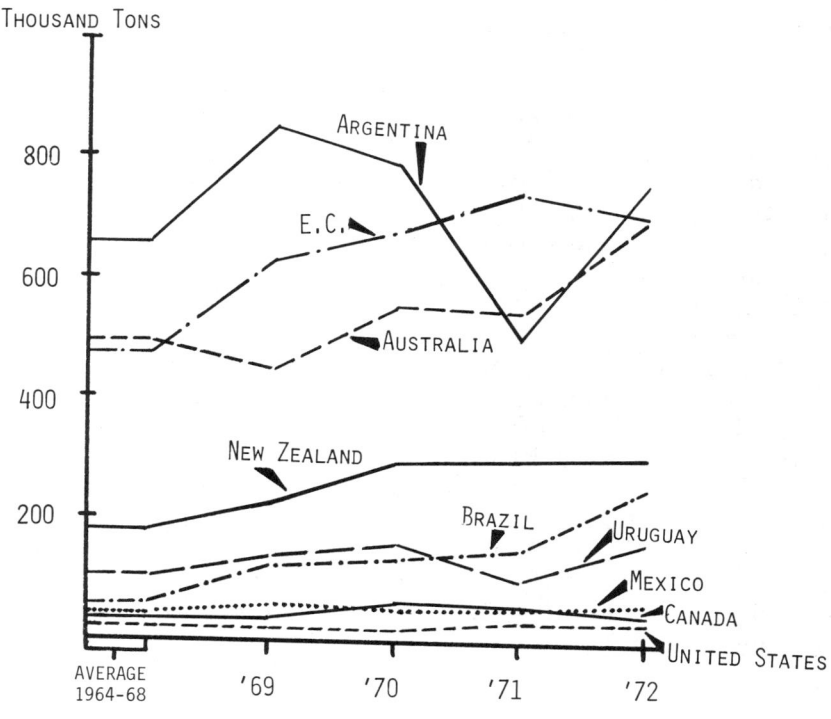

Fig. 2-17. Beef and veal exports of selected countries, average 1964-68, annual 1969-72. Carcass-weight equivalent basis; excludes fat, offals, and live animals. (Source: *Foreign Agriculture Circular*, Livestock and Meat, FLM 2-74, USDA, Foreign Agricultural Service, Washington, D.C., January 1974, pp. 10-11)

TABLE 2-8

MAJOR BEEF AND VEAL EXPORTING NATIONS
(1970 and 1985)

Country	Exports		Cattle Population (Including Buffaloes)	
	1970[1]	1985[2]	1970[3]	1985[2]
	(1,000 metric tons)		*(millions)*	
Argentina	650	2,067	52.0	57.9
Australia	320	547	20.7	23.8
Ireland	290	419	5.7	7.6
New Zealand	170	358	8.8	12.3
Mexico	113	293	24.6	36.0
Denmark	145	64	2.8	2.6
Uruguay	120	209	8.6	9.6
France	80	150	21.9	24.9
Brazil	70	241	91.9	112.5

[1]*Agricultural Commodity Projections, 1970-1980*, Volume I, Food and Agriculture Organization of the United Nations, Rome, 1971, Table 9, p. 144.

[2]1985 projections made by author, based on 1970-1980 trend projections from *Agricultural Commodity Projections, 1970-1980.* Their projections were extended by author to 1985.

[3]*Agricultural Commodity Projections, 1970-1980*, Volume II, Food and Agriculture Organization of the United Nations, Rome, 1971, Table 33, pp. 76-78.

Zealand third. Most of the beef exported by the EC is traded between member countries. Beef exports are along two major trade routes—Oceania (Australia and New Zealand) to North America and South America to the United Kingdom and the EC.

Beef exporters are divided into two foot-and-mouth disease groups—those that have it, and those that are free of the disease. Australia, New Zealand, Ireland, and the North American countries north of the Panama Canal are the primary foot-and-mouth disease-free exporters. Argentina, Brazil and Uruguay are the principal foot-and-mouth infected exporting countries.

All of the beef and veal exporting nations, with the possible exception of Denmark, are projected to increase their exports in the years ahead (Table 2-8). It is predicted that Argentina will triple its exports by 1985. It will achieve this primarily by higher calving percentages, lower mortality rates, marketing animals at younger ages, and finishing more cattle prior to slaughter, rather than greatly increasing cattle numbers. In an effort to keep up their exports in the early 1970s, Argentina, Uruguay, and Colombia had "meatless weeks" and "meatless days" to keep down domestic consumption.

It is also predicted that Australia will greatly increase its beef production and exports in the years ahead. Low wool and wheat prices gave impetus to the shift in the 1960s. No doubt, the sharp rise in wool and wheat prices in the early '70s will slow down this shift to beef, but it is unlikely that it will reverse it.

WORLD BEEF AND VEAL IMPORTS

The major beef and veal consuming nations are also the major importing and the richer nations (Fig. 2-18 and Table 2-9).

BEEF AND VEAL IMPORTS, 1964-72

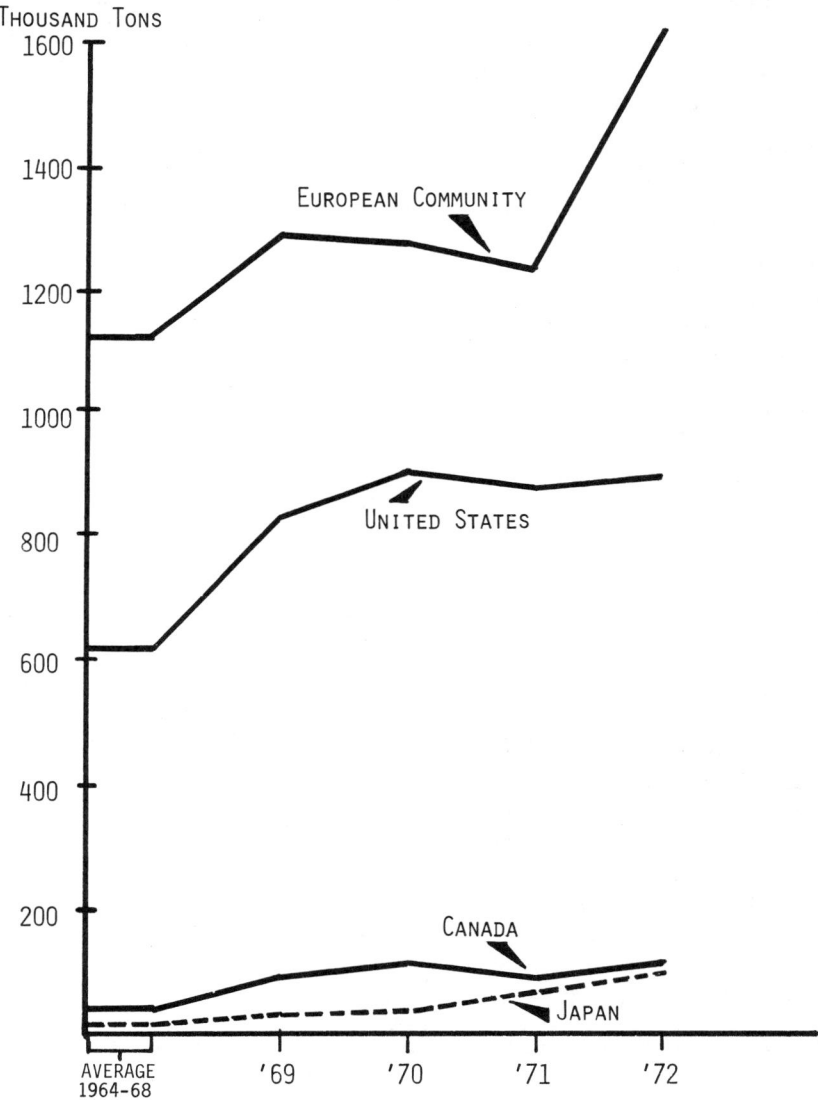

Fig. 2-18. Beef and veal imports of selected countries, average 1964-68, annual 1969-72. Carcass-weight equivalent basis; excludes fat, offals, and live animals. (Source: *Foreign Agriculture Circular*, Livestock and Meat, FLM 2-74, USDA, Foreign Agricultural Service, Washington, D.C., January 1974, pp. 10-11)

TABLE 2-9

MAJOR BEEF AND VEAL IMPORTING NATIONS
(1970 and 1985)

Country	Imports		Cattle Population (Including Buffaloes)	
	1970[1]	1985[2]	1970[3]	1985[2]
	(1,000 metric tons)		*(millions)*	
United States	719	1,534	112.3	177.6[4]
Italy	510	1,105	10.3	12.2
United Kingdom	480	278	12.7	14.2
Germany (West)	170	510	14.3	14.8
Spain	110	639	4.2	5.1
U.S.S.R.	115	3,071	102.0	115.8

[1]*Agricultural Commodity Projections, 1970-1980*, Volume I, Food and Agriculture Organization of the United Nations, Rome, 1971, Table 9, p. 144.

[2]1985 projections made by author, based on 1970-1980 trend projections from *Agricultural Commodity Projections, 1970-1980*. Their projections were extended by author to 1985.

[3]*Agricultural Commodity Projections, 1970-1980*, Volume II, Food and Agriculture Organization of the United Nations, Rome, 1971, Table 33, pp. 76-78.

[4]Projection made by author. (See Chapter 17, Table 17-1, of this book.)

The EC, the United States, and Japan accounted for 79 percent of all beef imports in 1972. The EC is the largest importer of beef, importing 47 percent of all beef traded in 1972. Imports into the EC countries reflect both increased imports from overseas and the growth of intra-Community trade as the Common Agricultural Policy integrated the European market in beef and veal. The United States suspended meat import quotas in June 1972, as a means of encouraging more imports. Japan has been a growing market for beef and veal, particularly from Oceania (Australia and New Zealand). Since the mid-1960s, internal demand pressures in Japan have resulted in expanded import beef quotas, followed by sharply increased imports, although the nation's beef imports are still comparatively small.

In the years ahead, it is expected that the U.S.S.R. will import considerable beef and veal, with the quantity determined by the success of its current emphasis on beef feeding.

Future U.S. beef imports will depend on the government's decision on the size of the quota or import restraint program.

BEEF TRADE PROBLEMS

The chief beef trade problems relate to the trade barriers of the main importing countries of North America and Western Europe and the quality and continuity of supplies from the exporting countries.

TRADE BARRIERS

It is particularly important to appreciate the reasons for trade barriers and the ways in which they operate. There are two main purposes of trade regulations: (1) to protect consumers from health risks; and (2) to protect the cattle industries of importing countries from both animal diseases and low priced imports.

● *Health and veterinary regulations*—In all the main meat importing countries, imports are subject to licensing. To obtain a license, the slaughtering facilities of the exporting country must satisfy the sanitary regulations of the importing country. This also applies to processing plants in the case of processed meat. Thus, importing countries try to protect their consumers from the possibility of food-borne diseases and food poisoning. Veterinary regulations are designed to prevent bringing in animal diseases which could have serious economic consequences for the importing country's own stockmen.

● *Tariffs and quotas*—In the developed countries, it is recognized that the income of farmers and ranchers tends to be lower than those of other professions. A prime objective of the agricultural policies, therefore, is to raise farmers' incomes. Generally this is attempted by introducing measures designed to raise the prices which domestic farmers receive above those prevailing in world markets. In the United States, quotas and tariffs exist.

Quotas limit the number of pounds of meat and the number of animals brought into a country. For example, U.S. Legislation of August, 1964, established a basic limit on meat plus an added factor based on the nation's beef and veal production.

Tariffs are duties, or charges (on a per pound basis in the United States), imposed by a government on imported meat, designed to raise the prices which cattlemen receive above those ruling in world markets.

Fig. 2-19 shows the effect of trade barriers on the prices received by farmers. The beef prices of a major exporter, Ireland, are compared with those of the main importing countries. As shown, the domestic prices of the

EFFECT OF TRADE BARRIERS ON DOMESTIC PRICES

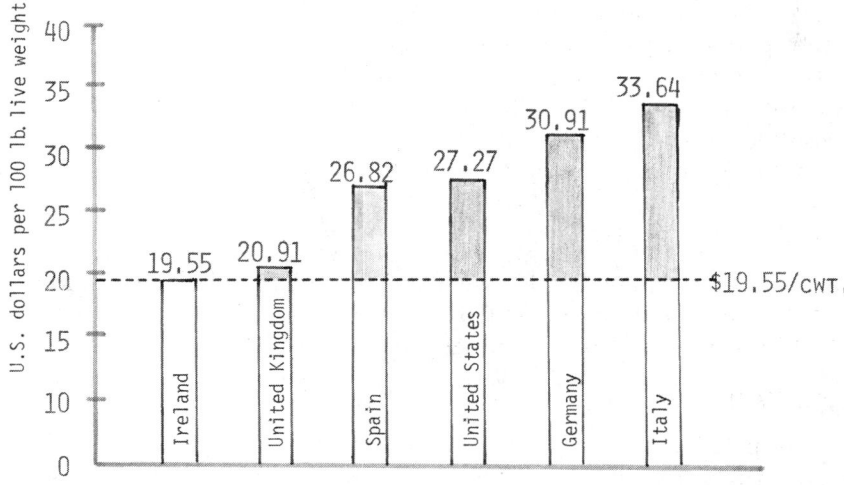

Fig. 2-19. Effect of import restrictions on the prices of beef in importing countries in comparison with the price of beef in an exporting nation, Ireland. (Source: *World Animal Review*, No. 4, FAO, 1972, p. 4)

latter are maintained at much higher levels than those prevailing in the world market.

BEEF SUPPLIES FOR EXPORT

Some of the main beef exporting countries in the southern hemisphere operate under two major handicaps:
1. Endemic cattle diseases, which are difficult and expensive to eradicate.
2. Serious periodic droughts, which not only prevent cattle fattening but deplete herds.

RECENT DEVELOPMENTS IN THE BEEF TRADE

Formerly, most beef sold on the world market was in the form of half or quarter carcasses. Processing into joints took place in the importing country. In recent years, more jointing has taken place before export. This development has been accelerated by importing countries as a result of (1) the spread of supermarkets selling prepackaged meats, and (2) the United Kingdom's ban, imposed in 1968, on imports of "bone in" beef from countries where foot-and-mouth disease is endemic (on the belief that the virus is most likely to be introduced in bone marrow, offal, and lymphatic glands).

In the main importing countries, the beef market is becoming increasingly split into two separate components: (1) high quality fresh beef for direct sale to consumers; and (2) beef for processing, known as manufacturing grade beef, suitable for soups, quick-frozen foods, and hamburgers. Intensive feeding techniques, such as the grain-finished cattle of the United States, result in beef which is unsuitable for manufacturing—it is too fat. On the other hand, beef from the developing nations is generally grass fed, lean, and ideal for manufacturing. Also, countries with cattle disease problems can supply manufacturing beef in cooked or frozen form, which is sterile and will not spread disease. Thus, in the United States, and to a lesser but increasing extent in other beef importing countries, most of the high quality fresh beef comes from domestic sources, and more and more of the manufacturing grades come from imports.

EUROPEAN COMMUNITY (EC)

The European Community (EC), or European Economic Community (EEC), built from the ruins of World War II, is one of the great trade centers of the world. It all began in 1950 when Belgium, Germany, Italy, Luxembourg, the Netherlands, and France pooled their coal and steel production under a common high authority, known as the European Coal and Steel Community (ECSC). This experiment was so successful that the six countries soon extended their cooperation to cover their economies as a whole, and the peaceful application of nuclear energy. They created the European Economic Community (EEC) or "Common Market" (commonly abbreviated Economic Community, or EC) and the European Atomic Energy Community (Euratom) in 1958.

Three other European nations—Britain, Ireland, and Denmark—joined the original six member nations on January 1, 1973, thereby bringing the total to nine—the EC-9.

The Common Agricultural Policy (CAP) has been described as the engine of the Common Market. The CAP seeks to ensure Community farmers a fair standard of living, stable markets, and improved methods. Its main features are free trade in farm produce (so that today one sees, for example, far more French goods in German shops, and vice versa, than before the EC); common price supports to raise and maintain farmers' incomes; variable levies at the Community frontier, which adjust the price of imported farm products to the general level of Community prices as set by the Council of Ministers; and a long-term, jointly financed plan to modernize farms and consolidate land-holdings that are too small to be efficient.

BEEF AND VEAL IN THE EC

The final step toward a common market in beef and veal was taken in August, 1968, with the fixing of a single guide price for all member countries, and a common external rate of duty on imports. The cornerstone of the program as it relates to beef and veal is the regulation of quantities coming onto the market in such a way that market prices approximate a predetermined guide price. This is done principally by influencing imports through the variable levy system, which discourages imports when market prices are low relative to the guide price, and allows freer access when they are above the guide price. In addition, imports are subject to a duty of 16 percent for live cattle and 20 percent for beef and veal.

When, despite the discouragement of imports through the levy system, producer prices fall significantly below the guide price, intervention purchases are made by government-sponsored intervention agencies, thus effectively providing a floor below which market prices will not fall.

Levies on beef and beef cattle are fixed every week, comprising the difference between the price at which the consignments are imported (including duties) and the guide price, but the proportion of the levy payable by the importer depends on the state of the home market. This is assessed on representative market prices, usually known as reference prices, which are calculated in each member country and then brought together as a Community reference price, weighted for the size of the cattle population in each country. Thus, depressed markets in one country do not have a disproportionate effect on the Community reference price.

Today, the EC is the most important importer of beef in the world, importing 47 percent of all beef traded in 1972. Most of the beef exported by the EC is traded between member countries. For example, France uses hindquarters of domestic cattle for steaks and roasts and exports fores to Germany and Belgium (for sausage) and to the United Kingdom, then imports hinds from these countries to supplement domestic production, and, in addition, buys boneless beef cuts from South America.

Cattle and Beef in the Global Perspective

Rising incomes around the world are making for higher standards of living and increased demands for the good things of life, including animal protein—and to most people this means beef. This, along with more people to feed and constraints in expanding cattle production, is causing supply to lag behind demand. Hence, we are witnessing world beef shortages and favorable prices—a situation which will probably prevail to the year 2000 A.D.—and perhaps beyond.

WORLD BEEF DEMAND AHEAD

Over a billion people—one-third of the world population—are seriously short of calories in their diets; two-thirds of the world population lack sufficient protein. In recent years, per capita real incomes have risen in almost every nation on the globe. More calories—merely satisfying hunger—have first call on income. Above this point, a part of this income is spent for increased quantities of animal protein. For this reason, it's the developed countries that are demanding more beef.

●*Japan*—The Japanese are turning away from their traditional fish diet, not only because of greater affluence, but also because of the mercury-contaminated fish scare. It is expected that Japan will continue to be a growing market for beef and veal, particularly from Australia and New Zealand, although nine other countries imported larger quantities of beef than Japan in 1971.

●*European Community (EC)*—The EC is the most important importer of beef, importing 47% of all beef traded in 1972. It is estimated that the real income in EC countries will increase about 4% yearly to 1985, and that the demand for meat will rise from 3 to 6% each year. Domestic production can support a rise of 2.2% of beef and veal. Increased beef production is being accomplished by shifting away from traditional calf and veal slaughter to older cattle, by feeding bulls, and by confinement finishing. However, the EC will always be a major beef importer, relying on South America, Australia, and New Zealand to fill the gap between its domestic supply and its demand.

● *The Soviet Union*—At the present time, the Soviet Union is attempting to develop confinement cattle feeding. To this end, they have launched a crash program to increase their grain production. Additionally, they have imported U.S. grains and technology. In time, they may decide that it is more economical to import meat than grain. More likely, they will do part of each—increase their cattle feeding and import a limited amount of grain, and import some meat.

● *United States*—The United States is the world's largest single nation beef importer (the EC embraces 9 nations), with 1.7 billion lb, or 31.6 percent of the world total, imported in 1971. It is predicted that U.S. beef consumption will rise to 140 lb per capita by 1985, making for a total requirement of 34.5 billion lb, or 61.2 percent more than we produced in 1973. Only 2 billion lb of this will be imported, unless we relax our import barriers and a remarkable change takes place in the world flow patterns.

It should be noted that U.S. beef production cannot respond quickly to demand and price changes in the years ahead. Seventy-five percent of slaughter cattle are now grain fed; thus, few more cattle are suitable for grain feeding. Calf slaughter has been reduced greatly, and dairy cow numbers have about bottomed out. Hence, for the most part, increased beef in the United States must come from more beef cows kept to produce more feeder calves. Cattlemen can produce more beef, provided it makes money. But remember that people do those things which are most profitable to them; and cattlemen are people. Remember, too, that there is a time lag in responding to prices—a fact that too many consumers fail to realize. For example, a heifer cannot be bred until she is about 1½ years of age, the pregnancy period takes another 9 months, and, finally, the young are usually grown 6 to 12 months before being sold to cattle feeders, who finish them from 4 months to a year. Thus, under the most favorable conditions, this manufacturing process, which is under biological control and cannot be speeded up, requires about 4 years in which to produce a new generation of market cattle.

MEETING DEMAND

As the world demand for beef increases, the major question is: How will this demand be met?

• *Developing nations*—Right off, it would appear that there is great potential for increasing beef production in many of the developing nations—in Africa, Latin America, and the Far East. Most of them have a surplus of forages. Also, they desperately need more high quality protein in the diet, and they need to improve their income. However, there are several bottlenecks which restrict fast increases in these areas; among them: (1) lack of infrastructure—railroads, roads, marketing facilities, slaughterhouses, and refrigeration; (2) periodic droughts; (3) diseases and parasites which either take huge tolls of the animal population and/or restrict expansion; (4) lack of soil testing facilities, irrigation, and fertilizers; (5) inadequate agricultural extension service and coordination between teaching, research, and extension; (6) insufficient transportation, storage (including refrigeration), and markets; and (7) insufficient investment capital.

It is difficult to make rapid increases in beef production in the developing countries. Due to the extremely low fertility and late maturity of the cattle in many of these countries, they are forced to slaughter some of the heifers in order to maintain an adequate offtake rate to meet demand rather than being able to keep these heifers to build up their cattle population. In the developing countries, the main efforts to increase beef production need to be directed so as to accomplish the following: (1) to increase the fertility of the breeding animals; (2) to decrease death losses; (3) to decrease the age at slaughter; and (4) to increase the average carcass weight. This calls for improvement in nutrition (better pastures, especially during dry seasons, and/or supplementation), management, health, and selection. Such improvements require both capital and knowledgeable personnel, both of which are generally in short supply in the developing nations.

•*Developed nations*—Quick mobilization for increased beef production can come about in the developed countries more readily than in the developing countries because they have the necessary infrastructure (transportation, marketing, slaughtering, and refrigeration), the capital for investment, and adequate-sized breeding herds, and the management support and knowledge necessary for bettering production. An example of this mobilization is the dramatic growth in cattle feeding in the United States. In 1947, only about 3.6 million head of cattle were grain fed, representing 35 percent of the slaughter cattle that year. In 1972, 26.8 million head of fed cattle were marketed, representing 75 percent of all cattle slaughtered that year.

•*Where will the beef come from?*—Perhaps, in the final analysis, increased beef will come from both the developing and the developed nations. Among the developing nations having high potential for increasing beef production are Argentina, Brazil, Uruguay, Paraguay, Colombia, and Mexico. However, it must be remembered that, due to the presence of foot-and-mouth disease in each of these countries, the major importing countries do not wish to take their beef unless it is deboned, cooked, or pasteurized. Consequently, the countries which have an immediate advantage of increasing exports are Australia, New Zealand, the Central American countries, Mexico, Ireland, the United States, and Canada.

WORLD PROTEIN SITUATION

Diogenes, the Greek philosopher, when asked about the proper time to eat, said: "If a rich man, when you will; if a poor man, when you can." This statement is particularly applicable to protein foods. Well-fed nations are scrambling to eat better, and hungry nations to eat at all.

Protein malnutrition is the most serious and common cause of infant mortality and general debility in developing countries, and among the poor in developed countries. The diseases kwashiorkor (primarily protein deficiency) and marasmus (severe undernourishment) directly or indirectly account for 3 to 10 times greater infant mortality and as much as 20 to 50 times higher death rate among 1- to 4-year-old children in certain African countries than in the industrialized regions of the world. Children below 5 years of age may account for 40 percent of the total mortality. The low protein reserves of the body and generally poor nutritional status cannot sustain those children who develop fevers from respiratory and gastrointestinal infections, which markedly increase the body losses of protein.

Food and Agriculture Organization (FAO) figures show that there are sufficient food supplies in the world to furnish 68 grams of protein per person per day, with 70% of it coming from plant foods and 30% from animal sources. This is enough to meet minimum needs if it were properly distributed according to individual needs of each person. But food supplies of different countries vary widely. In the United States, 93 grams of protein are available per capita daily, 70% of which comes from meat, milk, and eggs. By contrast, Nigeria has 59 grams of protein, India 50 grams, and Bolivia only 46 grams; and 74 to 91% of their protein is from plant sources. Not only that, these same countries are short on energy-rich foods, especially during the dry season, because of low production and limited food storage facilities.

Also, it is noteworthy that people in America are taller than their grandfathers. Perhaps the explanation lies in the 10-gram increase in per capita protein consumption which occurred between 1925 and 1950.

Proteins from animal sources (meat, milk, and eggs) have a higher biological value than plant sources because they contain more of the essential amino acids (protein building blocks) needed for growth. They are rich in the amino acids, lysine and methionine, in which vegetable sources are deficient. Also, they contain the essential vitamin B_{12}. Thus, the addition of even a small amount of animal protein will greatly improve the value of a diet rich in cereal grains, beans, or other plant foods. Soybean protein also has a high value, provided it is properly heat treated to improve its availability. But no soybean platter, no matter how well disguised, can ever inspire the toast or impart the status symbol of a roast of beef or a sizzling steak.

BEEF CATTLE AS A SOURCE OF PROTEIN

The primary reason for producing beef cattle is to obtain human food. Beef is good, and good for you. In addition to supplying the highest quality protein (along with milk and eggs), it is a rich source of the B vitamins, whereas some plant foods are deficient in these factors. Vitamin B_{12} does not occur in plant foods; only in animal sources and fermentation products. Beef is also a particularly rich source of iron, and the iron availability of beef is twice as high as in plants.

Some opinions to the contrary, there is as much protein in Prime beef as in the leanest piece. When finishing animals, fat merely replaces the moisture, without materially altering the protein content.

There has been much discussion about the efficiency of animals in converting feeds to food and their place in the economical production of human foods. Some plant scientists, economists, and others, claim that food needs should be met by plant sources entirely—that animals should be eliminated. This might be desirable if animals subsisted entirely on cereal grains and other edible foods, or if animals and feeds were grown on land needed to produce foods for human consumption. However, much of the land of the world is unsuited to the production of food crops. In temperate climates, this nonarable land is used to grow cattle, sheep, and goats. These ruminants can eat pasture and coarse roughages that have no value except as animal feed. Thus, they contribute high quality protein (meat), which is greatly prized for its nutritive qualities, plus its flavor and appetite appeal.

Cattle produce most of the meat consumed throughout the world.

NEW PROTEIN SOURCES

New protein sources are being used, and others will be discovered. However, people like beef, so they will pay more for it than for plant food, or even other kinds of meats. There is an old saying in the southwestern part of the United States that, "thin beef is better than fat beans." Nevertheless, there is no known nutritive essential in beef which cannot be provided from fish, vegetable sources, or by synthesis.

• *Fish protein*—The present world fish catch runs about 68 million metric tons per year, with an average protein content of about 15%. Fish fillets contain over 20% protein, hence, they are one of the richest sources of protein. Each year, some 25 million metric tons of fish are processed into fish meal for animal feed. With improved handling and processing techniques, some of this fish could be converted into fish protein concentrate (FPC) suitable for human consumption. FPC has great potential as an inexpensive source of high quality protein. It is produced from types of fish that are not popular in the usual channels of fresh fish trade. The fish are extracted to remove oil, dried, and ground to make a bland meal containing about 80% protein, 0.2% fat, and 13% mineral.

• *Soybean protein*—The protein of soybean, after suitable processing, has the highest nutritive value of any plant protein source. Foods made from soybeans include soy flour, soy milk, spun fiber, soy sauce, tofu and tempeh (fermented cheese and curd), and soy butter. Also, the soybeans may be cooked as green or dried beans, canned in sauce, or roasted (like peanuts).

Soy flour may be used in many baked products. Some six to eight percent of soy flour added to wheat flour used for bread and pastries will significantly improve the protein value of the product without making much change in texture and appearance.

Soy milk has been consumed for centuries in China and in some other countries of the Far East. A 200 ml bottle of soy milk will supply at least 6 grams of high quality protein, 50 percent of a child's daily requirement. Soy milk is made by soaking soybeans 4 to 5 hours, grinding them in a hot water slurry (1 part of soaked beans to 3 parts of water), then straining out the insoluble residue.

Isolated soybean protein can be spun into fibers, flavored, colored, and fabricated into meat-like products, including beef steaks, chicken, pork chops, lamb chops, ham, bacon, and sausage—all difficult to distinguish from the real products.

• *Amino acid fortification of food*—Experiments have shown improvement in growth and nitrogen retention of animals and children when diets high in cereals are supplemented with lysine, tryptophan, and methionine. These amino acids can be produced by the chemical industry in quantities suitable for supplementation of selected food items.

• *Single-cell protein (SCP)*—Some species of yeast, algae, and bacteria can be useful sources of protein and vitamins for human and animal feeding. The safety of these foods depends on the organisms selected, the quality of substrats used, and the conditions of growth. Of course, yeast and bacteria have been used for centuries in the baking, brewing, and distilling industries, in making cheese and other fermented foods, and in the storage and preservation of foods.

Dried brewers' yeast, a residue from the brewing industry, and Torula yeast, resulting from the fermentation of wood residues and other cellulose sources, have been marketed as animal feeds for years. With proper processing, they are also suitable for human foods.

Bacteria grow faster than yeasts, doubling their mass in a matter of minutes, rather than hours, under favorable conditions. Dried bacterial cells contain at least 55 percent protein.

Various bacteria and yeasts can be selected and cultured to grow on organic wastes. These include animal wastes, sewage, many different chemical residues from industrial plants, petroleum by-products, sawdust, and other fibrous residues. Petroleum companies in several countries are building factories to produce bacterial protein for the animal feed market. Also, considerable research is in progress to convert manure from poultry and other animals through bacterial fermentation into animal protein feed. This recycling process could produce much protein and help solve a pollution problem.

Algae are single-cell plants which may contain 20 to 60 percent protein on a dry basis. In Northern Nigeria, algae are dried and eaten for human food. Algae grow widely on the earth's water surfaces, but problems of harvesting and processing them into acceptable food products remain unsolved.

UNITED STATES CATTLE AND BEEF

It is important that U.S. producers and beef consumers consider cattle and beef in the global perspective (Table 2-10). Cattlemen need to know which countries are potential competitors and what's ahead—to 1985 and beyond. Beef consumers need to count their blessings.

Table 2-10 reveals that beef and veal demand will exceed production in 1985, by 1.5 million metric tons. This margin of deficit is small in relation to overall production and demand. Nevertheless, this would indicate that, despite some ups and downs, the beef industry is in a favorable position, on a worldwide basis and for many years to come.

Table 2-10 also shows that the United States—
1. Has 5.5% of the world's human population.
2. Has 9.7% of the world's cattle population.
3. Produces 24.2% of the world's beef and veal.
4. Consumes 26.4% of the world's beef and veal.

Think of it! Nine and seven-tenths percent of the world's cattle population producing a fantastic 24.2 percent of the world's beef! This points up the tremendous efficiency of the U.S. cattleman. The pounds of beef produced per year per inventory head of cattle offers a means of comparing the efficiency of cattlemen between countries. As shown in Fig. 2-2, the American cattleman ranks first.

Fig. 2-20 shows that U.S. cattle have become more efficient through the years. During the base period 1930-39 (Indexes, 1930-39 = 100), the nation had an average of 66.9 million head of cattle which produced 7.7 billion pounds of beef. Subsequently, beef and veal production rose faster than cattle numbers, with the gap widening each year. By 1974, cattle and calves on farms had increased by 191 percent over the 1930-39 base period while beef and veal production jumped by 278 percent over the base period. Without doubt, much of this increased beef production on a per head basis has come about as a result of increased cattle feeding and decreased calf slaughter. In

TABLE 2-10

UNITED STATES AND WORLD HUMAN AND CATTLE POPULATION
AND BEEF AND VEAL PRODUCTION AND CONSUMPTION

	Human Population	Cattle Population	Beef and Veal Production	Beef and Veal Consumption 1972-73, and Projected Demand 1985
	(millions)	(millions)	(1,000 metric tons)	(1,000 metric tons)
United States:				
Year as indicated	(1972) 209.2[1]	(1974) 127.5[2]	(1973) 9,787.7[6]	(1973) 10,692.4[9]
1985	246.3[1]	177.6[3]	15,649.1[7]	15,649.1[10]
World:				
Year as indicated	(1972) 3,782.0[1]	(1974) 1,311.5[4]	(1970-72 av.) 40,520.0[8]	(1970-72 av.) 40,520.0[8]
1985	4,933.0[1]	1,630.7[5]	61,008.0[8]	62,524.9[11]

[1]1972 World Population Data Sheet, Population Reference Bureau, Inc., Washington, D.C.
[2]Cattle, LvGb 1 (2-74), USDA, Statistical Reporting Service, Crop Reporting Board, February 1, 1974, p. 3.
[3]Projections made by author. (See Chapter 17, Table 17-1, of this book.)
[4]Foreign Agriculture Circular, Livestock and Meat, FLM 7-74, USDA, Foreign Agricultural Service, July 1974, p. 9.
[5]1985 projection made by author, based on estimated 2% yearly world increase.
[6]Foreign Agriculture Circular, Livestock and Meat, FLM 8-74, USDA, Foreign Agricultural Service, June 1974, p. 10.
[7]Projections made by the author. (See narrative in conjunction to Table 17-1.)
[8]FAO estimate, provided with a personal communication to author from Dr. R. S. Temple, FAO, United Nations, Rome.
[9]Based on 209.2 million people consuming 112.7 lb beef per capita. (See Table 17-1.)
[10]Based on author's projection of 140 lb per capita beef consumption in the United States in 1985, and 246.3 million people.
[11]1985 projections made by author, based on trend factors from Current Status of Livestock and Meat in World Commerce, Meeting the Challenge of World Commerce, Livestock Marketing Congress '73, Livestock Merchandising Institute, Kansas City, Missouri, p. 5.
    The United Nations projects (United Nations World Food Conference, Rome, Nov. 5-16, 1974, Table 22, p. 93) a deficit of 1.0 million metric tons of beef and veal in 1985, in comparison with the projected deficit of 1.5 million metric tons shown in Table 2-10.

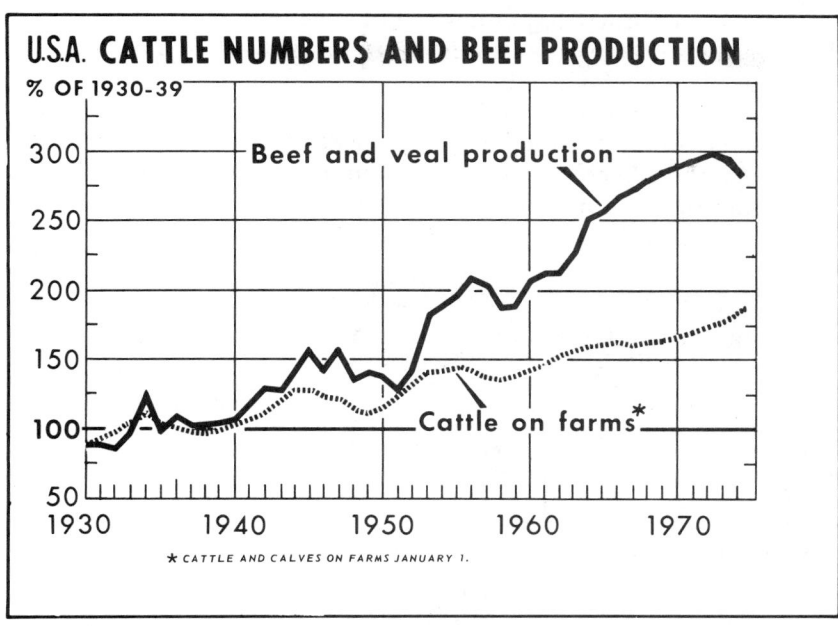

U.S.A. CATTLE NUMBERS AND BEEF PRODUCTION

Fig. 2-20. Beef production has gone up faster than cattle numbers, which shows that cattle have become more productive and efficient through the years. (Source: 1930-1973 data from *1973 Handbook of Agricultural Charts*, Agricultural Handbook No. 455, USDA, p. 75. 1974 figures computed by author based on data from *Cattle*, LvGb 1 (2-74), USDA, Statistical Reporting Service, Crop Reporting Board, February 1, 1974, p. 3, and from *Livestock Slaughter*, MtAn 1-2(1-74), USDA, Statistical Reporting Service, Crop Reporting Board, January 29, 1974, p. 2)

1947, we grain fed 3.6 million cattle; in 1972, 26.8 million head were grain fed.

But the United States is declining in its relative importance with regard to human population, cattle numbers, and beef and veal production and consumption. Other countries of the world are growing at a more rapid rate with regard to human population. Also, their per capita disposable income is rising, and with it have come newly found affluent life-styles and demand for more beef. As a result, people all over the world are demanding more beef, and they're able and willing to pay for it. Not only are beef prices in many countries higher than in the United States, but they have gone up faster in most countries. These forces will continue to operate at an accelerated rate in the years ahead. Translated into reality, world beef shortages in the years ahead mean decreased supplies and increased prices. But cattlemen throughout the world are demonstrating their willingness to respond to prices. As never before, they are applying science and technology; in genetics, in management, and in the use of humanly inedible feeds and nonprotein nitrogen.

U.S. Cattle Production

The present and future importance of beef cattle in the agriculture of the United States rests chiefly upon their ability to convert coarse forage, grass, and by-product feeds, along with a minimum of grain, into a palatable and nutritious food for human consumption. In 1973, 77 percent of U.S. beef production came from grain-fed beef, 23 percent came from nongrain-fed (grass) beef. As grain becomes scarcer and higher in price (due to increased human consumption), there will be less grain-fed beef and more short-fed cattle.

TABLE 2-11

NUMBER AND VALUE (VALUE/HEAD AND TOTAL VALUE) OF CATTLE IN UNITED STATES, JANUARY 1, 1974[1]

Class	Number	Farm Value	
		Total Value	Value
		($1,000)	(dollars/head)
All cattle	127,540,000	40,905,700	321.00
Cows and heifers that have calved:			
Beef cows	42,874,000		
Milk cows	11,284,000		

[1]*Cattle*, LvGb 1 (2-74), USDA, Statistical Reporting Service, Crop Reporting Board, February 1, 1974, pp. 5-7.

Table 2-11 shows the number and value of all cattle in the United States in 1974. As noted, there were 127,540,000 head on the nation's farms and ranches or in feedlots, of which 42,874,000 were beef cows that had calved and 11,284,000 were milk cows that had calved. Fig. 2-21 gives a breakdown of cattle into beef and dairy and shows the longtime trends of each. In 1950, there were a total of 77,963,000 cattle in the United States, of which 42,508,000 (or 55%) were classed as beef cattle and 35,455,000 (or 45%) were classed as dairy animals. Of the cows and heifers that had calved in 1974 (Table 2-11), 79% were beef cows and 21% were dairy cows. Hence, in the 25 year period, 1950 to 1974, beef cattle increased by 24% while dairy cattle decreased by 24%.

The production of beef cattle differs from that of most other classes of livestock in that the operation is frequently a two-phase proposition: (1) the production of stockers and feeders; and (2) the finishing of cattle. In general, each of these phases is distinctive to certain areas.

COW-CALF PRODUCTION

Cow-calf production refers to the breeding of cows and the raising of calves. In this system, the calves run with their dams, usually on pasture, until they are weaned, and the cows are not milked.

Fig. 2-22 shows the geographic location of the nation's beef brood cows, including (1) cows added in 1973, and (2) 1974 totals.

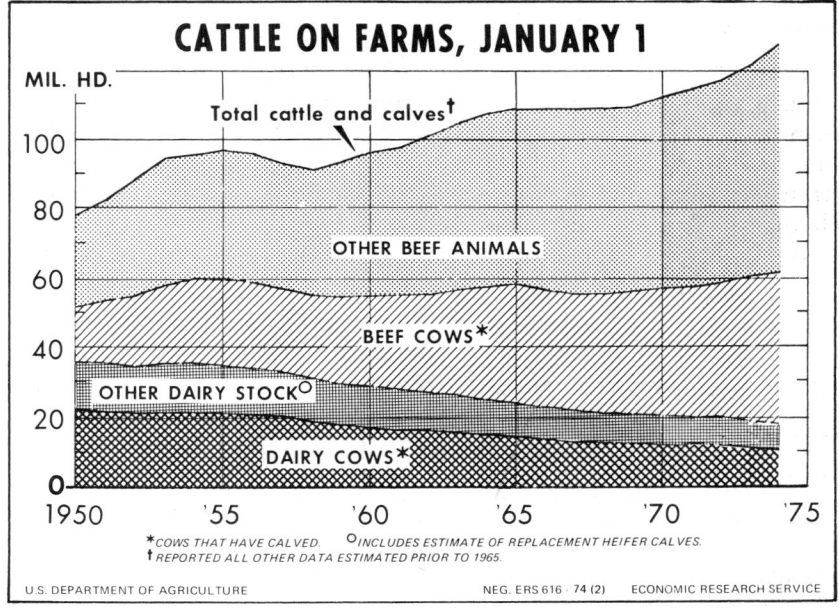

Fig. 2-21. Cattle on farms, by classes, 1950-1974. (Courtesy, U.S. Department of Agriculture)

GEOGRAPHIC LOCATION OF U. S. A. BEEF BROOD COWS

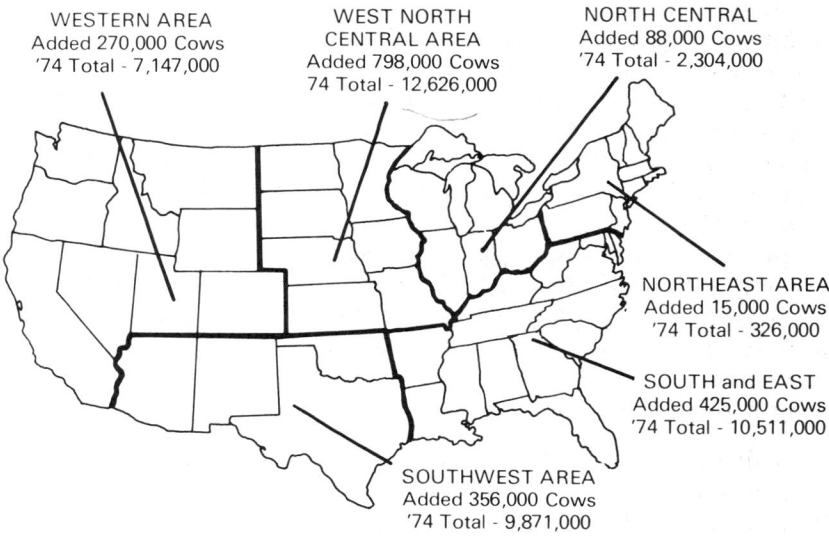

WESTERN AREA
Added 270,000 Cows
'74 Total - 7,147,000

WEST NORTH
CENTRAL AREA
Added 798,000 Cows
74 Total - 12,626,000

NORTH CENTRAL
Added 88,000 Cows
'74 Total - 2,304,000

NORTHEAST AREA
Added 15,000 Cows
'74 Total - 326,000

SOUTH and EAST
Added 425,000 Cows
'74 Total - 10,511,000

SOUTHWEST AREA
Added 356,000 Cows
'74 Total - 9,871,000

Fig. 2-22. The six areas of U.S. beef production. (Source: U.S. Department of Agriculture)

Some 54 percent of the U.S. beef cows are in the West North Central area (Missouri, Kansas, Nebraska, Iowa, South Dakota, North Dakota, and Minnesota) and the South and East (from Maryland, West Virginia, and Kentucky south and westward to and including Arkansas and Louisiana).

The greatest expansion since 1950 has been in the West North Central area, and in the South and East. Since 1950, there has been a 56 percent increase in beef cow numbers in the West North Central area, and a 54 percent increase in the South and East.

There are some rather characteristic production practices common to each cattle area. Some of these will follow.

● *West North Central area*—This 7-state area, which accounted for 29.5 percent of the nation's beef cows in 1974, embraces the 4 most westerly Corn Belt states (Missouri, Iowa, Kansas, and Nebraska) plus Minnesota and North and South Dakota. In many respects, it is a variable area. Iowa, north Missouri, eastern Kansas and South Dakota, and southern Minnesota are noted for fertile soil, medium-sized farms, high priced land, and corn. Here hogs compete with cattle and other classes of livestock for the available feeds. The western portion of the area—commonly referred to as the Plains area—is predominantly range and wheat land. Beef cow numbers in the Northern Plains area (Nebraska, and North and South Dakota) more than doubled during the period 1950-70, with the greatest increase in the humid, eastern portion of the area. Included in the West North Central area are two famous pasture areas—the Sand Hills of Nebraska, and the Flint Hills of Kansas.

Much of the growth in the West North Central area has come through (1) converting more marginal cropland to pasture, and (2) more effective utilization of relatively low quality roughages, like cornstalks, to winter feed. It is expected that this shift in land use from grain to forage production will continue, but the extent of this shift will depend upon the relative profitability of grain crops vs beef cattle.

Many of the farmers and ranchers of the West North Central area maintain small commercial breeding herds, the offspring of which are sold as stockers or feeders, or finished out on homegrown feeds. Pastures furnish practically all the feed for the breeding herds from May to November, and cornstalks and other roughages are utilized as winter feeds.

In addition to keeping small breeding units, many West North Central farmers make a regular practice of buying feeder cattle from the western ranges. These cattle are usually purchased in the fall of the year and obtained at auctions, through public stockyards, through dealers, or direct from the range. Some of these are roughed through the first winter by utilizing stubble or stalk fields and then pastured the next season. Others are carried on a program of winter feeding, then sold in the spring to go into feedlots.

The West North Central area is also noted for its excellent purebred herds. Because of its proximity to the western ranges and the demands of the ranch owners for bulls, there has always been a good market for superior breeding stock.

● *South and East area*—The South and East, with 10.5 million head, or 24.5 percent, of U.S. beef cows, is the second most populous beef cattle area.

Mild winters, adequate rainfall, and year-round grazing have encouraged many of the farmers of the area to turn to beef production. Other factors that have contributed greatly to the expansion of beef cattle in the area are: (1) the infusion of Brahman breeding, resulting in cattle better adapted to high summer temperatures and to resist the insects of the area; (2) improved pastures, suitable supplements for mineral deficiencies, and control of parasites; (3) adoption of better management practices; (4) increase in size of farms, with larger pasture acreages per farm; (5) increase in part-time farmers, who frequently choose beef cow enterprises because of the relatively small and flexible labor needs; and (6) more market outlets for feeder cattle to go into large finishing lots. All these improved conditions, together with year-round grazing, would indicate that the South and East area offers the greatest potential for future increase of beef cows of any area of the United States, and that the South and East will continue to expand in beef production on a sound basis. Some authorities estimate that this area could support 30 million beef cows—nearly three times present numbers, provided (1) current land were properly managed, and (2) idle and unproductive acres were put to work.

Factors that may have a restraining influence on beef cattle growth in the South and East are high cotton and soybean prices and shifts of land to nonfarm uses.

●Southwest area—The highest and best use for much of the land in the Southwest is for grazing. Hence, it has been a noted cow-calf area since the days of the Texas Longhorn. Except for some areas under irrigation, the units of operation are generally very large in size. In recent years, a noted cattle feeding area has developed in the High Plains area of Texas and Oklahoma. Future rate of expansion in beef cattle in the Southwest will be limited because it is already a well-stocked and highly specialized cattle area. Modest increases in cow-calf numbers will come from improvement in pasture production, and some increase in cattle feeding may be expected.

●Western area—The Western area, with 16.7 percent of U.S. beef cattle, ranks fourth in cattle numbers. It is characterized by great diversity of topography, soil, rainfall, and temperature. Accordingly, the amount of vegetation and the resulting carrying capacity are variable factors. Combinations of private and public grazing land often prevail.

In general, the western ranges supply an abundance of cheap grass, but only a limited amount of grain. Under these conditions, the cow-calf system is the dominant type of enterprise.

It is expected that beef cow numbers in the mountain region will expand at about the same rate as for the nation as a whole. Little change is expected in the arid portions of the West, because of current full utilization of existing forages. Likewise, little expansion of beef cattle numbers may be expected in the Pacific region because of competing land uses and rangeland limitations.

●North Central area—The North Central region accounts for only 5.4 percent of the nation's beef cows, despite the fact that it embraces the three Corn Belt states of Illinois, Indiana, and Ohio. The area does have considerable production potential for grass-legume forages; and it is well adapted to beef production. However, so long as dairying remains profitable in the Lakes States, shifts to beef will be minimal. Likewise, the most fertile soils of Illinois, Indiana, and Ohio will continue to be devoted to crops which yield greater economic returns, primarily corn and soybeans, with only the marginal cropland left for grass and beef cattle. However, it is expected that cornstalks will be more effectively utilized in the future.

● *Northeast area*—The production of beef cattle was formerly an important and highly developed industry through the Northeast. However, the opening up of the western ranges; the rapid increase in population of the East, with the resulting industrialization; the division of the eastern farming lands into smaller units; and the adoption of more intensive systems of farming caused a decline of beef production in the eastern states. With these changes in economic conditions, beef production was to a large extent supplanted by dairying. With the existence of these favorable conditions for milk production, beef cattle cannot compete with dairy cattle, particularly on small farms. On the other hand, where the distance to market is too great, or where labor difficulties exist, beef cattle have a place. But the beef industry in the Northeast can never regain its former magnitude nor hold the place that it does in the rest of the United States.

Today, the Northeast area accounts for only 0.8 percent of U.S. beef cattle. Although beef cow numbers in the area have increased slightly since 1950, the rise has not been sufficient to offset the decline in milk cow numbers. The beef herds of the Northeast are small and often operated as a supplementary enterprise by part-time farmers. Hence, the area is of minor importance in beef production.

LEADING STATES IN COW-CALF PRODUCTION

A ranking of the 10 leading states in beef cattle production, together with total numbers for the United States, is given in Table 2-12. As noted,

TABLE 2-12

TEN LEADING COW-CALF STATES, BY RANK, 1974

State	Beef Cows and Heifers That Have Calved[1]	Beef Heifer Replacements[2]
	(1,000 head)	*(1,000 head)*
Texas	6,470	1,115
Missouri	2,594	500
Oklahoma	2,379	460
Nebraska	2,248	415
South Dakota	2,058	345
Kansas	2,050	350
Iowa	1,790	268
Montana	1,746	355
Mississippi	1,285	261
Florida	1,282	224
Total United States	42,874	8,214

[1]*Cattle*, LvGb 1 (2-74), USDA, Statistical Reporting Service, Crop Reporting Board, February 1, 1974, p. 6.
[2]*Cattle*, LvGb 1 (2-74), USDA, Statistical Reporting Service, Crop Reporting Board, February 1, 1974, p. 8.

Texas is far in the lead. The large cattle numbers in the state of Texas may be attributed to the fact that it represents a truly great range-cattle country, to the immense size of the state, and to increased cattle finishing. Five of the top 10 states (Missouri, Nebraska, South Dakota, Kansas, and Iowa) are in the West North Central area, and two (Mississippi and Florida) are in the South and East. Montana, which ranks eighth, is a great range cattle state.

CATTLE FEEDING AREAS

The center of cattle feeding has shifted from the Corn Belt to the West and Southwest. In 1972, 70 percent of the fed cattle marketed in the 23 major producing states were fed in the 14 western states (Texas to North Dakota and westward). In the decade 1962-72, the greatest change occurred in Texas where 3.55 million more head were fed in 1972 than 10 years earlier, bringing the Texas total for 1972 to 4.31 million head. Colorado and California ranked fifth and sixth nationally in 1972, with 2.29 million and 2.06 million, respectively. (See Fig. 2-23.) Three Corn Belt states ranked second, third, and fourth, respectively; namely, Nebraska with 4.0 million, Iowa with 3.9 million, and Kansas with 2.4 million.

LEADING CATTLE FEEDING STATES 1972

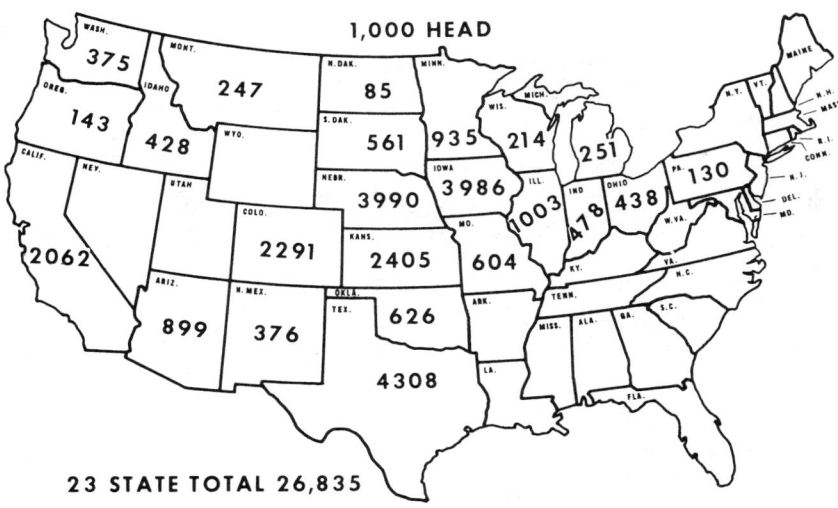

Fig. 2-23. Leading cattle feeding states, based on fed cattle marketings of the 23 leading feeding states in 1972. Note that the six leading states were: Texas, 1st; Nebraska, 2nd; Iowa, 3rd; Kansas, 4th; Colorado, 5th; and California, 6th. (Source: Using Information in Cattle Marketing Decisions, A Handbook *WEMC Pub. No. 5*, Feb., 1973, p. 37, Chart 14)

U.S. Beef Consumption

In 1973, the United States consumed 23.1 billion pounds of beef and veal, or an average of 111.1 pounds per person. The story of beef consumption is further presented in Table 2-13 and Fig. 2-24. From these, the following conclusions may be drawn:

1. Beef is the preferred red meat in the United States, having replaced pork in this position in the early 1950s.

2. The longtime trend in beef consumption is upward.

3. Beef consumption in 1973 fell in response to high beef prices created by a scarcity of beef following earlier beef controls.

TABLE 2-13

U.S. PER CAPITA MEAT CONSUMPTION[1]

Year	All Meats	Beef	Veal	Percent Beef and Veal of All Meats
	(lb)	*(lb)*	*(lb)*	*(%)*
1969	182.5	110.8	3.3	62.52
1970	186.3	113.7	2.9	62.59
1971	191.8	113.0	2.7	60.32
1972	188.9	116.0	2.2	62.57
1973	175.0	109.3	1.8	63.49

[1]1969-1972 figures from *Agricultural Statistics 1973*, USDA, Washington , D.C., Table 511, p. 349. 1973 figures from *National Food Situation*, NFS-147, USDA, Economic Research Service, February 1974, Table 6, p. 16.

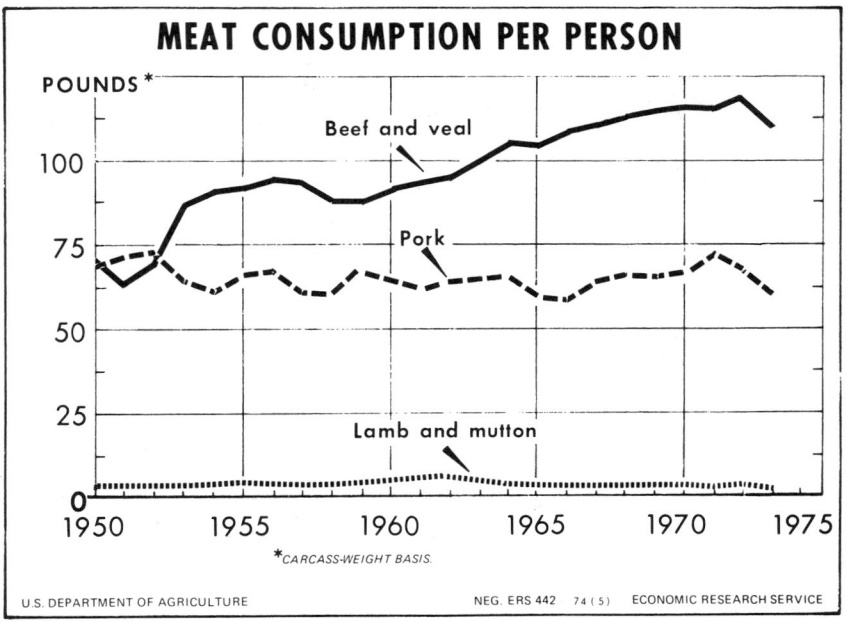

Fig. 2-24. Per capita meat consumption in the United States, by kind of meat, 1950-73. As noted, the amount of meat consumed in this country varies from year to year. Since the early 1950s, the average American has consumed more beef than any other kind of meat. (Courtesy, U.S. Department of Agriculture)

U.S. Beef Exports

Cattlemen are prone to ask why the United States, with a cattle population second only to India, buys cattle, and beef and veal, abroad. Conversely,
consumers sometimes wonder why we export beef and veal. Occasionally,
there is justification for such fears, on a temporary basis and in certain areas.

As shown in Table 2-14, (1) exports of beef have been negligible, and (2)
we have imported far more beef than we exported.

The amount of beef exported from this country is dependent upon: (1)
the volume of beef (and other meats) produced in the United States; (2) the
volume of beef (and other meats) produced abroad; and (3) the price of and
trade restrictions on beef abroad. With increased buying power abroad and
higher prices for beef than in the United States, it is probable that more beef
will be exported in the future.

Our exports of animals and animal products have consisted largely of
those by-products which we did not wish to use in the United States (Fig.
2-25). Normally, they were low value commodities; hence, from the
standpoint of dollar sales, they did not rank high.

As shown in Fig. 2-25, our exports of livestock products have consisted
primarily of lard and tallow, with hides and skins next in importance, followed by red meats, variety meats, live animals, and other livestock products
such as wool and mohair.

U.S.A. EXPORTS OF LIVESTOCK PRODUCTS

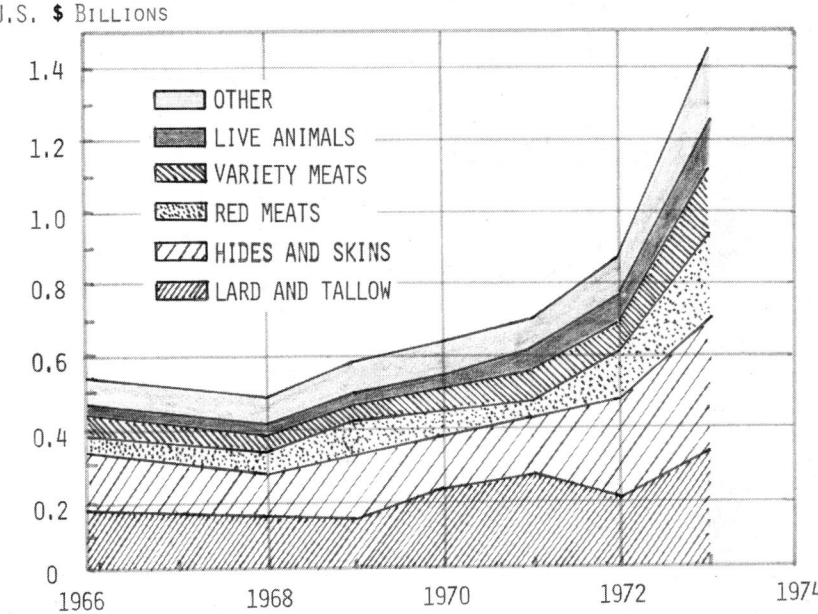

Fig. 2-25. U.S. exports of animal by-products and live animals, 1966-72.
(Source: *1973 Handbook of Agricultural Charts*, Agricultural Handbook No. 455,
USDA, p. 78)

TABLE 2-14

UNITED STATES IMPORTS AND EXPORTS OF BEEF AND OTHER MEATS[1]

Year	Imports			Exports		
	Beef and Veal	All Meats	Percent of Beef and Veal of All Meats	Beef and Veal	All Meats	Percent of Beef and Veal of All Meats
	(mil. lb)	(mil. lb)	(%)	(mil. lb)	(mil. lb)	(%)
1963	1,677	2,047	81.92	54	263	20.53
1964	1,085	1,432	75.77	96	315	30.48
1965	942	1,347	69.93	97	231	41.99
1966	1,204	1,721	69.96	88	233	37.77
1967	1,328	1,841	72.13	94	246	38.21
1968	1,518	2,081	72.95	94	288	32.64
1969	1,640	2,202	74.48	87	329	26.44
1970	1,816	2,387	76.08	104	288	36.11
1971	1,756	2,317	75.79	121	312	38.78
1972	1,996	2,653	75.24	124	354	35.03

[1]Agricultural Statistics 1973, USDA, Washington, D.C., Table 512, p. 350. All figures in carcass-weight equivalent.

Fig. 2-26 shows the leading U.S. agricultural exports, as a percentage of farm sales, 1973.

As shown, wheat was the leading agricultural export in 1973. Nearly 80% of farm sales of wheat entered the export market that year. Rice and soybeans ranked second and third, respectively. Tallow and cattle hides were in fourth and fifth positions in exports, with hides contributing more than 50%, and tallow more than 40%, of the farm sales moving into the export market.

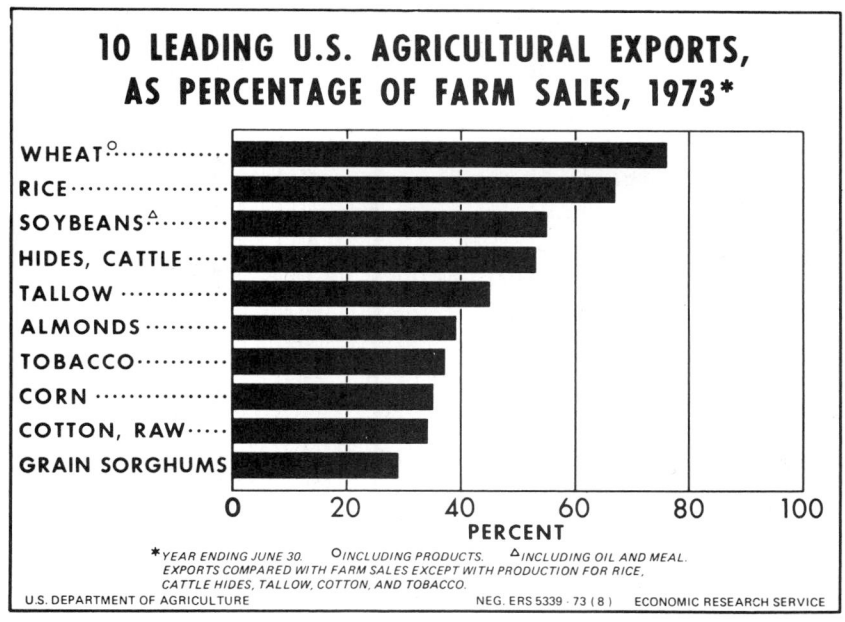

Fig. 2-26. Leading U.S. agricultural export products in 1973. (Source: *1973 Handbook of Agricultural Charts*, Agricultural Handbook No. 455, USDA, p. 49)

Although not large in dollar sales ($480.4 million in 1972), exports are extremely important to the cattle industry, even though most producers do not recognize it.

U.S. exports of live cattle, which have trended upward in recent years, totaled 106,000 head in 1972. Forty percent of these were for breeding purposes, with the principal destinations being Canada and Mexico.

U.S. Beef Imports

Table 2-14 reveals that the United States imports more beef than it exports. Table 2-15 places beef imports and production in perspective. As shown, total beef imports in the early 1970s constituted 8 to 9 percent of the available U.S. beef and veal. Fig. 2-27 shows that the longtime trend in beef and veal imports is upward.

TABLE 2-15

IMPORTS OF CATTLE, BEEF, AND VEAL
COMPARED WITH U.S. PRODUCTION

Year	Beef and Veal Imports[1]	Beef and Veal Production[2]	Beef Imports as a Percentage of Production
	(mil. lb)	(mil. lb)	(%)
1963	1,677	17,357	9.66
1964	1,085	19,442	5.58
1965	942	19,719	4.78
1966	1,204	20,604	5.84
1967	1,328	20,976	6.33
1968	1,518	21,580	7.03
1969	1,640	21,799	7.52
1970	1,816	22,240	8.17
1971	1,756	22,414	7.83
1972	1,996	22,839	8.74

[1]*Agricultural Statistics 1973*, USDA, Washington, D.C., Table 512, p. 350. All figures in carcass-weight equivalent.
[2]*Agricultural Statistics 1973*, USDA, Washington, D.C., Table 501, p. 342.

BEEF AND VEAL IMPORTS AS A PERCENTAGE
OF U.S.A. PRODUCTION

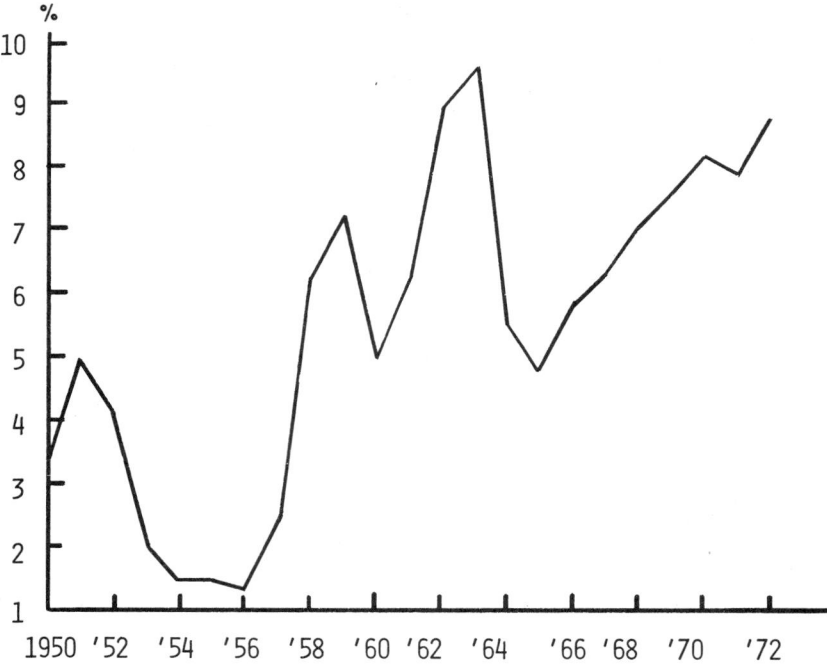

Fig. 2-27. Beef and veal imports as a percentage of U.S. production, 1950-1972.
As shown, the longtime trend is upward. Basic production and import figures are
expressed in carcass-weight equivalents. (Source: 1950-1962 figures from *Agricultural Statistics 1967*, USDA, Washington, D.C., Table 519, p. 418, and *Agricultural Statistics 1972*, USDA, Washington, D.C., Table 524, p. 419, and Table 513, p. 410.
1963-1972 figures from *Agricultural Statistics 1973*, USDA, Washington, D.C.,
Table 512, p. 350, and Table 501, p. 342. Also see Table 2-15 of this chapter)

The amount of beef imported from abroad depends to a substantial degree on (1) the level of U.S. beef (and meat) production, (2) consumer buying power, (3) cattle prices, (4) quotas and tariffs, and (5) need for manufacturing type beef.

Fig. 2-28 shows that beef and veal dominate beef imports. In 1972, the United States imported 1,996 million lb of beef and veal in comparison with 509 million lb of pork, 111 million lb of mutton and goat, and 37 million lb of lamb. Thus, 75 percent of all meats imported consisted of beef and veal. While livestock and meat products do not rank high in exports from the United States, they rank as the leading import commodity (Fig. 2-29).

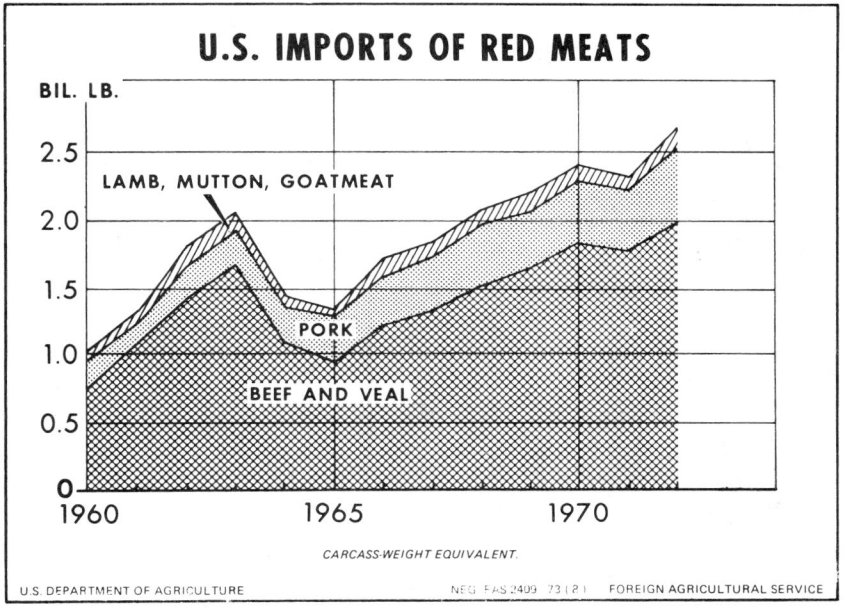

Fig. 2-28. U.S. imports of red meats. Note the dominant position of beef and veal. (Source: *1973 Handbook of Agricultural Charts*, Agricultural Handbook No. 455, USDA, p. 79)

But why are we importing so much beef and so many cattle into the United States? This question has two basic answers:

1. *To provide supplies when domestic beef is in short supply and high*—There may be some virtue in judiciously increasing imports of beef and cattle during times of scarcity and high prices, as an alternative to pricing beef out of the market. But, of course, beef imports can be overdone.

2. *To meet the demand for manufacturing meat*—Manufacturing-type beef is the kind that is boned and used in making hamburgers, franks, sausages, and bologna. The new generation of young people often prefer hamburgers to steaks. Thus, the demand for manufacturing meats began to exert itself in the early 1960s and was the primary factor in influencing the demand for imports. Hand in hand with this increase in demand, the domestic source of manufacturing meat—Utility and Canner dairy and beef cows, and

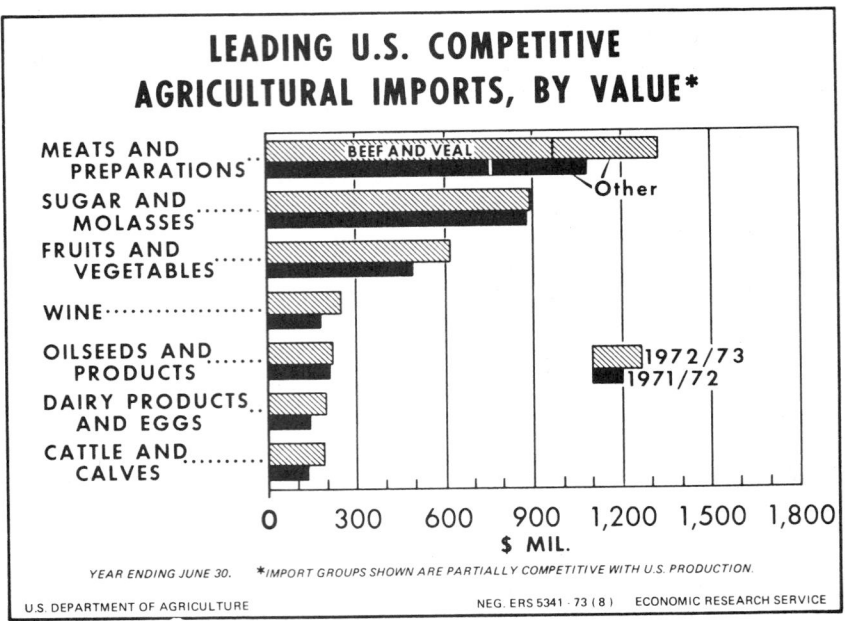

Fig. 2-29. Leading U.S. agricultural imports, by value. Note that beef and veal imports exceeded $1.3 billion in 1973. (Source: *1973 Handbook of Agricultural Charts*, Agricultural Handbook No. 455, USDA, p. 50)

older bulls—began to dwindle. Dairy cow numbers leveled off and fewer old bulls were needed in artificial insemination breeding. Per capita slaughter of cows tended downward from 40 to 50 pounds per capita (liveweight basis) during the early 1950s to around 30 pounds per capita in the late 1960s (Fig. 2-30). But steer and heifer beef production increased steadily from 1950-1971, with a slight tendency to vary cyclically. In 1962, of all the beef that was produced, only about 47% would have graded U.S. Choice. By 1971, 60% of it fell into the Choice grade (Fig. 2-31). However, Utility, Cutter, and Canner beef represented 17% of the total in 1962, but only 12% of the total in 1972.

To fill the gap for manufacturing beef, created by the increasing demand for hamburger (and other manufactured beef) and the decreasing domestic supply of Cutter and Canner cattle, the United States increased importations of frozen boneless beef, particularly from Australia and New Zealand.

3. *To provide more feeder cattle*—Beginning about 1950, the United States embarked upon an expansion program to produce more high quality beef, primarily in the Choice and Good grades. The result was a spectacular rise in grain-fed cattle; moving from 3.6 million head, representing 35 percent of the cattle slaughter that year, to 26.8 million head, representing 75 percent of the slaughter, in 1972. Simultaneously, we increased per capita beef and veal consumption, from 71.4 pounds in 1950 to 118.2 pounds in 1972. This increased the demand for feeder calves faster than could be satisfied by domestic cow-calf producers. To meet this demand, increased

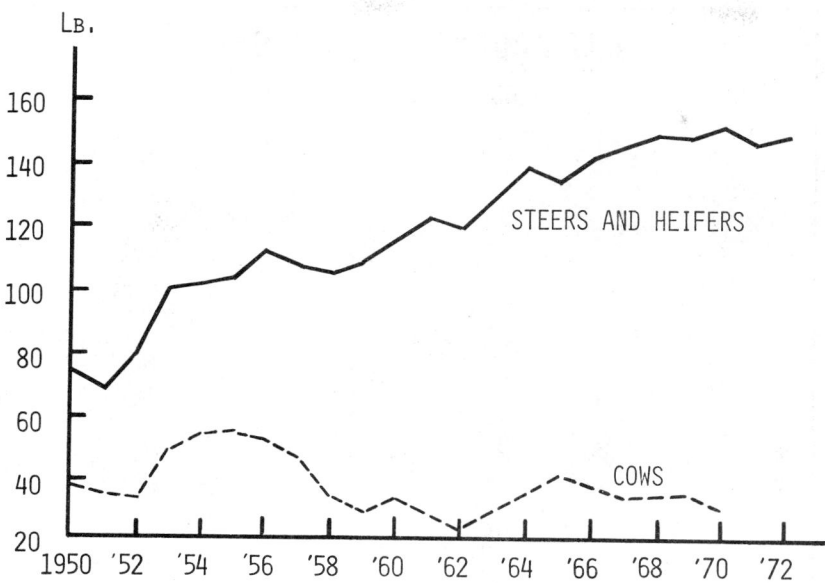

Fig. 2-30. Pounds per capita commercial slaughter of steers, heifers, and cows in U.S. Total per capita slaughter by classes was estimated by multiplying total commercial slaughter by the percentage of federally inspected slaughter that consisted of each class, multiplying by the average liveweight per animal and dividing by population. Average annual weights for live animals marketed at the Chicago terminal were used for all years except 1970-1972. Omaha weights, adjusted by historic relationship between weights at Chicago and Omaha, were used for 1970-1972. (Source: 1950-1970 figures from *Demand and Supply Functions for Beef Imports*, B 604, by Rollo L. Ehrich and Mohammad Usman, Division of Agricultural Economics, Agricultural Experiment Station, University of Wyoming, Laramie, Wyoming, January 1974, Chart 2, p. 5. 1971-1972 figures from *Livestock and Meat Statistics*, Statistical Bulletin No. 522, USDA, July 1973, Table 73, p. 96, Table 87, pp. 127-129, Table 105, p. 168; and from *Statistical Abstract of the United States 1973*, U.S. Department of Commerce, Bureau of the Census, 94th Edition, Washington, D.C., 1973, p. 5)

importations of feeder cattle were made from Mexico and Canada. In 1973, 669,200 head of cattle were imported from Mexico and 363,500 head from Canada.

U.S. beef and veal imports come from many countries. In 1972, 50% of the total came from Australia, and 20% from New Zealand. The remaining 30% came from the following countries, by rank: Mexico, Canada, Nicaragua, Costa Rica, Guatemala, Honduras, Ireland, and other countries. Because of restrictions designed to prevent the introduction of foot-and-mouth disease, neither fresh nor salted refrigerated beef can be imported to the United States from South America; beef importations from these countries are canned, cooked, or fully cured (i.e., corned beef). Most live cattle are imported for feeding and breeding purposes and relatively few are destined for immediate slaughter. The majority of feeders come from Mexico and Canada, but some breeding cattle enter the United States from a number of countries.

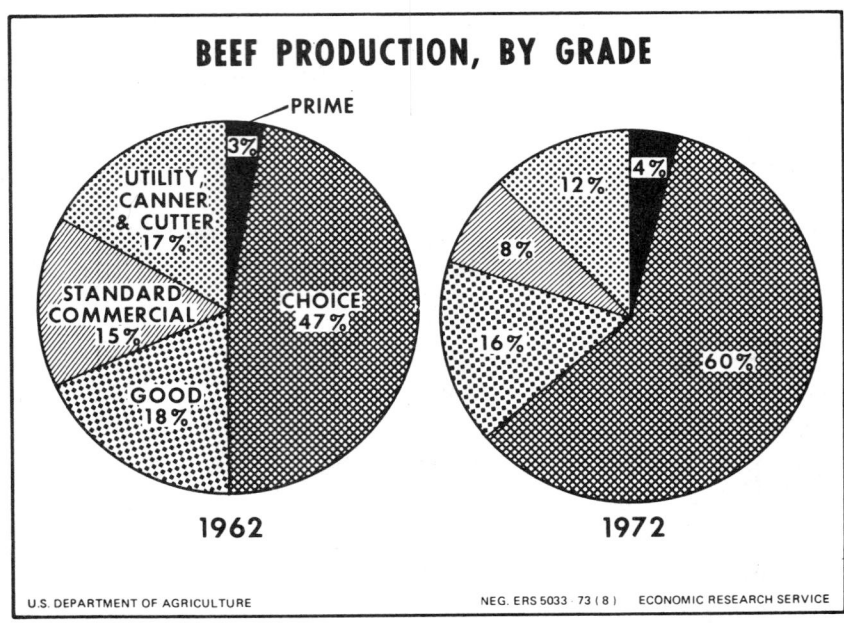

Fig. 2-31. U.S. beef production, by grade, 1962 and 1972.

QUOTAS AND TARIFFS

U.S. beef quotas and tariffs are for the purpose of encouraging domestic cattle production through protecting it from competition from foreign sources.

The United States passed a law (PL 88-482) in 1964 which provided for quotas on fresh and frozen meat imports, including beef and veal. Quotas are based on the relationship between imports and U.S. production of these meats during 1959-1963. Imports of fresh and frozen meat amounted to 4.6% of domestic production during those years. Thus, the "base quota" is equal to 4.6% of domestic commercial production of these meats. The actual quota or level of imports that would "trigger" imposition of quotas is 111% of the base quota.

The current quotas and tariffs on beef are given in Table 2-16.

As noted, quotas limit the number of pounds of meat and the number of animals that can be brought into the United States, whereas tariffs are duties, or charges, on a per pound basis imposed on imported meat for the purpose of raising prices which U.S. cattlemen receive above those prevailing in the world market.

It is noteworthy that the President of the United States is empowered to suspend quotas when he deems it in the best interest of the nation to do so. Thus, because of beef shortages and high prices, President Richard Nixon

TABLE 2-16

U.S. LIVESTOCK IMPORT TARIFF DUTIES

Import Item	Quotas[1] (No. of Head per Year)	Tariff (per pound)
Beef and Veal (fresh, frozen, or chilled)[2]	(See Footnote 2)	3 ¢
Cattle weighing: under 200 lb[3]	200,000	1½¢
between 200 and 700 lb	—	2½¢
Dairy cattle weighing: over 700 lb	—	7/10¢
Other cattle[4]	400,000	1½¢
Cattle for breeding (registered)	none	duty free

[1]Includes Canada, Mexico, and all other countries.
[2]Legislation of August, 1964, established a basic limit of .725.4 million lb plus an added factor based on U.S. production.
[3]For not over 200,000 head entered in the 12-month period beginning April 1 in any year (2½¢ per lb for any in excess of limitation).
[4]For not over 400,000 head entered in the 12-month period beginning April 1 in any year of which not over 120,000 shall be entered in any quarter beginning April 1, July 1, October 1, or January 1, and 2½¢ per lb for such "other" cattle entered in excess of any of the foregoing limitations.

suspended all beef quotas for the last half of 1972 through 1974. Actually, there was less need for U.S. quotas and tariffs on beef in the early 1970s, in a period of world beef shortages, than in previous years. Top grade beef was selling at a higher price in Europe and Japan than in the United States, with the result that the U.S. beef industry was eyeing exports; and we needed to import manufacturing beef, which was in short supply domestically.

U.S. Agricultural Trade and Balance of Payments

Trade balance, or balance of trade, is the difference in value over a period of time between exports and imports of commodities. When exports exceed imports, a favorable balance of trade is said to exist. Conversely, when imports exceed exports, there is an unfavorable balance of trade.

Agriculture has a long history of making important contributions to our international balance of payments. From the time of the first shipment of 2,500 pounds of tobacco from Jamestown, Virginia, to England in 1616 until 1916—300 years later—agricultural products accounted for the bulk of our exports. As late as 1900, agricultural products accounted for two-thirds of total merchandise exports.

Our agricultural exports dropped drastically in the 1920s and 1930s as European countries strived for self-sufficiency in agriculture. With the onset of World War II, our agricultural exports again rose significantly. They declined in the early 1950s, then moved upward quite steadily through the 1960s.

Beginning in 1973, the importance of U.S. agricultural products in world trade was hailed as the cure for the United States' nagging trade deficit which had persisted since 1970. Historically, the U.S. agricultural trade balance has been favorable. Further, its importance in recent years has been increasing significantly (see Fig. 2-32).

Normally, agricultural exports represent about 25 percent of total U.S. exports. In 1973, agricultural exports amounted to about $17 billion and made for a trade balance of $5.6 billion. But the nonagricultural sector does not fare so well. Even with exports of $52.5 billion in 1973, it did not make a positive contribution to the U.S. trade balance. So, agriculture has been responsible for solving the trade deficit problem.

AGRICULTURAL IMPORTS AND EXPORTS

FISCAL YEARS, 1965-74

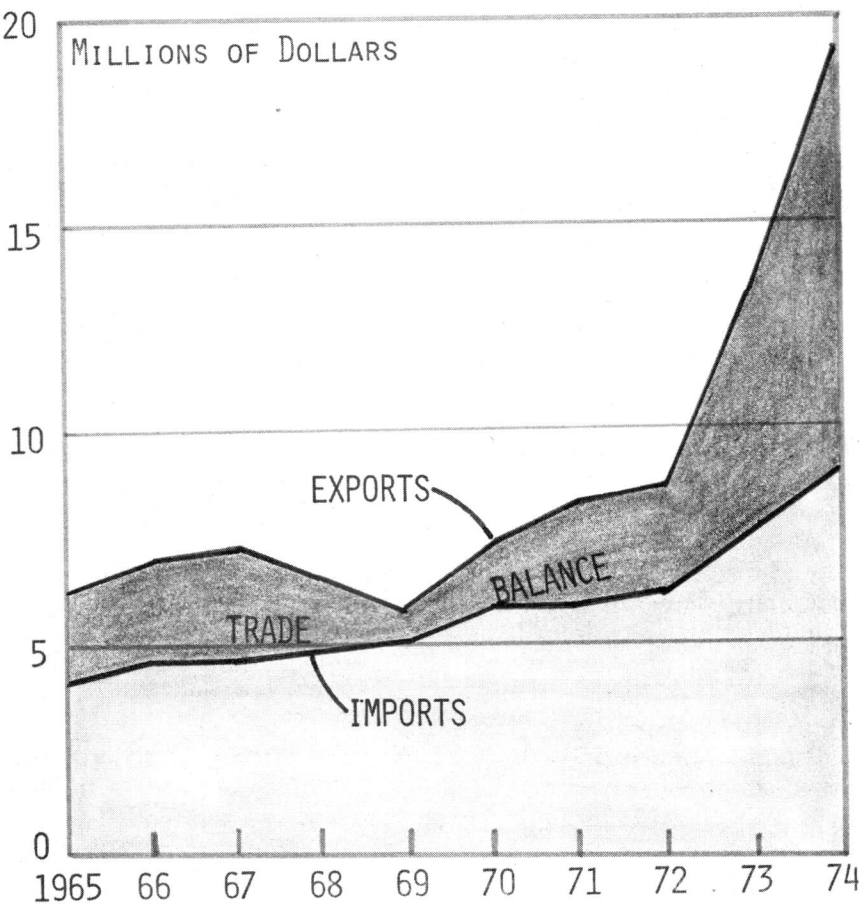

Fig. 2-32. Agricultural imports and exports, 1965-74. Note the increasing balance of trade. (Source: *Foreign Agriculture*, USDA, Foreign Agricultural Service, November 12, 1973, p. 2)

Although high prices abroad for agricultural products are one reason for agriculture's contribution to a favorable balance of trade, the following forces have also played a prominent role:

1. A very tight world grain supply situation, due to unfavorable growing conditions in many areas of the world.

2. Continued rapid improvement in world economic conditions.

3. Continued effort to upgrade diets, especially in regard to protein foods, are altering the eating habits—and demand patterns—of people all over the world.

4. Expansion of trade with the U.S.S.R. and the People's Republic of China.

5. The devaluation of the dollar, which made U.S. commodities less expensive in terms of most other currencies. For example, if beef in the United States cost a West German $1.00 per half pound in 1966, in 1974 it would cost him $0.68.

6. The availability of U.S. supplies and the capability to move large quantities of grains and oilseeds into the international market.

What's Ahead of the U.S. Beef Industry?

The author's crystal ball shows the following ahead for the U.S. beef industry.

● *50 million more cattle by 1985 (see Table 2-10)*—In 1974, there were 127.5 million cattle in the United States. By 1985, we shall need 177.6 million head—an increase of 50 million head.

● *Greatest cow-calf expansion will be in the North Central and Southeast*— The greatest future expansion of cow-calf operations will occur in the (1) West North Central, (2) North Central, and (3) South and East areas; primarily through more effective utilization of low quality roughages, such as cornstalks, in the first two areas, and through more improved pastures and year-round grazing in the South and East.

● *The Southern Plains and the Central Plains will be the growth areas in cattle feeding* —The growth areas in cattle feeding will be (1) the Southern Plains (Texas and Oklahoma), and (2) the Central Plains (Colorado, western Nebraska, and western Kansas). But future growth of cattle feeding in the Southern Plains will be slower than in the 1960s because of limitations in irrigation water and feed grains.

● *Beef imports will pose no great threat to U.S. cattle feeders*—Limited grain feeding potential in other countries precludes any real foreign threat to U.S. cattle feeders from the standpoint of importations of high quality beef. But we shall continue to import considerable quantities of lean, frozen, grass finished beef, known as manufacturing beef; and importation of any beef, regardless of quality, competes for the consumer's dollar.

● *World beef shortages and high prices will goad governments to apply all conceivable, and some inconceivable, means of increasing domestic supplies*—World beef shortages and rising prices have goaded, and will continue to goad, governments to invoke all the known methods, plus some new ones, in dealing with the situation, some of which will affect the U.S. beef industry. Governments, depending on their individual position in the world beef trade, have instituted consumer price ceilings and freezes—and higher producer price supports; consumption subsidies—and meatless days and weeks; lower tariffs—and higher export taxes; freer import quotas—and tighter export quotas; import subsidies—and export embargoes. All with the objective of increasing domestic supplies at lower consumer prices—and all counterproductive on a worldwide basis.

●*There will be pressure for liberalizing or removing U.S. quotas and tariffs*—American consumers will exert more and more pressure for liberalizing or removing quotas and tariffs. Also, in a time of world beef shortages, and when we are (1) eyeing exports for top grade beef, and (2) importing manufacturing beef suitable for hamburgers, hot dogs, and luncheon meats, there is less need for quotas and tariffs than in earlier years.

●*The United States will export more high quality beef*—The United States will continue to develop the high quality portion of the Japanese beef market, because American grain-fed beef closely resembles the highly marbled Kobe beef which is so coveted in Japan.

With increased buying power in Japan and Europe, and with the price of high quality beef in these countries higher than in the United States, it is inevitable that U.S. beef producers would like to export finished beef.

Finished beef, produced in U.S. feedlots, and fabricated, packaged, and frozen in U.S. packing plants, can be transported via refrigerated jet freight and marketed in Japan and Europe, at higher prices than can be secured at home.

All of the above could be realized on a free market. However, there are trade barriers in both Japan and Europe. Japan restricts U.S. beef by quotas; and the EC countries have a standard 20 percent duty on incoming beef in addition to a variable levy.

●*Ever-lengthening delivery times*—A growing proportion of world beef trade will be in chilled form as modern technology—vacuum packing and temperature-controlled containers—make possible ever-lengthening delivery times.

●*Agricultural exports will make for trade balance*—Agricultural exports will continue to be the great white hope for a favorable trade balance in the United States. Exporting a high value product, like grain-fed beef, would give a big assist.

●*Beef exporting countries will find it increasingly difficult to satisfy their own people*—Cattle expansion and development programs will continue around the world. But all people are demanding more beef. As a result, exporting countries will run into more and more difficulty satisfying their own people.

●*We shall import more manufacturing beef*—We shall import more and more manufacturing type beef—suitable for hamburgers, hot dogs, and luncheon meats—to meet the growing demands of the younger generation and those with moderate incomes.

●*Mexico and Canada will limit cattle and beef exports to the United*

States—Our neighbors will restrict exports of both feeder cattle and beef in order to meet their ever increasing domestic demands.

● *Trade barriers can make inefficient producers*—The more trade barriers (quotas, tariffs, subsidies, etc.)—the more protectionism—built around the beef industry(or any other agricultural economy), the more apt it is to become inefficient.

● *High income will make for beef eaters*—As personal incomes rise in countries around the world, and as people become more affluent, they will consume more meat—particularly beef. The two—beef consumption and affluency—will continue to rise together. Increasingly, beef will be a status symbol.

● *More animal protein, especially beef, will be demanded*—Population growth and rising per capita disposable income will make for a more affluent life-style and a greater demand for animal protein—especially beef, followed by beef shortages and higher prices.

● *Beef can be priced out of the market basket*—High prices for beef will do two things: (1) decrease demand; and (2) increase supply. This occurred in 1973.

● *World beef shortages and high prices will trigger passions and prejudices*—These will be expressed in the form of consumer boycotts, pickets, and hoarding; pressures from some for export controls, from others for import controls; and pressures from some to import more beef, and from others to export more beef.

● *Competition from simulated meats will increase*—The simulated meats (synthetic meats, or meat analogs) will likely become more competitive with beef in the future, as their price becomes relatively more favorable and their taste and texture are improved. In the early 1970s a product called "plus burger" (a combination of ground meat and soybeans) was being pushed by food chains. Not only was it selling at about 15 cents less per pound than hamburger, but it was extolled as containing less fat and being more nutritious than a beef burger. To meet this type of competition, the cattle industry of the future will place increasing emphasis on the palatability and nutritive qualities of beef.

● *Increasing health consciousness and warnings against unsaturated fats*—There will continue to be nationwide concern about nutrition, including warnings against consumption of foods high in unsaturated fats. Health faddists are telling people to eat fish instead of steak. This kind of thinking and publicity gives cause for cattlemen to keep a wary eye on long time beef consumption trends.

● *Inflation will increase demand for beef*—There will be continued inflation around the world, with the result that many countries will import more farm products to relieve the pressures of inflation.

● *The energy shortage will affect beef*—Energy shortages in many parts of the world will make for a decline in gross national product, which, in turn will lessen buying power and demand for beef.

● *More pollution control*—Environmentalists will force more and more pollution control.

● *Production and profits must go together*—The American cattleman can and will produce more beef, provided the business is profitable. Remember that people do those things which are most profitable to them. Remember, too, that cattlemen are people.

● *Increased human population*—The population of the United States continues to expand, even though it is at a slower rate. Therefore, it is reasonable to surmise that—as has happened in the older and more densely populated areas of the world—gradually less meat per capita will become available; and more and more grains will be consumed directly as human foods. This does not mean that the people of the United States are on the verge of going on an Asiatic grain diet. Rather, history often has an uncanny way of repeating itself—even though such changes come about ever so slowly. Certainly, these conditions would indicate the desirability of eliminating the less efficient animals.

● *Beef cattle will increasingly be "roughage burners"*—Beef cattle will increasingly be expected to rely upon their ability to convert coarse forage, grass, and by-product feeds, along with a minimum of grain, into palatable and nutritious food for human consumption; thereby not competing for humanly edible grains.

When put into the feedlot as calves or short yearlings and long fed on a high concentrate ration, their lifetime total feed (forage and grain combined) conversion is on the order of 10 to 1 liveweight.

In 1973, 77% of U.S. beef production was grain fed, and 23% was grass fed. There'll be less grain-fed and more grass-fed beef in the future.

● *There will be less emphasis on carcass quality*—In a "seller's market" for beef, there is the likely hazard that carcass quality will be relegated to a position of minor importance. Should this happen to any appreciable degree, per capita beef consumption could eventually suffer.

● *Improved genetics, management, and feed can make for more beef*—Our best hope in meeting beef shortages is through the proper application of our knowledge of genetics and management and the maximum use of coarse forages, grasses, by-product feeds, and nonprotein nitrogen.

● *Increased productivity per animal unit must come*—Increased productivity per animal unit to offset higher production costs, plus the continued output of a more desirable kind of beef, will be necessary for the prosperity and survival of America's number one agricultural industry—beef production.

● *"The tie that binds" will cross national boundaries*—There will always be political boundaries between nations. But cattlemen the world over will continue to work together and serve mankind through "the tie that binds"—the love for their cattle.

QUESTIONS FOR STUDY AND DISCUSSION

1. It is estimated that there will be 7.5 billion people in the world by the year 2007 A.D. Does this figure foretell the fulfillment of the "doomsday" prophesy of Thomas Malthus, made nearly two centuries ago, that world population grows faster than man's ability to increase food production?

2. Why do not cattle numbers reflect the carcass beef production of countries?

3. Discuss the beef production potential of Australia and New Zealand.

4. Why cannot most of the Economic Community (EC) countries make any marked transition from dairy to beef?

5. Discuss the increased beef production potential of South America.

6. How do you account for the fact that Australia and New Zealand lead the United States in *total* per capita meat consumption? How do you account for the fact that Argentina has the highest per capita beef consumption of any nation?

7. List the major beef and veal exporting nations. What is the significance of extensive land and low labor costs, which characterize these countries, from the standpoint of beef exports?

8. How do you account for the fact that the United States imports more beef and veal than any other country in the world?

9. What are the objects of trade barriers? Are such barriers good or bad so far as beef is concerned?

10. In the United States, most of our high quality fresh beef comes from domestic sources; and more and more of the manufacturing grades come from imports. Why is this so?

11. What do you foresee in world beef demand ahead in Japan, the European Community (EC), the Soviet Union, and the United States?

12. Can beef cattle be justified as a source of protein, or should beef cattle be eliminated?

13. List and discuss new protein sources.

14. Of the current total U.S. beef and veal production, about 77% comes from grain-fed cattle and 23% from nongrain-fed cattle. Will there be more grass finished beef in the United States in the future? Justify your answer.

15. The United States has 9.7% of the world's cattle population. Yet, it produces a fantastic 24.2% of the world's beef. What's the explanation of this situation?

16. The greatest expansion in U.S. cow-calf operations since 1950 has been in the West North Central area, and in the South and East. What's the explanation of the expansion in these two areas?

17. What's the explanation of the shift in the center of cattle feeding from the Corn Belt to the West and Southwest?

18. Why does the United States buy cattle, and beef and veal, from abroad?

19. What's the reason for U.S. beef quotas and tariffs?

20. Define trade balance, or balance of trade.

21. Will the United States export more high quality beef in the future? Justify your answer.

22. Will the United States import more or import less manufacturing type beef in the future—suitable for hamburgers, hot dogs, and luncheon meats?

23. Can beef be priced out of the market basket? Justify your answer.

24. Should cattlemen be concerned about increased competition from simulated meats?

25. Will more pollution control mitigate against beef production in the future?

26. Will U.S. beef cattle of the future increasingly be roughage burners, and will there be less emphasis on carcass quality?

27. How can U.S. cattlemen achieve increased productivity per cow unit?

SELECTED REFERENCES

Title of Publication	Author(s)	Publisher
Beef Cattle, Sixth Edition	A. L. Neumann R. R. Snapp	John Wiley & Sons, Inc., New York, N.Y., 1969
Beef Production and Distribution	H. Degraff	University of Oklahoma Press, Norman, Okla., 1960
World Cattle, Volumes I, II, and III	J. R. Rouse	University of Oklahoma Press, Norman, Okla., Volumes I and II, 1970; Volume III, 1973

CHAPTER 3

BREEDS OF CATTLE

Contents   **Page**

(Continued)

Contents   Page

A breed is a group of animals related by descent from common ances-tors and visibly similar in most characters. A breed may come about as a result of planned matings; or, as has been more frequently the case, it may be pure happenstance. Once a breed has evolved, a breed association is usu-ally organized.

One of the most frequently asked questions is: who approves a breed? The answer: no person or department has authority to approve a breed. The only legal basis for recognizing a breed is contained in the U.S. Tariff Act of 1930, which provides for the duty-free admission of purebred breeding ani-mals provided they are registered in the country of origin. But this applies to imported animals only. In this book, therefore, no official recognition of any breed is intended or implied. Rather, every effort has been made to present the factual story of the breeds. In particular, information about the new or less widely distributed breeds is needed and often difficult to obtain.

HISTORY OF CATTLE BREEDS

The first emphasis on ancestry occurred in human genealogies, where it is older than recorded history itself. But human pedigrees were emphasized for social purposes, to determine the inheritance of property, or for reasons of rank in a caste system, rather than because of any belief in the inheritance of physical and mental qualities. Human genealogies often recorded either the male or female line of descent (not both), as in the early chapters of Genesis, in Icelandic sagas, or in Maori legend.

More than a thousand years ago, the Arabs memorized the genealogies of their horses, tracing the pedigrees in the female line. We have no detailed knowledge of how these pedigrees were used, or if they were used at all, as guides in a breeding program.

The use of livestock pedigrees in the modern manner had its beginning in England late in the 18th Century, and the general formation of breed registry societies began around the middle of the 19th Century. The typical history of the formation of the various breeds of cattle may be summarized as follows:

1. A recognition of the existence of what was considered to be a more desirable and useful type.

2. The best animals of that type were gathered into one or a few herds which ceased to introduce outside blood.

3. Intense inbreeding to fix characters followed.

4. With greater numbers of animals, more herds were established.

5. When the breed became so numerous and the number of animal generations in the pedigrees increased until no man could remember all of the foundation animals far back in the pedigree, the necessity for a herd book arose. In addition to supplying knowledge of foundation animals, the herd book was designed to prevent, insofar as possible, unscrupulous traders from exporting grades or common stock as purebreds.

6. The breed association or society that published the stud book was also organized for the purposes of improving the breed and promoting the general interests of the breeders.

Naturally, not all breed histories were identical. Circumstances often varied the pattern that molded the breed, but the end results and objectives were similar. The first herd book of any breed or class of animals, known as "An Introduction to the General Stud Book," published in 1791, was for Thoroughbred horses. In it were included the pedigrees of horses winning the important races. Thus, it was really a record of performance. The Shorthorn herd book, first undertaken as the private venture of Mr. and Mrs. George Coates, of Great Britain, followed in 1822. Other societies and herd books for the various breeds were originated in due time.

It is noteworthy that some of the early breeders objected to furnishing pedigrees of their sale animals, fearing that they would thus give away valuable trade secrets.

THE GOLDEN PUREBRED ERA

The improvement that Robert Bakewell (1726-1795) and his followers, of England, had made in their breeding stock came to be known in other lands. Agriculture was on the move, and the golden age of stockbreeding was at hand. Animals possessing common characteristics were no longer to be confined to a small area and restricted to a few breeders.

Both in England and on the continent, the breeding of superior cattle engaged the interest of many persons of wealth and high position, even of royalty itself. Albert, Prince Consort of Queen Victoria, had a magnificent herd of 200 Shorthorns at the home farm near Windsor Castle, another herd of 90 Herefords at the Flemish Farm 2 miles distant, and 100 Devons on the Norfolk Farm. King William of Württemberg, as early as 1824, began to import and breed Shorthorns, as did his contemporaries, Nicholas II of Russia, Francis Joseph of Austria, and Louis Phillippe of France. Napoleon III was also a heavy buyer of English blooded stock; and if he could not buy the bulls that he desired, he often leased them—particularly was this true of sires from the Booth Herd at Warlaby. The King of Sardinia and the King of Spain became interested in stock breeding, crossing English Shorthorns on the long-horned white Tuscan cattle and black Spanish cows.

Early in the 19th Century, progressive U.S. cattle breeders, ever alert to their opportunity, proceeded to make large importations from England and Scotland. At this period in history, cattle were not valued solely for meat or milk. In 1817, Henry Clay of Kentucky, who made the first importation of Hereford cattle to the United States said of them: "My opinion is that the Herefords make better work cattle, are hardier, and will, upon being fattened, take themselves to market better than their rivals."

By 1850, the "battle of the breeds," which for many years had divided English stockmen into rival camps, was transferred across the Atlantic, where it has raged ever since. Except for the occasional emphasis on fancy points to the detriment of utility values, perhaps this breed competition has been a good thing. Undoubtedly, the extensive importations of Shorthorns, Herefords, Ayrshires, and Jerseys in the decades before the Civil War materially improved the quality of American cattle, both for beef and milk production.

In the early 1900s, the trend among U.S. beef cattlemen was toward purebreds or high grades. Breeding was a matter of pride, a status symbol, and usually profitable, too. The choice of the breed was determined by individual preference or prejudice; and, for the most part, it was limited to Shorthorns, Herefords, and Angus. The author recalls the case of one couple who came to his office for the purpose of discussing the purebred cattle venture which they were about to launch. When asked if they had decided on the choice of a breed, the lady spoke out: "Yes, we have white board fences and black Newfoundland dogs; I think black Angus would be beautiful." In the days when there were no production records, and when merely owning a cow was likely to turn a profit, perhaps that was as good a reason for choice of a breed as any.

CROSSBREEDING DATES TO THE LONGHORN

Crossbreeding is not new to American animal agriculture. An estimated 80 percent of the nation's hogs, sheep, and layers, and 95 percent of our broilers, are crossbreds. More remarkable yet, the latter are nearly all one cross—white Cornish roosters on White Plymouth Rock hens.

When we trace the history of the cattle industry of the United States, we become aware that the commercial cattleman has practiced crossbreeding and exploited the benefits of heterosis since the beginning of time. The commercial cattle industry of this country began with the Longhorn. Introduced by the Spaniards in the 16th Century, they were long on horns, hardiness, and longevity; and they had the ability to fight and fend for themselves as they reverted to the wild. In the movie, "The Rare Breed," Jimmie Stewart described them as meatless, milkless, and murderous. Yet, their tough, sinewy muscles made it possible for them to travel long distances to the sparsely located water holes on the early-day western ranges, and the even longer distances—often 1,000 miles or more—to the railheads.

As the living standards improved and the desire for better steaks developed, the aristocratic Shorthorns from Scotland came on the American scene. They were admirably promoted, and their assets that would improve the deficient traits of the Longhorn were exploited. The Shorthorn lived up to its name—it shortened the murderous weapons on the head of the Longhorn. Additionally, the Shorthorn X Longhorn crossbreds showed marked improvement in mothering ability, milk production, early maturity, weaning weight, beef qualities, and disposition. Yet, the hybrids retained the hardiness of the Longhorns. It was a perfect example of one breed complementing another.

Fig. 3-1. Longhorns, foundation of the American cattle industry. The Shorthorn, true to its name, shortened the murderous weapons on the head of the Longhorn.

Gradually, the Longhorn became extinct and the commercial herds of this country were dominated by Shorthorn blood. At the beginning of the present century, the grand champion steers at the Chicago International weighed 2,500 pounds and averaged 4 to 5 years of age. With the advent of USDA grading standards and the modern butcher, it became evident that carcass size of steers had to be reduced in keeping with consumer demand. Also, with the increase in Shorthorn blood and the decrease in heterosis, there was a decrease in the ability of the animals to withstand the rigors of the western range. At this point in history, the Hereford was lofted into prominence on the beef cattle scene. When crossed on the Shorthorn, they provided early maturity, smoothness of fleshing, hardiness, and ability to withstand severe winter conditions. The crossbreds retained good milk production from the Shorthorn, and, as a bonus, they got hybrid vigor.

Over the next 20 to 30 years, the cow herds of the beef producing areas of the nation became predominantly Hereford, and the Longhorn and Shorthorn were relegated to minor importance. Later, the Angus entered, and the crossbreeding and accompanying heterosis resulted in the "black baldies."

It is noteworthy that, for a considerable period of time, crossbreeding strengthened the position of purebred breeders and made for breed growth; first with the Shorthorn which was crossed on the Longhorn, second with the Hereford which was crossed on the Shorthorn, and third with the Angus which was crossed on the Hereford. Noteworthy, too, is the fact that, without a planned program to sustain heterosis, each breed brought on its own decline.

NEW BREEDS ARRIVE

From the time cattle of improved breeding began to make their influence evident in North America, the majority of the genetic material used was of British origin. During the early period, the most significant changes were in the proportions of the British breeds composing the national cow herd; first came the Shorthorns, then the Herefords, and finally the Angus.

The introduction of the Zebu type of cattle in the early part of the present century significantly improved the efficiency of beef production in subtropical and desert range areas of southern United States. Also, the Brahman was used in crossbreeding and in developing the Santa Gertrudis, Beefmaster, Charbray, and Braford breeds. But the gradual development and expansion of the Brahman and the new part-Brahman breeds were not considered a serious threat by supporters of the British breeds.

The beginning of the present exotic era must be credited to the Charolais, which found its way into the United States from Mexico in the late 1930s, thence spread north into Canada. In test stations, Charolais crosses demonstrated higher lean growth rate than straightbred or crossbred British breeds. Through active promotion based on performance facts, the enchantment of a new breed, and the momentum of a new registry, the Charolais was used widely enough to be regarded as a serious threat to the established breeds. Thus, the Charolais breed can rightfully be credited for accelerating genetic improvement programs in all other beef breeds, both old and new.

EXOTICS

Next came the exotics! According to Webster, the word "exotic" means, "from another country; not native to the place where found. Having the appeal of the unknown—mysterious, romantic, picturesque, glamorous. Strikingly unusual in color or design." Indeed, exotic cattle are all these things—and more.

Back of the rage for the exotics is the desire to use them in crossbreeding. Among commercial cattlemen, the tidal wave of exotics engendered enthusiasm and excitement such as had not been seen in recent times. But among some purebred breeders of the established breeds, it produced animosities reminiscent of the range wars in the days of intruding sheepmen and nesters. The "establishment" was riled because crossbreeding was being extolled, and bulls of these stark newcomers were being used. Arguments waxed hot; many oldtime purebred breeders became emotional and explosive. This was surprising because, as pointed out earlier in this discussion, the practice of crossbreeding has been common in this country since the birth of the cattle business. The commercial cattleman has always exploited the benefits of heterosis—first by using Shorthorns on Longhorns, thence followed by Herefords and Angus. Moreover, at one time or another, all breeds of cattle in North America, including the Longhorn and British breeds, were exotics—none of them were indigenous to this country. They are no more native than North American people, except the Indians. In the present century, the Brahman was the first exotic; thence the Charolais. But there were two great differences about the recent tidal wave of exotics: (1) they came in more quickly; and (2) they came in greater numbers, because of the following circumstances:

1. *Canada's quarantine station*—Under the leadership of Canada's astute and progressive Minister of Agriculture, Harry W. Hays (who recently released his own new breed—the Hays Converter), a quarantine station was established on Grosse Isle in the St. Lawrence River and opened in 1965. It was then possible to bring cattle from parts of Europe previously closed because of the disease situation. An influx soon followed.

2. *The time was ripe*—During much of the 1950s and 1960s, the cow-calf man was hurting financially. He recognized that profits could be increased by producing for less, as well as by selling for more. Crossbreeding with its demonstrated potential for producing for less through complementary genes and heterosis offered new hope—hope for survival. Many cattlemen had tried it with success in the '60s; and others were ready and anxious to do so.

3. *Artificial insemination (A.I.) facilitated it*—The use of A.I. in beef breeding herds increased substantially during the 1950s and '60s, with many beef producers becoming competent in managing such programs. This lessened the number of bulls that had to be maintained in conducting a crossbreeding program and made it possible to ship semen great distances.

4. *The dollar value of the Okie*—Cattle feeders came to a realization that crossbred cattle common to the south—the Okies—often made more rapid and efficient gains (hence, more dollars profit) than straightbreds.

5. *The Charolais inferiority complex*—A common reaction was: "I missed out on the Charolais, but I'm not going to miss out on the other new breeds."

6. *Promoters help usher them in*—Although most of the exotics were, and still are, in the hands of bonafide cattlemen intent on becoming constructive breeders, they attracted a considerable number of promoters whose main objective was to turn a big and quick profit. Their promotional know-how, along with money, fed the exotic boom.

The first Simmental was imported in 1967, and the first Limousin entered Canada in 1968. Other breeds followed, and the end is not in sight.

Indeed, the influx of exotics has disturbed the placid tranquillity of the pastoral scene. But a candid look at crossbreeding reveals the following:

1. The practice is as old as animal agriculture in North America—it has always been an economic necessity for commercial cattlemen.

2. The researcher hasn't discovered anything new; rather, he has evolved with new tools for applying, and new methods of measuring the merits of, a very old practice.

3. It is not the solution to all the commercial cattleman's problems. It will give him hybrid vigor, but—

a. The success of the resultant programs depends on the merits of the individual parents and the selection of breeds that are complementary.

b. He must sort out fact from fiction in the reports, both written and oral, extolling the virtues of both breeds and crossbreeding; and he must not forget the cattleman's old axiom that, "well bought is half sold." This double-barreled caution is prompted because the exotic boom is being ushered in by the greatest array of promoters, entrepreneurs, and "fast buck boys" ever to latch on to the cattle industry.

c. Without a longtime, planned breeding program, crossbreeding will almost inevitably end up with little to show for it, other than a motley collection of females and progeny varying in type and color.

d. The feeding and management of the exotics and crosses should satisfy their requirements for (1) increased growth rate, (2) greater milk production, (3) larger mature size, and/or (4) special needs because of differences in adaptability, winter hardiness, etc.

PUREBREDS CONTROL DESTINY

Despite the very considerable virtues of crossbreeding, there will always be purebreds, and they will control the destiny of cattle improvement. Not only are they needed to sustain hybrid vigor, but the only way to improve crossbreds above their present level of performance is through the development of superior purebreds. Commercial hybrids must depend on selection in the purebreds to improve those traits that are high in heritability, such as rapid and efficient gains and carcass merit. Master purebred breeders are to improved crossbreds what master designers are to better cars. Thus, crossbreeding is not a threat to the purebred breeder's existence; rather, it is an opportunity, *provided* he supplies animals of proven superior performance. Remember that bulls used in crossbreeding must be purebred if the breeder is to attain maximum hybrid vigor in the offspring. Remember, too, that the commercial man who crosses must return to the genetic well, or fountainhead, often to replenish his herd with identifiable unrelated genes. Hence, an everlasting bull market is assured. Not only that, heifers not needed as the purebred replacements may fit admirably into a program specializing in producing F_1 quality heifers. Under a system of using F_1 cows bred to a terminal breed, about half of the cattle must be purebreds.

Yet, breed partisans cannot be complacent. Not all breeds—from among the traditional beef breeds, dairy breeds, newly imported breeds, and newly synthesized breeds—are needed in a crossbreeding program, from the standpoint of complementary traits and adaption. Thus, only those breeds which provide the most of the best in desired qualities as reflected in net returns to the commercial cattleman will survive over a period of time; the rest will be relegated to a position of minor importance, or fall by the wayside. Thus, purebred breeders of today who wish to be in business tomorrow must produce superior animals backed by meaningful production records. Also, they must engineer (through selection) breeds that will not only grow fast, but which will calve annually with minimal assistance.

TYPES OF CATTLE

Type may be defined as an ideal or standard of perfection combining all the characters that contribute to the animal's usefulness for a specific purpose. It should be noted that this definition of type does not embrace breed fancy points. These have certain value as breed trademarks and for promotional purposes, but in no sense can it be said that they contribute to an animal's utility value. There are four distinct types of cattle in the world: beef type, dairy type, dual-purpose type, and draft type.

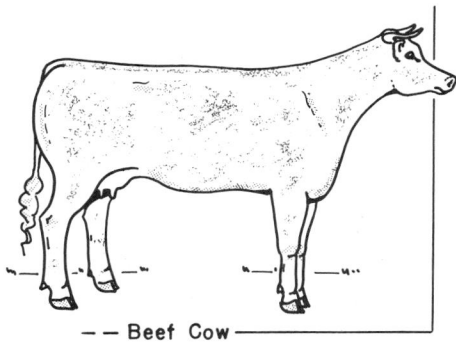

-- Beef Cow

--Dairy Cow

Fig. 3-2. Beef-type cow (above), characterized by bred-in meat qualities. Dairy-type cow (below), characterized by a lean, angular form and a well-developed mammary system. (Drawing by R. F. Johnson, Head, Dept. of Animal Science, California Polytechnic State University, San Luis Obispo, Calif.)

Dual Purpose Cow--

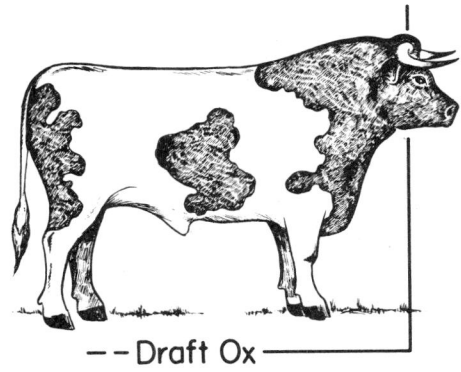

--Draft Ox

Fig. 3-3. Dual-purpose-type cow (above), intermediate between the beef type and dairy type in conformation. Draft-type ox (below), characterized by great size and ruggedness with considerable length of leg. (Drawing by R. F. Johnson)

Beef-type animals are characterized by meatiness. Their primary purpose is to convert feed efficiently into the maximum of high quality meat for human consumption.

Dairy-type animals are characterized by a lean, angular form and a well-developed mammary system. Their type is especially adapted to convert feed efficiently into the maximum of high quality milk.

Dual-purpose-type animals are intermediate between the beef type and dairy type in conformation, and also in the production of both meat and milk.

Although many breeders have the dual-purpose type clearly in mind, and although many fine specimens of the respective breeds have been produced, there is less uniformity in dual-purpose cattle than in strictly beef- or dairy-type animals. This is as one would expect when two important qualifications, beef and milk, are combined.

Draft-type animals, when true to form, are characterized by great size and ruggedness with considerable length of leg. Although oxen are seldom seen in the United States, except in the New England states, it must be remembered that these patient, steady, plodding beasts are still the chief source of power in many parts of the world.

Several distinct breeds of cattle of each of the types have been developed in different parts of the world. Although each of these breeds possesses one or more characteristics peculiar to the group (breed characteristics), in general the type of cow that will produce a large flow of milk is the same the world over, despite acknowledged differences in size, color, shape of head and horns, or in any other distinctive breed characteristic. Likewise, there is a general similarity between all the beef-type breeds.

BREEDS

Brief, but pertinent, information pertaining to each of the breeds of cattle follows. The breeds are listed in strictly alphabetical order, without regard to their more conventional division into beef, dairy, or dual-purpose type, because (1) beef type and dairy type have moved closer together in recent years, with the greater emphasis on red meat; and (2) more and more commercial cattlemen are crossing different types and breeds of cattle to achieve maximum hybrid vigor, milk production and weaning weights, and red meat.

The ultimate objective in beef production is the sale of beef over the block. But this is not enough. It is also imperative that feeds be *efficiently* converted into the maximum of high quality beef. How well the different breeds and crosses measure up to these requisites will determine their popularity in the future.

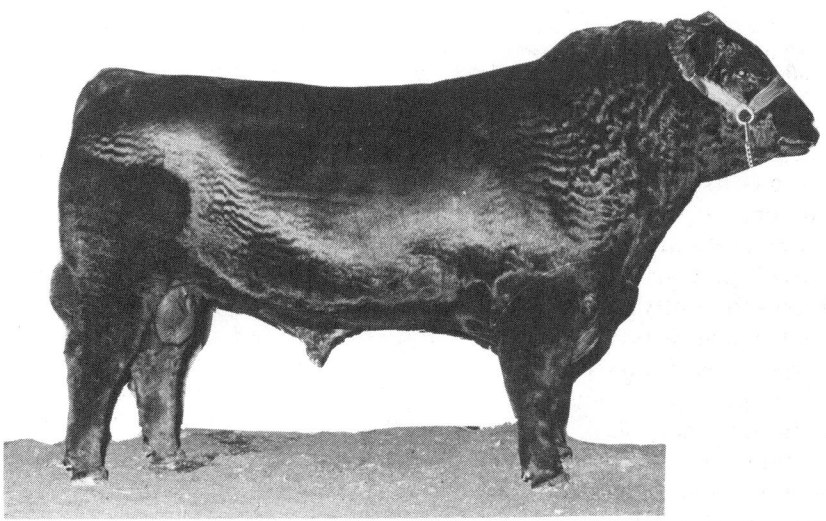

Fig. 3-4. Ankonian Dynamo, Grand Champion Angus bull at the 1973 National Western Stock Show, Denver, owned by Sayre Farms, Phelps, New York and Ankony Angus, Grand Junction, Colorado. (Courtesy, *Aberdeen Angus Journal*, Webster City, Iowa)

Angus

Origin—Scotland; in the northeastern counties of Aberdeen, Angus, Kincardine, and Forfar.

The origin of the breed is rather speculative. Some claim that it is a sport from an earlier black, horned breed of Scotland. Others are of the opinion that they sprang from the polled cattle of Britain.

Color—Black.

Distinguishing characteristics—Polled; comparatively smooth coat of hair; somewhat cylindrical body.

Disqualifications—Horns, scurs or buttons. Red color. A noticeable amount of white above the underline, or in front of the navel, or on one or more legs. Calves from females less than 18 months of age when calf was dropped, or from bulls less than 9 months of age at the time of service.

Comments—Sometimes the breed is referred to as "doddies." This is a Scotch term for polled or hornless.

In 1873, George Grant of Victoria, Kansas—a native of Banffshire, Scotland, and a retired London silk merchant—imported the first Angus bulls into the United States from Scotland, to use on his commercial range cattle. However, the first breeding herd—including animals of both sexes—to be imported into the New World was brought to Canada in 1876 by Professor Brown of Ontario Agricultural College.

Fig. 3-5. George Grant Memorial, Victoria, Kansas. Mr. Grant imported the first Angus bulls into the United States from Scotland in 1873. (Courtesy, American Angus Breeders' Association, St. Joseph, Mo.)

Ayrshire

Origin—County of Ayr, in southwestern Scotland.

Color—Light to deep cherry red, mahogany, brown, or a combination of these colors, with white; or white alone. Black or brindle are objectionable.

Distinguishing characteristics—Horns are widespread and tend to curve upward and outward. However, there is a polled strain. The udders are especially symmetrical and well attached to the body. The breed is noted for its style and animation, good feet and legs, and grazing ability.

Fig. 3-6. Hammond's Top Primrose, Ayrshire cow. Her lifetime production through 1966 was 149,754 lb of milk and 6,057 lb of fat, with an average butterfat test of 4.0. (Courtesy, Ayrshire Breeders' Assn., Brandon, Vt.)

Barzona

Origin—In the United States, on the Bard Ranches of Kirkland, Arizona; hence, the name Barzona (a contraction of Bard and Arizona). The development of the breed was started in 1942. The objective: to produce a breed of cattle adapted to the intermountain, desert, and mountain ranges of the Southwest and northern Mexico. The foundation consisted of an Afrikander-Hereford cross. The F1 females from this cross were divided into two herds; Santa Gertrudis bulls were used on one and Angus bulls on the other. Progeny were culled closely and crossed back with emphasis on fertility, mothering ability, and gains.

Color—Red.

Distinguishing characteristics—Well adapted to arid and semiarid ranges (the Bard Ranch, where the breed was developed, has an average annual rainfall of 12.76 inches); cows are regular producers, excellent mothers, and good milkers.

Comments—Breed enthusiasts recommend that Barzona bulls be used on commercial cows on southwestern ranges, so as to obtain a high degree of heterosis or hybrid vigor.

Fig. 3-7. Barzona bull; the breed that was developed from a planned crossbreeding program involving Afrikander, Hereford, Santa Gertrudis, and Angus. Adapted to the rigors of southwestern range conditions. (Courtesy, Curtiss Breeding Service, Cary, Ill.)

The breeding program followed by Bard Ranches to create the Barzona breed consisted of (1) forming a large genetic pool by crossing breeds, then breeding within the herd, and (2) using records to eliminate undesirable genes and retain desirable genes.

Beef Friesian

Origin—United States. In 1972, several U.S. cattlemen—headed by Premier Corporation, Fowlerville, Michigan, and C & J Cattle Company, Highmore, South Dakota—organized to import 25 Beef Friesian cattle from Ireland to the United States. In Europe, the Friesian has always been a dual-purpose animal, whereas the American descendant, the Holstein-Friesian, has been developed exclusively as a dairy breed. This explains the reason for the importation.

Two approaches are being followed in developing the Beef Friesian: (1) developing purebred Beef Friesians from Beef Friesian stock and through a grading up program by using purebred Beef Friesian bulls on selected U.S. Holstein females; and (2) developing a new breed from a Beef Friesian X Angus cross.

Color—The Beef Friesian is black and white. Beef Friesian X Angus are generally black.

Distinguishing characteristics—Rate and efficiency of gains comparable to the exotics; good marbling; little calving difficulty; good milking ability.

Comments—Some cattlemen will breed up to the purebred Friesian line, but most of those using Beef Friesians will likely stop with the first cross.

Fig. 3-8. "Admiral," one of five imported Beef Friesian bulls used as foundation sires for Beef Friesian development in the United States. Note his exceptional beefiness! These cattle consistently produce top carcasses in Ireland. (Courtesy, American Beef Friesian, Denver, Colo.)

Beefmaster (approximately ½ Brahman, and ¼ each Shorthorn and Hereford)

Origin—United States; on the Edward C. Lasater Ranch, in Fulfurrias, Texas, beginning in 1908. In 1931, Tom Lasater took over his father's herd and carried forward; in 1949, he moved the herd to its present location near Matheson, Colorado.

Color—Red is the dominant color, but color is variable and is disregarded in selection.

Distinguishing characteristics—The majority are horned, although a few are naturally polled. During the entire period of development, selection has been practiced for the following six essentials: disposition, fertility, weight, conformation, hardiness, and milk production. Beefmaster cows are good milk producers under range conditions and wean off heavy calves.

Comments—In order that each Beefmaster may be permanently identified with the breeder thereof, the breeder must use a prefix name, such as "Jones Beefmaster," "Smith Beefmaster," etc., to designate his cattle. Thus, in a unique way, the responsibility for the continued improvement of the breed is placed squarely upon the individual breeder.

Fig. 3-9. Lasater Beefmaster cow and calf, owned by Tom Lasater, The Lasater Ranch, Matheson, Colorado. (Unretouched range photograph by Darol Dickinson, Calhan, Colo.)

Belted Galloway

Origin—Scotland; in the southwestern district of Galloway. They were first imported to the U.S. in 1948.

Color—Black with a brownish tinge, or dun; with a white belt completely encircling the body between the shoulders and the hooks.

Distinguishing Characteristics—Polled; striking white belt; heavy coat of hair; hardiness; ability to graze rugged terrain; easy calving; good milkers.

Disqualifications—Red color, incomplete belt, other white marks, or scurs.

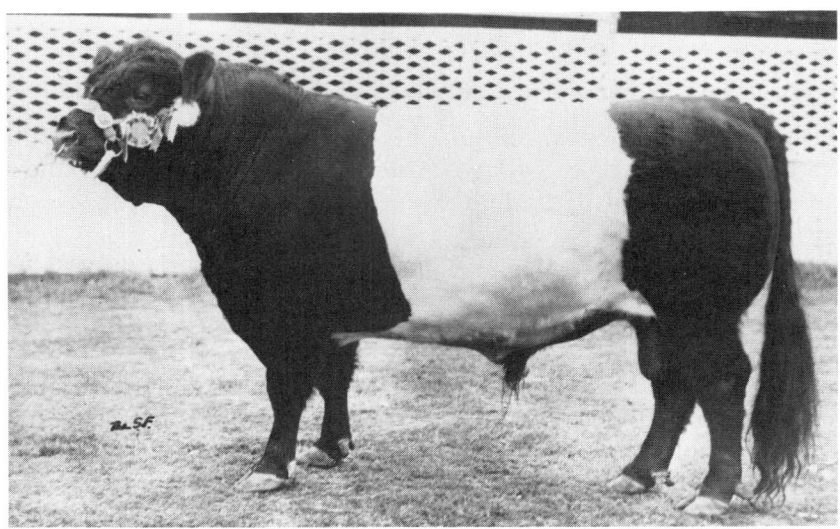

Fig. 3-10. Burnside Great Scot, Belted Galloway bull imported from Scotland. He was Supreme Champion at the Royal Highland Show in Aberdeen. (Courtesy, Aldermere Farm, Rockport, Maine)

Blonde d'Aquitaine (pronounced "blond dock-ee-tan")

Origin—The Blonde d'Aquitaine originated in southwest France in 1961, when three French strains of similar background—Garonne, Quercy, and Pyreneenne—combined. Today, there are about 500,000 cattle of this breed in France.

In 1972, Blonde d'Aquitaine semen in French mini-straws was released from the USDA quarantine to Blonde d'Aquitaine Breeders of America, a venture headed and managed by Interoceanic Cattle Company, Stillwater, Minnesota.

Color—Yellow brown, fawn, or wheat colored.

Distinguishing characteristics—The breed is similar to the Charolais in both physical appearance and size. Other characteristics are: relatively fine bone; high fertility; little calving difficulty, due to the width and shape of the pelvis; and growth rate comparable to the Charolais. Mature bulls weigh up to 2,600 pounds, and mature cows from 1,600 to 1,800 pounds.

Fig. 3-11. Flonflon, a Blonde d'Aquitaine bull. (Courtesy, Curtiss Breeding Service, Cary, Ill.)

Comments—Although Blonde d'Aquitaine beef cattle in France are far less numerous than either Charolais or Limousin, performance and progeny testing of the Blonde d'Aquitaine is considerably more intensive and extensive than that of the other breeds. Ordinarily, the top third of the bulls in a performance test are subsequently progeny tested. Of these, 5 to 10 will be retained and used further. Use of the Blonde d'Aquitaine as a terminal cross sire is suggested.

Braford

Origin—The Braford breed originated on Adams Ranches, near Fort Pierce, Florida. The foundation breeding and selection program was started about 1948; but a breed registry was not formed until 1973. The breed evolved from crossing Brahmans and Herefords. Today, Brafords are approximately ⅝ Hereford and ⅜ Brahman.

Color—Red or brindle, with white markings on the head and pigmentation around the eyes.

Distinguishing characteristics—Horned; short haired; heat tolerant; only a slight hump. Mature bulls weigh 1,500 to 2,000 pounds, and cows 1,000 to 1,500 pounds.

Fig. 3-12. Two-year-old Braford heifer and calf. (Courtesy, Adams Ranch, Inc., Fort Pierce, Fla.)

Disqualifications—Offspring cannot be registered if they are from cows that have not calved annually, or that required veterinary assistance at calving.

Comments—For registration, the International Braford Association requires pedigree information and performance records, and that the animal pass an inspection.

For many years, Adams Ranches have (1) culled nonbreeders, thereby obtaining good fertility, and (2) retained heavy weaning calves without creep feed, thereby getting good milk production and sound udders.

Brahman

Origin—United States, through the amalgamation of several Indian types, probably with a small infusion of European breeding.

Color—Gray or red preferred; either solid color, or a gradual blending of the two. However, there are brown, black, white, and spotted Brahmans.

Distinguishing characteristics—Drooping ears, and long face. Promi-

Fig. 3-13. IW's Rexcrata 204, Brahman bull bred and owned by 3W Ranches, Donie, Texas. This bull was the breed's National Grand Champion. He weighed 2,100 pounds when this photo was taken. (Courtesy, American Brahman Breeders Association, Houston, Tex.)

nent hump over the shoulders, which serves as a storehouse of energy and metabolic water for times of feed shortage and drought. An abundance of loose, pendulous skin under the throat and along the dewlap. A somewhat narrow body, an upstanding appearance due to somewhat longer legs, and a tucked up middle. A voice that resembles a grunt rather than a low.

Disqualifications—Brindle, grulla (a smutty or blackish red), or albino color. Cryptorchid bull. Freemartin heifer. Inherited lameness. Dwarf or midget characteristics.

Comments—The Brahman breed is well adapted to areas characterized by hot climates, heavy insect infestations, and sparse vegetation.

Brangus (⅜ Brahman X ⅝ Angus)

Origin—United States; on Clear Creek Ranch, Welch, Oklahoma, owned by Frank Buttram, beginning in 1942.

Color—Black.

Distinguishing characteristics—Polled; slight crest over the neck; smooth, sleek coat.

Disqualifications—Horns; any color other than black; white in front of navel; small for age; extremely nervous; too fine boned.

Fig. 3-14. Black Duke R 11657, Brangus bull, bred by Floyd Newcomer, Yuma Valley Cattle Co., Yuma, Arizona. (Courtesy, International Brangus Breeders Association, San Antonio, Tex.)

Brown Swiss (Dairy; Beef)

Origin—The Brown Swiss is a very old breed. It developed in the Alps of Switzerland, where they were triple-purpose cattle—used for meat, milk, and draft. The breed was first imported to America in 1869; and the Brown Swiss Cattle Breeders' Association of America was organized in 1880.

In answer to popular demand, in 1971, the Brown Swiss Cattle Breeders Association formed the Brown Swiss Beef International, Inc. Hence, the same registry association is now keeping separate Herd Books for each dairy-type and beef-type Brown Swiss. It keeps two complete sets of beef records: (1) the Brown Swiss Beef Purebred Registry; and (2) the Brown Swiss Beef Cross Record. The "Cross Record" covers all percentage females having less than ⅞ but no less than ½ Brown Swiss or Brown Swiss Beef blood, provided they are sired by registered Brown Swiss or a registered Brown Swiss Beef bull.

Color—Solid brown, varying from very light to dark. White markings are objectionable. The nose, hooves, and switch are black, and there is a characteristic light-colored band around the muzzle.

Distinguishing Characteristics—Medium length horns. Strong and rugged, with some tendency toward the heavy muscling characteristic of the beef breeds. Calm and unexcitable.

Brown Swiss females have good size, are very fertile, calve with ease, have mothering ability, and give plenty of milk for their calves.

Comments—The real advantage from crossbreeding with Brown Swiss bulls on commercial beef cows comes from using the F₁ heifers as brood cows.

Fig. 3-15. Schulte's Sunwise Pat, dairy-type Brown Swiss cow. Supreme Champion, National Dairy Cattle Congress 1971. Production record: 15,910 milk, 4.7%, 633 fat, 305-day, 2x. (Courtesy, The Brown Swiss Cattle Breeders' Association, Beloit, Wisc.)

Fig. 3-16. Ueli, imported beef-type Brown Swiss bull. Note his beef qualities. (Courtesy, Carnation-Genetics, Hughson, Calif.)

Charbray

Origin—United States, in the Rio Grande Valley of Texas, beginning in the late 1930s. It developed from Charolais X Brahman crosses. The Charbray is from ¾ to ⅞ Charolais.

Color—Light tan at birth, but usually change to a cream white in a few weeks.

Distinguishing characteristics—Horned; a slight hint of the Brahman dewlap remains. In appearance, the Charbray resembles the Charolais. But it is more rugged and less nervous than the Brahman. The Charbray has the growth thrust of the Charolais and the heat-insect tolerance of the Brahman.

Fig. 3-17. Charbray bull. (Courtesy, American International Charolais Association, Houston, Tex.)

Charolais (Usually spelled Charollais in France)

Origin—France; in the province of Charolles in Central France, and later in the province of Nivernais. The breed society was founded in France in 1887.

Color—White, golden wheat, or straw.

Distinguishing characteristics—Horned; pink skin and mucous membranes. The breed is noted for its large size, rapid gain, and bred-in red meat.

Disqualifications—The association disqualifies any animal that (1) has a black nose, (2) is spotted, or (3) has excessive dark skin pigmentation.

Fig. 3-18. 58's Miss Apollon Leesa, Grand Champion Charolais female of the 1972 American Royal, Kansas City, shown by Schuff Cattle Company, Sutherland, Nebraska. (Courtesy, American International Charolais Association, Houston, Tex.)

Chianina

Origin—In central Italy, in the Chiana Valley (from which they take their name), in the province of Tuscany. The Chianina is of very ancient origin, going back to the days of the Roman Empire, when they were used for draft. Shortly after the birth of Christ, the Latin poet Columella described the cattle of Rome as "boves albos et vastos," i.e., white cattle of great size. He went on to tell how they were used as sacrificial animals and for pulling the triumphal cart of the Emperor. Animal geneticists in Italy speculate that the breed is related to the *Bos indicus*, though it does not have the characteristic hump, excessive dewlap, or sweat glands. It is also possible that the breed was introduced into France during the first century A.D. by Roman colonizers, and subsequently served as a foundation breed for the present-day Charolais.

In 1972, Dr. Guccio Fortegueni, past president of the Italian Chianina Association, reported that the Chianina population in Italy was 400,000, of which 120,000 were registered with the ICA.

Color—Porcelain white hair, black switch, and dark skin, which give the Chianina resistance to heat—the white hair reflecting the sun's rays and the dark pigment preventing sunburn. Unlike the Charolais, the white of the Chianina is recessive in inheritance. Crosses of Chianina X Holstein are black; and crosses of Chianina X Brown Swiss are also dark, with Swiss markings on nose and ears.

Calves are born tan color, which gradually turns to white at about 60 days of age.

Distinguishing characteristics—The Chianina is probably the largest

Fig. 3-19. Ferrero, Chianina bull; bred by G. Fortegueni, Monteroni, Arbia, Italy; imported to the United States by American Breeders Service. (Courtesy, American Breeders Service, De Forest, Wisc.)

breed of cattle in the world. They are tall, long legged, long and round bodied, and heavy. Mature bulls stand about 6 feet high (18 hands) at the withers and weigh up to 4,000 pounds. Mature cows weigh up to 2,400 pounds. Other distinguishing characteristics of the breed are: uniformity of depth; trimness of middle; fineness of head, horn, and bone; absence of excessive dewlap and brisket; very strong feet and legs; poor milkers (Italian cattlemen creep feed Chianina calves beginning at about 4 months of age); and gentle disposition.

No serious genetic defects have been reported in the breed, despite the fact that the coefficient of inbreeding is between 13 and 24 percent.

Comments—Despite the large size of the breed, and large calves at birth (male calves average about 100 pounds at birth, and females about 85 pounds), calving difficulties are infrequent, perhaps due to the rather small heads and long, narrow bodies of the newborn.

The growth rate and leanness of the breed indicate that Chianina bulls may have an important place as a terminal breed in a crossbreeding program. Their low milk production and huge size would not seem to recommend them for use as brood cows. Their crossing potential with the large dairy breeds has been clearly established in Italy. Their greatest potential should accrue when Chianina bulls are crossed on large cows that can provide enough milk to allow the calf to express its growth potential.

Devon

Origin—England; in the counties of Devon and Somerset. Devons were probably the first purebred cattle to be brought to America. Edward Winslow, an agent of the Plymouth Colony, brought over a bull and three heifers in 1623. In 1817, the first registered Devons arrived in the United States, aboard the brig Margaretta.

Color—Red. A rich dark red is preferred; hence, the name "Ruby Reds."

Distinguishing characteristics—Creamy white horns with the black tips, but there are polled strains. Yellow skin. High fertility; cows calve easily without assistance; good milkers. Mature bulls in good working condition weigh from 1,800 to 2,400 pounds, and mature cows from 1,000 to 1,500 pounds.

Fig. 3-20. Devonacres Cascade Mr. Ideal, owned by Devonacres Ranch, Eagle Point, Oregon. (Courtesy, Edward T. Harrison, Jr., Devonacres Ranch, Eagle Point, Ore.)

Disqualifications—White other than in the switch or on small areas on the udder and belly.

Comments—The origin of the Devon is prehistoric; most writers claim that they descended from the aboriginal cattle in Britain. They are one of the oldest breeds in existence.

Dexter

Origin—Ireland in the southern and southwestern parts. They were named after their founder, a man by the name of Dexter.

Color—Black or red.

Distinguishing characteristics—Horned. Small size and short legs. Mature bulls should not exceed 1,000 pounds and mature cows 800 pounds. Some mature animals are less than 40 inches high.

Disqualifications—Animals having white other than on the belly, switch, udder, or scrotum are disqualified for registry.

Comments—The Dexter is the smallest American cattle breed; their

Fig. 3-21. Peerless Perfection II, 1,000-pound Dexter bull, owned by R. W. and Daisy Moore, Decorah, Iowa. (Courtesy, American Dexter Cattle Association, Decorah, Iowa)

smallness is accentuated by the shortness of their legs from the knees and hocks down.

"Bulldog" calves, a lethal condition, occurs in the breed. Such calves may be dropped or absorbed from the fourth to the eighth month of pregnancy.

Fleckvieh (German Simmental)

Fig. 3-22. Tattenhall Achilles, a Fleckvieh bull with excellent gaining ability. He had an adjusted weaning weight of 775 pounds and an adjusted yearling weight of 1,175 pounds. (Courtesy, Curtiss Breeding Service, Cary, Ill.)

Origin—In southern Germany, where it has been bred since 1895. It evolved from Simmental cattle, which originally came from Switzerland. The Fleckvieh is the German version of the Simmental breed. In Germany, the breed was first developed as a draft animal, with milk as a by-product. Today, there are approximately 2,000,000 Fleckvieh cows in Germany, counting both registered and unregistered.

Color—Generally red and white spotted, with a white face. The red varies from dark to a more common diluted, almost yellow shade.

Distinguishing characteristics—In Germany, it is now considered a dual-purpose animal, with emphasis on beef. The Fleckvieh is noted for good milk yield, fast growth rate, and excellent meat quality. Mature bulls average about 2,550 pounds, and cows about 1,550 pounds.

Comments—The progeny testing and selection program of the Fleckvieh breed in Bavaria is, without doubt, the best in the world. To be selected for licensing as a dam of a herd sire, a cow must be in the top 8 percent of the breed on milk production and classified by a committee for size and conformation. Also, they must meet rigid standards for calving intervals, calving ease, milking ease, disposition, and pedigree. In the testing program, 120 young sires resulting from special matings are selected each year for progeny testing. Then, the top 12 to 15 from these are selected, on the basis of their progeny, for A.I. service as proven sires.

Galloway

Origin—Scotland, in the southwestern province of Galloway. The breed descended from wild cattle native to the province of Galloway in southwestern Scotland.

Color—Black, sometimes with a brownish or reddish tint; or dun.

Fig. 3-23. Grange Bounty, Galloway bull imported from Scotland by Round Mountain Ranch, Round Mountain, California. (Courtesy, A. T. Carling-Smith, Round Mountain Ranch)

Distinguishing characteristics—Polled; long curly hair; hardiness and ability to rustle in cold weather.

Disqualifications—Scurs or horns; white marking on feet or legs or above the underline.

Comments—Galloway cattle are smaller and slower developing than most beef breeds.

Gelbvieh (German Yellow)

Origin—Germany, in Bavaria. The Gelbvieh is descended from the red-brown Keltic-German Landrace. This breed was then crossed with Simmental and Shorthorn in the early 1800s. The Gelbvieh actually came into being when four breeds of German cattle—Franconian, Glan-Donnersberg, Lahn, and Limpurg—amalgamated around 1920. Since then, the breed has been selected for solid color, growth, and, later, for carcass quality. In 1952, a planned program was started for selection for milk production. This program has been successful. Hence, the Gelbvieh claim to be the best solid-colored, dual-purpose breed.

The Gelbvieh breed was first imported to North America in 1972.

Color—Golden red to rust. Solid color.

Distinguishing characteristics—A large, long-bodied, well-muscled, fast-gaining, horned breed, with a reputation for high quality carcasses. Mature bulls weigh from 2,300 to 2,800 pounds (average 2,500), and cows 1,400

Fig. 3-24. Gelbvieh heifer in Germany. At the time photo was taken, she was 20 months old and weighed 1,369 pounds. (Courtesy, American Gelbvieh Association, Denver, Colo.)

to 1,800 pounds (average 1,500). The breed is of dual-purpose origin; hence, milk production is good. Cows average about 7,800 pounds of milk, with 4.07 percent fat.

Comments—In Germany, no Gelbvieh bull is put into general A.I. service until he is six years old and his progeny have proven him superior to his contemporaries.

The most logical crossbreeding use for the Gelbvieh appears to be as maternal sires, although their terminal sire capabilities should not be overlooked.

Guernsey

Origin—Isle of Guernsey.

Color—Fawn, with white markings clearly defined; preferably a clear (buff) muzzle.

Distinguishing characteristics—Good length of head; horns incline forward, are refined and medium in length, and taper toward the tips. The milk is especially yellow in color; golden yellow skin pigmentation; the unhaired portions of the body are light or pinkish in color (whereas in the Jersey they are near black); calves are relatively small at birth.

Fig. 3-25. Ideal Beacon's Nora, lifetime milk and butterfat champion of the Guernsey breed 4,328 days: 225,287 pounds of milk; 11,740 pounds of fat. Individual milk and fat record: 305 days-3x, 25,063 pounds of milk; 1,301 pounds of fat, AR. (Courtesy, American Guernsey Cattle Club, Peterborough, N.H.)

Hays Converter

Origin—In Canada, by the former Minister of Agriculture, Senator Harry Hays, Calgary, Alberta, beginning in 1957. The foundation breeds of the Hays Converters were Hereford, Brown Swiss, and Holstein. It is claimed that the breed converts feed into profit; hence, the name "Converter."

Every Hays Converter bull retained for service has been R.O.P. tested, with 3-pound minimum daily gain and over 1,100 pounds weight at 365 days of age.

It is the only entirely Canadian-developed breed of beef cattle.

Color—The predominant color is black with a white face, white feet, and a white tail. A few are red with white faces. But color is not a factor in selection.

Distinguishing characteristics—The traits upon which Senator Hays

Fig. 3-26. A two-year-old Hays Converter heifer and her calf. Heifers must calve without assistance and wean a 600-pound calf in 200 days to stay in the herd. Hays Converters are rugged; they can withstand 100° temperature in the summer and 40° below zero in the winter. (Courtesy, American Breeders Service, De Forest, Wisc.)

built the breed are: (1) growth; (2) fertility (replacement heifers must be from cows that settled on first service); (3) minimum calving problems; (4) well-attached udders; (5) abundant milk; (6) sound feet and legs; and (7) pigmentation. The Hays Converter has been called a "dual-purpose beef breed." Mature bulls weigh about 2,200 pounds and cows 1,400 pounds.

Senator Harry Hays, who developed the breed, gives the following pertinent facts about his herd: a 100% calf crop (the herd is pregnancy tested, and all drys are sent to the butcher); no female ever milked; no female requiring help at calving ever kept; and no foot ever trimmed.

Comments—The origin of the Hays Converter, and the system that produced the new breed, are important. Basically, it is the same technique that has been used so successfully for years in plant breeding. The steps: (1) select from a number of different breeds the important characteristics needed; (2) combine the genes from these selected individuals into one large breeding population; (3) select intensely for the important characteristics and cull ruthlessly for several generations; and (4) measure what you have after the hybrid vigor factor is no longer important, so that you are measuring the transmissible genetic superiority. A "multiple sire" breeding program was used in founding the breed.

The most logical crossbreeding use for the Hays Converter appears to be as maternal sires, although terminal sire capabilities must not be overlooked.

Hereford

Origin—England; in the county of Hereford.

Color—Red with white markings; white face and white on the under-

Fig. 3-27. "The Young Bull," from one of the great paintings of the world, by Paul Potter. Painted in 1647. Note the white-faced, red-bodied cow—an individual resembling many plain-looking Herefords of the past. (Courtesy, The Netherlands Information Bureau)

Fig. 3-28. 26 Lad F35, Hereford bull bred by 26 Bar Ranches (owned by John Wayne and Louis Johnson), Stanfield and Springerville, Arizona, sold in the November 1972 26 Bar sale to Rockbrook Farms (owned by Jim Stafford), Montalba, Texas, for $85,000 for ¾ interest and full possession for a valuation of a little over $113,000. "F35" was champion at the 1973 Register of Merit Arizona National Livestock Show, Phoenix, as well as the 1973 Southwestern Exposition and Fat Stock Show, Fort Worth, Texas. At the Arizona show, when "F35" was 28 months old, he weighed 1,975 pounds with a weight per day of age of 2.35 pounds. (Courtesy, 26 Bar Ranches)

line, flank, crest, switch, breast, and below the knees and hocks. White back of the crops, high on the flanks, or too high on the legs is objectionable. Likewise, dark or smutty noses and red necks are frowned upon.

Distinguishing characteristics—The white face is the distinct trademark of the breed. A thick coat of hair.

Disqualifications—Calves from females less than 21 months of age when calf was dropped, or from bulls less than 12 months of age when service producing the calf occurred, cannot be registered.

Comments—The earliest importation of Hereford cattle into the United States was made by Henry Clay of Kentucky, in 1817. He imported a bull and two females.

Holstein-Friesian

Origin—Netherlands and Northern Germany.

Color—Black and white markings, clearly defined.

Distinguishing characteristics—Clean-cut, broad muzzle, open nostrils, strong jaw, broad and moderately dished forehead, straight bridged nose. Large angular animal; females should weigh 1,500 pounds (mature); and males in breeding condition 2,200 pounds.

Disqualifications—The following colors bar registry: all black or all white, black in switch, black belly, black circling leg and touching hoof, black from hoof to knee or hock, black and white intermixed to give color other than distinct black and white.

Fig. 3-29. Princess Breezewood R A Patsy, Holstein-Friesian holder of U.S. butterfat production 305-day, twice-a-day milking. (Courtesy, Holstein-Friesian Association of America, Brattleboro, Vt.)

Indu Brazil (Zebu)

Origin—Brazil.

Color—Light gray to silver gray; dun to red.

Distinguishing characteristics—Indu Brazil animals are large, massive, well proportioned, and trim. The shape of the head, the type of the ear, and the shape and placement of the hump are distinct and different from all other types or breeds of Zebu or Brahman cattle.

Disqualifications—Brindle color combinations; white markings on the nose or switch; absence of loose, thick, mellow skin; weak and improperly formed hump.

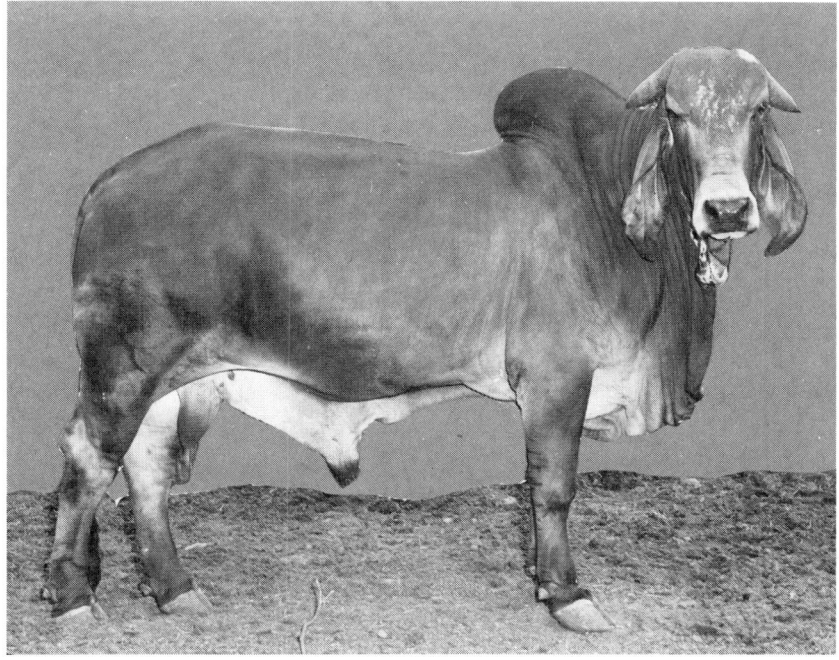

Fig. 3-30. Sahib, Grand Champion Indu Brazil bull at the 1974 Livestock Exposition, San Antonio, Texas. Bred and owned by Robert H. Coquat, Encinal, Texas. (Courtesy, Pan American Zebu Association)

Jersey

Origin—Island of Jersey.

Color—Jerseys vary greatly in color, but the characteristic color is some shade of fawn, with or without white markings.

Distinguishing characteristics—Head is clean-cut and proportionate to body. Forehead, broad and moderately dished with large, bright eyes. Body is very angular and refined. Jerseys are especially known for their well-shaped udders, strong udder attachments, and ease of calving.

Disqualifications—Total blindness, permanent lameness that interferes

Fig. 3-31. Beacon Bas Patience. Grand Champion female in 1964 All American Jersey Show. (Courtesy, The American Jersey Cattle Club, Columbus, Ohio)

with normal function, blind quarter, freemartin heifers, and animals showing signs of being operated upon or tampered with.

Limousin

Origin—In southwestern France, in the 19th Century. Presently, Limousins are found in largest numbers around Limoges. The breed takes its name from the Limousin Mountains. In France, there are approximately

Fig. 3-32. Limousin cow and calf. (Courtesy, North American Limousin Foundation, Denver, Colo.)

700,000 cattle of the Limousin breed, with about 12,000 head recorded in the Herd Book.

The first Limousin bull was imported to the North American Continent in 1967.

Color—Wheat to rust

Distinguishing characteristics—Horned. They are modern meat-type cattle—long, relatively shallow, with moderate to heavy muscling. Mature bulls average about 2,400 pounds, and cows about 1,300 pounds. The breed is noted for ease of calving and high carcass quality.

Comments—The Limousin is one of the new European breeds raised primarily for meat production, rather than the dual-purpose of meat and milk.

Lincoln Red

Origin—England, in Lincolnshire—the rugged east coast in England. They became the Lincoln Red in 1960.

Color—Deep cherry red, with occasional white markings.

Distinguishing characteristics—A long body; light birth weights and ease of calving; pigmentation; excellent milk production; fast growth rate; and good fertility. In England, they are a dual-purpose breed, with separate strains for beef and dairy; but milking herds of the breed are almost a rarity today. There are both horned and polled strains, with the polled predominating in their native land.

Comments—It appears that the main use for the breed in America will be in crossbreeding to produce F1 females since they make excellent brood cows.

Fig. 3-33. Cockerington Lord, a Lincoln Red bull. Weight, 2,275 pounds. (Courtesy, Carnation/Genetics, Hughson, Calif.)

Maine-Anjou

Origin—The breed originated in western France in the provinces of Maine and Anjou, from which it takes its name. Although cattle of the Maine-Anjou type (the result of crossing Durham cattle from England and Marcelle cows native to western France) were bred in this area from the middle of the 19th Century, the French Herd Book was not established until 1919. Today, some 50,000 cattle are registered therein.

Color—Dark red with white underline, often with small white patches on the body. Also, dark roans are found.

Distinguishing characteristics—Maine-Anjou cattle are the largest of the

Fig. 3-34. Maine-Anjou cow. (Courtesy, International Maine-Anjou Association, Kansas City, Mo.)

French breeds. They are considered a dual-purpose breed, with emphasis on beef. They are long, rather upstanding, have a particularly long rump, and are noted for rapid growth. Mature bulls weigh 2,500 pounds or above, and cows 2,000 pounds or more. Cows average about 5,000 pounds of milk per lactation, testing 3.7 percent.

Comments—The logical place for Maine-Anjou in American crossbreeding systems is as maternal sires, although use as terminal sires may occur to some extent. They appear to be a good addition to the beef herd that needs size and growth.

Marchigiana (pronounced Mar-key-jahna)

Origin—Italy, in the Marche region, around Rome. With the fall of the Roman Empire in the 5th Century, the Barbarians streamed into Rome and the surrounding provinces, bringing their cattle with them. These nomadic cattle were crossed with two native Italian breeds of the time—the Chianina and the Romagnola. Out of these crosses evolved the basic foundation stock for the Marchigiana.

Fig. 3-35. Lerico, Marchigiana bull. Weight, 2,500 lb. (Courtesy, Curtiss Breeding Service, Cary, Ill.)

The Herd Book was established in Italy in 1930. Originally, the Marchigiana was used for draft purposes. But the coming of the tractor relieved them of much of their draft duties. This led farmers to breed them for beef characteristics. Today, the Marchigiana makes up about 45 percent of the total cattle population of Italy.

Color—Grayish white, although bulls may be darker. Dark skin pigmentation, and dark muzzle, switch, and below or around the eyes. Calves are born tan but turn white at about two months of age.

Distinguishing characteristics—High growth rate; feed efficiency; mild disposition; and ability to do well under poor conditions. Mature bulls weigh 2,650 to 3,100 pounds. Mature cows weigh from 1,400 to 1,800 pounds.

Comments—In Italy, Marchigianas have been very popular in crossbreeding programs with dairy cattle.

Milking Shorthorn

Origin—England. The breed traces to a milking strain of Shorthorns developed by Thomas Bates, Kirklevington, Yorkshire, England.

Color—Red, white, or any combination of red and white.

Distinguishing characteristics—Fine horns that are rather short. Good milk production.

Disqualifications—No calf is eligible for registration unless its sire and dam were each at least 18 months of age at the birth day of the calf.

Comments—In 1949, the American Shorthorn Breeders Association split into (1) the American Milking Shorthorn Society, and (2) the American Shorthorn Association. Twenty-four years later, in 1973, the American Short-

Fig. 3-36. Clayside Cindy 2A, National Grand Champion Milking Shorthorn Cow of 1971. (Courtesy, American Milking Shorthorn Society, Springfield, Mo.)

horn Association made provision to accept Milking Shorthorn, or dual-purpose Shorthorn, blood in their Herd Book under two conditions:

1. That they trace back to the American Shorthorn Herd Book or Coates Herd Book.

2. That they be identified, and that their descendants be identified in the pedigree. This is to assure anyone looking at a registration certificate there *is*, or *is not*, any Milking Shorthorn or dual-purpose Shorthorn blood present.

Such registration is entirely voluntary. It does put Shorthorn (beef) breeders on the same basis as the Canadian Shorthorn registry. Also, Milking Shorthorn breeders may, if they so desire, register qualified animals in the American Shorthorn Association Herd Book, also; thereby double registering them.

There is no reciprocal arrangement for acceptance of any Shorthorn (beef) blood in the American Milking Shorthorn Herd Book.

Murray Grey

Origin—Australia. Had it not been for a Shorthorn cow that defied all genetic rules by failing to produce a black calf when mated to Angus bulls, the Murray Grey breed might never have been born. The breed descends from a mating, first made by the Sutherlands on "Thologolong," in the Murray Valley, near Wodonga, Victoria, Australia, in 1905, of a very light roan (almost white) Shorthorn cow and an Angus bull. This cow had 12 calves from similar matings during her lifetime, and all were dun grey. Later, with grey cattle acquired from Thologolong and from "Alberfeldy" (property of Mr. D. Ross of Holbrook, N.S.W.) Mervyn Gadd of Mt. Alfred, Victoria, set about to develop a breed. Because of the use of Angus bulls following the first cross, the Murray Grey is predominantly Angus. The Murray Grey Beef Cattle Society, of Australia, was formed in 1962; and the American Murray Grey Association, Inc., was organized in 1970.

Color—Silver grey color, which adapts them to sunny areas, as well as colder areas.

Distinguishing characteristics—Polled; ease of calving, because of small calves at birth; dark skin pigmentation, which lessens cancer eye;

Fig. 3-37. Murray Grey bull. (Courtesy, American Murray Grey Association, Inc., Billings, Mont.)

superior carcass; good dispositions. Bulls weigh around 2,000 pounds at maturity, and females from 1,100 to 1,300 pounds.

Comments—In the American Murray Grey Association, females with ⅞ Murray Grey blood can be registered. Bulls are eligible for registry with ¹⁵/₁₆ Murray Grey blood. In addition, Recordation Certificates can be obtained on any crosses of ½, or more, Murray Grey breeding. A ranch prefix (the owner's last name, the ranch name, or whatnot) is required of each breeder.

Normande

Origin—France, in the areas of Normandy, Brittany, and Maine. They are the dominant breed of cattle in France, where they comprise 60 percent of the cattle population.

Color—Primarily dark red and white. Colored patches around the eyes give them a "bespectacled" appearance and resistance to cancer eye and pinkeye; and dark pigmentation on the udder prevents sunburn.

Distinguishing characteristics—The Normande is known as a dual-purpose breed in France. It is very hardy, adaptable, and capable of surviving under a variety of environmental conditions. Mature bulls in good condition usually weigh 2,400 to 2,500 lb, although weights up to 2,800 lb have been reported. Cows weigh 1,500 to 1,600 lb and produce an average of 9,240 lb of milk per year. In France, the Normande is noted for exceptional mothering ability.

Fig. 3-38. Purebred Normande bull. (Courtesy, American Normande Association, Kearney, Mo.)

Norwegian Red (Norwegian Red-and-White)

Origin—Norway, where they are known as Norwegian Red-and-White. The first importation of the breed into the United States was made by the Southern Cattle Corporation, Memphis, Tennessee. It arrived in the United States on May 4, 1973.

Color—Red; red and white.

Distinguishing characteristics—Abundant milk production; excellent feed conversion; and good carcasses. In Norway, bulls are tested for growth rate, feed utilization, and carcass quality; and, among other things, cows are tested for fertility, ease of calving, and milking ability. Mature bulls weigh from 2,200 to 2,640 pounds, and mature cows from 1,210 to 1,430 pounds. Horned.

When the author of this book visited Norway in 1972, Professor Harold Skjervold, of the Agricultural College of Norway, advised him that the Norwegian Red-and-White cows at the College average 15,400 pounds of milk, with 4.0 percent fat.

In Norway, the Norwegian Red-and-White is a dual-purpose breed, kept for both milk and beef.

Comments—At one time, there were 8 different breeds of cattle in Nor-

Fig. 3-39. B. Nypan, Norwegian Red bull, imported from Norway by the Southern Cattle Corporation, Memphis, Tennessee. His performance record, made in Norway, follows: 360 day weight of 997 pounds; an average daily gain of 3.0 pounds from 90 to 360 days of age, on a high roughage ration. (Courtesy, Southern Cattle Corp., Memphis, Tenn.)

way. Today, the Norwegian Red-and-White has practically taken over; there are only 200 cattle of other breeds. Also, it is noteworthy that 58 percent of the cattle of Norway are registered purebreds.

Piedmont (Piemontese)

Origin—Italy, where they are the most popular breed; there are about 800,000 of them.

Color—White or pale grey with black points.

Distinguishing characteristics—About 80 percent of the bulls of the Piedmont breed are double muscled to some degree. In Italy, they report 9 percent higher dressing percentage and twice the steaks from double-muscled cattle over normal cattle. Piedmont cattle command a very considerable premium on the Italian market.

Scientific data relative to calving difficulties is lacking, but in personal visitations in Italy, the author gained the impression that this is not a serious problem. Originally, the Piedmont was considered to be a dual-purpose breed in Italy, but today it is selected and bred for beef qualities.

Comments—In Italy, the author got the impression that many of the disadvantages that U.S. cattlemen ascribe to double muscling have been minimized in the Piedmont breed. For example, they don't seem to complain too much about breeding or calving problems. Maybe they have overcome this through selection. Moreover, their system of handling many early-weaned calves—nurse-cowing (2 to 5 calves/cow) plus a starter ration, until 4 to 6 weeks of age—makes for a good start in life; even if double-muscled calves are less vigorous at birth.

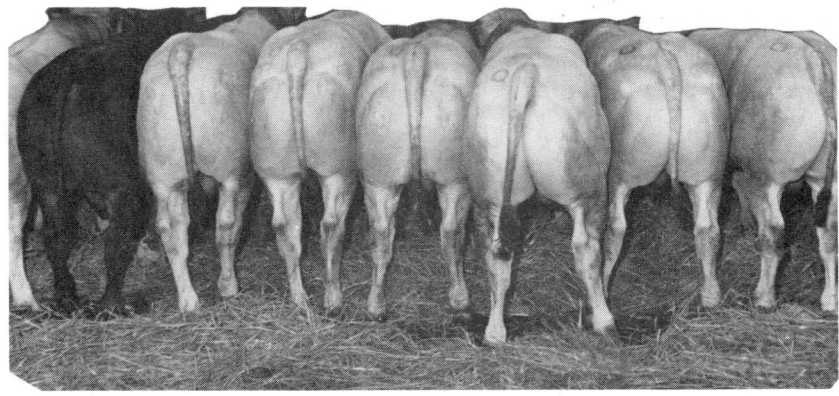

Fig. 3-40. Eight prize-winning double muscled bulls of the Piedmont breed, fed and exhibited by Ernest Gerbi, Asti, Italy. Age, 9 months; weight, 880 pounds; dressing percentage, 72% (vs 63% for normal cattle); steaks, 80% (vs 40% for normal cattle).

In Italy, double-muscled Piedmont cattle mean to the cattle industry what broad-breasted turkeys and Cornish cross broilers mean to the U.S. poultry industry. All are meat producers par excellence.

Pinzgauer (Pinzgau)

Origin—Austria and adjacent areas of Italy (South Tyrol) and Germany (Bavaria), in the Alpine region, where approximately 500,000 head of the breed are maintained.

Fig. 3-41. "Wohlmuth," 11 year old Pinzgauer cow. In 9 lactations, this cow averaged 14,392 pounds of milk with 4.28% fat. (Courtesy, Canadian Pinzgauer Association, Alberta, Canada)

Color—Chestnut brown sides with a white top line and underline, and usually, white feet. Deep orange pigment around eyes and on udder.

Distinguishing characteristics—A "beefy" breed; more so than most of the exotics, although it is classed as a dual-purpose breed in its native land. Horned. Mature bulls weigh 2,200 to 2,900 lb; and mature cows from 1,300 to 1,650 lb. The 1969 Herd Book (Germany) average milk production was 8,800 lb of milk, with 4.02% fat. This production was achieved with two-thirds of the cows on summer alpine pastures and no supplemental feed. The breed is noted for hardiness, longevity (the oldest cows and bulls reach 17 to 18 years of age), fertility, and foraging ability.

Comments—In Austria, all animals are subjected to and must pass a rigid conformation test before they can be registered. Additionally, performance of dams and daughters in milk production and butterfat content is a criterion in the selection of breeding bulls.

Genetically speaking, the breed is only distantly related to the older breeds in North America; hence, the first cross should produce maximum heterosis or hybrid vigor.

Polled Hereford

Origin—United States; in Iowa.

Color—Red with white markings, white face, and white on underline, flank, crest, switch, breast, and below the knees and hocks. White back of the crops, high on the flanks, or too high on the legs is objectionable. Likewise, dark or smutty noses are frowned upon.

Fig. 3-42. Warren Gammon (1846-1923), who, in 1902, assembled the 11 Hereford mutations from which the Polled Hereford breed developed. (Courtesy, B. O. Gammon)

Fig. 3-43. The Polled Hereford bull Giant 101740 AHR 1APHR, the sire that Warren Gammon used most extensively beginning 1901. The occurrence of the polled characteristic within the horned Hereford breed is an example of a mutation or "sport" of economic importance. Out of this gene change arose the Polled Hereford breed of cattle. (Courtesy, B. O. Gammon)

Fig. 3-44. Victorious, Polled Hereford bull. National Champion at the Fort Worth Show, January, 1973. Owned by Falklands Farm, Schelisburg, Pa. (Courtesy, Falklands Farm)

Distinguishing characteristics—Polled, with a white face. As would be expected, Polled Herefords are similar to Herefords except that they are without horns.

Disqualifications—Horned animals. No calf is eligible for registration unless its sire was at least 12 months of age at the time of conception, and its dam at least 21 months of age at time of calving.

Comments—Polled Herefords that are recorded in both the American Hereford Breeders' Association and the American Polled Hereford Breeders' Association are called double standard, whereas those that can be recorded only in the American Polled Hereford Breeders' Association are called single standard.

Polled Shorthorn

Origin—United States; in the north cental states, chiefly Ohio and Indiana.

Color—Red, white, or any combination of red and white.

Distinguishing characteristics—Polled. Other than being polled, they resemble horned Shorthorns, except there are more spotted animals among them.

Disqualifications—Horned animals. A "smutty nose," or dark nose, is objectionable.

Fig. 3-45. Kinnaber Mr. Thieman x (the "x" following the registered name means that the animal is a Polled Shorthorn), owned by SMIP Ranch, Woodside, California; Nold's Weston Shorthorns, Gettysburg, South Dakota; and Kinnaber Polled Shorthorns, Souris, Manitoba. One-third interest and possession was purchased by SMIP Ranch in 1972 for a record $30,000. This bull weaned at 635 pounds at 186 days. Weighed 1,100 pounds at 365 days, and 2,265 pounds as a 3-year-old. (Courtesy, American Shorthorn Assn., Omaha, Neb.)

Ranger

Origin—The breed originated in the United States, beginning in 1950, on the following three ranches: Barnes Livestock Company, Riverton, Wyoming; W. W. Ritchie and Family, Buffalo, Wyoming; and Watson Cattle Company, Cedarville, California. In the background of the Ranger are Hereford, Milking Shorthorn, Red Angus, Shorthorn, Beefmaster, Scotch Highland, and Brahman breeding. Educator 309 and Taryan 592 have been designated foundation sires of the Ranger breed. The name "Ranger" was selected because the breed was developed on, and is adapted to, the range areas of the West.

Color—They run the gamut of cattle colors, including both solid and broken colors.

Distinguishing characteristics—Medium size; hardy; fertile—animals have been selected to calve at an early age, at yearly intervals or less, and without assistance; adequate and persistent (over a long period of 250 to 300 days) milk production; heavy weaning weight; carcass quality.

Fig. 3-46. Educator 309, Ranger foundation sire. (Courtesy, Frank G. Watson, Ranger Cattle Company, Denver, Colo.)

Comments—The developers of the breed refer to it as, "a 'cow' breed for the cowman who must have a profitable commercial operation." Further, they claim that "the Ranger is the first and only beef breed to be created anywhere by utilizing, without arbitrary limits, the best genetic material available (from several breeds), combined with a complete performance evaluation program."

Red Angus

Origin—Scotland.
Color—Red.

Distinguishing characteristics—Polled. Similar to black Angus, except for recessive red color.

Fig. 3-47. Red Angus bull, Beckton Larkabelang 340, bred by Mrs. W. E. Forbes, Beckton Stock Farm, Sheridan, Wyoming. (Courtesy, Red Angus Association of America, Denton, Tex.)

Disqualifications—Any color other than red. White off the underline.

Comments—In 1972, the Red Angus Association expanded its enrolled Herd Book to include performance tested Angus (black Angus).

Red Brangus

Origin—United States, on Paleface Ranch, Spicewood, Texas, from Brahman X Angus cross (about 50% each), made in 1946. Registry chartered in 1956.

Color—Red. Pigmentation around the eyes.

Distinguishing characteristics—Polled; smooth, sleek coat; early sexual maturity; small calves; easy calving. Males have crest immediately forward of shoulders. Mature bulls weigh 1,800 to 2,200 pounds, and cows 1,200 to 1,400 pounds.

Disqualifications—Horns. White spotting other than on the underline, brindling or roan on the body, or black skin or mucus membrane. Long hair, or tight hide. Undersized; too rangy or too compact. Mature females with underdeveloped teats or udders. Mature males with an excessive or pendulous sheath, or the absence of a sheath.

Red Danish

Origin—Denmark, where Red Danish is the dominant breed; they ac-

Fig. 3-48. Typical Red Brangus bull. (Courtesy, American Red Brangus Association, Austin, Tex.)

Fig. 3-49. Sten, Red Danish bull at the A.I. Center at "Bellinge," in Denmark, which the author of this book visited in 1972. This bull, reputed to be the second largest in Denmark, weighs 2,970 pounds. (Photo by Audrey Ensminger)

count for 40 percent of the cows of the country. They were first imported to the United States in 1936, when Henry Wallace, who was then Secretary of Agriculture, brought over 2 bulls and 20 cows.

Color—Red, from light red to very deep mahogany. White spots on the underline are objectionable.

Distinguishing characteristics—Horned. Red Danish cattle are strikingly uniform in type, color, and production. Although they are strictly a dairy breed, the cows take on flesh readily and smooth up easily when dry. Also the bulls are meaty and well fleshed. Mature bulls weigh from 1,700 to 3,000 lb, and mature cows 1,100 to 1,600 lb. In a visit to Denmark in 1972, the author of this book gained the impression that they have larger Red Danish cattle than are being bred in the United States. The Red Danish cows of Denmark average about 10,000 lb of milk per lactation, testing 4.3 percent.

Comments—When used in beef crossbreeding programs, Red Danes will likely find their best use as a "dam breed."

Red Poll

Origin—England, in the eastern middle coastal counties of Norfolk and Suffolk.

Color—Red, varying from light to dark red.

Distinguishing characteristics—Polled; white in switch; limited white on underparts.

Disqualifications—White above underline, above switch of tail, or on legs. Bulls with white on underline forward of the navel region; or with only one testicle. Solid black or blue nose. Scurs or any horny growth. Total blindness.

Fig. 3-50. Pinpur Bonnie Bethel, with her 50-day-old bull calf. The cow was bred by Charles and Oneita Donohue, Earl Park, Indiana, and is owned by Pinney Purdue Agricultural Center, Wanatah, Indiana. (Courtesy, Red Poll Cattle Club of America, Lincoln, Neb.)

Romagnola

Origin—It is an ancient Italian breed, which originated in the lower Po Valley, from crossing the Podolic and native cattle.

Color—Solid white to light gray. The bulls have a characteristic darker color of hair about the shoulders, black color around the eyes, and a black switch.

Distinguishing characteristics—In Italy, the Romagnola is a dual purpose breed, used for both meat and milk.

Fig. 3-51. Monello Cra 3, Romagnola bull; an excellent representative of the breed. (Courtesy, American Breeders Service, De Forest, Wisc.)

They have horns which are longer and sharper than the other white breeds, and the horns grow upward and outward.

In Italy, the Romagnola has earned the title of "the rustic breed." Under harsh conditions, extreme heat or cold, and on the poorest of pastures, it is considered the hardiest of the Italian breeds.

The breed is of medium height and bone, well muscled, and early maturing. Mature bulls average 2,500 lb, and mature cows 1,500 lb. At birth, bull calves average about 110 lb, and heifers 95 lb.

Salers

Origin—In France, in the South Central area—a mountainous region.

Color—Solid, deep cherry red, with a white switch and sometimes white spots under the belly.

Distinguishing characteristics—In France, their native land, the breed is noted for rapid gain, hardiness, and adaptability. Mature bulls have an

Fig. 3-52. Vaillant, the first Salers bull on the North American continent. (Courtesy, Salers International, Calgary, Alberta, Canada)

average weight of 2,530 lb; mature cows average about 1,540 lb. The Salers was founded as a dual-purpose breed; hence, the cows are good milkers, with an average annual production of about 6,000 lb and a butterfat test of 3.7 percent.

Disqualifications—Salers International, the breed registry, makes the claim that no genetical defect and no double muscling has ever been reported in the Salers breed.

Comments—In the mountainous area, where the Salers breed originated, much cheese is produced. The name "Salers" was first applied to the cattle of this area in 1840, after a small town located in the center of the province. The Salers Herd Book was started in 1906.

Santa Gertrudis (⅝ Shorthorn and ⅜ Brahman)

Origin—United States; on the King Ranch in Texas.

Color—Red or cherry red.

Distinguishing characteristics—Hair should be short, straight, and slick. Hide should be loose, with surface area increased by neck folds and sheath or navel flap.

Disqualifications—White spots out of underline; fawn or cream color; brindling or roan condition; solid black.

Heredity deformities, such as hernia, cryptorchid, wry nose, wry tail, double muscling, malformed genitalia, undershot and overshot jaw, etc.

Comments—The Santa Gertrudis was the first new beef breed created in North America. Pertinent facts about the development of the breed are: de-

Fig. 3-53. WR Roccoco 576, Santa Gertrudis herd bull owned by Winrock Farms, Morrilton, Arkansas. (Courtesy, Jim Charlesworth, Cattle Manager, Winrock Farms)

veloped by famed King Ranch of Texas; named from the Santa Gertrudis Land Grant, granted by the Crown of Spain, on which the breed evolved, now the headquarters division of King Ranch; experimental crossing of Shorthorns and Brahmans initiated on King Ranch in 1910; in 1920, outstanding bull calf called "Monkey" was produced, and he became the foundation sire of the Santa Gertrudis breed; Santa Gertrudis Breeders International was organized in 1951.

Scotch Highland (or Highland)

Origin—Scotland.
Color—Red, yellow, silver, white, dun, black, or brindle.
Distinguishing characteristics—Long, shaggy hair; short head; long, widespread horns and heavy foretop; short legs.
Disqualifications—Polled. Mottled or spotted with white (white permissible on tip of tail or on udder).

Shorthorn

Origin—England; in the northeastern counties of Durham, Northumberland, York, and Lincoln. The Shorthorn was the first breed of cattle to have a Herd Book (the Coates Herd Book, founded in 1822 by George Coates, and one of the first of the beef breeds to be brought to America—in 1783). The name Shorthorn is derived from the fact that the early improvers of the breed shortened, through selection and breeding, the horns of the original long-horned cattle that were native to the district.

Fig. 3-54. LC Loretta's King, Scotch Highland Bull. Golden Certified Meat Sire. (Courtesy, American Scotch Highland Breeders' Association, Edgemont, S.D.)

Fig. 3-55. Sittyton, where Amos Cruickshank—the beloved herdsman of Abderdeenshire—developed the "Scotch" strain of Shorthorns. (Courtesy, Mr. Arnold Nicholson)

Color—Red, white, or any combination of red and white.

Distinguishing characteristics—Rather short, refined, incurving horns.

Disqualifications—No calf is eligible for registration unless its sire and dam were each at least 18 months of age at the birth date of the calf. A "smutty nose," or dark nose, is objectionable.

Fig. 3-56. Weston Romeo, Shorthorn bull owned by Dr. Martin Nold, Weston Shorthorns, Gettysburg, South Dakota. (Courtesy, American Shorthorn Association, Omaha, Neb.)

Simmental

Origin—The Simmental breed originated in the Simme Valley of western Switzerland, from which it derives its name. It is much older than the herd register, which was set up in Bern, Switzerland, in 1806. Slightly over 50 percent of the cattle of Switzerland are Simmental.

Color—Generally red-and-white spotted, although some are nearly solid in color; and a white face, which, like the Hereford, appears to be dominant in inheritance. The red varies from dark to a more common dilute, almost yellow, shade.

Distinguishing characteristics—The Simmental was first developed as a dual-purpose breed. They combine meat and milk to an unusually high degree, along with rapid growth rate. They are horned. Mature bulls average about 2,300 to 2,400 lb, and cows about 1,600 to 1,700 lb. The breed milk production average is about 8,000 lb, with a 4 percent butterfat test.

Disqualifications—Genetic unsoundnesses.

Comments—Size and weight gains of F₁ Simmental crossbreds in North America have been considerably in excess of most of the straightbred domes-

Fig. 3-57. A Simmental cow in Switzerland. This 6-year-old cow has had 4 calves. Two of her bull calves were exported. She has averaged 10,240 pounds of milk per lactation. (Courtesy, American Breeders Service, De Forest, Wisc.)

tic cattle with which they have been compared—enough to determine that the breed is contributing increased growth other than evidenced by heterosis alone.

The first Simmental bull was introduced in North America in 1967 by a group of southern Alberta cattlemen headed by Travers Smith of Cardston, Alberta, Canada.

South Devon

Origin—The breed originated in southern Devonshire, England, through infusion of Guernsey blood into the Devon breed. The South Devon has had its own Herd Book in England since 1891. Henry Wallace, former U.S. Vice President, brought the first shipment of South Devons to the United States in 1936; but they were not established as a breed in America at that time. Arthur V. Palmer, Big Beef Hybrids, Stillwater, Minnesota, reintroduced the breed in 1969.

Color—Medium light red color.

Distinguishing characteristics—The South Devon is a dual-purpose breed. It is of large size; mature bulls weigh 2,000 to 2,800 lb; and mature cows weigh 1,200 to 1,700 lb. Cows are heavy milkers; they average about 6,550 lb of milk per lactation, with 4.19 percent fat.

Comments—The South Devon is the only breed in England that both (1) receives a Milk Marketing Board premium for rich milk, and (2) qualifies for the British Beef Subsidy. The breed is being extolled as a "dam breed," superior in maternal traits.

Fig. 3-58. These South Devon cows are part of a milking herd in England. Multiple suckling enables 1 or 2 cows to raise the calves for the remainder of the herd. In some cases, a single cow raises 9 calves in 9 months—nursing calves from other cows that are being milked. (Courtesy, Big Beef Hybrids International Company, Stillwater, Minn.)

Sussex

Fig. 3-59. A Sussex heifer. (Courtesy, Sussex Cattle Association of America, Refugio, Tex.)

Origin—Sussex cattle are descended from the red cattle that inhabited Sussex and Kent Counties in England at the time of the Norman Conquest (1066). The first Sussex calves were registered in England in 1840, the first official Herd Book was published in 1879, and the Sussex Cattle Society was incorporated under the Companies Act in 1890. The first Sussex cattle were imported into the United States in 1883.

Color—Deep mahogany red.

Distinguishing characteristics—In England, the breed has earned the reputation as the "butcher's beast." Sussex cattle are characterized by evenness of fleshing, predominance of lean meat, and high dressing percentages. Good feet and legs enable the cattle to forage over vast areas without tiring. Sussex cattle are quiet and easy to handle. Quite a percentage of Sussex cattle are polled.

Comments—In the United States, the Sussex Cattle Association of America registers cattle in the English Herd Book. American entries in the English Herd Book commenced in 1967.

Tarentaise (Tarine)

Origin—France, in the Alps. The breed has been known as Tarentaise since 1863. The Herd Book was started in 1888. It is estimated that there are 220,000 Tarentaise cattle in France.

Color—Wheat to dark tan, resembling Jerseys in color.

Distinguishing characteristics—The breed is noted for easy calving, due to adequate pelvic capacity and small calves; vigorous calves at birth; hardiness; black hair around the eyes and pigmented udders and teats, thereby making cancer eye and sunburned teats rarities; good fertility; and milking

Fig. 3-60. Alpin, a Tarentaise bull imported from France, bred by G.A.E.C. D'Arcolliers, Yenne, Savori, France. This bull weighed 1,580 pounds at two years of age. (Courtesy, American Breeders Service, De Forest, Wisc.)

ability, with cows averaging about 8,000 pounds per lactation. The Tarentaise is smaller than most of the exotics, but mature bulls sometimes reach 2,200 pounds.

Disqualifications—In France, the breed registry advocates eliminating (although they do not "disqualify") widespread patches of white hairs or badger grey coloring, bright red or mahogany overall color, a stripe on the back lighter than the general coloring; very dark or black parts of the coat (cheeks, dewlap, shoulders, etc.); total absence of black pigmentation on mucous membranes and extremities; poor general conformation, particularly a crest-shaped tail.

Comments—The calves are small and vigorous at birth, which indicates that Tarentaise bulls are well suited for use on virgin heifers.

Texas Longhorn

Origin—The breed originated in the United States, from cattle of Spanish extraction. On his second voyage in 1493, Columbus brought Spanish cattle to Santa Domingo. In the two decades following the Civil War, an estimated 10 million head of Texas Longhorns were trailed north, either for fattening on midwestern pastures or for slaughter for the eastern market. But by 1900 the Texas Longhorn was driven to near extinction, replaced by the European breeds—the Shorthorn, Hereford, and Angus. Fi-

Fig. 3-61. A pair of Texas Longhorns. At the time the photo was taken, the bull, Sam Bass (TL #262), was 30 months of age; and the cow was 5 years of age. (Photo by Darol Dickinson, taken on his ranch at Calhan, Colo.)

nally, in 1927, the U.S. Congress appropriated $3,000 for the purpose of preserving the breed. Out of this, a nucleus herd was established by the Federal Government on the Wichita Mountain Wildlife Refuge in Oklahoma. From the Wichita herd, a second herd was started on the Fort Niobrara National Wildlife Refuge at Valentine, Nebraska. In the early 1930s, a small herd was donated to the State of Texas. The latter herd is still maintained.

The Texas Longhorn Breeders Association of America was organized in 1964. At that time, surveys showed that there were only approximately 1,500 head of genuine Texas Longhorn cattle in existence—about one-third of which were in the two Federal Refuges, the State of Texas herd, and in zoos and parks; the rest were in private herds.

Color—Texas Longhorns are characterized by a great variety of colors, in all degrees of richness, and in all possible combinations and patterns.

Distinguishing characteristics—The Texas Longhorn was shaped by nature. It is noted for fertility, ease of calving, hardiness, resistance to many common diseases, rustling ability, good feet and legs, longevity, and adaption to a wide variety of environmental conditions.

Welsh Black

Origin—In Wales, where they have long been bred as dual-purpose cattle. The Canadian Welsh Black Cattle Society was formed in 1970.

Color—Black.

Distinguishing characteristics—Horned (although there is a polled

Fig. 3-62. Harlec Einion 2nd, Welsh Black bull. (Courtesy, Curtiss Breeding Service, Cary, Ill.)

strain in Wales); high fertility; little calving difficulty; good milk production; adapted to harsh conditions of climate and forage; longevity; relative freedom from sunburned udders and cancer eye. Mature bulls weigh from 1,800 to 2,000 pounds, and cows from 1,000 to 1,300 pounds. Cows give 6,000 to 7,700 pounds of milk per lactation.

Comments—Until such time as a breed association is formed in the United States, the Canadian Society will accept registration and membership from American cattlemen.

The major impact of the breed is expected to be on brood cows—as maternal sires in a crossbreeding program.

Relative Popularity of Breeds of Cattle

Table 3-1 shows the 1973 and total registration to date of the breeds of cattle. In these changing times, the recent annual figures are more meaningful than the all-time registrations, although it is recognized that one year's data only fails to show trends. Further, it is realized that some of the exotics are so new that they have not yet established breed registry associations, or they have not had time to accumulate one year of registrations.

(Below: Herdsmen Tending Cattle. Mellon Collection. National Gallery of Art, Washington, D.C.)

TABLE 3-1

1973 AND TOTAL REGISTRATIONS OF CATTLE
IN U.S. BREED ASSOCIATIONS

Breed	1973 Registrations	Total Registrations
Angus	348,517	7,852,472
Holstein-Friesian	285,819	10,200,000
Hereford	260,676	16,570,592
Polled Hereford	168,746	3,500,000
Charolais[1]	115,003	1,025,000
Limousin	59,223	88,808
Simmental[2]	52,946	117,806
Jersey	33,104	3,565,996
Santa Gertrudis	32,360	317,442
Shorthorn & Polled Shorthorn	30,503	3,557,524
Guernsey	30,196	3,222,964
Brahman	22,750	517,737
Chianina	17,846	18,399
Beefmaster	16,572	39,596
Ayrshire	12,692	938,164
Brown Swiss	11,947	771,157
Brangus	11,740	88,248
Red Angus	7,452	54,128
Maine-Anjou	5,976	11,911
Milking Shorthorn	4,900	363,570
Gelbvieh	4,800	5,000
Murray Grey	3,000	4,000
Red Poll	2,442	249,931
Devon	1,334	49,625
Indu Brazil (Zebu)	1,019	34,084
Red Brangus	1,000	7,300
Texas Longhorn	821	7,870
South Devon	741	741
Barzona	391	3,788
Welsh Black[3]	370	1,241
Marchigiana	230	270
Blonde d'Aquitaine[4]	229	621
Sussex	50	493
Dexter	49	1,655

[1]Includes all Charbray figures.
[2]Includes ½ bloods, ¾ bloods, and purebreds.
[3]Includes ½ blood and ¾ blood cows, and purebred cows and bulls.
[4]Includes ½ blood heifers, purebred bulls, and crossbred bulls.

Comparative Breed Rating Chart

Table 3-2 points up (1) the large number of breeds of cattle available (53 breeds are listed in this table, and more new breeds will be brought in), (2) the number of economic traits that should be considered in evaluating breeds, and (3) breed differences in economic traits. As shown, there are wide differences between breeds. *But no attempt is made, or should be made, to rank the breeds by assuming that the traits are of equal value (which is not true) and averaging them.* It is important that the straightbred or purebred breeder—the breeder of seed stock—recognize the strong points and the weak points of the breed that he chooses. Armed with this informa-

TABLE 3-2—COMPARATIVE RATING (FOR BEEF PURPOSES)

Breed	Mature Size	Age of Puberty	Conception Rate	Gestation Period	Milk Production	Mothering Ability	Adaptation to Beef Management	Efficiency under Minimal Management	Calf Birth Weight	Hardiness
Angus	A	1	2	S	3	2	2	2	1	3
Ayrshire	A	2	3	A	1	3	4	5	2	3
Barzona	A	3	3	A	3	3	1	1	2	1
Beef Friesian	L	3	2	A	1	3	4	4	4	3
Beefmaster	A	3	2	A	3	1	1	1	2	1
Belted Galloway	A	2	2	S	2	2	2	2	1	2
Blonde d'Aquitaine	L	4	4	L	3	4	3	3	3	3
Braford	A	3	2	A	3	2	1	1	2	1
Brahman	A	5	5	L	3	2	2	2	2	2
Brangus	A	3	2	A	3	2	1	1	2	1
Brown Swiss	L	3	4	A	1	2	3	4	4	3
Charbray	L	4	4	L	3	3	3	3	3	2
Charolais	L	4	4	L	3	4	4	4	4	4
Chianina	L	5	3	L	3	4	4	4	5	3
Devon	A	3	2	A	3	2	2	2	1	2
Dexter	S	1	2	S	3	3	4	5	1	3
Fleckvieh	L	3	2	L	1	3	2	3	4	3
Galloway	A	2	2	S	3	2	2	2	1	2
Gelbvieh	A	3	3	A	2	3	3	3	3	3
Guernsey	S	2	3	S	1	3	4	4	2	4
Hays Converter	L	2	2	A	2	2	2	4	3	3
Hereford	A	3	2	A	4	3	2	2	2	2
Holstein-Friesian	L	3	2	A	1	3	4	5	4	3
Indu Brazil (Zebu)	A	5	5	L	3	2	2	2	2	2
Jersey	S	1	1	S	2	3	3	4	1	4
Limousin	L	3	3	L	3	3	2	3	3	3
Lincoln Red	A	3	2	A	2	2	2	3	1	2
Maine-Anjou	L	3	3	L	2	3	2	3	4	3
Marchigiana	L	4	3	L	3	3	3	3	3	3
Milking Shorthorn	A	2	4	A	2	3	3	4	2	3
Murray Grey	A	2	3	A	3	3	2	2	2	3
Normande	A	3	3	A	2	1	2	2	3	2
Norwegian Red	A	2	2	A	2	3	4	4	2	3
Piedmont	S	2	2	S	2	3	4	5	2	5
Pinzgauer	A	3	3	A	2	2	2	3	3	3
Polled Hereford	A	3	2	A	4	3	2	2	2	2
Polled Shorthorn	A	2	4	A	3	3	2	3	2	3
Ranger	A	3	3	A	3	3	2	2	3	3
Red Angus	A	1	2	S	3	2	2	2	1	3
Red Brangus	A	3	2	A	3	2	1	1	2	1
Red Danish	A	2	2	A	2	3	4	4	2	3
Red Poll	A	2	2	A	2	3	2	3	2	3
Romagnola	A	3	2	A	2	2	2	3	4	1
Salers	A	3	2	A	2	2	2	2	3	2
Santa Gertrudis	L	4	4	L	3	2	2	3	3	2
Scotch Highland	S	2	3	A	4	2	1	1	1	1
Shorthorn	A	2	4	A	3	3	2	3	2	3
Simmental	L	3	2	L	1	3	2	3	4	3
South Devon	L	3	3	A	2	3	2	3	3	3
Sussex	A	3	3	A	3	3	3	2	3	3
Tarentaise	S	1	1	S	2	2	3	3	1	1
Texas Longhorn	A	3	1	A	4	1	1	1	1	1
Welsh Black	A	3	2	A	2	2	2	2	2	1

[1]In the columns above, the following terms and values are used:

In the "Mature Size" column, A = average, L = large, and S = small.

In all the columns carrying numerical grades, grade 1 is the highest or most desirable, and grade 5 is the lowest or least desirable. For example, in the column headed "Age of Puberty" the number 1 indicates very early puberty, number 5 indicates very late puberty, and number 3 is intermediate.

In the "Gestation Period," A = average, L = long, and S = short.

OF ECONOMIC TRAITS OF 53 BREEDS OF CATTLE[1]

Calf Traits		Carcass			Bull Traits			Breed's Place in Crossbreeding		
Growth Rate	Optimum Slaughter Wt., Lb	Cutability (Muscle to Bone Ratio)	Marbling	Tenderness	Fertility	Freedom from Genital Defects	Calving Ease (Size Effect)	Maternal	Rotational	Terminal
3	1,000	2	1	2	2	2	2	X	X	
3	925	3	3	2	2	2	2	X		
3	1,000	3	3	3	2	2	2	X	X	
1	1,200	3	3	2	2	1	3	X	X	
2	1,100	2	3	3	2	2	1	X	X	
4	950	3	2	2	2	1	2	X	X	
1	1,225	1	4	2	3	2	3		X	X
3	1,100	2	3	3	3	2	1	X	X	
3	1,150	2	4	4	4	4	1	X	X	
3	1,050	2	2	3	3	2	1	X	X	
1	1,200	2	3	2	2	2	3	X	X	
2	1,200	1	4	3	3	3	4		X	X
1	1,250	1	4	2	3	2	5		X	X
1	1,350	2	3	3	3	2	2			X
3	1,050	3	2	2	2	1	1	X	X	
5	700	3	3	2	3	2	5			
1	1,250	2	3	2	2	1	5	X	X	X
4	950	3	2	2	2	1	2	X	X	
3	1,100	3	3	2	2	1	2	X	X	X
4	900	2	2	1	2	3	1	X		
2	1,150	2	3	2	2	1	3	X	X	
3	1,050	3	3	2	2	1	2	X	X	
1	1,200	3	3	2	2	1	3	X	X	
3	1,150	2	4	4	4	4	1	X	X	
5	850	2	2	1	1	3	1	X		
2	1,200	1	3	2	3	2	4		X	X
3	1,050	3	2	2	2	1	1	X	X	
1	1,250	4	2	2	3	1	2	X	X	X
3	1,150	3	3	3	2	1	3			X
3	975	4	2	2	3	1	2	X	X	
3	1,025	3	2	2	3	2	2	X	X	
3	1,100	3	3	3	3	3	3	X	X	
3	950	3	3	2	2	2	2	X	X	
3	875	1	3	2	2	2	3			X
3	1,075	3	3	2	2	2	3	X	X	
3	1,050	3	3	2	2	2	2	X	X	
3	1,000	4	2	2	3	2	2	X	X	
3	1,000	2	3	2	3	2	2	X	X	
3	1,000	2	1	2	2	2	2	X	X	
3	1,050	2	2	3	3	2	1	X	X	
3	950	3	3	2	2	2	2	X	X	
3	950	2	2	2	3	2	2	X	X	
3	1,075	3	3	3	3	3	3	X	X	
2	1,100	2	3	3	2	1	2	X	X	
2	1,150	2	3	3	4	3	2	X	X	X
4	900	3	2	2	3	1	2	X	X	
3	1,000	4	2	2	3	1	2	X	X	
1	1,250	2	3	2	2	1	5	X	X	X
2	1,125	3	2	2	3	2	3	X	X	
3	1,100	3	3	3	4	3	3	X	X	
4	900	2	2	1	1	2	1	X	X	
4	1,000	4	4	4	1	1	1	X	X	
3	1,000	3	3	2	2	2	2	X	X	

The check marks in the last three columns are the author's opinion as to the best use of the breed in selected systems of crossbreeding—as maternal foundation, rotational cross, or terminal cross.

No claim is made relative to the scientific accuracy of this table. Ratings are based on very limited experimental results (and it is difficult, if not impossible, to sample widely scattered breeds), opinions, and biases; hence, these values will be modified as more information becomes available. Also, more new breeds will be imported, and new breeds will be synthesized. In the meantime, it is hoped that this table will serve as a useful guide, give an assist in sorting fact from fiction, and stimulate the user to make further study.

tion, he is in a better position to bring about breed improvement. Also, he will know what the well-informed crossbreeding cattleman will be looking for in a particular breed. To the commercial cattleman, knowledge of such information as is presented in Table 3-2 makes it possible for him (1) more intelligently to select the breeds to use in a crossbreeding program which will complement each other, and (2) to identify those traits that should receive major attention when selecting individual breeding animals from a particular breed.

In using Table 3-2, it should be recognized that the characterization of breeds for the economically important traits that are controlled by many pairs of genes—such as weight, carcass quality, maternal ability, etc.—is more difficult than a classification based on characters controlled by only a few genes. This is so because wide variation exists within breeds. Nevertheless, breed differences do exist, and knowledge of these differences is important in deciding (1) which breed to raise in different areas of the country, and (2) which breeds to use in crossbreeding programs.

Some cattlemen may be confused because there are "too many breeds" from which to choose. However, a cattle breeding program is slow at best, and major mistakes are costly, simply because it takes so long to replace a cow herd by its own reproduction. It is far better, therefore, to consider all available alternatives at the beginning. Also, it is comforting to know that there is a large assortment of the world's most useful cattle germ plasm from which to choose for the divergent climates, objective, and economic conditions found in North America.

In addition to being acquainted with the breeds as such, the commercial cattleman who is following a crossbreeding program should select breeds that are complementary; that is, he should select those breeds that possess the favorable expression of traits which he desires in the crossbred cattle that he intends to produce.

This chart is presented as a service to cattlemen who are choosing breeds for crossbreeding programs. The author emphasizes that no claim is made relative to the scientific accuracy of this table. Ratings are based on very limited experimental results, along with the opinion, and perhaps even biases of the author; hence, these ideas will be modified as more information becomes available. In the meantime, it is hoped that it will serve as a useful guide, give an assist in sorting fact from fiction, and stimulate the user to make further study on his own. When using this table, cognizance needs to be taken of the fact that it is difficult, if not impossible, accurately to sample a widely scattered breed of cattle. Also, exceptions can be cited, both of individuals and herds, that excel or fall short of some of the values given in Table 3-2.

Table 3-2 should be taken for what it is, a statement of the author's best judgment of the relative strength of each of the breeds listed in various traits of economic importance, based on their relative performance as straightbreds in areas where they are most populous, and, presumably, well adapted. These values are not static. Rather, they will change as breeds change, as more experimental work becomes available, and as we learn more about some of the newer breeds. Moreover, additional new breeds will be imported and new breeds will be synthesized. In the meantime, Table 3-2 is presented for the guidance of cattlemen and students who are interested in studying the breeds from the standpoint of crossbreeding programs. Also, the table may serve as a challenge to purebred breeders regarding the traits that need improvement in their respective breeds.

There is no halo around any breed. During the next two decades, the competition among breeds will be intense, based on their performance as both straightbreds and crossbreds. The exotics must prove that the extra cost of their establishment is justified through (1) the value of broadening the germ plasm base, and (2) higher performance compared to local breeds. Hopefully, many breeds will meet the tests for total performance. Without doubt, some will fail and pass into oblivion. Still others may serve as part of the foundation material for developing entirely new breeds provided they possess one or two desirable traits that can be extracted for use through crossing and selection.

Choosing the Breed

No one breed of cattle can be said to excel all others in all points of beef production for all conditions. Hence, some choices must be made.

The purebred breeder can no longer choose a breed on the basis of breed preference based on (1) fancy points, (2) imparted status symbol, (3) the color of the cattle matching the color of the owner's dogs, and/or (4) numbers—because it is either very populous or very scarce. Today's cattle industry is too sophisticated and too profit-oriented to permit such luxury. Enlightened breed choice calls for anticipating the future needs of the beef cattle industry and choosing a breed for which there will be great demand (and profit). Breed choice calls for recognizing that few purebred breeders make money from selling cattle to each other; rather, that the vast majority of purebreds, especially bulls, must be sold to commercial cattlemen. Breed choice calls for choosing a breed that will be widely used by the commercial cattlemen of the area under consideration.

Fewer and fewer commercial cattle of the future will be straightbreds. Rather, they will be crossbreds in order to take advantage of complementary and heterosis. Once a decision to crossbreed has been reached, the most critical part of the plan is choosing the right parent breeds to fit the cattleman's objective and environment; then using the chosen breeds in a designed breeding system that will maximize the expression of their desirable traits and minimize the influence of their undesirable traits.

In addition to taking advantage of hybrid vigor or heterosis, crossbreeding should result in a desired combination of traits not available in any one breed. You cannot, for example, breed a bull from a cow family in which poor milkers are commonplace to a poor milking cow and get a heifer that is a superior milker. Neither can you cross breeds that are low gainers and get fast gaining offspring, although, due to hybrid vigor, the gaining ability of such offspring will be higher than the average of the parents. Those who are crossbreeding cattle also need to know which breeds to cross for a certain size animal, for improved carcass quality, and for every conceivable trait. Additionally, they need to consider all the points that follow:

● *Plan and follow a sound system of crossbreeding*—For maximum continuous high expression of heterosis, and maximum beef output per cow unit, it is necessary that the cattlemen plan and follow a sound system of crossbreeding. The systems of crossbreeding are fully covered in Chapter 5 of this book, Some Principles of Cattle Genetics; hence, repetition at this point is unnecessary.

● *Start with breed of present cow herd, then complement it*—Practical considerations favor utilizing the breed of the present cow herd (if the cattleman is already in business), with the additional breed choice made (or with additional breeds chosen) to complement the present breed so as to meet the objectives of the owner. Some cattlemen sell weaner calves, others market yearlings, and still others retain ownership through the feedlot (either their own lot or by custom feeding).

From the above, it may be concluded that the "best breed" is determined by the intended purpose. Smaller cattle have an advantage in efficiency for producing weaning calves, whereas larger cattle have the advantage if they are taken to heavier weights. Certain breeds are noted for superior carcasses. The cattle for use as a cow line in a hybrid operation are not the best for a sire line. Thus, different cattle are required for different markets, production systems, and breeding plans.

● *Cattle tend to grow and develop from conception to maturity in proportion to their mature size*—This means that—

1. As mature size increases, birth weight, weaning weight, yearling weight, and 18-month weight tend to increase; that is, at any given age, weight tends to increase.

2. As mature size increases, the age of puberty and size at any given level of finish increases; that is, at any given age, degree of maturity tends to decrease.

As a result of the above relationships, cattle of genetic potential for large size tend to gain faster and more efficiently (require less feed per pound of gain at any given weight). But they have to be carried to older ages and heavier weights in order to obtain market finish. Cows of the larger breeds tend to require more total feed for maintenance (nutrient requirements for maintenance are probably closely proportional to weight to the ¾ power).

● *Use high performing parent stock*—Both the selection of breeds and a planned program are important to the success of crossbreeding. However, the commercial cattleman must go further—he must use high performing parent stock.

● *Match or complement the breeds in crossbreeding*—In crossbreeding, the breeder can choose matching breeds. This is important because one breed simply cannot be all things. For example, a steer with a very high rate of gain simply cannot be expected from small cattle; likewise, a cow cannot be expected to have low feed intake for maintenance (efficient maintenance per cow unit) if she is very large. Characteristics which are especially desirable in the cow are: adaptation to area; small size and low maintenance requirement; high fertility; early sexual maturity; ease of calving; good milking ability; and longevity. Bulls should be selected for: large size and high rate of gain; high fertility, with libido; siring calves that make for ease in calving; lean and muscular, with high cutout; and tender palatable beef. Both cows and bulls should be sound and docile.

• *Traits low in heritability affected most by heterosis; highly heritable traits affected most by selection*—In using Table 3-2, it should be understood that the greatest amount of heterosis is expressed in those traits that are rather low in heritability, such as cow fertility, calf survivability, and weaning weight. Highly heritable traits—like rate and efficiency of gain, and carcass quality—are affected most by selection and reflect the average performance of the parent breeds.

• *Select less related breeds for maximum heterosis*—Crossing less closely related breeds will yield maximum heterosis. An example of choosing breeds to maximize heterosis among the British breeds can be found in the Hereford, Angus, and Shorthorn crossbred data. Since Shorthorns resemble Angus more than Herefords in several economic traits, the Hereford X Angus cross and the Hereford X Shorthorn cross calves have excelled Shorthorn X Angus calves in growth traits.

• *Adaption is important*—The ranking of each breed on traits of economic importance (in Table 3-2) is based upon its performance as straightbreds in regions where they are adapted and most frequently used. Hence, it would not be expected that Scotch Highland cattle would perform well in the South, or that Brahmans would perform so well in the North.

Different breeds or breed combinations for straightbreeding or crossbreeding will be desirable for different areas of the country and management systems. In the western range country, a medium-sized cow that is hardy, calves easily, has good mothering ability, and can utilize the maximum of range grass is desired. In the Corn Belt, where feed is more abundant and available with less walking, and cattle can be watched more closely, the larger, less active, and heavier milking breeds may be used. In three-breed fixed or static crossbreeding programs, maternal breeds may be selected for producing crossbred females and a sire of a larger breed with more growth potential and high carcass cutability can be used as a terminal cross for market animals.

• *Advantages of the exotics*—In general, the exotics differ from the standard British breeds in one or more of the following ways: they either grow faster; produce more milk; and/or have leaner carcasses.

• *Disadvantages of the exotics*—A major disadvantage shared by many of the exotics is the heavier calves at birth, which can lead to greater calving difficulty. Also, market discrimination, both as feeders and as slaughter animals, frequently occurs with any new breed or cross; but this generally disappears rather fast after satisfaction in the product is evident.

• *Recommended ways of using each breed in a crossbreeding program*—It is important that the choice of breeds be made with a designed breeding system in mind—one that will maximize the expression of their desirable traits and minimize the influence of their undesirable traits; thus, the last three columns of Table 3-2 indicate recommended ways of using each breed in a crossbreeding system. As noted, most of the breeds are recommended for use in both rotational crossbreeding and as a maternal component towards specific three-breed crossing. The latter are the breeds which have the genetic capacity for exceptional post-weaning muscle growth rate.

Feeding and Managing the New Breeds and Crosses

The feeding and managing of the new breeds and crosses should satisfy their requirements for (1) more rapid growth rate, (2) larger mature size, (3) increased milk production, and (4) special requirements because of difference in adaptability, winter hardiness, etc.

Bigger cattle produce faster growing calves that finish out at heavier weights. In order to meet current market requirements for finished cattle weighing 1,000 to 1,100 pounds, big, rapidly growing calves must receive more milk and be fed more liberally (with consideration given to creep feeding). Also, animals with more red meat and less fat likely require relatively more protein and less energy than the traditional British breeds.

A larger cow has larger maintenance requirements. If her minimum nutritional requirements are not met, poor conception and a lower calf crop will result. Such nutritional inadequacies are likely to become especially serious when rebreeding heifers that calved as two-year-olds.

The use of a dairy or dual-purpose breed in a crossbreeding program also calls for a special feeding and management program. Heavy-milking cows should be in good condition at calving; restricted somewhat immediately after calving, to lessen the milk flow (so as to avoid hand-milking and spoiled udders, and lessen calf scouring); flushed preceding breeding, so as to insure good conception; and fed rather liberally the latter part of the lactation period.

Breed Registry Associations

A breed registry association consists of a group of breeders banded together for the purposes of: (1) recording the lineage of their animals; (2) protecting the purity of the breed; (3) encouraging further improvement of the breed; and (4) promoting the interest of the breed. A list of the cattle breed registry associations is given in Table 3-3.

TABLE 3-3

BREED REGISTRY ASSOCIATIONS

Breed	Association	Address
ANGUS	American Angus Assn.	Lloyd Miller, Secy., 3201 Frederick Blvd, St. Joseph, Mo. 64501
AYRSHIRE	Ayrshire Breeders Assn.	David Gibson, Jr., Exec. Secy., Brandon, Vt. 05733
BARZONA	Barzona Breeders Assn. of America	Neil Hampton, Secy.-Treas., P.O. Box 1421, Carefree, Ariz. 85331
BEEF FRIESIAN	Beef Friesian Society	Maurice W. Boney, Admin. Dir., 210 Livestock Exchange Bldg., Denver, Colo. 80216
BEEFMASTER	Beefmaster Breeders Universal	Mrs. Richard E. Brown, Sec., G.P.M. South Tower, Suite 720, 800 N.W. Loop 410, San Antonio, Tex. 78216
	Foundation Beefmaster Assn.	Hal H. Nees, Exec. Secy., 201 Wyandot St., Denver, Colo. 80223
BELTED GALLOWAY	Belted Galloway Society, Inc.	Mrs. Meda McCord, Secy., P.O. Box 5, Summitville, Ohio 43962
BLONDE d'AQUITAINE	American Blonde d'Aquitaine Assn.	Ms. Lynne Dill, Exec. Dir. 217 Livestock Exchange Bldg., Denver, Colo. 80216
BRAFORD	International Braford Assn., Inc.	Hugh Whelchel, Secy., P.O. Box 1030, Fort Pierce, Fla. 33450
BRAHMAN	American Brahman Breeders Assn.	Wendell Schronk, Exec.-Secy., 1313 La Concha Lane, Houston, Tex. 77054
BRANGUS	International Brangus Breeders Assn., Inc.	Roy Lilley, Exec. Secy., 9500 Tioga Dr., San Antonio, Tex. 78230
BROWN SWISS	Brown Swiss Cattle Breeders' Assn.	Marvin L. Kruse, Secy., Box 1038, Beloit, Wisc. 53511
CHAROLAIS	American-International Charolais Assn.	Dr. J. W. Gossett, Exec. Secy., 1610 Old Spanish Trail, Houston, Tex. 77025
CHIANINA	American Chianina Assn.	Jack Phillips, Exec. Officer, P.O. Box 11537, Kansas City, Mo. 64138
DEVON	Devon Cattle Assn., Inc.	Dr. Stewart H. Fowler, Exec. Secy., P.O. Drawer 628, Uvalde, Tex. 78801
DEXTER	American Dexter Cattle Assn.	Mrs. Daisy Moore, Secy., 707 W. Water St., Decorah, Iowa 52101

(Continued)

TABLE 3-3 (Continued)

Breed	Association	Address
GALLOWAY	American Galloway Breeders Assn.	Elwood Marshall, Secy.-Treas., Room 302, Live-stock Exchange Bldg., Denver, Colo. 80216
	Galloway Cattle Society of America	Archie R. Minish, Secy., Springville, Iowa 52336
GELBVIEH	American Gelbvieh Assn.	Susan Carlson, Adm. Mgr., 202 Livestock Exchange Bldg., Denver, Colo. 80216
GUERNSEY	American Guernsey Cattle Club	Max L. Dawdy, Exec. Secy.-Treas., Box 126, Peter-borough, N.H. 03458
HEREFORD	American Hereford Assn.	H. H. Dickenson, Exec. V.P., 715 Hereford Dr., Kansas City, Mo. 64105
HOLSTEIN-FRIESIAN	Holstein-Friesian Assn. of America	Robert H. Rumler, Exec. Secy., P.O. Box 808, Brattleboro, Vt. 05301
JERSEY	American Jersey Cattle Club	J. F. Cavanaugh, Exec. Secy., 2105-J So. Hamilton Rd., Columbus, Ohio 43227
LIMOUSIN	North American Limousin Foundation	Robert H. Vantrease, Exec. V.P., 100 Livestock Exchange Bldg., Denver, Colo. 80216
MAINE-ANJOU	International Maine-Anjou Assn.	Ms. Annette Bennett, Performance Dir., 564 Livestock Exchange Bldg., Kansas City, Mo. 64102
MARCHIGIANA	American International Marchigiana Society	Bud Lester, Pres., R.R. 2, Box 65, Lindale, Tex. 75771
MILKING SHORTHORN	American Milking Shorthorn Society	Harry Clampitt, Exec. Secy., 313 So. Glenstone, Springfield, Mo. 65802
MURRAY GREY	American Murray Grey Assn., Inc.	Norman Warsinske, Exec. Secy., 1222 N. 27th, Billings, Mont. 59102
NORMANDE	American Normande Assn.	Jack Barr, Exec. V.P., P.O. Box 350, Kearney, Mo. 64060
NORWEGIAN RED	North American Norwegian Red Cattle Assn.	C/O Raymond McAnally, RFD 1, Box 346, Burns, Tenn. 37029
PINZGAUER	American Pinzgauer Assn.	Ben Livingston, Exec. Secy., 1415 Main St., Alamosa, Colo. 81101
POLLED HEREFORD	American Polled Hereford Assn.	Orville Sweet, Pres., 4700 E. 63rd St., Kansas City, Mo. 64130

(Continued)

TABLE 3-3 (Continued)

Breed	Association	Address
POLLED SHORTHORN	American Polled Shorthorn Soc.	Miss Charlotte Ekness, Secy., 8288 Hascall St., Omaha, Neb. 68124
RANGER	Ranger Cattle Co.	P.O. Box 21300, North Pecos Station, Denver, Colo. 80221
RED ANGUS	Red Angus Assn. of America	Box 776, Denton, Tex. 76201
RED BRANGUS	American Red Brangus Assn.	Bentley Syler, Office Mgr., P.O. Box 1326, Austin, Tex. 78767
RED POLL	American Red Poll Assn.	Wendell H. Severin, Secy.-Treas., 3275 Holdrege St., Lincoln, Neb. 68503
ROMAGNOLA	American Romagnola Assn.	Janna Neperud, Acting Secy., P.O. Box 8172 St. Paul, Minn. 55113
SALER	American Salers Assn.	Roy R. Moore, Jr., Secy., P.O. Box 30, Weiser, Ida. 83672
SANTA GERTRUDIS	Santa Gertrudis Breeders International	Tommy M. Cashion Exec. V.P., P.O. Box 1257, Kingsville, Tex. 78363
SCOTCH HIGHLAND	American Scotch Highland Breeders Assn.	Mrs. Margaret Manke, Secy., Edgemont, S.D. 57735
SHORTHORN	American Shorthorn Assn.	C. D. Swaffar, Secy., 8288 Hascall St., Omaha, Neb. 68124
SIMMENTAL	American Simmental Assn.	Don Vaniman, Exec. Secy., 1 Simmental Way, Bozeman, Mont. 59715
SOUTH DEVON	American South Devon Assn.	Arthur V. Palmer, Pres., P.O. Box 248, Stillwater, Minn. 55082
	International South Devon Assn.	Dr. T.E. Fitzpatrick, Exec. Secy., Lynnville, Iowa 50153
SUSSEX	Sussex Cattle Assn. of America	P.O. Drawer AA, Refugio, Tex. 78377
TARENTAISE	American Tarentaise Assn.	Floyd Tetreault, Exec. Secy., Box 1844, Fort Collins, Colo. 80522
TEXAS LONGHORN	Texas Longhorn Breeders Assn. of America	Manuel Gustamente, Jr., Exec. Secy., 204 Alamo Plaza (Menger Hotel), San Antonio, Tex. 78205
WELSH BLACK	U.S. Welsh Black Cattle Assn.	Max L. Allen, Secy., Route 1, Wahkon, Minn. 56386

Registering Animals Produced Through Artificial Insemination

Today, artificial insemination (A.I.) is more extensively practiced with dairy cattle than with any other class of farm animals. Because beef cattle are not kept under as close supervision as dairy cattle, and because of other differences in management, the practice does not lend itself as well to beef cattle production. Nevertheless, A.I. is increasing in beef cattle.

Table 3-4 summarizes the pertinent regulations relative to the registration of calves produced by artificial insemination. The three sections that follow Table 3-4—sections headed "Definition of Terms," "Blood Typing," and "Requirements Governing A.I. of Purebred Dairy Cattle"—detail certain matters pertinent to registering animals produced through A.I.

TABLE 3-4

RULES OF CATTLE REGISTRY ASSOCIATIONS RELATIVE TO
REGISTERING YOUNG PRODUCED ARTIFICIALLY[1]

Breeds	Registry Association	Procedure to Register A.I. Calves Assuming Breeder Is Not Owner of Sire, and Sire Is Approved for A.I. by the Breed Association
ANGUS	American Angus Assn. 3201 Frederick Blvd. St. Joseph, Mo. 64501	The owner of the dam must obtain one A.I. Service Certificate from the owner of the sire for each calf to be registered. Unlimited number of certificates available. When applying for A.I. Service Certificates, the sire owner must certify that to the best of his knowledge the bull has or has not transmitted any of the following genetic defects: red coat color, dwarfism, osteopetrosis, or double muscling. Any such defect will be listed on the A.I. Service Certificate, and, upon verification, such information will be made available to any member upon request. Calves may be registered after sire's death, without time limit, if his death and semen inventory are reported to the association within 90 days of death.
AYRSHIRE	Ayrshire Breeders' Assn. Brandon, Vt. 05733	*Requirements Governing A.I. of Purebred Dairy Cattle* (see section which follows Table 3-4)
BARZONA	Barzona Breeders Assn. of America P.O. Box 9 Kirkland, Ariz. 86332	The owner of the dam needs a standard Purebred Dairy Cattle Association (PDCA) type Breeding Receipt completed and signed by the inseminator. This must be attached to the application for registration or recordation. Percentage calves can be recorded in a grading-up program.

(Continued)

[1]The following authorities contributed to this table: William M. Durfey, Executive Secretary, National Association of Animal Breeders, Columbia, Mo.; Irving E. Nichols, Secretary, The Purebred Dairy Cattle Association, Inc., Peterborough, N.H.; and Keith Johnson, Beef Records Manager, Curtiss Breeding Service, Cary, Ill.

TABLE 3-4 (Continued)

Breeds	Registry Association	Procedure to Register A.I. Calves Assuming Breeder Is Not Owner of Sire, and Sire Is Approved for A.I. by the Breed Association
BEEF FRIESIAN	Beef Friesian Society 210 Livestock Exchange Bldg. Denver, Colo. 80216	Open A.I. rules. The sire must be recorded in the Beef Friesian Society Herd Books and must have a blood type recorded on file before any registration papers will be issued.
BEEFMASTER	Foundation Beefmaster Assn. 201 Wyandot St. Denver, Colo. 80223	No Distinction is made between A.I. and natural service.
	Beefmaster Breeders Universal G.P.M. Tower South, Suite 720 800 N. W. Loop 410 San Antonio, Tex. 78216	There are no A.I. rules relative to certified Beefmasters. However, in the BBU upgrading program, affidavits are required to the effect that semen from certified Beefmaster sires was used A.I. on listed base females.
BLONDE D'AQUITAINE	American Blonde D'Aquitaine Assn. RR 1, Box 14 Alta, Iowa 51002	Open A.I. policy. The owner of the dam obtains all necessary forms from the association and completes them. Percentage calves can be recorded in a grading-up program.
BRAFORD	International Braford Assn. Rt. 3, Box 490-U Fort Pierce, Fla. 33450	Open A.I. policy.
BRAHMAN	American Brahman Breeders Assn. 1313 La Concha Lane Houston, Tex. 77054	The owner of the dam must obtain one A.I. Service Certificate from the owner of the sire for each calf to be registered. This must be completed and attached to the application for registration. Unlimited number of certificates available. Calves may be registered after sire's death if his death and semen inventory are reported to the association within 90 days of death.
BRANGUS	International Brangus Breeders Assn. 9500 Tioga Drive San Antonio, Tex. 78230	The owner of the dam must obtain a completed PDCA type Breeding Receipt (or similar association form) signed by the inseminator. This inseminator must have a signature card on file with the association. Also, a between herds certificate must be obtained from the owner of the sire when the semen is purchased. Calves may be registered after sire's death if his death and semen inventory are reported to the association within 90 days of death.

(Continued)

TABLE 3-4 (Continued)

Breeds	Registry Association	Procedure to Register A.I. Calves Assuming Breeder Is Not Owner of Sire, and Sire Is Approved for A.I. by the Breed Association
BROWN SWISS (**Beef**)	Brown Swiss Cattle Breeders' Assn. Box 1038-A Beloit, Wisc. 53511	The owner of the dam must obtain a PDCA Semen Transfer from the owner of the sire when the semen is purchased. The owner of the dam obtains all other forms from the association and completes them. Calves, sired by registered bulls and out of either registered cows or grade cows in Identity Enrollment, can be registered as purebreds. Percentage calves can be recorded in a grading-up program.
BROWN SWISS (**Dairy**)	Brown Swiss Cattle Breeders' Assn. Box 1038 Beloit, Wisc. 53511	*Requirements Governing A.I. of Purebred Dairy Cattle* (see section which follows Table 3-4)
CHAROLAIS and **CHARBRAY**	American International Charolais Assn. 1610 Old Spanish Trail Houston, Tex. 77025	Open A.I. policy. The owner of the dam must obtain (1) a standard PDCA or AICA Breeding Receipt completed and signed by the inseminator, and (2) a Semen Transfer from the owner of the sire. As an alternate procedure, the owner of the sire can sign the application for registration. The Breeding Receipt must be attached to the application for registration or recordation. Percentage calves can be recorded in a grading-up program.
CHIANINA	American Chianina Assn. Box 11537 Kansas City, Mo. 64138	Open A.I. policy. The owner of dam obtains all necessary forms from the association and completes them, and he is required to be a member of the association. Percentage calves can be recorded in a grading-up program.
DEVON	Devon Cattle Assn. P.O. Drawer 628 Uvalde, Tex. 78801	Progeny of A.I. are registered without restriction provided (1) sire is licensed for A.I. by the Association, and (2) one of the following is furnished: (a) a statement from the sire owner identifying the dam and specifying the date she was bred, or (b) a breeding certificate from the inseminator identifying the licensed bull, the cow and the breeding date.
GALLOWAY	American Galloway Breeders Assn. Room 302, Livestock Exchange Bldg. Denver, Colo. 80216	Open A.I. rules. The owner of the dam obtains an application for registration from the association and has it signed by the owner of the sire. Calves may be registered after the death of their sire if his death and semen inventory are reported within 30 days of death.
GELBVIEH	American Gelbvieh Assn. 202 Livestock Exchange Bldg. Denver, Colo. 80216	Open A.I. policy. The owner of the dam obtains all necessary forms from the association and completes them. Percentage calves can be recorded in a grading-up program.

(Continued)

TABLE 3-4 (Continued)

Breeds	Registry Association	Procedure to Register A.I. Calves Assuming Breeder Is Not Owner of Sire, and Sire Is Approved for A.I. by the Breed Association
GUERNSEY	The American Guernsey Cattle Club Box 126 Peterborough, N.H. 03458	*Requirements Governing A.I. of Purebred Dairy Cattle* (see section which follows Table 3-4)
HEREFORD	American Hereford Assn. 715 Hereford Drive Kansas City, Mo. 64105	The owner of the dam must obtain (1) a completed PDCA type Breeding Receipt signed by the inseminator, and (2) one A.I. Service Certificate from the owner of the sire for each calf to be registered. These must be completed and attached to the application for registration. The number of A.I. Service Certificates issued is limited so it's wise to obtain the certificate before breeding the cow. Frozen semen may be used for 3 years after the death of the bull.
HOLSTEIN-FRIESIAN	Holstein-Friesian Assn. of America Box 808 Brattleboro, Vt. 05301	*Requirements Governing A.I. of Purebred Dairy Cattle* (see section which follows Table 3-4)
JERSEY	The American Jersey Cattle Club 2105-J South Hamilton Rd. Columbus, Ohio 43227	*Requirements Governing A.I. of Purebred Dairy Cattle* (see section which follows Table 3-4)
LIMOUSIN	North American Limousin Assn. 309 Livestock Exchange Bldg. Denver, Colo. 80216	Open A.I. policy. The owner of the dam must obtain a standard PDCA type Breeding Receipt completed and signed by the inseminator. This inseminator must have a signature card on file with the association. The Breeding Receipt must be attached to the application for registration or recordation. Percentage calves can be recorded in a grading-up program. Calves may be registered after sire's death if his death and semen inventory are reported to the Association within 90 days of death.
MAINE-ANJOU	American Maine-Anjou Assn. 564 Livestock Exchange Bldg. Kansas City, Mo. 64102	Open A.I. policy. The owner of the dam obtains all necessary forms from the association and completes them. Percentage calves can be recorded in a grading-up program.
MARCHIGIANA	American International Marchigiana Society P.O. Box 8103 St. Paul, Minn. 55113	Open A.I. policy. The owner of the dam obtains all necessary forms from the association and completes them, and he is required to be a member of the association. Percentage calves can be recorded in a grading-up program.

(Continued)

TABLE 3-4 (Continued)

Breeds	Registry Association	Procedure to Register A.I. Calves Assuming Breeder Is Not Owner of Sire, and Sire Is Approved for A.I. by the Breed Association
MILKING SHORTHORN	American Milking Shorthorn Society 313 S. Glenstone Avenue Springfield, Mo. 65802	*Requirements Governing A.I. of Purebred Dairy Cattle* (see section which follows Table 3-4)
MURRAY GREY	American Murray Grey Assn. Suite 105 1222 North 27 St. Billings, Mont. 59101	Open A.I. policy. The owner of the dam obtains all necessary forms from the association and completes them. Percentage calves can be recorded in a grading-up program.
POLLED HEREFORD	American Polled Hereford Assn. 4700 East 63 St. Kansas City, Mo. 64130	The owner of the dam must obtain one A.I. Service Certificate from the owner of the sire for each calf to be registered. This must be completed and attached to the application for registration. The number of A.I. Service Certificates issued is limited so it's wise to obtain the certificate before breeding the cow. Frozen semen can be used for one year following the bull's death.
POLLED SHORTHORN	American Shorthorn Assn. 8288 Hascall St. Omaha, Neb. 68124	The owner of the dam must obtain a standard PDCA type Breeding Receipt from the inseminator, and must send this receipt and a $10.00 per head permit fee to the association within 90 days of the service date. The owner of the dam must also obtain the signature of the owner of the sire on the application for registration. Frozen semen may be used after the death of a bull if his death and semen inventory are reported to the association within 30 days of his death.
RED ANGUS	Red Angus Assn. of America Box 776 Denton, Tex. 76201	An A.I. Bull Permit must first be completed and signed by the owner of the sire and the owner of the dam, and filed with the association. The owner of the dam then purchases a photostatic copy of this permit from the association, and attaches it to the application for registration. Calves can be registered for 15 years after the death of their sire if the sire's death and semen inventory were reported within 90 days of his death.
RED BRANGUS	American Red Brangus Assn. Box 1326 Austin, Tex. 78767	The owner of the dam must also be the owner of the sire or one of not more than 3 owners of the sire.
RED POLL	American Red Poll Assn. 3275 Holdredge St. Lincoln, Neb. 68500	The owner of the dam must obtain a standard PDCA type Breeding Receipt completed and signed by the inseminator. This inseminator must have a signature card on file with the association. The Breeding Receipt must be attached to the application for registration. Semen can be used after the death of the bull.

(Continued)

TABLE 3-4 (Continued)

Breeds	Registry Association	Procedure to Register A.I. Calves Assuming Breeder Is Not Owner of Sire, and Sire Is Approved for A.I. by the Breed Association
SANTA GERTRUDIS	Santa Gertrudis Breeders International P.O. Box 1257 Kingsville, Tex. 78363	The owner of the dam must obtain one A.I. Service Certificate from the owner of the sire for each calf to be registered. This must be completed and attached to the application for registration. The number of A.I. Service Certificates issued is limited so it's wise to obtain the certificate before breeding the cow. Percentage females can be recorded in a grading-up program without an A.I. Service Certificate. Frozen semen can be used after the death of the sire if his death and semen inventory are reported to the association within 90 days of his death.
SCOTCH HIGHLAND	American Scotch Highland Breeders' Assn. Edgemont, S.D. 57735	The owner of the dam must obtain a Breeders Certificate from the association which must be completed and signed by the inseminator on the date of service. This inseminator must have a signature card on file with the association. One copy of this form must be sent immediately to the Association. One copy is retained by the owner of the dam and is used as the "application for registration" of the resulting calf. A.I. calves must be registered before one year of age.
SHORTHORN	American Shorthorn Assn. 8288 Hascall St. Omaha, Neb. 68124	The owner of the dam must obtain a standard PDCA type Breeding Receipt from the inseminator, and must send this receipt and a $10.00 per head permit fee to the association within 90 days of the service date. The owner of the dam must also obtain the signature of the owner of the sire on the application for registration. Frozen semen may be used after the death of a bull if his death and semen inventory are reported to the association within 30 days of his death.
SIMMENTAL	American Simmental Assn. P.O. Box 24 Bozeman, Mont. 59715	Open A.I. policy. The owner of the dam obtains all necessary forms from the association, and completes them. Percentage calves can be recorded in a grading-up program.
SOUTH DEVON	American South Devon Assn. P.O. Box 248 Stillwater, Minn. 55082 International South Devon Assn. P.O. Box 1005 Newton, Iowa 50208	Open A.I. policy. The owner of the dam obtains all necessary forms from the association and completes them. Percentage calves can be recorded in a grading-up program. Calves resulting from A.I. or ova transplant accepted for registry. Part ownership of the sire not required. Blood type of sire must be filed with registry.

(Continued)

TABLE 3-4 (Continued)

Breeds	Registry Association	Procedure to Register A.I. Calves Assuming Breeder Is Not Owner of Sire, and Sire Is Approved for A.I. by the Breed Association
SUSSEX	Sussex Cattle Assn. of America, Refugio, Texas; with actual registration in the Sussex Cattle Society, 12 Lonsdale Gardens, Tunbridge Wells, Kent, England	Sussex cattle produced by means of A.I. are eligible for registration. Semen must be from a Sussex bull registered with either the American Sussex Cattle Association or the English Herd Book, or both. List of bulls used by A.I. must be filed with the Association by owner upon commencing A.I. To register an A.I. calf, a statement must be attached to the application giving the following information: identity of the approved bull, identity of the cow, date of breeding, and certification by the A.I. technician who inseminated the cow.
TARENTAISE	American Tarentaise Assn. Box 1844 Fort Collins, Colo. 80521	Open A.I. policy. A blood type record of all sires in A.I. must be on file with the association. For Tarentaise raised in the U.S. or Canada, it is recommended (1) that a sire used in A.I. have a yearling weight ratio of 110 or above, and (2) that any Tarentaise cross bull used in an upgrading program have a weaning weight ratio of 110 or above.
TEXAS LONGHORN	Texas Longhorn Breeders Assn. of America 204 Alamo Plaza San Antonio, Tex. 78205	A.I.-produced calves are eligible for registration provided: (1) semen is furnished by a reputable breeder, breeding service, or inseminator, and taken from approved bulls; (2) the cow bred by A.I. is properly identified; and (3) the application for registration is accompanied by a breeding receipt, signed by the inseminator, giving certain specified information.

DEFINITION OF TERMS[1]

The following terms apply to Table 3-4, and to A.I. in general:

• *Semen producing business*—An individual, or business entity, owning or leasing one or more dairy bulls from which the individual, or business entity, collects, processes, and distributes semen for use in the insemination of dairy cattle owned by others.

• *Semen freezing business (custom processing)*—An individual, or business entity, which for valuable consideration collects, processes and/or stores semen from dairy bulls owned by others when operating under an agreement with the Purebred Dairy Cattle Association.

• *Technicians*—

1. Affiliated Technicians are individuals authorized by a Semen Producing Business to issue breeding receipts in the name of the respective business showing use of semen from a Listed Bull.

   a. The respective Semen Producing Business is completely and solely responsible for the accuracy of each receipt issued in its name by an Affiliated Technician.

   b. Such Affiliated Technician is responsible only to the business in whose name the receipt is issued and such technician may affiliate with one or more Semen Producing Businesses.

   c. An Affiliated Technician may also be a Brown Swiss or Holstein-Friesian Technician.

2. Breed Certified Technicians are individuals who enter into an agreement with a Purebred Dairy Cattle Association (PDCA) member breed organization gaining authority to issue breeding receipts prescribed by or acceptable to the respective breed organization.

   a. Application for certification may be made with one or more PDCA member breed organizations that have such a program.

• *Standard breeding receipt*—

1. A breeding record form copyrighted by PDCA and issued in the name of, or identified as a document issued by, any Semen Producing Business operating under an agreement with PDCA.

2. A breed approved breeding record form issued only by a Breed Certified Technician in a suitable format.

• *Listed bull*—

1. A dairy bull from which semen is collected for use in the Artificial Insemination of dairy cattle.

   a. A bull must be listed each calendar year in which his semen is collected with the listing fee of $15 remitted to the respective breed organization by the owner or lessee.

   b. Once properly listed, breeding receipts with the name and registration number of such bull are continuously accepted by the respective breed organizations.

2. Failure properly to list a bull will result in his suspension as one for

---

[1]*Requirements Governing A.I. of Purebred Dairy Cattle Effective July 1, 1974*, adopted by The Purebred Cattle Association and The National Association of Animal Breeders.

which breeding receipts can be issued and accepted by the respective breed organization.

3. Standard and breed approved receipts can be accepted only when the bull shown thereon is a properly Listed Bull.

BLOOD TYPING[2]

One requirement that is common for all breed registry associations is that a record of the blood type of each bull used in artificial insemination be on file with the respective breed association. It is recommended that each bull be blood typed at the time that semen is first collected.

When having a bull blood typed for purposes of recording his blood type with the breed association, the blood should be sent to the blood typing laboratory with which that breed association is affiliated. Before collecting blood samples for blood typing, obtain the necessary instructions, report forms and blood tubes containing anti-coagulant solution from the respective breed registry association.

A list of the serology laboratories and the breeds affiliated with each one follows:

1. SEROLOGY LABORATORY
   UNIVERSITY OF CALIFORNIA
   DAVIS, CALIFORNIA 95616
   916/752-2211

Beef Friesian	Maine-Anjou
Blonde d'Aquitaine	Marchigiana
Devon	Murray Grey
Limousin	Simmental

2. CATTLE BLOOD TYPING LABORATORY
   OHIO STATE UNIVERSITY
   COLUMBUS, OHIO 43210
   614/422-6659

Angus	Red Angus
Chianina	Shorthorn
Polled Hereford	

3. IMMUNOGENETICS LABORATORY
   TEXAS A & M UNIVERSITY
   COLLEGE STATION, TEXAS 77843
   713/845-1543

Brahman	Gelbvieh
Brangus	Hereford
Brown Swiss	Santa Gertrudis
Charolais	

---

[2]A *Summary of A.I. Regulations of Beef Breed Associations*, compiled by National Association of Animal Breeders, p. 3.

REQUIREMENTS GOVERNING A.I. OF PUREBRED DAIRY CATTLE[3]

The Purebred Dairy Cattle Association (PDCA), a federation of dairy breed registry organizations (Ayrshire, Brown Swiss, Guernsey, Holstein-Friesian, Jersey, and Milking Shorthorn) adopted the following General Regulations and Operating Agreements effective July 1, 1974:

● *General requirements*—

1. All bulls from which semen is frozen must be blood typed along with his living unbloodtyped parents at the expense of the owner, with the fee paid the respective breed organization before offspring resulting from the use of frozen semen are eligible for registration.

2. Application for the registration of offspring resulting from the use of frozen semen must be accompanied by a properly completed standard or breed approved breeding receipt. If the bull whose semen was used is owned by the owner of the cow, no breeding receipt or artificial breeding form is required. This applies to all members of a syndicate ownership.

3. Calves must be blood typed with dam if in the opinion of the breed organization there is a question of parentage and the results of blood typing by the PDCA designated blood typing laboratory in all matters of identification and parentage will be accepted as official.

4. Each breed organization reserves the privilege of establishing separate conditions for the artificial insemination of animals within individual herds within its respective breed.

5. All exporters of semen will maintain complete and accurate records of all shipments of semen to foreign countries including registration name and number of bull, date of shipment, and number of inseminating units.

● *Operating agreement between Purebred Dairy Cattle Association (PDCA) and Semen Producing Business*—The Purebred Dairy Cattle Association and each of its members hereinafter known as the PDCA and_____, a Semen Producing Business meeting the terms of the definition set forth in the General Regulations hereinafter known as the SPB, do hereby enter into an agreement for the conduct of Artificial Insemination of dairy cattle in keeping with the following terms and conditions:

1. The terms of the agreement—

    a. The period ends December 31 of this calendar year.

    b. The operating agreement fee will be $100 annually and one-tenth of a cent per inseminating unit sold, payable quarterly.

    c. The operating agreement is automatically renewed effective January 1 each year providing conditions of the agreement have been met.

    d. The operating agreement may be terminated upon mutual agreement, or for cause by PDCA in keeping with the procedures for the conduct of investigations and hearings set forth in the Bylaws of one or more of the breed organization members of PDCA. The said SPB may terminate the operating agreement by giving 30 days prior written notice.

---

[3]*Requirements Governing A.I. of Purebred Dairy Cattle Effective July 1, 1974*, adopted by The Purebred Cattle Association and The National Association of Animal Breeders.

2. SPB responsibilities—

a. The said SPB agrees to assume complete responsibility for each Affiliated Technician to which semen is supplied. An Affiliated Technician is defined as one authorized by the SPB to issue breeding receipts in the name of the respective SPB, with semen from a Listed Bull.

Such technician is responsible only to the SPB in whose name the receipt is issued. Such technician may be affiliated with one or more Semen Producing Businesses, and may also be a Breed Certified Technician.

b. Said SPB agrees to—

(1) Provide each technician to which semen is supplied with pre-numbered Standard Breeding Receipt forms in triplicate described on page 12 of the Requirements.

(2) Maintain records that will permit producing complete individual herd breeding records for semen used by the Affiliated Technician for the preceding three years.

(3) Label each inseminating unit with the—

(a) Registered name and number of the bull.

(b) Collection Code.

(c) Code of the SPB assigned by USDA-PDCA.

(4) Provide the owner of each cow bred with a standard breeding receipt where the resulting progeny are to be considered eligible for registry or eligible for consideration in identification programs.

(5) Report the termination of each affiliated technician to each PDCA member organization, giving the date and reason with specific details of any irregularities or misconduct within 15 days of termination.

(6) Maintain records for six years showing all—

(a) Semen collections.

(b) Semen shipments.

(c) Semen inventory—central storage.

(d) Individual carbon receipts showing record of all inseminations.

(7) Submit within 15 days after the end of each quarter, to the respective breed organization, a list of bulls whose semen was collected, processed and stored for the first time during the current calendar year together with a fee of $15 for each bull so listed.

(8) Accept the results of blood typing by the Blood Typing Laboratory designated by the Association in matters of identification and parentage.

(9) Assume financial responsibility for clarifying questions of identification and parentage created by said SPB or its representative when proven at fault.

(10) Accept full and complete responsibility for all information shown on each breeding receipt issued in its name.

(11) Accept inspection by a representative of PDCA, or any one of its member breed organizations, making all records, except financial,

available for examination by the person making the inspection.

(12) Blood type all bulls whose semen is frozen and all living un-bloodtyped parents.

(13) Maintain complete and accurate record of all semen shipments to foreign countries including registered name and number of bull, date of shipment and number of inseminating units.

3. The PDCA and its members agree to—

a. Respect the right of said SPB to exercise complete jurisdiction over each Affiliated Technician for which it assumes complete responsibility.

b. Be responsible for, participate in, and/or aid in providing inspection service on a uniform basis for all SPB's.

c. Cooperate to the extent reasonable and practical in clarifying or resolving problems of parentage or identification.

Signed: _____

Secretary, Purebred Dairy Cattle Association

Date: _____

Signed: _____

Manager, Semen Producing Business

Date: _____

● *Operating agreement between Purebred Dairy Association (PDCA) and Semen Freezing Business*—The Purebred Dairy Cattle Association hereinafter known as PDCA and _____, a Semen Freezing Business meeting the terms of the definition set forth in the general regulations hereinafter known as the SFB do hereby enter into an agreement for the collecting, processing, storing and distributing of semen from dairy bulls owned by others in keeping with the following terms and conditions.

1. Terms of the agreement—

a. The period ends December 31 of this calendar year.

b. The operating agreement fee will be $100 annually and $2.00 per dairy bull (first collection annually) with payments made quarterly.

c. The operating agreement is automatically renewed effective January 1 each year provided conditions of the agreement have been met.

d. The operating agreement may be terminated upon mutual agreement, or for cause by PDCA in keeping with the procedures for the conduct of investigations and hearings set forth in the Bylaws of one or more of the breed organization members of PDCA. The SFB may terminate the operating agreement by giving 30 days prior written notice.

2. SFB responsibilities—

a. Label each inseminating unit with the—

(1) Registered name and number of the bull.

(2) Collection Code.

(3) Code of the SFB assigned by USDA-PDCA unless this frozen semen is to be used by a SPB whose code is printed with code of bull by request.

b. Maintain records for six years showing all—

(1) Semen collections.

(2) Semen shipments.

(3) Semen inventory—central storage.

c. Report each bull from which semen is collected at time of first collection each year to the respective breed organization on PDCA Form 4, Semen Freezing Report.

d. Identify each bull from which semen is collected with certificate of registration.

e. Accept inspection from a representative of the PDCA, or one of its member breed organizations, making all records, except financial, available for examination by the person making the inspection.

f. Inform owner of bull of blood typing requirement and communication with respective breed organization.

g. Assume financial responsibility for clarifying questions of identification and parentage created by said SFB when proven at fault.

3. The PDCA and its members agree to—

a. Be responsible for, participate in, and/or aid in providing inspection service on a uniform basis for all SFB's.

b. Cooperate to the extent reasonable and practical in clarifying or resolving problems of parentage or identification.

Signed: _____

Secretary, Purebred Dairy Cattle Association

Date: _____

Signed: _____

Manager, Semen Freezing Business

Date: _____

Cattle Magazines

The cattle magazines publish news items and informative articles of special interest to cattlemen. Also, many of them employ field representatives whose chief duty is to assist in the buying and selling of animals.

In the compilation of the list herewith presented (see Table 3-5), no attempt was made to list the general livestock magazines of which there are numerous outstanding ones. Only those magazines which are devoted to a specific class or breed of beef, dual-purpose, or dairy cattle are included.

TABLE 3-5

BREED MAGAZINES

Breed	Magazine	Address
ANGUS	Aberdeen-Angus Journal	808 Des Moines St., Webster City, Iowa 50595
AYRSHIRE	Ayrshire Digest	1 Union St. Brandon, Vt. 05733
BRAHMAN	The Brahman Journal	P.O. Box 389, Sealy, Tex. 77474

(Continued)

TABLE 3-5 (Continued)

Breed	Magazine	Address
BRANGUS	Brangus Journal	9500 Tioga Dr., San Antonio, Tex. 78230
BROWN SWISS	The Brown Swiss Bulletin	P.O. Box 1038, Beloit, Wisc. 53511
CHAROLAIS	Charolais Banner	Box 308, Shawnee Mission, Kan. 66201
	North American Charolais Journal	1120 Luke St., Irving, Tex. 75061
CHIANINA	American Chianina Journal	P.O. Box 11537, Kansas City, Mo. 64138
DEVON	Devon Cattle Roundup	P.O. Drawer 628, Uvalde, Tex. 78801
GELBVIEH	The Gelbvieh Magazine	Lyndon, Kan. 66451
GUERNSEY	Guernsey Breeders' Journal	70 Main St., Peterborough, N.H. 03458
HEREFORD	American Hereford Journal	715 Hereford Dr., Kansas City, Mo. 64105
HOLSTEIN-FRIESIAN	Holstein-Friesian World	Lacona, N.Y. 13083
JERSEY	Jersey Journal	2105-J South Hamilton Rd., Columbus, Ohio 43227
LIMOUSIN	International Limousin Journal	P.O. Box 2205, Fort Collins, Colo. 80522
	Limousin Journal	P.O. Box 2205, Fort Collins, Colo. 80522
MILKING SHORTHORN	Milking Shorthorn Journal	313 So. Glenstone, Springfield, Mo. 65802
POLLED HEREFORD	Polled Hereford World	4700 E. 63rd St., Kansas City, Mo. 64130
RED ANGUS	American Red Angus	Box 776, Denton, Tex. 76201
RED POLL	Red Poll News	3275 Holdrege St., Lincoln, Neb. 68503
SANTA GERTRUDIS	The Santa Gertrudis Journal	The Letz Co., P.O. Box 2386, Fort Worth, Tex. 76101
SHORTHORN	The Shorthorn Country	8288 Hascall St., Omaha, Neb. 68124
SIMMENTAL	Simmental Shield	P.O. Box 511, Lindsborg, Kan. 67456

QUESTIONS FOR STUDY AND DISCUSSION

1. Must a new breed of cattle be approved by someone, or can anyone start a new breed?
2. Give the step by step typical history of the formation of the various breeds of cattle.
3. Trace the crossbreeding of U.S. beef cattle, beginning with the Longhorn. Why didn't cattlemen continue to upgrade Longhorns with Shorthorns only; why did they turn to Herefords and eventually to Angus?
4. Why did supporters of the British breeds (Shorthorns, Herefords, and Angus) not consider the introduction of the Brahman a serious threat, whereas the introduction of the Charolais disturbed them greatly?

5. What breeds would you classify as among the exotics? What prompted the bringing in of the exotics? How do you account for the fact that the exotics came quickly and in large numbers?

6. Are purebreds important in an era of crossbreeding? If so, why?

7. Why are we currently bringing to North America so many dual-purpose type and draft-type cattle?

8. Why have beef-type and dairy-type (a) live animals, and (b) carcasses become more alike in recent years?

9. In outline form, list the (a) distinguishing characteristics, and (b) disqualifications of each breed of beef and dual-purpose cattle; then discuss the importance of these listings.

10. With what breed(s) is each of the following associated?
    a. Robert Bakewell.
    b. Thomas Bates.
    c. Henry Clay.
    d. Clear Creek Ranch.
    e. Amos Cruickshank.
    f. Warren Gammon.
    g. George Grant.
    h. King Ranch.
    i. Lasater Ranch.
    j. Monkey.
    k. Polled Durhams.
    l. Single Standard and Double Standard.

11. Justify any preference that you may have for one particular breed of beef or dual-purpose cattle.

12. Obtain breed registry association literature and a sample copy of a magazine of your favorite breed of cattle. (See Table 3-3 and 3-5 for addresses.) Evaluate the soundness and value of the material that you receive.

13. What factors should be considered in selecting the breeds for a crossbreeding program?

14. How should the feeding of the exotics differ from the feeding of Shorthorns, Herefords, and Angus?

15. Are the rules of cattle registry associations relative to registering young produced artificially becoming less rigid? If so, why?

16. Give the names and addresses of three different laboratories where a cattleman may send a blood sample for typing.

SELECTED REFERENCES

Title of Publication	Author(s)	Publisher
Aberdeen-Angus Breed: A History, The	J. R. Barclay A. Keith	The Aberdeen-Angus Cattle Society, Aberdeen, Scotland, U.K., 1958
Breeds of Livestock, The	C. W. Gay	The Macmillan Co., New York, N.Y., 1918
Breeds of Live Stock in America	H. W. Vaughan	R. G. Adams and Co., Columbus, Ohio, 1937
Dairy Cattle Breeds	R. B. Becker	University of Florida Press, Gainesville, Fla., 1973
Hereford in America, The	D. R. Ornduff	The author, Kansas City, Mo., 1957
Hereford Heritage	B. R. Taylor	The author, University of Arizona, Tucson, Ariz., 1953
History of Linebred Anxiety 4th Herefords, A	J. M. Hazelton	Assoc. Breeders of Anxiety 4th Herefords, Graphic Arts Bldg., Kansas City, Mo., 1939
Lasater Philosophy of Cattle Raising, The	L. M. Lasater	Texas Western Press, The University of Texas, El Paso, Tex., 1972
Modern Breeds of Livestock	H. M. Briggs	The Macmillan Co., New York, N.Y., 1969

(Continued)

Title of Publication	Author(s)	Publisher
Santa Gertrudis Breeders International Recorded Herds	R. J. Kleberg, Jr.	Santa Gertrudis International, Kingsville, Tex., 1953
Santa Gertrudis Breed, The	A. O. Rhoad	Inter-American Institute of Agric. Sciences, Turrialba, Costa Rica, 1949
Shorthorn Cattle	A. H. Sanders	Sanders Publishing Co., Chicago, Ill., 1918
Stockman's Handbook, The Fourth Edition	M. E. Ensminger	The Interstate Printers & Publishers, Inc., Danville, Ill., 1970
Study of Breeds in America, The	T. Shaw	Orange Judd Co., New York, N.Y., 1900
Story of the Herefords, The	A. H. Sanders	Breeders Gazette, Chicago, Ill., 1914
Types and Breeds of African Cattle	N. R. Joshi E. A. McLaughlin R. W. Phillips	Food and Agriculture Organization of the United Nations, Rome, Italy, 1957
Types and Breeds of Farm Animals	C. S. Plumb	Ginn and Company, Boston, Mass.,
World Cattle	J. E. Rouse	University of Oklahoma Press, Norman, Okla., Vols 1 & 2, 1970, Vol. 3, 1973
World Dictionary of Breeds Types and Varieties of Livestock, A	I. L. Mason	Commonwealth Agricultural Bureaux, Slough, Bucks, England, 1951
Zebu Cattle of India and Pakistan	N. R. Joshi R. W. Phillips	Food and Agriculture Organization of the United Nations, Rome, Italy, 1953

Also, breed literature pertaining to each breed may be secured by writing to the respective breed registry associations. (See Table 3-3 for the name and address of each association.)

SELECTING BEEF CATTLE[1]

Contents   **Page**

Virgil, great Roman poet who was born on a farm in Italy in 70 B.C., in his *Georgics*—Book III, a poem written in 37 to 30 B.C., dealing with the rearing of herds and flocks, had the following to say about the selection of cattle:[2]

> "Distinguish all betimes with branding fire,
> To note the tribe, the lineage and the sire;
> Whom to reserve for husband of the herd;
> Or who shall be to sacrifice preferred;
> Or whom thou shalt to turn thy glebe allow,
> To smooth the furrows, and sustain the plough:
> The rest, for whom no lot is yet decreed,
> May run in pastures, and at pleasure feed."

---

[1]The author expresses his sincere appreciation to the following persons who authoritatively reviewed this chapter: Dr. Harlan D. Ritchie, Department of Animal Husbandry, Michigan State University, East Lansing, Michigan; Dr. Clayton C. O'Mary, Department of Animal Science, Washington State University, Pullman, Washington; and Dr. John F. Lasley, Department of Animal Husbandry, University of Missouri, Columbia, Missouri.

Also, the author is most grateful to Professor Richard F. Johnson, Head, Department of Animal Science, California Polytechnic State University, San Luis Obispo, California, a fine artist and able animal scientist, who did all the drawings for this chapter.

[2]From *Georgics*—Book III, written by Virgil (Publius Vergilius Maro, 70-19 B.C.), Dryden's translation. Of *Georgics*, the poem of the land, it has been said, "The stateliest measure ever molded by the lips of man."

The foregoing verse gives conclusive evidence that the power of selection in cattle breeding was known and practiced long ago.

Whether establishing or maintaining a herd, cattlemen must constantly appraise or evaluate animals; they must buy, sell, retain, and cull. Where the beef cattle herd is neither being increased nor decreased in size, each year about 46 percent of the heifers, on the average, are retained in order to replace about 20 percent of the old cows.[3] In addition, bulls must be selected and culled, and steers and other surplus animals must be marketed. Thus, in normal operations, producers are constantly called upon to cull out animals, to select replacements, and to market surpluses. Each of these decisions calls for an evaluation or appraisal.

Cattlemen are ever aware of market demands as influenced by consumer preferences. Also, the great livestock shows throughout the land have exerted a powerful influence in molding cattle types.

It must be realized, however, that only a comparatively few animals on the farms and ranches are subjected annually to the scrutiny of market specialists or experienced show-ring judges. Rather, the vast majority of purebred animals and practically all commercial herds are evaluated by practical stockmen—men who select their own foundation or replacement stock and conduct their own culling operations. Such stockmen have no interest in the so-called breed fancy points. These practical operators may not be able to express fluently their reasons for selecting certain animals while culling others, but usually they become quite deft in their evaluations. Whether young animals are being raised for market or for breeding stock, successful livestock operators are generally good judges of livestock.

ESTABLISHING THE HERD

Except for the comparatively few persons who keep animals merely as a hobby, farmers and ranchers raise cattle because, over a period of years, they have been profitable, provided the production and marketing phases were conducted in an enlightened and intelligent manner. Therefore, after it has been ascertained that the feeds and available labor are adapted to cattle production, and that suitable potential markets exist, the next assignment is that of establishing a herd that is efficient from the standpoint of production and that meets market demands. This involves a number of considerations.

Purebred or Commercial Cattle

Broadly classified, cow-calf producers are either (1) purebred breeders, or (2) commercial cattlemen. Purebred breeders are a small, but select, group. The vast majority of cattlemen are commercial producers. An estimated 96 percent of the cattle of America are nonpurebreds.

---

[3]A herd of 100 cows will produce about 44 heifers each year where there is an 88% calf crop. Hence, 46% of the heifers will be needed to replace 20% of the cows ($44 \times .46 = 20$).

Purebred breeders produce "seedstock" for other purebred breeders, and both bulls and purebred females for F₁ heifer programs of commercial cattlemen. Purebred breeders need to be more than good cattlemen. They should be knowledgeable relative to breeding systems, pedigrees and registration, production testing, advertising, sales, and other special marketing methods, and perhaps fitting and showing. Also, they should be thoroughly knowledgeable relative to breeding commercial cattle the modern way in order that the needs of the commercial producer will be reflected in their breeding programs. Indeed, both types—purebred and commercial—are interdependent.

For the man with experience and adequate capital, the breeding of purebreds offers unlimited opportunities. It has been well said that honor, fame, and fortune are all within the realm of possible realization in the purebred business, but it should also be added that only a few achieve this high calling.

The goal of most commercial beef operations is to convert the production of the land—grass and crops—into dollars through the traditional cow-calf operation. Usually the product is marketed at the weaning stage, although some commercial cattlemen carry them to the yearling stage, or even finish them for market. More and more commercial herds will be crossbreds, simply because the economics favor crossbreeding accompanied by complementary genes and heterosis, and because the crossbreds can be used to produce beef according to specification.

As a group, commercial cattlemen are intensely practical. No animal meets with their favor unless it produces meat over the block at a profit. The commercial cattle business requires less outlay of cash than the purebred business on a per animal basis; and less knowledge relative to the many facets of the purebred business—pedigrees, promotion, etc.

Selection of the Breed or Cross

No one breed of cattle can be said to excel all others in all points of beef production for all conditions.

For the purebred breeder, the selection of a particular breed is most often a matter of personal preference, and usually the breed that he likes is the one with which he will have the greatest success. Where no definite preference exists, however, it is well to choose the breed that is most popular in the community—if any one breed predominates. If this procedure is followed, it is often possible to arrange for an exchange of animals, especially bulls. Moreover, if a given community is noted for producing good cattle of a particular breed, there are many advantages from the standpoints of advertising and sales.

Germ plasm choice for the commercial cattleman is becoming increasingly difficult because of the large number of breeds and breed cross combinations now available. With only 3 main breeds (Shorthorns, Herefords, and Angus), there are 3 single-cross combinations and 3 three-way cross combinations from which to choose. However, with 10 breeds there are 45 single-cross combinations and 360 possible three-way combinations from which to select. Of course, there are more than 10 breeds; 53 breeds are described in this book (Chapter 3, Breeds of Cattle), and more breeds will come.

Milking Ability

Weaning weight is the most important trait affecting net income in a cow-calf operation; and weaning weight of beef calves is influenced more by

Fig. 4-1. "Lotta" milk—"lotta" calf. Weaning weight of calves is influenced more by the dam's milk production than by any other single factor. (Courtesy, Red Angus Association of America, Denton, Tex.)

the dam's milk production than by any other single factor. For this reason, the pressure is on to increase the milk production of beef cows. For this reason, also, this subject "Milking Ability"—is fully covered herein.

Performance testing programs which emphasize weaning weight automatically result in selection for higher milk production. But, more rapidly to increase beef production, dairy breeding is being infused into many commercial beef herds.

Research has shown a strong correlation between the level of milk production of cows and the weaning weight of their calves. Also, conversion of milk to beef is rather efficient—on the order of 10 pounds of milk to one additional pound of weaned calf, although conversion may not be quite as efficient at higher levels of milk production.

● *Will more milk produce more beef?*—Beef showmen have long known that milk is an important ingredient in a calf's diet. As a result, they use nurse cows or milk replacer when fitting calves. Now there is unmistakable experimental evidence that more milk does, indeed, produce more beef.

Research at the Arkansas, Texas, Georgia, Oklahoma, Beltsville (USDA)

and Alberta (Canada) Stations has shown that 50 to 80 percent of the variåbil-
ity in the weaning weight of calves is due to the milk production of their
dams—that weaning weight is influenced by milk production more than all
other effects combined.

1. *What the Arkansas experiment showed*—In a study involving
Hereford cows, the Arkansas Station reported the following results:[4]

Average Daily Milk Production/Cow	Weaning Weight of Calves at 8 Months
(lb)	(lb)
5.0	354
15.6	475

Thus, a 10-pound per day difference in milk production made for 121
pounds greater weaning weight. With 50¢ calves, that's $60.50 per head
greater returns.

2. *What the Georgia Station found*—Using Hereford cows, the Georgia
Station found the following:[5]

Average Daily Milk Production/Cow	Weaning Weight of Calves at 205 days
(lb)	(lb)
4.9	350
12.2	475

Hence, 7.3 pounds more milk per day during the suckling period produced 125 pounds more calf at weaning time.

● *How much milk will a beef cow give?*—There's more difference
within than between breeds when it comes to milk production. Also, it's
difficult, if not impossible from a practical standpoint, to secure a truly representative sample of a widely scattered, populous breed of beef cattle.

Studies reveal that beef cattle range from less than 1 lb of milk per day
up to 25 lb, but it appears that most of them average around 10 lb per day. By
contrast, it's noteworthy that the nation's 12 million dairy cows average
about 40 lb per day. That's a big difference!

Table 4-1 shows what some of the experiment stations have found relative to the milk production of beef cows.

● *How many pounds of milk required to produce a pound of gain?*—
The pounds of milk required for a pound of gain vary. Breed, age of calf,
quantity of milk, and many other factors enter in. For example, the Texas
Station reported that Hereford calves required fewer pounds of milk per

---

[4]Gifford, Warren, Record of Performance Tests for Beef Cattle in Breeding Herds, *Bull. 531,*
Agricultural Experiment Station, University of Arkansas College of Agriculture, Fayetteville,
Ark., 1953, p. 28.

[5]Neville, W. E., Jr., To Spread Beef Gains You Need Milk, The University of Georgia,
reported in *Livestock Breeder Journal*, Sept., 1971, pp. 40, 42.

TABLE 4-1

MILK PRODUCTION OF BEEF COWS

Station	Breed	Average Daily Milk Production/Cow	Comments
		(lb)	
Arkansas[1]	Angus	8.5	The 8.5 figure was for a lactation period of 8 months. During the first 3 months, the Angus cows averaged 9.7 lb of milk per day.
Texas[2]	1. Hereford 2. Brahman 3. Brahman × Hereford	7.3 9.6 13.6	Crossbred dams yielded more milk and had faster growing calves than did Hereford or Brahman dams.
USDA, Beltsville[3]	Shorthorn	17.5	Lactation period of 252 days. Milk production estimated by weighing calves before and after nursing.
In coop. with Kansas State[3]	Shorthorn	13	Hand milked; complete lactation of 365 days or less.

[1]Gifford, Warren, Record of Performance Tests for Beef Cattle in Breeding Herds, *Bull. 531*, Agricultural Experiment Station, University of Arkansas College of Agriculture, Fayetteville, Ark., 1953, p. 21.
[2]Todd, J. C., H. A. Fitzhugh, Jr., and J. K. Riggs, Effect of Breed and Age of Dam on Milk Yield and Progeny Growth, *Beef Cattle Research in Texas*, Texas A & M University, College Station, Tex., 1969, p. 39.
[3]Dawson, W. M., A. C. Cook, and Bradford Knapp, Jr., Milk Production of Beef Shorthorn Cows, *Journal of Animal Science*, Vol. 19, No. 2, May, 1960, pp. 502-508.

pound of gain than Angus or Charolais.[6] But, based on a number of studies, it appears that, on the average, each additional 10 lb of milk will produce 1 lb of weaned calf. Hence, 1 gallon (8.6 lb) more milk per day may be expected to produce about a 150-lb heavier calf at weaning age.

● *How can you get more milk?*—Of course, if cows are not fed, either enough or properly, they will not produce. But, from a breeding standpoint, two main approaches may be used to get more milk: (1) selection; and (2) infusion of dairy breeding.

1. *Selection*—In dairy cattle, we know that milk yield is 25% heritable. Of course, few studies on milk production in beef cattle have been made. But recent work indicates that in beef cattle milk yield averages 32% heritable. So, for purposes of this discussion, let's use the 32% figure. This means that, if from a beef herd that averages 1,000 lb of milk in 205 days, you select top milk producing beef cows that average 2,000 lb, then mate them to a beef bull which you select in similar manner for 2,000-lb production, you may expect 32% apparent superiority of the parents to be expressed in the offspring. Since the selected parents in this case averaged 1,000 lb of milk higher than the herd, 32% of 1,000 = 320 lb. Thus, the offspring could be expected to average 1,320 lb of milk. This shows that you can soon improve the milking ability of a beef herd by giving proper attention to milk produc-

[6]Ibid., p. 38.

tion when selecting the herd bull. The best method is to select a bull whose sire and dam have produced good milking daughters. In beef cattle, where milk production is not actually measured, selecting individuals with superior weaning weights indirectly selects for higher milk production.

2. *Infuse dairy breeding, or higher milking strains of beef cattle*—Of course, it's possible rapidly to increase milk production to a much higher level by infusing dairy breeding than by selecting from within most existing beef breeds. The following results with four-year-old Hereford, Hereford X Holstein, and Holstein cows, reported by the Oklahoma Station, show the effect of infusing Holstein breeding:[7]

Breed	Av. Milk/Day; 205 Days	Av. Wt. of Calves; 240 Days
	(lb)	*(lb)*
Hereford	13.5	575
Hereford X Holstein	20.3	642
Holstein	27.2	708

As noted, the Holstein calves had a weight advantage of 133 pounds over the straightbred Herefords, whereas the Hereford X Holsteins enjoyed a weight advantage of 67 pounds over the Herefords. Clearly, milk and beef go hand in hand, and it doesn't make much difference whether the cow is red, magpie, black or polka dot in color.

● *What else do the experiments and experiences tell us?*—The experiments and experiences also tell us the following about milk and beef:

● The highest relationship of milk to calf weight gains is during the first 60 days of the calf's life. This is because the calf has need for a highly concentrated source of energy at a time when its consumption capacity is limited.

● Conversion of milk to beef may not be as efficient at high levels of milk production. Thus, according to the Georgia Station, instead of going beyond 3,000 lb of milk, in beef cows it may be more efficient to creep feed the calves.

● Maximum milk production of beef cows occurs during the first two months, then declines; whereas, with dairy cows, milk production increases up to the fourth to sixth month, then declines gradually. The level at which milk production is maintained as lactation progresses is what the dairyman calls "persistency."

● A good rule of thumb is that a beef calf will consume 1 lb of milk daily for each 10 lb body weight. Hence, a 300-lb calf should get 30 lb of milk daily, or approximately 3½ gallons—more than twice as much as most beef cows give. In this connection, it is noteworthy that the California Station reported that calves under 3 months of age consumed up to 50 lb of milk per day. If a cow produces more milk than her calf will take, the pressure will

---

[7]Lusby, K. S., *et al.*, Performance of Four-Year-Old Hereford, Hereford X Holstein and Holstein Females As Influenced by Level of Winter Supplementation Under Range Conditions, *Animal Sciences and Industry Research Report*, April 1974, p. 56.

build up in the udder and the drying-up process will usually take care of the situation; and there won't be as many spoiled udders as most cattlemen fear. So, unless there is an udder problem (like a sunburned udder or a big teat), it's not necessary to milk her out. Nevertheless, an excessive flow of milk in commercial cattle is undesirable, for no one likes to milk a wild cow.

• In selecting for milk, some valuable beef characteristics may be lessened because the higher the milk production the greater the dairy temperament, or angularity; they're built to convert a larger proportion of their feed to milk. But, of course, with the emphasis on red meat in recent years, beef type and dairy type have moved closer together.

• There is no indication that postweaning gains are affected by milk production of the dam up to 3,500 lb. Beyond this may be another story, because it is generally recognized that creep feeding results in slower and costlier gains during the feeding period following weaning.

• Bull calves suckle more frequently than steer or heifer calves. Consequently, cows with bull calves tend to give more milk.

• Cows nursing crossbred calves give more milk than cows suckling straightbred calves, perhaps due to the fact that the crossbreds are more vigorous nursers.

• Replacement heifers that become too fat as calves or yearlings will produce less milk during their first lactation, due to fat deposits in the mammary gland area. The solution: select replacement heifers when they're 3 to 4 months of age, then limit their preweaning and postweaning gains to 1.5 lb per head per day.

• Cows which tend to produce large amounts of milk may become very thin on poor pasture and may have a longer interval between calving and rebreeding than poorer milking cows on similar pasture.

• *Summary relative to milking ability*—Researchers are just now beginning carefully to study the milk production of beef cows. There may well be several answers to the two-pronged question, "How much milk should a beef cow produce, and what's the best way to increase milk production in a beef herd?" On a poor range where feed is sparse, a relatively low level of milk production may be necessary to allow good reproduction, while on improved pastures a very high level of milk production may be desirable. Also, more study needs to be given to selecting the high producing strains of beef cattle and to the use of "dam breeds." Reasonable goals for the more successful beef herds are: a cow averaging 20 pounds of milk per day (about double the present level); and weaning off a 600-pound calf. Cattlemen can be sure of one thing: "Little milk—little calf; lotta milk—lotta calf."

Size of the Herd

No minimum or maximum figures can be given as to the best size for the herd. Rather, each case is one for individual consideration. It is to be pointed out, however, that labor costs differ very little whether the herd numbers 100 or 300. The cost of purchasing and maintaining a herd bull also comes rather high when too few females are kept. Other efficiencies can be achieved through size, provided the operation is under competent management. For this reason, bigness in every kind of business, including the cattle business, is a sign of the times.

The extent and carrying capacity of the pasture, the amount of hay and other roughage produced, and the facilities for wintering stock are factors that should be considered in determining the size of herd for a particular farm unit. The system of disposing of the young stock will also be an influencing factor. For example, if the calves are disposed of at weaning time or finished as baby beef, practically no cattle other than the breeding herd are maintained. On the other hand, if the calves are carried over as stockers and feeders or are finished at an older age, more feed, pasture, and shelter are required.

Then, too, whether the beef herd is to be a major or minor enterprise will have to be decided upon. Here again, each case is one for individual consideration. In most instances, replacements should be made from heifers raised on the farm.

Uniformity

Uniformity in a herd has reference to the animals looking alike—"like peas in a pod," as a cattleman is prone to remark; particularly from the standpoints of size, type, and color.

Uniformity in color is still important in purebred herds. For the most part, however, the desire for uniformity in color of commercial herds went out as crossbreeding came in. So long as a quality product is produced efficiently, consumers and most cattle feeders have no interest in color of hair. It should be added for the benefit of the color conscious, however, that it is still possible to obtain uniformity of color even in crossbreds through making certain crosses; for example, uniform-colored animals with black bodies and white faces can be produced by crossing Herefords and Angus.

Fig. 4-2. Hereford X Angus crossbreds; uniform-colored animals with black bodies and white faces—"black baldies." (Courtesy, Dr. John F. Lasley, University of Missouri, Columbia, Mo.)

Size and type in any given lot of cattle are still important to cattle feeders and packer buyers. Cattle of uniform size and type feed better in a lot; and packers must provide their retail outlets with carcasses and cuts that meet their exacting specifications and grades. But buyer appeal, for both the feedlot and packer buyer, can be imparted amazingly well by sorting and grading prior to offering cattle for sale. Properly done, this practice can make for more uniformity faster in a few minutes than can be achieved through years of selective breeding.

Health

All animals selected should be in a thrifty, vigorous condition and free from diseases and parasites. They should give every evidence of a life of usefulness ahead of them. The cows should appear capable of producing good calves, and the bull should be able to withstand a normal breeding season. Tests should be made to make certain of freedom from both tuberculosis and contagious abortion, and perhaps certain other diseases in some areas. In fact, all purchases should be made subject to the animals being free from contagious diseases. With costly purebred animals, a health certificate should be furnished by a licensed veterinarian. Newly acquired animals should be isolated for several days before being turned with the rest of the herd.

(Also see Chapter 12, Beef Cattle Health, Disease Prevention, and Parasite Control.)

Condition

Although an extremely thin and emaciated condition, which may lower reproduction, is to be avoided, it must be remembered that an overfat condition may be equally harmful from the standpoint of reproduction.

It takes a unique ability to project the end result of feeding a few hundred pounds of grain or hay to a thin animal, and fortunate indeed is the cattleman who possesses this quality. This applies alike to both the purebred and the commercial producer. In fact, it is probably of greater importance with the commercial cattleman, for replacement females and stocker and feeder steers are usually in very average condition.

Age and Longevity

In establishing the herd, it is usually advisable to purchase a large proportion of mature cows (cows four to five years of age) that have a record of producing uniformly high quality calves. Perhaps it can be said that not over one-half of the newly founded herd should consist of untried heifers. Aside from the fact that some of the heifers may prove to be nonbreeders, they require more assistance during calving time than do older cows. Perhaps the best buy of all, when they are available, consists of buying cows with promising calves at side and rebred to a good bull—a three-in-one proposition.

Once the herd has been established, replacement females should come from the top heifers raised on the farm or ranch. Old cows, irregular breeders, and poor milkers sell to best advantage before they become thin and "shelly."

A sound practice in buying a bull is to seek one of serviceable age that is known to have sired desirable calves—a proved sire. However, with limited capital, it may be necessary to consider the purchase of a younger bull. Usually a wider selection is afforded with the latter procedure, and, also, such an individual has a longer life of usefulness ahead. Naturally, the time and number of services demanded of the bull will have considerable bearing on the age of the animal selected.

Since most beef females do not reproduce until they are two to three years of age, their regular and prolonged reproduction thereafter has an important bearing upon the overhead cost of developing breeding stock in relation to the number of calves produced. The longer the good, proved, producing cows can be kept without sacrifice of the calf crop or too much decrease in salvage value, the less the percentage replacement required. Moreover, the proportion of younger animals that can be marketed is correspondingly increased. Selection and improvement in longevity are possible in all breeds and should receive more attention.

In a survey made by Washington State University, it was found that old cows are culled or removed from the beef breeding herd at an average age of 9.6 years, and bulls at 6.3 years.[8] It is recognized that the severity of culling will vary somewhat from year to year, primarily on the basis of whether cattle numbers are expanding or declining; and that purebred cattle are usually retained longer than commercial cattle.

Adaptation

As has already been indicated—except in those localities where a certain breed predominates, thus making possible the exchange of breeding stock and joint benefits in selling surplus stock—one will usually do best to select that breed for which the producer may have a decided preference. On the other hand, there are certain areas and conditions wherein the adaptation of the breed or class of animals should be given consideration. For example, in the South, Brahman cattle and certain breeds with Brahman blood are able to thrive despite the extreme heat, heavy insect infestation, and less abundant vegetation common to the area. Because of this, Brahman blood has been added to many herds of the South and Southwest, and new strains of beef cattle have evolved.

The Missouri Station conducted some classical studies designed to show breed differences between Shorthorn, Santa Gertrudis (⅝ Shorthorn and ⅜ Brahman), and Brahman cattle.[9] The animals were housed in "climatic chambers," in which the temperature, humidity, and air movements were regulated as desired. The ability of representatives of the different breeds to withstand different temperatures was then determined by studying the respiration rate and body temperature, the feed consumption, and the productivity in growth, milk, beef, etc. Dr. Brody reported the following pertinent points:

---

[8]Ensminger, M. E., M. W. Galgan, and W. L. Slocum, Problems and Practices of American Cattlemen, *Wash. Agr. Exp. Sta. Bull.* 562, 1955.
[9]Brody, Samuel, *Jour. Series No. 1607*, Missouri Agr. Exp. Sta.

1. The most comfortable temperature for the Shorthorns was in the range of 30° to 60° F, while for the Brahmans it was 50° to 80° F (20° higher for the Brahmans), and for the Santa Gertrudis it was intermediate between the ideal temperatures for the Shorthorn and Brahman.

2. The Brahman cattle could tolerate more heat—they could withstand higher temperature better than Shorthorns, whereas the Santa Gertrudis approached the Brahman in heat tolerance.

3. The Shorthorn cattle could tolerate more cold—they could withstand a lower temperature better than the other two breeds, while the Santa Gertrudis were more cold-tolerant than Brahman cattle.

Dr. Brody attributed the higher heat tolerance of Brahman cattle to their lower heat production, greater surface area (their loose skin) per unit weight, shorter hair, and "other body-temperature regulating mechanisms not visually apparent."

Translated into practicality, the Missouri experiment proved what is generally suspected; namely, (1) that Brahman and Santa Gertrudis cattle are better equipped to withstand tropical and subtropical temperatures than the European breeds; (2) that mature breeding animals of acclimatized European breeds do not need expensive, warm barns; they merely need protection from wind, snow, and rain; and (3) that more attention needs to be given to providing summer shades and other devices to assure warm weather comfort for cattle.

The cattle producer must always breed for a strong constitution—the power to live and thrive under the adverse conditions to which most animals are subjected sometime during their lifetime. Under natural conditions, selection occurs for this characteristic by the elimination of the unfit. In domestic herds, however, the constitution of foundation or replacement animals should receive primary consideration.

Price

With a commercial herd, it is seldom necessary to pay much in excess of market prices for the cows. However, additional money paid for a superior bull, as compared to a mediocre sire, is always a good investment. In fact, a poor bull is high at any price.

With the purebred breeder, the matter of price for foundation stock is one of considerable importance. Though higher prices can be justified in the purebred business, sound judgment should always prevail.

BASES OF SELECTION

In simple terms, selection in cattle breeding is an attempt to secure or retain the best of those animals in the current generation as parents of the next generation. Obviously, the skill with which selections are made is all important in determining the future of the herd. It becomes perfectly clear, therefore, that the destiny of herd improvement is dependent upon the selection for breeding purposes of those animals which are genetically superior. Making the wrong selections and using genetically inferior animals for breeding purposes has ruined many a herd. Under the latter circumstances, the producer would be better off to let the cattle decide on the breeding program by random sampling.

The ultimate objective of beef production is selling beef over the block. Thus, fads or fancies in beef cattle selection that stray too far from this objective will, sooner or later, bring discredit and a penalty.

Strictly from the standpoint of the consumer, a beef animal should produce a carcass which has a high proportion of lean meat, no excess fat, and a minimum of bone—plus "eating quality," which includes tenderness, flavor, and juiciness. Additionally, for efficiency of production under practical farm or ranch conditions, the producer must have animals that produce regularly throughout a long life, and that utilize feed efficiently. From this it may be deducted that the profitability of any one animal, or of a herd, is determined by the following two factors:

1. *Individuality*—Which is based upon the ability of the animal to produce beef for a discriminating market.

2. *Performance or efficiency of production*—Which means the ability to reproduce regularly and utilize feed efficiently.

Regardless of the method of selection, the following points must be observed if maximum genetic progress is to be made:

● *Selection should be from among animals kept under an environment similar to that which you expect them and their offspring to perform*—This requisite applies to animals brought in from another herd, either foundation or replacement animals. For example, animals that are going into a range herd should be selected from among animals handled under range conditions, rather than from among stall-fed animals. This recommendation is based on the results of a long-time experiment conducted by the author and his colleagues at Washington State University.[10]

● *Selection should be for heritable traits of economic importance*—The traits upon which selection is based should be both highly heritable and of economic importance. (See Table 5-6, Economically Important Traits in Beef Cattle, and Their Heritability.) It stands to reason that the more highly heritable characters should receive higher priority in selection than those which are less heritable, for more progress can be made thereby.

By economic importance of traits is meant their dollars and cents value. Thus, those characteristics which have the greatest effect on profits should receive the most attention.

● *Selection must be accompanied by an orderly and accurate method of scoring or evaluating*—In order to determine whether progress is being made in a breeding program—in order to ascertain whether, through selection, each generation is actually better than the preceding one—a measure or yardstick must be applied to each trait to be evaluated. Moreover, individual measurements must be accurate. For example, weighing conditions must be alike; you cannot weigh some cattle "full" and others "shrunk." Likewise, if cutability is determined, it should be defined so that it can be repeated.

[10]*Wash. Agr. Exp. Sta. Bull. 34*, 1961.

Four methods of selection are at the disposal of the cattle producer: (1) selection based on individuality or appearance, which may involve either the traditional scorecard system or the new Functional Scoring System; (2) selection based on performance testing; (3) selection based on pedigree; and (4) selection based on show-ring winnings. Since each method of selection has its place, a cattleman, especially a purebred breeder, may make judicious use of more than one of them.

SELECTION BASED ON INDIVIDUALITY OR APPEARANCE

In starting a new herd certain matters pertinent thereto must be decided—like the breed or cross, whether to start with open heifers or bred cows, etc. Of equal importance to the success of the operation is the selection of the individuals—the choice of the cows and bulls that constitute the foundation herd, and the selection of replacement heifers, usually from within the herd.

Visual appearance (evaluation or appraisal) has been, and is, the basis of both feeder cattle and slaughter cattle trade. Likewise, individuality or appearance (scored or unscored) is the usual method followed by both commercial and purebred breeders. For the most part, it has been responsible for the transformation of the Texas Longhorn to the present-day bullock.

In making selections based on individuality, it must be borne in mind that the characteristics found in the parents are likely to be reflected in the offspring, for here, as in any breeding program, a fundamental principle is that "like tends to produce like." From a practical standpoint, this points up two things:

1. Only those animals which are at least average, or preferably better than average, should be used for breeding purposes.

2. A cow's inheritance will influence only one calf each year, whereas the herd bull may influence as many as 25 to 50 animals in a given season. Hence, in any selection based on individuality, the selection of the herd bull merits maximum attention.

Different methods of scoring individual animals have evolved; among them (1) the traditional scorecard system, and (2) the Functional Scoring System presented in this chapter. A discussion of each of these two methods of scoring will follow. But first it should be recognized that both methods are based on visual appraisal. This point bears emphasis because cattlemen and students often get the erroneous impression that, just because some visual scoring system (scoring systems based on visual appearance, in contrast to actual weights, measurements, etc.) is recommended for or used in conjunction with a production testing program, it must be more accurate than all other scoring systems. This isn't true. All are visual methods, and the score resulting from the use of any of them is no better than the person making it. Some method of selecting all animals by score, preferably on a systematic and written down basis, is the important thing. Any one of several methods may be used. It's the author's contention that the Functional Scoring System herein presented is the most complete and accurate, yet the most simple, of all methods of scoring, and that it is admirably adapted for use in all types of selection programs, including for use in scoring animals on performance test.

The Traditional Scorecard System

A scorecard is a listing of the different parts of an animal, with a numerical value assigned to each part according to its relative importance. It is a standard of excellence. The use of the scorecard involves studying each part (see Chapter 14, Fig. 14-30)—head, neck, back, loin, rump, round, fore and rear flanks, etc.—and assigning a rank or score to each.

Most of the breed registry associations have scorecards for their respective breeds, which are available on request. Also, a U.S. Department of Agriculture beef scorecard is herewith reproduced as Fig. 4-3.

BEEF CATTLE SCORING FORM

Animal No. ————Class ————Age ————Date ————

Weight ————Condition ————Station ————

Breed ————Sire ————Dam ————

Scorer ————

Points	Number 1	Number 2	Number 3	Number 4	Number 5
Scale					
Thickness of flesh					
Smoothness of body					
Head					
Chest					
Rear flank					
Back					
Loin					
Rump					
Round					
Body					

U.S. Government Printing Office   Total score ————

Fig. 4-3. U.S. Department of Agriculture beef cattle scorecard. A perfect score totals 100.

As noted, the scorecard lists several traits. These may either be ranked or given a value.

A scorecard is a valuable teaching aid for beginners. It systematizes judging and avoids any part of the animal being overlooked. However, a scorecard has the following limitations: (1) it is not adapted to scoring a great number of animals, or to comparative or show-ring judging, because of the time involved in using it; (2) a near worthless animal may score quite high—for example, an animal that is so structurally unsound that it can hardly walk may have a rather high total score; (3) it evaluates each part of an animal, rather than the systems—the skeletal system, the muscle system, etc.; (4) it is based almost entirely on consumer needs, on the end product—meat; and (5) it accords precious little consideration as to whether, or how, an animal can be changed better to conform to the needs and desires of man.

The Functional System, which follows, is a new and better method of making individual or appearance selection, on a systematic and recorded basis.

The Functional Scoring System

Many progressive cattlemen and scientists have long felt that there should be a better way of evaluating animals than either (1) unsystematic and unrecorded visual appraisal, or (2) the traditional scorecard system—that there is need for a system that relates structure to the function of producing quality beef more efficiently. Out of this thinking several related research findings and concepts based on anatomy have evolved in different parts of the world; and out of this thinking evolved the Functional Scoring System herein presented.

●*Basic Anatomy*—In order to select, score, or judge animals intelligently, it is first necessary to know (1) how they are built, (2) how they function, and (3) how we can change them better to fit our needs.

There are two ways of looking at the structure of anything, whether it be an automobile, a house, or an animal. We can either (1) look at each of the component parts or (2) look at the systems. Thus, in a car we can study each part—the engine, the dashboard, the seats, etc.; or we can study each system—the fuel system, the transmission system, the brake system, etc. In a house, we can either inspect each room; or we can study each system—the plumbing system, the electrical system, the heating system, etc. Likewise, when it comes to beef cattle, we can either look at each part of the animal— the head, neck, loin, etc.; or we can study each system—the skeletal (bone) system, the muscle system, etc. The old, traditional scorecard system uses the first approach, with each part of the animal evaluated. The new Functional Scoring System makes use of the systems approach—with primary consideration given to the bone, muscle, and fat. The author contends that, whether it be a car, a house, or a beef animal, a proper evaluation or score can be made only through use of the systems approach.

In order to use the systems approach effectively and intelligently, it is first necessary to have some basic knowledge of the three body (carcass) systems—bone, muscle, and fat; and to know how each of them makes for a desirable or undesirable carcass. Of course, there are several other animal systems, such as the nervous system, the digestive system, etc. Although of vital importance to the living animal, they are omitted in this discussion because they are mostly, or entirely, in the noncarcass part of the animal.

High cutability comes from carcasses which have a high proportion of lean meat, no excess fat, and a minimum of bone. Normally, as an animal grows, the proportions of lean, fat, and bone change. Soon after birth, the percentage of lean increases and later it decreases as the animal fattens. Fat percentage shows a continuous increase in animals fed for slaughter, and bone percentage slowly decreases.

Table 4-2 shows the composition of the three *major* tissues in carcasses.

TABLE 4-2

COMPOSITION OF THREE MAJOR TISSUES IN CARCASSES[1]

	Range in Values			Average Choice Grade Carcass	Ideal Carcass
	High	**Low**	**Variation**		
	(%)	(%)	(%)	(%)	(%)
Bone	17	9	8	12	10
Muscle	80	45	35	60	75
Fat	45	15	30	28	15

[1]Berg, R. T., The University of Alberta, Edmonton, Alberta, paper in *Beef Cattle Science Handbook*, Vol. 11, pub. by Agriservices Foundation and ed. by Dr. M. E. Ensminger.

As shown in Table 4-2, the amount of bone in carcasses is both smaller and less variable than the muscle and fat. Bone varies 8%, whereas muscle, and fat vary by 35% and 30%, respectively.

• *Bone*—The bones form the skeleton—the framework which provides support for the soft tissues, protection for the vital organs, and levers for the action of muscles. The skeleton varies very little insofar as number of bones is concerned or the location and size of the bones in relation to each other.

Animal species without a skeleton of some type have little or no regular form. The skeleton gives a basis for the external structure and appearance of most animals as we know them.

The bones of a newborn calf are well developed and relatively much longer than in the adult animal, thereby making it possible for it to run after its mother to nurse and to escape its enemies. The bones grow steadily at a very slow rate and become slightly heavier as the animal increases in weight.

When used in connection with evaluating cattle, (selecting, judging, buying or selling), the term "bone" refers to the circumference of the cannon bone about midway between the knee and ankle (front leg) or hock and ankle (back leg). Contrary to long-held opinion, bone is of little importance in a cattle selection program because (1) it is not a measure of bone alone—rather, it includes hair, hide, connective tissue, and tendons; (2) it tells nothing of either thickness of bone wall or bone density; (3) it varies little between animals of the same size; and (4) it is not necessarily indicative of the amount of muscle. Besides, bone breakage is not normally a problem in cattle. So, size of bone need not be considered in a selection program.

• *Muscle*—The muscles are the labor force of the body; they convert the energy taken in as feed into work. There are three types of muscle tissues: (1) striated voluntary (skeletal) muscles, which along with some connective tissue makes up the flesh of meat producing animals; (2) smooth (involuntary) muscles, which are found in the walls of the digestive tract, in the walls of the blood vessels, and in the walls of the urinary and reproductive organs; and (3) cardiac (involuntary striated) muscles, which are found in the heart. In cattle selection and evaluation, we are primarily interested in skeletal muscles, for this is the red meat that we eat.

Muscles contain fibers which are capable of contracting. Each skeletal muscle is attached to two or more points of the skeleton, so that the bones act as levers. These points of attachment are fixed; and they do not change from animal to animal; always a given muscle runs between the same points.

At birth, muscle is less well developed than bone. But it grows faster than bone. As a result, the heavier the animal becomes, the greater the difference in the weight of muscle and bone; i.e., the muscle:bone ratio becomes greater the heavier the animal becomes.

● *Fat*—Fat tissue is located in a variety of places around the body. It may be regarded as a liquid substance which flows around the body and locates in any region of low pressure. For example, in a lean animal it is unlikely to be deposited on the outside of the arm or the outside of the round so long as there are places of much less pressure, such as the flanks, cod or udder, and brisket. After these low pressure areas are filled, fat is deposited between muscles, inside bones, just under the skin, and within the muscles where it constitutes marbling. Fat does not have fixed boundaries like bones and muscles. Moreover, it is inclined to vary from animal to animal in the proportion deposited in each site.

Fat increases slowly from birth until the animal reaches a certain age and weight, then it commences to accumulate rapidly.

●*Relating Structure and Function to Usefulness*—We have three major animal components for use in scoring and evaluating animals:

1. *The very rigid skeleton*—which following slaughter provides bones, good only for soup.

2. *The softer muscle tissues*—which following slaughter provides the lean meat—a high quality animal protein, running 23 percent protein in a broiled sirloin.

3. *The liquid fat*—(which following slaughtering and chilling firms up), some of which is desirable because of the flavor and juiciness that it imparts, but much of which is considered waste.

How, then, can structure be related to usefulness? How can the skeleton be used as the basis of scoring or evaluating cattle for such a diverse threesome?

Remember that each bone is always located (1) in the same body part of an animal, and (2) in the same relationship to other bones. Remember, too, that the muscles are attached to the skeleton at the same points and in the same relation to each other, and that fat is deposited on the skeleton and musculature in certain low pressure areas. This clearly indicates that structure is related to function, and that the skeleton of cattle may be used for scoring or evaluating.

No longer do we want a beef animal that is "well let-down in its flanks, full in its brisket, and full in its twist." This is the wasty kind. Rather, the rear flank should be high and tight. The abdominal wall curves upward from the fore flank to the rear flank and attaches above the stifle joint.

A full brisket and loose hide in the twist are just as undesirable as loose hide in the flanks; hence, such animals should be scored down.

● *Use of the Functional Scoring System*—As previously indicated, this system relates structure (anatomy) to desired function and usefulness—*it is concerned with composition of gain* (with the quality and yield of the carcass), with eating value. It is limited to six traits—all of economic importance, and it is simple and easy to apply. It involves selecting animals for the following six traits:

1. Reproductive efficiency
2. Muscling
3. Size
4. Freedom from waste
5. Structural soundness
6. Breed type

Each animal is scored for each of the above six traits on a point basis; and with a total, or combined, score of 100.

Hence, the highest and lowest scores that an animal can receive are:

Trait	Highest Score	Lowest Score
Reproductive efficiency	20	1
Muscling	20	1
Size	15	1
Freedom from waste	15	1
Structural soundness	15	1
Breed type	15	1
Total	100	6

If desired, one can establish a minimum score for each character, then cull those animals that fall below this score for the particular character. For example, it might be decided to cull all animals that score less than 5 in muscling. When used in this manner, the Functional Scoring System is somewhat similar to the system of selection known as "Establishing minimum standards for each trait, and selecting simultaneously, but independently, for each trait," described in Chapter 5 of this book, except that the latter usually implies an objective measure (like pounds of feed, or pounds of gain) whereas the Functional Scoring System is visual. Of course, the minimum standards should increase slightly with each generation if progress is being made.

REPRODUCTIVE EFFICIENCY

Reproductive efficiency calls for females producing and weaning a calf each year, beginning at 2 years of age. A nonexistent or dead calf cannot be scored or evaluated. Thus, being born and born alive, or reproductive efficiency, is the most important trait affecting profit in the cattle business. Sometimes cattlemen lose sight of this simple fact for 2 reasons: (1) it is overshadowed by emphasis on such traits as rate and efficiency of gain; and (2) fertility in cattle is of low heritability—only about 10 percent. Most cattlemen lie awake nights trying to figure ways in which to improve weight gains, feed efficiency, and carcass quality; and they extol production records involving these traits. But similar concern over breeding efficiency is seldom evidenced, despite the fact that, when it comes to improving the dollar return on investment in beef cattle, no trait is so important as conception and getting a live calf.

Without a calf being born, and being born alive, the other economic traits are of academic interest only—they cannot be scored or evaluated because there is no calf. For example, size is economically important, but no matter how big they are, sterile cows and bulls are not producers. A "mating of the gods," involving the greatest genes in the world, is of no value unless those genes result in (1) the successful joining of the sperm and egg, and (2) the birth of a live calf.

Because reproductive efficiency is essential to successful beef production, it is important that heifers be selected from a herd with a consistent high calving percentage (certainly above 90%, and preferably above 95%). Also, heifers should be selected from a herd with a short calving season (calves coming in less than a 45-day period), and not to exceed a 12-month calving interval.

For the United States as a whole, out of each 100 cows bred, only about 88% drop calves annually.[11] The other 12% are nonproducers, either temporarily or permanently. In some herds, the calf crop percentage runs as low as 50%. With a 50% calf crop, it simply means that 2 cows are being maintained an entire year to produce 1 calf. As reproductive ability is fundamental to economical beef production, it can be readily understood that reproductive failure constitutes a major annual loss in the cattle business. In fact, cattlemen acknowledge that the calf crop percentage is the biggest single factor affecting profit in beef cattle production. Improper feeding and disease are the 2 most common causes of low percentage calf crops in cattle.

Overfeeding accompanied by extremely high condition, or underfeeding accompanied by an emaciated and run-down condition, usually results in temporary sterility that may persist until the condition is corrected. Lack of exercise, inflammation and infection of the reproductive tract resulting from retained afterbirth or other difficulties encountered at calving, and infections of various other kinds may also result in temporary sterility. The most common causes of permanent sterility in cattle are: old age; diseased reproductive organs, such as cystic ovaries; diseased Fallopian tubes; and heredity. The reproductive ability of an individual or an entire herd may also be

---

[11]Methods of computing calf crop percentages vary, but the three most common methods are: (1) number of calves born alive in comparison with the number of cows bred; (2) number of calves marketed or branded in relation to the cows bred; and (3) the number of calves that reach weaning age as compared with the number of cows bred. Perhaps the first method is the proper and most scientific one, but, for convenience reasons, some ranchers use the other two.

greatly affected, either temporarily or permanently, by the presence of brucellosis and certain other diseases.

Sterility may be present in either sex. Occasionally bulls are sterile even though sexually active. Differences in the fertility of bulls are especially revealed by the records kept and the semen studies made.

In addition to the above factors affecting reproduction, the percentage calf crop may be affected by the proportion of bulls to cows and their distribution on the range, by the season of breeding, and by diseases.

Of course, the most obvious indication of reproductive efficiency of a cow is a calf at side; and of a bull, it's a large number of calves from a season's service. However, breeding records are not always available and reliable; and they are nonexistent in animals that are not of breeding age. Hence, animals should be scored for reproductive efficiency, although it is recognized that fertility is a "zero" or "one" proposition so far as any one cow is concerned.

Fig. 4-4 illustrates and describes the difference in appearance between a highly fertile female and a lowly fertile female. Females should show femininity at all stages of development. Feminine females are trim in the jaw, throat, and dewlap, and they are smooth shouldered. Avoid coarse, heavy-fronted females that are excessively muscular and give an impression of masculinity rather than femininity.

Score

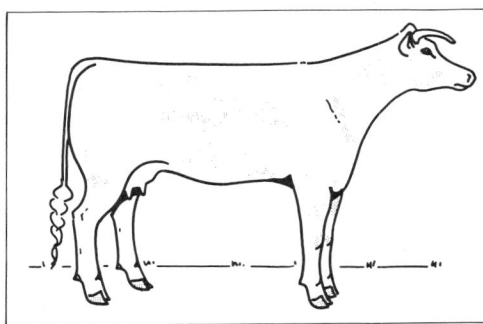

A. Feminine female— long body, lean, smooth muscled; refined, feminine head; lean cheek, jaw, neck, brisket, shoulder, and hindquarters; and a good functional udder (or promise of udder development in a heifer). Length of head in keeping with size of body.

20

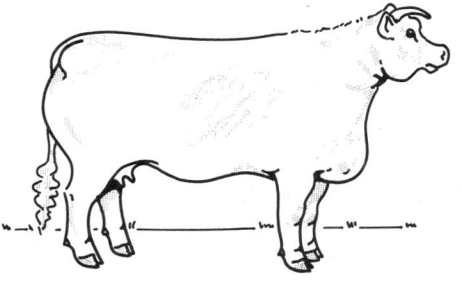

B. Steery female— coarse, heavy front, masculine rather than feminine; protruding brisket; bristly hair on neck and top of shoulders; rounded hindquarters; and fat deposits on the face, brisket, shoulders, hips, rump, pins, below the vulva, and in front of the udder.

1

Fig. 4-4. Fertility makes the difference! Highly fertile female (A) vs lowly fertile female (B).

Fig. 4-5 illustrates and describes the difference in appearance between a highly fertile bull and a lowly fertile bull. A mature bull should look masculine in front. But do not discriminate against a bull calf that lacks masculinity. Bull calves that show extreme masculinity at an early age are apt to mature too early and quit growing too soon. When selecting a bull, observe his testicles. Make certain that they are of normal size, and that they are well defined in the scrotum, rather than surrounded by excess fat.

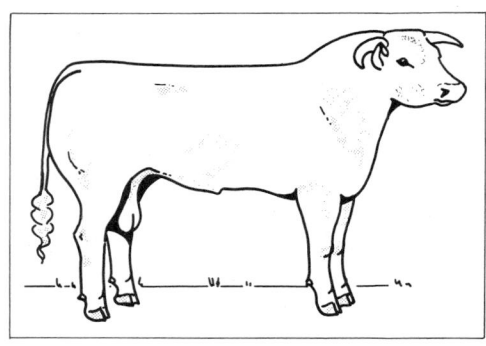

A. Bull with a masculine front and a sound pair of testicles behind; alert—he's "on the look," with head up and ears cocked; well-developed crest; muscles well developed and clearly defined, especially in the regions of the neck, loin, and thigh; and well-developed external genitalia, with testicles of equal size and well defined, and a proper neck to the scrotum. Length of head in keeping with size of body.

Score

20

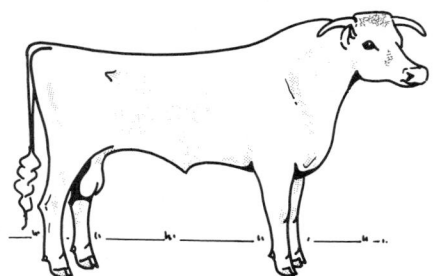

B. Bull lacking in masculinity; sleepy and droopy; ears not alert; undeveloped crest; muscles lacking development and not clearly defined; testicles may be small, unbalanced, or have one carried high; and the scrotum may be twisted or filled with fat.

1

Fig. 4-5. Fertility makes the difference! Highly fertile bull (A) vs lowly fertile bull (B).

Although masculinity in a yearling or older bull is an indication of fertility, and desirable, extreme masculinity in a bull under one year of age is undesirable. Development of the masculine sex characteristic at a young age is indicative of early maturity, which may result in reduced size and increased fat disposition.

MUSCLING

Consumers are demanding, and getting, beef with a maximum amount of lean and a minimum of fat. Fortunately, cattle can be bred for heavy muscling, even to the point that it can be overdone, resulting in double muscling.

Old-time draft horsemen always selected horses for heavy muscling by looking at those parts of the body where there was muscle alone, without fat—the arm, forearm, stifle, and gaskin. Shrewd cattlemen have done the same thing. But it remained for Butterfield of Australia to confirm the logic of this procedure experimentally; he showed a correlation of 0.93 to 0.99 between the weight of a single muscle or group of muscles and the total muscle in a carcass.[12] Thus, cattle that are heavily muscled in the regions of the arm, forearm, stifle, and gaskin are heavy muscled throughout the body. Butterfield also exploded a myth: he showed that the proportion of muscle in the high priced cuts in all breeds remains constant at 56 percent—that it is not possible to increase the proportion of muscle in the regions of the high priced cuts through selection.

A cattleman can increase total muscle per animal by selection, and thereby increase value based on cutout. A study conducted at the University of Wisconsin showed on-foot value differences of $4.67 per cwt on steers of comparable weight and finish—that's $50.00 on a 1,070-pound steer.[13] The U.S. Department of Agriculture reports (*Federal Register,* Vol. 39, No. 177, Sept. 11, 1974) that carcasses of the same carcass grade—Choice for example—can vary in value by $75.00 or more due to differences in cutability.

When scoring animals for muscling, it should be recognized that condition or fatness can create a false impression of muscling. For example, a very thin animal seldom has the appearance of muscularity, with the result that it is apt to be scored rather low in muscling. When the same animal is fattened, it is likely to score higher on muscling. Yet, when finish is overdone, the extra condition may cause the animal to lose its muscular appearance.

In addition to selecting cattle for heavy muscling, the producer has at his disposal many ways in which he can tailor-make his product so as to have a higher proportion of lean (muscling) to fat at slaughter time; among them—

● *Use of different breeds and crosses*—Breeds differ in the stage at which they begin to lay down fat. For the most part, the British breeds are early fatteners, whereas the exotics are late fatteners. Thus, the optimum slaughter weight for the British breeds is lighter than for the exotics. Also, the feedlot finishing programs should differ; the British breeds may be grown out (on a growing ration or handled as stockers) prior to fattening, whereas, the large, late-maturing exotics should be placed in the feedlot immediately following weaning and put on a high energy ration as soon as possible.

● *Delayed fattening*—Heavier slaughter weights without excess fat may be secured in types of cattle inclined to be early fatteners (1) by handling them as stockers, or backgrounding, before putting them on a "hot," finishing ration, (2) by feeding a low energy ration throughout the finishing period, or (3) by limiting feed intake.

● *Short feed*—Short-fed cattle (on feed less than 140 days) have less finish than long-fed cattle.

● *Lighter slaughter weight*—Marketing feedlot cattle at lighter slaughter

---

[12]Butterfield, R. M., The Relationship of Carcass Measurements and Dissection Data to Beef Carcass Composition, *Research in Veterinary Science*, Volume 6, No. 1, January 1965, Table III, p. 26.

[13]Kauffman, R. G., R. E. Smith, and R. A. Long, Bovine Topography and Its Relation to Composition. *Proceedings 23rd Annual Reciprocal Meat Conference of American Meat Science Association*, 1970.

weights will cut down on the number of overfinished, low cutability animals. Although this stage and weight will vary with the size of cattle and the ration fed, cattle should always be slaughtered when they have reached the optimum level of fatness for the market requirements.

●*Alter fat by energy of feed*—High energy rations can produce more fat at an early age. But *no amount or kind of feed will alter the amount of muscle.*

●*Heifers vs steers*—The muscles of heifers and steers grow much the same, but females start to lay down fat a little earlier in life than males. Hence, heifers should be slaughtered at lighter weights than steers. It is noteworthy that heifers finished following first calving produce very acceptable carcasses at carcass weights approaching those achieved in steers. Hence, if heavier heifer carcass weights are desired, a one-calf system followed by finishing may be considered.

●*Bulls vs steers*—In comparison with steers, bulls grow faster, are more efficient, and produce leaner carcasses of equal and acceptable quality. Also, they can be carried to heavier slaughter weights than steers without becoming excessively fat. Because of these several advantages, the feeding of young bulls is an accepted practice in many countries; and it will increase in North America.

●*Double muscling*—The proportion of muscle to fat and bone is greater in double muscled cattle, but there are many disadvantages of double muscling (see Chapter 5, section on Double Muscling). Moreover, selecting too strongly for heavy muscle development in any breed will, sooner or later, result in more double muscled cattle being produced, provided genes for double muscling are present in the breed.

In the Functional Scoring System, muscle is scored by looking at the points on the skeleton where there is little other than muscle—the arm, forearm, stifle, and gaskin. Look for the bulge in the muscle in these regions. When the animal walks, look for muscle movement and bulging in the shoulder and stifle regions. Fat just hangs and shakes on an animal, like it does on a fat man. Really muscular cattle are not smooth; rather, they show some creases and indentations between muscles. Also, they are slightly narrower through the heart girth and loin than through the shoulder and round.

Long, smooth muscling is preferred in calves, because such animals will usually grow for a longer period of time and get thicker with age.

Since muscling is a masculine trait, it is more important in bulls and steers than in heifers.

Breeding cattle with coarse shoulders (very heavy muscled shoulders) should be avoided since this condition is frequently associated with calving problems.

Fig. 4-6 shows four bulls that are the same in all traits except muscling. Note that E and F are front and rear views, respectively, of bulls A and D.[14]

---

[14]Four animals are pictured side view in this section, and under each subsequent trait, because this is the traditional number used in judging contests.

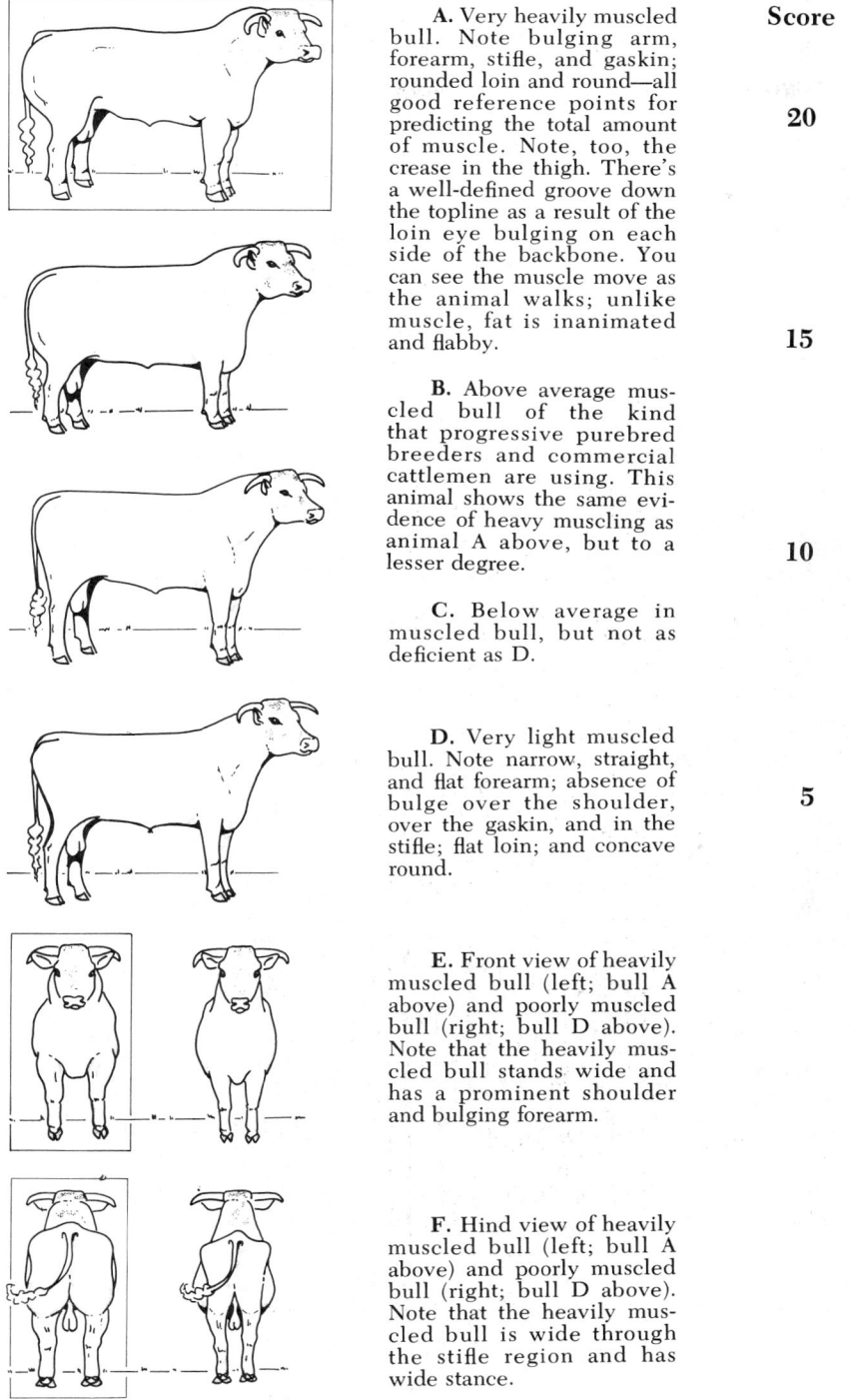

A. Very heavily muscled bull. Note bulging arm, forearm, stifle, and gaskin; rounded loin and round—all good reference points for predicting the total amount of muscle. Note, too, the crease in the thigh. There's a well-defined groove down the topline as a result of the loin eye bulging on each side of the backbone. You can see the muscle move as the animal walks; unlike muscle, fat is inanimated and flabby.

B. Above average muscled bull of the kind that progressive purebred breeders and commercial cattlemen are using. This animal shows the same evidence of heavy muscling as animal A above, but to a lesser degree.

C. Below average in muscled bull, but not as deficient as D.

D. Very light muscled bull. Note narrow, straight, and flat forearm; absence of bulge over the shoulder, over the gaskin, and in the stifle; flat loin; and concave round.

E. Front view of heavily muscled bull (left; bull A above) and poorly muscled bull (right; bull D above). Note that the heavily muscled bull stands wide and has a prominent shoulder and bulging forearm.

F. Hind view of heavily muscled bull (left; bull A above) and poorly muscled bull (right; bull D above). Note that the heavily muscled bull is wide through the stifle region and has wide stance.

Score

20

15

10

5

Fig. 4-6. Muscling makes the difference! Four bulls showing variations in muscling. Note that E and F are front and rear views of bulls A and D.

SIZE

The question of size of beef cattle has been a point of considerable controversy among producers. Generally speaking, arguments are advanced more in defense of a certain size (strain or breed) to which the cattleman is partial than on the basis of experimental results. Also, it is recognized that available feeds and markets are so variable as to preclude any standard size. Thus, the best size cattle for a given farm or ranch may not be the best size for another, even for a neighbor. Also, cow size must be considered as it relates to the production efficiency of the cow herd as well as the gain and finishing qualities of the progeny.

The following points are pertinent to the question of size in cattle:

● *How should size be measured or determined?*—To most cattlemen, size means weight per day of age (WPDA). This is understandable, simply because most cattle are sold by the pound. But weight per day of age (WPDA) is not the best measure of size of cattle. Within a breed, and at a given age (15 months of age in the case of Brungardt's work, University of Wisconsin), carcasses from cattle with the higher WPDA tend to be fatter and contain a lower amount of muscle.[15] At a given weight, however, the greater the weight per day of age, the less the total fat and the greater the amount of total muscling. Also, weight over the scales should not be considered when scoring animals because heavy weight may be due to higher finish, or more fat. For these reasons, size should be determined by height to the top of the shoulders and length from the nose to the tailhead.

● *Size has changed*—Fashions in beef cattle size have changed rather radically during the present century, moving from the big, rugged, beefy—but

Fig. 4-7. Cattle fashions have changed! "Firly," a prize ox of Britain in 1835, shown at 4 years and 8 months of age and weighing 3,000 pounds. The near animal is Royal Jupiter, the Shorthorn steer that was grand champion over all breeds at the 1946 Chicago International; shown by Oklahoma State University. (Courtesy, *The Shorthorn World*)

---

[15]Brungardt, V. H., *Research Report*, University of Wisconsin-Madison, R 2400, Efficiency and Profit Differences of Angus, Charolais, and Hereford Cattle Varying in Size and Growth, May 1972, p. 2.

oftentimes rough "farmer or rancher type" that our grandfathers produced up through the first third of the present century—to the smaller, earlier-maturing, blockier, smoother types in vogue from about 1935 to 1950. (The extremes in this era were known as "comprests" in Herefords and "compacts" in Short-horns.) Then, in the 1960s, the pendulum swung back to greater size. This time, the transition was not limited to selection within the existing breeds. It was speeded up through crossbreeding and the introduction of the exotics.

The reasons for these shifts in cattle size were many and varied. The smaller types evolved principally because of the demand on the part of the consumer for smaller cuts of meat. Likewise, the show-ring was an important factor, and show-ring fashions for both the finished product and breeding animals tended to follow consumer preferences.

Unfortunately, the tremendously important utility factor or economy of production was largely overlooked in this shift, and little information was available in regard to whether smaller cattle could be produced as economi-cally as larger cattle. Many producers felt that bigger cattle could be produced more economically, especially under conditions where operating costs are on a per-head rather than a per-pound basis (it requires no more labor to take care of a big cow than a little one), and where profit in a cattle enterprise depends primarily upon the ability of the animals to utilize efficiently large amounts of roughage. The opinion was also prevalent that show-ring fashions toward the low-set, blocky, earlier-maturing pony-type cattle went further than consumer demand justified. Additionally, there was the desire to get away from dwarfism, and to produce more meat during a period of world beef shortages and good prices. As a result, in recent years the pendulum has swung toward larger cattle; currently, medium or medium to large type cattle are favored by most breeders.

Yet, the right size cow to maintain in the beef breeding herd remains unresolved. To answer this question, cow size must be looked at in the total context of beef production; i.e. the production efficiency of the cow as well as the gain and finishing qualities of the progeny. On this basis, small cows appear to be more efficient producers than large cows.

●*What the experiments show*—Fortunately, some experimental work, designed to answer these and other questions, has been conducted. In a study involving large-, intermediate-, and small-type Hereford cattle, Stonaker of the Colorado Station reported that wide variations in mature size were not antagonistic with the market demands of the 1950s, efficiency of feed use, or carcass cutout values and grades.[16] However, the larger cattle produced a higher percentage calf crop and required some less fixed cost expenditure per pound of beef produced. In the Colorado experiments, all groups of steers were fed to about the same degree of finish. This is important because fat is high in energy; and the fatter an animal becomes the less efficiently it utilizes its feed. Thus, it stands to reason that smaller-type animals, when fed to the same weight as large type animals, will be fatter and, therefore, will require more feed per 100 pounds gain.

---

[16]Stonaker, H. H., *Colorado State University Bull. 501-S*, 1958.

In a study conducted cooperatively by the Kansas, Oklahoma, and Ohio Agricultural Experiment Stations, involving steers sired by small-, medium-, and large-size bulls, medium-size steers were favored. It was found that medium-size cattle tend to combine the gaining ability of large cattle and the finishing ability of small cattle without sacrifice of efficiency of gain.[17]

The New Mexico station compared the gains of, and carcasses produced from, compact, medium, and rangy steers.[18] They found that the rangy steers weighed more when put on feed, gained more, and yielded a higher dressing percentage than the compact steers; and that the medium type was intermediate in each case. There was no indication of any differences in economy of gain, however.

Brungardt, at the Wisconsin Station, made a study of efficiency and profit differences of Angus, Charolais, and Hereford cattle varying in size and growth. He concluded that (1) feed efficiency does not differ greatly among cattle of different sizes when they are fed to the same grade or degree of finish, and (2) at the same weight (but not at the same degree of finish) faster gaining cattle are more efficient than smaller and slower gaining cattle.[19]

Brown and Brown calculated maintenance costs for small, early-maturing cows and large, late-maturing cows. They concluded that the greater salvage value of larger cows may offset the greater yearly maintenance costs.[20]

Berg, of Canada, made calculations of 1,100-pound versus 900-pound cows.[21] Based on University of Alberta data, he projected that a cow with a 100-pound weight advantage would produce a 7-pound heavier calf at weaning. At the feed and calf prices that existed at the time he made the calculations, the Canadian investigator concluded: "The heavier cow hardly pays her way in extra calf produced at weaning. At lower feed prices and high calf prices, extra cow size would be more profitable." Then, Dr. Berg very wisely extended his projections to market weight, at which time his studies showed an advantage of 40 pounds for the calf from the larger cow. His final conclusion relative to the bigger cow at the prices of feed and cattle then prevailing, "definitely profitable."

An Oklahoma Station experiment, involving 14 calf crops for a total of 3,298 calves from 863 Hereford and Angus cows, showed that for each 100 pounds increase in the yearling weight of cows, a cattleman can expect 14 pounds increase in the weaning weight of their calves.[22]

● *Factors of importance in cattle size*—In the final analysis, therefore,

[17]Weber, A. D., A. E. Darlow, and Paul Gerlaugh, *American Hereford Journal*, Vol. 41, No. 22, 1951, p. 20.

[18]Knox, J. H. and Marvin Koger, *Journal of Animal Science*, Vol. 5, No. 4, 1946, p. 331.

[19]Brungardt, V. H., *Research Report*, University of Wisconsin-Madison, R 2400, Efficiency and Profit Differences of Angus, Charolais, and Hereford Cattle Varying in Size and Growth, May 1972, p. 2.

[20]Brown, C. J. and J. E. Brown, University of Arkansas, *Arkansas Research*, Vol. 20, May-June 1971, p. 3.

[21]Berg, R. T., How to Feed and Manage the "New Look" Beef Animal, *Beef Cattle Science Handbook*, Vol. 11, 1974, pub. by Agriservices Foundation and ed. by Dr. M. E. Ensminger.

[22]Selecting Replacement Females by Growth Boosts Weaning Performance of Calf Crop, *Better Beef Business*, Vol. 14, No. 9, July 1973, pp. 16-17.

the most practical size cattle will vary according to conditions. Among the factors that should be considered when choosing between large, medium, and small cattle are:

1. *Plane of nutrition and fertility*—Unless a cow's maintenance requirements are met, fertility (reproduction) will suffer. Thus, the genetic capabilities of the cow should match the available feed. Translated into practicality, this means that the better the pasture or range (or supplemental feeding), the bigger the cow; or conversely, the sparcer the range, the smaller the cow.

2. *Plane of nutrition and milk production*—Milking ability is positively correlated with weaning weight of calves. Hence, suckling calves should receive adequate milk. But cows milk to their genetic potential only after their nutritional needs for body maintenance are met. This means that the size cow should match the available feed on the pasture or range. Small cows will milk better on poorer ranges than big cows.

3. *Rapid gains of progeny*—Large cows pass greater growth potential, or more rapid gains, to their progeny than small cows. This is a major reason why the exotics have become so popular.

4. *Labor costs*—Under most conditions, the labor costs for handling larger cattle with their greater pounds are no greater than for the smaller ones.

5. *Relative price of feed and cattle*—Relatively low feed costs and high cattle prices tend to favor big cattle, whereas the reverse conditions favor small cattle.

6. *Weaning vs market weight of progeny*—Big cows and little cows differ very little on the basis of calf produced at weaning. But on the basis of progeny carried to market weight (steers 1,000 pounds or more), big cows are definitely more profitable than small cows.

7. *Market weight*—Heavy market weights (steers weighing around 1,200 pounds) favor big cattle. On the other hand, where the consumers desire high quality but smaller cuts of meat, the situation may favor the production of smaller cattle.

8. *Salvage value*—The greater salvage value of big cows than of small cows may offset their greater maintenance cost.

9. *Estimating production efficiency of cows of various sizes*—The two most important factors determining what size beef cow to keep are (a) the weight of the calf weaned, and (b) the total feed required to produce a weaner calf. Of course, the price of each of these is variable. To point up this situation, let us consider some estimates of production efficiency of cows of various sizes, based on specific prices of feed and weaned calves. Table 4-3 shows the annual feed cost for cows of different sizes, and the weights of calves that must be weaned to cover the additional cost of feed relative to the 900-lb cow weaning a 450-lb calf.

From these estimates, we can see that a 1,400-lb cow would have to wean a 518-lb calf at 40 cents per lb (518 − 450 = 68; then 68 × 40¢ = $27.20), or a 496-lb calf at 60 cents per pound (496 − 450 = 46; then 46 × 60¢ = $27.60), to cover the higher feed cost, compared to a 900-lb cow weaning a

TABLE 4-3

ESTIMATED MAINTENANCE AND PRODUCTION OF COWS
AT VARIOUS BODY WEIGHTS

Body Wt in Lb	Feed Cost ($/365 days)	Req'd Calf Wt @ 40¢/Lb	Req'd Calf Wt @ 60¢/Lb	Lb of Calf @ 60¢/Lb/$ Feed
900	70.00	450	450	6.43
1,000	76.30	466	461	6.04
1,100	81.90	480	470	5.74
1,200	86.80	492	478	5.51
1,300	92.40	506	487	5.27
1,400	97.30	518	496	5.10
1,600	108.50	546	514	4.74
1,800	118.30	571	531	4.49

450-lb calf. As feed prices increase, the necessary weaning weight, to cover the additional feed cost, also increases. When calves are selling at 60 cents per pound, the 1,400-lb cow is weaning 5.1 lb of calf per dollar of feed, while the 900-lb cow is weaning 6.43 lb. In order for a 1,400-lb cow to have the same production efficiency as the 900-lb cow (6.43 lb/$), she would have to wean a 626-lb calf. Similarly a 1,600-lb cow would have to wean a 700-lb calf. If it is possible for a 900-lb cow to wean a 500-lb calf, then the 1,400- and 1,600-lb cows would have to wean a 695- and 775-lb calf, respectively.

Also, bear in mind that light calves generally sell at a higher price per pound than heavy calves. This, too, favors small cows.

The above estimates indicate that small cows are more efficient producers than large cows.

More research on different size cows is needed, embracing (a) the production efficiency of the cows, and (b) the efficiency of their progeny when fed to a constant finish for slaughter.

Certainly animals can be too big. Huge size is usually accompanied by coarseness, poor fleshing qualities, loose rather than compact conformation, and slow maturity and finishing. Perhaps, under most conditions and in the final analysis, medium-type cattle are best from the standpoint of widest adaptability, general vigor, reproductive efficiency, milk production, longevity, and marketability. Yet, it is reasonable to conjecture that there are environmental and market conditions under which one or the other extremes in beef type (the large type or the small type) may be more profitable.

Size, as used in the Functional Scoring System, refers to height to top of the shoulders and length from the nose to the tailhead. Young breeding animals and steers should be long, tall, and not excessively fat—indications that they will continue to grow. Bulls should not be too masculine too early in life because the more masculine the animal, the more male hormones secreted. Male hormones, in addition to their reproductive function, inhibit long bone growth (and body growth). They cause a narrowing of the epiphyseal cartilage and a more rapid fusion or ossification of the epiphysis (breakjoint). It follows that bulls that show signs of early sexual maturity are not likely to make continued rapid growth and reach large mature size.

Fig. 4-8 shows four bulls that are the same in all major characteristics except size as determined by height and length.

Score

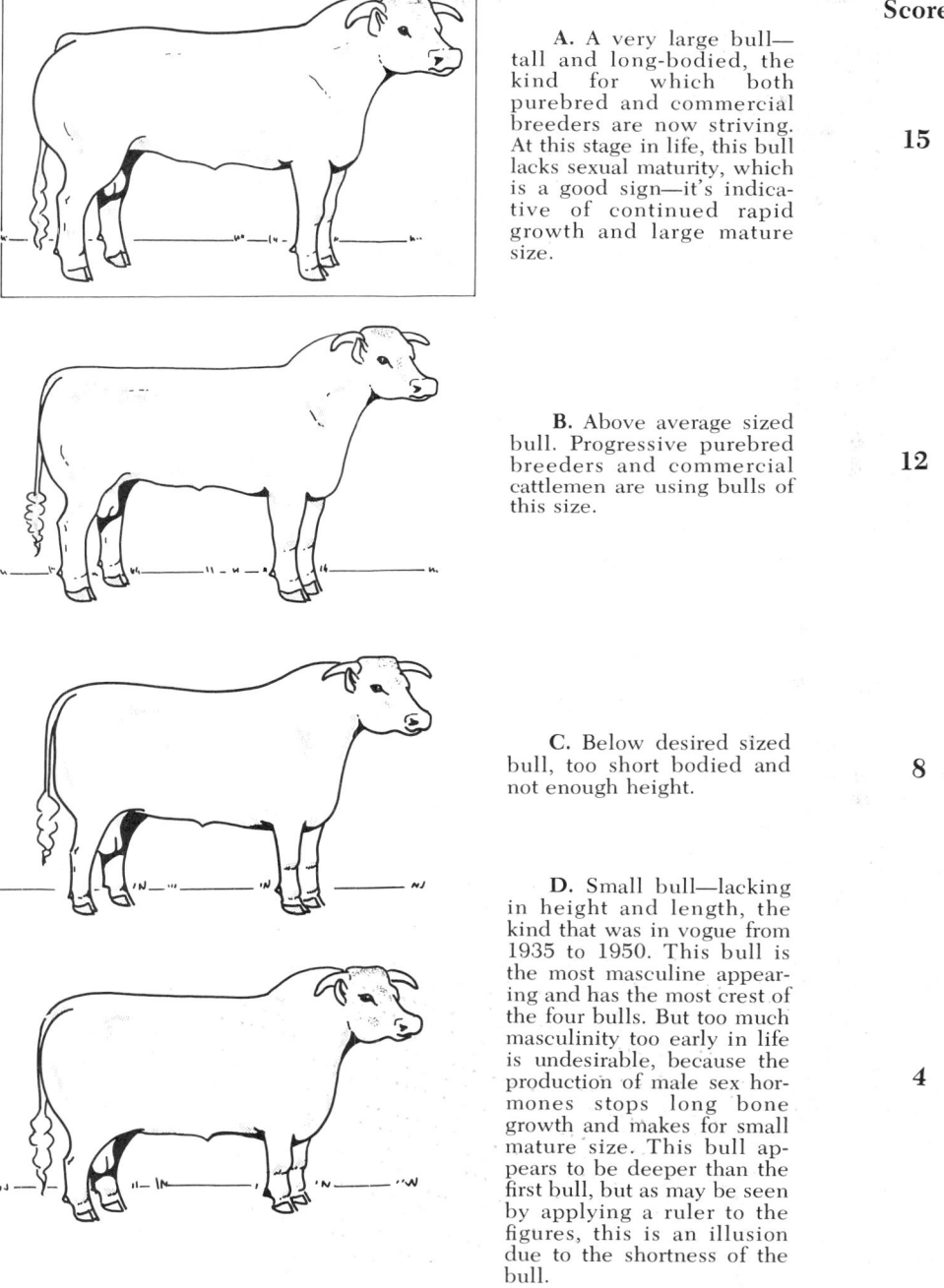

A. A very large bull—tall and long-bodied, the kind for which both purebred and commercial breeders are now striving. At this stage in life, this bull lacks sexual maturity, which is a good sign—it's indicative of continued rapid growth and large mature size.

15

B. Above average sized bull. Progressive purebred breeders and commercial cattlemen are using bulls of this size.

12

C. Below desired sized bull, too short bodied and not enough height.

8

D. Small bull—lacking in height and length, the kind that was in vogue from 1935 to 1950. This bull is the most masculine appearing and has the most crest of the four bulls. But too much masculinity too early in life is undesirable, because the production of male sex hormones stops long bone growth and makes for small mature size. This bull appears to be deeper than the first bull, but as may be seen by applying a ruler to the figures, this is an illusion due to the shortness of the bull.

4

Fig. 4-8. Size makes the difference! Four bulls of different sizes.

FREEDOM FROM WASTE

From the days of the Texas Longhorn through the 1960s, most cattlemen and scientists described their ideal beast about as follows: "The ideal beef animal should be low set and compact. The body should have great width and depth throughout, with good lines, and with all parts smoothly blended together."

How wrong we were! We now know that the smoother the loin, the squarer the rump, and the straighter the underline, the wastier the animal. We also know that high dressing percentage is *not* an indication of high carcass merit. Rather high dressing percentage is largely due to increase in carcass fat. Worse yet, as the fat increases there is a tendency for the greater proportion of it to be deposited on the cheaper cuts of meat (in the flanks, brisket, and along the underline) or in lumps where it will constitute fat trim. Thus, smoothness of form and straightness of lines should be penalized because they are indicative of fat deposition.

It is generally recognized that there is less fat deposition at early ages in late-maturing animals. This has been a factor in the rise in popularity of late-maturing Holsteins and exotics. This also indicates that when scoring weaner calves the too early maturing kind—those that look old for their age—should be scored down on "Freedom from Waste."

Also, it is noteworthy that excess fat has the greatest influence on cutability. Thus, estimating cutability involves estimating fatness. Basically, the Federal grading of beef in both the United States and Canada relies on measuring fatness on the carcass to estimate cutability.

Freedom from waste, or trimness, is important in both breeding and slaughter cattle. Excessively fat breeding cattle usually have lowered reproduction. Excessively fat slaughter cattle have reduced carcass value.

Look for fat over the following parts of the body: point of shoulder, back ribs, and along the topline directly above the backbone. No muscle should ever be found at these places, so if you feel something, it can be just one thing—fat.

Also, look for fat at those points where fat is deposited at a faster-than-average rate. Look for fat in the rear flank; and look for predisposition to waste as indicated by loose hide on the throat, dewlap, brisket, fore flank, navel or sheath, cod or udder, and twist.

Freedom from waste can be scored on the basis of (1) trimness of middle, and (2) freedom from fat or loose hide in the brisket and foreflank, for no muscle is ever found at these two points.

Of course, freedom from waste in young cattle does not necessarily measure how fat they are at the time of scoring. Rather, it's a score of their estimated, or projected, tendency to get wasty (1) following a normal feeding period, or (2) if fed to normal slaughter weights. Generally, calves that have large briskets, and that are deep in their flanks and twist will have a tendency to get overly fat as they mature.

Fig. 4-9 shows four bulls that are the same for all traits except waste. These bulls are the same length from nose to tailhead, and the same height from ground to top of shoulders (as can be determined by use of a ruler). The

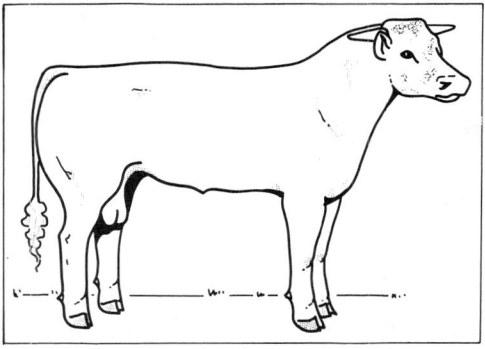

A. A very trim bull, free from waste fat. Note that the rear flank is high and tight, which is desirable. Note, too, that this bull appears to be longer and taller, and to have a longer neck, than the other three bulls. However, this is only an illusion, as can be proven by use of a ruler. All four bulls are the same in body length, from nose to tail; the same height, from ground to top of shoulders; and the same in length of neck. They differ only in depth of body and freedom from waste.

Score

15

B. A trim bull of the kind that most progressive purebred breeders and commercial cattlemen are using.

12

C. A slightly wasty bull, but not as wasty as D.

D. A very wasty bull. Note that the rear flank is loose, hangs low, and is likely filled with fat—a definite fault. Note, too, that this bull is smoother over the loin, squarer in the rump, straighter in the underline, and deeper in the body than the other three bulls. But the added depth of body is not due to a more capacious chest and middle; rather it is due to loose hide that is serving, or will serve, no more useful purpose than as a pouch for unwanted fat.

8

Also note that bull D appears to be shorter necked than the rest. However, this is only an illusion created by the loose skin in the dewlap and the added fat in the shoulder vein. As measurement with a ruler will reveal, the necks of all the bulls are the same length.

4

Fig. 4-9. Wastiness makes the difference! Four bulls varying in freedom from waste.

only difference is in the progressive increase in depth of body from bull A to bull D. However, this is not due to a more capacious chest and middle, which is desirable in all cattle, and which is obtained through long, well-sprung ribs. Rather, it is due to wastiness—to loose hide that is filled, or will fill, with fat.

STRUCTURAL SOUNDNESS

Correct skeletal structure is fundamental to a long and productive life in beef cattle, and to efficient beef production. Beef cattle must be able to travel freely and long distances over pasture or range, in order to harvest their feed—chiefly grass; and bulls must be able to follow and breed cows. This means that skeletal defects—particularly in the legs, feet, and joints—can shorten the productive life of any cow or bull. This calls for structural soundness, especially in the underpinning. The legs should be straight, true, and squarely set. The feet should be large, wide and deep at the heel, and have toes of equal size and shape that point straight ahead. The joints, particularly the hock and knee joints which are subject to great wear, should be large and correctly set, and should be clean—without tendency toward puffiness or swelling. Sickle-hocked, post-legged, back at the knees (calf-kneed), or over at the knees (buck-kneed) are faults, and such animals should be scored down.

Fig. 4-10 illustrates structural soundness, which is very important in breeding cattle, but only moderately so in slaughter steers.

Score

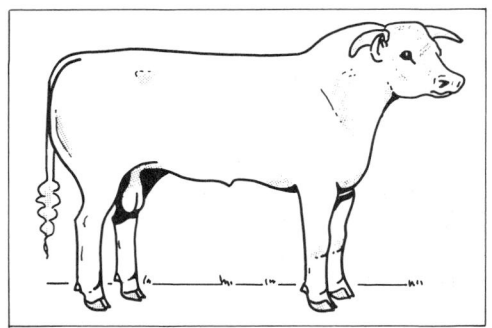

A. Correct skeletal structure. Note that the legs are squarely set, and that there is sufficient set or angle at the hocks to provide spring and flexion.

15

B. Sickle-hocked, which means that the hind feet are placed too far forward beneath the body; and back at the knees (calf-kneed). Sickle-hocked cattle frequently develop hock weaknesses. Cattle that are back at the knees stand and go uphill all the time.

4

Fig. 4-10. Soundness makes the difference!   (Continued next page)

C. Post-legged (too straight hock) and buck-kneed (over at the knees). Post-legged cattle have a short, stilted stride; and they're likely to be puffy in the hocks and predisposed to stifle injury. Buck-kneed cattle are less stable on their front legs.

4

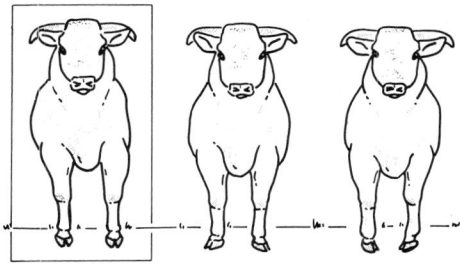

D. Left to right: (1) Correct front feet; (2) toes out; (3) toes in. Cattle that toe out swing the moving foot inward toward the supporting leg, whereas animals that toe in swing their feet outward when moving.

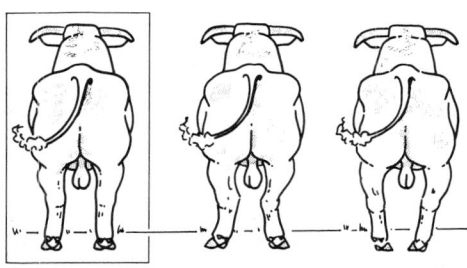

E. Left to right: (1) Correct hind feet; (2) cow-hocked; (3) wide at the hocks. Cow-hocks and wide at the hocks predispose to a twisting, rotating action when the animal is moving.

Fig. 4.10 continued.

BREED TYPE

Breed type may be defined as the distinctive characteristics which distinguish one breed from another. Usually, it involves color and markings, shape of head, presence or absence of horns (and shape of horns if present), set of ears, body shape, and size. The combination of these characteristics constitutes the breed "trademark." Generally they are highly valued by owners of purebred herds and used for promotional purposes.

So long as breed characteristics are not associated with undesirable traits, and so long as their selection does not lower the utility value of the animal, there is nothing wrong with them. Moreover, the choice of a breed is purely voluntary. Hence, when a breed is selected, the purebred breeder is morally obligated to comply with the rules of the registry association. Otherwise, he should either select another breed more compatible with his thinking or start another registry association.

First and foremost, the Functional Scoring System places emphasis on those traits which fit a beef animal for the functions that it is expected to perform—reproductive efficiency, size, muscling, freedom from waste, and structural soundness. Additionally, a registered animal (or a purebred which it is intended shall be registered) should be scored on breed type. Commercial cattlemen may disregard the latter trait. Likewise, it is unimportant in steers.

Fig. 4-11 shows some breed type differences as expressed in the heads of the Hereford, Angus, and Brahman breeds. Note the desirable versus the undesirable breed type heads of each sex of each of the three breeds.

(Below: Hunting the Prehistoric European Bison. After painting by Ernest Griset. Courtesy, Smithsonian Institution)

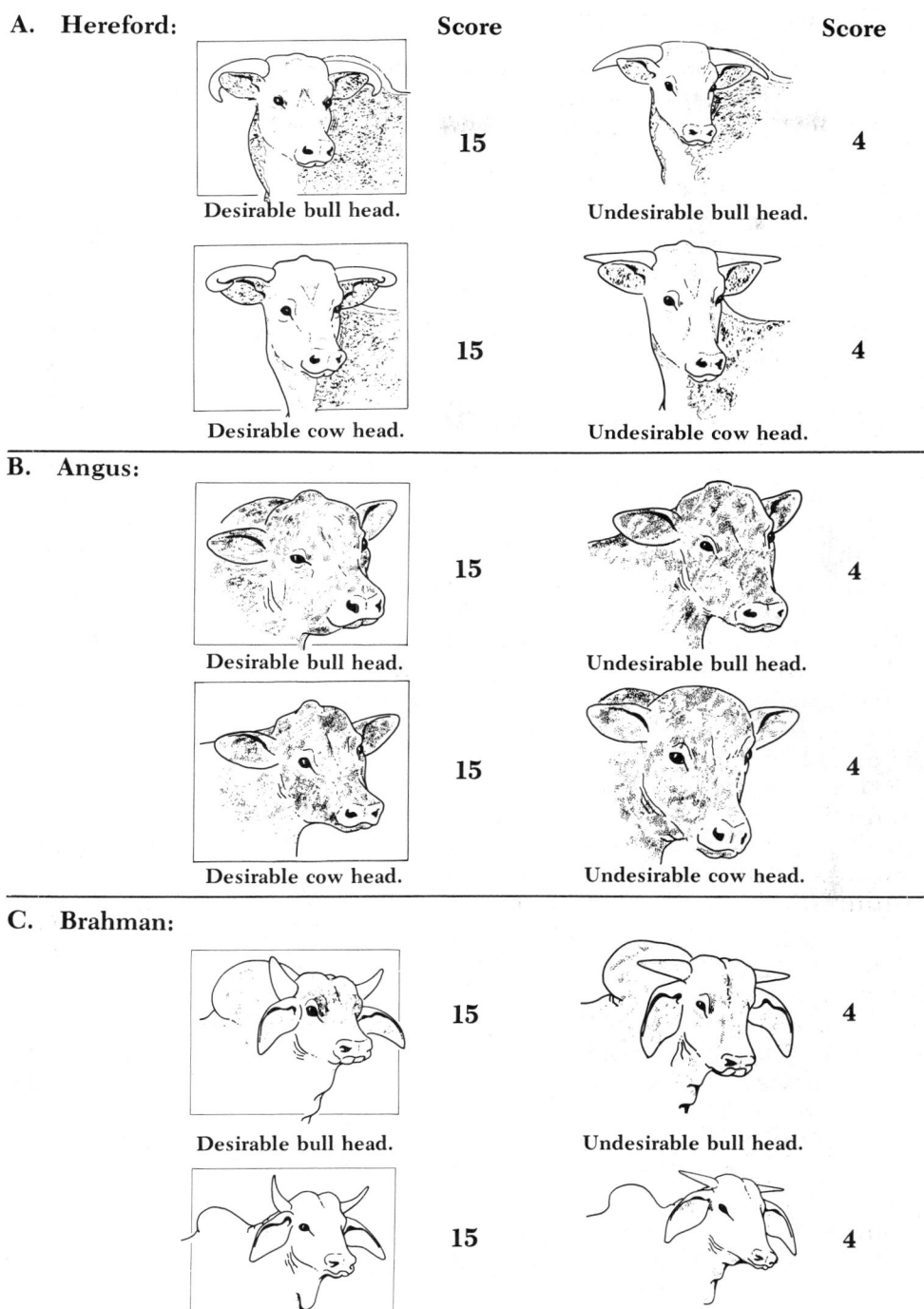

A. Hereford:

Score — 15 — Desirable bull head.
Score — 4 — Undesirable bull head.

Score — 15 — Desirable cow head.
Score — 4 — Undesirable cow head.

B. Angus:

15 — Desirable bull head.
4 — Undesirable bull head.

15 — Desirable cow head.
4 — Undesirable cow head.

C. Brahman:

15 — Desirable bull head.
4 — Undesirable bull head.

15 — Desirable cow head.
4 — Undesirable cow head.

Fig. 4-11. Breed type.

SUMMARY OF THE FUNCTIONAL SCORING SYSTEM

Like the computer, the Functional Scoring System is no better than the person back of it—the scorer. For best results, it is recommended (1) that the scorer completely disregard any facts about the animal that would color his thinking and affect his score—such as pedigree, purchase price, and show-ring record—and concentrate on the particular characteristic that he is scoring at the time; (2) that only one animal be scored at a time; and (3) that all cattle being scored be in the same corral or arena, on firm level footing, and at approximately the same location with reference to the scorer.

It is important that the same person score the animals of a given herd.

For convenience, the author has prepared the All-Breed Functional Scorecard which follows (Fig. 4-12).

SELECTION BASED ON PERFORMANCE TESTING

The modern era in beef cattle breeding was ushered in with the weighing of animals and the keeping of written production records. This led to performance testing programs, most of which were aimed at improvement of growth rate and feed efficiency. The ease of measuring these two characteristics made performance testing acceptable to producers.

Simply stated, performance testing is a record-keeping system for the purpose of collecting data to be used in selection. It has been an important selection tool in the hands of cattlemen. It brought an awareness that some animals are more efficient than others, and that such characteristics as rate of gain and feed efficiency are at least partially under genetic control and can be passed on to offspring.

However, a performance testing program based only on rate of gain and feed efficiency is not adequate, because composition of gain cannot be ignored and still meet consumer demand. We need to know if the animal is converting its feed to red meat or fat. We need to know the pounds of feed required to produce a pound of consumer accepted lean beef.

Rapid growth favors large size, but there is likely a point beyond which large size is not entirely compatible with other production traits. Also, within a breed, and at a given age (15 months of age in the case of Brungardt's work, University of Wisconsin), cattle with the higher weight per day of age tend to be fatter and contain a lower percent of muscle.[23] At a given weight, however, the greater the weight per day of age, the less the total fat and the greater the amount of total muscling.

Production testing programs that stop too early (i.e., under one year of age, or at less than 900 pounds weight) may lead to the selection of some animals which do not have the inheritance to continue to grow lean meat to heavier weights.

Thus, the goals of performance testing need to be reappraised, and measuring techniques need to be improved and become more sophisticated.

---

[23]Brungardt, V. H., *Research Report*, University of Wisconsin-Madison, Efficiency and Profit Differences of Angus, Charolais, and Hereford Cattle Varying in Size and Growth, R 2400, May 1972, p. 2.

	Perfect Score	Animal				
		No. 1	No. 2	No. 3	No. 4	Etc.
REPRODUCTIVE EFFICIENCY:	20					
Highly fertile female—Feminine—long body, lean, smooth muscled; refined, feminine head; lean cheek, jaw, neck, brisket, shoulder, and hindquarters; and a good functional udder (or promise of udder development in a heifer).						
Avoid lowly fertile female—Steery appearance—coarse, heavy front, masculine rather than feminine; protruding brisket; bristly hair on neck and top of shoulders; rounded hindquarters; and fat deposits on the face, brisket, shoulders, hips, rump, pins, below the vulva, and in front of the udder.						
Highly fertile bull—Masculine—"he's on the look," with head up and ears cocked; well-developed crest; muscles well developed and clearly defined especially in the regions of the neck, loin, and thigh; and well-developed genitalia, with testicles of equal size and well defined, and a proper neck to the scrotum.						
Avoid lowly fertile bull—Lacking masculinity—ears not alert; undeveloped crest; muscles lacking development and not clearly defined; testicles small, unbalanced, or with one carried high; scrotum that is twisted or filled with fat.						
MUSCLING:	20					
Well muscled—Bulging in those areas least affected by fatness—the arm, forearm, gaskin, and stifle, muscles move and bulge as animal walks. Look for curved loin and round; crease in thigh; well-defined groove down topline, with loin eye bulging on each side of backbone. Look for calves with long, smooth muscling, indicating continued growth. Since muscling is a masculine trait, it is more important in bulls and steers than in heifers.						
Avoid coarse shoulders in breeding cattle, because it is usually associated with calving problems.						
SIZE:	15					
Adequate size—As indicated by height to top of shoulders and length from nose to tailhead. Young breeding animals and steers should be long, tall, and not excessively fat—indications that they will continue to grow. *Avoid* bulls showing signs of early sexual maturity; they are not likely to make continued rapid growth and reach large mature size.						
FREEDOM FROM WASTE:	15					
Freedom from waste; trimness—In both breeding and slaughter cattle. Excessively fat breeding cattle usually have lowered reproduction. Excessively fat slaughter cattle have reduced carcass value.						
Avoid loose hide that is filled, or will fill, with fat. Look for loose hide on the throat, dewlap, brisket, fore flank, navel or sheath, and twist. Look for fat over back ribs, point of shoulder, and along backbone; since no muscle should be found at these places, if you feel something, it is fat.						
STRUCTURAL SOUNDNESS:	15					
Structurally sound—Legs straight, true, and squarely set; feet large, wide, and deep at the heel, with toes of equal size and shape that point straight ahead; hock and knee joints correctly set and clean.						
Avoid sickle-hocked, post-legged, back at the knees (calf-kneed), over at the knees (buck-kneed), or puffiness or swelling of knee or hock joints.						
BREED TYPE:	15					
Characteristics true to breed—Breed distinguished by color and markings, shape of head, presence or absence of horns (and shape of horns if present), set of ears, body shape, and size.						
Avoid breed characteristics associated with undesirable traits. Commercial cattlemen can disregard this trait. Likewise, it is unimportant in steers.						
TOTAL	100					

Fig. 4-12. All-Breed Functional Scorecard.

To growth rate and feed efficiency, we need to add composition of gain as determined by some such method as (1) the Functional Scoring System on live animals, and/or (2) carcass studies on slaughter animals.

The Beef Improvement Federation (BIF), whose membership is made up of United States organizations with an interest in performance testing, was formed on February 1, 1968. One of its main stated objectives was to establish accurate and uniform procedures for measuring and recording data concerning the performance of beef cattle.

Further details relative to production testing and traits of importance in beef cattle are presented in Chapter 5.

SELECTION BASED ON PEDIGREE

Selection based on pedigree refers to the selection of animals to be the parents of the next generation, based upon their ancestors. This method is used in most purebred herds, usually in combination with one or more of the other bases of selection: individuality, production test, or show-ring winnings.

Pedigree selection is of special importance (1) when progeny performance data are not available, or (2) when the animals are either so thin or so young that their individual merit cannot be ascertained with any degree of certainty. Also, when selection is being made between animals of comparable individual merit, the pedigree may be the determining factor.

In making use of pedigree selections, it must be remembered that the ancestors close up in the pedigree are much more important than those many generations removed. Also, pedigree fads as such should be avoided, especially if there has not also been rigid culling and selection based on utility value; and in all instances, poor individuals should be culled, regardless of the excellence of the relatives. Likewise, one should not be misled by or overestimate the value of such pedigree information as the following:

1. *Family names or favorite animals many generations removed*—The value of family names is generally overrated. Obviously, if the foundation animal (the one giving the family name) is very many generations removed, the genetic superiority of this family head is halved so many times by subsequent matings that there is little reason to think that one family is superior to another. Worse yet, some breeders extol a family name in a pedigree on the basis that there are few members of the elite family, little realizing that, in some cases at least, there may be unfortunate reasons for the scarcity in numbers.

2. *Percentage of blood*—With the need to build up populations of a new breed by upgrading, questionable pedigree schemes are sometimes devised by new breed associations. Among them:

a. *The use of "blood" fractions instead of percentages*—Instead of the use of such "blood" fractions as ½ blood, ¾ blood, ⅞ blood, etc., actual percentages should be calculated. To calculate the percentage of certain breeding in the offspring, add that of the sire and dam, then divide by 2. For example, if a 72% bull is mated to a 48% cow, the offspring will be 60% ($\frac{72 + 48}{2}$ = 60).

b. *Setting an arbitrary fraction for purebreds*—Some associations set arbitrary fractions, like ⅞ or ¹⁵/₁₆, and call such animals as reach this point purebred. Of course, there is nothing magical about reaching such a point, nor is any aura of superiority imparted. It would appear, therefore, that breeders and buyers should be allowed to make their own decisions on how much importance to place on different percentages. Simply list them, and let them decide.

● *Performance pedigree*—For maximum genetic progress, the pedigree should be more than a mere listing of birth dates and names of ancestors. A pedigree should combine genealogy and performance. Such a pedigree should contain a complete listing of an animal's performance record and its ancestor's performance and progeny records.

The Beef Improvement Federation (BIF) recommends that a performance pedigree contain the following basic information, with the format of the pedigree left to each recording organization:[24]

● *Animal's individual record*—
205 day adjusted weaning weight
weaning weight ratio[25]
number of contemporaries, weaning
365, 462, or 550 day adjusted yearling weight
yearling weight ratio[26]
number of contemporaries, yearling
● *Progeny of each individual in pedigree*—
Sons - number of calves or yearlings
average - 205 day adjusted weaning weight
weaning weight ratio[27]
number of contemporaries, weaning
365, 452, 550 day adjusted yearling weight
yearling weight ratio[28]
number of contemporaries, yearling
Daughters - number of calves or yearlings
average - 205 day adjusted weaning weight
weaning weight ratio[29]
number of contemporaries, weaning
365, 452, or 550 day adjusted yearling weight
yearling weight ratio[30]
number of contemporaries, yearling

[24]*Guidelines for Uniform Beef Improvement Programs*, Beef Improvement Federation Recommendation, October 1972, p. 32.
[25]The individual weight divided by the average weight of its contemporaries times 100.
[26]Ibid.
[27]Ibid.
[28]Ibid.
[29]Ibid.
[30]Ibid.

● *Progeny carcass information*—
number of steers, heifers, or bulls
carcass weight
average - USDA quality grade to ⅓
marbling score
percent cutability
fat thickness
loin eye area
pounds of trimmed retail cuts/day of age
● *Productivity of a sire's daughters*—This information will give a producer an idea of how the daughters of different bulls are producing or milking in his herd.

Average MPPA (Most Probable Producing Ability) for each sire's daughters as compared to her contemporaries or use average weaning weight ratios for this comparison.

● *Additional consideration*—The inclusion of breeding values for individuals is recommended when techniques and information for specific traits are available.

Fig. 4-13 suggests a layout for incorporating performance records with pedigree in a sale catalog.

SELECTION BASED ON SHOW-RING WINNINGS

For years, many cattlemen (purebred breeders and commercial cattlemen alike) looked favorably upon and used show-ring winnings as a basis of selection. Purebred breeders were quick to recognize this appeal and to extol their champions through advertising. In most instances, the selection of foundation and replacement cattle, and herd bulls, on the basis of show-ring winnings and standards was for the good. On some occasions, however, purebred and commercial breeders alike came to regret selection based on show-ring winnings. A case at point was the period from 1935 to 1946, when the smaller, earlier-maturing, blockier, and smoother types of cattle were winning. Among many, this debacle brought disrepute, from which livestock shows may never fully recover. This would indicate that some scrutiny should be exercised relative to the type of animals winning in the show, especially to ascertain whether they are the kind that are efficient from the standpoint of the producer, and whether, over a period of years, they will command a premium on a discriminating market.

Perhaps the principal value of selections based on show-ring winnings lies in the fact that shows direct the attention of new breeders to those types and strains of cattle that are meeting with the approval of the better breeders and judges.

(Also, see Chapter 14, Fitting, Showing, and Judging Cattle.)

HERD IMPROVEMENT THROUGH SELECTION

Once the herd has been established, the primary objective should be to

BREED   CONSIGNOR NAME AND ADDRESS   Name of Bull Calf   Reg.   Birth Date Mo. Da. Yr.   Tattoo

LOT

Sex	Check if Purebred	Percent if Not Purebred	Check if Polled	205 Day Wt.	205 Day Wt. Ratio	Weaning Grade	365 Day Wt.	Yearling Wt. Ratio

PEDIGREE

Sire Name

SIRE   REG.

GRAND SIRE   REG.

GRAND DAM   REG.

PROGENY RECORD

NO. CALVES ____ AVE 205 WT ____ 205 WT RATIO ____ AVE GRADE ____

NO. BULLS ____ AVE 365 WT ____ AVE 365 RATIO ____

PROGENY CARCASS DATA   No. ____ Ribeye Adl. ____ Carcass W/DA ____ Fat Thickness ____ Cutability ____ % Choice of Higher ____ Carcass Weight ____

DAM   REG.

GRAND SIRE   REG.

GRAND DAM   REG.

PROGENY RECORD

NO. CALVES ____ AVE 205 WT ____ 205 WT RATIO ____ AVE GRADE ____

NO. BULLS ____ AVE 365 WT ____ AVE 365 RATIO ____

Fig. 4-13. Suggested performance pedigree for a sale catalog.

improve it so as to obtain the maximum production of quality offspring. In order to accomplish this, there must be constant culling and careful selection of replacements. The breeders who have been most constructive in such a breeding program have usually used great breeding bulls and they have obtained their replacements by selecting some of the outstanding, early-maturing heifers from the more prolific families.

Improvements through selection are really twofold: (1) the immediate gain in increased calf production from the better animals that are retained; and (2) the genetic gain in the next generation. The first is important in all herds, whereas the second is of special importance in purebred herds and in all herds where replacement females are raised. Most of the immediate gain is attained in selecting the cows, which are more numerous than the bulls; whereas the majority of the genetic gain comes from the careful selection of bulls. The genetic gain is small, but it is permanent and can be considered a capital investment.

Many good cattle breeders consider it a sound practice to make about a 20 percent replacement each year. Under such a system of management, 46 percent of the heifer calves are retained each year.[31]

BULL GRADING

Bull grading programs have exerted a powerful influence in improving the commercial cattle on the western ranges of this country. Perhaps this movement received its greatest impetus in those areas where several owners run herds on unfenced public grazing lands. Formerly, those progressive ranchers who believed that only purebred beef bulls of high quality should be used could do nothing to prevent the presence of inferior bulls on these public ranges. The man who bought superior bulls got no more use from them than his neighbor who turned out scrubs because he could buy them cheaply. This problem was finally solved when groups of cattlemen using common ranges decided to have their bulls classified and to use only bulls meeting certain grades. Today, grazing permits are sometimes refused or delayed because of ranchers refusing to use graded bulls.

In some cases in the West, all animals consigned to range bull sales are graded and individual ranchers grade their young bulls before turning them with the cow herd. In some consignment sales, bulls must be of a certain specified minimum grade in order to be sold. Grading of sale bulls is especially popular with most buyers, but some sellers object to it.

Many different systems of bull grading have been used in the past. Today, most grading is on a numbering system, in which numbers from 0 to 17 are used.

---

[31]A herd of 100 cows will produce about 44 heifers each year (with an 88% calf crop—the national average). Hence, 46% of the heifers will be needed to replace 20% of the cows (44 × .46 = 20).

QUESTIONS FOR STUDY AND DISCUSSION

1. What is the object of selection (a) in breeding cattle, (b) in feeder cattle, and (c) slaughter cattle? Wherein does selection for each of these three purposes differ from the other two?

2. Select a certain farm or ranch (either your home farm or ranch, or one with which you are familiar). Assume that there are no beef cattle on this establishment at the present time. Outline, step by step, how you would go about (a) establishing a herd, and (b) selecting the individuals. Justify your decisions.

3. Should the selection of individuals for a purebred herd and for a commercial herd differ? If so, how?

4. How would you go about getting more milking ability (a) in a purebred herd, and (b) in a commercial herd?

5. Is uniformity of color in a herd important in a commercial herd where a crossbreeding program is being followed? If so, how could you obtain uniformity?

6. Discuss the economics of longevity in a cow herd. Use as the examples 2 cows dropping their first calves as 2-year-olds, each producing a calf each year thereafter—but one producing through age 8, and the other producing through age 10.

7. Discuss the practical importance of the Missouri Experiment Station study by Dr. Samuel Brody in which cattle were housed in "climatic chambers."

8. Why should selection be from among animals kept under an environment similar to that under which the cattleman expects them and their offspring to perform?

9. Under the Functional Scoring System, six traits are listed. Rank these traits in order of economic importance. Justify your ranking.

10. What size beef cattle do you feel is best for most farms and ranches of America—large, medium, or small? Justify your answer.

11. Is it possible through selection to increase (a) the overall, total muscling of cattle, and (b) the proportion of muscle in the high priced cuts?

12. With the current emphasis on selecting cattle for heavy muscling, will there be more and more double muscling? Is double muscling good or bad? Justify your answer.

13. In addition to selecting cattle that are heavy muscled genetically, a cattleman has at his disposal other ways through which he can tailor-make his product so as to have a higher proportion of lean to fat at slaughter time. List and discuss each of these.

14. Why does Europe feed bulls whereas America feeds steers?

15. If fat is unwanted, and largely waste, why are cattle finished out in a feedlot?

16. Should breed type be considered in a selection program?

17. What method of selection would you use, and why?

18. Do all methods of selection, including production testing, rely in part at least on visual appearance or "eyeballing"?

19. What are the advantages and disadvantages to the scorecard system of judging?

20. What are the advantages and disadvantages of the Functional Scoring System?

21. How would you improve a performance testing program that is based primarily upon rate and efficiency of gain?

22. How could pedigrees be made more useful from a selection standpoint?

23. Cite examples as proof of the fact that show-ring standards have not always been practical.

24. Under what conditions is bull grading particularly valuable? Why is bull grading generally more popular with buyers than sellers?

SELECTED REFERENCES

Title of Publication	Author(s)	Publisher
Animal Growth and Nutrition	Ed. by E. S. E. Hafez, I. A. Dyer	Lea & Febiger, Philadelphia, Penn., 1969
Beef Cattle, Sixth Edition	A. L. Neumann R. R. Snapp	John Wiley & Sons, Inc., New York, N.Y., 1969
Beef Production in the South	S. H. Fowler	The Interstate Printers & Publishers, Inc., Danville, Ill., 1969
Breeding and Improvement of Farm Animals, Sixth Ed.	V. A. Rice F. N. Andrews E. J. Warwick J. E. Legates	McGraw-Hill Book Co., New York, N.Y., 1967
Commercial Beef Cattle Production	Ed. by C. C. O'Mary, I. A. Dyer	Lea & Febiger, Philadelphia, Penn., 1972
Genetics of Livestock Improvement, Second Ed.	J. F. Lasley	Prentice-Hall, Inc., Englewood Cliffs, N.J., 1972
Guidelines for Uniform Beef Improvement Programs, Program Aid 1020	Beef Improvement Federation	U.S. Department of Agriculture, 1972
Improving Reproductive Efficiency in Beef Cattle	J. R. Beverly, *et al.*	Glidwell Printers, Bryan, Tex., 1972
Improvement of Livestock	R. Bogart	The Macmillan Company, New York, N.Y., 1959
Lasater Philosophy of Cattle Raising, The	L. M. Lasater	The Texas Western Press, University of Texas, El Paso, Tex., 1972
Proceedings—Second World Conference on Animal Production	R. E. Hodgson, *et al.*	American Dairy Science Association, Urbana, Ill., 1969
Stockman's Handbook, The, Fourth Edition	M. E. Ensminger	The Interstate Printers & Publishers, Inc., Danville, Ill., 1970

CHAPTER 5

SOME PRINCIPLES OF CATTLE GENETICS

Contents   **Page**

Cattle breeding dates back to the first book of the Bible. The thirtieth chapter of Genesis relates how the herdsman, Jacob, who was to receive his salary in the form of all the off-colored cattle that appeared in his father-in-law's herd, took action that marks him as an eager and crafty cattle breeder.

In an attempt to assure that speckled- and spotted-colored cattle appeared in goodly numbers, Jacob took green poles of poplar, hazel, and chestnut, peeled white rings and stripes around them, and set them around the watering troughs. When the cows came to drink and mate, Jacob theorized that these striped poles would have a prenatal effect on their offspring—resulting in off-colored calves. The whimsical story goes on to say that Jacob's herd "increased exceedingly."

Modern cattle breeders use more scientific approaches to change the specifications of their cattle. Nevertheless, the story of Jacob does show that cattle breeding is of ancient origin.

During the past century, remarkable progress has been made in breeding better beef cattle—animals that are more efficient, and that produce cuts of meat more nearly meeting the exacting requirements of the consuming public. The Texas Longhorn steer—built for stamina, ability to fight and to furnish its own transportation, but producing tough, stringy meat—has been replaced by the earlier-maturing, prime bullock.

Despite the remarkable progress of the past, much remains to be done. A casual glance at the daily receipts of any public market is convincing evidence of the need for further improvements. Also, cattle sterility, reproductive failures, and young calf losses are very costly. Moreover, the production of abnormal animals—even if they occur only to a limited extent—represents a considerable economic loss to the cattleman. In this age, there must be greater efficiency of production; this means more rapid growth, heavier weights, less feed to produce 100 pounds of beef, lifting the percentage calf crop well above the present U.S. average, and the production of tender beef with less exterior fat. The ultimate goal should be that of furnishing better animals for market and lowering the cost of production.

The laws of heredity apply to cattle breeding exactly as they do to all other classes of farm animals; but the breeding of cattle is less flexible in the hands of man because (1) cows seldom give birth to more than one offspring at a time, whereas sows are litter-bearing animals and ewes frequently produce twins or triplets, and (2) hogs and sheep breed at an earlier age, thus making for a shorter interval between generations. These and other breeding phenomena peculiar to beef cattle will be treated in this chapter.

With present, although incomplete, knowledge of genetics, progress should now be much more certain and rapid. In the past, cattle breeding has been an art; in the future, it should be both an art and a science.

MENDEL'S CONTRIBUTION TO GENETICS

Modern genetics was really founded by Gregor Johann Mendel, an Austrian monk, who smoked long, black cigars and gardened because of his obesity. He conducted breeding experiments with garden peas from 1857 to

1865, during the time of the Civil War in the United States. In his monastery at Brünn (now Brno, in Czechoslovakia), Mendel applied a powerful curiosity and a clear mind to reveal some of the basic principles of hereditary transmission. In 1866, he published in the proceedings of a local scientific society a report covering 8 years of his studies, but for 34 years his findings went unheralded and ignored. Finally, in 1900, 16 years after Mendel's

Fig. 5-1. Gregor Johann Mendel (1822-1884), a cigar-smoking Austrian monk, whose breeding experiments with garden peas founded modern genetics. (Courtesy, Bettman Archives)

death, three European biologists independently duplicated his findings. This led to the dusting off of the original paper published by the monk 34 years earlier.

The essence of Mendelism is that inheritance is by particles or units (called genes), that these genes are present in pairs—one member of each pair having come from each parent—and that each gene maintains its identity generation after generation. Thus, Mendel's work with peas laid the basis for two of the general laws of inheritance: (1) the law of segregation; and (2) the independent assortment of genes. Later genetic principles have been added; yet all the phenomena of inheritance, based upon the reactions of genes, are generally known under the collective term, Mendelism.

Thus, modern genetics is really unique in that it was founded by an amateur who was not trained as a geneticist and who did his work merely as a hobby. During the years since the rediscovery of Mendel's principles (in 1900), many additional genetic principles have been added, but the fundamentals as set forth by Mendel have been proved correct in every detail. It can be said, therefore, that inheritance in both plants and animals follows the biological laws discovered by Mendel.

SOME FUNDAMENTALS OF HEREDITY IN CATTLE

In the sections which follow, no attempt will be made to cover all of the diverse field of genetics. Rather, the author will present a condensation of a few of the known facts in regard to the field and briefly summarize their application to beef cattle.

The Gene as the Unit of Heredity

Genes determine all the hereditary characteristics of animals, from the body type to the color of the hair. They are truly the fundamental units of genetics.

The bodies of all animals are made up of millions or even billions of tiny cells, microscopic in size. Each cell contains a nucleus in which there are a number of pairs of bundles, called chromosomes. In turn, the chromosomes carry pairs of minute particles, called genes, which are the basic hereditary material. The nucleus of each body cell of cattle contains 30 pairs of chromosomes, or a total of 60[1], whereas there are perhaps thousands of pairs of genes. These genes determine all the hereditary characteristics of living animals. Thus, inheritance goes by units, rather than by the blending of two fluids as our grandfathers thought.

The modern breeder knows that the job of transmitting qualities from one generation to the next is performed by germ cells—a sperm from the male and an ovum or egg from the female. All animals, therefore, are the result of the union of two such tiny cells, one from each of its parents. These two germ cells contain all the anatomical, physiological, and psychological characters that the offspring will inherit. They determine whether a calf shall be polled instead of horned, black instead of white, a bull instead of a heifer, etc.

In the body cells of an animal, each of the chromosomes is duplicated; whereas in the formation of the sex cells, the egg and the sperm, a reduction division occurs and only one chromosome and one gene of each pair goes into a sex cell. This means that only half the number of chromosomes and genes present in the body cells of the animal go into each egg and sperm (thus each reproductive cell of cattle has 30 chromosomes), but each sperm or egg cell has genes for every characteristic of its species. As will be explained later, the particular half that any one germ cell gets is determined by chance. When mating and fertilization occur, the single chromosomes from the germ cell of each parent unite to form new pairs, and the genes are again present in duplicate in the body cells of the embryo.

With all possible combinations in 30 pairs of chromosomes (the species number in cattle) and the genes that they bear, any bull or cow can transmit over one billion different samples of its inheritance; and the combination from both parents makes possible one billion times one billion genetically different offspring. It is not strange, therefore, that no two animals within a given breed (except identical twins from a single egg split after fertilization)

---

[1]Horses also have 60 chromosomes; swine have 40; sheep have 54; and man has 46, not 48 as long believed. (The latter discovery was made by Dr. J. H. Tjio, and reported at the First International Congress of Human Genetics, Copenhagen.)

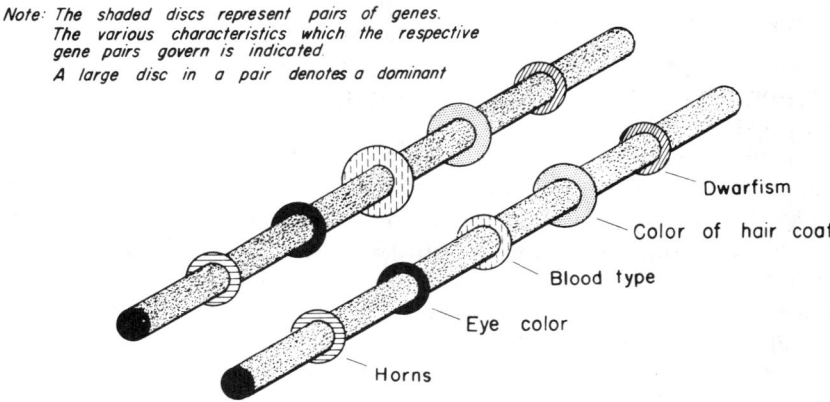

Note: The shaded discs represent pairs of genes. The various characteristics which the respective gene pairs govern is indicated. A large disc in a pair denotes a dominant

Schematic Drawing of a Pair of Chromosomes

Dwarfism
Color of hair coat
Blood type
Eye color
Horns

Fig. 5-2. A pair of bundles, called chromosomes, carrying minute particles, called genes. The genes determine all the hereditary characteristics of living animals, from length of leg to body size. (Drawing by R. F. Johnson, Head, Department of Animal Science, Calif. Polytechnic State University, San Luis Obispo, Calif.)

are exactly alike. Rather, we can marvel that the members of a given breed bear as much resemblance to each other as they do.

Even between such closely related individuals as full sisters, it is possible that there will be quite wide differences in size, growth rate, temperament, conformation, and in almost every conceivable character. Admitting that many of these differences may be due to undetected differences in environment, it is still true that in such animals much of the variation is due to hereditary differences. A bull, for example, will sometimes transmit to one offspring much better inheritance than he does to most of his get, simply as the result of chance differences in the genes that go to different sperm at the time of the reduction division. Such differences in inheritance in offspring have been called both the hope and the despair of the livestock breeder.

If an animal gets similar determiners or genes from each parent, it will produce uniform germ cells; because any half of its inheritance is just like any other half. For example, regardless of what combination of chromosomes goes into a particular germ cell, it will be just like any other egg or sperm from the same individual. Such animals are referred to as being homozygous. Few, if any, of our animals are in this hereditary state at the present time. Instead of being homozygous, they are heterozygous. This explains why there may be such wide variation within the offspring of any given sire or dam. The wise and progressive breeder recognizes this fact, and he insists on the production records of all get rather than that of just a few meritorious individuals.

Variation between the offspring of animals that are not pure or homozygous, to use the technical term, is not to be marveled at, but is rather to be expected. No one would expect to draw exactly 20 sound apples and 10 rotten ones every time he took a random sample of 30 from a barrel containing 40 sound ones and 20 rotten ones, although, on the average—if enough samples were drawn—he would expect to get about that proportion of each. Individual drawings would, of course, vary rather widely. Exactly the same situation applies to the relative numbers of "good" and "bad" genes that may be present in different germ cells from the same animal. Because of this situation, the mating of a cow with a fine show record to a bull that on the average transmits relatively good offspring will not always produce calves of merit equal to that of their parents. The calves could be markedly poorer than the parents or, happily, they could in some cases be better than either parent.

Selection and close breeding are the tools through which the cattle producer can obtain bulls and cows whose chromosomes and genes contain similar hereditary determiners—animals that are genetically more homozygous.

Actually, a completely homozygous state would be undesirable and unfortunate. This is so because economic and environmental changes are apt to dictate animal changes from time to time. With complete homozygosity, such shifts would be impossible except through the slow and uncertain process of mutations; thus, making it extremely difficult to effect a change in phenotype. Fortunately, enough heterozygosity exists in our improved breeds so that they are flexible in the hands of man; they can be molded in any desired direction. (In fact, enough heterozygosity exists that it is nigh impossible to fix complete homozygosity.) It may be said, therefore, that variation in the biological material permits the animal breeder to mold, change, and improve his stock.

Mutations

Gene changes are technically known as mutations. *A mutation may be defined as a sudden variation which is later passed on through inheritance and that results from changes in a gene or genes.* Mutations are not only rare, but they are prevailingly harmful, and most of them are recessive. Further, one cannot induce a particular kind of mutation. For all practical purposes, therefore, the genes can be thought of as unchanged from one generation to the next. The observed differences between animals are usually due to different combinations of genes being present rather than to mutations. Each gene probably changes only about once in each 100,000 to 1,000,000 animals produced.

Once in a great while a mutation occurs in a farm animal, and it produces a visible effect in the animal carrying it. These animals are commonly called "sports." Such sports are occasionally of practical value. The occurrence of the polled characteristic within the horned Hereford and Shorthorn breeds of cattle is an example of a mutation or sport of economic importance.[2] Out of this has arisen the Polled Hereford and Polled Shorthorn breeds.

Gene changes can be accelerated by exposure to X-rays, radium, mustard gas, and ultraviolet light rays. Such changes may eventually be observed in the offspring of both people and animals of Japan who were exposed to the atom bombs unleashed in World War II.

Although induced mutations have been used successfully in developing commercial varieties of plants, the technique does not appear very promising for the improvement of animals. This is so because (1) an enormous number of mutations would have to be induced in order to have much chance of getting one which had commercial value, and (2) its frequency would have to be increased by selection and as many of the concurrent undesirable mutations as possible eliminated from the stock.

Simple Gene Inheritance (Qualitative Traits)

In the simplest type of inheritance, only one pair of genes is involved. Thus, a pair of genes is responsible for the color of hair in Shorthorn cattle. This situation is illustrated by Fig. 5-3.

An animal having two genes for red (RR) is actually red in color, whereas an animal having two genes for white (rr) is white in color. On the other hand, a Shorthorn which has one gene for red (R) and one for white (r) is neither red nor white but roan (Rr), which is a mixture of red and white. Thus, red X white matings in Shorthorn cattle usually produce roan offspring. Likewise, white X white matings generally produce white offspring; but it must be remembered that white in Shorthorns is seldom pure, for the face bristles, eyelashes, and ears usually carry red hairs. Roans, having one gene for red and one for white on the paired chromosomes will never breed true and, if mated together, will produce calves in the proportion of one red, two roans, and one white. If one wishes to produce roans, the most certain way is to mate red cows with a white bull or vice versa, for then all the calves will be roan. If a roan animal is bred to a red one, one-half the offspring will be red, whereas the other half will be roan. Likewise when a roan animal is bred to a white one, approximately an equal number of roan and white calves will be produced.

This example illustrates the most important principles of inheritance; namely, (1) genes occur in animals in pairs because one member of each pair comes from each parent, and (2) each reproductive cell contains a sample half of the genes of that particular animal.

It should be borne in mind that there is no way to sort out the numerous genes so as to get the most desirable ones into the same reproductive cell except as it occurs by chance during the formation of eggs and sperm. Thus, it follows that the various gene combinations, such as referred to above, occur at random and that the various colors will appear in the offspring in the proportions indicated only when relatively large numbers are concerned.

---

[2]The horned gene mutates to the polled gene at a fairly high frequency; apparently at the rate of about 1 in 20,000.

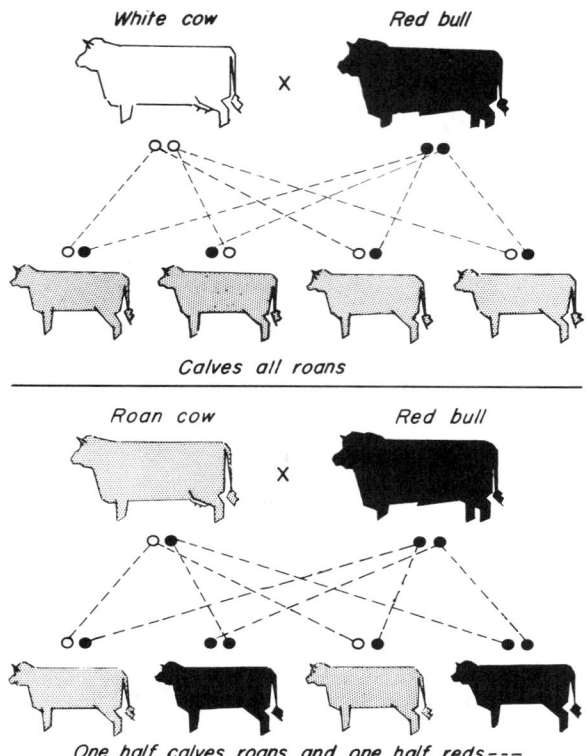

Fig. 5-3. Diagrammatic illustration of the inheritance of color in Shorthorn cattle. Red X white matings in Shorthorn cattle usually produce roan offspring, whereas roan X red matings produce one-half red offspring and one-half roan offspring. (Drawing by R. F. Johnson)

The possible gene combinations, therefore, are governed by the laws of chance, operating in much the same manner as the results obtained from flipping coins. For example, if a penny is flipped often enough, the number of heads and tails will come out about even. However, with the laws of chance in operation, it is possible that out of any 4 tosses one might get all heads, all tails, or even 3 to 1. In exactly the same manner, a Shorthorn breeder may be highly elated in obtaining 4 red calves from roan X roan matings only to be greatly depressed when the next 4 calves, from the same matings, are white in color.

In addition to color of hair, other examples of simple gene inheritance in animals (sometimes referred to as qualitative traits) include color of eyes, presence or absence of horns, type of blood, and lethals.

DOMINANT AND RECESSIVE FACTORS

In the example of Shorthorn colors, each gene of the pair (R and r) produced a visible effect, whether paired as identical genes (two red or two whites) or as two different genes (red and white).

This is not true of all genes; some of them have the ability to prevent or mask the expression of others, with the result that the genetic makeup of such animals cannot be recognized with perfect accuracy. This ability to cover up or mask the presence of one member of a set of genes is called dominance. The gene which masks the one is the dominant gene; the one which is masked is the recessive gene.

In cattle, the polled character is dominant to the horned character. Thus, if a *pure polled* bull is used on horned cows (or vice versa), the resulting progeny are not midway between two parents but are of polled character.[3] It must be remembered, however, that not all hornless animals are pure for the polled character; many of them carry a factor for horns in the hidden or recessive condition. In genetic terminology, animals that are pure for a certain character—for example the polled characteristic—are termed *homozygous*, whereas those that have one dominant and one recessive factor are termed *heterozygous*. A simple breeding test can be used in order to determine whether a polled bull is homozygous or heterozygous, but it is impossible to determine such purity or impurity through inspection. The breeding test consists of mating the polled sire with a number of horned females. If the bull is pure or homozygous for the polled character, all of the calves will be polled; whereas if he is impure or heterozygous, only half of the resulting offspring will, on the average, be polled and half will have horns like the horned parents. Many breeders of Polled Herefords or Polled Shorthorns test their herd sires in this manner, mating the prospective sire to several horned animals.

It is clear, therefore, that a dominant character will cover up a recessive. Hence, an animal's breeding performance cannot be recognized by its phenotype (how it looks), a fact which is of great significance in practical breeding.

Another example of dominance is that of the white face of Hereford cattle—the white face being dominant over the type of coloration in which the head and body are of the same color. Undoubtedly, this condition of dominance, which constitutes a trademark of the breed, has been of importance from a promotional standpoint.

As can be readily understood, dominance often makes the task of identifying and discarding all animals carrying an undesirable recessive factor a

---

[3]It is noteworthy, however, that when a homozygous polled animal is crossed with a homozygous horned animal, some "scurs" or small loosely attached horns usually appear. There are conflicting reports and opinions concerning the inheritance of scurs, with the following theories prevailing:

    a. That the gene for scurs is recessive and independent of the major genes for horns. According to this theory, scurs appear only in individuals homozygous for the scurred gene (sc sc).

    b. That scurs are a sex-influenced character. According to this theory, scurs will occur in males either homozygous (Sc Sc) or heterozygous (Sc sc) for the character, but only in females homozygous (Sc Sc) for the character; in other words, it acts as a dominant in polled males and a recessive in polled females.

    c. That the major gene (P) for polled condition is only partially dominant, with heterozygous individuals (Pp) tending to be scurred, especially in bulls.

    Horns prevent the expression of any genes an animal may have for scurs, and thus complicate studies designed to determine the exact mode of inheritance of scurs.

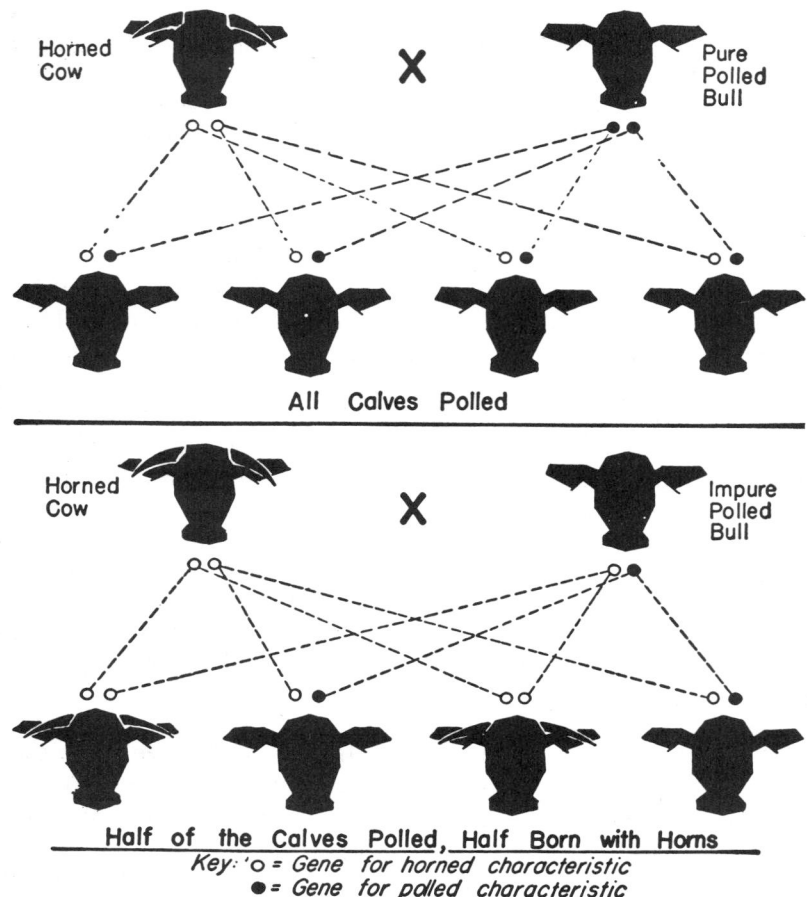

Fig. 5-4. Diagrammatic illustration of the inheritance of horns in cattle of European extraction. Although there may be a very occasional exception, if a bull that is considered pure or homozygous for the polled character is mated with a number of horned females, all of the calves will be polled; whereas if a bull that is impure or heterozygous for the polled character is mated with a number of horned females, only half of the calves will, on the average, be polled. (Drawing by R. F. Johnson)

difficult one. Recessive genes can be passed on from generation to generation, appearing only when two animals, both of which carry the recessive factor, happen to mate. Even then, only one out of four offspring produced will, on the average, be homozygous for the recessive factor and show it.

In Angus cattle, the red color is an example of such an undesirable recessive factor. Black polled cattle have been known in Scotland since 1523; and since the days of Hugh Watson, William McCombie, and George McPherson Grant, black has been the accepted color of the breed. Yet, down through the years, a recessive factor for red coat color has persisted in the breed. For this reason, a red calf occasionally and unexpectedly shows up in a purebred Angus herd (about 1 red calf appears in every 200 to 500 Angus calves dropped[4]). This occasional appearance of a red calf in the Angus breed does not signify any impurity of breeding but merely the outcropping of a long hidden recessive gene. When a red calf does appear, one can be very certain that both the sire and dam contributed equally to the condition and that each of them carried the recessive gene for red color. This fact should be given consideration in the culling program.

As the factor for red is recessive, the red animals are pure for color. The mating of two red animals, therefore, will always produce red calves.[5]

Other examples of undesirable recessives are red color in Holstein cattle and dwarfism in cattle.

Assuming that a hereditary defect or abnormality has occurred in a herd and that it is recessive in nature, the breeding program to be followed to prevent or minimize the possibility of its future occurrence will depend somewhat on the type of herd involved—especially on whether it is a commercial or purebred herd. In a commercial herd, the breeder can usually guard against further reappearance of the undesirable recessive simply by using an outcross (unrelated) sire within the same breed or by crossbreeding with a sire from another breed. With this system, the breeder is fully aware of the recessive being present, but he has taken action to keep it from showing up.

On the other hand, if such an undesirable recessive appears in a purebred herd, the action should be more drastic. A reputable purebred breeder has an obligation not only to himself but to his customers among both the purebred and commercial herds. Purebred animals must be purged of undesirable genes and lethals. This can be done by:

1. Eliminating those sires and dams that are known to have transmitted the undesirable recessive character.

2. Eliminating both the abnormal and normal offspring produced by these sires and dams (approximately half of the normal animals will carry the undesirable character in the recessive condition).

3. Breeding a prospective herd sire to a number of females known to carry the factor for the undesirable recessive, thus making sure that the new sire is free from the recessive.

Such action in a purebred herd is expensive, and it calls for considerable courage. Yet it is the only way in which the purebred livestock of the country can be freed from such undesirable genes.

INCOMPLETE OR PARTIAL DOMINANCE

The results of crossing polled with horned cattle are clear-cut because the polled character is completely dominant over its allele (horned). If, however, a cross is made between a red and a white Shorthorn, the result is a roan (mixture of red and white hairs) color pattern. In the latter cross, the

[4]In order to obtain one red calf out of 200, one parent out of every 7 must be a carrier of the red gene. Actually, to get $1/196$ red calves, there must be $1/14$ b (red gene) reproductive cells in both males and females because $1/14 \times 1/14 = 1/196$. Thus $1/7$ of the parents must be Bb (black in color, but carrying the red gene) while $6/7$ are BB (pure for black).

[5]A separate U.S. breed registry association for these Red Angus cattle was organized in 1954 (see Chapter 3, Table 3-3). To these breed enthusiasts, the recessive gene for red is desirable.

action of a gene is such that it does not cover the allele, which is known as incomplete dominance; or, stated differently, the roan color is the result of the action of a pair of genes (joint action) neither of which is dominant. This explains the futility of efforts to develop Shorthorns pure for roan.

The above discussion also indicates that there are varying degrees of dominance—from complete dominance to an entire lack of dominance. In the vast majority of cases, however, dominance is neither complete nor absent, but incomplete or partial. Also, it is now known that dominance is not the result of single-factor pairs but that the degree of dominance depends upon the animal's whole genetic make-up together with the environment to which it is exposed, and the various interactions between the genetic complex (genotype) and the environment.

Multiple Gene Inheritance (Quantitative Traits)

Relatively few characters of economic importance in farm animals are inherited in as simple a manner as the coat color or polled conditions described. Important characters—such as meat production and milk production—are due to many genes; thus, they are called multiple-factor characters or multiple-gene characters. Because such characters show all manner of gradation—from high to low performance, for example—they are sometimes referred to as quantitative traits.

In quantitative inheritance, the extremes (either good or bad) tend to swing back to the average. Thus, the offspring of a grand champion bull and a grand champion cow are not apt to be as good as either parent. Likewise, and happily so, the progeny of two very mediocre parents will likely be superior to either parent.

Estimates of the number of pairs of genes affecting each economically important characteristic vary greatly, but the majority of geneticists agree that for most such characters ten or more pairs of genes are involved. Growth rate in cattle, therefore, is affected by the following: (1) the animal's appetite; (2) feed consumption; (3) feed utilization—that is, the proportion of the feed eaten that is absorbed into the blood stream; (4) feed assimilation—the use to which the nutrients are put after absorption; and (5) feed conversion—whether used for muscle, fat or bone formation. This should indicate clearly enough that such a characteristic as growth rate is controlled by many genes and that it is difficult to determine the mode of inheritance of such characters.

Inheritance of Some Characters in Cattle

Mendelian characters are inherited in alternative pairs (or series). These alternative forms of a gene, which are located at the same point on each one of a pair of chromosomes, are called *alleles*—for example, horns (recessive) and polled (dominant).

An individual that is heterozygous with respect to one pair of allelic genes is a *monohybrid;* one that is heterozygous with respect to two pairs of allelic genes is a *dihybrid;* and one that is heterozygous with respect to three pairs of allelic genes is a *trihybrid.* Practical examples of the inheritance of each of these and other characters in cattle will follow.

A MONOHYBRID CROSS

When a homozygous polled animal (PP) is crossed with a homozygous horned animal (pp), the first cross (F_1) offspring is a monohybrid (Pp). This first cross animal is polled (see Fig. 5-4). If two of the monohybrids (Pp) are crossed, the second cross (F_2) offspring will be in the ratio of three polled to one horned. This phenomenon is explained and understood by assuming that one parent has the genes for the dominant polled characteristic and the other parent the genes for the recessive horned characteristic at similar loci in paired homologous chromosomes; that one pair of each of these genes is present in the monohybrid; but that these genes separate or segregate out when the monohybrid produces germ cells (sperms and eggs) with the half (haploid) number of chromosomes.

Also, it is noteworthy that the classical explanation for inheritance of the polled or horned condition in cattle, where (PP) and (Pp) are polled animals and (pp) are horned animals, apparently applies only to cattle of the European breeds. Where Brahman blood is introduced into crosses, the situation is considerably complicated and the exact mode of inheritance is not known. Thus, where polled animals of the European breeds of either sex are crossed with Brahman animals, it has been observed (1) that polled animals cannot be depended upon uniformly to dehorn the calves, but that the offspring from such a cross may be either polled, horned, scurred, or buttoned; (2) that the inheritance of horns may differ according to the way in which the cross is made—that is, as to whether the Brahman parent is the sire or the dam; and (3) that there is a greater tendency for bull calves of such crosses to be horned than for heifer calves. (The latter point applies to horns only; when horns, scurs, and buttons are considered collectively, there is apparently little or no difference between sexes.)

A DIHYBRID CROSS

Practical cattlemen are aware of the fact that the polled condition is dominant over the horned condition and that black color is dominant over red color. Angus cattle are homozygous for both the polled characteristic and black body color,[6] whereas Devon cattle are homozygous for the horned characteristic and red body color. Thus, if an Angus bull is mated to Devon cows, the results of the first cross (F_1) will be as follows:

```
polled, black    P  P  B  B   ×   p  P  b  b    horned, red
germ cells       P  B              p  b
F1 dihybrid         P  p  B  b              heterozygous polled, black
```

Apparently the genes for polled or horned and for black or red are located in different chromosomes, because, when we mate the dihybrids, Pb Bb, we get the following ratio in the second generation (F_2) crosses:

9 polled blacks	3 horned blacks
3 polled reds	1 horned red

[6]As already mentioned, a few Angus cattle carry a recessive for red color.

In tabular (checkerboard) form the results of crossing the F_1 dihybrids (Pp Bb X Pb Bb) are illustrated in Table 5-1.

TABLE 5-1

CHECKERBOARD OF THE F_2 (SECOND GENERATION) CROSS OF
ANGUS X DEVON CATTLE

Sperms:	Eggs:			
	P B	**P b**	**p B**	**p b**
P B	P P B B	P P B b	P p B B	P p B b
P b	P P B b	P P b b	P p B b	P p b b
p B	P p B B	P p B b	p p B B	p p B b
p b	P p B b	P p b b	p p B b	p p b b

As can be noted through studying the above chart, the nine polled black animals appearing in the second cross are of the same phenotype (they look alike), but they are of different genotype (they are different genetically and will transmit differently). Only the animal with the genetic make-up of PPBB is pure for both dominant characters and will breed pure, but this animal cannot be detected by appearance. Only matings to horned red animals will with certainty single this animal out. On the other hand, the one ppbb (horned red) animal can be identified both as to genotype and phenotype purely on the basis of appearance.

A TRIHYBRID CROSS

As already indicated, a trihybrid is one that is heterozygous for three pairs of allelic genes. A cross between Angus and Hereford cattle represents such a cross. Thus, the situation is as follows:

Angus (polled, black body, black face) = PP BB ww
Hereford (horned, red body, white face) = pp bb WW

All first-cross animals between the Angus and Hereford breeds will be polled, with black body and a white face—all three dominant characteristics being in evidence. However, when these F_1 animals are interbred, on the average and out of each 64 animals produced, the following results will be obtained:

27 *polled*, with a *black body* and a *white face*
 9 *polled*, with a *black body* and a *black face*
 9 *polled*, with a red body and a *white face*
 9 horned, with a *black body* and a *white face*
 3 *polled*, with a red body and a colored face
 3 horned, with a *black body* and a black face
 3 horned, with a red body and a *white face*
 1 horned, with a red body and a colored face

Table 5-2 indicates what one may expect to secure, on the average, with 64 animals resulting from a second generation cross (F_2) of Angus X Hereford. Careful analysis of this summary reveals that, of these 64 animals,

TABLE 5-2

CHECKERBOARD OF F_2 (SECOND GENERATION) CROSS BETWEEN AN ANGUS (*polled*, black-faced and *black-bodied*) AND A HEREFORD (horned, red-bodied and *white-faced*)

Eggs:	Sperms:							
	PWB	**PWb**	**PwB**	**Pwb**	**pWB**	**pWb**	**pwB**	**pwb**
PWB	PP WW BB	PP WW Bb	PP Ww BB	PP Ww Bb	Pp WW BB	Pp WW Bb	Pp Ww BB	Pp Ww Bb
PWb	PP WW Bb	PP WW bb	PP Ww Bb	PP Ww bb	Pp WW Bb	Pp WW bb	Pp Ww Bb	Pp Ww bb
PwB	PP Ww BB	PP Ww Bb	PP ww BB	PP ww Bb	Pp Ww BB	Pp Ww Bb	Pp ww BB	Pp ww Bb
Pwb	PP Ww Bb	PP Ww bb	PP ww Bb	PP ww bb	Pp Ww Bb	Pp Ww bb	Pp ww Bb	Pp ww bb
pWB	Pp WW BB	Pp WW Bb	Pp Ww BB	Pp Ww Bb	pp WW BB	pp WW Bb	pp Ww BB	pp Ww Bb
pWb	Pp WW Bb	Pp WW bb	Pp Ww Bb	Pp Ww bb	pp WW Bb	pp WW bb	pp Ww Bb	pp Ww bb
pwB	Pp Ww BB	Pp Ww Bb	Pp ww BB	Pp ww Bb	pp Ww BB	pp Ww Bb	pp ww BB	pp ww Bb
pwb	Pp Ww Bb	Pp Ww bb	Pp ww Bb	Pp ww bb	pp Ww Bb	pp Ww bb	pp ww Bb	pp ww bb

Fig. 5-5. Diagram showing a trihybrid cross; the inheritance of polled, white face, and black body characteristics in an Angus X Hereford cross. Note that all first cross (F_1) animals are polled, black bodied, and white faced; whereas, on the average, the $F_1 \times F_1$ cross results in the 27:9:9:9:3:3:3:1 ratio shown. (Drawing by R. F. Johnson)

27 different genetic combinations (genotypes) are listed. One and only one of these genotypes can be recognized on sight, namely, the last one involving all small letters or pure for the 3 recessive characters involved. Breeding tests would be necessary to reveal the genotypic differences between PP WW BB and Pp WW Bb individuals, etc. As is generally known, F_2 crosses of the above type are seldom produced in practical operations.

The ratios obtained in the cross with 1, 2, and 3 pairs of allelic genes are as follows:

the monohybrid cross = 3:1 ratio
the dihybrid cross = 9:3:3:1 ratio
the trihybrid cross = 27:9:9:9:3:3:3:1 ratio

Crosses involving 4 or more pairs of contrasted characters are possible, but these are unwieldy. For that reason, very few of these more complicated crosses have been worked out.

Multiple Births

Multiple births among cattle have been observed since their domestication.

A review of the literature reveals that, on the average, such multiple births occur at the frequences shown in Table 5-3.

Selection for natural twinning does not appear to hold much promise because the heritability of twinning is low. A herd in which selection for

TABLE 5-3

FREQUENCY OF TWINS IN CATTLE[1]

Breed	Total Number of Births	Percent of Twin Births
"Beef Cattle"		0.44
Angus	1,111	0.81
Grade Angus	586	1.71
Brown Swiss	14,111	2.70
Holstein	18,736	3.08
Jersey	87,926	1.02
Simmental	12,625	4.61

[1]Summary prepared by Dr. Clayton C. O'Mary, Department of Animal Science, Washington State University, Pullman, Wash. Provided in a personal communication to the author.

Fig. 5-6. Quadruplet purebred Angus heifers, at two weeks of age. Bred and owned by O. H. Delchamps, Point Clear, Ala. (Courtesy, Mr. Delchamps)

twinning was practiced for more than 20 years showed a twinning frequency of only 1.71% during the last 10 years.[7]

The repeatability of twinning is estimated to be three to four times higher than the average of the population, once a cow has given birth to the first set of twins.[8]

Twins may be produced in any of the following five ways:

1. By 2 eggs being produced at the same heat period, with both fertilized and carried to term.

2. By 2 eggs being shed at the same heat period, but the cow being bred to 2 different bulls with a sperm from each of the bulls uniting with an egg.

3. By a cow coming in heat and being bred, then 3 weeks later coming in heat again and being rebred; with both matings resulting in viable offspring.

4. By a single fertilized ovum splitting during the early stage of development.

5. By the use of hormones to induce superovulation.

Twins may be either fraternal (dizygotic) or identical (monozygotic). Fraternal twins are produced from two separate ova that were fertilized by two different sperm. Identical twins result when a single fertilized egg divides very early in its embryology, into two separate individuals.

In humans, nearly half of the like-sexed twins are identical, whereas in cattle only 5 to 12 percent of such births are identical. Such twins are always of the same sex, a pair of males or a pair of females, and alike genetically—their chromosomes and genes are alike; they are 100 percent related. When identical twins are not entirely separate, they are known as Siamese twins.

Genetically, fraternal twins are no more alike than full brothers and sisters born at different times; they are only 50 percent related. They usually resemble each other more, however, because they were subjected to the same intrauterine environment before birth and generally they are reared under much the same environment. Also, fraternal twins may be of different sexes.

Distinguishing between identical and fraternal twin calves is not easy, but the following characteristics of identical twins will be helpful:

1. Identical twins are usually born in rapid succession, and frequently there is only one placenta.

2. The calves are necessarily of the same sex.

3. The coat colors are identical; i.e., if there is a broken color, there must be a strong degree of resemblance in this respect.

4. There is little variation in birth weights, general conformation and, more particularly, the shape of the head, position of the horns and occurrence of skin pigmentation, rudimentary teats, etc.

5. Muzzle prints show a degree of resemblance.

6. The shape, twisting and position of the horns and behavior of the twins can be observed at a later stage. Identical twins are inclined to keep

---

[7]Summary prepared by Dr. Clayton C. O'Mary, Department of Animal Science, Washington State University, Pullman, Wash. Provided in a personal communication to the author.
[8]Ibid.

together when grazing, walking, lying down, or ruminating.

7. Identical twins have the same blood group.

Most cattlemen prefer single births to twins, for the following reasons:

1. The high incidence of stillbirths in twins. Herefords on the range show 3.6% stillbirths among singles vs 15.7% stillbirths among twins.[9] Despite this fact, twinning would result in more live calves per 100 cows calving; 96 live calves from singles (100 × 3.6% = 3.6; then, 100 − 3.6 = 96) vs 168 from twins (200 × 15.7% = 31.4; then, 200 − 31.4 = 168).

2. About 85 percent of all heifers born twin with a bull are apt to be freemartins (sterile heifers).[10]

3. Twin calves average 20 to 30% lighter weights at birth than singles.[11]

4. The tendency of cows that have produced twins to have a lowered conception rate following twinning.

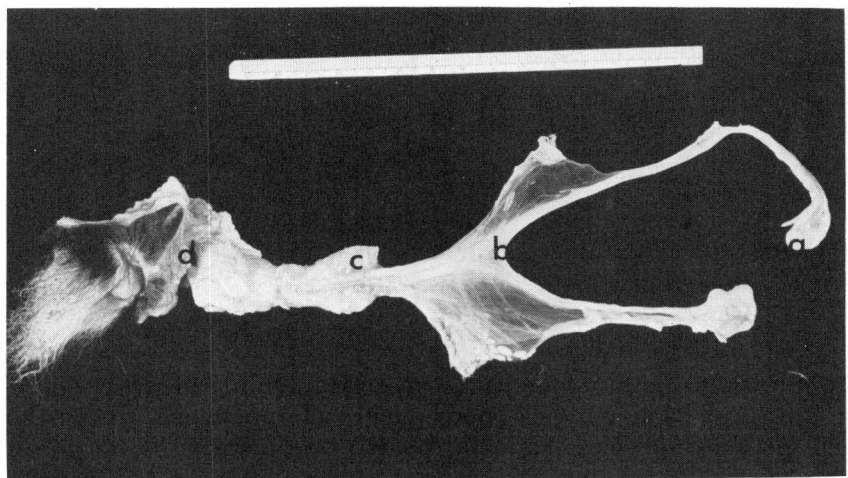

Fig. 5-7. Freemartin calf born co-twin to a male partner. Gonads (a), cordlike Mullerian ducts (b), seminal vesicles (c) and short vagina (d). (Courtesy, Professor H. W. Leipold, Kansas State University, Manhattan, Kan.)

How Sex is Determined

The possibility of sex determination and control has fascinated mankind since time immemorial. For example, in a book published in 1662,[12] studmasters were admonished as follows: "For a male colt you must bind back with a cord, or pull back his left stone, and for the female, bind back the right stone, and thus you may do unto all other kinds of cattle." The same book revealed that "if the rams be put with the ewes when the wind is in the

---

[9]Ibid.

[10]Ibid.

[11]Ibid.

[12]Mascal, Leonard, *The Government of Cattel*, London, printed for John Stafford and William Gilbertson, "and are to be sold at the George-Yard near Fleet-bridge; and at the Bible without New Gate," 1662.

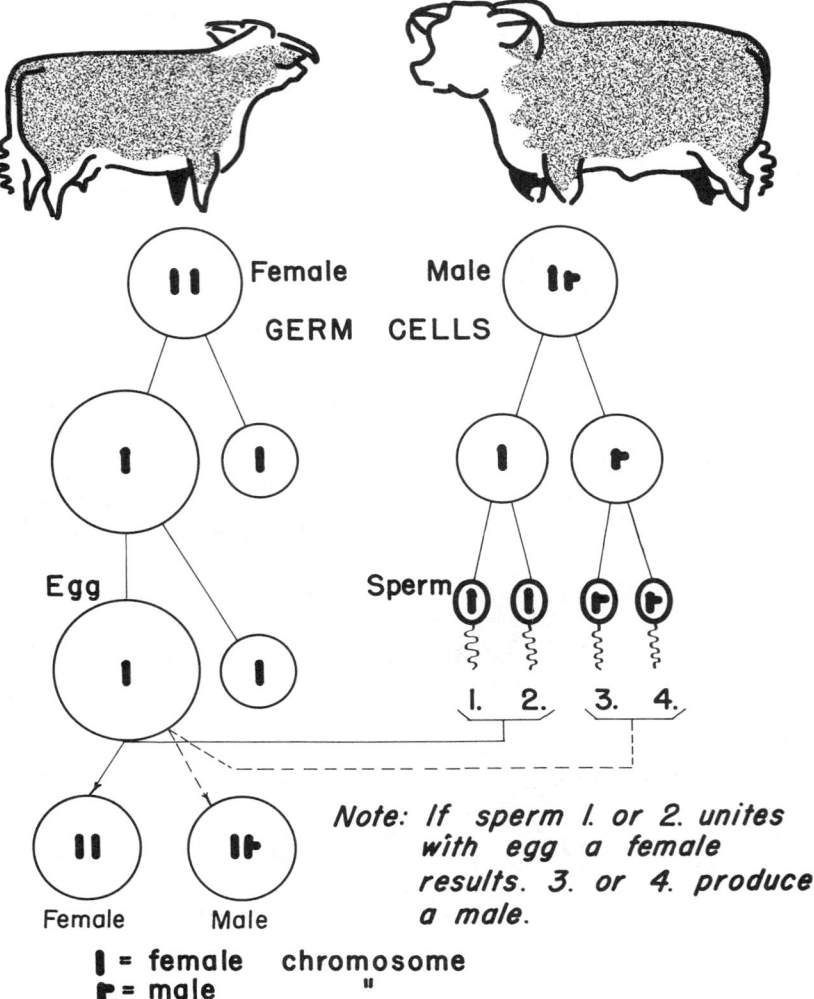

Note: If sperm 1. or 2. unites with egg a female results. 3. or 4. produce a male.

I = female chromosome
r = male

Fig. 5-8. Diagrammatic illustration of the mechanism of sex determination in cattle showing how sex is determined by the chromosomal makeup of the individual. The cow has a pair of like sex chromosomes, whereas the bull has a pair of unlike sex chromosomes. Thus, if an egg and sperm of like sex chromosomal makeup unite, the offspring will be female; whereas if an egg and sperm of unlike sex chromosomal makeup unite, the offspring will be a male. (Drawing by R. F. Johnson)

north, the ewes will bring males and if the wind be in the south when the ewes are covered, they will be females." These are but two of the hundreds of superstitions that have evolved concerning the control of sex.

On the average, and when considering a large population, approximately equal numbers of males and females are born in all common species of animals. To be sure, many notable exceptions can be found in individual herds. The history of the Washington State University Angus herd, for example, reads like a story book. The entire herd was built up from one foundation female purchased in 1910. She produced 7 daughters. In turn, her first daughter produced 6 females. Most remarkable yet and extremely fortunate from the standpoint of building up the cow herd, it was 4 years before any bull calves were dropped in the herd.[13]

Such unusual examples often erroneously lead the stockman to think that something peculiar to the treatment or management of a particular herd resulted in a preponderance of males or females, as the case may be. In brief, through such examples, the breeder may get the impression that variation in sex ratio is not random but that it is under the control of some unknown and mysterious influence. Under such conditions, it can be readily understood why the field of sex control is a fertile one. Certainly, any "foolproof" method of controlling sex would have tremendous commercial possibilities. For example, cattlemen wishing to build up a herd could then secure a high percentage of heifer calves. On the other hand, the commerical cattleman would then elect to produce only enough heifers for replacement purposes. From an economical standpoint, he would want a preponderance of bull calves for the reason that commercial steers sell for a higher price than do commercial heifers.

Sex is determined by the chromosomal makeup of the individual. One particular pair of the chromosomes is called the sex chromosomes. In farm animals, the female has a pair of similar chromosomes (usually called X chromosomes), whereas the male has a pair of unlike sex chromosomes (usually called X and Y chromosomes). In the bird, this condition is reversed, the females having the unlike pair and the male having the like pair.

The pairs of sex chromosomes separate out when the germ cells are formed. Thus, the ovum or egg produced by the cow contains the X chromosome; whereas the sperm of the bull are of two types, one half containing the X chromosome and the other half the Y chromosome. Since, on the average, the egg and sperm unite at random, it can be understood that half of the progeny will contain the chromosomal makeup XX (females) with the other one-half XY (males).[14]

Research workers have employed many techniques designed to change the sex ratio. Most efforts have been focused upon eliminating or altering the X or Y sperm in semen prior to artificial insemination. The three main approaches are: (1) sedementation, in which it is hoped that the heavy "X" sperm will settle to the bottom and the "Y" sperm float to the top, then draw off the "male" and/or "female" fraction of the semen; (2) electrophoresis, in which an anode and a cathode are placed in the semen to draw either X or Y sperm into its field of influence; and (3) treatment of sperm with a substance that is deadly to the chromosomes that determine maleness. Perhaps a successful method of separation will be found; thereby making sex determination possible.

_____

[13]The probability of getting 7 heifer calves in a row can be figured by multiplying ½ by itself 7 times. This is 1 out of 128, meaning that all 7 calves in a group of 7 will be heifers in less than 1 percent of the cases.

[14]The scientists' symbols for the male and female, respectively, are: ♂ (the sacred shield and spear of Mars, the Roman God of War), and ♀ (the looking glass of Venus, the Roman Goddess of Love and Beauty).

Abnormal Development of Sex in Cattle

Sex abnormalities occasionally occur in cattle; freemartins, intersexes, and hermaphrodites are the most common ones. Each of these is discussed in Chapter 7.

Lethals and Other Abnormalities in Cattle

Many abnormal animals are born on the nation's farms and ranches each year. Unfortunately, the purebred breeder, whose chief business is that of selling breeding stock, is likely to "keep mum" about the appearance of any defective animals in his herd because of the justifiable fear that it might hurt his sales. With the commercial producer, however, the appearance of such abnormalities is simply so much economic loss, with the result that he generally, openly and without embarrassment, admits the presence of such defects and seeks correction.

The embryological development—the development of the young from the time that the egg and the sperm unite until the animal is born—is very complicated. Thus, the oddity probably is that so many of the offspring develop normally rather than that a few develop abnormally.

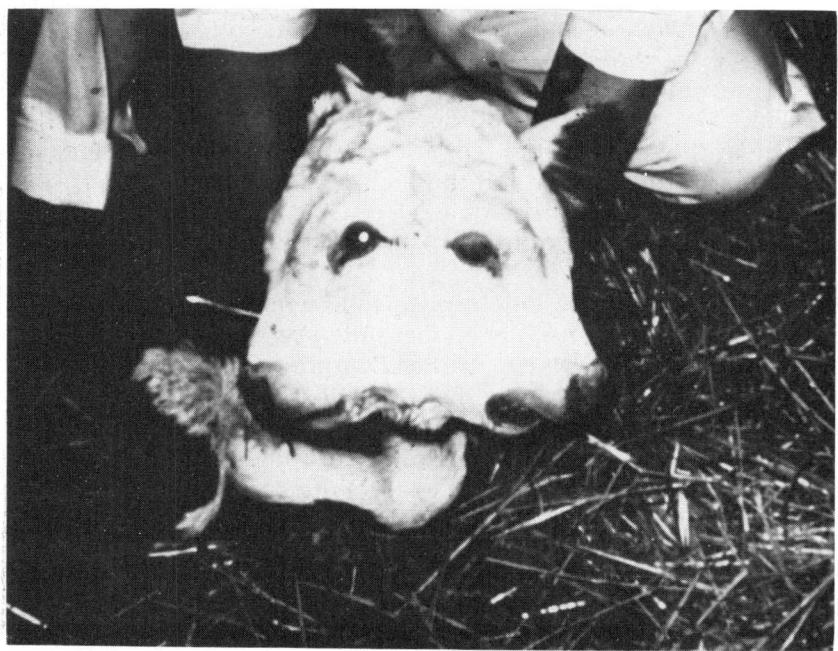

Fig. 5-9. Two-headed calf (Diprosopus). (Courtesy, Dr. Clair M. Hibbs, College of Agriculture, North Platte Station, The University of Nebraska, North Platte, Neb.)

Many such abnormalities (commonly known as monstrosities or freaks) are hereditary, being caused by certain "bad" genes. Most lethals are recessive and may, therefore, remain hidden for many generations. The prevention of such genetic abnormalities requires that the germ plasm be purged of the "bad" genes. This means that, where recessive lethals are involved, the stockman must be aware of the fact that both parents carry the gene. For the total removal of the lethals, test matings[15] and rigid selection must be practiced.

In addition to hereditary abnormalities, there are certain abnormalities that may be due to nutritional deficiencies, or to accidents of development—the latter including those which seem to occur sporadically and for which there is no well-defined reason. When only a few defective

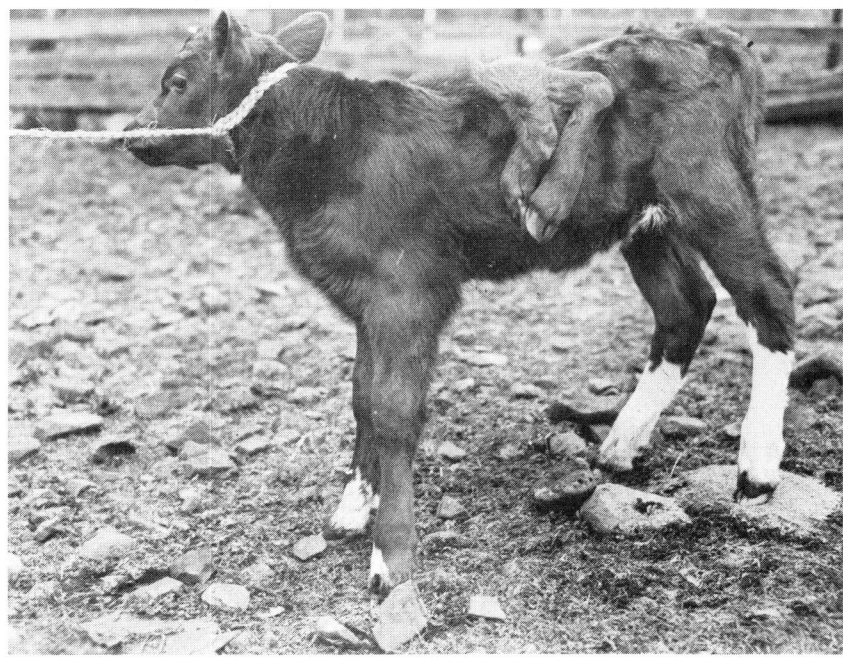

Fig. 5-10. Freak calf at two weeks of age. An accident of development, with three extra legs attached to the back. (Courtesy of the owner, C. M. Barker, White Salmon, Wash.)

individuals occur within a particular herd, it is often impossible to determine whether their occurrence is due to: (1) defective heredity; (2) defective nutrition; or (3) merely to accidents of development. If the same occurs in any appreciable number of animals, however, it is probably either hereditary or nutritional. In any event, the diagnosis of the condition is not always a simple matter.

The following conditions would tend to indicate an hereditary defect:
1. If the defect had previously been reported as hereditary in the same

[15]To test a bull at the .05 level of significance for heterozygosity for a single autosomal recessive gene, the following numbers of progeny are required: (1) 5 from homozygous females; (2) 11 from known heterozygous females; or (3) 23 from matings on daughters. (Source: Kidwell, J., *Journal of Heredity*, July-Aug. 1951)

Fig. 5-11. Any observed defect may be hereditary, nutritional, or a freak of nature (as a result of faulty development in the embryological life). Six types of defects sometimes observed in cattle are shown: (1) bulldog; (2) hairless streaks; (3) muscle contraction; (4) Siamese twins; (5) short spine; (6) dwarfism. (Drawing by Dr. E. S. E. Hafez)

breed of livestock.

2. If it occurred more frequently within certain families or when there had been inbreeding.

3. If it occurred in more than one season and when different rations had been fed.

The following conditions might be accepted as indications that the abnormality was due to a nutritional deficiency:

1. If previously it had been reliably reported to be due to a nutritional deficiency.

2. If it appeared to be restricted to a certain area.

3. If it occurred when the ration of the mother was known to be deficient.

4. If it disappeared when an improved ration was fed.

If there is suspicion that the ration is defective, it should be improved, not only from the standpoint of preventing such deformities, but from the standpoint of good and efficient management.

If there is good and sufficient evidence that the abnormal condition is hereditary, the steps to be followed in purging the herd of the undesirable gene are identical to those for ridding the herd of any other undesirable recessive factor. An inbreeding program, of course, is the most effective way in which to expose hereditary lethals in order that purging may follow.

Table 5-4 includes most of the well-authenticated abnormal conditions that have been reported in cattle. As noted, to date many of these defects are limited to specific breeds and certain countries. For convenience, these abnormalities are classified into five categories as follows:

1. *Lethals are genetic factors which cause death of the calves carrying them prior to or shortly after birth.* (See Table 5-4, Part I.) Except for Achondroplasia I, which is incompletely dominant, all lethals listed in Part I are recessive, either autosomal or sex-linked.

2. *Semi-lethals (or sub-lethals) are genetic factors which cause the death of the young after birth or sometime later in life if environmental situations aggravate the conditions.* (See Table 5-4, Part II.) Semi-lethals are probably far more serious than lethals because they are more likely to go unnoticed; yet they may be causing considerable damage.

3. *Delayed lethals are gene changes which are expressed later in life.* (See Table 5-4, Part III.) Delayed lethals interfere with normal reproduction of the cow or the bull.

4. *Harmful and defective genes cause a host of minor abnormalities which may afflict cattle, some of which are probably of hereditary nature.* (See Table 5-4, Part IV.) These genetic factors are not lethals, but they interfere with the usefulness of the animals. Genetic factors reducing body size (dwarfism) are representative cases. Other developmental defects such as notched ears and screw tail are hereditary, but they do not have marked effects on production and fertility.

5. *Congenital defects refer to qualities that an animal shows evidence of having at birth–natal conditions.* The calf has lived for several months before it is born. When a calf is born with an abnormality, the abnormality may be due to heredity; or it may be due to some freak of nature during prenatal development that had nothing whatsoever to do with chromosomes and genes. Congenital abnormalities may occur in any external or internal organ.

Dwarfism in Cattle

There are several different types of dwarfs, of which the short-headed, short-legged, pot-bellied dwarf—commonly referred to as the snorter dwarf—is the most frequent. The discussion that follows applies specifically to snorter dwarfism.

Though very small (usually weighing about half as much as normal calves), dwarf calves are exceedingly stocky and well-built. The eyes protrude, giving a characteristic pop-eyed appearance. Some dwarfs are weak and unsteady in gait at birth. Others appear to be strong enough, but soon develop a large stomach, heavy shoulders, crooked hind legs and sometimes labored breathing. Survival is somewhat lower than with normal calves, although most purebred breeders make no attempt to raise them.

There is complete agreement among scientists (1) that the dwarf condition is of genetic origin, and (2) that it is inherited as a simple autosomal

TABLE 5-4

LETHALS AND OTHER ABNORMALITIES IN CATTLE[1]

PART I. HEREDITARY LETHALS

(Except for Achondroplasia I, which is incompletely dominant, all lethals listed in Part I are recessive, either autosomal or sex-linked)

Lethal	Symptoms of Abnormality	Breed Reported In	Pertinent References
1. Achondroplasia I	Short legs and head; often have hernia and cleft palate. Most of them are absorbed from the fourth to eighth month of pregnancy. Delivery when close to term is usually difficult because of the extremely large head. About one-fourth of the Dexter X Dexter matings are "Bulldogs."	Dexter	Crew, 1923, *Proc. Royal Soc.,* (London) 95B, 228. Punnett, 1936, *J. Genetics,* 32, 65. Berger & Innes, 1948, *Vet. Record, 60,* 57.
2. Achondroplasia II	This type is similar to Achondroplasia I but less extreme. Calves are usually carried to term but die soon after.	Guernsey, Holstein, Jersey, Telemark	Mohr & Wriedt, 1925, *Z. fur Zuchtung,* B. 3, 223. Mohr, 1930, *Naturens Verden 14,* 1. Brandt, 1941, *J. Hered.,* 32, 183. Gregory *et al.,* 1942, *J. Hered.,* 33, 317.
3. Achondroplasia III	Less extreme than Bulldog II. Calves may live several hours.	Jersey, Swedish Red-and-White	Gregory *et al.,* 1942, *J. Hered.,* 33, 317. Johansson, 1953, *Hereditas,* 39, 75.
4. *Acroteriasis congenita*	Appendages short or absent. The calves are stillborn or die shortly after birth.	Swedish Holstein	Wriedt & Mohr, 1928, *J. Genetics,* 20, 187. Rieck & Buhr, 1967, *Deutsch, Tierazll, Wschs.*
5. Agnathia	Very short lower jaw (see Fig. 5-12).	Jersey	Annett, 1939, *J. Genetics,* 37, 301. Ely, et al., 1939, *J. Hered.,* 30, 105. Grant, 1956, *J. Hered.,* 47, 165.

(Continued)

Footnote on last page of table.

	Description	Breed	Reference
6. Amputated (See Acroteriasis congenita)			
7. Ankylosis	Ossification of joints. Legs rigid due to joints fusing; some cases of lower jaw fusions.	German, Norwegian Lyngdal	Mohr, 1930, *Naturens Verden*, *14*, 1. Eaton, 1937, *J. Hered.*, 28, 320. Stang, 1940, *Z. fur Zuchtung*, 36, 280. Tuff, 1948, *Skand. Vet. Tid.*, 38, 379.
8. Arthrogryposis	Front legs, hind legs, all four legs rigid. Cleft palate (see Figs. 5-13 and 5-14).	Charolais, Simmental, Maine-Anjou, Holstein, Hereford	Leipold, *et al.*, 1970, *J. Am. Vet. Res.* 31, 367. Greene, *et al.*, 1973, *J. Am. Vet. Res.* 34, 887.
9. Bulldog head	Short, broad skull; orbits large; upper jaw short; impaired vision (see Fig. 5-11).		Becker and Arnold, 1928, *J. Hered.* 10, 281.
10. Congenital dropsy	Water in tissues and cavities. The calves are born one or two months previous to term or at term.	Ayrshire, Swedish Lowland black-and-white	Johansson, 1941, *Proc. 7th Int. Cong. Genet.* (Edinburgh), 169. Donald *et al.*, 1952, *Brit. Vet. J.* 108, 227. Herrick & Eldridge, 1955, *J. Dairy Sci.* 38, 440.
11. Congenital ichthyosis	Scaly, cracked skin.	Norwegian Red Poll	Tuff & Gladitsch, 1949, *Nordisk Vet. Med.*, I, 619.
12. Digital abnormality	One toe shorter; toes spread; animal lame. Dominant.		Mead, *et al.*, 1949, *J. Hered.* 40, 151-155.
13. Edema	Watery enlargement of legs, muzzle, and belly.	Ayrshire	Eldridge and Atkinson, 1953, *J. Dairy Sc.* 35:598

Footnote on last page of table.

(Continued)

TABLE 5-4 Part I (Continued)

Lethal	Symptoms of Abnormality	Breed Reported In	Pertinent References
14. Epithelial defects	Defective formation of skin below knees, one or more claws undeveloped, deformed integument of muzzle and the mucus membranes of nostrils, tongue, palate, and cheek.	Holstein	Hadley & Cole, 1928, *Wisc. Agri. Exp. Sta. Bul. No. 86* Hutt & Frost, 1934, *J. Herd.*, 25, 41.
15. Epitheliogenesis imperfecta	Imperfect skin, partly hairless; septicemia. The calves are usually born at term but die as a result of bacterial invasion (See Fig. 5-15).	Holstein, Jersey	Hadley & Cole, 1928, *Wisc. Agri. Exp. Sta. Bul. No.* 86. Regan, *et al.*, 1935, *J. Hered.*, 26, 357. Hutt & Frost, 1948, *J. Hered.*, 39, 131. Leipold, *et al.*, 1973, *Can. Vet. J.* 14, 114.
16. Fetal resorption	Various stages from decomposed masses to only bones or dried mummies have been reported.	Dairy and beef breeds	
17. Hairless (See *Hypotrichosis congenita*)			
18. Hernia cerebri	Failure of frontal bones to fuse.		
19. *Hypotrichosis congenita*	A little hair is found on the muzzle, eyelids, ears, pasterns, and end of the tail. Most afflicted calves die shortly after birth; some live but grow slowly and never have normal hair.		Mohr & Wriedt, 1928, *J. Genetics*, 19, 314. Regan *et al.*, 1935, *J. Hered.*, 26, 357. Kidwell & Guilbert, 1950, *J. Hered.*, 41, 190. Surarrer, 1943, *J. Hered.*, 24, 175. Hutt & Saunders, *J. Hered.*, 44, 97.
20. Hydrocephalus	Affected animals have excess fluid in portions of the brain (see Fig. 5-16).		Houck, 1930, *Anat. Record*, 45, 83. Innes *et al.*, 1940, *J. Path & Bact*, 50, 456.

Footnote on last page of table.

(Continued)

			Cole & Moore, 1942, *J. Agr. Res.* 65, 483. Godgluck, 1942, *Monat, Fur Vet.*, 7, 250. Giannotti, 1952, *Mem. Soc. tosc. Sci. nat. B.* 59, 32. Kobozieff *et al.*, 1955, *Bul. biol.* 89, 189-210. Gilman, 1956, *Cornell Vet.*, 45, 487. Blackwell & Knox, 1959, *J. Hered.*, 50, 143-148. Baker *et al.*, 1961. *J. Hered.*, 52, 135-138. Leipold, *et al.*, 1971, *J. Am. Vet. Res.* 32, 1019. Greene, *et al.*, 1973, *Irish Vet. J.* 27, 37.
21. Impacted molars	Short lower jaw, impacted teeth; die within week.	Milking Shorthorn	Heizer & Hervey, 1937, *J. Hered.*, 28, 123.
22. Lameness in hind limbs .	Calves unable to stand.	Red Danish	Christensen & Christensen, 1952, *Norsk. Vet. Tid.*, 4, 861
23. Mannosidosis	Retardation of growth; aggressive behavior.	Angus	Jolly and Leipold, 1973 *New Zealand Vet. J.* in press)
24. Mummification	They have a short neck, stiff legs, and prominent joints. Fetuses die at about 8 months' gestation but carried to term.	Red Danish	Loje, 1930, *Tidssk. for Landok, 10*, 517.

Footnote on last page of table.

(Continued)

TABLE 5-4 Part I (Continued)

Lethal	Symptoms of Abnormality	Breed Reported In	Pertinent References
25. Muscle contracture	Head and legs drawn; joints stiff. Calves born at full term (see Fig. 5-11).	Holstein, Norwegian breeds	Hutt, 1934, *J. Hered.*, 25, 41. Nes, 1953, *Nordisk Vet. Med.*, 5, 869.
26. Night blindness	Animals see poorly in the twilight and at night, but have good daytime vision.		Craft, 1927, *J. Hered.* 15:255.
27. Osteopetrosis	Born 3 weeks premature; solid bones; no bone marrow cavities; short lower jaw (see Figs. 5-17 and 5-18).	Leipold, *et al.*, 1971, *An. Sel. Genet. Anim.* 3,245.	Leipold, *et al.*, 1971, *An. Sel. Genet. Anim.* 3,245.
28. Paralysis	Posterior paralysis—may have muscular tumors and blindness, die shortly after birth.		Tuff, 1948, *Vet.-Tid.* 38, 379. Cranek & Ralson, 1953, *J. Anim Sci.* 12, 892.
29. Paralyzed hindquarters ..	Calves appear normal except that they cannot stand on hindquarters.	Norwegian Red Poll, Red Danish	Loje, 1930, *Tidssk. for Landok.*, 10, 517.
30. Prolonged pregnancy I	The gestation period is from 311 to 403 days. Fetus grows in uterus almost as calves born normally, with the result that parturition is difficult or impossible.	Japanese breeds	Gregory *et al.*, 1951, *Portigualie Acta Biol. A*, 861. Jasper, 1951, *Cornell Vet.*, 40, 165.
31. Prolonged pregnancy II	Calves are carried up to 500 days. Calves show immature bone development, various degrees of hairlessness and lack of thyroid development; and they do not have pituitary gland.		Hallgren, 1951, *Nord. Vet. Med.*, 3, 1043.
32. Short limbs	Limbs short; hoofs undeveloped.	Russian breeds, Swiss breeds, Shorthorn	Leipold, *et al.*, 1970 *Can. Vet. J.* 11, 258.

Footnote on last page of table.

(Continued)

33. Short spine	Ribs and vertebrae fused; back bent down.	Norwegian, Mountain Angus	Mohr & Wriedt, 1930, *J. Genetics*, 22, 279. Leopold and Dennis, 1972, *Cornell Vet.* 62, 507.
34. Skinless (See Epitheliogenesis)			
35. Spasms	Calves first appear normal, but soon develop spasmodic muscular contractions. Affected animals die with a few weeks after birth.	Jersey	Shrode & Lush, 1947, *Advances in Genetics*, Academic Press, Inc., N.Y. Gregory, *et al.*, 1944. *J. Hered.*, 35, 195. Saunders, *et al.*, 1952, *Cornell Vet.*, 42, 559.
36. Streaked hairlessness ...	Gene carried only in females. Carrier females exhibit streaked hairlessness and produce a sex ratio of two females to one male (see Fig. 5-11).	Holstein	Eldridge & Atkeson, 1953, *J. Hered.*, 44, 265.
37. Syndactyly	Right front foot, both front feet, or 3 feet with a single toe. Rarely 4 feet have a single toe. Affected animals cannot withstand stress (see Fig. 5-19).	Holstein, Simmental, Chianina	Leipold, *et al.*, 1973, *Vet. Bull.* (in press)
38. Tendon contracture	Tendons pulled rigidly, calves are either born dead or die after birth.	Milking Shorthorn	Dale & Moxley, 1952, *Can. J. Comp. Med.*, 16, 399.

Footnote on last page of table.

(Continued)

TABLE 5-4 (Continued)

PART II. HEREDITARY SEMI-LETHALS

Semi-Lethal	Symptoms of Abnormality	Probable Mode of Inheritance	Breed Reported In	Pertinent References
1. *Atresia ani*	Closed anus. Calves do not survive corrective surgery.			Kuppuswami, 1937, IND. J. Vet. Sci. & Anim. Husb., 7, 305. Lerner, 1944, J. Hered., 35, 21.9.
2. Albinism	Complete absence of pigmentation in the skin, muzzle, and hoofs, and in the walls of the rumen. However, the eyes are not pink, due to a slight greyish coloration of the cortex (see Fig. 5-20).	Probably recessive	Hereford, Brown Swiss	Hafez et al., 1958, J. Hered., 49, 111. Greene, et al., 1973, J. Hered., (in press)
3. Albinism (partial)	White animals; spots in skin; blue eyes.	Dominant		Leipold, et al., 1968, J. Hered. 59, 2.
4. Imperforate Anus (See Atresia ani) ..				
5. Epilepsy	Cattle subject to epileptic type of attacks.	Dominant	Brown Swiss	Atkeson, et al., 1944, J. Hered., 35, 45.
6. Muscular dystrophy .	Genetic abnormality, develops over 2 years.	Dominant	Brown Swiss	Leipold, et al., 1968, Vet. Med. 68, 645.

Footnote on last page of table.

(Continued)

TABLE 5-4 (Continued)

PART III. HEREDITARY DELAYED LETHALS

Delayed lethal	Symptoms of Abnormality	Probable Mode of Inheritance	Breed Reported In	Pertinent References
1. Atrophy of testis	a. Gross microscopic changes to testes, including atrophy, calcification, degeneration of the seminiferous tubules, and varying degrees of fibriosis.	?	Swedish Highland	Rollinson, 1955, *Anim. Breed Abstr.*, 23, 215.
	b. Epithelium layers of seminal ducts are underdeveloped; in either one or in both testes.	?		
2. Cystic ovaries	Sterility; nymphomania.		Swedish Highland	Rollinson, 1955, *Anim. Breed. Abstr.*, 23, 215.
3. Female Sterility	Cows fail to settle.	Simple sex-limited		Gregory, *et al.*, 1945, *Genetics*, 30, 506. Gregory, *et al.*, 1951, *J. Dairy Sci.*, 34, 1047.
4. Gonad hypoplasis ..	Atrophy of one or two ovaries.	Autosomal		
5. Gonadless	Absence of ovaries.	?		
6. *Impotentia coeundia*	Bull does not possess ability to copulate due to failure of sigmoid curve of penis to straighten during coitus.	Autosomal recessive		Rollinson, 1955, *Anim. Breed. Abstr.*, 23, 215.
7. Knobbed spermatozoa	Abnormal formation of sperm. Formation vaculoes in developing sperm heads may be related to some unknown changes in the nucleic acid metabolism of the sperm head. Absence of any chromosomal aberration, quantitative or structural.	Autosomal sex-limited gene		Rollinson, 1955, *Anim. Breed. Abstr.*, 23, 215.

Footnote on last page of table.

(Continued)

TABLE 5-4. Part III (Continued)

Delayed lethal	Symptoms of Abnormality	Probable Mode of Inheritance	Breed Reported In	Pertinent References
8. *Turned sperm tails* .	Sperm tails turned back past the head.	?	Holstein	Warren & Atkeson, 1931, *J. Hered.*, 22, 345.
9. *Hernia, Umbilical* ..	It appers to be limited to males.	Dominant (?)	Holstein	Warren & Atkeson, 1931, *J. Hered.*, 22, 345.
10. *White heifer disease*	Persistent hymen or incomplete cervix; horns of uterus become distended with fluid. Most commonly found in white heifers of the Shorthorn breed, but it has been reported in roan and red Shorthorns and in colored animals of other breeds.	May be a double recessive.	Shorthorn, Other breeds	Gilmore, 1949, *J. Dairy Sci.*, 32, 71. Spriggs, *Et al.*, 1946, *Vet. Record*, 58, 405. Bennett & Olds, 1971, *Ky. An. Sci. Research Report* 196.

Footnote on last page of table.

(Continued)

PART IV. HARMFUL AND DEFECTIVE (NON-LETHALS) GENES

Type	Symptoms of Abnormality	Probable Mode of Inheritance	Breed Reported In	Pertinent References
1. Achondroplasia	Short legs and head, curly coat and thin.	Recessive	Shorthorn	Mead, et al., 1946, *J. Hered.*, 37, 183.
2. Cancer eye	Herefords without pigmented eyelids more susceptible than other cattle. Usually occurs in older animals.	?	Hereford	Frank, 1943, *J.A.V.M.A. 102*, 200. Guilbert, *et al.*, 1948, *J. Anim. Sci.*, 7, 426. Woodward & Knapp, 1950, *J. Anim. Sci.*, 9, 580. Anderson, *et al.*, 1957, *J. Anim. Sci.*, 16, 739.

Footnote on last page of table.

(Continued)

	Description	Inheritance	Breed	Reference
3. Cataract	Opaque condition of the lens of the eye.	Recessive	Jersey	Detlefson & Yapp, 1920, *Am. Naturalist*, 54, 277.
4. Cerebellar hypoplasia	Poor coordination. Cerebellum redimentary, excessive fluid. Some walk like ballet dancers.	Probably recessive	Holstein-Friesian, Jersey	Anderson & Davis, 1950, *J.A.V.M.A. 117*, 460. Saunders, *et al.*, 1952, *Cornell Vet.*, 42, 559.
5. Comprest	The animal is extremely bowlegged, often being unable to stand. The vertebrae do not have the spines which are pronounced in the "snorter dwarf"	Incomplete dominancy	Hereford	Lucas, 1950, Thesis (M.S.) Colo. State Univ. (Fort Collins). Stonaker, 1958, *Colo. Exp. Sta. Bul. 501-S.* Hafez, 1959, Unpublished data.
6. Crooked legs	Front legs crooked. Certain lines of breeding show this affliction whereas others do not, even in areas where it is prevalent. Those blood lines manifesting the condition in an area where it is prevalent are not afflicted in the areas where it is not found.	Nutritional-genetic interaction	Hereford	Stonaker, 1958, *Colo. Expt. Sta. Bull. 501-2.* Bogart, 1959, Personal Communication. Hafez, 1959, Unpublished data.
7. Cross eyes (see Strabismus)				
8. Curly hair	Hair in tight curls, viable.	Dominant	Ayrshire	Eldridge, *et al.*, 1949, *J. Hered.*, 40, 205.
9. Double ear	A thin, flat piece of cartilage lies parelel to the long axis of the ear, extends beyond the ear tip.	Dominant	Brahman	Lush, 1924, *J. Hered.*, 15, 93.

Footnote on last page of table.

(Continued)

Table 5-4. Part IV (Continued)

Type	Symptoms of Abnormality	Probable Mode of Inheritance	Breed Reported In	Pertinent Reference
10. Double muscled	Characterized by abnormally wide thighs, with this extreme width extending forward to include the loin. Deep grooves between the muscles are conspicuous externally. Little fat covering is present. (The "Dopplelander" calf is a mutation of doubtful value; the valuable muscles in the back and loin are doubled but the animals are sterile [Hammond, 1935, Emp. J. Exp. Agric. 3m Bi, 9]).	Single recessive gene, masked by dominant gene in heterozygous carriers.	Angus, Charolais, Hereford, Piedmont, Shorthorn.	Weber & Ibsen, 1934, *Proc. Am. Soc. Anim. Prod.*, p. 228. Kidwell *et al.*, 1952, *J. Hered.*, 43, 62. Oliver & Cartwright, 1969, *Bull 12, Texas Agric. Exp. Sta.*
11. Duck legged	Legs shorter than normal.	Dominant		Lush, 1927, *J. Hered.*, 21, 85.
12. Dwarfism	Animals small in size. Several kinds have been identified: snorter dwarfs, long-headed dwarfs, and comprest dwarfs are caused by 3 different genes. (See Fig. 5-11).	Recessive for some; partially dominant for others, (Comprest)	Angus, Hereford, Shorthorn	Mead, *et al.*, 1942, *J. Hered.*, 33, 411. Johnson, *et al.*, 1950, *J. Hered.*, 41, 177. Lindley, 1951, *J. Hered.* 42, 273. Gregory, *et al.*, 1953, *Hil-gardia*, 22, 407. Pahinsh, *et al.*, 1955, *J. Anim. Sci*, 14, 200, 1025.
13. Extra Toes (See Polydactylism)				
14. Flexed pasterns	Toes turned under.	Recessive	Jersey	Hable, 1948, *J. Vet. Res.*, 9, 131. Atkeson, *et at.*, 1943, *J. Hered.*, 34, 25. Mead, *et al.*, 1943, *J. Hered.*, 35, 367.

Footnote on last page of table.

(Continued)

	Description	Inheritance	Breed	References
15. Fused teats	The front and rear teats on the same side are fused.	Recessive	Hereford	Johnson, 1945, *J. Hered.*, 36, 317. Hiezer, 1932, *J. Hered.*, 23, 111.
16. Missing teat	One teat on left side.	Recessive	Holstein-Friesian	Rollinson, 1955, *Anim. Breed. Abstr.*, 23, 215.
17. Notched ears	Ears imperfect in shape.	Dominant		
18. Pink teeth (See Porphyrinuria)				
19. Polydactylism I	Affected animals have an extra toe on each front foot, with accompanying tenderness and lameness.	Dominant Autosomal	Hereford (males)	Roberts, 1921, *J. Hered.*, 12, 484. Morrill, 1945, *J. Hered.*, 36, 81. Shrode & Lush, 947, *Advances in Genetics*, Academic Press, N.Y.
20. Polydactylism II ...	A three-toed condition.	Dominant Autosomal	Holstein	
21. Porphyrinuria	a. Cattle are photosensitive and develop lesions in unpigmented areas of skin. b. Excessive coproporphyrin and uroporphyrin.	?	Shorthorn, Holstein-Friesian	Fourie, 1939, Onderst. *J. Vet. Sci.*, Anim. Indust., 13, 383. Jorenson & With, 1955, *Nature* (London), 176, 156. Clare, 1955, *Advan. Vet. Sci.*, 2, 191.
22. Screw tail	Tail appears to be broken due to fusion of two or more vertebrae.	Recessive	Red Poll	Knapp, 1936, *J. Hered.*, 27, 269
23. Semi-hairlessness ..	Hair coat thin; calves wild.	Recessive	Hereford, Polled Hereford	Craft & Blizzard, 1934, *J. Hered.*, 25, 385. Cole, 1919, *J. Hered.*, 22, 345.

Footnote on last page of table.

(Continued)

TABLE 5-4. Part IV (Continued)

Type	Symptoms of Abnormality	Probable Mode of Inheritance	Breed Reported In	Pertinent Reference
24. Spread hoofs	Hoofs spread greatly. Painful; animals often walk on knees.	Recessive	Jersey	Regan, *et al.*, 1944, *J. Hered.*, 35, 233.
25. Strabismus	Cattle have a cross-eyed condition. Not evident at birth but identified by 12 months of age.	Recessive		
26. Stumpy (See Achondroplasia) ..				
27. Syndactylism	One toe on each front foot instead of two.	Recessive	Holstein-Friesian	Eldridge, *et al.*, 1951, *J. Hered.*, 42, 241.

[1]This table was authoritatively reviewed by the following, who also made suggestions for its improvement: Prof. H. W. Leipold, College of Veterinary Medicine, Kansas State University, Manhattan, Kan.; Dr. Ralph Bogart, Director, Genetics Institute, Oregon State University, Corvallis, Ore.; and Dr. John F. Lasley, Animal Science Research Center, University of Missouri, Columbia, Mo.

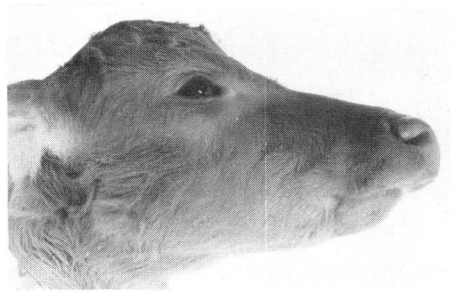

Fig. 5-12. Agnathia (short lower jaw) in a calf. (Courtesy, Professor H. W. Leipold, Kansas State University, Manhattan, Kan.)

Fig. 5-13. Arthrogryposis, showing rigid legs, in a Charolais calf. (Courtesy, Professor H. W. Leipold, Kansas State University, Manhattan, Kan.)

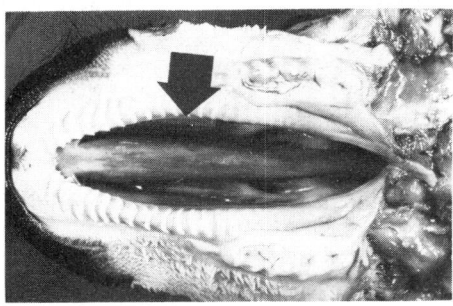

Fig. 5-14. Arthrogryposis, with related cleft palate (arrow) in a Charolais calf. (Courtesy, Professor H. W. Leipold, Kansas State University, Manhattan, Kan.)

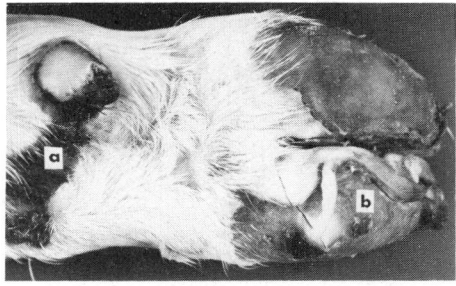

Fig. 5-15. Epithelogenesis imperfecta in a Holstein-Friesian calf. Note imperfectly developed dew claw (a) and defective claw (b). (Courtesy, Professor H. W. Leipold, Kansas State University, Manhattan, Kan.)

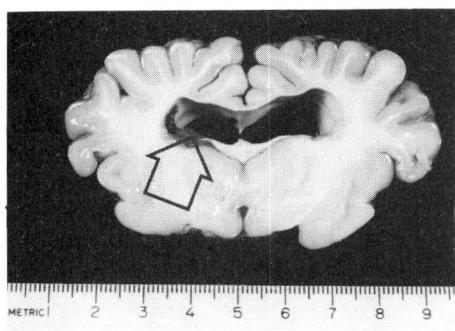

Fig. 5-16. Hydrocephalus (internal) in a Hereford calf. Note dilation of internal compartments of brain (arrow). (Courtesy, Professor H. W. Leipold, Kansas State University, Manhattan, Kan.)

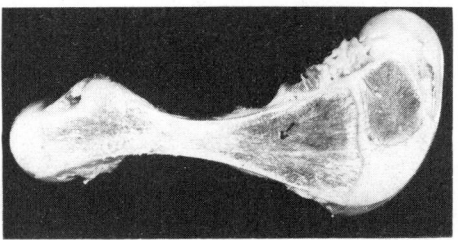

Fig. 5-17. Osteopetrosis bisectioned femur of an Angus affected with (marble-bone disease). Note dense bone without formation of bone marrow cavity (arrow). (Courtesy, Professor H. W. Leipold, Kansas State University, Manhattan, Kan.)

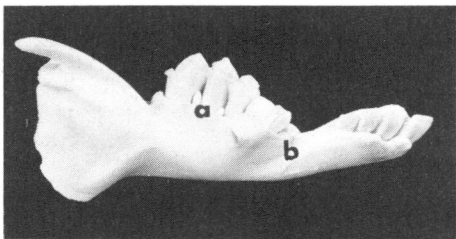

Fig. 5-18. Osteopetrosis, with related short lower jaw of a calf affected with osteopetrosis. Note impaction of molar teeth (a) and lack of the foramen mentale (b). (Courtesy, Professor H. W. Leipold, Kansas State University, Manhattan, Kan.)

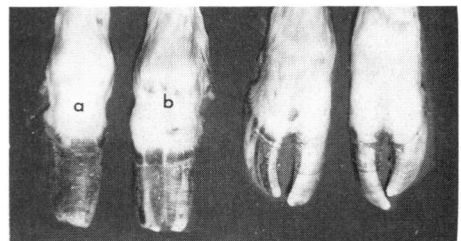

Fig. 5-19. Syndactyly in cattle. Note complete mulefoot of right front foot (a) and partial mulefoot of left front foot (b) compared to normal hind feet. (Courtesy, Professor H. W. Leipold, Kansas State University, Manhattan, Kan.)

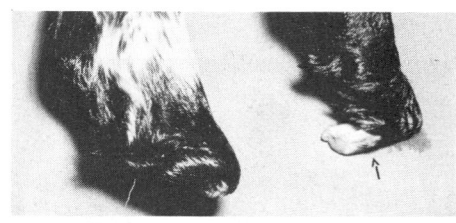

Fig. 5-21. Adactylia (absence of digital rays) in a Shorthorn calf. Note small uncloven hoof at distal end of leg (arrow). (Courtesy, Professor H. W. Leipold, Kansas State University, Manhattan, Kan.)

Fig. 5-20. Albino Hereford cattle. Note lack of pigment. (Courtesy, Professor H. W. Leipold, Kansas State University, Manhattan, Kan.)

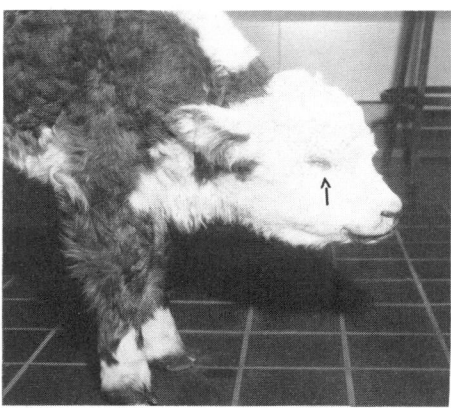

Fig. 5-22. Anophthalmia in a Hereford calf. Note lack of eye development (arrow). (Courtesy, Professor H. W. Leipold, Kansas State University, Manhattan, Kan.)

Fig. 5-23. Tailless cow and her normal-tailed calf sired by a tailless bull. (Courtesy, Professor H. W. Leipold, Kansas State University, Manhattan, Kan.)

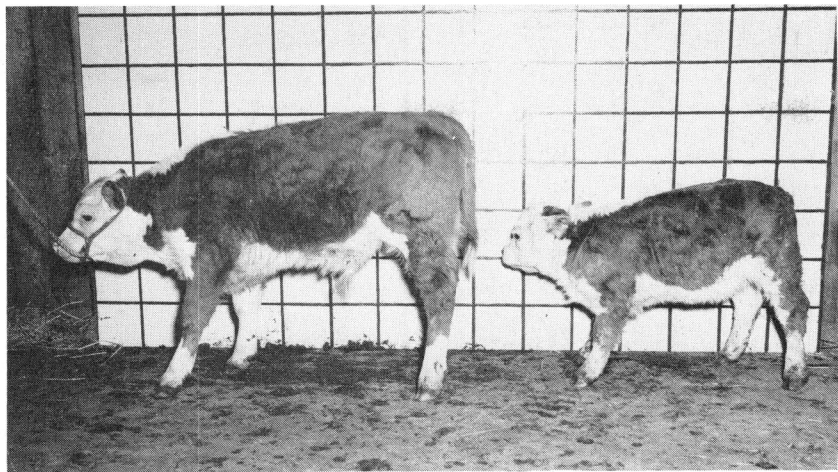

Fig. 5-24. Normal (left) and dwarf (right) Hereford calves. The normal calf is 2½ months of age while the dwarf is 3½ months of age. Note the smaller size of the dwarf and its stocky well built conformation. (Courtesy, Washington State University)

recessive (the word "autosomal" merely means that it is not carried on the sex chromosomes) and conditioned by at least two pairs of modifying genes. Thus, the birth of a dwarf calf identifies both the sire and the dam as carriers of the dwarf gene.

CONDITIONS PREVAILING IN DWARF-AFFLICTED HERDS

One of the conditions (or perhaps both conditions) shown in Figures 5-25 and 5-26 prevails in any herd of cattle in which dwarf-carrying animals are being used.

From Figure 5-25, it may be seen that 100 offspring from matings of carrier bulls X noncarrier cows will, on the average, possess the following genetic picture from the standpoint of dwarfism:[16]

50 carriers, although not dwarfs
50 noncarriers and nondwarfs
———
100 total.

From Figure 5-26, it may be seen that 100 offspring from matings of carrier bulls X carrier cows will, on the average, possess the following genetic picture from the standpoint of dwarfism:[16]

25 dwarfs
50 carriers, although not dwarfs
25 noncarriers and nondwarfs
———
100 total.

———
[16]All ratios are averages based on large numbers; thus they may not apply to any given herd.

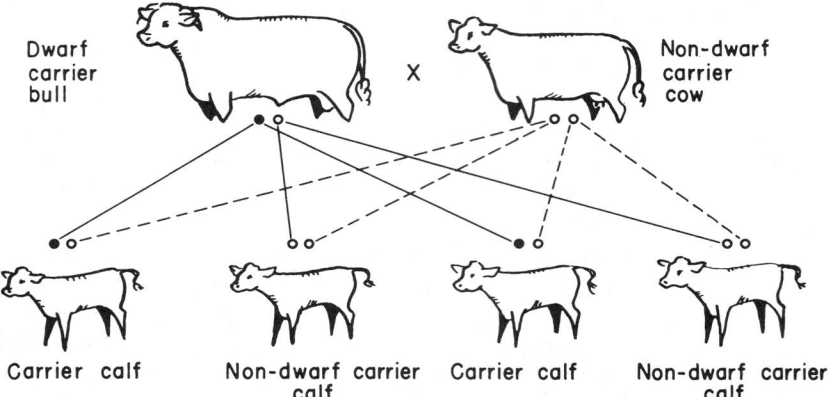

Fig. 5-25. Diagrammatic illustration of the inheritance of the most common type of dwarfism, showing what to expect when a carrier (heterozygous) bull(s) is mated to a noncarrier (homozygous normal) cow(s); or the sexes may be reversed. As shown, carrier X noncarrier matings will, *on the average*, produce calves of which (1) 50% are carriers, although not dwarfs, and (2) 50% are noncarriers and nondwarfs. Unfortunately, the two groups look alike and cannot be detected by sight. (Drawing by R. F. Johnson)

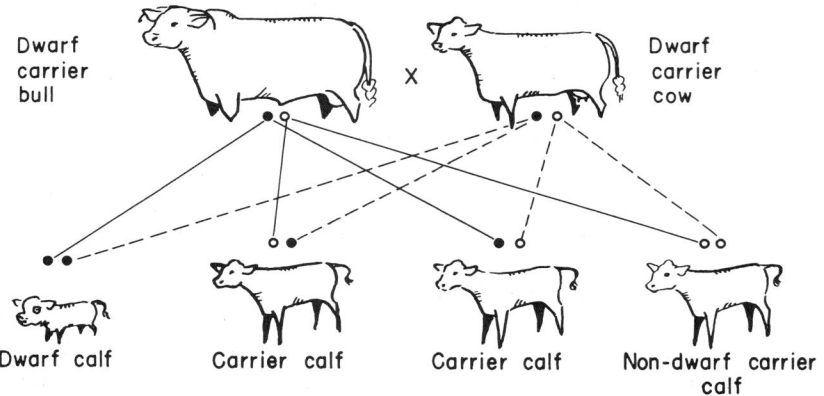

Fig. 5-26. Diagrammatic illustration of the inheritance of the most common type of dwarfism, showing what to expect when a carrier (heterozygous) bull(s) is mated to a carrier (heterozygous) cow(s); or the sexes may be reversed. As shown, carrier X carrier mating will, *on the average*, produce calves of which (1) 25% are dwarfs, (2) 50% are carriers, although not dwarfs, and (3) 25% are noncarriers and nondwarfs. Unfortunately, only the dwarfs can be detected by sight; the two nondwarf groups look alike and cannot be distinguished by sight. (Drawing by R. F. Johnson)

On the basis of these facts, it may be concluded that the following dwarfism genetic picture applies to any given calf having (1) one carrier parent, or (2) both carrier parents:

1. A calf out of parents one of which is a known carrier and the other a noncarrier has a 50 percent chance of being free of the dwarf factor.

2. A calf both of whose parents are carriers has only one chance in four of being free of the dwarf factor.

It is recognized that the percentage of carrier females in any given herd will vary. Obviously, where dwarf calves have appeared, there are both carrier bulls and carrier cows. Some breeders may remove carrier animals, especially cows, once they have dropped a dwarf calf, thus selecting away from the trait. Others may unwittingly select for, rather than against, animals of the carrier type, if such a carrier type exists and if it is associated with some much sought characteristic, such as a markedly dished face.

From Figures 5-25 and 5-26 the following deductions of value to practical cattlemen may be made: (1) where a carrier bull is mated to noncarrier cows, no dwarfs will be produced, but, on the average, one-half the calves will be carriers, and (2) the carrier heifers from this first cross can, and likely will, produce one-fourth dwarfs if they are mated back to a carrier bull. In other words, although the use of a carrier bull in a clean herd will not produce any dwarf calves, the seed for dwarfism is sown and it will crop out providing a second carrier bull is used in the herd.

Figures 5-25 and 5-26 also indicate the futility of continuing the use of carrier bulls or females. Also, it should be recognized that any animal producing a dwarf is a carrier, regardless of the number of dwarfs produced (one or several).

PURGING THE HERD OF DWARFS

The breeding program followed to remove or minimize the dwarf condition will depend somewhat on the type of herd involved—especially on whether it is a commercial or purebred herd.

In a commercial herd, the breeder may lessen the chances of obtaining dwarfs by using an outcross (unrelated) sire within the same breed or by crossbreeding with a sire from another breed. With this system, the dwarf carrying cows will remain, but—because of the recessive condition of the dwarf factor—it will be covered up.

In a purebred herd, the action taken in handling the dwarf situation should be more drastic. A reputable purebred breeder has an obligation, not only to himself, but to his customers among both the purebred and commercial herds. Purebred herds should be purged of the undesirable dwarf genes. This can be done through pursuing any one of the following three breeding systems.[17]

1. *Using sires of families free of the dwarf factor; pedigree-clean animals.* Within each breed where dwarfs have appeared, families exist that are free of the dwarf factor. Securing and using bulls from such families will "cover up" the dwarf situation.

Pedigree information is especially useful for early screening of prospective breeding animals and for small breeders who cannot afford the expense of progeny testing.

---

[17]It is recognized that it is practically impossible to eliminate completely the dwarf factor from a herd or breed of cattle once it has appeared. With a proper breeding program, however, the incidence of dwarfism will become so small as to be unimportant—much as the occurrence of the red color (caused by another recessive factor) has been minimized in the Angus and Holstein breeds of cattle.

2. *Testing bulls of present breeding in the present herd.* Continue to select bulls from within the herd or from herds from which purchases have been made previously, but select bulls that appear to be free from the dwarf factor as judged by pedigree and family background. Next, test these bulls by mating each of them to cows in the present herd that have produced dwarf calves. If each bull tested is mated to 15 known carrier cows and all the progeny are normal, there would be only 1.3 chances out of 100 (1.3%), or 1 chance in 75, that the bull is a dwarf-carrier and yet passed the test undetected.[18] A test of about the same validity can be secured by breeding a sire to 30 of his daughters. Accordingly, such a tested bull could be used in any herd with reasonable certainty that he is free of the dwarf-producing factor.

If rigidly adhered to, this system will eventually produce the desired results, but it has the following limitations: (a) It necessitates retaining carrier cows in the existing herd for bull testing purposes, thus giving doubting fellow breeders an opportunity to question the entire herd; and (b) there is always the temptation to retain outstanding calves although they are likely carriers.

3. *Using a commercial herd of test cows for the purpose of proving bulls.* If carefully followed, this is the most desirable of all the methods herein proposed, and over a period of years the one which will pay the most handsome dividends. At first glance, this method will appear drastic and expensive, but, in the end, the approach is the soundest of the three proposals. Under this system, it is suggested that the breeder assemble a herd of "tester cows" (from either purebreds or grades) each known to have dropped at least one dwarf calf, *with these animals operated strictly as a commercial herd.*

Prior to using any bull in the purebred herd, bulls that are otherwise desirable would be tested by mating each of them to approximately 15 of the dwarf-factor-carrying cows in this commercial herd. Then the top bulls from among those whose get are free of dwarfs could be used in the purebred herd with reasonable certainty.[19]

Carrier cows mated to bulls as indicated would, on the average, not produce in excess of 25 percent dwarfs if all the bulls were dwarf-carriers (see section relative to "Conditions Prevailing in Dwarf-Afflicted Herds"). Of

---

[18] A lesser number than 15 might be used, but the breeder could not be so sure of the results. For example, with 10 or 5 such matings, the chances of failing to detect a carrier bull are increased from 1.3% to 5.6% and 23.7%, respectively. Thus, with 10 such matings and all normal calves, there is only 1 chance in 18 that the bull is a dwarf-carrier and yet passed the test undetected.

This is computed as follows: the chance of obtaining a dwarf calf from a dwarf-carrier bull bred to known carrier cows is 1 in 4, or $\frac{1}{4}$. The chance of a dwarf not occurring from this type of mating is $1-\frac{1}{4}$, or $\frac{3}{4}$. Thus, if 5 such matings are made, $(\frac{3}{4})^5 = 23.7$, or the chances of failing to detect a carrier bull with 5 matings is 23.7%; with 10 matings it is $(\frac{3}{4})^{10}$, or 5.6%; and with 15 matings $(\frac{3}{4})^{15}$, or 1.3%.

If only a limited number of carrier cows are available, it may be desirable to breed each prospective herd sire to 4 to 6 carrier cows initially, followed by more thorough testing of those passing the initial screening.

If the factor becomes so rare that it is impossible to secure sufficient carrier cows for bull testing, the incidence of dwarfism will be so small as to be unimportant.

[19] The chances of avoiding a dwarf-factor-carrying bull through such testing on 15 carrier cows is covered in the section immediately preceding.

course, fewer dwarfs would be produced if some of the bulls were dwarf-free, as expected. Thus, there would be considerable remuneration from the sale of calves in the operation of such a commercial herd. Further, and most important, the merits of young sires, from the standpoint of type and efficiency of production, could be determined in the commercial herd prior to using them in the purebred herd—thus making it possible to select sires by modern record of performance methods.

Under any of the three systems herein proposed, it would be wise to eliminate those bulls and cows that are known to have produced dwarf calves as soon as desirable replacement animals proved to be free of the factor are available.

Providing one does not select a larger than normal percentage of carrier heifers as replacements for the herd, each generation of calves sired by dwarf-free bulls would halve (lessen by 50 percent) the number of carriers in the herd. True enough, there would always be some of the dwarf carriers present, for the incidence of carriers is halved with each generation of such matings, but not eliminated. Yet, after two generations of such matings, the incidence of dwarf carriers would be small. Also, it is noteworthy that the use of proved dwarf-factor-free bulls would give assurance that no more dwarf calves would be produced in the herd.

It is perfectly obvious that the elimination of the dwarf-producing factor is both slow and costly. Yet it is the only way in which cattle can be freed from dwarfs. Also it may require real courage to recognize openly the situation and discard outstanding animals that are known carriers.

Because dwarfs represent an almost complete economic loss, the problem deserves careful attention.

Double Muscling (Muscular Hypertrophy)

Double muscling refers to cattle characterized by bulging muscles of the shoulder and thigh, a very rounded rear end (as viewed from the side), a wide but shallow body throughout, appearance of intermuscular grooves, and fine bones.

Since beef cattle are produced primarily for their muscle, it's logical that selection should be centered around muscularity. It follows that cattle with "double muscles," or these tendencies, have appeal and have increased in frequency in the United States during recent years. But there are disadvantages as well as advantages to double muscling. Hence, cattlemen should be familiar with the characteristics and genetics of double muscling, and its side effects.

Double muscling is really a misnomer. Likewise, the scientific name, muscular hypertrophy, is incorrect because it implies increased size of fibers in each muscle, which is not the case. Rather, it has been shown that double muscled cattle have more fibers, not larger fibers.

In Germany, the trait is known as doppellender (double rump); in Italy, it's doppia (horse rump); and in France, it's culard.

Double muscling is a genetically controlled character. It appears to be caused by a single recessive gene, which tends to be "masked" by the dominant gene in the heterozygous carriers. Other examples of a character controlled by one pair of genes are: polledness and hornedness, and dwarfism. The genetics, therefore, are relatively simple. Since each animal has two genes for such characters, all cattle can be classified as follows:

DM DM—Homozygous normal; two dominant normal genes—a normal animal.

DM dm—Heterozygous; one dominant normal gene (DM) which tends to cover up the one recessive gene (dm)—these are called carriers. This coverup is not complete; hence, there is a tendency toward double muscling.

dm dm—Homozygous recessive; two recessive double muscle genes—a double muscled animal.

The progeny from a sire and dam of all these three genotypes are predictable, on the average, but not necessarily for any one offspring. The possible matings and progeny are:

Sire	Dam	Progeny
DM DM	DM DM	DM DM All normal.
DM DM or — DM dm	DM dm DM DM	½ DM DM, ½ DM dm All normal, but ½ carriers.
DM dm	DM dm	¼ DM DM, ½ DM dm, ¼ dm dm Of the ¾ normal, 2 out of 3 are carriers; the remaining ¼ are double muscled.
DM DM or — dm dm	dm dm DM DM	DM dm ALL carriers.
DM dm or — dm dm	dm dm DM dm	½ DM dm, ½ dm dm One-half carriers, ½ double muscled.
dm dm	dm dm	dm dm ALL double muscle.

APPEARANCE OF HOMOZYGOUS DOUBLE MUSCLED (dm dm) CATTLE

Obviously, the problem is to determine if an animal is DM dm (a carrier), rather than DM DM (a normal animal). There are two ways to do this: (1) appearance; and (2) breeding tests. Detection by appearance is not 100 percent sure, but the experienced observer doesn't make many mistakes.

The appearance of homozygous double muscle (dm dm) cattle follows, in summary form. But remember that the double muscle character is really a syndrome of many characteristics. Remember, too, that all of these traits may not be present to the same degree in any one animal.

BODY PART: TRAIT	APPEARANCE IN DOUBLE MUSCLED CATTLE
Rump and round	Protruding and rounding; definite grooves, or creases, between the thigh muscles.
Tail	Short. Attached far forward. Prominent tailhead.
Middle; heart girth; flank	Shallow bodied, light heart girth, tucked up flank; animal appears leggy and cylindrical.
Head	Small, long, carried lower than top of shoulders.
Shoulder	Large and bulging; grooves, or creases, evident in arm and forearm.
Cannon bone	Short and fine.
Stance	Animal stands camped out; forelegs extended to front and hindlegs stretched.
Vulva	The vulva of females is small, high, and far forward.
Testicles	Small and carried close to the stomach.
Lying down	Double muscled cattle spend a lot of time lying down; they're "muscle laden"; with a high proportion of flesh to bone.
Age	Double muscling is most conspicuous in young animals. It becomes less apparent with advancing age.
Sex	General lack of masculinity, other than muscularity, in bulls; and lack of a femininity in heifers and cows. At breeding age, double muscling is more marked in males than females.
Birth; early growth	Heavy birth weight. Calves often have enlarged tongues and crooked legs, and are weak. Good early growth. But growth markedly slower by one year of age; and small mature size.
Environment	Double muscling is more marked with superior environment; it shows up more in well-fed animals.

APPEARANCE OF CARRIER (DM dm) CATTLE

Generally speaking, homozygous, double muscled animals can be identified. But it isn't easy to pick out the heterozygotes—the carriers of the double muscle gene, due to the wide variation in expression. Some of the carriers look quite normal, others look like homozygous double muscled animals, and still others are intermediate between these two extremes. Also, identity is further complicated because few, if any, double muscled animals show all of the characteristics. Nevertheless, carriers are characterized by general overall trim appearance, thicker quarter with bulging, thicker round, and a higher tailhead setting than normal animals.

DOUBLE MUSCLED CATTLE—Good or Bad?

Pose the above question in Italy, and the answer is "good." Pose the same question in the United States, and the answer is "bad." Why the difference? Can both be right?

Here is what the author found in Italy:

Double muscled Piemontese (Piedmont) cattle mean to the Italian cattle industry what broad breasted turkeys and Cornish cross broilers mean to the U.S. poultry industry. All are meat producers par excellence.

Italian authorities estimate that 80 percent of the bulls of the Piemontese breed of cattle are double muscled to some degree. Moreover, producers select for the trait; and both feeders and slaughterers vie for double muscled bulls. (In Italy, they feed bulls; not steers.) The reason: All of them make more money from double muscled cattle than from normal cattle.

Piemontese cattle are the most popular breed in Italy. In 1964 (latest census available), there were 674,000 of them. The fact that 80 percent of the bulls are double muscled to some degree indicates that the character responds to selection. Knowledgeable Piemontese cattle breeders in Italy told the author that they expect the following results in their Piemontese breeding programs (see Fig. 5-27):

1. Phenotypically normal heifers mated to double muscled bulls will produce 80 percent double muscled bull calves.

2. By culling out the first calf heifers whose bull calves from the above mating were not double muscled, then mating only proved heterozygotes (or carriers) to double muscled bulls, 95 percent double muscled bull progeny will be produced.

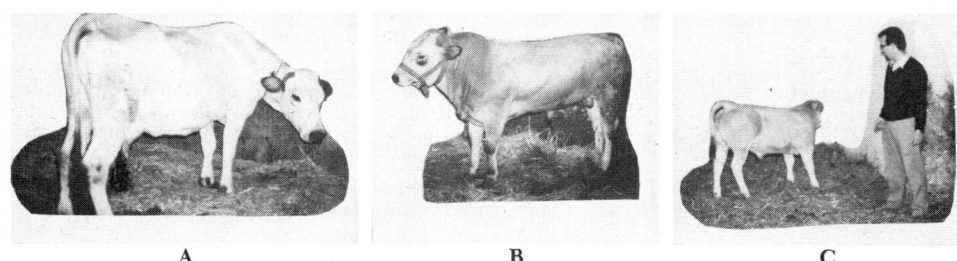

A   B   C

Fig. 5-27 A-B-C. Double muscling predictability in Piemontese (Piedmont) cattle, in Italy.

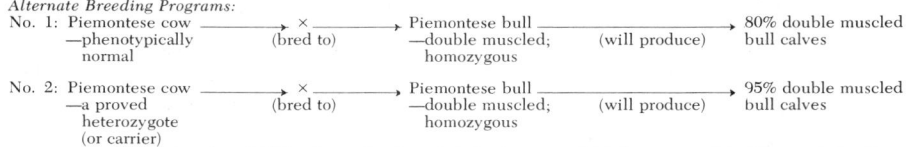

Alternate Breeding Programs:
No. 1: Piemontese cow ———— × ————→ Piemontese bull ———————————→ 80% double muscled
　　　　—phenotypically　　(bred to)　　　—double muscled;　　(will produce)　　bull calves
　　　　normal　　　　　　　　　　　　　　homozygous

No. 2: Piemontese cow ———— × ————→ Piemontese bull ———————————→ 95% double muscled
　　　　—a proved　　　　　(bred to)　　—double muscled;　　(will produce)　　bull calves
　　　　heterozygote　　　　　　　　　　homozygous
　　　　(or carrier)
The Piemontese breed differs from other breeds in having a very high frequency of double muscled cattle.

If double muscling is caused by a single recessive gene, as seems to be the case in the British breeds (Angus, Hereford, and Shorthorn), (1) mating phenotypically normal (noncarrier) cows to a homozygous, double muscled bull would produce 100% heterozygotes (carriers), none of which would be double muscled, and (2) mating known carrier cows (heterozygotes) to a homozygous, double muscled bull would produce 50% homozygous, double muscled calves and 50% heterozygous carriers. However, the Italians report breeding results from their Piemontese breed which suggest that modifier genes common in that breed tend to endow the double muscling gene with dominance (partial dominance). As a result of these modifier genes, the 100% heterozygotes (carriers) referred to could, in the Piemontese breed, easily be classified as 80% double muscled and 20% normal, thereby explaining Fig. 5-27, alternate breeding program No. 1. Likewise, in the

Piemontese breed, mating known carrier cows to double muscled bulls could result in 95% of the offspring being classed as double muscled (Fig. 5-27, alternate breeding program No. 2), as a result of the action of modifier genes causing a large part of the 50% heterozygotes to be classed as double muscled.

Also, it would seem reasonable to suspect that, if 80% of the bulls of the Piemontese breed are double muscled to some degree, a large part of the heifers are, also. Hence, many of the heifers classed as "phenotypically normal" in Fig. 5-27, alternate breeding program No. 1, would actually be carriers of the double muscled gene. However, due to the presence or absence of various modifier genes, perhaps they may appear completely normal or only moderately double muscled, with the result that they are classed as "normals."

In Italy, all members of the beef team are making money from double muscled cattle. Here are the facts:

1. Producers contract their double muscled bull calves to feeders at $360 per head at 5 days of age, for owner pickup at 1 month to 6 weeks of age. In Italy, that's about 2½ times the selling price of a normal calf of the same age and quality.

2. The cattle feeder has a ready market for finished, double muscled bulls. In 1973, double muscled Piemontese animals weighing 1,100 pounds were selling at $82/cwt on foot, which was $29/cwt more than could be obtained from Charolais cattle, or any other breed, of comparable finish and

Fig. 5-28. Double muscled bulls of the Piemontese (Piedmont) breed, fed in confinement by Ernest Gerbi, Asti, Italy. In Italy, cattle are never steered. During the author's visit, Mr. Gerbi sold this lot of bulls to a "country buyer." The vital statistics: Selling price, U.S. $82/cwt on foot, which was $29/cwt higher than Charolais cattle were selling at on the same market. Net profit to Mr. Gerbi, U.S. $242/head. Mr. Gerbi feeds out about 1,200 double muscled bulls each year. Double muscling—good or bad?

quality. Good cattle feeders were netting about $240 per double muscled animal.

3. The slaughterer and the retailer are happy, too. Double muscled cattle dress 72% (vs 63% for normal). Moreover, with their method of cutting, the retailer gets 80% steaks from double muscled carcasses vs 40% for normal carcasses. They literally cut steaks from end to end on a double muscled carcass.

ADVANTAGES OF DOUBLE MUSCLED CATTLE

Most cattlemen, in both Italy and in the United States, are in business to make money—that's as it should be. So, based on the figures already presented, it's obvious that all members of the beef team in Italy are profiting from double muscled cattle. But, they are quick to point out that there are other advantages, too. Here are some of the plusses that they list in favor of double muscled over normal cattle:

1. The calves grow more rapidly up to one year of age.
2. They convert feed more efficiently; it requires fewer pounds of feed to produce a pound of beef.
3. They (a) have a higher dressing percentage, and (b) yield a higher proportion of the more desirable cuts—more steaks. In comparison with normal cattle, double muscled cattle have a larger rib eye; produce less brisket, plate, flank; and produce less kidney and pelvic fat.

In summary: double muscled cattle are superior to normal cattle in (1) rate and efficiency of gain to one year of age, and (2) general carcass desirability.

DISADVANTAGES OF DOUBLE MUSCLED CATTLE

It is recognized that there may be breed differences when it comes to the advantages and disadvantages of double muscled cattle. Nevertheless, here are some of the disadvantages to double muscling that have been reported in different countries:

1. The conception rate is lower, due to (a) the infantile reproductive tracts or slow sexual maturity of some animals, and (b) the flat vulva, which makes copulation difficult.
2. The gestation period is about 10 days longer.
3. There is more calving difficulty (caesarian section; pulling calves; stillborn), due to heavier calves at birth (Piemontese double muscled calves average 108 pounds at birth vs 99 pounds for normal calves), along with the enlarged rump and round regions.
4. Double muscled calves are more difficult to raise; due to such things as (a) enlarged tongues (Macroglossia), and (b) greater susceptibility to disease.
5. Double muscled cows are poor milkers; they produce 30 to 50% less milk than normal cows.
6. Double muscled cattle must be fed a higher proportion of concentrate to roughage, simply because they cannot utilize roughage effectively.
7. There is less marbling; hence, carcasses of double muscled cattle are

penalized by the current U.S. Federal beef grades.

It's unlikely that all of the above disadvantages will occur in any one herd at any one time. Moreover, the degree to which they occur among breeds, and within breeds, will vary according to the "background" genes or modifying genes. Nevertheless, cattlemen should be apprised of the possibilities.

WHY ARE WE GETTING MORE DOUBLE MUSCLING?

Why are more double muscled cattle cropping up in recent years—in all breeds? The answer: In selecting breeding animals with more bred-in meat type, cattlemen are, unconsciously, selecting more carrier animals; simply because the carriers, or heterozygotes, are the heavier muscled ones. (They're the ones with the big rib eyes.)

Of course, double muscling has been around for a very long time—at least 200 years, in Europe, Australia, and the United States. Also, the condition has been reported in almost every breed. Hence, when selecting the heavier muscled animals for breeding purposes, more and more carriers are being used; and more and more double muscled cattle are showing up, and will continue to show up.

SUMMARY RELATIVE TO DOUBLE MUSCLING

Double muscled cattle—good or bad? Obviously, in Italy it's good—very good. If it were not so, no breed that produced 80% double muscled bull calves could survive. The main reason that double muscling in Italy is so good is that their slaughterers pay a premium of $29/cwt on foot for double muscled cattle at market time. With a 1,100-pound animal, that's a premium of $319/head. The reason for the premium: the reputed 9% higher dressing percentage (72 vs 63%) and twice the steaks (80 vs 40%) of double muscled cattle over normal cattle. Also, the Italians give the impression (without scientific proof) that many of the disadvantages that U.S. cattlemen attribute to double muscling have been minimized—they're less nettlesome—in the Piemontese breed and to Italian cattlemen. For example, they don't seem to complain too much about breeding or calving problems. Maybe they have overcome this through selection. Obviously, the expression of the trait in many approved heterozygote (carrier) Piemontese cows is nil (see Fig. 5-27A). Moreover, their system of handling early-weaned calves—nurse-cowing (2 to 5 calves/cow) plus a starter ration, until 4 to 6 weeks of age—makes for a good start in life; even if double muscled calves are less vigorous at birth.

Indeed, Italian cattlemen have something going for them. If the author were in Italy, he would breed and feed double muscled Piemontese cattle. In the United States, it's another story. Until, and unless, a premium is paid for double muscled beef over the butcher's block, there's no incentive; there's insufficient reason to risk the disadvantages of double muscled cattle—even if they could be minimized through selection. Yet, there are some aspects of double muscling in cattle that might be used in improving efficiency of beef production. Also, cattlemen may well emulate poultry breeders—the broad breasted turkey and Cornish chicken counterpart, by producing double muscled cattle (perhaps the heterozygotes) in a well-planned breeding program designed to minimize their production weaknesses and maximize their higher dressing percentage, more red meat, and more steaks; then promote and sell them at a premium. Experimental work conducted by Rollins *et al*, of the California Station, indicates this possibility. They found that, in comparison with normal cattle, calves from double muscled X normal parents (calves heterozygous for double muscling) had a 10 percent advantage in terms of pounds of trimmed retail cuts per day of age at marketing, with no undesirable side effects in either production or performance, and little or no reduction in carcass quality grade at marketing.[20] Broad breasted turkeys and chickens; double muscled cattle! Why not?

The Relative Importance of Sire and Dam

As a sire can have so many more offspring during a given season or a lifetime than a dam, he is from an hereditary standpoint a more important individual than any one female so far as the whole herd is concerned, although both the sire and the dam are of equal importance so far as concerns any one offspring. Because of their wider use, therefore, sires are usually culled more rigidly than females, and the breeder can well afford to pay more for an outstanding sire than for an equally outstanding female.

Experienced stockmen have long felt that sires often more closely resemble their daughters than their sons, whereas dams resemble their sons. Some sires and dams, therefore, enjoy a reputation based almost exclusively on the merit of their sons, whereas others owe their prestige to their daughters. Although this situation is likely to be exaggerated, any such phenomenon as may exist is due to sex-linked inheritance which may be explained as follows: the genes that determine sex are carried on one of the chromosomes. The other genes that are located on the same chromosome will be linked or associated with sex and will be transmitted to the next generation in combination with sex. Thus, because of sex linkage, there are more color-blind men than color-blind women. In poultry breeding, the sex-linked factor is used in a practical way for the purpose of distinguishing the pullets from the cockerels early in life, through the process known as "sexing" the chicks. When a black cock is crossed with barred hens, all the cocks come barred and all the hens come black. It should be emphasized, however, that under most conditions it appears that the influence of the sire and dam on any one offspring is about equal. Most breeders, therefore, will do well to seek excellence in both sexes of breeding animals.

Prepotency

Prepotency refers to the ability of the animal, either male or female, to stamp its own characteristics on its offspring. The offspring of a prepotent

---

[20]Rollins, W. C., R. B. Thiessen, and Moira Tanaka, "Usefulness of Market Calves Heterozygous for Double Muscling Gene," *California Agriculture*, Vol. 28, No. 3, March 1974, p. 8.

bull, for example, resemble both their sire and each other more closely than usual. The only conclusive and final test of prepotency consists of the inspection of the get.

From a genetic standpoint, there are two requisites that an animal must possess in order to be prepotent: (1) dominance, and (2) homozygosity. Every offspring that receives a dominant gene or genes will show the effect of that gene or genes in the particular character or characters which result therefrom. Moreover, a perfectly homozygous animal would transmit the same kind of genes to all of its offspring. Although entirely homozygous animals probably never exist, it is realized that a system of inbreeding is the only way to produce animals that are as nearly homozygous as possible.

Popular beliefs to the contrary, there is no evidence that prepotency can be predicted by the appearance of an animal. To be more specific, there is no reason why a vigorous, masculine-appearing sire will be any more prepotent than one less desirable in these respects.

It should also be emphasized that it is impossible to determine just how important prepotency may be in animal breeding, although many sires of the past have enjoyed a reputation of being extremely prepotent. Perhaps these animals were prepotent, but there is also the possibility that their reputation for producing outstanding animals may have rested upon the fact that they were mated to some of the best females of the breed.

In summary, it may be said that if a given sire or dam possesses a great number of genes that are completely dominant for desirable type and performance and if the animal is relatively homozygous, the offspring will closely resemble the parent and resemble each other, or be uniform. Fortunate, indeed, is the breeder who possesses such an animal.

Nicking

If the offspring of certain matings are especially outstanding and in general better than their parents, breeders are prone to say that the animals "nicked" well. For example, a cow may produce outstanding calves to the service of a certain bull, but when mated to another bull of apparent equal merit as a sire, the offspring may be disappointing. Or sometimes the mating of a rather average bull to an equally average cow will result in the production of a most outstanding individual both from the standpoint of type and performance.

So-called successful nicking is due, genetically speaking, to the fact that the right combination of genes for good characters is contributed by each parent, although each of the parents within itself may be lacking in certain genes necessary for excellence. In other words, the animals nicked well because their respective combinations of good genes were such as to complement each other.

The history of animal breeding includes records of several supposedly favorable nicks, one of the most famous of which was the favorable result secured from crossing sons of Anxiety 4th with daughters of North Pole in the Gudgell and Simpson herd of Hereford cattle. At this late date, it is impossible to determine whether these Anxiety 4th X North Pole matings were

Fig. 5-29. Anxiety 4th 9904, whose sons were alleged to "nick" exceedingly well on the daughters of North Pole in the Gudgell and Simpson herd of Hereford cattle. (Courtesy, *The American Hereford Journal*)

successful because of nicking or whether the good results should be more rightfully attributed to the fact that the sons of Anxiety 4th were great breeding bulls and that they merely happened to be mated, for the most part, with daughters of North Pole because the available females in the Gudgell and Simpson herd were of this particular breeding.

Because of the very nature of successful nicks, outstanding animals arising therefrom must be carefully scrutinized from a breeding standpoint, because, with their heterozygous origin, it is quite unlikely that they will breed true.

Family Names

In cattle, depending upon the breed, family names are traced through either the males or females. In Angus and Shorthorn cattle, the family names had their origin with certain great foundation females, whereas in Herefords the family names trace through the sires. Similar family names exist in horses, but in both hogs and sheep less importance is attached to them.

Unfortunately, the value of family names is generally grossly exaggerated. Obviously, if the foundation male or female, as the case may be, is very many generations removed, the genetic superiority of this head of a family is halved so many times by subsequent matings that there is little reason to think that one family is superior to another. For example, if a present-day Queen Mother (an old and well-known Angus family) is 18 generations removed from the founder, she would carry the following relationship to the head of the family: $(\frac{1}{2})$ 18 or $\frac{1}{262,144}$ or .0004%. Obviously, this Queen Mother may not have inherited a single gene from the foundation cow, and merely being a Queen Mother does not differentiate her much from other families which make up the breed.

The situation relative to family names is often further distorted by breeders placing a premium on family names of which there are few members, little realizing that, in at least some cases, there may be unfortunate reasons for the scarcity in numbers.

Such family names have about as much significance as human family names. Who would be so foolish as to think that the Joneses as a group are alike and different from the Smiths? Perhaps, if the truth were known, there have been many individuals with each of these family names who have been of no particular credit to the clan, and the same applies to all other family names.

Family names lend themselves readily to speculation. Because of this, the history of livestock breeding has often been blighted by instances of unwise pedigree selection on the basis of not too meaningful family names. The most classical example of a situation of this type occurred with the Duchess family of Shorthorn cattle, founded by the noted pioneer English Shorthorn breeder, Thomas Bates. Bates, and more especially those later breeders who emulated him, followed preferences in bloodlines within increasingly narrow limits, until ultimately they were breeding cattle solely according to fashionable pedigrees, using good, bad, and indifferent animals. Fad and fancy in pedigrees dominated the situation, and the fundamental importance of good individuality as the basis of selecting animals for breeding purposes was for the time largely ignored. The sole desire of these breeders was to concentrate the Duchess blood. The climax of the "Duchess boom" (or "Bates boom") came in September, 1873, when the New York Mills herd was sold at auction with English and American breeders competing for the offering. At this memorable event, 109 head of Duchess-bred cattle averaged $3,504 per head, with the seven-year-old 8th Duchess of Geneva selling at the world's record price of $40,600.

As with most booms, the New York Mills sale was followed by a rather critical reaction, and eventually the bottom dropped out of values. Even more tragic, the owners of Duchess Shorthorns suddenly came to a realization that indiscriminate inbreeding and a lack of selection had put the family name in disrepute. As a result, the strain became virtually extinct a few years later.

On the other hand, certain linebred families—linebred to a foundation sire or dam so that the family is kept highly related to it—do have genetic significance. Moreover, if the programs involved have been accompanied by rigid culling, many good individuals may have evolved, and the family name may be in good repute. The Anxiety 4th family of Hereford cattle is probably the best known family of this kind in meat animals. Even so, there is real danger in assuming that an "airtight" or "straightbred" Anxiety 4th pedigree is within itself meritorious and that this family is superior to that of any other family in Hereford cattle.

Heredity and Environment

A massive purebred bull, standing belly deep in straw and with a manger full of feed before him, is undeniably the result of two forces—heredity and environment (with the latter including training). If turned out on the range, an identical twin to the placid bull would present an entirely different appearance. By the same token, optimum environment could never make a champion out of a bull with scrub ancestry, but it might well be added that "fat and hair will cover up a multitude of sins."

Man-made treatments sometimes conceal undesirable traits to the extent that improvements in heredity are impeded. For example, the practice of cutting ties on cattle may be of immediate benefit in improving the appearance of treated animals, but the trait continues to be transmitted. Likewise, the use of nurse cows for developing young show animals in beef herds tends to favor the perpetuation of the genes for poor milkers.

These are extreme examples, but they do emphasize the fact that any particular animal is the product of heredity and environment. Stated differently, heredity may be thought of as the foundation, and environment as the structure. Heredity has already made its contribution at the time of fertilization, but environment works ceaselessly away until death.

Experimental work has long shown conclusively enough that the vigor and size of animals at birth is dependent upon the environment of the embryo from the minute the ovum or egg is fertilized by the sperm, and now we have evidence to indicate that newborn animals are affected by the environment of the egg and sperm long before fertilization has been accomplished. In other words, perhaps due to storage of nutritive factors, the kind and quality of the ration fed to young, growing females may later affect the quality of their progeny. Generally speaking, then, environment may inhibit the full expression of potentialities from a time preceding fertilization until physiological maturity has been attained.

Admittedly, after looking over an animal or studying its production record, a breeder cannot with certainty know whether it is genetically a high or a low producer. There can be no denying the fact that environment—including feeding, management, and disease—plays a tremendous part in determining the extent to which hereditary differences that are present will be expressed in animals. In general, however, the results of a longtime experiment conducted at Washington State University support the contention that selection of breeding animals should be carried on under the same environmental conditions as those under which commercial animals are produced.[21]

Within the pure breeds of livestock—managed under average or better than average conditions—it has been found that, in general, only 30 to 45 percent of the observed variation in a characteristic is actually brought about by hereditary variations. (See Table 5-6.) To be sure, if we contrast animals that differ very greatly in heredity—for example, a champion bull and a scrub—90 percent or more of the apparent differences in type may be due to heredity. The point is, however, that extreme cases such as the one just mentioned are not involved in the advancement within improved breeds of live-

---

[21]Fowler, Stewart H., and M. E. Ensminger, *Wash. Agr. Exp. Sta. Bull. 34*, January, 1961.

stock. Here the comparisons are between animals of average or better than average quality, and the observed differences are often very minor.

The problem of the progressive breeder is that of selecting the very best animals available genetically—these to be parents of the next generation of offspring in his herd. The fact that only 30 to 45 percent of the observed variation is due to differences in inheritance and that environmental differences can produce misleading variations makes mistakes in the selection of breeding animals inevitable. However, if the purebred breeder has clearly in mind a well-defined ideal and adheres rigidly to it in selecting his breeding stock, very definite progress can be made, especially if mild inbreeding is judiciously used as a tool through which to fix the hereditary material.

Hybrid Vigor or Heterosis

Heterosis, or hybrid vigor, is a name given to the biological phenomenon which causes crossbreds to outproduce the average of their parents. For numerous traits, the performance of the cross is superior to the average of the parental breeds. This phenomenon has been well known for years, and has been used in many breeding programs. The production of hybrid seed corn by developing inbred lines and then crossing them is probably the most important attempt by man to take advantage of hybrid vigor. Also, heterosis is being used extensively in commercial swine, sheep, layer, and broiler production today; an estimated 80 percent of market hogs, market lambs, and layers are crossbreds, and 95 percent of broilers are crosses.

The genetic explanation for the hybrid's extra vigor is basically the same, whether it be cattle, hogs, sheep, layers, broilers, hybrid corn, hybrid sorghum, or whatnot. Heterosis is produced by the fact that the dominant

Fig. 5-30. A growthy South Devon X Hereford crossbred calf with mother at the U.S. Meat Animal Research Center. (Courtesy, Arthur V. Palmer, Big Beef Hybrids International Company, Stillwater, Minn.)

gene of a parent is usually more favorable than its recessive partner. When the genetic groups differ in the frequency of genes they have and dominance exists, then heterosis will be produced.

Heterosis is measured by the amount the crossbred offspring exceeds the average of the two parent breeds or inbred lines for a particular trait, using the following formula for any one trait:

$$\frac{\text{Crossbred average} \quad (\text{minus}) \quad \text{Purebred average}}{\text{Purebred average}} \times 100 = \text{Percent Hybrid Vigor}$$

Thus, if the average of the two parent populations for weaning weight of calves at 205 days of age is 400 pounds and the average of their crossbred offspring is 420 pounds, application of the above formula shows that the amount of heterosis is 20 pounds, or 5 percent.

Traits high in heritability—like tenderness of rib eye—respond consistently to selection, but show little response in hybrid vigor. Traits low in heritability—like mothering ability, calving interval, and conception rate—usually show good response in hybrid vigor.

The level of hybrid vigor for all traits depends on the breeds crossed. The greater the genetic difference between two breeds, the greater the hybrid vigor expected. The genetic difference between a British breed and a breed of Indian origin is greater than the difference between one British breed and another British breed.

It is most important to have hybrid vigor in the cow herd where it results in increased fertility, survivability of the calves, milk production, growth rate of calves, and longevity of the cow—all factors that mean more profit to the cowman.

It is noteworthy that purebreds must be constantly tapped to renew the vigor of crossbreds; otherwise, the vigor is dissipated.

Complementary

Complementary refers to the advantage of a cross over another cross or a purebred resulting from the manner in which two or more characters combine or complement each other. It is a matching of breeds so that they compensate each other; the objective being to get the desirable traits of each. Thus, in a crossbreeding program, breeds that complement each other should be selected, thereby maximizing the desirable traits and minimizing the undesirable traits. Since breeds which are selected because they tend to express a maximum of some trait (e.g., high daily gain) will have some undesirable traits (e.g., large mature cow size and high maintenance cost), different breeds must be selected for different purposes. A well-known example of breed complementation for improving overall carcass desirability in the market animal is the Angus X Charolais cross, combining the higher carcass grade of the Angus with the higher cutability of the Charolais.

Fig. 5-31. Crossbred calf, out of an Angus cow and sired by a Charolais bull. (Courtesy, C & B Livestock Company, Baker, Ore.)

TRAITS OF BEEF CATTLE; THEIR (1) ECONOMIC VALUE, (2) HERITABILITY, AND (3) HETEROSIS

Selection and hybrid vigor combined can increase beef yield per cow maintained through continuous crossbreeding over straightbreds by 15 to 25 percent, depending on the choice of the breeds and the breeding system. The 15 to 25 percent increase is achieved in two ways: (1) through selection, based on production testing, of the purebreds used in the crossbreeding program, and (2) through heterosis increase of the crossbreds. This points up the interdependence of purebreds and crossbreds in modern beef production. For maximum beef over the block and increased profit, the best in both purebred breeding (selection) and crossbreeding (heterosis) must be utilized in cattle breeding programs.

In Table 5-5, the economically important traits of beef cattle are grouped into three broad classes—reproduction, production, and product; and the heritability and heterosis of each of these groups is given.

Calf crop percentage, average daily gain in the feedlot, and retail meat yield percentage in the carcass are examples of traits in the three classes of Table 5-5. Obviously, there is overlap. Based on average commercial cattle production conditions, the second column gives the relative economic value of the three classes of traits. Reproduction is at least 10 times as important as production in terms of net returns. This is so because being born and born alive are the two most important factors effecting profit in the cow business; without a live calf the other economic traits are of academic interest only— they cannot be evaluated because there is no calf. Production is twice as important today as product improvement. This stems from current buying

TABLE 5-5

ECONOMIC VALUE, SELECTION VARIATION, AND HETEROSIS
INCREASE OF THREE CLASSES OF TRAITS OF BEEF CATTLE[1]

Class of Traits	Relative Economic Values	Selection Variation	Heterosis Increases
		(%)	(%)
Reproduction (calf crop, etc.)	20	10	10
Production (av. da. gain, etc.)	2	40	5
Product (retail meat yield, etc.)	1	50	0

[1]*Beef Cattle Science Handbook*, Vol. 10, 1973, p. 194. From a paper on "Beef Breeding Programs," by Dr. Richard L. Willham, Professor of Animal Science, Iowa State University, Ames, Iowa.

procedures; hence, it could change rapidly with economic shifts. The latter point bears emphasis, because (1) cattle breeding programs cannot be shifted rapidly, and (2) history shows that livestock breeders have always paid dearly when they failed accurately to heed or project consumer preference on a longtime basis. The most notable example of this failure occurred in the swine industry. Pork lost its lead to beef in per capita consumption because it was too little and too late in shifting from a lard-type to a meat-type hog. So, cattlemen must continue to make and heed meat quality studies, despite the fact that consumers are unwilling to pay adequately for superior quality from time to time.

The third column of Table 5-5 gives the average heritabilities of the three classes of traits. These values reflect the percentage of variation among animals treated alike which is due to variation and is available for selective change. The greater the percent selection variation, the more effective the selection for increased performance. As noted, the heritability of reproductive traits is low—only 10 percent. Hence, little selection response can be expected even from intense culling on the reproductive traits. But production and product traits have relatively high heritability. Thus, selection for increased yearling weight (production) and retail meat yield (product) will be effective.

The last column gives the average percentage increase in performance that may be expected by crossing breeds of cattle—the heterosis. Reproductive traits show the biggest improvement from heterosis. Production traits exhibit some heterosis, while product traits, on the average, show little heterosis. The average of the parental performance usually predicts the response.

Table 5-5 is the key to breeding programs—both commercial and purebred. As shown, the column giving relative economic value is negatively related to the column giving available variation for selection. But the first column is positively related to the heterosis column. This indicates that some sort of commercial crossbreeding, including the crossbred cow, is the best way in which to utilize the reproductive heterosis. Further, the purebreds (usually bulls) used by commercial producers can be easily improved by selection for superior production and product traits, with this superiority passed on to the commercial producer.

To the commercial cattleman, this table suggests a crossbreeding system designed to maximize reproductive heterosis and put together desirable combination of traits in the market animal. To the purebred breeder, the table suggests that selection programs designed to improve production and product traits in the breeding stock herd would benefit the beef industry through direct transfer of superior parents (usually bulls) to the commercial industry.

BREEDING PROGRAMS

A breeding program is a complete system of management designed to bring about genetic change in a group of animals. Modern cattle breeding programs have the dual-objectives of (1) breeding better beef cattle, and (2) better beef cattle breeding. Two types of breeding programs exist—commercial and purebred. Generally speaking, the commercial cattleman crossbreeds—he mates animals of different breeds—although a gradually decreasing number still have straightbreds (males and females of the same genetic background, or breed). The seed stock producer is almost always a purebred breeder (straightbreds). All of them are more specialized and precise than formerly.

Most commercial breeding programs utilize crossbreeding because of hybrid vigor—which results in better phenotypic performance than straight breeding. But they rely on purebred breeders for seed stock (especially bulls), because the only way to improve crossbreds above their present level of performance is through the development of better straightbreds. Commercial programs usually market their end product (calves) on a per pound basis.

Purebred breeding programs are those which produce breeding stock. The cattle are merchandised on their breeding value rather than their actual performance. How the animals perform is not at issue; only as performance predicts breeding value at the commercial level is it of importance. This type of breeding program is restrictive in the kind of selection that can be practiced once the genetic groups have been chosen. Only traits having moderate to high heritability (production and product) can be improved appreciably.

The general form of beef breeding programs, both commercial and purebred, involves the production and measurement of a calf crop followed by the selection of the parents for the next generation. These matters are detailed in the section headed "Production Testing Beef Cattle," which follows later in this chapter.

In the future, most breeds will either be (1) propogated as straightbreds, or (2) used for crossbreeding. The vast majority of the straightbreds will be purebreds and production tested. Fewer and fewer grade herds will be maintained for commercial purposes, simply because the economics favor crossbreeding accompanied by complementary genes and heterosis. The straightbreds will be used primarily to produce (1) bulls and semen, and (2) purebred females for F_1 heifer programs. The crossbreds will be used to produce beef according to specification.

Since purebred breeding programs have as their function the supplying of breeding stock for commercial programs, the latter type will be presented first in order that its needs will be reflected in the purebred programs. However, both types are interdependent.

Commercial Breeding Programs

The goal of most commercial beef operations is to convert the production of the land—grass and/or crops—into dollars through the traditional cow-calf operation. Usually, the product is marketed at the weaning stage although some commercial cattlemen carry them to the yearling stage, or even finish them out for market. In any event, the goal of increased net returns must involve a critical analysis of current and future economics and the integration of this with the production potential available from the land. These are business and management aspects, rather than genetic; nevertheless, they must be analyzed and decisions must be reached. The level of genetic potential should be compatible with the resources available. Table

Fig. 5-32. Commercial cattleman Loren Lambert, Onida, South Dakota, and his Hereford X Friesian cross calves. (Courtesy, American Beef Friesian)

5-5 groups the classes of traits—reproduction, production, and product—and gives their relative economic values as they involve the goal of increased net returns. Improvement in reproductive output—a higher percent calf crop, born alive—is the key to increased net returns in most commercial cattle operations. Next in economic value comes improvement in pounds of beef produced per cow. Last is improvement in yield in retail beef of acceptable quality. In the long run, the latter may be more important economically than indicated by Table 5-5, based on what people are willing to pay for, simply because the future of beef production is only partially dependent upon economical protein production; it is more dependent upon palatability of the product and the utilization of roughage by the ruminant.

Choosing and keeping performance records in a commercial cattle operation is not easy. Yet, records are necessary. Specification of the product offered for sale (calves, in the case of the commercial cattleman) is becoming the rule. Thus, complete records, adequately analyzed and utilized, will help any commercial operation. But records cost money. A large commercial cow-calf operation (of 300 cows or more) may not be able to justify more than a simple feeder calf program involving the sampling of the product that it is offering for sale. On an every-other-year basis, this might involve a random sampling of calves which are fed out and slaughtered. The gain and carcass data are then used by the producer (1) in the development of his performance reputation (the production of reputation feeder cattle), and (2) in the selection of herd sires to improve his performance. Small commercial operations (with 50 to 300 cows) can well afford to keep more detailed records on their cowherds since this is a means by which they can more effectively compete. Their produce of dam records can be an aid in developing a high producing cow herd, and in adjusting the management to optimize production. Most state Beef Cattle Improvement Associations have such programs available at nominal cost. The use of such programs, even with multi-sire pastures, allows the producer at least to evaluate groups of sires purchased. In this manner, he can study the sources of breeding stock supply and be more critical in his future selection.

Germ plasm choice is becoming increasingly difficult because of the large number of breeds and breed cross combinations now available. With only 3 main breeds (as was once the case with Shorthorn, Hereford, and Angus), there are 3 single-cross combinations and 3 three-way cross combinations from which to select. However, with 10 breeds there are 45 single-cross combinations and 360 possible three-way combinations from which to select. Think for a moment of the combinations involved with the 53 breeds listed in Table 3-2. Remember, too, that more breeds will come. Thus, the commercial producer needs critically to evaluate the best information available in order to determine which breeds and breed crosses can be used in his particular program. Much money and effort is being, and will continue to be, put into promotion of breeds and their cross combinations. This promotional information must be carefully considered and evaluated by the commercial producer, especially if he has a long-standing, good herd of a traditional breed.

Purebred Breeding Programs

Purebred breeding programs are the key to continued genetic improvement of cattle. Thus, seed stock producers should establish and follow a modern, sophisticated program. Also, commercial cattlemen should be aware of exactly what constitutes a sound breeding program at the purebred level; otherwise, their choices among herds will likely be both limited and unwise.

The goal of the purebred cattleman is to provide stock with superior breeding values. Obviously, such a goal involves being paid according to the superiority of the breeding animals being offered for sale, along with the cost involved in promoting these animals. The goal of the purebred breeder must involve the improvement of the production and product traits (Table 5-5), along with at least maintaining the reproductive performance. Today, the emphasis is on improving production, especially yearling weight and efficiency of feed utilization.

Fig. 5-33. Purebred Shorthorn herd on pasture. (Courtesy, USDA)

In the past, breeding stock production could be accomplished in smaller units than commerical production. During the first half of the present century, it was not uncommon to find that many of the top purebred herds, in Europe as well as in America, only had 20 to 40 cows. Today, in order to take advantage of numbers in the development of selection schemes, purebred breeding is becoming centralized into larger units, including corporate structures. Since performance is fast becoming the rule in evaluating breeding stock, the use of animal breeding technology will be a necessary requisite to remaining in the purebred business. Fortunately, many of the breed registry associations are presently conducting cooperative sire evaluation programs within the breeds, thereby permitting the 100- to 200-head purebred breeder to participate in such a program and effectively compete with larger breeding units.

The various performance programs of the breeds and other organizations which are a part of the Beef Improvement Federation dictate a rather standard timing of records and a specification of the test. The important things in breeding stock production are that (1) they be as uniform from one calf crop to the next as is possible so that the complete set of records developed can be used in making selections, and (2) they conform as nearly as possible to the conditions under which the sale animals are to be used. This means that artificial conditions, other than those necessary to produce uniformity from year to year, should be avoided.

The record keeping system and the tests conducted should be as complete as possible. Current breed registry association programs, the state programs, and the Performance Registry International program are relatively complete in terms of providing a means of recording important measures from breeding to slaughter. However, the progressive breeder will measure additional traits that need improvement in his herd. Also, he will recognize that just keeping records is of little importance if they are not in a form in which they can be used in a creative selection program—and if they are not used.

Today, with the importation of numerous new European breeds, a wide choice in breeds is possible (see Table 3-2). Economics, timing, and promotion must be seriously considered before embarking on a program to breed one of the newly introduced breeds. Grading up to a new breed by the use of sires only is a long and costly process. It is rapid at first, but it becomes slower with each succeeding generation since purity is halved each time. Because of the long generation interval of cattle, the development of a seed stock herd through grading up will take approximately 20 years. Coupled with this, the heterosis produced in the initial single cross will be halved with each succeeding generation, thereby making it difficult to maintain the initial level of performance through selection alone. Such grading up to a purebred program will require perseverance, time, and money; and the reward may or may not be commensurate with it.

The choice of the selection program in the purebred herd is of great importance, since it determines the success of the program in producing genetic improvement. With heritability of 40 to 50 percent for production and product traits, selection based on own performance, coupled with sequential progeny testing among the selected individuals, provides near maximum selection advance. Using additional information on relatives, such as sibs, increases the accuracy of selection somewhat, provided the generation interval is not lengthened. Progeny testing is worthwhile only when used in combinations with own performance selection. The accuracy of progeny testing can more than compensate for the extra two years added to the generation interval of the bulls. The importance of thoroughly testing bulls is pointed up by the fact that over 90 percent of the selection advance will come from sire selection, with the remainder coming for the heifer selection. This results from the sex ratio being 50-50 while only one male to 20 females is required even under natural service. Of course, under A.I., the ratio can be 1 to 1,000 or greater.

Cooperative sire evaluation programs can be of distinct benefit to the small purebred breeder—the man with 100 to 300 cows. Such a program might consist of performance testing the bull crop, then selecting from these the top one or two bulls to go into a reference sire progeny test conducted by the breed.

If possible, the heifers in a purebred herd should also be tested—especially those that may become the dams of herd bulls. Such a program could make the difference between ordinary and superior. Testing heifer replacement for production traits (by saving more for breeding than is necessary) would enhance a breeding program. Also, a young cow herd—the product of the last superior sires—will increase the selection response per year.

In conclusion, it may be said that performance will be the rule in evaluating breeding stock of the future, and the use of animal breeding technology will be a necessary requisite to remaining in the business. The production and selling of breeding stock in the beef industry will become a specification business. But promotion will still be necessary. Central bull tests, correctly conducted, provide an excellent means by which to compete and promote. Also, breed sires evaluation programs will be a good promotional tool if the progeny of bulls compete with equal opportunity. Finally, it may be said that testing programs, whatever the kind, are no better than the accuracy of the records kept, the use made of the records in a selection program, and the honesty and integrity of the people back of the test.

SYSTEMS OF BREEDING

The many diverse types and breeds among each class of farm animals in existence today originated from only a few wild types within each species. These early domesticated animals possessed the pool of genes, which, through controlled matings and selection, proved flexible in the hands of man. In cattle, for example, through various systems of breeding, there evolved animals especially adapted to draft purposes, beef production, milk production, and dual-purpose needs.

Successful breeders follow a breeding system, the purpose of which is to give greater control of heredity than if selection alone is used. Thus, breeders need to know about the different breeding systems; what they can do well, and what they can do poorly or not at all.

Perhaps at the outset it should be stated that there is no one best system of breeding or secret of success for any and all conditions. Each breeding program is an individual case, requiring careful study. The choice of the system of breeding should be determined primarily by size and quality of herd, finances and skill of the operator, and the ultimate goal ahead.

Purebreeding

A purebred animal may be defined as a member of a breed, the animals of which possess a common ancestry and distinctive characteristics; and he is either registered or eligible for registry in the herd book of that breed.

It must be emphasized that pure breeding and homozygosity may bear very different connotations. The term "purebred" refers to animals whose entire lineage, regardless of the number of generations removed, traces back to the foundation animals accepted by the breed or to any animals which have been subsequently approved for infusion. On the other hand, homozygosity refers to the likeness of the genes.

Yet there is some interrelationship between purebreds and homozygosity. Because most breeds had a relatively small number of foundation animals, the unavoidable inbreeding and linebreeding during the formative stage resulted in a certain amount of homozygosity. Moreover, through the normal sequence of events, it is estimated that purebreds become more homozygous by from ¼ to ½ percent per animal generation. It should be emphasized that the word "purebred" does not necessarily guarantee superior type or high productivity. That is to say, the word "purebred" is not, within itself, magic, nor is it sacred. Many a person has found to his sorrow that there are such things as "purebred scrubs." Yet, on the average, purebred animals are superior to nonpurebreds.

For the man with experience and adequate capital, the breeding of purebreds may offer unlimited opportunities. It has been well said that honor, fame, and fortune are all within the realm of possible realization of the purebred breeder; but it should also be added that only a few achieve this high calling.

Purebred breeding is a highly specialized type of production. Generally speaking, only the experienced breeder should undertake the production of purebreds with the intention of furnishing foundation or replacement stock to other purebred breeders or purebred bulls to the producer of grades. Although we have had many constructive cattle breeders and great progress has been made, it must be remembered that only a few achieve sufficient success to classify as master breeders.

Inbreeding

Most scientists divide inbreeding into various categories, according to the closeness of the relationship of the animals mated and the purpose of the matings. There is considerable disagreement, however, as to both the terms used and the meanings that it is intended they should convey. For purposes of this book and the discussion which follows, the following definitions will be used.

Inbreeding is the mating of animals more closely related than the average of the population from which they came.

Closebreeding is the mating of closely related animals; such as sire to daughter, son to dam, and brother to sister.

Linebreeding is the mating of animals more distantly related than in

*closebreeding, and in which the matings are usually directed toward keep-
ing the offspring closely related to some highly admired ancestor; such as
half-brother and half-sister, female to grandsire, and cousins.*

CLOSEBREEDING

Closebreeding is rarely practiced among present-day cattlemen, though
it was common in the foundation animals of most of the breeds. For example,
it is interesting to note that Comet (155), an illustrious sire and noted as the
first Shorthorn to sell for $5,000, came from the mating of Favorite and
Young Phoenix, a heifer that had been produced from the union of Favorite
with his own dam. Such was the program of the Collings Brothers and many
other early-day beef cattle breeders, including those in all breeds.

Closebreeding is that system of breeding in which closely related animals are mated. In it there is a minimum number of different ancestors. In
the repeated mating of a brother with his full sister, for example, there are
only 2 grandparents instead of 4, only 2 great-grandparents instead of 8, and
only two different ancestors in each generation farther back—instead of the
theoretically possible 16, 32, 64, 128, etc. The most intensive form of inbreeding is self-fertilization. It occurs in some plants, such as wheat and
garden peas, and in some of the lower animals; but domestic animals are not
self-fertilized.

The reasons for practicing closebreeding are:

1. It increases the degree of homozygosity within animals, making the
resulting offspring pure or homozygous in a larger proportion of their gene
pairs than in the case of linebred or outcross animals. In so doing, the less
desirable recessive genes are brought to light so that they can be more readily culled. Thus, closebreeding together with rigid culling, affords the surest
and quickest method of fixing and perpetuating a desirable character or
group of characters.

2. If carried on for a period of time, it tends to create lines or strains of
animals that are uniform in type and other characteristics.

3. It keeps the relationship to a desirable ancestor highest.

4. Because of the greater homozygosity, it makes for greater prepotency.
That is, selected closebred animals are more homozygous for desirable
genes (genes which are often dominant), and they, therefore, transmit these
genes with greater uniformity.

5. Through the production of inbred lines or families by closebreeding
and the subsequent crossing of certain of these lines, it affords a modern
approach to livestock improvement. Moreover, the best of the inbred animals are likely to give superior results in outcrosses.

6. Where a breeder is in the unique position of having his herd so far
advanced that to go on the outside for seed stock would merely be a step
backward, it offers the only sound alternative for maintaining existing quality or making further improvement.

The precautions in closebreeding may be summarized as follows:

1. As closebreeding greatly enhances the chances that recessives will

appear during the early generations in obtaining homozygosity, it is almost certain to increase the proportion of worthless breeding stock produced. This may include such things as reduction in size, fertility, and general vigor. Also, lethals and other genetic abnormalities often appear with increased frequency in closebred animals.

2. Because of the rigid culling necessary to avoid the "fixing" of undesirable characters, especially in the first generations of a closebreeding program, it is almost imperative that this system of breeding be confined to a relatively large herd and to instances when the owner has sufficient finances to stand the rigid culling that must accompany such a program.

3. It requires skill in making planned matings and rigid selection, thus being most successful when applied by "master breeders."

4. It is not adapted for use by the man with average or below average stock because the very fact that his animals are average means that a goodly share of undesirable genes are present. Closebreeding would merely make the animals more homozygous for undesirable genes, and, therefore, worse.

Judging from outward manifestations alone, it might appear that closebreeding is predominantly harmful in its effects—often leading to the production of defective animals lacking in the vitality necessary for successful and profitable production. But this is by no means the whole story. Although closebreeding often leads to the production of animals of low value, the resulting superior animals can confidently be expected to be homozygous for a greater than average number of good genes and thus more valuable for breeding purposes. Figuratively speaking, therefore, closebreeding may be referred to as "trial by fire," and the breeder who practices it can expect to obtain many animals that fail to measure up and have to be culled. On the other hand, if closebreeding is properly handled, he can also expect to secure animals of exceptional value.

Although closebreeding has been practiced less during the past century than in the formative period of the different pure breeds of livestock, it has real merit when its principles and limitations are fully understood. Perhaps closebreeding had best be confined to use by the skilled master breeder who is in a sufficiently sound financial position to endure rigid and intelligent culling and delayed returns and whose herd is both large and above average in quality.

LINEBREEDING

From a biological standpoint, closebreeding and linebreeding are the same thing, differing merely in intensity. In general, closebreeding has been frowned upon by stockmen, but linebreeding (the less intensive form) has been looked upon with favor in many quarters.

In a linebreeding program, the degree of relationship is not closer than half-brother and half-sister or matings more distantly related; cousin matings, grandparent to grand offspring, etc.

Linebreeding is usually practiced in order to conserve and perpetuate the good traits of a certain outstanding sire or dam. Because such descen-

Fig. 5-34. Prince Domino 499611, calved September 13, 1914 and died April 4, 1930. Many great Herefords have been produced from linebreeding to this immortal sire. Prince Domino's final resting place at Wyoming Hereford Ranch, Cheyenne, Wyoming, is marked with the following epitaph: "He lived and died and won a lasting name." This is a rare tribute, indeed, to any beast—or man. (Courtesy, *The American Hereford Journal*)

dants are of similar lineage, they have the same general type of germ plasm and therefore exhibit a high degree of uniformity in type and performance. During the past five decades, for example, a great many Hereford herds have been linebred to Prince Domino, that immortal Gudgell and Simpson bred bull who, in the hands of Otto Fulscher and the Wyoming Hereford Ranch, contributed so much to the improvement of the Hereford breed.

In a more limited way, a linebreeding program has the same advantages and disadvantages of a closebreeding program. Stated differently, linebreeding offers fewer possibilities both for good and harm than closebreeding. It is a more conservative and safer type of program, offering less probability either to "hit the jackpot" or "sink the ship." It is a middle-of-the-road program that the vast majority of average and small breeders can safely follow to their advantage. Through it, reasonable progress can be made without taking any great risk. A degree of homozygosity of certain desirable genes can be secured without running too great a risk of intensifying undesirable ones.

Usually a linebreeding program is best accomplished through breeding to an outstanding sire rather than to an outstanding dam because of the greater number of offspring of the former. If a breeder found himself in possession of a great bull—proved great by the performance records of a large number of his get—a linebreeding program might be initiated in the following way: select two of the best sons of the noted bull and mate them to their half-sisters, balancing all possible defects in the subsequent matings. The next generation matings might well consist of breeding the daughters of one of the bulls to the son of the other, etc. If, in such a program, it seems wise to secure some outside blood (genes) to correct a common defect or defects in the herd, this may be done through selecting one or more outstanding proved cows from the outside—animals whose get are strong where the herd may be deficient—and then mating this female(s) to one of the linebred bulls with the hope of producing a son that may be used in the herd.

The owner of a small purebred herd with limited numbers can often follow a linebreeding program by buying all of his sires from a large breeder who follows such a program—thus in effect following the linebreeding program of the larger breeder.

Naturally, a linebreeding program may be achieved in other ways. Regardless of the actual matings used, the main objective in such a system of breeding is that of rendering the animals homozygous—in desired type and performance—to some great and highly regarded ancestor, while at the same time weeding out homozygous undesirable characteristics. The success of the program, therefore, is dependent upon having desirable genes with which to start and an intelligent intensification of these good genes.

It should be emphasized that there are some types of herds in which one should almost never closebreed or linebreed. These include grade or commercial herds and purebred herds of only average quality.

The owner of a grade or commercial herd runs the risk of undesirable results, and, even if successful, as a commercial breeder, he cannot sell his stock at increased prices for breeding purposes.

With purebred herds of only average quality, more rapid progress can usually be made by introducing superior outcross sires. Moreover, if the animals are of only average quality they must have a preponderance of "bad" genes that would only be intensified through a closebreeding or linebreeding program.

Outcrossing

Outcrossing is the mating of animals that are members of the same breed but which show no relationship close up in the pedigree (for at least the first four or six generations).

Most of our purebred animals of all classes of livestock are the result of outcrossing. It is a relatively safe system of breeding, for it is unlikely that two such unrelated animals will carry the same "undesirable" genes and pass them on to their offspring.

Perhaps it might well be added that the majority of purebred breeders with average or below average herds had best follow an outcrossing program, because, in such herds, the problem is that of retaining a heterozygous type of germ plasm with the hope that genes for undesirable characters will be counteracted by genes for desirable characters. With such average or below average herds, an inbreeding program would merely make the animals homozygous for the less desirable characters, the presence of which already makes for their mediocrity. In general, continued outcrossing offers neither the hope for improvement nor the hazard of retrogression of line-breeding or closebreeding programs.

Judicious and occasional outcrossing may well be an integral part of linebreeding or closebreeding programs. As closely inbred animals become increasingly homozygous with germ plasm for good characters, they may likewise become homozygous for certain undesirable characters even though their general overall type and performance remains well above the breed average. Such defects may best be remedied by introducing an outcross through an animal or animals known to be especially strong in the character or characters needing strengthening. This having been accomplished, the wise breeder will return to the original closebreeding or linebreeding program, realizing full well the limitations of an outcrossing program.

Grading Up

Grading up is that system of breeding in which purebred sires of a given pure breed are mated to native or grade females. Its purpose is to develop uniformity and quality and to increase performance in the offspring.

Many breeders will continue to produce purebred stock. However, the vast majority of animals in the United States—probably more than 97 percent—are not eligible for registry. In general, however, because of the obvious merit of using well-bred sires, farm animals are sired by purebreds. In comparison with the breeding of purebreds, such a system requires less outlay of cash, and less experience on the part of the producer.

Naturally, the greatest single step toward improved quality and performance occurs in the first cross. The first generation of such a program results in offspring carrying 50% of the hereditary material of the purebred parent (or 50% of the "blood" of the purebred parent, as many stockmen speak of it). The next generation gives offspring carrying 75% of the "blood" of the purebred breed, and in subsequent generations the proportion of inheritance remaining from the original scrub females is halved with each cross. Later crosses usually increase quality and performance still more, though in less marked degree. After the third or fourth cross, the offspring compare very favorably with purebred stock in conformation, and only exceptionally good sires can bring about further improvement. This is especially so if the males used in grading up successive generations are derived from the same strain within a breed. High grade animals that are the offspring of several generations of outstanding purebred sires can be and often are superior to average or inferior purebreds.

Crossbreeding

Crossbreeding is the mating of animals of different breeds. In a broad sense, crossbreeding also includes the mating of purebred sires of one breed with high grade females of another breed.

Today, there is great interest in crossbreeding cattle, and increased research is under way on the subject. Crossbreeding is being used by cattlemen to (1) increase productivity over straightbreds, because of the resulting hybrid vigor or heterosis, just as is being done by commercial corn and poultry producers; (2) produce commercial cattle with a desired combination of traits not available in any one breed; and (3) produce foundation stock for developing new breeds.

The motivating forces back of increased crossbreeding in cattle are: (1) more artificial insemination in beef cattle, thereby simplifying the rotation of bulls of different breeds; and (2) the necessity for cattlemen to become more efficient in order to meet their competition, from both inside and outside the cattle industry.

Fig. 5-35. Champion Red Angus X Hereford crossbred feeder steer calves. (Courtesy, Sally Forbes, Beckton Stock Farm, Sheridan, Wyo.)

Crossbreeding will play an increasing role in the production of market cattle in the future, because it offers the following advantages:

1. *Hybrid vigor in reproduction, survival, early growth rate, and maternal ability*—Crossbreeding results in hybrid vigor because the desirable genes from both breeds are combined and the undesirable genes from each tend to be overshadowed as recessives. That is to say, there has been an inevitable, though small, amount of inbreeding in all purebreds during the period of the last 100 to 150 years. This has been partly intentional and partly due to geographical limitations upon the free exchange of breeding stock from one part of the country to another. As a result of this slight degree of inbreeding, there has been a slow but rather constant increase in homozygosis within each of the pure breeds of livestock. Most of the factors fixed in the homozygous state are desirable; but inevitably some undesirable genes have probably been fixed, resulting in lowered vigor, slower growth rate, less ability to live, etc.

Theoretically, then, crossbreeding should be an aid in relegating these undesirable genes to a recessive position and in allowing more dominant genes to express themselves. Practical observation and experiments indicate that this does occur in crossbreeding.

An example of choosing a breed to maximize heterosis among the British breeds is found in the Hereford, Angus, and Shorthorn crossbred data. Since the Shorthorn resembles the Angus more than the Hereford in several traits, the Hereford X Angus and the Hereford X Shorthorn calves have excelled Shorthorn X Angus calves in growth traits.

2. *Opportunity to introduce new desired genes and incorporate them in the cowherd and market animal at a faster rate than by selection within a breed*—An example of introducing new genes at a fast rate for milk production would be crossing a dairy breed with a beef breed then selecting females from within the crossbred foundation for the future cowherd. Some of the American-created breeds were formed by using similar techniques; no doubt other new breeds will be developed in this manner in the future.

3. *Breed complementation or combining the desirable traits of two or more breeds to achieve a more desirable combination in the cow herd or the market animals than may be available in one breed*—The Angus X Charolais represents such a cross; the offspring combine the high carcass grade of the Angus and the high cutability of the Charolais.

4. *Opportunity to get hybrid vigor expressed in the cow*—Except for a two-breed cross, crossbreeding offers an opportunity to have hybrid vigor expressed by the cow. This enhances the cow's ability to "rough it," conceive, give birth, and nurse her calf well.

Many other examples of all factors could be cited. It should be noted, however, that the magnitude of the advantage of all these factors—achieving the 15 to 25 percent potential immediate increase in beef yield per cow unit through continuous crossbreeding compared to continuous straight breeding—depends upon the following:

1. *Making wide crosses*—The wider the cross, the greater the heterosis.

2. *Selecting breeds that are complementary*—A crossbreeding program should involve breeds that possess the favorable expression of traits desired in the crossbred offspring that will be produced.

3. *Using high performing stock*—Once a crossbreeding program is initiated, further genetic improvement is primarily dependent upon the use of high-indexing production-tested bulls.

4. *Following a sound crossbreeding system*—For a continuous high expression of heterosis and maximum beef output per cow, a sound system of crossbreeding must be followed. This should include the use of crossbred cows, for research clearly indicates that over one-half the higher profits from a crossbreeding program results therefrom.

CROSSBREEDING SYSTEMS

Without a planned breeding program, crossbreeding will almost inevitably end up with (1) a motley collection of females and progeny varying in type and color, and (2) minimum benefits from hybrid vigor or heterosis.

"Where do I go from here?" This is the question that many cattlemen frequently ask, almost frantically, after having heifers of breeding age sired by exotic bulls. Others get worried when they notice that calves out of their crossbred cows aren't doing so well as the first-cross calves. Of course, what these cattlemen really want to know is how they can maintain satisfactory hybrid vigor (heterosis) in animals when a herd is on a continuous crossbreeding program. They want to know how they can maintain 15 to 25 percent greater total efficiency in the crossbreds than the average of their parents; in production rate, calf livability, growth rate, and feed conversion.

Several different systems of crossbreeding may be used. Among them are the following:

1. *Two-breed cross*—This consists of mating purebred bulls to purebred or high-grade cows of another breed. An example would be using Angus bulls on Hereford cows, to give crossbred Angus X Hereford offspring— black baldies. This system of crossing has been used with success by cattlemen for many years.

TWO-BREED CROSS

ABERDEEN-ANGUS BULL

HEREFORD COW

F$_1$ GENERATION
ALL POLLED, BLACK BODIED
AND WHITE FACED

Fig. 5-36. Two-breed cross; Angus bull X Hereford cow, to produce black baldies.

In the two-breed cross, only the calves are crossbred—the breeding of the sires and dams remains the same. Hence, the two-breed cross imparts hybrid vigor only in the calf. On the average, it gives about an 8 to 10 percent increase in pounds of calf weaned per cow bred, plus another 2 to 3 percent advantage in rate of gain in the feedlot. In order to follow the two-breed cross indefinitely, the purebred females must be replaced with other purebreds sooner or later. They may either be purchased from another breeder or the breeder may want to produce his own purebred heifers within his own herd.

The two-breed cross is relatively simple. However, it has one major deficiency; it does not make use of the crossbred cow.

2. *Two-breed back cross or crisscross*—This system involves the use of bulls of breed A on cows of breed B, then back crossing the progeny to bulls of either breed A or B. The rotation is accomplished by using bulls of the breed least related to the particular set of cows. For example, if Charolais bulls are mated to Hereford cows, the crossbred Charolais X Hereford heifers could be retained and bred to either a Charolais or a Hereford bull. If Hereford bulls were used, the calves produced would be one-fourth Charolais and three-fourths Hereford. Later, if the heifers of this breeding are saved, they should be bred to a Charolais bull. The two-breed backcross results in about 67 percent of the maximum heterosis being attained in the crossbred calves. But since crossbred cows are used, overall performance should be a little better in pounds of calf weaned per cow bred than in the two-breed cross.

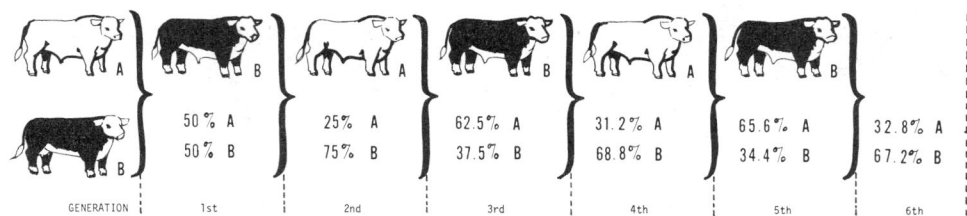

Fig. 5-37. Two-breed back cross or crisscross: Charolais bull X Hereford cow, thence female offspring bred to Hereford bull, thence female offspring bred to Charolais bull.

3. *Three-breed rotation cross*—This system calls for the selection of three breeds (e.g., breeds A, B, and C, which might represent Herefords, Brahmans, and Charolais), possessing the combination of maternal, growth, and carcass traits desired in the crossbred cows and the slaughter cattle produced. Crossbred females, selected for growth rate, are retained for breeding and bred to a purebred bull of one of the three breeds. Each new generation of crossbred females is retained for breeding and mated to a purebred bull until bulls of all three breeds have been used in rotation. Thus, such a system would operate as follows: Mate the existing B cow herd continuously to bulls of breed A; select crossbred heifers for growth rate and mate them

continuously to bulls of breed C; mate the selected C (AB) females to bulls of breed B. After the rotation of bulls from the three breeds is completed, the rotation of purebred sires begins all over again. Thus, mate the selected B X (ABC) females to bulls of breed A.

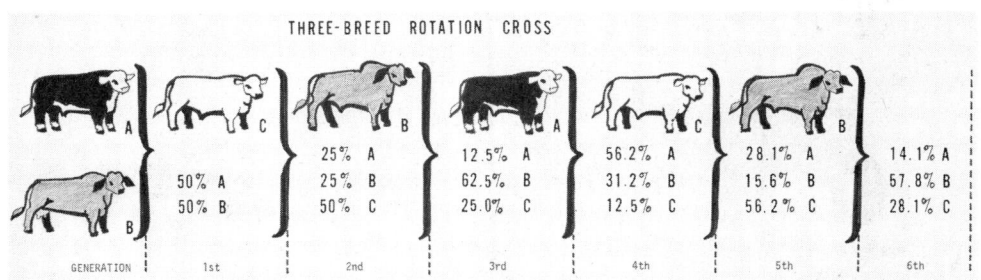

THREE-BREED ROTATION CROSS

50% A	25% A	12.5% A	56.2% A	28.1% A	14.1% A
50% B	25% B	62.5% B	31.2% B	15.6% B	57.8% B
	50% C	25.0% C	12.5% C	56.2% C	28.1% C
GENERATION 1st	2nd	3rd	4th	5th	6th

Fig. 5-38. Three-breed rotation cross; Hereford bull X Brahman female, thence female off-spring bred to Charolais bull, thence female offspring bred to Brahman bull.

Continue the same system indefinitely, always selecting the best performing crossbred females to be mated to the breed of sire in the program to which they are least related.

In addition to the genetic advantages of this system, commercial cattlemen select their own replacements; hence, the only outside cattle purchases are production tested bulls. The major disadvantage is that after the first four years it is necessary to maintain bulls of all three breeds simultaneously (unless A.I. is used).

A three-way rotation system results in about 87 percent of the maximum heterosis being attained.

4. *Three-breed fixed or static cross (terminal cross)*—In this system, crossbred cows from a two-breed cross (F_1s) are used as females and are mated to a bull of a third breed. All offspring from this cross are sold. When replacement females are needed, they are purchased. Thus, crossbred cows are used and crossbred calves with a fixed percentage of inheritance from three breeds are always produced.

In addition to realizing 100 percent of the maximum heterosis in each calf crop, this system allows the selection of maternal breeds to go into the production of the crossbred female and the selection of growthy breeds having desirable carcasses for the terminal cross sire breed. It allows the breeds to be used for their strong points without regard to some of their weaker points. A breeder can tailor-make the crossbred market animal, putting together in one animal desirable traits of several breeds. Such specification is not possible in the rotational system because all breeds contribute to maternal performance and calf performance.

The mechanics of this system consist in selecting three breeds for crossbreeding—two breeds (A and B) that will produce crossbred cows with

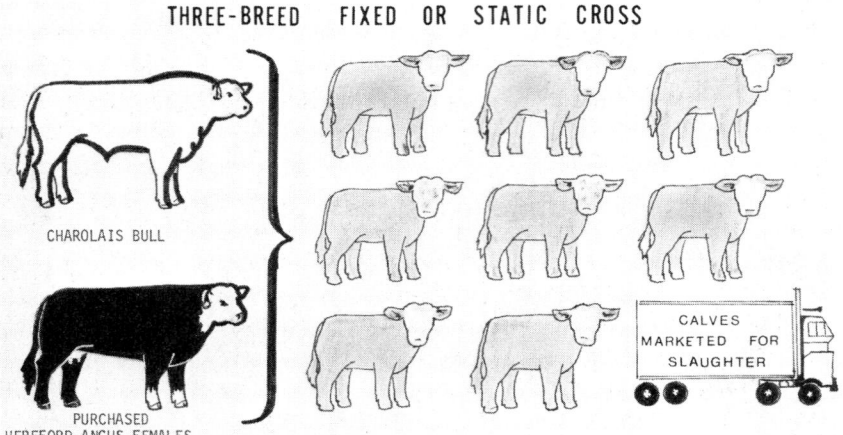

THREE-BREED FIXED OR STATIC CROSS

CHAROLAIS BULL

PURCHASED
HEREFORD-ANGUS FEMALES

CALVES
MARKETED FOR
SLAUGHTER

Fig. 5-39. Three-breed terminal cross; Charolais bull bred to crossbred Hereford-Angus female, with all the offspring marketed for slaughter.

outstanding maternal characteristics for fertility, milking ability, mothering ability, and adaptation. Select a third breed (breed C) with rapid, efficient, postweaning muscle growth rate. Breed C would be considered a "terminal" sire breed. All crossbred progeny of bull C are marketed for slaughter.

The problem with this system is the acquisition of production tested, crossbred (F_1) heifers for replacements in such a program, since all the three-way crosses are marketed. The system is perpetuated by having specialized multipliers produce crossbred (F_1) replacement females. Small operators (those with under 100 cows) might well use such a system where heifers are purchased along with the bulls. Large operators might well produce their own F_1 heifers in a specialized portion of their herd. Purebred breeders (seed stock breeders) would supply production selected terminal sires which the commercial cattleman would purchase for such a program.

Four, or more breeds may be used in a rotation crossbreeding system if the commercial producer so desires. However, the maximum hybrid vigor is usually realized with the three-breed cross.

Also, it is noteworthy that all of these crossbreeding systems rely upon the use of purebred bulls. Additionally, the two-breed cross relies on the use of purebred females, and the three-breed fixed or static cross relies on purebred females to produce the F_1 heifers necessary for the program.

Before going into a long-range crossbreeding program, the owner should know what is involved and what to expect. Plans should be developed before committing all available cattle and resources to a crossbreeding program. Consideration should be given to size of herd, markets, number of pastures, natural vs A.I. breeding, availability of breeding stock, etc. Sound management and sound selection of breeding stock based on performance, potential carcass characteristics, and overall productivity are just as important in crossbreeding as in any other breeding program.

Crossbreeding is no magic or "cure-all," but it will give a powerful assist to the pocketbook if properly used. Also—and this point bears emphasis—sound management and sound selection of breeding stock based on performance, potential carcass characteristics, and overall productivity are just as important in crossbreeding as in any other breeding program.

All crossbreeding programs involve some animal identification system so that (1) growth rate of heifers may be determined, with selection of replacements made on this basis; and (2) where more than one breed is involved, the cow herd can be sorted for assignment to specific sire breeds for mating. Unless A.I. is used, it is necessary to maintain bulls of whatever breeds are involved, along with separate breeding pastures for each sire breed. Also, it should be recognized that where bulls of two or three different breeds are used, the crossbred slaughter progeny will vary considerably in performance and carcass traits, because they will be sired by bulls of different breeds and be produced by cows of divergent breed backgrounds.

BULL SELECTION FOR CROSSBREEDING

Once the system of crossbreeding has been chosen, the most important recurring genetic decision is that of selecting bulls. It is just as important in commercial production to select superior performance bulls as it is in purebred breeding herds. This is so because the traits of production and product (see Table 5-5) being highly heritable can be transmitted directly from parent to commercial offspring. In order intelligently to buy bulls, the commercial cattleman must at least know where his herd stands in terms of average performance for the production and product traits. Then, he can select bulls strong in the weaker points of his herd and get bulls which, on the average, will improve his herd. The performance test of bulls to a year of age, plus possibly a sib carcass test, is an adequate estimate of breeding value for the commercial producer. On the average, the performance test will predict breeding value for the group, but for a particular individual it is not so good. Thus, the average performance of 10 yearling bulls will predict the average performance of their calves very well, but just which of the 10 is really the best bull is not well estimated. Of course, commercial producers using A.I. can afford to use progeny tested bulls available to them.

The selection of breeding herds from which to select commercial bulls is made difficult since only a fraction of herd differences are genetic—that is, the genetic differences between herds is probably less than 40 percent. On the other hand, research evidence suggests that for the production and product traits, the heritability of within herd differences among animals treated alike is between 40 and 50 percent. Thus, so long as the better herds are being patronized, selection of the best performing bulls within a herd is a safer bet than spending too much time deciding on which herd from which to buy bulls.

CROSSBRED BULLS

Cattlemen are sometimes tempted to use hybrid bulls as herd sires, particularly hybrids sired by high priced exotics. This inclination is under-standable, since the performance of hybrid bulls, as individuals, is often superior to either parent breed. But, before putting a hybrid bull with cows, some cautions are in order.

Unlike improved production resulting from selection, advantages from hybrid vigor are not transmitted from parent to progeny. Hence, to the extent to which superior performance is due to hybrid vigor, a hybrid bull will not breed true. If the hybrid bull is out of purebreds which have been selected for superior performance, this portion of his inherited superiority may be transmitted to his progeny. However, individual performance of hybrids is a less accurate indicator of their breeding value than is the performance of purebreds.

Work at the Experiment Station, Miles City, Montana indicates that there may be some gain from the use of crossbred bulls over straightbreds, in increased fertility, vigor, and livability of the calves. But the disadvantages of using a crossbred bull outweigh the advantages. Among the disadvantages are the following:

1. *Hybrid vigor may mask the true breeding worth of a hybrid bull*— The sires effect on profitability is basically indirect through the performance of his progeny. Thus, it is important that sires be accurately selected for the characters desired in their progeny. Hybrid vigor is not transmitted; hence, it may mask the true breeding worth of a hybrid bull. In other words, selection of purebreds is expected to be more effective.

2. *The progeny of hybrid sires tend to be more variable*—Since hybrid sires are less prepotent, their progeny will tend to be more variable in all measures of performance. Also, variation in color and conformation will tend to be more evident. As a result of this lack of uniformity, the market price of their progeny will be lower, especially when they are sold as feeders.

3. *Crossbred bulls have less effect on performance than crossbred cows*—Cows affect offspring through milk production and mothering ability; hence, they affect the performance of offspring more than the sire.

4. *The likelihood that crossbred bulls will be produced by breeders who haven't the best cattle*—A major problem in considering crossbred bulls is that there is a strong likelihood that they will be produced by breeders who haven't the best cattle from which to produce bulls.

Despite the above, there are special situations in which crossbred bulls may be considered. Two of these circumstances are:

1. *For the creation of a new breed*—The use of crossbred bulls is neces-sary in the creation of new breeds especially adapted to certain conditions. For example, the Santa Gertrudis breed of cattle, a breed derived from ⅝ Shorthorn and ⅜ Brahman, was developed to meet a need in the hot, dry, insect-infested area of the Southwest. Experienced cattlemen of the area will vouch for the fact that this is a practical example of a planned system of crossbreeding which has high utility value under the environmental condi-

tions common to the country. Also, crossbred Charolais bulls were used extensively and successfully for many years in the United States, during the formation of the breed by grading up. Still other examples of crossbreeding in the creation of breeds may be cited, including the breeding up to one of the new exotics in which purebred sires may be too expensive or scarce.

2. *For coping with harsh environmental conditions*—Under certain conditions, it may be desirable to incorporate hybrid vigor and adaptability in the bull, as well as in the brood cow, in order to cope with harsh environmental conditions. For this reason, Gulf Coast cattlemen often prefer bulls that have ⅛ to ¼ Brahman breeding.

From the above, it may be concluded that, except for special circumstances, crossbred bulls should not be used. They cannot be counted on to be herd improvers like comparable purebreds.

NOT ALL HYBRIDS EXCEL PUREBREDS

Cattlemen can learn from chickenmen when it comes to breeding, for the breeding of chickens has passed through the total presently known systems. In fact, each method has been, and still is, being used successfully.

The vast majority of chickens in America today are hybrids of one form or another—they're either strain crosses, breed crosses, or crosses between inbred lines. But they're exceptions!

Despite the fact that hybrids are widely used as commercial layers, it is noteworthy that egg-laying tests show that purebred Single Comb White Leghorns compete on even terms with hybrids under test conditions. Certainly, the hybrids are equal to the purebreds, but the point is that they do not excel them. The same principle applies to purebred vs crossbred cattle, *provided* the purebred cattle reach the pinnacle enjoyed by White Leghorns.

In broiler production, the main objective is the improvement of growth rate to eight weeks of age, although improvement in other economic factors is sought. Generally, growth rate and hybrid vigor are obtained by systematic matings that may involve crossing different breeds, different strains of the same breed, or the crossing of inbred lines. Most of the strains used as sires trace their ancestry to the broad-breasted Cornish breed. But there is some question whether heterosis, as obtained through hybrid breeding, contributes substantially to broiler weight. It is noteworthy, for example, that the best purebred New Hampshire strains generally equal the most rapid gaining crosses. The latter point is of great significance to purebred cattle breeders.

Body conformation is especially important in turkeys, because they are marketed at heavier weights than broilers and their carcass is usually left whole rather than cut up. Since conformation, size, and color of turkeys are highly heritable, they have responded well to simple methods of breeding and selection. As a result, most turkeys are bred as purebreds, rather than crossbreds. Also, it should be of more than passing interest to cattlemen to know that in turkey breeding programs (1) selections are largely based on physical appearances (phenotype), and (2) mass-matings (in which a number of males are allowed to run with the entire flock of hens) are the common practice.

BUFFALO X CATTLE HYBRIDS[22]

From time to time, American buffalo (*Bos bison*) and domestic beef cattle (*Bos taurus*) have been crossed, in Canada and the United States. Out of such crosses have evolved Cattalo (cattle of less than ½ bison parentage), Beefalo (⅜ buffalo, ⅜ Charolais, and ¼ Hereford), and the American Breed (⅛ buffalo, ½ Brahman, ¼ Charolais, 1/16 Durham, and 1/16 Hereford). These breeds are variously extolled, on the basis of their adaptability to cold, snowy climates; ability to thrive on weeds, shrubs, and other vegetation which domestic cattle pass up; small birth weights (straight buffalo calves weigh only about 25 pounds at birth); and leaner and more flavorful meat.

Fig. 5-40. Cattalo (¼ buffalo-¾ domestic cattle) cow. The initial Cattalo breeding experiment was started by the Dominion Experimental Station, Scott, Saskatchewan, Canada, in 1915. The foundation herd consisted of 16 female and 4 male hybrids. (Courtesy, Research Station, Canada Department of Agriculture, Lethbridge, Alberta, Canada)

Pertinent information relative to the reproductive ability of the American buffalo (*Bos bison*) X domestic cattle (*Bos taurus*) hybrids follows:
 1. Bison and domestic cattle interbreed.

---

[22]Buffalo breeders are banded together in the National Buffalo Association, Box 995, Pierre, South Dakota 57501.

2. Fewer maternal calving losses occur when domestic bulls are used on bison cows, although the reciprocal mating may be made.

3. Half-buffalo bull calves (F_1 hybrids) show normal sexual behavior, but they are always sterile. The scrotum is held close to the body cavity, as in the bison.

4. The half-buffalo heifers (F_1 hybrids) are fertile.

5. A few backcross bull hybrids have produced semen containing some sperm.

6. Reproductive ability improves in both sexes of further generations as the percentage of domestic blood increases.

It is possible that animals carrying a small percentage of buffalo breeding may have a place under certain conditions. However, more scientific research on the subject is needed.

PRODUCTION TESTING BEEF CATTLE[23]

Nearly every state now has an approved Beef Cattle Improvement Association (BCIA) program. Also, most breed registry associations have established programs. The Beef Improvement Federation (BIF) is a composite organization, responsible for standardization and uniformity in program systems.

Production testing embraces both (1) performance testing, and (2) progeny testing. The distinction between and the relationship of these terms is set forth in the following definitions:

1. **Performance testing**—*is the practice of evaluating and selecting animals on the basis of their individual merit.*

2. **Progeny testing**—*is the practice of selecting animals on the basis of the merit of their progeny.* It is usually costly and can be justified only for bulls of outstanding merit. Emphasis on progeny testing should be on traits not measurable in the bull himself—such as carcass traits and maternal ability of offspring. Generally speaking, the cost of progeny testing can be justified only for selecting bulls to be used extensively in artificial insemination or in very top seedstock herds.

3. **Production testing**—*is a more inclusive term, including performance testing and/or progeny testing.*

Production testing involves the taking of accurate records rather than casual observation. Also, in order to be most effective, the accompanying selection must be based on characteristics of economic importance and high heritability (see Table 5-6), and an objective measure or "yardstick," such as pounds, should be placed upon each of the traits to be measured. Finally, those breeding animals that fail to meet the high standards set forth must be removed from the herd promptly and unflinchingly.

In comparison with chickens or even swine, production testing of beef cattle is slow, and, like most investigational work with large animals, it is likely to be expensive. Even so, in realization that such testing is absolutely

---

[23]In the preparation of this section, the author drew heavily from *Guidelines for Uniform Beef Improvement Programs*, Beef Improvement Federation Recommendations, April 1972.

TABLE 5-6

ECONOMICALLY IMPORTANT TRAITS IN BEEF CATTLE, AND THEIR HERITABILITY[1]

Economically Important Traits	Approximate Heritability of Character (%)	Comments
Calving interval (fertility)	10	Fertility is economically the most important trait in beef cattle. Without a calf being born, and born alive, cattle are self eliminating.
Birth weight	40	Birth weight is associated with calf survival. Also, it has a positive correlation of .39 with growth rate. Selecting for increased birth weight is generally avoided because of likely increased calving difficulty.
Weaning weight	30	Heavy weaning weight is important because: 1. It is indicative of the milking ability of the cow. 2. Gains made before weaning are cheaper than those made after weaning. 3. Those who sell calves at weaning usually make more profit due to the heavier weight available to sell.
Cow maternal ability	40	Mothering ability is important in beef cows, because it contributes to calf survival and weaning weight.
Feedlot gain	45	Daily rate of gain is important because: 1. It is highly correlated with efficiency of gain. 2. It makes for a shorter time in reaching market weight and condition, thereby effecting a saving in labor and making for a more rapid turnover in capital.
Pasture gain	30	Most beef animals spend a good part of their lives on grass; hence, pasture gain is important.
Efficiency of gain	40	Efficiency of feed conversion is expressed as pounds of feed intake per 100 lb of gain. It is seldom measured in performance and progeny tests, because a positive relationship exists between rate and efficiency of gain. Hence, selection for rate of gain automatically selects for efficiency of gain.
Final feedlot weight	60	Final feedlot weight is usually referred to as *weight per day of age*. It is generally computed at one year of age or at the end of the performance test. It is probably the most important measurement of the estimated value of a beef bull. It is composed of birth weight, weaning weight, and postweaning gain.

(Continued)

Footnote on last page of table.

Table 5-6 (Continued)

Economically Important Traits	Approximate Heritability of Character (%)	Comments
Conformation score: Weaning Slaughter	25 40	This score should be based on skeletal soundness and indications of carcass desirability. Structural soundness, especially of the feet and legs, is most important in breeding animals.
Carcass traits:		Quality of product and quantity of edible portion are the basic factors of carcass merit.
		Where breeding animals are involved, and are not to be slaughtered, carcass quality may be evaluated by either (1) ultrasonic measurements, or (2) the K^{40} counter. Ultrasonics can be used to measure rib eye area and outside fat cover. The K^{40} counter evaluates the entire animal; it provides an effective method of measuring the total lean content of the live animal. (See Figs. 5-41 and 5-42.)
Carcass grade	40	High carcass grade is important because it determines selling and eating quality. BIF recommends that USDA Yield Grade (Nos. 1 to 5) also be used as a basis for evaluation of carcasses.
Rib eye area	70	The rib eye (the large muscle which lies in the angle of the rib and vertebra) is indicative of the bred-in muscling of the entire carcass. Thus, a large area of rib eye is much sought.
Tenderness	60	Warner-Bratzler shear test and taste panel test are recommended as methods of measuring tenderness.
Fat thickness	45	Fat thickness is taken at the twelfth rib.
Cancer eye susceptibility	30	There is indication that susceptibility to cancer eye is hereditary.

¹The heritability figures used herein are after: Gregory, *Beef Cattle Breeding*, USDA Agriculture Information Bulletin No. 286 (Revised).

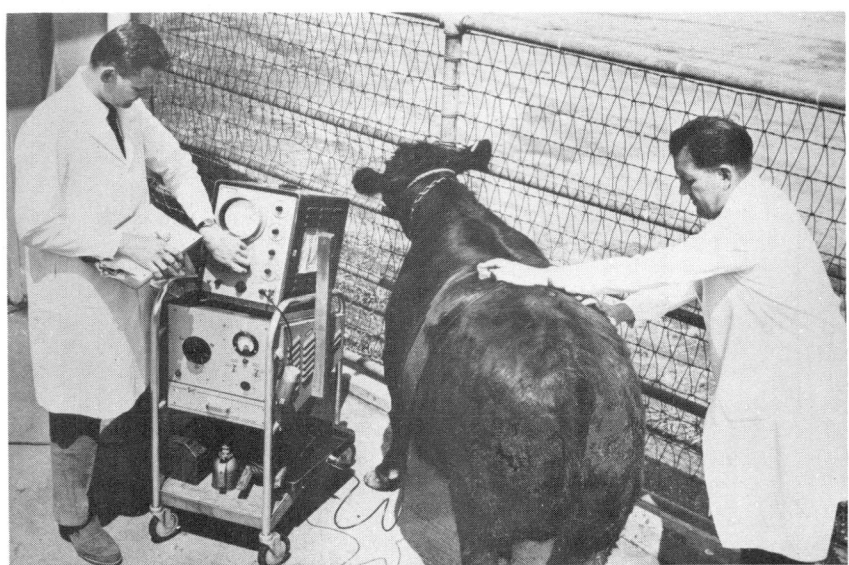

Fig. 5-41. High frequency sound (ultrasonics) being used to measure rib eye area and outside fat cover on live animal. (Courtesy, USDA)

Fig. 5-42. Taking K^{40} reading on a yearling bull. The K^{40} counter evaluates the entire live animal; it provides an effective method of measuring the total lean content of the live animal. (Courtesy, Martin Jorgensen, Jr., Ideal, S.D.)

necessary if maximum improvement is to be made, the progressive purebred beef cattle breeder will have his herd on production test.

Table 5-6 lists the economically important traits in beef cattle; those that contribute to both productive efficiency and desirability of product. Regularity of production, rapid growth, efficient use of feed, and carcass qualities preferred by consumers are economic traits of major importance. Performance testing offers beef cattle breeders a way of measuring differences among animals in heritable characters. Performance levels for these characters are related to ability to transmit desired traits to offspring.

Differences among animals in traits of economic value are, to a considerable entent, inherited differences. Thus, systematic measurement of these differences, the recording of the measurements, and the use of records in selection will increase the rate of genetic improvement in individual herds, and eventually in the breed and in the total cattle population.

Research has shown that when cattle are kept under nearly like conditions and their performance records are adjusted for known environmental differences—such as age, age of dam, and sex—genetically superior animals can be identified.

The rate of improvement in a herd, breed, and population is dependent on (1) the percentage of observed differences between animals that is due to heredity (heritability), (2) the difference between selected individuals and the average of the herd or group from which they come (selection differential), (3) the genetic association among traits upon which selection is based (genetic correlations), and (4) the average age of parents when the offspring are born (generation interval).

The essentials of effective record of performance programs are:

1. All animals of a given sex and age are given equal opportunity through uniform feeding and management.

2. Systematic written records are kept of important traits of economic value of all animals.

3. Records are adjusted for known sources of variation, such as age of dam, age of calf, and sex.

4. Records are used in selecting replacements (bulls and heifers) and in culling poor producers.

5. Nutritional program and management practices are practical and uniform for the entire herd and are similar to those where progeny of the herd are expected to perform.

• *Ratios*—Ratios—"Weight ratio," "gain ratio," and "conformation score ratio" are used to refer to the performance of an individual relative to the average of all animals of the same group. It is calculated as follows:

$$\frac{\text{Individual record}}{\text{Average of animals in group}} \times 100$$

It is a record of individual deviation from the group average expressed in terms of percentage. Thus, if an average test station gain was 3.00 lb per day, the gain ratio of a bull gaining 3.30 lb per day would be 110.

• *Farm or ranch preweaning testing program*—It is recommended that preweaning testing programs incorporate the following:

1. *Weaning weight and age*—It is recommended that the weaning weight be standardized to 205 days (with a maximum range of 160 to 250 days) of age and a mature dam equivalent.

It is recommended that 205-day weights be computed on the basis of average daily gains from birth to weaning. The following formula is used to provide an estimated 205-day weight, *unadjusted* for age of dam or sex of calf:

$$\text{Adjusted 205-day wt. (lb)} = \frac{\text{Actual weaning wt.} - 70}{\text{Weaning age in days}} \times 205 + 70$$

(The 70 is a constant; use actual birth wt. if available)

2. *Adjustment for age of dam*—To adjust for age of dam, the following adjustment factors are recommended:

> 2-year-old—multiply computed 205-day wt. by 1.15
> 3-year-old—multiply computed 205-day wt. by 1.10
> 4-year-old—multiply computed 205-day wt. by 1.05
> 5 through 10 year old—no adjustment
> 11-year-old and older—multiply computed 205-day wt. by 1.05

3. *Adjustment for sex of calf*—To adjust to a bull equivalent, the following adjustment factors are recommended:

Heifer wt.—multiply by 1.10
Steer wt.—multiply by 1.05

In a commercial herd where the majority of male calves are steers, records of heifer calves should be adjusted upward to a steer basis by multiplying by 1.05 and records of any bull calves should be adjusted to a steer basis by subtracting 5 percent or multiplying by .95.

Weaning weight ratios for sire, dam, and group should be calculated by dividing each individual's 205-day weight adjusted for (1) sex, and (2) age of dam by the average of the calves in its management code group, and expressing it as a percent of its management code group average. Sire, dam, and group summaries are made by averaging the weight ratios of the animals involved.

4. *Most Probable Producing Ability (MPPA)*—It is recommended that MPPA be included on Produce of Dam summaries and that ranking of dams be based on MPPA for 205-day weaning weight ratio. This is needed to compare dams which do not have the same number of calf records in their averages. For example, suppose six cows have the following records of production:

Cow	No. Calves	Average Weaning Weight Ratio	MPPA
A	1	85	94.0
B	2	88	93.2
C	4	90	92.7
D	3	110	106.7
E	4	112	108.8
F	1	115	106.0

MPPA is most helpful for identifying the lowest producing cows to be culled. In the example, cow A has the lowest lifetime average. However, this is for only a single calf for which environmental conditions or the calf's genetic potential for growth might have been below the average of what the cow would normally produce. One or more calves from cows B or C could also have had a record of 85 or less. All the 3 cows are probably low producers, but MPPA enables more accurate culling and in this example indicates that cows B and C are slightly lower producing cows than A.

MPPA for weaning weight ratio is computed by the following formula:

$$\text{MPPA} = \overline{H} + \frac{NR}{1 + (N\text{-}1)\,R}\,(\overline{C} - \overline{H})$$

Where $\overline{H}$ = 100, the herd average weaning weight ratio,

        N = the number of calves included in the cows average,

        R = .4, the repeatability factor for weaning weight ratio, and

        $\overline{C}$ = average for weaning weight ratio for all calves the cow has produced.

5. *Weaning conformation score*—The Beef Improvement Federation no longer recommends conformation scores. However, those breeders who desire to use conformation scores may well use the system formerly used by the BCIA, which is based on a scale of 3 to 17.

MPPA can also be computed for conformation score ratios using the same formula shown under point "4," Most Probable Producing Ability (MPPA) but with R = .3.

● *Farm or ranch postweaning program*—It is recommended that the postweaning program incorporate the following:

1. *Yearling weight and age*—The period between weaning and final weight should be at least 160 days. Yearling weights, taken at 365 days, or as long yearlings at 452 or 550 days of age, are particularly important because of their high heritability and high genetic association with efficiency of gain and pounds of retail trimmed boneless beef produced. Adjusted 365-day weights are recommended where animals on test are fed a high concentrate ration starting at weaning time. For bulls and heifers that are grown out more slowly, a long yearling weight may be used, with the test period being approximately 247 or 345 days postweaning.

Yearling weights should be computed and reported separately for each sex. Also, they should be adjusted for age of dam. For the latter reason, it is desirable to add postweaning gains in a 160-day postweaning period to 205-day weaning weight adjusted for age of dam to arrive at 365-day weight. The following formula is recommended:

$$\text{Adjusted 365 day wt.} = \frac{\text{actual final wt.} - \text{(minus) actual weaning wt.}}{\text{number of days between wts.}} \times 160$$
$$+ \quad \text{weaning wt. (205 days) adjusted for age of dam}$$

Adjusted long yearling weight (452 or 550 days) for each sex should be computed in the same manner as adjusted 365 day weight, using the following formula:

$$\text{Adjusted 550 day wt.} = \frac{\text{actual final wt.} - \text{(minus) actual weaning wt.}}{\text{number of days between weights}} \times 345$$
$$+ \quad \text{weaning wt. (205 days) adjusted for age of dam}$$

Of course, 247 would be substituted for 345 in the equation to compute 452-day weight.

2. *Weight ratios*—Weight ratios for either adjusted 365-day weight (yearling), adjusted 452-day weight or adjusted 550-day weight (long yearling) should be computed separately for each sex-management code group. Weight ratios should be reported for individual animals listed separately for each sex-management code group for ease of ranking individuals of each sex in making selections.

Sire and group summaries for yearling weight ratio should be computed as

$$\frac{\overline{W}_{ug} + \overline{P}_{sg}}{\overline{W}_{u} + \overline{P}_{s}} \times 100$$

Where W_{ug} = the sire progeny group average 205-day adjusted weight for all calves weaned, P_{sg} = the average 160 postweaning gain (247 or 345 for 452 or 550 day weights respectively) and W_u and P_s are the same as above.

3. *Final conformation score*—Final conformation scores taken at the end of the postweaning test can also be made on the conformation score based on a scale of 3 to 17.

● *Central testing stations—Central testing stations are locations where animals are assembled from several herds to evaluate differences in some performance traits under uniform conditions.* Central testing stations are used for: (1) comparing individual performance of potential seed-stock herd sires to similar animals from other herds; (2) comparing bulls being readied for sale to commercial producers; (3) finishing steers or heifers scheduled for slaughter as part of progeny test programs for growth and carcass traits; (4) as an educational tool to acquaint breeders with record of performance; and (5) estimating genetic differences between herds of between sire progenies in gaining ability, feed conversion, conformation, and carcass characteristics.

1. *Procedure*—The following procedures and policies are recommended for central testing of bulls:

a. Age of calves at time of delivery to test stations should be at least 180 days and not more than 305 days.

b. Herds from which bulls are consigned should be on herd testing programs for preweaning and postweaning performance. Calves should have completed the weaning phase of the performance records program and the following information should be submitted to the test station:

Sire, dam, birth date, actual weaning weight and date, adjusted 205-day weight, within herd weaning weight ratio (based on average of all bull calves in same weaning season and management group) and the number of calves making up this average.

c. There should be an adjustment or warm-up period of 21 days or *more* immediately prior to the test period.

d. The length of test should be 140 days or more.

e. Initial and final test weights should be an average of 2 full weights taken on different days.

f. All bulls sold in a test sale should be examined by a competent veterinarian for reproductive and structural soundness.

g. Test rations will vary according to locally available feeds and test objectives. Feeding should be ad lib. Rations between 60 and 70 percent total digestible nutrients (TDN) should be adequate for the expression of genetic differences in growth. The lower end of this range should result in few health problems and less excessive fattening.

h. Sire group testing of bulls is more desirable than individual testing because it provides more information to the breeder and to the prospective buyers.

2. *Test station reports*—Test station reports may include—

a. *140-day average daily gain and gain ratio*—These are the most important figures in test station results, because they measure growth during the period when the bulls are together under test conditions. The gain ratio is obtained by dividing the individual's gain by the test group average and multiplying by 100. A ratio of 100 means that the bull is average in his group, 115 means that he is 15% above the average, 90 means that he is 10% below the average.

b. *Weaning weights and within-herd weaning weight ratios*—This is the best available measure of the dam's milk production; hence, it is desirable to have a weaning weight above the average of the herd in which the calf was produced. Weaning weights provide good comparisons of bulls which come from the same herd but are less useful for comparing bulls from different herds.

c. *The 365-day adjusted weight and 365-day weight ratio*—This combines adjusted weaning weight and postweaning gain into one composite measurement. The 365-day weight ratio is the best measure for comparing growth of calves from the same herd. It is very highly heritable (around 60 percent). However, among bulls in a central test, care must be exercised in using this measurement, because the weaning weight portion was not made under comparable conditions.

d. *Efficiency of feed conversion*—This is expressed as pounds of feed per 100 pounds of gain. It is difficult to measure. Most tests do not attempt to get *individual* feed conversion because it would require individual feeding. Fortunately, growth rate and gain per unit of feed are highly correlated. It is estimated that selection for gain alone will result in 80 percent as much improvement in gain per unit of feed as selection directly for low feed requirement.

e. *Conformation score or grade*—This measure is optional among test stations. If it is made, it should be based strictly on skeletal soundness and indications of carcass desirability (including carcass weight *and* cutability).

f. *Rations*—Test rations among test stations vary widely, especially in level of energy. This variation causes some differences in the average daily gains of different tests.

● *Beef carcass evaluation—Carcass evaluation is the technique by which the components of quality and the components of quantity are measured.*

1. *Quality* refers to the overall palatability of the edible portion of the carcass. The USDA Quality Grades are Prime, Choice, Good, Standard, Commercial, Utility, Cutter and Canner. The grades are determined by visually evaluating certain carcass characteristics. These characteristics (excluding conformation) are maturity, marbling, texture of lean, color of lean, and firmness of lean. Once determined, the final grade should be reported by one third of a grade.

2. *Quantity* is the amount of salable meat the carcass will yield. It is recommended that USDA Yield Grades be used as a basis for evaluating carcass quality.

There are 5 USDA Yield Grades, numbered 1 through 5. Yield Grade 1 carcasses have the highest yields of retail cuts; Yield Grade 5 the lowest. The USDA Yield Grades are based on 4 factors:

a. Hot carcass weight.
b. Rib eye area at the twelfth rib.
c. Fat thickness at the twelfth rib.
d. Estimated percent kidney, pelvic, and heart fat.

The Yield Grade can be expressed in whole numbers from 1 to 5 or in tenths of the grade. The Yield Grade can also be expressed as a percentage. This percentage estimates the percent trimmed boneless retail cuts from the *round, loin, rib,* and *chuck.* For example, a carcass will have a Yield Grade 2.0 whether it is 2.0 or 2.9. A 3.9 Yield Grade indicates that a carcass is one-tenth better than a 4.0; however, it is still a Yield Grade 3.0. Yield Grades should be expressed to a tenth of a grade.

This percentage figure is commonly referred to as cutability. Various cutability figures correspond to Yield Grades, for example:

Yield Grade	Cutability
	(%)
1.0	54.6
1.5	53.5
2.0	52.3
2.5	51.2
3.0	50.0
3.5	48.9
4.0	47.7
4.5	46.6
5.0	45.4
5.5	44.3

The formula for calculating percent cutability is:

Percent cutability = 51.34 − 5.784 (single thickness of fat over *longissimus dorsi* in inches) − .462 (estimated percent kidney, pelvic and heart fat) + 0.740 (area *longissimus dorsi* in square inches) − 0.0093 (hot carcass weight in pounds).

Record Forms

A prerequisite for any production data is that each animal be positively identified—by means of ear notches, ear tags, or tattoos. For purebred breeders, who must use a system of animal identification anyway, this does not constitute an additional detail. But the taking of weights and grades does require additional time and labor—an expenditure which is highly worthwhile, however.

Fig. 5-43. Production records necessitate that each animal be individually identified. Note ear tag on this calf. (Courtesy, Martin Jorgensen, Jr., Ideal, S.D.)

In order not to be burdensome, the record forms should be relatively simple. Also, they should be in a form that will permit easy summarization—for example, the record of one cow should be on one sheet if possible. Suggested record forms are shown in Figs. 5-44 and 5-45.

Information on the productivity of *close relatives* (the sire and the dam and the brothers and sisters) can supplement that on the animal itself and thus be a distinct aid in selection. The production records of more distant relatives are of little significance, because, individually, due to the sampling nature of inheritance, they contribute only a few genes to an animal many generations removed.

GET OF SIRE RECORD

Calf Crop for Year of _____

Sex of Get[1] _____

Owner and Address _____

Sire's Name _____   Reg. No. _____

Date of Birth _____

		Calf Data						Yearling Data					Dam Data				Remarks
Herd No. of Calf	Date of Birth	Weaning Date	Weaning Age in Days	Weight in Lb	Daily Gain from Birth Weight, Lb	Adj. 205-Day Weaning Weight, Lb	Weanling Weight Ratio[2]	Confor-mation Score	Date Weighed	Weight, Lb	Wt. Adj. to days	Yr. Wt. Ratio[2]	Confor-mation Score	Herd No.	Age This Year	Mature Weight, Lb	Confor-mation Score	
Totals																		
Averages																		

[1]One sheet should be used to record all the bull calves and another sheet to record all the heifer calves by the same sire.

[2]Ratio calculated as follows:

$$\frac{\text{Individual record}}{\text{Av. of all calves on same farm and same season}} \times 100$$

Fig. 5-44. Get of Sire Record.

INDIVIDUAL COW RECORD

Tattoo _____ Reg. No. _____

Name _____

Bred by _____

Purchased from _____

Birth Date _____

Birth Wt., Lb _____

Sire _____

Address _____

Weaning Wt., Lb _____ Age _____ Conf. Score _____

Purchase Date _____ Price, $ _____

Yearling Wt., Lb _____ Age _____ Conf. Score _____

Disposition _____ Price, $ _____

Two Year Wt., Lb _____ Age _____ Conf. Score _____

Reason for Disposal _____

Av. Daily Gain Weaning to 1 yr., Lb _____

Feed Efficiency _____ lb feed/100 lb gain

Dam _____ Date _____

Temperament _____

Faults & Abnormalities _____

PRODUCE OF DAM RECORD

Calf Data

Birth Date	Sex	Tattoo	Sire	Birth Wt., Lb	Vigor at Birth[1]	Weaning Age Days	Weaning Wt., Lb Act.	Weaning Wt., Lb 205 day Adj.	Weaning Wt. Ratio[2]	Weaning Cond.	Conf. Score

Yearling Data / Production Testing

Date	Yr. Wt., Lb Adj.	Yr. Wt., Lb Days	Yearling Wt. Ratio[2]	Conf. Score[1]	Days on Feed	Av. Daily Gain, Lb	Gain Ratio[2]	Lb Feed /100 Lb Gain	Disposition; Price Remarks

[1]0=dead at birth; 1=definitely undersized at birth; 2=unthrifty, definite indications of disorders; 3=moderately thrifty, slight indications of disorders; 4=thrifty, no signs of disorders, dry hair coat; 5=thrifty, no signs of disorders, sleek hair coat; 6=very large, healthy, and vigorous

[2]Ratio calculated as follows:

$$\frac{\text{Individual record}}{\text{Av. of all calves on same farm and same season}} \times 100$$

Horn Brand or Neck Chain No. _____

Fig. 5-45A. Individual Cow Record (see Fig. 5-45B for reverse side of record form).

IMMUNIZATION AND TEST RECORD

Date[1]	Immunizations						Health Tests						Remarks
	Blklg.	M. Edema	Bangs	Misc.	TB-Bangs	Johnes	Lepto.	Anaplas.	Vib.	Trich.	Misc.		

[1]Indicate vaccinations by check in appropriate column opposite date given; indicate test results by P (positive), N (negative), or S (suspect) opposite date of test.

GENERAL INFORMATION

Record all facts pertinent to the history of this cow, viz.: veterinary treatment (except immunizations), udder condition, mothering instinct, calving peculiarities, etc.

Date	Remarks

Fig. 5-45B. Individual Cow Record. This is the reverse side of the record form shown in Fig. 5-45A.

Records by Computer

Beef production is becoming more sophisticated. Today, much of the pencil work normally involved in production testing is being eliminated by computer. A number of good computer record programs are available on a modest charge basis, including some sponsored by breed registry associations. The American Hereford Association's program, known as "Total Performance Records Program," is the world's largest beef cattle performance testing program. One of the several computer printouts of the Total Performance Record Program is herewith reproduced as Fig. 5-46.

T.P.R. BREEDING VALUE SELECTION WORK SHEET — AMERICAN HEREFORD ASSOCIATION — REPORT BASED ON WEANING DATA — DATE JUNE 21, 1973 PAGE 1 — BREEDER BEST HEREFORDS — FILE NUMBER B033016

HERD IDENTITY NUMBER	SEX	MTH	DAY	YEAR	SIRE HERD IDENTITY NUMBER	DAM HERD IDENTITY NUMBER	AGE OF DAM	ADJUSTED WEIGHT	WEIGHT RATIO	PAT HALF SIBS NUMBER HEAD	PAT HALF SIBS AVERAGE RATIO	MAT HALF SIBS NUMBER HEAD	MAT HALF SIBS AVERAGE RATIO	PROGENY NUMBER HEAD	PROGENY AVERAGE RATIO	BREEDING VALUE RATIO	SELECTION DECISIONS
BREEDING VALUES FOR ALL BULL CALVES																	
73370	B	4	27	73	451	70892	3	604	1102	19	1005	1	1123			1047	
73215	B	4	04	73	505	66392	7	540	985	29	1036	5	1042			1025	
73089	B	3	16	73	21Z	69716	4	574	1086	9	942	2	1070			1020	
73401	B	4	30	73	21Z	68982	5	546	996	9	952	3	1091			998	
73001	B	3	12	73	451	62333	11	476	912	19	1015	9	961			969	
BREEDING VALUES FOR ALL HEIFER CALVES																	
73789	C	5	01	73	505	71914	2	578	1187	29	1029					1080	
73178	C	4	01	73	505	64111	9	465	955	29	1037	7	980			1000	
73142	C	3	31	73	451	67781	6	468	961	19	1013	4	1016			996	
73095	C	3	25	73	21Z	63242	10	438	899	9	962	8	930			935	
BREEDING VALUES FOR ALL MOTHER COWS																	
71914	C					210		540	1120	30	1022	1	1150	1	1187	1097	
70892	C					218		520	1076	28	1013	4	1000	2	1112	1066	
66392	C					210		480	1100	30	1022	3	1041	6	1052	1065	
69716	C					210		505	1070	30	1023	5	1021	3	1075	1064	
68982	C					218		490	1062	28	1013	2	1010	4	1067	1056	
67781	C					218		492	1090	28	1012	2	1036	5	1005	1030	
64111	C							PURCHASED DAM						8	977	980	
62333	C							PURCHASED DAM						10	956	963	
63242	C							PURCHASED DAM						9	927	931	
BREEDING VALUES FOR ALL HERD SIRES																	
505	B					210		610	1120	30	1022	5	1182	30	1034	1080	
451	B					218		621	1143	28	1010	1	1096	20	1010	1041	
21Z	B							PURCHASED SIRE						10	956	953	

Fig. 5-46. *Breeding value selection work sheet* used by the American Hereford Association in its Total Performance Records Program.

With each set of computed data returned to the breeder, the American Hereford Association includes a calculated *breeding value selection work sheet* (Fig. 5-46). This form is designed especially for within herd selection. As noted, a *breeding value ratio* is computed on (1) each offspring by sex and management code, (2) on each cow, and (3) on each sire within the herd represented in either calf performance data or yearling data.

Calculated breeding values are based on the individual's own record, the individual's paternal and maternal half-sibs, and the individual's progeny. The combinations of these sources of information, using tested and proven mathematical formulas, will more accurately predict the genetic potential and rank within the herd and should be used along with the individual's record in the final decisions.

How To Use Herd Records in Selection

Herd records have little value unless they are intelligently used in culling operations and in deciding upon herd replacements. Also, most cattlemen can and should use production records for purposes of estimating the rate of progress and for determining the relative emphasis to place on each trait.

APPRAISING PERCENT OF CHANGE IN CHARACTERS DUE TO (1) HEREDITY, AND (2) ENVIRONMENT

Cattlemen are well aware that there are differences in birth weight, in weaning weight, in daily rate of gain, in body type, etc. If those animals which excel in the desired traits would, in turn, transmit without loss these same improved qualities to their offspring, progress would be simple and rapid. Unfortunately, this is not the case. Such economically important traits are greatly affected by environment (by feeding, care, management, type of birth [singles vs twins] age, diseases, etc.). Thus, only part of the apparent improvement in certain animals is hereditary, and can be transmitted on to the next generation.

As would be expected, improvements due to environment are not inherited. This means that if most of the improvement in an economically important trait is due to an improved environment, the heritability of that trait will be low and little progress can be made through selection. On the other hand, if the trait is highly heritable, marked progress can be made through selection. Thus, body color in cattle—e.g. red, roan, and white—is a highly heritable trait, for environment appears to have little or no part in determining the difference between animals that are red and those that are roan or white. On the other hand, such a trait as condition or degree of finish is of low heritability because, for the most part, it is affected by environment (feed, care, management, etc.).

There is need, therefore, to know the approximate amount or percentage of change in each economically important trait which is due to heredity and the amount which is due to environment. Table 5-6 gives this information for beef cattle in terms of the approximate percentage heritability of each of the economically important traits. The heritability figures given therein are averages based on large numbers; thus, some variations from these may be expected in individual herds. Even though the heritability of many of the economically important traits listed in Table 5-6 is disappointingly small, it is gratifying to know that much of it is cumulative and permanent.

ESTIMATING RATE OF PROGRESS

For purposes of illustrating the way in which the heritability figures in Table 5-6 may be used in practical breeding operations, the following example is given:

In a certain beef cattle herd, the calf crop in a given year averages a weaning weight of 500 pounds, with a range of 400 to 700 pounds. There are available sufficient of the heavier weaning calves weighing 600 lb from which to select replacement breeding stock. What amount of this heavier weaning weight (100 lb above the average) is likely to be transmitted to the offspring of these heavier weaning calves?

Step by step, the answer to this question is secured as follows:

1. 600 − 500 = 100 lb, the amount by which the selected calves exceed the average from which they arose.

2. By referring to Table 5-6, it is found that weaning weight is 30% heritable. This means that 30% of the 100 lb can be expected to be due to the superior heredity of the stock saved as breeders, and that the other 70% is due to environment (feed, care, management, etc.).

3. 100 × 30% = 30 lb; which means that for weaning weight the stock saved for the breeding herd is 30 lb superior, genetically, to the stock from which it was selected.

4. 500 + 30 = 530 lb weaning weight; which is the expected performance of the next generation.

It is to be emphasized that the 530 pounds weaning weight is merely the expected performance. The actual outcome may be altered by environment (feed, care, management, etc.) and by chance. Also, it should be recognized that where the heritability of a trait is lower less progress can be made. The latter point explains why the degree to which a character is heritable has a very definite influence in the effectiveness of mass selection.

Using the heritability figures given in Table 5-6, and assuming certain herd records, the progress to be expected from one generation of selection in a given beef cattle herd might appear somewhat as summarized in Table 5-7.

Appraising Factors Influencing Rate of Progress

Cattlemen need to be informed relative to the factors which influence the rate of progress that can be made through selection. They are:

1. *The heritability of the character*—When heritability is high, much of that which is selected for will appear in the next generation, and marked improvement will be evident.

2. *The number of characters selected for at the same time*—The greater the number of characters selected for at the same time, the slower the progress in each. In other words, greater progress can be attained in one character if selection is made for it alone. For example, if selection of equal intensity is practiced for 4 independent traits, the progress in any one will be only one-half of that which would occur if only one trait were considered; whereas selection for 9 traits will reduce the progress in any one to one-third. This emphasizes the importance of limiting the traits in selection to those which have greatest importance as determined by economic value and heritability. At the same time, it is recognized that it is rarely possible to select for one trait only, and that income is usually dependent upon several traits.

3. *The genotypic and phenotypic correlation between traits*—The effectiveness of selection is lessened by (a) negative correlation between two de-

TABLE 5-7

ESTIMATING RATE OF PROGRESS IN A BEEF CATTLE HERD

Economically Important Traits	Average of Herd	Selected Individuals for Replacements	Average Selection Advantage	Heritability Percent	Expected Performance Next Generation
Calving interval					
fertility, days	365	330	35	10	361.5
Birth weight, lb	70	80	10	40	74
Weaning weight, lb ...	500	600	100	30	530
Feedlot gain, lb	2.5	3.0	0.5	45	2.725
Pasture gain, lb	2.0	2.5	0.5	30	2.15
Efficiency of gain,					
lb feed/cwt gain	800	700	100	40	760
Final feedlot weight ...	1,000	1,100	100	60	1,060
Conformation score:					
Weaning	14	16	2	25	14.5
Slaughter	13	15	2	40	13.8

sirable traits, or (b) positive correlation of desirable with undesirable traits.

4. *The amount of heritable variation measured in such specific units as pounds, inches, numbers, etc.*—If the amount of heritable variation—measured in such specific units as pounds, inches, or numbers—is small, the animals selected cannot vary much above the average of the entire herd, and progress will be slow. For example, there is much less spread, in pounds, in the birth weights of calves than in weaning weights. Therefore, more marked progress in selection can be made in the older weights than in birth weights, when measurements at each stage are in pounds.

5. *The accuracy of records and adherence to an ideal*—It is a well-established fact that a breeder who maintains accurate records and consistently selects toward a certain ideal or goal can make more rapid progress than one whose records are inaccurate and whose ideals change with fads and fancies.

6. *The number of available animals*—The greater the number of animals available from which to select, the greater the progress that can be made. In other words, for maximum progress, enough animals must be born and raised to permit rigid culling. For this reason, more rapid progress can be made with swine than with animals that have only one offspring, and more rapid progress can be made when a herd is either being maintained at the same numbers or reduced than when it is being increased in size.

7. *The age at which selection is made*—Progress is more rapid if selection is practiced at an early age. This is so because more of the productive life is ahead of the animal, and the opportunity for gain is then greatest.

8. *The generation interval*—Generation interval refers to the period of time required for parents to be succeeded by their offspring, from the standpoint of reproduction. The minimum generation interval of farm animals is about as follows: horses, 4 years; cattle, 3 years; sheep, 2 years; and swine, 1 year. In actual practice, the generation intervals are somewhat longer. By way of comparison, it is noteworthy that the average length of a human generation is 33 years.

Shorter generation intervals will result in greater progress per year, provided the same proportion of animals is retained after selection.

Usually it is possible to reduce the generation interval of sires, but it is not considered practical to reduce materially the generation interval of females. Thus, if progress is being made, the best young males should be superior to their sires. Then the advantage of this superiority can be gained by changing to new generations as quickly as possible. To this end, it is recommended that the breeder change to younger sires whenever their records equal or excel those of the older sires. In considering this procedure, it should be recognized, however, that it is very difficult to compare records made in different years or at different ages.

9. *The calibre of the sires*—Since a much smaller proportion of males than of females is normally saved for replacements, it follows that selection among the males can be more rigorous and that most of the genetic progress in a herd will be made from selection of males. Thus, if 2% of the males and 50% of the females in a given herd become parents, then about 75% of the hereditary gain from selection will result from the selection of males and 25% from the selection of females, provided their generation lengths are equal. If the generation lengths of males are shorter than the generation lengths of females, the proportion of hereditary gain due to the selection of males will be even greater.

DETERMINING RELATIVE EMPHASIS TO PLACE ON EACH CHARACTER

A replacement animal seldom excels in all of the economically important traits. The cattleman must decide, therefore, how much importance shall be given to each factor. Thus, the beef cattle producer will have to decide how much emphasis shall be placed on birth weight, how much on weaning weight, how much on daily rate of gain, how much on efficiency of feed utilization, and how much on body type and carcass evaluation.

Perhaps the relative emphasis to place on each trait should vary according to the circumstances. Under certain conditions, some characters may even be ignored. Among the factors determining the emphasis to place on each trait are the following:

1. *The economic importance of the trait to the producer*—Table 5-6 lists the economically important traits in cattle, and summarizes (see comments column) their importance to the producer.

By economic importance is meant their dollars and cents value. Thus, those traits which have the greatest effect on profits should receive the most attention.

2. *The heritability of the trait*—It stands to reason that the more highly heritable traits should receive higher priority than those which are less heritable, for more progress can be made thereby.

3. *The genetic correlation between traits*—One trait may be so strongly correlated with another that selection for one automatically selects for the other. For example, rate of gain and economy of gain in beef cattle are correlated to the extent that selection for rate of gain tends to select for the most

economical gains as well; thus, economy of gain may be largely disregarded if rate of gain is given strong consideration. Conversely, one trait may be negatively correlated with another so that selection for one automatically selects against the other.

4. *The amount of variation in each trait*—Obviously, if all animals were exactly alike in a given trait, there could be no selection for that trait. Likewise, if the amount of variation in a given trait is small, the selected animals cannot be very much above the average of the entire herd, and progress will be slow.

5. *The level of performance already attained*—If a herd has reached a satisfactory level of performance for a certain trait, there is not much need for further selection for that trait.

It should be recognized, however, that sufficient selection pressure should be exerted to maintain the desired excellence of a given trait; for once selection for many of the economic quantitative traits is relaxed, there is a tendency for the trait to regress rather rapidly toward the average of the breed. For simple quantitative traits (controlled by a single pair of genes), it may be possible to rid the herd of the undesired gene, following which selection against the trait could be dropped, except when adding outside animals to the herd.

TYPES OF SELECTION

Finally, the cattleman needs to use a type of selection which will result in maximum total progress over a period of several years or animal generations. The three common types are:

1. *Tandem selection*—*This refers to that type of selection in which there is selection for only one trait at a time until the desired improvement in that particular trait is reached,* following which selection is made for another trait, etc. This system makes it possible to make rapid improvement in the trait for which selection is being practiced, but it has two major disadvantages: (a) usually it is not possible to select for one trait only; and (b) generally income is dependent on several traits.

Tandem selection is recommended only in those rare herds where one trait only is primarily in need of improvement.

2. *Establishing minimum standards for each trait, and selecting simultaneously, but independently, for each trait*—This system, in which several of the most important traits are selected for simultaneously, is without doubt the most common system of selection. It involves establishing minimum standards for each trait and culling animals which fall below these standards. For example, it might be decided to cull all calves weighing less than 55 lb at birth, or weighing less than 375 lb at weaning, or gaining less than 1¼ lb daily, or requiring more than 900 lb of feed per 100 lb gain, or grading 12 or less. Of course, the minimum standards may have to vary from year to year if environmental factors change markedly (for example, if calves average light at weaning time due to a severe drought and poor pasture).

The chief weakness of this system is that an individual may be culled because of being faulty in one character only, even though he is well nigh ideal otherwise.

3. *Selection index—A selection index combines all important traits into one overall value or index.* Theoretically, a selection index provides a more desirable way in which to select for several traits than either (1) the tandem type, or (2) the method of establishing minimum standards for each character and selecting simultaneously, but independently, for each character.

Selection indexes are designed to accomplish the following:

a. To give emphasis to the different traits in keeping with their relative importance.

b. To balance the strong points against the weak points of each animal.

c. To obtain an overall total score for each animal, following which all animals can be ranked from best to poorest.

d. To assure a constant and objective degree of emphasis on each trait being considered, without any shifting of ideals from year to year.

e. To provide a convenient way in which to correct for environmental effects, such as age of dam, etc.

Despite their acknowledged virtues, selection indexes are not perfect. Among their weaknesses are the following:

a. Their use may result in covering up or masking certain bad faults or defects.

b. They do not allow for year to year differences.

c. Their accuracy is dependent upon (1) the correct evaluation of the net worth of the economic traits considered, (2) the correctness of the estimate of heritability of the traits, and (3) the genetic correlation between the traits; and these estimates are often difficult to make.

In practice, the selection index is best used as a partial guide or tool in the selection program. For example, it may be used to select twice as many animals as are needed for herd or flock replacements, and this number may then be reduced through rigid culling on the basis of a thorough visual inspection for those traits that are not in the index, which may include such things as quality, freedom from defects, and market type.

QUESTIONS FOR STUDY AND DISCUSSION

1. Why is cattle breeding less flexible in the hands of man than swine or sheep breeding?
2. What unique circumstances surrounded the founding of genetics by Mendel?
3. Under what conditions might a theoretically completely homozygous state in cattle be undesirable and unfortunate?
4. How can you determine whether a polled bull is pure for the polled character?
5. Give examples of monohybrid, dihybrid, and trihybrid crosses in cattle; and describe each (a) the first cross animals, and (b) the second cross animals, and give the ratio of the second crosses.
6. Why do not fraternal (dizygotic) twins look more alike?
7. Why do most cattlemen prefer single births to twins?

8. Explain how sex is determined.

9. When abnormal animals are born, what conditions tend to indicate each: (a) an hereditary defect, or (b) a nutritional deficiency?

10. Give the expected genetic picture of dwarfism of 100 offspring from matings of (a) carrier bulls X noncarrier cows, and (b) carrier bulls X carrier cows. What steps can be taken to get rid of dwarfism?

11. Is double muscling good or bad? Justify your answer.

12. The "sire is half the herd!" Is this an understatement or overstatement?

13. In order to make intelligent selections and breed progress, is it necessary to fit, stall-feed, or place animals in show condition; or may they be selected in their "work clothes" off the farm or ranch?

14. Define the word "complementary," and give an example of it.

15. What cattle traits respond well to selection? What traits usually show good response in hybrid vigor?

16. Challenge the following statements: In the future most breeds of cattle will either be (a) propagated as straightbreds, or (b) used for crossbreeding. The vast majority of straightbreds will be purebreds and production tested.

17. What system of breeding do you consider to be best adapted to your herd, or to a herd with which you are familiar? Justify your choice.

18. Crossbreeding will play an increasing role in the production of market cattle in the future because of its several advantages. List these advantages.

19. Under what circumstances would you recommend a three-breed fixed or static cross?

20. List the disadvantages of using a crossbred bull.

21. Cite examples in poultry that show that not all hybrids excel purebreds.

22. How would you go about performance testing a herd of beef cattle? List and discuss each step.

23. List the factors which should be considered in determining the relative emphasis to place on each trait in a beef cattle selection program.

24. If an average test station gain was 3.00 pounds per day, what would be the gain ratio of a bull gaining 3.30 pounds per day?

25. In beef carcass evaluation, what method and terms would you use in arriving at (a) quality, and (b) quantity?

26. What type of selection—(a) tandem, (b) establishing minimum culling levels, or (c) selection index—would you recommend and why?

SELECTED REFERENCES

Title of Publication	Author(s)	Publisher
Animal Breeding	A. L. Hagedoorn	Crosby Lockwood & Son, Ltd., London, England, 1950
Animal Breeding	L. M. Winters	John Wiley & Sons, Inc., New York N.Y., 1948
Animal Breeding Plans	J. L. Lush	Collegiate Press, Inc., Ames, Iowa, 1965
Breeding and Improvement of Farm Animals	V. A. Rice F. N. Andrews E. J. Warwick J. E. Legates	McGraw-Hill Book Co., Inc., New York, N.Y., 1967
Breeding Better Livestock	V. A. Rice F. N. Andrews E. J. Warwick	McGraw-Hill Book Co., Inc., New York, N.Y., 1953

(Continued)

Title of Publication	Author(s)	Publisher
Breeding Livestock Adapted to Unfavorable Environments	R. W. Phillips	Food and Agriculture Organization of the United Nations, Washington, D.C., 1949
Crossbreeding Beef Cattle, Series 2	M. Koger T. J. Cunha A. B. Warnick	University of Florida Press, Gainsville, Fla., 1973
Elements of Genetics, The	C. D. Carlington K. Mather	The Macmillan Co., New York, N.Y., 1950
Farm Animals	J. Hammond	Edward Arnold & Co., London, England, 1952
Genetic Resistance to Disease in Domestic Animals	F. B. Hutt	Comstock Publishing Assn., Cornell University Press, Ithaca, N.Y., 1958
Genetics and Animal Breeding	I. Johansson J. Rendel	Oliver and Boyd, Ltd., Edinburgh, Scotland, 1968
Genetics Is Easy, Fourth Revised Edition	P. Goldstein	Lantern Press, Inc., New York, N.Y., 1967
Genetics of Livestock Improvement	J. F. Lasley	Prentice Hall, Englewood Cliffs, N.J., 1972
How Life Begins	J. Power	Simon and Schuster, Inc., New York, N.Y., 1965
Improvement of Livestock	R. Bogart	The Macmillan Co., New York, N.Y., 1959
Lasater Philosophy of Cattle Raising, The	L. M. Lasater	Texas Western Press, The University of Texas, El Paso, Tex., 1972
Livestock Improvement, Fourth Edition	J. E. Nichols	Oliver and Boyd, Ltd., Edinburgh, Scotland, 1957
Modern Developments in Animal Breeding	I. Lerner H. P. Donald	Academic Press Inc., New York, N.Y., 1966
Principles of Genetics	I. H. Herskowitz	The Macmillan Company, New York, N.Y., 1973
Problems and Practices of American Cattlemen, Wash. Agr. Exp. Sta. Bull. 562	M. E. Ensminger M. W. Galgan W. L. Slocum	Washington State University, Pullman, Wash., 1955
Robert Bakewell— Pioneer Livestock Breeder	H. C. Pawson	Crosby Lockwood & Son, Ltd., London, England, 1957

CHAPTER 6

PHYSIOLOGY OF REPRODUCTION IN CATTLE

Contents   **Page**

Reproduction is the first and most important requisite of cattle breeding. Without a calf being born and born alive, the other economic traits are of academic interest only. Yet, 12 percent of the nation's cows never calve, and there is an appalling calf loss of 6 percent at birth.

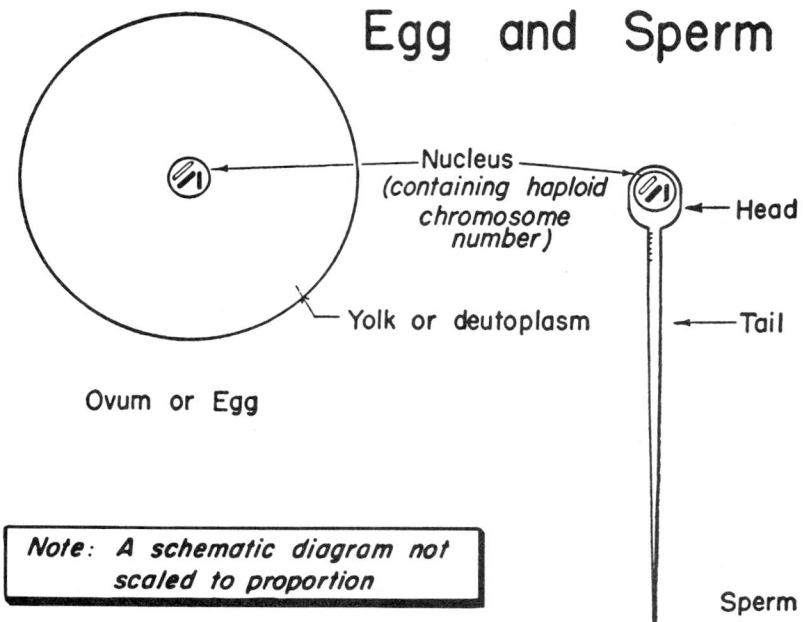

Fig. 6-1. Egg and sperm. The parent germ cells, the egg from the female and the sperm from the male, unite and transmit to the offspring all the characters that it will inherit. (Drawing by R. F. Johnson, Head, Dept. of Animal Science, Calif. Polytechnic State University, San Luis Obispo, Calif.)

Many outstanding individuals, and even whole families, are disappointments because they are either sterile or reproduce poorly. The subject of physiology of reproduction is, therefore, of great importance.

THE REPRODUCTIVE ORGANS OF THE BULL

The bull's functions in reproduction are: (1) to produce the male reproductive cells, the *sperm* or *spermatozoa;* and (2) to introduce sperm into the female reproductive tract at the proper time. In order that these functions may be fulfilled, cattlemen should have a clear understanding of the anatomy of the reproductive system of the bull and of the functions of each of its parts. Fig. 6-2 shows the reproductive organs of the bull. A description of each part follows:

1. *Scrotum*—This is a diverticulum of the abdomen, which encloses the testicles. Its chief function is thermoregulatory; to maintain the testicles at temperatures several degrees lower than that of the body proper.

2. *Testicles*—The testicles of the mature bull measure 4 to 5 inches in length and 2 to 3 inches in width. Their primary functions are the production of sperm and the male hormone, testosterone.

Once the animal reaches sexual maturity, sperm production in the seminiferous tubules—the glandular portion of the testicles, in which are situated the spermatogonia (sperm-producing cells)—is a continuous process. Around and between the seminiferous tubules are the interstitial cells which produce testosterone or androgen.

A sperm is a small (less than 1/500 inch in length), tadpole-shaped living entity, in which the head contains the unit of inheritance and the tail provides the means of locomotion.

Testosterone is essential for the development and function of male reproductive organs, male characteristics, and sexual drive.

Cryptorchids are males one or both of whose testicles have not descended to the scrotum. The undescended testicle(s) is usually sterile because of the high temperature in the abdomen.

The testicles communicate through the inguinal canal with the pelvic cavity, where accessory organs and glands are located. A weakness of the inguinal canal sometimes allows part of the vicera to pass out into the scrotum—a condition called *scrotal hernia.*

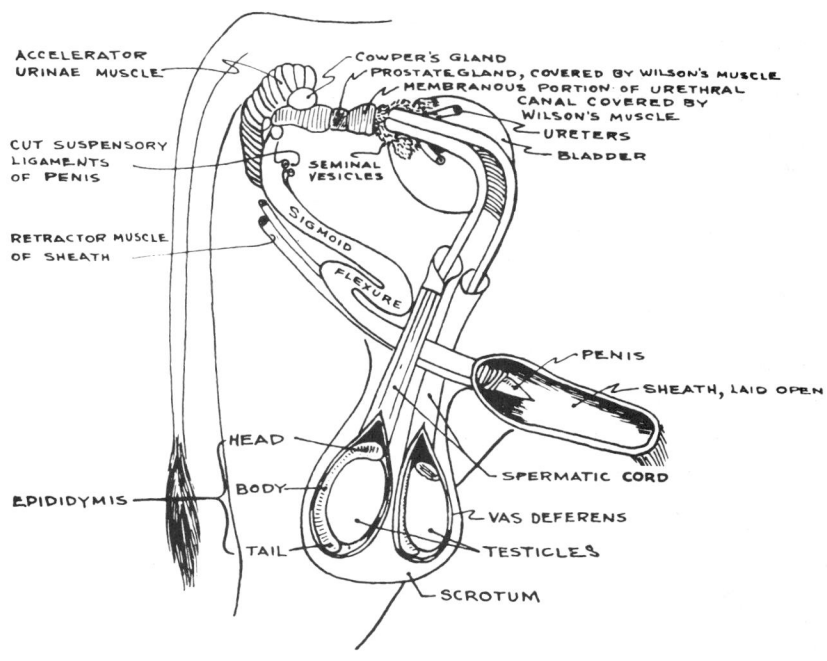

Fig. 6-2. Diagram of the reproductive organs of the bull.

3. *Epididymis*—The efferent ducts of each testis unite into one duct, thus forming the epidiymis. This long and greatly coiled tube consists of three parts:

a. *The head*—Consisting of several tubules which are grouped into lobules.

b. *The body*—The part of the epididymis which passes down along the sides of the testis.

c. *The tail*—The part located at the bottom of the testis.

The epididymis has four functions; namely, (1) as a passage way for sperm from the seminiferous tubules; (2) the storage of sperm; (3) the secretion of a fluid which probably nourishes the sperm; and (4) the maturation or ripening of the sperm.

4. *Vas deferens (Ductus deferens)*—This slender tube, which is lined with ciliated cells, leads from the tail of the epididymis to the pelvic part of the urethra. Its primary function is to move sperm into the urethra at the time of ejaculation.

The cutting or closing off of the vas deferens, known as *vasectomy,* is the most usual operation performed to produce sterility, where sterility without castration is desired.

5. *Spermatic cord*—The vas deferens—together with the longitudinal strands of smooth muscle, blood vessels, and nerves; all encased in a fibrous sheath—make up the spermatic cord (two of them) which pass up through an opening in the abdominal wall, the inguinal canal, into the pelvic cavity.

6. *Seminal vesicles (or vesicula seminalis)*—These compact glandular organs with a lobulated surface flank the vas deferens near its point of termination. They are the largest of the accessory glands of reproduction in the male. In the mature bull, they measure 4 to 5 inches in length and 2 inches in width at their largest part and are located in the pelvic cavity.

7. *Prostate gland*—This gland is located at the neck of the bladder, surrounding or nearly surrounding the urethra and ventral to the rectum. The secretion of the prostate gland is thick and rich in proteins and salts. It is alkaline, and it has a characteristic odor.

It cleanses the urethra prior to and during ejaculation, and provides bulk and a suitable medium for the transport of sperm.

8. *Cowper's gland (Bulbo-urethral gland)*—These two glands, which are deeply inbedded in muscular tissue in the bull, are located on either side of the urethra in the pelvic region. They communicate with the urethra by means of a number of small ducts.

It is thought that these glands produce an alkaline secretion for the purpose of neutralizing or cleansing the urethra prior to the passage of semen.

9. *Urethra*— This is a long tube which extends from the bladder to the glans penis. The vas deferens and seminal vesicle open to the urethra close to its point of origin.

The urethra serves for the passage of both urine and semen.

10. *Penis*—This is the bull's organ of copulation. Also, it conveys urine to the exterior. It is composed essentially of erectile tissue, which, at the

times of erection, becomes gorged with blood. Just behind the scrotum it forms an S-shaped curve, known as the sigmoid flexure, which allows for extension of the penis during erection. In the mature bull, the erected penis is about 3 feet long.

In total, the reproductive organs of the bull are designed to produce semen and to convey it to the female at the time of mating. The semen consists of two parts; namely (1) the sperm which are produced by the testes, and (2) the liquid portion, or semen plasma, which is secreted by the seminiferous tubules, the epididymis, the vas deferens, the seminal vesicles, the prostate, and the Cowper's glands. Actually, the sperm make up only a small portion of the ejaculate. On the average, at the time of each service, a bull ejaculates 4 to 7 cubic centimeters of semen, containing about 6 to 10 billion sperm. The sperm concentration is about 1½ billion per cubic centimeter.

THE REPRODUCTIVE ORGANS OF THE COW

The cow's function in reproduction are: (1) to produce the female reproductive cells, the *eggs* or *ova*; (2) to develop the new individual, the *embryo*, in the uterus; (3) to expel the fully developed young at time of *birth* or *parturition;* and (4) to produce milk for the nourishment of the young. Actually, the part played by the cow in the generative process is much more complicated than that of the bull. It is imperative, therefore, that the modern cattle producer have a full understanding of the anatomy of the reproductive organs of the cow and the functions of each part. Fig. 6-3 shows the reproductive organs of the cow, and a description of each part follows:

1. *Ovaries*—The two irregular-shaped ovaries of the cow are supported by a structure called the broad ligament, and lie rather loosely in the abdominal cavity 16 to 18 inches from the vulvar orifice. They average about 1½ inches in length, 1 inch in width, and ½ inch in thickness.

The ovaries have three functions: (1) to produce the female reproductive cells, the *eggs* or *ova*; (2) to secrete the female sex hormones, *estrogen* and *progesterone* (the latter is the hormone of the corpus luteum); and (3) to form the *corpora lutea*. The ovaries may alternate somewhat irregularly in the performance of these functions.

The ovaries differ from the testes in that generally only one egg is produced at intervals, toward the end of the heat period or after heat symptoms have passed. Each miniature egg is contained in a sac, called *Graafian follicle*, a large number of which are scattered throughout the ovary. (It has been estimated that there are more than 75,000 follicles in the ovaries of a heifer calf at birth.) Generally, the follicles remain in an unchanged state until the advent of puberty, at which time some of them begin to enlarge through an increase in the follicular liquid within. Toward the end of heat, a follicle ruptures and discharges an egg, which process is known as *ovulation*. As soon as the egg is released, the corpus luteum makes its appearance. This corpus luteum secretes a hormone called progesterone, which (1) acts on the uterus so that it implants and nourishes the embryo, (2) prevents other eggs

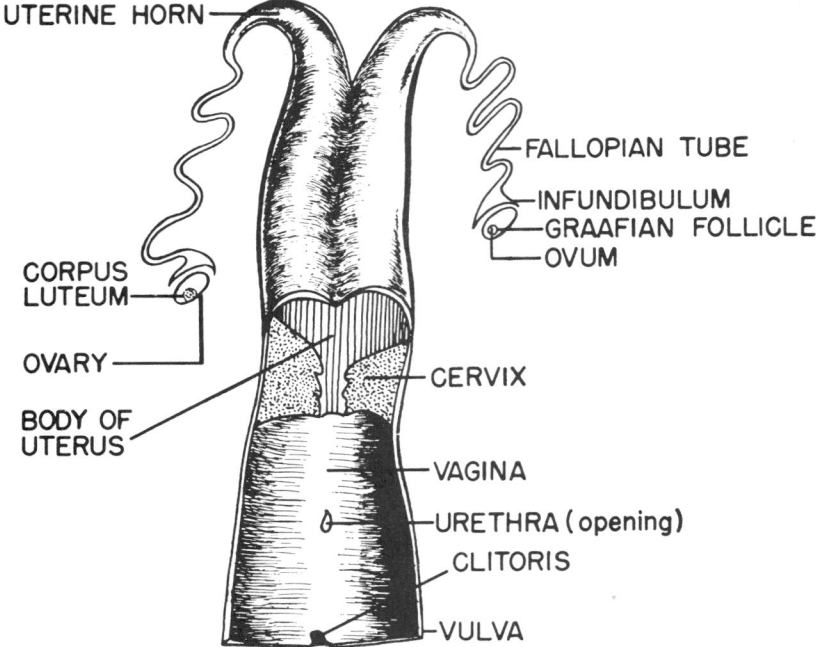

UTERINE HORN

FALLOPIAN TUBE

INFUNDIBULUM
GRAAFIAN FOLLICLE
OVUM

CORPUS
LUTEUM

OVARY

CERVIX

BODY OF
UTERUS

VAGINA

URETHRA (opening)

CLITORIS

VULVA

Fig. 6-3. The reproductive organs of the cow.

from maturing and keeps the animal from coming in heat during pregnancy, (3) maintains the animal in a pregnant condition, and (4) assists the female hormone in the development of the mammary glands. If the egg is not fertilized, however, the corpus luteum atrophies and allows a new follicle to ripen and a new heat to appear. Occasionally the corpus luteum fails to atrophy at the normal time, thus inducing temporary sterility. This persistent corpus luteum can be squeezed out.

The egg-containing follicles also secrete into the blood the female sex hormone, estrogen. Estrogen is necessary for the development of the female reproductive system, for the mating behavior or heat of the female, for the development of the mammary glands, and for the development of the secondary sex characteristics, or femininity, in the cow.

From the standpoint of the practical cattle breeder, the ripening of the first Graafian follicle in a heifer generally coincides with puberty, and this marks the beginning of reproduction.

2. *Fallopian tubes (or oviducts)*—These small, cilia-lined tubes or ducts lead from the ovaries to the horns of the uterus. They are about 5 to 6 inches long in the cow, and the end of each tube nearest the ovary, called *infundibulum,* flares out like a funnel. They are not attached to the ovaries but lie so close to them that they seldom fail to catch the released eggs.

At ovulation, the egg passes into the infundibulum where, within a few minutes, the ciliary movement within the tube, assisted by the muscular movements of the tube itself, carries it down into the oviduct. If mating has

Ovary of Cow During Heat

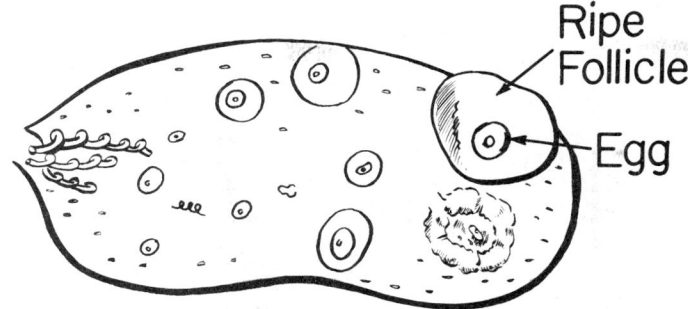

Ovary of Cow Not in Heat

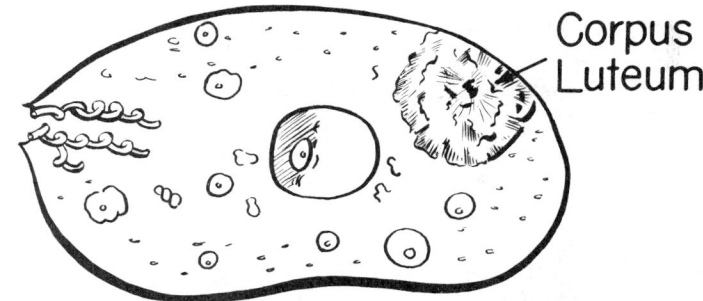

Fig. 6-4. Diagrammatic illustration of the ovary of the cow. Top: ovary of cow in heat. The ripe follicle secretes the estrogenic hormone responsible for the symptoms of heat. Bottom: ovary of cow not in heat. The corpus luteum is a glandular structure and secretes the hormone progesterone which maintains pregnancy in the pregnant cow. (Drawing by Steve Allured)

taken place, the union of the sperm and egg usually takes place in the upper third of the Fallopian tube. Thence the fertilized egg moves into the uterine horn. All this movement from ovary to the uterine horn takes place in 3 to 4 days.

3. *Uterus*—The uterus is the muscular sac, connecting the Fallopian tubes and the vagina, in which the fertilized egg attaches itself and develops until expelled from the body of the cow at the time of parturition. The uterus consists of the 2 horns (cornua), the body, and the neck (or cervix) of the womb. In the cow, the horns are about 15 inches long, the body about 1 ½ inches long, and the cervix about 4 inches long. In the mature cow, the uterus lies almost entirely within the abdominal cavity.

In the cow, the fetal membranes that surround the developing embryo are in contact with the lining of the uterus through buttons or cotyledons.

The thick, muscular, fold-containing portion of the uterus, known as the

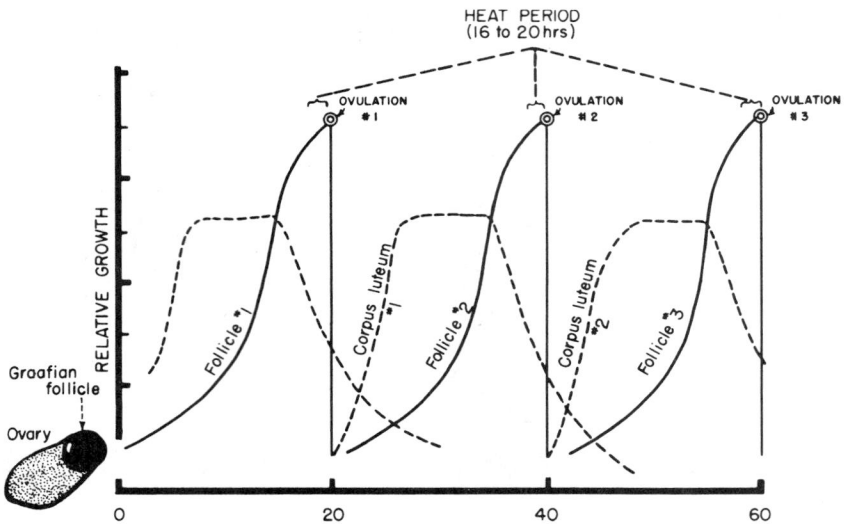

Fig. 6-5. Diagrammatic illustration of recurrence of the estrus cycle and the chain of events which takes place in the ovary of the nonpregnant cow. Continuous line shows the relative growth of the Graafian follicle. When maximum growth of the follicle is attained, rupture takes place and the ova is released (see double circles on top of diagram). At rupture, the size of the follicle reaches zero and the follicle is replaced by the corpus luteum (dotted line) which increases in size, reaches a maximum, then declines before the subsequent follicle starts to develop. Note that the rupture of the follicles coincides with the end of heat symptoms. The interval between two ovulations is 19 to 20 days on an average.

cervix, forms an effective seal at the posterior end of the uterus. Cells within the cervix secrete copious amounts of mucus, forming the cervical plug. This gelatine-like material is discharged just prior to the occurrence of a normal heat period.

4. *Vagina*—The vagina admits the penis of the bull at the time of service and receives the semen. At the time of birth, it expands and serves as the final passageway for the fetus. In the nonpregnant cow, the vagina is 10 to 12 inches in length, but it is somewhat longer in the pregnant animal.

5. *Clitoris*—The clitoris is the erectile and sensory organ of the female, which is homologous to the penis in the male. It is situated just inside the portion of the vulva farthest removed from the anus.

6. *Urethra*—The urine makes its exit through the opening of the urethra.

7. *Vulva (or urogenital sinus)*—The vulva is the external opening of both the urinary and genital tracts.

The reproductive system of the cow is regulated by a complex endocrine system. The functions of the reproductive organs and the occurrence of estrus, conception, pregnancy, parturition and lactation are all regulated and coordinated by the hormones of the pituitary, the ovarian follicle, the corpus luteum, and the placenta. The cyclic nature of these phenomena is shown in graphic form in Fig.6-5.

FERTILIZATION

Fertilization is the union of the male and female germ cells, sperm and ovum. The sperm are deposited in the vagina at the time of service and from there ascend the female reproductive tract. Under favorable conditions, they meet the egg and one of them fertilizes it in the upper part of the oviduct near the ovary.

In cows, fertilization is an all or none phenomenon, since only one ovum is ordinarily involved. Thus, the breeder's problem is to synchronize ovulation and insemination; to ensure that large numbers of vigorous, fresh

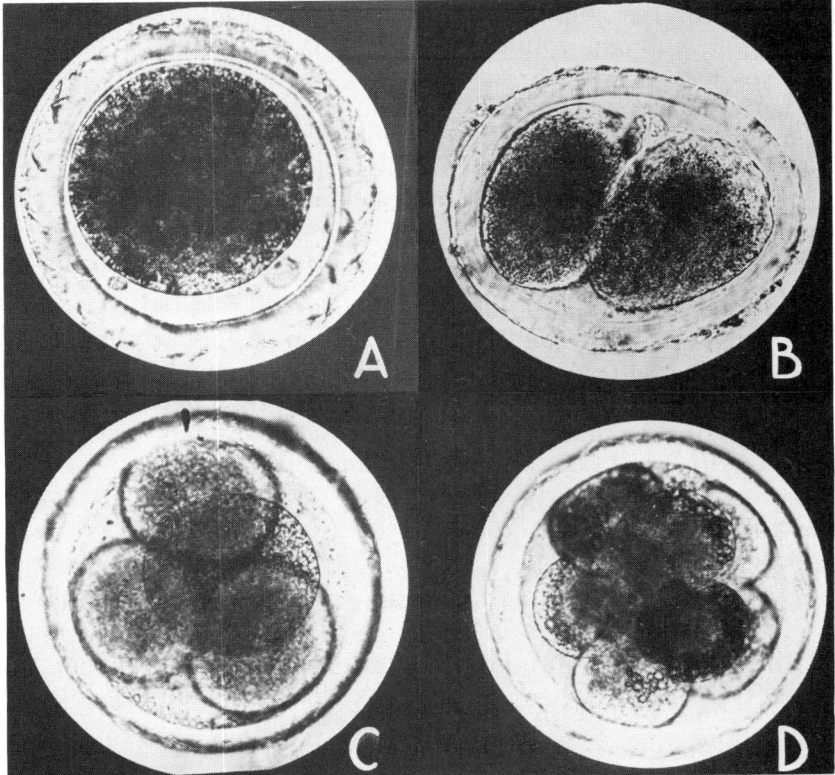

Fig. 6-6. Developmental changes in the egg at and after fertilization (these four stages all occur in the Fallopian tube).

A. A healthy egg in the one-cell stage. Several spermatozoa attempt to fertilize the egg, but only one sperm will be able to do so. A few may succeed but die in the zona pellucida (outer circle) where the tails of the dead sperms may be seen.

B. Fertilization is followed by the multiplication of cells. An egg in the 2-cell stage; this stage in the cow takes place 24 to 30 hours after fertilization, i.e. some 46 hours after the last symptoms of estrus.

C. The egg cleaves further (an increase in number of cells) without any increase in cellular mass. An egg in the 4-cell stage; this stage in the cow takes place 10 hours following stage "B."

D. An egg in the 8-cell stage; this stage in the cow takes place 16 hours after stage "C." After this stage the developing ovum migrates to the uterus where it will increase in cellular and protoplasmic mass.

(Courtesy, Dr. J. A. Laig, University of Bristol, Bristol, England)

sperm will be present in the Fallopian tubes at the time of ovulation. This is very difficult, because (1) there is no reliable way of predicting the length of heat or the time of ovulation (it is known that ovulation generally takes place toward the end of or following the heat period; however, it may occur during the heat period or as late as 36 hours after it); (2) like all biological phenomena, there is considerable individual variation; (3) the sperm cells of the bull live only 24 to 30 hours in the reproductive tract of the female; (4) an unfertilized egg will not live over about 6 hours, and (5) it may require less than 1 minute for sperm cells to ascend the female reproductive tract of a cow.

From the above, it is perfectly clear that a series of delicate time relationships must be met; that breeding must take place at the right time. For the maximum rate of conception, therefore, it is recommended that breeding be done the latter part of the heat period; but, since the duration of heat in cattle is very short (seldom exceeding 20 hours), to delay too long may result in the cow being out of heat when mating is attempted. Cows that are detected in heat in the morning should be bred during the afternoon of the same day, and those detected in the afternoon, should be bred late that evening or early the next morning.

NORMAL BREEDING HABITS OF COWS

In general, cattle that are bred when out on pasture or range are mated under environmental conditions approaching those which existed in nature prior to domestication. Less breeding trouble is generally encountered among such animals than among beef or dairy animals that are kept in confined conditions and under forced production.

Age of Puberty

The normal age of puberty of cattle is 8 to 12 months. It is recognized, however, that the age at which puberty is attained varies according to: (1) breeds, with the smaller breeds having an earlier onset of puberty than the larger, slower maturing ones; and (2) nutritional and environmental factors, with puberty occurring when animals have reached about one-third of their adult size.

Heat Periods

The period of duration of heat—that is, the time during which the cow will take the bull—is very short, usually not over 16 to 20 hours, although it may vary from about 6 to 30 hours. Cows tend to have a characteristic pattern of estrus behavior; for example, they come in heat during the morning hours, go out of heat in the evening or early part of night, and then ovulate approximately fourteen hours after the end of heat.

Females of all species bred near the end of the heat period are much more likely to conceive than if bred at any other time. The heat period recurs approximately at 21 day intervals, but it may vary from 19 to 23 days. In most cases, cows do not show signs of estrus until some 6 to 8 weeks after parturition, or in some instances even longer. Occasionally, an abnormal condition develops in cows that makes them remain in heat constantly. Such animals are known as nymphomaniacs.

SIGNS OF ESTRUS

It has been well said that, "The cow that stands is the cow in heat," for this is the best single indicator of the heat period. Also, cows in heat usually exhibit one or more of the following characteristic symptoms: (1) rough hair on tailhead; (2) mud marks on side when the ground is wet; (3) nervousness; (4) bawling; (5) frequent urination; (6) mucus on rump and tail; and (7) a moist and swollen vulva. Dry cows and heifers usually show a noticeable swelling or enlargement of the udder during estrus, whereas in lactating cows a rather sharp decrease in milk production is often noted. A day or two following estrus, a bloody discharge is sometimes present.

The subject of "Heat Detection Methods and Devices" is presented in Chapter 10 of this book.

Gestation Period

The average gestation period of cows is 283 days, or roughly about 9½ months. Though there may be considerable breed and individual variation in the length of the gestation period, it is estimated that ⅔ of all cows will calve between 278 and 288 days after breeding.

INDUCED CALVING (Shortened Gestation)

Instead of letting "nature take its course," scientists are now artificially shortening gestation. The objectives of induced early calving are: (1) lowering birth weight of calves, thereby lessening parturition difficulty; (2) predicting calving dates in order to pool labor and concentrate watching; and (3) gaining a longer period from calving until rebreeding.

Females that have passed the 269th day of pregnancy will calve within 24 to 72 hours if injected intramuscularly with an adrenal steroid. Experimental work indicates that such induced calving will result in (1) 5 to 8 days earlier than normal calving, and (2) 6 to 8 pounds lighter birth weight than calves carried to term. However, a higher incidence of retained placentas and lowered milk production accompany early calving. Antibiotics are indicated where the fetal membrane remains attached to the uterus longer than normal. Failure to expel the membranes after induced calving appears to have little effect on fertility as cows suffering this problem usually have no trouble breeding back.

FERTILITY IN BEEF CATTLE

Fertility refers to the ability of the male or female to produce viable germ cells capable of uniting with the germ cells of the opposite sex and of producing vigorous, living offspring. Fertility is lacking in very young animals, manifests itself first at puberty, increases for a time, then levels out, and finally recedes with the onset of senility. In cattle, as with other classes of farm animals, fertility is determined by heredity and environment.

In the wild state, when a female was served several times during her heat period, an annual conception rate of 90 to 100 percent was common rather than the exception. Aside from frequency of service, the outdoor exercise, vigor, good nutrition, and regular breeding habits and lack of contamination were conducive to conception. On the other hand, when handled under unnatural conditions in confinement, when the female is generally bred as soon as she starts to show signs of heat, it is not surprising that the conception rate of cattle is rarely higher than 88 percent, and is frequently much less.

Fig. 6-7. The goal of every cattleman is to have a 100 percent calf crop. (Courtesy, Tom Lasater, Lasater Ranch, Mattheson, Colo.)

Of course, the final test for fertility is whether young are produced, but unfortunately this test is both slow and expensive. Through evaluation of the quality of semen, it is possible to make a fairly satisfactory appraisal of the male's fertility; but no comparable measure of the female's relative fertility has yet been devised, although a pregnancy test may be made.

METHODS OF MATING

The two methods of mating beef cattle by natural service are: (1) hand mating; and (2) pasture mating. It has been found that 98 percent of the commercial cattlemen and 68 percent of the purebred breeders use pasture mating; the rest use hand mating.[1]

Hand Mating

In hand mating, the bull is kept separate from the cows at all times, except when an individual cow is to be bred and is turned in with him for

---

[1]Ensminger, M. E., M. W. Galgan, Problems and Practices of American Cattlemen, *Wash. Agr. Exp. Sta. Bull.* 562, 1955.

this purpose. As a rule, in hand mating, only a single service is allowed, the cow being removed immediately after service. In the breeding of purebred cattle, when breeding records are so important, this method is frequently followed. Hand mating allows for a more accurate check on whether the bull is settling the cows. It also permits a larger number of cows to be served by a bull, an especially important consideration with a proved sire.

Pasture Mating

In this system the bull is turned in with the herd, either throughout the entire year or during the breeding season. Even with pasture breeding, when it is desired to have the calves all come within a few weeks of each other thereby assuring more uniformity in size and offspring, the herd bull should be separated from the cows except during the breeding season. Uniformity in size is very important from the standpoint of marketing the calves advantageously. Also, by having the calves come as nearly as possible at one time, closer observation may be given the herd at the time of parturition.

Pasture breeding is most often followed with a commercial herd. As a rule, this system requires less labor, and there is less danger of missing cows when they are in heat. However, the convenience of pasture mating should not result in neglect to check whether the cows are being settled during the breeding season.

PREGNANCY TEST

Pregnancy tests can cut the wintering bill and make for increased profits. It is expensive business to over-winter a cow that will not produce a calf; necessitating feed, interest, labor, and other costs. Table 6-1 shows the additional income which may reasonably be expected from pregnancy testing a 300-cow herd.

Barren cows can usually be marketed satisfactorily following testing; most feeders and packers will actually pay a premium for cows known to be open. Where valuable purebred animals are involved, bred cows can be sold with a more certain guarantee of being safely in calf, and barren cows may be accorded special care or hormone treatment.

Absence of heat is not always a sign of pregnancy, but a positive diagnosis can be made. By about the second month in heifers and the third month in cows, the uterus becomes enlarged, especially in the pregnant horn, and drops into the abdominal cavity. An experienced technician can ascertain this sign of pregnancy by *feeling with the gloved hand through the rectum wall.* Application of this method depends upon the recognition of changes in tone, size, and location of the uterine horns and changes in the uterine arteries. This is the most common test of pregnancy. It is popular because it affords early diagnosis, and there is little hazard when performed by experienced operators. It is recommended that cows be pregnancy tested, by this method, about two months after the bulls have been removed. Palpation takes only a few seconds. The speed with which pregnancy is determined depends largely on management of the cows as they come through the chutes, stage of pregnancy, and the experience of the tester. As many as 800 head of cows, or more, can be palpated in a normal working day under ideal conditions.

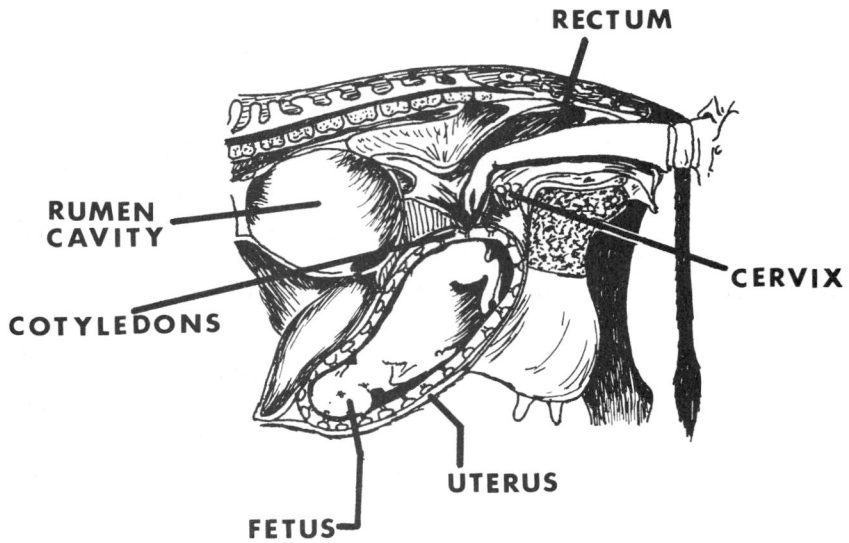

Fig. 6-8. Rectal method for determining pregnancy in the cow.

TABLE 6-1

VALUE OF PREGNANCY TESTING[1]

	With Pregnancy Testing			Without Pregnancy Testing		
	%	No.	Income	%	No.	Income
Calf crop	90	270		80	240	
Replacement heifers saved	17	51		12	36	
Calves for sale		219	$44,347.50[2]		204	$41,310.00[2]
Cull cows		45	11,340.00[3]		30	7,560.00[3]
Total income			$55.687.50			$48,870.00
Additional income from testing						$ 6,817.50

[1]300 cow herd, 2% death loss.
[2]450 lb x 45ᶜ.
[3]900 lb x 28ᶜ.

Other less-used tests for pregnancy in cows are:

1. *Abdominal ballottement* may be used from the fifth to the seventh months of pregnancy. This consists in feeling the fetus by the following techniques: (a) place the hand or fist against the abdomen in the lower right flank region; and (b) execute a short, vigorous, inward-upward thrust in this region and retain the hand in place. The hard fetus may be felt. Because of the amniotic fluid, the technique described above will make the fetus re-

cede, but it will fall back in place almost immediately.

2. *The fetal heart beat* can sometimes be detected after the sixth month of pregnancy, though this method is not as certain in the cow as in other classes of farm animals. Use of a stethoscope is preferred, though good results are sometimes secured by merely placing the ear against the right lower abdominal region and listening. The fetal heart beat can be distinguished from that of the mother because of its greater frequency and lighter and higher pitch.

3. *Fetal movements* can sometimes be observed through the abdominal wall during the latter half of pregnancy. This method of detecting pregnancy requires much patience. The observer simply must wait until voluntary movement of the fetus on the right side of the cow is observed. The practice of trying to induce movement of the fetus by allowing a very thirsty cow to take on a fill of cold water is cruel and is to be condemned.

CARE OF THE PREGNANT COW

The nutritive requirements of the pregnant cow are less rigorous than those during lactation. In general, pregnant cows should be provided as nearly year-round pasture as possible. During times of inclement weather or when deep snows or droughts make supplemental feeding necessary, dry roughages and silage are the common feeds. If produced on fertile soils, such forage will usually provide all the needed nutrients for reproduction. Further discussion of the nutritive needs of pregnant cows is contained in Chapters 8 and 19.

No shelter is necessary except during periods of inclement weather. Normally, the cows will prefer to run outdoors. This desire is to be encouraged—in order to provide exercise, fresh air, and sunshine. Where and when shelter is necessary, it should be neither elaborate nor expensive. An open shed facing away from the direction of prevailing winds is quite as satisfactory for the protection of dry cows as a warm bank barn with individual box stalls—and it is far less expensive. The chief requirements are that the shelter be tight overhead, that it be sufficiently deep to afford protection from inclement weather and remain dry (depths of 34 to 36 feet are preferred), that it is well drained, and that it is of sufficient size to allow the animals to move about and lie down in comfort.

CARE OF THE COW AT CALVING TIME

The careful and observant caretaker will be ever alert and make definite preparations for calving in ample time. It is especially important that first-calf heifers be watched at calving time, for frequently they will need some assistance. Older cows that habitually have trouble in parturition may well be culled from the herd.

Signs of Approaching Parturition

Perhaps the first sign of approaching parturition is a distended udder, which may be observed some weeks before calving time. Near the end of the gestation period, the content of the udder changes from a watery secretion to a thick, milky colostrum. As parturition approaches, there generally will be a marked shrinkage or falling away of the muscular parts in the region of the tail head and pin bones, together with a noticeable enlargement and swelling of the vulva.

The immediate indications that the cow is about to calve are extreme nervousness and uneasiness, separation from the rest of the herd, and muscular exertion and distress.

Preparation for Calving

At the time the signs of approaching parturition seem to indicate that the calf may be expected within a short time, arrangements for the place of calving should be completed.

During the seasons of the year when the weather is warm, the most natural and ideal place for calving is a clean open pasture away from other livestock. Hogs should not be allowed in the same place with the cow, for they are likely to injure or kill the young calf. They have even been known to injure the cow.

Under pasture conditions, there is decidedly less danger of either infection or mechanical injury to the cow and calf. In commercial range operations, it is common practice to ride the range more frequently at calving time. A better procedure consists of having a smaller pasture adjoining headquarters into which heavy springing cows are placed a few days before calving. With the added convenience of such an arrangement, the animals can be given more careful attention.

During inclement weather, the cow should be placed in a roomy (10 or 12 feet square), well-lighted, well-ventilated, comfortable box stall or maternity pen which should first be carefully cleaned, disinfected, and bedded for the occasion.

Normal Presentation[2]

Labor pains in a mild form usually start some hours before actual parturition. After a time, the water bag appears on the outside, usually increasing in size until it ruptures from the weight of its own contents. This is closely followed by the appearance of the amniotic bladder (the second water bag), with the fetus. With the rupture of the second water bag, the straining becomes more violent, and presentation soon follows. Most commonly in presentation, the front feet come first followed by the nose which is resting on them, then the shoulders, the middle, the hips, and finally the hind legs and feet.

With posterior presentation (hind feet first), there is likely to be difficulty in calving. Moreover, there is considerably more danger of having the calf suffocate through rupture of the umbilical cord and strangulation.

---

[2]Figs. 6-9, 6-10, and 6-11 reviewed by Dr. Frank Bracken DVM, Washington State University.

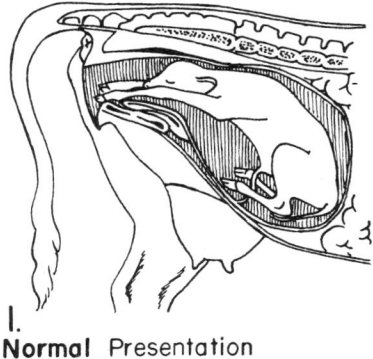

I.
Normal Presentation

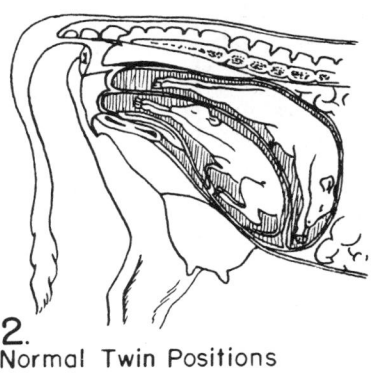

2.
Normal Twin Positions

Fig. 6-9. 1. Normal single presentation; the back of the fetus is directly toward that of the mother, the forelegs are extended toward the vulva, and the head rests between the forelegs. If it is necessary to render assistance, apply ropes above the ankle joints and pull alternately downward on each leg as the cow strains.

2. Normal twin positions. If delivery does not proceed normally, this is a case for a veterinarian.

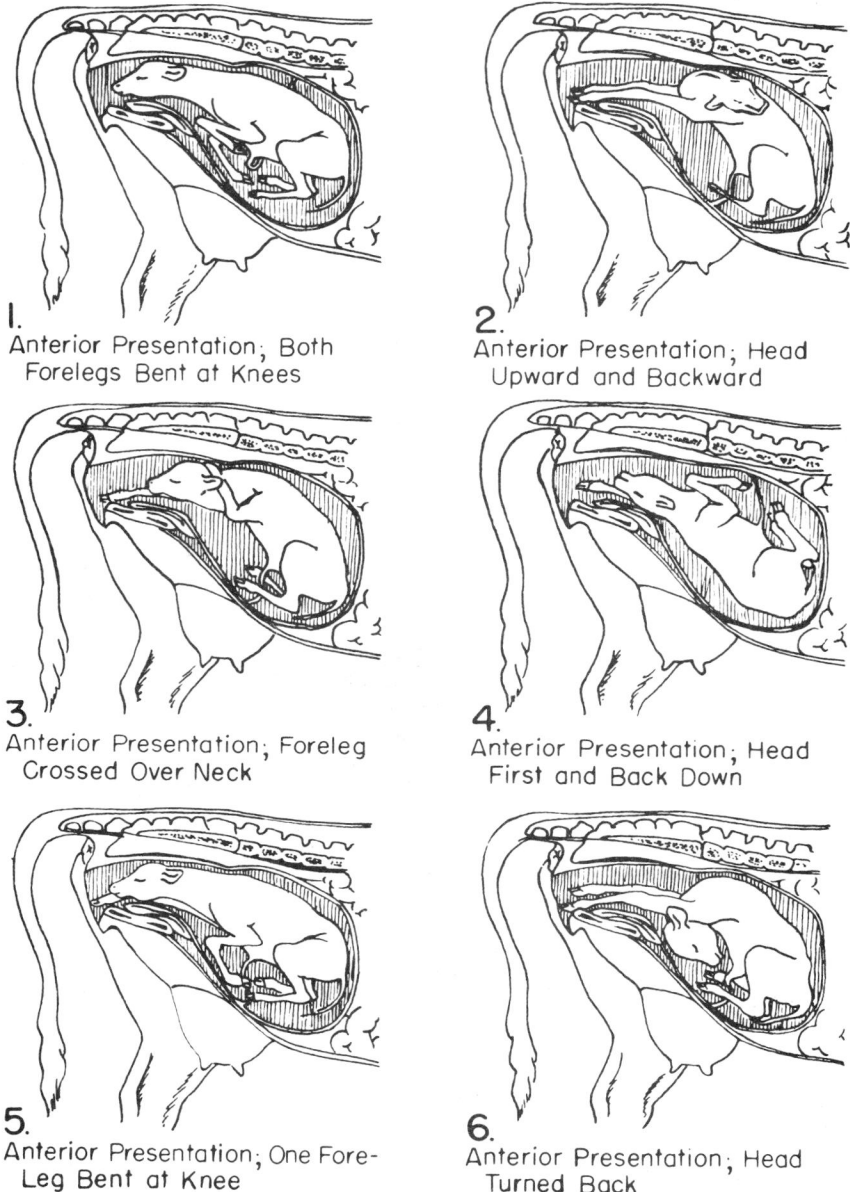

1.
Anterior Presentation; Both
Forelegs Bent at Knees

2.
Anterior Presentation; Head
Upward and Backward

3.
Anterior Presentation; Foreleg
Crossed Over Neck

4.
Anterior Presentation; Head
First and Back Down

5.
Anterior Presentation; One Fore-
Leg Bent at Knee

6.
Anterior Presentation; Head
Turned Back

Fig. 6-10. Some abnormal presentations with suggestions for correction:
1. Extend the legs so that delivery can be accomplished.
2. Push back the fetus, which will often bring the head into its normal position.
3. Grasp the crossed leg a little above the ankle, raise it, draw it to the proper side, and extend it in the genital canal.
4. Rotate the fetus, extend the forelegs and deliver by traction.
5. Lift the head and draw up and rope the leg so that it does not slip back again.
6. Rope the forelegs and then push them forward; place the head in normal position.

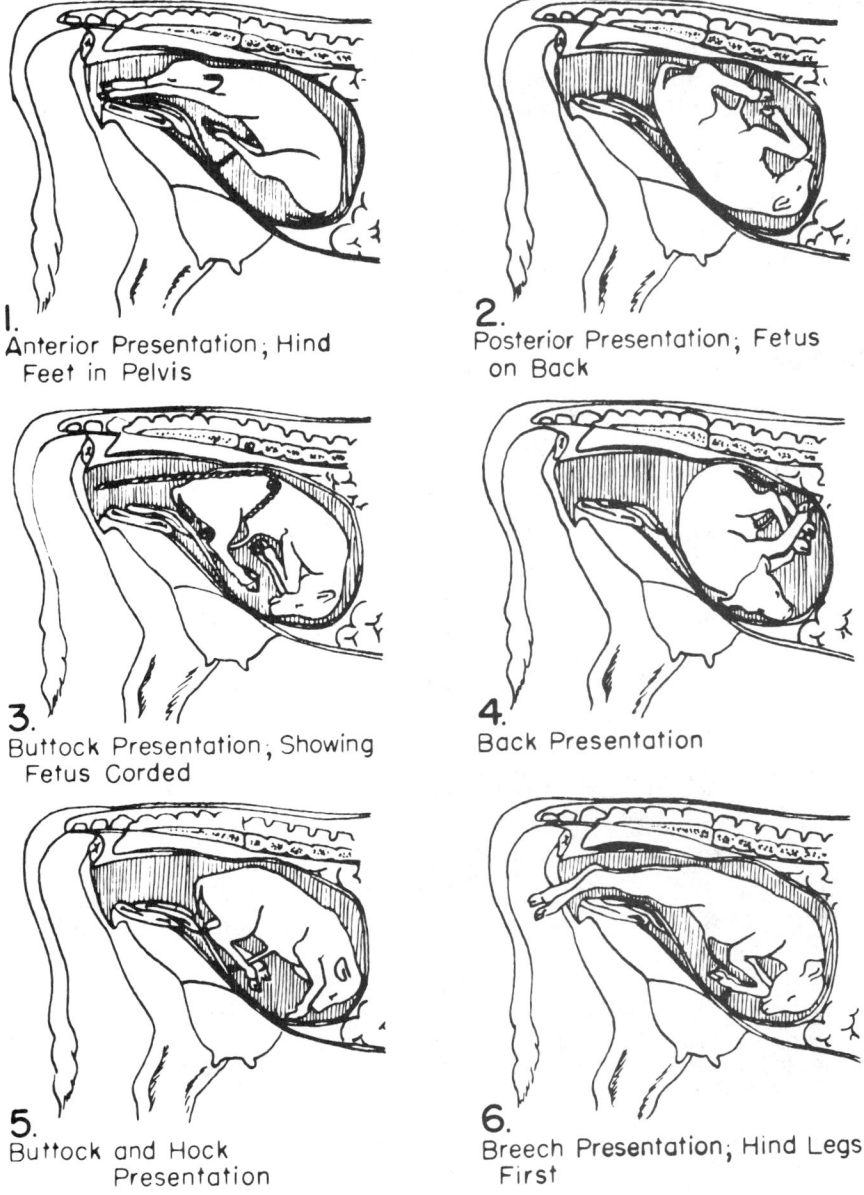

1.
Anterior Presentation; Hind
  Feet in Pelvis

2.
Posterior Presentation; Fetus
  on Back

3.
Buttock Presentation; Showing
  Fetus Corded

4.
Back Presentation

5.
Buttock and Hock
  Presentation

6.
Breech Presentation; Hind Legs
  First

Fig. 6-11. Some abnormal presentations with suggestions for correction:
  1. Force back the hind feet. This is a very serious malpresentation, in which it is generally impossible to save the fetus if delivery is far advanced.
  2. Rotate the fetus, extend the rear legs, and deliver by traction.
  3. Push the fetus forward and bring the legs properly into the genital passage.
  4. Turn the fetus so that either the head and forelegs or the rear legs can be started through the pelvis.
  5. Push the fetus forward and bring the legs properly into the genital passage.
  6. Usually delivery is normal but traction may facilitate; beware of prolonged labor because calf may suffocate due to rupture of the navel cord.

Rendering Assistance

A good rule for the attendant is to be near but not in sight. If presentation is normal and within an hour or two after the onset of signs of calving, no assistance will be necessary. On the other hand, if the cow has labored for some time with little progress or is laboring rather infrequently, it is usually time to give assistance. Such aid will usually consist of fastening small ropes around the pasterns and pulling the young outward and downward as the cow strains. This should be done by an experienced caretaker or a competent veterinarian. It is always well to be reminded that rough, careless, or unsanitary methods at such a time may do more harm than good.

The Newborn Calf

If parturition has been normal, the cow can usually take care of the newborn calf, and it is best not to interfere. However, in unusual cases, it may be necessary to wipe the mucus from the nostrils to permit breathing; or, more rarely yet, artificial respiration methods may have to be applied to some calves. This may be done by blowing into the mouth, working the ribs, rubbing the body rather vigorously, and permitting the calf to fall gently. The cow should be permitted to lick the calf dry.

With calves born in sanitary quarters or out on clean pastures, there is little likelihood of navel infection. To lessen the danger of such infection, the navel cord of the newborn calf should be treated at once with a two percent solution of tincture of iodine.

Fig. 6-12. A good start in life. When the weather is warm, the most natural and ideal place for calving is a clean, open pasture, away from other livestock. Under pasture conditions, there is less danger of either infection or mechanical injury to the cow and calf.

A vigorous calf will attempt to rise in about 15 minutes and usually will be nursing in half an hour to an hour. The weaker the calf, the longer the time before it will be able to be up and nursing. Sometimes it may even become necessary to assist the calf by holding it up to the cow's udder.

The colostrum (the milk yielded by the mother for a short period following the birth of the young) is most important for the well-being of the newborn calf. Experiments have shown that it is almost impossible to raise a calf that has not received any colostrum. Aside from the difference in chemical composition, compared with later milk, the colostrum seems to have the following functions:

1. It contains antibodies which temporarily protect the calf against certain infections, especially those of the digestive tract.

2. It serves as a natural purgative, removing fecal matter which has accumulated in the digestive tract.

3. It contains a very high content of vitamin A, from 10 to 100 times that of ordinary milk. This provides the young calf, which is born with little body storage of this vitamin, with as much vitamin A on the first day as it would secure in some weeks from normal milk.

Usually it is best to keep the cow and calf in a small pasture for a few days. After this, they may be turned back with the main herd. Nothing is better for the cow at calving time than plenty of grass, and both the cow and calf will be helped by an abundance of fresh air and sunshine. The cow may deliberately hide the calf for the first few days, and the job may be so thoroughly done as to require considerable cleverness on the part of the caretaker to find it.

The Afterbirth

Under normal conditions, the fetal membranes (placenta or afterbirth) are expelled from 3 to 6 hours after parturition. Should they remain as long as 24 hours after calving, competent assistance should be given by an experienced caretaker or a licensed veterinarian. The operation of removing a retained afterbirth requires skill and experience; and, if improperly done, the cow may be made a non-breeder. Before attempting to do so, the fingernails should be trimmed closely, and the hands and arms should be thoroughly washed with soap and warm water, disinfected, and then lubricated with vaseline or linseed oil. In no case should a weight be tied to placenta in an attempt to force removal.

As soon as the afterbirth is ejected, it should be removed and burned or buried in lime, thereby preventing the development of bacteria and foul odors. This step is less necessary on the open range, where animals travel over a wide area.

CARE AND MANAGEMENT OF THE BULL

Outdoor exercise throughout the year is one of the first essentials in keeping the bull virile and in a thrifty, natural condition. The finest and easiest method of providing such exercise is to arrange for a well-fenced, grassy paddock (about two acres is a good size for one bull). Many valuable sires have been ruined through close confinement in a small stall—or more likely yet—through being kept knee deep in mud within a small filthy enclosure. In addition to the valuable exercise obtained in the grassy paddock, the animal gets succulent pasture, an ideal feed for the herd bull.

A satisfactory and inexpensive shelter should be provided for the bull. The most convenient arrangement is to have this within or adjacent to the paddock, so that the bull may run in and out at will. Sufficient storage space for feed and conveniences for caring for the bull should be provided in this building. Normally, purebred bulls are kept in separate stalls and enclosures, though some successful purebred breeders regularly run several valuable bulls in one enclosure. Bulls used in commercial herds are usually run together, both on the range and when separated out from the cows. Because of their scuffling and fighting, there is more injury hazard when bulls are handled in a group.

Under range conditions, it is rather difficult to give the bulls much attention during the breeding season. Usually the proper number of bulls is simply turned with the cow herd. During the balance of the year, however, the bulls are usually kept separate. Thus, if the producer desires calves that are dropped from February 1 to June 1, the bulls are turned with the cows about May 1 and are removed September 1.

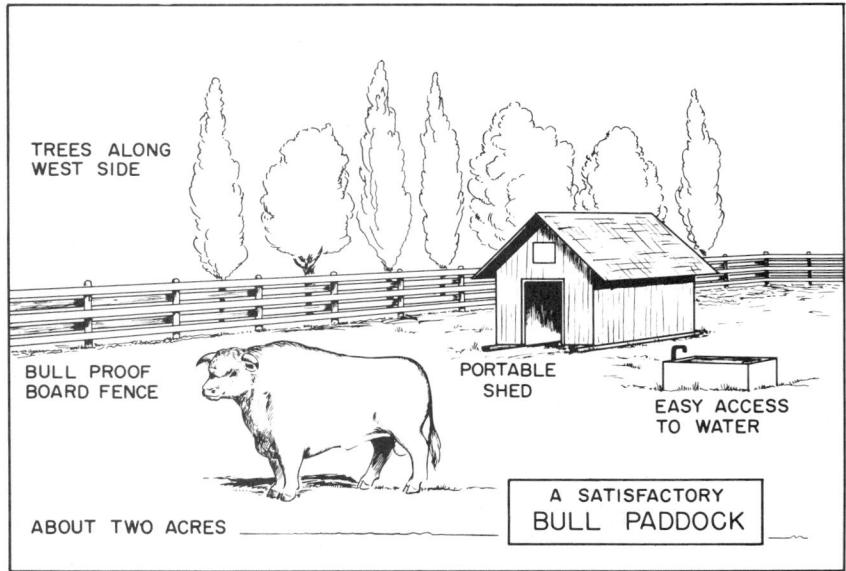

Fig. 6-13. This arrangement has proved most satisfactory at Washington State University. Note the size of the pasture paddock, the individual portable shelter, and the water tank, shade, and bull proof board fence. (Drawing by R. F. Johnson)

The feeding of the herd bull is fully covered in Chapters 8 and 20. In brief, it may be said that the feeding program should be such as to keep the bull in a thrifty, vigorous condition at all times.

AGE AND SERVICE OF THE BULL

The number and quality of calves that a bull sires in a given season is more important than the total number of services. The number of services allowed will vary with the age, development, temperament, health, breeding condition, distribution of services, and system of mating (pasture or hand mating). With pasture mating, size of area, carrying capacity of the range, and the size of the herd are important factors. Therefore, no definite best number of services can be recommended for any and all conditions, and yet the practices followed by good cattlemen do not differ greatly. For best results, a bull should be at least fifteen months old and well grown for his age before being put into service. Even then, it is best to follow a system of hand mating until the bull is two years of age.

Table 6-2 gives pertinent information relative to the use of the bull, including consideration that should be given to age and method of mating.

TABLE 6-2

HANDY BULL MATING GUIDE

Age	No. of Cows/Yr.		Comments
	Hand-mating	Pasture-mating	
Yearling	10-12	8-10	Most western ranchers use 1 bull to about 25 cows.
Two-yr.-old ...	25-30	20-25	A bull should remain a vigorous and reliable breeder up to 10 years or older; up to 6 to 7 years under range conditions.
Three-yr.-old or over	40-50	25-40	

Fig. 6-14. Breeding cows on pasture at Flat Top Ranch, Walnut Springs, Texas. Pregnant cows should be provided as much year-round pasture as possible. Also, shade is necessary in the southern states. (Courtesy, National Cottonseed Products Association, Inc.)

In a survey conducted by Washington State University, it was found that one bull was used for every 21.5 cows and heifers bred.[3]

Should the bull prove to be an uncertain breeder, he should be given rest from service, forced to take plenty of exercise, and then placed in proper condition—neither fat nor thin. Sometimes a bull that is being let down in condition following showing will be temporarily sterile during the reducing process. Even though this lack of fertility may last for a year, usually such animals bounce back.

NORMAL BREEDING SEASON AND TIME OF CALVING

The season at which the cows are bred depends primarily on the facilities at hand, taking into consideration the feed supply, pasture, equipment, labor, and weather conditions; and whether the cattle are being produced for commercial or for purebred purposes.

The purebred breeder who exhibits cattle should plan the breeding program so that maximum advantage will be taken of various age groups. In most livestock shows throughout the country, the classifications are based upon the dates of January 1, May 1, and September 1. Further information relative to show classifications is presented in Chapter 14.

In commercial herds of beef cattle, two systems of breeding are commonly practiced in regard to the season of the year. In one system, the bulls are allowed to run with the cows throughout the year so that calving is on a year-round basis. This system results in greater use of the bull, and there is less delay in the first breeding of the heifers as soon as they are sufficiently mature. On the other hand, often the calves arrive at undesired and poorly adapted times; the breeding system is without order and regularity; and the calves usually lack uniformity. This system is frequently followed in the central and southern states.

The other system of breeding followed in commercial herds, and the most widely used system on the western range, is that of having all of the breeding done within a restricted season (of about three months) so that the calves arrive within a short spread of time—usually in the spring. Having the calves born about the same time, whether it be fall or spring, results in greater uniformity. Thus, it is easier to care for (brand, dehorn, castrate, vaccinate, etc.) and market such animals. Each farm has its individual problems, and the decision must be made accordingly.

Figs. 6-15, 6-16, 6-17, and 6-18, are based on a survey made by Washington State University.[4] These figures show that there is a marked area difference in the breeding season followed by commercial cattlemen, due to weather conditions. Thus, in the South about a third of the cattlemen leave the bulls with the cow herd the year around; where this is not done, the breeding season is much more prolonged and the peak of the breeding season is in January. In each of the other areas, however, the breeding season is

---

[3]Ensminger, M. E., M. W. Galgan, and W. L. Slocum, Problems and Practices of American Cattlemen, *Wash. Agr. Exp. Sta. Bull.* 562, 1955.
[4]Ibid.

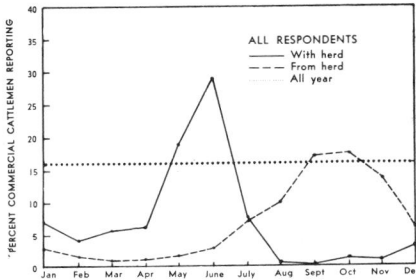

Fig. 6-15. The breeding season followed over most of the U.S. The solid line shows the period during which the bulls are put with the herd, whereas the broken line shows the dates for their removal. These curves show that most cattlemen breed within a restricted season of 3 to 4 months. The dotted line shows that about 16 percent of the cattlemen leave the bulls with the herd the year around.

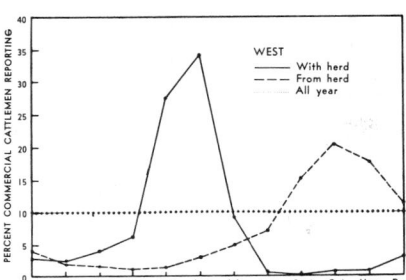

Fig. 6-16. The breeding season followed in the 11 western states. Only about 10 percent of the cattlemen of this area breed on the year-round basis (see dotted line); most of the rest breed from May to October.

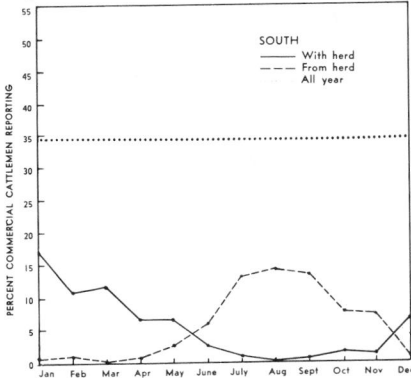

Fig. 6-17. The breeding season followed in the southern states. As noted, a third of the cattlemen leave the bulls with the cows the year around (see dotted line); the balance extend the breeding season over several months, reaching a peak in January.

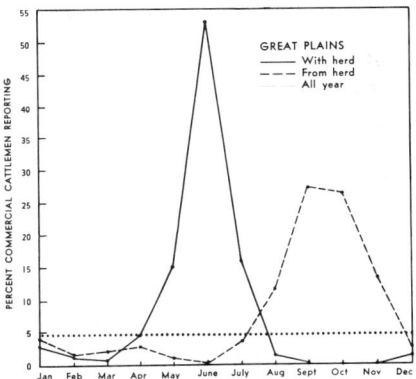

Fig. 6-18. The breeding season followed in the Great Plains. As shown, less than 5 percent of the bulls are left with the cows the year around, and the vast majority of the cattlemen breed during a relatively short season extending from June to October.

later and more restricted; most cattlemen do all of the breeding within a season of three to four months.

ARTIFICIAL INSEMINATION[5]

Artificial insemination is, by definition, the deposition of spermatozoa in the female genitalia by artificial rather than by natural means.

Legend has it that artificial insemination had its origin in 1322, at which time an Arab chieftain used artificial methods to impregnate a prized mare

---

[5]This section was authoritatively reviewed by Dr. H. A. Herman, Executive Secretary Emeritus, National Association of Animal Breeders, Columbia, Mo.; and Mr. Melvin L. Kenley, Consultant, and Dr. David E. Bartlett, DVM, Vice President-Production and Veterinarian, American Breeders Service, De Forest, Wisc; and Mr. Clarke A. Nelson, Vice President, Carnation Company, Los Angeles, Calif.

with semen stealthily collected by night from the sheath of a stallion belonging to an enemy tribe. There is no substantial evidence, however, to indicate
that the Arabs practiced artificial insemination to any appreciable degree.

The first scientific research in artificial insemination of domestic animals
was conducted with dogs by the Italian physiologist, Lazarro Spallanzani, in
1780. A century later, American veterinarians employed artificial means to
get mares in foal that persistently had failed to settle to natural service. They
noticed that because of obstructions the semen was often found in the vagina
and not in the uterus following natural service. By collecting the semen into
a syringe from the floor of the vagina and injecting it into the uterus, they
were able to impregnate mares with these anatomical difficulties.

The Russian physiologist, Ivanoff, began a study of artificial insemination of farm animals, particularly horses, in 1899; and in 1922, he was called
upon by the Russian government to apply his findings in an effort to reestab-
lish the livestock industry following its depletion during World War I. Crude
as his methods were, his work with horses must be considered the foundation upon which the success of the more recent work is based.

The shifting of the large-scale use of artificial insemination to cattle and
sheep, two decades after it was first introduced for horses, was not caused by
the fading importance of the horse and the increased demand for cattle and
sheep. Rather, it was found that progress was quicker and more easily
achieved with these animals, because the exact time of ovulation in relation
to signs of heat is more easily detected in the cow and ewe than in the mare.
It was also discovered that the sperm of bulls and rams survive better in
storage than stallion sperm.

Following World War II, British scientists were called upon to make
wide use of artificial insemination in reestablishing the livestock industry of
England. They looked upon it as: (1) a way in which to increase more
rapidly the efficiency and utility value of their animals through making
wider use of outstanding sires; (2) a means of controlling certain diseases;
and (3) the best way in which to increase breeding efficiency.

Today, artificial insemination is more extensively practiced with dairy
cattle than with any other class of farm animals. In 1938, only about 7,000
cows were bred by this means in organized groups in the United States;
whereas in 1974, 7,500,000 head, or 49.7 percent, of the dairy cows and heifers were bred by artificial insemination. Additionally, in that same year,
2,700,000 head, or 5 percent, of beef cattle were bred by artificial insemination. Also, it is noteworthy that the average number of cows bred per bull
increased from 228 in 1939—when the first artificial breeding association
was organized in this country—to 3,620 in 1971. (That's more than 100 times
greater than by natural service.)

It is recognized that—due primarily to their more frequent handling (in
milking) and greater accessibility—artificial insemination is vastly easier to
apply to dairy cattle than to beef cattle; and, therefore, more common in
dairy cattle enterprises. However, the use of artificial insemination in beef
cattle is growing. During 1974, over 2,700,000 beef cows were inseminated
by A.I. breeding organizations and several hundred thousand registered

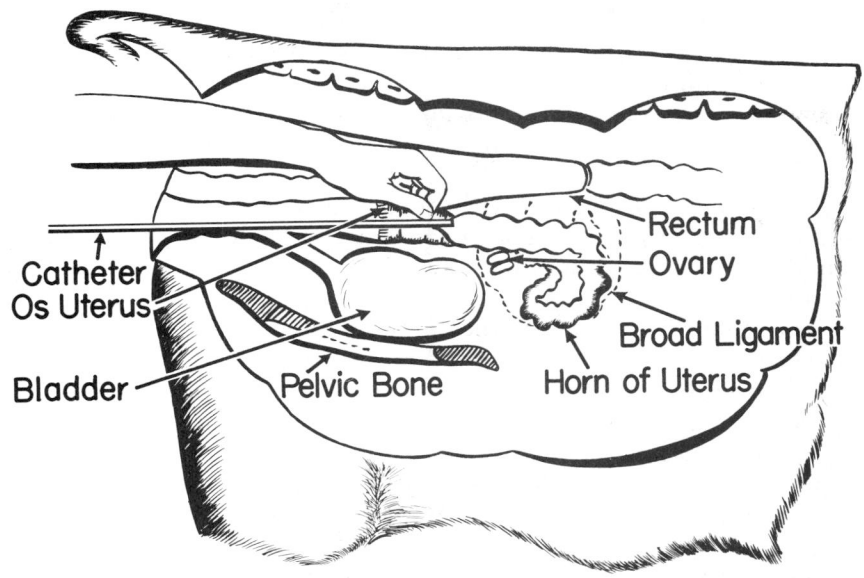

Fig. 6-19. Deep uterine insemination of the cow. The cervix is grasped *per rectum*, and the inseminating tube is carefully worked into and through the cervical canal. (Courtesy, Dr. H. A. Herman, Executive Secretary Emeritus, National Association of Artificial Breeders)

cows were serviced by A.I. in private herds, utilizing key sires, or partnership sires, in addition to the number reported by bull stud operations.

Subsequent to a Consent Decree between the American Angus Association and the United States Government, which was signed on July 13, 1970, there have been major changes liberalizing the rules for registration of offspring from A.I. by most purebred beef associations. The current attitudes and rulings of the cattle registry associations relative to artificial insemination are summarized in Chapter 3 of this book.

Frozen Semen

The freezing of semen, particularly slow freezing such as might occur from exposure to ordinary sub-freezing weather conditions or from placing a vial of semen in the freezing compartment of a refrigerator, will kill all spermatozoa immediately. However, spermatozoa can be preserved for several years provided glycerine is first added and it is frozen at a certain rate for storage in liquid nitrogen at $-196°$ C, or $-320°$ F. Liquid nitrogen is the universally used refrigerant in the United States because uniform temperatures may be maintained for long periods of time and the method is more convenient in shipping and storing frozen semen. Dry ice at $-110°$ F isn't cold enough, and mechanical refrigerators proved unreliable. Although the frozen semen technique is now being used extensively in a practical way, more research on the subject is needed; we still need to know more about the exact nature of freezing and thawing damages, the mechanism of action of glycerol, the optimum diluting media, freezing temperatures and techniques, storage conditions, and the length of survival of frozen semen.

Frozen bull semen has been stored as long as 20 years and conception obtained. There is a small reduction in fertility of semen stored long periods, but it is usual to find semen stored one year or longer in routine use. Because changes in recommendations are constantly being made in light of new findings, no attempt is made herein to detail semen freezing techniques.

Frozen semen, refrigerated with liquid nitrogen, can be shipped to all parts of the world. The thawing of semen is accomplished by placing the vial in a container of water containing thawing ice (34-38° F) immediately prior to insemination.

Frozen semen is potentially the most valuable breeding technique yet known. Through it, the following may be achieved:

1. The usefulness of outstanding bulls can be extended far beyond their lifetime; also, it insures the proven sire should he die.

2. Outstanding bulls can be used, nationwide and worldwide.

3. A multi-herd progeny test can be completed at a much earlier age.

4. A stock of semen can be built up while waiting for a progeny record assessment.

5. Long-term storage of semen lessens semen wastage, and facilitates long-distance transport.

6. Semen from valuable sires may be fully utilized.

7. A herd owner can usually obtain the sire of his choice at any time.

Advantages of Artificial Insemination

Some of the advantages of artificial insemination are:

1. *It increases the use of outstanding sires*—Through artificial insemination, many breeders can avail themselves of the use of an outstanding sire, whereas the services of such an animal were formerly limited to one owner, or, at the most, to a partnership.

2. *It alleviates the danger and bother of keeping a sire*—Some hazard and bother is usually involved in keeping a bull. The ready availability of semen rules out this necessity.

3. *It makes it possible to overcome certain physical handicaps to mating*—Artificial insemination is of value (a) in mating animals of greatly different sizes; for example, in using heavy mature bulls on yearling heifers, and (b) in using stifled or otherwise crippled sires that are unable to perform natural service.

4. *It makes it possible to use a bull that is not alive at the time*—Since frozen semen can be stored for many years, it is possible to use a bull far beyond his lifetime.

5. *It lessens sire costs*—In smaller herds (in herds with fewer than 50 cows), artificial insemination is less expensive than the ownership of a worthwhile sire together with the accompanying building, feed, and labor costs.

6. *It reduces the likelihood of costly delays through using sterile sires*—Because the breeding efficiency of sires used artificially is constantly checked, it reduces the likelihood of breeding females to a sterile sire for an

extended period of time.

7. *It makes it feasible to prove more sires*—Because of the small size of the herds in which they are used in natural service, many sires are never proved. Still others are destroyed before their true breeding worth is known. Through artificial insemination, it is possible to determine the genetic worth of a sire at an earlier age and with more certainty than in natural service.

8. *It creates large families of animals*—The use of artificial insemination makes possible the development of large numbers of animals within a superior family, thus providing uniformity and giving a better basis for a constructive breeding program.

9. *It increases pride of ownership*—The ownership of progeny of outstanding sires inevitably makes for pride of ownership, with accompanying improved feeding and management.

10. *It may lessen and control certain diseases*—Artificial insemination has proven to be equally valuable as a means of preventing and controlling the spread of certain types of cattle diseases, especially those associated with the organs of reproduction, such as vibriosis and trichomoniasis. However, when improperly practiced, it may be an added means of spreading disease. Therefore, it is most essential (a) that all males be carefully examined for symptoms of transmissible diseases, (b) that bacterial contamination be avoided during the collection and storage of semen, and (c) that clean, sterile equipment be used in the insemination. Artificial insemination organizations that are members of the National Association of Animal Breeders, Inc., are expected to follow the rigid Sire Health Code approved by the American Veterinary Medical Association and adopted by the National Association of Animal Breeders.

11. *It increases profits*—The offspring of outstanding sires are usually higher and more efficient producers, and thus more profitable. Artificial insemination provides a means of using such sires more widely.

Limitations of Artificial Insemination

Like many other wonderful techniques, artificial insemination is not without its limitations. A full understanding of such limitations, however, will merely accentuate and extend its usefulness. Some of the limitations of artificial insemination are:

1. *It must conform to physiological principles*—One would naturally expect that the practice of artificial insemination must conform to certain physiological principles. Unfortunately, some false information concerning the usefulness of artificial insemination has been encountered—for example, the belief that females will conceive if artificially inseminated at any time during the estrus cycle. Others have even accepted exaggerated claims that the quality of semen may be improved through such handling, only to be disappointed.

2. *It requires skilled technicians*—In order to be successful, artificial insemination must be carried out by skilled technicians who have had considerable training and experience.

3. *It necessitates considerable capital to initiate and operate an artifi-*

cial insemination organization—Considerable money is necessary to initiate an artificial insemination enterprise, and still more is needed to expand and develop it properly. It is noteworthy, however, that 50 percent of all cows bred artificially in the United States in 1971 were serviced with semen provided by privately owned establishments and 50 percent by cooperatives.

4. *It may accentuate the damage of a poor sire*—It must be realized that when a male sires the wrong type of offspring, his damage is accentuated because of the increased number of progeny possible.

5. *It may restrict the sire market*—The fact that the market demand for poor or average sires will decrease if artificial insemination is widely adopted should be considered an attribute rather than a limitation. Also, it is noteworthy that a sizeable number of the nation's beef cattle are still bred to scrub and nondescript bulls.

6. *It may increase the spread of disease*—As previously indicated, the careful and intelligent use of artificial insemination will lessen the spread of disease. To date, no outbreaks of disease traceable to the use of artificial insemination have been reported in the United States. However, it must be recognized that carelessness and ignorance may result in the rapid spread of disease. Semen should always be obtained from a source known to observe recommended health and sanitation procedures. The organized A.I. Studs have adopted a rigid Sire Health Code.

7. *It may be subject to certain abuses*—If semen is transported from farm to farm, the character of the technician must be above reproach. Trained workers can detect differences in the spermatozoa of the bull, ram, boar, stallion, or cock; but even the most skilled scientist is unable to differentiate between the semen of a Hereford and a Shorthorn, to say nothing of the difference between 2 bulls of the same breed. However, it appears that such abuse is more suspicioned than real. In a blood type study[6] with cattle, Rendel found 4.2 percent family records in error out of 615 animals by natural service, compared to 4 percent family records in error out of 199 sired by artificial insemination.

Of course, with skilled workers performing the techniques required in artificial insemination, there usually is more check on the operations and perhaps less likelihood of dishonesty than when only the owner is involved, such as is usually the situation with natural service. Also, bulls used in artificial insemination are blood typed in order to make possible investigation of suspected errors.

Some Practical Considerations

Based on present knowledge, gained through research and practical observation, it may be concluded that cattlemen can make artificial insemination more successful through the following:

1. Give the female a reasonable rest following parturition and before rebreeding; in cows this should be about 60 days.

---

[6]Rendel, J., Studies of cattle blood groups, II. Parentage Tests, *Acta. Agric. Scand.*, 1958, 8:131, page 140.

2. Keep record of heat periods and note irregularities.

3. Watch carefully for heat signs, especially at the approximate time.

4. Notify the insemination technician promptly when an animal comes in heat.

5. Avoid breeding diseased females or females showing pus in their mucus. The latter condition indicates an infection somewhere in the reproductive tract.

6. Have the veterinarian examine females that have been bred three times without conception or that show other reproductive abnormalities.

7. Have a pregnancy diagnosis made.

HORMONAL CONTROL OF HEAT IN COWS[7]

Planned parenthood is not new. It has long been practiced among females of all species, women included. For several years, a product has been available and used by veterinarians to defer heat in dogs. Back in the 1930s, much progress was made in the isolation and identification of progesterone from the corpus luteum. In the 1940s, researchers at the University of Wisconsin injected progesterone into various species of farm animals and successfully synchronized estrus. They found that, during the treatment, the corpus luteum regressed; and, upon discontinuing it, heat followed. In the 1950s, certain pharmaceutical houses chemically synthesized progestogens (progesterone-like compounds)—and the race was on. Many new drugs have been administered in attempts to control the estrus cycle of cows. Although more research needs to be done, it appears that prostaglandins, hormone-like substances which regress the corpus luteum, may be the answer to controlled breeding.

Sly Controls of Heat Have Been Used a Long Time

Stockmen have long "tampered with" the breeding and parturition season that was common in the wild state. Prior to domestication, animals brought forth their young in the fields and glens, inhibited only by age and feed, and influenced somewhat by seasons. But man changed all this—even without the use of hormones. Sly controls have been exercised over breeding for a very long time. For example, farm flock owners controlled reproduction in chickens by the simple act of putting eggs under an old setting hen—unless she hid out. Today, modern poultry producers regulate chick hatchings by controlling when, and how many, eggs go into the incubator. It's more difficult to accomplish the same thing in four-footed animals.

The motivating reasons back of man-made changes vary somewhat by species. Horsemen, especially those who race or show, want their mares to foal as soon after January 1 as possible, because a horse's age is computed on a January 1 basis, regardless of how late in the year he may have been born. Sheepmen and hogmen strive for two crops of offspring per year, and for

[7]This section was authoritatively reviewed by the following: Dr. J. N. Wiltbank, Texas A & M University, College Station, Tex.; Dr. Charles K. Vincent, Codding Embryological Sciences, Inc., Foraker, Okla.; Dr. David E. Bartlett, DVM, Vice President-Production and Veterinarian, American Breeders Service, De Forest, Wisc.; and Dr. H. A. Herman, Executive Secretary Emeritus, National Association of Animal Breeders, Columbia, Mo.

multiple births. Purebred cattlemen who show, plan their breeding programs to take maximum advantage of show classifications; commercial cattlemen are concerned with weather and feed supply; and dairymen want the largest flow of milk at a time when the product is likely to bring the highest price. Also, cattlemen recognize that controlled estrus would greatly facilitate both artificial insemination and ova transplantation.

Stockmen have altered nature's way in farm animals (1) by confining the male at certain times, or hand mating, (2) by emulating spring conditions—through providing better feed, shelter, and/or blankets when breeding at other times of the year, (3) by flushing—through feeding females more liberally two to three weeks ahead of the breeding season, and (4) by artificially controlling the hours of light per day—through use of ordinary electric lights, which activate hormone production. Each of these methods has been used with varying degrees of success. All have fallen short of achieving the hoped-for goal—that of bringing females in heat at will, followed by a high conception rate. Hormonal control appears to be the answer.

Advantages to Accrue from Bringing Cows in Heat at Will

Many obvious advantages would accrue from a sure method of controlling the breeding dates of cows, and, consequently, the birth dates of calves; among them, the following:

1. More cows would calve early in the breeding season and wean off heavier calves.

2. Cattlemen could have their calves come within a restricted period of time. They could even swap help with the neighbors, according to a predetermined schedule.

3. Calves of uniform age would also be more uniform in size, thereby making them easier to handle, feed, and sell.

4. It would greatly facilitate such management practices as pregnancy testing, dehorning, castrating, and vaccinating; all of which could be done at one time, rather than piecemeal.

5. Cattlemen could plan their breeding programs so that calves would be ready for market when seasonal prices are highest. They could contract a certain number of calves of the same age on a specified date.

6. Artificial insemination would be simplified if the time of ovulation could be controlled to the extent that the hour and date of insemination of a breeding herd could be precisely determined, thereby alleviating the problem of heat detection and lowering the cost of semen distribution. This would greatly facilitate breeding range cattle by A.I.

7. Ova transfer would be facilitated. One of the major hurdles to overcome in ova transfer is the synchronization of the estrus cycles of both the donor and the recipient females. Both must be in similar stages in their reproductive cycles to allow a successful transfer to take place.

Drugs for Estrus Control

Many different drugs have been administered (either orally, by injection, or by implantation) in attempts to control the estrus cycle of cattle, with progestins and prostaglandins heading the list. Unfortunately, low conception rates have been a problem with most drugs; prostaglandin appears to be an exception.

● *Progestins*—are compounds that mimic the hormone progesterone, which is produced naturally by the corpus luteum on the cow's ovary. Following release of the egg (ovulation) from the mature follicle on the ovary shortly after heat, the follicle changes into a mass of yellow tissue called the corpus luteum (CL).

The CL produces the hormone progesterone, which prepares the reproductive organs for the growth of the embryo. Progesterone also prevents the maturation of new follicles (containing eggs) on the ovary and, through this action, prevents the return of the animal to heat.

If pregnancy does not occur, the CL begins to regress on day 15 after heat. The regression of the CL results in declining progesterone in the blood and, by day 20 or 21, a new follicle matures; the animal returns to heat; and ovulation is repeated, with the formation of a new CL.

If pregnancy occurs, the CL does not regress, but continues to release progesterone which, basically, prevents recurrent heats and maintains the reproductive organs in a state conducive to embryo growth.

Feeding, injecting, or implanting nonpregnant cattle with progestins for a sustained period of 10 to 20 days, followed by withdrawal of the dose, will result in animals exhibiting heat 2 to 7 days after withdrawal. Essentially, this method of control places animals in a type of false pregnancy; they are fooled by the high levels of the progesterone-mimicking progestin.

Upon withdrawal of the compound, progestin levels in the blood drop and new follicles mature, bringing animals similarly treated into heat synchronously.

Three problems have been encountered with progestins: (1) heats and ovulations in animals so treated are not sufficiently synchronized to breed them successfully at a prescribed time; (2) conception at the first synchronized heat is subnormal; and (3) the progestin must be administered over several days, which involves much labor.

● *Prostaglandins*—A completely new era of research is underway with prostaglandins. These are hormone-like substances, found in almost every cell and tissue, that are believed to play a key role in regulating cellular metabolism. The name "prostaglandin," which is a misnomer, was given to these substances because they were believed to have originated in the male's prostate gland. Later research revealed that they actually come from the seminal vesicles (another accessory sex gland of the male).

Over 14 natural compounds have, to date, been identified as prostaglandins. The marine animal—the sea whip—which is found off the Florida coast, is a rich source. Also, all 14 prostaglandins are now being synthesized from commercially available materials.

These highly potent substances have been called local hormones or tissue hormones because they do their work in the immediate area in which they are produced, as distinguished from circulating hormones which aim at distant targets. In cattle, prostaglandins are being used to regress the corpus luteum (the growth on the ovary that prevents ovulation). This allows the natural estrus cycle to begin again. Research indicates that the time interval between injection with a single prostaglandin dosage and the onset of estrus and ovulation is very short—about 90 hours after injection, and is predictable.

Pregnancy rates were similar between control and prostaglandin-treated cattle in 7 independent studies. Pregnancy rates for data pooled within control and within prostagladin-treated cattle for the studies were 58.8 percent and 59.1 percent, respectively, based on 216 control cattle and 286 prostaglandin-treated cattle. These data demonstrate no significant alternation in fertility of cattle inseminated at estrus following prostaglandin treatment.

In summary, it may be said that research indicates that cows treated with a single dose of prostaglandin may be bred at a predetermined time, and over a short period, and still have satisfactory fertility. More research needs to be done on this promising new drug. We need to know what effects, if any, prostaglandins have on fertility. We need to know what dosage levels to administer. We need to know if prostaglandins have any side effects on animals. Also, they must be approved by the U.S. Food and Drug Administration prior to commercial release. Prostaglandins may be the answer to future controlled breeding in cattle.

What Estrus Synchronization Will and Will Not Do

With any new program that offers improvement, hope springs eternal. Progressive cattlemen are always on the alert for something that will increase efficiency still further. On the other hand, submarginal cattlemen are usually hunting for a panacea for all their toubles—for a "crutch," or an easy way out. A heat-grouping program is not likely to do all that is expected of it, but it will do more for the good producer than for the poor one. Here are some of the things that it will and will not do:

It will:

1. Add a management tool that will better the opportunity for improved fertility.
2. Shorten the breeding and calving seasons.
3. Give as good, and likely better, conception rate than obtained without hormone synchronization.
4. Make for earlier breeding.
5. Improve calf survival.
6. Increase weaning weight of calves.
7. Help improve the management program.
8. Make for a more uniform calf crop.
9. Facilitate A.I. and ova transplantation.

It will not:

1. Start a cow or heifer cycling.
2. Improve fertility.
3. Make it possible to breed sooner than 60 days following calving.

4. Lower feed requirements.

5. Overcome sterility or delayed breeding due to (a) genital infections and disease, (b) poor management and feeding, (c) inherited abnormalities, (d) anatomical defects and injuries, and (e) shy or difficult breeders.

6. Assure good weather at a time when all of the calves are coming, for it pretty much "puts all the eggs in one basket."

Results of Hormone-Induced Estrus

The major criteria for measuring the success of hormone-induced estrus are: (1) the percentage of the cows that come in heat; and (2) the percentage of conception.

Admittedly, the effectiveness of hormone-induced heat varies widely from herd to herd. This is expected, for the results are affected by reproductive abnormalities and management. For any such product to be most effective, the cows must be having normal estrus prior to treatment; the caretaker must be capable of detecting cows in heat in order to breed them; and the plane and kind of nutrition must be satisfactory. In fact, the better the level of the herd fertility without hormone treatment, the better the results with treatment. It follows, therefore, that hormone-planned parenthood in the cow business is not for the poor manager.

In view of the above statement, prior to undertaking any estrus synchronization program, the following questions should be answered for each herd, honestly and unflinchingly:

1. Are the females having normal heat periods, and are they reproducing normally?

2. Is the management program good?

3. Is the nutrition satisfactory?

4. Is the herd health good?

5. Has at least 60 days elapsed since calving?

6. Are heifers of proper age and size?

7. Will it be possible to administer the treatment according to directions?

8. Is the caretaker capable of detecting the cows in heat?

9. Are the facilities adequate to allow breeding (natural or artificial insemination) on the same day that cows are in heat?

10. Is the semen of good quality; and, if artificial insemination is used, is the inseminator experienced and properly trained?

If the answer to each of the above questions is affirmative, you are ready to proceed and the results will be good. If the answer to one or more of these questions is negative or in doubt, the results will be just as questionable, or even disappointing.

Summary Relative to Hormone-Controlled Estrus

Researchers in both colleges and industries are in general agreement that hormone-controlled estrus synchronization will work, and that it offers promise of good returns when properly used in a well-managed cow herd.

Scientists also realize that we don't know all the answers; that further research work is necessary. Among other things:

1. **We need to know which hormone(s) to use, and how to give it;** whether to feed, inject, or implant it; what dosage to give; and when to give it.

2. **We need to know how to lower costs exclusive of the drugs—** primarily, added labor and perhaps feed, when administering hormones.

3. **We need to know how to obtain a higher conception rate following hormone-induced heat—**although it is recognized that conception is as good, or better, in hormone-treated cows than in untreated cows.

4. **We need to know the effect of the stage of the estrus cycle on treatment—**For example, when treatment is started at or near estrus, some animals show estrus and/or ovulate during treatment.

5. **We need to know more about the effect of lactation and suckling,** both on sexual behavior and on the endocrine control of the ovary.

6. **We need to know more about the influence of location, season, nutrition, social factors, and time of day;** on suppression, synchronization, and fertility.

7. **We need to lower costs or increase the gross—**for regardless of product, it is net returns that count.

It appears that planned parenthood in the cow business is here to stay; that its wide use only awaits getting the technique perfected and lowering costs, both of which will come. In the meantime, cattlemen are admonished to keep abreast of developments and to rely on well-informed advisers.

SUPEROVULATION[8]

The bull is capable of producing from several thousand to millions of sperm daily whereas the cow normally produces one ovum (occasionally two ova) every 17 to 21 days. Now it is possible, through the administration of hormones, to obtain up to 100 ova from a cow at one estrus cycle. It is also feasible to obtain a large number of eggs from very young calves, by injection of hormones.

Superovulation begins with the selection of healthy heifers or cows which are cycling normally. The animal is ready to be injected with hormone, starting about 16 days after the last heat; although this time may be varied by using prostaglandins or other compounds which cause degeneration of the corpus luteum, or yellow body. (See section on "Drugs for Estrus Control.")

Eggs which are shed from the ovaries are stored in large follicles. The basic principle of superovulation is to stimulate extensive follicular development through the use of a hormone preparation, given intramuscularly or subcutaneously, with follicle-stimulating hormone (FSH) activity. The

---

[8]This section was authoritatively reviewed by Dr. Charles K. Vincent, Codding Embryological Sciences, Inc., Foraker, Okla.; Dr. J. N. Wiltbank, Texas A & M University, College Station, Tex.; Dr. David E. Bartlett, DVM, Vice President-Production and Veterinarian, American Breeders Service, De Forest, Wisc.; and Dr. H. A. Herman, Executive Secretary Emeritus, National Association of Animal Breeders, Columbia, Mo.

most common sources of such a hormone are pregnant mares' serum (PMSG) and FSH extracts from pituitaries of slaughtered animals. Many animals so treated will come into estrus about five days after initiation of treatment and ovulate, through release of their own luteinizing hormone (LH). However, to help assure that multiple ovulations occur, the ovulating LH from pituitaries or in human chorionic gonadotropin (HCG) is injected. The multiple ovulations occur at about the same time the cow would have normally ovulated one egg (21 days after the previous ovulation).

Early studies have confirmed that FSH should be administered twice daily over a period of about five days. PMSG has a longer biological life and a single subcutaneous injection is normally used. Five or six days after the original FSH or PMSG "shot," LH or HCG is given intravenously.

The heifers should ovulate by the seventh day after starting hormone treatment.

An example of a superovulated ovary is shown in Fig. 6-20.

Since ovulation occurs over a period of time, not all the eggs are fertilized unless the donor is inseminated repeatedly. A yield of four or five good fertilized eggs per donor is about average.

Of course, the real economic value of superovulation lies in the successful transfer of excess eggs from more valuable donor cows to less valuable recipient cows. As a result of this technique, someday stockmen may refer to litter-bearing cows.

OVA TRANSPLANTATION[9]

Artificial insemination has given a means for the widespread distribution of desirable genes via the sperm. Similar genetic selection through high quality females has, however, been limited since, normally, one cow will produce one calf per year and the average number of offspring per female will seldom exceed five in a lifetime. Out of the latter arose the idea that a marked increase in the production of offspring from desirable cows might be effected by superovulation, followed by transfer of the fertilized ova to less desirable cows, with the latter serving as host-mothers or foster-mothers to the developing embryo. Ova transplantation is a seven-step process as follows:

1. *Synchronize heat-cycles of donor and recipient cows*—The first step is to determine the donor cow's heat-cycle timing and to synchronize a group of potential recipient cows so that all are releasing eggs at about the same time.

2. *Obtain a large number of ova (superovulation)*—A drug is given the donor cow so that she "superovulates," resulting in multiple egg release. Without a superovulation drug, the cow would normally release only one egg for fertilization. (See section entitled "Superovulation.")

[9]This section was authoritatively reviewed by Dr. Charles K. Vincent, Codding Embryological Sciences, Inc., Foraker, Okla.; Dr. J. N. Wiltbank, Texas A & M University, College Station, Tex.; Dr. David E. Bartlett, DVM, Vice President-Production and Veterinarian, American Breeders Service, De Forest, Wisc.; and Dr. H. A. Herman, Executive Secretary Emeritus, National Association of Animal Breeders, Columbia, Mo.

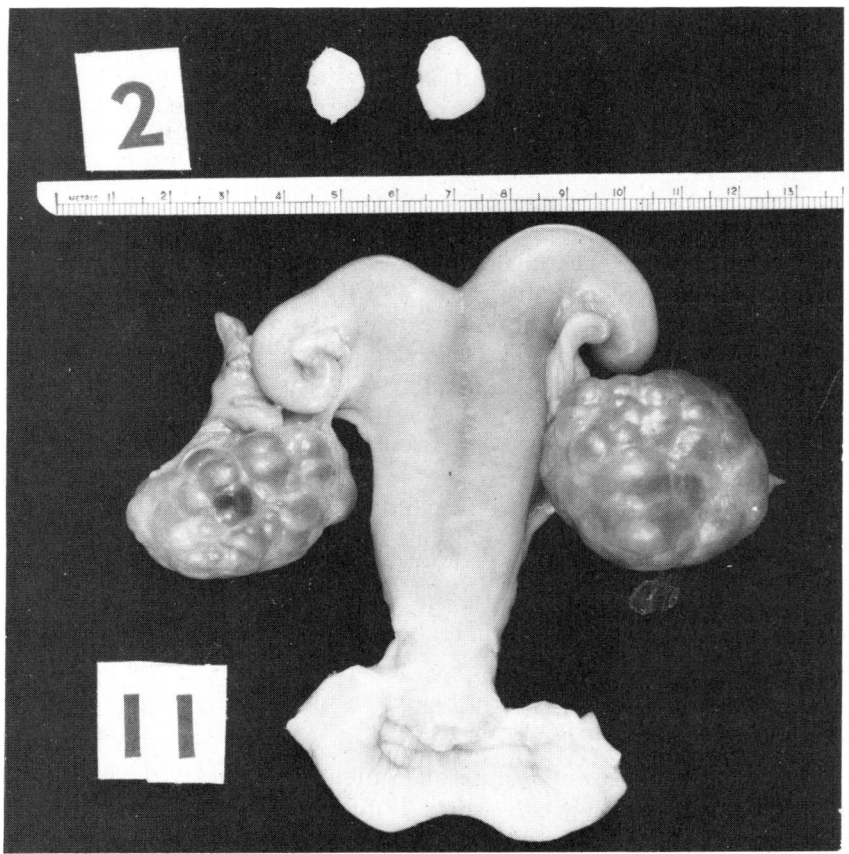

Fig. 6-20. Superovulation of the calf:
   Calf No. 2—Ovaries of a control 4-month-old calf. The ovaries contain thousands of dormant Graafian follicles.
   Calf No. 11—Ovaries and genital organs of a superovulated 4-month-old calf. The calf was not showing any signs of heat, since it had not reached sexual maturity. However, it was injected with 10 rat units of the gonadotropic hormone "Vetrophin" on each of three successive days and slaughtered 5 days after the last injection. The two ovaries contained 97 ripe follicles. Note the size of the superovulated ovaries in relation to the immature uterus. (Courtesy, Dr. E. S. E. Hafez; photo, Washington State University)

   3. *Breed donor cow*—Breed (usually by A.I.) donor cow to bull of choice. Five days later, at which time the fertilized ova have developed to the 8- to 32-cell stage, the donor cow is ready for ova transfer.
   4. *Collect ova*—For obvious reasons, only recovery of eggs by surgical or nonsurgical means from live animals (rather than slaughter donors) is practical. Japanese workers have had fairly good success inserting a special tube with several channels through the cervix nonsurgically, and flushing eggs from the uterus. However, at the present time only surgical recovery of eggs is used because more eggs are recovered this way.
   Beginning about 3 days after insemination, the donor cow is fasted for 2 days in preparation for surgery; on day 5, she enters surgery, under very

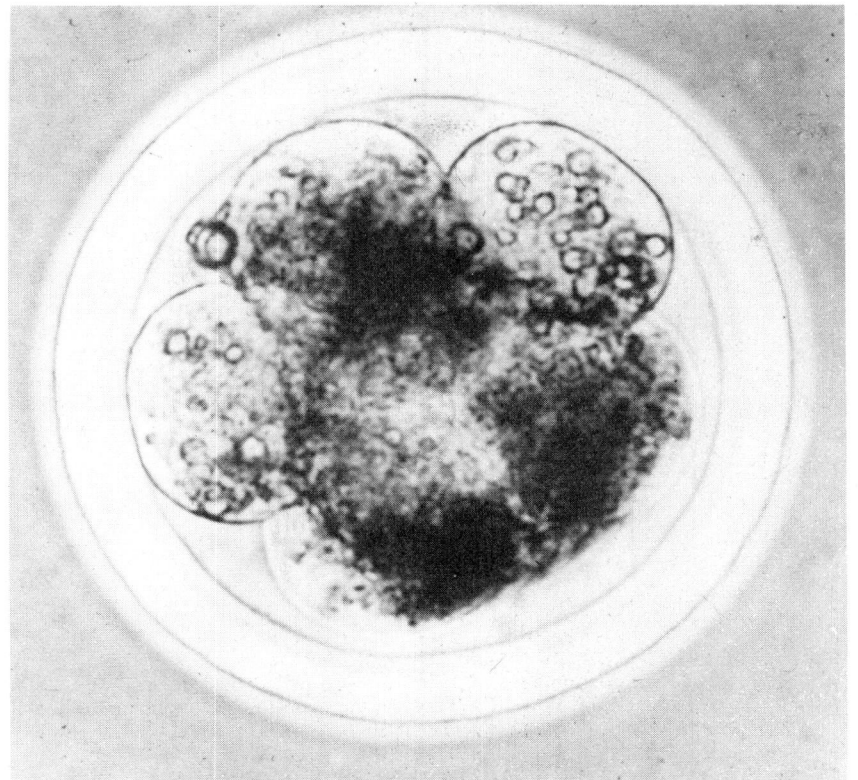

Fig. 6-21. Fertilized ovum in dividing stage, ready to be flushed from donor cow's oviduct and transferred to recipient cow. (Courtesy, A. H. J. Rajamannan, International Cryo-Biological Services, Inc., St. Paul, Minn.)

sanitary conditions; with sterile equipment, a small incision is made in the abdomen just in front of the udder; the reproductive tract is exposed; each uterine horn is flushed (Fig. 6-22); the fertilized eggs are collected into a glass container; the donor cow's reproductive organs are put back in the abdominal cavity; and the incision is closed. Thereafter, the donor cow can continue to have normal estrus cycles, and can be reused.

5. *Examine eggs*—Immediately take the eggs into the laboratory and examine them microscopically for normal appearance and fertilization.

6. *Prepare foster mothers*—Foster mothers should have ovulated at the same time as the donor, but should not have been mated. The animals selected as recipients can be synchronized with the donor in either of two ways:

a. *Natural synchronization*—With natural synchronization one must have a large number of cows from which to choose several that are in natural heat the same day as the donor.

b. Treat recipient cows so that they will be synchronized, using progestins or prostaglandins. (See section entitled, "Hormonal Control of Heat In Cows," in this chapter.)

OVUM FLUSHING TECHNIQUE

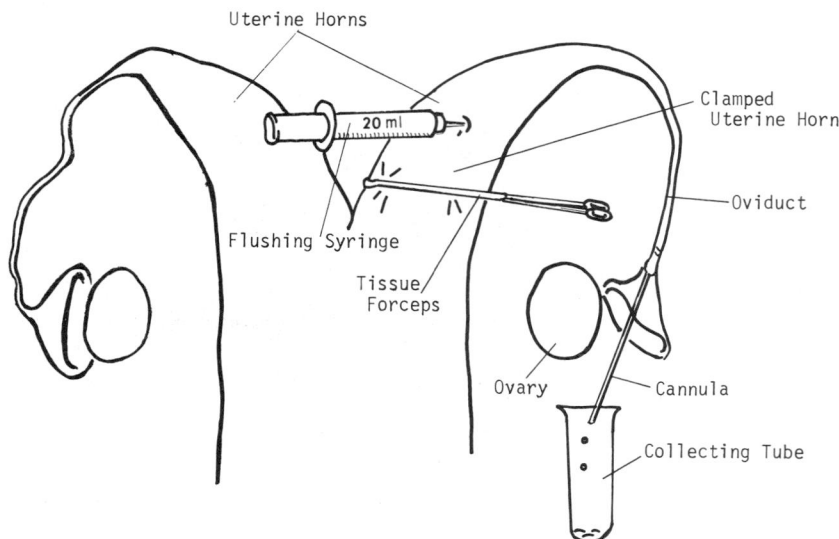

Fig. 6-22. Diagram showing the technique of recovering fertilized eggs from cattle. The collecting tube is held in the oviduct with the fingers or a small soft clamp. (Courtesy, *The Shorthorn World*, Brandon, S.D., February 1974, p. 38, and Dr. Robert H. Foote, Cornell University, Ithaca, N.Y.)

7. *Transfer eggs to recipients*—The recipients are surgically prepared in the same way as the donor. Then one good egg is transferred to a small pipette under the microscope; the pipette is inserted through a small puncture in the uterus near the oviduct; the egg in its nourishing medium is carefully expelled into the uterus; and the animal is sewn up. One-egg transfers are made into the uterine horn on the same side that the animal ovulated on and has a corpus luteum on the ovary.

Pregnancy in the recipients can be diagnosed in about 30 days. Full-term pregnancies result in full sibs (brothers and sisters) with the genetic traits of the donor cow and the bull to which she was bred. Recipients have no genetic influence on the calves they carry—they merely serve as "incubators."

The following advantages would accrue from extensive use of ova transplantation:

1. A dozen calves might be obtained from a valuable cow during a year's time.

2. The rate of progress in genetic improvement would be speeded up, because of the increased number of progeny from valuable cows.

3. Valuable cows that produce normal ova but fail to conceive due to some hormonal or anatomic defects would not need to be culled because of sterility; such animals could be used as donors for supplying ova for transplantation.

Fig. 6-23. Eight purebred calves (6 bulls and 2 heifers), all from the Maine-Anjou cow, Cetlela; the result of ova transplantation to 7 Jersey and Holstein cows (one cow had twins). All embryos transferred in this operation resulted in live calves. (Courtesy, the owner—Harold Biensch, Nilburg, Saskatchewan, Canada)

4. Heifers could be effectively progeny tested at an early age. If large numbers of fertilized eggs could be procured from calves and transplanted to sexually mature recipients, the generation time of cattle could be reduced by one year or more.

5. It would be possible to produce calves of the beef breed of preference from dairy cows.

Because of recent advances, ova transplantation in cows is being employed commercially by several groups, and in several countries. However, several problems must be overcome before it can be employed on a wide-scale basis.

We need to lower costs; understandably, surgical recovery and transfer of embryos is costly.

We need to develop nonsurgical techniques of flushing eggs from the donor cow and passing them through the cervix of the recipient, somewhat like sperm are deposited in artificial breeding.

We need to overcome the difficulties of recovering eggs from small calves, thereby obtaining more eggs per female during her lifetime.

We need to remove eggs (oocytes) directly from the ovarian follicle, thereby increasing the egg yield; then devise a system of fertilizing them in a test tube.

We need to develop methods of storing embryos, so as to facilitate collection at one time and transfer at another time.

We need to be able to sex eggs at the time of storage, thereby permitting eggs of known sex to be transferred.

In summary, it may be said that egg transfer can be done very successfully by skilled teams. But the high cost of present techniques limit the application to the most elite stock. With more research, techniques will become more efficient, simple, and economical, and embryo transfer will be more widely used.

BLOOD TYPING CATTLE

Cattle blood typing was developed at the University of Wisconsin during the decade 1940-50. It involves a study of the components of the blood, which are inherited according to strict genetic rules that have been established in the research laboratory. By determining the genetic "markers" in each sample and then applying the rules of inheritance, parentage can be determined. To qualify as the offspring of a given cow and bull, an animal must not possess any genetic markers not present in his alleged parents. If he does, it constitutes grounds for illegitimacy.

Blood typing is used for the following purposes:

• *To verify parentage*—The test is used in instances where the offspring may bear some unusual color or markings or carry some undesirable recessive characteristic. It may also be used to verify a registration certificate. When one considers that about 5% of all registered cattle in the United States are illegitimate, there is need to use blood typing much more extensively as a bulwark of breed integrity. Through blood typing, parentage can be verified with 90% accuracy.[10] Although this means that 10% of the cases can't be settled, it's not possible to do any better than that in human blood typing.

• *To determine which of two bulls*—When a cow has been served by two or more bulls during one breeding season.

• *To provide a permanent blood type record for identification purposes*—Two samples of blood are required for each animal to be studied; and the samples must be taken in tubes and in keeping with detailed instructions provided by the laboratory. In parentage cases, this calls for blood samples from the calf and both parents; in paternity cases, samples must be taken from the calf, the cow, and all the bulls.

• *To detect fertile heifers born co-twin with bulls*—About 15 percent of all heifers born twin with a bull are potentially fertile; the other 85 percent are sterile, or freemartins. You need not wait until such heifers reach breeding age in order to ascertain their breeding potentialities. Instead, you can submit blood samples from each of the twins (the bull and the heifer) to a service-typing laboratory and request a diagnosis. If the bull and heifer have *like* blood types (except possible differences in the J system), the heifer is diagnosed as a freemartin and nonbreeder. If the bull and heifer have unlike blood types (except possible differences in the J system alone), the heifer is diagnosed as potentially fertile.

The basis for this remarkable method of diagnosing the breeding poten-

---

[10]In a personal communication to the author, Dr. Clyde Stormont, Professor of Immunogenetics, Department of Reproduction, School of Veterinary Medicine, University of California, Davis, reported that in the California Laboratory they have been able to solve approximately 91 percent of all the cattle parentage cases.

tialities of a heifer born twin with a bull goes back to the early events in the embryology of cattle twins. In about 85 percent of the cattle embryos, some of the chorionic blood vessels become anastomosed, or joined together. This results in a communal blood vascular system. Hence, the twins come to share each other's blood forming tissues. As a result, they have like blood types.

● *To substitute for fingerprinting*—Much attention is now being given to the idea of utilizing blood typing as a positive means of identification of stolen animals, through proving their parentage.

The following laboratories are capable of determining bull parentage:

1. Department of Reproduction, School of Veterinary Medicine, University of California, Davis, California 95612.

2. Immunogenetics Laboratory, Department of Animal Science, Texas A & M University, College Station, Texas 77843.

3. Cattle Blood Typing Laboratory, Department of Dairy Science, 625 Stadium Drive, Columbus, Ohio 43210.

4. Animal Disease Research Institute (E), Health of Animals Branch, Agriculture Canada, P.O. Box 11300, Postal Station "H," Ottawa, Ontario, Canada K2H 8P9.

FACTORS AFFECTING REPRODUCTION

Two major problems plague the cow-calf man: (1) the long calving season; and (2) a poor calf crop. Both can be lessened very materially.

Nature ordained that the gestation period of a beef cow be about 283 days. This means that a cow must be pregnant again within 80 days of calving if she is going to produce a calf each year (283 + 80 = 363). This is not easy, primarily because the interval from calving to first estrus is long. Warnick reported that the average interval from calving to first estrus is 61 days.[11] But, as shown in Table 6-3, it is longer in heifers than in older cows.

Table 6-3 reveals that, 50 days after calving, 53% of the mature cows had shown heat, in comparison with 24% of the heifers; hence, more than twice as many of the old cows were in heat. By 70 days after calving, 82% of the

TABLE 6-3

INTERVAL FROM CALVING TO FIRST ESTRUS;
FIRST-CALF HEIFERS VS MATURE COWS[1]

No. of Days After Calving	Percent of Cows in Heat at the Time	
	3-Year-Old Heifers	Cows 5 Years Old or Older
40	15	30
50	24	53
60	47	72
70	62	82
80	68	89
90	79	94

[1]Wiltbank, J. N. (1970), Research Needs in Beef Cattle Reproduction, *Journal of Animal Science*, Vol. 31, p. 755.

---

[11]Warnick, A. C., Factors Associated with the Interval from Parturition to First Estrus in Beef Cattle, *Journal of Animal Science*, Vol. 14, No. 4, 1955, p. 1003.

old cows had shown heat, in comparison with 62% of the first-calf heifers. Clearly, this shows that the interval from calving to first heat is too long, particularly in first-calf heifers.

This long interval from calving to first heat makes for reproductive problems such as (1) a high proportion of 2- and 3-year-old cows which are open or calving late, and (2) calving intervals of 13 to 14 months.

Since the pattern for a heifer's entire reproductive life is established with her first calf, it is important to get her off on the right start. To do so, 50 percent more replacement heifers should be selected than are needed as brood cows in order to get those which will produce early calves. Next, heifers should be kept separate from the brood cows, because younger animals don't compete well with older animals. From this stage on, getting heifers to calve early is a matter of condition, nutrition, and planning.

A poor calf crop is the other major problem affecting profits. Calving losses at birth have been estimated to run more than six percent of the potential calf crop, with these losses ranking second in magnitude only to cows failing to conceive. Fortunately, over 50 percent of these losses could be prevented by improved management.

Since calving difficulty is so important, a study was conducted at the Miles City Station to determine some of the causes of dystocia in first-calf heifers, and the relative importance of each. The study involved 95 Hereford and 103 Angus first-calf heifers. All heifers were bred artificially to one Angus or one Hereford bull to produce reciprocal crossbred calves. Heifers were held in feedlots throughout the study and gained 0.7 pound per day during gestation. During the calving season, they were observed 24 hours daily and calving difficulty was scored from 1 (none) to 4 (extreme). Calving difficulty was considered to depend on gestation length, gestation weight gain, sex of calf, pelvic area of dam, precalving weight of dam, and birth weight of calf. The results are shown in bar-graph form (Fig. 6-24), separately for Hereford and Angus heifers. The bars extending above the dotted line were statistically significant.

Fig. 6-24 shows that gestation length and gestation weight gain did not cause calving difficulty. The effect of sex of calf was significant in Hereford dams, but not in Angus.Pelvic area of dam was a factor in both breeds. Precalving weight of dam was associated with calving difficulty in Angus heifers, but not in Herefords. The fact that larger Angus heifers had more calving difficulty is perplexing and awaits additional experiments. Birth weight of the calf was the most important factor in both breeds. This is understandable. Two-year-old heifers are about 75 percent of their mature size, yet they produce calves 90 percent as large as those of older cows; hence, calving difficulty is caused by calves which are large in relation to the size of the dam. In this study, the effect on birth weight was the result of only one sire being used within each breed. Of course, there are many experiments showing that the bull or sire used has a marked effect on birth weight and calving difficulty.

Because of the economic importance of reproduction, a review of the experiments dealing therewith follows.

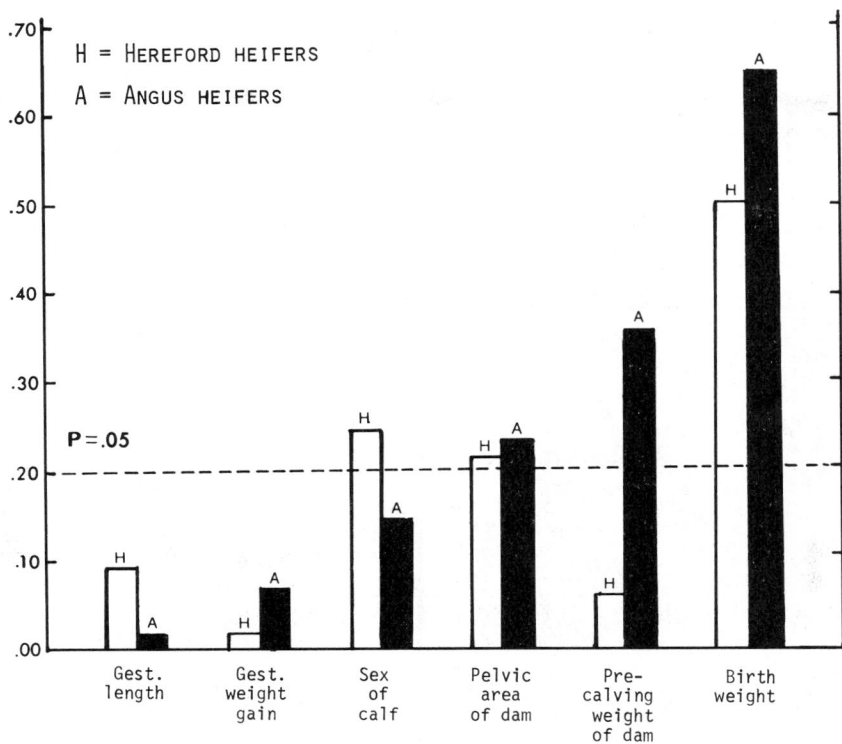

Fig. 6-24. Relative Importance of Factors Associated with Calving Difficulty of First-Calf Hereford and Angus Heifers. Bars extending above the line were statistically significant. (From: Short, R. E. and R. A. Bellows, U.S. Range Livestock Experiment Station, Miles City, Montana, Factors Affecting Calf Losses and Calving Difficulty, *Beef Cattle Science Handbook*, Vol. 11, p. 260, edited by M. E. Ensminger and published by Agriservices Foundation.)

Heifers vs Older Cows

As every cowman knows, calving difficulty and losses, known technically as dystocia, are higher in 2-year-old heifers than in older cows. Surveys indicate that approximately 50% of calving 2-year-olds require help at calving and 15% of the calves and 5% of the heifers are lost at parturition.[12]

At the U.S. Range Livestock Experiment Station, Miles City, Montana, where there is better than average care, 10% of the calves born to first-calf heifers during the 6-year period 1956-61 were lost, compared to a 2% to 4% loss in older cows. Table 6-4 shows the calving difficulty and calving assistance in cows of different ages at the Miles City Station.

---

[12]Moore, D. G., D. Chambers, J. A. Whatley, Jr., and W. D. Campbell, Some Factors Affecting Difficulty at Parturition of Two-Year-Old Hereford Heifers, *Journal of Animal Science*, Vol. 15, No. 4, p. 1225, 1956.

TABLE 6-4

CALVING DIFFICULTY AND CALVING ASSISTANCE
IN COWS OF DIFFERENT AGES[1]

Age of Dam	No. Calving	Av. Birth Wt.	Assisted Deliveries		
			Normal Presentations	Abnormal Presentations	Total
(yr.)		(lb)	(%)	(%)	(%)
2[2]	287	72	43	3	46
3[2]	158	78	16	6	22
4	465	83	1	2	3

[1]Short, R. E. and R. A. Bellows, U.S. Range Livestock Experiment Station, Miles City, Montana, Factors Affecting Calf Losses and Calving Difficulty, *Beef Cattle Science Handbook*, Vol. 11, edited by M. E. Ensminger and published by Agriservices Foundation, p. 259.
[2]First-calf heifers.

Fig. 6-25. Two-year-old heifers have a much higher incidence of calving difficulty than older cows. (Courtesy, American Breeders Service, De Forest, Wisc.)

Nutritional Factors Affecting Reproduction

Improper feeding—either (1) uncommonly high or low feed intake, or (2) a deficiency of specific nutrients—can affect reproduction. Experiments pertaining thereto are summarized in the sections that follow.

(Also see Chapter 19, section entitled, "Nutritional Reproductive Failure in Cows.")

LIBERAL FEEDING MAKES FOR EARLY SEXUAL MATURITY

The onset of heat and ovulation in young heifers is definitely and positively correlated with the level of nutrient intake. Thus, if feed intake is too low for normal rate of growth and development, the onset of reproductive function is delayed.

● *A long-term experiment to determine the effect of plane of nutrition on sexual maturity*—Cornell University raised 3 groups of Holstein heifers from birth to first calving on 3 different nutritive levels—*62%* (low), *100%* (medium), and *146%* (high)—of the standard amount of total digestible nutrients (TDN).[13] An extremely important finding in the Cornell study was that regardless of plane of nutrition, the heifers (Holsteins) came into heat at about 600 pounds body weight; showing that size and weight, not age, determine the time of sexual maturity. Thus, most researchers and knowledgeable cattlemen recommend breeding according to size and not age.

● *The effect of three winter feeding levels on early breeding of heifers*—The U.S. Range Experiment Station, Miles City, Montana, studied the effects of 3 winter feeding levels—*low*, average daily gain 0.6 lb; *medium*, average daily gain 1.0 lb; and *high*, average daily gain 1.5 lb. The effect on the number of heifers bred and pregnant early in the breeding season is shown in Table 6-5.

TABLE 6-5

EFFECT OF THREE WINTERING LEVELS OF ANGUS × HEREFORD CROSSBRED HEIFERS ON NUMBER BRED AND PREGNANT[1]

	(1) Low Level	(2) Medium Level	(3) High Level
Number of heifers	30	29	30
Winter gain, daily, lb	0.6	1.0	1.5
Summer gain, daily, lb	1.3	1.2	0.9
Weight at start of breeding (June 15), lb	458	527	584
Age at puberty (days)	434	412	388
Percent in heat by—			
Start of breeding (June 15)	7	31	83
End of breeding (August 15)	80	97	100
Not bred	20	3	0
Percent pregnant by—			
July 5 (20 days of breeding)	30	62	60
July 25 (40 days of breeding)	40	83	80
August 15 (60 days of breeding)	50	86	87

[1]Bellows, R. A., *Improving Reproductive Performance of Heifers*, U.S. Range Experiment Station, Miles City, Montana, 1971.

In the Montana study, the heifers on the low level winter ration averaged 458 pounds at the beginning of breeding. Only 7 percent of them were cycling, and only 30 percent of them conceived during the first 20 days of breeding. There was no significant difference between the medium level and high level groups in the percent of heifers that conceived.

Both the Cornell and Montana experiments showed that well-fed heifers developed faster sexually than those raised on a lower plane of nutrition, and that size and weight, not age, determine the time of sexual maturity.

---

[13]Reid, J. T., J. K. Loosli, G. W. Trimberger, K. L. Turk, S. A. Asdell, and S. E. Smith, Causes and Prevention of Reproductive Failures in Dairy Cattle, IV. Effect of Plane of Nutrition During Early Life on Growth, Reproduction, Production, Health, and Longevity of Holstein Cows,1. Birth to Fifth Calving, *Cornell University Agr. Exp. Sta. Bull.* 987, February 1964.

FLUSHING

Flushing refers to the practice of feeding thin cows and heifers to gain approximately 1.5 pounds per head daily beginning 20 days before the start of and continuing throughout the breeding season. It may be accomplished either by providing more lush pasture or range or by feeding grain.

By 30-day grain (corn) flushing during the spring breeding season, the Louisiana Station increased the calf crop by 11 percent.[14]

Although it is not likely that all of the benefits ascribed to flushing will be fully realized under all conditions, the general feeling persists that the practice will cause more cows to come into heat, to breed early in the season, and to conceive at first service. Certainly, following calving a cow should maintain her weight and/or make small gains.

ENERGY (FEED) LEVEL

Restricted rations often occur during periods of drought, when pastures or ranges are overstocked, or when winter rations are skimpy. When such feed shortages are extreme, there may be lowered reproductive efficiency. Likewise, too liberal feeding and high condition may cause sterility. Summaries of several experiments pertaining to the energy (feed) level of cows follow.

ENERGY (FEED) LEVEL DURING GESTATION

The best calf crop is produced by cows that are kept in vigorous breeding condition—that are neither overfat nor in a thin, run-down condition.

● *Effect of feed (energy) level of heifers during gestation on calf birth weight, calving difficulty, and when the dam comes back in heat after calving*—The U.S. Range Livestock Experiment Station, Miles City, Montana, conducted three studies designed to answer this question. The design of the experiment and the results are given in Table 6-6.

As shown in Table 6-6, reciprocal crossbred Angus-Hereford heifers were used each year. In 1969, one Angus bull was used; in 1970, one Hereford bull; and in 1971, 2 different Charolais bulls—one whose calves had a moderate weight of 81 pounds at birth, and the other whose progeny had a high birth weight of 97 pounds. Ninety days prior to the predicted calving date, all heifers were placed in a feedlot on either a low or a high feed level until calving. The pounds TDN of the low group were: 7.5, 7.0, and 8.0, respectively, for the years 1969, 1970, and 1971. The high level groups received nearly twice as many pounds of TDN per head daily: 13.8, 14.0, and 15.0, respectively, for 1969, 1970, and 1971.

All heifers were observed 24 hours per day during calving. Each calving was given a numerical score, ranging from "1" where there was no difficulty to "4" where there was extreme difficulty, including Caesarean delivery. Calf birth weights were taken immediately after calving.

---

[14]Loyacano, A. F., Grain Flushing Ups Calfing Percentage, *Better Beef Business*, May 1973, p. 22.

TABLE 6-6

EFFECTS OF GESTATION FEED LEVEL OF HEIFERS ON CALF BIRTH WEIGHT,
CALVING DIFFICULTY, AND WHEN HEIFERS COME BACK IN HEAT[1]

Year	Dam	No. Cows	Sire	Feed Level Last 90 Days of Gestation	Pre-calving Wt.	Calf Birth Wt.	Calving Difficulty Per-cent	Calving Difficulty Avg Sc.[2]	Cows in Heat by 6/15	Oct. Preg.
					(lb)	(lb)			(%)	(%)
1969	Reciprocal crossbreds Angus-Hereford	30 / 32	Angus	Low (7.5 lb TDN)	725	59	37	1.5	47	60
				High (13.8 lb TDN)	828	63	37	1.6	78	79
				Diff.	103	4	0	0.1	31	19
1970	Reciprocal crossbreds Angus-Hereford	11 / 12	Hereford	Low (7.0 lb TDN)	793	63	60	1.9	27	78
				High (14.0 lb TDN)	888	72	58	2.0	58	85
				Diff.	95	9	2	0.1	31	7
1971	Reciprocal crossbreds Angus-Hereford	14 / 14	Char. #1 (Moderate birth Wt. sire—81 lb)	Low (8.0 lb TDN)	827	73	86	2.9	-g	-g
				High (15.0 lb TDN)	900	72	78	2.4	-g	-g
				Diff.	73	1	8	0.5	—	—
		10 / 10	Char. #2 (High birth Wt. sire—97 lb)	Low (8.0 lb TDN)	801	79	100	2.8	17	58
				High (15.0 lb TDN)	900	76	90	2.7	36	84
				Diff.	99	3	10	0.1	19	26

[1]Short, R. E. and R. A. Bellows, U.S. Range Livestock Experiment Station, Miles City, Montana, Factors Affecting Calf Losses and Calving Difficulty, *Beef Cattle Science Handbook*, Vol. 11, 1974, edited by M. E. Ensminger and published by Agriservices Foundation, p. 263.

[2]Scores range: 1 = no difficulty to 4 = extreme.

It is noteworthy that the feed level during the last 90 days of gestation made quite a difference in precalving weights. The heifers on the high feed level weighed 103 and 95 pounds more than their counterparts on the low level in 1969 and 1970, respectively. In 1971, the differences were 73 and 99 pounds for the heifers bred to the moderate and high birth weight sires.

These studies showed that cutting the feed level prior to calving had little effect on either calf birth weight or calving difficulty, but it did have a marked depressing effect on subsequent reproduction. The heifers which were on the low levels were not showing heat (cycling) as well as the high level group at the beginning of the breeding season (June 15) and fewer became pregnant. The latter finding is very important. If cows are not cycling before the breeding season starts, they tend to breed late or not at all.

A word of caution relative to the above experiment appears to be in order. If the gestation feed level is very high and heifers become excessively fat, the effects on calving difficulty will be drastically different. Studies conducted by the U.S. Department of Agriculture at Beltsville, Maryland, showed that extreme calving difficulty and high calf and dam losses resulted when heifers were too fat at calving. Also, excessively fat heifers do not milk as well as heifers in medium flesh, and the depressing effect on milk production of a high plane of nutrition becomes more evident with each lactation.

● *Effect of two levels of energy from 120 days prior to calving until calving as two-year-olds*—In a study conducted at the Fort Robinson Beef Cattle Research Station, with the U.S. Department of Agriculture and the University of Nebraska cooperating, the effect of 2 levels of energy— *moderate* and *low*— on calving difficulty and calf losses of 2-year-old heifers was determined. Beginning 120 days prior to calving, heifers were fed 2 levels of energy. The moderate level energy group received 8 lb of TDN and gained 1 to 1.25 lb per day, while the low level group received 4.3 lb of TDN and gained 0.1 to 0.3 lb per day. The results of this study are summarized in Table 6-7.

At calving time, the heifers on the moderate level of feed weighed approximately 100 lb more than the ones on the low level of feed. At calving,

TABLE 6-7

EFFECT OF LEVEL OF ENERGY LAST 120 DAYS OF GESTATION ON CALVING DIFFICULTY AND CALF LOSSES OF TWO-YEAR-OLD ANGUS AND HEREFORD HEIFERS[1]

	First Year		Second Year	
	Moderate (8 Lb TDN)	Low (4.3 Lb TDN)	Moderate (8 Lb TDN)	Low (4.3 Lb TDN)
Number of cows	140	94	123	111
120-Day weight gain	150	35	120	13
Cow weight change[2]	+6	−130	−10	−125
Birth weight of calf, lb	70	63	70	64
Difficult births, %	37	34	36	20

[1]Wiltbank, J. N., *Proceedings of the Fifth Conference on Artificial Insemination of Beef Cattle*, National Association of Animal Breeders, pp. 18-24, 1971.
[2]Weight one day after calving minus weight 120 days before calving.

the moderate plane heifers lost 137 lb weight in comparison with 151 lb loss for the low level group. Calf birth weight was 7 lb heavier for the heifers on the moderate level feed than those on the low level of feed. There was no significant difference between the 2 groups in calves alive at birth and 2 weeks after birth.

Although the Fort Robinson Station reduced birth weight of calves and calving difficulty by feeding low levels of energy, calving difficulties were not eliminated. Also, and most important, the heifers receiving the low energy diet did not return to heat and conceive as readily as heifers on the moderate level of feed.

As a result of this study, the investigators (1) concluded that, "you cannot starve calving losses out of a group of heifers"; and (2) recommended that heifers be fed so as to gain approximately one pound per head per day for the last 100 to 120 days before calving.

• *Effect of energy level before calving on interval from calving to first estrus*—The Fort Robinson Beef Research Station, Crawford, Nebraska, found that energy level prior to calving had a marked effect on the interval from calving to first estrus.(See Table 6-8)

TABLE 6-8

EFFECT OF ENERGY LEVEL BEFORE CALVING ON POSTPARTUM
INTERVAL TO FIRST ESTRUS IN TWO-YEAR-OLD HEIFERS[1]

No. of Days After Calving	Percent 2-Year-Old Heifers Showing Estrus at the Time	
	4.3 Lb TDN Daily	8.0 Lb TDN DAILY
40	7	22
60	49	81
80	73	92
100	88	97

[1]Dunn, T. G., J. E. Ingalls, D. R. Zimmerman, and J. N. Wiltbank, Reproductive Performance of 2-Year-Old Hereford and Angus Heifers as Influenced by Pre- and Post-Calving Energy Intake, *Journal of Animal Science*, 1969, Vol. 29, p. 719.

Table 6-8 shows that more heifers receiving 8.0 pounds of total digestible nutrients (TDN) prior to calving were in heat by 40, 60, 80, or 100 days after calving than heifers receiving 4.3 pounds of TDN.

CONDITION OF HEIFERS AT CALVING

Excessively fat heifers have calving problems, with heavy calf losses at and soon after birth. However, the losses at birth are not the result of increased birth weight. Rather, they are due to the pelvic area being filled with fat and the calves being presented backward.

• *Effect of 3 levels of energy on calf losses of heifers*—Table 6-9 summarizes an experiment in which 3 levels of energy were fed from weaning until calving as 2-year-olds.

As shown in Table 6-9, within 24 hours after birth, 12 calves were dead in the high energy, overly fat, heifer group compared to 3 calves in each of the other groups; and the fat heifers weaned a 48 percent calf crop. This

TABLE 6-9

EFFECT OF LEVEL OF ENERGY ON CALVING OF TWO-YEAR-OLD HEIFERS[1]

| | Level of Energy | | |
	Low	Medium	High
Number of heifers calving ...	30	34	33
Gestation length	279	279	277
Condition at calving	thin	good flesh	extremely fat
Calves living:			
When born	27	31	27
24 hours after birth	27	31	21
2 weeks after birth	27	31	18
At weaning time	27	29	16
Average birth weight	45.1	59.5	56.2
Heifer weight before calving	698	921	1,104
Heifer weight after calving ..	618	831	1,012

[1]Wiltbank, J. N., E. J. Warwick, R. E. Davis, A. C. Cook, W. L. Reynolds, and M. W. Hazen, Influence of Total Feed and Protein Intake on Reproductive Performance of the Beef Female Through Second Calving, *USDA Technical Bulletin 1314*, 1965, pp. 20-21.

Fig. 6-26. Very fat heifers often have calving difficulty, because of excess fat decreasing the size of the birth canal. (Courtesy, American Breeders Service, De Forest, Wisc.)

means that 2 heifers were kept to produce one weaner calf. This does not indicate that a cattleman should attempt to alleviate calving difficulties by starving heifers. Rather, they should be in medium condition, neither too fat nor too thin.

ENERGY (FEED) LEVEL AFTER CALVING

Cows with calves at side should be fed for the production of milk, which requirements are more rigorous than those during pregnancy. The energy requirements of a cow nursing a calf are about 50 percent higher than that of a dry pregnant cow.

● *Effect of energy level following calving on estrus*—Energy level after calving affects the proportion of cows showing heat (see Table 6-10).

Table 6-10 shows that 19% (100 - 81 = 19) of the heifers on a low energy ration of 7 lb TDN daily failed to show heat during a 100-day breeding season in comparison with 3% and 2%, respectively, of the heifers receiving 13 and 22 lb of TDN. Also, fewer heifers receiving 7 lb of TDN showed heat 40, 60, and 80 days than ones receiving the two higher levels of energy.

TABLE 6-10

EFFECT OF ENERGY LEVEL AFTER CALVING ON POSTPARTUM
INTERVAL TO FIRST ESTRUS IN TWO-YEAR-OLD HEIFERS[1]

No. of Days After Calving	Percent 2-Year-Old Heifers Showing Estrus at the Time		
	7 Lb TDN Daily	13 Lb TDN Daily	22 Lb TDN Daily
40	17	22	36
60	62	64	81
80	81	92	93
100	81	97	98

[1]Dunn, T. G., J. E. Ingalls, D. R. Zimmerman, and J. N. Wiltbank, Reproductive Performance of 2-Year-Old Hereford and Angus Heifers As Influenced by Pre- and Post-Calving Energy Intake, *Journal of Animal Science*, 1969, Vol. 29, p. 719.

● *Effect of level of energy following calving on early pregnancy*—Table 6-11 summarizes the results of a series of experiments designed to determine the effect of energy following calving on reproductive performance.

Table 6-11 shows that the low level of energy after calving had the following marked effects on both older cows and heifers: (1) a much greater number of females did not show heat during the breeding season; and (2) a sharp decrease in the number of females pregnant at the end of 20 days of breeding, from first service, and at the end of the breeding season.

In experiment No. 1, of the older cows which received 8 lb of TDN (16 to 18 lb of hay) following calving, 34 percent were pregnant after 20 days of breeding compared to 60 percent of cows receiving 16 lb of TDN (32 to 35 lb of hay). Similar results were obtained in 2-year-old heifers (experiments 2 and 3), although pregnancy rates were lower than in the older cows. The latter situation is understandable because the older cows were bred for 90 days, whereas the heifers were bred for 60 days only. Also, breeding on both groups started 60 days after calving, which disadvantaged the heifers.

Table 6-11 also reveals that there are two reasons for the poor reproductive performance in cows which are on low levels of feed and losing weight following calving: (1) some cows do not show heat during the breeding; and (2) the conception rate at first service is low.

TABLE 6-11

EFFECT OF TWO LEVELS OF ENERGY FOLLOWING CALVING ON EARLY PREGNANCY[1]

Exp. No.	Level of Feed		No. Cows	Did Not Show Heat During Breeding Season (%)	Pregnant		
	Before Calving TDN (lb)	After Calving TDN (lb)			At End of 20 Days of Breeding (%)	From First Service (%)	At End of Breeding Season (%)
1				Cows 5 Years or Older			
	9	16	21	0	60	67	95
	9	8	22	14	34	42	77
2				Cows 2 Years of Age			
	8	13	37	3	54	63	71
	8	7	42	19	33	53	64
3				Cows 2 Years of Age			
	8	13	24	0	54	50	79
	8	7	13	8	23	37	76

[1]Wiltbank, J. N., W. W. Rowden, J. E. Ingalls, K. E. Gregory, and R. M. Koch, Effect of Energy Level on Reproductive Phenomena of Mature Hereford Cows, *Journal of Animal Science*, 1962, Vol. 21, p. 219.

Wiltbank, J. N., W. W. Rowden, J. E. Ingalls, and D. R. Zimmerman, Influence of Post-Partum Energy Intake on Reproductive Performance of Hereford Cows Restricted in Energy Intake Prior to Calving, *Journal of Animal Science*, 1962, Vol. 21, p. 658.

Dunn, T.G., J. E. Ingalls, D. R. Zimmerman, and J. N. Wiltbank, Reproductive Performance of 2-Year-Old Hereford and Angus Heifers As Influenced by Pre- and Post-Calving Energy Intake, *Journal of Animal Science*, 1969, Vol. 29, p. 719.

PROTEIN

Cattlemen are prone to skimp on protein, especially (1) when cows are grazed on dried grass cured on the stalk in times of drought or during the winter months, or (2) when protein supplements are high in price. Sometimes both producers and those who counsel with them suspicion that such practice may cause cows to abort, have calving difficulty, drop dead or weak calves, fail to rebreed, and/or not milk well. Fortunately, experiments designed to answer these questions have been conducted.

● *Effect of protein level on number of calves born, calf birth weight, and calf losses*—Three levels of protein were fed to cows from conception through calving: *High,* 1.38 lb of digestible protein; *medium,* 0.89 lb of digestible protein; and *low,* 0.26 lb of digestible protein. The results are summarized in Table 6-12.

TABLE 6-12

EFFECT OF PROTEIN LEVEL ON NUMBER OF CALVES BORN,
CALF BIRTH WEIGHT, AND CALF LOSSES[1]

	Level of Digestible Protein Consumed Per Head Daily, Conception Through Weaning		
	Low (0.26 lb)	Medium (0.89 lb)	High (1.38 lb)
Number of heifers calving	35	32	30
Gestation length	277	278	280
Condition of calving	thin	good flesh	good flesh
Calves living:			
When born	29	30	26
24 hours after birth	29	26	24
2 weeks after birth	28	26	22
At weaning time	25	26	21
Average birth weight	49.4	55.0	56.4
Heifer weight before calving	840	948	935
Heifer weight after calving	758	856	848

[1]Wiltbank, J. N., E. J. Warwick, R. E. Davis, A. C. Cook, W. L. Reynolds, and M. W. Hazen, Influence of Total Feed and Protein Intake on Reproductive Performance in the Beef Female Through Second Calving, *USDA Technical Bulletin 1314,* 1965, pp. 22 & 23.

Except for the lower birth weight of the calves born to cows on a low level of protein, there were no adverse effects from a low protein ration on the calf crop—in number of calves born and calf survival.

● *Effect of protein level on rebreeding and on calf gains of first-calf heifers*—Among cattlemen the feeling persists that a shortage of protein following calving will have an adverse effect on rebreeding and on calf gains. This question was pursued by the same investigators who conducted the experiment summarized in Table 6-12.

Cows suckling their first calves received 3 levels of digestible protein: high, 2.0 lb; medium, 1.3 lb; and low, 0.4 lb. The results are summarized in Table 6-13.

The onset of heat was delayed in the cows on the low level of protein but conception was not adversely affected. The weight gains on calves were lowered by the low level of protein. The medium level of protein was as satisfactory as the high level; hence, there is no need to go to the added cost of providing more protein than is needed.

TABLE 6-13

EFFECT OF PROTEIN LEVEL ON REBREEDING, AND CALF GAINS OF
COWS SUCKLING THEIR FIRST CALVES[1]

| | Digestible Protein Consumed Per Head Daily During Lactation | | |
	Low (0.4 lb)	Medium (1.3 lb)	High (2.0 lb)
Number of cows	21	21	21
Proportion in heat:			
60 days after calving, %	36	48	71
90 days after calving, %	54	90	81
Pregnant, %	91	90	100
Average daily gains of calves:			
First 60 days of life, lb	1.1	1.7	1.7
First 150 days of life, lb	1.3	1.7	1.8

[1]Adapted from: Wiltbank, J. N., E. J. Warwick, R. E. Davis, A. C. Cook, W. L. Reynolds, and M. W. Hazen, Influence of Total Feed and Protein Intake on Reproductive Performance in the Beef Female Through Second Calving, USDA Technical Bulletin 1314, 1965, pp. 24-27. Low energy data not included.

PHOSPHORUS

Generally speaking, range forages are low in phosphorus. This is especially true of dried grass cured on the stalk, such as exists during droughts and in the winter.

•Effect of phosphorus on reproduction—Black and coworkers showed an increase in reproductive efficiency from phosphorus supplementation of cows in a five-year study on King Ranch in Texas, where the range is deficient in phosphorus (Table 6-14).

In the King Ranch study, phosphorus was supplied by either (1) bone meal, self-fed, (2) dicalcium phosphate in drinking water, or (3) fertilized pasture. As shown, phosphorus supplementation markedly increased the percentage of calves dropped (by 26%) and weaned (by 28%), the weaning weight of calves (54 lb), and the pounds calf weaned per cow (by 179 lb).

TABLE 6-14

EFFECT OF PHOSPHORUS ON REPRODUCTION
AT KING RANCH IN TEXAS[1]

Group	No. of Cows	Cows Calving	Calves Weaned	Calf Weaning Weight	Calf Weaned /Cow
		(%)	(%)	(lb)	(lb)
No phosphorus supplement	42	69	64	489	319
Three phosphorus supplemented groups:	141	95	92	543	498
Bone meal, self fed	42	92	88	535	468
Disodium phosphate in drinking water	42	96	92	542	500
Fertilized pasture	57	98	96	551	527

[1]Black, W. H., L. H. Tash, J. M. Jones, and R. J. Klebert, Jr., Comparison of Methods of Supplying Phosphorus to Range Cattle, USDA Technical Bulletin 981, 1949.

Other experiments have also shown that beef cows generally respond favorably to phosphorus supplementation; among them, studies conducted by the New Mexico Station and the Oklahoma Station. Also, there is general agreement that phosphorus should be available to beef cows throughout the year.

Calving Difficulty

Difficult births may cause pathological conditions resulting in uterine tissue damage. Such conditions may or may not affect milk production, but they very likely will result in greater difficulty in getting affected cows settled.

• *Effect of calving difficulty on rebreeding fertility*—In a study of 1,889 Angus and Hereford cows and heifers at the Meat Animal Research Center (MARC), Clay Center, Nebraska, it was found that cows experiencing difficult calving had significantly lower rebreeding fertility (Table 6-15).

Each of the measures of fertility shown in Table 6-15 indicates that calving difficulty increases the percentage of cows that do not resume their normal estrus cycle before the end of the breeding season.

TABLE 6-15

EFFECT OF CALVING DIFFICULTY ON REBREEDING[1]

	Cows Experiencing	
	Calving Difficulty	No Calving Difficulty
Number of cows	466	1,423
Detected in heat (45 days A.I.), % ..	60	74
Producing an A.I. calf, %	54	69
Producing a calf (70-day breeding season), %	69	85

[1]*The Profit Brand*, Vol. 5, No. 6, American Breeders Service, De Forest, Wisc., 1973.

Suckling Stimulus

Studies have shown that the interval from calving to first heat is 15 to 52 days longer in cows suckling calves than in cows which are milked. In a herd of Milking Shorthorns, Wiltbank and Cook found that the interval from calving to first estrus was 30 days longer, and the interval from first service to conception was 27 days longer, in nursed cows than in cows milked twice a day.[15] This indicates that early weaning—around 45 to 60 days of age—would give a big assist in bringing cows in heat following calving.

Size of Pelvic Opening

The calf must come through the birth canal. Consequently, when the calf is too big or the pelvic opening is too small, calving difficulties result. It would appear, therefore, that measurement of the pelvic area should provide

___

[15]Wiltbank, J. N., and A. C. Cook, The Comparative Reproductive Performance of Nursed Cows and Milked Cows, *Journal of Animal Science*, Vol. 17, No. 3, 1958, p. 640.

a helpful index to use in decreasing calving difficulty.

The pelvic opening measurement technique consists of rectally insert-ing the hand containing the instrument (either a pelvimeter or a pair of slid-ing calipers) and determining the height and width of the birth canal. These results are multiplied together to determine the area.

● *The size of the pelvic opening in two-year-old heifers is small com-pared to that in older cows*—Studies conducted at the U.S. Range Livestock Experiment Station, Miles City, Montana, show that the pelvic opening of first-calf heifers is small (Table 6-16).

TABLE 6-16

PELVIC OPENING OF COWS OF DIFFERENT AGES[1]

Age	No.	Pelvic Area		Body Weight
(yr.)		*(sq. in.)*	*(sq. cm)*	*(lb)*
2[2]	198	38.75	250	813
3[2]	251	45.26	292	1,023
4 and 5	75	51.46	332	1,045

[1]Short, R. E. and R. A. Bellows, U. S. Range Livestock Experiment Station, Miles City, Montana, Factors Affecting Calf Losses and Calving Difficulty, *Beef Cattle Science Handbook*, Vol. 11, edited by M. E. Ensminger and published by Agriservices Foundation, 1974, p. 259.
[2]First-calf heifers.

Table 6-16 shows that the precalving pelvic area of 2-year-old heifers studied was 6.51 sq. in. (or 42 sq. cm) smaller than the pelvic area of the first-calf 3-year-olds, and 12.71 sq. in. (or 82 sq. cm) smaller than the pelvic area of the 4- and 5-year-old cows. The Montana Station workers point out that (1) numbers were small in the group recorded as 4- and 5-year-old cows, and (2) the measurements of the older cows were not obtained from the same animals represented in the 2- and 3-year-old age groups. Nevertheless, there is strong indication that the size of the birth canal increases with age to a certain stage of maturity; hence, it suggests a reason for less calving difficulty in older animals.

Experiments show that the percentage of heifers experiencing calving difficulty decreases steadily as the pelvic area increases, particularly where the calf is presented normally. The only very difficult births encountered are in those heifers having a pelvic area less than 34.10 sq. in. (220 sq. cm). This would indicate that it would be possible to cull heifers with pelvic openings of less than 34.10 sq. in. (220 sq. cm) and decrease calving difficulty mark-edly. However, in some herds this would result in culling half, or more, of the heifers, which would be impossible. So, two approaches are suggested: (1) Breed heifers with pelvic areas smaller than 34.10 sq. in. (220 sq. cm) to a bull that will sire calves with low birth weights; and (2) include in the re-placement heifer selection program data on pelvic area, then cull those with small pelvic openings.

Sire

Most cattlemen assume that the bull is a major factor contributing to calving difficulty, and many of them will tell you that the way to reduce calving difficulty in heifers calving as two-year-olds is to use Angus bulls, because, so they say, they sire small calves at birth. The first assumption is correct, but the second is not necessarily so.

● *Effect of sire breed on calving difficulty*—Many experiments have shown that the sire breed used has a marked effect on birth weight and calving difficulty. The results of one such study, conducted at the U.S. Range Livestock Experiment Station, Miles City, Montana, are reported in Table 6-17.

TABLE 6-17

EFFECTS OF SIRE BREED ON BIRTH WEIGHT AND CALF LOSSES[1]

| Sire Breed | Birth Wt. | | | Calf Losses | |
	Straight-bred Matings	Crossbred Matings	Brown Swiss Matings	Heifers[2]	Cows
	(lb)	(lb)	(lb)	(%)	(%)
Hereford	77	78	87	3.6	1.1
Angus	70	78	90	3.1	3.5
Charolais	90	89	109	6.8	3.9

[1]Short, R. E. and R. A. Bellows, U. S. Range Livestock Experiment Station, Miles City, Montana, Factors Affecting Calf Losses and Calving Difficulty, *Beef Cattle Science Handbook*, Vol. 11, 1974, edited by Dr. M. E. Ensminger, and published by Agriservices Foundation, p. 262.
[2]First-calf 3-year-olds.

The Miles City data (Table 6-17), show that the birth weights of calves sired by Charolais bulls was higher than calves sired by Hereford or Angus bulls. Likewise, the Charolais gave the highest calf losses.

● *Effect of sire breed on birth weight and calving difficulty*—Data on 793 calves from 2-year-old Hereford and Angus heifers and 1,802 calves from Hereford and Angus 3- , 4- , and 5-year-old cows at the Meat Animal Research Center, Clay Center, Nebraska, show the effect of sire breed on birth weight and calving difficulty (Table 6-18).

TABLE 6-18

EFFECT OF SIRE BREED ON BIRTH WEIGHT AND CALVING DIFFICULTY[1]

| Breed of Sire | Average Birth Wt. | Percent Assisted | |
		2-Year-Old Heifers	3- , 4- , 5-Year-Old Cows
	(lb)	(%)	(%)
Jersey	73	17	1
Angus	79	41	7
Hereford	82	46	6
South Devon	86	55	17
Limousin	87	75	13
Simmental	91	74	17
Charolais	92	77	24

[1]*The Profit Brand*, Vol. 5, No. 6, published by American Breeders Service, De Forest, Wisc., 1973.

The Clay Center data (Table 6-18) also show that there are large differences between sire breeds in birth weight and calving difficulty. The four exotic breeds listed gave far more problems than Angus or Herefords; Jerseys gave the least problem. This would indicate that, in some cases, a Jersey sire should be seriously considered for use on heifers.

It is noteworthy that the correlation between average birth weight and percent requiring assistance is .96 for the 2-year-old heifers and .93 for the 3- to 5-year-old cows, indicating that almost all differences in calving difficulty caused by these breeds of sires are due to the birth weight of their calves.

The Clay Center data also indicate that, all other factors being equal, each one-pound increase in average birth weight will increase the percent of difficult calving by 3.2 percent for 2-year-old Hereford and Angus heifers and by 1.1 percent for 3- , 4- , and 5-year-old Hereford and Angus cows.

Clearly, therefore, choosing sires of breeds with lower birth weights will reduce the percentage of difficult births.

● *Angus bulls do not necessarily sire small calves*—In an experiment in which Angus and Hereford heifers were bred to Angus and Hereford bulls, there was no decrease in calving difficulties when Hereford heifers were bred to Angus bulls (Table 6-19).

TABLE 6-19

CALVING DIFFICULTY IN ANGUS AND HEREFORD HEIFERS BRED TO ANGUS AND HEREFORD BULLS TO CALVE AS TWO-YEAR-OLDS[1]

Breed of Sire	Breed of Dam	1963		1964	
		Number Calves Born	% Calving Difficulty	Number Calves Born	% Calving Difficulty
Hereford	Hereford	84	24	64	52
Angus	Hereford	71	30	61	56
Hereford	Angus	67	29	56	64
Angus	Angus	55	26	55	59

[1]Wiltbank, J. N., Relationship of Energy, Cow Size and Sire to Calving Difficulty, *Proceedings of the Fifth Conference on Artificial Insemination of Beef Cattle*, National Association of Animal Breeders, 1971, p. 22.

The study reported in Table 6-19 involved four Angus and four Hereford bulls each year. The investigators found more difference in calving difficulty within each sire breed than between breeds. Also, the bulls causing the greatest calving difficulty tended to sire calves which had the highest birth weights.

● *Within breed differences of bulls in weight of calves sired*—Even though Tables 6-18 and 6-19 show wide differences in birth weights and calving difficulty by breed of sire, the author wishes to emphasize that sires should not be selected by breed alone; rather, they should be selected on the basis of individual records. Although there are breed differences in weight of calves sired, much variation exists between sires within a breed. This situation is pointed up in Table 6-20.

Table 6-20 shows that birth weights range from 68 to 80 pounds from 5 different sire lines within the Hereford breed. Other breeds show similar

TABLE 6-20

EFFECT OF DIFFERENT HEREFORD LINES ON
BIRTH WEIGHT OF CALVES[1]

Sire Line	No. Calves	Birth Weight
		(%)
9	20	68
6	18	70
4	23	77
1	19	80
10	22	80

[1]Short, R. E. and R. A. Bellows, U. S. Range Livestock Experiment Station, Miles City, Montana, Factors Affecting Calf Losses and Calving Difficulty, *Beef Cattle Science Handbook*, L/O Vol. II, 1974, Edited by M. E. Ensminger and published by Agriservices Foundation, p. 262.

differences. Thus, to lessen calving difficulty, the important thing is to select a bull known to throw light calves, or in case of an untried sire, select one that had a small birth weight.

Induced Calving and Exercise

Experiments show that the longer the gestation period the higher the calf birth weight. Since birth weight is the most important cause of calving difficulty, it would appear that calves from shorter gestation periods would encounter less difficulty at calving. Recent research has shown that treating pregnant cows with adrenal steroids will induce calving. Thus, a method is available to shorten gestation and presumably get lighter calves.

Most cattlemen feel that adequate exercise in the pregnant female is of great importance in preventing calving difficulty. Their reasoning: Without sufficient exercise, muscle tone is not maintained, with the result that the muscles involved in the abdominal press exerted during calving do not have sufficient force to expel the calf.

● *Effect of induced calving and exercise on birth weight, calving difficulty, and retained placentas*—The U.S. Range Livestock Experiment Station, Miles City, Montana, conducted a study of the effects of (1) shortened gestation (induced calving), and (2) exercise on calving difficulty.

Angus-Hereford crossbred cows and heifers were used, and they were bred to the same Charolais sire. Ninety days prior to calving, pregnant dams were divided into 2 groups. One group was confined to a small feedlot, where activity was restricted, and fed at the rate of 10 pounds of TDN daily. The second group was maintained under range conditions and fed the same type of feed as the drylot group in amounts necessary to obtain the same weight gains as their counterparts in the drylot. The pasture animals were fed 1 mile from the only water source, thereby forcing them to walk at least 2 miles per day.

All animals were calved in corrals. Those treated for induced calving were injected intramuscularly with adrenal steroid (flumethazone) at the rate of 10 mg per head on approximately day 269 of gestation.

The results (preliminary) are summarized in Table 6-21.

Calving was successfully induced with an adrenal steroid, and gestation

TABLE 6-21

EFFECT OF INDUCED CALVING AND EXERCISE ON BIRTH WEIGHT,
CALVING DIFFICULTY, AND RETAINED PLACENTAS (PRELIMINARY)[1]

Age of Dam, Activity, and Type of Calving	No.	Induction Time[2]	Gest. Length	Birth Wt.	Calf. Diff.		Retained Placentas
		(hr.)	(days)	(lb)	(%)[3]	(sc.)[4]	(%)[5]
Heifers							
Forced Activity[6]							
Natural Calving	9	—	280.3	70.4	44	1.56	22
Induced calving	12	43.2	272.7	61.4	50	1.92	50
Restricted activity[7]							
Natural calving	13	—	279.2	66.9	46	1.69	8
Induced calving	10	45.4	271.4	65.9	70	2.10	80
Cows							
Forced activity[6]							
Natural calving	21	—	280.6	77.4	5	1.05	5
Induced calving	16	47.6	272.2	67.8	12	1.19	69
Restricted activity[7]							
Natural calving	18	—	281.5	75.6	6	1.22	11
Induced calving	19	48.6	272.7	72.5	10	1.26	68

[1]Short, R. E. and R. A. Bellows, U. S. Range Livestock Experiment Station, Miles City, Montana, Factors Affecting Calf Losses and Calving Difficulty, *Beef Cattle Science Handbook*, Vol. 11, 1974, edited by M. E. Ensminger and published by Agriservices Foundation, p. 264.
[2]Hours elapsing from injection of sterioid until calf was born.
[3]Represents all dams given assistance.
[4]Average score including abnormal presentations.
[5]Fetal membranes retained longer than 12 hours.
[6]Forced walk of 2 miles daily last 90 days of gestation.
[7]Held in feedlots last 90 days of gestation.

was shortened from 280.4 days (controls) to 272.2 days (treated). Also, the shortened gestation period resulted in an average of 5.7 pounds lighter birth weight. However, induced calving increased both calving difficulty and retained placentas. It is noteworthy, however, that the vigor and livability of the induced calves were as good as the noninduced ones. Noteworthy, too, is the fact that induced calving has labor-concentrating and saving possibilities in herds where breeding dates are known.

Forced or restricted activity during gestation had little or no effect on gestation length, calf birth weight, incidence of calving difficulty, or calving difficulty score.

Late Calving Heifers Usually Remain Late Calvers

Late calving heifers have little chance to calve early the next year because early in the breeding season in subsequent years (1) too few of them show heat, and (2) too few of them conceive at first service. Also, it is recognized that the time for young cows to show heat after calving is longer than that required by older cows.

Nature ordained that the uterus must return to normal for good conception rates. Available data indicate that 50 to 60 days after calving are required for conception rates to reach their maximum level.

If a 90 to 95 percent calf crop is expected each year, with 75 to 80 percent of the calves born early in the season, the management system must allow each cow sufficient time for the uterus to return to normal after calving and some time to come back in heat. The only feasible way to provide this time is to shorten the calving season. And since it takes a longer time for heifers to show heat after calving than older cows, the only feasible way of providing this added time is to breed heifers earlier than the rest of the herd, so that they will drop their calves earlier and be ready to rebreed on schedule each year thereafter.

The solution to the above problems appears to be as follows: Start with about 50 percent more heifers than needed, breed them 21 days earlier than the old cows, and shorten their breeding season to 45 days.

A Management Program for Improved Reproduction

Any management program for improved reproduction must begin with heifers, for the pattern of a cow's entire productive life is established with her first calf. Based on experiments and experiences, the following management system is recommended for replacement heifers:

• Select 50 percent more heifers than you actually need as replacements, thereby permitting culling the late breeders—those that fail to conceive in a 45-day breeding season.

• Separate heifers from older cows.

• Feed heifers so that they weigh 650 to 750 lb at breeding time. (Remember that there are breed differences. Remember, too, that size, not weight, determines the onset of estrus. For example, Angus heifers reach puberty at about 650 lb, and Herefords at 750.) Thus, gauge the feeding program accordingly. If a Hereford heifer weighs 500 lb at weaning (7 months), and it is planned to breed her at 14 months of age, she needs to gain 250 lb—that's 1.2 lb per day (250 ÷ 210 days = 1.2).

• Breed heifers 21 days earlier than older cows and limit their first breeding season to 45 days. Cull those that do not conceive.

• Feed pregnant heifers to gain about one lb per day during the 120 days prior to calving. That way they will retain their normal weight after calving. This is important because a cow's nutritional requirements double after calving, making it difficult to put weight on a thin first-calf heifer.

A management system like this has been designed and tested experimentally by the Colorado Station in a five-year study, the results of which are summarized in Tables 6-22, 6-23, 6-24, 6-25, and 6-26.[16]

The Colorado workers studied the effect of early calving of heifers on their future reproductive performance and pounds of calf weaned. One hundred and forty registered yearling heifers were divided into a New Management (NM) group designed to ensure that the heifers would calve early, and a control group (C) in which there was no attempt to get the heifers to calve early. On the New Management group, 170 percent more heifers than needed as replacements were bred; breeding lasted only 45 days; breeding started 21 days earlier than the cow herd; and estrus synchronization was

[16]Spitzer, John C., James N. Wiltbank, and Donald G. LeFever, Improve Reproduction—Wean More Pounds of Calf, Research Highlights of the Animal Science Department, Colorado State University, Experiment Station, Fort Collins, Colo. General Series 931, pp. 20-22.

TABLE 6-22

EXPERIMENTAL DESIGN

Group	New Management	Control
No. pregnant replacements needed	50	50
No. exposed for breeding	85	54
Breeding season started—		
As heifers	4-22	5-12
3-year-olds and greater	5-12	5-12
Length of breeding (days)	45	90
Estrus synchronization	yes	no
No. cycling 4/22/68	54 (64%)	35 (63%)

TABLE 6-23

MORE HEIFERS BECAME PREGNANT EARLY IN THE 1ST YEAR (1968)

	New Management		Control	
	In Heat	Pregnant	In Heat	Pregnant
By 4/22/68[1]	54 (64%)	—	35 (63%)	—
After 4 days of breeding	53 (62%)	30 (35%)	9 (17%)	1 (2%)
After 21 days of breeding ...	62 (73%)	45 (50%)	38 (70%)	26 (48%)
After 45 days of breeding ...	68 (80%)	56 (66%)	45 (83%)	32 (56%)
After 90 days of breeding ...	—	—	52 (96%)	48 (89%)

[1]The start of the breeding season for the NM group.

TABLE 6-24

COWS CONTINUE TO BECOME PREGNANT EARLY IN
THE BREEDING SEASON

	NM	Control	Diff.
	Pregnant during 1st 21 days of breeding season		
	(%)	(%)	(%)
2nd year (1969)	54[1]	38[1]	16
3rd year (1970)	68	58	10
4th year (1971)	86	48	38
5th year (1972)	72	40	32
	Pregnant after 45 days of breeding		
	(%)	(%)	(%)
2nd year (1969)	74[1]	51[1]	23
3rd year (1970)	90	80	10
4th year (1971)	94	79	15
5th year (1972)	91	89	3
	Pregnant at end of breeding season		
	(%)	(%)	
2nd year (1969)	74 (45 days)[1]	77 (90 days)[1]	
3rd year (1970)	90	93	
4th year (1971)	94	92	
5th year (1972)	91	98	

[1]A poor collection of semen resulted in these lowered pregnancy rates.

TABLE 6-25

COWS BECOME PREGNANT EARLY BECAUSE THEY SHOWED HEAT
EARLY IN THE BREEDING SEASON

	NM	Control	Diff.
	In heat after 1st 21 days of breeding season		
	(%)	(%)	(%)
2nd year (1969)	96	67	29
3rd year (1970)	93	73	20
4th year (1971)	92	77	15
5th year (1972)	100	92	8
	In heat after 45 days of breeding		
	(%)	(%)	(%)
2nd year (1969)	100	95	5
3rd year (1970)	100	93	7
4th year (1971)	100	94	6
5th year (1972)	100	100	0
	In heat at end of breeding season		
	(%)	(%)	
2nd year (1969)	100 (45 days)	100 (90 days)	
3rd year (1970)	100	100	
4th year (1971)	100	100	
5th year (1972)	100	100	

TABLE 6-26

WEANING WEIGHTS OF CALVES (ADJUSTED TO BULL WEIGHTS)

	NM	Control	Diff.
	(lb)	(lb)	(lb)
1st calf (1969)	409	342	67
2nd calf (1970)	436	421	15
3rd calf (1971)	434	398	36
4th calf (1972)	452	423	29
4-year average			37

used. In the C group approximately the same number of heifers as needed for replacements were bred; breeding lasted 90 days and started the same day as the cow herd (Table 6-22). The replacement heifers came from their respective groups and were handled the same as their dams had been the first year. The criteria for selection of NM replacements was early pregnancy while C replacements were selected on the basis of 205-day adjusted weights and conformation. All heifers and cows were bred by artificial insemination to one bull each year.

Results of the Colorado study show that the New Management (NM) group of replacement heifers continued to be "early calvers" in their subsequent calving seasons. This was largely the result of exposing more heifers in the NM group than needed as replacements. This becomes evident by noting that the percent in heat and the percent pregnant at 21 and 45 days were very similar in both groups (Table 6-23). The differences in pregnancy rates after 21 days of breeding were 16%, 10%, 38%, and 32%; and after 45 days of breeding were 23%, 10%, 15%, and 3% in the second through the fifth years, respectively (Table 6-24). The pregnancy rates were comparable in the 2 groups at the end of the breeding season each year, but the NM group was only bred for 45 days while the C group was bred for 90 days. These differences in pregnancy rates were the result of more of the NM cows being in heat early in the breeding season (Table 6-25). The increase in reproductive performance also resulted in an average increase of 37 pounds weaning weight per cow per year in the NM group (Table 6-26).

From this study, it may be concluded that a management system, including proper nutrition, in which a surplus of heifers is bred 21 days earlier than the cows, and in which the first breeding season is limited to 45 days, will result in early heat and conception throughout life and more total pounds of calf weaned.

It is emphasized that a 45-day breeding season may be disastrous unless it is accompanied by the proper nutrition level so that cows will show estrus and have a high conception rate at first service. However, experiments and experiences indicate that under optimum conditions, a 45-day breeding season will not decrease the number of cows calving.

Summary of Factors Affecting Reproduction

Difficult calving is a major cause of calf losses and a contributing factor to rebreeding difficulty.

Any factor which increases the size of the calf, and/or decreases the size of the birth canal of the mother, causes an increase in the percentage of difficult births.

Breeds differ widely in the incidence of calving difficulty. Apparently bulls of the same breed do not cause large differences in the incidence of difficult births.

The following conclusions appear to be justified based on experiments and experiences:

1. *Magnitude of calf losses*—Calf losses at birth have been estimated to run more than 6% of the potential calf crop, with these losses ranking second in magnitude only to cows failing to conceive. Fortunately, over 50% of these losses could be prevented by improved management.

2. *Age of Cow*—Difficult births and calf losses from heifers are higher than in older cows; 10% of the calves from first-calf heifers may be lost, in comparison with 6% for cows of all ages in the United States.

3. *Liberal feeding of growing heifers*—Liberal feeding makes for early sexual maturity. This is important because size and weight, not age, determine the time of sexual maturity.

4. *Flushing*—Flushing by feeding cows and heifers to gain approximately 1.5 pounds per head daily beginning 20 days before the start of and continuing through the breeding season will result in an increase in the calf crop and in the cows breeding both earlier and more nearly at the same time.

5. *Large calves*—Large calves, as determined by birth weight, are the leading cause of calving difficulty.

6. *Sex of calf*—Two-thirds of all difficult births involve bull calves. This appears to be largely due to birth weight, as bull calves average 5 to 7 pounds heavier at birth than heifers.

7. *Gestation length*—Length of gestation is positively correlated (.30) with birth weight; hence, it is a cause of larger calves at birth. However, short gestations do not automatically mean small calves at birth, as evidenced by the Holstein breed, which has a short gestation period (average of 277 days), but relatively heavy birth weights.

8. *Low levels of energy during gestation*—Low levels of energy the last 90 to 120 days of gestation may reduce calf birth weight and calving difficulty. But it will have a marked adverse effect on rebreeding—fewer heifers will be in heat at the beginning of the breeding season, and fewer will become pregnant—unless they are fed high energy diets immediately after calving and until breeding.

9. *Low levels of energy after calving*—Both older cows and heifers that receive a low level of energy following calving show a marked decrease in the number pregnant after 20 days of breeding and at the end of the breeding season. There are 2 reasons for the poor reproductive performance in cows on low levels of feed and losing weight following calving: (a) Some of them do not show heat during the breeding season, and (b) the conception rate at first service is low.

10. *Low level of protein*—A low level of protein during gestation results in lighter birth weight of calves, but it has no adverse effect on number of calves born and calf survival.

A low level of protein following calving delays heat and results in lowered calf weight gains.

No beneficial effect is derived from feeding a high level of protein over a medium level; hence, there is no need to go to the added expense of providing more protein than is needed.

11. *Low phosphorus*—Low phosphorus will markedly decrease the percentage of calves dropped and weaned, the weaning weight of calves, and the pounds calf weaned per cow.

It is recommended that phosphorus supplementation be provided free-choice throughout the year, and that supplementation of other minerals be provided if local conditions and feeds indicate a deficiency.

12. *Over-fed and over-fat*—Excessively fat heifers often have difficult calvings, with heavy calf losses at and soon after birth, apparently as a result of excess fat decreasing the size of the birth canal.

13. *Exercise during gestation*—Exercise during the last third of gestation has little or no effect on gestation length, calf birth weight, or calving difficulty.

14. *Induced Calving*—Induced calving has a major labor-concentrating and saving possibilities in herds where breeding dates are known. However, it increases both calving difficulty and retained placentas.

15. *Small cows*—Small cows tend to experience more calving difficulty

than large cows of the same breed, as indicated by the average correlation of -.16 between cow weight and calving difficulty scores from 4 studies.[17]

16. *Pelvic opening*—The pelvic opening in 2-year-old heifers is small compared to that in older cows. Heifers with pelvic areas smaller than 34.10 sq. in. (220 sq. cm) should be bred to a bull that sires calves with small birth weights. Also, data on pelvic area should be included in the replacement heifer selection program.

17. *Difference between sire breeds*—Choosing sire breeds with lower birth weights will reduce the percentage of difficult births.

18. *Differences within the same sire breed*—Choosing among sires of the same breed can affect calving difficulty, but to a far lesser extent than choosing among breeds.

19. *Abnormal presentation*—Calves which enter the birth canal abnormally (backward, with the head or leg bent backward, etc.) often experience difficult births. Fortunately, abnormal presentations occur at relatively low frequency (1 to 5%) from all sires and in all breeds.

20. *Calf shape*—Shape of calf has very little effect on difficulty of calving, except for double muscled calves.

21. *Double muscling*—Double muscling appears to make difficult calving, despite moderate birth weights.

22. *Crossbreeding*—Crossbreeding causes little or no increase in calving difficulty when the breeds that are crossed are similar in size. But significant increase in calving difficulty occurs when cows of small breeds are mated to bulls of breeds with heavy birth weights.

23. *The interval from calving to first estrus is long in first-calf heifers*—Time must be allowed for a cow to return to heat after calving—especially heifers. It requires 50 to 60 days for the uterus to return to normal and for the animal to come back in heat, but it is longer in heifers than in cows. Hence, the only feasible way to give heifers more time to return to heat after calving is to calve them early.

24. *Suckling*—The interval from calving to first heat is 15 to 52 days longer in cows suckling calves than in cows which are milked.

25. *The interval from calving to first estrus can be shortened by high energy levels and early weaning*—It can be shortened by more liberal feeding, both before and after calving, and by early weaning.

26. *Late calving cows usually remain late throughout life*—Cows that calve late have low conception rates at first service when they are bred early in the breeding season; hence, late calving cows usually remain late calving throughout life.

27. *Management system for first-calf heifers*—A management system, including proper nutrition, in which 50% more replacement heifers than needed are bred 21 days earlier than the cows, and in which the first breeding season is limited to 45 days, will result in early heat and conception throughout life and more total pounds of calf weaned.

28. *Cows experiencing difficult calving*—Such animals should either be culled or bred to a bull of known small birth weights.

---

[17]*The Profit Brand*, Vol. 5, No. 6, published by American Breeders Service De Forest, Wisc., 1973.

QUESTIONS FOR STUDY AND DISCUSSION

1. Twelve percent of the nation's cows never calve, and there is an appalling calf loss of 6 percent between birth and weaning. Discuss the causes and economics of this situation.

2. Diagram and label the reproductive organs of the bull.

3. Diagram and label the reproductive organs of the cow.

4. In order to synchronize ovulation and insemination, when should cows be bred with relation to the heat period?

5. What are the objectives of induced calving (shortened gestation)? What are the hazards?

6. How do you account for (a) so much pasture mating among purebred breeders, and (b) so little use of artificial insemination in purebred beef cattle?

7. Discuss the economic aspects of pregnancy testing cows.

8. For your home farm or ranch (or one with which you are familiar), what do you consider to be the most desirable breeding season and time of calving? Justify your answer.

9. List the advantages to accrue from bringing cows in heat at will. Discuss each of them.

10. What advantages could accrue from the practical and extensive use of (a) superovulation, and (b) ova transplantation in beef cattle?

11. For what purposes may blood typing be used?

12. Why are more calving difficulties encountered in 2-year-old heifers than in older cows?

13. Discuss the effect of each of the following nutritional factors on reproduction: (a) liberal feeding, (b) flushing, (c) energy level, (d) protein, and (e) phosphorus.

14. Discuss the effect of each of the following factors on reproduction: (a) suckling stimulus, (b) size of pelvic opening, and (c) sire.

15. Experiments clearly show that late calving heifers usually remain late calvers throughout life. Outline a program to lessen late calvers.

SELECTED REFERENCES

Title of Publication	Author(s)	Publisher
Anatomy and Physiology of Farm Animals	R. D. Frandson	Lea & Febiger, Philadelphia, Penn, 1965
Anatomy of the Domestic Animals, The, Fourth Edition	S. Sisson J. D. Grossman	W. B. Saunders Company, Philadelphia, Penn, 1953
Animal Breeding	A. L. Hagedoorn	Crosby Lockwood & Son, Ltd., London, England, 1950
Animal Breeding Plans	J. L. Lush	Collegiate Press, Inc., Ames, Iowa, 1965
Artificial Insemination of Dairy and Beef Cattle, The, Fourth Edition	H. A. Herman F. W. Madden	Lucas Brothers Publishers, Columbia, Mo., 1972
Artificial Insemination of Farm Animals, The, Fourth Revised Edition	Ed. by E. J. Perry	Rutgers University Press, New Brunswick, N.J., 1968

(Continued)

Title of Publication	Author(s)	Publisher
Breeding Beef Cattle for Unfavorable Environments	Ed. by A. O. Rhoad	University of Texas Press, Austin, Tex., 1955
Breeding Better Livestock	V. A. Rice F. N. Andrews E. J. Warwick	McGraw-Hill Book Co., Inc., New York, N.Y., 1953
Breeding and Improvement of Farm Animals	V. A. Rice F. N. Andrews E. J. Warwick J. E. Legates	McGraw-Hill Book Co., Inc., New York, N.Y., 1967
Developmental Anatomy, Fourth Edition	L. B. Arey	W. B. Saunders Company, Philadelphia, Penn., 1940
Dukes' Physiology of Domestic Animals, Eighth Edition	Ed. by M. J. Swenson	Cornell University Press, Ithaca, N.Y., 1970
Factors Affecting Calf Crop	Ed. by T. Cunha, A. C. Warnick, M. Koger	University of Florida Press, Gainesville, Fla., 1967
Farm Animals	J. Hammond	Edward Arnold & Co., London, England, 1952
Hammond's Farm Animals, Fourth Edition	J. Hammond, Jr. I. L. Mason T. J. Robinson	Edward Arnold (Publishers) Ltd. London, England, 1971
How Life Begins	J. Power	Simon and Schuster, Inc., New York, N.Y., 1965
Improvement of Livestock	R. Bogart	The Macmillan Co., New York, N.Y., 1959
Livestock Improvement	J. E. Nichols	Oliver and Boyd, Ltd., London, England, 1957
Modern Developments in Animal Breeding	I. M. Lerner H. P. Donald	Academic Press, Inc., New York, N.Y., 1966
Prenatal and Postnatal Mortality in Cattle	Subcommittee on Prenatal and Postnatal Mortality in Bovines, Committee on Animal Health, National Research Council	National Academy of Sciences, Washington, D.C., 1968
Problems and Practices of American Cattlemen, Wash. Agr. Exp. Sta. Bull. 562	M. E. Ensminger M. W. Galgan W. L. Slocum	Washington State University, Pullman, Wash., 1955
Progress in the Physiology of Farm Animals, Vols. I, II, and III	Ed. by J. Hammond	Butterworths Scientific Publications Ltd., London, England, 1954-1957
Reproduction in Farm Animals, Second Edition	Ed by E. S. E. Hafez	Lea & Febiger, Philadelphia, Penn., 1968
Reproductive Physiology	A. V. Nalbandov	W. H. Freeman & Co., San Francisco, Calif., 1958
Veterinary Endocrinology and Reproduction	L. E. McDonald	Lea & Febiger, Philadelphia, Penn., 1969

CHAPTER 7

STERILITY AND DELAYED BREEDING IN BEEF CATTLE[1]

Contents	Page

---

[1]In the preparation of this chapter, the author had the authoritative help of Dr. Robert F. Behlow, Professor and Extension Veterinarian, North Carolina State University, Raleigh, N.C.; and Dr. Durward Olds, Professor of Physiology, University of Kentucky, Lexington, Ky.

Sterility (infertility or barrenness) may be defined as temporary or permanent reproductive failure; resulting from anestrus (lack of heat), failure to conceive, or abortion. Animals are not simply fertile or sterile; rather, all degrees of fertility exist in both sexes.

In practical operations the breeding efficiency of most beef cattle is expressed in terms of the annual calf crop[2]—the most fertile herds being those in which the highest percentage of all cows conceive on a schedule which results in the spacing of calves each 12 months. Based on number of calves born in comparison with number of cows bred, the United States calf crop for all cattle (beef and dairy) is 88 percent. This means that the other 12 percent are nonproducers, either temporarily or permanently. Since reproductive ability is fundamental to economical beef production, it can be readily understood that sterility constitutes a major annual loss in the cattle business. In fact, most cattlemen acknowledge that the calf crop percentage is the biggest single factor affecting profit in beef cattle production.

Fig. 7-1. Every cattleman is concerned with getting a live, healthy calf from each cow every 12 months. (Courtesy, American Breeders Service, De Forest, Wisc.)

---

[2]Methods of computing calf crop percentages vary. But the three most common methods are: (1) number of calves born in comparison with the number of cows bred; (2) the number of calves marketed or branded in relation to the cows bred; and (3) the number of calves that reach weaning age as compared with the number of cows bred. Perhaps the first method is the proper and most scientific one, but, for convenience reasons, some ranchers use the other two.

It is recognized that the problem of sterility is difficult to study, especially under range conditions where the majority of beef cattle are found. For this reason, most of the experimental studies on this problem have been conducted with dairy cattle. It is believed, however, that most principles apply to all breeds of *Bos taurus* and *Bos indicus,* regardless of type.

The incidence of sterility varies greatly from herd to herd, and within the same herd from year to year. Despite this fact, cattlemen should establish arbitrary standards by which breeding performance may be gauged. To this end, the following reasonable averages are proposed: not more than 10 percent breeding difficulty in the cows at any one time[3]; not more than an average of 1.85 service per conception[4], and not lower than a 94 percent calf crop. Of course, the better managed and the more fortunate herds will do better. If this standard is accepted, however, there should be reason for concern if performance falls below these averages—it should then be assumed that something is wrong and that investigation is needed.

Fortunately, comparatively few barren cows and sterile bulls are totally and permanently infertile. Those that are should be sold for slaughter without further delay or expense. Most of the others will regain their breeding abilities with good care and management and appropriate treatment. Mating at the proper stage of estrus and the correct training and use of bulls will do much to maintain a high conception rate. Care in the selection of disease-free breeding stock, isolation of newly purchased animals, and periodic health examinations are effective preventive measures. When breeding irregularities are noted or disease strikes, however, treatment should be prompt. In general, diagnosis and treatment should be left to a veterinarian who possesses training, experience, and skill in handling reproductive failures; and since infertility constitutes one of the major problems with which the veterinarian must deal, he will wish to be well informed.

Cattlemen should also be well informed relative to reproductive failures, because the enlightened producer will (1) encounter less trouble as a result of the application of preventive measures, (2) more readily recognize serious trouble when it is encountered, and (3) be more competent in carrying out the treatment prescribed by the veterinarian.

Table 7-1 lists the most common causes of infertility in cattle and shows the relative importance of each.

Table 7-1 accounts for 75.6 percent of all infertility and reveals that infectious diseases and poor management are the most common causes of infertility in cattle. Additional causes are listed in the discussion that follows, and, of course, it is recognized that the calf crop may be affected by the proportion of bulls to cows, their distribution on the range, the season of breeding, and other factors.

---

[3]These were the average figures obtained in a survey of dairy cattle in New York, as reported in *Northeastern States Regional Bulletin 32 (Cornell Univ. Agr. Exp. Sta. Bull. 924),* p. 5.

[4]Ibid.

TABLE 7—1

COMMON CAUSES OF INFERTILITY IN CATTLE AND THE
RELATIVE IMPORTANCE OF EACH[1]

Cause	Percentage of Infertile Cattle Affected by	Percentage of Affected Animals Having Lowered Fertility	Percentage of All Infertility Which It Accounts for
Vibriosis	25	60	15
Purulent metritis	10	100	10[2]
Glandular vaginitis	50	10	5
Leptospirosis	15	20	3
Brucellosis	3	50	1.5
Trichomoniasis	1	100	1.0
Silent heats	20	36	7.2
Nymphomania	5	100	5.0
Anestrus	10	30	3.0
Ovulation failure	3	30	0.9
Genetic defects			10
Nutritional deficiencies			10
Improper care at calving			10[2]
Lack of observation			2
Breeding too soon after calving			2
Total			75.6[2]

[1]Estimates by Durward Olds, University of Kentucky, Lexington, Ky. The percentage figures in the first column total more than 100 because of the fact that simultaneous infections occur in many cases. It should be noted that bovine virus diarrhea, infectious bovine rhinotracheitis, and epizootic bovine abortion, all of which cause abortion, are not listed in Table 7-1 because they did not become prevalent in the United States until recent years.
[2]The 10 percent due to purulent metritis is also listed as improper care at calving.

STERILITY IN THE COW

Usually the failure of a female to have a heat period before 18 months of age, or to come in heat within 3 months after calving, or to conceive after 3 matings should constitute sufficient basis for assuming that an abnormal condition exists and that the services of a veterinarian should be obtained for diagnosis and possible treatment. Occasionally such conditions will correct themselves without treatment; in other cases, they subside for a time only to recur later—they become irregular breeders.

Repeat breeders—cows which exhibit regular or irregular heat periods, but fail to conceive—are most perplexing. The condition may be due to failure of fertilization or to early embryonic death.

When a cow fails to come in heat, she should first be checked for pregnancy (see section entitled "Pregnancy Test," Chapter 6); approximately 1 cow in 20 thought to be sterile will be found safely in calf.

For convenience, the common causes of sterility and delayed breeding in beef cows are herein classified as (1) genital infections and diseases, (2) poor management and feeding, (3) physiological and endocrine disturbances, (4) inherited (genetic) abnormalities, (5) anatomical defects and injuries, and

(6) miscellaneous and unknown causes. Of course, at the outset it is recognized that no definite demarcation exists between these classifications; that many of the causes of sterility may be, and are by some authorities, listed under other classifications than those given herein; and that there may be interaction between two or more forces. Also, whatever the cause of sterility, there are no "cure-alls"; rather, each individual case requires careful diagnosis and specific treatment for whatever is wrong.

Genital Infections and Diseases

Table 7-1 shows that specific genital diseases account for 35.5 percent of all sterility in cattle. In addition, nonspecific infections of the cervix and/or uterus are common causes of sterility.

NONSPECIFIC INFECTIONS OF THE GENITAL TRACT

Studies reveal that the invasion of the cervix and uterus by a variety of microorganisms normally follows parturition, but that such infections usually clear up, without treatment, within 40 to 50 days after calving. The return of estrual cycles of normal length and duration is, therefore, a reasonably good indication that the reproductive tract is again normal.

Because bacteria are likely to be present in the reproductive tract immediately following calving, and since breeding at this time may interfere with the normal breeding process and extend the period of infertility, it is recommended that cows not be rebred within 60 days after calving. This will give the cow sufficient time to recover. If the placenta was retained or other calving difficulties were encountered, the cow should be allowed to go through at least one normal heat period before rebreeding.

The veterinarian-prescribed treatment for nonspecific genital infections will depend upon their nature and the extent to which they have invaded the reproductive tract; antiseptics, antibiotics, and other drugs may be indicated for local or systemic use, and/or sexual rest may be recommended.

SPECIFIC GENITAL DISEASES

Bovine virus diarrhea (BVD, mucosal-disease complex), brucellosis, epizootic bovine abortion (EBA, foothill abortion), infectious bovine rhinotracheitis (IBR), leptospirosis, metritis, trichomoniasis, vaginitis, and vibriosis are the most troublesome specific genital diseases of cattle.

• *Bovine virus diarrhea (BVD, mucosal-disease complex)* is an infectious disease of cattle caused by a myxovirus, characterized by diarrhea and dehydration. Also, the BVD virus may cause abortions in pregnant cows. Such abortions generally occur three to six weeks after infection, with all stages of gestation affected. A modified live-virus vaccine is available which can provide immunity for a certain time, the length of which has not been established. However, there is some question as to the economic justification for vaccination, along with concern lest the vaccine precipitate the clinical disease in certain circumstances. For these reasons, vaccination should be limited to those herds in which the disease is a constant problem.

• *Brucellosis* in cattle is a serious genital disease caused primarily by the *Brucella abortus* bacteria, although the suis and melitensis types are also seen in cattle. It is characterized by (1) abortions at any stage of pregnancy, but most commonly between the fifth and eighth months, (2) above normal incidence of retained placenta, and (3) lowered conception rate. Brucellosis can be readily and accurately detected by both blood and milk tests, or by the new rapid card test.

• *Epizootic bovine abortion (EBA, foothill abortion)* is an infectious disease of cattle, epizootic to California, where it is known as "foothill abortion" because of the high incidence in cows which are pastured on foothill terrain. It is caused by *Chlamydia psittaci*. Infected cows usually abort between the fifth and seventh months of gestation. In epizootic areas, only first-calf heifers and new cattle introduced from areas free of the disease are affected. The abortion rate varies from 25 to 75 percent. Aborted animals appear to immune and should be retained in a herd. Under experimental conditions, it has been shown that feeding 2 grams of chlortetracycline will prevent pregnant cows from aborting, but field application of this preventive measure has been limited because of cost and the problem of maintaining adequate dosage levels.

• *Infectious bovine rhinotracheitis (IBR)* is an acute contagious viral infection characterized by inflammation of the upper respiratory tract. It was first diagnosed in the United States in 1950. IBR is most prevalent where there are large concentrations of cattle under confinement. The virus may invade the placenta and fetus via the maternal bloodstream, causing abortion or stillbirth from two to three months subsequent to the respiratory infection.

• *Leptospirosis* is caused by several species of corkscrew-shaped organisms of the spirochete group. Among other symptoms, it is apt to produce a large number of abortions anywhere from the sixth month of pregnancy to term. In newly infected herds, abortions may approach 30 percent. The disease can be diagnosed by a blood test. For control, cattle owners rely on annual vaccinations.

• *Metritis* is caused by various types of bacteria. Lacerations at the time of calving, wounds inflicted by well-meaning but inexperienced operators, and/or retention of afterbirth are the principal predisposing factors. Metritis usually develops soon after parturition. It is characterized by a foul-smelling discharge from the vulva that may be brownish or blood-stained and finally becomes thick and yellow. An acute infection may develop into the chronic form, producing sterility. Treatment should be left to the veterinarian. Most cases are treated by the introduction of an antibiotic or sulfa into the uterus.

• *Trichomoniasis* is a genital-tract infection caused by the protozoan organism, *Trichomonas foetus*. The disease is characterized by (1) irregular sexual cycles, (2) early abortions, usually between 60 and 120 days, (3) a whitish vaginal discharge, and (4) resorption of the fetus, while the uterus becomes filled with a thin grayish fluid. When these symptoms are observed in a herd known to be free of brucellosis, trichomonad infection should be suspected. The diagnosis can be confirmed microscopically by finding the organism. Trichomoniasis may be eliminated from the herd by adoption of a hygienic breeding program. Such a program depends upon using semen from bulls free of trichomoniasis, which can be accomplished by artificial insemination. In cows, the disease appears to be self-limiting; that is, cows appear to acquire an immunity after about three months' sexual rest. If natural service must be used, infected bulls must either be treated or replaced with clean animals. Trichomonad-infected bulls can be successfully treated, provided the time-consuming prescribed treatment is followed. The preferred treatment consists of the local application of Bovoflavin ointment.

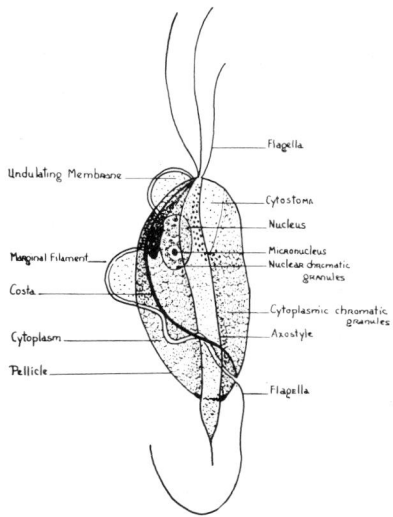

Fig. 7-2. Protozoan *Trichomonas foetus*. (Source: Dr. H. A. Herman, *The Artificial Insemination of Dairy Cattle*, Lucas Brothers, Columbia, Mo., p. 69; courtesy, Dr. Herman)

● *Vaginitis* is an infection of the vagina and vulva which causes an inflammation of varying intensity and results in difficult breeders. It may be caused by bruising or laceration of the vagina and vulva at parturition, service from a large and vigorous bull, or prolapse of the vagina. The inflamed vagina is painful, swollen, and often there is an offensive-smelling discharge, indicating infection. The veterinarian may place antibiotics in the uterus and vagina, or he may treat by injecting antibiotics or sulfa drugs.

Glandular venereal disease (glandular vaginitis) is characterized by small spherical nodules on the vulva mucosa of cattle. A similar condition may occur in the lymphatic follicles of the bull's penis. The condition is a response of the lymphatic tissue in the affected area to an irritant or antigen; it is not a disease in the classic sense. Losses in females are in terms of lower percent calf crop and decreased milk production. Bulls may refuse to breed.

Treatment of females is not recommended. It will clear up spontaneously, although it may take several weeks. The condition in the bull is more persistent and should be treated. The prolapsed penis and sheath should be massaged with a suitable antibiotic ointment, repeated sufficiently often to assure elimination of any existing infection.

● *Vibriosis* is caused by the microorganism *Vibrio fetus*, which is transmitted at the time of breeding. The disease is characterized by (1) several (4 or more) services per conception, (2) cows exhibiting irregular heat periods, but finally settling without much difficulty and carrying calves to normal term, and (3) 3-5 percent abortions, usually between the fifth and seventh month of pregnancy. For positive diagnosis, laboratory methods must be used. Infected cows may be treated by injecting drugs into the uterus and/or by allowing sexual rest. Artificial insemination with semen (1) from known noninfected bulls, or (2) which has been antibiotic-treated, is a rapid and practical method of stopping the transmission of infection from cow to cow. The disease tends to be self-limiting in the cow; a cow seems to be free of the disease once she has had a calf following an infection. Vaccination is effective in controlling the disease, especially in beef herds kept under range conditions. Usually vaccination is done at least two weeks before breeding and is repeated annually. Bulls may be vaccinated but the value is unknown.

(Also see Chapter 12, Beef Cattle Health, Disease Prevention, and Parasite Control.)

Poor Management and Feeding

The term "management" is somewhat elusive and all inclusive. As used in the discussion that follows, reference is made only to those beef cattle management practices pertinent to breeding efficiency. It is noteworthy, however, that management is very important in determining breeding efficiency, and that the breeder largely determines his own destiny in this regard.

OVULATION AND BREEDING NOT PROPERLY SYNCHRONIZED

If the sperm are introduced in the female reproductive tract much in advance of the egg's release, the chances of fertilization are greatly reduced. It has been shown that the best conception rates are obtained when cows are bred during the final 10 hours of standing heat, or during the first 10 hours after the end of standing heat. To meet these timing relationships, cows that are detected in heat in the morning should be bred during the afternoon of the same day, and those detected in the afternoon should be bred late that evening or early the next morning.

Where hand mating is practiced, this means that failure to detect accurately when a cow is in heat may account for some breeding failures. Experience and careful observation are the answers to this problem. In addition to recognizing the signs of estrus (see Chapter 6, section on "Signs of Estrus"), where hand mating is to be followed, the herdsman should keep a record of breeding dates and observed heat periods for each cow.

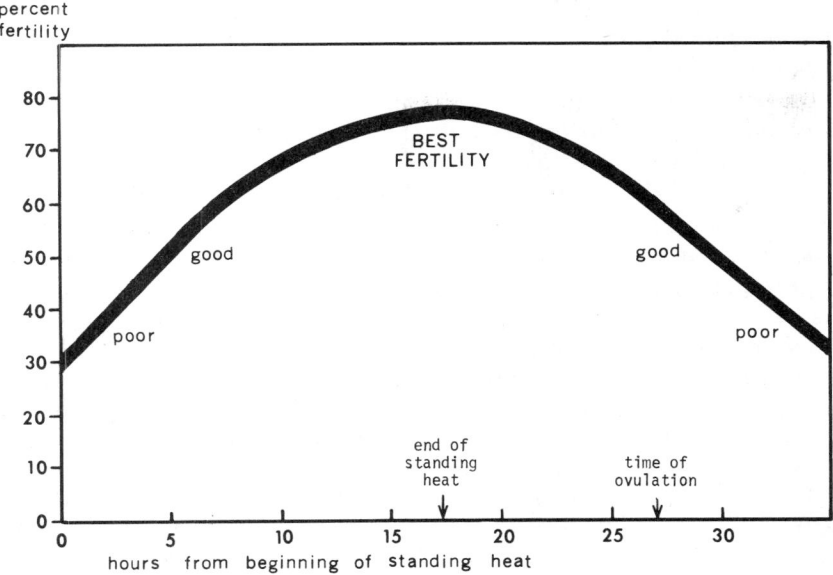

Fig. 7-3. For best conception, breed near the end of standing heat.

IMPROPER FEEDING

Improper feeding may imply (1) uncommonly high or low feed intake, or (2) a deficiency of specific nutrients.

Restricted rations often occur during periods of drought, when pastures or ranges are overstocked, or when winter rations are skimpy. When such deprivations are extreme, there may be lowered reproductive efficiency on a temporary basis. It is noteworthy, however, that experimental results to date fail to show that the fertility of the germ cells is seriously impaired by uncommonly high or low feed and nutrient intakes.[5] In contrast to these findings, it is equally clear that the onset of heat and ovulation in young heifers is definitely and positively correlated with the level of nutrient intake; thus, if feed or nutrient intake is too low for normal rates of growth and development, the onset of reproductive function is delayed.[6]

Likewise, too liberal feeding and high condition may cause sterility. Sometimes the presence of an excessive amount of fat in the pelvic region leads to a partial protrusion of the vagina and eventual inflammation. Also, where a female remains in an excessively high condition for an extended period of time, degeneration of the ovarian follicles may occur, thereby producing a prolonged state of sterility.

Under natural conditions, a deficiency of vitamin A is the only vitamin

[5]*Northeastern States Regional Bulletin 32 (Cornell Univ. Agr. Exp. Sta. Bull. 924)*, p. 30.
[6]Ibid.

likely to be lacking in cattle rations; and they obtain plenty of the precursor (carotene) when they are on green pasture or receive reasonable amounts of green hay not over one year old. A severe deficiency of vitamin A may result in a low conception rate, a small calf crop—with many calves weak or stillborn and with some calves born blind or without eyeballs, but estrus may remain normal.

Under practical conditions, it has been observed that when cows are on a phosphorus-deficient ration, there is a marked inhibition of estrus and a tendency to reproduce every other year. Also, it is believed that heat periods are suppressed when there is a copper deficiency. However, experimental results have failed to show that the fertility of cows is seriously impaired by low trace-mineral content rations.[7]

Although an adequate supply of minerals and vitamins is essential for normal growth and health, adding an excess of these nutrients to a well-balanced beef cattle ration fed according to recommended practices has no known value in curing breeding troubles.

(Also see Chapter 19, section on "Nutritional Reproductive Failure in Cows.")

EXERCISE

Although heat periods are more easily detected when cows are out in the open, exercise is not essential for normal reproduction. It is recognized, however, that exercise is necessary for the normal well-being of the individual.

SEXUAL REST

For maximum reproductive efficiency, cows should not be rebred too soon after calving. Although the reproductive tract usually returns to normal within about 6 weeks, barring infection or other abnormalities, it is inadvisable to rebreed a cow earlier than 60 days following parturition. This will still allow time for a second service if needed without exceeding the 12-month calving interval. A New York study revealed that the highest conception rate was obtained when breeding was between 70 and 90 days after calving.[8] In case of retained placenta or other calving difficulties, the cow should be allowed to go through at least one normal heat period before rebreeding.

SEASON AND LIGHT

Seasonal variations in conception rate have been observed in artificial insemination associations; poorest conception is obtained in the winter and maximum in the spring. From this it may be concluded that the amount of

[7]Ibid.
[8]Ibid., p. 5.

daylight (and perhaps the temperature) has an influence on fertility—as the amount of daylight increases, breeding efficiency increases proportionately until temperatures become too hot. In the South, the poorest conception rates occur in the hot summer months.

Physiological and Endocrine Disturbances

The development of the reproductive organs, the production of ova, sexual behavior, the attachment and development of the fetus, parturition, and lactation are primarily regulated by hormones. Many cases of reproductive failure, particularly of conception and early development, may be due to hormone imbalance.

If neither infection of the genital tract nor any unusual condition is observed, the administration of hormones may be indicated. A wide variety of both natural and synthetic hormones is available, but none should be used except under the direction of a well-informed practitioner. The temptation is always strong to administer a mixture of hormones in the hope that one of them will correct the trouble. A wiser plan is to prescribe the specific hormone which it is believed will produce the desired results; however, this is often difficult or impossible, with the result that "trial and error" methods may be the only alternative.

ANESTRUS (Failure to Come in Heat)

Anestrus is the prolonged period of sexual quiescence between the mating seasons of animals. The term is used to describe a cow which is not showing any external signs of heat, which may or may not be associated with ovarian inactivity. This condition is normal in cows immediately following calving. It may also occur in cows in the late winter and early spring when nutritional levels are low. Estrus can be readily induced in such animals by the administration of one of several natural or synthetic estrogens, such as estrone or diethylstilbestrol. Although it is unlikely that ovulation will accompany the induced heat, normal cycles are often reestablished thereby and conception may occur at the next heat period.

In some cases of anestrus, it may be preferable to administer follicle-stimulating hormones. Frequently, these will induce both estrus and ovulation, but there is a likelihood of multiple ovulations from such treatment.

DISTURBED ESTRUS CYCLES

Numerous variations of the normal estrus cycle occur, all of which are explainable in terms of improper gonadotropin-estrogen-progesterone relationships. Among such conditions are the following:

1. Ovulation without estrus; silent heats; or estrus of such low intensity that recognition is difficult. It has been estimated that 20 percent of ovulations in cows are not accompanied by external signs of heat.

2. Animals showing estrus cycles, but with a delay in ovulation.

3. Long or short heat periods.

4. Abnormal intervals between heat periods.
5. Estrus without ovulation, known as an anovulatory cycle.

If it is definitely determined that cows with disturbed estrus cycles are not pregnant, and if infections have been ruled out, the judicious and careful use of hormones may be appropriate. Such treatment may restore the endocrine balance, which, in turn, will condition the reproductive system for estrus, conception, and pregnancy.

SEXUAL INFANTILISM

In this condition, the entire reproductive tract remains small. Ovulation may not occur; if it does, affected heifers may exhibit silent heat periods or be irregular in their sexual cycles. Sometimes sexual development is delayed but reproduction is normal after puberty. Heifers with this condition may become excessively fat and resemble spayed heifers or steers.

Sexual infantilism appears to be due to lack of gonadotropic hormone secretion by the anterior pituitary gland, but, unfortunately, treatment with gonadotropins is not often successful. If malnutrition does not appear to account for the condition, the possibility of a heritable factor should be suspected.

RETAINED CORPUS LUTEUM (Retained Yellow Body)

After the ovarian follicle ruptures and the egg is released, the cells within the follicular cavity change in character and function, forming a corpus luteum or yellow body in the cavity of the ruptured follicle. If the corpus luteum persists, subsequent heat periods do not usually occur. The corpus luteum produces a hormone (progesterone) which suppresses the pituitary output of the follicle-stimulating hormone (FSH). Thus, future follicular development and ripening is inhibited and the estrogens which would induce heat periods are not produced.

A retained corpus luteum may be suspected if a cow is not seen in heat within 60 days after calving. A few years ago, these were removed manually, but this procedure usually results in some bleeding. Injections of FSH have replaced the old method.

CYSTIC OVARIES

Cystic ovaries may result when the ovarian follicle fails to rupture. The follicle persists, increases in size, and forms a cyst. It is believed that this condition is due to a derangement of gonadotropic-hormone secretion. An excessive secretion of FSH without adequate luteinizing hormone (LH) to produce ovulation causes continued follicular development and estrogen production.

When the condition is allowed to persist in cows, they frequently become chronic "bullers" or nymphomaniacs. Such individuals may show pronounced anatomic and psychologic changes; the pelvic ligaments relax so that there is a sagging of the loin region and elevation of the tailhead. Af-

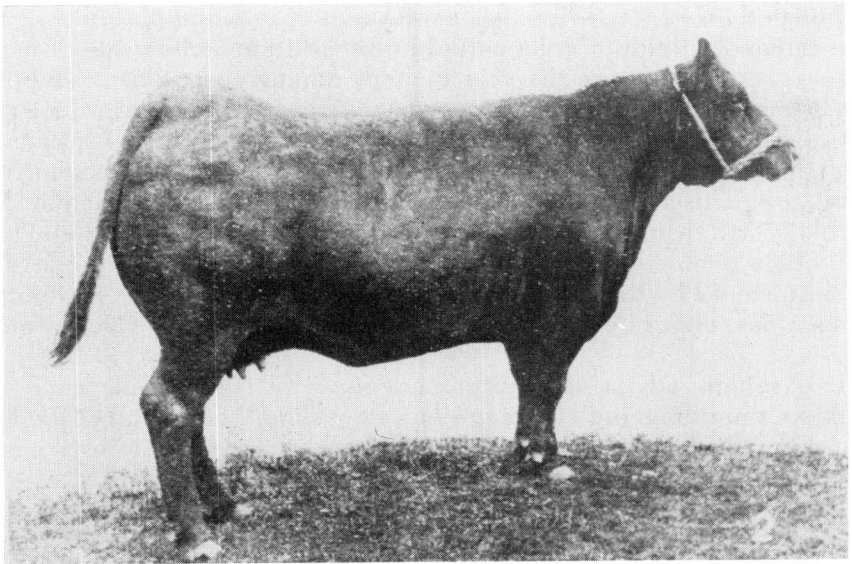

Fig. 7-4. A chronic "buller" or nymphomaniac cow, showing the characteristic sagging of the loin region and elevation of the tailhead. (Source: *Physiology of Reproduction*, by Marshall, p. 667; courtesy, the Royal Society of Edinburgh)

fected cows may acquire such male characteristics as thickened forequarters and may bellow and behave like a bull.

Recommended treatment for cystic ovaries consists in the administration of a gonadotropin preparation rich in the LH fraction, made either from anterior pituitary glands or from human pregnancy urine. The use of progesterone for 14 consecutive days will clear up this condition. Rupturing the cyst manually damages the ovary.

There is evidence that the tendency toward cystic ovaries is inherited in cattle. Also, there is a higher incidence in the dairy breeds than in the beef breeds.

RETAINED PLACENTA (Retained Afterbirth)

Normally, the placenta is expelled within 3 to 6 hours after parturition. If it is retained as long as 12 hours after calving, competent assistance should be rendered.

Retained placenta occurs in about 5 percent of the parturitions in beef cattle and 10 percent in dairy cattle. It is more common following abnormally short or abnormally long pregnancies, among older cows, and following twinning. Experimentally, it has been found that a high incidence of retained afterbirth occurs when premature calving is induced by the administration of glucocorticoid drugs.

While infections such as brucellosis, vibriosis, and others have been associated with abortion and retained afterbirth, these are by no means the only causes. Nutritionally, deficiencies of carotene or vitamin A have been incriminated. Also, it appears that fewer cases of retained placenta occur (1) when calves stay with their dams and nurse for 12 to 24 hours, and (2) when cows are kept on pasture the year around. Among cows which have previously retained the placenta, 20 percent are likely to do so again.

Calves born when the placenta is retained are likely to be weak. A retained placenta may cause pathological conditions resulting in uterine tissue destruction. This condition may or may not affect milk production, but it very likely will result in 5 to 10 percent lower fertility than for normal cows.

When a retained placenta is encountered, appropriate treatment should be administered by the veterinarian. He will likely use either antibiotics or sulfonamides, either by direct infusion into the uterus or by other routes (or both).

It is seldom advisable to attempt removal of retained placenta. If the membranes are dragging on the ground, they should be cut off at the hocks. But never, never tie bricks or other objects to it. In most instances, the membranes will fall out by themselves in one to two weeks.

It is desirable to have all cows which have had retained afterbirth examined at about 30 days after calving. If pus is present, they may be treated with estrogenic hormones to induce heat and then the uterus can be infused with an antibiotic solution or perhaps with a dilute Lugol's (iodine) solution. Such examination and treatment may save considerable time with regard to the onset of normal cycles and may result in a higher conception rate.

INTERSEXES AND HERMAPHRODITES

Stedman's Medical Dictionary gives the following definitions of intersexes and hermaphrodites:

• **Intersex**—*An individual showing both maleness and femaleness, in which the sex differences are not confined to clearly demarcated parts of the body but blend more or less with one another.*

• **Hermaphrodite**—*An individual whose genital organs have the characters of both male and female in greater or less degree.*

These conditions which occur only rarely in cattle are a result of (1) imbalance in the maternal fetal endocrine system, or (2) genetic factors.

Inherited (Genetic) Abnormalities

The development of breeds or families within breeds which differ in prolificacy is good evidence that fertility may have a genetic basis. Thus, it is common knowledge that some once-popular families have become extinct because of the high incidence of infertility. A classical example of a situation of this type occurred in the Duchess family of Shorthorn cattle, founded by the noted pioneer English Shorthorn breeder, Thomas Bates. Bates, and more especially those later breeders who emulated him, followed preferences in bloodlines until, ultimately, they were selecting cattle solely on the basis of fashionable pedigrees, without regard to fertility. Ironically, during their heyday, the scarcity of this strain of cattle contributed to their value.

Eventually, but all too late, the owners of Duchess Shorthorns suddenly came to a realization that indiscriminate inbreeding and lack of selection had increased sterility to the point that the family name was in disrepute; a high incidence of sterility had actually contributed to their scarcity. As a result, Duchess Shorthorns became virtually extinct a few years later.

Of course, reproductive disorders of a heritable nature should not knowingly be perpetuated.

LETHAL GENES

Lethal genes, and other recessive genes causing sterility, belong in the group of highly heritable characters. Among such hereditary lethals are mummification, cystic ovaries, and gonad hypoplasia or gonadless. (See Chapter 5, Table 5-4, for a summary relative to lethal characters.) Generally these abnormalities are easily recognized by their actions, and their mode of transmission can often be analyzed and determined. Although lethals are of interest to the geneticist, they are generally no great problem to the cattleman because their distribution remains under control. Usually recessive, their gene frequency is kept low by the self-destruction of the double recessive; thus, they are self-selective.

WHITE HEIFER DISEASE

This name is a misnomer, for, although the condition is most commonly found in white heifers of the Shorthorn breed, it has been reported in roan and red Shorthorns and in colored animals of other breeds. It appears to be due to faulty development of the Mullerian ducts. Some of the more com-

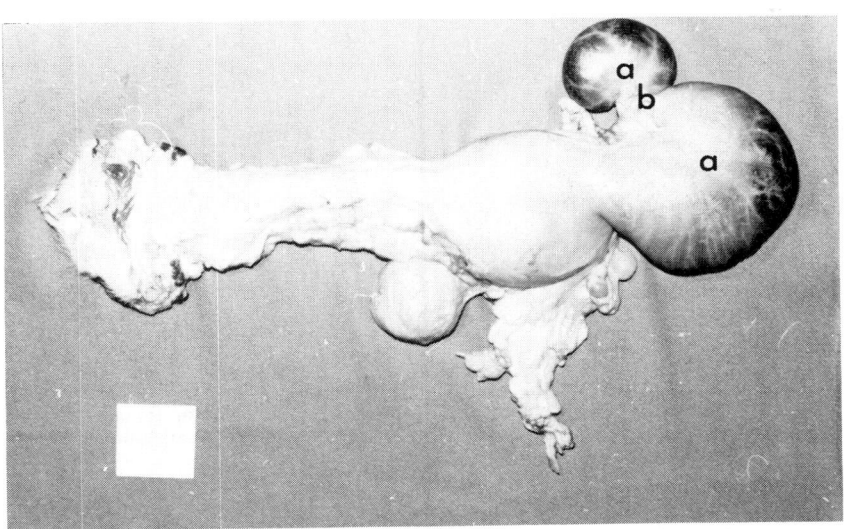

Fig. 7-5. White heifer disease. Note (1) balloon-like distention of uterine horns, and (2) segmental blockage. (Courtesy, Prof. H. W. Leipold, Kansas State University, Manhattan, Kan.)

mon characteristics are: closed hymen or hymen persisting in varying degrees, distention of one or both uterine horns, and uterine body present in rudimentary form, complete absence of cervix, and anterior vagina.

Anatomical Defects and Injuries

A long list of anatomical defects and injuries to the genital organs has been reported. Some of these are so severe as to cause sterility; others affect the degree of fertility. A Pennsylvania study of repeat breeders—cows which had failed to conceive after four services—revealed that 13 percent were anatomically abnormal.[9] A brief account of some of the more general anatomical defects follows.

FREEMARTIN HEIFERS

Sterile heifers that are born twin with a bull are known as freemartins. This condition prevails in about 9 out of 10 twin births when a calf of each sex is involved. The fetal circulations fuse, and the male hormones get into the circulation of the unborn female where they interfere with the normal development of sex and modify the female embryo in the direction of the male. In approximately 15 percent of twin births of unlike sexes, fusion of the circulation does not occur, and the animal is normal and fertile.

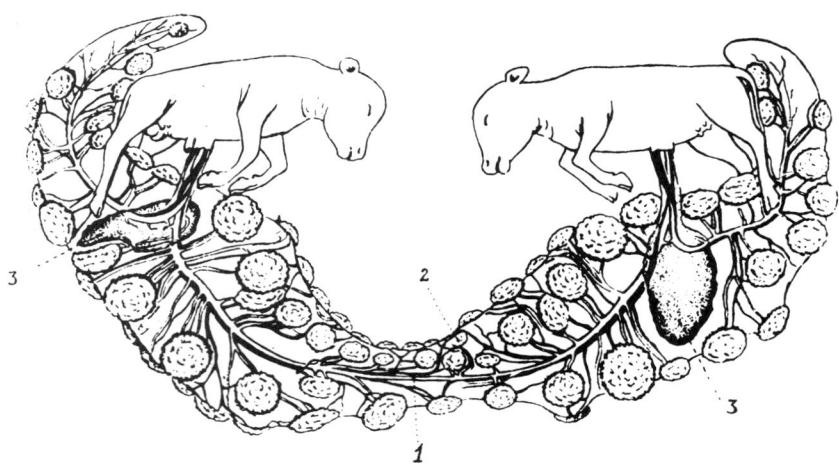

Fig. 7-6. Diagram showing fused fetal circulation of twin calves of opposite sex. Note (1) the fetal circulation of the male fused with that of the female, (2) fetal cotyledon free yolk sac, and (3) normal bull on the left and freemartin heifer on the right. (Source: *Physiology of Reproduction*, by Marshall; courtesy, the publisher, Longmans, Green and Co., Ltd., London, England)

---

[9]*Northeastern States Regional Bulletin 32 (Cornell Univ. Agr. Exp. Sta. Bull. 924),* p. 21.

Since only about 15 percent of such heifers are fertile, it is usually best to assume that they are sterile and market them, unless (1) an experienced person determined at the time of birth that their circulatory systems were not fused, (2) an examination of the vagina reveals that the animal is normal (in freemartin heifers, the vagina is usually about one-third normal length), or (3) skin-grafting[10] or blood-typing[11] techniques show that they are not freemartins and that they may, therefore, be regarded as reproductively normal.

MECHANICAL INJURIES TO THE GENITAL ORGANS

Mechanical injuries may occur in the female at service and at parturition. A large vigorous male may inflict injury when breeding; and complicated parturition may result in the loss of the offspring, permanent damage to the reproductive organs, or even death of the cow herself. Perforation of the uterine or vaginal walls, laceration of the cervix, eversion of the vagina, cervix, and uterus or of the rectum may all be sequelae of complicated birth. In some cases, adhesions or secondary infections of the genitalia cause tissue damage which prevent further reproduction.

Miscellaneous and Unknown Causes

Unfortunately, many of the causes of sterility are unrecognized and unknown. A discussion of one of these follows.

EMBRYONIC MORTALITY (Fetal Death or Prenatal Mortality)

It is well known that early embryonic mortality occurs normally in the pig and the rabbit, where there is a surplus production of female gametes. Although it is not so common in the cow, it appears that 20 to 30 percent of the ova fertilized may meet embryonic death in 2 to 6 weeks. In such cases, the embryo may be absorbed or be expelled unobserved from the female reproductive tract. The cow may assume normal sexual cycles with the conclusion the fertilization did not take place.

Fetal death, followed by resorption or abortion, may occur at any stage of pregnancy. Such fetuses may range from decomposed masses to bones or dried mummies.

The cause or causes of prenatal death are not known.

STERILITY IN THE BULL

Any bull of breeding age that is purchased should be a guaranteed breeder; in fact, this is usually understood among reputable cattlemen.

The most reliable and obvious indication of fertility in a bull is a large number of healthy calves from a season's service. However, a good evaluation of a bull's fertility may be obtained through a microscopic examination

---

[10]Billingham, R. E. and G. H. Lampkin, *J. Embryol. Exp. Morph.*, Dec. 1957, Vol. 5, part 4, pp. 351-367.
[11]Stormont, Clyde, *Journal of Animal Science*, Feb. 1954, Vol. 13, No. 1, pp. 94-98.

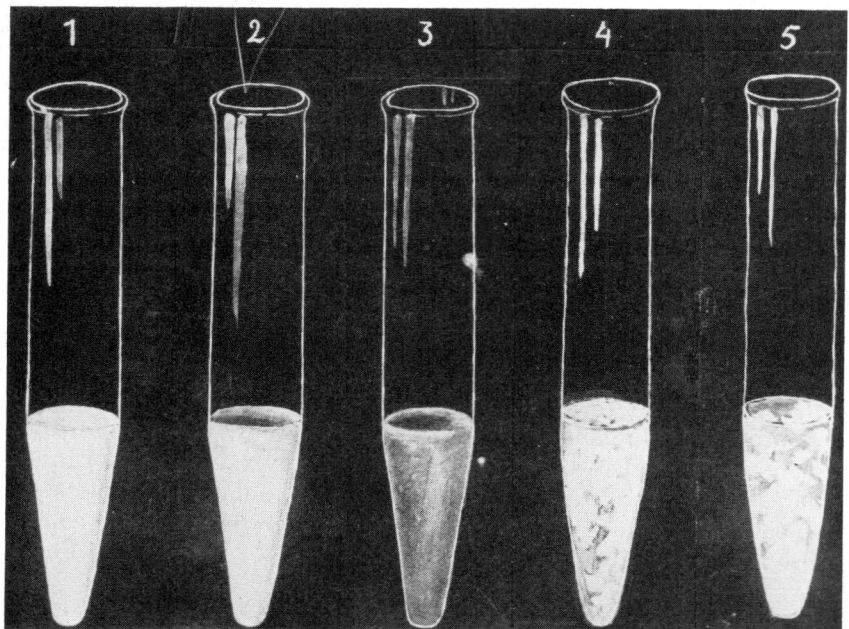

Fig. 7-7. Semen samples of different kinds. 1. Semen with normal appearance, about 1,000,000 spermatozoa per cubic mm. 2. & 3. Semen from a bull with hypoplastic testicles. Sample 2 contains about 200,000 spermatozoa per cubic mm. Sample 3, which is almost transparent contains about 25,000 spermatozoa per cubic mm. 4. & 5. Semen from a bull with inflammation in the seminal vesicles. In the semen, which is almost transparent, there are big, purulent flocci. (Courtesy, Prof. Nils Lagerlof, Department of Obstetrics and Gynecology, Royal Veterinary College, Stockholm, Sweden)

of the semen made by an experienced person. It is recommended that all bulls be semen tested prior to the breeding season; and, where valuable purebred bulls are involved, periodic tests during the breeding season are desirable. Such procedure may alleviate much loss in time, feed, and labor, and avoid delayed and small calf crops.

For purposes of convenience, the common causes of sterility and delayed breeding in bulls are herein classified as (1) poor semen, (2) physical defects and injuries, (3) psychological, (4) genital infections and disease, (5) poor management and feeding, (6) physiological and endocrine disturbances, and (7) inherited (genetic) abnormalities. Also, a section on "'Bull Testing' to Evaluate Breeding Soundness" follows.

Poor Semen

It is always well to obtain a sample of semen and to make a laboratory examination of the number and condition of the sperm. The four main criteria of semen quality are (1) volume, (2) sperm count, (3) progressive movement, and (4) morphology (shape). Although this technique is not infallible, an experienced person can predict with reasonable accuracy the relative fertility of bulls so examined.

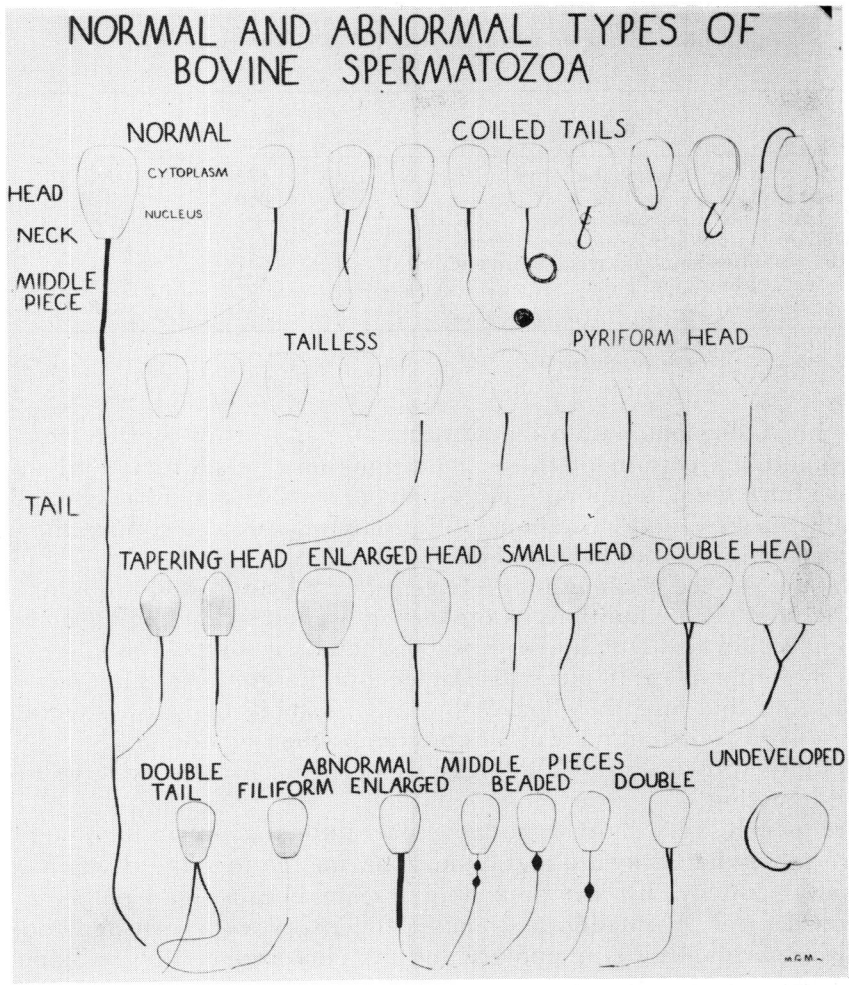

Fig. 7-8. Normal and morphologically abnormal spermatozoa of the bull. (Source: Dr. H. A. Herman, *The Artificial Insemination of Dairy Cattle*, Lucas Brothers, Columbia, Mo., p. 27; courtesy, Dr. Herman)

Physical Defects and Injuries

Based on a study of 10,940 bulls, Colorado State University found 2,266, or 20 percent, questionable or unsatisfactory for breeding purposes, for the reasons shown in Table 7-2.[12]

Some bulls possessed more than one breeding deficiency, which accounts for the total number of defects recorded exceeding the 2,266 figure.

The most common defects involving the reproductive system are: degenerating testes, abscessed testes, fibromas of the penis, broken penis,

---

[12]Breeding soundness in bulls—a summary of 10,940 examinations, *Journal of American Veterinary Medical Association*, Vol. 142, No. 10, pp. 1105-1111.

TABLE 7-2

REASONS FOR BULL BEING QUESTIONABLE OR
UNSATISFACTORY[1]

Reasons	No. of Bulls	Percent
Semen quality questionable or unsatisfactory	1,832	16.7[2]
Defects of penis	388	3.5
Occurrence of testicular defects	1,931	17.6
Defects of epididymis and vas deferens	112	1.0
Abnormalities found by rectal examination	547	5.0
Defects of limbs and feet	597	5.5
Eye defects	421	3.8

[1]10,940 bulls tested; 2,266 found unsatisfactory.
[2]1,832 is 16.7% of 10,940 bulls tested.

hematoma, adhesions within the sheath, and paralysis of the retractor muscle. Sometimes certain of these conditions can be corrected by skilled surgery. Infrequent, temporary, or permanent sterility may result from bruising, inflammation, and lacerations of the scrotum and testes. Also, the penis may become bruised or lacerated during service.

Natural service is sometimes interfered with by unsound limbs and feet. This includes such conditions as broken limbs, bad sickle hocks, sore, overgrown or malformed feet, and arthritic or rheumatic joints. The latter condition is more prevalent in old bulls. Defects of the hind limbs are especially troublesome, because they make the bull unstable or cause severe pain when he shifts much of his weight onto them when mounting.

Keratitis (pinkeye) and blindness, especially if in both eyes, will interfere with mating.

Frequently, very short-legged, compact bulls that are in high condition and are somewhat awkward and clumsy are unable to serve cows. Also, an extremely paunchy bull may be unable to mate, because the "potbelly" acts as a mechanical obstruction and causes the penis to be directed too low. Usually such inability is temporary and may be corrected by reducing the ration and increasing the exercise.

Improvements can be effected in some of these physical defects and injuries, while others are irreparable.

Psychological

Sterile bulls are frequently victims of psychological sterility. Usually, this condition expresses itself in either (1) absence of or lowered sex drive, or (2) faulty reflex behavior during mating or ejaculation.

As is true in people, there is a wide individual difference in the reaction of bulls to their environment; and, likewise, these factors are, in part, hereditary. Thus, there are four major types of temperament in bulls: the nervous, the sulky, the placid, and the treacherous. The treacherous types should always be culled at an early age. Then, in selecting from among the other three types, consideration should be given to the herd requirements—whether a commercial or purebred herd is involved, whether pasture- or hand-mating is to be used, etc. Also, it is generally recognized that bulls lacking in masculinity or secondary sexual characteristics (bulls that are very docile, very fine boned, and lacking development of the crest, etc.) are likely to possess poor psycho-physiological sexual activity.

Young, inexperienced bulls mating for the first time are usually awkward to handle. They approach the cow hesitantly, spend a long time exploring the genitalia, mount hesitantly without erection, descend and try to mount again. Extreme patience and careful handling should be exercised during this critical first service; otherwise, difficult breeding habits may be established.

CAUSES OF PSYCHOLOGICAL STERILITY

Some common causes of psychological sterility are:

1. *Excitement*—shouting, noises, distractions during mounting, and the presence of strangers may cause low sex drive. When restrained with a dog, a highly fertile bull may become sexually impotent. Bulls show evidence of such excitement when they urinate more frequently, as they do following the visit of strangers or the introduction of new animals.

Also, it is important to keep sires as quiet as possible in the nonbreeding season. Undue excitement causes sexual impulses and results in the flow of semen into the ampullae.

2. *Transportation*—Psychological sterility is not uncommon in bulls that have been transported long distances by truck or rail.

3. *Animal management*—Young bulls that are isolated from all cows for long periods of time manifest homosexual reflexes and may become impotent.

Also, inadequate sex drive may be due to an attempt to use a young bull on a cow that is too large for him to mount successfully. As a result of his failure to copulate, he develops a mild sense of frustration. Frequently, such a bull can be restored to a high state of breeding efficiency by giving him assistance and trying him on a heifer selected for her small size and willingness to stand quietly; or the same effect may be obtained by standing a large cow in a pit.

Where hand-mating is practiced, the bull should not be used immediately following feeding.

4. *Wrong technique during semen collection*—Certain inhibitory reactions may develop from improper use of the artificial vagina, including wrong timing in applying it, too hot or too cold water, or holding it at a wrong angle.

Genital Infections and Diseases

The presence of bacteria in the semen of bulls will not only tend to decrease the viability of the sperm cells but may very easily infect the cows to which they are mated, thereby preventing their conception even when bred to other bulls. If this condition is due to inflammation of the seminal vesicles or prostate, the systematic use of the appropriate antibiotic or sulfa drug may be recommended by the veterinarian. Occasionally, rectal massage of an infected prostate or seminal vesicle will aid the passage of the purulent exudate into the urethra and thus hasten recovery.

Brucellosis, trichomoniasis, glandular venereal disease (glandular vaginitis) and vibriosis are the specific genital diseases of most concern in bulls. Each of these diseases is fully covered in Chapter 12; and, likewise, a pertinent discussion of each disease from the standpoint of the cow appears earlier in this chapter. Thus, the ensuing comments will be limited to effects on the bull.

The Brucella organism may localize in the testes, seminal vesicles, or vas deferens, and bulls may spread the disease by copulation. Sex drive is generally reduced if testicular involvement occurs.

In trichomoniasis, the bull is the source of infection. Positive diagnosis of infection in the bull may be made by means of (1) a microscopic examination of smears taken from the prepuce, or (2) mating him with a virgin heifer and checking her cervical and uterine smears for the organism as she approaches the next heat period.

It is believed that glandular venereal disease is commonly transmitted by the bull at the time of service, but this is not the only means of transmission since virgin heifers may be infected. Infected bulls should either be treated or sold for slaughter. Treatment consists in massaging the prolapsed penis and sheath with a suitable antibiotic ointment, repeated sufficiently often to assure elimination of infection.

In vibriosis, no clinical lesions are observed in bulls. However, V. *fetus* appears to persist indefinitely in the genital tract of carrier bulls and the bull may transmit the infection. For diagnosis of vibriosis in individual bulls, one should (1) culture the semen, and, where possible, (2) breed to one or more virgin heifers and collect vaginal mucus for culture 10 to 20 days later. Effective control can be obtained by (1) adding antibiotics to semen from infected bulls and breeding artificially, or (2) establishing a new herd of sexually immature animals.

Poor Management and Feeding

In all too many cases, little thought is given to the management and feeding of the bull, other than during the breeding season. Instead, the program throughout the entire year should be such as to keep the bull in a vigorous, thrifty condition at all times. Also, lack of fertility in the bull may often be traced back to his early care and feeding. He may have been small and weak at birth, he may have been improperly fed during the first year of his life, or he may have been prey to infection.

IMPROPER FEEDING

Overfat, heavy bulls should be regarded with suspicion, for they may be uncertain breeders. On the other hand, a poor, thin, run-down condition is also to be avoided.

Fitting bulls for show and sale results in a variable effect upon their semen producing ability; in some it has no detectable detrimental effect, while others are severely affected by such practices. Many highly fitted bulls show a complete lack of sperm cells; others produce semen comparable in quality to a bull exhibiting testicular degeneration. Many, perhaps most, such fat bulls eventually reach normal breeding efficiency if they are properly let down—primarily by reducing the grain and by increasing the exercise.

Sterility in bulls has occasionally been traced to a lack of vitamin C. This can be determined by an analysis of a blood or semen sample. Should the analysis disclose a deficiency of vitamin C, the required amount can be supplied by injecting ascorbic acid under the skin of the bull. The usual treatment consists of injecting 1 gram of ascorbic acid dissolved in sodium-potassium phosphate buffer, physiologic saline, or sterile water, twice a week.

EXERCISE

Exercise is necessary for the normal well-being of the bull. Although it is difficult to show precisely to what extent lack of exercise induces low fertility, there seems to be ample reason for concluding that it constitutes a highly important factor.

Generally speaking, beef bulls should be exercised by allowing them the run of the pasture or large corral. However, it may be necessary to lead fat show bulls from 2 to 4 miles daily following the show season.

RETARDED SEXUAL MATURITY

As stated earlier in this chapter, the normal age of puberty in cattle is 12 months. But there is wide variation in the age of sexual maturity of bulls as measured by semen quality. Some yearling bulls are fully equipped to produce a healthy percentage of calves; others are not. At the present time, it is not known whether such retarded sexual development is due to hereditary, nutritional, and/or management factors. There is need, therefore, for additional research on this subject.

THE OVERWORKED BULL

Low fertility of the bull is frequently caused by overservice. Table 6-2, of Chapter 6 may be used as a bull mating guide for different age animals, under both hand-mating and pasture-mating practices.

In pasture breeding, the bull may copulate four or five times in succession, thereby correspondingly reducing his powers and lessening the size of the herd on which he should be used. However, it is recognized that this situation is compensated for, in part at least, by the fact that bulls on pasture have sexual vigor not possessed by stall-fed bulls.

Physiological and Endocrine Disturbances

The development of the reproductive organs, the production of sperm, and sexual behavior are primarily regulated by hormones; thus, the possibilities of endocrine disturbances are endless. Examples of some of the more common abnormalities follow.

SEXUAL INFANTILISM

In this condition, the entire reproductive tract remains small and the testes are visibly reduced in size. Affected animals lack sex drive. If a low plane of nutrition does not appear to account for the condition, the possibility of a genetic factor should be suspected.

SEX DRIVE

If sperm are being produced, but the bull is unwilling to perform service due to low sex drive, the administration of an androgen, such as testosterone, may restore mating.

Inherited (Genetic) Abnormalities

Lack of fertility in the bull may often be traced right back to his own sire and dam. He may have been sired by a bull of low vigor and fertility and out of a cow of equally low fertility.

Reproductive disorders of a heritable nature should not knowingly be perpetuated. In addition to those which follow, Table 5-4 of Chapter 5 contains a summary of some fertility-affecting hereditary abnormalities in bulls, including impotentia, atropied testes, knobbed spermatozoa, and turned sperm tails.

CRYPTORCHIDISM

When one or both of the testicles of a bull have not descended to the scrotum, the animal is known as a cryptorchid. The undescended testicle(s) is usually sterile because of the high temperature in the abdomen. Since this condition may be heritable, it is recommended that animals so affected not be retained for breeding purposes.

SCROTAL HERNIA

When a weakness of the inguinal canal allows part of the viscera to pass out into the scrotum, the condition is called scrotal hernia. This abnormality may interfere with the circulation in the testes and result in their atrophy.

UMBILICAL HERNIA

This condition, which may interfere with breeding efficiency, has been reported as due to (1) a sex-limited dominant gene, or (2) one or more pairs of autosomal recessive factors.

"Bull Testing" to Evaluate Breeding Soundness

The term "bull testing" as used herein refers to a method for evaluating the breeding soundness of beef bulls; actually, it is a method for detecting infertility rather than determining fertility. An extensive program of this type was first initiated by Colorado State University. Since then, it has spread to several states. Essentially, such testing embodies the following three evaluations, all of which have a bearing on a bull from the standpoint of the efficiency of his reproductive performance:

1. Checking for physical defects which might impair breeding capacity.

2. Collecting a representative semen sample by means of the electric ejaculator and examining it under the microscope. Semen quality is based on four criteria: (a) degree of vigor; (b) percent living sperm as determined by the live-dead stain technique; (c) concentration of sperm cells; and (d) morphology.

3. Evaluating and determining the breeding history of a nonvirgin bull whose status, based on points 1 and 2, is questionable.

By use of these criteria,[13] Colorado State University classifies each bull examined to denote his relative breeding capacity, using the following classification:

1. *Satisfactory*—This group includes bulls (a) that have no major physical defects which will impair their ability to travel and to serve females, and (b) that produce semen within the range of quality standards necessary for conception in natural mating. With normal females, conception rates with such bulls should be above 60 percent with one service.

2. *Questionable*—These bulls show some faults affecting ability to settle cows; many cows will not conceive to their services. Some bulls in this class may, in time, improve; but later calves and a low percentage calf crop will likely result. Such bulls should be rechecked before the breeding season. Conception rates would be expected to fall below 50 percent on one service.

3. *Unsatisfactory*—These bulls are, as the term indicates, unsatisfactory for breeding purposes and, unless very valuable, should be sold for slaughter without delay.

Out of 3,400 bulls tested, the Colorado workers found 6 percent of them questionable and 6 percent unsatisfactory; or 12 percent that gave reason for concern as breeders. This points up the need for a systematic and regular examination of breeding bulls.

A PROGRAM OF IMPROVED FERTILITY AND BREEDING EFFICIENCY

A program designed to give improved fertility and breeding efficiency in beef cattle follows, in summary form.[14]

---

[13]Actually, the third criterion is not invoked unless the first two criteria show that the bull is questionable.

[14]In this section, special emphasis is placed on increased fertility and breeding efficiency; thus, there is some repetition and there are some additions to the section entitled, "A Program of Beef Cattle Health, Disease Prevention, and Parasite Control," as given in Chapter 12 of this book.

1. Keep complete breeding records. Maintain complete fertility records on each animal, examine them periodically, and cull low producers. Where hand-mating is followed, keep a record of dates bred and observed heat periods of each cow; calculate the expected estrus cycle by adding 21 days to the date of last estrus.

2. Before the breeding season, check the bull for physical defects and quality of semen.

3. Breed only healthy cows to healthy bulls.

4. Avoid either an overfat or a thin, emaciated condition in all breeding animals; and feed balanced rations.

5. Provide plenty of exercise for bulls and pregnant cows, preferably by allowing them to graze in well-fenced pastures in which plenty of shade and water are available.

6. Do not breed young animals until they are sufficiently mature.

7. Provide an ample rest period between pregnancies; do not rebreed within 60 days after calving. Where the placenta was retained or other calving difficulties encountered, allow the cow to go through at least one normal heat period before rebreeding.

8. Do not overwork the bull.

9. Observe breeding females carefully during the breeding season; otherwise heat periods of short duration may be missed. Also, keep a close watch for shy breeders; expose them to the bull often.

10. Diagnose cows for pregnancy.

11. Handle the newborn so that its health shall be assured, and in order that it may have uninterrupted development.

12. Retain as future replacements only those animals which are the progeny of healthy parents, that were carried in utero for a normal gestation period of from 279 to 288 days, and that were born without difficult calving, retained afterbirth, or metritis.

13. Isolate newly acquired animals for a minimum of 3 weeks, during which time they should be tested for brucellosis, leptospirosis, trichomoniasis, and vibriosis. However, first make every reasonable effort to ascertain that they came from herds which are known to be free from these and other diseases.

14. When possible, purchase virgin heifers and bulls. Isolate nonvirgin bulls for a period of 3 weeks, and then turn them with a limited number of virgin heifers; observe these heifers for 30 to 60 days after breeding as an aid in preventing the introduction of breeding diseases.

15. When sterility is encountered, promptly call upon the veterinarian for treatment; do not delay action until the condition is of long standing.

QUESTIONS FOR STUDY AND DISCUSSION

1. Compute the following for your beef cattle herd, or for a herd with which you are familiar:
   a. How much is being lost in annual gross sales of calves at weaning time, when considering the current calf crop percentage vs a 100 percent calf crop?
   b. How much does it cost to maintain all of the barren cows for a year?

2. If a certain cattleman is experiencing (a) 15 percent breeding difficulty in his cow herd at a given time, (b) 2.5 services per conception, and (c) a 75 percent calf crop, outline, step by step, your recommendations for determining the difficulties and improving the situation.

3. What precautions should a cattleman take to avoid the introduction into the herd of genital infections and diseases?

4. What symptoms characterize each of the following specific genital diseases in cattle: bovine virus diarrhea, brucellosis, epizootic bovine abortion, infectious bovine rhinotracheitis, leptospirosis, metritis, trichomoniasis, vaginitis, and vibriosis? What positive diagnosis, if any, can be made of each? What control program should be initiated when the presence of each disease is known?

5. For best results, (a) how many days should be allowed to elapse following parturition before rebreeding, and (b) how many hours should elapse following the known onset of heat before breeding?

6. May sterility be caused by (a) lack of feed as sometimes occurs during droughts, or (b) high condition as when fitted for show?

7. Why should the cattleman call on the veterinarian if hormone injections are to be given to cattle?

8. On the basis of experimental evidence, should the corpus luteum be removed by an experienced technician by pressure applied through the rectal wall (rectal palpation) if the cow does not come in heat within 60 days after calving?

9. How should a cattleman handle a case of retained placenta?

10. Define the following: (a) nymphomania; (b) freemartin; (c) hermaphrodite; and (d) intersex.

11. How can you tell whether a heifer is a freemartin?

12. What precautions should be taken when purchasing a bull to avoid sterility and delayed breeding?

13. When bringing in a new bull, what precautions should be taken to avoid psychological sterility.

14. Prepare a recommended breeding schedule for bulls of different ages showing the interval between services.

15. How would you go about testing a bull for breeding soundness?

16. Write out a "program of improved fertility and breeding efficiency" for your herd or for a herd with which you are familiar.

SELECTED REFERENCES

Title of Publication	Author(s)	Publisher
Breeding Difficulties in Dairy Cattle, Cornell Univ. Agr. Exp. Sta. Bull. 924	S. A. Asdell	Cornell University, Ithaca, N.Y., 1957
Breeding Difficulties of Cattle	C. Staff	General Mills, Larro Feeds, Chicago, Ill.
Cattle Fertility and Sterility	S. A. Asdell	Little, Brown and Co., Boston, Mass., 1968
Dairy Cattle Sterility	H. D. Hays L. J. Boyd	*Hoard's Dairyman,* W. E. Hoard and Sons, Company, Fort Atkinson, Wisc.

(Continued)

Title of Publication	Author(s)	Publisher
Factors Affecting Reproductive Efficiency in Dairy Cattle, Kentucky Agr. Exp. Sta. Bull.	D. Olds D. M. Seath	University of Kentucky, Lexington, Ky., 1954
Farm Animals	J. Hammond	Edward Arnold & Co., London, England, 1952
Problems and Practices of American Cattlemen, Wash. Agr. Exp. Sta. Bull. 562	M. E. Ensminger M. W. Galgan W. L. Slocum	Washington State University, Pullman, Wash., 1955
Reproduction and Infertility	Centennial Symposium	Agricultural Experiment Station, Michigan State University, Mich., 1955
Reproduction and Infertility III Symposium	Ed. by F. X. Gassner	Pergamon Press, New York, N.Y.
Reproduction in Dairy Cattle, Ext. Bull. 115	C. H. Boynton	University of New Hampshire, Durham, N.H.
Reproductive Physiology	A. V. Nalbandov	W. H. Freeman & Co., San Francisco, Calif., 1958

FEEDING BEEF CATTLE[1]

Contents   **Page**

---

[1]The author gratefully acknowledges the helpful suggestions of Dr. Wilton W. Heinemann, Washington State University, Prosser, Wash., who reviewed this chapter.

    Although Webster defines the noun *ration* as "the amount of food supplied to an animal for a definite period, usually for a day," to most stockmen the word implies the feeds fed to an animal or animals without limitation to the time in which they are consumed. In this and other chapters of *Beef Cattle Science*, the author accedes to the common usage of the word rather than to dictionary correctness.

Cattle inherit certain genetic potentialities, but how well they develop depends upon the environment to which they are subjected; and the most important factor in the environment is the feed. The feeding of cattle also constitutes the greatest single cost item of their production. It is important, therefore, that the feeding practices be as satisfactory and economical as possible.

Fig. 8-1. Under the old system of unforced production and marketing at three to five years of age, reasonably good pasture and hay sufficed. Not so today! Meeting consumer demand for smaller cuts, more lean, less fat, and greater tenderness, calls for the production of baby beef—animals in forced production, growing and finishing simultaneously, and frequently subjected to stress and confinement. Consequently, the nutritive requirements are more critical than formerly, especially from the standpoint of proteins, minerals, and vitamins.

Pastures and other roughages, preferably with a maximum of the former, are the very foundation of successful beef cattle production. In fact, it may be said that the principal function of beef cattle is to harvest vast acreages of forages, and, with or without supplementation, to convert these feeds into more nutritious and palatable products for human consumption. It is estimated (1) that 73.8 percent of the total feed of beef cattle is derived from roughages (see Table 8-1), and (2) that 49.4 percent of the land area of the United States is pastured all or part of the year, and much of this area is utilized by beef cattle. If produced on well-fertilized soils, green grass and well-cured, green, leafy hay can supply all of the nutrient requirements of beef cattle, except the need for common salt and whatever energy-rich feeds may be necessary for additional conditioning or drylot finishing.

DIGESTIVE SYSTEM

An understanding of the principal parts and functions of the digestive system is essential to intelligent feeding of cattle. Fig. 8-2 shows the location

TABLE 8-1

PERCENTAGE OF FEED FOR DIFFERENT CLASSES OF
LIVESTOCK DERIVED FROM (1) CONCENTRATES, AND
(2) ROUGHAGES, INCLUDING PASTURE, 1970-1972[1]

Class	Concentrates	Roughages
	(%)	*(%)*
Beef cattle	26.2	73.8
Dairy cattle	35.8	64.2
Sheep and goats	11.0	89.0
Swine	85.2	14.8
Horses and mules	20.6	79.4
Poultry	97.4	2.6
All livestock	45.9	54.1

[1]Unpublished data provided by Mr. Earl F. Hodges, Leader—Feed Livestock Group, Production Research Branch, Economic Research Service, USDA, Washington, D.C.

of the parts of the ruminant's stomach and the route of digestion followed by most feed. Table 8-2 gives capacity figures.

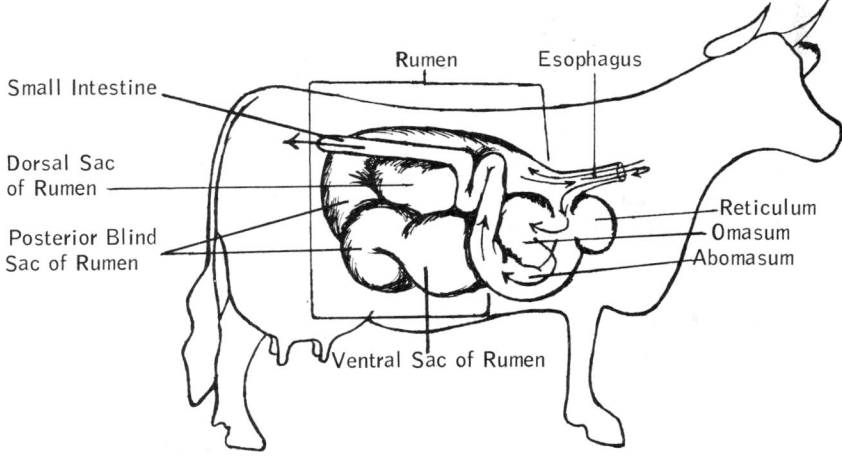

Fig. 8-2. Location and parts of the ruminant stomach (four compartments), with pathway of feeds indicated by arrows.

TABLE 8-2

PARTS AND CAPACITIES OF DIGESTIVE TRACTS (QUARTS)

Parts	Cow	Horse	Pig
	— *(qt capacity)* —		
Stomach:	(200)	8-16	6-8
Rumen (paunch)	160		
Reticulum (honeycomb)	10		
Omasum (manyplies)	15		
Abomasum (true stomach)	15		
Small intestine	62	48	9
Cecum		28-32	
Large intestine	40	80	10

To be useful to animals, nutrients must enter the bloodstream for transport to various parts of the body. The process whereby the animal releases feed nutrients from feed is termed digestion. As commonly used, the process also includes absorption of food from the digestive tract into the bloodstream. Most of the unused portion of the feed is eliminated in the feces, although a considerable proportion is also given off as gas through the mouth and nose.

Cattle belong to the ruminant or cud-chewing group of animals (which includes cattle and sheep), whereas swine and chickens are monogastric animals (those having only one stomach).

The primary nutritional difference between cows, or ruminants, and simple-stomached animals comes about through the normal functioning of the rumen and its billions of microorganisms—the bacteria and protozoa. These microorganisms live on the food material supplied in the cow's diet. The rumen provides a highly desirable environment in which they can grow and reproduce themselves as rapidly as every half hour. In turn, they release to the cow during their life and from their bodies as protoplasm upon death, many of the cow's required nutrients. This is how the cow gets many nutrients needed for production that are not supplied in the diet.

Parts and Functions of the Ruminant Stomach

The feed taken in by cows is mixed with a heavy flow of saliva, which is needed to help in chewing and swallowing of dry materials. This saliva of ruminants, unlike that of nonruminants, does not contain enzymes to aid in the digestion of starches. However, the saliva of cows, estimated to be about 120 pounds per day in a mature animal, does have enough buffers (sodium bicarbonate) to neutralize the fatty acids produced in the rumen and maintain the rumen contents at approximately a neutral pH (7.0).

The ruminant stomach consists of four distinct compartments: (1) the rumen, or "paunch," as it is commonly called; (2) the reticulum, or "honeycomb"; (3) the omasum, or "manyplies" (so called because of the plies or folds); and (4) the abomasum, or "true stomach." When feeds are ingested, the normal pathway they follow is in the order just listed, with portions being returned to the mouth for chewing before they enter the omasum. These four compartments do not lie in a straight arrangement; rather, they are bunched and joined together to form a compact structure. A discussion of each of the parts of the ruminant's stomach follows:

1. *Rumen*—In large, mature cows, the rumen has a capacity of about 40 gallons, or as much as 300 pounds of material. The functions of the rumen are: (a) to act as a storage place (it holds the feed which the cow gathers during her feeding period; later, when she rests, she regurgitates the larger particles back to the mouth, to grind them more completely, so that they pass from the rumen more quickly and make room for more feed); and (b) to provide a place for fermentation—there is a continual flow of feed material into and out of the rumen, constant introduction of saliva which controls the pH, absorption of the end products of the microbial action, and a population of

microorganisms—bacteria and yeasts, both of which are plants, and one-celled animals called protozoa—which develop in accordance with the amount and type of feed consumed. These organisms in the paunch (a) help digest the crude fiber, (b) form essential amino acids and build up complete proteins, and (c) manufacture the B-complex vitamins.

For satisfactory rumen function in the digestive tract and for synthesis of B vitamins and proteins, rumen bacteria and protozoa require adequate nutrients from beef cattle rations at all times, including (a) energy, involving small amounts of available energy, such as sugars and starches; (b) ammonia-bearing ingredients, such as proteins, urea, and ammonium salts; (c) major minerals, especially sodium, potassium, and phosphorus; (d) cobalt and possibly other trace minerals; and (e) miscellaneous growth factors (branched-chain volatile fatty acids) arising from or found in certain natural feeds rich in protein or nonprotein nitrogenous constituents.

The fermentation process produces large amounts of carbon dioxide, methane, and ammonia; smaller amounts of hydrogen, hydrogen sulfide, and carbon monoxide; and probably trace amounts of other gases. Normally, these gases are passed off by the reflex action of belching. However, sometimes a cow is unable to get rid of this gas, and bloating results.

2. *Reticulum*—The reticulum lies directly in front of the rumen. Actually, the rumen and the reticulum are not completely separated; hence, food particles pass freely from one to the other. The interior of the reticulum is lined like a honeycomb; hence, the popular name—"the honeycomb."

The main function of the reticulum is its action as a screening device. Heavy objects, such as nails or wire eaten with the feed, have a tendency to settle out into this compartment. Therefore, it is sometimes referred to by cattlemen as the "hardware stomach."

3. *Omasum*—The third compartment is lined with plies or folds of tissue; hence, it is sometimes called the "manyplies." Less is known about the functions of the omasum than of any of the other compartments of the ruminant stomach. It is generally believed, however, that the primary functions of this compartment are (a) to reduce the water content of the feedstuffs, and (b) to exert a regrinding and squeezing action on the material.

4. *Abomasum*—The abomasum is often referred to as the "true stomach" because its action is similar to the stomach action in monogastric animals. As in the monogastric stomach, digestive juices are added and the moisture content of the feed material is increased. The digestive juices contain enzymes, resulting in protein digestion in the abomasum. Little or no digestion of fat, cellulose, or starch occurs in this organ.

The feed material leaving the abomasum is highly fluid in nature. It is then passed out of the stomach through the small intestine, where additional digestion occurs, and the unabsorbed material is then excreted by way of the large intestine.

Rumination

The process known as rumination, or chewing the cud, occupies about eight hours of the cow's time each day. When the cow regurgitates, a soft mass of coarse feed particles, called a bolus, passes from the rumen through the esophagus in a fraction of a second. She chews each bolus for about one minute, then swallows the entire mass again. A placid cow lying under a tree slowly chewing her cud conveys a special sense of contentment, symbolic of the tranquility of the countryside. But this activity is most significant to the cow. Originally, it was thought that the regrinding which occurred during rechewing helped the digestion by exposing a greater surface area to fiber-digesting microflora. But recent experiments indicate that rechewing does not improve digestibility. Instead, rumination has an important effect on the amount of feed the animal can utilize. Feed particle size must be reduced to allow passage of the material from the rumen. It follows that high quality forages require much less rechewing and pass out of the rumen at a faster rate; hence, they allow a cow to eat more. This concept is very important to the production of beef and milk because a cow will eat only as much coarse material as she can grind up by ruminating not more than nine hours per day.

Calf's Stomach

When the calf is born, the rumen is small and the fourth stomach is by far the largest of the compartments. Thus, digestion in the young calf is more like that of a single-stomached animal than that of a ruminant. The milk which the calf normally consumes bypasses the first two compartments and goes almost directly to the fourth stomach in which the rennin and other compounds for the digestion of milk are produced. If the calf gulps too rapidly, or gorges itself, the milk may go into the rumen where it is not digested properly and may cause upset of the calf's digestive system. As the calf nibbles at hay, small amounts of material get into the rumen. When certain bacteria become established, the rumen develops and the calf gradually becomes a full-fledged ruminant.

FEED CONSUMPTION AND RATE OF GAIN

If they don't eat it, they won't gain. But feed consumption and rate of gain are affected by many things.

The daily consumption of dry matter (feed) by cattle is primarily dependent upon the following:

1. *Size*—Large feedlot cattle consume more feed, animal for animal, than small cattle. But they may or may not be more efficient than small cattle when consideration is given to (a) the production efficiency of their dams, and (b) carrying them to the same degree of finish.

2. *Age and condition*—Older and more fleshy feedlot cattle consume less feed per unit of liveweight than do younger, leaner animals. Mature animals in good condition may be expected to consume amounts of dry matter equal to two percent or more of their liveweight, whereas thin animals eating high quality roughage should eat amounts equal to three percent of their liveweight per day.

3. *Digestible nutrient content (energy density)*—As digestible nutrient content (energy density) increases, consumption of feed dry matter is usually reduced. It follows that feed efficiency is improved in high- and all-concentrate rations, due to their high energy. In the final analysis, however, the comparative price of concentrates and roughages—the economics of the situation—will be a major determining factor.

4. *Environmental stress*—Cattlemen have long known that environmental stress caused by high and low temperatures, mud, and other adverse environmental factors can affect the voluntary consumption of feed. For example, feedlot cattle consume less during very hot weather than in cold weather.

The rate of gain of feedlot cattle is influenced by the following factors:

1. *Sex*—Under feedlot conditions, at comparable weight and finish, bulls can be expected to make about 10 percent greater gains than steers, and steers can be expected to make about 10 percent greater gains than heifers.

2. *Implants and growth stimulants*—The use of certain implants and growth stimulants in finishing steers and heifers usually increases gains by 8 to 12 percent.

Elucidation and application of the subject "Feed Consumption and Rate of Gain" as it pertains to each of the respective classes of cattle is presented in the following chapters of this book: Chapter 19, Feeding and Managing Brood Cows; Chapter 20, Bulls; Chapter 21, Feeding and Handling Calves; Chapter 22, Replacement Heifers; and Chapter 30, Feeding Finishing (Fattening) Cattle.

NUTRITIVE NEEDS OF BEEF CATTLE

The nutritive requirements of beef cattle have become more critical with the shift in beef production practices. Steers were formerly permitted to make their growth primarily on roughages—pastures in the summertime and hay and other forages in the winter. After making moderate and unforced growth for two to four years, usually the animals were either turned into the feedlot or placed on more lush pastures for a reasonable degree of finishing. With this system, the growth and finishing requirements of cattle came largely at two separate periods in the life of the animal.

Under the old system of moderate growth rate, reasonably good pastures and good quality hay fully met the protein requirements, as well as the mineral and vitamin needs. As the feeding period was not so long with these older cattle, and the stress was not so great, in comparison with the period required in the finishing of calves or yearlings in a drylot, there also was less tendency for vitamin deficiencies to develop in the feedlot; and the protein requirements were less important during the finishing period.

The preference of the consumer for smaller cuts of beef—meats that are more tender and have less fat—has caused a shift in management and marketing. Today, increasing numbers of cattle are finished at younger ages, in a shorter period than formerly, in bigger feedlots, and marketed as baby beef. Such animals are in forced production. Their bodies are simultaneously laying on fat and growing rapidly in protein tissues and skeleton. Consequently, the nutritive requirements are more critical than those of older cattle, especially from the standpoint of proteins, minerals, and vitamins.

In recent years, the introduction of crossbreeding and the exotic breeds has produced faster gaining calves, later maturing cattle, and heavier milking cows. Also, more and more heifers are being bred to calve as two-year-olds. In this revision, provision has been made for the nutritive needs created by these changes.

As feeds represent by far the greatest cost item in beef production, it is important that there be a basic understanding of the nutritive requirements. For convenience, these needs will be discussed under the following groups: (1) energy, (2) protein, (3) mineral, (4) vitamin, and (5) water.

Energy

Carbohydrates, which constitute about 75 percent of all the dry matter of plants, are the chief sources of energy of cattle feeds. In the usual chemical analysis of feeds, carbohydrates are divided into nitrogen-free extract (NFE) and fiber. The NFE, which is the more soluble part, includes sugars, starches, organic acids such as lactic and acetic acid (which are present in silage), and other more complex carbohydrates. Fiber includes the relatively insoluble carbohydrates, such as cellulose.

Next to carbohydrates, fats are important as energy sources. Because of their larger proportion of carbon and hydrogen than the carbohydrates, they liberate more energy; furnishing approximately 2.25 times as much heat or energy per pound as do the carbohydrates.

In addition to supplying nitrogen, natural plant protein compounds also supply a certain amount of energy.

A relatively large portion of the feeds consumed by beef cattle is used in meeting the energy needs, regardless of whether the animals are merely being maintained (as in wintering) or fed for growth, finishing, or reproduction.

The first and most important function of feeds is that of meeting the maintenance needs. If there is not sufficient feed, as is frequently true during periods of droughts or when winter rations are skimpy, the energy needs of the body are met by the breakdown of tissue. This results in loss of condition and body weight.

After the energy needs for body maintenance have been met, any surplus energy may be used for growth, finishing, reproduction, or lactation. With the present practice of finishing cattle at early ages, growth and finishing are in most instances simultaneous, and, therefore, not easily separated.

In the finishing process, the percentage of protein, ash, and water steadily decreases as the animal matures and fattens, whereas the percentage of fat increases. Thus the body of a calf at birth may contain about 70% water and 4% fat; whereas the body of a fat two-year-old steer may contain only 45 to 50% water but from 30 to 35% fat. This storage of fat requires a liberal allowance of energy feeds.

Through bacterial action in the rumen, cattle are able to utilize a considerable portion of roughages as sources of energy. Yet it must be realized that with extremely bulky rations the animal cannot consume sufficient quantities to produce the maximum amount of fat. For this reason, finishing rations contain a considerable proportion of concentrated feeds, mostly cereal grains. On the other hand, when the energy requirements are primarily for maintenance, roughages are usually the most economical sources of energy for beef cattle.

At times, fats may be cheap enough to merit consideration as partial substitutes for standard energy feeds. Also, it is probable that very small amounts of fatty acids are essential for beef cattle, as is true in certain other species, but no requirements have thus far been established.

SYMPTOMS OF ENERGY DEFICIENCY (Underfeeding)

Many cattle throughout the world are underfed all or some part of the year. In fact, lack of sufficient total feed is probably the most common deficiency suffered by beef cattle, although it is recognized that underfeeding is frequently complicated by concomitant shortage of protein and other nutrients. Restricted rations often occur during periods of drought, when pastures or ranges are overstocked, or when winter rations are skimpy. Also, many range cattlemen regularly plan that cows in good flesh should lose some condition during the winter months; they feel that it is uneconomical to feed sufficient to retain the fleshy condition. Fortunately, during such times of restricted feed intake, animals have nutritive reserves upon which they can draw. Although they may survive for a considerable period of time under these conditions, there is an inevitable loss in body weight and condition; and, varying with the degree of underfeeding, there may be a slowing or cessation of growth (including skeletal growth), failure to conceive, and increased mortality. Low feed intake also commonly results in increased deaths from toxic plants and from lowered resistance to parasites and diseases.

Research workers of the U.S. Department of Agriculture conducted an experiment to determine some of the economic effects and possible harm to animals of limited rations.[2] Identical twin calves were used and the following planes of nutrition were studied: (1) full feed—gains of more than 1.5 lb daily, (2) 75% of full feed—gains of 1.0 lb per day, (3) 62% of full feed—gains of 0.5 lb a day, and (4) a maintenance ration of about 50% of full feed—they neither gained nor lost in weight.

All animals—including those on the low energy rations—received ample protein, vitamins, minerals, and other nutrients. At the end of the period of retarded feeding, the steers were fed liberally until they reached a slaughter weight of 1,000 pounds.

Although the low-plane-of-nutrition animals reached slaughter weight

---

[2]Winchester, C. F., and Paul E. Howe, *USDA Tech. Bull. No. 1108.*

from 10 to 20 weeks later than did their twins, the former attained their weight on approximately the same total feed intake as the latter; which means that, after limited feeding ended, the retarded animals made more economical gains than did their twins. Carcass quality, amount of lean meat, and grade were not affected.

This experiment showed that, under conditions of feed scarcity, beef cattle between the ages of 6 and 12 months can be carried on a maintenance ration—so they will neither gain nor lose in weight—provided the nutrient needs other than energy are supplied—without subsequent loss in feed efficiency, carcass quality, or quantity of lean meat. Also, it shows that compensatory gains occur following a low plane of nutrition; it shows why feedlot finishers prefer feeder cattle that have not been backgrounded at a high rate of gain.

METHODS OF MEASURING ENERGY[3]

Scientists generally agree that the units used to measure the nutritive requirements of animals and to evaluate feeds should be one and the same. But, there is considerable disagreement, throughout the world, as to what units, or system, to use.

The earliest attempt to put feed comparisons on a quantitative basis was made by Thaer, a German scientist, in 1810. He took meadow hay as his standard, compared the extractable nutrients of other feeds to it, then assigned them "hay values." Other systems followed. In 1859, Grouven made use of analyses of protein, fat, and carbohydrate to formulate the first feeding standard for farm animals. In 1864, Wolff devised a standard based on digestible nutrients obtained from feeding trials. In 1897, Lehmann modified Wolff's standards. Others followed. Among them were the starch values of Kellner, a German scientist, in 1907 (starch values are still used in Europe); the Scandinavian feed-unit system (Woll, 1912); the dairy cow standards of Haecker of Minnesota, in 1914; the net-energy values by Armsby, in 1915; the productive units developed by Mollgaard, of Denmark, in 1939; the productive-energy values computed by Fraps of the Texas station (1937, 1941); and the Morrison standards, based upon total digestible nutrients (TDN).

Beginning about the time of World War II, research on the energy values of feeds gave way to the more popular studies on hormones, antibiotics, and other ration ingredients. Today, the pendulum is swinging back, and there is renewed interest in methods of measuring energy. As would be expected, there is much disagreement among scientists and stockmen as to the best system, for none of them is perfect. Generally speaking, there are two schools of thought; and each is inclined to be militant and uncompromising. On the one hand, there are those who'll fight for a continuation of the total digestible nutrient system (TDN), rather than switch to the calorie system. Most of them recognize the weaknesses of the TDN system, but they favor a

[3]The author gratefully acknowledges the authoritative help of Dr. Lorin E. Harris, Utah State University, Logan, Utah, in the preparation of this entire section.

"let well enough alone" policy because, so they argue, folks are generally using it, whereas a change would confuse them. On the other hand, those who champion the adoption of the calorie system, are usually impatient; they want to dump the TDN system and move on.

The author favors a transition from the TDN system to the calorie system. For this reason, both systems are presented in this book. In Table I-3 of the Appendix, metabolizable energy (ME), net energy for maintenance (NEm), net energy for gain (NEgain), and TDN values of feeds are presented.

The discussion that follows will, of necessity, be brief and simplified.

TOTAL DIGESTIBLE NUTRIENTS (TDN) SYSTEM

Total digestible nutrients (TDN) is the sum of the digestible protein, fiber, nitrogen-free extract, and fat × 2.25. Back of TDN values are the following steps:

1. *Digestibility*—The digestibility of a particular feed for a specific class of stock is determined by a digestion trial. It is made by determining the percentage of each nutrient in the feed through chemical analysis; giving the feed to the test animal for a preliminary period, so that all residues of former feeds will pass out of the digestive tract; giving weighed amounts of the feed during the test period; collecting, weighing, and analyzing the feces; determining the difference between the amount of the nutrient fed and the amount found in the feces; and computing the percentage of each nutrient digested. The latter figure is known as the *digestion coefficient* for that nutrient in the feed.

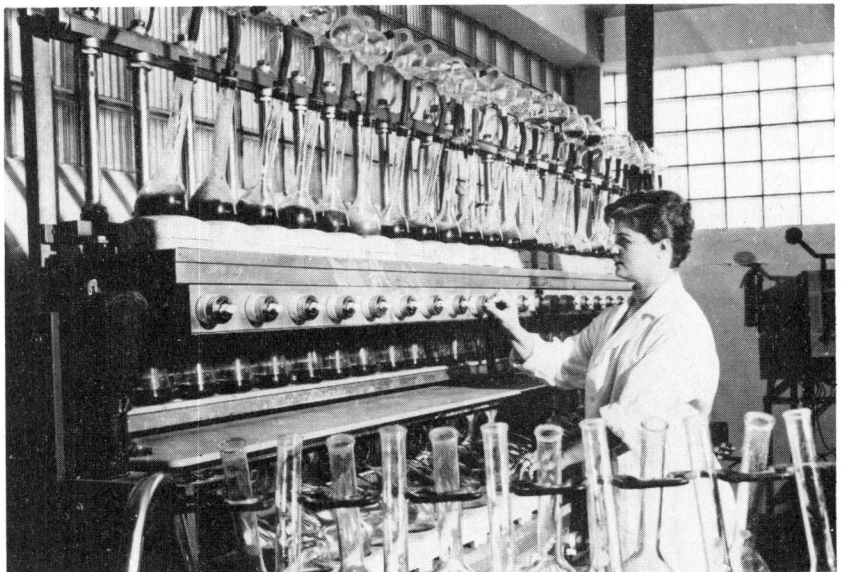

Fig. 8-3. A chemical analysis is the first step in determining the total digestible nutrients (TDN) of a feed. (Courtesy, Beacon Division of Textron, Inc., Cayuga, N.Y.)

2. *Computation of digestible nutrients*—Digestible nutrients are computed by multiplying the percentage of each nutrient in the feed (protein, fiber, nitrogen-free extract [NFE], and fat) by its digestion coefficient. The result is expressed as digestible protein, digestible fiber, digestible NFE, and digestible fat. Thus, for corn of the composition shown in the left-hand column that follows, the digestible nutrients could be estimated as:

Total % of nutrient in feed		×	$\dfrac{\text{Digestion Coefficient}}{100}$	=	% Digestible nutrient	
					Digestible Nutrient	
					(%)	(lb)
Protein:	9.3%	×	$\dfrac{67}{100}$	=	6.2 (protein)	6.2
Fiber:	1.9%	×	$\dfrac{39}{100}$	=	0.7 (fiber)	0.7
Nitrogen-free extract (NFE)	70.1%	×	$\dfrac{85}{100}$	=	59.6 (NFE)	59.6
Ether extract: (fat)	3.9%	×	$\dfrac{85}{100}$	=	3.3 (fat)	3.3

In this example, corn grain has 6.2 percent digestible protein. Every feed has a digestible protein value which was determined in this manner. (See Appendix, Table I-3.)

3. *Computation of total digestible nutrients*—To approximate the greater caloric value of fat, the percentage of digestible fat is multiplied by 2.25. Hence, for the above sample of corn, the TDN may be calculated as follows:

	%				%
Digestible protein	6.2	×	1	=	6.2
Digestible crude fiber	0.7	×	1	=	0.7
Digestible NFE	59.6	×	1	=	59.6
Digestible either extract (fat)	3.3	×	2.25	=	7.4

73.9% TDN, or 73.9 lb TDN/100 lb (73.9 kg TDN/100 kg)

4. *Animal requirements or feeding standards*—In the TDN system, the feed requirements (energy) of farm animals are given as pounds of total digestible nutrients; and, in addition, the pounds of dry matter are given also to insure that the stated amount of TDN is fed.

Advantages and Disadvantages of the TDN System

The main *advantage* of the TDN system is:
It has been used a long time and many people are acquainted with it.

The main *disadvantages* of the TDN system are:

1. It is based on physiological fuel values for humans and dogs. These do not apply to ruminants. The factors of 1 for protein, crude fiber, and nitrogen free extract, and 2.25 for fat, are not always constant.

2. It overevaluates high fiber feeds (roughages) in relation to low fiber feeds (concentrates) when fed for high rates of production, due to the higher heat loss per pound of TDN in the case of the high fiber feeds.

3. It does not measure energy in energy units.

4. It does not measure all losses of energy from the body.

CALORIE SYSTEM

Energy is used in many forms—as light, electricity, atomic force, work, or heat; and it is measured by several units such as candle power, kilowatts, foot pounds, joules, and calories. In animals, energy is expended as work and/or heat or stored as products. It would appear, therefore, that it should be measured in units suitable for these purposes. Thus, a heat unit is an excellent way in which to measure the potential energy of feeds, the energy of animal products, and the heat that results from body processes. The heat unit used by animal nutritionists is the calorie.

A calorie (cal always written with a small c) is the amount of heat required to raise the temperature of one gram of water one degree centigrade. To measure this heat, an instrument known as the bomb calorimeter is used, in which the feed (or other substance) tested is placed and burned in the presence of oxygen. (See Fig. 8-4.)

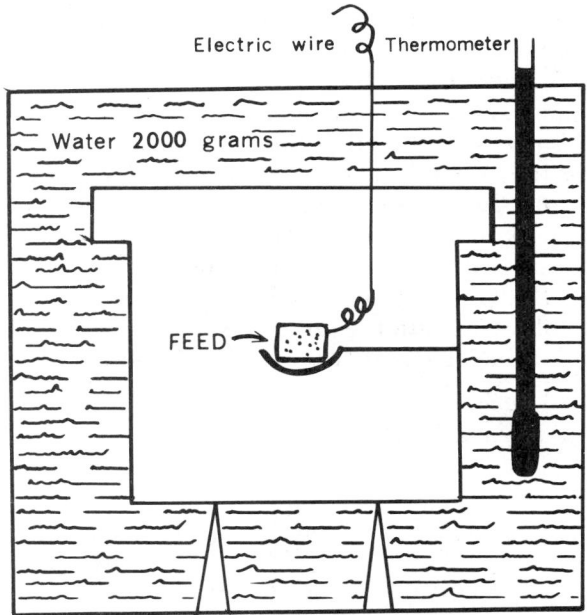

Fig. 8-4. Diagrammatic sketch of a bomb calorimeter used for the determination of the gross energy value (caloric content) of various materials.

It is noteworthy that the determination of the heat of combustion with a bomb calorimeter is not as difficult or time consuming as the chemical analyses used in arriving at TDN values. Briefly stated, the procedure is as follows: An electric wire is attached to the material being tested, so that it can be ignited by remote control; 2,000 g of water are poured around the bomb; 25 to 30 atmospheres of oxygen are added to the bomb; the material is ignited; the heat given off from the burned material warms the water; and a thermometer registers the change in temperature of the water. For example, if one g of material is burned and the temperature of the water is raised one degree centigrade, 2,000 cal are given off. Hence, the material contains 2,000 cal per g, or 907,200 cal per lb (2 million cal per kg), or 907 kilocalories (kcal) per lb, or 0.907 megacalories (Mcal) per lb. This value is known as the gross energy (GE) content of the material. Thus, one kcal is equivalent to 1,000 cal while one Mcal is equivalent to 1,000,000 cal or 1,000 kcal.

The energy requirements of beef cattle are expressed in terms of metabolizable energy (ME), net energy for maintenance (NEm), net energy for gain (NEg), and total digestible nutrients (TDN). Definitions of these and other energy terms are given in Section I of the Appendix of this book.

A major advantage of separate net energy requirements for maintenance and gain is that animal requirements stated in this way do not vary when different roughage-concentrate ratios are fed.

How to Use the Multiple Net Energy System

The following examples will illustrate two important uses of the multiple NE system:[4]

• *To calculate the quantity of feed necessary to meet an animal's energy needs and to compound a ration to supply the needed concentration of energy per unit of dry matter*—For example, a 770-lb (350 kg) steer gaining 2.86 lb (1.3 kg) per day requires 6.24 Mcal of NEm for maintenance and 6.48 Mcal of NEg for body weight gain (Tables 8-18 or 8-20). For estimating purposes we can assume a good quality roughage will support a gain of 1.1 lb (0.5 kg) per day and the maximum gain likely on a high concentrate ration would be 3.1 lb (1.4 kg). With these assumptions it can be estimated that the widest roughage-concentrate ratio likely to support a gain of 2.86 lb (1.3 kg) per day is 15:85. If alfalfa hay were fed as the roughage, and barley as the concentrate, the following ration would satisfy the energy requirements (see Appendix Table I-3 for NEm and NEg for alfalfa and barley):

	Lb or Kg	NEm (Mcal /lb)	NEm (Mcal /kg)	NEg (Mcal /lb)	NEg (Mcal /kg)
Alfalfa					
(s.c. midbloom)	0.15	.08	0.17	.04	0.08
Barley	0.85	.74	1.62	.48	1.06
Total	1.00	.82	1.79	.52	1.14

[4]Adapted by the author from *Nutrient Requirements of Beef Cattle*, fifth revised edition, National Academy of Sciences, Washington, D.C., 1976, pp. 4-5.

Fig. 8-5. A cow in an open circuit respiration chamber. The gas meter to the left of the chamber is used to measure the respiratory exchange of the cow. These data, plus the gas composition, provide the information needed to calculate the Heat Production (HP) of the cow. The HP of an animal consuming feed in a ther-moneutral environment is composed of the heat increment (heat of fermentation plus heat of nutrient metabolism) plus heat used for maintenance (basal metabolism plus voluntary activity). It can be estimated by three procedures: (1) by measuring the amount of oxygen an animal consumes (open circuit method), shown above; (2) by measuring directly the amount of heat produced by the animal (direct method); and (3) the comparative slaughter technique, in which two comparable animals are slaughtered, one at the beginning of the test period, and the other at the end of the test period, and the energy content of each is determined, then the difference between these two values represents the amount of energy gained. (Courtesy, USDA, Beltsville, Md.)

This ration contains 0.82 Mcal of NE_m and 0.52 Mcal NE_g per lb (1.79 Mcal of NE_m and 1.14 Mcal NE_g per kg) of dry matter. It would require 6.24 ÷ .82, or 7.61 lb (6.24 ÷ 1.79, or 3.49 kg) of this mixture for maintenance. To gain 2.9 lb (1.3 kg)/day, the 770-lb (350-kg) steer would require 6.48 Mcal NE_g (Table 8-18). Thus, 12.46 lb of feed would be necessary to supply 6.48 Mcal NE_g (6.48 ÷ .52). The total amount of feed required for 2.9 lb (1.3 kg) gain would be 20.07 lb (9.12 kg)—7.61 (3.46 kg) for maintenance and 12.46 (5.66 kg) for gain. Table 8-18 indicates that the minimum amount of feed containing 15 percent roughage and 85 percent concentrate a 770-lb (350-kg) steer must eat to deposit 6.48 Mcal of NE_g is 17.6 lb (8.0 kg). The procedure illustrated in this example can easily be modified for conditions where expected gains may differ from the assumptions used. Also different combinations of feeds may require small changes in the trial ration to give the exact energy concentration desired.

● *To predict weight gains and to determine whether cattle have gained weight in accordance with expectations*—This information can be helpful in evaluating feeding and management programs and in estimating the future weight of a given group of cattle if the type of ration plus the intake are known. An example illustrating this latter point follows: A pen of steers weighing 495 lb (225 kg) is to be fed a ration containing 0.83 Mcal of NE_m and 0.51 Mcal of NE_g per lb (1.83 Mcal of NE_m and 1.13 Mcal of NE_g per kg) and marketed subsequently at 1,045 lb (475 kg). The mean steer weight for the feeding period is 770 lb (350 kg), so the NE_m requirements (from Table 8-20) will average 6.24 Mcal NE_m per day, or 7.5 lb (6.24 ÷ 0.83), or 3.4 kg (6.24 ÷ 1.83), of feed for maintenance. Total feed consumption of this ration should be about 17.6 (8.0 kg) per day (Table 8-18), which leaves 10.1 lb (17.6 − 7.5), or 4.5 kg (8.0 − 3.5), of feed available for weight gain. Ten and one-tenth lb (4.59 kg) of this ration will contain 5.19 Mcal NE_g (10.1 × 0.514, or 4.59 × 1.13). A group of steers averaging 770 lb (350 kg) during a feeding period and having available 5.19 Mcal NE_g should make a weight gain of about 2.3 lb (1.05 kg) per day. Thus, this group of steers should reach the desired market weight in about 240 days.

Protein

Proteins are composed of different nitrogenous compounds called amino acids. At least 24 amino acids have been identified and may occur in combinations to form an almost limitless number of proteins.

In a chemical analysis, crude protein refers to all the nitrogenous compounds in a feed. It is determined by finding the nitrogen content and multiplying the result by 6.25. The nitrogen content of proteins averages about 16 percent (100 ÷ 16 = 6.25).

The protein allowance for beef cattle, regardless of age or system of production, should be ample to replace the daily breakdown of the tissues of the body including the growth of hair, horns, and hoofs. In general, the protein needs are greatest for the growth of the young calf and for the gestating-lactating cow.

The protein requirements listed in Tables 8-18 and 8-19 are estimated needs for optimal production. They can be exceeded without toxicity or reduced animal performance. As noted, the requirements are expressed on the basis of both total and digestible protein. Nitrogen values were converted to digestible protein values by multiplying by the factor 6.25 and using an average biological value of 77.5. Cattle fed these levels of protein have gained and reproduced at optimum rates. Methods of feeding, feed preparation, and various feed additives do not appear to alter protein requirements. Feed consumption is reduced when all-concentrate rations are fed. As consumption declines, the percentage of protein in such rations should be increased proportionally.

The protein supplements are regularly in shorter supply and higher priced than the cereal grains and other high energy feeds used in livestock feeding. Normally, the United States produces about 19 million tons of protein supplements, exclusive of urea (Fig. 8-6). But it is estimated that an

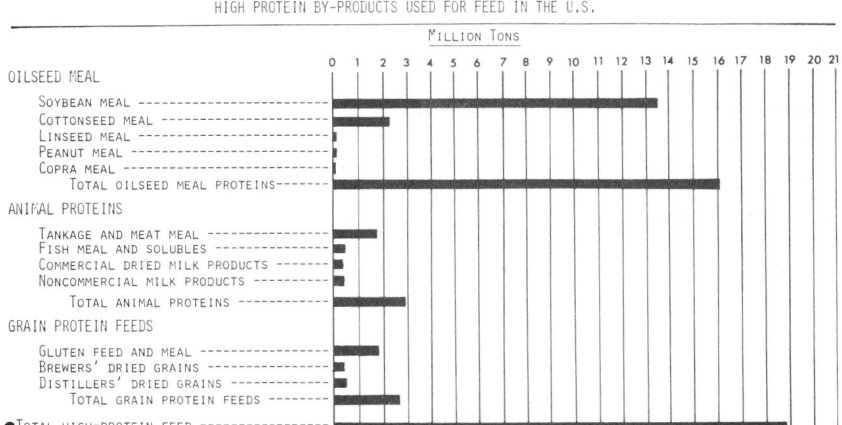

Fig. 8-6. Tonnage of different protein supplements used for feed in the United States, 1973. (Source: *Feed Situation*, FdS 253, USDA, May 1974, Table 23, p. 6)

additional 2 to 5 million tons of these products could be used advantageously if all animals were supplied an adequate amount of protein.

As protein supplements ordinarily cost more per ton than grains, normally beef cattle should not be fed larger quantities of these supplements than actually needed to balance the ration.

With stocker cattle, or in the maintenance of the beef breeding herd, it usually does not pay to add a protein supplement when a legume hay is fed. With feedlot cattle on high concentrate rations, or when the breeding herd is being wintered on a nonlegume roughage, sufficient protein supplement— usually one to two pounds daily—should be added to the ration.

Because of rumen synthesis of essential amino acids by microorganisms, the quality of proteins (or balance of essential amino acids) is of less importance in the feeding of beef cattle than in feeding some other classes of stock. Proteins from plant sources, therefore, are quite satisfactory. Also, these microorganisms—which are a low form of plant life and are able to use inorganic compounds such as ammonia just as plants utilize chemical fertilizers—build body proteins of high quality in their cells from sources of inorganic nitrogen that nonruminants cannot use. Since the life span of these bacteria is short, further on in the digestive tract, the ruminant digests the bacteria and obtains good protein therefrom. In ruminant nutrition, therefore, even such nonprotein sources of combined nitrogen as urea and ammonia have a protein replacement value. An exception is the very young ruminant in which the rumen and its ability to synthesize are not yet well developed. For such an animal, high quality proteins in the diet are requisite to normal development.

SYMPTOMS OF PROTEIN DEFICIENCY

Depressed appetite is the primary symptom of protein deficiency in beef cattle rations. Depressed appetite may, in turn, lead to an inadequate intake of energy; hence, protein deficiency and energy deficiency often occur together.

Other symptoms of protein deficiency are loss of weight, poor growth, irregular or delayed estrus, and reduced milk production.

NONPROTEIN NITROGEN SOURCES

Certain nonprotein nitrogen sources may be substituted for all or much of the supplemental protein required in most beef cattle rations, provided such rations are adequate in minerals and readily available carbohydrates. Among such products are urea, ammoniated molasses, ammoniated beet pulp, ammoniated cottonseed meal, ammoniated citrus pulp, and ammoniated rice hulls. The possibility exists that other products will be forthcoming. Each such product should be evaluated by controlled feeding trials.

UREA

Approximately 837,000 tons of urea are fed annually in the United States, as a source of protein for cattle, sheep, and goats.[5] In recent years there has been increased interest in feeding urea to cattle, due primarily to the following circumstances:

1. *Shortage of oil meal proteins*—The scarcity and high price of normal supplies of oil meal protein feeds is well known.

2. *Progress in fundamental ruminant nutrition*—Through basic studies, scientists have established many of the nutrient requirements of rumen microorganisms, thereby permitting the preparation of balanced supplements designed to enable animals to get the most out of the roughages they consume. This knowledge has led to the extensive use of such low grade roughages as corncobs, straws, and poor-grade hays—many of which had been wasted previously.

These factors, plus meeting the needs of a rapidly expanding human population, are likely to continue to accentuate the interest of feed manufacturers and cattle producers in utilizing urea and other nonprotein nitrogen sources.

When properly used, urea is a valuable cattle feed; when improperly used, it may be a needless expense, or worse yet, a hazard. To the end that the first condition may more frequently prevail, the following summary is presented:

1. *Feed grade urea*—Initially, the protein equivalent value of feed grade urea was 42 (% nitrogen) times 6.25 (common protein factor), or 262% protein. Today, more concentrated 45% nitrogen (45 × 6.25 ÷ 281) urea has replaced most of the 42% grade, at a lower unit cost.

---

[5]The 1973 U.S. consumption of urea was 3,805,000 tons, with use as follows: liquid fertilizer, 37%; dry fertilizer, 24%; feed industry, 22%; and industrial users, 12%. Hence, 3,805,000 × 22% = 837,000 tons of urea used for feed. (Source: John Chohlis, Ralston Purina Company, St. Louis, Mo.)

2. *Feeding value of urea*—Attempts have been made to equate urea to oil meals by various thumb rules. One such thumb rule is that 1 pound of urea plus 6 pounds of corn equal 7 pounds of soybean (or cottonseed) meal. This combination of corn and urea supplies as much nitrogen as does soybean meal and, thus, could be considered equal to it in crude protein content. This is true if the rumen microorganisms can convert the urea nitrogen to protein. But this doesn't tell the whole story! Table 8-3 shows the inequalities in mineral content of these two feeds.

TABLE 8-3

COMPARISON OF A CORN:UREA (6:1) MIXTURE WITH SOYBEAN MEAL

Nutrient	Corn:Urea (6:1) as a % of Soybean Meal
TDN	88.9
N	100.0
Ca	13.5
P	38.1
Mg	36.0
K	14.7
S	9.1
Mn	14.6
Co	95.2
Cu	20.0
Fe	13.1
Zn	63.9

As noted in Table 8-3, the corn-urea combination is a little low in TDN, or energy. Additionally, the combination supplies only 13.5% as much calcium and 38.1% as much phosphorus as does soybean meal; and only in the case of cobalt does the combination supply as high a mineral level as does the soybean meal. Unless these deficiencies are met, poor utilization of the urea-containing ration can be expected. Of course, minerals can be added, but all ingredients cost money.

3. *Utilization of urea by ruminants*—Nature has provided cattle and other ruminants with a remarkable digestive system centering around the first stomach or rumen. The rumen serves as a large fermentation vat where billions of bacteria and other microorganisms live and multiply, if properly nourished. Rumen bacteria are actually tiny plants, and, like the bacteria in field soils, they are able to break down tough, coarse, fibrous material into simpler carbohydrates and to use these carbohydrates along with the sugars and starches, protein, nonprotein nitrogen (such as urea), and minerals in forming their own bodies.

Rumen bacteria have first access to all the feeds consumed by the ruminating animals. In carrying out their own living processes, they convert much of the feeds into simple substances which are absorbed by the animals and used for maintenance and for meat and milk production. As the bacteria grow and multiply they produce bacterial protein. They digest the proteins in the feeds and also make protein from simple sources such as urea. While scientists do not yet completely understand all aspects of protein utilization in the rumen, there is considerable evidence that the bacteria change a substantial amount of all proteins into bacterial protein, which is of good quality. This development enables ruminants to have good quality protein available and makes them comparatively independent of the quality of the protein in the feed. This is in marked contrast to the situation with man, dogs, swine, and poultry where considerable care must be exercised to provide all the essential amino acids in proper proportions for efficient and economical production. Young calves are like other mammals and do need the high quality protein, such as that in their mother's milk, until the rumen has developed enough to function efficiently.

The rumen bacteria are very small, and they are constantly being carried down into the true stomach and small intestines of the animal where digestive juices produced by the animal break the bacterial protein down into the simple amino acids which are then absorbed and used by the host.

4. *Urea is best utilized in well-balanced high energy rations*—Urea is not well utilized for supplementing low quality roughages. The explanation is that the carbohydrates in grasses and hays appear to be so slowly available that the bacteria have difficulty in using the energy from roughages to make use of urea in preparing bacterial protein. It is generally held that some preformed protein should be present in the feed also. Part of this will be provided by the grains, and frequently some oil meals are used in preparing the formula feeds.

Other components of a balanced feed include calcium, phosphorus, iron, copper, cobalt, manganese, iodine, and perhaps zinc and sulfur and magnesium. The need for these minerals as well as for vitamin A will depend upon local conditions with respect to the types of roughages produced and the influence of weather upon the quality of such roughages.

Other factors influencing the formulation include the purpose such as for the breeding herd, creep-feeding, or for use in the finishing or feedlot.

5. *Factors essential for optimum use of urea*—Urea can be successfully and effectively used, or it can be abused. Observance of the following pointers will assure optimum use of urea:

a. Mix the urea thoroughly.

b. Feed urea only to cattle, sheep, or other ruminants. Never feed it to swine, poultry, or horses.[6]

c. Limit the intake of urea to recommended maximum levels. (See Table 8-4.)

d. Provide a readily available energy source, such as molasses or grain.

e. Supply adequate and balanced levels of minerals, including calcium, phosphorus, and trace minerals (especially cobalt and zinc).

f. Achieve a nitrogen-sulfur ratio not wider than 15:1.

g. Incorporate alfalfa meal as a source of unidentified factors to stimulate the microbial synthesis of protein.

---

[6]Most state laws restrict the use of urea to ruminant rations; hence, it is illegal to add it to swine, poultry, or horse rations. Urea may be toxic to foals.

h. Include adequate salt for palatability; 0.5 percent in complete rations and 3.5 percent in protein supplements.

i. Provide the proper level of vitamin A, and of such other vitamins, hormones, antibiotics, and additives as desired.

j. Use a free-flowing urea, and mix it thoroughly; avoid sifting or sorting of ingredients. Never use high urea-containing supplements as a top dressing.

k. Accustom animals gradually to urea-containing feeds (over a period of 5 to 7 days), exercise caution in feeding very hungry animals, and feed at frequent intervals.

l. Never use raw soybeans or beans of any kind, lespedeza seed, alfalfa seed, or wild mustard seed in a grain mixture containing urea for the reason that an enzyme (urease) in the beans will break down urea into ammonia and carbon dioxide. The liberated ammonia may be strong enough to be objectionable to cattle. Eventually, the animals will eat the feed, but the protein level will be reduced.

6. *Quantity of urea that may be fed*—Urea may constitute up to 33⅓ percent[7] of the total protein of growing-finishing rations and 25 percent of the total for pregnant and lactating cows, provided additional energy is added in the form of molasses or grain to compensate for the lack of energy in the urea,[8] in order to feed the rumen bacteria properly. By total protein is meant the protein intake of the entire ration—including forage, grain, and protein supplements.

Common guidelines relative to the use of urea are given in Table 8-4.

Less urea is recommended in range cubes or pellets because (a) of the more limited grain and the poor quality roughage usually fed and (b) of the uncertainty of feeds being consumed regularly under adverse weather conditions. Thus, it is recommended that urea be limited to five percent, by weight, of range cubes or pellets used primarily to supplement dried range grass cured on the stalk.[9] Also, when feeding on the range, it is important that the supplement be spread out evenly and in such manner that the gluttonous animals do not get more than their share and the weak ones, that need help the most, are denied the benefits of the supplementary feed.

7. *Toxicity*—The symptoms of animals reacting from high urea intake include uneasiness, muscular incoordination, bloat, prostration, convulsions, and even death.

The veterinarian should be called to treat cases of urea toxicity. As an emergency measure, one gallon of vinegar may be administered to cattle as a drench. The acetic acid furnished by the vinegar lowers rumen pH and neutralizes ammonia, thus preventing further absorption of ammonia into the bloodstream.

---

[7]Some authorities place the upper limit at 50%. Some state laws limit the urea level to not more than ⅓ of the total nitrogen of the ration.

[8]For every pound of urea added to the ration, 5 to 6 pounds of a cereal grain or molasses should be added in order to replace the energy lost.

[9]The balance of the ingredients in range cubes or pellets usually consists of ground grain, molasses, oil meal proteins, and, under certain conditions, minerals (including trace minerals) and vitamin A supplements. Generally, the urea and molasses are first mixed, and then added to the rest of the concentrate.

TABLE 8-4

COMMON GUIDELINES TO THE USE OF UREA FOR CATTLE[1]

	For Finishing Cattle	For Grower (Stocker) Cattle	For Wintering Pregnant and Lactating Cows
Percent of total protein in ration from urea .... (%)	33⅓	25.0	25.0
Maximum urea/animal/day .... (lb)	0.22 (100 g)	0.15 (68 g)	—
Percent of urea, by weight of total air-dry feed consumed .... (%)	1.0	1.0	1.0
Percent of urea, by weight, of concentrate mix (grain plus protein supplement)[2] .... (%)	2.0-3.0	3.0	3.0
Percent of urea, by weight, of the protein supplement .... (%)	20-30[3]	10.0[4]	10.0
Percent of supplemental nitrogen in high protein supplement from urea[5] .... (%)	60-90[6]	30.0	30.0
Pounds of urea added/ton of corn silage at ensiling time[7] .... (lb)	10.0 (4.5 kg)	10.0 (4.5 kg)	10.0 (4.5 kg)

[1]In the preparation of this table, the author had the authoritative help of Dr. W. M. Beeson, Department of Animal Sciences, Purdue University, Lafayette, Ind.; Dr. William H. Hale, Department of Animal Science, The University of Arizona, Tucson, Ariz.; and Dr. W. E. Dinusson, Department of Animal Science, North Dakota State University, Fargo, N.D.

[2]Feed intake may be depressed if over 1% is used. Yet, many beef men are successfully using 2%.

[3]This means that as much as 60 to 90% of the protein value of the supplement may come from nonprotein sources. However, since such a supplement will constitute only 2 to 5% of the total ration fed, the first rule of thumb given in Table 8-4 still applies; namely, only ¼ to ⅓ of the total protein in the ration will be supplied from a nonprotein source.

[4]A protein supplement containing 10% urea provides 28.1% of protein equivalent (281% × .10) from nonprotein nitrogen.

[5]High urea supplements are best fed in complete mixed rations, which are *thoroughly* mixed. *Supplements containing 20-30% urea require extreme caution when being hand fed.*

[6]In a feedlot ration, this may be equivalent to 25 to 40% of the total nitrogen from all sources.

[7]On a dry matter basis, corn silage ensiled at the well-dented stage runs about 8% protein. The addition of 10 pounds of urea per ton (or 5 kg/1,000 kg) of silage increases the protein content to 13%. However, there is loss of flexibility in feeding such a ration, and the rate of gain will be less than can be secured from higher, more dense rations. Also, it is extremely important that the urea be well mixed in the silage, otherwise there is hazard of toxicity.

8. *Palatability*—Although various opinions exist relative to the palatability of urea and urea-containing feeds, most feeders feel that urea is not palatable and, therefore, that feed consumption may be lowered in comparison with rations in which oil meal protein supplements are used entirely. For this reason, care should be exercised in selecting an appetizing urea-supplemented mixture.

In contrast to the above opinion, it should be noted that, occasionally, cattle will consume straight fertilizer urea or ammonium nitrate in sufficient amounts to poison themselves.

Sometimes cattle will consume a urea-containing feed for a few days or weeks and then refuse it. This has occurred in drought areas where farmers have tried to extend their roughage supplies by feeding straw and other mineral-poor, low quality roughages. Appropriately increasing phosphorus and trace minerals have corrected the latter problem.

9. *How to compute how much urea is in a feed*—The level of urea in a feed may be noted in the following ways:

a. *Percent of urea in the feed*—When the percent of urea is given, one can calculate the amount of protein furnished by urea by multiplying the percent urea by 281 (the protein equivalent of urea). For example, if a 40% supplement contains 5% urea, then 14% protein is furnished by urea (281 × 5% = 14%). To determine the percent of the total protein furnished by urea, divide the percent of protein as urea by the percent of protein in the supplement (14 ÷ 40% = 35%). In this case, slightly more than one-third of the protein in the supplement is furnished by urea.

b. *Percent protein as urea*—When the urea in the supplement is expressed in percent protein as urea, one can determine the amount of urea by dividing this value by 281%. For example, if a 36% protein supplement has 12% protein as urea, it contains 4.3% urea (12 ÷ 281 = 4.3%). One-third of the protein in the supplement is furnished by urea (12 ÷ 36% = 33.33%).

Dry vs Liquid High-Urea Supplements

Liquid urea supplements are available in most areas; and some large feedlots are mixing them for their own use. Such supplements normally contain molasses, urea, phosphoric acid, vitamins, and trace minerals. Also, some of them contain alcohol, stilbestrol, and other ingredients.

TABLE 8-5

DRY AND LIQUID SUPPLEMENTS[1]

Ingredient	Purdue Dry 64 (64% protein)	Purdue Liquid 64 (64% protein)
	(lb)	*(lb)*
Urea (45% N)	200	—
Liquid urea (32% N)[2]	—	290
Cane molasses	140	385
Dehydrated alfalfa meal (17%)[3]	510	—
Ammoniated polyphos (10-34-0)	—	90
Distillers' solubles (27% dry matter)	—	93
Dicalcium phosphate	105	—
Iodized salt	35	—
Salt solution (28% + 72% water)	—	90
Calcium chloride	—	12
Sodium sulfate	—	10
Premix	10[4]	30[5]
	1,000	1,000

[1]Beeson, W. M., Purdue University, Lafayette, Ind.; paper in *Beef Cattle Science Handbook*, Vol. II, published by Agriservices Foundation and edited by Dr. M. E. Ensminger, 1974, p. 396.

[2]Liquid urea—70% urea and 30% water.

[3]Distillers' dried grains with solubles will effectively replace 50 to 100% of the dehydrated alfalfa meal to supply urea protein factors (UPF).

[4]Premix: 20 million IU vitamin A; 1,250 g zinc oxide; 4 g cobalt carbonate; 7 lb (3.2 kg) dehydrated alfalfa meal.

[5]Liquid supplement premix: 20 million IU vitamin A (soluble); 9.5 g cobalt sulfate (21% Co); 4,350 g zinc sulfate (23% Zn); 19 lb (8.6 kg) water (H2O).

Table 8-5 gives the formulas of the dry and liquid urea supplements used at Purdue University.

Liquid supplements are usually fed as a top dressing. Occasionally, they are self-fed, but the latter is not recommended due to the tendency of cattle to overeat such supplements, thereby increasing the cost of gains.

Feedlot tests with beef steers have shown rather conclusively that there is no significant difference in the nutritional value or cattle response to high-urea dry or liquid supplements provided the supplement and/or ration contains the same essential nutrients in proper balance. In other words, cattle do not distinguish between the same nutrients fed in dry or liquid form. There is no nutritional advantage in liquid supplements or liquid feeds; it is just another way to balance the ration for growing or finishing cattle.

Several experiments have shown that for optimum performance with cattle, urea supplements must contain some source of unidentified urea factors; hence, it is recommended that high-urea dry supplements contain a ratio of 2 parts of either dehydrated alfalfa meal or distillers' dried grain solubles to 1 part of urea; and that liquid supplements contain distillers' solubles at a level of 2.5 percent on a dry matter basis.

In summary, there is no difference in the nutritional value of liquid and dry supplements built around urea if the supplements contain the same basic nutrients. Thus, it is a matter of personal choice, convenience, and ingredient costs as to which is used by cattle feeders.

Slow Released Nonprotein Nitrogen Products

Although products such as urea have many advantages, compounds that are nonprotein in nature possess the ability to liberate free ammonia. When fed to cattle in excess of their ability to utilize this free ammonia, elevated blood ammonia levels will occur. Should the levels be sufficiently high that the normal metabolic processes cannot detoxify and eliminate by way of the kidneys (urine), death may result. Although this hazard does not exist with the oil seed proteins, the latter have become very high in price on the world market. This stems mainly from the fact that refined products derived from these sources are (1) increasingly finding their way into channels for human consumption, and (2) in great demand for monogastric animals. This has spurred researchers on in an effort to improve nonprotein nitrogen sources. Much of this research has dealt basically with slowing down the release of ammonia from these nonprotein nitrogen sources so that the animal's metabolic system is not overworked or overloaded with ammonia at any one given point in time. This lower, more constant supply of ammonia can thereby be more efficiently utilized by the rumen microflora and subsequently the animal. Thus, the protection of protein and the coating of amino acids can improve growth and feed efficiency.

Most of the effort to improve slow-release nonprotein nitrogen products has utilized a chemical approach. Basically, the chemical method involves utilizing nonprotein nitrogen ingredients in conjunction with a source of carbohydrates. Today, controlled ammonia-released products are on the market utilizing three different carbohydrate sources: (1) simple sugars (molasses); (2) starch (grain); and (3) cellulose (forage). Since cellulose is digested more slowly than starch or sugar, it adapts itself well to reducing ammonia release and is the most desirable carbohydrate source.

With protein in the world market in such a critical demand, cattle feeders must find more and better ways to utilize nonprotein nitrogen in their rations. The following are among such products that are on the market now: Starea, a combination of urea and gelatinized starch, developed by Bartley at Kansas State University, in 1969; and Golden Pro, a gelatinized corn combined with urea.

Minerals[10] [11]

Beef cattle are liable to the usual inefficiencies and ailments when exposed to (1) prolonged and severe mineral deficiencies, or (2) excesses of fluorine, selenium, or molybdenum. (See Table 8-17 for a summary of Nutritional Diseases and Ailments of Cattle.)

Although acute mineral deficiency diseases and actual death losses are relatively rare, inadequate supplies of any one of the essential mineral elements may result in lack of thrift, poor gains, inefficient feed utilization, lowered reproduction, and decreased production of milk. Thus, like a thief in the night, subacute mineral deficiencies in beef cattle each year steal away millions of dollars in profits from the farmers and ranchers of America and, for the most part, go unnoticed. Only when the mineral deficiency reaches such proportions that it results in excess emaciation, reproductive failure, or death, is it likely to be detected.

Fig. 8-7. Mineral self-feeder on a South Dakota range being refilled. (Courtesy, Martin Jorgensen, Jr., Ideal, S.D.)

[10]In the preparation of this section, including the accompanying tables, the author had the authoritative review and help of Dr. O. O. Thomas, Professor, Montana State University, Bozeman, Mont.; Dr. D. J. Thompson, Manager, Technical Service, Animal Health and Nutrition Division, International Minerals and Chemical Corp., Libertyville, Ill.; Dr. R. F. Klay, Nutrition Coordinator, MoorMan Manufacturing Company, Quincy, Ill.; and Dr. Vernon R. Heaton, Director of Research and Development, Calcium Carbonate Company, Quincy, Ill.

[11]See Table 8-17 for a summary of Nutritional Diseases and Ailments of Cattle.

Needed minerals may be incorporated in beef cattle rations or in the water. In addition, it is recommended that all classes and ages of cattle be allowed free access to a two-compartment mineral box, with (1) salt (iodized salt in iodine-deficient areas) in one side, and (2) a suitable mineral mixture in the other side. Free-choice feeding is in the nature of cheap insurance, with the animals consuming the minerals if they are needed.

The 15 mineral elements which, up to the present time, have been shown to be essential for most animals are: calcium, phosphorus, sodium, chlorine, iodine, iron, copper, manganese, magnesium, sulfur, zinc, potassium, cobalt, selenium, and molybdenum. This list is based on experiments with one or more species, for all elements have not been tested with all species. This does not mean that all 15 of these minerals must always be included in the mineral supplement. Rather, only the specific minerals that are deficient in the feed—and in the quantities necessary—should be supplied. Excesses and mineral imbalances are to be avoided.

● *Salt*—Salt should be available at all times. It may be fed in the form of granulated, half ground, or block salt; but because of weathering losses, flake salt is not satisfactory for feeding in the open. If block salt is used, the softer types should be selected.

Most ranchers compute the yearly salt requirements on the basis of about 25 pounds for each cow.[12] Mature animals will consume 3 to 5 pounds of salt per month when pastures are lush and succulent, and 1 to 1½ pounds per month during the balance of the season.

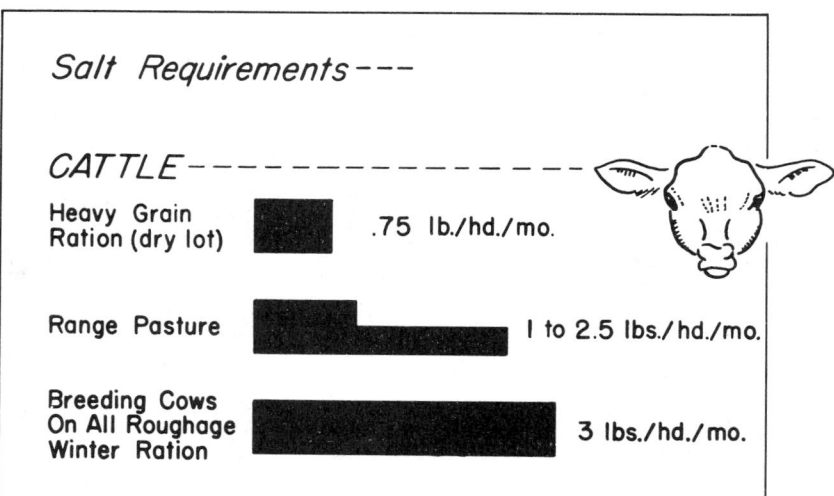

Fig. 8-8. The average salt requirements of cattle.

[12]Ensminger, M. E., M. W. Galgan, and W. L. Slocum, Problems and Practices of American Cattlemen, *Wash. Agr. Exp. Sta. Bull.* 562, 1955, p. 43.

The careful location of the salt supply is recognized as an important adjunct in proper range management. Through judicious scattering of the salt supply and the moving of it at proper intervals, the animals can be distributed more properly; and overgrazing of certain areas can be minimized.

● *Calcium*—In contrast to phosphorus deficiency, calcium deficiency in beef cattle is relatively rare and mild and the symptoms much less conspicuous. In general, when the forage of cattle consists of at least one-third legume (legume hay, pasture, or silage), ample calcium will be provided. But even nonlegume forages contain more calcium than cereal grains. This indicates that a mineral source of calcium is less necessary when large quantities of roughage are being consumed. Also, plants grown on calcium-rich soils contain a higher content of this element.

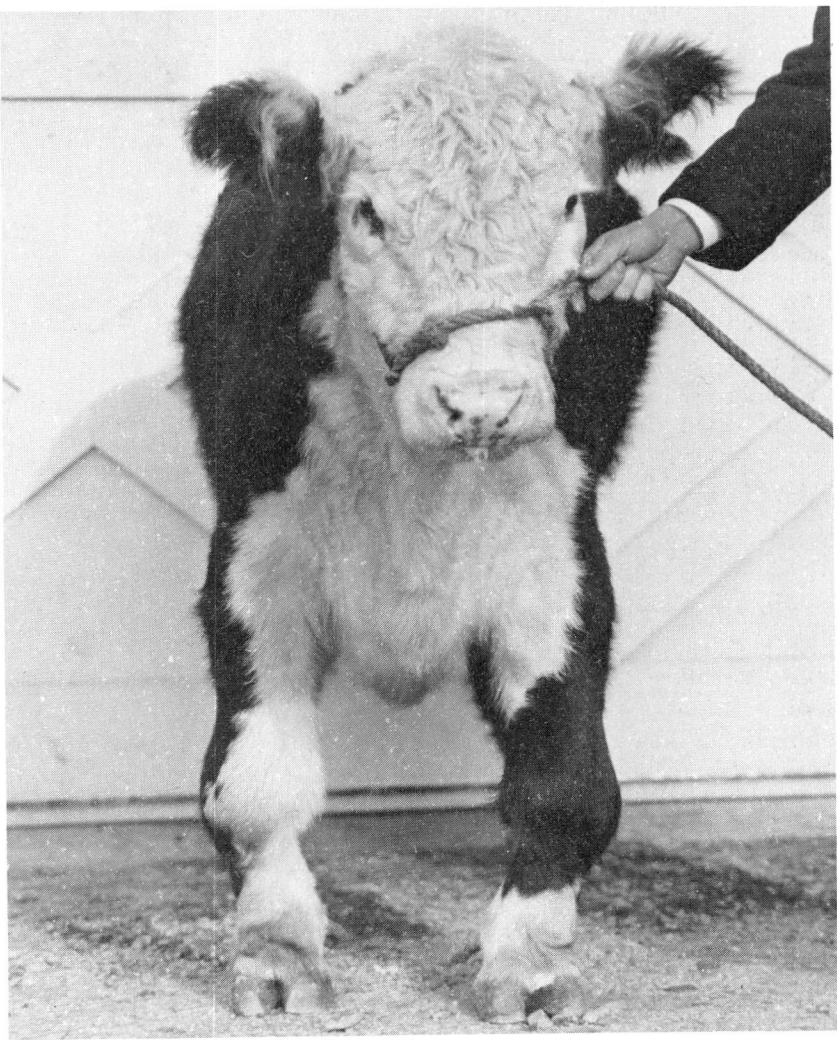

Fig. 8-9. This cow developed rickets early in life. Note the bowed front legs and enlarged joints. (Courtesy, USDA, Washington, D.C.)

As finishing cattle consume a high proportion of grains to roughages—and the grains are low in calcium—they have a greater need for a calcium supplement than do beef cattle that are being fed largely on roughages. This is especially true of cattle of the younger ages and where a long feeding period is involved.

When the ration of beef cattle is suspected of being low in calcium, the animals should be given free access to a calcium supplement, with salt provided separately; or a calcium supplement may be added to the daily ration in keeping with nutrient requirements (See Table 8-6.)

TABLE 8-6

MINERAL REQUIREMENTS OF BEEF CATTLE
(IN PERCENTAGE OF DIET DRY MATTER OR AMOUNT
PER KILOGRAM OF DRY DIET)[1]

Nutrient	Growing and Finishing Steers and Heifers	Dry Pregnant Cows	Breeding Bulls and Lactating Cows	Possible Toxic Levels[5] (mg/kg diet)
Minerals				
Sodium .... %	0.06	0.06	0.06	—
Calcium[2] .... %	0.18-1.04	0.18	0.18-0.44	—
Phosphorus[2] .... %	0.18-0.70	0.18	0.18-0.39	—
Magnesium .... %	0.04-0.10	—[4]	0.18	—
Potassium .... %	0.6-0.8	—[4]	—[4]	—
Sulfur .... %	0.1	—[4]	—[4]	—
Iodine .... mcg	—[3]	50-100	50-100	100
Iron .... mg	10	—[4]	—[4]	400
Copper .... mg	4	—[4]	—[4]	115
Cobalt .... mg	0.05-0.10	0.05-0.10	0.05-0.10	10-15
Manganese .... mg	1.0-10.0	20.0	—[4]	150
Zinc .... mg	20-30	—[4]	—[4]	900
Selenium .... mg	0.10	0.05-0.10	0.05-0.10	5

[1]From: *Nutrient Requirements of Beef Cattle*, fifth revised edition, National Academy of Sciences, Washington, D.C., 1976, p. 35, Table 5.
[2]See Tables 8-18 and 8-19 for more detailed data on requirements.
[3]Very small, but unknown.
[4]Unknown. It is suggested that the level for the growing and finishing animal be used.
[5]The level of mineral that is toxic is at best an estimate, and is dependent upon such factors as length of intake, availability of the mineral in the feedstuff or compound, and other mineral levels.

Table 8-7 gives the composition of some common calcium and phosphorus supplements.

• *Phosphorus*—In some sections of the United States and other countries, the soils are so deficient in phosphorus that the feeds produced thereon do not provide enough of this mineral for cattle or other classes of stock. As a result, the cattle produced in these areas may have depraved appetites, may fail to breed regularly, and may produce markedly less milk. Growth and development are slow, and the animals become emaciated and fail to reach normal adult size. Death losses are abnormally high.

In range areas where the soils are either known or suspected to be deficient in phosphorus, cattle should always be given free access to a suitable phosphorus supplement.

To be on the safe side, the general recommendation for beef cattle on both the range and in the finishing lot is to allow free choice of a suitable

TABLE 8-7

COMPOSITION OF CALCIUM AND PHOSPHORUS SUPPLEMENTS[1]

Mineral Supplement	Calcium			Phosphorus		
	%	Grams		%	Grams	
		Per Lb	Per Kg		Per Lb	Per Kg
Oyster shells, ground	38.0	172	380	—	—	—
Limestone, ground (or calcium carbonate, 38-40% Ca)	33.8	153	338	—	—	—
Bone black, spent	30.1	137	302	14.14	64	141
Bone meal, steamed	30.0	136	300	14.31	65	143
Dicalcium phosphate	23.13	105	231	18.65	85	187
Defluorinated phosphate	33.0	150	330	18.0	82	181
Monosodium phosphate	—	—	—	22.4	102	225
Diammonium phosphate (N = 18%)	—	—	—	20.0	91	201

[1]Other calcium supplements are calcium chloride (often used in liquid supplements) and calcium sulfate. Other phosphorus supplements include disodium phosphate, sodium tripoly phosphate, ammonium polyphosphate, and phosphoric acid.

Fig. 8-10. Bone chewing by cattle is a common sign of phosphorus deficiency. (From *Tex. Sta. Bull. 344*, courtesy of The Fertilizer Institute, Washington, D.C.)

phosphorus supplement in a mineral box, or to add a phosphorus supplement to the ration in keeping with nutrient requirements. (See Tables 8-6 and 8-8.)

Where phosphorus is added to the water, either of the following methods may be employed:

1. If added by hand, add ¼ ounce of monosodium phosphate per 8 gallons of water, or ¼ ounce per head daily.

2. If added by automatic dispenser, a stock solution of 2½ pounds of monosodium phosphate per gallon of water (or 100 lb to 40 gallons of water) is recommended. The machine automatically proportions the mineral to the water.

• *Cobalt*—Deficiencies of cobalt in cattle are costly, for the affected animals become weak and emaciated and eventually die. Florida is without doubt the most serious cobalt-deficient area in the United States, but similar deficiencies of a lesser order have been observed in Michigan, Wisconsin, Massachusetts, New Hampshire, Pennsylvania, and New York. Cattle in these affected areas should have access to a cobaltized mineral mixture, made by mixing 0.2 ounce of cobalt chloride, cobalt sulfate, cobalt oxide, or

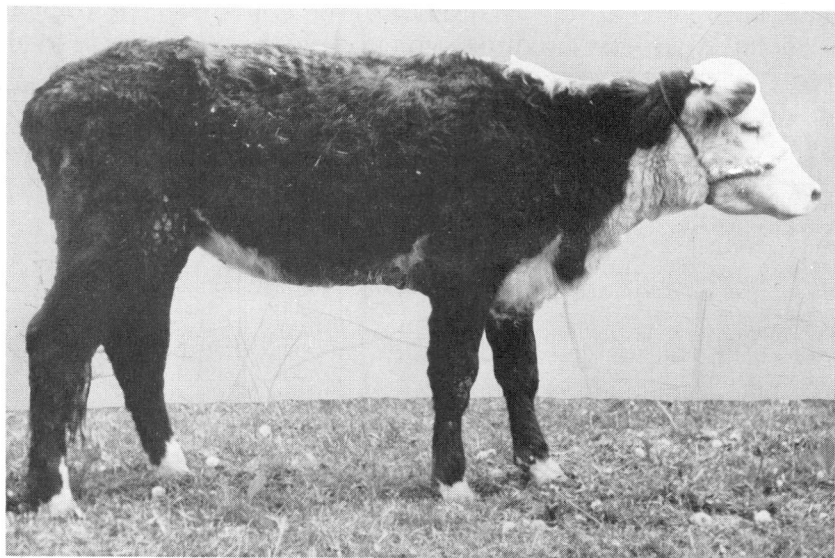

Fig. 8-11. Cobalt deficiency:
    The upper picture shows a heifer suffering from a cobalt deficiency. Anemia, loss of appetite, and roughness of hair coat characterize the malady.
    The lower picture illustrates the remarkable recovery in the same animal brought about by the administration of cobalt.
    (Courtesy, Michigan State University, East Lansing, Mich.)

cobalt carbonate per 100 pounds of either (1) salt, or (2) mineral mix.

In other areas of the world, cobalt deficiency is known as Denmark disease, coastal disease, enzootic marasmus, bush sickness, salt sickness, nakuritis, and pining disease.

● *Copper*—This is sometimes deficient in the soils of certain areas, notably in the state of Florida. In such areas, 0.25 to 0.5 percent of copper sulfate or copper oxide should be incorporated in the salt or mineral mixture.

In addition to being an area disease, copper deficiencies have occurred in beef calves kept on nurse cows for periods extending beyond normal weaning age.

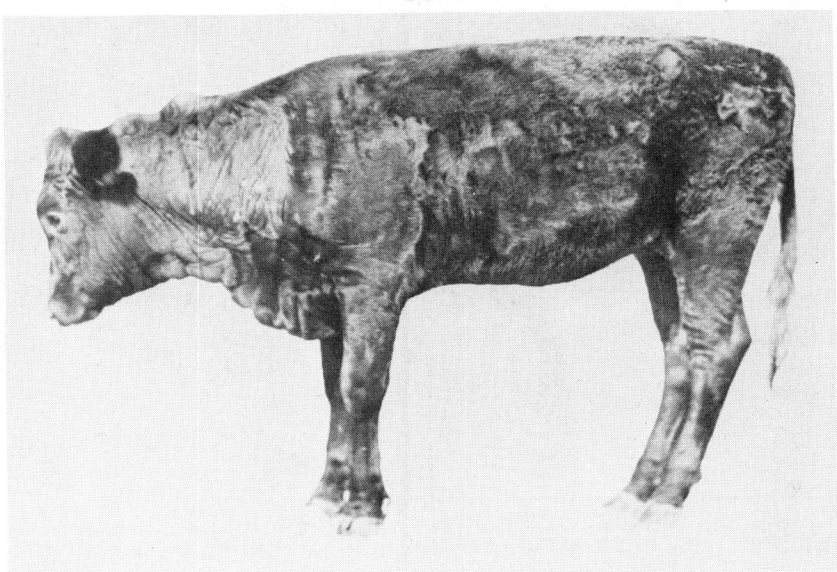

Fig. 8-12. Copper deficiency in calf. Note rough coat and bleaching of hair. (Courtesy, H. L. Chapman, Jr., University of Florida, Ona, Fla.)

● *Iodine*—Iodized salt should always be fed to cattle in iodine-deficient areas (such as the northwestern United States and the Great Lakes region). This can be easily and cheaply accomplished by providing stabilized iodized salt containing 0.01 percent potassium iodide (0.0076 percent iodine). Under some conditions, organic iodine appears to be an effective aid in the prevention and treatment of foot rot and lumpy jaw (soft tissues) in cattle. For prevention, add to the ration 50 mg of ethylenediamine dihydriodide (EDDI) per head daily; for treatment, add 250 to 500 mg per head daily for two to three weeks.

● *Magnesium*—Certain pastures in early spring are inadequate in magnesium, with the result that grass staggers may occur in cattle grazed on such pastures. Lactating cows are most commonly affected. In problem areas, as high as 20 grams of supplemental magnesium per head daily may be required to prevent this malady.

● *Manganese*—A deficiency of manganese exists in some areas of the northwestern United States, where it has been shown to be one cause of "crooked calves"—calves born with enlarged joints, stiffness, twisted legs, "over-knuckling," and weak and shortened bones. (See Fig. 8-13.)

● *Selenium*—Cows grazing on low selenium pastures may produce calves with nutritional muscular dystrophy (white muscle disease), the symptoms of which are similar to those of vitamin E deficiency. Also, it has been shown that the performance of feedlot cattle fed a selenium-deficient

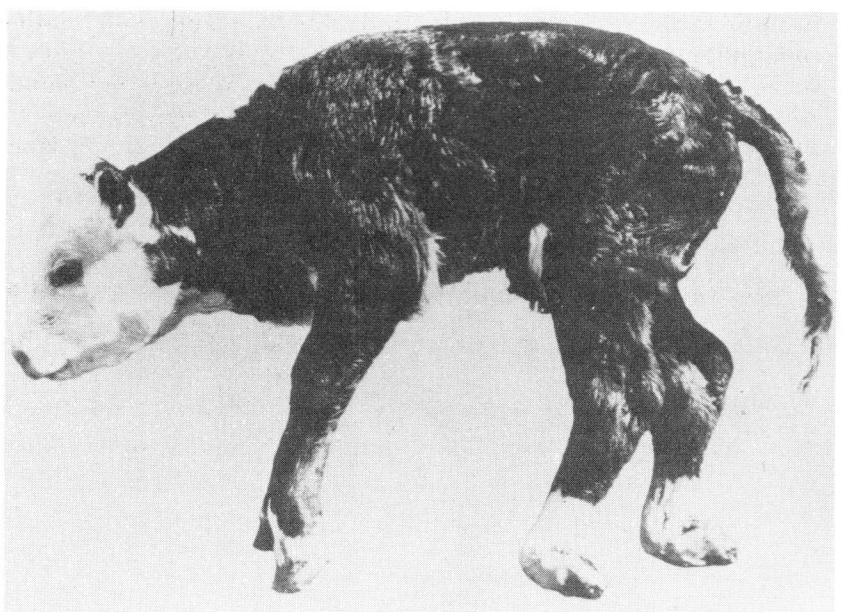

Fig. 8-13. Manganese deficiency in a newborn calf. Note weak legs and over-knuckling. (Courtesy, I. A. Dyer, Washington State University, Pullman, Wash.)

Fig. 8-14. Selenium toxicity in cow grazing on forage produced on alkali soil containing excessive selenium. Note emaciated condition, curvature of back, and deformed hoofs. (Courtesy, Wyoming Agricultural Experiment Station, Laramie, Wyo.)

ration is improved by injecting selenium, the only approved method of selenium supplementation. (See Fig. 8-14.)

• *Zinc*—Added zinc intake has been shown to increase the rate and efficiency of gains by feedlot cattle in certain areas. This may be due to the relationship between (1) phytic acid and zinc, and (2) calcium and zinc, the improper ratios of which may create a need for supplemental zinc.

MINERAL REQUIREMENTS OF BEEF CATTLE

The National Research Council mineral requirements of beef cattle are given in Table 8-6. Table 8-8 is a beef cattle mineral summary.

TABLE 8-8—BEEF CATT

Minerals Which May Be Deficient Under Normal Conditions	Conditions Usually Prevailing Where Deficiencies Are Reported	Function of Mineral	Some Deficiency Symptom
Salt (sodium and chlorine). The requirements for sodium and chlorine are commonly expressed as salt requirements because salt is an effective, economical way of supplementing diets with these elements.	Negligence; for salt is inexpensive.	Sodium chloride helps maintain osmotic pressure in body cells, upon which depends the transfer of nutrients to the cells and the removal of waste materials and the maintenance of water balance among the tissues. Also sodium is important in making bile, which aids in the digestion of fats and carbohydrates, and chlorine is required for the formation of hydrochloric acid in the gastric juice so vital to protein digestion. It is noteworthy that when salt is omitted, sodium expresses its deficiency first.	Intensive craving of salt, manif by the animals chewing and ing various objects. Prolon deficiency results in lack of a tite, unthrifty appearance, decreased production. High ducing milk cows may coll and die when salt deficiency been of long duration. Excessive salt intake can resu toxicity. But as much as 3 lb be consumed per cow daily out harm provided animals free access to plenty of water
Calcium	When finishing cattle are fed heavily on concentrates and limited quantities of nonlegume roughage, especially young cattle on a long feed. Adding calcium to such a ration increases the rate of gain, improves feed utilization, results in heavier, stronger bones, and enhances market grades. When the diet consists chiefly of dried mature grasses or cereal straws. When cows are in heavy lactation.	Essential for development and maintenance of normal bones and teeth. Important in blood coagulation and lactation. Enables heart, nerves, and muscles to function. Regulates permeability of tissue cells. Affects availability of phosphorus and zinc.	Calcium deficiency in beef ca rare and mild; the symptom inconspicuous. With severe p tion, there may be fract bones, poor gains and bone velopment, and lower ma grade.

Footnote on last page of table.

NERAL CHART

ily Nutrients/ Animal	Nutrient Requirements[1] Percentage of Ration	Recommended Allowances[1]	Practical Sources of the Mineral	Comments
young, growing nals: 2-3 g of ium, and less 5 g of chlorine. lactating cows: of sodium, and of chlorine.	*0.10% salt in the total ration dry matter.	Cows on pasture or on high-roughage winter rations will consume from 1 to 3 lb (.45 to 1.36 kg) salt per head per month; finishing steers on heavy grain rations in drylot will consume 1 to 3.5 lb (.45 to 1.59 kg) per head per month; a wide range due to differences in age, rations, form of salt (rock vs block), and weather losses. Most ranchers compute the yearly salt requirements on the basis of 25 lb per cow. The careful location of the salt supply is an important adjunct in range management.	Salt should be available at all times. It should be both (1) self-fed, free-choice, and (2) mixed with other ration ingredients. Free access to salt in the form of loose-rock, or block-salt. Cattle prefer loose salt to block salt, since it can be eaten more rapidly and with less effort. However, experiments with growing dairy heifers and lactating cows have shown fully as good results with block salt as with loose salt even though smaller quantities were consumed. This means that the additional intake of loose salt over block salt does not appear to benefit cattle. Commercial mineral mixes (in block, or loose form) may contain one-third or more salt.	The salt requirements of cattle differ (1) between individuals, (2) according to whether milk is produced (being higher for lactating cows than for dry cows, because of the salt in the milk), (3) from season to season, (4) according to the weathering losses to which the salt is subjected (being higher on pasture than in the drylot; exposed block salt loses about 15% per month), (5) between block and loose salt (animals often consuming twice as much easy-to-get loose salt as block salt), and (6) according to the salt content of the soil, feed, and water (being higher when vegetable proteins are fed, than when animal proteins are fed, higher on predominantly forage rations than on predominantly concentrate rations, and higher on lush early pasture than on more mature grasses). These are some of the reasons why free-choice feeding of salt is advocated.
iable, according class, age, and ght of cattle. e Table 8-18.)	*Variable, according to class, age, and weight of cattle. (See Table 8-19.)	Free access to a calcium supplement, or 0.1 lb (45 g) of a calcium supplement added to the daily ration. Calf rations should contain a minimum of .4% calcium, .3% phosphorus, and 200 IU of vitamin D per lb (440 IU per kg).	Ground limestone, steamed bone meal, oyster shell flour, dicalcium phosphate, or defluorinated phosphate; free-choice, or 0.1 lb (45 g) per head daily added to the ration. Where both calcium and phosphorus need to be supplemented, they should be provided in a readily available and palatable form such as dicalcium phosphate, defluorinated phosphate, or bone meal.	In addition to an adequate supply of calcium, proper utilization is dependent upon (1) a highly available source of the mineral, (2) a suitable ratio between calcium and phosphorus (somewhere between 1 to 2 parts of calcium to 1 part of phosphorus). Calcium-phosphorus ratios of 2:1 have been shown to be beneficial in reducing urinary calculi. When calculi are encountered, even higher levels of calcium may be advisable. Ratios between calcium and phosphorus of 7:1 have been reported to be satisfactory for cattle. Generally when cattle receive at least 1/3 of a legume forage, ample calcium will be provided. But even nonlegume forages contain more calcium than cereal grains. Plants grown on calcium-rich soils are high in calcium. Calcium availability of 70% is generally assumed for all feedstuffs.

(Continued)

TABLE

Minerals Which May Be Deficient Under Normal Conditions	Conditions Usually Prevailing Where Deficiencies Are Reported	Function of Mineral	Some Deficiency Symptom
Phosphorus	Semiarid regions are commonly associated with soils deficient in phosphorus. The phosphorus content of plants generally decreases markedly with maturity, with the result that deficiencies often occur in cattle subsisting for long periods on mature dried forage.	Essential for sound bones and teeth, and for the assimilation of carbohydrates and fats. A vital ingredient of the proteins in all body cells. Necessary for enzyme activation. Acts as a buffer in blood and tissue. Occupies a key position in biologic oxidation, and reactions requiring energy.	Loss of appetite, poor gains creased milk production creased feed efficiency, dep appetite—with special cravi chewing bones and eating Lameness and stiffness of j broken bones. Rickets in young animals an teomalacia, osteoporosis, ar teitis fibrosa in mature anim Breeding problems, milk feve tained afterbirth, and blinc (Phosphorus is necessary to vert carotene to vitamin A.)
Cobalt	In cobalt-deficient areas (soils) where this element is not provided (in Fla., Mich., Wisc., Mass., N.H., Penn., and N.Y.).	Cobalt is an integral component of the vitamin B$_{12}$ molecule, and vitamin B$_{12}$ is synthesized by microorganisms in the reticulum.	Affected animals become emaciated, and eventually Other symptoms include l appetite, craving for hai wood, scaliness of skin, and times diarrhea.
Copper	In copper-deficient areas (soils), as in Florida and the Coastal Plain region. On peat and muck soils. Deficiencies have occurred in calves kept on an exclusive milk diet for long periods.	Copper, along with iron and vitamin B$_{12}$, is necessary for hemoglobin formation, although it forms no part of the hemoglobin molecule (or red blood cells). Copper is essential in enzyme systems, hair development and pigmentation, bone development, reproduction, and lactation.	Emaciation, depigmentation turn yellowish) and loss of stunted growth, anemia, an tle and malformed bones heat periods are suppresse there may be depraved ap and diarrhea. Young calve have straight pasterns and forward on their toes.
Fluorine	Feeding rock phosphate which has not been defluorinated and which may contain 3.5 to 4.0% fluorine, a toxic level.	No essential known function in beef cattle. But excessive levels of fluorine are harmful to cattle.	Teeth may erode and the e may become mottled; the become thickened and sof their breaking strength decr appetite is decreased, and growth results.

Footnote on last page of table.

ntinued)

Nutrient Requirements[1]		Recommended Allowances[1]	Practical Sources of the Mineral	Comments
aily Nutrients/ Animal	Percentage of Ration			
iable, according class, age, and ght of cattle. e Table 8-18.)	*High energy rations should contain at least 0.22% phosphorus; other rations should contain at least 0.18%. (See Table 8-19.)	Free access to a phosphorus supplement, or 0.1 lb (45 g) of a phosphorus supplement added to the daily ration. Where phosphorus is added to water, either of the following methods may be employed: 1. Added by hand at rate of ¼ oz of monosodium phosphate/8 gal water, or ¼ oz/ head/day. 2. Added by dispenser, using stock solution of 2½ lb of monosodium phosphate/gal water (or 100 lb/40 gal water).	Dicalcium phosphate, defluorinated phosphate, monosodium phosphate, diammonium phosphate, steamed bone meal; free-choice, or 0.1 lb (45 g) per head daily added to the ration.	Grains, grain by-products, and high-protein supplements are fairly high in phosphorus; hence, rations high in such ingredients require little or no phosphorus supplementation. Calcium-phosphorus ratios of 2:1 are beneficial in reducing urinary calculi; and even higher levels of calcium may be necessary when urinary calculi are encountered. Ratios between calcium and phosphorus of 7:1 have been reported to be satisfactory for cattle.
227 to 0.045 mg · lb (or 0.05 to 0 mg per kg) of t dry matter.	*0.05 to 0.10% of diet dry matter.	Free access to a cobaltized mineral mixture in cobalt-deficient areas.	A cobaltized mineral mixture may be prepared by adding cobalt at the rate of 0.2 oz per 100 lb (1.25 mg/kg) of salt as cobalt chloride or cobalt sulfate, cobalt carbonate, cobalt oxide, or a good commercial mineral mixture or salt product may be used.	Several good commercial cobalt-containing minerals are on the market. A vitamin B_{12} injection will relieve a cobalt deficiency. Toxicity in calves produced by feeding 1.2 mg/kg body wt.
	*1.8 mg/lb (4 mg/kg) of total ration dry matter when rations contain low levels of molybdenum and sulfate.	*Copper deficiency can be prevented by adding 0.5% copper sulfate to salt fed free-choice. *Copper (Cu) added to total feed (dry basis) 4.00 ppm. Copper may also be injected as glycinate to meet the nutritional needs for the mineral.	*Salt containing 0.5% copper sulfate.	Copper deficient cattle can be returned to normal by feeding 3 g of copper sulfate or blue vitriol every 10 days. An interesting interrelation exists between copper and molybdenum. An excess of molybdenum (in the presence of sulfate) causes a condition which can be cured only by administering copper. Excess copper is toxic; it accumulates in the liver, and death may result.
	*Safe fluorine levels: No more than 45 mg/lb (100 mg/kg) of diet for finishing cattle, and no more than 18.2 mg/lb (40 mg/kg) for animals to be kept for breeding.			Fluorine is a cumulative poison; hence, the toxic effects may not be noticed for some time.

(Continued)

TABLE

Minerals Which May Be Deficient Under Normal Conditions	Conditions Usually Prevailing Where Deficiencies Are Reported	Function of Mineral	Some Deficiency Symptoms
Iodine	In iodine-deficient areas (soils) where iodized salt is not fed (in northwestern U.S. and in the Great Lakes region). Where feeds come from iodine-deficient areas.	Iodine is needed by the thyroid gland in making thyroxin (an iodine-containing hormone which controls the rate of body metabolism or heat production).	Production of weak, goitrous dead calves. Occasional borderline cases survive; in these, the mode thyroid enlargement disapp in a few weeks.
Iron	Calves on an exclusive milk diet.	Necessary constituent of hemoglobin (oxygen carrying system of the blood). Deficiencies result in anemia. Up to 20 weeks of age, supplemental iron contributes to improved weight gain in calves fed milk diets.	Anemia and decreased growt calves on exclusive milk diet. Excessive amounts of iron are t
Magnesium	When milk feeding of calves is prolonged without grain or hay. (Milk is rather low in magnesium.) Certain pastures early in the spring. Lactating cows are most commonly affected.	Essential for the bones and teeth, and required for various body processes. Aids in maintaining acid base equilibrium and in activating many enzyme systems. In cells, magnesium is present in far greater concentration than calcium.	Grass tetany or grass stagg characterized by anore hyperemia, hyperirritability, vulsions, and death.
Manganese	In northwestern U.S. All-concentrate diets based on corn supplemented with non-protein nitrogen.		Delayed estrus, reduced ferti abortions, and deformed yo Calves born to magnesi deficient cows may exhibit formed legs (enlarged joints, ness, twisted legs, "o knuckling"), weak and shorte bones, and poor growth.
Molybdenum	Molybdenum toxicity occurs only occasionally in cattle and appears to be an area problem.	Molybdenum is a constituent of the enzyme xanthine oxidase, which plays an essential role in purine metabolism and is found in the liver, intestinal tissue, and milk.	Molybdenum toxicity results in vere scours and loss of condit
Potassium	When drylot finishing cattle receive high or all-concentrate rations.	Essential for proper enzyme, muscle and nerve function, rumen microorganism activity, and appetite.	Poor appetite and feed conver slow growth, stiffness, and en ation.

Footnote on last page of table.

tinued)

Nutrient Requirements[1]		Recommended Allowances[1]	Practical Sources of the Mineral	Comments
y Nutrients/ Animal	Percentage of Ration			
o 800 micro- iodine/day egnant-lactat- ef cows.	*Iodized salt at rate of 0.10% of dry diet.	Free access to stabilized iodized salt contain- ing 0.01% potassium iodide (0.0076% iodine).	Stablized iodized salt containing 0.01% potassium iodide. Calcium iodate. Ethylenediamine di- hydriodide (EDDI).	The enlargement of the thyroid gland (goiter) is nature's way of trying to make enough thyroxin, when there is in- sufficient iodine in the feed. Toxicity can occur, resulting in depressed appetite, dull list- less appearance, difficulty in swallowing, hacking cough, and weepy eyes.
g daily during to 8 weeks, or ons of 500 mg rth and at 8 of age.	*10 ppm.	Iron (Fe) 40 mg daily.	Levels of iron in feed believed to be ample, since feeds contain 36 to 45 mg/lb (80-100 mg/kg) in most re- gions. Iron sulfate.	After calves are past 20 weeks of age, iron does not seem to be beneficial. About 30% of all calves are af- fected by prenatal iron defi- ciency.
13.6 mg/lb (12 mg/kg) body t per day. In em areas, up g of supple- al magnesi- ad daily may quired to pre- grass tetany.	*For lactating cows, 0.18% of ration dry matter.		Commonly fed rough- ages and concentrates usually contain ample magnesium, but it may not be present in an available form. Magnesium sulfate or oxide may be used as a supplement.	Although grass tetany is attrib- uted to magnesium defi- ciency, uncomplicated cases have been produced only on purified diets or by prolonged feeding of calves on milk. However, supplemental feed- ing of magnesium (20 g/day) reduces the incidence of grass tetany in many out- breaks.
	*0.45 to 4.5 mg/lb (1 to 10 mg/kg) of diet.	An intake of 20 ppm will prevent deformi- ties in the fetus.	Most roughages contain over 13.6 mg of manganese/lb (30 mg/kg) of dry matter, and most grains con- tain about half this amount. Manganous oxide, sul- fate, and carbonate.	A deficiency of manganese exists in northwestern U.S., where it has been shown to cause "crooked calves." Where animals are heavily parasitized, iron is beneficial. 20 ppm prevents deformities in the fetus.
	.0045 mg/lb (0.01 mg/kg) of dry diet.	As a feed additive, molybdenum is not cleared by Food and Drug Administration.	Many feeds contain 6.8 to 13.6 mg/lb of ration dry matter.	Toxic levels of molybdenum in- terfere with copper metabo- lism and thus increase copper requirements. Increasing copper level in diet to one g/head daily is effec- tive in overcoming molyb- denum toxicity in beef cattle. Phosphorus, manganese, potassium, zinc, and sulfur have also been reported to af- fect the degree of molyb- denum toxicity.
	*0.6 to 0.8% of the total ration dry mat- ter.	*0.8 to 1.0% of the total ration dry matter.	Roughages usually con- tain ample potassium. Potassium chloride is the supplement of choice.	Grains often contain less than 0.5% potassium. Excessive levels of potassium have been found to interfere with magnesium absorption. Also, excessive levels of potassium along with high levels of phosphorus, in- crease the incidence of phos- phatic urinary calculi.

(Continued)

TABL

Minerals Which May Be Deficient Under Normal Conditions	Conditions Usually Prevailing Where Deficiencies Are Reported	Function of Mineral	Some Deficiency Sympto
Selenium	Low selenium forage and low vitamin E. It is an area problem, but it occurs in many parts of the U.S.	Vitamin E fills a vital role in the functioning of every cell of the body. It serves as a powerful biological antioxidant in the body cells and in the digestive tract. It has a sparing effect on the selenium requirement. Toxic levels reported in S.D., N.D., Mont., Wyo., Utah, Neb., Kan., and Colo.	White muscle disease; c terized by white muscle, failure, and paralysis. Hol swayed back. Often a dys tongue.
Sulfur	In high urea rations.	Essential for the synthesis of methionine.	Depressed appetite, loss of w poor growth, irregular or d estrus, and reduced milk p tion.
Zinc		Essential in skin, hair, and bone development.	Parakeratosis in young calve denced by inflamed nos mouth, unthrifty appea roughened haircoat, and st of joints.

[1]As used herein, the distinction between "nutrient requirements" and "recommended allowances" is as follows: In nutrient re ments, no margins of safety are included intentionally; whereas in "recommended allowances," margins of safety are provided to co sate for variations in feed composition, environment, and possible losses during storage or processing. *Where preceded by an asteri nutrient requirements and recommended allowances listed herein were taken from *Nutrient Requirements of Beef Cattle*, fifth edition, National Academy of Sciences, Washington, D.C., 1976.

tinued)

Nutrient Requirements[1]		Recommended Allowances[1]	Practical Sources of the Mineral	Comments
ly Nutrients/ Animal	Percentage of Ration			
	0.045 mg/lb (0.1 mg/kg) of dry ration. Cows grazing on pastures with less than this quantity of selenium produce calves with white muscle disease.		Any of following 4 procedures are effective against white muscle disease: (1) selenium as a drench; (2) subcutaneous or intramuscular injections; (3) selenium as a feed additive; or (4) adding selenium to fertilizer applied to pasture.	Selenium toxicity may occur when cattle consume feeds containing 10-30 ppm of selenium on a dry matter basis for an extended period. It has been shown that the performance of feedlot cattle fed a selenium-deficient ration is improved by injecting selenium, the only approved method of selenium supplementation.
	*As low as 0.1% of ration dry matter.	*3 g of inorganic sulfur to 100 g urea, or 1 part of inorganic sulfur to 15 parts of nonprotein nitrogen.	Organic forms of sulfur are most readily utilized, elemental sulfur is least so, and sulfates are intermediate in this respect.	Sulfur requirements are primarily those involving amino acid nutrition.
	*9.1 to 13.6 mg/lb (20 to 30 mg/kg) of ration dry matter.	20 to 30 ppm zinc in the total feed (air-dry basis).	Many feeds contain 6.8 to 13.6 mg/lb (15-30 mg/kg) of ration dry matter; adding zinc to fertilizer applied to pasture.	Mild zinc deficiency in feedlot cattle results in lowered weight gains without the development of a specific syndrome.

Mineral recommendations for all classes and ages of cattle: Provide free access to a two-compartment mineral box, with (1) salt (iodized in iodine-deficient areas) in one side and (2) dicalcium phosphate, defluorinated phosphate, or a mixture of ⅓ salt (salt added for ses of palatability) and ⅔ steamed bone meal in the other side. The mineral requirements may be met by using a good commercial ral, in either block or loose form. If desired, the mineral supplement may be incorporated in the ration in keeping with the recom- ed allowances given in this Table.

COMMERCIAL MINERAL MIXTURES

Commercial mineral mixtures are just what the name implies—minerals mixed by manufacturers who specialize in the commercial mineral business, either handling minerals alone or a combination feed and mineral business. Most commercial minerals are very good.

The commercial mineral manufacturer has the distinct advantages over farm- or ranch-mixing of (1) purchase of minerals in quantity lots, thereby obtaining price advantages, (2) economical and controlled mixing, (3) the hiring of scientifically trained personnel for use in determining the formulations, and (4) quality control. Additionally, most farmers and ranchers do not have the equipment with which to mix minerals properly. Besides, mineral mixes have become more complicated with recognition of the increasing importance of trace elements and interrelationships. For these reasons, commercial minerals are finding a place of increasing importance in all livestock feeding.

Good mineral mixtures supply only the specific minerals that are deficient, and in the quantities necessary. Excesses and mineral imbalances are avoided. Thus, the value of any mineral mixture can easily be determined by how well it meets the needs.

How to Select and Buy Commercial Mineral Mixes

The informed cattleman will know what constitutes the best commercial mineral mix for his needs, and how to determine the best buy. Here are the factors to consider when buying a commercial mineral:

1. *The reputation of the manufacturer*—This can be determined by (a) checking on who is back of it, (b) conferring with other cattlemen who have used the particular product, and (c) checking on whether the product under consideration has consistently met its guarantees. The latter can be determined by reading the bulletins or reports published by the respective state departments in charge of enforcing feed laws.

2. *Determining your needs*—The mineral requirements of cattle are much the same everywhere, although it is recognized that age, pregnancy, and lactation make for differences in mineral needs within a given herd. Additionally, there are some area differences. For example, the northern Great Plains and the Southwest are generally recognized as phosphorus-deficient areas—their grasses and hays are usually low in phosphorus. Accordingly, a high phosphorus mineral is needed for such areas—one containing 10 to 15 percent phosphorus. Also, unless there is a concomitant deficiency of calcium (which is not likely), the calcium:phosphorus ratio of a mineral should not be wider than 2:1.

The minimum daily phosphorus need for dry, pregnant, mature cows, established by the National Research Council, is .18 percent of the total ration or about .5 ounces (15 g) per head. In phosphorus-deficient areas, these minimum recommendations should be exceeded; thus, the daily recommended phosphorus allowances will approximate .21 percent of the total ration, or .6 ounces (18 g) per head. However, under most conditions it is not necessary to supply more than one-third of the daily phosphorus need in supplemental form because cattle will get the rest of it from available feeds.

3. *Choose method of supplying minerals*—The daily phosphorus allowance recommended above, .6 ounces (18 g)/head/day, can be met in any of the following ways:

a. *Self-feeding a 10 percent phosphorus supplement*—If the monthly consumption of such a supplement is 3 to 4 pounds per head, the average daily intake of phosphorus will equal one-third of the requirement. Actually, in phosphorus-deficient areas, the use of a 10 to 15 percent phosphorus mineral for cows is in the nature of good insurance; thereby affording protection during droughts and other periods of low feed consumption.

b. *Feeding a high phosphorus range supplement*—A daily consumption of 1 to 2 pounds of a supplement containing 1 to 1.5 percent phosphorus will meet the needs.

c. *Adding phosphorus to the drinking water*—If cattle drink an average of 10 gallons of phosphorus-treated water per head daily, adding one-half pound of monosodium phosphate (22% phosphorus) per 100 gallons of water will meet the requirements.

4. *What's on the tag?*—Cattlemen should study, and be able to interpret what's on the tag. Does it contain what you need?

5. *Determine the best buy*—When buying a mineral, the cattleman should check price against value received. For example, let's assume that the main need is for phosphorus and that we wish to compare two minerals, which we shall call brands "X" and "Y." Brand "X" contains 12 percent phosphorus and sells at $340.00 per ton or $17.00/cwt; whereas brand "Y" contains 10 percent phosphorus and sells at $320.00 per ton or $16.00/cwt. Which is the better buy?

COMPARATIVE VALUE OF BRANDS "X" AND "Y"

(Based on Phosphorus Content Alone)

Brand	Phosphorus	Price/Cwt	Cost/Lb Phosphorus
	(%)	($)	($)
"X"	12	17.00	1.41
"Y"	10	16.00	1.60

Hence, brand "X" is the better buy, even though it costs $1.00 more per hundred, or $20.00 more per ton.

One other thing is important. As a usual thing, the more scientifically formulated mineral mixes will have plus values in terms of (a) trace mineral (needs and balance), and (b) palatability. (Cattle will eat just the right amount of a good mineral, but they won't overdo it—due to appetizers, rather than needs.)

Commercial mineral mixtures costing $1.40 to $1.60 per pound of phosphorus are not excessively priced. If the average consumption per head per month of a mineral mix costing $17.00/cwt is 3 pounds, the monthly per head cost will be about $.51, or less than $.02 per cow per day.

HOME-MIXED MINERALS FOR CATTLE

When buying and home-mixing minerals, as when buying commercial mineral mixes, the stockman should first determine his needs, based on (1) available feeds, (2) area (for example, the Northern Great Plains and the Southwest are phosphorus-deficient areas), and (3) the age and reproduction status (pregnancy and lactation make a difference) of the animals for which the mineral mix is intended. Of course, the available feeds, and the age and reproduction status of animals, on a given farm or ranch vary from time to time; and usually not all animals on a given establishment for which the mineral is to be used are of the same age and reproduction status. For the ultimate in exactness, therefore, there would have to be many changes and many different mineral mixes. Fortunately, a reasonable range in the allowances of the different mineral elements is permissible. As a result, the selection of the mineral mix, or mixes, for a particular farm or ranch usually involves a compromise, reached primarily on the basis of what will meet the needs of most of the animals, particularly during most critical stages, for the specific farm or ranch under consideration.

One of the mineral mixes given in Table 8-9 will usually suffice for free-choice feeding most cattle. Moreover, where special circumstances necessitate some other formulation, Table 8-9 will serve as a useful guide. As noted, provision is made for cattle receiving rations containing different roughages and different levels of cereal grains. For example, a ration consisting of 15 lb of shelled corn and 5 lb of alfalfa-brome hay contains 75% cereal grain; hence, the mineral mix should contain 20% ground limestone and 80% trace mineralized salt. If the 5 lb of alfalfa-brome hay were replaced with 5 lb of grass hay (brome or timothy, for example), the mineral mix should be 50% ground limestone and 50% trace mineralized salt. If the 5 lb of alfalfa-brome were replaced with 15 lb of silage (wet basis), the mineral mix should be 60% ground limestone and 40% trace mineralized salt. As noted, the Table 8-9 mineral mixes for silage are on a wet basis; hence, a ration of 15 lb of wet silage and 15 lb of corn is considered as having 50% cereal grain.

TABLE 8-9

MINERAL MIXTURES FOR CATTLE FOR FREE-CHOICE FEEDING
WITH DIFFERENT RATIONS

Mineral Supplement	Type of Roughage		
	Alfalfa, or Alfalfa-Grass Hay	Grass Hay	Corn Silage
	Rations containing up to 20% cereal grains[1]		
Dicalcium phosphate or bone meal[2]	—	40%	30%
Ground limestone	—	—	—
Trace mineralized salt	100%	60%	70%
	Rations containing 20% to 40% cereal grains[1]		
Dicalcium phosphate or bone meal[2]	—	25%	—
Ground limestone	—	25%	50%
Trace mineralized salt ,...................	100%	50%	50%
	Rations containing 40% to 60% cereal grains[1]		
Dicalcium phosphate or bone meal[2]	—	20%	—
Ground limestone	—	40%	60%
Trace mineralized salt	100%	40%	40%
	Rations containing 60% to 80% cereal grains[1]		
Dicalcium phosphate or bone meal[2]	—	—	—
Ground limestone	20%	50%	80%
Trace mineralized salt	80%	50%	20%
	Rations containing 80% to 100% cereal grains[1]		
Dicalcium phosphate or bone meal[2]	—	—	—
Ground limestone	50%	70%	80%
Trace mineralized salt	50%	30%	20%

[1]Percents on an "as-fed" basis. Examples: Ration composed of 15 lb shelled corn and 5 lb hay contains 75% cereal grains. Ration composed of 30 lb corn silage and 10 lb shelled corn contains 25% cereal grain.
[2]Defluorinated rock phosphate may also be used to replace bone meal and dicalcium phosphate.

Cattle pastured on native grass should be offered a free-choice mineral mix consisting of 40% dicalcium phosphate or bone meal and 60% trace mineralized salt. Minerals that are self-fed on pastures or in corrals should be in boxes protected from the weather.

Where a high proportion of urea is used as the protein source, it is necessary to add minerals to the ration unless the remainder of the ration contains adequate minerals to make up for the complete lack of minerals in the urea.

Where there is need to increase, or assure, the palatability of a mineral mix, usually molasses is added; and where there may be wind losses, a bland oil is usually added.

Salt should always be available on a free-choice basis in addition to whatever mineral mix is provided.

Vitamins[13] [14]

The absence of one or more vitamins in the ration may lead to a failure in growth or reproduction, or to characteristic disorders known as deficiency diseases. In severe cases, death itself may follow. Although the occasional deficiency symptoms are the most striking result of vitamin deficiencies, it must be emphasized that in practice, mild deficiencies probably cause higher total economic losses than do severe deficiencies. It is relatively uncommon for a ration, or diet, to contain so little of a vitamin that obvious symptoms of a deficiency occur. When one such case does appear, it is reasonable to suppose that there must be several cases that are too mild to produce characteristic symptoms but which are sufficiently severe to lower the state of health and the efficiency of production.

It is also recognized that vitamin deficiencies may occur as a result of lack of availability of vitamins or because of the presence of antimetabolites. Both are important concepts. For example, analyses show corn to be adequate in niacin. Yet, due either to an antimetabolite or unavailability, there may be niacin deficiencies when corn is fed—deficiencies which can be remedied by niacin supplementation.

● *Vitamin A*—The vitamin most likely to be deficient in beef cattle rations is vitamin A. True vitamin A is a chemically formed compound, which does not occur in plants. It is furnished in most beef cattle rations in the form of its precursor, carotene. However, plants are a variable, and often undependable, source of carotene due to oxidation. Also, cattle are relatively inefficient converters of carotene to vitamin A. The latter fact was taken into consideration in the development of international standards for vitamin A, which are based on the rate at which the rat converts beta-carotene to vitamin A. The conversion rate for the rat is one mg of beta-carotene to 1,667 IU of vitamin A, whereas it is estimated that one mg of beta-carotene is equal to 400 IU of vitamin A in cattle. Moreover, the conversion rate for cattle varies

---

[13]In the preparation of this section and the Beef Cattle Vitamin Chart herein, the author had the authoritative review and help of Dr. W. M. Beeson, Department of Animal Sciences, Purdue University, Lafayette, Ind.; Dr. William H. Hale, Animal Science Department, The University of Arizona, Tucson, Ariz.; Dr. W. E. Dinusson, North Dakota State University, Fargo, N.D.; Dr. C. R. Adams, Technical Services, Roche Chemical Division, Hoffman-La Roche, Inc., Nutley, N.J.; and Dr. J. L. Turk, Manager, Developmental Research, Pfizer Agricultural Division, Terre Haute, Ind.

[14]See Table 8-17 for a summary of Nutritional Diseases and Ailments of Cattle.

under different conditions; it is influenced by type of cartenoid, breed, individual differences in animals, and level of carotene intake. Stress
conditions—such as extremely hot weather, viral infections, and altered
thyroid function—have also been suggested as causes for reduced conversion.

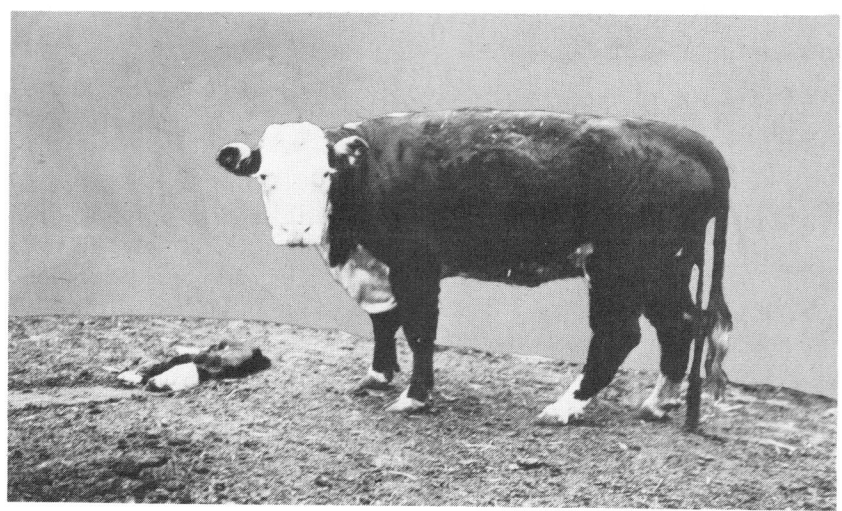

Fig. 8-15. Effect of vitamin A deficiency on reproduction.
  The heifer in the upper picture received a ration deficient in vitamin A, but
otherwise complete. She became night blind and aborted during the last month of
pregnancy; also, note the retained placenta.
  The heifer in the lower picture received the same ration, but during the latter
part of the gestation period, a supplement of one pound daily of dehydrated alfalfa
meal containing 50 mg of carotene was added. She produced a normal vigorous
calf.
  (Courtesy, California Agricultural Experiment Station)

Under practical feeding conditions, cattlemen should consider (1) previous feeding as it influences body stores of vitamin A; (2) vitamin A destruction during processing or when mixed with oxidizing materials; and (3) carotene destruction in feeds during storage.

Table 8-10 gives the estimated carotene content of feeds in relation to appearance and method of conservation.

TABLE 8-10

ESTIMATED CAROTENE CONTENT OF FEEDS IN RELATION
TO APPEARANCE AND METHODS OF CONSERVATION[1]

Feedstuff	Carotene	
	(mg/lb)	*(mg/kg)*
Fresh green legumes and grasses, immature	15-40	33-88
Dehydrated alfalfa meal, fresh, dehydrated without field curing, very bright green color[2]	110-135	242-298
Dehydrated alfalfa meal after considerable time in storage, bright green color	50-70	110-154
Alfalfa leaf meal, bright green color	60-80	132-176
Legume hays, including alfalfa, very quickly cured, with minimum sun exposure, bright green color, leafy	35-40	77-88
Legume hays, including alfalfa, good green color, leafy	18-27	40-60
Legume hays, including alfalfa, partly bleached, moderate amount of green color	9-14	20-31
Legume hays, including alfalfa, badly bleached or discolored, traces of green color	4-8	9-18
Nonlegume hays, including timothy, cereal, and prairie hays, well cured, good green color	9-14	20-31
Nonlegume hays, average quality, bleached, some green color	4-8	9-18
Legume silage	5-20	11-44
Corn and sorghum silages, medium to good green color	2-10	4.4-22
Grains, mill feeds, protein concentrates, and by-product concentrates, except yellow corn and its by-products	.01-0.2	0.2-0.4

[1]From *Nutrient Requirements of Beef Cattle*, Pub. No. 579, National Academy of Sciences, Washington, D.C., with metric system added by the author.
[2]Green color is not uniformly indicative of high carotene content.

● *Vitamin D*—When exposed to enough direct sunlight, beef cattle normally acquire their vitamin D needs, for the ultraviolet rays in sunlight penetrate the skin and produce vitamin D from traces of sterols in the tissues. Also, cattle obtain vitamin D from sun-cured roughages. However, the addition of vitamin D to the ration is important where cattle, especially calves, are kept in a barn most of the day, where there is limited sunshine, where the calcium:phosphorus ratio leaves much to be desired, and/or where little or no sun-cured hay is fed. Vitamin D helps build strong bones and sturdy frames. It is usually added at a level of about one-seventh the level of added vitamin A.

● *Vitamin E*—Added vitamin E may be necessary under certain conditions because of its relationship to vitamin A utilization and the prevention of white muscle disease. (See Fig. 8-16, page 470.)

● *B vitamins*—Dietary requirements for the B vitamins (thiamine, biotin, niacin, pyridoxine, pantothenic acid, riboflavin, and vitamin B12) have

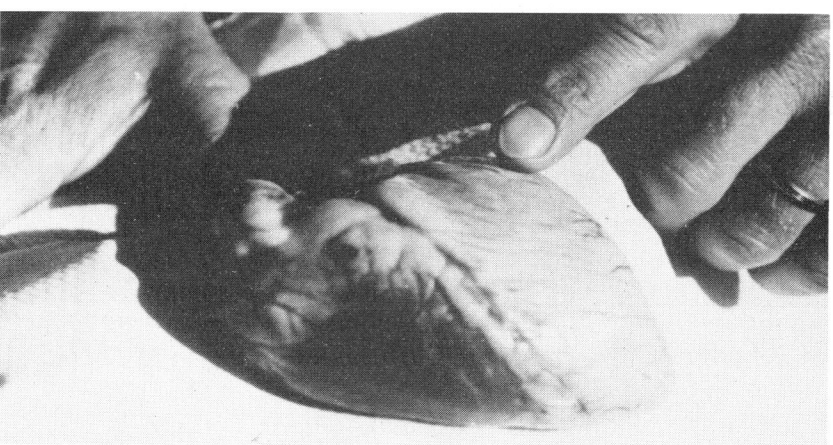

Fig. 8-16. White muscle disease in a calf.
Upper picture shows the generalized weakness of muscles, lameness, and difficulty in locomotion of an afflicted calf. Calf is about three months old.
Lower picture shows abnormal white areas in the heart muscles of a six-week-old calf afflicted with white muscle disease.
(Courtesy, Oregon Agricultural Experiment Station, Corvallis, Ore.)

been demonstrated experimentally for the young calf during the first eight weeks of life, prior to the development of the functioning rumen. At this stage in life, these requirements are usually met by the milk of the dam. Later, the B vitamins appear to be synthesized in sufficient quantities by rumen bacterial fermentation. However, inadequacy of protein or other nutrients in the ration may impair rumen fermentation, with the result that sufficient quantities of the B vitamins will not be synthesized.

● *Vitamin K*—Under normal conditions, adequate vitamin K is synthesized in the rumen of cattle. However, symptoms of inadequacy (a bleeding syndrome known as "sweetclover disease") occur when moldy sweetclover hay, high in dicoumarol content, is fed.

Appendix Table I-3 gives the provitamin A (carotene), vitamin E, and vitamin D of feeds commonly used in beef cattle rations.

VITAMIN REQUIREMENTS OF BEEF CATTLE

The National Research Council vitamin requirements of beef cattle are given in Table 8-11. Table 8-12 (see page 472) is a beef cattle vitamin summary.

TABLE 8-11

VITAMIN REQUIREMENTS OF BEEF CATTLE
(IN PERCENTAGE OR AMOUNT PER KILOGRAM OF DRY DIET)[1]

Nutrient	Growing and Finishing Steers and Heifers	Dry Pregnant Cows	Breeding Bulls and Lactating Cows
Vitamin A activity .... IU[2][3]	2200	2800	3900
Vitamin D .... IU	275	275	275
Vitamin E .... IU	15-60	——	15-60

[1]From *Nutrient Requirements of Beef Cattle*, fifth revised edition, National Academy of Sciences, Washington, D.C., 1976, p. 35, Table 5.
[2]May be vitamin A or provitamin A equivalent.
[3]See Tables 8-18, 8-18a, and 8-18b for more detailed data on requirements.

Water

Water is the most vital of all nutrients. It is needed for all the essential processes of the body, such as the digestion and absorption of food nutrients, the removal of waste, and in regulating body temperature. Animals can survive for a longer period without feed than they can without water. Yet, under ordinary conditions, it can be readily provided in abundance and at little cost.

Beef cattle should have an abundant supply of water before them at all times. Mature cattle will consume an average of about 11 gallons of water per head daily, with younger animals requiring proportionally less. The water requirement is influenced by several factors, including rate and composition of gain, pregnancy, lactation, activity, type of ration, feed intake, and environmental temperature.

Saline water containing one percent soluble salts may be toxic. Excessive nitrates or alkalinity may make water unsatisfactory for cattle.

In the northern latitudes, heaters must be provided to make the water available, but they are not needed to warm the water further.

TABLE 8-12—BEEF CATTL

Vitamins Which May Be Deficient Under Normal Conditions	Conditions Usually Prevailing Where Deficiencies Are Reported	Function of Vitamin	Some Deficiency Symptoms
A (Vitamin A is found only in animals; plants contain the precursor—carotene.	Vitamin A deficiencies may occur when—(1) extended drought results in dry, bleached pastures; (2) winter feeding on bleached hays (especially over-ripe cereal hays or straws) with little or no green hay or silage; (3) drylot finishing on rations with little or no green forage or yellow corn, especially for feeding periods longer than 2 to 3 months; and (4) there is high nitrate intake, in either water or feed.	Vitamin A—(1) promotes growth and stimulates appetite; (2) assists in reproduction and lactation; (3) helps keep the mucous membranes of respiratory and other tracts in healthy condition; and (4) makes for normal vision.	**Mild deficiency:** Lowered feed consumption and we gains. **Severe deficiency:** Night blindness, muscular incoord tion, staggering gait, and convulsive zures. Total and permanent blindnes young animals. Other localized paral may occur. Excessive watering of eyes (rather than Xerophthalmia) ally occurs; the corneas of the eyes come keratinized and may, upon in tion, develop ulceration. Severe and termittent diarrhea at advanced stage deficiency is characteristic. **Finishing cattle:** Generalized edema or anasarca may cur, with symptoms of lameness in hock and knee joints and swelling in brisket area. Pulmonary complicati culminating in pneumonia have b reported. **Bulls of breeding age:** Decline in sexual activity. Spermatc decrease in numbers and motility, there is a marked increase in abnor forms. **Breeding cows:** Estrus may continue, but concept rate may be low. Pregnant cows r abort or give birth at term of dead, we or blind calves. Retained placentas common. **Vitamin A deficiency can be detected carotene and Vitamin A analysis blood and liver tissue of cattle.**
D	Young calves kept indoors, especially in the winter-time. Finishing cattle in northern U.S. on high silage and grain rations and a minimum of sun-cured hay.	Aids in assimilation and utilization of calcium and phosphorus, and necessary in the normal bone development of animals—including the bone of the fetus.	Rickets in young calves, the symptom. which are: decreased appetite, lowe growth rate, digestive disturbanc stiffness in gait, labored breathing, ritability, weakness, and, occasiona tetany and convulsions. Later, enlar ment of the joints, slight arching of back, bowing of the legs, and the sion of the joint surfaces cause diffic in locomotion. Posterior paralysis r follow fracture of vertebrae. Sympto develop more slowly in older animal Vitamin D deficiency in the pregn animal may result in dead, weak or formed calves at birth.
E (Also see Table 8-17.)	Abnormally high levels of nitrites may produce vitamin E deficiencies. Where soils are very low in selenium.	Serves as a physiological antioxidant, facilitating the absorption and storage of vitamin A. Its other biochemical roles in the animal body appear to be related to its antioxidant capability, including the protection of vitamin A.	Muscular dystrophy (commonly cal white muscle disease) in calves 2 to weeks of age; characterized by he failure and paralysis varying in seve from slight lameness to inability stand. Also, a dystrophic tongue is of seen in affected animals.

Footnote on last page of table.

ʃAMIN CHART

Nutrient Requirements[1]				
ʃily Nutrients/ Animal (or Injection)	Amount/Lb (or kg) of Feed	Recommended Allowances[1]	Practical Sources of the Vitamin	Comments
ʃriable according to class, age, ʃ weight of catʃ. (See Tables 8, 8-18a, and ʃ8b.) ʃection of 1 ʃlion IU of viʃnin A intraʃscularly will ʃvent deficienʃsymptoms for ʃ months in ʃwing or breedʃcattle.	*Variable according to class, age, and weight of cattle. (See Table 8-11.) On a dry diet basis, the vitamin A requirements are about as follows: *1. Growing-finishing steers and heifers, 1,000 IU/lb (2,200 IU/kg). *2. Pregnant heifers and cows, 1,270 IU/lb (2,800 IU/kg). *3. Lactating cows and breeding bulls, 1,770 IU/lb (3,900 IU/kg).	See section on "Vitamins." Inject newborn calves (at birth) with 250,000 to 1,000,000 IU of vitamin A (use the higher level under confinement production or where scours may be a problem).	Stabilized vitamin A. Green pasture. Grass or legume silages. Yellow corn. Green hay not over 1 yr. old The average carotene content of some common feeds is as follows: mg. Carotene/ lb Legume hays (including alfalfa), average quality . . . . .9-14 Nonlegume hays, average quality . . . . . .4-8 Dehydrated alfalfa meal, average quality . . . . .50-70 Yellow corn . . . .0.8-1.0 Silages, corn or sorghum 2-10	Hay over 1 yr. old, regardless of green color, is usually not an adequate source of carotene or vitamin A activity. The younger the animal, the quicker vitamin A deficiencies will show up. Mature animals may store sufficient vitamin A to last 6 months. When deficiency symptoms appear, it is recommended that there be added to the ration either (1) a stabilized vitamin A product, or (2) dehydrated alfalfa or grass. Corn and sorghum silage may contain a substance which destroys carotene and/or vitamin A.
	*125 IU/lb (275 IU/kg) of dry diet.	Normally, beef cattle receive sufficient vitamin D from exposure to direct sunlight or from sun-cured hay.	Exposure to direct sunlight. Sun-cured hay. Irradiated yeast.	Sun-cured alfalfa hay contains 300 to 1,000 IU/lb (661 to 2,204/kg). Vitamin D is usually added at level of about 1/7 level of vitamin A.
	*dl-alpha-tocopherol acetate added to dry diet at level of 6.8 to 27.3 IU/lb (15 to 60 IU/kg).	Generally natural feeds supply adequate quantities of alpha-tocopherol for mature cattle, although muscular dystrophy in calves occurs in certain areas.	Alpha-tocopherol, added to the diet or injected intramuscularly. Commercial vitamin E supplements. Grains contain 6-15 mg vitamin E/lb (13-33 mg/kg).	The incidence of white muscle disease appears to be lower where the cows receive 2 to 3 lb (.91-1.36 kg) of grain during last 60 days of pregnancy. Where supplemental vitamin E is needed, it may be added to the ration or injected intramuscularly.

(Continued)

TABLE 8

Vitamins Which May Be Deficient Under Normal Conditions	Conditions Usually Prevailing Where Deficiencies Are Reported	Function of Vitamin	Some Deficiency Symptoms
B Vitamins: thiamine biotin niacin pyridoxine pantothenic acid riboflavin vitamin B_{12}	Severe inadequacy of protein or other nutrients in the diet may impair rumen fermentation to such an extent that sufficient quantities of B vitamins will not be synthesized.		
K	When moldy sweetclover hay high in dicoumarol content is fed, resulting in a bleeding syndrome called sweet clover poisoning or bleeding disease.		

[1]As used herein, the distinction between "nutrient requirements" and "recommended allowances" is as follows: In nutrient require ments, no margins of safety are included intentionally; whereas in nutrient allowances, margins of safety are provided in order to comp sate for variations in feed composition, environment, and possible losses during storage or processing. *Where preceded by an asterisk

Fig. 8-17. Typical watering facility on a western range—well, windmill, and tank. (Courtesy, Tom Lasater, Lasater Ranch, Matheson, Colo.)

(continued)

Nutrient Requirements[1]		Recommended Allowances[1]	Practical Sources of the Vitamin	Comments
Daily Nutrients/ Animal (or Injection)	Amount/Lb (or kg) of Feed			
		Usually, no dietary B vitamins need be supplied to cattle.	Milk supplied by the cow during early lactation.	During the first 8 weeks of life of the calf, the dietary requirements for the B vitamins are usually adequately met by milk from the dam; after this, these vitamins are usually synthesized by the rumen bacteria.
				Except when the dicoumarol content of hay is excessively high (as in moldy sweet clover hay) sufficient vitamin K is synthesized in the rumen of cattle.

nt requirements listed herein were taken from *Nutrient Requirements of Beef Cattle*, fifth revised edition, National Academy of ces, Washington, D.C., 1976, p. 35, Table 5.

FEEDS FOR BEEF CATTLE

Beef cattle feeding practices vary according to the relative availability of grasses, dry roughages, and grains. Where roughages are abundant and grain is limited, as in the western range states, cattle are primarily grown out or finished on roughages. On the other hand, where grain is relatively more abundant, as in the Corn Belt and in the High Plains area of Texas and Oklahoma, finishing with more concentrates is common.

Pastures

Good pasture is the cornerstone of successful beef cattle production. In fact, there has never been a great beef cattle country or area which did not produce good grass. It has been said that a good farmer or rancher can be recognized by the character of his pastures and that good cattle graze good pastures. Thus, the three go hand in hand—good farmers, good pastures, and good cattle. The relationship and importance of cattle and pastures has been further extolled in an old Flemish proverb which says, "No grass, no cattle; no cattle, no manure; no manure, no crops."

Approximately 49.4 percent of the total land area of the United States is devoted to pasture and grazing lands. Much of this area, especially in the far West, can be utilized only by beef cattle or sheep. Although the term pasture usually suggests growing plants, it is correct to speak of pasturing stalk and stubble fields. In fact, in the broad sense, pastures include all crops that are harvested directly by animals.

The type of pasture, as well as its carrying capacity and seasonable use, varies according to topography, soil, and climate. Because of the hundreds of species of grasses and legumes that are used as beef cattle pastures, each with its own best adaptation, no attempt is made to discuss the respective virtues of each variety. Instead, it is recommended that the farmer or rancher seek the advice of his local county agricultural agent, or write to his state agricultural college.

Fig. 8-18. Cows grazing on tall fescue fall regrowth and field-stored round bales, in southeastern Ohio. (Courtesy, Dr. Robert W. Van Keuren, Ohio Agricultural Research and Development Center, Wooster, Ohio)

No method of harvesting has yet been devised that is as cheap as that which can be accomplished through grazing by animals. Accordingly, successful beef cattle management necessitates as nearly year-round grazing as possible. In the northern latitudes of the United States, the grazing season is usually of about six months' duration, whereas in the deep South, yearlong grazing is approached. In many range areas of the West, the breeding herds obtain practically all their forage the year-round from the range, being given supplemental roughage only if the grass or browse is buried deep in snow.

During the winter months, and in periods of drought, the pasture utilized by beef cattle may consist of dried grass cured on the stalk. On a dry basis, the crude protein content of mature, weathered grasses may be three percent or less. To supplement such feed, cattlemen commonly feed cake or cubes. The use of cake or cubes instead of meal reduces losses from wind blowing, an especially important factor on the range.

In some instances, cattle on pasture fail to make the proper growth or gain in condition because the soil is seriously deficient in fertility or the pasture has not been well managed. In such instances, striking improvement will result from proper fertilization and management.

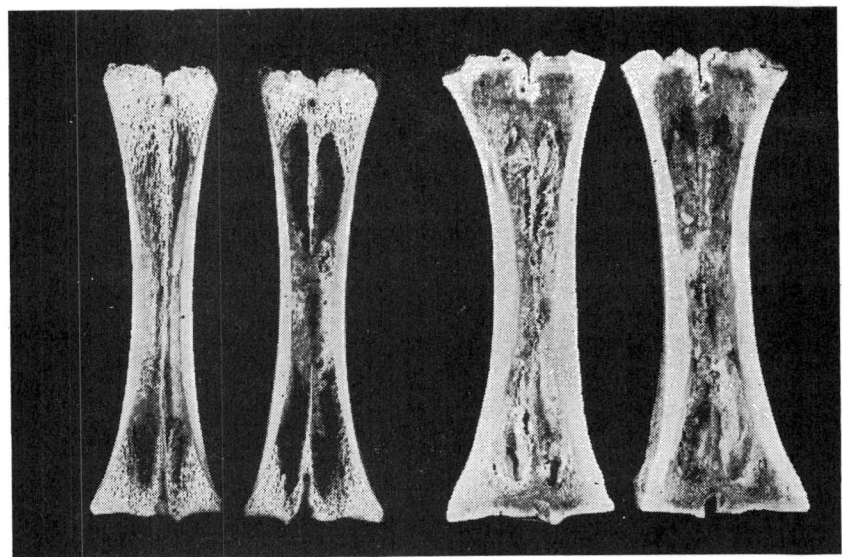

Fig. 8-19. Shin bones from two calves of similar breeding and age. Small, fragile, pitted bones (left pair) obtained from calf pastured on "belly deep" grasses grown on highly weathered but untreated soil. Heavy bones (right pair) obtained from calf pastured on grasses grown on moderately weathered but fertilized soil, including "trace" elements. (Courtesy, University of Missouri, Columbia, Mo.)

Hay and Other Dry Roughages

Hay is the most important harvested roughage fed to beef cattle, although many other dry roughages can be and are utilized.

The dry roughages are all high in fiber and, therefore, lower than concentrates in total digestible nutrients. Hay averages about 28% fiber and straw approximately 38%, whereas such concentrates as corn and wheat contain only 2 to 3% fiber. Fortunately, cattle are equipped to handle large quantities of roughages. In the first place, the paunch of a mature cow has a capacity of 3 to 5 bushels, thus providing ample storage for large quantities of less concentrated feeds. Secondly, the billions of microorganisms in the rumen attack the cellulose and pentosans of the fibrous roughages, such as hay, breaking them down into available and useful nutrients. In addition to providing nutrients at low cost, the roughages add needed bulk to cattle rations.

Roughages, like concentrates, may be classified as carbonaceous or nitrogenous, depending on their protein content. The principal dry carbonaceous roughages used by cattle include hay from the grasses, the straws and hays from cereal grains, corncobs, and the stalks and leaves of corn and the grain sorghums. Cured nitrogenous roughages include the various legume hays such as alfalfa, the clover hays, peanut hay, soybean hay, cowpea hay, and velvet bean hay.

Although leguminous roughages are preferable, weather conditions and soils often make it more practical to produce the nonlegumes. Also, in many

Fig. 8-20. Haying has gone modern! This shows a tightly packed, round-topped, virtually weatherproof, movable hay stack. It lessens labor and makes it possible to deposit the stacks in the field, fence them, and feed cattle right from the stack. (Courtesy, Deere & Company, Moline, Ill.)

areas, such feeds as dry grass cured on the stalk, cereal straws, corncobs, and cottonseed hulls are abundantly available and cheap. Under such circumstances, these feeds had best be used as part of the ration for wintering beef cows, for wintering stockers that are more than one year of age, or for finishing beef cattle.

In comparison with good quality legume hays, the carbonaceous roughages are lower in protein content and in quality of proteins, lower in calcium, and generally deficient in carotene (provitamin A). Thus, where nonlegume roughages are used for extended periods, these nutritive deficiencies should be corrected; this is especially true with the gestating-lactating cow or the young, growing calf. To the end that the feeding value of some of the common nonlegumes may be enhanced for beef cattle, the following facts are pertinent:

1. The feeding value of nonlegume hays can be increased by cutting them at an early stage of maturity and curing so as to retain as much of the carotene content as possible.

2. Where dry and bleached pastures are grazed for an extended period of time, or where there is an unusually long winter, it is important that at least part of the roughage be a legume, either silage or hay, or that vitamin A either be added to the ration or injected intramuscularly; and that suitable energy, protein, and mineral supplements be provided.

3. Potentially, corncobs—which were formerly considered a waste product and of little worth—have a feeding value approaching that of hay. However, their energy cannot be utilized unless they are fortified with certain nutrients which help the bacteria and other organisms of the rumen

break them down into a form which can be digested. Also, corncobs are low in palatability.

4. Cereal straws and cottonseed hulls may be incorporated in the wintering ration of pregnant cows or in the ration of finishing cattle provided their fundamental characteristics and nutritional limitations are recognized and corrected.

Silages and Roots

Silage is an important adjunct to pastures in beef cattle production, it being possible to use a combination of the two forages in furnishing green, succulent feeds on a year-round basis. Extensive use of silage for beef cattle dates back only to about 1910. Prior to that time, it was generally thought of as a feed for dairy cows. Even today, only a relatively small percentage of the beef cattle of the United States is fed silage.

Where silage has been used, it has proved very popular. Some of the more important reasons advanced in favor of silage are as follows:

1. On most beef cattle producing farms, silage is the cheapest form in which a good succulent winter feed can be provided.

2. It is the cheapest form in which the whole stalk of an acre of corn or sorghum can be processed and stored.

3. Good silage can be made during times of rainy weather when it would be impossible to cure properly hay or fodder.

4. It helps to control weeds, which are often spread through hay or fodder.

5. Grass silage is a better source of vitamins, especially carotene and perhaps some of the unknown factors, than dried forages.

6. There is no danger of fire loss to silage.

7. Silage is a very palatable feed and a mild laxative.

8. Converting the crop into silage clears the land earlier than would otherwise be possible.

9. Silage makes for less waste, the entire plant being eaten with relish.

10. The removal of cornstalks, as is required in making silage, is one of the best methods of controlling the European corn borer.

11. Silage increases the number of animals that can be fed from the produce of a given area of land.

Corn was the first and still remains the principal crop used in the making of silage, but many other crops are ensiled in various sections of the country. The sorghums are the leading ensilage crop in the Southwest, and grasses and legumes are the leading ensilage crops in the Northeast. Also, in different sections of the country to which they are adapted, the following feeds are ensiled: cereal grains, field peas, cowpeas, soybeans, potatoes, and numerous fruit and vegetable refuse products. A rule of thumb is that crops that are palatable and nutritious to animals as pasture, as freshly harvested feed, or as dry forage, also make palatable and nutritious silage. Conversely, crops that are unpalatable and nonnutritious as pasture, as green feed, or as dry forage, also make unpalatable and nonnutritious silage.

Grass silage can be produced in those areas where the climate is too

Fig. 8-21. Mechanized feeding of corn silage in an Illinois feedlot. (Courtesy, James Mfg. Co., Fort Atkinson, Wisc.)

cool and the growing season too short for corn or sorghum silage. It is generally higher in protein and carotene, but lower in total digestible nutrients and vitamin D than corn or sorghum silage. Generally grass silage contains about 90 percent as much total digestible nutrients (TDN) as corn silage, but it is equal in TDN when 150 pounds of grain per ton have been added as a preservative. Thus, grass silage generally requires the addition to the ration of less protein supplement but more total concentrates than corn or sorghum silage. This would indicate that corn or sorghum silage would be slightly preferable to grass silage in high roughage finishing rations for beef cattle, whereas grass silage would be preferable in high roughage rations for young, growing beef cattle.

When silage is fed to cattle, it must be remembered that, because of its high moisture content, about 3 lb of silage are generally considered equivalent to 1 lb of dry roughage of comparable quality. A ration of 55 to 60 lb of corn silage plus ½ to ¾ lb of a protein concentrate daily will carry a dry cow through the winter. The ration may be improved, however, by replacing ⅓ to ½ of the silage with an equivalent amount of a dry roughage, adding 1 lb of dry roughage for each 3 lb of silage replaced.

Silage may be successfully used for finishing steers. Long yearling steers will eat 25 to 35 lb a day at the beginning of the feeding period, the larger amounts being consumed when no hay is fed with it. Better results are obtained, however, if hay is included in the ration. The amount of silage is gradually decreased as the concentrates are increased. At the end of the feeding period, the cattle should be getting around 4 to 6 lb of silage and 1 or 2 lb of hay. Because of the more limited digestive capacity, the allowance of silage fed to calves should be correspondingly less.

Usually, silage provides a much cheaper succulent feed for beef cattle than roots. For this reason, the use of roots for beef cattle is very limited, being confined almost entirely to the northern areas.

Concentrates

The concentrates include those feeds which are low in fiber and high in energy. For purposes of convenience, concentrates are often further classified as (1) carbonaceous feeds, and (2) nitrogenous feeds.

In general, the use of concentrates for beef cattle is limited to (1) the finishing of cattle, (2) the development of young stock, and (3) use as limited supplements in the winter ration. Over most of the United States, the cereal grains are the chief concentrates fed to beef cattle—these grains being combined, if necessary, with protein supplements to balance the ration.

The chief carbonaceous concentrates used for beef cattle are the cereal grains and such processed feeds as hominy feed, beet pulp, and molasses. The choice of the particular feeds is usually determined primarily by price and availability.

For best results, the feeder should correct the nutritive deficiencies of the cereal grains. All of them are low in protein, low in calcium, and lacking in vitamin D. All except yellow corn are also deficient in carotene. Regardless of whether the cereal grains are fed to growing, breeding, or finishing animals, their nutritive deficiencies can be corrected in a very effective and practical way either by adding (1) a good quality legume hay to the ration, or (2) a protein concentrate plus suitable minerals and vitamins.

Preparation of Feeds

The physical preparation of cereal grains for cattle by soaking and cooking has been practiced by cattle showmen for a very long time. In recent years, many sophisticated techniques for the processing of grains have been developed, especially for feedlot cattle. (See Chapter 30, section entitled, "Feed Preparation.") Basically, however, grain is either soaked, cooked,

Fig. 8-22. Cattle feed processing facilities at the Benedict Feeding Co., Casa Grande, Arizona. (Courtesy, Benedict Bros.)

ground, or rolled (wet or dry), and hay is either pelleted, cut, shredded, ground, or cubed.

A pertinent summary relative to the preparation of feeds for beef cattle follows:

• *Soaking or cooking (steaming)*—Professional herdsmen often cook (or steam) grains, in order to increase palatability and feed consumption. However, soaking or cooking has the following limitations:

1. It is apt to decrease the digestibility of the proteins, even though it may slightly increase the digestibility of the starches.

2. It does not improve feeding value.

• *Grinding or rolling grains*—Among the factors to consider when deciding on the grain processing method are the size of kernel, percentage of moisture, and percentage of grain in the ration. Also, when any of the dry processing methods are used, it is important that the kernel be broken, but that there be coarseness and relative freedom from "fines." Additionally, the following points are pertinent to grinding or rolling grains:

1. It may not be profitable to grind corn for calves under 9 to 12 months of age, because young calves masticate grains well.

2. Professional herdsmen and feedlot operators prefer steam rolled grains, because they are lighter and fewer digestive disturbances are encountered.

3. Based on University of Arizona studies, it is recommended that milo, barley, and wheat (and perhaps other cereal grains) for finishing cattle be processed under 20 pounds of steam pressure for 20 minutes, at approximately 205° F; then, at 18 to 20 percent moisture content, rolled thinly (flaked).

Fig. 8-23. Flaked wheat. (Courtesy, Benedict Feeding Co., Casa Grande, Ariz.)

● *Chopping, shredding, or grinding roughage*—In the West, hay is frequently chopped because (1) it is easier to handle, (2) it can be stored in a smaller area at less cost, (3) it is fed with less waste, and (4) it may be han-

Fig. 8-24. Field-chopped hay stored in long, rectangular stacks at C & B Livestock, Inc., a modern cattle feedyard, at Hermiston, Oregon. (Courtesy, Ron Baker, President, C & B Livestock, Inc.)

dled mechanically. Also, cutting or shredding such coarse forages as corn fodder or stover makes them easier to handle and store, and results in less waste. Additional points pertinent to roughage preparation follow:

1. Chopping forages for cattle is more common in areas where forages are relatively more abundant and cheaper than the grains, with the result that a higher proportion of them is fed. Also, it follows that there is apt to be greater waste of forage under liberal feeding, unless precautions are taken to alleviate it.

2. The preparation of roughages does not increase the value of the initial product.

3. For cattle, roughages should be coarsely chopped, not less than 2 inches in length.

4. In preparing forages, avoid (a) processing those with high moisture, which may heat and produce spontaneous combustion, and (b) processing those in which there are foreign objects (wire and other hardware), which the animals may not be able to select out and which may ignite a fire when being processed.

• *Pelleting*—Roughages alone, concentrates alone, or a combination of roughages and concentrates may be, and are, pelleted. Many cattlemen prefer to feed pellets or cubes on the range, primarily for reasons of convenience and reducing losses from wind blowing. The practice of pelleting will increase for the following reasons:

1. Pelleted feeds are less bulky and easier to store and handle, thus lessening storage and labor costs.

2. Pelleting prevents animals from selectively wasting ingredients likely to be high in certain dietary essentials; each bite is a balanced feed.

3. Pelleting alleviates wastage of relatively unpalatable feeds, such as rye and ground alfalfa.

The biggest deterrent to pelleting forages is that of being able to process chopped forage which is coarse enough to allow for optimum cellulose digestion in the rumen, and which will not increase the incidence of bloat. As a rule of thumb, one would be on the safe side if the forage were not chopped more finely than silage. Also, there is the cost factor; processors charge about $6.00 to $7.00 per ton to prepare an all-pelleted ration (grain and hay combined), and $10.00 per ton for an all-roughage pellet.

Of course, the increased value of pellets should be appraised against the increased cost of pelleting.

• *Hay cubes*—This refers to the practice of compressing long or coarsely cut hay in cubes or wafers, which are larger and coarser than pellets. Most cubes are about 1¼" square and 2" long, with a bulk density of 30 to 32 pounds per cubic foot. Cubing costs about $5.00 per ton more than baling.

This method of haymaking is increasing, because it offers most of the advantages of pelleted forages, with few of the disadvantages. Cubed or wafered forage (1) simplifies haymaking, (2) alleviates fine grinding, (3) facilitates automation in both haymaking and feeding, (4) lessens transportation costs and storage space—cubed roughages require about one-third as much space as when the forage is baled and stacked, (5) reduces labor, (6) decreases nutrient losses, and (7) eliminates dust.

● *All-pelleted rations (grain and forage combined)*—Among the virtues ascribed to an all-pelleted (grain and forage combined) ration are: (1) it prevents selective eating—if properly formulated, each mouthful is a balanced diet; (2) it alleviates waste; (3) it eliminates dust; (4) it lessens labor and equipment—just fill self-feeders; (5) it lessens storage; and (6) it lends itself to automation. In all-pelleted rations, the ratio of roughage to concentrate should be higher than where long hay is fed.

With pelleting or cubing, the spread between high and low quality roughage is narrowed; that is, the poorer the quality of the roughage, the greater the advantage from pelleting or cubing. This is so because such preparation assures complete consumption of the roughage.

Feed Substitution Tables[15]

The successful cattleman is a keen student of values. He recognizes that feeds of similar nutritive properties can and should be interchanged in the ration as price relationships warrant, thus making it possible at all times to obtain a balanced ration at the lowest cost.

Table 8-13, Handy Feed Substitution Table for Beef Cattle, is a summary of the comparative values of the most common U.S. feeds. In arriving at these values, two primary factors besides chemical composition and feeding value have been considered—namely, palatability and carcass quality.

In using this feed substitution table, the following facts should be recognized:

1. That, for best results, different ages and groups of animals within classes should be fed differently.

2. That individual feeds differ widely in feeding value. Barley and oats, for example, vary widely in feeding value according to the hull content and the test weight per bushel, and forages vary widely according to the stage of maturity at which they are cut and how well they are cured and stored.

3. That nonlegume forages may have a higher relative value to legumes than herein indicated provided the chief need of the animal is for additional energy rather than for supplemented protein. Thus, the nonlegume forages of low value can be used to better advantage for wintering mature, dry beef cows than for young calves.

On the other hand, legumes may actually have a higher value relative to nonlegumes than herein indicated provided the chief need is for additional

[15]The author gratefully acknowledges the helpful suggestions of the following authorities who reviewed this section and table 8-13: Dr. W. W. Heinemann, Washington State University, Pullman, Wash.; Dr. Edward L. Stephenson, Head, Department of Animal Science, University of Arkansas, Fayetteville, Ark.; Dr. George L. Robertson, Head, Department of Animal Science, Louisiana State University, Baton Rouge, La.; Dr. O. O. Thomas, Nutritionist, Montana State University, Bozeman, Mont.; Dr. J. Matsushima, Department of Animal Science, Colorado State University, Fort Collins, Colo.; Prof. L. B. Embry, Animal Nutritionist, South Dakota State University, Brookings, S.D.; and Dr. T. J. Cunha, Dean, School of Agriculture, California State Polytechnic University, Pomona, Calif.

protein rather than for added energy. Thus, no protein supplement is necessary for breeding beef cows provided a good quality legume forage is fed.

4. That, based primarily on available supply and price, certain feeds—especially those of medium protein content, such as brewers' dried grains, corn gluten feed (gluten feed), distillers' dried grains, distillers' dried solubles, peanuts, and peas (dried)—may be used interchangeably as (a) grains and by-product feeds, and/or (b) protein supplements.

5. That the feeding value of certain feeds is materially affected by preparation. Thus, wheat must be coarsely ground or rolled for cattle. The values herein reported are based on proper feed preparation in each case.

For the reasons noted above, the comparative values of feeds shown in the feed substitution table are not absolute. Rather, they are reasonably accurate approximations based on average quality feeds, together with experiences and experiments.

(See Table 8-13, beginning on p. 487.)

Home-Mixed vs Commercial Feeds

The value of farm-grown grains—plus the cost of ingredients which need to be purchased to balance the ration, and the cost of grinding and mixing—as compared to the cost of commercial ready-mixed feeds laid down on the farm, should determine whether it is best to mix feeds at home or depend on ready-mixed feeds.

Although there is nothing about the mixing of feeds which is beyond the capacity of the intelligent farmer or rancher, under many conditions a commercial mixed feed supplied by a reputable dealer may be the most economical and the least irksome. The commercial dealer has the distinct advantages of (1) purchase of feeds in quantity lots, making possible price advantages, (2) economical and controlled mixing, and (3) the hiring of scientifically trained personnel for use in determining the rations. Because of these advantages, commercial feeds are finding a place of increasing importance in American agriculture.

Also, it is to the everlasting credit of reputable feed dealers that they have been good teachers, often getting stockmen started in the feeding of balanced rations, a habit which is likely to remain with them whether or not they continue to buy commercial feeds.

HOW TO SELECT COMMERCIAL FEEDS

There is a difference in commercial feeds! That is, there is a difference from the standpoint of what a stockman can purchase with his feed dollars. The smart operator will know how to determine what constitutes the best in commercial feeds for his specific needs. He will not rely solely on how the feed looks and smells or on the feed salesman. The most important factors to consider or look for in buying a commercial feed are listed on page 497.

TABLE 8-13

HANDY FEED SUBSTITUTION TABLE FOR BEEF CATTLE, AS FED BASIS

Feedstuff	Relative Feeding Value (Lb for Lb or Kg for Kg) in Comparison with the Designated (Underlined) Base Feed Which = 100	Maximum Percentage of Base Feed (or Comparable Feed or Feeds) Which It Can Replace for Best Results	Remarks
GRAINS, BY-PRODUCT FEEDS, ROOTS AND TUBERS:[1] (Low and Medium Protein Feeds)			
Corn, No. 2	**100**	**100**	The most important concentrate for finishing cattle in the United States. Grind coarsely or roll.
Almond hulls, dried, no shells	70-75	15-30	
Almond hulls and shell meal	35	15-20	
Apple pomace, air-dry basis	78	33⅓	
Bakery products, dried	100	15-30	
Bakery waste, not dried (30% water) ...	75	15-30	
Barley	90	25-100	The heavier the barley and the smaller the proportion of the hulls, the higher the feeding value. Grind coarsely or roll for cattle. In Canada, where considerable barley is fed, it is often used as the only basal feed in the ration once animals are accustomed to it.
Beans (cull)	80	10	Best when cooked, but can also be fed raw. Beans should be ground. Scouring may occur if constitutes more than 15% of total ration.
Beet pulp, dried	90	50	
Beet pulp, molasses, dried	90-95	50	
Beet pulp, wet	25	40	May compose 40% of ration on dry matter basis. 50% the value of corn silage.
Brewers' grains, dried	80	33⅓	Not very palatable. Fed chiefly to dairy cattle. Somewhat bulky.

Footnotes on last page of table.

(Continued)

TABLE 8-13 (Continued)

Feedstuff	Relative Feeding Value (Lb for Lb or Kg for Kg) in Comparison with the Designated (Underlined) Base Feed Which = 100	Maximum Percentage of Base Feed (or Comparable Feed or Feeds) Which It Can Replace for Best Results	Remarks
Brewers' grains (wet)	13-15	33⅓	Grains usually come from barley. Best to haul and feed directly. Can be stored in silo if salt is added at rate of 25 lb per ton of grains.
Buckwheat	55-75	33⅓	Should be ground and mixed with other grains.
Carrots (cull)	10-15	20-25	Store 3 to 4 weeks before using; fresh carrots cause scouring. Feed whole or sliced.
Citrus pulp, dried	80-88	25-50	
Corn and cob meal	85-90	100	
Corn gluten feed (gluten feed)	85-90	50	
Distillers' dried grains	73-90	33⅓	Rye distillers' dried grains are of lower value than similar products made from corn or wheat. Distillers' dried grains are used chiefly for dairy cattle.
Distillers' dried solubles	73-90	33⅓	The chief difference between distillers' dried grains and distillers' dried solubles is the higher B-vitamin content of the latter. Normally this is not important for cattle.
Fat (animal or vegetable)	225	5	Fat has 203 megacalories energy/100 lb for maintenance and 127 megacalories for weight gain, as compared to 92 and 60, respectively, for corn.
Grape pomace	35	25	
Hominy feed	100	50	
Manure, cattle	75	50	Approximately 80% of the total nutrients of feeds is excreted as animal manure. However, the feeding value of manure will vary according to (1) the nutritive value of the feeds initially fed, (2) the class, age, and individuality of the animal to which the feeds were initially fed, and (3) the handling and processing of the manure.

Feed			Comments
Manure, poultry (see poultry house litter)			
Molasses, beet	75	10-40	Value is highest when used as an appetizer. May be laxative if fed at levels above 6 lb daily.
Molasses, cane	75	10-40	Value is highest when used as an appetizer.
Molasses, citrus	65-75	10-40	
Molasses, wood	26-30	10-20	Rather unpalatable.
Oats	70-90	10-100	Valuable for young stock, for breeding stock and for getting animals on feed. Oats have lowest value for finishing cattle and should be limited to ⅓ of such rations. Also, the feeding value of oats varies according to the test weight per bushel. Grind or roll for cattle.
Paunch, dried (also see "paunch-blood" under Protein Supplements of this table)	90	5-10	Dried paunch is not too palatable, with the result that it depresses appetite. Rate of gain is not affected, but feed efficiency is slightly lowered.
Peas (cull), dried	88	40	Peas appear to be unpalatable to certain individuals. Also, there is bloat hazard if they exceed 40% of the ration.
Pear waste, air-dry	75	40	When fed with alfalfa hay, they are worth about 80% as much per ton as corn silage.
Potato, (Irish), wet	20-25	85	Do not feed frozen. Sunburned, decomposed, or sprouted potatoes should not make up more than 10% of potatoes fed. Keep steers' heads down while eating to prevent choking.
Potato, (Irish) dehydrated	88	50	Excellent source of energy, but deficient in protein, minerals, and vitamins.
Potato, (sweet)	25	85	Dehydrated sweet potatoes are more palatable than dehydrated Irish potatoes.
Potato, (sweet) dehydrated	95-100	50	
Poultry house litter	10-40	15-25	Poultry house litter may also be used as a protein source (see Protein Supplement this table).
Prunes	62	15	Because of the laxative quality of prunes, they should be limited to 7% of the total ration.

(Continued)

Footnotes on last page of table.

TABLE 8-13 (Continued)

Feedstuff	Relative Feeding Value (Lb for Lb or Kg for Kg) in Comparison with the Designated (Underlined) Base Feed Which=100	Maximum Percentage of Base Feed (or Comparable Feed or Feeds) Which It Can Replace for Best Results	Remarks
Raisins, (cull)	70	33⅓	
Raisin pulp	53	25	
Rice, (rough rice)	80	100	
Rice bran	66⅔-75	33⅓	Not palatable when fed in larger amounts.
Rice polishings	88	25	
Rye	96	33⅓	Should be finely ground to kill noxious weed seeds.
Screenings, refuse	62-70	25-35	Quality varies; good quality screenings are equal to oats, whereas poor quality screenings resemble straw.
Sorghum (milo, Kafir), grain ...	90-95	100	Varieties vary in protein content. Grind or roll for cattle.
Spelt and emmer	70-90	30-100	Similar to oats.
Wheat	100-105	50	Grind coarsely or roll.
Wheat bran	70-90	25-33⅓	Because of its bulk and fiber, bran is not desirable for finishing rations. Bran is valuable for young animals, for breeding animals, and for starting animals on feed.
Wheat-mixed feed (mill run) ...	95	33⅓	Sometimes fed to the breeding herd, to young calves, and to finishing cattle being started on feed.
Wheat screenings	85	50	Wood products, which are largely cellulose and lignin, must be cooked before animals can digest them.
Wood (cooked)	75-80	70	

PROTEIN SUPPLEMENTS:

Soybean meal (41%)	**100**	**100**	Slightly laxative.
Alfalfa or clover screenings	70-75	50	Grind finely to destroy weed seeds.
Brewers' dried grains	55-65	50	Not very palatable. Fed chiefly to dairy cattle.
Copra meal (coconut oil meal)	90-100	50	
Corn gluten feed (gluten feed)	65-75	50-100	
Corn gluten meal (gluten meal)	90-100	50	Somewhat unpalatable.
Cottonseed meal (41%)	100	100	Among practical cattlemen, the feeling persists that cottonseed meal has a constipating effect; some experimental work to the contrary. Although it may be fed as the only protein supplement, best results are obtained when it is fed with linseed meal for finishing cattle.
Distillers' dried grains	65-70	100	Rye distillers' dried grains are about 10% lower in protein than similar products made from corn or wheat.
Distillers' dried solubles	70	100	Low in palatability.
Feather meal (hydrolyzed; 84% protein)	175	50	Feather meal is unpalatable; hence, cattle must be accustomed to it gradually and it must be limited in quantity. It is best used for wintering brood cows and stocker cattle.
Legume screenings	75	75	Satisfactory, but less palatable than soybean or cottonseed meal.
Linseed meal (35%)			
For other than finishing cattle	95	100	Linseed meal has laxative effect. Some cattle will not tolerate more than 5 to 8% linseed meal in the ration.
For finishing cattle	115	100	Higher value for finishing cattle due to both greater efficiency and higher selling price of the cattle because of the increased bloom.
Paunch-blood feed (also see "paunch, dried" under Grains section of this table)	100	100	At slaughter, each cow yields about 20 lb of paunch and 20 lb of blood. Dried paunch runs around 10% protein, dried blood around 80%, and a 50-50 mixture of the 2 products around 45%.
Peanut meal (43%)	100	100	Peanut meal may become rancid if stored too long, especially in warm, moist climates.
Peas (cull), dried	65-75	50	
Poultry house litter	50-55	25	Poultry house litter may also be used as an energy source (see Grains section of this table).

Footnotes on last page of table.

(Continued)

TABLE 8-13 (Continued)

Feedstuff	Relative Feeding Value (Lb for Lb or Kg for Kg) in Comparison with the Designated (Underlined) Base Feed Which = 100	Maximum Percentage of Base Feed (or Comparable Feed or Feeds) Which It Can Replace for Best Results	Remarks
Rapeseed meal (36%)	88	75	Rapeseed meal should be limited to not more than 2 lb per cow.
Safflower meal, well hulled	92	100	Safflower meal with hulls is unpalatable. Thus, it should
Safflower meal, with hulls	40-45	100	be mixed with more palatable feeds.
Sesame meal	90-95	25	Not satisfactory for finishing calves.
Soybeans, whole	95-100	95	Soybean allowance should be limited to amount necessary to balance the ration. Larger amounts may be unduly laxative and throw cattle "off feed."
Sunflower meal (39%)	95-100	100	If poorly hulled and lower protein content than 39%, feeding value will be lowered accordingly. It is well liked by cattle and keeps well in storage.
DRY FORAGES AND SILAGES:[2]			
Alfalfa hay, all analyses	100	100	All the dry nonlegume forages listed herein are satisfactory when needed minerals and either a limited amount of legume hay or a protein supplement are supplied to balance the ration.
Alfalfa silage	33⅓-50	50-85	Does away with or lessens protein supplement requirements.
			When alfalfa silage replaces corn silage, more energy feed must be provided, but less protein.
Alfalfa straw	37	50	Feed with good hay.
Apple pomace silage	17-25	50-85	Usually fed as a substitute for corn or grass silage. 50% the value of corn silage. Sometimes fed out of a stack or trench silo.

Feed			Notes
Apples	17-25	50-85	Do not feed more than 25 lb/cow. Not recommended for finishing cattle. Danger of choking when fed whole. Relatively high handling cost.
Bagasse, dried; sugarcane or sorghum	10-20	5-10	Has negative protein value.
Barley hay	70	100	Avoid bearded varieties.
Barley straw	63	70	Of the cereal straws, barley ranks next to oat straw in feeding value. Use for dry pregnant cows. Supplement daily with 5-6 lb alfalfa hay or 1-2 lb of 30-40% protein supplement.
Bean straw	34	50	Feed with good hay.
Beet tops, fresh	20	33⅓-50	In the West, large acreages of fresh beet tops are pastured off by cattle and sheep. Bloat may be problem when tops are frozen. Tops are laxative. Add 2½ lb of ground limestone/ton of feed.
Beet top silage, sugar	17-25	33⅓-50	Feed 2 oz of finely ground limestone or chalk with each 100 lb of tops, as calcium changes the oxalic acid to insoluble calcium oxalate.
Clover hay, crimson	90-100	100	Crimson clover hay has a considerable lower value if not cut at an early stage.
Clover hay, red	90-100	100	If the rest of the ration is adequate in protein, clover hay will be equal to alfalfa in feeding value; otherwise, it will be lower.
Clover straw	37	50	Feed with good hay.
Clover-timothy hay	80-90	100	Value of clover-timothy mixed hay depends on the proportion of clover present and the stage of maturity at which it is cut.
Corncobs, ground	70	90	Ground corncobs can be used as the only roughage for beef cattle if properly supplemented with proteins, minerals, and vitamins.
Corn fodder	75	80-90	Highest and best use is for dry pregnant cows. It is slightly higher in energy and more palatable than corn stover.
Corn husklage (shucklage)	50	80-90	

(Continued)

Footnotes on last page of table.

TABLE 8-13 (Continued)

Feedstuff	Relative Feeding Value (Lb for Lb or Kg for Kg) in Comparison with the Designated (Underlined) Base Feed Which=100	Maximum Percentage of Base Feed (or Comparable Feed or Feeds) Which It Can Replace for Best Results	Remarks
Corn silage	33⅓-50	50-85	
Corn (sweet) silage, cannery waste	26-40	50-85	
Corn stover	45	70-90	Corn stover will meet the energy needs of dry pregnant cows, but is deficient in protein and low in phosphorus and vitamin A. Two acres of cornstalks will carry a cow 100-120 days.
Corn (sweet) stover	50	100	Use for dry pregnant cows. Supplement daily with 4-6 lb of good legume hay or 1-2 lb of 30-40% protein supplement.
Cottonseed hulls	66⅔	75	
Cowpea hay	90-100	100	Add molasses to improve palatability; low in TDN and protein; high in fiber.
Gin trash, cotton	75	75	
Grape pomace or meal	15-30	10-25	
Grass-legume mixed hay	80-90	100	Value depends on the proportion of legume present and the stage of maturity at which it is cut.
Grass-legume silage	32-47	50-85	Unless grain is added as a preservative, grass silage requires more energy feed, but less protein supplement than corn silage when fed to finishing cattle.
Grass silage	30-45	50-85	For finishing cattle, grass silage must be supplemented with additional energy feeds, such as cereal grain or molasses, to be of the same value as corn silage.
Hop vine silage	20	50-75	Should be chopped when placed in the silo.
Hops, spent, dehydrated	80	50-65	Devoid of carotene; feed with legume hay.
Johnson grass hay	70	100	
Lespedeza hay	80-100	100	Feeding value of lespedeza hay varies considerably with stage of maturity at which it is cut.

Feed			Comments
Mint hay	70-80	75	Cattle tire of mint hay when it is fed as the only roughage for extended periods.
Oat hay	75	100	
Oat silage	32-47	50-85	Must be chopped finely to exclude air from silo.
Oat straw	66⅔	75	Oat straw is the best of the cereal straws. Use for dry pregnant cows. Supplement daily with 4-6 lb of good legume hay or 1-2 lb of 30-40% protein supplement.
Paper (newspaper; waste paper)	66⅔	50	Paper varies in feeding value in proportion to the cellulose (most paper is 60-90% cellulose) and lignin content. Magazine and bookstock papers are higher in cellulose and lower in lignin than newspapers; hence, of higher feeding value. Pelleting or cubing may increase the value of paper. *Caution:* Some newspapers contain heavy metals (boron, lead, barium, and antimony), sometimes used as a dye carrier in printer's ink, which may be toxic to animals.
Pea Straw	45-75	60-75	Can constitute the only roughage for finishing cattle.
Pea-vine hay	100-110	75-90	
Pea-vine silage	33⅓-50	50-85	Unless grain is added as a preservative, pea-vine silage requires more energy feed, but less protein supplement than corn silage when fed to finishing cattle.
Potato silage	25-30	50-75	About 75% the value of corn silage.
Prairie hay	65-70	100	
Reed canarygrass hay	70	100	
Rice straw	47	70	High levels of rice straw can be used for wintering cattle if the straw is properly fortified.
Sawdust	75-80	70	Digestibility is increased by cooking and other treatments. There are indications that the presence of sawdust will reduce liver abscesses in feedlot cattle.
Sorghum fodder	70	100	
Sorghum silage (grain varieties)	32-47	50-85	For finishing cattle, sorghum silage is 85 to 90% as valuable as corn silage and must be supplemented in the same manner as corn silage.

(Continued)

TABLE 8-13 (Continued)

Feedstuff	Relative Feeding Value (Lb for Lb or Kg for Kg) in Comparison with the Designated (Underlined) Base Feed Which=100	Maximum Percentage of Base Feed (or Comparable Feed or Feeds) Which It Can Replace for Best Results	Remarks
Sorghum silage (sweet varieties)	25-30	50-85	Nearly equal to grain varieties in value per acre because of greater yield.
Sorghum (milo) stover	35	70-90	Can be grazed or harvested and stored either as dry feed or silage. About 2% higher in protein, but less palatable, than corn stover.
Soybean hay	85-90	50-75	Lower value than alfalfa hay, largely due to greater wastage in feeding.
Sudan grass hay	70	100	It may cause scouring when fed alone.
Sunflower silage	25-35	50-85	65 to 75% value of corn silage. Somewhat unpalatable and may cause constipation. Harvest for silage when ½ to ⅔ of heads are in bloom.
Sweet clover hay	100	100	Value of sweet clover hay varies widely. Second year crop is less desirable than first year and is more apt to cause sweet clover disease.
Timothy hay	70	100	The higher the proportion of vetch, the higher the value.
Vetch-oat hay	80-90	100	
Wheat hay	70	100	Of the cereal straws, wheat ranks third in nutritive value, behind oat straw and barley straw. Highest and best use is for dry pregnant cows. Supplement daily with 6 lb of alfalfa or 2 lb of a 30-40% protein supplement.
Wheat straw	60	65	

[1]Roots and tubers are of lower value than the grain and by-product feeds due to their higher moisture content.
[2]Silages are of lower value than dry forages due to their higher moisture content.

1. *The reputation of the manufacturer*—This should be determined by (a) conferring with other stockmen who have used the particular products, and (b) checking on whether or not the commercial feed under consideration has consistently met its guarantees. The latter can be determined by reading the bulletins or reports published by the respective state departments in charge of enforcing feed laws.

2. *The specific needs*—Feed needs vary according to (a) the class, age, and productivity of the animals, and (b) whether the animals are fed primarily for maintenance, growth, finishing (or show-ring fitting), reproduction, lactation, or work. The wise operator will buy different formula feeds for different needs.

3. *The feed tag*—Most states require that mixed feeds carry a tag that guarantees the ingredients and the chemical makeup of the feed. Feeds with more protein and fat are better, and feeds with less fiber are better.

In general, if the fiber content is less than 8%, the feed may be considered as top quality; if the fiber is more than 8 but less than 12% the feed may be considered as medium quality; while feeds containing more than 12% fiber should be considered carefully. Of course, many feeds are high in fiber simply because they contain generous quantities of alfalfa; yet they may be perfectly good feeds for the purpose intended. On the other hand, if oat hulls and similar types of high fiber ingredients are responsible for the high fiber content of the feed, the quality should be questioned. The latter type of fiber is poorly digested and does not provide the nutrients required to stimulate the digestion of the fiber in roughages.

4. *Flexible formulas*—Feeds with flexible formulas are usually the best buy. This is because the price of feed ingredients in different source feeds varies considerably from time to time. Thus, a good feed manufacturer will shift his formulas as prices change, so as to give the stockman the most for his money. This is as it should be, for (a) there is no one best ration, and (b) if substitutions are made wisely, the price of the feed can be kept down and the feeder will continue to get equally good results.

BALANCED RATIONS FOR BEEF CATTLE

A balanced ration is one which provides an animal the proper proportions and amounts of all the required nutrients for a period of 24 hours.

Several suggested rations for different classes of cattle are listed in Table 8-15 of this chapter. Generally these rations will suffice, but it is recognized that rations should vary with conditions, and that many times they should be formulated to meet the conditions of a specific farm or ranch, or to meet the practices common to an area. Thus, where cattle are on pasture, or are receiving forage in the drylot, the added feed (generally grains, by-product feeds, and/or protein supplements), if any, should be formulated so as to meet the nutritive requirements not already provided by the forage.

Rations may be formulated by the methods which follow, but first the following pointers are noteworthy:

1. In computing rations, more than simple arithmetic should be considered, for no set of figures can substitute for experience. Compounding rations is both an art and a science—the art comes from cattle know-how and experience, and keen observation; the science is largely founded on chemistry, physiology, and bacteriology. Both are essential for success.

Also, a good cattleman should know how to balance a ration. Then, if the occasion demands, he can do it. Perhaps of even greater importance, he will then be able to select and buy rations with informed appraisal; to check on how well his manufacturer, dealer, or consultant is meeting his needs; and to evaluate the results.

2. Before attempting to balance a ration for cattle, the following major points should be considered:

a. *Availability and cost of the different feed ingredients*—Preferably, cost of ingredients should be based on delivery to the mill and after processing—because delivery and processing costs are quite variable. A simple method of evaluating feeds is presented in the section headed, "How to Determine the Best Buy in Feeds," beginning on page 518.

b. *Moisture content*—When considering costs and balancing rations, feeds should be placed on a comparable moisture basis; usually, an air-dry basis, or 10 percent moisture content, is used. This is especially important in the case of silage. Here's how silage may be converted to an air-dry (10 percent moisture) basis:

If silage has a moisture content of—	Divide by
(%)	
75	3.6
70	3.0
65	2.6
60	2.25

c. *Composition of the feeds under consideration*—Feed composition tables ("book values"), or average analysis, should be considered only as guides, because of wide variations in the composition of feeds. For example, the protein and moisture contents of milo, hay, and silages are quite variable. Whenever possible, especially with large operations, it is best to take a representative sample of each major feed ingredient and have a chemical analysis made of it for the more common constituents—protein, fat, fiber, nitrogen-free extract, and moisture; and often calcium, phosphorus, and carotene. Such ingredients as oil meals and prepared supplements, which must meet specific standards, need not be analyzed so often, except as quality control measures.

Despite the recognized value of a chemical analysis, it is not the total answer. It does not provide information on the availability of nutrients to the animal; it varies from sample to sample, because feeds vary and a representative sample is not always easily obtained, and it does not tell anything about the associated effects of feedstuffs—for example, the apparent way in which beet pulp enhances the value of ground milo. Nor does a chemical analysis tell anything about taste, palatability, texture, or undesirable physiological effects such as bloat and laxativeness.

However, a chemical analysis does give a solid foundation on which to start the evaluation of feeds. Also, with chemical analysis at hand, and bearing in mind that it's the composition of the total feed (the finished ration) that counts, the person formulating the ration can more intelligently determine the quantity of protein to buy, and the kind and amounts of minerals and vitamins to add.

d. *Soil analysis*—If the origin of a given feed ingredient is known, a soil analysis or knowledge of the soils of the area can be very helpful; for example, (1) the phosphorus content of soils affects plant composition, (2) soils high in molybdenum and selenium affect the composition of the feeds produced, (3) iodine- and cobalt-deficient areas are important in animal nutrition, and (4) other similar soil-plant-animal relationships exist.

e. *The nutrient allowances*—This should be known for the particular class of cattle for which a ration is to be formulated; and, preferably, it should be based on controlled feeding experiments. Also, it must be recognized that nutrient requirements and allowances must be changed from time to time, as a result of new experimental findings.

3. In addition to providing a proper quantity of feed and to meeting the protein and energy requirements, a well-balanced and satisfactory ration should be:

a. Palatable and digestible.

b. Economical. Generally speaking, this calls for the maximum use of feeds available in the area, especially forages.

c. Adequate in protein content, but not higher than is actually needed. Generally speaking, medium and high protein feeds are in scarcer supply and higher in price than high energy feeds. In this connection, it is noteworthy that the newer findings in nutrition indicate (1) that much of the value formerly attributed to proteins, as such, was probably due to the vitamins and minerals which they furnished, and (2) that lower protein content rations may be used successfully provided they are fortified properly with the needed vitamins and minerals.

d. Well fortified with the needed minerals, or free access to suitable minerals should be provided; but mineral imbalances should be avoided.

e. Well fortified with the needed vitamins.

f. So formulated as to nourish the billions of bacteria in the paunch of ruminants that there will be satisfactory (1) digestion of roughages, (2) utilization of lower quality and cheaper proteins and other nitrogenous products (thus, it is possible to use urea to constitute up to one-third of the total protein of the ration of ruminants, provided care is taken to supply enough carbohydrates and other nutrients to assure adequate nutrition for rumen bacteria), and (3) synthesis of B vitamins.

This means that rumen microorganisms must be supplied adequate (1) energy, including small amounts of readily available energy such as sugars or starches; (2) ammonia-bearing ingredients such as proteins, urea, and ammonium salts; (3) major minerals, especially sodium, potassium, and phosphorus; (4) cobalt and possibly other trace minerals; and (5) unidentified factors found in certain natural feeds rich in protein or nonprotein nitrogenous constituents.

g. One that will enhance, rather than impair, the quality of meat produced.

The above points are pertinent to the balancing of rations, regardless of the mechanics of computation used. In the sections that follow, three different methods of ration formulation are presented: (1) the square method; (2) the trial and error method; and (3) the computer method. Despite the sometimes confusing mechanics of each system, if done properly, the end result of all three methods is the same—a ration that provides the desired allowances of nutrients in correct proportions economically (or at least cost), but, more important, so as to achieve the greatest net returns—for it's net profit rather than cost per bag that counts. Since feed represents by far the greatest cost item in beef production (about 80 percent of the cost of finishing feedlot cattle, exclusive of the purchase price of the animals) the importance of balanced rations is evident.

How to Balance a Ration by the Square Method

The so-called "square method" (or the Pearson Square Method) is one of several methods that may be employed to balance rations.

The square method is simple, direct, and easy. Also, it permits quick substitution of feed ingredients in keeping with market fluctuations, without disturbing the protein content.

In balancing rations by the square method, it is recognized that protein content alone receives major consideration. Correctly speaking, therefore, it is a method of balancing the protein requirement, with only incidental consideration given to the vitamin, mineral, and other nutritive requirements.

With the instructions given herein, the square method may be employed to balance rations.

To compute balanced rations by the square method, or by any other method, it is first necessary to have available both feeding standards and feed composition tables. Several feeding standards can be and are used, and there is practically no limit to the number of nutrients that can be listed in feed composition tables.

For purposes of simplification, the author has prepared Table 8-14, Handy Beef Cattle Feeding Recommendations. Then, the crude protein content of most common feeds can be obtained from Table I-3 of the Appendix. These two tables are adequate for balancing most rations by the square method.

TABLE 8-14

HANDY BEEF CATTLE FEEDING RECOMMENDATIONS[1]

Description of Animals (1)	Recommendations[2] (2)	In Drylot, with Following Types of Forages:			On Pasture of the Following Grades:		
		Legume and/or Legume-Nonlegume Mixed Forages of High Quality; Consisting of Dry Forages and/or Silage (High Protein Forages) (3)	Legume and Non-legume Forages Mixed; Consisting of Dry Forages and/or Silage (Medium Protein Forages) (4)	Nonlegume Forage; Consisting of Dry Forages and/or Silage (Low Protein Forages) (5)	Excellent (6)	Fair to Good (7)	Poor; Including Winter Pasture Consisting of Dry Grass Cured on the Stalk[3] (8)
Mature pregnant beef breeding cows (av. wt. 1,100 lb; 500 kg). Medium and low protein forages may be used for pregnant cows.	Forage per head daily, in lb.	18-20 (8.2-9.1 kg)	18-20 (8.2-9.1 kg)	18-20 (8.2-9.1 kg)			
	Concentrate: (1) Supplement allowance of soybean meal (or equivalent 41-45% crude protein) per head daily, in lb.[4]			½-1½ (.23-.68 kg)			½-1½ (.23-.68 kg)
Mature lactating beef breeding cows (av. wt. 1,100 lb; 500 kg). When possible, use high quality, high protein forage for nursing cows.	Forage per head daily, in lb.	26 (11.8 kg)	24 (10.9 kg)	22 (10 kg)			
	Concentrate: (1) Total concentrate allowance per head daily, including protein supplement, in lb.		2½ (1.1 kg)	5 (2.3 kg)		2½ (1.1 kg)	5 (2.3 kg)
	(2) Supplement allowance of soybean meal (or equivalent 41-45% crude protein) per head daily, in lb.[4,5]		1½ (.68 kg)	3 (1.4 kg)		1½ (.68 kg)	3 (1.4 kg)
	(3) Crude protein composition of total concentrate, in %.	10-14	14-18	18-20	10-14	14-18	18-20

Footnotes on last page of table.

(Continued)

TABLE 8-14 (Continued)

Description of Animals (1)	Recommendations[2] (2)	In Drylot, with Following Types of Forages:			On Pasture of the Following Grades:		
		Legume and/or Legume-Nonlegume Mixed Forages of High Quality; Consisting of Dry Forages and/or Silage (High Protein Forages) (3)	Legume and Nonlegume Forages Mixed; Consisting of Dry Forages and/or Silage (Medium Protein Forages) (4)	Nonlegume Forage; Consisting of Dry Forages and/or Silage (Low Protein Forages) (5)	Excellent (6)	Fair to Good (7)	Poor; Including Winter Pasture Consisting of Dry Grass Cured on the Stalk[3] (8)
Replacement heifers (weighing 400-500 lb; 181-227 kg); to be bred to calve as 2-year-olds. Heifers bred to calve as 3-year-olds can be wintered at a lower level.	Forage per head daily, in lb.	12-18 (5.4-8.2 kg)	12-18 (5.4-8.2 kg)	12-18 (5.4-8.2 kg)			
	Concentrate: (1) Total concentrate allowance per head daily, including protein supplement, in lb.	2-4 (.91-1.8 kg)	2½-4 (1.1-1.8 kg)	2½-4½ (1.1-2.0 kg)			2½-4½ (1.1-2.0 kg)
	(2) Supplement allowance of soybean meal (or equivalent 41-45% crude protein) per head daily, in lb.[4, 5]		½-1 (.23-.45 kg)	1¼-1½ (.57-.68 kg)			1¼-1½ (.57-.68 kg)
	(3) Crude protein composition of total concentrate in %.	9-13 (Cereal grains only will suffice)	14-18	17-22			17-22
Stocker calves: roughed through the winter and generally grazed the following summer. Fed for winter gains of ¾ to 1 lb per head daily (weighing 400-500 lb, or 181-227 kg, start of period).	Forage per head daily, in lb.	12-18 (5.4-8.2 kg)	12-18 (5.4-8.2 kg)	12-18 (5.4-8.2 kg)			
	Concentrate: (1) Supplement allowance of soybean meal (or equivalent 41-45% crude protein) per head daily, in lb.[4]		¼-1 (.1-.45 kg)	1¼-1½ (.57-.68 kg)			1¼-1½ (.57-.68 kg)

Finishing calves (weighing 400-500 lb, or 181-227 kg, start of feeding, and 750-850 lb, or 340-386 kg, at marketing).	Forage per head daily, in lb.	2-6 (.9-2.7 kg)	2-6 (.9-2.7 kg)	2-5 (.9-2.3 kg)			
	Concentrate: (1) Total concentrate allowance per head daily, including protein supplement, in lb.	12-15 (5.4-6.8 kg)	12-15 (5.4-6.8 kg)	12-15 (5.4-6.8 kg)	10-12 (4.5-5.4 kg)	11-13 (5-5.9 kg)	12-14 (5.4-6.4 kg)
	(2) Supplement allowance of soybean meal (or equivalent 41-45% crude protein) per head daily, in lb.[4, 5]	1-1½ (.45-.68 kg)	1½-1¾ (.68-.8 kg)	1¾-2¼ (.8-1.0 kg)		1½-1¾ (.68-.8 kg)	1¾-2¼ (.8-1.0 kg)
	(3) Crude protein composition of total concentrate, in %.	9-11 (Cereal grains only will suffice)	12-13	13-15	9-11 (Cereal grains only will suffice)	12-13	13-15
Yearlings: roughed through the winter, and pasture finished the following summer. Fed for winter gains of 1 to 1¼ lb, or 0.45 to 0.57 kg, per head daily (weighing about 600 lb, or 272 kg, start of wintering).	Forage per head daily, in lb.	16-24 (7.3-10.9 kg)	16-24 (7.3-10.9 kg)	16-24 (7.3-10.9 kg)			
	Concentrate: (1) Supplement allowance of soybean meal (or equivalent 41-45% crude protein) per head daily, in lb.[4]		1-1½ (.45-.68 kg)	1½-1¾ (.68-.8 kg)			1½-1¾ (.68-.8 kg)
Finishing yearlings (weighing about 600 lb, or 272 kg, start of feeding, and 850 to 1,050 lb, or 386 to 476 kg, at marketing).	Forage per head daily, in lb.	2-8 (.9-3.6 kg)	2-8 (.9-3.6 kg)	2-8 (.9-3.6 kg)			
	Concentrate: (1) Total concentrate allowance per head daily, including protein supplement, in lb.	15-19½ (6.8-8.9 kg)	15-19¾ (6.8-9.0 kg)	15-20 (6.8-9.1 kg)	12-18 (5.4-8.2 kg)	13-19 (5.9-8.6 kg)	14-20 (8.4-9.1 kg)
	(2) Supplement allowance of soybean meal (or equivalent 41-45% crude protein) per head daily, in lb.[4, 5]	1-1½ (.45-.68 kg)	1¼-1¾ (.57-.79 kg)	1½-2½ (.68-1.1 kg)		1¼-1¾ (.57-.79 kg)	1½-2½ (.68-1.1 kg)
	(3) Crude protein composition of total concentrate, in %.	8-10 (Cereal grains only will suffice)	11-12	12-13	8-10 (Cereal grains only will suffice)	11-12	12-13

Footnotes on last page of table.

(Continued)

TABLE 8-14 (Continued)

Description of Animals (1)	Recommendations[2] (2)	In Drylot, with Following Types of Forages:			On Pasture of the Following Grades:		
		Legume and/or Legume-Nonlegume Mixed Forages of High Quality; Consisting of Dry Forages and/or Silage (High Protein Forages) (3)	Legume and Non-legume Forages Mixed; Consisting of Dry Forages and/or Silage (Medium Protein Forages) (4)	Nonlegume Forage; Consisting of Dry Forages and/or Silage (Low Protein Forages) (5)	Excellent (6)	Fair to Good (7)	Poor, Including Winter Pasture Consisting of Dry Grass Cured on the Stalk[3] (8)
Finishing long-yearling steers (weighing about 800 lb, or 363 kg, start of feeding and 1,000 to 1,100 lb, or 454 to 499 kg, at marketing).	Forage per head daily, in lb.	2-12 (.9-5.4 kg)	2-12 (.9-5.4 kg)	2-12 (.9-5.4 kg)			
	Concentrate:						
	(1) Total concentrate allowance per head daily, including protein supplement, in lb.	16-22 (7.3-10 kg)	16-22 (7.3-10 kg)	16¾-22¾ (7.5-10.3 kg)	13-19 (5.9-8.6 kg)	14-20 (6.4-9.1 kg)	15-21 (6.8-9.5 kg)
	(2) Supplement allowance of soybean meal (or equivalent 41-45% crude protein) per head daily, in lb.[4,5]	½-¾ (.23-.34 kg)	½-¾ (.23-.34 kg)	1½-1¾ (.68-.79 kg)		½-¾ (.23-.3 kg)	1½-1¾ (.68-.79 kg)
	(3) Crude protein composition of total concentrate, in %.	9-12 (Cereal grains only will suffice)	10-11	11-12	9-10 (Cereal grains only will suffice)	10-11	11-12

[1]This table was authoritatively reviewed by Dr. Robert Totusek, Department of Animal Sciences and Industry, Oklahoma State University, Stillwater, Okla.; Dr. W. M. Warren, Head, Department of Animal and Dairy Sciences, Auburn University, Auburn, Ala.; Dr. A. T. Ralston, Department of Animal Science, Oregon State University, Corvallis, Ore.; Dr. E. R. Barrick, Department of Animal Science, North Carolina State University, Raleigh, N.C.; Dr. H. B. Geurin, Director of Feed Research, W. R. Grace and Co., St. Louis, Mo.; Dr. R. I. Pick, Walnut Grove Products, Atlantic, Iowa; and Dr. W. P. Lehrer, Jr., Director, Nutrition and Research, Albers Milling Company, Los Angeles, Calif.

[2]The daily forage recommendations given herein are based on dry forage. When silage is included in the ration, figure 3 lb of silage equivalent to 1 lb of dry forage, due to the higher moisture content of silage. Many cattlemen do not winter feed as liberally as herein recommended. In general, these operators feel that it is more profitable (1) to let cattle "hold their own" or even lose in condition during the winter months (so long as they remain healthy), to keep winter feed and labor costs at a minimum, and (2) to make all or most of the gains on grass.

[3]On a dry basis, the crude protein content of mature, weathered grasses may be 3% or less. The upper limit of the concentrate allowance recommended in column 8 should be fed on winter range when (1) the grass is less abundant, and/or (2) the grass is relatively low in protein.

[4]Soybean meal, which usually ranges from 41 to 45% protein content, is herein used as a standard merely because it is the leading U.S. protein supplement. It is to be emphasized, however, (1) that other protein supplements, including numerous commercial products, may be used, (2) that, in general, those supplements should be purchased which provide a unit of protein at the lowest cost, and those feeds which are highest in protein content are usually the most economical, and (3) that where other protein feeds are substituted for the soybean meal recommended herein (41-45% protein), an equivalent amount of crude protein should be provided—for example, approximately 2 lb of a 20% crude protein supplement should be provided to replace each 1 lb of soybean meal (although it is recognized that 2 lb of a 20% protein feed will generally provide more energy, and may supply more of certain other important nutrients, than 1 lb of soybean meal).

[5]The recommended supplement allowance is based on the assumptions (1) that cereal grains, averaging 9 to 13% crude protein content, comprise the major part of the concentrate mix, and (2) that the forage is not comprised entirely or predominantly of nonlegume silage. Naturally, less protein supplement will need to be added where feeds of higher protein content than the cereal grains predominate. Also less protein supplement is required to balance a ration consisting predominantly of barley (of 12.7% crude protein content) than one consisting mostly of corn (of 8.7% crude protein content). Likewise, the upper limit of protein supplement recommended herein (or even a higher figure) is required to balance a ration where the forage is comprised entirely or largely of very low protein forages such as those that are mature and weathered.

In using Table 8-14 and Appendix Table I-3, the following points should be noted:[16]

1. Under "Description of Animals"—column 1 of Table 8-14—are sufficient groups to cover the vast majority of cattle found on the nation's farms and ranches.

2. Columns 2 to 8 give pertinent recommendations relative to both forages and concentrates. These recommendations are in keeping with those advocated by scientists, and with the actual practices followed by successful operators.

In particular, it should be noted that all protein recommendations are in terms of *crude protein* content,[17] rather than digestible protein. This was decided upon because (a) this is what the feed manufacturer wants to know as he plans a feed formula, and (b) this is what the stockman sees on the feed tag when he purchases feed.

3. It is recognized that most farmers and ranchers generally grow their own forages, and purchase part or all of the concentrates. Thus, they generally wish to know what crude protein content of concentrate alone (including grains, by-product feeds, and/or protein supplements) they need to feed to balance out the forage which is available. Likewise, feed manufacturers have need for this information in compounding mixes. For these reasons, harvested forages in Table 8-14 are classified as (a) high protein forages, (b) medium protein forages, and (c) low protein forages; and specific recommendations are made for each. Similar classifications and recommendations are made for (a) excellent, (b) fair to good, and (c) poor pastures.

4. It is often hazardous to formulate rations for excellent pastures that are different from those for poor pastures, because (a) cattlemen may be in error in appraising the quality of their pastures, and (b) pastures are generally excellent in the early spring, but become progressively poorer as the season advances unless they are irrigated and fertilized.

For purposes of illustration, let us refer to Table 8-14. Under column 5, it is noted that a mature beef breeding cow (av. wt. 1,100 lb) that is being fed a daily ration of somewhere between 18 to 20 lb of grass hay or other non-legume dry roughage should receive, in addition, ½ to 1½ lb daily of a protein supplement of soybean meal (or some other protein supplement which will provide an amount equivalent to 41 to 45 percent crude protein). To be sure, it is entirely proper to meet this recommended crude protein content of concentrate by feeding double the allowance of some protein supplement with approximately 20 percent crude protein content. Many times the latter may be more economical, and even advisable—for example, when the forage is of poor quality and added energy feed is needed. In general, however, those feeds should be purchased which furnish a unit of protein at the low-

---

[16]In addition, see pertinent footnotes which accompany Table 8-14.

[17]Also, it is recognized (1) that beef cattle consume a large proportion of forage, and (2) that the percentage digestibility of protein of forages differs tremendously—for example, the percent digestibility of protein of wheat straw is 11, whereas for alfalfa hay it is 71. On the other hand, the grains do not differ greatly in percent digestibility of protein. The National Research Council expresses digestible protein as 77.5 percent of total protein.

est cost, and those feeds which supply the protein in the most concentrated form are usually the most economical.

Under column 2 of Table 8-14, additional information, of value to both the feed manufacturer and the stockman who mixes his own rations, is given. For example, in Table 8-14, under "Finishing long yearling steers . . . ," recommendations are given relative to the following:

"(3) Crude protein composition of total concentrate, in %."

The application of the square method will be illustrated by solving some practical problems.

Problem No. 1:

A cattleman wishes to compute a balanced ration for 800-pound yearling finishing steers in drylot. Grass hay is on hand, and corn (No. 2 grade) and soybean meal (solvent process) are the cheapest concentrate feeds available. The cattleman wishes to know (1) the pounds each of forage and of concentrate to feed daily, and (2) the proportions of corn and of soybean meal to put in the concentrate mixture.

Step by step, the answers may be calculated as follows:

1. Table 8-14, Handy Beef Cattle Feeding Recommendations (column 5, Nonlegume Forage), gives the following requirements for 800-lb finishing steers in drylot:

    a. Forage per head per day = 2 to 8 lb.

    b. Concentrate per head per day = 15 to 20 lb.

    c. Crude protein content of the concentrate alone where a grass hay is fed = 12 to 13%.

2. Thus, when on full feed the steers should receive daily feed allowances of somewhere between 2 and 8 lb of the grass hay, and between 15 and 20 lb of the concentrate mixture. A range is given, because (a) individual animals and different lots of cattle differ in feed capacity, (b) feeds differ in composition and feeding value, and (c) the proportion of forage should decrease whereas the proportion of concentrate should increase as the finishing period advances.

3. The proportions of corn and of soybean meal to put in the concentrate mixture may be obtained by the square method as follows:

    a. Place in the center of the square the percentage of crude protein needed in the mixture; in this case 13% (using the upper limit).

    b. Place at the upper left-hand corner of the square the percentage of crude protein in the soybean meal; in this case 46.6 as-fed basis. (See Appendix Table I-3.)

    c. Place at the lower left-hand corner of the square the percentage of protein in the corn (maize); in this case 8.9% as-fed basis. (See Appendix Table I-3.)

    d. Connect the diagonal corners of the square with lines, and subtract, diagonally across the square, the smaller (46.6 − 13 = 33.6 and 13 − 8.9 = 4.1) figure from the larger. Place the answers at the opposite corners. This gives the following:

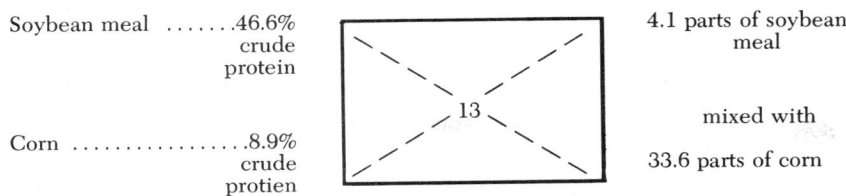

Soybean meal ......46.6% crude protein

4.1 parts of soybean meal

13

mixed with

Corn ....8.9% crude protien

33.6 parts of corn

e. Thus, a mixture of 4.1 parts of soybean meal and 33.6 parts of corn (37.7 parts total) will provide a concentrate mix with the desired 13% crude protein content.

f. The proportions of soybean meal and corn can be converted to 100 lb of mixture as follows (or to a ton basis by using 2,000 instead of 100):

$$\frac{4.1 \text{ parts soybean meal}}{37.7 \text{ (total of soybean meal and corn)}} \times 100 = 10.88 \text{ lb soybean meal}$$

$$\frac{33.6 \text{ parts corn}}{37.7 \text{ (total of soybean meal and corn)}} \times 100 = 89.12 \text{ lb corn}$$

Total . . . . . . 100.00 lb mix

g. Thus, to make a 13% crude protein concentrate mixture from corn and soybean meal, the cattleman will need to mix 10.88 lb of soybean meal and 89.12 lb of corn for each 100 lb of mix.

Problem No. 2:

A cattle feeder is planning to finish long yearling steers in the drylot. He has on hand corn silage and alfalfa hay, but wishes to purchase the entire concentrate mixture from a commercial feed manufacturer. What percent crude protein content (as shown on the feed tag) is needed in the commercial concentrate?

This can be readily and directly answered simply by referring to Table 8-14, Handy Beef Cattle Feeding Recommendations. As noted in Table 8-14 (column 4), where corn silage and a legume hay constitute the roughage ration, 10 to 11 percent crude protein concentrate is recommended.

How to Balance a Ration by the Trial and Error Method

As stated, balancing rations by the square method is simple, direct, and easy, but protein alone receives major consideration. Balancing by the trial and error method goes further. In it, consideration is given to meeting whatever allowances are decided upon of each of the nutrients that one cares to list and consider. Here are the steps in the trial and error method:

Set down the allowances—Refer to Tables 8-18 and 8-19, but bear in mind that they list requirements, and not allowances. Requirements do not provide for margins of safety to compensate for variations in feed composition, and possible losses during storage and processing. Nor do they recognize that the needs of cattle do not necessarily remain the same from day to day or from period to period; that age and size of animal, stage of gestation and lactation, the kind and degree of activity, and the amount of stress; the system of management; the health, condition, and temperament of the animal; and the kind, quality, and amount of feed are all exerting a powerful influence in determining nutritive needs. How well the stockman understands, anticipates, interprets, and meets these variable requirements usually determines the success or failure of the ration and the results obtained. Because the effects of each of these factors on nutritive needs vary in degree, and so little is known about many of them, the author suggests that each cattleman or feed company arrive at the nutritive allowances for a specific class of cattle and set of conditions as follows:

1. *Base protein and energy allowances on Tables 8-18 and 8-19*, bearing in mind that the figures given therein are requirements, and not allowances. Considering all the factors affecting nutritive needs, come to a decision as to what margin of safety, if any, to provide over and above the protein and energy requirements given in Tables 8-18 and 8-19; then add it to the requirements, whatever it may be—0, 10 percent, 20 percent, or some other level.

2. *Obtain mineral allowances from Table 8-8.*

3. *Obtain vitamin allowances from Table 8-12.*

4. *Set down the allowances*—Simply list them, as a result of following steps 1, 2, and 3 above. Thus, for finishing 660-pound yearling steers in drylot, gaining 2.9 lb/head/day, the feed allowances are:

Daily Feed	Proportion Concentrate and Roughage[18]	Crude Protein	TDN	Calcium	Phosphorus
(lb)		*(%)*	*(%)*	*(%)*	*(%)*
15.6	85% Concentrate 15% Roughage	11.7[19]	83	0.41	0.32

5. *Apply the trial and error method*—Considering (a) available feeds, and (b) common feeding practices, the next step is arbitrarily to set down a ration and see how well it measures up to the desired allowances. The approximate composition of the available feeds may be arrived at from feed composition tables (Appendix Table I-3) if an actual chemical analysis is not available. Where commercial supplements are used, the guarantee on the feed tag may be used.

Let's try the following ration[20]:

Ingredients	Percent	Per Ton
	(%)	*(lb)*
Corn (No. 2)	69.7	1,394
Alfalfa hay (mid bloom)	5.0	100
Cottonseed hulls	10.0	200
Molasses (sugarcane)	10.0	200
Soybean meal (46.6%)	5.0	100
Salt	.3	6
TOTAL	100.0	2,000

[18]The proportion of roughage to concentrate is an arbitrary decision, which cattlemen must make.

[19]See Table 8-19. 0.4% added to protein requirements, as a margin of safety.

[20]All calculations are on as fed basis.

Here's a listing of the desired allowances (step No. 4), followed by the composition of the proposed ration[21]:

	Daily Feed	Conc. to Roughage	Crude Protein	TDN	Calcium	Phosphorus
	(lb)		(%)	(%)	(%)	(%)
Desired allowances	15.6	85% Conc. 15% Rough.	11.7	83	0.41	0.32
Approx. analysis of proposed ration	15.6	85% Conc. 15% Rough.	10.15	73.61	0.19	0.28

Thus, the proposed ration is low in protein (1.55% under), low in TDN (9.39% under), low in calcium (0.22% under), and slightly low in phosphorus. To correct these deficiencies, let's go to a 90% concentrate and 10% roughage ration, increase the corn, alfalfa, and soybean meal, delete the cottonseed hulls and molasses, and add fat and ground limestone. Thus, our second trial ration is:

Ingredients	Percent	Per Ton
	(%)	(%)
Corn (No. 2)	77	1,540
Alfalfa hay (mid bloom)	10	200
Fat (TDN = 233)	5	100
Soybean meal (46.6%)	7	140
Dicalcium phosphate	0.7	10
Salt	0.3	10
TOTAL	100.0	2,000

	Daily Feed	Con. to Roughage	Crude Protein	TDN	Calcium	Phosphorus
	(lb)		(%)	(%)	(%)	(%)
Desired allowances	15.6	90% Conc. 10% Rough.	11.7	83	0.41	0.32
Approx. comp. of 2nd. prop. allowances	15.6	90% Conc. 10% Rough.	11.7	84	0.40	0.39

This ration approximates the desired allowances and may be considered satisfactory.

How to Balance a Ration by the Computer Method[22]

Many large cattle feedlots, and most feed companies, now use computers for ration formulation as well as for other purposes; and their use will increase.

Despite their sophistication, there is nothing magical or mysterious about balancing rations by computer. Although they can alleviate many human errors in calculations, the data which come out of a machine are no

---

[21]Same as footnote 20.

[22]This section was authoritatively reviewed by IBM's specialist in the beef cattle area, namely, Mr. M. E. Smith, International Business Machine Corporation, Chicago, Ill.; and Professor Clyde L. Hart, Computer Science Technology, Texas State Technical Institute, Waco, Tex.

better than those which went into it; without a man, they don't know the difference between a Doberman and a Hereford. The men back of the computer—the cattleman and his nutritionist who prepare the data that go into it, and who evaluate and apply the results that come out of it—become more important than ever. This is so because an electronic computer doesn't know anything about (1) feed palatability, (2) bloat prevention, (3) limitations that must be imposed on certain feeds to obtain maximum utilization, (4) the goals in the feeding program—such as growing or finishing, (5) home-grown feeds for which there may not be a suitable market, (6) feed processing and storage facilities, (7) the health, environment, and stress of the cattle, and (8) the men responsible for actual feed preparation and feeding. Additionally, it must be recognized that a computer may even reflect, without challenge, the prejudices and whims of those who prepare the data for it.

Hand in hand with the use of computers in balancing rations, the term "least cost ration formulation" evolved. In some respects this designation was unfortunate, for the use of least cost rations does not necessarily assure the highest net returns—and net profit is more important than cost per ton. For example, the least cost ration may not produce the desired daily gain or carcass quality.

An electronic computer can do little more than a good mathematician can do, but it can do it a lot faster and it can check all possible combinations. It alleviates the endless calculations and many hours of time required for hand calculations. For example, it is estimated that there may be as many as 500 practical solutions when 6 quality specifications and 10 feedstuffs are considered for a ration.

Generally speaking, electronic feed formulation (1) effects a greater saving when first applied to a ration than in subsequent applications, and (2) is of most use where a wide selection of feed ingredients is available and/or prices shift rather rapidly.

STEPS IN BALANCING A RATION BY COMPUTER

The information needed and the procedure followed in formulating rations by computer are exactly the same as in the hand-method of ration formulation; namely, (1) the nutritive requirements for the particular class and kind of animal, (2) nutritive content of the feeds, and (3) ingredient costs. Sometimes this simple fact is overlooked because of the awesomeness of the computer, and the jargon used by those who wish to impress fellow scientists. Step by step, the procedure in formulating rations by computer is:

1. *List available feed ingredients, and the cost of each*—It is necessary that all of the available feeds be listed along with the unit cost (usually/ton) of each; preferably, ingredient cost should be based on market price plus delivery, storage, and processing cost.

2. *Record quality of feed*—The more that is known about the quality of feed the better. This is so because of the wide variation in composition and feeding value within ingredients; for example, between two samples of al-

falfa hay. See Appendix of this book for a qualitative description of each feed ingredient.

Whenever possible, an actual chemical analysis of a representative sample of each ingredient under consideration, rather than "book values," should be available and used. However, the imperfections of a chemical analysis of a feedstuff should be recognized; chiefly, (a) it does not provide information on the availability of nutrients to animals, and (b) there are variations between samples.

3. *Establish ration specifications*—Set down the ration specifications— the nutrients and the levels of each that are to be met. This is exactly the same procedure as is followed in the hand method. (See "How to Balance a Ration by the Trial and Error Method," step 4.) In arriving at ration specifications, the nutritionist considers the following: (a) age, weight, and grade of cattle; (b) length of feeding period; (c) the probable market; (d) season of year; (e) background and stress of animals; and (f) other similar factors.

4. *Give restrictions*—Usually it is necessary to establish certain limitations on the use of ingredients; for example, (a) the maximum amount of roughage, (b) the maximum amount of urea, (c) the minimum and maximum amounts of fat, (d) the proportion of cottonseed hulls to alfalfa hay, (e) the proportion of one grain to another—such as 60% barley and 40% milo, (f) an upper limit of some ingredients—such as 20% rye, (g) the exact amount of the premix, and (h) the lower and upper limits of molasses, as between 5 and 10%.

It must be recognized that the more narrow the limitations imposed on the computer, the less the choice it will have in ration formulation and the higher the cost.

5. *Stipulate feed additives*—Generally speaking, the nutritionist makes rigid stipulations as to amounts of these ingredients, much as he does with added vitamins and minerals. All of them cost money, and many of them must be used in compliance with the Food and Drug Administration regulations.

6. *Obtain program*—Take data, ration specifications, and restrictions to an experienced computer programmer or systems analyst. He will either (a) "tailor-make" a program for a given situation, or (b) suggest one of the "canned" linear programs available from the larger computer companies. The canned programs are, by necessity, general in nature, because they are written for a wide variety of applications; but they cost less than a tailor-made program.

7. *Put data on cards*—The data obtained from steps 1 through 5 (above) must be punched into standard punch cards (unless key-to-tape or key-to-disk equipment is available), and verified by competent keypunchers or verifiers.

8. *"Feed" the punched cards into the computer*—When the punched cards or other input media are fed into the computer, it treats the data as one gigantic algebra problem and arrives at the ration formulation in a matter of minutes. Based on available feeds, analysis, and price, the computer evolves with the mix that will meet the desired nutritive allowances at the least possible cost.

9. *Formulate as necessary*—All rations should be reviewed at frequent intervals, and reformulated when there are shifts in (a) availability of ingredients (certain ingredients may no longer be available, but new ones may have evolved), (b) price, and/or (c) chemical composition.

10. *Validate the restrictions*—That is, test or confirm them.

USE OF COMPUTER AS FEED BUYING AND SELLING AID

When some computer programs formulate a ration, they also give a complete set of "shadow prices," which may be used as follows:

1. If a certain ingredient does not enter the formula due to its chemical analysis as related to price, the shadow price will indicate how much the market price of this ingredient must go down in order for it to enter the formula. For example, if soybean meal is selling for $175 per ton and the shadow price is $143, soybean meal will not be included in the formula until the price goes down to, or falls below, $143; unless, of course, a minimum amount of soybean meal is specified.

2. If an ingredient is home-grown and on hand, and the feeder desires to use it, despite the fact that its market value is out of line, the shadow price will indicate to him the penalty that he will pay for using it. Sometimes, it may become obvious that it is good business to sell a certain home-grown product and buy something else to replace it.

3. The shadow price provides a technique for determining the value of each ingredient based on its chemical analysis, thereby making it possible to determine which ingredient is the best buy. As a result, it is an excellent management tool for buying and selling feed ingredients.

4. The shadow price is used in price mapping (the range of prices over which an ingredient will stay in the formula). By considering one ingredient at a time, this is an excellent buying guide.

5. The shadow price is used in determining the cost of restrictions; that is, the decrease in price that will occur if a restriction is released.

FEED ALLOWANCE AND SOME SUGGESTED RATIONS[23]

Some general rules of feeding may be given, but it must be remembered that *"the eye of the master fattens his cattle."* Nevertheless, the beginner may well profit from the experience of successful feeders. It is with this hope that the suggested rations are herewith presented.

Table 8-15 (see pages 514 and 515) is a handy beef cattle feeding guide for different classes and ages of cattle.[24] All of these are merely intended as general guides. Variations can and should be made in the rations used. The feeder should give consideration to (1) the supply of home-grown feeds, (2)

---

[23]Insofar as possible, these rations were computed from the requirements as reported by the National Research Council and applied by the author.

[24]Recommendations relative to feeding show animals are included in Chapter 14 of this book.

the availability and price of purchased feeds, (3) the class and age of cattle, (4) the health and condition of the animals, and (5) the length of the grazing season.

In using Table 8-15 as a guide, it is to be recognized that feeds of similar nutritive properties can and should be interchanged as price relationships warrant. Thus, (1) the cereal grains may consist of corn, barley, wheat, oats, and/or sorghum; (2) the protein supplement may consist of soybean, cottonseed, peanut, sunflower, and/or linseed meal; (3) the roughage may include many varieties of hays and silages; and (4) a vast array of by-product feeds may be utilized.

FEED TO FOOD EFFICIENCY

Table 8-16 points up the continued need for greater efficiency of animal production, the urgency of which becomes altogether too apparent in light of population increases and impending world food shortages. Even the dairy cow, the most efficient of all animal producers, is only (1) 26 percent efficient in converting the energy of feed to food, and (2) 34 percent efficient in converting protein of feed to food. However, there is a wide gap between milk cows and layers, with the former having a 33.4 point higher total score than the latter. Table 8-16 also reveals that the beef cow is the least efficient of all feed converters (requiring 10 lb of feed to produce 1 lb of meat), and that only sheep are lower on the totem pole from the standpoint of protein conversion. Granted, feed is only one factor responsible for the productivity rating given in Table 8-16; other environmental factors and heredity play a part. Nevertheless, feed is a costly and important item, and, historically, hungry people throughout the world have been forced to eliminate some animals and to consume grains directly to avoid famine.

In recent years, science and technology have teamed up and made for great strides in improving the productive efficiency of poultry. No doubt, further progress with all animals lies ahead. Also, as increasing quantities of cereal grains are needed for human consumption, ruminants will utilize higher proportions of roughages to concentrates.

TAI

HANDY BEEF CATT

SUGGESTED RATIONS With all rations and for all classes and ages of cattle, provide free access in separate containers to (1) salt (iodized salt in iodine-deficient areas), and (2) a suitable mineral mixture	Wintering mature pregnant beef breeding cows (av. wt. 1,100 lb or 499 kg)		Wintering mature lactating beef breeding cows (av. wt. 1,100 lb or 499 kg)		Wintering repl ment heifers (we ing 400 to 500 l 181 to 227 kg sta wintering)	
	Per Day		Per Day		Per Day	
	(lb)	(kg)	(lb)	(kg)	(lb)	(kg)
1. Legume hay or grass-legume mixed hay, good quality	18 -20	8.2 - 9.1	30	13.6	13 -15[4]	5.9 -
Grain					2 - 3	.91-
Protein supplement						
2. Grass hay or other nonlegume dry roughage	18 -20	8.2 - 9.1	24-26	10.9-11.8	12 -18[4]	5.4 -
Grain			2	.91	2½- 4½	1.13-
Protein supplement	½- 1	.23- .45	3	1.36	1¼- 1½	.57-
3. Legume hay or grass-legume mixed hay, good quality	7 -11	3.2 - 5.0	26-28	11.8-12.7	8 -12[4]	3.6 -
Grass hay or other nonlegume dry roughage	9 -11	4.1 - 5.0			4 - 6	1.8 -
Grain			1	.45	2½- 4	1.13-
Protein supplement			1	.45	½- 1	.23-
4. Corn or sorghum silage	50 -55	22.7 -25	55	25	25 -40	11.3 -18
Grain			2	.91		
Protein	0 - ½	0 - .22	3	1.36	1½- 1¾	.68-
5. Grass silage, half or more legume	50	22.7	50	22.7	25 -40	11.4 -18
Grain			4	1.81	3 - 4	1.36- 1
Protein supplement					½	.23
6. Silage (corn or sorghum silage fed with legume hay or legume silage fed with grass hay)	35	15.9	40	18.1	15 -30	6.8 -13
Hay	5 - 6	2.3 - 2.7	10	4.5	3 - 4	1.4 - 1
Grain					1 - 2	.45- 0
Protein supplement	0 - ½	.23			½- 1	.22-

[1]This table was authoritatively reviewed by Dr. Robert Totusek, Department of Animal Sciences and Industry, Oklahoma State University, Stillwater, Okla.; Dr. W. M. Warren, Head, Department of Animal and Dairy Sciences, Auburn University, Auburn, Ala.; Dr A. T. Ralston, Department of Animal Science, Oregon State University, Corvallis, Ore.; Dr. E. R. Barrick, Department of Animal Science, North Carolina State University, Raleigh, N.C.; Dr. H. B. Geurin, Director of Feed Research, W. R. Grace and Co., St. Louis, Mo.; Dr. R. I. Pick, Walnut Grove Products, Atlantic, Iowa; and Dr. W. P. Lehrer, Jr., Director, Nutrition and Research, Albers Milling Company, Los Angeles, Calif.

[2]If stocker calves are late or the roughage is fair to poor in quality, it may be desirable to add 2-4 lb (.91-1.81 kg) of grain per head daily. If farm scales are available, monthly weights may be used as the criterion for grain feeding. Keep in mind that the calves should

:DING GUIDE[1]

ring stocker calves ed through winter razed the following ...er. Fed for winter of ¾ to 1 lb (.34 to g) per head daily hing 400 to 500 lb to 227 kg start of ring)[2]		Finishing calves in drylot, generally in winter. (Weighing 400 to 500 lb or 181 to 227 kg start of feeding and 750 to 850 lb or 340 to 386 kg at marketing)[3]		Wintering yearlings; roughed through the winter, and generally pasture finished the following summer. Fed for winter gains of 1 to 1¼ lb or .45 to .57 kg per head daily (weighing about 600 lb or 272 kg start of wintering)		Finishing yearlings in drylot, generally in winter (weighing about 600 lb or 272 kg start of feeding, and 900 to 1,050 lb or 409 to 477 kg at marketing)[3]		Finishing long-yearling steers in drylot generally about 850 lb or 386 kg start of feeding and 1,000 to 1,100 lb or 454 to 499 kg at marketing)[3]	
Per Day		Per Day		Per Day		Per Day		Per Day	
(lb)	(kg)	(lb)	(kg)	(lb)	(kg)	(lb)	(kg)	(lb)	(kg)
-18[4]	5.4 - 8.2	4 -6	1.8 -2.7	16 -24	7.2 -10.9	4 - 8	1.8 - 3.6	6 -12	2.7 - 5.4
		12 -15	5.4 -6.8			15 -19½	6.8 - 8.8	16 -22	7.2 -10.0
		1 - 1½	.45- .68			1 - 1½	.45- .68		
-18[4]	5.4 - 8.2	4 - 5	1.8 -2.3	16 -24	7.2 -10.9	4 - 8	1.8 - 3.6	6 -12	2.7 - 5.4
- 1½	.57- .68	12 -15	5.4 -6.8	1½- 1¾	.68- .79	15 -20	6.8 - 9.1	16½-22¾	7.5 -10.3
		1¾- 2	.79- .91			1½- 2½	.68- 1.1	1½- 1¾	.68- .79
-12[4]	5.4 - 8.2	2 - 3	.91-1.36	6 - 8	2.7 - 3.6	2 - 8	.91- 1.81	3 - 6	1.4 - 2.7
- 6	1.8 - 2.7	2 - 3	.91-1.36	10 -16	4.5 - 7.2	2 - 4	.91- 1.81	3 - 6	1.4 - 2.7
		12 -15	5.4 -6.8	1 - 1½	.45- .68	15 -19¾	6.8 - 9.0	16 -22	7.2 -10.0
- 1	.11- .45	1½- 1¾	.68- .79			1¼- 1¾	.57- .79	½- ¾	.23- .34
-40	11.3 -18.2	6 -16	2.7 -7.2	40 -55	18.2 -24.9	6 -25	2.7 -11.4	6 -35	2.7 - 5.9
		8 -12	3.6 -5.4			11 -16	5.0 - 7.2	15 -21	6.8 - 9.5
- 1¼	.45- .57	2	.91	1¼- 1½	.57- .68	2	.91	1¼- 1½	.57- .68
-40	11.3 -18.2	6 -16	2.7 -7.2	40-55	18.1 -24.9	6 -25	2.7 -11.3	6 -35	2.7 -15.9
- 3	.90- 1.36	8 -12	3.6 -5.4	4 - 5	1.8 - 2.3	11 -16	5.0 - 7.3	15 -21	6.8 - 9.5
½	.23	1 - 2	.45- .91	½	.23	1 - 1½	.45- .68	1	.45
-30	6.8 -13.6	3 - 8	1.4 -3.6	20 -35	9.1 -15.9	3 -15	1.4 - 6.8	3 -15	1.4 - 6.8
- 4	1.4 - 1.8	1 - 3	.45-1.4	7	3.2	1 - 4	.45- 1.8	1 - 7	.45- 3.2
- 2	.45- .91	8 -12	2.6 -5.4			11 -16	5.0 - 7.2	15 -21	6.8 - 9.5
½	.23	1 - 2	.45- .91	½- ¾	.23- .34	1 - 1¾	.45- .79	1 - 1¼	.45- .57

₂ ¾-1 lb (.34-.45 kg) daily.

[3]In general, the experienced feeder plans that cattle on full feed shall consume (1) feeds in amounts (daily: air-dry basis) equal to it 2.5-3.0% of their liveweight, (2) 70-90% concentrates, and (3) a minimum of 2.0-4 lb (.9-1.8 kg) roughage for each 100 lb (45 kg) weight. In areas where roughage is more abundant and comparatively cheaper than grain, the proportions of roughage to grain ild be somewhat higher than indicated. In computing roughage consumption, 3 lb (1.36 kg) of silage are considered equivalent to lb (.45 kg) of hay.

[4]With calves (both replacement heifers and stockers) an extra 2 lb (.91 kg) of hay daily, over and above requirements, are herewith cated to allow for wastage. Practical operators generally feed stemmy or other hay left over by claves to the cow herd.

TABLE 8-16

FEED TO FOOD EFFICIENCY RATING BY SPECIES OF ANIMALS
(Based on Energy as TDN or DE and Crude Protein in Feed Eaten by Various Kinds of Animals Converted into Calories and Protein Content of Ready-to-Eat Human Food)

Species	Unit of Production (on foot)	Feed Required to Produce One Production Unit[1]				Dressing Yield		Ready-to-eat; Yield of Edible Product (meat and fish deboned and after cooking)				Efficiency Rating		Total Score	Rank
		Pounds	TDN	DE[2]	Protein	Percent	Net Left	As % of Raw Product (carcass)	Amount Remaining from One Unit of Production	Calorie[3]	Protein[3]	Calorie Efficiency[4]	Protein Efficiency[5]	(Calories + Protein)	(Calories & Protein)
		(lb)	(lb)	(kcal)	(lb)	(%)	(lb)	(%)	(lb)	(kcal)	(lb)	(%)	(%)		
Dairy cow .... 1 lb milk		.85[6]	.60	1,200	.11	100	1.0	100	1.0	309	.037	25.8	33.6	59.4	1
Fish .... 1 lb fish		1.67	.90	1,800	.57	65[8]	.65	57[9]	.37	285	.093	15.8	16.3	32.1	2
Layer .... 1 lb eggs (8 eggs)		4.0[10]	2.96	5,920	.68	100	1.0	100[11]	1.0[11]	616	.106	10.4	15.6	26.0	3
Broiler .... 1 lb chicken		3.0[10]	2.37	4,740	.66	72[12]	.72	54[13]	.39	274	.11	5.8	16.7	22.5	4
Turkey .... 1 lb turkey		5.2[10]	3.95	7,900	1.19	79.7[12]	.797	57[14]	.45	446	.146	5.6	12.3	17.9	5
Hog (birth to 200 lb) ...... 1 lb pork		4.9[6]	3.67	7,340	.69	70[15]	.70	44[16]	.31	341	.088	4.6	12.7	17.3	6
Beef steer (yearling finishing period in feedlot) .... 1 lb beef		10.0[7]	6.50	13,000	1.00	58[15]	.58	49[16]	.28	342	.085	2.6	8.5	11.1	7
Lamb (finishing period in feedlot) .... 1 lb lamb		9.0[8]	5.58	11,160	.96	47[15]	.47	40[16]	.19	225	.052	2.1	5.4	7.5	8

[1]Includes provision for body maintenance, for the reason that feed energy must be expended in body maintenance before there can be any production; for example, the dairy cow must be maintained before there can be any milk production.

[2]Digestible Energy (DE) in this column given in kcal, which is 1 Calorie (written with a capital C), or 1,000 calories (written with a small c). Kilocalories computed from TDN values in column to immediate left as follows: 1 lb TDN=2,000 kcal.

[3]From Lessons on Meat, 1965. National Live Stock and Meat Board.

[4]Kilocalories in ready-to-eat food=kilocalories in feed consumed, converted to percentage.

[5]Protein in ready-to-eat food=protein in feed consumed, converted to percentage.

[6]Computations made by the author.

[7]Data from Feedstuffs, April 15, 1967, report by Dr. Phillip J. Schaible, Michigan State University.

[8]Industrial Fishery Technology, ed. by Maurice E. Stansby, Reinhold Pub. Corp., 1963, Ch. 26, Table 26-1.

[9]Ibid. Reports that, "Dressed fish averages about 73% flesh, 21% bone, and 6% skin." In limited experiments conducted by A. Ensminger, it was found that there was a 22% cooking loss on filet of sole. Hence, these values—73% flesh from dressed fish, plus 22% cooking losses—give 57% yield of edible fish after cooking, as a percent of the raw, dressed product.

[10]Handbook of Agriculture Charts 1965, Agric. Handbook No. 300, p. 58, USDA, Oct., 1965.

[11]Calories and protein computed basis per egg; hence, the values herein are 100% and 1.0 lb, respectively.

[12]Marketing Poultry Products, p. 147.

[13]Factors Affecting Poultry Meat Yields, Univ. of Minn. Sta. Bul. 476, Table 11 (fricassee), p. 29, 1964.

[14]Ibid. Table 10, p. 28, 1964.

[15]The Stockman's Handbook, 4th Ed., Sec. XII

[16]Allowance made for both cutting and cooking losses following dressing. Thus, values are on a cooked, ready-to-eat basis of lean and marbled meat, exclusive of bone, gristle, and fat. Values provided by National Live Stock and Meat Board (personal communication of June 5, 1967, from Dr. Wm. C. Sherman, Director, Nutrition Research to the author; and based on data from The Nutritive Value of Cooked Meat, by Ruth M. Leverton and George V. Odell, Misc. Pub. MP-49, Appendix C, March 1958).

[17]Beef Cattle Science, 4th Ed., Ch. 14, Table 14-21.

[18]Sheep and Wool Science, 3rd Ed., p. 250.

STATE COMMERCIAL FEED LAWS

Nearly all the states have laws regulating the sale of commercial feeds. These benefit both cattlemen and reputable feed manufacturers. In most states the laws require that every brand of commercial feed sold in the state be licensed, and that the chemical composition be guaranteed.

Samples of each commercial feed are taken each year, and analyzed chemically in the state's laboratory to determine if the manufacturer lived up to his guarantee. Additionally, skilled microscopists examine the sample to ascertain that the ingredients present are the same as those guaranteed. Flagrant violations on the latter point may be prosecuted.

Results of these examinations are generally published, annually, by the state department in charge of such regulatory work. Usually, the publication of the guarantee alongside any "short-changing" is sufficient to cause the manufacturer promptly to rectify the situation, for such public information soon becomes known to both users and competitors.

The sections that follow brief some of the pertinent information required on the label of commercial feeds by most states.

• *Medicated Feed Tags and Labels*—Medicated feeds (those which contain drug ingredients intended or represented for the cure, mitigation, treatment, or prevention of diseases of animals) must also carry the following information in their labeling: (1) the purpose of the medication; (2) directions for the use of the feed; (3) the names and amounts of all active drug ingredients; (4) a warning or caution statement for a withdrawal period prior to marketing when required for a particular drug; and (5) warnings against misuse.

• *Vitamin Product Labels*—When a product is marketed as a vitamin supplement *per se*, the quantitative guarantees (unit/lb) of vitamins A and D are expressed in USP units; of E in IU; and of other vitamins in milligrams per pound.

• *Mineral Product Labels*—Some states require that all minerals except salt (NaCl) be quantitatively guaranteed in terms of percentage of the element(s); others require milligrams per pound.

• *Other Rules and Regulations*—Generally, the following rules and regulations also apply in the different states:

1. The brand or product name must not be misleading.

2. The sliding scale or range (for example, 15% to 18% crude protein) method of expressing guarantees is prohibited.

3. Ingredient names are those adopted by the Association of American Feed Control Officials.

4. The term "dehydrated" may precede the name of any product that has been artificially dried.

5. Urea and other nonprotein nitrogen products are acceptable ingredients for ruminant animals only.

• *Terms Used in Analyses and Guarantees*—Knowledge of the following terms is requisite to understanding analyses and guarantees:

Dry matter is found by determining the percentage of water and subtracting the water content from 100 percent.

Crude protein is used to designate the nitrogenous constituents of a feed. The percentage is obtained by multiplying the percentage of total nitrogen by the factor 6.25. The nitrogen is derived chiefly from complex chemical compounds called amino acids.

Crude fat is the material that is extracted from moisture-free feeds by ether. It consists largely of fats and oils with small amounts of waxes, resins, and coloring matter. In calculating the heat and energy value of the feed, the fat is considered 2.25 times that of either nitrogen-free extract or protein.

Crude fiber is the relatively insoluble carbohydrate portion of a feed consisting chiefly of cellulose. It is determined by its insolubility in dilute acids and alkalies.

Ash is the mineral matter of a feed. It is the residue remaining after complete burning of the organic matter.

Nitrogen-free extract consists principally of sugars, starches, pentoses and nonnitrogenous organic acids. The percentage is determined by subtracting the sum of the percentages of moisture, crude protein, crude fat, crude fiber, and ash from 100.

Carbohydrates represent the sum of the crude fiber and nitrogen-free extract.

Calcium and phosphorus are essential mineral elements that are present in feeds in varying quantities. Mineral feeds are usually high in source materials of these elements.

TDN—The digestible nutrients of any ingredient are obtained by multiplying the percentage of each nutrient by the digestion coefficient. For example, dent corn contains 10 percent protein of which 77 percent is digestible. Therefore, the percent of digestible protein is 7.7.

The TDN is the sum of all the digestible organic nutrients—protein, fiber, nitrogen-free extract, and fat (the latter multiplied by 2.25).

In buying feeds, the cattleman should check prices against values.

HOW TO DETERMINE THE BEST BUY IN FEEDS

Feed prices vary widely. For profitable production, therefore, feeds with similar nutritive properties should be interchanged as price relationships warrant.

In buying feeds, the cattleman should check prices against values received. This may be done by computing the cost per pound of protein and TDN. The use of this method can best be illustrated by the examples that follow:

If 44% protein (crude) soybean meal is selling at $6.00 per 100 lb whereas 35% protein (crude) linseed meal sells for $5.00 per 100 lb, which is the better buy? Divide $6.00 by 44 to get 13.6c per lb of crude protein for the soybean meal. Then divide $5.00 by 35 and get 14.3c per lb for the linseed meal. Thus, at these prices soybean meal is the better buy—by 0.7c per lb of crude protein.

When buying energy feed, one can compare the cost per lb of total digestible nutrients (TDN). For example, if corn is priced at $4.00 per 100 lb and has a TDN of 91%, divide $4.00 by 91 and the result is 4.44c per pound of TDN.

If barley with 83% TDN sells for $3.85 per 100 lb, divide $3.85 by 83, and the price is 4.64c per lb of TDN. Thus, corn would be the better buy by 0.2c per lb of TDN.

Of course, it is recognized that many other factors affect the actual feeding value of each feed, such as (1) palatability, (2) grade of feed, (3) preparation of feed, (4) ingredients with which each feed is combined, and (5) quantities of each feed fed.

NUTRITIONAL DISEASES AND AILMENTS

More animals (and people) throughout the world suffer from hunger—from just plain lack of sufficient feed—than from the lack of one or more specific nutrients. Therefore, it is recognized that nutritional deficiencies may be brought about by either (1) too little feed, or (2) rations that are too low in one or more nutrients.

Also, forced production (such as very high milk yields and finishing animals at early ages) and the feeding of forages and grains which are often produced on leached or depleted soils have created many problems in nutrition. This condition has been further aggravated through the increased confinement of stock, many animals being confined to lots or buildings all or a large part of the year. Under these unnatural conditions, nutritional diseases and ailments have become increasingly common.

Although the cause, prevention, and treatment of most of these nutritional diseases and ailments are known, they continue to reduce profits in the livestock industry simply because the available knowledge is not put into practice. Moreover, those widespread nutritional deficiencies which are not of sufficient proportions to produce clear-cut deficiency symptoms cause even greater economic losses because they go unnoticed and unrectified. Table 8-17 contains a summary of the important nutritional diseases and ailments affecting cattle.

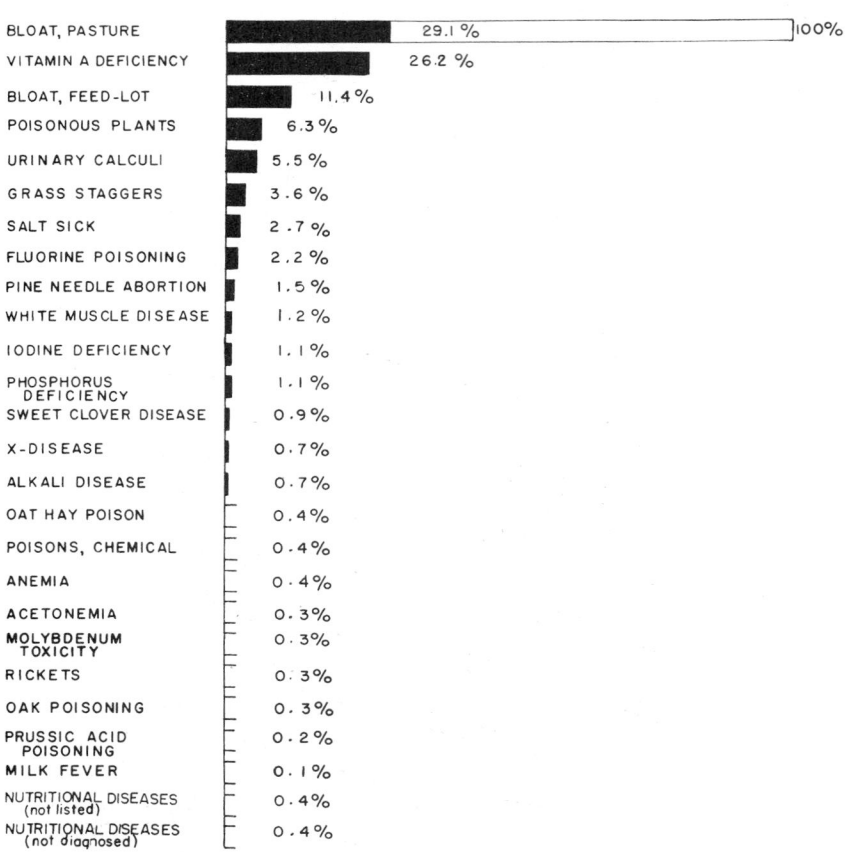

Fig. 8-25. Relative frequency of beef cattle nutritional diseases and ailments. This figure, taken from Wash. Agr. Exp. Sta. Bull. 562, is based on a 24-state survey made by the author which covered over a half million cattle. It is recognized that cattlemen may not always have been accurate in their diagnosis; nevertheless, this is the most complete record of its type available.

Bloat accounted for 40.5% of all nutritional diseases and ailments; vitamin A deficiency ranked second. The survey also brought out some area difference: vitamin A deficiency and poisonous plant losses were relatively higher in the West; bloat was higher in the South; urinary calculi was higher in the Great Plains; and white muscle disease was higher in the Pacific Northwest.

During the course of a year, 1.33% of the nation's cattle were afflicted by nutritional deficiency diseases and ailments, of which 0.32% died therefrom. Of the death losses due to nutritional deficiency diseases and ailments, 36% were attributed to bloat, 22% to poisonous plants, and 13% to urinary calculi.

Fig. 8-26. Bloat. There is complete absence of bloat in identical twin on right given 10 g of poloxalene before being fed freshly cut alfalfa. Twin mate on the left received no poloxalene and is bloating. (Courtesy, Dr. Erle E. Bartley, Department of Dairy and Poultry Science, Kansas State University, Manhattan, Kan.)

Fig. 8-27. Bloat. When the cap on the rumen plug is removed, frothy rumen contents are evacuated with considerable force. (Courtesy, Dr. Erle E. Bartley, Department of Dairy and Poultry Science, Kansas State University, Manhattan, Kan.)

TABLE 8-17—NUTRITIONAL DISEA

Disease	Cause	Symptoms (and Age or Group Most Affected)	Distribution a Losses Caused
Acetonemia in Cattle (See Ketosis)			
Alkali Disease (See Selenium poisoning)			
Anemia, nutritional	Commonly an iron deficiency, but it may be caused by a deficiency of copper, cobalt, and/or certain vitamins.	Loss of appetite, progressive emaciation, and death. Most prevalent in suckling young.	Worldwide. Losses consist of and inefficient and deaths.
Aphosphorosis	Low phosphorus in feed.	Depraved appetite; chewing bones, wood, hair, rags, etc. Stiff joints and fragile bones. Breeding problems and a high incidence of milk fever in dairy cattle.	Worldwide in United States.
Bloat—Feedlot	High concentrate rations increase numbers of slime-producing bacteria in rumen. Slime traps fermentation gas and produces bloat.	Symptoms same as pasture bloat. Occurs when cattle have been fed high-concentrate low-roughage rations for approximately 60 days or longer.	Survey[2] of Kansas lots showed died of bloat, bloated severely 0.6% mildly to erately. Mild probably affect imal performanc
Bloat—Pasture	Most common on lush legume pastures. Incidence on wheat pasture has been increasing in recent years. Pasture bloat is a frothy bloat caused by interaction of several factors—plant, animal, and microbial. Soluble plant proteins play a prominent role in permitting stable froth formation.	First observed as distention of paunch on left side in front of hip bone. This is followed by distention of right side, protrusion of anus, respiratory distress, cyanosis of tongue, struggling, and death if not treated.	Widespread, alth some areas app have more bloat others. Often results in d 36% of all mort due to nutrit diseases and ments is attribu bloat.[3] Causes average a losses in beef dairy cattle (in ing milk) $104,904,000.
Crooked Calves	Manganese deficiency.	Calves born with crooked necks and legs.	Northwestern U.S
Fluorine Poisoning (Fluorosis)	Ingesting excessive quantities of fluorine through either the feed or water.	Abnormal teeth (especially mottled enamel) and bones, (bones become thickened and softened), stiffness of joints, loss of appetite, emaciation, reduction in milk flow, diarrhea, and salt hunger.	The water in par Arkansas, Calif South Carolina, Texas has bee ported to contai cess fluorine. C sionally throug the U.S., fluorine phosp are used in mi mixtures.
Founder	Overeating, overdrinking or from inflammation of the uterus following parturition. Also intestinal inflammation. Too rapid a change in the ration.	Extreme pain, fever (103° to 106° F) and reluctance to move. If neglected, chronic laminitis will develop, resulting in a dropping of the hoof soles and a turning up of the toe walls.	Worldwide. Actual death lo from founder ar very great.

Footnotes on last page of table.

D AILMENTS OF CATTLE[1]

Treatment	Control and Eradication	Prevention	Remarks
...ide dietary sources of the nu-[e]nt or nutrients, the deficiency which is known to cause the ...ndition.	When nutritional anemia is encountered, it can usually be brought under control by supplying dietary sources of the nutrient or nutrients, the deficiency of which is known to cause the condition.	Supply dietary sources of iron, copper, cobalt, and certain vitamins. Levels of iron in feed believed to be ample, since feeds contain 40 to 400 milligrams/lb.	Anemia is a condition in which the blood is either deficient in quality or quantity. (A deficient quality refers to a deficiency in hemoglobin and/or red cells.)
	Controlled by feeding phosphorus, either free-choice or added to the ration.	Feed phosphorus in feed and/or as mineral supplement (free-choice). Keep the Ca:P ratio within the range 2:1 or 1:1.	Generally caused by lack of phosphorus in the pasture. Phosphorus fertilizing may help.
...nch with 1 to 2 oz poloxalene ...herabloat®), and then relieve ...e gas with stomach tube 10 mi-...tes after treatment.	If feasible, increase proportion of roughage in ration. However, good quality legume hay may increase incidence of feedlot bloat. In this instance poloxalene (Bloat Guard®) is effective preventive. Use according to manufacturer's directions.	No effective preventive drug available.	Feedlot bloat may occur during any month of year; however, more common during hot, humid weather.
...e permitting, severe cases of ...oat should be treated by a veterinarian. Puncturing of the ...unch should be a last resort. ...l cases may be home-treated by ...keeping the animal on its feet ...d moving, and (2) drenching ...ther with (a) ½ to 1 quart min-...al oil or (b) 1 to 2 oz poloxalene ...herabloat®). Mineral oil will ...use cattle to go off feed whereas ...loxalene will not.	When there is high incidence of bloat, it may be desirable to change the feed. Where legume bloat is encountered, use poloxalene (Bloat Guard®) according to manufacturer's directions.	The incidence is lessened by (1) avoiding straight legume pastures, (2) feeding dry forage along with pasture, (3) avoiding a rapid fill from an empty start, (4) keeping animals continuously on pasture after they are once turned out, (5) keeping salt and water conveniently accessible at all times, and (6) avoiding frosted pastures. Use poloxalene (Bloat Guard®), a nonionic surfactant, according to manufacturer's directions, for the control of legume bloat.	Legume pastures, alfalfa hay, and barley appear to be associated with a higher incidence of bloat than many other feeds. Legume pastures are particularly hazardous when moist, after a light rain or dew.
		Feed manganese; 30 ppm of total feed.	The Utah station has also produced crooked calves by feeding lupine.
...damage may be permanent, but ...imals which have not de-...loped severe symptoms may be ...lped to some extent, if the ...urces of excess fluorine are ...iminated.	Discontinue the use of feeds, water, or mineral supplements containing excessive fluorine.	Avoid the use of feeds, water, or mineral supplements containing excessive fluorine. 100 ppm (0.01%) fluorine of the total dry ration is the borderline in toxicity for cattle. At levels of 25-100 ppm, some mottling of the teeth may occur over periods of 3-5 years. In breeding animals, therefore, the permissible level is 40 ppm of the total dry ration. Not more than 65-100 ppm fluorine should be present in dry matter of rations when rock phosphate is fed.	Fluorine is a cumulative poison. Undefluorinated rock phosphate often contains 3.5 to 4.0 percent fluorine.
...ding arrival of the veterinarian, ...e attendant should stand the ...nimal's feet in a cold water bath.	Alleviate the causes; namely, (1) overeating, (2) overdrinking, and/or (3) inflammation of the uterus following parturition.	Avoid overeating and overdrinking (especially when hot).	Unless foundered animals are quite valuable, it is usually desirable to dispose of them following a case of severe founder.

(Continued)

TABLE 8

Disease	Cause	Symptoms (and Age or Group Most Affected)	Distribution ar Losses Caused
Goiter (See Iodine Deficiency)			
Grass Tetany (Grass Staggers)	Magnesium deficiency.	Generally occurs during first 2 weeks of pasture season. Nervousness, twitching of muscles (usually of head and neck), head held high, accelerated respiration, high temperature, gnashing of the teeth, and abundant salivation. Slight stimulus may precipitate a crash to the ground, and finally death.	Reported in Nebr Kentucky, Miss Iowa, Washin and other states found in New land, England Holland. Highly fatal if treated quickly. Causes average a losses in beef dairy cattle (in ing milk) $662,000.
Iodine Deficiency (Goiter)	A failure of the body to obtain sufficient iodine from which the thyroid gland can form thyroxin (an iodine-containing compound).	Goiter (big neck) is the most characteristic symptom in humans, calves, lambs, and kids. Also, there may be reproductive failures and weak offspring that fail to survive. Pigs may be born hairless, and foals may be weak.	Northwestern U.S the Great Lakes ion.
Ketosis (Acetonemia)	A metabolic disorder, thought to be a disturbance in the carbohydrate metabolism.	In cows, ketosis or acetonemia is usually observed within first 1-6 weeks after calving. Affected animals show loss in appetite and condition, a marked decline in milk production, and the production of a peculiar, sweetish chloroform-like odor of acetone that may be present in the milk and pervade the barn.	Worldwide. Ketosis or aceton affects dairy c throughout the Causes annual l in dairy cattle cluding milk $1,899,000.
Milk Fever	Low blood calcium concentration. Too much calcium in the ration can cause this condition. In milking cows, the Ca:P ratio should not exceed 2:1.	Commonly occurs soon after calving and in high-producing cows. Rarely occurs at first calving. First symptoms are loss of appetite, constipation, and general depression. This is followed by nervousness and finally collapse and complete loss of consciousness. The head is usually turned back.	A common, widesp disease of cows. Losses are not gre though untre animals are like die. Causes average a losses in dairy (including mil $10,619,000.
Molybdenum Toxicity (Commonly called teartness)	As little as 10 to 20 ppm in forages result in toxic symptoms.	Toxic levels of molybdenum interfere with copper metabolism, thereby increasing the copper requirement and producing typical copper deficiency symptoms. The physical symptoms are anemia and extreme diarrhea, with consequent loss in weight and milk yield.	England, and Florida, Califo and Manitoba.

Footnotes on last page of table.

ontinued)

Treatment	Control and Eradication	Prevention	Remarks
venous injection of a solution calcium and/or magnesium salt a veterinarian.	(See Prevention.)	Grass tetany can be prevented by not turning animals to pasture, but this is not practical. Feeding hay at night during the first 2 weeks of the pasture season is helpful. A salt lick of 10 parts each of magnesium sulfate and calcium diphosphate with 80 parts of salt will aid in prevention.	Affected animals show low blood magnesium, often low serum calcium. Treated cattle may be aggressive on arising; so watch out!
sionally borderline cases may rvive; in these the moderate yroid enlargement disappears a few weeks.	At the first signs of iodine-deficiency, iodized salt should be fed to all farm animals.	In iodine-deficient areas, feed iodized salt to all farm animals throughout the year. Stabilized iodized salt containing 0.01% potassium iodide is recommended.	The enlarged thyroid gland (goiter) is nature's way of attempting to make sufficient thyroxine under conditions where a deficiency exists.
affected animal ½ to 1 lb of ther propylene glycol or sodium opionate per day, with the dose vided into 2 administrations per y. Treat for 5 to 10 days. Put atment on grain if cow is eat-g; otherwise, give as drench. avenous injection of glucose 00 ml of a 50% glucose solution) a rapid way of getting outside pply of sugar in blood.		The incidence of ketosis can be lessened by avoiding excessively fat cows at calving; (2) increasing the level of concentrates rapidly after calving; (3) feeding good quality roughage after calving, and avoiding abrupt changes in roughage; (4) feeding adequate proteins, minerals, and vitamins, and (5) providing comfort, exercise, and ventilation. In problem herds, feeding ¼ lb daily of propylene glycol or sodium propionate may be helpful.	The clinical findings are similar in the case of affected cattle and sheep, but it usually strikes ewes just before lambing, whereas cows are usually affected within the first 1 to 6 weeks after calving.
tment consists of having the terinarian give an intravenous jection of a calcium salt.	(See Prevention.)	Each of the following measures will lessen the incidence of milk fever: 1. Calcium-phosphorus, ratio and amounts—Approximately a 2.3:1 Ca:P ratio. Feed a ration that contains 0.5 to 0.7% Ca and 0.3 to 0.4% P. 2. Calcium shock treatment—10 to 14 days before calving, feed a Ca deficient ration with a Ca:P ratio of 1:2. This activates the cow's calcium-mobilizing mechanism for drawing calcium from the bones, with the result that it is functioning before calving and, milk fever is avoided. 3. High vitamin D—This consists in feeding 20 million units of vitamin D/cow/day starting about 5 days before calving and continuing through the first day postpartum, with a maximum dosage period of 7 days.	The name "milk fever" is a misnomer, because the disease is not accompanied by fever; the temperature really being below normal.
gram of copper sulfate per ad daily will cure symptoms of olybdenum toxicity.		One gram of copper sulfate per head daily will prevent molybdenum toxicity.	When feeds are high in sulfate, toxic symptoms will be produced on lower levels of molybdenum and, conversely, higher levels of molybdenum can be tolerated with low levels of sulfate.

(Continued)

TABLE 8

Disease	Cause	Symptoms (and Age or Group Most Affected)	Distribution an Losses Caused
Nitrate (Oat hay poisoning, corn stalk poisoning)	1. The forages (seeds do not appear to accumulate nitrate nitrogen) of most grain crops (oats, wheat, barley, rye, corn, and sorghum), Sudan grass, and numerous weeds; especially (a) under stress —drought, insufficient sunlight, following spraying by weed killers, or after frost, or (b) following the application of high soil nitrate nitrogen (through nitrogen fertilizer, green manure crops, or barnyard manure) which may boost the nitrate nitrogen of plants to dangerous levels. Also, sometimes nitrate appears to be formed after forage is stacked. 2. Inorganic salts of nitrate or nitrite (including fertilizers) carelessly applied to fields or left where animals have access to them. Sometimes these chemicals are also mistakenly used in place of common salt. 3. Pond or shallow well water into which heavy rains have washed a high concentration of nitrate from (a) fertilizer from heavily fertilized fields, or (b) feedlot drainage (ammonium nitrate).	Accelerated respiration and pulse rate; diarrhea; frequent urination; loss of appetite; general weakness, trembling, and a staggering gait; frothing from the mouth; lowered milk production; abortion; blue color of the mucous membrane, muzzle, and udder due to lack of oxygen; and death in 4½ to 9 hrs. after eating lethal doses of nitrate. A rapid and accurate diagnosis of nitrate poisoning may be made by drawing and examining a venous (jugular) blood sample. Normal blood is red and becomes brighter on standing. Brown-colored blood, due to the formation of methemoglobin, is characteristic of animals suffering from nitrate poisoning; chemically, the nitrate oxidizes the ferrous hemoglobin (oxyhemoglobin) to ferric hemoglobin (methemoglobin) which cannot transport oxygen, with the result that death due to nitrate poisoning may be compared to asphyxiation or strangulation. Death occurs when about ¾ of the oxyhemoglobin (the oxygen carrier in the blood) has been converted to methemoglobin.	Excessive nitrate tent of feeds is creasingly impo cause of poisoni farm animals, primarily to and more high gen fertilization nitrate toxicity new, having reported as ear 1850, and havin curred in sem regions of this other countrie years.
Oat Hay Poisoning (See Nitrate poisoning)			
Osteomalacia	Inadequate phosphorus (sometimes inadequate calcium). Lack of vitamin D in confined cattle. Incorrect ratio of calcium to phosphorus.	Phosphorus deficiency symptoms are: depraved appetite (gnawing on bones, wood, or other objects, or eating dirt); lack of appetite, stiffness of joints, failure to breed regularly, decreased milk production, and an emaciated appearance. Calcium deficiency symptoms are: fragile bones, reproductive failures, and lowered lactations. Mature animals most affected. Most of the acute cases occur during pregnancy and lactation.	Southwestern U. classed as a p phorus-deficie area, whereas cium-deficient have been rep in parts of Flo Louisiana, Nebr Virginia, and Virginia.
Pine Needle Abortion	Needles of Yellow Pine (*Pinus ponderosa*); commonly called Yellow Pine, or Jack Pine (*Pinus banksiana*). It is suspected that the high turpentine content of Yellow Pine needles actually causes the abortion, for there is evidence that turpentine can cause abortion in the human female.	Pregnant cows, free of brucellosis, abort.	British Columbia, ada, and in the s of Washington, ho, and Oregon
Rickets	Lack of either calcium, phosphorus, or vitamin D; or an incorrect ratio of the two minerals.	Enlargement of the knee and hock joints, and the animal may exhibit great pain when moving about. Irregular bulges (beaded ribs) at juncture of ribs with breastbone, and bowed legs. Rickets is a disease of young animals—calves, foals, pigs, lambs, kids, pups, and chicks.	Worldwide. It is seldom fatal.

Footnotes on last page of table.

ntinued)

Treatment	Control and Eradication	Prevention	Remarks
solution of methylene blue (in % glucose or a 1.8% sodium ate solution) administered by eterinarian intravenously at the of 100 cc/1,000 lb liveweight.	(See Prevention.)	Regard any amount of nitrate nitrogen over 0.5% of the total ration (moisture-free basis) as a potential source of trouble. When in doubt, have the feed analyzed (first make a rapid, qualitative field test, using a commercial test kit according to direction; then, if high-nitrate samples are spotted, follow with a quantitative laboratory chemical test). Nitrate poisoning may be lowered by (1) feeding high levels of carbohydrates or energy feeds (grain or molasses) and vitamin A, (2) feeding limited amounts of high nitrate forage, (3) alternating or mixing high and low nitrate forages, and (4) ensiling forages high in nitrates, since fermentation reduces some of the nitrates to gas (but beware of nitric oxide and nitrogen dioxide gas, which is released as yellow-red fumes in the early stages of fermentation and may cause silo gas poisoning to both humans and animals). After 3-4 weeks the silage has usually lost most of its nitrates and is safe to feed.	Nitrate form nitrogen does not appear to cause the actual toxicity. During digestion, the nitrate is reduced to nitrite, a far more toxic form (10 to 15 times more toxic than nitrates). In cows and sheep, this conversion takes place in the rumen (paunch). Lethal dose varies with (1) nutritional state, size and type of animal; and (2) the consumption of feed other than nitrate-containing material. Nitrate over 5% of total ration is a potential source of trouble; 0.75% content nitrate forages must be fed with caution, and milk production will be lowered; and at 1.5% death will likely occur. Where nitrate troubles are suspected, consult the local veterinarian or county agent.
ase the calcium and phos-rus content of feed through lizing the soils. t natural feeds that contain ficient quantities of calcium phosphorus. a special mineral supplement upplements. is disease is far advanced, tment will not be successful.	(See Treatment.)	Feed balanced rations, and allow animals free access to a suitable phosphorus and calcium supplement.	Calcium deficiencies are much more rare than phosphorus deficiencies in cattle.
eatment known.		Keep pregnant cows away from Yellow Pine trees.	Lodgepole Pine (Pinus contorta)—commonly called Black Pine, Jack Pine, Western Jack Pine, White Pine (Pinus monticola) or Cypress (Cypressus arizonica)—does not appear to cause abortion in cattle. Pregnant cows will consume quantities of pine needles even though fed an adequate ration.
disease has not advanced too treatment may be successful supplying adequate amounts vitamin D, calcium, and phos-rus, and/or adjusting the ratio alcium to phosphorus.	(See Prevention.)	Provide (1) sufficient calcium, phosphorus, and vitamin D, and (2) a correct ratio of the two minerals.	Rickets is characterized by a failure of growing bone to ossify, or harden, properly.

(Continued)

TABLE 8

Disease	Cause	Symptoms (and Age or Group Most Affected)	Distribution a Losses Caused
Salt Deficiency (Sodium chloride)	Lack of salt (sodium chloride).	Loss of appetite, retarded growth, loss of weight, a rough coat, lowered production of milk, and a ravenous appetite for salt.	Worldwide.
Salt Sick (Cobalt Deficiency)	Cobalt deficiency, associated with copper and perhaps iron deficiencies.	Loss of appetite, depraved appetite, scaliness of skin, listlessness, and lack of thrift.	Florida; on sandy
Selenium Poisoning (Alkali Disease)	Consumption of plants grown on soils containing selenium.	Loss of hair from the tail in cattle. In severe cases, the hoofs slough off, lameness occurs, feed consumption decreases, and death may occur by starvation.	In certain regic western U especially c areas in Dakota, Mor Wyoming, Neb Kansas, and pe areas in other in the Great I and Rocky M tains. Also in Canada.
Sweet Clover Disease	Usually produced only by moldy or spoiled sweet clover, hay or silage. Caused by presence of dicoumarol which interferes with vitamin K in blood clotting.	Loss of clotting power of the blood. As a result, blood forms soft swellings beneath skin on different parts of body. Serious or fatal bleeding may occur at time of dehorning, castration, parturition, or following injury. All ages affected. A newborn animal may also have the condition at birth.	Wherever sweet is grown.
Urinary Calculi (Gravel, Stones, Water Belly)	Unknown, but it does seem to be nutritional. Experiments have shown a higher incidence of urinary calculi when there is (1) a high potassium intake, (2) an incorrect Ca:P ratio, or (3) a high proportion of beet pulp or grain sorghum in the ration.	Frequent attempts to urinate, dribbling or stoppage of the urine, pain and renal colic. Usually only males affected, the females being able to pass the concretions. Bladder may rupture, with death following. Otherwise, uremic poisoning may set in.	Worldwide. Affected anima dom recover pletely. Causes estimated age annual l beef cattle $4,052,000.
Vitamin A Deficiency (Night Blindness and Xerophthalmia)	Vitamin A deficiency. High levels of nitrate in hay or silage.	Night blindness, the first symptom of vitamin A deficiency, is characterized by faulty vision, especially noticeable when the affected animal is forced to move about in twilight in strange surroundings. Xerophthalmia develops in the advanced stages of vitamin A deficiency. The eyes become severely affected, and blindness may follow. Severe diarrhea in young calves and intermittent diarrhea in advanced stages in adults. In finishing cattle, generalized edema or anasarca with lameness in hock and knee joints and swelling in the brisket area.	Worldwide. Espe prevalent in w U.S. where o the following tions frequently vails: (1) ext drought, and winter feedi bleached grass on the stalk bleached hay.

Footnotes on last page of table.

ntinued)

Treatment	Control and Eradication	Prevention	Remarks
starved animals should be dually accustomed to salt, wly increasing the hand-fed alvance until the animals may be ely allowed free access to it.	(See Treatment and Prevention.)	Provide plenty of salt at all times, preferably by free-choice feeding.	Common salt is one of the most essential minerals for grass-eating animals, and one of the easiest and cheapest to provide. Excessive salt intake can result in toxicity if animals are deprived of water.
		Mix 0.2 oz of cobalt chloride, cobalt sulfate, or cobalt carbonate/100 lb of either (1) salt, or (2) the mineral other than salt.	
use of salt containing 37.5 ppm arsenic may reduce the incince of chronic selenium isoning in cattle on selenifers range. ure rotation and use of supplental feeds from nonselenifers areas are practical solutions to e problem. re is no known treatment for ute selenium poisoning.	(Control measures based on Prevention.)	Abandon areas where soils contain excess selenium, because crops produced on such soils constitute a menace to both animals and man.	Selenium toxicity may occur when cattle consume feeds containing 10-30 ppm of selenium on a dry matter basis for an extended period. Chronic cases of selenium poisoning occur when cattle consume feeds containing 8.5 ppm of selenium over an extended period; acute cases occur on 500 to 1,000 ppm.
ove the offending materials and minister menadione (vitamin). veterinarian usually gives the ected animal an injection of asma or whole blood from a rmal animal that was not fed on e same feed.	When a case of sweet clover disease is observed in the herd, either (1) discontinue feeding the damaged product or (2) alternate it with a better quality hay, especially alfalfa.	Properly cure any sweet clover hay or ensilage.	The disease has also been produced from feeding moldy lespedeza hay and from sweet clover pasture.
e calculi develop, dietary atment appears to be of little lue. oth muscle relaxants may allow ssage of calculi if used before pture of bladder. Surgery may ve the animal, but such treatent will result in bulls becomg nonbreeders. n the condition strikes in feed-t cattle, increase the salt content the diet if the animals are not ady to market.	If severe outbreaks of urinary calculi occur in finishing steers, it is usually well to dispose of them if they are carrying acceptable finishing. When an outbreak occurs in cattle, one of the following treatments is recommended: 1. Add ammonium chloride at the rate of 1¼ to 1½ oz/head/day. 2. Increase the salt content of the total ration to a level of 1½%. Too high levels will lower feed consumption. 3. Incorporate 20% alfalfa in the ration.	Good feed and management appear to lessen the incidence. Delayed castration (castration of bull calves at 4-5 mo. of age) and high salt diets of feedlot cattle (1-3% salt in the grain ration, using the upper limits in the winter months) in order to induce more water consumption are effective preventive measures. Avoid high phosphorus and low calcium. Provide adequate vitamin A, salt and water.	Calculi are stonelike concretions in the urinary tract which almost always originate in the kidneys. These stones block the passage of urine. Ammonium chloride (see Control and Eradication) appears to be the product of choice. However, ammonium sulfate may be used, at the rate of 1.7 to 2.0 oz/head/day. Add it to the ration when an outbreak occurs.
tment consists of correcting the etary deficiencies and (1) adng vitamin A to the ration, or (2) jecting intramuscularly or inruminal 500,000 to 1,000,000 of vitamin A.	(See Prevention and Treatment.)	Provide good sources of carotene (vitamin A) through green, leafy hays; silage; lush, green pastures; yellow corn; or add stabilized vitamin A to the ration.	High levels of nitrates interfere with the conversion of carotene to vitamin A.

(Continued)

Disease	Cause	Symptoms (and Age or Group Most Affected)	Distribution a Losses Caused
White Muscle Disease (muscular dystrophy)	A deficiency of selenium and/or vitamin E.	Symptoms range from mild "founderlike" stiffness to sudden death. Calves continue to nurse as long as they can reach the cow's teats. Many calves stand or lie with protruded tongue, fighting for breath against a severe pulmonary edema. It seems that more calves than lambs develop fatal heart damage. Affected calves show pathological lesions similar to those of "stiff lambs" (white muscle disease in lambs); namely, whitish areas or streaks in the heart and other muscles. Affects calves from birth to 3 months of age.	Throughout the but the inci appears to be h in the intermo area, betwee Rocky and Ca mountains.

[1]This table was authoritatively reviewed by the following: Dr. Erle E. Bartley, Department of Dairy and Poultry Science, Kansas University, Manhattan, Kansas; Dr. C. Brent Theurer, Department of Animal Science, The University of Arizona, Tucson, Ariz.; Dr Huber, Department of Dairy Science, Michigan State University, East Lansing, Mich.; and Dr. T. H. Blosser, Department of A Sciences, Washington State University, Pullman, Wash.

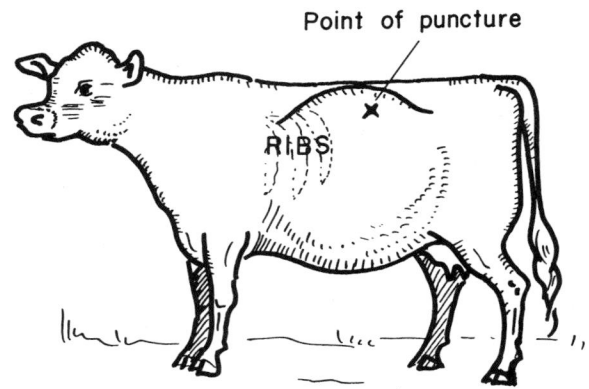

Fig. 8-28. When bloat is encountered, puncturing of the paunch should be a last resort. However, if such treatment must be administered to save the life of an animal, the stockman should know where to make the puncture. As noted in the above drawing, it should be made on the left side at the location shown. (Drawing by R. F. Johnson)

ontinued)

Treatment	Control and Eradication	Prevention	Remarks
fine affected animals to a stall, d give plenty of rest. veterinarian may administer phatocopherol.	(See Prevention and Treatment.)	Feed 1¼ lb linseed meal per cow daily during last 2 months of gestation. It is thought that linseed meal contains sufficient selenium to serve as a preventative. Alpha-tocopherol, added to the diet or injected intramuscularly. Commercial vitamin E supplements.	White muscle disease is often overlooked in calves.

Unless otherwise indicated, the estimated average annual loss figures (in dollars) given in column 4 of this table were taken from *Losses Agriculture*, Agr. Hdbk. No. 291, U.S. Department of Agriculture, Washington, D.C., 1965.
[2]Meyer, R. M., *27th Kansas Formula Feed Conference Proceedings, P. Hl*, Kansas State University, Manhattan, Kan., 1972.
[3]Ensminger, M. E., M. W. Galgan, and W. L. Slocum, Problems and Practices of American Cattlemen, *Wash. Agr. Exp. Sta. Bull. 562*, 55, p. 18.

Fig. 8-29. Sweet clover disease. Note the collection of blood at the point of the left shoulder. (Courtesy, Dept. of Veterinary Pathology and Hygiene, College of Veterinary Medicine, University of Illinois, Urbana, Ill.)

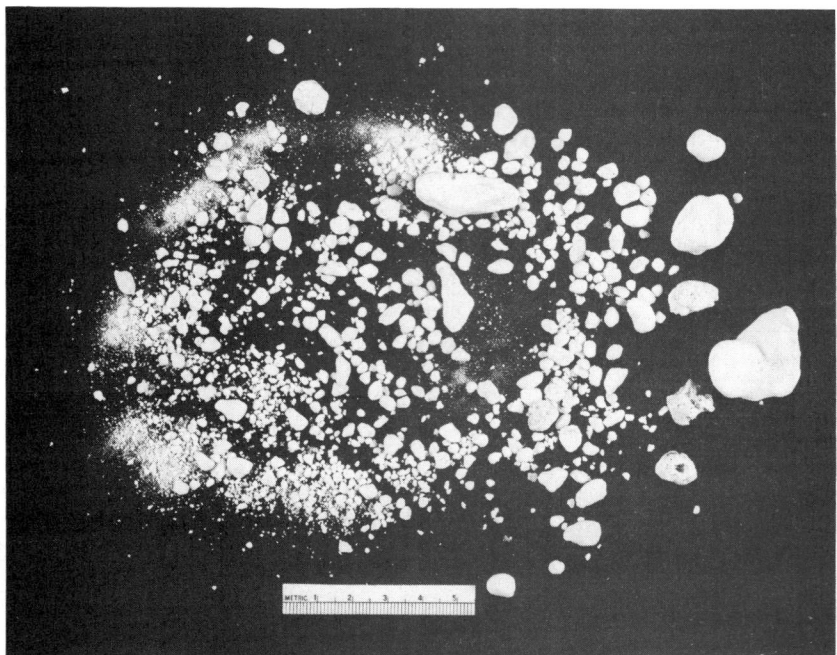

Fig. 8-30. Urinary calculi. Concretions or stones obtained from a $6,500 imported Shorthorn bull that died from urinary calculi. The cause of urinary calculi is unknown, but it does seem to be tied up with nutrition. (Courtesy, Washington State University, Pullman, Wash.)

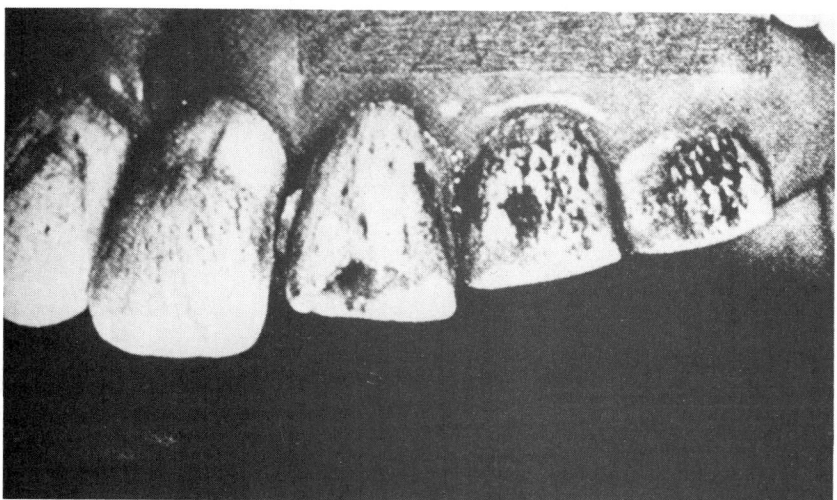

Fig. 8-31. Fluorosis in cattle. Incisors of 5-year-old cow fed ration averaging 37 ppm fluorine content. Note mottled enamel. (Courtesy, American Institute of Nutrition, Bethesda, Md.)

NUTRIENT REQUIREMENTS OF BEEF CATTLE

The nutrient requirements of beef cattle for which there are reasonably reliable data have been summarized by a subcommittee of the National Research Council (*Nutrient Requirements of Beef Cattle*, Fifth Revised Edition, 1976) and are herewith reproduced at the end of this chapter in Tables 8-18 to 8-20. The figures in these tables are, for the most part, requirements (rather than allowances); hence, they do not provide for margins of safety to compensate for variations in feed composition, environment, and possible losses of nutrients during storage or processing.

FEED COMPOSITION

Nutrient compositions of feedstuffs are necessary for intelligent ration preparation, animal health, and feed efficiency. Appendix, Table I-3, contains the most commonly used feeds for beef cattle.

TABLE 8-18

NUTRIENT REQUIREMENTS FOR GROWING-FINISHING STEER CALVES AND YEARLINGS (DAILY NUTRIENTS PER ANIMAL)*

Weight[1]		Daily Gain		Minimum Dry Matter Consumption[2]		Roughage[2]	Total Protein	Dig. Protein	NEm	NEg	ME[2]	TDN[2,3]		Ca	P	Vitamin A (Thousands IU)
kg	(lb)	kg	(lb)	kg	(lb)	%	kg	kg	Mcal	Mcal	Mcal	kg	(lb)	g	g	
100	(220)	0	0	2.1	4.6	100	0.18	0.10	2.43	0	4.2	1.2	2.6	4	4	5
		.5	1.1	2.9	6.4	70-80	0.36	0.24	2.43	.89	6.6	1.8	4.0	14	11	6
		.7	1.5	2.7	6.0	50-60	0.40	0.28	2.43	1.27	7.1	2.0	4.4	19	13	6
		.9	2.0	2.8	6.2	25-30	0.46	0.33	2.43	1.68	7.7	2.1	4.6	24	16	7
		1.1	2.4	2.7	6.0	15	0.49	0.36	2.43	2.10	8.4	2.3	5.1	28	19	7
150	(331)	0	0	2.8	6.2	100	0.23	0.13	3.30	0	5.6	1.6	3.5	5	5	6
		.5	1.1	4.0	8.8	70-80	0.44	0.28	3.30	1.20	9.0	2.5	5.5	14	12	9
		.7	1.5	3.9	8.6	50-60	0.49	0.33	3.30	1.73	9.6	2.7	6.0	18	14	9
		.9	2.0	3.8	8.4	25-30	0.54	0.37	3.30	2.27	10.7	3.0	6.6	23	17	9
		1.1	2.4	3.7	8.2	15	0.58	0.41	3.30	2.84	11.3	3.1	6.8	28	20	9
200	(441)	0	0	3.5	7.7	100	0.30	0.17	4.10	0	7.0	1.9	4.2	6	6	8
		.5	1.1	5.8	12.8	80-90	0.57	0.35	4.10	1.49	12.1	3.4	7.5	14	13	12
		.7	1.5	5.7	12.6	70-80	0.61	0.39	4.10	2.14	13.0	3.6	7.9	18	16	13
		.9	2.0	4.9	10.8	35-45	0.61	0.40	4.10	2.82	13.3	3.7	8.2	23	18	13
		1.1	2.4	4.6	10.1	15	0.63	0.43	4.10	3.52	14.1	3.9	8.6	27	20	13
250	(551)	0	0	4.4	9.7	100	0.35	0.20	4.84	0	8.2	2.3	5.1	8	8	9
		.7	1.5	5.8	12.8	55-65	0.62	0.39	4.84	2.53	14.4	4.0	8.8	18	16	14
		.9	2.0	6.2	13.7	45-50	0.69	0.44	4.84	3.33	16.2	4.5	9.9	22	19	14
		1.1	2.4	6.0	13.2	20-25	0.73	0.48	4.84	4.17	17.0	4.7	10.4	26	21	14
		1.3	2.9	6.0	13.2	15	0.76	0.51	4.84	5.04	18.6	5.2	11.5	30	23	14
300	(661)	0	0	4.7	10.4	100	0.40	0.23	5.55	0	9.4	2.6	5.7	9	9	10
		.9	2.0	8.1	17.9	55-65	0.81	0.50	5.55	3.82	19.5	5.4	11.9	22	19	16
		1.1	2.4	7.6	16.8	20-25	0.82	0.52	5.55	4.78	20.4	5.6	12.3	25	22	16
		1.3	2.9	7.1	15.6	15	0.83	0.54	5.55	5.77	21.6	6.0	13.2	29	23	16
		1.4[4]	3.1	7.3	16.1	15	0.87	0.57	5.55	6.29	22.5	6.2	13.7	31	25	16
350	(772)	0	0	5.3	11.7	100	0.46	0.26	6.24	0	10.6	2.9	6.4	10	10	12
		.9	2.0	8.0	17.6	45-55	0.80	0.49	6.24	4.29	20.8	5.8	12.8	20	18	18
		1.1	2.4	8.0	17.6	20-25	0.83	0.52	6.24	5.36	22.4	6.2	13.7	23	20	18
		1.3	2.9	8.0	17.6	15	0.87	0.55	6.24	6.48	24.2	6.8	15.0	26	22	18
		1.4[4]	3.1	8.2	18.1	15	0.90	0.57	6.24	7.06	25.3	7.0	15.4	28	24	18
400	(882)	0	0	5.9	13.0	100	0.51	0.29	6.89	0	11.8	3.3	7.3	11	11	13
		1.0	2.2	9.4	20.7	45-55	0.87	0.54	6.89	5.33	24.5	6.8	15.0	21	20	19
		1.2	2.6	8.5	18.7	20-25	0.87	0.54	6.89	6.54	25.4	7.0	15.4	23	21	19
		1.3	2.9	8.6	19.0	15	0.90	0.56	6.89	7.16	26.5	7.3	16.1	25	22	19
		1.4[4]	3.1	9.0	19.8	15	0.94	0.59	6.89	7.80	28.0	7.7	17.0	26	23	19

TABLE 8-18 (Continued)

Weight[1] kg	(lb)	Daily Gain kg	(lb)	Minimum Dry Matter Consumption[2] kg	(lb)	Roughage[2] %	Total Protein kg	Dig. Protein kg	NEm Mcal	NEg Mcal	ME[2] Mcal	TDN[2][3] kg	(lb)	Ca g	P g	Vitamin A (Thousands IU)
450	(992)	0	0	6.4	14.1	100	0.54	0.31	7.52	0	12.8	3.6	–	12	12	14
		1.0	2.2	10.3	22.7	45-55	0.96	0.57	7.52	5.82	26.7	7.4	16.3	20	20	20
		1.2	2.6	10.2	22.5	20-25	0.97	0.58	7.52	7.14	28.6	7.9	17.4	23	22	20
		1.3	2.9	9.3	20.5	15	0.97	0.59	7.52	7.83	29.0	8.0	17.6	24	23	20
		1.4[4]	3.1	9.8	21.6	15	0.98	0.60	7.52	8.52	30.5	8.4	18.5	25	23	20
500	(1102)	0	0	7.0	15.4	100	0.60	0.34	8.14	0	13.9	3.8	8.4	13	13	15
		.9	2.0	10.5	23.1	45-55	0.95	0.56	8.14	5.60	27.1	7.5	16.5	19	19	23
		1.1	2.4	10.4	22.9	20-25	0.96	0.57	8.14	7.01	29.2	8.1	17.8	20	20	23
		1.2	2.6	9.6	21.2	15	0.96	0.58	8.14	7.73	29.7	8.2	18.1	21	21	23
		1.3[4]	2.9	10.0	22.0	15	0.97	0.60	8.14	8.47	31.4	8.7	19.2	22	22	23

*Source: Tables 8-18, 8-18a, and 8-18b from *Nutrient Requirements of Beef Cattle*, fifth revised edition, National Academy of Sciences, Washington, D.C., 1976, pp. 22-27.

[1] Average weight for a feeding period.

[2] Dry matter consumption, ME and TDN allowances are based on NE requirements and the general types of ration indicated in the Roughage column. Most roughages will contain 1.9-2.2 Mcal of ME/kg dry matter and 90-100% concentrate rations are expected to contain 3.1-3.3 Mcal of ME/kg.

[3] TDN was calculated by assuming 3.6155 Mcal of ME per kg of TDN.

[4] Most steers of the weight indicated and not exhibiting compensatory growth, will not sustain an energy intake necessary to maintain this rate of gain for an extended period.

TABLE 8-18a

NUTRIENT REQUIREMENTS FOR GROWING-FINISHING HEIFER CALVES AND YEARLINGS (DAILY NUTRIENTS PER ANIMAL)

Weight[1]		Daily Gain		Minimum Dry Matter Consumption[2]		Roughage[2]	Total Protein	Dig. Protein	NEm	NEg	ME[2]	TDN[2 3]		Ca	P	Vitamin A (Thousands IU)
kg	(lb)	kg	(lb)	kg	(lb)	%	kg	kg	Mcal	Mcal	Mcal	kg	(lb)	g	g	
100	(220)	0	0	2.1	4.6	100	0.18	0.10	2.43	0	4.2	1.2	2.6	4	4	5
		.5	1.1	3.0	6.6	70-80	0.37	0.25	2.43	.99	6.9	1.9	4.2	14	11	6
		.7	1.5	2.9	6.4	50-60	0.42	0.29	2.43	1.44	7.5	2.1	4.6	19	14	6
		.9	2.0	3.0	6.6	25-30	0.48	0.34	2.43	1.92	8.3	2.3	5.1	24	17	7
		1.1	2.4	3.0	6.6	<15	0.53	0.39	2.43	2.43	9.2	2.5	5.5	29	19	7
150	(331)	0	0	2.8	6.2	100	0.24	0.14	3.30	0	5.6	1.6	3.5	5	5	6
		.5	1.1	4.1	9.0	70-80	0.45	0.29	3.30	1.34	9.4	2.6	5.7	14	12	9
		.7	1.5	4.0	8.8	50-60	0.50	0.33	3.30	1.95	10.4	2.8	6.2	18	14	9
		.9	2.0	4.0	8.8	25-30	0.54	0.37	3.30	2.60	11.3	3.1	6.8	23	17	9
		1.1	2.4	4.0	8.8	<15	0.60	0.42	3.30	3.30	12.4	3.4	7.5	28	20	9
200	(441)	0	0	3.5	7.7	100	0.30	0.17	4.10	0	7.0	1.9	4.2	6	6	8
		.3	0.7	5.4	11.9	100	0.49	0.29	4.10	0.95	10.8	3.0	6.6	10	10	12
		.5	1.1	6.0	13.2	80-90	0.58	0.35	4.10	1.66	12.7	3.5	7.7	14	13	13
		.7	1.5	6.0	13.2	70-80	0.61	0.39	4.10	2.42	13.8	3.8	8.4	18	16	13
		.9	2.0	5.3	11.7	35-45	0.62	0.40	4.10	3.23	14.3	4.0	8.8	22	17	13
		1.1	2.4	5.0	11.0	<15	0.64	0.43	4.10	4.09	15.4	4.3	9.5	25	19	13
250	(551)	0	0	4.1	9.0	100	0.35	0.20	4.84	0	8.3	2.3	5.1	7	7	9
		.3	.7	6.4	14.1	100	0.57	0.33	4.84	1.13	12.8	3.5	7.8	12	12	14
		.5	1.1	6.5	14.3	80-90	0.62	0.37	4.84	1.96	14.2	3.9	8.6	13	13	14
		.7	1.5	5.8	12.8	55-65	0.62	0.38	4.84	2.86	15.0	4.1	9.1	17	15	14
		.9	2.0	5.9	13.0	35-45	0.65	0.42	4.84	3.81	16.5	4.6	10.1	21	17	14
		1.1	2.4	6.5	14.3	20-25	0.74	0.48	4.84	4.84	18.7	5.2	11.5	25	20	14
		1.2	2.6	6.3	13.9	<15	0.75	0.49	4.84	5.37	19.4	5.4	11.9	27	21	14
300	(661)	0	0	4.7	10.4	100	0.40	0.23	5.55	0	9.5	2.6	5.7	9	9	10
		.3	.7	7.4	16.3	100	0.63	0.36	5.55	1.29	14.5	4.0	8.4	13	13	16
		.5	1.1	7.4	16.3	80-90	0.67	0.40	5.55	2.25	16.3	4.5	9.9	14	14	16
		.7	1.5	6.6	14.6	55-65	0.67	0.40	5.55	3.37	17.1	4.7	10.4	16	15	16
		.9	2.0	6.8	15.0	35-45	0.70	0.44	5.55	4.37	19.0	5.2	11.5	19	17	16
		1.1	2.4	7.5	16.5	20-25	0.78	0.49	5.55	5.55	21.5	6.0	13.2	23	20	16
		1.2	2.6	7.2	15.9	<15	0.79	0.50	5.55	6.16	22.3	6.2	13.7	24	20	16
350	(772)	0	0	5.3	11.7	100	0.46	0.26	6.24	0	10.6	2.9	6.4	10	10	12
		.3	.7	8.2	18.1	100	0.69	0.39	6.24	1.45	16.5	4.6	10.0	15	15	18
		.5	1.1	8.3	18.3	80-90	0.73	0.42	6.24	2.52	18.3	5.1	11.2	15	15	18
		.7	1.5	7.9	17.4	55-65	0.73	0.43	6.24	3.68	19.7	5.4	11.9	16	15	18
		.9	2.0	8.1	17.9	35-45	0.77	0.46	6.24	4.91	21.8	6.0	13.2	17	17	18
		1.1	2.4	8.3	18.3	20-25	0.81	0.50	6.24	6.23	24.0	6.6	14.5	20	19	18
		1.2[4]	2.6	8.1	17.9	<15	0.81	0.50	6.24	6.91	25.0	6.9	15.2	21	20	18

TABLE 8-18a (Continued)

Weight[1]		Daily Gain		Minimum Dry Matter Consumption[2]		Roughage[2]	Total Protein	Dig. Protein	NEm	NEg	ME[2]	TDN[2][3]		Ca	P	Vitamin A (Thousands IU)
kg	(lb)	kg	(lb)	kg	(lb)	%	kg	kg	Mcal	Mcal	Mcal	kg	(lb)	g	g	
400	(882)	0	0	5.9	13.0	100	0.51	0.29	6.89	0	11.8	3.3	7.3	11	11	13
		.3	.7	9.1	20.0	100	0.76	0.43	6.89	1.61	18.2	5.0	11.1	16	16	19
		.5	1.1	8.5	18.7	70-80	0.78	0.43	6.89	2.79	19.5	5.4	11.9	15	15	19
		.7	1.5	8.7	19.2	55-65	0.79	0.46	6.89	4.06	21.7	6.0	13.2	16	16	19
		.9	2.0	8.4	18.5	20-25	0.79	0.47	6.89	5.43	23.5	6.5	14.3	17	17	19
		1.1[4]	2.4	8.3	18.3	<15	0.81	0.49	6.89	6.88	25.9	7.2	15.9	19	18	19
450	(992)	0	0	6.4	14.1	100	0.55	0.31	7.52	0	12.9	3.6	7.9	12	12	14
		.2	.4	8.7	19.2	100	0.74	0.41	7.52	1.14	17.4	4.8	10.6	16	16	19
		.5	1.1	9.3	20.5	70-80	0.80	0.46	7.52	3.05	21.3	5.9	13.0	17	17	20
		.8	1.8	9.1	20.1	35-45	0.82	0.48	7.52	5.17	24.5	6.8	15.0	16	16	20
		1.0[4]	2.2	8.5	18.7	<15	0.83	0.48	7.52	6.71	26.8	7.4	16.3	19	19	20

[1]Average weight for a feeding period.
[2]Dry matter consumption, ME and TDN allowances are based on NE requirements and the general type of ration indicated in the Roughage column. Most roughages will contain 1.9-2.2 Mcal of ME/kg dry matter and 90-100% concentrate rations are expected to have 3.1 to 3.3 Mcal of ME/kg.
[3]TDN was calculated by assuming 3.6155 kcal of ME per g of TDN.
[4]Most heifers of the weight indicated and not exhibiting compensatory growth will not sustain the energy intake necessary to maintain this rate of gain for an extended period.

TABLE 8-18b

NUTRIENT REQUIREMENTS FOR BEEF CATTLE BREEDING HERD (DAILY NUTRIENTS PER ANIMAL)

Weight[1] kg	(lb)	Daily Gain kg	(lb)	Minimum Dry Matter Consumption[2] kg	(lb)	Roughage[2] %	Total Protein kg	Dig. Protein kg	NEm Mcal	NEg Mcal	ME[2] Mcal	TDN[2][3] kg	(lb)	Ca g	P g	Vitamin A (Thousands IU)
						Pregnant yearling heifers - Last 3-4 months of pregnancy										
325	(716)	.4[3]	.9	6.6	14.5	100[4]	0.58	0.34	5.89	0.62	12.6	3.5	7.7	15	15	19
		.6	1.3	8.5	18.7	100	0.75	0.42	5.89	1.52	16.2	4.5	9.9	18	18	23
		.8	1.8	9.4	20.7	85-100	0.85	0.50	5.89	2.49	20.1	5.6	12.3	22	20	26
350	(772)	.4[3]	.9	6.9	15.2	100	0.61	0.35	6.23	0.65	13.2	3.7	8.1	15	15	19
		.6	1.3	8.9	19.6	100	0.78	0.45	6.23	1.60	16.9	4.7	10.3	19	19	25
		.8	1.8	10.0	22.0	85-100	0.88	0.51	6.24	2.63	21.1	5.8	12.9	22	21	28
375	(827)	.4[3]	.9	7.2	15.9	100	0.63	0.36	6.56	0.68	13.7	3.8	8.4	15	15	20
		.6	1.3	9.3	20.5	100	0.81	0.46	6.56	1.68	17.7	4.9	10.8	19	19	26
		.8	1.8	11.0	24.2	85-100	0.96	0.55	6.56	2.76	22.1	6.1	13.5	22	22	31
400	(882)	.4[3]	.9	7.5	16.5	100	0.65	0.38	6.89	0.71	14.2	3.9	8.7	16	16	21
		.6	1.3	9.7	21.4	100	0.84	0.48	6.89	1.76	18.5	5.1	11.3	19	19	27
		.8	1.8	11.6	25.6	85-100	1.01	0.57	6.89	2.90	23.0	6.4	14.0	22	22	33
425	(937)	.4[3]	.9	7.8	17.2	100	0.69	0.40	7.21	0.74	14.8	4.1	9.0	16	16	22
		.6	1.3	10.1	22.3	100	0.88	0.50	7.21	1.84	19.2	5.3	11.7	19	19	28
		.8	1.8	12.1	26.7	85-100	1.05	0.60	7.21	3.03	24.0	6.6	14.6	22	22	34
						Dry pregnant mature cows—Middle third of pregnancy										
350	(772)			5.5	12.2	100[4]	.32	.15	6.23		10.8	3.0	6.6	10	10	15
400	(882)			6.1	13.4	100	.36	.17	6.89		11.9	3.3	7.3	11	11	17
450	(992)			6.7	14.8	100	.39	.19	7.52		13.0	3.6	7.9	12	12	19
500	(1102)			7.2	15.9	100	.42	.20	8.14		14.1	3.9	8.6	13	13	20
550	(1213)			7.7	17.0	100	.45	.22	8.75		15.1	4.2	9.2	14	14	22
600	(1323)			8.3	18.3	100	.49	.23	9.33		16.1	4.4	9.8	15	15	23
650	(1433)			8.8	19.4	100	.52	.25	9.91		17.1	4.7	10.4	16	16	25
						Dry pregnant mature cows—Last 2-3 months of pregnancy										
350	(772)	.4[3]	.9	6.9	13.9	100[4]	.41	.19	7.8		13.2	3.6	8.0	12	12	19
400	(882)	.4	.9	7.5	15.4	100	.44	.21	8.4		14.3	4.0	8.7	14	14	21
450	(992)	.4	.9	8.1	16.5	100	.48	.23	9.1		15.4	4.2	9.4	15	15	23
500	(1102)	.4	.9	8.6	17.9	100	.51	.24	9.7		16.4	4.5	10.0	15	15	24
550	(1213)	.4	.9	9.1	19.0	100	.54	.25	10.3		17.5	4.8	10.7	16	16	26
600	(1323)	.4	.9	9.7	20.3	100	.57	.27	10.9		18.5	5.1	11.2	17	17	27
650	(1433)	.4	.9	10.2	22.4	100	.60	.29	11.5		19.6	5.4	11.9	18	18	29

TABLE 8-18b (Continued)

Weight[1] kg	(lb)	Daily Gain kg	(lb)	Minimum Dry Matter Consumption[2] kg	(lb)	Roughage[2] %	Total Protein kg	Dig. Protein kg	NEm Mcal	NEg Mcal	ME[2] Mcal	TDN[2,3] kg	(lb)	Ca g	P g	Vitamin A (Thousands IU)
Cows nursing calves—Average milking ability[5]—First 3-4 months post-partum																
350	(772)			8.2	18.1	100[4]	.75	.44	9.2		15.9	4.4	9.7	24	24	19
400	(882)			8.8	19.4	100	.81	.48	9.9		17.0	4.7	10.4	25	25	21
450	(992)			9.3	20.5	100	.86	.50	10.5		18.1	5.0	11.0	26	26	23
500	(1102)			9.8	21.6	100	.90	.53	11.1		19.2	5.3	11.7	27	27	24
550	(1213)			10.5	23.1	100	.97	.57	11.9		20.3	5.6	12.3	28	28	26
600	(1323)			11.0	24.2	100	1.01	.59	12.3		21.3	5.9	13.0	28	28	27
650	(1433)			11.4	25.1	100	1.05	.62	12.9		22.3	6.2	13.7	29	29	29
Cows nursing calves—Superior milking ability[6]—First 3-4 months post-partum																
350	(772)			10.2	22.4	100[7]	1.11	0.65	12.3		21.0	5.8	12.8	45	40	32
400	(882)			10.8	23.8	100	1.17	0.69	13.0		22.1	6.1	13.5	45	41	34
450	(992)			11.3	24.9	100	1.23	0.72	13.6		23.2	6.4	14.1	45	42	36
500	(1102)			11.8	26.0	100	1.29	0.76	14.2		24.3	6.7	14.8	46	43	38
550	(1213)			12.4	27.3	100	1.35	0.79	14.9		25.3	7.0	15.4	46	44	41
600	(1323)			12.9	28.4	100	1.41	0.83	15.5		26.4	7.3	16.1	46	44	43
650	(1433)			13.4	29.5	100	1.46	0.86	16.2		27.5	7.6	16.8	47	45	45
Bulls, growth and maintenance (moderate activity)																
300	(661)	1.00	2.2	8.8	19.4	70-75	0.90	0.55	5.6	3.8	20.4	5.6	12.3	27	23	34
400	(882)	.90	2.0	11.0	24.2	70-75	1.03	0.62	6.9	4.1	25.2	7.0	15.4	23	23	43
500	(1102)	.70	1.5	12.2	26.9	80-85	1.07	0.62	8.5	3.7	27.0	7.5	16.5	22	22	48
600	(1323)	.50	1.1	12.0	26.4	80-85	1.02	0.60	9.8	3.0	26.4	7.3	16.1	22	22	48
700	(1543)	.30	0.7	12.9	28.4	90-100[7]	1.08	0.60	11.0	2.0	27.7	7.7	17.0	23	23	50
800	(1764)	0.0	0.0	10.5	23.1	100[7]	0.89	0.50	12.2	0	21.0	5.8	12.8	19	19	41
900	(1984)	0.0	0.0	11.4	25.1	100[7]	0.99	0.55	13.3	0	22.8	6.3	13.9	21	21	44
1000	(2205)	0.0	0.0	12.4	27.3	100[7]	1.05	0.60	14.4	0	24.8	6.9	15.2	22	22	48

[1]Average weight for a feeding period.
[2]Dry matter consumption, ME and TDN requirements are based on the general type of ration indicated in the Roughage column.
[3]Approximately .4 ± .1 kg of weight gain/day over the last third of pregnancy is accounted for by the products of conception.
[4]Average quality roughage containing about 1.9-2.0 Mcal ME/kg dry matter.
[5]5.0 ± .5 kg of milk/day.
[6]10 ± 1 kg of milk/day.
[7]Good quality roughage containing at least 2.0 Mcal ME/kg dry matter.

TABLE 8-19

NUTRIENT REQUIREMENTS FOR GROWING-FINISHING STEER CALVES AND YEARLINGS*
(NUTRIENT CONCENTRATION IN RATION DRY MATTER)

Weight[1]		Daily Gain[2]		Minimum Dry Matter Consumption[2]		Roughage[2]	Total Protein	Dig. Protein	NEm[5]		NEg[5]		ME[5]		TDN[5]	Ca	P
kg	(lb)	kg	(lb)	kg	(lb)	%	%	%	Mcal/kg	Mcal/lb	Mcal/kg	Mcal/lb	Mcal/kg	Mcal/lb	%	%	%
100	(220)	0	0	2.1	4.6	100	8.7	5.0	1.17	0.53	—	—	2.0	0.91	55	.18	.18
		.5	1.1	2.9	6.4	70-80	12.4	8.3	1.35	0.60	0.75	0.23	2.2	1.00	62	.48	.38
		.7	1.5	2.7	6.0	50-60	14.8	10.7	1.60	0.71	1.00	0.43	2.5	1.13	70	.70	.48
		.9	2.0	2.8	6.2	25-30	16.4	11.8	1.81	0.82	1.18	0.54	2.8	1.27	77	.86	.57
		1.1	2.4	2.7	6.0	<15	18.2	13.3	2.07	0.94	1.37	0.62	3.1	1.41	86	1.04	.70
150	(331)	0	0	2.8	6.2	100	8.7	5.0	1.17	0.53	—	—	2.0	0.91	55	.18	.18
		.5	1.1	4.0	8.8	70-80	11.0	7.0	1.35	0.60	0.75	0.23	2.2	1.00	62	.35	.32
		.7	1.5	3.9	8.6	50-60	12.6	8.5	1.60	0.71	1.00	0.43	2.5	1.13	70	.46	.36
		.9	2.0	3.8	8.4	25-30	14.1	9.7	1.81	0.82	1.18	0.54	2.8	1.27	77	.61	.45
		1.1	2.4	3.7	8.2	<15	15.6	11.1	2.07	0.94	1.37	0.62	3.1	1.41	86	.76	.54
200	(441)	0	0	3.5	7.7	100	8.5	4.8	1.17	0.53	—	—	2.0	0.91	55	.18	.18
		.5	1.1	5.8	12.8	80-90	9.9	6.0	1.25	0.56	0.60	0.27	2.1	0.95	58	.24	.22
		.7	1.5	5.7	12.6	70-80	10.8	6.8	1.40	0.64	0.78	0.35	2.3	1.04	64	.32	.28
		.9	2.0	4.9	10.8	35-45	12.3	8.2	1.70	0.78	1.10	0.50	2.7	1.22	75	.47	.37
		1.1	2.4	4.6	10.1	<15	13.6	9.3	2.07	0.94	1.37	0.62	3.1	1.41	86	.59	.43
250	(551)	0	0	4.1	9.7	100	8.5	4.8	1.17	0.53	—	—	2.0	0.91	55	.18	.18
		.7	1.5	5.8	12.8	55-65	10.7	6.7	1.56	0.71	0.95	0.43	2.5	1.13	70	.31	.28
		.9	2.0	6.2	13.7	45-50	11.1	7.1	1.64	0.74	1.02	0.46	2.6	1.18	72	.35	.31
		1.1	2.4	6.0	13.2	20-25	12.1	8.0	1.81	0.82	1.18	0.54	2.8	1.27	77	.43	.35
		1.3	2.9	6.0	13.2	<15	12.7	8.5	2.07	0.94	1.37	0.62	3.1	1.41	86	.50	.38
300	(661)	0	0	4.7	10.4	100	8.6	4.8	1.17	0.53	—	—	2.0	0.91	55	.18	.18
		.9	2.0	8.1	17.9	55-65	10.0	6.2	1.56	0.71	0.95	0.43	2.5	1.18	70	.27	.23
		1.1	2.4	7.6	16.8	20-25	10.8	6.8	1.81	0.82	1.18	0.54	2.8	1.27	77	.33	.29
		1.3	2.9	7.1	15.6	<15	11.7	7.6	1.98	0.90	1.31	0.59	3.0	1.36	83	.41	.32
		1.4[4]	3.1	7.3	16.1	<15	11.9	7.8	2.07	0.94	1.37	0.62	3.1	1.41	86	.42	.34
350	(772)	0	0	5.3	11.7	100	8.5	4.8	1.17	0.53	—	—	2.0	0.91	55	.18	.18
		.9	2.0	8.0	17.6	45-55	10.0	6.1	1.64	0.74	1.02	0.46	2.6	1.18	72	.25	.22
		1.1	2.4	8.0	17.6	20-25	10.4	6.5	1.81	0.82	1.18	0.54	2.8	1.27	80	.29	.25
		1.3	2.9	8.0	17.6	<15	10.8	6.9	1.98	0.90	1.31	0.59	3.0	1.36	83	.32	.28
		1.4[4]	3.1	8.2	18.1	<15	10.9	7.0	2.07	0.98	1.37	0.62	3.1	1.41	86	.34	.29
400	(882)	0	0	5.9	13.0	100	8.5	4.8	1.17	0.53	—	—	2.0	0.91	55	.18	.18
		1.0	2.2	9.4	20.7	45-55	9.4	5.7	1.64	0.74	1.02	0.46	2.6	1.18	72	.22	.21
		1.2	2.6	8.5	18.7	20-25	10.2	6.3	1.81	0.82	1.18	0.54	2.8	1.27	80	.27	.25
		1.3	2.9	8.6	19.0	<15	10.4	6.5	2.07	0.98	1.37	0.62	3.1	1.41	86	.29	.26
		1.4[4]	3.1	9.0	19.8	<15	10.5	6.6	2.07	0.98	1.37	0.62	3.1	1.41	86	.29	.26

TABLE 8-19 (Continued)

Weight[1]		Daily Gain[2]		Minimum Dry Matter Consumption[2]		Roughage[2]	Total Protein	Dig. Protein	NEm[5]		NEg[5]		ME[5]		TDN[5]	Ca	P
kg	(lb)	kg	(lb)	kg	(lb)	%	%	%	Mcal/kg	Mcal/lb	Mcal/kg	Mcal/lb	Mcal/kg	Mcal/lb	%	%	%
450	(992)	0	0	6.4	14.1	100	8.5	4.8	1.17	0.53	—	—	2.0	0.91	55	.18	.18
		1.0	2.2	10.3	22.7	45-55	9.3	5.5	1.64	0.74	1.02	0.46	2.6	1.18	72	.19	.19
		1.2	2.6	10.2	22.5	20-25	9.5	5.7	1.81	0.82	1.18	0.54	2.8	1.27	80	.23	.22
		1.3	2.9	9.3	20.5	<15	10.4	6.3	2.07	0.98	1.31	0.62	3.1	1.41	86	.26	.25
		1.4[4]	3.1	9.8	21.6	<15	10.0	6.1	2.07	0.98	1.37	0.62	3.1	1.41	86	.26	.23
500	(1102)	0	0	7.0	15.4	100	8.5	4.8	1.17	0.53	—	—	2.0	0.91	55	.18	.18
		.9	2.0	10.5	23.1	45-55	9.1	5.3	1.64	0.74	1.02	0.46	2.6	1.18	72	.18	.18
		1.1	2.4	10.4	22.9	20-25	9.2	5.5	1.81	0.82	1.18	0.54	2.8	1.27	80	.19	.19
		1.2	2.6	9.6	21.2	<15	10.0	6.0	2.07	0.98	1.31	0.62	3.1	1.41	86	.22	.22
		1.3[4]	2.9	10.0	22.0	<15	9.7	6.0	2.07	0.98	1.37	0.62	3.1	1.41	86	.22	.22

*Tables 8-19, 8-19a, and 8-19b from *Nutrient Requirements of Beef Cattle*, fifth revised edition, National Academy of Sciences, Washington, D.C., 1976, pp. 28-33.
The concentration of vitamin A in all rations for finishing steers and heifers is 2,200 IU/kg of dry ration; pregnant heifers and cows is 2,800 IU/kg dry ration; and lactating cows and breeding bulls is 3,900 IU/kg.

[1] Average weight for a feeding period.
[2] Dry matter consumption, ME and TDN allowances are based on NE requirements and the general types of ration indicated in the Roughage column. Most roughages will contain 1.9-2.2 Mcal of ME/kg dry matter and 90-100% concentrate rations are expected to contain 3.1-3.3 Mcal of ME/kg.
[3] TDN was calculated by assuming 3.6155 Mcal of ME per kg of TDN.
[4] Most steers of the weight indicated and not exhibiting compensatory growth will not sustain an energy intake necessary to maintain this rate of gain for an extended period.
[5] Due to conversion and rounding variation, the figures in these columns may not be in exact agreement with a similar energy concentration figure calculated from the data of Table 8-18.

TABLE 8-19a

NUTRIENT REQUIREMENTS FOR GROWING—FINISHING HEIFER CALVES AND YEARLINGS*
(Nutrient Concentration in Ration Dry Matter)

Weight[1]		Daily Gain		Minimum Dry Matter Consumption[2]		Roughage[2]	Total Protein	Dig. Protein	NEm[5]		NEg[6]		ME[5]		TDN[5]	Ca	P
kg	(lb)	kg	(lb)	kg	(lb)	%	%	%	Mcal/kg	Mcal/lb	Mcal/kg	Mcal/lb	Mcal/kg	Mcal/lb	%	%	%
100	(220)	0	0	2.1	4.6	100	8.7	5.0	1.17	0.53	—	—	2.0	0.91	55	.18	.18
		.5	1.1	3.0	6.6	70-80	12.4	8.3	1.32	0.60	0.70	0.32	2.2	1.00	61	.47	.37
		.7	1.5	2.9	6.4	50-60	14.4	10.0	1.56	0.71	0.95	0.43	2.5	1.13	69	.66	.48
		.9	2.0	3.0	6.6	25-30	15.9	11.3	1.81	0.82	1.18	0.54	2.8	1.27	77	.80	.57
		1.1	2.4	3.0	6.6	<15	17.8	13.0	2.07	0.94	1.37	0.62	3.1	1.41	86	.97	.63
150	(331)	0	0	2.8	6.2	100	8.7	5.0	1.17	0.53	—	—	2.0	0.91	55	.18	.18
		.5	1.1	4.1	9.0	70-80	11.0	7.1	1.32	0.60	0.70	0.32	2.2	1.00	61	.34	.29
		.7	1.5	4.0	8.8	50-60	12.4	8.2	1.56	0.71	0.95	0.43	2.5	1.13	69	.45	.35
		.9	2.0	4.0	8.8	25-30	13.5	9.2	1.81	0.82	1.18	0.54	2.8	1.27	77	.57	.42
		1.1	2.4	4.0	8.8	<15	15.0	10.5	2.07	0.94	1.37	0.62	3.1	1.41	86	.70	.50
200	(441)	0	0	3.5	7.7	100	8.5	4.9	1.17	0.53	—	—	2.0	0.91	55	.18	.18
		.3	.7	5.4	11.9	100	9.1	5.4	1.17	0.53	0.50	0.23	2.0	0.91	55	.18	.18
		.5	1.1	6.0	13.2	80-90	9.6	5.8	1.24	0.56	0.60	0.27	2.1	0.95	58	.23	.22
		.7	1.5	6.0	13.2	70-80	10.2	6.5	1.40	0.64	0.87	0.39	2.3	1.04	64	.30	.27
		.9	2.0	5.3	11.7	35-45	11.7	7.5	1.72	0.78	1.10	0.50	2.7	1.22	75	.41	.32
		1.1	2.4	5.0	11.0	<15	12.8	8.6	2.07	0.94	1.37	0.62	3.1	1.41	86	.50	.38
250	(551)	0	0	4.1	9.0	100	8.5	4.9	1.17	0.53	—	—	2.0	0.91	55	.18	.18
		.3	.7	6.4	14.1	100	8.9	5.2	1.17	0.53	0.50	0.23	2.1	0.91	55	.18	.18
		.5	1.1	6.5	14.3	80-90	9.5	5.7	1.24	0.56	0.60	0.27	2.1	0.95	58	.20	.20
		.7	1.5	5.8	12.8	55-65	10.5	6.5	1.64	0.74	1.02	0.46	2.6	1.18	72	.29	.26
		.9	2.0	5.9	13.0	35-45	11.1	7.1	1.81	0.82	1.18	0.54	2.8	1.27	77	.36	.29
		1.1	2.4	6.5	14.3	20-25	11.4	7.4	1.89	0.86	1.25	0.57	2.9	1.31	80	.38	.31
		1.2	2.6	6.3	13.9	<15	11.9	7.8	2.07	0.94	1.37	0.62	3.1	1.41	86	.43	.33
300	(661)	0	0	4.7	10.4	100	8.6	4.9	1.17	0.53	—	—	2.0	0.91	55	.18	.18
		.3	.7	7.4	16.3	100	8.5	4.9	1.17	0.53	0.50	0.23	2.0	0.91	55	.18	.18
		.5	1.1	7.4	16.3	80-90	9.2	5.4	1.32	0.60	0.70	0.32	2.2	1.00	61	.19	.19
		.7	1.5	6.6	14.6	55-65	10.1	6.1	1.64	0.74	1.02	0.46	2.6	1.18	72	.24	.23
		.9	2.0	6.8	15.0	35-45	10.4	6.5	1.81	0.82	1.18	0.54	2.8	1.27	77	.28	.25
		1.1	2.4	7.5	16.5	20-25	10.4	6.5	1.89	0.86	1.25	0.57	2.9	1.31	80	.31	.27
		1.2	2.6	7.2	15.9	<15	10.9	6.9	2.07	0.94	1.37	0.62	3.1	1.41	86	.33	.28
350	(772)	0	0	5.3	11.7	100	8.5	4.8	1.17	0.53	—	—	2.0	0.91	55	.18	.18
		.3	.7	8.2	18.1	100	8.5	4.8	1.17	0.53	0.50	0.23	2.0	0.91	55	.18	.18
		.5	1.1	8.3	18.3	80-90	8.7	5.1	1.32	0.60	0.70	0.32	2.2	1.00	61	.18	.18
		.7	1.5	7.9	17.4	55-65	9.2	5.4	1.56	0.71	0.95	0.43	2.5	1.13	69	.19	.19
		.9	2.0	8.1	17.9	35-45	9.5	5.7	1.72	0.78	1.10	0.50	2.7	1.22	75	.21	.21
		1.1	2.4	8.3	18.3	20-25	9.9	6.0	1.89	0.86	1.25	0.57	2.9	1.31	80	.24	.23
		1.2[4]	2.6	8.1	17.9	<15	10.0	6.2	2.07	0.94	1.37	0.62	3.1	1.41	86	.26	.25

TABLE 8-19a (Continued)

Weight[1]		Daily Gain		Minimum Dry Matter Consumption[2]		Roughage[2]	Total Protein	Dig. Protein	NEm[5]		NEg[5]		ME[5]		TDN[5]	Ca	P
kg	(lb)	kg	(lb)	kg	(lb)	%	%	%	Mcal/kg	Mcal/lb	Mcal/kg	Mcal/lb	Mcal/kg	Mcal/lb	%	%	%
400	(882)	0	0	5.9	13.0	100	8.5	4.8	1.17	0.53	—	—	2.0	0.91	55	.18	.18
		.3	.7	9.1	20.0	100	8.5	4.8	1.17	0.53	0.50	0.23	2.0	0.91	55	.18	.18
		.5	1.1	8.5	18.7	70-80	8.8	5.1	1.40	0.64	0.78	0.35	2.3	1.04	64	.18	.18
		.7	1.5	8.7	19.2	55-65	9.0	5.3	1.56	0.71	0.95	0.43	2.5	1.09	66	.18	.18
		.9	2.0	8.4	18.5	20-25	9.4	5.6	1.81	0.82	1.18	0.54	2.8	1.27	77	.20	.20
		1.1[4]	2.4	8.3	18.3	<15	9.7	5.9	2.07	0.94	1.37	0.62	3.1	1.41	86	.23	.22
450	(992)	0	0	6.4	14.1	100	8.5	4.8	1.17	0.53	—	—	2.0	0.91	55	.18	.18
		.2	.4	8.7	19.2	100	8.5	4.7	1.17	0.53	0.50	0.23	2.0	0.91	55	.18	.18
		.5	1.1	9.3	20.5	70-80	8.6	4.9	1.40	0.64	0.78	0.35	2.3	1.04	64	.18	.18
		.8	1.8	9.1	20.1	35-45	9.0	5.3	1.72	0.78	1.10	0.50	2.7	1.22	75	.18	.18
		1.0[4]	2.2	8.5	18.7	<15	9.5		2.07	0.94	1.37	0.62	3.1	1.41	86	.22	.22

*The concentration of vitamin A in all rations for finishing steers and heifers is 2,200 IU/kg of dry ration; pregnant heifers and cows is 2,800 IU/kg dry ration; and lactating cows and breeding bulls is 3,900 IU/kg.

[1] Average weight for a feeding period.

[2] Dry matter consumption, ME and TDN allowances are based on NE requirements and the general type of ration indicated in the Roughage column. Most roughages will contain 1.9-2.2 Mcal of ME/kg dry matter and 90-100% concentrate rations are expected to have 3.1 to 3.3 Mcal of ME/kg.

[3] TDN was calculated by assuming 3.6155 kcal of ME per g of TDN.

[4] Most heifers of the weight indicated and not exhibiting compensatory growth will not sustain the energy intake necessary to maintain this rate of gain for an extended period.

[5] Due to conversion and rounding variation, the figure in these columns may not be in exact agreement with a similar energy concentration figure calculated from the data of Table 8-18a.

TABLE 8-19b

NUTRIENT REQUIREMENTS FOR BEEF CATTLE BREEDING HERD*
(Nutrient Concentration in Ration Dry Matter)

Weight[1]		Daily Gain		Dry Matter[2] Consumption		Roughage[2]	Total Protein	Dig. Protein	NEm[8]		NEg[8]		ME[8]		TDN[8]	Ca	P
kg	(lb)	kg	(lb)	kg	(lb)	%	%	%	Mcal/kg	Mcal/lb	Mcal/kg	Mcal/lb	Mcal/kg	Mcal/lb	%	%	%
															%	%	%
Pregnant yearling heifers—Last 3-4 months of pregnancy
325	(716)	.4[3]	.9	6.6	14.5	100[4]	8.8	5.1	1.09	0.49	0.38	0.17	1.9	0.86	52	.23	.23
		.6	1.3	8.5	18.7	100	8.8	5.1	1.09	0.49	0.38	0.17	1.9	0.86	52	.21	.21
		.8	1.8	9.4	20.7	85-100	9.0	5.3	1.24	0.56	0.60	0.27	2.1	0.95	58	.23	.21
350	(772)	.4[3]	.9	6.9	15.2	100	8.8	5.1	1.09	0.49	0.38	0.17	1.9	0.86	52	.22	.22
		.6	1.3	8.9	19.6	100	8.8	5.1	1.09	0.49	0.38	0.17	1.9	0.86	52	.21	.21
		.8	1.8	10.0	22.0	85-100	8.8	5.1	1.24	0.56	0.60	0.27	2.1	0.95	58	.22	.21
375	(827)	.4[3]	.9	7.2	15.9	100	8.7	5.0	1.09	0.49	0.38	0.17	1.9	0.86	52	.21	.21
		.6	1.3	9.3	20.5	100	8.7	5.0	1.09	0.49	0.38	0.17	1.9	0.86	52	.20	.20
		.8	1.8	11.0	24.2	85-100	8.7	5.0	1.17	0.53	0.50	0.23	2.0	0.91	55	.20	.20
400	(882)	.4[3]	.9	7.5	16.5	100	8.7	5.0	1.09	0.49	0.38	0.17	1.9	0.86	52	.21	.21
		.6	1.3	9.7	21.4	100	8.7	5.0	1.09	0.49	0.38	0.17	1.9	0.86	52	.20	.20
		.8	1.8	11.6	25.6	85-100	8.7	5.0	1.17	0.53	0.50	0.23	2.0	0.91	55	.19	.19
425	(937)	.4[3]	.9	7.8	17.2	100	8.8	5.1	1.09	0.49	0.38	0.17	1.9	0.86	52	.20	.20
		.6	1.3	10.1	22.3	100	8.7	5.0	1.09	0.49	0.38	0.17	1.9	0.86	52	.19	.19
		.8	1.8	12.1	26.7	85-100	8.7	5.0	1.17	0.53	0.50	0.23	2.0	0.91	55	.18	.18
Dry pregnant mature cows—Middle third of pregnancy
350	(772)			5.5	12.2	100[4]	5.9	2.8	1.09	0.49	—	—	1.9	0.86	52	.18	.18
400	(882)			6.1	13.4	100	5.9	2.8	1.09	0.49	—	—	1.9	0.86	52	.18	.18
450	(992)			6.7	14.8	100	5.9	2.8	1.09	0.49	—	—	1.9	0.86	52	.18	.18
500	(1102)			7.2	15.9	100	5.9	2.8	1.09	0.49	—	—	1.9	0.86	52	.18	.18
550	(1213)			7.7	17.0	100	5.9	2.8	1.09	0.49	—	—	1.9	0.86	52	.18	.18
600	(1323)			8.3	18.3	100	5.9	2.8	1.09	0.49	—	—	1.9	0.86	52	.18	.18
650	(1433)			8.8	19.4	100	5.9	2.8	1.09	0.49	—	—	1.9	0.86	52	.18	.18
Dry pregnant mature cows—Last 2-3 months of pregnancy
350	(772)	.4[3]	.9	6.9	13.9	100[4]	5.9	2.8	1.09	0.49	—	—	1.9	0.86	52	.18	.18
400	(882)	.4	.9	7.5	15.4	100	5.9	2.8	1.09	0.49	—	—	1.9	0.86	52	.18	.18
450	(992)	.4	.9	8.1	16.5	100	5.9	2.8	1.09	0.49	—	—	1.9	0.86	52	.18	.18
500	(1102)	.4	.9	8.6	17.9	100	5.9	2.8	1.09	0.49	—	—	1.9	0.86	52	.18	.18
550	(1213)	.4	.9	9.1	19.0	100	5.9	2.8	1.09	0.49	—	—	1.9	0.86	52	.18	.18
600	(1323)	.4	.9	9.7	20.3	100	5.9	2.8	1.09	0.49	—	—	1.9	0.86	52	.18	.18
650	(1433)	.4	.9	10.2	22.4	100	5.9	2.8	1.09	0.49	—	—	1.9	0.86	52	.18	.18

TABLE 8-19b (Continued)

Weight[1]		Daily Gain		Dry Matter[2] Consumption		Roughage[2]	Total Protein	Dig. Protein	NEm*		NEg*		ME*		TDN*	Ca	P
kg	(lb)	kg	(lb)	kg	(lb)	%	%	%	Mcal/kg	Mcal/lb	Mcal/kg	Mcal/lb	Mcal/kg	Mcal/lb	%	%	%
						Cows nursing calves—Average milking ability[5]—First 3-4 months post-partum											
350	(772)			8.2	18.1	100[4]	9.2	5.4	1.09	0.49	—	—	1.9	0.86	52	0.29	0.29
400	(882)			8.8	19.4	100	9.2	5.4	1.09	0.49	—	—	1.9	0.86	52	0.28	0.28
450	(992)			9.3	20.5	100	9.2	5.4	1.09	0.49	—	—	1.9	0.86	52	0.28	0.28
500	(1102)			9.8	21.6	100	9.2	5.4	1.09	0.49	—	—	1.9	0.86	52	0.28	0.28
550	(1213)			10.5	23.1	100	9.2	5.4	1.09	0.49	—	—	1.9	0.86	52	0.27	0.27
600	(1323)			11.0	24.2	100	9.2	5.4	1.09	0.49	—	—	1.9	0.86	52	0.25	0.25
650	(1433)			11.4	25.1	100	9.2	5.4	1.09	0.49	—	—	1.9	0.86	52	0.25	0.25
						Cows nursing calves—Superior milking ability[5]—First 3-4 months post-partum											
350	(772)			10.2	22.4	100[7]	10.9	6.4	1.17	0.53	—	—	2.0	0.91	55	0.44	0.39
400	(882)			10.8	23.8	100	10.9	6.4	1.17	0.53	—	—	2.0	0.91	55	0.42	0.38
450	(992)			11.3	24.9	100	10.9	6.4	1.17	0.53	—	—	2.0	0.91	55	0.40	0.37
500	(1102)			11.8	26.0	100	10.9	6.4	1.17	0.53	—	—	2.0	0.91	55	0.39	0.36
550	(1213)			12.4	27.3	100	10.9	6.4	1.17	0.53	—	—	2.0	0.91	55	0.37	0.35
600	(1323)			12.9	28.4	100	10.9	6.4	1.17	0.53	—	—	2.0	0.91	55	0.36	0.34
650	(1433)			13.4	29.5	100	10.9	6.4	1.17	0.53	—	—	2.0	0.91	55	0.35	0.33
						Bulls, growth and maintenance (moderate activity)											
300	(661)	1.00	2.2	8.8	19.4	70-75	10.2	6.3	1.40	0.64	0.78	0.35	2.3	1.04	64	0.31	0.26
400	(882)	.90	2.0	11.0	24.2	70-75	9.4	5.6	1.40	0.64	0.78	0.35	2.3	1.04	64	0.21	0.21
500	(1102)	.70	1.5	12.2	26.9	80-85	8.8	5.1	1.32	0.60	0.70	0.32	2.2	1.00	61	0.18	0.18
600	(1323)	.50	1.1	12.0	26.4	80-85	8.8	5.0	1.32	0.60	0.70	0.32	2.2	1.00	61	0.18	0.18
700	(1543)	.30	0.7	12.9	28.4	90-100[7]	8.5	4.8	1.17	0.53	0.50	0.17	2.0	0.91	55	0.18	0.18
800	(1764)	0.0	0.0	10.5	23.1	100[7]	8.5	4.8	1.17	0.53	—	—	2.0	0.91	55	0.18	0.18
900	(1984)	0.0	0.0	11.4	25.1	100[7]	8.5	4.8	1.17	0.53	—	—	2.0	0.91	55	0.18	0.18
1000	(2205)	0.0	0.0	12.4	27.3	100[7]	8.5	4.8	1.17	0.53	—	—	2.0	0.91	55	0.18	0.18

*The concentration of vitamin A in all rations for finishing steers and heifers is 2,200 IU/kg of dry ration; pregnant heifers and cows is 2,800 IU/kg dry ration; and lactating cows and breeding bulls is 3,900 IU/kg.

[1] Average weight for a feeding period.

[2] Dry matter consumption, ME and TDN requirements are based on the general type of ration indicated in the Roughage column.

[3] Approximately .4 ± .1 kg of weight gain/day over the last third of pregnancy is accounted for by the products of conception.

[4] Average quality roughage containing about 1.9-2.0 Mcal ME/kg dry matter.

[5] 5.0 ± .5 kg of milk/day.

[6] 10 ± 1 kg of milk/day.

[7] Good quality roughage containing 2.0 Mcal ME/kg dry matter.

[8] Due to conversion and rounding variation, the figures in these columns may not be in exact agreement with a similar figure calculated from the data in Table 8-18b.

TABLE 8-20

Net Energy Requirements of Growing and Finishing Beef Cattle*
(Megacalories per Animal per Day)

Total Weight (kg):	100	150	200	250	300	350	400	450	500
NE Required:	2.43	3.30	4.10	4.84	5.55	6.24	6.89	7.52	8.14
Daily Gain (kg)	NEgain Required								
STEERS									
0.1	0.17	0.23	0.28	0.34	0.39	0.43	0.48	0.52	0.56
0.2	0.34	0.46	0.57	0.68	0.78	0.88	0.97	1.06	1.14
0.3	0.52	0.70	0.87	1.03	1.18	1.33	1.47	1.61	1.74
0.4	0.70	0.95	1.18	1.40	1.60	1.80	1.99	2.17	2.34
0.5	0.89	1.20	1.49	1.77	2.02	2.27	2.51	2.74	2.97
0.6	1.08	1.46	1.81	2.15	2.46	2.76	3.05	3.33	3.60
0.7	1.27	1.73	2.14	2.53	2.90	3.26	3.60	3.93	4.25
0.8	1.47	2.00	2.47	2.93	3.36	3.77	4.17	4.55	4.92
0.9	1.68	2.27	2.81	3.33	3.82	4.29	4.74	5.18	5.60
1.0	1.88	2.55	3.16	3.75	4.29	4.82	5.33	5.82	6.29
1.1	2.10	2.84	3.52	4.17	4.78	5.36	5.93	6.47	7.01
1.2	2.31	3.13	3.88	4.60	5.27	5.92	6.54	7.14	7.73
1.3	2.53	3.43	4.26	5.04	5.77	6.48	7.16	7.83	8.47
1.4	2.76	3.74	4.63	5.49	6.29	7.06	7.80	8.52	9.22
1.5	2.99	4.05	5.02	5.95	6.81	7.65	8.46	9.23	9.98
HEIFERS									
0.1	0.18	0.25	0.30	0.36	0.41	0.46	0.51	0.56	0.61
0.2	0.37	0.50	0.62	0.74	0.84	0.95	1.05	1.14	1.24
0.3	0.57	0.77	0.95	1.13	1.29	1.45	1.61	1.75	1.90
0.4	0.77	1.05	1.30	1.54	1.76	1.98	2.18	2.39	2.58
0.5	0.99	1.34	1.66	1.96	2.25	2.52	2.79	3.05	3.30
0.6	1.21	1.64	2.03	2.40	2.75	3.09	3.41	3.73	4.03
0.7	1.44	1.95	2.42	2.86	3.27	3.68	4.06	4.44	4.80
0.8	1.67	2.28	2.81	3.33	3.82	4.28	4.73	5.17	5.59
0.9	1.92	2.60	3.23	3.81	4.37	4.91	5.43	5.93	6.41
1.0	2.17	2.94	3.65	4.32	4.95	5.56	6.14	6.71	7.26
1.1	2.43	3.30	4.09	4.84	5.55	6.23	6.88	7.52	8.13
1.2	2.70	3.66	4.55	5.37	6.16	6.91	7.64	8.35	9.03
1.3	2.98	4.04	5.01	5.92	6.79	7.63	8.42	9.21	9.96
1.4	3.26	4.42	5.49	6.49	7.44	8.36	9.23	10.09	10.91
1.5	3.56	4.82	5.98	7.07	8.11	9.11	10.06	11.00	11.90

*Source: *Nutrient Requirements of Beef Cattle*, fifth revised edition, National Academy of Sciences, Washington, D.C., 1976, p. 34.

QUESTIONS FOR STUDY AND DISCUSSION

1. Why is knowledge of beef cattle feeding so important?

2. Compare the nutritive needs of beef cattle with those of simple-stomached animals.

3. Sketch the ruminant's stomach and show the route of digestion followed by most feeds.

4. List and discuss the factors upon which the daily consumption of dry matter (feed) is primarily dependent.

5. In recent years, crossbreeding and the exotic breeds have produced faster-gaining calves, later-maturing cattle, and heavier-milking cows. How has this changed nutritive needs?

6. Describe the symptoms of (a) energy deficiency, and (b) protein deficiency in cattle.

7. Discuss the advantages and disadvantages of each (a) the TDN system, and (b) the calorie system.

8. Discuss the impact of the expanding world human population on the use of urea and other nonprotein sources.

9. Explain how you would go about determining mineral needs and buying a mineral mix.

10. For beef cattle, list the vitamins most apt to be deficient; then (a) list some of the deficiency symptoms, and (b) give practical sources of each vitamin for use on a farm or ranch.

11. Will the proportion of concentrates and roughages used in beef cattle rations change in the years ahead? Justify your answer.

12. What recent (within the past decade) nutritional discovery has had the greatest impact on the beef cattle industry? Justify your answer.

13. How would you go about selecting a commercial feed for beef cattle?

14. Which of the three methods listed for balancing rations do you prefer? Justify your choice.

15. Table 8-16 shows that beef cattle are near the bottom of the "animal totem pole" in "feed to food efficiency." In light of this fact, will beef cattle decline in numbers throughout the world?

16. In order to feed beef cattle efficiently and economically, one must thoroughly understand the nutrients furnished by the available feeds, the extent to which cattle can utilize each feed, and the actual feeding value of these feeds. This can be accomplished only through careful and thorough study of the different feeds.

    Table 8-13 of this chapter is a summary of the comparative values of the most common United States feeds. However, the following exercises are designed better to acquaint the student with the feeds commonly fed to beef cattle:

    a. *A study of available roughages*–Table 8-1 shows that 73.8 percent of the feed of beef cattle is derived from roughages. Also, it is well known that a relatively large portion of the feed consumed by beef cattle is used in meeting the energy needs. Thus, a convenient and reasonably accurate way of determining which roughages are most economical under the conditions existing in a particular area at any given time is to compute the cost at which each of the available roughages furnishes 100 pounds of total digestible nutrients. This is a measure of the economy with which the various feeds furnish fuel or energy. Refer to Table I-3 in the Appendix of this book for analyses, and obtain prices of available roughages locally. Then fill out the following table:

TABLE I— AVAILABLE ROUGHAGES

Feed	Farm Price per Ton	Total Protein per 100 Lb	Digestible Protein per 100 Lb	TDN per 100 Lb	Cost per 100 lb TDN	Carotene Mg per Lb	Calcium Content (%)	Phosphorus Content (%)

The student should also become familiar with the protein, calcium, phosphorus, and carotene content of the different roughages. The above table is so designed.

  b. *A study of available grains and by-products*—At least one of the cereal grains is grown in almost every section of the country, and all of them are used quite widely as beef cattle feeds. As a group, the cereals and their by-products are high in energy. However, they possess certain nutritive deficiencies which may prove to be quite limiting if they are not properly used. Refer to Table I-3 in the Appendix of this book for analyses, and obtain prices of available grain and by-product feeds locally. Then fill out the following table:

TABLE II—AVAILABLE GRAIN AND BY-PRODUCT FEEDS

Feed	Retail Price per Cwt	Total Protein per 100 Lb	Digestible Protein per 100 Lb	TDN per 100 Lb	Cost per 100 Lb Digestible Protein	Cost per 100 Lb TDN	Carotene mg per lb	Calcium Content (%)	Phosphorus Content (%)

In addition to considering the cost per 100 pounds of each TDN and digestible protein, give consideration to calcium, phosphorus, and carotene content of each of these feeds. Also, in studying the by-product feeds, be sure to understand the source of the feed and just what part of the original grain or seed goes into the by-product.

17. Select a specific class of beef cattle and prepare a balanced ration, using those feeds that are available at the lowest cost.

SELECTED REFERENCES

Title of Publication	Author(s)	Publisher
AFMA Liquid Feed Symposium Proceedings	American Feed Manufacturers Association	AFMA, Chicago, Ill., 1971
Alternative Sources of Protein for Animal Production	National Research Council	National Academy of Sciences, Washington, D.C., 1973
Animal Feeds	M. Gutcho	Noyes Data Corporation, Park Ridge, N.J., 1970
Animal Nutrition	L. A. Maynard J. K. Loosli	McGraw-Hill Book Co., New York, N.Y., 1969
Antibiotics in Nutrition	Thomas H. Jukes	Medical Encyclopedia, Inc., New York, N.Y., 1955

(Continued)

Title of Publication	Author(s)	Publisher
Applied Animal Feeding and Nutrition	M. H. Jurgens	Kendall/Hunt Publishing Company, Dubuque, Iowa, 1972
Applied Animal Nutrition	E. W. Crampton L. E. Harris	W. H. Freeman and Co., San Francisco, Calif., 1969
Association of American Feed Control Officials Incorporated, Official Publication	Association of American Feed Control Officials, Inc.	Association of American Feed Control Officials, Inc., Annual
Atlas of Nutritional Data on United States and Canadian Feeds	National Research Council, U.S.A.; Committee on Feed Composition, Research Branch, Canada Dept. of Agriculture	National Academy of Sciences, Washington, D.C., 1971
Body Composition in Animals and Man, Pub. 1598	National Research Council	National Academy of Sciences, Washington, D.C., 1968
Cereal Processing and Digestion	U.S. Feed Grains Council USDA Foreign Agricultural Service	U.S. Feed Grains Council, London, England, 1972
Composition of Cereal Grains and Forages, Pub. 585	National Research Council	National Academy of Sciences, Washington, D.C.
Composition of Concentrate By-Product Feeding Stuffs, Pub. 449	National Research Council	National Academy of Sciences, Washington, D.C., 1956
Digestive Physiology and Nutrition of the Ruminant	Ed. by D. Lewis	Butterworth & Co., Ltd., London, England, 1961
Digestive Physiology and Nutrition of Ruminants, Volumes 1 and 3	D. C. Church	D. C. Church, Dept. of Animal Science, Oregon State University, Corvallis, Ore., 1969 and 1972
Digest of Research on Urea and Ruminant Nutrition	E. I. Du Pont De Nemours & Company (Inc.), Polychemicals Department	E. I. Du Pont De Nemours & Company, Wilmington, Del., 1958
Effect of Processing on the Nutritional Value of Feeds	National Research Council	National Academy of Sciences, Washington, D.C., 1973
Energy Metabolism of Ruminants	K. L. Blaxter	Hutchinson & Co., Ltd., London, England, 1962
1974 Feed Additive Compendium	Ed. by D. Natz, The Animal Health Institute, Washington, D.C.	The Miller Publishing Company, Minneapolis, Minn., Annual

(Continued)

Title of Publication	Author(s)	Publisher
Feed Composition, Tables of, Pub. 1232	National Research Council	National Academy of Sciences, Washington, D.C., 1964
Feed Flavor and Animal Nutrition	T. B. Tribble	Agriaids, Inc., Chicago, Ill., 1962
Feed Formulations	T. W. Perry	The Interstate Printers & Publishers, Inc., Danville, Ill., 1975
Feeds and Feeding, Abridged	F. B. Morrison	The Morrison Publishing Company, Ithaca, N.Y., 1956
Feeds and Feeding, 22nd Edition	F. B. Morrison	The Morrison Publishing Company, Ithaca, N.Y., 1956
Feeds for Livestock, Poultry and Pets	M. H. Gutcho	Noyes Data Corporation, Park Ridge, N.J., 1973
Feeds of the World	B. H. Schneider	Agricultural Experiment Station, West Virginia University, Morgantown, W. Va., 1947
Fundamentals of Nutrition	E. W. Crampton L. E. Lloyd	W. H. Freeman and Company, San Francisco, Calif., 1959
Handbook of Feedstuffs, The	R. Seiden W. H. Pfander	Springer Publishing Company. Inc., New York, N.Y., 1957
International Conference on the Use of Antibiotics in Agriculture, Proceedings	American Cyanamid Company, Scientific Sessions	American Cyanamid Company, Wayne, N.J.
International Feed Nomenclature and Methods for Summarizing and Using Feed Data to Calculate Diets, An, Bull. 479	L. E. Harris J. M. Asplund E. W. Crampton	Agricultural Experiment Station, Utah State University, Logan, Utah, 1968
Lysine in Animal Nutrition, Annotated Bibliography	Merck Sharp & Dohme Research Laboratories	Merck & Co., Inc., Rahway, N.J., 1960
Manual of Clinical Nutrition	R. S. Goodhart M. G. Wohl	Lea & Febiger Philadelphia, Penn., 1964
Manual for the Computer Formulation of Livestock Feed Mixtures	W. K. McPherson	M. L. McPherson, Gainesville, Fla., 1971
Mineral Metabolism, An Advanced Treatise, Volume I, Part A, Volume II, Part A, and Volume III	Ed. by C. L. Comar, Felix Bronner	Academic Press Inc., New York, N.Y., 1960, 1964, 1969
Mineral Nutrition of Livestock, The	E. J. Underwood	Food and Agriculture Organization of the United Nations, Commonwealth Agricultural Bureaux, Rome, Italy, 1966
Mineral Nutrition of Plants and Animals	F. A. Gilbert	University of Oklahoma Press, Norman, Okla., 1948

(Continued)

Title of Publication	Author(s)	Publisher
Nonprotein Nitrogen in the Nutrition of Ruminants	J. K. Loosli I. W. McDonald	Food and Agriculture Organization of the United Nations, Rome, Italy, 1968
Nutrient Requirements of Beef Cattle, Fifth Revised Edition	National Research Council	National Academy of Sciences, Washington, D.C., 1976
Nutrient Requirements of Farm Livestock, The, No. 3, Pigs	Agricultural Research Council	Agricultural Research Council, London, England, 1967
Nutritional Data	H. A. Wooster, Jr. F. C. Blanck	H. J. Heinz Company, Pittsburgh, Penn., 1950
Nutritional Deficiencies in Livestock	R. T. Allman T. S. Hamilton	Food and Agriculture Organization of the United Nations, Rome, Italy, 1952
Nutrition of Animals of Agricultural Importance, Parts 1 and 2	Ed. by D. Cuthbertson	Pergamon Press, London, England, 1969
Nutrition of Plants, Animals, Man	College of Agriculture, Michigan State University	Michigan State University, East Lansing, Mich., 1955
Nutrition Research Techniques for Domestic and Wild Animals, Volume 1	L. E. Harris	Lorin E. Harris, Logan, Utah, 1970
Physiology of Digestion in the Ruminant	Dougherty, *et al.*	Butterworth, Inc., Washington, D.C., 1964
Proceedings of the Fifth International Congress on Nutrition	National Research Council	Waverly Press, Inc., Baltimore, Md., 1961
Processed Plant Protein Foodstuffs	Ed. by A. M. Altschul	Academic Press Inc., New York, N.Y., 1958
Processing and Utilization of Animal By-Products	I. Mann	Food and Agriculture Organization of the United Nations, Rome, Italy, 1962
Proteins - Their Chemistry and Politics	A. M. Altschul	Basic Books, Inc., New York, N.Y., 1965
Rations for Livestock, Bull. No. 48	R. E. Evans	Her Majesty's Stationery Office, London, England, 1960
Rumen and Its Microbes, The	R. E. Hungate	Academic Press Inc., New York, N.Y., 1966
Selenium in Nutrition	National Research Council	National Academy of Sciences, Washington, D.C., 1971
Single-Cell Protein	Ed. by R. I. Mateles, S. R. Tannenbaum	The M.I.T. Press, Cambridge, Mass., 1968

(Continued)

Title of Publication	Author(s)	Publisher
Stockman's Hand-book, The Fourth Edition	M. E. Ensminger	The Interstate Printers & Publishers, Inc., Danville, Ill., 1970
Trace Elements in Agriculture	V. Sauchelli	Van Nostrand Reinhold Company, New York, N.Y., 1969
Urea and Non-Protein Nitrogen in Ruminant Nutrition, Second Edition	Ed. by H. J. Stangel	Nitrogen Division, Allied Chemical Corporation, Morristown, N.J., 1963
Urea as a Protein Supplement	Ed. by M. H. Briggs	Pergamon Press, New York, N.Y., 1967
Use of Drugs in Animal Feeds, The, Pub. 1679	National Academy of Sciences	National Academy of Sciences, Washington, D.C., 1969
Vitamin B12 in Animal Nutrition	Merck & Co., Inc., Chemical Division	Merck & Co., Inc., Rahway, N.J., 1957
Vitamin B12, Selected Annotated Bibliography	Merck & Co., Inc., Chemical Division	Merck & Co., Inc., Rahway, N.J., 1954
Vitamins, The - Chemistry, Physiology, Pathology, Methods, Volumes I-III, Second Edition	Ed. by W. H. Sebrell, Jr., R. S. Harris	Academic Press Inc., New York, N.Y., 1967, 1968, 1971
Vitamins in Feeds for Livestock	F. C. Aitken R. G. Hankin	Commonwealth Agricultural Bureaux, Farnham Royal, Bucks, England, 1970
Vitamins and Hormones, Volume XV	Ed. by R. S. Harris, G. F. Marrian, K. V. Thimann	Academic Press Inc., New York, N.Y., 1957

CHAPTER 9

PASTURES AND RANGES

Contents   **Page**

The great cattle producing areas of the world are characterized by good pastures. Good cattlemen, good cattle, and good pastures go hand in hand. Indeed, pasture is the cornerstone of successful beef cattle production. The economic importance of pastures is further attested by the following facts:

553

. A total of 26.7 percent of the total land area of the United States (50 _) is used solely for grassland. (See Fig. 9-3.)

2. A total of 73.8 percent of the feed supply of all United States beef cattle is derived from forage; in season this means pasture.

3. Good pasture alone will produce 200 to 400 pounds of beef per acre annually (in weight of calves weaned, or in added weight of older cattle); superior pastures will do much better.

4. No method of harvesting has yet been devised which is as cheap as that which can be accomplished by animals, although nutrient yields are generally less when highly productive forage crops are pastured rather than harvested as hay or silage. The difference in yield is generally more than offset by the added expense of harvesting, storing, and feeding.

Fig. 9-1. On pasture. (Courtesy, American Shorthorn Association, Omaha, Neb.)

Fig. 9-2. On the range. (Courtesy, Arizona Cattle Growers' Association, Phoenix, Ariz.)

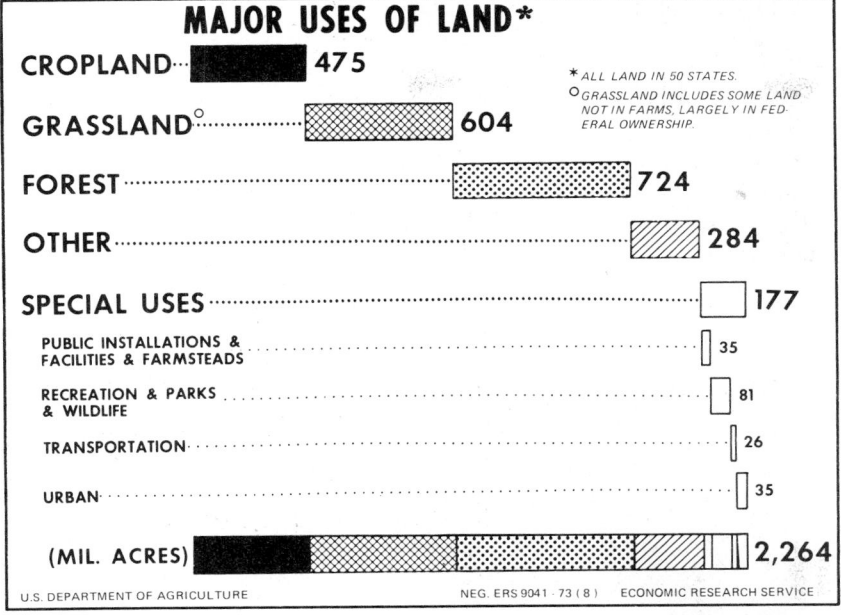

MAJOR USES OF LAND*

CROPLAND··· 475

GRASSLAND° ············· 604

FOREST ················· 724

OTHER ················· 284

SPECIAL USES ················· 177

PUBLIC INSTALLATIONS &
FACILITIES & FARMSTEADS ·················· 35

RECREATION & PARKS
& WILDLIFE ·················· 81

TRANSPORTATION· · · · · · · · · 26

URBAN· · · · · · · · · · · 35

(MIL. ACRES) 2,264

*ALL LAND IN 50 STATES.
°GRASSLAND INCLUDES SOME LAND
NOT IN FARMS, LARGELY IN FED-
ERAL OWNERSHIP.

U.S. DEPARTMENT OF AGRICULTURE   NEG. ERS 9041 - 73 (8)   ECONOMIC RESEARCH SERVICE

Fig. 9-3. About 604 million acres of pasture and range area in the United States are used solely as grazing land. (Courtesy, USDA)

5. Pasture gains are generally cheaper than drylot gains because (a) less labor is required, (b) grass is the cheapest of all roughages, (c) less expensive protein supplement is required, (d) the animals scatter their own droppings, thus alleviating hauling manure, and (e) fewer buildings and less equipment are necessary.

But grass—the nation's largest crop—should not be taken for granted. Again and again, scientists and practical farmers and ranchers have demonstrated that the following desired goals in pasture production are well within the realm of possibility:

● To produce higher yields of palatable and nutritious forage.

● To extend the grazing season from as early in the spring to as late in the fall as possible.

● To provide a fairly uniform supply of feed throughout the entire season.

At the outset, it should be recognized that no one plant embodies all the desirable characteristics necessary to meet the above goals. None of them will grow year-round, or during extremely cold or dry weather. Each of them has a period of peak growth which must be conserved for periods of little growth. Consequently, the progressive cattleman will find it desirable (1) to grow more than one species, and (2) to plan pastures for each season of the year. In general, a combination of permanent, rotation, and temporary pastures—accompanied by scientific management—will best achieve these ends.

Broadly speaking, all United States pastures may be classified as either (1) tame (seeded) pastures, or (2) native pastures. (See Fig. 9-4.) Although no sharp line of demarcation exists between the two groups, tame pastures include those which either receive more than approximately 20 inches of rainfall annually or are irrigated, whereas the latter group includes those range pastures which receive less than 20 inches of rainfall annually. The general principles and the objectives sought are the same, but, as will be discussed later, there are considerable differences in the recommended seeding and management practices for the two groups.

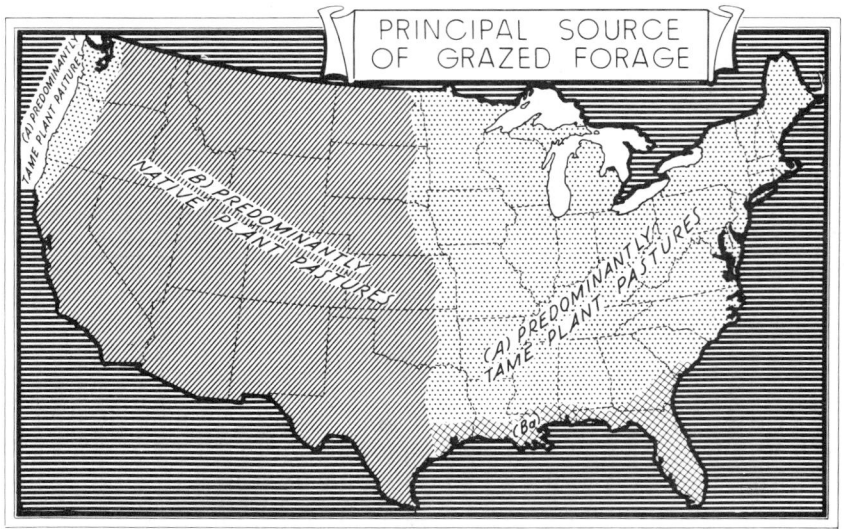

Fig. 9-4. The two major United States pasture areas—(A) tame (seeded), and (B) native (range)—about equally divide the 48 contiguous states into east and west halves. (Courtesy, USDA)

Pastures may be further classified as follows:

1. *Permanent pastures*—Those which, with proper care, last for many years. They are most commonly found on land that cannot be used profitably for cultivated crops, mainly because of topography, moisture, or fertility. The vast majority of the farms of the United States have one or more permanent pastures, and most range areas come under this classification.

2. *Rotation pastures*—Those that are used as a part of the established crop rotation. They are generally used for two to seven years before plowing.

3. *Temporary and supplemental pastures*—Those that are used for a short period, usually annuals such as Sudan grass, millet, rye, wheat, oats, rape, or soybeans. They are seeded for the purpose of providing supplemental grazing during the season when the regular permanent or rotation pastures are relatively unproductive.

PART I. TAME (SEEDED) PASTURES

Tame pastures—those which either receive over 20 inches of rainfall annually or are irrigated—include the seeded (cultivated) pastures of the Corn Belt, the South, the East, and the irrigated areas, and smaller and scattered moderate- to high-rainfall areas throughout the West.

ADAPTED VARIETIES AND SUITABLE MIXTURES

The specific grass or grass-legume mixture will vary from area to area, according to differences in soil, temperature, and rainfall. A complete listing of all adapted and recommended grasses and legumes for cattle pastures would be too lengthy for this book. However, Table 9-1 shows the most important ones for each of the 10 generally recognized United States pasture areas (Fig. 9-5). In using this chart, bear in mind that many species of forages have wide geographic adaption, but subspecies or varieties often have rather specific adaptation. Alfalfa, for example, is represented by many varieties which give this species adaptation to nearly all states. Variety then, within species, make many forages adapted to the widely varying climate and geographic areas. The county agricultural agent or state agricultural college can furnish recommendations for the area that they serve.

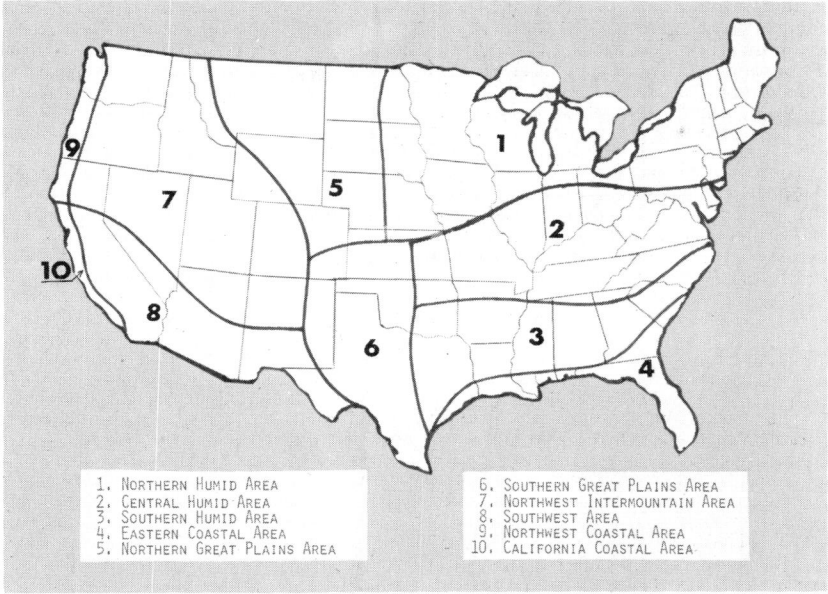

1. NORTHERN HUMID AREA
2. CENTRAL HUMID AREA
3. SOUTHERN HUMID AREA
4. EASTERN COASTAL AREA
5. NORTHERN GREAT PLAINS AREA
6. SOUTHERN GREAT PLAINS AREA
7. NORTHWEST INTERMOUNTAIN AREA
8. SOUTHWEST AREA
9. NORTHWEST COASTAL AREA
10. CALIFORNIA COASTAL AREA

Fig. 9-5. The 10 generally recognized United States pasture areas.

TABLE 9-1

GRASSES AND LEGUMES ADAPTED TO THE 10 AREAS OF THE 48 CONTIGUOUS STATES[1]

	Regions of Adaptation (See Fig. 9-5)									
	1	2	3	4	5	6	7	8	9	10
GRASSES										
Bahia Grass			X	X						
Bermuda Grass		X	X	X				X		X
Big Bluestem	X				X	X		X		
Blue Grama						X				
Blue Panicum						X				
Buffalo Grass					X	X				
Carpet Grass			X	X						
Chewings Fescue								X		
Common Ryegrass	X	X	X	X				X	X	X
Creeping Red Fescue								X		
Dallis Grass			X	X				X		X
Harding Grass								X		X
Indian Grass					X	X				
Kentucky Bluegrass	X	X						X		
Little Bluestem	X				X	X				
Lovegrasses					X	X	X			
Meadow Fescue	X	X					X		X	
Meadow Foxtail	X						X		X	
Millets	X				X	X				
Mountain Bromegrass							X			
Orchardgrass	X	X	X	X			X	X	X	X
Pangola				X						
Perennial Ryegrass	X	X	X	X				X	X	X
Redtop	X	X	X					X		
Reed Canary Grass	X	X							X	X
Rescuegrass			X			X		X		X
Rhodes Grass						X		X		X
Sand Dropseed					X	X				
Side Oats Grama					X	X				
Smooth Bromegrass	X					X	X	X		X
Soft Chess										X
Tall Fescue	X	X	X	X			X	X	X	X
Tall Oatgrass	X	X					X	X	X	
Timothy	X	X			X		X	X	X	
Veldt Grass										X
Wheatgrasses						X	X	X		
Yellow Bluestem						X				
LEGUMES										
Alfalfa	X	X	X	X	X	X	X	X	X	X
Alsike Clover	X	X			X		X		X	
Birdsfoot Trefoil	X	X					X	X	X	
Bur Clover			X	X		X			X	X
Crimson Clover			X	X	X					X
Hop Clover			X							
Kudzu		X	X	X						
Ladino Clover	X		X	X	X	X	X	X	X	X
Lespedeza		X	X	X						
Lupine			X							
Persian Clover			X							
Red Clover	X	X	X		X	X	X	X	X	X
Sweet Clover	X	X	X		X	X	X	X	X	X
Vetches	X	X	X	X	X	X	X	X	X	
White Clover	X	X	X	X	X	X	X	X	X	X

[1]Adapted from Keystone Steel and Wire, Peoria, Ill.

Five grass species—orchardgrass, reed canary grass, fescue, smooth bromegrass, and Bermuda grass—account for the major portion of tame grasses in the United States. The leading legumes are alfalfa, trefoil, lupine, sweet clover, Kudzu, and clover.

Most cattle pastures can be improved by fertilizing and management. Also, cattlemen need to give attention to balancing pastures nutritionally. Early-in-the-season grasses are of high water content and lack energy. Mature weathered grass is almost always deficient in protein (being as low as 3% or less) and low in carotene. But these deficiencies can be corrected by proper supplemental feeding.

ESTABLISHING A NEW PASTURE

The following practices are usually adhered to in successfully establishing a new pasture in the tame pasture area:

1. *Adapted varieties and suitable mixtures are selected*—The first requisite of successful pastures is that adapted varieties of grasses and/or legumes shall be selected for the area and for the purposes intended. Table 9-1 gives the general recommendations. For more specific recommendations for a particular farm or ranch, the cattleman should consult such local authorities as the county agent, vocational agriculture instructor, or successful neighbors.

Where grass-legume mixtures are to be grown, a 50-50 mixture is satisfactory for most purposes and conditions.

2. *The soil is tested and fertilized*—The soil is tested and fertilized (and limed if necessary) according to needs. The three elements required by all grasses and legumes in greatest abundance are nitrogen, phosphorus, and potassium. In addition, where legumes are grown, acid soils need lime. The pH of the soil should be about 6.5. It is best to work lime into the soil considerably in advance of seeding, but commercial fertilizers should be applied at seeding time.

A thin, uniform mulch of barnyard manure is especially valuable in establishing a new seeding.

3. *High quality seed is purchased*—The seed should be of good quality, of high germination and purity as indicated on the tag, and free of noxious weeds. Also, proof of origin is of prime importance when an imported variety is secured. Certified seed carries a little more assurance of being high quality than noncertified seed, and gives proof of its origin much as a registration certificate does on a purebred animal.

4. *Scarified legume seed is used*—In the purchase of certain legume seed, it is important that it be scarified, which breaks the seed coat and allows faster moisture penetration—thus assuring quicker and more uniform germination and a better stand the first year.

5. *Legume seed is inoculated*—Since legumes can use nitrogen from the air provided they are inoculated with the proper bacteria, it is important that legume seed be inoculated.

Inoculant comes in several different forms, usually in a can with directions given thereon. It is important that the seed not be treated more than a few hours before seeding because these nitrogen-fixing bacteria are easily killed by drying, heat, sunlight, or by chemical seed treatment.

6. *A good seed bed is prepared*—A good seed bed is free from weeds, fine-textured, firm and moist.

Weeds are usually destroyed by growing row crops or a small grain the year preceding seeding to pasture and by cultivating frequently following the harvesting of this crop.

There are many different ways in which to prepare a good seed bed. Perhaps as good a method as any consists in (a) plowing as far in advance of seeding as possible, (b) disking, (c) harrowing one or more times to level up the field and smooth down the surface, and (d) cultipacking or rolling. A properly prepared seed bed should be so firm that one barely leaves a footprint when walking across it; the firmer the better from the standpoint of moisture conservation of small seeds.

7. *The seeding operation is timed and carried out properly*—The seeding time will vary, being determined primarily by the area and by the species or mixture used.

The actual seeding operation may be (a) by broadcasting, with a whirlwind seeder or by hand, or (b) by drilling, with any one of several types of conventional seeders. Drilling is the preferred method, for it insures more uniform placement of seed in both depth and amount of seed per acre and results in a more uniform stand.

Since most grass and legume seeds are very small, they should not be covered deeply. A good rule of thumb is that they should not be covered more than four or five times the width of the seed; usually this means not more than one-fourth inch.

8. *A companion or nurse crop may or may not be included*—The value of planting a "companion" or nurse crop—usually consisting of annuals—with new seed crops is controversial.

The advantages are: (a) It furnishes a crop of value while the new seeding is being established; (b) it lessens erosion; and (c) it reduces the weed population.

The disadvantages are: (a) It may retard the growth of the seedlings for whose protection it is grown; and (b) it may rob the new seeding of so much moisture that it kills them during dry spells unless the companion crop is harvested early as pasture, hay or silage.

IMPROVING OR RENOVATING AN OLD PASTURE

In altogether too many cases old permanent pastures are merely gymnasiums for livestock. Generally this condition exists because the least productive areas are used for pastures and because little attention is given to fertility and pasture management.

Permanent pastures in the tame pasture area that are run-down may be brought back into production by either of the following methods:

1. *By reseeding without growing a crop in the interim*—Poor, run-down

permanent pastures are frequently renovated by reseeding without growing a crop in the interim; in other words, pasture follows pasture. This kind of renovation is designed to increase pasture yields without subjecting the soil to excessive erosion and without keeping the area out of pasture production any longer than necessary. The actual operations involved in renovating will vary from area to area, and from field to field. In general, it involves (a) cultivating (preferably by plowing, but by disking or other methods in unplowable areas) so as to destroy all existing vegetation, (b) fertilizing and liming, and (c) preparing the seed bed and seeding with an adapted high-yielding pasture mixture.

2. *By fertilizing, overseeding and managing*—Where a fair but unproductive permanent pasture stand exists, pasture improvement or renovation may consist of (a) fertilizing (and liming where needed), (b) seeding (overseeding) with desirable and adapted varieties, and (c) managing in accordance with the outline which follows. Usually the fertilizer and the seed are worked into the soil with a disk and spring-tooth harrow, but a minimum of the existing sod is destroyed.

MANAGEMENT OF PASTURES

Many good horse pastures have been established only to be lost through careless management. Good pasture management in the tame pasture area involves the following practices.

Controlled Grazing

Nothing contributes more to good pasture management than controlled grazing. At its best, it embraces the following:

1. *Protection of first year seedings*—First year seedings should be grazed lightly or not at all in order that they may get a good start in life. Where practical, instead of grazing, it is preferable to mow a new first year seeding about 3 inches above the ground and to utilize it as hay, provided there is sufficient growth to justify this procedure.

2. *Rotation or alternate grazing*—Rotation or alternate grazing is accomplished by dividing a pasture into fields (usually 2 to 4) of approximately equal size, so that one field can be grazed while the others are allowed to make new growth. This results in increased pasture yields, more uniform grazing, and higher quality forage.

Generally speaking, rotation or alternate grazing is (a) more practical and profitable on rotation and supplemental pastures than on permanent pastures, and (b) more beneficial where parasite infestations are heavy than where little or no parasitic problems are involved.

3. *Shifting the location of salt, shade, and water*—Where portable salt containers are used, more uniform grazing and scattering of the droppings may be obtained simply by the practice of shifting the location of the salt to the less grazed areas of the pasture. Where possible and practical, the shade and the water should likewise be shifted.

4. *Deferred spring grazing*—Allow 6 to 8 inches of growth before turn-

ing cattle out to pasture in the spring, thus giving grass a needed start. Anyway, the early spring growth of pastures is washy and high in moisture.

5. *Avoiding close late fall grazing*—Pastures that are grazed closely late in the fall start late in the spring. With most pastures, 3 to 5 inches of growth should be left for winter cover.

6. *Avoiding overgrazing*—Never graze more closely than 2 to 3 inches during the pasture season. Continued close grazing reduces the yield, weakens the plants, allows weeds to invade, and increases soil erosion. The use of temporary and supplemental pastures, may "spell off" regular pastures through seasons of drought and other pasture shortages and thus alleviate overgrazing.

7. *Avoiding undergrazing*—Undergrazing seeded pastures should also be avoided, because (a) mature forage is unpalatable and of low nutritive value, (b) tall-growing grasses may drive out such low-growing plants due to shading, and (c) weeds, brush, and coarse grasses are more apt to gain a foothold when the pasture is grazed insufficiently. It is a good rule, therefore, to graze the pasture fairly close at least once each year.

Clipping Pastures and Controlling Weeds

Pastures should be clipped at such intervals as necessary to control weeds (and brush) and to get rid of uneaten clumps and other unpalatable coarse growth left after incomplete grazing. Pastures that are grazed continuously may be clipped at or just preceding the usual haymaking time; rotated pastures may be clipped at the close of the grazing period.

Topdressing

Like animals, for best results grasses and legumes must be fed properly throughout a lifetime. It is not sufficient that they be fertilized (and limed if necessary) at or prior to seeding time. In addition, in most areas it is desirable and profitable to topdress pastures with fertilizer annually, and, at less frequent intervals, with lime (lime to maintain a pH of about 6.5). Such treatments should be based on soil tests, and are usually applied in the spring or fall.

Scattering Droppings

The droppings should be scattered three or four times each year and at the end of each grazing season in order to prevent animals from leaving ungrazed clumps and to help them fertilize a larger area. This can best be done by the use of a brush harrow or chain harrow.

Grazing by More Than One Class of Animals

Grazing by two or more classes of animals makes for more uniform pasture utilization and fewer weeds and parasites, provided the area is not overstocked. Different kinds of livestock have different habits of grazing; they show preference for different plants and graze to different heights.

Irrigating Where Practical and Feasible

Where irrigation is practical and feasible, it alleviates the necessity of depending on the weather.

EXTENDING THE GRAZING SEASON

In the South and in Hawaii, year-round grazing is a reality on many a successful cattle establishment. By careful planning and by selecting the proper combination of crops, other areas can approach this desired goal.

In addition to lengthening the grazing season through the selection of species, earlier spring pastures can be secured by avoiding grazing too late in the fall and by the application of a nitrogen fertilizer in the fall or early spring. Nitrogen fertilizers will often stimulate the growth of grass so that it will be ready for grazing 10 days to 2 weeks earlier than unfertilized areas.

PART II. THE WESTERN RANGE[1]

Various geographical divisions are assumed in referring to the western range area—the native pasture area. Sometimes reference is made to the 17 range states, embracing a land area of approximately 1.16 billion acres. At other times this larger division is broken down, chiefly on the basis of topography, into: (1) the Great Plains area (the 6 states of North Dakota, South Dakota, Nebraska, Kansas, Oklahoma, and Texas); and (2) the 11 western states (Arizona, California, Colorado, Idaho, Montana, Nevada, New Mexico, Oregon, Utah, Washington, and Wyoming). In addition to these major and commonly referred to geographical divisions, there are numerous other groupings. These are of importance to the cattleman in that they affect the type of management and, to some extent, the kind of animals kept.

Almost half (47.8%) of the land area in the 11 western states is Federally owned. Domestic livestock graze on 73% of this area. Federal land is estimated to supply 12% of all grazing resources in the region and to provide the equivalent of the feed required *yearlong* for 1.7 million head of cattle and 1.0 million sheep.

Because of the magnitude of the range beef cattle industry and the fact that it is a highly specialized type of operation, considerable discussion will be devoted to the range area and the care and management of cattle in the range method.

The carrying capacity of much of the western range is low, and little of it provides yearlong grazing. Moreover, variation in vegetative types, climate,

[1]The author is very grateful to the following specialists who reviewed this section: Frank J. Smith, Director of Range Management, Forest Service, U.S. Department of Agriculture, Washington, D.C.; Mr. Charles H. Stoddard, Director, Bureau of Land Management, U.S. Department of the Interior, Washington, D.C. (now retired and serving as Resource Consultant, Wolf Springs Forest, Minong, Wisconsin); Dr. Leo B. Merrill, In Charge, Texas A & M University Agricultural Research Station at Sonora, Tex.; and Dr. Grant A. Harris, Chairman, Forestry and Range Management, Washington State University, Pullman, Wash.

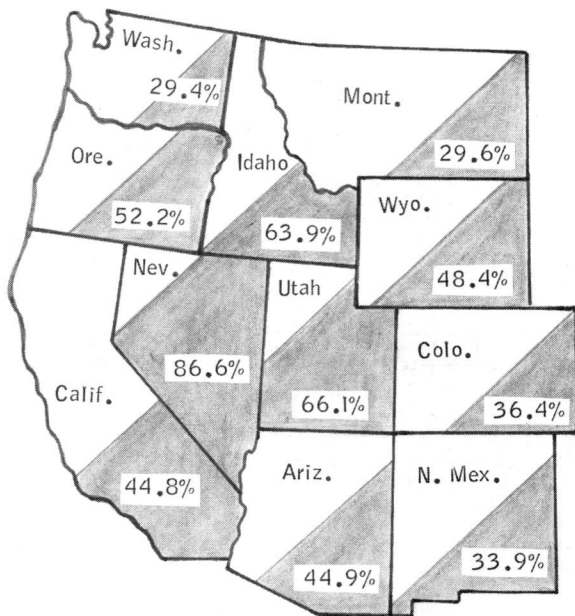

Fig. 9-6. A map showing the 11 western states and the proportion of land in each of these states that is owned by the U.S. Government. (Source: *Congressional Record*, Vol. 120, No. 41, March 26, 1974, p. S4433)

and topography in the range country is accompanied by great diversity in the seasonal use made of it. As a result, rangelands are usually grazed during different parts of the year, and the herds and flocks migrate with the season, moving to the mountains and higher elevations in summer and returning to the lower ranges in winter.

From the standpoint of vegetation and utilization by livestock, ranges differ from tame, or cultivated, pastures as follows:

1. *They are less productive*—Generally, their productive capacity is lower. This is as one would expect, for they are largely made up of the residue remaining after the usable agricultural lands have been taken up. Also, plant growth on range lands frequently is limited by low and undependable rainfall (even drought), short growing seasons, shallow or rocky soil, alkali or salt accumulations, steep topography, etc. Under such conditions, forage plants are usually less resistant to grazing damage than those growing under a more favorable environment.

2. *They are more likely to progress to less palatable plants*—Range vegetation consists of a mixture of native and introduced plants, varying greatly in palatability, nutritive value, and productive ability. Grazing animals select the most palatable plants first. Thus, unless careful management is practiced, the best plants are crowded out through a combination of grazing injury and competition from the ungrazed, low value plants. Continued poor management can result in good forage plants being almost completely replaced by low value annual, weedy, or shrubby vegetation, or left denuded and subject to severe erosion.

3. *They are more difficult to restore when depleted*—Once a range becomes depleted, it is a slow process to rebuild it. Plowing and drilling are impractical on most rangelands; thus, very often the only feasible way of restoring a range to good condition is to stock it conservatively and manage it well.

4. *They often serve multiple uses*—Rangelands often have other uses in addition to grazing values. Among such uses are: water production, timber production, mineral production, wildlife production, and recreation (camping, hiking, picnicking, etc.).

Thus, many people, in addition to the livestock producer, have an interest in the grazing management practiced on ranges. This is part of the justification given for Federal government ownership of large tracts of rangeland.

Wild Animals of the Range

While stockmen were expanding to the West, they had to compete for forage with great herds of wild grazing animals, the most numerous of which were buffalo, wild horses, and elk. Contrary to many present-day opinions, these animals were so numerous as to be a major factor in the utilization of the available forage. For example, it has been estimated, by those competent to judge, that there were 15 million buffalo in the West in 1864 and that the combined number of buffalo, wild horses, and elk in 1873 amounted to not less than 100 million. It is reasonable to assume, therefore, that the forage consumed annually by wild animals exceeded the amount utilized by the cattle and sheep grazed on these same ranges today. Except for the heavy and destructive grazing which occurred in the vicinity of strategic watering places and near salt licks, these wild animals alone were not particularly

Fig. 9-7. Herd of elk at feed. It has been estimated by those competent to judge that the combined number of buffalo, wild horses, and elk on the western ranges in 1873 amounted to not less than 100 million. (Courtesy, U.S. Forest Service)

active to the virgin vegetation, primarily because their normal seasonal migrations permitted recovery of the range. But when cattle and sheep were added, overgrazing and range deterioration occurred for a time.

History of the Range Cattle Industry

Animals of Spanish extraction served as sturdy foundation stock for the great cattle herds which were eventually to populate the western range. Although cattle have grazed intermittently on the southern plains of the United States since 1540, at which time they were introduced by Coronado, the period of continuous grazing accompanied the establishment of several Jesuit missions in Arizona, New Mexico, and Texas in the period from 1670 to 1690.

The growth of the industrial East and the subsequent development and extension of the railroads provided the necessary stimulus for further expansion of the range cattle industry. The grass supply of the vast ranges seemed unlimited, and the region was regarded as a permanent paradise for cattle. About 1880, the lure of the grass bonanza fired the imagination of investors, big and little. Cowboys, lawyers, farmers, merchants, laborers, and bankers—many of them English and Scotch investors in great companies—rushed in to seek their fortunes. The number of cattle increased rapidly, and soon the range was overstocked. Regulations were few, and the guiding philosophy was "to get what you can while the getting is good, and let the devil take the hindmost."

Then, suddenly, it became apparent that greed had taken its toll. The supply of tall grass was exhausted. Even more tragic, the winter of 1886-87 was unusually severe, and few owners had made provisions for winter feed. Cattle perished by the thousands. In some herds, 85 percent of the animals starved. And this was not all! The prolonged drought of 1886-95 brought further losses to the cattle companies, and the inroads of the homesteaders on the range and the growth of the range sheep industry contributed other difficulties. These circumstances marked the beginning of the end of the large cattle companies, the gradual growth of smaller operators, and increased attention to management.

Types of Range Vegetation

Cattlemen and students alike—whether they reside in the East, West, North, or South—should be well informed concerning range grasslands, the very foundation of the range livestock industry. This is so because this vast area, comprising 37 percent of the total land area and two-thirds of the total grazing lands of the United States, is one of the greatest cattle countries in the world and a potential competitor of every American cattle producer. Since cattle and sheep compete successfully with each other in utilizing most range forages, both classes of animals necessarily will be mentioned in the discussion which follows relative to types and uses of range vegetation.

Chiefly because of climate, topography, and soil, the character and composition of native range vegetation is quite variable. Ten broad types of veg-

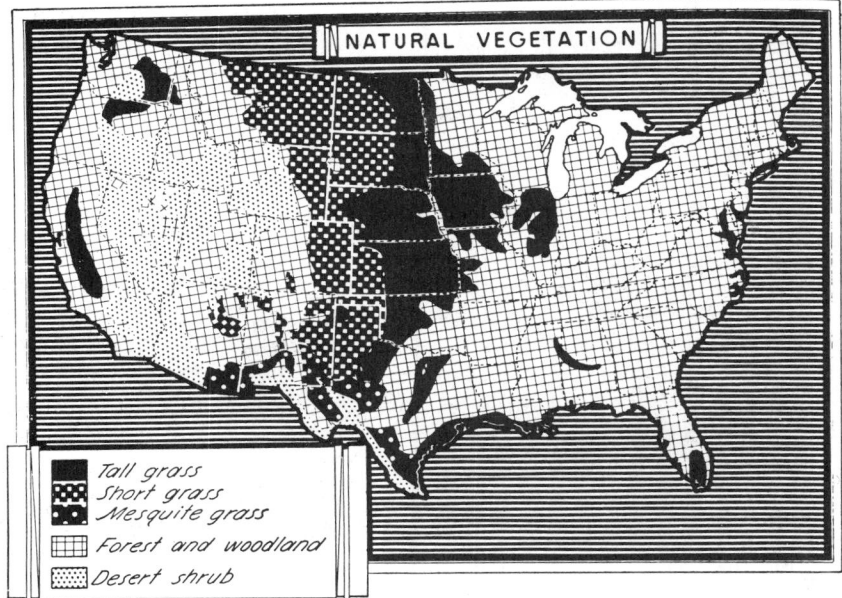

Fig. 9-8. Five major types of natural vegetation of importance to United States beef cattle production. (Courtesy, USDA)

etation native to the western ranges of the United States are discussed in this chapter.

TALL-GRASS TYPE

It is estimated that about 20 million acres of tall-grass range remain, most of which is in the eastern Great Plains region, in a rainfall area varying from 20 to 40 inches. Although differing with the soil, topography, and rainfall, the dominant native tall-grass species include the bluestem (big and little), Indiangrass, switchgrass, side oats grama, and slough grass. Such famous grazing areas as the Flint Hills of Kansas, the Osage Pastures of Oklahoma, and the Sand Hills of Nebraska belong to the tall-grass type of vegetation. For the most part, this type of range is utilized by cattle, although sheep graze some of it. Each fall, thousands of fat cattle are marketed after being finished in tall-grass vegetation without a grain supplement. The carrying capacity of these ranges is very high.

Fig. 9-9. Steers on tall-grass type range in the famous Flint Hills area of Kansas.

SHORT-GRASS TYPE

The short-grass range, which is the largest and most important grassland type in the United States, embraces an area of approximately 280 million acres. It extends from the Texas Panhandle to the Canadian border and from the foothills of the Rocky Mountains eastward midway into the Dakotas. The common grasses of this area include the grama grasses, buffalograss, and western wheatgrass, all of which are well adapted to making their growth during the time of favorable moisture conditions in the late spring and early summer. Although they become bleached and cured on the stalk, because of the small amount of leaching in the fall and winter months, these plants retain sufficient nutrients to furnish valuable winter grazing. Because the forages in the short-grass area are dry during much of the year and droughts are rather frequent, cow-calf operations predominate in the cattle industry of the area; and most of the calves and older steers are finished in Corn Belt feedlots or in irrigated valleys prior to slaughter. The smaller fine-wool breeds of sheep are also most numerous, and most of the lambs go the feeder route.

Fig. 9-10. Short-grass type range in Montana. Cow-calf operations predominate throughout this area, which represents the largest and most important grassland type in the U.S. (Courtesy, U.S. Forest Service)

SEMIDESERT-GRASS TYPE

The semidesert-grass type—which predominates in an area characterized by low rainfall, frequent droughts, and mild winters—embraces about 93 million acres of grasslands in central and southwestern Texas, Arizona, and New Mexico. It provides year-round grazing. Because of great differences in climate and soil, the vegetation is quite variable. The most common grasses are grama, curly-mesquite, and black grama. Scattered among the more or less sparse grasses are many scraggly shrubs, dwarf trees, yuccas, and cacti. Some of these—especially saltbush, mesquite, ratany, and

Fig. 9-11. Semidesert-grass type range in New Mexico, showing dry cows on excellent grama grass. Generally, this type of range provides year-round grazing for commercial cow-calf enterprises, although some of it is utilized by sheep and goats. (Courtesy, U.S. Forest Service)

scrub oak—are rather palatable and are browsed effectively by goats. For the most part, the semidesert-grass area is utilized by commercial cattle as a cow-calf proposition. But bands of breeding sheep are found throughout the area. Both sheep and goats are common in southwestern Texas. The sheep of this area are kept primarily for wool production, and production of feeder lambs is secondary.

PACIFIC BUNCHGRASS TYPE

The Pacific bunchgrass area embraces about 60 million acres in western Montana, eastern Washington and Oregon, northern and southwestern Idaho, and central California. Much of the original bunchgrass area, including the famous Palouse area of eastern Washington and northern Idaho, is now devoted to the production of wheat and peas. Though well adapted to the dry summers and moist winters of the area, the native tall bunch or tuft-growing grasses of this area—bluebunch wheatgrass, Idaho fescue, Sandberg bluegrass, and California needlegrass—did not withstand overgrazing and have largely been replaced by such annuals as alfileria, bur-clover, and cheatgrass (in the Northwest), and wild oatgrasses (in California). These ranges furnish excellent grazing in the spring and fall months but are too dry for summer use. The Pacific bunchgrass area is best adapted for spring, fall, and winter grazing by cattle and sheep.

SAGEBRUSH-GRASS TYPE

The sagebrush area, which is the third largest of all range types, em-

Fig. 9-12. Sagebrush-grass type range in Idaho, showing big sagebrush with excellent understory of bluebunch wheatgrass, bluegrass, and palatable perennial weeds. Generally, this type of range is used for early spring and late fall grazing for cattle and sheep. (Courtesy, U.S. Forest Service)

braces between 90 and 100 million acres extending from northern New Mexico and Arizona northwestward into Montana and to the east slope of the Cascades in the Pacific Northwest. This type of vegetation is characteristic of low rainfall areas where most of the meager precipitation occurs during the winter and spring seasons. Interspersed among the ever-present sagebrush, of which there are several kinds, are many species of native grasses among which are bluebunch and western wheat-grasses, needle-and-thread, Indian ricegrass, Sandberg bluegrass, and numerous species of weeds. The sagebrush, which varies from 2 to 7 feet in height, provides little forage except when winter snows blanket the grasses. For the most part, the sagebrush type of vegetation is used for early spring and late fall grazing for cattle and sheep. It furnishes interim pasture until more distant summer and winter grazing areas may be used. In recent years, studies have shown that the carrying capacity of sagebrush areas may be increased by destruction of the sagebrush, which encourages greater growth of the grasses.

SALT-DESERT-SHRUB TYPE

About 40 million acres in central Nevada, Utah, southwestern Wyoming, western Colorado, and southern Idaho are covered with a mixture of low shrubs and scattered grasses. The common browse species of the area are shadscale, saltbrush, black sagebrush, winterfat, rabbit-brush, greasewood, spiny hop-sage, and horsebrush; and the rather sparse grass species include blue grama, sand dropseed, galleta, and Indian ricegrass. Because there is not any dependable source of water and because of the high temperature and dryness during the summer months, the use of much of this area by cattle and sheep is restricted to the winter months when there is snow. Other areas cannot be grazed because of the high alkali content of the soil.

SOUTHERN-DESERT-SHRUB TYPE

Approximately 50 million acres, located chiefly in southeastern California, southern Nevada, and southwestern Arizona, are classed as southern-desert-shrub vegetation. The common shrubs are the creosotebush and different kinds of cacti. Normally, the scant rainfall, extremely high temperatures, and sparse vegetation of this area make it rather poor grazing for cattle or sheep. However, when moisture conditions are favorable, there is growth of such annuals as alfileria, Indian-wheat, bur clover, black grama, tobosa, dropseed, and six weeks' fescue. When forage and water are available, nearby ranchers make use of the southern-desert-shrub area, primarily for winter grazing, although it is used for spring and fall grazing and in a few cases throughout the year.

PINYON-JUNIPER TYPE

The pinyon-juniper type of vegetation forms the transition zone from the shrub and grass areas of the lower elevations to the forests of the mountains. The 76 million acres in this general type area extend all the way from southwest Texas to south central Oregon. As the name would indicate, pinyon and juniper trees are common to the area. These scattered trees range in height from 20 feet to 40 feet, and interspersed among them are such low-growing shrubs as sagebrush, bitterbrush, mountain-mahogany and cliff-rose, and grasses like the gramas, bluebunch and bluestem wheatgrass, and galleta. For the most part, this area is used for spring and fall grazing by cattle and sheep; but in the Southwest, where the forage cures on the ground and retains much of its nutritive value through the winter, yearlong grazing is prevalent.

Fig. 9-13. Pinyon-juniper type range in Arizona, showing young juniper trees encroaching on the area. For the most part, this type of range is used for spring and fall grazing by cattle and sheep, although in the Southwest yearlong grazing is prevalent. (Courtesy, U.S. Forest Service)

WOODLAND-CHAPARRAL TYPE

This type of vegetation is characteristic of parts of California and Arizona. It varies all the way from an open forest of parklike oak and other hardwood trees with an undergrowth of herbaceous plants and shrubs to dense "chaparral" thickets of no value to animals. Alfileria, slender oatgrass, and bur clover have been introduced in the more open areas. Though somewhat restricted, woodland-chaparral is used for fall, winter, and spring grazing by cattle and sheep.

OPEN-FOREST TYPE

The 130 million acres of open forests, found scattered in practically all the mountain ranges, constitute the second largest range-type vegetation. This is the summer range of the West, which provides grazing for large numbers of cattle, sheep, and big game. Many grass-fat cattle and lambs are sent to market directly off these cool, lush, high-altitude ranges. For the most part, the tree growth common to the area consists of pine, fir, and spruce;

Fig. 9-14. Open forest type range in ponderosa pine timber of Oregon. (Courtesy, U.S. Forest Service)

and the grasses include blue grama, fescues, bluestem, wheatgrasses, timothy, bluegrasses, sedges, and many others. More than half of these mountain ranges are Federally owned as national forest lands. In addition to serving as valuable grazing areas, the open forests are important for lumbering and recreational purposes.

Grazing Publicly Owned Lands

The ownership of United States land is summarized in Table 9-2.

TABLE 9-2

OWNERSHIP OF U.S. LAND (50 STATES)[1]

Ownership	Acreage	Percentage of Total
	(million acres)	(%)
1. Private ownership	1,317	58.2
2. Indian land	50	2.2
3. Public ownership	879	39.6
(a) Federal	763	33.7
(b) State and local governments	134	5.9

[1]*Statistical Abstract of the United States*, 1973, Table 313, p. 197.

About half the public lands are in Alaska. Because of its remoteness and northern location, land development has been slow in this state. As a result, the Federal Government still owns over 95 percent of all the lands in Alaska.

The other half of the public lands is located in the 48 contiguous states, but it is not evenly distributed across the country. Over 90 percent of the Federal lands outside Alaska are in the 11 western states.

Today, in the 11 western public land states, the Federal Government owns and administers approximately 273 million acres on which grazing is allowed. At one time or another during the year, domestic cattle and sheep graze on about half of these public lands. More of the public lands are used for this purpose than for any other economic activity. In 1971, lands in the 11 western states administered by the Bureau of Land Management and the U.S. Forest Service provided grazing for all or part of the year for 4,592,793 cattle and horses and 6,763,835 sheep and goats, or a total of 11,356,628 head of all classes, or a total of 18,308,758 animal unit months.

AGENCIES ADMINISTERING PUBLIC LANDS

Because much of the grazing land that ranchers rely upon to maintain their cattle and sheep enterprises is built up into operating units by leasing or by obtaining use permits from seveal Federal and state agencies, private corporations, and individuals, it is imperative that the owner have a working knowledge of the most important of these agencies. Some range operators are placed in the position of using range rented from as many as six landlords; either private, state, and/or Federal.

The bulk of Federal land is administered by the following six agencies: the Bureau of Land Management, the U.S. Forest Service, the Bureau of Indian Affairs, the Department of Defense, the National Park Service, and the Bureau of Reclamation. The most important Federal lands from the standpoint of grazing permits and utilization of grazing areas by animals are administered by the first three of these agencies.

1. *Bureau of Land Management*—The Bureau of Land Management of the U.S. Department of the Interior administers over 60 percent of all Federal lands. Almost two-thirds of the lands it manages are in Alaska. The remainder is almost entirely in the 11 western states.

From the standpoint of the stockman, the most important function of the Bureau of Land Management is its administration of the grazing districts established under the Taylor Grazing Act of 1934 and of the unreserved public land situated outside of these districts which are subject to grazing lease under Section 15 of the Act. This Federal act and its amendments authorize the withdrawal[2] of public domain from homestead entry and its organization into grazing districts administered by the Department of the Interior. Also, this legislation, as amended, allows the Bureau of Land Management to administer state and privately owned lands under a cooperative arrangement.

In 1972, the Bureau of Land Management had 52 grazing districts,

---

[2]On May 28, 1954, a bill was signed by President Eisenhower lifting the 142 million acre limitation on public domain lands that can be included in Taylor Grazing Act districts.

operating in the 10 western states and totaling 157.3 million acres of public lands. In these districts, 13,811 operators were granted privileges to graze 5,619,485 head of livestock for an average of about 4½ months each year. These operators paid the United States, as grazing fees for this range use, a total of $6,575,904. In addition to this livestock use, in 1972 the grazing districts supported, for approximately 5 months of the year, an estimated 1,439,191 big game animals, most of which were deer.

Fig. 9-15. Hereford cattle on the ranch of a grazing permittee of the Bureau of Land Management. (Courtesy, Bureau of Land Management, U.S. Department of the Interior, Washington, D.C.)

In addition to, and outside of, the grazing districts, in 1972 the Bureau of Land Management supervised 17.7 million acres of public domain in the western states, most of which were leased to 8,000 stockmen for 3,257,875 head of livestock for about 1¾ months. These operators paid rentals in the amount of $1,034,873 for the use of these lands.

Each grazing district is administered by a District Manager, who is a technically trained employee of the Bureau of Land Management. He is responsible to the state bureau office for the proper use, management, and welfare of the public land resources of his district. In turn, the state office is responsible to an area office, and the area office to the Director's office in Washington, D.C.

In each of the grazing districts, local groups of stockmen, elected by the users, and a wildlife representative, serve on an advisory board in allocating grazing privileges and supervising details of administration.

Grazing privileges are allocated to individual operators, associations, and corporations on the basis of: (1) priority of use; (2) ownership or control of base property dependent on grazing district land for forage during certain seasons of the year, or control of permanent water needed to graze district land; (3) proximity of home ranch to the grazing district; and (4) adequate property to supply the feed needed along with grazing privileges, to maintain throughout the year the livestock permitted on public range. All of these lands are subject to classification and disposal under Sections 7 and 14 of the Taylor Grazing Act, for any higher use or other appropriate purpose. Grazing privileges may, therefore, be cancelled whenever such lands are determined to be more suitable for other purposes.

A fee is charged for grazing privileges. In 1975 the basic fee was equivalent to $1.51 per animal unit month; or $1.51 per cow month, $1.51 per horse month, and $0.30 per sheep or goat month.

The Taylor Grazing Act has been responsible for many changes, not all of which have been popular. Some stockmen complain about the loss of their ranges; others tell of increased costs; and there are those who resent government controls, and, above all, the confusion which results from dealing with several agencies. Without doubt, many of these criticisms are justified, and some errors in administration should be rectified; but those who would be fair are agreed that the ranges as a whole have improved under the supervision of the Bureau of Land Management and that further improvements are in the offing.

2. *U.S. Forest Service*—Almost one-fourth of the Federal lands are administered by the Forest Service. Over 100 million acres of the national forests are used for grazing under a system of permits issued to local farmers and ranchers by the Forest Service of the U.S. Department of Agriculture. In 1973 about 1,598,000 mature sheep and goats and 1,396,000 mature cattle and horses (mostly cattle), owned by over 17,000 paid permit operators, were grazed on national forests for some part of the year. In addition, there were many calves and lambs for which no fee is charged and additional stock that were grazed under free permits to local settlers. This made an estimated total of 6 million domestic animals grazed on the national forest ranges in 1973.

The Forest Service issues 10-year term permits to stockmen who hold preferences and annual permits to those who hold temporary use. Among other things, the permit prescribes the boundaries of the range which they may use, the maximum number of animals allowed, and the season when grazing is permitted.

Preferences may be acquired through prior use, through a grant, or through purchase of land or livestock, or both, of a user who already holds a preference.

The requisites in order to qualify for a permit are:

a. *Ownership*—The ownership of both the livestock and commensurate ranch property.

b. *Dependency*—The need for forest range in order to round out an operation to obtain proper and practical use of commensurate property.

c. *Commensurability*—The ability of the land to support livestock during the period when not on forest land.

Fig. 9-16. Cows and calves on the Santa Rita Range Reserve, in Arizona. (Courtesy, U.S. Forest Service)

A grazing preference is not a property right. Rather, it is approved for the exclusive use and benefit of a person to whom allowed. Preferences or permits may be revoked in whole or in part for a clearly established violation of the terms of the permit, the regulations upon which it is based, or the instructions of forest officers issued thereunder.

A ranger administers the grazing use on each National Forest Ranger District. Several districts (usually three to six or more) comprise a national forest. A forest supervisor, with his staff, administers the national forest. Several national forests, under the direction of a regional forester and staff, comprise a forest region. The Chief administers the Forest Service from Washington, D.C., under the supervision of the Secretary of Agriculture.

As is true in the administration of Taylor grazing districts, local farmers and ranchers act in an advisory capacity in the allocation of grazing privileges and in details of administration of the national forests. About 840 such livestock associations and advisory boards are recognized and in operation.

Forest Service grazing fees are based on a formula which takes into account livestock prices over the past 10 years, the quality of forage on the allotment, and the cost of ranch operation. In 1976, average charges were $1.60 per animal unit month (AUM); or $1.60 per cow month, $1.60 per horse month, and $0.32 per sheep or goat month.

Although shortcomings exist in the management of the national forests, it is generally agreed that these ranges have been vastly improved under the administration of the Forest Service. Many of them now approach the quality that existed in their virgin state. Perhaps the most heated arguments between stockmen and the Forest Service arise over the relative importance attached to the multiple use of big game and other wildlife, recreation, etc. For example, it was estimated that in 1974 there were 4.6 million big game animals (85% of which were deer) in the national forests. As would be expected, these wild animals compete with domestic animals for use of the range, thus creating a most difficult problem.

3. *Bureau of Indian Affairs*—Most Indian lands, comprising nearly 50 million acres, are really not public lands. Rather, these lands are held in trust for the benefit or use of the Indians and are merely administered by the Bureau of Indian Affairs of the Department of the Interior. Because over 80 percent of Indian lands are in the range area of the West, they are suited primarily to livestock. Thus, it is noteworthy that the sale of livestock and animal by-products regularly account for two-thirds of the total Indian agricultural income. Although the Indians themselves own most of the stock grazed on these lands, animals owned by non-Indians utilize one-fourth of the Indian lands devoted to grazing. Provisions for such use is handled under lease agreement jointly approved by the Indian owners and the Bureau of Indian Affairs.

Many of the Indian lands have suffered serious vegetative depletion, but a concerted effort is now being made to decrease livestock numbers in keeping with available feed supplies and to improve the quality of animals produced. However, overstocking continues to be a difficult problem on the Navajo, Hopi, and Papago Reservations.

4. *State and local government-owned lands*—A total of 134 million acres are owned by state and local governments. For the most part, the management of these areas is diverse and confused, each state and local government having established different regulations relative to the lands under its ownership. In general, however, such lands are operated on a stipulated lease arrangement. On many such areas, range depletion has been severe.

5. *Railroad-owned lands*—Recognizing that the main deterrent to rapid settlement and development of the West was the lack of adequate transportation facilities, the Federal Government very early encouraged the construction and westward extension of the railroads by means of large grants of land. It was intended that the railroads should sell or otherwise utilize these lands in financing their costs of construction. These initial grants, totaling 94,355,739 acres, consisted of alternate sections extending in a checkerboard fashion for a distance of from 10 to 40 miles on each side of the right-of-way. Today, less than 20 million acres of these lands are held by railroads. Many of these holdings are leased to stockmen, but because of inconvenience, past abuses, or other reasons, some of these lands are considered worthless for grazing. In general, railroad lease agreements do not restrict the number of stock to be grazed or the season during which the land may be so used.

ENVIRONMENTAL EFFECTS OF GRAZING PUBLIC LANDS

Grazing influences the environment on Federal lands. Under poor range management, the environment is affected adversely; under good range management, such as exists on most ranges today, grazing actually improves the environment.

Eating of plant materials by animals is a natural process in earthly and aquatic systems. Thus, the coming of the white man to what is now the United States, along with the introduction of domestic animals, did not constitute

THE WHEEL OF ECOLOGY

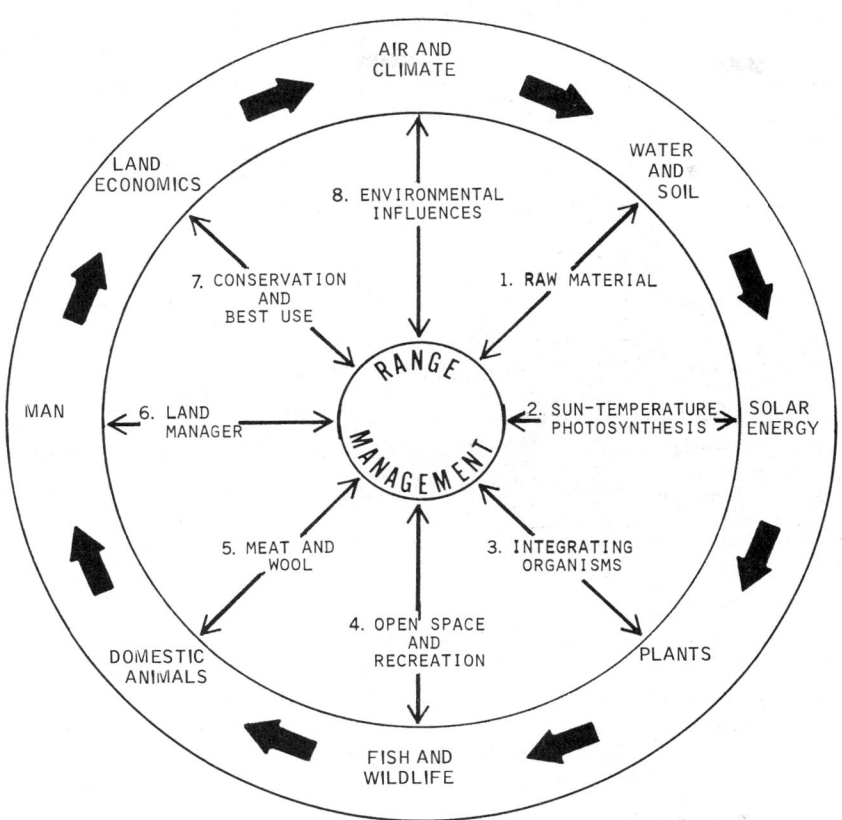

Fig. 9-17. The wheel of ecology. Cattlemen share today's increasing national concern for the quality of our environment, and, through scientific range management, they are doing much to improve it.

an entirely new component in the environment. Rather, domestic animals replaced, or added to, the wild animals that were already there.

Mistakes in grazing practices have occurred in the United States in the past, the most significant of which was the exploitative grazing practices between 1865 and the 1930s. The effects were almost catastrophic. Nevertheless, they were not the result of grazing ranges that had never been grazed before. Rather, they resulted from several decades of grazing the western ranges with too many animals for too long, and often at the wrong season of the year. Most range livestock operators of that period were not aware of the benefits that could accrue to them from improved range management.

Scientific management of rangeland began at the turn of the century. Range managers and livestock operators found that controlling grazing improved both range conditions and livestock production. Development of this

Fig. 9-18. Aerial topdressing. Scientific range management benefits the environment.

new concept marked the beginning of the end of the exploitative period of grazing and the introduction of managed grazing on the western ranges.

The environmental effects of grazing depend upon the kind of range, the intensity of grazing, and the kind of management employed to control livestock on the range. It is generally recognized that unregulated heavy grazing results in loss of desirable forage plants, increased runoff and erosion, and other indications of range deterioration. On the other hand, planned seasonal grazing and controlled animal distribution foster rapid vegetational growth. Most grazing experiments have shown that ranges may be improved more rapidly under proper grazing management than with no grazing at all.

There is no evidence that well-managed grazing of domestic livestock is incompatible with a high quality environment. But there is ample evidence that managed grazing by livestock enhances certain uses and that poor management detracts from them. Properly managed grazing is a reasonable and beneficial use of the range.

Ecologists tell us that good range management will support more wildlife than the wilderness. This explains why big game numbers on Federal lands have increased during recent years, and why wildlife production is an increasingly important use of rangelands.

Indeed, ranges actually improve while being properly utilized by domestic livestock. The benefits which accrue to the range include increased vegetation cover, improved plant species composition, improved soil fertility and soil structure, and greater yield of high quality water. When cattle and sheep go, rank underbrush takes over, and fire becomes a real hazard.

Both upland game birds and big game animals are benefited by grazing that promotes good cover for mating sites and enhances food supply and other habitat requirements.

On ranges with mixed types of vegetation, herbaceous species increase and browse species decline when grazed only by game. The converse is true when cattle graze the land. The combined grazing by two groups of animals maintains a better balance of browse species, preferred by game animals, and of herbaceous species, preferred by cattle.

Heavy livestock grazing is beneficial to irrigated pastures used by geese and other migratory waterfowl. Unless the vegetation is closely cropped, these areas are unattractive to the birds.

Thus, livestock grazing of the public lands is contributing to improved wildlife habitat conditions and increased numbers of game animals. Range development programs, particularly livestock water developments, have made more public land usable by game animals and is partly responsible for the vast increase in game numbers over the years.

On many grass-shrub ranges, livestock grazing reduces the danger of fire by preventing a buildup of dry grass, which is highly inflammable.

Grazing systems and manipulation of vegetation can create contrast in vegetation color and pattern, thereby improving the aesthetical value of the landscape. Also, the livestock industry is traditional to the West; hence, a well-managed range with its cattle herd and roundup, or with its sheep camp, has recreational values. Indeed, cattle and sheep on the landscape are pleasing to tourists who come to view the "Old West."

Ranges properly grazed by hoofed animals produce safe water. Counts of fecal coliform organisms, as indicators of water pollution by warm-blooded animals, relate more closely to the quantity of fecal material than to the kind of animal. Investigations have shown that the count of harmful bacteria in streams is no greater in areas grazed by livestock than in areas grazed by wild animals alone, and that modern livestock grazing has little effect upon the chemical and physical quality of the water.

It is noteworthy, too, that few western ranges are ever in a stable, natural condition, whether or not they are grazed by domestic animals. Rather, most of them are in a stage of vegetational development following disturbances by such phenomena as drought, flood, avalanche, frost, or fire. Also, cyclic phenomena, such as large numbers of deer, rodent epidemics, or insect plagues, temporarily change the natural ecosystems. Thus, an absolutely stable rangeland is seldom attained or maintained.

Significantly, the greatest diversity of animal and plant species and the highest rates of reproduction occur when the landscape supports many stages of ecosystem development. Fire, grazing, and drought stimulate plants and animals to new growth. Each stage of vegetational development is more productive of certain animal species than of others.

Finally, in an era of world food shortages, the contribution of properly managed Federal lands in terms of food and fiber production needs to be recognized. More and more grains will be used for direct human consumption. As a result, there will be an increased reliance on ranges for meat and wool production. It just makes sense to preserve all the natural food and fiber that we can. Remember that petroleum is not needed to make wool. Remember, too, that cattle and sheep are completely recyclable. It takes thousands of years to create coal, oil, and natural gas; and when they're gone, they're gone forever. But animals produce a new crop each year and perpetuate themselves through their offspring.

Approximately 273 million acres of Federal land, plus about 4.5 million acres of Indian land, are administered for livestock grazing. Some 55,300 leases, licenses, or permits are issued to about 60,000 farmer and rancher families to graze approximately 5 million head of cattle and 11.4 million head of sheep on these lands. In terms of dollars, grazing on public lands represents a gross product income of approximately $585 million annually to the United States and about $525 million annually to the 11 western states.

Both stockmen and environmentalists need to recognize (1) that forage is a renewable natural resource, which regrows each year and is wasted unless it is utilized annually; (2) that grazing on Federal rangelands helps to keep the natural environmental systems active and productive; (3) that we cannot allow overgrazing by domestic livestock, bison, deer, or wild horses; and (4) that grazing must be scientifically controlled and responsive to the needs of all users.

Indeed, it may be said that man's influence on and use of the environment will determine how well we live—and how long we live.

Seasonal Use of the Ranges

A prime requisite of successful range management for both cattle and sheep is that there shall be as nearly year-round grazing as possible and that both animals and the range shall thrive. In some areas, especially in the southwestern Great Plains region, these conditions are met without necessitating extensive migration of animals. The winter climate is mild, and the native forages cure well on the stalk, thus providing nutritious dry feed at times when green vegetation is not available. Generally speaking, however, most of the cattle and sheep from such areas are marketed via the feeder route rather than as grass-finished slaughter animals.

In general, the most desirable management, both from the standpoint of the animals and the vegetation, consists of the proper seasonal use of the range. Although there is wide variety in the customs and requirements for seasonal use of the range—because of the spread in climate, topography, and vegetative types included in the vast expanse of range country—seasonal-use ranges are usually placed in four major classes: (1) spring-fall, (2) winter, (3) spring-fall-winter, and (4) summer.

Because a range band of sheep can be moved and herded on unenclosed areas with greater ease than a herd of cattle and because investigations in range livestock management have been conducted more extensively with sheep, greater seasonal use of ranges is made with sheep. On the other hand, the more progressive cattlemen are finding ways and means of adopting many of the same methods.

Despite the values of yearlong grazing, it is recognized that the prevalence of severe winters in some parts of the West precludes winter grazing except to a limited degree, and stock must be fed during at least a part of the winter season. Where these conditions prevail, cattle and sheep are usually wintered in the irrigated valleys, close to the feed supply, especially a supply of alfalfa or meadow hay.

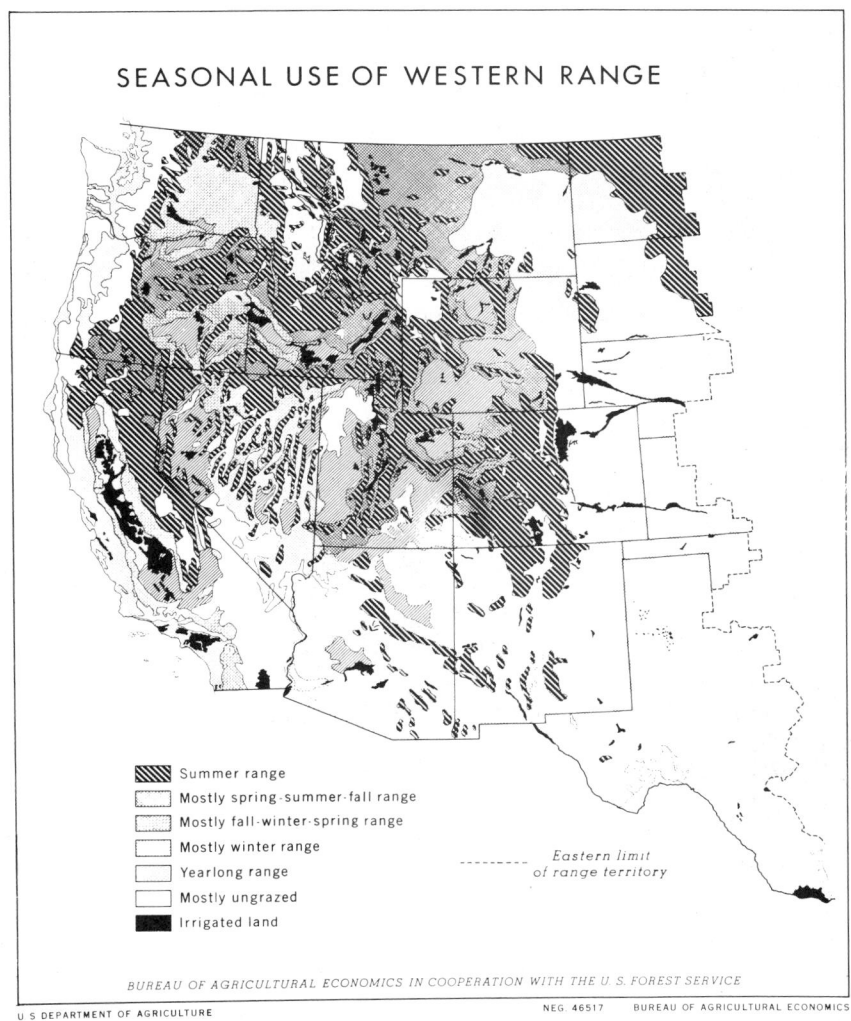

SEASONAL USE OF WESTERN RANGE

▨ Summer range
☐ Mostly spring-summer-fall range
☐ Mostly fall-winter-spring range
☐ Mostly winter range
☐ Yearlong range
☐ Mostly ungrazed
■ Irrigated land

--------- Eastern limit
of range territory

BUREAU OF AGRICULTURAL ECONOMICS IN COOPERATION WITH THE U. S. FOREST SERVICE

U S DEPARTMENT OF AGRICULTURE   NEG. 46517   BUREAU OF AGRICULTURAL ECONOMICS

Fig. 9-19. Seasonal use of the western range. In general, the most desirable management, both from the standpoint of animals and vegetation, consists of the proper seasonal use of the range. (Courtesy, USDA)

Some pertinent points in determining the proper season of use of the range follow:

1. *Elevation*—Generally speaking, vegetative development is delayed 10 or 15 days by each 1,000-foot increase in elevation. Also, severe storms occur later in the spring and earlier in the fall at higher altitudes than at low, desert locations.

Fig. 9-20. Trailing cattle from a winter range in Utah. Calves are from one day to several weeks of age. (Courtesy, U.S. Forest Service)

2. *Availability of water*—Certain desert areas are so poorly watered that only the occurrence of winter snows makes their use practical.

3. *Early forage "washy"*—Early spring forage is extremely "washy," and may be incapable of supporting stock.

4. *Soil tramping*—Soil tramping may be serious in early spring.

5. *Poisonous plants grow early*—Most poisonous plants are very early growers and cause their greatest damage when animals are turned out too early. Larkspur, which affects cattle, and death camas, which affects sheep, are two examples. Poisoning losses from these two plants are usually negligible if stock are detained until the best forage plants have made suitable growth.

6. *Winter range should be saved*—If stock are allowed to remain on winter ranges too long after spring growth begins, the next winter's feed will be reduced, because the forage produced on these ranges grows mainly during spring and early summer.

SPRING-FALL RANGE

The foothills and plateaus, lying between the plains or desert sections and mountains, furnish desirable spring and fall grazing for both cattle and sheep. Areas of this type are, therefore, commonly referred to as spring and fall range. The sagebrush-grass type is the leading range vegetation used for spring and fall range. Compared with the vegetation of the plains and desert regions, that growing on the foothills and plateaus in season is more luxuriant and succulent and better suited to milk production and finishing. Although spring and fall ranges are accessible most of the year—because such areas are not too extensive—usually they are not grazed for a very extended period. Sometimes fall grazing for the herd or flock is provided through using owned or leased stubble fields and meadow aftermath.

Fig. 9-21. Cattle on rich fall range in Oregon. (Courtesy, Bureau of Land Management, Department of the Interior)

WINTER RANGE

Areas adapted to yearlong grazing possess desirable climatic and vegetative conditions for winter range. Many such areas are used only during the winter season because of their proximity to more desirable spring-fall and summer ranges. Desert areas, such as the salt-desert-shrub type, which would otherwise be suited to yearlong grazing, must be restricted to use as winter range. This is due to the absence of any reliable water supply except that secured from the snow or rain during the winter months. In some sections, winter wheat is used for late fall and winter grazing.

Fig. 9-22. Cattle on winter range in New Mexico. (Courtesy, U.S. Forest Service)

SPRING-FALL-WINTER RANGE

In general, spring-fall-winter-type ranges are adapted to year-round grazing. But because of their proximity to more lush mountain ranges, they are not used during the summer months. On typical spring-fall-winter ranges, the annual grass species dry up during the summer months.

SUMMER RANGE

The most desirable summer ranges are located in the cooler and higher altitudes of the mountains where the vegetation is lush, palatable, and nutritious during the 3 to 6 summer months. There is usually a bountiful supply of water. Much of this range is of the open-forest type on national forest lands, where the average grazing period is 5.6 months for cattle and 3.1

Fig. 9-23. A summer range for cattle in the Sawtooth National Forest of Idaho. (Courtesy, U.S. Forest Service)

months for sheep. Although summer range is usually very rugged, the conditions are generally ideal for finishing, so that year after year it is the most consistent type of range area for the production of grass-finished cattle and lambs. Sheep are sometimes trailed or shipped a distance of 200 miles in order to have access to a desirable summer range, but cattle are usually not moved so far. These summer ranges supplement the lower spring, fall, and winter ranges (either publicly or privately owned). Thus, they hold a key position in the yearlong operations of many livestock producers. Sometimes seeded grass-legumes or irrigated pastures are used for summer grazing instead of mountain ranges.

THE RANGE CATTLE HERD

The range herd may be owned by an individual, a cattle company, or a corporation. Usually the ranch headquarter consists of a ranch-type house, bunk house, sheds, corrals, and a water supply and there is ownership of sufficient adjacent land for the production of winter forage and a limited amount of grain. Additional and more distant rangeland is either owned or leased.

Fig. 9-24. A range herd. (Courtesy, Arizona Cattle Growers' Association, Phoenix, Ariz.)

Size of Herd and Carrying Capacity

Range herds vary in size from about a hundred head up to several thousands. In general, the most important single criterion in determining the size of herd is the number of animals that the unit will support each season over a period of years without injury to the range. When the range is stocked more heavily than its true grazing capacity, usually three conditions become evident: (1) the animals fail to thrive as they should; (2) the vegetation gets thinner and of less desirable species; and (3) erosion and soil losses occur. Practical observations and controlled experiments would indicate that somewhere between 20 and 30 per cent of the palatable growth of the more important forage species should be left ungrazed each year. Additional considerations in arriving at the size of herd and carrying capacity are: (1) available reserve feed supply for drought and winter feeding, (2) the long-range economic conditions, and (3) the topography, water supply, and poisonous plants of the area.

Stocking Rate

The key to successful long-term operation of rangeland lies in a good estimate of the grazing capacity of the individual units. Stocking too lightly wastes forage, while stocking too heavily results in reduced plant vigor and less forage produced per plant, as well as a change of forage plant cover from an abundance of valuable forage plants to an abundance of worthless plants.

Of course, the stocking rate for any given unit may vary widely from year to year, depending on the forage production as affected by weather and other factors. For this reason, stocking should either be adjusted to forage yield each year, or set at a constant rate that will assure a sustained yield of the most valuable forage plants. (Constant stocking at about 25% below average capacity will usually achieve the latter.)

Recognition must also be given to the fact that animals do not graze uniformly over a range unit—that certain areas are more attractive to them. Consequently, some areas produce most of the grazed forage, while others may go practically unused. Cattle tend to congregate on fairly level creek bottoms, ridge tops, and around water and shade; whereas sheep, if herded, can be moved more uniformly over a unit. But even sheep graze some areas more heavily than others if not herded properly. For the purpose of determining grazing capacity, the key areas—those rather extensive parts of the range which are most heavily grazed—must be given greatest consideration. If preferred or key areas are maintained in good condition, the whole unit will generally remain in good condition. Conversely, if key areas are allowed to deteriorate, the grazing capacity of the whole unit will be endangered.

Grazing capacity determinations are relatively complex and require careful study over a period of several years. They are arrived at most simply and accurately by observing soil conditions and changes in plant cover. If the best plants are being destroyed and soil movement is observed, numbers of animals should be reduced or season of use changed; conversely, if excessive forage remains at the end of the grazing season, numbers should be slowly increased until a balance is struck.

The following rule of thumb, applied to the more heavily grazed key areas, may be used in arriving at the proper stocking rate: "Use half and save half, and the half you save will grow bigger and bigger." The rule refers to half the weight, which is concentrated at the bottom of the plant, and not to half the height. Thus, when the 50% rule of thumb is applied to bluebunch wheatgrass, a common range plant, it means that approximately 75% of the bunches have been grazed to an average stubble height of about 4 inches, and the remaining 25% of the plants left relatively ungrazed.

In arriving at grazing capacity determinations, it is generally wise to seek assistance from county agents, soil conservation service technicians, or other trained specialists.

The commonly used terms for describing range condition are: (1) excellent, (2) good, (3) fair, and (4) poor. If the range is covered with 75% or more high value forage plants, it is classified as being in excellent condition. If the best plants constitute less than 25% of the total cover, it is classified as being in poor condition. Good and fair classifications are intermediate. The trend in condition is also important; if the range condition is improving, the trend is upward, and vice versa. Actually, the range condition reflects the kind of management practiced in the past.

Water Supply

Cattle should always be provided with a dependable and adequate supply of good water. On some ranges, this problem presents a difficult, costly, or even impossible situation. The source(s) of water vary between ranges and areas, with reservoirs or ponds, streams, deep wells, and springs being utilized.

Some cattlemen haul water to the range. Although there is considerable variation in the equipment, the better outfits use the following: (1) a main storage tank that will hold at least one week's supply; (2) a 1,000-gallon tank truck; and (3) galvanized rectangular troughs, approximately 8 feet long and

Fig. 9-25. A cattle water supply on TO Ranch, Raton, N.M. Deep wells and concrete tanks are becoming more common on the western range. (Courtesy, National Cottonseed Products Association, Inc., Dallas, Tex.)

Fig. 9-26. A cattle water hole in New Mexico. (Courtesy, Bureau of Land Management, Department of the Interior)

2 feet wide, with sloping sides so as to permit nesting and easy transportation. On unwatered ranges, hauling can make the difference between grazing and no grazing. Also, hauling helps obtain more uniform use of forage, reduces trailing damage, permits grazing at the most appropriate time, and makes for more animal weight gains. It generally costs $1.50 to $3.00 per animal per month to truck water where hauling distances do not exceed 15 miles for the round trip.

For cattle, the ideal arrangement consists of having sufficient watering places distributed over the range so that the animals never have to travel over one-half mile, and not to exceed two miles at the most. Sheep can travel about twice as far to water as cattle. If cattle must travel great distances to water, the amount of walking will use up much of the energy which would otherwise be available for the production of meat or milk. They also will water infrequently and overgraze and tramp out the grass near the water supply.

Range Grazing Systems

Consciously or unconsciously, cattle ranchers follow one or various combinations of the following grazing systems: (1) continuous grazing, (2) rotation grazing, (3) rotation-deferred grazing, or (4) rest-rotation grazing.

Currently, most progressive ranchers of the northwestern United States use combinations of the four grazing systems, adjusting the system to fit their range needs. In the Great Plains and other regions of good summer rainfall, continuous grazing is most widely used.

CONTINUOUS GRAZING

Perhaps most ranges of the West are grazed more or less continuously, although some rest is given them through the use of seasonal ranges (such as in migrating to summer ranges in the mountains). Where continuous grazing is **moderate**, it is perhaps more suitable and practical than rotation-deferred or rest-rotation grazing under the following conditions: (1) when the construction of fences or barriers is very costly; (2) when the important forage species are not dependent upon reseeding for reproduction; and (3) when seasonal ranges are available and used.

ROTATION GRAZING

Rotation grazing is that system in which the grazing of areas is alternated at intervals throughout the season. A heavy concentration of animals is placed on a given area for a few weeks, after which all the stock are moved on to another area or areas and are finally returned to the first field when the growth is sufficient to withstand another period of grazing. This system is best adapted to the utilization of cultivated pastures in the irrigated valleys of the West or to the humid regions of the United States. However, if a high intensity-low frequency system is followed, it will work on the arid ranges of the West.

ROTATION-DEFERRED GRAZING

In rotation-deferred grazing, the range usually is divided into three to five or more units. The grazing on at least one unit is deferred each year until after the seed crop has matured, or through a complete growing season or period. The next year a second area is deferred, and the grazing on the first area is delayed as late as possible to afford opportunity for the young seedlings to become established. By so treating a new unit each year, the entire area is rested, allowed to reseed itself, and grazed in rotation.

Sometimes rotation-deferred grazing is used as a part of rotation grazing and continuous grazing to improve plant vigor, insure natural seed production and establishment, or in conjunction with practical range reseeding and brush control.

REST-ROTATION GRAZING

This is a relatively new range grazing system. It consists of resting one subunit (range or area), while grazing the others.

The Texas Station reports that their studies show conclusively that, from the standpoint of range improvement and economic return, rest-rotation grazing is superior to continuous grazing.[3]

---

[3]In a personal communication to the author from Dr. Leo B. Merrill, In Charge, Texas A & M University Agricultural Research Station at Sonora, Tex.

Range specialists are agreed that a rest-rotation grazing system has much to offer. However, there are two schools of thought as to the best way in which to apply it, with the controversy centering around the length of the rest period and the intensity of grazing. The two systems are:

● *Rest-rotation grazing, with the alternate resting of each pasture for one year*—Essentially, this system consists of dividing a grazing area into 3 pastures and using them in a 3-year rotation. One pasture is rested while the other 2 are grazed. Here is how it works:

1. Rest one pasture completely for one year, to allow establishment of valuable seedlings. During this rest period, both low-value and high-value plants will recover; but it is hoped that the latter will be favored.

2. Crowd all the livestock on the two remaining, unrested pastures, thereby forcing the grazing of low-value plants and unpopular sites. However, the following differences will prevail in the grazing of these two pastures:

   a. One pasture will be grazed continuously all season.
   b. The other pasture will be allowed to mature seeds before turning stock in, following which it will be grazed heavily so as to tramp the seeds into the ground.

3. Rotate the use of the pastures the next year. Thus, during a given three-year period, each of three pastures will be (a) completely rested one year, (b) grazed continuously all season one year, and (c) allowed to make seed, followed by heavy grazing, one year.

● *Rest-rotation grazing, using a high intensity-low frequency system*—This system uses one herd for five or more pastures. Each pasture is grazed for several weeks (but not utilized much over half), then rested for at least four months. Advocates of the high intensive-low frequency system claim that, from the standpoint of utilizing less palatable forage, it is far superior to rest-rotation, with each pasture rested one year.

Distribution of Animals on the Range

Next to the proper rate of stocking and proper seasonal use, distribution of the animals on the range is the most important feature in range management. Proper distribution of animals is reflected in more even utilization of the forage. This assignment is more difficult with cattle than with sheep, especially on rough or mountainous land. Cattle have more of a tendency to utilize the flatter areas and to congregate around watering places. Better distribution of cattle on the range may be accomplished through: (1) fencing or riding the range (or herding); (2) providing water at short intervals (under ideal conditions, the distance between water in rough country should not exceed one-half mile for cattle, in level country 2 to 2½ miles); (3) systematically locating salt grounds away from watering areas and salting at the proper intervals and in the right quantities, and (4) building trails into inaccessible parts of the range.

Fig. 9-27. Showing how systematically locating salt grounds away from watering areas may be used as a means of obtaining better distribution of cattle on the range. (Courtesy, U.S. Forest Service)

Range Riding

On a cattle spread, the range rider is the counterpart of the sheep herder. Riders prevent straying, force better distribution of animals on the range and more even utilization of forage, provide salt, service watering facilities, give minor repairs to fences, herd the animals away from areas infested with poisonous plants, see that "bogged down" animals are removed and cared for, dispose of dead carcasses, administer such assistance as may be required when animals are injured and at calving time, and warn the owner of any unusual parasite or disease problems. The number of cattle cared for by one rider and his saddle and pack animals will vary from 150 to 1,200 head. The number will be determined chiefly by the character of the country and the carrying capacity of the range.

Fencing on the Range

Some fences are essential to the improvement of both cattle and ranges. Cattle fences are less costly than sheep fences. Moreover, cattle are less well adapted to herding methods than sheep. For these reasons, cattle ranges are more frequently fenced than sheep ranges.

The cattle rancher commonly refers to two types of fences: (1) drift fences, and (2) division fences. Drift fences are those that are not intended as enclosures, but which serve as barriers to retain animals within a certain area and to prevent drifting into an area where it is not desired that the animals shall travel. Usually drift fences extend between such natural barriers as steep ridges, ravines, etc. Frequently drift fences are useful for such things as preventing animals from following the snow line back in the

Fig. 9-28. A drift fence on the Uncompahgre National Forest of Colorado's western slope. Drift fences are not intended as enclosures, but serve as barriers to retain animals within a certain area and to prevent drifting into an area where it is not desired that the animals shall travel. (Courtesy, *The Record Stockman*, Denver, Colo.)

spring, avoiding poison infested areas, confining animals to the area owned or leased by the operator, and holding stock during the roundup.

Division fences are those which enclose the boundaries of the range or field. They are usually used in keeping animals on the area owned or leased, in segregating different age and sex groups, and in conducting deferred or alternate systems of grazing.

Breeding Season and Calving Time

The breeding season on the range is usually timed so that the calves will be dropped in the spring of the year, with the coming of mild weather and green forage. Quite naturally, therefore, the calendar date of calving will vary in different sections of the United States. It is earlier on the ranges of the Southwest (February to April) and later in the Northwest (April to June).

Some range producers let the bulls run with the cows throughout the year, reasoning that a calf born out of season is better than no calf at all. The vast majority of the better producers, however, prefer to have the calves born in the shortest possible time. They make every effort to get calving over within two to three months.

Winter Feeding Cattle on the Range

Cattle must be given rather close supervision when on the winter range, and usually some supplemental feeding is desirable. Unless they are on fenced range or herded, they are inclined to drift great distances during blizzards and in cold weather. Unlike horses, they seldom learn to paw snow off the ground in search of forage.

Fig. 9-29. Winter feeding cattle on the range. (Courtesy, U. P. Railroad Company)

In the early days of the range cattle industry, the animals were usually moved to the lower winter ranges and turned loose to get their feed as best they could. There was precious little feeding of supplemental forage or concentrates. If the winter happened to be mild and if a reasonable amount of grass had cured on the stalk, the herd came through in pretty good shape. During an exceedingly cold winter, however, the losses were severe and often disastrous. Today, the practical and successful rancher winter feeds. He generally provides 1 to 2 pounds daily of a protein supplement and has an adequate supply of fodder or hay for those periods when the range is covered with snow. Young animals and cows with calves are more liberally fed than mature animals.

The progressive rancher is also equipped to meet emergency feeding periods, of which droughts are the most common in the West. Concentrates and roughages should be available for such emergencies.

The Roundup

Because of the large territory over which range cattle graze, it is common practice to gather them together at least twice each year, in the spring and again in the fall, for the purpose of carrying out certain routine assignments.

Fig. 9-30. Cattle roundup on a Nevada ranch. Note animals trailing in from several directions. (Photo by Bennett, Reno, Nev., through the courtesy of *The Western Horseman*)

The spring roundup takes place between April and June, the exact time depending upon the earliness of the forage and the time of calving. The objects of this roundup are: (1) to get an accurate count on the stock; (2) to castrate, brand, and vaccinate calves against blackleg; and (3) to separate the breeding animals from the steers and heifers.

The fall roundup, which usually takes place in September and October, is for the purpose of: (1) castrating, branding, and blackleg-vaccinating calves born since the spring roundup; (2) culling out and marketing barren, old, or otherwise undesirable breeding stock which it is not desired to winter; and (3) weaning the calves.

QUESTIONS FOR STUDY AND DISCUSSION

1. Give facts and figures pointing up the economics of pastures in the United States.
2. What are the primary differences between (a) tame (seeded) pastures, and (b) native pastures?
3. How would you go about determining what grass and/or legume to seed on a particular farm or ranch?

4. Outline, step by step, for tame pastures, the procedure (a) for establishing a new pasture and (b) for managing a new pasture.

5. Discuss the economic importance of the western range for cattle production.

6. Why is so much of the range area of the West publicly owned and unenclosed? Is it good or bad to have so much public domain?

7. Do you concur in the policy which permits a sizable number of wild animals to feed on privately owned land?

8. If you live in the West, classify the range grass of your ranch or area as to (1) type (which of the 10 broad types), and (2) dominant native species.

9. What similarities and differences characterize the various agencies administering public lands?

10. Some environmentalists are agitating for a ban on grazing rights of public lands. What are the pros and cons for such action, and what is your recommendation?

11. Discuss the seasonal use of western ranges.

12. How can you detect if a given range has been overstocked? If overstocking is apparent, how would you rectify the situation?

13. Would you change the grazing system followed on your ranch, or in your area? Justify your answer.

14. Wherein does a modern cattle roundup differ from the roundup portrayed in most western movies?

15. How has the handling of cattle on the western range been mechanized? What further automation and mechanization of the western range can you suggest?

SELECTED REFERENCES

Title of Publication	Author(s)	Publisher
Commercial Beef Cattle Production	Ed. by C. C. O'Mary, I. A. Dyer	Lea & Febiger, Philadelphia, Penn., 1972
Crop Production, Third Edition	R. J. Delorit H. L. Ahlgren	Prentice-Hall, Inc., Englewood Cliffs, N.J., 1967
Forage and Pasture Crops	W. A. Wheeler	D. Van Nostrand Company, Inc., New York, N.Y., 1950
Forages, The Science of Grassland Agriculture, Third Edition	M. E. Heath D. S. Metcalfe R. F. Barnes	The Iowa State University Press, Ames, Iowa, 1973
Grass, The Yearbook of Agriculture 1948	U. S. Department of Agriculture	U. S. Government Printing Office, Washington, D.C., 1948
Livestock Husbandry on Range and Pasture	A. W. Sampson	John Wiley & Sons, Inc., New York, N.Y., 1928
Manual of the Grasses of the United States, Second Edition	A. S. Hitchcock, Rev. by A. Chase	U. S. Government Printing Office, Washington, D.C., 1950
One Third of the Nation's Land	Public Land Law Review Commission	U. S. Government Printing Office, Washington, D.C., 1970
Pasture and Range Plants	Phillips Petroleum Company	Phillips Petroleum Company, Bartlesville, Okla., 1963
Pasture Book, The	W. R. Thompson	W. R. Thompson, State College, Miss., 1950

(Continued)

Title of Publication	Author(s)	Publisher
Practical Grassland Management	B. W. Allred, Ed. by H. M. Phillips	*Sheep and Goat Raiser Magazine,* San Angelo, Tex., 1950
Problems and Practices of American Cattlemen	M. E. Ensminger M. W. Galgan W. L. Slocum	Wash. Agr. Exp. Sta. Bull. 562, Washington State University, Pullman, Wash., 1955
Range Management, Principles and Practices	A. W. Sampson	John Wiley & Sons, Inc., New York, N.Y., 1952
Stockman's Handbook, The, Fourth Edition	M. E. Ensminger	The Interstate Printers & Publishers, Inc., Danville, Ill., 1970
Western Range Livestock Industry, The	M. Clawson	McGraw-Hill Book Company, Inc., New York, N.Y., 1950

CHAPTER 10

BEEF CATTLE MANAGEMENT

Contents   **Page**

According to Webster, management is *"the act, or art, of managing, handling, controlling or directing."*

Three major ingredients are essential to success in the cattle business: (1) good cattle; (2) a sound feeding program; and (3) good and aggressive management. The author called upon selected cattlemen and beef cattle specialists to rank these three factors in order of importance. Over 80 percent of the respondents put management at the top of the list. Further, their consensus was that a poor cow-calf manager can half the calf crop and lower weaning weights by a third, and a poor cattle feedlot manager can lower average daily gains by as much as three-fourths of a pound, and increase costs as much as nine cents per pound. Indeed, a manager can make or break a cattle outfit. Unfortunately, this fact is often overlooked in the present era, primarily because the accent is on scientific findings, automation, and new products.

Management gives point and purpose to everything else. The skill of the manager materially affects how well cattle are bought and sold, the health of

Fig. 10-1. Cattle management has gone modern! On Tequesquite Ranch, Albert, New Mexico, this helicopter is used for observation, transporting crews, and rounding up cattle in large pastures. (Courtesy, A. J. Mitchell, Albert, N.M.)

the animals, the results of the ration, the stress of the cattle, the percent calf crop and the weaning weights of the calves, the rate of gain and feed efficiency, the performance of labor, the public relations of the outfit, and even the expression of the genetic potential of the cattle. Indeed, a cattle manager must wear many hats—and he must wear each of them well.

The bigger and the more complicated the cattle operation, the more competent the management required. This point merits emphasis because, currently, (1) bigness is a sign of the times, and (2) the most common method of attempting to "bail out" of an unprofitable cattle venture is to increase its size. Although it's easier to achieve efficiency of equipment, labor, purchases, and marketing in big operations, bigness alone will not make for greater efficiency as some owners have discovered to their sorrow, and others will experience. Management is still the key to success. When in financial trouble, owners should have no illusions on this point.

In manufacturing and commerce, the importance and scarcity of top managers are generally recognized and reflected in the salaries paid to persons in such positions. Unfortunately, agriculture as a whole has lagged; and altogether too many owners still subscribe to the philosophy that the way to make money out of the cattle business is to hire a manager cheap, with the result that they usually get what they pay for—a "cheap" manager.

Without attempting to cover all management practices, facts relative to—and methods of accomplishing—some simple beef cattle management practices will follow.

HEAT DETECTION METHODS AND DEVICES

The problem of heat detection becomes more important as herds get larger, good hired help is more difficult to come by, cows produce more milk, and animal value increases.

Under ordinary farm conditions, herdsmen miss an estimated 25 to 50 percent of the heat periods. On the average, a missed heat period prolongs the calving interval by 30 to 40 days and means a loss of more than $20 in a dairy herd and $10 in a beef herd. Some owners pay their employees a bonus for catching a cow in heat. For these reasons, cattlemen are interested in heat detection methods. Among them are the following:

1. *Chin-Ball Marker*—This device was developed in New Zealand. It is similar to a ball-point pen attached to a halter under the chin of a surgically modified, teaser bull, often called a "Gomer." (One of the first ranches in North America to use the Chin-Ball Marker gave this name to the bull on which it was used.) During preservice sex play, it is usual for a bull to place his head over the shoulders, back, and rump of the cow. This causes a smearing of the colored ink from the ball-point onto the cow.

Fig. 10-2. Chin-Ball marking device. (Courtesy, American Breeders Service, De Forest, Wisc.)

One filling of the stainless steel container is sufficient to mark 15 to 25 cows. Experience indicates that one Gomer bull can work approximately 80 cows. In large pastures and in larger sized herds, it is best to have two bulls. This method of heat detection is a most dependable management tool.

2. *The KaMaR Heat-Mount Detector*—The heat-mount detector is a 2″ × 4″ fabric base to which is attached a white plastic capsule. Inside the capsule is a small plastic tube containing red dye. The tube is constructed so the dye is released slowly by moderate pressure. When enough dye is re-

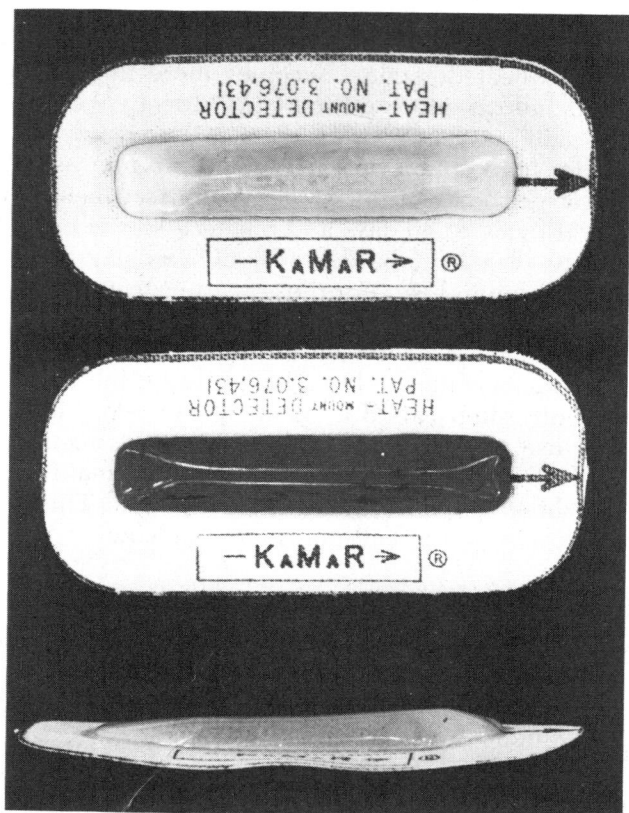

Fig. 10-3. Device for heat detection, as an aid in the artificial insemination of beef and dairy cows. At the top, the KaMaR Heat-Mount Detector is shown before activation. Center shows detector bright red after activation, indicating that cow is in heat. Lower view shows side or profile view of the device, which is applied to cow by an adhesive. (Courtesy, KaMaR, Inc., Steamboat Springs, Colo.)

leased from the tube (after about four to five seconds of pressure), it spreads over the inner lining of the capsule, causing it to turn red.

The detector relies on the natural bovine instinct of "bulling" or mounting during estrus. The pressure from the brisket of a mounting animal causes the dye to be released and the detector to turn red. If the cow does not stand for the mounting animal, there will not be enough pressure to release the dye and turn the detector red. This device has resulted in catching 95 percent of the heat periods.

3. *Pen-O-Block*—The Pen-O-Block is a plastic tube placed within the bull's sheath and held in place with a stainless steel pin. The bull can detect cows in heat and mount them in a normal way, but the device mechanically prevents him from making contact with the cow.

The Pen-O-Block consists of a white plastic tube, the pin or cannula, two washers, and a cotter pin. The device is inserted within the bull's sheath and held in place by the cannula. The procedure is best carried out by a veterinarian, as it requires skill.

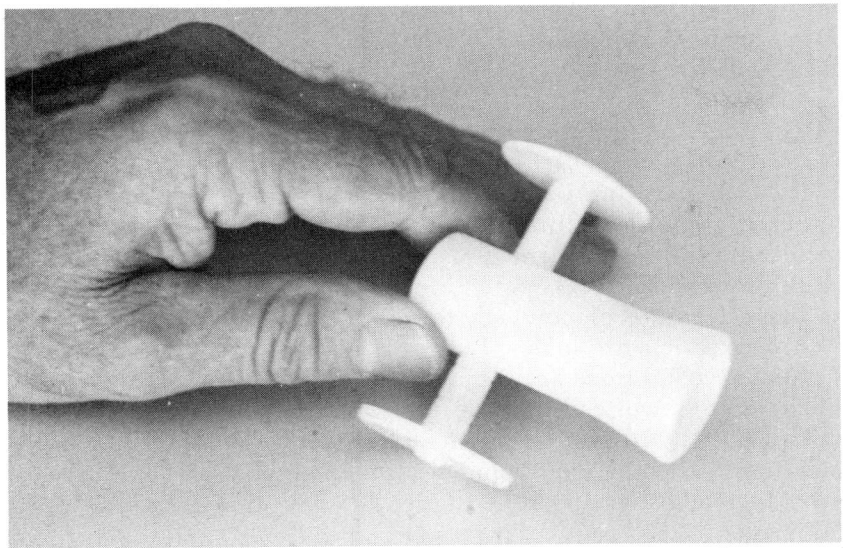

Fig. 10-4. Pen-O-Block marking device. (Courtesy, American Breeders Service, De Forest, Wisc.)

Properly used, these three aids will improve heat detection. They are by no means replacements for visual heat detection; nor will they solve all the problems in breeding a beef herd artificially. Other factors that need attention are:

1. *Nutrition*—Cows must have adequate nutrition to cycle at a satisfactory rate for successful breeding.

2. *Rest interval*—This is very important, as cows must have calved at least 60 days prior to breeding for satisfactory performance.

3. *A.I. facilities*—Facilities should be adequate for handling and breeding the cow herd. Locate them where the cows tend to gather, such as the watering hole.

4. *Personnel*—Trained personnel are needed to do heat detection, gather the in-heat cows, and inseminate the herd.

PREGNANCY TESTING

(See Chapter 6, Physiology of Reproduction in Cattle, section on "Pregnancy Test.")

CALVING SEASON

Most cattlemen favor spring calves over fall calves because they involve less intensive and expensive management practices than at any other season. Spring calving is more "in tune" with mother nature, because forage production is at its best and cows are generally more fertile.

Cows can usually drop spring calves outdoors, away from buildings, which helps reduce health problems in newborn calves. Also, pregnant cows can be wintered more economically than lactating cows with fall calves. Then, too, spring-born calves are ready for sale in the fall when demand for feeder calves is greatest.

The advantages and disadvantages of spring and fall calves follow:

• *Advantages of spring calves*—The production of spring calves has the following advantages:

1. The cows are bred during the most natural breeding season—at a time when they are on pasture, gaining in flesh, and most likely to conceive. As a result, the calving percentage is usually higher with a system of spring calving.

2. The calves are old enough to use the cow's abundant milk supply when spring pastures are lush.

3. Weather conditions are usually favorable.

4. There are fewer calf diseases than among fall calves.

5. The calves will be in shape to sell directly from the cows in the fall, at which time there is a good demand for feeder calves.

6. Spring calves usually have higher 205-day adjusted weaning weights than fall calves.

7. If the calves are to be sold as yearlings, one wintering is saved; or if they are to be sold at weaning time, no wintering is required.

8. Because of greater utilization of cheap roughage, dry cows may be wintered more cheaply.

9. Spring calves require little or no supplemental feeding and utilize the maximum amount of pasture and roughage if marketed at weaning.

10. Less labor and attention is required in caring for the calves the first winter.

• *Disadvantages of spring calves*—Labor is needed for calving when it is least available, due to spring work.

• *Advantages of fall calves*—The production of fall calves has the following advantages:

1. The cows are in better condition at calving time.

2. Labor is more readily available for late fall calving than for spring calving.

3. The cows give more milk for a longer period.

4. The calves make better use of the grass during their first summer.

5. The calves escape flies, screwworms, and heat while they are small. (This is especially important in the South.)

6. Upon being weaned the following spring, the calves can be placed directly on pasture instead of in a drylot; or, if it is desired to sell, they usually find a ready market ahead of the influx of fall feeder calves from the range area.

7. When it is intended to sell market milk from dual-purpose cows, fall calves are usually best. The greater flow of milk is obtained during the period of highest prices.

• *Disadvantages of fall calves*—The two main disadvantages of fall calves are:

1. Supplemental feed for the lactating cow must be provided in the drylot.

2. Although calves are heavier the next fall than spring calves sold at weaning, the price per hundredweight is often lower.

DEHORNING

Although the presence of well-trained and properly polished horns may add to the attractiveness of the horned breeds in the show-ring or of the purebred herd, horns are objectionable on animals in the commerical herd and should always be removed.

● *The chief reasons for dehorning are—*

1. Less shed and feeding space is required for dehorned cattle.

2. Dehorned cattle are less likely to inflict injury upon other cattle or upon the attendant.

3. Dehorned cattle are quieter and easier to handle.

4. Feeders prefer to buy dehorned cattle.

5. Dehorned cattle suffer fewer bruised carcasses and damaged hides in shipment to market, thus commanding a premium of 50 to 75 cents per hundredweight more than horned cattle of similar market class and grade.

Although the advantages in dehorning commercial cattle far outweigh the disadvantages, and progressive farmers and ranchers regularly dehorn their calf crop, there are certain unfavorable aspects. The disadvantages are:

1. Dehorning gives animals a setback, especially when the operation is performed on older animals. With yearling steers, the South Dakota station found[1] (a) that about two weeks were needed for dehorned steers to equal their initial weight, and (b) that dehorned steers failed to catch up with horned animals; that, due to shrink at dehorning time, dehorned steers weighed slightly less at marketing time than horned steers.

2. Labor and equipment are required in dehorning.

3. There are some death losses which result from excessive bleeding, screwworms, or infection.

4. Scurs may result if the operation is not carefully done.

5. Diseases may be spread unless equipment (except the hot iron) is disinfected between animals.

Age and Season to Dehorn

Cattle should be dehorned early in life, preferably before they are two months old, so as to minimize shock. Young calves are easier to handle, lose less blood, and suffer less setback. The danger of screwworm trouble and infection is also reduced when calves are dehorned while young.

In a study[2] which included more than a half million cattle in a 24-state area, it was found that, on the average, calves are dehorned at 5.2 months of age, which is considerably later in life than is desirable.

Less insect trouble is encountered when dehorning is done in the early

---

[1]Luther, Richard M., *South Dakota Farm and Home Research*, Vol. IX, No. 2, Feb. 1958, pp. 16-19.

[2]Ensminger, M. E., M. W. Galgan, and W. L. Slocum, Problems and Practices of American Cattlemen, *Wash. Agr. Exp. Sta. Bull.* 562.

spring or late fall—an especially important consideration in dehorning animals in an area where screwworms exist.

Methods of Dehorning

Dehorning is accomplished in the ways shown in Fig. 10-5, which also shows the percentage of cattlemen using each method. As noted, most cattle are dehorned by means of saws and clippers, with the use of chemicals and polled bulls almost tied for second position. Because of the screwworm problem, the use of polled bulls is especially popular in the South.

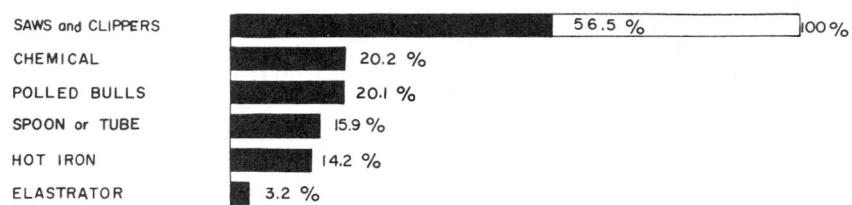

SAWS and CLIPPERS   56.5 %   100%
CHEMICAL   20.2 %
POLLED BULLS   20.1 %
SPOON or TUBE   15.9 %
HOT IRON   14.2 %
ELASTRATOR   3.2 %

Fig. 10-5. Methods of dehorning and the percentage of cattlemen using each. (Source: *Wash. Agr. Exp. Sta. Bull. 562*, courtesy, Washington State University)

When using a mechanical method to dehorn, most stockmen (96%)[3] routinely do their own work; others (4%)[4] call upon the veterinarian. Perhaps the most important thing is that it be done at the proper time.

BREEDING POLLED CATTLE

The use of polled bulls is the most humane method of securing cattle without horns. If such a bull is "pure" polled (carrying in his blood no tendency to produce horns), practically all of his calves will be polled, even though their dams have horns.[5] This system of securing hornless calves saves labor and avoids pain and possible setback to the calves. Without doubt, the breeding of polled calves will increase in popularity.

USE OF CHEMICALS

With small herds kept under close supervision, newborn calves may be dehorned satisfactorily by the use of caustic potash (potassium hydroxide) or caustic soda (sodium hydroxide). These chemicals can be purchased at almost any drug store and come in the form of a stick, paste, or a lacquer base (the accompanying use of petroleum jelly is not necessary with the lacquer base). This method really prevents horn growth and does not actually remove the horns. The treatment should be applied when the calf is from 3 to

---

[3]Ibid.
[4]Ibid.
[5]This applies to cattle of European extraction. See section in Chapter 5 of this book relative to "Dominant and Recessive Factors" and the footnote therewith, for a more complete treatment of this subject.

10 days old and when only small buttons are present. After the hair around the buttons has been clipped or sheared closely, smear a ring of heavy grease or petroleum jelly around the clipped area to keep the caustic from running into the calf's eyes. Then rub the caustic material over the button or little horn until the blood appears. This should be done carefully, for otherwise some of the horn cells may not be destroyed and a scur may develop.

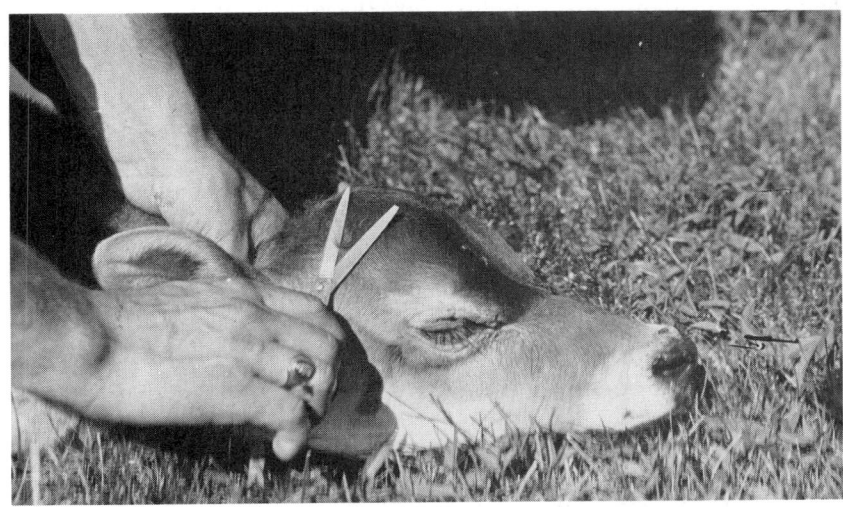

Fig. 10-6. Use of chemicals in dehorning. First the hair around the "button" is clipped (upper), following which caustic material should be rubbed over the little horn until blood appears (lower). (Courtesy, USDA)

The caustic should be wrapped in a cloth or paper to protect the operator's hands from serious burns. Within a week or 10 days, the thick scab that appears over the horn buttons will drop off, and the calf will suffer little inconvenience. Calves treated with caustic should be protected from rain for a day following the application, for the caustic may wash down and injure the side of the face. Also, it is best not to turn the calves back with the dams for a few hours following the application of the caustic.

SAWS; CLIPPERS (Barnes-Type Dehorner)

Saws or various forms of shears and clippers (including the Barnes-type dehorner) are used almost exclusively for dehorning in the range country, and for dehorning calves over four months old. Even on the general live-stock farms of the central and eastern states, mechanical methods are more generally used than chemicals. Whatever the instrument used (saws or clippers), it is necessary to remove the horn with about ¼ to ½ inch of the skin around its base to make certain that the horn-forming cells are destroyed. The skin should then be allowed to grow over the wound. Dehorning of young animals can be done with greater ease, and there is less shock to the animals. However, some attention must be given to the season.

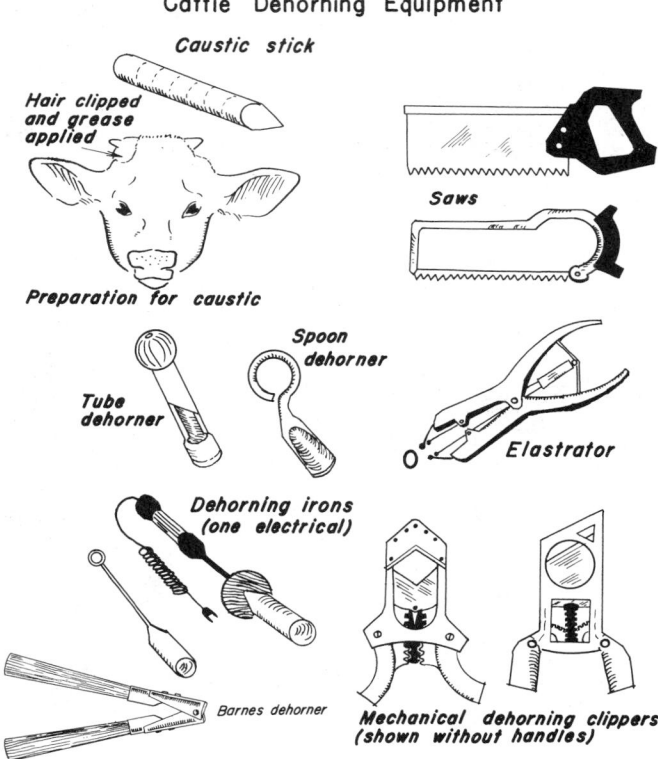

Fig. 10-7. Common instruments used for dehorning cattle. (Drawing by R. F. Johnson)

Ordinarily, clippers are satisfactory for removing the horns of younger cattle; but the hard, brittle horns of mature cattle can best be removed with a saw. With older animals, clippers are likely to sliver or crack the bone that forms the horn core. Moreover, the saw results in less loss of blood, for the action of the saw blade produces a lacerating of the blood vessels rather than a clean-cut cross section. On the other hand, the ragged wound made by a saw heals more slowly, and the operation is less rapid.

While the dehorning operation with older cattle is being performed, it is necessary to have some device for confining or restraining animals. For this purpose, various types and arrangements of dehorning chutes, pinch gates, squeeze pens, and cattle stocks have been devised. Calves may be handled by throwing or snubbing them to a fence post and tying one side of the body against a strong fence or solid wall. Such methods are more difficult, however, for both man and beast.

THE HOT IRON (Fire Irons; Electrically Heated Irons)

The hot-iron method of dehorning consists of the application of a specially designed hot iron to the horn of young calves. The iron is fashioned with one end cupped out (bell-shaped) so that it fits over the small horn button. Some ranchers first cut out the horn with a sharp knife before applying the hot iron; whereas others use a larger iron and fit it over the small horn. This system of dehorning is bloodless and may be used any time of the year, but it can be used on young calves only—preferably calves under five months of age.

Where electricity is available, the electric hot iron may be used. It keeps an even temperature, without getting too hot or too cold.

THE DEHORNING SPOON AND DEHORNING TUBE

The dehorning spoon (or gouge) is a small instrument with which the horns of young calves can be gouged out. In the hands of an experienced operator, it is both fast and effective. The use of the spoon leaves the head slightly rounded, and very seldom do scurs occur.

The dehorning tube is a newer instrument than the dehorning spoon. In comparison with the spoon, the tube is easier, faster, and less tiresome to use, and more certain to avoid regrowth.

Dehorning tubes come in four sizes, varying in diameter from ¾ of an inch for the smallest to 1⅛ inches for the largest. All four sizes should be available.

The steps and directions for using the dehorning tube are as follows:

1. Restrain the calf.

2. Select a sharp tube of proper size to fit over the base of the horn and include about ⅛ of an inch of skin all the way around.

3. Place the cutting edge straight down over the horn and then push and twist, first one way and then the other, until the skin has been cut through. A cut from ⅛ to ⅜ of an inch deep is required, the greater depth being neces-

sary with calves about 3 months of age. Going deeper than necessary will cause excessive bleeding.

4. Turn the tube to about a 45 degree angle and rapidly shove and turn the cutting edge until the button comes off.

Most ranchers who use the dehorning spoon or tube do so at the time of branding, thus avoiding extra handling. Either instrument can be used on calves up to 60 days of age.

Treatment After Dehorning

If dehorning is done in cool weather (spring or fall), when there are no flies, no treatment is required. On the other hand, if the operation is performed when flies are present, it is important that a good repellent be applied to the wound. Smear 62 or EQ335 (or some equivalent) is very satisfactory for this purpose. As a rule, there will be no danger from excess bleeding. The danger of infection will be materially reduced if cleanliness is practiced and the instruments (except hot irons) are disinfected at intervals.

CASTRATING

Castrating is the unsexing of a male animal. The practice of castrating males in animal species used for food production purposes is universally practiced and is one of the oldest surgical operations known to man.

Most male calves not intended for breeding purposes are castrated. It makes animals quieter and easier to handle.

The California Station measured the effects on healthy calves of castration by removal of 1/3 of the scrotum and pulling of testicle with cord. The study extended 28 days following castration. In comparison with the noncastrated animals, the castrated calves gained 18 pounds less per head, consumed 12 percent less feed, and required 22 percent more feed to produce a pound of body weight.[6]

Age and Season for Castrating

Bull calves are generally castrated at about 2.5 months of age,[7] although they can be desexed at any age. The older the animal at the time of the operation, the greater the shock and risk. Moreover, if a bull calf is not castrated before he is 10 to 12 months old, he may become "staggy"—a very objectionable characteristic in the feeder or finished steer.

Currently, there is a trend toward castrating at slightly older ages than formerly, primarily (1) to take advantage of the higher gains and greater efficiency of bulls, and (2) to lessen the hazard of urinary calculi.

As in dehorning, it is best to perform this operation in the early spring or late fall so as to avoid infestation from flies. Moreover, it is unwise to castrate during periods of inclement weather. If castration is performed in an area

[6]Addis, D. G., *et al, Research Reports*, University of California, December 3, 1973.
[7]Ensminger, M. E., M. W. Galgan, and W. L. Slocum, Problems and Practices of American Cattlemen, *Wash. Agr. Exp. Sta. Bull. 562.*

where screwworms exist, a fly repellent (Smear 62, EQ335, or some equivalent) should be applied to the wound and the animal should be kept under close observation until the wound has healed over; or a bloodless method of castration should be used.

Methods of Castrating

Fig. 10-8 shows the common methods of castrating and the percentage of cattlemen using each. As noted, the vast majority of cattlemen use the knife.

As is true in dehorning, most stockmen[8] routinely castrate their calves (94.9%); others call upon the veterinarian (5.1%). Perhaps the most important thing is that it be done at the proper time.

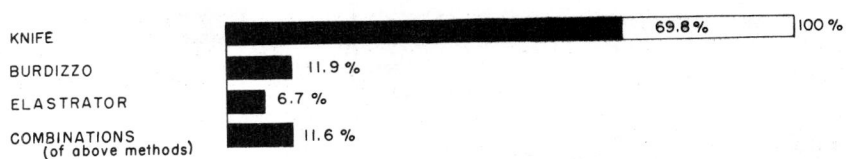

Fig. 10-8. Methods of castrating and the percentage of cattlemen using each. (Source: *Wash. Agr. Exp. Sta. Bull. 562;* courtesy, Washington State University)

REMOVAL OF LOWER END OF SCROTUM

In this method, approximately the lower one-third of the scrotum is removed (exposing the testicles from below); the membrane covering each testicle is slit (if desired, the membrane need not be slit; simply remove it along with the testicle), and the testicles are removed by pulling them out. In older cattle, excessive bleeding may be prevented through severing the partially withdrawn cord by scraping with a knife or clamping with an emasculator. Although the removal of the end of the scrotum allows for excellent drainage, this method is not recommended for calves intended for show purposes because the cod will not be so large and shapely. Most ranchers castrate calves by this method.

SLITTING SCROTUM DOWN THE SIDES

In this method, one testicle is pulled down at a time and is held firmly to the outside so that the skin of the scrotum is tight over the testicle. With a sharp knife, an incision is then made on the outside of the scrotum next to the leg. It is important that the incision extend well down to the end of the scrotum to allow for proper drainage and that it extend both through the scrotum and membrane (if desired, the membrane need not be slit; simply remove it along with the testicle). The testicle may be removed as previously indicated.

---

[8]Ibid.

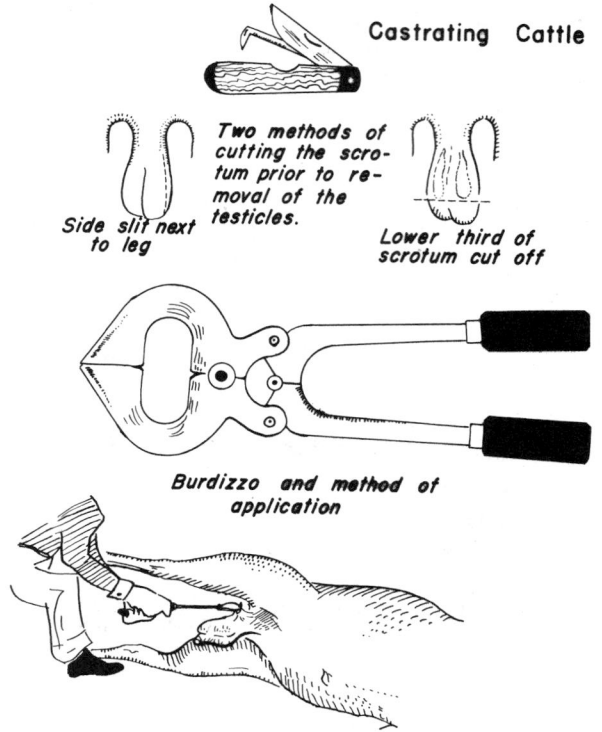

Fig. 10-9. Two common pieces of castrating equipment (the knife and the Bur-dizzo), and method(s) of using each. (Drawing by R. F. Johnson)

THE BURDIZZO PINCERS

"Burdizzo" pincers (named after their inventor, Dr. Burdizzo, and man-ufactured in Italy) are sometimes used in making a "bloodless castration." In this method, the cords and associated blood vessels are crushed or severed so completely that the testicles waste away from want of circulation. After the animal is thrown, the cord is worked to the side of the scrotum, and the Burdizzo is clamped on about 1½ to 2 inches above the testicle, where it is held for a few seconds. Then, this operation is repeated on the same cord at a location about ¼ inch removed from the first one. The same procedure is then followed on the other testicle. In using the Burdizzo, it is important that the cord not slip out, that only one cord be clamped at a time, and that there be no interference with the circulation of the blood through the central por-tion of the scrotum.

This method is a satisfactory means of castration if done properly and by an experienced operator. But if the operation is not performed correctly, the cord may be incompletely crushed and the animal may develop stagginess later. This is especially disturbing when discovered in experimental or show animals or steers that are nearly ready for market or show. Because there is no break in the skin with this type of castration, there is no external bleed-ing. Nor can there be any trouble from screwworms—an important consideration in the South. Furthermore, steers so castrated usually develop very large and shapely cods, a characteristic that is very desirable in well-finished steers.

THE ELASTRATOR

The elastrator (developed in New Zealand) may be used in stretching a specially made rubber ring over the scrotum to castrate young calves. It works best on calves under two months of age. The directions are as follows: (1) hold the calf in either a sitting or lying position; and (2) press both testicles through the ring and to the lower end of the scrotum, then release the rubber ring.

SHORT SCROTUM BULLS

This new method of rendering the intact male infertile was developed by the New Mexico Station. The animal is really a pseudocryptorchid in that the scrotum is shortened by bringing it through a distended rubber band with an elastrator when the calf is one to three months of age. Before the band is released, the testicles are moved near the abdominal wall. The scrotum below the rubber band sloughs off after three or four weeks.

Shortening the scrotum requires considerably less time than castration. It can be done in 15 to 30 seconds. Another advantage over castration is that there is no weight loss. As a result of the shortened scrotum, the testicles lie close to the abdominal wall and the animal is sterile. The testicles then develop to about half the weight of those from fertile bulls of the same age and weight. This new technique eliminates indiscriminate breeding while retaining and even enhancing the testosterone level in the pseudocryptorchid male. The short scrotum treatment does not change either the temperament or the urge of the animal; hence, there will be riding if sexes are not kept separate.

Experimental work to date indicates that the rate and efficiency of gains of short scrotum bulls and intact bulls are about the same; and that the carcasses of short scrotum bulls are leaner than the carcasses of steers.

Position and Treatment in Castrating

Young animals are usually thrown to be castrated, whereas animals eight months of age or older may be more easily operated on in a standing position.

Before making any incision, the hands of the operator and the knife should be thoroughly cleaned and washed in a good disinfectant. With these precautions, there is usually little danger of infection in castrating under range conditions; for normally the calves are turned back on the range where there is plenty of sunshine, fresh air, and no contamination. If the operation is performed in fly season and in an area where screwworms are prevalent, a good repellent should be applied. Smear 62 or EQ335 (or some equivalent) is commonly used for the latter purpose.

SPAYING

In females, the operation corresponding to castration is known as spaying. Under most conditions, desexing of the heifers is not recommended because: (1) the operation is more complicated and difficult, requiring a very experienced man; (2) spaying is attended with more danger than castration; (3) it eliminates the heifers for possible replacement purposes or sale as breeding stock; and (4) experiments and practical operations with spayed heifers have generally shown that the selling price obtained is not sufficiently higher to compensate for the lower and less efficient gains plus the attendant risk of the operation. On the other hand, spaying does prevent the possibility of heifers becoming pregnant, and eliminates the necessity of separating heifers from bulls or steers.

In a survey[9] made by the author, it was found that 9% of cattlemen spayed heifers, that they spayed an average of 67.5 heifers each, and that they spayed at an average of 8.9 months of age. Also, it was reported that 80.5% of the owners did their own spaying, while 19.5% had the work done by the veterinarian.

MARKING OR IDENTIFYING

It is of historical interest to note that one of the first uses of branding in many countries of the world was permanently to identify human criminals.

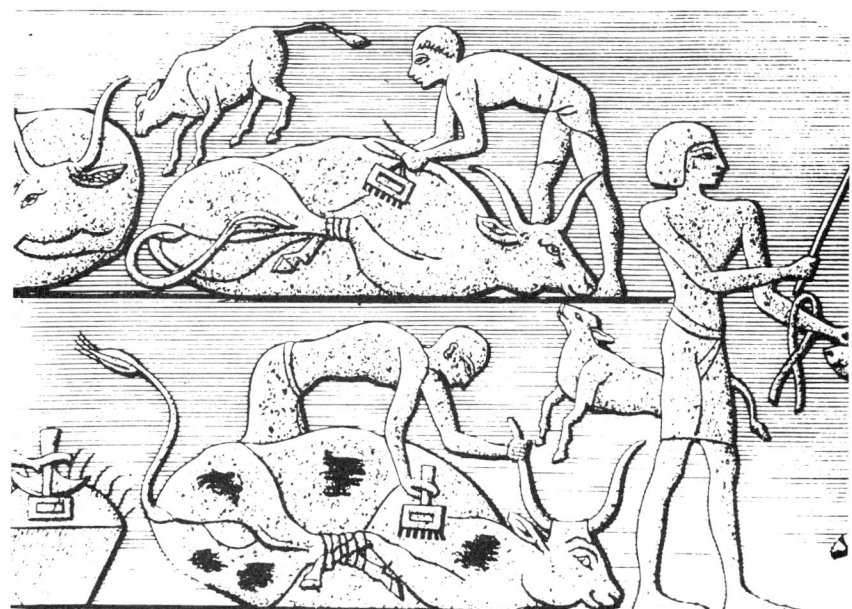

Fig. 10-10. Branding is a very old management practice. This picture made from an Egyptian tomb shows Egyptian drivers branding their cattle. (Courtesy, The Bettmann Archive)

---

[9]Ibid.

Also, of interest is the fact that the noun "Maverick," meaning unbranded cattle, and now a recognized part of the American lexicon, originated with lawyer-cattleman Samuel A. Maverick (1803-1870). Maverick, who accepted 600 head of cattle as an attorney's fee, failed to brand his young stock. As a result, year after year, his unbranded yearlings fell into the hands of other cattlemen who promptly placed their brands on them. After 10 discouraging years of such operation, Maverick sold his depleted herd for the amount of the original fee.

Ranchers take pride in their brands, for to them it is much more than a sign of ownership; it is a symbol of service—a pledge of integrity of the man behind it and a mark of courage, character, and wisdom. It is indicative of the quality of stock that the owner raises, the class of bulls used, the condition of the range, and the kind of cowhands connected with the outfit.

On the western range, marking or branding is primarily a method of establishing ownership and/or age. In the small herd, particularly in the purebred herd, it is a means of ascertaining ancestry or pedigree. The method of marking employed will depend primarily upon the objective sought and the area.

The common methods of identification are listed in Table 10-1 and pictured in the sections that follow. As indicated, the three leading methods of identification, by rank are: hide brands, ear marks, and tattoos. However, there are some differences between areas and between purebred and commercial herds (Table 10-1). Hide brands and ear marks are popular in the West and in commercial herds; and tattooing is a requisite to registration in most beef cattle registry associations. In this same study,[10] it was found that

TABLE 10-1

METHOD OF IDENTIFICATION

Method of Identification	Cattlemen Using the Method[1]					Purebred vs Commercial	
	In West	In South	In Great Plains	In Pacific Northwest	All Respondents	Purebred Herds	Commercial Herds
	(%)	(%)	(%)	(%)	(%)	(%)	(%)
Hide Brands	78.8	66.7	77.2	69.4	73.4	30.0	86.0
Ear Marks	52.6	39.8	38.6	52.0	44.7	5.7	52.2
Tattooing	22.0	26.5	24.7	29.5	24.6	84.9	6.0
Metal Ear Tags	14.3	5.4	13.9	15.0	11.1	10.8	9.4
Neck Chains or Straps	3.6	11.0	5.0	2.9	7.1	14.2	2.3
Horn Brands	5.5	4.8	7.3	5.8	5.6	12.6	0.0
Knows Cows	4.7	7.1	1.5	4.6	4.9	2.7	5.6
Others	2.3	0.4	1.2	2.9	1.3	0.0	1.5

[1]Ensminger, M. E., M. W. Galgan, and W. L. Slocum, Problems and Practices of American Cattlemen, *Wash. Agr. Exp. Sta. Bull. 562.* The columns will total over 100 percent because several cattlemen reported using more than one method of identification.

---

[10]Ibid.

42.8 percent of the cows were individually identified and 36.2 percent of the calves were individually identified before weaning.

Hide Brands

When properly applied, hide brands are permanent. Throughout the range country, the hide brand is recognized as the cattleman's trademark. Most of the western states require that each brand be recorded as to both type and location in order to avoid duplication. When stock are run close to a state boundary, the same brand may be recorded in two states.

In addition to the regular brand, many ranchers identify the age of the females by adding the last number of the year in which they were born (usually at a different location). Thus, heifer calves born in 1976 might be identified by adding the number 6 to the regular brand. At the end of 10 years, the numbers are used over again, for there is seldom any difficulty in determining ages where there is a 10-year spread. In those states where brands are recorded, these added numbers or brands must also be approved by and recorded with the Registrar of Brands.

Hide brands have the disadvantage of being unsightly, and hot iron brands lower the market value of the hide. For these reasons, they are not recommended except when necessary for identification purposes. Even then, it is desirable that their size be as small as possible, consistent with

Fig. 10-11. Branding time at Tequesquite Ranch, Albert, New Mexico. With this size crew, 700 calves are branded in one day. (Photo by H. D. Dolcater, Amarillo, Tex. Courtesy, A. J. Mitchell, Albert, N.M.)

serving the primary objective of the brand. The pertinent facts relative to branding are:

1. *Time*—In the range country, the usual practice is to brand calves at the same time they are castrated and vaccinated against blackleg.

2. *Location on animal*—The brand is located on a body area where it may be easily seen and where it will do the least possible damage.[11] Hips and thighs are favorite body areas for brands.

3. *Preparation*—Usually calves are thrown for branding—roped by the hind legs and dragged to the place of branding. Older cattle, however, are restrained in a chute. Some ranchers now prefer to use the specially designed branding chutes for calves.

4. *Four methods of applying brands are:*

a. *The hot iron*—To date, this has been the preferred method. Ranchers heat the irons to a temperature that will burn sufficiently deep to make the scab peel, but which will not leave deep scar tissue. The proper temperature of the hot iron is indicated by a yellowish color. Branding is accomplished by placing the heated branding iron firmly against the body area which it is desired to mark and by not allowing it to slip for the few seconds when the hide is burned. The branding iron should be kept free from dirt and adhering hair at all times.

The California Station measured the effect of hot iron branding (in one group they branded the rib; in the other they branded the hip). They found that branding had no effect on 28-day weight gains, pounds of feed consumed daily, or pounds of feed required to produce a pound of gain.[12]

Other methods of identification do not provide the excitement of roping and branding cattle with a hot iron. Hot branding, like the TV western, is here to stay.

Where electricity is available, the electric iron may be used; it keeps an even temperature, and if properly used, makes a clear, uniform brand.

b. *Branding fluids*—Branding fluids, which are less widely used in making hide brands, consist of caustic material applied by means of a cold iron. Best results are secured if the area is first clipped. The chemical method of producing hide brands is slower; the results are generally less satisfactory, particularly if the operator is inexperienced with the method; and the resulting brand is less permanent.

c. *Freeze branding*—This new and promising method, developed by the U.S. Department of Agriculture, at Washington State University, makes use of a super-chilled (by dry ice or liquid nitrogen) copper branding "iron" which is applied to the closely clipped surface for about 20 seconds, thereby depigmenting the hair follicles, following which the

---

[11]In arriving at both the kind and location of the brand, the owner should first check with the brand inspector or the local county agent, to determine if any part of the animal is reserved for state or Federal disease control programs; for example, the cheek of cattle is used for brucellosis reactor identification.

[12]Addis, D. G., *et al*, *Research Reports*, University of California, December 3, 1973.

hair grows out white. When properly done, this method is painless, permanent, and there is no hide damage. (Hot-iron branding costs the livestock industry and tanning trade an estimated $20 million annually.)

Fig. 10-12. Freeze branded replacement heifers. (Courtesy, Martin Jorgensen, Jr., Ideal, S.D.)

On white cattle, deliberate overbranding (30 seconds or more) will produce a bald brand suitable for identification after clipping.

To date, freeze branding has not been widely accepted in the United States. Moreover, it will not be legal until many state laws, built around hot-iron branding, are changed.

d. *Laser brand,* which is permanent, but which needs further development and experimental study.

5. *Characteristics of a good brand*—A good brand is one that is easily read, that is of simple design and yet cannot be easily changed or tampered with, that has no welds or thick points in the iron, and that interferes with the circulation as little as possible. Thick points mean deeper burning and slower healing; whereas small enclosed areas, such as a small "0," will slough out entirely.

Earmarks

Earmarks are permanent and easily recognized, but they're unsightly. They may be administered with either a sharp knife or a regular ear notcher. Sometimes polled animals are individually identified through ear notches. In such instances a definite value is assigned to each area location. When earmarks are used in commercial operations, however, they are uniform and recorded for any given ranch. Some of the more common earmarks are "crops," "swallow forks," "bobs," "over-bits," "under-bits," and "splits."

Metal or Plastic Earmarkers (Tags and Buttons)

Metal or plastic earmarkers are easily attached, but sometimes they are easily pulled out and lost, although the latter problem has been reduced by the new plastic types which cannot cut through the ears as easily as the sharper metal varieties. Also, eartags frequently rub and scratch the skin, thus making openings for screwworms infestation—an important consideration in the South.

Fig. 10-13. A show of numbers at Ankony Hyland! Like mother, like calf. Also, ear tags of calves are color-coded by sires. Hence, a calf and its mother can be readily paired, and the calf's sire identified by color of tag. (Courtesy, Ankony Angus, Highmore, S.D.)

Neck Chains or Straps

Neck chains or straps are the most frequently used means of identifying polled cattle. Occasionally, chains or straps may be lost, but this is not particularly serious if the caretaker is on the alert and immediately replaces each one that is lost, without allowing several losses to accumulate before taking action. In rare instances, an animal will hang itself by the chain.

Neck chains or straps must be adjusted, for young animals grow, or animals change in condition.

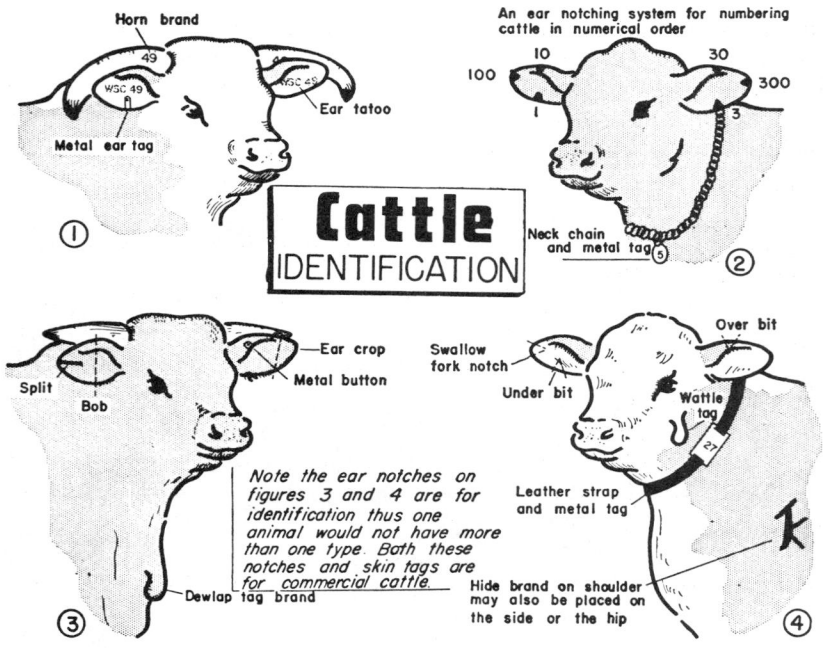

Fig. 10-14. Composite drawings showing a number of methods of cattle identification. It is unlikely that any individual animal will carry more than one or two of these methods of identification. (Drawing by R. F. Johnson)

Horn Brands

Horn branding for individual identification is commonly used among breeding or sale animals of the horned breeds. Usually horn brands are made by heating small copper numbers with a blow torch or charcoal burner. On mature animals, this method of branding works fairly well, but it cannot be used on young animals while the horns are still growing, unless it is repeated at intervals.

Tattoos

Most purebred beef cattle registry associations require that registered animals be individually tattooed. This method of marking consists of piercing the skin with instruments equipped with needle points which form letters or numbers. This operation is followed by rubbing indelible ink into the freshly pierced area. It is well to disinfect the tattooing instrument carefully between each operation in order to alleviate the hazard of spreading warts to the pierced area, for warts make it impossible to read the tattoo.

Cattle must be confined in order to read tattoo numbers. Even then, tattoos are difficult to decipher on dark-skinned animals.

Other Identifications

Other identification marks used on the range include: (1) "buds" formed by making a strip incision through the nose; (2) "wattles" made by cutting down a strip of skin on the jaw bone; and (3) "dewlaps" formed by cutting down a strip of skin on the brisket.

The U.S. Department of Agriculture requires that most cattle two years of age or older be backtagged or eartagged to identify the animals to their herd of origin before they are shipped across state lines.

Various electronic devices are in different stages of research and development; among them—

1. *Radio transmitter in the second stomach*—The animal swallows a small radio transmitter enclosed in a ¾" x 2½" plastic capsule, which lodges in the second stomach. From there, it transmits a coded number when signaled by a receiving unit to do so. The transmitter can be retrieved at slaughter and reused.

2. *Implant behind the poll*—A ¼" cube device, coded to give specific information about the animal when read by a "transponder," is implanted behind the poll. This device may include such information as birth date, original owner, state of origin, year of implant, and temperature of animal. This method holds great promise as a means of combatting cattle theft.

MARKING PUREBRED CATTLE

Table 10-2, Handy Marking or Identifying Guide for Registered Cattle, summarizes the pertinent regulations of the beef and dual-purpose cattle registry associations relative to marking or identifying.

TABLE 10-2

HANDY MARKING OR IDENTIFYING GUIDE FOR REGISTERED CATTLE

Breed	Association Rules Relative to Marking
Angus	Each animal, for which application for registry is submitted, must be tattooed alike in both ears.
	Each breeder may devise his own tattooing system, using a series of numbers or letters, or a combination of numbers and letters. Tattoo marks are limited to four units in each ear; and only standard numerals or letters are acceptable.
	Each animal of the same sex to be registered by any one breeder must be tattooed differently.
Ayrshire	All calves must be tattooed before leaving individual pens or ties.
	Both ears may be used and the letters and numbers in

(Continued)

TABLE 10-2 (Continued)

Breed	Association Rules Relative to Marking
	the ears must be stated on the application for registration. Tattoos must include at least one letter and one number. Duplicate tattoos are not allowed in the same herd. The letters "I," "O," "Q," and "V" may be used only if accompanied by one or more other letters in the same ear. Tattoos may not exceed a total of five letters and numbers per line in each ear. The number shall be followed by the year letter designated by the Association.
Beefmaster	*Foundation Beefmaster Association:* The Association accepts any permanent marking system, such as fire brand or tattoo. The Association does not specify any particular type of marking for certified animals.
Brahman	A holding brand (a symbol, letter, combination of letters and/or symbols, numerals, replica of some object, etc., to denote ownership or breeder) and private herd number (both branded by fire) are required on a calf before it may be registered.
Brangus	Each animal for which application is submitted must be fire branded on the body with the owner's holding brand and a private herd number, and the year brand. The application for registration or enrollment must show where this brand is located on the body. The Association suggests starting with the number "1" on the private herd number and numbering consecutively. The holding brand is any mark, initial, or number, or combination of all three which the breeder chooses to use, and which is approved by the breed registry (IBBA).
Brown Swiss	Each animal must be tattooed in ear with indelible ink with such letters and numbers as the owner may select. No two animals in the same herd, of the same sex, shall have the same number. Both ears may be used. If only one ear is used, it is recommended that it be the left ear. All calves must be tattooed before leaving individual pens or ties.
Charbray	Ear tattoo or fire brand. The breeder may devise his own system.
Charolais	Ear tattoo or fire brand. Breeder may devise his own system.
Chianina	All animals must be individually tattooed.
Devon	Each breeder is assigned a herd tattoo code of three letters, which must be applied to the right ear. The individual herd number plus the letter code for the year of birth must be tattooed in the left ear.
Dexter	Tattoo in either or both ears.
Galloway	Must have ear tattoo.
Gelbvieh	Each calf must be permanently identified with ear tattoo, freeze brand, or fire brand, showing breeder's 3-letter herd prefix, a 3 or 4 digit number, and the international year code letter suffix. Example: RFR 3136 E; where RFR stands for Rocky Ford Ranch, the 3136 is the number within the individual herd; and the E the 1973 year code.
Guernsey	The animal must be plainly tattooed in the ear with indelible ink or paste before application for registration is made. Both ears may be used, but it is not required.

(Continued)

TABLE 10-2 (Continued)

Breed	Association Rules Relative to Marking
Hereford	The tattoo may consist of either a series of numbers or a combination of letters and numbers selected by the owner but may not exceed a total of six numbers and letters. No two animals in the same herd and of the same sex can have the same tattoo. Vaccination tattoos are not acceptable as identification for registration. Each animal, for which application for registry is submitted, must be tattooed in one ear. Tattooing in both ears is recommended. The Association recommends (1) starting with number "1" in each ear and proceeding upward in regular order to 999; (2) preferably, limiting numbers to 3 digits; and (3) using the tattoo number as the horn brand number, also.
Holstein-Friesian	Eartag identification may be used but is not considered official for registration by the Association.
Jersey	All calves must be tattooed before leaving individual pens or ties. Both ears may be used and the letters and numbers in the ears must be stated on the application for registration. Tattoos must include at least one letter and one number. The letters "I," "O," "Q," and "V" may be used only if accompanied by an additional letter in the same ear. Tattoos may not exceed a total of seven letters and numbers in each ear. The Association recommends that calves be tattooed in both ears with the same tattoo.
Limousin	All animals must have the first owner's herd prefix in one ear (4 letters). Each individual tattoo number must also have a letter at the end indicating the year of birth: A=1969, B=1970, C=1971, D=1972, E=1973, F=1974, etc.
Milking Shorthorn	Milking Shorthorn cattle cannot be registered unless they have been tattooed in the ear with an individual identification number. An indentification letter or initial may precede the number if desired. The application for registry of a calf must show whether the tattoo number appears in the right or left ear. Duplication of numbers for calves of same sex in the same herd is not permissible. Use of initial letters without a number is not sufficient—a number is required.
Murray Grey	Each animal must be tattooed in the left ear. If member so desires, a corresponding tattoo may also be placed in the right ear.
Polled Hereford	Each animal, for which application for registry is submitted, must be tattooed with a number in one or both ears (the Association recommends tattooing in both ears). Application for registry must (1) give the tattoo number, and (2) indicate in which ear or both ears, as the case may be, that it appears. Each animal of the same sex registered by any one breeder should be tattooed differently.
Red Angus	Calves must be tattooed, not exceeding 4 digits in one ear as follows: Right ear to have owner's assigned (by Sec.-Treas.) letters and last digit of year of animal's

(Continued)

TABLE 10-2 (Continued)

Breed	Association Rules Relative to Marking
	date of birth. Left ear to have herd identification number of owner's own system; but the animal must be definitely identified and without duplication in the herd. Special symbols, diagonals, brands, bars, joined letters, etc. cannot be recorded.
Red Brangus	Permanently identified by fire-brand, including the owner's or breeder's brand and the animal's private herd number.
Red Poll	Breeders and first owners select their own tattoo marking system. However, identifying numbers must be tattooed in both ears, but either the same or different marks may be used in each ear (if different marks are used, they must be so specified on application for registry). Animals of the same sex and near the same age must be tattooed with a different number, but animals of different sex and widely different age may have same number.
Santa Gertrudis	Each animal must be previously numbered by fire-brand or freezebrand so that it can be individually identified. Two or more animals of the same sex may not bear the same number in a given herd for a minimum period of 10 years.
Scotch Highland	Owners must submit herd designation, which must be tattooed in ear or branded on animal; and herd designation must have prior approval in registry office so that no two breeders use the same designation. The Association recommends (1) that the left ear carry the two letters for herd designation, followed by the year (thus, a calf in 1976 on the Double X ranch might be marked XX76), and (2) that individual animal numbers either appear in the right ear or be branded on the animal.
Shorthorn	Each animal, for which application for registry is submitted, must be tattooed with a number in one ear. A letter or initial may or may not precede the number. The application for registry must show whether the tattoo appears in the calf's right or left ear. Duplication of numbers in the same sex and herd is not permissible.
Simmental	Each animal to have a private herd number (brand or tattoo), which shall include the International Year/Letter Designation: A=1969, B=1970, C=1972, D=1973, etc. Duplicate numbers cannot be used in the same herd the same year.
South Devon	Each animal must be tattooed or branded.

CREEP FEEDING

(See Chapter 21, Feeding and Handling Calves, section on "Creep Feeding.")

WEANING

Early spring calves should be weaned in the fall, preferably before the forage becomes dry or just before moving the breeding herd to the winter range or into winter quarters. Calves are usually weaned when they are 6 to 8 months of age.[13] Usually a cow will wean her calf by the time it reaches 10 to 11 months of age, even if the 2 animals are not separated.

(See Chapter 21, Feeding and Handling Calves, section on "Weaning.")

HARDWARE DISEASE PREVENTION

The term "hardware disease" (traumatic gastritis) is used to describe the condition that results from swallowing foreign materials, usually metal (nails, wire, screws, pins, etc.). Cattle are involved more than other classes of animals. In most cases, the metal is found only in the reticulum (second stomach).

Nearly 7,000 cattle are condemned each year by the Federal Meat Inspection Service as unfit for food because of hardware disease. Clinical reports indicate that the problem is increasing due to the use of more chopped feeds and more contamination. Sharp objects will injure the lining of the stomach and cause infection and inflammation, a condition known as traumatic gastritis.

Hardware disease is a problem in cattle because of their eating habits and stomach arrangement. The usual source of metals is the feed. The animals eat rapidly and are not able to sort foreign objects from their feed.

• *Prevention*—Avoid foreign objects getting into the feed through good management. Also, install strong magnets, in keeping with the manufacturer's directions, (1) at the outlets of mechanical silo unloaders, and (2) in feed processing equipment.

• *Symptoms*—The most common symptoms are: loss of appetite and digestive disturbance; slow and stiff movement and arched back; elbows that bow outward; decreased rumen movement and chewing; possible diarrhea; tendency to stand with the front feet elevated so as to lessen the pressure of the viscera on the inflamed area; rise in body temperature; and swellings under the jaw, at the brisket, and at the hock joints. Bulls may be reluctant to mate.

• *Treatment*—Powerful magnets may be permanently placed in the cow's second stomach, for the purpose of holding objects that have not penetrated the stomach wall. However, the only sure cure for traumatic gastritis is veterinary surgery. Surgery will be successful only if performed before the condition has progressed to the point that damage has been done to the heart or other organs.

BEHAVIOR OF CATTLE

A knowledge of cattle behavior ("cow sense") is necessary in order to manage and handle them successfully.

• *Agonism*—Agonistic behavior encompasses a behavior associated with conflict, including escape and submissive behavior. When there is stress on

---

[13]On the average, beef calves are weaned at 7.3 months of age. (Source: *Wash. Agr. Exp. Sta. Bull. 562*)

an animal, its first reaction is to run away until it reaches a point where running won't help; then, and then only, will it turn and fight.

• *Care seeking and care giving*—A newborn calf cannot see too well, but it can smell, touch, and taste. It associates everything that is good and that gives it care with its mother. This is the beginning of herd instinct.

Also, the young calf associates everything good or bad with humans. Unfortunately, this is the period in life during which calves are dehorned, castrated, branded, and vaccinated; hence, it's no wonder some cattle are hard to handle. In order to minimize this problem, calves should be worked as little as possible, with all jobs done at one time.

After weaning, the calf looks for care and shelter from the herd itself. Thus, if you separate an animal from the herd, it is stressed. It will jump fences because of its strong instinct to rejoin the herd.

• *Dominance order*—The outcome of an agonistic interaction (or a fight) determines the dominance-subordination relationship of two animals, usually as long as they remain in the herd. Once the relationship is firmly established, it is recognized by each and further combat is superfluous; the subordinate animal retreats from the dominant at the slighest threat.

• *Leadership*—The leader is an individual which frequently is at the head of a moving column and often seems to initiate a new activity. The leader in a herd is not always the dominant animal, contrary to common opinion.

• *Shelter seeking*—Cattle seek shelter in the center of the herd, because those on the outside may get caught. This is also the normal reaction of wild game herds. Thus, the weaker (more submissive) animals and the ones that fight back are usually on the outside of the herd. So, in separating cattle, you may be putting the weak and the aggressive together.

• *Sight*—A cow does not see what a person sees. Her eyes are on each side of her head, giving her a panoramic view. She doesn't see in color either; she sees in shades of grays and blacks. If a cow sees movement, her instinct is to escape; hence, movements around cattle should be made very quietly and slowly.

• *Smell*—Cattle can smell at a greater distance than people. On a day with a 5 mile wind and a humidity of 75 percent, a cow can smell up to 6 miles away; as wind and humidity increase, she can smell even further.

• *Touch and taste*—A cow touches and tastes her offspring immediately after birth. Touch and taste are also important in the sex act, especially for the bull.

The most natural position for feeding cattle is at ground level. With the head in a downward position, the cow produces more saliva; and saliva aids rumen digestion. Cattle fed from their natural downward position convert their feed better. The lower the head, the better.

BEDDING CATTLE

Bedding or litter is used primarily for the purposes of keeping animals clean and comfortable. But bedding has the following added values from the standpoint of the manure:

1. It soaks up the urine which contains about one-half the total plant food of manure.

2. It makes manure easier to handle.

3. It absorbs plant nutrients, fixing both ammonia and potash in relatively insoluble forms that protects them against losses by leaching. This characteristic of bedding is especially important in peat moss, but of little significance with sawdust and shavings.

Kind and Amount of Bedding

The kind of bedding material selected should be determined primarily by (1) availability and price, (2) absorptive capacity, and (3) plant nutrient content. In addition, a desirable bedding should not be dusty, should not be excessively coarse, and should remain well in place and not be too readily kicked aside. Table 10-3 summarizes the characteristics of some common bedding materials.

TABLE 10-3
WATER ABSORPTION CAPACITY OF BEDDING MATERIALS

Material	Lb of Water Per Lb of Bedding	Material	Lb of Water Per Lb of Bedding
Barley straw	2.10	Sand	.25
Cocoa shells	2.70	Sugar cane bagasse	2.20
Corn stover (shredded)	2.50	Vermiculite[1]	3.50
Corncobs (crushed or ground)	2.10	Wheat straw (long)	2.20
Cottonseed hulls	2.50	(chopped)	2.95
Flax straw	2.60	Wood	
Hay (mature, chopped)	3.00	Dry fine bark	2.50
Leaves (broadleaf)	2.00	Tanning bark	4.00
(pine needles)	1.00	Pine chips	3.00
Oat hulls	2.00	sawdust	2.50
Oat straw (long)	2.80	shavings	2.00
(chopped)	3.75	needles	1.00
Peanut hulls	2.50	Hardwood chips	1.50
Peat moss	10.00	shavings	1.50
Rye straw	2.10	sawdust	1.50

[1]This is a mica-like mineral mined chiefly in South Carolina and Montana.

In addition to the bedding materials listed in Table 10-3, many other products can be and are successfully used for this purpose, including leaves of many kinds, tobacco stalks, buckwheat hulls, processed manure (made by separating solid fibers from the liquid and water-soluble material in animal wastes), and shredded paper.

Naturally the availability and price per ton of various bedding materials vary from area to area, and from year to year. Thus, in the New England states shavings and sawdust are available, whereas other forms of bedding are scarce, and straws are more plentiful in the central and western states.

Table 10-3 shows that bedding materials differ considerably in their relative capacities to absorb liquid. Also, it is noteworthy that chopped straw will absorb more liquid than long straw. But there are disadvantages to chopping; chopped straws do not stay in place, and they may be dusty.

The suspicion that sawdust or shavings will hurt the land is rather widespread, but unfounded. It is true that these products decompose slowly. But this process can be expedited by the addition of nitrogen fertilizers.

The minimum desirable amount of bedding to use is the amount necessary to absorb completely the liquids in manure. Some helpful guides to the end that this may be accomplished follow:

1. With 24-hour stabling, the minimum daily bedding requirements, based on uncut wheat or oats straw, of cattle is as follows: cows, 9 pounds; steers, 7 to 10 pounds. With other bedding materials, these quantities will vary according to their respective absorptive capacities (see Table 10-3). Also, more than these minimum quantities of bedding may be desirable where cleanliness and comfort of the animal are important. Comfortable animals lie down more and utilize a higher proportion of the energy of the feed for productive purposes. (Cattle require 9 percent less energy when lying down than when standing.)

2. Under average conditions, about 500 pounds of bedding are used for each ton of excrement.

3. Where the liquid excrement is collected separately in a cistern or tank—as is the common practice in Denmark, Germany, and France, and on some dairy farms in the United States—less bedding is required than where the liquid and solid excrement are kept together.

Many of today's improved crops are "shorty" varieties, with the result that smaller quantities of stalks and straws are available for bedding. Hand in hand with this transition, confinement animal production has increased. As a result, bedding materials have become scarce and high in price. This situation is prompting interest in slatted (slotted) floors and other means of lessening or alleviating bedding.

CATTLE MANURE[14]

The term "manure" refers to a mixture of animal excrements (consisting of undigested feeds plus certain body wastes) and bedding. Increased confinement production has made for manure disposal problems. This has been particularly acute in most cattle feedlots. As a result, in many areas manure is unwanted. Worse yet, it may be looked upon as a foul-smelling, fly-breeding, dusty, unattractive centerpiece in a feedlot. In due time, however, science and technology will evolve with new methods of using and handling manure; and we shall learn to live with it. From the standpoint of soils and crops, barnyard manure contains the following valuable ingredients:

1. *Organic matter*—It supplies valuable organic matter which cannot be secured in commercial fertilizers. Organic matter—which constitutes three

---

[14]The author gratefully acknowledges the helpful suggestions of Dr. H. M. Reisenauer, Soil Specialist, U. of Cal., who very kindly reviewed this material.

Fig. 10-15. Colorado feedlot showing heavy manure production. (Courtesy, The Great Western Sugar Co.)

to six percent, by weight, of most soils—improves soil tilth, increases water-holding capacity, lessens water and wind erosion, improves aeration, and has a beneficial effect of soil microorganisms. It is the "lifeblood" of the land.

2. *Plant food*—It supplies plant food or fertility—especially nitrogen, phosphorus, and potassium. In addition to these three nutrients, manure contains organic matter, calcium, and trace elements such as boron, manganese, copper, and zinc. A ton of well-preserved manure, including bedding, contains plant food nutrients equal to about 100 pounds of 10-5-10 fertilizer. Thus, when manure is spread at the rate of 8 tons per acre, it is like applying 800 pounds of 10-5-10 commercial fertilizer.

Amount and Composition of Manure

The quantity, composition, and value of manure produced vary according to species, weight, kind and amount of feed, and kind and amount of bedding. The author's computations are on a fresh manure (exclusive of bedding) basis. Table 10-4 presents data by species per 1,000 pounds liveweight, whereas Table 10-5 gives yearly tonnage and value.

The data in Table 10-4 and Fig. 10-16 are based on animals confined to stalls the year around. Actually, the manure recovered and available to spread where desired is considerably less than indicated because (1) animals are kept on pasture and along roads and lanes much of the year, where the manure is dropped, and (2) losses in weight often run as high as 60 percent when manure is exposed to the weather for a considerable time.

As shown in Fig. 10-17, about 75% of the nitrogen, 80% of the phosphorus, and 85% of the potassium contained in animal feeds are returned as

TABLE 10-4

QUANTITY, COMPOSITION, AND VALUE OF FRESH MANURE
(FREE OF BEDDING) EXCRETED BY 1,000 POUNDS LIVEWEIGHT
OF VARIOUS KINDS OF FARM ANIMALS

(1)	(2)	Composition and Value of Manure on a Tonnage Basis[2]						
		(3)	(4)	(5)	(6)	(7)	(8)	(9)
Animal	Tons Excreted/ Year/1,000 lb Liveweight[1]	Excre- ment	Lb/ ton[3]	Water	N	P[4]	K[4]	Value/ Ton[5]
				(%)	(lb)	(lb)	(lb)	($)
Cow	12	Liquid Solid Total	600 1,400 2,000	79	11.2	2.0	10.0	1.79
Steer (finishing cattle)	8.5	Liquid Solid Total	600 1,400 2,000	80	14.0	4.0	9.0	2.27
Sheep	6	Liquid Solid Total	660 1,340 2,000	65	28.0	4.2	20.0	4.11
Swine	16	Liquid Solid Total	800 1,200 2,000	75	10.0	2.8	7.6	1.67
Horse and Foal	8	Liquid Solid Total	400 1,600 2,000	60	13.8	2.0	12.0	2.12
Chicken	4.5	Total	2,000	54	31.2	8.0	7.0	4.27

[1]Manure Is Worth Money—It Deserves Good Care, U. of Illinois Circ. 595, 1953, p. 4.
[2]Columns 5, 6, 7, and 8 from: Farm Manures, U. of Kentucky Circ. 593, 1964, Table 2, p. 5.
[3]From: Reference Material for 1951 Saddle and Sirloin Essay Contest, p. 43, compiled by M. E. Ensminger, data from Fertilizers and Crop Production, by Van Slyke, published by Orange Judd Publishing Co.
[4]Phosphorus (P) can be converted to P_2O_5 by multiplying the figure given above by 2.29, and potassium (K) can be converted to K_2O by multiplying by 1.2.
[5]Calculated on the assumption that nitrogen (N) retails at 9¢, phosphorus (P) at 14¢, and potassium (K) at 5¢ per pound in commercial fertilizers.

TABLE 10-5

TONNAGE AND VALUE OF MANURE (EXCLUSIVE OF BEDDING)
EXCRETED IN 1972 BY U.S. LIVESTOCK[1]

Class of Livestock	No. of Animals on Farms[2]	Average Liveweight	Tons Manure Excreted/Year/ 1,000 Lb Liveweight[3]	Total Manure Production	Total Value of Manure[4]
		(lb)	(tons)	(tons)	($)
Cattle (beef and dairy; including steers)	130,655,000	900	11	1,293,484,500	2,315,337,255
Sheep	17,726,000	100	6	10,635,600	43,712,316
Swine	61,502,000	200	16	196,806,400	328,666,688
Chickens					
Layers	406,528,000	4.5	4.5	8,232,192	35,151,460
Broilers	599,211,600	3.5	4.5	9,437,583	40,298,479
Turkeys	132,302,000	22	4.5	13,097,898	55,928,024
Horses	7,000,000	1,000	8	56,000,000	118,720,000
				1,587,694,173	$2,937,814,222

[1]In these computations, no provision was made for animals that died or were slaughtered during the year. Rather, it was assumed that their places were taken by younger animals, and that the population of each specie was stable throughout the year.
[2]From USDA, Statistical Reporting Service; and assumed as average throughout the year.
[3]Manure Is Worth Money—It Deserves Good Care, U. of Illinois Circ. 595, 1953, p. 4.
[4]Computed on the basis of the value per ton given in the right-hand column of Table 10-4.

Fig. 10-16. Manure is a big "crop"! On the average, each class of stall-confined animals produces per year per 1,000 pounds weight the tonnages shown above. (Drawing by R. F. Johnson)

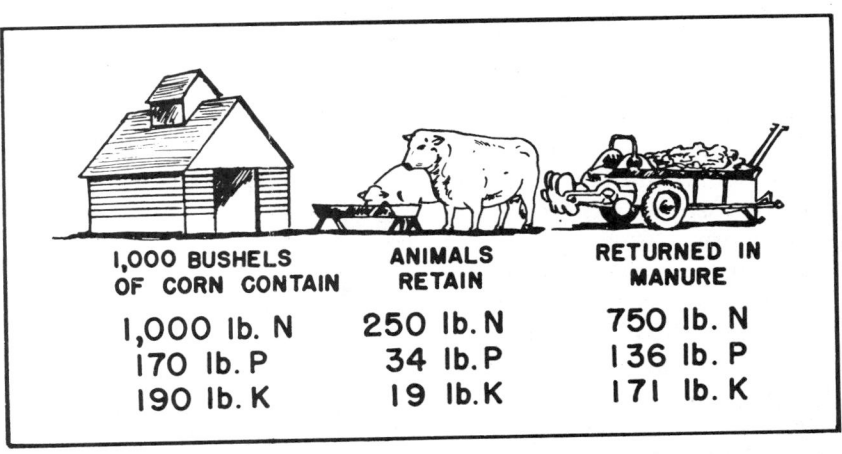

Fig. 10-17. Animals retain about 20 percent of the nutrients in feed; the rest is excreted in manure. (Drawing by Steve Allured)

manure. In addition, about 40% of the organic matter in feeds is excreted as manure. As a rule of thumb, it is commonly estimated that 80% of the total nutrients in feeds are excreted by animals as manure.

Naturally, it follows that the manure from well-fed animals is higher in nutrients and worth more than that from poorly fed ones. For example, the manure produced from steers liberally fed on nutritious concentrates is more valuable than that produced from cattle wintered on hay.

Although varying with (1) the kind of feed, (2) the class of animal, (3) the age, condition, and individuality of the animal, and (4) the kind and amount of bedding used, a ton of *fresh* barnyard manure has approximately the composition shown in Figure 10-18.

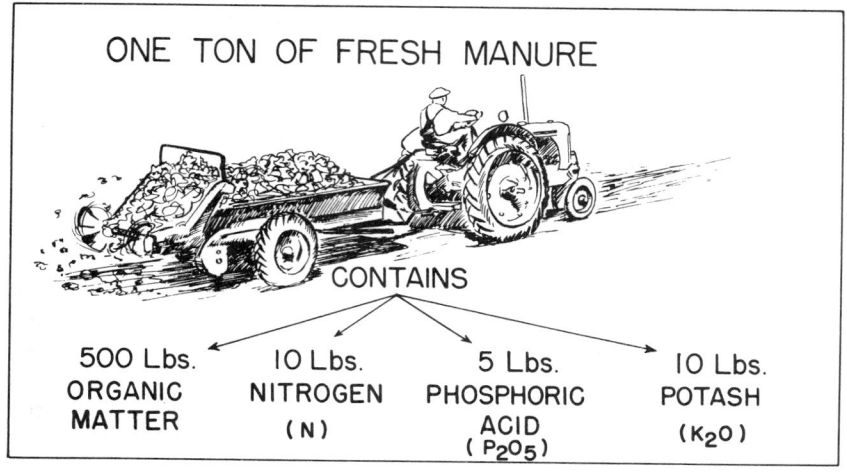

Fig. 10-18. The contents of one ton of average *fresh* manure. Cattle feedlot manure is much more nutritious; on a per ton basis, it contains about 27 pounds of nitrogen, 24 pounds of P_2O_5, and 36 pounds of K_2O.

According to the U.S. Department of Agriculture, a 1,000-lb steer will produce 17 lb of urine and 43 lb of semisolids each day. Of this 60 lb, there are 5 lb of dry organic and inorganic materials. If this figure is multiplied by 365 days, about 1 ton of dry manure is produced annually by each steer. A ton of *dry* beef manure contains about 70 lb of nitrogen, 20 lb of phosphorus, 45 lb of potassium, about 0.8 to 0.9 ton of organic material, and numerous trace elements.

Table 10-4 gives the nutrients, by classes of animals, in one ton of fresh manure. As shown, in terms of nutrients contained, chicken manure is the most valuable. Sheep manure ranks second, and steer manure third. Cow manure ranks lower than steer manure, because cows are fed largely on roughages, rather than concentrates.

The urine makes up 20% of the total weight of the excrement of horses, and 40% of that of hogs; these figures represent the two extremes in farm animals. Yet the urine, or liquid manure, contains nearly 50% of the nitrogen, 6% of the phosphorus, and 60% of the potassium of average manure; roughly one-half of the total plant food of manure (see Figure 10-19). Also, it is noteworthy that the nutrients in liquid manure are more readily available to plants than the nutrients in the solid excrement. These are the reasons why it is important to conserve the urine.

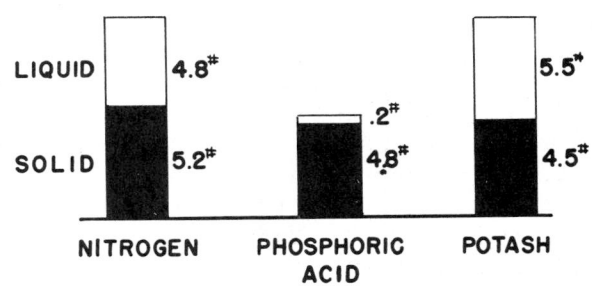

Fig. 10-19. Distribution of plant nutrients between liquid and solid portions of a ton of average farm manure. As noted, the urine contains about half the fertility value of manure. (Drawing by Steve Allured)

Manure Gases

When stored inside a building, gases from liquid wastes create a hazard and undesirable odors. Most (95% or more) of the gas produced by manure decomposition is methane, ammonia, hydrogen sulfide, and carbon dioxide. Several have undesirable odors or possible animal toxicity, and some promote corrosion of equipment. Table 10-6 gives some properties of the more abundant gases.

TABLE 10-6

PROPERTIES OF THE MORE ABUNDANT MANURE GASES[1]

Gas		Weight Air=1	Physiologic Affect	Other Properties
CH_4	Methane	½	Anesthetic	Odorless, explosive
NH_3	Ammonia	⅔	Irritant	Strong odor, corrosive
H_2S	Hydrogen Sulfide	1+	Poison	Rotten-egg odor, corrosive
CO_2	Carbon Dioxide	1⅓	Asphyxiant	Odorless, mildly corrosive

[1]*Beef Housing and Equipment Handbook*, MWPS-6, Midwest Plan Service, Ames, Iowa, 1968, p. 10.

The primary hazard to animal health occurs only with inadequate ventilation. Animals asphyxiate when methane and carbon dioxide displace oxygen. Ammonia can irritate the eyes and respiratory tracts and make them more susceptible to disease.

Value of Manure as a Fertilizer

The actual monetary value of manure can and should be based on (1) equivalent cost of a like amount of commercial fertilizer, and (2) increased crop yields. Tables 10-4 and 10-5 give the equivalent cost of a like amount of commercial fertilizer. Numerous experiments and practical observations have shown the measurable monetary value of manure in increased crop yields.

When the author was a boy on a Missouri farm, we fed livestock to produce manure, to grow more crops to feed more livestock, to produce more manure. But things changed! The use of chemical fertilizer expanded many fold; labor costs rose to the point where it was costly to conserve and spread manure on the land; more animals were raised in confinement; and a predominantly urban population didn't appreciate what they referred to as "foul-smelling, fly-breeding stuff." As a result, what to do with manure became a major problem on many livestock establishments.

We are producing manure (exclusive of bedding) at the rate of 1.6 billion tons annually (see Table 10-5). That's sufficient manure to add nearly one ton each year to every acre of the total land area (1.9 billion acres) of the 48 contiguous states of the United States.

Based on equivalent fertilizer prices (see Table 10-4, right-hand column), the yearly manure crop is worth $3 billion (Table 10-5). That's a potential income of more than $1,000 for each of the nation's 2.8 million farms.

The value of manure varies according to (1) the class of animals, (2) the kind of feed consumed and the kind of bedding used, (3) the method of handling, (4) the rate and method of application, and (5) the kind of soil and crops on which it is used.

Numerous experiments and experiences have shown the measurable monetary value of manure in increased crop yields.

At the Iowa Experiment Station, beginning in 1915, 8 tons of manure per acre were applied to the sod and plowed under for corn in a 4-year rotation of corn-corn-oats-meadow. For the 10-year period, 1942 to 1951, the results shown in Table 10-7 were obtained.

Since 8 tons of manure were used, this means an average of $7.84 worth of increased crops from each ton of manure. But this was based on 1942-1951

TABLE 10-7

INCREASED YIELD AND VALUE FROM MANURE

Crop	Increased Yield from Manure	Prevailing Prices	Increased Value from Manure
Corn	15 bu	$ 1.50/bu	$22.50
Corn	15 bu	1.50/bu	22.50
Oats	9 bu	0.80/bu	7.20
Clover	0.7 ton	15.00/ton	10.50
Total increased value from manure			$62.70

crop prices. At today's feed prices, the value of manure in increased crop yields would be more than double the Iowa figure of $7.84 per ton.

At the Ohio Agricultural Experiment Station, in 3 experiments in which barnyard manure was the only source of applied nutrients for a period of 50 years, manure showed an average value of $9.50 per ton, based on increased crop yields at 1952 prices.

At the Illinois Experiment Station, extensive experiments with manure were conducted at 20 different locations over the state during the 4-year period 1948 to 1951. The crop rotations varied somewhat, but corn and hay were common to all fields. The manured fields received an application of one ton of manure for each ton of crops removed from the land, or an average of 2.36 tons per acre annually. In comparison with fields which were not manured, the increased crop yields obtained on the manured plots were sufficient to give manure a value of $10.11 per ton, when the value of the increased yields was computed on the basis of the average prices received by farmers during the same 4-year period (1948 to 1951).

Of course, the value of manure cannot be measured alone in terms of increased crop yields and equivalent cost of a like amount of commercial fertilizer. It has additional value for the organic matter which it contains, which almost all soils need, and which farmers and ranchers cannot buy in a sack.

Also, it is noteworthy that, due to the slower availability of its nitrogen and to its contribution to the soil humus, manure produces rather lasting benefits, which may continue for many years. Approximately one-half of the plant nutrients in manure are available to and effective upon the crops in the immediate cycle of the rotation to which the application is made. Of the unused remainder, about one-half, in turn, is taken up by the crops in the second cycle of the rotation; one-half of the remainder in the third cycle, etc. Likewise, the continuous use of manure through several rounds of a rotation builds up a backlog which brings additional benefits, and a measurable climb in yield levels.

Stockmen sometimes fail to recognize the value of this barnyard crop because (1) it is produced whether or not it is wanted, and (2) it is available without cost. Most of all, no one is selling it. Who ever heard of a traveling manure salesman?

Modern Ways of Handling Manure

Modern handling of manure involves maximum automation and a minimum loss of nutrients. Among the methods being used, with varying degrees of success, are: slotted floors emptying or pumping into irrigation systems; storage vats; spreaders (including those designed to handle liquids alone or liquids and solids together); dehydration; power loaders; conveyors; industrial-type vacuums; lagoons; and oxidation ditches. Actually, there is no one best manure management system for all situations; rather, it is a matter of designing and using that system which will be most practical for a particular set of conditions.

PRECAUTIONS WHEN USING MANURE AS FERTILIZER

The following precautions should be observed when returning manure to the land:

1. Avoid applying waste closer than 100 feet to waterways, streams, lakes, wells, springs, or ponds.

2. Do not apply where downward movement of water is not good, or where irrigation water is very salty or inadequate to move salts down.

3. Incorporate (preferably by plowing or disking) manure into the soil as quickly as possible after application. This will maximize nutrient conservation, reduce odors, and minimize runoff pollution.

4. Distribute the waste as uniformly as possible on the area to be covered.

5. Irrigate thoroughly to leach excess salts below the root zone.

6. Allow about a month after irrigation before planting, to enable soil microorganisms to begin decomposition of manure.

7. Minimize odor problems by—

a. Spreading raw manure frequently, especially during the summer.

b. Spread early in the day as the air is warming up, rather than late in the day when the air is cooling.

c. Do not spread on days when the wind is blowing toward populated areas.

How Much Manure Can Be Applied to the Land?

With today's heavy animal concentration in one location, the question is being asked: how high rates of manure can be applied to the land without depressing crop yields, making for salt problems in the soil, making for nitrate problems in the feed, or contributing excess nitrate to groundwater or surface streams.

Based on earlier studies in the Midwest, before the rise of commercial fertilizers, it would appear that one can apply from 5 to 20 tons of manure per acre, year after year, with benefit.

Heavier applications can be made, but probably should not be repeated every year. With higher rates than 20 tons per year annually, there may be excess salt and nitrate buildup. Excess nitrate from manure can pollute streams and groundwater and result in toxic levels of nitrate in crops. Without doubt the maximum rate at which manure can be applied to the land will vary widely according to soil type, rainfall, and temperature.

State regulations differ in limiting the rate of manure application. Missouri draws the line at 30 tons per acre on pasture, and 40 tons per acre on cropland. Indiana limits manure application according to the amount of nitrogen applied, with the maximum limit set at 225 pounds per acre per year. Nebraska requires only one-half acre of land for liquid manure disposal per acre of feedlot, which appears to be the least acreage for manure disposal required by any state.

A total of 26.8 million tons of cattle feedlot manure were produced in 1972, or one ton per animal. Assuming 2¼ turnover of cattle per feedlot per year, and the production of one ton of manure per animal, it would take 6 animals a whole year to produce enough manure to apply on one acre at the rate of 13.5 tons. But it takes the crop production of 18 acres of land to feed 6 feedlot steers for a year; hence, 17 acres would not receive any manure.

One of the big problems in applying animal waste to the land as a fertilizer is knowing the plant nutrient content of the material. If this is known, the amount of manure necessary to supply the needed nutrients can be added. So, representative samples of the manure should be analyzed for nitrogen, phosphorus, potash, and moisture content. Then the application rates should be based on soil tests, crop requirements, and composition of the manure sample. This relatively inexpensive procedure will avoid errors in application rate.

The amount of manure to be applied can usually be geared to the amount of nitrogen that the crop needs. Thus, if 150 pounds of available nitrogen would be adequate for maximum crop production, the manure containing 300 pounds of total nitrogen should be applied for the first year of use (twice the amount of nitrogen needed the first year is applied, because only half the nitrogen is available the first year).

When a farmer has sufficient land, he should use rates of manure which supply only the nutrients needed by the crop rather than the maximum possible amounts suggested for pollution control.

Objectionable Features of Manure as a Fertilizer

Despite the recognized virtues of manure as a fertilizer, it does possess the following objectionable features:

1. *It is costly to haul and apply*—In the past, a main deterrent to the use of manure as a fertilizer has been the high cost of hauling and applying. But high priced chemical fertilizers have made manure more attractive.

2. *It can create salt (sodium chloride) problems*—Because manure contains appreciable quantities of salt, excessive applications of animal waste can result in salt accumulations in the soil, which will harm crops.

3. *It can create nitrate problems*—Excessive nitrogen in manure can result in excessive nitrates in the groundwater, which can be hazardous for animal or human consumption. Normally, the nitrate content of manure is low, but microorganisms rapidly convert ammonium to nitrate when conditions are favorable for crop growth. Nitrate is the mobile form of nitrogen—the form usually associated with pollution.

4. *It can create phosphorus problems*—Excessive phosphorus can interfere with zinc nutrition of crops and pollute streams.

5. *It can create potassium problems*—The high potassium content of manure tends to reduce the absorption of magnesium and calcium from the soil by plants and results in a greater possibility of grass tetany because of the lowered magnesium content of the forage.

6. *It may propagate insects*—Unless precautions are taken, often manure is the preferred breeding place for flies and other insects. It is noteworthy, however, that comparatively few houseflies are reared in cow manure.

7. *It may spread diseases and parasites*—Where animals are allowed to come in contact with their own excrement there is always danger of infections from diseases and parasites.

8. *It may produce undesirable odors and dust*—Where manure is stored improperly, there may be a nuisance from odors and dust.

9. *It may scatter weed seeds*—Even when fermented, manure usually contains a certain quantity of viable weed seeds which may be scattered over the land.

Manure Uses Other Than Fertilizer

Recycling manure as a livestock feed is the most promising of the non-fertilizer uses. Various processing methods are being employed; some are even feeding manure without processing. Further experimentation and Food and Drug Administration approval will be required before the use of manure-feeds becomes widespread; but some researchers predict that eventually wastes may supply up to 20 percent of the nation's livestock feed, thereby freeing an equivalent amount of grain for human consumption.

Manure may also serve as a source of energy, which, of course, is not new. The pioneers burned dried bison dung, which they dubbed "buffalo chips," to heat their sod shanties. In this century, methane from manure has been used for power in European farm hamlets when natural gas was hard to get. While the costs of constructing plants to produce energy from manure on a large scale basis may be high, some energy specialists feel that a prolonged fuel shortage will make such plants economical. India now has about 10,000 anaerobic digestion plants in operation. In 1974, Monfort of Colorado, operator of 2 of the world's largest cattle feedlots, announced that it had granted an option for the construction of a facility to produce 4,000,000 cubic feet of methane gas per day from the 225,000 tons of dry weight manure produced yearly in one of its feedlots. The announcement further stated that the process (anaerobic digestion) reduces the odor associated with manure handling and improves the residue as a fertilizer. The methane, of course, will be usable like natural gas. There is nothing new or mysterious about this process. Sanitary engineers have long known that a family of bacteria produces methane when they ferment organic matter under strictly anaerobic conditions. (Grandad called it swamp gas; his city cousin called it sewer gas.) However, it should be added that due to the capital and technical resources needed, for sometime to come, the production of methane gas by anaerobic digestion will likely be limited to municipal or corporate industries. If all animal manure were converted to energy, it has been estimated that it would produce energy equal to 10 percent of the petroleum requirements or 12½ percent of our natural gas requirements.

One researcher has come up with a way in which to combine manure with broken glass to produce bricks, decorative and roofing tiles, wall core material, and garden stones.

QUESTIONS FOR STUDY AND DISCUSSION

1. Under ordinary conditions, herdsmen fail to detect 25 to 50 percent of the heat periods. What are the consequences of this failure, and how would you recommend that it be rectified?

2. What are the advantages and disadvantages of each (a) spring calving, and (b) fall calving?

3. By what method and at what age and season would you dehorn cattle in your area? Give reasons for your answers.

4. By what method and at what age and season would you castrate cattle in your area? Give reasons for your answers.

5. What did the California Station find relative to the effects of each (a) hot-iron branding, and (b) castrating healthy calves, on the basis of a 28-day study?

6. Do you feel that the short scrotum bull technique will be widely used in the future? Justify your answer.

7. Are there any circumstances under which you would recommend that heifers be spayed? Justify your answer.

8. Under your conditions, what method of marking or identifying cattle would you select? Justify your selection and tell how you would apply this method.

9. Why is there more and more hardware disease? What can be done to lessen it?

10. How will a knowledge of cattle behavior be of value in managing and handling cattle?

11. What type of cattle bedding is commonly used on your farm or ranch (or on a farm or ranch with which you are familiar)? Would some other type of bedding be more practical? If so, why?

12. For your farm or ranch (or one with which you are familiar) is it preferable and practical to apply manure to the land, or should commercial fertilizers be used instead?

13. How would you recommend that manure be handled?

14. What precautions should be taken when using manure as a fertilizer?

15. What tonnage per acre of manure can be applied to the land?

16. What nonfertilizer use for manure offers the most profit potential?

17. India has 10,000 anaerobic digestion plants in operation to produce methane gas, whereas there are few such facilities in the United States. Why the difference?

SELECTED REFERENCES

Title of Publication	Author(s)	Publisher
Beef Cattle, Sixth Edition	A. L. Neumann R. R. Snapp	John Wiley & Sons, Inc., New York, N.Y., 1969
Beef Cattle Production	K. A. Wagnon R. Albaugh G. H. Hart	The MacMillan Company, New York, N.Y., 1960
Beef Production in the South	S. H. Fowler	The Interstate Printers & Publishers, Inc., Danville, Ill., 1969
Beef Cattle Science Handbook	Ed. by M. E. Ensminger	Agriservices Foundation, Clovis, Calif., pub. annually since 1964
Commercial Beef Cattle Production	Ed. by C. C. O'Mary, I. A. Dyer	Lea & Febiger, Philadelphia, Penn., 1972
Stockman's Handbook, The, Fourth Edition	M. E. Ensminger	The Interstate Printers & Publishers, Inc., Danville, Ill., 1970

(Continued)

Title of Publication	Author(s)	Publisher
Problems and Practices of American Cattlemen, Wash. Agr. Exp. Sta. Bull. 562	M. E. Ensminger M. W. Galgan W. L. Slocum	Washington State University, Pullman, Wash., 1955

CHAPTER 11

BUILDINGS AND EQUIPMENT FOR BEEF CATTLE

The economical production of beef cattle in most sections of the United States depends largely upon the investment in practical, durable, and convenient buildings and equipment, as well as upon the care, feeding, and management of the herd. As would be expected in a country so large and diverse as the United States, there are wide differences in the system of beef production. In a broad general way, a major difference in management exists between the farm herd method and the range cattle method. In addition, further management differences exist within each area according to whether the enterprise is commercial or purebred, whether it is a cow-calf proposition or devoted to one of the many methods of growing stockers and feeders or cattle for finishing, or whether it is a combination of two or more of these systems of beef production. Climatic differences also vary, all the way from nearly year-round grazing in the deep South to a long winter-feeding period in the northern part of the United States. Then, too, the size of the herd may vary all the way from a few animals up to an operation involving many thousands of head. Finally, there is the matter of availability of materials and labor and individual preferences.

Fig. 11-1. Attractive farmstead on an Iowa beef cattle establishment. (Courtesy, *The Corn Belt Farm Dailies*, Chicago, Ill.)

Except for the classic experiment conducted by the University of Missouri, little experimental work has been done on the basic building requirements of beef cattle.

Brody, of the Missouri Station[1], placed cattle in "climatic chambers," in which the temperature, humidity, and air movements were regulated as desired. The ability of the animals to withstand different temperatures was then determined by studying the respiration rate and body temperature, the feed consumption, and the productivity in growth, milk, beef, etc. Out of these experiments came the following pertinent findings: The most comfortable temperature for Shorthorns was within the range of 30° to 60° F; for the Brahmans it was 50° to 80° F (20° higher); and for the Santa Gertrudis (⅝ Shorthorn X ⅜ Brahman) it was intermediate between the ideal temperatures given above. Translated into practicality, the Missouri experiment proved what is generally suspected; namely:

1. That there are breed differences; that Brahman and Santa Gertrudis

---

[1]Brody, Samuel, Mo. Agr. Exp. Sta. Journal Series No. 1607.

cattle can tolerate more heat than the European breeds—thus, they are better equipped to withstand tropical and subtropical temperatures.

2. That acclimated European cattle do not need expensive, warm barns—they merely need protection from wind, snow, and rain.

3. That more attention needs to be given to providing summer shades and other devices to assure warm weather comfort for cattle.

The basic building and equipment requirements for beef cattle include: (1) some type of protection from weather; (2) adequate water supply and watering equipment; (3) feed storage, and possibly some equipment for feed processing and conveying; (4) feeding equipment; and (5) waste handling.

No standard set of buildings and equipment can be expected to be adapted to all the diverse conditions and systems of beef production. In presenting the discussion and illustrations that follow, it is intended, therefore, that they be considered as guides only. Detailed plans and specifications for buildings and equipment can usually be obtained through the local county agricultural agent, FFA instructor, or lumber dealer, or through writing the college of agriculture in the state.

The right kind of beef cattle buildings and equipment can materially lower the work required to do the job. Although beef cattle normally require a relatively small amount of labor, a great deal can be done to lower costs and shorten the hours of work.

WEATHER INFLUENCE ON BEEF CATTLE PERFORMANCE

Extreme weather can cause wide fluctuations in cattle performance. The research data clearly show that winter shelters and summer shades almost always improve gains and feed efficiency. However, the additional costs incurred by shelters have frequently exceeded the benefits gained by improved performance. For this reason, except for extreme weather stress of long duration, it becomes a question of whether to shelter cattle at all.

The effects of winter shelter have been widely tested in the Midwest. Table 11-1 shows (1) the reductions in performance, and (2) the feed cost increases resulting from lack of shelter in 13 separate tests at various research stations.

For the 13 trials summarized in Table 11-1, no shelter decreased the average daily gain by 12% and increased feed cost by 14%. The spreads of 4% to 22% decrease in daily gain and 4% to 28% increase in feed cost are typical of fluctuations that occur from one year to the next. Of course, the dollars and cents benefits to accrue from cattle shelters should be based on the added value of the cattle gains and the decreased feed cost.

Summer shade has resulted in less improvement in performance of cattle than winter shelter. Table 11-2 shows the results of eight trials.

For the Plains and the Midwest, it is doubtful that summer shade will pay.

Studies in the desert Southwest demonstrate a greater advantage to shade than the results presented in Table 11-2. In that section of the country, it would appear that shade is not only a paying proposition but a necessity.

TABLE 11-1

EFFECT OF NO SHELTER ON WINTER CATTLE PERFORMANCE[1]

Station	Date	Decrease in Average Daily Gain	Increase in Feed Cost
		(%)	(%)
Iowa	1962	18	21
Iowa	1963	15	17
Iowa	1965	15	15
Iowa	1966	11	8
Iowa	1967	13	14
Iowa	1968	9	7
Ohio	1959	2	8
Michigan	1968	14	23
Michigan	1968	10	14
Kansas	1961	22	28
South Dakota	1968	4	4
Connecticut	1963	7	12
Saskatchewan	1958	15	14
Average of 13 Trials		12	14

[1]*Doane's Agricultural Report*, Nov. 20, 1970, Vol. 33, No. 45-5.

TABLE 11-2

EFFECT OF NO SUMMER SHADE ON CATTLE PERFORMANCE[1]

Station	Date	Decrease in Average Daily Gain	Increase in Feed Cost
		(%)	(%)
Iowa	1962	6	4
Iowa	1964	7	7
Iowa	1965	7	5
Iowa	1966	4	1
Iowa	1967	3	3
Kansas	1961	7	8
South Dakota	1967	2	5
Ohio	1964	7	2
Average of 8 Trials		5	4

[1]*Doane's Agricultural Report*, Nov. 20, 1970, Vol. 33, No. 45-5.

ENVIRONMENTAL CONTROL OF BEEF CATTLE[2]

Environment may be defined as all the conditions, circumstances, and influences surrounding and affecting the growth, development, and production of a living thing. In beef cattle, this includes the air temperature, relative humidity, air velocity, wet bedding, dust, light, ammonia buildup, odors, and space requirements. Control or modification of these factors offer possibilities for improving cattle performance. There is still much to be learned about environmental control, but the gap between awareness and application is becoming smaller.

Cattlemen were little concerned with the effect of environment on beef cattle so long as they grazed on pastures or ranges. Space requirements, wet bedding, ammonia buildup, odors, and manure disposal were no problem. But the concentration of cattle into smaller spaces changed all this. With the shift to confinement structures and high density production operations, building design became more critical.

In addition to improved performance, the primary reasons cited for increased confinement housing of cattle are: (1) saving in land cost, and (2) saving in labor.

Before an environmental system can be designed for beef cattle, it is important to know (1) their heat production, (2) vapor production, and (3) space requirements. This information is as pertinent to designing beef cattle buildings as nutrient requirements are to balancing rations.

The tables accompanying the three sections that follow cover all classes of livestock (1) for comparative purposes, and (2) because most cattlemen have more than one species.

Heat Production of Animals

The heat production of cattle and calves is given in Table 11-3, along with the heat production of other classes of animals for comparative purposes. Table 11-3 may be used as a guide, but in doing so, consideration should be given to the fact that heat production varies with age, body weight, ration, breed, activity, house temperature, and humidity at high temperatures. As noted, Table 11-3 gives both total heat production and sensible heat production. Total heat production includes both sensible heat and latent heat combined. Latent heat refers to the energy involved in a change

---

[2]The four tables in this section—Tables 11-3, 11-4, 11-5, and 11-6—were authoritatively reviewed by the following agricultural engineers: Prof. H. D. Bartlett, The Pennsylvania State University; Dr. William H. Brown, Mississippi State University; Mr. W. H. Collins, Virginia Polytechnic Institute and State University; Mr. Spencer H. Daines, Utah State University; Mr. L. Bynum Driggers, North Carolina State University; Dr. Robert E. Graves, University of Wisconsin; Dr. LeRoy Hahn, Dr. Neil Meador, and Dr. R. E. Phillips, University of Missouri; Mr. Ralph W. Hansen, Colorado State University; Mr. Charles M. Hohn, New Mexico State University; Prof. D. G. Jedele, University of Illinois; Dr. Kenneth A. Jordan, Prof. Donald W. Bates, and Mr. Martin Hellickson, University of Minnesota; Dr. James A. Lindley, The University of Connecticut; Mr. Robert L. Maddex, Michigan State University; Mr. Michael A. McNamee, University of Wyoming; Prof. June Roberts, Washington State University; Dr. Jorge H. Rodriguez-Arias, University of Puerto Rico; Dr. Charles K. Spillman, Kansas State University; Dr. Bill R. Stewart, Texas A & M University; Mr. Roy E. Taylor, University of Idaho; Mr. Norman C. Teter, University of Nebraska; and Mr. Ernest W. Walpole, University of Delaware.

TABLE 11-3

HEAT PRODUCTION OF ANIMALS[1]

Heat Source	Unit (lb)	Unit (kg)	Heat Production, Btu/hr Temperature (°F)	Total	Sensible	Heat Production, Kcal/hr Temperature (°C)	Total	Sensible
Cow	1,000	453.6	40	3,600	2,640	4	907.2	665.3
			70	3,000	1,550	21	756.0	390.6
Calves (6-10 months)	—	—	60	780	660	16	196.6	166.3
			80	720	420	27	181.4	105.8
Hog: Sow & litter (3 weeks after farrowing)	400	181.4	—	2,000	1,000	—	504.0	252.0
Finishing	200	90.7	35	860	740	2	216.7	186.5
			70	610	435	21	153.7	109.6
Layer Hen	4.5	2.04	50	40	28	10	10.1	7.1
Sheep	100	45.4	0.039 in. fleece length 45	560	500	0.099 cm fleece length 7	141.1	126.0
			70	320	245	21	80.6	61.7
			3.937 in. fleece length 45	245	185	10.000 cm fleece length 7	61.7	46.6
			70	260	125	21	65.5	31.5
Horse	1,000	454	70	1,800-2,500[2]	—	21	453.6-630	—

[1]Adapted by the author from 1974 Agricultural Engineers Yearbook, St. Joseph, Michigan, ASAE Data Sheet D-249.2, page 424, except for horse. Data for horse from Farm Buildings, by John C. Wooley, McGraw-Hill Book Company, Inc., 1946, Table 24, p. 140.

[2]Armsby and Kriss, in a paper entitled, "Some Fundamentals of Stable Ventilation," published in the Journal of Agricultural Research, Vol. 21, June 1921, p. 343, list the total heat output as follows: a 1,000 lb horse, 1,500 Btu per hour; a 1,500 lb horse, 2,450 Btu per hour.

of state and cannot be measured with a thermometer; evaporation of water or respired moisture from the lungs are examples. Sensible heat is that portion of the total heat, measurable with a thermometer, that can be used for warming air, compensating for building losses, etc. Heat is measured in British thermal units (Btu). One Btu is the amount of heat required to raise the temperature of one pound of water one degree Fahrenheit.

Vapor Production of Animals

Cattle give off moisture during normal respiration; and the higher the temperature the greater the moisture. This moisture should be removed from buildings through the ventilation system. Most building designers govern the amount of winter ventilation by the need for moisture removal. Also, cognizance is taken of the fact that moisture removal in the winter is lower than in the summer; hence, less air is needed. However, lack of heat makes moisture removal more difficult in the winter time. Table 11-4 gives the information necessary for determining the approximate amount of moisture to be removed.

Since ventilation also involves a transfer of heat, it is important to conserve heat in the building to maintain desired temperatures and reduce the need for supplemental heat. In a well-insulated building, mature animals may produce sufficient heat to provide a desirable balance between heat and moisture; but young animals will usually require supplemental heat. The major requirement of summer ventilation is temperature control, which requires moving more air than in the winter.

Recommended Environmental Conditions for Animals

The comfort of animals (or man) is a function of temperature, humidity, and air movement. Likewise, the heat loss from animals is a function of these three items.

The prime function of the winter ventilation system is to control moisture, whereas the summer ventilation system is primarily for temperature control. If air in beef cattle barns is supplied at a rate sufficient to control moisture—that is, to keep the inside relative humidity in winter below 75 percent—then this will usually provide the needed fresh air, help suppress odors, and prevent an ammonia buildup.

Some typical temperature, humidity, and ventilation recommendations for different classes of livestock are given in Table 11-5. This table will be helpful in obtaining a satisfactory environment in confinement livestock buildings, which require careful planning and design.

Space Requirements of Buildings and Equipment for Beef Cattle

One of the first and frequently one of the most difficult problems confronting the farmer or rancher who wishes to construct a building or item of equipment for beef cattle is that of arriving at the proper size or dimensions. Table 11-6 contains some conservative average figures which, it is hoped,

TABLE 11-4

VAPOR PRODUCTION OF ANIMALS[1]

Vapor Source	Unit		Temperature		Vapor Production		Vapor Production	
	(lb)	(kg)	(°F)	(°C)	(lb/hr)	(Btu/hr)	(kg/hr)	(kcal/hr)
Cow	1,000	453.6	40	4	0.92	960	0.42	241.9
			70	21	1.38	1,450	0.63	365.4
Calves (6-10 months)	—	—	60	16	0.11	120	0.05	30.2
			80	27	0.29	300	0.13	75.6
Hog:								
Sow & litter (3 weeks after farrowing)	400	181.4	—	—	0.97	1,020	0.44	257.0
			35	2	0.11	120	0.05	30.2
Finishing	200	90.7	70	21	0.16	175	0.07	44.1
Layer hen	4.5	2.04	50	10	0.012	12	0.005	3.0
Sheep	100	45.4	{ 45 70	7 21	0.039 in. fleece length 0.06 0.07	60 75	0.099 cm fleece length 0.03 0.03	15.1 18.9
			{ 45 70	7 21	3.937 in. fleece length 0.06 0.13	60 135	10.000 cm fleece length 0.03 0.06	15.1 34.0
Horse	1,000	454	70	21	0.729	—	0.33	—

[1]Adapted by the author from 1974 Agricultural Engineers Yearbook, St. Joseph, Michigan, ASAE Data Sheet D-249.2, p. 424, except for horse. Data for horse from Farm Buildings, by John C. Wooley, McGraw-Hill Book Company, Inc., 1946, Table 25, p. 141.

TABLE 11-5

RECOMMENDED ENVIRONMENTAL CONDITIONS FOR ANIMALS

Class of Animal	Temperature Comfort Zone (°F)	Comfort Zone (°C)	Optimum (°F)	Optimum (°C)	Acceptable Humidity (%)	Commonly Used Ventilation Rates[1] Basis	Winter[2] (cfm)	Winter (m³/min.)	Summer (cfm)	Summer (m³/min.)	Drinking Water Winter (°F)	Winter (°C)	Summer (°F)	Summer (°C)
Beef cow	40-70	5-21	50-60	10-15	50-75	1,000 lb (or 454 kg)	100	2.8	200	5.7	50	10	60-75	15-24
Steer, enclosed bldg. on slotted floor	40-70	5-21	50-60	10-15	50-75	1,000 lb (or 454 kg)	100	2.1-2.3	200	14.2	50	10	60-75	15-24
Dairy cow	40-70	5-21	50-60	10-15	50-75	1,000 lb (or 454 kg)	100	2.8	200	5.7	50	10	60-75	15-24
Dairy calves	50-75	10-24	65	17		per 100 lb (45 kg)	10		25					
Sheep:														
Ewe	45-75	7-24	55	13	50-75		20-25	.6-.7	40-50	1.1-1.4	40-45	5-8	60-75	15-24
Feeder lamb	40-70	5-21	50-60	10-15	50-75		15	.3	30	.65	40-45	5-8	60-75	15-24
Newborn lamb	75-80	24-27												
Swine:														
Sow, farrowing house	60-70	15-20	65	17	60-85	Sow and litter	80	1.4	210	2.8	50	10	60-75	15-24
Newborn pigs (brooder area)	80-90	27-32	85	29	60-85									
Growing-finishing hogs	60-65	15-17	60	15	60-85	125 lb (or 57 kg)	15	.7	75	2.1	50	10	60-75	15-24
Horse	45-75	7-24	55	13	50-75	1,000 lb (or 454 kg)	60	1.7	160	4.5	40-45	5-8	60-75	15-24
Newborn foal	75-80	24-27												
Poultry:														
Layers	50-75	10-24	55-70	13-20	50-75	per bird	2		5		50	10	60-75	15-24
Broilers	85-95	21-27	70	24	50-75	per lb body weight	½		1		50	10	60-75	15-24
Turkeys	95-100 (beginning poults)	35-38				per lb body weight	½		1		50	10	60-75	15-24

[1]Generally two different ventilating systems are provided; one for winter, and an additional one for summer. Hence, as shown in Table 11-5, the winter ventilating system in a beef cow barn should be designed to provide 100 cfm (cubic feet/minute) for each 1,000-pound cow. Then, the summer system should be designed to provide an added 100 cfm, thereby providing a total of 200 cfm for summer ventilation.

In practice, in many buildings, added summer ventilation is provided by opening (1) barn doors, and (2) high-up hinged walls.

[2]Provide approximately ¼ the winter rate continuously for moisture removal.

TABLE 11-6

SPACE REQUIREMENTS OF BUILDINGS AND EQUIPMENT FOR BEEF CATTLE

Class, Age and Size of Animal	Barn or Shed		Shades		Feedlots[1]		Hay or Silage Manger, or Rack					Feed Bunk or Trough for Hand-Feeding Grain[2]				Self-Feeder	Water	
	Floor Area Per Animal	Height of Ceiling[3]	Shade Per Animal	Shade Height	Area If Ordinary Dirt Lot	Area If Paved Lot	Length Per Animal[4]	Width If Feeds from 1 Side	Width If Feeds from 2 Sides	Width If Attached Side of Barn	Height at Throat	Length Per Animal	Width If Feeds from 1 Side	Width If Feeds from 2 Sides	Height at Throat	Trough Length If Feeder Is Kept Filled	Water Per Animal Per Day	Water Trough
	(sq ft)	(ft)	(sq ft)	(ft)	(sq ft)	(sq ft)	(in.)	(in.)	(in.)	(in.)	(in.)	(in.)	(in.)	(in.)	(in.)	(in.)	(gal)	
Cows, 2 years or over	40-50	8½-10	30-40	10-12	300[5]	50-100	24-30	30	48-60	30	24	24-30	18-30	48	24	6-12 per animal	12	Allow one linear ft of open water tank space for each 10 cattle; or one automatic watering bowl for each 25 cattle. A satisfactory water temperature range in winter is 40-45° F; in summer, 60-80° F.
Yearling finishing cattle	Solid floor: 30-40 Slotted floor: 20-25	''	25-35	''	125-200[5]	30-50	20	''	''	''	20	18-24	18	''	22	6-9	10	
Calves, 350 to 500 lb	20-30	''	15-25	''	130-175[5]	20-50	18	''	''	''	20	18	''	''	18	6-8	8	
Cows in maternity stall	100-120	''	35-40	''	1-2 acre pasture paddock	---	30	---	---	---	26	30	---	---	30	9-12	15	
Herd bulls	100-150	''	35-45	''	''	---	''	30	36-40	30	''	''	30	36-40	''	''	''	

[1]Allow slope of ⅜ in./ft in paved lots, and ½ in. or more in dirt lots (depending on soil and climate conditions).
[2]Feed bunks should be about 8 in. deep for calves and 12 in. for older cattle.
[3]Minimum ceiling height of 9 ft necessary where a power-operated manure loader is to be used.
[4]With liberal grain or other concentrate feeding, half the recommended space given herein. With bunker or self-feeder silos, allow 6 in./animal.
[5]More space is desirable under some soil and climatic conditions.

Remarks:

Animals with horns require about one linear ft more manger or trough space per animal than the figures given in this table. Movable hayracks or feed bunks are usually 12 to 16 ft in length.

Provide a paved area of at least 10 ft around waterers, feed bunks, and roughage racks.

For specifications on slotted floors see Chapter 28, section entitled, "Design Requirements for Cattle Confinement Buildings and Slotted Floors."

Re: Water per animal per day. A minimum of 20 gal/day is needed for continuous flow to keep the water clean, and to keep it from freezing in the winter months.

will prove helpful. In general, less space than indicated may jeopardize the health and well-being of the animals, whereas more space may make the buildings and equipment more expensive than necessary.

BEEF CATTLE BUILDINGS

Beef cattle are not so sensitive to extremes in temperature—heat and cold—as are dairy cattle or swine. In fact, mature beef animals will withstand extremely cold weather if kept dry.

It is especially noteworthy that finishing steers, whose bodies generate considerable heat from the digestion and assimilation of their rations, do not need a warm shelter even during the cold winter months. Their chief need is for a dry bed and protection against cold winds, rains, and snow. About the same thing can be said about the sheltering of dry cows and stockers and feeders. Young stock require more protection. A shelter also permits the feeder to do his work in greater comfort. In the deep South and in the Southwest, barns are not necessary for cattle, except on rare occasions.

Beef cattle shelters are of two kinds, natural and artificial. The former includes hills and valleys, timber, and other natural windbreaks. The artificial shelters include those man-made structures (solid fences, stacks, barns, and sheds) designed to protect cattle against the elements—heat, cold, wind, rains, and snows.

It is with beef cattle barns and sheds that this discussion will deal.

In addition to protecting the animals during severe cold and stormy weather and at winter calving time, these structures should (1) provide a reasonably dry bed for the animals, (2) simplify feeding and management, (3) provide storage for feed and bedding when necessary, and (4) protect young calves.

Although the discussion describes beef cattle barns and sheds, it must be recognized that on small farms, which have a limited number of beef cattle, the animals are usually housed in a general-purpose barn or shed or in extensions to other barns rather than in separate and specially designed beef cattle structures.

Beef Cattle Barns

Barns are more substantial structures than sheds and provide more complete protection for stock in the colder areas. In addition to housing the animals, such structures usually provide adequate facilities for all of the roughage and bedding needed during the winter season and for a considerable proportion of the concentrates. Stalls, pens, and storerooms may also be included—additions which are especially important where a breeding herd is to be served. In general, beef cattle barns effect a saving of labor and time in feeding, and save feed.

Fig. 11-2. Beef cattle can withstand extremely cold weather if kept dry. Cows grazing cornstalks covered with a foot of snow. (Courtesy, The University of Nebraska, Lincoln, Neb.)

Fig. 11-3. Cattle shelter and equipment on an Iowa farm. With finishing cattle, whose bodies generate considerable heat from the digestion and assimilation of their rations, an open shed may suffice. (Courtesy, *The Corn Belt Farm Dailies*, Chicago, Ill.)

KINDS OF BEEF CATTLE BARNS

The type of barn is determined by the kind of stock and the method of handling and management. In general, the following two types of beef cattle barns, or modifications thereof, are in common use:

1. *The pen, stall, or general utility barn*—These barns range up to 40 feet in width and may be either one- or two-story, the only essential difference between the two being the overhead loft for hay and bedding storage in the latter. If feeding is done inside, mangers are placed along the alley and hayracks along the side walls; or the location of the racks and bunks may be reversed. Frequently, especially in a feeder cattle operation, hay is fed inside the barn and grain and silage outside. If this type of barn is to be used for breeding stock, the plan usually provides for grain bins and stalls and pens for calves, cows, and bulls. The barn may be an oblong structure; or it may have wings or extensions to form an L, T, or U shape.

2. *The central storage type with attached sheds or livestock sections*—This is a popular and economical type of cattle barn. These barns vary in width, but perhaps the average structure is about 60 feet wide and consists of a 24-foot center and two 18-foot wings. Sometimes the wings may extend around three sides of the building, thus providing more shelter space in relation to storage. The general floor plan for this type of barn consists of ground-to-roof storage for hay and bedding in the center and cattle sheds on each side. Racks for hay are adjacent to the central storage area. Troughs for grain and silage may be along the outside wall of the barn or outside in the lot.

CENTRAL STORAGE WITH CATTLE SHEDS ATTACHED ON SIDES

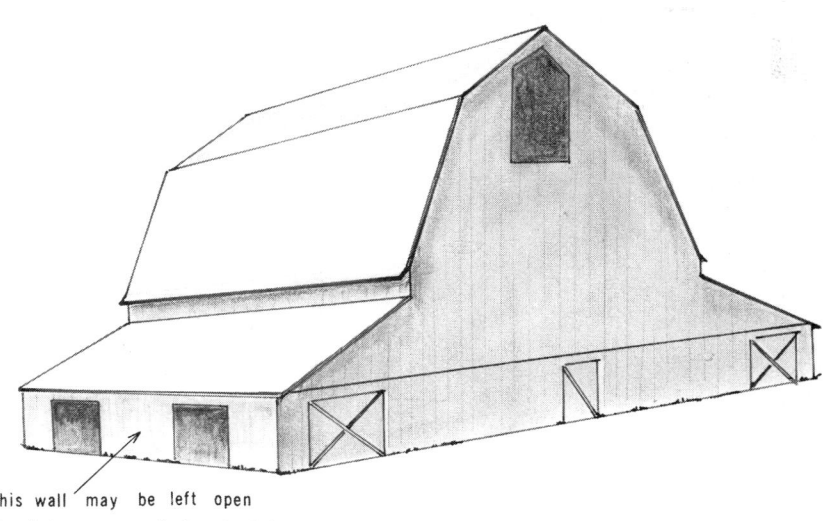

This wall may be left open
if winters are not too severe

Fig. 11-4. A central storage barn with attached sheds. Where considerable hay storage is desired, the central storage barn with attached sheds is popular.

Beef Cattle Sheds

Sheds are the most versatile and widely used beef cattle shelters throughout the United States. They are used for cattle in the feedlot, as a range shelter for dry cows with calves, and for housing young stock. They usually open to the south or east, preferably opposite to the direction of the prevailing winds and toward the sun. They are enclosed on the ends and sides. Sometimes the front is partially closed, and in severe weather drop-doors may be used. The latter arrangement is especially desirable when the ceiling height is sufficient to accommodate a power manure loader.

Fig. 11-5. An open shed. Sheds are the most versatile and widely used beef cattle shelters throughout the United States. (Courtesy, USDA)

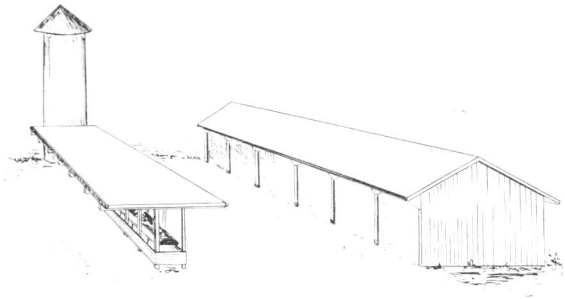

Fig. 11-6. A modern low-cost, labor-saving (1) feed bunker (which may be equipped with a chain or belt feeder) and (2) open shed.

So that the bedding be kept reasonably dry, it is important that sheds be located on high, well-drained ground; that eave troughs and down spouts drain into suitable tile lines, or surface drains; and that the structures have sufficient width to prevent rain and snow from blowing to the back end. Sheds should be a minimum of 24 feet in depth, front to back, with depths up to 36 feet preferable. As a height of 8½ feet is necessary to accommodate some power-operated manure loaders, when this type of equipment is to be used in the shed, a minimum ceiling height of 9 feet is recommended. The extra 6 inches allow for the accumulation of manure. Lower ceiling heights are satisfactory when it is intended to use a blade or pitchfork in cleaning the building.

The length of the shed can be varied according to needed capacity. Likewise, the shape may be either a single long shed or in the form of an L or T. The long arrangement permits more corral space. When an open shed is contemplated, thought should be given to feed storage and feeding problems.

Sometimes hayracks are built along the back wall of sheds, or next to an alley, if the shed is very wide or if there is some hay storage overhead. Most generally, however, hayracks, feed bunks, and watering troughs are placed outside the structure.

A Modern Multiple-Use Barn

The two-story red barn, long a traditional American trademark, is fast giving way to cheaper one-story structures of more flexible design and lower operating costs.

Figures 11-7 and 11-8 show a modern low-cost, labor-saving, multiple-use barn. It may be used for beef cattle, for dairy cattle (as a loafing barn), for sheep, for swine, for horses, and/or for storage of feed, seed, fertilizer, and machinery. This barn is flexible and versatile. Figure 11-9 shows how this barn can be adapted to hillside construction, thereby obtaining a gravity feeding arrangement.

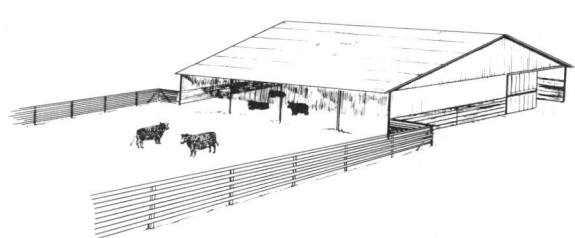

Fig. 11-7. A modern multiple-use barn. It is flexible and versatile. (Drawing by Steve Allured)

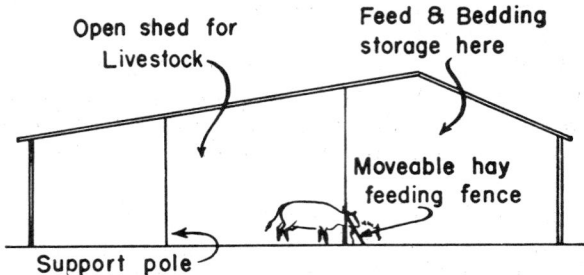

Fig. 11-8. Cross section of the modern multiple-use barn shown in Fig. 11-7. Note the movable hay feeding fence which makes it possible (1) to decrease the feed and bedding storage area and to increase the animal area as winter advances, and (2) to keep the feed and bedding in close proximity to the supply, thereby lessening labor and drudgery. (Drawing by Steve Allured)

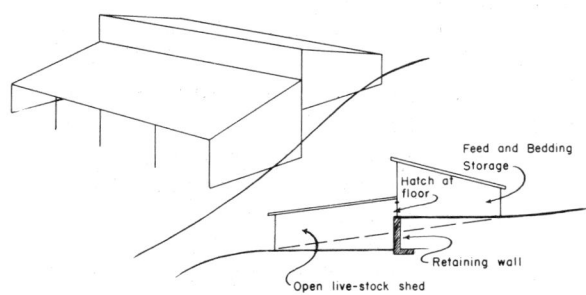

Fig. 11-9. Modern multiple-use barn adapted to hillside construction. This design alleviates costly leveling of the building site and provides for easy gravity handling of the feed and bedding. (Drawing by Steve Allured)

IT HAS DESIRABLE FEATURES

If properly designed and constructed, this modern multiple-use barn should possess the following desirable features:

1. *Meet needed animal space and feed storage requirements*—It should meet the specific animal space and feed storage requirements of the farm or ranch.

2. *Face in the right direction*—Except in the deep South, it should open to the south or east, preferably opposite the direction of the prevailing winds and toward the sun.

3. *Be relatively dry*—It should be constructed to assist in providing a reasonably dry bed for animals.

4. *Be of proper height*—A minimum ceiling height of 9 feet is recommended in order to accommodate a power-operated manure loader.

5. *Possess movable equipment*—Wherever possible, barn equipment should be as movable as the furniture in a home. Thus, where the entire

floor of the barn is on the same ground level, movable feed facilities make it possible (a) to decrease the feed and bedding storage area and to increase the animal area as winter advances, and (b) to keep the feed and bedding in close proximity to the supply, thereby lessening labor and drudgery.

6. *Promote animal health*—This is most important, for healthy animals are the profitable and efficient ones. Today, it is recognized (a) that open sheds provide the cheapest type of ventilation, and usually the best, and (b) that, except for newborn or sick animals, the inside temperature of the barn should be as close to the outside temperature as possible. The modern multiple-use type of barn shown in Figure 11-7 meets these requisites.

7. *Be flexible*—The need for flexibility is best illustrated by referring to the great number of obsolete draft horse and mule barns throughout the country, usually two-story structures with built-in stalls and permanent feed mangers. With the passing of the draft animal, many of these old barns have either remained unused or have been modernized at high cost. But this is not the end of such changes: Who can predict, with certainty for example, what the future holds relative to methods of harvesting, curing, and storing hay? Thus, it is important that buildings be of such flexible design that they can be easily and inexpensively modernized to meet changes in a changing world.

This desired flexibility is best obtained by constructing a single-story building with movable equipment—features of the modern barn shown in Figure 11-8.

8. *Gravity feed if on a hillside*—In a hilly area, it is recommended that the design of the barn be adapted as shown in Figure 11-9. This alleviates costly leveling of the building site and provides for easy gravity handling of the feed and bedding.

9. *Possess such added rooms and stalls as are needed*—Maternity stalls, feed, seed, or fertilizer rooms, a milking room, and/or a tack room may be incorporated in this type of barn if they are needed.

10. *Possess adjacent corrals and a loading chute*—Suitable corrals and a loading chute should be provided adjacent to the building.

IT IS REASONABLE IN COST

On many livestock establishments, the cost of the improvements is about equal to the value of the land alone. Thus, it is important that every consideration should be given to affecting savings in the construction of buildings and equipment. Among the most important reasons why the modern type barn shown in Figure 11-7 is reasonable in cost, are the following:

1. *It is of simple pole-frame construction*—It is supported by poles chemically treated under pressure and set like fence posts 4 to 5 feet in the ground and spaced at least 12 feet apart. This barn (a) eliminates scaffolding, for even the highest points can be reached from ladders, trucks, or wagons; (b) lessens bracing; and (c) saves labor, for this simple type of construction can be built with any farm labor capable of building a movable hog house.

2. *It uses the ground to support feed and bedding*—This method alleviates the necessity of heavy construction to support this weight overhead.

3. *It is built low to the ground*—This makes for less wind pressure to resist, and requires less bracing.

4. *It leaves one side open*—This saves on material, and desirable ventilation is obtained without cost.

IT SAVES LABOR

The importance of designing farm buildings for efficiency of operation becomes apparent when it is realized that the average stockman spends more than half his working hours in and around the farm buildings.

The type of barn shown in Figure 11-7 is labor-saving in comparison with conventional two-story barns for the following reasons:

1. *Feed is stored on the ground level*—The truck, or other vehicle, may be driven directly into the ground-level storage area and unloaded without hoisting or elevating.

2. *Feed and bedding are stored where used*—A movable hay and feed bunk can be so designed that it can be moved back as the animals eat the feed and use the bedding. This eliminates the necessity of climbing into mows, poking feed and bedding down a chute, and carrying it some distance to where it is to be used. Instead, the feed may be tossed directly and easily into the rack or bunk and the bedding into the area where it is used.

3. *Overflow feed and bedding storage is convenient*—If the farm production of feed and bedding is higher than normal, the added tonnage may be temporarily stored in the animal area. Then as it is used, more and more of the animal space becomes available as winter advances. With a conventional two-story barn, such flexibility is not easily obtained.

4. *Manure may be removed by power loader*—The ceiling height and open shed arrangement make for ease in operating a manure loader and in getting in and out with a spreader. No hand labor is required.

BEEF CATTLE EQUIPMENT

It is not proposed that all of the numerous types of beef cattle equipment will be described herein; rather, only those articles that are most common. Suitable equipment saves feed and labor, conserves manure, and makes for increased production.

Hayracks

Various sizes and designs of racks are used in feeding hay and other forages. In general, these structures are of two types: overhead racks, and low mangers. Overhead racks are easily moved, but they are difficult to clean; and often there is considerable wastage, especially when cows have horns and do considerable fighting. Low mangers result in less wastage of the leaves and other fine particles than overhead racks because the cattle must work down from the top. Low mangers may be satisfactorily used in the feeding of kaffir and other coarse forages, also.

Fig. 11-10. Beef cattle equipment has gone modern! (Upper) The way it used to be done—with a bushel basket. (Lower) A self-unloading feed truck. (Courtesy: Upper—*Wallaces' Farmer and Iowa Homestead*, Des Moines, Iowa; lower—Massey-Ferguson, Inc., Des Moines, Iowa)

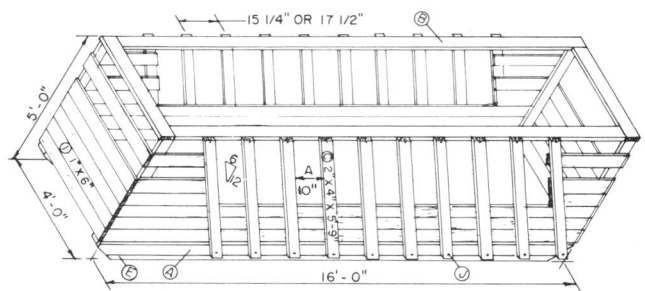

Fig. 11-11. Portable hayrack for cattle. (From *Beef Equipment Plans*, MWPS-6, Midwest Plan Service, Iowa State University, Ames, Iowa, 1968, p. 29.)

It is preferable that racks be of sufficient size so that one filling will last several days, thus lessening the labor requirements. It is desirable also that the rack be mounted on runners or wheels, thus making it convenient to move it from place to place.

Feed Troughs or Bunks

Feed troughs or bunks are used for feeding both grain and silage. Figure 11-12 shows a common and desirable type of cattle bunk that is well braced, free from sharp corners, portable, and made so as to prevent cattle from throwing feed out. It could be made stationary by extending the posts into the ground or by setting the posts on concrete foundations. Stationary troughs should be located on a well-drained site, or, preferably, placed on a concrete or other hard-surfaced platform. The dimensions of the feed bunk or trough should be in keeping with those given in Table 11-6. Bunks of a height for mature cattle may be used for calves by digging holes for the legs or runners. In many large cattle feeding yards, especially in the West, a "manger-type" trough forms a part of the enclosure and is filled from outside the corral along a service lane or road.

Stationary troughs in barns should be provided with adjustments for height, thus making it possible to adjust for cattle of different ages and for the accumulation of manure.

Self-Feeders

Most cattle are hand-fed, but the use of self-feeders for finishing cattle is increasing. A self-feeder does a better job of feeding than a careless man. Self-feeders may either be stationary or portable. The latter are usually equipped with runners or are mounted on low wheels. The most desirable cattle self-feeders have feed troughs along both sides, have a capacity adequate to hold a feed supply that will last 10 days to 2 weeks, and are easily filled.

Fig. 11-12. Cattle feed bunk. Note pole-type open shed at back of lot. (Courtesy, Ralston Purina Company, St. Louis, Mo.)

Fig. 11-13. Homemade self-feeder, showing adjustable feed opening arrangement which is 1″ × 12″ with wing-nut bolts in 2″ slots. (Courtesy, Dr. Wilton W. Heinemann, Irrigated Agriculture Research and Extension Center, Washington State University, Prosser, Wash.)

Watering Facilities

Water is more essential than feed, for animals will subsist longer without feed than without water. As indicated in Table 11-6, on the average, mature cattle consume approximately 12 gallons of water per head daily, with variations according to size of animal, season, and type of feed. Cattle are frequently watered from reservoirs, springs, lakes, and streams; but if surface water is not available, labor can be saved by having reliable power for pumping and by piping water under pressure to tanks or troughs where it will be available at all times. When wells are used as a source of supply, windmills are the most commonly used power unit, although electric motors and gas engines are sometimes used. In cold areas, outside tanks are generally provided with tank heaters or covers during the winter months.

Fig. 11-14. A good concrete water tank for cattle. Note fence over top of tank to keep animals out. (Courtesy, USDA)

Regardless of the source of the water supply—wells, springs, streams, or surface-storage supplies—it is important that it be abundantly available at all times. It should also be fresh and well protected. When watering tanks and troughs are used, they should be of adequate size, and there should be provision for keeping animals out of the tank. In areas in which mud is a problem, it is desirable that tanks be surrounded by a pavement at least 6 feet in width, and provision should be made to pipe the overflow away from the vicinity of the tank.

Shades

Cattle should be provided with suitable shade during the hot, summer months. An unshaded cow standing in an air temperature of 100° F has to dispose of enough heat in a 10-hour period to bring 9 gallons of ice water to the boiling point. At the Imperial Valley Field Station, El Centro, California, it was found (1) that the difference in heat load in an animal under a shade 10 or 12 feet high at 100° F, and one in the sun, was 1,334 Btu per hour, and (2) that to make 100 pounds gain during midsummer required 200 to 300 pounds more feed without shade than with shade.

Fig. 11-15. A 10' × 20' shade. (Courtesy, Kindred P. Caskey, Jr., Bar K Ranch, Weslaco, Tex.)

The most satisfactory cattle shades are (1) oriented with a north-south placement, as such shades are drier underneath than those with east-west orientation, because the sun can get underneath to dry out the manure and urine; (2) at least 10 to 12 feet in height (in addition to being cooler, high shades allow a truck to be driven under and cattle to be worked by a man on a horse); and (3) open all around.

Corrals

No equipment adds more to the ease and pleasure of handling cattle than a convenient system of well-constructed corrals. In addition, such a system saves money by reducing the shrinkage resulting from sorting and handling cattle. On the western range, this type of equipment is considered a virtual necessity. To be sure, it is not presumed that each operator will have need for the same size, number, and arrangement of lots. These will vary according to the size of the herd, management practices, and individual preferences. There are, however, certain salient features that should be observed in planning a system of corrals: (1) They should be large enough to accommodate easily the number of cattle involved; (2) they should be conveniently located with respect to the pasture or range and a nearby water supply; and (3) they should include at least one large pen for holding the herd, a chute or alley to be used for separating or crowding work, and two smaller pens in which to put the separated cattle.

Fig. 11-16. A paved corral lot used for finishing steers. Note that the cattle are clean and free from mud and manure. (Courtesy, Portland Cement Association)

Loading Chute

The extensive use of trucks makes it desirable that the stock farm and ranch be equipped with a chute for loading and unloading stock. Such equipment may be either portable or stationary. In the latter case, it is usually desirable to attach the chute to the corrals or feedlot. A loading chute of sufficiently durable construction for cattle is equally satisfactory for sheep and swine. The main essentials are: (1) that the chute have proper height for the truck commonly served (or preferably have an adjustable height arrangement); (2) that it have adequate width to accommodate animals; (3) that it have sufficient slope and cleating to the platform approach to prevent slipping. Most chutes are about 28 to 30 inches in width and 46 inches high.

Squeeze

A squeeze for handling cattle can be profitably used on any farm or ranch for dehorning, branding, castrating, testing for tuberculosis, vaccinating, or in performing minor surgical operations. Lack of such equipment usually entails a great deal of labor in catching and throwing animals and is hard on both man and beast. Numerous designs of homemade and commercial cattle squeezes are available, but the essential features of all of them are: (1) durability, (2) thorough restraint of the animal, and (3) convenience

Fig. 11-17. Portable loading chute. (Courtesy, Pioneer Hi-Bred Corn Company, Des Moines, Iowa)

Fig. 11-18. Cattle move through a curved alley better than through a straight alley because they cannot see what is ahead of them. A swinging gate is used to guide cattle into the head catch or up the loading chute. (Courtesy, University of Illinois, Urbana, Ill.)

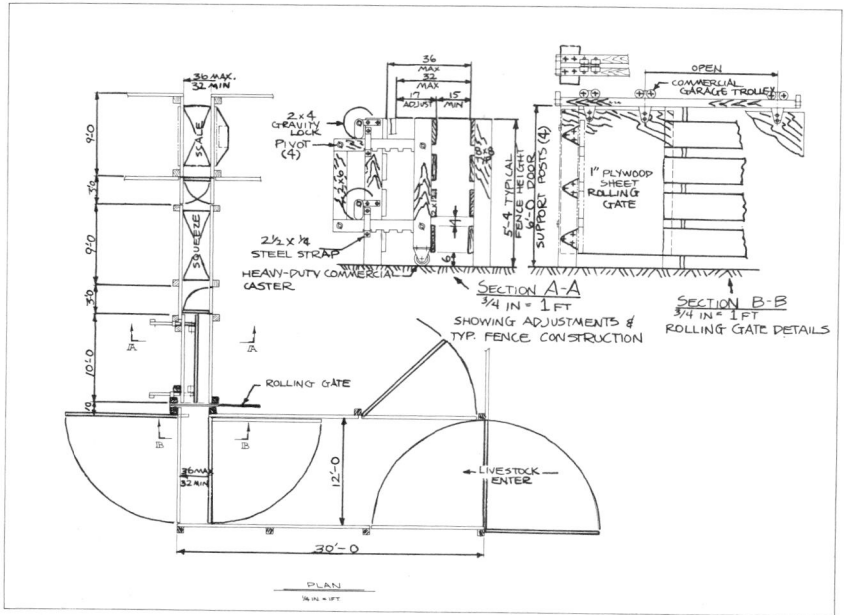

Fig. 11-19. An excellent chute, squeeze, and scale arrangement which can be adapted to any cattle corral (new or old). As shown, it features (1) a very short approach chute, with vertical sides that are adjustable for width, and (2) a gate located so that animals have no choice other than to go into the chute. The squeeze should (1) have a stanchion-type head gate, and (2) be designed so that the sides close in on the back of the animal. With this arrangement, animals will stand still without struggling. Where the narrow portion of the squeeze is at the bottom, squeezing tends to lift the animal off its feet and causes it to fight. (Drawing prepared under the direction of Dr. A. T. Ralston, Department of Animal Science, Oregon State University)

for the operator. When cattle corrals are constructed, the cattle squeeze is normally a part of the pen arrangement.

Stocks

Cattle stocks are primarily an item of equipment for the purebred herd and the feedlot. Usually they are so located that animals must be led into them rather than driven, thus limiting their use to cattle that may be rather easily handled. Cattle stocks may be used for trimming and treating hoofs, dehorning, horn branding, drenching, ringing bulls, swinging injured animals, and restraining animals during surgical operations. The essential features of cattle stocks are: (1) durability; (2) thorough restraint of the animals; (3) convenience for the operator; (4) a canvas sling to place under the animal to prevent it from lying down while in the stocks (the swing may be wound up on side rollers by means of turning rods); and (5) wooden sills that extend along either side at a height of 15 inches from the floor and on which the feet may be rested and tied while being trimmed or treated.

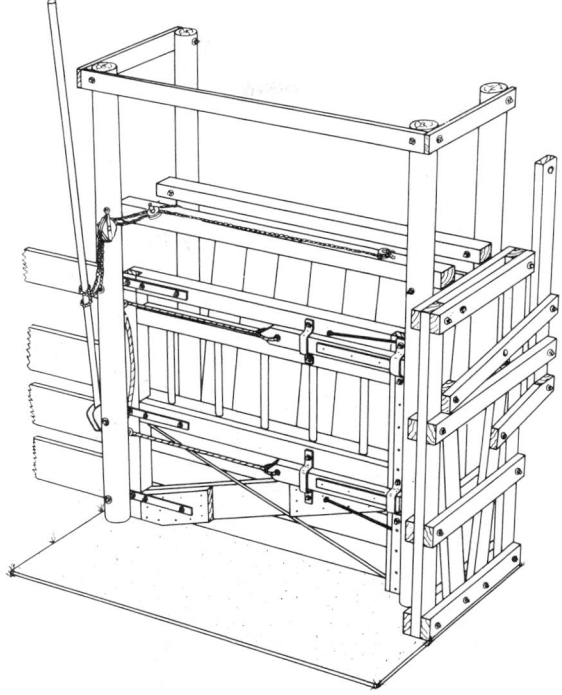

Fig. 11-20. A squeeze. Detailed plans and specifications for such equipment can usually be obtained through the local lumber dealer, county agricultural agent or FFA instructor, or through writing the college of agriculture in the state. (Courtesy, University of Wyoming, Laramie, Wyo.)

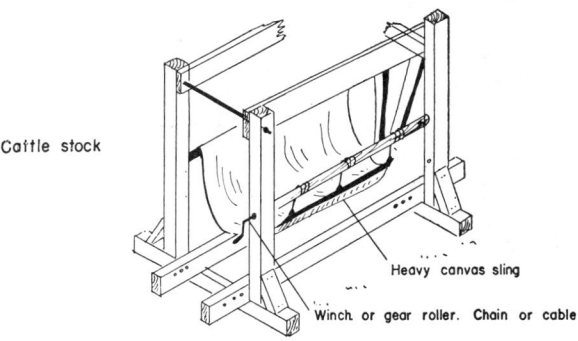

Cattle stock

Heavy canvas sling

Winch or gear roller. Chain or cable

Fig. 11-21. Cattle stock. Detailed plans and specifications for such equipment can usually be obtained through the local lumber dealer, county agricultural agent or FFA instructor, or through writing the college of agriculture in the state. (Drawing by R. F. Johnson)

Breeding Rack

Breeding racks are sometimes used by purebred operators who desire to breed young heifers to mature, heavy bulls. Figure 11-22 shows a very satisfactory type of breeding rack.

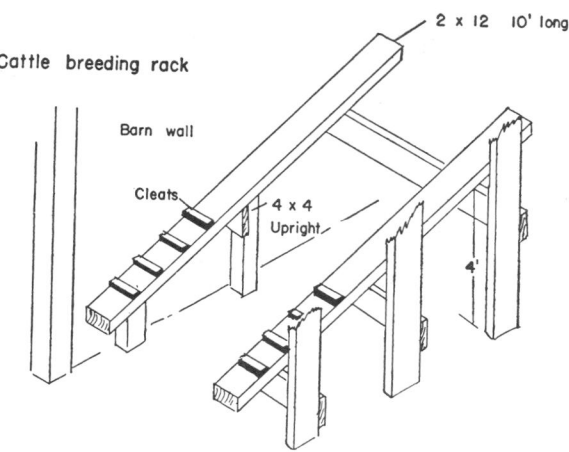

Fig. 11-22. Cattle breeding rack. Detailed plans and specifications for such equipment can usually be obtained through the local lumber dealer, county agricultural agent or FFA instructor, or through writing the college of agriculture in the state. (Drawing by R. F. Johnson)

Dipping Vat and Spraying Equipment

Dipping vats have long been used successfully and rather extensively on the western range in treating cattle, sheep, and sometimes horses, for external parasites. The vat is usually built at one side of the corral system. The chief virtue of the dipping vat lies in the fact that animals so treated are thoroughly covered. On the other hand, the vats are costly to construct and lack mobility and flexibility; and there is always considerable leftover dip at the finish of the operation. Frequently, dipping vats are built as a cooperative enterprise by a group of producers rather than by an individual operator. With the development of modern insecticides and improved spraying equipment, it appears probable that in the future spraying operations will increase and that fewer expensive dipping vats will be constructed.

Other Beef Cattle Equipment

There is hardly any limit to the number of different articles of beef cattle equipment, and the design of each. In addition to those already listed, Figures 11-23, 11-24, and 11-25, show a calf creep, a salt-mineral feeder, and scales, respectively.

Fig. 11-23. A calf creep. Note the small entrance for the calves. (Courtesy, Kindred P. Caskey, Jr., Bar K Ranch, Weslaco, Tex.)

Fig. 11-24. A salt-mineral feeder on skids. Note that it is a three-compartment arrangement. (Courtesy, Kindred P. Caskey, Bar K Ranch, Weslaco, Tex.)

Fig. 11-25. Scales take the "guess" out of many farm and ranch operations. (Courtesy, Swift & Company, Chicago, Ill.)

SILOS

The general kinds of silos are: tower silos, pit silos, trench silos, self-feeder or bunker silos, above ground temporary silos, and plastic silos. The kind of silo decided upon and the choice of construction material should be determined primarily by the cost and by the suitability to the particular needs of the farm or ranch.

RECOMMENDED MINIMUM WIDTH OF SERVICE PASSAGES

In general, the requirements for service passages are similar, regardless of the kind of animals. Accordingly, the suggestions contained in Table 11-7 are equally applicable to cattle, sheep, swine, and horse barns.

TABLE 11-7

RECOMMENDED MINIMUM WIDTHS FOR SERVICE PASSAGES

Kind of Passage	Use	Minimum Width
Feed alley	For feed cart	4' - 0"
Driveway	For wagon, spreader, or truck	9' - 0"
Doors and gate	Drive-through	9' - 0"
Doors and gate	To small pens	4' - 0"

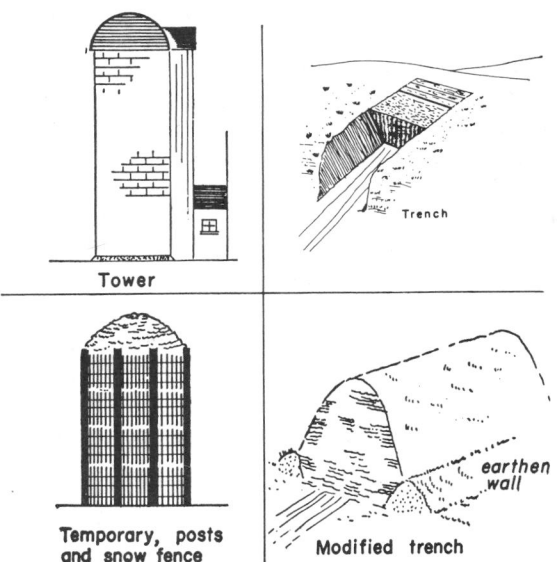

Fig. 11-26. Four kinds of silos: (1) tower silo (upper left); (2) trench silo (upper right); (3) enclosed stack silo (lower left); and (4) modified trench-stack silo (lower right). The latter two are both above-ground temporary silos. (Drawing by R. F. Johnson)

Fig. 11-27. A laborsaving self-feeding silo. (Courtesy, Republic Steel Corporation, Cleveland, Ohio)

STORAGE SPACE REQUIREMENTS FOR FEED AND BEDDING

The space requirements for feed storage for the livestock enterprise— whether it be for cattle, hogs, sheep, or horses, or as is more frequently the case, a combination of these—vary so widely that it is difficult to provide a suggested method of calculating space requirements applicable to such diverse conditions. The amount of feed to be stored depends primarily upon (1) length of pasture season, (2) method of feeding and management, (3) kind of feed, (4) climate, and (5) the proportion of feeds produced on the farm or ranch in comparison with those purchased. Normally, the storage capacity should be sufficient to handle all feed grain and silage grown on the farm and to hold purchased supplies. Forage and bedding may or may not be stored under cover. In those areas where weather conditions permit, hay and straw are frequently stacked in the fields or near the barns in loose, baled, or chopped form. Sometimes poled framed sheds or a cheap cover of waterproof paper or wild grass is used for protection. Other forms of low-cost storage include temporary upright silos, trench silos, and temporary grain bins.

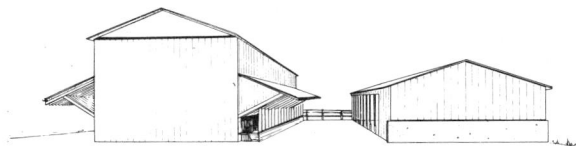

Fig. 11-28. A modern low-cost, laborsaving (1) hay feeding shed and (2) shelter. The shelter may have some boxed-in maternity stalls if desired, and the corral may be paved.

Table 11-8 gives the storage space requirements for feed and bedding. This information may be helpful to the individual operator who desires to compute the barn space required for a specific livestock enterprise. This table also provides a convenient means of estimating the amount of feed or bedding in storage.

HOW TO DETERMINE THE SIZE BARN TO BUILD

The length and depth of the barn may be varied according to needs. The size barn to build for any given farm or ranch may be determined as follows:

1. Estimate the number and kind of animals to be quartered and compute their total animal space requirements. (See Table 11-6.)

2. Compute the yearly feed requirements of the animals to be fed and quartered by referring to Table 8-15, giving consideration to the length of the pasture season and the quantity and quality of the grass.

3. Estimate the farm production of feeds and bedding to be stored in the barn. In most operations this should coincide reasonably close to the total animal requirements (point 2), but there may be circumstances where the feed and bedding storage requirements are more or less than the animal feed requirements.

TABLE 11-8

STORAGE SPACE REQUIREMENTS FOR FEED AND BEDDING[1]

Kind of Feed or Bedding	Pounds Per Cubic Foot	Cubic Feet Per Ton	Pounds Per Bushel of Grain
Hay-Straw:			
1. Loose			
Alfalfa	4.4-4.0	450-500	
Nonlegume	4.4-3.3	450-600	
Straw	3.0-2.0	670-1000	
2. Baled			
Alfalfa	10.0-6.0	200-330	
Nonlegume	8.0-6.0	250-330	
Straw	5.0-4.0	400-500	
3. Chopped			
Alfalfa	7.0-5.5	285-360	
Nonlegume	6.7-5.0	300-400	
Straw	8.0-5.7	250-350	
Corn:			
15½% moisture			
Shelled	44.8		56
Ear	28.0		70
Shelled, ground	38.0		48
Ear, ground	36.0		45
30% moisture			
Shelled	54.0		67.5
Ear, ground	35.8		89.6
Barley, 15%	38.4		48.0
ground	28.0		37.0
Flax, 11% moisture	44.8		56.0
Oats, 16% moisture	25.6		32.0
ground	18.0		23.0
Rye, 16% moisture	44.8		56.0
ground	38.0		48.0
Sorghum, grain 15% moisture	44.8		56.0
Soybeans, 14% moisture	48.0		60.0
Wheat, 14% moisture	48.0		60.0
ground	43.0		50.0

[1]*Beef Housing and Equipment Handbook*, MWPS-6, Midwest Plan Service, Iowa State University, Ames, Iowa, 1968, p. 62.

4. Estimate the total tonnage of feed and bedding to be stored by correlating the animal feed needs and the farm or ranch production. (Correlate the results of points 2 and 3.) Then determine the total storage space requirements for feed and bedding from Table 11-8.

5. Determine the size of barn to build from the total animal space requirements and the total yearly feed and bedding storage requirements (points 1 and 4).

In general, modern multiple-use barns of the type shown in Figure 11-7 are 52 feet deep, with 26 feet devoted to feed and bedding storage and 26 feet to animals; although both the depth and the length can be varied to meet specific needs.

HOW TO DETERMINE THE SIZE SILO TO BUILD

The size of silo to build should be determined by needs. With tower type and pit silos, this means (1) that the diameter should be determined by quantity of silage to be fed daily, and (2) that the height (depth in a pit silo) should be determined by the length of the silage feeding period. Similar consideration should be accorded with trench silos.

Size of Tower Silo

If the diameter is too great, the silage will be exposed too long before it is fed; and unless a quantity is thrown away each day, spoiled silage will be fed.

The minimum recommended rate of removal of silage varies with the temperature. In most sections of the United States, it is desirable that a minimum of 1½ inches of silage be removed from tower silos daily during the winter feeding period, with the quantity increased to a minimum of 3 inches when summer feeding is practiced. Of course, the total daily silage consumption on any given farm or ranch will be determined by (1) the class and size of animals, (2) the number of animals, and (3) the rate of silage feeding. Some suggestions on how much silage to feed cattle are found in Table 8-15.

Silo height should be determined primarily by the length of the intended feeding period. In general, however, the height should not be less than twice, nor more than three and one-half times the diameter. The greater the depth, the greater the unit capacity. Extreme height is to be avoided because of (1) the excessive power required to elevate the cut silage material, and (2) the heavier construction material required. Also, it is noteworthy that, with silos of the larger diameters, more labor is required in carrying the silage to the silo door for removal.

Table 11-9 may be used as a guide in computing the proper diameter of tower silo for any given farm or ranch.

TABLE 11-9

MAXIMUM DIAMETER OF TOWER SILO TO BUILD IF SILAGE
IS TO BE KEPT FRESH

Inches of Silage Removed Daily	Total Silage Removed Daily with an Inside Silo Diameter of—					
	10 feet	12 feet	14 feet	16 feet	18 feet	20 feet
	(lb)	(lb)	(lb)	(lb)	(lb)	(lb)
Summer: 3″ daily will remove[1]	786	1,312	1,539	2,010	2,545	3,142
Winter: 1½″ daily will remove[1]	393	656	770	1,005	1,272	1,571

[1]The pounds listed in each of the columns to the right are approximations based on an average constant weight of 40 lb of silage per cubic foot.

Figure 11-29 shows capacities of tower silos of different heights and diameters. It is based on well-eared corn silage harvested in the early dent stage, cut in ¼″ lengths, well tramped when filled, and with the silo refilled once after settling for a day.

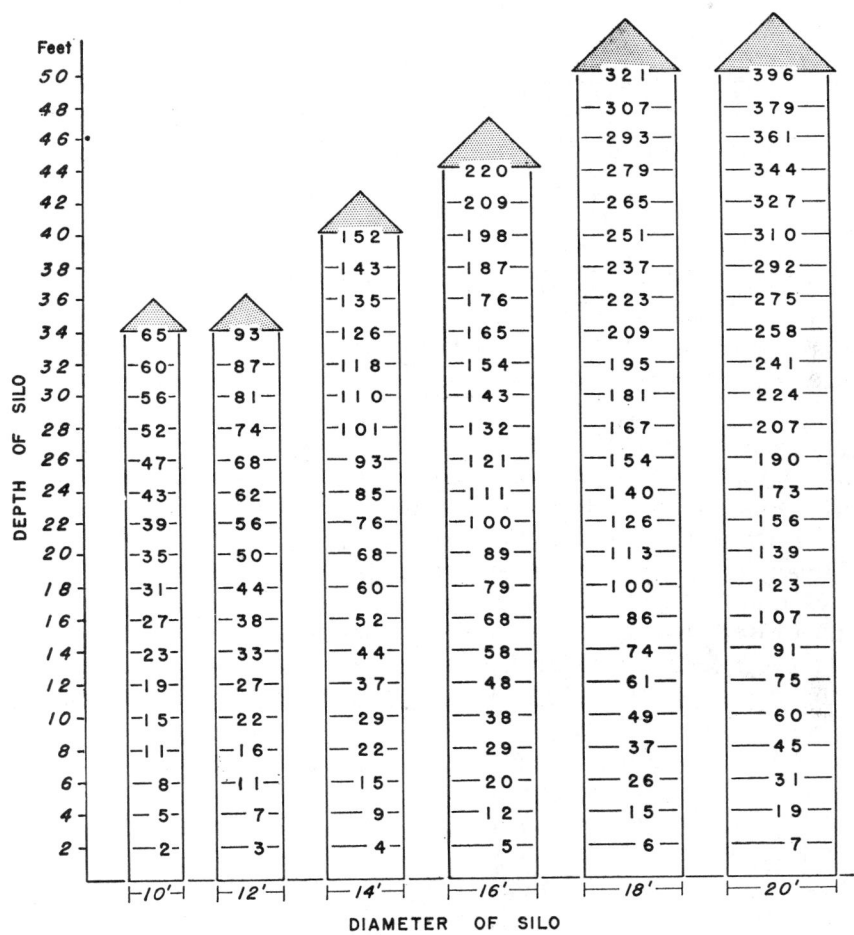

Fig. 11-29. Capacity in tons of settled corn silage in tower silos of varying sizes (based on data reported in USDA Circ. 603; drawing by R. F. Johnson)

Figure 11-29 can be adapted for corn silage of different stages of maturity and grain content, and for other kinds of silage, by applying the following rules of thumb:

Kind of Silage:	Changes to Be Made in the Number of Tons Shown in Fig. 11-29
1. For corn silage ensiled when less mature than usual	Add 5 to 10%
2. For corn ensiled when dry or overripe	Deduct 5 to 10%
3. For corn very rich in grain	Add 5 to 10%
4. For corn with very little grain	Deduct 5 to 10%
5. For sorghum silage	Use the same weights as used for corn silage of comparable grain and maturity
6. For sunflower silage	Add 5 to 10%
7. For grass silage	Add 10 to 15%[1]

[1]For this reason, a stronger structure is necessary where grass silage is stored.

The following example will serve to illustrate how to determine the size tower silo to build:

Over a period of years, a farmer plans to winter 34 head of 425-pound stocker calves on a ration of corn silage and protein supplement. There is a 240-day wintering period. No increase in the herd is planned. What size tower silo should be built?

The answer is obtained as follows:

1. First, here are the silage requirements:

a. Table 8-15, Handy Beef Cattle Feeding Guide, indicates that 425-pound stocker calves on a ration of corn silage and protein supplement should receive about 30 lb of silage per head per day.

b. 34 × 30=1,020 lb of silage required daily for the 34 calves.

c. 1,020 × 240=244,800 lb, or 122.4 tons, of silage required for the 240-day wintering period for the 34 calves.

2. Next, here is the size silo to build:

a. Table 11-9 shows that in order to remove 1,005 lb of silage daily (which is only slightly less than the 1,020 lb needed daily), with 1½ in. removed from the top of the silo each day, diameter of the silo should not be greater than 16 ft.

b. Fig. 11-29 can now be used as a guide in determining both the proper height (or depth) and diameter of the silo. Fig. 11-29 shows that a silo 16 ft in diameter and 27 ft high will hold 127 tons of silage, which would allow for 4.6 tons spoilage in excess of the required 122.4 tons. However, the height of a silo should not be less than twice the diameter. It appears best, therefore, to plan on a 14-ft diameter silo. As noted in Fig. 11-29, 34 ft of settled silage in a 14-ft diameter silo will provide 126 tons of silage, which would allow for 3.6 tons spoilage in excess of the required 122.4 tons. To allow for settling, an additional 4 to 6 ft should be added to the height, thus making a 38- to 40-ft height.

c. The size silo to build to meet the needs outlined in this example, therefore, is one that is 14 ft in diameter and 38 to 40 ft high.

Size of Trench Silo

As in an upright silo, the cross sectional area of a trench silo should be determined by the quantity of silage to be fed daily. The length is determined by the number of days of the silage feeding period. The only difference is that generally greater allowance for spoilage is made in the case of trench silos, though this factor varies rather widely.

Under most conditions, it is recommended that a minimum 4-inch slice be fed daily from the face (from the top to the bottom of the trench) of a trench silo during the winter months, with a somewhat thicker slice preferable during the summer months.

The dimensions, areas, and capacities given in Table 11-10 are based on the assumption that the silage weighs 35[3] lb per cubic foot, which is an average figure for corn or sorghum silage. Thus, a trench silo 8 feet deep, 6 feet wide at the bottom, and 10 feet wide at the top has a cross sectional area of 64 feet. This size silo will hold 747 lb of silage for each 4-inch slice, or 2,240 lb of silage for each 1-foot slice, or 112 tons in a trench 100 feet long.

TABLE 11-10

DIMENSIONS, CROSS SECTION AREA OF TRENCH SILO, AND
WEIGHT OF SILAGE IN 4-INCH SLICE AND PER LINEAL FOOT[1]

Side Slope Per Foot of Depth	Depth	Bottom Width	Top Width		Cross Sectional Area	Weight of Silage	
						4-inch slice	1-foot slice
(in.)	(ft)	(ft)	(ft)	(in.)	(sq ft)	(lb)	(lb)
3	4	5	7	0	24	280	840
4	4	6	8	8	29	338	1,015
5	4	7	10	4	33	385	1,155
3	6	6	9	0	45	525	1,575
4	6	7	11	0	54	630	1,890
5	6	8	13	0	63	735	2,205
3	8	6	10	0	64	747	2,240
4	8	7	12	4	77	898	2,695
5	8	8	14	8	91	1,062	3,185
3	10	6	11	0	85	992	2,975
4	10	8	14	8	113	1,318	3,955
5	10	10	18	4	142	1,657	4,970

[1]*Silos, Types and Construction*, USDA, Farmers' Bulletin No. 1820, Revised 1948, p. 55.

For illustrative purposes, let us use the same example and silage requirements as given on page 676, but this time determine the size trench silo to build rather than the size tower silo. Briefly, the requirements are for 1,020 lb of silage daily for a 240-day wintering period. As noted in Table 11-10, one day's feed or 1,020 lb of silage (1,062 lb to be exact) can be obtained in each 4-inch slice of a trench silo 8 feet wide at the bottom, 14 feet 8 inches wide at the top, and 8 feet deep; or a 91 square feet cross sectional

---

[3]Because the silage in trench silos is generally not so deep and well packed as the silage in tower silos, an average figure of 35 lb per cubic foot is used herein for trench silos and 40 lb for upright silos. With all types of silos—including above-ground and below-ground types—the weight of a cubic foot of silage varies with the kind and maturity of the material, moisture content, length of cut, rate of filling, and depth of the silo. Corn silage harvested when about 74% of the grain has passed the milk stage and containing approximately 70% moisture is considered average silage. Volume for volume, sorghum silage weighs about the same as corn silage. Grass or grass-legume silage is 10 to 15% heavier than corn silage.

area. The cross sectional area should not be larger than this if a 4-inch slice is to be removed daily in order to alleviate spoilage.

In order to obtain a 240-day feed supply, the filled trench must be 80 feet long (⅓ of 240; the ⅓ represents ⅓ foot or 4 inches).

The size trench silo to build to meet the specified needs, therefore, is one that is 8 feet wide at the bottom, 14 feet 8 inches wide at the top, 8 feet deep, and 80 feet long. In order to take care of spoilage and to provide a measure of safety, it is recommended that the actual length be from 85 to 90 feet.

About 8 feet for a trench silo is the most economical depth from the standpoint of cost and feeding. Of course, in filling it is desirable to pile silage 3 feet higher over the center of the trench and round it off. This provides for settlement.

FENCES FOR CATTLE

Good fences (1) maintain farm boundaries, (2) make livestock operations possible, (3) reduce losses to both animals and crops, (4) increase land values, (5) promote better relationships between neighbors, (6) lessen accidents from animals getting on roads, and (7) add to the attractiveness and distinctiveness of the premises.

Fig. 11-30. Woven wire cattle fence. (Courtesy, Keystone Steel and Wire, Peoria, Ill.)

The discussion which follows will be limited primarily to wire fencing, although it is recognized that such materials as rails, poles, boards, stone, and hedge have a place and are used under certain circumstances. Also, where there is a heavy concentration of animals, such as in corrals and feed yards, there is need for a more rigid type of fencing material than wire. Moreover, certain fencing materials have more artistic appeal than others; and this is an especially important consideration on the purebred establishment.

The kind of wire to purchase should be determined primarily by the class of animals to be confined. Tables 11-11 and 11-12 are suggested guides.

The following additional points are pertinent in the selection of wire:

1. *Styles of woven wire*—The standard styles of woven wire fences are designated by numbers as 958, *1155*, 849, *1047*, 741, *939*, 832, and *726*. (The figures in italic are most common.) The first one or two digits represent the number of line (horizontal) wires; the last two the height in inches; i.e., 1155 has 11 horizontal wires and is 55 inches in height. Each style can be obtained in either (a) 12-inch spacing of stays (or mesh), or (b) 6-inch spacing of stays.

2. *Mesh*—Generally, a close-spaced fence with stay or vertical wires 6 inches apart (6-in. mesh) will give better service than a wide-spaced (12-in. mesh) fence. However, some fence manufacturers believe that 12-inch spacing with No. 9 wire is superior to a 6-inch spacing with No. 11 filler wire. (About the same amount of material is involved in each case.)

3. *Weight of wire*—A fence made of heavier weight wires will usually last longer and prove cheaper than one made of light wires. Heavier or larger size wire is designated by a smaller gauge number. Thus, No. 9 gauge wire is heavier and larger than No. 11 gauge. Woven wire fencing comes in Nos. 9, 11, 12½ and 14½ gauges—which refer to the gauge of the wires other than the top and bottom line wires. Barbed wire is usually 12½ gauge.

Heavier or larger wire than normal should be used in those areas subject to (a) salty air from the ocean, (b) smoke from close proximity industries which may give off chemical fumes into the atmosphere, (c) rapid temperature changes, or (d) overflow or flood.

Heavier or larger wire than normal should be used in fencing (a) small areas, (b) where a dense concentration of animals is involved, and (c) where animals have already learned to get out.

4. *Styles of barbed wire*—Styles of barbed wire differ in the shape and number of the points of the barb, and the spacing of the barbs on the line wires. The 2-point barbs are commonly spaced 4 inches apart while 4-point barbs are generally spaced 5 inches apart. Since any style is satisfactory, selection is a matter of personal preference.

5. *Standard size rolls or spools*—Woven wire comes in 20 and 40 rod rolls; barbed wire in 80 rod spools.

6. *Wire coating*—The kind and amount of coating on wire definitely affects its lasting qualities.

Three kinds of material are commonly used for fence posts: wood, metal, and concrete. The selection of the particular kind of posts should be determined by (a) the availability and cost of each, (b) the length of service desired (posts should last as long as the fencing material attached to it, or the

TABLE 11-11

HANDY WOVEN WIRE FENCE CHART

Kind of Stock	Recommended Woven Wire Height	Recommended Weight of Stay Wire	Recommended Mesh or Spacing Between Stays	Number Recommended Strands of Barbed Wire to Add to Woven Wire	Comments
	(inches)	(gauge)	(inches)		
Cattle	47, 48 or 55	9 or 11	12	1 strand on top (2" to 3" above woven wire), with points 4 or 5 in. apart.	Also satisfactory for all farm animals, except young pigs. Fences for cattle feedlots should be constructed of wood, cable, pipe, or other strong material, and should be 60 in. high.
Sheep	32	11 or 12½	12	2 strands on top.	Sheep fences should total 39 in. in height. 12 in. mesh is best for sheep as they will not get their heads caught if they attempt to reach through. With a heavy concentration of feeder lambs, use woven fence 39 in. high.
Swine	26, 32, or 39	9 or 11	6	1 strand on bottom.	Barbed wire on bottom prevents rooting.
Horses	55 or 58	9 or 11	12	2 strands on top; top strand with points 4 or 5 in. apart.	Also satisfactory for all farm animals, except young pigs.
All farm animals	26 or 32	9 or 11	6	3 strands on top; 1 strand on bottom.	Cyclone wood, pole or other durable and attractive materials are usually used around the headquarters.
	32	9 or 11	6	2 strands on top; 1 strand on bottom.	

TABLE 11-12

HANDY BARBED WIRE FENCE CHART

Kind of Stock	Recommended Number of Points	Recommended Spacing Between Points	Recommended Weight of Strands	Recommended Number of Lines of Barbed Wire to Install	Comments
		(inches)	(gauge)		
Cattle or horses; in farm pastures	2 or 4	4 or 5	12½	4 or 5	2-point barbs are generally 4 in. apart; 4-point are 5 in. apart.
Cattle or horses; on the range	2 or 4	4 or 5	12½	2 or 3	Not all animals will be restrained by 2 or 3 strands.
Sheep	Barbed wire is not considered suitable for sheep because it tears the fleece.				
Swine	2 or 4	4 or 5	12½	6	A 6-strand barbed wire fence for swine may cost more to build and maintain than a woven wire fence.

maintenance cost may be too high), (c) the kind and amount of livestock to be confined, and (d) the cost of installation.

1. *Wood posts*—Osage orange, black locust, chestnut, red cedar, black walnut, mulberry, and catalpa—each with an average life of 15 to 30 years without treatment—are the most durable wood posts, but they are not available in all sections. Untreated posts of the other and less durable woods will last 3 to 8 years only, but they are satisfactory if properly butt treated (to 6 to 8 in. above the ground line) with a good wood preservative.

The proper size of wood posts varies considerably with the strength and durability of the species used. In general, however, large posts last longer than small ones. Satisfactory line posts of osage orange may be as small as 2½ inches in diameter; whereas line posts of other woods should be 4 to 8 inches in diameter at the smaller end. Split posts should be a minimum of 5 inches in diameter. Line posts are generally 7 to 8 feet in length, depending on the height of the fence to be constructed.

Wood corner, end, and gate posts should be substantial, usually not less than 10 to 12 inches in diameter. Also they should be long enough so that they can be set in the ground to a depth of at least 36 inches.

The less durable types of fence posts will last about five times longer when treated than when untreated. This effects yearly savings in two ways: (a) in the cost of posts, and (b) in the labor involved in fence construction.

Although the relative durability of posts does not materially affect initial fencing costs, the length of life of the posts is the greatest single factor in determining the cost of a fence on an annual basis.

Some recommended preservatives are: creosote, pentachlorophenol, zinc chloride, and chromated zinc chloride. Creosote and "penta" should be used only on dry seasoned posts; the others are effective on green posts with the bark left on.

2. *Metal posts*—Metal posts (made of steel or wrought iron) last longer, require less storage space when not in use, and require less labor in setting than wood posts. Also they may give protection against lightning by grounding the current. However, such protection is questionable in dry weather or in areas with a low water table. But they are usually higher in price than wood posts.

Metal line posts are made in different styles and cross sections. Heavier studded "Y" section posts are most popular for livestock, although lighter channel posts may be used for temporary and movable fences. Line posts are available in lengths of 5 to 8 feet in increments of 6 inches. Metal corner, end, and gate posts are commonly made from angle sections, and come in 7 to 9 feet lengths.

3. *Concrete posts*—When properly made, concrete posts give excellent service over many years. In general, however, they are expensive.

Electric Fences

Where a temporary enclosure is desired or where existing fences need bolstering from roguish or breachy animals, it may be desirable to install an electric fence, which can be done at minimum cost.

The following points are pertinent in the construction of an electric fence:

1. *Safety*—If an electric fence is to be installed and used, (a) necessary safety precautions against accidents to both persons and animals should be taken, and (b) the farmer or rancher should first check into the regulations of his own state relative to the installation and use of electric fences. *Remember that an electric fence can be dangerous.* Fence controllers should be purchased from a reliable manufacturer; homemade controllers may be dangerous.

2. *Charger*—The charger should be safe and effective. (Purchase one made by a reputable manufacturer.) There are four types of chargers: (a) *the battery charger*, which uses a 6-volt hot shot battery; (b) *the inductive discharge system*, in which the current is fed to an interrupter device called a circuit breaker or chopper which energizes a current limiting transformer; (c) *the capacitor discharge system*, in which the power line is rectified to direct current and the current is stored in the capacitor; and (d) *the continuous current type*, in which a transformer regulates the flow of current from the powerline to the fence.

3. *Wire height*—As a rule of thumb, the correct wire height for an electric fence is about three-fourths the height of the animal; with two wires provided for sheep and swine. Following are average fence heights above the ground for cattle and calves:

Cattle: 30 to 40 inches
Calves: 12 to 18 inches
Mixed Livestock,
    three wires—8, 12, and 32 inches

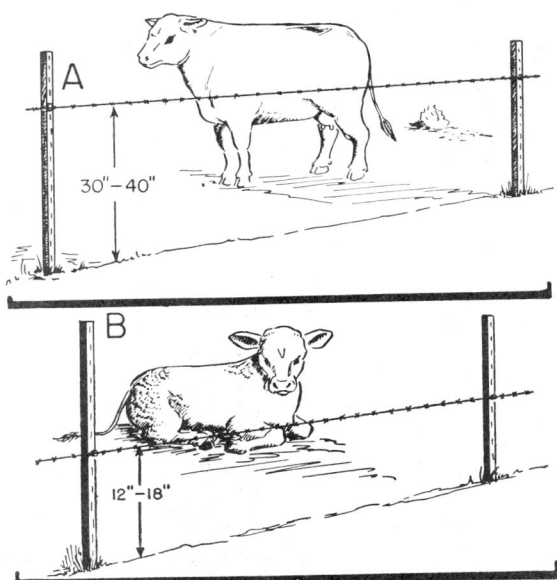

Fig. 11-31. Recommended height for electric fence for: (A), cattle, and (B), calves. (Drawing by R. F. Johnson)

4. *Posts*—Either wood or steel posts may be used for electric fencing. Corner posts should be as firmly set and well braced as required for any nonelectric fence so as to stand the pull necessary to stretch the wire tight. Line posts (a) need only be heavy enough to support the wire and withstand the elements, and (b) may be spaced 40 to 50 feet apart for cattle.

5. *Wire*—In those states where barbed wire is legal, new four point 12½ gauge hog wire is preferred. Barbed wire is recommended, because the barbs will penetrate the hair of animals and touch the skin, but smooth wire can be used satisfactorily. Rusty wire should never be used, because rust is an insulator.

6. *Insulators*—Wire should be fastened to the posts by insulators and should not come into direct contact with posts, weeds, or the ground. Inexpensive solid glass, porcelain, or plastic insulators should be used, rather than old rubber or necks of bottles.

7. *Grounding*—One lead from the controller should be grounded to a pipe driven into the moist earth. *An electric fence should never be grounded to a water pipe, because it could carry lightning directly to connecting buildings.* A lightning arrester should be installed on the ground wire.

Guards or Bumper Gates

Cattle guards or bumper gates set in a fence permit convenient passage of automobiles and trucks but deter cattle, hogs, sheep, and most horses and mules. Where cattle guards are installed, gates should be constructed nearby in order to allow for the movement of animals.

QUESTIONS FOR STUDY AND DISCUSSION

1. Brody, of the Missouri Station, studied the basic building requirements of beef cattle. What was the nature of this classic study, what was found, and what is the practical application of this experiment?

2. Except for extremely cold or hot weather, does it normally pay to provide shelter for cattle?

3. Define environment. In beef cattle, what environmental factors are involved?

4. Define the following terms: total heat, sensible heat, latent heat, Btu, and vapor.

5. The major requirement of winter ventilation is moisture removal, whereas the major requirement of summer ventilation is temperature control. Why the difference?

6. One of the first and frequently one of the most difficult problems confronting the cattleman who wishes to construct a building or item of equipment is that of arriving at the proper size or dimensions. In planning to construct new buildings and equipment for beef cattle, what factors and measurements for buildings and equipment should be considered?

7. Which is more important in beef cattle buildings and equipment—initial cost or laborsaving in operation?

8. Make a critical evaluation of your own beef cattle barn(s), or one with which you are very familiar. Determine its (a) desirable and (b) undesirable features.

9. List and discuss the factors determining the type and size of beef cattle buildings.

10. Make a critical evaluation of your own beef cattle equipment, or of the equipment on a beef cattle establishment with which you are familiar. Determine the (a) desirable and (b) undesirable features of each item.

11. List and discuss the steps in determining the size beef cattle barn to build on any given farm or ranch.

12. List and discuss the steps in determining the size silo to build on any given farm or ranch.

13. List and discuss the needed and desirable features of a modern multiple-use barn such as shown in Figures 11-7 to 11-9.

14. In the selection of woven wire fence, what is meant by the number 1155? What other factors should be considered in the selection of woven wire? What specifications may be used in ordering barbed wire?

15. Under what circumstances would you use electric fences for cattle?

SELECTED REFERENCES

Title of Publication	Author(s)	Publisher
Agricultural Engineers Yearbook	Ed. by R. H. Hahn, Jr.	American Society of Agricultural Engineers, St. Joseph, Mich., 1973
Beef Equipment Plans	Midwest Plan Service	Midwest Plan Service, Iowa State University, Ames, Iowa, 1963
Bibliography of Livestock Waste Management	J. R. Miner D. Bundy G. Christenbury	Office of Research and Monitoring, U.S. Environmental Protection Agency, Washington, D.C., 1972
Facts & Figures for Farmers	Doane Agricultural Service, Inc.	Doane Agricultural Service, Inc., St. Louis, Mo., 1972
Farm Builder's Handbook, Second Edition	R. J. Lytle	Structures Publishing Company, Farmington, Mich., 1973
Farm Building Design	L. W. Neubauer H. B. Walker	Prentice-Hall, Inc., Englewood Cliffs, N.J., 1961
Farm Buildings, Third Edition	D. G. Carter W. A. Foster	John Wiley & Sons, Inc., New York, N.Y., 1947
Farm Buildings, Second Edition	J. C. Wooley	McGraw-Hill Book Company, Inc., New York, N.Y., 1946
Farm Service Buildings	H. E. Gray	McGraw-Hill Book Company, Inc., New York, N.Y., 1955
Farm Structures	H. J. Barre L. L. Sammet	John Wiley & Sons, Inc., New York, N.Y., 1950
Handbook of Livestock Equipment	E. M. Juergenson	The Interstate Printers & Publishers, Inc., Danville, Ill., 1971
Livestock Waste Management System Design Conference for Consulting and SCS Engineers		University of Nebraska Cooperative Extension Service, Lincoln, Neb., 1973
Livestock Waste Management and Pollution Abatement	American Society of Agricultural Engineers	American Society of Agricultural Engineers, St. Joseph, Mich., 1971
Machines for Power Farming, Second Edition	A. A. Stone H. E. Gulvin	John Wiley & Sons, Inc., New York, N.Y., 1967

(Continued)

Title of Publication	Author(s)	Publisher
Practical Farm Buildings	J. S. Boyd	The Interstate Printers & Publishers, Inc., Danville, Ill., 1973
Principles of Animal Environment	M. L. Esmay	The Avi Publishing Company, Inc., Westport, Conn., 1969
The Stockman's Handbook, Fourth Edition	M. E. Ensminger	The Interstate Printers & Publishers, Inc., Danville, Ill., 1970
Structures and Environment Handbook	Midwest Plan Service	Midwest Plan Service, Iowa State University, Ames, Iowa, 1972

* * * * *

Plans and specifications for beef cattle buildings and equipment can also be obtained from the local county agricultural agent, your state college of agriculture, and materials and equipment manufacturers and dealers.

CHAPTER 12

BEEF CATTLE HEALTH, DISEASE PREVENTION, AND PARASITE CONTROL[1]

By
Dr. Robert F. Behlow, DVM
Professor and Extension Veterinarian
North Carolina State University
Raleigh, North Carolina
and
Dr. M. E. Ensminger, Ph.D.
Distinguished Professor, University of Wisconsin-River Falls
and
Adjunct Professor, California State University-Fresno

Contents

(Continued)

---

[1]The authors are indebted to a number of veterinarians and other specialists who reviewed the portions of this chapter indicated. It is emphasized, however, that the review accorded by these eminent authorities does not constitute either full approval or full agreement of the reviewers and the authors on the contents of this chapter.

The material presented in this chapter is based on factual information believed to be accurate, but it is not guaranteed. When the instruction and precautions given herein are in disagreement with those of competent local authorities or manufacturers, always follow the latter two.

688   BEEF CATTLE SCIENCE

Contents   **Page**

Without doubt, one of the most serious menaces threatening the livestock industry is animal ill-health, of which the largest loss is a result of the diseases that are due to a common factor transmitted from animal to animal.

Fig. 12-1. The cow doctor. From a woodcut, 1875. (Courtesy, The Bettmann Archive, Inc., New York, N.Y.)

Today, with modern rapid transportation facilities and the dense livestock population centers, the opportunities for animals to become infected are greatly increased compared with a generation ago.

Each year, cattlemen suffer staggering losses from diseases and parasites—internal and external. Death takes a tremendous toll. Even greater economic losses—hidden losses—result from failure to reproduce living young, and from losses due to retarded growth and poor feed efficiency, carcass condemnations and decreases in meat quality, and labor and drug costs. Also, considerable cost is involved in keeping out diseases that do not exist in the United States; and quarantine of a diseased area may cause depreciation of land values or even restrict whole agricultural programs. It has been conservatively estimated that annual U.S. losses from the more important diseases, parasites, and pests of livestock and poultry aggregate $4.1 billion.[2]

---

[2]*Losses in Agriculture*, Agriculture Handbook No. 291, 1965, pp. 73-82, Tables 26-32, reported average annual losses of $2.8 billion for the period 1951-60. The author updated the latter figure by applying 1974 livestock and poultry numbers ($\times$ 1.13) and prices ($\times$ 1.3) to give the figure of $4.1 billion. Estimated losses from certain specific diseases and parasites have been updated in like manner in this chapter. In doing so, it is recognized that the incidence of specific diseases and parasites may have changed since the 1951-60 study; for example, a new and effective vaccine or wormer may have evolved subsequently. Nevertheless, these figures will serve to underscore the staggering losses from diseases and parasites.

Thus, the potential for increasing beef cattle profits through disease prevention is great.

Cattlemen should also be well informed relative to the relationship of cattle diseases and parasites to other classes of animals and to man, because many of them are transmissible between species. For example, approximately 200 different types of infectious and parasitic diseases can be transmitted from animals to human beings.[3] Of most concern in the latter respect are such animal diseases as brucellosis (undulant fever), leptospirosis, anthrax, Q fever, rabies, trichinosis, tuberculosis, and tularemia. Thus, rigid meat and milk inspection is necessary for the protection of human health. This is an added expense which the producer, processor, and consumer must share.

Fortunately for the producer, beef animals are out in the open much of the time, with a minimum of confinement in shelters and small enclosures and close contact with each other. On the other hand, the very fact that beef herds are usually inspected infrequently may result in serious and widespread losses before the ravages of diseases or parasites are observed and control measures instituted.

NORMAL TEMPERATURE, PULSE RATE, AND RESPIRATION RATE OF FARM ANIMALS

Table 12-1 gives the normal temperature, pulse rate, and breathing rate of farm animals. In general, any marked and persistent deviations from these normals may be looked upon as a sign of animal ill health.

Every stockman should provide himself with an animal thermometer, which is heavier and more rugged than the ordinary human thermometer. The temperature is measured by inserting the thermometer full length in the rectum, where it should be left a minimum of three minutes. Prior to inserting the thermometer, a long string should be tied to the end. Where a large number of animals is involved, a more rapid electronic thermometer may be used.

TABLE 12-1

NORMAL TEMPERATURE, PULSE RATE, AND RESPIRATION RATE
OF FARM ANIMALS

| Animal | Normal Rectal Temperature | | Normal Pulse Rate | Normal Respiration Rate |
	Average	Range		
	(degrees F)	(degrees F)	(rate/minute)	(rate/minute)
Cattle	101.5	100.4-102.8	60-70	10-30
Sheep	102.3	100.9-103.8	70-80	12-20
Goats	103.8	101.7-105.3	70-80	12-20
Swine	102.6	102 -103.6	60-80	8-13
Horses	100.5	99 -100.8	32-44	8-16

[3]Hull, Thomas G., *Diseases Transmitted from Animals to Man*, Charles C Thomas, Publisher, Springfield, Ill., 1963, Table 84, p. 915.

In general, infectious diseases are ushered in with a rise in body temperature, but it must be remembered that body temperature is affected by stable or outside temperature, exercise, excitement, age, feed, etc. It is lower in cold weather, in older animals and at night.

The pulse rate indicates the rapidity of the heart action. The pulse of cattle is taken either on the outside of the jaw just above its lower border, on the soft area immediately above the inner dewclaw, or just above the hock joint. It should be pointed out that the younger, the smaller, and the more nervous the animal, the higher the pulse rate. Also, the pulse rate increases with exercise, excitement, digestion, and high outside temperature.

The respiration rate can be determined by placing the hand on the flank, by observing the rise and fall of the flanks, or, in the winter, by watching the breath condensate in coming from the nostrils. Rapid breathing due to recent exercise, excitement, hot weather, or poorly ventilated buildings should not be confused with disease. Respiration is accelerated in pain and in febrile conditions.

A PROGRAM OF BEEF CATTLE HEALTH, DISEASE PREVENTION, AND PARASITE CONTROL

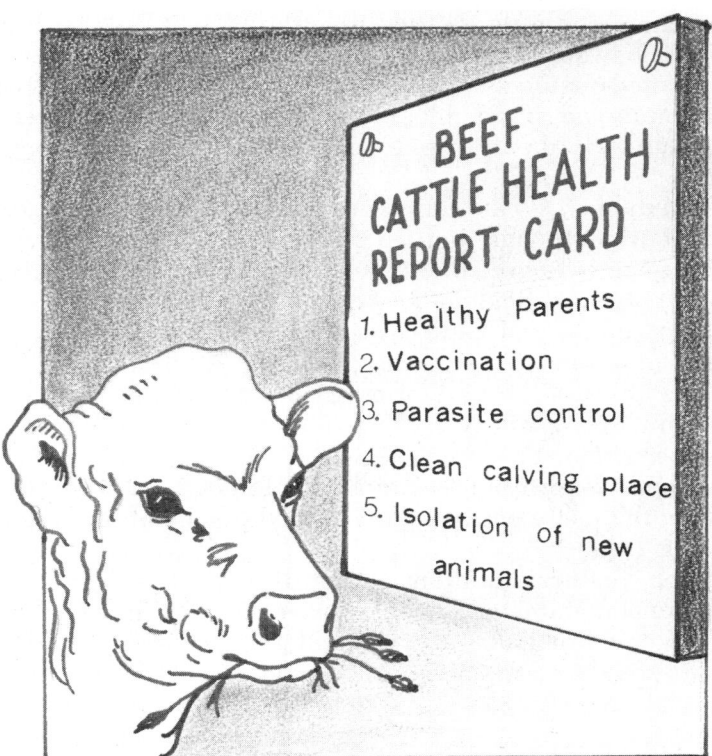

Fig. 12-2. Beef cattle health, disease prevention, and parasite control.

Although the exact program will and should vary according to the specific conditions existing on each individual farm or ranch, the basic principles will remain the same. With this thought in mind, the following program of beef cattle health, disease prevention, and parasite control is presented with the hope that the beef cattle producer will use it (1) as a yardstick with which to compare his existing program, and (2) as a guidepost so that he and his local veterinarian, and other advisors, may develop a similar and specific program for his own enterprise:

1. *General Beef Cattle Program*

    a. Breed only healthy cows to healthy bulls.

    b. Avoid either an overfat or a thin, emaciated condition in all breeding animals.

    c. Provide plenty of exercise for bulls and pregnant cows, preferably by allowing them to graze in well-fenced pastures in which plenty of shade and water are available.

    d. Keep lots and corrals well drained and as dry as practical to prevent breeding places for foot rot, other diseases, and parasites. Fence cattle out of pasture mudholes for the same reason.

    e. If possible, divert drainage from adjacent infected premises and avoid across-the-fence contact with the neighbors' cattle unless they are definitely disease free. Do not visit farms where infectious diseases exist, as the germs may be brought home on shoes, clothing, or vehicles. For the same reason, feeds should not be bought from such farms, and one should beware of used feed bags.

    f. If rented pastures must be used, avoid areas on which cattle have overwintered; and, preferably, use only those rented pastures that have not had cattle on them for one year or that have been plowed in the interim.

    g. Test the entire herd for tuberculosis and brucellosis each fall or at the time they are brought in from pasture and placed in winter quarters.

    h. Eliminate the breeding ground of parasites as far as practical and use the proper insecticide or anthelmintic for their control.

    i. Keep commercial cattle—such as stocker and feeder cattle, and finishing cattle—in isolated areas away from breeding animals.

    j. Have all cows checked for pregnancy 45 to 60 days after the breeding season is over. When problems are encountered, immediately consult a veterinarian.

    k. When disease troubles are encountered, isolate affected animals and follow the instructions and prescribed treatment of a veterinarian.

2. *Calving Time*

    a. When weather conditions permit, allow parturient cows to calve in a clean, uncontaminated, open pasture. During inclement weather, place the cows in isolated, roomy, light, well-ventilated maternity stalls—which should first be carefully cleaned, thoroughly disinfected, and provided with clean bedding for the occasion. After calving, all wet, stained, or soiled bedding should be removed and the floor sprinkled with lime; the afterbirth should be burned or buried deep in lime; and, if there has been trouble, the cows should be kept isolated until all discharges have ceased.

b. Unless the calves are born on a clean pasture away from possible infection, treat the navel cord of each newborn animal with tincture of iodine.

c. See that the newborn calf gets colostrum milk as soon as possible. But bear in mind that the antibodies of colostrum depend upon the dam's disease history, either directly or through vaccinations.

3. *Suckling Calves*

a. If the baby calves are confined to stalls, scrub stalls thoroughly twice each week with warm soap solution and disinfect the walls and feed bunks and/or mangers.

b. Vaccinate calves with blackleg and malignant edema bacterin at 2 to 3 months of age (at 1 month of age in endemic areas).

4. *New Stock*

a. Vaccinate calves against blackleg and malignant edema in areas that are endemic for these diseases.

b. Isolate newly acquired animals for a minimum of three weeks, during which time they should be cared for by a separate caretaker.

c. While in isolation, test all newly acquired breeding animals for tuberculosis, brucellosis, leptospirosis, anaplasmosis, vibriosis, and Johne's disease; first, however, make every reasonable effort to ascertain that they come from herds which are known to be free from these and other diseases.

d. Spray newly acquired animals for lice control; and check them for internal parasites, and treat where indicated.

e. When possible, it is preferable to purchase virgin heifers and bulls, from a disease control standpoint. Isolate "tried" (nonvirgin) bulls for a period of 3 weeks, and then turn them with a limited number of virgin heifers; observe these heifers for 30 to 60 days after breeding, as an aid in preventing the introduction of breeding diseases.

f. Thoroughly clean and disinfect the isolation stall after each animal(s) is removed and before a new animal(s) is placed therein.

A CATTLE FEEDLOT DISEASE AND PARASITE CONTROL PROGRAM

Cattle feedlot health is fully covered in Chapter 33 of this book; hence, the reader is referred thereto.

DISEASES OF BEEF CATTLE[4]

Disease may be defined as an illness, sickness, or any deviation from a state of health.

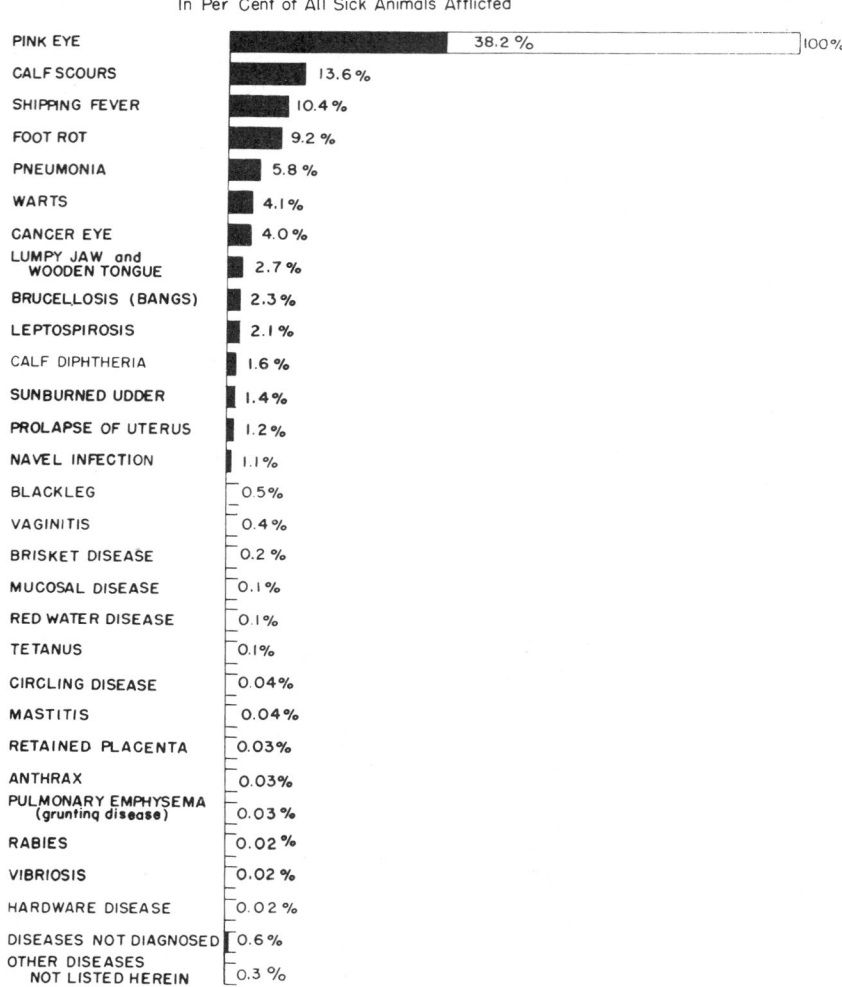

Fig. 12-3. Relative frequency of nonnutritional beef cattle diseases and ailments. This figure, from *Wash. Agr. Exp. Sta. Bull. 562*, p. 6, is based on a 24-state survey made by the author of this book, which covered over a half million cattle. It is recognized that cattlemen may not always have been accurate in their diagnosis; nevertheless, this is the most complete record of its type available. The 5 most common beef cattle diseases, in order of frequency—pinkeye, calf scours, shipping fever, foot rot, and pneumonia—accounted for 77.2% of all sick cattle (nonnutritional). During the course of a year, 8% of the nation's cattle were afflicted by one or more nonnutritional diseases, and 0.59% of all cattle died from such diseases. By rank, 4 diseases—pneumonia, calf scours, shipping fever, and blackleg—accounted for 60% of all death losses from nonnutritional diseases.

―――――――――
[4](See footnote on next page.)

Seventy-five percent of the losses from beef cattle diseases are due to common, well-known diseases. Modern science makes it possible effectively to prevent many of these diseases through a combination of good management, proper nutrition, vaccination, and other disease prevention practices.

It is not intended that this book shall serve as a source of home remedies. Rather, the enlightened cattleman will institute a program designed to assure herd health, disease prevention, and parasite control. When animal disease troubles are encountered, he will not attempt to diagnose or treat but will call upon his local veterinarian in exactly the same manner as he calls upon the family doctor when human ill health is encountered.

This chapter is limited to nonnutritional diseases and ailments; the nutritional diseases and ailments of cattle are covered in Table 8-17, Chapter 8.

Anthrax (Charbon, or Splenic Fever)

Anthrax, also referred to as splenic fever or charbon, is an acute, infectious disease affecting all warm-blooded animals and man; but mature cattle are most susceptible. It usually occurs as scattered outbreaks or cases, but hundreds of animals may be involved. Certain sections are known as anthrax districts because of the repeated appearance of the disease. Grazing animals are particularly subject to anthrax, especially when pasturing closely following a drought or on land that has been recently flooded. In the United States, most human infections of anthrax result from handling diseased or dead animals on the farm or from handling hides, hair, and wool in factories.

Historically, anthrax is of great importance. It was one of the first scourges to be described in ancient and Biblical literature; it marked the beginning of modern bacteriology, being described by Koch in 1876; and it was the first disease in which immunization was effected by means of an attentuated culture, Pasteur having immunized animals against anthrax in 1881. Anthrax causes estimated average annual beef cattle losses of $31,122.[5]

SYMPTOMS AND SIGNS[6]

The mortality is usually quite high. It runs a very short course and is characterized by a blood poisoning (septicemia). The first indication of the disease may be the presence of severe symptoms of colic accompanied by high temperature, loss of appetite, muscular weakness, depression, and the passage of blood-stained feces. Swellings may be observed over the body,

---

[4]This entire section on diseases was reviewed by the following eminent veterinarians: Dr. R. E. Watts, DVM, Professor, Veterinary Science, The University of Arizona, Tucson, Ariz.; Dr. Charles S. Crane, DVM, Porterville, Calif.; Mr. Cecil F. Metzger, Manager, Special Products, Norden Laboratories, Lincoln, Neb.; Dr. William T. Davis, Director, Animal Health Clinical Research, E. R. Squibb & Sons, Three Bridges, N. J.; and Dr. John B. Parks, DVM, Director of Biological Research, Diamond Laboratories, Des Moines, Iowa.

[5]*Losses in Agriculture*, Agriculture Handbook No. 291, 1965, p. 73, reported average annual losses of $19,000 for the period 1951-60. The author updated the latter figure by applying 1974 cattle numbers ($\times$ 1.17) and prices ($\times$ 1.4), to give the figure of $31,122.

[6]Currently, many veterinarians prefer the word "signs" rather than "symptoms," but throughout this chapter the author accedes to the more commonly accepted terminology among cattlemen and includes the word "symptoms."

especially around the neck region. Milk secretion may turn bloody or cease entirely; and there may be a bloody discharge from all body openings.

CAUSE, PREVENTION, AND TREATMENT

The disease is identified by a microscopic examination of the blood in which will be found the typical large, rod-shaped organisms (*Bacillus anthracus*) causing anthrax. The bacillus can survive for years in a spore stage, resisting most destructive agents. As a result, it may remain in the soil for extremely long periods.

This disease is one that can largely be prevented by immunization. In the so-called anthrax regions, vaccination should be performed annually, usually in the spring, and well in advance of the time when the disease normally makes its appearance. At least nine types of biologics (serums, bacterins, and vaccines) are now available for use in anthrax prevention. The choice of the biologic is dependent upon the local situation and should be left to the veterinarian or state livestock sanitary officials. Also, in the infected areas, adequate fly control should be obtained by spraying animals during the insect season.

Herds that are infected should be quarantined, and all milk and other products should be withheld from the market until the danger of disease transmission is past. The farmer or rancher should never open the carcass of a dead animal suspected of having died from anthrax, because the organism is infectious to man; instead, the veterinarian should be summoned at the first sign of an outbreak.

When the presence of anthrax is suspected or proved, all carcasses and contaminated material should be completely burned or deeply buried and covered with quicklime, preferably on the spot. This precaution is important because the disease can be spread by dogs, coyotes, buzzards, and other flesh eaters, and by flies and other insects.

When an outbreak of anthrax is discovered, all sick animals should be isolated promptly and treated. All exposed healthy animals should be vaccinated, pastures should be rotated, and a rigid program of sanitation should be initiated. Anthrax is a reportable disease, requiring quarantine. Hence, control measures will be carried out under the supervision of state or Federal regulatory officials.

Early treatment of affected animals, with massive doses of penicillin, may be effective if given soon enough.

Bacillary Hemoglobinuria (or Red Water Disease)

Bacillary hemoglobinuria, which is an acute, infectious disease, is often confused with other cattle diseases in which blood-colored urine is seen. The disease usually occurs in cattle that are pastured on meadows or irrigated lands where drainage is poor, and during the summer and early fall months. Sheep are also affected, but to a lesser degree than cattle. A mortality rate up to 100 percent occurs in untreated cases.

SYMPTOMS AND SIGNS

All ages are affected, but most losses occur in cows over one year of age.

The course of the disease is usually two days or less. Appetite, rumination, and milk flow suddenly cease. The animal hesitates to move and stands apart from the herd. The eyes are sunken, bloodshot, and may appear yellow. Breathing is rapid, and the temperature is high. The urine and feces are usually both blood tinged from destroyed red cells. It must be understood, however, that bloody urine (red water) may also be one of the symptoms in such conditions and diseases as lack of phosphorus, leptospirosis, Texas fever, plant poisoning, and anthrax. As red water disease can easily be confused with other conditions producing bloody urine, laboratory assistance is usually indicated in the event of unknown hemoglobinuria (bloody urine).

CAUSE, PREVENTION, AND TREATMENT

An anaerobic bacterium called *Clostridium hemolyticum* is the primary causative agent, and its toxin causes the blood breakdown. It is found in moist alkaline soils, especially in the low lying valley land of the Sierra Nevada, and the Coast and Cascade ranges in Nevada, California, and Oregon; and in Washington, Louisiana, Florida, Montana, Idaho, and Texas.

Inoculations with a bacterin (inactivated bacteria) or toxoid (inactivated toxin) to stimulate immunity are valuable in communities where annual losses occur. This vaccination should occur two weeks prior to the time of the previous annual outbreak. Unless unavoidable, cattle should not be pastured on areas of known infection. Pools of stagnant water should be drained as such areas provide a favorable environment for the growth of the causative agent.

Antiserum from hyper-immunized animals and blood transfusions, in conjunction with antibiotics, is the only known treatment for animals showing symptoms of the disease, but suitable medical treatment in the way of stimulants may be a valuable adjunct. Since the disease may be confused with shipping fever, leptospirosis, and plant poisoning, it is essential that a veterinarian make the diagnosis.

Blackleg (Black Quarter, Emphysematous Gangrene, Quarter-Ill, or Symptomatic Anthrax)

This is a very infectious, highly fatal disease of cattle, and less frequently of sheep and goats. The disease is widespread, especially in the western range states. It occurs at almost any season, predominating in the spring and fall months among pastured cattle; but it may occur in the winter in stabled cattle. Once prevalent in a community, the disease remains there as a permanent hazard, the infected territory being referred to as a "hot area." It is seen most frequently in cattle ranging in age from three months to two years, but it may occur in older animals.

SYMPTOMS AND SIGNS

The incubation period is from 1 to 5 days, and its course is from 1 to 3 days. The first symptom noted is lameness, usually accompanied by or followed by swellings of gas under the skin in the areas of the neck, shoulder, flanks, thighs, and breast, which crackle under pressure. High fever, loss of appetite, and severe depression accompany the symptoms. Although there are a few recoveries, death is the usual termination, occurring within three days of the onset of symptoms.

Fig. 12-4. Heifer with blackleg, six hours before death. Note the lameness and the swelling over the neck and shoulder. (Courtesy, Veterinary Research Laboratory, Montana State University)

CAUSE, PREVENTION, AND TREATMENT

This disease is caused by an anaerobic bacterium, called *Clostridium chauvoei*, although it is often accompanied by other of the *Clostridia* genus. Infection is usually the result of wound contamination or ingestion of the organisms.

Prevention consists of vaccination (currently, most vaccines contain both blackleg and malignant edema) of all animals at approximately 3 to 4 months of age, followed by a second, or booster, vaccination at about 12 months of age, using one of the approved vaccines. In endemic areas, the first vaccination should be given at one month of age. A natural resistance tends to develop when the animal is about 2 years of age.

Animals that die from blackleg should not be cut open unless under the direction of a qualified veterinarian. The carcasses should be burned or deeply buried and the contaminated area disinfected. Eradication of blackleg from pastures is difficult if not impossible.

In the early stages of the disease, massive doses of antibiotics will sometimes save an animal. But a good immunization program is the key to preventing losses due to blackleg.

Bovine Pulmonary Emphysema

This disease is also known as cow asthma, panters, lungers, fog fever, skyline fever, summer pneumonia, green grass poisoning, and grunts.

Bovine pulmonary emphysema usually occurs when cattle are abruptly changed from dry, mature feed to green, immature pasture. Typically, cases develop when cattle are moved from dry range to green mountain or irrigated pastures in the fall or late summer. It may also occur when cattle are first turned to lush pasture in the spring.

SYMPTOMS AND SIGNS

The disease is characterized by rapid and labored breathing (the animal forces air from the lungs, and many grunt with each breath). Affected animals may breathe through and froth at the mouth. The temperature remains normal or only slightly elevated, and the appetite is good.

CAUSE, PREVENTION, AND TREATMENT

The cause is unknown. It is noteworthy, however, that workers at Washington State University have produced the condition experimentally by giving cattle the essential amino acid, tryptophan, leading to speculation that rapidly growing pastures may be high in tryptophan.

Prevention consists in avoiding sudden changes from dry or poor pasture to immature, green feed. Some dry hay should be fed while making the transition.

When bovine pulmonary emphysema strikes, call a veterinarian. Atropine and antihistamines are useful in treating the disease.

Bovine Virus Diarrhea (BVD, or Mucosal Disease)

Bovine virus diarrhea is not new, having first been described in 1946; but, in different periods of time and areas, it has been known as New York Virus Diarrhea, Indiana Virus Diarrhea, and Mucosal Disease. Improved methods of diagnosis and laboratory techniques now indicate that most, if not all, of the bovine virus diarrhea-mucosal disease outbreaks are caused by the same organism.

The disease is widespread in the United States. The greatest losses are in weight, condition, and feed. Mortality is low, rarely exceeding five percent.

SYMPTOMS AND SIGNS

The incubation period is 7 to 9 days following exposure to the virus. The disease is characterized by high temperature (104° to 107° F) for 2 to 5 days, nasal discharge, rapid breathing, depression, and loss of appetite. Some animals make a prompt recovery. In other cases, signs persist, including nasal discharge and diarrhea. Sometimes blood flecks occur in the feces. Coughing, eye lesions, and lameness may affect 10 percent of the herd. In pregnant cows, abortions generally appear 3 to 6 weeks after infection; and, in lactating cows, a marked loss in milk production occurs.

CAUSE, PREVENTION, AND TREATMENT

As indicated by the name, the disease is caused by a virus.

The most effective preventive measures consist in avoiding contact with affected animals and in keeping away from contaminated feed and water. Also, all incoming animals should be isolated for at least 30 days. Once the disease makes its appearance, sick animals should be isolated and rigid sanitary measures should be initiated.

Where virus diarrhea is a constant problem, cows and feedlot cattle should be vaccinated. The BVD vaccine, which is a modified live virus vaccine, is one of the most effective cattle vaccines available. But two "don'ts" should be observed: (1) don't use the vaccine on pregnant cows because of possible abortions and birth defects; and (2) don't vaccinate calves under six months of age because it may be ineffective due to the temporary immunity from colostrum of immune dams. One vaccination should last a lifetime.

Antibiotics or sulfonamides effectively combat the secondary bacterial invaders that accompany the disease.

Brucellosis (Bang's Disease)

Brucellosis, which occurs throughout the world, is an insidious (hidden) disease in which the lesions frequently are not evident. Although the medical term "brucellosis" is used in a collective way to designate the disease caused by the three different but closely related *Brucella* organisms, the specie names further differentiate the organisms as: (1) *Br. abortus;* (2) *Br. suis;* and (3) *Br. melitensis.*

The disease is known as brucellosis, Bang's disease (after Professor Bang, noted Danish research worker, who, in 1896, first discovered the organism responsible for bovine brucellosis), or contagious abortion in cattle, caused by *Brucella abortus.* In swine, it's Traum's disease, or infectious abortion, caused by *Brucella suis.* In goats, it's Malta fever, or abortion, caused by *Brucella melitensis.* In man, it's Mediterranean fever, or undulant fever. The causative organism is also associated with fistuous withers and poll-evil in horses.

Brucellosis is one of the most serious and widespread diseases affecting the livestock industry. For the nation as a whole, fewer than 1.0 percent (0.7 percent in 1972) of all cattle tested (including both beef and dairy animals) react. Control and eradication of the disease are important for two reasons:

(1) the danger of human infection, it being one of the most important U.S. animal-human diseases; and (2) the economic cattle loss in the form of fewer live calves, more retained placentas, more breeding trouble, more arthritis, more mastitis, and lowered milk production—losses totaling $75,926,214 annually.[7] A number of biological tests are used in the diagnosis of brucellosis in cattle.

The blood (agglutination) test is a safe, reliable (though not perfect), and practical method for the diagnosis of brucellosis in all farm animals. Either the tube test method, the rapid (stained antigen) plate method, or the card test is satisfactory when conducted by an experienced technician.

There is nothing mysterious about the blood test. It is simply based on the phenomenon that the bloodstream of an infected animal contains an antibody, known as agglutinin. When blood serum containing this substance is brought in contact with a suspension of *Brucella* organisms (called an antigen), it causes the organisms to adhere to one another and form clumps. This action, known as agglutination, constitutes a simple test for diagnosing brucellosis in the living animal.

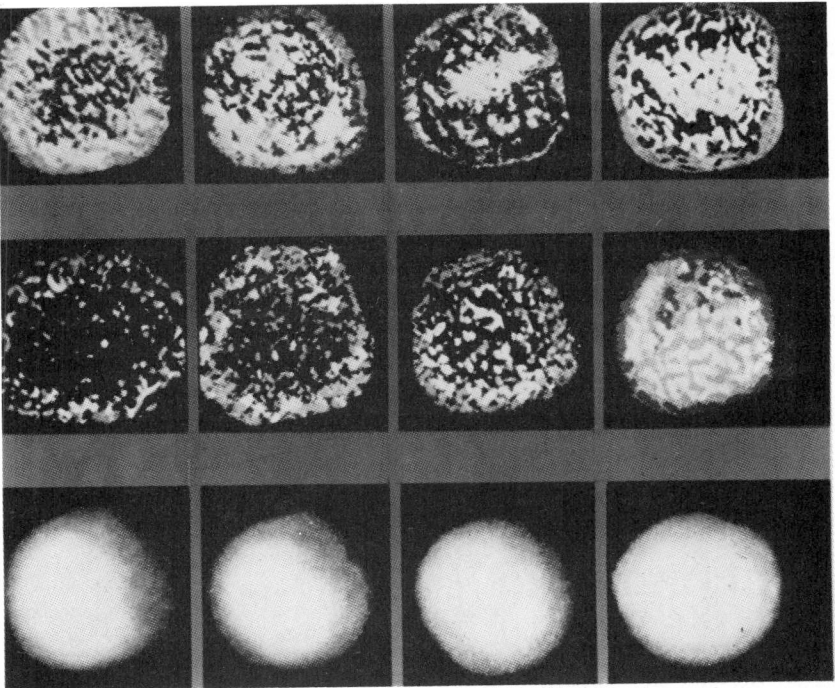

Fig. 12-5. Microscopic picture showing the blood (agglutination) test for diagnosis of brucellosis. Top row shows clumping (agglutination), indicating brucellosis. Center row shows complete clumping in the first 3 dilutions and partial clumping in the 1:200 dilution. Bottom row shows a negative test, indicating brucellosis free. (Courtesy, Lederle Laboratories)

[7]*Losses in Agriculture*, Agriculture Handbook No. 291, 1965, p. 73, reported average annual losses of $46,353,000 for the period 1951-60. The author updated the latter figure by applying the 1974 cattle numbers (× 1.17) and prices (× 1.4), to give the figure of $75,926,214.

In 1966, the Brucellosis Committee of the United States Livestock Sanitary Association recommended the Brucellosis Card Test for brucellosis testing. It detects infected animals (as verified by isolation of *Brucella abortus*) that may be missed by the plate and tube agglutination tests. Also, it detects infected animals earlier, thereby making it possible to clear up an infected herd with fewer tests; it does not show nearly so many false reactions caused by nonspecific agglutinins as do the older tests; and it is more likely to show Strain 19 vaccinated animals as negative (unless an actual infection is present), rather than as reactors or suspects as the older tests frequently do.

The milk fat or ring test is a satisfactory method of detecting brucellosis-infected dairy herds.

SYMPTOMS AND SIGNS

Unfortunately, the symptoms of brucellosis are often rather indefinite. While abortion is the most readily observed symptom in cows, it should be borne in mind that not all animals that abort are affected with brucellosis and that not all animals affected with brucellosis will necessarily abort. On the other hand, every case of abortion should be regarded with suspicion until proved noninfectious.

Fig. 12-6. Cow with aborted fetus. Every case of abortion should be regarded with suspicion until proved noninfectious. (Courtesy, USDA)

The infected animal may prematurely give birth to a dead fetus, usually during the last third of pregnancy. On the other hand, the birth may be entirely normal; but the calf may be weak, or there may be retention of the afterbirth, inflammation of the uterus, and/or difficulty in future conception. The milk production is usually reduced. There may be abscess formation in the testicles of the male and swelling of the joints (arthritis). The observed symptoms in man include weakness, joint pains, undulating (varying) fever, and occasionally orchitis (inflammation of the testes).

CAUSE, PREVENTION, AND TREATMENT

The disease is caused by a bacteria called *Brucella abortus* in cattle, *Brucella suis* in swine, and *Brucella melitensis* in goats. The suis and melitensis types are seen in cattle, but the incidence is rare; swine are infected with both the suis and melitensis types; and horses may become infected with all three types.

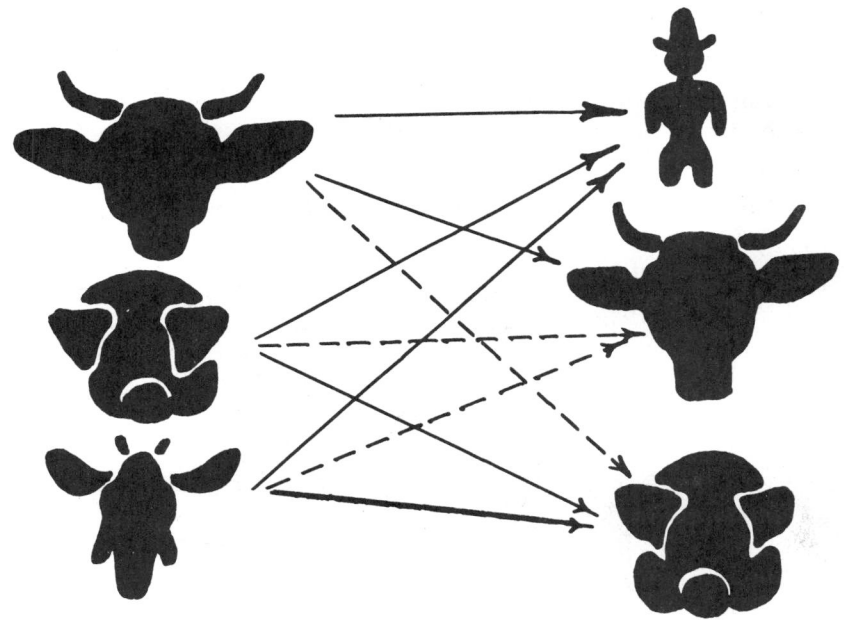

SOURCES OF INFECTION
Dotted lines indicate sometimes a source.

Fig. 12-7. Sources of brucellosis infection. (Drawing by R. F. Johnson)

Man is susceptible to all three species of brucellosis. In most areas, the vast majority of undulant fever cases in man are due to *Brucella suis*. The swine organism causes a more severe disease in human beings than the cattle organism, although not so severe as that induced by the goat type (*Brucella melitensis*). Fortunately, far fewer people are exposed to the latter, simply because of the limited number of goats and the rarity of the disease in goats in the United States. Stockmen are aware of the possibility that human beings may contact undulant fever from handling affected animals, especially at the time of parturition; from slaughtering operations or handling raw meats from affected animals; or from consuming raw milk or other raw by-products from cows or goats, and eating uncooked meats infected with brucellosis organisms. The simple precautions of pasteurizing milk and cooking meat, however, make these foods safe for human consumption.

The *Brucella* organism is quite resistant to drying but is killed by the common disinfectants and by pasteurization. It is found in immense numbers in the various tissues of the aborted young and in the discharges and membranes from the aborted animal. It is harbored indefinitely in the udder and may also be found in the sex glands, spleen, liver, kidneys, bloodstream, joints, and lymph nodes.

Brucellosis appears to be commonly acquired through the mouth in feed and water contaminated with the bacteria, or by licking infected animals, contaminated feeders, or other objects to which the bacteria may adhere. Venereal transmission by infected bulls to susceptible cows through natural service may occur, but it is rare.

ROUTE OF BRUCELLOSIS GERMS IN THEIR ATTACK ON CATTLE

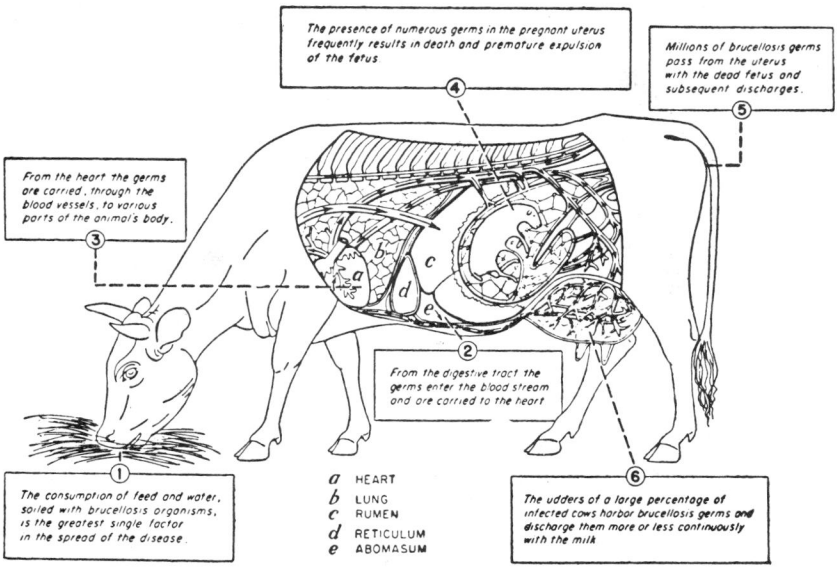

Fig. 12-8. Diagram showing how cattle become infected with the *Brucella* organism and its route of attack. (Courtesy, USDA)

Freedom from disease should be the goal of all control programs. An annual blood test (and more frequently if the disease is encountered), the removal of infected animals, strict sanitation, proper and liberal use of disinfectants, isolation at the time of parturition, and the control of animals, feed, and water brought into the premises are the key to the successful control or eradication of brucellosis.

Sound management practices, which include either buying replacement animals that are free of the disease or raising all females, are a necessary adjunct in prevention. Drainage from infected areas should be diverted or fenced off, and visitors (man and animal) should be kept away from animal barns and feed lots. Feeds should not be bought from farms that have infected animals, and one should beware of used feed bags. Animals taken to livestock shows and fairs should be isolated on their return and tested 30 days later.

In 1934, a brucellosis control and eradication program was initiated in connection with the cattle-reduction program necessitated by the drought of that year. This program has continued to operate, with provision for the slaughter of animals that react positively to the test. Under the slaughter provision, several states pay partial indemnity to farmers whose animals are condemned under the program. During the 40-year period, 1934-1974, $700 million in tax funds were spent on brucellosis control.

Brucellosis cannot be eradicated by the use of Strain 19 alone, because (1) it is a live vaccine, and (2) it is only 65 to 70 percent effective under normal field conditions. Nevertheless, in problem herds and problem areas, it is recommended that dairy heifer calves be vaccinated with Strain 19 at 2 through 6 months of age, and that beef heifers be vaccinated at 2 to 10 months of age. In 1972, 4.3 million calves were vaccinated for brucellosis in the United States.

Although great progress has been made, and although 99 percent of the nation's cattle are free of brucellosis, it is disturbing to know that 23,588 infected herds, or a total of 97,585 head of cattle, were found on farms and ranches in 1972 (see Fig. 12-9).

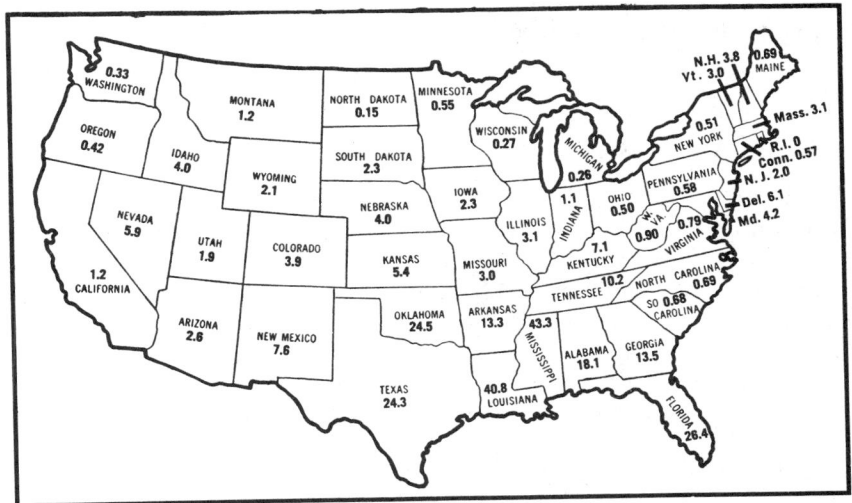

Fig. 12-9. Rate of brucellosis infection by states for the period July 1, 1972, to June 30, 1973. The figure in each state shows the infection rate per 1,000 herds. Thus, the 40.8 in Louisiana means that there were 40.8 infected herds per 1,000 herds, or that 4.08 percent of the herds were infected. Note that infection is highest in the South and Southeast. (Courtesy, USDA)

Slaughter or quarantine of infected cattle, together with rigid sanitation, must be a part of any successful eradication program once the disease has made its appearance.

To date, there is no known medicinal agent that is completely effective in the treatment of brucellosis in any class of farm animals. Thus, the farmer and rancher should not waste valuable time and money on so-called cures that are advocated by fraudulent operators.

The Federal brucellosis regulations were completely revised in 1975, with the program detailed in a report entitled *Brucellosis Eradication, Recommended Uniform Methods and Rules (APHIS 91-1)*, which may be obtained from the Animal Plant Health Inspection Service, U.S. Department of Agriculture, Hyattsville, Maryland 20782. Also, the Federal regulations with respect to brucellosis were amended in Title 9 of the Code of Federal Regulations in April 1975, the principle provisions of which are:

Part 78. (Brucellosis) requires that reactors be identified and moved for slaughter; exposed cattle must be identified and may be moved for slaughter or to a quarantined feedlot. Other cattle must be tested, or vaccinated, and so certified, depending on the status of the state and herd of origin.

(Also see later section in this chapter entitled, "Federal and State Regulations Relative to Disease Control.")

Calf Diphtheria

Calf diphtheria is an acute, infectious disease of housed suckling calves and young feedlot cattle. The disease sometimes attacks these young animals as early as the third or fourth day after birth. If untreated, the mortality rate is very high. The disease is not to be confused with the sore mouth virus of sheep or the diphtheria of humans, with which it has no relationship.

SYMPTOMS AND SIGNS

The affected animal shows difficulty in breathing, eating, and drinking. Drooling and swallowing movements may also be noted. Inspection of the mouth reveals yellowish crumbling masses and patches of dead tissue (diphtheritic membranes) on the borders of the tongue, adjacent to the molar teeth, and in the throat. Once established and unchecked, it will spread rapidly, eventually causing the death of the animal.

CAUSE, PREVENTION, AND TREATMENT

The alleged cause of this malady is the soil organism *Spherophorus necrophorus*, the same organism that is often found in foot rot. The organism gains entrance to the tissues through wounds or eruptions in the mouth, and within five days after entrance will develop the symptoms noted.

Prevention consists of segregating the sick animals from the healthy ones and cleaning and disinfecting the quarters, not only after infection breaks out but before the calf is born. After outbreaks, all well animals should be checked daily.

Treatment consists of using sulfa drugs or broad spectrum antibiotics. The local application of a proteolytic enzyme for removal of the dead tissue is indicated.

Calf Scours (Infectious Diarrhea, or White-Scours)[8]

This is an acute, contagious, and often rapidly fatal disease of young or newborn calves. Most affected calves are less than two days of age. Outbreaks of calf scours are most common during fall, winter, and early spring.

Scours is the cause of more calf deaths than all other diseases combined. It is estimated that 10% of all calves in the United States are affected by the disease, and that 8% of beef calves and 18% of dairy calves so affected die.[9] It is further estimated that calf scours cost the cattle industry $200 million annually.[10]

SYMPTOMS AND SIGNS

Calf scours can vary from a mild to a severe disease. In the mild form, the main symptom is softer than normal feces. The severely affected calf initially appears depressed and has a lack of appetite. Then begins a severe diarrhea which consists of yellowish, foul smelling, watery or foamy feces. These calves can have a rough hair coat, sunken eyes, and appear emaciated. In very acute cases, death can occur before diarrhea is observed; however, death usually occurs two to three days after the onset of diarrhea. Some degree of associated pneumonia occurs more frequently in stabled dairy calves than in beef calves on the range.

It is sometimes difficult to distinguish the infectious disease from diarrhea caused by other factors—such as overfeeding, irregular feeding, use of unclean utensils, too rapid changes in feed, or exposure to drafts and cold, damp floors. With the infectious type of scours, however, several calves are usually affected; and some animals may die quickly.

---

[8]This section was reviewed by Dr. Charles A. Mebus, DVM, Department of Veterinary Science, The University of Nebraska, Lincoln.
[9]*Better Beef Business*, April, 1973.
[10]*Successful Farming*, May, 1973.

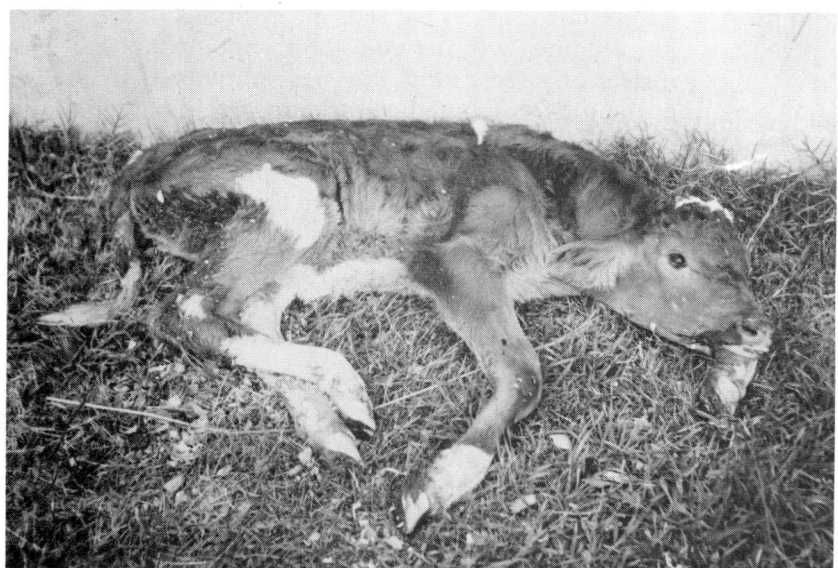

Fig. 12-10. Young calf with severe calf scours. Note sunken eyes, profuse salivation, and watery bowel discharge; which are characteristic symptoms. (Courtesy, Professor G. H. Weise, University of North Carolina)

CAUSE, PREVENTION, AND TREATMENT

Infectious diarrhea in calves under two weeks of age has been shown
experimentally to be caused by both viruses and bacteria. Viral infections
alter the lining of the intestinal tract so that the calf cannot normally digest
and absorb milk. This altered intestine allows an overgrowth of bacteria in
the intestine and escape of bacteria into the bloodstream. Mortality in these
viral infections is dependent on the severity of the secondary bacterial infections and quality of treatment. Certain strains of the colon bacillus (*E. coli*)
have also been shown to cause diarrhea.

In 1968, Nebraska Station workers isolated a virus, called a reo-virus
(short for respiratory-enteric orphan virus), which actually triggers the disease. This virus seems to be responsible for calf scours that occur from 12
hours to 5 days of age. A second virus, called corona virus, is responsible for
scours in calves from 5 to 10 days of age.

To keep the disease away from the herd, one must prevent primary infection of the newborn. This rests on strict sanitary measures and isolation.
The disease can be introduced by adding calves or adult animals from
another herd. Calf diarrhea frequently occurs when a newly assembled herd
begins to calf.

Weather conditions permitting, birth should preferably take place in the
open on an uncontaminated, sun-exposed pasture. Otherwise, a clean, disinfected maternity stall should be provided, and the navel cord of the newborn
calf should be treated with tincture of iodine. The newborn animal should
be segregated from other animals and the contaminated quarters thoroughly

cleaned and disinfected. Prevention should also include proper feeding of pregnant cows and giving colostrum to calves that are subsequently to be raised on a milk replacer. When calf scours appear, infected animals should be segregated and the premises and feed containers should be thoroughly cleaned and disinfected.

If the virus causing calf diarrhea in a herd can be identified, vaccination of the cows before parturition or of the calves at birth with that specific agent will reduce the incidence and severity of the disease. Vaccination of cows before parturition with bacterins has been less successful because of the variation between strains of bacteria.

The most effective preventive measure of calf scours is a new modified live virus vaccine, developed by the University of Nebraska and released in 1973 by Norden Laboratories under the name Scourvax-Reo (after the reovirus that the University of Nebraska identified as the primary cause of scours in young calves). Studies show that the product is about 80 percent effective in preventing the early, or neonatal, form of the disease that scientists have found is caused by a reo-virus. The vaccine is not 100 percent effective because it was developed to counteract a specific virus—reovirus—while there are a number of other agents that can also cause diarrhea. Scourvax-Reo is sprayed into the back of the calf's mouth in a 4.0 ml dose 12-24 hours after birth. The modified live virus is incapable of causing disease, but produces an immune response when given orally to a susceptible calf.

Treatment of severely affected diarrheic calves should include: discontinuing feeding milk for 24 to 48 hours; giving fluids orally and by injection to combat dehydration; administering gastrointestinal protectants; and giving antibiotics orally and by injection. The choice of the antibiotic should be made by the veterinarian for in many areas the bacteria associated with calf diarrhea are resistant to many of the available drugs.

Circling Disease (Encephalitis, or Listeriosis)

Circling disease, also called Listerellosis, Listeriosis, or Encephalitis, is an infectious disease affecting mainly cattle, sheep, and goats; but it has been reported in swine, foals, and other animals, and man. Cattle of all ages are susceptible. One to seven percent of the herd may be infected, and the mortality rate of affected animals is extremely high. The disease causes an estimated annual beef cattle loss of $88,452.[11]

SYMPTOMS AND SIGNS

This disease affects the nervous system. Depression, staggering, circling, and strange awkward movements are noted. One eye and one ear may be paralyzed. The animal may be seen holding a mouthful of hay for hours.

---

[11]*Losses in Agriculture*, Agriculture Handbook No. 291, 1965, p. 73, reported average annual losses of $54,000 for the period 1951-60. The author updated the latter figure by applying 1974 cattle numbers ($\times$ 1.17) and prices ($\times$ 1.4), to give the figure of $88,452.

There may be inflammation around the eye, and abortion may occur. The course of the disease is very short, with paralysis and death the usual termination. Positive diagnosis can be made only by laboratory examination of the brain.

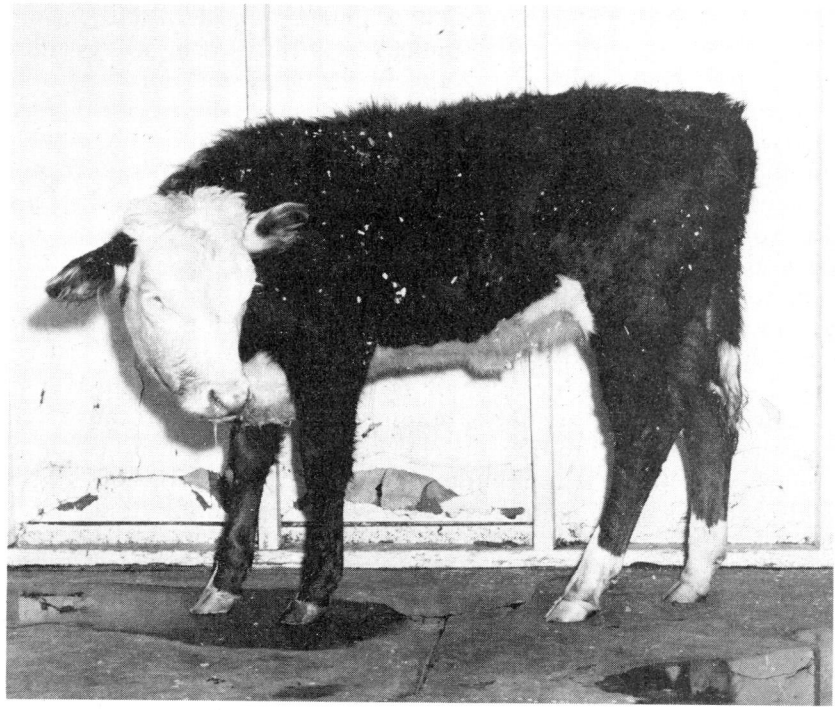

Fig. 12-11. Calf with circling disease. Note the drooping ear, slobbering, and tendency to turn to the left. (Courtesy, Department of Veterinary Pathology and Hygiene, College of Veterinary Medicine, University of Illinois)

CAUSE, PREVENTION, AND TREATMENT

Circling disease results from the invasion of the central nervous system by bacteria called *Listeria monocytogenes*. The method of transmission is unknown. In an outbreak, affected animals should be segregated. If silage is being fed, discontinue that particular silage on a trial basis. Spoiled silage should be avoided, routinely.

Various sulfa derivatives, alone and in combination with antibiotics, have shown beneficial results if given early.

Fescue Foot (Fescue Toxicity)

Fescue is a valuable pasture grass that grows best in cool and cold weather. Under some conditions, which are not completely understood, a fescue pasture may become toxic. Cattle that graze on such pastures do not perform well. In cold climates, they occasionally develop a crippling disease, known as fescue foot or fescue toxicity. Both beef and dairy cattle are susceptible; and this disease has been reported in sheep in Australia.

Most cases of fescue toxicity occur among cattle that graze pure stands of fescue during late fall and winter; and most toxic stands of fescue pasture are several years old. Fescue toxicity is more prevalent in animals suffering from malnutrition or parasitism.

SYMPTOMS AND SIGNS

There are variations in the severity of symptoms in cattle on toxic pastures. Some animals show no apparent lameness, whereas others show varying degrees of sloughing (necrosis) on the ends of their tails.

During the summer, cattle grazing toxic pastures show a poor growth rate, increased temperatures, and increased pulse and respiratory rates. The only complaint that cattlemen make is the fact that the cattle are not doing as well as in previous years. In some herds, the weaning weights decline for two or three years before cattlemen realize that they have a problem.

CAUSE, PREVENTION, AND TREATMENT

The exact cause of this condition is still unknown. It is either a mycotoxin on the fescue or some change in the fescue plant itself that makes it toxic under some conditions. The toxin is a vaso constrictor that affects the blood vessels. Cold weather is also a constrictor of blood vessels, which explains the extreme lameness found in the winter months. Occasionally, the circulation is closed to a degree that causes the entire foot to slough off. Such cattle walk on stumps of bone.

Proper management of fescue pastures is the best way to prevent fescue toxicity. Toxic pastures should be renovated and some legume should be seeded with the fescue. It requires good pasture management, along with fertilization, to maintain a good fescue-legume pasture.

No medication is effective for cattle with fescue foot. In severe cases where sloughing has occurred, the animal should be destroyed for humane reasons.

Cattle usually recover completely if they are removed from fescue pasture or fescue hay and are given other feed or pasture as soon as the first signs of the disease appears.

Foot-and-Mouth Disease

This is a highly contagious disease of cloven-footed animals (mainly cattle, sheep, and swine) characterized by the appearance of watery blisters in the mouth (and in the snout in the case of hogs), on the skin between and around the claws of the hoof, and on the teats and udder. Fever, diminished rumination and reduced appetite are other signs of the disease.

Man is mildly susceptible but very rarely infected, whereas the horse is immune.

Unfortunately, one attack does not render the animal permanently immune, but the disease has a tendency to recur, perhaps because there are several strains of the causative virus. The disease is not present in the United States, but there were at least 9 outbreaks (some authorities claim 10) in this country between 1870 and 1929, each of which was stamped out by the prompt slaughter of every affected and exposed animal. No U.S. outbreak has occurred since 1929, but the disease is greatly feared. Drastic measures are exercised in preventing the introduction of the disease into the United States, or, in the case of actual outbreak, in eradicating it.

In December, 1946, an outbreak of foot-and-mouth disease was confirmed in Mexico, and the border was closed from that date to September 1, 1952. Then the border remained open from September 1, 1952, until May 23, 1953, at which time another outbreak occurred and it was again closed. The U.S. Secretary of Agriculture again opened the border on January 1, 1955.

On February 25, 1952, an outbreak of foot-and-mouth disease was diagnosed in Saskatchewan, Canada. This resulted in the U.S.-Canada border being closed from this date until March, 1953.

Foot-and-mouth disease is constantly present in Europe, Asia, Japan, the Philippines, Africa, and South America. It has not been reported in New Zealand or Australia.

SYMPTOMS AND SIGNS

The disease is characterized by the formation of blisters (vesicles) and a moderate fever three to six days following exposure. These blisters are found on the mucus membranes of the tongue, lips, palate, cheeks, and on the skin around the claws of the feet, and on the teats and udder.

Presence of these vesicles, especially in the mouth of cattle, stimulates a profuse flow of saliva that hangs from the lips in strings. Complicating or secondary factors are infected feet, caked udder, abortion, and great loss of weight. The mortality of adult animals is not ordinarily high, but the usefulness and productivity of affected animals is likely to be greatly damaged, thus causing great economic loss.

CAUSE, PREVENTION, AND TREATMENT

The infective agent of this disease is one of the smallest of the filtrable viruses. In fact, it now appears that there are at least six strains of the virus. Infection with one strain does not protect against the other strains.

The virus is present in the fluid and coverings of the blisters, in the blood, meat, milk, saliva, urine, and other secretions of the infected animal. The virus may be excreted in the urine for over 200 days following experimental inoculation. The virus can also be spread through infected biological products, such as smallpox vaccine and hog cholera virus and serum, and by the cattle fever tick.

Except for the nine outbreaks mentioned, the disease has been kept out of the United States by extreme precautions, such as quarantine at ports of entry and assistance with eradication in neighboring lands when introduc-

Fig. 12-12. Cow with foot-and-mouth disease. The animal is reluctant to stand because of sore feet. The characteristic profuse flow of saliva is caused by blisters in the mouth. (Courtesy, USDA)

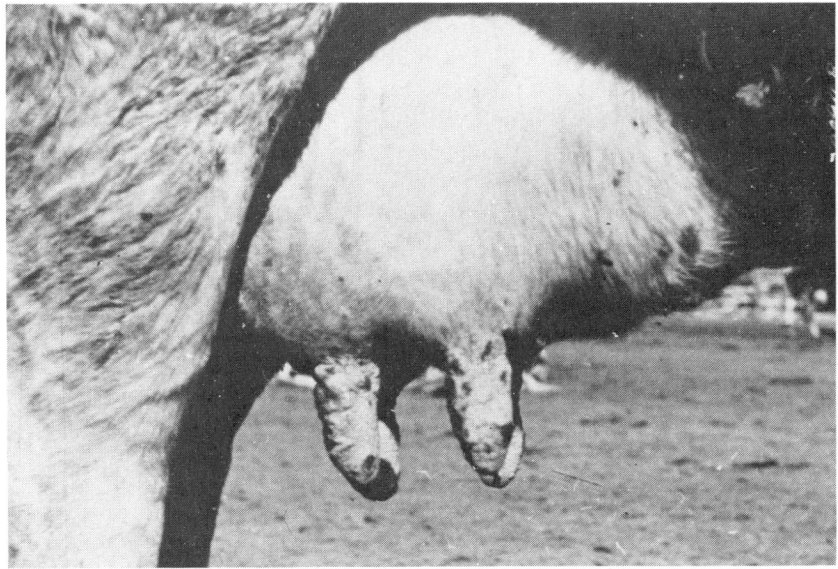

Fig. 12-13. Blisters (vesicles) on the teats of a cow with foot-and-mouth disease. (Courtesy, USDA)

tion appears imminent. Neither live cloven-hoofed animals nor their fresh, frozen, or chilled meats can be imported from any country in which it has been determined that foot-and-mouth disease exists (meat imports from these countries must be canned or fully cured).[12]

Two methods have been applied in control: the slaughter method, and the quarantine procedure. Then, if the existence of the disease is confirmed by diagnosis, the area is immediately placed under strict quarantine; infected and exposed animals are slaughtered and buried, with owners being paid indemnities based on their appraised value. Everything is cleaned and thoroughly disinfected.

In some countries, a third method, the use of a tissue culture vaccine which contains all the strains of virus, is being used. Vaccines have not been used in the outbreaks in the United States because they have not been regarded as favorable to rapid, complete eradication of the infection.

Fortunately, the foot-and-mouth disease virus is quickly destroyed by a solution of the cheap and common chemical, sodium hydroxide (lye). Because quick control action is necessary, state or Federal authorities must be notified the very moment the presence of the disease is suspected.

No effective treatment is known.

Foot-Rot (or Foul-Foot)

This disease is an inflammation of the hoofs of cattle, sheep, and goats; but cross-infections of foot-rot between cattle and sheep do not occur. It is a potential hazard wherever animals of these species are kept; especially in wet muddy areas.

Foot-rot in beef cattle causes estimated average annual losses of $4,863,222.[13]

SYMPTOMS AND SIGNS

A shrewd observer will first notice a reddening and swelling of the skin just above the hoof, between the toes, or in the bulb of the heel. As the infection progresses, lameness will be noted. If not arrested, the infection will invade the soft tissue and cause a discharge of pus from the infected breaks in the skin. At this stage, a characteristic foul odor is present. Later, the joint cavities may be involved, and the animal may show fever and depression characteristic of a general infection. Affected animals lose weight, and, if lactating, produce less milk; and they may die if unattended.

---

[12]Effective April 11, 1974, this rule was altered to permit dependent territories or possessions to be determined free of foot-and-mouth disease and rinderpest, regardless of the disease status of the mother country. Thus, dependent territories or possessions that are geographically separated from their mother country, such as colonies or former colonies, can be judged as to their livestock disease status by the same criteria previously applied to politically separate countries.

[13]*Losses in Agriculture*, Agriculture Handbook No. 291, 1965, p. 73, reported average annual losses of $2,969,000 for the period 1951-60. The author updated the latter figure by applying 1974 cattle numbers (× 1.17) and prices (× 1.4), to give the figure of $4,863,222.

CAUSE, PREVENTION, AND TREATMENT

Because the feet of animals are continually being exposed to all types of filth containing millions of microorganisms, it is difficult to incriminate the causative agent or agents. The soil organism, *Spherophorus necrophorus*, is most frequently recovered from cases of foot-rot in cattle; but pus-forming bacteria, colon bacilli, and others may lend support to the destructive process. Similar types of infection may have different causative agents depending on area, soil, and other factors.

The prevention of this disease is much more effective than the treatment, because once established it is difficult to control the spread. Draining of muddy corrals and the segregation of infected and new animals is recommended. If the disease appears, a good cleaning is in order, and unaffected animals should be moved to clean quarters and pastures if possible. Also, effective prevention may be obtained through subjecting animals to (1) a foot bath of 2 to 5 percent bluestone (copper sulfate), or (2) a walk way of air-slaked lime. In some cases, the inclusion of an organic iodide compound (ethylene diamine dihydriodide; EDDI) in the feed or salt has markedly reduced the incidence of foot-rot in problem herds. For prevention, use 50 mg of EDDI per head daily on a continuous basis; for treatment, once cattle have foot-rot, use 500 mg per head daily for 2 to 3 weeks.

The success of treatment seems to depend on the stage of infection and perhaps on the causative agent. The usual treatment consists of the following:

1. Place in a clean, dry place.
2. If necessary, trim away the affected part of the foot. Also, check for, and if necessary eliminate, foreign bodies in the hoof, or wire that might be wrapped around just above the hoof.

If these steps fail, the veterinarian may use sulfonamide or antibiotic therapy—accompanied by cleaning, disinfecting, and packing the affected area. Broad spectrum antibiotics have also been used to good effect. In advanced stages, best results are obtained by surgical amputation of the affected claw. Animals so treated soon walk as before on the one remaining healthy claw.

Infectious Bovine Rhinotracheitis (IBR, or Red Nose)

Infectious bovine rhinotracheitis was first found in a Colorado feedlot in 1950. Since then, it has occurred throughout the United States. The main economic losses from the disease are in time, weight, milk production, and drugs.

SYMPTOMS AND SIGNS

Affected animals go off feed and lose weight; generally cough; may show pain in swallowing; usually slobber and show a nasal discharge; breathe rapidly, with difficulty, and in severe cases through the mouth; show severe inflammation of the nostrils and trachea; have a high fever, 104° to

107°F; and may remain sick for as long as a week. When the disease breaks out, 25 to 100 percent of the animals are affected. Death loss rarely exceeds 5 percent.

Although IBR is usually thought of as a respiratory disease, it may cause inflammation of the eyes and/or vagina. Also, it may cause abortion.

CAUSE, PREVENTION, AND TREATMENT

The disease is caused by a virus.

Infectious bovine rhinotracheitis can be prevented by the use of a vaccine, of which there are two types. The modified live virus vaccine provides lasting immunity, but it should not be used on pregnant cows or calves under six months of age. Killed virus vaccines must be repeated.

There is no known treatment, but sulfonamides and antibiotics effectively combat the secondary bacterial invaders that accompany the disease.

Infectious Embolic Meningoencephalitis (Thromboembolic Meningoencephalitis)

Infectious embolic meningoencephalitis is an acute, febrile disease of feedlot cattle, characterized by incoordination and coma. Only one to two cases develop in a lot at a time, but 10 percent of the cattle may be affected before the disease runs its course.

Twelve outbreaks of the disease were reported in feedlot cattle in Colorado from 1949 to 1956. Subsequently, it has been reported throughout the West.

SYMPTOMS AND SIGNS

The disease occurs in feedlot cattle, in the fall and winter months. It affects both sexes, usually animals one to two years of age. It is characterized by incoordination, coma, sometimes blindness, and always fever (near 107°F). Death usually follows in two to four days. Positive diagnosis can be made upon autopsy by the inflamed areas of infection observed in the brain.

The disease should not be confused with *Polioencephalomalacia*, which also affects feedlot cattle and causes incoordination, but with which fever is rarely associated.

CAUSE, PREVENTION, AND TREATMENT

It is thought to be caused by a *Haemophilus*-like gram-negative organism.

Application of the usual sanitary measures may help control the disease. Also, for the first 28 days after arrival, fortify the ration with the following per head per day: 350 mg of Aureomycin plus 350 mg of sulfamethiazine.

If identified sufficiently early, treatment with one of the broad spectrum antibiotics or with a sulfonamide may help affected animals. If the situation is serious enough, it may be desirable to (1) change diet (to at least 50%

roughage) and pens, and (2) scatter animals out on pasture, despite the reluctance of feeders to move cattle and change rations.

Johne's Disease (Chronic Bacterial Dysentery, or Paratuberculosis)

This is a chronic, incurable, infectious disease seen chiefly in cattle; also found in sheep and goats, and more rarely in swine and horses. It resembles tuberculosis in many respects. The disease is very widespread, having been observed in practically every country where cattle are raised on a large scale. Apparently, it is increasing in the United States. It is one of the most difficult diseases to eradicate from a herd.

Johne's disease in beef and dairy cattle causes an estimated annual loss of $22,604,400.[14]

SYMPTOMS AND SIGNS

The disease seems to involve calfhood exposure with no evidence of infection for 6 to 18 months. At the end of this time, the animal loses flesh and displays intermittent diarrhea and constipation, the former becoming more prevalent. Affected animals may retain a good appetite and normal temperature. The feces are watery but contain no blood and have a normal

Fig. 12-14. Cow with Johne's disease. Note marked loss of flesh and tucked up flank. Diarrhea is a common symptom of the disease. (Courtesy, Department of Veterinary Pathology and Hygiene, College of Veterinary Medicine, University of Illinois)

[14]*Losses in Agriculture*, Agriculture Handbook No. 291, 1965, p. 73, reported average annual losses of $13,800,000 for the period 1951-60. The author updated the latter figure by applying 1974 cattle numbers (× 1.17) and prices (× 1.4), to give the figure of $22,604,400.

odor. The disease is almost always fatal, but with the animal living from a month to two years.

Upon autopsy, the thickening of the infected part of the intestines, covered by a slimy discharge, is all that is evident. This thickening prevents the proper digestion and absorption of food and explains the emaciation.

CAUSE, PREVENTION, AND TREATMENT

The disease is caused by the ingestion of a bacterium, *Mycobacterium paratuberculosis.* Inasmuch as this organism is acid-fast (that is, it retains certain dyes during a staining procedure), it resembles tuberculosis.

Effective prevention is accomplished by keeping the herd away from infected animals. If it is necessary to introduce new animals into a herd, they should be purchased from reputable breeders; and the owner should be questioned regarding the history of his herd.

It must be borne in mind that apparently healthy animals can spread the disease. Testing at regular intervals of three to six months with "Johnin," removing reactors, disinfecting quarters, and raising young stock away from mature animals should be practiced in infected herds. In using the Johnin test, however, it should be realized that it is not entirely accurate as a diagnostic agent. Some affected animals fail to react to the test.

No satisfactory treatment for Johne's disease has yet been found.

Leptospirosis

Leptospirosis was first observed in man in 1915-16, in dogs in 1931, and in cattle in 1934. It has also been reported in hogs, horses, and sheep.

It was first reported in cattle in the United States in 1944, although it had been found in dogs in the United States since 1939. Bovine leptospirosis has been reported in Europe, Australia, and the United States.

Human infections may be contracted through skin abrasions when handling or slaughtering infected animals, by swimming in contaminated water, through consuming raw beef or other uncooked foods that are contaminated, or through drinking unpasteurized milk.

Today, leptospirosis costs $100 million annually.[15]

SYMPTOMS AND SIGNS

In most herds, leptospirosis is a mild disease. However, the symptoms may vary from herd to herd, or even within a herd. In general, the symptoms noted in cattle are: (1) high fever; (2) poor appetite; (3) abortion at any time; (4) bloody urine; (5) anemia; and (6) ropy milk.

All ages of cattle, and both sexes (including steers) are affected.

CAUSE, PREVENTION, AND TREATMENT

The disease is caused by several species of corkscrew-shaped organisms

---

[15]USDA, The Yearbook of Agriculture 1956, *Animal Diseases,* p. 12.

of the spirochete group; primarily *Leptospira pomona* in cattle, although *L. hardjo* and *L. grippotyphosa* are becoming more common.

The following preventive measures are recommended:

1. Blood test animals prior to purchase, isolate for 30 days and then retest prior to adding them to the herd.

2. Do not allow animals to consume contaminated feed or water, or to breathe contaminated urinal mist.

3. Keep premises clean, and avoid use of stagnant water.

4. Vaccinate susceptible animals annually if the disease is present in the area.

Where a herd is infected, the following control measures should be initiated:

1. Blood test the herd and dispose of reactors either through (a) strict isolation, or (b) sale for slaughter only. Then retest every 30 days until two consecutive clean tests are obtained.

The same blood sample used in a brucellosis test may also be used for a leptospirosis test, by simply dividing the serum.

2. Spread the cattle over a large area; avoid congestion in a corral or barn.

3. Do not let animals drink from ponds, swamps, or slow-running streams, and avoid contaminated feed.

4. Clean and disinfect the premises; exterminate the rodents.

It should be recognized that carrier animals—animals that have had leptospirosis and survived—may spread the infection by shedding the organism in the urine. The infected urine may then either (1) be breathed as a mist in cow barns, or (2) contaminate feed and/or water and thus spread the infection. It is known that such recovered animals may remain carriers for two to three months or longer after getting over the marked symptoms. Fortunately, the organisms seldom survive for more than 30 days outside the animal. However, stagnant water and mild temperatures favor their survival.

Treatment, which should be prescribed by a veterinarian, may include blood transfusions, administration of selected antibiotics, and good care.

Lumpy Jaw and Wooden Tongue

These two infections are chronic diseases affecting mainly the head of cattle—hence, the name "big head." They occur most frequently in young cattle during the period of changing teeth. At one time, both of the conditions were referred to as actinomycosis, but now this term is used only for lumpy jaw. Actinobacillosis is the synonym for wooden tongue and soft tissue lesions.

Each year, about 200,000 parts of Federally inspected carcasses are condemned from lumpy jaw.[16]

---

[16]Ensminger, M. E., M. W. Galgan, and W. L. Slocum, *Problems and Practices of American Cattlemen*, Wash. Agr. Exp. Sta. Bull. 562, p. 15.

SYMPTOMS AND SIGNS

Because of the area involved in these diseases, there is usually emaciation resulting from the difficulty encountered in chewing and swallowing.

Lumpy jaw only rarely attacks the soft tissue. It is usually confined to the bones of the lower jaw, although the upper jaw and nasal bones may be involved. The affected bone becomes enlarged and spongy and filled with creamy pus. As the disease progresses, inflamed cauliflower masses of tissue spread out and may appear on the surface, discharging pus of foul odor. The surrounding flesh will also show inflammation, and the teeth may become loosened.

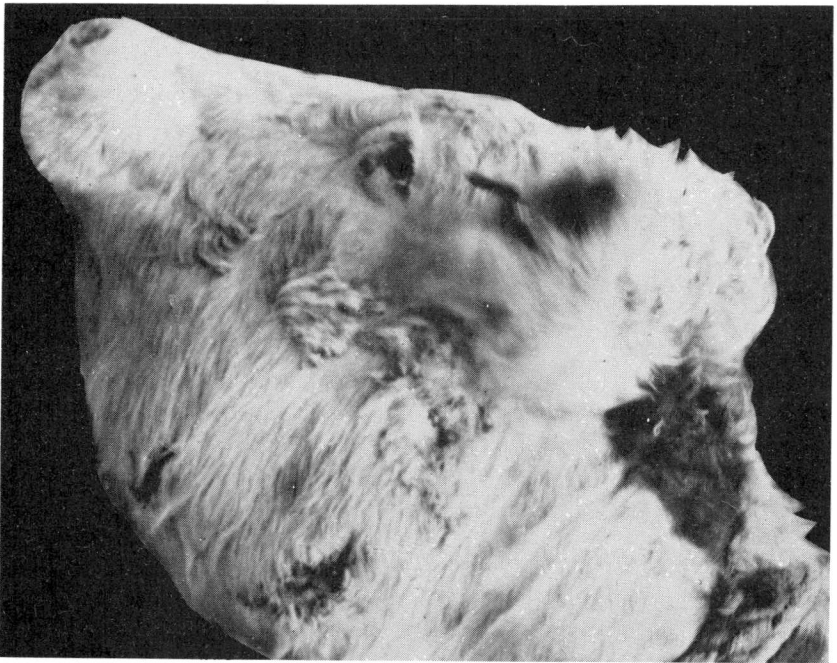

Fig. 12-15. Head of steer showing a bad case of lumpy jaw. It is usually confined to the bones of the lower jaw, although the upper jaw and nasal bones may be involved. (Courtesy, USDA)

The same organism that causes lumpy jaw occasionally attacks the udder of sows, where it is characterized by many small abscesses filled with calcified granules. There may be fistulous tracts to the outside of the udder, discharging pus. On rare occasions, the organism causing the disease has also been found in fistulous withers of the horse, in conjunction with *Brucella* organisms.

Wooden tongue attacks chiefly the tissue in the throat area of cattle, but is also often seen in the tongue, stomach, lungs, and lymph glands. The first lesion usually observed is a movable, tumor-like swelling about the size of a small egg under the skin in the infected area. The enlargements usually break open and discharge a light colored and very sticky pus. An involved tongue may or may not be ulcerated but will show an increase in size and hardness. The tongue may become quite immobile and may protrude from the mouth.

With this wooden tongue condition, there will be constant drooling, and the animal will lose weight and condition through inability to take feed. Although any chronic swelling in the region of the head should lead one to suspect the presence of this infection, a positive diagnosis depends upon a microscopic examination of the yellowish, granular pus material that will eventually discharge from the swelling.

The same organism that causes wooden tongue occasionally attacks the lips and face of sheep.

CAUSE, PREVENTION, AND TREATMENT

The organism of actinobacillosis (wooden tongue) is called *Actinobacillus lignieresei,* and the one causing actinomycosis (lumpy jaw) is called *Actinomyces bovis.* In each case, they may be assisted by secondary invaders. Both organisms lack invasive power, often being found in a normal oral cavity. They are thought to enter the tissue only by wound infections—for example, they may be carried in by the sharp awns of fox tail, barley, rye, bearded wheat, or oats.

Prevention consists of segregation and proper treatment or elimination of infected animals and the restricted feeding of material having sharp awns that might injure the animal's mouth. The latter precaution is important as the organism is a normal inhabitant of the mucous membranes of the mouth and nasal cavity of animals and man.

Under some conditions, organic iodine appears to be an effective aid in the prevention and treatment of lumpy jaw in cattle. For prevention, add to the ration or salt 50 mg of ethylenediamine dihydriodide (EDDI) per head daily; for treatment, add 250 to 500 mg per head daily for 2 to 3 weeks.

For treatment, the veterinarian may, under certain conditions, (1) administer a water solution of an iodine salt of sodium or potassium, (2) prescribe an antibiotic, (3) resort to surgery, or (4) use X-ray therapy.

Treatment with iodine salt of sodium or potassium consists of daily internal administration of a water solution of an iodine salt until iodism appears. Iodism is characterized by flow of tears, roughness of hair coat, and loss of appetite. Administration should be discontinued until these symptoms disappear, and treatment should then be resumed. Such treatment causes tainted milk and is not recommended for animals in advanced pregnancy.

Treatment of lumpy jaw is not very satisfactory, but most cases of wooden tongue yield readily to treatment.

Superficial abcesses should be opened, drained, and swabbed with tincture of iodine.

Malignant Edema (Gas Gangrene)

This is an acute infectious, but noncontagious, disease characterized by gangrene and emphysema around a wound. The incidence in a single herd may be high following castration, dehorning, or accidental wounds.

SYMPTOMS AND SIGNS

The affected animal goes off feed, breathes rapidly, and is profoundly depressed. A swelling forms around the wound. A gaseous and malodorous fluid exudes from the wound. In advanced stages of the disease, the animal is prostrated and often disoriented. There may or may not be a rise in temperature. Death occurs after a course of 12 to 48 hours. The mortality rate is high.

CAUSE, PREVENTION, AND TREATMENT

Malignant edema is caused by *Clostridium septicum* and related bacteria.

Since malignant edema is associated with contamination of wounds, the disease can be partially prevented by minimizing wounds and by castrating and dehorning under hygienic conditions.

Vaccination of young cattle with a vaccine containing *Cl. septicum* (for malignant edema) along with *Cl. chauvoei* (for blackleg) at the time of the blackleg vaccination(s) will give some protection against malignant edema. Also, antibiotics may be administered four to five days following surgery.

In the early stages of the disease, treatment with massive doses of antibiotics may be effective.

Metritis

Metritis is an inflammation of the uterus, which affects cattle, horses, sheep and swine.

SYMPTOMS AND SIGNS

Metritis usually develops soon after the animal has given birth. It is characterized by a foul smelling discharge from the vulva that becomes thick and yellow or white, and finally brownish or blood-stained. Also, there is chilling, high temperature, rapid breathing, marked thirst, loss of appetite, and lowered milk production. Pressure on the right flank may produce pain. The animal may lie down and refuse to get up. Affected animals may die in one to two days; or the acute infection may develop into a chronic form, producing sterility.

CAUSE, PREVENTION, AND TREATMENT

Metritis is caused by various types of bacteria. Laceration at the time of calving, wounds caused by inexperienced operators and/or retention of the afterbirth are the principal predisposing causes.

Preventive measures consist in alleviating as many of the predisposing factors as possible, including bruises and tears while giving birth, exposure to wet and cold, and the actual introduction of disease-causing bacteria during delivery or the manual removal of the afterbirth. Clean maternity stalls should be provided. If assistance at calving time becomes necessary, the caretaker should first disinfect his hands and arms as well as the animal's external genitals.

Difficult parturition should be left to the veterinarian. Nothing is so distressing to the veterinarian as a history of long labor and well meaning but ill guided attempts to remove a calf. Most cases are treated by introducing (in solution or tablets) an antibiotic or sulfa into the uterus.

Navel Infection (Joint-Ill, or Navel-Ill)

Navel infection is an infectious disease of newborn calves, foals, and lambs. It occurs less frequently in calves and lambs than in foals.

SYMPTOMS AND SIGNS

Navel infection is characterized by loss of appetite, by swelling, soreness and stiffness in the joints, by umbilical swelling and discharge, and by general listlessness.

CAUSE, PREVENTION, AND TREATMENT

Navel infection is caused by several kinds of bacteria.

The recommended preventive measures are: sanitation and hygiene at mating and parturition, painting the navel cord of the newborn animal with iodine, and the administration of bacterins.

For treatment, the veterinarian may give a blood transfusion, or he may administer an antibiotic.

Pinkeye (Keratitis)

This is the common name for an infectious disease that affects the eyes of cattle. It may be caused by several different infectious agents. Of the two most common forms of the disease, one is caused by a bacteria and the other by a virus. It attacks animals of any age, but is more common in young animals. It seems to become more virulent in certain years and in certain communities. The disease is widespread throughout the United States, especially among range and feedlot cattle. Pinkeye is encountered in nearly half of U.S. beef cattle herds and affects three percent of all beef cattle.[17]

SYMPTOMS AND SIGNS

The first thing one may notice in bacterial pinkeye is the liberal flow of tears and the tendency to keep the eyes closed. There will be redness and

---

[17]Ibid., p. 12.

swelling of the lining membrane of the eyelids and sometimes of the visible part of the eye. There may also be a discharge of pus. Ulcers may form on the cornea. If unchecked, they may cause blindness and even loss of the eye. The attack may also be marked by slight fever, reduction in milk flow, and slight digestive upset.

Fig. 12-16. Cow with pinkeye. Note eye discharge and the cloudiness or milkiness of the cornea or covering of the eyeball. (Courtesy, Department of Veterinary Pathology and Hygiene, College of Veterinary Medicine, University of Illinois)

About one-half the animals in a herd become infected regardless of the treatment or control measures employed.

One record of steers on pasture showed that affected steers gained an average of 50 pounds less during the grazing season than those not affected.[18]

In viral pinkeye, the eyeball itself is only slightly affected. Infectious bovine rhinotracheitis (IBR), a virus infection of the eyes of cattle, mainly affects the eyelids and the tissues surrounding the eyes. It causes a severe swelling of the lining of the lids.

CAUSE, PREVENTION, AND TREATMENT

• *Bacterial pinkeye*—The most prevalent bacterial form of the disease is caused by *Moraxella bovis*. This organism produces a toxin which irritates and erodes the covering of the eye. Bacterial pinkeye occurs mainly during warm weather. Bright sunlight, wind, and dust may contribute to the cause of the disease. Cattle with white faces or lack of pigment around the eyes are

---

[18]Ibid., p. 15.

rather susceptible. Transmission is mainly by flies and other insects that feed on eye discharges of infected animals and then carry the infection to susceptible animals. Also, the disease is spread by direct contact, from animal to animal.

Prevention of bacterial pinkeye consists of the following: controlling face flies and other insects that feed around the eyes; good nutrition, including adequate vitamin A; and isolation of affected animals.

The most common treatment for bacterial pinkeye is the application of antibiotics or sulfa drugs to the affected eye as ointments, powders, or sprays; preferably, with treatment made twice daily. Foreign protein therapy, which is the subcutaneous or intramuscular injection of such things as sterile milk, has been used to treat pinkeye for years, with some success. Cortisone is sometimes combined with antibiotics and injected under the covering (at the outer edge) of the eyeball. The cortisone aids in reducing inflammation and pain and lessens the tears. Recovery is speeded up by keeping the infected animals in a dark barn. A commercially produced protective eye patch is now available. It completely covers the infected eye, holding the medication in place, protects the eye from insects and bright sunlight, and reduces the work and expense of handling and isolation. Held in place by a special adhesive, the eye patch drops off and decomposes after about 7 to 10 days.

Fig. 12-17. An eye patch may be used in pinkeye treatment to protect the inflamed eye from sunlight. Periodic treatment should be used with the patch. (Courtesy, University of Illinois, Urbana, Ill.)

● *Virus pinkeye*—The best known virus infection of the eyes of cattle is caused by the "red nose"or infectious bovine rhinotracheitis (IBR) virus. It is much less common than bacterial pinkeye. When this organism infects the eyes of cattle, there may or may not be other signs of disease, such as respiratory infection, vaginitis, or abortion commonly associated with IBR. IBR conjunctivitis occurs most frequently in the winter, but it may be seen at other times of the year. The disease is highly contagious by direct and indirect contact of infected animals with susceptible animals.

IBR conjunctivitis may be prevented by proper vaccination of animals prior to onset of the disease. The herd should not be vaccinated once the disease appears; nor should pregnant cows be vaccinated. Affected animals should be isolated.

Treatment of IBR conjunctivitis is seldom of value, although antibiotics sometimes help reduce the secondary bacterial infection.

Pneumonia

Pneumonia is an inflammation of the lungs in which the alveoli (air sacs) fill up with an inflammatory exudate or discharge. The disease is often secondary to many other conditions. It is difficult to describe and classify for the lung is subject to more forms of inflammation than any other organ in the body. It affects all animals. In cattle, it is seen most commonly as calf pneumonia, and frequently it accompanies shipping fever. If untreated, 50 to 75 percent of affected animals die. Pneumonia causes one-fifth of all nonnutritional mortality in U.S. beef cattle.[19]

SYMPTOMS AND SIGNS

The disease is ushered in by a chill, followed by elevated temperature. There is quick, shallow respiration, with discharge from the nostrils and perhaps from the eyes. A cough may be present. The animal appears distressed, stands with legs wide apart, drops in milk production, shows no appetite, and is constipated. There may be crackling noises with breathing, and gasping for breath may be noted. If the disease terminates favorably, the cough loosens and the appetite picks up.

CAUSE, PREVENTION, AND TREATMENT

The causes are numerous. Many microorganisms found in other acute and chronic diseases, such as mastitis and metritis, have been incriminated; and pneumonia can be caused by a number of different viruses. One common cause that should be stressed is the inhalation of water or medicines that well-meaning but untrained persons give to animals in drenches. Also, it is generally recognized that changeable weather during the spring and fall, and damp barns, are conducive to pneumonia.

Prevention includes providing good hygienic surroundings and practicing good, sound husbandry.

---

[19]Ibid., p. 12.

Fig. 12-18. Calf with pneumonia. Note characteristic spread of front legs in an effort to ease breathing. (Courtesy, Department of Veterinary Pathology and Hygiene, College of Veterinary Medicine, University of Illinois)

Sick animals should be segregated and placed in quiet, clean quarters away from drafts.

Calves can be treated with a broad-spectrum antibiotic for four to five days. Secondary bacterial pneumonia may also be treated with sulfonamides or an antibiotic.

Polioencephalomalacia (Cerebrocortical Necrosis, or Forage Poisoning)

This is a noninfectious disease of pasture and feedlot cattle, affecting animals between 3 months and 4 years of age. The highest incidence in feedlot cattle occurs in animals 12 to 18 months of age.

SYMPTOMS AND SIGNS

Some sudden deaths occur in the feedlots. Sick animals are excitable, incoordinated, and have impaired vision. On driving, these animals go down into a convulsion.

CAUSE, PREVENTION, AND TREATMENT

This disease is believed to be due to a thiamine (B1) deficiency. It is possible that enzymes or toxins of fungi or other microbes in the rumen may destroy the vitamin before absorption can take place.

Intramuscular or intravenous thiamin injections should be administered to sick animals. Supplementary fluids should be given by way of the stomach tube.

Rabies (Hydrophobia, or Madness)

Rabies (hydrophobia or "madness") is an acute infectious disease of all warm-blooded animals and man. It is characterized by deranged consciousness and paralysis, and terminates fatally. This disease is one that is far too prevalent, and, if present knowledge were applied, it could be controlled and even eradicated.

When a human being is bitten by a dog that is suspected of being rabid, the first impulse is to kill the dog immediately. This is a mistake. Instead, it is important to confine the animal under the observation of a veterinarian until the disease, if it is present, has a chance to develop and run its course. If no recognizable symptoms appear in the animal within a period of two weeks after it inflicted the bite, it is safe to assume that there was no rabies at the time. Death occurs within a few days after the symptoms appear, and the dog's brain can be examined for specific evidence of rabies.

With this procedure, unless the bite is in the region of the neck or head, there will usually be ample time in which to administer the Pasteur treatment to exposed human beings. As the virus has been found in the saliva of a dog at least five days before the appearance of the clinically recognizable symptoms, the bite of a dog should always be considered potentially dangerous until proved otherwise. In any event, when a human being is bitten by a dog, it is recommended that a physician be consulted immediately. Each year about 30,000 persons in the United States undergo the Pasteur treatment.

But not all animals that have bitten humans should be held for observation. Wild animals (skunks, racoons, foxes, etc.) should be killed immediately and examined for evidence of rabies infection, because the signs of rabies in wild animals are variable and the duration of the virus excretion before clinical rabies develops may be longer than in dogs.

Rabies cause an estimated average annual loss in beef cattle of only $44,226,[20] which is minor. However, the fact that man is susceptible to the disease makes it more important than the economic losses would indicate.

SYMPTOMS AND SIGNS

Less than 10 percent of the rabies cases appear in cattle, horses, swine, or sheep. The disease usually manifests itself in two forms: The furious, irritable, or violent form, or the dumb or paralytic form. It is often difficult to distinguish between the two forms, however. The furious type usually merges into the dumb form because paralysis always occurs just before death.

The furious form is seen most often in cattle. In its early stages, the

---

[20]*Losses in Agriculture*, Agriculture Handbook No. 291, 1965, p. 73, reported average annual losses of $27,000 for the period 1951-60. The author updated the latter figure by applying 1974 cattle numbers (× 1.17) and prices (× 1.4), to give the figure of $44,226.

Fig. 12-19. Cow with rabies. Note the violent butting with the head; a characteristic of the furious form which is seen most often in cattle. At this stage, the animal is insane and is very dangerous, for it may attack and bite itself, other animals, or man. (Courtesy, Pitman-Moore, Indianapolis, Ind.)

disease is marked by loss of appetite, cessation in milk secretion, anxiety, restlessness, and a change in disposition. This initial phase is followed by a stage of madness and extreme excitation indicated by a loud bellowing marked by a change in the voice, pawing of the ground, inability to swallow, and violent butting with the head. In all respects, the animal is insane and is very dangerous, for it may attack and bite itself or other animals and man. On the fourth or fifth day, the animal becomes quieter and unsteady. This indicates approach of posterior paralysis. Loss of flesh is already very evident. On the sixth day, the animal may go into a coma and die.

CAUSE, PREVENTION, AND TREATMENT

Rabies is caused by a filtrable virus which is usually carried into a bite wound by the infected saliva of a rabid animal. The malady is generally transmitted to farm animals by dogs and certain wild animals, such as the fox and skunk.

Rabies can best be prevented by attacking it at its chief source, the dog. With the advent of an improved anti-rabies vaccine for the dog, it should be a requirement that all dogs be immunized. This should be supplemented by regulations governing the licensing, quarantine, and transportation of dogs. For understandable reasons, the control of rabies in wild animals and bats is extremely difficult. In areas where rabies is present, all cattle should be vaccinated.

After the disease is fully developed in cattle, there is no known treatment. However, immediate use of either HEP Flury vaccine or the phenolized nerve tissue vaccine is recommended following known exposure of cattle.

Persons bitten by a rabid animal should immediately report to the family doctor who usually administers Semple type vaccine, although irradiated vaccines are used to some extent. With severe bites, especially those around the head, antiserum is particularly indicated.

Shipping Fever (Hemorrhagic Septicemia, Para-Influenza, or Stockyard Fever)

Shipping fever is an acute respiratory disease of cattle. It is most common in calves and following shipment. It occurs widely throughout the world, expecially among thin and poorly nourished young animals that are subjected to shipment by truck or rail during periods of inclement weather, though it may occur in animals in good condition. The disease is a serious problem to both shippers and receivers of cattle.

Based on an extensive survey, it was estimated that shipping fever caused out-of-pocket losses of $100 million in 1972. Also, it is conjectured that the losses due to sickness and poor performance probably were twice as great as the actual death-and-drug losses. That's another $200 million, making for a total loss of $300 million due to shipping fever in 1972.[21]

SYMPTOMS AND SIGNS

The first sign of the disease (which may appear within 2 to 21 days after moving cattle) is a tired appearance and reduced appetite. The affected animal may show signs of depression, watery to slimelike nasal discharge, increased body temperature, occasional soft or hacking cough, rapid breathing, loss of appetite, followed by loss of body weight. In very acute forms, animals may die showing no symptoms. Death losses may be high in untreated cases.

CAUSE, PREVENTION, AND TREATMENT

Shipping fever is caused from multiple infection due to the interaction of viruses and bacteria, accentuated by environmental conditions creating physical tension or stress. Change in weather and feed, overcrowding, hard driving, lack of rest, and improper shelter all help usher in the disease.

As a preventive measure, one should eliminate as many as possible of the predisposing factors that lower the animal's vitality. Also, newly purchased animals should be isolated for two to three weeks before being placed in the herd.

Several vaccines, both modified and inactivated, are available. Vaccination should be done three to four weeks before exposure.

Where cattle have been subjected to great stress—long shipment, extensive handling, and/or exposure to severe weather conditions—it is recommended that they be handled as follows: Adult cattle should be given long grass or oat hay, rolled oats, and/or wheat bran during the first week. Newly

---

[21]*Beef,* July, 1973; study made by Livestock Conservation, Inc., with Dr. Don Gill, Oklahoma State University, serving as Chairman of the Committee.

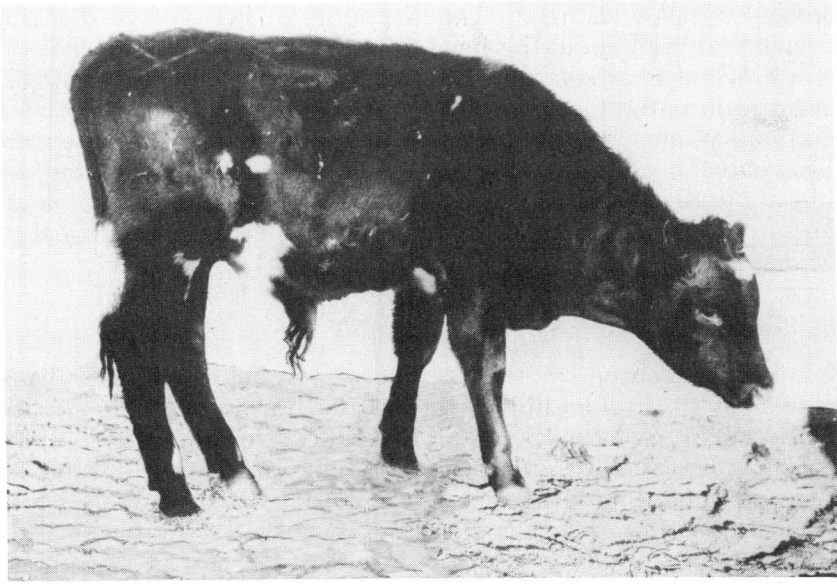

Fig. 12-20. A steer with shipping fever. The disease is most frequently associated with animals whose resistance has been lowered due to travel; hence, the name shipping fever. (Courtesy, USDA)

weaned calves can be given a calf starter ration in addition. All animals should have access to plenty of clear fresh water at all times. For the first 28 days after arrival, fortify the ration with the following per head per day: 350 mg of Aureomycin plus 350 mg of sulfamethiazine. Newly arrived cattle should also receive 50,000 IU of vitamin A per head daily and have free access to a good mineral mixture.

Antibiotics and sulfa drugs are effective treatments if given early in the course of the disease. Treatment after pneumonia develops is often disappointing.

Tetanus (Lockjaw)

Tetanus is chiefly a wound infection disease that attacks the nervous system of horses (and other equines) and man, although it does occur in cattle, swine, sheep and goats. In the Southwest, it is quite common in sheep after shearing, docking, and castrating. In the Central States, tetanus frequently affects calves, lambs, and pigs, following castration or other wounds. It is generally referred to as lockjaw.

In the United States, the disease occurs most frequently in the South, where precautions against tetanus are an essential part of the routine treatment of wounds. The disease is worldwide in distribution.

SYMPTOMS AND SIGNS

The incubation period of tetanus varies from one to two weeks, but may be from one day to many months. It is usually associated with a wound but may not directly follow an injury. The first noticeable sign of the disease is a stiffness first observed about the head. The animal often chews slowly and weakly and swallows awkwardly. The third eyelid is seen protruding over the forward surface of the eyeball (called "haws"). Violent spasm or contractions of groups of muscles may be brought on by the slightest movement or noise. The animal usually attempts to remain standing throughout the course of the disease until close to death. If recovery occurs, it will take a month or more. In over 80 percent of the cases, however, death ensues, usually because of sheer exhaustion or paralysis of vital organs.

CAUSE, PREVENTION, AND TREATMENT

The disease is caused by an exceedingly powerful toxin (more than 100 times as toxic as strychnine) liberated by the tetanus organism, *Clostridium tetani*. This organism is an anaerobe (lives in absence of oxygen) which forms the most hardy spores known. It may be found in certain soils, horse dung, and sometimes in human excreta. The organism usually causes trouble when it gets into a wound that rapidly heals or closes over it. In the absence of oxygen, it then grows and liberates the toxin which follows up nerve trunks. Upon reaching the spinal cord, the toxin excites the symptoms noted above.

The disease can largely be prevented by reducing the probability of wounds, by general cleanliness, by proper wound treatment, and by vaccination with tetanus toxoid in the so-called hot areas. When an animal has received a wound from which tetanus may result, short term immunity can be conferred immediately by use of tetanus antitoxin, but is of little or no value after the symptoms have developed. All valuable animals should be protected with tetanus toxoid.

All perceptible wounds should be properly treated, and the animal should be kept quiet and preferably should be placed in a dark quiet corner free from flies. Supportive treatment is of great importance and will contribute towards a favorable course. This may entail artificial feeding. The animal should be placed under the care of a veterinarian.

Tuberculosis

Tuberculosis is a chronic infectious disease of man and animals, which occurs worldwide. It is characterized by the development of nodules (tubercules) that may calcify or turn into abscesses. The disease spreads very slowly, and affects mainly the lymph nodes. There are three kinds of tuberculosis bacilli—the human, the bovine, and the avian (bird) types. Practically every species of animal is subject to one or more of the three kinds, as shown in Table 12-2.

In general, the incidence of tuberculosis is steadily declining in the United States, both in animals and humans. In 1917, when a thorough nationwide eradication campaign was first initiated, 1 cow in 20 had the disease; whereas today the number is about 1 in 1,250. Meanwhile, human mortality from tuberculosis dropped from 150 per 100,000 population in 1918 to

TABLE 12-2

RELATIVE SUSCEPTIBILITY OF MAN AND ANIMALS TO 3 DIFFERENT KINDS OF TUBERCULOSIS BACILLI

Species	Susceptibility to Three Kinds of Tuberculosis Germs			Comments
	Human Type	Bovine Type	Avian (Bird) Type	
Humans	Susceptible	Moderately susceptible	Questionable	Pathogenicity of avian type for humans is practically nil.
Cattle	Slightly susceptible	Susceptible	Slightly susceptible	
Swine	Moderately susceptible	Susceptible	Susceptible	Ninety percent of all swine cases are due to the avian type.
Chickens	Resistant	Resistant	Very susceptible	Chickens only have the avian type.
Horses and mules	Relatively resistant	Moderately susceptible	Relatively resistant	Rarely seen in these animals in the U.S.
Sheep	Fairly resistant	Susceptible	Susceptible	Rarely seen in these animals.
Goats	Marked resistance	Highly susceptible	Susceptible	Rarely seen in these animals in the U.S.
Dogs	Susceptible	Susceptible	Resistant	Highly resistant.
Cats	Quite resistant	Susceptible	Quite resistant	Usually obtained from milk of tubercular cows.

2.1 per 100,000 in 1971. Some decline in the incidence of the disease in poultry and swine has also been noted, but the reduction among these species has been far less marked.

Tuberculosis still makes for an estimated annual loss in beef cattle of $1,018,836,[22]

SYMPTOMS AND SIGNS

Tuberculosis may take one or more of several forms. Human beings get tuberculosis of the skin (lupus), of the lymph nodes (scrofula), of the bones and joints, of the lining of the brain (tuberculous meningitis), and of the lungs. For the most part, tuberculosis in animals involves the lungs and lymph nodes. In cows, the udder becomes infected in chronic cases.

---

[22]*Losses in Agriculture*, Agriculture Handbook No. 291, 1965, p. 73, reported average annual losses of $622,000 for the period 1951-60. The author updated the latter figure by applying 1974 cattle numbers (× 1.17) and prices (× 1.4), to give the figure of $1,018, 836.

Fig. 12-21. Cow in the last stages of tuberculosis. Cattle are susceptible to all three kinds of tuberculosis. (Courtesy, USDA)

Many times an infected animal will show no outward physical signs of the disease. There may be a gradual loss of weight and condition and swelling of joints, especially older animals. If the respiratory system is affected, there may be a chronic cough and labored breathing. Next to the lungs and lymph nodes, the udder is most frequently affected, showing increased size and swelling of the supra mammary lymph gland. Other seats of infection are genitals, central nervous system, and the digestive system.

CAUSE, PREVENTION, AND TREATMENT

The causative agent is a rod-shaped organism belonging to the acid-fast group known as *Mycobacterium tuberculosis*. The disease is usually contracted by eating feed or drinking fluids contaminated by the discharges of infected animals. Hogs may also contract the disease by eating part of a tubercular chicken.

With cattle, periodic testing and removal of the reactors is the only effective method of control. Also, avoid housing or pasturing cattle with chickens. It is well to abide by the old adage, "once a reactor, always a reactor."

The test consists of the introduction of tuberculin—a standardized solution of the products of the tubercle bacillus—into an approved location on the animal.

There are three principal methods of tuberculin testing—the intradermic, subcutaneous, and ophthalmic. The first of these is the method now principally used. It consists of the injection of tuberculin into the dermis (the true skin).

Upon injection into an infected animal, tuberculin will set in the body a reaction characterized by a swelling at the site of injection. In human beings, the X-ray is usually used for purposes of detecting the presence of the disease.

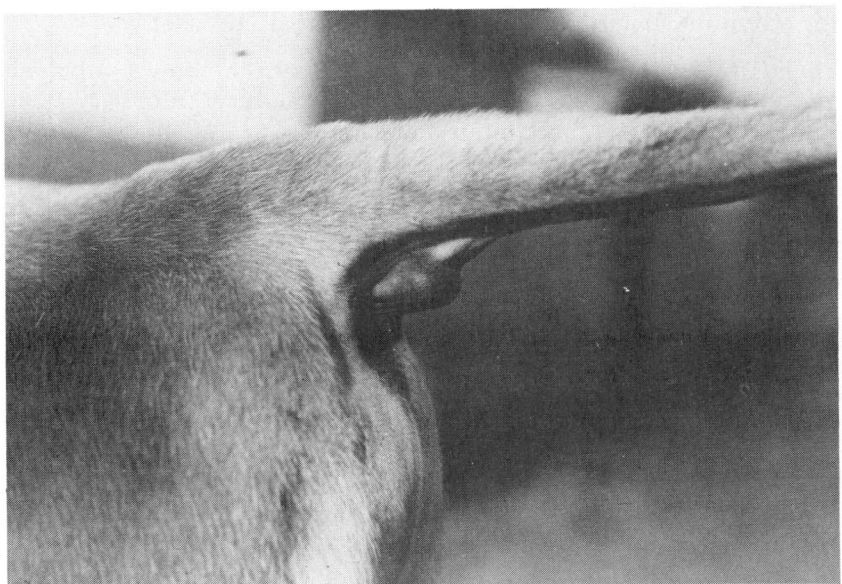

Fig. 12-22. A positive reaction (indicating the presence of tuberculosis) to the intradermic (into the true skin) tuberculin test in a cow. Reactors show a noticeable swelling, varying from the size of a pea to the size of a walnut, at the point of injection. The reading is made approximately 72 hours after injection. (Courtesy, Department of Veterinary Pathology and Hygiene, College of Veterinary Medicine, University of Illinois)

As a part of the Federal-state tuberculosis eradication campaign of 1917, provision was made for indemnity payments on animals slaughtered.

Preventive treatment for both humans and animals consists of pasteurization of milk and creamery by-products and the removal and supervised slaughter of reactor animals. Vaccination against human tuberculosis, using B.C.G., is now being generally practiced in some foreign countries and is being experimentally tried on some groups in this country. B.C.G. is the abbreviated designation of Bacille' Calmette-Guerin, after two French scientists who first prepared the vaccine and who are its chief advocates.

In human beings, tuberculosis can be arrested by hospitalization and complete rest, along with drug therapy. But in animals this method of treatment is neither effective nor practical. Infected animals should be sent to slaughter.

All 50 states will accept for entry cattle meeting either of the following T.B. tests: (1) accredited herds tested within 12 months, or (2) individual negative test within 30 days. Title 9 of the Code of Federal Regulations, Part 77, Tuberculosis of Cattle, requires the branding of reactors and tuberculin testing cattle which originate in areas which are not modified accredited.

Cattle to be exported must be tested for tuberculosis and found free of the disease within 90 days of shipment.

(Also see later section in this chapter entitled, "Federal and State Regulations Relative to Disease Control.")

Vaginitis (Granular Vaginitis, or Granular Venereal Disease)

This is an infectious disease which localizes in the cow's vulva and on the penis and prepuce of the bull, causing an inflammation of varying intensity. It occurs throughout the United States.

SYMPTOMS AND SIGNS

The tissue of the vagina is reddish, roughened, and granular in appearance. Infected animals are usually difficult breeders. Economic losses are in terms of lower percentage calf crop and decreased milk production.

CAUSE, PREVENTION, AND TREATMENT

This condition is considered to be a response of the lymphatic tissue to an unknown irritant or antigen. It is nonspecific, and not a disease in the classic sense.

Prevention consists in purchasing clean animals from clean herds and avoiding the use of bulls that have been exposed to the infection. Artificial insemination can be effectively employed in a control program. Also, in problem herds, vaccination 30 to 60 days prior to the breeding season should be considered.

Treatment of females is not indicated. The condition will clear up by itself in several weeks. Vaginitis in females is not directly related to fertility. However, the predisposing agents (viruses, bacteria, or fungi) may affect fertility.

The condition in bulls tends to be more persistent; and affected bulls may refuse to breed. They should be treated by massage of the anesthetized prolapsed penis and sheath with a suitable antibiotic ointment.

Vibrionic Abortion (Vibrio Fetus, or Vibriosis)

This is an infectious venereal disease of cattle, which causes infertility and abortion. For diagnosis, laboratory methods must be used.

SYMPTOMS AND SIGNS

Infected herds are characterized by (1) abortions in the middle third of pregnancy, (2) several services per conception, and (3) irregular heat periods. In total, vibrionic abortion costs U.S. beef cattlemen $170,352,000 annually.[23]

CAUSE, PREVENTION, AND TREATMENT

The disease is caused by the microorganism *Vibrio fetus*, which is transmitted at the time of coitus.

---

[23]*Losses in Agriculture*, Agriculture Handbook No. 291, 1965, p. 73, reported average annual losses of $104,000,000 for the period 1951-60. The author updated the latter figure by applying 1974 cattle numbers ($\times$ 1.17) and prices ($\times$ 1.4), to give the figure of $170,352,000.

Prevention consists in avoiding contact with diseased animals and contaminated feed, water, and materials. Also, vaccination, repeated annually, is effective in controlling the disease, especially in beef herds kept under range conditions. Artificial insemination is a rapid and practical method of stopping infection from cow to cow.

Aborting cows should be isolated, and aborted fetuses and membranes should be burned or buried. Contaminated quarters should be thoroughly cleaned and disinfected.

Infected cows are treated by injecting drugs into the uterus and/or by allowing sexual rest.

Warts (Papillomatosis)

Warts, which are small tumors, are an infectious disease of cattle and other animals and man. Young animals, under two years of age, are most often affected.

SYMPTOMS AND SIGNS

Warts are protruding growths on the skin, varying from very small to quite large, pendulous growths weighing several pounds. They may appear anywhere on the body, but are especially common on the teats and/or around the head.

Although warts are a nuisance, their presence does not normally inter-

Fig. 12-23. An extreme case of warts. (Courtesy, Fort Dodge Laboratories, Inc., Fort Dodge, Iowa)

fere with the animal's health. However, they damage the hide, making the leather derived therefrom weak in the affected area.

CAUSE, PREVENTION, AND TREATMENT

Warts are caused by a virus. It appears that each species of animals is attacked by a specific virus.

The following preventive measures are recommended:
1. Segregate "warty" cattle.
2. Clean and disinfect all exposed pens, stables, chutes, and rubbing posts.
3. Administer wart vaccine.

The most common treatment among cattlemen consists in softening the wart with oil for several days, and then tying off the growth with thread or a rubber band or snipping it off with sterile scissors. The stump should then be treated with tincture of iodine. Wart vaccines help in some cases, but, generally speaking, they are more effective in prevention than in treatment. The veterinarian may resort to surgical removal of extremely large warts.

Winter Dysentery (Winter Scours)

This is an acute infectious disease of stabled cattle, both dairy and beef, most frequently occurring between the months of November and March.

SYMPTOMS AND SIGNS

It causes few death losses, but afflicted animals lose in condition; and, in lactating animals, there is a sharp reduction in milk flow.

The period of incubation is extremely short, varying from three to five days. A profuse watery diarrhea is the main symptom. Often the feces are dark brown in color, and tend to become darker when intestinal hemorrhages occur. Usually the temperature remains normal, and the appetite is unchanged. Calves and young animals are least susceptible, but animals of all ages are affected. The seasonal incidence of the disease, the age and number of animals affected, together with the suddenness of the onset, are helpful in arriving at a correct diagnosis.

CAUSE, PREVENTION, AND TREATMENT

The disease is caused by an organism, *Vibrio jejuni*, and perhaps a virus.

Prevention consists in isolating new or replacement animals. Also, any animal suffering from an acute attack of dysentery should be separated from the herd. Where the disease is encountered, rigid sanitation should be practiced.

The veterinarian may administer antibiotics and intestinal astringents. For relief of dehydration, intravenous injections of physiological salt solution, accompanied by glucose, are useful.

X-Disease (Hyperkeratosis)

Following the discovery that X-disease is caused from the ingestion of, or contact with, highly chlorinated naphthalenes found in certain lubricants, the disease was almost completely eliminated. At the present time, therefore, X-disease is primarily of historic interest. It is noteworthy, however, that the disease once caused annual losses of more than $1¾ million.[24]

SYMPTOMS AND SIGNS

X-disease is characterized by a watering of the eyes, wrinkling of the thickened skin, loss of hair, diarrhea and emaciation. All ages are affected, but animals under two years of age are most susceptible. The death loss may be as high as 80 percent in young calves.

CAUSE, PREVENTION, AND TREATMENT

Chlorinated naphthalene compounds, such as are contained in certain oils and greases used on machinery, were identified as the cause.

Fig. 12-24. Animal with X-disease. Note wrinkling of the thickened skin. (Courtesy, College of Veterinary Medicine, University of Illinois)

Prevention consists in keeping cattle away from tractors, combines, old oil drums, drainage from grease racks, and roofing asphalt; or other places where it may be possible for them to contact grease or oil. If pelleted feeds are used, make sure that the lubricant used in the pelleting machinery contains no highly chlorinated naphthalene compounds.

Treatments are not too satisfactory, but administering vitamin A appears to help.

---

[24]*Losses in Agriculture*, Agriculture Handbook No. 291, 1965, p. 73, reported average annual losses of $1,791,000 for the period 1951-60.

PARASITES OF BEEF CATTLE[25]

Beef cattle are attacked by a wide variety of internal and external parasites.

Fig. 12-25. Calf with "bottle jaw." Generally this condition is indicative of a heavy infestation of several species of internal parasites. (Courtesy, School of Veterinary Medicine, Alabama Polytechnic Institute)

The prevention and control of parasites is one of the quickest, cheapest, and most dependable methods of increasing beef and milk production with no extra cattle, no additional feed, and little more labor. This is important, for, after all, the farmer or rancher bears the brunt of this reduced meat and milk production, wasted feed, and damaged hides. It is hoped that the discussion that follows may be helpful in (1) preventing the propagation of parasites, and (2) causing the destruction of parasites through the use of the most effective anthelmintic or insecticide.

Recommended anthelmintics for the control of internal parasites are given in the sections that follow. But no specific suggestions concerning insecticides are given because of (1) the diversity of environments and management practices under which they occur, (2) the varying restrictions on the use of insecticides from area to area, and (3) the fact that registered uses of insecticides change from time to time. Information about what insecticide is available and registered for use in a specific area can be obtained from the County Agent, Extension Entomologist, or Agricultural Consultant.

---

[25]This section was authoritatively reviewed by the following: Dr. A. C. Todd, Department of Veterinary Science, University of Wisconsin, Madison, Wisc.; Dr. George T. Edds, Department of Veterinary Science, University of Florida, Gainsville, Fla.; Dr. J. F. Landram, Parasitologist, Dr. W. S. McGregor, Entomologist, and Dr. L. L. Wade, Entomologist, The Dow Chemical Company, Freeport, Tex.; Dr. R. J. Boisvenue, Parasitologist, Eli Lilly and Company, Greenfield, Iowa; and Dr. R. L. Harris, U.S. Department of Agriculture, Agriculture Research Service, Kerrville, Tex.

From time to time, new insecticides and vermifuges are approved and old ones are banned or dropped. When parasitism is encountered, therefore, it is suggested that the stockman obtain from local authorities the current recommendation relative to the choice and concentration of the insecticide and vermifuge to use.

Anaplasmosis

Anaplasmosis is an infectious disease whose etiology and symptomatology are similar to cattle tick fever, except that more carriers are involved. It is caused by a minute parasite, *Anaplasma marginale,* which invades the red blood cells. The parasite is transmitted from infected to healthy animals by ticks, horseflies, stable flies, mosquitoes, deer flies, and probably by other biting insects.

DISTRIBUTION AND LOSSES CAUSED BY ANAPLASMOSIS

The disease is widely distributed in warm climates throughout the world. In the United States, it has been prevalent throughout the southern states, but it is slowly spreading to the northern states.

The mortality rate may vary from 2 to 5 percent to 50 to 60 percent. The most severe losses are found in older animals and in hot weather.

Anaplasmosis in beef and dairy cattle causes the United States annual losses of approximately $100 million.[26]

LIFE HISTORY AND HABITS

In infected animals, the causative parasite, *Anaplasma marginale,* live in the red blood cells. The parasite and, consequently, the disease may be transmitted from animal to animal by means of biting insects and by such mechanical agencies as needles, dehorning instruments, etc. Any animal that has once contracted the disease permanently retains the parasite in the blood, though no signs of ill health may be evident. Such animals are "carriers," and are potential sources of danger to others.

In addition to carrier animals, there is another reservoir of anaplasmosis infection in the western range states—the wood tick. This insect is a biological vector, since the disease will overwinter in its body.

DAMAGE INFLICTED; SYMPTOMS AND SIGNS OF AFFECTED ANIMALS

The symptoms may be those of a mild, acute, or chronic condition. Calves usually have the mild type of infection, simply becoming "dumpy" for a few days and then apparently recovering, though their blood remains the permanent abode of the parasite.

The more characteristic symptoms in mature animals include rapid, pounding heart action, labored and difficult breathing, rise in temperature (up to 107°F), dry muzzle, marked depression, tremors of the muscles, loss of appetite, and a great reduction in the milk flow. Animals usually show yellowing of the eye and other mucous membranes and of the skin, as in jaundice. Depraved appetite, evidenced by the eating of bones or dirt, is not uncommon. Sick animals may also show brain symptoms and an inclination

---

[26]U.S. Department of Agriculture estimate, made by Dr. B. R. McCallon, Chief, Veterinary Staff, Viral and Parasitic Diseases, Animal Health Division. Reported in *Arkansas Cattle Business,* May, 1974, p. 8.

Fig. 12-26. Cow exhibiting typical symptoms of acute anaplasmosis. (Courtesy, USDA)

to fight. Unlike cattle tick fever, bloody urine is not common in anaplasmosis. In severe acute cases, death may follow in one to a few days. Recovery is usually very slow, and although no clinical symptoms remain, such animals continue as permanent carriers of the parasite.

PREVENTION, CONTROL, AND TREATMENT

Once an animal is infected with anaplasmosis, it usually becomes a carrier, harboring the disease agent in its bloodstream for life. This animal then becomes a continuous source of infection in the herd.

Importing infected carrier cattle into clean areas and clean herds is the most common method of spreading the disease. Once an infected animal has been introduced, the disease is spread within the herd by insect vectors, primarily the wood tick and biting flies, or by mechanical means such as castrating, ear tagging, dehorning, and vaccination. Infection can be transferred any time fresh blood is transferred from infected to a noninfected animal.

Once an animal becomes a carrier, it is immune to subsequent exposure to the disease agent. However, carrier animals that have been treated and freed of infection become susceptible again.

Prevention in lightly infected areas consists in testing the herd and finding infected animals. Then, either removing the infected animals by culling them for slaughter or feeding an antibiotic. (Either Terramycin or

Aureomycin, fed in a supplement, at a level of 5 mg of the antibiotic/lb body weight daily for 45 days.) Also, new additions to the herd should be tested to avoid reinfection. A new rapid card test, conducted at cowside, can be quickly made in the field to detect the presence of anaplasmosis. This test is approved by the U.S. Department of Agriculture.

The U.S. Department of Agriculture, in cooperation with several states, is now certifying beef cow herds as anaplasmosis free.

In heavily infected areas, where there has been a high rate of infection for several years, the procedure, at present, is to live with the disease. Calves raised in such areas and herds will have a high rate of infection, develop a degree of immunity, and become carriers for life.

Vaccination with a killed vaccine (first reported by the Oklahoma Station in 1965) does not prevent the animal from getting the disease, but it does reduce the severity of the infection. Vaccination of bulls is particularly recommended. The killed vaccine should be administered in two doses separated by an interval of six weeks, followed by a single annual booster injection.

In 1973, the University of Illinois developed an attenuated anaplasmosis vaccine. But, to date, the U.S. Department of Agriculture has not licensed the product, because an attenuated vaccine interferes in tests of virulent anaplasmosis. The product is manufactured in Mexico.

Cattle that are acutely ill with anaplasmosis are often treated with an antibiotic (Terramycin or Aureomycin), but blood transfusion remains the most effective treatment. The chronically affected carrier animal can be treated effectively by the prolonged, high level feeding of an antibiotic with feed or water.

Blowfly (Fleece Worms, Fly Strike, or Wool Maggots)

The flies of the blowfly group include a number of species that find their principal breeding ground in dead and putrifying flesh, although they sometimes infest wounds or unhealthy tissues of live animals and fresh or cooked meat. Black blowfly larvae frequently infest dehorning wounds during winter months and occasionally the navel of newborn animals. All the important species of blowflies except the flesh flies, which are grayish and have three dark stripes on their backs, have a more or less metallic luster.

DISTRIBUTION AND LOSSES CAUSED BY BLOWFLIES

Although blowflies are widespread, they present the greatest problem in the Pacific Northwest and in the South and southwestern states. Death losses from blowflies are not excessive, but they cause much discomfort to affected animals and lower production.

LIFE HISTORY AND HABITS

With the exception of the group known as gray flesh flies, which deposit tiny living maggots instead of eggs, the blowflies have a similar life cycle to

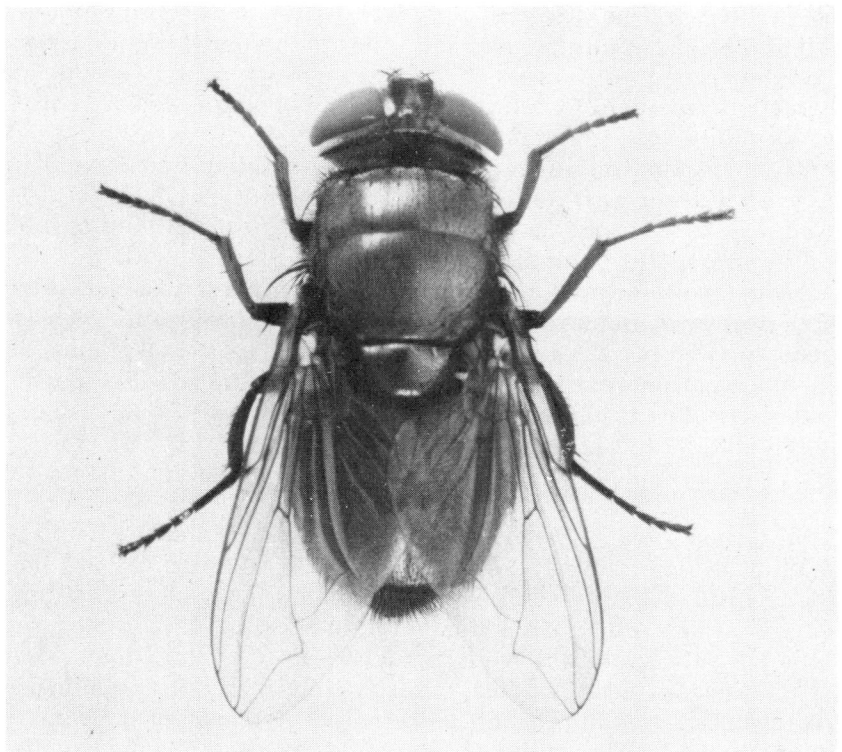

Fig. 12-27. The black blowfly (*Phormia regina*). These flies are characterized by a metallic luster. (Courtesy, USDA)

the screwworm, except that the cycle is completed in about one-half the time.

DAMAGE INFLICTED; SYMPTOMS AND SIGNS OF AFFECTED ANIMALS

The blowfly causes its greatest damage by infesting wounds and the soiled hair or fleece of living animals. Such damage, which is largely limited to the black blowfly (or wool-maggot fly), is similar to that caused by screwworms. Sheep are especially susceptible to attacks of blowflies, because their wool frequently becomes soiled or moistened by rain and accumulations of feces and urine. The maggots spread over the body, feeding on the skin surface, where they produce a severe irritation and destroy the ability of the skin to function. Infested animals rapidly become weak and fevered; and, although they recover, they may remain in an unthrifty condition for a long period.

Because blowflies infest fresh or cooked meat, they are often a problem of major importance around packing houses or farm homes.

PREVENTION, CONTROL, AND TREATMENT

Prevention of blowfly damage consists of eliminating the pest and decreasing the susceptibility of animals to infestation.

As blowflies breed principally in dead carcasses, the most effective control is effected by promptly destroying all dead animals by burning or deep burial. The use of traps, poisoned baits, and electrified screens is also helpful in reducing trouble from blowflies. Suitable repellents, such as pine tar oil, help prevent the fly from depositing its eggs.

When animals become infested with blowfly maggots, their wounds should be treated twice weekly with a smear, dust, or pressurized spray of the proper insecticide, used according to manufacturer's directions.

Bovine Trichomoniasis

This is a protozoan venereal disease of cattle characterized by early abortions (usually between the second and fourth months of pregnancy) and temporary sterility. The protozoa that cause the disease, known as *Trichomonas foetus*, are one-celled, microscopic in size, and capable of movement. They are found in aborted fetuses, fetal membranes and fluids, vaginal secretions of infected animals, and the sheaths of infected bulls. Diagnosis can be confirmed microscopically. The infected bull is the source of the infection. On the other hand, the disease appears to be self-limiting in the cow.

DISTRIBUTION AND LOSSES CAUSED BY BOVINE TRICHOMONIASIS

This disease is being reported with increasing frequency throughout the United States and has become a serious problem in many herds. The economic loss is primarily due to the low percentage calf crops in infected herds.

Trichomoniasis causes estimated average annual losses in beef and dairy cattle of $13,169,520.[27]

LIFE HISTORY AND HABITS

The protozoa that cause the disease are a one-celled microscopic organism with three threadlike whips (flagella) at the front and one at the rear. The evidence indicates that the disease is spread from the infected to the clean cow by an infected bull at the time of service and that other types of contact infection do not occur. Following one or perhaps two abortions, cows appear to be immune to reinfection. Further than these facts, little is known of the life history and habits of *Trichomonas foetus*.

---

[27]*Losses in Agriculture*, Agriculture Handbook No. 291, 1965, p. 79, reported average annual losses of $8,040,000 for the period 1951-60. The author updated the latter figure by applying 1974 cattle numbers ($\times$ 1.17) and prices ($\times$ 1.4), to give the figure of $13,169,520.

DAMAGE INFLICTED; SYMPTOMS AND SIGNS OF AFFECTED ANIMALS

There is no systemic disturbance manifested by the infected bull. There may be some mucus discharged from the sheath, and the latter may be slightly inflamed. The only clinical evidence of infection is the transmission of the disease to the cattle serviced.

Fig. 12-28. Bull with trichomoniasis. This animal appeared normal, but spread the disease during the breeding act. (Courtesy, Department of Veterinary Pathology and Hygiene, College of Veterinary Medicine, University of Illinois)

Infected cows frequently show a whitish vaginal discharge, and the following characteristic conditions usually exist when a herd is infected: (1) abortions in the first third of pregnancy; (2) uterine infections; (3) irregular heat periods; and (4) several services per conception.

Early abortions or erratic heat periods in individuals or herds that are known to be free of Bang's disease should lead one to suspect the presence of the Trichomonad infection. Definite diagnosis of infection in the bull is made by means of a microscopic examination of smears taken from the prepuce of the bull or the vagina of the cow.

PREVENTION, CONTROL, AND TREATMENT

Prevention lies in the use of clean bulls or artificial insemination. If practical, infected animals should be sold for slaughter or allowed 90 days of sexual rest. Otherwise, treatment should be attempted.

Effective control consists of exercising great precaution in introducing new animals into the herd, in breeding outside cows, and in taking cows outside the herd for breeding purposes.

Treatment of the bull consists in massaging the penis and the entire mucous membrane of the prepuce for 30 minutes with either 250 gram of Bovoflavin Ointment, or a 0.5% acriflavine ointment; repeated in 10-14 days. Concurrent with this local treatment, the animal should be given 5 gram of sodium iodide per 100 lb of body weight; intravenous for 5 consecutive doses at 48-hour intervals.

Cattle Tick Fever (Splenetic Fever, Texas Fever, etc.)

This is an infectious protozoan disease of adult cattle caused by *Babesia bigemina,* which depends upon the tick, chiefly *Boophilus annulatus,* for its survival and transmission.

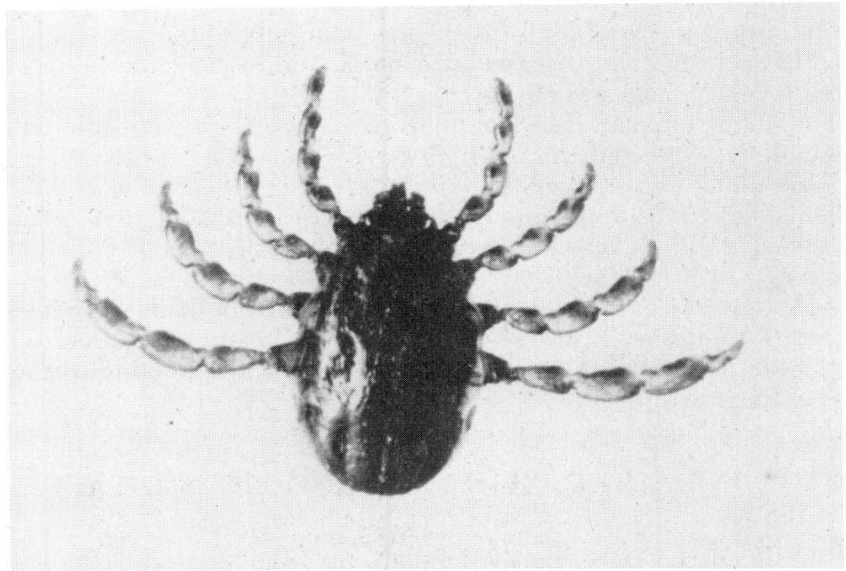

Fig. 12-29. An adult cattle tick (*Boophilus annulatus*), chief transmitter of cattle tick fever. (Courtesy, USDA)

DISTRIBUTION AND LOSSES CAUSED BY CATTLE TICK FEVER

Prior to 1906, at which time a concerted effort was initiated to eradicate the cattle fever tick, this infectious disease of cattle and the parasite which transmits it were the most serious obstacles faced by the cattle industry in the fifteen southern and southwestern states, representing a combined area of nearly one-fourth of the United States. At that time, conservative estimates placed the yearly losses at $40,000,000. Today, 99 percent of the formerly infested area has been freed (for the most part, the tick is confined to the Texas-Mexico border), and the once appalling losses have been practically eliminated. Cattle tick fever is not uncommon in Central and South America.

In addition to the serious death losses encountered in infected herds, the loss of blood—the only food of the cattle fever tick—results in serious damage. Infected young animals are stunted, mature animals are emaciated, and the milk flow of infected dairy animals is greatly reduced. Death occurs in about 10 percent of the chronic and 90 percent of the acute cases.

LIFE HISTORY AND HABITS

In 1889 and 1890, investigators of the USDA, Bureau of Animal Industry, established (1) that intracellular one-celled parasites, or protozoa, known as *Babesia* are the direct causative agents of the disease, and (2) that cattle tick infestation is necessary in the transmission of the disease. Thus, for the first time in either human or veterinary medicine, the discovery was made that an intermediate biological carrier may transmit a disease. It is noteworthy that this pioneer work opened up an entirely new field in medical science, pointing the way for studies that later solved the problems of the spread of such dreaded diseases as malaria, yellow fever, Rocky Mountain spotted fever, typhus, and others.

The life history and habits of the protozoa which causes cattle tick fever are as follows: infected ticks, which have sucked blood from an infected cow, pass along the protozoa *Babesia (Piroplasma) bigemina* to their eggs. The female tick falls to the ground and deposits from 2,000 to 4,000 eggs. In 2 to 3 weeks, these eggs hatch into young ticks or larvae. The larvae climb on nearby vegetation to await the passing of cattle to which they attach themselves, biting and sucking blood from the host. In the latter process, the protozoa (*Piroplasma*) is passed into the blood of cattle—the protozoa of infected ticks having been passed into the eggs of the tick and through all stages of its growth.

DAMAGE INFLICTED; SYMPTOMS AND SIGNS OF AFFECTED ANIMALS

The incubation period of cattle tick fever is about 10 days. The disease is characterized by high temperature, rapid breathing, enlarged spleen, engorged liver, pale and yellow membranes, and red to black urine. Sometimes the symptoms subside only to reoccur at another time. Although immune, the recovered animals are permanent carriers of the disease. In infected areas, native cattle are either immune or only slightly affected.

PREVENTION, CONTROL, AND TREATMENT

Prevention of the disease consists of avoiding contact with the cattle fever tick, the only natural agent by which cattle tick fever is transmitted from animal to animal.

The most effective control measures are directed at the eradication of the fever ticks, either by killing them on the pastures or on the cattle. Pastures may be rendered tick-free by excluding all the host animals—cattle, horses, and mules—until all the ticks have died of starvation in 8 to 10 months.

Fig. 12-30. Cattle with cattle tick fever. The disease is characterized by high temperature, rapid breathing, enlarged spleen, engorged liver, pale and yellow membranes, and red to black urine. (Courtesy, Dr. B. T. Simms, USDA)

Dipping or spraying with an approved insecticide is the most effective method of control.

Successful treatment of sick animals depends upon early recognition of the disease and prompt treatment. Agents traditionally used include trypan blue, trypaflavine, and quinuronium sulfate.

Although immune, recovered animals are permanent carriers of the disease.

Title 9 of the Code of Federal Regulations, part 72, Texas fever in cattle, indicates areas under quarantine and requires dipping in a permitted dip prior to movement from an infected area.

(Also see later section in this chapter entitled, "Federal and State Regulations Relative to Disease Control.")

Coccidiosis

Coccidiosis—a parasitic disease affecting cattle, sheep, goats, swine, pet stock, and poultry—is caused by microscopic protozoan organisms known as Coccidia, which live in the cells of the intestinal lining. Each class of domestic livestock harbors its own species of Coccidia, thus there is no cross-infection between animals.

Cattle are known to be affected by 21 species of coccidia. But only

Eimeria bovis and *E. zuerni* are important in the United States, with *E. zuerni* tending to cause the most serious infections.

DISTRIBUTION AND LOSSES CAUSED BY COCCIDIOSIS

The distribution of the disease is worldwide. Except in very severe infections, or where a secondary bacterial invasion develops, infested farm animals usually recover. The chief economic loss, which is estimated at $47 million in the United States and $472 million worldwide (both beef and dairy animals),[28] is in lowered gain and production. It is most severe in feeder cattle and young dairy calves.

LIFE HISTORY AND HABITS

Infected animals may eliminate daily with their droppings thousands of coccidia organisms (in the resistant öocyst stage). Under favorable conditions of temperature and moisture, Coccidia sporulate to maturity in 3 to 5 days, and each öocyst contains eight infective sporozoites. The öocyst then gains entrance into an animal by being swallowed with contaminated feed or water. In the host's intestine, the outer membrane of the öocyst, acted on by the digestive juices, ruptures and liberates the 8 sporozoites within. Each sporozoite then attacks and penetrates an epithelial cell, ultimately destroying it. While destroying the cell, however, the parasite undergoes sexual multiplication and fertilization with the formation of new öocysts. The parasite (öocyst) is then expelled with the feces and is again in a position to reinfest a new host.

The coccidia parasite abounds in wet, filthy surroundings; resists freezing and ordinary disinfectants; and can be carried long distances in streams.

DAMAGE INFLICTED; SYMPTOMS AND SIGNS OF AFFECTED ANIMALS

A severe infection with Coccidia produces diarrhea, and the feces may be bloody. The bloody discharge is due to the destruction of the epithelial cells lining the intestines. Ensuing exposure and rupture of the blood vessels then produces hemorrhage into the intestinal lumen.

In addition to a bloody diarrhea, affected animals usually show pronounced unthriftiness and weakness.

PREVENTION, CONTROL, AND TREATMENT

Coccidiosis can be prevented by protecting animals from feed or water that is contaminated with the protozoa that causes the disease. Prompt segregation of affected animals is important and should be done if practical. Manure and contaminated bedding should be removed daily. Low, wet areas should be drained. If possible, segregation and isolation of animals by age should be used in controlling the disease. All precautions should be

---

[28]*Calf News*, June, 1974.

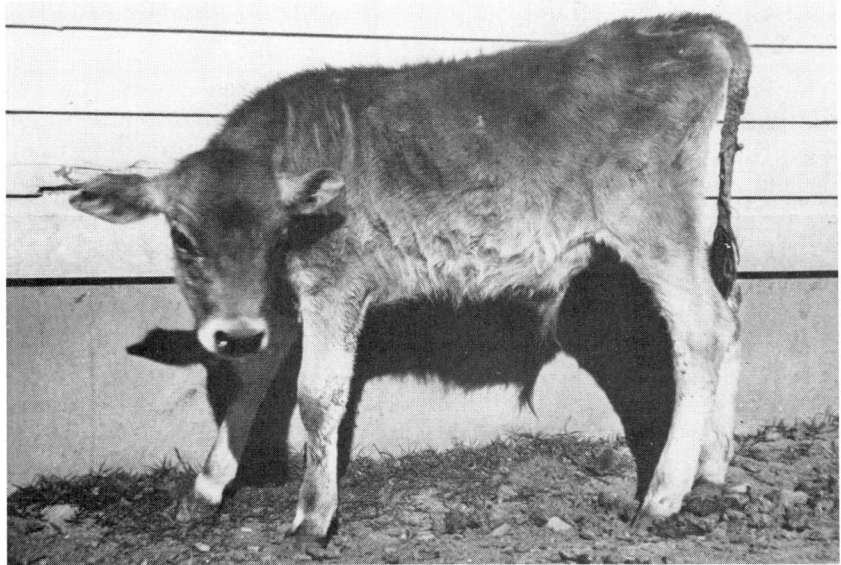

Fig. 12-31. Calf suffering from coccidiosis. The soiled tail is a typical symptom. (Courtesy, USDA, Regional Laboratory An. Dis. Res., Auburn, Ala.)

undertaken to keep droppings from contaminating the feed. Although the öocysts resist freezing and certain disinfectants and may remain viable outside the body for one or two years, they are readily destroyed by direct sunlight and complete drying.

Some of the sulfonamides, antibiotics, astringents, and antidiarrhetic compounds may be given to support affected animals. Also, amprolium, a drug long used against coccidiosis of chickens, appears to be very effective.

If coccidiosis is suspected, the veterinarian should be consulted for treatment of affected animals and for advice on the management steps necessary to prevent further losses in the herd.

Face Fly

The face fly, *Musca autumnalis*, was first found in this country in New York in 1953. It is a close relative of and similar in appearance to the housefly.

The face fly is primarily a pest of cattle—particularly animals in open fields and away from shade, although it will attack horses and sheep.

DISTRIBUTION AND LOSSES CAUSED BY FACE FLIES

By 1970, face flies had been reported in most of the states. Only a few of the states in southern United States have remained free of face flies.

It is estimated that the face fly causes annual losses of $111,384,000.[29]

---

[29]*Losses in Agriculture*, Agriculture Handbook No. 291, 1965, p. 82, reported average annual losses of $68,000,000 for the period 1951-60. The author updated the latter figure by applying 1974 cattle numbers (× 1.17) and prices (× 1.4), to give the figure of $111,384,000.

DAMAGE INFLICTED; SYMPTOMS AND SIGNS OF AFFECTED ANIMALS

The face fly does not bite. It feeds on moist areas around the eyes and muzzle and on blood that oozes from cuts and wounds. It is extremely annoying to animals; interfering with vision and breathing, and preventing normal grazing. Large populations force animals to leave pastures and seek relief in wooded areas and shelter.

When cattle enter a barn or darkened area, the fly leaves the animal's face and rests on fence posts, gates, sides of barns, etc. Adult flies hibernate in attics and other protected places during the winter.

The life cycle of the face fly closely resembles that of the horn fly in that fresh manure is required for the development of each; both lay their eggs in fresh manure, where the larvae develop.

PREVENTION, CONTROL, AND TREATMENT

Prevention consists in scattering or removing fresh cow manure. Face flies on cattle may be controlled by regular spraying or dusting or by self-treatment with dust bags or backrubbers.

Gastro-Intestinal Nematode Worms

More than 20 species of harmful parasites inhabit the fourth stomach or abomasum, the small intestine, the large intestine, the lungs, and the liver of cattle; among them, the following: stomach worms, bankrupt worms, medium brown stomach worms, nematodirus, threadworms, hookworms, cooperia, nodular worms, whipworms, lungworms, tapeworms, deer liver fluke, and coccidia.

When their presence is suspected, their diagnosis should be confirmed by microscopic examination by the veterinarian of fresh manure samples. Such examination of the samples will show if parasite eggs or öocysts are present, and in what numbers.

DISTRIBUTION AND LOSSES CAUSED BY GASTRO-INTESTINAL NEMATODE WORMS

One or more species of internal parasites of cattle are found throughout the United States, but they may be especially severe in the South and on irrigated and permanent pastures.

The losses, which are estimated at $163,875,340 annually for all U.S. cattle and calves (both beef and dairy animals)[30] are in terms of lowered feed efficiency caused by disturbed digestion, lowered meat and milk production, and some death losses. Young animals are more severely affected than mature cattle.

---

[30]*Losses in Agriculture*, Agriculture Handbook No. 291, 1965, p. 79, reported average annual losses of $100,046,000 for the period 1951-60. The author updated the latter figure by applying 1974 cattle numbers ($\times$ 1.17) and prices ($\times$ 1.4), to give the figure of $163,875,340.

LIFE HISTORY AND HABITS

Most of the parasites listed above have similar life cycles, of which the common stomach worm is typical. Infected cattle carry the mature worms in the fourth stomach. Eggs from these worms are expelled with manure and develop on the pasture into infective larvae. Then cattle become infected by eating grass infected with the larvae. The latter develop into mature worms in the stomach and these again produce eggs that recontaminate the pasture.

DAMAGE INFLICTED; OR SYMPTOMS AND SIGNS OF AFFECTED ANIMALS

An animal with a light infection rarely shows any outward symptoms, and the symptoms are not specific. But infected animals generally show loss of weight, retarded growth, anemia, diarrhea, and/or lowered resistance to other diseases. With a heavy infection there may be a swelling under the jaw (bottle jaw); and the parasites may even cause the death of the animals.

Fig. 12-32. Calf with stomach worms. (Courtesy, Department of Veterinary Pathology and Hygiene, College of Veterinary Medicine, University of Illinois)

PREVENTION, CONTROL, AND TREATMENT

Preventive and control measures include (1) rotating pastures, (2) segregating calves from mature animals, (3) cross-grazing with cattle and horses, (4) avoiding overstocking or overgrazing pastures since most of the infective larvae are on the bottom inch of grass, and (5) keeping feeders and waterers sanitary.

Treatment consists of therapeutic doses in the spring and again in the fall (or at such other times as necessary) of one of the following drugs, given according to manufacturers' directions: phenothiazine, thiabendazole (Thibenzole or T.B.Z.), Tox-I-Ton (a copper sulfate drench), Baymix (an organophosphate), Haloxen (an organophosphate), or tetramisole (Levamisole, Tramisol, or Ripercol). Tetramisole is also effective on lung worms. From this long list, the cattleman and his veterinarian can (1) rotate drugs to prevent parasitic resistance (for example, alternating treatment with such anthelmintics as Phenothiazine, Thibenzole, and Tramisol would be preferable to repeated use of any one of them), and (2) select the method of administration easiest to follow (although varying according to drug, they may be given as a drench, bolus, feed or mineral mix, paste, or injection). Phenothiazine at the rate of 1 gram per day mixed in the salt affects the reproduction of parasites not removed by treatment.

Grubs (Heel Fly, or Warbles)

Cattle grubs are the maggot stage of insects known as heel flies, warble flies, or gadflies. Two species of cattle grubs are present in the United States.

Fig. 12-33. Adult female heel fly (*Hypoderma lineatum*) whose maggot stage is responsible for the common cattlegrub. (Courtesy, USDA)

The northern cattle grub (*Hypoderma bovis*) occurs mainly in the north though it is found as far south as southern California, northern Arizona, Oklahoma, Tennessee, South Carolina, and Hawaii. The common cattle grub (*Hypoderma lineatum*) occurs throughout the 48 contiguous states and in Hawaii and Alaska. The cattle grub or heel fly is probably the most destructive insect attacking beef and dairy animals.

DISTRIBUTION AND LOSSES CAUSED BY CATTLE GRUBS

The species *Hypoderma lineatum* is widely distributed throughout the United States, whereas *Hypoderma bovis* is chiefly confined to the northern states.

The total annual economic loss from cattle grubs in the United States, including both beef and dairy animals, is estimated at $314,496,000,[31] and it may even exceed this figure when decreased gains and milk production are taken into consideration. The damage inflicted by cattle grubs affects cattlemen, packers, tanners, and, finally, consumers. The kinds of losses include the following:

1. *Decreased gains or milk production, mechanical injury, or even death*—Though the fly does not bite or sting, when it lays its eggs on the lower leg, it usually terrifies the animal, causing it to run with tail hoisted, seeking relief. It may run through fences, over cliffs, or become hopelessly bogged down in a mudhole or swamp. Beef animals suffer weight losses, and milk production from dairy cows may be reduced from 10 to 25 percent during the period heel flies are laying their eggs. Livestock Conservation, Inc. is authority for the statement that grub treatment can mean the following to cattlemen: 20 to 25 lb more per head in the feedlot, 30- to 40-lb heavier weaning weights, 50-lb heavier yearlings, and $25 per head more at slaughter.

Fig. 12-34. Heifers running away from heel flies. Though the fly does not bite or sting, when it lays its eggs on the lower leg, it usually terrifies the animal, causing it to run with tail hoisted, seeking relief. (Courtesy, Livestock Conservation, Inc.)

2. *Carcass damage*—According to meat packers, about 35 percent of all beef carcasses are damaged by grubs. The yellowish, watery patches caused

[31]*Losses in Agriculture*, Agriculture Handbook No. 291, 1965, p. 82, reported average annual losses of $192,000,000 for the period 1951-60. The author updated the latter figure by applying 1974 cattle numbers (× 1.17) and prices (× 1.4), to give the figure of $314,496,000.

by the migration of the larvae under the skin are referred to by butchers as "pilled" or "licked beef." Two to three pounds of "jellied" beef must be trimmed from the loins and ribs of each "grubby" animal, and the damaged cut of meat is devalued two cents per pound because of the ragged and unattractive appearance.

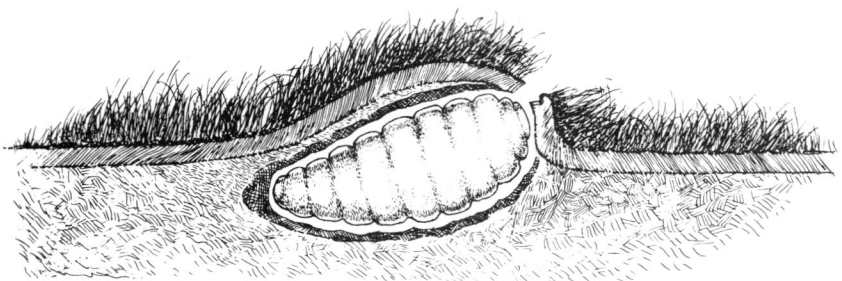

Fig. 12-35. Hole in the hide on the back of an animal, with a grub inside. Note hide opening through which the grub obtains air and finally escapes. (Courtesy, USDA)

3. *Injury of hides*—Approximately one-third of all cattle hides produced in the United states are damaged by grubs. This loss is caused by the migration of the grub through the back, which leaves a scar in the most valuable part of the hide. According to trade custom, if a hide has as many as five grub holes, it is classed as grade No. 2 and is subject to a discount of one cent per pound. Commonly as many as 40, and occasionally 100 or more, grub holes are found in a single hide. Hides of the latter quality are not considered worth tanning and are sold for by-products.

4. *Shock to animal*—In certain older animals that have been previously sensitized, the breaking of a grub under its skin may cause a terrific reaction (anaphylaxis or allergic reaction). The area may be greatly swollen and form an abscess, and there may be such a general reaction that the animal may die from shock. At the first signs of shock, the local veterinarian should be summoned quickly, to administer appropriate stimulants.

LIFE HISTORY AND HABITS

Basically the two species have a similar life cycle. The female flies, called heel flies, attach their eggs to the hairs of the legs and bodies of cattle. The eggs hatch into larvae after about 3 days and enter the animals at the bases of the hairs. Once inside, the common cattle grub migrates from the point of entry to the gullet; the northern cattle grub migrates to the spinal column (both migrations take 2 to 4 months). After some additional months in the gullet or spinal column, grubs of both species migrate to the animals' backs, cut breathing holes in the hide, and remain there for about 6 weeks while they increase greatly in size. The resultant swellings are often called wolves or warbles. Fully grown grubs leave the hide through the breathing holes and drop to the ground where they pupate. Then, in a few weeks they

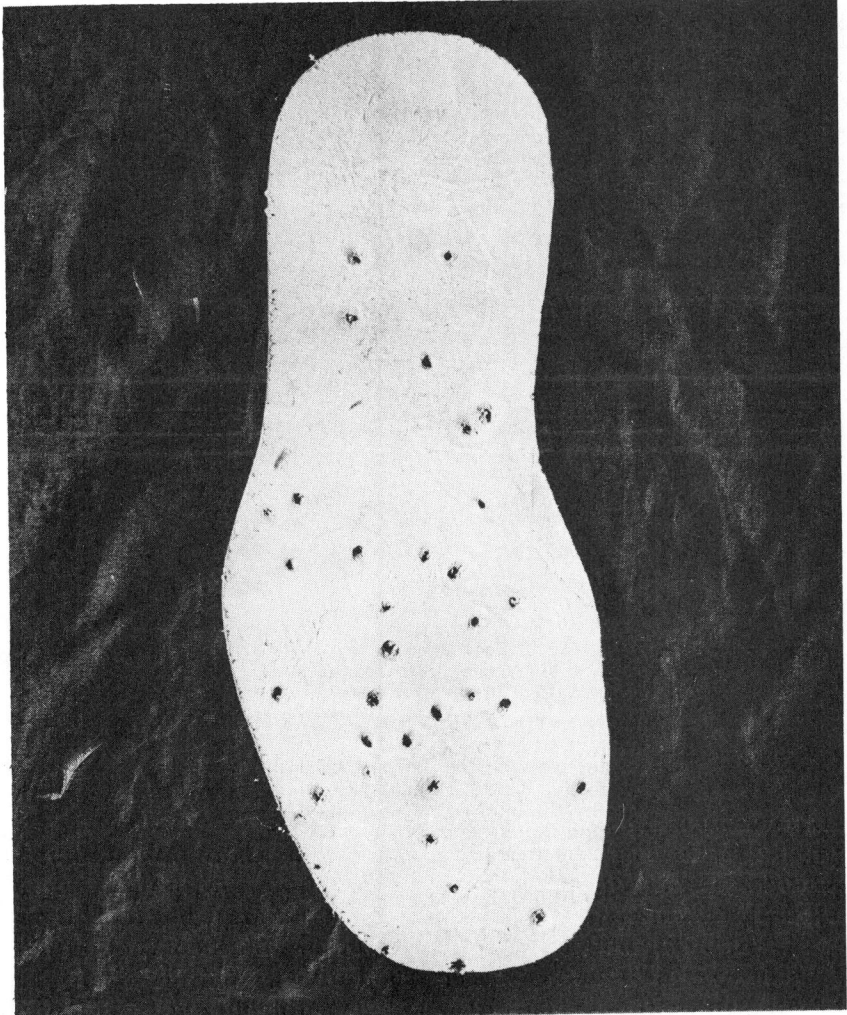

Fig. 12-36. Shoe sole damaged by cattle grubs. About one-third of all cattle hides produced in the U.S. are damaged by grubs. (Courtesy, Livestock Conservation, Inc.)

transform to nonfeeding adult flies that emerge and mate. On bright sunny days the female then seeks cattle for egg laying, which causes gadding. The entire life cycle takes about 1 year, and the same stages are usually found at about the same time each year in any given area.

DAMAGE INFLICTED; SYMPTOMS AND SIGNS OF AFFECTED ANIMALS

The attack of the heel fly is unmistakable when, in the spring or early summer, cattle are seen madly running with their tails hoisted high over their backs in an attempt to escape. The presence of the grub (larva) in the

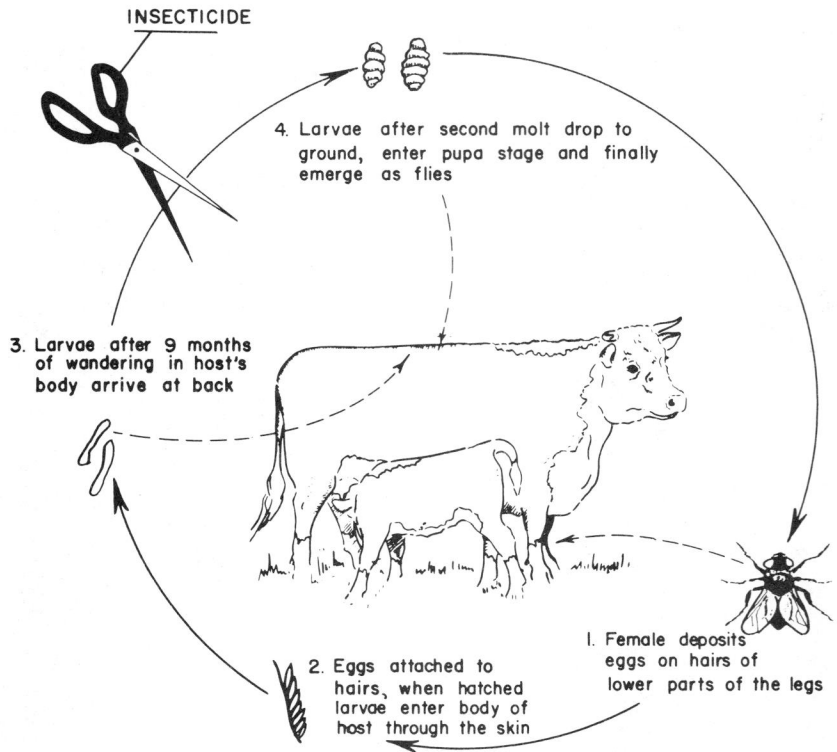

INSECTICIDE

4. Larvae after second molt drop to ground, enter pupa stage and finally emerge as flies

3. Larvae after 9 months of wandering in host's body arrive at back

2. Eggs attached to hairs, when hatched larvae enter body of host through the skin

1. Female deposits eggs on hairs of lower parts of the legs

Fig. 12-37. Diagram showing the life history and habits of cattle grubs. As noted (see scissors) effective control and treatment (cutting the cycle of the parasite) may be obtained by the application of a suitable insecticide (spray, dip, or dust). The first application should be made 25 to 30 days after grubs first appear in the back, with subsequent treatments, at 30-day intervals thereafter as long as grubs are present in the back. Control may also be secured by the use of a systemic insecticide. (Drawing by R. F. Johnson)

back, usually from December to May, causes a characteristic swelling (and an opening in the skin, from which pus is discharged), which usually becomes conspicuous, so that a grubby back has a lumpy appearance.

PREVENTION, CONTROL, AND TREATMENT

Complete prevention of cattle grub damage within any given herd cannot be obtained unless all cattle grubs throughout the country are exterminated. This means a nationwide campaign in which all cooperate to eradicate the menace, farm by farm, county by county, and state by state.

Cattle grubs are most effectively controlled by applying a systemic insecticide to cattle as soon as possible after the activity of the heel flies ceases since these insecticides kill the young larvae in the animal's body. When the grubs are near the back or located in the back, treatments are less effective, and possible side effects are more likely. Side effects may also occur when there is a concentration of grubs in the gullet or spinal cord of treated cattle. A single treatment with a systemic insecticide should give excellent control of cattle grubs. For the correct timing, each owner is advised to check with his local County Agent or Consultant. Systemics may be administered as sprays, dips, pourons, or as feed additives. Never use more than one systemic insecticide at a time, and always use a systemic in keeping with manufacturers' directions.

Horn Fly (Cattle Fly, Cow Fly, or Stock Fly)

This fly is one of the most numerous and worst annoyances of cattle. It is often found resting at the base of the horn; hence, the name horn fly. But by no means does it confine itself to this location on the animal. Horn flies may congregate by the hundreds or even thousands on the backs, shoulders, and bellies of cattle.

Fig. 12-38. Cattle heavily infested with horn flies at the base of the horn, hence the name horn fly. (Courtesy, USDA)

DISTRIBUTION AND LOSSES CAUSED BY THE HORN FLY

The horn fly is widely distributed throughout the United States. It is one of the most numerous of the biting flies affecting cattle in North America. An average of 4,000 to 5,000 flies per animal is not uncommon, and individual animals may support as many as 10,000 to 20,000 flies. Horn flies are more of a problem on grazing cattle than of cattle held in lots. The presence of this insect produces irritation and worry, loss of blood, reduced vitality, and, in the South, sores that may become infested with screwworms. From an

Fig. 12-39. Enlargement of an adult female horn fly. It is about one-half the size of an ordinary housefly. (Courtesy, USDA)

economic standpoint, beef cattle gains are sharply reduced, and dairy cattle suffer lowered milk production. Studies indicate that the weight gains of beef animals may be reduced by as much as 50 pounds per animal during fly season when high populations of horn flies go uncontrolled. Losses, to both beef and dairy animals, are estimated at $293,202,000 annually.[32]

LIFE HISTORY AND HABITS

This fly, *Haematobia irritans,* is about one-half the size of an ordinary housefly or the stablefly and possesses a piercing beak. Unlike the housefly and stablefly, it remains on cattle throughout the day and night. Horn flies usually feed twice a day, sometimes more frequently.

The life cycle and habits of the horn fly are as follows:

1. The adult fly leaves the cow only for a brief five- to ten-minute period

---

[32]*Losses in Agriculture,* Agriculture Handbook No. 291, 1965, p. 82, reported average annual losses of $179,000,000 for the period 1951-60. The author updated the latter figure by applying 1974 cattle numbers (× 1.17) and prices (× 1.4), to give the figure of $293,202,000.

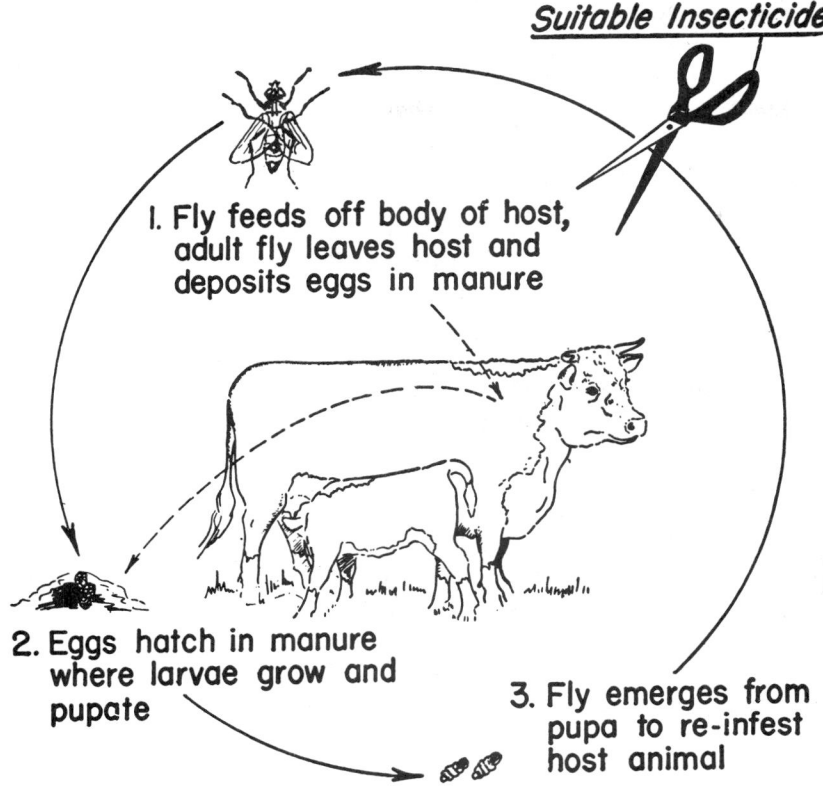

Suitable Insecticide

1. Fly feeds off body of host, adult fly leaves host and deposits eggs in manure

2. Eggs hatch in manure where larvae grow and pupate

3. Fly emerges from pupa to re-infest host animal

Fig. 12-40. Diagram showing the life history and habits of the horn fly. As noted (see scissors) effective control (cutting the life cycle of the parasite) consists of thorough spraying of animals with a suitable insecticide throughout the fly season. First and foremost, however, it is important that fresh droppings be spread with a spring tooth harrow to hasten their drying and that accumulations of cattle manure around barns be hauled out at frequent intervals and spread thinly on the land. (Drawing by R. F. Johnson)

to lay eggs. The eggs are laid beneath the droppings, where they are protected from the sun and rain and hatch out into tiny white maggots in about 16 hours. Each female fly is capable of laying 375 to 400 eggs during a lifetime.

2. The maggots crawl into the droppings where they feed and grow. They become full-grown in another four days; then they crawl down into the lower part of the droppings or into the soil to pupate and later emerge as a fly.

3. About an hour after the fly emerges from its pupal case, it seeks the nearest cow where it settles and starts feeding. The fly may mate as early as the second day after emergence and may deposit eggs on the third day.

The entire life cycle of the horn fly averages about 9 to 12 days during the summer months, and the adult fly lives about 7 weeks. Thus, it is small wonder that such hordes of this insect exist.

DAMAGE INFLICTED; SYMPTOMS AND SIGNS OF AFFECTED ANIMALS

The tormented cattle often refuse to graze during the day and seek protection by hiding in dark buildings, brush, or tall grass. Heavily infested cattle may also have a rough, sore skin, and they suffer an inevitable loss in condition.

PREVENTION, CONTROL, AND TREATMENT

Prevention and control rest chiefly in disturbing the main breeding ground of the horn fly. This is best accomplished by spreading fresh droppings with a spring tooth harrow in order to hasten their drying. Running pigs with cattle will accomplish the same purpose. These methods of control are not practical on extensive grazing areas, but they may be used in small pastures.

The horn fly, because of its dependency on the host animal, can be very effectively controlled in the adult stage by treating cattle. When sprays are used, only a small deposit of insecticide on the hair is sufficient. Also, complete coverage is not necessary because horn flies move about on the animal sufficiently to come in contact with insecticide deposited on almost any part of the animal's body. Thus, protection of beef cattle from attacks of horn flies can be accomplished by use of sprays, dusts, dust bags, dips, backrubbers, and pourons.

Horse Flies and Deer Flies

An annoying group of biting flies that attack cattle, the Tabanidae, has two particularly troublesome genera, *Tabanus* (horse flies) and *Chrysops* (deer flies).

DISTRIBUTION AND LOSSES CAUSED BY HORSE FLIES AND DEER FLIES

Tabanids are found in all parts of the United States, and large numbers may be expected wherever there are extended areas of permanently wet, undeveloped land and a mild climate. Generally, horse flies are more of a problem to livestock than deer flies, but deer flies are often extremely annoying in the coastal areas of the south and the mountain areas of the west.

LIFE HISTORY AND HABITS

Females of both genera lay masses of eggs (as many as 1,000 eggs/mass) on foliage or on other objects that project over water or moist ground. The eggs hatch in 5 to 7 days, and the larvae drop into the water or upon the moist soil. Then they burrow into soil and feed on organic matter that may include the juices of other insect larvae, other tabanid larvae, and earthworms. Larvae can ordinarily be found in the top 2 or 3 inches of soil in swamps and around lakes, ponds, and permanent streams; however, recent studies indicate that they may also be widely distributed in drier soils such as forest floors and decaying wood. Generally, tabanids remain in the larval stage during the summer, fall, and winter months; then in the spring, they move to drier soil where they pupate. The pupal stage lasts 2 to 3 weeks. Most species complete one generation a year.

DAMAGE INFLICTED; SYMPTOMS AND SIGNS OF AFFECTED ANIMALS

Male horse flies and deer flies feed on vegetable sap, and some may suck juices of soft-bodied insects, but they do not attack warmblooded animals. Female horse flies and deer flies require a blood meal soon after emergence. They feed primarily during the daytime, but some feed at dusk and dawn, and a few species will attack animals in complete darkness. The bite from the slashing mouthparts of these insects is very painful, and animals generally try to dislodge the fly with their tail or tongue or by stamping their feet. Heavily attacked animals stop grazing and tend to bunch together or seek shelter. Severe outbreaks can seriously affect weight gain. Moreover, tabanids are also implicated in disease transmission because their habit of feeding on one animal and immediately attacking another can result in the direct mechanical transfer of pathogenic organisms that live in blood.

PREVENTION, CONTROL, AND TREATMENT

No really satisfactory method exists of controlling horse flies or deer flies. If possible, avoid pasturing cattle near swampy wooded areas when these flies are numerous. Also, sheltering animals is often beneficial since tabanids do not ordinarily enter enclosures.

Houseflies

Houseflies (*Musca domestica*) are nonbiting flies that are common around barns and lots.

DISTRIBUTION AND LOSSES CAUSED BY HOUSFLIES

Houseflies become numerous both inside and outside barns and farm buildings. Perhaps they are the most abundant insect pest of feedlots. Houseflies are annoying to cattle and people, and they can spread human and animal diseases.

LIFE HISTORY AND HABITS

Houseflies breed in manure, garbage, and decaying vegetable matter. The eggs hatch after an incubation period of 12 to 36 hours. The larvae feed on the organic medium and grow to full size in 6 to 11 days.

DAMAGE INFLICTED; SYMPTOMS AND SIGNS OF AFFECTED ANIMALS

Although houseflies are nonbiting, they cause serious economic losses through annoyance of cattle and by disease transmission. Also, they create public health problems.

PREVENTION, CONTROL, AND TREATMENT

Insecticides alone will not control houseflies. Adequate sanitary measures, including proper disposition or handling of manure, are necessary to eliminate fly breeding areas. Spread manure thinly in fields so fly eggs and larvae will be killed by drying and heat.

Several insecticides in spray or bait forms may be used to control adult flies in barns.

Lice

The louse is a small, flattened, wingless insect parasite of which there are several species.

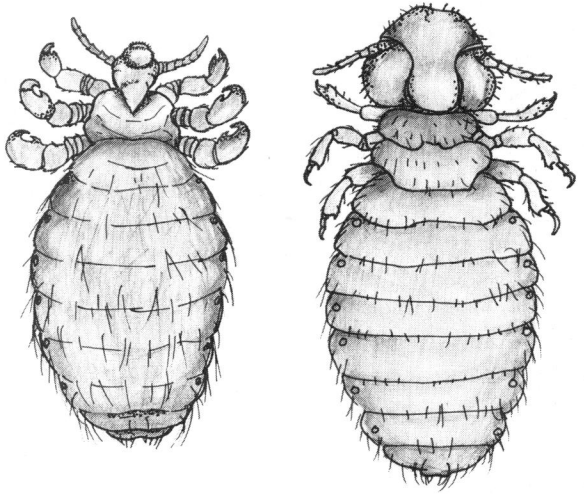

Fig. 12-41. Two species of cattle lice: (1) (left) hairy cattle louse—a blood-sucking louse; and (2) (right) chewing, or biting, cattle louse.

Cattle are attacked by four species of bloodsucking lice—the short-nosed cattle louse (*Haematopinus eurysternus*), the cattle tail louse (*Haematopinus quadripertusus*), the long-nosed cattle louse (*Linognathus vituli*), and the little blue louse (*Solenopotes capillatus*), and by one species of biting louse, the cattle biting louse (*Bovicola bovis*). Cattle lice will not remain on other farm animals, nor will lice from other animals infest cattle. Lice are always more abundant on weak, unthrifty animals and are more troublesome during the winter months than during the rest of the year.

DISTRIBUTION AND LOSSES CAUSED BY LICE

The presence of lice upon animals is almost universal, but the degree of infestation depends largely upon the state of animal nutrition and the extent to which the owner will tolerate parasites. The irritation caused by the presence of lice retards growth, gains, and/or production of milk. The combined losses from cattle lice, including both beef and dairy animals, are estimated at $76,986,000, based on 1974 figures.[33]

LIFE HISTORY AND HABITS

Lice spend their entire life cycle on the host's body. They attach their eggs or nits to the hair near the skin where they hatch in about 2 weeks. Two weeks later the young females begin laying eggs, and after reproduction they die on the host. Lice do not survive more than a week when separated from the host, but, under favorable conditions, eggs clinging to detached hairs may continue to hatch for 2 or 3 weeks.

DAMAGE INFLICTED; SYMPTOMS AND SIGNS OF AFFECTED ANIMALS

Infestation shows up most commonly in winter in ill-nourished and neglected animals. There is intense irritation, restlessness, and loss of condition. As many lice are blood suckers, they devitalize their host. There may be

Fig. 12-42. Cow's neck heavily infested with short-nosed sucking lice. These pests seek the sheltered parts of the body on which to feed. (Courtesy, Department of Veterinary Pathology and Hygiene, College of Veterinary Medicine, University of Illinois)

---

[33]*Losses in Agriculture*, Agriculture Handbook No. 291, 1965, p. 82, reported average annual losses of $47,000,000 for the period 1951-60. The author updated the latter figure by applying 1974 cattle numbers (× 1.17) and prices (× 1.4), to give the figure of $76,986,000.

severe itching and the animal may be seen scratching, rubbing, and gnawing at the skin. The hair may be rough, thin, and lack luster; and scabs may be evident. In cattle, favorite locations for lice are the root of the tail, on the inside of the thighs, over the fetlock region, and along the neck and shoulders. In some cases, the symptoms may resemble that of mange and it must be kept in mind that the two may occur simultaneously.

With the coming of spring, when the hair sheds and the animals go to pasture, the problem of lice is greatly diminished.

PREVENTION, CONTROL, AND TREATMENT

Because of the close contact of domesticated animals, especially during the winter months, it is practically impossible to prevent entire herds from

Fig. 12-43. Spraying cattle with a power sprayer. Two- to four-hundred pounds pressure is adequate. (Courtesy, Food Machinery and Chemical Corporation)

becoming slightly infested with the pests. Nevertheless, lice can be kept under control.

For effective control, all members of the herd must be treated simultaneously at intervals, and this is especially necessary during the fall months about the time they are placed in winter quarters. Cattle should be inspected for lice periodically throughout the winter and spring and retreated when necessary. Insecticides applied by spraying or dipping are the most effective against lice, but some control may be obtained by dusting.

Liver Fluke

The liver fluke, *Fasciola hepatica,* is a flattened, leaf-like, brown worm, usually about an inch long. It affects cattle, sheep, goats, and other animals.

DISTRIBUTION AND LOSSES CAUSED BY LIVER FLUKE

The liver fluke is distributed throughout the world, wherever there are low-lying wet areas and suitable snails. In the United States, it is most common in some of the areas of the western and southwestern range country and in Florida.

Lowered gains and milk production and feed inefficiency are the chief losses. In addition, vast quantities of liver are condemned each year at the time of slaughter—an estimated 1,200 tons annually. In packinghouses, such livers are referred to as "fluky livers" or "rotten livers." In total, the annual U.S. losses from liver fluke are estimated at $4,950,036, for beef and dairy animals combined.[34]

LIFE HISTORY AND HABITS

Flukes reproduce by means of eggs which, after passing from the host, hatch into embryos equipped with cilia that enable them to move about. Upon encountering certain kinds of snails, they penetrate into the body of the intermediate host and develop into cercariae (flukes in the larval stage), which leave the snails and become encysted on the nearby vegetation. The encysted cercariae are then ingested by animals during grazing. The fluke is liberated from the cyst, penetrates the intestinal wall, migrates about the abdominal cavity, and finally reaches the liver where maturity is attained two or three months after infestation.

DAMAGE INFLICTED; SYMPTOMS AND SIGNS OF AFFECTED ANIMALS

Infested cattle show anemia, as indicated by pale mucous membranes, digestive disturbances, loss of weight and general weakness. As with most parasites, positive diagnosis consists of finding eggs in the feces by microscopic examination.

---

[34]*Losses in Agriculture*, Agriculture Handbook No. 291, 1965, p. 73, reported average annual losses of $3,022,000 for the period 1951-60. The author updated the latter figure by applying 1974 cattle numbers (× 1.17) and prices (× 1.4), to give the figure of $4,950,036.

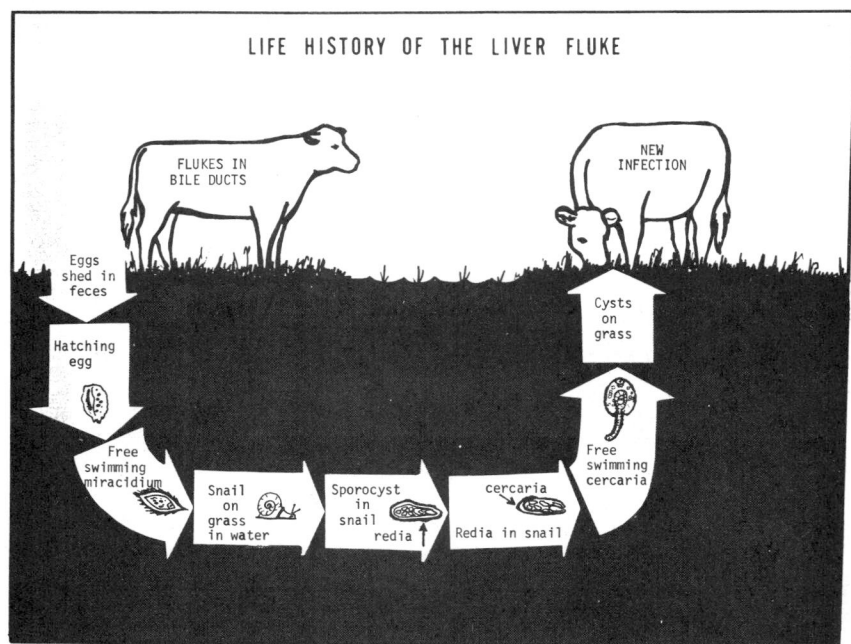

Fig. 12-44. Diagram showing the life history and habits of the liver fluke, *Fasciola hepatica*. When the eggs of the liver fluke hatch, they liberate larva called miracidium. The miracidium is ciliated and swims about in the water until it comes in contact with a suitable snail, into which it bores. Then it changes a number of times, finally becoming what is known as a redia. Large numbers of minute stages, shaped like tadpoles and known as cercariae, are produced in the redia. These eventually escape from the snail, swim about, and become encysted on grass or other vegetation.

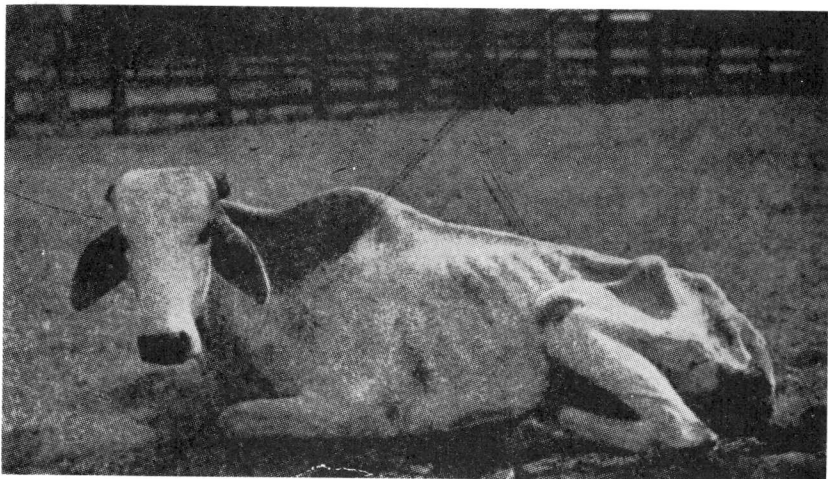

Fig. 12-45. Brahman bull dying from fluke infestation. (Courtesy, Pitman-Moore Co., Indianapolis, Ind.)

PREVENTION, CONTROL, AND TREATMENT

The following measures are recommended for the prevention and control of liver fluke:

1. Drainage or avoidance of wet pastures.
2. Where relatively small snail-infested areas are involved, it may be practical to destroy the snail (carrier of liver fluke), preferably in the spring season, through—

    a. Applying 3 to 6 lb of copper sulfate (bluestone or blue vitrol) per acre of grassland, mixing and applying the small quantity of copper sulfate with a suitable carrier (such as a mixture of 1 part of the copper sulfate to 4 to 8 parts of either sand or lime), and

    b. Treating ponds or sloughs with 1 part of copper sulfate to 500,000 parts of water.

When copper sulfate is used in the dilutions indicated, it is not injurious to grasses and will not poison farm animals, but it may kill fish.

Snail-infested pastures should not be used for making hay.

The recommended treatment, which should be under the direction of a veterinarian, consists in drenching with hexachlorethane. It kills the common bile duct fluke, *Fasciola hepatica*, with variable results. But it is not effective against the large American liver fluke, *Fascioloides magna*.

Lungworm

The lungworm, *Dictyocaulus viviparus*, is a white, threadlike worm 1½ to 3 inches long, found in the trachea and bronchi of cattle—especially calves.

DISTRIBUTION AND LOSSES CAUSED BY LUNGWORMS

Lungworms are distributed throughout the United States, especially on wet pastures.

The losses are chiefly in lowered feed efficiency, milk, and meat production. With a heavy infestation, there may be death losses.

LIFE HISTORY AND HABITS

In the bronchial tubes of the lungs, female lungworms produce large numbers of eggs. Usually, these hatch in the air passages and liberate larvae that are coughed up, swallowed, and eliminated in the feces. Sometimes the coughed-up eggs hatch in the stomach or intestines, but they may pass unhatched from the host, particularly when there is severe diarrhea.

Under favorable conditions, the larvae eliminated with the feces develop into the infective stage in about a week. Then they crawl upon blades of grass, where they are ingested by grazing cattle. Thence they penetrate the intestinal wall and reach the lymph glands, from which they are eventually carried to the lungs.

Cattle lungworms mature in 3 to 4 weeks, at which time larvae appear in the feces. The worms live from 2 to 4 months in the host.

DAMAGE INFLICTED; SYMPTOMS AND SIGNS OF AFFECTED ANIMALS

Typical symptoms include coughing, labored breathing, loss of appetite, unthriftiness, and intermittent diarrhea. Death may follow, probably from suffocation or pneumonia.

PREVENTION, CONTROL, AND TREATMENT

Where lungworms are found, the following prevention and control measures are recommended:
1. Practice rigid sanitation.
2. Where practical, segregate calves from older animals.
3. Keep calves on a good ration.
4. Do not spread infested manure on pastures.
5. Utilize dry pasture if possible.

Tramisol, which is also effective against gastro-intestinal worms, is the drug of choice for lungworm treatment. Like all drugs, it should be given according to the manufacturer's directions.

Measles (Cysticercosis, or Measly Beef)

A parasitic disease of cattle, cysticercosis is an invasion of the musculature and viscera by larvae, *Cysticercus bovis*, of *Taenia saginata,* the beef tapeworm of man. Cattle are the intermediate host and people the definite host. The name is a misnomer, for it has no relationship, and little resemblance, to human measles. In humans, the disease is caused by a virus; in cattle, the disease is caused by a tapeworm cyst which lodges and grows in the muscle tissue.

DISTRIBUTION AND LOSSES CAUSED BY MEASLES

Beef measles is worldwide. However, the incidence is highest in Africa, the Middle East, Asia, and South America. In the United States, the measles problem is largely confined to the Southwest.

At the time of slaughter, losses result from extensive trimming and prolonged storing of mildly infected carcasses and from condemning heavily infected carcasses. In 1967, the total loss, including cost of inspection, was estimated at $740,000.

LIFE HISTORY AND HABITS

Beef measles is transmitted to cattle by contamination of their feed and water with viable tapeworm eggs. The eggs are ingested by cattle. Within the bovine intestine, the embryo hatches, penetrates the intestinal wall, enters the bloodstream, wends its way to muscle tissue where it develops into "measles" or cysticerci. Each cysticercus appears like a small (when fully developed, they are about ¼ inch in diameter) white balloon filled with fluid. Within the cyst is the head of a new tapeworm. People become infected by eating rare beef.

Humans are the sole host for the adult tapeworm, and cattle are the only intermediate host. No other animals are involved in the life cycle. Thus, control consists of disposing of human excrement in such manner that it cannot come in contact with cattle.

DAMAGE INFLICTED; SYMPTOMS AND SIGNS OF AFFECTED ANIMALS

Most cases of beef measles produce few signs in live cattle. In the carcass, the cysticerci (cysts) are readily discernible. Carcasses that are excessively infested are unsatisfactory for food and should be condemned. If only a few cysticerci are found, the entire carcass is frozen sufficiently long to insure that the cysticerci are killed.

PREVENTION, CONTROL, AND TREATMENT

In endemic areas, workers employed in and around feedlots, cattle pastures, and dairies should be medically examined for beef tapeworm parasitism, and infested individuals should be treated for removal of the worms. Sanitary latrines should be provided for caretakers, and people should be forbidden to defecate in feedlots or pastures where cattle feed. At slaughter, cysticercus-infested meat should be disposed of in a manner which avoids the inclusion of viable cysticerci in human food. Meats should be thoroughly cooked to destroy viable cysticerci.

No effective treatment for bovine measles is known.

Mites (Mange, or Scabies)

Scabies in cattle, also known as scab, mange, or itch, is caused by mites living on or in the skin. Three of the more important species are the psoroptic or common scab mite (*Psoroptes equi* var. *bovis*), the sarcoptic mange mite (*Sarcoptes scabei* var. *bovis*), and the chorioptic or symbiotic scab mite (*Chorioptes bovis*).

DISTRIBUTION AND LOSSES CAUSED BY MITES

Injury from mites is caused by blood sucking and the formation of scabs and other skin affections. In a severe attack, the skins may be much less valuable for leather. Growth is retarded, and production of meat and milk is lowered.

Mites cause estimated average annual losses to all cattle and calves of $4,914,000.[35]

---

[35]*Losses in Agriculture*, Agriculture Handbook No. 291, 1965, p. 82, reported average annual losses of $3,000,000 for the period 1951-60. The author updated the latter figure by applying 1974 cattle numbers ($\times$ 1.17) and prices ($\times$ 1.4), to give the figure of $4,914,000.

LIFE HISTORY AND HABITS

The mites that attack cattle breed exclusively on the bodies of their hosts, and will live for only 2 or 3 weeks when removed therefrom. The female mite which produces sarcoptic mange—the most severe form of scabies—lays from 10 to 25 eggs during the egg-laying period, which lasts about 2 weeks. At the end of another 2 weeks, the eggs have hatched and the mites have reached maturity. A new generation of mites may be produced every 15 days.

Mites are more prevalent during the winter months, when animals are confined and in close contact with each other.

DAMAGE INFLICTED; SYMPTOMS AND SIGNS OF AFFECTED ANIMALS

When the mite pierces the skin to feed on cells and lymph, there is marked irritation, itching, and scratching. Exudate forms on the surface, and this coagulates, crusting over the surface. The crusting is often accompanied or followed by the formation of a thick, tough, wrinkled skin. Often there are secondary skin infections. The only certain method of diagnosis is to demonstrate the presence of the mites.

Fig. 12-46. A severe case of mange on a cow, caused by sarcoptic mites. Note the rough and wrinkled condition of the skin, with crusting over the surface. (J. W. McManigal, Agricultural Photographer, Horton, Kan.)

PREVENTION, CONTROL, AND TREATMENT

Prevention consists of avoiding contact with diseased animals or infested premises. Mange is a reportable disease; hence, infestations should be reported to the proper livestock inspection authorities. The presence of certain species of cattle scab mites will result in the herd being quarantined. This prohibits their movement until they are inspected and found free of scabies.

Mites can be controlled by spraying or dipping infested animals with suitable insecticidal solutions, and by quarantine of affected herds.

Title 9 of the Code of Federal Regulations, part 73, scabies in cattle, requires that affected cattle be dipped in a permitted dip prior to interstate movement.

(Also see later section in this chapter entitled, "Federal and State Regulations Relative to Disease Control.")

Fig. 12-47. Dipping cattle. The chief virtue of dipping vats lies in the fact that animals so treated are thoroughly covered. (Courtesy, Washington State University)

Mosquitoes

Mosquitoes, particularly species of the genera *Aedes, Psorophora,* and *Culex,* are a severe nuisance to cattle in many areas, especially in swampy regions that have permanent pools of water or that are exposed to frequent flooding.

DISTRIBUTION AND LOSSES CAUSED BY MOSQUITOES

Mosquitoes are rather widely distributed, but they are most numerous in southeastern United States. Sometimes they kill cattle, although this is rare.

LIFE HISTORY AND HABITS

Almost all female mosquitoes must take a blood meal before they can lay eggs. (The males do not suck blood, but feed on nectar and other plant juices.) Eggs are laid singly or in rafts on the surface of the water or on the ground in depressions that are flooded by tidal waters, seepage, overflow, or rainwater. The larvae and pupae are aquatic.

DAMAGE INFLICTED; SYMPTOMS AND SIGNS OF AFFECTED ANIMALS

Mosquitoes may occur in such abundance that cattle refuse to graze. Instead, they bunch together or stand neck deep in water to protect themselves from attack. Moreover, mosquitoes will annoy cattle day and night, so they can cause serious losses in meat production—or even death in extreme cases. Also, they may be disease carriers.

PREVENTION, CONTROL, AND TREATMENT

Mosquitoes can be controlled in several ways: (1) by elimination of breeding places, through providing fills, ditches, impoundments, improved irrigation methods, and other means of water manipulation; (2) by chemical destruction of larvae, by treating the relatively restricted breeding areas with proper larvicides; and (3) by chemical destruction of adults. Elimination of breeding sites is by far the most satisfactory and effective method of control. However, either this method or chemical destruction of larvae may not be economically practical if the breeding area is extensive. When the latter is the case, control can best be accomplished through group action, such as mosquito abatement districts. The cattle producer can achieve some control by fogging or spraying the pasture or rangeland, and some relief can be obtained by spraying the animals with insecticides.

Ringworm

Ringworm, or barn itch, is a contagious disease of the outer layers of skin. It is caused by certain microscopic molds or fungi (*Trichophyton, Achorion,* or *Microsporon*). All animals and man are susceptible.

DISTRIBUTION AND LOSSES CAUSED BY RINGWORM

Ringworm is widespread throughout the United States. It is a contagious disease of the outer layer of skin caused by certain microscopic molds or fungi, which affects all animals and man. Though it may appear among animals on pasture, it is far more prevalent as a stable disease. It is unsightly, and affected animals may experience considerable discomfort; but the actual economic losses attributed to the disease are not too great.

LIFE HISTORY AND HABITS

The period of incubation for this disease is about one week. The fungi form seed or spores that may live eighteen months or longer in barns or elsewhere.

Ringworm is usually a winter disease, with recovery the following summer when the animals are on pasture.

DAMAGE INFLICTED; SYMPTOMS AND SIGNS OF AFFECTED ANIMALS

Round, scaly areas almost devoid of hair appear mainly in the vicinity of the eyes, ears, side of the neck, or the root of the tail. Crusts may form, and the skin may have a gray, powdery, asbestos-like appearance. The infected patches, if not checked, gradually increase in size. Mild itching usually accompanies the disease.

Fig. 12-48. Heifer with ringworm. The fungi causing these raised circular areas may be transmitted to man. (Courtesy, Department of Veterinary Pathology and Hygiene, College of Veterinary Medicine, University of Illinois)

PREVENTION, CONTROL, AND TREATMENT

The organisms are spread from animal to animal or through the medium of contaminated fence posts, curry combs, and brushes. Thus prevention and control consists of disinfecting everything that has been in contact with infected animals. The infected animals should also be isolated. Strict sanitation is an essential in the control of ringworm.

The hair should be clipped, the scabs removed, and the area brushed and washed with soap. The diseased parts should be painted with tincture of iodine or salicylic acid and alcohol (one part in ten) every three days, until cleared up. Certain proprietary remedies available only from veterinarians have proved very effective in treatment.

Screwworm

The screwworm (*Cochliomyia hominivorax*) was eradicated from the southeastern United States in the late 1950s. Now, many millions of sterile flies are being released to eradicate completely this pest from Mexico and adjoining states of the United States. Since the screwworm fly may infest an animal through wounds and also through lesions caused by ticks, horse flies, or even horn flies, there is an added impetus to control these parasites. They are not found in cold-blooded animals such as turtles, snakes, and lizards.

Man-made wounds resulting from branding, castrating, and dehorning afford a breeding ground for this parasite. Add to this the wounds from some types of vegetation, from fighting, and from blood-sucking insects and ample places for propagation are provided.

DISTRIBUTION AND LOSSES CAUSED BY SCREWWORMS

Screwworm infestations occur occasionally in certain areas of Arizona, New Mexico, and Texas, and occur often in many areas of Mexico and other Latin American countries to the south.

LIFE HISTORY AND HABITS

The primary screwworm fly is bluish green in color, with three dark stripes on its back and reddish or orange color below the eyes. The fly generally deposits its eggs in shingle-like masses on the edges or the dry portion of wounds. From 50 to 300 eggs are laid at one time, with a single female being capable of laying about 3,000 eggs in a lifetime. Hatching of the eggs occurs in eleven hours, and the young whitish worms (larvae or maggots) immediately burrow into the living flesh. There they feed and grow for a period of 4 to 7 days, shedding their skin twice during this period.

When these worms have reached their full growth, they assume a pinkish color, leave the wound, and drop to the ground, where they dig beneath the surface of the soil and undergo a transformation to the hard-skinned, dark-brown, motionless pupa. It is during the pupa stage that the maggot changes to the adult fly.

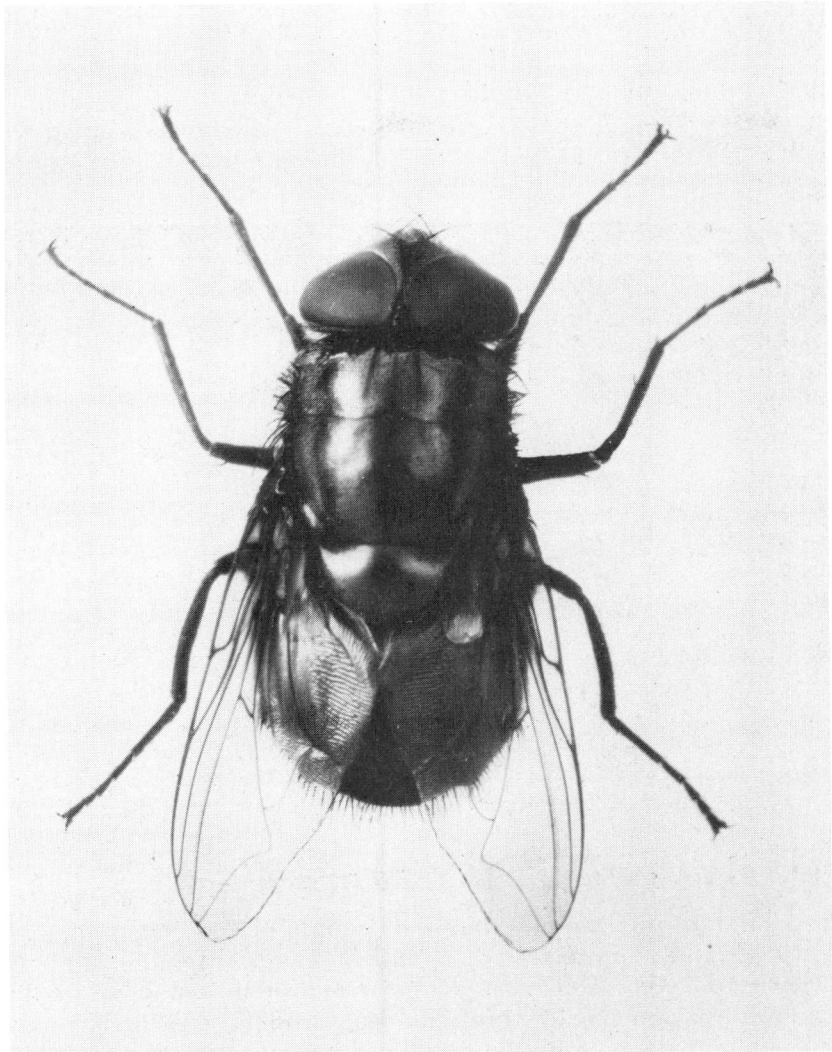

Fig. 12-49. The screwworm fly (*Cochliomyia americana*). This fly is bluish green in color, with three dark stripes on its back and reddish or orange color below the eyes. (Courtesy, USDA)

After the pupa has been in the soil from seven to sixty days, the fly emerges from it, works its way to the surface of the ground, and crawls up on some nearby object (bush, weed, etc.) to allow its wings to unfold and otherwise to mature. Under favorable conditions, the newly emerged female fly becomes sexually mature and will lay eggs five days later. During warm weather, the entire life cycle is usually completed in 21 days, but under cold, unfavorable conditions the cycle may take as many as 80 days or longer.

DAMAGE INFLICTED; SYMPTOMS AND SIGNS OF AFFECTED ANIMALS

The injury caused by this parasite is inflicted chiefly by the maggots. The early symptoms in affected animals are loss of appetite and condition, and listlessness. Unless proper treatment is administered, the great destruction of many tissues kills the host in a few days.

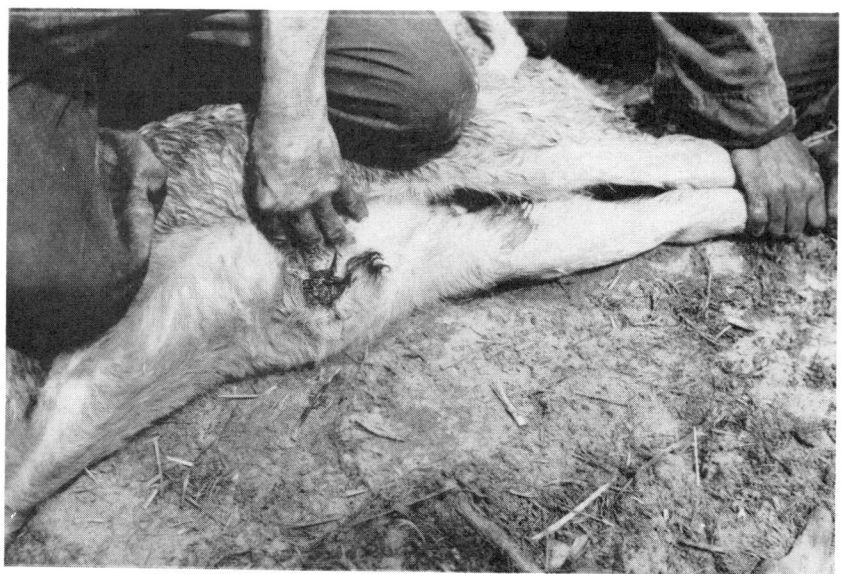

Fig. 12-50. Screwworm in navel of calf. (Courtesy, USDA)

PREVENTION, CONTROL, AND TREATMENT

Prevention in infested areas consists mainly of keeping animal wounds to a minimum and of protecting those that do materialize.

In 1958, the U.S. Department of Agriculture initiated an eradication program. Screwworm larvae were reared on artificial media. Two days before fly emergence, the pupae were exposed to gamma irradiation at a dosage which caused sexual sterility but no other deleterious effects. Sterile flies were distributed over the entire screwworm-infested region in sufficient quantity to outnumber the native flies, at an average rate of 400 males per square mile per week. The female mates only once and, therefore, when mated with a sterile male does not reproduce. There was a decline in the native population each generation until the native males were so outnumbered by sterile males that no fertile matings occurred and the native flies were eliminated. This program has virtually eliminated all the losses caused by screwworms in the United States. Unfortunately, the states bordering on Mexico are periodically reinfested by mated female flies from Mexico. For this reason, permanent elimination of screwworms from the United States by the sterile-male technique cannot be hoped for until they are also eradicated in Mexico.

When maggots (larvae) are found in an animal, they should be removed and sent to the proper authorities for identification, and the animal should be treated with a proper insecticide or smear.

Title 9 of the Code of Federal Regulations, part 83, screwworms, gives the screwworm control zone, the areas of recurring infestation, and the inspection and treatment requirements for movement from these areas.

(Also see later section in this chapter entitled, "Federal and State Regulations Relative to Disease Control.")

Stablefly (Biting Housefly, Dog Fly, or Stock Fly)

The stablefly, *Stomoxys calcitrans*, which is about the size of a housefly, is usually found in the vicinity of animals and derives its nourishment from sucking blood. It attacks all classes of farm animals and man.

DISTRIBUTION AND LOSSES CAUSED BY THE STABLEFLY

Stableflies are found in all parts of the U.S.A. and in many other countries. They are especially numerous in the central and southeastern states and some coastal states. They are not usually a problem on the open range, but large numbers can occur around barns where livestock congregate and where there are accumulations of decaying plant material. The chief economic loss from the stablefly is in terms of decreased gains in beef cattle and lowered milk production in dairy cattle, amounting to as much as 50 percent in seasons when the numbers of flies become large. It is estimated that U.S. cattle losses, including both beef and dairy animals, due to the stablefly, approach $232,596,000 annually.[36] In addition to causing such lowered production losses, the stablefly has been incriminated in the mechanical transmission of anthrax, swamp fever, surra, and even infantile paralysis of man. It is also known to be an intermediate host for the peritoneal roundworm of cattle and the small-mouth stomach worm of horses. Along certain beaches, the annoyance to man has been so great that resort interests have been affected and real estate values have been lowered.

LIFE HISTORY AND HABITS

Like other true flies, the stablefly has four stages—egg, larva, pupa, and adult fly.

Although cattle feces are a suitable habitat for the development of stablefly larvae, they do not seem to be attractive to adult flies for egg laying unless they are mixed with straw, feed, or similar materials. Therefore, the eggs are commonly deposited in soggy hay or grain in the bottoms of and underneath feed racks, in piles of moist fermenting weeds and grass cuttings, in moist piles of fermenting peanut litter, in deposits of grass along beaches (especially in northwest Florida), in straw in calf pens, and in rot-

---

[36]*Losses in Agriculture*, Agriculture Handbook No. 291, 1965, p. 82, reported average annual losses of $142,000,000 for the period 1951-60. The author updated the latter figure by applying 1974 cattle numbers ($\times$ 1.17) and prices ($\times$ 1.4), to give the figure of $232,596,000.

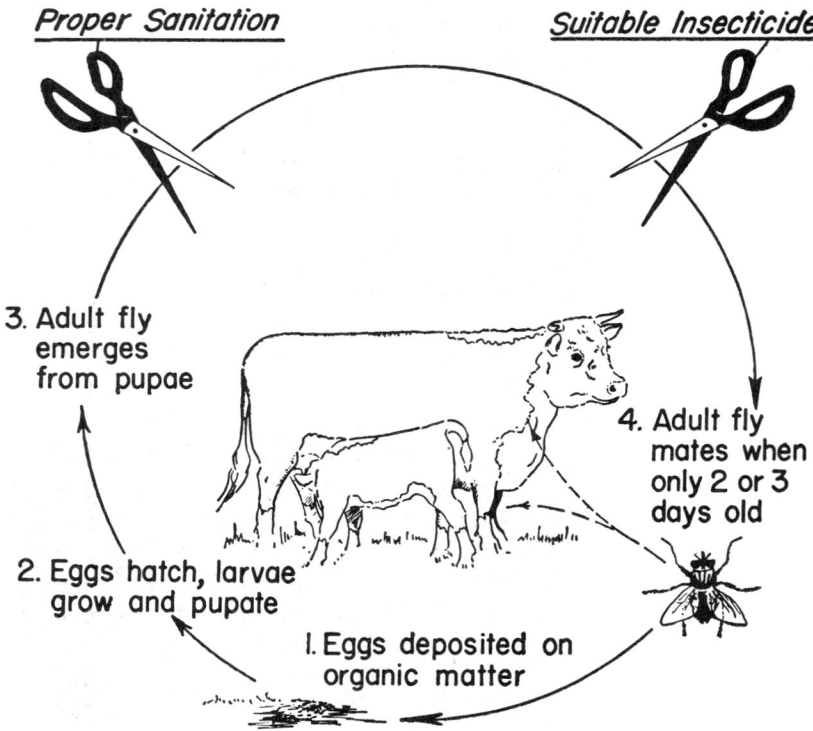

Fig. 12-51. Diagram showing the life history and habits of the stablefly. As noted (see scissors) effective control (cutting the life cycle of the parasite) consists of thorough spraying—especially the legs and lower part of the body—with a suitable insecticide throughout the fly season. First and foremost, however, it is important that there be proper disposal of manure and elimination of all fermenting or decaying organic material. (Drawing by R. F. Johnson)

ting hay at the edge of hay stacks. The eggs hatch in 2 days, and larvae complete development in about 10 days. The pupal stage lasts 4 to 5 days. The entire life cycle takes about 3 weeks so several generations can develop during the summer.

In northern climates, stableflies overwinter in either the larval or pupal stage. In the warmer southern regions, the length of each developmental period may be increased in the winter months, but flies may be present during warmer days, and breeding is generally continuous throughout the year.

DAMAGE INFLICTED; SYMPTOMS AND SIGNS OF AFFECTED ANIMALS

Stableflies commonly rest on the vertical surfaces of fences, buildings, trees, or other structures near cattle. They visit the animals only long enough to obtain a blood meal that they generally prefer to take from the legs and lower parts of the animal's body. The fly usually takes 5 or 10 minutes to engorge on blood, but it may puncture the skin several times with its proboscis in the process. The severe pain caused by this piercing and probing action causes cattle to fight the flies constantly by licking themselves and stomping their feet. As the fly feeds, its abdomen swells to almost twice the original size and turns a reddish color. The fully engorged fly then leaves the animal and finds a suitable resting place to digest the blood meal. Both sexes take one or more such feeds daily.

PREVENTION, CONTROL, AND TREATMENT

Control of stableflies by direct application of insecticides to cattle is usually not satisfactory. They are best controlled by sanitation and by application of insecticides to the resting surfaces. Sanitation is undoubtedly the most effective method of controlling stableflies in such areas as feedlots and barnyards because it breaks the life cycle by removing the breeding sites. Barnyards and feedlots should be well drained; manure and decaying organic matter should be removed from inside and outside buildings and disposed of weekly or more often, if possible, by spreading it out to dry (this kills developing larvae). If manure cannot be spread, it should be placed in compact piles where the surface will dry quickly and become unattractive to females.

With stableflies, insecticides should be used as a supplement to good sanitation, rather than as the principal method of control, because alone they may not do a satisfactory job. Residual sprays are the most effective method of treatment. They should be applied inside and outside barns and other farm structures where stableflies rest. The spray may be applied by spray gun or by fogging or misting devices. Care should be taken to prevent contamination of feed and drinking water, and animals should be removed from the area during the spraying. Application of insecticides to the cattle may afford only temporary relief. The preferred method of application is spraying since the insecticide should be applied to the legs and lower body of the animal.

Although the animals may be annoyed by only 5 or more flies each, control measures should be initiated when even these few flies are present because the number of stableflies biting an animal at any one time may represent only 2 to 3 percent of the total number of stableflies in the area. As a blood-feeding fly, the stablefly is implicated in carrying disease, and high populations cause reduced weight gains in cattle.

Ticks

The lone star tick (*Amblyomma americanum*), the Gulf Coast tick (*Amblyomma maculatum*), the Rocky Mountain wood tick (*Dermacentor andersoni*), the Pacific Coast tick (*Dermacentor occidentalis*), and the American dog tick (*Dermacentor variabilis*) are 3-host species that attack cattle during the summer months. The black-legged tick (*Ixodes scapularis*) is a 3-host tick that is common in late winter and early spring. The winter tick (*Dermacentor albipictus*) is a 1-host species found on cattle and horses in the fall and winter. In addition, larvae and nymphs of the so-called "spinose" ear tick (*Octobius megnini*), a 1-host species, attach deep in the ears of cattle and feed there for several months.

DISTRIBUTION AND LOSSES CAUSED BY TICKS

Ticks are widely distributed, especially throughout the southern part of the United States. But they are usually seasonal in their activities.

Ticks suck blood. They cause economic losses by transmitting diseases; by restlessness, anemia, and inefficient feed utilization; and by necessitating expensive treatments. Among the diseases transmitted to or produced in cattle by ticks are Texas fever, anaplasmosis, Q fever, tick paralysis, and piroplasmosis.

It is estimated that ticks cause annual losses to the United States cattle industry amounting to $60 million. Additional, but unmeasured, losses result from incapacitation of people affected with diseases transmitted by ticks of cattle.

LIFE HISTORY AND HABITS

Generally, all species of ticks have similar stages of development. The females lay eggs that hatch into 6-legged larvae (seed ticks). The larvae attach to a host, engorge on blood, and molt to 8-legged nymphs. The nymphs attach to a host, engorge on blood, and molt to 8-legged adults. Mating usually occurs on the host. The female then engorges fully, drops off the host, lays several thousand eggs, and dies.

DAMAGE INFLICTED; SYMPTOMS AND SIGNS OF AFFECTED ANIMALS

Generally speaking, injury to cattle from tick parasitism varies directly with numbers of parasites. Ticks feed exclusively on blood. Thus, when several hundred ticks feed, the host becomes anemic, unthrifty, and loses weight. In addition, some female ticks generate a paralyzing toxin. The spinose ear tick, commonly called the "ear tick," takes up residence along the inner surfaces of the ears and in the external ear canals, where they are extremely annoying. Cattle heavily parasitized by spinose ear ticks droop their heads, rub and shake their ears, and turn the head to one side.

PREVENTION, CONTROL, AND TREATMENT

Because most species of ticks, except the ear tick, attach to the external surfaces of cattle, dipping and spraying are the most effective methods of control; however, dusts may be used. To treat animals for ear ticks, the chemical should be applied into the ears of the cattle.

Control of External Parasites of Beef Cattle

Losses to the beef cattle industry due to insect pests can be considerably reduced by treating animals with insecticides that give economical control.

The cost of controlling external parasites on beef cattle with insecticides is usually very small compared with losses incurred when the infestations go uncontrolled. A good program for controlling external parasites of cattle should (1) be initiated during the early stages of infestation, (2) include use of good sanitation practices and manipulation of standing water in addition to the application of insecticides (not all external parasites can be effectively controlled with insecticides), and (3) require use of insecticides in complete accordance with the labels and instructions.

No suggestions concerning the specific insecticides that should be used are presented in this book because of (1) the diversity of environments and management practices represented by this group, (2) the varying restrictions on the use of insecticides from area to area, and (3) the fact that registered uses of insecticides change from time to time. Information about what insecticide is available and registered for use in a specific area can be obtained from the County Agent, Extension Entomologist, or Agricultural Consultant. Rather, a review follows relative to some of the types of formulations that are available, the type of treatments or applications with chemical that may be used to control external parasites of beef cattle, and some of the restrictions and precautions to be observed when treating animals.

FORMS OF INSECTICIDES

Insecticides for use on beef cattle may be purchased in several forms. The most common are emulsifiable concentrates, dusts, wettable powders, and oil solutions. When treating cattle, be sure to use only insecticide formulations that are prepared specifically for livestock.

● *Emulsifiable concentrates* (EC)—Emulsifiable concentrates, which are probably the most common type of formulation, are solutions of insecticides in petroleum oils or other solvents. An emulsifier has been added so that the solution will mix well with water. On occasion, usually after extended storage, an EC may separate into its various parts; in that case, it should be discarded. An emulsion may also separate if it is allowed to stand after the concentrate has been added to the water; periodic agitation will help prevent this.

● *Dusts*—Dusts are applied directly to animals in the dry form and cannot be used as sprays.

● *Wettable powders*—Wettable powders are also dry, but the addition of a dispersing and wetting agent allows them to be suspended in water for application to cattle. Continuous agitation of the mixture is important when treating with wettable powders.

● *Oil solutions*—Oil solutions are insecticides dissolved in oil; no emulsifier is added. These materials are usually ready for use and should not be added to water.

APPLICATION OF INSECTICIDES

The availability of an insecticide and the type of application(s) for which it was formulated are of prime importance, but the treatment of a herd is dictated pretty much by the size of herd, available handling facilities, time or season, the target pest, management practices, and cost.

● *Spraying*—When done properly, spraying gives adequate control of many external parasites. This method of application is quite economical because only the amount of material needed at the time of spraying is prepared. The sprayer should agitate the mixture adequately and have a pressure of at least 200 psi. A long narrow pen is ideal because it allows the operator to spray one side of the cattle as they move to one end of the pen and the other side as they return to the first end. It is very difficult to treat each animal uniformly when attempting to spray large numbers of cattle in a large pen, even when the equipment provides several hundred pounds of pressure. There is waste, and there is the danger of excessive drift that may be harmful to people, plants, or nontarget animals. Sprays for livestock may also be administered by specialized devices such as mist machines and aerial applicators that deliver low volumes of insecticide; or by automatic devices that apply small quantities of insecticide on individual animals when they move through a narrow passage.

● *Dipping*—Animals dipped in a vat are wetted thoroughly, and good coverage is assured. In operations such as feedlots and intensified grazing units where the herd is very large and concentrated and the vat can be used several times during the season, such treatment is quite economical. However, in many grazing operations where cattle are distributed in smaller herds through several pastures of a farm or ranch, dipping is very expensive compared with other methods of applying insecticides. There are, however, certain quarantinable parasites in some localities that must be controlled by dipping.

● *Dusting*—When beef cattle must pass through gates to obtain water, mineral, or feed, or when they can be trained to use dust bags in loafing areas, dust bags provide a very effective and economical means of treatment. Hang the bags with the bottom about 12 to 24 inches below the average height of the cattle or just high enough to prevent straddling, and maintain an adequate quantity of the dust in the bags because the added weight provides resistance to movement to insure good treatment. A number of manufacturers provide durable weatherproof bags for this purpose. Dust may also be applied with shakers or mechanical dusters.

● *Backrubbers*—Backrubbers can be obtained from several manufacturers; or they can be homemade by suspending wire, cable, rope, or chain between posts, covering it with burlap, and saturating it with the insecticide as directed by the label. Backrubbers may be located at the entrance to pens that contain water, salt, or feed, in alleys, or in loafing areas. When a backrubber is located in a gate or alley, it should sag low enough so that it will drape over the animal as it passes under it but high enough to prevent the animals from straddling it. All self-treating devices should be recharged regularly with insecticide.

● *Pouron*—When good pens, chutes, and labor are available, the pouron technique is quick and easy. A small quantity of mixture, usually ½ to 1 oz/100 pounds body weight up to a maximum of 800 pounds, is poured along the back of each animal.

● *Feed additives*—Certain external parasites of beef cattle can be controlled by adding insecticide either to the mineral mix or the feed mix. There may be a problem in treating all the animals uniformly by this method because of variation in consumption by individual animals.

PRECAUTIONS ON THE USE OF INSECTICIDES

Certain basic precautions must be observed when insecticides are to be used because, used improperly, they can be injurious to man, domestic animals, wildlife, and beneficial insects. Follow the directions and heed all the precautions on the labels. Other pertinent information follows.

● *Selecting insecticides*—Always select the formulation and insecticide labeled for the purpose for which it is to be used.

● *Storing insecticides*—Always store insecticides in original containers. Never transfer them to unlabeled containers or to feed or beverage containers. Store insecticides in a dry place out of reach of children, animals, or unauthorized persons.

● *Disposing of empty containers and unused insecticides*—Properly and promptly dispose of all empty insecticide containers. Do not reuse. Break and bury glass containers. Chop holes in, crush, and bury metal containers. Bury containers and unused insecticides at least 18 inches deep in the soil in a sanitary landfill or dump, or dump in a level isolated place where water supplies will not be contaminated. Check with local authorities to determine specific procedures for your area.

● *Mixing and handling*—Mix and prepare insecticides in the open or in a well-ventilated place. Wear rubber gloves and clean dry clothing (respirator device may be necessary with some products). If any insecticide is spilled on you or your clothing, wash with soap and water immediately and change clothing. Avoid prolonged inhalation. Do not smoke, eat, or drink when mixing and handling insecticides.

● *Applying*—Use only amounts recommended. Apply at the correct time to avoid unlawful residues in meat. Avoid treating calves younger than specified on the label. Avoid retreating more often than label restrictions. Avoid drift on nearby crops, pastures, livestock, or other nontarget areas. Avoid prolonged contact with all insecticides. Do not eat, drink, or smoke until all operations have ceased and hands and face are thoroughly washed. Change and launder clothing after each day's work.

● *In case of an emergency*—If you accidently swallow an insecticide, induce vomiting by taking one tablespoonful of salt in a glass of water. Repeat if necessary. Call a doctor.

DISINFECTANTS[37]

A disinfectant is a bactericidal or microbicidal agent that frees from infection (usually a chemical agent which destroys disease germs or other microorganisms, or inactivates viruses).

---

[37]The author had the benefit of the authoritative review and suggestions of Mr. C. H. Shaffer, Jr., Microbiologist in Charge, United States Environmental Protection Agency, Washington, D.C., in the preparation of this section and Table 12-3.

TABLE 12-3

HANDY DISINFECTANT GUIDE

Kind of Disinfectant	Usefulness	Strength	Limitations and Comments
Alcohol (ethyl-ethanol, Isopropyl, Methanol)	Primarily as skin disinfectants and for emergency purposes on instruments.	70% alcohol—the content usually found in rubbing alcohol.	They are too costly for general disinfection. They are ineffective against bacterial spores.
Boric Acid[1]	As a wash for eyes, and other sensitive parts of the body.	1 oz in 1 pt water (about 6% solution).	It is a weak antiseptic. It may cause harm to the nervous system if absorbed into the body in large amounts. For this and other reasons, antibiotic solutions and saline solutions are fast replacing it.
Chlorines (sodium hypochlorate, chlormine-T)	Used for dairy equipment and as deodorants. They will kill all kinds of bacteria, fungi, and viruses, providing the concentration is sufficiently high.	Generally used at about 200 ppm for dairy equipment and as a deodorant.	They are corrosive to metals and neutralized by organic materials.
Cresols (Many commercial products available)	A generally reliable class of disinfectant. Effective against brucellosis, shipping fever, swine erysipelas, and tuberculosis.	Cresol is usually used as a 2 to 4% solution (1 cup to 2 gallons of water makes a 4% solution).	Cannot be used where odor may be absorbed, and, therefore, not suited for use around milk and meat.
Formaldehyde (gaseous disinfectant)	Formaldehyde will kill anthrax spores, TB organisms, and animal viruses in a 1 to 2% solution. It is often used to disinfect buildings following a disease outbreak. A 1 to 2% solution may be used as a foot bath to control foot rot.	As a liquid disinfectant, it is usually used as a 1 to 2% solution. As a gaseous disinfectant (fumigant), use 1½ lb of potassium permanganate plus 3 pt of formaldehyde. Also, gas may be released by heating paraformaldehyde.	It has a disagreeable odor, destroys living tissue, and can be extremely poisonous. The bactericidal effectiveness of the gas is dependent upon having the proper relative humidity (above 75%) and temperature (above 30°C and preferably near 60°C).
Heat (by steam, hot water, burning, or boiling)	In the burning of rubbish or articles of little value, and in disposing of infected body discharges. The steam "Jenny" is effective for disinfection (example: poultry equipment) if properly employed, particularly if used in conjunction with a phenolic germicide.	10 minutes exposure to boiling water is usually sufficient.	Exposure to boiling water will destroy all ordinary disease germs but sometimes fails to kill the spores of such diseases as anthrax and tetanus. Moist heat is preferred to dry heat, and steam under pressure is the most effective. Heat may be impractical or too expensive.
Iodine[1] (tincture)	Extensively used as skin disinfectant, for minor cuts and bruises.	Generally used as tincture of iodine, either 2% or 7%.	Never cover with a bandage. Clean skin before applying iodine. It is corrosive to metals.
Idophore (tamed iodine)	Primarily used for dairy utensils. Effective against all bacteria (both gram-negative and gram-positive), fungi, and most viruses.	Usually used as disinfectants at concentrations of 50-75 ppm titratable iodine, and as sanitizers at levels of 12.5 to 25 ppm. At 12.5 ppm titratable iodine, they can be used as an antiseptic in drinking water.	They are inhibited in their activity by organic matter. They are quite expensive.

(Continued)

TABLE 12-3 (Continued)

Kind of Disinfectant	Usefulness	Strength	Limitations and Comments
Lime (quicklime; burnt lime; calcium oxide)	As a deodorant when sprinkled on manure and animal discharges; or as a disinfectant when sprinkled on the floor or used as a newly-made "milk of lime" or as a whitewash.	Use as a dust; as "milk of lime"; or as a whitewash, but **use fresh.**	Not effective against anthrax or tetanus spores. Wear goggles, when adding water to quicklime.
Lye (sodium hydroxide; caustic soda)	On concrete floors; in milk houses because there is no odor; against microorganisms of brucellosis and the viruses of foot-and-mouth disease, hog cholera, and vesicular exanthema. In strong solution (5%), effective against anthrax and blackleg.	Lye is usually used as either a 2% or 5% solution. To prepare a 2% solution, add 1 can of lye to 5 gallons of water. To prepare a 5% solution, add 1 can of lye to 2 gallons of water. A 2% solution will destroy the organisms causing foot-and-mouth disease, but a 5% solution is necessary to destroy the spores of anthrax.	Damages fabrics, aluminum, and painted surfaces. Be careful, for it will burn the hands and face. Not effective against organism of T.B. or Johne's disease. Lye solutions are most effective when used hot. **Diluted vinegar can be used to neutralize lye.**
Lysol (the brand name of a product of cresol plus soap)	For disinfecting surgical instruments and instruments used in dehorning, castrating, and tattooing. Useful as a skin disinfectant before surgery, and for use on the hands before castrating.	0.5 to 2.0%	Has a disagreeable odor. Does not mix well with hard water. Less costly than phenol.
Phenol (carbolic acid): 1. Phenolics—coal tar derivatives. 2. Synthetic phenols.	They are ideal general-purpose disinfectants. Effective and inexpensive. They're very resistant to the inhibiting effects of organic residue; hence, they are suitable for barn disinfection, and foot and wheel dip-baths.	Both phenolics (coal tar) and synthetic phenols vary widely in efficacy from one compound to another. So, note and follow manufacturers' directions. Generally used in a 5% solution.	They are corrosive, and they're toxic to animals and humans. Ineffective on fungi and viruses.
Quaternary Ammonium Compounds (QAC)	Very water soluble, ultra-rapid kill rate, effective deodorizing properties, and moderately priced. Good detergent characteristics and harmless to skin.	Follow manufacturers directions.	They can corrode metal. Not very potent in combating viruses. Adversely affected by organic matter.
Sal Soda	It may be used in place of lye against foot-and-mouth disease and vesicular exanthema.	10½% solution (13½ oz to 1 gal water)	
Sal Soda and Soda Ash (or sodium carbonate)	They may be used in place of lye against foot-and-mouth disease and vesicular exanthema.	4% solution (1 lb to 3 gal water). Most effective in hot solution.	Commonly used as cleansing agents, but have disinfectant properties, especially when used as a hot solution.
Soap	Its power to kill germs is very limited. Greatest usefulness is in cleansing and dissolving coatings from various surfaces, including the skin, prior to application of a good disinfectant.	As commercially prepared.	Although indispensable to sanitizing surfaces, soaps should not be used as disinfectants. They are not regularly effective, staphylococci and the organisms which cause diarrheal disease are resistant.

[1]Sometimes loosely classed as disinfectant but actually an antiseptic and practically useful only on living tissue.

The high concentration of animals and continuous use of modern live-stock buildings often results in a condition referred to as disease buildup. As disease-producing organisms—viruses, bacteria, fungi, and parasite eggs—accumulate in the environment, disease problems can become more severe and be transmitted to each succeeding group of animals raised on the same premises. Under these circumstances, cleaning and disinfection become extremely important in breaking the life cycle. Also, in the case of a disease outbreak, the premises must be disinfected.

Under ordinary conditions, proper cleaning of barns removes most of the microorganisms, along with the filth, thus eliminating the necessity of disinfection.

Effective disinfection depends on five things:

1. Thorough cleaning before application.

2. The phenol coefficient of the disinfectant, which indicates the killing strength of a disinfectant as compared to phenol (carbolic acid). It is determined by a standard laboratory test in which the typhoid fever germ often is used as the test organism.

3. The dilution at which the disinfectant is used.

4. The temperature; most disinfectants are much more effective if applied hot.

5. Thoroughness of application, and time of exposure.

Disinfection must in all cases be preceded by a very thorough cleaning, for organic matter serves to protect disease germs and otherwise interferes with the activity of the disinfecting agent.

Sunlight possesses disinfecting properties, but it is variable and superficial in its action. Heat and some of the chemical disinfectants are more effective.

The application of heat by steam, by hot water, by burning, or by boiling is an effective method of disinfection. In many cases, however, it may not be practical to use heat.

A good disinfectant should (1) have the power to kill disease-producing organisms, (2) remain stable in the presence of organic matter (manure, hair, soil), (3) dissolve readily in water and remain in solution, (4) be nontoxic to animals and humans, (5) penetrate organic matter rapidly, (6) remove dirt and grease, and (7) be economical to use.

The number of available disinfectants is large because the ideal universally applicable disinfectant does not exist. Table 12-3 gives a summary of the limitations, usefulness, and strength of some common disinfectants.

When using a disinfectant, *always read and follow the manufacturer's directions.*

POISONOUS PLANTS[38]

Poisonous plants have been known to man since time immemorial. Bib-

---

[38]In the preparation of this section, the author had the benefit of the authoritative review and suggestions of Dr. Wayne Binns, Director of the Poisonous Plant Research Laboratory, U.S. Department of Agriculture, Logan, Utah.

lical literature alludes to the poisonous properties of certain plants, and history records that hemlock (a poison made from the plant from which it takes its name) was administered by the Greeks to Socrates and other state prisoners.

No section of the United States is entirely free of poisonous plants, for there are hundreds of them. But the heaviest livestock losses from them occur on the western ranges because (1) there has been less cultivation and destruction of poisonous plants in range areas, and (2) the frequent overgrazing on some of the western ranges has resulted in the elimination of some of the more nutritious and desirable plants, and these have been replaced by increased numbers of the less desirable and poisonous species. It is estimated that poisonous plants account for 8 to 10 percent of all range animal losses each year; and even more in some areas. For the nation as a whole, beef cattle losses attributed to poisonous plants total $28,342,314 annually.[39]

Diagnosis of Plant Poisoning

The diagnosis of plant poisoning in animals is not an easy or precise procedure. Any case of sudden illness or death with no apparent cause is commonly considered to be a poisoning. This may not always be correct. When large numbers of animals are suddenly affected, however, a suspicion of poisoning is justified until it has been proven otherwise.

Symptoms or signs induced by eating poisonous plants may include: (1) sudden death; (2) transitory illness; (3) general body weakness; (4) disturbance of the central nervous, vascular, and endocrine systems; (5) photosensitization; (6) frequent urination; (7) diarrhea; (8) bloating; (9) chronic debilitation and death; (10) embryonic death; (11) fetal death; (12) abortion; (13) extensive liver necrosis and/or cirrhosis; (14) edema and/or abdominal dropsy; (15) tumor growths in tissues; (16) congenital deformities; (17) metabolic deficiencies; and (18) physical injury.

No general set of symptoms and signs *per se* irrefutably provides all the information necessary to make a diagnosis of plant poisonings. Nevertheless, a careful description of the toxic signs coupled with information pertaining to available plants provides a meaningful basis for a tentative diagnosis. Additional information essential to a poisonous plant diagnosis includes: (1) type of feed, site grazed, and availability of water; (2) identification and relative abundance of all poisonous plants available to animals; (3) amount and stage of growth of various poisonous plants being grazed; (4) the toxicity and palatability of the plants in relation to their stage of growth; (5) time from eating the plants until onset of toxic signs; (6) species, age, and sex of animals affected; (7) clinical signs of toxic reactions; (8) chemical analysis of plants; and (9) a careful evaluation of all the information relative to the etiology of the disease.

---

[39]*Losses in Agriculture,* Agriculture Handbook No. 291, 1965, p. 73, reported average annual losses of $17,303,000 for the period 1951-60. The author updated the latter figure by applying 1974 cattle numbers (× 1.17) and prices (× 1.4), to give the figure of $28,342,314.

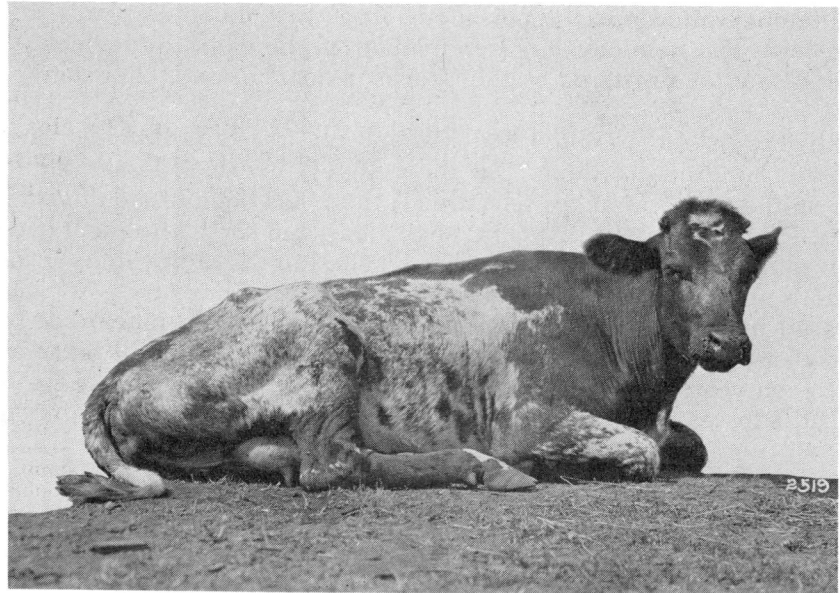

Fig. 12-52. Cow with white snakeroot poisoning. Marked weakness results in the "tremble" characteristic of this condition. (Courtesy, Dept. of Veterinary Pathology and Hygiene, College of Veterinary Medicine, University of Illinois)

Why Animals Eat Poisonous Plants

A frequently asked question is: why do animals eat poisonous plants? The answer is not simple, but among the reasons are the following: (1) total lack of sufficient palatable forage—the animals are hungry; (2) decrease in palatability and nutrients of mature, weathered range grasses, with the result that poisonous plants become more appealing, comparatively speaking; (3) insufficient spring grass; (4) rain, melting snow, and heavy dew may enhance the palatability of some poisonous plants; (5) going without water too long, which results in a reduction in feed intake, then, after watering, they develop a ravenous appetite and eat anything in sight—including less palatable poisonous plants; and (6) poisonous plants vary in palatability—between species, and within species, plants at different stages of growth. For example, poison hemlock is never palatable and is eaten only as a last resort—when palatable forage is not available. Locoweed and black nightshade are eaten at any stage of growth or when mixed with hay. Others, such as lupines, horsebrush, and death camas may be eaten only at certain stages of growth. Still others, such as milk vetch, larkspur, and halogeton, are highly palatable to livestock at any and all times, with the result that if they're present animals will seek them out and there will be losses. Then, too, certain plants are poisonous to cattle but not to sheep (and vice versa), as shown in Table 12-4.

Some Common Poisonous Plants

The list of poisonous plants is so extensive that no attempt is made herein to discuss them. Nevertheless, both the stockman and the veterinarian should have a working knowledge of the principal poisonous species in the area in which they operate. The common poisonous plants of the intermountain ranges to which cattle and/or sheep are susceptible at certain times of the grazing season are listed in Table 12-4.

TABLE 12-4

TYPE OF RANGE ANIMAL SUSCEPTIBLE TO POISONOUS PLANTS AT DEFINITE SEASONS

Poisonous to cattle	Time of year	Poisonous to sheep	Time of year	Poisonous to cattle & sheep	Time of year
Low larkspur	Spring	Death Camas	Spring	Broomweed	Spring and Summer
Oak	Spring	Greasewood	Fall	Chokecherry	Spring
Tall larkspur	Early Summer & early Fall	Horsebrush	Spring	Copperweed	Summer
Timber Milk Vetch	Spring	Rubberweed	Summer	Desert Parsley	Spring
Water Hemlock	Spring	Sneezeweed	Summer	Halogeton	All year
				Loco	Spring
				Lupine	Summer and Fall
				Milkweeds	Summer
				Veratrum	Summer

Preventing Losses From Poisonous Plants

With poisonous plants, the emphasis should be on prevention of losses rather than on treatment, no matter how successful the latter. The following are effective preventive measures:

1. *Follow good pasture or range management* in order to improve the quality of the pasture or range. Plant poisoning is nature's sign of a "sick" pasture or range, usually resulting from misuse. When a sufficient supply of desirable forage is available, poisonous plants may not be eaten, for they are usually less palatable. On the other hand, when overgrazing reduces the available supply of the more palatable and safe vegetation, animals may, through sheer hunger, consume the toxic plants.

2. *Know the poisonous plants common to the area.* This can usually be accomplished through (a) studying drawings, photographs, and/or descriptions, (b) checking with local authorities, or (c) sending two or three fresh whole plants (if possible, include the roots, stems, leaves, and flowers) to the state agricultural college—first wrapping the plants in several thicknesses of moist paper.

By knowing the poisonous plants common to the area, it will be possible—

a. To avoid areas heavily infested with poisonous plants which, due to animal concentration and overgrazing, usually include waterholes, salt grounds, bed grounds, and trails.

b. To control and eradicate the poisonous plants effectively, by mechanical or chemical means (as recommended by local authorities) or by fencing off.

c. To recognize more surely and readily the particular kind of plant poisoning when it strikes, for time is important.

d. To know what first aid, if any, to apply, especially when death is imminent or where a veterinarian is not readily available.

e. To graze with a class of livestock not harmed by the particular poisonous plant or plants, where this is possible. Many plants seriously poisonous to one kind of livestock are not poisonous to another, at least under practical conditions.

f. To shift the grazing season to a time when the plant is not dangerous, where this is possible. That is, some plants are poisonous at certain seasons of the year, but comparatively harmless at other seasons.

g. To avoid cutting poison-infested meadows for hay when it is known that the dried cured plant is poisonous. Some plants are poisonous in either green or dry form, whereas others are harmless when dry. When poisonous plants (or seeds) become mixed with hay (or grain), it is difficult for animals to separate the safe from the toxic material.

3. *Know the symptoms that generally indicate plant poisoning*, thus making for early action.

4. *Avoid turning to pasture in very early spring.* Nature has ordained most poisonous plants as early growers—earlier than the desirable forage. For this reason, as well as from the standpoint of desirable pasture management, animals should not be turned to pasture in the early spring before the usual forage has become plentiful.

5. *Provide supplemental feed during droughts, after plants become mature, and after early frost.* Otherwise, hungry animals may eat poisonous plants in an effort to survive.

6. *Avoid turning very hungry animals where there are poisonous plants,* especially those that have been in corrals for branding, etc., that have been recently shipped or trailed long distances, or that have been wintered on dry forage. First feed the animals to satisfy their hunger or allow a fill on an area known to be free from poisonous plants.

7. *Avoid driving animals too fast when trailing.* On long drives, either allow them to graze along the way or stop frequently and provide supplemental feed.

8. *Remove promptly all animals from infested areas when plant poisoning strikes.* Hopefully, this will check further losses.

9. *Treat promptly,* preferably by a veterinarian.

Treatment

Unfortunately, plant-poisoned animals are not generally discovered in sufficient time to prevent loss. Thus, prevention is decidedly superior to treatment.

When trouble is encountered, the owner or caretaker should *promptly* call a veterinarian. In the meantime, the animal should be (1) placed where adequate care and treatment can be given, (2) protected from excessive heat and cold, and (3) allowed to eat only feeds known to be safe.

The veterinarian may determine the kind of poisonous plant involved (1) by observing the symptoms, and/or (2) by finding out exactly what poisonous plant was eaten through looking over the pasture and/or hay and identifying leaves or other plant parts found in the animal's digestive tract at the time of autopsy.

It is to be emphasized, however, that many poisoned animals that would have recovered had they been left undisturbed, have been killed by attempts to administer home remedies by well-meaning but untrained persons.

FEDERAL AND STATE REGULATIONS RELATIVE TO DISEASE CONTROL[40]

Certain diseases are so devastating that no individual livestock producer could long protect his herd and flock against their invasion. Moreover, where human health is involved, the problem is much too important to be entrusted to individual action. In the United States, therefore, certain regulatory activities in animal-disease control are under the supervision of various Federal and state organizations. Federally, this responsibility is entrusted to the following agency:

Veterinary Services
Animal and Plant Health Inspection Service
U.S. Department of Agriculture
Federal Center Building
Hyattsville, Md. 20782

In addition to the Federal interstate regulations, each of the states has requirements for the entry of livestock. Generally, these requirements include compliance with interstate regulations. States usually require a certificate of health or a permit, or both, and additional testing requirements depending upon the class of livestock involved.

Detailed information relative to animal disease control can be obtained from Federal and state animal health officials, or from accredited veterinarians in all states. Shippers are urged to obtain such information prior to making interstate shipments of livestock.

Many highly infectious diseases are prevented by quarantine from (1) gaining a foothold in this country, or (2) spreading. *By quarantine is meant (1) segregation and confinement of one or more animals in the smallest possible area to prevent any direct or indirect contact with animals not so restrained; or (2) regulating movement of animals at points of entry.* When an infectious disease outbreak occurs, drastic quarantine must be imposed to restrict movement out of an area or within areas. The type of quarantine

---

[40]This section was reviewed by the Administrator, Animal and Plant Health Inspection Service, U.S. Department of Agriculture, Washington, D.C.

varies from one involving a mere physical examination and movement under proper certification to the complete prohibition against the movement of animals, produce, vehicles, and even human beings.

Where certain animal diseases are involved, the stockman can obtain financial assistance in eradication programs through Federal and state sources.

Federal Quarantine Center

A Federal Quarantine Center was authorized in Public Law 91-239, signed by the President on May 6, 1970. A 16.1-acre site for the Center has been selected at Fleming Key, near Key West, Florida; and preliminary planning has been done. As soon as the necessary funds are secured, the provisions of the act will be carried out.

The Quarantine Center is designed to hold some 400 head of cattle or other animals at one time, for a five-month quarantine period. This maximum security station will enable American livestock producers to import breeding animals from all parts of the world, while at the same time safeguarding our domestic beef cattle herds from such diseases as foot-and-mouth disease, rinderpest, piroplasmosis, and others.

QUESTIONS FOR STUDY AND DISCUSSION

1. Why are good beef cattle health books of value to livestock producers?
2. What is normal temperature, pulse rate, and respiration rate of beef cattle, and how would you determine each?
3. Select a specific farm or ranch (either your own or one with which you are familiar) and outline (in 1,2,3, order) a program of beef cattle health, disease prevention, and parasite control.
4. Assume that a specific contagious disease (you name it) has broken out in your herd. What steps would you take to meet the situation (list in 1,2,3, order; be specific)?
5. Since 1934, the United States has conducted a brucellosis control program at a cost of $700 million in tax funds. Yet, in 1972, after nearly 40 years and an expenditure of ¾ billion dollars, 23,588 herds, representing nearly one percent of the nation's cattle, were still infected with the disease, and over 4 million calves were vaccinated with Strain 19 during the year. Has the program been a success or a failure?
6. Outline a program for the prevention and treatment of calf scours.
7. Outline a program for the prevention of shipping fever in purchased feeder calves, beginning with preconditioning.
8. Assume that a specific parasite (you name it) has become troublesome in your herd. What steps would you take to meet the situation (list in 1,2,3, order; be specific)?
9. Outline a treatment program for the control of gastro-intestinal nematode worms, including (a) drug(s), (b) time of year to treat, and (c) preferred method of administering.
10. Explain (a) how screwworm flies are sterilized, and (b) how the screwworm control program works.
11. When mange is confirmed in a herd, why is that herd quarantined? Who does the quarantining?
12. Assume that you have, during a period of a year, encountered cattle death losses from three different diseases (you name them). What kind of disinfectant would you use in each case?

13. Assume that you have encountered death losses from a certain poisonous plant (you name it). What steps would you take to meet the situation (list in 1,2,3, order; be specific)?

14. When people are ill, they call the family doctor. Isn't it just as logical that a veterinarian be called when cattle are sick?

SELECTED REFERENCES

Title of Publication	Author(s)	Publisher
Abortion Diseases of Livestock	Ed. by L. C. Faulkner	Charles C Thomas, Springfield, Ill., 1968
Animal Agents and Vectors of Human Disease	E. C. Faust	Lea & Febiger, Philadelphia, Penn., 1955
Animal Diseases	Yearbook of Agriculture, 1956	U.S. Department of Agriculture, Washington, D.C., 1956
Animal Parasitism	C. P. Read	Prentice-Hall, Inc., Englewood Cliffs, N.J., 1972
Animal Sanitation and Disease Control, Sixth Edition	R. R. Dykstra	The Interstate Printers & Publishers, Inc., Danville, Ill., 1961
Bovine Medicine and Surgery	Ed. by W. J. Gibbons, E. J. Catcott, J. F. Smithcors	American Veterinary Publications, Inc., Wheaton, Ill., 1970
Brucellosis	National Institutes of Health of the Public Health Service Federal Security Agency U.S. Department of Agriculture National Research Council	American Association for the Advancement of Science, Washington, D.C., 1950
Diseases of Cattle, Revised Second Edition	Ed. by W. J. Gibbons	American Veterinary Publications, Inc., Wheaton, Ill., 1963
Diseases of Feedlot Cattle, Second Edition	R. Jensen D. R. Mackey	Lea & Febiger, Philadelphia, Penn., 1971
Diseases of Livestock, Sixth Edition	T. G. Hungerford	Angus and Robertson Ltd., Sydney, Australia, 1967
Diseases Transmitted from Animals to Man, Fifth Edition	Ed. by T. G. Hull	Charles C Thomas, Springfield, Ill., 1963
Disinfection, Sterilization, and Preservation	C. A. Lawrence S. S. Block	Lea & Febiger, Philadelphia, Penn., 1968
Farm Animal Health and Disease Control	J. H. Galloway	Lea & Febiger, Philadelphia, Penn., 1972
Farmer's Veterinary Handbook	J. J. Haberman	Prentice-Hall, Inc., New York, N.Y., 1953
Foreign Animal Diseases, Their Prevention, Diagnosis and Control	Official Report of the United States Livestock Sanitary Association	U.S. Livestock Sanitary Association, Trenton, N.J., 1954

(Continued)

Title of Publication	Author(s)	Publisher
Hagan's Infectious Diseases of Domestic Animals, Sixth Edition	D. W. Bruner J. H. Gillespie	Comstock Publishing Associates, Ithaca, N.Y., 1973
Home Veterinarian's Handbook, The	E. T. Baker	The Macmillan Company, New York, N.Y., 1949
Immunity to Animal Parasites	Ed. by E. J. L. Soulsby	Academic Press, Inc., New York, N.Y., 1972
Keeping Livestock Healthy	Yearbook of Agriculture, 1942	U.S. Department of Agriculture, Washington, D.C., 1942
Lecture Outline of Preventive Veterinary Medicine for Animal Science Students, Third Edition	I. A. Schipper	Burgess Publishing Company, Minneapolis, Minn., 1962
Livestock Health Encyclopedia	Ed. by R. Seiden	Springer Publishing Co., Inc., New York, N.Y., 1951
Losses in Agriculture	Agriculture Handbook No. 291	U.S. Department of Agriculture, Washington, D.C., 1965
Merck Veterinary Manual, The, Fourth Edition	Ed. by O. H. Siegmund	Merck & Co., Inc., Rahway, N.J., 1973
New Zealand Farmer's Veterinary Guide, The	D. G. Edgar D. C. Dodd T. J. McClure E. L. Owens T. Wallace	The New Zealand Dairy Exporter, Wellington, New Zealand, 1962
Pathology of Domestic Animals, Second Edition, Volumes 1 and 2	K. V. F. Jubb P. C. Kennedy	Academic Press, Inc., New York, N.Y., 1970
Principles of Veterinary Science, Fourth Edition	F. B. Hadley	W. B. Saunders Company, Philadelphia, Penn., 1949
Some Important Animal Diseases in Europe, FAO Agricultural Studies No. 10	Ed. by K. V. L. Kesteven	Food and Agriculture Organization of the United Nations, Rome, Italy, 1948
Special Report on Diseases of Cattle	V. T. Atkinson W. Dickson W. H. Harbaugh J. Law W. H. Lowe J. R. Mohler A. J. Murray L. Pearson B. H. Ransom M. R. Trumbrower	U.S. Department of Agriculture, Washington, D.C., 1942
Stockman's Handbook, The, Fourth Edition	M. E. Ensminger	The Interstate Printers & Publishers, Inc., Danville, Ill., 1970

(Continued)

Title of Publication	Author(s)	Publisher
Veterinary Guide for Farmers	G. W. Stamm, Ed. by D. S. Burch	Windsor Press, Chicago, Ill., 1950
Veterinary Handbook for Cattlemen, Second Edition	J. W. Bailey	Springer Publishing Company, Inc., New York, N.Y., 1958
Veterinary Medicine	D. C. Blood J. A. Henderson	The Williams and Wilkins Company, Baltimore, Md., 1960
Veterinary Parasitology, Second Revised Edition	G. Lapage	Charles C Thomas, Publisher, Springfield, Ill., 1968

* * * * *

In addition to the above selected references, valuable publications on different subjects pertaining to beef cattle diseases, parasites, disinfectants, and poisonous plants can be obtained from the following sources:

1. Division of Publications
   Office of Information
   U.S. Department of Agriculture
   Washington, D.C. 20250

2. Your state agricultural college.

3. Several biological, pharmaceutical, and chemical companies.

CHAPTER 13

BUSINESS ASPECTS OF BEEF PRODUCTION[1]

---

[1]In recognition of the great importance of this chapter, the author obtained the review and suggestions of the following knowledgeable persons in its preparation: Mr. E. A. Jaenke, Governor, Farm Credit Administration, Washington, D. C.; Mr. G. W. Hickman, Vice President, Bank of America, San Francisco, Calif.; Mr. Malcolm E. Adams, President, Bank of Livingston, Livingston, Mont.; Mr. Oliver A. Hansen, Superintendent, State of Iowa Department of Banking, Des Moines, Iowa; and Mr. Arthur Solomon, Manager and President, Howard County Production Credit Association, Howard, Kan.

Contents   **Page**

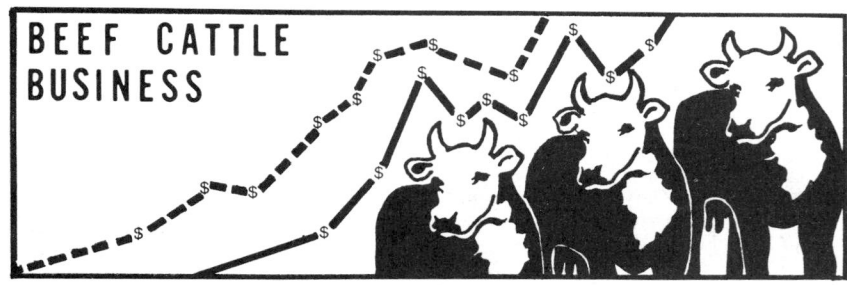

Fig. 13-1. Beef cattle business.

The cattleman of the future will be a good businessman as well as a good cattleman. He also will need an operation that is large enough to provide his family an adequate standard of living and generate enough capital to keep expanding. Since profit margins will likely decline further, there will be greater stress on business and financial management skills.

Generating both equity and debt capital, or risk and borrowed capital, will be one of the main concerns of the future cattleman. The large investment, plus the need to keep competitive by utilizing new and usually expensive technological advances, will cause capital to be very important.

To obtain capital several things will be necessary. The cattleman-businessman will have to prepare (1) profit and loss statements to show that his operation is profitable, (2) financial statements to show that progress is being made, and (3) cash-flow projections to show loan repayability. Then, and then only, will the cattleman be ready to go looking for funds.

Skill in capital budgeting and analyzing alternative investment opportunities will be needed to see that the limited capital is invested where payoff will be the greatest. The cattleman will also have to exercise budget and cost controls of his business. Skill in building sound credit will be needed.

The greatest payoffs in the future are likely to come from the efforts devoted to improving cattlemen's skills in business and financial management.

TYPES OF BUSINESS ORGANIZATION

The success of today's cattle business is very dependent on the type of business organization. No one type of organization is superior under all circumstances; rather, each situation must be considered individually. The size of the operation, the family situation, the enterprises, the objectives—all these, and more, are important in determining the best way in which to organize the cattle business.

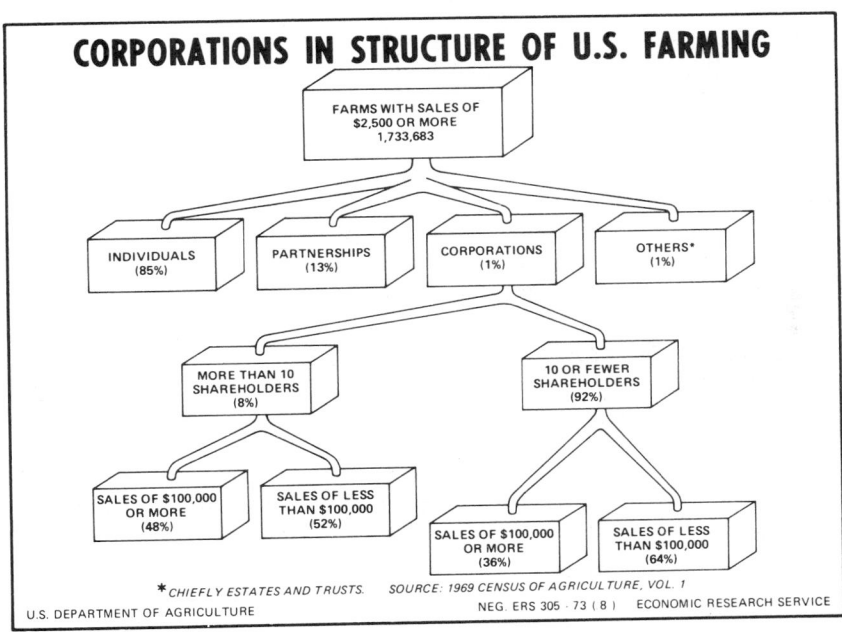

Fig. 13-2. (1) Types of business organizations in United States farming; and (2) corporate structure.

Three major types of business organizations are commonly found among cattle enterprises: (1) the sole proprietorship; (2) the partnership; and (3) the corporation. Additionally, there are agency services. As shown in Fig. 13-2, among U.S. farms with $2,500 or more in annual income, the breakdown in types of business organizations is as follows: individual, 85%; partnership, 13%; and corporations, 1%. Among the factors which should be considered when deciding which business form best fits a given set of circumstances are the following:

1. Which type of organization is most likely to be looked upon favorably from the standpoint of more credit and capital?
2. How much capital will be required of each individual involved?
3. Are there tax advantages to be gained from the business organization?
4. Is expansion of the business feasible and facilitated?
5. Which type of organization reduces risks and liability most?
6. Which type of organization can be terminated most easily and readily?

7. Which type of ownership provides for the most continuity and ease of transfer?

8. What costs for legal and accounting fees are involved, in setting up the organization and in the preparation of the annual reports required by law?

9. Who will manage the business?

Most cow-calf enterprises are operated as sole proprietorships; not necessarily because this is the best type of organization, but with no effort to form some other type of organization it naturally results. Both the partnership and the corporation, which require special planning and effort to bring about, are well suited to the operation of large commercial cattle feedlots.

Proprietorship (Individual)

This is the most common type of business organization in U.S. farming—85 percent of the nation's farms are individually owned. (See Fig. 13-2.) Under the sole proprietorship, or individual ownership, one man controls the business. He may not provide all the capital used in the business; in fact, he usually does not. However, he has sole management and control of the operation, although this may be modified and delegated somewhat through lease agreements, contracts, etc. Basically, the sole proprietor gets all the profits of the business. Likewise, he must absorb all the losses.

In comparison with other forms of organization, the sole proprietorship has two major limitations: (1) it may be more difficult to acquire new capital for expansion; and (2) not much can be done to provide for continuity and to keep the present business going as a unit, with the result that it usually goes out of existence with the passing of the owner.

Partnership (General Partnership)

A partnership is an association of two or more persons who, as co-owners, operate the business. About 13 percent of U.S. farms are partnerships. (See Fig. 13-2.)

The basic idea of two or more persons joining together to carry out a business venture can be traced back to the syndicates that were used in major trading centers in Western Europe in the Middle Ages. Many of the early efforts to colonize the New World were also partnerships, or "companies" which provided venture capital, ships, provisions, and trade goods to induce settlement of large land grants.

Most farm partnerships involve family members who have pooled land, machinery, working capital, and often their labor and management to operate a larger business than would be possible if each member limited his operation to his own resources. It is a good way in which to bring a son, who is usually short on capital, into the business, yet keep the father in active participation. Although there are financial risks to each member of such a partnership, and potential conflicts in management decisions, the existence of family ties tends to minimize such problems.

In order for a partnership to be successful, the enterprise must be sufficiently large to utilize the abilities and skills of the partners and to compensate each adequately in keeping with his contribution to the business.

A partnership has the following *advantages*:

1. *Combining resources*—A partnership often increases returns from the operation due to combining resources. For example, one partner may contribute his labor and management skills, whereas another may provide the capital. Under such an arrangement, it is very important that the partners agree on the value of each person's contribution to the business, and that this be clearly spelled out in the partnership agreement.

2. *Equitable management*—Unless otherwise agreed upon, all partners have equal rights, regardless of financial interest. Any limitations, such as voting rights proportionate to investments, should be a written part of the agreement.

3. *Tax savings*—A partnership does not pay any tax on its income, but it must file an informational return. The tax is paid as part of the individual tax returns of the respective partners, usually at lower tax rates.

4. *Flexibility*—Usually, the partnership does not need outside approval to change its structure or operation—the vote of the partners suffices.

Partnerships may have the following *disadvantages*:

1. *Liability for debts and obligations of the partnership*—In a partnership, each partner is liable for all the debts and obligations of the partnership.

2. *Uncertainty of length of agreement*—A partnership ceases with the death or withdrawal of any partner, unless the agreement provides for continuation by the remaining partners.

3. *Difficulty of determining value of partner's interest*—Since a partner owns a share of every individual item involved in the partnership, it is often very difficult to judge value. This tends to make transfer of a partnership difficult. This disadvantage may be lessened by determining market values regularly.

The above is what is known as a partnership or general partnership. It is characterized by (1) management of the business being shared by the partners, and (2) each partner being responsible for the activities and liabilities of all of the partners, in addition to his own activities within the partnership.

LIMITED PARTNERSHIP

A *limited partnership is an arrangement in which two or more parties supply the capital, but only one partner is involved in the management.* This is a special type of partnership with one or more "general" partners and one or more "limited" partners.

The limited partnership avoids many of the problems inherent in a partnership (general partnership) and has become the chief legal device for attracting outside investor capital into farm and ranch ventures. Although this device has been widely used in the oil and gas industry, and for acquiring income-producing urban real estate for a number of years, its application to agricultural ventures on a national scale is quite new. As the term implies, the financial liability of each partner is limited to his original investment, and the partnership does not require, and in fact prohibits, direct involvement of the limited partners in management. In many ways, a limited partner is in a similar position to a stockholder in a corporation.

A limited partnership must have at least one general partner who is responsible for managing the business and who is fully liable for all obligations.

Like most other business organizations, the limited partnership has both advantages and disadvantages.

The *advantages* of a limited partnership are:

1. *It facilitates bringing in outside capital*—A limited partnership provides a way in which to bring outside capital into an agricultural operation, without giving up any control of the operation.

2. *It need not dissolve with the loss of a partner*—Unlike a regular partnership, a limited partnership is not necessarily dissolved upon the death, bankruptcy, or withdrawal of a limited partner.

3. *Interests may be sold or transferred*—A limited partner may sell or transfer his interests without disrupting the partnership, so long as it is done in keeping with the agreement that is drawn up at the time of organizing. Usually, it is specified that a limited partner who wishes to sell his interest in the business must give the other partners first refusal before offering it to outsiders.

4. *The business is taxed as a partnership*—This allows profits and expenses to be passed on to the investor on a pro rata share based on his investment.

5. *Liability is limited*—The investor's liability is limited to the amount of his investment.

6. *It may be used as a tax shelter*—It makes possible the deferral of taxable income into the next year through expending of the normal feeding expenses in the year paid, along with the purchase of grain and other storable commodities in the current tax year for use in the next year. These expenses can be used to offset other income which in many cases includes profits from cattle fed during the prior year. The Internal Revenue Service may remove this advantage in the future. But so long as it exists, cattlemen should take advantage of it.

The *disadvantages* of a limited partnership are:

1. *The general partner has unlimited liability*—Although this fact should be recognized, it need not be reason for concern *provided* the operation is sound and well managed.

2. *The limited partners have no voice in management*—This is not an unmixed blessing, for it alleviates the hazard of compromise and weakness, which can, and often does, mitigate against a business.

Corporations

A corporation is a device for carrying out a farming or ranching enter-

prise as an entity entirely distinct from the persons who are interested in and control it. Each state authorizes the existence of corporations. As long as the corporation complies with the provisions of the law, it continues to exist—irrespective of changes in its membership.

Until about 1960, few farms and ranches were operated as corporations. In recent years, however, there has been increased interest in the use of corporations for the conduct of farm and ranch business. Even so, only about one percent of U.S. farms use the corporate structure. (See Fig. 13-2.) In 1974, the U.S. Department of Agriculture reported that, based on a 1969 study, there were 21,513 farm corporations in the United States. But 90 percent of these were "closely held," which means that they were mostly family owned. When closely held corporations were deleted from the total, corporate farms totaled only 1,797 and accounted for only 2.9 percent of farm sales.

From an operational standpoint, a corporation possesses many of the privileges and responsibilities of a real person. It can own property; it can hire labor; it can sue and be sued; and it pays taxes.

Separation of ownership and management is a unique feature of corporations. The owners' interest in a corporation is represented by shares of stock. The shareholders elect the board of directors who, in turn, elect the officers. The officers are responsible for the day to day operation of the business. Of course, in a close family corporation, shareholders, directors, and officers can be the same persons.

The major *advantages* of a corporate structure are:

1. *Continuity*—It provides for continuity of the business despite the death of a stockholder.

2. *It facilitates transfer of ownership*—Since stock (rather than physical property) is sold, exchanged, or given, transfer of stock is easy.

3. *The liability of shareholders is limited to the value of their stock*—In an incorporated business, the shareholders are liable for the debts of the corporation and for any liability caused by negligent employees who injure others in the course of their employment, but only to the extent of their investment in the corporation.

4. *There may be some savings in income taxes*—In particular, this may be so if the enterprise produces taxable income of $10,000 or more. A corporation pays a Federal corporate income tax of 30% on the first $25,000 net profit (before dividends are paid), and 52% on profits above $25,000. Moreover, salaries paid to owners who also are employees of the corporation, along with costs of insurance, health, and retirement plans, can be deducted as business expenses, thereby reducing the amount of taxable income.

The *disadvantages* of a corporation are:

1. *Restricted to charter*—Corporations are restricted to doing only what is specified in their charter.

2. *Must register in each state*—They must register in each state in which they do business and abide by the regulations of each state.

3. *Must comply with regulations*—They must comply with specific reg-

ulations regarding meetings, records, and reorganizations; and they are subject to certain legal fees, payroll taxes, accumulated earnings taxes on profits not distributed to stockholders, and other fees. Thus, a corporation involves considerable paper work and expense.

4. *The hazard of higher taxes*—A corporation is always a ripe target for higher income taxes, particularly Federal income taxes.

5. *Control can be lost*—Any person or group owning 51 percent, or more, of the stock can elect the Board of Directors and gain control.

Still another type of corporation is family owned (privately owned). It enjoys most of the advantages of its generally larger outside-investor counterpart, with few of the disadvantages. The chief *advantages* of the family owned corporation over a partnership arrangement are:

1. *It alleviates unlimited liability*—For this reason, a lawsuit cannot destroy the entire business and all the individual partners with it.

2. *It facilitates estate planning and ownership transfer*—It makes it possible to handle the estate and keep the business in the family and going if one of the partners should die. Each of the heirs can be given shares of stock—which are easy to sell or transfer and can be used as collateral to borrow money—while leaving the management of the enterprise to those heirs interested in operating it, or even outsiders.

TAX-OPTION CORPORATION (Subchapter S Corporation)

Instead of paying a corporate tax, a corporation with no more than 10 stockholders may elect to be taxed as a partnership, with the income or losses passed directly to the shareholders each of whom pays taxes on his share of the profits. This special type of corporation, which Congress created in 1958 by amending the Internal Revenue Code, is variously referred to as a "tax-option" corporation, "subchapter S" corporation, pseudo-corporation, or elective corporation.

For income tax purposes, the owners of a tax-option corporation are taxed as if they were a partnership. That is, income earned by the corporation passes through the corporation to the personal income tax returns of the individual shareholders. Thus, the corporation does not pay any income tax. Instead, each shareholder pays tax on his share of corporate income at his individual tax rate, and each shareholder reports his share of long-term capital gains and receives his deductions therefor. Although each shareholder's portion of any corporate losses from current operations is deducted from his personal return, capital losses incurred by the corporation cannot be passed through to the shareholders.

Thus, there are some very real advantages to be gained from a "subchapter S" or "tax-option" corporation. However, in order to qualify as a Subchapter S Corporation, the following requisites must be met:

1. There cannot be more than 10 stockholders.
2. All stockholders must agree to be taxed as a partnership.
3. Nonresident aliens cannot own stock.
4. There can be only one class of stock.

5. Not more than 20 percent of the gross receipts of the corporation can be from royalties, rents, dividends, interest, or annuities plus gains from sale or exchange of stock and securities; and not more than 80 percent of gross receipts can be from sources outside the United States.

General Advantages of Limited Partnerships and Corporations

In addition to the advantages peculiar to (1) limited partnerships, and (2) corporations, and covered under each, limited partnerships and corporations have the following advantages over individual ownership in the acquisition of capital:

1. They make it possible for several producers to pool their resources and develop an economic-sized operation, which might be too large for any one of them to finance individually.

2. They make it possible for persons outside agriculture to invest through purchase of shares or stock in the business.

3. They can generally borrow money easier because the strength of the loan is not dependent on the financial and managerial capability of one person.

4. They give assurance that the business will continue, even if one of the owners should die or decide to sell his interest.

5. They provide built-in management, with continuity; and, generally speaking, they attract very able management.

Thus, cattlemen—both cow-calf men and cattle feeders—can and do use either of these two business organizations—a limited partnership or a corporation—to develop and maintain an economically sound operation. Actually, no one type of business organization is best suited for all purposes. Rather, each case must be analyzed, with the assistance of qualified specialists, to determine whether there is an advantage to using one of these types of organizations, and, if so, which organization is best suited to the proposed business.

Agency Services

As an alternative to entering into the limited partnership arrangement or owning stock in a corporation, nonfarm investors wishing to engage in cattle feeding can utilize the services of several firms which specialize in purchasing feeders, contracting with feedlots, and selling market cattle. Also, similar services are available for acquiring and managing commercial breeding herds. Under the agency arrangement, the investor establishes a drawing account for the agent, arranges his own financing if he wishes, and can withdraw any profits realized. Because he obtains legal title to specific lots of cattle, he may use them as collateral for loans.

Firms offering such services charge a flat fee per head, or a percentage of gross sales. They usually do not have financial interests in the feedlots or ranches with whom they contract.

CAPITAL

In 1974, U.S. farm investment in real estate, livestock, machinery and motor vehicles, crops stored on and off farms, and household equipment and furnishings totaled $431.2 billion,[2] which (1) ranked it as America's biggest single industry, and (2) is equivalent to about three-fifths the value of all the stocks of all corporations represented on the New York Stock Exchange. Another noteworthy statistic is that it takes about $13 in farm assets to produce $1 of net farm income.[3]

There is no available breakdown of total capital investment (including land, buildings and equipment, as well as animals) by classes of livestock, but it is reasonable to assume that beef cattle head the list.

In comparison with most other businesses, cow-calf operations are characterized by slow turnover. Nature ordained that heifers not reach puberty until 8 to 12 months of age, that they not be bred before about 15 months of age, and that there be a pregnancy period of 9 months. This means that the minimum generation interval of cattle is about 3 years. The producer can do very little to speed it up.

A faster capital turnover exists in cattle finishing than in cow-calf operations, but there is also higher risk.

Sources of Capital (Where Farmers Borrow)

Traditionally, agriculture has been financed by two kinds of capital, known as equity capital and credit (debt) capital. Formerly, equity capital came only from farmers—those who operated farms and ranches; debt capital came from a variety of sources. This was one of a number of characteristics of farm businesses that differentiated this sector from the rest of the economy.

Table 13-1 shows where farmers borrow, the amount of loans from each source, and the percent of the total held by each type of lender. Also, it shows borrowing trends. (See last column.)

But, agricultural financing is changing, and it will continue to change even more in the years ahead. Today, cattlemen are tapping the vast supply of farm equity or risk capital that is constantly seeking investment opportunities—nonfarm equity capital is being used in the cattle business.

Sometime or other most cattlemen find it necessary to borrow money to buy land; to construct buildings and other improvements; to purchase equipment, seed, and cattle; and/or to pay for seasonal labor. They should know something, therefore, about the lending organizations available to them in order that they may determine which one will best serve their needs. The leading sources of farm credit are:

1. *Commercial banks*—Commercial bank funds for loans are provided by bank deposit. Since these fluctuate rather widely, the number of long-term loans banks can make is limited.

Progressive bankers are willing to lend to people who are beginning or

---

[2]Source: *Agricultural Finance Outlook*, USDA, March, 1974, p. 19.
[3]Source: *Agricultural Finance Statistics*, USDA, June, 1974, p. 78.

TABLE 13-1
WHERE FARMERS BORROW (1974)[1]

Type and Source of Loan	Amount of Loan	Percent of Total	Percent of Change from '73
	(million dollars)	(%)	(%)
Real Estate Mortgage Loans:			
Individuals and others	13,880	35.3	+ 14.2
Federal Land Banks	10,950	27.9	+ 21.0
Insurance companies	5,992	15.2	+ 6.2
Commercial banks	5,458	13.9	+ 13.9
Farmers Home Administration	3,040	7.7	+ 7.2
Total	39,320	100.0	+ 14.1
Nonreal Estate Loans:			
Commercial banks	17,328	40.2	+ 21.0
Dealers, merchants & others	15,900	36.9	+ 3.5
Production Credit Association	7,829	18.2	+ 18.5
Farmers Home Administration	877	2.0	+ 12.3
Commodity Credit Corp	800	1.9	− 55.6
Federal Intermediate Credit Banks	331	0.8	+ 31.9
Total	43,065	100.0	+ 10.1
Total loans	82,385		
Percent Real Estate	47.7		
Percent Nonreal Estate	52.3		

[1]Data provided with a personal communication to the author from Mr. E. A. Jaenke, Governor, Farm Credit Administration, Washington, D.C.

expanding their cow herds. Most commonly, they will write a note for 6 to 12 months; however, some of them will make intermediate-term loans on breeding animals. A 6- or 12-month note is reviewed annually. Terms of the loan usually call for payment of interest due and some payment on the principal. In most cases, the note is renewed for the smaller principal amount.

Notes written for a year or less enable both the banker and the borrower to evaluate the progress being made by the borrower and permit the lender to update the interest rate being charged. A financial statement is required by bank examiners. This statement enables both the borrower and the lender to visualize how financing the cow loan may affect the financial status of a total operation.

Loans may be more easily obtained by the cattleman who already has a cash income from land, cash grain, or other livestock. Progressive commercial bankers also are more willing to lend to the producer who is using management techniques that will increase his level of efficiency. Good record keeping on cow herds, use of performance tested bulls, proper range and/or pasture management that could increase carrying capacity are but a few of the management techniques presently viewed as profit boosters.

Some commercial banks have special agricultural representatives who are qualified to assist the borrower in many ways.

2. *Farm credit system*—The Farm Credit System includes three Federal lending units—Federal Land Banks, Production Credit Associations (PCAs), and Banks for Cooperatives—all under the supervision of the Farm Credit Administration, an independent Federal agency.

The Farm Credit System obtains its loan funds primarily through the sale of its bonds in the nation's private money market.

The Federal Land Banks and PCAs supply nearly one-fourth of the credit used by farmers, and the Banks for Cooperatives provide nearly two-thirds of the borrowed capital used by farmer cooperatives.

a. *Federal Land Banks (FLB)* are long-term farm real estate mortgage lenders. Loans are made through local Federal Land Bank Associations which are local cooperative lending organizations that operate under the supervision of 12 Federal Land Banks and the Farm Credit Administration. Their loan funds come from private investors who buy FLB bonds and from interest received from mortgage loans. Since the Federal Land Bank is a farmer cooperative, borrowers are required to purchase stock in the local land bank equal to 5 percent of their loan.

By law, Federal Land Banks may make loans up to 40 years for an amount not to exceed 85 percent of the appraised market value. However, traditionally, they have been conservative. In 1973, the average Land Bank loan was for a term of 27 years at 59 percent of the present market value. Loans are on first mortgage only. Consideration is also given to the income and management ability of the borrower. The Federal Land Banks and local associations appraise the farms or ranches and supervise and administer the loans.

In addition to providing loans for the purchase of farmland, farmers may borrow from the Federal Land Bank to build farm buildings, drain or level land, or to refinance existing debts. Also, the FLB can provide loans for beef breeding herds if a first mortgage is taken on the farm or ranch real estate.

Land Bank loans may be paid ahead of schedule without penalty. The interest rate on FLB loans varies directly with the cost of money.

Loans and information about loans may be obtained from the manager of the local Federal Land Bank Association nearest you or from a local association representative. The address of the latter may be obtained from the local county agricultural agent or from the Farm Credit Administration, Washington, D.C. 20578.

b. *Production Credit Associations (PCAs)* are local cooperative lending organizations that operate under the supervision of the 12 Intermediate Credit Banks and the Farm Credit Administration.

The PCA is very willing to provide financing for beef herds for qualified cow-calf operators. This financing may be done in conjunction with other ranch or farm enterprises. Most beef cow financing is on a 3-year basis, although it may be up to 7 years. Depending upon the managerial financial ability of the borrower, the PCA may loan up to 100 percent of the breeding herd cost. However, a 70-30 share is most common with the borrower providing a 30 percent margin. The loan limit varies with individual cases. Also, the PCA may make short-term loans for feed, seed, fertilizer, chemicals, fuel, building supplies, fencing, machinery, labor and professional services, rent, taxes, and property insurance.

All borrowers become stockholders in the PCA. The PCA requires all borrowers to purchase PCA stock in an amount equal to 10 percent of the loan. The investment in the stock is redeemed when the loan is repaid. Interest rates vary depending on the cost of money on the open market.

Obtaining a PCA loan begins with evaluating a prospective buyer's operating statement and operating ability. PCA managers encourage cowmen to build their herd over a period of time, rather than buying into the business at full capacity. This allows for both management and financial success.

c. *Banks for Cooperatives* provide the majority of financing for the nation's farm supply, marketing, and business service cooperatives.

3. *Farmers Home Administration (FHA)*—The Farm Home Administration is a Federal agency which makes loans to farmers and ranchers who are unable to obtain adequate credit from other sources at reasonable rates and terms. Applicants who are veterans and have farm experience receive preference. FHA helps applicants determine their credit needs, work out debt repayment schedules, and solve other financial problems even though they may not be eligible for a loan.

Farm Home Administration loans are made to farmers who carry on farming operations on a scale large enough to support their families and to farmers on small farms who obtain income from off-farm employment. Each farm-ownership loan is based on a plan that when followed will provide enough income from the farm and other sources to enable the family to have a reasonable standard of living and make payments on its debts when due.

The county supervisors of Farm Home Administration help borrowers prepare the plan and provide on-the-farm assistance with management problems. Farm-ownership loans are scheduled for repayment according to the borrowers' ability to repay, over a period not exceeding 40 years.

In addition to farm-ownership loans, Farm Home Administration makes loans to purchase cows, operating loans, housing, soil and water conservation, and emergency. Generally, cow loans are drawn up when the cows are purchased. The number of cows purchased is based on the grass and feed available on the farm or ranch; and the loan is set up on a 7-year note. Normally, the upper limit on such a loan will be 100 percent of the cost of the cows or up to $50,000. Repayments must be made at least annually, right after calves are sold.

In 1974, the U.S. Congress enacted legislation providing guaranteed emergency loans to financially distressed livestock producers, primarily cattle feeders, to help them recover from recent financial hardships caused primarily by high feed costs and low market prices. Loans were made available only to farmers and ranchers whose time and income were mainly associated with livestock raising and who needed Federal leverage to stay in business. The law set a maximum of $250,000 for any one producer of beef or dairy cattle, hogs, sheep, goats, chickens, or turkeys. (Subsequently, the upper limit on such loans was increased to $350,000.) It also limited the guarantees to 80 percent of the loans, which had to be made by banks or other private sources. The loans were for up to 3 years, with renewal up to 2 additional years in appropriate cases. The Farmers Home Administration (FHA) conducted the program.

To obtain loans or detailed information about loans, apply to the Farm Home Administration. The address of this office can be obtained from the county agricultural agent or the Farm Home Administration, Washington, D.C. 20250.

4. *Individuals*—Individuals with money, including retired farmers, landlords, and relatives, are a very important source of real estate credit.

Today, a number of sellers loan the amount of equity they have in land which they sell to a friend, associate, or member of the family. A retiring farmer, for example, may sell his farm to a young man he knows will run it properly, but who can't make a down payment or obtain a large enough mortgage. The seller may be willing to take a second mortgage on part of the purchase price at a reasonable rate of interest. Frequently, the selling farmer will do this under an "installment land contract," in which the down payment may vary from nothing to 29 percent of the purchase price. He then retains title until a certain agreed-to part of the price has been paid. By staying at 30 percent or under on the down payments, the sales qualify as installment sales and the seller(s) doesn't have to pay the capital gains tax except as proration when he gets the money.

Under a land contract, the selling farmer usually gets a favorable price, along with a tax break on the profits of the sale. The buying farmer obtains the advantage of using the land which he is otherwise unable to finance through any other credit source.

After a land contract is run for some years and the new farm operator has improved his financial condition, he will often refinance his land contract with an institutional lender. Close to 40 percent of all transfers of farm property and 45 percent of the dollar volume of farm land credit purchases occur in this manner—either by land contract or seller-held mortgage.

One disadvantage of a loan from an individual is that the arrangement may be complicated by his death unless adequate provision has been made for this eventuality. Individuals also make some nonreal estate loans to farmers. Here again, the terms and repayment provisions vary widely depending upon the agreement reached by the individual parties.

5. *Life Insurance Companies*—Many life insurance companies invest a part of their policyholders' reserve funds in loans secured by first mortgages on farm real estate. Terms usually range between 5 and 40 years and amortized repayment plans are now common. They usually offer a competitive rate of interest and amortized repayment plans. Generally insurance companies will make loans up to 60 percent of the appraised value of the farm or up to 50 percent of the sale value.

Life insurance activity in farm and ranch loans varies from region to region. In some states, they hold close to one-half of the farm and ranch mortgage debt. In others, they have less than one-half of one percent. Their contacts are through branch offices or other local agents. Realtors can almost always provide information concerning insurance agencies actively making loans in the area.

6. *Merchants and dealers*—Many cattlemen buy machinery, feed, and supplies on time. This is usually the easiest type of credit to get, but it may be the most costly. If farmers and ranchers use this type of credit, they should try to find out how much it costs and compare it with costs with that of other sources of credit.

It must be realized that dealers and merchants extend credit to farmers and ranchers primarily for the purpose of promoting the sale of products and services. To a machinery dealer, the interest return from a loan is less important than the sale of a tractor. This is so because his profits come from both sales and interest, and he is selling "hard goods" that can be repossessed; hence, he can afford to make relatively high risk loans. Installments over the expected productive life of the equipment, often several years, are the general practice.

The seller of "soft goods," like feed and fertilizer, must look at credit differently. Their products, once used, cannot be repossessed. Thus, the debt for feed and fertilizer customarily must be paid from the income of a single year.

7. *Farm mortgage companies (Finance companies)*—These companies make loans for various purposes. In comparison to local bank rates, their interest rates are much higher—2 percent per month or 30 to 36 percent per year are common. If a borrower gets only $7 to use out of each $10 that he borrows, he will need to make very high profits on his business. Few farmers can afford loans from mortgage or finance companies if other sources are available to them.

8. *Rural Development Act (FHA 90% guaranteed)*—The passage of the Rural Development Act of 1972 made a great variety of farm and rural community development loans available. In most cases, the loan is made through and serviced by the local banker or Production Credit Association, and the Farmers Home Administration (FHA) guarantees the loan up to about 90 percent. There are 2 types of loans:

   a. *Guaranteed loans*—Loans made and serviced by a private lender, with FHA guaranteeing to limit any loss to a specific percentage (up to 90 percent).

   b. *Insured loans*—Loans made, as in the past, by FHA personnel to individuals or agencies who cannot obtain loans elsewhere at reasonable rates.

Loans of up to $100,000 may be made and serviced by private lenders, with an FHA guarantee up to 90 percent of any losses. Loans of this nature have been made previously by FHA itself, or with participation of another lender. With local lending agency involvement, the program is expected to gain new momentum.

Rural Development Act loans may be used to buy or enlarge farms; to construct, improve, or repair farm homes and farm service buildings; to drill wells and otherwise improve water supply systems for home use, livestock, and irrigation; to refinance debts; to develop and improve farmland, clear and level land, establish and improve farm forests, provide drainage systems, carry out basic land treatment practices such as liming, and to make other improvements; and to finance, on family farms, enterprises involving such things as camping and swimming facilities, tennis courts, riding stables, vacation rental cottages, lakes and ponds for boating and fishing, docks, nature trails and picnic grounds, repair shops, roadside markets, souvenir shops, wood or metal working facilities, and small grocery stores or service station facilities.

To be eligible, the loan applicant must manage and operate the farm, and the farm must produce a substantial amount of his total income.

Borrowers pay five percent per year on the unpaid principal, with a maximum term of 40 years. FHA makes loan subsidy payments to private lenders if necessary, to bring the lender's return up to the lender's usual rate for such loans or the FHA established rate.

Community facility loans are also available for such purposes as to relocate roads, bridges, and utilities; to construct libraries, hospitals, industrial parks, and other public buildings; and to develop recreation, transportation, and related projects.

Business and industrial loans are available for nonagricultural ventures in a rural community, for the purpose of providing employment in small towns.

Rural single-family housing loans are available, also.

NONFARM EQUITY CAPITAL IN AGRICULTURE[4]

Is a cow-calf operation or feeding cattle any different in terms of needs for capital than financing any other production process, or building a shopping center, or drilling for oil or gas? There are risks in all such ventures that go beyond those that most suppliers of debt capital are willing to assume. Perhaps, too, the chances of return to equity capital in certain farming ventures are competitive with the opportunities available in nonfarm businesses.

Among the various arrangements now being offered nonfarm investors to enable them to participate in agricultural ventures are the following:

1. *Limited partnership*—Unlimited legal liability, the need for active participation in management by all members, and the problem of termination or reorganization if one member withdraws makes the general partnership inappropriate for outside investors. The limited partnership alleviates these disadvantages.

The main attraction of a limited partnership from the investor's standpoint is that it provides a direct conduit for tax deferral and pass-through of profits and losses. The partnership pays no taxes on its earnings because profits and losses are allocated to each partner in proportion to his respective interest and are reported on his individual return. These tax fea-

---

[4]In the preparation of this section, the author drew heavily from the paper entitled, "Nonfarm Equity Capital in Agriculture," by William H. Scofield, Economic Research Service, USDA, Washington, D.C., which appeared in *Agricultural Finance Review*, Vol. 33, July 1972.

tures are not available to holders of common stock because a corporation is a taxable entity and losses, as well as income and capital gains, are lodged within the corporation and are not passed directly to shareholders for tax purposes.

The organizational aspects of creating and selling limited partnership interests are relatively simple. First, an existing firm, such as an incorporated feedlot, forms a subsidiary corporation to act as the general partners. The subsidiary usually acquires its capitalization and often additional borrowing capacity from its parent company.[5] After registration with the appropriate State agency, the general partner then files a registration statement with the SEC if an interstate offering is planned. Following clearance, the general partner proceeds to offer limited partnership interests to the public, either directly or through security dealers.

A study of recent registration statements pertaining to agricultural ventures filed with the Securities and Exchange Commission (SEC) shows that more than one-half were offering limited partnership interests. Until 1974, when cattle feeders suffered disastrous losses, limited partnerships to raise capital for cattle feeding were the most numerous of agricultural ventures. Also, there were offerings of beef breeding herds. Most of the latter were purebreds, of the "exotic" breeds, rather than commercial cow herds. Prices asked were substantially above prevailing market prices of similar animals available directly from purebred breeders or public auctions. The purchase of a herd, usually 5 or 10 cows with breeding arrangements, in most cases also involved maintenance contracts at an annual cost of $250-$400 per head. Most of the firms offering breeding herds owned or leased range lands and had other facilities needed to care for the herds. Some firms planned to place the cattle on ranches under contractual arrangements.

Apart from the satisfaction that investors may derive from owning a herd of cattle, the prime consideration is the tax shelter such an investment offers to persons in high tax brackets. Although certain changes were made in the Tax Reform Act of 1969 to reduce the tax incentives in such investments, it is still possible to convert ordinary income into capital gains via cattle breeding herds. The extension of investment tax credit to breeding animals permitted under the 1971 amendments offset part of the restrictions imposed in the earlier tax provision. As a result, more offerings of breeding herds will be made in the future.

2. *Common stock*—One of the potential advantages of incorporation of existing farm businesses is that a mechanism is provided for the sale of stock to persons outside the firm. Additional equity capital can thus be brought into the firm for expansion, or to replace debt capital. In actual practice, however, relatively few incorporated farms are sufficiently large or have the financial history of profitability needed to attract investors. Also, the subsequent earnings and the behavior of the stocks of some of those that have incorporated and gone public have been disappointing to investors.

---

[5]Internal Revenue Service regulations require that the capitalization of the general partnership must be 10 to 15 percent of the amount of capital contributed by the limited partners, depending upon the total amount involved.

In recent years, obtaining equity capital via public offerings of common stock has been used to a limited extent in the cattle business. Typically, the technique involves establishing a new business by merging several existing privately held firms. Stock in the new firm is exchanged for the stock of each acquired firm. The financial statements of the new firm reflect the combined assets, debts, and pro forma earnings based on records of the acquired firms. A public offering of stock in the new firm can then be made, as well as new debt offerings secured by assets of the firms acquired.

Several benefits accrue to the firms that are consolidated into the new firm. There is usually a pooling of management talents from the acquired firms into the new firm, and possibly new outside management may be brought in. If the merger is of a horizontal or vertical nature, there can be economies in purchasing, sales, and in distribution costs. However, perhaps the most important future benefit is the liquidity afforded the shareholders of the acquired firms as a result of the exchange of stock which now has a public market. Liquidity can also be achieved by sale of a business through the exchange of stock with a firm that already has an established market for its stock. Lack of liquidity is a perennial problem of closely held firms that hampers growth and often creates estate settlement and tax problems.

There are no apparent reasons why the corporation technique cannot be used to consolidate several existing farm businesses that are well managed and can show an attractive pro forma earnings statement. It is feasible to make an intrastate offering of common stock in a consolidated farm business which is of sufficient scale to support capable management.

3. *Private placements*—In addition to the public equity capital market widely used by nonfarm firms, a substantial volume of capital flows directly from private investors, insurance companies, and specialized firms to new business ventures that could not otherwise obtain such risk capital. For example, so-called letter stocks have become a favorite type of investment for growth-oriented mutual funds, and for specialized investment capital firms offering both financial and managerial support.[6]

Firms seeking venture capital may deal directly with the primary source of such funds, or they may utilize the services of agents who bring the two parties together. A number of major corporations in recent years have established separate departments to evaluate new products or services developed by other firms and will advance seed capital to those that appear most promising.

Because of the unorganized and informal nature of the venture capital market, it is difficult to judge its importance as a source of capital for the farm sector. It is known, however, that cattle feeding firms have tapped this market in the past, and even now may be receiving more equity capital via this avenue than from the sale of limited partnership interests. Also, it is noteworthy that a mutual fund based in Luxembourg has as its stated pur-

---

[6]Letter stocks derive their name from a written agreement between the firm issuing unregistered securities and the firm making the purchase that the purchaser will not offer such stock for sale to the public. The issuing company may, of course, later file a registration statement and qualify the stock for sale. This is often a condition imposed by the firm acquiring the stock.

pose the acquisition of equity interests in various types of agricultural ventures in this country, including ranches.[7] This fund was said to hold about $23 million of such interests in 1970.

SUMMARY OF NONFARM EQUITY CAPITAL

Individuals in upper tax brackets represent a potential source of equity capital that can be directed to the farm sector if appropriate legal arrangements can be developed and channels established. However, opportunities in the farm sector would need to compare favorably with the alternatives available to such investors. More specifically, such investors evaluate alternative risk capital ventures with respect to the following considerations:

1. Potential for leveraging equity capital.
2. Potential for deferral of taxes on ordinary income.
3. Opportunities for converting ordinary income into capital gains.
4. Effectiveness of the investment as a hedge against inflation.
5. Potential after-tax returns in comparison with other venture capital investment opportunities available.

Also, some weight may be given to noneconomic considerations such as the novelty or uniqueness of the venture and the personal satisfaction and prestige to accrue from being associated with it. Thus, ownership of land, or cattle may be more important to some investors than prospective monetary returns.

Innovative arrangements to attract nonfarm equity capital into the agricultural sector hold promise for alleviating the shortage of equity capital that exists in certain kinds of farm businesses. The techniques in present use have been borrowed from the nonfarm sector, but further modifications may be necessary to adjust them to the special characteristics of farms, ranches, and feedlots. Thus far, the most significant penetration of nonfarm equity capital has been in cattle feeding, which has paralleled the rapid growth of large-scale commercial feedlots. Despite the attention given to nonfarm investments in cattle breeding herds, the total amount of capital attracted to such ventures is quite small.

In the years ahead, there likely will be more opportunities for joint venture arrangements among farmers, ranchers, and feedlots than are presently being realized. But judicious use of equity capital from nonfarm investors will be needed to finance further growth of production or for forward integration into processing and marketing. Conversely, there may also be opportunities for firms now engaged in providing inputs, or in processing and marketing to develop mutually beneficial arrangements with producers through mergers or the investment of equity capital in production facilities. If a more effective and efficient food production and distribution system is to be realized in coming decades, major attention will need to be given to wider use of financial arrangements that now exist only in rare instances.

---

[7]Barrons, Feb. 23, 1970.

CREDIT IN THE CATTLE BUSINESS[8]

Credit, upon which the whole vast structure of our commercial world rests, has been around for a very long time. According to the Greek philosopher Plato, the last words of Socrates were: "I owe a cock to Asclepius. Will you remember to pay the debt?" But it was the Romans who coined the word "credit," which comes from the Latin "credere," meaning to believe or trust.

Without credit, great businesses, including the cattle business, wouldn't make it, for few could supply all the capital needed.

Total physical farm assets as of January 1, 1974, were estimated at $431.2 billion. Additionally, there were financial assets of $28.6 billion, making for total assets of $459.8 billion[9].

Total farm debt was estimated at $81.7 billion. This means that, in the aggregate, farmers had an 82 percent equity in their business, and 18 percent borrowed capital. Perhaps they have been too conservative, for it is estimated that one-fourth to one-third of American farmers could profit from the use of more credit in their operations.

Credit is an integral part of today's cattle business. Wise use of it can be profitable, but unwise use of it can be disastrous. Accordingly, cattlemen should know more about it. They need to know something about the lending agencies available to them, the types of credit, and how to go about obtaining a loan.

Types of Credit or Loans

Getting the needed credit through the right kind of loan is an important part of sound financial farm management. The following three general types of agricultural credit are available, based on length of life and type of collateral needed:

● *Short-term loans*—This type of loan is made for operating expenses and is usually for one year or less. It is used for the purchase of feeders, feed, seed, fertilizer, gasoline, and family living expenses. Security such as a chattel mortgage on the feeders or crop may be required by the lender; and the loan is repaid when the animals or crop are sold.

● *Intermediate-term loans*—These loans are used to buy equipment and breeding stock, for making land improvements, and for remodeling existing buildings. They are paid back in one to seven years. Generally, they are secured by a chattel mortgage on livestock and machinery.

● *Long-term loans*—These loans are secured by mortgage on real estate and are used to buy land or make major improvements to farm land and buildings or to finance construction of new buildings. They may be for as long as 40 years. Usually they are paid off in regular annual or semiannual

[8]This section was reviewed by, and helpful suggestions were received from, Mr. Douglas D. Kleist, Vice President-Loan Administrator, National Bank of Agriculture (now American National Bank), Fresno, Calif.

[9]*Agricultural Finance Outlook*, USDA, March, 1974, p. 77. Data for 48 states only.

payments. The best sources of long-term loans are: an insurance company, the Federal Land Bank, the Farm Home Administration, or an individual.

Categories of Cattle Loans

There are three categories of cattle loans: (1) cow-calf loans; (2) cattle feeder loans (the financing of stocker cattle and finishing cattle is much the same); and (3) cattle on pasture loans.

COW-CALF LOANS

Many cow-calf men operate on considerable borrowed capital. Their loans are for three purposes:

1. *Purchase of farm or ranch*—This is a long-term loan, with repayments made over several years.

2. *Production loans*—This refers primarily to initial loans for purchase of the breeding herd, or to loans made for the production of a crop. It also includes loans made against the breeding herd to refinance other debts, to purchase machinery and equipment, and to make minor ranch improvements. Usually, these loans are set up on an intermediate-term basis, mostly for 1 to 5 years, although they may range from 1 to 7 years. This means that the cash flow should be such that the operator can reduce his intermediate credit loan by $1/5$ or $1/7$ each year, depending on the term.

3. *Operating budget loans*—These loans, which usually do not exceed a 12-month period, are for the purpose of financing the recurring expenses during the year's operation. Barring disaster (drought, insects), the borrower should be able to pay back operating loans from current income.

CATTLE FEEDER LOANS

Adequate financing to carry on a cattle feeding program is extremely important. Thus, it behooves the feeder to have credit well established before jumping into cattle feeding. Consideration should be given to (1) the cash requirements for plant and equipment, for purchase of feeder cattle and feed, and for operating; and (2) the program to be followed—including kind and source of feeder cattle and feed, and market outlets. Both the financial needs and the program should be carefully conceived and put in writing. Then they should be reviewed by and discussed with a lender who has an understanding of cattle feeding and in whom the feeder has confidence. Once agreed upon, both the feeder and the lender should adhere to the program.

Many cattle feeders make use of long-term and intermediate credit for the purchase of land, the construction of the feedlot and mill, and the purchase of equipment. Additionally, cattle feeder loans fall into the following four categories:

1. *Loans to farmer-feeders*—These feeders generally grow most of their feeds. Frequently, they also produce part, or all, of their calves. They gener-

ally buy protein supplements, minerals, and feed additives. Usually, loans to farmer-feeders are modest in size and on a short-term basis.

2. *Revolving loans to year-round feeders*—This is the type of feeder who maintains cattle in the feedlot the year round. As he sells finished cattle, he brings in feeders. Revolving loans (also referred to as line of credit) are needed for this type of operation.

3. *Loans to short-term feeders*—This is the kind that buys a certain number of cattle, feeds them out, and then pays his loan in full. Short-term loans serve this need.

4. *Loans to grower-feeders*—These are generally calf-type programs, in which the cattle are grown to 600 to 700 pounds weight, on relatively high roughage rations, then sold to finishers. This kind of feeder either produces or buys his roughages, then buys calves to utilize the feed. Some grower-feeders operate on a year-round basis and require revolving loans.

Buying feeder calves involves less risk than finishing yearlings, simply because less money is tied up in cattle and more in feed. When purchase price of animals is included, feed costs make up about 40 percent of the total cost of calves and about 35 percent of the total cost of yearlings.

The following pointers are pertinent to each of the above four categories of cattle feeder loans:

● Many feedyard operators make the mistake of overcapitalizing in feedlot facilities and equipment, with the result that they have insufficient working capital to operate. Generally speaking, financial institutions require a margin of 20 to 25 percent of the feeder cattle and feed. This would appear as follows:

700-lb yearling steer	@40¢ =	$280.00
Feed to put on 350 lb gain	@32¢ =	112.00
		$392.00
20% margin	=	78.40
Amount to borrow	=	$313.60

A sound operator will generally maintain a safe margin of 30 to 40 percent so that he can keep going after a market reversal, which could cost more than $100 per head. He will be careful not to dip into the safe margin to finance long-term fixed assets; rather, he will finance such fixed assets through either (1) current margins in excess of his safe margin, or (2) intermediate or long-term loans.

● Lenders normally lend up to 100 percent of the purchase price of cattle if the borrower can cover feed and operating expenses.

● Lenders look upon the following as highly important factors when considering a feedlot loan: the ability, experience, and past record of the feeder; feeder's knowledge of cattle markets, and of buying and selling cattle; adequacy of the feedlot and facilities; availability and source of cattle and feed; and adequacy of margin, and the liquid assets of the operation.

● Feedlot expansion and improvement should be tied directly to the amount of the margin in the livestock and to the net worth of the feeder. In some cases, cattle equities are used for new homes, cars, and travel (for what is known as consumption purposes), with the result that insufficient money is available to operate on or to meet loans, particularly when some adversity strikes. As a result, the lender has no alternative but to stop financing the feeders.

CATTLE ON PASTURE LOANS

Pasture loans are very similar to cattle feeder loans. They are usually made in the spring and paid off in the fall.

Lenders will usually advance 75 to 80 percent of the cost of stockers for going on pasture provided the borrower (1) can put up the rest, and (2) has ample pasture and capital for operating expenses.

Fig. 13-3. Cattlemen can usually borrow up to 75 to 80 percent of the cost of stockers for pasture finishing. (Courtesy, USDA)

Credit Factors Considered and Evaluated by Lenders

Potential money borrowers sometimes make their first big mistake by going in "cold" to see a lender, without adequate facts and figures, with the result that they already have two strikes against them.

When considering and reviewing cattle loan requests, the lender tries to arrive at the repayment ability of the potential borrower. Likewise, the borrower has no reason to obtain money unless it will make money.

Lenders need certain basic information in order to evaluate the soundness of a loan request. To this end, the following information should be submitted:

1. *Analysis and feasibility study*—Lenders are impressed with a borrower who has a written-down program; showing where he is now, where

he's going, and how he expects to get there. In addition to spelling out the goals, this should give assurance of the necessary management skills to achieve them. Such an analysis of the present and projection into the future is imperative in big operations.

2. *About the applicant*—The applicant should furnish the following information:

    a. Name of applicant and wife; age of applicant.
    b. Number of children (minors, legal age).
    c. Partners in business, if any.
    d. Years in area.
    e. Experience—practical and educational.
    f. Estate planning (will, trust, etc.)
    g. References.

3. *About the present farm or ranch, if any*—If a loan is being obtained for the operation or expansion of the present ranch, the following information should be provided:

    a. Owner or tenant.
    b. Location; legal description and county, and direction and distance from nearest town.
    c. Type of ranch enterprise; cow-calf, feedlot, etc.

4. *Financial statement*—This document indicates the borrower's financial record and current financial position, his potential ahead, and his liability to others. The net worth statement records the financial status of a business at a particular point in time, whereas the profit and loss statement (P & L) measures a flow-through time. The borrower should always have sufficient solvency to absorb reasonable losses due to such unforeseen happenstances as storms, droughts, diseases, and poor markets; thereby permitting the lender to stay with him in adversity and to give him a chance to recoup his losses in the future.

Fig. 13-4 shows the Financial Statement used by the Bank of America, the world's largest bank.

5. *Profit and Loss (P & L) statement—The profit and loss statement (P & L) is a measure of the income generated, and the cost incurred, during a specific period of time—usually one year.* It is completed at the end of the farm business year to arrive at actual returns for the year. To prepare an estimated profit and loss statement for the coming year, you need to estimate your farm sales, operating expenses, depreciation, and net inventory change. It is much better to know what is expected and change plans if necessary, than to wait until the end of the year and discover that your farm business did not make a satisfactory profit.

The P & L statement serves as a valuable guide to the potential ahead; hence, it is important to the lender. Preferably, it should cover the previous three years. Also, most lenders prefer that the P & L statement be on an accrual basis (even if the cattleman is on a cash basis in reporting to the Internal Revenue Service).

BANK OF AMERICA

SPECIAL PURPOSE FINANCIAL STATEMENT
FARMER OR STOCKMAN

NAME _____

STREET ADDRESS _____

MAILING ADDRESS _____

RECEIVED AT _____ OFFICE

AT CLOSE OF BUSINESS _____ 19____

TAX I.D. NUMBER _____

BUSINESS STRUCTURE _____

ASSETS			LIABILITIES		
CURRENT ASSETS			**CURRENT LIABILITIES**		
Cash: Checking Acct. _____	$ _____		Notes & Accts. Payable (Sched. I)	$ _____	
Cash: Savings Acct. _____	_____		Long-Term Debt due within 12 months (Sched. H)	_____	
Cash: Other Banks _____	_____		Federal & State Income Taxes for year(s) _____	_____	
Current Receivables (Sched. A)	_____				
Feeder Livestock (Sched. B)	_____		Cash Rent Accrued	_____	
Harvested Crops/Feed on Hand (Sched. C)	_____		Taxes Due & Unpaid	_____	
Investment in Growing Crops (Sched. C)	_____		Interest Accrued & Unpaid	_____	
Listed Stocks & Bonds (Sched. D)	_____		Other Debt due within		
Farm Supplies on Hand	_____		12 Months _____	_____	
Other: _____	_____				
Total Current Assets	$ _____		Total Current Liabilities	$ _____	
FIXED ASSETS			**LONG-TERM LIABILITIES**		
Equipment & Vehicles (Sched. E)	$ _____		Long-Term Debt less current portion (Sched. H)	$ _____	
Breeding Stock (Sched. B)	_____		Other Long-Term Liabilities:		
Receivables (Over 1 yr. or Past Due) (Sched. A)	_____		_____	_____	
Unlisted Stocks (Sched. D)	_____		_____	_____	
Farm Land & Improvements (Sched. F)	_____		_____	_____	
Non-Farm Real Estate (Sched. G)	_____		Total Long-Term Liabilities	$ _____	
Other Personal Property	_____				
Life Insurance Cash Value	_____				
Other: _____	_____		**TOTAL LIABILITIES**	$ _____	
Total Fixed Assets	$ _____		**NET WORTH** (Assets less Liabilities)	$ _____	
TOTAL ASSETS	$ _____		**LIABILITIES PLUS NET WORTH**	$ _____	

WORKING CAPITAL
(Current Assets less Current Liabilities) $ _____

MEMO: Net change in value of Farm Land & Improvements (from page 2) $ _____

CONTINGENT LIABILITIES (Not included above)

As Guarantor or Endorser	$ _____
Accounts, Notes, or Trade Acceptances discounted or pledged	_____
Surety on Bonds, or other Contingent Liability	_____
Agreements to Repurchase	_____
Servicing Agreements or Guaranties of Collection	_____
Lease Liabilities	_____
Lease Equipment (Monthly Payments)	_____
Unsatisfied Tax Liens, or Notices from the Federal or State Governments of Intention to Assess Deficiencies.	$ _____

The undersigned (BORROWER) delivers the foregoing (and attached, if any) financial statements, certifies that they fully, truly and correctly represent BORROWER'S financial condition as of the effective date indicated and furnishes them to Bank of America National Trust and Savings Association (BANK) in order to induce BANK to establish and/or continue credit for the benefit of BORROWER.

CHANGE IN CONDITION. There has been no material change in BORROWER'S financial condition since the effective date. BORROWER agrees to immediately notify BANK of (1) any material change in BORROWER'S financial condition, (2) any change known to BORROWER in the financial condition of any endorser, co-maker or guarantor of BORROWER'S present and/or future obligations to BANK, and (3) the death of any such endorser, co-maker or guarantor. Unless and until BANK receives such a notice, BANK is requested and authorized to rely upon this financial statement upon each and every transaction in and by which BORROWER becomes indebted to BANK.

DEFAULT AND ACCELERATION. Upon the occurrence of any of the following events, all of BORROWER'S obligations to or held by BANK shall, at the option of BANK, become immediately due and payable without prior demand or notice: (1) Death of BORROWER; (2) Insolvency of, or the commission of an act of bankruptcy by BORROWER or any endorser, co-maker or guarantor of BORROWER'S indebtedness to or held by BANK; (3) Sale or transfer of a business owned or controlled by BORROWER; (4) institution of legal proceedings, issuance of a writ of attachment, garnishment, execution or other legal process against BORROWER'S property; (5) assessment of taxes (other than taxes on real property) by the federal, state or local government; (6) failure to give the notices as agreed in the preceding paragraph; or (7) any of the representations in the above financial statement prove to be untrue.

STATEMENT OF BANK OFFICER:
Insofar as our records reveal, this financial statement is accurate and true. The foregoing statement is (a copy of) (the original) signed by the maker, in the credit files of this bank. The average daily balance for the past year has been:

Coml. A-C $_____ Sav. A-C $_____

_____ Asst. Cashier-Manager

Signature _____

CR-10AB 11-72

Fig. 13-4a. Financial Statement for farmer or stockman used by Bank of America—page 1. (Continued)

PROFIT AND LOSS STATEMENT
For Current Period

Date Prepared: _____ 19___

Period Covered: _____ 19___ through _____ 19___

SALES: (Commodities sold during the period)

Dairy Products	$ _____
Livestock	_____
Poultry or Eggs	_____
Other Farm Products	_____
Total Sales	$ _____

OPERATING EXPENSES: (Incurred during the period)

Feed	$ _____
Labor	_____
Cash Rent	_____
Other Operating Costs	_____
Total Operating Expenses	$ _____

OPERATING MARGIN (Sales less Expense) $ _____

ADJUSTMENTS FOR INVENTORY CHANGES:

	FEEDER LIVESTOCK	HARVESTED CROPS	GROWING CROPS	FARM SUPPLIES	
Ending Inventory	_____	_____	_____	_____	
Less: Beginning Inventory	_____	_____	_____	_____	
Net Change (±)	_____	_____	_____	_____	(±) _____

(PLUS IF INCREASED, MINUS IF DECREASED)

ADJUSTED OPERATING MARGIN $ _____

OTHER FARM INCOME: (describe) _____ (+) _____

OTHER FARM EXPENSES: (describe) _____
_____ (−) _____

ADJUSTMENTS FOR CAPITAL ITEM CHANGES:

	EQUIPMENT & VEHICLES	BREEDING & DAIRY STOCK	FARM LAND & IMPROVEMENTS	
Ending Inventory	_____	_____	_____	
Plus: Sales	_____	_____	_____	
Total	_____	_____	_____	
Less: Beginning Inventory	_____	_____	_____	
Less: Purchases	_____	_____	_____	
Net Change (±)	_____	_____	_____	(±) _____

NET PROFIT BEFORE INCOME TAXES	$ _____
Less: Federal & State Income Taxes	_____
NET PROFIT	$ _____
Add: Net Worth at beginning of period	_____

OTHER ADDITIONS: (describe) _____

TOTAL	$ _____

LESS: Other Deductions & Withdrawals _____

NET WORTH ON THIS FINANCIAL STATEMENT $ _____

INSURANCE	

Farm Products $ _____
Livestock $ _____
Merchandise $ _____
Machinery $ _____
Buildings $ _____
Is fallen building clause waived? _____
Or extended coverage endorsement included? _____
Life Insurance $ _____
What Company? _____
Name of beneficiary _____

Automobiles and Trucks:
Public Liability $ _____ M/ $ _____ M
Collision $ _____
Property Damage $ _____
Earthquake $ _____

Do you carry Workmen's Compensation Insurance? _____
BankAmericard Number _____
Are you borrowing from
  any other office of this bank? _____
Have you applied for
  credit at any other source? _____
Have you ever filed a petition in bankruptcy
  or entered into composition with creditors? _____ If so, attach
  a complete explanation & state basis of settlement.
Amount for which you are
  liable for partnership obligations _____
Have you made a will? _____

Fig. 13-4b. Financial Statement, page 2. (Continued)

SUPPORTING SCHEDULES

A. NOTES & ACCOUNTS RECEIVABLE:

FROM WHOM	DUE DATE	AMOUNT DUE WITHIN 1 YEAR	PAST DUE OR DUE AFTER 1 YEAR	TERMS AND COLLATERAL
		$	$	
	TOTAL	$	$	(INCLUDE SUBSIDIES & CO-OP RETAINS)

B. LIVESTOCK

NO.	DESCRIPTION	CURRENT UNIT VALUE	TOTAL VALUE FEEDER STOCK HELD FOR SALE	BREEDING AND DAIRY STOCK
			$	$
		TOTAL	$	$

C. CROPS & FEED

DESCRIPTION	HARVESTED CROPS/ FEED ON HAND UNITS	PRICE	VALUE	GROWING CROPS ACRES	INVESTMENT
			$		$
	TOTAL		$		$

D. STOCKS & BONDS Describe fully. Stocks without readily available price quotes are considered Restricted.

NO. OF SHARES AMT. OF BONDS	NAME AND ISSUE	COST	VALUE AS OF STATEMENT DATE MARKETABLE	RESTRICTED	HOW ARRIVED AT
			$	$	
		TOTAL	$	$	

E. EQUIPMENT & VEHICLES Attach list if necessary. Enter book values.

ARTICLE								TOTAL VALUE
NUMBER								
VALUE	$	$	$	$	$	$	$	$

Fig. 13-4c. Financial Statement, page 3. (Continued)

F. FARMLAND & IMPROVEMENTS:

DESCRIPTION	TITLE IN NAMES OF	FAIR VALUE	MEMO: BOOK VALUE	MORTGAGE
		$	$	$
	TOTAL	$	X X X X X X X X X X X X X X	

G. NON-FARM REAL ESTATE:

DESCRIPTION	TITLE IN NAMES OF	FAIR VALUE	MEMO: BOOK VALUE	MORTGAGE
		$	$	$
	TOTAL	$	X X X X X X X X X X X X X X	

H. LONG-TERM LIABILITIES: (Enter mortgages shown in schedules F & G first.)

YEAR ORIGINATED AND PURPOSE	HOLDER	ORIGINAL AMOUNT	BALANCE OWED	CURRENT PORTION	REPAYMENT ARRANGEMENTS
		$	$	$	
	TOTAL	$	$	$	←TOTAL – Due Within 12 Months

I. NOTES & ACCOUNTS PAYABLE: (Enter notes first and identify)

DATE ORIGINATED	DATE DUE	AMOUNT OWED	TO WHOM	SECURED BY (IF ANY)	REPAYMENT ARRANGEMENTS
		$			
	TOTAL	$			

Fig. 13-4d. Financial Statement, page 4.

6. *The potential borrower*—Most lenders agree that the potential borrower is the most important part of the loan.

Lenders consider his—

a. Character.
b. Honesty and integrity.
c. Experience and ability.
d. Moral and credit rating.
e. Age and health.
f. Family cooperation.
g. Continuity, or line of succession.

Lenders are quick to sense the "high liver"—the fellow who lives beyond his means; the poor manager—the kind who would have made it except for hard luck, and to whom the hard luck happened many times; and the dishonest, lazy, and incompetent.

In recognition of the importance of the man back of the loan, "key man" insurance on the owner or manager should be considered by both the lender and the borrower.

7. *Production records*—This refers to a good set of records showing efficiency of production. Such records should show weight and price of products sold, calf-crop percentage and weaning weight, efficiency of feed utilization and rate of gain on feedlot cattle, age of livestock, heifer replacement program, depreciation schedule, average crop yield, and other pertinent information. Lenders will increasingly insist on good records.

8. *Progress with previous loans*—Has the borrower paid back previous loans plus interest; has he reduced the amount of the loan, thereby giving evidence of progress?

9. *Physical plant*—

a. Is it an economic unit?
b. Does it have adequate water, and is it well balanced in feed and livestock?
c. Is there adequate diversification?
d. Is the right kind of livestock being produced?
e. Are the right crops and varieties grown; and are approved methods of tillage and fertilizer practices being followed?
f. Is the farmstead neat and well kept?

10. *Collateral (or security)*—

a. Adequate to cover loan, with margin.
b. Quality of security:
 (1) Grade and age of livestock.
 (2) Type and condition of machinery.
 (3) If grain storage is involved, adequacy to protect from moisture and rodents.
 (4) Government participation.
c. Identification of security:
 (1) Brands, ear tags, tattoo marks of livestock.
 (2) Serial numbers on machinery.

THE LOAN REQUEST

Farmers and ranchers are in competition for money from urban businessmen. Hence, it is important that their request for a loan be well presented and supported. The potential borrower should tell the purpose of the loan; how much money is needed, when it's needed, and what it's needed for; the soundness of the venture; and the repayment schedule.

Here is a hypothetical example of the kind of loan request that a lender appreciates:

In 1974, the Bar-None Ranch consisted of 2,000 acres and a 200-head cow herd, all unencumbered. Beginning in 1975, it was desired to expand the cow herd to 500 head, and to establish a 1,000-head cattle finishing operation, with a master plan for the feedlot to go to 2,000 head should the economics of the operation so warrant. Except for replacement heifers, it was proposed that the annual calf crop would be wintered over as stockers, grazed and grained on irrigated pastures from May 15 through September, then put in the dry lot October 1 and finished for market before Christmas. Further, it was proposed that the purchased stockers would be handled similar to the stockers raised—that is, purchased in October and marketed the next December. How sound is the proposal, and how much capital will be needed, for what will it be needed, and when will it be needed?

Step by step, here is the answer:

1. *Soundness*—This requires an analysis of each (a) the cow-calf enterprise (see Table 13-2) and the feedlot operation (see Table 13-3). They are analyzed separately because each enterprise should stand on its own.

TABLE 13-2
COW-CALF OPERATIONS
BAR-NONE RANCH

(Per cow basis; total for year)

Basis: 500-head cow herd	
Investment/acre/animal unit (one mature cow) in land and improvement (real estate)	$ 1,200.00
Cost to produce a 400-lb calf to weaning age	$ 160.00
One calf, 400 lb, sold @ 50¢	$ 200.00
Management income/calf	$ 40.00
A total of 450 calves @ $40.00	
Total management income	$ 18,000.00

TABLE 13-3
FEEDLOT OPERATION
BAR-NONE RANCH

(Per head basis; total for year)	On calves raised	On calves bought
Expenses:		
Initial cost	$ 160.00	$ 200.00[1]
Feed and pasture as stocker	50.00	50.00
Feed for finishing	120.00	120.00
Nonfeed costs	19.95	19.95
	$ 349.95	$ 389.95
Miscellaneous costs	5.00	5.00
Total cost to produce	$ 354.95	$ 394.95
Income/steer[2]	$ 420.00	$ 420.00
For management	$ 65.05	$ 25.05
Total for management:		
360 head @ $65.05	$ 23,418.00	
640 head @ $25.05		$ 16,032.00

[1] 400-lb calves @ 50¢/lb.
[2] 1,050 lb @ 40¢/lb.

2. *Milling considerations*—Before feed milling and storage facilities are purchased, it is necessary to arrive at the quantity of feed needed and the method of processing it (see Table 13-4).

TABLE 13-4
MILLING CONSIDERATIONS
BAR-NONE RANCH

For stockers:	
Two lb daily, 7 mo., 500 lb/animal	
for 1,000 animals	250 tons
For finishing:	
80% grain and 20% hay + protein supplement,	
or 21 lb grain/head/day, 75 days,	
1,575 lb (rounded ¾ tons)/animal	
for 1,00 animals	750 tons
Total	1,000 tons
If proper steam roller mill increases feeding	
value $5.00/ton, that's	$ 5,000/year
Daily capacity:	
For 1,000 stockers & 1,000 finishers	12 tons/day
For 2,000 stockers & 2,000 finishers	24 tons/day
Storage for 50 tons *protein supplement*/year (but storage for	
two months will be adequate)	10 tons

3. *Feedlot, mill, and equipment costs*—Next, the feedlot, mill, and equipment costs should be determined (see Table 13-5).

TABLE 13-5
FEEDLOT, MILL, AND EQUIPMENT COSTS
BAR-NONE RANCH

Mill	$15,000
Feedlot (corrals, bunks, apron,	
scales, and waterers)	13,000
Storage	12,000
Truck and box	12,000
Tractor and loader	12,000
Total	$64,000

4. *Annual cash expense*—This is shown in Table 13-6. Actually, a more detailed cash expense budget than Table 13-6 is arrived at for use by the operator (see Table 13-11). But since the banker is not particularly interested in a detailed breakdown of costs for labor, grease, insurance, utilities, and other items, Table 13-6 will suffice.

5. *Annual cash income budget*—This is shown in Table 13-7.

6. *Annual cash expense and income budget (cash flow)*—This is given in Table 13-8. This shows, by months, the estimated expenses and income, and the particular months when money will need to be borrowed—and how much. Further, it shows that the loan will be reduced to $246,804 at the beginning of 1976. Table 13-8 is a money flow chart.

Credit Factors Considered by Borrowers

Credit is a two-way street; it must be good for both the borrower and the lender. If a borrower is the right kind of person and on a sound basis, more than one lender will want his business. Thus, it is usually well that a borrower shop around a bit; that he be familiar with several sources of credit and see what they have to offer. There are basic differences in length and type of loan, repayment schedules, services provided with the loan, interest rate, and the ability and willingness of lenders to stick by the borrower in emergencies and times of adversity. Thus, interest rates and willingness to loan are only two of the several factors to consider. If at all possible, all borrowing should be done from one source; a one-source lender will know more about the borrower's operations and be in a better position to help him. Lenders strongly discourage "split borrowings," and, in most cases, will not lend under such conditions.

Helpful Hints for Building and Maintaining a Good Credit Rating

Cattlemen who wish to build up and maintain good credit are admonished to do the following:

1. *Keep credit in one place, or in few places*—Generally, lenders frown upon "split financing." Shop around for a creditor (a) who is able, willing, and interested in extending the kind and amount of credit needed, and (b) who will lend at a reasonable rate of interest; then stay with him.

2. *Get the right kind of credit*—Don't obtain short-term credit to finance

TABLE 13-6
ANNUAL CASH EXPENSE BUDGET
BAR-NONE RANCH FOR 1974 AND 1975
(dollars)

ITEM	Total	Jan.	Feb.	March	April	May	June	July	Aug.	Sept.	Oct.	Nov.	Dec.	
1974														
300 cows & calves bought + 200 on hand	155,336				102,000	6,667[1]	6,667	6,667	6,667	6,667	6,667	6,667	6,667	Remark # 1
640 feeder calves	149,600										128,000	10,800[2]	10,800	Remark # 2
Feedlot	15,000						3,750[3]	3,750[3]	3,750[3]	3,750[3]				Remark # 3
Total Expense	319,936				102,000	6,667	10,417	10,417	10,417	10,417	134,667	17,467	17,467	
1975														
500 cows	26,668	6,667	6,667	6,667	6,667									
Stockers-finishing 1,000 head	121,600	10,800	10,800	10,800	10,800						31,200[4]	31,200	16,000	
Stockers 640	198,600						30,000[3]	19,000[3]			128,000	10,800	10,800	
Total Expense	346,868	17,467	17,467	17,467	17,467		30,000	19,000			159,200	42,000	26,800	

[1]Based on $160/cow, 500 cows; total of $80,000 for the year or $6,667/mo. 500 total includes 300 purchased and 200 on hand.

[2]Based on 36¢/head/day, or $360/day, or $10,800/mo. on 1,000 head.

[3]Feed mill, storage, and handling equipment purchased and installed over two month period.

[4]Based on 850-lb steer, consuming 26 lb/head/day, feed cost of 4¢/lb, $1.04/head/day, or $1,040/day; or $31,200/mo. (Only $16,000 for December because marketed before Christmas.)

Remark # 1. 300 cows and calves purchased April @ $340/pair or total cost of $102,000—200 cows on hand.

Remark # 2. 640 stockers purchased Oct. @ 50¢, wt 400 lb, $200/head, or total of $128,000.

Remark # 3. Feedlot, but not feed mill, installed at a total cost of $15,000, but distributed over 4 months, June to Sept., $3,750 each month.

TABLE 13-7
ANNUAL CASH INCOME BUDGET
BAR-NONE RANCH FOR 1974 and 1975

(dollars)

(No income in 1974, because the first finished cattle will not
be ready to market until Dec., 1975)

Item	Total	Jan.	Feb.	Mar.	Apr.	May	June	July	Aug.	Sept.	Oct.	Nov.	Dec.
1974													
Total	none	none	none	none	none	none	none	none	none	none	none	none	none
1975 1,000 finished steers & heifers	420,000												420,000
Total 1,000 head	420,000												420,000

TABLE 13-8

ANNUAL CASH EXPENSE AND INCOME BUDGET (CASH FLOW)
BAR-NONE RANCH FOR 1974 AND 1975

Item	Total	Jan.	Feb.	Mar	Apr.	May	June	July	Aug.	Sept.	Oct.	Nov.	Dec.
					(dollars)								
1974													
Gross Income	None												
Gross Expense	319,936				102,000	6,667	10,417	10,417	10,417	10,417	134,667	17,467	17,467
Difference	319,936				102,000	6,667	10,417	10,417	10,417	10,417	134,667	17,467	17,467
Surplus (+) or Deficit (−)	—	—	—	—	—	—	—	—	—	—	—	—	—
1975													
Gross Income	420,000												420,000
Gross Expense	346,868	17,467	17,467	17,467	17,467		30,000	19,000			159,200	42,000	26,800
Difference	73,132	17,467	17,467	17,467	17,467		30,000	19,000			159,200	42,000	26,800
Surplus (+) or Deficit (−)	+	—	—		—		—	—		—	—	—	+
1976													
Surplus (+) or Deficit (−)	246,804	—											

long-term improvements or other capital investments. Also, use the credit for the purpose intended.

3. *Be frank with the lender*—Be completely open and above board. Mutual confidence and esteem should prevail between borrower and lender.

4. *Keep complete and accurate records*—Complete and accurate records should be kept by enterprises. By knowing the cost of doing business, decision-making can be on a sound basis.

5. *Keep annual inventory*—Take an annual inventory for the purpose of showing progress made during the year.

6. *Repay loans when due*—Borrowers should work out a repayment schedule on each loan, then meet payments when due. Sale proceeds should be promptly applied on loans.

7. *Plan ahead*—Analyze the next year's operation and project ahead.

Borrow Money to Make Money

Cattlemen should never borrow money unless they are reasonably certain that it will make or save money. With this in mind, borrowers should ask, "How much should I borrow?" rather than, "How much will you lend me?"

Cost of Credit (Actual Rate of Interest)

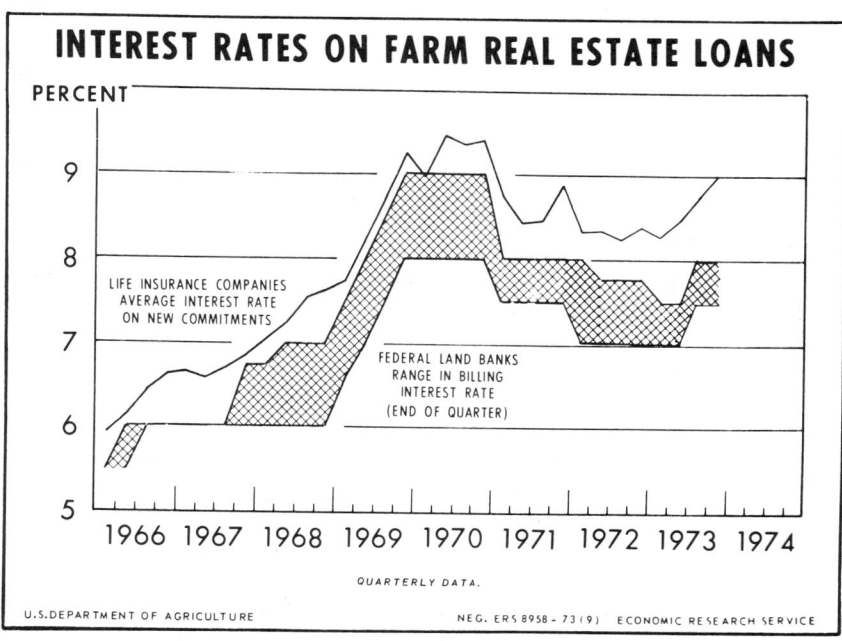

Fig. 13-5. Interest rates vary. Hence, it pays to shop around when buying a farm or ranch. This figure shows the interest rates on farm loans charged by life insurance companies and the Federal Land Banks since 1966. (Courtesy, USDA, Washington, D.C.)

The charge for the use of money is called interest. The basic charge is strongly influenced by the following:

1. The *basic cost* of money in the money market.
2. The *servicing costs* of making, handling, collecting, and keeping necessary records on loans.
3. The *risk* of loss.

Interest rates vary among lenders and can be quoted and applied in several different ways. The quoted rate is not always the basis for proper comparison and analysis of credit costs. Even though several lenders may quote the same interest rate, the effective or simple annual rate of interest may vary widely. The more common procedures for determining the actual annual interest rate, or the equivalent of simple interest on the unpaid balance, follow.

1. *Simple or true annual interest on the unpaid balance*—A $1,200 note payable at maturity (12 months) with 8% interest:

Interest Paid $.08 \times \$1,200 = \$96$

Average use of the money $1,200 for the entire year

Actual rate of interest

$$\frac{96 \text{ (interest)}}{\$1,200 \text{ (Used for one year)}} = 8\%$$

2. *Installment loan (with interest on unpaid balance)*[10]—A $1,200 note payable on 12 monthly installments with 8% interest on the unpaid balance:

Interest paid ranges from:

First month ....

$$\frac{.08 \times \$1,200}{12} = \$8.00$$

to

Twelfth month ....

$$\frac{.08 \times 100}{12} = \$.67$$

Total for 12 months is $52.00

Average use of the money ranges from $1,200 for the first month down to $100 for the twelfth month, an average of $650 for 12 months.

Effective rate of interest ....

$$\frac{\$52}{\$650} = 8\%$$

3. *Add-on installment loan (with interest on face amount)*—A $1,200 note payable in 12 monthly installments with 8% interest on face amount of loan:

Interest paid ....

$.08 \times \$1,200 = \$96$

Average use of the money ranges from $1,200 for the first month down to $100 for the twelfth month, an average of $650 for 12 months.

---

[10]This method is used for amortized loans.

Effective rate of interest .... $\dfrac{\$\ 96}{\$650}$ = 14.77%

4. *If interest is not stated, use this formula to determine the effective annual interest rate:*

Effective rate of interest = $\dfrac{2\ \times\ \begin{matrix}\text{Number of}\\\text{payment periods}\\\text{in one year}\end{matrix}^{11}\ \times\ \begin{matrix}\text{Finance}\\\text{charges}\end{matrix}^{12}}{\text{Balance owed}^{13}\ \times\ \begin{matrix}\text{Number of pay-}\\\text{ments in contract}\\\text{plus one}\end{matrix}}$

For example, a store advertises a refrigerator for $250. It can be purchased on the installment plan for $30 down and monthly payments of $20 for 12 months. What is the actual rate of interest if you buy on the time payment plan?

Effective rate of interest .... $\dfrac{2 \times 12 \times \$20}{\$220 \times (12\ \text{plus}\ 1)} = \dfrac{\$480}{\$2,860} = 16.8\%$

HIDDEN CHARGES

Before using credit, the borrower should thoroughly understand *all* the costs of borrowing. Many of these costs are hidden; hence, it is important to investigate, ask questions, and consider every expense connected with the use of credit. Some lenders, for example, require the purchase of stock based on a percentage of the loan; others require minimum balances; still others pay commissions or "finder's fees" for loans. All these things, and more, point up the importance of comparing actual dollar charges when shopping for credit.

MANAGEMENT

Fortunes have been made and lost in the beef cattle industry. Although it is not possible to arrive at any overall, certain formula for success, in general those operators who have made money have paid close attention to the details of management—they have been good managers.

The almost innumerable beef cattle management practices vary widely between both areas and individual farmers and ranchers. In a general sort of way, however, the principles of good management are much alike. The main differences arise from the sheer size of big cattle enterprise, which means that things must be done in a big way.

---

[11]Regardless of the total number of payments you will make, use 12 if the payments are monthly, use 6 if payments are every other month, or use 2 if payments are semi-annual.

[12]Use either the time payment price less the cash price, or the amount you pay the lender less the amount you received if negotiating for a loan.

[13]Use cash price less down payment or, if negotiating for a loan, the amount you receive.

Traits of a Good Cattle Manager

There are established bases for evaluating many articles of trade, including cattle, hay, and grain. They are graded according to well-defined standards. Additionally, we chemically analyze feeds and conduct feeding trials. But no such standard or system of evaluation has evolved for cattle managers, despite their acknowledged importance.

The author has prepared the Cattle Manager Check List, given in Table 13-9, which (1) employers may find useful when selecting or evaluating a manager, and (2) managers may apply to themselves for self-improvement purposes. No attempt has been made to assign a percentage score to each trait, because this will vary among cattle establishments. Rather, it is hoped that this check list will serve as a useful guide (1) to the traits of a good manager, and (2) to what the boss wants.

TABLE 13-9
CATTLE MANAGER CHECK LIST

☐CHARACTER—
Absolute sincerity, honesty, integrity, and loyalty; ethical.

☐INDUSTRY—
Work, work, work; enthusiasm, initiative, and aggressiveness.

☐ABILITY—
Cattle know-how and experience, business acumen—including ability systematically to arrive at the financial aspects and convert this information into sound and timely management decisions, knowledge of how to automate and cut costs, common sense, organized, growth potential.

☐PLANS—
Sets goals, prepares organization chart and job description, plans work, and works plans.

☐ANALYZES—
Identifies the problem, determines pros and cons, then comes to a decision.

☐COURAGE—
To accept responsibility, to innovate, and to keep on keeping on.

☐PROMPTNESS AND DEPENDABILITY—
A self-starter; has "T.N.T.," which means that he does it "today, not tomorrow."

☐LEADERSHIP—
Stimulates subordinates, and delegates responsibility.

☐PERSONALITY—
Cheerful, not a complainer.

Organization Chart and Job Description

It is important that every worker know to whom he is responsible and for what he is responsible; and the bigger and the more complex the operation, the more important this becomes. This should be written down in an organization chart and a job description.

Here's how they look:

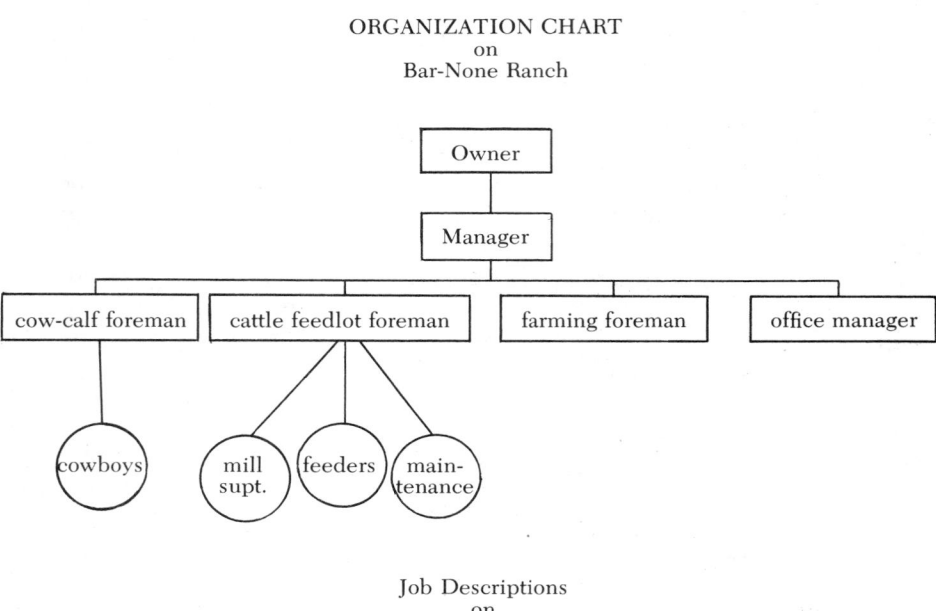

ORGANIZATION CHART
on
Bar-None Ranch

Job Descriptions
on
Bar-None Ranch

Owner	Manager	Cattle Feedlot Foreman	Cattle Feeder No. 1
Responsible for:	**Responsible for:**	**Responsible for:**	**Responsible for:**
1. Selecting management.	1. Supervising all staff.	1. Directing feedlot staff.	1. Morning and evening feedings.
2. Making policy decisions.	2. Preparing proposed long-time plan.	2. Buying and selling cattle.	2. Cleaning water troughs.
3. Borrowing capital.	3. Budgets.	3. Processing incoming cattle.	3. (List others)
4. (List others)	4. (List others)	4. Animal health.	
		5. Feedlot rations.	
		6. (List others)	

An Incentive Basis for the Help

Big farms and ranches must rely on hired labor, all or in part. Good help—the kind that everyone wants—is hard to come by; it's scarce, in strong demand, and difficult to keep. And the farm manpower situation is going to become more difficult in the years ahead. There is need, therefore, for some system that will (1) give a big assist in getting and holding top-flight help, and (2) cut costs and boost profits. An incentive basis that makes hired help partners in profit is the answer.

Many manufacturers have long had an incentive basis. Executives are frequently accorded stock option privileges, through which they prosper as the business prospers. Common laborers may receive bonuses based on piecework or quotas (number of units, pounds produced). Also, most factory

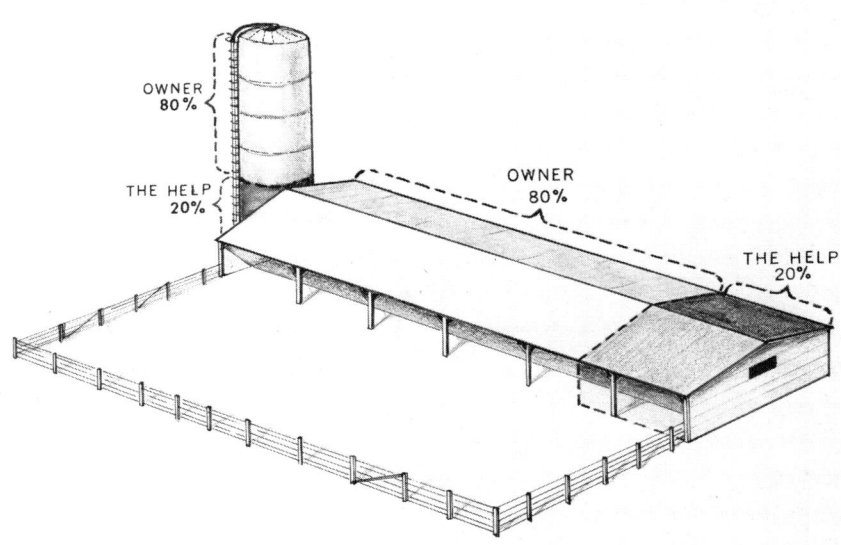

Fig. 13-6. The author favors an 80:20 percent split of the net between the owner and the help.

workers get overtime pay and have group insurance and a retirement plan. A few industries have a true profit-sharing arrangement based on net profits as such, a specified percentage of which is divided among employees. No two systems are alike. Yet, each is designed to pay more for labor, provided labor improves production and efficiency. In this way, both owners and laborers benefit from better performance.

Family-owned and family-operated farms have a built-in incentive basis; there is pride of ownership, and all members of the family are fully cognizant that they prosper as the business prospers.

Many different incentive plans can be, and are, used. There is no best one for all operations. The various plans given in Table 13-10 are intended as guides only.

The incentive basis chosen should be tailored to fit the specific operation; with consideration given to kind and size of operation, extent of owner's supervision, present and projected productivity levels, mechanization, and other factors.

For most cow-calf and cattle feedlot operations, the author favors a "production sharing and prevailing price" type of incentive.

HOW MUCH INCENTIVE PAY?

After (1) reaching a decision to go on an incentive basis, and (2) deciding on the kind of incentive, it is necessary to arrive at how much to pay. Here are some guidelines that may be helpful in determining this:

1. Pay the going base, or guaranteed, salary; then add the incentive pay above this.

TABLE 13-10

INCENTIVE PLANS FOR SOME BEEF CATTLE ESTABLISHMENTS

Types of Incentives	Pertinent Provisions of Some Known Incentive Systems in Use	Advantages	Disadvantages	Comments
1. Bonuses	A flat, arbitrary bonus; at Christmas time, year-end or quarterly or other intervals. A tenure bonus such as (1) 5 to 10% of the base wage or 2 to 4 weeks additional salary paid at Christmas time or year-end, (2) 2 to 4 weeks vacation with pay, depending on length and quality of service, or (3) $3.00 to $20/week set aside and to be paid if employee stays on the job a specified time.	It's simple and direct.	Not very effective in increasing production and profits.	
2. Equity-building plan	Employee is allowed to own a certain number of animals. In cow-calf operations, these are usually fed without charge.	It imparts pride of ownership to the employee.	The hazard that the owner may feel that employee accords his animals preferential treatment; suspicioned if not proved. It is difficult to control animal numbers, feed, etc.	
3. Production sharing	$1.00 to $3.00/calf weaned 50¢/cwt of gain on feeder cattle. 25¢ to $1.00/head on fed cattle marketed.	It's an effective way to achieve higher production.	Net returns may suffer. For example, a higher rate of gain than is economical may be achieved by feeding stockers more concentrated and expensive feeds than are practical. This can be alleviated by (1) specifying the ration, and (2) setting an upper limit on the gains to which the incentive will apply. If a high performance level already exists, further gains or improvements may be hard to come by.	Incentive payments for production above certain levels—for example, above 450 lb calf weaned/cow bred—are more effective than paying for all units produced.
4. Profit sharing: a. Percent of gross income	1% to 2% of the gross.			

(Continued)

TABLE 13-10 (Continued)

Types of Incentives	Pertinent Provisions of Some Known Incentive Systems in Use	Advantages	Disadvantages	Comments
b. Percent of net income	10 to 20% of the net after deducting all costs.	Net income sharing works better for managers, supervisors, and foremen than for common laborers, because fewer hazards are involved to opening up the books to them. It's an effective way to get hired help to cut costs. It's a good plan for a hustler.	Percent of gross does not impart cost of production consciousness. Both (1) percent of gross income, and (2) percent of net income expose the books and accounts to workers, who may not understand accounting principles. This can lead to suspicion and distrust. Controversy may arise (1) over accounting procedure—for example, from the standpoint of the owner a fast tax write-off may be desirable on new equipment, but this reduces the net shared with the worker; and (2) because some owners are prone to overbuild and over-equip, thereby decreasing net. It is difficult to define what constitutes net income.	There must be, prior agreement on what constitutes gross or net receipts, as the case may be, and how it is figured.
5. Production sharing and prevailing price	See Chapter 24, section entitled, "An Incentive Basis for the Help." See Chapter 27 section entitled, "An Incentive Basis for the Help."	It embraces the best features of both production sharing and profit sharing, without the major disadvantages of each. It (1) encourages high productivity and likely profits, (2) is tied in with prevailing prices, (3) does not necessitate opening the books, and (4) is flexible—it can be split between owner and employee on any basis desired, and the production part can be adapted to a sliding scale or escalator arrangement—for example, the incentive basis can be higher for the quarter pound of feedlot gain made in excess of 2% pounds than for a quarter pound gain in excess of 2¼ pounds.	It is a bit more complicated than some other plans, and it requires more complete records.	When properly done, and all factors considered, this is the most satisfactory incentive basis for a cattle enterprise, for both cow-calf and feedlot.

2. Determine the total stipend (the base salary plus incentive) to which you are willing to go.

3. Before making any offers, always check the plan on paper to see (a) how it would have worked out in past years based on your records, and (b) how it will work out as you achieve the future projected production.

Let's take the following example:

A foreman of a 500-cow herd is now producing an average of 400 pounds of calf weaned per cow bred. He is receiving a base salary of $500/month plus house, garden, and 600 pounds of dressed beef/year. The owner prefers a "production sharing and prevailing price" type of incentive.

Step by step, here is the procedure for arriving at an incentive arrangement based on increased production:

1. By checking with local sources, it is determined that the present salary of $500/month plus "extras" is the going wage; and, of course, the foreman receives this regardless of what the year's calf production or price turns out to be—it's guaranteed.

2. A study of the cow-calf records reveals that with a little extra care on the part of the foreman—particularly in pregnancy testing, at calving time, and in rotating pastures—the average weaning weight of calves per cow bred can be boosted enough to permit paying him $700/month, or $200/month more than he's now getting. That's $2,400 more per year. This can be fitted into the incentive plan.

3. An average increase of 50 pounds of calf weaned per cow bred at 50 cents/pound would mean $25.00/cow, or $12,500 on a 500-cow herd. With an 80:20 split between owner and manager, the foreman would get $2,500 or slightly over $200/month.

REQUISITES OF AN INCENTIVE BASIS

Owners who have not previously had experience with an incentive basis are admonished not to start with any plan until they are sure of both their plan and their help. Also, it is well to start with a conservative plan; then a change can be made to a more inclusive and sophisticated plan after experience is acquired.

Regardless of the incentive plan adopted for a specific operation, it should encompass the following essential features:

1. Good owner (or manager) and good workers. No incentive basis can overcome a poor manager. He must be a good supervisor and fair to his help. Also, on big establishments, he must prepare a written-down organization chart and job description so the help knows (a) to whom they are responsible, and (b) for what they are responsible. Likewise, no incentive basis can spur employees who are not able, interested, and/or willing. This necessitates that employees must be selected with special care where they will be on an incentive basis. Hence, the three—good owner (manager), good employees, and good incentive—go hand in hand.

2. It must be fair to both employer and employees.

3. It must be based on and make for mutual trust and esteem.

4. It must compensate for extra performance, rather than substitute for a reasonable base salary and other considerations (house, utilities, and certain provision).

5. It must be as simple, direct, and easily understood as possible.

6. It should compensate all members of the team; from cowboys to manager on a cow-calf outfit, and from feeders and feed processors to manager in a cattle feedlot.

7. It must be put in writing, so that there will be no misunderstanding. If some production-sharing plan is used in a cattle feedlot, it should stipulate the ration (or who is responsible for ration formulation), the maximum gain of stocker cattle, and the grade to which finishing cattle are to be carried. On a cow-calf outfit, it should stipulate the ration, the culling of cows, and other pertinent factors.

8. It is preferable, although not essential, that workers receive incentive payments (a) at rather frequent intervals, rather than annually, and (b) immediately after accomplishing the extra performance.

9. It should give the hired help a certain amount of responsibility, from the wise exercise of which they will benefit through the incentive arrangements.

10. It must be backed up by good records; otherwise, there is nothing on which to base incentive payments.

11. It should be a two-way street. If employees are compensated for superior performance, they should be penalized (or, under most circumstances, fired) for poor performance. It serves no useful purpose to reward the unwilling, the incompetent, and the stupid. For example, no overtime pay should be given to an employee who must work longer because of slowness or correcting mistakes of his own making. Likewise, if the reasonable break-even point on a cow-calf operation is an average of a 400-pound calf weaned per cow bred, and this production level is not reached because of obvious neglect (for example, not being on the job at calving time), the employee(s) should be penalized (or fired).

INDIRECT INCENTIVES

Normally, we think of incentives as monetary in nature—as direct payments or bonuses for extra production or efficiency. However, there are other ways of encouraging employees to do a better job. The latter are known as indirect incentives. Among them are: (1) good wages; (2) good labor relations; (3) adequate house plus such privileges as the use of the farm truck or car, payment of electric bill, use of a swimming pool, hunting and fishing, use of horse, and furnishing meat, milk, and eggs; (4) good buildings and equipment; (5) vacation time with pay, time off, and sick leave; (6) group health; (7) security; (8) the opportunity for self-improvement that can accrue from working for a top man; (9) the right to invest in the business; (10) an all-expense-paid trip to a short course, show, or convention; and (11) year-end bonus for staying all year. These indirect incentives will be accorded to the help of more and more establishments, especially the big ones.

ANNUAL INVENTORY

The annual inventory is the most valuable record that a cattleman can keep. It gives the most complete statement of financial conditions. A complete farm inventory is usually necessary before any other kind of records or accounts can be kept and analyzed; it should be considered the first and most important record to be assembled and used by all cattlemen. The inventory should include a list and value of real estate, livestock, equipment, feed, supplies, and all other property, including cash on hand, notes, bills receivable and growing crops. Also, it should include a list of mortgages, notes, and bills payable. It shows the farmer what he owns and what he owes; whether he is getting ahead or going behind. The following pointers may be helpful relative to the annual inventory.

1. *Time to take inventory*—The inventory should be taken at the beginning of the account year; usually this means Dec. 31 or Jan. 1.

2. *Proper and complete listing*—It is important that each item be properly and separately listed.

3. *Method of arriving at inventory values*—It is difficult to set up any hard and fast rule to follow in estimating values when taking inventories. Perhaps the following guides are as good as any:
     a. *Real estate*—Estimating the value of farm real estate is, without doubt, the most difficult of all. It is suggested that the farmer use either (1) the cost of the farm, (2) the present sale value of the farm, or (3) the capitalized rent value according to its productive ability with an average operator.
     b. *Buildings*—Buildings are generally inventoried on the basis of cost less observed depreciation and obsolescence. Once the original value of a building is arrived at, it is usually best to take depreciation on a straight line basis by dividing the original value by the estimated life in terms of years. Usually four percent or more depreciation is charged off each year for income tax purposes (which means that buildings are normally depreciated in 25 years).
     c. *Livestock*—Animals are usually not too difficult to inventory because there are generally sufficient current sales to serve as a reliable estimate of value.
     d. *Machinery*—The inventory value of machinery is usually arrived at by either of two methods: (1) the original cost less a reasonable allowance for depreciation each year; or (2) the probable price that it would bring at a well-attended auction.
       Under conditions of ordinary wear and reasonable care, it can be assumed that the general run of farm machinery (except trucks and autos) will last about 10 years. Thus, with new machinery, the annual depreciation will be the original cost divided by 10.
     e. *Feed and supplies*—The value of feed and supplies can be based on market price.
Two further points are important. Whatever method is used in arriving at inventory value (1) should be followed at both the beginning and the end

of the year, and (2) should reflect the operator's opinion of the value of the property involved.

BUDGETS IN THE CATTLE BUSINESS

A budget is a projection of records and accounts and a plan for organizing and operating ahead for a specific period of time. A short-time budget is usually for one year, whereas a long-time budget is for a period of years. The principal value of a farm budget is that it provides a working plan through which the operation can be coordinated. Changes in prices, droughts, and other factors make adjustments necessary. But these adjustments are more simply and wisely made if there is a written budget to use as a reference.

How to Set Up a Budget

It's unimportant whether a printed form (of which there are many good ones) is used or a form made up on an ordinary ruled 8½"×11" sheet placed sidewise. The important things are: (1) that a budget is kept; (2) that it be on a monthly basis; and (3) that the operator be "comfortable" with whatever forms or system with which he evolves.

No budget is perfect. But it should be as good an estimate as can be made—despite the fact that it will be affected by such things as droughts, diseases, markets, and many other unpredictables.

A simple, easily kept, and adequate budget can be prepared by using the following three types of budget forms, all of which are shown earlier in this chapter in the section entitled, "The Loan Request":

1. Annual cash expense budget (see Table 13-6).
2. Annual cash income budget (see Table 13-7).
3. Annual cash expense and income budget—cash flow (see Table 13-8).

The annual cash expense budget should show the monthly breakdown of various recurring items—everything except the initial loan and capital improvements. It includes labor, feed, supplies, fertilizer, taxes, interest, utilities, etc.

The annual cash income budget is just what the name implies—an estimated cash income by months.

The annual cash expense and income budget is a cash flow budget, obtained from the first two forms. It's a money "flow" summary by months. From this, it can be ascertained when, and how much, money will need to be borrowed, and the length of the loan along with a repayment schedule. It makes it possible to avoid tying up capital unnecessarily, and to avoid unnecessary interest.

How to Figure Net Income

Table 13-8 shows a gross income statement. There are other expenses that must be taken care of before net profit is determined; namely—

1. *Depreciation on buildings and equipment*—It is suggested that the "useful life" of buildings and equipment be as follows:

Buildings .... 25 years
Machinery and equipment .... 10 years

Sometimes a higher depreciation, or amortization, is desirable because it produces tax savings, and for protection against obsolescence due to scientific and technological developments.

2. *Interest on owner's money invested in farm and equipment*—This should be computed at the going rate in the area, say 9 percent.

Here's an example of how the above works: Let's assume that on a given farm there was a gross income of $125,000 and a gross expense of $75,000, or a surplus of $50,000. Let's further assume that there are $40,000 worth of machinery, $30,000 worth of buildings, and $175,000 of the owner's money invested in farm and equipment. Let's further assume that buildings are being depreciated in 25 years and machinery in 10 years. Here is the result:

Gross profit $50,000
Depreciation:
Machinery, $ 40,000 @ 10%   = $ 4,000
Buildings, $ 30,000 @ 4%   =   1,200
                                         $ 5,200
Interest, $175,000 @ 9%   =   15,750
                                                    20,950

Return to labor and management $29,050

Some folks prefer to measure management by return on invested capital, and not wages. This approach may be accomplished by paying management wages first, then figuring return on investment.

Enterprise Accounts

Where a cattle enterprise is diversified—for example, a farm or ranch having a cow-calf operation, a feedlot, and crops—enterprise accounts should be kept; in this case three different accounts for three different enterprises. The reasons for keeping enterprise accounts are:

1. It makes it possible to determine which enterprises have been most profitable, and which least profitable.

2. It makes it possible to compare a given enterprise with competing enterprises of like kind, from the standpoint of ascertaining comparative performance.

3. It makes it possible to determine the profitableness of an enterprise at the margin (the last unit of production). This will give an indication as to whether to increase the size of a certain enterprise at the expense of an alternative existing enterprise when both enterprises are profitable in total.

COMPUTERS IN THE CATTLE BUSINESS[14]

Accurate and up-to-the-minute records and controls have taken on increasing importance in all agriculture, including the cattle business, as the investment required to engage in farming and ranching has risen and profit margins have narrowed. Today's successful farmers and ranchers must have, and use, as complete records as any other business. Also, records must be kept current; it no longer suffices merely to know the bank balance at the end of the year.

Big and complex enterprises have outgrown hand record keeping. It's too time consuming, with the result that it doesn't allow management enough time for planning and decision making. Additionally, it does not permit an all-at-once consideration of the complex interrelationships which affect the economic success of the business. This has prompted a new computer technique known as linear programming.

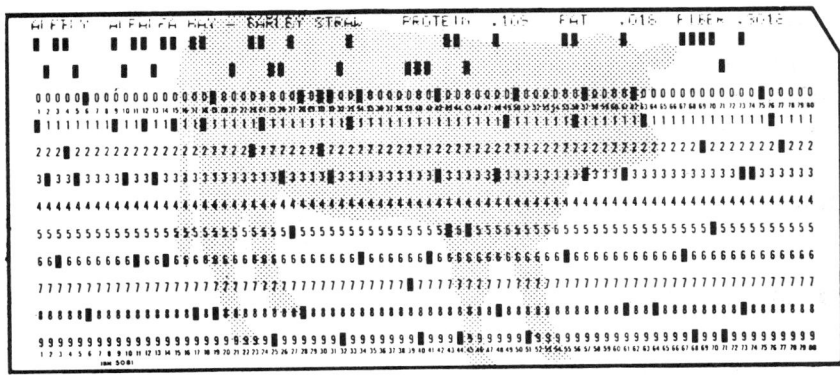

Fig. 13-7. Computers in the cow business.

Linear programming is similar to budgeting, in that it compares several plans simultaneously and chooses from among them the one likely to yield the greatest overall profit. It is a way in which to analyze a great mass of data and consider many alternatives. It is not a managerial genie; nor will it replace decision-making managers. However, it is a modern and effective tool in the present age, when just a few dollars per head can spell the difference between profit and loss.

There is hardly any limit to what computers can do if fed the proper information. Among the difficult questions that they can answer for a specific farm or ranch are:

1. *How is the entire operation doing so far?* It is preferable to obtain quarterly or monthly progress reports; often making it possible to spot trouble before it's too late.

---

[14]This section was authoritatively reviewed by the following: Mr. M. E. Smith, Beef Cattle Specialist, International Business Machine Corporation, Chicago, Ill.; Professor Clyde L. Hart, Computer Science Technology, Texas State Technical Institute, Waco, Tex.; and Dr. R. W. Sammons, Executive Vice President, Management Computer Network, New Hampton, Iowa.

2. *What farm enterprises are making money; which ones are freeloading or losing?* By keeping records by enterprises—cow-calf, cattle feedlot, wheat, corn—it is possible to determine strengths and weaknesses; then either to rectify the situation or shift labor and capital to a more profitable operation. Through "enterprise analysis" some operators have discovered that one part of the farm business may earn $5, or more, per hour for labor and management, whereas another may earn only $1 per hour, and still another may lose money.

3. *Is each enterprise yielding maximum returns?* By having profit or performance indicators in each enterprise (see Chapter 24, Table 24-4, for Cow-Calf Profit Indicators; and see Chapter 27, Table 27-3, for Cattle Finishing Profit Indicators), it is possible to compare these (1) with the historical average of the same farm or ranch, or (2) with the same indicators of other farms or ranches.

4. *How does this ranch stack up with its competition?* Without revealing names, the computing center (local, state, area, or national) can determine how a given ranch compares with others—either the average, or the top (say 5 percent).

5. *How can you plan ahead?* By using projected prices and costs, computers can show what moves to make for the future—they can be a powerful planning tool. They can be used in determining when to plant, when to schedule farm machine use, etc.

6. *How can income taxes be cut to the legal minimum?* By keeping accurate record of expenses and figuring depreciations accurately, computers make for a saving in income taxes on most farms and ranches.

For providing answers to the above questions, and many more, computer accounting costs an average of about 1 percent of the gross farm income. By comparison, it is noteworthy that city businesses pay double this amount.

There are three requisites for linear programming a farm or ranch; namely—

1. Access to a computer.

2. Computer know-how, so as to set the program up properly and be able to analyze and interpret the results.

3. Good records; preferably, covering a period of years.

The pioneering computer services available to farmers and ranchers were operated by universities, trade associations, and government—most of them were on an experimental basis. Subsequently, others have entered the field, including commercial data processing firms, banks, machinery companies, feed and fertilizer companies, and farm suppliers. They are using it as a "service sell," as a replacement for the days of "hard sell."

Programmed cattle business is here to stay, and it will increase.

ANALYZING A CATTLE BUSINESS: IS IT PROFITABLE?

Most people are in business to make money—and cattlemen are people. In some areas, particularly near cities and where the population is dense, land values may appreciate so as to be a very considerable profit factor. Also, a tax angle may be important. But neither of these should be counted upon.

The cattle operation should make a reasonable return on the investment; otherwise, the owner should not be in the cattle business.

A cattle owner or manager needs to analyze his business—to determine how well he's doing. With big operations, it's no longer possible to base such an analysis on the bank balance at the end of the year. In the first place, once a year is not frequent enough, for it is possible to go broke, without really knowing it, in that period of time. Secondly, a balance statement gives no basis for analyzing an operation—for ferreting out its strengths and weaknesses. In large cattle feedlots, it is strongly recommended that progress be charted by means of monthly or quarterly closings of financial records.

Also, a cattleman must not only compete with other cattlemen down the road, but he must compete with himself—with his record last year and the year before. He must work ceaselessly at making progress, improving the end product, and lowering costs of production.

To analyze a cattle business, two things are essential: (1) good records, and (2) yardsticks, or profit indicators, with which to measure an operation.

A profit indicator is a gauge for measuring the primary factors contributing to profit. In order for a cattleman to determine how well he's doing, he must be able to compare his own operations with something else; for example, (1) his own historical five-year average, (2) the average for the United States or for his particular area, or (3) the top five percent. The author favors the latter, for high goals have a tendency to spur superior achievements.

Like most profit indicators, the ones presented in this book (see Chapter 24, Table 24-4 for Cow-Calf Profit Indicators; and see Chapter 27, Table 27-3, for Cattle Finishing Profit Indicators) are not perfect. But they will serve as useful guides. Also, on some establishments, there may be reason for adding or deleting some of the indicators; and this can be done. The important thing is that each cattle operation have adequate profit indicators, and that these be applied as frequently as possible; in a cattle feedlot, this may be done monthly with some indicators.

GUIDELINES RELATIVE TO FACILITY AND EQUIPMENT COSTS

Overinvestment is a rather common mistake. Cow-calf men are prone to invest more in land and improvements than reasonably can be expected to

Fig. 13-8. Haymaking; pitchfork vs automation. How much can you afford to invest in equipment to replace hired labor? (Left, picture courtesy USDA; right, picture courtesy Allis-Chalmers Mfg. Co.)

make a satisfactory return; and cattle feedlot operators frequently invest too much in feed mills and equipment. Sometimes small cattle feeders fail to recognize that it may cost half as much to mechanize to feed 500 head as it costs to mechanize to feed 2,000 head.

In order to lessen overinvestment by the uninformed, guidelines are useful. Here are some:

1. **Guideline No. 1—The break-even point on how much you can afford to invest in equipment to replace hired labor can be arrived at by the following formula:**

$$\frac{\text{Annual saving in hired labor from new equipment}}{.15} = \text{amount you can afford to invest}$$

Example:

If saving in hired labor costs is $3,600 per year, this becomes—

$$\frac{\$3,600}{.15} = \$24,000, \text{ the break-even point on new equipment.}$$

Since labor costs are going up faster than machinery and equipment costs, it may be good business to exceed this limitation under some circumstances. Nevertheless, the break-even point, $24,000 in this case, is probably the maximum expenditure that can be economically justified at the time.

2. **Guideline No. 2—Assuming an annual cost plus operation of power machinery and equipment equal to 20 percent of new cost, the break-even point to justify replacement of one hired man is as follows:**

If annual cost of one hired man is—		The break-even point on new investment is—
$4,000 (20%) × 5	....	$20,000
5,000 (20%) × 5	....	25,000
6,000 (20%) × 5	....	30,000

In the above figures, it is assumed that the productivity of men at different salaries is the same, which may or may not be the case.

Example:
Assume that the new cost of added equipment comes to $5,000, that the annual cost is 20 percent of this amount, and that the new equipment would save 2 hours of labor per day for 6 months of the year. Here's how to figure the value of labor to justify an expenditure of $5,000 for this item:

$5,000 (new cost) × 20%=$1,000, which is the annual ownership, use cost.
$1,000 ÷ 360 hours (labor saved)=$2.78 hour.

So, if labor costs less than $2.78/hour, you probably shouldn't buy the new item.

3. **Guideline No. 3**—The annual cost for cattle feedlot and equipment should not run over 12 to 21 percent of the new (initial) cost, with a breakdown as follows:

	Minimum (%)	Maximum (%)
Depreciation	5	10[15]
Interest[16]	4	5
Repairs & Upkeep[16]	2	4
Taxes & Insurance[16]	1	2
	12%	21%

Example #1: Let's assume that the total feedlot expenditure for 1,000 head is $80,000. That's $80/head. Twenty-one (21) percent (taking the maximum annual feedlot and equipment cost) of this is $16.80/animal.

Example #2: Let's assume that the total feedlot expenditure for 2,000 head is $140,000. That's $70/head. Twenty-one (21) percent (the maximum annual feedlot and equipment cost) of this is $14.70/animal; which is $2.10 less/animal than shown in Example 1.

4. **Guideline No. 4**—Do not spend over 4 to 5 percent (average of 4.5 percent) of the annual product sold for annual cattle feedlot and equipment costs.

Example #1: 1,000 steers sold @ $420 each, for a total of $420,000
$420,000 × 4.5% (av.) = $18,900
$18,900 ÷ 1,000 = $18.90/head/year, the maximum that may be spent on feedlot and equipment.

Example #2: 2,000 steers sold @ $420 each, for a total of $840,000
$840,000 × 4.5% (the av.) = $37,800
$37,800 ÷ 2,000 = $18.90/head/year.

CATTLE INSURANCE

The ownership of cattle (or any kind of animal) constitutes a risk and a chance of financial loss. Unless the owner is in such strong financial position that he can assume this financial risk, the animal should be insured.

Several good companies write livestock insurance. In general, their policies and rates do not differ greatly. Conditions and rates for *cattle—beef or dairy*—are about as follows:

Transportation, exhibition and all normal usage of the animal is included in each policy.

Loss, directly or indirectly, due to animal(s) giving birth prematurely or otherwise, shall not be included unless insured on a policy written for a period in **excess of two months.**

---

[15]Up to 20% depreciation may be advisable in the case of certain feedlot equipment subject to heavy use and obsolescence due to new technology, such as milling and conveying equipment.

[16]About one-half of normal rate because, during the life of the item, its value averages about half of new cost.

Loss due to **bloat** is **excluded** in nonregistered or feeder cattle **if they are fed on leguminous pasture.**

CATTLE—BEEF OR DAIRY

AGE LIMITS 3 MONTHS TO 7 YEARS

15-Day Term .... $1.50 per $100.00
 (A 15-Day policy may be endorsed "15-day cover to have its
 effective starting date at time of actual shipment.")
1-Month Term .... $2.50 per $100.00
2-Month Term .... $3.00 per $100.00
3-Month Term .... $3.50 per $100.00
6-Month Term .... $4.00 per $100.00
1-Year Term .... $6.00 per $100.00

AGE EXCEPTIONS—NOTE ADDED PREMIUM

Calves, 2 weeks to 7 weeks old .... $4.00 per $100.00
Calves, 7 weeks to 3 months old .... $2.00 per $100.00
 (This is additional premium to be added to period coverage and
 considered earned in entirety when written.)

Bulls past 6, up to eighth birthday, eligible for insurance after amount of cover has been confirmed by company. One dollar per hundred additional premium charged for each year or part over seventh birthday. **At ninth birthday insurance not available.**

Cows past seven may be covered in same manner. **At tenth birthday annual insurance not available.**

Generally, special stipulations and rates apply to (1) group (herd) insurance for cattle, and (2) 4-H and FFA calves. For information relative to these, or other special types of coverage, the owner should make inquiry of a livestock insurance agent.

In order to obtain insurance, the following information pertaining to the animal is generally required: name, registry or tattoo number, breed, sex, birth date, and a statement of health examination by a veterinarian—certifying that, at the time of applying for insurance, the animal was in a state of good physical health and condition.

LIABILITY INSURANCE; WORKMEN'S COMPENSATION INSURANCE

Most farmers are in such financial position that they are vulnerable to damage suits. Moreover, the number of damage suits arising each year is increasing at an almost alarming rate, and astronomical damages are being claimed. Studies reveal that about 95 percent of the court cases involving injury result in damages being awarded.

Several types of liability insurance offer a safeguard against liability suits brought as a result of injuries suffered by other persons or damage to their property.

Comprehensive personal liability insurance protects a farm operator who is sued for alleged damages suffered from an accident involving his property or family. The kinds of situations from which a claim might arise are quite broad, including suits for injuries caused by animals, equipment, or personal acts.

Both workmen's compensation insurance and employer's liability insurance protect farmers against claims or court awards resulting from injury to hired help. Workmen's compensation usually costs slightly more than straight employer's liability insurance, but it carries more benefits to the worker. An injured employee must prove negligence by his employer before the company will pay a claim under employer's liability insurance, whereas workmen's compensation benefits are established by state law, and settlements are made by the insurance company without regard to whom was negligent in causing the injury. Conditions governing participation in workmen's compensation insurance vary among the states.

CATTLE AND INCOME TAXES (Tax Shelters)[17]

Since Federal income tax was introduced in 1913, numerous changes have been made in the law, usually in the form of special deductions, credits, exclusions, exemptions, and special rates. Occasionally, these changes allow some taxpayers to escape paying as much income tax as their critics think they should. Such provisions are popularly called "tax loopholes" or "tax shelters." Attacking the cattle industry as a tax haven makes good newspaper copy.

Special farm tax rules, when combined with high nonfarm income, can permit the deferral of income tax on nonfarm income. In addition, some farming activities allow the conversion of ordinary income into capital gains income, which has a lower tax rate. Also, if a farming investment produces a "real" loss, there is little likelihood that the loss will be disallowed for tax purposes. These are the characteristics of an ideal tax shelter.

Some high tax bracket nonagricultural men and women are investing in cattle programs to shelter nonfarm earnings. Such "tax loss" cattlemen are not "typical" in United States agriculture. But neither are they a rarity.

Tax Sheltered Cattle Investments

What, then, is the investment and tax appeal of cattle enterprises? How do they work?

Special treatment for "farmers" under provisions of the Internal Revenue Code arises primarily from the following four sources:

1. *Cash accounting*—Farmers are permitted to choose between the accrual method and the cash receipts and disbursement method (the cash method). As a result, the cash method may be used as a means of minimizing and postponing income tax payments. For example, an investor may buy feeder cattle near the end of his tax year, and prepay feed and interest costs,

---

[17]This section was authoritatively reviewed by an eminent tax specialist, Mr. S. P. Kurth, of the law firm of Kurth, Felt, Speare and Lalonde, P.C., Billings, Mont.

which would be deductible as a business expense and cause a loss on farm operations, which could be offset against his ordinary income.

2. *Current deduction of capital expenditures*—Investment spending (the cost of acquiring and developing capital assets) is generally not deductible from income as a current expense for income tax purposes. Instead, these costs are required to be capitalized and recovered through depreciation over the useful life of the asset. Livestock breeders are allowed to deviate from this general rule. In a breeding program, they are allowed to use their expenses to offset ordinary income while the herd is being built up and treated as a capital asset.

Additional provisions allow taxpayers reporting farming operations to deduct certain expenses which, if incurred in other businesses, would be capitalized. These include (a) soil and water conservation expenditures (not to exceed 25 percent of gross income from farming for the year), (b) land clearing expenses or expenditures incurred in making land suitable for farming (limited to the lesser of $5,000 or 25 percent of taxable income from farming during the year) and (c) expenses for fertilizer applied on land used in farming.

3. *Livestock as a capital asset*—Livestock held for draft, breeding, dairy, or sporting purposes is treated as property used in a trade or business. This means that livestock (except poultry) held for the above purposes is entitled to capital gain treatment upon sale provided holding period requirements are met. These holding periods are 24 months for cattle and horses and 12 months for all other qualifying livestock.

When cash accounting is used, the benefit of this provision is multiplied. Expenses of raising the animal are deductible currently and the entire sales price is taxed as a capital gain.

Of course, purchased livestock held for draft, breeding, dairy, or sporting purposes can be depreciated just as any other purchased capital asset. But there is recapture of depreciation upon sale.

4. *Using farm losses to offset, or shelter from tax, the nonfarm income*—Taxpayers with large nonfarm incomes can make farm investments and, using the special farm tax rules, generate artificial losses which are then used to offset, or shelter from tax, the nonfarm income, provided they use the special accounting procedure for "excess" farm losses. Under this procedure, taxpayers with nonfarm adjusted gross income over $50,000, must place farm losses in excess of $25,000 in a special excess-deductions account (EDA). Gain on the subsequent sale of "farm recapture property" must be treated as ordinary income to the extent of EDA balances. Amounts in an EDA are reduced by farm income in subsequent years and to the extent that they are used to offset capital gains on the sale of farm property. For example, suppose a taxpayer with a nonfarm adjusted gross income of $50,000 reports a farm loss of $30,000. He would claim the $30,000 farm loss but credit his EDA account for $5,000. Thus, in a subsequent year, the first $5,000 realized for the sale of farm recapture property would be treated as ordinary income. The remainder of the sale, assuming no subsequent additions to EDA, would be handled as capital gains. This provision continues to allow tax

payment to be delayed but limits the tax-reduction feature for taxpayers meeting its tests.

Cattle Tax Shelters—Good or Bad?

Some tax experts feel that special tax rules, instead of helping, actually foster unfair competition for the farmer who makes his living from the land. Real dirt farmers must compete with tax-loss farmers who are mostly rich or well-off city people more interested in saving taxes on their nonfarm incomes than on making money on their investments in cattle, orchards, or whatever. This charge is based on the fact that many of the benefits of special farm tax rules are available primarily to the taxpayer who has (1) tax losses which are not economic losses, and (2) substantial nonfarm income against which to offset the farm "tax loss." As a result of these tax benefits, the argument goes, high-bracket taxpayers may bid up the price of farmland beyond its productivity value and beyond the value which would have prevailed in the absence of tax benefits.

In beef cattle (both cow-calf and cattle feeding), there is some evidence that feeding margins may have been "squeezed" by the presence of funds not requiring as high a before-tax return as more traditional cattle feeding funds.

It is probable that "tax shelter" capital may have contributed to the generally adequate supply of investment capital available to agriculture in recent years. To the extent that real farmers and feeders (as opposed to tax loss farmers) have had access to these tax-motivated funds, capital may have been available at a lower cost than capital from more traditional sources. However, when adversity struck the cattle feeding industry in 1974, "tax shelter" cattlemen fled in droves, leaving a capital void which further aggravated and depressed the cattle market.

Many countries of the world accord special tax consideration to farmers, as a means of (1) encouraging the production of more food, and (2) keeping consumer prices down. In the present era of world food shortages, this policy will likely be expanded on a worldwide basis. In the United States, however, it appears probable that tax rules will be modified so as to (1) advantage the real farmer, and (2) disadvantage the tax shelter farmer.

ESTATE PLANNING[18]

Human nature being what it is, most cattlemen shy away from suggestions that someone help plan the disposition of their property and other assets after they're gone. Also, they have a long-standing distrust of lawyers, legal terms, and trusts; and, to them, the subject of taxes seldom makes for pleasant conversation.

If a cattleman has prepared a will or placed his property in joint tenancy, his estate will be distributed as he has specified. If not, it goes to his heirs,

---

[18]This entire section was authoritatively reviewed by the author's son, Mr. John J. Ensminger, Attorney, New York, N.Y.; and by Mr. Robert L. Hobson, Assistant Vice President-Trust, Bank of America, Bank of America Center, San Francisco, California.

according to the laws governing intestate (without a will) succession of property. His heirs are those persons whom the law appoints to succeed to his property in the event of intestacy, and are not necessarily the persons the cattleman would want to have his property.

Every cattleman should consider how, and more importantly *when*, he wishes to dispose of his property. Sometimes it is preferable to release ownership of a portion of his land during his lifetime. Thereafter, as everyone else, he has certain objectives he feels are most important—such as providing adequately for his wife, and minimizing taxes and, consequently, leaving more to his family.

If no plans are made, estate taxes and settlement costs often run considerably higher than if proper estate planning is done and a trust or will is made to carry out these plans. Today, cattle raising is big business; many cattlemen have more than $500,000 invested in land, cattle, and equipment. Thus, it is not a satisfying thought to one who has worked hard to build and maintain a good enterprise during his lifetime to feel that his heirs will have to sell the business to raise enough cash to pay federal estate and inheritance taxes. By using a good estate planning service, a cattleman can generally save thousands of dollars for his family in estate and inheritance taxes and in estate settlement costs. For assistance, farmers should go to an estate planning specialist—an individual or company specializing in this work, or the trust department of a commercial bank.

Tools of Estate Planning

Several tools may be used in estate planning, regardless of the size of the estate. The cattleman's banker and accountant can determine his financial needs, and his lawyer can determine which tools should be used and on what portions of the estate each tool should be used.

WILLS

A will is a set of instructions drawn up by or for an individual which details how he wishes his estate to be handled after his passing. Despite the importance of a will in distributing property in keeping with the individual's wishes, about 50 percent of farmers and ranchers pass away without having written a will. This means that state law determines property distribution in such cases.

Every cattleman should have a will. By so doing, (1) the property will be distributed in keeping with his wishes, (2) he can name the executor of the estate, and (3) sizable tax savings can be made by the way in which the property is distributed. Because technical and legal rules govern the preparation, validity, and execution of a will, it should be drawn up by an attorney. Wills can and should be changed and updated from time to time. This can be done either by (1) a properly drawn-up codicil (formal amendment to a will), or (2) a completely new will which revokes the old one.

The same attorney should prepare both the husband's and wife's wills so that a common disaster clause can be incorporated and the estate planning of each can be coordinated.

JOINT TENANCY (Co-Ownership of Property)

When property is owned concurrently by two or more persons, title may be held in several ways: (1) as "tenants in common"; (2) as "joint tenants"; (3) as "community property" (in 8 states); or (4) as "tenancy by the entirety" (in approximately 20 other states). The latter two forms of co-ownership apply only to married couples.

In the above four forms, each named person is considered owner of an undivided share of the entire property. Upon the death of one of the "tenants in common," his undivided ownership passes as he willed it, or, in the absence of a will, according to the state laws of intestate succession. When a "joint tenant" dies, his undivided ownership passes to the surviving joint tenant or tenants. The latter provision is called right of survivorship. Each spouse's community property interest is subject to disposition by will.

Tenancy by the entirety operates much like joint tenancy; the death of one spouse vests the entire estate in the survivor. There are a variety of consequences in the forms of ownership and transfer of property on death, with variation from state to state. Hence, uniform advice on a nationwide basis is impossible. Nevertheless, certain effects, particularly of Federal taxation, should be noted.

There are two main differences between transferring property by will and by joint tenancy: (1) a will takes effect only when the maker of that document dies, whereas a transfer into joint tenancy is effective the moment it is made; and (2) a will is not necessarily final in the assignment of property, since the maker can amend his will at anytime, but once property is placed in joint tenancy, that transfer is final unless the joint tenants agree to make further transfers.

If the transfer to joint tenancy is gratuitous and there is an intent to transfer a present interest to the new joint tenant, the transfer will be subject to the gift tax laws. There are other tax effects to holding property in joint tenancy which should be fully explored with your attorney, estate planner, and/or bank trust officer.

The following points are noteworthy about joint tenancy:

1. If an owner plans to put property in joint tenancy, he should keep in mind that he has no assurance that his property will return to him upon the death of the other joint tenant, for a joint tenant can destroy the right of survivorship at anytime—by sale, mortgage, partition, or gift. If the right of survivorship is destroyed, the original owner of that property is left with only his own share upon the other joint tenant's death.

2. Property in joint tenancy passes to the survivor without probate proceedings or without administration, except as the property is subject to state inheritance and Federal death taxes. Even when tax returns are required on property in joint tenancy, probate costs are usually less because only a termination of joint tenancy is required which is generally much less complicated than estate administration.

3. There are possible disadvantages from joint tenancy when large estates are involved. Federal tax laws assume that the deceased owned all the joint tenancy property himself, and the entire property is taxed to him, unless the surviving joint tenant can prove that he or she contributed to the purchase price or that the property was originally taken in joint tenancy by gift or inheritance from someone other than the deceased. If a husband and wife have joint tenancy property amounting to more than $120,000 in value, there may be a double Federal tax on part of it, unless the disposal of the estate has been carefully planned. The following alternatives point up this situation:[19]

Alternative No. 1—Let's assume a rancher dies leaving a total net estate of $300,000, which is held in joint tenancy with his wife. (Administration and other expenses and obligations, including state inheritance taxes, are not taken into consideration in the computations.)

Husband's Estate		$300,000
Federal Estate Tax (after marital deduction)		$17,900
Wife's subsequent Estate	$282,100	
Federal Estate Tax (no marital deduction)		$57,330
Reduction by Federal Estate Tax		$75,230

Alternative No. 2—Now let's assume the same rancher dies leaving a total net estate of $300,000, but not in joint tenancy, and he leaves only ½ of it to his wife (for marital deduction purposes). The other ½ is placed in trust for the *benefit of the wife for her lifetime*, and then passes to their children. (Administration and other expenses, not taken into consideration as in above example.)

Husband's Estate		$300,000
Federal Estate Tax (after marital deducation)		$17,900
Wife's subsequent Estate	$150,000	
Federal Estate Tax (no marital deduction)		$17,900
Reduction by Federal Estate Tax		$35,800
Savings to Estate by use of trust rather than joint tenancy		$39,430

GIFTS

Nothing pleases parents more than seeing their children succeed, and, generally speaking, having them take over the family farm, ranch, or feedlot

---

[19]Tax rates in both alternatives were computed from the Internal Revenue Code of 1954.

makes for the ultimate in parental pride and satisfaction. Moreover, such an arrangement can make for a fine financial start in life for the young man who desires to carry on, provided, while the parents are still living, advantage is taken of the very considerable savings in Federal estate taxes, as provided by law.

The Federal gift tax laws were enacted to compliment the Federal estate tax and to prevent the circumvention of the estate tax by lifetime gifts. The two taxes are closely intertwined and with proper estate planning and a thorough understanding of the two taxes, thousands of dollars in tax money can be saved.

Under the Federal law, a person may give up to $3,000 in any year to any particular person without incurring a gift tax or filing a gift tax return. Thus, a father can give $3,000 to each of his children in any year without gift tax consequences. A married person, whose spouse will consent to have one-half of the gift treated as if it came from him or her, may make gifts to individuals up to $6,000 per year—$3,000 to each person from the husband and $3,000 from the wife. So, if both parents of the children are still living, $6,000, not subject to Federal gift taxes, can be given each child each year.

In addition to the annual exclusion discussed above, each donor has a lifetime exemption of $30,000. Thus, a husband and wife, who never before gave more than their annual exclusions, may in one year give to one child $66,000 without tax consequences, or to 2 children $72,000. This combines the annual exclusion of husband and wife of $6,000 for each child, plus the $30,000 each to exhaust their lifetime exemptions. The year they exceed their annual exclusions they must file information gift tax returns, although no tax is payable until the lifetime exemption is exhausted.

Once the lifetime exemption ($30,000 from each parent) has been used, the $3,000 annual exclusion is still available for future gifts, but a tax is payable upon any excess at a cumulative and increasing rate. The tax is assessed against the donor, but if he does not pay, the recipient of the gift can be assessed.

The gift(s) may be in the form of cash, or it may consist of income producing property—land, cattle, corporate stocks, savings bonds, mutual funds, or whole life and annuity life insurance. Also, group or term life insurance can be gifted where the policy itself and state law permit. Most such policies and state laws do permit this.

Frequently, even where it is the full intent and desire of the parents and the children that the latter continue with the cattle operation, the gift tax provision is not considered. Then, upon death of the parent, the heir(s) may be required to raise such a large amount of cash to pay the inheritance taxes that a part or all of the operation may have to be liquidated.

Most states also have gift tax laws which generally work in conjunction with the state inheritance tax, to prevent loss of revenue through circumvention of the inheritance tax. These laws, like the various inheritance tax laws, vary from state to state.

PARTNERSHIP CONTRACT

A second logical step in this transfer is a partnership contract between the parents and their heir(s). Appropriate counsel should be consulted in the preparation of this agreement. Where the partnership contract is between the father and the heir, a provision should be included permitting the heir to purchase the father's share of the partnership for a fixed amount, through a buy-sell agreement. Funding for such a buy-sell agreement is commonly done with life insurance. This will provide for proper and uninterrupted operation of the beef cattle enterprise, because at the father's death, the partnership is legally terminated.

TRUSTS

A trust is a legal arrangement by which the management, control, and legal title of property is placed in the hands of a trustee (bank, corporation, or individual) for the benefit of specified beneficiaries. Almost any kind of asset can be included in a trust. It is important that the trust property be large enough so that it will more than pay the management fee out of current earned income. Also, management skill should be considered carefully in choosing a trustee.

● *Testamentary trust*—A testamentary trust is created by will and becomes effective upon the death of the trustor. It can accomplish several things, such as (1) providing income for life of the spouse, (2) providing income to children up to a certain age, following which the principal is turned over to them, or (3) providing income to the children, with the grandchildren receiving the principal upon their parents' death. Because business and investment conditions change, as well as the needs of the beneficiaries, it is recommended that broad discretionary powers be written into trusts.

● *Living trusts*—Living trusts are those that are commenced during the lifetime of the trustor, and which may be either revocable or irrevocable. Revocable trusts can be amended, altered, or totally revoked at any time the grantor chooses. Irrevocable trusts are permanent according to their terms, and those terms cannot be changed by the grantor. The living trust provides the cattleman a means of passing along property to others upon his passing, yet having the income payable to him during his lifetime. Also, through it property may be passed along to others after the owner's death without either probate expense, or loss of control during lifetime. It is not subject to probate administration because the trust is a separate and distinct entity from the grantor (creator) of the trust, and its existence survives the death of the grantor. Therefore, the trustee himself transfers the title to property to persons taking it after the grantor's death, not the executor of the decedent's grantor's estate.

LIFE INSURANCE

Life insurance is gaining increasing acceptance among farmers and ranchers. It can be used to solve liquidity problems, to pay estate and inheritance taxes, and to increase the total estate. Proper estate planning with life insurance policies can also save thousands of dollars in estate and inheritance taxes.

INHERITANCE AND ESTATE TAXES

The Federal estate tax is *not* an important source of Federal revenue. It probably accounts for no more than 2%-3% of that annual revenue. Rather, it is more in the nature of a "social policy" tax for restraining the accumulation of large fortunes in the hands of those who have not earned them by their own labors and endeavors.

Whether the revenue raised by the estate tax is large or not, in the total revenue picture, and whether it is basically "social" or monetary, the Federal estate tax is a large bite which can generally be substantially reduced through proper estate planning.

The Federal estate tax is a tax on the entire estate of the decedent (after taking exemptions and deductions into account) and is basically a transfer tax on the privilege of transferring property at one's death. The Federal gift tax is merely a companion tax to prevent circumvention of the estate tax laws.

The inheritance tax (an estate tax in some states) is an additional tax imposed by the various states and is imposed on the individual amounts of inheritances (as opposed to a tax on the entire estate). Contrary to the Federal estate tax, most inheritance taxes are primarily enacted for revenue purposes, and again, are a tax on the privilege of transferring property at one's death. Generally, state gift taxes are companion taxes to prevent circumvention of the inheritance tax laws.

Estate and inheritance taxes are imposed on transfers at death. They are the same whether property passes in joint tenancy, by the laws of intestate succession, or by will. State death taxes are usually small compared to Federal estate taxes. However, they often involve smaller estates and more people.

The Federal estate tax is payable from the estate. A charitable deduction is allowed for bequests to qualified charitable organizations. Up to one-half of the estate may be given to a surviving husband or wife without being taxed (except with community property estates, where the surviving spouse's community interest is not included); this is called the estate tax marital deduction. Finally, there is a $60,000 specific exemption allowable against the remaining property. Thus, a deceased person with an estate of $300,000 after payment of debts and expenses may give one-half to his surviving spouse, the remainder to one or more of his children with a tax imposed only upon $90,000 ($300,000 less $150,000 less $60,000). That $90,000 is subject to a Federal estate tax amounting to $17,900.

The Federal estate tax may range from 3 percent to 77 percent. It is not computed on the total estate, but on what remains after certain exemptions and deductions are taken.

The first $60,000 of the estate is not subject to tax. This is known as the specific exemption. The law also allows deductions from the remainder of the estate for such items as bequests to qualified charitable organizations;

attorneys, accountants, and executor's fees; other administration costs connected with the estate; and funeral expenses.

In the case of a married couple, the estate of the first decedent may qualify for a marital deduction, which amounts to half of the adjusted gross estate. Thus, it is possible for either the husband or wife to leave the other an estate amounting to as much as $120,000 tax-free, by taking advantage of the $60,000 specific exemption and the marital deduction.

It should be remembered that a sound estate plan should consider the estate of both marriage partners. Often the planning of the surviving spouse's estate requires more care because the marital deduction is not available to this estate. Here the major objective is to insulate a substantial portion of the first deceased's estate from the estate of the surviving spouse. This can be accomplished by a variety of methods, including (1) gifts and/or (2) trusts. Such procedure requires careful planning and legal assistance.

Estate planning involves expense and time, but the importance of a good estate plan far outweighs these. After taking an accurate inventory of your assets and liabilities, get together with your lawyer, accountant, and life insurance representative. Their assistance on this important plan is essential in preventing legal and financial problems.

PREDICTING WHAT'S AHEAD

Most cattlemen are doing a good job when it comes to feeding and management. But they do little or nothing about scientifically projecting what's ahead. Worse yet, they are prone to flock with other cattlemen, in Judas fashion, simply because they have no watchdog or barometer. As a result, they buy and sell when everyone else does. Altogether too often they buy when cattle are up and sell when they're down.

The trouble with most existing cattle forecasts is that they come too late. Information on numbers and weights of cattle on feed, for example, only confirm what has already happened. Additionally, even the best prediction may be upset by wars, droughts, and politics.

Ideally, the cattleman needs to spot a zig or zag in the graphs that may mark the beginning of a trend. He needs signs that indicate trouble in advance of an actual downturn. Then he can shift his program to take advantage of the situation. The author feels that this can be achieved by applying a new sophistication to certain indicators. By using computers, the time lag for most information gathering can be reduced. To be sure of what the machine-produced information means requires experience in analysis and interpretation. Also, the relative importance of the different factors must be weighed and evaluated—there is need for an index figure.

The author predicts that, eventually, progressive individual cattlemen, partnerships, and corporations will do it by hiring economists or professional analysts. They will utilize such public information figures as are available, but they will supplement them as necessary and evolve with their own barometer. Personnel for such an assignment will cost money, for it requires great competence; and there will be added charges for computers. But it will pay handsome dividends. For example, a saving of $7/10$ cents per pound on a 14,000-capacity feedlot, handling 30,000 cattle annually and adding an average of 425 pounds gain per animal, would mean $89,250 annually. That's a sizable sum! Yet, increased profits of this amount, or more, are well within the realm of possibility through the system herein proposed.

Here are some of the harbingers—the indicators of what's ahead in the cattle business, that should be considered:

1. *Cattle numbers*—Total cattle numbers, and number and weight of cattle on feed, are very important indicators. When cattle production begins to outrun beef consumption, it's time to slow down.

2. *Beef imports*—Beef imports are a factor in determining domestic cattle and beef prices, although opinions differ as to their importance. According to one U.S. Department of Agriculture study, when imports equal about 10% of total domestic beef production, which they did in 1962 and 1963, a further increase of 10% in imports would cause about a 1% drop in Choice steer prices. But there are those who feel that the effect of imports is more marked. Certainly, psychological reactions are difficult to measure.

3. *Other meat competition*—Consideration must be given to the available supply of meats other than beef. This is so because when the housewife walks along the counter, she, and she alone, determines what she shall buy. If she feels that beef is too high, she simply does not buy it. Instead, she may move down the counter a few feet and select some broilers, or perhaps pork. This simply means that the price of beef must be competitive—that relative price is an important factor. Thus, the cattleman's barometer needs to ferret out the quantity and effect of competitive meats. As is generally known, the relative importance of competitive meats changes from time to time. In particular, the broiler industry has enjoyed a tremendous growth, largely due to technological developments.

4. *Employment and wages*—This indicator is of particular importance to the cattle feeders. This is so because in periods of prosperity—when employment is high and wages are good—consumers place a premium on choice cuts and top grades of beef.

5. *Feed grain; production, prices, and Federal programs*—Since about 80 percent of the cost of finishing cattle (exclusive of the initial purchase price of animals) is for feed, it is obvious that feed prices are a major item in determining profits and losses. The cattleman's barometer needs to keep tab on the factors which influence feed prices—acreages and yields, droughts, Federal programs, and carry-over stocks.

6. *The feeder's margin* (also see Chapter 27, section on "Computing Break-Even Prices for Cattle")—The wise feeder projects the necessary break-even margin before laying in cattle; keeping in mind that (a) the better the grade and the younger the cattle, the smaller the necessary margin; (b) the higher the cost of feed, the greater the necessary margin; (c) the heavier the initial weight of the feeder, the smaller the necessary margin; and (d) the longer the feeding period and the fatter the cattle, the greater the margin. Generally speaking, those feeders who ignore these facts and simply play for a rising slaughter cattle market don't stay in business very long.

7. *The buying power of the dollar*—This, of course, affects the price of all commodities. Consideration of it is a must in predicting what's ahead in the cattle business.

8. *The beef futures market, including both live cattle and carcass beef*—This brings into focus both hedgers and speculators. Since their money is involved, it reflects their thinking relative to what lies ahead. As such, it merits careful study.

Despite all the above, no one can consistently outguess the market. Thus, cattlemen are admonished to adopt a strategy that will minimize high losses.

QUESTIONS FOR STUDY AND DISCUSSION

1. Why have the business aspects of beef production become so important in recent years?

2. How can one acquire the needed training and experience in the business aspects of beef production?

3. Will the 100-head beef cow owner be able to compete 20 years hence? Will the farmer-feeder with 50 to 100 head of steers on feed be able to compete?

4. Between 1935 and 1973—within a span of 38 years—59 percent of U.S. farms disappeared, and the average size farm increased from 155 to 385 acres. How do you account for these changes?

5. A mere 1.4 percent of the nation's cattle feedlots finish and market two-thirds of the fed cattle. Is it good or bad to have such concentration of cattle feeding? Justify your answer.

6. Discuss the advantages and the disadvantages of each of the 3 major types of business organizations common to cattle enterprises.

7. It takes about $13 in farm assets to generate $1.00 of net farm income. How does this ratio compare with other businesses? (Check with the managers of 2 successful businesses not in agriculture.)

8. List the lending types of business organizations that are available to farmers and ranchers, and tell about each.

9. What arrangements are being used to attract nonfarm equity capital in agriculture? Which of these would you recommend for each (a) cow-calf operation, and (b) a cattle feeding operation?

10. Compare the types and sources of credit for each (a) cow-calf loans, (b) cattle feeder loans, and (c) cattle on pasture loans.

11. Assume a certain kind and size of cattle operation—either cow-calf, or cattle feedlot, or a combination of the 2—then prepare a detailed report and request for credit.

12. Use Table 13-9 to rate yourself as a cattle manager. How do you stack up?

13. Take your own farm or ranch, or one with which you are familiar, and develop a workable incentive basis for the help.

14. Using Tables 13-6, 13-7, and 13-8 as guides, develop a budget for the year ahead for your own farm or ranch, or for one with which you are familiar.

15. How may computers be used, on a practical basis in the cattle business?

16. Why does a cattleman need profit indicators?

17. Cite one guideline relative to facility and equipment costs. Give an example to show how it works.

18. What tax sheltered cattle investments are available; and how do they work?

19. Are cattle tax shelters good or bad? Justify your opinion.

20. What tools of estate planning are available to a cattleman; and how should each be used?

SELECTED REFERENCES

Title of Publication	Author(s)	Publisher
Agricultural Finance Statistics	AFS-2	Economic Research Service, U.S. Department of Agriculture, Washington, D.C., 1974
Complete Guide to Making a Public Stock Offering, A	E. L. Winter	Prentice-Hall, Inc., Englewood Cliffs, N.J., 1962
Concepts Involved in Defining and Identifying Farms	ERS 448	Economic Research Service, U.S. Department of Agriculture, Washington, D.C., 1970
Contract Farming, U.S.A.	E. P. Roy	The Interstate Printers & Publishers, Inc., Danville, Ill., 1963
Cooperatives: Today and Tomorrow, Second Edition	E. P. Roy	The Interstate Printers & Publishers, Inc., Danville, Ill., 1969
Cowboy Arithmetic, Cattle as an Investment	H. L. Oppenheimer	The Interstate Printers & Publishers, Inc., Danville, Ill., 1961
Cowboy Economics, Rural Land as an Investment	H. L. Oppenheimer	The Interstate Printers & Publishers, Inc., Danville, Ill., 1966
Cowboy Litigation, Cattle and the Income Tax	H. L. Oppenheimer J. D. Keast	The Interstate Printers & Publishers, Inc., Danville, Ill., 1968
Doane's Tax Management Guide	J. C. O'Byrne C. Davenport J. D. Keast	Doane Agricultural Service, Inc., St. Louis, Mo., 1973
Do It Right the First Time	J. D. Keast J. W. Looney D. K. Banner L. R. Miller R. C. Matthews C. L. Pilmer R. J. McTigue	Doane Agricultural Service, Inc., St. Louis, Mo., 1973
Exploring Agribusiness	E. P. Roy	The Interstate Printers & Publishers, Inc., Danville, Ill., 1967
Farm Accounting	D. R. Mitchell	McGraw-Hill Book Company, Inc., New York, N.Y., 1941
Farm Management Economics	E. O. Heady H. R. Jensen	Prentice-Hall, Inc., Englewood Cliffs, N.J., 1954
Farm Management Handbook, Third Edition	Queensland Department of Primary Industries	Queensland Department of Primary Industries, Brisbane, Queensland, Australia, 1970
Farm Records and Accounts	J. N. Efferson	John Wiley & Sons, Inc., New York N.Y., 1949
Financing the Farm Business	I. W. Duggan R. U. Battles	John Wiley & Sons, Inc., New York N.Y., 1950
Going Public	Corplan Associates	Corplan Associates, IIT Research Institute, Chicago, Ill.
How To Do a Private Offering–Using Venture Capital	A. A. Sommer, Jr.	Practicing Law Institute, New York, N.Y., 1970
Introduction to Agri-Business Management, An	W. J. Wills	The Interstate Printers & Publishers, Inc., Danville, Ill., 1972

(Continued)

Title of Publication	Author(s)	Publisher
Land Speculation, An Evaluation and Analysis	H. L. Oppenheimer	The Interstate Printers & Publishers, Inc., Danville, Ill., 1972
Law on the Farm	H. W. Hannah	The Macmillan Company, New York, N.Y., 1950
Law and the Farmer	J. H. Beuscher	Springer Publishing Company, Inc., New York, N.Y., 1953
Law for the Veterinarian and Livestock Owner	H. W. Hannah D. F. Storm	The Interstate Printers & Publishers, Inc., Danville, Ill., 1959
Life Insurance Company Farm-Mortgage Loans	ERS 439	Economic Research Service, U.S. Department of Agriculture, Washington, D.C., 1970
Stockman's Handbook, The, Fourth Edition	M. E. Ensminger	The Interstate Printers & Publishers, Inc., Danville, Ill., 1970
Tax Loss Farming	ERS-546	Economic Research Service, U.S. Department of Agriculture, Washington, D.C., 1974
1971 U.S. Master Tax Guide		Commerce Clearing House, Inc., Chicago, Ill., 1970

CHAPTER 14

FITTING, SHOWING, AND JUDGING BEEF CATTLE[1]

The show-ring has long been a major force in shaping beef cattle type. The first American livestock show was held at Pittsfield, Massachusetts in 1810, but livestock exhibitions had been initiated in Europe many years earlier.

The mere mention of a livestock show causes the author to become as nostalgic as an old fire horse hearing a siren. He grew up in an era when the show-ring was as traditional as mom's apple pie. It was the way—the only way—to establish livestock standards.

---

[1]The author expresses his sincere appreciation to the following persons who authoritatively reviewed this chapter: Dr. G. L. Minish, Department of Animal Science, Virginia State University, Blacksburg, Va.; Dr. Harlan D. Ritchie, Department of Animal Husbandry, Michigan State University, East Lansing, Mich.; and Mr. William Bennett, B B Cattle Co., Connell, Wash.

Fig. 14-1. Cody, a Maine-Anjou X Angus, Grand Champion Steer at the 1974 Houston Livestock Show, weighed 1,283 pounds and sold for a world record price of $40,000. Cody was owned, fitted, and shown by Theresa Scott, Rising Star, Texas, a 13-year-old 4-H Club girl. (Courtesy, Theresa Scott)

Then came the cattlemen's mid-century wrangle relative to the merits or demerits of livestock shows—eyeball vs science, which has raged ever since. Those who would do away with livestock shows argue that we know how they should look. What we need to know, they add, is how they will perform on the pasture or range, in the feedlot, and, finally, on the grill or in the roaster. Production records, bolstered by computers and son-o-ray or K-40 readings, are the answer, according to this school of thought. Then they clinch their argument by citing the debacle which brought disrepute to cattle shows from which they may never fully recover. In the period 1935 to 1946, small, early-maturing, blocky, smooth cattle were sweeping shows from one end of the country to the other, with this beauty contest (it wasn't a beef contest) reaching the ultimate in the "comprests" in Herefords and the "compacts" in Shorthorns, and culminating in dwarfism.

It's not all that bad, counter the traditionalists. Individual selection, or eyeballing, is the usual method followed by both commercial and purebred cattlemen when selecting foundation animals, when adding new blood, and when selecting replacements from within the herd. Also, it's the basis of evaluating the vast majority of feeder and slaughter cattle. Then they add this unassailable fact: eyeballing and livestock shows have been largely responsible for the transformation of the Texas Longhorn to the present-day bullock.

Fig. 14-2. Show arena with fitted Hereford show cattle. (Courtesy, The American Hereford Association, Kansas City, Mo.)

Both schools of thought have a point. Livestock shows need to change; progress in the beef industry bypassed them. To something old, they need to add something new. The show-ring has been, and can continue to be, an important vehicle for getting cattle and people together in one place and at one time to compare, design, and engineer the most economically productive and desirable beef model.

In recent years, shows have aided in the genetic improvement of beef cattle. The cattle winning in the major shows have been the growthy, meaty, productive, useful kind that fit the mold and standards of today's producers and consumers. Research has clearly shown that well-trained personnel can estimate lean to fat ratio (cutability) very well. With accurate performance and carcass information available, shows will be a major force in identifying cattle that have the genetic potential to improve the pounds of edible Choice grade lean beef produced per day of age, at a minimum cost.

ADVANTAGES AND DISADVANTAGES OF SHOWING

Livestock shows have both advantages and disadvantages.

Human nature being what it is, not all exhibitors share equally in the many advantages that may accrue from showing. In general, however, cattle shows offer the following *advantages*:

1. They afford the best medium yet discovered for molding breed type. For this reason, it behooves the breed registry associations and the purebred breeders alike to accept their rightful responsibility in seeing that the animals winning top honors are those which most nearly meet the efficiency of production requirements of the producer and the demands of the consumer.

2. They provide an incentive to breed better cattle, for the breeder can

determine how well he is keeping pace with his competitors only after securing an impartial appraisal of his entries in comparison with others.

3. They offer an opportunity to study the progress being made within other breeds and classes of livestock.

4. They serve as one of the very best advertising or promotional mediums for both the breed and the breeder.

5. They give breeders an opportunity to exchange ideas, thus serving as an educational event.

6. They offer an opportunity to sell a limited number of breeding animals.

7. They set sale values for the animals back home, for such values are based on the sale of show animals.

8. They direct the attention of new breeders to those types and strains of cattle that are meeting with the approval of the better breeders and judges.

9. They are an effective means of distributing new breeds to new geographic locations, throughout the United States and the world.

10. They provide instruction, training, and encouragement for young people to join the expanding and dynamic beef industry.

11. They attract new producers and their dollars, both of which are needed if the industry is to grow in numbers and economically.

Like many good things in life, livestock shows also have some *disadvantages*, among which are the following:

1. Breed fancy points—involving such things as color and markings,

Fig. 14-3. A Grand Champion Hereford bull, beautifully fitted, perfectly groomed, and properly shown by one of the best professionals in the game. The bull: BBC 1 Domino 158. The owner: William Bennett, BB Cattle Co., Connell, Wash., at the halter. (Courtesy, William Bennett)

shape of head, shape of horns (if present), and set of ears—may overshadow utility value.

2. They may mislead, as occurred during the period 1935 to 1946—the era of the comprests and compacts. Even now, the author's crystal ball tells him that our current emphasis on heavy muscling in the show-ring may presage an era of double muscling.

3. The desire to win sometimes causes exhibitors to resort to "surgical means" and "filling" in order to correct defects. Admittedly, such man-made corrections are not hereditary, and their effects are often not too durable—as is belatedly discovered by some innocent purchaser.

4. Valuable animals are sometimes kept out of reproduction in order to enhance their likelihood of winning in the show-ring.

5. Their educational value has not been exploited. Many people who attend a cattle show for the first time are bewildered by the procedure and the many breeds and classes that they see. Others have never gone to a cattle show because they feel that their lack of knowledge of showing and judging cattle would prevent them from enjoying it.

Recent Improvements in Livestock Shows

Recently, there has been a concerted effort to improve livestock shows; to maximize their advantages and to minimize their disadvantages. To this end, there has been a trend toward the following changes:

1. To provide breeding classes with a maximum of a one-month age span per class.

2. To divide the cattle into evenly sized classes of 15 or less after they arrive at the show. For example, the show committee might decide to have 2 April heifer classes of 15 animals each instead of one class of 30.

3. To line up by age, from oldest to youngest.

4. To eliminate the older heifer classes (beyond early junior yearling), because of the feeling that senior yearling and 2-year-old females should be left home to raise their first calves.

5. To weigh all cattle after they arrive at the show.

6. To provide the judge and the spectators with the following information pertaining to each animal:

   a. Birth date.
   b. Weight per day of age.
   c. Son-o-ray fat thickness and rib eye measurements (or K-40 counter readings).
   d. Linear height measurements taken at the withers.

7. To combine live animal and carcass evaluation for steers, by requiring that the top 3 to 5 head of every class be slaughtered with carcass information obtained on these cattle.

8. To specify minimum carcass weight in carcass show classes, and to calculate and consider pounds of carcass per day of age.

SELECTING SHOW CATTLE

The cattleman who is planning to show has the difficult job of selecting animals which, following fitting, will meet the show-ring ideal. Requiring as it does a projection into the future, no judging assignment is quite so difficult. But the success or failure attained in the show-ring depends to a very great extent on the animals selected. Inasmuch as several months are required to bring an animal into the peak of show condition, the selection of animals for the show should be made as early in the season as possible, and from among animals that are healthy and vigorous and have good dispositions.

Fig. 14-4. No judging assignment is quite so difficult as that of selecting show animals for further development, for it requires a projection into the future.

When selecting steers for fitting and showing, look for growthiness, skeletal size, muscling, and trimness. The primary objective is to identify steers that will efficiently produce the greatest quantity of Choice lean beef per day of age. Purchasing calves from purebred or commercial herds that emphasize growth rate and use fast gaining, performance-tested bulls will give added assurance of rapid and efficient gains.

The ideal show steer should meet the following specifications:
1. Weigh 550 lb at weaning (205 days).
2. Make an average daily gain of 3.0 lb from weaning to slaughter.
3. Have feed efficiency of less than 6.0 lb of feed per pound of gain.
4. Weigh over 950 lb at 12 months.
5. Grade Choice when at slaughter weight of 1,000 to 1,350 lb.
6. Yield grade less than 2.9.

The selection of show bulls and females differs from the selection of show steers primarily because sex character, along with greater emphasis on structural soundness, is important in breeding animals. Like steers, they should possess growthiness, skeletal size, muscling, and trimness. Additionally, when selecting bulls, add masculinity and structural soundness to the list; and when selecting heifers, add femininity and structural soundness.

Fig. 14-5. Uniform, growthy steer calves. Excellent show prospects. Champion carload of Hereford feeder calves at the National Western Stock Show, Denver, Colo., exhibited by Fawn Lake Ranch, Gordon, Nebraska. (Courtesy, The American Hereford Association, Kansas City, Mo.)

The essential factors to be considered when selecting individual show prospects are: (1) type; (2) breeding and performance; and (3) show classification.

(Also, see Chapter 4, Selecting Beef Cattle.)

Type

When they make their placings, livestock judges endeavor to select cattle that are most efficient at producing quality beef. Steers that are growthy, muscular, and lack excessive finish (external fat) are usually placed at the top. Judges look for the same things in breeding classes, but, in addition, they look for structural soundess and sex character—traits that indicate that they will be efficient producers of offspring in the future.

Both breeding cattle and steers should be growthy. They should be long, tall, and not excessively fat if they are to continue to grow. Above all, avoid selecting a young animal that looks mature (that looks old for its age) and is overfat, because such characteristics are indicative of an animal that will not grow enough to be competitive in the show-ring.

Muscling, being a masculine trait, is more important in bulls and steers than in heifers. Long smooth muscling is preferred in young animals because they will usually grow for a longer period of time and get thicker as they get older. Truly muscular cattle are not smooth all over. They show some creases and indentations between the muscles; they are prominent in the forearm and stifle; and they are slightly narrower through the heart girth and loin than through the shoulder and round. Breeding cattle with coarse shoulders (very heavy muscled shoulders) should be avoided since this condition is frequently associated with calving problems.

Trimness, or freedom from predisposition to waste, is important in both breeding and slaughter cattle. Excessively fat breeding cattle will usually have poorer reproduction than trimmer ones; and very fat slaughter cattle will have reduced carcass value. Usually, prospective calves are not fat to begin with; hence, the showman must estimate whether they are the kind that are likely to get overfat. Generally, calves that have large briskets and that are deep in the flanks and twist will have a tendency to be overfat when fitted to mature weight, and should be avoided.

Structural soundness is especially important in breeding cattle, but only moderately so in show steers. Nevertheless, the skeletal structure of steers should be sufficiently correct that they will not develop any serious unsoundnesses as fitting progresses. Breeding cattle must be structurally sound so that they will calve easily and be able to travel well on the pasture or range. Proper structural soundness involves a big foot, a deep heel, toes that are the same size and point straight ahead, clean joints, and correctly set legs. Avoid cattle that are sickle-hocked or post-legged. Breeding cattle should be wide through the pins.

Signs of fertility and reproductive efficiency are extremely important in breeding cattle. Bulls should have two testicles that are of normal size, and that are well defined in the scrotum, rather than surrounded by excess fat. A mature bull should be masculine in front. But bull calves should not show extreme masculinity at a young age, because such animals are apt to mature too early and stop growing.

When selecting females for show, it is well to look for femininity at all ages. Feminine females are trim in the jaw, throat, and dewlap, and smooth shouldered. Avoid masculine females—the kind that are coarse, heavy fronted, and excessively muscular.

In summary, when selecting prospective show steers, emphasize growthiness, skeletal size, muscling, and trimness. Add masculinity and soundness to the list when selecting bulls; and add femininity and structural soundness when selecting females.

(Also see Chapter 4, Selecting Beef Cattle, section entitled, "The Functional Scoring System.")

Breeding and Performance

Animals selected for show should always be from good ancestry, for this is added assurance of satisfactory future development.

Projecting the growthiness of cattle is more certain if they have been performance tested. If performance records are available, the showman should select cattle that have weight ratios of 100 or above, which indicates above average growth. However, it should be recognized that cattle with weight ratios below 100 may be very growthy if the herd has been selected for superior performing cattle for a number of years.

(Also see Chapter 5, Some Principles of Cattle Genetics.)

Show Classifications

Distinct and separate show classifications are provided for breeding animals and for steers. Also, there is a breakdown of several classes within each category. These follow:

● *Breeding beef cattle*—Breeding classes vary according to breed. Most breeds make available to livestock shows, and to breeders, a "Standard Show Classification," setting forth their recommended classes and ages.

Show classifications are usually divided into three basic classes within a calendar year. Those calves born between January 1 and April 30 are known as *junior calves*; those born between May 1 and August 31 are called *summer calves*; and those dropped between September 1 and December 31 are known as *senior calves*. These divisions may be divided further to provide for classes of calves, yearlings, and two-year-olds. The standard Classification of the American Hereford Association is given in Table 14-1. Some breed differences exist; hence, the showman should secure from the breed registry and/or show(s) the classification for the breed that he is exhibiting.

TABLE 14-1

TYPICAL SHOW CLASSIFICATION FOR BREEDING BEEF CATTLE

Class	Age of Bulls	Age of Females
Junior calves	Born after Jan. 1 of the current show year. In shows between Jan. 1 and May 1, this class for bulls is divided into spring calves, and those born after April 1.	Same as bulls.
Winter calves	Born between Nov. 1 and Dec. 31 of the previous year.	Same as bulls.
Senior calves	Born between Sept. 1 and Oct. 31 of the previous year.	Same as bulls.
Summer yearlings	Born between May 1 and Aug. 31 of the previous year.	Same as bulls.
Spring yearlings	Born between March 1 and April 30 of the previous year.	Same as bulls.
Junior yearlings	Born between Jan. 1 and Feb. 28 of the previous year.	No class.
Senior yearlings	Born between Sept. 1 and Dec. 31 2 years prior to the show.	No class.
Two-year-olds	Born between March 1 and Aug. 31 2 years prior to the show.	No class.

Table 14-1 shows that there are no classes for breeding females older than spring yearlings. Older female classes have been eliminated from the classification because it is believed that the place for females of such age should be at home where they can start production, rather than continue under fitted show condition with the possible impairment in future productivity and possible loss to the breed.

It should be noted that there is a trend in the major cattle shows throughout the country to have breeding classes with a maximum two- to three-month age span per class.

In addition to providing for individual classifications for each sex as shown in Table 14-1, the major shows also make provision for championships and for various group classes. Since these differ somewhat between both shows and breeds, no attempt will be made to list them herein. Instead, the showman is admonished to study the premium list of the show or shows in which he plans to exhibit. Entries must be made for both individual and group classifications, but no entries for championship classes are required.

● *Steers*—The steer classification varies considerably; with some shows following age divisions, others following weight divisions, and still others following a combination of both age and weight. Each system has its advocates. The 1973 Chicago International used the following on-foot and carcass divisions:

● *On-foot steer division*—The 1973 Chicago International classes were as follows:

1. *Senior calf*—Born between September 1 and December 31, 1972. Minimum weight 900 pounds.

2. *Summer yearling*—Born between June 1 and August 31, 1972. Minimum weight 1,025 pounds.

3. *Junior yearling*—Born between March 1 and May 31, 1972. Minimum weight 1,125 pounds. Junior yearlings could have permanent incisors showing, but none in wear.

4. *Champion steer*—Any age.

5. *Group of three*—Any age from above classes owned by exhibitor.

The above classes were provided for (1) Angus, (2) Hereford, (3) Shorthorn, and (4) purebreds of other breeds and crossbreds. Then the Champion and Reserve Champion in each of these divisions competed for the Grand Champion Steer and the Reserve Grand Champion Steer of the show.

All steers placing fifth and higher in the on-foot competition, including the Reserve Grand Champion, were slaughtered for carcass competition. The Grand Champion could be slaughtered and could compete in the carcass competition at the option of the buyer.

● *Beef Carcass Division*—The 1973 Chicago International classes were as follows:

1. *Senior calf*—Calved between September 1 and December 31, 1972. Minimum hot carcass weight 540 pounds.

2. *Summer yearling*—Calved between June 1 and August 31, 1972. Minimum hot carcass weight 650 pounds.

3. *Junior yearling*—Calved between March 1 and May 31, 1972. Minimum hot carcass weight 708 pounds. Junior yearlings could have permanent incisors showing, but none in wear.

4. *Champion*—Selected from the first place winners in each of the above classes.

The above classes were provided for (1) Angus, (2) Hereford, (3) Shorthorn, and (4) other breeds and crossbreds. Also, the Champions in each of the above classes competed for Grand Champion Carcass Over All Breeds.

Additional rules pertinent to the Beef Carcass Division of the 1973 Chicago International follow:

1. All entries were required to provide the date of birth (verified by the breed registry association representative, by agricultural extension personnel, or by a local vocational agricultural instructor), identity of the sire, and the breed or predominant breeds of the dam.

2. All entries were weighed and rate of gain per day of age was calculated and made available to the judge of the live competition.

3. Awards were made only to beef carcasses with average or better USDA Choice quality and 3.5 or better yield grade (USDA).

FEEDING FOR THE SHOW

The subject of nutrition and feeding is fully covered in Chapter 8 of this book, so needless repetition will be avoided at this point; only the application of feeding for the show will be discussed.

All animals intended for show purposes should be placed in the proper state of condition—they should be neither too fat nor too thin. Steers should be fed to a degree of finish that will help ensure their ability to grade at least low Choice on the rail. Requirements differ slightly between breeds and lines of cattle. For example, the larger, leaner continental European breeds do not mature as quickly nor fatten as readily as the British breeds. The essentials in feeding cattle for the show might be described as similar to those in feeding steers for market, except that more attention must be given to the smallest details. A suitable ration must be selected and the animal or animals must be fed with care over a sufficiently long period.

At the beginning of show preparation, check for both internal or external parasites. If any are present, apply the recommended treatment. (See Chapter 12, Beef Cattle Health, Disease Prevention, and Parasite Control.)

Rules of Feeding Show Cattle

Some general rules of feeding may be given, but it must be remembered that "the eye of the master conditions his cattle." The most successful cattle fitters have worked out systems of their own through years of practical experience and close observation—they do not follow set rules. Nevertheless, the beginner may well profit by the experience of successful fitters, and it is with this hope that the following general rules of feeding show cattle are presented:

1. *Practice economy, but avoid false economy*—Although the ration should be as economical as possible, it must be remembered that rapid growth and ideal condition are primary objectives, even at somewhat additional expense. It is beneficial to start feeding prospective show calves for a short time prior to weaning so that they are accustomed to eating grain when they are eventually weaned from their dams. A suitable creep ration is given in Chapter 21.

2. *Use care in getting animals on feed*—In starting animals on feed, use extreme caution to see that they do not get digestive disturbances. Animals

that are not accustomed to grain or other concentrates must be started on feed gradually, or digestive trouble may result. Until the animal gets on full feed, at no time should it be given more grain than can be cleaned up in about 30 to 60 minutes' time. In starting, a safe plan is to feed not more than *1 lb of grain at the first feed, or 2 lb for the day. This may be increased approximately by ¼ to ½ lb daily until the animal is on full feed about 3 weeks later.* From the beginning, it is safe to full-feed grass hay or the hay to which the animal is accustomed. Oats are the best concentrate for the beginning ration. As the grain feed is increased according to the directions given, gradually (1) replace the oats with the mixed ration selected, and (2) decrease the hay. Whenever the calf does not seem to have a good appetite and does not clean up the grain within an hour's time, the allowance should be reduced for the next few days and then gradually worked up to a full feed again. With careful observation and good judgment, it is possible to have an animal on full feed in 3 weeks' time. After full feeding is reached, the animal may be fed according to either of 2 plans: (a) by governing the allowance according to the amount of feed that the animal will clean up in about 30 to 60 minutes' time; or (b) by more or less self-feeding, with some feed being kept before the animal most of the time. Perhaps, in the final analysis, the method of feeding decided upon should vary according to the individual feeder.

Ordinarily it is best to start the animal on feed with grass hays and then gradually change to legumes, if legumes are to be included in the ration. By this method, the animal may be allowed a full feed of hay at the beginning of the feeding period with no danger of scouring. When forced feeding is being sought, it is usually preferable to use a grass hay or a mixed hay with a limited amount of legume in it because of (a) the laxative effect of a straight legume under heavy feeding, and (b) the possible bloat hazard of legumes.

3. *Provide a variety of feeds*—A good variety of feeds increases the palatability of the ration and makes it easier to supply the proper balance of nutrients. Furthermore, a finishing ration consisting of only one or two feeds may lose its palatability during a long feeding period.

4. *Feed a balanced ration*—A balanced ration will be more economical and will result in better gains. That is to say, the ration should contain the proper balance of energy, protein, minerals, and vitamins. It must also be remembered that growing animals require a higher proportion of grain than mature ones. Then, too, because of its high cost, it is not economical under most conditions to feed more protein than is required.

5. *Do not overfeed*—Feed plenty but do not overfeed. Overfeeding is usually caused by the desire of the inexperienced caretaker to push the animal too rapidly. After the animal has reached full feed, it may be given all the grain that it will clean up, provided that the appetite and well-being seem to so warrant and that the droppings are of the proper consistency.

6. *Keep the feed box clean*—Never leave uneaten feed in the box or trough. It may become sour and cause the animal to go off feed.

7. *Do not underfeed*—It never pays to starve an animal. Gains and profits result from feeds consumed in excess of the maintenance require-

ment. A common expression among cattle feeders is: "Get every bit of feed under their hides that you can." When animals are consuming too little grain to grow-finish the feeder may look for several causes, such as the consumption of too much roughage, unpalatability of the ration, or discontentment on the part of the animal—that is, if the caretaker is not deliberately withholding grain (false economy).

8. *Full feed for economical finishing*—When on full feed, the average animal will eat from 2 to 2½ pounds of grain for each 100 pounds of liveweight. The exact amount will depend primarily upon the age, size, and individuality of the animal; the bulkiness and palatability of the ration; and the amount of fat that the animal is carrying. A full feed of grain is the amount that an animal will clean up nicely in ½ to 1 hour's time at each feeding period.

9. *Supply palatable feeds*—In order to consume the maximum amount, the animal must relish the feed. Unpalatable feeds may be fed in limited quantities, provided that they are mixed with more palatable ingredients. Blackstrap molasses is relished by animals and is excellent for increasing the palatability of the ration (although blackstrap molasses is preferable, beet molasses is satisfactory). Usually the molasses is added by diluting with water (warm water in winter) and mixing it with the grain ration just before feeding. One-half to one pint of molasses diluted with an equal volume of water will be entirely satisfactory for this purpose. Most commercial feeds contain some molasses in the mixture. Cooking certain feeds, especially barley, also makes for increased palatability.

10. *Provide succulent feeds*—Succulence is provided in such feeds as silage, root crops, and grasses. These feeds have a beneficial effect in the ration. They increase the palatability and produce a laxative effect on the animal's digestive system.

11. *The ration must provide the correct amount of bulk*—The beef animal is a ruminant, and, therefore, requires some bulk in order to distend the digestive tract. Mature animals can handle more roughage than calves. Furthermore, more bulk may be fed at the beginning of the finishing period than at the end. As the grain ration is increased, the animal will consume less roughage. When on full feed, three to six pounds of hay daily are ample. Consumption of too much bulk will cause the animal to become paunchy.

12. *Do not feed damaged feeds*—Moldy, musty, or spoiled feeds may cause digestive disturbances and should not be fed to animals being fitted for show or sale.

13. *Prepare grains*—The grain ration of cattle intended for show purposes is almost always coarsely ground or rolled. Most herdsmen prefer steamed flaked grains. The preparation of hay is neither necessary nor advisable.

14. *Feed regularly*—Animals intended for show should be fed with exacting regularity. In the earlier part of the feeding period, two feedings per day may be adequate. Later, the animals may be fed three times a day, particularly if they are rather thin and rapid improvement in condition is desired.

15. *Avoid sudden changes*—Sudden changes in either the kind or amount of feed are apt to cause digestive disturbances. Any necessary changes should be gradual.

16. *Provide minerals*—All animals should be given free access to salt at all times. For feeding where the salt is not exposed to the weather, loose salt, rock salt, and block salt are all satisfactory. Loose salt, however, is preferred if kept under shelter. Stabilized iodized salt should be used in iodine-deficient areas; and trace mineralized salt should be used in areas where other trace mineral deficiencies exist.

When a nonlegume roughage is provided, it is especially likely that there will be a deficiency of calcium and phosphorus. Under these conditions and even when a legume roughage is used, the addition of calcium and phosphorus gives protection at very little cost. Probably as satisfactory and inexpensive a mineral as can be provided for cattle is steamed bone meal or dicalcium phosphate. The mineral may be placed in a box to which the animals have free access. The best arrangement is to provide a double-compartment mineral box with salt in one side and a mixture of one-third salt and either steamed bone meal or dicalcium phosphate in the other.

17. *Keep the animal quiet and contented*—Quiet and contentment are necessary for profit in feeding. The restless animal rarely makes good gains, whereas the quiet animal that will "eat and lie down" will show superior gains. This is not due to differences in digestive or assimilative powers, but rather to the fact that the quiet animal is putting on weight while the wild, nervous animal is using surplus energy for nonproductive purposes. Uncomfortable quarters, isolation from other animals, annoyance by parasites, sudden changes in quarters or feeds, improper handling, and unnecessary noise are the most common causes of discontentment.

18. *Provide exercise*—A certain amount of exercise is necessary in order to promote good circulation and to increase the thrift and vigor of the animal. Exercise also tends to stimulate the appetite and makes for greater feed consumption. Animals can usually be kept in condition by turning them in a paddock at night.

19. *Avoid scouring*—If the droppings are too thin or there is scouring, (a) decrease the grain allowance, and (b) clean up the quarters. If trouble still persists, decrease the legume roughage and the protein supplement (especially linseed meal).

20. *Avoid sudden water changes*—Frequently, show cattle fail to drink enough water while at the fair. As a result, they become gaunt and show at a disadvantage. Usually this situation is caused by (a) the sudden change in drinking from a trough or tank at home to drinking out of a bucket at the show, and (b) the different taste of the water, primarily due to chlorine or mineral content. This problem can be alleviated by (a) getting the animal accustomed to drinking from the same bucket that will be used at the fair, beginning 7 to 12 days before leaving home, and (b) adding a tiny bit of molasses to each bucket of water, from the time bucket watering is started until the show is over, thus avoiding any flavor or taste change in the water.

Some Suggested Rations

Variations can and should be made in the rations, depending upon the individual animal, the relative prices of feeds, and the supply of home-grown feeds. To secure the correct state of condition, a suitable ration must be selected and the animal or animals must be fed with care over a sufficiently long period. The rations that follow are ones that have been used, and are being used, by successful showmen. In general, when show animals are being forced fed on any one of these concentrate mixtures, experienced herdsmen prefer to feed a grass hay or a grass-legume mixed hay to a straight legume, because of the laxative effect and possible bloat hazard of the latter.

Ration No. 1
Rolled barley .... 50 lb
Rolled oats .... 20 lb
Wheat bran .... 20 lb
Protein supplement[2] .... 10 lb

Ration No. 2
Rolled corn .... 20 lb
Rolled barley .... 30 lb
Rolled oats .... 20 lb
Wheat bran .... 20 lb
Protein supplement[2] .... 10 lb

Ration No. 3
Rolled oats .... 30 lb
Rolled barley .... 30 lb
Rolled corn .... 10 lb
Wheat bran .... 20 lb
Protein supplement[2] .... 10 lb

Ration No. 4
Rolled oats .... 30 lb
Rolled corn .... 60 lb
Protein supplement[2] .... 10 lb

Ration No. 5
Rolled corn .... 40 lb
Rolled oats .... 30 lb
Wheat bran .... 20 lb
Protein supplement[2] .... 10 lb

Ration No. 6
Rolled corn or sorghum
    chop .... 50 lb
Rolled barley .... 40 lb
Protein supplement[2] .... 10 lb

Ration No. 7
Rolled corn .... 60 lb
Rolled oats .... 20 lb
Dry beet pulp .... 10 lb
Protein supplement[2] .... 10 lb

Ration No. 8
Rolled corn .... 40 lb
Rolled barley .... 20 lb
Rolled oats .... 10 lb
Dry beet pulp .... 10 lb
Wheat bran .... 10 lb
Protein supplement[2] .... 10 lb

Ration No. 9
Rolled oats .... 25 lb
Rolled barley .... 20 lb
Rolled wheat .... 20 lb
Rolled corn .... 20 lb
Wheat bran .... 10 lb
Protein supplement[2] .... 5 lb

Ration No. 10
Rolled barley .... 35 lb
Rolled oats .... 20 lb
Rolled wheat .... 20 lb
Dry beet pulp .... 15 lb
Protein supplement[2] .... 10 lb

Ration No. 11
Rolled barley .... 20 lb
Rolled corn .... 20 lb
Rolled oats .... 20 lb
Whole barley (dry wt.
    basis, but cooked
    before feeding) .... 13 lb
Beet pulp, dried
    molasses .... 4 lb
Wheat bran .... 6 lb
Commercial supplement .... 8 lb
Linseed oil meal
    (pellets) .... 8 lb
Salt .... 1 lb

---

[2]The protein supplement may consist of linseed, cottonseed, peanut, or soybean meal, or a commercial supplement may be used. With most herdsmen, linseed meal is the preferred protein supplement. It gives the animal a sleek haircoat and a pliable hide. However, it is a laxative feed. Caution should be exercised in feeding it.

Although it is true that an animal getting good clover or alfalfa hay needs less protein supplement than does one eating nonleguminous roughage, it is not possible to supply all the needed protein with hay and still get enough grain into young animals to finish them quickly.

Rations 1 to 5 are bulky. They are recommended for use in starting prospective show animals on feed. Rations 6 to 11 are less bulky and more fattening. They are recommended for use during the latter part of the fitting period.

Ration 11 is the one which the author uses in fitting show steers. The cooked barley is prepared by (1) adding water in the proportion of 2 to 2½ gallons to each gallon of dry barley, and (2) cooking until the kernels are thoroughly swelled and can be easily crushed between the thumb and forefinger. Each young steer also receives 4 lb daily of a supplement high in milk by-products. As the animal approaches show finish, the ration is changed by decreasing the rolled barley by 7 lb and increasing the rolled oats by 5 lb and the wheat bran by 2 lb.

EQUIPMENT FOR FITTING AND SHOWING CATTLE

Every showman should have a durable and attractive box in which to keep the necessary equipment. The attractiveness of the exhibit can be enhanced by the presence of a nicely painted box on which the name of the exhibitor is neatly printed.

The equipment for the show may include the following: brushes, liners, curry combs, Scotch combs, material and cloth for polishing horns and hoofs, soap or shampoo, bluing, saddle soap and/or butch wax, rasp, knife, rope halters, leather show halters, show sticks, hair oil, clippers, thinning shears, water buckets, feed boxes, and clean clothes and other articles for the personal use of the caretaker. It may also be well to include a small amount of epsom salts, raw linseed oil, mineral oil, and iodine. The size of the box and the amount of equipment one carries will depend upon the number of cattle being shown and the length of the show circuit.

A Rope Halter and How to Make It

Attractive leather halters with lead straps and chains are ideal for showing cattle, but they are rather expensive. Rope halters are much more practical for everyday use.

MATERIALS NEEDED

The following materials are needed for making a rope halter:

1. *Rope*—Thirteen feet (15 feet when making halters to use in breaking cattle to lead, thereby having a longer lead) of 3-strand manila or nylon rope (nylon will not draw up when it gets wet). Use ½-inch rope for cattle over 6 months of age and ⅜-inch rope for calves under 6 months.

2. *Marlinspike*—This is needed for opening up the strands of rope. It may be made by taking a piece of ¾-inch round, hardwood stick 6 inches long, and tapering it to a point on one end; or a small pointed piece of iron may be used.

3. *Measure and pencil*—A rule or tape and a pencil.

DIRECTIONS FOR MAKING

Here are the directions for making an eye-loop rope halter (see Fig. 14-6), which is adjustable in every respect except for the nose band. This type of halter keeps its adjustments. By contrast, the double-loop rope halter is objectionable for use in tying because it adjusts too easily with the result that many tied calves free themselves.

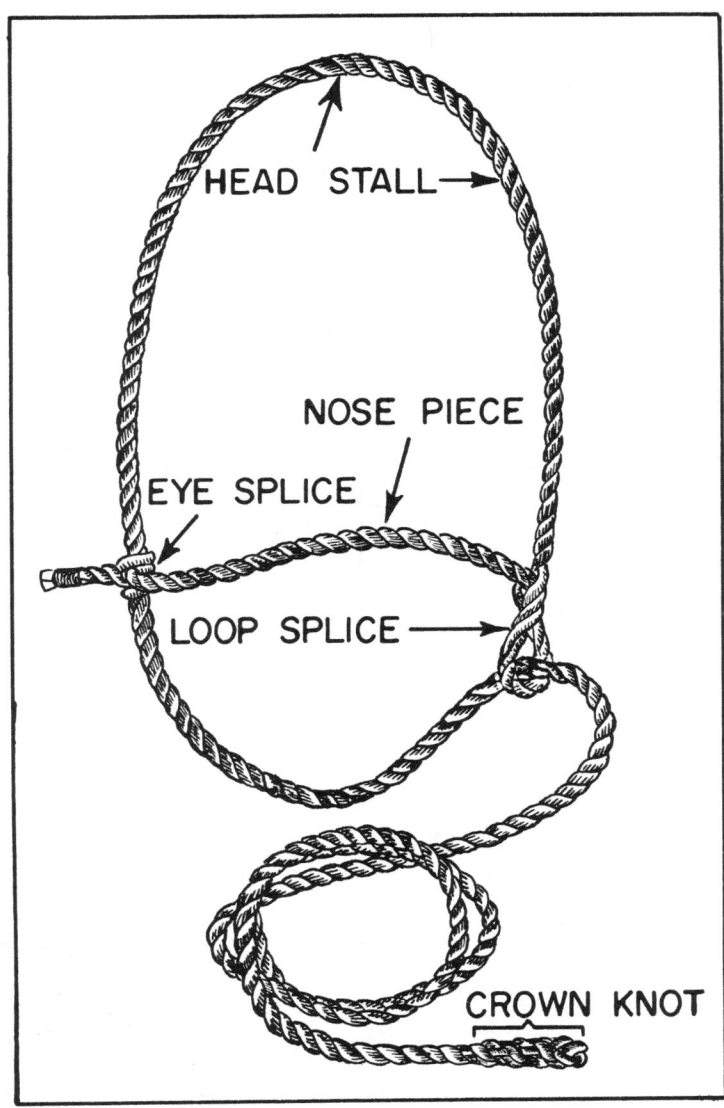

HEAD STALL →

NOSE PIECE

EYE SPLICE

LOOP SPLICE →

CROWN KNOT

Fig. 14-6. Completed eye-loop rope halter.

The steps in making an eye-loop rope halter are:

1. *Whip-slice one end*—The end which is whip-sliced, to prevent it from unraveling, is known as the short end. For this purpose, use a waxed cord, fish line, or strong piece of string about 40 inches long.

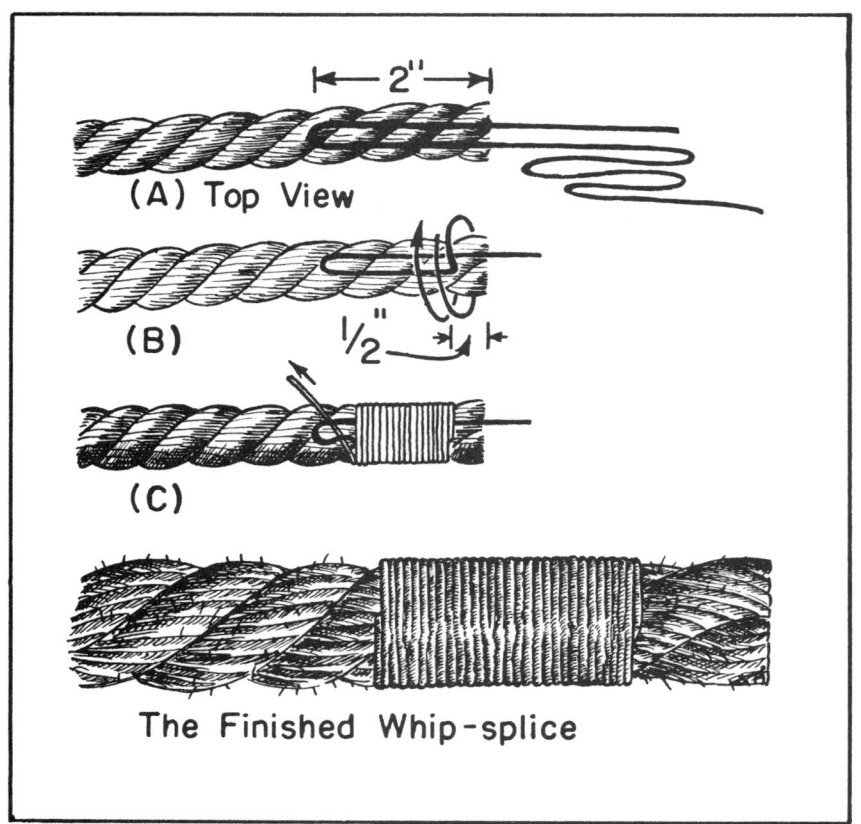

Fig. 14-7. Proper method of whip-splicing.

Double the whipping cord from one end to form a loop approximately 6 inches long. Lay this loop on top of the rope (Fig. 14-7, Step A), which is held in the left hand, with the end of the loop about 2 inches from the end of the rope. Now with the thumb holding the cord about ½ inch from the end of the rope (Fig. 14-7, Step B), make the first turn from front to back or clockwise around the rope, and be sure that this first turn locks the first wind in place.

Continue to wind tightly and neatly for a length of one inch. Then run the last wind up through the loop (Fig. 14-7, Step C) and draw up tightly against the last turn and hold in place. Next grasp the cord out over the end of the rope and pull the loop under the windings for about ½ inch or half the length of the windings. These will now stay in place. With a knife, cut the remaining cord off even with the first and last windings. Soaking the whipping in waterproof cement makes the job more secure.

2. *Make the loop splice*—The loop splice is made as follows (see **Fig. 14-8**): The short end of the rope will form the nosepiece of the halter and will be its only permanent dimension. For average size cattle, use 15 to 16 inches for the short end (the future nosepiece); for large bulls use 18 to 20 inches. Measure off this amount (the 15 to 16 inches, or up to 18 to 20

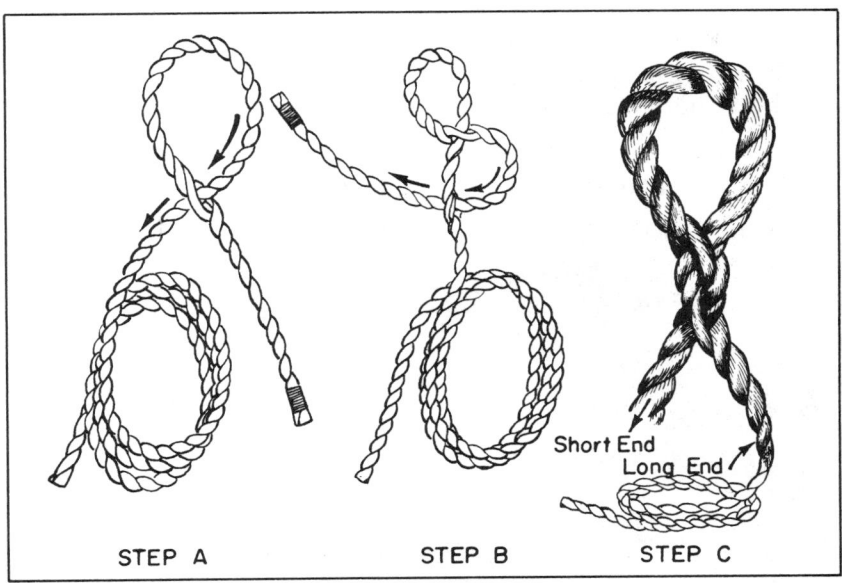

Fig. 14-8. Steps in making the loop splice.

inches—as decided upon) from the whipping on the short end that you have just completed whip-splicing, and pencil-mark the two strands to be raised. Lay the rope in front of you, short end to the right. Bend the long end up or away from you clockwise. With the aid of the marlinspike, raise the two strands on the top of the short end at the 15- to 16-inch mark and pull the long end down toward you and under the strands. Draw this long end through until the loop is about 1½ to 2 inches (make a 2-inch loop where the halter may get wet) inside diameter (see Fig. 14-8, Step A). A rule of thumb is that the inside diameter of the loop should be at least twice the thickness of the rope; a loop that is too small will close too tightly when the halter becomes wet and shrinks.

Next, take the short end and, with the help of the marlinspike, pass it under a top strand on the long rope as close to the loop as possible (see **Fig. 14-8, Step B**).

Pull the lead end snug, and the loop splice is complete as shown in **Fig. 14-8, Step C**. This makes for a loop with an equal number of strands on each side of the splice and leaves the inside of the splice fairly smooth where it bears against the jaw.

3. *Make the eye splice*—Grasp the whip-spliced end in your left hand with the left thumb a couple of inches from the end and on top of the rope.

Grasp the nosepiece (short end) with your right-hand thumb on the bottom of the rope, a couple of inches from the left hand. Twist your right hand away from you (Fig. 14-9, Step A), pushing while you twist, so that the strands separate and finally kink (Fig. 14-9, Step B). Take each kink on the same side of the rope and in line. Place the marlinspike through the three kinks (as shown in Fig. 14-9, Step C or 14-9, Step D—either of which is correct), and then follow with the long rope through all three (Fig. 14-9, Step E). Draw the long rope through the eye splice until it is free of kinks and then adjust it back until it is approximately the size of the headpiece. Arrange the kinks neatly on the rope.

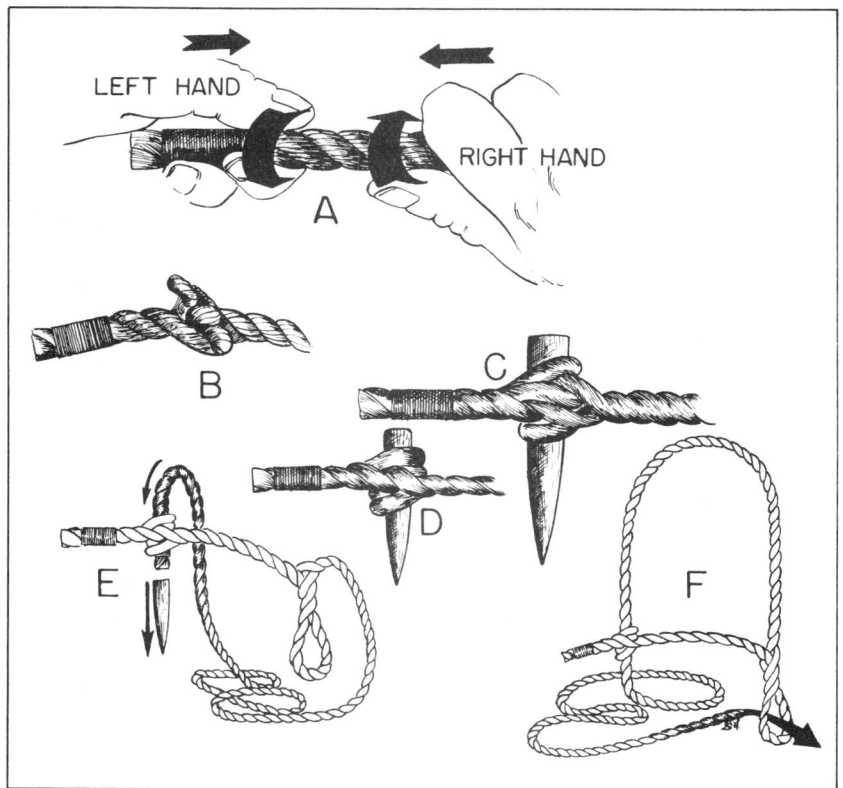

Fig. 14-9. Steps in making the eye splice.

Pass the long rope through the loop splice (Fig. 14-9, Step F) and the halter is complete except for preparing the end of the lead rope. The halter is placed on the animal so that the loop splice is at the left of the jaw. Needed adjustments may be readily made after fitting it on the animal. (When properly fitted on the animal, the nosepiece should fall about two-thirds the distance from the muzzle to the eyes.)

4. *Prepare the end of the rope*—The end of the lead rope may be prepared in either of three ways: (a) by whipping; (b) by making a crown knot; or (c) by making a wall knot.

Whipping leaves the end of the lead sufficiently small that it can be passed easily through tie rings. On the other hand, both the crown knot and the wall knot make it easier to hold on to an animal that is trying to get away. The procedure for preparing the end of the rope by each of these methods follows:

a. *Whipping the end of the rope*—If the lead end of the rope is to be whipped, this step is generally done at the very beginning as already directed for the short end of the rope (see Fig. 14-7).

b. *Make the crown knot*—Make the crown knot as follows (see Fig. 14-10):

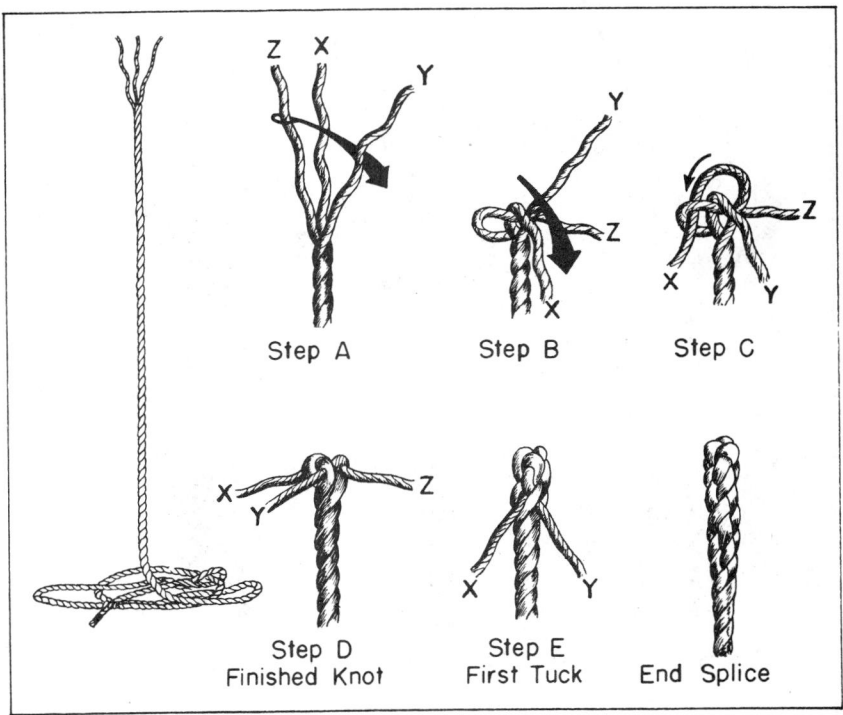

Fig. 14-10. Steps in making the crown knot.

(1) Unlay the rope for five to seven turns, depending upon the length of finish desired (Fig. 14-10, Step A) and throw up a bight in z between x and y and lock with the forefinger (see Fig. 14-10, Step A).

(2) Pull x over z and lock with the thumb (see Fig. 14-10, Step B).

(3) Pass y over x and linewards through bight z (see Fig. 14-10, Step C).

(4) Hold the formation in the tips of the fingers and set by pulling on x, y, and z in turn until they are uniformly snug (see Fig. 14-10, Step D).

(5) Tuck by passing each strand over one strand and under one strand, straight down the line, removing one or more yarns of the

strand at each tuck after the first, according to the taper desired. (See Fig. 14-10).

(6) Moisten (barely) the completed crown knot and roll the splice under the foot.

c. *Make the wall knot*—If desired, a wall knot may be made as follows (see Fig. 14-11):

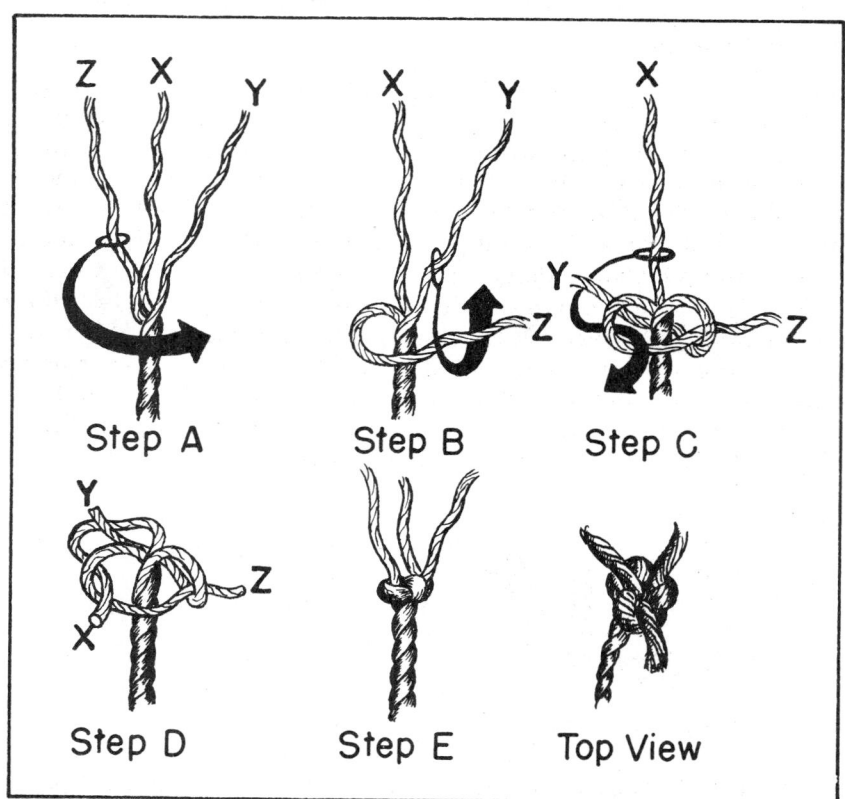

Fig. 14-11. Steps in making the wall knot.

(1) Unlay the rope from five to seven turns (Fig. 14-11, Step A) and throw down a bight in z and lock with the thumb (see Fig. 14-11, Step B).

(2) Throw a bight in y around z and lock under the first finger (see Fig. 14-11, Step B).

(3) Throw a bight in x around y, extending the end through the bight in z (see Fig. 14-11, Steps C and D).

(4) Hold in the tips of the fingers and set by pulling alternately on each strand until they are uniformly snug (see Fig. 14-11, Step E).

TRAINING AND GROOMING FOR THE SHOW

Assuming that the animal has been carefully selected and properly fed, there yet remains the assignment of parading before the judge. In order to present a pleasing appearance in the show-ring, the animal must be well trained and thoroughly groomed. Competition is keen, and often the winner will be selected by a very narrow margin. Close attention to details may, therefore, be a determining factor in the decision.

Gentling and Posing

Proper training of the animal requires time, and one must be patient, firm, and persistent. Such schooling makes it possible for the judge to see the animal at its best. First, the animal should be gentled by petting and brushing, and then it may be haltered. After the animal has become accustomed to the halter, it should be taught to lead. This should be correctly done from the left side and with the halter strap or rope neatly coiled in the right hand. Although most dairy showmen prefer to walk backward, beef showmen follow the custom of walking forward, glancing back over the right shoulder at frequent intervals. The rope halter is preferable when starting the training program, but it is very important that the animal become accustomed to being led with the leather show halter well in advance of the show. The latter precaution is important because the animal reacts differently when led with the show halter than when led with the rope halter.

Fig. 14-12. Gentling and posing. (Drawing by R. F. Johnson, Head, Dept. of Animal Science, California Polytechnic State University, San Luis Obispo, Calif.)

The next step is that of teaching the animal to stand or "pose" properly, so that the judge may have an opportunity to examine it carefully. For correct posing the animal must stand squarely on all 4 feet (preferably with the forefeet on slightly higher ground than the rear feet). The back should be held perfectly straight and the head held on a level with the top of the back. At first, this position may be quite strained and unnatural for the animal. For this reason, it should not be required to hold this position too long. Later, it should be possible to "pose" the animal for 15 or 20 minutes at a time. In "posing," the showman should hold the strap in the left hand and face toward the animal. A show stick is usually used in placing the hind feet but the showman's foot is best used in obtaining correct placement of the front feet.

Trimming, Cleaning, and Oiling the Feet

So that the animal will stand squarely and walk properly, the feet should be trimmed regularly. Long toes or unevenly worn hoofs are unsightly in

Fig. 14-13. Trimming the feet. (Drawing by R. F. Johnson)

appearance. Trimming can best be done with the animal either in a set of stocks or on a trimming table. With this method, it is possible to square up the sole and the sides of the feet as well as to cut back the ends of the toes. The practice of merely shortening the toes by standing the animal on a hard surface and cutting off the ends with a hammer and chisel gives only temporary relief, very often not really correcting the difficulty. If a set of stocks or a table cannot be made available, it may be advisable to throw the animal as shown in Fig. 14-14, thereby making it possible to work on the bottoms of the feet.

Fig. 14-14. A simple method of throwing an animal is illustrated here using a rope halter and a rope about 40 feet long. With the animal haltered, tie the halter shank or lead to a stout post, tree, etc. Then, with one end of the rope around the neck, tie a bowline (nonslip) knot. Next, circle the calf's body just behind the shoulder and a half-hitch at the withers; continue the rope back to the loin and make a second half-hitch and circle the rope around the body at the flanks. Make sure the second half-hitch is just in front of the hooks. A strong backward pull on the rope will cause the animal to sink, and a shift in the pull to the side on which the animal is to fall will result in an easy, soft fall to the ground. Maintaining the strong pull on the rope will keep the animal lying on the ground, making it possible to do minor "doctoring," foot trimming, etc.

The feet of some animals should be trimmed regularly as often as every two months. Too much trimming at any one time, however, may result in lameness. For this reason, it is not advisable to work on the feet immediately before the show.

Among the tools that may be used for trimming are the electric sander, chisel, nippers, farrier's knife, and rasp. However, not all these need be available.

Quite often, when cattle are kept constantly in stables, the feet may become dry and brittle. This condition can usually be corrected by turning the animals out in a pasture paddock at night when there is dew on the grass. Packing the hoofs with wet clay, or applying neat's-foot oil will also be helpful in such cases. If the animal gets sore feet from standing in a filthy stable, the soreness should first be corrected. Following this, the feet should be washed and disinfected.

Before the animal enters the show-ring, its hoofs and dew claws should be thoroughly cleaned and then oiled or sprayed with colored spray paint.

Training and Polishing the Horns

Often the horns will grow in an improper shape unless they are weighted or scraped to correct the faults. A well-curved set of horns will command the admiration of the judge, but poorly shaped horns will give the head a coarse, unattractive appearance. It must also be remembered that there is a difference in the desired shape with different breeds. The horns of the Hereford should curve downward, whereas the horns of the Shorthorn should curve slightly forward and inward.

Fig. 14-15. Note the properly (left) and improperly shaped horns (right). (Courtesy, USDA)

As soon as the horns are long enough (3 to 4 inches long) and sufficiently strong to bear the weight, it is time to begin training. For this purpose one-half pound weights (the correct size can best be determined by study and experience) are usually used. Care should be taken to prevent making a sharp turn in the horn by using a weight that is too heavy or by allowing the weights to remain on for too long a time. If the horns yield too readily, it is best to remove the weights and give the horns a rest of from 10 days to a month, the length of time depending upon their condition. Then replace the weights until the desired effect is obtained. Weights should be removed when the horns become level with the top of the head, or not more than one inch below this level. Leaving the weights on longer tends to cause the

Fig. 14-16. Horn weights may be used in order to obtain symmetrical, properly curved and attractive horns. As soon as the horns are long enough and sufficiently strong, it is time to begin with a light weight. There is a variation in the degree of training that is desired in the different breeds. (Courtesy, USDA)

horns to curve inward too much, thus causing a problem later. If the screw type of fastener is used, one should be careful not to force the screw into the horn so deeply that the depression cannot be removed. Horn weight losses may be reduced by tying a strong cord around the screws on the two weights; then if one weight is knocked off, it will not be lost. Horns may be pulled forward when they are 3 to 4 inches long by using a suitable spring or strap device for the purpose.

Extremely long horns may appear out of proportion and unsightly. In such cases, they can often be cut back as much as 2 or 3 inches, provided that not more than half an inch is removed at any one time and at no more frequent intervals than a month or 6 weeks. As a rule, most of the black tip can be removed without harming the sensitive part.

After the horns have been properly shaped, the next job is that of trimming and polishing. Usually it is best to smooth them down a week or two before the show. The rough surface may be smoothed with a sharp knife, a rasp, or a steel scraper; always scrape from the base toward the top. The final smoothing or finishing touches may then be given by using sandpaper, fine emery cloth, steel wool, or a flannel cloth and emery dust.

Horns are usually polished just before the show. An excellent polish that will not collect dust can be obtained as follows: Apply a paste which is made by mixing olive oil or sweet oil with pumice stone or tripoli. Polish by rubbing briskly with a flannel cloth. A quick and more simple polish can be obtained by the use of glycerine, linseed oil, or mineral oil. However, a polish obtained in this manner is rather temporary, and the oil will collect dust quickly.

Grooming

Animals intended for show should be brushed at least 15 minutes twice daily. Vigorous brushing stimulates the circulation in the hide and keeps it in a loose, pliable condition. It also brings out the natural oil in the hair; removes dandruff, dirt, and dead hair; and induces hair to grow. Short-haired animals should be brushed downward and to the rear with the lay of the hair; whereas long-haired animals should first be brushed downward and then upward in the opposite direction of the lay of the hair in order to make

Fig. 14-17. Grooming with rice root grooming brush and Scotch comb. (Courtesy, American Shorthorn Association, Omaha, Neb.)

it loose and fluffy. A woolen cloth may also be used to advantage for removal of dirt. Hold the brush in one hand and the cloth in the other.

It is well to use the curry comb sparingly at all times, especially in sensitive regions where the bone is near the surface. The primary purpose of the curry comb is to remove clinging particles that cannot be taken out with the brush. Furthermore, the curry comb should never be used on the tail switch, for it will pull out too many hairs and leave the switch light and skimpy.

Clipping

Before clipping an animal, have a mental picture of the ideal counterpart; then strive to emulate it.

Custom decrees certain breed differences in clipping ("haircuts") and "hairdos," and differences between steers and breeding animals. Clipping may involve the back and sides, the underline, the tail, and/or the head.

Clipping is best done with electric clippers. With the exceptions noted in the discussion which follows, it should be done about a week before the show, so that the clipped hair will lose its stubby, "fresh haircut" appearance.

Animals with long hair usually look better if they are "blocked," or clipped, over the top and down the sides. This blocking should be done at home, but the finishing touches should be administered at the show after the final washing.

Most animals need to have their underlines clipped, so that they will

Fig. 14-18. Trimming the underline and blocking the animal along the sides and back is practiced by most showmen to help complement the good points and diminish the bad points of an animal. (Courtesy, American Angus Association, St. Joseph, Mo.)

appear trimmer. Such clipping should be done three or four weeks in advance of the show in order to avoid a "bald belly" look. When clipping the underline, gradually blend the clipped area with the long hair on the sides of the animal, thereby avoiding any unartistic clipper line. This requires a steady hand. Also, the blending can be facilitated by use of a clipper guard (a plastic comb-like attachment) or by use of thinning shears.

One of the main objects in clipping the tail is to show the fullness of the twist and the thickness or beefiness of the hindquarters. In order to do this to best advantage with each individual, good judgment should be exercised in determining the extreme points of clipping that will show these characteristics to advantage. A general guide is to clip the tail from the high point of the twist to the tailhead. At the tailhead, the hair should be gradually tapered off near the body so that the tail blends nicely with the rump. To avoid leaving ridges of long hair extending down the center of the tail, the clipper should be run across the tail after it has been run upward to the tailhead.

Sometimes the head is clipped so that it will appear cleaner cut, more shapely, and the poll more clearly defined. Such heads are clipped from a point just back of the jawbone and one-half to one inch behind the ears; leaving the hair on the ears, eyelashes, and end of the muzzle.

Generally speaking, custom decrees that clipping of heads and tails shall be as follows:

STEERS

    1. All tails clipped.
    2. All heads clipped, but do not clip the ears.

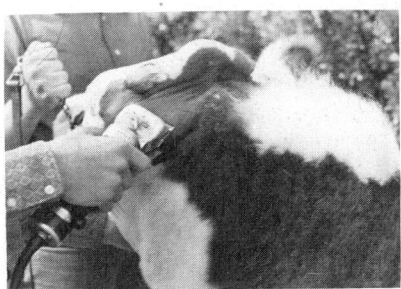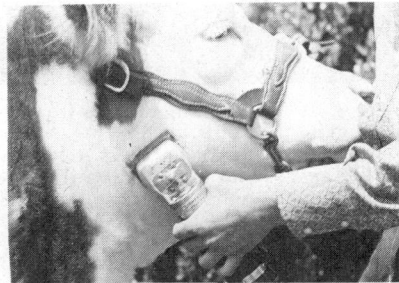

Fig. 14-19. Clipping a steer's head. Clipping should begin at the top of the neck just behind the halter head strap (left). From this point, the head is clipped straight down the neck on both sides to the underside (right). Once this line is established, everything in front is clipped to the skin, except the ears and eyelashes. Most people prefer that hair be left on the ears, so trim only that which may be unsightly. (Courtesy, The American Hereford Association, Kansas City, Mo.)

BREEDING CATTLE

    1. Angus—Heads (but do not clip either the inside or the outside of the ears) and tails clipped.
    2. Shorthorns (both horned and polled)—Neither tails nor head clipped.
    3. Herefords (both horned and polled)—Tails clipped; heads not clipped. (The poll is trimmed on Polled Herefords.)

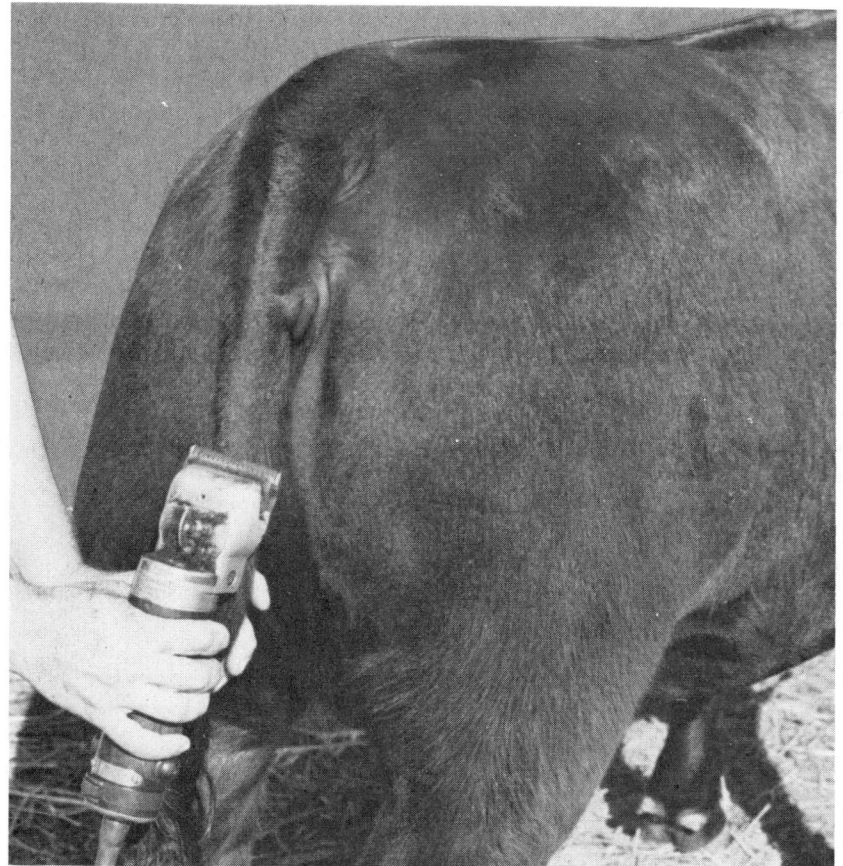

Fig. 14-20. Clipping the hair from the tail of an Angus. One of the main objects in clipping the tail is to show the thickness or beefiness of the hindquarters. Clipping should begin above the switch of the tail opposite the point where the twist is fullest; and extend to the tailhead where it gradually tapers off, giving a blended effect with the rump. (Courtesy, American Angus Association, St. Joseph, Mo.)

Washing

Frequent washing keeps the animal clean; stimulates a heavy growth of loose, fluffy hair; and keeps the skin smooth and mellow. Wash monthly until six weeks before the show, wash weekly thereafter. In preparation for washing, place a chain about the neck. Never use a rope about the animal's neck when washing, for a wet rope cannot be easily loosened should the animal fall or otherwise get into trouble. Immediately prior to washing, remove all possible dirt and dandruff before wetting the animal. The latter may be accomplished either (1) by use of a vacuum blower, or (2) by vigorous brushing with a stiff brush.

Most showmen now use a special commercial detergent, as a cleaner and hair conditioner for washing. However, if desired, an excellent preparation may be made as follows: mix one to two cups of good concentrated liquid coconut oil shampoo in one and one-half gallons of lukewarm water (or high-grade castile soap may be used). With a bristle wash brush, thoroughly wet the dry animal with the soap solution. Work the soap into a good lather, making sure that all parts of the body are well scrubbed and clean. Some parts of the animal that are frequently neglected in washing are the head, tail, legs, brisket, and belly.

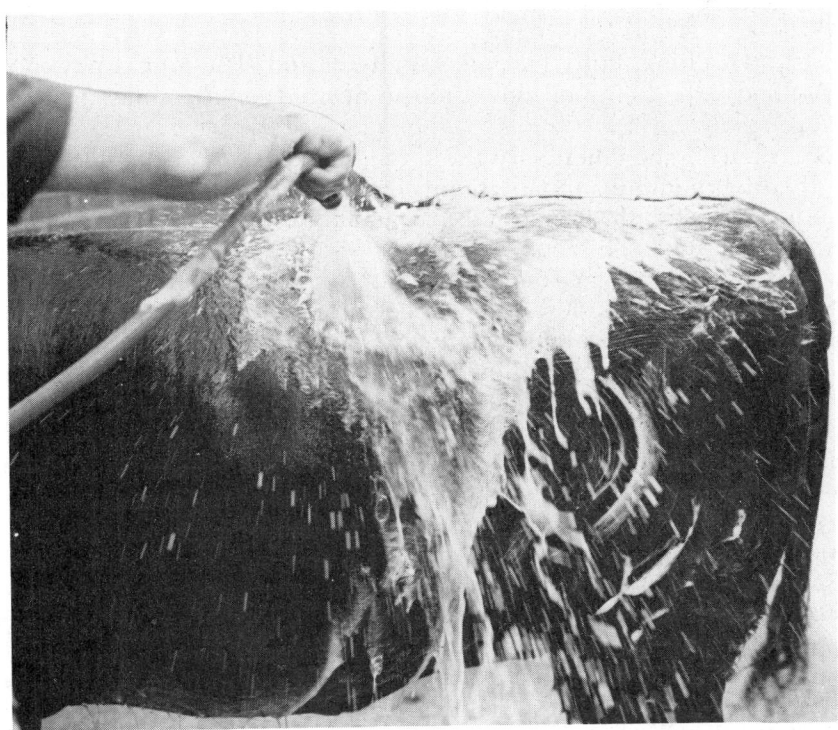

Fig. 14-21. Washing. Frequent washing guarantees a clean coat and stimulates hair growth. Show animals should be washed at least once a month until six weeks before the show and at least once a week thereafter until show day. Use a mild, cold water soap followed by a thorough lathering, rinsing, and brushing to remove all excess soap, as the latter can cause dandruff. (Courtesy, American Angus Association, St. Joseph, Mo.)

In washing the head, avoid getting soap or water in the eyes, ears, nostrils, and mouth. Cattle do not like to have their heads washed. A precaution commonly used in washing the head safely is to wash one side at a time while firmly holding the ear on the side being washed. Death of the animal may result from getting water into the lungs through the nostrils or mouth.

Following washing, the animal should be rinsed off very thoroughly in order to remove all traces of soap from the hair and skin, because soap left on the animal causes dandruff. For animals with light parts, a little bluing added to the last rinse water will improve the results. At this stage (after scraping off the surplus water), some showmen spray the animal lightly with a solution made by adding one tablespoon creosote dip to one gallon of water, because the dip will help the curl. Others may use a vinegar rinse to soften the hair—a quart of vinegar to three gallons of water. Do not rinse after applying the dip or vinegar. After drying to the stage of dampness, the hair is pulled up and the animal may be dried with an electric blower.

The final washing should be done the day before the show, although white areas may need to be washed the day of the show.

Curling

Cattle with long, fluffy hair are usually curled. The hair is pulled up all over the body, then clipped down to the desired length, so as to shape the animal properly.

Experienced showmen vary the method or type of curling according to the individuality of the animal; giving consideration to the length of hair and to whether it is straight or naturally curly, and to the conformation and condition of the animal. The most skillful showmen are able to produce a more natural effect rather than something that is quite artificial in appearance. Also, there are some differences between breeds; for example, in Angus breeding classes it is common practice to curl the hair in the regions of the neck, the forearm, and on the lower part of the thighs—leaving the hair on the other parts of the body uncurled.

Regardless of the type of curl desired, in preparation therefor, the animal is always either (1) washed, rinsed, and scraped—free of surplus water—as described under washing, or (2) dry-brushed, wetted down all over (preferably by means of a hand sprayer, although satisfactory wetting down may be accomplished by dipping a stiff brush into the water or creosote solution and by brushing the hair smoothly against the animal) with water alone or a solution made by adding one tablespoonful of creosote to each gallon of water, and scraped free of any surplus water. Naturally, the latter procedure is followed when complete washing is not necessary or desired, as is frequently the case when animals are curled daily or twice daily when on the show circuit or when curled one to two hours before showing.

In general, the following distinct types of curls are used:

1. *The fluffy curl*—This is really a misnomer, for it is not a curl in the true sense of the word. Instead, the object is to produce a hair effect very much like that of a fluffy teddy bear or a fluffy ball of fur. When the art is plied by a master showman to a beast which possesses a long dense coat of hair, but which does not possess a tight natural curl (for example, most Shorthorns are straight-haired), the fluffy curl produces the most natural effect of all methods. Step by step, the fluffy curl is produced as follows:

a. Either wash or wet down the hair of the animal as previously described.

b. Beginning along the top of the animal, make waves ½ inch to 1 inch apart (and about 2 inches wide) by drawing the back tip of a round curry comb in the direction of the natural lay of the hair in a short, wavy manner. Make waves on all body parts except the head and rear end. Proceed to step c, thus allowing the necessary 2 to 3 minutes for the curl to set before applying step d.

Fig. 14-22. The fluffy curl of a Shorthorn. (Courtesy, American Shorthorn Association, Omaha, Neb.)

c. Comb (with a Scotch comb) and/or brush the hair on the head forward and toward the muzzle on horned animals (generally the head of a polled animal is clipped), outward from the median line at the top of the neck and withers, and outward from the median line of the rear end.

d. In order to tip back the ends of the hairs, while the coat is drying, brush it lightly (with a dry rice brush) opposite the direction of its natural lay (upward on all body parts except the top line; forward on the top line from the tailhead to a point just back of the withers). Such brushing should begin at the tops of the hoofs and extend to the top of the animal.

As the hair becomes fairly dry (the drying time varying with the weather and density of hair), discontinue brushing and draw it (with a Scotch comb) opposite the direction of its natural lay.

After the hair of straight-haired animals is well trained, one may start this step (step d) with the Scotch comb, without first brushing as indicated above. But the procedure as indicated should always be followed with animals that are naturally curly haired.

e. Continue to comb the dry hair outward and upward (with a large, coarse, hard rubber comb).

2. *The wave curl*—This is sometimes referred to as the Hereford curl, because it is especially popular with Hereford showmen. This type of curl may be produced on any animal which possesses a rather tight natural curl, such as characterizes most Herefords, whereas the fluffy curl is produced on straight-haired animals. Step by step, the wave curl is produced in exactly the same manner as previously outlined for the fluffy curl. Thus, the reader is referred to the previous section. The chief difference is the end result obtained from applying step d; namely in brushing and combing the hair in the opposite direction of its natural lay. In straight-haired animals (such as most Shorthorns), the application of this step results in tipping the hair

Fig. 14-23. The wave curl. The round comb may be used in curling the hair of the Hereford to produce a wave curl. This may be made by drawing the back side of a round curry comb downward in a short, wavy manner. The hair is then brushed or combed upward. (In the above picture, a Scotch comb is being used for combing upward.) Likewise, a wave curl may be made by catching the tips of the hair and combing upward with the back side of the round curry comb, followed by brushing. (Courtesy, The American Hereford Association, Kansas City, Mo.)

back—thus producing the fluffy curl; whereas in naturally curly-haired animals (such as most Herefords), the application of this step results in denser, tighter curls—thus producing the wave curl.

In producing the fluffy curl and the wave curl, one further difference may be observed. In the wave curl, step e may either be omitted or applied lightly, depending on how tight a curl is desired.

3. *The parallel curl*—In this type of curl, parallel lines spaced about one inch apart are marked along the body.

Step by step, the parallel curl is produced as follows:

a. Either wash or wet down the hair of the animal as previously described.

b. Mark off (with a "liner") parallel lines about one inch apart along the sides of the body, from in front of the shoulders to the rear edge of the thigh and from the point where the back and the sides blend to the knees and hocks. Either line both sides from rear to front or vice versa.

c. From the tops of the hoofs to the point where the back and the sides blend, draw (with a Scotch comb) the hair upward, and then in order to curl the ends of the hairs, brush upward with a dry brush while it is drying.

d. Comb (with a Scotch comb) the hair on the head forward and toward the muzzle on horned animals, from the median line across the withers, from the center top line to where the back and sides blend, and from the median line of the rear end.

Fig. 14-24. The parallel curl on a Shorthorn.

4. *Diamond curl*—In this the animal is first lined (with a liner) in the shape of diamonds.

Like the parallel curl, this method is easily mastered by the amateur and is valuable in training the hair in early stages. However, like the parallel curl, it leaves marks that still show on the animal after further brushing, combing, and drying, and produces a distinctly artificial effect instead of the much sought natural effect.

"Boning" the Legs

Many judges still try to select heavy boned cattle in breeding classes, despite the fact that studies reveal (1) that bone is of little importance in a selection program, (2) that bone varies little between animals of the same age, and (3) that bone cannot be determined by "eyeballing," because what appears to be bone is really a combination of hair, hide, connective tissue, tendon, and bone (see Chapter 4, under the section entitled, "The Functional Scoring System"). Nevertheless, the showman is exhibiting before the judge; he is not educating him. Hence, he should, to the extent possible, give him what he wants. Since many judges insist on evaluating bone in breeding classes, and since no judge will place an animal down for appearing to have heavy bone (even though he is well aware of the facts stated above), it is good business to accentuate the bone. This is done by "boning the legs."

Animals are boned by rubbing a limited amount (just enough to make the hair stand out when combed) of either saddle soap (first soak the bar in water until it is soft) or butch wax on the cannon bone (between the knee and ankle of the front legs, and between the hock and the ankle of the hind legs), then combing the hair so that it stands straight out from the legs. When done throughout the length of the cannon bone, heaviness of bone is imparted. Of course, not all animals should be made to appear heavy boned. For example, a short-legged animal with a lot of hair on its legs may be made to look taller by clipping much of the hair on the cannon bones.

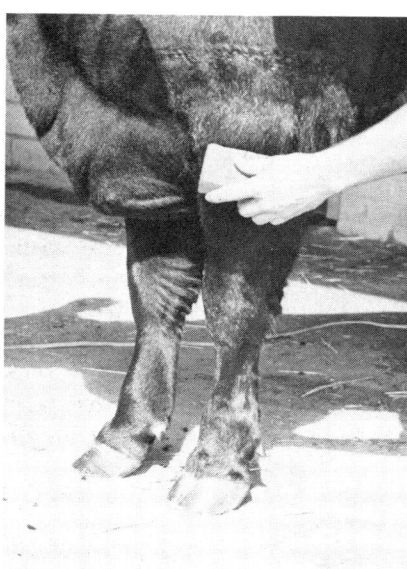 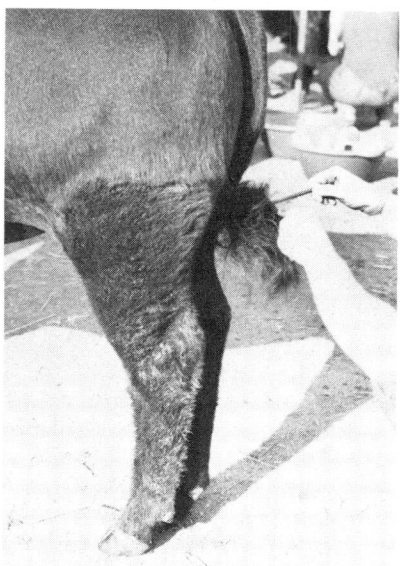

Fig. 14-25. By using a bar of glycerine saddle soap (as shown above), or wax, and a Scotch comb, an animal can be made to appear to be heavy boned. The bar is rubbed on the animal from the hoof to a point about 10 inches into the forearm (left), followed by use of the Scotch comb to pull the hair on the legs. This practice is also applied to the hind legs to show more bone and quarter in the rear of the animal (right). (Courtesy, American Angus Association, St. Joseph, Mo.)

But legs are boned for an added reason! Crooked legs may be made to appear straight(er) through the art of boning. For example, cattle that are sickle-hocked (their hind legs extend too far forward) can be made to look straighter by making the hair just above the dew claw (the lower cannon) stand out while gradually tapering the hair toward the hock, with the hair on the back of the hock flat. Post-legged cattle (cattle that are too straight in the hocks) will look better with the hair on the back of the hock standing out, then gradually tapering the hair so that it will be flat just about the dew claws. Through application of this same technique (by varying the length of the hair to rectify a fault or weakness), both crooked front legs and crooked hind legs can be made to appear straighter.

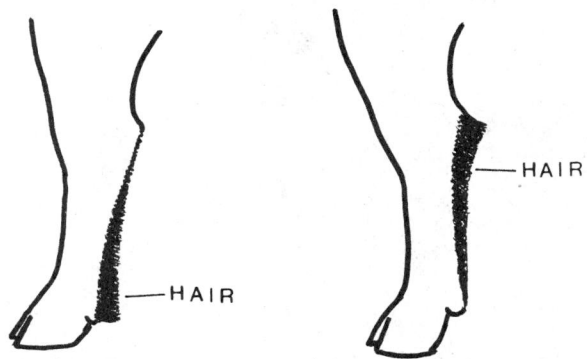

Fig. 14-26. "Boned" hind legs to achieve proper stance. A naturally sickle-hocked animal properly boned to make the leg appear straighter (left); and a naturally post-legged animal properly boned to make the leg appear to have more set (right).

Cleaning and Fluffing the Switch

The switch can usually be made most attractive by the following procedure: the evening before the show, wash the switch thoroughly with soap and warm water. If it is white, it may be made brighter by adding a little bluing to the water. Rinse it in water to which a small amount of alum has been added. (The alum will make the switch more fluffy when combed out.) Then, about an hour or two before the show, (1) fluff the switch by brushing the hair upward a few strands at a time, and (2) "tease" the hair, just like girls do. Hair spray can be used to help hold the switch (see Fig. 14-27).

Oiling

After "boning" the legs and "teasing" the switch, the animal should be oiled. If oiled before boning, oil may get on the legs and make boning more difficult. Also, if applied too early, the oil will tend to dry out and catch dirt.

Either a hair dressing or a homemade mixture of equal parts of glycerine, sweet oil, and rubbing alcohol may be used. After curling, boning, and teasing, apply lightly and evenly (except on any white) with a small hand sprayer, woolen cloth, or turkish towel, then brush the animal.

It is to be remembered, however, that an animal should not be shown when the hair is either shiny or gummy from too much oil. Likewise, an animal should not be shown when the hair is wet.

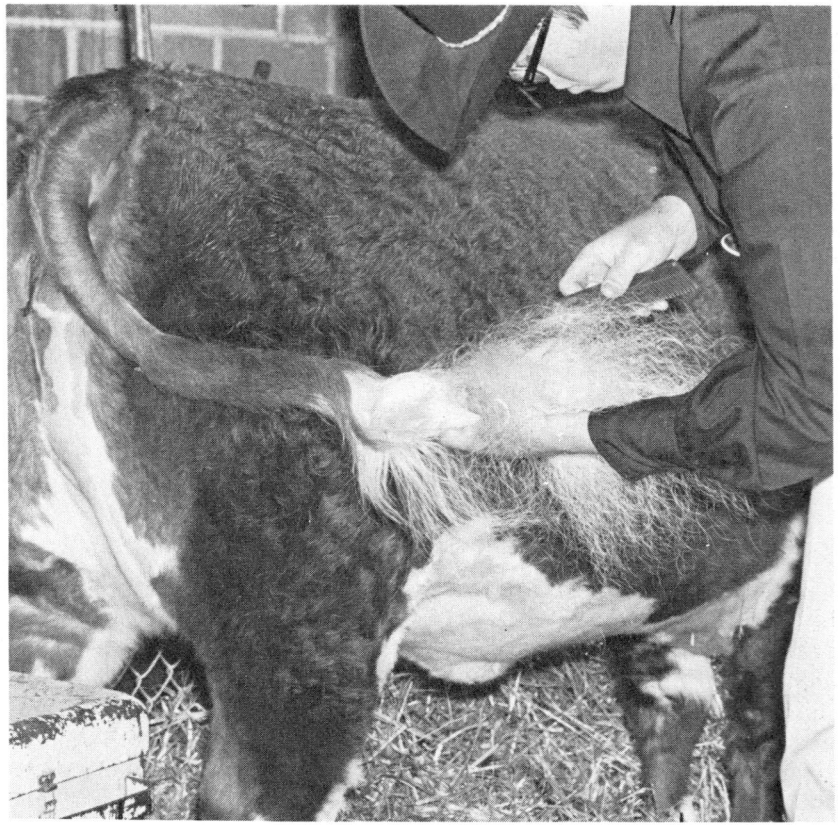

Fig. 14-27. Fluffing the switch. The final step in preparing an animal for the show-ring is ratting the tail, or combing and teasing the hair back on the tail. This helps give the animal the appearance of a thicker quarter. Use a comb or push the hair back on the switch with your fingers. (Courtesy, The American Hereford Association, Kansas City, Mo.)

Other Fitting and Grooming Tips

In addition to the training and grooming points already covered, the following fitting and showing tips are pertinent:

• *Hair*—A good coat of hair can, in the hands of an experienced showman, cover up a multitude of defects. Hair can be grown even in hot weather. Keep animals inside during the day, then wet with a fine mist of water and turn them out at night. The combination of wet hair and cool nights awakens nature's instinct to provide more protection—hair.

• *Short and unruly hair*—If the hair is short and unruly over the body, curling may be limited to the neck, forearm, and lower quarter. This type of grooming is increasing.

• *Fill before showing*—Give the animal a final fill of feed and water about an hour before show time.

MAKING FAIR ENTRIES

Well in advance of the show, the exhibitor should request that the show manager or secretary provide him with a premium list and entry blanks. All rules and regulations of the show should be studied carefully and followed to the letter—including requirements relative to entrance, registration certificates, vaccinations, health certificates, stall fees, exhibitor's and helper's tickets, and other matters pertaining to the show.

Generally, entries must be filed with the show about 30 days in advance of the opening date. Most shows specify that entries be made out on printed forms and in accordance with instructions thereon. The class, age, breed, registry number, and usually the name and registry number of the sire and dam must be given. Entries must be made in all individual and group classes in which it is intended to show, but no entries are made in the championship classes, the first place winners being eligible for the latter.

PROVIDING HEALTH CERTIFICATES

Health certificates, signed by an accredited veterinarian of the state of origin, are always required for show animals. For cattle, most shows specify that this certificate indicate that within 30 days prior to entry and within 90 days prior to the date of exhibiting, the veterinarian has examined each animal offered for entry and has found it free of tuberculosis, brucellosis, and other infectious or contagious diseases. This provides reasonable assurance that diseases are not being spread. In addition, some states require that a special permit issued by the state veterinarian must accompany cattle on their trip home from the show.

SHIPPING TO THE FAIR

Show animals are usually shipped via truck; rail shipments are seldom made anymore. It is important that the following details receive consideration:

1. Schedule the transportation so that the cattle will arrive within the limitations imposed by the show and at least two to three days in advance of the date that they vie for awards.

2. Before using, thoroughly clean and disinfect any public conveyances.

3. Use long, clean, bright straw for bedding in order not to soil the hair or introduce foreign matter into it. It is also a good plan to sand the floor so that cattle will not slip.

4. In transporting by truck, cattle are generally stood crosswise of the truck, with the largest animal near the cab and tied facing to one side. The direction of facing the remaining animals is alternated; the second animal is faced in the opposite direction from the first, and so on. Some prefer to tie the animals so that all of them face the same direction. It takes more space, but they stay cleaner that way. In either case, it is best to tie animals fairly short and near enough together so that they will not lie down.

5. If space is at a premium, place the feed supply, bedding, and show equipment on a deck or platform in the truck, preferably at least 5½ feet above the floor. Allow for air circulation and tying of smaller animals under the deck.

6. When mixed feeds are used, as is usually the case in fitting rations, a supply adequate for the entire trip should be taken along in the truck. This will reduce the hazard of animals going off feed because of feed changes.

7. Limit show cattle to a half feed at the last feeding before loading out and while in transit.

8. In transit, the animals should be handled quietly and should not be allowed to become hot nor to be in a draft.

STALL SPACE, FEEDING, AND MANAGEMENT AT THE FAIR

As soon as the show is reached, the animals should be unloaded and placed in clean stalls that are freshly bedded with clean straw. The cattle should be arranged in order of size so as to make the exhibit as attractive as possible.

While at the show, it is preferable that the cattle receive the same ration to which they were accustomed at home. Usually only a half ration is allowed for the first 24 hours after arrival at the show, and a normal ration is provided thereafter. So that the animals will maintain their appetite, however, it is necessary that they receive exercise while at the show. It is usually best to exercise the animals one-half hour or more in the cool of the evening and morning, when the animals are being led to and from their nightly tie-outs. This also is a convenient time to clean out the stalls.

It is customary for each exhibitor to identify his exhibit by means of a neat and attractive sign, the size of which must be within the limitations imposed by the show. This sign usually gives the name of the breed of cattle and the name and address of the exhibitor.

SHOWING BEEF CATTLE

Expert showmanship cannot be achieved through reading any set of instructions. Each show and each ring will be found to present unusual circumstances. However, there are certain guiding principles which are always adhered to by the most successful cattle showmen. Some of these are:

1. Train the animal long before entering the ring.

2. Have the animal carefully groomed and ready for the parade before the judge.

3. Dress neatly for the occasion.

4. Enter the ring promptly when the class is called.

5. Lead the animal from the left side (walking near the left shoulder), with the halter strap in the right hand.

6. When asked to line up, go quickly but not brashly.

7. When stopped, pose the animal correctly, and so as to minimize faults. Take the strap in the left hand and set the animal up with a leg under each corner. Generally, it is best to set the hind feet before setting the front

Fig. 14-28. Some of the guiding principles observed by the most successful showmen. (Drawing by R. F. Johnson)

feet. Keep the animal's head up and his back straight. A firm pressure near the navel, applied with the show stick, will help keep the weak-backed animal straighter. Animals with high loins can be pinched down with your fingers to straighten their tops. Cow-hocked animals can be made to look straight by pulling on the hocks with your hands.

8. Stroke the animal along the back or under the belly while posing. It will calm him.

9. When the judge handles your animal, react properly. If you feel that your animal may be slightly overdone, or too soft, turn its head away from the judge—thereby imparting firmness to the touch. If you think your animal is too bare, turn his head toward the judge—thereby imparting softness. After the judge handles the animal, comb the hair up where he handled it.

10. Keep one eye on the judge and the other on the animal. Center your

attention entirely on showing the animal. The animal may be under the observation of the judge at a time when you least suspect it.

11. Let the animal stand "at ease" if you are in a big class and the judge is working at the other end of the ring. Calm him by scratching him with the show stick.

12. Never stand so that you block the judge's view; he's interested in seeing your animal—not you.

13. If you find that you are hemmed in and that the judge cannot see your animal, move to another location of vantage, unless, of course, the judge has asked you to hold your position.

14. Keep calm and collected. Remember that the nervous showman creates an unfavorable impression.

15. Work in close partnership with the animal.

16. Be courteous and respect the rights of other exhibitors.

17. Do not enter into conversation with the judge. Speak to him only when he asks you a question; and never question the judge's placings.

18. Be a good sport. *Win without bragging and lose without squealing.*

Fig. 14-29. Posing the animal. A good showman never gets excited, but is constantly doing three things: (1) watching the judge and showing the animal; (2) combing hair into place; and (3) getting the animal and feet into position. (Courtesy, The American Hereford Association, Kansas City, Mo.)

JUDGING BEEF CATTLE

Livestock judging is the art of visual appraisal, or the making of a subjective evaluation, of an animal. Cattlemen use it every day to select herd sires and replacement females for their breeding programs and to determine when animals are ready for market.

In addition to individual merit, the word judging implies the comparative appraisal or placing of several animals. In most judging contests, four animals are used in each class; and they are numbered 1, 2, 3, and 4, left to right as viewed from the rear. In livestock shows, a great number of animals may be ranked, or placed, in each class.

Judging beef cattle is an art, the rudiments of which must be obtained through patient study and long practice. The master breeders throughout the years have been competent livestock judges.

The Judge

The judge(s) is the person chosen by the show management to determine the relative merits of the animals entered in the show. Judging is hard work and a great responsibility. Not only does the judge pick the winners, but he leads or misleads many people. For better or worse, he may be the cause of changing breeding programs and affecting the traits of the entire breed.

The essential qualifications that a good judge of beef cattle must possess, and the recommended procedure to follow in selecting or judging are as follows:

1. *Knowledge of the parts of cattle and the relative importance of each*—This consists in mastering the language that describes and locates the

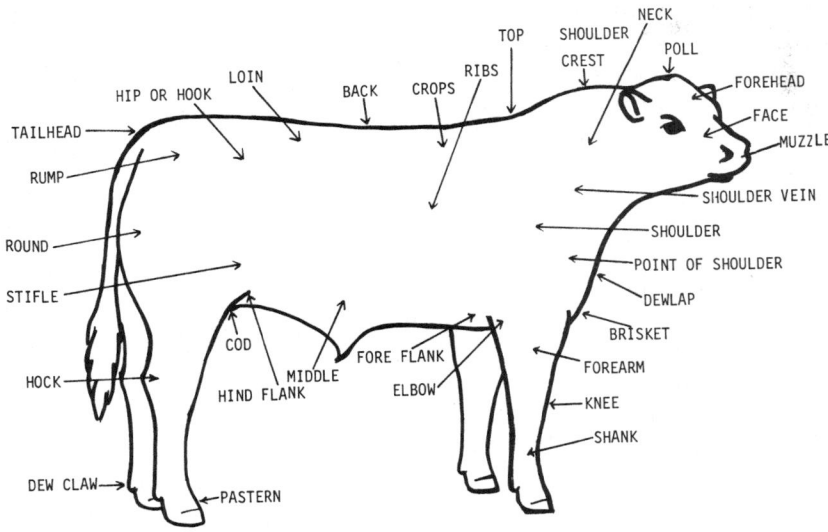

Fig. 14-30. Parts of a steer. The first step in preparation for judging beef cattle consists of mastering the language that describes and locates the different parts of the animal.

different parts of cattle (see Fig. 14-30). In addition, it is necessary to know which of these parts are of major importance in terms of form relating to function. In a slaughter animal, the latter necessitates knowledge of cutability and quality grades. In breeding animals, the Functional Scoring System (see Chapter 4 for pictures and description) may serve as a useful guide.

2. *A clearly defined ideal or standard of perfection*—The successful cattle judge must know for what he is looking; that is, he must have in mind an ideal or standard of perfection based on a combination of (a) the efficient performance of the animal from the standpoint of the producer, and (b) the desirable carcass characteristics of market animals as determined by the consumer. The major factors to consider are:

a. *Beef type*—This refers to the sum total of those economic traits included in the makeup of a commonly accepted ideal bull, female, or steer.

b. *Conformation*—This refers to the parts which make up the animal structure. Each part should be in proper form and proportion to lend balance in body conformation. Faulty conformation affects animal structure, form, and utility.

c. *Substance*—This refers to muscle and bone. Beef cattle need muscling to yield high cutability carcasses, and they need well-placed bone for skeletal soundness.

d. *Constitution*—This refers to capacity of chest and middle. Beef cattle should have ample capacity to house the digestive, respiratory, and reproductive systems.

e. *Size*—This refers to the skeletal frame and weight for age. The frame should be large enough to make possible rapid growth and efficiency in production and reproduction.

f. *Quality*—This is indicated by hide, hair, bone, and general bloom of the animal. Quality is not of significant importance, but it does add to animal appeal and attractiveness.

g. *Breed Character*—This relates to the features of an animal which identify it with a particular breed.

h. *Sex Character*—This refers to masculinity and femininity in breeding animals. Bulls should be strong, burly, and well developed in male sex features, with much importance placed on testicle development and position. Females should be feminine, indicating good mothering and milking ability.

i. *Disposition and temperament*—This is important in handling cattle on the range or in the feedlot. Good disposition is of economic value in beef production.

All of the above factors are important. The trained judge can evaluate them in a matter of seconds, using an objectively developed mental index.

It must be recognized, however, that the perfect specimen has never been produced. Each animal possesses one or several faults. In appraising an individual animal, therefore, its good points and its faults must be recognized, weighed, and evaluated in terms of an ideal. In comparative judging—that is, in judging a class of animals—the good points and the faults

Modern Type

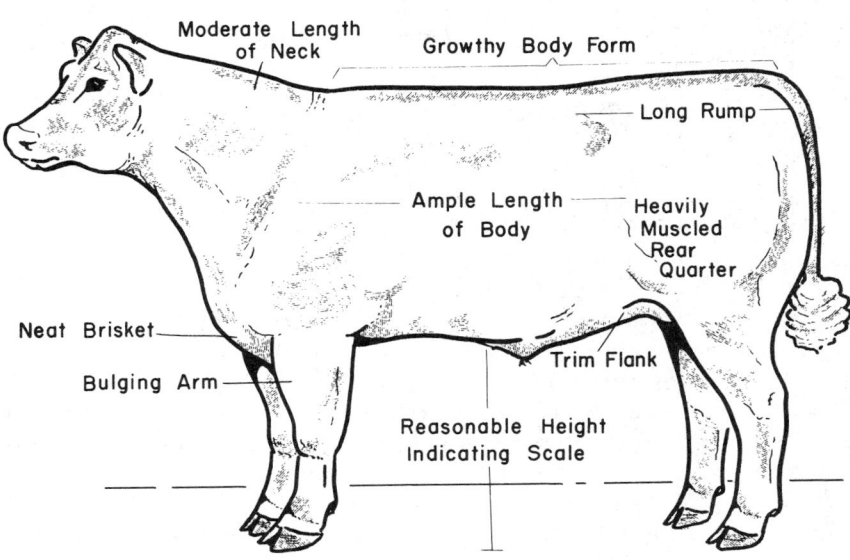

Moderate Length of Neck

Growthy Body Form

Long Rump

Ample Length of Body

Heavily Muscled Rear Quarter

Neat Brisket

Bulging Arm

Trim Flank

Reasonable Height Indicating Scale

Faults

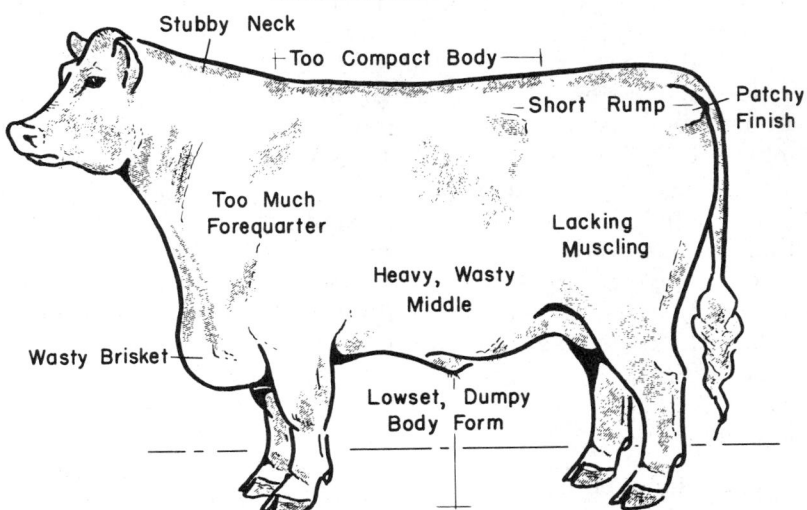

Stubby Neck

Too Compact Body

Short Rump

Patchy Finish

Too Much Forequarter

Lacking Muscling

Heavy, Wasty Middle

Wasty Brisket

Lowset, Dumpy Body Form

Fig. 14-31. Ideal beef type vs common faults. After mastering the language that describes and locates the different parts of an animal, the next requisite in judging is to have clearly in mind a standard or ideal. In brief, the successful beef cattle judge must know what he is looking for, and be able to recognize and appraise the common faults. (Drawing by R. F. Johnson)

of each animal must be compared with the good points and the faults of every other animal in the class. In no other manner can they be ranked.

In addition to recognizing the strong and weak points in an animal, it is necessary that the successful judge recognize the degree to which the given points are good or bad. A sound evaluation of this kind requires patient study and long experience.

3. *Keen observation and sound judgment*—The good judge possesses the ability to observe both good conformation and defects, and to weigh and evaluate the relative importance of the various good and bad features.

4. *Honesty and courage*—The good judge of any class of livestock must possess honesty and courage, whether it be in making a show-ring placing or conducting a breeding and marketing program. For example, it often requires considerable courage to place a class of cattle without regard to: (a) placings in previous shows; (b) ownership; and (c) public applause. It may take even greater courage and honesty with oneself to discard from the herd a costly animal whose progeny has failed to measure up.

5. *Tact*—In discussing either (a) a show-ring class, or (b) animals on a cattleman's farm or ranch, it is important that the judge be tactful. The owner is likely to resent any remarks that imply that his animal(s) is inferior.

Having acquired the above referred to knowledge, long hours must be spent in patient study and practice in comparing animals. Even this will not make expert and proficient judges in all instances, for there may be a grain of truth in the statement that "the best judges are born and not made." Nevertheless, training in judging and selecting cattle is effective when directed by a competent instructor or experienced cattleman.

In the major shows, usually there is a separate judge(s) for each breed; in small shows, one judge may place all breeds. Each show must select its own judge or judges. The judge is usually selected because of his expertise. He knows modern beef type, and in breeding classes he is familiar with the breed type, or character, of the particular breed being judged. Breed registry associations usually conduct judging clinics to train judges and provide literature setting forth their breed standards, generally detailing the relative importance of each trait and listing faults and disqualifications. Some breed registry associations compile a certified list of judges with proven ability and make their names available to cattle show officials, as a recommended list from which to choose a judge(s).

Shows vary in the compensation of judges; generally speaking, the bigger the show the better the compensation. Most shows pay a professional fee plus expenses.

Placing the Class

When placing a class, the judge picks for his top animal the one which most nearly approaches his ideal, all traits considered. Then, he ranks the rest of the class on the same basis, or he places as many animals as there are premium monies or ribbons.

Of course, opinions differ. As a result, two equally competent judges may come up with different placings on the same animals exhibited in successive shows. The more faults animals have, the more judges' opinions will differ. This is because they evaluate faults differently.

When officiating, most judges follow a sequence. Actually, it makes little difference as to the order of the views in inspecting cattle; what is important is that the same procedure be followed each time. With some slight variations, most judges proceed about as follows:

1. The judge starts working a class after the ringmaster has lined them up, side by side, and advised him that it is ready.

2. The animals in a class are examined as follows:

a. Observed at a distance, in order to obtain a panoramic view. Also, the animals are viewed from all directions—(1) rear view, (2) front view, and (3) side view, thereby avoiding overlooking anything and making it easier to retain the observations that are made.

b. Handled (with the hands), in order to determine the degree and quality of finish.

c. Moved, in order to study their action—to see how they walk. The judge usually stands in the center of the ring, and the ringmaster asks that the exhibitors circle clockwise of the ring, thereby avoiding the necessity of the showmen who are leading the animals (from the left side) from being between the animal and the judge.

3. The judge evolves with his ranking of the class, based on the nearness with which each approaches the breed standard of perfection, along with each animal's merits and demerits.

4. The judge signals the animals to line up, in the order of his placings.

5. The judge may walk the class "Indian file" according to his placings, to double check his decision and let the crowd see his ranking.

6. The judge may give oral reasons on the class, defending his placings.

7. The ringmaster hands out the ribbons.

8. The judge goes over to the judge's table and signs the judge's books, in which the clerk has recorded his placings.

9. The ringmaster announces the next class. Then, class by class, this same procedure is followed until the class winners in each the male and female classes are judged, followed by the selection of champions.

Composite Judging System

The Composite Judging System is now being introduced in beef cattle shows. Because of its several advantages, it seems destined to increase. It is designed to reduce the inconsistencies inherent in traditional show-ring judging. It practically eliminates the personal prejudice, bias, and abrupt changes in type sometimes found in one-man judging; and it alleviates the dominant personality of a committee placing.

Three judges are used in the Composite Judging System. Ideally, each should represent a different segment of the industry, i.e., breeder, commercial producer, and university personnel. At the beginning of each class, judges draw for their designated title in that class—*Judge A, Judge B,* and *Referee.* Then, regardless of title, each judge independently places each class, with no communication between them.

Show-ring procedure is much the same as in traditional one-man judging. The animals are lined up by age; then observed, walked, and judged. The Referee instructs the Ringman on when and where to walk the cattle and line up. The cattle are always lined up according to age until the final composite placing is announced; judges are not permitted to make pair comparisons, side by side.

Each judge selects enough top animals to fill the monied or ribbon placings. Thus, if there are 8 places, each judge selects his 8 top animals for further consideration. These selections are written on the judge's card, which is turned in. Animals not picked by any one of the 3 judges may either be excused from the ring or left standing at one end. Of course, in small classes where all animals are in the money, the latter procedure is not necessary.

Next, each judge places, or ranks, the "top cut" animals on his scorecard, and turns his card into the Ringman. Then the composite placing is calculated and posted by the Ring Secretary as the final placing.

In compiling the composite placing, only the highest placing animal on each Judge A's and Judge B's card is considered. If these 2 animals are one and the same, that animal wins first place. If they are different animals, the Referee's card is consulted to see which of these 2 animals he ranked higher. The higher placing animal of the 2 on the Referee's card then goes into the final placing. Once an animal is ranked in the final placing, his number is crossed off of Judge A's and Judge B's cards.

This same procedure is repeated until all monied placings have been filled. Table 14-2, Composite Compilation Sheet, shows the placings on one class of cattle and illustrates how the Composite Judging System works.

TABLE 14-2

COMPOSITE COMPILATION SHEET

Placing	Judge A	Judge B	Referee	Final
1st	106	106	104	106
2nd	111	110	106	110
3rd	104	104	110	104
4th	110	111	111	111
5th	108	108	109	108
6th	105	105	108	105
7th	107	109	105	109
8th	109	107	107	107

BCS, Chapter 14, MSp 1153.

As shown in Table 14-2, Judges A and B agree that entry 106 is first, so he wins the class. For second place, Judge A has 111 and Judge B has 110. They disagree, so the Referee's card is consulted, with consideration given only to 111 and 110. The Referee placed 110 over 111, so 110 places second in the class.

Using the highest placing animals on each card that have not been placed, consideration for third place is between 111 from Judge A's card and

104 from Judge B's card. Since they differ, the Referee's card is again consulted; and because he placed 104 higher, 104 goes in third place.

For fourth place, 111 is the highest unplaced animal on both Judge A's and Judge B's cards, so 111 wins fourth place. The fifth and sixth place animals are the same on both A's and B's cards, so they win fifth and sixth places, respectively, regardless of how the Referee placed them. The two judges differ on seventh and eighth places, so the final placing is resolved by the same procedure previously followed—that of using the Referee's placings.

In calculating the final placing, two basic principles are followed:

1. Only the top animal on each A's and B's card is considered.

2. The Referee's card is considered only when disagreement exists between Judges A and B; then only those two animals are considered.

After the final composite placing is completed, the Ringman lines up the cattle in the order of their final placing. Then the results are announced for exhibitor and spectator information.

The Composite Judging System does not require any longer than a committee or two-judge system, once it is understood. There is no possibility of a deadlock. If the judges disagree, the Referee's card can be used to make the final placing.

The Composite Judging System is a foolproof system of combining the judgment of three judges.

HOW TO DETERMINE THE AGE OF CATTLE BY THE TEETH

Many shows are mouthing cattle, as a means of assuring "honest" show classification according to age. Thus, it is important that the showman know how to determine the age of cattle by the teeth. Also, because the life span of cattle is relatively short, the age of cattle is of practical importance to the breeder, the seller, and the buyer.

The approximate age of cattle can be determined by the teeth as described and illustrated herewith. There is nothing mysterious about this procedure. It is simply a matter of noting the time of appearance and the degree of wear of the temporary and of the permanent teeth. The temporary or milk teeth are readily distinguished from the permanent ones by their smaller size and whiter color.

It should be realized, however, that theoretical knowledge is not sufficient and that anyone who would become proficient must also have practical experience. The best way to learn how to recognize the age of cattle is to examine the teeth of individuals of known ages.

Of course, age determination of cattle is an art, and not a science. "Mouthing cattle" is subject to human judgment, much like a human medical diagnosis based on observation. Also, it should be noted that variation in an animal's dental development can result from several causes, but primarily from nutritional variation or stress from sickness and environmental conditions. It is affected little by genetics or breeds.

At maturity cattle have 32 teeth, of which 8 are incisors in the lower jaw. The 2 central incisors are known as pinchers; the next 2 are called first in-

termediates; the third pair is called second intermediates or laterals; and the
outer pair is known as the corners. There are no upper incisor teeth; only the
thick, hard dental pad.

Table 14-3 illustrates and describes how to determine the age of cattle
by the teeth.

TABLE 14-3

HANDY GUIDE TO DETERMINING THE AGE OF CATTLE
BY THE TEETH

Drawing of Teeth	Age of Animal	Description of Teeth
Fig. 14-32	At birth to 1 month	Two or more of the temporary incisor teeth present. Within first month, entire 8 temporary incisors appear.
Fig 14-33	2 years	As a long-yearling, the central pair of temporary incisor teeth or pinchers is replaced by the permanent pinchers. At 2 years, the central permanent incisors attain full development.
Fig. 14-34	2½ years	Permanent first intermediates, one on each side of the pinchers, are cut. Usually these are fully developed at 3 years.
Fig. 14-35	3½ years	The second intermediates or laterals are cut. They are on a level with the first intermediates and begin to wear at 4 years.
Fig. 14-36	4½ years	The corner teeth are replaced. At 5 years the animal usually has the full complement of incisors with the corners fully developed.
Fig. 14-37	5 or 6 years	The permanent pinchers are leveled, both pairs of intermediates are partially leveled, and the corner incisors show wear.
Fig. 14-38	7 to 10 years	At 7 or 8 years the pinchers show noticeable wear; at 8 or 9 years the middle pairs show noticeable wear; and at 10 years, the corner teeth show noticeable wear.
Fig. 14-39	12 years	After the animal passes the 6th year, the arch gradually loses its rounded contour and becomes nearly straight by the 12th year. In the meantime, the teeth gradually become triangular in shape, distinctly separated, and show progressive wearing to stubs. These conditions become more marked with increasing age.

AFTER THE FAIR IS OVER

Most shows have regulations requiring that all exhibits remain on the grounds until a specified time, after which signed releases must be secured from the superintendent of the show. Because most exhibitors are anxious to travel when the show is over and there is considerable confusion at this time, it is usually advisable to load all equipment, leftover feed, and other articles before the release of animals is secured. Then all that remains to be done is to load out the animals.

Upon returning to the farm or ranch, it is usually good policy to isolate the show herd for a period of three weeks. This procedure reduces the possibility of spreading diseases or parasites to the balance of the herd.

It is important that young stock to be developed for show purposes the following year continue to receive an adequate, though lighter, grain ration.

Where the herd is being exhibited on a circuit, the herdsman must use great care in keeping the animals in show condition at all times. The peak condition should be reached at the strongest show. In order to be successful, showing on the circuit requires great skill on the part of the caretaker, especially from the standpoint of feeding and exercising the cattle.

QUESTIONS FOR STUDY AND DISCUSSION

1. Under what circumstances would you recommend that each a purebred, a commercial, and a 4-H club member (a) should show, and (b) should not show beef cattle?

2. Take and defend either the affirmative or the negative position of each of the following statements:

    a. Fitting and showing does not harm cattle.
    b. Livestock shows have been a powerful force in beef cattle improvement.
    c. Too much money is spent on livestock shows.
    d. Unless all cattle are fitted, groomed, and shown to the same degree of perfection, show-ring winnings are not indicative of the comparative quality of animals.

3. In recent years, most major shows have augmented visual appraisal judging with performance and carcass information. When conducted in this manner, will cattle shows make for genetic improvement?

4. How would you improve upon the typical show classification for breeding beef cattle given in Table 14-1 of this chapter?

5. How would you improve upon the steer classification (both the on-foot and the carcass divisions) of the Chicago International as presented in this chapter?

6. Wherein does feeding cattle for the show differ from feeding cattle in a commercial cattle feedlot?

7. Do you favor (a) curling the hair of beef cattle, or (b) showing them smooth? Justify your answer.

8. "Boning" the legs is a deceptive practice, designed for the purposes of (a) making the cannon bone appear larger, and/or (b) improving stance. Is this good or bad? Justify your answer.

9. Are livestock shows making for more double muscling? Justify your opinion.

10. Some livestock shows have an "expert" mouth cattle, in an attempt to keep ages "honest." How accurate is such an age determination?

SELECTED REFERENCES

Title of Publication	Author(s)	Publisher
Beef Cattle, Sixth Edition	A. L. Neumann R. R. Snapp	John Wiley & Sons, Inc., New York, N.Y., 1969
Beef Production	R. V. Diggins C. E. Bundy	Prentice-Hall, Inc., Englewood Cliffs, N.J., 1958
Beef Production in the South	S. H. Fowler	The Interstate Printers & Publishers, Inc., Danville, Ill., 1969
Selecting, Fitting and Showing Beef Cattle, Sixth Edition	J. E. Nordby H. E. Lattig	The Interstate Printers & Publishers, Inc., Danville, Ill., 1956
Stockman's Handbook, The, Fourth Edition	M. E. Ensminger	The Interstate Printers & Publishers, Inc., Danville, Ill., 1967

CHAPTER 15

MARKETING AND SLAUGHTERING CATTLE AND CALVES

Contents   **Page**

Marketing is an integral part of modern cattle production. It is the end of the line; that part which gives point and purpose to all that has gone before. The importance of cattle marketing is further attested to by the following facts:

1. A total of 119.1 million head of cattle and calves were marketed in 1973, in comparison with 99.7 million hogs and 18.1 million sheep and lambs. (See Fig. 15-2, which also gives 1971 and 1972 figures.)

On the average, cattle and calves had a marketing turnover rate of 2.02

Fig. 15-1. Two Egyptian cattlemen taking an ox to market; from Bas Relief found in Sakara in the tomb of King Ephto Stoptep. At first, meat animals were bartered for articles made by craftsmen. Eventually, bartering gave way to cash sales as coined money began to circulate. (Courtesy, The Bettmann Archive)

VOLUME OF LIVESTOCK
MARKETED BY SPECIES

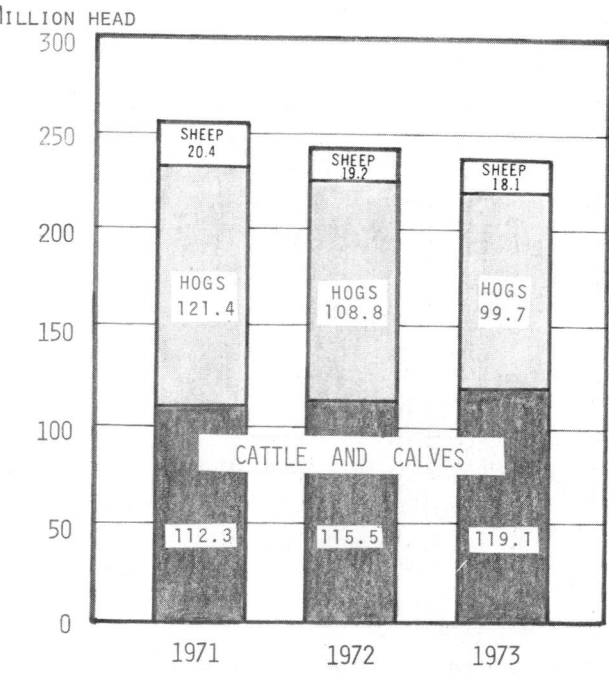

Fig. 15-2. Volume of livestock marketed by species, 1971-73. (Courtesy, Livestock Merchandising Institute, Kansas City, Mo.)

times in 1973, which means that they were marketed twice, or changed ownership two times, during the course of the year.

2. Cattle and calf transactions in 1973 had a total value of $34.6 billion, in comparison with $7.7 billion for hogs. (See Fig. 15-3, which also gives 1971 and 1972 figures.)

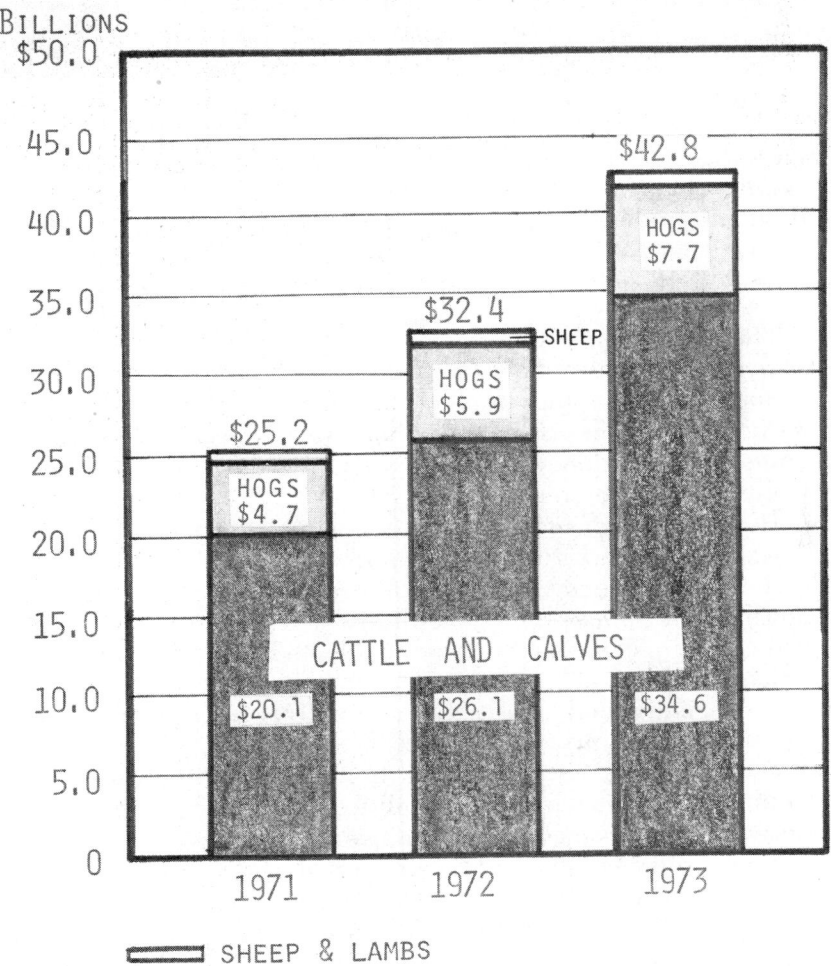

Fig. 15-3. Value of livestock marketed by species, 1971-73. (Courtesy, Livestock Merchandising Institute, Kansas City, Mo.)

3. In 1972, United States farmers and ranchers received 30 percent of the total cash income which they derived from livestock and livestock products *plus* crops from marketing cattle and calves.[1]

4. In 1972, U.S. farmers and ranchers received 51 percent of the cash income which they derived from livestock and livestock products (exclusive of crops) from marketing cattle and calves.[2]

5. In 1973, 33.7 million cattle and 2.2 million calves were slaughtered in the United States.[3]

6. Livestock markets establish values of all animals, including those on the farm or ranch. On January 1, 1974, there were 127,540,000 head of cattle in the United States, with an aggregate value of $40.9 billion or $321 per head.[4]

It is important, therefore, that the cattleman know and follow good marketing practices.

MARKET PRICES

Market prices of cattle are determined by the following four factors:

1. *Supply of cattle*—This refers to the number of cattle that are available and which producers are willing to place on the market at the prices prevailing during a particular period of time.

2. *Consumer demand*—The demand for beef is influenced by five factors:

   a. The population within the market area.
   b. Income of consumers.
   c. Substitution of other products for beef. For example, if pork chops are selling at $.75 per pound and T-bone steaks are $2.50 per pound, the consumer has a tendency to buy pork chops rather than beef steak.
   d. Tastes and preferences.
   e. Habits, culture, and other environmental characteristics.

3. *Variation in marketing costs*—Marketing costs vary. They are affected by the market agency, transportation costs, etc.

4. *Government policy*—The government is deeply involved in policies favoring consumers. Thus, beef prices are affected by beef imports, government purchases of beef, etc.

In terms of individual sales, other factors that affect cattle prices are: size of lot, uniformity, quality, and record of desirability. The latter point refers to such things as feeder animals from a production tested herd, or finished animals with known high cutability.

In a private enterprise economy, the beef business competes with other enterprises for resources—for land, labor, capital, and management. Thus, if beef cattle are more profitable than alternate enterprises, resources will be allocated to cattle. However, if more money can be made raising vegetables, the resources may be switched to vegetable crops.

---

[1]*Agricultural Statistics 1973*, USDA, p. 468.
[2]Ibid.
[3]Data from *Livestock Slaughter*, USDA, Mt. An. 1-2-1 (74), April, 1974, p. 3.
[4]*Cattle*, USDA, LvGb 1 (2-74), p. 3.

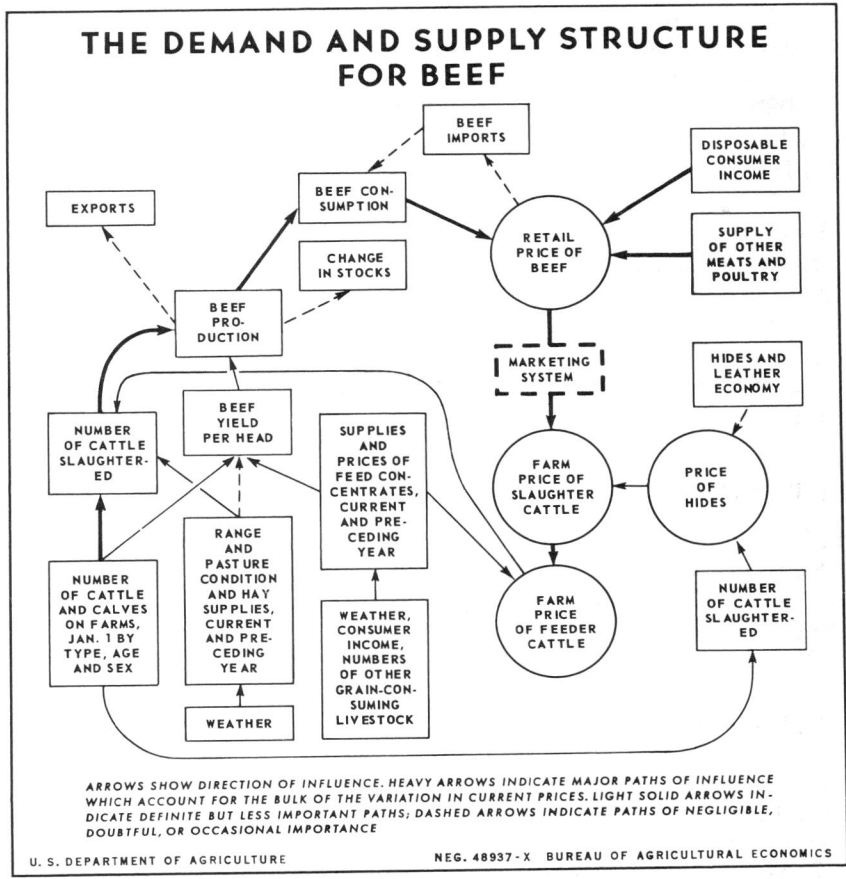

THE DEMAND AND SUPPLY STRUCTURE FOR BEEF

ARROWS SHOW DIRECTION OF INFLUENCE. HEAVY ARROWS INDICATE MAJOR PATHS OF INFLUENCE WHICH ACCOUNT FOR THE BULK OF THE VARIATION IN CURRENT PRICES. LIGHT SOLID ARROWS INDICATE DEFINITE BUT LESS IMPORTANT PATHS; DASHED ARROWS INDICATE PATHS OF NEGLIGIBLE, DOUBTFUL, OR OCCASIONAL IMPORTANCE

U. S. DEPARTMENT OF AGRICULTURE   NEG. 48937-X BUREAU OF AGRICULTURAL ECONOMICS

Fig. 15-4. Cattle prices are determined by two basic groups of factors—supply and demand. (Courtesy, USDA)

When cattle prices are depressed, the following psychological, and less tangible, factors come into play and tend to keep prices down.

1. *Pessimism*—Cattlemen tend to be either incurable optimists or unshakable pessimists. When the market price is up, bidding for cattle is lively. When the market goes down, buyers are as scarce as "hen's teeth." This tends to exaggerate price fluctuations and extremes.

2. *Banker's mood*—Bankers are prone to wide fluctuations in mood—between mild pessimism (never optimism) and extreme pessimism. When prices are down, they get very nervous.

3. *Low equity levels*—Commercial cattle feeders normally operate with only 30 to 40 percent equity. Thus, a 35 percent drop in slaughter cattle prices may wipe out or leave equity levels perilously low (which bankers view dimly) and limit the feeder's ability to restock his feedlot.

4. *The commercial feeder lacks the stability of the farmer-feeder*—The farmer-feeder, who produces most of his feed, is a tremendous stabilizing

influence. He is able and willing to accept market risks. If the price is down, he may forego buying a new car and the wife may go without a new dress. But he continues to feed cattle. It's another story with his counterpart—the commercial feeder!

5. *Retail beef prices don't usually fall proportionally*—Packers and retailers may maintain or increase their margins during a depressed cattle market. As a result, there may be consumer resistance at the meat counter, further stymieing demand.

METHODS OF MARKETING CATTLE

The cattleman is confronted with the perplexing problem of determining where and how to market his animals. Usually there is a choice of market outlets, and the one selected often varies between classes and grades of cattle and among sections of the country. Thus, the method of marketing usually differs between slaughter and feeder cattle, and both of these differ from the marketing of purebreds.

Most market cattle are sold through the following channels: (1) public stockyards; (2) livestock auction markets; or (3) country selling (direct)—including sales direct to packers, local dealers, and farmers.

In 1972, the marketing channels for slaughter cattle and calves were:[5]

	Cattle	Calves
	(%)	(%)
Direct, Country Sales, etc.	72.2	31.6
Auctions	14.6	60.7
Public stockyards	13.2	7.7

More cattle (72.2%) were sold direct; more calves (60.7%) were sold through auctions.

Public Stockyards

Public stockyards (also referred to as terminal public markets, central public markets, or terminal markets) are livestock trading centers, where livestock is consigned to commission firms for selling. Two or more commission firms must operate at a public stockyards. The stockyards company owns and maintains the physical facilities, such as yards, alleys, scales, loading and unloading docks, office buildings, and facilities for feeding and watering livestock. Individuals, partnerships, corporations, and cooperatives operate as commission agencies at public stockyards.

The first public stockyards was established at Chicago in 1865. Most of the larger public stockyards operating today were organized in the latter half of the 19th Century. With the development of the Corn Belt and western range cattle industry, the markets established at various points on the Great

[5]*Packers and Stockyards Resume*, USDA, Vol. XI, No. 13, pp. 15 and 17.

Lakes and along the Mississippi and Missouri Rivers became important. It is logical that these should have become leading cattle markets because of their proximity to both producers and consumers.

Up through World War I, the majority of slaughter livestock in the United States was sold through public stockyards directly by farmers or by local buyers shipping to them. Since then, the importance of these markets has declined in relation to other outlets. In 1923, Federally inspected slaughterers purchased 90% of their cattle in public stockyards; in 1951, 73%. During this same period, the percentage of Federally inspected slaughter of calves sold through public stockyards declined from 86% to 56%. By 1972, the percentage of Federally inspected slaughter cattle bought at public stockyards dropped to about 13.2% and calves to 7.7%.[6] But these figures do not tell the whole story; when the total slaughter at all commercial slaughtering establishments is considered, the percentage bought at public stockyards is somewhat less. The latter situation is so because nonfederally inspected slaughterers tend to buy larger proportions of their livestock at markets other than public stockyards.

Fig. 15-5. Terminal market cattle buying scene. (Courtesy, *Corn Belt Farm Dailies*, Chicago, Ill.)

---

[6]Ibid., p. 8.

Although public stockyards vary from year to year in total receipts, Table 15-1 shows the largest public stockyards in cattle receipts. It is noteworthy that Omaha leads.

TABLE 15-1

CATTLE RECEIPTS (EXCLUDING CALVES) OF 10 LEADING
PUBLIC STOCKYARDS, BY RANK, 1972[1]

Market	1972
Omaha, Neb.	962,289
So. St. Paul, Minn.	829,057
Oklahoma City, Okla.	791,312
Sioux City, Iowa	680,534
Amarillo, Tex.	544,062
Springfield, Mo.	444,651
LaJunta, Colo.	391,949
Kansas City, Mo.	381,222
Sioux Falls, S.D.	370,947
St. Joseph, Mo.	346,938
Total of 46 public markets	11,334,272

[1]*Livestock and Meat Statistics*, USDA Statistical Bull. 522, p. 51.

At public stockyards, cattle under approximately 400 pounds in weight are designated as calves. As would be expected, many of these animals are of dairy breeding, especially the surplus bull calves that are not needed for breeding purposes. Of the remainder, a considerable number are culled out from beef herds because of undesirable type or breeding from the standpoint of future development. It can be expected, therefore, that the leading calf markets would not coincide with the leading cattle markets.

Table 15-2 shows the 10 leading public stockyards in calf receipts in the United States. Milwaukee and South St. Paul, the two largest calf markets, are primarily outlets for veal calves from neighboring dairy areas.

TABLE 15-2

CALF RECEIPTS OF 10 LEADING PUBLIC STOCKYARDS, BY RANK, 1972[1]

Market	1972
Milwaukee, Wisc.	130,387
So. St. Paul, Minn.	125,826
Oklahoma City, Okla.	118,725
Fort Worth, Tex.	94,105
Kansas City, Mo.	59,489
Louisville, Ky.	37,087
St. Joseph, Mo.	29,168
Springfield, Mo.	27,075
Sioux City, Iowa	25,959
Memphis, Tenn.	22,589
Total of 46 markets	834,153

[1]*Livestock and Meat Statistics*, USDA Statistical Bull. No. 522, p. 53.

It is natural that the leading feeder cattle markets should be conveniently located between the cow-calf areas, and the feeding areas. Table 15-3
gives the 10 leading public stockyards in feeder cattle out-shipments, by
rank. As noted, Oklahoma City has a commanding lead, due primarily to its
proximity to the High Plains feeding area.

TABLE 15-3

FEEDER CATTLE SHIPPED FROM LEADING FEDERALLY
INSPECTED PUBLIC STOCKYARDS, BY RANK, 1972[1]

Market	1972
Oklahoma City, Okla.	565,794
Kansas City, Mo.	258,437
Sioux City, Iowa	175,039
So. St. Paul, Minn.	148,500
So. St. Joseph, Mo.	126,399
Omaha, Neb.	126,383
E. St. Louis, Ill.	84,328
Sioux Falls, S.D.	75,169

[1]*Livestock and Meat Statistics*, USDA Statistical Bull. No. 522, pp. 60-66.

In recent years, there has been an increasing tendency to market feeder
cattle direct, without passing them through a public stockyards. Also, an increasing number of producers are arranging to have their feeders custom fed.

Cattlemen need to be acquainted with livestock marketing costs. Although public stockyards charges vary slightly (1) according to size of consignment, and (2) between markets, out-of-pocket costs to producers average
about $3.15 per head—for commission, yardage, and feed,[7] with these
charges on a per head basis rather than a percentage of the proceeds.

Livestock Auction Markets

*Livestock auctions (also referred to as sales barns, livestock auction
agencies, community sales, and community auctions) are trading centers
where animals are sold by public bidding to the buyer who offers the highest price per hundredweight or per head.* Auctions may be owned by individuals, partnerships, corporations, or cooperative associations.

This method of selling livestock in this country is very old, apparently
being copied from Great Britain where auction sales date back many centuries.

The auction method of selling was used in many of the colonies as a
means of disposing of property, imported goods, secondhand household furnishings, farm utensils, and animals.

According to available records, the first public livestock auction sale was
held in Ohio in 1836 by the Ohio Company, whose business was importing
English cattle. This event also marked the first sale of purebred cattle ever
held in America.

Although there are some records of occasional livestock auction sales

---

[7]Johnson, Ralph D., Agricultural Economist, Economic Research Service, U.S. Department
of Agriculture, stationed at the University of Nebraska.

Fig. 15-6. A livestock auction. (Courtesy, Michigan State University, E. Lansing, Mich.)

during the 19th Century, there is no indication of any auction market that continued operation throughout the period of the greatest development of public stockyards markets. It is within the current century that present auction development had its beginnings. In fact, livestock auction markets had their greatest growth from 1930 to 1952.

About 200 auctions were operating in 1930; by 1937, this number had increased to 1,345. The peak in numbers was reached in 1952, when over 2,500 different livestock auctions were holding sales. Subsequently, auctions have grown larger in volume of business and declined in numbers. In 1973, there were an estimated 1,000 active, regularly conducted livestock auction markets in the United States.[8]

Several factors contributed to the phenomenal growth in auction markets during the 1930s, chief of which were the following:

1. The decentralization of markets. Associated with this was the improvement and extension of hard surfaced roads accompanied by the increased use of trucks as a means of transporting livestock to and from the market place. Use of trucks increased flexibility in the handling of various sized lots of livestock and in the direction of movement; and, with the advent of better roads, trucks could be used for transporting livestock moderate distances both quickly and economically. Also, growing numbers of small packers, located at distant points from public stockyards, were able to procure livestock more efficiently at auction markets.

2. The development of more uniform class and grade classifications for livestock.

---

[8]Estimate given in a personal communication to the author, from Competitive Livestock Marketing Association, Kansas City, Mo.

3. Improvements made by the Federal Government in providing more extensive collection and dissemination of market news.

4. The greater convenience afforded in disposing of small lots of livestock and in purchasing stockers, feeders, and breeding animals.

5. The recognized educational value of these nearby markets, which enabled producers to keep currently informed of local market conditions and livestock prices.

6. The depression of 1930-33. When livestock prices are low—such as was true during the depression years—transportation and other marketing expenses compose a greater part of the total gross value received from the sale of livestock. Since at this time the commission charges at most auctions were based on a percentage of the gross sale value of the animal, marketing expenses at auctions tended to be low when prices were low. Also, with the proximity of auctions to producing areas, out-of-pocket transportation expenses were less for livestock sent to local auction than for shipments to more distant market outlets.

7. The abnormal feed distribution caused by the droughts of 1934 and 1936 in the western Corn Belt and range states created conditions favoring increased sales at auctions. Some immature and unfinished stock in these areas were sold at nearby auctions to farmers having a plentiful supply of feed on hand. Also, stocker and feeder cattle were shipped out of the drought sections to auctions located in areas where feed supplies were more abundant.

8. The desire to sell near home. By contrast to large public stockyards which receive some livestock from considerable distances, auction markets draw their supplies largely from the communities in which they are located.

Prior to the advent of community livestock auctions, the small livestock operator had two main market outlets for his animals: (1) shipping them to the nearest public stockyards; or (2) selling them to buyers who came to his farm or ranch. Generally, the first method was too expensive because of the transportation distance involved and the greater expense in shipping small lots. The second method pretty much put the producer at the mercy of the buyer, because he had no good alternative to taking the price he was offered, and often he did not know the value of his animals. By contrast, the big operator is not particularly concerned about these things. Because of his large scale, usually he can take advantage of any of several public stockyards, and he knows enough about values that he can deal satisfactorily with buyers who come to his feedlot, farm, or ranch. Thus, livestock auctions are really of greatest importance to the small operator.

Rates charged for marketing livestock vary at different auctions. Services for which charges are levied may include selling, yardage, weighing, insurance, brand inspection, and health inspection. Many auctions do not provide all these services. A commission or selling fee, however, is charged at all markets and is the primary source of income to auction operators. At some auctions, the commission covers yardage and weighing in addition to the selling service. Some operators levy a separate charge for each service provided, while others charge a single rate to cover all services.

Auction operators levy their charge on a per head basis, on a percentage of gross value, or by a combination of the two methods.

Straight per head charges on cattle and calves vary considerably. However, the most usual charges for cattle range from $2.50 to $4.50 per head. Charges on calves are less than on cattle, usually ranging from $1.50 to $2.50 per head.

For auctions reporting straight percentage charges, the most usual charge on cattle and calves is three percent.

FEEDER CALF AUCTIONS

Feeder calves are sold through the following market channels: auctions, direct, public stockyards, and dealers. In recent years, feeder calf sales, held auction style, have been especially popular in many states. Generally such sales are organized on a statewide or area basis, with the state agricultural extension service cooperating.

In advance of the auction, all entries are usually "sifted" by a committee, whose duty it is to assure the desired quality. The calves that pass the initial inspection are then delivered to the auction market a day ahead of the sale, at which time they are (1) tagged, (2) weighed, and (3) graded. Following this, each calf is penned with other calves of similar breed, sex, grade, and weight; thereby providing uniform lots for buyers. Usually, the pens are marked, giving the grade, sex, breed, and average weight of each lot of calves. The buyers are then given an opportunity to inspect the pens of calves prior to the sale. At the appointed time, each uniform lot of calves is offered for sale by the auction method.

For successful feeder calf auctions, it is necessary that there be both volume and quality. Large numbers attract more buyers and make it possible to sell calves in larger and more uniform lots.

The better managed feeder calf sales serve as an excellent education medium. Producers observe the grading demonstration, see how the weights of their calves compare with those of the neighbors, and realize the price spread between grades.

Two of the most important advantages to accrue from feeder calf sales are: (1) producers with a few head are provided market advantages comparable to large operators; and (2) cattle feeders are given an opportunity to purchase feeder calves of specified quality and in small or large lots, without the time and expense of shopping around.

Country Selling (Direct Marketing)

Country selling, or direct marketing, refers to producers' sales of livestock directly to packers, local dealers, or farmers without the support of commission men, selling agents, buying agents, or brokers.

Prior to the advent of public stockyards in 1865, country selling accounted for virtually all sales of livestock. Sales of livestock in the country declined with the growth of public stockyards until the latter method reached its peak of selling at the time of World War I. Country selling was accelerated by the large nationwide packers following World War I in order to meet the increased buying competition of the small interior packers.

Like auction selling, direct selling has a certain appeal, inasmuch as it permits producers to observe and exercise some control over selling while it takes place; whereas consignment to distant public stockyards at times represents an irreversible commitment to sell. Larger and more specialized livestock farmers feel competent to sell their livestock direct.

Improved highways and trucking facilitated the growth of country selling. Farmers were no longer tied to outlets located at important railroad terminals or river crossings. Livestock could move in any direction. Improved communications, such as the radio and telephone, and an expanded market information service, also aided in the development of country selling of livestock, especially in sales direct to packers.

Direct selling to meat packers is the most important outlet for slaughter cattle. Some packers buy cattle direct from producers at the plant; others send their buyers into the country, from farm to farm or feedlot to feedlot, where they make bids on the livestock that they inspect.

Local dealers operate in all parts of the country. These include country buying operations by local buyers, by contract buyers for later delivery, by buyers purchasing on orders for others, and by a variety of speculative buyers. Speculative buyers are known by various names in different parts of the country; they are sometimes called livestock buyers, traders, scalpers, truck buyers, stock buyers, pin-hookers, and scavengers. Some of these country buyers purchase livestock at fixed establishments similar to packer-owned country buying points.

Other Livestock Market Channels

In addition to public stockyards, livestock auctions, and country selling, producers in most sections of the United States have access to several other market channels; among them, the following:

● *Local markets; concentration yards*—These are sometimes referred to as local stockyards or union stockyards. At such markets, animals are purchased from farmers on a lot or graded basis, following which they are usually resorted and sold to slaughterers, order buyers, or to other markets. Local markets have fixed facilities for handling livestock, such as chutes and pens.

● *Packing plants; packer buying station*—Cattle may be sold by a producer to the slaughtering plant or to yards owned and operated some distance away from the slaughtering plant. In some states, packer buying stations are called concentration yards.

● *Order buyers*—Order buyers serve as agents of livestock buyers in the procurement of livestock. Usually, they buy through public stockyards or auctions, or from dealers in local markets. However, they occasionally act as agents of the buyer in purchase of livestock directly from producers.

● *Locker plants and retailers*—Occasionally, farmers sell a few head of livestock to a local locker plant, or to a store which retails meat.

• *Cooperative shipping associations*—These associations, which are owned and operated by producers, assemble livestock from farmers, load the livestock and ship cooperatively by truck or rail to a market, usually where the selling function is performed by commission men. Thus, the primary function of cooperative shipping associations is assembling and forwarding livestock.

• *Cooperative selling associations*—These are cooperatives which operate much like the cooperative shipping association, except that they perform more services such as obtaining bids on livestock, selecting outlets for livestock, and providing information to producers.

Commercial feedlot operations, country commission men, and producers' bargaining associations can be added to the above market channels.

Admittedly, there is duplication in the above listing. For example, order buyers operate at public stockyards and auctions, as well as carrying out free-lance country operations. Many public stockyards have incorporated auctions in their operations. Packer buyers operate at public stockyards and auctions, and buy directly from feeders in the country.

Choice of Market Channel

Marketing is dynamic; thus, changes are inevitable in types of market channels, market structures, and market services. Some outlets have gained in importance; others have declined.

The choice of a market channel represents the seller's evaluation of the most favorable market among the number of alternatives available to him. No simple and brief statement of criteria can be given as a guide to the choice of the most favorable market channel. Rather, an evaluation is required of the contributions made by alternative markets in terms of available services offered, selling costs, the competitive nature of the pricing process, and ultimately the producer's net return. Thus, an accurate appraisal is not simple.

From time to time, producers can be expected to shift from one type of market outlet to another. Because price changes at different market outlets do not take place simultaneously, nor in the same amount, nor even in the same direction, one market may be the most advantageous outlet for a particular class and grade of cattle at one time, but another may be more advantageous at some other time. The situation may differ for different classes and kinds of livestock and may vary from one area to another.

Regardless of the channel through which the producer markets his cattle, in one way or another, he pays or bears, either in the price received from the livestock or otherwise, the entire cost of marketing. Because of this, he should never choose a market because of convenience of habit, or because of personal acquaintance with the market and its operator. Rather, the choice should be determined strictly by the net returns from the sale of his livestock; effective selling and net returns are more important than selling costs. To arrive at net returns, the seller must deduct such indirect marketing costs as transportation and shrinkage. Likewise, it costs a packer money to keep a buyer at the market.

PREPARING AND SHIPPING CATTLE

Cattle are transported by truck, rail, and air. Regardless of the type of conveyance, proper preparing and shipping are much the same.

Improper handling of cattle immediately prior to and during shipment may result in excess shrinkage; high death, bruise, and crippling losses; disappointing sales; and dissatisfied buyers. Unfortunately, many stockmen who do a superb job of producing cattle, dissipate all the good things that have gone before by doing a poor job of preparing and shipping. Generally speaking, such omissions are due to lack of know-how, rather than any deliberate attempt to take advantage of anyone. Even if the sale is consummated prior to delivery, negligence at shipping time will make for a dissatisfied customer. Buyers soon learn what to expect from various producers and place their bids accordingly.

Fig. 15-7. Choice slaughter steers moving through curved steel loading chutes onto a cattle truck at Farr Feeders, Inc., near Greeley, Colo., which recently opened its $4 million, 80,000 head per year capacity, commercial feedlot. This cattle shipping area is specially designed to insure easy mobility through the chutes with loading platforms constructed at the same level as the truck gates, thereby reducing the chance of injury. Two loading chutes and one unloading chute are located in this shipping area. (Courtesy, Farr Feeders, Inc., Greeley, Colo.)

The following general considerations should be accorded in preparing cattle for shipment and in transporting them to market:

1. *Select the best suited method of transportation*—The cattleman should decide between truck, rail, and air transportation on the basis of which method best suits his particular situation.

The railroad system set the pattern for concentrated and centralized livestock marketing and meat packing locations. Then came the motor truck and highway system which were instrumental in reversal of the organizational structure—causing decentralization.

Many improvements have been and are being made within each system (rail and truck) which sustain a continuous competitive contest between these two types of carriers. Trucks virtually have taken over all short hauls of livestock. It is estimated that more than 80 percent of all livestock is moved to market by truck and trailer. In short hauls, trucks have the advantages of faster time, convenience, and flexibility—including pickup and delivery at destinations. Trucks are also vying for long hauls of livestock by increasing capacity, controlling temperature, and improving designs that reduce bruising and crippling and expedite loading and unloading. Double-deck cattle trucks have been in use for a number of years.

Railroads, too, have developed multi-deck cars and improved construction for greater animal comfort and reduced bruising. These include such things as improved ventilation, suspension, coupling, and operational procedures that reduce starting and stopping jerks. Most railroads now operate fast "express" livestock trains to speed delivery. Some years ago railroads introduced "piggyback" hauling of loaded trucks as a countermeasure to truck competition. Piggyback trucks are used to some extent for live animals.

Fig. 15-8. Registered Hereford calves bred by Tequesquite Ranch, Albert, New Mexico, being loaded on a jet freighter at Amarillo, Texas, for air shipment to Parker Ranch, Hawaii. A total of 160 450-lb calves were loaded on this DC-8. (Courtesy, Albert J. Mitchell, Tequesquite Ranch, Albert, N.M.)

Truck rates are usually less than rail rates on hauls of up to 200 to 300 miles. However, the rate for interstate trucks is usually greater than that for rail on both feeder and slaughter cattle on distances above 300-400 miles. Rates are important, but other factors must also be considered. Usually, some trucking is required to bring livestock to a rail loading point and take them away from the unloading point. Time in transit also is important. The adequacy of rest stops is a factor in shrinkage. On the latter score, railroads have had better facilities than were available for trucks. However, in recent years improvements have been made in privately-owned rest stops for trucks.

Plane shipments are a specialty, the details of which had best be left in the hands of an experienced person or agency, such as the livestock airlift specialist of an airline or an import-export company. At the present time, cattle shipments via air are largely confined to foreign movements, especially valuable breeding animals.

2. *Feed properly prior to loading out*—Never ship cattle on an excess fill. Instead, withhold grain feeding 12 hours before loading (omit one feed), and do not allow access to water within 2 to 3 hours of shipment. Cattle may be allowed free access to dry, well-cured grass hay up to loading time, but more laxative-type hays, such as alfalfa or clover, should not be fed within 12 hours of shipment even if the animals were accustomed to them previously. Likewise, cattle on green or succulent feed should be conditioned to dry feeds prior to shipment.

Cattle that are too full of concentrated or succulent feeds or full of water at the time of loading will scour and urinate excessively. As a result, the floors become dirty and slippery and the animals befoul themselves. Such cattle undergo a heavy shrink and present an unattractive appearance when unloaded.

Abrupt ration changes of any kind prior to shipment should be avoided. Occasionally, a misinformed cattleman withholds water, but gives a liberal feeding of salt prior to shipment, to obtain maximum water consumption and fill on the market. This "sharp" practice cannot be condemned too strongly; it is cruel to animals, and experienced buyers are never deceived.

3. *Keep cattle quiet*—Prior to and during shipment, cattle should be handled carefully. Hot, excited animals are subject to more shrinkage, more disease and injury, and more dark cutting if slaughtered following shipment.[9]

If the animals are trailed on-foot to the shipping point, they should be moved slowly and allowed to rest and to drink moderately prior to loading. Although loading may be exasperating at times, take it easy; never lose your temper. Avoid hurrying, loud hollering, and striking. Never beat an animal with such objects as pipes, sticks, canes, or forks; instead, use either (a) a flat, wide canvas slapper with a handle, or (b) an electric prod (the latter judiciously).

[9]Brady, D. E. and H. B. Hedrick, *Journal of Animal Science*, Vol. 15, No. 4, Nov. 1956, p. 1290; and Hedrick, H. B., D. E. Brady, and C. W. Turner, *Proceedings of the Ninth Research Conference*, sponsored by the Council on Research, American Meat Institute, at the Univ. of Chicago, March 22-23, 1957.

4. *Consider health certificates, permits, and brand inspection in interstate shipments*—When cattle are to be shipped into another state, the shipper should check into and comply with the state regulations relative to health certificates, permits, and brand inspection. Usually, the local veterinarian, railroad agent, or trucker will have this information. Should there be any question about the health regulations, however, the state livestock sanitary board (usually located at the state capital) of the state of destination should be consulted. Knowledge of and compliance with such regulations well in advance of shipment will avoid frustrations and costly delays.

5. *Comply with the 28-hour law in rail shipments*—Actually, the shipper has no alternative to taking advantage of feed and rest stops during long hauls by rail; by Federal law, passed in 1873, livestock cannot be transported by rail for a longer period than 28 consecutive hours without unloading for the purpose of giving feed, water, and rest for a period of at least 5 consecutive hours before resuming transportation. The period may be extended to 36 hours upon written request from the owner of the animals; and most experienced cattle shippers routinely so request. With less than carload lots (LCL shipments) the owner may provide feed and water in the car with instructions that the animals be fed and watered en route.

The shipper may instruct the railway company on the kind and amount of feed to be given in transit, with these instructions written on the waybill or on the livestock contract which each shipper signs. If no such instructions are given by the owner of the cattle, the amount of feed prescribed by the USDA is given at the livestock feeding yards; namely, per carload of beef cattle or range calves (a) 200 pounds of hay at the first feeding station, and (b) 300 pounds of hay at the second and subsequent feed stations. Usually grass hay of good quality is fed. The feeding is done by the railway company crew, and charge is made to the shipper for the amount of feed consumed.

The need for the 28-hour law is considerably diminished now, compared to earlier years. Improved, faster freight service makes it possible for many shipments to reach their destination within 36 hours.

Trucks were not included in the provisions of the original 28-hour law, simply because there were no trucks at the time. But most buyers of livestock insist on a rest stop in an extended truck shipment, just as is required by law with rail shipments.

6. *Use partitions when necessary*—When mixed loads (consisting of cattle, sheep, and/or hogs) are placed in the same truck or car, partition each class off separately. Also, partition calves from cattle, and separate out cripples and stags; tie bulls.

7. *Avoid shipping during extremes in weather*—Whenever possible, avoid shipping when the weather is either very hot or very cold. During such times, shrinkage and death losses are higher than normal. During warm weather, avoid transporting animals during the heat of the day; travel at night or in the evening or early morning.

Additional points pertinent to proper preparing and shipping cattle are covered in the four sections which follow.

How to Prevent Bruises, Crippling, and Death Losses

Losses from bruising, crippling, and death that occur during the marketing process represent a part of the cost of marketing livestock; and, indirectly, the producer foots most of the bill.

The following precautions are suggested as a means of reducing cattle market losses from bruises, crippling, and death:

1. Dehorn cattle, preferably when young.
2. Remove projecting nails, splinters, and broken boards in feed racks and fences.
3. Keep feedlots free from old machinery, trash, and any obstacle that may bruise.
4. Do not feed grain heavily just prior to loading.
5. Use good loading chutes; not too steep.
6. Bed with sand free from stones, to prevent slipping.
7. For calves, cover sand with straw in cold weather.
8. Provide covers for trucks to protect from sun in summer and cold in winter.
9. Always partition mixed loads into separate classes, and partition calves from cattle.
10. Have upper deck of truck high enough to prevent back bruises on animals below.
11. Remove protruding nails, bolts, or any sharp objects in truck or car.
12. Load slowly to prevent crowding against sharp corners and to avoid excitement. Do not overload.
13. Use canvas slappers instead of clubs or canes.
14. Tie all bulls in truck or car, and partition stags and cripples.
15. Place bull board in position and secure before car door is closed on loaded cattle.
16. Drive trucks carefully; slow down on sharp turns and avoid sudden stops.
17. Inspect load en route to prevent trampling of animals that may be down.
18. Back truck slowly and squarely against unloading dock.
19. Unload slowly. Do not drop animals from upper to lower deck; use cleated inclines.

All these precautions are simple to apply; yet all are violated every day of the year.

Number of Cattle in a Truck and in a Railroad Car

Overcrowding of market animals causes heavy losses. Sometimes a truck or a railroad car is overloaded in an attempt to effect a saving in hauling charges. More frequently, however, it is simply the result of not knowing space requirements.

The suggested number of animals, Tables 15-4 and 15-5, should be tempered by such factors as distance of haul, class of cattle, weather, and road conditions.

TABLE 15-4

NUMBER OF CATTLE FOR SAFE LOADING IN A TRUCK[1]

Floor Length	Average Weight of Cattle, Lbs					
	450	600	800	1000	1200	1400
(ft)	(no.)	(no.)	(no.)	(no.)	(no.)	(no.)
8	8	7	5	4	4	3
10	10	8	7	6	5	4
12	13	10	8	7	6	5
15	16	13	10	9	8	7
18	20	16	13	11	9	8
20	22	18	14	12	10	9
24	27	22	17	15	13	11
28	31	25	20	17	15	13
30	34	27	22	19	16	14
32	36	29	23	20	17	15
36	41	33	26	22	19	17
42	48	39	31	28	22	20

[1]Table 15-4 is taken from the authoritative recommendations of Livestock Conservation, Inc.

1. *By truck*—Table 15-4 shows the number of cattle for safe trucking.
2. *By rail*—Normally, railroad cars are either 36 or 40 feet in length. The size of the car and the class and size of animals determine the number of head that can be loaded in a given car. For comfort in shipping, the car should be loaded heavily enough so that the animals stand close together, but overcrowding is to be avoided. Table 15-5 gives some indication as to the number of cattle that may be loaded in a railroad car.

TABLE 15-5

CATTLE PER RAILROAD CAR

Car Size	Cattle, Wt. 1,000 Pounds
36-foot car	26
40-foot car	28

Kind of Bedding to Use for Cattle in Transit

Among the several factors affecting livestock losses, perhaps none is more important than proper bedding and footing in transit. This applies to both truck and rail shipments, and to all classes of animals.

Footing, such as sand, is required at all times of the year, to prevent the car or truck floor from becoming wet and slick, thus predisposing to injury of animals by slipping or falling. Bedding, such as straw, is recommended for warmth in the shipment of calves during extremely cold weather, and as cushioning for dairy cows, breeding stock, or other animals loaded lightly enough to permit their lying down. Recommended kinds and amounts of bedding and footing materials are given in Table 15-6. (Because many loads are mixed, information relative to each class of animals is provided in Table 15-6.)

TABLE 15-6

HANDY GUIDE RELATIVE TO BEDDING AND FOOTING MATERIAL
WHEN TRANSPORTING LIVESTOCK[1]

Class of Livestock	Kind of Bedding for Moderate or Warm Weather; Above 50° F	Kind of Bedding for Cool or Cold Weather Below 50° F
Cattle	Sand, 2 in.	Sand; for calves use sand covered with straw
Sheep and goats	Sand	Sand covered with straw
Swine	Sand, ½ in. to 2 in.	Sand covered with straw
Horses and mules	Sand	Sand

[1]Straw or other suitable bedding (covered over sand) should be used for protecting and for cushioning breeding stock that are loaded lightly enough to permit their lying down in the car or truck.
Sand should be clean and medium-fine, and free from brick, stones, coarse gravel, dirt, or dust.
Fine cinders may be used as footing for cattle, horses and mules, but not for sheep or hogs. They are picked up by and damage the wool of sheep, and they damage hog casings.
In hot weather, wet sand down before loading.

Shrinkage in Marketing Cattle

The shrinkage (or drift) refers to the weight loss encountered from the time animals leave the farm, ranch, or feedlot until they are weighed over the scales at their destination. Thus, if a steer weighed 1,000 lb at the feedlot and had a market weight of 970 lb, the shrinkage would be 30 lb or 3.0 percent. Shrink is usually expressed in terms of percentage. Most of this weight loss is due to excretion, in the form of feces and urine and the moisture in the expired air. On the other hand, there is some tissue shrinkage, which results from metabolic or breakdown changes.

The most important factors affecting shrinkage are:

1. *The fill*—Naturally, the larger the fill animals take upon their arrival at the market, the smaller the shrinkage.

2. *Time in transit*—The longer the animals are in transit and the greater the distance, the higher the total shrinkage.

3. *Truck vs rail transportation*—Based on practical experience and observation, most stockmen are of the opinion (a) that truck shipments result in less shrinkage than rail shipments for short hauls, and (b) that rail shipments result in less shrinkage than truck shipments for long hauls. This latter situation may be due to the fact that cattle hauled by rail have a feed and rest stop while those moved by truck may not.

4. *Season*—Extremes in temperature, either very hot or very cold weather, result in higher shrinkage.

5. *Age and weight*—Young animals of all species shrink proportionally more than older animals.

6. *Overloading*—Overloading always results in abnormally high shrinkage.

7. *Rough ride, abnormal feeding and mixed loads*—Each of these factors will increase shrinkage.

8. *Auction sale or ranch origin*—Iowa State University reported an average shrink of 9.1 percent for cattle originating at auction sales compared with 7.2 percent shrink in cattle purchased directly from ranches. On the average, market cattle shrink from 3 to 10 percent, with younger animals

shrinking more than older and fatter animals. Experienced cattle feeders report that it takes an average of 7 days after arrival at the feedyard to regain a shrink of 10 percent on feeder cattle.

SHRINKAGE TABLES

Cattle are sometimes bought or sold with a certain percentage shrink from actual weights. This is known as pencil shrinkage. Both cattle sellers and buyers must give consideration to such shrinkage. For example, if a buyer offers $40.00 per 100 with a 4% shrink allowance, the producer will want to know how much he will receive. The answer can be quickly and easily obtained from Table 15-7 as follows: look at $40.00 under column 1, headed "Offer." Go across to column headed "4%." As shown, the producer will receive $38.40 for his cattle.

TABLE 15-7

SELLING CATTLE
NET PRICES AFTER ALLOWING FOR SHRINKAGE
(Prices, per Cwt)

Offer	Shrink				
	2%	3%	4%	6%	8%
($)	($)	($)	($)	($)	($)
80.00	78.40	77.60	76.80	75.20	73.60
79.00	77.42	76.63	75.84	74.26	72.68
78.00	76.44	75.66	74.88	73.32	71.76
77.00	75.46	74.69	73.92	72.38	70.84
76.00	74.48	73.72	72.96	71.44	69.92
75.00	73.50	72.75	72.00	70.50	69.00
74.00	72.52	71.78	71.04	69.56	68.08
73.00	71.54	70.81	70.08	68.62	67.16
72.00	70.56	69.84	69.12	67.68	66.24
71.00	69.58	68.87	68.16	66.74	65.32
70.00	68.60	67.90	67.20	65.80	64.40
69.00	67.62	66.93	66.24	64.86	63.48
68.00	66.64	65.96	65.28	63.92	62.56
67.00	65.66	64.99	64.32	62.98	61.64
66.00	64.68	64.02	63.36	62.04	60.72
65.00	63.70	63.05	62.40	61.10	59.80
64.00	62.72	62.08	61.44	60.16	58.88
63.00	61.74	61.11	60.48	59.22	57.96
62.00	60.76	60.14	59.52	58.28	57.04
61.00	59.78	59.17	58.56	57.34	56.12
60.00	58.80	58.20	57.60	56.40	55.20
59.00	57.82	57.23	56.64	55.46	54.28
58.00	56.84	56.26	55.68	54.52	53.36
57.00	55.86	55.29	54.72	53.58	52.44
56.00	54.88	54.32	53.76	52.64	51.52
55.00	53.90	53.35	52.80	51.70	50.60
54.00	52.92	52.38	51.84	50.76	49.68
53.00	51.94	51.41	50.88	49.82	48.76
52.00	50.96	50.44	49.92	48.88	47.84
51.00	49.98	49.47	48.96	47.94	46.92
50.00	49.00	48.50	48.00	47.00	46.00

(Continued)

Table 15-7 (Continued)

Offer	Shrink				
	2%	3%	4%	6%	8%
($)	($)	($)	($)	($)	($)
49.00	48.02	47.53	47.04	46.06	45.08
48.00	47.04	46.56	46.08	45.12	44.16
47.00	46.06	45.59	45.12	44.18	43.24
46.00	45.08	44.62	44.16	43.24	42.32
45.00	44.10	43.65	43.20	42.30	41.40
44.00	43.12	42.68	42.24	41.36	40.48
43.00	42.14	41.71	41.28	40.42	39.56
42.00	41.16	40.74	40.32	39.48	38.64
41.00	40.18	39.77	39.36	38.54	37.72
40.00	39.20	38.80	38.40	37.60	36.80
39.00	38.22	37.83	37.44	36.66	35.88
38.00	37.24	36.86	36.48	35.72	34.96
37.00	36.26	35.89	35.52	34.78	34.04
36.00	35.28	34.92	34.56	33.84	33.12
35.00	34.30	33.95	33.60	32.90	32.20
34.00	33.32	32.98	32.64	31.96	31.28
33.00	32.34	32.01	31.68	31.02	30.36
32.00	31.36	31.04	30.72	30.08	29.44
31.00	30.38	30.07	29.76	29.14	28.52
30.00	29.40	29.10	28.80	28.20	27.60
29.00	28.42	28.13	27.84	27.26	26.68
28.00	27.44	27.16	26.88	26.32	25.76
27.00	26.46	26.19	25.92	25.38	24.84
26.00	25.48	25.22	24.96	24.44	23.92

If the producer has decided that $40.00 is his minimum asking price, he may refuse the offer and ship his cattle to market. Then, he will wish to know how much he will have to receive in order to compensate for shrinkage. The answer can be obtained from Table 15-8, as follows: Look at $40.00 under column 1, headed "Asking"; then read under the proper column to the right. Thus, if the animals shrink 4 percent during marketing, the price will have to be $41.67 in order to compensate for shrinkage.

TABLE 15-8

BUYING CATTLE
CHANGE IN PRICE TO COMPENSATE FOR SHRINKAGE
(Prices, per Cwt)

Asking	Shrink				
	2%	3%	4%	6%	8%
($)	($)	($)	($)	($)	($)
80.00	81.63	82.47	83.33	85.11	86.96
79.00	80.61	81.44	82.29	84.04	85.87
78.00	79.59	80.41	81.25	82.98	84.78
77.00	78.57	79.38	80.21	81.91	83.70
76.00	77.55	78.35	79.17	80.85	82.61
75.00	76.53	77.32	78.13	79.79	81.52
74.00	75.51	76.29	77.08	78.72	80.43
73.00	74.49	75.26	76.04	77.66	79.35

(Continued)

Table 15-8 (Continued)

Asking	Shrink				
	2%	3%	4%	6%	8%
($)	($)	($)	($)	($)	($)
72.00	73.47	74.23	75.00	76.60	78.26
71.00	72.45	73.20	73.96	75.53	77.17
70.00	71.43	72.16	72.92	74.47	76.09
69.00	70.41	71.13	71.88	73.40	75.00
68.00	69.39	70.10	70.83	72.34	73.91
67.00	68.37	69.07	69.79	71.28	72.83
66.00	67.35	68.04	68.75	70.21	71.74
65.00	66.33	67.01	67.71	69.15	70.65
64.00	65.31	65.98	66.67	68.09	69.57
63.00	64.21	64.95	65.63	67.02	68.48
62.00	63.27	63.92	64.58	65.96	67.39
61.00	62.24	62.89	63.54	64.89	66.30
60.00	61.22	61.86	62.50	63.83	65.22
59.00	60.20	60.82	61.46	62.77	64.13
58.00	59.18	59.79	60.42	61.70	63.04
57.00	58.16	58.76	59.38	60.64	61.96
56.00	57.14	57.73	58.33	59.57	60.87
55.00	56.12	56.70	57.29	58.51	59.78
54.00	55.10	55.67	56.25	57.45	58.70
53.00	54.08	54.64	55.21	56.38	57.61
52.00	53.06	53.61	54.17	55.32	56.52
51.00	52.04	52.58	53.13	54.26	55.43
50.00	51.02	51.55	52.08	53.19	54.35
49.00	50.00	50.52	51.04	52.13	53.26
48.00	48.98	49.48	50.00	51.06	52.17
47.00	47.96	48.45	48.96	50.00	51.09
46.00	46.94	47.42	47.92	48.94	50.00
45.00	45.92	46.39	46.88	47.87	48.91
44.00	44.90	45.36	45.83	46.81	47.83
43.00	43.88	44.33	44.79	45.74	46.74
42.00	42.86	43.30	43.75	44.68	45.65
41.00	41.84	42.27	42.71	43.62	44.57
40.00	40.82	41.24	41.67	42.55	43.48
39.00	39.80	40.21	40.63	41.49	42.39
38.00	38.78	39.81	39.58	40.43	41.30
37.00	37.76	38.14	38.54	39.36	40.22
36.00	36.73	37.11	37.50	38.30	39.13
35.00	35.71	36.08	36.46	37.23	38.04
34.00	34.69	35.05	35.42	36.17	36.96
33.00	33.67	34.02	34.37	35.11	35.87
32.00	32.66	32.99	33.33	34.04	34.78
31.00	31.63	31.96	32.29	32.98	33.70
30.00	30.61	30.93	31.25	31.91	32.61

MARKET CLASSES AND GRADES OF CATTLE[10]

The generally accepted market classes and grades of live cattle are summarized in Table 15-9. The first five divisions and subdivisions include those factors that determine the class of the animal or the use to which it will be put. The grades, which are a combination of both their quality and yield

---

[10]This section was authoritatively reviewed by Mr. John C. Pierce, Director, Livestock Division, Agricultural Marketing Service, USDA.

grades (except that slaughter bulls are yield graded only), indicate how well the cattle fulfill the requirements to which they are put.

TABLE 15-9

THE MARKET CLASSES AND QUALITY GRADES OF CATTLE

Cattle or Calves	Use Selection	Sex Classes	Age	Weight Divisions Wt. (Group)	(lb)	Commonly Used Quality Grades[1]
Cattle	Slaughter cattle[1]	Steers	Yearlings	Light Medium Heavy	750 down 750-950 950 up	Prime, Choice, Good, Standard, Utility, Cutter, Canner
			2-year-old and over	Light Medium Heavy	1,100 down 1,100-1,300 1,300 up	Prime, Choice, Good, Standard, Commercial, Utility, Cutter, Canner
		Heifers	Yearlings	Light Medium Heavy	750 down 750-900 900 up	Prime, Choice, Good, Standard, Utility, Cutter, Canner
			2-year-old and over	Light Medium Heavy	900 down 900-1,050 1,050 up	Prime, Choice, Good, Standard, Commercial, Utility, Cutter, Canner
		Cows	All ages	All Weights		Choice, Good, Standard, Commercial, Utility, Cutter, Canner
		Bullocks	24 mo. & under	All Weights		Prime, Choice, Good, Standard, Utility
		Bulls		All Weights		None (yield graded only)
	Feeder cattle	Steers	Yearlings	Light Medium Heavy Mixed		Prime, Choice, Good, Standard, Utility, Inferior
			2-year-old and over	Light Medium Heavy Mixed		Prime, Choice, Good, Standard, Commercial, Utility, Inferior
		Heifers	Yearlings	Light Medium Heavy Mixed		Prime, Choice, Good, Standard, Utility, Inferior
			2-year-old and over	Light Medium Heavy Mixed		Prime, Choice, Good, Standard, Commercial, Utility, Inferior
		Cows	All ages	All Weights		Choice, Good, Standard, Commercial, Utility, Inferior
		Bullocks	24 mo. & under	All Weights		Prime, Choice, Good, Standard, Utility, Inferior
		Bulls	24 mo. & over	All Weights		None
	Milkers & springers	Cows (milkers or springers)	All ages	All Weights		None
Calves	Vealers	No Sex Class (Sex characteristics of no importance at this age)	Under 3 mo.	Light Medium Heavy	110 down 110-180 180 up	Prime, Choice, Good, Standard, Utility
	Slaughter calves	Steers Heifers Bulls	3 mo. to 1 year	Light Medium Heavy	200 down 200-300 300 up	Prime, Choice, Good, Standard, Utility
	Feeder calves	Steers Heifers Bulls	Usually 6 mo. to 1 year	Light Medium Heavy Mixed		Prime, Choice, Good, Standard, Utility, Inferior

[1]In addition to the quality grades, there are the following yield grades for all slaughter cattle, except bulls: Yield Grade 1, Yield Grade 2, Yield Grade 3, Yield Grade 4, and Yield Grade 5; with Yield Grade 1 representing the highest cutability, and Yield Grade 5 the lowest. Thus, slaughter cattle are graded for both quality and yield grade.

Factors Determining Market Classes of Cattle

The market class of cattle is determined by (1) use selection, (2) sex, (3) age, and (4) weight (see Table 15-9).

CATTLE AND CALVES

All members of the bovine family are designated as calves until they are one year of age, after which they are known as cattle. On the average, about 93 percent of market animals are cattle and only 7 percent are calves.

BABY BEEF

From the 1920s to about 1950, the term "baby beef" was well known and widely used in cattle production and marketing circles. The ultimate in baby beef was the 4-H Club steer of the era. It referred to well-finished young animals, marketed at 12 to 18 months of age, weighing 700 to 1,200 pounds, grading Good to Prime. With the coming of large feedlots, and the marketing of more grain-fed animals and animals at slightly older ages, the term was no longer used.

In 1974, baby beef was again being promoted, primarily as a means of reducing cattle numbers and lessening grain feeding. But it took on a new look! Today, baby beef connotes heavy calves that are fat enough for slaughter at weaning time, weighing 400 to 700 pounds on foot, the meat of which is lean and tender.

USE SELECTION OF CATTLE AND CALVES

The cattle group is further divided into three use divisions, each indicating something of the purpose to which the animals will be put. These divisions are: (1) slaughter cattle; (2) feeder cattle; and (3) milkers and springers. Slaughter cattle include those which are considered suitable for immediate slaughter; feeders include those which are to be taken back to the country and grown for a time or fattened; and milkers and springers include those cows recently freshened or soon due to calve and which are sold for milk purposes.

The calf group is also subdivided into three classes: (1) vealers, including milk-fat animals under three months of age which are sold for immediate slaughter; (2) slaughter calves that are between the ages of three months and one year, which have usually received grain in addition to milk and which are fat enough for slaughter; and (3) feeder calves which are of weaning age and are sold to go back into the country for further growing or finishing.

In the selection of feeder cattle or calves, the sex, age, weight, and grade are of importance. In addition, consideration should be given to the following factors: (1) constitution and thrift; (2) natural fleshing; (3) breeding; (4) uniformity; (5) absence of horns; and (6) temperament and disposition.

As can be readily understood, the use to which animals are put is not always clear-cut and definite. Thus, when feed is abundant and factors are favorable for cattle finishing, feeders may outbid packer buyers for some of the animals that would normally go for slaughter purposes. On the other hand, slaughterers frequently outbid feeders for some of those animals that would normally go the feeder route.

THE SEX CLASSES

Cattle are divided into five sex classes: steers, heifers, cows, bullocks, and bulls. Each of these five groups has rather definite and easily distinguishable characteristics that are related to the commercial value of the carcass—especially in the cattle group—and which are important in determining the suitability of animals as stockers and feeders. In older cattle, sex is an important factor affecting carcass quality, finish, and conformation. The definition of each sex class follows:

1. *Steer*—A male bovine castrated when young and which has not begun to develop the secondary physical characteristics of a bull.

2. *Heifer*—An immature female bovine that has not developed the physical characteristics typical of cows.

3. *Cow*—A female bovine that has developed through reproduction or with age, the relatively prominent hips, large middle, and other physical characteristics typical of mature females.

4. *Bullock*—A young (under approximately 24 months of age) male bovine (castrated or uncastrated) that has developed or begun to develop the secondary physical characteristics of a bull.

5. *Bull*—A mature (approximately 24 months of age or older) uncastrated, male bovine. However, for the purpose of these standards, any mature, castrated, male bovine which has developed or begun to develop the secondary physical characteristics of an uncastrated male also will be considered a bull.

Calves are merely divided into three sex classes: steers, heifers, or bulls. Because the secondary sex characteristics are not very pronounced in this group, the sex classes are of less importance for slaughter purposes than in older cattle. On the other hand, bull calves are not preferred as feeders because castration involves extra trouble and risk of loss.

AGE GROUPS

Because the age of cattle does affect certain carcass characteristics, it is logical that age groups should exist in market classifications. The terms used to indicate approximate age ranges for cattle are: vealers, calves, yearlings, 2-year-old and older cattle. As previously indicated, vealers are under 3 months of age,[11] whereas calves are young cattle between the vealer and yearling stage. Yearlings range from 12 to 24 months in age, and 2-year-olds from 24 to 36 months. Older cattle are usually grouped along with the 2-year-olds as "2-year-old and over."

---

[11]Vealers are generally over 21 days of age at time of slaughter, although Federal and most state regulations governing meat inspection do not specify a minimum age. Rather, it is a matter of maturity. A big Holstein calf may, for example, make satisfactory veal at two weeks of age, whereas calves of some of the smaller dairy breeds might not pass inspection before 4 or 5 weeks of age. Animals that are too young or immature are generally considered unsuited for human consumption primarily for esthetic reasons, rather than because of any harmful effect of the meat. On the market, underage veal calves are called "deacons" or "bob veal."

WEIGHT DIVISIONS

It is common to have three weight divisions: light, medium, and heavy. When several weight divisions are included together, they are referred to as "mixed weight." The usual practice is to group animals by rather narrow weight divisions because purchasers are frequently rather "choosey" about weights, and market values often vary quite sharply with variations in weights.

The Market Grades of Cattle

While no official grading of live animals is done by the U.S. Department of Agriculture, market grades do form a basis for the uniform reporting of livestock marketed. The grade is the final step in classifying any kind of market livestock. It indicates the relative degree of excellence of an animal or group of animals. Slaughter cattle quality grades are based on an evaluation of factors related to the palatability of the lean, referred to as "quality." The yield grades, which estimate the amount of salable meat a carcass will yield, are based on evaluations of muscling and fatness.

Table 15-9 lists the commonly used quality grades of cattle by classes. There also are five yield grades (see Table 15-9 footnote), designated by the numbers 1 through 5, which are applicable to all classes of slaughter cattle, with Yield Grade 1 representing the highest cutability and Yield Grade 5 the lowest. As noted, the number of quality grades varies somewhat between classes chiefly because certain groups of animals present a wider range of variations in conformation, finish, and quality than do other groups. Slaughter steers and heifers are divided into eight grades: Prime, Choice, Good, Standard, Commercial, Utility, Cutter, and Canner. As shown in Table 15-9, seven grades apply to slaughter cows, and five grades to slaughter bullocks. The Prime grade is not applied to cows, chiefly because of deficient conformation, finish, and quality in this class. Further, as in carcasses, bullocks on foot are always designated as "slaughter bullocks" since meat obtained from this class is never interchanged with meat carrying the same grade name from steers, heifers, and cows.

The terms Cutter or Canner are applied to the two lowest grades of slaughter cattle. Cutter cattle are so poor in form and lacking in muscle and fat covering that only such wholesale cuts as the loin and round are cut out and sold over the block. The balance of the carcass is boned out and used in sausage and canned-meat products. Canners are almost entirely processed as ground and canned meats.

The grades of feeder cattle are: Prime, Choice, Good, Standard, Commercial, Utility, and Inferior. These grades are based on two value-determining characteristics—logical slaughter potential and thriftiness. The logical slaughter potential of an animal is the slaughter grade at the time the animal's carcass quality grade and carcass conformation grade become equal. Thriftiness in feeder cattle refers to the ability of the animal to gain weight and finish rapidly and efficiently.

SLAUGHTER STEERS
U.S. GRADES
(QUALITY)

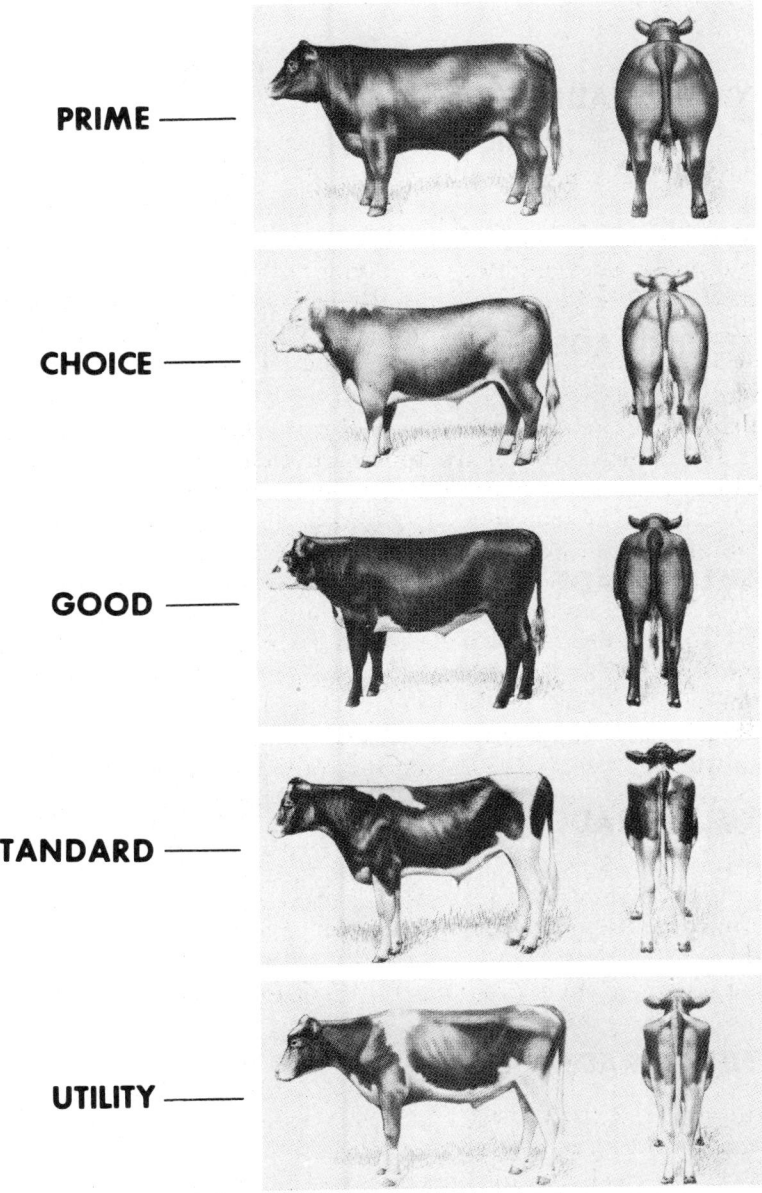

PRIME

CHOICE

GOOD

STANDARD

UTILITY

Fig. 15-9. U.S. quality grades of slaughter steers. (Courtesy, USDA)

SLAUGHTER STEERS
U.S. GRADES
(YIELD)

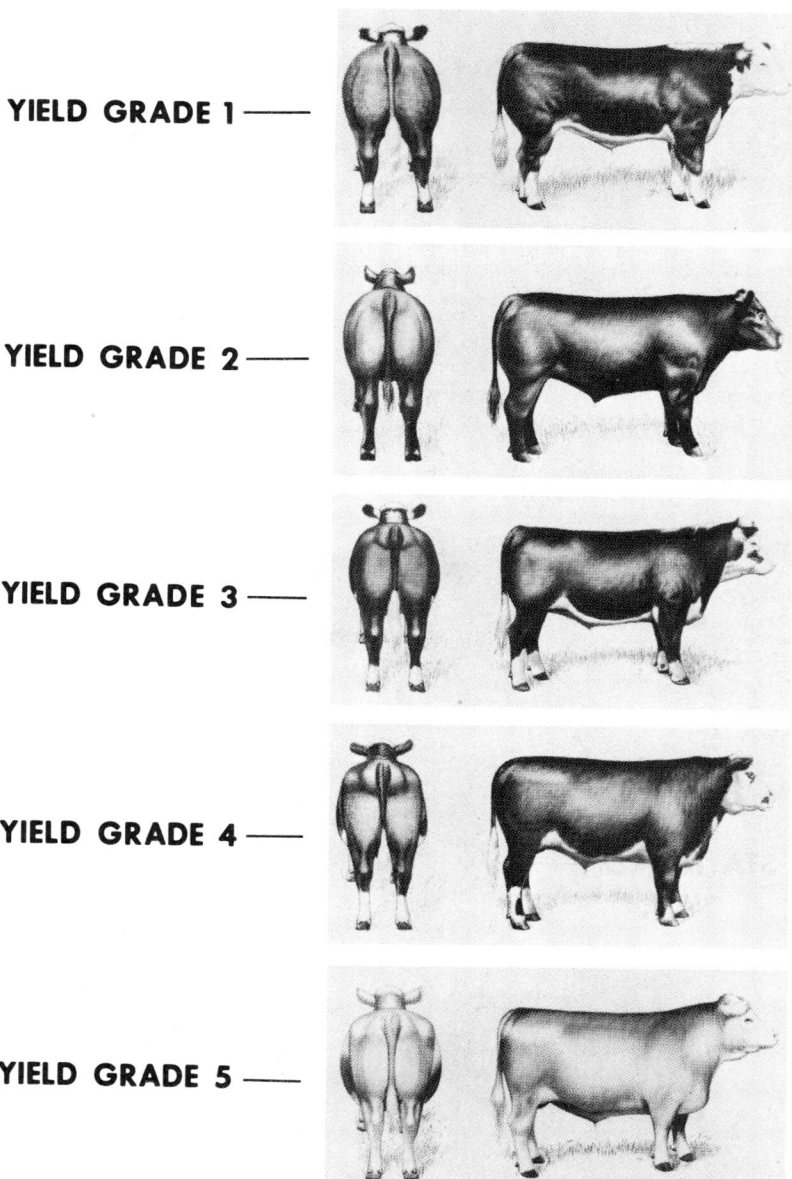

YIELD GRADE 1 ———

YIELD GRADE 2 ———

YIELD GRADE 3 ———

YIELD GRADE 4 ———

YIELD GRADE 5 ———

Fig. 15-10. U.S. yield grades of slaughter steers. (Courtesy, USDA)

FEEDER STEERS
U.S. GRADES

Fig. 15-11. U.S. grades of feeder steers. (Courtesy, USDA)

FEEDER STEERS
(CALVES)
U. S. GRADES

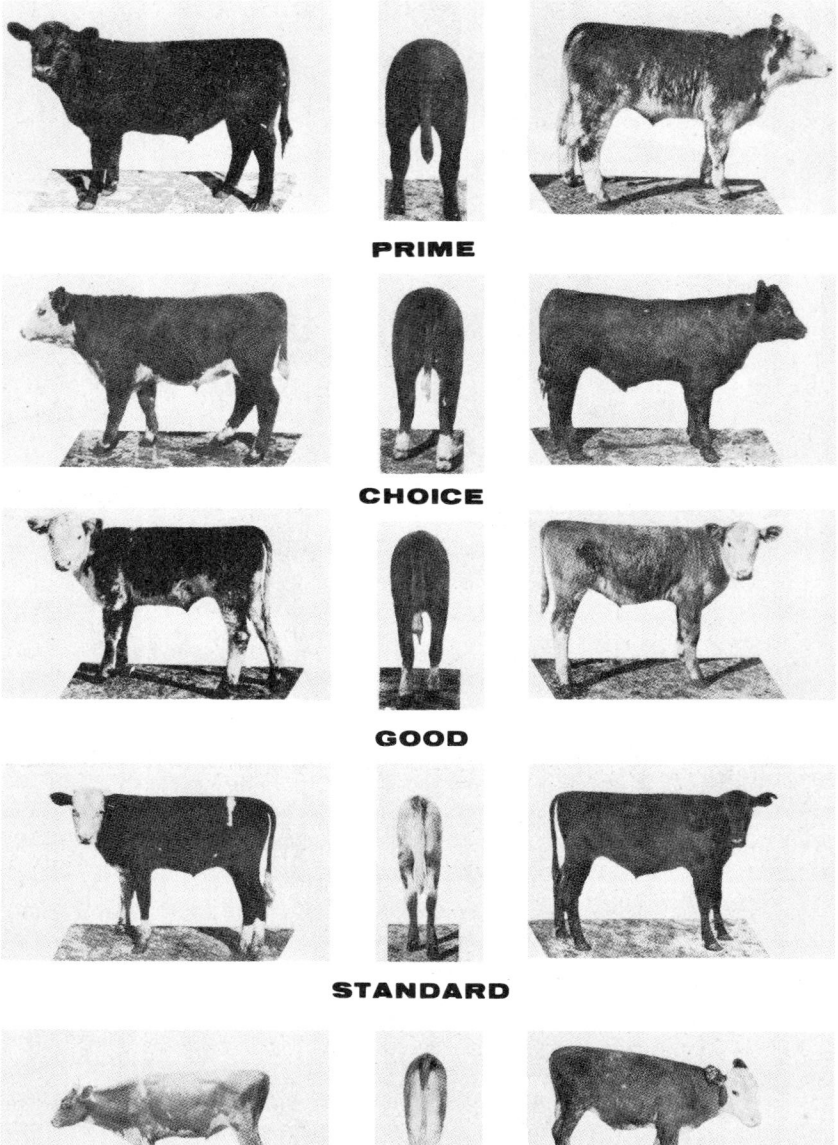

PRIME

CHOICE

GOOD

STANDARD

UTILITY

Fig. 15-12. U.S. grades of feeder steer calves. (Courtesy, USDA)

The grades of vealers and slaughter calves are: Prime, Choice, Good, Standard, and Utility.

As would be expected, the higher grades (Prime and Choice) of finished cattle usually carry more weight, and the lower grades are lighter and usually underfinished. Fig. 15-13 shows the relationship of the weight of beef steers to the grade.

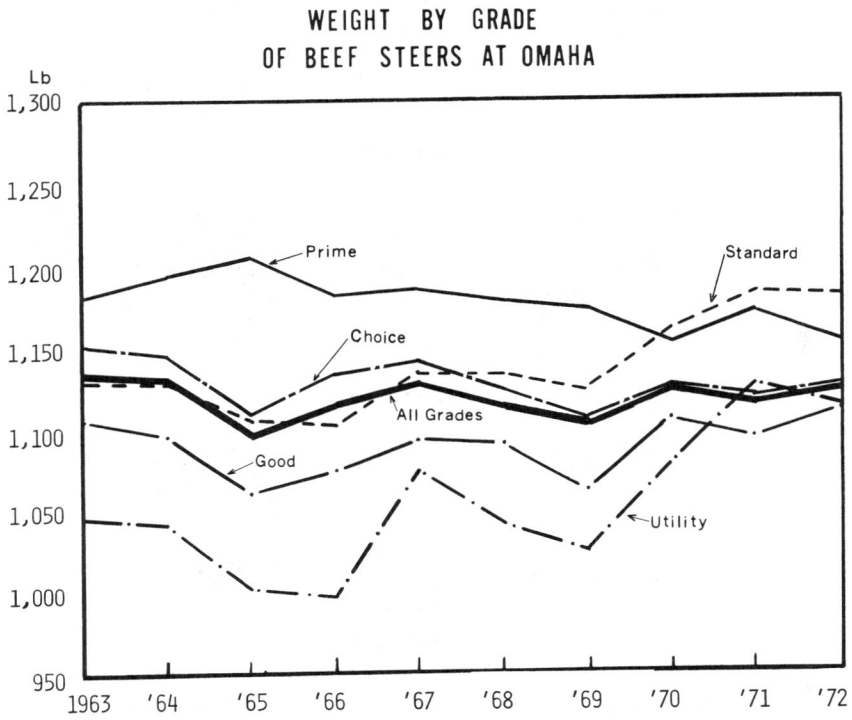

Fig. 15-13. Weight by grade of beef steers sold at Omaha, 1963-1972. This shows that the Prime and Choice grades of finished cattle usually carry more weight, and the lower grades are lighter and usually underfinished. (Data provided by John C. Pierce, Director, Livestock Division, Consumer and Marketing Service, USDA)

Because the production of the better grades of cattle usually involves more expenditure in the breeding operations (due to the need for superior animals) and feeding to a higher degree of finish, there must be a spread in market grades in order to make the production of the top grades profitable. Figure 15-14 shows the 10-year average price per hundredweight of beef steers, by grades, on the Chicago market from 1963 to 1972.

Grades of slaughter (live) cattle are intended to be directly related to the grades of the carcasses that they will produce. To accomplish this, slaughter grade standards are based on factors which are related to the quality grade and the yield grade of beef carcasses.

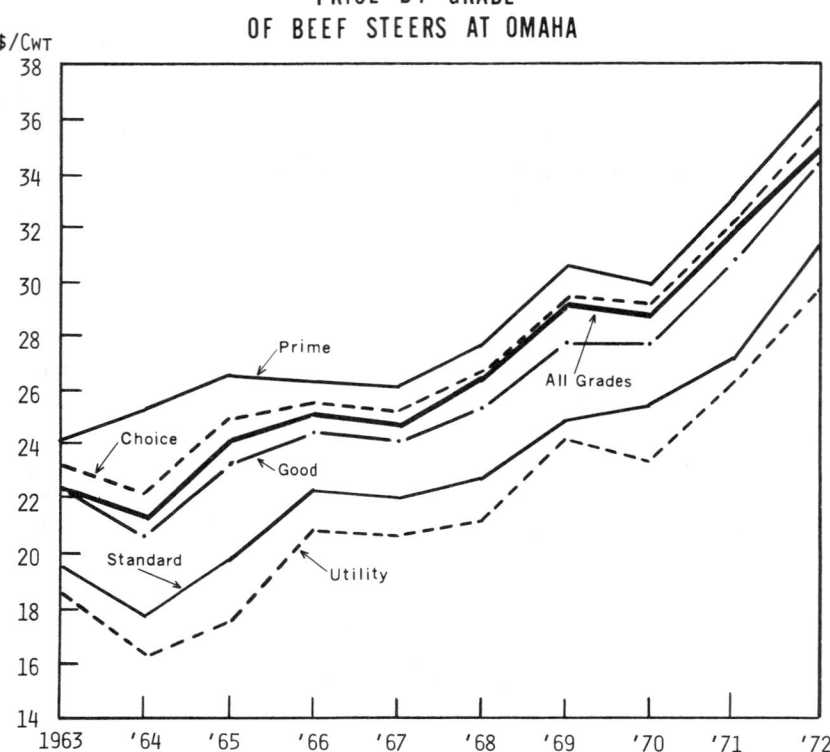

Fig. 15-14. Prices of beef steers sold out of first hand, by grades, at Omaha, 1963-1972. This shows that there is a rather uniform difference in the selling price of the different grades of cattle, with a slightly greater spread between the lower grades. (Data provided by John C. Pierce, Director, Livestock Division, Consumer and Marketing Service, USDA)

QUALITY GRADES

Slaughter cattle quality grades are based on an evaluation of factors related to the palatability of the lean, herein referred to as "quality." Quality in slaughter cattle is evaluated primarily by the amount and distribution of finish, the firmness of muscling, and the physical characteristics of the animal associated with maturity. Progressive changes in maturity past 30 months of age and in the amount and distribution of finish and firmness of muscling have opposite effects on quality. Therefore, for cattle over 30 months of age in each grade, the standards require a progressively greater development of the other quality-indicating factors. In cattle under about 30 months of age, a progressively greater development of the other quality-indicating characteristics is not required.

Since carcass indices of quality are not directly evident in slaughter cattle, some other factors in which differences can be noted must be used to evaluate quality in on-foot cattle. Therefore, the amount of external finish is included as a major grade factor, even though cattle with a specific degree of fatness may have widely varying degrees of quality. Differentiation between the quality of cattle with the same degree of fatness is based on distribution of finish and firmness of muscling.

Approximate maximum age limitations for the specific grades of steers, heifers, and cows are as follows: (1) Prime, Choice, Good, and Standard grades of steers, heifers, and cows—42 months; (2) Commercial grade for steers, heifers, and cows—only cattle over approximately 42 months; (3) no age limitations for the Utility, Cutter, and Canner grades; and (4) all grades of bullocks—approximately 24 months.

YIELD GRADES

There are five yield grades, which are applicable to all classes of slaughter cattle. They are designated by numbers 1 through 5, with yield grade 1 representing the highest degree of cutability.

The yield grades, which estimate the amount of salable meat a carcass will yield, are based on evaluations of muscling and fatness. Those factors and the change in each which is required to make a full yield grade change are summarized in Table 15-10.

TABLE 15-10

FACTORS, AND CHANGES IN EACH, REQUIRED
TO MAKE A FULL YIELD GRADE CHANGE

Factor	Effect of Increase on Yield Grade[1]	Approximate Change in Each Factor Required to Make a Full Yield Grade Change[2]
Thickness of fat over rib eye ..	Decreases	4/10 of an inch
Percent of kidney, pelvic and heart fat	Decreases	5 percent
Carcass weight	Decreases	260 pounds
Area of rib eye	Increases	3 square inches

[1]The yield grades are denoted by numbers 1 through 5 with yield grade 1 representing the highest cutability or yield of closely trimmed retail cuts. Thus, an "increase" in cutability means a smaller yield grade number while a "decrease" in cutability means a larger yield grade number.
[2]This assumes no change in the other factors.

The most practical method of appraising on-foot cattle for yield grade is to use only two factors—muscling and fatness. In this approach, evaluation of the thickness and fullness of muscling in relation to skeletal size largely accounts for the effects of two of the factors—area of rib eye and carcass weight. By the same token, an appraisal of the degree of external fatness largely accounts for the effects of thickness of fat over the rib eye and the percent of kidney, pelvic, and heart fat.

These fatness and muscling evaluations can best be made simultaneously. This is accomplished by considering the development of the various parts based on an understanding of how each part is affected by variations in muscling and fatness. While muscling of most cattle develops uniformly, fat is normally deposited at a considerably faster rate on some parts of the body than on others. Therefore, muscling can be appraised best by giving primary consideration to the parts least affected by fatness, such as the round and the forearm. Differences in thickness and fullness of these parts—with appropriate adjustments for the effects of variations in fatness—are the best indicators of the overall degree of muscling in live cattle.

The overall fatness of an animal can be determined best by observing those parts on which fat is deposited at a faster-than-average rate. These include the back, loin, rump, flank, cod or udder, twist, and brisket. As cattle increase in fatness, these parts appear progressively fuller, thicker, and more distended in relation to the thickness and fullness of the other parts, particularly the round. In thinly muscled cattle with a low degree of finish, the width of the back usually will be greater than the width through the center of the round. The back on either side of the backbone also will be flat or slightly sunken. Conversely, in thickly muscled cattle with a similar degree of finish, the thickness through the rounds will be greater than through the back and the back will appear full and rounded. At an intermediate degree of fatness, cattle which are thickly muscled will be about the same width through the round and back and the back will appear only slightly rounded. Thinly muscled cattle with an intermediate degree of finish will be considerably wider through the back than through the round and will be nearly flat across the back. Very fat cattle will be wider through the back than through the round, but this difference will be greater in thinly muscled cattle than in those that are thickly muscled. Such cattle with thin muscling also will have a distinct break from the back into the sides, while those with thick muscling will be nearly flat on top, but will have a less distinct break into the sides. As cattle increase in fatness, they also become deeper bodied because of large deposits of fat in the flanks and brisket and along the underline. Fullness of the twist and cod or udder and the bulge of the flanks, best observed when an animal walks, are other indications of fatness.

In determining yield grade, variations in fatness are much more important than variations in muscling.

SOME CATTLE MARKETING CONSIDERATIONS

Enlightened marketing practices generally characterize the successful cattle enterprise. Among the considerations of importance in marketing cattle, are those which follow.

Cyclical Trends in Market Cattle

The price cycle as it applies to livestock may be defined as that period of time during which the price for a certain kind of livestock advances from a low point to a high point and then declines to a low point again. In reality, it is a change in animal numbers that represents the stockman's response to prices. Although there is considerable variation in the length of the cycle within any given class of stock, in recent years it has been observed that the price cycle of cattle is about 10 years. (See Fig. 15-15) Also, it has been observed that cyclical trends have become less pronounced.

PRICE CYCLES

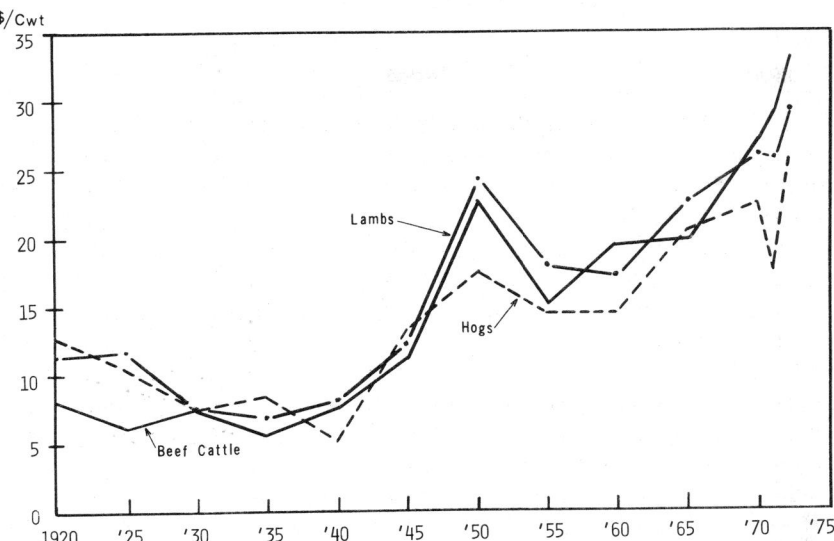

Fig. 15-15. Average price received by U.S. farmers for each class of livestock, 1920-1972. In general, this shows that the price cycle of each animal is approximately as follows: hogs, 3 to 5 years; sheep, 9 to 10 years; and cattle, 10 years. (Data provided by John C. Pierce, Director, Livestock Division, Consumer and Marketing Service, USDA)

The specie cycles are a direct reflection of the rapidity with which the numbers of each class of farm animals can be shifted under practical conditions to meet consumer meat demands. Thus, litter-bearing and early-producing swine can be increased in numbers much more rapidly than either cattle or sheep.

When market cattle prices are favorable, established cattle enterprises are expanded, and new herds are founded, so that about every 10 years, on the average, the market is glutted and prices fall, only to rise again because too few cattle are being produced to take care of the demand for beef. Normal cycles are disturbed by droughts, wars, general periods of depression or inflation, and Federal controls.

Seasonal Changes in Market Cattle

Cattle prices vary by seasons, by classes, and by grades, as shown in Table 15-11, and Figures 15-16, 15-17, 15-18, and 15-19. Consideration is given herein to slaughter steers, slaughter heifers, cows, and feeder steers.

1. *Slaughter steers*—Fig. 15-16 shows the seasonal variation in slaughter steer prices, by grades. From this and other market information, the following conclusions are drawn:

a. The cattle feeder will usually hit the highest market in May to August, and the lowest market from September to March.

b. Prices of steers of various grades parallel each other.

TABLE 15-11

WHEN TO BUY AND SELL CATTLE[1]

Kind of Class of Cattle	Lowest Prices (When to Buy)	Highest Prices (When to Sell)
Slaughter Steers	Sept. to March	May to Aug.
Slaughter Heifers	Sept. to Nov., and Jan.	May to July
Cows	Oct. to Feb.	March to Sept.
Feeder Steers	Aug.	March to June, and Oct. through Dec.

[1]Based on averages.

PRICE OF SLAUGHTER STEERS BY GRADE

PRICE/Cwt

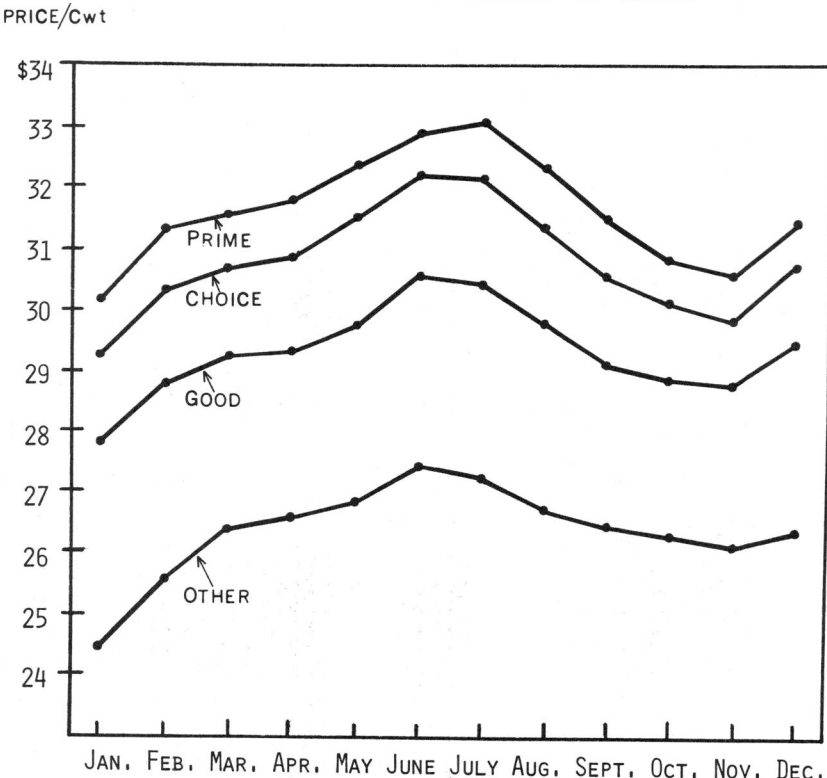

Fig. 15-16. Omaha monthly average price of slaughter steers, by grades, for the 5-year period, 1968-1972. (Based on data from *Livestock and Meat Statistics*, USDA, Statistical Bull. No. 522, pp. 224-227).

c. There is less spread in price between the two top grades, Prime and Choice, than between the other grades.

2. *Slaughter heifers*—Fig. 15-17 shows the seasonal variation in slaughter heifer prices. From this and other market information, the following conclusions are drawn:

PRICE OF SLAUGHTER HEIFERS

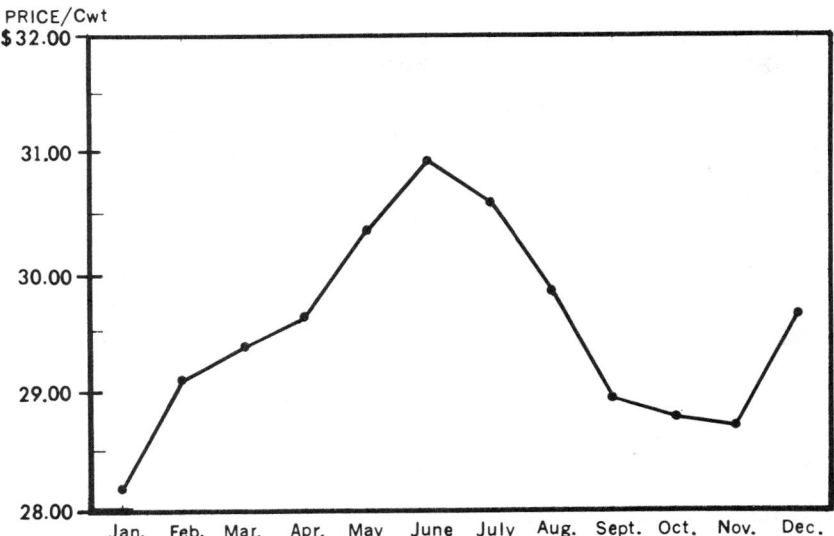

Fig. 15-17. Omaha top price of slaughter heifers for the 5-year period, 1968-1972. This shows the seasonal variation in fat heifer price. (Based on data from *Livestock and Meat Statistics*, USDA, Statistical Bull. 522, p. 233)

a. The price of slaughter heifers shows greatest strength in May to July, and greatest weakness in September to November, and in January.

b. It is more difficult for the producer to take advantage of high market seasons with heifers than with steers, because the former are usually discounted in price after they pass the 750- to 900-pound mark and, therefore, must be sold.

3. *Cows*—Fig. 15-18 shows the seasonal variation in cow prices. From this and other market information, the following conclusions are drawn:

a. Usually, dry cows and cows that are to be culled from the herd for one reason or another had best be marketed from March to September.

b. The heavy run of grass cattle in the late fall causes lower prices for cows from October to February.

c. The lower prices in the fall will likely nullify the value of any increased weight put on cows during the summer and fall. Thus, unless there is an overabundance of grass, or unless the cows can be raised a grade through increased weight gains, it is usually best to market them from March to September.

4. *Feeder cattle*—Fig. 15-19 shows the seasonal variation in feeder steer prices. To cattle feeders, this graph is a buying guide; to cattle producers, it is a selling guide. From this figure and other market information, the following conclusions are drawn:

a. Feeder steer prices are usually lowest in August, and highest from March to June, and from October through December.

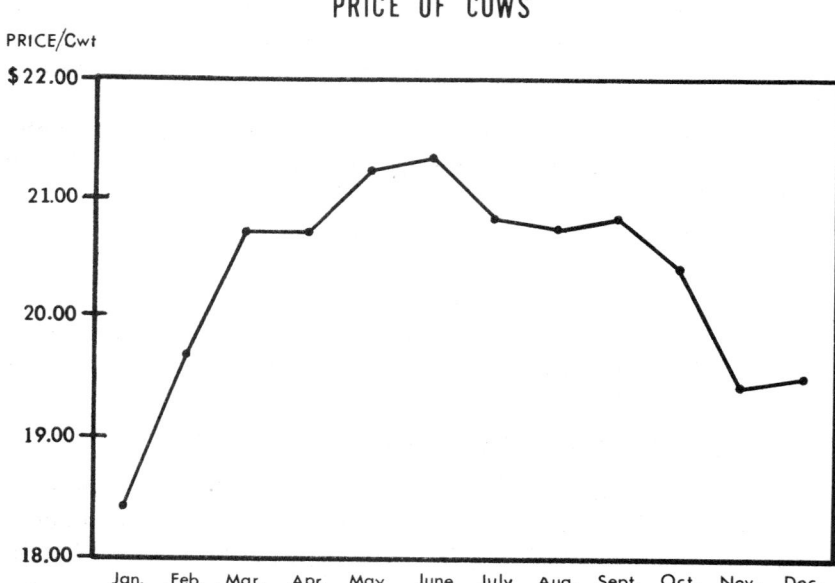

Fig. 15-18. Omaha monthly average price of cows for the 5-year period, 1968-1972. This shows the seasonal variation in cow prices. (*Livestock and Meat Statistics*, USDA, Statistical Bull. No. 522, pp. 243 and 244)

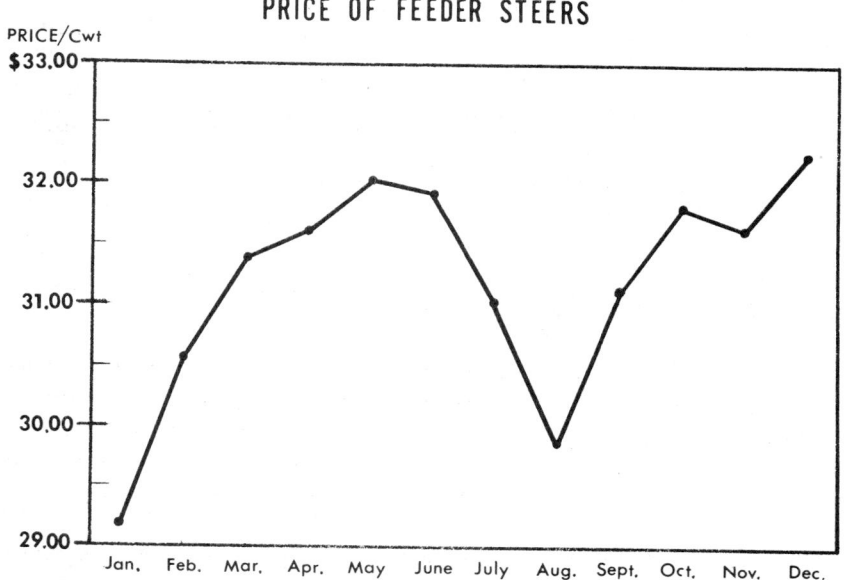

Fig. 15-19. Kansas City monthly average price of feeder steers (of all weights and grades) for the 5-year period, 1968-1972. This shows the seasonal variation in feeder steer prices. (*Livestock and Meat Statistics*, USDA, Statistical Bull. No. 522, p. 217)

b. Feeder cattle receipts are greatest in the fall because this is the season when farmers and ranchers sell their surplus cattle off grass in preparation for winter, and when feeders want to fill their feedlots in order to utilize the new crop and available labor.

c. It is difficult for the producer of feeder cattle to take advantage of seasonal trends because pasture and range conditions often determine, or actually dictate, the time of sale from the range.

Dockage

The value of some market animals is low because dressing losses are high, or because part of the product is of low quality. Cattle with lumpy jaw are usually bought subject to the amount of wastage.

PACKER SLAUGHTERING AND DRESSING OF CATTLE

Table 15-12 shows the proportion of cattle and calves slaughtered commercially (meaning that they were slaughtered in Federally inspected and other wholesale and retail establishments), and the proportion slaughtered on farms. The total figure refers to the number dressed in all establishments and on farms. As shown, farm slaughter is negligible, with a slightly higher percentage of calves farm-slaughtered than of cattle.

TABLE 15-12

PROPORTION OF CATTLE AND CALVES SLAUGHTERED
COMMERCIALLY, 1968 TO 1972[1]

Year	Cattle			Calves		
	Total Number Slaughtered (Commercial and Farm)	Number Slaughtered Commercially	Percent Slaughtered Commercially	Total Number Slaughtered (Commercial and Farm)	Number Slaughtered Commercially	Percent Slaughtered Commercially
	(1,000 head)	*(1,000 head)*	*(%)*	*(1,000 head)*	*(1,000 head)*	*(%)*
1968	35,418	35,026	98.90	5,616	5,443	96.97
1969	35,573	35,237	99.05	5,011	4,863	97.07
1970	35,354	35,025	99.07	4,203	4,072	96.88
1971	35,895	35,585	99.13	3,821	3,689	96.55
1972	36,074	35,779	99.18	3,180	3,053	96.00

[1]*Livestock and Meat Statistics*, USDA Statistical Bull. No. 522, p. 95.

Table 15-13 shows the region, or area, where most of the cattle and calves are slaughtered under Federal inspection. This table points up that by far the greatest number of cattle are slaughtered in the North Central states of North Dakota, South Dakota, Minnesota, Iowa, and Nebraska. But the highest calf slaughter is in the North Atlantic region, where dairy cattle are most concentrated.

Cattle intended for slaughter purposes are bought primarily on the basis of projected quality and yield of carcasses. Upon reaching the packinghouse, they rapidly pass through the operations of killing and dressing. Unlike most manufacturing, meat packing is primarily a disassembly process wherein the manufacturing operation starts with a complete unit that is progressively

TABLE 15-13

REGIONAL LIVESTOCK SLAUGHTER:
NUMBER UNDER FEDERAL INSPECTION, 1972[1]

Region	Cattle	Calves
	(1,000)	*(1,000)*
North Central-East (Ohio, Ind., Ill., Mich., Wisc.)	3,607	431
North Central-N.W. (N.D., S.D., Minn., Iowa, Neb.)	11,408	173
North Central-S.W. (Mo. and Kan.)	3,218	—
North Atlantic (Maine, N.H., Vt., Mass., R.I., Conn., N.Y., N.J., Penn.)	1,148	1,254
South Atlantic (Md., Del., Va., W.Va., N.C., S.C., Ga., Fla.)	729	302
South Central (Ky., Tenn., Ala., Miss., Ark., La., Okla., Tex.)	4,619	114
Mountain (Mont., Ida., Wyo., Colo., N.M., Ariz., Utah, Nev.)	4,009	—
Pacific (Wash., Ore., Calif.)	3,526	131

[1]*Livestock and Meat Statistics*, USDA, Statistical Bull. No. 522, pp. 110-117.

broken down into its component parts. The various parts then are subjected to divergent processing operations. In most of the larger and newer slaughtering plants, cattle are processed with the endless chain method of dressing, similar to that used for dressing of hogs, calves, and sheep.

Steps in Slaughtering and Dressing Cattle

Although the procedure differs between plants, in general the endless chain method of slaughtering and dressing cattle involves the following steps, carried out in rapid succession:

1. *Rendering insensible*—The cattle are rendered insensible.[12] The following methods are accepted as humane for cattle and calves: captive bolt stunners, gun shot, or electric current. Also, carbon dioxide may be used for calves.

2. *Shackling, hoisting, sticking, and bleeding*—The animal is next shackled, hoisted by the hind legs, stuck, and bled. The head is then skinned and removed.

---

[12]By Federal law (known as the Humane Slaughter Act), passed in 1958 and effective June 30, 1960, unless a packer uses humane slaughter methods, he forfeits the right to sell meat to the government. The law lists the following two methods as humane:

1. By rendering insensible to pain by a single blow or gunshot or an electrical, chemical, or other means that is rapid and effective, before being shackled, hoisted, thrown, cast or cut.

2. By slaughtering in accordance with the ritual requirements of the Jewish faith or any other religious faith.

3. *Skinning*—The shanks are skinned and removed at the knees and hocks; beef hooks are inserted on the gam cord; the hide is opened along the median line of the belly and is removed from the belly and sides; then hide pullers are used for removing the rest of the hide in most of the newer plants. The breast and aitch (rump) bones are split by sawing.

4. *Removing viscera*—All internal organs are removed except the kidneys. If the plant is under Federal inspection, the carcass and viscera are examined at this stage in the slaughtering process.

5. *Splitting carcass and removing tail*—The carcass is then split through the center of the backbone and the tail is removed.

6. *Washing and drying*—The split carcasses or sides are washed with warm water under pressure.

7. *Shrouding*—The better carcasses are shrouded tightly with cloth so that they may have a smoother appearance following chilling.

8. *Sending to coolers*—Following slaughtering, the sides are sent to the coolers where they are kept at a temperature of about 34°F for a minimum period of 24 hours before ribbing.

How Slaughter of Veal Calves Differs

Because of their smaller size, calves are almost always dressed by the endless-chain method. A wheel hoist is used in lifting the shackled calves to the rail. They are then stuck, bled, dressed, and washed. Because of the high moisture content of veal, the hide is usually left on for the purpose of reducing evaporation. This also produces a more desirable carcass color. When the hide is left on, it is thoroughly washed before dressing.

Kosher Slaughter

Meat for the Jewish trade must come from animals slaughtered according to the rules of *Shehitah* (the ancient dietary rules). Although we usually think in terms of cattle when kosher slaughtering is mentioned, calves, sheep, lambs, goats, and poultry are slaughtered in a similar manner.

The killing is performed by a rabbi of the Jewish church or a specially trained representative; a person called the *shohet* or *shochet*, meaning slaughterer. In kosher slaughter, the animal is hoisted without stunning and is cut across the throat with a special razor-sharp knife, known as a *chalaf*. With one quick, clean stroke the throat is cut, through the jugular vein and other large vessels, together with the gullet and windpipe. Two reasons are given for using this method of killing instead of the more conventional method of stunning and sticking; namely, (1) it produces more instant death with less pain, and (2) it results in more rapid and complete bleeding, which Orthodox Hebrews consider essential for sanitary reasons.

The shohet also makes an inspection of the lungs, stomach and other organs while dressing. If the carcass is acceptable, it is marked on the brisket with a cross inside a circle. The mark also gives the date of slaughter and the name of the inspector.

Since neither packers nor meat retailers can hold kosher meat longer than 216 hours (and even then it must be washed at 72-hour intervals), rapid handling is imperative. This fact, plus the heavy concentration of Jewish folks in the eastern cities results in large numbers of live cattle being shipped from the markets farther west to be slaughtered in or near the eastern consuming areas.

The Dressing Percentage

Dressing percentage may be defined as the percentage yield of hot carcass in relation to the weight of the animal on foot. For example, a steer which weighed 1,200 pounds on foot and yielded a hot carcass weighing 720 pounds may be said to have a dressing percentage of 60. The offal—so-called because formerly (with the exception of the hide, tallow, and tongue) the offal (waste) was thrown away—consists of the blood, head, shanks, tail, hide, viscera, and loose fat.

A high carcass yield is desirable because the carcass is much more valuable than the by-products. Although the packers have done a marvelous job in utilizing by-products, about 90.8 percent of the income from cattle and calves is derived from the sale of the carcass and only 9.2 percent from the by-products. Thus, the estimated dressing percentage of slaughter cattle is justifiably a major factor in determining the price or value of the live animal.

The chief factors determining the dressing percentage of cattle are: (1) the amount of fill; (2) the finish or degree of fatness; (3) the thickness of muscling; (4) the general quality and refinement (refinement of head, bone, hide, etc); and (5) the size of udder. The better grades of steers have the highest dressing percentage, with thin canner cows showing the lowest yield. Table 15-14 and 15-15 give the dressing percentages that may be expected for different grades of cattle and calves.

TABLE 15-14

RANGE AND AVERAGE IN DRESSING PERCENTAGES,
BASED ON HOT CARCASS WEIGHTS, OF CATTLE, BY GRADES[1]

Grade	Dressing Percentage	
	Range	Average
	(%)	(%)
CATTLE:		
Prime	62-67	64
Choice	59-65	62
Good	58-62	60
Standard	55-60	57
Commercial	54-62	57
Utility	49-57	53
Cutter	45-54	49
Canner	40-48	45

[1]Dressing percentages provided by John C. Pierce, Director, Livestock Division, Consumer and Marketing Service, USDA.

TABLE 15-15

RANGE AND AVERAGE IN DRESSING PERCENTAGES BASED ON
HOT CARCASS WEIGHTS, OF CALVES AND VEALERS, BY GRADES[1]

Grade	Dressing Percentage	
	Range	Average
	(%)	(%)
CALVES AND VEALERS:		
Prime	59-65	62
Choice	56-60	58
Good	52-57	55
Standard	47-54	51
Utility	40-48	46

[1]Dressing percentages provided by John C. Pierce, Director, Livestock Division, Consumer and Marketing Service, USDA.

The highest dressing percentage on record was a yield of 76¾ percent made by a spayed Angus heifer at the Smithfield Fat Stock Show in England. The average liveweights of cattle and calves dressed by Federally inspected meat packing plants, and their percentage yield in meats, for the year 1971 are given in Table 15-16.

TABLE 15-16

AVERAGE LIVEWEIGHT, DRESSED WEIGHT AND DRESSING
PERCENTAGES OF ALL CATTLE AND CALVES COMMERCIALLY
SLAUGHTERED IN THE U.S. IN 1971[1]

Percentage	Average Liveweight	Average Dressed Weight	Dressing Percentage
	(lb)	(lb)	(%)
Cattle	1,028	611	59.4
Calves	249	140	56.2

[1]Data from *Livestock and Meat Statistics*, Statistical Bull. No. 333, USDA.

Fabricated, Boxed Beef

Traditionally, beef is shipped in exposed halves, quarters, or wholesale cuts and divided into retail cuts in the backrooms of 250,000 supermarkets. This procedure leaves much to be desired from the standpoints of efficiency, sanitation, shrink, spoilage, and discoloration. To improve this situation, more and more packers are fabricating and boxing beef in their plants. In the Iowa Beef Processors' ultramodern plant at Dakota City, Nebraska, beef is fabricated and handled as follows: after chilling, the carcass is subjected to a disassembly process, in which it is fabricated or broken into subprimal cuts; vacuum-sealed; boxed; moved into storage by a fully automated, computer-controlled system; loaded by computer-automation into trailers; and shipped to retailers across the nation. The fabrication and boxing of beef at the packing plant will increase in the future.

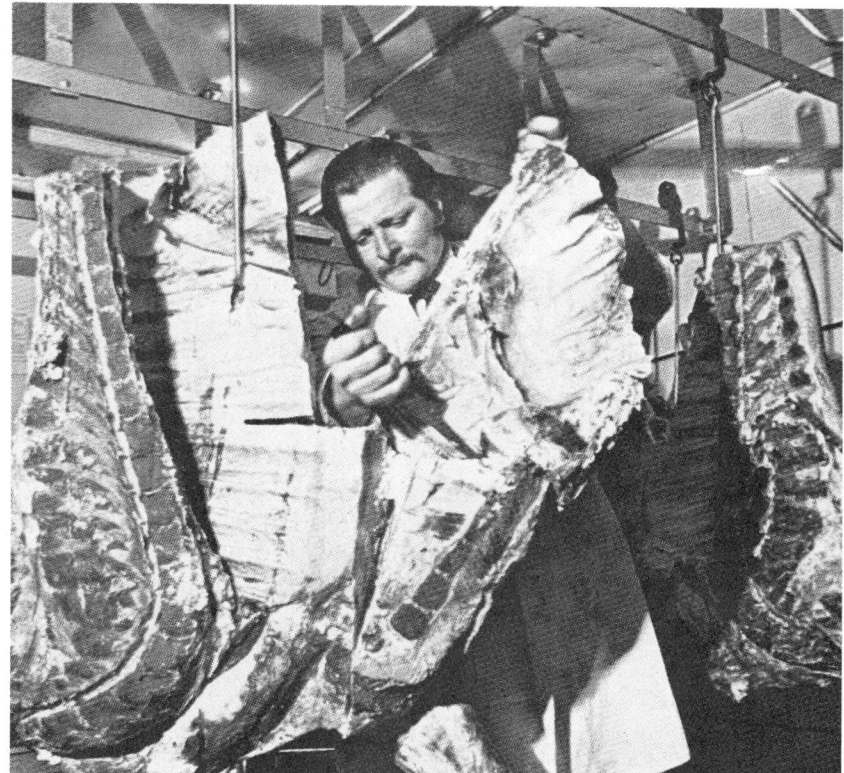

Fig. 15-20. The traditional: Breaking beef carcasses at the store level. This necessitates moving excess fat and bone 1,000 miles or more and incurring excessive shrink. (Courtesy, Iowa Beef Processors, Inc., Dakota City, Neb.)

Fig. 15-21. The modern: (Left) Each cut is trimmed to specification, placed in a vacuum-sealed bag, and boxed. (Center) Each box is code-labeled and moved by computer to storage. (Right) Boxed beef is stored at 28°F prior to shipment. (Courtesy, Iowa Beef Processors, Inc., Dakota City, Neb.)

Fig. 15-22. Efficiency: By fabrication at the packing plant, approximately 100 pounds of fat (left) and bone (right) per carcass remain at the plant. No longer is it necessary to ship unwanted fat and bone across the country. (Courtesy, Iowa Beef Processors, Inc., Dakota City, Neb.)

MOVEMENT OF FEEDER CATTLE

In the past, the traditional sources of feeder cattle for the Corn Belt and the West have been the cow-calf operations in the Southwest. However, in recent years, the number of cattle moving between these areas has decreased as the Southwest has kept back feeder cattle to supply its own ever-increasing and larger cattle feeding program. Texas and Kansas now feed more cattle than are available as feeders from their own yearly calf crop.

In the meantime, the southern and southeastern states have become more and more important as a source of feeder cattle. This transition is the result of (1) increases in the percent calf crop produced, and (2) a sizable reduction in veal calf slaughter.

In the future, the Corn Belt will produce more of its feeder cattle, almost out of necessity, and a number of feeder cattle deficit states will depend on feeder cattle from the southeastern states. Many of these may be grown on pastures in Kansas, Oklahoma, and Texas before reaching feedlots, as in the past. A predicted direction of feeder cattle movement to 1985 from their source to feedlots for finishing is shown in Fig. 15-23.

FEEDER CATTLE MOVEMENT

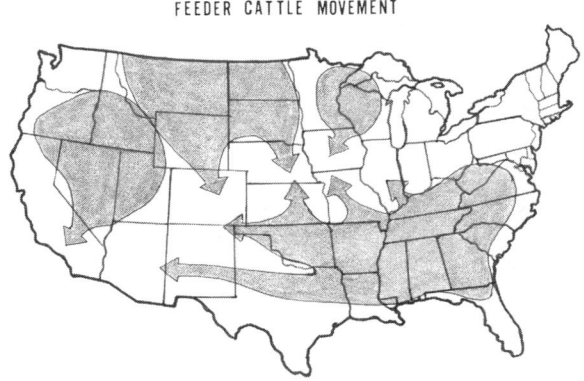

Fig. 15-23. Predicted directions of feeder cattle movements to 1985.

SELLING PUREBRED CATTLE

Selling purebred animals is a highly specialized and scientific business. Purebred animals are usually sold at private treaty directly to other purebred breeders or commercial producers or through auctions which may either be sponsored by one or a few breeders (joint sales or consignment sales).

Fig. 15-24. The success of a cattle auction depends upon many things in addition to the quality of the cattle; among them, (1) the number of buyers present, rather than just spectators; (2) the presence of a specialty consignment(s) which will attract buyers, who may bid on other consignments as well; and (3) the reputation of and homework done by the sales manager and consignors.

Altogether too often, a good or bad sale is a reflection of the above factors, rather than the quality of the cattle. (Courtesy, *Bull-O-Gram*, published by Litton Charolais Ranch, Chillicothe, Mo.)

In general, the vast majority of bulls saved for breeding purposes go into commercial herds. Only the elite sires are retained with the hope of affecting further breed improvement in purebred herds. On the other hand, the sale of purebred females is fairly well restricted to meeting the requirements for replacement purposes in existing purebred herds or for establishing new purebred herds.

Most consignment sales are sponsored by a breed association, either local, statewide, or national in character. Such auctions, therefore, are usually limited to one breed. Purebred auction sales are conducted by highly specialized auctioneers. In addition to being good salesmen, such auctioneers must have a keen knowledge of values and be familiar with the blood lines of the breeding stock.

QUESTIONS FOR STUDY AND DISCUSSION

1. Discuss the magnitude of cattle marketing.

2. Many top authorities feel that marketing, not production, is the cattleman's number one problem. Do you agree? Justify your answer.

3. U.S. farmers and ranchers receive a major portion of their cash income from cattle (a fantastic 30% of their total income from livestock and products *plus* crops; 51% of livestock and products—exclusive of crops). Is this good or bad from the standpoint of cattlemen? Is it good or bad from the standpoint of consumers?

4. Since World War I, public stockyards have declined in importance while country sales (direct marketing) has increased. Why has this happened?

5. Why are most calves sold through auctions, whereas most cattle are sold direct?

6. What market channel(s) would you recommend for the sale of cattle off your farm or ranch, or off a farm or ranch with which you are familiar?

7. Does each market channel give adequate assurance of honesty, of sanitation, and of humane treatment of animals?

8. Why is it important that a cattleman know the leading markets for each class of cattle?

9. How do you account for the fact that the leading terminal public markets for finished cattle, for calves, and for feeder cattle do not coincide?

10. Which is the more important to the livestock seller: (a) low marketing costs; or (b) effective selling and net returns?

11. Outline, step by step, how you would prepare and ship cattle.

12. Under what circumstances would you recommend the use of each of the following methods of transporting market cattle: (a) truck; (b) rail; and (c) air?

13. Assume that you have two offers for a 1,000-pound steer: one buyer is offering $40.00/cwt with a 4% pencil shrink. The other buyer is offering $39.00/cwt with no pencil shrink. Which offer would you take? Show your computations.

14. Define on-foot market (a) classes, and (b) grades of cattle, and tell of their value.

15. Why was the "bullock" sex class added? Will we feed out more bulls in the future?

16. What's the difference between (a) quality grades and (b) yield grades?

17. Why is it important that a cattleman know the market classes and grades of cattle and what each implies?

18. Why are grade names and specifications changed from time to time?

19. Since there is a rather uniform difference in the selling price of the different grades of cattle, with the top grades bringing the higher prices, why do not more cattlemen produce the top grades?

20. How may a cattleman take advantage of cyclical trends and seasonal changes?

21. Discuss the importance of each: (a) proper bedding for cattle in transit, and (b) shrinkage.

22. Do packers control market cattle prices?

23. How do you account for the commanding lead of the North Central region in number of cattle slaughtered under Federal inspection?

24. Discuss the movement of feeder cattle, including (a) production, (b) the stocker stage, (c) feedlot, and (d) slaughter.

25. In what ways does the selling of purebred cattle differ from the selling of commercial cattle?

SELECTED REFERENCES

Title of Publication	Author(s)	Publisher
Animal Science, Sixth Edition	M. E. Ensminger	The Interstate Printers & Publishers, Inc., Danville, Ill., 1969
Essentials of Marketing Livestock	R. C. Ashby	R. C. Ashby, Sioux City, Iowa, 1953
Lessons on Meat, Second Edition	National Live Stock and Meat Board	National Live Stock and Meat Board, Chicago, Ill., 1972
Livestock Marketing	A. A. Dowell, K. Bjorka	McGraw-Hill Book Company, Inc., New York, N.Y., 1941
Livestock and Meat Marketing	J. H. McCoy	The Avi Publishing Company, Inc., Westport, Conn., 1972
Marketing, The Yearbook of Agriculture 1954	Ed. by A. Stefferud	U.S. Department of Agriculture, Washington, D.C., 1954
Marketing of Livestock and Meat, The, Second Edition	S. H. Fowler	The Interstate Printers & Publishers, Inc., Danville, Ill., 1961
Meat Handbook, The, Second Edition	A. Levie	The Avi Publishing Company, Inc., Westport, Conn., 1967
Problems and Practices of American Cattlemen, Wash. Agr. Exp. Sta. Bull. 562	M. E. Ensminger, M. W. Galgan, W. L. Slocum	Washington State University, Pullman, Wash., 1955
Processed Meats	W. E. Kramlich, A. M. Pearson, F. W. Tauber	The Avi Publishing Company, Inc., Westport, Conn., 1973
Stockman's Handbook, The, Fourth Edition	M. E. Ensminger	The Interstate Printers & Publishers, Inc., Danville, Ill., 1970
Uniform Retail Meat Identity Standards	Industrywide Cooperative Meat Identification Standards Committee	National Live Stock and Meat Board, Chicago, Ill., 1973

CHAPTER 16

BEEF AND VEAL, AND BY-PRODUCTS
FROM CATTLE SLAUGHTER

Beef over the block is the ultimate objective in producing cattle; it is the end product of all breeding, feeding, care and management, marketing, and processing. It is imperative, therefore, that the progressive cattleman, the student, and the beef cattle scientist have a working knowledge of beef and veal and of the by-products from cattle slaughter. Such knowledge will be of value in selecting animals and in determining policies relative to their handling. To this end, this chapter is presented.

Of course, the type of animals best adapted to the production of meat over the block has changed in a changing world. Thus, in the early history of this country, the very survival of animals was often dependent upon their speed, hardiness, and ability to fight. Moreover, long legs and plenty of bone were important attributes when it came time for animals to trail hundreds of miles as drovers took them to market. The Texas Longhorn was adapted to these conditions.

Fig. 16-1. A modern meat market. Meat over the block is the ultimate objective in producing cattle, sheep, and swine. Providing America's meat requires about 300,000 retail meat dealers. (Courtesy, Tyler Refrigeration Corporation, Niles, Mich.)

With the advent of rail transportation and improved care and feeding methods, the ability of animals to travel and fight diminished in importance. It was then possible, through selection and breeding, to produce meat animals better suited to the needs of more critical consumers. With the development of large cities, artisans and craftsmen and their successors in industry required fewer calories than those who were engaged in the more arduous tasks of logging, building railroads, etc. Simultaneously, the American family decreased in size. The demand shifted, therefore, to smaller and less fatty cuts of meats; and, with greater prosperity, high quality steaks and roasts were in demand. To meet the needs of the consumer, the producer gradually shifted to the breeding and marketing of younger animals with maximum cutout value, instead of marketing large, ponderous, fat, three- to five-year-old steers.

Thus, through the years, consumer demand has exerted a powerful influence upon the type of cattle produced. To be sure, it is necessary that such production factors as prolificacy, economy of feed utilization, rapidity of gains, size, and longevity, receive due consideration along with consumer demands. But once these production factors have received due weight, cattle producers—whether they be purebred or commercial operators—must remember that meat over the block is the ultimate objective.

Now, and in the future, beef producers need to select and feed so as to obtain increased quality and cutout value, without excess fat.

QUALITIES IN BEEF DESIRED BY THE CONSUMER

Because consumer preference is such an important item in the production of beef, all members of the beef team—the farmer and rancher, the packer, and the meat retailer—should be familiar with these qualities, which are summarized as follows:

1. *Palatability*—First and foremost, people eat meat because they like it. Palatability is influenced by the tenderness, juiciness, and flavor of the fat and lean.

2. *Attractiveness*—The general attractiveness is an important factor in selling meats to the housewife. The color of the lean, the degree of fatness, and the marbling are leading factors in determining buyer appeal. Most consumers prefer a white fat and a light or medium red color in the lean.

3. *Moderate amount of fat*—Middle- and low-income groups discriminate against too much fat, especially when it must be trimmed heavily.

4. *Tenderness*—Consumers want fine-grained, tender meat, in contrast to coarse-grained, less tender meat.

5. *Small cuts*—Most purchasers prefer to buy cuts of meat that are of a proper size to meet the needs of their respective families. Because the American family has decreased in size, this has meant smaller cuts. In turn, this has had a profound influence on the type of animals and on market age and weight.

6. *Repeatability*—The housewife wants a cut of meat just like the one that she purchased last time, which calls for repeatability.

7. *Ease of preparation*—In general, the housewife prefers to select those cuts of beef that will give her the greatest amount of leisure time. Steaks of beef can be prepared with greater ease and in less time than can roasts or stews. Hamburger and sausage are also easy to prepare.

THE FEDERAL GRADES OF BEEF[1]

In 1926, a producer group, known as The Better Beef Association, petitioned the U.S. Department of Agriculture to set up a service for grading and stamping beef. Such a program was activated in 1927, and continues to this day. Since then, federal beef grades have been changed many times, in response to changes in consumer tastes and preferences, as well as changes in technology.

On March 6, 1975, the U.S. Department of Agriculture announced beef grading changes to become effective April 14, 1975. Although subsequent court action halted the implementation of the new grade standards for nearly a year, they became effective on February 23, 1976. Basically, the revised standards are designed to accomplish the following:

1. Reduce slightly the marbling required up to 30 months of age for Prime and Choice beef, thereby making it possible for leaner and younger beef to qualify for the 2 top grades.

---

[1]This section was authoritatively reviewed by Mr. John C. Pierce, Director, Livestock Division, Agricultural Marketing Service, USDA.

2. Establish a more restrictive (narrower) Good grade.

3. Eliminate conformation as a factor when carcasses are graded for quality.

4. Make yield grading mandatory; that is, require that all graded beef be dual graded—graded for both quality and yield (percentage of retail cuts).

Unlike meat inspection, government meat grading is purely voluntary, on a charge basis. In 1974, a government meat grader cost $16 per hour, plus the use of an hourly paid man to roll the beef.

The value of beef carcass depends chiefly upon two factors—the quality of the meat and the amount of salable meat the carcass will yield, particularly the yield of the high value, preferred, retail cuts. To reflect these factors, all USDA graded carcasses are identified for both quality grade and yield grade.

Because the USDA beef grades have become so widely used in marketing, it is important that everyone from producer to consumer have an understanding of the Federal standards for grades of beef. Producers need to understand grades in order to make wise production and marketing plans. Meat packers, wholesalers, and retailers also must be familiar with grades to facilitate their buying and selling operations. Consumers need to have an understanding of the grades in order to select the quality they desire. USDA yield grades also are important to consumers when they buy carcasses or wholesale cuts for the freezer.

Pertinent facts about each the quality grade and the yield grade follow:

1. *Quality grade*—Quality refers to the palatability-indicating characteristics of the lean and is evaluated by considering the marbling (flecks of fat within the lean) and firmness of the lean as observed in a cut surface in relation to the apparent maturity of the animal from which the carcass was produced. The maturity of the carcass is determined by evaluating the size, shape, and ossification of the bones and cartilages—especially the split chine bones—and the color and texture of the lean flesh. Superior quality implies firm, well-marbled lean that is fine in texture and has a light red, youthful color.

USDA quality grades for beef—Prime, Choice, Good, Standard, Commercial, Utility, Cutter, and Canner—have for many years served as nationally reliable guides to the eating quality of beef—its tenderness, juiciness, and flavor.

Research has confirmed that marbling and maturity are the two most important factors that can be used in grading to identify differences in palatability.

In general, increases in marbling improve the palatability or eating quality of beef, but increases in maturity have the opposite effect. However, some recent research has indicated that for younger cattle (under 30 months of age) maturity changes do not have an appreciable effect upon palatability. Therefore, within each of the Prime, Choice, Good, and Standard grades the minimum requirement for marbling is the same for all carcasses from animals under 30 months of age. This reduction in marbling requirement, which was promulgated by the USDA in 1976, resulted in slightly leaner beef with less excess fat, particularly in the Prime and Choice grades, and less grain required to produce cattle that would qualify for the top grades.

For cattle over 30 months of age, increased marbling is required with increasing maturity within each grade.

There is no information which indicates that variations in conformation are related to differences in beef's palatability. Therefore, in 1976, the USDA eliminated consideration of conformation as a factor in determining the quality grade.

2. *Yield grade*—It is recognized that variations in conformation (shape) due to differences in muscling do affect yields of lean—and carcass value. This is reflected by the yield grades. The significance of yield grades becomes evident when tests reveal that carcasses of the same quality grade— Choice for example—can vary in value by $75 or more due to differences in cutability.

The yield grade of a beef carcass is determined by considering four characteristics: (1) the amount of external fat; (2) the amount of kidney, pelvic, and heart fat; (3) the area of the ribeye muscle; and (4) the carcass weight.

Thus, in the same way that quality grades identify beef for differences in palatability, USDA yield grades for beef provide a nationally uniform method of identifying cutability differences among carcasses. Specifically, they are based on the percentage yields of boneless, closely trimmed, retail cuts from the high value parts of the carcass—the round, loin, rib, and chuck—which account for more than 80 percent of its value. However, they also reflect differences in total yields of retail cuts.

There are five USDA yield grades numbered 1 through 5. Yield Grade 1 carcasses have the highest yields of retail cuts; yield Grade 5 the lowest. A carcass which is typical of its yield grade would be expected to yield about 4.6 percent more in retail cuts than the next lower yield grade, when USDA cutting and trimming methods are followed. (Other differences in yield may result from different cutting procedures, but these should result in similar differences between yield grades.) When used in conjunction with quality grade, yield grades can be of benefit to all segments of the industry.

A discussion of each of the factors affecting yield grade follows:

a. *External fat*—The amount of fat over the outside of a carcass is the most important yield grade factor since it is a good indication of the amount of fat that is trimmed in making retail cuts. The less trimmable fat, the higher the yield grade and the more the carcass is worth. A single fat thickness measurement over the ribeye muscle has been found to be the most practical indicator of external fatness for use in a grading program. Four-tenths of an inch variation in thickness of fat over the ribeye makes a full yield grade change.

b. *Kidney, pelvic, and heart fat*—Fat deposits on the inside of the carcass around the kidney and in the pelvic and heart areas also affect yields. Since practically all of this fat is removed in trimming, increases in these fats decrease the yields of retail cuts. A change of five percent in these fats makes a full yield grade change.

c. *Area of ribeye*—The ribeye muscle lies on each side of the backbone and runs the full length of the back. It is the largest muscle in the carcass and one of the most palatable. When the side is separated into a hindquarter and a forequarter, a cross section of the ribeye muscle is exposed. Its area (in square inches) at this point is another factor used in determining the yield grade. Among carcasses of the same fatness and weight, an increase in the ribeye area indicates an increase in the yield of retail cuts. A change of 3 square inches in ribeye area make almost a full yield grade change.

d. *Carcass weight*—The warm carcass weight is the weight used in yield grading. When carcass weight is used in conjunction with the other three yield factors, an increase in weight indicates a decrease in yield of retail cuts. Weight is the least important yield grade factor. It takes a change of about 250 pounds in carcass weight to make a full yield grade change.

Since yield grades essentially measure proportions of lean and fat, USDA meat graders can determine the correct yield grade of most carcasses after a simple, rapid, visual appraisal of the fatness and muscling of the carcass. But objective measurements can be made when necessary and the yield grade also can be calculated from an equation which gives each factor its proper weight. Very little extra time is required to grade carcasses for both quality and yield over that required to grade them only for quality.

Federally graded meats are so stamped (with an edible vegetable dye) that the grade will appear on the retail cuts as well as on the carcass and wholesale cuts. These are summarized in Table 16-1 for beef and veal.

TABLE 16-1

FEDERAL QUALITY AND YIELD GRADES OF BEEF AND VEAL[1]

Beef		Calf and Veal
Quality Grades	**Cutability Grades**	**(Quality Grades Only)**
1. Prime[2]	1. Yield Grade 1	1. Prime
2. Choice	2. Yield Grade 2	2. Choice
3. Good	3. Yield Grade 3	3. Good
4. Standard	4. Yield Grade 4	4. Standard
5. Commercial	5. Yield Grade 5	5. Utility
6. Utility		
7. Cutter		
8. Canner		

[1]In rolling meat, the letters "USDA" are included in a shield with each Federal Grade name. This is important, as only government-graded meat can be so marked. For convenience, however, the letters "USDA" are not used in this table or in the discussion which follows.
[2]Cow beef is not eligible for the prime grade. The quality grade designations for bullock carcasses are Prime, Choice, Good, Standard, and Utility. Bull carcasses are eligible for yield grade only.

Unlike meat inspection, government grading is purely voluntary, on a charge basis. In 1971, 64.3% of the total U.S. commercial production of beef was Federally graded for quality, and 13.4% of the veal and calf. In addition, 18.2 % of the total beef production was yield graded.

Fig. 16-2 shows the percentage distribution of beef by classes and grades. American consumers are eating better grade beef than ever before. As shown, about 80 percent of all beef eaten in 1972 was in the three top grades—Prime, Choice, and Good.

BEEF PRODUCTION BY CLASS AND BY GRADE, 1972

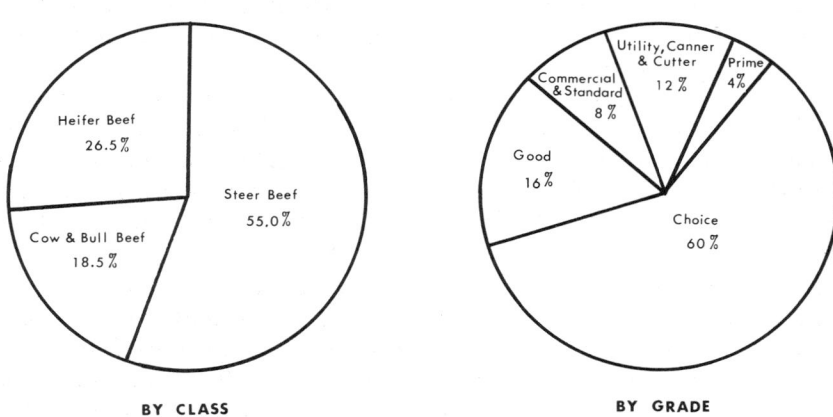

BY CLASS   BY GRADE

Fig. 16-2. 81.5% of all beef produced is steer and heifer beef; most of the rest is cow beef. Because the largest part of steer and heifer is the equivalent of Good grade or better, 80% of the total beef production is of the three top grades. (Source: *Livestock and Meat Statistics*, USDA, Statistical Bull. No. 522, pp. 127-130)

Some additional and pertinent facts relative to Federal grades of beef are:

1. *There is no sex differentiation between steer, heifer and cow beef*—Federal grades make no distinction between steer, heifer and cow beef. It is not intended that this should be construed to imply that there is no carcass difference between these sex classes. Rather, there is no evidence that beef from steers, heifers, and cows with the same combination of quality-indicating characteristics is different in palatability. This step in simplification was taken so that it might be easier for the buyer or consumer to purchase meat on the basis of quality, without the added confusion of a more complicated system.

2. *Bull and bullock beef are identified*—Bull and bullock beef are identified by class as "bull" beef and "bullock" beef, respectively. Within the bullock class, there are the following five grades: (1) Prime; (2) Choice; (3) Good; (4) Standard; and (5) Utility. However, no designated grade of bullock beef is necessarily comparable in quality to a similarly designated grade of beef obtained from steers, heifers, or cows. Neither is the yield in a designated yield grade of bull or bullock beef comparable to a similarly designated yield grade of steer, heifer, or cow beef.

3. *Lower grades seldom sold as retail cuts*—It is seldom that the lower grades—Cutter and Canner beef, and Utility veal—are sold as retail cuts. The consumer, therefore, only needs to become familiar with the upper grades of each kind of meat.

4. *Yield grading wholesale cuts*—Only loins, short loins, and ribs can be graded as individual cuts, because these are the only cuts which contain a cross section of the ribeye muscle at the twelfth rib—a requirement in determining the yield grade. However, rounds, chucks, and other wholesale cuts can be graded as cuts if they remain attached to a rib, short loin, or loin.

5. *Yield grading will reduce excess fat*—The real potential for reducing excess fat on beef is dependent upon the greater use of the yield grading system in reflecting appropriate price differentials.

6. *Use of quality and yield grades will be reflected from consumers to producers*—Grading beef carcasses for both quality and yield, rather than for one or the other, increases the effectiveness of the grades as a tool for reflecting consumer preferences back through marketing channels to producers. Also, if the market for beef and cattle reflects the full retail sales value differences associated with variations in both quality and cutability, producers will respond by increasing the production of high quality, high cutability beef. This would be advantageous to all segments of the industry and to consumers by providing leaner beef with less waste in keeping with consumer tastes.

AGING BEEF

Except for veal, fresh beef is not at optimum tenderness immediately after slaughter. It must undergo an aging or ripening process before it really becomes tender. This process consists of the dissolution of the connective tissue (collagen) by the action of enzymes. Beef should be aged from 2 to 6 weeks at temperatures ranging from 34° to 38°F, but only the better grades

Fig. 16-3. Aging beef in cooler. It should be aged from 2 to 6 weeks at temperatures ranging from 34° to 38° F. (Courtesy, *Meat Magazine*)

can be aged for the longer periods. Beef must have a fat covering to protect the meat from bacterial action by sealing it from the air. With well-finished beef, some trimming is necessary anyway and the removal of the mold does not constitute any additional loss. The aging process may be hastened by the use of ultraviolet lights in high temperature coolers with controlled humidity.

DISPOSITION OF THE BEEF CARCASS

Beef carcasses are of two types, namely those suitable for (1) block beef, or (2) processed meats.

1. *Block beef*—Block beef refers to beef that is suitable for sale over the block. Such beef is purchased by the retailer in sides, quarters, wholesale cuts, or retail cuts. Block beef may enter regular channels of trade either as "fresh chilled" or "fresh frozen." Fresh chilled beef is chilled at temperatures ranging from 34° to 38° F for a minimum of 24 hours before moving out to the retail trade, or it may be held longer if aging is desired. Frozen beef is subjected to temperatures of 0° F or below and is frozen solid, in which form it can be kept for a period of several months. U.S. consumers prefer "fresh chilled" beef, although frozen beef is increasing in acceptance and quantity. On the other hand, the bulk of imported beef is frozen, rather than chilled.

2. *Processed meats*—Beef that is not suitable for sale over the block is (a) boned out and disposed of as boneless cuts, (b) canned, (c) made into sausage, or (d) cured by drying and smoking. It is estimated that about one-fifth of all slaughter cattle are disposed of as processed meats.

THE BEEF AND VEAL CARCASSES AND WHOLESALE CUTS

Fig. 16-4 shows the location of the wholesale cuts on a live animal.

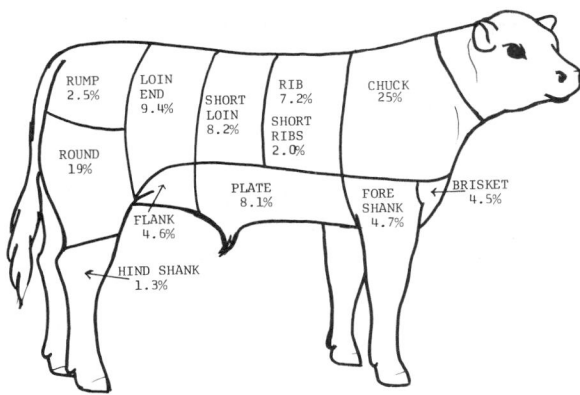

Fig. 16-4. Location of the wholesale cuts commonly derived from a beef animal.

There was a time when each area of the United States had its traditional cuts of beef and veal. However, increased central processing and boxed beef prompted the need for greater uniformity in cutting and labeling, among both packers and retailers. Out of this need arose a new nationwide standardized identification-labeling system, coordinated by the National Live Stock and Meat Board and adopted by the industry in 1973. The names for various cuts of beef, pork, and lamb sold in U.S. food stores were reduced from more than 1,000 to about 300. As a result of this system of uniform labeling, a ribeye steak is a ribeye steak—not a Delmonico steak at one place, a filet steak someplace else, or a spencer steak or beauty steak in still other stores, depending on where you live in the United States—or even where you shop in the same city. Figures 16-5 and 16-6 show for beef and veal, respectively, the primal area (or wholesale cuts) from which the retail cuts are derived.

PRIMAL (WHOLESALE) CUTS AND BONE STRUCTURE OF BEEF

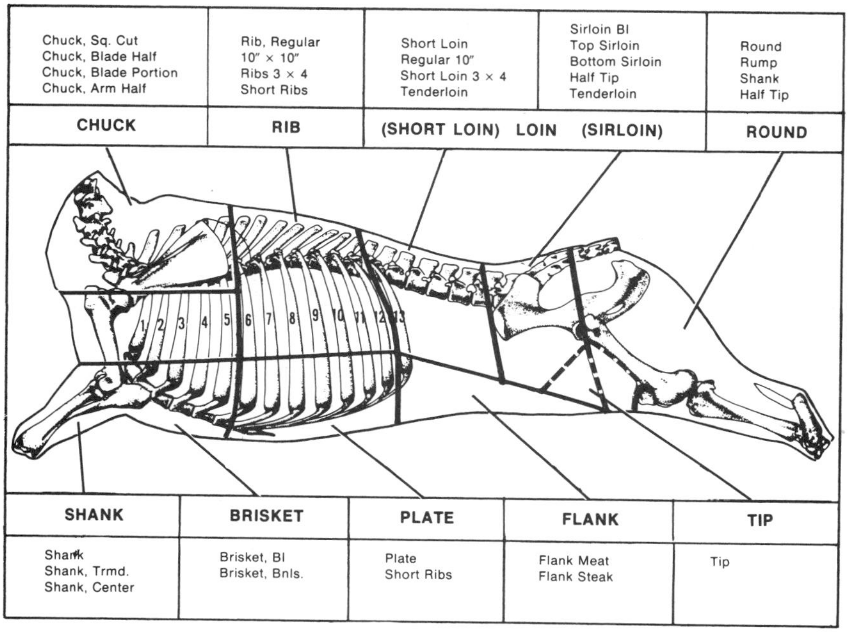

Chuck, Sq. Cut Chuck, Blade Half Chuck, Blade Portion Chuck, Arm Half	Rib, Regular 10" x 10" Ribs 3 x 4 Short Ribs	Short Loin Regular 10" Short Loin 3 x 4 Tenderloin	Sirloin Bl Top Sirloin Bottom Sirloin Half Tip Tenderloin	Round Rump Shank Half Tip
CHUCK	**RIB**	**(SHORT LOIN) LOIN (SIRLOIN)**		**ROUND**
SHANK	**BRISKET**	**PLATE**	**FLANK**	**TIP**
Shank Shank, Trmd. Shank, Center	Brisket, Bl Brisket, Bnls.	Plate Short Ribs	Flank Meat Flank Steak	Tip

Fig. 16-5. Beef primal (wholesale) and subprimal cuts, and bone structure. (Courtesy, National Live Stock and Meat Board, Chicago, Ill.)

PRIMAL (WHOLESALE) CUTS AND BONE STRUCTURE OF VEAL

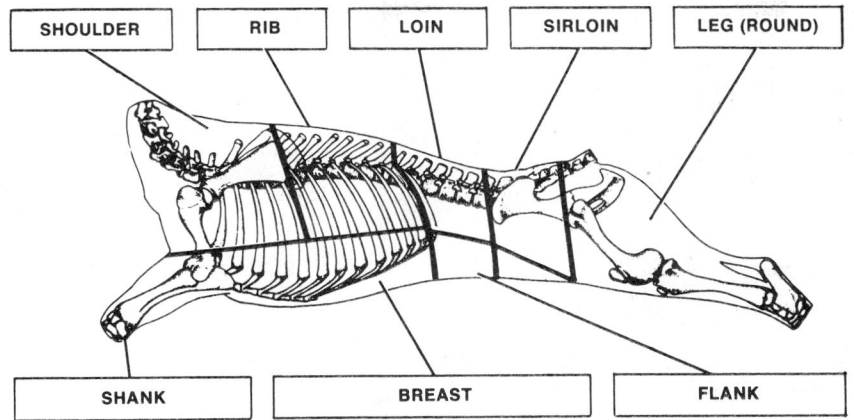

| SHOULDER | RIB | LOIN | SIRLOIN | LEG (ROUND) |

| SHANK | BREAST | FLANK |

Fig. 16-6. Veal primal (wholesale) and subprimal cuts, and bone structure. (Courtesy, National Live Stock and Meat Board, Chicago, Ill.)

BEEF AND VEAL RETAIL CUTS AND HOW TO COOK THEM

Whether a beef carcass is cut up in the home or by an expert, it should always be cut across the grain of the muscle tissue and the thick cuts should be separated from the thin cuts and the tender cuts from the less tender cuts.

Fig. 16-7 shows the wholesale and retail cuts of beef and gives the recommended method(s) of cooking each. Fig. 16-8 presents similar information for veal.

In order to buy and/or process beef and veal wisely, and to make the best use of each part of the carcass, the consumer should be familiar with the types of cuts and how each should be processed.

Every grade and cut of meat can be made tender and palatable provided it is cooked by the proper method. Also, it is important that meat be cooked at low to moderate temperatures, usually between 300° and 325° F for roasting. At this temperature, it cooks slowly, and as a result is juicier, shrinks less, and has a better flavor than when cooked at high temperatures.

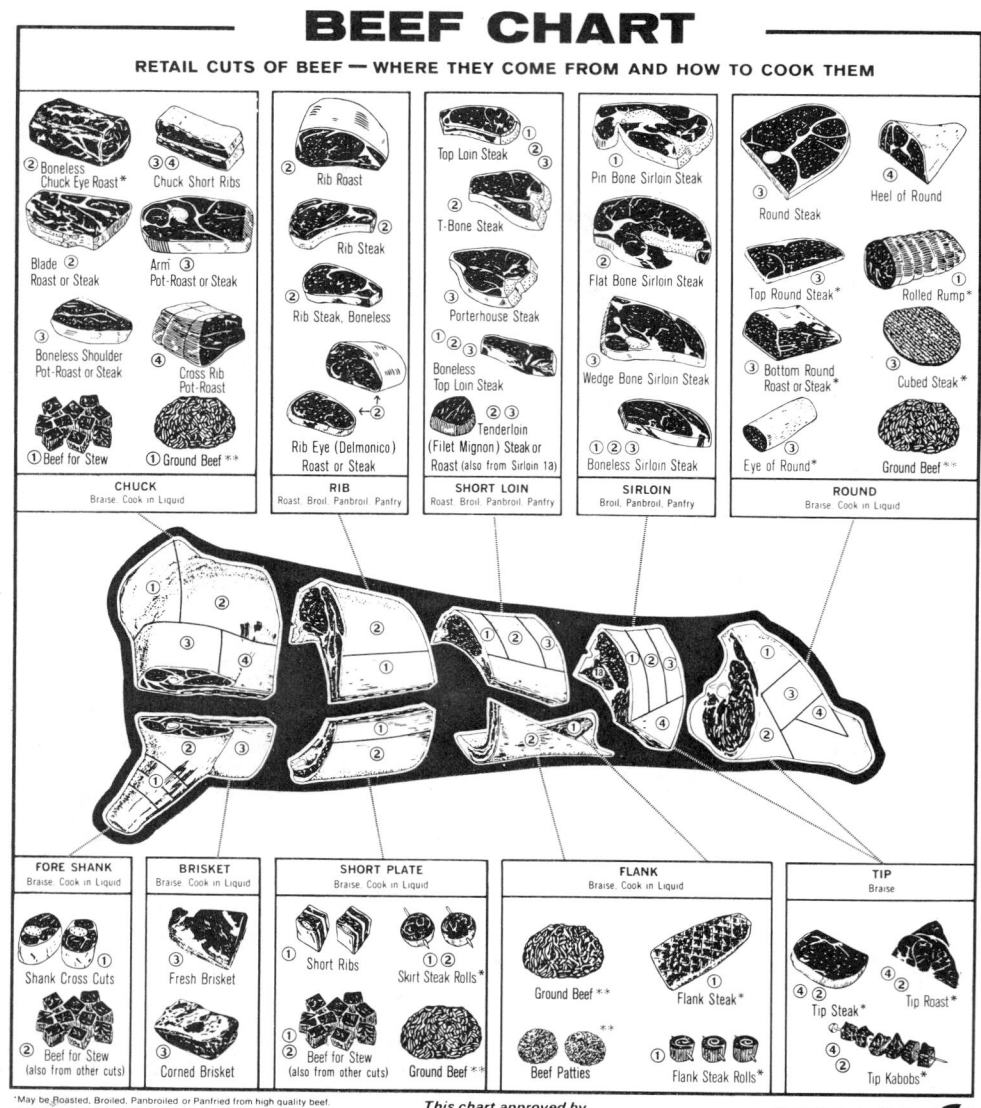

Fig. 16-7. The wholesale and retail cuts of beef, and the recommended method(s) of cooking each. (Courtesy, National Live Stock and Meat Board, Chicago, Ill.)

VEAL CHART

RETAIL CUTS OF VEAL — WHERE THEY COME FROM AND HOW TO COOK THEM

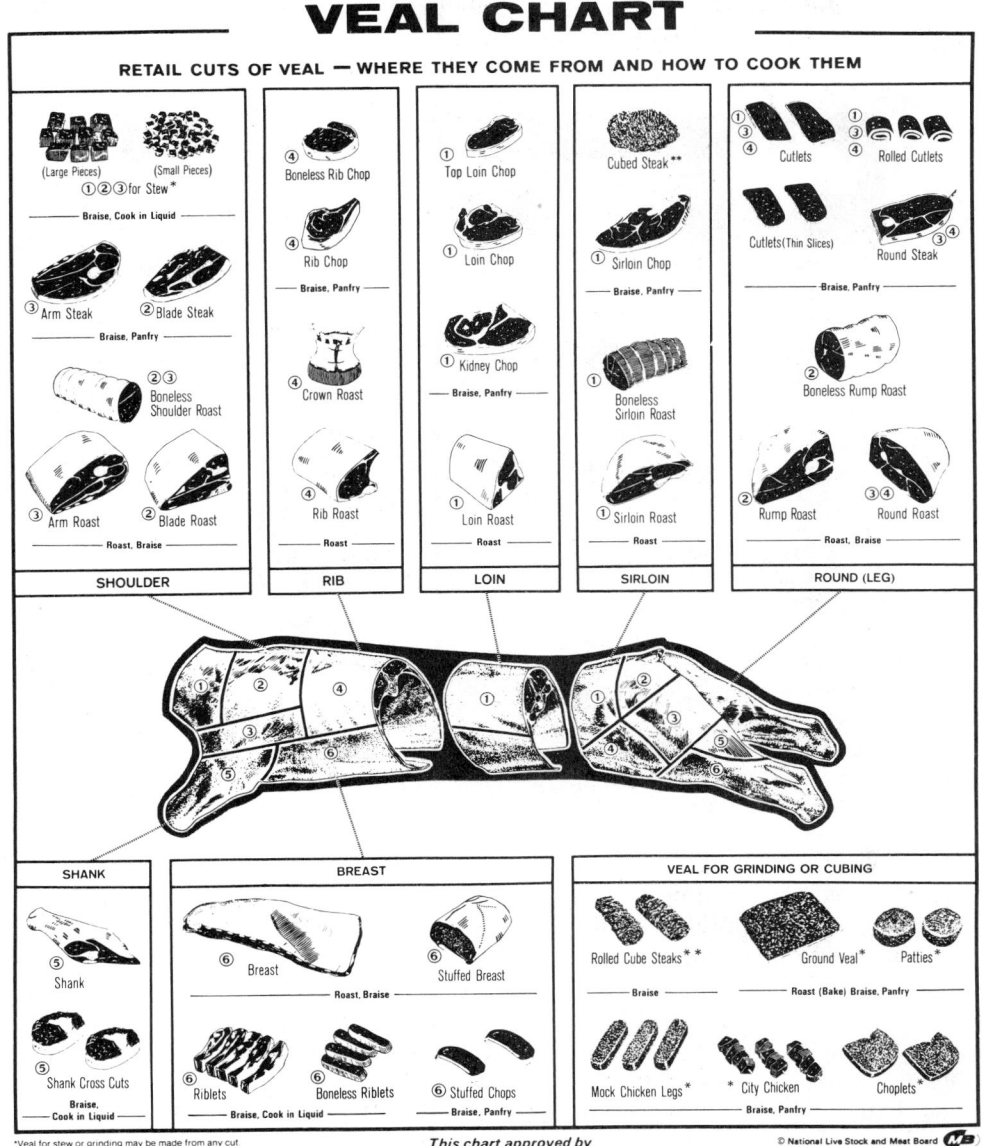

Fig. 16-8. The wholesale and retail cuts of veal, and the recommended method(s) of cooking each. (Courtesy, National Live Stock and Meat Board, Chicago, Ill.)

THE NUTRITIVE QUALITIES OF BEEF

Perhaps most people eat beef simply because they like it. They derive a rich enjoyment and satisfaction therefrom.

Fig. 16-9. Rib of beef, prepared by roasting. (Courtesy, National Live Stock and Meat Board, Chicago, Ill.)

But beef is far more than just a very tempting and delicious food. From a nutritional standpoint, it contains certain essentials of an adequate diet; high quality proteins, calories, minerals, and vitamins. This is important, for how we live and how long we live are determined in large part by our diet.

It is estimated that the average American gets the percentages of his daily nutrient allowances shown in Fig. 16-10 from a 3½-ounce serving of meat.

Effective beef promotion necessitates full knowledge of the nutritive qualities of meats, the pertinent facts of which follow:

1. *Proteins*—The word protein is derived from the Greek word *protos*, meaning "in first place." Protein is recognized as a most important body builder. Fortunately, meat contains the proper quantity and quality of protein for the building and repair of body tissues. On a fresh basis, it contains 15 to 20 percent protein. Also, it contains all of the amino acids, or building blocks, which are necessary for the making of new tissue; and the proportion of amino acids in meat almost exactly parallels that in human protein.

2. *Calories*—Meat is a rich source of energy, the energy value being

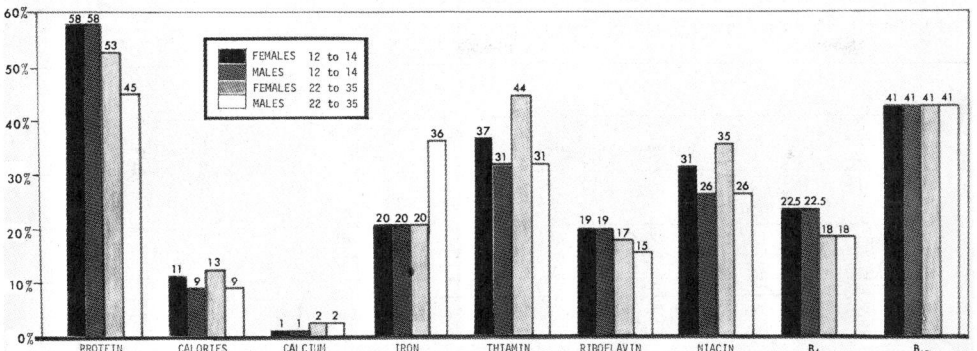

PERCENTAGE OF DAILY RECOMMENDED ALLOWANCES
FURNISHED BY A SERVING OF MEAT (3½ OZ COOKED)

Fig. 16-10. Percentage of daily recommended nutrient allowances furnished by a 3½-ounce serving of meat. Note that, from left to right, each nutrient is given for the following sex-age groups: females, 12-14 years of age; males, 12-14 years of age; females, 22-35 years of age; and males, 22-35 years of age. (Source: Recommended allowances from Food and Nutrition Board, National Research Council, [NRC], Recommended Daily Dietary Allowances, Washington, D.C. Nutrients in meat from National Live Stock and Meat Board, Chicago, Ill.)

dependent largely upon the amount of fat it contains. A pound of moderately fat beef provides about 1,350 calories, or nearly half the daily energy requirement of the average adult.

3. *Minerals*—Minerals are necessary in order to build and maintain the body skeleton and tissues and to regulate body functions. Meat is a rich source of several minerals, but is especially good as a source of phosphorus and iron. Phosphorus combines with calcium in building the bones and teeth. Phosphorus also enters into the structure of every body cell, helps to buffer the blood, is involved in the output of nervous energy, and has other important duties.

Iron is necessary for the building of blood, and its presence protects against nutritional anemia. It is a constituent of the hemoglobin or red pigment of the blood corpuscles. Thus, it helps to carry the life-giving oxygen to every part of the body. The average adult male would be assured of an adequate supply of iron if two servings of meat were taken daily along with one serving of liver each week.

4. *Vitamins*—Thousands of years ago, people knew that certain foods possessed unique nutritive properties. For example, as early as 1500 B.C., the Egyptians and Chinese hit upon the discovery that eating livers would improve one's vision in dim light. We now know that liver contains vitamin A, a very important factor for night vision. In fact, medical authorities recognize that night blindness, glare blindness, and poor vision in dim light are all common signs pointing to the fact that the person so affected is not getting enough vitamin A in his diet.

Meat is one of the richest sources of the important B group of vitamins (see Table 16-2). The B vitamins are now being used to reinforce certain

TABLE 16-2

VITAMIN B CONTENT OF 100 GRAMS OF ROUND STEAK[1]

	Raw	Cooked
	(mg)	*(mg)*
Thiamin	0.09	0.08
Riboflavin	0.18	0.22
Niacin	4.8	5.6
Pantothenic acid	1.0	0.63
Vitamin B6	0.37	0.45
Folic acid	0.03	
	(μg)	*(μg)*
Vitamin B12	2.0	1.7
Biotin	4.6	

[1]Leverton and Odell (1958); Watt and Merrill (1963); Kiernat *et al* (1964).

foods and are indispensable in our daily diet. Thiamin stimulates appetite, prevents beriberi, aids in the utilization of carbohydrates, and promotes the health of the body cells. Riboflavin is necessary for cell oxidation and protects against nervous disorders and also liver disturbances. Niacin prevents and cures pellagra, a diet deficiency disease which formerly afflicted hundreds of thousands of persons in this country, especially in the South. Vitamin B12 stimulates the appetite, increases the rate of growth and the efficiency of food utilization, and is necessary for normal reproduction. Indeed, one of the reasons for the rapid decline in B vitamin deficiencies in America may well be the increased amount of meat and other B vitamin-containing foods in the daily diet.

5. *Digestibility*—Finally in considering the nutritive qualities of meats, it should be noted that this food is highly digestible. About 97 percent of meat proteins and 95 percent of meat fats are digested.

We come to realize, therefore, the important part that beef is playing in the nutrition of the nation.

MEAT MYTHS

Much has been spoken and written linking the consumption of meat to certain diseases. Also, the sustained annual increase in per capita and total beef consumption over a period of years may have lulled the industry into a sense of false security. A summary of three incorrect statements along with the correct facts follows:

• *Myth: Meat fats cause coronary heart disease*—Much has been said about the relationship of meat fats to heart disease, prompted by the extremely high mortality from coronary heart disease in the United States in recent years. In 1965, deaths from cardiovascular disease in the United States rose to more than one million, over half the total deaths from all causes. Some have conjectured that cholesterol and saturated fat, found in animal fat, raise blood cholesterol and cause a high incidence of coronary heart disease.

Fact—It is tempting to ascribe the high incidence of coronary heart disease to a single factor; for example, to blame either saturated fatty acids, cholesterol, calories, stress, heredity, hypertension, or lack of exercise. It seems far more prudent, however, to recognize that many factors are involved, and perhaps they're involved to different degrees under various sets of conditions.

Researchers at the University of Georgia have found that only one saturated fatty acid—caproic acid—raises serum cholesterol; but there is no caproic acid in meat fat. The Georgia studies also revealed that stearic acid, one of the main saturated fatty acids in meat fats, lowers both cholesterol and blood pressure.

Dr. Michael E. DeBakey, noted heart transplant surgeon and President of Baylor College of Medicine, Houston, Texas, reports that only about 30% of heart patients have any form of abnormality in their cholesterol. While Dr. DeBakey has not ruled out the importance of diet, he has concluded that it is not the specific cause of heart disease. Said he, "We don't know the cause and we need to take a much saner attitude towards diet in relation to the disease since it is obvious that diet, so far as 65% to 70% of the patients are concerned, has not been related to or associated with the disease in our experience."

Dr. T. C. Huang, Director of Research, Timken Mercy Hospital, Canton, Ohio, (and a former student of the author of this book) found that rats fed on diets containing saturated animal fats supplemented with cholesterol grew better and lived longer than rats given unsaturated fats.

Dr. Raymond Reiser, Distinguished Professor of Biochemistry and Biophysics at Texas A&M University, in a review of all pertinent research studies which have been the basis for the popular concept that saturated fats (largely of animal origin) raise blood cholesterol and cause heart disease, concluded that saturated fats do not, by any criterion, elevate cholesterol to high risk levels, "if indeed they raise it at all."

In a study of the Masai tribesmen of Africa, Dr. George V. Mann, Vanderbilt University School of Medicine, Nashville, Tennessee, found that these nomadic people, whose diet is made up almost entirely of meat and milk, had unusually low blood cholesterol levels and were virtually free of coronary heart disease.

Based on the above, it may be concluded that until more is known relative to the causes of heart disease, restraint should be exercised in making unsubstantiated recommendations relative to prevention and treatment lest there be created more problems than those for which attempt is being made to solve.

● *Myth: Beef causes bowel cancer*—In 1973, a "scare story" appeared on the front pages of a number of metropolitan newspapers, quoting the National Cancer Institute to the effect that, "There is now substantial evidence that beef consumption is a key factor in determining bowel cancer."

Fact—The study to which reference is made pertained to Hawaiian residents of Japanese descent. The truth of the matter is that string beans, peas, and macaroni all had more of a statistical relationship to colon cancer than beef. Also, the Japanese in Hawaii changed more than their eating habits when they left Japan. They changed their lifestyles too. Eating more beef wasn't the only change. They also ate less of some Japanese foods which could have been functioning as cancer supressants.

Others have conjectured that colon cancer is predisposed by (1) lack of crude fiber or roughage in the diet, or (2) high amounts of fat. But the truth of the matter is that the real culprit is unknown.

In each myth—that meat fats cause heart disease, and that beef causes bowel cancer—the facts support the thesis that the primary role of the researcher should remain that of helping people live longer and healthier, based on research facts, rather than advancing phobias.

● *Myth: U.S. beef consumption will continue unabated*—The optimistic view is that, *come what may,* U.S. beef consumption, both in total and on a per capita basis, is elastic and unlimited.

Fact—This is not necessarily so. The slump in beef consumption and prices in 1973 had a sobering effect on the cattle industry. Per capita beef consumption had increased annually since 1958—for 16 years. Then, suddenly, it slumped 5.1 pounds per capita. Simultaneously, cattle prices plummeted and cattle feeders lost millions of dollars. Two forces appear to have triggered the situation: (1) removal of government price ceilings on beef; and (2) consumer resistance to higher prices. The message was loud and clear: neither beef consumption nor production is elastic *at any price.* They are subject to the old law of supply and demand. Cattlemen can, and will, produce enough beef if it is profitable. Price at the meat counter and consumption are, and will continue to be, governed by what consumers as a group are able and willing to pay.

Also, remember that the rate of U.S. human population increase has slowed; hence, there will be fewer beef eaters in the years ahead. Remember, too, that, on a worldwide basis, there will be pressure to consume grains directly, and to lessen the animal stage; and that the continued energy crisis will mitigate against transportation—getting animals and feed together.

WHY $2.00 STEAKS FROM 40¢ CATTLE?

Some folks are prone to compare what the packer is paying for cattle on-the-hoof to what they are paying for a pound of steak over the meat counter. Vent to such feelings is sometimes manifested in political campaign propaganda, consumer boycotts, supermarket pickets, sensational news stories, and the chain-type telephone meat strike. "Why," they scream, "$2.00 steaks from 40¢ cattle?"

Is there any real justification for this often-vicious criticism? Who or what is to blame for high meat prices and for the spread?

If good public relations are to be maintained, it is imperative that each member of the beef team—the cattleman, the packer, and the retailer—be fully armed with documented facts and figures with which to answer such questions and to refute such criticisms. Also, the consumer should know the truth of the situation.

Who Controls Beef Prices; Are Profits Excessive?

On the surface, the above heading appears to pose a very controversial subject. Actually, much of it is due to misunderstanding; most folks know too little about the other fellow's business. When meat prices are high, there is a tendency on the part of the consumer to blame any or all of the following: (1) the farmer or rancher; (2) the meat packer; or (3) the meat retailer. Also, these three may blame each other. The following sections are designed to present the facts relative to beef prices and profits.

BEEF PRICES AND CATTLEMEN'S PROFITS

It is preposterous to think that U.S. cattle producers could control prices, for cattlemen are well-known individualists and the competition between them is too great. However, some consumers feel that cattlemen, as a group, make excess profits—that they are responsible for the high price of beef. They hear that ranchers are getting 50¢ a pound for feeder calves weighing 500 lb at weaning time, and that a cattle feeder can buy a 500-lb calf for $250 and sell it 5 to 6 months later as a 1,000-lb steer at $400; so they decide that it is, indeed, a lucrative business.

Unfortunately, all too many meat consumers fail to realize that a heifer cannot be bred until she is about 1½ years of age, that the pregnancy period requires another 9 months, and, finally, that the young are usually grown 6 to 8 months before being sold to cattle feeders. Then follows 5 to 6 months in the feedlot during which time they receive about 2,400 lb grain, 300 lb protein supplement, and 1,350 lb of hay or equivalent roughage, after which they are sent to slaughter. Thus, under the most favorable conditions, this manufacturing process, which is under biological control and cannot be speeded up, requires about 3 to 4 years in which to produce a new generation of market cattle. Most of these critics also fail to realize that, for various reasons, only an average of 81 out of each 100 cows bred in the United States wean off young; that, in addition to cattle and feed costs, there are shipping charges, interest on borrowed money, death losses, marketing charges, taxes and numerous other costs, before the steer finally reaches the packer.

For the cattle, and all the expenses and services that go into their production, the cattleman (including both the cow-calf man and the feeder) gets 64 cents of the consumer's dollar (see Fig. 16-11). But this figure does not tell the whole story! It does not relate the shivering hours spent in the barn or shed as numb fingers attempt to bring life into a newborn calf; and it does not tell of droughts or of the sweat of a long work week which begins at 40 hours. Nor does this figure tell that during the 10-year period, 1963-1972, farmers and ranchers averaged only 3.9 percent return on the equity of their capital investment in farming. How many folks would invest in such a business, knowing this? Also, lost in the furor that some consumers stir up over beef prices is the fact that, in terms of disposable income (income after taxes), the average income of farm people lags 17 percent behind the average income of nonfarm people; in 1972, per capita farm income was $3,182 vs $3,847 nonfarm. And that's not all! Nearly half the income of farm people

WHERE THE CONSUMER'S BEEF DOLLAR GOES

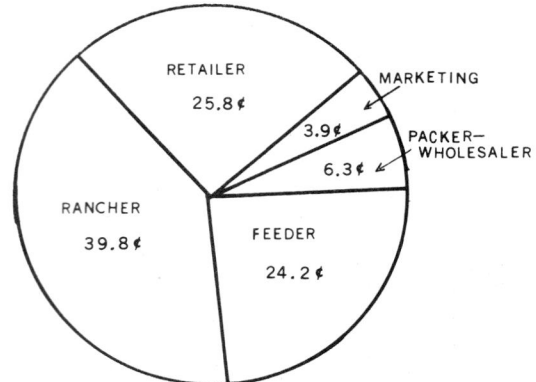

Fig. 16-11. Farmer's share of the consumer's beef dollar—the portion of the retail price of beef that the farmer receives. On the average, on Choice grade beef, the rancher and the feeder share 64 percent of the consumer's dollar. (Source: *Price Spreads for Beef and Pork*, Misc. Pub. No. 1174, Economic Research Service, USDA, May 1970, with adjustments by the author)

comes from off-farm sources. If farmers had to rely solely on income from farming, the average income of farm people would be less than half of the average income of nonfarm people.[2]

Thus, it is perfectly clear that the farmer or rancher is not receiving excessive profits in meat production. With knowledge of the above facts, and considering the long-time risks involved, it is doubtful if very many people will object to the occasional good years when the cattleman has a chance to recoup the losses he suffers during the lean years.

BEEF PRICES AND MEAT PACKERS' PROFITS

Of course, meat packing companies, like other businesses, are owned and operated by people; and all people want to be paid for their work. Therefore, they are entitled to a fair and reasonable profit; otherwise, they would not stay in business. The only question is—do they control prices and are they making too lush profits? Here are the pertinent facts on which to base an answer to this question:

1. Packers do not and cannot control prices because there is too intensive competition between them; there are more than 5,800 livestock slaughtering establishments in the United States.

2. From 1965-1972, meat packers' net earnings amounted to 0.94 cent per dollar of sales, or less than 1 cent per dollar of sales; which is not enough. By comparison, it is noteworthy that the net earnings per dollar of

---

[2]Source: The figures used in this paragraph came from *What's Happened to Food Prices?* USDA, Office of Communications, April, 1973.

sales of 24 individual companies[3] in the United States averaged 7.4 cents in 1972.[4]

3. From 1970-1972, the net profit of packing companies per 100 lb of live animal averaged 45 cents; and for 100 lb of dressed meat, it averaged 69 cents.[5] Thus, on the average for each 1,000-lb steer purchased, the packer netted about $4.50 in profit. Certainly this is a reasonable and legitimate earning. Of course, the volume of sales (the number of animals processed in a year), the efficiency of operations, and the utilization of by-products makes it possible for the industry to operate on these comparatively small margins.

Actually, the packers' profits are so small that, were they eliminated entirely, they would have practically no effect on the ultimate selling price of

EARNING RATIOS OF MEAT PACKING INDUSTRY, 1949-72

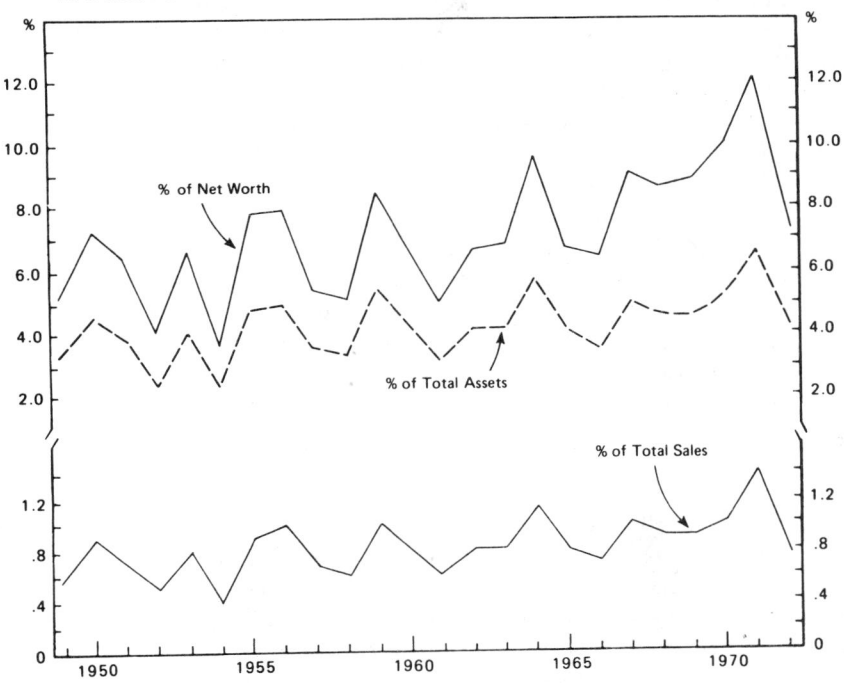

Fig. 16-12. On the average, from 1965 to 1972, meat packers' net earnings amounted to about 9/10 cents per dollar of sales, or about 45 cents per 100 lb liveweight of meat animals handled. (*Financial Facts About the Meat Packing Industry*, 1972, American Meat Institute, Washington, D.C.)

---

[3]In 1972 each of 24 individual companies in the United States made more money than the entire meat packing industry. As a whole, the net earnings of these 24 companies amounted to $16,070.7 million, compared to the net earnings of $207 million made by all the meat packers; hence, the net income of these 24 companies was 78 times greater than the meat industry's earnings.

[4]*Financial Facts About the Meat Packing Industry*, 1972, American Meat Institute, Washington, D.C.

[5]Ibid.

the retail cuts—only about 1 cent per pound. The truth of the matter is that the meat packer is in the middle of an impossible situation; on the one hand, the nation's livestock producers want high prices for all the animals that they can sell, while, on the other, more than 200 million consumers desire to buy as much meat as they wish at low prices.

BEEF PRICES AND MEAT RETAILERS' PROFITS

Finally, let us consider the meat retailer. Like the cattleman and the packer, he, too, is in business to make money; thus, he buys carcass beef at as low a price as possible and he sells the retail cuts at as high a price as consumers will pay. But does he control meat prices, and is he making excessive profits? Here are the facts:

1. There are an estimated 300,000 meat outlets in the United States, about 45,000 of which are supermarkets that handle about 75 percent of all meat sales.[6]

2. On the average, year after year, the meat retailer nets less than 1.5¢ after taxes on each $1.00 of sales.[7]

3. There is much competition between the 300,000 U.S. retail outlets that sell meat. As a result, no meat retailer can keep his prices out of line for very long; otherwise, consumers will just quit patronizing his market.

4. There is an old axium in the meat business which says "you sell it or you smell it." This simply refers to the fact that the vast majority of meats (about 75 percent) are sold in fresh form, and that, as a perishable product, it must be moved promptly into consumption channels and sold for whatever it will bring. If the retailer attempted to get a higher price than the market afforded, the amount of the product demanded would diminish, the meat would not be sold, and it would soon spoil.

CONSUMERS CONTROL BEEF PRICES

From the above it should be obvious that neither the cattleman, the packer, nor the meat retailer controls meat prices. Unknowingly, the people who really dictate the price of cattle on-foot, the price of dressed beef, and the price of retail cuts are the consumers. They actually put the price tag on the retail cuts. For purposes of illustrating their impact on meat and cattle prices, let us assume the following vital statistics: that the cattleman sells the packer a Choice grade steer at 40¢ per pound, that the packer sells the retailer the carcass (wholesale dressed meat) at 80¢ per pound, and that the retailer prices steaks at $2.00 per pound.

When consumers walk along the counter, they, and they alone, determine what they will buy. If they feel that $2.00 a pound is too much to pay for the steak, they simply do not buy it. They may buy hamburger, chuck

---

[6]Data on supermarkets provided in a personal communication to the author by Dr. S. Kent Christensen, Vice President and Agricultural Counsel, National Association of Food Chains, Washington, D.C.

[7]Information provided, in personal communication to the author by Dr. S. Kent Christensen (see footnote 6).

Fig. 16-13. Consumers cast their "vote" at the meat counter. They really determine meat prices. (Courtesy, National Live Stock and Meat Board, Chicago, Ill.)

pot-roast, or perhaps move down the counter a few feet and buy some pork chops, or a little further down the counter they will find some broilers or perhaps some fish. They do not say anything to the retailer about his price of $2.00 being too high for the steak. Generally speaking, there is no organized boycott on the part of the consumers, but it does not take the retailer long to discover that his price on steak is out of line.

If steak will not sell at $2.00, the retailer may lower the price to $1.80 a pound. He probably will try to raise the prices on some other beef cuts to take care of the loss on steak. If he is unable to make up the deficit, there is only one thing to do and that is to tell the packer that he cannot afford to pay 80¢ a pound for wholesale beef. He may tell the packer that he will pay 78¢ a pound, and in all probability the packer will refuse this offer; so the retailer will buy one side of beef instead of two; he will buy 400 pounds of pork loins instead of 300 pounds, and he may double his order for poultry. Then, in two or three days, as beef begins to back up in his coolers, the packer will probably call the retailer and make a deal with him. They may compromise on 79¢ a pound instead of 78¢ or 80¢, and the retailer starts buying his normal quantity of beef.

If the packer cannot get 80¢ a pound for wholesale beef, it does not take him very long to tell his livestock buyers that he cannot afford to pay 40¢ a pound for live steers and that they will have to cut the price to 39¢. If the run is heavy, the packer may be able to buy all the cattle he needs at 39¢, but if the run is light, he may be forced to pay 40¢, realizing that he will lose less money by paying 40¢ for steers than by not having enough work to keep his employees busy. The packer faces a dilemma; he must have enough live-stock coming into the plant to keep it in operation, but he must also buy livestock at low enough prices so that the dressed meat can be sold at a profit.

From the above, it is apparent that, as in the case of all other com-modities on a free market, meat prices are determined by the laws of supply and demand; by what the consumers as a group are able and willing to pay for the available supply. In plain simple terms, this means that what you and your neighbor and all America eat tonight will determine tomorrow's beef prices.

What Determines Beef Spread?

When a cattleman receives a check for $400 for a 1,000-pound steer—40¢ a pound—and on the way home stops at a retail meat market and buys a steak at $2 per pound, he is prone to think that he is on the wrong side of the counter; that he ought to be a meat packer or meat retailer.

Why is there so much spread between the price of a steer on-foot and the price of a pound of steak? This is a straightforward question which de-serves a straightforward answer. Here are the facts.

A STEER IS NOT ALL STEAK

Cattle are not all beef, and beef is not all steak. It is important, therefore, that those who produce and slaughter animals and those who purchase wholesale and/or retail cuts know the approximate (1) percentage yield of chilled carcass in relation to the weight of the animal on foot, and (2) yield of different retail cuts. For example, the average steer weighing 1,000 lb on foot and grading Choice will yield only 432 lb of retail cuts. (The balance consists of hide, internal organs, etc.) Thus, less than half of a live beef ani-mal can be sold as retail cuts of beef. In other words, the price of beef at retail would have to be more than double the live cost even if there were no processing and marketing charges at all. Secondly, the higher priced cuts make up only a small part of the carcass. Thus, this 432 lb will cut out only about 75 lb of porterhouse, T-bone, club, and sirloin steak. The other cuts retail at lower prices than do these choice cuts; also, there are bones, fat, and cutting losses which must be considered.

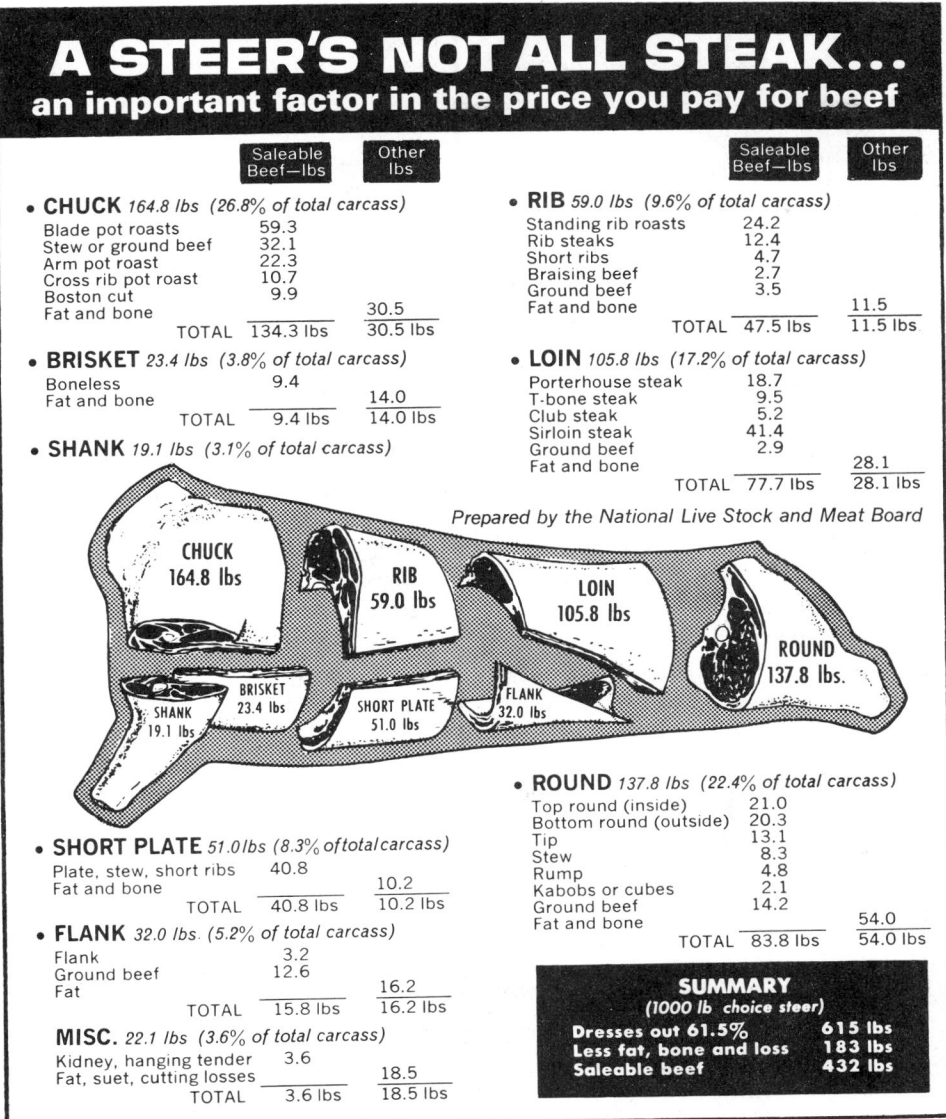

A STEER'S NOT ALL STEAK...
an important factor in the price you pay for beef

CHUCK *164.8 lbs (26.8% of total carcass)*

	Saleable Beef—lbs	Other lbs
Blade pot roasts	59.3	
Stew or ground beef	32.1	
Arm pot roast	22.3	
Cross rib pot roast	10.7	
Boston cut	9.9	
Fat and bone		30.5
TOTAL	134.3 lbs	30.5 lbs

BRISKET *23.4 lbs (3.8% of total carcass)*

	Saleable Beef—lbs	Other lbs
Boneless	9.4	
Fat and bone		14.0
TOTAL	9.4 lbs	14.0 lbs

SHANK *19.1 lbs (3.1% of total carcass)*

RIB *59.0 lbs (9.6% of total carcass)*

	Saleable Beef—lbs	Other lbs
Standing rib roasts	24.2	
Rib steaks	12.4	
Short ribs	4.7	
Braising beef	2.7	
Ground beef	3.5	
Fat and bone		11.5
TOTAL	47.5 lbs	11.5 lbs

LOIN *105.8 lbs (17.2% of total carcass)*

	Saleable Beef—lbs	Other lbs
Porterhouse steak	18.7	
T-bone steak	9.5	
Club steak	5.2	
Sirloin steak	41.4	
Ground beef	2.9	
Fat and bone		28.1
TOTAL	77.7 lbs	28.1 lbs

Prepared by the National Live Stock and Meat Board

CHUCK 164.8 lbs
RIB 59.0 lbs
LOIN 105.8 lbs
ROUND 137.8 lbs.
BRISKET 23.4 lbs
SHANK 19.1 lbs
SHORT PLATE 51.0 lbs
FLANK 32.0 lbs

ROUND *137.8 lbs (22.4% of total carcass)*

	Saleable Beef—lbs	Other lbs
Top round (inside)	21.0	
Bottom round (outside)	20.3	
Tip	13.1	
Stew	8.3	
Rump	4.8	
Kabobs or cubes	2.1	
Ground beef	14.2	
Fat and bone		54.0
TOTAL	83.8 lbs	54.0 lbs

SHORT PLATE *51.0 lbs (8.3% of total carcass)*

	Saleable Beef—lbs	Other lbs
Plate, stew, short ribs	40.8	
Fat and bone		10.2
TOTAL	40.8 lbs	10.2 lbs

FLANK *32.0 lbs. (5.2% of total carcass)*

	Saleable Beef—lbs	Other lbs
Flank	3.2	
Ground beef	12.6	
Fat		16.2
TOTAL	15.8 lbs	16.2 lbs

MISC. *22.1 lbs (3.6% of total carcass)*

	Saleable Beef—lbs	Other lbs
Kidney, hanging tender	3.6	
Fat, suet, cutting losses		18.5
TOTAL	3.6 lbs	18.5 lbs

SUMMARY
(1000 lb choice steer)

Dresses out 61.5%	615 lbs
Less fat, bone and loss	183 lbs
Saleable beef	432 lbs

Supply and Demand are not the only factors in the price you pay for beef. For instance, today's modern-type 1,000 lb choice steer produces an approximate 615 lb carcass which the packer sells to a retailer who trims away 183 lbs of fat, bone and waste ... ending up with only 432 lbs of beef that he cuts, wraps and sells to customers.

Of that a surprisingly small amount is steak and a much larger quantity is roasts as shown in the chart above. Retail stores put a higher price on steak and a lower price on pot-roasts and ground beef so that they sell it all ... not end up with only less-in-demand cuts like pot-roasts and short ribs left in the cooler.

Fig. 16-14. This shows that, on the average, a Choice 1,000-lb steer will yield about 615 lb of carcass or 432 lb of salable beef. (Courtesy, National Live Stock and Meat Board, Chicago, Ill.)

INCREASED SERVICES AND ATTRACTIVENESS

Since about 48 percent of the nation's women are working and the other 52 percent are spending more time at the club, it is understandable why they want more convenience. They desire that food purchases be largely prepared for immediate cooking, for their kitchen time is limited. Thus, when the housewife buys meat, she also buys many unseen services such as trimming, boning, packaging, tenderizing and freezing. All of these services have increased the farm-to-consumer spread.

MARKETING AND PROCESSING CHARGES AND PROFITS

Everyone and everything connected with the meat industry influences the spread between on-foot and retail prices. Investment capital is not free; it must yield returns comparable with other industries which compete for the use of capital. In addition, there are costs for labor, rent, supplies, transportation, and equipment. Over and above these costs, there should be a reasonable and fair profit.

What about decreasing margins by reducing the profits of the marketing agencies? The two major organizations involved are the meat packing industry and the retail stores. As previously stated, the average net profit of each—the meat packer and the meat retailer—amounting to less than 1 and 2 percent, respectively, on each $1.00 of sales—is very small. Were profits from both businesses eliminated entirely—and if producers received all of this additional amount—cattle prices would be raised about 1 cent a pound.

Cattlemen and consumers also need to recognize that when the demand or the supply changes for beef, there is of necessity much more change in the price of live cattle than there is in the price of beef over the counter. This is so because of the tendency for marketing margins to be more fixed— that is, the costs of labor, rent, supplies, transportation and equipment do not fluctuate rapidly.

Other Factors Affecting Beef Prices and Spread

This includes those forces other than supply and demand, per se, which affect beef prices and help to explain why a steak may cost five times the price of cattle on foot.

CONSUMERS WITH HIGH INCOMES DEMAND CHOICE CUTS AND TOP GRADES

In periods of prosperity—when incomes are rising—consumers place a premium on the preferred kind, cut, and quality of meat, rather than any marked increase in total meat consumption. Also, people on higher incomes eat more beef, veal, and lamb, and less pork, and they eat more of the expensive cuts of beef, such as steaks and roasts, and less stewing beef and sausage.

Also, due to the increased money available and shorter working hours, there is a desire for more leisure time, which, in turn, increases the demand for those meat cuts or products which require the minimum time in preparation. In many respects these two factors operate together; in other words, with high buying power, people hunt the choicer and easier prepared cuts of meat—porterhouse and T-bone steaks, and hamburger (the latter because of the ease of preparation).

All this suggests that producers of meat animals and processors and distributors have much to gain in periods of good times by taking steps to provide the desired kind and quality of products; by breeding for increased carcass quality, by feeding cattle to enhance grade, and by processing and preparing a higher quality and more attractive product.

NATURE MAKES FEW CHOICE CUTS AND TOP GRADES

But the novice may wonder why these choice cuts are so scarce, even though people are able and willing to pay a premium for them. The answer is simple; nature does not make many choice cuts or top grades, regardless of price. Cattle are born with only two loins—a right and left one. The balance of the cuts are equally wholesome and nutritious, and, though there are other steaks, many of the cuts are better adapted for use as roasts, stews, and soup bones. To make bad matters worse, not all cattle are of a quality suitable for the production of steaks—for example, the meat from most "worn-out" dairy animals and thin cattle of beef breeding is not sold over the block. To be sure, the lower grades are equally wholesome and nutritious; they are simply graded down because the carcass is somewhat less desirable in conformation, finish, and quality.

Thus, when the national income is exceedingly high, there is a demand for the choicest but limited cuts from the very top but limited grades of meats. This is bound to make for extremely high prices for such grades—for the supply is limited, but the demand is great.

CONSUMER TREND TO MORE BEEF AND LESS PORK

Another underlying trend which accentuates the price of beef is the shift in consumer preference from pork to beef. Expenditures for pork dropped from around 3.25% of the consumer income in the early 1930s to 1.4% in 1972; whereas, during this same period, consumer expenditures for beef increased from 2.3% to 2.6%.[8]

COMPETITION WITH OTHER FOOD PRODUCTS AND SERVICES

Meat must compete with other products for the consumer's dollar. Thus, in addition to preference, relative price is an important factor.

Also, meat must compete with certain nonfood items, for there are people who would go hungry in order to be able to spend their money for other purposes. On the average, consumers spend about 4% of their disposable income, or about 25% of their food budget, for meat.

Finally, sometimes the consumer is prone to blame his budget troubles on food, and meat in particular, simply because he forgets that he is spending part of his income on things which he did not have before—including such things as TV sets, automatic dryers, and two automobiles.

---

[8]*Meat Facts 1973*, American Meat Institute, Washington, D.C.

BEEF IS A GOOD BUY

TABLE 16-3

MINUTES OF FACTORY WORK REQUIRED TO PURCHASE ONE POUND
OF SIRLOIN STEAK[1]

Country	Work Time
	(Minutes)
United States	24
Canada	26
Brazil	60
France	90
U.S.S.R.	95
Japan	120

[1]Source: *The Stockman's Guide*, January 1974, syndicated column by Dr. M. E. Ensminger.

One of the best ways in which to evaluate whether a given product is a good or poor buy is in terms of the work hours required to purchase it. Table 16-3 presents convincing evidence that, in comparison with the other countries listed, beef is a good buy in the United States.

PACKINGHOUSE BY-PRODUCTS FROM CATTLE SLAUGHTER

The meat or flesh of animals is the primary object of slaughtering. The numerous other products are obtained incidentally. Thus, all products other than the carcass meat are designated as by-products, even though many of them are wholesome and highly nutritious articles of the human diet. Yet it must be realized that, upon slaughter, cattle yield an average of 40 percent of products other than carcass meat. When a meat packer buys a steer, he buys far more than the cuts of meat that will eventually be obtained from the carcass; that is, only about 60 percent of a steer is meat.

In the early days of the meat packing industry, the only salvaged cattle by-products were hides, tallow, and tongue. The remainder of the offal was usually carted away and dumped into the river or burned or buried. In some instances, packers even paid for having the offal taken away. In due time, factories for the manufacture of glue, fertilizer, soap, buttons, and numerous other by-products sprang up in the vicinity of the packing plants. Some factories were company-owned; others were independent industries. Soon much of the former waste material was being converted into materials of value.

Naturally, the relative value of carcass meat and by-products varies both according to the class of livestock and from year to year. It is estimated that packers retrieved 9.2 percent of the live cost of slaughter cattle from the value of the by-products (edible and inedible) in 1972. Hides are the most important by-product from cattle slaughter.

In contrast to the three early-day by-products—hide, tallow, and tongue—modern cattle slaughter alone produces approximately 80 by-products which have a great variety of uses. Although many of the by-products from cattle, sheep and hogs are utilized in a like manner, there are a few special products which are peculiar to the class of animals (e.g., wool and "catgut" from sheep).

The complete utilization of by-products is one of the chief reasons why large packers are able to compete so successfully with local butchers. Were it not for this conversion of waste material into salable form, the price of meat would be much higher than under existing conditions. In fact, under normal conditions, the wholesale value of the carcass is about the same as the cost of the animal on foot. The returns from the sale of by-products cover all operating costs and return a reasonable profit.

It is not intended that this book should describe all of the by-products obtained from cattle slaughter. Rather, only a few of the more important ones will be listed and discussed briefly (see Figure 16-15).

Fig. 16-15. It is not within the scope of this book to picture and describe all of the products which are manufactured from by-products obtained from slaughter. Instead, Fig. 16-15 is presented in order to show some of the more important items for which by-products are used, items which contribute to the convenience, enjoyment and health of people in all walks of life. (Courtesy, American Meat Institute, Washington, D.C.)

1. Bone for bone china.
2. Horn and bone handles for carving sets.
3. Hides and skins for leather goods.
4. Rennet for cheese making.
5. Gelatin for marshmallows, photographic film, printers' rollers.
6. Stearin for making chewing gum and candies.
7. Glycerin for explosives used in mining and blasting.
8. Lanolin for cosmetics.
9. Chemicals for tires that run cooler.
10. Binders for asphalt paving.
11. Medicines such as various hormones and glandular extracts, insulin, pepsin, epinephrine, ACTH, cortisone; and surgical sutures.
12. Drumheads and violin strings.
13. Animal fats for soap.
14. Wool for clothing.
15. Camel's hair (actually from cattle ears) for artists' brushes.
16. Cutting oils and other special industrial lubricants.
17. Bone charcoal for high-grade steel, such as ball bearings.
18. Special glues for marine plywoods, paper, matches, window shades.
19. Curled hair for upholstery. Leather for covering fine furniture.
20. High-protein livestock feeds.

1. *Hides*—Hides are particularly valuable as a by-product of cattle slaughter. Thus, most of the discussion in this section will be especially applicable to cattle hides.

Cattle hides have been used by man since the dawn of time; and leather, particularly cowhide, has held an important place in commerce throughout recorded history. It was an important part of the clothing and armor of ancient and medieval times. Today, it has hundreds of industrial uses.

On the average, the hide represents 2.5 to 3.0 percent of the total on-foot value of steers. There are two great classes of cattle hides, based on their place of origin; packer hides and country hides. Packer hides are the most valuable of the two because they are more uniform in shape, cure, and handling; much freer from cuts and gashes; and uniformly graded and available in larger lots.

The presence of needlessly large brands lowers the value of the hide. Cattle grubs (ox-warbles) also damage hides. It is estimated that one-third of all cattle hides produced in the United States are damaged by grubs. If there are five or more grub holes in the hide, it is classed as No. 2 and is discounted one cent per pound. Because of the larger throat cut, hides from kosher-killed cattle are less valuable.

The leather from animal hides is used for shoes, harness and saddles, belting, traveling bags, razor strops, footballs, baseball mitts, "sheepskins" for diplomas, sweat bands for hats, gloves, and numerous other leather goods.

2. *The fats*—Next to hides and pelts, the fats are the most valuable by-products derived from slaughtering. Products rendered from them are used in the manufacture of oleomargarine, soaps, animal feeds, lubricants, leather dressing, candles, fertilizer, etc.

Oleomargarine, which is perhaps the best known of the products in which rendered animal fat is incorporated, is usually a mixture of vegetable oils and select animal fat.[9] Oleo oil, one of the chief animal fats of this product, is obtained from beef and mutton or lamb.

3. *Variety meats*—The heart, liver, brains, kidneys, tongue, cheek meat, tail, feet, sweetbreads (thymus and pancreatic glands), and tripe (pickled rumen of cattle and sheep) are sold over the counter as variety meats or fancy meats.

4. *Hair*—Artist and camel-hair brushes are made from the fine hair on the inside of the ears of cattle. Other hair from cattle and hogs is used in toothbrushes; paintbrushes; mattresses; upholstery for furniture, automobiles, and passenger planes; air filters; baseball mitts; parachute seat pads, etc.

5. *The horns and hoofs*—At one time considered a nuisance, horns and hoofs are now converted into napkin rings, goblets, tobacco boxes, knife and umbrella handles, combs, buttons, etc.

---

[9]Oleomargarine is of two kinds: (1) a mixture of 50 to 80% animal fat and 20 to 50% vegetable oil, churned with pasteurized skimmed milk, or (2) 100% vegetable oil, churned with pasteurized skimmed milk. Oleomargarine was first perfected in 1869 by the Frenchman, Mege, who won a prize offered by Napoleon III for a palatable table fat which would be cheaper than butter, keep better, and be less subject to rancidity.

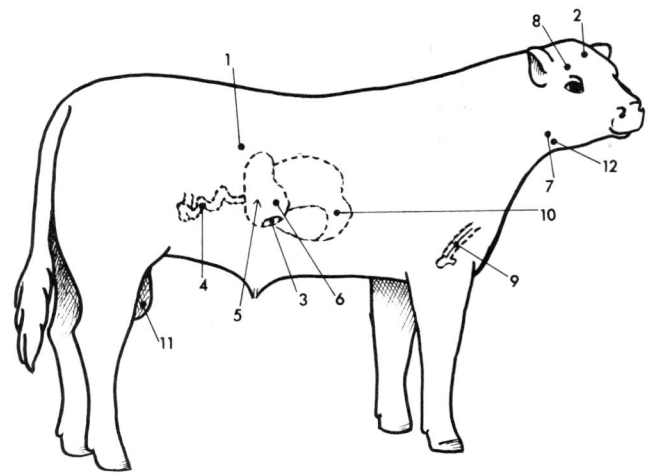

Fig. 16-16. Meat animals are the source of more than 100 medicines and medical products, which doctors and veterinarians administer daily to millions of people and animals to save lives, battle disease, relieve pain, and restore health. This figure shows the approximate location of a few of the glands and other tissues used in the manufacture of some of the pharmaceutical products of human and veterinary medicine.

1. *Adrenal (suprarenals)*—Source of (a) epinephrine (used for asthma, hay fever, allergies and shock), and (b) adrenal cortex extract (used for Addison's Disease, and in postsurgical and burn shock).

2. *Brain*—Source of kephalin (or cephalin), used on oozing surfaces to check bleeding.

3. *Gallbladder*—Source of (a) bile salts and (b) dehydrocholic acid—used for gallbladder disturbances and abnormalities of fat digestion, and (c) cortisone (used for rheumatic fever, arthritis, various allergies, inflammatory eye diseases, etc.)

4. *Intestines*—Lamb intestines are used for surgical sutures.

5. *Liver*—Source of (a) liver extract (used for pernicious anemia), and (b) heparin (used to delay clotting of shed blood of ulcers and following surgery).

6. *Pancreas*—Source of (a) insulin (the only substance known to medical science which can control diabetes), (b) trypsin (the protein-digesting enzyme), (c) amalase (the starch-splitting enzyme), and (d) lipase (the fat-splitting enzyme). Each enzyme is used for digestion of these respective nutrients; and trypsin is also used to soften scar tissue or digest necrotic tissue in wounds and ulcers.

7. *Parathyroid*—Parathyroid extract is used for tetany, which follows removal of these glands.

8. *Pituitary*—Source of (a) posterior pituitary extract (used to increase blood pressure during shock, to promote uterine contraction during and after childbirth, and to control excessive urination of diabetes insipidus), and (b) ACTH (used for rheumatic fever, arthritis, acute inflammation of eyes and skin, acute alcoholism, severe asthma, and hay fever and other allergy conditions).

9. *Red bone marrow*—Bone marrow concentrates used in treatment of various blood disorders.

10. *Stomach*—Source of rennin, used to aid milk digestion.

11. *Testes*—Source of the enzyme hyaluronidase.

12. *Thyroid*—Thyroid extract is used for malfunctions of the thyroid gland (some goiters, cretinism and myxedemal).

13. *Blood*—Source of thrombin, applied locally to wounds to stop bleeding.

14. *Bones and hides*—Source of gelatin, used as a plasma extender.

15. *Lard*—With 1% benzoin added, it produces Benzoinated Lard, which is used as a medical ointment base.

6. *Blood*—The blood is used in the refining of sugar, in making blood sausage and stock feeds, in making buttons, and in making shoe polish, etc.

7. *Meat scraps and muscle tissue*—After the grease is removed from meat scraps and muscle tissue, they are made into meat-meal or tankage.

8. *Bones*—The bones and cartilage are converted into stock feed, fertilizer, glue, crochet needles, dice, knife handles, buttons, teething rings, toothbrush handles, and numerous other articles.

9. *Intestines and bladders*—Intestines and bladders are used as sausage, lard, cheese, snuff, and putty containers. Lamb casings are used in making surgical sutures, strings for various musical instruments, and strings for tennis rackets.

10. *Glands*—Various glands of the body are used in the manufacture of numerous pharmaceutical preparations (see Figure 16-16).

Proper preparation of glands requires quick chilling and skillful handling. Moreover, a very large number of glands must be collected in order to obtain any appreciable amount of most of these pharmaceutical products. For example, it takes pancreas glands from 1,500 cattle to produce one precious ounce of insulin. But, fortunately, only minute amounts of insulin are required.

11. *Collagen*—The collagen of the connective tissues—sinews, lips, head, knuckles, feet, and bones—is made into glue and gelatin. The most important use for glue is in the woodworking industry. Gelatin is used in baking, ice cream making, capsules for medicine, coating for pills, photography, culture media for bacteria, etc. About 50 percent of the U.S. production of gelatin comes from veal.

12. *Contents of the stomach*—Contents of the stomach are used in making fertilizer and feed.

Thus, in a modern packing plant, there is no waste; literally speaking, "everything but the squeal" is saved. These by-products benefit the human race in many ways. Moreover, their utilization makes it possible to slaughter and process beef at a much lower cost. But this is not the end of accomplishment! Scientists are continually striving to find new and better uses for packinghouse by-products in an effort to increase their value.

BEEF CHECK-OFF

In 1975, legislation officially known as the Beef Research and Consumer Education Act was introduced in the U.S. Senate. Its stated purpose:

> To establish, finance, and carry out a coordinated program of research, producer and consumer education, and promotion to improve, maintain, and develop markets for cattle, beef and beef products, and to provide an adequate, steady supply of high-quality beef and beef products readily available to consumers at reasonable prices.

The bill is modeled after current cotton and egg check-off programs. It is based on a value added formula at the contribution rate of 0.3 of 1%. In other words, when Seller "A" sells an animal to Buyer "B" for $200, he actually receives $199.40 ($200.00 minus 0.3%). If Buyer "B" then fattens the animal and sells to Buyer "C" for $400.00, he actually receives $398.80. Not until the final sale before slaughter is the money actually collected and sent to the Beef Board, which will then determine exactly how to spend it.

After the first year of operation, the assessment rate may be set anywhere from 0.1% to 0.3% of the sale price of the cattle sold, and the assessment can be raised above 0.5% only with the approval of the producers in a new referendum.

The legislation was signed into law by President Gerald R. Ford on May 28, 1976.

The entire program is voluntary, so at any step in the selling process the seller can request his contribution back.

It is projected that the program will generate $30 to $40 million annually once it is fully operational. The Beef Board—which will administer the funds—will be composed of 68 cattlemen appointed by the Secretary of Agriculture from recommendations made by cattle organizations.

QUESTIONS FOR STUDY AND DISCUSSION

1. Why should the cattleman have a reasonable working knowledge of the end products, beef and veal?
2. Consumer studies indicate the following transition in preferences relative to beef:
   a. Preference for more red meat and less fat.
   b. Preference for more highly processed meat; that is, meat that is boned-out, trimmed, etc., prior to purchase.
   c. Preference for more frozen meat.
   Discuss the impact of each of these trends from the standpoint of the producer, the processor, and the consumer.
3. What factors account for the favored position of beef in world meat consumption?
4. What factors contributed to beef taking the lead over pork in U.S. per capital consumption?
5. What qualities do you desire in beef? Are these qualities reflected adequately in the top Federal grades of beef?
6. Do you approve of Federal grading of beef? Justify your answer.
7. What's the difference between quality grades and yield grades? Which is the most important?
8. Why are separate (from steers, heifers, and cows) grade indentifications used for bull and for bullock beef?
9. Why do most consumers prefer beef that has been properly aged?
10. Discuss the relationship of choice of retail beef cuts and method of cookery.
11. What facts relative to the nutritive qualities of beef are important in effective beef promotion?
12. Choose and debate either the affirmative or negative of each of the following questions:
    a. Beef prices are controlled by (1) the cattleman, (2) the packer, or (3) the retailer.
    b. Excessive profits are made by (1) the cattleman, (2) the packer, or (3) the retailer.
    c. Beef causes coronary heart disease.
    d. Beef causes bowel cancer.
13. Discuss the factors that determine the spread between on-foot cattle prices and retail beef prices.

14. How can small local slaughterers without the benefit of modern by-product processing facilities compete with larger packers?

15. Why are consumers prone to complain about the high price of beef more than the high price of a TV set, of a swimming pool, or of coffee, beer, or soft drinks?

SELECTED REFERENCES

Title of Publication	Author(s)	Publisher
Animal Science, Sixth Edition	M. E. Ensminger	The Interstate Printers & Publishers, Inc., Danville, Ill., 1969
Adventures in Diet	V. Stefansson	Reprinted from *Harper's Magazine* by American Meat Institute, Washington, D.C.
Beef Production and Distribution	H. DeGraff	University of Oklahoma Press, Norman, Okla., 1960
By-Products in the Packing Industry	R. A. Clemen	University of Chicago Press, Chicago, Ill., 1927
Food from Farmer to Consumer	National Commission on Food Marketing	U.S. Government Printing Office, Washington, D.C., 1966
Hides and Skins	National Hide Association	Jacobsen Publishing Co., Chicago, Ill., 1970
Marketing of Livestock and Meat, The, Second Edition	S. H. Fowler	The Interstate Printers & Publishers, Inc., Danville, Ill., 1961
Meat for the Table	S. Bull	McGraw-Hill Book Co., New York, N.Y., 1951
Meat Reference Book		American Meat Institute, Washington, D.C.
Meat We Eat, The, Tenth Edition	J. R. Romans P. T. Ziegler	The Interstate Printers & Publishers, Inc., Danville, Ill., 1974
Stockman's Handbook, The, Fourth Edition	M. E. Ensminger	The Interstate Printers & Publishers, Inc., Danville, Ill., 1970
Lessons on Meat, Second Edition	National Live Stock and Meat Board	National Live Stock and Meat Board, Chicago, Ill., 1972
Using Information in Cattle Marketing Decisions, A Handbook	WEMC Publication No. 5	Western Extension Marketing Committee Task Force on the Economics of Marketing Livestock, Fort Collins, Colo., 1973

Also, literature on meats may be secured by writing to meat packers and processors and trade organizations; in particular, the following two trade organizations:

American Meat Institute
P. O. Box 3556
Washington, D.C. 20007

National Live Stock and Meat Board
444 N. Michigan Avenue
Chicago, Illinois 60611

PART II

COW-CALF SYSTEM; STOCKERS

The cow-calf system refers to the breeding of cows and the raising of calves. In this system, the calves run with their dams, usually on pasture, until they are weaned, and the cows are not milked. It is the very foundation of beef production. Without the cow-calf system, there would be no stocker programs and no finishing operations, for there wouldn't be any raw material—calves and feeders. The importance of the cow-calf system in the agriculture of the nation rests chiefly upon the conversion of coarse forage and grass into palatable and nutritious food for human consumption. It is especially adapted to regions where pasture is plentiful and land is cheap; hence, it is the standard system of beef production on the western range. Reproduction—the production of calves—is the first and most important requisite of the cow-calf system, for if animals fail to reproduce, the breeder will

Fig. II-1. A new "breed" of cattlemen has changed the cow-calf industry of America more in the last two decades than it has changed since the days of the Texas Longhorn. No cattleman has wrought more of these changes than Tom Lasater of Beefmaster fame. Pictured: Lane Lasater, son of Tom and Mary Lasater, at age four, and yearling Beefmaster heifers. (Courtesy, Tom Lasater, Tom Lasater Ranch, Matheson, Colo.)

soon be out of business. Nationally, cattlemen get an 88 percent calf crop, which is not good enough. It can and should be 95 percent.

Stockers are young cattle that are fed and cared for so as to promote growth rather than finish. It includes steers and heifers that will either be (1) marketed as feeders, or (2) finished out by the owner who produced them. Strictly speaking, the term "stocker" does not include heifers intended for breeding purposes; the latter are more correctly designated as "replacement heifers."

Since the 1960s, the cow-calf business of America has changed more than it has since the days of the Texas Longhorn. One noteworthy evidence of this tremendous transition appears in the listing of breeds of beef cattle in college textbooks. A popular textbook published in 1920 listed 16 breeds of cattle, whereas the author's book, *Beef Cattle Science*, Fourth Edition, published in 1968, listed 21 breeds. Hence, only 5 more breeds were added in the 48-year period. This edition (Fifth) of *Beef Cattle Science*, lists 53 breeds—that's 32 new breeds, or more than double the number listed 8 years earlier in the preceding edition of this same textbook.

Hand in hand with changes in the cow herd, stocker programs have changed. Many stockers no longer look like "peas in a pod"—and on the small side. Instead, more and more of them are multi-colored crossbreds, with production tested ancestry, weaned at heavy weights off milk and grass. Consequently, the trend is to shorten the finishing stage.

Indeed, mankind makes constant progress and nature undergoes constant change—they never remain the same. Therefore the beef industry must go on researching, discovering, creating, and advancing. Chapters 17 through 25, which follow, will give a big assist in charting the course of the cow-calf system.

CATTLE RAISING[1]

Cattle raising must continue to improve and expand if consumers are to keep getting high quality beef at reasonable prices. Changes within the beef industry, improved forage production and utilization, and favorable prices are major factors in encouraging cattle raising.

---

[1]In the preparation of this chapter, the author drew from *Cattle Raising in the United States*, USDA, Agr. Econ. Report No. 235.

CONSIDERATIONS WHEN ESTABLISHING-EXPANDING THE BEEF ENTERPRISE

Whether or not cows and calves should be produced on a particular farm or ranch should be determined by a careful analysis of (1) the available resources (land, labor, capital, and managerial skills), and (2) the relative profitability of beef cattle in comparison with alternative enterprises.

Choices of alternate enterprises are limited on many ranches of the West and Southwest. Much of the land is suited only for grazing cattle or sheep. If beef cattle are selected rather than sheep, the only decision that remains is whether it should be strictly cow-calf, with all calves except replacements marketed at weaning time; a combination of cow-calf and stockers; or stockers only.

Fig. 17-1. Partners in the cattle business.

Farmers in the central and eastern United States have more options; hence, the choice of the enterprise becomes more difficult. They must first decide between grain and animal production, or some combination of the two. Additionally, if they decide to go the animal route, they must decide between beef cattle, dairy cattle, hogs, and sheep, or some combination of them.

Beef cattle do have certain advantages. Likewise, they have their disadvantages. These follow.

Factors Favorable To Cow-Calf Production

Some of the special advantages of cow-calf production as compared to other kinds of livestock on the farm or ranch are:

1. *It utilizes land not suited for grain production*—Beef cattle are well adapted to the use of the millions of acres of land unsuited for the production of grains or for any other type of farming, including humid areas for which the highest and best use is pasture, the arid and semiarid grazing lands of the West and Southwest, and the brush, forest, cut-over, and swamplands found in various sections of the United States. For the 50 states of the United States, there are 604 million acres of grassland, involving 27 percent of U.S. land, plus additional acreage that is grazed part of the year (meadow aftermath following a hay crop, cornstalks following harvest, cotton fields after harvest, winter wheat, etc.).

2. *It utilizes low quality roughages*—Beef cattle efficiently utilize large quantities of coarse, relatively low quality roughages produced on farms and ranches, including cornstalks, straw, and coarse or low-grade hays.

Fig. 17-2. Cows winter grazing cornstalks. Beef cows can utilize large quantities of low quality roughages. (Courtesy, The University of Nebraska, Lincoln, Neb.)

3. *It provides a profitable outlet for by-product feeds*—Beef cattle provide a profitable outlet for many by-product feeds, including corncobs, cottonseed hulls and gin trash, the oilseed meals, beet pulp, citrus pulp, molasses (cane, beet, citrus, and wood), wood by-products (sawdust), rice bran and hulls, and fruit, nut, and vegetable refuse.

4. *It uses homegrown feeds*—Cattle can use the total homegrown production of grains and roughages, with or without the purchase of other feeds, more efficiently than any other class of livestock.

5. *It provides an elastic outlet for grain*—Beef cattle production provides an elastic outlet for grain. When plentiful, more grain can go into beef. When scarce, less grain and more grass and roughage will still produce beef.

6. *It maintains fertility*—Beef cattle provide an excellent way in which to maintain fertility on cultivated land. In addition to returning to the soil approximately 80 percent of the fertilizing constituents in the feeds, they offer a profitable way in which to utilize soil-building legumes that are usually a part of improved crop rotations.

7. *It requires a minimum of labor*—A beef cattle enterprise requires less labor than most other animal enterprises, with the result that it is relatively free from labor problems and adapted where labor is scarce and costly. Under average commercial range conditions, one man can care for approximately 300 cows.

8. *It distributes labor*—Beef cattle help to distribute the labor requirements throughout the year; they require but little attention except during the winter months.

9. *It requires small investment in buildings and equipment*—Beef cattle require a comparatively small investment in buildings and equipment.

10. *It entails little death risk*—Cattle past weaning age entail little death risk, as they are susceptible to comparatively few diseases and parasites. Death losses average about as follows: the mature beef herd, 2%; replacement heifers, 1%; and feedlot cattle, 1%.

11. *It is not normally a source of pollution*—Most cow-calf operations are not a source of pollution, because a minimum of confinement is involved and, for the most part, the animals defecate on the pasture or range—as wild animals do, and as nature intended.

12. *It produces the maximum amount of meat from milk and grass*—Beef cattle are adapted to the production of a maximum amount of meat from milk and grass, through (a) heavy weaning weights, or (b) finishing on pasture.

13. *It provides flexibility*—Cow-calf operations are flexible. Based primarily upon the availability and price of feed and the price of cattle, the options are: (a) sell all calves as weaners; (b) sell some calves as weaners and hold the balance over for sale as short or long yearlings; (c) buy additional calves and sell them as yearlings; or (d) feed home-raised calves to slaughter weight.

14. *It is suited to part-time farming*—Beef cattle are suited, as a supplementary enterprise, for part-time farming where the owner has off-farm employment or is semiretired, because of the relatively small and flexible labor needs.

15. *It produces a preferred food*—Beef is a preferred food, with the result that, generally speaking, the demand is good and the price favorable. From 1958 to 1972, the per capita consumption of beef rose nearly 50%,

while that of cereals and potatoes declined about 5%. Of the total per capita consumption of red meat (beef, veal, lamb, and pork) in 1972, 61% was beef.

16. *It results in more than 60% of the consumer's beef dollar going to the producer*—The farmer gets a larger share of the retail dollar spent for beef than for most products, simply because the cost of processing beef is much lower than the cost of processing products like a loaf of bread. In 1972, for example, farmers received an average of 64¢ of each dollar spent for Choice beef at the store, in comparison with an average of 15¢ for white bread and 40¢ for all farm-produced foods.

17. *Its product—beef—will increase in demand*—Demand for beef and affluency go hand in hand; and people around the world are becoming more affluent. It follows, therefore, that the demand for beef will increase worldwide in the years ahead.

18. *It will be profitable in the years ahead*—World beef shortages, along with increased per capita beef consumption at home and abroad, indicate that, generally speaking, beef prices and profits will be favorable for several years ahead.

19. *It need not compete with people for grains*—World food shortages favor the retention of beef cattle because they are adapted to the use of a maximum of such humanly inedible feeds as pasture and hay. Hence, cattle can adjust to the increased competition for grains for human consumption in the years ahead.

20. *It is in a favorable export situation*—U.S. beef cattle production and prices are in a favorable position due to world beef shortages and high beef prices abroad, factors which (a) lessen the consequences of beef imports through changing controls (quotas, tariffs, and embargoes), and (b) create increased markets through exports.

21. *It imparts pride of ownership*—Ownership of beef cattle imparts pride of ownership and serves as a status symbol more than any other class of livestock, which is an important consideration for many people, farmers and nonfarm investors alike.

22. *It makes for a favorable balance of trade*—Through expanded exports, beef cattle can play a leading role in bringing about a favorable balance of trade, a more stable dollar, and a healthier world.

Factors Unfavorable to Cow-Calf Production

Factors which, under certain conditions, may be unfavorable to cow-calf production are:

1. *It requires considerable land*—A cow-calf operation as normally conducted necessitates considerable land, which requires much capital. Land cost an estimated $900 to $1,100 per commercial cow unit in 1973. In recent years, land values have risen sharply because of the appeal of land to investors (a) as a means of getting away from the losses that many of them suffered on the stock market in the 1960s and early 1970s, and (b) as a hedge against inflation. These same forces will continue to push land values still higher, to a projected (by the author) $2,000 to $2,500 per commercial cow unit by 1985.

2. *It requires much capital for equipment, cattle, and operation*—In addition to the capital required for real estate (land and buildings), a cattle operation necessitates a considerable investment for equipment, cattle, and operation.

3. *It requires fencing and water*—Beef cattle require fencing and water—factors which have limited their expansion in the Corn Belt and Lake States.

4. *It requires considerable knowledge and management ability*—A cow-calf operation requires know-how and managerial experience and ability commensurate with the size and sophistication of the operation.

5. *It isn't easy to comply with the grazing regulations on public lands*—The grazing regulations on the various Federal- and state-controlled lands are often difficult with which to comply.

6. *It makes for high sire costs in boom periods*—During boom periods, high quality purebred bulls are usually high in price and difficult to obtain.

7. *It is characterized by great spread between classes and grades*—The spread in price in market cattle is usually greater than is encountered in any other class of livestock. Shelly old cows that have outlived their usefulness in the breeding herd will bring comparatively less than an old brood sow. In other words, a greater spread usually exists between the price of Choice steers and Utility cows than between barrows and sows. In 1972, the comparative figures were: on seven Corn Belt markets, the spread between barrows and sows was $3.41 per hundred. On the Omaha market, the spread between Choice steers and Utility cows in 1972 was $10.62 per hundred.

8. *It cannot be expanded or liquidated quickly*—Breeding herds cannot be expanded quickly, and the response to prices is slow. Ordinarily, a heifer does not calve until she is about 2 years of age, the pregnancy period requires another 9 months, and, finally, the young are usually grown 6 to 12 months before being sold to cattle feeders, who finish them for 4 months to a year; in total, involving about 4 years to produce a new generation.

Neither can breeding herds be liquidated quickly without great loss to the operator and much waste in terms of resources.

9. *It propogates slowly*—Cattle are neither as early breeders nor as prolific as hogs or sheep. Normally, a bovine female is not bred until she is 1½ years of age, gives birth only once per year, and produces only one young at a time. (See Fig. 17-3.)

10. *It is subject to the hazard of such dreaded foreign diseases as foot-and-mouth disease*—The presence of foot-and-mouth disease, and of certain other diseases, in other countries always constitutes a hazard.

11. *It is inefficient in converting feed*—The beef cow is the least efficient of all feed converters, in pounds of feed required to produce one pound of product. (See Fig. 17-4.)

12. *It is inefficient in converting protein*—Beef cattle are not efficient in converting protein in feed to ready-to-eat food. Only sheep are less efficient. (See Fig. 17-5.)

PROLIFICACY

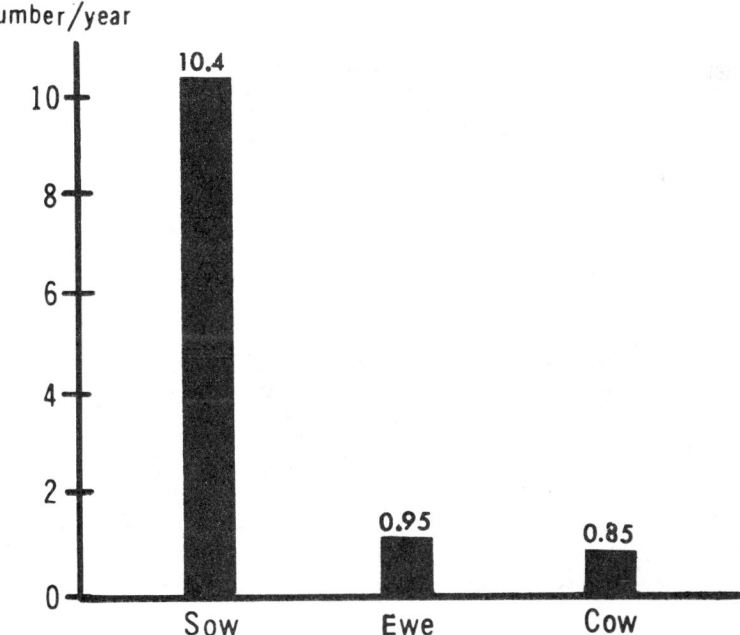

Fig. 17-3. Number of young raised to weaning age/year/breeding female.

FEED EFFICIENCY

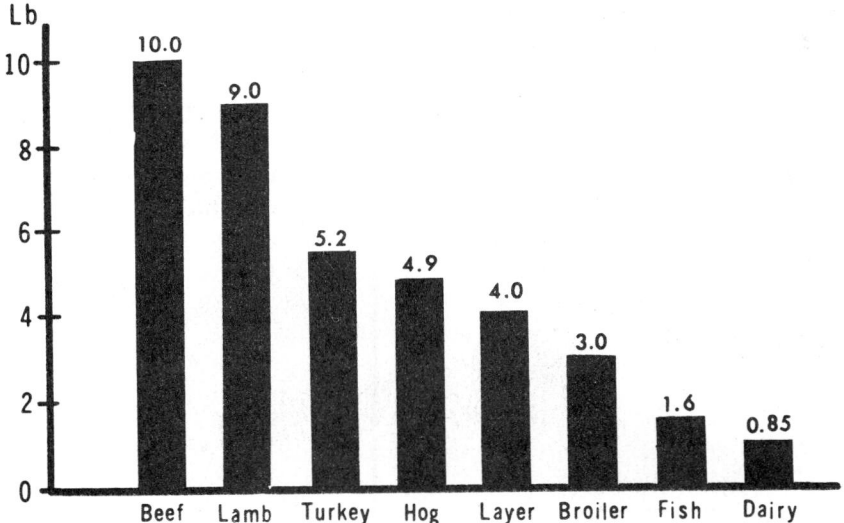

Fig. 17-4. Pounds of feed required to produce one pound of product—meat, milk, or eggs.

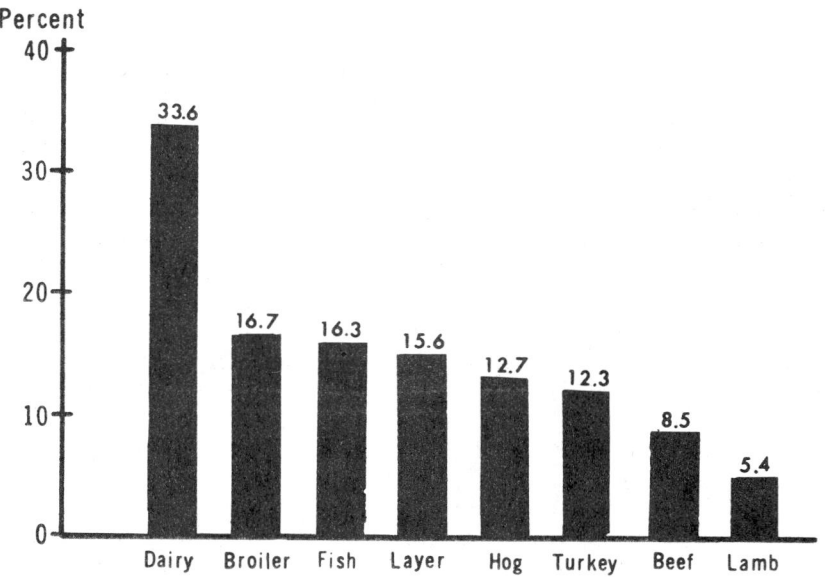

Fig. 17-5. Protein efficiency rating—protein in feed consumed converted to protein in ready-to-eat food, in percentage.

13. *Its tax shelter incentive could be lost*—Current Federal income tax regulations provide an incentive (tax shelter) to own beef cattle—especially for investors in high tax brackets through (a) the rapid depreciation allowances, and (b) the possibility of investment credit. But these could change with the flick of a pen.

14. *It is costly to assemble feeder cattle*—Assembling uniform lots of feeder cattle from many small producers for movement to a few large feedlots presents structural problems.

15. *It is not very responsive to price and cost changes*—Because of the preponderance of small herds, the beef industry as a whole does not respond too well to price and cost changes, with the result that small herds tend to depress prices further and longer during times of overproduction.

16. *It is not very responsive to technological advances*—Because of the preponderance of small herds, the beef industry as a whole does not respond too well to technological advances, simply because most small operators cannot justify the cost of the technique. As a result, there is a tendency for the industry not to be as progressive as some.

17. *It overreacts to the economy*—Beef is the "Cadillac" of foods. Hence, like the "Cadillac" of cars, its sales slump faster and lower in times of a recession or depression than economy foods.

18. *It will constantly be threatened by soybeans and other meat substitutes*—Beef is somewhat vulnerable from substitutes by soybean and

other proteins, provided the latter secure consumer acceptance and are low cost, especially if incomes fall.

Basic Resources for Cow-Calf Production

Basically, farmers and ranchers have four resources at their disposal with which to change cow numbers, i.e., land, labor, capital, and managerial skills. In the years ahead, the most successful cattlemen will put these together in such manner as to maximize profits, followed by increased cow numbers. The cow-calf resource requirements, compared to alternate enterprises, are: (1) large acreages of land for which the highest and best use is pasture, along with considerable quantities of comparatively low quality winter roughage; (2) available labor during the calving season (preferably with a liking for and knowledge of cattle); (3) adequate capital; and (4) able management commensurate with the size and sophistication of the operation.

Kinds of Cow-Calf Operations

Usually, the cow-calf operator has several options open to him. He may choose between (1) a farm or ranch herd, (2) running commercial or purebred cattle, (3) selling weaners or stockers, and (4) dual-purpose production.

FARM OR RANCH HERD

In general, beef cattle production in the farm states is merely part of a diversified type of farming. Grain and pasture crops are produced; and, on the same farm, beef cattle may compete with dairy cattle, hogs, and sheep for the available feeds. This applies to practically all the farms located to the east of the 17 western range states. In general, farm beef cattle herds are much smaller than range herds of the West, and many of them lack the uniformity which prevails in range cattle.

Fig. 17-6. A farm herd of beef cattle. In general, farm herds are much smaller than range herds. Moreover, beef production in the farm states is usually part of a diversified type of farming.

More than half of all U.S. beef cattle are produced on the western range. Because a considerable portion of the range area is not suited to the production of grains, and because sheep, and in some areas big game, offer the only other major use of the grasses, it seems evident that range beef cattle production will continue to hold a place of prominence in American agriculture.

Fig. 17-7. Hereford cattle on a range in Oregon. South Sister Peak in the background. (Courtesy, U.S. Forest Service)

From the foregoing, it should be concluded that geographic location largely determines whether a herd of cattle shall be operated as a farm herd or range herd. Thus, the majority of the beef herds in the West and Southwest, except for relatively small herds in irrigated areas, are operated as range herds, whereas the vast majority of the herds in the central and eastern parts of the United States are operated as relatively small farm herds.

PUREBRED OR COMMERCIAL

A purebred animal is a member of a breed, the animals of which possess a common ancestry and distinctive characteristics; and it is either registered or eligible for registry in the herd book of the breed.

Based on number of calves raised vs number of calves registered, the author's computations show that only about 4.0 percent of U.S. cattle (beef and dairy) are registered purebreds. Hence, purebreds are a small, but mighty, minority.

There is nothing sacred about purebreds, nor does the word itself imply any magic. It is generally agreed, however, that purebreds have been the major factor in the beef improvement of the past, and that they will continue to exert a powerful influence in the future. Although limited in numbers, purebred herds are scattered throughout the United States and include both farm and ranch herds.

Fig. 17-8. Purebred Polled Hereford heifers at Morlunda Farms, Lewisburg, West Virginia. (Courtesy, Oscar Nelson, Jr., Morlunda Farms)

The purebred cattle business is a specialized type of production. Generally speaking, few cattlemen should undertake the production of purebreds with the intention of furnishing foundation or replacement stock to other purebred breeders, or purebred bulls to the commercial producer. Although there have been many constructive purebred beef cattle breeders, and great progress has been made, it must be remembered that only a few master breeders such as Bakewell, Cruickshank, Gudgell and Simpson, and Congdon and Battles are among the immortals. Few breeders achieve the success that was theirs.

All nonpurebred cattle are known as commercial cattle. This includes the vast majority of the beef cattle of the United States—probably 96 percent. In general, however, because of the obvious merit of using well-bred bulls, most commercial calves are sired by purebreds. Beef over the block is the ultimate article of commerce of the commercial cow-calf man; although in the process of getting from pasture to packer the calf may be subjected to one or two intermediate stages—as stockers, and in the feedlot.

The commercial cattleman is intensely practical. No cow meets with his favor unless she regularly produces the right kind of calf. Experience, industry, and good judgment are requisites to success in the commercial cow-calf business. Additionally, the commercial cattleman is a key man in the nation's economy.

Fig. 17-9. Commercial beef cattle production on the Black Ranch, Brewster, Nebraska. In this system, the calves are run with their dams until weaned and the cows are not milked. It is the most common system of beef production on the western range. (Courtesy, C.B. & Q. Railroad Co.)

WEANERS OR STOCKERS

The commercial cattleman seldom adheres strictly to a cow-calf operation as such. Rather, based primarily upon the price of feed and the price of cattle, he may option to (1) sell weaners, except for replacements, (2) carry all or part of his calf crop over to the yearling stage, (3) buy additional calves, and/or (4) finish out his home-produced calves, in his own feedlot or in a custom lot.

DUAL-PURPOSE PRODUCTION

For the most part, dual-purpose production has been confined to the small farmer who lives upon the land and who makes his living therefrom. Cows of dual-purpose breeding are often referred to as the "farmer's cow." In this type of production, an attempt is made to obtain, simultaneously, as much beef and milk as possible. That is to say, in its truest form, this type of management cannot be classified as either beef or dairy production.

One of the chief virtues of dual-purpose production is the flexibility which it affords. When labor is available and dairy products are high in price, the herd may be managed for market milk production. On the other hand, when labor is scarce and dairy products are low in price, calves may be left running with their dams, and emphasis may be placed on beef production.

Because of the very nature of dual-purpose production, it is not adapted to the extensive ranches of western and southwestern United States. Rather, it is practiced on a limited number of small farms scattered throughout the humid area of central and eastern United States. It is noteworthy, however, that many of the exotic breeds recently introduced into the United States

Fig. 17-10. Red Poll steers on feed. Dual-purpose cattle provide considerable flexibility. Thus, when labor is scarce and dairy products are low in price, the emphasis may be placed on beef production. (Courtesy, Red Poll Cattle Club of America)

and Canada, primarily for crossbreeding purposes in an effort to secure cows that will produce more milk for their calves, are known as dual-purpose breeds in the countries of their origin. Noteworthy, too, is the fact that these breeds are being used for crossbreeding in both farm and ranch herds.

Beef Needs and Supplies

Dramatic changes have occurred in the beef industry, but more changes lie ahead. Pertinent information pertaining to past, present, and future cattle raising is given in Table 17-1.

An analysis of Table 17-1, especially the author's projections for 1985, follows:

• *Beef eaters*—In 1972, we had 209.2 million mouths to feed. By 1985, we shall have 246.3 million, an increase of 37.1 million people.

• *Per capita beef and veal consumption*—Per capita beef and veal consumption rose from 71.4 lb in 1950 to 112.7 lb in 1973, an increase of 41.3 lb in 23 years, for an average per capita increase of 1.8 lb per year. It is projected that per capita consumption will be 140 lb in 1985, which calls for an increase of 27.3 lb, or an average yearly per capita increase of 2.28 lb.

TABLE 17-1

THE U.S. PEOPLE AND BEEF AND VEAL SITUATION[1]

Year	No. of People	Per Capita Beef and Veal Consumption	No. of Cattle and Calves
	(million)	*(lb)*	*(million)*
1950	151.0	71.4	77.96
1960	180.0	91.2	96.24
1973 (or as indicated)	209.2[2] (1972)	112.7	121.5
1985 (estimated)	246.3[2]	140.0[3]	177.6[3]

[1]Except where otherwise noted, the figures in this table were obtained from various U.S. Department of Agriculture sources.
[2]Source: 1972 *World Population Data Sheet,* Population Reference Bureau, Inc., Washington, D. C.
[3]Author's projections. The 1985 projection of 177.6 million cattle and calves takes into consideration both increased human population and increased per capita beef and veal consumption.

• *Cattle and calf numbers*—From 1950 to 1973, cattle and calves increased by 43.5 million head, or at an average rate of 1.9 million head or 2.4 percent per year. As indicated above, it is estimated that we shall have 246.3 million people to feed in 1985, and that the average beef and veal consumption (demand) at that time will be 140 lb. To meet this need, we must produce 10.9 billion pounds more beef in 1985, than in 1973, or a total of 34.5 billion pounds (15,649,097 metric tons) of beef.

In 1973, 121.5 million cattle and calves produced 23.6 billion pounds of beef and veal. Hence, 177.6 million cattle and calves will be required to produce the 34.5 billion pounds of beef needed in 1985. That's 56.1 million, or 46.2 percent, more cattle! This is based on the assumption that carcass weights will remain the same (622 lb), no increase in beef imports, and beef cows not becoming more efficient.

To increase U.S. cattle and calves by 56.1 million head over a 12-year period (1973 to 1985) calls for an average increase of 4.7 million head per year. This is a projected expansion of 3.9 percent per year (4.7 ÷ 121.5 × 100). This seems feasible.

Table 17-1 shows that more people and higher per capita beef consumption created the demand for more beef cows. Other forces, not shown in Table 17-1, also operated to cause changes in the beef cattle industry, particularly rising income and change in consumer preference. As a result, an increasing number of animals were fed for slaughter and cattle feedlots went commercial and got big. Hand in hand with this development, there was a reduction in dairy cows, veal calves, and sheep. More recently, cow-calf production changed in breeds and breeding.

In the years ahead, more beef calves must be born, and each cow must produce more beef. Small, but insignificant, increases in beef may be expected from carrying slaughter animals to heavier weights and from imports. But there are practical limits on big and overweight cattle, and imports will be limited by world beef shortages accompanied by higher beef prices in many competing countries.

● *Veal slaughter; grain-fed slaughter*—In 1947, 13.7 million head of calves were slaughtered in the United States, representing 38% of all cattle and calves slaughtered that year. In 1972, only 3.2 million head of calves were slaughtered, representing a mere 8% of all cattle and calves slaughtered that year. Hence, in terms of numbers, the relative importance of calf slaughter to total slaughter fell from 38% to 8% in the 25-year period from 1947 to 1972. As would be expected, per capita veal consumption declined sharply in this same period of time; it went from 10.8 pounds in 1947 to 2.2 pounds in 1972.

In 1947, only 6.9 million head of market cattle were grain fed, representing 30% of the slaughter cattle that year. In 1972, 26.8 million head of grain-fed cattle were marketed, representing nearly 75% of all cattle marketed that year. Hence, in terms of numbers, the relative importance of fed cattle slaughter to total cattle slaughter rose from 30% to 75% in the 25-year period from 1947 to 1972. In 1973, 77% of the total beef produced in the United States came from grain-fed cattle, 23% from "grass"-fed cattle. (This means that the 75% grain-fed cattle accounted for 77% of the beef produced. This is as one would expect, because grain-fed cattle are generally heavier than nongrain-fed cattle.)

Today, few nonfed steers and heifers are being marketed, and the beef cattle industry has expanded nearly to the maximum by grain feeding cattle to slaughter weights of 1,000 to 1,100 pounds, rather than slaughtering them as calves or lightweight animals off pasture.

● *Specialization in cattle feeding*—Cattle feeding has become increasingly specialized. It has changed from that of many small farmers feeding during the winter months to fewer and larger year-round commercial feeding operations. The percentage of cattle marketed from feedlots of greater than 1,000-head capacity increased from 36 to 62 percent between 1962 and 1972. Percentage of cattle placed on feed by calendar year quarters went from a 21-16-21-42 distribution in 1960 to 21-22-25-32 in 1970. This shows the movement away from winter operations to a demand for feeder cattle more evenly distributed throughout the year.

● *Shift from dairy and sheep to beef*—Milk cow numbers decreased from 23.9 million in 1950 to 11.6 million in 1973. This decline was caused by (1) a decrease in the per capita consumption of dairy products, on a total equivalent basis, from 653 pounds in 1960 to 570 pounds in 1970; and (2) an increased production per cow, from 7,029 pounds in 1960 to 9,609 pounds in 1971.

From 1950 to 1973, breeding sheep on farms decreased from 26.2 million head to 14.8 million.

Past increases in beef cow numbers have been possible, in part, because of a decline in dairy cattle and sheep and a shift to beef. But it appears that decline in dairy cattle numbers has about bottomed out, with a decrease of only 3 million head projected for the period 1973 to 1985. Suitability of land grazed by sheep and the profitability of sheep will determine the future of sheep-to-beef shifts.

On a feed-weighted animal unit basis, milk cows and beef cows substitute at a 1:1 ratio, whereas sheep and cattle substitute at a 5:1 ratio. That is, it takes 5 sheep to equal 1 cow. Thus, the reduction of 11.4 million head of sheep (as occurred from 1950 to 1973) would only account for an increase of 2.3 million beef cows, even if all the land released from sheep production were switched to beef cattle. Thus, the drop in numbers of breeding sheep, which has occurred in all regions of the United States, does not go very far toward explaining the increase in beef cattle.

From the above, it may be concluded that (1) calf slaughter has declined to the point where little slack remains in the number available for feeding instead of vealing, (2) the percentage of cattle grain fed is near the practical ceiling, and (3) the shift from dairy and sheep to beef has about leveled out.

Certainly, much of the increase in beef supplies during the period 1960-70 resulted from dairy-to-beef shifts and from grain feeding young beef animals. But neither of these sources holds much potential for further additions to the beef supply. Under these circumstances, can supplies of beef continue to be forthcoming in amounts that will meet a sustained growth in demand, yet be available at a price that consumers are able and willing to pay? Two possibilities exist: (1) to increase beef cow numbers; and (2) to increase production per cow. Both of these possibilities will be pursued, but first let us consider trends—trends geographically and trends in herd size and numbers of herds.

TRENDS IN COW-CALF OPERATIONS

Prior to 1950, there were more milk cows than beef cows in the United States. Beef cow numbers first exceeded milk cow numbers in 1954, and the gap has been widening ever since. (See Fig. 17-11.)

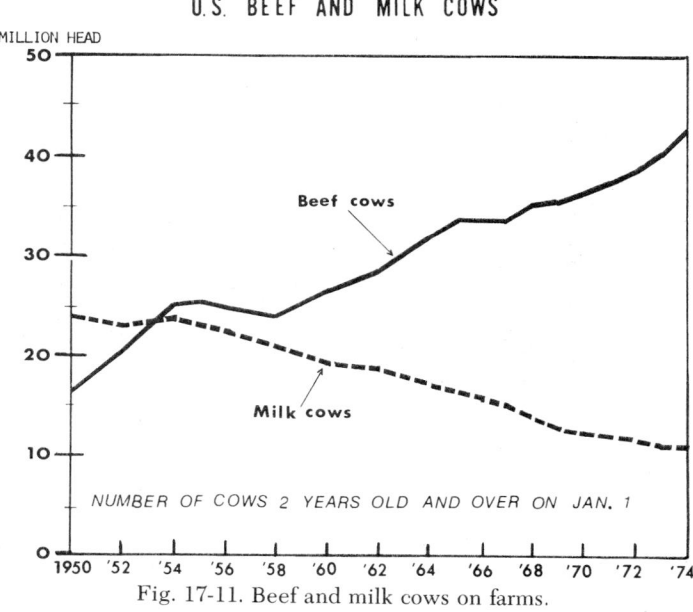

Fig. 17-11. Beef and milk cows on farms.

Beef cow numbers grew from 16.7 million in 1950 to 42.9 million in 1974 (see Fig. 17-11), and are expected to reach 51 million in 1985. The expansion ahead will need to be accomplished with no more total land to be had, increased competition for grains for human consumption, more pollution control, and more labor problems. The cattlemen of America can and will meet this challenge.

Geographic Trends

The number of beef cows is expected to increase in all states and regions except the Northeast. However, great differences among states and regions may be expected. Generally, the largest relative growth is expected in the humid regions, where crop-livestock farms are dominant, and in those crop farming areas or regions where ranching is the major type of agriculture. The Western States area, with vast acreages of public land and open range, is well suited to beef raising and will continue as an important beef production center. The land's aridity and rather limited productivity, except where irrigated, tend to limit herd expansion, however. Region by region (see Fig. 17-12), here is the situation—past, present, and future—as the author sees it:

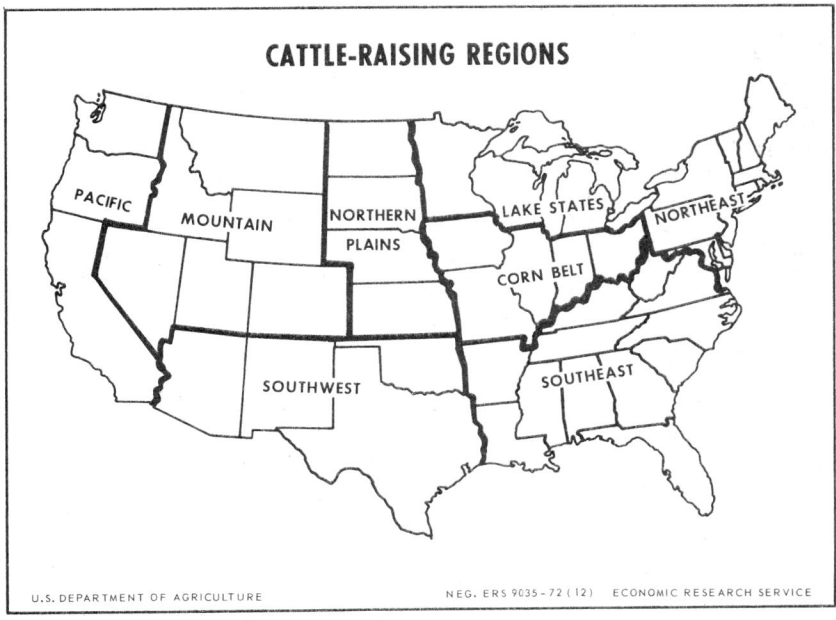

Fig. 17-12. Geographical regions which will be discussed. (Courtesy, USDA)

NORTHEAST

● *Present status*—Beef cow numbers in the Northeast increased from 1950 to 1970, but the rise was not sufficient to offset the decline in milk cow numbers. The beef herds of the Northeast are small and are often operated as a supplementary enterprise by part-time farmers. Hence, the area is of minor importance in beef production.

● *Future growth*—No significant increase in beef cow numbers is expected in the Northeast. Dairy cattle will continue to be the dominant animal enterprise, to supply the highly concentrated human population with fresh milk.

CORN BELT

● *Present status*—Beef cow numbers in the Corn Belt and Lake States, combined, increased by 2.9 times from 1950 to 1970; yet, the 3.75 million-head increase of beef cows in this area during this period just about offset the 3.97 million-head decrease in milk cows.

Even with its relative abundance of forage from crop residue, especially from corn, the Corn Belt has not expanded to its potential. Among the explanations as to why beef cattle have not increased more in this area are: (1) pasture acreage decreased (by 9% in the Corn Belt and Lake States from 1958-67); (2) the inability of beef cows to compete for land resources with crops which yield greater economic returns; (3) the beef cow herds are small on most farms, and small, fragmented landholdings do not lend themselves to enlarging the beef enterprise; and (4) lack of fencing, water, and managerial experience in beef production.

● *Future growth*—Increases in the Corn Belt area are expected to come largely through (1) converting more marginal crop land to pasture, and (2) more effective utilization of relatively low quality roughages, like cornstalks, for winter feed. The greatest expansion will occur in Missouri and Iowa, and growth will come largely by expansion of cow-calf enterprises on farms that already have beef cow herds.

LAKE STATES

● *Present status*—From 1950-1970, beef cow numbers in the Lake States increased at about the same rate as they did in the Corn Belt in the same period—about 2.9 times.

● *Future growth*—So long as dairying remains profitable in the Lake States, shifts to beef will be minimal. The area does have considerable production potential for grass-legume forages; and it is well adapted to beef production. Beef gains in this region will be greater in Wisconsin and Minnesota than in Michigan.

SOUTHEAST

● *Present status*—Although beef cow numbers rose in all regions, the 12 states comprising the Southeast Region had the most dramatic increase of any area during the period 1950-70—about 6.4 million head. The factors which contributed most to the growth in beef cows in this area were: (1) the introduction and promotion of improved pastures, especially tall fescue; (2) development of crossbred animals that can withstand hot humid weather, diseases, and insects; (3) the increase in size of farms, with larger pasture acreage per farm; and (4) the increase in part-time farmers (those who took off-farm employment while continuing to farm) who frequently chose beef cow enterprises because of the relatively small and flexible labor needs.

Fig. 17-13. Cows on winter fescue pasture, with round bales; on marginal crop-land that was converted to pasture. (Courtesy, University of Illinois, Urbana, Ill.)

The upward trend in calving percentages in the Southeast, from 78 percent in 1950 to 86 percent in 1970, reflects improvement in the quality of the cows and adoption of better management practices.

● *Future growth*—The Southeast region offers the greatest potential for future increase of beef cows of any area of the United States. Adequate rainfall and a long growing season contribute to good forage production, thus sustaining and encouraging cattle grazing. Generally speaking, the factors favorable to past beef cattle growth will continue to influence expansion in the Southeast; namely, reduction in tobacco acreage, farm consolidation, scarce and costly farm labor, technological improvements in forage production and utilization, and a continuing trend toward part-time farming.

Factors having a restraining effect on beef cattle growth in the Southeast are: rising land costs, high cotton and soybean prices, shifts of land to non-farm uses, and increase in calf grow-out (stockers) and grain finishing cattle on grass—which limits the pasture available for brood cows. All in all, beef cattle numbers in the area are expected to increase at a slower rate than in earlier years.

NORTHERN PLAINS

● *Present status*—Beef cow numbers more than doubled in the Northern Plains during the period 1950-70, with the greatest increases in the humid, eastern portion of the area. Improved pasture and forage crops, decline in milk cow numbers, and increased farm and ranch size, along with larger acreages of pasture available per farm, seem to account for most of the beef cow expansion in the Northern Plains.

● *Future growth*—Most of the expansion ahead in the Northern Plains will continue to occur in the eastern part of the area, and most of it will come about as a result of improvement in the production and utilization of forage crops. Shifts in land use from grain to forage production will depend upon the relative profitability of grain crops vs beef cattle.

SOUTHWEST

● *Present status*—The highest and best use for much of the land in the Southwest, where combinations of private and public land often prevail, is for grazing. Hence, it has been a noted beef cattle area for a very long time. Beef cow numbers in the area increased more than 3.9 million head from 1950 to 1970, primarily as a result of increase in acreage of pasture and range, but partly because of a substitution of beef cows for milk cows in the humid areas.

● *Future growth*—Rate of expansion in beef cows in the Southwest will be limited, partly because it is already a well-stocked and highly specialized beef area. Modest increases will come from improvement in forage production and expansion of cattle raising in the cropping areas. A shift from sheep and goats to beef will allow for some growth in beef cows.

MOUNTAIN

● *Present status*—Beef cow numbers almost doubled in the Mountain region from 1950-70. This area is characterized by three different kinds of cattle ranches: (1) mountain ranches, which, typically, combine irrigated meadows with range areas and private land with grazing permits to public lands; (2) desert ranches, involving public and private land, perhaps some irrigated hayland, and year-long grazing; and (3) mixed crop-livestock farms. The latter type of operation has shown the greatest expansion in beef cattle numbers, primarily as a result of the consolidation of farms and ranches into units capable of supporting a profitable beef herd.

● *Future growth*—Beef cow numbers in the Mountain region will expand at about the same rate as for the nation as a whole. Little change is expected in Utah and Nevada because of the arid climate and full utilization of existing forages. Increases in the region will come primarily from technological advances in forage production on rangelands and in the crop farming sections.

PACIFIC STATES

● *Present status*—This area showed an 80 percent increase in beef cow numbers during the 20-year period, 1950-70. It differs widely in climate and topography. Cow-calf operations are under mountain ranches, intermountain deserts, and mixed crop-livestock situations. In comparison with other regions, this region held more stable in milk cow numbers, which probably reflected increased human populations. So, beef cattle increases have come from increased production and utilization of forage.

●*Future growth*—Little expansion of beef cattle numbers may be expected in the Pacific region because of competing land uses and rangeland limitations.

A graphic picture of increase in beef cow numbers by states from 1950 to 1970 is presented in Fig. 17-14. As shown, the increases were greatest in the Plains, Central, and Southern states. The following states had increases of more than 500,000 head during this period: Texas, Oklahoma, Kansas, Nebraska, South Dakota, North Dakota, Montana, Iowa, Missouri, Arkansas, Kentucky, Tennessee, Mississippi, Alabama, and Georgia.

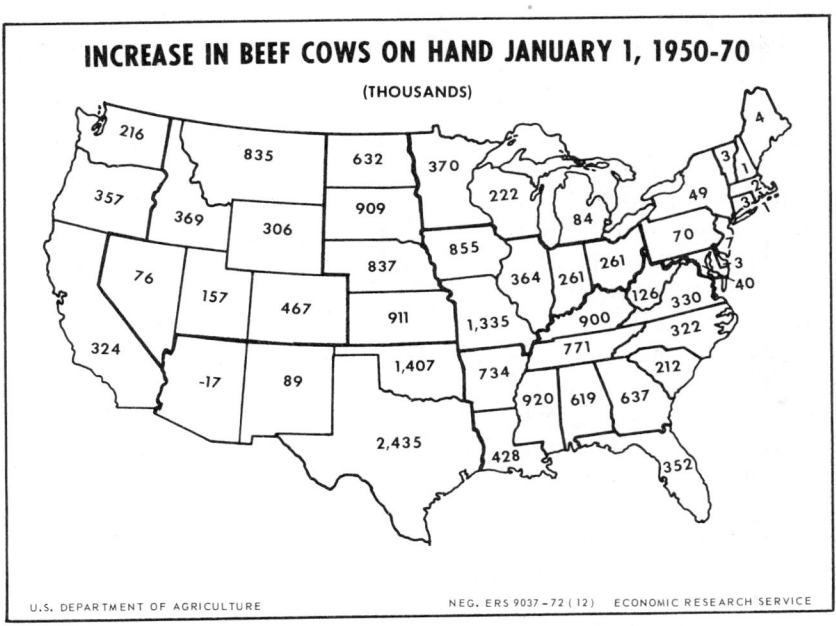

Fig. 17-14. Twenty-year trend (1950-70) in beef cow numbers, by states. As noted, the largest increases have occurred in the Northern Plains, Oklahoma, Texas, Iowa, Missouri, and the Southeast. (Courtesy, USDA)

Trends in Numbers of Herds and Herd Size

About two U.S. farmers in three own cows and heifers.

Contrary to popular belief, the average U.S. beef cattle herd is relatively small. In 1964, 1,322,783 U.S. farms (including part-time and part-retirement farms) had 32,639,630 beef cows, or an average of only 25 cows per herd. That year, two-thirds of U.S. farms reporting beef cows had herds with fewer than 20 cows and almost 90 percent had herds with fewer than 50.

In 1969 (latest figures available), 739,076 commercial farms (*not* including part-time and part-retirement farms, figures for which were not available) had 31,022,389 beef cows, for an average of 42 cows per herd. Thus, even when excluding the small part-time and retirement farmers, of which there are a very considerable number, beef cow herds are very small, on the average.

The above figures pertain to the size of beef cattle herds, exclusive of dairy. There are more recent figures on size of cattle herd, of *all cattle* (beef and dairy combined). On January 1, 1973, 1,830,810 U.S. farms had 121,534,000 cattle and calves (beef and dairy), for an average of 66 cows and calves per herd.[2] Here again, this points up the fact that, on the average, cattle herds are small.

In the 1960s, the most common cow herd sizes by areas were: Corn Belt, 38% under 20 head; Lake States, 38% under 20 head; and Southeast, 44% under 50 or more head. The largest herds (with the figures in parentheses indicating the percent of herds with 100 or more cows) were in the Mountain (72.7%), Pacific (70.1%), Southwest (53.9%), and Northern Plains (41.9%) regions.

It is noteworthy that, despite the rapid increases in beef cow inventory, the typical beef cow herd tends to be rather small and supplementary. Noteworthy, too, is the fact that further increases in part-time farming activities and continued farm consolidation may both be important in shaping future increases in beef supplies. Part-time farming favors a beef cattle enterprise which can be managed on a labor-extensive basis. As farms consolidate, small holdings of pasture are brought under one management, which makes it possible to support a profitable beef enterprise.

The preponderance of small beef cow herds is likely to affect the beef industry of the future in two ways: (1) They won't respond dramatically to price and cost changes—that is, part-time farmers are less responsive to prices than commercial farmers; and (2) they may limit technological change, simply because they cannot justify the cost of the technique.

Also, this means that (1) many producers will continue to lack sufficient volume to justify economically the use of many available technologies, and (2) assembling uniform lots of feeder cattle from many small producers for movement to a few large feedlots will continue to present structural problems for the entire beef industry. Nevertheless, the rising number of producers with larger herds can be expected to lead the way in the adoption of improved management and technology.

FACTORS INFLUENCING THE NUMBER OF BEEF COWS

Over a long period of time, economics determine the number of beef cows. In recent years, economic conditions have favored cattle production. As a result, beef cow numbers have increased greatly in all areas of the

[2]*Cattle*, Feb. 1, 1974, USDA, Statistical Reporting Service, Crop Reporting Board, LvGb 1 (2-74), pp. 3 and 13, Washington, D.C.

United States. In 1973, beef cows totaled nearly 2.5 times the number 20 years ago.

Beef cattle numbers are expected to continue to grow substantially in the years ahead, accounting for 87 percent of the increased beef needs from 1972 to 1985. A discussion of the influencing factors and the relative importance of each will follow.

Shifts in Crop Production

Traditionally, land for which the highest and best use is for pasture and hay has been used for beef cows. This practice will continue in the years ahead. However, forage crops produced with known technology will become more competitive with grain production. Forage for beef cows grown on land now occupied by wheat and cotton is uncertain, and largely dependent upon the world market for food and fiber. Also, should such acreage be released, the potential for soybeans and newly introduced cultural practices for raising corn will act as a counter force.

Shifts in Livestock Enterprises

Beef cow numbers already occupy much of the land formerly used by dairy cows, particularly in the humid regions. By 1985, it is estimated that U.S. per capita milk consumption in all forms will be about 470 lb, down from 570 lb in 1970. But milk production per cow will rise, from just under 10,000 lb in 1973 to 13,300 lb in 1985. Thus, even with 245 million people in 1985 (up from 208 million in 1972), the total milk needs will be met by 8,657,895 milk cows, a decrease of 3 million cows from 1973 to 1985. Hence, this will free sufficient land resources to feed 3 million more beef cows by 1985. Also, it is expected that more dairy enterprises will be moved into year-round confinement that utilizes only harvested feeds, thereby freeing some permanent pastures.

Nationally, beef cows have been substituted for sheep through the years. Further decreases in sheep numbers in the West and the Southwest are anticipated, due primarily to difficulty in obtaining labor, followed by a conversion from sheep to beef.

Traditionally, few cattle enterprises are strictly cow-calf operations as such. Instead, the cattleman exercises one or more of the options open to him, based primarily upon the availability and price of feed and the price of cattle. Some cow-calf outfits sell some calves and hold the remainder over for sale as short or long yearlings. Others buy additional calves and sell them as yearlings. Still others feed their calves to slaughter weight. Additionally, some beef cattle operators buy and sell stocker cattle only; they have no cows. Once weaned, a calf competes with beef cows for feed and forage and, consequently, acts as a constraint on production of additional calves.

In the past, the Corn Belt cattleman who has fed out his calves in his own small feedlot has found the finishing phase relatively unprofitable. Higher grain prices and smaller net profits per head in finishing operations will cause more Midwest farmers to discontinue feedlot operations, increase

ow numbers, and push for heavier weaned calves. The resources thus freed
rom cattle finishing offer the greatest single potential for increased feeder
calf production in this region.

Changes in the Southeast will curtail cow-calf expansion. Here the trend
is toward more calf grow-out and grain feeding on pasture.

Southwestern cattlemen engage in three distinct cattle raising programs:
(1) cow-calf; (2) cow-yearling; and (3) yearling-stocker. Except in Arizona,
straight cow-calf programs are expected to decline somewhat throughout the
Southwest. The slack will be taken up by cow-yearling programs in Texas
and New Mexico and yearling-stocker programs in Oklahoma, where considerable wheat grazing will be used. This shift among cattle-raising systems
will be prompted by the desire of cattlemen to capture returns from additional gains on the calves they produce, rather than raise more calves. The
movement in this direction will be more noticeable in areas in close proximity to large feedlots. Of course, the price of weaner calves will determine the
extent of this trend, for, given an option, cattlemen will choose that system
which is most profitable to them.

No significant changes in the existing cattle raising programs are expected in the Northern Plains, Mountain, and Pacific regions.

Forage Production, Harvesting, and Utilization

BEEF COWS PER 1000 ACRES

UNITED STATES AVERAGE 28.2

Fig. 17-15. Density of beef cows—number of beef cows per 1,000 acres. Most
pastures and ranges are fully stocked. But carrying capacity can be improved by the
application of forage technology and by irrigation. (Source: *Using Information in
Cattle Marketing Decisions—A Handbook*, WEMC Pub. No. 5, Feb., 1973, p. 20,
prepared by the Western Extension Marketing Committee, Task Force on the
Economics of Marketing Livestock, Fort Collins, Colo.)

Most ranges are now fully stocked, some are overgrazed. Hence, increased beef cow numbers are dependent upon increased forage production. Available forage will be increased by—

1. Application of forage technology, including fertilization of pasture· and hay crops, improvement of forage plant mixes, controlled grazing, renovation and reseeding of existing pastures and ranges, and use of herbicides to control undesirable plants.

2. Irrigation of forage crops in some of the Western dryland areas.

3. Improvements in methods of salvaging crop residues, especially "husklage" and "stalklage" from corn.

Fig. 17-16. Harvesting corn stalklage with flail attachment. (Courtesy, Koehring Farm Division, Appleton, Wisc.)

Public Policies

Public policies will change from time to time, but, for the most part, it is anticipated that they will favor expansion in cow numbers. The author does not foresee any long-time depressing effect on beef cattle numbers to accrue from crop control programs; from banning drugs (such as diethylstilbestrol, or DES); from environmental control; from livestock wastes; from imports and exports of beef; or from changes in Federal grades of beef. However, cow numbers may be affected by the following:

● *Income tax regulations*—Current Federal income tax regulations provide an incentive to own beef cows, especially ownership by investors in high tax brackets because of (1) the rapid depreciation allowances, and (2)

he possibility of investment credit. Such tax shelter possibilities encourage
outside investment from persons with access to large amounts of capital.

• *Substitutes for beef*—Current public standards allow beef products to
be extended by a specified amount with soybean and other proteins. If consumer acceptance is achieved and costs are lowered, extenders will likely be
used in increasing amounts. However, unless incomes fall, neither meat extenders nor pork, nor other red meats are expected to have much bearing on
beef consumption.

Cattle Prices

Anticipation of favorable feeder cattle prices to 1985 is one of the
strongest reasons for projecting expanding numbers of beef cows. People
usually do those things which are most profitable to them—and cattlemen
are people. World beef shortages, along with increased per capita beef consumption at home and abroad, indicate that, generally speaking, beef prices
and profits will be favorable for several years ahead.

Popularity of Cattle Raising

Being a cattleman serves as a status symbol for many people, farmers
and nonfarm investors alike. Considerable romanticism has been attached to
cattle raising through the years; and it has been enhanced by the recent
publicity on crossbreeding and the exotic breeds. Ownership of cattle carries
more prestige than ownership of any other kind of livestock.

Shifts in Land Use

There will be increased demands for grazing lands for industrial, residential, and recreational uses, all of which will mitigate against the expansion of cow-calf operations. In particular, pressures for recreational development and from the environmentalists will increase in the Western
range area.

Elderly and Part-Time Farmers

The increasing age of farmers and the growth of part-time farming favor
cow-calf operations. Farmers, like other people, are living longer. Many of
them choose to pull up the reins gradually by shifting to less labor-intensive
enterprises. A beef cow herd is often the enterprise of choice. Rising cost
and scarcity of hired labor also favor beef cattle. More part-time cattle farming seems likely.

CHANGES IN PRODUCTION PER COW

The major addition to the supply of beef in the years ahead is expected
to come from increased cow numbers (87% of it), which will produce a larger
number of calves to move into feedlots. But the cow of tomorrow will also

produce more beef as an individual—she will be a product of research. The author's projections show that 13 percent of the increased beef from 1972 to 1985 will come from more beef produced per cow.

Several potential changes in productivity per cow will directly affect the supply of beef obtainable from a given inventory of cows, among them those which follow.

Increased Percent Calf Crop

The U.S. Department of Agriculture reports a 94.3% calf crop in 1972 and a 93% calf crop in 1973, nationally. This is based on calves born as a percentage of cows and heifers (beef and dairy cows combined) 2 years old and over January 1.[3] This figure is higher than the actual rate, primarily because (1) more heifers are now calving at less than two years of age, with the result that the calves are counted but the dams are not; and (2) cows that are pregnancy tested, found barren, and marketed for slaughter are usually not counted as having been bred. Nevertheless, the USDA figures are the best that we have; hence, they serve as a useful guide. The author estimates that the actual percent calf crop for the United States is about 88%. No great improvement is seen in calf crop percentage ahead, primarily because gains become more difficult as calving percentages approach perfection—or 100%. It is estimated that the figure will move up only 2% by 1985, becoming about 90%. The largest gains are expected in the Southeast and Southwest, where rates have been the lowest.

Calf crop percentage on some cow-calf operations will be increased through selection; management practices, such as controlled breeding seasons (by use of hormones), fertility testing of bulls, and pregnancy testing of cows; improved nutrition, including improved pastures and ranges; and calving heifers at two years of age, instead of three-year-olds.

Superovulation (twinning and multiple births) could dramatically improve production per cow, if some of the obstacles could be overcome.

But the increased application of several practices that have an adverse effect on calving rate will partially offset the gains. This includes artificial insemination; the use of the larger breeds of cattle, which will make for more calving difficulty; shifts in the calving period; and more confinement cow operations.

Reduced Calf Losses

U.S. calf death losses from birth to weaning averaged 8.0 percent of the total calf crop dropped in 1972 and 1973, according to the USDA. Of course, there is room for improvement. Young calf losses will be reduced to some extent through the hybrid vigor of crossbreds and improved nutrition and management. However, these gains will be partially offset by increased

---

[3]Based on USDA statistics the figures are: in 1972, 94.29% calf crop and 7.04% calf death loss. In 1973, 92.95% calf crop and 8.97% calf loss. Sources: *Agricultural Statistics 1973*, USDA, Table 439, p. 299; and *Meat Animals*, Farm Production, Disposition, Income 1972-1973, USDA, MtAn-1 (74), April 1974, Table 1, p. 4 (1972 data), and Table 2, p. 6 (1973 data).

confinement calving, which will make for more calf scours and other diseases. Also, any severe restriction in the use of antibiotics could have a negative effect. Although there are many unknowns, it would seem that a calf death loss of 4.0 percent from birth to weaning would be within the realm of possibility by 1985.

Use of Dairy Calves

The average milk cow breaks down or is sold because of poor production after being in the milk string 4 years. To maintain a 100-cow dairy, therefore, 25 first-calf heifers must replace their elders on the milk line each year. But not all dairy heifer calves become tomorrow's cows! There are calf losses, and others must be culled for one reason or another. To maintain *status quo* in a milking herd, therefore—with no provision whatsoever for expansion—each year a dairyman must start with a minimum of 3 heifer calves for every 10 cows in the milking string. That's a 30 percent replacement. Of course, cull dairy cows and replacement heifers end up as beef, along with dairy steers and bulls.

In recent years, part of the demand for feeder calves has been met by increasing the proportion of the dairy calf crop fed out as dairy beef rather than vealed or kept for replacements. (Of course, fewer replacements, percentagewise, are kept when dairy cow numbers are being reduced instead of expanded.) Also, the need for dairy bulls has declined with increased artificial insemination. Although milk cow numbers halved in the 1950s and '60s, and some modest declines will occur in the years ahead, it is estimated that dairy beef, from cull dairy cows and dairy calves that are fed out, will provide about 15 percent of the nation's total annual beef supply to 1985.

In 1970, an estimated 40% of the nation's dairy calf crop was used for veal. This will change drastically, due to a continuing gain in demand for feeder calves relative to that for veal. By 1985, the author predicts that of all dairy calves born, 30% will be retained as replacements, 60% will be finished out as dairy beef, and only 10% will be vealed. Each calf sent to the feedlot means about 800 pounds more liveweight at slaughter than would have been obtained had the animal been vealed; hence, such a shift in disposition of dairy calves will increase production per cow.

Weights and Ages of Cull Breeding Stock

Cull cows contribute significantly to the total supply of beef, and account for nearly all nonfed beef produced in the United States. About 20 percent of beef cows and 25 percent of milk cows are culled each year and sent to slaughter. Beef cows are culled at an average of 7 to 9 years of age. No significant changes in age of culling are expected in the years ahead. Milk cows average only 4 years on the production line; they have a shorter life of usefulness than beef cows.

The trend is toward heavier weights of cull cows. Today, cull cows in most areas average around 1,000 lb in weight, except in the Southeast and Southwest where they are 100 to 150 lb lighter. By 1985, the author predicts an average of 75 lb heavier cow culling weights, as a result of (1) favorable beef prices serving as an incentive to put a little more weight on them before marketing, (2) improved nutrition, (3) increasing numbers of the larger exotics, and (4) conversion of more dairies to Holsteins, with lessening of the lighter breeds.

Production Testing and Crossbreeding

Production testing and crossbreeding, combined, which are out of the experimental and in the practical realm, can increase beef yield per cow maintained over straightbreds by 15 to 25 percent, depending on the choice of breeds and the breeding system. The 15 to 25 percent is achieved in two ways: (1) through selection, based on production testing, of the purebreds used in the crossbreeding program; and (2) through heterosis increase of the crossbreds.

Through production testing, it is possible to achieve (1) heavier weaning weights (29% heritability)—and young gains off milk and grass are very efficient; (2) higher daily gains from weaning to marketing, making for a shorter time in reaching market weight and condition, thereby effecting a saving in labor and making for a more rapid turnover in capital; and (3) greater efficiency of feed utilization, thereby making it possible to feed more cows and calves on a given quantity of feed.

A 2-breed cross (in which only the calves are crossbred) gives an 8 to 10 percent increase in pounds of calf weaned per cow bred. Through a 3-breed cross, it is possible to achieve even greater production per cow.

Calf Weights

Cow-calf men will produce heavier calves in their efforts to increase efficiency and obtain a larger share of the total returns to the beef industry. Of course, heavier calves at weaning add to the total beef supply only if they are eventually carried to heavier slaughter weights. When calves are sold at around 400 lb weight, there is need for a stocker or backgrounding stage because most cattle feeders prefer cattle weighing 600 to 700 lb. Weaning heavier calves off milk and grass, especially if weights of 500 to 600 lb are achieved, will have a tendency to eliminate or shorten the stocker stage.

Larger cattle and heavier milking strains will result in heavier weaning weights in the years ahead. Much of this transition will come about as a result of exotic crosses and dairy crosses. Also, increased calf weights will result from selection of British breeding stock for more size and more milk, from improved nutrition of cows and suckling calves, and from greater use of production records to select breeding stock with superior growth rate potential.

The author predicts that the combination of breeding (improved British breeds, along with infusion of exotic and dairy breeding) plus improved nutrition, will result in an average of 480-lb weaning weights at 7 to 8 months of age by 1985, vs about 407 lb in 1970. This means that the better herds will be weaning off calves weighing 600 lb or better. The greatest increase will occur in the Southeast, but it should be added that they have the most room for improvement, because their weaning weights have always been lower than other sections of the country.

Yearling Weights

Some cattlemen have long grown calves beyond weaning weight, with the prevalence of the practice varied by years according to available resources—particularly feed. It is expected that weaning calves at heavier weights will lessen the practice of holding them over as stockers except in the Southwest, where yearling cattle will remain important.

Short yearlings are weaner calves held over because they are too light to sell at weaning time and/or because adequate forage is available. Usually, they are sold at just under one year of age, although they may range up to 14 months in Arizona and New Mexico. Long yearlings average about 16 months of age.

Market weights of each group will gradually move upward, due primarily to improved forage supplies and selecting breeding stock for more size and more milk. The author predicts that, by 1985, without much change in age, short yearlings will weigh around 675 lb, 87 lb more than in 1970; and long yearlings will weigh around 750 lb, 60 lb more than in 1970.

QUESTIONS FOR STUDY AND DISCUSSION

1. Discuss the characteristics of, the relative importance of, and the relationship between (a) the cow-calf system, (b) the growing of stockers, and (c) drylot or pasture finishing.

2. Will the three phases of beef production—cow-calf, growing stockers, and drylot finishing—be more integrated or less integrated in the future?

3. What major changes have occurred in the cow-calf business since 1960? What forces have caused these changes?

4. How and why have stocker programs changed in recent years?

5. What factors should determine whether cows-calves will be produced on a particular farm or ranch?

6. Why have beef cattle increased while sheep have decreased on the western range?

7. What factors determine whether a beef operation shall be (a) strictly cow-calf, with all calves except replacements marketed at weaning time, (b) a combination cow-calf and stocker operation, or (c) stockers only?

8. Why do farmers in central and eastern United States have more options in the choice of enterprises than western ranchers?

9. List and discuss what you consider to be the six *most favorable* factors to cow-calf production.

10. List and discuss what you consider to be the *most unfavorable* factors to cow-calf production.

11. Basically, farmers and ranchers have four resources at their disposal with which to change cow numbers—land, labor, capital, and managerial skills. In the years ahead, how will the most successful cattlemen put these together to maximize profits and increase cow numbers?

    How do the cow-calf resource requirements compare to alternate enterprises?

12. What are the pros, what are the cons, and what is your choice between (a) a farm or ranch herd, (b) purebred or commercial cattle, (c) selling weaners or stockers, and (d) dual-purpose production?

13. How do you account for the fact that probably fewer than 4.0 percent of U.S. cattle are registered purebreds?

14. What changes do you foresee in cattle raising in the years ahead from the standpoints of (a) beef eaters and per capita consumption, (b) beef cow numbers, (c) veal slaughter, (d) percent of slaughter cattle that have been grain fed, (e) specialized vs farmer feeders, (f) shift from dairy to beef, and (g) shift from sheep to beef?

15. Prior to 1950, there were more milk cows in the United States. But beef cows passed milk cows in 1954. By 1974, there were 3.8 times more beef cows than dairy cows in the United States. Why did such a shift from dairy cows to beef cows occur?

16. The Corn Belt has an abundance of crop residue, especially from corn, most of which is left to rot in the field. Why, then, have beef cattle not increased more in this area?

17. What factors have contributed to the dramatic growth of the beef industry in the Southeast?

18. What factors will have a restraining effect on future beef cattle growth in the Southeast?

19. Fig. 17-14 shows that over the 20-year period, 1950-70, the largest increases in beef cattle numbers occurred in the Northern Plains, Oklahoma, Texas, Iowa, Missouri, and the Southeast. Why were the increases greater in these areas and states than in the rest of the United States?

20. If you were starting a beef cattle enterprise, and if you had the flexibility of choice, in what area would you locate, and why would you locate there?

21. Why are the vast majority of United States beef cattle herds small—with fewer than 50 cows?

22. How do you account for the fact that the largest beef cattle herds are in the Mountain and Pacific areas?

23. How will the preponderance of small beef cow herds likely affect the beef cattle industry?

24. How do you account for the fact that, in 1973, beef cows totaled nearly 2.5 times the number 20 years earlier?

25. Do you feel that beef cattle will be able to compete with wheat, cotton, corn, and soybeans for the use of the land in the United States? Justify your answer.

26. What shifts in land use do you see in the years ahead: (a) from dairy to beef; (b) from sheep to beef; (c) from Corn Belt farmer-feeder to cow-calf man; (d) from Southeast cow-calf expansion to growing stockers and pasture finishing; and (e) from Southwest cow-calf programs to cow-yearling programs?

27. Discuss the possibility of increasing beef production through increased pasture and range production by (a) fertilization, (b) improved forage plant mixes, (c) controlled grazing, (d) renovation, (e) reseeding, (f) use of herbicides, and (g) irrigation.

28. Describe the modern methods that are being employed to salvage husklage and stalklage.

29. Discuss the impact on cow numbers of each of the following: (a) banning drugs; (b) environmental control; (c) imports and exports of beef; (d) changes in Federal grades of beef; (e) the rapid depreciation allowance; (f) investment credit; (g) substitutes for beef; (h) prices; (i) cattle serving as a status symbol; (j) increased demands for grazing lands for industrial, residential, and recreational use; and (k) part-time farmers.

30. The author of this book estimates that the major addition to the supply of beef in the years ahead will come from the following sources: (a) increased cow numbers, 87 percent; and (b) more beef produced per cow, 13 percent. Do you agree or disagree? Justify your answer.

31. How will the following changes in productivity per cow affect the supply of beef obtainable from a given inventory of cows: (a) percent calf crop; (b) reduced calf losses; (c) use of dairy calves; (d) weights and ages of cull breeding stock; (e) production testing and crossbreeding; (f) calf weights; and (g) yearling weights?

SELECTED REFERENCES

Title of Publication	Author(s)	Publisher
Animal Science, Sixth Edition	M. E. Ensminger	The Interstate Printers & Publishers, Inc., Danville, Ill., 1969
Beef Cattle, Sixth Edition	A. L. Neumann R. R. Snapp	John Wiley & Sons, Inc., New York, N. Y., 1969
Beef Cattle Production	K. A. Wagnon R. Albaugh G. H. Hart	The Macmillan Company, New York, N. Y., 1960
Beef Cattle Science Handbook	Ed. by M. E. Ensminger	Agriservices Foundation, Clovis, Calif., pub. annually since 1964
Beef Production in the South	S. H. Fowler	The Interstate Printers & Publishers, Inc., Danville, Ill., 1969
Cattle Raising in the United States	R. N. Van Arsdall M. D. Skold	Economic Research Service, U.S. Department of Agriculture, Washington, D.C., 1973
Commercial Beef Cattle Production	Ed. by C. C. O'Mary, I. A. Dyer	Lea & Febiger, Philadelphia, Penn., 1972
Stockman's Handbook, The, Fourth Edition	M. E. Ensminger	The Interstate Printers & Publishers, Inc., Danville, Ill., 1970

CHAPTER 18

THE CATTLE FARM OR RANCH[1]

Contents   **Page**

---

[1]This chapter was authoritatively reviewed by the following: Mr. H. G. E. Fick, Accredited Rural Appraiser, Vice Chairman, Board of Directors, Doane, St. Louis, Mo.; Mr. Richard Thallmann, Manager, Myers Real Estate Co., Corpus Christi, Tex.; and Mr. Melvin L. Fox, Arthur G. Bishop & Co., Realtor, Denver, Colo.

"The past is prologue," according to a sign at the entrance to the National Archives Building in Washington, D.C. This gem of wisdom speaks loudly and clearly to cattlemen as they travel to the year 2000 A.D. The more successful operators will be those who recognize that the past is just the beginning (prologue)—those who focus their eyes on the future, then prepare for it. Indeed, the beef cattle industry is in the era of its greatest development as it explores and applies new genetic, nutrition, and marketing technology to meet an unprecedented worldwide demand for more beef.

Fig. 18-1. Attractive farmstead, enhanced by a fine herd of Polled Herefords and a good pasture, of Morlunda Farms, Lewisburg, West Va., owned by Oscar Nelson, Jr. (Courtesy, *Progressive Farmer*, Birmingham, Ala.)

People of all walks of life are eager to get into the cattle business, causing a rush reminiscent of the opening up of the western range from the 1860s to 1880s. But then, as now, ownership of cows and a potential market are not enough. Then, as now, land, feed, water, and management are necessary. Lest cattlemen forget the story of 1886, it should be told again and again. It was a severe winter of the type that is the bane of the cattleman's existence. With the melting of the snow in the spring of 1887, thousands of cattle skeletons lay weathering on the western range, a grim reminder of overstocking and inadequate feed supplies. Many ranchers went broke, and the cattle industry of the West suffered a crippling blow that plagued it for the next two decades. Out of this disaster, however, ranchers learned the never-to-be-forgotten lessons of (1) avoiding overexpansion and too close grazing, and (2) the necessity of an adequate winter feed supply. This story is not retold for pessimistic reasons. Rather, it is repeated for purposes of emphasizing that in the future, as in the past, the farm or ranch is the most important basic resource to the success of a cow-calf operation.

Many people are interested in buying a cattle spread. Tenants are climbing the ladder to ownership. Present owners are mechanizing their holdings and borrowing more money, with the result that they want more acres. Farms and ranches are being bought for sons and daughters. City folks want to fulfill their dreams by semiretiring on a cattle ranch. Investors, who have been disillusioned with the stock market, who are concerned about continued inflation, and who feel that manufacturing profits are on the wane, are looking for cattle farms and ranches. Some of these buyers will be happy with their purchases. Others will rue the day that they made the decision. How well the farm or ranch is selected, bought, and managed will make the difference. Indeed, no decision in the cattle business has greater consequences for the individual than selecting and purchasing the farm or ranch.

Although a cattle farm or ranch may be resold, most cattlemen plan to operate the place that they acquire for a lifetime. This means that the fields and buildings are to be the purchaser's workshop, and that the alternatives open thereafter are greatly reduced. He can no longer consider operating in another area, even if the climate, feed, and market are more favorable. In many cases, he is committed to a narrow range of alternatives, such as (1) cattle only, like on some holdings of the Southwest; (2) the particular combination of livestock, for example, hogs cannot be run on the western range; and (3) adapted crops and grasses. Moreover, except for an absentee landlord, the farm or ranch selected will be the family home. As such, the family will develop community ties that may last for years or even for generations, since many cattle farms and ranches are passed on from one generation to the next.

Indeed, the ultimate success of a cattle operation is determined by the careful selection of a farm or ranch; the proper combination of cattle, feed, and sometimes other livestock; the weather, disease, price fluctuations, and market demands; and the well-being and happiness of the family.

WHY OWN LAND?

In addition to providing self-employment, making a living, and obtaining a reasonable return on equity invested, ownership of land is attractive to investors for the following reasons:

1. *It is expected that land values will go up and up*—During the 20-year period 1953-1973, farm land values increased at an average annual compounded rate of 5.6 percent, nationally (48 mainland states). There is every reason to expect that the rise in land values in the years ahead will at least match this long-term trend.

Three big economic forces are at work to raise land values: (a) the population is growing while land area remains about the same; (b) our rising standard of living requires land for many purposes; and (c) inflation raises the money value of land.

2. *It provides an opportunity to obtain relatively large, long-term capital gains*—In land, the major increase is in its value, rather than from operating income. In 1945, United States land (farm real estate, exclusive of livestock, bank deposits, etc.) was valued at $53.9 billion; by 1974 it was $309.7 billion. That's a 5.75-fold increase. It is expected that this trend will con-

tinue. There is a tremendous advantage in getting long-term capital gain from land, because it is not taxed as heavily as ordinary income. For individuals (husband and wife) with up to $50,000 gains in any one tax year, the maximum tax is 25%; above $50,000, it is 35%. For corporations, the capital gains tax is 30%. So, in either case, there is a tremendous advantage in getting long-term capital gain from land.

3. *It provides an effective hedge against inflation*—If what you can purchase with your salary diminishes each year and your savings account is shrinking, while, at the same time, the price of groceries is climbing, what do you do? The truth of the matter is that most folks don't do anything before it's too late. This is primarily because (a) they are too busy "grubbing" out a living, and (b) they do not recognize the erosive nature of inflation. However, those with perception have frequently read the signs and found the answer in land. Historically, land has increased in value faster than other prices. For example, from 1960 to 1973 the value of United States farmland increased from $116 to $247 per acre, for an increase of 113%, whereas, during this same period of time, consumer prices went up only 46%.

Also, it is noteworthy that real estate is the traditional hedge against inflation, and that real estate values from 1972 to 1973 climbed 3.6 times (land: 12.8%; Consumer Price Index, 3.6%) more rapidly than the Government's Consumer Price Index.

Most economists agree that long-term inflation will not disappear and that land will not go down in value. Thus, as inflation continues, people are likely to bid up land prices more and more in an effort to protect themselves against the declining purchasing power of their dollars. Indeed, assets in underdeveloped land bid fair to replace "growth" stocks in the years ahead.

4. *It makes a good alternative to a jittery stock market*—The 1960-69 composite return on common stocks reported by Standard and Poor was 3.19%. For 1970-72, it was 3.27%.

5. *It affords investment opportunity in America's biggest single industry*—On January 1, 1974, farm assets (physical and financial) totaled $459.8 billion, which is equivalent to three-fifths the value of all stocks of all the corporations represented on the New York Stock Exchange.

6. *It's a business where the greatest need and action of the future lie*—The greatest need and action of the future will be that of providing food and fiber for the world's exploding human population—3.8 billion in 1972; and projected to be 4.9 billion by 1985, and 7.5 billion by 2007.

7. *It provides a way through which to benefit from increased farm exports*—Expanded export opportunities for United States farm products assure a bright future for United States agriculture.

8. *It offers an escape from some of our environmental and social problems*—There is growing concern with environmental pollution, poverty, urban congestion, and social unrest. As a result, these forces are giving a boost to interest in rural living.

At the turn of the century, 75% of the nation's population lived in rural areas—mostly on farms. Today, fewer than 30% are in rural areas and only 3.8% live on farms.

9. *It furnishes a recreational and vacation area of which the investor is part owner*—Such an arrangement (a) assures greater privacy than public recreational areas, and (b) makes it possible to combine business with pleasure.

10. *It provides good "leverage"*—In most cases, a person may borrow 65 to 90% of the cost of property, thereby tying up only a small amount of his own capital.

11. *It satisfies the psychological desire of people wanting to own some land*—Ownership of land—the good earth—imparts pride of ownership. Also, there is no security like the security of owning land.

TYPES OF CATTLE FARMS OR RANCHES

When selecting a cattle farm or ranch, the first major consideration should be the purpose of the ranch. For what use will it be put? Is it intended to be used as a full- or part-time cattle operation? Will it be used for cattle only; will it be used for cattle and one or more other classes of farm animals; or will it be a combination cattle and crop operation? Is it flexible enough to permit some choices should the economic conditions so indicate? Is it limited to a certain type of operation, like the Southwest where the choices, for the most part, are cattle or sheep? All these alternatives, and more, should be considered prior to purchasing a ranch. Of course, alternatives will also be considered, from time to time, in an established ranching operation. Nevertheless, the type of farm or ranch is important, for greatest success is usually achieved when the operator does those things which he likes best.

Full-Time Cattle Farm or Ranch

A full-time cattle farm or ranch is one in which the operator devotes his full time to the enterprise and depends entirely upon cattle for his income. As indicated later in this chapter (see section headed "Carrying Capacity; Size of Herd or Ranch"), (1) carrying capacity, not acreage, determines whether a unit is big enough to constitute a full-time operation, and (2) the author estimates that a farm or ranch with a minimum carrying capacity of 300 head of brood cows (or equivalent; for example, two yearling stockers may be substituted for one cow) is necessary to be an economic unit for a full-time cattle operation. Without doubt, exceptions can be cited where operators with smaller herds are making a good living, with no other source of income. However, most operators with fewer than 300 brood cows are either employed part-time off the farm or ranch or have another sizable farming enterprise in combination with the cattle.

In full-time cattle operations, it is important that the labor force be utilized efficiently throughout the year. This means that the size of the family, or the amount of hired help, should be considered. Generally speaking, an efficient full-time cattle operation should not have in excess of one man to 300 cows. When a cattleman improves on that, he can make money; when his labor force is in excess of that, there may be trouble. Of course, the more specialized the cattle operation is, the more difficult it is to distribute labor properly throughout the year. For example, it is very difficult for a cattleman who, during the grazing season, pasture finishes steers in the Flint Hills of Kansas to use his labor throughout the year unless he has a cow-calf operation in addition. Likewise, it is easier to use labor effectively throughout the year where a good part or all of the winter feed is homegrown than where the cattleman relies entirely on purchased winter feeds.

On some cattle ranches, none of the area is suitable for cultivation, or even for hay production, with the result that all supplemental feeds must be purchased. On still others, nearly year-round grazing is possible, with the result that little supplemental feed is needed. On the vast majority of the

Fig. 18-2. Hilly and wooded farm not suited to crop production. (Courtesy, University of Missouri, Columbia, Mo.)

nation's cattle farms and ranches, however, the operator has the option of either (1) pasture only, with all supplemental feed purchased, or (2) pasture, with homegrown supplemental feed. The advantages and disadvantages of each of these systems follow:

PASTURE ONLY, WITH ALL SUPPLEMENTAL FEED PURCHASED

In comparison with an operation where supplemental feed is homegrown, the advantages and disadvantages of purchasing supplemental feed are:

● *Advantages*

1. More cattle can be kept.
2. Less machinery and equipment necessary.
3. More time can be spent caring for the cattle, especially when they need it—as when calving, when there is a disease outbreak, etc.

4. Purchase of additional feed can give flexibility to the operation. When cattle are down in price and feeds are reasonable, the cattleman may buy feed and finish his own calves or even buy more calves.

● *Disadvantages*

1. It is more difficult to distribute the labor throughout the year.

2. The peak pasture growth is usually not efficiently utilized—that is, pasture may go to waste during the lush growing season.

3. The feed supply is often uncertain and is subject to high prices during periods of drought or other times of scarcity.

4. It may be difficult to borrow money to buy feed when it is needed.

PASTURE, WITH HOMEGROWN SUPPLEMENTAL FEED

In comparison with a farm or ranch that purchases all supplemental feed, the advantages and disadvantages of home-growing supplemental feeds are:

Fig. 18-3. Cattle and homegrown hay on a Corn Belt farm. (Courtesy, *The Corn Belt Farm Dailies*, Chicago, Ill.)

● *Advantages*

1. Labor can be used more effectively and efficiently throughout the year.

2. Land can be used for its highest and best use; that is, sometimes the highest and best use for land is to produce winter feed rather than to produce pasture.

3. It evens out costs, because it avoids the necessity of buying feed when it is scarce and high in price.

4. It makes it possible to rotate the use of land.

● *Disadvantages*

1. It limits the number of animals, by the amount of land that must be used for supplemental feed.

2. It requires more able management to produce both cattle and supplemental feed than to produce cattle alone, simply because the knowledge of each type of operation is necessary.

3. It requires more machinery and equipment.

4. The labor force may be divided at a time when it is urgently needed for the cattle, as at calving time, when dehorning and castrating, or when there is a disease outbreak.

When selecting a cattle ranch, the above alternative management systems should be considered. What are the pros and cons for each system, then what is the decision?

More and more full-time cattle farms and ranches of the future will specialize in cattle only, without diversifying in cash crops or another class of farm animals. Also, cow herds will get larger; and more and more of them will be under multiple ownerships and have holdings in different sections of the United States and even abroad.

Part-Time Cattle Farm or Ranch

Most cattle operators with fewer than 200 brood cows are either employed part-time off the farm or ranch or semiretired and have another source of income. Most of these have fewer than 50 cows.[2] In total, they account for nearly half the beef cattle numbers of the United States.

Many part-time cattle farmers or ranchers are individuals who have always wanted to own and operate a cattle ranch but who, because of limited funds, cannot acquire big enough spreads to make a living therefrom. The partial fulfillment of their dreams is realized by buying a small place on which they live, then they supplement their income from another job. They take care of their cattle before and after hours, and on weekends and holidays. Some of these part-time farmers hope eventually to farm full time. They plan to continue their nonfarm work until they accumulate enough capital for a full-time farm business, gaining valuable farming experience in the meantime on a part-time farm.

Many senior citizens and semiretired people find beef production a rewarding and renumerative experience. Folks are living longer. Many of them choose to pull up the reins gradually by shifting to less labor-intensive enterprises. Beef cattle fit their need. These retired or semiretired senior citizens are less dependent upon cattle prices and profits than full-time operators.

---

[2]In 1964 (last year for which complete data is available), almost 90% of U.S. farms reporting beef cows had fewer than 50 cows. But, in total, they had 46% of the nation's beef cows.

The above figures are for beef cattle, exclusive of milk cows. On January 1, 1973, 1,830,810 U.S. farms had 121,534,000 cattle and calves (beef and dairy), for an average of 66 cows and calves per herd (see Chapter 17 section entitled, "Trends in Numbers of Herds and Herd Size"). Obviously, therefore, many cattlemen secure a good deal of their income from sources other than cattle.

When it is planned that the operator seek off-farm employment, it is important that the ranch be selected with this in mind. This means, that, in addition to all the other factors that should be weighed when selecting a cattle operation, the off-farm worker must consider the availability of additional work and the distance thereto. Likewise, he must think how his cattle operation and the off-farm labor fit together. For example, it is very difficult to operate a part-time cattle farm or ranch when the operator must be away for three or four days each week, including overnight. Ideally, the part-time cattleman should be home each night, so that he can look after the cows night and morning. Likewise, he can use his weekends and holidays for such things as mending fences, putting up hay, branding the calves, and so forth.

Part-time cattle enterprises will increase in the future. More and more folks will own a little cattle farm or ranch, as a source of some income and as a way of life. But they will derive most of their income from off-farm employment. An additional, and growing, number of part-time cattle farmers and ranchers will consist of senior citizens and semiretired folks. They are more interested in an enterprise that is relatively free from labor problems, and in the good life, than in monetary gain, for most of them have already made it.

Kind of Beef Farm or Ranch

Traditionally, few beef cattle farms have been strictly cow-calf enterprises, limited to selling calves at weaning time. Some cow-calf producers sell part of the calf crop and hold the remainder over for sale as short or long yearlings, depending on the forage supply. Others buy additional calves and sell them as yearlings. Still others finish their home-raised calves out to slaughter weight. Additionally, there are cattle operations that own no cows; instead, they buy and sell stocker cattle.

In selecting a cattle farm or ranch, the above alternatives should be considered. Additionally, projected future changes in cattle raising in different regions should be considered. For example, because cattle feeding in the Corn Belt hasn't been too profitable in the past, it is expected that this area will go more to cow-calf production and push for heavier calves to increase returns. In the Southeast, more calves will be grown out to the yearling stage and grain-finished on pasture. Cow-calf systems are expected to decrease somewhat in some of the southwest states and to be replaced by more cow-yearling programs. In the Northern Plains the trend will be for more cow-calf programs and fewer cow-yearling and stocker programs.

Cattle and Cash Crop Combinations

Cattle and cash crop combination farm and ranches are preferred by some operators. It makes for desirable diversification, and, hopefully, cattle and the cash crop(s) will not be down in price at the same time.

When selecting a cattle-cash crop combination ranch, geographic location and crop adaptation must be considered. (See Fig. 17-12 for geographical regions, such as Corn Belt, Lake States, etc. to which reference will be made in this chapter.) In the Corn Belt and Lake States, for example, the cash crops might consist of either corn and/or soybeans. In the Northern Plains and the northwest, small grain crops, like wheat, barley, and milo, might be considered. In the south, cotton, rice, milo, soybeans, and peanuts are possibilities.

Generally speaking, the most successful cattle and cash crop combination farms and ranches utilize those areas for which the highest and best use is pasture and hay for cattle. Then, the amount of cash crop acreage in the rest of the farm is determined primarily by prices. When cash crops are high in price, the maximum amount of area can be devoted to their production. On the other hand, when crop prices are down, more of the land can be converted to cattle feeds.

Still another type of flexibility exists on this type of operation. When crop prices are down and cattle prices are up, it may be advantageous to feed out the home-raised calves or even buy more feeders. When grain prices are high, however, it may be advantageous to sell the cash crop and market the calves as weaners. Thus, the cash crop can be expanded or contracted as determined by price. Cattle-crop combinations also offer a fine opportunity to utilize land and crops to the highest level of efficiency. Small grain crops, like wheat and rye, can provide winter and spring grazing and after-harvest stubble grazing. Cornstalks can serve as a source of cheap roughage during the winter months.

Cattle-crop combinations provide an opportunity for specialization of labor. For example, in a family operation, in a partnership, or in a corporation, one or more persons may have expertise in cattle, whereas other members may have primary interest in machinery and crops.

Government crop programs may also play a part in cattle-crop combination farms and ranches. Like taxes, however, such programs are subject to frequent changes; hence, a farm or ranch should not be purchased on the basis of the existing government program. Rather, like tax shelters, the farm or ranch owner should take advantage of whatever government crop programs exist at the time.

Cattle and Other Crop Combinations

Although they are not likely to be primary factors in determining the selection of a ranch, or how much will be paid for it, certain other cattle-crop combinations should not be overlooked. This includes both by-product feeds and specialty crops.

Innumerable by-products—both roughages and concentrates—from plant and animal processing, and from industrial manufacturing, are available and used as cattle feeds in different areas. Mention has already been made of small grain stubble fields and cornstalks. Cotton fields may also be pastured following harvest. Then, there are such additional by-product feeds as cull potatoes, cottonseed hulls, corncobs, cull citrus, cannery refuse, beet tops, and a host of other similar products. Also, on many farms and ranches there are either low, wet areas, or rough broken areas, which cannot be used profitably in crop production. Such areas can be fenced and made available to cattle.

Cotton and tobacco prices are subject to rather wide fluctuations. Also, an increasingly large proportion of these crops in the Southeast will be grown on fewer and larger farms. Forage for cows will be grown on some of the released cropland. But, of course, soybean and corn will be competing for the released acreage. Cattle and hay ranching make a good combination. Most always, some poor quality hay is produced as a result of unfavorable haying conditions at harvest time, or during a wet season silage may be made instead of hay. Rain-damaged hay and/or silage may be fed to beef cows. In some areas, cattle and timber make a good combination. This is true on much of the leased land under the supervision of the Bureau of Land Management and the U.S. Forest Service. Likewise, some small farms, particularly in the Southeast, are adapted to pine trees on the rough areas and cattle on the more level areas.

Cattle and Other Livestock Combinations

In many areas, and on many farms and ranches, beef cattle and one or more other classes of livestock may be combined to advantage, thereby using the resources more efficiently and increasing income.

Farmers in the central states long ago recognized the advantages of combining beef cattle and hogs. Regardless of the system of beef production—cow and calf proposition, the growing of stockers and feeders, finishing steers, dual-purpose production, or a combination of two or more of these systems—beef cattle and swine enterprises complement each other in balanced feeding. The cattle are able to utilize effectively great quantities of roughages, both dry forages (hay, cornstalks, etc.) and pastures; whereas pigs are fed primarily on concentrates. In brief, the beef cattle-hog combinations makes it possible to market efficiently all the forages and grains through livestock, with the manure being available for application back on the land. Such a combination makes for excellent distribution of labor. The largest labor requirements for both beef cattle and hogs comes in the winter and early spring. During the growing and harvesting seasons, therefore, most of the labor is released for attention to the crops.

In the early history of the range livestock industry of both the United States and Canada, the cattle-sheep feuds frequently waxed hot. Each group warned the other away from its range. For the most part, however, the hatchet has long since been buried and only the legendary stories linger on. Today, many cattlemen would do well seriously to consider adding sheep to their enterprises. Limited experiments, along with observations, indicate that it is more efficient to graze sheep and cattle together than to graze either species alone. Joint grazing results in (1) the production of more total pounds of meat, or greater carrying capacity, per acre, and (2) more complete and uniform grazing than pasturing by either species alone. This is attributed to the difference in the grazing habits of the two species. Cattle tend to leave patches of forage almost untouched, especially areas around urine spots and manure droppings. Also, cattle take larger bites and are less selective in their eating habits than sheep. Sheep tend to be selective of plant parts and will strip the leaves of plants. Also, sheep will graze many common weeds, even when good quality grasses and legumes are abundant.

Beef cattle and dairy cattle compete for about the same feeds. Thus, a beef and market milk enterprise is seldom conducted in combination, with both as major enterprises on the same farm or ranch. In the East, cow-calf programs sometimes fit in where dairy farms have more feed than is needed in the dairy program. Also, some dairymen, particularly small operators, have always used beef bulls as "cow fresheners" on lower producing cows, then marketed the calves as feeders. But a new type of beef-dairy combination has developed in recent years. It is known as "dairy beef." Dairy beef is just what the term implies—beef derived from cattle of dairy breeding, or from dairy X beef crossbreds. Today, it is extolled with pride. The shift in consumer preference, along with rapidly expanding population and increased per capita consumption of beef, will result in the production of increased quantities of dairy beef.

Both commercial cattle feeders and dairymen are interested in producing dairy beef, with the result that there is competition between them. Many dairy steers and cull heifers will continue to be finished out in commercial cattle feedlots. However, an increasing number of these animals will be finished out by dairymen as a means of augmenting their income and diversifying.

Cattle and horses are found on many farms and ranches. In addition to being of use in working cattle, horses may contribute to the income of the operation. Also, it is noteworthy that cattle and horses are rotated on the pastures in the great horse breeding centers of the world (including the blue-grass area of Kentucky and the lush pastures of Ireland) for parasite control.

Combinations of beef cattle and poultry are occasionally seen, primarily as a means of disposing of poultry manure. Modern poultry operations require little land. However, manure disposal is a problem. Some poultrymen have solved their problem by having extensive pastures adjacent to the poultry operation on which the manure is spread. Such pastures produce an abundance of grass and are utilized by beef cattle.

Irrigated Farm or Ranch

Rising land and labor costs favor more irrigation. Where the cost per cow-calf carrying capacity is cheaper under irrigation than a dryland operation, irrigation will increase. Likewise, intensive cow-calf operations under irrigation usually require less labor per cow-calf unit than more extensive dryland operations. Of course, once an area is irrigated, new crop alternatives are opened up, with the result that a determination will have to be made as to which will be the most profitable—cattle or crops.

Attached Government and Private Leases

In the West, much of the grazing land that ranchers rely upon to maintain their cattle is built up into operating units by leasing or by obtaining use permits from several Federal and state agencies. Overall, about half the range area in the 11 western states is Federal or state land. This land is made available through permit or lease to nearby ranch operators, usually at a fixed annual fee per head of livestock. Although most of these leases are subject to ready cancellation, it is noteworthy that many of them have continued for years, even through two or three generations of the same family. Nevertheless, now and in the future, increasing pressure is being brought to use such public lands for recreation, wildlife, and environmental control. As a result, the future of such leases is less secure than in the past, and ranches made up principally of deeded land go for substantially more than those made up chiefly of government-leased land; and this gap will widen. Of course, the type of government lease, as well as the way the area lies in relationship to deeded land, can make a tremendous variation in land value.

The bulk of Federal land is administered by the Bureau of Land Management and the U.S. Forest Service. Thus, where a cattle ranch is being acquired in the West, it is important that the cattleman have knowledge of these particular agencies.

BUREAU OF LAND MANAGEMENT

The Bureau of Land Management of the U.S. Department of the Interior administers about 60 percent of Federally owned land. From the standpoint of the cattleman, the most important function of the Bureau of Land Management is its administration of the grazing districts established under the Taylor Grazing Act of 1934 and of the unreserved public land situated outside of these districts which are subject to grazing lease under Section 15 of the act.

Fig. 18-4. Hereford breeding herd on the home ranch of a grazing permittee. These cattle summer in Colorado and winter in Utah. (Courtesy, Bureau of Land Management, U.S. Department of the Interior, Washington, D.C.)

Grazing privileges are allocated to individual operators, associations, and corporations, and a fee is charged for grazing privileges. In 1976, the Bureau of Land Management charged $1.51 per cow per month. (Also see Chapter 9, section entitled, "Agencies Administering Public Lands.")

U.S. FOREST SERVICE

Approximately 106 million acres of the national forests are used for grazing under a system of permits issued to local farmers and ranchers by the Forest Service of the U.S. Department of Agriculture. These grazing allotments provide grazing for about 6 million livestock and about 4.6 million head of big game animals.

The Forest Service issues 10-year term permits to stockmen who hold preferences and annual permits to those who hold temporary use. Among other things, the permit prescribes the boundaries of the range which they may use, the maximum number of animals allowed, and the season when grazing is permitted.

Forest Service grazing fees are based on a formula which takes into account livestock prices over the past 10 years, the quality of forage on the allotment, and the cost of ranch operation. In 1976, the charge came to $1.60 per animal unit month. (Also see Chapter 9, section entitled, "Agencies Administering Public Lands.")

Hobby Farm

In recent years, many men of wealth have established outstanding cattle herds, especially purebred herds. Most of these folks operate such cattle ranches as moneymaking enterprises. They are just as "money hungry" in the cattle business as they were in the industry from which they made their initial wealth. Also, conducting the cattle operation as a business is more of a challenge to them than if it were a hobby. Besides, the income tax regulations today are such as to make it impossible for many to afford not to operate a cattle farm or ranch as a business.

Internal Revenue Service agents are prone to attack the "farmer" status of absentee owners. Most challenges are raised where the taxpayer earns substantial off-farm or ranch income and is showing farm or ranch losses for a particular year or over a period of years. Though the concept is similar to the "hobby farm" challenge, the distinction exists in that the disallowance or challenge relates to capitalization of expenses. In hobby farm situations, the expenses are considered personal, and thus not deductible; nor can they be capitalized.

Of course, if a ranch is to be purchased and run as a hobby, the net return in terms of investment is likely to be of less concern to that individual.

SELECTING AND BUYING A BEEF CATTLE FARM OR RANCH

The first and most important requisite for success in the cattle business is proper selecting and buying of the beef cattle farm or ranch. The fundamental considerations will be discussed in the sections that follow.

Carrying Capacity: Size of Herd or Ranch

Carrying capacity is defined as the number of animal units (one cow, plus suckling calf—if there is a calf; or one heifer two years old or over) a property will carry on a year-round basis. This includes the land grazed plus the land necessary to produce the winter feed. Thus, if a 3,000-acre ranch provides all the pasture, along with winter feed, for a 300-cow herd plus their suckling calves, it has a carrying capacity of 10 acres per cow (3,000 ÷ 300 = 10). Two yearlings are considered equivalent to one cow-calf unit. The carrying capacity may vary anywhere from a productive irrigated farm with a carrying capacity of one acre per cow-calf to some ranges of the Southwest where grass and browse species are so sparse that it takes 60 acres to support one cow-calf. Most areas west of the 100th meridian require an average of about 30 acres of grazing land per animal unit.

Thus, when a realtor glibly refers to a certain farm or ranch as having a carrying capacity of 500 head, it is well to pin him down by asking, "head of what?" Is he basing his figures on the definition given above, or is he counting calves, also; and is he counting each yearling as one-half animal unit?

Admittedly, the carrying capacity of a ranch is difficult to determine. Past range management is a factor. If a range has been grazed too closely, carrying capacity is apt to be overrated. To continue stocking too heavily means lighter calf weaning weights, thinner cows, and perhaps a smaller calf crop. To rest and improve the range takes time and money.

From the above it should be deducted that carrying capacity, rather than number of acres, determines the number of cows that can be run on a given unit. No minimum or maximum figures can be given as to the best size of herd. Rather, each case is one for individual consideration. It is noteworthy, however, that labor costs differ very little whether the herd numbers 10 or 80. Noteworthy, too, is the fact that the cost of purchasing and maintaining a herd bull comes rather high when too few females are kept. The extent and carrying capacity of the pasture, the amount of hay and other roughage produced, and the facilities for wintering stock are factors that should be considered in determining the size of a herd for a particular farm or ranch unit. The system of disposing of the young stock will also be an influencing factor. For example, if the calves are disposed of at weaning time and replacement heifers are bought, practically no animals other than the brood cows and herd bulls are kept. On the other hand, if the calves are carried over as stockers or finished at an older age, more feed, pasture, and shelter are required.

Then, too, whether the beef herd is to be a major or minor enterprise will have to be decided. Here again, each case is one for individual consideration. Big operations are getting bigger, with an increasing number of them involving multiple ownership. At the same time, there are an increasing number of small part-time herds, owned by individuals who have off-farm employment or are semiretired. Movement in both directions will continue.

Table 18-1 shows the average acreage and number of cattle on selected cattle ranches of the Northern Plains and Rocky Mountain Areas.

TABLE 18-1

ACREAGE AND CATTLE ON NORTHERN PLAINS AND ROCKY MOUNTAIN
CATTLE RANCHES WITH 200 to 500 BROOD COWS[1]

Item	Northern Plains[2]			Rocky Mountain[3]		
	Average 1960-64	1971	1972	Average 1960-64	1971	1972
	(acres)	(acres)	(acres)	(acres)	(acres)	(acres)
Total land operated	11,500	11,500	11,500	5,900	5,900	5,900
Land Owned	8,940	8,940	8,940	4,000	4,000	4,000
Land Rented	2,560	2,560	2,560	1,900	1,900	1,900
		(number)			(number)	
All Cattle	432	487	492	363	431	435
Cows & Heifers, 2 yrs old or over	292	311	316	277	315	315

[1]Goodsell, Wylie D., and Macie J. Belfield, *Cost and Returns, Northwest Cattle Ranches, 1972*, USDA, ERS-525, Table 1, p. 4, and Table 4, p. 6.
[2]The Northern Plains survey included southeastern Montana, northeastern Wyoming, and western South Dakota.
[3]The Rocky Mountain survey included southwestern Montana and east-central Idaho.

Table 18-1 is based on a survey of the larger cattle ranches with 200 to 500 brood cows; those that produce 35 percent of the cattle in the Northern Plains and 30 percent of the cattle in the Rocky Mountains. Thus, it included most of the larger outfits of the two areas surveyed, and most of the operations on which cattle were the major enterprise. However, it did not include approximately two-thirds of the ranches having cattle—those with fewer than 200 brood cows, on many of which cattle are a part-time business.

Gray *et al* reported that, in 1971, Southwest cattle ranches in Arizona, New Mexico, and western Texas averaged 17,720 acres of owned and leased land, with an average of 280 cows and heifers 2 years old and over plus 107 head of other cattle.[3] The latter study was also based on a survey of ranches with 200-500 head of brood cows—ranches which produce nearly two-fifths of the area's calves.

Based on the above information, plus observation, the author estimates that a farm or ranch with a carrying capacity of 300 to 500 head of brood cows (or equivalent; for example, 2 yearling stockers may be substituted for one cow) is necessary to be an economic unit for a cattle rancher who devotes his full time to the operation and depends entirely upon his cattle for his income.

Since carrying capacity is the most important estimate in the entire appraisal, when buying a ranch it should be done by an expert who is com-

[3]Gray, J. R., W. D. Goodsell, and J. M. Belfield, *Costs and Returns from Southwest Cattle Ranches, 1971*, USDA, ERS-83.

pletely familiar with the area. Generally, the seller is prone to overestimate carrying capacity; his tendency is to estimate carrying capacity on the basis of his best season in 20 years, without including the other 19 years in the average.

Capital Requirements

Those thinking of becoming cow-calf operators inevitably ask, "how much money will it take, and what can I make?" The operator needs to take stock of his financial situation. How much money does he have available now, how much credit does he have, and how much debt is he willing to assume?

Cattle ranching is big business, requiring large amounts of capital, ranging from $350,000 to well over $1,000,000 per ranch for a 300-cow, or larger, outfit. The capital requirements have risen as buyers have bid up the price of land and cattle, and as costs of machinery, equipment, and related items have advanced with the general price level. Investment per animal unit provides a suitable basis for comparing areas and measuring increases in capital requirements. (See Table 18-2.)

TABLE 18-2

INVESTMENT PER ANIMAL UNIT, SELECTED WESTERN LIVESTOCK RANCHES, 1960, 1965, AND 1972[1]

| | Cattle Ranches | | | | | | | | |
| | Northern Plain[2] | | | Rocky Mountain[3] | | | Southwest[4] | | |
Item	1960	1965	1972	1960	1965	1972	1960	1965	1972
	Number			Number			Number		
Animal Units[5]	403	430	454	344	361	411	NA	348	341
	Dollars			Dollars			Dollars		
Investment in									
Land & buildings ...	484	581	796	400	526	563	NA	908	1,327
Livestock	162	133	245	161	141	247	NA	124	268
Machinery & equipment	35	37	49	40	43	51	NA	30	38
Crops	7	10	15	24	30	32	NA	0	0
Total	688	761	1,105	625	740	893	NA	1,062	1,633

[1]Goodsell, Wylie D., and Macie J. Belfield, *Cost and Returns, Northwest Cattle Ranches, 1972*, USDA, ERS-525, Table 7, p.9. The survey was limited to 200- to 500- cow outfits.
[2]Consists of 15 counties in Montana, 8 counties in Wyoming, and 9 counties in South Dakota.
[3]Consists of 12 counties in Montana and 7 counties in Idaho.
[4]Consists of 20 counties in Texas, 11 counties in New Mexico, and 3 counties in Arizona.
[5]An animal unit consists of 1.0 cow or heifer 2 years old and over, 1.33 steers or heifers 1 year old, 0.83 bull of breeding age.
NA = not available.

It is noteworthy that the total investment per cattle ranch increased at the following average yearly rate during the period (1960 to 1972) reported in Table 18-2: Northern Plains, 6.2%; Northern Rocky Mountain, 5.5%; and Southwest, 6.4%.

Table 18-3 shows the capital requirements in two different areas of Nebraska: (1) the Sandhills ranching area, and (2) the general farming area of eastern Nebraska.

TABLE 18-3

ESTIMATED CAPITAL REQUIREMENT ON A PER COW BASIS IN TWO
AREAS OF NEBRASKA, EARLY IN 1973[1]

	Sandhills Ranches in Nebraska		Crop and Livestock Farms in Eastern Nebraska	
Cow		$350[2]		$350[2]
Heifers		50[2]		50[2]
Share of Bull	(1/25)	30[2]	(1/30)	25[2]
Pastureland	(15-16 acres)	930	(3-4 acres)	700
Hayland	(1¼-1¾ acres)	188	(.5 acre)	250
Buildings and Equipment		10		35
Horses		5		—
Total		$1,563		$1,410

[1]*Opportunities for Cow-Calf Operations in Nebraska, A Prospectus,* University of Nebraska, College of Agriculture, Extension Service, EC 73-254, Table 2, p. 21.
[2]Average values while in herd based on early 1973 prices.

The figures reported in Tables 18-1, 18-2, and 18-3 are excellent, as far as they go. But, they cover only a restricted area and a limited number of ranches. Similar surveys have been made in certain other states or areas, but most of them are out-of-date. No scientific, nationwide study of this type has ever been made. Thus, there is a paucity of information on the subject of cow-calf capital needs, despite the fact that it is sorely needed by both investors and lenders. To fill this gap for this revision, the author resorted to a consensus of knowledgeable Chairmen of Animal Science Departments in the Land Grant colleges and universities throughout the United States. A summary of the results of this study is presented in Chapter 24, "Business Aspects of a Cow-Calf System," Table 24-1.

Table 24-1 reveals that, on the average and on a nationwide basis, for a commercial cow-calf operation, (1) an investment of $900 to $1,100 per mature cow for land and improvements is required, plus machinery and equipment; (2) the cost to produce one beef calf to weaning age varies from $209.90 to $257.90; and (3) the larger the herd, the lower the investment per animal unit and the cost to produce a weaner calf. Similar figures are presented for purebred herds.

It should be noted that ranches made up principally of deeded land sell for substantially more than those made up mainly of government lease land. A rule of thumb is that, on a per animal unit basis, all deeded land is worth twice as much as comparable land of which only one-half is deeded. Of course, the type of government lease, as well as the way the leased area lies in relationship to the deeded land, can make a tremendous difference in value. Yet, no matter what the history of the lease or what the old timers say, a government lease can be cancelled quickly and without compensation.

WHAT RETURN?

All cow-calf enterprises are owned by people, and most people want to be paid for their investment and work. This is as it should be. The only question is—how much return?

As shown in Table 24-1, it costs from $209.90 to $257.90 to produce a commercial calf to weaning age, depending on the size cow herd. As previously indicated, the author feels that a commercial herd of 300 to 500 cows is necessary for an economic unit; hence, as shown in Table 24-1, it costs $208.40 to produce a beef calf to weaning age in such a herd. Assuming a weaning weight of 500 pounds (U.S. average in 1970 was 407 pounds), and a price of 50 cents per pound (the going price for calves in the West in 1973), such a calf would bring $250.00. Thus, $250.00 − $208.40 = $41.60, the net return per calf. The 300-499 cow outfit has a capital investment of $1,395 per cow unit (land $950 + machinery and equipment $125 + cow $320). Thus, with a net of $41.60 per calf, the return to ranch capital is 3.0%. In terms of operator labor and management, a 400-cow herd would return $16,640. Over and above this, added incentives may accrue from tax shelter and increased land value, but these should not be counted upon, or used as reasons for investing in a cattle farm or ranch. Remember, that tax laws can and do change, and that land values may not continue to go up at current rates (although the author feels that they will). Remember, too, that increases in land values are unrealized income, or "paper profits" until the property is sold. They cannot be used as earnings with which to buy groceries.

Goodsell and Belfield reported returns from cattle ranch capital for the period 1960 to 1972. (See Table 18-4.)

TABLE 18-4

RETURNS TO CATTLE RANCH CAPITAL, 1960-72[1]

Year	Cattle Ranches		
	Southwest	Northern Plains	Rocky Mountain
	(%)[2]	(%)[2]	(%)[2]
1960	NA	2.8	3.1
1961	NA	2.3	4.1
1962	NA	4.5	5.7
1963	NA	3.7	5.0
1964	NA	2.0	2.5
1965	1.6	2.1	3.5
1966	1.9	3.3	4.8
1967	1.2	3.2	4.9
1968	1.4	3.5	5.5
1969	1.6	4.1	7.0
1970	1.1	4.7	6.5
1971	1.0	5.4	7.4
1972	3.6	7.6	10.8

[1]Goodsell, Wylie D., and Macie J. Belfield, *Cost and Returns, Northwest Cattle Ranches, 1972*, USDA, ERS-525, Table 8, p.9.

[2]Net ranch income less a nominal charge (annual wage to year-round hands × 1.25) for operator's labor and management divided by total ranch investment.

NA = Not available.

As shown in Table 18-4, during the years for which data are available on southwest cattle ranches, returns were much lower than on cattle ranches in the Northern Plains and Rocky Mountains. There are two reasons for this: (1) investment runs high on southwest ranches; and (2) the Southwest has been plagued with droughts during much of the time reported. For the three areas, the returns on ranch capital averaged 3.92 percent for the period 1960 to 1972, which is not enough. Hence, it prompts the following three questions, to which the author gives his answers:

1. Q. *Would an investor do better in the stock market?*

A. Not likely. The 1960-69 composite return on common stocks reported by Standard and Poor was 3.19%. For 1970-72, it was 3.27%. As noted above, the average return on cattle ranches was 3.92% for the 13-year period, which was slightly higher than the return from common stocks.

2. Q. *Would an investor make more money in some agricultural enterprise other than cattle?*

A. Not likely. In the 10-year period 1963 to 1972, farmers as a whole averaged only a 3.9 percent return on the equity of their capital investment in farming—the same as cattlemen averaged during this period.

3. Q. *Would an investor put money in a cattle ranch if he knew in advance that he would get an average return of only 3.92 percent?*

A. What's better or more sure than an investment in the good earth? Marshall Field put it this way: "Buying real estate is not only the best way, the quickest way, and the safest way, but the only way to become wealthy." Andrew Carnegie said: "Ninety percent of all millionaires became so through owning real estate. More money has been made in real estate than in all industrial investments. The wise young man or wage earner of today invests his money in real estate."

RISKS

The ownership of a farm or ranch does involve some risks, which are not pleasantly recalled by most current owners, and of which all prospective owners should be aware. Among them are the following:

1. *Drought*—A long dry period will shorten the grazing season; cut down on crops to be harvested for winter feed; result in loss in weight of cows; make for a smaller calf crop percentage; lower weaning weights; necessitate the purchase of feed; and/or even cause liquidation of all or part of the herd at unfavorable prices.

2. *Floods, storms, blizzards, fires*—Historically, some cattle are lost as a result of sudden rain storms and blizzards. Also, some are lost by fire.

3. *Disease outbreaks*—Traditionally, a cattleman vaccinates against the diseases most common to his area, provided such a preventive exists. Then, he takes a calculated risk relative to a long list of other diseases. But costly disease outbreaks do occur, resulting in death losses and inefficiency among the living.

Selecting the Location and the Farm or Ranch

Choosing the right location and the right farm or ranch are very important requisites for success. Nearness to friends and relatives is a major, but unmeasurable, factor in determining location, although it is less important than formerly with the development of more rapid transportation (improved highways and air travel) between areas. However, many nonsentimental factors should be considered in the selection of the particular location and the specific farm or ranch; among them, those which follow.

AREA AND CLIMATE

Area and climate (rain, snow, heat, cold, and wind) affect land use, cattle, and people; hence, they may determine the type of cattle farming or ranching of the particular area. In addition to soil, it takes moisture and reasonably warm weather to grow grass; and the longer the growing season, the more grass. Also, winter feeding is always more costly than summer grazing; hence, an area with mild winters has considerable advantage. So, the official weather bureau rainfall and temperature records—minimum, maximum, and average, by months and over a period of years—should be studied. Also, it is well to check the number of frost-free days, and even day lengths.

In the Corn Belt and Northern Plains, more than half of the land is cropland. In the southern states and mountain states, pasture and rangeland predominate. In the Northeast, forest is the major use of land, as it is in the Appalachian, Southeast, Delta states, and the Pacific Coast north of San Francisco.

Across the 48 contiguous states, there has been no appreciable overall change in land use in the past 20 years. Cropland still totals ¼ of the land; forest land, ⅓; grassland, pasture and range, ⅓; and wasteland, 3%. Urban land has doubled, mounting to nearly 2% of the total land area, and transportation land remains less than 1.5% of total acreage.

When all 50 states, including Alaska and Hawaii, are considered, there has been a significant change since 1950 in the 20-million acre increase in land used for recreation and wildlife areas, now occupying 4 percent of our land. Much of this increase was in Alaska.

Fig. 18-5 shows today's farmland and how it may be used by the year 2000 A.D.

Area and climate also affect both cattle and people. For example, Brahman, or part Brahman cattle, are better adapted to hot, insect-infested areas than the European breeds. Climate affects people, too; hence, it must be considered by the cattleman and his family. Some folks want to avoid hurricanes and tornadoes, no matter how lush the grass or how long the grazing season. Others want no part of shoveling snow and an area that may have to rely on emergency haylift by helicopter or airplane. Still others object to dust storms or to a hot, humid climate. Thus, in locating a farm or ranch, these factors are very important.

A fairly good idea of the prosperity of an area and of the kind of people

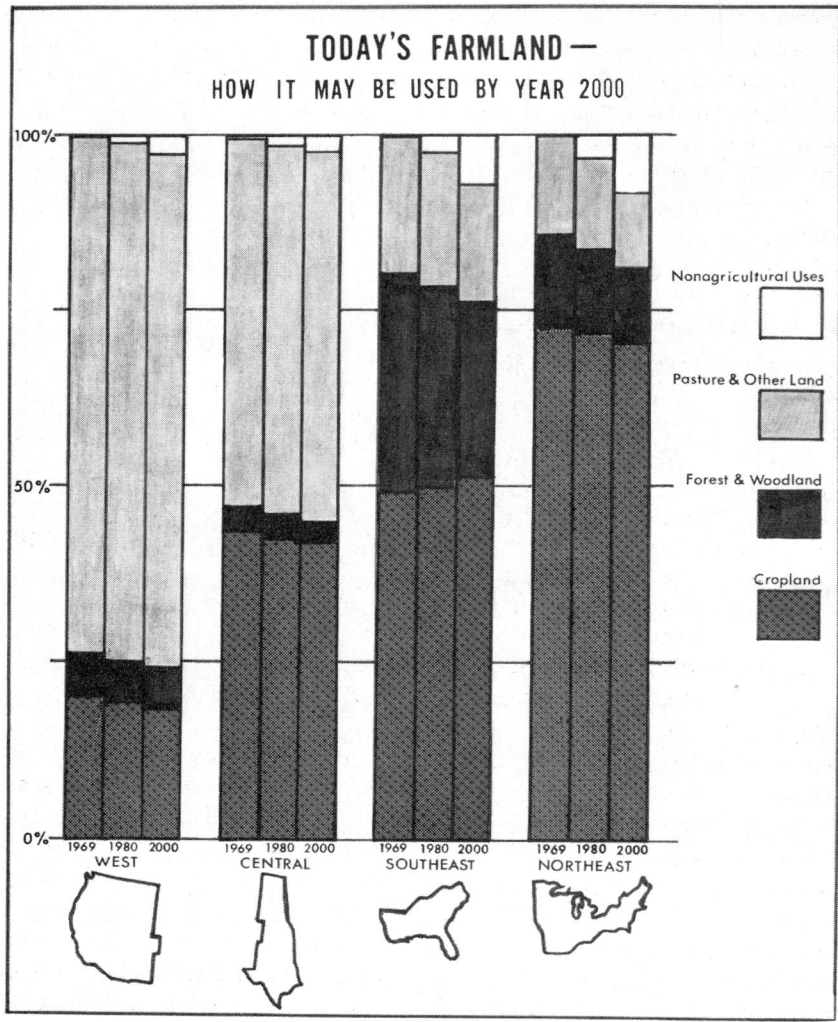

Fig. 18-5. Land in farms is the predominant land use in the 48 contiguous states, occupying slightly over 1 billion acres in 1969—more than 55 percent of total land area.

who live there can be obtained by driving around and looking at the neighborhood. Painted and well-manicured buildings and surroundings indicate that the owners are doing all right. Poorly kept premises, with debris scattered about, give evidence of the "dry rot" of the occupants.

SOIL AND TOPOGRAPHY

Capacity to produce is the main thing for which to look when buying a cattle farm or ranch. Soil is the most basic thing affecting this ability. But soils differ, and their ability to produce is dependent not only on the soil itself but the use to be made of it. Thus, the prospective purchaser should learn about the main soil types in the area where he is planning to locate, so that he can recognize most of them. He should find out what kinds of yield of grass or crops can be expected on typical soils of each type. Do they have any special problems, such as alkali or poor drainage? If drainage is a problem, for example, he will want to find out if provisions have been made for tile or open ditch drainage on the farm or ranch, and if these are working satisfactorily.

Remember that soil is "life blood" of the farm or ranch. To a very great extent, it sets the degree of profitability of a farm or ranch. It is normally cheaper to buy good soil than to purchase poor land and try to build it up.

There are several things a buyer can check on that will help determine the quality of soil. Study records of yields per acre, growth of crops or weeds on the land, soil tests, and production capacity maps. Note the depth of the soil, along road cuts or ditch banks or as determined by a soil auger. Land use capability maps are helpful to the buyer. Also, the seller or the Soil Conservation Service office may be able to provide complete plans showing soil types and classes, recommended crop use, and long-term development plans.

It has long been said that poor soils and poor people go together. Thus, one of the guidelines in estimating the worth of a farm or ranch is the prosperity of the entire community. If houses are modern, if buildings are painted, and if fences are in good repair, the area is likely productive and prosperous.

Pasture and hay crops—hence, cattle—are well adapted to hilly, rolling land. However, it must be realized that type of topography is a factor in erosion; and that erosion control measures are costly. Also, the cost of seedbed preparation and putting steep, barren slopes into grass is sometimes prohibitive, although initial costs might make the farm or ranch look attractive.

It is well to have an aerial photo of a farm or ranch at the time it is inspected in detail. This can usually be obtained from the ASC or SCS office in the county. The aerial photo will give a clue of the undesirable features of the property. It will show (1) how ditches cut through the place, (2) inaccessible or poorly watered pastures, (3) "point" rows due to curving roads or railroads, and (4) soils low in organic matter as indicated by light spots.

Grass is the cornerstone of successful cattle production. It is estimated that 73.8% of the total feed supply of United States beef cattle is derived from forage; in season, this means pasture. In purchasing a farm or ranch, therefore, it is important that there be abundant grass, preferably with a long grazing season.

Also, it is well to confer with the neighbors concerning the productivity of the land. They have the advantage of having seen it in all seasons, and over a period of years, whereas the prospective purchaser may be seeing it at a particularly favorable or unfavorable time. Belly deep grass makes quite a different impression than dry, parched pasture or a snow-covered range, yet neither represents average conditions.

Leasing a place with the option to purchase is the surest of all ways in which to "try it out," or evaluate it. Thus, this possibility should not be overlooked, although it is recognized that it is seldom possible.

IMPROVEMENTS

The improvements—buildings, fences, corrals, scales, windmills, stock tanks, and irrigation ditches—are important on a cattle farm or ranch. Hence, they should be carefully considered when buying a place. They should be adequate, without being elaborate, and they should meet the needs of the type of farm or ranch operation contemplated. Consider their suitability, flexibility, and state of repair. Old barns that cannot easily be adapted may be completely useless; worse yet, they may be a detriment. Most people find it better to purchase a farm or ranch with adequate buildings, fences, and corrals, rather than to attempt to build them. Construction costs are high. Moreover, if improvements have to be added immediately after purchase, they should be considered as part of the cost of the establishment; and it should be recognized that they will reduce working capital.

One should avoid overinvestment in buildings. The amount of barn rent for which a cow can pay is limited. There is an old saying in the industry that, "Fancy white houses and big red barns won't put fat on cattle. You just need grass and water." Yet, a cattleman needs adequate facilities. In addition to their adequacy for shelter, cattle barns and shelters should be evaluated from the standpoint of storage of feed, saving of labor, and handling of manure.

Family happiness is important to the success of any cattle venture. Thus, the home should come in for its share of investigation, also. If it is not completely modern so far as plumbing, heating, and lighting are concerned, it will need to be made so as soon as possible. Of course, if these things have already been done, this can alleviate a sizable future expense.

Fences need to be considered. Repairing fencing or adding new fencing is expensive. For holding cattle, especially where calves and yearlings are involved, four barbed wires are minimum, and five would be better. It should be kept in mind that the life of the fence depends to great extent on the life of the posts, and the stability of the corner posts.

An essential part of a beef cow operation is a well-planned, workable corral system for handling cattle. Corrals should be located where they are accessible and, if practical, should contain a water supply. A loading chute and a two-foot wide working chute should be included in the facility. Also, it would be helpful if a platform type livestock scale (with panels), of 10,000 lb capacity, were a part of the corral. A workable set of corrals need not be elaborate, yet such things cost money. It is preferable, therefore, that a suitable corral be on the farm or ranch at the time of purchase. It will avoid time and money in planning and developing such handling facilities.

If the farm or ranch is in an area where drainage is a problem, get a tile map of the farm. Check the outlets and see if the system has been working. If there is a county drainage ditch involved, take a look to see if it will have to be cleaned out soon; this could mean a special assessment for the farm.

WIND DIRECTION, WINDBREAKS, AND NATURAL SHELTERS

While most farm families are conditioned to farm aromas, there is no use making for an unpleasant situation when it can be avoided. Thus, the house should be located on the windward side of the headquarters, with special consideration given to summer winds.

Unless hills form a natural windbreak, it is desirable to arrange a suitable tree planting for the farm or ranch headquarters. Usually, a tree windbreak is located 75 to 150 feet from the buildings to be protected, with 3 to 7 rows of trees 20 to 75 feet wide.

Natural shelters for beef cattle may consist of hills and valleys, timber, and other natural windbreaks. If natural windbreaks are adequate, it may avoid the necessity of constructing shelter.

Wind direction is also important from the standpoint of the location of open sheds. They should face away from the direction of the prevailing winds.

SERVICE FACILITIES, COMMUNITY, AND MARKETS

The service facilities, community, and markets are very important. It is important that the headquarters be near an all-weather road or highway that is well maintained. In some areas, dirt and gravel roads seem to be bottomless three to four months out of the year. In the northern states, snowplows are necessary at times. Nothing is more disturbing than trying to make delivery on contracted calves, only to find that your truck is hopelessly mired in a mudhole or stuck in a snowbank, with the animals shrinking all the while.

Normally, a location along an all-weather road has better access to electric and telephone lines, the school bus, the mail, religious and recreational facilities, and other services. Also, in irrigated areas, the irrigation turnout is usually near an all-weather road. Always make sure who is responsible for the maintenance of access roads, and how well the maintenance is done. The construction and maintenance of a road can be a considerable expense. When along dirt and gravel roads, the headquarters should be far enough away, taking into consideration the direction of the prevailing winds, to keep the dust from becoming a nuisance.

If possible, the headquarters should be near well-maintained telephone and electric lines. Farming or ranching is a business, and it is difficult to conduct any kind of business without access to a telephone. Likewise, electricity is essential for the operation of most modern facilities and automated equipment. Hence, power should be available at each area of main operation and not just at the headquarters.

The farm or ranch should have convenient access to an established mail route, a school bus, and delivery services (milk, laundry, bread, etc.) The availability of various social institutions is a good criterion of the community. Easy access to a good school, to adequate hospital facilities, to the church of choice, to recreational facilities of interest, and to farm, home, and youth organizations should not be overlooked when selecting a farm or ranch.

Cattle must be bought and sold; hence, distance from and kind(s) of market are important. Preferably, there should be a choice of markets for the kind and quality of animals that it is proposed to produce. Remember that raising cattle is a business, and that buying and selling cattle is a large part of the total operation; hence, it must be well done in order to be successful. Remember, too, that the greater the distance to market, the higher the trucking cost and shrinkage. Where supplemental feeds must be purchased, nearness to where they are produced cuts down on trucking expense.

WATER; WATER RIGHTS

Only seeing people without water is more disturbing than seeing cattle dying of thirst. Both people and cattle can survive longer without food than without water. Rainfall, wells, rivers, snow, streams, creeks, springs, lakes, ponds, or any other source of water should be considered. Because of the importance of water, the official weather rainfall records, by months and over a period of years, should be obtained.

In most states, domestic users have the first right to water. Agricultural uses usually rank second. In some instances, where industry is highly centralized and promotes the welfare of the public, manufacturing use is given priority to water over agriculture.

Plenty of good drinking water for human use is a high-priority item. Generally, this involves well water. Wells should be checked to make sure that they meet county health department standards, and that there is adequate water supply, for drilling is costly. The only accurate test of a well is to have it pumped dry and see how fast it fills up, or to see how much water can be pumped out of it and how fast, without substantially changing the water level. Of course, there is variation from season to season, and from year to year. In most western states, the general underlying water table is going down every year; hence, it is impossible to predict what it might be 10 years from date of purchase. Artesian wells are the most risky of all; nobody can predict when they're going to stop flowing.

From the standpoint of the water supply for cattle, consideration should be given to the distribution of rainfall, to snowfall in the northern areas and to river frontage wherever streams are found. Where running water is not available, man-made lakes, ponds, developed springs, and wells may be relied upon, but all these cost money. Hence, they should be considered at the time of purchase.

Where water is to be used for irrigation, both availability and cost must be considered. A cattleman with free riparian water rights has considerable advantage over his neighbor who must pay an average of $20 per acre per season for the same amount of water.

Water rights have been of prime importance to the development of civilization. Nearly every society had its own system of regulating water. According to the Bedouin "code of the desert," a traveler might drink of a well, but "should the well bear the *wasm* (camel brand) of a local tribe, and should the traveler, without permission, water his flocks and camels not bearing this brand, then should he be slain, and his body left to be devoured by the birds of the air and the beasts of the field."

Water rights have always been essential in arid countries where there is limited water supply. In western and southwestern United States, where the use of water is essential to the productivity of the land and to all living things on the land, the use of waterways is usually written into the deed of the farm or ranch. This water right becomes part and parcel of the land, meaning that it cannot be separated from the land. This is because much of the value of the property depends on accessibility to water. The rights are vested, and thus are considered private property.

Basically, there are two types of recognized doctrines, or water rights—the riparian right, based on English law, and the appropriative right.

● *Riparian Right*—A riparian right is the right of an owner who owns land adjacent to a body of water. Under riparian law, the property owner who has land lying next to a stream, or having a stream running through it, is entitled to water that is required for his domestic consumption and for his livestock. The English Common Law, from which the riparian doctrine stems, further states that the owner is entitled to have this stream flow un-diminished in quantity or quality.

The Americanized version of the riparian right has a reasonable use clause which allows the riparian owner to make beneficial use of water so long as the quantity and quality of stream flow are not materially reduced. In other words, a downstream riparian landowner enjoys the same rights as an upstream owner. If, through unreasonable use, an upstream owner infringes upon the rights of a downstream owner and deprives him of water, the latter could sue and possibly collect damages and halt the excessive use.

Groundwater may be a different story. Straight riparian rights apply to well-defined underground streams. But percolating ground water (water moving downward through the soil) is defined as real property in some states. As such, it is owned by the overlying landowner. Use of such water, even when it deprives a neighbor, cannot be contested.

There are many state variations and interpretations of riparian water rights. Beyond the right to use water for domestic consumption, riparian rules are usually vague. Ordinarily, this does not create a problem in the normally high rainfall areas of eastern United States. Yet, the irrigator does not really have adequate protection. Because of the irrigation boon, most riparian states have a special legislative committee studying water rights. They have established, or they will establish, priority among users.

In new irrigation areas, a land owner should consult an attorney on how to protect his rights and investment. Records of the date irrigation started, the amount of water used, the acreage irrigated, and the return from the crops grown are information that can be valuable later if beneficial use must be proved.

When a person diverts water that other landowners have legal rights to, without being stopped, he gains the rights to continued use. This is known as prescriptive rights. To gain prescriptive rights, the diversion must have been for a period of time designated by state statute, and the water must have been used openly.

● *Appropriative rights*—An appropriative right is the right for a certain amount of water at a given place. It is not necessary for the water source to lie next to the land where the water is to be used. Rather, it is transported by such means as an irrigation ditch.

The appropriative rights for regulating water use are found mainly in the arid western states, although certain other states, including Minnesota and Mississippi, have adopted this type of law. Where states recognize both riparian and appropriative rights, the appropriative rights are dominant. Under the appropriative right, the individual may acquire the right to use water for beneficial purposes on a given tract of land by fulfilling certain requirements of written law. The appropriative right fully recognizes the public ownership of water. A landowner must apply to the controlling state agency to obtain the right to use water; and the right can be lost through nonuse after a stated period of time. In the appropriation states, rules vary on groundwater.

In appropriative rights, the right to use water depends upon prior claims made against the water sources. Anyone who filed to use water before you is entitled to his water needs before yours can be filled.

A vested right protects the rights of persons putting water to beneficial use before the appropriative law was passed. Usually, vested rights also apply to water applied beneficially within three years after passage of the law. Although a vested right has three years priority, it can be lost through nonuse.

Where to Go for Water Right Help

It is very important that both prospective and present landowners obtain authoritative information relative to water rights.

In some riparian states, there are agencies that are studying water legislation and are in charge of regulating present water laws.

In states that use the appropriative doctrine, there is an authoritative agency, usually a chief engineer, to whom applications must be placed to obtain a right to use water beneficially. This official can give information concerning what must be done to abide by the statutes of his state.

You should request a copy of the water laws from either the state agency, your state senator or representative, or your state agricultural college. Where water rights may be an individual legal question, a competent attorney should be consulted.

MINERAL RIGHTS

Mineral rights are usually involved in the buying and selling of land, especially in oil and gas producing areas. Buyers of land should always have the title checked to see if the mineral rights have been severed. Generally, they are broken down into two broad classifications—surface and subsurface.

In the United States, the following two major theories prevail concerning the actual ownership of any subsurface wealth:

1. The ownership-in-place theory generally recognizes that any mineral deposit is actually a part of the land and is owned by the individual holding title to the land.

2. The nonownership theory in which the landowner does not have outright title to underlying mineral deposits but has the right to explore and retain any deposit developed.

Regardless of which theory is recognized in a particular state, the landowner has the following mineral rights:

1. Rights may be transferred to others.

2. Mineral deposits are part of the land.

3. Landowner has right to withdraw minerals, and, in the case of gas and oil, be free of liability for drainage (the pumping of oil or gas from under adjoining, nonowned property).

SEPARATING MINERAL RIGHTS

Surface and subsurface rights can be separated. In "ownership-in-place" states, you dispose of the minerals, while in "nonownership-in-place" states, it is simply the right to explore and retain any minerals recovered.

It should be noted that the term "minerals" refers to gas and oil, unless specifically stated otherwise. Moreover, any conveyance of a named mineral does not include other minerals unless so stated.

Surface and subsurface rights may be severed in any one of the following six basic ways:

1. By deed conveying all or part interest in the minerals.

2. By deed conveying land but retaining mineral rights.

3. By land contract excepting the minerals.

4. By mineral lease in an ownership-in-place jurisdiction. (Conveys present undivided ⅞ interest to the lessee on any mineral developed, except in Kansas.)

5. Court judgments setting aside mineral rights or part interest in a lawsuit.

6. Formation of a mining partnership.

Separating mineral interests can create problems, especially if interest is divided among many parties. For example, if the property has been through three or four different hands, possibly ⅛ of the minerals might be left for the new purchaser. Not only is such a small fraction of the mineral interest unattractive to the buyer, but widespread breakdown of the interest runs up cost of bringing land abstracts up to date and often discourages leasing and well development. Separation of mineral interest reduces loan values and usually increases difficulty in obtaining credit.

As a general rule, it is not considered good business for a surface rights owner to dispose of over fifty percent of the subsurface rights.

OIL AND GAS LEASES

Oil and gas leases have become fairly standard. Nevertheless, one should check any proposed lease with an attorney before signing.

A general division on any developed oil and gas is ⅛ for the lessor (the mineral owner) and ⅞ for the lessee (the persons taking a lease on the land).

Most leases are perpetual as long as certain qualifications are met. Leases are generally subject to termination by the lessee anytime within the base period by failure to begin a well or pay the delay rental by a stated date.

In some areas, mineral rights are more valuable than the surface.

TIMBER

Trees are pretty, but they may or may not have monetary value on a farm or ranch. Evaluation of timber is a job for an expert. The value of timber delivered to market may have little relationship to its quality as it stands. In some cases, the cost of cutting and transporting trees is exorbitant.

EASEMENTS; PROPERTY LINES

An easement is the right to go on and use the land of another in a particular manner. Two common types of easements are: (1) the grant of one landowner to another of the right to build or use a roadway across the land to provide access to another tract; and (2) where a power company purchases an easement to string an electric line across land. Before purchasing a farm or ranch, all easements should be checked because an easement (1) limits the landowner's use of his property, and (2) is valid against the purchaser. The author recalls one near-sale of a ranch that was being bought primarily because of a beautiful sight on which to build the new owner's dream house. Just as the deal was about to be closed, it was discovered that the county had a permanent easement for a 25-yard strip for a road right through the intended house location.

Three types of legal descriptions of farms and ranches are commonly used: (1) the rectangular survey, based on meridians and parallels; (2) metes and bounds (metes are measures of length—feet, inches, or perches—a perch equals 16½ feet; bounds are artificial boundaries such as roads, streams, adjoining farms); or (3) monuments (iron pin, blazed tree, lake, stream). Disagreements over property lines have led to feuds and lawsuits. Accordingly, before purchasing land, the property lines should be determined. Usually, it is wise to engage the services of a professional surveyor. The author knows of one case where failure to do this resulted in the cattleman building a new ranch home only to discover, some years later, that the house was on an adjacent property—not his own.

NEARNESS TO FACTORIES OR CITY

Nearness to factories or city has both advantages and disadvantages from the standpoint of farm or ranch location. Some of these will be discussed.

1. *Off-farm employment*—Unless a cattle farm or ranch is large enough to support a 300-brood-cow herd, it is well to keep in mind that the operator will likely find it necessary to have a part-time off-farm employment in order to make a go of it. This fact should be kept in mind at the time of purchase, even though the one making the purchase has sufficient income from other sources not to require part-time employment. This is so for resale reasons.

In a study conducted by the University of Arkansas, it was found that off-farm work was an important source of income for all of the upland areas of the state, with over 50 percent of the cattle producers working off-farm.[4] This means that a farm or ranch which will not support 300 brood cows should be located within a 50-mile commuting radius of a city or factory payroll.

2. *Big city attractions*—The new generation of cattlemen has gone modern. Although they may be perfectly content to live on an isolated farm or ranch and near a small town, most of them still like to be fairly near a large city. The big city is important to them from the standpoints of shopping, concerts, junior colleges, airports, etc.

3. *Air and noise pollution*—An increasingly important consideration in selecting a cattle farm or ranch is air and noise pollution. Thus, if the cattle operation being considered is located near a factory, an effort should be made to determine if the factory is likely to cause either water, air, or noise pollution. Although a national effort is being made to reduce such pollution, there continue to be claims, imagined or real, that the contamination from certain factories has polluted the air or water to the extent that the growth of grass and the well-being of cattle have been affected adversely. Also, there are reports where streams have been polluted to the extent of affecting certain species of fish, thereby reducing the value of the ranch from a recreational standpoint. Airports and highways are also sources of noise. Although animals adjust to usual noises, they may be excited and become nervous by unusual noises.

EXPANSION POSSIBILITIES

Sooner or later, most cattle operators try to increase the size of their holdings. When that time comes, it is a tremendous advantage if they are not "hemmed in"—if there are expansion possibilities. Expansion possibilities usually exist where a farm or ranch is surrounded by other farms or ranches. Then, when a neighbor retires, first refusal may be obtained.

Where a farm or ranch is adjacent to an airport, factory, golf course, and the like, expansion possibilities are always limited.

In an Arkansas study, the typical beef cattle producer indicated that he would like to have more forage land for his herd, but that lack of land was the most limiting factor.[5]

Things to Do When Buying a Farm

After the prospective purchaser has found the farm or ranch that he likes and has determined that the price is right, the following things should be done:

---

[4]*Production, Financing, and Off-Farm Employment Aspects of Beef Farming in Arkansas, 1969,* Agriculture Experiment Station, University of Arkansas, Fayetteville, Bull. 785, November 1973, p. 24.

[5]Ibid., p. 20.

1. *Have it appraised*—Buying a farm or ranch is a big financial transaction. Thus, it is good business to have an accredited rural appraiser make a detailed appraisal of the property. The appraiser's fee will vary depending upon the size of the ranch to be inspected and the time required to document income and expense items, search out and view comparable sales, and prepare a confidential written appraisal report. Appraisal fees normally run from $1.00 per acre on 1,000 acre ranches down to $0.35 per acre on much larger ranches.

There are three basic approaches to appraisal of land, or estimating its value. These are: (a) market value, or what similar land has sold for recently; (b) productive value, or net income the land will produce; and (c) present value of useful improvements.

The appraisal should show the fair market value of the property. This is frequently defined as "the price at which a willing seller would sell and a willing buyer would buy, neither being under abnormal pressure." This definition assumes that both buyer and seller are fully informed as to the property and as to the state of the market for that type of property, and that the property has been exposed in the open market for a reasonable time.

The appraisal should include maps of the farm, showing the physical features and the various soils. There should be a summary sheet listing all the buildings and their size and description. The appraiser should also allocate a value to buildings, fences, tiling, wells, pipe lines, and other depreciable items for income tax purposes. This will give the buyer a reliable value from which to set up a depreciation schedule.

The appraiser should evaluate improvements in terms of the owner's intended use. For example, a $20,000 turkey shed is worth only scrap lumber to an owner who is going to run cows. Worse yet, it may have a negative value because of taxes.

Preferably, the appraiser should be familiar with cow-calf operations. Such an appraiser can point out the highest and best use of the property for the intended purpose. Also, he may be able to indicate factors which would alter considerably the plans of the potential buyer.

2. *Check government programs*—Government programs change from time to time, but usually one or more government agencies is involved in most farms and ranches. Thus, it is well to check into the situation.

The county Agricultural Stabilization and Conservation Service (ASCS) offices advise on and administer commodity programs, including allotments and marketing quotas for the basic commodities. These offices can also supply information about the soil, water, timber, and wildlife conservation practices that the Agricultural Conservation Program helps carry out on individual farms. ASCS offices also are charged with the local administration of price support commodity loans made available through the Commodity Credit Corporation, certain emergency programs in designated areas affected by drought or floods, the feed grain program, and other farm programs.

The Soil Conservation Service (SCS) has offices in nearly every county. They provide technical assistance and information on soil and water conservation, land-use alternatives, soil surveys, and resource use.

3. *Check courthouse records*—A courthouse check is standard proce-dure for most appraisers. However, the prospective owner should make sure that it is not overlooked.

The courthouse records should show what the property tax has been running on the farm or ranch in question, and if the property is subject to special levies for drainage or irrigation districts. The plat books should be checked to make certain of the boundaries of the property and how many acres are actually involved. What mortgages are on record against the farm? Check the recorder's office also for any special agreements in regard to property-line fences. The latter is especially important if there is a "water gap" where fences must frequently be rebuilt. Also, important water and mineral rights should be checked.

4. *Check mortgages*—The prospective purchaser should check the mortgage situation. Is there a mortgage on the property? If so, how much, at what interest rate; and can it be assumed without penalty? These questions are particularly important during times of scarce money and high interest rates.

Things to Avoid When Buying a Farm

In the purchase of a farm or ranch, there are certain pitfalls which should be avoided; among them the following:

1. *Avoid legal problems*—Regardless of whether a farm is purchased di-rect from the seller on a first hand negotiated basis or through a real estate broker, the buyer should have an attorney check the details, thereby lessen-ing the hazard of legal problems.

2. *Beware of the glamor states*—Much land in California, Florida, and Arizona is priced so high that it is difficult to show a profit from the operation of a cattle farm or ranch. In these states, either a higher and better use must be considered, or the land must be purchased on the basis of speculation—its potential for recreational development, housing, etc. Beautiful mountains, trees, streams, and sunsets all make for a heap of living and enjoyment, but they don't feed a cow. Thus, they should be secondary in the selection of a farm or ranch.

3. *Avoid city suburbs and high taxes*—Where rapid-growth cities are involved, it is generally wise to stay at least 50 miles away if one wishes to develop a cattle farm or ranch. Of course, there are many small towns or cities that are not subject to rapid growth where it is possible to be closer in without the hazard of subdivisions or high taxes.

4. *Avoid overelaborate improvements*—Improvements are always ex-pensive to maintain and they are subject to taxes. Hence, they should have utility value. No matter how attractive they may be, unless improvements contribute to the income of the farm or ranch, they have a negative value.

Purchase Contract

After a prospective buyer has found the particular property that he wants and has agreed upon a price with the owner, he should have an attorney draft an agreement covering the terms of the purchase. Then he should sign it and submit it as an offer to buy. It does not become a binding contract until the seller signs it, also. After both parties sign the contract, there is little bargaining power left; hence, the buyer should get all stipulations covered in the original contract.

The purchase contract should be relatively simple, amounting to a mere memorandum signed by buyer and seller, but it should clearly specify the following:

1. Amount and method of paying purchase price.

2. The amount of deposit or down payment, and the method of handling it. Is it to be applied to the total purchase price; when is it to be forfeited or returned; and is payment made to a responsible person?

3. Method of financing the purchase. Does purchaser assume and agree to pay existing mortgage? Will purchaser obtain a new mortgage loan?

4. Is seller to furnish an abstract of title brought up to date or a good and clear title that can be insured?

5. Date that possession can be taken.

6. A list of items that go with the property.

7. Who pays accrued and current taxes.

8. The legal rights of any tenant on the property.

It is important to remember that the buyer assumes risk of loss as soon as the contract is signed, even though the deed has not been delivered. If a barn burns and the seller has no insurance, the buyer could be forced to pay the full purchase price agreed upon, even though the barn has burned. So it is important that the buyer make certain that he has insurance on the buildings during the interim period.

It is customary to prorate annual taxes, with the buyer and seller assuming responsibility for the number of months that each has actual possession of the property.

The purchase contract should require the seller to deliver to the buyer an abstract of title for the property, certified to the date of sale. If the seller cannot produce a clear title or satisfactory title insurance by a certain date, the contract should call for a refund of the buyer's down payment.

Of course, the purchase contract should include the price being paid for the property and the date on which the seller guarantees possession. There should be a complete legal description of the holdings. If it is understood that certain portable buildings go with the farm, they should be itemized in the purchase contract.

The contract should detail how payment is to be made and the time of settlement. If the seller is to carry a mortgage on the farm or ranch for part of the purchase price, the buyer will need to work out the usual arrangement as to interest rates, prepayment privileges, etc.

If the buyer wishes to go on the place to do certain work in advance of taking actual possession—like repairing buildings, or reseeding a pasture—this should be spelled out in the contract.

Nothing should be left to oral agreements, because most such agreements are unenforceable. Also, the contract should be binding upon the heirs and assigns of the seller.

In summary, when completing the purchase of a farm or ranch, give attention to the following details:

1. Do you have satisfactory evidence that the seller has complete title to the property and can convey it to the purchaser by deed?

2. Examine the seller's deed. Does the wife release her dower; does seller warrant free and clear of all encumbrances; is the description of boundaries and acreage correct; are the easements or right-of-way for or against the farm; are mineral rights or water rights reserved; and has seller attached U.S. transfer stamps?

3. Examine mortgage and note before signing.

4. Immediately record deed with the Register of Deeds office in the county where the property is located.

5. Insure all uninsured buildings.

6. Make sure that expenses incurred in the last 60 days have been paid for materials or work done on buildings, for land clearing or leveling, and for wells or pipe lines.

Ways of Acquiring a Farm or Ranch

Getting a suitable farm is a big problem, particularly for the beginner. He must compete for available farms with established farmers as well as with other beginners. Many established farmers need more land to enlarge their operations. Others move during the year, getting a better or more suitable farm. Some simply move to a new locality for personal reasons.

Farms or ranches may be acquired through gift or inheritance, by marriage, by renting or leasing, or by purchasing.

1. *Gift or inheritance*—Many farms and ranches are inherited and stay in the family for several generations. Certainly, inheriting a farm or ranch is a real advantage to anyone desiring to stay in the business and having the know-how and ability to operate the enterprise. Farmland and ranchland, and livestock (as well as corporate stocks, U.S. savings bonds, mutual funds, money, whole life and annuity life insurance, and commercial real estate) can be transferred to relatives and friends, with certain tax savings, provided certain well-established rules are observed. As evidence of the legitimate savings that may be affected through gift or inheritance, the following example is cited, in which the author served as consultant: Three owners, all in their 70s, with a cattle ranch valued at approximately $4 million, had done no estate planning. By having the property appraised and following a carefully planned gift program, $¾ million in inheritance taxes will be saved—all legally.

The basic rule in giving land or animals to another member of the family is that the donor (person giving the property) must give up control. Of course, if it actually or constructively passes through the hands of the donor, it is taxed to him.

A transfer of property either in trust, to a custodian, or outright, may incur Federal gift tax. However, you can give away property valued at $3,000 each year ($6,000 if married and the wife or husband consents) to any number of donees without incurring a gift tax obligation. Additionally, the law provides for a total lifetime exemption of $30,000 ($60,000 if married and the wife or husband consents).

You can give relatives or friends income-producing property and the income will be taxed to the person receiving the gift (donee). But the property must actually be transferred. Any strings attached to the ownership whereby the donor can have control over the property or get it back at some future time will not meet the requirements of the law. The basic elements are:

a. There must be intention to make a gift.

b. Transfer of legal title and control.

c. The donee accepts the gift.

d. No consideration (money or property) to be exchanged for the property.

If a person is expecting to inherit a farm or ranch, he should become fully cognizant of the inheritance tax laws. Also, it is most important that he have clear title to the land. If the title of ownership of a farm or ranch is left unsettled through two or three generations, the value of the land may almost be expended in settling the estate.

Where gift or inheritance money or property are involved, always seek the advice of a tax accountant and/or tax attorney.

(Also, see Chapter 13, Business Aspects of Beef Production, sections entitled "Cattle and Income Tax (Tax Shelters)," and "Estate Planning").

2. *Marry it*—Although young men wanting to enter farming or ranching are frequently admonished, facetiously, that they should "marry for love, but love a woman with plenty of money," there is more than a little bit of truth in the advice.

In some countries, the social orders call for the parents to arrange the marriages of sons and daughters, primarily to keep them within the same class strata, thereby not dividing property with those who "have not." However, this method is fast giving way to the new social order, which, like that in the United States, results in marriages between individuals of vastly different amounts of wealth.

Attitudes are important in any marriage, but they are doubly important in those marriages in which the wife contributes most of the wealth. The ability of the young man marrying the wealth, along with the attitude of the wife and the in-laws, can have a big part in contributing to the success or failure of both the marriage and the cattle farming or ranching.

3. *Rent or lease*—The main advantage in renting over buying is that less capital is required and less financial risk is involved. The main disadvantages are insecurity of tenure and that the farming enterprise may be limited in size or kind because the landowner is reluctant to make needed additional investments in buildings and facilities. These disadvantages can be minimized and sometimes eliminated, by a suitable lease—an agreement between landlord and tenant under which the farm or ranch is rented and operated. Such a lease should always be in writing.

The most common types of cattle farm and ranch leases are:

a. *Cash lease*—This is a good type of lease for (1) the small farm or where the landlord lives at a distance, and (2) a tenant who has adequate livestock, equipment, and working capital. It encourages livestock farming because all of the crop can easily be fed on the farm. Also, it is simple, with little chance for controversy.

There are two types of cash leases: (1) that type in which a fixed rent per acre is agreed upon when the lease is drawn; and (2) that type in which the rent is adjusted to prices of farm products which prevail during the lease year. Under the second plan, the landlord bears part of the risk of price changes; however, it is difficult to keep cash rent in line with farm product prices. If product prices are used as a basis for rent changes, the products, markets, and dates should be specified.

The landlord may prefer a cash lease because (1) the amount paid is definite, and (2) it requires less supervision by the owner. On the other hand, it may not always be desirable from the standpoint of the landlord because (1) it generally makes for lower income, (2) it gives him less control of the farm, and (3) it is difficult to collect rent if crops fail.

The tenant may prefer a cash lease because (1) it will make for more profit if he is a successful manager, (2) it makes for more independence in the operation, and (3) it makes for more profit in the good years.

b. *Livestock share lease*—Cow-calf share leases vary considerably. But most of them provide for 50-50 ownership of the herd; 50-50 sharing of the costs of production—especially feed and veterinary expenses; and 50-50 division of the income from the sale of animals. Buildings are generally a cost borne by the landowner. Labor is the responsibility of the operator.

A livestock share lease fits the tenant who wants to raise livestock, but cannot finance a program. It is especially suited where tenant and landlord get along well and where the landlord can make a good contribution in management.

In order for this type of lease to work best, the landlord should live close to the farm, and either give it his personal attention or arrange for adequate management help such as can be provided through a professional farm management service.

The landlord may prefer a livestock share lease because (1) it encourages more livestock and more manure, (2) low quality crops can be utilized more easily, (3) he retains an active interest in management, and (4) it generally makes for more profit.

The tenant may prefer a livestock share lease because (1) the risk is less since rent is based on net income on the farm, (2) it requires less tenant capital, (3) the landlord is more willing to make improvements, and (4) he can gain experience from the guidance of a successful owner.

A careful determination of lease provisions and putting them in writing will result in a lease that is more equitable to both tenant and landowner, and will avert later misunderstandings and friction between the two parties. Standard lease forms are available, so the detailed provisions of a lease need not be spelled out in this book.

Renting or leasing might be a desirable way to start in the cattle business even if funds are available for purchase. This is particularly true where there is an option-to-buy clause. This gives the renter an opportunity to study the ranch more carefully and gain additional management experience without committing his entire assets.

4. *Purchase*—Purchase of a cattle farm or ranch has the advantages of security of tenure and freedom to make management decisions. Earnings from the operator's equity capital may be added to labor and management earnings for living expenses, reinvestment in the business, or other uses. Also, the value of the land may rise over a period of time. On the other hand, ownership may involve substantial indebtedness. Also, risks of financial loss are greater than in renting.

Of course, ownership brings with it financial responsibility that is both greater and longer lasting than the financial responsibility which renting entails. Few persons buy more than one farm in a lifetime; moves are time-consuming and expensive.

Some part-time farmers, working at nonfarm jobs, use their off-farm income to move gradually into full-time farming. They use their initial savings to make a down payment on a small farm and to buy enough cattle and equipment to permit limited farming or ranching operations for the first few years. The off-farm income makes them better credit risks for lenders than if they were wholly dependent on farm earnings. They can continue to borrow to build up their farm business to a point where it will support their family to pay off previous loans. Such a gradual shift into full-time farming can usually be made with less sacrifice in family living standards and better chances of eventual success than an abrupt change to full-time farming.

Generally speaking, purchase of land is either by (a) land purchase contract, or (b) mortgage contract.

Purchase of land by use of land purchase contract has become much more important in recent years. These contracts allow the use of lower buyer down payments (usually from nothing to 29 percent) with the balance paid over a long period of years in annual payments. For the buyer, this offers a way in which to buy land without having to make a big down payment. For the seller, it has certain capital gain tax advantages and it usually attracts more prospective buyers and makes for a higher sale price. In order to qualify for special treatment on capital gains for Federal income tax purposes, the seller must not receive more than 30 percent of the purchase price in the year of sale.

Mortgage contracts differ from land purchase contracts in the following ways: (1) they are of longer duration—usually 20 to 30 years, or up to 40 years, whereas land purchase contracts are commonly for 10 years or less; (2) the law provides for specified grace periods after default in payments before the seller can foreclose the mortgage; and (3) larger down payments are normally required—frequently 40 or 50 percent of the purchase price.

What's a Farm or Ranch Worth?

The above question is best answered by still another question: How much will it make? The most logical answer to the latter question is that the farm or ranch should be expected to return to the prospective buyer as much on his investment as he could earn were his money invested in the best alternative enterprise of comparable risk. Thus, if an alternative enterprise of comparable risk will return seven percent on investment, the farm or ranch should do likewise.

The above income-productivity approach does not take into consideration a possible tax shelter or increase in land values. However, neither of these should be counted upon. Although the prospective land buyer should have no objection to Uncle Sam being lenient on his taxes, or to striking oil, or to having a fashionable summer resort go up on the adjacent property, none of these possibilities can be counted upon. The tax structure may change; and land values may not increase. Besides, the cattleman living upon the land and depending upon his earnings from the operation to pay interest on borrowed money and buy groceries for his family cannot rely on "paper profits" (unrealized profits).

The income-productivity approach, which calls for calculating receipts and expenses, should be based on projected longtime productivity levels, prices, and costs, without either undue optimism or pessimism.

In plain simple terms, the income-productivity approach is based on the capacity of the farm or ranch to make money—the more money it will make, the more it is worth. Step by step, it is determined as follows:

1. Record expected receipts and expenses from the operation of the cattle farm or ranch.

2. Determine the net returns to the land by subtracting all expenses from gross receipts, including a return to the operator for his labor and management.

3. Divide the dollar returns to land by the rate of interest that you could get were your money invested in the best alternative enterprise of comparable risk to arrive at the "income-productivity value" of the farm or ranch. This value is the maximum that a buyer can pay for a farm or ranch based on its "productivity value." If the price is below this figure, it's a good buy. If it's above, watch out.

The above procedure is given in Table 18-5, based on an actual ranch for which the author serves as consultant.

Thus, based on income productivity, a buyer who wishes to allow himself $20,000 per year for labor and management, and who wishes to realize 7 percent on his investment, could consider the Bar-None Ranch—a 500-cow, 5,000-acre ranch—a good buy at up to $500,000, or $100 per acre. On the basis of a 500-cow carrying capacity, that's $1,000 per cow for the land and improvements.

Historically, the "income productivity" value of cattle farms and ranches has been below current market prices. This means that either land is too high or cattle are too cheap. Moreover, a prospective purchaser must recognize that he will likely have to pay the going market price for a farm or ranch. That is, he will have to pay a price close to that for which comparable land is selling. Except when buying from relatives, there are precious few really good buys. However, there are two exceptions: (1) when an existing

TABLE 18-5

INCOME PRODUCTIVITY VALUE OF
BAR-NONE RANCH
A 500-COW OPERATION ON A 5,000-ACRE RANCH[1]

	Amount
Receipts:	
Cattle sales	$ 101,250
Expenses:	
Cash expenses:	
Feed purchases	12,115
Hired labor	5,000
Machinery	7,785
Property tax	4,025
Other	1,900
Noncash expenses:	
Depreciation	11,425
Interest on operating expenses	4,000
Operator labor and management	20,000
Total expenses	66,250
Net return to land (ranch)	
Income productivity value at 7%	
rate of interest ($35,000 ÷ .07)	$ 500,000

[1]This is an abbreviated form. A competent appraiser will detail this. For example, under "cattle sales," there should be a breakdown as to number, weight, and price of steer calves; number, weight, and price of heifer calves; number, weight, and price of yearlings, by sex; number, weight, and price of cows; and number, weight, and price of bulls. Also, hay sales, crop sales, and pasture rents would be included—if income from these sources is expected.
Similarly, under expenses there should be a more detailed breakdown of costs.

cattleman enlarges his present holdings by acquiring nearby land, thereby lowering the cost per cow expenses—primarily through more efficient use of labor, management, and equipment; and (2) when unimproved land can be developed to where its "income productivity" exceeds its cost plus development.

Fig. 18-6 shows farmland value 1960-1974, based on actual sales.

U.S. FARMLAND VALUE $ PER ACRE, MARCH 1

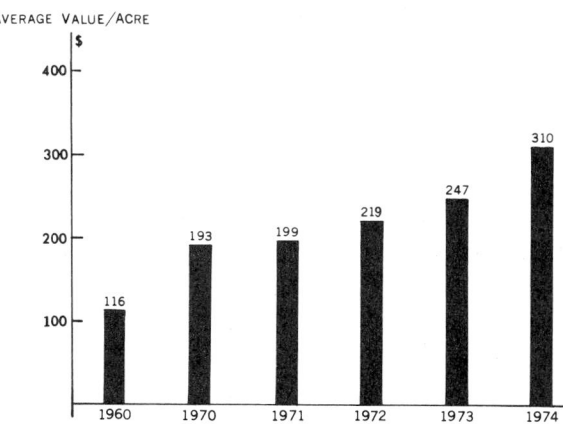

Fig. 18-6. United States farmland value in dollars per acre, 1960 to 1974.

Financing the Farm or Ranch

In 1972, the average U.S. farmer or rancher had nearly $100,000 invested in land, machinery, livestock, working capital, and farm buildings other than his house. Many cattlemen have much larger amounts invested. As the author has previously pointed out, a minimum of a 300-cow outfit is necessary for a profitable unit where the operator expects all of his income from the farm or ranch. Further, as noted in Chapter 24, Table 24-1, it takes about $1,395 per cow to cover land and improvements, machinery and equipment, and the brood cow. This means that for a 300-cow unit, a total of $418,500 is necessary. Of course, many cattle operations have more than 300 cows. Likewise, many of them have more than $1,395 invested per cow unit. As a result, cattle investments of over $1,000,000 per farm or ranch are not uncommon. Hence, a big cattle spread necessitates both money and knowledge of financing.

CREDIT

Credit may be defined as "belief in the truth of a statement, or in the sincerity of a person." In farming and ranching, or in any other business transaction, credit means confidence that men will take care of their future obligations. Credit is the life blood of the cattle business. Without it, few large cattle operations would be possible, for not many people are able to provide all of the capital that they need.

Most commercial lenders have guides and standards that set upper limits on the amount they will lend. Usually, to get credit on a mortgage for buying a farm, the borrower is expected to make a down payment of 40 to 50 percent of the purchase price. Lenders usually will make loans on livestock and on new machinery for up to 80 percent of the purchase price.

Total farm investment in land, buildings, livestock and equipment has more than doubled (2.6 times) in 22 years, rising from $167 billion in 1952 to $431.2 billion on January 1, 1974. Farm debts have increased even more—they are 5.6 times larger than 22 years ago. The amount of debt owed by farmers and ranchers has risen from $14.7 billion in 1952 to $81.7 billion in 1974; and the trend shows no signs of letting up.

TYPES OF CREDIT OR LOANS

(See Chapter 13, Business Aspects of Beef Production, section entitled, "Types of Credit or Loans.")

SOURCES OF LOANS

Hand in hand with getting the right kind of loan, it is important that the best available source of the loan be secured. Table 18-6 shows the main sources of the three main kinds of loans.

TABLE 18-6

PRINCIPAL SOURCES OF THREE MAIN KINDS OF FARM LOANS

Credit Source	Kind of Loan		
	Long Term	Intermediate Term	Short Term
Commercial banks	X	X	X
Dealers and merchants		X	X
Farm mortgage companies	X		
Farmers home administration	X	X	X
Federal land bank associations	X	X	
Individual lenders	X	X	X
Insurance companies	X		
Production credit associations		X	X

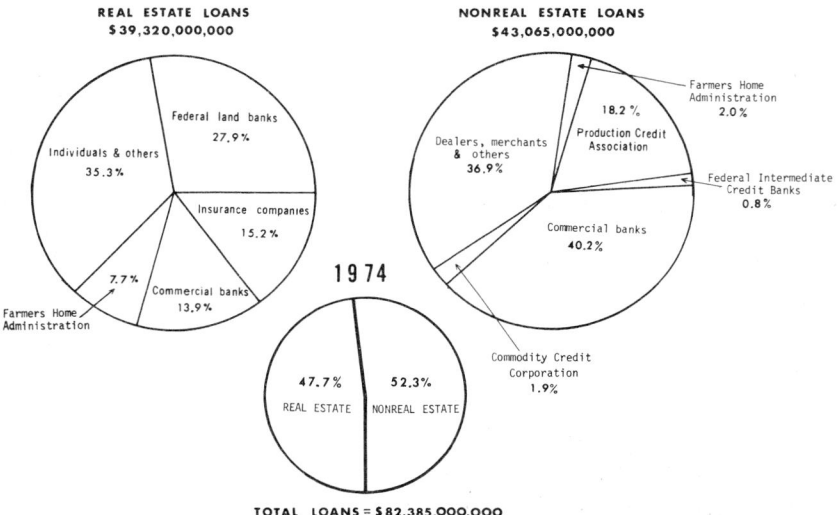

Fig. 18-7. Where farmers borrow (1974). (Source: Data provided in a personal communication to the author from Mr. E. A. Jaenke, Governor, Farm Credit Administration, Washington, D.C.)

Banks, merchants, and individuals provide 80 percent of the farm credit.

In seeking a loan, it usually pays to "shop around" in advance of actual need to see which source is best under the circumstances. Compare cost of credit, length of loan, loan fees, repayment privileges, and security required. Also, for long-term loans check the reputation of the lenders for staying with worthy borrowers in times of adversity.

In comparing costs of credit, look at the total dollar amounts, not just interest rates. Lenders figure charges in different ways. For example, if you buy feed or farm machinery on the time-purchase plan, you may find that you will pay more in interest than you would if you borrowed the money to pay for the purchase outright.

To obtain loans or information about loans from commercial banks, dealers and merchants, farm mortgage companies, individuals, and insurance companies, apply directly to these sources or to their local representatives. Local banks and farm real estate dealers often serve as loan correspondents for life insurance or farm mortgage companies; they can tell you about loan requirements, terms, conditions, and interest rates, and arrange for loans.

(Also see Chapter 13, Business Aspects of Beef Production, section entitled, "Sources of Capital.")

QUESTIONS FOR STUDY AND DISCUSSION

1. Why is the ultimate success of a cattle operation so dependent upon the selection of the proper farm or ranch?
2. Why is the ownership of land attractive to investors?
3. Explain how land provides a means of obtaining relatively large capital gains.
4. Explain how land provides an effective hedge against inflation.
5. Why are there so many part-time beef cattle ranches? From the standpoint of the beef cattle industry as a whole, is the preponderance of part-time farmers good or bad?
6. What factors are likely to favor an irrigated cattle farm or ranch over a dryland operation?
7. Why have ranches made up of deeded land increased in value more than those made up chiefly of government-leased land?
8. List and discuss the fundamental considerations when selecting the location of and the specific farm or ranch.
9. Define "carrying capacity." How important is a high carrying capacity?
10. Define "investment per animal unit." How important is it? What maximum dollars per animal unit can be justified in a commercial cattle operation at this time?
11. Cattlemen as a whole averaged only 3.9 percent return on their equity for the 10-year period 1963-1972. Would they do better in the stock market?
12. How important are water and mineral rights to a land buyer?
13. List and discuss the important "things to do" when buying a farm.
14. When buying a farm, why should a purchaser avoid overelaborate improvements?
15. What are the advantages and disadvantages of acquiring a farm by (a) gift or inheritance, (b) marrying it, (c) rent or lease, and (d) purchase?
16. Show how the income-productivity approach may be used to determine what a farm or ranch is worth?
17. How would you go about selecting and buying a cattle farm or ranch?
18. Can a young man become a ranch owner today? If your answer is in the affimative, explain how.

SELECTED REFERENCES

Title of Publication	Author(s)	Publisher
Cattle Raising in the United States	R. N. Van Arsdall M.D. Skold	Economic Research Service, U.S. Department of Agriculture, Washington, D.C., 1973

(Continued)

Title of Publication	Author(s)	Publisher
Cowboy Arithmetic, Cattle as an Investment, Second Edition	H. L. Oppen-heimer	The Interstate Printers & Publishers, Inc., Danville, Ill., 1971
Cowboy Economics, Rural Land as an Investment	H. L. Oppen-heimer	The Interstate Printers & Publishers, Inc., Danville, Ill., 1976
Cowboy Litigation, Cattle and the Income Tax	H. L. Oppen-heimer J. D. Keast	The Interstate Printers & Publishers, Inc., Danville, Ill., 1972
Do It Right the First Time	J. D. Keast J. W. Looney D. K. Banner L. R. Miller R. C. Matthews C. L. Pilmer R. J. McTigue	Doane Agricultural Service, Inc., St. Louis, Mo., 1973
Farm Management Economics	E. O. Heady H. R. Jensen	Prentice-Hall, Inc., Englewood Cliffs, N.J., 1954
Farm Records and Accounts	J. N. Efferson	John Wiley & Sons, Inc., New York, N.Y., 1949
Financing the Farm Business	I. W. Duggan R. U. Battles	John Wiley & Sons, Inc., New York, N.Y., 1950
Land Speculation, An Evaluation and Analysis	H. L. Oppen-heimer	The Interstate Printers & Publishers, Inc., Danville, Ill., 1972
Law on the Farm	H. W. Hannah	The Macmillan Company, New York, N.Y., 1948
Law and the Farmer	J. H. Beuscher	Springer Publishing Company, Inc., New York, N.Y., 1953

CHAPTER 19

FEEDING AND MANAGING BROOD COWS[1]

Contents **Page**

[1]The author is grateful to Dr. Wilton W. Heinemann, Washington State University, for authoritatively reviewing this chapter.

Feed affects total profit and cow productivity. It accounts for 65 to 75 percent of the total cost of keeping cows, and it exerts a powerful influence on cow fertility and calf weaning weight—the two biggest success factors in the cattle business.

Fig. 19-1. "Producing calves" is the name of the cow-calf game. Brood cows should be fed and managed so that they approach a 100 percent calf crop and wean off heavy calves. (Courtesy, University of Illinois, Urbana, Ill.)

NUTRITIONAL REQUIREMENTS OF BROOD COWS (Also see Chapter 8 of this book)

Experiments and practical observations reveal that the period during which calf crop percentage is affected most by nutrition extends from 30 days before calving until 70 days after calving—until after rebreeding; a period of approximately 100 days. This, then, is the most critical period in the cow-calf business. It's when life begins—that period within which one calf is born and another is conceived. The needs for the cow during this most critical production period are approximately equal to her needs for the remainder of the year. This fact is pointed up in Fig. 19-2, showing the estimated energy requirements of a 1,000-pound (454 kg) beef cow during her 12 months' reproductive cycle.

The average daily energy requirement is about 14.5 lb TDN (24 megacalories). However, as shown in Fig. 19-2, the requirements are above the average for nearly 6½ months of the year. This means that, for reasons of economy, the calving season should be timed so that much of the feed can be supplied by pasture and other economical sources of homegrown energy and protein.

A second important requisite of a sound beef cattle nutrition program is to feed animals according to their requirements. It is impossible to feed the herd properly where calving occurs the year around, or when dry pregnant cows, replacement heifers, and cows nursing calves are run together. This point becomes apparent in Figs. 19-3 to 19-5, which show certain nutritive

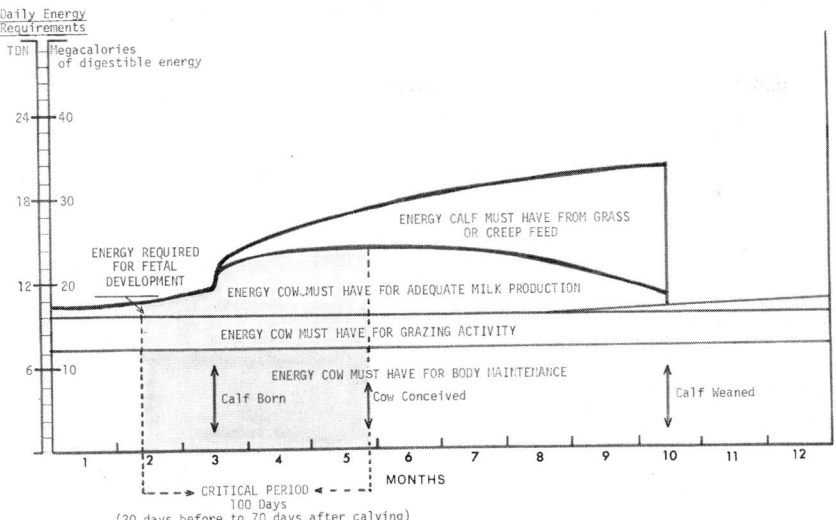

Fig. 19-2. Estimated energy requirements of a mature 1,000-pound (454 kg) beef cow during her 12-month reproductive cycle; based on a 90-day calving season and 500-pound calf at 7 months of age. (Adapted by the author from *Nutrient Requirements of the Cow and Calf,* Texas A&M University, B-1044, p. 7, Fig. 2)

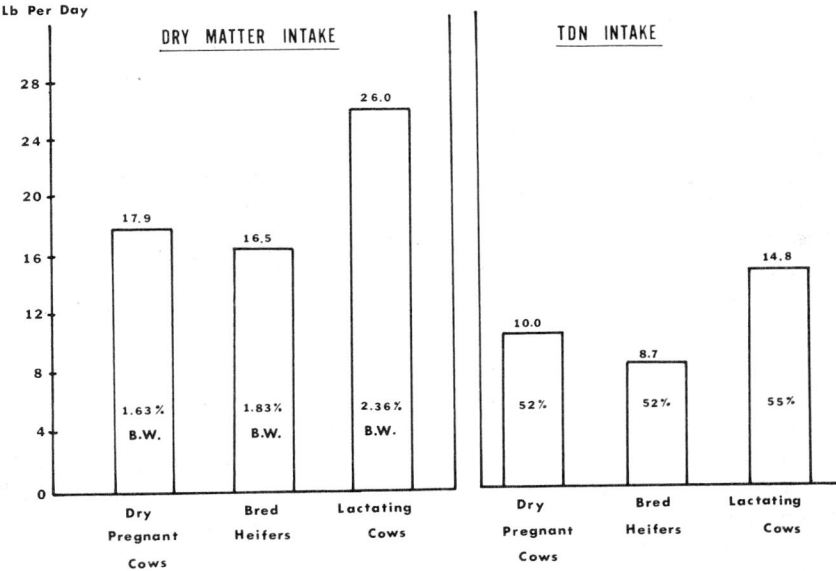

Fig. 19-3. Dry matter (in lb daily and in % of body weight) and TDN requirements (in lb daily and in % of the ration) of dry pregnant cows, bred heifers, and lactating cows. As shown, lactating cows require considerably larger quantities of feed (26 vs 17.9 lb/day), along with a slightly higher energy ration (3% more), than dry pregnant cows. Replacement heifers, which must provide for the growth of their own bodies as well as for the growth of the fetus, consume more feed in relation to body weight (B.W.). (Based on data from *Nutrient Requirements of Beef Cattle,* Fifth Revised Edition, National Research Council, National Academy of Sciences, Washington, D.C., 1976)

requirements for the following classes of cattle: (1) dry mature cows, last 2-3 months of pregnancy, weighing 1,100 lb; (2) yearling heifers, last 3-4 months of pregnancy, weighing 900 lb; and (3) cows nursing calves, superior milking ability, first 3-4 months after calving, weighing 1,100 lb.

Figs. 19-3 to 19-5 show that the lactation requirements are the highest and most critical of all; in total feed consumed, in TDN and crude protein of the ration, and in the major minerals—calcium and phosphorus. After a cow calves, her energy needs jump about 50 percent, her protein needs double, and her calcium and phosphorus needs triple. These figures also show that the requirements of growing heifers are higher than those of mature cows.

Weight also makes a difference, as shown in Table 19-1, which gives the daily nutrient requirements at various weights of (1) dry pregnant cows and (2) cows nursing calves.

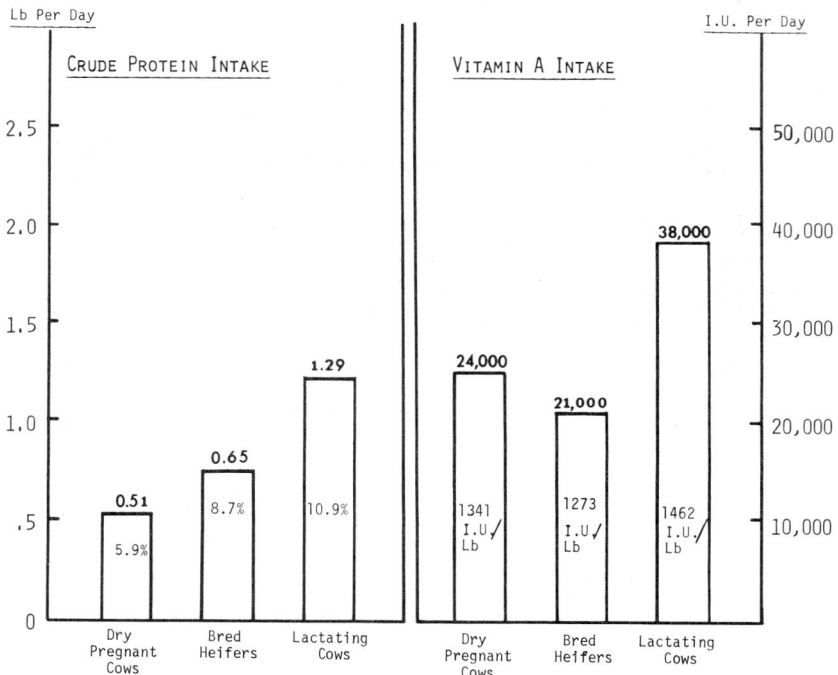

Fig. 19-4. Crude protein and vitamin A requirements of dry pregnant cows, bred heifers, and lactating cows. Milk is high in protein; hence, it follows that for superior milk production the protein requirements of lactating cows are high, in terms of both crude protein intake and percent of the ration. Also, in order to meet the simultaneous protein requirements for body growth and fetal development, heifers require a greater daily protein intake and a higher protein ration than dry cows.

Lactating cows have the highest vitamin A requirement of the three groups, followed, in order, by dry pregnant cows and bred heifers. (Based on data from *Nutrient Requirements of Beef Cattle*, Fifth Revised Edition, National Research Council, National Academy of Sciences, Washington, D.C., 1976)

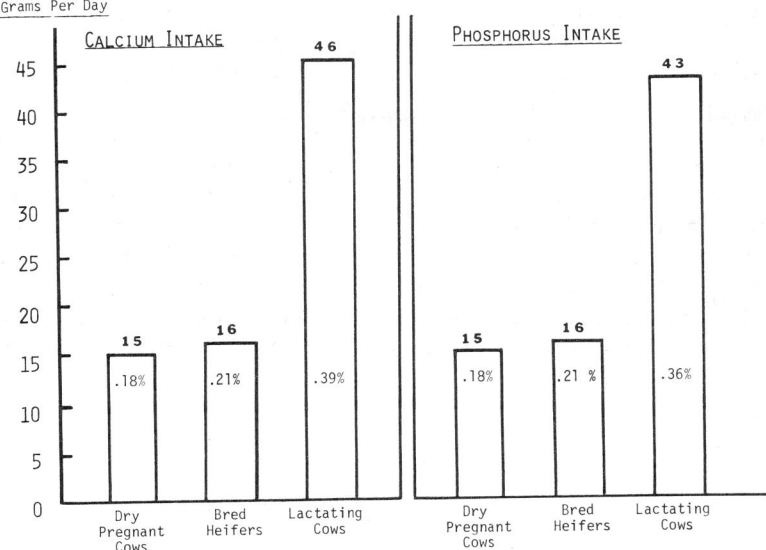

Fig. 19-5. Calcium and phosphorus requirements of dry pregnant cows, bred heifers, and lactating cows. Milk is a rich source of calcium and phosphorus; hence, it is no surprise to find that the requirements for each of these minerals is markedly higher for lactating cows than for either of the other groups. Also, the calcium and phosphorus requirements of growing heifers are slightly higher than those of mature dry cows. (Based on data from *Nutrient Requirements of Beef Cattle,* Fifth Revised Edition, National Research Council, National Academy of Sciences, Washington, D.C., 1976)

TABLE 19-1

DAILY NUTRIENT REQUIREMENTS OF BEEF COWS[1]

Body Weight		Crude Protein		TDN		Calcium	Phosphorus
Dry pregnant mature cows (middle third of pregnancy)							
(lb)	*(kg)*	*(lb)*	*(kg)*	*(lb)*	*(kg)*	*(g)*	*(g)*
800	363	.70	.32	6.6	3.0	10	10
900	408	.79	.36	7.3	3.3	11	11
1,000	454	.86	.39	7.9	3.6	12	12
1,100	499	.93	.42	8.6	3.9	13	13
1,200	544	.99	.45	9.2	4.2	14	14
1,300	590	1.08	.49	9.8	4.4	15	15
1,400	635	1.15	.52	10.4	4.7	16	16
Cows nursing calves, first 3 to 4 months after calving (superior milking ability)							
(lb)	*(kg)*	*(lb)*	*(kg)*	*(lb)*	*(kg)*	*(g)*	*(g)*
800	363	2.45	1.11	12.8	5.8	45	40
900	408	2.60	1.17	13.5	6.1	45	41
1,000	454	2.71	1.23	14.1	6.4	45	42
1,100	499	2.84	1.29	14.8	6.7	46	43
1,200	544	2.98	1.35	15.4	7.0	46	44
1,300	590	3.11	1.41	16.1	7.3	46	44
1,400	635	3.22	1.46	16.8	7.6	47	45

[1]Adapted by the author from *Nutrient Requirements of Beef Cattle,* Fifth Revised Edition, National Research Council, National Academy of Sciences, Washington, D.C., 1976, pp. 26, 27; with U.S. Customary added by the author.

Of course, these are minimum requirements. Hence, it would be well to add one percentage unit to the crude protein requirement and 3.0 percentage units to the TDN requirement, and to self-feed the minerals. This would take care of variations in feedstuffs and differences in requirements between individual cows within a herd.

Both nutritional requirements and recommended allowances (the latter provide for margins of safety) are given in Chapter 8, Feeding Beef Cattle, of this book; hence, the reader is referred thereto.

Heavy grain feeding is uneconomical and unnecessary for the beef breeding herd. The nutrient requirements should be adequate merely to provide for maintenance, growth (if the animals are immature), and reproduction. Fortunately, these requirements can largely be met through feeding roughages.

NUTRITIONAL REPRODUCTIVE FAILURE IN COWS

Since cattlemen largely determine their own destiny when it comes to feeding, it is important that they know the causes of nutritional reproductive failure and how to rectify them.

A review of the literature clearly points to three important reproductive difficulties: (1) the small number of cows in heat and bred the first 21 days of the breeding season, (2) the low conception rate at first service, and (3) the excessive calf losses at birth or within the first 2 weeks of age. Also, it is noteworthy that each of the causes is more marked in young cows (first-calf heifers) than in mature cows.

Research throughout the country gives ample evidence that the real cause of most beef cow reproductive failure is a deficiency of one or more nutrients just before and immediately following calving—nutritive deficiencies during that critical 100-day period when life begins—a deficiency of energy, protein, minerals and/or vitamins. In the sections that follow, reproductive failures attributed to deficiencies of specific nutrients are summarized.

(Also see Chapter 6, section entitled, "Nutritional Factors Affecting Reproduction.")

Energy Reproductive Failures

Energy is essential for the normal life processes of the cow, including body maintenance, reproduction, and lactation. Because energy is necessary for life itself, it is the most important nutrient.

● *New Mexico workers showed* that inadequate energy intake, rather than a shortage of protein or vitamin A, was the major cause of lowered reproductive performance in cows grazing on semi-desert grassland during drought years (Table 19-2). Supplementation started approximately one month (February 10) before calving and continued until an appreciable amount of green forage was available (June 1). Cows received approximately one lb of supplement before calving and approximately 2 lb after calving. The New Mexico Station was able to increase the percentage calf crop by 9%

(from 78% to 87%) by providing 1 to 2 lb of supplemental energy feed (ground corn) to cows on semi-desert grassland, starting about a month before calving and continuing until adequate green grass was available.

TABLE 19-2

RANGE CONDITIONS AND RESPONSE TO SUPPLEMENTAL
FEEDING IN NEW MEXICO[1]

	Control	Ground Corn	Cottonseed Cake	Cottonseed Cake + Dehy. Alfalfa
Weight of calves:				
1. Years of average rainfall, lb	428	424	430	436
2. Drought years, lb .	351	366	366	367
Percent of cows calving:				
1. After years of average rainfall, %	93	92	91	95
2. After drought years, %	78	87	84	84

[1]Knox, J. H., and W. E. Watkins, Supplements for Range Cows, *New Mexico Agricultural Experiment Station Bulletin 425*, 1958.

• *In a Montana study,*[2] cows on heavily grazed (23.1 acres per cow yearly) ranges averaged 15 percent fewer calves born than those on moderately (30.5 acres per cow yearly) or lightly grazed (38.8 acres per cow yearly) ranges. A shortage of forage appeared to be responsible for the results obtained.

• *In an Oklahoma study,* the effect of winter plane of nutrition on the performance of spring-calving 3-year-old beef cows was studied (Table 19-3). The heifers were started on winter feed in early November of each year and fed each winter (approximately 160 days to mid-April) according to the following plan:

TABLE 19-3

EFFECT OF FOUR WIDELY DIFFERENT WINTER FEED LEVELS ON
PERFORMANCE OF THREE-YEAR-OLD BEEF COWS[1]

	Low	Medium	High	Very High
Winter feed/head, lb:				
Cottonseed cake	32	202	206	—
Ground milo	—	88	815	—
50% concentrate mix	—	—	—	4,660
Winter feed cost/cow, 3 years .	$ 74.62	$107.29	$147.87	$ 281.71
% calf crop weaned	71.4	85.7	92.9	84.6
Average birth weight, lb ...	69.5	73.8	75.3	77.0
Average weaning weight, lb	361	455	512	455

[1]Pinney, D., L. S. Pope, C. V. Cotthem, and K. Urban, Effect of Winter Plane of Nutrition on the Performance of Three- and Four-Year-Old Beef Cows, *36th Annual Livestock Feeders' Day, Oklahoma State University*, April 21, 1962, Table 2, p. 53.

[2]Marsh, H., K. F. Swingle, R. R. Woodward, G. F. Payne, E. E. Frahm, L. H. Johnson, and J. C. Hide, Nutrition of Cattle on an Eastern Montana Range As Related to Weather, Soil, and Forage, *Montana Agricultural Experiment Station Bulletin 549*, 1959.

1. *Low plane*—No gain during the first winter as weaner calves, with a loss of at least 20% of fall weight during subsequent winters as bred females.

2. *Moderate plane*—Gains of 0.5 pound per head daily the first winter as weaner calves, with a loss of nearly 10% of fall weight during subsequent winters as bred females.

3. *High plane*—Gains of approximately one pound per head daily during the first winter, with no loss in weight during subsequent winters.

4. *Very high plane*—Self-fed a 50% concentrate mixture to gain as rapidly as possible both as weaner calves and in subsequent winters.

The results obtained in this study point up the danger of underfeeding—low calf crop, smaller calves at birth and weaning, and depressed milk production. However, overfeeding resulted in more calving difficulty, depressed milk production, and a tremendous increase in feed cost. From the economic standpoint, a medium level of feeding appeared to be most desirable.

● The National Research Council lists "failure to conceive" as one of the symptoms of energy deficiency in beef cattle.[3]

Protein Reproductive Failures

Protein is essential for maintenance and building of muscle tissue and bone, including growth of hair and hoofs; for the development of the fetus; for the growth of young stock; and for milk production.

● *Workers at the Montana Station conducted* an experiment to compare the feeding of one pound per head per day of either a 10 or 20% protein supplement during the winter on the production of range cows. Hereford cows fed one lb of the 20% protein supplement daily weaned calves which, on a 205-day corrected weaning weight basis, averaged 22 lb heavier than those receiving one lb of the 10% supplement daily (Table 19-4).

TABLE 19-4

PERFORMANCE OF MATURE HEREFORD COWS
FED ONE POUND PER HEAD PER DAY OF EITHER
A 10 OR 20 PERCENT PROTEIN SUPPLEMENT[1]

	Protein Content of Supplement	
	10%	20%
Number of head	24	24
Cows calving	24	24
Average birth weight, lb	78	82
Average weaning weight, lb	406	416
Average adjusted weaning weight, lb (205 days)	478	500

[1]Thomas, O. O., J. L. Van Horn, and F. S. Willson, Effect of Level of Protein in Winter Rations and Added Feed at Calving Upon Production of Range Cows, *8th Cattle Feeders' Day, Montana State College*, March 20, 1964.

[3]*Nutrient Requirements of Beef Cattle*, Fifth Revised Edition, National Research Council, National Academy of Sciences, Washington, D.C., 1976.

• *The National Research Council* reports that irregular or delayed estrus is the major symptom of protein shortage in rations for breeding females.[4]

Combined Energy and Protein Reproductive Failures

Lack of energy in cows results in poor reproduction: failure of some cows to show heat, more services per conception, lowered calf crops, and lightweight calves.

A severe shortage of protein causes depressed appetite, poor growth, loss of weight, reduced milk production, irregular estrus, and lowered calf crops.

• *California workers found* that cows receiving supplemental feed during periods when forage was scarce had better reproductive performance than cows not receiving a supplement (Table 19-5). The supplementation consisted of one lb of 43 percent cottonseed cake beginning in August; 2 lb of cottonseed cake from calving time (October, November, and December) to first rains (usually January or February); and 2 lb of cottonseed cake plus one lb of barley from time of first rain until new forage growth was available. This study was conducted in the Sierra Nevada foothills where rainfall averaged 19.2 inches yearly for the 12 years of the experiment.

TABLE 19-5

REPRODUCTION IN CALIFORNIA[1]

	Number of Cow Years	Cows Calving	Calf Crop	Weaning Weight	Weaning Age	Lb of Calf Weaned Per Cow Bred
		(%)	*(%)*	*(lb)*	*(days)*	*(lb)*
Supplemented	478	90	83	464	240	385
Unsupplemented	469	75	66	406	230	270

[1]Wagnon, K. A., H. R. Guilbert, and G. H. Hart, Beef Cattle Investigations on the San Joaquin Experimental Range, *California Agricultural Experiment Station Bulletin 765,* 1959.

• *Bond and Wiltbank found* that there was a wide variation in reproductive performance in heifers fed different levels of energy and protein. (Table 19-6). Angus heifers were placed on the experimental rations shortly after weaning and calved at approximately two years of age. Results observed in the heifers receiving low levels of protein were confounded because of inadequate intake. This low level intake resulted in an energy deficiency as well as a protein deficiency. It appeared that adequate energy was more important than adequate protein on reproductive performance.

[4]*Nutrient Requirements of Beef Cattle,* Fifth Revised Edition, National Research Council, National Academy of Sciences, Washington, D.C., 1976.

TABLE 19-6

EFFECT OF VARIOUS LEVELS OF PROTEIN AND ENERGY ON REPRODUCTIVE
PERFORMANCE—MARYLAND AND LOUISIANA[1]

Energy Intake		Intake of DP		Showing Heat	Calves Lost at Birth	Interval from Calving to First Heat	Milk Production at 60 Days	Wt. at First Heat
(lb)		(lb)		(%)	(%)	(days)	(lb)	(lb)
(H)	11	(H)	1.11	100	73	51	17.2	599
(H)	10	(M)	0.65	100	45	80	14.6	586
(H)	6	(L)	0.24	50	10	66	9.3	412
(M)	6	(H)	1.11	100	0	63	15.0	506
(M)	6	(M)	0.68	100	9	68	16.8	552
(M)	5	(L)	0.24	67	17	118	13.0	427
(L)	3.5	(H)	1.06	25	25	—	6.8	440
(L)	3.4	(M)	0.66	58	9	163	8.7	436
(L)	3.3	(L)	0.22	33	17	130	8.0	433

(H) = High
(M) = Medium
(L) = Low
DP = Digestible Protein

[1]*Beef Cattle Science Handbook*, Volume 2, edited by M. E. Ensminger and published by Agriservices Foundation, 1965, p. 135. Table 19-6 is based on work conducted by J. Bond and J. N. Wiltbank.

● *Nelson and coworkers in Oklahoma presented* the results of a study on wintering 2 groups of 3-year-old fall-calving beef cows (Table 19-7). During the summer, all cattle grazed the native grass pasture. Cows in the low group were fed 1.43 lb of pelleted cottonseed meal per head daily in addition to prairie hay starting October 27. The high group of cows was self-fed a cottonseed meal-milo-salt mixture (average consumption—1.80 lb cottonseed meal, 5.32 lb ground milo, 1.84 lb salt) as supplement to prairie hay. The level of winter feeding had a marked effect on weaning weight of calves. It is difficult to determine whether the additional protein or the energy accounted for the greater net return from the high group of cows.

TABLE 19-7

SUPPLEMENTAL WINTER FEEDING OF THREE-YEAR-OLD BEEF COWS[1]

	Low	High
Winter loss in wt., lb	158	144
Birth wt. of calves, lb	71	72
Spring wt. of calves, lb	214	264
Weaning wt. of calves, lb	382	473
Feed cost/cow	$ 57.73	$ 68.38
Selling value of calves	$100.42	$121.22
Selling value minus feed cost	$ 42.69	$ 52.84

[1]Nelson, A. B., R. D. Furr, and G. R. Waller, Level of Wintering Fall-Calving Beef Cows, *36th Annual Livestock Feeders' Day*, Oklahoma State University, April 21, 1962, Table 1, p. 73.

Phosphorus Reproductive Failures

Cattle produced in phosphorus-deficient areas may have depraved appetites, may fail to breed regularly, and may produce markedly less milk.

Growth and development are slow, and the animals become emaciated and fail to reach normal adult size. Death losses are abnormally high.

● *Montana workers reported* the results of a study designed to determine the phosphorus requirements of breeding cows on winter range (Table 19-8). Heifers fed the supplement containing 1.0 percent phosphorus weaned the heaviest calves.

TABLE 19-8

CALVING RECORDS OF TWO-YEAR-OLD HEIFERS FED DIFFERENT LEVELS OF PHOSPHORUS IN SUPPLEMENTS[1]

	2 Lb of a 20% Protein Supplement/Day			
	0.5%P	1.0%P	1.5%P	2.0%P
Number of head	14	14	13	14
Cows calving	11	11	13	13
Average birth wt., lb	71	70	64	64
Average weaning wt., lb	308	324	307	314
Average adjusted weaning wt., lb (205 days)	444	466	445	435

[1]*Beef Cattle Science Handbook,* Volume 2, edited by M. E. Ensminger and published by Agriservices Foundation, 1965, p. 138. Table 19-8 is based on work by O. O. Thomas, J. L. Van Horn, and F. S. Willson.

Vitamin A Reproductive Failures

A severe deficiency of vitamin A may result in a low conception rate; a small calf crop; many calves weak or stillborn, with some born blind or without eyeballs; more cows with retained placentas; low gains; greater susceptibility to calf scours; and more respiratory troubles.

Lane summarized the results[5] of work in Arizona with range cows for two years (1962 and 1963). Yearling heifers receiving a rumen injection of 1,000,000 IU of vitamin A plus 100,000 units of vitamin D_1 per cow and calving as 2-year-olds produced calves that averaged 20 pounds heavier than those from untreated heifers. When treatment was prior to the breeding season, the percent calf crop increased 10 to 11% (65% to 76%). The heifers were on ranches that were extremely droughty at the time of injection. Range conditions on each individual ranch should determine whether this injection should be given in the spring or fall.

● *Verdugo, of California State University-Fresno,* summarized[6] data collected for over 20 years from a fall-calving Hereford herd in the foothills of the Sierra Nevada mountains near Porterville, California (Fig. 19-6). Rainfall, which was highly variable and occurred in the winter, averaged slightly over 14 inches per year. Changes in a number of management and nutritional practices contributed to the increased calf crop over the years. The important

[5]Lane, A., Vitamin A Injections in Range Cows, *Arizona Cattle Feeders' Day, University of Arizona,* May 7, 1964, p. 28.

[6]Verdugo, W. R., Recorded Research and Management Problems for 22 years, AceHi Polled Hereford Ranch, Porterville, California, *Beef Cattle Science Handbook,* Volume 2, edited by M. E. Ensminger and published by Agriservices Foundation, 1965, p. 137.

change, particularly in the heifers, appeared to be the addition, beginning in 1948, of 10,000 IU of vitamin A per head daily to the supplement of alfalfa meal, cottonseed meal, and salt which was fed in the late summer or early fall until the dry grass season was over. Upon charting yearly calf crop and yearly rainfall, a striking parallel was noted between heifer calf crop and yearly rainfall up to 1948 when vitamin A supplementation was initiated (Fig. 19-6).

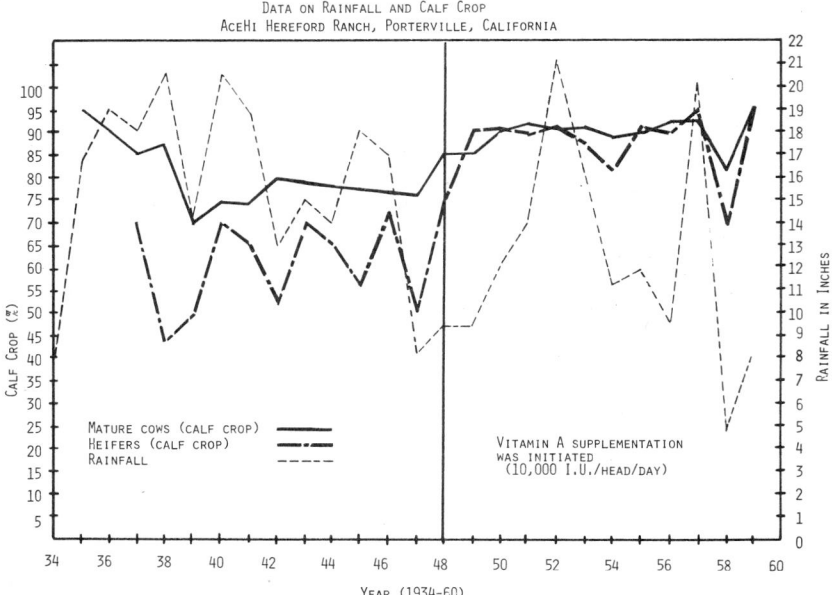

Fig. 19-6. Data on rainfall and calf crop, AceHi Hereford Ranch, Porterville, California.

• *Temple, Southern Regional Beef Cattle Research, reported*[7] that studies at Front Royal showed that (1) vitamin A injections during pregnancy lowered the incidence of stillborn calves by 2.2 percent, and (2) vitamin A supplementation at birth lowered newborn calf mortality by 8.4 percent.

• *Hentges of the Florida Station, in a review of the literature, reported*[8] that deficiencies of vitamin A in cows will cause delayed conception, retained placentas, and dead, weak, or malformed calves; in bulls, delayed puberty, testicular tissue degeneration, and pituitary cysts. Also, Hentges reported that a deficiency of vitamin D will cause infrequent heat periods and lowered reproduction.

Based on experiments plus experiences, it is recommended that vitamin

---

[7]Summarized by the author in *The Stockman's Guide*, February, 1970, syndicated monthly livestock column.
[8]Ibid.

A be supplied to cows and calves after periods of four months on dry, weathered forage—either winter pasture or harvested forage. This can be done either by injection of one million IU of vitamin A or by feeding a supplement.

Summary of Nutritional Reproductive Failure in Cows

From the above review of literature, the following conclusions seem justified:

1. Energy is more important than protein in reproduction.

2. Beef cows receiving inadequate energy reproduce at a low level.

3. Phosphorus supplementation of cows on range areas deficient in phosphorus increases the calf crop.

4. Administering additional vitamin A to heifers grazing dry forage increases the calf crop.

5. The level and kind of feed before and after calving will determine how many cows will show heat—and conceive. After calving, feed requirements increase tremendously because of milk production; hence, when a cow is suckling a calf, she needs approximately 50 percent greater feed allowance than during the pregnancy period (see Fig. 19-2). Otherwise, she will suffer a serious loss in weight and fail to come in heat and conceive.

6. Cows in average condition should gain a minimum of 100 lb during the pregnancy period, followed by a gain of ½ to ¾ lb daily after calving and extending through the breeding season. If they are on the thin side at calving time, they should gain 1½ to 2 lb daily after they drop calves. This calls for 7 to 12 lb of TDN daily before calving (which can be provided by feeding 14 to 22 lb of average quality hay), and 10 to 17 lb of TDN after calving (which can be provided by feeding 14 to 28 lb of hay plus 4 lb grain), with the lactating requirement dependent on both cow weight and milking ability. Additionally, there must be adequate protein, minerals, and vitamins.

WINTER FEEDING BROOD COWS

In a country as large and diverse as the United States, wide variations exist in both the length of the winter season and the available feeds. But the same principles are applicable to all areas and enterprises, and the chief objective remains the same—economically to produce high percentage calf crops with heavy birth and weaning weights.

Winter feeding is the most expensive time in cow-calf operations. From an economic standpoint, therefore, it is important that wintering practices be both knowledgeable and wise. The cheaper homegrown roughages should constitute the bulk of the winter ration for dry pregnant cows. Most of the grain and the higher class roughages may be used for other classes of livestock. A practical ration may consist of silage and/or dry roughages (legume or grass hays) combined with a small quantity of protein-rich concentrates (such as soybean meal or cottonseed meal). With the use of a leguminous roughage, the protein-rich concentrate may be omitted. Dusty or moldy feed and frozen silage should be avoided in feeding all cattle—especially in the

Fig. 19-7. Winter feed for brood cows. Laborsaving, tightly-packaged, loaf-like stacks of corn stover. Such stacks may be deposited in the feeding area, then fenced, and cattle may be fed right from the stack. (Courtesy, Hesston Corporation, Hesston, Kan.)

case of the pregnant cow, for such feed may produce complications and possible abortion.

Except during the winter months, pastures constitute most of the feed of beef cattle. By fall, however, grass is usually in short supply and relatively poor as a source of protein, certain minerals (especially phosphorus), and carotene (provitamin A). To overcome these deficiencies, the cattleman must resort either to (1) supplemental feeding on pasture, or (2) drylot feeding. At no other time in the operations is a possible profit so likely to be dissipated and replaced by a loss.

Fall feeding should not be delayed so long that animals begin to lose weight. The reason cattle often eat and get poor on dry, weathered grass is that it is low in energy, protein, carotene, and phosphorus and perhaps certain other minerals. These deficiencies become more acute and increase in severity as winter advances. Cattle simply cannot consume sufficient quantities of such bulky, low quality roughage to meet their needs; and the younger the animal the more acute the problem. Under such circumstances, the maintenance needs are met by the breakdown of body tissues, accompanied by the observed loss in weight and condition. Young animals fail to grow; it makes for lightweight calves. Also, reproduction is affected adversely; serious underfeeding results in lowered calf crops. Supplementing fall grass with a concentrated type of supplement is the practical and ideal way in which to alleviate such nutritive deficiencies.

Likewise, spring feeding should be continued until grass has attained sufficient growth and sustenance. The calf crop will be adversely affected in pregnant cows and milk production will be lowered in lactating cows if they subsist for prolonged periods on insufficient amounts of forage or on frosted pastures that are low in protein, vitamin A, and minerals.

When roughage is scarce and high in price, feed less of it and more concentrate; conversely, feed more concentrate and less hay when grain is plentiful. Generally speaking, one pound of grain can replace two pounds of dry roughage, providing they are of comparable quality.

The best calf crop is produced by cows that are kept in vigorous breeding condition—that are neither overfat nor in thin, run-down condition. Generally speaking, this calls for winter feeding, with the maximum use of roughage. The kind and amount of concentrate needed will depend upon (1) the amount and kind of roughage given, (2) the age and condition of the cattle, and (3) whether the cows are dry or suckling calves. In total, it is important that the ration provides the kinds and amounts of nutrients needed, along with sufficient bulk to satisfy the appetite reasonably well. On a dry-feed basis, the daily requirement of dry, pregnant cows is about as follows: thin cows, 2¼% of their liveweight; cows in average flesh, 2% of liveweight; and cows in good condition, 1¾% of liveweight. Cows suckling calves should receive approximately 50% more feed than dry cows of comparable weight and condition. From this it should be concluded that, unless the herd is so small as to make it impractical, dry cows should be wintered separate from those that are suckling calves. This makes it possible to limit the feed of dry cows and to effect certain other economies in their handling.

Dry pregnant cows in average condition should gain in weight sufficient to account for the growth of the fetus (60 to 90 lb) plus sufficient increase in weight and condition to carry them through the suckling period. In total, they should gain 100 to 150 lb during the pregnancy period, or at the rate of approximately one-half lb daily. Of course, the size and condition of the cow is the best gauge as to the feed allowance and desired gain. As previously noted, dry cows require less supplementation than cows suckling calves. Nevertheless, they should not be permitted to lose too much flesh, unless, of course, they are overfat. Also, it is recognized that it requires less feed to keep cattle from losing flesh than it does to restore them to proper condition after they have become thin. Thus, it is good economy to start feeding before cows show any signs of malnutrition. Unless a good quality legume roughage is fed, the concentrate should provide protein, energy, and needed minerals and vitamins.

Noteworthy, too, is the makeup of a calf at birth. An average 70-lb calf at birth is about 75% water, 20% protein, and 5% ash. The calf's 70-lb weight represents about 17.5 lb dry matter. From this, it is apparent that the dry gestation period does not create a heavy nutritional drain. Thus, this is a period when a cattleman may economize by utilizing crop residues and winter pasture.

Cows with calves at side should be fed for the production of milk, which requirements are more rigorous than those during pregnancy. This is important because, until weaning time, the growth of the calf is determined chiefly by the nourishment available through the milk of its dam. The principal part of the calf's ration, therefore, may be cheaply and safely provided by giving its mother the proper feed for the production of milk. To stimulate milk flow, most beef cows need a concentrate during the winter months, and the poorer or the more limited the roughage the higher the supplement requirement. On the average, cows that calve in the fall should be fed a minimum of 4 to 6 pounds concentrate daily throughout the winter.

Purebred breeders recognize that good condition and attractive appearance of breeding animals are important assets from the standpoint of favorably impressing prospective purchasers. Thus, a certain amount of the feed consumed by purebred cows—that which is above the amount needed for good health and vigor—can rightfully be charged to advertising costs. This is so because no more effective means of advertising exists than well-conditioned animals, regardless of the season of the year. Also, supplemental feeding of purebred cows that are suckling is favorably reflected in their calves—in greater growth, development, and bloom. In short, purebred breeders usually find it advantageous to feed a concentrate to cows during the winter months. Any purebred breeder who is content with less than the maximum potential should not be in the purebred business in the first place; he should confine his efforts to commercial production.

Rations for Dry Pregnant Cows

When winter grazing is not possible, the rations in Table 19-9 may be used to meet the daily needs for energy and protein of a 1,100-pound, dry, pregnant cow. A combination of legume roughage with lower quality roughage (such as stalklage, straw, corncobs, or cottonseed hulls) will meet both the energy and protein requirements without the use of a supplement.

TABLE 19-9

WINTERING RATIONS FOR A 1,100-POUND, DRY, PREGNANT COW

	Rations				
	1	2	3	4	5
	(Pounds per day)				
Legume-grass hay	18				10
Legume-grass haylage[1]		30			
Corn or grain sorghum silage			35		
Stalklage or husklage				45	
Straw, cobs, or cottonseed hulls					10
Supplement[2]				.5	1

[1]Haylage figured at 55% dry matter, corn or grain sorghum silage at 35% dry matter, stalklage-husklage at 45% dry matter.

[2]Supplement figured at 48% crude protein. Quantity to be adjusted in keeping with the protein content of the supplement. For example, if a 24% crude protein supplement is fed, the quantity of supplement should be doubled.

Rations for Cows Nursing Calves

The energy requirement of a cow nursing a calf is about 50 percent higher than that of a dry pregnant cow; and the protein, calcium, and phosphorus requirements are about double. Since the vast majority of the nation's cows with calves at side are on pasture most, if not all, of the lactation period, the only supplemental need is for salt and other minerals, unless the pasture is insufficient in quantity or quality of feed to support adequate milk production. The rations in Table 19-10 may be used for drylot feeding of beef cows nursing calves. Of course, the daily levels shown in Table 19-10 should be approached gradually so that nutritional scours will not develop in baby calves.

TABLE 19-10

WINTERING RATION FOR A 1,100-POUND COW NURSING A CALF

	Rations				
	1	2	3	4	5
	(Pounds per day)				
Legume-grass hay	30			20	10
Legume-grass haylage[1]		50			
Corn or grain sorghum silage[1]			60		40
Grain				5	
Supplement[2]		1.5			

[1]Haylage figured at 55% dry matter; corn or grain sorghum silage figured at 35% dry matter.
[2]Supplement figured at 48% crude protein. Quantity to be adjusted in keeping with the content of the supplement. For example, if a 24% crude protein supplement is fed, the quantity of the supplement should be doubled.

Crop Residues and Winter Pastures for Brood Cows

Two requisites are important in wintering the cow herd: (1) bringing them through the winter in proper condition for calving, and (2) keeping feeding costs to the minimum consistent with nutritional demands. Meeting these requirements has prompted increased use of crop residues and winter pastures for brood cows. As the ever increasing human population of the world consumes a higher proportion of grains and seeds, and their by-products, directly, cattle will utilize a maximum of crop residues and pastures and a minimum of products suitable for human consumption. Thus, more and more farmers with crops will include a beef herd in their operations and realize a fair return from feeds which would otherwise be wasted.

CROP RESIDUES

Generally speaking, crop residues may either be grazed, processed as dry feed, or made into silage. The important thing to remember is their relatively low value, in comparison with grains, necessitates low cost harvesting, storing, and feeding. Also, they must be fed to the right class of animals, and they must be properly supplemented. Remember, too, that there is a marked difference between economical wintering and deficient wintering.

In addition to the crop residues discussed in the sections that follow, good quality roughages—such as alfalfa, brome, and corn or grass silage—can be used to meet the nutritional requirements of the brood cow. But such high quality feeds must be limited, otherwise animals will get too fat.

CORN RESIDUES

Of all crop residues, the residue of corn is produced in greatest abundance and offers the greatest potential for expansion in cow numbers. In 1972, 57,289,000 acres of corn, yielding 96.9 bushels per acre, were harvested in this country. For the most part, over and above the grain, 2½ to 3 tons of dry matter produced per acre (40 to 50 percent of the energy value of the total corn plant) were left to rot in the field. That's 157 million tons of potential cow feed wasted, enough to winter 118 million dry pregnant cows consuming an average of 22 pounds of corn refuse per head per day during a 4-month period. Moreover, mature cows are physiologically well adapted to utilizing such roughage. And that's not all! When corn residue is used to the maximum as cow feed, acreage which would otherwise be used to pasture the herd is liberated to produce more corn and other crops. Also, there are many other crop residues, which, if properly utilized, could increase the 118 million figure given above.

Although corn refuse offers tremendous potential as a cow feed, there are difficulties in harvesting and storing it. But science and technology have teamed up and are working ceaselessly away at solving these problems.

Fig. 19-8. Fox equipment with flail attachment harvesting corn stalklage. (Courtesy, Koehring Farm Division, Appleton, Wisc.)

Broadly speaking, three alternate methods of salvaging corn refuse are being used: (1) grazing, (2) harvesting and dry feeding, and (3) ensiling; with different ways of accomplishing each. The choice of the method should be determined primarily by (1) cost, (2) the proportion of refuse utilized, and (3) how well it meshes in with other farm enterprises—for example, in some cases the need for fall plowing will necessitate removal of the material from the land and eliminate grazing as an alternative. Costs and the proportion of the refuse salvaged vary widely, as shown in Table 19-11.

TABLE 19-11

HARVESTING COST PER TON OF DRY MATTER OF CORN FORAGE
BY VARIOUS HARVESTING SYSTEMS[1]

Harvesting System	Forage Yield /Acre[2]	Harvesting Cost/Ton[3]		
		80 Acres	160 Acres	240 Acres
	(tons)	($)	($)	($)
Corn silage, with forage harvester	6.0	5.30	(100 Acres) 4.75	—
Corn stalklage, with flail-type forage harvester	1.5	10.85	8.36	7.59
Corn stalklage, with 1 ton stacker	1.5	6.75	5.49	5.20
Corn stalklage, with 3 ton stacker	1.5	8.86	6.48	5.83
Corn husklage, with straw-buncher behind	1.0	6.22	3.21	2.22

[1]Ayres, George E., Harvesting and Handling Forage For Beef Cows, paper presented at Second Annual Corn Belt Cow-Calf Conference, Ottumwa, Iowa, February 24, 1973.
[2]Tons of dry matter.
[3]Includes cost of all harvesting machines and labor, but does not include storage costs.

Table 19-11 shows that the cost of each method of harvesting is lowered as the acreage increases.

Grazing (Pasturing)

This refers to turning the animals directly into the stalk field—the traditional way of utilizing cornstalks. Letting the animals do their own harvesting is the simplest and least expensive method devised for utilizing a crop. However, there is considerable wastage, and it is not possible to prolong the winter feeding period. In an open fall and winter, 2 acres of cornstalks will carry a pregnant cow for 100 to 120 days. But the following problems are associated with this method of harvesting:

1. *Selective grazing*—Cows are selective grazers. They will consume the more palatable portions of corn refuse first, in the following order: corn ears, husks, leaves, and stalk.

2. *Waste*—Only an average of one-third of the stover is actually used, with the amount varying from 15 to 40 percent, depending primarily on weather conditions.

3. *Fencing*—Many cornfields are unfenced; hence, fence must be constructed in order to confine the animals. Also, strip grazing (grazing a part of the field at a time) will improve the utilization of stalks in large fields by making more uniform nutrition available throughout the grazing period. It prevents selective grazing over the entire field, with the result that the animals consume the more palatable portions of the plant first, and leave the bare stalks until last.

Fig. 19-9. Angus cows grazing cornstalks in the Corn Belt. (Courtesy, *Wallaces Farmer*, Des Moines, Iowa)

The fencing problem may be solved economically with electric fence.

4. *Snow cover; fall plowing; soil puddling; stock water*—In the northern part of the United States, snow cover prevents grazing for part of the winter and necessitates a reserve feed supply.

Another drawback is that grazing prevents fall plowing, a recommended practice on heavy soils. Also, cattle may puddle and pack the soil, which lowers crop yields.

Frequently, supplying the herd with drinking water is costly. It may necessitate drilling a well or piping water from a distance.

Stalklage

Stalklage refers to all the residue remaining after harvesting corn with a combine or picker. It may either be stored as dry stalklage or ensiled.

● *Dry Stalklage*—Stalklage is more difficult to collect than husklage; and more expensive, since it involves more equipment and another trip across the field. A number of different machines for harvesting stalklage are being used; among them, forage harvesters, balers, stackers with flail pickups, and choppers and stackers. By operating the machine a few inches above the ground so as to prevent excess soil pickup, a yield of 1 to 3 tons of residue per acre may be obtained with the moisture content ranging from 20 to 55 percent, depending on the time of harvest. Stacked or baled cornstalks should be at the low end of this moisture range to reduce heating and spoilage.

Cows like dry stover. Self-feeders around a stack make feeding convenient. Leftover material may be used as bedding.

● *Stalklage ensilage (stover silage)*—Stalklage may also be ensiled, producing a product known as corn stover silage or cornstalk silage. When this is done, the use of a forage harvester equipped with a screen or a recutter-blower at the silo is necessary in order to chop the material finely. Fine chopping will insure good packing and improve consumption by avoiding selectivity.

Where corn stover silage is to be made, the residue should be harvested as soon as possible after the grain is taken off, before the residue loses any moisture. At that time, the grain moisture will generally be under 30% and the refuse moisture will be above 48%. In an airtight silo, 40 to 45% moisture will suffice. In an unsealed or bunker silo, the moisture content should be 48 to 55% for proper lactic acid formation. Water may be added at the silo if necessary. As a precaution, some authorities recommend the addition of 56 pounds of corn meal (or other finely ground grain) per ton of corn stover silage, as a means of providing carbohydrates from which acids will form and act as a preservative. With husklage, the latter precaution is not necessary since there is sufficient grain remaining in the husk and cob.

The biggest deterrent to harvesting stalklage, in either dry or ensiled form, is the cost—primarily for the equipment. Rather than own such expensive equipment, which is used for a short period only, custom harvesting of stalklage is likely cheapest for most operators.

Husklage (Shucklage)

Husklage refers to the forage discharged from the rear of a combine when harvesting corn. It consists of the husks, cobs, and any grain carried over the combine, collected in a wagon or straw buncher pulled behind the combine. This operation minimizes labor and does not slow the grain harvest, because the husklage piles can be dumped at the end of the field for supplemental feeding or later pickup by a front-end loader and moved to another location for stacking or ensiling. The moisture content of this material will usually run between 30 and 40 percent, and the yields will be between 1 and 1.5 tons per acre.

The greatest difficulty encountered in feeding husklage dumps at the end of the field is waste. Depending on weather conditions, as much as 50

Fig. 19-10. Cows on Circle S Ranch, Rockwood, Ontario, Canada, on a winter ration of corn husklage—the husks, cobs, and waste grain; supplemented with 2 pounds of grain and one pound of a 32 percent protein concentrate daily, beginning 6 weeks before calving. (Courtesy, *Country Guide*, Winnipeg, Manitoba, Canada)

percent of the material may be wasted. But wastage of husklage dumps can be materially lessened by controlling access to them.

Stacking of husklage has been satisfactory for some producers.

Ensiling husklage, along with recutting and adding water, results in increased cow consumption and less rejection of cobs.

Since grazing stalk fields is widely practiced in the fall and winter, the feeding of baled, piled or stacked corn residues in the field permits feeding cows on stalks most of the winter without supplemental feeding of hay or silage. Some molds may develop in the collected material, but usually they are not sufficient to affect the feed intake or health of mature cows.

Feeding Value of and Supplements for Corn Residues

Table 19-12 lists the daily nutritive requirements of a dry, pregnant cow weighing 1,000 pounds. Table 19-13 gives the nutritive composition of air-dry corn stover and husklage.

TABLE 19-12

NUTRITIVE REQUIREMENTS OF A DRY PREGNANT COW
(MIDDLE THIRD OF PREGNANCY) WEIGHING 1,000 POUNDS[1]

Dry matter, daily	14.8 lb
TDN, daily	7.9 lb
Crude protein, % of ration	5.9%
Calcium, % of ration	0.18%
Phosphorus, % of ration	0.18%
Vitamin A	19,000 IU

[1]*Nutrient Requirements of Beef Cattle*, Fifth Revised Edition, National Research Council, National Academy of Sciences, Washington, D.C., 1976.

TABLE 19-13

ANALYSIS OF AIR-DRY CORN STOVER AND HUSKLAGE[1]

	Corn Stover	Husklage
	(%)	(%)
TDN	48	57
Crude Protein	4.5	3.4
Calcium	0.4	0.02
Phosphorus	0.07	0.05
Vitamin A	—	—

[1]*Cow-Calf Information Roundup*, University of Illinois, 1971, Table 2, p. 10.

Not all corn refuse will be of the same composition as Table 19-13; some will be better, some will be poorer. The quality declines with the passing of time following grain harvest; and the more severe the weather, the greater the decline. Also, cultural practices during the growing season may alter corn residue quality.

Studies show that a 1,000-lb cow will eat approximately 18 to 20 lb per day of palatable, air-dry stover, or about 2 lb of air-dry stover per cwt per day. She will eat slightly larger amounts of husklage. This consumption, along with the information presented in Tables 19-12 and 19-13, suggests that stover and/or husklage rations will meet the daily energy (TDN) needs of dry pregnant cows, but such rations will be slightly deficient in protein, and low in phosphorus and vitamin A. Nevertheless, the highest and best use for corn residue is for dry pregnant cows for the period following conception to about 30 days before calving.

For corn refuse feeding, mature cows should be in medium to good condition at the start of the winter feeding period; and they should not be permitted to lose over 10 to 15 percent of their weight from fall through calving. Heifer weight losses should be under 5 percent. When weight loss approaches this limit, it's time to feed some grain or silage.

The following additional information is pertinent to the feeding value and supplementation of corn residues for cattle:

1. *Digestibility*—The components of corn residue rank as follows in digestibility, in descending order: remaining grain, husk, leaf, cob, and stalks.

2. *Energy*—Corn residues provide adequate energy to maintain dry

pregnant cows, but they must be supplemented with additional energy when fed to cows nursing calves or to young, growing animals.

3. *Protein*—The crude protein content of corn stover is on the low side for dry pregnant cows. It runs 4.5% (Table 19-13), whereas a dry pregnant cow requires 5.9% crude protein (Table 19-12). Thus, it is recommended that ½ lb per head per day of a 30 to 40% crude protein equivalent (CPE) supplement be provided.

It follows that the protein content of corn refuse is much too low to support either productivity or growth. For example, a 1,000-lb lactating cow requires a daily allowance of 2.0 lb of crude protein. However, a daily consumption of 20 lb of stover will provide only about 0.9 lb—less than half the need.

For nursing cows, the protein deficiency of stover and/or husklage may be corrected by supplementation with the following, on a per head per day basis: 2 lb of a 40 percent protein supplement, or 6 lb of a good legume hay. If desired, the protein supplement may be provided in the form of protein blocks, with one block provided for each 15 cows. Where hay is fed, it should be taken to the field, rather than fed in a feedlot, as this will encourage the cows to stay in the field and graze the cornstalks.

4. *Minerals*—Phosphorus should be provided to all cattle fed corn residue. Calcium may be deficient, especially for lactating cows. Also, some of the trace elements may be deficient. Hence, it is recommended that all cattle on high corn refuse have free access to a complete mineral. A mineral mixture with a Ca:P ratio of 1:2 is recommended for gestating cows. Lactating cows might perform better on a 1:1 ratio.

5. *Vitamin A*—Corn residue, along with other crop residues, is deficient in vitamin A. Hence, it must be supplemented. The precalving and postcalving (heavy milking) needs of approximately 24,000 and 38,000 IU per head per day, respectively (NRC-1976), may be met by feeding vitamin A supplement, intramuscular injection of vitamin A solution, or by feeding adequate levels of green-leafy hay.

It is important that corn residue be tailored to match the cow's nutritional needs. This is relatively simple with dry pregnant cows, where supplementation with a high phosphorus mineral and vitamin A will usually suffice. Beginning four to six weeks before calving and continuing through the lactation period, much heavier supplementation is necessary; in addition to phosphorus and vitamin A, protein must be added, and preferably some energy and calcium for nursing cows.

OTHER CROP RESIDUES

A host of crop residues, other than corn residue, can be used for feeding cows. Some of these follow.

Sorghum (Milo)

Cows will make good use of sorghum as a winter feed. It can be grazed or harvested and stored either as dry feed or silage. The sorghum plant stays green late in the fall; hence, good sorghum stover silage can be made without additional water. In comparison with corn residue, sorghum residue (1) is higher in protein content (corn residue averages 4.5% crude protein, whereas sorghum residue averages 6.5%); (2) is less palatable (if given a choice, cows will select corn refuse in preference to sorghum refuse); (3) comprises a lower percentage of the total plant dry matter than corn (40% of the total plant dry matter of sorghum is residue compared with 40 to 50% for corn); and (4) is lower yielding.

After harvesting, sorghum will send up new shoots if moisture permits. The prussic acid content of these shoots may be harmful to grazing animals; hence, cattlemen should be aware of this possible poisoning. These shoots can be grazed safely four to six days after a hard killing frost.

Soybean Refuse

The stems and pods of soybean refuse available for feeding yield approximately one-fourth ton per acre, with a ratio of stems to pods of about 2:1. The digestibility of stems is low—25 to 35%—due to their high lignin content (18 to 20% for the stalk portion). The digestibility of pods is much higher, ranging from 58 to 63%.

Small Grain Refuse

This refers to (1) straw and (2) tailings—the chaff and grain behind the combine. In the days of binders and threshing machines, straw stacks were commonplace; and they were extensively used for winter cattle feed. With the advent of combines, much of the straw was left to rot in the field. During periods of hay scarcity and high-priced hay, straw is frequently used as either a "hay-stretcher" or "hay-replacer."

Of the common cereal straws, oat straw is the most palatable and nutritious. Barley straw ranks second, and wheat straw is third.

Straw is a bulky feed; and it must be properly supplemented. It is low in protein (wheat straw averages about 3.6% crude protein), low in phosphorus, and low in vitamin A. Dry pregnant cows can be wintered on straw plus a daily allowance of either 5 to 6 pounds of good quality alfalfa hay or 1 to 2 pounds of a 30 to 40% protein supplement, along with free access to a high phosphorus mineral (one containing at least 12% phosphorus). If no legume hay is fed, vitamin A should be fed or injected. When oilseed meals are scarce and high in price, a slow-release nonprotein nitrogen may be used.

The tailings—the chaff and grain behind the combine—are generally used by farmers and ranchers, either as dry feed or mixed with silage.

Legume and Grass Seed Straws

In addition to the cereal straws, other low-cost roughages available in certain sections of the United States are lentil straw, field pea straw, bean straw, clover straw, and bluegrass straw.

Cottonseed Hulls

Cottonseed hulls are one of the most important roughages in the South, especially for cattle. They supply 43.7% TDN, which is about as much as is furnished by late-cut grass hay or by oat straw. They are low in protein (3.9%)—and practically none of it is digestible—low in calcium (0.13%), very low in phosphorus (0.06%), and lacking in carotene. To correct these deficiencies when fed to dry pregnant cows, hulls should be supplemented with a daily allowance of either (1) 6 pounds of a good quality legume hay, or (2) 2 pounds of a 30 to 40% protein supplement, along with free access to a complete mineral, high in phosphorus unless a phosphorus-rich supplement such as cottonseed meal is fed. If no legume is fed, vitamin A should be fed or injected.

When properly fed, cottonseed hulls are about equal in quality to fair quality grass hay and are worth more per ton than corn or sorghum stover, straw, or poor hay. Also, they can be fed without further processing—there is no chopping; and they are well liked by cattle, even when fed as the only roughage.

Pelleted hulls are now on the market. In comparison with regular hulls, they are more digestible, require less transportation and storage space—because of their high density—and are easier to handle.

TREATING CROP RESIDUES TO INCREASE DIGESTIBILITY

Crop residues are inefficiently utilized by animals because of the high content and poor digestibility of the fibrous fraction. This poor digestibility is related to the extent of lignification of the cell wall component of these low quality forages. Although crop residues provide a satisfactory ration for dry gestating cows, they do not provide sufficient energy for either young ruminants or lactating cows—they simply cannot hold enough of these low quality roughages to provide adequate energy. This prompts interest in increasing the digestibility of these crop residues.

There are many approaches to delignifying and increasing the digestibility of crop residues; among them, treatment with sodium hydroxide, potassium hydroxide, ammonium hydroxide, and pressurized heating. The potential of such treatments becomes apparent when it is realized that straw, for example, is only 30 to 40% digestible before treatment. When pressure heated with water, it becomes 50 to 60% digestible; and digestibility increases to 70 to 80% when sodium hydroxide is added prior to cooking. By treating corn husklage and milo residue, the Nebraska Station was able to increase their energy value to 90% that of corn silage.

Lowering the cost of treating crop residues to increase digestibility is the primary area which must be researched before these procedures can be applied in practical operations.

WINTER PASTURE

Where feasible, winter pasture offers cattlemen a means of reducing costs. By accumulating the feed in the field, rather than harvesting, storing, and handling the forage, the cost and labor of winter feeding can be substantially reduced. Also, costs of bedding and manure hauling can be eliminated.

Tall fescue is used as a winter pasture in the area to which it is adapted—Missouri, Illinois, Indiana, and Ohio. Usually, the new regrowth is baled in late June into round bales and left in the field. The round bales shed rain and snow and, together with the regrowth, make excellent late fall and winter grazing. Experience shows that field-stored forage has adequate quality to maintain beef cows in good condition.

Fig. 19-11. Cows on winter fescue pasture, supplemented with round bales of fescue harvested the previous June and left in the field. (Courtesy, University of Illinois, Urbana, Ill.)

Fescue is a cool season grass; it actually grows some during the winter in the area to which it is adapted; and it is more palatable during the fall and winter than any other season because of the high concentration of soluble sugars. Trampling during the fall, winter, and spring months does not injure the turf.

The Ohio Station reported that tall fescue winter pasture—including both standing growth and baled hay—carried 2 cows per acre for a 4-month period. The use of electric fence to strip graze the bales and regrowth increases carrying capacity by 50 to 60 percent over permitting the herd access to the entire field.

RANGE CATTLE SUPPLEMENTATION

Improved range should be the first goal of cattlemen, without using supplemental feeding as a substitute for good grass or as a crutch for poor range. Instead, the two—good range and proper supplemental feeding—go hand in hand.

Range Nutrient Deficiencies

Growing grasses provide adequate nutrients for beef cattle in **unforced** production when (1) produced on fertile soils, (2) available in sufficient quantities, (3) not washy, and (4) not weathered, leached, or bleached. However, the simultaneous fulfillment of all these conditions is the exception, rather than the rule. Every cattleman worthy of the name, forces his young stock for an early market; most soils are deficient in certain nutrients, which, in turn, affect the plants and the animals feeding thereon; during droughts and early and late in the season, feed may be in short supply (thereby limiting energy and other nutrients); early spring pastures are washy and lacking in energy; during droughts and late in the season, grasses become mature, leached and bleached—they increase in fiber and decrease in protein, phosphorus, and carotene. To meet these conditions, a supplemental source of energy, protein, phosphorus, and vitamin A is necessary.

ENERGY

Hunger, due to lack of feed, is the most common deficiency on the western range. Thus, the first and most important range pasture need is that there be sufficient feed for the animal—to provide the necessary energy required to maintain the body. Over and above these needs, any surplus energy may be used for growth, reproduction, and conditioning.

With bulky, low quality roughages—such as dry grass cured on the stalk, common to drought periods and late in the season—animals may not be able to consume sufficient quantities to meet their energy needs; and the younger the animal the more acute this problem. Also, very early spring grass is washy and lacking in energy. Under such energy-deficient circumstances, the maintenance needs of animals are met by the breakdown of body tissues, accompanied by the observed loss in weight and condition and the failure of young animals to grow. Also, reproduction is adversely affected; serious underfeeding results in the failure of some cows to show heat, more services per conception, lowered calf crops, and lightweight calves. Supplemental feeding is the practical way in which to alleviate such energy deficiencies.

PROTEIN

There is adequate evidence that a deficiency of protein results in depressed appetite, poor growth, loss of weight, reduced milk production, irregular estrus, and lowered calf crops.

The protein allowance of beef cattle, regardless of age, should be ample to replace the daily breakdown of tissues of the body, including the growth of hair, horns, and hoofs. In general, this need is most critical for the growth of the young calf and for the gestating-lactating cow.

Mature, weathered native range grass is almost always deficient in protein—being as low as 3 percent, or less. Protein leaching losses due to fall and winter rains may range from 37 to 73 percent.

Because protein supplements ordinarily cost more per ton than grains, beef cattle should not be fed larger quantities of them than are actually needed to balance the ration. But the temptation is to feed too little of them. When on mature, weathered grass, cows should receive about two pounds of concentrate supplement daily—the exact amount depending upon the nutrient content of the supplement and other factors.

The protein and energy requirements are closely interdependent; hence, it follows that energy rather than feed intake should be the dietary component relative to which the nutrient needs are adjusted.

MINERALS

Growth, reproduction, and lactation require adequate minerals. Although the mineral requirements of dry pregnant cows and of lactating cows are much the same everywhere, it is recognized that age and individuality make a difference. Additionally, there are area and feed differences. Thus, the informed cattleman will supply the specific minerals that are deficient in the ration, and in the quantities necessary. Excesses and mineral imbalances will be avoided.

Salt should be available at all times, on the basis of about 25 pounds per range cow annually.

Phosphorus deficiencies are rather common among range beef cattle. A severe phosphorus deficiency will result in depraved appetite, emaciation, retarded growth and development, failure to reach normal adult size, failure to breed regularly, lowered calf crop, lowered milk production, and high death losses.

The New Mexico station reported (1) phosphorus losses in grasses of 49 to 83 percent during the winter period, (2) increased average annual calf production per range cow of 53 pounds through proper mineral supplementation, and (3) that the phosphorus supply should be continous throughout the year, and not limited to the winter months.

Iodine, copper, and cobalt supplements should be added in those areas where deficiencies of these minerals are known to exist.

Other mineral elements are thought to be essential, but the picture is somewhat confused; and new findings clearly indicate that we have reason to be less certain of our mineral recommendations than heretofore. For the latter reason, the judicious use of certain trace minerals may be good insurance, even though not all the requirements are known.

VITAMINS

Under normal conditions, vitamin A is the vitamin most likely to be deficient in cattle rations, because dry, bleached range grass is very low in carotene (the precursor of vitamin A).

Inadequate amounts of vitamin A (carotene) during pregnancy may cause cows to abort or give birth to dead or weak calves. Extreme deficiencies may also impair the ability of cows to conceive. Bulls receiving insufficient vitamin A show a decline in sexual activity and semen quality.

In low sunshine areas, especially during the winter months, it is recommended that vitamin D be added to the ration.

INDIRECT DEFICIENCY LOSSES

In addition to the conditions given above, nutritional deficiencies on the range are accompanied by lowered resistance to parasites and diseases, and increased mortality of both breeding stock and calves. Also, where feed is scarce there are usually increased deaths from toxic plants.

Supplementing Early Spring Range

Turning out on the range when the first blades of green grass appear will usually make for a temporary deficiency of energy, due to (1) washy (high water content) grasses and (2) inadequate forage for the animal to consume. As a result, stockmen are often disappointed by the poor gains made during this period.

If there is good reason why grazing cannot be delayed until there is adequate spring growth, it is recommended that early pastures be supplemented with grass hay or straw (a legume hay will accentuate looseness, which usually exists under such circumstances), preferably placed in a rack; perhaps with an added high energy concentrate.

Supplementing Dry Range

Dry, mature, weathered, bleached grass characterizes (1) drought periods and (2) fall-winter range. Such cured-on-the-stalk grasses are low in energy, protein, carotene, and phosphorus, and perhaps certain other minerals. These deficiencies in range plants become more acute following frost and increase in severity as winter advances. This explains the often severe shrinkage encountered on the range following the first fall freeze.

In addition to the deficiencies which normally characterize whatever plants are available, dry ranges may be plagued by a short supply of feed.

Generally speaking, a concentrated type of supplement is best used during droughts or on fall-winter ranges. However, when there is an acute shortage of forage, hay or other roughage may be used with or without a concentrate.

SUPPLEMENTING THE PUREBRED HERD ON PASTURE

Purebred cattle breeders and commercial cattlemen, alike, recognize the following as important profit factors: (1) percent calf crop, (2) weaning weight, and (3) number of calves a cow produces in her lifetime. But, to be successful, the purebred man must go further. He must also give attention to (1) saleability—sleek, bloomy, well-conditioned animals attract buyers and sell better; and (2) maximum development of genetic potential in characteristics of economic importance (rate of gain, feed efficiency, etc.), otherwise he cannot make intelligent selections and make breed progress.

Further, the acid test of the competence of the purebred breeder is that these objectives shall be achieved without jeopardizing the breeding performance of the animal—and that's not easy.

Good Pastures and Good Purebreds Go Together

Pastures are the foundation of successful purebred production. It has been well said that good purebred breeders, good pastures that are properly supplemented, and good cattle go hand in hand.

Fig. 19-12. Good purebred breeders, good cattle, and good pastures go hand in hand. Picture shows a purebred South Devon cow and calf on summer pasture in Minnesota. (Courtesy, Arthur V. Palmer, Big Beef Hybrids International Company, Stillwater, Minn.)

Under the old system of moderate growth, good pastures produced on well-fertilized soils met the nutritive requirements of most beef cattle. However, the nutritive requirements of animals—especially purebreds—have become more critical in recent years. Today, more and more purebreds are being production tested. Among other traits, they are being selected for more rapid and efficient gains—they are in forced production. Consequently, their nutritive requirements are more critical—especially from the standpoints of energy, protein, minerals, and vitamins.

Eye Appeal Important in Purebreds

Successful purebred breeders recognize the importance of maintaining animals in good condition and attractive surroundings where they may be seen and admired by potential buyers. Certainly, a lush pasture is ideal from the standpoint of presenting beef cattle. Obtaining proper condition without overfatness and lowered reproduction is not easy. However, a combination of outdoor exercise, good grass, and proper supplemental feeding is the answer. Above all, replacement heifers should not be ruined by overfatness; it's expensive advertising and poor public relations.

Fig. 19-13. Eye appealing registered Herefords on Tequesquite Ranch, Albert, New Mexico. (Courtesy, T. E. Mitchell & Son, Tequesquite Ranch, Albert, N.M.)

Basic Considerations in Supplementing Purebreds on Pasture

First and foremost, it must be recognized that, no matter how excellent pastures may be, they are roughages and not concentrates. Therefore, for the purebred herd, judicious supplemental feeding on grass may be warranted and profitable. In doing so, the following "to do" list is noteworthy:

1. *Practice economy, but avoid false economy*—Although the ration should be as economical as possible, condition and results are the primary objective, even at somewhat added expense.

2. *Supplement to achieve a balanced ration*—The supplement(s) should balance out the available grass, keeping in mind the varied requirements of (a) cattle of different ages, (b) lactating vs. dry animals, and (c) other differences.

3. *Give an assist to early pastures*—If at all possible, allow 6 to 8 inches of growth before turning to pasture. If there are good reasons to turn out earlier, remember that young, tender grass is washy (high in moisture) and low in energy. Thus, it should be supplemented with carbohydrate feeds, preferably grass hay plus high energy concentrates.

4. *Increase carrying capacity*—Supplemental feeding offers a practical way to step up the carrying capacity of pastures, especially prior to and following the period of peak pasture growth.

5. *Extend the grazing season*—Since there is no finer place for pure-

breds, the pasture season should be extended as much as possible; year-round grazing being the ultimate where the climate will permit it. Experiments show that it is possible in many areas to lengthen the grazing season by nearly two months by supplemental feeding on pasture.

6. *Prevent overgrazing or undergrazing*—Pastures should never be grazed more closely than 2 to 3 inches during the growing season, and 3 to 5 inches should be left for winter cover. Also, undergrazing should be avoided because (a) mature forage is unpalatable and of lower nutritive value, (b) tall-growing plants may drive out low-growing plants by shading, and (c) weeds, brush, and coarse grasses are apt to gain a foothold. Supplemental feeding is the answer to governing this situation.

Separating Purebreds by Classes and Ages

It is important that different classes and ages of purebred cattle be kept in separate pastures and supplemented according to needs. The following groups should be sorted out:

1. *Heifers*—Because they are growing, heifers require more liberal feeding, especially from the standpoint of energy and proteins. However, this word of caution is in order: *Do not overfeed heifers* to the point that reproduction is adversely affected. A combination of grass, exercise, and proper supplementation offers the ideal way in which to condition and grow heifers.

2. *Pregnant cows*—Pregnant cows should be in healthy, vigorous condition. Mature cows that are overfat—such as show cows—should be let down through a happy combination of exercise and limited, but balanced, rations.

Most of the ration of pregnant cows should consist of pasture plus such supplements as required—with emphasis on proteins, minerals, and vitamin A; and the kind and level of supplementation should be varied according to the quality and quantity of grass available.

3. *Lactating cows*—Immediately before and after calving, cows should be fed lightly and with laxative feeds. At this time, the amount of supplementation should be governed by the milk flow, the condition of the udder, the demands of the calf, and the appetite and condition of the cow.

Until weaning time, the growth of the calf is chiefly determined by the amount of milk available from its dam, plus whatever assist is given through creep feeding. With purebreds, where maximum early development of the calf is so important, it is generally good business to give the mother the proper feed for production of milk. A combination of pasture plus supplement is the practical way to bring this about.

End Result in Purebreds Important

In purebreds, the end result is all-important, even at somewhat added expense. In addition to supplying proper balance of nutrients, the supplement should provide variety and palatability. Also, it should not make for bloat, scours, or other digestive disturbances.

PASTURE AND RANGE SUPPLEMENTATION

Where dried grass cured on the stalk is grazed, or where insufficient pasture is available—perhaps due to drought or overstocking—supplemental feeding is necessary. Also, supplemental feeding is a way in which to extend the grazing season, both early and late.

Sorting Pasture and Range Cattle

When supplemental feeding is planned, it is strongly recommended that cattle first be sorted by age and condition groups. Heifers should not be supplemented at the same levels as older cattle. They are growing; thus they must be fed more liberally (see Fig. 19-3). Also, heifers have need for more protein (see Fig. 19-4), and they must be fed for a longer period. Likewise, thin cows should be placed where they can be given extra feed and special care. More energy, proteins, and minerals, are required for a cow suckling a calf than for a pregnant or dry cow. The nutritional requirement of cows nursing calves is approximately 50 percent higher than for pregnant cows.

How to Choose a Pasture or Range Supplement

Every cowman faces the question of what supplement to use, when to feed, and how much to feed under his conditions.

In supplying a supplement to range cattle, the following requisites should be observed:

1. It should balance the diet of the animals to which it is fed, which means that it should supply all the nutrients needed by the animal which are missing in the forage.

2. It should be fed in such a way that each animal gets its proper portion.

3. It should be fed in a form that is convenient and practical from the standpoint of the feeder, and that will least disturb the animal.

The net profit resulting from the use of the supplement, rather than the cost per ton, should determine the choice of the supplement. This philosophy is the same as that which normally prevails in livestock marketing; where net receipts, rather than charges for marketing services, should be the determining factor.

Types and Systems of Pasture and Range Supplementation

There is no one best and most practical pasture or range supplement for any and all conditions. Many different feeds may be, and are, used; among them (1) ranch- or locally-produced hay, (2) alfalfa pellets or cubes, with or without fortification, and (3) supplements of various kinds.

Also, cattlemen can lessen the labor attendant to the daily feeding of a pasture or range supplement by (1) hand-feeding cubes at intervals, rather than daily, (2) use of protein blocks, (3) use of liquid protein supplements, or (4) self-feeding salt-feed mixtures. Where these feeding systems do not result in the neglect of the herd, there is no effect upon the health and weight of the cows, percent calf crop, or weaning weight of calves.

RANGE CUBES OR PELLETS

Traditionally, cattle have been supplemented either once or twice daily on pasture or range. Where this practice is followed, a urea-containing range cube or pellet, similar to the formulation shown in Table 19-14, may be used. Cubes may be scattered on the ground.

Fig. 19-14. Range cubes fed on pasture or range. Many cattlemen prefer this method of supplementation, primarily for reasons of convenience and reducing losses from wind blowing. (Courtesy, Ralston Purina Company, St. Louis, Mo.)

Urea-containing supplements, particularly those containing high levels of urea, should not be fed at intervals on the range because (1) range forages are relatively low in energy, and (2) urea is extremely soluble and its nitrogen becomes available very quickly in the rumen. Where nonprotein nitrogen is used in a range cube or pellet, a slow-release product is safest.

TABLE 19-14

RANGE CUBE OR PELLET, WITH UREA

Ingredient	Percent	Per Ton
	(%)	(lb)
Alfalfa meal	15.0	300
Soybean, cottonseed, linseed[1]		
and/or peanut meal	32.5	650
Urea, 45% grade	4.0	80
Corn, barley, wheat, oats, and/or		
milo	34.7	694
Molasses	10.0	200
Salt	1.0	20
Dical., or equivalent	2.0	40
Trace minerals	.5	10
Vitamin A[2] (30,000 IU/gram potency)	.3	6
TOTAL	100.0	2,000

Calculated Analysis:

	(%)
Crude protein[3]	32.2
Fat	2.0
Fiber	6.4
Calcium	.9
Phosphorus	.8
TDN	67.2

[1]If linseed is used, limit to 6% of the ration.
[2]In low sunshine areas, also add 6 million IU of vitamin D/ton of finished feed.
[3]This includes not more than 11.24% equivalent protein from nonprotein nitrogen; 34.9% of the total protein is furnished by urea.

HAND-FEEDING AT INTERVALS, RATHER THAN DAILY

The Texas Station compared (1) daily feeding, (2) twice weekly feeding, and (3) three times weekly feeding; with each group hand-fed. Cottonseed meal was used as the supplement at two levels: (1) 14 pounds/head/week, and (2) 21 pounds/head/week.

As a result of a 4-year study,[9] it was reported that—

1. The group fed twice weekly had a slight advantage over feeding daily and 3 times weekly in (a) weight change of cows, (b) percent calf crop weaned, and (c) weaning weight of calves.

2. Twice weekly feeding saved 60 percent in time and equipment over daily feeding.

3. Twice weekly feeding did not cause any digestive disturbances, even when fed 10½ lb at one time (one-half the allowance of the 21 lb/week level). It took these cows about 2 hours to consume their 10½ lb share of cottonseed cake, which gave the timid and the slow eaters an opportunity to get their share.

4. The cows fed twice weekly grazed more widely over their range than those fed daily.

Based on the above study, plus observations and experiences, the author

---

[9]*The Hereford Journal,* Jan. 1, 1966, pp. 62, 63, and 150.

recommends feeding nonurea range supplement twice weekly; allocating in each of the two feedings one-half as much supplement as would have been fed in a week on a daily feeding basis.

Protein cubes may be scattered on the ground—two or three times a week. This offers a method of checking the animals because they are attracted by the sight or sound of the vehicle when they know that there is something to eat.

A suggested pasture-range supplement without urea is given in Table 19-15.

TABLE 19-15

RANGE CUBE OR PELLET, WITHOUT UREA

Ingredient	Percent	Per Ton
	(%)	(lb)
Soybean or cottonseed meal (41%)	72.7	1,454
Alfalfa meal	15.0	300
Molasses	8.5	170
Salt	1.0	20
Dical., or equivalent	2.0	40
Trace minerals	.5	10
Vitamin A[1] (30,000 IU/gram potency)	.3	6
TOTAL	100.0	2,000

Calculated Analysis:

	(%)
Crude protein	32.6
Fat	1.0
Fiber	8.3
Calcium	.9
Phosphorus	.9
TDN	70.9

[1]In low sunshine areas, also add 6 million IU of vitamin D/ton of finished feed.

Twice weekly feeding has two distinct advantages over the use of salt-feed mixes; (1) It alleviates the cost of using excess salt, which has no nutritive value when so used; and (2) it forces inspection of the herd two times per week, which is as infrequent as is desirable.

PROTEIN BLOCKS

Protein blocks are just what the designation implies. They are compressed protein blocks, generally weighing from 30 to 50 pounds each.

Blocks may be placed in grazing areas where cattle have frequent access to them, with one block provided to 15 cows. Intake will vary with the feed supply and the type of block. Generally, it is planned to limit feed consumption to about two pounds per head per day by hardness of block and salt and/or fat content.

Fig. 19-15. Protein block in use on pasture—a means of lessening the labor attendant to the daily feeding of a protein supplement on pasture or range. (Courtesy, Moorman Manufacturing Co., Quincy, Ill.)

LIQUID PROTEIN SUPPLEMENTS

Liquid supplements in a "lick" tank can be offered free-choice. This is a convenient and satisfactory way in which to supply protein, energy, and other nutrients, so long as the cattle do not consume more than they need.

In the United States, the vast majority of the liquid protein supplements are patented. Also, they are difficult to home mix. As a result, the universities have done little experimental work on them. Moreover, most such supplements are fed to finishing (fattening) cattle in drylot; only a small quantity is fed to dairy cattle, and relatively little is used to supplement dry grass for cow-calf operations. However, Australia has long fed a great deal of molasses-urea on the range, due to its frequent droughts. Also, daily feeding on its vast stations (ranches), many of which stretch over many miles, isn't practical—it requires too much labor and travel. Hence, the Australians are more experienced than we are when it comes to using molasses-urea to supplement grass, and as a means of saving labor. Accordingly, the author has come to know their formulations and techniques, firsthand, in three visitations that he has made "down under." Their feeding directions appear to be about as follows:

1. Do not feed the urea-containing supplement to cattle that have been without feed for 36 hours until they have had a chance to fill the rumen with grass.

2. Since some animals tend to overeat, it is desirable to restrict consumption to a desired level (1 to 2 lb/head/day) by physical means. This may be accomplished by a free-turning plastic or wood wheel dipped in a tank or other similar equipment.

Fig. 19-16. Liquid protein supplement free-choice fed in a "lick" tank, with consumption limited by a tongue-turning plastic wheel; used by Sr. Guillermo Osuna, Infante Ranch, Muzquiz, Coahuila, Mexico. (Photo by Audrey Ensminger)

3. It may take several days for cattle on dry grass to get accustomed to a liquid protein supplement.

4. Once you start feeding the liquid protein supplement, never let the animals run out of it until you discontinue feeding it entirely.

5. Self-feed regular minerals in addition.

SELF-FEEDING SALT-FEED MIXTURES

The practice of using salt as a governor to limit feed consumption on pasture or range has been around a very long time. It was ushered in as a laborsaving device for cattle and sheep in inaccessible and rough areas.

The proportion of salt to feed may vary anywhere from 5 to 40% (with 30 to 33$\frac{1}{3}$% salt content being most common), with the actual intake of feed supplement limited to 1 to 2½ pounds daily. By varying the proportion of salt in the mixture, it is possible to hold the consumption of feed supplement to any level desired. In some range areas, a reduction of the salt level from 33$\frac{1}{3}$ to 24% will increase consumption by about 50%. When a liberal feeding of grain on pasture is desired, 5% salt may be sufficient.

Two suggested salt-meal supplements follow:

SALT-COTTONSEED OR SOYBEAN MEAL, 41% (Do Not Pellet)

1 part salt-2 parts meal (plus 4 lb vitamin A, 30,000 IU/gram potency, per ton of mix)—Guarantee: 27% crude protein, maximum of 35% salt, and 18,000 IU of vitamin A per pound. Cattle will generally consume about 1½ lb daily of such a mix.

1 part salt-3 parts meal (plus 3 lb vitamin A, 30,000 IU/gram potency, per ton of mix)—Guarantee: 30% crude protein, maximum of 27% salt, and 13,000 IU of vitamin A per pound. Cattle will generally consume about 2 lb daily of such a mix.

Fig. 19-17. Cattle self-fed a salt-feed mixture, with consumption governed by
the amount of salt; on Infante Ranch, Muzquiz, Coahuila, Mexico, owned by Sr.
Guillermo Osuna. (Photo by Audrey Ensminger)

Based on experiments and experiences, the following points are pertinent to self-feeding salt-feed mixtures to range cattle:

1. The practice need not be limited to any specific protein supplement
or feed.

2. It is best that salt mixes be in meal form, rather than pelleted. If
pellets are small and soft, they will work satisfactorily. However, there is
always the hazard that they will be hard enough to permit cows to swallow
them without the salt being fully effective as an inhibitor, with the resulting
overeating.

3. The proportion of salt and feed may vary anywhere from 5 to 40%
(with 30 to 33⅓% salt content being most common), with the actual intake of
feed supplement limited to 1 to 2½ lb daily. By varying the proportion of salt
in the mixture, it is possible to hold the consumption of feed supplement to
any level desired. In some range areas, a reduction of the salt level from
33⅓ to 24% will increase consumption by about 50%. When a liberal feeding of grain on pasture is desired, 5% salt may be sufficient.[10]

4. The quantity of salt and the proportion of salt to supplement required
to govern supplement consumption varies according to (a) the daily rate of
feed consumption desired, (b) the age and weight of animals (higher quantities of salt are required in the case of older animals), (c) the fineness of the
salt grind (fine grinding lowers the salt requirement), (d) the salinity of the
water, (e) the severity of the weather, (f) the quality and quantity of forage,

---

[10]At the Irrigation Experiment Station, Prosser, Wash., it was found that 7½% salt limited
grain consumption by yearling steers on pasture to 10 to 12 lb daily; 5% salt limited grain
consumption to 12 to 14 lb.

and (g) the length of the feeding period (as animals become accustomed to the mixture, it may be necessary to increase the proportion of salt).

5. It is common practice to prepare the starting feed by mixing 1 lb of salt to 4 lb of feed supplement, and to increase the proportion of salt in the mixture as the animals become accustomed to the feed.

6. It lessens the difficulty in starting cattle on a supplement, for sprinkling a little salt on the meal makes it more palatable.

7. It is recommended that animals be hand-fed a week or so before allowing free-choice to a salt-feed mixture; thus, getting them on feed gradually.

8. It is necessary to regulate or limit (by hand-feeding for a few days) the supply of salt-feed mixture when it is desired to shift animals from a straight feed supplement (such as cottonseed meal alone) to a salt-feed mixture. Otherwise, hungry animals may consume too much.

9. It is estimated that the practice increases the total salt consumption to 8 to 10 times that required in conventional salt feeding, and doubles or triples the water consumption.

10. If the salt-feed mixture is placed in close proximity to the water supply, it will make for restricted grazing distribution on the range, because of the greater intake of water on a high salt diet. On the other hand, if the salt-feed mixture is shifted about on the range, it will make for desirable distribution, because of the animals following the feed supply.

11. It reduces the labor required in feeding, promotes more uniform feed consumption (among the greedy and the timid), and permits animals to eat at their leisure with less disturbance during blizzards or cold weather.

12. It lessens the space required for feed equipment (bunks or feeders) to 20% of that required in conventional hand-feeding, but makes it desirable that the feeder be constructed so as to protect the mixture from the weather (especially wind and rain).

13. It is equally applicable to feeding during droughts, on dry summer range, and in the winter months.

14. It is commonly believed that under conditions of short feed supply (submaintenance) and relatively inaccessible water supply, animals may consume sufficient salt in this manner to produce toxic effects, especially during the winter months when low temperatures tend to lessen the water intake.

15. The practice of self-feeding salt-feed mixtures is well adapted to inaccessible and rougher areas, where daily feeding is difficult. In no case, however, should it be an excuse to neglect animals, for herds need to be checked often.

16. It reduces the consumption of minerals other than salt to practically nothing, with the result that mineral deficiencies must be considered.

FATTEN CULL COWS FOR MARKET

There is a new profit potential in the cow business—that of fattening cull cows for market. Where herd numbers are being held fairly constant, about 20 percent of the cows are culled each year, because of (1) poor calves, (2) being barren, (3) spoiled udders, (4) disease, (5) old age, and (6) miscellaneous reasons. Traditionally, these culls were sent to market—and the sooner the better. Today, good money can be made by holding and fattening these culls prior to slaughter.

Cutter and canner cows are in demand; and the price spread between them and Choice steers has narrowed, both on foot and on the rail—especially the latter. The reasons: (1) the increased demand for hamburger and other ground and prepared meats, for which cow beef is admirably suited; (2) fewer cows being slaughtered, because decrease in dairy cows has slowed and beef cow numbers are expanding; (3) imports of manufacturing-type beef having held at rather stable levels; and (4) better quality cow carcasses available than formerly.

All this suggests the following to cow-calf producers:

1. More than ever, they cannot afford to keep marginal or barren cows. They should keep records, perform pregnancy tests, and remove loafers from the herd and fatten them for slaughter.

2. They should fatten cull cows on cheap, high roughage rations, using a maximum of such feeds as silage, haylage, green chop, wheat pasture, irrigated pasture, etc. If they have good teeth and they're healthy, they will make remarkable gains.

3. They should compare selling on foot vs selling on the rail. The latter may be best.

Also, buying cull cows and finishing them commercially offers good opportunity, but numbers are limited. But buying such cows is a problem. They're scattered; and it is important that they have good teeth, and that they be healthy. Remember, too, that cows are usually culled because of feed shortages, breeding problems, disease, internal parasites, or age. In short, somebody had a problem with them.

FACILITIES

The investment in beef cattle buildings and equipment should be kept to a minimum. The farmstead should be neat and attractive, particularly where purebred cattle are sold for breeding purposes.

Buildings and yards should be located on a well-drained area; and plans should be made for the efficient feeding and management of the herd.

● *Housing*—Housing for beef cows need not be elaborate or expensive. Allow the herd to be outside during the grazing season and most other times—even in winter. Beef cattle naturally grow long thick hair coats in the fall. Except where it is extremely cold and windy, the most shelter they need is a wooded area or a hill for a windbreak. Generally speaking, the cattleman should give more attention to feed storage and to saving labor in feeding and manure handling than to the necessity of getting the cows inside. Also, it should be recognized that a combination of drafts, dampness, poor ventilation, and lack of sunlight create hazards.

If a shed or barn is used, have it open to the south or east (away from the direction of prevailing winds), with an adjoining lot to permit them to stay indoors or run out at will. Mature brood cows require 50 to 60 square feet of shelter, yearlings 35 to 40 square feet, and weaned calves 25 to 30 square feet. If the cattle are fed roughage under the shelter, more space will be needed—60 to 75 square feet per cow, and a little more for young stock than the figures given.

Sheds more than 20 feet deep are preferred. The greater the depth, the warmer and drier the building.

Pole-type barns or sheds are excellent for beef cattle. They should be built on high ground, so that there will be good drainage away from them; they should have dirt floors; and they should be built high enough for convenience in removing manure—10 feet from the floor to the plate is sufficient.

Pole-type barns can be arranged so that roughage can be stored in the back of the shed, and so that, by use of movable racks, the cattle are permitted to eat their way back during the winter.

• *Fencing*—All corral fences should be built of heavy board material, pole, or rail type construction.

For holding cattle, 3 or 4 barbed wires are sufficient. Four wires should be used along roads and boundary lines, and 3 wires are sufficient for other areas. Woven wire is satisfactory, but more expensive. The life of the fence depends to a great extent on the life of the posts and the stability of the corner posts.

Electric fences are satisfactory for temporary fencing or rotation grazing.

• *Cattle-handling facilities*—Lack of adequate facilities for handling beef cattle prevents cattlemen from carrying out practices which would otherwise be routine and would increase their returns from the beef cattle operation. Time and money spent in planning and developing handling facilities for cattle will return dividends in terms of added profits and greater efficiency. Cattle-handling facilities should include the following:

1. A permanent corral or holding pen located near the main livestock buildings, with a working alley and several attached smaller catch pens to help in sorting cattle.

2. A headgate, with a chute leading thereto.

3. A portable corral for use in pastures that are considerable distance from headquarters, or a permanent corral constructed in such pastures.

4. A stationary or portable loading chute.

5. A scale, or ready access to a scale.

6. Cattle stocks where cattle are fitted for shows and sales, and for use in trimming feet.

• *Other buildings and equipment*—Other types of buildings and equipment for the beef cattle establishment are covered in Chapter 11, Buildings and Equipment for Beef Cattle, of this book; hence, the reader is referred thereto.

QUESTIONS FOR STUDY AND DISCUSSION

1. How may feed costs, which account for 65 to 75 percent of the total cost of keeping cows, be lowered?

2. How may a practical cattleman meet the added energy requirements of a brood cow during the critical 100 days, extending from 30 days before calving until 70 days after calving?

3. Compare and discuss the nutritive requirements of (a) dry pregnant cows, (b) bred heifers, and (c) lactating cows. From a practical standpoint does this mean that a cattleman should separate different classes and ages of cattle, then feed them according to needs?

4. Cite experimental studies showing reproductive failure in cows due to a deficiency of each of the following: (a) energy, (b) protein, (c) phosphorus, and (d) vitamin A. How would you rectify each of these deficiencies?

5. It is estimated that the United States annual production of corn residue totals 157 million tons, sufficient to winter 118 million pregnant cows. Yet, much of this potential cow feed is not utilized. Why is not more corn residue fed to cattle?

6. List and discuss the feeding value of crop residues, other than corn residue, that can be used for feeding cows.

7. Discuss nutrient deficiencies of cattle that are frequently encountered on United States ranges. How would you rectify each one?

8. Wherein does the supplementation of purebreds on pastures and ranges differ from the supplementation of commercial cattle?

9. What type and system of pasture and range supplementation would you recommend? Justify your answer.

10. Considering current feed and fat cow prices, present figures to show whether or not it will pay to fatten cull cows for market.

11. Select a particular area, then determine what facilities would be necessary and practical for the operation of a 300-cow herd.

SELECTED REFERENCES

Title of Publication	Author(s)	Publisher
Beef Cattle, Sixth Edition	A. L. Neumann R. R. Snapp	John Wiley & Sons, Inc., New York, N.Y., 1969
Beef Cattle Production	K. A. Wagnon R. Albaugh G. H. Hart	The Macmillan Company, New York, N.Y., 1960
Beef Cattle Science Handbook	Ed. by M. E. Ensminger	Agriservices Foundation, Clovis Calif., pub. annually since 1964
Beef Production in the South	S. H. Fowler	The Interstate Printers & Publishers, Inc., Danville, Ill., 1969
Commercial Beef Cattle Production	C. C. O'Mary I. A. Dyer	Lea & Febiger, Philadelphia, Penn., 1972
Stockman's Handbook, The, Fourth Edition	M. E. Ensminger	The Interstate Printers & Publishers, Inc., Danville, Ill., 1970

BULLS

Contents

There are an estimated 2 million beef bulls of breeding age in the United States, and about 800,000 new bulls are selected each year. But numbers alone do not portray their importance. They are the most influential part of any beef cattle breeding program. Their selection can make or break any herd, either purebred or commercial. Since a much smaller proportion of males than of females is normally saved for replacements, it follows that selection among males can be more rigorous and that most of the genetic progress in a herd will be made from selection of males. Thus, if 2% of the males and 50% of the females in a given herd become parents, about 75% of the hereditary gain from selection will result from the selection of males and 25% from the selection of females, provided their generation lengths are equal. If the generation lengths of males are shorter than the generation lengths of females, as is true in most beef herds, the proportion of hereditary gain due to the selection of males will be even greater. Thus, it is estimated that between 80% and 90% of the progress realized in most beef herds comes about through the bulls used. There is little doubt, therefore, that proper bull selection is the most effective way in which to make genetic progress in a beef herd.

Fig. 20-1. W.S.U. Cornerstone 704 X, impressive Polled Shorthorn bull, bred by Washington State University, shown in breeding condition on the John F. Shuman & Sons' Ranch, Byers, Colorado. When fitted, this bull weighed 2,300 lb. (Courtesy, American Shorthorn Association, Omaha, Neb.)

The importance of bull selection is sometimes confused by the statement that the bull is half the herd. This is true in the sense that half of the genetic material making up the calf crop comes from the sires used. However, each bull's contribution to the calf crop is 30 times greater than the average cow if we assume 30 cows per bull and a 100% calf crop from the cow herd.

Also, it is noteworthy that the bull's influence extends beyond the immediate calf crop in that his daughters, and perhaps sons, too, remain in production in the herd long after he is gone—they're constant "reminders."

Another factor which causes the bull to play such an important role in herd improvement is the much greater selection differential that can be realized in selecting bulls compared to selecting females. This means that the bulls selected can be much better than the herd average compared to the heifer replacements selected. This is primarily because of the relative numbers of each that are required as replacements.

SOURCE OF BULLS

Generally speaking, the cattleman has three sources of herd sires: (1) purchase; (2) raising his own; or (3) artificial insemination service. Additionally, there is a choice of sources when it comes to purchase or A.I. service.

Purchase of Bulls

Bulls may be purchased from a breeder, at a central test station, or at a consignment sale.

• *Purchase from a breeder*—Purchase directly from a breeder, either at private treaty or in a production sale, is the most common source of bulls. From the standpoint of the bull buyer, purchase from the breeder has the following advantages: (1) it provides an opportunity to buy a number of bulls of similar breeding, thereby leading to greater uniformity in the herd in which they are used; (2) it makes it possible to select from among animals that have been exposed to the same environment from birth, and that can be compared on the basis of their performance without environmental differences; (3) it is usually possible to inspect their sires, dams, and near relatives; and (4) there is adequate time for evaluation. The only real disadvantage in this method of buying is that there is no opportunity to make direct comparison with cattle bred by other breeders; hence, the bull buyer must be an astute cattleman.

• *Purchase at a central test station sale*—In recent years, there has been a trend toward evaluating bulls in central test stations, followed by an auction sale at the close of the test period. At these stations, bulls are entered for postweaning gain comparisons with bulls from other herds. The chief advantages of purchasing bulls from a test station are: (1) the chance for buyers to get acquainted with performance breeders who are willing to compete with each other on the basis of the performance of their breeding stock when placed in a comparable environment; (2) the opportunity to make gain-on-test comparisons between herds; (3) the likelihood that breeders have sent their best performance prospects to the test station for the promotional benefit, and that purchases can be made therefrom; and (4) the bulls are seldom heavily fitted. The disadvantages are that weight per day of age during the test period is affected to some degree by preweaning environment; hence, valid comparisons are possible only between bulls of the same herd. However, most test stations now require an adjustment period of three to four weeks prior to the actual test as a means of alleviating much of the difference due to compensatory gain.

• *Consignment sales*—Traditionally, consignment sales have been (1) a market place for establishing values back home, (2) a source of revenue of breed associations, (3) a meeting place for purebred breeders and commercial cattlemen, and (4) a stage for creating and promoting change in the kind of cattle produced. Normally, bulls in consignment sales are sifted, graded and/or placed, and those with obvious defects or unsoundnesses are culled.

The chief advantages of a consignment sale from the standpoint of a bull buyer are (1) convenience, and (2) high quality animals, due to (a) sifting, and (b) sellers using them to establish values back home. The disadvantages: (1) frequently bulls sold in consignment sales are overfitted; and (2) they provide little basis to compare growth rates of the consigned bulls, except within breeder groups.

Raising Your Own Bulls

Large purebred cattle establishments commonly raise a good part of their own replacement bulls. Some large commercial establishments also find it desirable to raise their own bulls. The main advantages from raising your own replacements are: (1) the bulls are raised and selected in the environment in which they are expected to perform; (2) management has complete control of the breeding program, with the result that they can provide specific lines and select for certain traits; and (3) the cost is usually less, because it alleviates transportation and selling charges. The disadvantages are that (1) a good commercial cattleman may not necessarily be a good seed stock producer, and (2) it requires more skilled labor, more accurate records, and more facilities and fencing.

Artificial Insemination (A.I.) Bulls

Selection of artificial insemination sires involves essentially the same considerations as the selection of bulls for natural service. Additionally, they are usually progeny tested—their calves have been evaluated for performance. This means that for an A.I. bull it is usually possible to get information on birth weight of his calves, ease of calving and mothering ability of his daughters, and carcass yield and grade of his progeny.

Unfortunately, such complete information is not available on many of the newly imported breeds. Only calves under nine months of age have been eligible for importation through the quarantine procedures and facilities established by the Canadian Department of Agriculture. However, this situation is changing very rapidly as calves born on this continent become eligible for testing and selection.

BULL SELECTION—FACTORS TO CONSIDER

The cow-calf producer is striving for genetic improvement. He is hoping that the bulls which he selects are genetically superior to his cows—that they will transmit their apparent superiority to their offspring, thereby continuing to improve his herd. But how can he be sure?

Although there's no foolproof method through which herd-improving sires can be assured, application of the selection procedure and criteria which follow will come as near to doing so as is scientifically possible today.

Before starting out to buy a new herd sire, evaluate your cow herd and current calf crop. Where do they need the most improvement? Is it size, muscling, soundness, gaining ability, or some other trait? Then look for a bull to correct those weaknesses. But do not lose existing valuable traits to improve a weak one.

Decide which herds you should visit or which sales you should attend. Buy from reputable breeders who are known to be doing a good job of production. Take time in making a selection. Start well ahead of the time you need a bull, so that you will have a better selection of bulls from which to choose.

Many bulls are selected on visual appraisal and age alone. But more and more cattlemen, both purebred and commercial, are selecting sires only after studying detailed performance records on not only the individual bulls but their near relatives. Since sire selection may be the most important factor in determining the success of a beef cattle breeding program, every breeder should use all the information at his disposal when selecting herd sires. At the outset, however, it should be recognized that a performance tested bull will seldom excel in all the economically important traits. The cattleman must decide, therefore, how much importance shall be accorded each trait—how much emphasis will be placed on yearling weight, rate of gain, and carcass evaluation, etc. This will vary from herd to herd, depending primarily on the level of performance in the trait already attained in the particular herd.

For maximum improvement in the herd, the factors which follow should be considered in bull selection.

Breed

The commercial cattleman has a wide germ plasm choice available to him, especially if he can use artificial insemination. Over 300 breeds are available, along with all combinations of crosses. But the choice of the breed(s) or cross is not easy. Available experiment station, U.S. Department of Agriculture, and even field trial data must be interpreted and inferences made by the producer to his own management. Just how well will this new breed fit into beef production in the United States and on the particular farm or ranch? If replacement females are to be retained, care must be taken to evaluate the reproductive and maternal potentials of the breed. Will the breed have a terminal sire market only, or will it be a maternal breed?

Selection of the breed and planning the breeding program must go hand in hand. Remember that grading up takes a long time. Remember, too, that half the heterosis will be lost with each successive back cross—and this may result in disappointment.

Age

The use of well-grown bulls, 14 to 15 months of age, is increasing. Performance tested bulls are generally large enough to use and sexually mature by the end of their test period. When you hold a bull out of service for 10 to 12 months longer than necessary, you've got nothing to show for it but a feed bill.

Where a breeder is making substantial progress in his herd, consideration may well be given to securing bulls out of young sires and young dams even though there is not as much data on which to make a decision.

An older bull that has been proven on the basis of his progeny to be a superior breeding sire, and that is free of reproductive diseases, is usually a sound investment. Such a bull may be available from a neighbor, on an exchange arrangement. Also, proven bulls should be available through artificial insemination.

Herd Merit

The really difficult problem is the selection of the herd or source, once the breed has been decided upon. Whenever possible, bulls should be obtained from herds having the highest average merit. But it is difficult to estimate the average merit of a herd in relation to other herds and the rest of the breed. Performance test comparisons between herds are not very meaningful, for the reason that environmental factors, especially feeding and herdsmanship, exert a powerful influence. However, by determining the production level of the herd and by observing how animals produced in the herd have performed for other breeders, an evaluation can be made.

Of course, not all animals produced in a herd will be breeding bull prospects. So, relative rank of a bull prospect among other bulls produced in the same calf crop and managed in the same way is important. The bull buyer should take the "top cut" of the performance tested bulls from a given, outstanding herd that is on production test. Weight ratios computed in most performance records are a convenient device to determine how an animal ranks among other animals raised at the same time in a herd.

Fig. 20-2. Two-year-old production tested Angus bull. (Courtesy, Martin Jorgensen, Jr., Ideal, S.D.)

Some important questions to ask of a seller of breeding stock as a means of determining the merit of his herd are:

1. How many years has this herd been selected for performance?
2. What is the average level of performance in your herd for the relevant traits?
3. What is your breeding program?
4. What is your management system?

Weaning Weight (205-Day Weight)

A bull's own weaning weight (adjusted for age of dam) is both an indication of his genetic capability for growth and of his dam's mothering (including milking) ability. A weaning weight ratio will indicate how a particular bull's weaning weight compared to the average of other bull calves raised at the same time under the same conditions.

Postweaning Growth Rate

A measure of gain from weaning time to one year of age is a good indication of a bull's genetic capability for gain. The gain period may be in a central test station in comparison with bulls from other herds or it may be on the farm or ranch in comparison with other bulls raised in the same herd. Either test is satisfactory. But the rate of gain itself must be evaluated in light of the feeding level and management conditions.

Adjusted 365-day weight or weight per day of age up to a year are good overall measures of growth rate as they include both weaning weight and postweaning gain, and they are about 60 percent heritable.

Carcass Merit

The two main considerations in carcass merit are (1) quality, and (2) ratio of lean to fat and bone. At the present time, carcass quality can only be determined by slaughtering an animal and measuring the various things that contribute to quality in the carcass. This means that carcass quality information on prospective sires can only be obtained by slaughtering relatives, such as half-sibs from the same sires. Of course, once a bull is selected, he can be progeny tested by slaughtering some of his offspring and obtaining carcass information, but this takes time. Few breeders can supply meaningful records of carcass quality of their herds.

Various methods can be used to estimate muscle content of prospective breeding animals directly. A trained judge can exhibit considerable accuracy by visual appraisal. Fat thickness can be obtained by mechanical means such as probing or by the use of ultrasonic devices. Knowing fat thickness helps in estimating muscle content of animals. Ultrasonic devices can be used to estimate the size of certain muscles (such as the loin eye muscle) which are related to total muscle content of the animal.

The K-40 whole body counter is being used to estimate the total muscle content of live beef cattle. By using this method, prospective breeding animals can be evaluated for muscle content and the better animals selected for breeding use or further testing. The K-40 count is highly related to total muscle content if the variables are fairly constant. Animals to be evaluated and compared should be of about the same age and weight and should come from the same management background. Also, there must be meaningful comparisons between animals. For example, it would do little good to obtain a K-40 count on a single sire in range condition.

The results of K-40 counter-evaluation are usually expressed as percent lean of liveweight, or as lean per day of age. To be good breeding prospects, bulls should grow rapidly to a year of age and still have a high percent lean of liveweight.

Sire's and Dam's Performances

The records of the sire and dam of each prospective herd bull should be evaluated if possible. The dam's record may show her regularity of production, her mothering ability as reflected by the weaning weight of her calves, and the postweaning gain and carcass merit of her calves.

If sires have produced several calves, their records of weaning weight, postweaning gain, yearling weight, and carcass merit should be evaluated, particularly relative to the records of other sires used in the same herd.

Individuality

Only bulls that are better than average should be used for breeding purposes.

The bull being considered for purchase should be well developed for his age, of good type, masculine, and structurally sound. Young bulls should be long, tall, and not excessively fat—indications that they will continue to grow. Bulls that show signs of early sexual maturity are not good prospects for continued rapid growth and large mature size.

The bull should have a strong, straight top line, well-sprung foreribs, and be well muscled. When viewed from the rear, the back and loin should slope gently downward, like a Quonset roof. Extreme flatness over the top indicates excess fatness, and is undesirable. The bull should be deeper in the fore flank than in the rear flank.

Look for a bulging arm, forearm, gaskin, and stifle; a rounded loin and round; and creases in the thighs. There should be a well-defined groove down the top line as a result of the loin eye bulging on each side of the backbone; and you should see the muscles when the animal walks.

If a bull is very smooth, he is likely either too fat or muscle deficient, or both. The bull should be masculine in front and have a sound pair of testicles behind; he should be alert—"on the look," with head up and ears cocked; he should have a bold, masculine head, with reasonably prominent eyes, broad muzzle, and large nostrils; he should have a well-developed crest; he should have well-developed and clearly defined muscles, especially in the regions of the neck, loin, and thigh; and he should have well-developed external genitalia, with testicles of equal size and well defined, and a proper neck to the scrotum.

Avoid heavy, pendulous sheaths, which tend to be a problem in certain breeds. Look for a sheath that is of medium size and neatly attached. It appears that bulls with pendulous sheaths transmit excessive development of sheath (and perhaps navel) to many of their offspring. Excessive sheath development often leads to injury to the sheath lining, and sometimes to the penis, especially when the bull is run on rough, brushy range.

The bull should be able to follow and breed cows on the pasture and range. This means that structural defects—particularly in the legs, feet, and joints—can shorten the productive life of any bull. This calls for structural soundness, especially in the underpinning. The legs should be straight, true, and squarely set. The feet should be large, deep at the heel, and have toes of equal size and shape that point straight ahead. The joints, particularly the hock and knee joints which are subject to great wear, should be correctly set, and without tendency toward puffiness or swelling. Sickle-hocked, post-legged, back at the knees (calf-kneed), or over at the knees (buck-kneed) are faults, and such animals should be scored down.

Where a purebred herd is involved, the bull should be true to breed type and color, factors of importance in merchandising cattle.

Semen and Fertility Evaluation

Semen quality is based on evaluating the ejaculate for (1) density of concentration, (2) rate of movement, (3) motility, and (4) morphology.

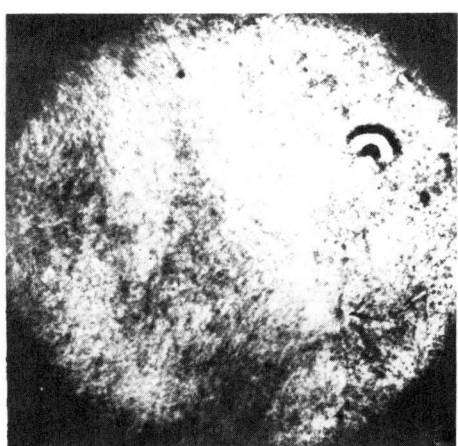 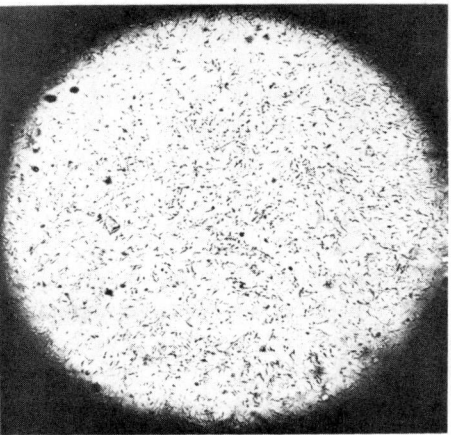

Fig. 20-3. Characteristics of semen in relation to fertility in the bull. Microscopic appearance of semen from two beef bulls (X100).
Left—Highly fertile bull gives semen of high density showing wave motion of active sperms; individual sperms cannot be distinguished.
Right—Low fertile bulls give ejaculates of low density; sperms can be seen individually with sluggish nonprogressive movements.
(Courtesy, Dr. J. A. Laing, University of Bristol, Langford, England)

It is advisable to have a semen or fertility evaluation made before you buy a bull, especially when purchasing for a single sire herd. The penalty for an infertile multisire herd is not so great. Although a semen test will not provide absolute assurance that the bull will settle the females to which he is exposed, it is a strong indicator. Bulls that are not producing sperm cells, that are producing a high percentage of nonmotile or abnormal sperm cells, and those that have infections in the reproductive tract should be avoided. No matter how high the growth rate, and no matter how admirable all the other qualities, a bull that cannot sire a calf crop is useless.

Pedigree

In addition to recording lineage, pedigrees are useful in selecting against undesirable recessive genes and such traits as vaginal or uterine prolapse.

The pedigree of a bull is not as important as it was once believed. When verbal descriptions were in vogue, names and numbers were sufficient. Today, selection pressure is being directed toward improving highly heritable traits, for which the bull's own performance is a good indicator of breeding value. Hence, the pedigree is less important.

In addition to identification of immediate ancestors, performance pedigrees, which list performance records and ratios of all immediate ancestors, are coming into use. Such pedigrees are an important aid to accurate sire selection and should be used when available.

Although the format varies, more and more pedigrees contain information like the one in Fig. 20-4.

Hereditary Defects (Undesirable Recessives)

Bull buyers should always be on the alert for hereditary defects. Even though the bull selected shows no evidence of defects, he may be a "carrier" if his relatives have a history of such defects. Among the more than 200 hereditary abnormalities in cattle to guard against are: dwarfism, double muscling, hernia, cryptorchidism, hydrocephalus, and arthrogryposis. Cattlemen commonly refer to such defects as "undesirable recessives" or "freaks." Such traits range all the way from lethals, which cause the death of the affected animal, to those that cause only an economic loss.

In addition to the more simply inherited hereditary defects, there is now ample evidence that the predisposition to such undesirable characteristics as bloat, vaginal and uterine prolapse, and cancer eye is also heritable. Hence, one should not buy a bull whose relatives exhibit a high incidence of these problems.

Price

Once you have found him, what's a good bull worth?

One commercial cow-calf producer of the author's acquaintance pays an average of $1,800 for his bulls. He weans a 97 percent calf crop, with an average calf weight of 590 pounds, and his reputation calves bring $2.00/cwt premium. Currently, he's netting $86 per cow. And that's not all! After this commercial cattleman is finished with his bulls, he sells them for an average of $1,200 per head. Many of them go into purebred herds, at a premium price because they have been progeny tested. This cattleman's success story can be attributed to 2 things: (1) the use of outstanding sires; and (2) the production of reputation calves.

Cowboy arithmetic shows that a cattleman can afford to pay as much as $1,250 more for a good bull than for an average bull. Here is how most producers figure the situation: Calves sired by a superior bull will usually aver-

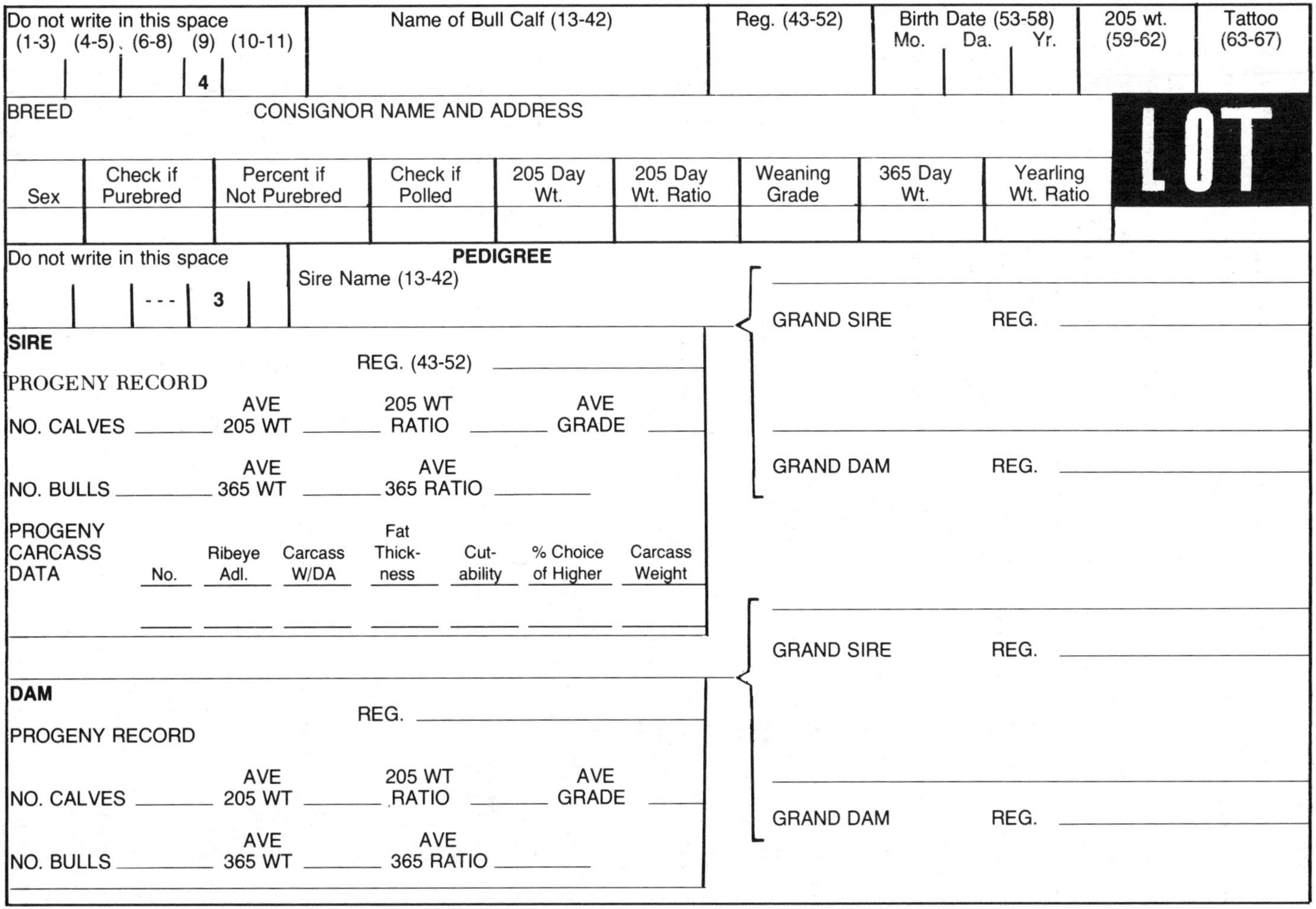

BREED

Do not write in this space (1-3) (4-5) (6-8) (9) (10-11) 4

CONSIGNOR NAME AND ADDRESS

Name of Bull Calf (13-42)

Reg. (43-52)

Birth Date (53-58) Mo. Da. Yr.

205 wt. (59-62)

Tattoo (63-67)

LOT

Sex	Check if Purebred	Percent if Not Purebred	Check if Polled	205 Day Wt.	205 Day Wt. Ratio	Weaning Grade	365 Day Wt.	Yearling Wt. Ratio

Do not write in this space - - - 3

PEDIGREE

Sire Name (13-42)

SIRE REG. (43-52)

PROGENY RECORD

NO. CALVES _____ AVE 205 WT _____ 205 WT RATIO _____ AVE GRADE _____

NO. BULLS _____ AVE 365 WT _____ AVE 365 RATIO _____

PROGENY CARCASS DATA

No.	Ribeye Adl.	Carcass W/DA	Fat Thickness	Cutability	% Choice of Higher	Carcass Weight

GRAND SIRE _____ REG.

GRAND DAM _____ REG.

DAM REG.

PROGENY RECORD

NO. CALVES _____ AVE 205 WT _____ 205 WT RATIO _____ AVE GRADE _____

NO. BULLS _____ AVE 365 WT _____ AVE 365 RATIO _____

GRAND SIRE _____ REG.

GRAND DAM _____ REG.

Fig. 20-4. Performance pedigree. (Source: *Guidelines for Uniform Beef Improvement Programs*, Beef Improvement Federation Recommendations, April 1972, p. 69)

age 25 pounds more at weaning time than calves by an ordinary bull. Let's assume that these calves bring 50 cents per pound, and that 25 calves are weaned annually over the 4 years the bull is in service. That's an extra $12.50 a calf, $312.50 a year, and $1,250 more for the 4 years. Usually, the superior bull will be more growthy than an ordinary bull by perhaps 300 pounds. At $33 per cwt, the bull will bring an extra $100 when he has passed his useful breeding age and is marketed for slaughter, bringing the total added value of the superior bull to $1,350. Thus, if a cattleman is willing to pay $600 for an ordinary bull, he can afford to pay up to $1,950 for a superior bull. Besides, if he is saving replacement heifers, a further bonus will accrue to the superior sire through his daughters.

A superior sire can improve immediate net profit through his calves, and make for capital improvement through his daughters. It follows, therefore, that the monetary value of these improvements represents the added worth of a superior bull over an ordinary bull. This may be accurately and scientifically computed as follows:

1. *Net profit from using a superior sire is the product of* (a) the number of calves produced per year, (b) the average pounds sold per calf, (c) the net value obtained per pound of calf, and (d) the number of years used, *less* this same product from an ordinary sire.

2. *Capital improvement through the daughters is the product of* (a) the number of daughters saved, (b) the average pounds sold per calf (maternal plus growth), (c) the net value obtained per pound of calf, and (d) the average number of calves produced per daughter, *less* the same product for daughters out of ordinary sires.

Based on the above, if a superior sire produces 100 calves that wean at 50 lb more at 10¢ net per pound, that's $500 more profit than from an ordinary sire whose calves are 50 lb lighter. Additionally, if 20 daughters are saved that produce 25 lb more calf because of improved maternal ability plus 25 lb more expected in growth and they average 5 calves, that's another $500 profit. Thus, a cattleman could afford to pay up to $1,000 more for a superior sire of this caliber than for an ordinary bull. Then, there is an added bonus. Bulls that increase net value per pound 1¢ on a 1,000-lb steer return $1,000 per 100 calves.

Of course, the purebred seed stock producer has even more to gain from using a top sire, since his worth is multiplied through both his sons and daughters.

Another formula that is sometimes used in pricing a bull is based on worth as determined by performance test. An example follows:

> Let's assume that as a result of a performance test, it is found that *bull A* has a weight per day of age advantage of .5 lb over *bull B*. Then calves from *bull A* would be expected to gain .15 lb per day faster up to a year of age than calves from *bull B* (.5 × .6 heritability estimate for weight per day of age ÷ 2 for sire effect only). Let's further assume that 20 steer calves from each bull sell at 50¢ per pound at one year of age. The increased annual income from *bull A* over *bull B* would be $547.50. Let's further assume that these bulls are used 3 years. Thus, *bull A* is worth $1,642.50 more than *bull B*. The advantage from the heifer calves, many of which would likely be used as replacements, would be additional—and would be cumulative.

From the above, it may be concluded that the value of a bull is determined by his ability to produce calves that have enough quality to command the top dollar and enough pounds to make it pay.

Guarantee

Traditionally, beef bulls are "guaranteed breeders," meaning that they are guaranteed to be fertile and to sire live calves. Liability is limited to an option to select a replacement of equal value. High priced bulls are sometimes sold under more detailed contracts, which may specify percent conception and name the committee that must certify to the fertility of the bull.

Where a bull is bought for an artificial insemination stud, the sale contract may stipulate that payment is contingent upon the bull producing semen in specified quantity and quality useful in A.I.

Warranty for frozen semen is restricted to quality control characteristics, with liability limited to replacement in kind.

Of course, sellers of beef bulls cannot, and should not, be held responsible for infertility due to such things as infectious diseases, nutritional deficiencies, or mismanagement of the cow herd, or to injury to the bull after purchase. Likewise, those who sell frozen semen cannot be responsible for the cow herd's infertility for reasons such as those given above or for improper handling of semen or insemination procedure.

Bull Selection Guidelines

One of the most important management decisions a cattleman makes is the selection of a herd sire. Careful thought and planning are required. After deciding on the breed, evaluate your cow herd and current calf crop. Where do they need improvement the most? Is it size, muscling, soundness, gaining ability, or some other trait?

Next, decide what herds you plan to visit or which sales you wish to attend. Buy from reputable breeders who are doing a good job of production and who will give a breeder's guarantee (and preferably a semen test) with the animals they sell. Patronize those who are cattle breeders in the truest sense—those who are making real progress in improving the quality and performance of their own cattle, and who are not just "multipliers."

Take time in making a selection. Start well in advance of the time you need a bull. The earlier you start, the greater the number of bulls from which you may choose.

Select a bull that meets the following criteria:

● Performance tested, with a good 205-day weight (adjusted for age of dam) and a good 365-day adjusted weight.

● Large framed, with plenty of size for his age.

● Well muscled.

● Structurally sound, including the feet and legs, and free from hereditary defects (and not a "carrier" of hereditary defects).

● Good disposition. Avoid flighty, nervous bulls; heifers sired by such a bull will likely exhibit the same traits.

• Sired by a bull that has been doing a good job of settling cows and of siring large-framed, fast-gaining calves.

• Out of a cow that consistently ranks in the top half of the herd in terms of production.

• Normal in testicular development—both testicles are present and they are fully descended, sound, and approximately equal in size.

• Superior carcasses of half-sibs of a young herd bull prospect, as evidenced by records of ribeye area, fat cover over the twelfth rib, cutability (yield grade), and overall quality grade.

• Free of reproductive diseases, as determined by blood tests and verified by health papers.

Sire Selection Is a Continuous Process

The sire selection process continues even after the bull has settled his cows. His progeny should be evaluated as soon as possible. Then, on the basis of progeny performance, the breeder should make the decision either to retain or reject the bull for future use. If the progeny performance is acceptable to the breeder and the sire is retained, the selection procedure dealing with semen and libido checks begin again during the next breeding season and should continue each breeding season that the sire is used. Thus, sire selection is truly a continuous process.

MANAGING AND FEEDING BULLS[1]

Frequently, little thought is given to the management and feeding of bulls except during the breeding season. Instead, the feeding program for herd bulls should be such as to keep them in a thrifty, vigorous condition at all times. They should neither be overfitted nor in thin, run-down condition. Also, exercise is necessary for the normal well-being of the bull.

Periodic management practices for bulls should include:

1. External parasite control of lice and flies.
2. Fecal examination for internal parasites.
3. Foot examination and care if needed—trimmed, corns removed, treatment of foot infections and other injuries. Neglected foot trimming has probably ruined more good bulls than deficiency in any other single management practice.
4. Eye examination, and treatment if necessary.
5. Reproductive examination, and semen evaluation if needed.

The feeding and management of bulls differ according to age and condition. For this reason, sale bulls, young bulls, and mature bulls are treated separately in the sections that follow:

---

[1]The nutritive requirements of bulls for growth and maintenance, at different weights, are given in Chapter 8, Tables 8-18 and 8-19.

Fig. 20-5. Registered Brahman bulls on burr clover pasture. (Courtesy, Soil Conservation Service, USDA)

Sale Bulls

Most bull sales are held in late winter and early spring, at which time mostly yearling and two-year-old bulls are sold. In order to attract buyers, they have usually been grain-fed since calfhood. Most bull buyers—especially commercial cattlemen in rougher range areas—would rather have their new bulls in less than fitted sale condition. They find that such bulls are more fertile and more apt to range with the cows when turned to pasture during the breeding season.

Sale and show bulls should be acquired 2 to 3 months ahead of the breeding season, so that they may be conditioned, or let down. Also, bear in mind that it takes about 40 days from the time a sperm cell is formed until it is ready to be ejaculated. Since the stress of handling and hauling a bull can reduce his fertility for about 40 days, the rest period lets his body overcome these problems.

Handling highly conditioned sale bulls during the critical period—after the sale is over, and just ahead of the breeding season—is all-important. Experienced cattlemen "let them down" and yet retain strong vigorous animals. They do this successfully by (1) providing plenty of exercise, (2) increasing the amount of bulky feeds, such as oats, in the ration, (3) cutting down gradually on the grain allowance, and (4) retaining the succulent feeds and increasing the pasture and hay.

Exercise is most important during this period, and the more excessive the finish the more vital the exercise. Heavily fitted bulls can best be exercised by leading as much as two miles daily. Moderately fitted bulls can be force-exercised by placing the feed and water on opposite sides of the field.

In summary, therefore, conditioning highly fitted sale bulls for breeding consists of the gradual elimination of high energy rations along with forced exercise.

Young Bulls

Lack of fertility in a bull may often be traced back to his early care and feeding. From weaning to three years of age, bulls should be kept separate by age groups. Young bulls should be fed more liberally than mature bulls because their growth requirements must be met before any improvement in condition can take place.

Following weaning, bulls should be fed and developed sufficiently to show their inherited characteristics, but without excessive finishing. Simultaneously, they should be given plenty of exercise. Overfeeding and lack of exercise are apt to result in infertility, low quality sperm, and unsound feet and legs.

To achieve proper development, young bulls should gain at least 2½ pounds daily from weaning to 12 to 15 months of age. This will necessitate a daily feed allowance equal to about 2½% of their body weight, with a ration comprised of 50% or more concentrate. From 15 months to 3 years old, they should make a daily gain of 2 to 2¼ pounds and receive a feed allowance equal to 2 to 2¼% of their body weight, with the proportion of roughage increased after the first year.

If desired, the roughage may be chopped and mixed with the concentrate; and the ration may be hand-fed or self-fed. If self-fed, the feed consumption may be held at the desired level by using salt as a regulator. Other cattlemen prefer to feed the roughage and concentrate separately; they usually free-choice the roughage and either self-feed a salt-concentrate mix or hand-feed the concentrate alone. When grain is fed separate from the roughage, about 1½ pounds per 100 pounds of body weight can be fed at the beginning, gradually decreasing the grain and increasing the roughage as the animal grows older.

Without doubt, the least laborious and most convenient management arrangement in handling young bulls consists in allowing a group of not to exceed 10 to 15 head of uniform size and age the run of a pasture or enclosure of ample size, thereby providing (1) exercise, and (2) pasture in season. Of course, wherever possible, bulls should be performance tested while being developed. Ideally, this calls for individual feed and body weight records, although group feeding plus individual weight records will suffice.

At intervals, young bulls should be thoroughly checked for feet and leg defects, such as sickle-hocked, post-legged, cow-hocked, bow-legged, toeing out, or toeing in; and corrective hoof trimming should be administered as needed.

Bulls handled as recommended above will generally attain half their mature weight by the time they are 14 to 15 months of age and may be used in limited service.

During the breeding season, young bulls should be fed a grain ration consistent with pasture quality and number of cows to be bred in order to promote proper growth and development. Drought, over pasturing, and poor

Fig. 20-6. Weaned bull calves on performance test being fed on a high roughage ration in a self-feeder. (Courtesy, University of Illinois, Urbana, Ill.)

quality pastures are situations in which grain supplementation is particularly needed. Heavy service and poor pasture with no supplemental feeding may shorten the breeding career of a young bull.

After the breeding season, yearling bulls generally need 5 to 6 pounds of grain along with good roughage.

Mature Bulls

Winter is the proper time to condition bulls for the next breeding season. Bulls that have been running on pasture with the cows are likely to be thin; thus, they require sufficient concentrate to put them in proper flesh. Mature bulls will consume daily amounts of feeds equal to 1½ to 3 percent of their liveweight, depending upon condition and individuality.

Outdoor exercise is also essential in keeping bulls virile and thrifty. The finest and easiest method of providing such exercise is to allow them the run of a well-fenced pasture. About two acres is a good size for one bull, with a larger enclosure where several bulls are run together.

The importance of having bulls in proper condition at the opening of the breeding season cannot be overemphasized. Nothing is quite so disheartening or costly as a small calf crop, with many of the calves coming late. Lack of fertility in the bull may often be traced back to his care and feeding.

Feed mature bulls all the legume hay they will eat plus 3 to 5 pounds of ground or rolled grain and 1 pound of a 32 percent protein supplement (or equivalent) per head per day. Also, provide free access to a suitable mineral mixture. About 60 days before the bulls are turned with the cows, increase the concentrate allowance by 25 to 50 percent, with the amount of the increase determined by the condition of the bulls.

The mature herd bull needs no additional feed when running with the cow herd on good summer pasture.

Managing Bulls During the Breeding Season

Even though a 100 percent calf crop is difficult to obtain, a herd owner should strive to approach this figure as closely as possible. Poor management of the herd at breeding time can greatly reduce the calving percentage.

Fig. 20-7. Yearling Hereford bull in hard breeding condition on the range of Sr. Guillermo Finan, Ganaderia Valle Colombia, Muzquiz, Coahuila, Mexico. (Photo by Audrey Ensminger)

The following management practices during the breeding season are recommended:

• *Purchase the bull early*—Bulls should be purchased early in the season when selection is best. Also, they should be on the farm or ranch where they are to be used 60 days prior to breeding. This will provide a period of isolation and give them a chance to adjust to the "bugs" on the new premises. Purchasing well in advance of the breeding season also gives bulls a chance to overcome any fertility problems encountered from a sale or shipment.

• *Evaluate semen and reproductive soundness*—Perform a semen and reproductive soundness evaluation on all bulls two weeks ahead of the breeding season. This practice will detect sterile bulls and those with obvious low fertility prior to their use and will allow time to replace them.

If foot trimming is required, do it at this time—at least two weeks before breeding.

• *Provide adequate bull power*—The proportion of bulls to cows is dependent on (1) the age of the bulls—young or old bulls cannot carry their share of the breeding responsibility; (2) the topography and feed conditions—rough areas and sparse vegetation require more bulls; (3) condition of bulls—excessively fat or excessively thin bulls will handle fewer cows; and (4) the length of the breeding season—the shorter the breeding season, the greater the stress on the herd bull battery. Under average conditions, one yearling bull should be provided for each 15 to 20 cows and one mature bull for each 25 to 35 cows. With hand mating, more cows may be bred.

• *Shorten the breeding season*—Gradually shorten the breeding season each year until it is no longer than 60 to 75 days.

• *Check the breeding herd frequently*—Watch for and record cows in heat; and see that the bull is finding and breeding these cows. Bulls that have low libido and are reluctant to mate even though cows are showing visible signs of estrus should be culled immediately. Also, if a high percentage of the cows have not been settled after two heat periods, the bull should be replaced.

• *Modify hot weather*—Breeding during hot weather often reduces the quality of sperm and lowers conception rates. Because of this, it is important that cattle have access to plenty of fresh water, shelter from the sun, and protection from flies.

BULL COSTS; A.I. COSTS

With the increase in artificial insemination in recent years, a frequently asked question is: What's the cost of A.I. vs natural service?

Many cost figures seen in various publications show bull costs by natural service ranging from $3 to $6 per cow. The Nebraska Station reported that the cost of keeping a bull is much higher than this, even if the bull is depreciated over a 4-year period. (See Table 20-1.)

As shown in Table 20-1, one of the largest bull cost items is depreciation. These figures are based on a purchase price of $1,120 and a selling price of $480. On this basis, bull depreciation alone would amount to more than $5 per cow. Direct costs, excluding pasture, amount to $10 to $15 per cow (depending on how long the bull is kept), and $12 to $17 per cow if the charge for summer pasture is included.

No allowance was made for investment credit in the bull costs shown. According to tax laws in effect in 1973, this would be a consideration and would reduce the bull costs slightly. In most cases, it would amount to no more than $1 to $2 per cow in the herd, however.

Artificial insemination eliminates the expense and problems associated with maintaining bulls. But, of course, it involves some different expenses, primarily for the cost of semen and the added labor.

Based on a study of 37 commercial ranches in Wyoming, the Wyoming

TABLE 20-1

ESTIMATED ANNUAL COST OF OWNERSHIP AND MAINTENANCE OF A BULL[1]

	1973 Price Conditions			
	If Depreciated Over 2 Years		If Depreciated Over 4 Years	
	Total Costs	Direct Costs	Total Costs	Direct Costs
Hay, 1.5 tons	$ 45.00	$ 45.00	$ 45.00	$ 45.00
Winter pasture	16.88	——	16.88	——
Summer pasture	63.00	——	63.00	——
Salt & mineral	2.00	2.00	2.00	2.00
Veterinary & medicine	7.50	7.50	7.50	7.50
Death loss	5.60	5.60	5.60	5.60
Depreciation (purchase price $1,120; selling price $480)	320.00	320.00	160.00	160.00
Taxes	6.45	6.45	6.45	6.45
Interest on bull	54.40	54.40	54.40	54.40
Interest on feed & operating expense	5.80	2.37	5.08	2.37
Labor @ $2.50/hr.	25.00	——	25.00	——
Use of bldgs. & equip	9.30	3.72	9.30	3.72
Misc. cattle expense	3.00	3.00	3.00	3.00
TOTAL	$563.21	$450.04	$403.21	$290.04
Per Cow (based on 30 cows)	18.77	15.00	13.44	9.67

[1]Opportunities For Cow-Calf Operations—a prospectus, EC 73-254, University of Nebraska, June 1973, p. 29.

Experiment Station reported $1.87 greater cost per calf from A.I. than from natural service.[2] However, the A.I.-sired calves had a $7.05 per head greater value than the calves sired by natural service, leaving $5.18 net per calf in favor of A.I. This comparison included all charges related to breeding the cow herds.

QUESTIONS FOR STUDY AND DISCUSSION

1. Why does 80 to 90 percent of the hereditary gain in a given beef herd result from the selection of the bull?

2. What source of bulls do you favor—purchase from a breeder at private treaty or in a production sale, purchase at a central test station sale, raising your own, or artificial insemination; and why do you prefer it over other sources?

3. In a bull selection program, would you buy a young bull or a proved sire? Why?

4. How would you determine what herd has the highest merit—hence, the place for you to buy?

5. If you were limited to selection for just one of the following 3 production traits, which one would you choose, and why would you choose it: weaning weight (205-day weight), postweaning growth weight, and carcass merit?

6. How important is "individuality" in selecting a bull?

7. Is a semen test 100 percent accurate in determining bull fertility?

8. Of what value is a regular pedigree in bull selection? Of what value is a performance pedigree?

[2]Stevens, D. M. and T. Mohr (1969), Artificial Insemination of Range Cattle in Wyoming: an economic analysis, *Wyoming Bulletin 496.*

9. Detail by example how you would arrive at the price that you would be willing to pay for a particular bull?

10. When buying a bull, what kind of a guarantee would you expect?

11. Briefly and concisely, list the criteria that you recommend using in selecting a bull.

12. How would you handle a heavily fitted sale bull from auction time to breeding season, which we shall assume to be a period of 60 days?

13. Wherein does the management of a young bull 13 to 14 months old differ from the management of a mature bull?

14. List and discuss the most important bull management practices during the breeding season.

15. Which will cost the most, and which will make for the highest net return—bulls used in natural service or artificial insemination?

16 Select a particular farm or ranch, either your own or one with which you are familiar. Would you buy bulls or buy A.I. semen for this particular operation?

SELECTED REFERENCES

Title of Publication	Author(s)	Publisher
Commercial Beef Cattle Production	Ed. by C. C. O'Mary, I. A. Dyer	Lea & Febiger, Phildelphia, Penn., 1972
Beef Cattle	A. L. Neumann R. R. Snapp	John Wiley & Sons, Inc., New York, N.Y., 1969
Beef Production in the South	S. H. Fowler	The Interstate Printers & Publishers, Inc., Danville, Ill., 1969

FEEDING AND HANDLING CALVES[1]

Contents   **Page**

Cattlemen, as a whole, have lagged in applying much of what we know about feeding and handling calves. They're inclined to let mother cows and mother nature fend for the calves. Indeed, more proved good practices, based on both successful operations and research, need to be put to use in feeding and handling calves.

(The nutritive requirements of all cattle, including calves, are presented in Chapter 8 of this book; hence, the reader is referred thereto.)

MANAGEMENT AT BIRTH

Losing a calf means losing the profit on the cow for a whole year. Proper management at birth can make the difference.

Close observation of the cow herd at calving time is essential so that assistance can be provided when necessary. First-calf heifers will usually have more calving difficulty than older cows; hence, they should receive very close attention during the calving season.

Recommended calving-time management practices follow:

1. Have cows in an area that can be checked easily.

2. Keep the cows in a clean place during calving—it's the best way to avoid scours and other diseases. In season, a clean pasture (one that has

---

[1]The author gratefully acknowledges the helpful suggestions of Dr. Wilton W. Heinemann, Washington State University, who reviewed this chapter.

Fig. 21-1. Calving on clean ground makes for a good start in life and lessens calf scours. (Courtesy, University of Illinois, Urbana, Ill.)

been idle for a time) is ideal. When calving in confinement in winter, early spring, or late fall, provide a clean, freshly bedded shed.

3. If a cow is in true labor for more than 2 hours or in unusually severe labor, she should be examined. In normal presentation, the head is between and slightly above the front feet. The experienced caretaker may correct a minor problem, such as a front foot bent back or the calf's head turned back. For a difficult position problem, however, the veterinarian should be called. Difficult calving may be caused by any of the following conditions:

    a. The calf is very large, especially in the shoulders or hips—or both.

    b. The cow has a small pelvic area, or the pelvic area is filled with fat (as happens in excessively fat animals).

    c. The cow fails to dilate.

    d. The calf is presented backward (breech birth).

    e. One or both front legs are bent back.

    f. The head is bent back.

4. As soon as possible after birth, remove mucus from the calf's nose and mouth. If the calf does not start to breathe normally, hold it by its hind legs and shake it vigorously, or apply artificial respiration by alternate pressure and release on the rib cage.

5. Disinfect the navel with a 2-percent tincture of iodine solution; and inject the newborn calf with 250,000 to 1,000,000 IU of vitamin A (use the higher level in confinement production or where scours may be a problem).

6. See that the newborn calf nurses within 2 hours after birth. It is essential that it receive colostrum. The caretaker may have to assist a calf to nurse a dam that has very large teats or an udder that hangs very low. Also, weak calves should be helped to nurse.

7. If a cow does not claim her calf, put them in a separate pen and do not disturb them for a few hours.

8. If a cow fails to clean properly (has a retained placenta) within 12 hours after calving, call your veterinarian.

9. Keep the cows that have calved separate from those that are still to calve.

Fig. 21-2. It is usually best to designate a small pasture adjoining headquarters as the calving area. The cow and calf should remain in this area for a few days, following which they may be turned with the main herd, like the above group of Angus. Note the snow on the ground. (Courtesy, Martin Jorgensen, Jr., Ideal, S.D.)

10. Keep a close watch for signs of mastitis or injury to udders. It may be necessary to milk out a few cows for the first two or three days after calving.

11. Be sure cows have access to plenty of clean, fresh water.

12. Ear tag or tattoo the calf, and record pertinent information on your calving record, including date of birth, ease of calving, abnormalities, availability of colostrum, strength of calf, birth weight of calf, and temperament of the cow.

(Also see Chapter 6, Physiology of Reproduction in Cattle.)

RAISING ORPHAN AND MULTIPLE BIRTH CALVES

Occasionally a cow dies during or immediately after parturition, leaving an orphan calf to be raised. Also, there are times when cows fail to give a sufficient quantity of milk for the newborn calf. Sometimes, there are multiple births.

If there are only a few orphans, usually they can be grafted onto another cow (or adopted)—either one that has lost her calf or one that gives sufficient milk to raise two calves. Where such calves cannot be grafted, they must be raised by artificial methods—without a cow.

Regardless of whether orphans are grafted or raised artificially, the problem will be simplified if the calf receives colostrum, the first milk produced by a cow after giving birth to a calf, during the first 24 hours, and preferably for the first three days, of its life—from its mother, from another fresh cow, or from frozen-stored colostrum. Colostrum is higher than normal milk in dry matter, protein, vitamins, and minerals. Also, it contains antibodies (found in the gamma globulins) that give newborn calves a passive immunity against common calfhood diseases.

Because colostrum is so important for the newborn calf, cattlemen should store a surplus of it from time to time. It can be frozen and stored for a period of one year or longer, then, as needed, thawed and warmed to 100° to 105° F, and fed.

Fortunately, orphan calves can now be raised successfully on a milk replacer and calf starter ration, using them as directed. The milk replacer may be fed by using a bottle or pail equipped with a rubber nipple, or the calf may be taught to drink from a pail. It is important that all receptacles be kept absolutely clean and sanitary (clean and scald each time) and that feeding be at regular intervals. Dry feed should be started at the earliest possible time; not later than one week of age. With proper management, healthy calves may be switched entirely to a suitable dry feed at four to five weeks of age.

Basically, calves are fed according to one of 3 systems: (1) the whole milk system, (2) the combination whole milk-milk replacer system, or (3) the combination whole milk-calf starter system. Of course, various combinations of these 3 systems are used, also. A suggested schedule for each of these 3 systems is given in Table 21-1.

The whole milk method costs the most, but it produces the fastest gains, the best appearing calves, and requires the least skill of any system. Milk replacers can be fed as the only feed, immediately following the colostrum period; or, as shown in Table 21-1, they may replace whole milk beginning on about the seventh day.

There is hardly any limit to the number of calf starters on the market. Most of them are mixed commercially. Because of the difficulty in formulating a home-mixed calf starter ration, the purchase of a good commercial feed usually represents a wise investment.

Two suggested calf starter rations are given in Table 21-2.

Starter Ration A, of Table 21-2, is designed for feeding anytime after the first day following birth. Starter Ration B is designed for feeding beginning about 45 days of age. As is true in any ration change, the transition from Ration A to Ration B should be made gradually by blending the feeds over a period of 2 to 3 days.

Good quality hay for young calves is essential; it provides an economical source of nutrients, helps maintain rate of gain, and speeds up the development of the rumen.

Many cattlemen make the mistake of placing calves on pasture at too early an age. Unless pastures are properly supplemented, young calves simply cannot hold enough grass, or other pasturage, to obtain sufficient nutrients for their growing bodies. Accordingly, growth will be retarded.

TABLE 21-1

SCHEDULE FOR FEEDING CALVES BY THREE DIFFERENT SYSTEMS

Age of Calf	Whole Milk System	Whole Milk-Milk Replacer System	Whole Milk-Calf Starter System
0 to 3 days	Calf should receive colostrum during first 3 days.	Calf should receive colostrum during first 3 days.	Calf should receive colostrum during first 3 days.
3 days	Start feeding whole milk at the rate of 1 lb milk to 10 lb body weight.[1]	Start feeding whole milk at the rate of 1 lb milk to 10 lb body weight.[1]	Start feeding whole milk at the rate of 1 lb milk to 10 lb body weight.[1]
7 days	Make grain available in box in pen (see Table 21-2).	Make calf starter available in box in pen.	Make calf starter available in box in pen.
7 to 10 days		Start replacing whole milk with fluid milk replacer. Replace 1 to 2 lb milk daily with fluid milk replacer until change is completed.	
14 days		Transition to milk replacer should be completed.	
21 days	Make good quality hay available in rack in pen.	Make good quality hay available in rack in pen.	Make good quality hay available in rack in pen.
60 days		Discontinue milk replacer.	Discontinue feeding whole milk. Larger more vigorous calves may have whole milk stopped as early as 42 days.
60 to 120 days	Permit calves to consume grain free-choice, up to 4 or 5 lb daily. Rest of nourishment should be obtained from hay.	Permit calves to consume calf starter-free-choice, up to 4 to 5 lb daily. Rest of nourishment should be obtained from hay.	Permit calves to consume calf starter-free-choice, up to 4 or 5 lb daily. Rest of nourishment should be obtained from hay.
90 days	Discontinue whole milk.		

[1]For economic reasons, it is never advisable to feed calves more than 12 pounds whole milk daily during the entire whole milk feeding period.

TABLE 21-2

CALF STARTER RATIONS

Ingredients	Starter Ration A[1] (for feeding first 45 days, along with liquid skim milk)	Starter Ration B[1] (for feeding after first 45 days, with dry skim milk therein)
	(lb)	(lb)
Dried skim milk		400
Soybean or cottonseed meal (41%)	560	450
Barley	1,000	750
Wheat bran	200	150
Dicalcium phosphate	20	20
Trace mineralized salt	20	20
Antibiotic (follow mfr.'s directions)	10	10
Vitamin A	10,000 IU/lb	2,000 IU/lb
Vitamin D (not needed if calf is in sunlight)	2,000 IU/lb	400 IU/lb
Molasses	200	200
Total	2,010	2,000
PROXIMATE ANALYSIS:	(%)	(%)
Crude protein	18.8	19.8
Fiber	6.6	4.1
Fat	2.3	1.4
Calcium	.43	.69
Phosphorus	.77	.74
TDN	69.9	74.7

[1]In ⅛- or ³/₁₆-inch pellets.

EARLY WEANING

Early weaning refers to the practice of weaning calves earlier than the usual weaning age of about 7 months, usually within the range of 45 days to 5 months of age. Although it is not common practice among United States beef cattlemen, dairymen have been weaning 3-day-old calves for years. Also, early weaning has long been an integral part of many of the beef programs of Europe.

Currently, there is much interest in early weaning because (1) it fits into a drylot cow-calf management system, and (2) it can give a big assist in getting females, especially two-year-old heifers, to rebreed in a short period of time.

The current interest in increasing the number of cows in the Corn Belt is largely predicated on more efficient use of crop residues, especially corn and sorghum residues. With heavy cropping, there is little or no pasture as such; hence, drylot management systems are evolving. When using crop residues as a basic feed source, the lactating cow is likely to need supplemental feed. Considering the low efficiency involved in converting supplemental energy to milk and in converting milk to meat, it is apparent that a more efficient use of feed could be achieved by giving the supplemental feed directly to the calf. A lactating cow requires about 50 percent more feed than a dry cow. So, rather than give her that additional feed, it is more efficient to give feed directly to the calf.

Weaning calves early from two-year-old, first-calf heifers reduces the stress of milking and raising a calf. As a result, they recycle and rebreed earlier and grow out more rapidly. As heifers are bred for higher milk production, this reason for early weaning takes on greater importance, for the more milk they give, the slower they are to cycle. It appears doubtful that any level of nutrition will have the same effect on reproduction efficiency of the heifer as early weaning.

In addition to fitting into a drylot cow-calf system and facilitating a program of calving two-year-old heifers, early weaning may be desirable for the following reasons:

- It may be the answer to getting one calf per cow every 12 months in intensive management systems.
- It may be the key to the most efficient feed utilization during times of droughts and other periods of feed shortages. Under such conditions, it might be advantageous to wean the calves early and to provide them with the highest quality feed available, while restricting the quality and quantity of feed fed to the cows.
- It fits in with fall calving where heavy winter feeding is required. As soon as the calves are weaned, cows may be turned to stalk or stubble fields to winter on cheap feed.
- It may make it possible to keep a particularly valuable old purebred cow in production longer.
- Young gains are cheap gains, due to (1) the higher water and lower fat content of young animals in comparison to older animals, and (2) the higher feed consumption per unit weight of young animals. Thus, the feed efficiency of early weaned calves is excellent, ranging from three to four pounds of TDN per pound of gain.
- Lactating cows decline in milk production after about one to two months following parturition.
- Parasite problems are minimized in an early weaning program.

Where early weaning is successful, the only responsibility of the beef cow is to produce a calf and give it a good start in life for a brief period, then go on a maintenance ration the rest of the year.

Like many good things in life, early weaning does have some disadvantages. To be successful, superior nutrition and management are essential; and the earlier the weaning age the more exacting these requirements.

How Early Is Early Weaning?

Experiments and experiences indicate that it is practical to wean calves as early as 45 days of age. This is long enough to stress the cow a bit, but short enough to get her to recycling and rebred so that a calf will be produced each 12 months. Also, it allows the cow to function as a lactating animal. Weaning at 3 to 5 months of age doesn't make for early recycling; hence, it doesn't contribute to getting one calf per cow every 12 months.

Fig. 21-3. A vigorous, growthy Angus calf at 45 days of age, old enough for early weaning. (Courtesy, Martin Jorgensen, Jr., Ideal, S.D.)

What To Feed Early Weaned Calves

From 45 days of age on, early weaned calves can be fed any good starter ration, most of which contain dry skim milk. One such ration is given in Table 21-2, Ration B. Most commercial feed companies manufacture a starter ration. Of course, the starter ration should be made available to the calves well ahead of weaning in order that they will be accustomed to it, thereby avoiding any setback.

MILK REPLACERS

Several reputable commercial companies now produce and sell milk replacers, which are composed of sizable amounts of milk by-products, such as dry skim milk, buttermilk, and/or whey, along with additives.

Although scientists have not yet learned how to compound a synthetic product that will alleviate the necessity of colostrum, in certain other respects they have been able to improve upon nature's product, milk. For example, it has long been known that milk is deficient in iron and copper, thus resulting in anemia in suckling young if proper precautions are not taken. In addition to correcting these deficiencies, synthetic milks are fortified with vitamins, minerals, and antibiotics. They can be fed as the only feed immediately following the colostrum period; or, as shown in Table 21-1, they may replace whole milk beginning about the seventh day.

From the standpoint of the beef cattle producer, a milk replacer is of primary interest for two uses; namely, (1) for raising orphaned and early-weaned calves, and (2) for replacing nurse cows. Also, it is a valuable adjunct in certain disease control programs, especially those diseases that may be transmitted from dam to offspring.

CREEP FEEDING

Creep feeding is the supplementation of calves while they are nursing their dams. It increases weaning weight. The basis for this response is related to the lactation curve of beef cows, the increasing nutrient requirements of the calf during the nursing period, and the decline in feed quality and quantity typical of most pastures or ranges which support the cows and calves during lactation. Studies reveal that milk production of dairy cows increases up to the fourth or sixth month following freshening, then declines gradually. By contrast, maximum milk production of beef cows occurs during the first two months after calving, then declines.

Fig. 21-4 shows why creep feeding is important. From birth to weaning, the protein and energy requirements of a growing calf increase well beyond the ability of most beef cows to meet those needs. For example, to meet the protein and energy requirements for growth, a 100-lb calf needs 10 lb of milk, whereas a 500-lb calf needs 50 lb of milk. Since the average beef cow gives only 13 lb of milk per day throughout a 7-month suckling period, a 500-lb calf lacks 40 lb of getting enough milk from its dam at this stage of lactation to meet its needs—that's the "hungry calf gap."

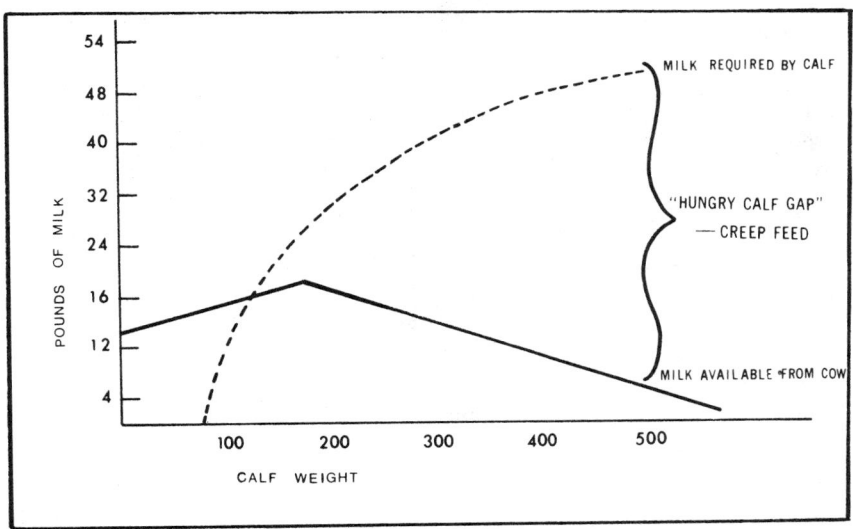

Fig. 21-4. Milk yield of a typical beef cow vs nutrient requirements of a nursing calf. This points up the need for creep feeding.

To fill the "hungry calf gap"—the nutrient requirements over and above that provided by 13 lb of milk—would require the consumption of 50 lb of green grass daily. Of course, that's a physical impossibility, because a 500-lb calf simply cannot hold that much bulk. So, the best way to fill the "hungry calf gap" is to creep feed.

Creep feeding is no longer primarily an emergency program to supplement drought-stricken grasses and other conditions resulting in poor pastures. Rather, continuous feeding of calves from birth to weaning is on the increase because milk in quantity, and pasture in quality, are not normally available season-long to supply the necessary nutrition (1) to produce calves that meet today's market demands, and (2) to realize the maximum genetic potential from improved breeding.

Most calves will continue to be raised on their mother's milk plus whatever pasture or other feed they share with their dams. However, more and more of them will be creep fed in addition.

The Creep

A creep is an enclosure or feeder for feeding purposes which is accessible to the calves but through which the cows cannot pass. It allows for the feeding of the calves but not their dams. For best results, the creep should be built at a spot where the herd is inclined to loiter; on high ground, in the shade, and near watering and salting grounds. The enclosing fence may be

Fig. 21-5. A calf creep. Note available shade, and calves at feed. (Courtesy, Purdue University, Lafayette, Ind.)

of board, pole, or metal construction, with an entrance 16 to 20 inches wide and 3 to 3½ feet high. Self-feeders, troughs, or racks may serve as feed containers; allowing 4 to 5 inches of space per calf for self-feeding and 8 to 12 inches for hand-feeding. Also, there are on the market portable, metal creep feeders consisting of a self-feeder to which the enclosing fence is firmly attached. These are especially suited for use on large range pastures, where frequent moving is necessary.

Creep Rations; Feeding Directions

Creep-fed calves need special rations. They are bovine babies; and they are both in forced production and finishing. They are expected simultaneously to lay on fat and grow in protein tissues and skeleton. Consequently, their ration requirements are for feed high in protein; rich in readily available energy; fortified with vitamins, minerals, and unidentified factors; and with all the nutrients in proper balance. Also, the ration must be very palatable. This calls for an exacting ration. To meet these needs, more and more cattlemen are finding it practical to buy a commercially prepared complete creep feed, or a well fortified and highly concentrated supplement to add to locally available feeds, rather than purchase individual ingredients and mix from the ground up.

Tables 21-3 and 21-4 show two creep rations, formulated by the author, that have been widely and successfully used.

TABLE 21-3

CALF CREEP RATION NO. 1[1]

Ingredient	Lb/Ton Mix	Calculated Analysis:	(%)
Oats	800	Crude protein	14.1
Corn	300	Fat	2.9
Barley	200	Fiber	8.4
Wheat bran	200	TDN	69.9
Dried molasses		Calcium	.37
beet pulp	200	Phosphorus	.50
Soybean meal	200		
Salt	10		
Dical	10		
Trace minerals[2]	1		
Vitamin A (30,000 IU/g)	1.5		
Molasses	100		
TOTAL	2,022.5		

[1]*Feed Preparation:* Preferably ⅛ or 3/16-inch pellets. Otherwise, steam roll grains, or grind grains coarsely.
[2]See Chapter 8, Table 8-8, of this book for recommended trace mineral levels. Follow manufacturer's directions.

TABLE 21-4

CALF CREEP RATION NO. 2

Ingredient	Lb/Ton Mix	Calculated Analysis:	(%)
Alfalfa meal	450	Crude Protein	13.8
Soybean meal	124	Fat	3.2
Linseed meal	100	Fiber	12.8
Corncobs or grass hay	200	TDN	60.5
Oats	400	Calcium	1.01
Corn	485	Phosphorus	.71
Bran	100		
Dical	40		
Vitamin A (325,000 IU units/g)	84 g[1]		
Tra-min[2]	1		
Molasses	100		
TOTAL	2,000		

[1]When 4 lb/head/day of the Calf Creep Ration is consumed, 54,600 IU units of vitamin A will be obtained in the feed.
[2]See Chapter 8, Table 8-8, of this book for recommended trace mineral levels. Follow manufacturer's directions.

It takes considerable effort and patience to start calves on feed. Also, a little "calf psychology" helps; remember that calves do not go for the privilege of eating, they need to be persuaded. One or more of the following techniques will usually prove helpful: Shut a gentle cow or a few calves in the creep, to serve as a decoy(s); scatter a little feed near the creep so that the cows will loiter nearby; and/or spread a little feed near and extending through the creep opening. It is also recognized that fall and early spring calves take to creep feeding better than late spring calves, simply because they have less grass and milk available. (Grass stimulates milk flow.)

When 3 to 4 weeks of age, calves should be started on feed very gradually. For the first 3 to 5 days, only about one-fourth pound of feed per calf should be placed in the container(s) each day, and any leftover feed should be removed and given to the cows. In this manner, the feed will be kept clean and fresh. When calves are on lush pasture and their mothers are milking well, difficulty may be experienced in getting them to eat; but time and patience will pay off, and the results will become evident in 2 to 3 months.

After 5 to 7 days of hand-feeding, the creep ration can be left before the calves safely. Once they are on full feed, never let the feeder become empty; and avoid sudden changes. During the first 30 days, they will consume about a pound per head daily. By the end of the fifth month, they should be up to 8 pounds daily. Of course, with good pastures and plenty of milk, these consumption figures may be halved. Calves will consume approximately 500 pounds of creep feed per head from one month of age to weaning. In years of lush pasture, it will be less; in dry years more.

Why Creep Feed?

Unquestionably, the best yardstick for measuring performance in a beef breeding herd is pounds of calf weaned per cow bred. This fact, along with the demand for healthy, "gain-ready" feeder calves and the prices being paid, is causing cow-calf operators to take a new "pencil-pushing" look at the immediate and residual benefits of creep feeding.

Among the reasons for or the benefits from creep feeding are the following:

1. *It provides a way to fill the "hungry calf gap"*—Creep feeding provides a logical and practical way to compensate for insufficient milk which usually characterizes the following conditions: (a) the normal falloff in milk production of beef cows about two months after freshening; (b) periods of unfavorable feed conditions—droughts, overgrazing, early and late in the season, and fall-calving herds that are maintained on a low plane of winter nutrition—when the calf is shortchanged on both milk and pasture; (c) first-calf heifers, whose milk production is generally lower than mature cows; (d) shelly and poor-doing cows; and (e) poor milkers. Under such conditions, creep feeding makes for heavier weights.

2. *It makes for heavier weaning weights*—Creep feeding results in 50 to 75 lb heavier weaning weight per calf, at no extra cost for the capital investment in land and cows.

3. *It facilitates fall calving*—Normally, fall-calving cows do not milk as well as spring-calving cows, because of the lack of lush grass to stimulate milk flow. As a result, fall calves neither gain as well nor have as heavy weaning weights as spring calves. It follows that creep feeding is more effective with fall calves than with spring calves. One study showed that creep-fed fall calves made 20 pounds more gain than creep-fed spring calves.[2]

4. *The calves are more uniform*—Creep-fed calves are more uniform, because those that are getting less milk eat more feed in order to meet their nutritive requirements.

5. *Calves achieve full genetic growth potential*—Today, the emphasis in beef cattle is on size and growth, brought about through selection, introduction of the exotics, and crossbreeding. But most cattlemen have improved their cattle faster than they have improved their pastures. As a result, many calves are not getting sufficient nutrients from milk and grass to achieve the full genetic growth potential that is bred into them. Under these circumstances, creep feeding is the answer.

6. *Young gains are cheap gains*—Creep feeding makes for efficient and cheap gains, and profit.

The Virginia Station reported[3] that drylot calves which only had access to cow's milk gained only .33 lb per day. Later, when creep fed (at 3 to 4 months of age), these same calves gained 2.2 lb per day. More remarkable yet, the creep-fed calves made one lb of gain for each 3 lb of creep feed.

7. *It is more efficient to feed calves directly than to feed cows too liberally*—This is so because when beef cows are fed above a certain level, they have a tendency to put tallow on their backs instead of milk in their bags.

8. *It makes for attractive purebred calves*—Creep feeding is the ideal way in which to obtain the important development and bloom in purebred calves; for fat is a very pretty color. This is especially desirable if they are to be sold or shown at young ages.

9. *It makes it easy to reinforce and improve milk*—Creep feeding affords a convenient way in which to improve upon milk, chiefly by reinforcing it with certain vitamins and minerals; and in which to add such additives as desired.

10. *It controls parasites*—Creep feeding lessens parasites, simply because well-nourished animals have fewer parasites.

11. *It simplifies weaning*—Creep-fed calves rely less on the dam for nourishment and develop independence, with the result that there is less bawling, stress, and sickness when they are separated from the cows.

12. *It facilitates early weaning*—It makes it possible to wean calves at early ages—1½ to 5 months instead of the normal 7—a practice which is increasing.

13. *It makes for marketing flexibility*—Creep-fed calves may go the

[2]Hodge, Dr. Dean E., Director, Beef Cattle Research, Ralston Purina, St. Louis, Missouri. Quoted in article entitled, "Creep Feeding," *The Cattleman*, March, 1974, p. 154.

[3]Fontenot, J. P., Virginia Polytechnic Institute and State University, Blacksburg, Va. Quoted in article entitled, "Creep Feeding," *The Cattleman*, March, 1974, p. 154.

slaughter route—in addition to going into feedlots or as stockers; with the particular market avenue selected being determined by price.

14. *It narrows the price between heifers and steers*—It permits heifers to be marketed for slaughter at a weight when they will bring about as much as steers.

15. *It makes for better lifetime reproductive performance of heifers*—Creep-fed heifers are heavier at weaning, reach puberty at an earlier age, and tend to breed earlier in the season throughout life, than noncreep-fed heifers.

16. *They're "bunk-broke"*—It provides calves that are accustomed to feed, with the result that they will continue on feed in the feedlot with a minimum of stress, shrink, digestive disturbances, and death loss, and without the normal period of 3 weeks to get them on full feed. This, plus their higher initial condition, results in their being ready for market approximately 40 days sooner than noncreep-fed calves. In turn, the shorter feeding period makes for a saving in interest payment, and less exposure to disease, injuries, and weather.

17. *It makes for earlier cycling and conception*—Calves that are creep fed have a source of nutrients other than from their dams; hence, they nurse less and put less stress on their mothers than calves that are not creep fed. As a result, dams of creep-fed calves lose less weight and show heat and breed back earlier than the dams of noncreep-fed calves.

18. *Cows are in better condition*—It leaves the dams less suckled down and in better condition (25 to 50 lb heavier) at weaning time because the calves partially wean themselves, an important consideration relative to cows from the standpoint of either sale or the wintering period ahead.

19. *It gives first-calf heifers a needed assist*—Calves from first-calf heifers tend to be lighter at birth than calves from older cows; consequently, there is more potential for good conversion of creep feed to additional gains on them. Also, the milk production of two-year-old heifers is only about 70 percent of their mature production.[4] Further, it is generally recognized that some of the stresses on first-calf heifers need to be lessened in order to get them rebred. It follows that calves from first-calf heifers generally give more response to creep feeding than calves from older cows, and that the heifers whose calves are creep fed show greater response in rebreeding than mature cows whose progeny are creep fed.

20. *It usually pays*—The potential profitability of creep feeding depends upon (1) the price of cattle and (2) the price of feed.

The following rule of thumb may be used to determine whether or not it will pay to creep feed: It pays to creep feed when the selling price per hundred pounds of calf is greater than the cost of one-fourth ton (500 lb) of feed.

One cattleman of the author's acquaintance summed up the economics of creep feeding in this way: "If your cost of gains runs 35 cents per pound, and you can sell the calves at 50 cents per pound, it's almost as good as

---

[4]Ensminger, M. E., *Dairy Cattle Science*, The Interstate Printers & Publishers, Inc., Danville, Ill., 1971, p. 51.

finding a money tree." Assuming a $1.00/cwt selling advantage for noncreep-fed calves, because they're lighter and thinner, most cattlemen feel that from creep feeding, they'll net, $3.00 to $5.00 per calf more after feed and other costs are deducted.

Limitations of Creep Feeding

Like many good things, creep feeding does have its limitations; among them, the following:

1. *It isn't always profitable*—Creep feeding may not be profitable because of the cost of the creep ration, low response, and/or price discrimination against fleshy weaner calves.

2. *It lowers feedlot gains and efficiency*—Fleshy creep-fed calves make slightly less rapid and efficient gains than calves not creep fed when they are (a) moved directly into the feedlot following weaning and (b) long-fed, because creep feeding alleviates compensatory gains. Also, it may cause small type cattle to get to fleshy—to "stall" or stop growing before weaning. However, if the latter occurs, it's a sure sign that the wrong kind of cattle are being bred. However, these disadvantages may be compensated for, in part at least, by the shorter feedlot feeding period and more desirable market weights of creep-fed calves.

3. *It makes for less desirable stockers*—Creep-fed calves do not make as desirable stockers as calves that have not been creep fed, simply because the latter are normally placed on less nutritious growing rations consisting predominantly of roughages. This may be further explained in this way: One of the basic rules in feeding slaughter cattle is always to proceed to a higher plane of nutrition; never go down.

4. *It mitigates against selecting cows for milk production*—Creep feeding makes it difficult to put selection pressure on cows for good milk production since creep feeding cancels differences in weaning weight due to lactation differences in dams. This may be an important consideration in production testing programs.

5. *It is difficult in less remote areas*—Creep feeding is difficult on less accessible ranges, because the very nature of creep feeding calls for close attention.

6. *It cannot be done where there are hogs, sheep, and goats*—These animals can enter any creep opening that is big enough for a calf.

GROWTH STIMULANTS FOR CALVES

Growth-promoting implants generally increase rate of gain and weaning weight of creep-fed calves. But there is little or no benefit from implanting noncreep-fed calves because the function of a growth stimulant is to improve utilization of energy, especially concentrates.

(Also see Chapter 30, Feeding Finishing Cattle.)

IDENTIFICATION

All calves should be identified as soon after birth as possible, and not later than three days of age. A combination of flexible plastic ear tags and tattoo numbers is recommended, thereby securing both ease of reading and permanency.

(Also see Chapter 10, Beef Cattle Management.)

Fig. 21-6. Roping calves for tattooing and branding on Tequesquite Ranch, Albert, New Mexico. (Courtesy, Albert J. Mitchell, Tequesquite Ranch, Albert, N.M.)

DEHORNING

Dehorning is an economic necessity, because horned calves usually bring lower prices. In addition, dehorned and naturally polled animals do less damage to facilities and other animals than cattle that have horns.

All naturally horned animals should be dehorned, preferably before they are two months old to minimize the "shock effects" of the operation. At that time, the blood vessels in the horn area are very small, which means less blood loss and minimum shock.

Dehorning may be accomplished by any of the following means:

1. *Mechanical*—A tube, spoon, or Barnes-type dehorner.

2. *Electrical*—Electrically heated irons have been used with good results.

3. *Chemical*—Liquids, pastes, or caustic sticks can be used for dehorning, but all of them should be used with caution and according to directions. Chemical dehorning materials should be allowed to dry before a calf is returned to its mother. Also, this type of dehorning must be done when calves are very young (less than two weeks of age).

If dehorning is done during the fly season, a good fly repellent should be applied.

(Also see Chapter 10, Beef Cattle Management.)

CASTRATING

Castration is recommended for all bull calves destined to be sold as feeders or finished in the feedlot. Bull calves and staggy-looking steer calves will not be accepted in many feeder calf sales. If sold as feeder animals, they usually bring a reduced price.

Castration time will vary according to method employed and management program, and it will be different for a commercial than for a purebred operation. Some cattlemen use elastrator bands when calves are only a few days old. Others use a knife or burdizzo (clamp) when the calves are three to four months old. Still others wait until weaning time to castrate. Bull calves will weigh more at weaning than steer calves; however, younger calves are easier to restrain for castration and suffer less shock therefrom than older animals.

The commercial cow-calf operator should castrate all bull calves before weaning, and preferably before they are four months old.

Most purebred breeders who raise bulls for sale may wish to wait until after weaning to castrate bull calves, so that they can evaluate weaning weights and use them in their bull selections. However, poor quality purebred bull calves should be castrated earlier when it is evident that they are not bull prospects.

The use of a knife, preferably by removing the lower one-fourth of the scrotum—thence the testicles, assures complete removal of the testicles and good drainage. There is, however, some risk in hemorrhage and postcastration infection. Poor technique with the burdizzo or elastrator results in incomplete castration and price discount. With the elastrator it is necessary that both testicles be below the band. With the burdizzo both cords must be completely crushed.

Whenever castration is by knife and during the fly season, a fly repellent should be used to keep flies away and help prevent infection.

(Also see Chapter 10, Beef Cattle Management.)

WEANING

Weaning is a traumatic experience for a calf. It represents environmental, nutritional, psychological, and, altogether too often, vaccination-castration-dehorning changes—all of which make for great stress. Generally, at weaning time the calf is moved into a strange environment to which it must adjust—a new pasture, corral, or shelter; its food supply—milk—is suddenly removed; and its association with and protection by its mother is lost. Under such circumstances, it's small wonder that calves lose weight and become more susceptible to disease. The marvel is that they survive such mistreatment so well.

Calves should be weaned when they are seven to eight months old. Weaning at this age fits in well with the weight record-keeping requirements of most performance testing programs. Also, calves will be about the right age and weight for fall feeder calf sales.

Weaning earlier than seven or eight months may be necessary in years when pastures are short or when calves are from first-calf heifers.

The best way to wean is to remove the calves from their dams and keep them out of sight of each other. Cows and calves should never be turned together once the separation has been made. Such a practice will only prolong the weaning process, and it may also cause digestive disorders in the calf. Provide calves with plenty of water, free-choice hay, and three to four pounds of grain per head per day. If calves were creep fed, continue their rations during the weaning period.

During the weaning process, calves should be confined to a small area to cut down on walking and shrinkage. In bad weather, protection should be provided from cold wind and rain; they should have access to a shed, wooded area, gorge, or other protection.

●*How to minimize stress and weight loss*—Weaning calves is more a matter of preparation than of absolute separation from the dam. Minimizing stress and weight loss depends largely upon the thoroughness of the preparation. The following procedure is recommended:

1. Dehorn and castrate well ahead of weaning time.

2. Bring the cows and calves into a small pasture paddock two to three weeks before weaning. If weaning paddocks are not available, use small pastures. Avoid dusty pens in order to minimize respiratory diseases and pneumonia.

3. Creep feed the calves, so that they learn to eat. Use the same feed as the preconditioning ration (such as corn or sorghum silage plus a protein supplement; or a combination of a high quality grass-legume hay plus a concentrate supplement; or a combination of these feeds); or if they have been on a creep ration all along, continue with the creep ration to which they are accustomed.

Keep the feed clean and free from mold.

4. Give the first immunizations while the calves are still nursing, and follow with the booster shots as recommended by the manufacturer or the veterinarian.

5. Treat for parasites, internal and/or external, before weaning.

6. Wean the calves by moving the cows out of the weaning paddock(s) so that the calves remain in familiar surroundings.

7. Check calves two or three times daily. Remove sick ones and take them to the hospital area for treatment.

8. Administer TLC (tender loving care) to calves being weaned. Always handle gently; never rope them.

9. Consider (confer with your veterinarian) the use of tranquilizers during the weaning process, as a means of calming calves and minimizing weight loss.

●*Drying up the cow*—With higher milking strains of beef cattle, when drying up cows, cattlemen will have the same concerns as dairymen—that of avoiding "spoiled udders." To alleviate this problem, the following procedure is recommended:

1. Do not feed milk-stimulating feeds at weaning time. Either put the cows on poorer pastures or feed a nonlegume forage.

2. Let "back pressure" in the udder build up. Examine the udder at intervals, *but do not milk it out.* If the bag fills up and gets tight, rub an oil preparation (such as camphorated oil or a mixture of lard and spirits of camphor) on it, *but do not milk it out.* At the end of five to seven days, when the bag is soft and flabby, what little secretion remains (perhaps not more than half a cup) may be milked out if so desired.

PRECONDITIONING

Preconditioning is a way of preparing the calf to withstand the stress and rigors of leaving its mother, learning to eat new kinds of feed, and shipping from the farm or ranch to the feedlot. To the cow-calf producer, it is a program of management, nutrition, and immunization.

A study conducted by Washington State University revealed that preconditioning of calves increased profits by $10.56 per head on steers and $7.67 on heifers, after deducting all preconditioning costs. This explains why some feedlot operators are willing to pay $1.00 to $2.00 per cwt more for preconditioned calves, and why others are lowering the price if it is not done.

The term "preconditioning" is new, but the concept has long been recommended. Stated simply, preconditioning is the schedule of practices used in preparing feeder calves to withstand the stress of leaving their mothers, shipping, and adapting to feedlot conditions. It consists of administering generous amounts of TLC (tender, loving care) along with immunological practices and treatment for parasites.

Fig. 21-7. Preconditioning, along with production testing, will be the trademark of the producer of reputation feeder calves of the future. (Courtesy, Ralston Purina Company, St. Louis, Mo.)

Changed environment; excitement of sorting, loading, and shipping; long periods without feed and/or water; movement through one or more assembly points; change of feed; and exposure to disease—all add up to *fatigue, stress, shrink,* and *lowered disease resistance.*

Preconditioning is the answer. The steps used in preconditioning may, and should, vary somewhat among areas, farms, and ranches. The important thing is that the program be written down, adhered to rigidly, then certified to by both the owner and the veterinarian. The producer should take the lead in developing such a program, but he should seek the counsel of his veterinarian and potential buyers.

Preconditioning is often confused with handling newly arrived feedlot cattle, and backgrounding. This is understandable because all three of them are important phases between weaning and finishing. Yet, each of them is a distinct and separate phase.

● *Where does stress take place?*—Stress takes place when a calf is weaned, at which time the social structure is disrupted. Also, there is stress when an animal is placed in a different environment, whether it be in a pen on the home ranch or on a loaded truck on the way to market. If stress is to be lessened, therefore, it is obvious that measures must be taken before any of these steps happen.

● *Losses attributed to prefeedlot stress*—It is estimated that nearly 1.5 million head of cattle die in feedlots each year, making for an estimated annual loss of more than $500 million. Losses from sickness are even greater—they run 2 to 5 times as great as the actual death loss. Sickness losses accrue from the expense and treatment plus the resulting inefficiency. Thus, the combined losses—death losses, and losses due to sickness—add $20 to $25 per head onto the cost of every feedlot-finished animal.

● *What are preconditioned cattle?*—Opinions differ rather widely as to what constitutes properly preconditioned cattle. However, the following preconditioning program is presented with the hope that the beef producer will use it (1) as a yardstick with which to compare his existing program, or (2) as a guidepost so that he and his local veterinarian, and other advisers, may develop a similar and specific program for his own enterprise:

1. *Handle quietly*—Calves should be handled quietly, with a minimum of excitement.

2. *Dehorn and castrate*—All calves that will eventually go into feedlots should be dehorned (although tipping of horns is acceptable), and they should be castrated unless they are to be fed out as bulls. There is far less stress if calves are dehorned and castrated well ahead of weaning—about 2 months of age is best.

3. *Wean*—Calves should be weaned 30 days ahead of shipment.

4. *Start on feed*—Adjust to feed bunks and water troughs and start on a ration similar to that which they will get in the feedlot. For the first 3 days following weaning, calves should have access to loose grass hay. Additionally, they should be started on a ration of about the following composition:

Crude protein, minimum %	12.0
Calcium, %	.5
Phosphorus, %	.3
Vitamin A, IU/lb	5,000
Net energy for production, (NE$_p$), Mcal	38
Roughage: concentrate ratio, approx	40:60

If weaning is totally impractical, calves should be started on a creep feed similar to the above ration.

This type of ration will be very similar to the starting ration that calves will receive when they arrive in the feedlot.

Use medicated feed only on the recommendation of your veterinarian.

5. *Vaccinate*—Vaccinate 2 weeks after weaning. If calves were vaccinated for blackleg, malignant edema, and leptospirosis before 3 months of age, revaccinate. Simultaneously, vaccinate for "red nose" (infectious bovine rhinotracheitis, or IBR), bovine virus diarrhea (BVD), and para-influenza 3 (PI-3). In some instances, clostridial toxoids for types C and D are needed. Follow your veterinarian's advice for vaccination procedures. If a direct sale to a feedlot is involved, the calves should be vaccinated in keeping with the regular program of the feedlot.

6. *Treat for parasites*—At the time of weaning, and prior to shipment, calves should be checked for both internal and external parasites, and treated as necessary. Usually this involves (a) treating for grubs, through either spray, pour-on, or feed; (b) spraying for lice; and (c) checking for worm eggs, and worming if necessary.

7. *Reduce time from farm or ranch to feedlot*—Every effort should be made to reduce the total time between the moment calves leave the farm or ranch and when they arrive at the feedlot.

Where either truck or rail shipments are longer than 36 hours (the 28-hour law governing rail shipments may be extended to 36 hours upon written request of the owner), unload en route for the purpose of giving feed, water, and rest for a period of at least 5 consecutive hours before resuming transportation.

8. *Reduce stress and exposure to infection*—The stress and exposure to infection during the marketing and transportation periods should be reduced to a minimum.

9. *Preconditioning certificate*—It is extremely important that records be kept of all husbandry, nutritional, and medical histories, and that the man who sells the feeder cattle should provide the man receiving them with a written record of all of them. This will help the feedlot operator fit the cattle to his program and minimize costly and unnecessary procedures. A suggested preconditioning certificate is herewith presented as Fig. 21-8.

• *Cost of preconditioning*—The cost of preconditioning will run about $10 per head, the amount depending primarily upon the number and kind of vaccinations administered and the cost of feed. Some or all of this cost will be recovered by gains made during the preconditioning period.

• *Marketing agencies can lessen stress and exposure*—The above discussion pertains to preconditioning at the farm or ranch level. It is recog-

PRECONDITIONING CERTIFICATE

Date _____
Number of Cattle _____
Steers _____ Heifers _____ Bulls _____ Breed _____ Age _____ Brand _____

PRACTICES, TREATMENT, AND VACCINATION

	Date	Product Brand	Signature of Responsible Person
Castrated			
Dehorned			
Blackleg & Malignant Edema Vaccination			
Shipping Fever Vacc.			
Lepto Vaccine			
IBR Vaccine			
BVD Vaccine			
Other Vaccines			
Weaned			
Wormed			
Grub Treated			
Lice (Treatment) Spray or Dip			
Vitamin A.D.E. Inj.			
Medication, Antibiotics, Sulfa, electrolytes			

Ration during preconditioning period _____

Loading Point _____

       The undersigned hereby declares and certifies that the practices, treatments and vaccinations indicated above have been carried out and administered to all the cattle described and identified by this certificate.

Seller:

_____
(Signature of owner or authorized representative)   (date)

Cattle Producer:

_____
(Signature)   (date)

Seller's Veterinarian:

_____
(Signature)   (date)

Fig. 21-8. Preconditioning certificate.

nized, however, that not all cattle are moved directly from the cow-calf producer to the feedlot. Many of them pass through auction markets, terminal markets, and other similar intermediate locations. Hence, the marketing agencies handling feeder cattle can do much to cut down on stress and exposure to disease. Recommendations to this end are:

1. Refuse to accept sick animals, or at least isolate them. This is important because one sick animal may cause a disease break in hundreds of other cattle.

2. Isolate cattle from individual producers whenever possible. Here again, this is recommended in order to cut down the spread of infectious diseases.

3. Keep dust to a minimum.

4. Feed animals well-balanced rations similar to the rations to which they have been accustomed.

5. Move cattle to their final destination as expeditiously as possible.

6. Prevent bruises.

7. Improve sanitation in all handling facilities.

8. Reduce weather stress.

9. Keep records which will help pinpoint problem areas and unscrupulous individuals.

• *Preconditioning will increase*—On an industry-wide basis, preconditioning feeder cattle could save millions of dollars now lost in sickness, shrink, and death.

Feeder buyers will determine how quickly preconditioned cattle become generally available. They will speed their availability when they ask the seller:

1. When were they weaned (if calves are involved)?

2. Are they accustomed to bunk feeding and trough watering?

3. Have they been treated for grubs?

4. Have they been examined for internal and external parasites, and treated if necessary?

5. Have they been vaccinated for IBR, BVD, shipping fever, blackleg?

6. Is certification for the above procedures available?

Buyers will increasingly favor those producers who follow such a program. Preconditioning, along with improved breeding based on production testing, will be the trademark of the producer of reputation feeder calves of the future.

MARKETING ALTERNATIVES

Marketing the calf is the end of the line; it is that part of a cow-calf program which gives point and purpose to all that has gone before. After breeding, feeding, managing, and preconditioning the calf, it is still necessary to market it in such manner as to receive the greatest net income.

In recent years, the marketing of feeder calves has greatly benefited from numerous outlets, improved truck transportation, and improved market reporting. However, it requires more skill on the part of the cow-calf producer to benefit from today's modern marketing system than it did a generation ago. The producer must be aware of the numerous market channels, market news services, the value of his calves, and the economic and political factors affecting markets in order to sell his calf crop to the best advantage.

The following marketing alternatives are available to most cow-calf producers:

• *Selling feeder calves*—Spring-dropped calves may be weaned and sold as feeder calves in the fall when they are 7 to 8 months old. Steer calves normally weigh 400 to 500 pounds. Usually heifer calves weigh about 5 percent less than steer calves.

Many feeder calf sales are held each fall; and a few are held in the spring. These provide a good outlet for locally produced calves fresh off the farm or ranch. Usually calves are sorted into uniform lots by weight and grade. The smaller producer benefits by being able to "pool" his calves with those of similar weight and quality, thus making a larger lot and bringing a higher price per cwt.

Cow-calf producers may also sell directly to cattle feeders, order buyers, or cattle dealers; or they may consign to terminal markets or decentralized buying stations.

• *Selling yearling feeders*—Some cattlemen have always grown calves beyond weaning weight, with the prevalence of the practice varied by years according to feed. This practice is most common in southwestern United States.

Short yearlings are weaner calves held over because they are too light to sell at weaning time and/or because adequate forage is available. Usually, they are sold at just under one year of age, although they may range up to 14 months in Arizona and New Mexico. Long yearlings average about 16 months of age when sold to go into feedlots.

• *Finishing cattle*—This alternative is open to all cow-calf producers; either in their own facilities if they have them, or in custom feedlots. Calves may go directly into the feedlot at weaning to be marketed as finished cattle at 12 to 16 months old; or they may be grown out as stockers, then finished out at slightly older ages and heavier weights.

The producer who has the choice of selling his calves at weaning or as finished cattle can generally make more profit per cow by feeding his own calves, especially if they are sired by performance-tested bulls with genetic potential for rapid growth. He may as well reap the full benefits from his breeding program, rather than sell his calves and let someone else have a sizable share of the profits.

• *Replacement heifers*—The best source of replacement heifers is from the breeder's own herd, regardless of whether the operation is purebred or commercial. This is so because more is known about them—their age, immunization history, health status, and performance—than replacements from any other source. Thus, at weaning time, the top performing (heaviest) heifer calves should be retained—keeping 50 percent more than will actually be needed.

There is one exception where saving replacements would not be practical—that's a crossbreeding program where specialized F₁ females are mated to a terminal sire breed. In such a program, all calves are marketed at weaning or after finishing. Thus, it is necessary to purchase F₁ heifers from an outside source.

QUESTIONS FOR STUDY AND DISCUSSION

1. It is estimated that there is an appalling calf loss of seven percent between birth and weaning. What can cattlemen do to lessen these losses?

2. Outline a management program for calving time.

3. How would you raise orphan calves or multiple birth calves? Give the feeds and the schedule of feeding for the first 120 days of a calf's life.

4. What are the advantages and disadvantages of early weaning? Outline a program for early weaning, giving the age of weaning, the feed, and the feeding schedule.

5. What are the advantages and disadvantages of creep feeding? Under what conditions would you recommend creep feeding; under what conditions would you recommend against creep feeding?

6. Should growth-promoting stimulants be used on calves that are not creep fed? Justify your answer.

7. How would you go about weaning calves, from the standpoint of both the cows and calves?

8. Who benefits the most from a preconditioning program, the cow-calf man or the cattle feeder?

9. Detail a preconditioning program.

10. What marketing alternative—selling feeder calves, selling yearling feeders, or finishing cattle—should a cow-calf producer select?

SELECTED REFERENCES

Title of Publication	Author(s)	Publisher
Beef Cattle, Sixth Edition	A. L. Neumann R. R. Snapp	John Wiley & Sons, Inc., New York, N.Y., 1969
Beef Cattle Production	K. A. Wagnon R. Albaugh G. H. Hart	The Macmillan Company, New York, N.Y., 1960
Beef Cattle Science Handbook	Ed. by M. E. Ensminger	Agriservices Foundation, Clovis, Calif., pub. annually since 1964
Beef Production in the South	S. H. Fowler	The Interstate Printers & Publishers, Inc., Danville, Ill., 1969
Commercial Beef Cattle Production	Ed. by C. C. O'Mary, I. A. Dyer	Lea & Febiger, Philadelphia, Penn., 1972
The Stockman's Handbook, Fourth Edition	M. E. Ensminger	The Interstate Printers & Publishers, Inc., Danville, Ill., 1970

CHAPTER 22

REPLACEMENT HEIFERS

There is no better or quicker way to improve the reproductive performance of a herd than through the selection and proper development of replacement heifers.

Where the beef cattle herd is neither being increased nor decreased in size, each year about 20 percent of the heifers, on the average, are retained as replacements, and about the same percentage of old cows is culled. That means that there is a complete turnover in the cow herd every five years.

But not all of today's heifers become tomorrow's cows! Statistics show that about seven percent of all calves born die before reaching weaning age. Still others must be culled for one reason or another, either before or after weaning. Thus, to maintain *status quo* in a beef herd—with no provision for expansion whatsoever—each year the cattleman should start with 50 percent more weaner heifers than he actually needs. This means that for every 100 cows in the herd, 20 replacement heifers are actually needed to maintain the same size herd. However, 30 weaner replacement heifer prospects (50 percent more than actually needed) should be held, simply because, based on averages, 10 of these will fall by the wayside—they will either die or have to be culled before they replace their elders.

1173

Fig. 22-1. Replacement heifers on pasture. Good pasture in season, water, shade, and minerals will make for proper growth and development of heifers. (Courtesy, American Hereford Association, Kansas City, Mo.)

CRITERIA FOR SELECTION OF REPLACEMENT HEIFERS

Selection of replacement heifers is an attempt to retain the best of those animals in the current generation as parents of the next generation. Obviously, the skill with which these selections are made is all important in determining the future of the herd.

The cattleman who raises his replacements is in a better position to evaluate the animal genetically than the operator who buys replacements, simply because their close relatives are available in the same herd under similar feed and management conditions.

The type and performance of each individual heifer and of her close relatives are the criteria to use in selecting replacement heifers. Since replacement heifers are not old enough to have progeny, the closest relatives are the parents and half-sibs (half-brothers and half-sisters).

Selection should be based on as accurate and complete records as possible, with consideration given to the criteria in the sections which follow.

Individuality

Each replacement heifer should be scored for conformation at the weaning stage, and again at the yearling stage. This score should relate the structure of the heifer to her function—that of producing calves, which, in turn, will produce quality beef efficiently. Thus, scoring had best be done by the Functional Scoring System, which involves selecting animals for the following six traits:

1. Reproductive efficiency
2. Size
3. Muscling
4. Freedom from waste
5. Structural soundness
6. Breed type

The Functional Scoring System is detailed in Chapter 4 of this book.

Performance Test

Performance testing of heifers should include a record of (1) weaning weight, and (2) yearling weight, taken over scales. Heavy weaning weight is indicative of milking ability of the dam. Heavy yearling weight is indicative of size and growthiness and is correlated with feed efficiency. Basically, performance testing is for the purposes of improving growth rate, or rate of gain (which is 45% heritable), and feed efficiency (which is 40% heritable).

Traditionally, beef cattle performance testing is conducted as follows: Each animal is individually identified by means of an ear tattoo, ear tag, ear notches, brand, or neck strap. Soon after weaning, and following an adjustment period of 2 to 3 weeks, animals are individually weighed and individually full-fed on weighed amounts of a high energy feed for a period of 140 days, followed by individual weighing at the end of this period; thereby obtaining an individual record of both (1) rate of gain and (2) feed efficiency—the pounds of feed required to make 100 pounds of gain.

Under practical conditions, performance testing of cattle is usually limited to bulls for the following reasons: (1) a bull produces in his lifetime many more offspring than a female, which means that his influence on the total genetic progress of the herd is greater; (2) many more replacement females than herd bulls are selected, with the result that the facilities, labor, and cost of individually testing them in the traditional manner would be overwhelming; and (3) if heifers are production tested in the traditional manner by full-feeding on a high energy ration, they may become so fat that it will impair their reproductive performance. Yet, where possible, heifers should be performance tested, with some modification of the above procedure, because greater genetic progress can be made thereby.

From the above, it may be concluded that the three major problems encountered in performance testing heifers in the traditional manner are:

1. The very considerable labor and expense involved in individually feeding large numbers of replacement heifers in order to measure feed efficiency.

2. The hazard of lowered reproductive performance as a result of heifers becoming too fat when full-fed on a high energy ration.

3. At the close of such a test, fat heifers must be placed on reducing rations, which means that much of the feed that went into making gains during the test period is lost.

To alleviate these three problems, it is recommended that heifers be performance tested somewhat differently than bulls; that their performance test be modified as follows:

1. *Measure rate of gain only*—Fortunately, there is a significant correlation between rate and efficiency of gain (r = .40 to .60); hence, selection for rapid gains should also improve efficiency of gains. Thus, it is recommended that heifers be tested only for rate of gain, which can be accomplished by individually weighing at the beginning and end of the test period.

By measuring rate of gain only and eliminating feed efficiency, heifers may be group fed; thereby alleviating the very considerable labor and facilities of individual feeding. Heifers gaining the fastest may then be selected for replacements.

2. *Full-feed a high roughage ration*—In order that heifers will not become too fat when full-fed during the test period, they should be fed a high roughage-low energy ration. They may even be performance tested on pasture, as pasture gain has an estimated heritability of 30 percent (in comparison with 45 percent for feedlot gain).

Fig. 22-2. Yearling Angus heifers on summer range. Heifers may be performance tested on pasture, as pasture gain has a heritability of 30 percent—in comparison with 45 percent for feedlot gain. (Courtesy, Martin Jorgensen, Jr., Ideal, S.D.)

By a high roughage drylot ration is meant one in which the ratio of roughage to concentrate is somewhere between (a) one part of roughage to one part of concentrate, and (b) two parts of roughage to one part of concentrate.

In order to avoid selective eating, high roughage rations should be fed as a complete mixed feed, pelleted or unpelleted. Also, the ration should be full-fed, allowing each heifer to eat according to her appetite, because a good appetite is inherited and conducive to faster gains—a desirable characteristic, especially in feedlot cattle.

3. *Test for 140 days*—The performance test should cover a minimum of 140 days. A period of this length allows each heifer to express her potential

growth rate and tends to average some of the environmental effects on growth rate during the feeding period—such as the condition (fatness or thinness) of each heifer when she is placed on test.

Usually, animals are performance tested soon after weaning, primarily for convenience reasons. It is noteworthy, however, that selection for faster growth rate during one period of life should improve growth rate in another period of life.

4. *365-day or 550-day weights may be used*—Often it is not practical to performance test heifers by full-feeding them in a drylot for 140 days or more, with weights taken at the beginning and the end of the period. It is generally more practical to feed them under normal conditions, then obtain either a 365-day or a 550-day weight. Heifers gaining the fastest may be selected for replacements.

The 365-day and 550-day weights may be calculated as follows:

a.
$$\text{adjusted 365-day weight} = \frac{\text{actual final weight} - \text{(minus) actual 205-day weaning weight}}{\text{number of days between weights}} \times 160 + \text{205-day weaning weight adjusted for age of dam}$$

b.
$$\text{adjusted 550-day weight} = \frac{\text{actual final weight} - \text{(minus) actual 205-day weaning weight}}{\text{number of days between weights}} \times 345 + \text{205-day weaning weight adjusted for age of dam}$$

Yearling (365-day) weights or 550-day weights are 30 to 35 percent heritable, whereas rate of gain on full feed in a drylot is 45 percent heritable. Nevertheless, selection of heifers on the basis of either 365-day or 550-day weights should result in some progress and is generally much more practical than performance testing on a full feed.

Record of the Sire and Dam

Every animal receives one-half of its inheritance from each of its parents. Hence, the more superior the sire and dam are genetically, the more likely they are to transmit genetic superiority to their offspring. But even genetically superior parents will produce offstring which range from good to poor in type and performance. Neverthelesss, the average of the progeny will be higher than the average of progeny from less superior parents.

The data presented in Table 22-1 show that the individual's own record is more important than the records of its parents or grandparents. These data also show that the individual's own records plus those of the parents increase the accuracy of selection very little over that attained when selection is based only on the individual's record.

TABLE 22-1

ACCURACY OF SELECTION[1] BASED ON THE RECORDS OF THE PARENTS
AND GRANDPARENTS AS COMPARED TO SELECTION ON THE BASIS
OF THE INDIVIDUAL'S OWN RECORDS[2]

Trait	Fertility	Weaning Wt. & Conf. Scores[3]	Yearling Weight
Heritability of Each Trait	.10	.30	.40
Accuracy of selection based on the individual's records	.32	.55	.63
Accuracy of selection based on records of the individual plus records of one parent	.35	.58	.66
Accuracy of selection based on records of the sire, dam, and all four grandparents but not the individual's own records	.27	.43	.48

[1]Perfect accuracy of selection would be 1.0.
[2]Lasley, John F., Selection and Acquisition of Foundation Cows and Replacement Heifers, *Commercial Beef Cattle Production*, ed. by C. C. O'Mary and Irwin A. Dyer, pub. by Lea & Febiger, Philadelphia, Penn., 1972, Chapter 6, Table 6-2.
[3]Weaning weight, weaning conformation score, and yearling conformation score are heritable to about the same degree.

Record of the Half-Sibs

An average beef animal has very few full brothers and sisters during its lifetime, simply because only one offspring is normally produced at each birth. Hoever, it may have many half-sibs born in the same herd and in the same season because a bull sires many calves in a given year. An animal is related to its half-sibs by about 25 percent if no inbreeding has been practiced.

The data presented in Table 22-2 show the relative accuracy of selection based on the records of the individual's half-sibs as compared to selection on its own records only.

Table 22-2 reveals the following: (1) the larger the number of half-sib records available, the more accurate the selection; and (2) selection on the basis of half-sib records is more accurate than selection only on the basis of the individual's own records (a) when the trait is lowly heritable, and (b) when records are available on 30 or more half-sibs.

Whenever possible, selection should be based on the individual animal's own records. Otherwise, selecting superior individuals from superior families is the method of choice.

TABLE 22-2

THE RELATIVE ACCURACY OF SELECTION BASED ON THE RECORDS
OF THE INDIVIDUAL'S HALF-SIBS AS COMPARED TO SELECTION
ON THE BASIS OF THE INDIVIDUAL'S OWN RECORDS[1]

Number of Half-sibs	Fertility with Heritability of .10	Weaning Weight and Conformation Scores with Heritability of .30	Yearling Weight with Heritability of about .40
1	.25	.25	.25
2	.35	.34	.34
3	.42	.40	.40
4	.48	.45	.44
5	.53	.49	.47
6	.58	.52	.50
7	.62	.55	.52
8	.65	.57	.54
9	.69	.59	.56
10	.71	.61	.57
20	.92	.72	.66
30	1.04	.77	.69
40	1.13	.80	.72
50	1.19	.82	.73
100	1.34	.86	.76

[1]Lasley, John F., Selection and Acquisition of Foundation Cows and Replacement Heifers, *Commercial Beef Cattle Production*, ed. by C. C. O'Mary and Irwin A. Dyer, pub. by Lea & Febiger, Philadelphia, Penn., 1972, Chapter 6, Table 6-3.

Calculations consider many factors. A relative accuracy of 1.00 would mean that selection on the basis of half-sibs would be as accurate as selection based on the individual animal's own record. A relative accuracy of .25 would mean that selection on the basis of half-sib records would be only one-fourth as accurate as selection on the basis of the individual's own records.

Select Against Excessively Fat Heifers

A certain amount of condition improves the type and conformation of an animal, and makes for a pleasing appearance. But excessive feeding increases feed costs and produces too much fat, which is harmful from the standpoints of reproduction and milk production. Hence, in selecting replacement heifers, those which are excessively fat should be avoided. Among the reasons for selecting against excessive fatness are those which follow:

• *Excessive fatness has an adverse effect on reproduction*—A high plane of nutrition in growing heifers results in early sexual maturity, but it can be overdone. Excessive fatness has a detrimental effect on the reproductive process. It makes for calving problems, with heavy calf losses at and soon after birth as a result of excess fat decreasing the size of the birth canal and the calves being presented backward.

• *Excessive fatness has a detrimental effect on udder development and milk production*—Research with beef, dairy, and laboratory animals has demonstrated the adverse effect of very high planes of nutrition on mammary development and milk yield. Thus, in addition to increased calving difficulty and drastically increased feed costs, feeding replacement heifers excessively high amounts of feed results in decreased milk production. This was clearly demonstrated by the Oklahoma Station, in a study in which replacement Hereford heifers were allotted to four planes of winter feed: (1) low, (2) moderate, (3) high, and (4) very high. The "very high" group was self-fed a 50 percent concentrate mixture to gain as rapidly as possible, both as weanling calves and in subsequent winters. The depression in milk yield is shown in Fig. 22-3.

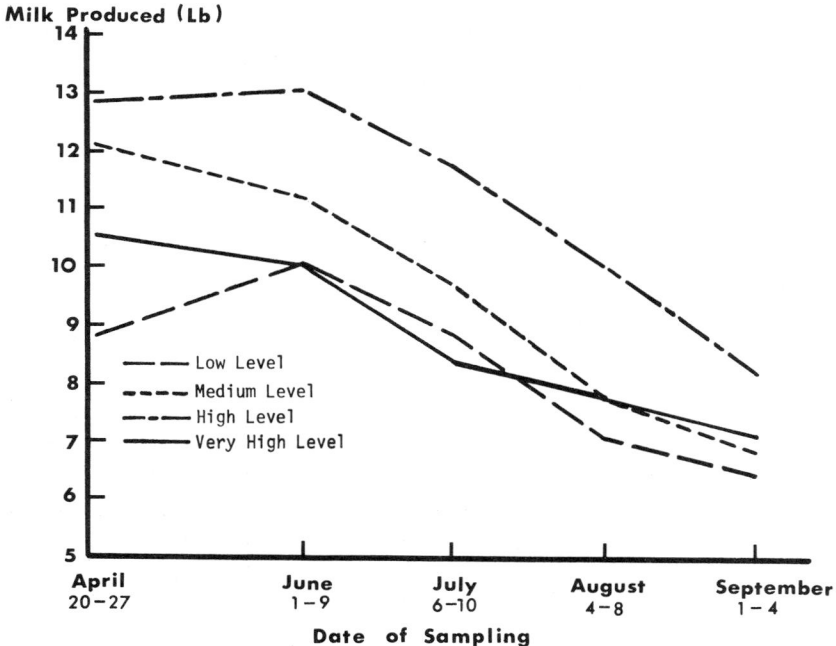

Fig. 22-3. Milk production of three-year-old Hereford cows wintered on four different levels of feed—low, medium, high, and very high. (Source: *Feeding and Breeding Tests*, 36th Annual Livestock Feeder's Day, MP-67, Oklahoma Agr. Exp. Sta., Oklahoma State University, Stillwater, Okla., April 21, 1962, p. 55)

The depressed milk production of the very high level group is clearly shown in Fig. 22-3.

The detrimental effects on milk production seem to increase from the first to subsequent lactations, according to the Cornell Station.[1] It would appear, therefore, that early overfatness has a detrimental residual effect on the development of secretory tissue which increases with time.

Select Against Underfed Heifers

Restricted rations often occur during periods of drought, when pastures or ranges are overstocked, or when winter rations are skimpy. When such deprivations are extreme, there may be lowered reproductive efficiency on a

---

[1]Reid, J. T., J. K. Loosli, G. W. Trimberger, K. L. Turk, S. A. Asdell, and S. E. Smith, Causes and Prevention of Reproductive Failures in Dairy Cattle, *Cornell Agr. Exp. Sta. Bull. 987*, Feb. 1964.

temporary basis. It is noteworthy, however, that experimental results to date fail to show that the fertility of the germ cells is seriously impaired by uncommonly high or low feed and nutrient intakes.[2] In contrast to these findings, it is equally clear that the onset of heat and ovulation in young heifers is definitely and positively correlated with the level of nutrient intake; thus, if feed or nutrient intake is too low for normal rates of growth and development, the onset of reproductive function is delayed.[3]

Although healthy, undernourished heifers will usually recover when the level of feeding is returned to normal, it is not always possible to determine if poor condition is due to lack of feed, parasitism, or disease. Thus, there is reason to discriminate against poorly developed heifers.

Culling Replacement Heifers

As previously indicated, when neither increasing nor decreasing the size of the herd, cattlemen should start with 50% more replacement heifers than they actually need. This means that for each 100 cows in the herd, 30 weaner heifer replacement prospects will be held; 20 head of which will actually be needed to replace their elders (20% replacement per year), and 10 of which will be culled.

Selection and culling is a continuous process. Some replacement heifers will be culled because they do not measure up from the standpoint of individuality. For example, when scoring (grading) them under the Functional Scoring System (Chapter 4), as weanlings and again as yearlings, it might be decided to cull all animals that are not in the upper 20% in each of the 6 traits—reproductive efficiency, size, muscling, freedom from waste, structural soundness, and breed type. Further, it might be decided to cull those in the lower 10% of the performance test—in rate of gain. Others may be culled because of sickness or injuries or because of such abnormalities as chronic bloat, founder, poor eyesight, or poor teeth. Still others may be culled because of such things as failing to conceive readily, calving difficulty, giving birth to a dead calf, or giving insufficient milk for the calf. A few heifers may even be carried until they wean their first calves, then culled because of the light weight of their calves, which is about 50% repeatable.

BREEDING TWO-YEAR-OLD HEIFERS

Surveys indicate that approximately 50% of 2-year-old heifers require help at calving, and that 15% of the calves and 5% of the heifers are lost at parturition.[4] Thus, optimum reproductive performance of 2-year-old heifers can be achieved only if there is application of the best in breeding, feeding, and management, along with knowledge of the factors affecting reproduc-

[2]*Northeastern States Regional Bull. 32*, Cornell University Agr. Exp. Sta. Bull. 924, p. 30.
[3]Ibid.
[4]Moore, D. G., D. Chambers, J. A. Whatley, Jr., and W. D. Campbell, Some Factors Affecting Difficulty at Parturition of Two-Year-Old Hereford Heifers, *Journal of Animal Science*, Vol. 15, No. 4, 1956, p. 1225.

tion. (See Chapter 6 for "Factors Affecting Reproduction," much of which pertains to calving 2-year-old heifers.)

Age of Puberty

Puberty may be defined as the age of sexual maturity, or the first ovulation estrus. The normal age of puberty in cattle is 8 to 12 months, but it may range from 4 to 14 months. The age at which puberty is attained varies according to (1) breed, with the smaller, earlier maturing breeds having an earlier onset of puberty than the larger, slower maturing ones; and (2) nutritional and environmental factors, with puberty occurring when animals have reached about one-third of their adult size.

Jersey heifers usually reach puberty at an early age. Brahman heifers may reach puberty 6 to 12 months later than heifers of the European breeds. Crossbred heifers reach puberty at lighter weights than straightbreds.

Age to Breed Heifers

The age at which to breed heifers will vary with their growth and development. However, when heifers are reasonably well grown and weigh 600 to 700 pounds, a safe rule is to breed at the first breeding season after they are 13 to 14 months old. Some breed registry associations will not register a calf born to the heifer under a certain stipulated age; thus, purebred breeders need to be informed relative to such rules.

There appears to be an increasing trend among commercial cattlemen to breed heifers to calve as two-year-olds. The more successful early-breeding advocates feed such heifers rather liberally, either (1) by grazing them on the choicest range, and/or (2) by feeding added concentrates during the winter months. Also, more and more cattlemen who calve out two-year-old heifers use a bull known to sire small calves at birth. Certainly, if the dam and the calf are not adversely affected, breeding at an early age is advantageous from the standpoint of cutting production costs.

Added Beef from Two-Year-Olds

Population experts agree that there will be 7.5 billion people in the world by the year 2007—double the 3.8 billion today. Right here in the United States we shall have need for 69 percent more beef by 1985; with no further assist from either high energy rations or dairy beef (from the culling of dairy cows and the growing out of dairy calves rather than vealing). How are we going to meet it?

More beef can, and will, be produced in two ways: (1) by increasing cow numbers; and (2) by stepping up the production per cow. In the latter category, breeding heifers to calve as two-year-olds offers one of the most practical approaches—and profitable, too.

On January 1, 1974, there were 8,214,000 replacement heifers in the United States.[5] If 80 percent of these heifers were bred to calve as 2-year-olds rather than as threes; if they weaned an 80 percent calf crop, or

6,571,200 calves, averaging 400 lb; if all these calves were finished off to an average on-foot weight of 1,048 lb and an average carcass weight of 620 lb (same as the 1972 average on-foot and carcass weights of fed cattle); that would make for a total of 4,074 million pounds more beef produced than would have been obtained had all these heifers calved as 3-year-olds. That's enough beef to feed 37 million people a whole year at our 1973 per capita beef consumption level of 109.3 lb.

Also, and most important, producing more beef by calving out 2-year-olds instead of 3-year-olds doesn't require more cow numbers. Hence, to produce the 6,571,200 head of 400-pound weaner calves would require few more acres and little more feed and equipment. This means that the capital investment is less than when cow numbers are increased. True enough, more feed will be required to finish the calves from weaning to slaughter. But the fact remains that greater efficiency can be achieved by putting a heifer to work early in life.

The Choice—Calve Twos or Threes

For practical reasons, most commercial cattlemen must either calve out as twos or threes. They have no in-between choice. This is so because of the common practice of one-season calving, spring or fall. But the choice makes a difference of 12 months—one whole year. In terms of the productive lifetime of an 8-year-old cow, a heifer calving at 2 will begin paying her keep after only 25 percent of her lifetime, whereas one calving at 3 will "freeload" 37.5 percent of her lifetime.

FACTORS FAVORABLE TO CALVING TWO-YEAR-OLDS

Based on experiments and experiences, the primary factors *favorable* to calving 2-year-olds are:

1. On a lifetime basis, this will result in the production of about one more calf and an added calf weight of approximately 400 pounds.

2. It is a way in which to increase beef production with no more cow numbers and little more feed and equipment.

3. Cow cost per 100 pounds of weaned calf will be lower.

4. Increasing each animal's lifetime production efficiency provides a way in which to combat increased fixed costs of maintaining a beef animal.

FACTORS UNFAVORABLE TO CALVING TWO-YEAR-OLDS

The chief factors *unfavorable* to calving 2-year-olds are:

1. The conception rate of young heifers bred when just reaching puberty is lower than in older ones. This may result in spreading the calving season over a longer period, with accompanying greater inconvenience and expense.

---

[5]U. S. Department of Agriculture, Statistical Reporting Service, Crop Reporting Board, LvGb1, February 1, 1974

2. The percentage calf crop of heifers calving as twos will be about 10 percent lower than of older cows.

3. More calving troubles, and a higher death loss of both dams and calves will be encountered.

4. Early calving heifers will likely be somewhat undersized until they reach 4 to 5 years of age.

5. The calves will wean at 25 to 50 pounds lighter weights than calves from older cows.

6. Heifers calving at 2 years of age will not breed back to have their second calves as 3-year-olds if they do not receive enough energy during their first gestation and lactation periods, simply because they are developing and providing milk for a calf while still growing themselves.

7. It may not be desirable for the purebred breeder from the standpoints of (a) compliance with breed registry rules relative to minimum age at calving time (check them out; different breed registries have different rules), (b) selling of open heifers, (c) having well-grown, young cows for visitors to see, and (d) distribution of birth dates so as to fill more show classifications; but, of course, these points do not affect commercial cattlemen.

FEEDING AND MANAGING REPLACEMENT HEIFERS

The feed and management program of replacement heifers will have a lifelong effect on their productivity. It will determine how young they may be bred, whether they calve early or late, whether they are good milkers or poor milkers, the weaning weight of their calves, and how long they remain in the herd. Also, feed accounts for 40 to 70 percent of the cost of raising replacement heifers; hence, it is important to know whether it is possible to effect savings on feed during the growing period without affecting reproduction adversely. It is even more important to know whether by suitable nutrition and management of heifers, we can enhance their performance as adult animals.

The pregnancy requirements are really not too great. The body of an 80-lb, newborn calf contains only about 12 lb of protein, 3.0 lb of fat, and 3.6 lb of mineral matter. But the lactation requirements are much more rigorous. If a 2-year-old heifer gives her calf an average of 1¾ gallons of milk per day over a 7-month suckling period, she will produce in that milk a total of 93 lb of protein, 107 lb of fat, 133 lb of sugar, and 20 lb of minerals.

Hence, the comparison: 12 lb of protein in the fetus vs 93 lb in the milk during the suckling period. This means that nearly 8 times more protein is required for 7 months of lactation than for 9 months of pregnancy.

Also, when breeding yearlings to calve as two-year-olds, cattlemen should be aware that nature has ordained that the growth of the fetus, and the lactation which follows, shall take priority over the maternal requirements. Hence, when there is a nutritive deficiency, the young mother's body will be deprived, or even stunted, before the developing fetus or milk production will be materially affected.

Nutrient Requirements of Replacement Heifers

Meeting the nutrient requirements of heifers from weaning to first calving is of great importance. The requirements of heifers of different body weights and growth rates are given in Table 22-3.

TABLE 22-3

DAILY NUTRIENT REQUIREMENTS OF GROWING HEIFERS[1]

Body Weight		Daily Gain		Crude Protein	TDN	Calcium	Phosphorus
(lb)	(kg)	(lb)	(kg)	(lb)	(lb)	(g)	(g)
300	136	1.50	0.7	1.08	6.0	18	14
400	182	1.50	0.7	1.25	7.52	18	16
500	227	1.50	0.7	1.36	8.8	18	16
600	273	2.00	0.9	1.47	10.7	19	17
700	318	2.00	0.9	1.60	12.2	19	17
800	364	2.00	0.9	1.72	13.6	19	17

[1]Adapted by the author from *Nutrient Requirements of Beef Cattle*, National Academy of Sciences, Washington, D.C., 1976, pp. 24, 25, Table 1A.

Rations for Replacement Heifers

In season, good pasture plus mineral supplements fed free-choice will meet the nutrient requirements for proper growth and development of heifers.

On the winter range, when dry forage is of low quality, and sometimes not too abundant, one to two pounds of a protein supplement should be provided in the form of cubes, blocks, meal-salt, or liquid. When consumed at the intended level, the supplement should contain sufficient vitamin A to meet the requirements. Mineral supplements should also be provided, preferably free-choice.

Where winter grazing is not available, heifers must be drylotted and fed a complete ration. Sufficient nutrients should be provided to meet the requirements and to keep heifers in a thrifty condition, neither too fat nor too thin.

The wintering rations in Table 22-4 for 500-pound heifer calves should result in a rate of gain of 1 to 1.5 pounds per day.

The wintering rations in Table 22-5 for 800- to 900-pound bred yearling heifers should allow a gain of 0.75 to 1 pound per day during the wintering period prior to calving.

Replacement heifers should be fed rather liberally—more so than stocker cattle which are being grown for the feedlot, to the end that they will acquire most of their growth and development before calving. With limited feeding, they will not have enough weight for age to breed when they are 15 months old; and it is best not to have them calve until they are 30 months of age. Cost of the ration is important, but too limited a ration may actually be costly. This is to say that it is usually cheaper in the long run to grow heifers out well rather than to delay their development and run into management problems, higher death losses, greater pasture costs, high interest on invest-

Fig. 22-4. A large group of replacement heifers in a drylot, well grown and thrifty. (Courtesy, Martin Jorgensen, Jr., Ideal, S.D.)

TABLE 22-4

DAILY RATIONS FOR HEIFER CALVES (500 POUNDS)

	Rations				
	1	2	3	4	5
	(lb/day)				
Legume-grass haylage	25				
Legume-grass hay		10	10		5
Corn or sorghum silage				30	20
Ground ear corn		4			
Corn, grain sorghum, or barley			3		
Supplement[1]				1	

[1]Supplement contains 48 percent crude protein.

TABLE 22-5

RATIONS FOR BRED YEARLING HEIFERS (800 TO 900 POUNDS)

	Rations				
	1	2	3	4	5
	(lb/day)				
Corn or sorghum silage	45	25			
Legume-grass hay		10	20		15
Legume-grass haylage				35	
Corn, grain sorghum, or barley					3
Supplement[1]	1.5				

[1]Supplement contains 48 percent crude protein.

ment, and increased labor costs. Also, where breeding animals are sold, it is recognized that prospective purchasers are impressed by well-grown and well-conditioned young stock, whereas they are seldom interested in stunted animals at any price. The latter situation is accentuated in purebred heifers, hence it is important that they be more liberally fed than replacement heifers for the commercial herd.

Occasionally, a replacement animal is injured by overfeeding or by fitting for the show, but such losses are insignificant compared with those resulting from the thousands of undersized, poorly developed animals that are grossly underfed.

During the winter months, the feeding of a ration containing adequate protein and energy, plus the required vitamins and minerals, is necessary to keep young stock growing and gaining in weight.

During their second winter, heifers bred to calve as two-year-olds should be fed more liberally than mature cows of comparable condition. The added feed is necessary because, in addition to maintenance and development of the fetus, provision must be made for body growth. Even then, these heifers should have close supervision at calving time and the calves should be weaned at an early age in order to alleviate the strain of lactation.

Separate Heifers by Ages

The nutritive requirements for heifers differ according to body weight and expected daily gain (Table 22-3). Consequently, the recommended ration for a 500-pound heifer calf (Table 22-4) differs from that of an 800- to 900-pound bred heifer (Table 22-5). It is important, therefore, that replacement heifers be separated by ages for wintering, with coming yearlings in one group and coming twos in another.

SUMMARY RELATIVE TO CALVING TWO-YEAR-OLDS

From the above, it may be concluded that more cattlemen can, and should, breed yearling heifers to calve as two-year-olds. But, in doing so, the following practices should be observed in order to lessen calving difficulties:

1. Select the heaviest and highest scoring individual heifers at weaning. Weight at weaning is a means of evaluating the dam's milking ability.

2. Keep heifers separate from older cows.

3. Start with 50% more weaner replacement heifers than needed if it is the intent to maintain the same size herd—with no provision for expansion whatsoever. This means that for every 100 cows in the herd, 20 replacement heifers are actually needed each year in order to maintain the same size herd. (There is about a 20% replacement in each herd each year.) However, 30 weaner replacement prospects (50% more than actually needed) should be held, simply because, based on averages, 10 of them will either die or have to be culled before they replace older cows.

4. Replacement heifers should be fed for gains of approximately 1.0 lb per head per day from weaning to calving. From weaning to mid-pregnancy, 1¼ lb gain per day is about right. From mid-pregnancy until calving, the

Fig. 22-5. Replacement heifers. There is no better or quicker way in which to improve the reproductive performance of a herd than through the selection and proper development of replacement heifers. In the future, more and more heifers will be bred to calve as two-year-olds. (Courtesy, Ralston Purina Company, St. Louis, Mo.)

gain may be lowered to 0.75 lb per day.

5. Select yearlings and coming 2-year-old heifers on the basis of individuality and rate of gain. Also, cull heifers with small pelvic openings; those with large pelvic openings (above 34 sq in., or 220 sq cm) have less calving difficulty. Avoid excessively fat heifers.

6. Breed only well-developed heifers, weighing 600 to 750 lb (depending on breed) at 13 to 14 months of age. Size at breeding is more important than age. Also, some breeds come in heat and mature a little earlier than others.

7. Breed heifers 20 days earlier than the cow herd and restrict the breeding season to 45 days. This gives a short concentrated calving period; therefore, proper attention and help can be given heifers at calving time.

8. "Flush" feed heifers to gain approximately 1.5 lb per head daily beginning 20 days before the start of and continuing through the breeding season.

9. Breed heifers to a bull known to sire small calves at birth.

10. Feed a well-balanced ration, and feed for continuous gain of 0.75 to 1 lb during the pregnancy period; but don't get them too fat.

11. Feed heifers to weigh at least 775 lb by 120 days before calving.

12. Feed heifers to gain 100 to 120 lb from 120 days prior to calving. Heifers should weigh at least 875 lb just before calving and approximately 775 lb shortly after calving.

13. Give heifers special care at calving time. This should include—

a. Providing adequate facilities, including (1) a pull stall, and (2) small pens each suitable for confining a heifer and her calf for approximately 24 hours of "mothering up."

    b. Move each heifer into the calving area approximately 2 weeks before the expected calving date.

    c. Check heifers for calving at 2-hour intervals.

    d. Render assistance quickly and expertly when it is needed.

    e. Remove heifer and calf from calving area within 24 hours after birth and put them into a clean, dry pasture or other similiar area.

    14. Provide superior nutrition—well balanced, and rather liberal—during the lactation period, because a heifer's nutritional requirements double after calving. This requires a good ration—one containing adequate energy and proteins, and fortified with the necessary vitamins and minerals. In season, usually this can be accomplished by keeping these heifers on good pastures, with or without supplemental feeding, both during pregnancy and lactation. When good grass is not available—in the winter, early and late, or during droughts—proper feeding must be relied upon.

    15. If practical, wean early; at 2 to 6 months of age, rather than the normal 7 months.

    16. Run heifers that calved as 2-year-olds in a separate herd until after they have had their second calf.

    17. Try it out on half of your replacement heifers to start with; make sure that you know what is involved before going all out.

    Of course, the below average breeder—the fellow who has lightweight, poorly developed heifers, and who wouldn't think of staying up nights and having cold, numb fingers while being nursemaid to a heifer and a newborn calf—should take another year and stick to calving out three-year-olds. But progressive, commercial cattlemen should calve out more two-year-olds from the standpoint of cutting production costs and increasing profits.

QUESTIONS FOR STUDY AND DISCUSSION

1. If a cow herd is being built up through increasing numbers by 10 percent per year, how many replacement heifers at weaning time per 100 mature cows should be selected?

2. What consideration should be given to each of the following criteria for selection of replacement heifers: (a) individuality; (b) performance test; (c) record of the sire and dam; (d) record of half-sibs; (e) overly fat heifers; and (f) underfed heifers?

3. Discuss each of the following major problems that are frequently encountered in performance testing heifers in the usual manner: (a) the labor and expense of individual feeding; (b) excessive fatness; and (c) letting heifers down in condition at the close of the test.

4. How may the problems listed in question 3 (above) be alleviated, and yet have the animals performance tested?

5. How do you account for the fact that, in a selection program, a heifer's own record is more important than the records of her parents or grandparents?

6. Under what circumstances would you recommend breeding heifers to calve as 2-year-olds? Under what circumstances would you recommend against calving heifers as 2-year-olds?

7. Do you agree or disagree with the author's added beef projections to result from breeding 2-year-old heifers? Justify your answer.

8. List and discuss each of the factors *favorable* to calving 2-year-old heifers.

9. List and discuss each of the factors *unfavorable* to calving 2-year-old heifers.

10. How do the nutrient requirements of heifers calving as 2-year-olds differ from the nutrient requirements of heifers calving as 3-year-olds?

11. Why should heifers be separated according to age, body weight, and expected daily gain?

12. Under what circumstances would you recommend buying replacement heifers, rather than raising them?

SELECTED REFERENCES

Title of Publication	Author(s)	Publisher
Beef Cattle, Sixth Edition	A. L. Neumann R. R. Snapp	John Wiley & Sons, Inc., New York, N.Y., 1969
Beef Cattle Production	K. A. Wagnon R. Albaugh G. H. Hart	The Macmillan Company, New York, N.Y., 1960
Beef Production in the South	S. H. Fowler	The Interstate Printers & Publishers, Inc., Danville, Ill., 1969
Commercial Beef Cattle Production	Ed. by C. C. O'Mary, I. A. Dyer	Lea & Febiger, Philadelphia, Penn., 1972
Nutrient Requirements of Beef Cattle, Fifth Revised Edition	National Research Council	National Academy of Sciences, Washington, D.C., 1976

CHAPTER 23

CONFINEMENT (DRYLOT) BEEF COWS

Contents   Page

Confinement of cattle is not new. The pioneers planted trees and shrubs, such as osage orange, or built rail or stone fences, to hold their animals. Then, in 1873, Joseph F. Glidden of DeKalb, Illinois, invented barbed wire, or the "devil's rope," as it was dubbed by those who considered the barbs inhumane or who opposed its use because it marked the beginning of the end of the open range and free grazing. But, for the most part, the wire that fenced the West confined cattle to large areas—pastures.

Today, experimentally and on a very limited commercial basis, beef cows are being confined to small quarters—to drylots, all or part of the year. The practice will increase. More and more feeder calves will be produced under an intensive system.

Fig. 23-1. Part of a herd of 80 brood cows in confinement in a deep, open shed on Circle S Ranch, Rockwood, Ontario, Canada. (Courtesy, *Cattlemen*, Winnipeg, Manitoba)

Historically, as countries become more densely populated, land values and tax rates rise, necessitating that the highest and best use of land be made in order to pay taxes and yield a satisfactory return on investment. It's the same principle that spawns multistory buildings—as urban land prices mount, skyscrapers appear.

The land area available for beef production is slowly decreasing due to urban development, highways, airports, and recreational areas; and precious little new land can be brought into production. Hand in hand with this transition, the price of grassland on a per animal unit basis has increased to the point where it no longer provides cheap animal feed for cattle. On most cow-calf operations, the largest single cost item is interest on the money invested in land. At $1,000 per cow unit and 8 percent interest, the yearly interest tab is $80 per cow. Under confinement, at least 60 cows can be handled on an acre. That's a land cost of $16.67 per cow ($1,000 ÷ 60), or an interest charge of $1.34 per year—a saving of $78.66 per cow on interest on land. This saving can be applied toward the increased feed and labor cost of a confinement operation. This situation has prompted interest in more intensive use of productive lands—in the production of more high yielding crops such as corn and sorghum. Although intensive production requires considerable investment in facilities, equipment, and labor, the point has been reached where these added costs are largely offset by the lower investment in land. With the demand for more beef and world beef shortages, this gap will widen in the years ahead. As a result, intensive drylot cow-calf production will increase on the more productive lands.

Although confinement production as a management technique in beef cow-calf production is new in the United States, the concept is very old. To gain efficiency and reduce cost of production, American dairy, poultry, and swine producers have, of necessity, invested in intensive production units. Also, confinement cattle production has long been traditional in China, Japan, and Europe—in the more densely populated areas of the world, where land is scarce and high in price.

WHAT DRYLOT COW-CALF EXPERIMENTS SHOW

Fortunately, several longtime research studies have been conducted on confinement cow-calf production. Thus, a considerable body of data and experience is available. A brief review of the experimental work involving drylotting cows the year-round follows:

● *Alabama Station, Lower Coastal Plain Station, Camden, Alabama*[1]

Two confinement systems and a pasture system were compared over a 5-year period by the Alabama Station. Cows under the pasture system were full-fed Coastal bermudagrass hay plus 2 lb of cottonseed meal per head daily during the winter months—from November 1 until spring grazing. The balance of the year, they grazed Coastal bermudagrass pasture at the rate of one cow-calf unit per acre. Cows in continuous confinement were fed either

---

[1]Harris, R. R., V. L. Brown, W. B. Anthony, and C. C. King, Jr., Confined Feeding of Beef Brood Cows, *Alabama Agr. Exp. Sta. Bull. 411*, 1970.

(1) sorghum silage plus 1.5 lb of a 65 percent protein supplement per head daily for the first 180 days after calving, then one lb per day for the rest of the year; or (2) Coastal bermudagrass hay plus one lb of a 65 percent protein supplement per head daily for the first 180 days post-calving and none for the rest of the year. Hereford cows calving in the spring were used the first 3 years and Angus-Hereford crossbreds calving in the fall were used the last 2 years. Hence, none of the cows were kept on the treatments for the duration of the study—5 years. Natural breeding was practiced.

Results of the 5-year trial are shown in Table 23-1.

TABLE 23-1

COMPARATIVE PERFORMANCE OF BEEF COWS MANAGED UNDER TWO
CONFINEMENT FEEDING SYSTEMS AND A CONVENTIONAL
GRAZING SYSTEM FOR FIVE YEARS, IN ALABAMA

Item	Management System		
	Summer Grazed, Drylot Wintered	Confined, Fed Silage	Confined, Fed Hay
Hay or silage, tons/cow	1.58	12.81	4.31
Hay or silage production, tons/acre	1.68	17.15	7.49
Land requirements; cow-calf units/acre (feed)	1.06	1.34	1.74
Calving rate, %	95.0	92.0	82.0[1]
Calf crop weaned, %[2]	92.0	88.0	84.0[1]
Adjusted calf weaning weight, lb	453.0	551.0	511.0
Cow-calf units/acre (feed)	1.06	1.34	1.74

[1]The 82% is based on 1964-68 data, whereas the 84% is based on 1963-67 data.
[2]Calves weaned based on cows bred.

The Alabama experiment showed the following:

1. The confined cows performed satisfactorily in every respect except that their average calving percentages were lower than for the pasture group.

2. One acre of Coastal bermudagrass fertilized with 200 pounds of nitrogen provided grazing and hay for one conventionally managed cow-calf unit annually. One acre devoted to the production of sorghum silage or Coastal bermudagrass for hay supported 1.34 and 1.74 cow units, respectively. Thus, only 0.58 acre was required to produce sufficient hay for one cow-calf unit kept in confinement.

3. A cow-calf unit in confinement the year around required 12.8 tons of sorghum silage (35% dry matter) or 4.3 tons of Coastal bermudagrass annually.

4. Cows fed conventionally weaned a larger percent calf crop (92%) than did either of the confinement groups (88 and 84%, respectively).

5. The cows in confinement weaned calves that averaged 78 pounds heavier than the pastured cows. However, their calves were creep-fed and consumed an average of 1,422 pounds of feed per calf; that's 18 pounds of feed per pound of gain.

●*Arizona Station*[2]—Total and partial confinement are being studied at the Arizona Station. The partially confined cows are on irrigated pastures during a long grazing season, with oats and barley providing winter grazing and Sudan grass summer grazing. Alfalfa and fescue were added so as to extend the grazing season to nine months. The totally confined cows are fed a mixture of alfalfa hay and bermudagrass straw, with adjustments in proportions and quantities fed as well as in the use of supplements at different stages of the reproductive cycle. The calves in both groups are creep-fed.

Annual feed costs have been approximately $10 higher per cow and calf for the pairs totally confined than those partially confined. The Arizona investigators suggest that the cost of drylot cows could be lowered by one cent per cow per day by feeding on alternate days, based on their studies showing that feeding every other day is satisfactory.

●*Illinois Station*[3]—In a 4-year trial, initiated in 1963, using spring-calving Shorthorn cows, the Illinois Station compared a year-round drylot cow-calf system with a summer grazing winter drylot system. The pasture, which consisted of bromegrass, orchardgrass, and alfalfa, was grazed from May 15 to October 15. The cows in drylot were fed (1) cornstalk silage with 25 pounds of dried molasses and 75 pounds of corn meal added per ton of forage at ensiling time, from November 1 to February 15; (2) corn silage from February 16 to June 30; and (3) haylage from July 1 until October 31. Urea, dicalcium phosphate, and trace mineralized salt were added to both corn forages at ensiling to provide a balanced ration of approximately 11 percent crude protein on an air-dry basis. Cows were bred by natural service. Calves in both groups were creep-fed.

The Illinois experiment showed that (1) cows in drylot required fewer services per conception than those on pasture, (2) the drylot cows weaned slightly heavier calves than the pasture cows, and (3) the carrying capacity of the land was increased 47 percent by harvesting and feeding haylage rather than grazing the forage as pasture.

●*Iowa Station*[4]—In a study initiated in 1958, involving cows nursing calves, the Iowa Station compared (1) drylot cows fed 114 lb of alfalfa-grass mixture as green chop, and (2) cows on pasture. The cows on green chop gained an average of 7 lb during the summer period and weaned calves weighing 405 lb, whereas the cows on pasture lost an average of 70 lb and weaned calves weighing 370 lb. Thus, the green-chop cows were 47 lb heavier and weaned off calves that were 35 lb heavier than the drylot cows. But the heavier calf weights were not sufficient to compensate for the added charges of harvesting and feeding the green chop; the cows on pasture returned $9.34 more per calf than the drylot cows fed green chop.

---

[2]McGintry, D. D., H. I. Essign, and E. Hussmann, Production of Feeder Calves in Intensive Management Systems, *Cattle Feeders' Day*, Department of Animal Science, The University of Arizona, May 7, 1970, p. 11.

[3]Albert, W. W., Good Performance from Beef Cows Confined to Drylot the Year-Round, *Ill. Agr. Exp. Sta. Ill. Res.*, Winter, 1969, p. 8.

[4]Woods, W., B. Taylor, and W. Burroughs, Feeder Calf Production by Intensive Methods on Iowa Corn Land, *Animal Husbandry Leaflet 230*, Iowa State University, Ames, Iowa, February, 1958.

● *Minnesota Station*[5]—In a 5-year study, Angus and Hereford cows were maintained under two systems; one group was kept on legume-grass pasture during the grazing season and confined to drylot during the rest of the year, and the other group was confined to a drylot continuously for five years where they were fed legume-grass forage in season, similar to the forage grazed by the pastured herd. All cows were bred by natural service, and all calves were creep-fed.

Table 23-2 shows the performance of the cows and calves.

TABLE 23-2

COMPARATIVE PERFORMANCE OF BEEF COWS MAINTAINED
IN PASTURE AND DRYLOT HERDS FOR FIVE YEARS, IN MINNESOTA[1]

Treatment	Angus		Hereford	
	Pasture	Drylot	Pasture	Drylot
Conception rate, %	96	96	88	88
Calf birth weight, lb	64	63	68	67
Av. adjusted weaning weight 1963-1967, lb	470	450	490	426

[1]Meiske, J. C., and R. D. Goodrich, Drylot vs Conventional Cow-Calf Production, *Minn. Res. Rep. B-118,* University of Minnesota, 1968.

The Minnesota investigators concluded that (1) about twice as many cows could be maintained per unit of land if they were confined instead of grazed in the conventional manner; (2) conception rates, calf crop percentages, and birth weights were similar; and (3) the drylot group had fewer losses, but lower weaning weights.

● *South Dakota Station*[6]—In 1968, the South Dakota Station initiated a study to compare the performance of cows (1) maintained continuously in drylot, vs (2) cows summered on native pasture and wintered in drylot. They used Hereford-Angus crossbred cows, which were mated to Angus, Charolais, and Hereford bulls.

Results during the first 5 years of the study showed that the pasture cows have a higher percentage of calves born (4% more) and weaned (5% more), and heavier weaning weights (18 lb more) than cows maintained continuously in drylot.

● *Texas A & M Agricultural Research Station, Spur, Texas*[7]—In 1959, workers at the Rolling Plains Livestock Research Station at Spur, Texas, initiated a lifetime performance study of cows in total confinement compared with cows maintained continuously on native pasture. The experiment was started with 72 yearling Hereford heifers, and involved 2 systems—(1) pasture, and (2) drylot. Each system was further subdivided into 3 groups of 12

[5]Meiske, J. C., and R. D. Goodrich, Drylot vs Conventional Cow-Calf Production, *Minn. Res. Rep. B-118,* University of Minnesota, 1968.

[6]Slyter, A. L., Influence of Mating and Management Systems on the Performance of Beef Cows and Calves, *A. S. Series 72-5,* South Dakota State University, August 25, 1972.

[7]Marion, P. T., J. K. Riggs, and J. L. Arnold, Calving Performance of Drylot and Pasture Cows over an 11-Year Period, *Beef Cattle Research in Texas, Consolidated PR-2963-2999,* June, 1971.

heifers each, which were fed rations of cottonseed meal, sorghum grain, and silage to provide 3 levels of energy designated as low, medium, or high during the winter period, November through April. Table 23-3 shows the digestible protein and metabolizable energy of the rations fed, and Table 23-4 shows the performance of the cows over an 11-year period. All animals, both drylot and pasture, had free access to plain block stock salt at all times; and during the period 4 to 6 weeks after calving, all cows had access to a 50/50 mixture of bone meal and salt. No vitamin supplements were used. Breeding was by natural service.

As shown in Table 23-4, the data was broken down to reflect the trouble years—the 3-year period, 1967-1969, when birth weights and calf survival were adversely affected by what subsequently appeared to be a protein deficiency.

TABLE 23-3

DIGESTIBLE PROTEIN (DP) AND METABOLIZABLE ENERGY (ME) FOR RATIONS FED 1959-68 and 1969-70[1]

| | Winter Rations for Dry Cows, Level of Energy | | | | | | Summer Ration, | |
| | Low | | Medium | | High | | Lactating Cows | |
Year	DP	ME	DP	ME	DP	ME	DP	ME
	(lb)	(Mcal)	(lb)	(Mcal)	(lb)	(Mcal)	(lb)	(Mcal)
1959-68	0.57	8.70	0.56	11.88	0.53	15.90	0.70	15.80
1969-70	0.95	10.1	0.85	13.00	0.77	16.42	0.91	17.10
Requirements for 1,000-pound cow[2]					0.42	12.40	1.17	20.4

[1]Beef Cattle Research in Texas, *Tex. Agr. Exp. Sta. Consolidated PR-2963-2999*, Texas A&M University, June 1971, Table 1, p. 6.
[2]*Nutrient Requirements of Beef Cattle*, Fourth Rev. Ed., National Academy of Sciences, Washington, D.C., 1970.

TABLE 23-4

PERFORMANCE OF PASTURE VS DRYLOT COWS (1960-1970)[1]

| | Pasture | | | | Drylot | | | |
| | Calf Crop | | Birth | Weaning | Calf | | Birth | Weaning |
Period	Born	Weaned	Weight	Weight	Crop	Weaned	Weight	Weight
	(%)	(%)	(lb)	(lb)	(%)	(%)	(lb)	(lb)
1960 to 1966 — 7 yr. av.	94.8	88	72.9	468	96.9	88	73.0	476
1967 to 1969 — 3 yr. av.	93.3	89	75.4	492	91.4	69	57.4	455
1970	100.0	90	87.4	618	90.4	86	68.7	527

[1]Beef Cattle Research in Texas, *Tex. Agr. Exp. Sta. Consolidated PR-2963-2999*, Texas A&M University, June, 1971, Table 2, p. 6.

Through the first 7 years of the Texas study, there was no significant difference in the average percent calf crop and weaning weights of the calves produced in drylot and on pasture. However, large differences between the two systems occurred in 1967 and continued through 1969. It was conjectured that the adverse effects of drylotting during this 3-year period—as evidenced in percentage calves weaned and lighter calves at birth and weaning—was caused by a cumulative nutritional deficiency) or deficiencies) over the period of years in confinement. So, a change was made in the rations during the winter of 1968-69, by increasing the protein and phosphorus intake. The change in protein was prompted so that the rations would meet National Research Council (NRC) requirements. The rations fed from 1959 to 1968 met the NRC digestible protein (DP) standards for dry pregnant cows, from November through March; but during lactation, from April through October, the cows received 0.5 lb less DP than the NRC recommendation. The investigators reasoned that this protein deficiency for the first 5 to 6 months of gestation while the cows were lactating, during the period of greatest stress on the cows, must have been a contributing factor in the birth of small, weak calves. The 1969-70 ration change consisted in increasing the DP in the ration by 0.21 lb during lactation and 0.24 to 0.38 lb during the dry period (Table 23-3). Although this ration change did not affect the 1969 calf crop, it apparently greatly improved the birth weights and calf survival of the 1970 crop (Table 23-4).

The Texas experiment showed the following:

1. There were no significant differences in the average percent calf crop and weaning weight of calves produced by drylot and pasture cows through the seventh calf crop.

2. In drylot cow-calf production, it is important that the nutrient requirements be established and met, with increases during the lactating period when the requirements are most rigorous. Otherwise, a cumulative deficiency (or deficiencies) is likely to show up on cows confined for a number of years. In the Texas study, a protein deficiency became evident after 7 years.

3. Overfeeding and excess fatness of drylot cows is costly and detrimental.

4. The teeth of the cows on pasture showed severe wear, whereas those of the drylot cows remained in good condition.

5. Eighteen cows—an equal number—in each group, drylot and pasture, had to be removed during the 12-year study. It is noteworthy, however, that 5 of the drylot cows were removed for being open 2 successive years, whereas no pasture cows were removed for this reason.

6. The average cow-calf consumption of the drylot cows was 4,403 pounds of TDN.

7. Costs favored the pasture cows, primarily because summer feed for the drylot cows was expensive. The Texas workers suggested that if drylot cows could be grazed on more economical temporary pasture for 90 to 120 days during the summer, an additional $10 to $15 could be added to the net return per cow.

8. There is need for longtime studies (lifetime, and even generation after generation) where a cumulative nutritional deficiency, or deficiencies, may affect reproduction. In this study, it took 7 years for the protein deficiency to become evident.

ADVANTAGES AND DISADVANTAGES OF CONFINING BEEF COWS

The logical procedure to follow in arriving at any important business decision, including whether or not to switch to or start a confinement cow-calf business, involves three steps:

1. What are the pros?
2. What are the cons?
3. What is the decision?

Based on experiments and experiences, the following advantages and disadvantages are inherent in most confinement cow-calf operations in comparison with conventional pasture systems.

The advantages:

1. They require less investment in land per cow unit.
2. Cow numbers can be increased without obtaining more land.
3. They maximize feed production per acre of land, through utilizing harvested forages instead of grazing.
4. Hazards of drought and adverse weather conditions may be minimized through storing adequate feed reserves for future use.
5. Precise breeding programs can be designed and carried out more easily.
6. Control of estrus and use of artificial insemination programs become more practical.

Fig. 23-2. Sorting confinement cows for artificial insemination on the Tom Brothers Ranch of Texas. Confinement cow-calf operations facilitate A.I. breeding because of the close observation accorded. (Courtesy, Lytle Tom, Jr., Tom Brothers Ranch, Campbellton, Tex.)

7. Fewer services per conception are required in drylot than on pasture.

8. Individual performance records can be kept more easily.

9. Selection and culling are easier, because cattle can be observed more closely.

10. Plane of nutrition can be accurately known and controlled in accordance with the age and production needs of the cow; for example, nutrients can be increased during the critical 100 days, extending from 30 days before calving to 70 days after calving. After a calf is born, the cow's nutritional requirements rise approximately 50 percent over maintenance alone.

11. They make it easier to flush cows prior to the breeding season.

12. By-products and low quality feeds may be utilized to advantage.

13. Partial environmental control may be achieved.

14. They permit close observation of cattle. Hence, illness and injuries are quickly detected and may be promptly treated.

15. Calving at various times of the year is more practical than in pasture handling.

16. They make for maximum flexibility in creep feeding. For example, steers and heifers can be fed separately.

17. Confinement-produced calves wean more easily and go on finishing rations more readily.

The disadvantages:

1. They usually require a higher investment in buildings and equipment.

2. They require more labor.

3. Labor must be provided seven days per week. Hence, it is more confining to the operator, unless he has additional help.

4. Disease problems, especially among calves, are generally more acute.

5. All feed must be harvested and moved to the feedlot, rather than being harvested directly by the cows.

6. An assured, adequate, and economic feed supply must be available. The capital tied up in stored feeds may be quite large.

7. More able management is required.

8. More knowledge of beef cow nutrition and ration formulation is needed.

9. Risks due to blizzards, hurricanes, tornadoes, floods, and lightning are increased due to the heavy concentration of animals.

KIND OF COW FOR CONFINEMENT (DRYLOT) PRODUCTION

What will the confinement cow of tomorrow look like? What kind of cow is best suited for confinement (drylot) production? How will she differ from her counterpart on the farm or range?

Under confinement, cows will be completely dependent upon man. They will no longer need to fight and fend for themselves, as did the Longhorn—or even the cow on the range. Will they, under such conditions, become timid and completely defenseless like sheep? Will they lose their independence of behavior and be unable ever to return to the wild state, or become feral? These traits in sheep are clearly the result of selection by man; and their dependence on man is a logical final result of domestication.

What size cow is best suited to confinement production, where man harvests the feed for her? Generally speaking, the European breeds of beef cattle which evolved under confined or semiconfined conditions were larger beasts than their counterparts that roamed the hills and pastures, gleaning the feed provided by nature.

Will the desirability of adaptation to environment be eliminated? Will heat and cold tolerance be necessary under confinement where partial environmental control is provided by shelters and shades, and where perishing in the winter and overheating in the summer are no longer hazards?

Answers to these questions, and more, will determine the ideal cow for confinement production. The author's specifications of her follow:

1. She will be an F_1 (first cross) cow, which is the most productive of all cows; and she will be estrus-controlled and bred A.I. to a bull of a third breed to tailor her calves to command top price. Such a precise breeding program can be designed and carried out under confinement more easily than on pasture or range.

2. She will be a medium-sized cow, weighing around 1,000 pounds, rather than a huge beast, because:

   a. The maintenance requirement for energy is proportional to body weight to the three-fourths power ($W^{0.75}$). This favors a small cow where feed costs are high.

   b. By breeding a medium-sized cow to a large bull (without accentuating calving problems), a rapid-gaining calf which will have large mature size can be secured without the added maintenance cost of a big cow.

3. Behavior-wise, she should (a) be docile and easily handled, (b) exhibit a minimum of independence and dominance, and (c) be able to tolerate crowding.

4. She should breed at an early age, and be highly fertile and prolific—twinning would be desirable if it becomes practical.

5. She may calve when desired, because calving at various times of the year is more practical under confinement than in pasture handling.

6. Heavy and persistent (over a long period of time) milking will not be essential, because early weaning at about 2 months of age will be practiced. It is more efficient to feed calves directly than it is to feed cows to produce milk to feed calves. Using Hereford and Charolais cows, Melton *et al* showed that calves required an average of 2.1 lb TDN (including the TDN in both the dam's milk and the creep feed) per pound of gain from birth to 210 days of age, and 3.4 lb of TDN per pound of gain from birth to 365 days of age. When the 365-day TDN requirement of the dam was included, the TDN requirements per pound of calf produced increased to 8.5 and 7.2 lb at 210 and 365 days of age, respectively.[8]

Also, it is easier to get cows to rebreed when they are not nursing a calf.

---

[8]Melton, A. A., T. C. Cartwright, and W. E. Kruse, Cow Size as Related to Efficiency of Beef Production, *Tex. Agr. Exp. Sta. P.R. 2485*, 1967, p. 11.

7. She will be of higher quality than pasture cows, because selection and culling are easier and cattle can be observed more closely under confinement.

8. She should be a very efficient converter of roughages and by-product feeds.

9. She should be adapted to semiconfinement. In season, and if desired, she should be able to graze stalkage, or native or irrigated pasture.

10. She should not have horns—she should either be naturally polled or dehorned.

11. She should be more resistant to diseases than her less exposed counterpart on pastures and ranges.

FACILITIES FOR DRYLOTTING BEEF COWS

Facilities for drylotting beef cows must be adequate and functional, but they need not be elaborate. The requirements change somewhat according to the stage of the reproductive cycle and the area. For example, dry pregnant cows require less space than cows suckling calves. Dry and well-drained areas require less space than high rainfall, poorly-drained areas. Cold areas require shelter, and hot areas require shade. However, the basic facility needs, subject to some adaptation, for a confinement cow-calf operation are as follows:

• *Site*—Pollution is a most critical factor in site selection of a confinement cow-calf operation. Remoteness from urban development is recommended because of dust and odor. Also, before constructing facilities, the owner should familiarize himself with both state and Federal regulations, then comply therewith.

• *Corral space*—Under average conditions, and an unsurfaced lot, it is recommended that 300 sq ft of corral space be provided per cow-calf. More or less space may be provided, depending primarily on rainfall, drainage, dust control, pollution control, etc. Up to 400 sq ft may be required in a wet, muddy area. The Arizona Station provided 355 sq ft per cow and calf; the Texas Station provided 200 sq ft for a dry pregnant cow and 300 sq ft per cow-calf after calving; and a successful commercial operator near Toronto, Canada, provides 2,700 sq ft of corral space, or $1/16$ acre, per cow for his 400-head operation.

In addition to corral space, provision must be made for feed storage, alleys, and working areas. So, under average conditions the total space requirements for confinement (drylot) beef cows is on the order of one acre for 60 to 65 cows.

• *Fencing*—The specifications for a desirable fence are: 54 to 60 inches high; posts of 3- to 4-inch diameter pipe, or treated wood with 5-inch top diameter, set in concrete and spaced 8 to 12 feet apart; enclosed (except for the feed bunk area) by 4 strands of ⅜-inch steel cable spaced at intervals (from ground up) of 18 inches, 12 inches, 12 inches, and 13 inches. Four strands of the cable should be placed above the feed bunk, with these spaced equal distances apart. The cable above the bunks should be so

Fig. 23-3. Confinement cow-calf feed bunk equipped with protective bar to keep cows from butting calves into the bunk. If such a bar or wire is not provided, some calves will be turned upside down in the bunk and die if the caretaker does not rescue them very soon. (Courtesy, Lytle Tom, Jr., Tom Brothers Ranch, Campbellton, Tex.)

spaced as to keep a cow from butting a calf into the trough. Pipe or wooden rails may be used in place of steel cable if desired.

The above arrangement will confine the cows, but not the calves. Thirty-nine inch woven wire, with number 9 top and bottom wires and 6-inch mesh, should be placed around the entire corral. Also, calves must be kept out of the feed bunks.

• *Cows per pen*—Except at calving time, 50 to 100 cows may be run together in one group.

• *Shelter and shade*—Where winters are severe and snowfall is heavy, a shelter should be provided. This is especially important for newborn calves. A high and dry, deep, pole-type shed, opening away from the direction of prevailing winds is excellent. In areas with mild winters, a windbreak may be provided by hills, trees, or board fences.

In hot climates, a shade 12 feet or more high, oriented north-south (so that the sun will shine under the shade early in the morning and late in the evening), should be provided, with approximately 40 sq ft of area per cow-calf.

• *Bunk and concrete apron*—Feed bunks, which form part of the pen fence, should provide 24 to 30 inches of space per cow. Bunks should be constructed of concrete with the outside (alley side) of the bunk 22 to 36 inches high, and the inside (the pen side) 22 to 24 inches high. The bottom of the bunk should be rounded and about 18 inches wide.

An 8-foot concrete apron (platform), 4 to 6 inches thick, should extend from the feed bunk into the pen to provide solid footing for the cattle. The apron should have a slope of one inch per foot, which will make it nearly self-cleaning.

• *Water troughs*—Confinement cows should have access to water at all times. One linear foot of trough space for each 8 cows is sufficient. There should be sufficient water supply to provide 20 gallons per head daily. Shallow, low capacity troughs are preferable, since frequent drainage is necessary to keep them clean. Continuous-flow troughs are excellent and help to keep clean water. Also, in most areas, continuous-flow troughs will not freeze during the winter months; hence, heat is not required. Heated and insulated troughs are necessary in cold areas.

• *Maternity stalls*—Where cows calve in confinement, maternity stalls 100 to 120 sq ft in size should be provided for occupancy during calving and continuing for 1 to 2 days thereafter. This is particularly true of first-calf heifers. When several heifers calve the same day in a large corral, they frequently claim the wrong calf, or no calf at all—with the result that they end up as dogies. Thus, it is best that a heifer and her calf be confined to a maternity stall until they pair up.

When calving during the winter months in cold areas, the maternity stalls should be in a barn or shed; and heat lamps should be provided. In warm areas, uncovered pens will suffice. All maternity stalls should be cleaned, disinfected, and freshly bedded for each birth.

• *Calf creep*—Calves should be fed separately from the cows, in a creep.

• *Hospital area*—A special hospital area should be provided, with individual pens in which to treat and isolate sick animals. It should be equipped with feed facilities for the sick animals and a chute for handling cows. In areas subject to severe weather, shelter should be provided for the hospitalized animals. Also, there should be a room equipped with refrigerator, running water, medicine, and equipment storage.

• *Loading and unloading facilities, working chutes, squeeze, and scales*—An area should be provided for receiving and shipping cows, working animals, and weighing. The size and facilities of this area will be determined by the number of cows.

RATIONS FOR DRYLOT COWS

Rations for drylot cows generally consist of cheap roughages—such as crop refuse, straw, cottonseed hulls, and gin trash—supplemented with protein, grain, vitamins, and minerals as required. Where available, higher quality roughages—such as silages, hays, and haylages—may be used, especially (1) during the critical 100 days, beginning 30 days before calving and extending 70 days after calving, and (2) for heifers calving as 2-year-olds. Also, during the summer and fall, green chop is frequently fed. Cows in partial confinement may, in season, graze such forages as cornstalks, or irrigated or native pastures.

Phase-feeding according to stage of production and age of animals is recommended.

It is relatively easy to meet the nutritive requirements of a dry pregnant cow. Generally speaking, low quality roughages or a combination of low quality roughages and high quality roughages, properly supplemented, will suffice. If only high quality roughages are fed, they must be limited; otherwise, the cows will get too fat. However, such limited feeding does not meet the maximum fill requirements of the rumen, with the result that the cows nibble at the manure and pick up small feed particles scattered about the corral. This may make for disease and parasite problems. A simple solution to the weight and scavenger problems is to use a combination of low quality and high quality roughages.

The lactation requirements are much more rigorous than the dry pregnancy requirements; and the higher the milk production, the higher the nutritive requirements. For the latter reason, requirements for two levels of milk production for lactating cows—(1) average milk production, and (2) superior milk production—are presented in Tables 8-18 and 8-19 of this book.

Under a drylot system, heifers are commonly calved as two-year-olds. Such heifers are still growing. Thus, during lactation they have a nutritional requirement for both growth and lactation. The nutritional requirements for these animals will be large, particularly if they are crossbreds and bred for rapid growth, considerable size, and high milk production.

The mineral needs of confinement cows may be met either by incorporating the needed minerals in the supplement which is fed, or by feeding the required minerals free-choice.

Vitamin A supplementation is extremely important for drylot cows. The carotene content of the dry forage should be disregarded and the total vitamin A requirement met by supplementation. This can be done by feeding a supplement of 2 pounds of mill waste containing 1 million IU of vitamin A per animal—feeding this vitamin A supplement once a month to heifers, and every other month to older cows. With older cows receiving high levels of dry forages containing normal amounts of carotene, it is probable that the vitamin A requirements are being met. However, it has been demonstrated under range conditions that (1) percent calf crop is markedly increased by supplementing with vitamin A during drought years, and (2) calves respond to vitamin A treatments given their dams 90 days prior to calving.

MANAGEMENT OF DRYLOT COWS

In order for a confinement cow-calf operation to be profitable, it is essential that it have a high percentage calf crop, heavy weaning weights, a low incidence of disease, and a low cost ration. To these ends, successful confinement operators pay close attention to the details of management. Among the good management techniques which are being used in intensified beef production are the following:

1. *Dehorning*—Cows kept in a drylot should either be naturally polled, dehorned, or have their horns tipped. (When the latter is done, leave only a 3- or 4-inch stub.)

2. *Flushing*—Females (both cows and heifers) should be flushed 2 to 3 weeks before breeding. This may be accomplished by replacing low quality roughage with good alfalfa hay, or by adding about 4 pounds of grain per head per day.

Semiconfined cows can be put on more lush pastures as a means of flushing.

3. *Breed yearlings; calve 2-year-olds*—All heifers should be bred as yearlings to calve as 2-year-olds. Because they are still growing, they require extra feed and attention, particularly if they are to be rebred to calve as 3-year-olds. Also, they must be watched more closely, and assisted when necessary, at calving time.

4. *Pregnancy testing*—Pregnancy testing is essential following the breeding season so that barren cows will not be kept.

5. *Day of birth*—At birth, each calf should be weighed, tattoed, ear-tagged, injected with 1 million IU of vitamin A, and the navel should be treated with iodine.

6. *Excess milk*—With heavy milking cows, their feed should be limited until the calf is 10 days to 2 weeks old.

7. *Scours*—Calf scours is the bane of the confinement cow-calf operator. A drylot aggravates scours and favors a buildup of the scour problem over the years.

Wherever possible, it is strongly recommended that cows be removed from the drylot immediately before calving and placed on a sizable clean pasture (one that has been idle for a period of time) for calving out, thereby alleviating most, if not all, scouring. Also, a good pasture will stimulate milk flow and make for a good nutritional start in life for calves.

Where calving on clean pasture is not practical, the following precautions against scours should be taken:

    a. Clean, disinfect, and bed the maternity stall after each birth.

    b. Inject the newborn calf with 1 million IU of vitamin A.

    c. Limit the feed of heavy producing cows until the calf is 10 days to 2 weeks old.

    d. Where scours develop, see your local veterinarian for recommended treatment.

8. *Consider early weaning*—Weaning calves at 2 months of age will save feed and result in getting the cows rebred more quickly; hence, early weaning should be considered.

9. *Feed cost*—The key to a successful confinement cow-calf operation is to locate where a sure supply of inexpensive roughage (or by-product feed) is available, with a minimum transportation cost.

10. *Amount of feed per cow*—An average cow will consume about 3¾ tons of air-dry feed per year. That's 20 lb per day. Many cows will stay in condition with only 14 lb of feed during the dry period, but they will need around 24 lb per day during the nursing period.

11. *Amount of creep feed*—Each calf will consume 200 to 500 lb of creep feed while nursing.

12. *Mineral-soil box*—A 3-compartment mineral box should be provided

for each pen. In it, place (a) salt, (b) a mixture of equal parts of salt and dicalcium phosphate (or bone meal), and (c) soil.

13. *Alternate day feeding*—Cows kept in confinement may be fed every other day without altering performance, thereby effecting a saving in labor.

14. *Parasites*—Both internal and external parasites should be controlled.

15. *Dust control*—In dry, windy areas, pens should be equipped with sprinklers for dust control.

SEMICONFINEMENT (OR PARTIAL CONFINEMENT) COW HERDS

A semiconfinement (or partial confinement) operation is one which takes advantage of grazing during part of the year, such as winter grazing of corn or sorghum stalks or seasonal grazing of pastures. In addition to providing low cost feed and allowing the animals to do their own harvesting, breeding may be timed so that the calves will be dropped on clean pasture as a means of (1) preventing calf scours, and (2) stimulating milk flow.

Grazing crop residue and pastures is covered in Chapter 19 of this book; hence, the reader is referred thereto.

Fig. 23-4. By going to semiconfinement, the Tom Brothers Ranch was able to increase cow numbers from 600 to 1,075 head, with no added acreage. (Courtesy, Lytle Tom, Jr., Tom Brothers Ranch, Campbellton, Tex.)

QUESTIONS FOR STUDY AND DISCUSSION

1. What factors favor increased confinement production in the future?
2. Summarize what drylot cow-calf experiments show.
3. List the advantages of confinement cow-calf production.
4. List the disadvantages of confinement cow-calf production.

5. Describe the ideal cow for confinement (drylot) production.

6. List and describe the basic facility needs for a confinement cow-calf operation.

7. Discuss phase-feeding confinement cows according to stage of production and age of animals.

8. Outline a management program for a confinement cow-calf operation.

9. Discuss the advantages and disadvantages of (a) a semiconfinement vs (b) a year-round (total) confinement cow-calf operation.

10. Do confinement cow-calf operations lend themselves to (a) artificial insemination, and (b) early weaning? Justify your answer.

11. Which cow-calf system necessitates the most able management—(a) conventional pasture system, or (b) confinement?

12. Why has confinement cow-calf production existed for so much longer in China, Japan, and Europe than in the United States?

SELECTED REFERENCES

Title of Publication	Author(s)	Publisher
Beef Cattle Science Handbook	Ed. by M. E. Ensminger	Agriservices Foundation, Clovis Calif., pub. annually since 1964
Commercial Beef Cattle Production	Ed. by C. C. O'Mary, I. A. Dyer	Lea & Febiger, Philadelphia, Penn., 1972

BUSINESS ASPECTS OF A COW-CALF SYSTEM[1]

The cow business has changed from a way of life to a way of making a living. It's big and important business, and intricate too; and it is destined to get bigger and more complicated.

Fig. 24-1. Business aspects of cow-calf production.

---

[1]Tables 24-1 and 24-4 are a consensus (or judgment) of several Chairmen of University Animal Science Departments and Agricultural Consultants, from throughout the United States, based on a survey made by the author. A consensus was resorted to for the reasons that (1) no extensive, nationwide, scientific sampling of cattlemen on these matters has ever been made, and (2) this information is much needed by cattlemen and those who counsel with them. No claim is made relative to the scientific accuracy of the data; rather, it is presented (1) because it is the best, if not the only, information of its kind presently available on a nationwide basis, and (2) with the hope that it will stimulate needed research along these lines.

In 1935, there were 6,814,000 farms in the United States with an average size of 155 acres. By 1973, there were 2,831,290 farms, with an average size of 385 acres. Thus, within a span of 38 years, 3,982,710 farms—59 percent of our farms—disappeared from American agriculture; and the average size farm increased from 155 acres to 385 acres—more than double. With this transition, cow herds increased in size.

In the 21-year period 1952 to 1973, farm investment in land, buildings, livestock, and equipment more than doubled, rising from $167 billion to $384 billion. During this same period of time, farm debts increased by 5-fold, going from $14.7 billion to $73.6 billion.

Another noteworthy statistic is that, during the 10-year period 1963 to 1973, U.S. farmers averaged only 3.9 percent return on the equity of their capital investment in farming. This clearly indicates that the business aspects of agriculture merit much greater attention than they have received in the past.

The above facts and figures point up the two main deterrents to bigness in the cow business: (1) the inability to bring together the amount of capital required; and (2) the scarcity of competent management.

COW-CALF CAPITAL NEEDS

Those thinking of becoming cow-calf operators inevitably ask, "How much money will it take, and what can I make?" Unfortunately, there is a paucity of information on this subject on which to base an answer, despite the fact that it is sorely needed by both investors and lenders. To fill this gap, the author made a nationwide survey, the results of which are reported in Table 24-1. This reveals that, on the average, (1) it requires an investment of $1,000 to $1,700 per mature cow for land, improvements, machinery and equipment; (2) physical plant and cattle are the main costs; (3) the larger the herd, the lower the per animal unit investment and the feed and labor costs; and (4) purebred operations are much more costly than commercial.

What Return?

All cow-calf enterprises are owned by people, and most people want to be paid for their investment and work. This is as it should be. The only question is—how much return?

The figures, taken from and following Table 24-1, for a commercial herd within the range of 300 to 499 cows, show (1) the cost (exclusive of depreciation) to produce one calf to weaning age, and (2) the possible net income:

TABLE 24-1

AVERAGE INVESTMENT IN U.S. COW-CALF ENTERPRISES[1]

Investment—	In a *Commercial* cow-calf operation with the following number of cows of breeding age				In a *Purebred* cow-calf operation with the following number of cows of breeding age			
	Under 99 $	*100-299* $	*300-499* $	*500 or more* $	*Under 99* $	*100-299* $	*300-499* $	*500 or more* $
1. Investment/animal unit (one mature cow):								
a. In land and improvements (real estate)	1,100	1,025	950	900	1,400	1,325	1,250	1,200
b. In machinery and equipment	200	175	125	100	300	275	225	200
Total	1,300	1,200	1,075	1,000	1,700	1,600	1,475	1,400
2. Investment to produce one beef calf to weaning age (with provision made for barren cows):								
a. Land (investment/cow unit × going rate of interest)[2]	68.00	68.00	68.00	88.00	88.00	88.00	88.00	88.00
b. Improvements (buildings, etc.; depreciation plus interest on a per calf basis)[3]	30.00	21.00	12.00	6.00	36.00	27.00	18.00	12.00
c. Machinery and equipment (depreciation plus interest on a per calf basis)[4]	36.00	31.50	22.50	18.00	54.00	49.50	40.50	36.00
d. Cattle (depreciation plus interest on a per calf basis)[5]	49.90	49.90	49.90	49.90	72.15	72.15	72.15	72.15
e. Feed (including feed for dam)	50.00	45.00	40.00	35.00	75.00	70.00	65.00	60.00
f. Labor (including labor for dam)	18.00	15.00	12.00	10.00	36.00	30.00	24.00	20.00
g. Other	6.00	5.00	4.00	3.00	12.00	10.00	8.00	6.00
Total	257.90	235.40	208.40	209.90	373.15	346.65	315.65	294.15

[1]1973 survey.
[2]Land for commercial cattle at $850/animal unit; land for purebred cattle at $1,100/animal unit. Interest 8%.
[3]Buildings depreciated 25 years. Interest 8%.
[4]Machinery and equipment depreciated 10 years. Interest 8%.
[5]Commercial cows $320. Purebred cows $420. Cattle depreciated 7 years. Salvage value $150/head. Interest 8%.

1. Cash expense—

Feed	$	40.00
Labor		12.00
Other		4.00
Total	$	56.00

2. Investment in—

Land	850.00
Improvements	100.00
Machinery and equipment	125.00
Cattle	320.00
Total	$1,395.00
@ 8% interest	.08
Yearly interest	$ 111.60

3. Total cost—

Cash expense for feed, labor, and other	$	56.00
Interest		111.60
Total cost	$	167.60

4. Gross income, basis 450-pound calf, at 50¢
   450 × .50 = $225.00

5. Net income
   $225 − 167.60 = $57.40

Thus, based on the above figures, a commercial cattleman would need a 300-cow herd to make $17,000 per year (300 × $57.40 = $17,220). Of course, efficient operators will do much better. High quality feeder calves will bring $1.00/cwt more than the average kind—that's $4.50 per head, or $1,215 per year on a 300-cow herd with a 90% calf crop. A 5% increase in the calf crop means 15 more calves, or $3,375 in a 300-cow herd, with calves at 50¢ per pound. Fifty pounds more weaning weight per calf can mean $6,750 more in a 300-cow herd, with a 90% calf crop and 50¢ calves.

For land and cattle to be profitable, they should yield a return sufficient to the owner to (1) meet the interest payment on the investment, (2) retire a reasonable portion of the loan, and (3) provide satisfactory management return. But, there is no more reason why large land and cattle holdings should be debt free than there is for General Motors, or any other big corporation, to be debt free. Some land is overpriced, particularly in the more populous areas. Yet, there are compensating factors in many of the latter regions—their greater appreciation as potential building sites and recreational areas.

Other Plus Values to Land and Cattle Ownership

Some buy land to balance operations and cut costs; some to keep the children down on the farm; others because it adjoins their present property; and still others because of pride of ownership. There are also other important plus values to land and cattle ownership. Among them:

1. Appreciation in land values. From 1960 to 1974, U.S. farmland values rose from $116 per acre to $310 per acre, an increase of 2.7 times. They soared a record 25 percent for the year ending March 1, 1974. The latter

increase surpassed the largest previous rise of 22 percent scored in 1920. And all indications point to continued rises in land price in the years ahead.

2. As an effective hedge against inflation.

3. As a desirable alternative to a jittery stock market.

4. As a tax shelter.

5. A boom down on the farm in the years ahead brought about by (a) expanded export opportunities for U.S. farm products which have shifted the expectations of many farmland buyers and their lenders to a new plateau regarding the future well-being of U.S. agriculture, with the result that more people are interested in buying farmland; and (b) spreading roads and suburbs, and precious little new land that can be brought under production.

There is little doubt that farm real estate values will continue to climb. The only question is "how much."

Land and Cattle Investment Companies Will Come

Big land-cattle operations will get bigger, demanding more and more capital and top management. In the years ahead, many investors will become part owners in land and cattle, much as they now do through corporate stocks and bonds; and they will leave the management of the holdings to the professionals. Such an arrangement will also make it possible to (1) diversify in countries and types of investments—in different areas of the United States, and in Australia, Canada, South America, and other areas where there

Fig. 24-2. The major capital needs for a cow-calf enterprise are for land and cattle. Capital turnover is slow. (Courtesy, California State Polytechnic University, San Luis Obispo)

are vast acreages of rangeland with great potential for improvement that can be secured at reasonable prices; (2) minimize risks of loss from droughts and local depressions; (3) obtain for investors, big and little, the benefits that accrue to bigness, such as lower investment per cow unit, and lower feed and labor costs; (4) furnish recreational and vacation areas on farms and ranches of which they are part owner; and (5) provide know-how, continuity, and able management.

Part ownership in land investment companies affords a modern way in which to spread investments and minimize risks—as is done with stocks and bonds, grain and livestock futures, and syndicated sires.

RECORDS AND MONEY MANAGEMENT

In times of money scarcity and narrow profit margins, there will be greater stress on business and financial management skills. Also, cattlemen must realize that they are competing with many other users of credit, including retail merchants, manufacturers, home buyers, and professional people. Many of these borrowers can and do provide the lender with profit and loss statements and net worth statements, prepared by a CPA. Also, they usually submit annual budgets. As cattlemen increase the amount of borrowed capital in their operations, they, too, will be required to furnish adequate records and budgets if they are to compete successfully for the available capital.

Types of Records

Three general types of records are involved in a cow-calf operation. All are equally important to a well-run operation. They are:

• *Daily and weekly report*—A farm or ranch manager should keep a daily record, like the daily-weekly report shown in Fig. 24-3. It takes little time to keep it. Nevertheless, a certain time should be set aside daily for this purpose, thereby assuring that it will be kept—and that it will be kept accurately. For the manager, such a report provides an invaluable record of the day-by-day operations. For the owner, it's a quick and easy way to keep informed. This record should be filed, where it can be referred to as needed.

• *Production records*—Production records include such important profit indicators as percent calf crop dropped, percent calf crop weaned, weaning weight of calves at seven months, the pounds of beef per cow or per acre, the pounds of feed required to produce a pound of beef, death losses, etc.

• *Financial records*—The name of the game is "profit." Thus, it is necessary that production be translated into dollars and weighed against costs involved in achieving that production.

Many different kinds of financial records are used by cattlemen; among the most common ones are the following: inventory, budgets, income, expenses, cash flow, depreciation schedule, annual net income, enterprise accounts, profit and loss statement, and net worth statement. (Also, see Chapter 13, Business Aspects of Beef Production.)

"BAR-NONE RANCH," FOR WEEK BEGINNING _____,
(month, day, year)

Prepared by _____

	Mon.	Tue.	Wed.	Thur.	Fri.	Sat.	Sun.	Comments for day or week
WEATHER Rain or snow, inch .... Temp.—high .... low								
LABOR Accident or sick .... Who, what, how .... Changes								
EQUIPMENT BREAKDOWN Kind and make .... Cause .... Cost to repair								
ANIMAL HEALTH PROGRAM (indicate cow-calf/cattle feedlot) Vaccination .... Treatment								
ANIMAL LOSSES (indicate cow-calf/cattle feedlot) No. .... Kind .... Cause								
ANIMALS RECEIVED (indicate cow-calf/feedlot) No. .... Kind .... From .... Price or custom feed								
ANIMALS SOLD (indicate cow-calf/cattle feedlot) No. .... Kind .... To .... Price								
CROP HARVESTED Kind .... Field No. .... Acreage .... Yield/acre								
VISITORS Name and address								

Fig. 24-3. A good type of daily and weekly report.

Why Keep Financial Records?

There are many reasons for keeping good financial records, the most important of which follow:

1. *To save on income taxes*—Good tax management and reporting consists in complying with the law, but in paying no more tax than is required. The cardinal principles of good tax management are: (a) maintenance of adequate records so as to assure payment of taxes and amounts no less or no more than required by law, and (b) conduct of business affairs to the end that the tax required by law is no greater than necessary.

Also, cattlemen need to recognize that good tax management and good farm management do not necessarily go hand in hand. In fact, they may be in conflict. When the latter condition prevails, the advantages of one must be balanced against the disadvantages of the other to the end that there shall be the greatest net return.

2. *To obtain needed credit*—Lenders need certain basic information in order to evaluate the soundness of a loan request. The financial record will show the borrower's current financial position, his potential ahead, and his liability to others.

3. *To guide changes in enterprises*—Farm and ranch records should provide information from which the farm business may be analyzed, with its strong and its weak points ascertained. From the facts thus determined, the manager may adjust current operations and develop a more profitable organization. The enterprise should be above average before the owner borrows to expand it. Is the ranch too small? Is it more profitable to sell weaners or yearlings? Should the farm or ranch produce or buy hay?

4. *To serve as a guide for current income and expenses*—Records of cost and returns of previous years are very valuable as guides, and as a means of spotting trouble. Items which deviate substantially from the historical record should be studied with care.

5. *To provide for continuity of the business*—"Barn door" and "memory" records are insufficient. They mitigate against continuity of the business. The sudden passing of a manager places a severe stress on a business even under the most favorable circumstances. However, a good set of financial records gives a big assist to those who must take over during such times.

Kind of Record and Account Book

A farmer can make his own record book by ruling off the pages of a bound notebook to fit his specific needs, but the saving is negligible. Instead, it is recommended that he obtain a copy of a farm record book prepared for and adapted to his area. Such a book may usually be obtained at a nominal cost from the agricultural economics department of each state college of agriculture. Also, certain commercial companies distribute very acceptable farm record and account books at no cost.

Kind of Financial Records to Keep

At the outset it should be recognized that farm or ranch records should give the information desired to make a valuable analysis of the business. In general, the functions enumerated under the earlier section entitled, "Why Keep Financial Records?" can be met by the following kinds of records:

1. Annual inventory.
2. Budget.
3. Profit and Loss (P & L) statement.
4. Net Worth Statement.
5. Enterprise accounts.

Each of these records is discussed in Chapter 13; hence, the reader is referred thereto.

Who Shall Keep the Records?

The records may be kept either by someone in the farm or ranch business, or by someone hired to perform this service—a professional.

●*Farm or ranch help*—Very frequently, farm or ranch records are kept by the wife. If she has adequate time, there is no reason why she cannot keep good records.

Most farmers are not good record keepers, primarily because the operation of an ordinary farm requires large amounts of physical labor. As a result, most farmers and ranchers are physically exhausted at the end of the day's work and have neither the time nor the ambition to record in at least four different places each transaction which occurred during the day, as would be necessary of the usual double-entry bookkeeping system.

●*A professional*—There are, of course, individuals and firms who make a business of keeping farm or ranch records. Usually, they are accountants or farm management specialists.

How Shall Records Be Kept?

Records may be kept either by hand or by computer. Accurate and up-to-the-minute records and controls have taken on increasing importance in all agriculture, including the cattle business, as the investment required to engage in farming and ranching has risen and profit margins have narrowed. Today's successful farmers and ranchers must have, and use, as complete records as any other business. Also, records must be kept current; it no longer suffices merely to know the bank balance at the end of the year.

●*By hand*—The hand system can be used, but is slow and tedious. This service is usually performed by a member of the family or by an accountant living in the community. Nevertheless, after learning what is wanted, a first-rate accounting system can be kept by hand.

●*By computer*—Big and complex cattle enterprises have outgrown hand record keeping. It's too time consuming, with the result that it doesn't allow management enough time for planning and decision making. Additionally, it does not permit an all-at-once consideration of the complex interrelationships which affect the economic success of the business. This has prompted a new computer technique known as linear programming.

In the past, the biggest deterrent to production testing on cow-calf operations, both purebred and commercial, has been the voluminous and time-consuming record keeping involved. Keeping records *per se* does not change what an animal will transmit, but records must be used to locate and propagate the genetically superior animals if genetic improvement is to be accomplished.

Production testing has been covered elsewhere in this book (see Chapter 5); hence, the reader is referred thereto. As has been pointed out, two factors contribute to optimum beef production; namely (1) heredity, and (2) environment. Records of traits must be adjusted for such well-known sources of variation as age of dam, age of calf, and sex. This is tedious and time consuming when records are kept and analyzed by hand. However, punched card equipment, or a computer, can handle this assignment efficiently and with less risks of errors and omissions than the hand method. Several processing centers—universities, state associations, performance registries, breed associations, cooperatives, and private firms—now offer this service. Their goals are similar, but their methods of accomplishing them differ slightly. The important thing is to use one of them; generally speaking, this should be the one in use in the area which best fits the individual needs.

In addition to their use in production testing, computerized records can be, and are, used for herd record purposes—as a means of keeping management up-to-date and serving as an alert on problems or work to be done. Each animal must be individually identified. Reports can be obtained at such intervals as desired, usually monthly or every two weeks. Also, the owner can keep as complete or as few records as desired, and the system may be adapted to either purebred or commercial herds. Of course, commercial operators do not need pedigrees. Here are several of the records that can be kept by computer:

1. Pedigrees.
2. Animals that need attention such as—
    a. Animals 4 months old that are unregistered.
    b. Females 6 months old that are not vaccinated.
    c. Bulls that are 6 months old, and which should either be marked to keep or be castrated.
    d. Heifers 18 months old that haven't been bred.
    e. Cows that have been bred 2 consecutive times.
    f. Cows not rebred 3 months after calving.
    g. Cows that have reached 9 years of age, and that may be getting shelly and should be culled.
    h. Cows due to calve in 30 days.
    i. Calves 7 months of age that haven't been weaned.
    j. Calves 7 months of age that haven't been scored and weighed.
3. A running, or cumulative, inventory of the herd, by sex; including calves dropped, calves due, and purchases and sales—in number of animals and dollars.
4. The depreciation of purchased animals according to the accounting method of choice.

It is predicted that "card pokes" will come to be almost as well-known in the future as "cow-pokes" are at present.

Whether a cattleman should keep his own records or hire a professional, and whether records should be kept by hand or by computer, each individual must decide for himself. For the most part, the decision should be based on weighing the usefulness of the information each provides against the cost of obtaining it.

Tax Planning[2]

There are some things that a cattleman can do to lessen the tax bite. Some of these will be discussed.

CLAIM AS MUCH DEPRECIATION AS POSSIBLE

This means claiming the full amount of extra first-year 20 percent depreciation, then taking the regular depreciation on the remaining value.

The farmer or rancher will want to consider using fast methods of depreciation on new machinery and equipment having a useful life of three years or more (used items are not eligible). Breeding stock may be eligible if they meet the useful life test and are not "used."

There are two fast depreciation methods—the double declining balance, and the sum-of-the-year's digits. Of the two, the double declining balance method is the simplest. All it involves is doubling the straight-line rate and applying that factor each year against the undepreciated balance.

With the sum-of-the-year's-digits method, the annual depreciation allowance is figured by applying a declining fraction to the cost of the property, reduced by the salvage value. In the fraction, the numerator is the remaining years of useful life and the denominator is the sum of the digits of each of these years of useful life. For an item with a 10-year life, first-year depreciation would be 10/55 of cost. The last year would be 1/55.

However, the fast method cannot be used on most depreciable real property. Only new residential rental housing is eligible. For other new depreciable real property, depreciation may be claimed no faster than the 150 percent declining balance will permit. For used depreciable real property (buildings on a newly purchased farm, for example), depreciation is limited to the straight-line method which involves an equal deduction each year of the property's useful life. Used residential rental housing can go to a 125 percent declining balance if it has a useful life of 20 years or more.

Table 24-2 shows how the methods compare on the purchase of a new $10,000 item having a useful life of 10 years.

MAKE USE OF INVESTMENT CREDIT

Investment credit can be a massive tax saver. It permits the farmer or rancher to deduct 7 percent of the cost of purchased property off the tax he would otherwise have to pay. The effect is to reduce his tax by 7 percent on most items.

---

[2]This section was authoritatively reviewed by an eminent tax specialist, Mr. S. P. Kurth, of the law firm of Kurth, Felt, Speare and Lalonde, P.C., Billings, Mont.

TABLE 24-2

ANNUAL DEPRECIATION[1]

Year	Straight Line Method	Declining Balance Method	Sum of Year's Digits
First year additional depreciation	$ 2,000	$2,000	$ 2,000
First year ordinary depreciation	800	1,600	1,455
Total depreciation for first year	$ 2,800	$3,600	$ 3,455
Second year	$ 800	$1,280	$ 1,309
Third year	800	1,024	1,164
Fourth year	800	819	1,018
Fifth year	800	655	873
Sixth year	800	524	727
Seventh year	800	420	582
Eighth year	800	336	436
Ninth year	800	268	291
Tenth year	800	215	145
TOTAL	$10,000	$9,141	$10,000
Unrecovered cost		$ 859	....

[1]Source: *Farmer's Tax Guide*, Internal Revenue Service.

On eligible new or used property that the cattleman purchases, he can take 7 percent of the "qualified investment" off the tax he would otherwise have to pay. This means that if he bought a tractor for $10,000, the investment credit could save him $700 in taxes. That's worth as much as $2,800 in added depreciation deduction for a taxpayer in the 25 percent bracket.

A lot of different kinds of farm property are eligible for the investment credit. Along with the usual array of machinery and equipment, breeding stock now qualify. Even tile drains, fences, feeding floors, outside electrical installations, and water systems are eligible.

Buildings don't usually qualify, but the credit can be claimed on silos and grain bins. In recent years, structures and equipment used in livestock confinement systems have qualified. The test is whether the structure houses property "used as an integral part of production, manufacturing, or extractive activity." Also, whether the structure is so closely related to use of the equipment that it can be expected to be replaced when the equipment is changed.

OTHER TAX SAVERS

Other tax saving ideas that the cattleman may apply are:

1. *Buy ahead on certain inputs*—The cattleman might consider buying ahead feed, fertilizer, seed, chemicals, or other items. In doing so, however, he should pay for it in the year for which claim is being made, and, if possible, take delivery. A mere deposit for future delivery on verbal contract isn't enough to back up a deduction as far as IRS is concerned.

2. *Cull the herd*—The cattleman may sell additional breeding stock

rather than replacement stock. Remember that only half of the income from the sale of home-raised breeding animals is taxable. Remember, too, that cattle must be kept two years to qualify as capital gain breeding stock.

3. *Pay children for farm work*—Payments made to children for farm work are business expenses. But the cattleman must be able to show that a true employer-employee relationship exists. To do so, children should be assigned definite jobs at agreed-upon wages and paid regularly.

Wages paid to children under 21 years of age are not subject to Social Security tax. However, if a child nets more than $400 on a 4-H or FFA cattle project, he is liable for Social Security tax as a self-employed person.

There are other ways to save on taxes. But there are also some "traps" into which the cattleman may fall. That's why it's important to consult with a good farm or ranch tax advisor—one who has had considerable experience in working out tax plans involving breeding animals and other unique aspects of farm taxes.

Cow-Calf Profit Indicators

Many factors determine the profitableness of a cow-calf enterprise. Certainly, a favorable per animal unit capital investment in land and improvements is a first requisite. Additionally, percent calf crop weaned and weaning weight are exceedingly important, as shown in Table 24-3.

TABLE 24-3

SIZE AND WEIGHT OF CALF CROP WEANED ARE IMPORTANT

Calf Crop	Yearly Operating Cost—$150.00 Per Cow Calf Weights				
(%)	(450 lb)	(425 lb)	(400 lb)	(375 lb)	(350 lb)
90%	405 37.0¢	382 39.3¢	360 41.7¢	337 44.5¢	315 47.6¢
85%	382 39.3¢	361 41.5¢	340 44.1¢	319 47.0¢	298 50.3¢
80%	360 41.7¢	340 44.0¢	320 46.8¢	300 50.0¢	280 53.6¢
75%	337 44.5¢	319 47.0¢	300 50.0¢	281 53.4¢	263 57.0¢
Break-even point @ 44.0 cents/lb.					

In Table 24-3, it is assumed that the yearly operating cost is $150 per cow. Then, the effect of size and weight of calf crop is computed. As shown, with a 90% calf crop and an average weaning weight per cow bred of 405 lb (450 × 90% = 405), a selling price of 37¢/lb will meet the break-even cost of $150. With a 75% calf crop, because of fewer pounds per cow bred (450 × 75% = 337 lb), the calves would have to bring 44.5¢/lb in order to break even.

Table 24-4 will serve as a yardstick for determining (1) how you stack up with the nation's (a) average, and (b) top five percent of cow-calf operators, and (c) where you are falling down in your cow-calf enterprise.

TABLE 24-4

COW-CALF PROFIT INDICATORS

	Average for U.S. Cow-Calf Operations			
	Commercial		Purebred	
	Average	Top 5%	Average	Top 5%
Investment/animal unit (one mature cow) in land and improvements (real estate), $	1,025	1,075	1,325	1,400
Percent calf crop dropped (based on no. cows bred), %	88	93	89	94
Percent of calf crop weaned (based on no. cows bred), %	81	89	82	91
Weaning weight of calf at 7 mo., lb .	400	500	425	520
Age and longevity of cows:				
Age when removed from herd, yrs.	9	10	9.5	10.5
No. calves produced in lifetime of cow, no.	6.0	8.0	7.0	8.5
Labor/cow/year, hrs	10.0	9.5	16.0	13.50
Net return per cow to management,[1] $	27.25	41.50	38.00	95.00

[1]Net return to management after deducting from gross receipts all costs, including depreciation on machinery, buildings and cattle, and interest on investment.

It is noteworthy that Table 24-4 reveals that the top five percent operators have a higher investment/animal unit in land and improvements than their average counterparts. Obviously, better operators have better land and improvements; their savings are made in the handling of the herd. It's not unlike selecting a ration—where it's net returns, rather than cost per ton, that counts.

No claim is made that Table 24-4 is perfect. Admittedly, there are wide area differences, and no two farms or ranches are alike. Also, there are seasonal differences; for example, a drought will materially affect weaning weight of calves. Yet, Table 24-4 will serve as a useful guide.

An Incentive Basis for the Help

On cow-calf enterprises, there is need for some system which will encourage caretakers to be good nursemaids to newborn calves, though it may mean the loss of sleep, and working with cold, numb fingers. Additionally, there is need to do all those things which make for the maximum percent calf crop weaned at a heavy weight.

From the standpoint of the owner of a cow-calf enterprise, production expenses remain practically unchanged regardless of the efficiency of the operation. Thus, the investment in land, buildings and equipment, cows, feed, and labor differs very little with a change (up or down) in the percent calf crop or the weaning weight of calves; and income above a certain break-even point is largely net profit. Yet, it must be remembered that owners take all the risks; hence, they should benefit most from profits.

On a cow-calf operation, the author recommends that profits beyond the break-even point (after deducting all expenses) be split on an 80:20 basis. This means that every dollar made above a certain level is split, with the owner taking 80 cents and the employees getting 20 cents. Also, there is merit in an escalator arrangement; with the split changed to 70:30, for example, when a certain plateau of efficiency is reached. Moreover, that which goes to the employees should be divided on the basis of their respective contributions, all the way down the line; for example, 25% of it might go to the manager, 25% might be divided among the foremen, and 50% of it divided among the rest of the help.

A true profit-sharing system on a cow-calf outfit based on net profit has the disadvantages of (1) employees not benefiting when there are losses, as frequently happens in the cattle business, and (2) management opening up the books, which may lead to gossip, misinterpretation, and misunderstanding. An incentive system based on major profit factors alleviates these disadvantages.

Gross income in cow-calf operations is determined primarly by (1) percent calf crop weaned, (2) weaning weight of calves, and (3) price. The first two factors can easily be determined. Usually, enough calves are sold to establish price; otherwise, the going price can be used.

The incentive basis proposed in Table 24-5 for cow-calf operations is simple, direct, and easily applied. As noted, it is based on average pounds of calf weaned per cow, which factor encompasses both percent calf crop and weaning weight.

TABLE 24-5

A PROPOSED INCENTIVE BASIS FOR COW-CALF OPERATIONS

Average Pounds of Calf Weaned/Cow Bred	Here's How It Works
(lb)	
350	On this particular operation, the break-even point is assumed to be an average of 400 pounds of calf weaned per cow bred; and, of course, this is arrived at after including all costs of production factors.
375	
400 (break-even	
425 point)	
450	Pounds of calf weaned per cow bred in excess of the break-even point are sold or evaluated at the going price.
475	
500	
525	
550	If an average of 450 pounds of calf per cow bred is weaned, and if mixed steers and heifers of this quality are worth 50 cents per pound, that's $25 more net profit per cow. In a 500-cow herd, that's $12,500. With an 80:20 division, $10,000 would go to the owner, and $2,500 would be distributed among the employees.
575	
600	
	Or, if desired, and if there is an escalator arrangement, there might be an 80:20 split at 425 lb, a 70:30 split at 450 lb, and a 65:35 split at 475 lb.

QUESTIONS FOR STUDY AND DISCUSSION

1. Take your own cattle farm or ranch, or one with which you are familiar, and see how you compare in investment, breakdown, and total to the figures given in Table 24-1.

2. What net return per calf weaned should a cattleman reasonably expect?

3. Land values have risen rather sharply in recent years. What *plus* values do you see ahead to land and cattle ownership?

4. How can business and financial management skills help a cattleman in times of money scarcity, high interest rates, and narrow profit margins?

5. How would you improve on Fig. 24-3, daily and weekly report?

6. How will good financial records give an assist in (a) saving on income taxes, (b) obtaining needed credit, (c) guiding changes in enterprises, (d) guiding current income and expenses, and (e) providing for continuity of the business?

7. What's the value of an annual inventory?

8. How would you go about setting up a budget for a cow-calf enterprise?

9. Who should keep the farm and ranch records—the farm or ranch help, or a professional? Justify your answer.

10. How should records be kept—by hand, or by computer?

11. What kind of cow-calf records can be kept by computer? What kind of questions can be answered by computer?

12. How may a cattleman use each of the following to lessen the tax bite: (a) depreciation; (b) the seven percent investment credit; (c) buying feed ahead; (d) culling; and (e) paying children for work?

13. Of what value are cow-calf profit indicators? Apply the Table 24-4 Cow-Calf Profit Indicators to your beef herd, or a herd with which you are familiar. Determine (a) how well you are doing, and (b) where you are falling down.

14. Evaluate the Table 24-5, a Proposed Incentive Basis for Cow-Calf Operations. How would you improve it?

SELECTED REFERENCES

Title of Publication	Author(s)	Publisher
Analysis of Beef Costs and Returns in California, An, AXT-258	A. D. Reed	Agricultural Extension Service, University of California, Davis, Calif., 1967
Beef Cattle Economics Series, AXT-378	J. H. Cothern	Agricultural Extension Service, University of California, Davis, Calif., 1973
Beef Cattle Production, Bull. 527	A. L. Barr	Agricultural Experiment Station, West Virginia University, Morgantown, W.Va., 1966
Beef Cow Costs and Returns in Northern Wisconsin, Res. Rep. 22	R. E. Rieck G. C. Pulver W. Henquinet	Experiment Station, University of Wisconsin, Madison, Wisc., 1966
Cattle Raising in the United States	R. N. Van Arsdall M. D. Skold	Economic Research Service, U.S. Department of Agriculture, Washington, D.C., 1973
Commercial Beef Cattle Production	Ed. by C. C. O'Mary, I. A. Dyer	Lea & Febiger, Philadelphia, Penn., 1972
Complete Guide to Making a Public Stock Offering, A	E. L. Winter	Prentice-Hall, Inc., Englewood Cliffs, N.J., 1962

(Continued)

Title of Publication	Author(s)	Publisher
Corporation Guide		Prentice-Hall, Inc., Englewood Cliffs, N.J., 1968
Corporations with Farming Operations	G. W. Coffman	Economic Research Service, U.S. Department of Agriculture, Washington, D.C., 1971
Costs and Returns of Cow-Calf Operations, MP-947	T. E. Prater	Texas Agricultural Extension Service, Texas A & M University, College Station, Tex., 1970
Costs and Returns, Northwest Cattle Ranches, 1972, ERS-525	W. D. Goodsell M. J. Belfield	Economic Research Service, U.S. Department of Agriculture, Washington, D.C., 1973
Costs and Returns, Southwest Cattle Ranches, 1971, FCR-83	J. R. Gray W. D. Goodsell M. J. Belfield	Economic Research Service, U.S. Department of Agriculture, Washington, D.C., 1972
Credit for Farming, Bull. 175	G. A. Stevens L. D. Rhoades	Extension Service, University of Rhode Island, Kingston, R.I., 1960
Economic Analysis of the Beef Cattle Enterprise at the Southwest Georgia Branch Experiment Station for a 5-Year Period, 1967-71, Res. Rep. 155	R. B. Moss F. B. Saunders W. C. McCormick	College of Agriculture Experiment Station, University of Georgia, Athens, Ga., 1973
Economic Aspects of Registered Cattle Enterprises in New Mexico, Bull. 538	K. H. Wendland J. R. Gray	Agricultural Experiment Station, New Mexico State University, Las Cruces, N.M., 1968
Economic Comparison of Cow-Calf and Stocker Calf Production Systems In the Georgia Piedmont Area, Res. Rep. 151	G. C. Lance W. A. Griffey G. V. Calvert	College of Agriculture Experiment Station, University of Georgia, Athens, Ga., 1973
Economic Effects of Changes in Production Practices in Beef Cattle Production in the Southwest Louisiana Rice Area	A. R. Gerlow W. F. Woolf	Agricultural Experiment Station, Louisiana State University, Baton Rouge, La.; Economic Research Service, U.S. Department of Agriculture, Washington, D.C., 1971
Economic and Operational Characteristics of Beef Cattle Ranches in West Central Florida	H. D. Brodnax, Jr. B. R. Eddleman	Florida Agricultural Experiment Stations, Gainsville, Fla.; Economic Research Service, U.S. Department of Agriculture, Washington, D.C., 1969
Economic Strategies for Foothill Beef Cattle Ranchers, Bull. 824	G. W. Dean A. J. Finch J. A. Petit, Jr.	California Agricultural Experiment Station, University of California, Davis, Calif., 1966
Economics of Retained Ownership of Calves on Eastern Colorado Cattle Ranches, The, Bull. 551S	C. K. Gee J. K. Pursley	Experiment Station, Colorado State University, Fort Collins, Colo.; Economic Research Service, U.S. Department of Agriculture, Washington, D.C., 1972

(Continued)

Title of Publication	Author(s)	Publisher
Estimated Annual Costs, Production, and Income for Selected Livestock and Crop Enterprises, Eastern West Virginia, Bull. 594T	R. A. Layton A. L. Barr P. E. Nesselroad	Agricultural Experiment Station, West Virginia University, Morgantown, W.Va., 1970.
Farm Corporation, The, Pm-273	J. C. O'Byrne N. G. P. Krausz N. E. Harl H. Jurgenson	Cooperative Extension Service, Iowa State University, Ames, Iowa, 1970
General Partnership for Agricultural Producers, Bull. E-731	R. E. Hepp M. Kelsey	Cooperative Extension Service, Michigan State University, E. Lansing, Mich., 1972
Going Public	Corplan Associates	Corplan Associates, IIT Research Institute, Chicago, Ill.
Guide for Estimating Annual Return to Labor, Management and Capital, Cow-Calf Operation–Central and East Texas, MP-398	T. E. Prater L. A. Maddox	Texas Agricultural Extension Service, Texas A & M University, College Station, Tex., 1967
Investment Requirements for an Approximate $3,000 Return to Labor-Management, MP-380	T. E. Prater	Texas Agricultural Extension Service, Texas A & M University, College Station, Tex., 1968
Investment Requirements for Full Employment–One-Man Equivalent, MP-931	T. E. Prater	Texas Agricultural Extension Service, Texas A & M University, College Station, Tex., 1970
Making and Evaluating Farm Partnership Agreements, MF-223	F. L. Overley	Cooperative Extension Service, Kansas State University, Manhattan, Kan., 1970
Mechanics of Farm Financial Planning, Circ. 1042	R. B. Schwart J. M. Holcomb A. G. Mueller	Cooperative Extension Service, University of Illinois, Urbana, Ill., 1972
Optimum Enterprise Combinations and Resource Use on Mountain Cattle Ranches in Colorado, Bull. 546S	C. K. Gee M. D. Skold	Colorado State University Experiment Station, Fort Collins, Colo.; Economic Research Service, U.S. Department of Agriculture, Washington, D.C., 1970
Organization, Costs, and Incomes of Western Cattle and Sheep Ranches, Bull. 587	J. R. Gray	Agricultural Experiment Station, New Mexico State University, Las Cruces, N. Mex., 1971
Organization, Costs, and Returns, Northwest Cattle Ranches, 1960-71	W. D. Goodsell	Economic Research Service, U.S. Department of Agriculture, Washington, D.C., 1972

(Continued)

Title of Publication	Author(s)	Publisher
Production of Beef Cattle in the Southeast: Selected Characteristics and Trends, EIR-16	H. C. Gilliam, Jr.	Agricultural Experiment Station, North Carolina State University, Raleigh, N.C.; Economic Research Service, U.S. Department of Agriculture, Washington, D.C., 1970
Production, Financing, and Off-Farm Employment Aspects of Beef Farming in Arkansas, 1969, Bull.785	W. A. Halbrook A. E. Spooner M. L. Ray	Agricultural Experiment Station, University of Arkansas, Fayetteville, Ark., 1973
Requirements and Returns for 1000-Cow Beef Herds on Flatwood Soils in Florida, Circ. 385	C. L. Anderson T. S. Hipp	Cooperative Extension Service, University of Florida, Gainsville, Fla., 1973
Resources, Production and Income on Eastern West Virginia Beef Cattle Farms, Bull. 546	A. L. Barr G. E. Toben C. C. Wilson, Jr.	Agricultural Experiment Station, West Virginia University, Morgantown, W. Va., 1967
Southwest Cattle Ranches, Organization Costs, and Returns 1964-72	W. D. Goodsell J. R. Gray M. J. Belfield	Economic Research Service, U.S. Department of Agriculture, Washington, D.C., 1974
Stockman's Handbook, The, Fourth Edition	M. E. Ensminger	The Interstate Printers & Publishers, Inc., Danville, Ill., 1970
Stretching Your Resources Through Borrowing Money for Your Farm and Home Business, C-365	W. B. Thomas J. L. Treat J. R. Urich	Extension Service, Kansas State University, Manhattan, Kan., 1966
Summary of Illinois Farm Business Records, 1971, Circ. 1058	D. F. Wilken R. P. Kesler	Cooperative Extension Service, University of Illinois, Urbana, Ill., 1972
Tax-Sheltered Investments	W. J. Casey	Institute for Business Planning, Inc., New York, N.Y., 1973
Using Agricultural Credit, Circ. 890	J. M. Holcomb R. B. Schwart	Cooperative Extension Service, University of Illinois, Urbana, Ill., 1964

CHAPTER 25

STOCKER (FEEDER) CATTLE

Contents   **Page**

Stocker cattle and stocker cattle programs have changed with the passing of time. Until the early 1900s, stockers involved growing purchased or homegrown calves or yearlings on grass and hay until they were 3 to 4 years of age. As calf weaning weights increased and finished slaughter cattle weights decreased, the amount of time and gain required to grow calves

Fig. 25-1. Stocker calves on pasture.

1229

from weaning until the beginning of the finishing period was substantially shortened. The stocker cattle industry became a calf-yearling industry, usually starting with 300- to 500-pound calves and ending with a yearling sold to a feeder at 600 to 700 pounds.

The development of large feedlots and year-round feeding increased the demand for feeders ready to go on high concentrate rations. The main reason that the larger feedlots like to purchase feeders ready to go on high energy rations is that roughage is used more efficiently in growing cattle, whereas it usually is an expensive item to use in large feedlots.

Today, the stocker stage is changing again, as a result of forces working in opposite directions, as follows, with one force favoring lengthening of the stocker stage and the other favoring shortening it:

1. Scarce and high priced grains favor more roughage feeding and less grain feeding, resulting in carrying stockers to older ages and heavier weights, followed by a shorter feedlot period.

2. Heavier milking cows and heavier weaning weights, coupled with high priced land, favor shortening the stocker stage, or even eliminating it, as 600-pound, or heavier, weaning weights are achieved.

In the future, both types of stocker operations will prevail, with the choice determined primarily by the price of grain and the weaning weight of the calves. Heavy weaned calves will likely go directly into the feedlot or for slaughter. Calves with light to average weaning weights will likely be carried as stockers to 700- to 800-pound weights, thereby shortening the feedlot period and lessening grain feeding.

Thus, the growing of calves from weaning until placing on finishing rations is not new. However, in recent years some new "wrinkles" have been added to the methods of conducting it. Today, stockers are grown according to two systems: (1) calves or light yearlings are either roughed through the winter followed by grazing, or grazed only, then sold as feeders in late summer and fall; or (2) calves or yearlings are fed harvested roughage and grain in drylot, and then transferred to another location for finishing. Also, some new terms have evolved, definitions of which follow.

● *Stockers* are calves and yearlings, both steers and heifers, that are intended for eventual finishing and slaughtering and which are being fed and cared for in such manner that growth rather than finishing will be realized. They are generally younger and thinner than feeder cattle.

● *Feeders* are calves and yearlings, both steers and heifers, carrying more weight and/or finish than stockers, which are ready to be placed on high energy rations for finishing and slaughtering.

● *Replacement heifers* are the top end of the heifer calves selected to replace the older cows that are culled from the herd.

● *Preconditioning* refers to preparing the calf to withstand the stress and rigors of leaving its mother, learning to eat new kinds of feeds, and shipment from the farm or ranch to the feedlot or stocker grower.

● *Backgrounding* is an old practice with a new emphasis and a new name. Actually backgrounding and the stocker stage are one and the same. Both refer to that period in the life of a calf from weaning to around 800 pounds weight, when they are ready to go on a high energy finishing ration. However, the term "backgrounding," which was ushered in with the development of large feedlots, indicates a shift in emphasis. The term "stocker stage" connotes emphasis on marketing roughages through thin cattle, whereas "backgrounding" connotes emphasis on growing out feeder calves ready to go on a high energy finishing ration. Backgrounding may be done on pasture or in the drylot, or some combination of both. At its best, the animals should be in good health, bunk broke, and ready to go on full feed. (Also, see section of "Backgrounding" in Chapter 31, Management of Feedlot Cattle.)

From the above, it may be concluded that in the variable period of a calf's life between weaning and finishing, it is usually classed as either a stocker, a feeder, or a replacement heifer. Prior to weaning, calves may or may not be preconditioned.

The dividing line between stockers and feeders is not always as clear cut as the above definitions would indicate. That is, not all thin cattle are suitable for stockers. For example, very large yearlings and most heifers are usually sold as feeders, to be placed on high energy feeds. Also, "Okie" type cattle are usually backgrounded for 50 to 60 days, then placed on a finishing ration.

TYPES OF STOCKER PROGRAMS

Sometimes the stocker operation is the only cattle enterprise on the farm or ranch. More frequently, however, it is conducted in conjunction with a cow-calf operation or it precedes the finishing program.

When the stocker enterprise is the only cattle enterprise on a farm or ranch, it is usually conducted in one of the following ways:

1. Calves or light yearlings are bought in the fall to be wintered on high roughage rations in drylot and sold in the spring to buyers either (a) to go on grass for the summer, or (b) to go on a drylot finishing program.

2. Lightweight calves are bought in the fall to be wintered on roughage

Fig. 25-2. Yearling stockers on pasture. (Courtesy, USDA)

rations, then, under the same ownership, grazed throughout the following pasture season and sold in the fall. Under this plan, usually lighter weight calves are acquired and they are wintered at a lower rate of gain than in plan 1.

3. In Kansas, Oklahoma, and Texas, calves or light yearlings are bought in the fall and grazed on winter small grains, chiefly wheat. Good wheat pastures will produce very acceptable stocker gains. The main disadvantage to the program is that, due to weather conditions, winter wheat pasture cannot always be counted upon. When it fails, the stockers must either be sold or fed a higher cost roughage.

4. In the southeastern states, which is primarily a cow-calf area, winter oats and fescue are used extensively in stocker programs. This area is turning to stocker programs in order to utilize profitably winter pastures, and to satisfy the demand for 600- to 800-pound feeder steers as a result of the expansion of feedlots.

There is a trend for more and more calves (not yearlings) to be handled according to plan 1,—that is, bought in the fall, wintered on roughage, and sold directly into a finishing program. This trend will be accelerated because of heavier calves being weaned in the fall, and because it is more profitable either to use presently available pasture areas for brood cows to produce more calves or for crop production.

The most common type of operation is a combination stocker-feeder program, typical of the Corn Belt and the irrigated sections of the West, where high yielding corn and sorghum crops are produced for silage. In these areas, cattle feeders usually purchase steer calves or light yearlings in the fall or late winter; fall-graze stalk fields and small-grain stubble where available; move into the drylot for the winter and feed corn or sorghum silage, supplemented with a legume hay or protein supplement; then finish on a high energy ration either in the drylot or on pasture and sell for slaughter in the summer or fall.

An increasing number of feeders are grown on contract for and delivered to a feedlot for finishing. This trend has been prompted by the competition between feedlots. It is their way of assuring a continuous supply of feeders of the desired weights and quality. As a further inducement, many of the feedlots finance the grower (backgrounding) operation.

ADVANTAGES AND DISADVANTAGES OF A STOCKER PROGRAM

A stocker enterprise has both advantages and disadvantages in comparison with a cow-calf or a cattle finishing operation. These should always be weighed and balanced, especially where there is a choice.

A stocker program has the following *advantages* over other types of cattle programs:

1. *Flexibility*—A stocker operation is more flexible than a cow-calf operation from the standpoint of adjusting to feed supplies and cost, labor, and economic outlook. The number of stockers purchased each year may be altered accordingly.

2. *Efficient gains*—Stocker operators have the cattle when they make the most efficient gains.

3. *Low labor requirement*—Stocker cattle have a lower labor requirement than a cow-calf operation conducted on the same amount of land.

4. *Distribution of labor*—The peak labor requirement in wintering stockers in the drylot is completed ahead of spring and summer farm work.

5. *Quick returns*—Where a stocker program is limited either to wintering or pasturing only, returns come quickly, within 4 to 6 months. In some cases, this quick turnover permits handling 2 or 3 droves of stockers per year.

6. *Adapted to areas lacking accessibility to fat cattle markets or sources of grain*—A stocker operation is better adapted than a cattle finishing operation to areas lacking accessibility to slaughter cattle markets or sources of grain. Of course, such areas are also well suited to cow-calf operations; hence, a choice must be made.

7. *Utilize roughage and salvage feeds*—Stockers are adapted to the use of roughage and salvage feeds.

8. *Investment in buildings and equipment may be less*—If the stocker program is limited to grazing winter or summer pastures, a minimum of buildings and equipment is required.

9. *Contract basis requires little capital*—If a grower contract is arranged with a feedlot, as is sometimes possible, little capital is required.

There are also *disadvantages* to a stocker program, including the following:

1. *High risk*—It is a high risk venture. High risk results from seasonal and yearly price fluctuations in feeder cattle, and the fact that total gains are not large in proportion to the weight purchased. Also, stocker operators who rely on winter wheat or oat pastures or summer grazing may have to purchase cattle when the price is high and sell them when the market is flooded with similar cattle. Moreover, lack of rain may make for small gains and high costs.

2. *Cost of gain may have to offset negative margins*—Cost of gain must be kept down to offset negative margins that may prevail.

3. *High buying and selling skills are required*—Buying and selling skills are extremely important because (a) there is no established market as exists with slaughter cattle, and (b) the original weight purchased is a high percentage of the weight sold. It follows that any mistake made in buying and selling, such as mistakes in judging the quality and health of the stockers, has a greater influence on profits or losses than in a finishing program. Moreover, the entire livestock inventory is bought and sold at least once per year.

4. *Buying, selling, and shrink costs must be absorbed by a limited amount of gain*—Shrink on both ends and buying and selling costs can wipe out any economical, but relatively small, gains that can be made.

5. *High land, labor, and interest costs mitigate against stocker programs*—Because of the relatively small gains made by stockers, high land, labor, and interest costs mitigate against such programs. For this

reason, growing calves at rates of less than one pound per head daily becomes increasingly difficult to justify.

6. *Disease can make for heavy losses*—The weanling calf is at the most susceptible stage in life to contagious diseases. Also, the limited weight gains in growing operations leaves little opportunity to recover severe losses.

7. *High transportation costs*—Often the transportation cost is high because of long distance from the sources of supply and/or feedlots.

FACTS PERTINENT TO STOCKER PROGRAMS

The following points are pertinent to stocker programs:

● *Stocker programs are used by large cattle feedlots to insure continuous supply of feeder replacements*—Many large feedlots, which feed on a year-round basis, are effectively using stocker programs to insure a continuous supply of feeder replacements. They usually accomplish this by buying both calves and yearlings when they are available at favorable prices, then putting them on different stocker programs, often on a contract basis, designed to stagger their readiness to move to the feedlot at the weight and time desired.

● *Stocker stage will be either lengthened or shortened, depending on the circumstances*—High priced grains and cheap roughages, comparatively speaking, favor older and heavier stockers. So, yearling stockers will increase where and when pastures and dry roughages are relatively cheap. However, the following factors favor younger and lighter stockers—calves: (1) heavier weaning weight; (2) greater efficiency of feed utilization of younger cattle; (3) the need to place larger strains of cattle (the exotics) on finishing rations at younger ages—otherwise their carcasses become too heavy; (4) the demand for leaner beef; (5) higher remuneration from cow-calf operations; and (6) high land, labor, and interest costs mitigating against the small gains common to stocker operations.

● *Diseases are a problem*—Diseases can take a tremendous toll in a stocker program. The weanling calf is at the most susceptible stage of life to contagious diseases. This coupled with the limited weight gains in growing operations leaves little opportunity to recover severe losses. Plans should be made for routine immunization, starting rations, and veterinary treatment. Also, the operator will be well paid for providing TLC (tender loving care) and spotting sick animals early.

As every cowman knows, losses due to diseases and accidents are higher in stocker calves than in yearlings. Studies show that death losses of calves usually run between 2.0 and 2.5%; yearlings about 0.75 to 1.00%; and 2-year-old and older cattle 0.4 to 0.5%. Most of these losses are due to stress and to viruses to which the animals are subjected between the farm or ranch and the final destination of the stockers. Shipping fever takes the heaviest toll. It and other diseases are fully covered in Chapter 12.

● *Lightweight calves gain more efficiently*—The feed requirement for calf gain is directly related to the animal's weight. Because of low maintenance requirements, lightweight calves gain more efficiently than heavyweight calves. This is illustrated in Fig. 25-3 which shows the megacalories of digestible energy required for each 100-lb gain when calves are gaining at the rate of 1½ lb per day. All things being equal (health, genetics, etc.), calves will gain 100 lb on less feed at the lighter weights, which is an advantage for which most stocker buyers are looking.

EFFECT OF CALF WEIGHT ON FEED EFFICIENCY

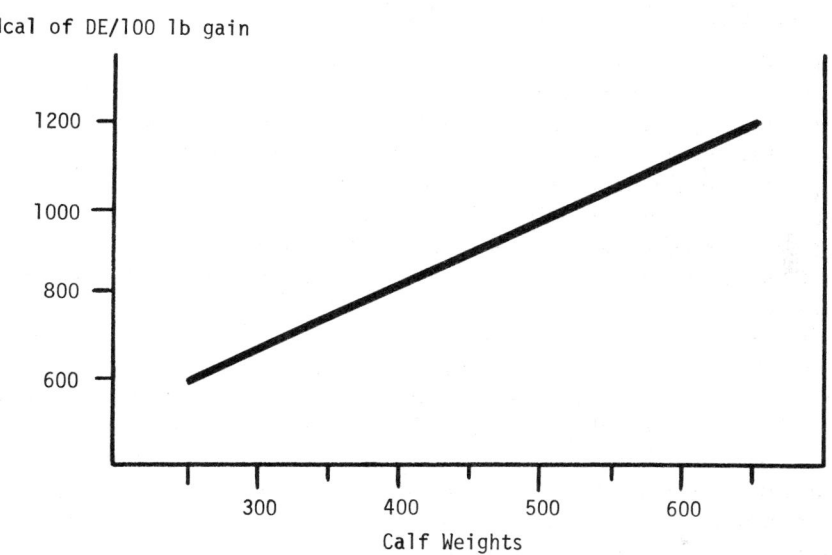

Fig. 25-3. Feed required per 100-lb gain for calves of different weights gaining 1½ lb per day. (Source: *Keys to Profitable Stocker Calf Operations*, MP-964, Texas A&M Univ. Agr. Ext. Service, 1970, p. 5)

● *Lightweight calves require higher quality feed*—Although lightweight calves gain more efficiently than heavier calves, they require better quality feed for the same gain. For this reason, forage (pasture, hay, or silage) which will merely furnish a 300-lb calf sufficient feed to maintain body weight will allow a mature animal to gain weight. Fig. 25-4 illustrates this point. Note that 300-lb calves gaining 1½ lb per day would need forage containing 1.3 megacalories of digestible energy per pound whereas 600-lb calves would gain at the same rate on forage with only 1.1 megacalories of energy per pound. This points up the fact that stocker operators should buy calves of the right age and weight to match their feed.

EFFECT OF CALF WEIGHT ON QUALITY OF FEED

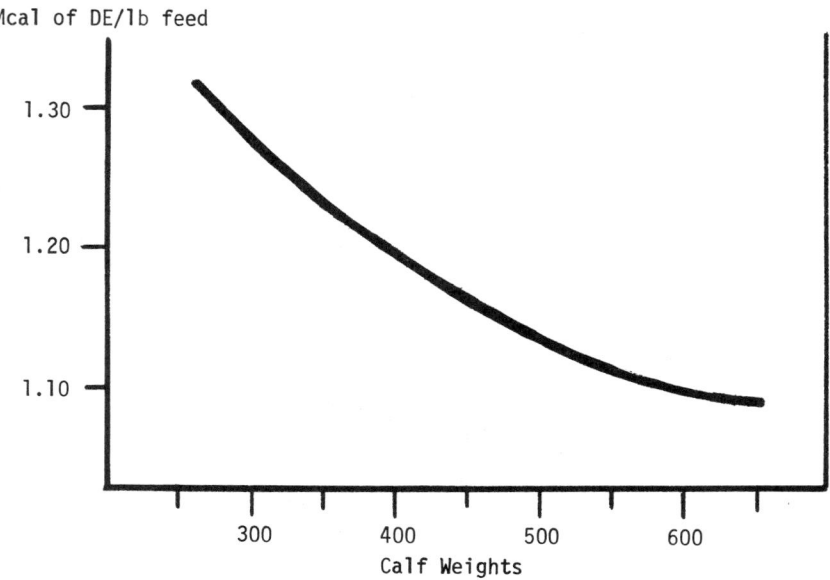

Fig. 25-4. Quality of feed required for calves of different weights to gain 1½ lb per day. (Source: *Keys to Profitable Stocker Calf Operations*, MP-964, Texas A&M Univ. Agr. Ext. Service, 1970, p. 5)

● *Faster gains are cheaper*—Faster gains are cheaper because the maintenance requirement is the same regardless of the daily gain; hence, the maintenance requirement is a smaller part of the total feed consumed as gains increase. Fig. 25-5 illustrates this point. A calf gaining 2 lb per day requires only approximately 800 megacalories of digestible energy per 100-lb gain, whereas the energy requirements per pound of gain for a calf gaining .6 lb per day are doubled—to 1,600 megacalories. With interest rates, taxes, labor costs, etc., rising as they have, growing calves at rates of less than one lb per head daily becomes increasingly difficult to justify.

● *Compensatory growth*—Compensatory growth is making up for a bad start in life. It is common practice for stocker cattle to be "roughed through" the winter as cheaply as possible, with limited daily gains. Then, in the spring, the animals are turned to lush spring pasture or put in a feedlot on a high energy ration. Animals so managed exhibit the phenomena of "compensatory growth"; that is, on the high energy diet they gain faster and more efficiently than similar cattle which were fed more liberally during the wintering period. Feedlot operators were quick to sense this situation, and to take advantage of it. This is the chief reason for the popularity of Okie-type cattle. They are animals whose growth has been held back to less than their genetic potential. When fed more liberally, they exhibit a surge in growth rate and feed efficiency. Large compensatory growth usually indicates that

EFFECT OF RATE OF GAIN ON FEED EFFICIENCY

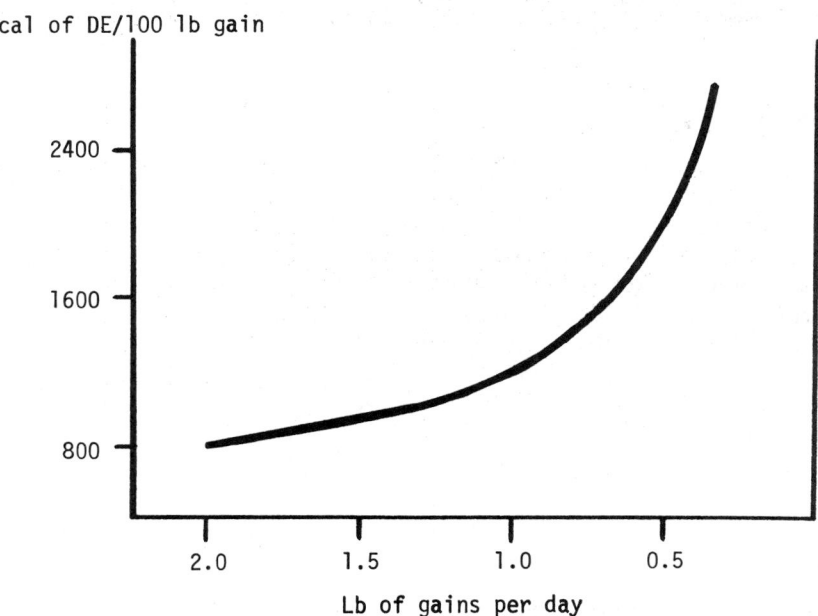

Fig. 25-5. Feed required per 100-lb gain for a 450-lb calf gaining at different rates per day. (Source: *Keys to Profitable Stocker Calf Operations*, MP-964, Texas A&M Univ. Agr. Ext. Service, 1970, p. 5)

someone (the stocker operator) has lost money while someone else (the feeder) has made money. It is noteworthy that Holsteins and the larger exotics should never be handled so as to exhibit compensatory gains. If they're held back in the winter, they're too heavy when they finish.

• *Avoid excess feeder condition*—If cattle get too fleshy as feeders, (1) they may reach market finish before they attain desirable market weight, and (2) they will tend to gain more slowly than desirable during the feedlot finishing period. The planned rate of gain should be determined by feeder finish, growth potential, sex, and the beginning weight of the feeder calves. It will also depend upon how much condition is acceptable to the buyer. British breed or crossbred steer calves gaining at the rate of 1.25 lb daily will about maintain their condition; at 1.5 lb daily they will add some condition; at 2.0 lb daily they may be too fleshy by the time they reach 700- to 750-lb feeder weight. British breed or crossbred yearlings making daily gains of 0.9 lb will maintain growth without fattening. Larger cattle of some of the exotic breeds may make larger gains without fattening.

• *Buy and sell carefully*—Because the original weight purchased in a stocker program is a high percentage of the weight sold by the grower, any mistake made in buying or selling has a greater impact on profits or losses than it would in a finishing program. If you pay too much, buy with too much fill, pay for quality that the feeder is not willing to pay for, or sell too low, profits will be rapidly eroded away.

• *Steers are heavier*—Steer calves normally weigh about five percent more than heifer calves.

MARKET CLASSES AND GRADES OF STOCKERS

Stocker cattle are of many kinds, displaying a wide range of combinations of the various characteristics such as breeding, sex, age, weight, size, conformation, and condition. Fortunately, there is a market for each kind, with variation in prices reflecting both the supply of and the demand for each kind and the degree of suitability for the intended purpose. The use of uniform terms and descriptions of all classes and grades of cattle by all members of the beef team—cow-calf producers, feeders, packers, and buyers and sellers—throughout the country has contributed much to the orderly marketing of all cattle, including stockers. The market classes and grades of stocker cattle are fully covered in Chapter 15 of this book; hence, the reader is referred thereto.

SOURCES OF STOCKERS

Sources of stockers vary from area to area. Nationwide, most stockers are secured through auctions and direct purchase from cow-calf producers. In the Corn Belt, however, most stocker cattle are secured through public stockyards, followed by local dealers, producers, and auction markets.

Stockers may be purchased to advantage in an auction, but there are problems. They have already been stressed by being loaded, hauled, unloaded, and handled; in most cases, the buyer knows little or nothing of the origin of the cattle; and they may either have come straight off native grass (which is good), or they may be making their third sale in the last 10 days (which is very bad).

The best doing stockers are usually those coming directly from cow-calf producers—preconditioned, and without passing through a market facility.

Native (Homegrown or Local) or Western Stockers

Native stockers are those coming from the farms of the Corn Belt, Great Lakes, and Southeast, whereas western stockers are the branded animals coming from the western ranges. For the most part, native cattle come from comparatively small farm herds. Calves produced in these herds are often fed out on the farms where they are produced. Others are offered for sale locally. There are both advantages and disadvantages to buying native stocker cattle.

Native stockers usually have the following *advantages*:

1. *Fewer disease problems*—Native stockers are subjected to less exposure and stress than western stockers, with the result that they have fewer

diseases such as shipping fever. This is especially true if they are bought directly from the farm where produced.

2. *Lower freight and buying costs*—Usually freight cost is much less because of shorter distances; and buying costs may be less, especially where they are bought directly from the neighbors.

3. *Lower shrink*—Shrink, especially tissue shrink—the decrease in carcass weight, which occurs on long extended hauls and during long periods of fast—is usually much lower.

4. *They're acclimated*—Native stockers do not have to adjust to the weather and altitude of the area.

5. *They're accustomed to the feed*—Native cattle are usually accustomed to the feeds of the area.

Native stockers usually have the following disadvantages:

1. *Lack uniformity*—They are usually less uniform than western stockers in breeding, quality, condition, and age.

2. *May be of lower quality*—On the average, there is little to choose between the quality of native and western cattle. However, quality is much more variable between native herds, with the result that they range from very high quality to very low quality.

3. *They're fleshier*—Native stockers are likely to be fleshier than western stockers, with the result that they are not apt to make as good gains when placed on high roughage rations.

4. *Small lots must be combined*—Where a number of stockers is desired, it is usually necessary to buy and put together several small lots of cattle.

5. *Availability*—Native stockers may not be available in large enough numbers when desired.

METHOD OF BUYING STOCKERS

Stocker cattle may be purchased by stocker operators themselves, or by salaried buyers, order buyers, commission firms, or cattle dealers. Although the owner or manager of a stocker operation should be knowledgeable enough to buy stocker cattle, if he is a large operator, he is generally too busy to do so if he is doing a proper job of running his outfit. Accordingly, the trend is to shift the buying responsibility to specialists—primarily order buyers.

TRANSPORTATION AND SHRINK OF STOCKER CATTLE

Improper handling of stocker cattle immediately prior to and during shipment may result in (1) excess shrinkage; (2) high death, bruise, and crippling losses; (3) disappointing sales; and (4) dissatisfied buyers. Unfortunately, many cow-calf men who do a superb job of producing stockers dissipate all the good things that have gone before by doing a poor job of preparing and shipping. The subject of "Preparing and Shipping Cattle" is fully covered in Chapter 15 of this book; hence, the reader is referred thereto.

PRECONDITIONING; HANDLING NEWLY ARRIVED CATTLE

These important subjects are fully covered under similar headings in Chapters 21 and 33, respectively, of this book; hence, the reader is referred thereto.

FEEDING STOCKERS

For a stocker operation to be profitable, the grower must be ever aware of the following reasons back of it and feed stockers accordingly: (1) to provide a supply of the kind of cattle desired by finishing lots at the time needed, (2) to utilize roughages and other low cost feeds, and (3) to "cheapen down" the cattle.

Because of the very nature of the operation, the successful feeding of stockers requires the maximum of economy consistent with normal growth and development. This necessitates cheap feed—either pasture or range grazing or such cheap harvested roughage as hay, straw, fodder, and silage. In general, the winter feeds for stockers consist of the less desirable and less marketable roughages. It is important, therefore, that the high roughage rations of young stockers be properly supplemented from the standpoints of proteins, minerals, and vitamins.

The feed consumption of stockers will vary somewhat with the quality of the roughage available, the age of the cattle, and the rate of gain desired. As far as practical, the stocker ration should prepare the cattle for making maximum use of the feed which follows—either the finishing ration or grass. The rate and efficiency of gain in the feedlot or on pasture varies inversely with the amount of gain made during the stocker stage—that is, the smaller the stocker gains, the higher the finishing gains; and the higher the stocker gains, the lower the finishing gains. This phenomenon is known as compensatory gain.

Of course, too small gains may be unprofitable to the grower. Besides, young animals can be stunted. To make maximum growth without fattening—just to maintain condition—calves of the British breeds and crossbreds should gain 1.25 pounds daily, and yearlings should gain 0.9 pound daily.

The amount of gain desired in a grower program depends largely on the way the cattle are to be handled in the next stage. For example, stockers that are to be grain fed on grass or go directly to the finishing lot can be wintered more liberally, and make a higher rate of gain, than cattle that are to be turned to pasture only. Also, winter and summer gains are not exactly inversely proportional. For example, if one lot of steers gains twice as much during the winter as a second lot, its summer gains won't be limited to half those of the second lot. Rather, they will likely be 70 to 90 percent as much. Thus, where the stocker grower retains ownership of the cattle through the finishing stage, he may well find that the cattle that make the largest stocker gains also make the most economical and largest total gain for the entire period—from weaning to slaughter.

Tables 25-1 and 25-2 contain some recommended rations for stocker cat-

TABLE 25-1

DAILY RATION FOR STOCKER CALVES (400-500 Lb)

Rations (Fed for Gains of 1.25 Lb/Head/Day)

	1		2		3		4		5		6		7		8		9		10	
	(lb)	(kg)	(lb)	(kg)	(lb)	(kg)	(lb)	(kg)	(lb)	(kg)	(lb)	(kg)	(lb)	(kg)	(lb)	(kg)	(lb)	(kg)	(lb)	(kg)
Legume hay or grass-legume mixed hay	12-18	5.4-8.2	12-18	5.4-8.2	8-12	3.6-5.4							8-10	3.6-4.5						
Grass hay					4-6	1.8-2.7			2-4	0.9-1.8	2-4	0.9-1.8					10-12	4.5-5.4		
Straw, corncobs, cornstalks, stalklage, cottonseed hulls													2-4	.9-1.8	2-3	.9-1.4			2	.9
Corn or sorghum silage							25-40	11.4-18.2	20-30	9.1-13.6	20-30	9.1-13.6			20-25	9.1-11.4				
Legume-grass silage, or oat silage																				
Legume-grass haylage, or oat haylage																			20-25	9.1-11.4
Grain (corn, sorghum, barley, or oats)													4-5	1.8-2.3			4-5	1.8-2.3	4-5	1.8-2.3
Protein supplement (41% or equivalent)			1¼-1½	0.6-0.7	¼-1	0.1-0.5	1-1¼	0.5-0.6	¾-1	0.3-0.5	1¼-1½	0.6-0.7			1-1½	0.5-0.7	1-1½	0.5-0.7		

With all rations, provide suitable minerals (see Tables 8-7 and 8-8).

TABLE 25-2

DAILY RATION FOR YEARLING STOCKERS (600-700 Lb)

Rations (Fed for Gains of 0.9 Lb/Head/Day)

	1		2		3		4		5		6		7		8		9		10	
	(lb)	(kg)	(lb)	(kg)	(lb)	(kg)	(lb)	(kg)	(lb)	(kg)	(lb)	(kg)	(lb)	(kg)	(lb)	(kg)	(lb)	(kg)	(lb)	(kg)
Legume hay or grass-legume mixed hay	16-24	7.3-10.9			6-8	2.7-3.6							6-8	2.7-3.6						
Grass hay			16-24	7.3-10.9	10-16	4.5-7.3			2-4	0.9-1.8	2-4	0.9-1.8					16-20	7.3-9.1		
Straw, corncobs, cornstalks, stalkage, cottonseed hulls													12-15	5.4-6.8	10-12	4.5-5.4			2	0.9
Corn or sorghum silage							45-55	20.4-25.0	40-50	18.2-22.7	40-50	18.2-22.7								
Legume-grass silage, or oat silage															20	9.1			35-40	15.9-18.2
Legume-grass haylage, or oat haylage																				
Grain (corn, sorghum, barley, or oats)													5-6	2.3-2.7			5-6	2.3-2.7	5-6	2.3-2.7
Protein supplement (41% or equivalent)			1½-1¾	0.7-0.8	1-1½	0.5-0.7	1¼-1½	0.6-0.7	¾-1	0.3-0.5	1¼-1½	0.6-0.7			1	0.5	1-1½	0.5-0.7		

With all rations, provide suitable minerals (See Tables 8-7 and 8-8).

tle. Variations can and should be made in the rations used. The grower should give consideration to (1) the supply of homegrown feeds, (2) the availability and price of purchased feeds, (3) the class and age of cattle, (4) the health and condition of animals, and (5) the kind of feeder cattle in demand by feedlots.

In using Tables 25-1 and 25-2 as guides, it is to be recognized that feeds of similar nutritive properties can and should be interchanged as price relationships warrant. Thus, (1) the cereal grains may consist of corn, barley, wheat, oats, and/or sorghum; (2) the protein supplement may consist of soybean, cottonseed, peanut, linseed, safflower, and/or sunflower meal; (3) the roughage may include many varieties of hays and silages; and (4) a vast array of by-product feeds may be utilized.

The following points are pertinent to the success of a stocker operation and should be kept in mind:

• *Recommended nutrient allowances*—Where grower rations are formulated on the basis of percentage of nutrients in the ration, the following allowances are recommended:

Protein
For up to 1.5 lb daily gain 10.0%
For 1.5 lb daily gain or more 10.5%

Calcium and Phosphorus
For up to 500 lb liveweight 3.0%
For over 500 lb liveweight .... .25%

Vitamin A
Air-Dry Feed (10% moisture) .... 800 to 1,000 IU per lb
.... 10,000 IU daily per head

Implant
Gains of more than 1.5 lb
per head daily .... Include growth stimulant implant

• *Protein*—Calves have a higher protein requirement per 100 lb of liveweight than older cattle and are more apt to be deficient on low quality roughage. Extra energy will not be efficiently used unless protein intake is adequate.

Calves wintered on range will require .5 to .7 lb of crude protein from supplements to gain .5 lb daily. Calves wintered on good grass hay or meadow hay will require less supplemental protein and phosphorus than when wintered on range.

One lb of 41 percent supplement will supply about half the total protein requirement of a 500-lb calf. Three lb of alfalfa hay or 2½ lb of dehydrated alfalfa will furnish about the same amount of protein but more energy than one lb of a 41 percent protein concentrate.

Urea is not well utilized as a protein supplement in high roughage rations. Because of this, usually it is best to use a plant protein supplement or a slow release urea product in growing rations.

Work at Missouri and other stations indicates no advantage to feeding protein supplements on pasture which contains legumes or grasses in lush condition.

● *Energy*—Satisfactory and efficient winter gains of weanling calves depend upon sufficient energy, along with a proper balance of protein and energy in the ration.

There is much information that indicates stocker cattle should not be winter-fed over one lb of grain per 100 lb of body weight if they are to make best use of pasture the following summer. Also, delayed grain feeding on pasture until after peak pasture growth is recommended.

● *Vitamins and minerals*—All rations should provide adequate carotene or vitamin A, calcium, and phosphorus.

Most stockers are bought in the fall to be wintered on low quality roughage or on winter wheat or other cool-season pasture. Some are bought in the spring to be placed on bluestem, Bermuda grass, or other native pastures. Such calves usually do not have a high store of vitamin A; hence, an intraruminal or intramuscular injection of 500,000 to 1,000,000 IU/head of vitamin A upon their arrival at the place where they will go on the stocker program could be helpful, especially if a nitrate problem is anticipated on wheat pastures.

● *Roughage*—Feeding good quality roughage in a drylot may be more desirable for weanling calves than wintering on the range. Winter range is most efficiently used by yearling steers and mature cattle. Meadow hay, a mixture of alfalfa and grass hay, good quality upland hay, or silage provide a good basal diet for wintering calves.

Corn or sorghum silage does not need to be supplemented with additional grain or dry roughage unless it was put up too wet or had little grain as the result of drought. Sorghum silage or grass-legume silage that has more than 65 to 70 percent moisture will require more grain supplementation than drier silage of comparable quality.

Grass-legume silage cut at the proper stage of maturity and carefully ensiled is an excellent feed for stocker cattle. Such silage contains adequate protein and need not be supplemented with a protein concentrate. However, it is much lower in energy than corn or sorghum silage; hence, the gains will be smaller unless (1) 150 to 200 pounds of grain per ton are added as a preservative, or (2) it is supplemented at feeding time with some grain or other energy source.

Grass-legume hay of good quality is excellent for stocker cattle. It may or may not be supplemented with grain.

Haylage, made from grass-legume or straight legume forage, wilted to about 50 percent moisture content, and commonly stored in an oxygen-free silo, is increasing as a stocker feed. Except for the difference in moisture content, haylage has a feeding value comparable to silage or hay made from a similar crop.

Grass hay, such as prairie hay, Bermuda, timothy, Sudan, Johnson grass, etc., may make up most of the ration of stocker cattle. Energy supplementation is needed if improvement in the condition of stockers is desired. Also, such hays must be properly supplemented with protein, minerals, and vitamins.

Stubble and stalk fields furnish much feed for stocker cattle, especially yearlings, in the late fall and early winter. Unless there is access to a good winter pasture, cattle on stalk fields should be fed 4 to 6 pounds of legume hay or 1½ pounds of protein concentrate daily. In addition, minerals should be provided.

Occasionally, roughages are high in nitrates. Actually, calves can consume very high levels of nitrate provided (1) they are fed with regularity, (2) changes in feed are gradual, (3) feeds are mixed uniformly, and (4) ample water is provided at all times. Where nitrate is a problem, cattle may die of oxygen insufficiency. Others in the lot will show discoloration of nonpigmented areas of the epithelium and chocolate-colored blood.

● *Keep roughage waste low*—Roughage need not be processed. But it should be fed so that there will be a minimum of waste.

● *Grain*—Calves are unable to consume enough dry roughage to gain more than a pound a day. Thus, grain should be added in the quantity necessary to achieve the desired gains. Bear in mind that with calves of the British breeds and crossbreds it takes a gain of about 1.25 pounds daily to maintain condition; yearlings of the British breeds or crossbreds will maintain condition on a gain of about 0.9 pounds daily.

Some grain should be included in the ration of stocker cattle when (1) they are to be finished immediately after the wintering period, (2) they weigh less than 350 pounds when started on winter feeding, and (3) heifers are to be bred when they are 13 to 15 months old.

Calves that are full-fed corn or sorghum silage high in grain content, plus one pound of protein concentrate or 4 to 5 pounds of legume hay, need not be fed grain.

● *Level of wintering*—The level of wintering stockers affects the gains in the next stage. Thus, calves gaining the most during the winter make the least gains on pasture the following summer. This is clearly shown in Table 25-3.

Calves wintered to gain 1.0 pound daily make satisfactory summer pasture gains. This level is recommended for calves to be grazed season-long the following summer, provided the same ownership is retained all the way through. One to 2 pounds daily gain during the winter is usually desirable if calves (1) are to be sold in the spring, (2) will be on full feed 2 to 3 months after going to grass, (3) will be receiving a limited feed of grain on grass, or (4) are replacement heifers that are to be bred at 13 to 15 months of age.

Since yearlings are not growing as rapidly as calves, they may be fed for smaller gains than calves, and yet show comparable condition. Thus, for maximum growth without fattening (for just holding their condition) calves should gain approximately 1.25 pounds daily, whereas yearlings need to gain only 0.9 pound daily.

TABLE 25-3

CALVES TO YEARLINGS—

EFFECT OF WINTER GAINS ON SUBSEQUENT GAINS
THE FOLLOWING SUMMER AS YEARLINGS[1]

Lot No.	Winter	Summer	Total
	(lb)	(lb)	(lb)
	Valentine, Nebraska		
1	115	205	320
2	120	181	301
3	127	183	310
4	129	183	312
5	150	170	321
6	157	143	300
7	164	170	304
8	179	160	339
9	184	150	324
10	186	152	338
11	186	162	348
	Fort Robinson, Nebraska		
1	67	226	309
2	83	218	308
3	87	231	298
4	90	221	308
5	104	202	306
6	136	183	319

[1]*Beef-Forage Notebook*, Cattle Management, 5-C-1, Univ. of Neb. Ext. Services, December, 1967.

Winter Pastures

Wherever possible, stocker calf operations are planned around a winter pasture program. Weanling calves or lightweight, thin yearlings are purchased in the fall. In some cases, homegrown calves are retained and developed under this system for sale as yearlings. As would be expected, winter pasturing of stockers is largely limited to the southern part of the United States, with the kind of pasture varying from area to area.

● *Winter wheat pastures*—Winter wheat pastures are widely used for stocker cattle in Kansas, Oklahoma, and Texas. When such pastures are good, cattle make very acceptable gains on them. However, wet weather or droughts make winter wheat pastures unreliable, with the result that it is important that there be flexibility in the stocker program, both in numbers and season of use.

● *Other cool-season pastures*—In the southern states, extensive use is made of oats, rye, ryegrass, vetch, and fescue—a perennial grass that remains green throughout the winter. This area is turning more and more to winter grazing, as a means of making profitable year-round use of their land and labor and providing 600- to 800-pound feeder cattle in greatest demand by feedlots.

GRASS TETANY

Grass tetany (grass staggers or wheat pasture poisoning), a highly fatal nutritional ailment, is one of the hazards of winter pastures. Although it is more common among lactating cows, stocker cattle are affected. It generally occurs during the first 2 weeks of the pasture season, particularly in cattle grazing wheat or other cereal crops. The characteristic symptoms and signs are: nervousness, flickering of the third eyelid, twitching of the muscles (usually of the head and neck), head held high, accelerated respiration, high temperature, gnashing of the teeth, and abundant salivation. A slight stimulus may precipitate a crash to the ground and, finally, death. The condition follows a rapid course, with usually a lapse of only 2 to 6 hours between onset and death.

The exact cause of grass tetany is not completely understood. It is known that hypomagnesemia (low level of magnesium in the blood serum) is associated with grass tetany. But the causes of low levels of blood magnesium cannot be explained in all cases by rations being deficient in magnesium. Nevertheless, it has been established that supplemental magnesium will increase blood levels and alleviate at least in part the grass tetany problem. Approximately 6 grams of magnesium per day (magnesium may be in the form of magnesium oxide, magnesium sulfate, or magnesium carbonate; since magnesium oxide is approximately 60 percent magnesium, it will require 10 grams of it daily to provide 6 grams of magnesium) added to the salt, mineral, or concentrate supplement will suffice. Commercial magnesium supplements are available in wheat pasture areas. These should be fed according to manufacturers' directions. Also, magnesium alloy "bullets" have been developed for cattle to give a slow release of magnesium in the rumen.

In addition to magnesium supplementation, access to some hay or dried mature grass during the first two weeks after stockers are first placed on winter pasture (wheat pasture or whatever kind) is also helpful in preventing grass tetany.

If grass tetany is suspected, a veterinarian should be called immediately for proper diagnosis and prescribed treatment. Normally, the prescribed treatment consists of an intravenous injection of at least 500 cc of a dextrose solution containing both magnesium and calcium.

STOCKERS AND GROWER CONTRACTS

Hand in hand with the development of big feedlots and year-round feeding came the need for an assured supply of feeder cattle of the desired kind on a continuous basis. To meet this need, more and more feedlots have turned to contractual arrangements with stocker growers, with numerous kinds of contracts. Usually, the cattle are owned by the feedlot, most of which are large and in a stronger financial position than the majority of stocker growers. The two most common kinds of contracts are based on either (1) a fixed cost for the gain, or (2) an agreed feed cost plus an extra charge for labor and lot rental. Usually, there is provision for adjusting for death loss. Such contracts should always be in writing, with all provisions, including weighing conditions, spelled out.

Although the use of stocker and grower contracts has increased in recent years, the concept is not new. Many of the Kansas bluestem pasture owners have long grown out yearlings owned by Iowa and other Corn Belt feeders.

Today, many corn farmers in the fertile irrigated area around Greeley, Colorado, make corn silage and feed cattle on a contract basis to stockers owned by one of several large feedlots in the vicinity. Stocker cattle are also being grown under contract on the wheat pastures of Kansas, Oklahoma, and Texas; on hay and other roughages in the irrigated valleys of the West; and on sorghum silage and stalk fields throughout the Southwest.

QUESTIONS FOR STUDY AND DISCUSSION

1. How and why have stocker programs changed through the years?
2. Define the following terms: (a) stockers; (b) feeders; (c) replacement heifers; (d) preconditioning; and (e) backgrounding.
3. What are the common types of stocker programs, and what are the characteristics of each?
4. What are the advantages and disadvantages of a stocker program?
5. Do you feel that, for the most part, the stocker stage will be lengthened or shortened? Justify your answer.
6. Why do lightweight calves (a) gain more efficiently, (b) require higher quality feed, and (c) make faster and cheaper gains than heavyweight calves?
7. For the man who produces his own stocker calves, and who finishes them out on a custom basis, is compensatory growth good or bad? Justify your answer.
8. What are the market classes and grades of stocker cattle? How do these relate to finished cattle grades?
9. How would you go about buying stocker cattle?
10. What are the advantages and disadvantages of (a) native stockers, and (b) western stockers?
11. Discuss rations and rate of gain of (a) stocker calves, and (b) stocker yearlings.
12. What are the primary differences between stocker and finishing rations?
13. Tell how the level of wintering stockers affects the next stage.
14. Discuss the favorable and unfavorable factors of winter wheat pastures.
15. Discuss the favorable and unfavorable factors of the common cool-season pastures of the Southeast.
16. How would you lessen the hazard of grass tetany?
17. What provisions should be incorporated in a stocker and grower contract?

SELECTED REFERENCES

Title of Publication	Author(s)	Publisher
Beef Cattle, Sixth Edition	A. L. Neumann R. R. Snapp	John Wiley & Sons, Inc., New York, N.Y., 1969
Beef Cattle Science Handbook	Ed. by M. E. Ensminger	Agriservices Foundation, Clovis, Calif., pub. annually since 1964

PART III

CATTLE FEEDLOTS; PASTURE FINISHING

The finishing of cattle is what the name implies, the laying on of fat. The ultimate aim of the finishing process is to produce beef that will best answer the requirements and desires of the consumer. This is accomplished through an improvement in the flavor, tenderness, and quality of the lean beef which results from marbling.

In a general way, there are two methods of finishing cattle for market: (1) cattle feedlots, including confinement (sheltered) finishing; and (2) pasture finishing. Prior to 1900, the majority of fat cattle sent to market were four- to six-year-old steers that had been finished primarily on grass. Even today, the utilization of pastures continues to play an important role in all types of cattle finishing operations.

Cattle finishing of the 1950s and 1960s was characterized by a dramatic increase in grain feeding. In 1947, only about 3.6 million head of cattle were grain-fed, representing 35 percent of the slaughter cattle that year. In 1972, 26.8 million head of fed cattle were marketed, representing nearly 75 percent of all the cattle slaughtered that year.

In the "good old days" of the horse-drawn wagon, scoop shovel, wicker basket, and processing ear corn by breaking the nubbins (ears) on the edge of the feed bunk, one man fed 100 head of fattening cattle. Today, in a modern cattle feedlot, one man feeds 1,500 cattle, and in the more mechanized lots, one man feeds as many as 2,500.

Today, cattle feedlots with more than 1,000 capacity dominate cattle feeding. In 1972, there were 2,107 feedlots with capacity of 1,000 head and over; and these relatively few lots marketed 62% of the nation's 26,835,000 fed cattle. That same year, there were 152,429 feedlots with capacity less than 1,000 head; and they marketed 38% of the fed cattle. Thus, 1.4% of the nation's feedlots marketed nearly ⅔ of the fed cattle.

In 1973, beef prices spiraled to unprecedented heights, in response to increased costs of production brought on by scarce and high priced grain. The reaction was loud and clear; neither beef consumption nor production is elastic at any price. They are subject to the old law of supply and demand. Cattle feeders suffered staggering losses, and cattle investors fled. Out of this chaos, feedlot cattle increasingly became roughage burners. Cattle feeders came more and more to rely upon the ability of the ruminant to convert coarse forage, grass, and by-product feeds, along with a minimum of grain, into palatable and nutritious food for human consumption; thereby not competing so much for humanly edible grains.

In 1973, 77 percent of the beef produced in the United States was grain-fed, 23 percent was grass-fed. There'll be less grain-fed and more grass-fed beef in the future.

Chapters 26 to 35 will (1) give cattle feeders, and those who counsel with them, a big assist in the years ahead; and (2) provide students with scientific and technological information relative to modern cattle feeding.

CHAPTER 26

MODERN CATTLE FEEDING

The finishing of cattle is what the term implies, the laying on of fat. The ultimate aim of the finishing process is to produce meat that will best answer the requirements and desires of the consumer. This is accomplished through an improvement in the flavor, tenderness, and quality of the lean meat which results from marbling. Fat also adds to the digestibility and nutritive value of the product.

Fig. 26-1. The modern, shaded cattle feedlot of McElhaney Feeding Co., Yuma County, Arizona. (Courtesy, Sam McElhaney, Wellton, Ariz.)

In a general way, there are two methods of finishing cattle for market: (1) open lot and confinement (sheltered) finishing, and (2) pasture finishing. Prior to 1900, the majority of fat cattle sent to the market were four- to six-year-old steers that had been finished primarily on grass. Even today, the utilization of pastures continues to play an important part in all types of cattle feeding operations.

Cattle feeders are commonly classed as either (1) commercial feeders or (2) farmer-feeders, based largely on numbers. From the standpoint of statistical reporting, the U.S. Department of Agriculture commonly draws the line at 1,000 head. A commercial cattle feeding operation is defined as one having a capacity of 1,000 head or more, at any one time.

The traditional farmer-feeder evolved with Corn Belt farming, in the north central region of the United States. Generally speaking, he markets his crop, usually corn, through cattle (or hogs, or lambs), and spreads the manure on the land. The purchase of feeder cattle for this type of enterprise is generally in the fall, with the actual feeding done during the winter months when labor is available due to limited field work. This traditional farmer-feeder type of operation has persisted to the present time, although it has been modernized through the years.

In addition to being larger, commercial cattle feeders generally differ from farmer-feeders in the following respects: (1) they usually feed cattle on a year-round basis, rather than during the winter months only; (2) they may grow little, or none, of their feed; (3) they are highly mechanized; and (4) they are knowledgeable of costs and returns, skillful buyers and sellers, and aware of market trends. Today, commercial feedlots with more than 1,000 head capacity dominate cattle feeding.

The trend is for larger but fewer feedlots. In 1972, there were 2,107 feedlots with capacity of 1,000 head and over; and these relatively few lots marketed 62% of the nation's 26,835,000 fed cattle. That same year, there were 152,429 feedlots with capacity less than 1,000 head; and they marketed 38% of the fed cattle. Thus, 1.4% of the nation's feedlots marketed nearly ⅔ of the fed cattle. Noteworthy, too, is the fact that feedlots with capacity of over 32,000 head showed the largest percentage gain in 1972, both in

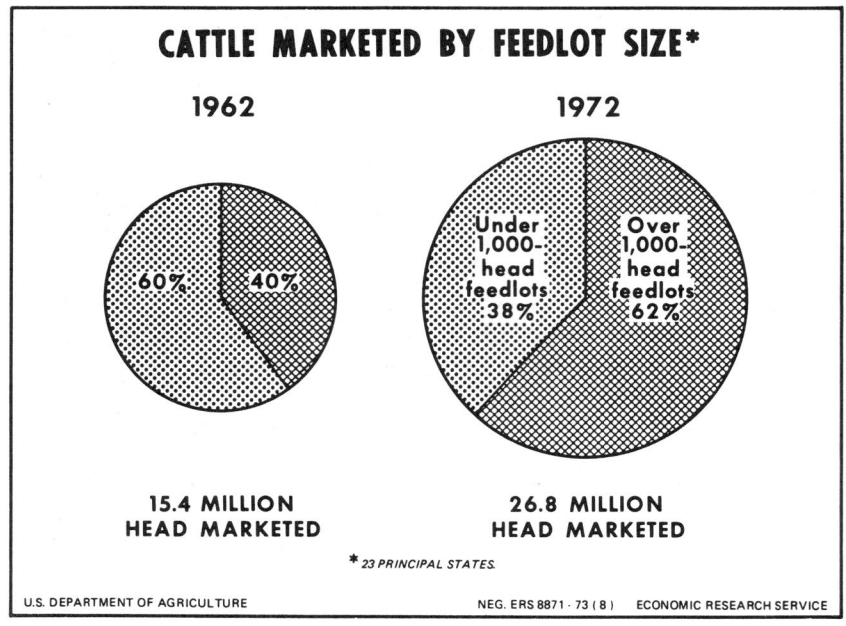

Fig. 26-2. Feedlots have gotten bigger! In 1962, feedlots with more than 1,000 head marketed 40% of the fed cattle. By 1972, 10 years later, feedlots with 1,000 head and over marketed 62% of the nation's fed cattle. (Courtesy, USDA)

number of lots and fed cattle marketings. There were a total of 59 of these big lots with 32,000 head or more, up 30% from the previous year.

Feedlots exceeding 100,000 capacity are now in operation in Colorado, Arizona, and the northern Texas Panhandle. Even in the Corn Belt, feedlots are getting bigger.

Cattle feeding in the national perspective is herewith detailed in terms of growth and shift in geography.

It is important that cattlemen, both cow-calf operators and cattle feeders, be knowledgeable relative to the transition and impact of cattle feeding. More important still, this story needs to be told to consumers again and again in order to supplant their altogether too common image that beef is sold to the packer directly from the range. Actually, when estimating profits, they are prone to compute from range (off grass) to range (stove), with no intermediate steps between.

GROWTH AND COMPOSITION OF CATTLE FEEDING

The growth of the cattle feeding industry in the United States has been spectacular since World War II. In 1947, only 6.9 million head of market cattle were grain-fed, representing 30.1 percent of the slaughter cattle that year. In 1972, 26.8 million head of fed cattle were marketed, representing

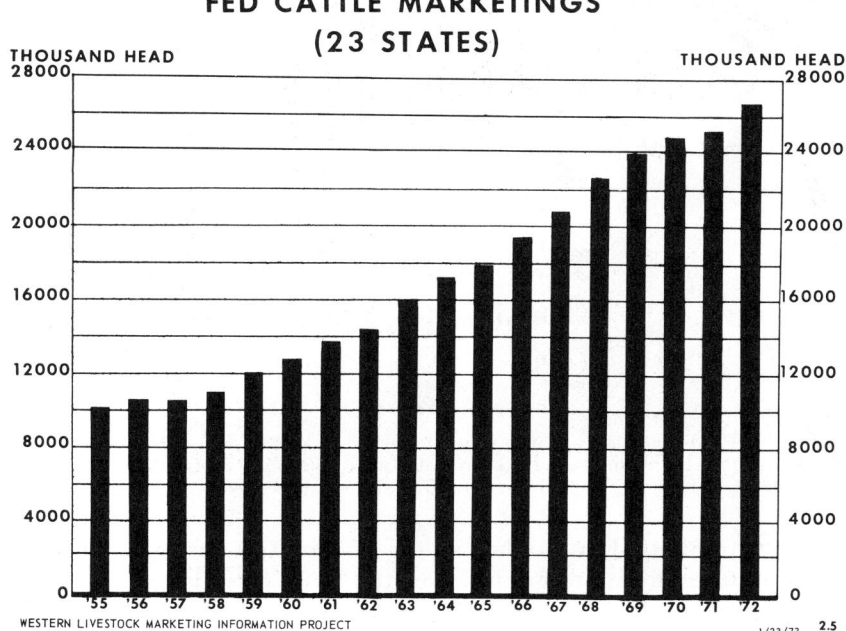

Fig. 26-3. Growth of cattle feeding. In 1955, 10 million head of fed cattle were marketed. In 1972, 26.8 million head of fed cattle were marketed. (Source: Using Information in Cattle Marketing Decisions, A Handbook, *WEMC Pub. No. 5*, Feb., 1973, p. 36, Chart 13)

nearly 75 percent of all cattle slaughtered that year. Thus, the upswing in feedlot feeding has averaged 800,000 head per year since 1947.

Fig. 26-4 shows the composition of cattle slaughter, 1965 to 1972.

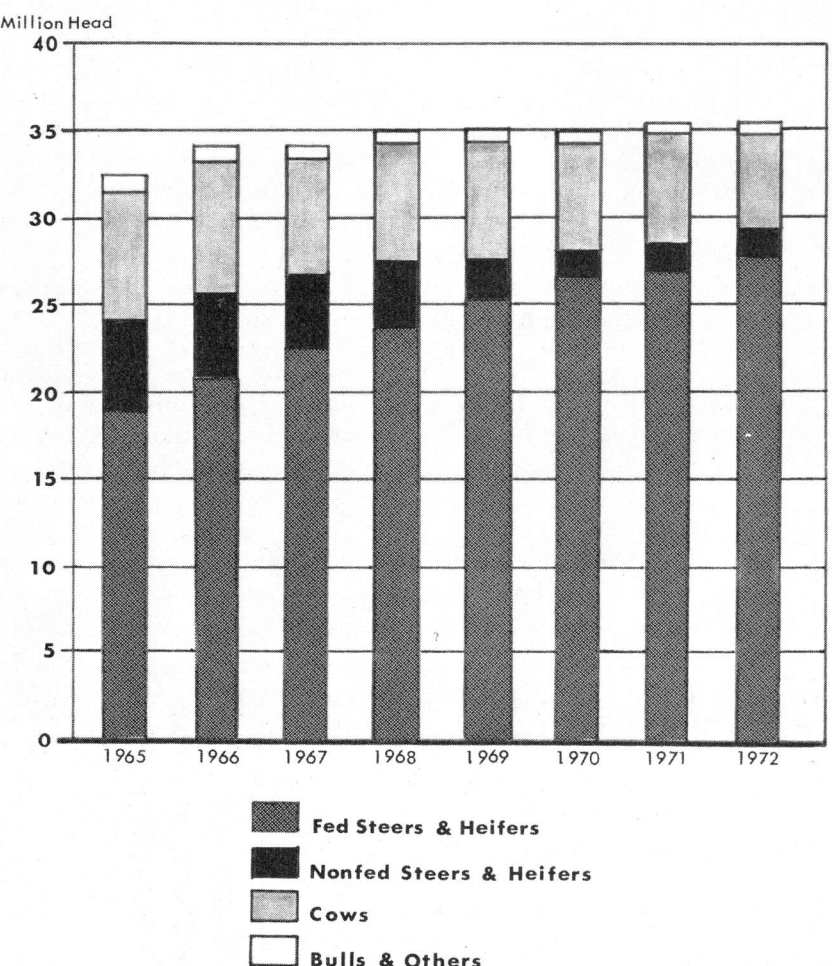

Fig. 26-4. Composition of cattle slaughter, 1965 to 1972. Few nonfed steers and heifers are now being marketed. (Source: Using Information in Cattle Marketing Decisions, A Handbook, *WEMC Pub. No. 5*, Feb., 1973, p. 65)

As noted, precious few nonfed steers and heifers are now being marketed; hence, to secure more grain-fed cattle, more calves must be born.

The primary reasons for the phenomenal growth of the cattle feeding industry since World War II are:

• *Increased population*—The U.S. population—the number of meat eaters (except for a few vegetarians)—increased from 144 million in 1947 to 208 million in 1972.

● *Increased beef consumption*—Per capita beef consumption increased from 69.6 pounds in 1947 to 115.9 pounds in 1972; hence, it nearly doubled in this period.

● *Beef the preferred meat*—Beef is the preferred red meat in the United States, having replaced pork in this position in 1953; and the gap between beef and pork has widened every year since. In 1953, the per capita figures were: beef 77.6 pounds; pork, 63.5 pounds. In 1972 the comparable figures were: beef, 115.9 pounds; pork, 67.4 pounds.

● *Increased disposable income*—U.S. disposable income per person increased from $1,264 in 1949 to $3,807 in 1972, more than a three-fold increase. During this period, the percent of disposable income spent for beef remained nearly the same; in 1949 it was 2.7 percent, in 1972 it was 2.6 percent. Of course, in 1972 the percentage figure was applied to more than 3 times as many dollars as in 1949.

● *Cattle feeding was profitable*—Most people are in business to make money, which is as it should be; and cattle feeders are no exception. Through the 1950s and 1960s, cattle feeding was, generally speaking, highly profitable. This caused rather large quantities of money to be poured into the expansion of the industry.

● *Consumer demand for beef grading U.S. Choice, with repeatability (the housewife wanting the same quality of meat as she purchased last week)*—For the most part, meeting this need calls for grain-fed cattle. This, then, was the motivating force back of the growth of the cattle feeding industry. To meet consumer needs in their mass merchandising, self-service retail stores, both chain stores (which merchandise about one-half of all the beef sold today) and independent affiliated retailers rely on rigid buying specifications when dealing with packers. In turn, the packers reflect this demand back to the producer when buying animals on foot. Thus, in a very real sense, consumer demand set off the "chain reaction" that spawned the cattle feeding industry.

● *Modern, commercial cattle feeders were attuned to consumer demands*—Farmer-feeders, who generally feed cattle on a part-time basis and grow crops, are not usually as acutely attuned to consumer demands as the commercial feeder who specializes in the business. Following World War II, the latter sensed that consumers were ready and willing to pay for more grain-fed beef, and they set about to fill this need.

● *Finishing of cattle on a year-round basis*—With the advent of large, commercial lots, cattle were fed the year-round, hence, grain-fed beef became available to consumers throughout the year. This made for growth in cattle feeding.

● *Dual grading improved the product*—Dual grading of carcasses, which emphasized lean yield, resulted in beef better meeting consumer preference.

● *Increased mechanization and labor efficiency*—The mechanization and labor efficiency of commercial feedlots made them competitive with farmer-feeders, and stimulated their growth. When the author was a boy on a Missouri farm in the Corn Belt, one man—conveying feed in a mule-drawn wagon and a bushel wicker basket, and processing ear corn by breaking the nubbins (ears) on the edge of the feed bunk—fed 100 head of fattening cattle. Today, in a modern feedlot, one man feeds 1,500 cattle, and in the more mechanized lots, one man feeds as many as 2,500.

• *Additives and hormones improved performance*—Antibiotics, hormones, and other additives improved the rate of gain and feed efficiency of feedlot cattle, thereby making cattle feeding more profitable and resulting in lower beef prices to the consumer than would otherwise have been possible.

• *New methods of grain processing improved efficiency*—The several new methods of grain processing developed since World War II greatly improved feed efficiency, particularly with grain sorghum.

• *Marketing of finished cattle became more efficient*—With increased direct selling of cattle to packers, the buyers went to the cattle, rather than the cattle to the buyers (as in the days of terminal markets), thereby lowering marketing costs and shrinkage.

• *Beef packing became more efficient*—The moving of the beef packing plants from the traditional river markets of the Midwest near the cattle feeding centers lessened transportation costs, shrinkage, and bruises. Also, the new packing plants that evolved were very modern and efficient.

• *The advent of futures trading*—Futures trading gave producers a basis for market decisions and provided a means to lessen risks.

SHIFT IN GEOGRAPHY OF CATTLE FEEDING

Regional or geographical shift in cattle feeding are a matter of history. Fig. 26-5 records the developments in the 3 major cattle feeding areas of the nation during the 13-year period 1960-1972. The most marked change was in the Southern Plains states, where cattle feeding increased by 504 percent from 1960 to 1972. The percentage increase in the other 2 areas have been much smaller; in the case of the Corn Belt, this was partially because the 1960 level of marketings was relatively high. It is predicted that future growth in cattle feeding in the Southern Plains will be slowed because of limitations in irrigation water and feed grains.

The primary reasons for the shift in geography of cattle feeding were as follows:

• *Commercial cattle feeders were free to choose their locations*—Most farmer-feeders are "tied" to a particular farm; they cannot locate their feedyards elsewhere. However, most commercial cattle feeders are free to choose an area. As a result, they located where they felt that they had the best combination of factors favorable to their success, with consideration given to availability and price of feed, supply of feeder cattle, weather, slaughter plants, etc. This made for a shift in the geography of cattle feeding.

• *Large grain supplies, especially sorghum, in the Plains area*—Hand in hand with the increased production of sorghum, more cattle feeding emerged. The economics of this close correlation between the area of sorghum production and cattle feeding becomes apparent when it is realized that it requires approximately 8 pounds of feed to produce 1 pound of on-foot

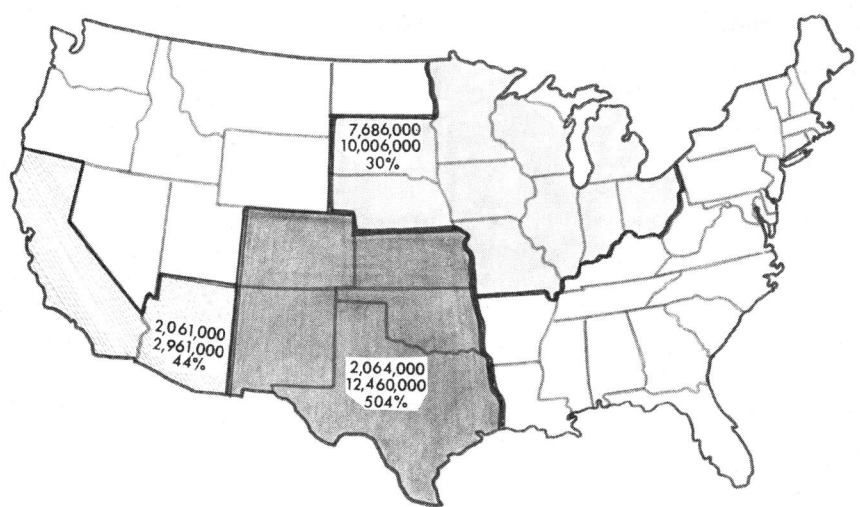

FED CATTLE MARKETINGS IN THE MAJOR FEEDING AREAS,
1960 AND 1972 WITH PERCENT INCREASES 1960 TO 1972

Fig. 26-5. Geographical shifts in cattle feeding, based on fed cattle marketings. Legend: top figure—1960 fed cattle marketings; middle figure—1972 fed cattle marketings; bottom figure—percent increase, 1960-72. (Source: 1960 figures from *Livestock and Meat Statistics 1962*, Statistical Bulletin No. 333, USDA, July 1963, and Supplement for 1969 to Statistical Bulletin No. 333. 1972 figures from Using Information in Cattle Marketing Decisions, A Handbook, *WEMC Pub. No. 5*, February 1973, Chart 14, p. 36)

gain. Moreover, each $1 increase in freight cost per ton of feed increases feed cost per pound of gain by approximately 0.4¢. Thus, considerable saving in cost can be effected by moving the feeder cattle near the feed source, rather than moving the feed to the cattle.

● *Nearness of large feeder calf supplies in Texas and the Southeast—* For many years, the cow-calf population has been concentrated west of the Mississippi River. It was logical, therefore, that proximity to feeder cattle supplies (as well as feed) should be considered in locating cattle feedlots. This was a factor in the development of the Plains feeding area.

● *Decentralization of livestock marketing and meat packing—* Following World War II, terminal markets declined and meat packers decentralized. Both auction markets (especially feeder cattle auctions) and beef slaughterers located near concentrated cattle feeding areas.

IMPACT OF MODERN CATTLE FEEDING

Cattle feeding is a "beef factory." More than 10 million tons of cattle, feed, beef, and other supplies must be transported annually to service this factory. In particular, feed grain growers and cow-calf operators benefit from cattle feedlots; and the nearer they are to one or more big feedlots, the greater the benefits.

The impact of one large commercial feedlot is enormous. For example, a

40,000-head capacity feedlot with a turnover of 2.5 times its capacity yearly, will turn out 100,000 finished animals, or 60 million pounds of carcass beef each year. Large quantities of raw materials are required to keep this "beef factory" going. It will need the feeder calf production from 142,857 cows if 70 percent of the cows produce calves for the feedlot, or 1,428 herds each averaging 100 head of brood cows. To maintain these cows on the western range will require 1.5 million acres of average grazing land. To feed the cattle in this 40,000-head capacity lot (100,000 animals per year) will require 100,000 tons of corn and/or sorghum, or the production of nearly 100,000 acres averaging 60 bushels per acre. And that's not all!

The Texas cattle feeding industry alone purchases annually $7 million in pharmaceuticals, uses $1.7 million in electrical power, and pays more than $1 million for vehicles and related equipment.[1] Additionally, the clothing merchant, the doctor, the lawyer, the banker, and the barber all benefit from the increased economic activity of the region. The local, state, and Federal governments obtain added taxes. Indeed, the economic impact of the cattle feeding industry is enormous.

CHARACTERISTICS OF MODERN CATTLE FEEDING

The characteristics of the modern cattle feeding industry are:

● *It is unique to the United States*—Finishing (fattening) cattle by feeding grain, and other concentrates, is not practiced to any great extent elsewhere in the world, although interest is evolving in certain countries.

● *It is highly concentrated in a relatively few states*—Six states—Texas, Nebraska, Iowa, Kansas, Colorado, and California—account for nearly 70 percent of all the cattle fed in the United States.

● *It gave rise to a new "beef belt"*—Beginning in the early 1950s, a massive "beef belt" formed, extending from the southern Great Plains through Nebraska. It was ushered in with increased irrigated feed crops, particularly sorghum, of the area, along with new feedlot technology and mechanization. Today, this area has the largest concentration of feedlot cattle in the world.

● *It is a "high risk" business*—In some years, the feeder may not even recover his feed costs. Thus, cattle feeders should be in a relatively strong financial position.

● *Facilities and man-hours required for feeding vary*—Facilities vary from a dirt lot on a general farm, where a farmer may spend as much as 25 man-hours of labor to finish one steer, to elaborate, highly specialized facilities, costing almost as much per square foot as a modern home, where labor needs may drop to about one man-hour per head for the entire feeding period.

● *Feed costs alone account for about 80 percent of the variable costs*—Variable costs include feed, interest on feed and feeder cattle, labor, death

---

[1]*Meat Prices and the Public Interest*, A Study by the Subcommittee on Livestock and Grains of the Committee on Agriculture, House of Representatives, together with additional views, July 27, 1972, p. 18.

loss, and veterinary and medicine costs. Feed costs, which are the largest item, account for about 80 percent of these variable costs.

• *Several methods of financing are being used*—Because of the large amounts of capital required, big feedlots are being financed by outside capital, much like any other big business. Each of the major methods of financing is detailed in Chapter 27 of this book.

• *Type of legal ownership varies with size*—Although there is no set pattern, small commercial lots are generally single-proprietor owned; medium-sized lots are partnership owned; and large lots, with capacities of 10,000 head or more, are incorporated.

• *Large amounts of investor money are involved*—Doctors, lawyers, merchants, industrialists, are investing in cattle feeding; through consignment feeding (ownership) of cattle that are custom fed, limited partnerships, shareholders in a corporation, and other arrangements. Prior to the severe cattle feedlot losses of 1974, it was estimated that these "Wall Street cowboys" were financing up to one-half the cattle on feed.

• *There is considerable vertical integration of lots, but little horizontal integration*—Vertical integration refers to control or ownership by other levels of the functional system; i.e., a cow-calf producer, cattle dealer, or packer who feeds cattle, either in his own lot or on a custom basis. Horizontal integration is attained by ownership or control of similar functional levels; i.e., a feedlot that merges with another feedlot. Some vertical integration of feedlots exists, but horizontal integration is limited. In 1971, only 4.5 percent of the nation's fed cattle in 39 states were fed by packers.[2]

• *Larger lots do more custom feeding*—Most farmer-feeders and small commercial feeders own their cattle. Large commercial lots do a great deal of custom feeding; and, generally speaking, the proportion of cattle custom fed varies almost directly with size.

A Texas Station study showed the following types of ownership of custom fed cattle on the southern plains; over 50 percent owned by farmers and ranchers; about one-third owned by cattle buyers, cattle dealers, and other types of investors; and 10 to 11 percent owned by packers.[3]

• *Most feeder cattle are bought through auctions or by direct purchase*—The two primary sources of feeders are: (1) auctions, and (2) direct purchase from producers. Feeder cattle are purchased by feedlot operators, salaried buyers of feedlots, order buyers, commission firms, and cattle dealers.

• *Most finished cattle are sold directly to packing plants*—Approximately 77 percent of feedlot cattle are sold directly to packers. The majority of them are sold FOB the feedlot, on foot. Some (usually relatively small lots) are sold to packers on a grade and carcass weight basis or on a carcass weight basis. In many areas, relatively small lots of finished cattle are sold through auctions and terminal markets.

• *Most big lots buy every week; sell every week*—Few experienced

[2]*Packers and Stockyards Résumé*, Vol. X, No. 13, Dec. 15, 1972, p. 4.

[3]Dietrich, Raymond A., *Cost and Economies of Size in Texas-Oklahoma Cattle Feedlot Operations*, B-1083, May 1969, Texas A&M University.

commercial cattle feeders are "in-and-outers." They assume that future price levels cannot be accurately predicted. Therefore, they buy feeder cattle and sell finished cattle on the market each week.

●*Nutrition and animal health consultants are used*—The larger commercial feedlots usually rely on two different consultants: (1) a nutritionist for ration formulation and feed purchase advice, and (2) a veterinarian for animal health, disease, and sanitation decisions.

QUESTIONS FOR STUDY AND DISCUSSION

1. What is your prediction relative to the future of cattle finishing from the standpoints of —
   a. Open lot vs confinement (sheltered) finishing?
   b. Pasture finishing?
   c. Commercial feeders vs farmer-feeders?
   d. Larger feedlots?
   e. Shifts in geography?

2. During the period of a depressed fed cattle market and/or high feed prices, which is hurt most—a farmer-feeder or a commercial feedlot? Justify your answer.

3. When estimating profits, consumers are prone to compute from range (off grass) to range (stove), with no cattle finishing phase between. Why is this so?

4. Detail the reasons back of the phenomenal growth of cattle feeding—a growth which took it from 30.1% of all slaughter cattle grain-fed in 1947, to 75% of all slaughter cattle grain-fed in 1972. Which force was most important in this growth?

5. List the six leading cattle feeding states of the nation. What factors are favorable to cattle feeding in each of them?

6. What were the primary reasons for the shift in the center of the geography of cattle feeding from the Corn Belt to the West? Which force was most important in this shift?

7. Cattle feeding is a "beef factory." Hence, establishing a big cattle feedlot in a community makes for increased business. Will a realization of this fact make feedlot flies, dust, and odors less obnoxious to the nearby urban population?

8. List and discuss the characteristics of modern cattle feeding.

9. Are the "Wall Street cowboys" (the doctors, lawyers, merchants, and industrialists) who invest in cattle feeding good or bad for the cattle feeding industry? Justify your answer.

10. Would you recommend ownership of a commercial cattle feedlot as an investment?

SELECTED REFERENCES

Title of Publication	Author(s)	Publisher
Beef Cattle Science Handbook	Ed. by M. E. Ensminger	Agriservices Foundation, Clovis, Calif., pub. annually since 1964
Feedlot, The	Ed. by I. A. Dyer, C. C. O'Mary	Lea & Febiger, Philadelphia, Penn., 1972

BUSINESS ASPECTS OF CATTLE FEEDING

Contents

Cattle feeding is big and important business, and intricate, too; and it is destined to get bigger and more complicated. Today, cattle feedlots with more than 1,000 capacity dominate cattle feeding. In 1972, there were 2,107 feedlots with capacity of 1,000 head and over; and these relatively few lots marketed 62 percent of the nation's 26,835,000 fed cattle. That same year, there were 152,429 feedlots with capacity less than 1,000 head; and they marketed 38 percent of the fed cattle. Thus, 1.4 percent of the nation's feedlots marketed nearly ⅔ of the fed cattle.

Cattle feeders are businessmen and, like other businessmen, they hope to obtain a reasonably good return for the use of their capital, labor, and management. To this end, their business aspects must become more sophisticated and efficient; they must—

1. Compute break-even prices prior to buying feeder cattle, especially if they do not buy and sell each week.
2. Buy feeder cattle of the right size, quality, and price.
3. Sell the cattle to the best advantage.
4. Integrate when possible.
5. Feed cattle to weight and grade.
6. Evaluate performance.
7. Obtain economies with size.
8. Finance the feedlot and cattle properly and adequately.

Of course, the above eight points represent a great oversimplification of a complex business, but they do clearly set forth the main requisites for profitable cattle feeding. Anyone who wishes to make money feeding cattle must have expertise in these eight areas.

A new "breed" of cattle feeder is providing business acumen at a highly professional level, with the same degree of confidence that exists in any other big business. These people are attracting large amounts of new capital from sources outside of agriculture. Business aspects outweigh all other factors—feed additives, crossbreds, pollution control, etc.—producing change in cattle feeding. It is important, therefore, that feeders and those who counsel with them be thoroughly grounded in each of these areas. To this end, each of these business aspects is covered more completely in a section that follows.

COMPUTING BREAK-EVEN PRICES FOR CATTLE

Those who feed on a large scale and on a continuous basis try to build in some insurance against the consequences of price changes through their buying programs. When finished cattle are sold, they try to replace them with feeders bought at a price which would allow a suitable profit if they were sold at the same time as the finished cattle they replace. To the extent that prices of finished cattle and feeder cattle move together (both in direction and magnitude), this work reasonably well. But they don't always move together.

Except for big cattle feeders who buy and sell cattle each week as a means of hedging, cattle feeders should compute break-even prices before buying feeder cattle. Even the big operators who buy and sell weekly do not keep the same number of cattle on feed from year to year, or throughout the year. As a result, the demand for feeders tends to be buoyant following a period of good profits from cattle feeding and depressed following a period of low returns or losses.

A nomograph (a graph or chart) may be used in computing break-even prices for cattle. It can give a quick, preliminary idea of cost, price, and investment relationships. But a nomograph should not replace more detailed budgeting which should precede all major buying, selling, and investment decisions. Also, one should realize that a nomograph will give erroneous and misleading information unless based upon accurate and realistic cost and return data from the problem at hand.

Fig. 27-1 is a nomograph for use in making quick calculation of break-even buying and selling prices for cattle. The section that follows gives an example and step-by-step instructions on how to use it.

Instructions for Using Nomograph

This nomograph provides a quick method of computing the "break-even prices" a feeder can afford to pay for incoming cattle, based upon assumptions relative to costs, amount of gain, and the estimated selling price for finished cattle. Likewise, the break-even selling price for finished cattle can be determined, based upon assumptions concerning costs, amount of gain, and the prices of feeder cattle.

The following step-by-step procedure illustrates the use of the nomograph:

DATA NEEDED		EXAMPLE	YOUR FARM
1. Purchase weight of feeder cattle	(W_1)	400#	_____
2. Selling weight of finished cattle	(W_2)	1000#	_____
3. Feed costs per cwt of gain	(L)	$32.00	_____
4. Price per cwt of feeder cattle	(P_1)	$50.00	_____
5. Price per cwt of finished cattle	(P_2)	$39.00	_____

• *Procedure for determining the break-even price for feeder cattle*–

Step 1. On Chart 1, locate the purchased weight of feeder cattle on scale W_1 and the final weight on scale W_2 and connect the two with a straight line.

Step 2. Read the W_1/W_2 weight-gain ratio on the center scale.

Step 3. On Chart 2, locate the same weight-gain ratio *vertical line*, W_1/W_2, on the scale to the immediate right.

Step 4. Follow this line straight down to the point where it intersects the *diagonal* line which represents the cost per cwt of gain—*mark this point.*

Step 5. On the vertical scale P_2 (break-even price), locate the expected selling price per cwt of finished cattle—*mark this point.*

Step 6. Draw a straight line to connect the two points located above. The point of intersection with the vertical scale P_1 on the left indicates

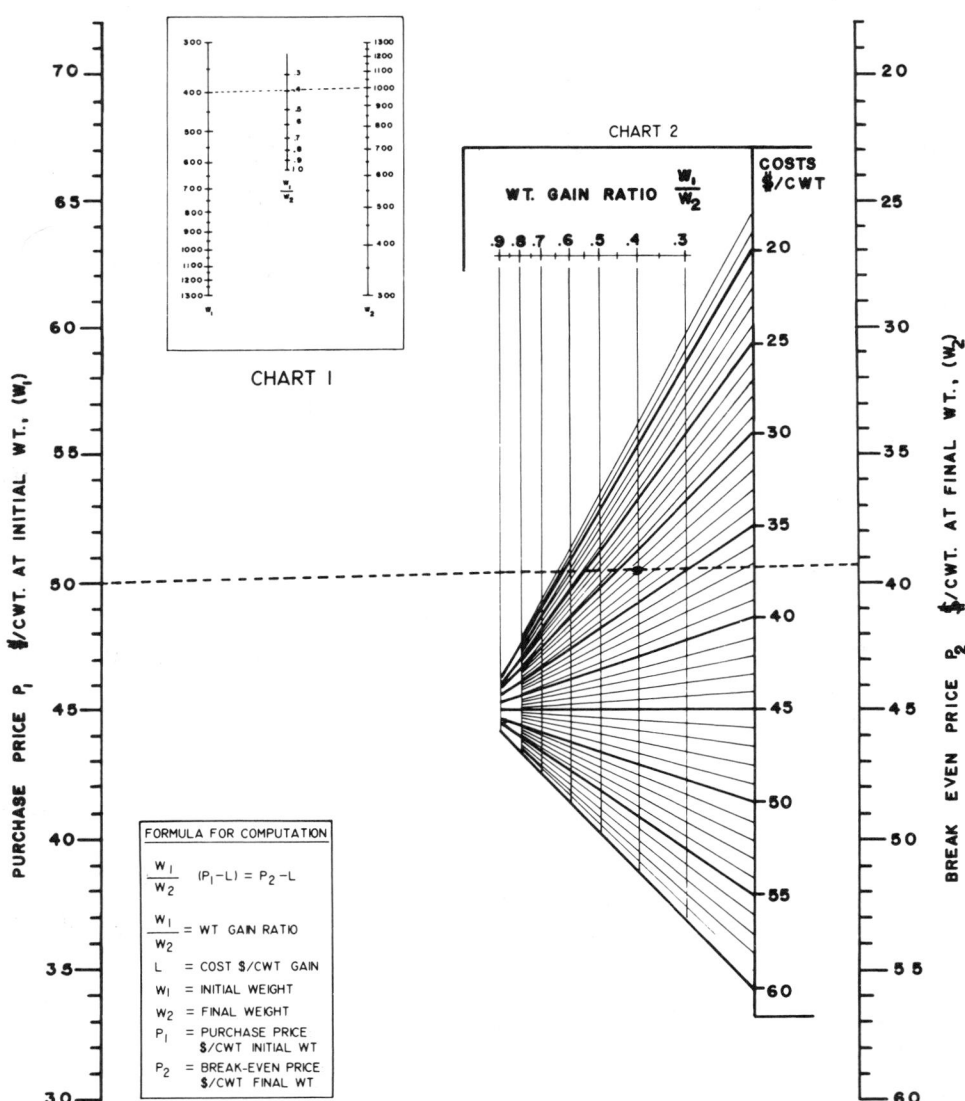

Fig. 27-1. Nomograph for computing the "break-even prices" for cattle. Beef cattle break-even price is determined by purchase price, feed cost ($/cwt), and weight-gain ratio. (This ingenious nomograph, along with instructions on how to use it, was prepared by Robert M. George and Albert R. Hagan, University of Missouri, Columbia, Mo.)

the break-even price per unit for feeder cattle—the price which will allow recovery of costs.

NOTE: In Step 4, a "diagonal cost line" may be selected to represent any cost per cwt to be recovered—feed costs only, total variable costs, total variable plus fixed costs, or total costs plus some desired profit per cwt of gain.

● *Procedure for determining the break-even price for finished cattle—* To determine the break-even price one must receive for finished cattle, to recover the costs of feed, feeder cattle, etc., follow the same procedure down to Step 5, then locate the purchase price for feeder cattle on the vertical scale P_1, and, with a straight line, locate the break-even price for finished cattle on scale P_2.

Other factors (than the cost of feed, amount of gain, and price of slaughter cattle—the factors considered in the nomograph) affecting the price that a feeder can afford to pay for feeder cattle are:

1. Condition of the cattle. Thin cattle, if in good health, will make faster gains than fleshy cattle.

2. Growthy cattle—cattle that are big framed and on the rangy order—make better gains than the little, compact kind; and they may be carried to heavier weights. If the feeder cannot obtain cattle backed by production records, "eyeballing" will help.

3. Younger, lighter weight cattle tend to make more efficient gains.

4. Cattle of known, superior ancestry with gaining ability are worth more.

5. Higher grade cattle are worth more. This is so because better grades generally bring a higher selling price, and, therefore, a higher price is obtained on their gains made in the feedlot.

6. The higher the cost of feed, the greater the necessary margin between the cost of feeder cattle and the selling price of finished animals. This is so because of the high cost of gains as compared to their selling price.

7. Feeder steers are generally worth approximately $3 to $4 per cwt more than heifers. This is because they gain about 10 percent faster, require 5 to 10 percent less feed, and bring from $.50 to $1.50 per cwt more than heifers when finished. Additionally, there is no pregnancy problem.

8. Good crossbreds will make 2 to 4 percent more rapid and efficient gains than the average of the parent breeds.

Margin

A positive margin exists when feeder cattle cost less than finished cattle. A negative margin exists when feeder cattle cost more than finished cattle. Cattle feeders will pay more for feeder cattle than they expect to receive for them as finished animals when there appears to be a favorable margin on the gain in weight. That is, when they can sell the gain in weight for considerably more than it cost to produce it.

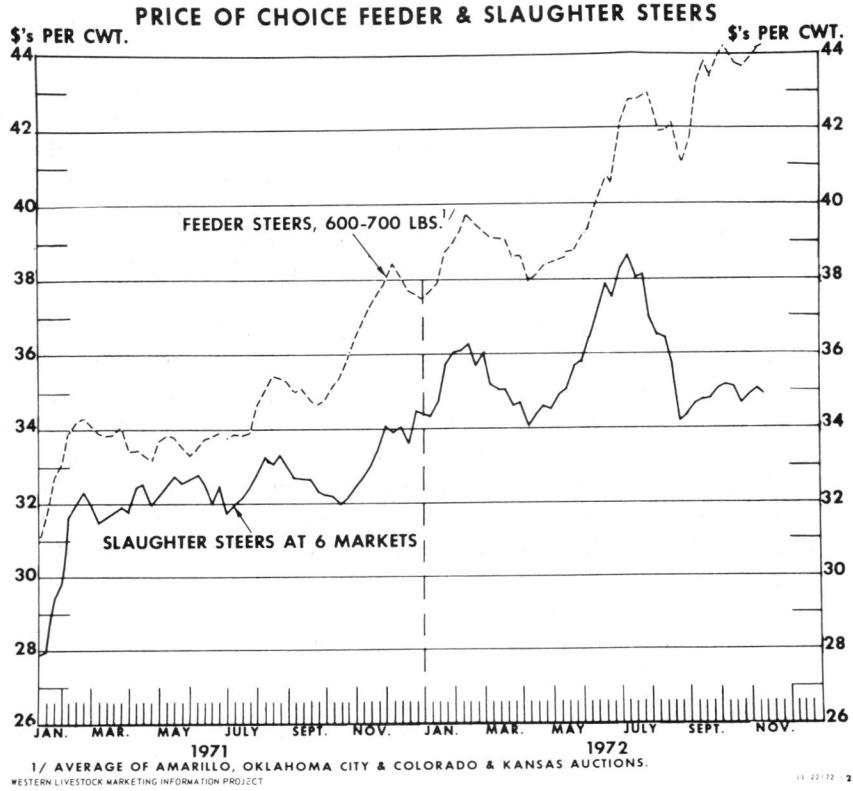

Fig. 27-2. A negative feeder's margin, 1971-1972. A negative margin may exist, which means that feeder cattle cost more than finished cattle. (Source: Using Information in Cattle Marketing Decisions, A Handbook, *WEMC Pub. No. 5*, Feb., 1973, p. 26, Chart 7)

Profits in cattle feeding come from two different kinds of margins—price margin and feeding margin.

Price margin is the difference between the cost per cwt of the feeder animal and the selling price per cwt of the same animal when finished.

For example, if a feeder pays $45 per cwt for a 600-pound steer and sells him for $40 per cwt, the price margin is a negative $5. This means that the cattle feeder would take a $30 loss on the original 600 pounds he bought.

Feeding margin is the difference between the cost of putting on 100 pounds of gain and the selling price per cwt of the same animal when finished. Thus, if it costs $30 per cwt to put gain on yearling steers, and if a cattle feeder could sell his cattle for $40 per cwt, he would have a feeding margin of $10 per cwt. Assuming a market weight of 1,000 pounds, or of 400 pounds gain, he could expect to make about $40 on feeding margin.

The amount a cattle feeder makes as a result of a good feeding margin can more than offset the losses accruing from a negative price margin, but it doesn't always work that way. It depends on many different things—the selling price, the cost of gain, the price paid for feeder animals, and other factors.

In the example just cited, the feeding margin amounted to $40 per animal. This is not to suggest, however, that cattle feeders should always put more gain on yearling steers. How much gain a cattle feeder should put on depends upon the kind of feeding program being followed, the kind and condition of the feeder cattle when they go into the lot, rate of gain, and several other factors. Research and experiences have clearly demonstrated that costs of gain go up pretty fast if cattle are fed much beyond Choice slaughter grade.

The principle of profits from price margin and feeding margin applies to feeder calves, also. But the relative importance of price margin vs feeding margin is not quite the same for calves as for yearlings. Let us analyze the situation further: the feeder who is feeding yearlings buys 600 pounds of the 1,000 pounds he finally sells. Getting cattle bought right is pretty important to him. If he pays $1 per cwt too much for his feeders, that takes $6 off the potential profit from price margin and from total profits. On the other hand, the feeder who buys calves is more interested in costs of gain and feeding margin than in price margin because about 60 percent of the weight he sells is from the gain he puts on in the feedlot. Thus, if he pays $1 per cwt too much for a 400-pound feeder calf, it hurts, but not quite so much—$4 compared to $6 per cwt per head.

If a farmer-feeder just manages to balance his gains from feeding margin with his losses from price margin, this does not necessarily mean that he should not feed cattle. Actually, he isn't in too bad shape. He's getting paid market price for his feed, a going wage for his labor, around eight percent on his own capital, and he's getting enough to cover all fixed costs like depreciation, taxes, etc., on his lot and equipment. With the commercial feeder, it's another story. He usually buys most of his feed, and he operates on borrowed capital. Thus, to stay in business, he must turn in a profit over and above these costs.

BUYING FEEDER CATTLE

When buying feeder cattle, it is well to keep in mind the old, but true, cliché: *cattle well bought are half sold.* Indeed, wise buying will increase profits or minimize losses on any lot of cattle. The following factors are pertinent to buying feeder cattle:

• *Decide on kind of finished cattle*—The feeder should first decide on the kind of finished cattle that he wishes to market. Then, he is in a position to decide on the kind of feeder animal required. For example, a certain feeder may decide that he wishes to produce a finished steer weighing around 1,100 pounds, grading 85 percent Choice, and yielding about 62 percent (cold) in the packing plant. Next, it may be decided that to meet this goal the feeder will need to start with a steer weighing 650 to 725 pounds, capable of reaching the desired weight and grade as quickly and as efficiently as possible. This weight of feeder steer may be dictated by the fact that it takes about 400 pounds of gain in the feedlot to get such an animal to grade Choice.

● *Market sources of feeder cattle*—Most feeder cattle are secured through auctions and direct purchase, rather than through terminal markets.

The best cattle from the standpoint of feedlot performance are usually those feeders coming directly from cow-calf producers—preconditioned and without passing through a market facility.

Feeder cattle may be purchased to advantage in an auction, but the problems are greater; among them, (1) they have already been stressed by being loaded, hauled, unloaded, and handled; (2) the buyer knows little or nothing of the origin of the cattle in most cases; and (3) they may have come straight off native grass (which is good), or they may be making their third sale in the last 10 days (which is very bad).

● *Method of buying*—Feeder cattle may be purchased by feedlot operators themselves, or by salaried buyers, order buyers, commission firms, or cattle dealers. Although the owner or manager of a feedlot should be knowledgeable enough to buy feeder cattle, he is generally too busy to do so if he is doing a proper job of running the lot. Accordingly, the trend is to shift the buying responsibility to specialists—primarily order buyers. Specialized buyers purchase about three-fourths of all feeder cattle in California, Colorado, and the High Plains of Texas and Oklahoma.[1]

Specialized buyers offer several advantages. They are constantly on the market for feeder cattle; hence, they are knowledgeable relative to the availability of cattle, prices, and feeding and slaughter potential of various lots of cattle. Evaluating and pricing feeder cattle is not easy, for it involves a projection ahead relative to the finished product. Of course, much of the guess work is taken out of the projection where one can obtain and feed cattle with known performance records back of them.

● *Shrinkage*—There are two types of weight loss or shrink: (1) excretory shrink, or loss of bellyfill; and (2) tissue shrink, or a decrease in the carcass weight of the animal. In the early part of shipment only excretory shrink occurs. Tissue shrink occurs on long extended hauls or during long periods of fast. It takes longer for animals to recover from tissue shrink than from excretory shrink. At an undefined stage in movement, both excretory and tissue shrinkage occur simultaneously. During the latter part in transit, tissue shrink is relatively more important.

In a study in Iowa involving 4,685 feeder cattle, it was found that an average shrink of 7.2% occurred with cattle purchased from a rancher as contrasted to 9.1% from those purchased from a sale yard. The cattle were shipped varying distances from 150 to 1,133 miles. It was found that there was a .61% shrink for each 100 miles in transit.

The cost of regaining weight loss during movement to the feedlot and getting feeder cattle back on a normal rate of gain is called "interruption cost." Older feeders and preconditioned feeders have a lower interruption cost than younger feeders.

---

[1]*Cattle Feeding in the United States*, Economic Research Service, U.S. Department of Agriculture, Agricultural Economic Report No. 186, p. 47.

Factors affecting shrinkage, along with shrinkage tables designed to make adjustments to compensate for different shrinkage conditions, are presented in Chapter 15 of this book; hence, the reader is referred thereto.

• *Preconditioning*—Preconditioning is not well defined in terms of standardized practices and procedures. In fact, it is often confused with backgrounding. The term "preconditioning" as used herein refers to dehorning and castrating well in advance of weaning, providing necessary vaccinations, treatment for control of parasites, and weaning at least three weeks before shipment. In addition, calves should be familiarized with drylot confinement, feed bunks, and water troughs.

The transition of calves from a cow herd in an open range environment to closely confined quarters, perhaps in a different climate hundreds of miles away, is a shock. This shock is a costly problem for cattle feeders.

Calves are under stress from a combination of factors such as weaning, medical treatment, rough handling, lack of feed and water during transit, strange environment, and change in feeds. Reduction or elimination of these factors lowers death loss, shrinkage, and time required to get cattle started on feed. Cattle feeders want cattle to reach their feedlots in the best possible condition. Thus, preconditioning is important to them.

SELLING FINISHED (FED) CATTLE

The feeding and marketing phases of cattle are closely related. So much so that selling should never end at the point of reaching an agreed price. Rather, the feeder should follow through on the performance of the cattle. He should learn the final weight of the cattle at the plant, their grade and yield, their cutability, and how the retailer liked them from the standpoint of tenderness and salability. This type of total marketing knowledge is necessary for the feeder to improve his performance in the feedlot.

• *When to sell fed cattle*—Cattle should be sold when they are properly finished. This usually means that they should be sold as finished cattle when they reach the same grade they had as feeder cattle. In some cases, fed cattle may be carried to one grade higher. However, the hazard of overweight cattle should be avoided.

• *Market channels for fed cattle*—Although there are area differences, fed cattle are marketed as follows: liveweight at the feedlot, carcass grade and weight basis, by auction, or through terminal markets.

During the 1950s, when cattle feeding was just entering its period of rapid growth, terminal public markets were the most widely used outlet for slaughter cattle in the United States. In 1955, over 38% of all fed cattle moved through terminal markets, and only 12% were sold direct to packers. Today, direct selling (country; private treaty) is most common. In 1972, the nation's packers purchased their fed cattle (steers and heifers) through the following market channels: country (direct), 81.1%; terminals, 11.9%; and auction, 7.0%.[2]

---

[2]Figures from "Packers and Stockyards Resúmé", Vol. XI, No. 3, Dec. 14, 1973, pp. 2 and 4.

Carcass grade and weight selling of cattle is increasing somewhat in those areas where direct-liveweight marketing is dominant, with 22.6 percent of all cattle sold on this basis in 1972.[3]

Direct-liveweight, and grade- and carcass-weight sales increase as sales lot capacity increases. Commercial feedlots, with their large volume of fed cattle marketing, can attract buyers to the feedlot. Farmer-feeders, particularly the smaller ones, cannot attract buyers to the feedlot; hence, most of them sell through auctions and terminal markets.

Each cattle feeder should study his market options carefully to determine which may be most desirable for him, with consideration given to (1) price, (2) marketing costs, and (3) convenience. Additionally, cattle feeders who sell their cattle at the lot will need to keep well informed about market conditions if they are to be on an equal bargaining basis with most cattle buyers.

● *Factors influencing value of fed cattle*—Three major factors influence the value of slaughter cattle. They are:
1. Grade.
2. Dressing percent (carcass weight in relation to liveweight).
3. Retail yield (weight of retail cuts in relation to carcass weight).

In the past, packers bought cattle on a liveweight basis and sold beef on a carcass weight basis. As a result, only grade and dressing percent were considered in determining value. Additionally, variations in retail yield between carcasses should be considered to price live cattle accurately.

Research studies have indicated that differences in value due to retail yield (or cutout) may be as much, or more, than $50 per 1,000-pound live animal. At the present time, however, only small differentials, if any, are being paid for the higher retail yielding cattle. It is important that the market system develop price differences that accurately reflect cutout value, because producing cattle that have a high percentage of retail cuts will contribute to a more efficient beef industry.

● *Seasonality of prices*—The seasonality price pattern for fed cattle has not been consistent in recent years. This is because cattle feeding is now a year-round business, and adjustments can be made to increase or decrease marketings in any particular period. Consequently, it is expected that the seasonal pattern for fed slaughter cattle will continue unstable in the future.

● *Shrinkage*—Shrinkage in market cattle is influenced by a number of factors, especially (1) time off feed and water, and (2) stresses of being hauled and handled.

Several studies have been made of the relationship of shrink to distance hauled. All of these studies show that shrinkage rises rapidly in the first few miles cattle are hauled and continues to increase, but at a decreasing rate as mileage accumulates. The results of one such study are given in Table 27-1.

Tables 15-7 and 15-8 of Chapter 15 are designed to make price adjustments to compensate for different shrinkage conditions.

In some cases cattle are purchased or sold with a certain percentage shrink from actual weights. This is known as pencil shrinkage.

[3]Ibid.

TABLE 27-1

AVERAGE PERCENTAGE SHRINK OF SLAUGHTER CATTLE
IN RELATION TO DISTANCE HAULED[1]

Distance Hauled	Shrink
(mi)	*(%)*
0 - 34	1.36
35 - 70	1.43
71 - 104	2.07
105 - 144	2.47
145 & over	2.83

[1]Henning, G. F., and P. R. Thomas, *Factors Influencing the Shrinkage of Livestock from the Farm to First Market*, Ohio Agricultural Experiment Station, Columbus, Ohio, AES Bull. 925, October, 1962.

● *Recommended handling of fed cattle prior to and during shipment*—Market cattle should be handled as follows prior to and during shipment:

1. Stop feeding all feed additives and drugs that require a certain time interval between use and slaughter. *Check directions and recommendations on containers and follow closely.*

2. If cattle are to be weighed at lot or off trucks, continue normal feeding schedule until they are weighed, with the exception of withdrawing additives or drugs as required.

3. If selling on a market where cattle will have access to feed and water before being weighed, as is the case at terminal markets and auctions—

    a. Withdraw protein from ration at least 48 hours prior to shipment.

    b. Reduce grain ration to half feed the day before shipment.

    c. Provide grass hay free-choice during the last two or three days prior to shipment.

    d. Don't allow animals to eat grain for 12 hours before shipment.

    e. Allow animals to have free access to water until shipment.

4. If selling on a carcass grade and weight basis, all feed can be withdrawn 48 hours prior to slaughter. It should not be withdrawn more than 48 hours because there may be some tissue shrink. Animals sold on a carcass grade and weight basis can have free access to water until slaughter.

5. Move slowly and quietly when loading and handling cattle. Don't overcrowd trucks. Gate cattle into compartments on big trucks to take pressure off end animals when truck starts and stops.

INTEGRATE WHEN POSSIBLE

There are two types of integration—horizontal and vertical. Horizontal integration is the combining of one or more firms engaged in the same enterprise into a larger firm; for example, when a cattle feeder in California acquires a feedlot in Texas. Vertical integration refers to combining two or more stages of production, processing, and distribution into a single entity; for example, a cow-calf producer entering the feedlot business, or a feedlot operator entering the packing business.

There has been, and will continue to be, some of each type of integration. But the greatest economic advantage can accrue from vertical integration; hence, it will be the dominant type of integration in the future, and it will be discussed herein.

In order to check what it saw as a trend toward "monopolistic conditions" in the meat industry, USDA's Packers and Stockyards Administration ruled that effective July 1, 1974, (1) packers cannot own, operate, or control *custom* feedlots, and (2) *custom* feedlot owners cannot own, operate, or control packing plants. However, packers are not prohibited from feeding their own cattle for their own slaughter needs.

● *Why integrate?* –
. . . *and the cow-calf man sells to the stocker operator.*
. . . *and the stocker operator sells to the feedlot operator.*
. . . *and the feedlot operator sells to the packer.*
. . . *and commission and transportation costs multiply.*
. . . *and the calves shrink and get sick.*
. . . *and they're given shots on top of shots.*
. . . *and cattle die!*

Indeed, the several ownerships characteristic of the traditional buy-and-sell routine makes for a vicious cycle; and it's costly, too. There must be a better way—better for the cattle, and better for the pocketbooks of everybody along the production line. Vertical integration appears to be the answer.

Three Kansas State University economists estimated that the added costs due to lack of vertical integration—costs for commission, freight, pencil shrink, and extra veterinary and medicine costs due to stress and change in environment—may add nearly $40 to the overhead cost of a steer during his lifetime. Table 27-2 gives a breakdown of their estimate.

Table 27-2 points up the following:

1. If the cow-calf operator would retain ownership of the steer all the way through the feedlot, he could eliminate some $20 worth of trading costs, which would make for a nice profit within itself. Bear in mind that this $20 profit would accrue even if the steer were to be moved from the ranch to the backgrounder, thence on to the feedlot.

2. Nearly $17 of extra costs are added to the animal by moving him from one location to another; of course, transportation figures are highly variable, dependent upon the distance involved. On long-haul cattle, the figure would be considerably higher.

Table 27-2 also explains why many single phase operations are not profitable; for example—

1. The feedlot owner who buys calves and does his own backgrounding at the feedlot location could eliminate $22.25 from the list of costs given in the chart.

2. The rancher who does the backgrounding job on his own ranch then sells to the feedlot owner, could eliminate nearly $19 worth of costs from the table.

Of course, the cost figures shown in Table 27-2 will vary considerably, from one bunch of cattle to the next, and from one year to the next. But, regardless of costs, it can be concluded that the longer a calf is under one

TABLE 27-2

SUMMARY OF NORMAL EXPENSES FROM COW TO PACKER

Production-Marketing Phase	Type of Expense	Estimated Cost[1]	
		Item	Total
Cow-calf man—	Freight to sale	$1.00	
	Sale commission	3.50	$ 4.50
The calf buyer spends—	Commission (time spent)	1.00	
	Trucking home	1.00	
	Cost to recover shrink	2.00	
	Vet expense—due to stress, change of feed & environment	1.00	5.00
After six months wintering— Calf sold to grass man—	Trucking to sale	1.25	
	Commission	4.00	5.25
The grass man spends—	Trucking home	1.25	
	Commission	1.00	
	Recover shrink	2.00	4.25
Steer is sold in fall—	Truck to feeder sale	1.25	
	Commission	4.00	5.25
Feedlot owner buys him—	Freight	1.50	
	Commission	4.00	
	Shrink	2.00	7.50
Feedlot owner sells to packer—	Commission	5.00	
	Freight	1.50	6.50
		Total	$38.25

[1]Estimates prepared by Larry R. Kepley, Bill D. Collins, and Hobart Frederick, Kansas State University.

ownership, the better the chances of profit. This explains why a small farmer-feeder with a cow herd and a finishing lot is able to compete with a large commercial feedlot. According to most economists, they're doing it all wrong. Instead of specializing, they are a trinity—(1) cow-calf man, (2) backgrounder, and (3) feeder. They turn only one crop of cattle through the feedlot each year (instead of the 2¼ to 2½ common to commercial feedlots); and they aren't big enough to get any economy of size. Yet, they keep on making money—even when the big operators are "losing their shirts." The explanation: they alleviate all the buying and selling that goes on in the cattle business, which, as shown in Table 27-2, can come to nearly $40 per head.

● *Who will do the integrating*—The integration operation could be set up in a number of ways. It could be done—

1. By the cow-calf man, who would retain ownership of his calves all the way through the feedlot.

2. By the feedlot, who would buy the calves or get control of them by contract at weaning time.

3. By the stocker operator, who would buy the calves and eventually have them finished on a custom basis in a commercial lot.

FEEDING CATTLE

Rations as such are covered later, in Chapter 30; thus, they will not be repeated herein. Rather, points pertaining to feeds and feeding not normally considered "nutritional," but which affect profits and losses—the business aspects—will be considered here:

1. Consider the moisture level of feed when buying and selling, formulating and mixing the ration, and in evaluating feed consumption.

2. Check the accuracy of scales used to weigh feed.

3. Where split loads are used in feeding two or more lots of cattle, make certain that each lot gets the intended amount of feed daily.

4. Feed cattle when they want to eat, rather than at the convenience of the feeder. For example, in a very hot climate, in the summertime, cattle may do 80 percent of their eating at night. In such areas, it is desirable to do feeding late in the evening and early in the morning.

5. The fixed, nonfeed costs associated with the ownership of cattle—including interest, labor, equipment, drugs, veterinary expenses, death losses, and taxes—will add 15 to 20 cents/lb to the cost of low gains of 1¼ to 1¾ lb/day, in comparison with no more than 5 to 6 cents/lb where gains of 3 lb/head/day are being obtained.

EVALUATING THE PERFORMANCE

A feeder cannot tell where he is going if he does not know where he is or where he has been. Thus, he must keep adequate records. Every feedlot manager should have access to records that show him the cost of gain, rate of gain, and feed conversion of each pen of cattle that leaves his lot; the grade and yield of each lot of cattle slaughtered; inventory figures showing his cattle and feed inventory on hand every day; death loss by pen or group, and the cause of death; and the air-dry feed consumption per head per day for each group or pen of cattle, as well as an idea of their average for the period.

Also, it is recommended that cattle feeders compute their profits on a per head per day basis. This is important whether they own only the cattle, only the feedlot, or both. In any case, the important thing is to make the most profit per head of capacity per day. For example, it is better to make $10 per head on an animal that is fed 110 days ($.0909/head/day) than to make $30 per head on an animal that is fed a whole year ($.0822/head/day). The feedlot can also be evaluated on a profit per head per day basis to determine performance. Comparisons can be made in this way in the same lot with different numbers of cattle on feed, and between several lots of different size to determine relative performance.

Cattle Feedlot Profit Indicators

Cattle feedlot operators need to keep good records and make frequent analyses (at least once monthly) to determine how well they are doing. A determination of assets and liabilities at the end of the fiscal year is not good enough, primarily because it is available only once per year. Cattle feeding requires much capital; hence, records should be kept as current as possible at all times.

Operators are primarily interested in two questions; namely, (1) how well am I doing—profitwise, and (2) how do I compare with other feedlots? Table 27-3 gives some guidelines for answering these questions. It may be used in making an analysis of a specific feedlot; for determining (1) the strengths and weaknesses within the feedlot, and (2) how it stacks up with the top 5 percent of the nation's feedlots. It is noteworthy that the top 5 percent operators invest more in land and equipment than their average counterparts. Obviously, they effect savings in operation rather than in physical plant.

Admittedly, profit indicators, such as those given in Table 27-3, are not perfect; simply because no two feedlots are the same. Nationally, there are wide area differences in climate, feeds, land costs, salaries and wages, and other factors. Nevertheless, indicators *per se* serve as a valuable yardstick. Through them, it is possible to measure how well a given feedlot is doing— to ascertain if it is out of line in any one category, and, if so, the extent of same.

After a few years of operation, it is desirable that a feedlot evolve with its own yardstick and profit indicators, based on its own historical records and averages. Even with this, there will be year to year fluctuations due to

TABLE 27-3

CATTLE FINISHING PROFIT INDICATORS[1]

	Calves		Yearlings	
	Av. for U.S. Feedlots	Top 5% of U.S. Feedlots	Av. for U.S. Feedlots	Top 5% of U.S. Feedlots
Initial (new) land, feedlot and equipment cost basis/animal capacity, $	45.00	47.00	47.00	49.00
Feedlot and equipment cost charged off/ animal finished out, $	2.25	2.00	1.75	1.25
Daily nonfeed costs/animal,[2] ¢	3.00	2.5	2.5	2.25
Salaries and wages/head/day, ¢	2.5	2.25	2.25	2.0
Death, losses, %	3.7	2.0	1.8	0.8
Vet. fees and medicine/head/day, ¢	2.5	1.8	1.6	1.2
Feed/lb gain: Steers, lb	7.3	6.9	8.0	7.6
Heifers, lb	7.2	7.0	8.6	8.0
Daily rate of gain: Steers, lb	2.3	2.5	2.7	2.85
Heifers, lb	2.0	2.15	2.45	2.6
Net return per head to management,[3] $	7.00	8.50	7.50	9.00

[1]This is a consensus (or judgment) of several knowledgeable heads of university animal science departments and agricultural consultants, based on a survey made by the author. A consensus was resorted to for the reasons that (a) no extensive, nationwide, scientific sampling of cattlemen on these matters has ever been made, and (b) this information is much needed by cattlemen and those who counsel with them. No claim is made relative to the scientific accuracy of the data; rather, it is presented (a) because it is the best, if not the only, information of its kind presently available on a nationwide basis, and (b) with the hope that it will stimulate needed research along these lines.

[2]This embraces all costs other than cattle and feed. It includes salaries and wages; taxes, interest, insurance; utilities; gasoline, oil, grease; depreciation; repairs; vet. and medical; consultant and legal service; trucking; promotion; and other costs.

[3]This means just what it says; net return to management after deducting from gross receipts all costs, including depreciation on machinery and buildings, and interest on investment.

seasonal differences, cattle and feed price changes, disease outbreaks, changes in managers, wars and inflation, and other happenstances.

OBTAIN ECONOMIES WITH SIZE

Big feedlots have become bigger. This trend will continue. The economies of size associated with cattle feedlots come about through the ability of the larger lots to obtain higher prices and buy at lower costs. Also, in some cases management simply wants a larger operation in order to increase total income.

Studies and practical observation show that larger feedlots have the following advantages in efficiency: they operate more nearly to capacity, with the feedlot fuller; there is greater specialization of labor; they are more highly mechanized; they can outdo smaller lots in buying and selling cattle, feeds, equipment, and supplies; and they have more opportunities to coordinate vertically with firms engaged in feed processing, by-product feeds, transportation, meat packing, and retail outlets; and they attract more custom feeding.

Some inefficiencies arise with extremely large feedlots, chief among these are: a weakening of coordination among the many feedlot activities; a higher proportion of time needed to travel within the feedlot complex; a tendency to outgrow managerial capacities; and increased problems in manure disposal and pollution control.

FINANCING FEEDLOT CATTLE[4]

Big commercial cattle feedlots are now in the "era of money management," where success comes to those who obtain and use most effectively a quantity of borrowed capital. A 20,000-head lot requires about $7 million for cattle and feed for the typical 6-month feeding period. If the firm is able to borrow $3 for each dollar of equity, it will need about $1.75 million of equity capital. That's a far cry from the financial needs for feeding 20 to 100 head on home-grown feeds. Indeed, big cattle feedlots are "beef factories."

The acquisition and use of money are discussed in the sections that follow. The tables and projections therein are based on a 20,000-head capacity feedlot. Note that these are 1971 figures based on costs and returns in Oklahoma at the time. They will serve as useful guides, but like any other figures of similar nature, they must be adapted to a particular area and updated in keeping with the economy at the time of use.

Fixed Costs (Feedlot and Equipment)

Fixed assets include land, feed mill, bunks, waterers, office, shop, etc. Needs vary, depending on size of feedlot and region. Most studies show that

---

[4]Except where otherwise noted, the tables used in this section, and its subsections, are from *Beef Cattle Science Handbook*, Vol. 8, 1971, published by Agriservices Foundation and edited by Dr. M. E. Ensminger, pp. 398-409. These tables were provided by Paul H. Hitch, Henry R. Hitch Ranch, Inc., Guymon, Okla., a large and successful cattle feeder.

the larger the feedlot the lower the investment per head capacity. On the other hand, extremely large commercial feedlots generally require (1) rather elaborate types of equipment, and (2) some duplication in equipment. As a consequence, fixed investment costs per head capacity gradually decrease up to about 10,000-head capacity, then increase slightly in larger feedlots.

Table 27-4 shows the fixed costs for a 20,000-head feedlot, starting from scratch. Note that this represents a per head cost of $41.25.

TABLE 27-4

CAPITAL INVESTMENTS FOR A 20,000-HEAD COMMERCIAL FEEDLOT

Feed mill-stream flaking	$ 300,000.00
Feedyard:	
1. Land	64,000.00
2. Bunks and aprons	130,000.00
3. Fences, waterers, sick pens, working chutes, etc.	150,000.00
Total Feedyard	$ 344,000.00
Office (equipped)	25,000.00
Shop (equipped)	20,000.00
Trench Silos	36,000.00
Feeding and maintenance equipment	100,000.00
TOTAL	$ 825,000.00
Per Head Capacity—$41.25	

In a study made in 1970, involving feedlots in Colorado, Kansas, Nebraska, Wyoming, most of Oklahoma and New Mexico, and the western part of Missouri, the Federal Reserve Bank of Kansas City found that the capital investment per head for items other than the animal itself varied from less than $30 to more than $90, with a median of $50 on real estate and equipment.[5] The capital investment in most California feedlots is higher than the figures cited because of being located near suburban or industrial areas on expensive land.

A distinctive characteristic of large cattle feeding operations is mechanization. It enables them to (1) substitute more capital per head (which is relatively cheaper than labor) for labor, and (2) process feed so that it can be utilized more efficiently. The result is that labor used per animal fed is almost negligible (less than one hour per animal) in large lots, and better gains are obtained per unit of feed fed. These gains, however, are achieved at a substantial investment in facilities and other operating costs.

DEPRECIATION SCHEDULE

Table 27-5 shows the depreciation schedule for this 20,000-head installation, as well as fixed costs. Note that the total depreciation each year is $97,850.00. Note, too, that taxes, insurance, and management costs bring the annual fixed charges up to $130,150.00. Of course, land is not depreciable.

---

[5]Doll, Raymond J., "Cattle Feeding in the Tenth District: Financing," *Monthly Review*, Federal Reserve Bank of Kansas City, July-August, 1970, p. 11.

The feed mill, bunks, fences, silos, etc., have been depreciated over a 10-year period, whereas office equipment, feeding equipment, and shop, which become obsolete rather quickly, have been depreciated over a 4-year period.

TABLE 27-5

TOTAL ESTIMATED FIXED COSTS FOR A 20,000-HEAD FEEDLOT

	Cost	Annual Depreciation
Land	$ 64,000.00	$
Feed Mill	300,000.00	30,000.00
Bunks and aprons	130,000.00	13,000.00
Fences, waterers, sick pens, working chutes, etc.	150,000.00	15,000.00
Trench silo	36,000.00	3,600.00
Office (equipped)[1]	25,000.00	6,250.00
Feeding and maintenance equipment[1]	100,000.00	25,000.00
Shop (equipped)[1]	20,000.00	5,000.00
Total	$ 825,000.00	$ 97,850.00
Annual depreciation	$	97,850.00
Taxes		3,500.00
Insurance		3,800.00
Management		25,000.00
Total fixed charges		$ 130,150.00

[1]Office, feeding equipment, and shop are depreciated straight line on a 4-year life; all other depreciations are straight line on a 10-year life.

Variable Cost (Nonfeed Costs)

Variable costs are those which change with the volume of cattle handled by the feedlot; they increase in total with the size of the feedlot, but they decrease on a per head basis with increased size. They include labor, veterinary supplies, insecticides, dues, trucking, equipment maintenance and repairs, electricity, and fuel. These are the costs that make the outfit run. Table 27-6 shows that it costs $277,500.00 to keep this 20,000-head feedlot operating during the year. As noted, salaries and wages are the largest cost item, accounting for 59 percent.

TABLE 27-6

ANNUAL VARIABLE COSTS OF OPERATING A 20,000-HEAD FEEDLOT

Salaries and wages	$ 165,000.00
Vet. supplies	30,000.00
Insecticides	2,000.00
Dues, fees, and subscriptions	500.00
Trucking—other than cattle	20,000.00
Equipment maintenance and repairs	35,000.00
Electricity	10,000.00
Fuel	15,000.00
Total	$ 277,500.00

Feed Costs

Feed costs alone account for about 80 percent of the total costs of feeding cattle exclusive of the purchase of animals. They vary considerably by size of feedlot and location of the feeding enterprise. Also, the relative importance of feed as a cost item is directly affected by the price level for the major feed ingredients. Of course, feed costs per pound of gain are greatly affected by the type of cattle fed, the feeding and management practices employed, and the location of the feedlot.

After determining feedlot and equipment costs, next the feed costs should be determined. First, the feeder will need to have a ration in mind and some ideas relative to consumption. For purposes of this illustration, a very simple ration consisting of corn silage, dehydrated alfalfa pellets, corn, and protein supplement has been chosen. (See Table 27-7.)

Of course, other feed ingredients can and should be used in different locations, depending primarily on feed availability and price.

For purposes of this illustration, it is further assumed that 700-lb feeder cattle will be fed; that they will gain 3 lb per day; that they will be fed 135 days to a weight of 1,100 lb; and that there will be 4 different ration changes during the feeding period.

Next, feed costs need to be determined on the basis of (1) per pound of feed (Table 27-7), (2) per head per day (Table 27-8), (3) for the feeding period (135 in this case) (Table 27-9), and (4) for the year. (Table 27-10).

● *Feed costs per pound of feed—*

TABLE 27-7

PRICES FOR FEED INPUTS

		Cost/Lb.
Corn silage	$ 10.00/ton	$ .0050
Dehy	40.00/ton	.0200
Corn	2.20/cwt	.0220
Protein supplement	84.00/ton	.0420

● *Feed costs per head per day—*

TABLE 27-8

APPROXIMATE RATION IN INGREDIENTS
NUMBER OF POUNDS OF FEED AND COST, PER HEAD, PER DAY

	%	Lb of Feed	Cost/Hd/Day
			($)
Ration 1 (7 days):			
Supplement	3	.87	.03654
Dehy.	2	.58	.01160
Silage	52	15.08	.07540
Corn	43	12.47	.27434
		29.00	.39788

(Continued)

TABLE 27-8 (Continued)

	%	Lb of Feed	Cost/Hd/Day
Ration 2 (7 days):			
Supplement	3	.90	.0378
Dehy.	2	.60	.0120
Silage	40	12.00	.0600
Corn	55	16.50	.3630
		30.00	.4728
Ration 3 (46 days):			
Supplement	3	.93	.03906
Dehy.	2	.62	.01240
Silage	30	9.30	.04650
Corn	65	20.15	.44330
		31.00	.54126
Ration 4 (75 days):			
Supplement	3	.93	.03906
Dehy.	2	.62	.01240
Silage	24	7.44	.03720
Corn	71	22.01	.48422
		31.00	.57288

● *Feed costs for 135 days*–For one steer, for the 135-day feeding period, the average feed cost per day not including the markup will be $.54784. Yardage will increase this cost to $.59784. The total feed and yardage costs for the 135-day period will be $80.71. (See Table 27-9.)

TABLE 27-9

COST OF RATION FOR ONE ANIMAL FOR 135 DAYS ON FEED

Ration	No. Days	Feed $/Day	Total $/Day	Total Cost
1	7	.39788	.44788	$ 3.13516
2	7	.47280	.52280	3.65960
3	46	.54126	.59126	27.19796
4	75	.57288	.62288	46.71600
	135	.54784	.59784	$ 80.70872
Average feed cost/lb of gain	$ .18261			
Average total cost/lb of gain	.19928			

● *Feed costs for 20,000 head per year with turnover of 2.5 and 90 percent full*–It is further assumed that the feedlot will turnover about 2½ times in the course of the year, or that the 20,000-head feedlot will turn out about 45,000 head of cattle, provided the lot is kept 90 percent full all the time. To feed these cattle will require nearly $3½ million in feedstuffs. Included in this cost is a shrinkage factor for each ingredient since these losses will occur. (See Table 27-10.)

TABLE 27-10

TOTAL NUMBER OF POUNDS OF FEED & COST TO OPERATE A
20,000-HEAD CAPACITY FEEDLOT ONE YEAR, INCLUDING SHRINKAGES

	Shrinkage	Total Lb of Feed		Total Cost
	(%)	(lb)		($)
Supplement ...	3	5,793,750	2,897 (tons)	243,348.00
Dehy.	5	3,972,456	1,986 (tons)	79,440.00
Silage	15	62,128,125	31,064 (tons)	310,640.00
Corn	3	128,853,000	1,288,530 (cwt)	2,834,766.00
				3,468,194.00

Flow Chart For Feed

Table 27-11 is a flow analysis for feed. It shows, by months, the tons of feed bought, the total cost of the feed for the month, and the tons used. This covers each of the ingredients making up the ration in Table 27-8. In Table 27-11, Month 1 is September. Thus, as shown, silage is received over the 3 months of September, October, and November, but it is fed out over 12 months. Hence, the total cost of silage purchase is much higher in those 3 months than in the rest of the year. Unless the silage growers, from whom silage has been contracted, will carry the feedlot for awhile additional funds will be needed to start the crop year beginning in September.

Cash Flow Chart All Expenses and Income

Finally, Table 27-12 is a summary of the whole lot in operation for a year, with all expenditures and receipts shown. It is a complete cash flow chart.

Table 27-12 takes into account all the costs which have to be paid to keep the lot running. The facilities are to be paid out on a 10-year schedule. The money borrowed for the construction of the lot costs 9 percent. As shown, the initial purchase of feedstuffs (in this case silage) creates a cash deficit during the first three months of operation. Although this deficit declines thereafter, it is not until the eighth month of the year that cash on hand actually comes into the black. Obviously, therefore, this effect will need to be taken into consideration in planning the year's fund requirement.

Cattle Cost

If the lot owns its cattle, it will be concerned with cattle financing. If the lot is entirely on a custom feeding basis, financing of the cattle may be only of secondary consideration. But, assuming that the feedlot owns its cattle, an analysis of costs and returns will be necessary. Thus, the example given in Table 27-13 is presented.

Further study of these figures reveals that this animal has a break-even price of about $29.22 per cwt. Whether that price is likely to occur is a matter for the individual to decide. No one can be sure.

TABLE 27-11

CASH FLOW ANALYSIS FOR FEED
20,000-HEAD FEEDYARD @ 90% CAPACITY

MONTH	1	2	3	4	5	6	7	8	9	10	11	12	TOTAL[1]
Total feed cost	$ 325,259	449,510	325,254	263,130	263,129	263,130	263,129	263,130	263,129	263,130	263,129	263,130	$ 3,468,194
Supplement													
Buy ton	241	242	241	242	241	242	241	242	241	242	241	242	2,897
Cost	$ 20,279	20,279	20,279	20,279	20,279	20,279	20,279	20,279	20,279	20,279	20,279	20,279	$ 243,348
Use ton	234	235	234	235	234	235	234	235	234	235	234	235	2,813
Dehy.													
Buy ton	165	166	165	166	165	166	165	166	165	166	165	166	1,986
Cost	$ 6,620	6,620	6,620	6,620	6,620	6,620	6,620	6,620	6,620	6,620	6,620	6,620	$ 79,440
Use ton	156	156	156	156	156	156	156	156	156	156	156	156	1,874
Silage													
Buy ton	6,213	18,638	6,213										31,064
Cost	$ 62,130	186,380	62,130										$ 310,640
Use ton	2,203	2,203	2,203	2,203	2,203	2,203	2,203	2,203	2,203	2,203	2,203	2,203	26,438
Corn													
Buy ton	107,377	107,378	107,377	107,378	107,377	107,378	107,377	107,378	107,377	107,378	107,377	107,378	1,288,530
Cost	$ 236,230	236,231	236,230	236,231	236,230	236,231	236,230	236,231	236,230	236,231	236,230	236,231	$ 2,834,766
Use cwt	104,250	104,250	104,250	104,250	104,250	104,250	104,250	104,250	104,250	104,250	104,250	104,250	1,251,000

[1]Numbers may not add to total because of rounding.

TABLE 27-12

CASH FLOW ANALYSIS FOR A 20,000-HEAD FEEDLOT OPERATION ONLY @ 90% CAPACITY

MONTH	1	2	3	4	5	6	7	8	9	10	11	12	TOTAL[1]
Facilities repayment	$ 6,875	6,875	6,875	6,875	6,875	6,875	6,875	6,875	6,875	6,875	6,875	6,875	$ 82,500
Interest—1st yr. 9%	$ 6,188	6,188	6,188	6,188	6,188	6,188	6,188	6,188	6,188	6,188	6,188	6,188	$ 74,250
Feed costs	$ 325,259	449,510	325,259	263,130	263,129	263,130	263,129	263,130	263,129	263,130	263,129	263,130	$ 3,468,194
TOTALS:	$ 338,322	462,573	338,322	276,193	276,193	276,193	276,192	276,193	276,192	276,193	276,192	276,193	$ 3,624,944
Fixed costs	$ 10,846	10,846	10,846	10,846	10,846	10,846	10,846	10,846	10,846	10,846	10,846	10,846	$ 130,150
Variable costs	$ 23,125	23,125	23,125	23,125	23,125	23,125	23,125	23,125	23,125	23,125	23,125	23,125	$ 277,500
Costs— total feedyard operation	$ 372,293	496,544	372,293	310,164	310,163	310,164	310,163	310,164	310,163	310,164	310,163	310,164	$ 4,032,594
Charge for feed per head per day $8.602624	$ 325,417	325,417	325,417	325,417	325,417	325,417	325,417	325,417	325,417	325,417	325,417	325,417	$ 3,905,004
.05¢/hd. /day	$ 27,000	27,000	27,000	27,000	27,000	27,000	27,000	27,000	27,000	27,000	27,000	27,000	$ 324,000
TOTAL INCOME	$ 352,417	352,417	352,417	352,417	352,417	352,417	352,417	352,417	352,417	352,417	352,417	352,417	$ 4,229,004
Cash on hand	$ -19,876	-164,003	-183,879	-141,626	-99,372	-57,119	-14,865	27,388	69,642	111,895	154,149	196,402	$ 196,410

[1]The sum of each line does not equal the total because the numbers are rounded off.

TABLE 27-13

PER STEER COST, SELLING PRICE, NET RETURN

Buy	700-lb steer @ $31/cwt	$ 217.00
Add[1]	400-lb of gain	80.00
Add[1]	Interest @ 9.5%	9.03
Add[1]	Death loss @ 1%	2.57
	Cost of the animal	$ 308.60
Sell[2]	1,056 lb @ $31/cwt	327.36
Net return/head—based on the above prices		$ 18.76

[1]Cost of gain including interest and death loss is $22.90/cwt.
[2]In selling the animal, there was a 4% pencil-shrink figured on the liveweight of 1,100 pounds.

Sources of Finances

The dynamic cattle feeding industry needs different kinds of credit involving different sources. Large scale feedlot operators have been innovative and aggressive in meeting these needs. Commercial banks are the primary source of capital. Other important sources of money for feedlots are the Federal Land Banks, Production Credit Associations, and private individuals.

• *Financing land and plant (real estate and equipment)*—The main sources of credit for financing the land and physical plant are private individuals, commercial banks, Federal Land Banks, Production Credit Associations (PCAs), the Small Business Administration (SBA), and life insurance companies.

• *Financing feed and operating expenses*—Although the amount of funds needed for financing feed inventories and operating expenses is usually relatively small, it can be sizable if a full year's supply of grain and forage is purchased at harvest time and stored. Also, the security available is frequently less distinct and less familiar to lenders than a mortgage on cattle or facilities. Thus, financing this phase of a feedlot operation sometimes requires special treatment.

Financing of feed inventories and operating expenses varies widely, but banks, PCAs, and individuals are the main sources. Other methods of financing feeds include (1) deferred payments, with interest, from the farmer who produced the feed, and (2) warehouse receipts.

• *Financing cattle*—A major portion of the funds needed for operating a feedlot is required for financing the feeder cattle. The two major sources of credit for financing cattle are commercial banks and PCAs. Most banks and PCAs do not require a fixed margin when lending to a farmer-feeder; rather, they consider the various resources and financial position of the borrower. Where a commercial feeder is involved, however, they generally limit the loan to 75 to 80 percent of the purchase price of the cattle (thus, they require a margin equivalent to 20 to 25 percent of the value of the feeder animals), plus all the feed costs. In some areas, banks do almost all of this kind of financing, while PCAs do almost all of it in other areas. In some cases, both lending organizations are involved; that is, the feedlot operator may use the local bank for credit up to its loan limit, then use the PCA for overlines which frequently are large. The lending agency usually retains first mortgage on the cattle.

In addition to the common sources of credit already mentioned—commercial banks, Federal Land Banks, Production Credit Associations, private individuals, Small Business Administration, and life insurance companies—cattle feeders sometimes use the following sources of credit:

1. *Bank acceptance financing*—This method involves the following: The owner puts up equity money for the cattle (usually 20 to 30%), signs a note for the balance showing that the bank accepts (guarantees) the note, then the note is sold on the open market just like any other commercial paper.

2. *Warehouse receipts*—Cattle are sometimes "warehoused," in the same manner as grain is sometimes warehoused. The warehouse receipt merely provides a guarantee of collateral by a third party. The firm (third party) does not lend money; hence, the owner of the cattle or the feedlot must negotiate credit lines.

3. *Leasing equipment*—Automobiles, trucks, and heavy equipment may be leased.

4. *Selling the feedlot and leasing back*—This method works particularly well if the plant and equipment are worth considerably more than their tax basis.

5. *Cattle clubs*—Essentially, these are corporations whose job it is to own cattle in custom feedlots.

Outside Capital in Cattle Feeding

The best facilities and equipment, the best cattle, and the best management will not make for a successful operation unless the cattle feeder establishes a "line of credit" that will provide the basic needs in sufficient quantities to promote efficiencies. Generally speaking, this calls for outside capital, from some source.

It is difficult to determine how much investor money has come into the cattle industry during the last few years. However, it is estimated that between 25 and 50 percent of the nation's cattle are owned by outside investors (investors not actively engaged in the cattle business).

During the period of rapid growth of commercial feedlots in the 1960s and early 1970s, the agricultural industry, with the help of banks, agricultural credit organizations, and insurance companies, was able to provide the facilities and equipment for expansion of cattle feeding. But they could not provide the equity capital for cattle ownership without help from outside the industry itself. Nor can they do so in the years ahead. Indeed, the financial requirements of cattle feedlots will get bigger. As a result, the capital needed to establish a new enterprise or to expand an old one will likely be greater than an individual has access to or is able to borrow. Increasingly, therefore, feedlot operators will be seeking outside capital.

The credit needs of agriculture in the United States by 1985 may be double the 1975 figure. In particular, the credit needs for larger and more complex cattle feedlots will be enormous. This need will be met through new and ingenious concepts. With it, lenders will place different requirements on borrowers. Cattle feeders will need better records, cash flow statements in greater detail, performance records for a period of years (ideally five), and future plans for proving their ability to repay comfortably.

Among the business arrangements for providing outside capital are the several that follow. (Also, see Chapter 13, section on Types of Business Organization.)

LIMITED PARTNERSHIPS

A limited partnership is an arrangement in which two or more parties supply the capital, but only one partner is involved in the management. In 1969, following the format of oil and gas drilling programs, the first limited partnership to engage solely in commercial cattle feeding was filed with the Securities and Exchange Commission (SEC). At first the brokerage community was slow to accept cattle feeding as an investment opportunity for their clients.

Although a few of the offerings of interests in feeding funds have been made by new firms without feeding facilities, most of the offerings have been sponsored by firms with substantial feeding capacity. The general partner organized by the feeding firm is responsible for making credit arrangements, buying feeder cattle, the selection and placement in feedlots and marketing of finished cattle. The accounts of the limited partners are charged a fee of $5 or $6 per head, or a percentage of gross sales, for these services. Office and administrative expenses of the general partner are also charged to partnership accounts.

During the late 1960s, when many of the large lots were being built and feeding margins were often above $20 per head, such firms apparently could obtain equity capital without tapping the national market. But with further expansion of capacity and less attractive margins, a more permanent source of equity capital than from the in-and-out investor became necessary. This need to lock in blocks of capital from outside investors appears to have been at least partly met by limited partnerships, commonly known as the feeding fund.

Although feeding funds do not offer the same kind of tax shelter as breeding herds, they do provide a mechanism for tax deferral from high to lower tax bracket years. This is possible through the prepayment of expenses for feed and the possibility of a buildup in capital invested in successive lots of cattle which can postpone taxable income for several years. The practice of obtaining equity leverage through borrowing for the purchase of both cattle and feed also makes participation in feeding funds attractive in terms of potential returns to equity. When gross margins are favorable, returns of 20 percent or more on equity can be realized. Conversely, of course, substantial losses can be incurred when margins are less favorable because of the high proportion of debt capital involved, as some investors discovered in 1974 when heavy cattle feeding losses were sustained.

In addition to normal market risks associated with fluctuations in prices of feeder cattle, feed, and market cattle, the returns to investors are directly affected by both the quality of management and the practices that management follows.

Because most general partners are wholly owned subsidiaries of firms operating feedlots, any profit from the usual markup of feed costs, yardage fees, and charges for veterinary services will accrue to the parent company. Thus, potential exists for conflicts of interest between the general partner, the limited partners they serve, and the parent company. Because these costs are not specified in advance, the investor may find it difficult to determine the probable net returns from participation in a feeding fund in comparison with contractual arrangements that he could enter into directly with a custom feedlot, or through an agent.

Most offerings of limited partnership interests specify that the investor's capital be committed for about 3 years. Several of the offerings are designed to carry out a monthly quota of placements, with the investor participating in each lot. Under this arrangement his investment builds up over a 6-month period, then if he were to terminate the plan at that point, an additional 6-7 months would be required before he could withdraw all of his capital.

Because no organized market exists for limited partnership interests, and because general partners are not required to purchase such interests, the liquidation value of each interest cannot be determined in advance. In general, the value of such interests is increased from the original investment by retained feeding profits, and reduced by losses. Thus, while a book value may be determinable at any time, the liquidation value would usually be less than this amount.

Shrewd brokers and investors are aware of the following two great appeals to cattle feeding limited partnerships:

1. *Good returns on equity capital*—On a continuous basis, for the 8 years preceding 1972, returns ran between 20 to 25 percent in well-managed lots.

2. *Tax advantage*—Tax advantages are available by use of the cash basis of accounting, which allows the deduction of expenses when paid. This provides an opportunity to create a tax loss during the first year, and to defer recapturing the tax loss as income until some future date.

Today, practically every brokerage house has a tax shelter department which offers cattle feeding, along with oil and real estate, investment programs. Also, investment advisory firms have sprung up, who, for a fee, will advise a potential investor and assist him in setting up a cattle feeding, oil and gas, or real estate tax shelter program.

A cattle feeding limited partnership, like any other limited partnership, has two kinds of partners: (1) a general partner (an individual, another partnership, or a corporation), who is the manager of the operation and has unlimited liability; and (2) the limited partner (or partners), who cannot have any voice in the management of the operation and whose liability is no greater than his investment.

Cattle feeding funds organized as limited partnerships are a very successful means of obtaining investment capital for feeding cattle. Generally speaking, the organizers and promoters of cattle feeding funds—the general partners—are of two kinds: (1) management companies that do not own feed-lot facilities; and (2) feedlot owners. Of the two, management companies must charge the higher management fee, because they place the cattle in custom feedlots from which they derive no fees. Yet, they have flexibility relative to the area or lot in which they place the cattle, which can be very advantageous. A general partner who is a feedlot owner can settle for a management fee from the partnership only large enough to cover his cost of managing the limited partnership and pay him for his risks, for his main compensation will come from custom feeding the cattle.

Limited partnerships in cattle feeding fall into three basic classifications:

1. SEC registered offerings.
2. Regulation A offerings.
3. Intrastate offerings.

SEC REGISTERED OFFERING

Where a public offering of limited partner interests in a cattle feeding fund in excess of $500,000 is to be marketed, a registration statement providing full and fair disclosure of the character of the securities must be filed with the Securities and Exchange Commission (SEC), in keeping with the Securities Act of 1933. The SEC considers selling interests in a cattle feeding fund similar to selling stock in a corporation. Thus, a prospectus, which reveals all pertinent facts of the securities offered, must be printed. The latter is used as informational matter to explain partnership operations to potential investors. Also, the services of a lawyer knowledgeable in the area of public securities offerings is necessary in preparing the prospectus, filing the registration statement, and negotiating with the underwriter. Generally speaking, an investment banker, or underwriter, is needed to market the offering—as a middleman—to bring buyers and sellers together, for which he charges a commission. Although there is nothing to keep the person or persons offering securities from marketing them privately, they usually do not have the necessary time, staff, or expertise. In selling limited partnership interests in cattle feeding funds, underwriters generally work on a best effort basis. This means that they are not obliged to market all the securities. The limited partnership interests are sold to the public at a price previously established by the feedlot company. The time involved and the cost of registering a cattle feeding fund are very considerable. Normally, registration will require 4 to 8 months; and the total cost for legal, accounting, and printing, will run anywhere from $50,000 to $150,000, depending upon the size of the offering, legal fees, and underwriter fees. Other costs of organizing a limited partnership include registering with the state in which business will be conducted. While these costs may be passed on to the purchasers of the limited partners interests, such intention must be stated in the prospectus. Moreover, there is no assurance that the offering will sell, in which case the general partner must stand all costs.

The question of whether or not a registration fee must be filed with the SEC should be answered by an attorney as there are a number of exemptions for which provision is made in the SEC regulations. It is possible that the sale of the securities may be exempt from the registration provisions of the SEC if the offering does not involve a "public offering," if the aggregate amount of the offering to the public does not exceed $500,000, or if the issue is to be sold on an intrastate basis.

REGULATION A OFFERING

Regulation A, issued by the Commission, provides for the exemption of certain classes of domestic and Canadian securities where the aggregate offering to the public does not exceed $500,000. While a registration statement need not be filed with the SEC under Regulation A, notification and reports are required. Also, the regulation requires that offering circulars containing information prescribed by the Commission must be furnished to buyers. Although a filing under Regulation A is not as difficult, time consuming, or expensive as a registration, it still involves expense and labor.

INTRASTATE OFFERING

Some persons or companies may wish to sell their stock within the confines of a state since costs, fees, and time can be saved. Quite often the savings in filing fees, printing, etc. may bring the cost down to 25 percent of the expenditures that are incurred in an SEC registration. Certain states are in a position to qualify an issue within a week if it is presented properly, as compared to the six weeks or longer that are normally required in filing with the SEC.

It is necessary to determine whether an issue is exempt from registration with the SEC under the provisions of Section 3A (11) of the Securities Exchange Act of 1933 which provides:

> Any security which is a part of an issue offered and sold only to persons resident within a single State or Territory, where the issuer of such security is a person resident and doing business within, or, if a corporation, incorporated by and doing business within such State and Territory.

State laws differ as to registration of intrastate issues. Thus, if the offering can be sold within the confines of a state, it would be well to consider with an attorney the advantages and disadvantages of this type of registration.

WHY SHOULD AN INVESTOR BE INTERESTED IN A LIMITED PARTNERSHIP?

Why should a doctor, lawyer, or other wage earner be interested in investing in a limited partnership? The following two points are pertinent to answering this question:

1. A farmer operating a farm for profit is allowed to deduct production expenses from his gross income, ultimately to arrive at the income which is taxable.

2. A wage earner or a person receiving most of his income from a salary does not have production expenses to deduct from his income. Consequently, many salaried people receive a very high income, only to have a large

amount of it taxed away. Such individuals look for types of investments where their net income position can be improved by being taxed in a lower bracket.

The following example will serve to illustrate how a limited partnership arrangement might work: individual A invests $5,000 as a limited partner in a cattle feeding operation with general partner B. In turn, general partner B invests the $5,000 for purchase of cattle and feed for the limited partner. This initial collateral enables general partner B to borrow an additional $15,000 so that A's working capital is actually $20,000, although he is not liable for the additional borrowed money. This means that limited partner A can actually have tax deduction a great deal in excess of the total initial amount of money invested. In fact, approximately 150 percent of the total initial amount invested in such an operation could be tax deductible, simply because the leveraged investment represents an immediately deductible expense to limited partner A.

If limited partner A did invest $5,000 in such a business and received a $7,500 deduction, and was 50 percent taxable, he could actually save $3,750 in taxes he would otherwise have to pay. This simply means that the $5,000 additional investment, less the tax saving of $3,750, would limit partner A's actual hard cash investment to $1,250. This situation would actually defer payment of taxes since investor A would have additional income once the cattle were sold.

Investor A might well consider, concurrently with the cattle feeding program, the investment of any profits accruing from the feeding enterprise into a breeding program. This would convert ordinary income into capital gains income due to income tax treatment of breeding stock. In addition, investor A could enter a new cattle feeding partnership in a subsequent year and invest money in this new partnership in the same manner as the previous one.

Thus, the incentive for limited partner A lies in the possibility of (1) making a profit from the cattle feeding enterprise, and (2) gaining tax advantages through deferral of income in the short run and conversion of profits into capital gains income taxed at a lower rate in the long run.

The ultimate success of limited partnership rests on the fact that it will have to make a profit in the conventional economic sense rather than rely on tax savings in order to survive.

MONEY MATTERS PERTINENT TO LIMITED PARTNERSHIPS

The following money matters are pertinent to the success of organizing and operating a limited partnership to feed cattle:

• *There should be profits, as opposed to tax deferrals only*—The Internal Revenue Service may, sooner or later, rule against allowing prepayment for storable feed as an expense in the year paid. Also, Congress may even go so far as to limit further or disallow the use of expenses incurred in an agricultural enterprise to offset nonagricultural income. Thus, cattle feeding funds should make money for investors, and should not use tax shelter as a crutch.

● *Cattle feeding funds should be of at least three years duration*—This extended period will spread the cost of organization and registration over more cattle to be fed, and greatly lessen the possibility of an investor being "soured" by one bad year.

● *Liquidity*—The liquidity must be available whereby an investor can get out early in an emergency, but it should be with enough discount to discourage investors that would get in late in one year and get out in the next year.

● *Leverage*—The capital contributed by the partners in feeding funds is usually leveraged by as much as three times or more, through obtaining additional debt financing.

EFFECTS OF LIMITED PARTNERSHIPS ON THE CATTLE INDUSTRY

Over and above providing funds for expansion of the cattle industry itself, limited partnership funds, largely from outside the industry, have had the following effects:

1. *They make for responsibility similar to that of any other publicly owned company*—The books are open to the world. Thus, it places a responsibility for effective management and results that compare favorably with other like companies.

2. *They call for business acumen*—This includes sophistication and knowledge in financing, accounting and income tax and finance reporting, and for a knowledge of the securities rules and regulations of SEC and various states and the liabilities which may be incurred if these rules and regulations are violated.

3. *They may make for higher feeder cattle and grain prices*—It appears that these public moneys have made for higher feeder cattle and feed ingredient prices, especially toward the end of the year.

4. *They may make for overproduction*—It seems reasonable that available capital and promoters may, at times, make for overproduction, particularly if a tax shelter motive, rather than a profit motive, dominates the situation.

5. *They may have prompted the Internal Revenue Service to scrutinize these tax deductions*—The promotion and use of cattle feedings funds for tax shelter purposes has caused the Internal Revenue Service to scrutinize the tax shelter advantage that goes with the use of the cash basis method of accounting, which allows the purchase of feed ahead of the time that it is actually used.

Like any other funds, limited partnership funds will be available so long as those using them produce a reasonable return on investment, and, failing to do so, they will dry up just as they did when heavy losses were encountered in 1974.

CORPORATIONS

A corporation is a device for carrying out a cattle feeding enterprise as an entity entirely distinct from the persons who are interested in and control it.

The advantages and disadvantages of corporations are presented in Chapter 13; hence, the reader is referred thereto.

Cattle feeders, and other agriculturalists, can and do use either a limited partnership or a corporation to develop and maintain an economically sound operation. Actually, no one type of business organization is best suited for all feedlots. Rather, each case must be analyzed, with the assistance of qualified specialists, to determine whether there is an advantage to using one of these types of organizations, and, if so, which organization is best suited to the proposed business.

CENTRAL MONEY MARKETS

This refers to financing cattle feeding operations by using central money markets through the medium of commercial paper in competition with other lending institutions. This concept was initiated by the Central National Bank of Chicago, through the bank's Central Ag-Finance Corp., on March 1, 1972. According to bank officials, their use of Central Money Markets was prompted because of what they saw as a developing situation in which the public agencies—Federal Intermediate Credit Banks, Production Credit Associations, and other federal and quasi-federal agencies—may not be able to provide the necessary financing for the burgeoning needs of cattle feeders. Also, commercial banks have been somewhat stymied in the past by the limit of their deposits. The answer, according to this school of thought, is to put agricultural financing on a par with commercial financing—into the Central Money Marts with commercial paper as the basis for securing money.

Central Ag-Finance Corp. (a subsidiary of Central National Bank of Chicago) is selling commercial paper in the open market, at rates competitive to the market, and then using the proceeds to make loans generally of one year or less maturity to agricultural producers. This is a new concept in agricultural financing. It may well prove to be the most desirable method of financing large commercial cattle feedlots in the future.

CONDOMINIUMS

The condominium concept in feedlots which started in Iowa in 1972 is new. It refers to a feedlot in which there is separate ownership and management of lots, but a joint ownership of part of the facilities (perhaps the feed mill, feed trucks, etc.). The concept is the same as that of condominium apartment houses; in which the owners have full and complete possession of their living quarters, but in which they may share in the heating unit, swimming pool, outside maintenance, etc.

Property taxes are leveled on the individual condominium units. By-laws are required in which procedures are specified for purposes of the overall administration and maintenance of a condominium feedlot.

The *advantages* of condominium cattle feedlots over individually owned and smaller lots are:

1. *Lower feeding costs*—Iowa State University studies show that feed-

ing 600 head of cattle instead of 100 head lowers the feeding costs $3.25 per 100 pounds of gain. Feeding 1,000 head instead of 600 lowers the cost another $.75 per 100 pounds gain.

2. *More efficient waste management*—Waste products can be managed more efficiently in larger units.

3. *Easier to adopt new feeding technology*—A larger lot can adopt new cattle feeding technology more easily than a smaller operation.

4. *They attract more buyers*—Cattle buyers will come regularly to a large feedlot, which means more competition from buyers—and could mean a higher price.

But there are *disadvantages* to a condominium! Among them, are—

1. *Capable management difficult to come by*—It takes a more able manager to conduct a big feedlot than a small one.

2. *Disease problems increased*—Disease problems are increased by having different groups of cattle in nearby pens.

3. *Group decision making can be troublesome*—Group decisions sometimes make for compromise and weakness and cause other problems among condominium feedlot owners.

More time and research is needed in order to evaluate the condominium concept. Such feedlots will not be the answer for all feeders, but large feedlots can spell extra profits if the necessary conditions exist.

Coop Owned Feedlots

Most large scale cattle feedlots are owned by individuals, partnerships, or corporations. It is expected that this will continue to be the dominant type of ownership. In 1968, cooperative feedlots fed less than one percent of the nation's cattle. However, there is increasing interest in the concept, as a means of (1) accommodating relatively small operators who wish to feed out their own cattle, without necessitating that they have cattle feedlot managerial ability or large capital investment in a feedlot and equipment; and (2) attracting outside investment capital.

ALTERNATE WAYS TO ORGANIZE A COOP FEEDLOT

Two alternatives generally exist in setting up a cooperative feedlot; namely, (1) organize it as a separate department of an existing cooperative, or (2) organize a new and distinct cooperative feedlot—start from scratch.

The first method generally reduces the capital outlay, particularly when existing feed milling and storage facilities may be used. On the other hand, such facilities may be so far removed from the feedlot that higher feed transportation costs offset any capital outlay advantage.

The organization of an entirely new cooperative feedlot allows (1) more flexibility, and (2) the opportunity to design modern feed milling and storage facilities tailor-made for the particular lot.

MEMBERSHIP

Certain requisites for membership should be established; among them

(1) that the member provide a stipulated number of cattle, thereby assuring that the feedlot will be kept full, and (2) a relatively high initial investment, thereby assuring continued support.

OPERATING POLICIES

The general principles and practices of buying, handling, and marketing cattle are much the same in both individually and cooperatively owned feed-lots; hence, only the operating policies unique to coops will be covered in the points that follow:

1. *Management*—In addition to meeting the traits of a good manager (see Chapter 13), a successful coop manager must (1) be able to work harmoniously with his members and board, and (2) be a strong leader, relying on his board for broad general policy decisions, but avoiding the hazardous pitfall of committee action, compromise, and weakness.

2. *Pooling vs individual ownership of cattle*—A decision must be reached on whether to operate the yard by pooling the cattle or by maintaining individual ownership.

The advantages of operating the feedlot on a pool basis are: (a) it requires fewer pens and eliminates under-utilization and overutilization of pens; (b) it lessens record keeping; (c) it makes it easier to group and handle small incoming lots of cattle; (d) it facilitates grouping of cattle according to weight, quality, and grade for marketing; (e) it spreads risks for members in that they become partners in a year-round marketing program, rather than seasonal; and (f) it provides greater potential bargaining power when marketing. However, pooling (a) requires that coops paying cash for pooled cattle have more capital than where individual ownership is retained; (b) may make for disgruntled members, because cattlemen are prone to feel that their cattle are better than the evaluation that the manager is willing to give them; (c) disadvantages the producer of high performing cattle; and (d) takes away pride of ownership.

For a pooling arrangement to work at its best, the cattle owned by the members must be fairly uniform in quality and performance, and the coop must be in a strong financial position if cash is to be paid at delivery time (in contrast to the book entry method). Also, some "refusal" arrangement is important. Usually, the latter works as follows: when the cattle are delivered to the feedyard, they are evaluated according to the current price. The owner is then given a choice: (1) if he accepts the price, the cattle are pooled and he receives either cash or a book entry; (2) if he refuses the price, his cattle are fed separately. Usually, a higher charge is levied where individual ownership is retained than where animals are pooled, so as to cover less efficient utilization of pen space, and other slightly higher costs in care and marketing.

3. *Charges*—Coops charge their members on different bases, just as custom feeders do. Basically, the methods of making charges are: (a) daily yardage; (b) on a straight tonnage of feed markup above feed costs; (c) a combination of feed markup and yardage; and (d) per pound of gain.

METHOD OF FINANCING

Three basic sources of capital for financing a cooperative feedlot are available; namely, (1) coop members, (2) nonmember investors, and (3) lending agencies. Sometimes funds are obtained from all three sources.

Membership fees set at a high level raise capital and encourage membership participation in the enterprise. Certificates of indebtedness and preferred stock are often attractive investments in a cooperative for both members and nonmembers. Other means of membership financing include common stock, deferred patronage refunds, and revolving funds. Local banks and banks for cooperatives are also potential sources for financing cooperatives.

CUSTOM (CONTRACT) FEEDING[6]

Custom cattle feeding is the feeding of cattle for a fee, usually without taking ownership of the animals.

Contract feeding is not new. It made rapid development after 1929, and there was much of it during the severe drought of 1934. From this time to World War II, contract feeding decreased in importance—a decline attributed to improved feed conditions on the western range, higher prices for feeder animals, and the availability of more credit through federal and private loan agencies.

Custom cattle feeding as we know it today paralleled the development of commercial feedlots. California pioneered in it; thence, it spread to Arizona, the Northwest and other areas of the West, Nebraska, Texas, the Oklahoma Panhandle, and western Kansas. Even today, these are the principal custom feeding areas. Custom feeding provided a means of financing the rapid growth of cattle numbers needed to utilize the highly mechanized, large volume feeding operations. Individuals who did not have sufficient capital to build and operate a feedlot large enough to perform economically could acquire the necessary capital and volume by custom feeding cattle for others. In this way, part of the burden of providing capital for efficient operation of a large feedlot was shifted to outside interests.

Capital requirements, periods of severe economic conditions (like scarce money and high interest), times of depressed feeder cattle prices, and adverse pasture conditions caused custom feeding to grow following World War II. These same forces, along with the need for high occupancy (full feedlots) and increased integration, have resulted in further expansion of custom feeding.

Most custom feeders have developed large, highly mechanized, and very efficient plants. Usually, they have on their staffs highly trained nutritionists who are charged with the responsibility of formulating rations and of obtaining maximum gains and feed efficiency at the lowest possible cost.

---

[6]This section was authoritatively reviewed by, and helpful suggestions were received from, the following: Dr. Willard F. Williams, Department of Agricultural Economics, Texas Tech University, Lubbock, Tex.; and Mr. Ronald R. Baker, President, C & B Livestock, Inc., Hermiston, Ore.

Through custom feeding, they sell the use of their facilities, services, and know-how to cattle owners, usually with profit to each party.

The Packers and Stockyards Administration, of the U.S. Department of Agriculture, ruled that, effective July 1, 1974, (1) packers could not own, operate, or control *custom* feedlots, and (2) *custom* feedlot owners could not own, operate, or control packing plants. This action was taken in order to avoid monopolistic conditions. This ruling does not prohibit packers from feeding their own cattle for their own slaughter needs.

The proportion of custom fed cattle to cattle owned by the feedlot varies (1) *in period of time*—it increases in times of financial stress (when cattle feeding is not profitable, money is scarce, and interest is high); (2) *according to area*—for example, there is more custom feeding in California than Colorado; (3) *according to size of feedlot*—generally speaking, the larger the feedlot, the greater the percentage of custom feeding. Some feedlots do not do any custom feeding whatsoever; others are almost wholly on a custom basis; but most lots have part of each. Feedlots that do both—those in the dual role of custom feeding and owning cattle—vary in the proportion of cattle in each category, but most of them seem to prefer about ⅔ custom fed cattle and ⅓ ownership. It's a good bread-and-butter division; in times when fed cattle lose money, such a feedlot has sufficient assured income to pay its bills.

The ownership of custom fed cattle is diverse. It includes (1) cow-calf men (farmers and ranchers) who wish to retain ownership of the cattle that they produce through the feedlot phase, (2) packers, and (3) investors, including limited partnerships, corporations, cattle buyers, cattle dealers, and others.

Provisions of a Custom Feeding Contract

Custom feeding contracts should always be detailed and in writing, for a good understanding is the best way to avoid a misunderstanding. Also, contracts should be fair to both parties—to both the feedlot owner and the cattle owner.

The experience of feedlot owners and cattle owners, and the difficulties encountered, suggest that custom feeding contracts should include provision for the following:

● *Ration*—Some basics about the rations—such as the different rations to be used when getting the cattle on full feed, proportion of concentration to roughage, and energy and protein content—should be spelled out in the contract. Yet, the feedlot operator should be permitted flexibility, so that he can take advantage of price changes in ingredients, etc.

● *Veterinary expenses*—Veterinary expenses, medication, dehorning, castrating, and dipping charges should be specified. They will vary with the age of the cattle (calves require more attention than older cattle), the preconditioning, if any, etc. Usually, the assessment for veterinary expenses runs from $1.50 to $3.00 per head.

● *Responsibility for death losses*—The contract should specify responsibility for death losses. The cattle owner normally assumes all losses prior to the arrival of the cattle at the feedlot. After a reasonable time at the feedlot, losses may be the fault of the feedlot; hence, it may logically be expected to assume a share of the loss. Usually the partitioning of death losses is accomplished by (1) the cattle owner standing loss of the initial cost of the feeder cattle delivered to the lot, and (2) the feedlot operator canceling out all or part of his charges for yardage, feed, and/or cost of gain from the time the custom fed cattle arrive in the feedlot until the time any death losses may occur.

● *Buying and selling services*—The buying and selling services provided by the feedlot are important to the success of the operation. In most cases, the feedlot is in the best position to sell the finished cattle. But this responsibility should be spelled out in the contract, along with any fee to be charged for the services.

● *Right to reject poor-doing cattle*—A feedlot using a payment-for-gain contract should reserve the right to reject poor-doing cattle. Of course, where the feedlot purchases the cattle for the owner, he accepts the responsibility to get good-doing cattle.

● *Power of attorney for feedyard operator*—From a legal standpoint, feedyards should have power of attorney for buying, selling, and borrowing for customers.

● *Arbitration*—In cases of dispute and disagreement over a custom feeding contract that cannot be resolved by the owner of the cattle and the feedlot, the contract should provide for arbitration to be conducted by a committee of three—each party to the contract choosing a representative and these two then choosing a third party, to study the case and recommend settlement.

Types of Custom Feeding Contracts

The services rendered vary from feedlot to feedlot and according to the type of contract. In some instances, the services may be so complete that the customer never sees the cattle. The feedlot operator may buy the feeder cattle, feed them, market them, and send the customer (his client) a check for the balance, after deducting input costs, interest charges, and custom-feeding charges. Less complete services are usually available to suit the customer.

Both the feedlot owner and the cattle owner should analyze different types of contracts and determine which best fits their respective circumstances. Some feedlots offer several types of contracts, thereby according the cattle owner a choice.

Competition may dictate the type of contract and the changes made. But by knowing the variables and managing them correctly, the feedlot owner can write and carry out a contract that will be fair to himself and his customer.

Generally speaking, contracts with fixed charges are the most satisfactory and the most common, primarily because there is less room for misunderstanding.

Although there are many types of custom cattle feeding contracts, and many variations of each kind exist, most of them can be classified under one of the following types:

• *Feed cost plus daily yardage fee per head*—This type of contract is based on the cost of feed plus an additional 5 to 9 cents per head per day to cover handling, yardage, feed grinding, and similar expenses. Generally, an additional charge of $1.50 to $3.00 per head is made to cover medication, vaccination, branding, dehorning, and dipping. The customer finances purchase of the cattle.

With this type of contract, the feedlot does not assume any risk whatsoever. It is merely their intent to sell feed, facilities, and services at an agreed price. However, it does not permit the feedlot owner to participate in the greater revenue generated by high performing animals as does a margin based on markup on feed costs.

• *Feed cost plus markup*—This type of contract calls for reimbursement on cost of feed plus a feed markup on either (1) a flat rate, (2) a percent of cost, or (3) a percent of moisture added in steam processing.

With a flat rate markup, a $4 to $11 charge per ton above feed cost is made to cover feed handling, grinding, and labor costs. An additional assessment is made to cover veterinary services and medication; and the customer finances the purchase of cattle. Since actual feed milling costs (for labor, power, insurance, mill maintenance, etc.) run $3 to $6 per ton, profit to the feedlot accrues from having a higher markup than the milling cost. It should be noted that markup on such high moisture feeds as silage and wet beet pulp is generally computed on a dry basis. On a wet basis, the markup on silage is usually figured at about $3 per ton. A flat markup per ton of feed favors the feedlot when prices fall, and the cattle owner when feed prices rise.

Also, markup on feed may be on a percentage of cost basis. With this arrangement, higher feed costs favor the feedlot, whereas lower feed costs reduce the actual return to the feedlot.

Markup on feed is sometimes through the addition of steam in processing. Where this method is used, the feedlot must either add enough steam to generate returns comparable to other markup systems or increase the daily yardage fee.

Any system of feed markup will be more profitable to the feedlot with heavy, high-performing cattle than with light slow-gaining cattle, simply because the former eat more.

With the "feed cost plus markup contract," the feedlot is essentially a feed manufacturer processing and delivering feed to its customers—the owners of the cattle.

• *Feed cost plus (1) daily yardage fee per head and (2) markup per ton of feed*—This is a combination of the first two types. Those feedlots that charge the higher yardage rates per head daily add a smaller markup per ton of feed above actual ingredient prices; conversely, those that charge the lower yardage rates per head per day make a higher feed charge over and above actual ingredient cost. As a rule of thumb, feed markup is generally lowered by $1 per ton for each one cent per head per day of yardage charged. At the present time, it appears that the lower yardage cost and the higher feed markup is the favored basis; primarily because (1) the owner of the cattle is less inclined to object to such charges, and (2) increased competition has driven custom feeders to make their charges on the least conspicuous basis.

• *Agreement to purchase contract*—In this plan the cattle feedlot operator buys the feeder cattle in his own name, usually with the client required to make a down payment of 20 to 30 percent of the purchase price. The client then executes an agreement to buy the cattle when they are ready for slaughter, including the original purchase price of the feeder cattle (less any down payment made) and all feeding, handling, and interest charges. (Interest charges are tax deductible.)

There are several variations of this type of contract. But all of them are much like buying commodities on the Chicago Mercantile Exchange.

• *Payment for weight gained*—This plan is based on a charge per pound of gain. In this arrangement, the feeder is reimbursed on the basis of the gain in weight put on the cattle, at an agreed price of so many dollars per hundred. This type of contract has decreased in importance in recent years, because it frequently results in poor owner-feeder relations—due primarily to the following reasons: (1) it is impossible in advance of feeding to detect those lots of animals that will be "poor doers" because of such factors as nervousness, nutritional deficiencies, diseases, and/or parasites; (2) the longer the feeding period, the greater the cost of gains, and the length of the feeding period is seldom stipulated in such contracts; (3) because weather and disease, which are uncontrollable, affect rate of gain; and (4) the amount of fill or shrinkage when weighing animals in and out the feedlot is of great importance to both the owner and the feeder, and a source of argument. Also, a major disadvantage of using a payment-for-gain system is that it does not adjust for feed costs; the feeder must absorb any increase in feed cost. However, it does allow a feedlot to take advantage of opportunity feeds— such as down corn, or grass which cannot be delivered to the feedlot. Also, the small feedlot operator who does not have scales may use this type of custom contract.

Payment for weight gained may be used in growing and backgrounding operations, where it may be desirable to specify both the minimum and maximum rate of gain. Also, it may be used on cattle that are pastured for a time before being sent to the feedlot.

• *The incentive basis contract*—This is another system of charging on a payment-for-gain basis that some cattle owners like, because it gives an incentive for the feedlot to produce rapid daily gains. It consists in paying the feedlot for all feed plus a charge arrived at by "multiplying the average daily gain times itself, or times some factor." Thus, if the cattle being finished should average 2½ pounds gain daily over the entire feeding period, the per-head per-day basis of payment to the feeder would be as follows:

2½ (gain) × 2½[7]=6¼ cents per head per day. If the average daily gain

---

[7]This factor might be varied in keeping with economic conditions. There is no reason why it must be identical to the average daily gains. It may be higher or lower according to economic conditions.

per head over the entire period is 2 pounds, then the basis of payment would be as follows:

2 (gain) $\times$ 2^8=4 cents per head per day.

The incentive basis isn't desirable from the standpoint of the feedlot owner if the cattle are "poor doers" or if, because of disease or other factors beyond his control, poor gains are obtained. For understandable reasons, however, this type of contract does have a very strong appeal to the cattle owner.

Operation Procedures

The operation procedures followed by feedlots that custom feed vary. In particular there should be an understanding between the feedlot and the customer relative to billing and payment of feed costs, selling arrangement, and payment for slaughter cattle.

• *Billing for feed or gain costs*—Most feedlots bill their clients for feed or gain costs either on the 1st and 15th of each month, at the end of the month, or at the end of the feeding period. The more frequent the billing, the smaller the amount of short-term capital required by the feedlot. Where feedlots carry feed costs longer than a month, they usually charge interest.

The vast majority of custom feedlots worthy of the name are equipped with platform or hopper scales and/or scale trucks with electric or mechanical scales mounted under the feeding box; hence, billings are based on actual weights. Occasionally, a farmer without scales will feed a few cattle on a custom basis. In the latter case, the feedlot owner and the cattle owner may agree to (1) compute feed costs on the basis of 3 percent of the incoming body weight of the cattle plus 7 pounds per day (for example, the daily feed consumption of a steer weighing 700 pounds at the time of delivery to the yard would be estimated at 28 pounds); or (2) feed on a cost-of-gain basis, with payment delayed until the cattle are weighed and marketed at the end of the feeding period.

• *Selling arrangement*—Feedlot managers generally handle the selling of custom fed cattle in the same manner, and through the same channels, as cattle which the feedlot owns. Feedlot managers, or their representatives, are usually in a better position to estimate the weights and grades of cattle on feed than are their clients. Also, they are more familiar with the type and quality of fed cattle desired by various packer buyers.

• *Payment for custom fed cattle*—Market payments for custom finished cattle may be made either directly to the owner of the cattle or to the feedlot, depending on prior arrangements between the feedlot and the client. However, commercial banks and other lending institutions generally retain a first lien on the client's cattle; and this must be satisfied. Likewise, finance agencies generally provide the necessary financing for feed and other custom feeding charges; hence, the feedlot operator ordinarily is assured of receiving full payment for feed bills and other services. In the event the client has outstanding bills with the custom feeder, feedlots handling payments for

---

[8]Same as footnote 7.

their clients are permitted to retain sufficient funds to satisfy these debts, after satisfying the first mortgage holder.

What Cattle Owners Expect of Custom Feedlots

The owner who entrusts his valuable cattle to a feedlot for custom feeding rightfully expects certain things of the feedlot, chief of which are:

● *Profits*—Cattle owners assume considerable risk; hence, they expect profits commensurate therewith. They will contract with the feedlot which consistently returns the most on their investment.

● *Cattle feeding know-how, honesty, and integrity*—Cattle owners will not, knowingly, place cattle in a custom feedlot under inexperienced management. Also, because cattle owners are not involved in day-to-day management, the honesty and integrity of the custom feedlot management are very important to them.

● *Progress reports*—Cattle owners wish to keep informed of how well their investment is doing. To this end, they expect monthly progress reports giving such pertinent information as (1) rate of gain, (2) feed efficiency, (3) cost of gains, and (4) sickness and death losses.

● *Courteous customer treatment*—Like any other customer, a custom cattle owner likes to feel that the feedlot wants and appreciates his business.

● *Competitiveness*—The cattle owner expects the feedlot in which he places cattle for custom feeding to be competitive in charges and performance with other feedlots of the area.

● *Satisfactory financial position*—The customer expects that the feedlot with which he does business be in sufficiently strong financial position that it buys feeds when it is most advantageous (usually at harvest), and that it pays its bills regularly.

● *Knowledge of the financial position of packer buyers*—The cattle owner expects that the feedlot owner know the financial position of the packer buyer to whom his cattle are sold, so that there can be no concern relative to payment. A few cattle owners who custom feed even require that the feedlot handling the sale of the finished cattle guarantee payment for them.

How to Attract Custom Feeders

Success in attracting customers depends on reputation and performance. Reputation will attract new customers, but only performance will hold them.

If a feedlot relies on custom fed cattle for a certain percentage of its capacity, it's important that there be enough clients with sufficient cattle to keep the lots filled, or nearly so, at all times.

● *Be successful*—The best way to recruit and retain customers is to be successful. After a customer has used the services of a given feedlot, he will know what it can do. Until then, the feedlot manager will have to convince the prospective customer of his ability. Among the tools which the manager may use in proving his ability are records and computers.

● *A complete set of records*—A good set of records will allow the feedlot manager to predict with confidence what he can do for the prospective custom feeding client in rate, efficiency, and cost of gain; in disease and death losses; and in buying and selling cattle.

• *A computerized system*—Prospective clients will be impressed with the sophistication of a computerized accounting system and a computerized closeout statement. It indicates the feedlot's capability of complete, accurate, and prompt records. In addition to being a good means of keeping customers well informed, a computerized system may be used to provide them with an accurate and complete set of records for accounting purposes. Also, records are useful to the feedlot, because it provides periodic analysis of progress and allows the manager to monitor the feeding program for needed changes.

• *Financing cattle and feed fed on a custom basis*—Financing helps to attract clients. Since feedlot operators are under pressure to keep their feedlots filled, financing of cattle and feed by custom operators will probably continue to be an important source of funds to those who place cattle in these lots.

Commercial banks are the primary source of financing for cattle fed on a custom basis. They generally require a margin equivalent to 25 to 30 percent of the value of the feeder cattle. In addition, banks make loans to cover feeding charges. It is not uncommon for banks to finance 80 percent of the feeder cattle price plus all of the feeding charges. Depending on the reputation of the feedlot operator and the custom feeder, banks and other lending institutions may secure only the cattle as collateral for the loan. They may also specify that feeder cattle be hedged on a futures market before negotiating loans, although this has not been general practice to date. Some lending agencies will even advance funds for the margin on cattle futures contracts, as part of a feeder cattle loan.

Also, some custom feedlots finance cattle purchases in their own names, with the client executing an "agreement to purchase" the cattle when they are sold at a cost equal to the initial purchase price of the feeder plus all feeding, handling, and interest charges.

• *Bring prospective feeding customers together*—The feedlot manager may attract some customers by bringing them together, especially those who have insufficient funds to carry on a continuous feeding program by themselves. The feedlot manager can provide a real service by assisting them in the formation of clubs or corporations whose pooled resources are large enough to feed cattle continuously.

COMPUTERS IN CATTLE FEEDLOTS

Ration formulation is only one use of computers in cattle feedlots. As every cattleman knows, many management decisions are involved in obtaining the highest net returns from a given lot of cattle; and the bigger and the more complex the feedlot, the more important it is that the decisions be right.

Problems can be solved without the aid of a computer. But the machine has the distinct advantages of (1) speed, (2) coming up with answers to each of several problems simultaneously, and (3) offering the best single alternative, all factors considered. Among the possible uses of computers in cattle feedlots, particularly the largest operations, are:

1. Ration formulation (see Chapter 8).
2. How to determine the best ingredient buy.
3. The most profitable ration—the one that costs the least per pound of gain produced. In many cases, the cheapest ration will actually increase cost of gain. Also, rations that produce the most rapid gains may be too costly; for example, silage rations will not produce as high gains as an all-concentrate ration, yet the net profit from their use may be greater.
4. The most profitable kind of cattle—age, weight, sex, and grade—to feed in relation to available feeds and feed prices. (See Chapter 29, Kind of Cattle to Feed.)
5. Seasonal differences in performance of cattle in a given feedlot.
6. As a means of forecasting profits or losses; with all-at-once consideration given to feeder prices, slaughter cattle outlook, probable rate of gain, and interest and overhead costs.
7. As a means of keeping and updating the voluminous daily feed transfers from feed inventories, to mixed rations, to records for each lot of cattle. Accurate and current feed inventories are necessary for wise ingredient buying and for financing feed inventories; and accurate and current feed records by lots are important for both privately owned and custom fed cattle.

AN INCENTIVE BASIS FOR THE HELP

An incentive basis for cattle feedlot help is needed for motivation purposes, just as it is in cow-calf operations. It is the most effective way in which to lessen absenteeism, poor processing and mixing of feeds, irregular and careless feeding, unsanitary troughs and water, sickness, shrinkage, and other profit-sapping factors.

The incentive basis for cattle feedlots shown in Table 27-14 is simple, direct, and easily applied.

Whenever possible, the break-even points—(1) pounds feed/pound gain, and (2) daily rate of gain—should be arrived at from actual records accumulated by the specific feedlot, preferably over a period of years. Perhaps, too, they should be moving averages, based on 5 to 10 years, with older years dropped out and more recent years added from time to time; thereby reflecting improvements in efficiency and rate of gain due primarily to changing technology, rather than to the efforts of the caretakers.

With a new feedlot, on which there are no historical records from which to arrive at break-even points of feed efficiency and rate of gain, it is recommended that the figures of other similar feedlots be used at the outset. These can be revised as actual records on the specific feedlot become available. It is important, however, that the new feedlot start an incentive basis, even though the break-even points must be arbitrarily assumed at the time.

Because of the high correlation between feed efficiency and rate of gain, the incentive basis recommended in Table 27-14 does result in an overlapping of measures. Nevertheless, both efficiency and rate of gain are impor-

TABLE 27-14

A PROPOSED INCENTIVE BASIS FOR CATTLE FEEDLOTS

Feed/Lb Gain	Daily Rate of Gain	Here's How It Works
(lb)	*(lb)*	
12.0	0.5	On this particular cattle finishing operation, the break-
11.5	0.75	even points are assumed to be (1) 8.5 lb of feed/lb gain,
11.0	1.00	and (2) 2.25 lb daily gain. Of course, for cattle on
10.5	1.25	grower rations, the break-even points would be differ-
10.0	1.50	ent; for example, feed efficiency might be 10.0 and the
9.5	1.75	daily rate of gain 1 lb; also, on grower rations an upper
9.0	2.00	limit on daily gains could be set, beyond which there
8.5 (Break-even point)	2.25	would be no bonus benefits.
———	———	Feed saved and gains made in excess of the break-even
8.0	2.50	points are computed at going prices.
7.5	2.75	
7.0	3.00	If the feed efficiency drops to 8.0 and the gain increases
6.5	3.25	to 2.5; and if feed costs $80 per ton and cattle are worth
6.0	3.50	40 cents per lb, then these feed savings and increased
5.5	3.75	gains are worth——

Feed saved (lb)	Cost of feed/lb	Value of feed saved/ lb gain	Value of feed saved on 2.5 lb gain
0.5	4.0¢	2.0¢	5.0¢

Gains made (lb)	Per lb mkt. value of gains	Value of incr. daily gain
0.25	50¢	12.5¢

Increased profit/head/day:
(cents)

Feed	5.0
Gain	12.5
Total	17.5

On steers fed for 150 days, that's $26.25/head. With 500 steers, that's $13,125 total. When divided on an 80:20 basis, that's $10,500 for the owner and $2,625 for the employees.

tant profit indicators to cattlemen. Because of the overlapping, however, some may prefer to choose one or the other of the measures, rather than both.

Another incentive basis followed in a few large feedlots consists of the following: a certain percent (say 15 percent) of the net earnings set aside in a trust account, which is divided among and applied to the account of each employee according to salary and/or length of service, and paid to employees upon retirement or after a specified period of years. The main disadvantages to this incentive basis are that there may not be any net some years, that some employees do not want to wait that long for their added compensation, and that it opens up the books of the business.

FUTURES TRADING IN FINISHED CATTLE, FEEDER CATTLE, AND FEED[9]

The three big uncertainties in the cattle feeding business, any one of which can cause a cattle feeder to suffer heavy losses, are prices of (1) feeder cattle, (2) feed, and (3) finished cattle. Through futures contracts, the cattle feeder can now hedge all three. In advance of feeding, he can lock in his price of feeder cattle, feed, and finished cattle.

Fig. 27-3. Over-all view of the trading floor of the Chicago Mercantile Exchange. (Courtesy, Chicago Mercantile Exchange)

This discussion is devoted primarily to live (slaughter) beef cattle futures as they apply to cattle feedlot operators, because it is the highest risk phase of the cattle business, as well as the least flexible. Unless a feeder contracts ahead, he has no assurance of what his finished cattle will bring when they are ready to go. Moreover, there is little flexibility in market time, for the reason that excess finish is costly and unwanted by the consumer. As a result of this uncertainty of market price, and in realization of the high risks involved, sleepless nights are rather commonplace among cattle feeders; they find it difficult to concentrate on the business at hand—the efficient feeding and management of cattle. Live (slaughter) beef cattle futures provide a means through which a cattle feeder can fix his selling price at the outset of the feeding period.

The second major item of the triumvirate making for uncertainties in cattle feeding is the price of feeder cattle. Only by contracting ahead, can the

---

[9]This entire section on beef futures was authoritatively reviewed by and helpful suggestions were received from the following: Mr. Everett B. Harris, President, Chicago Mercantile Exchange, Chicago, Ill.; Mr. Owen S. Wirak, Extension Economist, Washington State University, Pullman, Wash.; and Mr. W. Fred Benson, Cattle Analyst, Merrill Lynch, Pierce, Fenner & Smith Inc., New York, N.Y. Also, most of the examples of how a futures contract works were taken from literature provided by the Chicago Mercantile Exchange.

cattle feeder be sure of the price that he will have to pay when he is ready to lay in feeder cattle. For many years, a fairly effective, albeit unorganized, cash contracting system has been operating relative to feeder cattle. Feeder cattle futures now offer, on an organized basis, a method for cattle feeders to lock in the price of feeder cattle well ahead of taking delivery, thereby alleviating possible heavy losses due to sharp price rises of feeder cattle. Without feeder cattle, a feedlot is not in business. Yet, much of the overhead cost for facilities and staff continues. Hence, a full feedlot is important. The cow-calf man—the producer of feeder cattle—has more flexibility, and is less dependent on contracting ahead, than the cattle feeder. If the feeder cattle market isn't good, he can hold his calf crop for a time; he may even carry them over for another year—to the yearling stage. Also, rather than accept what he considers to be an unfavorable price for his stockers, he can have them custom fed, or he can feed them out himself. Certainly, there are risks in the cow-calf business, but, in comparison with cattle feeding, there is more flexibility, and the timing is not so exacting.

Since feed represents such a large proportion of the cost of feeding cattle (amounting to approximately 80 percent of the costs exclusive of the purchase price of the feeder cattle), it is wise to set the price months in advance whenever possible. Usually, feed can be bought most advantageously at harvest time. Thus, the cattle feedlot owner who has adequate storage and finances generally buys his main feed ingredients at that time. By so doing, he can project with reasonable accuracy what it will cost him to feed cattle. Grain sorghum (milo), corn, and soybean meal futures permit the cattle feeder to accomplish the same thing without actually taking delivery on the feed and incurring storage costs and risks of physical deterioration. The cattle feeder can use such futures to protect against increases in feed prices.

It should be noted that beef carcass futures are also traded. But these are of primary interest to packers and wholesale meat dealers, rather than cattlemen; hence, they will not be covered herein. Nevertheless, the same principles of other commodities apply to carcass futures.

What Is Futures Trading?

Futures trading is not new. It is a well-accepted, century-old procedure used in many commodities; for protecting profits, stabilizing prices, and smoothing out the flow of merchandise. For example, it has long been an integral part of the grain industry; grain elevators, flour millers, feed manufacturers, and others, have used it to protect themselves against losses due to price fluctuations. Also, a number of livestock products—hides, tallow, frozen pork bellies, and hams—were traded on the futures market before the advent of beef futures. Many of these operators prefer to forego the possibility of making a high speculative profit in favor of earning a normal margin or service charge through efficient operation of their business. They look to futures markets to provide (1) an insurance medium in the marketing field, and (2) the facilities and machinery for underwriting price risks.

A commodity exchange is a place where buyers and sellers meet on an organized market and transact business on paper, without the physical presence of the commodity. The exchange neither buys nor sells; rather, it provides the facilities, establishes rules, serves as a clearing house, holds the margin money deposited by both buyers and sellers, and guarantees delivery on all contracts. Buyers and sellers either trade on their own account or are represented by brokerage firms.

The unique characteristics of futures markets is that trading is in terms of contracts to deliver or to take delivery, rather than on the immediate transfer of the physical commodity. In practice, however, very few contracts are held until the delivery date. The vast majority of them are cancelled by offsetting transactions made before the delivery date.

Many cow-calf men have long forward contracted their calves for future delivery without the medium of an exchange. They contract to sell and deliver to a buyer a certain number and kind of calves at an agreed upon price and place. Hence, the risk of loss from a decrease in price after the contract is shifted to the buyer; and, by the same token, the seller foregoes the possibility of a price rise. In reality, such contracting is a form of futures trading. Unlike futures trading on an exchange, however, actual delivery of the cattle is a must. Also, such privately arranged contracts are not always available, the terms may not be acceptable, and the only recourse to default on the contract is a lawsuit. By contract, futures contracts are readily available and easily disposed.

Magnitude of Livestock Futures Trading

Livestock futures trading is relatively new. It was not until November 1964 that trading in live cattle futures opened on the Chicago Mercantile Exchange. Trading in live hogs began 15 months later. Since then, livestock futures trading has grown enormously. Between 1965 and 1972, the volume of cattle futures increased from 2.7 to 38.6 billion pounds. Trading in hog futures went from 71 million pounds in 1965 to almost 11 billion pounds in 1972. In terms of dollar volume, livestock futures trading is now more than $16 billion annual business, with no sign of a letup. Despite the growth and magnitude of futures, it represents only a small fraction of fed cattle in the U.S.; in 1969, it represented less than 5 percent of the cattle on feed. As feedlots increase in size and margins become smaller, futures trading will increase.

What Is a Futures Contract?

A futures contract is a standardized, legal, binding paper transaction in which the seller promises to make delivery and the buyer promises to take delivery of a specified quantity and type of a commodity at a specified location(s) during a specified future month. The buying and selling are done through a third party (the exchange clearing member) so that the buyer and seller remain anonymous; the validity of the contract is guaranteed by reputable and well-financed exchange clearing members; and either buyer or seller can readily liquidate his position by simply offsetting sale or purchase.

The Chicago Mercantile Exchange specifications of finished cattle, feeder cattle, and grains sorghum contracts follow:

● *Specifications for a live (slaughter) cattle contract are*—Delivery and acceptance of 40,000 pounds of steers (approximately 37 head) Choice grade and yield grades 1, 2, or 3, within the weight range of 1,050 to 1,125.5 pounds, and hot weight yielding 62 percent, or within the weight range of 1,125.5 to 1,200 pounds, and hot weight yielding 63 percent; stated discounts and tolerances including substitutions in estimated grade, weight, yield, fat thickness, and other details; and delivery to Peoria, Omaha, and Sioux City, with Guymon, Oklahoma as an alternative at $1 less per cwt.

● *Specifications for a feeder cattle contract are*—Delivery unit of 42,000 pounds of feeder steers averaging 650 pounds (approximately 65 head) of which 80 percent must be Choice or better and 20 percent Good, within a weight range of 50 pounds above and 50 pounds below the average weight of unit; set allowances including substitutions on estimated grade and weight; and delivery at Omaha, Nebraska and Sioux City, Iowa, with a $50 discount at St. Paul, Minnesota; Springfield, Missouri; Greeley, Colorado; Kansas City, Missouri; St. Joseph, Missouri; Emporia, Kansas; Wichita, Kansas; Oklahoma City, Oklahoma; and Amarillo, Texas; and with $.75 discount at Billings, Montana.

● *Specifications for a grain sorghum (milo) contract are*—Trading unit is 400,000 pounds (7142.9 bushels) of No. 2 yellow sorghum (milo) as described by the Official Grain Standards of the United States. A one cent per cwt discount is made on delivery units that are graded No. 3 yellow sorghum plus further specified discounts for test weight and moisture content.

Commission Fees and Margin Requirements on Cattle Contracts

The commission fee on both finished cattle and feeder cattle contracts, covering both purchase and sale (called a round turn), is $40. In August, 1974, the minimum margin on live cattle was $900; the minimum hedge margin was $1,000; and the speculative margin was $1,200. On feeders, the speculative margin is $900 and the hedge margin is $600. The margin deposit may be increased by the broker if the value of the contract should change unfavorably.

The $40/contract commission fee on 37 head makes for $1.08 commission charge per head on the live (slaughter) cattle contract, and $0.62 per head on the feeder cattle contract.

How Does a Futures Contract Work?

Traditionally, futures contracts have been used for two purposes: (1) hedging; and (2) speculating.

Hedging is an offsetting transaction by which purchases or sales of a commodity in the cash market are counterbalanced by sales or purchases of an equivalent quantity of futures contracts in the same commodity.

Speculating is risk-taking by anyone who hopes to make a profit in the advances or declines in the price of the futures contract.

Examples of futures hedging by each (1) a cattle feeder, (2) a cow-calf man, (3) a packer, (4) a cattle feeder using a long hedge to protect the price of feeder cattle replacements at the time he forward contracts finished cattle, and (5) a cattle feeder hedging on grain sorghum follow. These illustrate hedging procedures, although it must be borne in mind that in actual application the hedges may not work out as perfectly as these.

● *Example I—A cattle feeder hedging to lock in price (see Table 27-15)*—It is now November, and the cattle feeder has just purchased his feeder cattle to place in the feedlot. Based on past experience, he is quite confident that these cattle should be ready for market the following April. Through good record keeping, he is also quite confident that his production (including labor) and marketing costs should be about $40.50/cwt.

TABLE 27-15

EXAMPLE OF A CATTLE FEEDER USING A SHORT HEDGE TO LOCK IN A PRICE

Cash Market		Futures Market		Basis
	Per Cwt		*Per Cwt*	*Per Cwt*
Nov. 15:				
Expected production costs of	$40.50	Sells April futures at or localized price of $41.25/cwt	$43.25	$2.75
April 10:				
Sells cattle on cash market at	$37.50	Buys April futures at or localized price of $37.50/cwt	$39.50	$2.00
Loss	$ 3.00	Gain	$ 3.75	Profit $0.75

He decides to hedge his cattle with the April futures contract which at the time is selling for $43.25/cwt. He has also estimated his hedging costs and basis—that is, costs of transportation, shrinkage, brokerage fees, interest, and other marketing expenses—to be about $2/cwt. He subtracts this figure from the April futures price and gets his localized futures price of $41.25/cwt or an estimated $.75/cwt profit; hence, he sells April futures.

In this example, the estimated profit of $.75/cwt was realized by the feeder because he had correctly calculated the basis.

The cattle feeder sold his finished cattle on the cash market for $37.50/cwt which, after subtracting his production costs of $40.50/cwt gives him a loss of $3/cwt. However, he realized a profit of $3.75/cwt on his futures transaction, so that his total profit was $.75/cwt.

This example illustrates what could happen on a declining market. The feeder still showed a profit, even though he had to sell his cattle in the cash market for a price lower than his production costs because this loss was offset by a larger profit in the futures market. This is true because, as the cash price declined, the futures prices also declined.

If, however, the cash and futures prices had risen, he still could have made a profit; this time in the cash market. But because of a loss in the futures market, his total profit would have been less than had he not hedged. Nevertheless, he still received the price protection he desired, which was his main purpose in hedging.

● *Example II—A cow-calf man using a short hedge (see Table 27-16)—*

During April, as his calves are being born, a rancher decides to hedge these calves on a feeder cattle contract. Most of the calves will be sold as feeders during October. Through experience he has estimated that it costs him $48/cwt to produce these feeders. The futures market is showing October feeder cattle at $51.25/cwt or at a localized price of $50.75/cwt. The rancher feels that this assures him of a reasonable profit; hence, he sells October futures.

TABLE 27-16

EXAMPLE OF A COW-CALF MAN HEDGING TO LOCK IN A PRICE

Cash Market		Futures Market		Basis	
	Per Cwt		Per Cwt		Per Cwt
April 25:					
Expected production costs of	$48.00	Sells October futures at or localized price of $50.75/cwt	$51.25		$3.25
Oct. 10:					
Sells feeder cattle on cash market at	$48.50	Buys Oct. futures at or localized price of $48.50/cwt	$49.00		$0.50
Gain	$ 0.50	Gain	$ 2.25	Profit	$2.75

Even though the cash market was not as strong as the rancher had hoped, he was still able to realize the profit he wanted because of his hedge.

● *Example III—A meat packer using a long hedge (see Table 27-17)*— The above examples are illustrations of short hedges. The following example will be of a long hedge, where the futures contract is bought.

The meat packer has determined, from his basis charts, that the February cattle futures are normally $1/cwt above the local cash price in February. This then assures him of the maximum amount that he will have to pay for cattle in February—that is, if the basis does narrow to $1 during February, the cost of the slaughter cattle will be $1 below the futures price in February.

The meat packer has also determined that the most he can pay for the slaughter cattle and still make a profit is $38/cwt. He is confident that the basis will be $1 in February.

TABLE 27-17

EXAMPLE OF A PACKER USING A LONG HEDGE

Cash Market		Futures Market		Basis	
	Per Cwt		Per Cwt		Per Cwt
Sept. 27:					
Expected cash price based on past basis relationships	$38.00	Buys amount needed of Feb. futures at	$37.80	−	$0.20
Feb. 20:					
Buys slaughter cattle at	$44.00	Sells Feb. futures at	$45.00	+	$1.00
Loss	$ 6.00	Gain	$ 7.20	Profit	$1.20

The profit in the futures market of $7.20/cwt, when applied to the higher than expected cash prices, assured the meat packer that he could purchase the slaughter cattle at a price that allowed him to protect his profit margin.

●*Example IV—A cattle feeder hedging on feeder replacements (see Table 27-18)*—It is not uncommon for a feeder to contract slaughter cattle for future delivery at a set price to a packer before he has acquired the necessary feeder cattle. If by the time of purchase the price of feeders increases beyond the feeder's expectation, his feeding margin may be substantially reduced or he may even suffer a loss on the contract.

The futures contract in feeder cattle can be used in a long hedge to reduce or eliminate the risk involved in an adverse movement of feeder cattle prices. At the time that he negotiates a forward contract with the packer, the feeder would buy feeder futures contracts, preferably for the month in which he actually planned to buy the feeders. If the price of the futures followed local market prices, the feeder would not care what happened to the level of feeder cattle prices because any loss or gain in the cash market would be offset by an opposite outcome in futures.

The arithmetic of this particular long hedge is illustrated in Table 27-18.

TABLE 27-18

A LONG HEDGE TO PROTECT FEEDING MARGIN
ON FORWARD CONTRACTED FED CATTLE

Cash Market		Futures Market		Basis
	Per Cwt		*Per Cwt*	*Per Cwt*
Jan. 15: Contracts slaughter cattle based on feeder cost of	$37.60	Buys Feb. futures	$36.75	$0.85 over
Feb. 15: Buys feeder cattle	$39.50	Sells Feb. futures	$39.15	$0.35 over
Loss	$ 1.90	Gain	$ 2.40	Profit $0.50

If the feeder had not hedged, he would have had to pay $1.90/cwt more for his feeder cattle than he had planned. By hedging he paid $37.10/cwt ($37.60 - $.50 basis gain; or $39.50 - $2.40 profit on futures transaction), plus hedging costs, or about what he based his contract price on in January.

It is advisable to place such a hedge in the futures contract month in which the cattle will be purchased because the cash-futures relationship is more predictable at this time. Also, in the event that local feeder cattle prices advanced substantially more than futures, the feeder could stand for delivery of the feeder cattle at the Chicago futures price quotation.

A long hedge to fix the price of feeder cattle may be used in lieu of forward contracting to fix or cheapen the price of feeder cattle needed for replacements several weeks or months in advance of actual purchase. However, the same businesslike procedures are required to get a good buy in feeder futures as in forward contracting. The most likely time to buy futures is when the particular contract in which the feeder is interested is favorably priced relative to local feeder prices after allowing for costs of actually taking delivery.

If futures and cash prices move together closely, the feeder can buy cattle locally when he needs them at about the price prevailing when he placed the long hedge. If futures advances relative to local cash, the feeder can buy cattle as needed and cheapen their cost by the gain in the futures transaction. If the futures declines relative to cash, the feeder has the option of taking delivery of his cattle under the futures contract at the original purchase price.

• *Example V—A cattle feeder hedging on grain sorghum (see Tables 27-19 and 27-20)*—Grain sorghum is herein used as the example of hedging a feed ingredient because 85 percent of it is used in the production of meat, primarily beef. It should be noted, however, that the same principles given relative to sorghum apply to corn and soybean meal. This example is an illustration of a long hedge—a hedge in which futures contracts are bought rather than sold.

This particular cattle feeder does not have the storage capacity to store more than a four-month supply of feed, as is the case with most commercial cattle feeders. He is short of grain for feeding requirements, so he decides to buy futures contracts in the amount of his expected feed requirements.

In December, the cattle feeder has enough feed to last until March. Therefore, he buys March futures. Upon checking his basis charts, he determines that the March futures should be about 10 cents above the local cash price in March. This, then, assures him of the maximum amount that he will have to pay for the cash grain in March—that is, if the basis does in fact narrow to 10 cents during March, the maximum cost of the grain will be 10 cents below the futures price in March.

Let us assume that this feeder feels that the most he can pay for grain, and still make a profit on his feeding operation, is $1.85. He is confident that the basis will be 10 cents in March.

TABLE 27-19

A LONG HEDGE ON GRAIN SORGHUM ON A DECLINING MARKET

Cash Market		Futures Market		Basis
	Per Cwt		*Per Cwt*	*Per Cwt*
Dec. 10: Expected cash price	$ 1.85	Buys amount needed of March futures at	$ 1.95	$0.10
March 15: Buys cash grain at	$ 1.80	Sells March futures at	$ 1.90	$0.10
Gain	$ 0.05	Loss	$ 0.05	

Thus, the feedlot operator—the hedger—has broken even on the grain sorghum transaction despite a declining market.

If the market should rise, the situation given in Table 27-20 would apply.

TABLE 27-20

A LONG HEDGE ON GRAIN SORGHUM ON A RISING MARKET

Cash Market		Futures Market		Basis
	Per Cwt		*Per Cwt*	*Per Cwt*
Dec. 10: Expected cash price	$ 1.85	Buys amount needed of March futures at	$ 1.95	$0.10
March 15: Buys cash grain at	$ 1.90	Sells March futures at	$ 2.00	$0.10
Loss	$ 0.05	Gain	$ 0.05	

Again the feedlot operator—the hedger—has broken even on the grain sorghum transaction on a rising market.

Delivery Against the Contract

Although very few contracts, usually fewer than three percent, are consummated by actual delivery of the commodity, a hedger should consider delivery as one of his alternatives, particularly when the cash and futures prices are out of line with each other. However, due consideration must be given to the costs of delivering or receiving delivery, since such costs may be of such magnitude as to offset the differences between the cash and futures prices.

It is not the function of the futures market to provide an alternative source of supply nor an alternate means of disposal of surplus commodities. The purpose of delivery is merely to serve as a safeguard to be used when all else fails.

Facts About Futures Contracts

A cardinal feature of any workable futures contract—whether it be steers, grain, or any other commodity—is that there shall be maintained a solid connection with the commodity; that is, cash and futures must be tied together.

Any contract held until maturity must be delivered. This keeps the futures price in line with the cash price at the livestock market.

During the delivery month the cash and futures market tend to come together at the point of delivery. If this were not so, traders would quickly take advantage of the situation. For example, if prior to the termination of trading on August futures, the price of U.S. slaughter steers on the terminal market was $5/cwt below August futures, traders could buy cattle and sell futures, then deliver on the contract for a profit of $5/cwt (less marketing and brokerage fees).

It is not good business to use the futures market as a selling hedge when the futures price is below the estimated break-even price. The most desirable hedging situation is when the futures price is higher than the break-even price plus an adequate profit margin. The wider the spread, the better the hedge.

A feedlot operator who buys feeder cattle and sells finished cattle on a regularly scheduled basis—weekly, each two weeks, or monthly—throughout the year is doing his own hedging, provided feeder prices roughly parallel fed cattle prices.

Advantages and Limitations of Live (Slaughter) Cattle Futures

In this section, only live (slaughter) cattle futures will be discussed, simply because they constitute the greatest uncertainty, or risk, in the cattle business; hence, they will always dominate the futures market insofar as cattle feeding is concerned. Nevertheless, many of the same advantages and limitations apply to feeder cattle and feeds.

Live (slaughter) cattle futures are serving a useful purpose; and they are here to stay. Before using them, however, a cattleman should understand what they will and will not do for him.

Among the *advantages* of beef cattle futures are:

1. They serve as a price barometer for several months ahead, thereby increasing the range of information and judgments brought to bear on finished cattle prices and making it easier for feeders to choose a preferred course of action.

2. Through hedging, they can provide price protection or insurance to cattlemen against major breaks in the market.

3. They permit prices to be "locked in" anytime during the feeding period. Thus, they allow selectivity of the market time over the entire feeding period, rather than limit it to the one day that cattle are ready to go to market.

4. They make it possible for cattlemen to obtain credit more easily and to increase financial leverage. For example, let's assume that without hedging, a particular cattleman is able to borrow 70 percent of the cost of feeder cattle. If he has $90,000 of his own capital to invest, this will enable him to purchase $300,000 worth of feeder cattle. However, if he hedges the cattle he buys, the lender may be willing to lend up to 90 percent of their cost. His $90,000 of capital will then permit him to purchase $900,000 worth of feeder cattle. In this case, therefore, hedging tripled the number of cattle he could purchase and, likewise, his profit potential.

Some lenders will advance funds for the margin on cattle futures contracts.

5. They make for a more stable market, with fewer peaks and valleys of price movements.

6. They make it possible for a meat packer to protect himself when he contracts with a feeder for delivery of finished cattle, (a) for a few months ahead, and (b) at the futures market price at the date of specified delivery. Thereupon, the packer initiates a hedge by selling futures to offset his purchase contract.

Like many good things in life, live beef cattle futures are not perfect. They will not solve all the cattlemen's price problems, they will not raise longtime price levels, nor will they cause people to eat more beef. But these are not disadvantages; they're facts.

Among the *limitations* of live beef cattle futures are the following:

1. During an extended period of rising finished cattle prices, the cattle feeder is disadvantaged when he fixes a price for his cattle in advance. One study revealed that over a 6-year period the consistent hedger would have sacrificed 23 percent in profits to attain a 74 percent reduction in profit variability.[10]

2. No provision for heifers is available; only steers may be traded.

3. Only Choice grade and yield grades 1, 2, or 3 may be traded (although there is provision for substituting yield grade 4's at a 4-cent discount,

---

[10]*Livestock Breeder Journal*, August 1973, p. 12, article entitled, "Beef Futures Trading Hit $16 Billion; No Let-Up Is in Sight," by Carol E. Curtis, Economic Research Service, USDA.

but no more than 18 head of 4's can be delivered in any single unit), whereas the feeder may have lower grade animals. Also, it is recognized that (a) grades of live cattle are more difficult to standardize than grades of grain, and some other commodities, and (b) price relationships between grades of cattle in the cash market vary from time to time.

4. The cattle inventory (the cattle in a particular feedlot, for example) may be at a location which would require costly transportation to one of the specified delivery points.

5. Some delivery months may not move exactly as the cash market does.

6. A change in the basis (the spread between the cash price and the price of the futures) can mean a hedging loss as well as a hedging profit.

7. There is a relatively narrow range of time during which it is practical to hold slaughter cattle while waiting for a change in the basis.

8. The feeder must not forget to offset by purchase of another contract at the proper time; otherwise, he may find it necessary to deliver.

9. If the feeder sells futures for a greater amount than the finished weight of his cattle, he is engaged in speculation for the amount of the excess.

10. There are some costs in futures which must be considered; namely, commission and interest on margin capital. These should be considered as costs of doing business; for the protection secured, the cattleman must pay a commission—much as he does for a life insurance policy.

11. Unless a feeder has maintained good and accurate records, and can project his costs with reasonable accuracy, he cannot intelligently determine if a futures price is favorable for placing a hedge.

How to Go About Hedging Beef Futures

Here is the "how and where" that a cattleman interested in hedging must follow:

1. Have good and accurate records of costs.

2. Contact a brokerage house that holds a membership in the commodity exchange.

3. Open up a trading account with the broker, by signing an agreement with him authorizing him to execute trades.

4. Deposit with the broker the necessary margin money ($900 on live [slaughter] cattle and $600 on feeder contracts) for each contract desired. He will then maintain a separate count for the cattleman. The $40/contract commission fee is due when the contract is fulfilled by either delivery or offsetting purchase or sale of another contract.

5. Maintain basis charts; showing the relationship between (a) local prices of feeders and slaughter cattle, and (b) live beef futures.

Glossary of Futures Market Terms

Futures markets have a jargon and language of their own. It is not necessary that cattlemen dealing in futures master all of them, but it will facilitate matters if they at least have a working knowledge of the following:

Basis—The difference or spread between the cash price of live beef cattle at a particular market and the price of a futures contract. This spread differs from one market to another and changes with time.

Basis movement—The change which occurs in a particular cash-futures price relationship. It is the change in basis that determines the success or failure of a hedge, rather than changes in market price. One should always hedge according to basis rather than price.

Bear market—A downward moving or lower market is considered "bearish," because the bear strikes down its victim.

Bid—A bid subject to immediate acceptance made on the floor of an exchange to buy a definite quantity of a commodity future at a specified price.

Break—A more or less sharp price decline.

Broker—An agent who handles the execution of all trades. He may also represent a clearing house member.

Bull market—An upward moving or higher market is considered "bullish," because the bull tosses his victim upward on impaled horns.

Cash market—Cattle bought and sold for immediate delivery. Also known as spot market.

Cash (spot)—The cash price refers to the price of live animals and not futures contract. Also, known as spot commodity.

Chicago Board of Trade—It was founded in 1848. The Chicago Board of Trade handles futures trading in such commodities as beef carcasses, lard, wheat, corn, oats, rye, soybeans, soybean oil and meal, and grain sorghum.

Chicago Mercantile Exchange—It was founded in 1919. It provides a market place for trading in futures commodities, including finished and feeder cattle.

Commission—The charge made by a broker for buying or selling a cattle futures contract.

Commission house—A firm which buys and sells actual commodities or futures contracts for the accounts of its customers.

Delivery—The tender and receipt of the actual commodity, or warehouse receipts covering such commodity, in settlement of a futures contract.

Delivery points—Those points designated by futures exchanges at which the physical commodity covered by futures contract may be delivered in fulfillment of such a contract.

Discount to futures—When the cash price is under the futures price.

Forward contract—A forward contract calls for delivery at sometime in the future. In a forward contract, a cattleman might make a deal with a buyer during the summer months that calls for delivery of cattle in the fall at the price agreed upon in the contract.

Futures—A term used to designate any and all contracts which are made or established subject to the rules for delivery at a later date.

Hedge—A sale of any commodity for future delivery on or subject to the rules of any futures market to the extent that such sales are offset in quantity by the ownership or purchase of the same cash commodity or, conversely, purchase of any commodity for future delivery on or subject to the rules of any futures market to the extent that such purchases are offset by sales of the same cash commodity. "Hedging" means to safeguard oneself by making compensatory arrangements.

Hedgers—Persons who desire to avoid risks, and who try to increase their normal profit margins through buying and selling futures contracts. They are feeders, packers, and others actually involved in production, processing, or marketing of beef. Their primary objective is to establish future prices and costs so that operational decisions can be made on the basis of known relationships.

Limit order—Placing price limitations on orders given the brokerage firm.

Long—The buying side of an open futures contract. A trader whose net position in the futures market shows an excess of open purchases over open sales is said to be "long."

Long hedge—Buying on the futures market contracts against anticipated need in the future in order to protect against a rise in the market price. Thus, futures contracts in feeder cattle can be used in a long hedge to reduce or eliminate the risk involved in a rise of feeder cattle prices. At the time the feeder negotiates a forward contract with the packer, he would buy feeder futures contracts, preferably for the month in which he actually planned to buy the feeders.

Margin—Cash or equivalent posted as guarantee of fulfillment of a futures contract (not a payment or purchase).

Margin call—If the market price of a futures contract changes after the cattleman has sold or purchased a futures contract he will either make a profit or lose money. If the price moves in such a direction so that he loses money his broker will deduct the losses from his original "margin" and call for additional funds in order to bring the "margin" back up to the original amount.

For example, a feeder might have his broker sell a futures contract at $40/cwt. He would deposit $800 margin with the broker. If the price of futures were to advance to $40.50, the feeder would have lost 50 cents/cwt or $200. His broker would deduct the $200 from his original margin of $600, leaving $400. The broker would issue a "margin call" for an additional $200.

Pit—An octagonal platform on the trading floor of an exchange consisting of steps upon which traders and brokers stand while executing futures trades.

Premium—When the cash price is above the futures.

Rally—Quick advance in prices following a decline.

Ring—A circular platform on the trading floor of an exchange, consisting of steps on which traders and brokers stand while executing futures trades.

Round turn—A purchase and its liquidating sale, or a sale and its liquidating purchase.

Ticker—A teletype machine which sends and receives futures market and cash market information.

Trend—The direction prices are taking.

Short—The selling of an open futures contract. A trader whose net position in the futures market shows an excess of open sales over open purchases is said to be "short."

Short hedge—When one owns an inventory of a commodity and hedges by selling an equivalent amount of futures contracts, he has sold short or is short futures and has what is called a short hedge. An example of a short hedge is selling on the futures markets contracts of live cattle which represent cattle that are on feed in the feedlot in order to protect the enterprise against a severe decline in the market.

Speculators—Persons who are willing to accept the risks associated with price changes in the hope of profiting from increases or decreases in futures prices.

Spot commodity—The actual physical commodity such as live cattle as distinguished from the futures. Also, known as cash commodity.

Straddle—The purchase in one market and the simultaneous sale of the same commodity in some other market. It may also refer to the purchase of one commodity against the sale of a different commodity, both of which should normally be closely allied in price movements.

Washout—Offsetting a position taken earlier.

QUESTIONS FOR STUDY AND DISCUSSION

1. Use the nomograph (Fig. 27-1) to compute the break-even price that you can afford to pay for feeder cattle where the following circumstances prevail: purchase weight of feeder cattle, 500 lb; selling weight of fat cattle, 1,050 lb; feed cost per cwt gain, $32.00; price per cwt of fat cattle, $40.

2. Define (a) price margin, (b) feeding margin, (c) positive margin, and (d) negative margin. Why isn't the relative importance of price margin vs feeding margin the same for calves as for yearlings?

3. List and discuss each of the factors pertinent to buying feeder cattle.

4. Discuss each of the following points pertinent to selling finished (fed) cattle: (a) when to sell; (b) market channels; (c) factors influencing value; (d) seasonality of prices; (e) shrinkage; and (f) handling prior to and during shipment.

5. Define (a) horizontal integration, and (b) vertical integration. What are the advantages and disadvantages of each?

6. Using Table 27-3, Cattle Finishing Profit Indicators, and taking either your own cattle finishing enterprise or one with which you are familiar, determine how well you're doing.

7. Discuss the importance of each of the following from the standpoint of financing feedlot cattle: (a) fixed costs; (b) variable costs; (c) depreciation schedule; (d) feed costs; (e) flow chart for feed; and (f) cattle costs.

8. What kinds and sources of credit are available to the cattle feeder?

9. Discuss the pros and cons of each of the following sources of outside capital for cattle feeding: (a) limited partnership; (b) corporations; (c) central money markets; and (d) condominiums. Which source of outside capital would you recommend for cattle feeding?

10. What is involved in an SEC registered public offering?

11. Develop plans for a coop-owned cattle feedyard to serve 50-to 300-head cow owners who desire to finish their own calves.

12. Discuss each of the following points pertinent to custom feeding; (a) provisions of a contract; (b) types of contracts; (c) operation procedures; (d) what cattle owners expect of custom feedlots; and (e) how to attract custom feeders.

13. Why did USDA's Packers and Stockyards Administration rule that packers could not own, operate, or control *custom* feedlots? Why did they rule that *custom* feedlots cannot own packing plants?

14. What practical use can cattle feeders make of computers?

15. Detail a desirable and workable incentive basis for cattle feedlot help.

16. Discuss each of the following points pertinent to futures trading in finished cattle, feeder cattle, and feed: (a) magnitude; (b) contracts; (c) commission fees and margin requirements; (d) how a futures contract works; (e) delivery against contract; (f) facts about futures contracts; (g) advantages and limitations of live (slaughter) cattle futures; and (h) how to go about hedging beef futures.

17. Under what circumstances would you recommend that a cattleman use beef futures? Under what circumstances would you recommend that a cattleman not use beef futures?

SELECTED REFERENCES

Title of Publication	Author(s)	Publisher
Analysis of Beef Costs and Returns in California, An, AXT-258	A. D. Reed	Agricultural Extension Service, University of California, 1971
Beef Cattle Feeding and Slaughtering in California, Bull. 826	S. H. Logan G. A. King	Agricultural Experiment Station, University of California, Davis, Calif., 1966
Beef Cattle Futures: A Marketing Management Tool, Bull. 663	R. F. Bucher G. L. Cramer	Agricultural Experiment Station, Montana State University, Bozeman, Mont., 1972
Budgeted Costs and Returns of Fifteen Cattle Feeding Systems in Four Areas of Texas, MP-1022	E. Williams D. E. Farris	Agricultural Experiment Station, Texas A&M University, College Station, Tex., 1972
California's Competitive Position in Cattle Feeding and Poultry Production: A Review of Interregional Competition Studies, No. 72-1	H. F. Carman	Agricultural Extension Service, University of California, Davis, Calif., 1972
Capital Structure and Financial Management Practices of the Texas Cattle Feeding Industry, The, B-1128	R. A. Dietrich J. R. Martin P. W. Ljungdahl	Agricultural Experiment Station, Texas A & M University, College Station, Tex.; Economic Research Service, U.S. Department of Agriculture, Washington, D.C., 1972
Cattle Feeding in California	J. A. Hopkin R. C. Kramer	Bank of America N.T. & S.A., San Francisco, Calif., 1965
Cattle Feeding Industry's Place in the Economy of Imperial County, The	J. R. Dunbar A. D. Reed	Agricultural Extension Service, University of California, El Centro, Calif., 1972
Cattle Feeding in the United States	R. A. Gustafson R. N. Van Arsdall	Economic Research Service, U.S. Department of Agriculture, Washington, D.C., 1970

(Continued)

Title of Publication	Author(s)	Publisher
Cattle Futures Handbook	J. Sampier J. March	J. Sampier, J. March, Chicago, Ill., 1966
Chicago Mercantile Exchange Yearbook 1969-1970		Market News Department, Chicago Mercantile Exchange, Chicago, Ill.
Commodity Futures Statistics	Stat. Bull. No. 444	U. S. Department of Agriculture, Washington, D.C., 1970
Costs and Economies of Size in Texas–Oklahoma Cattle Feedlot Operations, B-1083	R. A. Dietrich	Agricultural Experiment Station, Texas A&M University, College, Station, Tex., 1969
Economic Comparison of Confinement, Conventional Drylot and Open-lot Beef Feeding Systems, AE-4250	D. E. Erickson	Cooperative Extension Service, University of Illinois, Urbana, Ill., 1970
Economics of Cattle Feeding on Imperial Valley Field Crop Farms, Bull. 813	H. O. Carter G. W. Dean P. H. Maxwell	Agricultural Experiment Station, University of California, Davis, Calif., 1965
Economics of Futures Trading for Commercial and Personal Profit	T. A. Hieronymus	Commodity Research Bureau, Inc., New York, N.Y., 1971
Evolution of Futures Trading	H. S. Irwin	Mimir Publishers, Inc., Madison, Wisc., 1954
Factors Affecting Cattle Prices	J. Ferris	Cooperative Extension Service, Michigan State University, East Lansing, Mich., 1969
Facts You Should Know About the Chicago Mercantile Exchange and Its Commodities		Chicago Mercantile Exchange, Chicago, Ill.
Feedlot, The	Ed. by I. A. Dyer, C. C. O'Mary	Lea & Febiger, Philadelphia, Penn., 1972
Futures Markets for Livestock–Value as Marketing and Management Tools, Res. Rep. 63	W. D. Dobson	Research Division, The University of Wisconsin, Madison, Wisc., 1970
Futures Trading, Direct Marketing, and Efficiency of the Cattle Marketing System, Res. Jrnl. 9	R. L. Ehrich	Agricultural Experiment Station, University of Wyoming, Laramie, Wyo., 1967
Futures Trading in Live Beef Cattle, E.M. 2587	O. S. Wirak	Cooperative Extension Service, Washington State University, Pullman, Wash., 1965
Futures Trading in Livestock–Origins and Concepts	Ed. by H. H. Bakken	Mimir Publishers, Inc., Madison, Wisc., 1970

(Continued)

Title of Publication	Author(s)	Publisher
Futures Trading Seminar, Influence of Environmental Factors, Volume II	H. B. Arthur H. E. Erdman O. B. Jesness M. Ezekiel	Mimir Publishers, Inc., Madison, Wisc., 1963
Hedging as a Marketing Tool for Western Cattle Feeders Bull. 203	E. L. Menzie T. F. Archer	Agricultural Experiment Station, The University of Arizona, Tucson, Ariz., 1973
Hedging Crops and Livestock	M. J. Oster	Professional Farmers of America, Cedar Falls, Iowa, 1973
Hedging Feeder Cattle in Washington, Bull. 601	O. S. Wirak	Cooperative Extension Service, Washington State University, Pullman, Wash., 1970
Hedging Potential in Grain Storage and Livestock Feeding	R. G. Heifner	Economic Research Service, U.S. Department of Agriculture, Washington, D.C., 1973
How to Buy and Sell Commodities	Merrill Lynch, Pierce, Fenner & Smith Inc.	Merrill Lynch, Pierce, Fenner & Smith Inc., New York, N.Y., 1966
How to Hedge Commodities	Merrill Lynch, Pierce, Fenner & Smith Inc.	Merrill Lynch, Pierce, Fenner & Smith Inc., New York, N.Y., 1963
Interregional Competition in the Cattle Feeding Economy with Special Emphasis on Economies of Size, B-1115	R. A. Dietrich	Agricultural Experiment Station, Texas A&M University, College Station, Tex., 1971
Stockman's Handbook, The, Fourth Edition	M. E. Ensminger	The Interstate Printers & Publishers, Inc., Danville, Ill., 1970
Summary of Illinois Farm Business Records, 1971, Circ. 1058	D. F. Wilken R. P. Kesler	Cooperative Extension Service, University of Illinois, Urbana, Ill., 1972
Texas–Oklahoma Cattle Feeding Industry, The, B-1079	R. A. Dietrich	Agricultural Experiment Station, Texas A&M University, College Station, Tex., 1968
Trading in Live Beef Cattle Futures	R. P. Shiner D. G. Nash D. Schambach	Commodity Exchange Authority, U.S. Department of Agriculture, Washington, D.C., 1970
Trends in Cattle and Calf Numbers and Income (Selected States), AXT-378	J. H. Cothern	Agricultural Extension Service, University of California, Davis, Calif., 1973
Understanding the Commodity Futures Markets	Commodity Research Bureau, Inc.	Commodity Research Bureau, Inc., New York, N.Y., 1963

CHAPTER 28

FEEDLOT FACILITIES AND EQUIPMENT

Contents   **Page**

Cattle feeding facilities and equipment are a manufacturing plant, wherein animate objects (cattle) convert feed into beef. Hence, they merit the same level of competence in planning and design as any other sophisticated manufacturing plant.

Because of variations in climatic conditions, number of cattle to be fed, and factors prevalent at the location where it is desired to construct the feedlot, no attempt will be made herein to present detailed facility and equipment plans and specifications. Rather, it is proposed to convey suggestions regarding the desirable features of cattle feeding facilities in various sections of the country. For detailed plans and specifications the feeder should (1) study facilities and equipment in other feedlots, and (2) engage the services of a consultant(s) with expertise in cattle feeding facilities and equipment.

Fig. 28-1. Cattle feeding has gone modern. (Upper) Yesterday. (Lower) Today. (Upper picture—courtesy, Iowa Beef Processors, Dakota City, Neb.; lower picture—Pioneer Hi-Bred Corn Co., Des Moines, Iowa)

Some preliminary feedlot planning suggestions follow:

1. Decide on the number of cattle and the feed and storage requirements with provision for expansion.

2. Determine the justifiable investment in cattle feeding facilities. (See the section that follows.)

3. Select the facilities, equipment, and arrangement that best fit the management program you have chosen; for example, (a) fence-line bunks and a central feed processing plant, (b) upright storage with distributors and bunks, or (c) self-feeders.

4. Decide on the type of facilities: (a) feedlot (open pen); (b) cold confinement; or (c) warm confinement.

5. Design a system that is practical, laborsaving, environmentally suitable for economical gains of cattle, and attractive.

JUSTIFIABLE INVESTMENT IN CATTLE FEEDING FACILITIES

Cattle feeders need to know the size investment that they can justify in cattle feeding facilities. A nomograph may be used for this purpose. It can give a quick, preliminary idea of cost, gross profit, returns, and investment relationships. But a nomograph should not replace more detailed figuring which should precede all major investment decisions. Also, one should realize that a nomograph will give erroneous and misleading information unless based upon accurate and realistic cost and return data from the problem at hand.

Fig. 28-2 is a nomograph for use in making quick calculations of justifiable investments in beef feeding systems. The section that follows gives an example and "step-by-step" instructions for using the nomograph. By working through an example, you will soon discover how quickly the nomograph can help in evaluating capital investments for cattle feeding enterprises.

Instructions for Using Nomograph

This nomograph provides a quick method of computing the investment in cattle feeding facilities one can justify on the basis of three factors: (1) the percent of the total investment to "charge off" as annual costs each year; (2) the gross profit (G.P.) expected per head; and (3) the return per head (R/C & L) desired for labor, management, and interest on the capital required for cattle, feed, and miscellaneous equipment (*excluding* the investment in the feedlot facilities under consideration).

Determining "% Annual Costs"—Annual fixed costs include such items as interest, insurance, and taxes (usually from 4 to 6%) plus an allowance for depreciation or annual principal repayments required, if making a cash-flow analysis (usually from 5 to 15%). These two percents should be combined to get the proper "total annual costs, percent of investment" for this analysis.

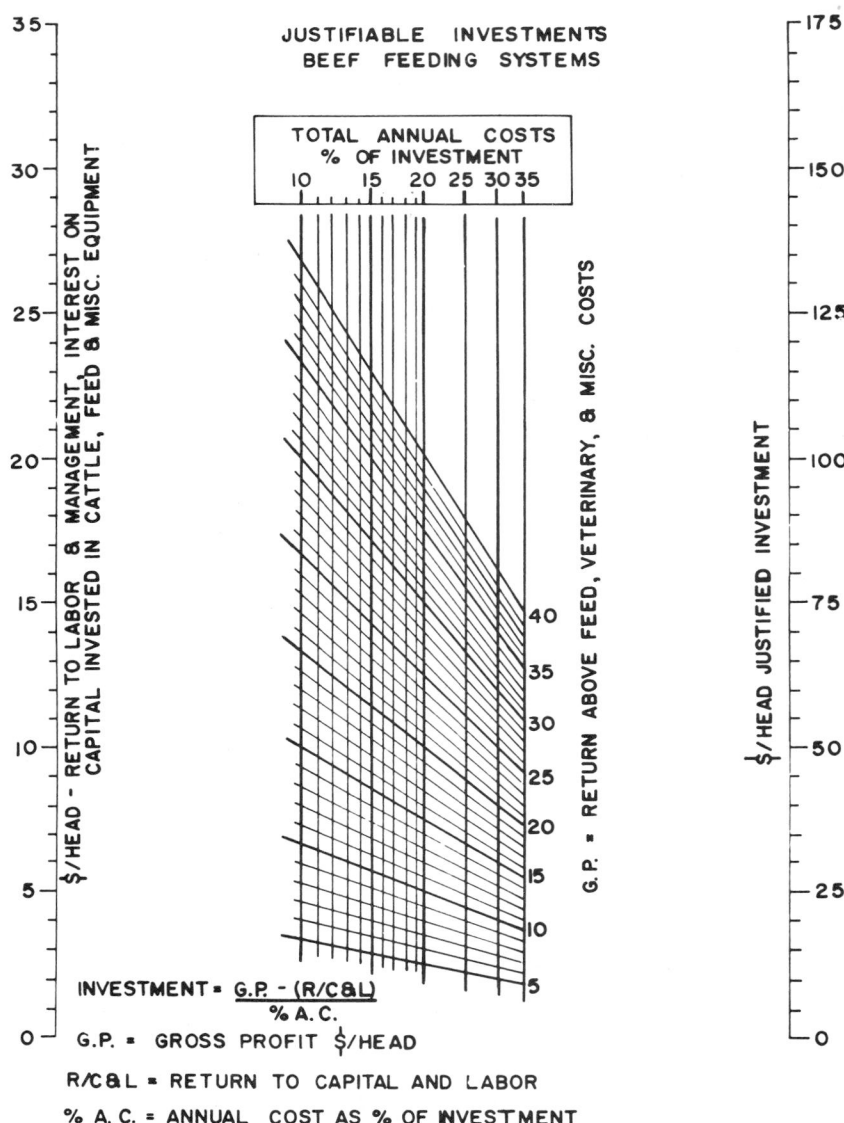

Fig. 28-2. Nomograph for computing "justifiable investments in beef feeding systems." (Nomograph, and instructions for using it, prepared by Robert M. George, University of Missouri, Columbia, Mo.)

DATA NEEDED FOR COMPUTATIONS	EXAMPLE	YOUR FARM
1. Percent (%) annual costs	15%	_____
2. Gross profit (return over variable costs) per head	$20.00	_____
3. Return to labor, mgt., and capital (R/C & L)	$14.00	_____
4. Investment per head justified	$40.00	_____

● *Procedure for determining the investment justified in cattle feeding facilities—*

1. Locate the *vertical line* representing the "total annual costs, % of investment" (15% in the example).

2. Follow this line straight down to the point where it intersects with the diagonal line representing the expected gross profit (returns over variable costs) per head ($20.00 in the example)—*Mark this point;*

3. Locate the point on the left-hand vertical scale which represents the desired return ($/head) for labor, management, and capital invested in cattle, feed, and miscellaneous equipment ($14.00/head in example)—*Mark this point;*

4. Draw a straight line to connect the two points located above. The point of intersection with the right-hand vertical scale indicates the justified dollar-per-head investment in facilities ($40.00/head in example).

● *Procedure for determining the probable return ($/head) with a known, or contemplated, investment per head in facilities—*Follow the above procedure through point 2. Then, locate the dollar investment per head on the right-hand vertical scale ($40.00 per head in example). Connect a straight line from this point through the one located in point 2. The point of intersection with the left-hand vertical scale gives the probable $ return/head ($14.00 in example).

Other variations in the use of the nomograph are obvious. For example, if the investment per head in facilities is $50.00, a return of $15/head is desired, and the gross profit per head is $20.00—10% annual costs would be indicated.

FEEDLOT (Open Pen)

The vast majority of feedlot cattle are fed in open pens, without shelter. Some are provided wind protection (trees, hills, or wind fences). Still others are provided an open-front building, for shelter from wind, sun, rain, and snow.

The sections that follow present information pertinent to feedlot design and construction.

Cost

Before constructing a feedlot, cost must be considered for two reasons: (1) capital must be secured; and (2) cost must be amortized. The usual basis of computing cost is on a "per animal unit capacity." This will run about the same whether calves or yearlings are involved, because per unit capacity

Fig. 28-3. Feedlot windbreak and cable fence. Upright fence boards are spiked to horizontal nail ties (20-inch lumber) for a windbreak in mounded feedlot. Windbreak, supported by treated poles, is vertical to the ground so it spans curved ground without leaning. Cable fence in foreground is braced with pipes. Pipes have "feet" welded to them that are spiked to the posts. Cables pass through holes drilled in posts. A cable brace with turnbuckle is anchored diagonally from gatepost to second post. (Photo by A. M. Wettach, Mount Pleasant, Iowa)

must consider carrying the animals to market time. Chapter 27, Table 27-3, Cattle Finishing Profit Indicators, gives the estimated initial (new) land, feedlot, and equipment cost on the basis of per animal unit capacity.

The area affects cost from the standpoint of shelter requirements and land values. Thus, because of the necessity for winter protection and shelters, feedlot costs are higher in the northern tier of states than in the South. Land values are higher in California than most areas of the United States, with the result that land costs become a factor.

Size of feedlot affects per animal cost. Most studies reveal that investment savings do accrue to the larger feedlots. Thus, the cost per animal usually decreases up to about 10,000-head capacity, then it increases slightly with larger lots. The slightly higher cost per head capacity of the larger lots appears to be due to duplication in equipment and the tendency to become more highly mechanized and elaborate.

An open lot without shelter is the cheapest type of feedlot construction of any. In the Southern Plains area, where the weather is mild and shelters are unnecessary, investment costs range from $30 to $50 per head of capacity.[1]

---

[1]Gill, Don, and Myron D. Paine, (Oklahoma State University), "Feedlot Design and Construction," *Feedlot Management*, Nov. 1972, p. 84.

Housing increases costs, and the more elaborate the housing the greater the cost. University of Minnesota studies showed the following costs relative to three types of cattle feedlot facilities.[2]

Type of Facility	Sq. Ft. per Animal	Cost per Animal Capacity
		($)
Open Shed	20	70
Cold confinement	17	115
Warm confinement (heated)	17	170

Location

In the present day and age, pollution control is the first and most important consideration in locating a cattle feedlot. The location should avoid (1) neighbors complaining about odors, flies, and dust, and (2) pollution of surface and underground water. Also, feedlots should be located on a well-

Fig. 28-4. In the present era, pollution control is most important. Besides, when a truck is bed deep in mud, it's difficult for a caretaker to remember that he's supposed to feed cattle, and it's difficult for the cattle to know that they are supposed to eat. Few present-day cattle feeders permit such conditions as pictured above. (Courtesy, Iowa Beef Processors, Dakota City, Neb.)

drained site, with area available for expansion. Whenever possible, they should be built on a slope, preferably at the top of it. There should be a minimum amount of runoff from areas above lots (a diversion terrace can be used if necessary); and there should be ample space below feedlots for necessary water pollution control measures. Also, feedlots should be located where there is ample space for expansion, if and when desired. Of course,

---

[2]*Feedlot Management*, November 1972, p. 44.

the space requirements will vary with the type of facility. Open lots require the most space; confinement housing the least. Minimum space requirements for an open feedlot—including lots, mill, office, etc.—are approximately 8/10 acre per 100 head or 7 acres per 1,000 head. In order to allow for expansion to double this size, it is recommended that there be 1.4 acres per 100 head, or 12 acres per 1,000 head.

Layout

Prior to starting construction, anyone contemplating a feedlot may avoid much subsequent difficulty and expense by doing some paper and pencil planning at the outset. First, decide on the size of the enterprise and the management system. Then, establish traffic routes for animals, feed, cleaning equipment, supply trucks, etc. Next, sketch out the facilities and equipment required to meet these needs in the most efficient and economical manner, including pens, mill, scales, office, etc. Where the area permits, the ideal feedlot should be U-shaped, with the following arrangement: the facilities for receiving and loading out cattle, scales, milling feed, office, and equipment barn should be located near the center of the U. Pen facilities should be located on the three closed sides of the U. The open end of the U should be connected to a public road. Also, trench silos should be located at the mouth of the U. The mouth of the U should be kept open in order to allow for the flow of livestock, feed and visitors. The smallest pens should be located as near the feed mill as possible, in order to minimize travel time in feeding. Feed alleys should parallel the legs and closed end of the U, forming a semicircle around the feed mill area. The corners of the pens should be rounded to allow feed trucks to turn at all intersections.

The above information, constituting the layout of operations, should first be put on paper in sketch form, by the cattleman. From this, the engineer or consultant can design facilities and equipment which most effectively and economically meet the production requirements of the specific enterprise.

Pens

Pens are the working end of the business; they have much to do with the well-being and performance of the cattle. Hence, their design is most important; and the more severe the climate, the more important the design becomes. Consideration should be given to the following points when designing cattle pens:

1. *Drainage*—Muddy pens reduce cattle performance more than wind and rain; a muddy lot may reduce daily gains and feed efficiency by 25 to 35%. To supplement the natural drainage, surface grading should be done prior to starting construction. A grade of 4 to 6% should be established. Excessive slopes (above 10%) should be avoided, as they make for difficult footing and are subject to erosion.

2. *Mounds*—Mounds of dirt in each pen will provide a drier resting area for cattle. Preferably, mounds should be pushed up before the lot fences are installed. They should be 6 to 8 feet high, with a 4:1 or 5:1 (horizontal to

vertical) slope on the sides. The top of the mound should be fairly narrow (about 10 feet wide) and crowned for good drainage. The size of the mound will vary with the size of the lot and the number of cattle for which it is intended that it shall provide a rest area. Each mound should be large enough so that the cattle can rest on the upper half of it; 10 to 15 square feet of mound per animal will accomplish this, although not all animals will use mounds. The orientation of earth mounds is unimportant, so long as they do not block feedlot drainage. The mound should be built parallel with the general lot drainage to assure that liquids can readily drain from the mound area.

Fig. 28-5. This Caterpillar 613 self-loader works year-round in mounding pens at C & B Livestock Co. (Courtesy C & B Livestock Co., Hermiston, Ore.)

Mounds will require some maintenance. A logical time for this work is when manure is cleaned from the pen.

3. *Surface*—When cattle are fed in outside pens during favorable weather, the type of pen surface (concrete vs dirt) is probably of little importance. Where good drainage cannot be provided, concrete surfaced pens should be considered. Cattle can be finished successfully in outside pens surfaced with concrete as evidenced by the feedlots in which such footing is provided.

4. *Size*—Climate and the amount of paving are the main factors determining pen size. It will vary anywhere from a minimum of 30 square feet per animal with a surface lot and open housing to 400 square feet per head for an unsurfaced lot in a wet, muddy area. With an open, dirt lot, 75 square feet of pen space per head is adequate in a dry climate, whereas up to 400 square feet per animal may be required in a wet climate. On the average, a pen space allowance of 125 to 200 square feet per head is recommended if the lot is unpaved.

The number of animals per pen is variable. Investment may be reduced by using fewer but larger pens. However, visual inspection of cattle for sickness is easier in smaller pens, as are the separation and removal of individual animals. Custom feedlots require a variety of pen sizes in order to accommodate each customer's cattle in separate pens. In custom feedlots, the majority of pens with a capacity for 120 head appear to be desirable, because of customer convenience. From the standpoint of trucking requirements, pens should be sized to hold multiples of 60 heads. In most feedlots, where some custom feeding is done, it is recommended that feedlot sizes vary from 120-head capacity to 300 head.

Generally, most feedlots have too many large pens to permit them to feed the numerous small lots of cattle that they receive.

5. *Shape*—Pens should be rectangular in shape, preferably with rounded corners to allow the feed truck to turn at all intersections. The depth of the pen will depend upon the length of the bunk; there must be sufficient bunk space to accommodate the cattle in any given pen; then there should be sufficient depth to provide the number of square feet per animal intended.

6. *Fences*—A corral system is no more dependable than its fences. The specifications for the most desirable feedlot fence are: 54 to 60 inches high; posts of 3- to 4-inch diameter pipe, or treated wood with 5-inch top diameter, set in concrete and spaced 8 to 12 feet apart; enclosed (except for the feed bunk area) by 4 strands of ⅜-inch steel cable spaced at intervals (from ground up) of 18 inches, 12 inches, 12 inches, and 13 inches. Three strands of the cable should be placed above the feed bunk, with these spaced equal distances apart. Steel or aluminum pipe, wood, or wire may be used in place of steel cable if desired.

An outside fence, enclosing the entire cattle feedlot complex, is also recommended. A 60-inch high chain link fence, with steel posts set in concrete, is very satisfactory for this purpose. Such a fence is effective in keeping prowlers, dogs, and predators out; and in holding cattle which, for one reason or another, have escaped from their pens or handling facilities.

Working corral fences should be higher and stronger than ordinary feedlot fences. They may be constructed with treated wood posts, 6 inches top diameter, spaced 8 feet apart and set in concrete, enclosed by 2-inch dimension lumber; or 4-inch diameter pipe and ⅞-inch sucker rod or equivalent may be used.

7. *Feed bunks and aprons*—Six to nine inches of bunk space per head are adequate when cattle have access to feed at all times. The bunk forms part of the fence for the pen. It should be constructed of concrete with the outside (alley side) of the bunk 22 to 36 inches high, and the inside (the pen side) 18 to 22 inches high. The bottom of the bunk should be rounded and about 18 inches wide.

Fig. 28-6. Concrete feed bunk. (Courtesy, C & B Livestock Co., Hermiston, Ore.)

A 6- to 8-foot concrete apron (platform), 4 to 6 inches thick should extend from the feed bunk into the pen to provide solid footing for the cattle. The apron should have a minimum slope of ½- to 1-inch per foot (a 1-inch per foot slope will be nearly self-cleaning). A lip at the low end of the concrete apron, to divert the water along the length of the slab and out the pen, rather than merely off the slab area and into the pen dirt area, will alleviate the low muddy spot that usually develops at the juncture of the concrete and dirt.

Fig. 28-7. A concrete apron adjacent to the bunk and a mound, connected by a gravel pad, avoids muddy conditions in the feedlot. (Courtesy, C & B Livestock Co., Hermiston, Ore.)

8. *Water troughs*—Feedlot cattle should have access to clean fresh water at all times. One linear foot of trough space for each 10 head of cattle is sufficient. Shallow, low capacity troughs are preferable, since frequent drainage is necessary to keep them clean. There should be sufficient water supply to provide 20 gallons per head per day. Continuous-flow troughs are excellent and help to keep clean water. In most areas, continuous-flow troughs will not freeze during the winter months; hence, heat is not required. However, heated and insulated troughs are necessary in cold areas. Heat is usually provided by a 250-watt infrared reflector-type lamp, which is set to start when the temperature goes down to 33°F and to cut off when the temperature rises above 33°F.

9. *Windbreaks and shades*—In those sections of the country where snow and cold winds are a problem, windbreaks may be provided by hills, trees, or board fences. Where board fences are used, they are generally constructed of 1-inch lumber and are 7 feet high.

In hot climates, shades should be provided. (See Section on "Shades" later in this chapter.)

10. *Gates*—Gates should be of the same height as the fences. The length of most gates should be coordinated with the width of alleyways and crowding areas. Thus, gates along a 12-foot alley should be 12 feet long and capable of swinging either way. All gates should have a substantial lock and catch to prevent breaking open under pressure. The easier operation, less sagging, and longer life of metal gates in comparison with wooden gates will usually more than compensate for their additional cost.

Most managers of big feedlots seem to agree that the most desirable arrangement consists in having two gates to each pen, with both gates offset to provide easy access for cattle and equipment. One of these gates should be located in the corner of the alley nearest the scales. The other should be located adjacent to the feed bunk, but in the side fence. The latter gate will facilitate the use of equipment in cleaning manure off aprons, and is useful for cowboys or equipment operators in moving from pen to pen without entering the cattle alley.

Alleys

Two types of alleys are common to most feedlots—feed alleys and drive alleys. In feedlots below approximately 5,000-head capacity, feed and drive alleys should be combined, for cost reasons. In larger feedlots, it is recommended that feed alleys be separated from drive alleys, so that feeding and cattle movement can be done without excessive interference. Feed alleys, and combined feed and drive alleys, should be at least 20 feet wide, so as to permit the passing of trucks. Working and drive alleys which do not handle truck traffic should be 12 feet wide. Alleyways should not occupy more than 20 percent of the total corral area unless their space can also be used effectively as pens.

Cattle Loading and Unloading Facilities

The facilities for loading and unloading cattle are the connecting link between the corrals and the various types of vehicles used to transport cattle in and out. They should include a truck dock (with loading chute), scales, working alley, and holding pens. These facilities should be located centrally so that a smooth flow of cattle trucks in and out can be maintained. However, they should be somewhat removed from the feed-mill area, in order to separate the traffic flow of feed and cattle and alleviate congestion of traffic.

The truck dock generally consists of a platform, chute, and chute pen. Most loading platforms are 46 inches high, although they may range from 30 to 50 inches in order to accommodate various sizes of trucks. The ramp in a loading chute should provide a safe gradual slope for the cattle to climb. Ten feet is considered a minimum length, while many feedlot operators prefer loading chutes up to 16 feet long. Most chutes are 30 to 42 inches wide, with 48- to 54-inch high fences.

Fig. 28-8. Cattle loading facilities. Note the curved chute, which works well. (Courtesy, Benedict Feeding Co., Casa Grande, Ariz.)

In order to have favorable weighing conditions for sale cattle, it is necessary that they be moved from their pens to the scale expeditiously and with a minimum of stress. Larger feedlots—those of over 30,000-head capacity—usually accomplish this by having two shipping areas. In such lots, a good design allows for the movement of cattle to and from the cattle scale at the same time other cattle are being loaded or unloaded through the loading chute.

Also, there is need for holding pens convenient to the loading and unloading facilities. The number of such pens will be determined by the size of the feedlot. In general, each pen should be of sufficient size to hold two loads of cattle. The shipping pens—those for outgoing cattle—need not contain feed and water facilities, since they are used only to hold cattle temporarily that are to be loaded onto trucks. However, receiving pens—those for incoming cattle—should have provision for fresh feed and water. This is especially true for cattle received during the night, which might otherwise be left in a pen or alley without feed or water until the next morning.

Scales

Smaller feedlots can use a combination cattle and commodity scale. With larger feedlots, however, it is recommended that there be separate scales for each of these uses.

The scale should have sufficient capacity to meet the largest anticipated volume demand, without the necessity of dividing a lot of cattle and weighing in 2 or more drafts. In the larger feedlots, this calls for a scale approximately 10 feet wide and 60 feet long, with a capacity of 100,000 pounds. To avoid exceeding the weighing capacity of a scale, provide one square foot or less of platform space for each 110 pounds of rated capacity.

Fig. 28-9. A load of finished cattle being weighed over the scales. (Courtesy, Benedict Feeding Co., Casa Grande, Ariz.)

The scale should be equipped with a ticket printer, in order to verify weight records and avoid human error in market transactions.

Cattle Processing Facilities

A separate area should be provided where cattle can be branded, dehorned, castrated, vaccinated, and sprayed or dipped. This should include crowding areas and a curved chute 18 to 30 feet long to aid in the movement of the cattle to the squeeze chute. Generally cattle will work best in a chute constructed with a modest curvature; a sharp curve may spook them. A manually operated squeeze chute is satisfactory for a small feedlot, but larger feedlots should use a hydraulic squeeze chute for reasons of efficiency and ease of operation.

Where large numbers of cattle are to be processed daily, the processing pens should be equipped with feed and water facilities. Such provisions lessen stress and make for better performance of the cattle.

In areas subject to severe weather conditions—either very hot or very cold—the processing facilities should be covered. This makes for greater comfort of both the help and the cattle, thereby assuring better processing.

Hospital Areas

Hospital pens are for holding sick or injured animals. From 2 to 5 percent of the feedlot area should be allocated for this purpose in lots feeding mostly yearling cattle. Lots feeding calves will need more intensive care areas. Larger feedlots prefer to disperse the hospital areas into sections of the feedlot, with one such hospital area provided for about 6,000 head of cattle. This alleviates the necessity of moving sick cattle excessive distances for treatment. Each hospital area should be equipped with a squeeze chute, refrigerator, running water, medicine, equipment storage, and feed facilities for the sick animals. Two or three small pens at the hospital for cattle at various stages of recovery are recommended. In areas subject to severe weather, shelter should be provided for hospitalized animals.

Feed Mill

The feed mill should be located convenient to the feeding area, and away from cattle traffic patterns. Mill type will be determined by the number of cattle to be fed and the feeds to be used. Generally speaking, it is recommended that feedlots with fewer than 5,000 head of cattle use a self-mixing, self-unloading truck, especially if milling of the grain is not necessary immediately prior to feeding. For lots above 5,000-head capacity, a mixing mill and self-unloading feed truck are more economical. Mill type will be determined by the feeds that are to be used. If whole corn is fed, a minimum of mill facilities is needed. Where sorghum grain is the primary feed, steam-flaking equipment or other similar processing equipment, must be considered. Steamflake type mills, along with storage facilities, make for added costs of from $17 to $20 per head of capacity.

Mill capacity is also important. For a 5,000-head feedlot, the mill should be able to process 20 tons of feed per hour; for a 10,000-head feedlot, it should be able to process 35 to 40 tons per hour. Modern mills automatically mix the feed, with operation from a central control panel.

Fig. 28-10. Cattle feed processing facilities at the Benedict Feeding Co., Casa Grande, Arizona. (Courtesy, Benedict Feeding Co.)

Fig. 28-11. A central computerized system housed in the feed mill controls all feed ingredient distribution in Farr Feeders' new $4 million, 80,000 head per year capacity, feedlot near Greeley, Colorado. Here, the manager of operations is mixing and loading feed. Each truck load is assigned a specific pen in the feedlot, and that pen receives a precise ration to meet daily nutritional requirements. (Courtesy, Farr Feeders, Inc., Greeley, Colo.)

Office and Parking

Since the office is the headquarters of the cattle feedlot business, its location is of importance. It should be located near the main access road in order to minimize nonessential traffic in the feedlot proper. In most feedlots, the truck scales are adjacent to the office, and weighings are made in the office. Where this arrangement exists, the location of the office should be such as to provide for a good traffic pattern for both cattle and feed trucks. The size of the office will vary. Generally speaking, custom feedlots need more office space than feedlots that own all of the cattle. Automobile parking space is generally provided near the office. As a rule of thumb, most commercial feedlots plan on having 2 to 3 parking spaces for every 1,000-head animal capacity.

Equipment Storage and Repair Building

Most commercial feedlots have a separate building in which they store trucks, tractors, silage loaders, spare parts, and miscellaneous supplies. Also, this building is used for repair work. The size of the building will vary with the size of the cattle feeding operation. Usually, such buildings are 30 feet deep and 30 to 100 feet long.

Lights

Feedlot lights serve three purposes: (1) they have a calming and quieting effect on the cattle, although there is no conclusive experimental work to support the claim that lights will improve rate of gain or feed efficiency; (2) they prevent prowlers and pilfering; and (3) they make for convenience in working or loading cattle at night. Mercury vapor lights, equipped with a photo eye (which turn on automatically), are recommended.

Pollution Control

Pollution control is a most critical factor in site selection and operation of a cattle feedlot. Remoteness from urban development is recommended because of dust and odor. Also, before constructing a cattle feedlot, the owner should familiarize himself with both Federal and state regulations. Federal regulations apply to feedlots (1) with 1,000, or more, feeder cattle; or (2) in which animal wastes either empty directly into a stream that crosses the feedlot, or are conveyed directly into a nearby waterway by a pipe, ditch, or other man-made means. The state regulations can be secured from the state water board. They differ from state to state, but most states require a catch basin (detention pond) sufficient to contain the runoff from a storm of the magnitude of the largest rainfall during a 48-hour period of the most recent 10 years. A feedlot may minimize runoff by locating near the top of the slope and, if necessary, by using diversion embankments to divert runoff from other areas.

Various methods of handling manure are being studied, including recycling it for feed, the production of methane gas, the production of garden fertilizer, etc. Yet, today, and perhaps for sometime to come, the vast majority of manure will be spread on farmland for use as a fertilizer. Thus, it is imperative that adequate farmland be available, and that a suitable location be made for stockpiling waste until it is used. The amount of farmland necessary will vary. For example, less manure can be applied to farmland that is alkaline than to land that is acidic. A common rule of thumb, however, is that the feedlot should plan to have one acre of farmland for manure for each three- to five-head of cattle capacity. In order to lessen hauling costs, this land should be as close as possible to the feedlot.

Dust from feedlots in arid areas is a problem. Likewise, dust from feed mills can cause damage to crops and discomfort to people. Also, dust can (1) aggravate respiratory diseases in cattle, and (2) create poor visibility and make for possible accidents on highways. Sprinkling is the most effective method of preventing feedlot dust.

Fig. 28-12. Water wagon for dust control in alleys and pens. (Courtesy, Benedict Feeding Co., Casa Grande, Ariz.)

Before building a new feedlot or expanding an old one, the owner should be fully aware of the following:

1. A feedlot may be required by its neighbors to control odor. Both large units and periodic field distribution or large amounts of waste contribute to high odor levels.

2. Pollution of surface and underground water can occur, and should be avoided.

3. Runoff from lots must be held in detention ponds before disposal by hauling and spreading or by distribution in an irrigation system.

4. Runoff from fields may pollute surface water if wastes are spread on snow or frozen ground.

5. Leached nitrates from decomposing manure may pollute nearby shallow wells.

6. Wastes must be field-spread and absorbed, digested in an adequate lagoon, or otherwise handled to minimize pollution.

CONFINEMENT FEEDING; SLOTTED (OR SLATTED) FLOORS

Currently, there is much interest in cattle confinement feeding and slotted floors. The main deterrent is cost; construction costs vary with type of structure and may range up to $200 per steer space.

Fig. 28-13. Confinement cattle feeding facility in Nebraska. (Courtesy, University of Nebraska, Lincoln, Neb.)

Confinement cattle feeding refers to feeding in limited quarters, generally 20 to 25 square feet per yearling animal, which is about ⅛ the space normally allotted to a yearling in an unsurfaced lot and ⅓ that of a paved lot. The confinement is usually on slotted floors.

Slotted floors are floors with slots through which the feces and urine pass to a storage area immediately below or nearby.

Interest in confinement feeding and slotted floors is prompted in an effort to (1) automate and save labor, (2) cut down on bedding and facilitate manure handling, (3) lessen mud, dust, odor, and fly problems, (4) increase gains and save feed, (5) require less land, and (6) lessen pollution.

Fig. 28-14. Cattle on slotted floors. (Courtesy, Pioneer Hi-Bred Corn Company, Des Moines, Iowa)

Research has shown conclusively that cattle fed during the winter months in cold areas gain faster and more efficiently if they are sheltered. However, as pointed out earlier in this chapter under the section headed "Feedlot (Open Pen)—Cost," the per head cost is much higher for confined or sheltered cattle. Thus, the decision on whether or not cattle confinement can be justified, even in the northern part of the United States, should be determined by economics. Will the cattle in confinement quarters gain sufficiently more rapidly and efficiently to justify the added cost? Of course, manure disposal and pollution control should also be considered.

Cold Confinement[3]

Cold confinement refers to a more or less open shed for confining cattle; hence, winter temperatures therein are within a few degrees of outdoor temperatures. Open sheds should be faced away from the direction of the prevailing winds. Additionally, doors or other openings in the closed walls should be provided for summer ventilation.

Warm Confinement[4]

Warm confinement refers to a confinement building for cattle which is sufficiently insulated and ventilated to maintain inside winter conditions

---

[3]The terms "cold confinement" and "warm confinement" refer to winter conditions. Without mechanical cooling, both systems are "warm" during the summer months.
[4]Ibid.

Fig. 28-15. A cold confinement cattle feeding facility in Iowa, open to the south. (Courtesy, Pioneer Hi-Bred Corn Company, Des Moines, Iowa)

above 35° F in severe weather, and in the range of 50° to 60° F most of the time.

Design Requirements for Cattle Confinement Buildings and Slotted Floors

The figures that follow, based on information and experiences presently available, may be used as guides. (Also see Chapter 11, Table 11-6, Space Requirements of Buildings and Equipment for Beef Cattle.)

●*Floor space*–Allot 15 to 30 square feet per animal exclusive of the bunk and alley, with an average of 20 to 25 square feet for a 1,000-pound animal.

●*Animals per pen*–25 to 100 head per pen, with 25 to 30 being most common.

●*Bunk*–Allow 6 to 18 inches of linear bunk space per animal, with the amount of feeding space determined by frequency of feeding and size of animal.

●*Waterers*–Locate one waterer per 25 head at the back (opposite feed bunk) of each pen, preferably in the center.

●*Slats*–Reinforced concrete or steel may be used. Most concrete slats are 5 to 6 inches wide across the top, 6 to 7 inches deep and tapered to 3 to 4 inches wide at the bottom; and placed so as to provide a slot width of 1½ to 1¾ inches.

●*Manure production and storage*–Manure production will vary with size of animal and kind of feed, but it will be approximately as follows:

Animal	Cu Ft/Day Solids & Liquid	% Water	Gallons/Day
1,000-lb steer	1-1½	80-90	7½-10¾

Here is how to determine how much manure will need to be stored:

Storage capacity =Number of animals × daily manure production
× desired storage time (days) + extra water.

A rule of thumb is that when the pit occupies the entire area beneath the cattle, it will fill at a rate of 8 to 10 inches per month.

Fig. 28-16. Cattle on concrete slats, over a 3-foot pit. The excreta solids, which run about 9 percent, are evacuated by pumps into irrigation water. The cattle are confined to 20 square feet per animal. On either side of the slats is a 4-foot concrete apron with a 5 percent slope. The cattle move excreta into the pit by their natural movement. (Courtesy, E. S. Erwin and Associates, Tolleson, Ariz.)

SHADES

A hot steer is a poor doer! Hence, providing adequate shade to protect cattle from the sun is among the more important and widely used devices for improving the environment of cattle in hot climates. Tests conducted by the University of California, at the Imperial Valley Field Station, showed that it required 200 to 300 pounds more feed to make 100 pounds gain during mid-summer without shade than with shade. Based on work conducted at the Yuma, Arizona Station, where summer temperatures of 110° F are not uncommon, University of Arizona workers found that good shades can increase feedlot gains by 20 to 25 percent and improve feed efficiency by 14 to 20 percent.

Shades should be 12 feet or more high; provide 20 to 25 square feet per animal; and be oriented north and south, so that the sun will shine under the shade early in the morning and late in the evening. Shades may be run east-west in the hot deserts of the southwest to take fullest advantage of the cooler north sky.

Fig. 28-17. Shades are necessary in a hot climate. (Courtesy, Benedict Feeding Co., Casa Grande, Ariz.)

Fans, sprinklers, and other cooling devices increase the effectiveness of shades. At Yuma, Arizona, where August temperatures reach daily highs of 110° F, the Arizona Station reported that yearling steers in pens equipped with shades and evaporative coolers gained at the rate of 3.01 lb per head daily, compared with 2.66 lb daily for steers equipped with shades and sprinklers, and 2.62 lb per day in the control pens—shades only. The steers exposed to the evaporative coolers ate 17% more feed and gained 15% faster than the controls (shades only). Steers penned in the corrals with sprinklers ate 2% more feed and gained 29% *faster than the animals penned with shades only.*

● *Temperature and humidity*–University of Missouri researchers found that Shorthorn calves gained about ½ pound more per head per day and required 0.49 fewer pound of TND per pound of gain at 50° F than at 80° F, and that the heat-tolerant Brahmans were more efficient at 80° F than at 50° F.[5] At the University of California Station in the Imperial Valley, Herefords gained 0.36 pound more per head per day when they had access to shelter cooled 7° F below the high outside temperature.[6]

Somewhere between 50 and 75 percent relative humidity is considered optimum.

● *Insulation*–In cold climates, provide for R = 14 in the walls and R = 23 in the ceiling. Protect the insulation with 4-ml polyethylene plastic or equivalent.

● *Ventilation*–In most areas, summer and winter ventilation require-

[5]Brody, Samuel, *Climate Physiology of Cattle*, Journal Series No. 1607, Mo. Agr. Exp. Sta.
[6]Confinement Rearing of Cattle, USDA, ARS 22-89, October, 1964, p. 4.

ments are so different as to require two different systems. For a completely enclosed slotted-floor cattle barn, the fans should provide—

Winter—75 to 80 cfm/1,000-pound animal
Summer—225 cfm/1,000-pound animal

But the summer requirements in a completely enclosed building can be lessened by providing doors and/or hinged walls (high up), which can be opened during warm weather. Besides, this is good insurance when, and if, the electricity fails. (Also see Table 11-5, Recommended Enviromental Conditions for Animals.)

FEED DELIVERY EQUIPMENT

Most feedlots transport the feed from the mill to the feedlot by truck, using either (1) a self-unloading truck, or (2) a self-mixing self-unloading truck.

● *Self-unloading trucks*–These trucks do not mix the feed; they merely convey and unload it. The feed is mixed at the mill, loaded into the truck bed, driven to the lots, and augered into the bunks. Then, the truck returns to the feed mill or silage pit for another load. Self-unloading trucks are designed to convey the feed to one end of the bed and discharge it to one side. Some of these trucks, used by the larger lots, are very large—up to 20 tons capacity.

Fig. 28-18. Self-unloading truck in operation, conveying feed through the outlet spout into the feed bunk. (Courtesy, Benedict Feeding Co., Casa Grande, Ariz.)

● *Self-mixing self-unloading trucks*–These trucks, which are usually limited to 6 to 8 tons capacity, are well suited for use in feedlots with fewer than 5,000 cattle. They are equipped with their own bed mixing mechanism, which makes it possible to alleviate mixing at the mill. The separate ingredients are loaded at the mill, delivered to the lots, mixed in the truck bed, and conveyed through the outlet spout into the feed bunk. Self-mixing self-unloading trucks are also well suited to adding silage to the concentrate mix.

Scales may be installed on either type of truck. Generally speaking, pen size and ownership of the cattle determine whether on not trucks are equipped with scales. Where cattle are custom fed and/or small lots of cattle are involved, scales are important. For big lots of cattle, and where the cattle are owned by the feedlot, scales are not so necessary.

Two-way radios are standard equipment on most southwest feedlots.

Some small to medium-sized feedlots use conveyors (auger or belt-type) to move feed from the mill to the cattle. Unless the mill and the cattle are in close proximity (not more than 1,000 to 1,200 feet apart), conveyors are likely to be more costly than trucks. Also, when it comes to changing the feed formula of cattle, they are less flexible than trucks. Confinement systems, with a great concentration of cattle and short runs, make conveyors more practical.

One California feedlot of the author's acquaintance uses self-powered rail cars to deliver feed. Two men feed 23,000 cattle with ease. In comparison with truck delivery, a railroad has a higher initial cost. But it is more durable, and the maintenance cost on the mangers is low. One of San Francisco's old, but famed, cable cars is used to take buyers and visitors through this particular lot. Although this railroad has successfully operated in this feedlot for a number of years, the author's analysis is that only a "railroader" would be happy in using a rail system to feed cattle.

Among other feed delivery systems that are being, or have been, used are the following: a slurry system, in which the feed and water are mixed together and moved to the cattle; and pneumatic delivery.

MANURE CLEANING EQUIPMENT

Pens are generally cleaned after the cattle have been removed and before a new lot is brought in. Sometimes, a tractor skiploader is used to push manure from the concrete slab adjacent to the bunks and from around waterers, even when cattle are in the pen. However, when conditions are bad enough to require machine work in lots occupied by cattle, the corral is usually so wet that the equipment cannot operate efficiently. Thus, it is best to avoid such a condition by reducing the concentration of cattle in wet weather.

The most common manure cleaning equipment consists of a front-end loader attached to a tractor and a dump truck. The front-end loader is a scoop 4 to 5 feet wide, and 15 inches deep. The truck is a 4-wheel dump. Typically, feedlots with 1,000-head capacity use one tractor-loader and one truck; 5,000-head capacity lots use one tractor and two trucks; and 10,000-head capacity lots use two tractor-loaders and four trucks. The number of trucks needed is dependent on the distance that the manure must be hauled. The average load hauled from the pens is 7,000 pounds.

Feedlots with large pens sometimes use self-loading wheel scrapers.

Fig. 28-19. Renovating (loosening) manure prior to removal from pens. (Courtesy, Benedict Feeding Co., Casa Grande, Ariz.)

SPRAYING PEN; DIPPING VAT

Periodic and regular treatment for the control of lice, grubs, summer flies, and other parasites is essential to good management. For this purpose, a spray pen or a dipping vat will be needed.

A spray pen should not be over 15 feet wide, and it should have solid sides, good drainage, and a rough-paved or gravel surface. This type of pen will keep the animals close enough to the spray nozzle to give good spray penetration and reduce drift of the spray materials.

Dipping vats are increasing in popularity. A properly constructed system of chutes and pens with a dipping vat can provide fast, positive control. A metal vat can be purchased or a concrete vat can be built. The vat is usually 28 to 32 feet long, plus an entry chute at one end and a drip pen on the other.

BACK SCRATCHERS

Some feedlots provide back scratchers in their feeding pens. A back scratcher is a horizontal suspended arm with a burlap type cloth attached at cattle back level. This device is so arranged that the cloth is always saturated with insect repellent. The repellent is placed on the animal's back through contact with the scratcher, thereby preventing flies and other insects from molesting the cattle.

QUESTIONS FOR STUDY AND DISCUSSION

1. Select a certain area for feeding cattle. Then, give for that particular area the pros and the cons for each (a) an open feedlot, (b) cold confinement, and (c) warm confinement. Finally, give your recommendation.

2. What can a feedlot do to lessen pollution?

3. Describe a desirable feedlot mound.

4. Give the specifications for (a) a feedline bunk, and (b) an apron.

5. Why are so many cattle chutes curved?

6. What purposes do feedlot lights serve?

7. In confinement, how much ventilation should be provided (a) in the winter, and (b) in the summer?

8. Give the specifications for desirable cattle shades.

9. Under what circumstances would you recommend each of the following feed delivery systems: (a) self-unloading truck, (b) self-mixing self-unloading truck, and (c) conveyors (auger; belt)?

10. How, and how often, would you remove the manure from an open feedlot?

11. What are the advantages and disadvantages of slotted floors?

12. What are the advantages and what are the disadvantages of each (a) dipping, and (b) spraying?

13. Of what value are back scratchers?

14. What major changes do you foresee in cattle feedlot facilities and equipment from now until 2000 A.D.?

SELECTED REFERENCES

Title of Publication	Author(s)	Publisher
Beef Housing and Equipment Handbook	Midwest Plan Service	Midwest Plan Service, Iowa State University, Ames, Iowa, 1975
Cattle Feeding in the United States	Agricultural Economic Report No. 186	Economic Research Service, USDA, 1970
Feedlot, The	Ed. by I. A. Dyer, C. C. O'Mary	Lea & Febiger, Philadelphia, Penn., 1972
Improved Methods and Facilities for Commercial Cattle Feedlots	Marketing Research Report No. 517	Agricultural Marketing Service, USDA, 1962
Stockman's Handbook, The, Fourth Edition	M. E. Ensminger	The Interstate Printers and Publishers, Inc., Danville, Ill., 1970
Structures and Environment Handbook	Midwest Plan Service	Midwest Plan Service, Iowa State University, Ames, Iowa, 1972

KIND OF CATTLE TO FEED[1]

All kinds of cattle may be, and are, fed. But, for maximum success, it is imperative that the right kind of cattle be selected for a particular feedlot. The cattle should match the operator's available feed, labor, shelter, and credit. Also, it is imperative that there be a suitable market outlet following finishing; for example, it would be unwise to feed lightweight heifers in an area where the strongest slaughter market is for heavy steers; nor should one finish out heavy Holstein steers where the primary interest of packers is for Choice baby beef. But, assuming that a satisfactory slaughter outlet exists for different kinds of cattle, the following general guides will be helpful in determining what kind of cattle to feed in a given lot.

Fig. 29-1. Hereford feeder steers in feedlot of Master Feeders, Hooker, Oklahoma. (Courtesy, The American Hereford Association, Kansas City, Mo.)

---

[1]Tables 29-1 to 29-5 are a consensus (or judgment) of several knowledgeable feedlot consultants and Chairmen of Animal Science Departments, based on a survey made by the author. A consensus was resorted to for the reasons that (1) no extensive, nationwide, scientific study of this sort has ever been made, and (2) this information is much needed by cattlemen and those who counsel with them. No claim is made relative to the scientific accuracy of the data; rather, it is presented (1) because it is the best, if not the only, information of the kind presently available, and (2) with the hope that it will stimulate needed research along these lines.

Fig. 29-2. Mexican cattle in an Arizona feedlot. (Courtesy, Benedict Feeding Co., Casa Grande, Ariz.)

Fig. 29-3. Holstein steers on feed in Wisconsin. (Courtesy, Ralston Purina Co., St. Louis, Mo.)

AGE AND WEIGHT OF CATTLE

A generation ago the term "feeder steer" signified to both the rancher and the Corn Belt feeder a 2½- to 3-year-old animal weighing approximately 1,000 pounds. Today, cattle are referred to by ages as calves, yearlings, and 2-year-olds. This shift to younger cattle has been brought about primarily by consumer demand for smaller and lighter cuts of meat and improved feeding and management practices.

The age of cattle to feed is one of the most important questions to be decided upon by every practical cattleman. The following factors should be considered in reaching an intelligent decision on this point:

• *Rate of gain*—When cattle are fed liberally from the time they are calves, the daily gains will reach their maximum the first year and decline with each succeeding year thereafter. On the other hand, when in comparable condition, thin but healthy two-year-old steers will make more rapid gains in the feedlot than yearlings; likewise yearlings will make more rapid gains than calves. Table 29-1 illustrates this situation.

TABLE 29-1

EFFECT OF AGE OF CATTLE ON DAILY GAINS

	Daily Gains	
Age	Av. of U.S. Feedlots	Top 5% of U.S. Feedlots
	(lb)	*(lb)*
Calves	2.2	2.4
Yearlings	2.7	2.9
Two-year-olds	2.9	3.2

• *Economy of gain*—Calves require less feed to produce 100 pounds of beef than do older cattle. This may be explained as follows:

1. The increase in body weight of older cattle is largely due to the deposition of high energy fat, whereas the increase in body weight of young animals is due mostly to the growth of muscles, bones, and organs. Thus, the body of a calf at birth usually consists of more than 70 percent water, whereas the body of a fat 2-year-old steer will contain only 45 percent moisture. In the latter case, a considerable part of the water has been replaced by fat.

In passing, it is noteworthy that there is much room for improvement in cattle feedlot efficiency—even with calves, for the 8.0:1 feed efficiency of cattle must compete with 2.3:1 feed efficiency of the modern broiler industry.

For feedlot efficiency figures to be very meaningful, it is necessary to know what kind of ration was fed; otherwise, on a poundage basis, it is not unlike comparing steaks and carrots in the human diet. The more concentrated and the higher the energy value of the ration, the fewer the pounds of feed required to produce 100 pounds of gain. Yet, many times it is in the nature of good business to feed rations that necessitate more pounds of feed to produce 100 pounds of gain simply because lower cost gains can be produced thereby. This applies to feeding corn silage in many areas.

2. Calves consume a larger proportion of feed in proportion to their body weight than do older cattle.

3. Calves masticate and digest their feed more thoroughly than older cattle. Despite the fact that calves require less feed per 100 pounds gain—because of the high energy value of fat—older cattle store as much energy in their bodies for each 100 pounds of total digestible nutrients consumed as do younger animals.

From the above, it is apparent that age of cattle affects the pounds of feed required to produce 100 pounds of gain—that the younger the cattle the greater the feed efficiency.

Table 29-2 points up the effect of age of cattle on feedlot efficiency.

TABLE 29-2

EFFECT OF AGE OF STEERS ON FEED EFFICIENCY

	Av. of U.S. Feedlots			Top 5% of U.S. Feedlots		
	Feed/Lb Gain	TDN/Lb Gain	Mcal/Lb Gain[1]	Feed/Lb Gain	TDN/Lb Gain	Mcal/Lb Gain[1]
	(lb)	(lb)		(lb)	(lb)	
Calves	8.0	6.0	9.86	7.5	5.8	9.53
Yearlings	8.3	6.2	10.19	7.9	6.0	9.86
Two-year-olds and over	8.5	6.4	10.52	8.1	6.2	10.19

[1]Mcal was calculated by assuming 1.6434 Mcal = 1 TDN.

A more accurate measure of feed efficiency than pounds feed per pound gain can be obtained through the use of energy conversion; the TDN or Mcal required to produce a pound of beef. It alleviates much of the inevitable disadvantage to which a relatively low energy, bulky ration (such as a high silage ration) is put when it is compared on a poundage basis to a more highly concentrated feed (such as an all-concentrate ration). Thus, energy requirements (in both TDN and Mcal per pound of gain) are given in Table 29-2.

• *Flexibility in marketing*—Calves will continue to make satisfactory gains at the end of the ordinary feeding period, whereas the efficiency of feed utilization decreases very sharply when mature steers are held past the time that they are finished. Therefore, under unfavorable market conditions, calves can be successfully held for a reasonable length of time, whereas prolonging the finishing period of older cattle is usually unprofitable.

• *Length of feeding period*—Calves require a somewhat longer feeding period than older cattle to reach comparable finish. To reach Choice condition, steer calves are usually full-fed about 8 to 9 months; yearlings 5 to 6 months; and 2-year-olds only about 4 months. Table 29-3 points up this situation. The longer finishing period required for calves is due to the fact that they are growing as well as finishing.

TABLE 29-3

EFFECT OF AGE OF CATTLE ON LENGTH OF FEEDING PERIOD

Age	Av. for U.S. Feedlots	Top 5% of U.S. Feedlots
	(days)	(days)
Steer calves	250	230
Yearlings	160	135
Two-year-olds	120	110

● *Total gain required to finish*—In general, calves must put on more total gains in the feedlot than older animals to attain the same degree of finish. In terms of initial weight, calves practically double their weight in the feedlot. On the average, yearlings increase in weight about 450 pounds, and 2-year-olds increase their initial feedlot weight about 355 pounds. Table 29-4 illustrates this situation.

TABLE 29-4

EFFECT OF AGE OF CATTLE ON TOTAL GAIN REQUIRED TO FINISH

Age	Av. for U.S. Feedlots	Top 5% of U.S. Feedlots
	(lb)	(lb)
Calves	590	575
Yearlings	450	430
Two-year-olds	355	340

● *Total feed consumed*—Because of their smaller size, the daily feed consumption of calves is considerably less than for older cattle. However, as calves must be fed a longer feeding period, the total feed requirement for the entire finishing period is approximately the same for cattle of different ages.

● *Experience of the feeder*—Young cattle are bovine "babies." As such, they must be fed more expertly. Thus, the inexperienced feeder had best feed older cattle.

● *Kind and quality of feed*—Because calves are growing, it is necessary that they have more protein in the ration. Since protein supplements are higher in price than carbonaceous feeds, the younger the cattle the more expensive the ration. Also, because of smaller digestive capacity, calves cannot utilize as much coarse roughage, pasture, or cheap by-product feeds as older cattle.

Calves also are more likely to develop peculiar eating habits than older cattle. They may reject coarse, stemmy roughages or moldy or damaged feeds that would be eaten readily by older cattle. Calves also require more elaborate preparation of the ration and attention to other small details designed to increase their appetite.

● *Comparative costs and selling price*—Calves cost more per 100 pounds as feeders than do older cattle. They also sell for a higher price per hundredweight as finished cattle.

● *Dressing percentage and quality of beef*—Older cattle have a slightly higher dressing percentage than calves or baby beef. Moreover, many consumers have a decided preference for the greater flavor of beef obtained from older animals.

From the above discussion, it should be perfectly clear that there is no best age of cattle to feed under any and all conditions. Rather, each situation requires individual study and all factors must be weighed and balanced.

Baby Beef

Today, baby beef refers to heavy calves that are fat enough for slaughter at weaning time, weighing 400 to 700 pounds on foot. In its truest form, the production of baby beeves involves the breeding, rearing, and finishing of calves on the same farm. The first requirement of baby beef production is superior breeding. Secondly, calves intended for the baby beef route should never be allowed to lose their baby fat. This calls for dams that milk well and/or creep feeding the calves.

Baby beef production is well suited to the production of the maximum amount of beef from milk and grass, with a minimum of grain. Hence, it is predicted that it will increase in the future.

SEX OF CATTLE

More steers than heifers are fed, simply because more of them are available. A portion of the heifers is held back for replacement purposes. In the future, more young bulls will be fed. Thus, the feedlot operator must give consideration to the sex of cattle fed. First, and foremost, he should consider his market outlets.

TABLE 29-5

EFFECT OF SEX OF YEARLING CATTLE ON GAIN

Sex	Av. for U.S. Feedlots		Top 5% U.S. Feedlots	
	Daily Gain	Feed/Lb Gain	Daily Gain	Feed/Lb Gain
	(lb)	*(lb)*	*(lb)*	*(lb)*
Heifers	2.4	7.5	2.6	6.9
Steers	2.8	6.9	3.0	6.6
Bulls	3.0	6.7	3.2	6.4

Table 29-5 shows the effect of sex on rate of gain and feed efficiency. It is noteworthy that bulls gain more rapidly on less feed than steers, and that steers gain more rapidly on less feed than heifers.

Steers vs Heifers

On the market, cattle are divided into five sex classes: steers, heifers, cows, bullocks, and bulls. The sex of feeder cattle is important to the producer from the standpoint of cost and selling price (or margin), the contemplated length of feeding period, quality of feeds available, and ease of handling. The consumer is conscious of sex differences in cattle and is of the impression that it affects the quality, finish, and conformation of the carcass.

Steers are by far the most important of any of the sex classes on the market, both from the standpoint of numbers and their availability throughout the year, whereas heifers are second.

The relative merits of steers vs heifers, both from the standpoint of feedlot performance and the quality of carcass produced, has long been a controversial issue. Based on experiments[2] and practical observations, the following conclusions and deductions seem to be warranted relative to this question:

• *Length of feeding periods*—Heifers mature earlier than steers and finish sooner, thus making for a shorter feeding period. In general, heifers may be ready for the market 30 to 40 days earlier than steers of the same age started on feed at the same time.

• *Market weight*—The most attractive heifer carcasses are obtained from animals weighing 650 to 900 pounds on foot, showing good condition and finish but not patchy and wasty.

• *Rate and economy of gain*—Because of their slower daily gains and lower feed efficiency, the feedlot gains made by heifers are usually somewhat more costly than those made by steers of the same age.

• *Price*—Because of existing prejudices, feeder heifers can be purchased at a lower price per pound than steers, but they also bring a lower price when marketed. Thus, the net return per head may or may not be greater with heifers.

• *Carcass quality*—In England, there is no discrimination in price against well-finished heifers. In fact, the English argue that the grain of meat in heifer carcasses is finer and the quality superior. On the other hand, the hotels, clubs, and elite butcher shops in the United States hold a prejudice against heifer beef.

Carefully controlled experiments have now shown conclusively that when heifers are marketed at the proper weight and degree of finish, sex makes no appreciable difference in the dressing percentage, in the retail value of the carcasses, or in the color, tenderness, and palability of the meat.

• *Ease of handling in the feedlot*—Because of disturbances at heat periods, many feeders do not like to handle heifers in the feedlot. Of course, the incidence of estrus can be lowered by feeding the additive MGA (see Table 30-6).

• *Flexibility in marketing*—If the market is unfavorable, it is usually less advisable to carry heifers on feed for a longer period than planned because (1) of possible pregnancies, and (2) they become too patchy and wasty.

---

[2]Bull, Sleeter, F. C. Olson, and John H. Longwell, *Ill. Agr. Exp. Sta. Bull. No. 355*, 1930.

• *Effect of pregnancy*—Packer buyers have long insisted that they are justified in buying finished heifers at a lower price than steers of comparable quality and finish because: (1) most heifers are pregnant and have a lower dressing percentage; and (2) pregnant heifers yield less desirable carcasses. In realization that the packer will lower the price anyway, many feeders make it a regular practice to turn a bull with heifers about three to four months before the market period. Such feeders contend that the animals are then quieter and will make better feedlot gains.

In a carefully controlled experiment designed to ferret out the facts of this controversy, the Illinois Experiment Station[3] compared open and bred heifers in a five-month period with the following pertinent results:

1. The bred heifers were quieter and easier to handle in the feedlot.
2. The bred heifers possessed keener appetites.
3. The pregnant heifers grew less and put on more finish.
4. When kept at the same level of feed consumption, pregnancy had no appreciable effect on average daily gain.
5. Pregnancy did not affect dressing percentage. This was probably due to the higher finish and the lighter hides, heads, shanks, stomach and smaller fill of the pregnant heifers. These factors were sufficient to overcome the weight of the 3 to 4 months' fetus which averaged 30 pounds.
6. The carcasses of the bred heifers were noticeably better finished.
7. The flank, loin-end, and round were lighter in the bred than in the open heifers, but the other wholesale cuts were practically the same in each group.

The results of the Illinois experiment would indicate that the producer is entirely justified in breeding full-fed heifers three to four months prior to the time of marketing. It would also seem that packers are not justified in buying such bred heifers at a lower price because of any alleged lower dressing percentage or difference in carcass quality.

The trade in feeder cows and heifers assumes considerable volume only in the fall and early winter—at the close of the grazing season when the farmer or rancher is culling his herd and prior to the start of the wintering operations. When market conditions are favorable and an abundance of cheap roughage is available, cows may often be fed at a profit.

When there is considerable demand for cheap meats, the feeder may find it profitable to finish old bulls and stags. Usually it is difficult to purchase such animals in large numbers. Here, as with the finishing of old cows, the feeder should plan to utilize the maximum of cheap roughage.

Spayed Heifers

Spaying prevents heifers from becoming pregnant and eliminates the necessity of separating heifers from bulls or steers. Also, some buyers pay a slight premium for spayed heifers. However, most experiments have shown that spayed heifers make less rapid gains and require more feed per 100

---

[3]Snapp, R. R., and Sleeter Bull, *Ill. Agr. Exp. Sta. Bull. No. 508*, 1944.

pounds gain than open (control) heifers.[4] The latter facts, plus the attendant danger of the operation, generally do not justify spaying unless the selling price is sufficiently higher.

Bulls

The feeding of bulls (uncastrated males) instead of steers has been standard practice throughout Europe for many years. For example, since about 1954 Germany has, for the most part, fed out and slaughtered bulls as yearlings, instead of steers; because they obtain 10 to 15 percent greater rate of gain and feed efficiency thereby. The practice will increase in the United States now that carcasses from young bulls are Federally graded as "bullock beef" (the use of the term "bullock beef" to identify meat from young bulls became effective July 1, 1973) rather than "bull beef," thereby removing the connotation that the meat is inferior to or different from steer or heifer beef.

The carcasses from older bulls are still labeled "bull beef," to differentiate them from the carcasses of younger bulls. Bullock beef from young bulls is graded according to the same quality standards as beef from steers and heifers.

Also, the economics of the situation favor the feeding of bulls instead of steers. The male hormones secreted by the testicles are excellent growth stimulants and will improve gain and feed efficiency by 10 to 15 percent. Also, bulls will produce more lean meat than steers, and research has shown that bull meat is equal in value, quality, and palatability to steer meat.

Now that the carcasses from young bulls are differentiated from older bulls, only consumer acceptance remains. Without doubt, beef shortages will speed acceptance of bullock beef.

The following guidelines are recommended in the feeding of bulls:

1. Start young bulls on full feed at weaning age (6 to 7 months) and feed out as rapidly as possible to a market weight of 1,100 pounds.

2. Use high-energy rations for bull feeding, because they tend to grow rapidly and lay down less fat than steers.

3. Feed out bulls so that they are finished for market before 18 months of age.

4. Do not add new bulls to the pen after the weanling bulls are started on feed, because this tends to encourage fighting and riding, and results in reduced gains.

5. Keep bulls separate from other cattle when marketing them. If possible, do not permit the bulls to stand in the pen overnight before slaughter.

6. Bulls of beef breeding are less nervous and ride less than bulls of dairy breeding.

---

[4]Dinussion, W. E., F. N. Andrews, and W. M. Beeson, *Journal of Animal Science*, Vol. 9, 1950, p. 321; Gramlich, H. J, and R. R. Thalman, *Neb. Agr. Exp. Sta. Bull. 252*, 1930; Hart, G. H., H. R. Guilbert, and H. H. Cole, *Calif. Agr. Exp. Sta. Bull. 645*, 1940; Langford, L. H., R. J. Douglas, and M. L. Buchanan, *N.D. Agr. Exp. Sta. Bimonthly Bull.*, Vol. XVIII, No. 2, 1955.

GRADE OF CATTLE

The most profitable grade of cattle to feed will generally be that kind of cattle in which there is the greatest spread of margin between their purchase price as feeders and their selling price as fat cattle. As can be readily understood, one cannot arrive at this decision by merely comparing the existing price between the various grades at the time of purchase. Rather, it is necessary to project the differences that will probably exist, based on past records, when the animals are finished and ready for market.

As fewer grain-fed cattle are marketed in the summer and fall, the spread in price between Good and Choice fed cattle and those of the lower grades is usually the greatest during this season. On the other hand, the spread between these grades is likely to be least in late winter and early spring, when a large number of well-finished cattle are coming to market from the feedlots.

The length of the feeding period and the type of feed available should also receive consideration in determining the grade of cattle to feed. Thus, for a long feed and when a liberal allowance of grain is to be fed, only the better grades of feeders should be purchased. On the other hand, when a maximum quantity of coarse roughage is to be utilized and a short feed is planned, cattle of the medium or lower grades are most suitable. Thus, successful cattle feeders match the quality of the cattle selected with the quality of the available feed; the better the feed the higher the grade of cattle.

Cattle of the lower grades should be selected with very special care to make certain that only thrifty animals are bought. Ordinarily, death losses are much higher among low-grade feeder cattle, especially when the low-grade animals are calves. The death loss in handling average or high-grade feeders seldom exceeds one to two percent; whereas with "cull" or "dogie" cattle, it frequently is twice or three times this amount. Many low-grade cattle are horned, and dehorning further increases the death risk—in addition to the added labor and shrinkage resulting therefrom.

No given set of rules is applicable under any and all conditions in arriving at the particular grade of cattle to feed, but the following factors should receive consideration.

1. The feeding of high-grade cattle is favored when:
   a. The feeder is more experienced.
   b. A long feed with a maximum of grain in the ration is planned.
   c. Conditions point to a wide spread in price between grades at marketing time. Such conditions normally prevail in the late summer or early fall.

2. The feeding of average or low-grade cattle is favored when:
   a. The feeder is less experienced.[5]
   b. A short feed with a maximum of roughage or cheap by-products is planned.

[5]In general, the inexperienced feeder should stick to the middle kind and leave the extremes—the fancy and the plain cattle—to the man with experience.

c. Conditions point to a narrow spread in price between grades at marketing time. Such conditions normally prevail in the spring.

3. In addition to the profit factors enumerated above, it should be pointed out that with well-bred cattle the following conditions prevail:

a. Well-bred cattle possess greater capacity for consuming large quantities of feed than steers of a more common grade, especially during the latter part of the feeding period.

b. The higher the grade of cattle, the higher the dressing percentage and the greater the proportionate development of the high-priced cuts.

c. The higher the grade of the cattle, the greater the opportunities for both profit and loss.

d. There is a great sense of pride and satisfaction in feeding well-bred cattle.

Certainly the producer who raises his own feeder cattle should always strive to breed high quality cattle, regardless of whether he finishes them himself or sells them as feeders. On the other hand, the purchaser of feeder steers can well afford to appraise the situation fully prior to purchasing any particular grade.

BREEDING AND TYPE OF CATTLE

Although the supporting data are rather limited, it is fully realized that there is considerable difference between individual animals insofar as rate and economy of gain is concerned. It is to be emphasized that these differences are greater within breeds than between breeds.

The University of California has compared the feedlot performance and carcass quality of Okie and Hereford calves and yearlings.[6] Contrary to the implication of the name, Okie cattle do not necessarily originate in Oklahoma. Rather, they are cattle of nondescript breeding, including both beef and dairy background, that originate on small farms in the southern part of the United States.

The California studies showed that—

1. Herefords gained faster, utilized their feeds more efficiently, and graded higher than Okies.

2. Okies were upgraded from feeder to slaughter stage, whereas Herefords maintained their grade.

No claim can be made that the animals used in the California studies were representative of all cattle of the breeds and strains studied, for, when limited to small numbers, it is impossible accurately to select a representative sample of a widely scattered breed or strain.

Crossbreds

Good crossbreds will likely show two to four percent improvement over the average of the parent breeds for rate and efficiency of gains. Additional-

---

[6]California Feeders' Day reports of 1965 and 1966.

ly, even larger advantages accrue to the cow-calf producer. Thus, it is inevitable that an increasing number of crossbred feedlot cattle will be seen. The primary characteristics desired in feedlot cattle are the same, whether they be crossbreds or straightbreds; namely, (1) high rate of gain, (2) efficient feed conversion, (3) high cutout percent, and (4) tender-palatable beef.

Fig. 29-4. Crossbred cattle are becoming increasingly popular among producers feeding their own cattle. (Courtesy, Ron Baker, C & B Livestock, Hermiston, Ore.)

Dairy Beef

Dairy beef accounts for about one-fourth of the beef consumed in this country, with these animals marketed as veal calves, cull dairy cows and bulls, and finished dairy heifers and steers. Improvements in the science and technology of feeding and processing favor growing and finishing dairy beef, and minimum slaughter of veal calves.

"Dairy beef" is just what the term implies—beef derived from cattle of dairy breeding, or from dairy × beef crossbreds. Today, it's extolled with pride.

But dairy beef hasn't always enjoyed status. Prior to about 1960, few self-respecting beef cattlemen would admit to finishing out dairy cattle. Given a choice between (1) topping the market with a uniform load of well-bred beef steers, even if they were fed at a loss, or (2) making money by feeding cattle of dairy breeding, most cattle feeders would have elected the first alternative—that is, they would have done so until recent years. They derived much satisfaction from topping the market, and they took pride in telling their neighbors about it. Likewise, meat packers were reluctant to have visitors see yellow-finished carcasses in their coolers, because of the yellow fat being indicative of dairy breeding or grass finish. Only the presence, suspicioned if not real, of goat carcasses was more humiliating to a packer. The near-contempt formerly evoked by cattle of dairy breeding was further evidenced on the nation's terminal markets by the names that were applied to them. Holsteins were known as "magpies," and Jerseys were known as "yellow hammers"—terms which were neither endearing nor appetizing.

Fig. 29-5. Dairy beef in the making. Holstein steers on feed in a commercial feedlot near Corona, Calif. (Courtesy, Albers Milling Co., Los Angeles, Calif.)

But time was! Today's cattle feeders are primarily concerned with rate and efficiency of gains, and net returns. As a result, most of them would just as soon feed steers of dairy breeding, either purebreds or crossbreds; some actually prefer them. Consumers demand beef that has a maximum of lean, with a minimum amount of waste fat, and which is tender and flavorful; and they couldn't care less whether it comes from a critter that was black, white faced, roan, pink, yellow, or polka dot. As a result, more and more steers of dairy breeding are going the feedlot route, rather than as veal. As evidence of this transition, during the 30-year period 1942-1972, U.S. per capita veal consumption declined from 8.2 to 2.2 pounds, while per capita beef consumption increased from 61.2 to 115.9 pounds in this same period of time. Also, the shift in consumer demand to more lean and less fat has been reflected in the changed Federal grades of beef. As a result when properly fed, Holstein steers will make Choice, Good, or Standard grades.

● *Britains' "barley beef"*—The popularity of dairy beef traces to Europe. It was pioneered by Dr. T. R. Preston, well-known Scottish animal scientist, in Scotland and throughout the United Kingdom, where it became known as "barley beef." Holstein Friesian bull calves (not steers) were fed on all-concentrate rations consisting chiefly of barley. The British reported gains of 2½ to 3 pounds per day, slaughter weights of 900 pounds in less than one year of age, and lifetime feed conversions under 5 to 1.

●*High growth thrust essential*—For a dairy beef program to be most successful, scientists and cattlemen in both Britain and the United States are agreed that the animals should have a high growth potential, as evidenced by heavy birth weight and heavy weight at maturity. Since Holsteins are heavy at birth and mature out at around 1,400 pounds, in comparison with mature weights of 1,000 to 1,200 pounds of the European beef breeds, it can be readily understood that Holsteins are ideal when it comes to producing dairy beef.

●*Decide on feeding program*—There is, of course, no one best system of producing dairy beef for any and all conditions. As is true in any type of cattle feeding program, the operator should make the best use of those feeds that are readily available at the lowest possible cost. Then, these feeds must be combined into satisfactory rations, with consideration given to both economy and probable market price of finished cattle of various weights, grades, and degrees of finish.

●*High energy rations; light market weights*—If dairy steers are to be slaughtered at young ages and light weights, high energy (low roughage) rations are imperative. Under this system, usually young calves of either dairy or dairy × beef breeding are fed in confinement—in barns; and fed milk replacers from 1 to 4 days of age to 200 to 300 lb; and are full-fed a high-concentrate ration from about 300 lb to market weight of 750 to 950 lb. Essentially, this is the "barley beef" program of Europe.

Crowding for market at an early age takes advantage of the fact that growth is generally most economical when most rapid, and that young gains are cheap gains. Also, experience shows that when Holstein calves are started on super energy rations at around 300 to 350 pounds weight and marketed under 950 pounds, (1) there's excellent marbling with very little bark (outside fat), and (2) many of these animals will grade Choice.

●*High roughage rations; heavy market weights*—If roughages are relatively more abundant and cheaper than concentrates, then it may be more remunerative to feed dairy beef more roughage and market at heavier weights—and with it to expect slower and less efficient gains. In any event, it's net returns that count, rather than rate of gain and pounds of feed required per pound of gain.

Under the high roughage system, steers of dairy breeding are grown on maximum roughage to 600 to 750 pounds weight, following which the ratio of concentrate to roughage is increased. Most dairy steers fed according to this system are marketed at weights of 1,050 to 1,200 pounds, grading Good or Commercial. (Most of them are too old to grade Standard, and lack the necessary marbling to grade Choice.)

●*Dairy beef has good potential*—There is ample evidence that male calves of the larger dairy breeds (Holsteins and Brown Swiss), along with dairy beef crosses, have the potential for producing acceptable beef with good feed efficiency, under a system of either (1) full-feeding from an early age on a high energy ration, or (2) growing and finishing on a maximum of roughage and marketing at older ages and heavier weights. In the final analysis, therefore, the system selected should be determined by net returns.

Both methods necessitate the rearing of young, one- to four-day-old calves to weights of around 300 pounds, with such early rearing done by either a calf-raising specialist or by the cattle feeder who will do the ultimate finishing. Such calves must usually be obtained from over a wide area, and of variable ages and sizes; hence, they are difficult to come by. Also, death losses are frequently high and discouraging.

Both commercial cattle feeders and dairymen are showing increased interest in producing dairy beef, with the result that there is competition between them. More dairy beef will be produced in the future.

QUESTIONS FOR STUDY AND DISCUSSION

1. Why is it important that the kind of cattle to feed match the operator's available feed, labor, shelter, and credit?

2. Discuss how each of the following enter into the choice of the kind of cattle to feed in a given lot:

   a. Age and weight of cattle.
   b. Sex of cattle.
   c. Grade of cattle.
   d. Breeding and type of cattle.

3. Should cattle feeders achieve fewer pounds of feed per 100 pounds gain by going to high energy rations, like broiler rations?

4. Will more young bulls be fed in the United States in the future? Justify your answer.

5. Why do the English prefer heifer beef, whereas Americans prefer steer beef?

6. Would you recommend that a cattle feeder breed all heifers about three to four months before marketing? Justify your answer.

7. Discuss the advantages and the disadvantages of preventing heifers from coming in heat (a) by spaying, or (b) by feeding MGA.

8. Will more bulls be fed in the United States in the future, as is done in Europe?

9. What are Okie Steers? Under what circumstances would you buy and feed Okie steers instead of Hereford steers?

10. What advantages may accrue from feeding crossbreds rather than straightbreds?

11. Will more dairy beef be fed in the future? Who will feed the dairy beef of the future—the dairyman, or the specialized feedlot feeder?

12. What's "barley beef"?

13. What breeds of cattle are particularly suited to the production of dairy beef?

14. When feeding dairy beef, what will determine the choice between—(a) high energy rations and light market weight; vs (b) high roughage rations and heavy market weight?

SELECTED REFERENCES

Title of Publication	Author(s)	Publisher
Beef Cattle, Sixth Edition	A. L. Neumann R. R. Snapp	John Wiley & Sons, Inc., New York, N.Y., 1969
Beef Cattle Science Handbook	Ed. by M. E. Ensminger	Agriservices Foundation, Clovis, Calif., pub. annually since 1964
Feedlot, The	Ed. by I. A. Dyer, C. C. O'Mary	Lea & Febiger, Philadelphia, Penn., 1972

FEEDING FINISHING (FATTENING) CATTLE[1]

An understanding of digestion is essential to intelligent feeding of finishing cattle. Fig. 30-2 shows the location of the parts of the ruminant stomach and the processes that occur in energy digestion and metabolism.

Digestion of feedstuffs in ruminants is primarily a fermentation process that occurs in the rumen. This allows ruminant animals to use both roughages and grains as sources of carbohydrates for energy. Part of the carbohydrates pass through the rumen and are digested in the abomasum and small intestine. Most carbohydrates in feeds are converted to either acetic, propionic, or butyric acid by rumen bacteria and protozoa. These short-chain fatty acids are then absorbed through the rumen wall into the bloodstream and are eventually used for energy in body tissue.

The major nutritional requirements of finishing cattle are: energy, protein, minerals, vitamins, and water.

[1]The author is grateful to the following scientists for their authoritative review of this chapter: Dr. G. P. Lofgreen, Nutritionist, University of California, Imperial Valley Field Station, El Centro, Calif., and Dr. Wilton W. Heinemann, Animal Scientist, Irrigated Agriculture Research and Extension Center, Washington State University, Prosser, Wash.

The principles of beef cattle nutrition are covered in Chapter 8; hence, they will not be repeated in this chapter. Instead, the application of nutrition to cattle finishing will be covered in this chapter.

Fig. 30-1. Feeding cattle the modern way. Feed truck, equipped with automatic augering system, is shown conveying feed through the outlet spout into the feed bunk. This is part of the facilities and equipment at Farr Feeders, Inc., Greeley, Colo., a new $4 million, fully computerized commercial cattle feedlot capable of producing for market more than 80,000 head of cattle a year. (Courtesy, Farr Feeders, Inc., Greeley, Colo.)

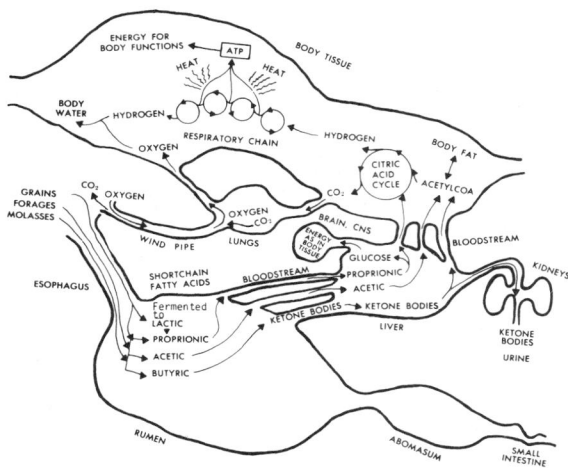

Fig. 30-2. Energy digestion and metabolism in ruminants. (Source: *Great Plains Beef Cattle Feeding Handbook*, GPE-1000)

About 80 percent of the cost of finishing cattle, exclusive of the purchase price of the feeders, is feedstuffs—grain, hay, silage, and miscellaneous wastes and by-products. The greatest need is for energy. Of course, net profit depends on how much of that energy can be converted to pounds of gain— and how efficiently.

EXPRESSING ENERGY VALUES OF FEEDS

Energy is the largest cost item in the ration. Thus, the value of a feed ingredient is based primarily upon the energy which it will provide the ani-

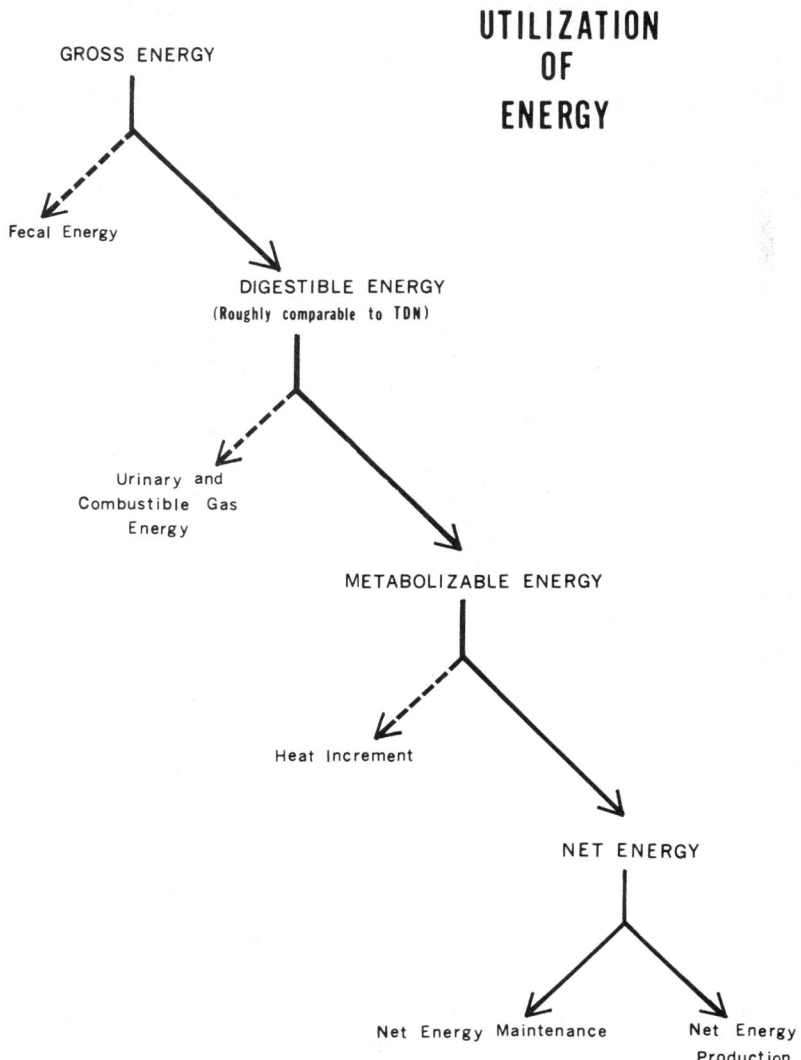

Fig. 30-3. Utilization of energy.

mal, assuming it is not used specifically as a supplement to provide protein, minerals, vitamins, or additives. Cereal grains are higher than roughages in energy. Although grains usually cost more on a weight basis than roughages, they are often a cheaper source of energy for beef rations.

Broadly speaking, two methods of measuring energy are used in this country—the total digestible nutrient system (TDN), and the calorie system. Both systems are discussed in Chapter 8 of this book, under the heading "Methods of Measuring Energy."

Calories are used to express the energy value of feedstuffs. One calorie is the amount of heat required to raise the temperature of one gram of water 1°C. One kcal = 1,000 calories. One mcal or therm = 1,000,000 calories.

Through various digestive and metabolic processes, much of the energy in feed is dissipated as it passes through the animal's digestive system. About 60 percent of the total combustible energy in grain and about 80 percent of the total combustible energy in roughage is lost as feces, urine, gases, and heat. These losses are illustrated in Fig. 30-3.

As shown in Fig. 30-3, energy losses occur in the digestion and metabolism of feed. Measures that are used to express animal requirements and the energy content of feeds differ primarily in the digestive and metabolic losses that are included in their determination. Thus, the following terms are used to express energy value of feeds:

● *Gross energy (GE)*—Gross energy represents the total combustible energy in a feedstuff. It does not differ greatly between feeds, except for those high in fat. For example, one pound of corncobs contains about the same amount of GE as one pound of shelled corn. Therefore, GE does little to describe the useful energy in feeds for finishing cattle.

● *Digestible energy (DE)*—Digestible energy is that portion of the GE in a feed that is not excreted in the feces.

● *Metabolizable energy (ME)*—Metabolizable energy represents that portion of the GE that is not lost in the feces, urine, and gas. Although ME more accurately describes the useful energy in the feed than does GE or DE, it does not take into account the energy lost as heat.

● *Net energy (NE)*—Net energy represents the energy fraction in a feed that is left after the fecal, urinary, gas, and heat losses are deducted from the GE. The net energy, because of its greater accuracy, is being used increasingly in feedlot ration formulations, especially in computerized formulations for large feedlots.

Although net energy is a more precise measure of the real value of the feed than other energy values, it is much more difficult to determine. Heat increment must be determined with the whole body or respiration calorimeter. California workers have recently made effective use of a comparative slaughter technique and balance trials to calculate net energy values of feeds and determine animal requirements.

The net energy values of feeds are different for maintenance and production. Roughages compare more favorably with grain for maintenance than for production.

•*Net energy for maintenance (NEm)*—This is the fraction of the net energy that keeps the animal in energy equilibrium.

•*Net energy for gain (NEgain)*—This is the fraction of the net energy available for gain in weight.

RATION FORMULATION

Feedlot cattle have access only to the rations provided by the caretaker. It is important, therefore, that cattle feedlot rations be balanced, and that they make for maximum net returns.

In addition to considering changes in availability of feeds and feed prices, ration formulation should be altered at stages to correspond to weight increases in the cattle.

Some suggested rations that may serve as useful guides are given in Chapter 8, Table 8-15.

Ration formulation consists of combining feeds to make a ration that will be eaten in the amount needed to supply the daily nutrient requirements of the animal. The nutritive requirements of all classes and ages of beef cattle, including finishing cattle, are given in Tables 8-18 and 8-19 of Chapter 8. The nutrient compositions of feeds commonly fed to beef cattle are given in the Appendix, Table I-3. In these tables, the net energy of each feed is partitioned into energy used for body maintenance and for production; thus, the net energy values in megacalories (mcal) per pound are given for each feed for maintenance (NEm) and for gain (NEgain).

Table 30-2 shows the net energy requirements of both steers and heifers, for growing and finishing.

Using Net Energy to Calculate Rations

Any system of expressing energy requirements must lend itself to use in ration formulation. Net energy values are no exception. The following examples will show how to use net energy values to calculate rations.

•*Example 1—Using net energy values to predict number of pounds of the ration that must be consumed to produce a specific gain*—In the following example, net energy values of feeds are used to predict the number of pounds of a given ration that a 775-pound yearling steer would need to consume to gain 2.9 pounds daily. Bear in mind that the ration (in this example, the ration in Table 30-1) must be balanced for protein, minerals, and vitamins, in order for these net energy values to have validity for predicting gain.

Step 1—First calculate the net energy for maintenance (NEm) and gain (NEgain) values for a pound of the ration shown in Table 30-1.

Use Table I-3, of the Appendix, which gives composition values of feeds on a dry basis (moisture free). One pound of the Table 30-1 ration supplies 0.8998 megacalories of net energy for maintenance (mcal NEm) and 0.5561 megacalories of net energy for gain (mcal NEgain).

TABLE 30-1

RATION FOR FINISHING CATTLE

Ration Ingredient	Lb	Composition of Ingredient NEm[1] (Mcal/lb)[3]	Ration Supplies NEm[1] (Mcal)[3]	Composition of Ingredients NEgain[2] (Mcal/lb)[3]	Ration Supplies NEgain[2] (Mcal)[3]
Shelled corn, No. 2	68.60	1.04	71.34[4]	0.67	45.96[5]
Soybean meal (solvent)	4.00	0.88	3.52	0.59	2.36
Alfalfa hay (midbloom)	27.00	0.56	15.12	0.27	7.29
Salt	0.40	——	——	——	——
Total	100.00		89.98		55.61

[1]NEm = net energy for maintenance.
[2]NEgain = net energy for gain.
[3]Mcal stands for Megacalorie.
[4]68.60 lb × 1.04 = 71.34.
[5]68.60 lb × 0.67 mcal = 45.96.

Step 2—From Table 30-2 find the requirement for a 775-pound yearling steer to gain 2.9 pounds daily. This follows:

	Mcal/day
NEm	6.24
NEgain	6.48

Step 3—Pounds of feed (dry basis, moisture free) to meet the daily maintenance requirement:

$$6.24 \text{ mcal} \div .8998 \text{ mcal} = 6.92 \text{ lb}$$

Step 4—Pounds of feed (dry basis, moisture free) to meet the requirement for 2.9 pounds daily gain:

$$6.48 \text{ mcal} \div .5561 \text{ mcal} = 11.65 \text{ lb}$$

Step 5—Total pounds of feed (dry basis, moisture free) steer must eat daily to gain 2.9 pounds:

$$6.92 \text{ lb} + 11.65 \text{ lb} = 18.57 \text{ lb}$$

Step 6—Convert to an "as fed" basis. Bear in mind that the 18.57 pounds is on a dry basis, moisture free. It can be converted to an "as fed" basis by following the procedure outlined in the Appendix, Section I, under "Formulating Rations." Or if, for example, the ration contained 90% DM, the daily ration (on an "as fed" basis) can be calculated by dividing 18.57 pounds by .90:

$$18.57 \div .90 = 20.63$$

● *Example 2—Using net energy to predict the average daily gain of a 775-pound steer that is consuming a certain number of pounds of a specified*

TABLE 30-2

NET ENERGY REQUIREMENTS[1]

FOR GROWING-FINISHING STEERS

Body Weight:								
lb		441	551	661	772	882	992	1,102
kg		200	250	300	350	400	450	500
Daily Gain:		**NEm[2] Required, Mcal/day[3]**						
lb	*kg*							
0	0	4.1	4.84	5.55	6.24	6.89	7.52	8.14
		NEg[4] Required, Mcal/day[3]						
1.1	.5	1.49	1.77	2.02	2.27	2.51	2.74	2.97
1.5	.7	2.14	2.53	2.90	3.26	3.60	3.93	4.25
2.0	.9	2.82	3.33	3.82	4.29	4.74	5.18	5.60
2.2	1.0	3.16	3.75	4.29	4.82	5.33	5.82	6.29
2.4	1.1	3.52	4.17	4.78	5.36	5.93	6.47	7.01
2.6	1.2	3.88	4.60	5.27	5.92	6.54	7.14	7.73
2.9	1.3	4.26	5.04	5.77	6.48	7.16	7.83	8.47
3.1	1.4	4.63	5.49	6.29	7.06	7.80	8.52	9.22

FOR GROWING-FINISHING HEIFERS

Daily Gain:		**NEm[2] Required, Mcal/day[3]**						
0	0	4.1	4.84	5.55	6.24	6.89	7.52	8.14
		NEg[4] Required, Mcal/day[3]						
.4	.2	.62	.74	.84	.95	1.05	1.14	1.24
.7	.3	.95	1.13	1.29	1.45	1.61	1.75	1.90
1.1	.5	1.66	1.96	2.25	2.52	2.79	3.05	3.30
1.5	.7	2.42	2.86	3.27	3.68	4.06	4.44	4.80
1.8	.8	2.81	3.33	3.82	4.28	4.73	5.17	5.59
2.0	.9	3.23	3.81	4.37	4.91	5.43	5.93	6.41
2.2	1.0	3.65	4.32	4.95	5.56	6.14	6.71	7.26
2.4	1.1	4.09	4.84	5.55	6.23	6.88	7.52	8.13
2.6	1.2	4.55	5.37	6.16	6.91	7.64	8.35	9.03

[1]Adapted by the author from *Nutrient Requirements of Beef Cattle*, fifth revised edition, 1976, National Academy of Sciences, Washington, D.C., p. 34.
[2]Net energy for maintenance.
[3]Megacalories per day.
[4]Net energy for gain.

ration–Let's assume that we have a 775-pound steer that is consuming 18 pounds of the ration shown in Table 30-1. What daily gain should be expected?

Step 1—Pounds of feed (dry basis, moisture free) to meet the daily maintenance requirement = 6.92 pounds (see prior example).

Step 2—Pounds of feed left for gain:

$$18 \text{ lb} - 6.92 \text{ lb} = 11.08 \text{ lb}$$

Step 3—Mcal of NE_{gain} supplied by remaining feed:

$$11.08 \text{ lb} \times .5561 \text{ mcal} = 6.16 \text{ mcal}$$

Step 4—Daily gain expected from 6.16 Mcal of NE_{gain} (Table 30-2).

$$2.6 \text{ lb gain} = 5.92 \text{ Mcal}$$

Thus, slightly over 2.6 pounds daily gain should be produced.

FEEDS

The growth of the cattle feeding industry of America has gone hand in hand with the production and feeding of more grains and by-product feeds. Such feeds, which are high in energy and low in fiber, are usually the most economical part of a finishing ration. Hence, their availability influences the location and type of feeding program. Roughages, which require relatively more energy to digest and metabolize than grains, are used at low levels in most finishing rations. However, they are important in growing programs or in warm-up rations and they are even more important for maintenance of the breeding herd.

For convenience, the commonly used beef cattle finishing feeds are herein classified as (1) concentrates (2) by-product feeds, (3) protein supplements, (4) roughages, (5) minerals, (6) vitamins, and (7) additives.

Concentrates

Concentrate feeds are those which are high in energy and low in fiber. Many different kinds of concentrate feeds can be, and are, used in beef cattle finishing. Availability and price are the two most important factors determining the choice of concentrates. Consideration of the latter factor—price—necessitates that the cattle feeder be a keen student of values. He must change the formulations of his ration(s) in keeping with comparative feed prices.

Corn is the most common and the most desirable grain used in finishing cattle. In 1972, the United States produced 5.5 billion bushels of corn, in comparison with 0.8 billion bushels of grain sorghum that same year. Corn is palatable and rich in the energy-producing carbohydrates and fats, and low in fiber. Also, corn is easily stored, only moisture and carotene being lost over a period of time. However, corn has certain very definite limitations—it is low in protein and calcium.

The grain sorghums are assuming an increasingly important role in cattle feeding, particularly in the fringe areas of the Corn Belt, and in the South and Southwest where moisture conditions are less favorable. New and high-yielding varieties have been developed and become popular. As a result, more and more grain sorghums are being fed to cattle. The chemical composition of sorghum (milo) is similar to corn except that the protein content is generally higher and more variable. Its feeding value is greatly enhanced by steam processing and other similar methods of preparation.

Although corn and sorghum are by far the most common grains used in finishing steers, such grains as barley, rye, oats, and wheat are used in many sections of the United States and Canada. The small grains are excellent for finishing cattle when properly used. In comparison with corn feeding: (1) barley-fed cattle are more susceptible to bloat (for this reason, it is best not to use a straight legume hay along with a grain ration high in barley; a mixture of barley and dried beet pulp is commonly used in the West); (2) barley- or wheat-fed animals are more apt to tire of their ration during a long feeding period; (3) rye should not constitute more than one-third of the grain ration because it is unpalatable; (4) more care is necessary to prevent wheat-fed cattle from going off feed; and (5) oats should not constitute more than one-half of the ration, and preferably not more than one-third, because of its bulk. Fortunately, these limitations can be lessened considerably by mixing these feeds together, or by mixing the cereal grain with beet pulp, silage, or chopped hay. Also, it is important that the small grains be coarsely ground or properly rolled. It is recognized that wheat and oats are frequently too expensive to include in cattle finishing rations.

HIGH-MOISTURE (EARLY HARVESTED) GRAIN

High moisture grain, especially corn and sorghum, is grain that is harvested at 24 to 30% moisture content and stored in an airtight silo. Experimental tests show that the feed efficiency of a high-moisture milo is improved by 8 to 15%, although there is little increase in daily gain. There is less improvement from high-moisture processing of shelled corn, and the results have been most variable. High-moisture milo and corn should be ground or rolled before it is fed. It is questionable, however, if it pays to process high-moisture corn in rations that have less than 15% roughage.

● *Moisture is important when buying feeds*—When buying grains, a feeder should never lose sight of how much water he may be purchasing. Table 30-3 illustrates the relative value (dry matter purchased) when paying for corn on a 15.5% moisture basis while actually receiving corn of another moisture content. Thus, if the feeder were receiving 19% moisture corn and paying for 15.5% moisture, he would receive only 95.86% of the dry matter for which he paid. On the other hand, if corn is delivered with 7% moisture, while paying on a 15.5% moisture basis, the feeder would receive 110.06% of what he paid for.

● *Moisture is important in formulating rations*—A careful feeder must constantly watch the moisture content of the feeds he buys, and the effect of moisture on his nutritional quality control. Most good feeders will readjust

TABLE 30-3

RELATIVE VALUE OF U.S. NO. 2 CORN (15.5% MOISTURE)
AS EFFECTED BY CHANGES IN MOISTURE[1]

Moisture	DM Basis Multiplier	Moisture	DM Basis Multiplier
(%)		(%)	
0	1.1834		
1	1.1716	19	.9586
2	1.1598	20	.9467
3	1.1479	21	.9349
4	1.1361	22	.9231
5	1.1243	23	.9112
6	1.1124	24	.8994
7	1.1006	25	.8876
8	1.0888	26	.8757
9	1.0769	27	.8639
10	1.0651	28	.8521
11	1.0533	29	.8402
12	1.0414	30	.8284
13	1.0296	31	.8166
14	1.0178	32	.8047
15	1.0059	33	.7929
16	.9941	34	.7811
17	.9822	35	.7691
18	.9704	36	.7574

[1]If 15.5% moisture corn is the purchase basis, it will require 1.1834 units of purchase base corn to make 1 unit of 100 dry matter base corn. (Source: *Great Plains Beef Cattle Feeding Handbook*, GPE-1602)

feeding formulas whenever moisture in a leading ingredient changes over one percent.

The way in which moisture changes cause imbalances is pointed up in the following example:

Let's assume that a feeder is using a ration which has as one of its main ingredients corn silage with 68% moisture content, and that this ration requires 1.9% supplement on an "as fed" basis. Now assume that the moisture of the silage suddenly decreased to 55%, and with it the necessary supplement to balance the ration increased to 2.62%. Obviously, if the feeder did not adjust the feeding formula, a serious shortage of protein could result. In this case, the cattle would receive only 72.5% as much supplement as they should have since the mixing formula was not recalculated.

The multipliers in Table 30-4 may be used to determine the price per unit of dry matter simply by multiplying price times the appropriate factor for the indicated moisture.

● *Acid treated high-moisture grain*—In the past, high-moisture grain has been either (1) artificially dried, or (2) stored in an airtight silo, to prevent spoilage. Now there is a third alternative—the use of naturally occurring acetic and propionic acids for reducing mold growth and other deterioration of high-moisture grain. Tests with acid treated (sprayed on at the time of storage) high-moisture grain suggests that rate of gain and dry matter conversion are at least equal, if not superior, to artificially dried or airtight stored grain.

TABLE 30-4

CORRECTION FACTORS TO USE WHEN CONVERTING FEEDS OF
VARIOUS MOISTURE CONTENTS TO A 100% DRY
MATTER BASIS (0% MOISTURE)[1]

Moisture	100% DM Basis Multiplier	Moisture	100% DM Basis Multiplier	Moisture	100% DM Basis Multiplier
0	1.0000	29	1.4084	58	2.3809
1	1.0101	30	1.4285	59	2.4390
2	1.0204	31	1.4492	60	2.5000
3	1.0309	32	1.4705	61	2.5641
4	1.0416	33	1.4925	62	2.6315
5	1.0526	34	1.5151	63	2.7020
6	1.0638	35	1.5384	64	2.7777
7	1.0752	36	1.5625	65	2.8571
8	1.0869	37	1.5873	66	2.9411
9	1.0989	38	1.6129	67	3,0303
10	1.1111	39	1.6393	68	3.1250
11	1.1235	40	1.6666	69	3.2258
12	1.1363	41	1.6949	70	3.3333
13	1.1494	42	1.7241	71	3.4482
14	1.1627	43	1.7543	72	3.5714
15	1.1765	44	1.7857	73	3.7037
16	1.1904	45	1.8181	74	3.8461
17	1.2048	46	1.8518	75	4.0000
18	1.2195	47	1.8867	76	4.1666
19	1.2345	48	1.9231	77	4.3478
20	1.2500	49	1.9607	78	4.5454
21	1.2658	50	2.0000	79	4.7619
22	1.2820	51	2.0408	80	5.0000
23	1.2987	52	2.0833	81	5.2631
24	1.3157	53	2.1276	82	5.5555
25	1.3333	54	2.1739	83	5.8824
26	1.3513	55	2.2222	84	6.2500
27	1.3698	56	2.2727	85	6.6666
28	1.3889	57	2.3255		

[1]Source: *Great Plains Beef Cattle Feeding Handbook*, GPE-1602.

MOLASSES

Molasses—cane, beet, corn, and wood—is palatable (wood molasses is the least palatable) and a good source of energy; and cane molasses is also a good source of certain trace minerals. All molasses are low in protein. As a rule of thumb, molasses should not cost more than three-fourths as much as cereal grain, pound for pound, to be economical. Generally speaking, molasses is limited to 5 to 10 percent of the ration, although up to 15, or even 20, percent can be added if the mixing facilities will handle it and the price is favorable.

FATS

Feeding of fats was prompted in an effort to find a profitable outlet for surplus packing house fats. For the most part fats were formerly used for soapmaking, but they are not used extensively in detergents. Thus, with the rise in the use of detergents in recent years, they became a "drug on the market."

Animal and vegetable fats seem to be equally effective additions to feed-lot rations; thus, selection should be determined solely by comparative price. Ordinarily, animal fats are much cheaper than such vegetable fats as soybean or cottonseed oil.

Several different fat products are used as cattle feed. Each of them should be bought by specifications and guarantees. Pertinent specifications follow:[2]

1. *Acidulated soap stock (foots)*—This end product may be composed of acidulated soap stock from all vegetable and animal sources which have been washed free of mineral acids.

2. *Tallows*—This end product shall be composed of rendered, clean, disease-free, filtered animal fats.

3. *Greases (white and yellow)*—This end product shall be composed of rendered, clean, disease-free, filtered animal fats and/or restaurant greases.

4. *Blended feeding fat*—This end product may be composed of rendered animal fat, animal grease, restaurant grease, vegetable oil, or acidulated soap stock in any combination.

5. *Other materials*—House grease, brown grease, sewer grease, modified yellow grease, and other low-grade materials are sometimes used.

Fat serves the following three practical functions when added to cattle rations:

1. *It increases the caloric density of the ration*—This appears to be very important with cattle on feed in hot climates and during the summer months. Since fat contains approximately 2¼ times as much energy as soluble carbohydrate, it is possible to increase the energy content with little increase of the bulk of the ration. Thus, with the same feed intake, energy intake is higher.

2. *It controls dust*—Thus, the addition of 1 to 2 percent fat materially lessens the dust involved in hay grinding. Also, it is well known that dusty rations are not consumed readily by cattle; hence, fat enhances consumption from this standpoint.

3. *It lessens the wear and tear on feed mixing equipment*—This is important, because both breakdowns and new equipment are costly.

It is recommended that 2 to 5% fat be added to high-concentrate rations in which milo, barley, and/or wheat are the chief grain sources. Higher levels of fat usually result in drastically lowered feed consumption. When fed at a 2 to 5% level, the energy value of fat is approximately 2¼ times that of the grains. When corn is the major source of grain, fat additions can be

---

[2]Cattle feeders' fat trading standards of California Cattle Feeders' Association, Oct. 11, 1972.

expected to be less useful than with the small grains. This is understandable when it is realized that corn contains approximately 4% fat as compared to 1 to 1½ for the other feed grains.

When fat is added, the calcium and phosphorus levels of the ration should be 0.55% and 0.33%, respectively.

One of the more exciting areas, which is still in the experimental stage, is the so-called rumen protected fat system. In this system, the fat is emulsified with a protein, then the protein is treated with formaldehyde. The product cannot be digested in the rumen due to the formaldehyde linkage on the protein. However, the pH of the abomasum is such that the formaldehyde linkage of the protein is broken, with the result that the protein in fat will then be digested in the small intestine (similar to that of the monogastric animal). Studies from Australia and the University of Arizona have shown that the utilization of fat by the ruminant can be improved by the use of this system. Also, and most significant, by the use of the protein protected fat system it is possible to alter the ratio of the saturated to the unsaturated fatty acids in beef depot fat. The same is true of butterfat. A beef fat which contains high levels of unsaturated fatty acids is especially attractive to people with certain types of heart conditions. Thus, in the future we may see some beef cattle fed rations containing protein protected fat for the express purpose of producing Choice beef in which the fat contains a high level of unsaturated fatty acids.

ALL-CONCENTRATE AND HIGH-CONCENTRATE RATIONS

Based on experiments and experiences, the following conclusions relative to all-concentrated and high-concentrate rations appear to be justified:

1. Ruminants need some "roughness factor" or "scratch factor" to stimulate the rumen papillae for normal functioning. In high- or all-concentrate rations, this can be achieved partially by rolling or coarse grinding.

2. With the possible exception of whole shelled corn, and such high fiber feeds as oats and barley, a high level of management is needed to make an all-concentrate system work under feedlot conditions. Problems associated with high-concentrate rations include acidosis, founder, and liver abscesses.

3. Some research indicates that continued high levels of performance can be better maintained by including 10-15 percent roughage in high concentrate rations.

4. Rations having a concentrate content of 90 percent or more should be self-fed, and a liberal amount of feed should be available at all times.

5. Feed efficiency is improved in high- and all-concentrate rations, due to their high energy; but rate of gain is not materially affected.

6. The energy of a ration may be increased without eliminating much of the roughage by adding 4 to 5 percent fat.

7. Feed formulation and balance of nutrients become more critical on high- and all-concentrate rations; specifically—

   a. Higher levels of vitamin A must be added—50,000 to 100,000 IU/head/day.

b. The ration must be fortified with the calcium, phosphorus, and trace elements that are normally provided through the roughage.

c. The unidentified growth factors are usually reduced with a reduction of the roughage. To compensate therefor, 5 percent dehydrated alfalfa meal may be added to the ration.

8. Cattle on high- and all-concentrate rations stall and go off feed more frequently.

9. Pelleting high- or all-concentrate rations lowers daily gains.

10. In the final analysis, the comparative price of concentrates and roughages—the economics of the situation—along with management practices, will be the major determining factors.

By-Product Feeds

Prior to 1900, by-product feeds were unwanted, and practically unused. Wheat bran was dumped into the Mississippi River, because nobody wanted to buy it; cottonseed meal was used as a fertilizer; most of the linseed meal was shipped to Europe; and tankage had not been processed. Today, these are standard and valuable animal feeds. Innumerable other by-products—both roughages and concentrates—from plant and animal processing, and from industrial manufacturing, are available and used as cattle feeds, in different areas, including the following: potatoes and potato pulp, pea vines and corn refuse silage from the canning industry, and by-products from numerous fruits and nuts.

As is true of any ration ingredient, the requisites to effective and profitable use of each by-product feed in cattle feeding are: (1) that it be bought at a favorable price, nutritive composition considered; (2) that its proximate composition be known, and that it be incorporated in a balanced ration; (3) that it be palatable and consumed in adequate quantity; and (4) that it not adversely affect carcass quality, particularly from the standpoint of harmful chemical residues from pesticides applied to crops. Generally speaking, the use of by-product feeds calls for ingenuity and experience in handling them, special knowledge relative to their nutritive qualities and use in balanced rations, and relatively high labor costs. As a result, many cattle feeders are not interested in using them, whereas others find it a lucrative business.

The feeding value and the maximum amount that can be fed to cattle of several by-product feeds are given in Chapter 8, Table 8-13, Handy Feed Substitution Table for Beef Cattle.

Roughages

Roughages are used in feedlot rations to supply bulk, physical properties, energy, protein, minerals, and vitamins. They contain considerable fiber (cellulose, hemicellulose, and lignin); consequently, they have a lower available energy content than concentrates. For this reason, only limited amounts of roughages are incorporated in finishing rations, particularly toward the end of the feeding period. They are, however, used extensively in growing programs and in warm-up rations.

Fig. 30-4. Drylot steers eating field cured dry beet tops. (Courtesy, The Great Western Sugar Co., Denver, Colo.)

The amount of roughage in feedlot rations varies over a wide range—from roughage alone in some grower rations to all-concentrate rations in some finishing rations, and many roughage proportions between these two extremes. Each of these roughage to concentrate ratios may be highly successful and very practical under certain conditions. In the final analysis, therefore, the roughage to concentrate ratio for a given feedlot should be determined by (1) the available feeds and comparative prices, (2) feed processing facilities, (3) the feed handling charges, (4) the age and quality of cattle, (5) the stage in the feeding period (i.e., starting vs finishing period), (6) temperature (decrease roughage and increase concentrate in hot summer months, because high-roughage rations produce more body heat), (7) the feeder, (8) the troubles encountered (off feed, founder, scours, bloat), and (9) the results obtained by previous experience.

Also, step-wise reductions in the amount of roughage are usually made at least three times during the finishing period; for example—

Starter ration—70% roughage, 30% concentrate

Intermediate ration—30% roughage, 70% concentrate

Final ration—10% roughage, 90% concentrate

Such step changes in roughage to concentrate ratios should be made gradually, by blending the two mixes for two to four days.

In drylot finishing, the kind of roughage fed varies from area to area. This is so because, normally, it is not practical to move roughages great distances. Thus, generally speaking, cattle feeders utilize those roughages that are most readily available and lowest in price.

● *Hay*—Hay is the most important harvested roughage fed to feedlot cattle, although many other roughages can be, and are, utilized.

High quality legume hays are superior from the standpoint of cattle performance—and, when they are fed, supplemental feed requirements are lower. However, lower quality hays give satisfactory results when properly supplemented.

Alfalfa is the roughage of choice in commercial feedlots. It averages 17.1% protein on a dry (moisture free) basis and contains 58% TDN. On a net energy basis for production (NE$_p$), it is estimated to contain 27 Mcal per cwt, in comparison with a value of 56 Mcal per cwt for milo.

● *Silage*—Increasing quantities of corn and milo silage are being fed to finishing cattle. At the present time, it is estimated that 90 percent of the nation's silage is made from corn and sorghum and the other 10 percent from hay and pasture crops, small grains, by-products, and other feeds. The following silage pointers, based on experiments and practical observation, are generally observed by successful cattle feeders:

1. Where cattle are given a liberal amount of grain, any silage that is fed is considered a part of the roughage ration; hence, the silage should be fed in accordance with the well-recognized rules for feeding roughages. Also, it can be assumed that about two pounds of good corn silage may be substituted for one pound of hay in cattle rations.

2. The maximum use of silage is best obtained early in the finishing period or with more mature steers that possess a larger digestive capacity.

3. Low grade cattle may be fed either (a) entirely on silage or (b) liberally on silage throughout the finishing period, but cattle that grade Good or better may return more profit if they are full-fed on grain during the last half of the feeding period.

4. Cattle can be placed on full feed of silage from the beginning of the feeding period without any detrimental effects.

5. When fed on silage alone, cattle on full feed will consume 6 to 7 pounds per 100 pounds weight.

6. Sorghum silage has 60 to 90 percent of the value of corn silage and should be supplemented in the same manner as corn silage.

7. Grass silage should be supplemented with additional energy feeds, such as cereal grain or molasses, to be of the same value as corn silage.

● *Haylage*—Tests to date have shown that haylage, which generally contains about 40 percent moisture, is an excellent feed for finishing cattle. More research is needed, but indications are that costs of gains may be a little higher than for corn silage.

A fairly accurate rule of thumb is that it takes 1.5 pounds of 40 percent moisture haylage to equal one pound of hay.

● *Other roughages*—Among the other roughages (other than hays and silages) used for feedlot cattle are: cottonseed hulls, corncobs, cereal straws, milo stover, Bermuda straw, Sudan hay, sawdust and other wood products, oat hulls, beet tops, peanut hay, newspapers, and a host of others. When properly (1) combined with a high quality legume roughage, and/or (2) supplemented with the necessary protein, minerals, and vitamins, all of them are excellent feeds. Availability, costs, and results should be the determining factors in their use, just as the economics of the situation should determine the use of any other feed ingredient.

(Also see Chapter 8, Table 8-13, Handy Feed Substitution Table for Beef Cattle, for the feeding value and the maximum amount that can be fed to cattle of each of several roughages.)

ROUGHAGE SUBSTITUTES

It appears that the rumen animal requires a minimum amount of roughage factor for normal rumen function. All-concentrate rations show a response to small amounts of roughage factor. As little as one pound per day of some low quality roughage such as alfalfa stems, rice hulls, or sorghum straw have given improvements in efficiency ranging up to 15 percent of all-concentrate rations.

To meet the "roughage factor" need, different roughage substitutes have been developed and used with varying degrees of success.

Protein Supplements

The daily protein requirements of cattle and the percentage of protein needed in the ration are given in Tables 8-18 and 8-19 of Chapter 8. It should be noted, however, that a larger percentage of protein is needed in rations with higher energy density. This is because fewer pounds of the high energy ration are needed daily to meet the animal's energy requirement but the protein needs stay the same. Other factors that must be considered in formulating feedlot rations are the variations in protein content of feedstuffs, the digestibility of the protein, metabolic efficiency of protein utilization, and the previous nutritional treatment of the cattle. Age, genetic background, and health may influence efficiency of protein utilization. Because of the many factors that affect protein requirements, it is advisable to include a safety factor when balancing for protein. Although excess protein in the ration can be partly utilized for energy, each one percent increase in protein above the required level may increase the cost of gain ¼ to ½ cent per pound. However, underfeeding protein can cost much more than overfeeding protein due to the slow gains and poor feed efficiency. There appears to be a trend among research workers and nutritionists to recommend higher protein levels than were believed necessary in recent years, especially on high-concentrate rations. Of course, price levels of protein supplements are a factor.

KIND OF PROTEIN

Quality of protein, or balance of essential amino acids, is not a critical factor in most beef cattle finishing rations, because bacteria in the rumen "manufacture" proteins that are used by cattle. For this reason, it makes little difference to a steer whether his protein comes from one source or several. Yet, practical feeders recognize that protein mixtures may be more palatable than a single ingredient and cause animals to eat a little more feed, and that, on long-fed cattle, fed to high Choice or Prime grade, linseed meal produces extra "bloom" which makes it a little more valuable than its protein content would indicate.

The choice of a protein supplement should usually be determined by the comparative price of a pound of protein in the available supplements. (See Chapter 8, section on "How to Determine the Best Buy in Feeds.") The leading protein supplements for finishing cattle are soybean meal, cotton-seed meal, linseed meal, urea, and slow released nonprotein nitrogen products.

AMOUNT OF PROTEIN SUPPLEMENT TO FEED

The percent protein supplement to add to the ration will depend upon the age of the cattle, the kind and amount of roughage, and the protein content of the grain(s), or other carbonaceous concentrate being fed. Also, more protein is needed in rations with higher energy density. Thus, the amount of supplement should be determined for each lot. On a percentage of total ration basis, it decreases as the cattle grow older. Here are the recommendations for crude protein in the total ration, on an air dry basis (which is about 10% moisture):

Stage in Feedlot		Crude Protein (%)
Calves	First 60-90 days	11-12
	Next 100-200 days	10.5-11
	200 days to market	10.0-10.5
Yearlings	First 60-90 days	10.5-11.0
	100 days to market	10.0-10.5

Based on recent studies, the Ohio Station workers recommend that protein supplements be deleted from feedlot rations after cattle are past 750 pounds weight. After removing the protein supplement, their test ration contained from 8.2 to 8.6 percent protein, a good 25 percent below the usual recommendations. They do warn that when lowering the protein the feeder must not decrease the levels of minerals and vitamins in the ration.

For cattle on relatively high concentrate rations in the drylot, the following rules of thumb may be used:

1. Where no legume hay is fed, add two pounds of oilseed protein supplement per head daily.

2. Where half the roughage consists of legume hay, add one pound of oilseed protein supplement per head daily.

Where cattle are getting a full feed of good quality legume hay, haylage, or silage, and limited grain, it is not necessary to add a protein supplement.

Because protein supplements are usually expensive, normally one should not add more than is required to balance the ration. Neither should they be shorted, for digestion of roughage is lowered if there is a lack of protein in the ration.

UREA

Urea is not a protein. It is a simple nitrogen compound, $NH_2-\overset{\overset{\displaystyle O}{\|}}{C}-NH_2$, from which the microorganisms can obtain nitrogen, synthesize amino acids, and finally bacterial protein—provided that all the nutrients essential for protein synthesis are present.

Each year, urea is replacing a larger percent of the supplementary protein in cattle feedlot rations. Because of the high price of oilseed proteins, due to their increasing use for human consumption and for monogastric animals, eventually urea and/or other nonprotein nitrogen compounds will be used as a major source of supplementary protein for feedlot cattle.

Urea or other nonprotein nitrogen compounds can furnish 33 percent of the total protein requirement of feedlot cattle; higher levels cause a depression in gain and feed efficiency. It is noteworthy, however, that, even in supplements in which 90 percent of the protein equivalent is from urea, only about one-third of the total protein in the ration is supplied from nonprotein nitrogen—the remainder is supplied from grain and roughages.

The following conditions are essential for proper urea utilization by feedlot cattle:

1. High level of bacteria population in the rumen. Cattle off feed or sick cattle do not utilize urea very effectively.

2. Slow release of ammonia from urea.

3. Two to four weeks' adjustment period.

4. Low level of natural protein in the diet. Urea utilization is depressed when used with increasing levels of natural protein. Microorganisms prefer the nitrogen from natural protein to nonprotein nitrogen because natural proteins, such as soybean meal, furnish other nutrients which are beneficial to bacterial and protozoan life.

5. High quality ingredients are required in high-urea supplements. For this reason, high-fiber filler feeds (such as ground corncobs, oat hulls, rice hulls, cottonseed hulls, cellulose, paper, and sawdust) should not be used with urea.

6. Response to urea is greater on high energy and limited roughage rations than on low energy and high roughage diets.

7. Urea supplements should be well mixed (homogenous).

8. High-urea dry supplements should be protected from rain and kept as dry as possible, because urea is hygroscopic. (It picks up moisture.)

9. Essential nutritional factors including (a) readily available source of energy (such as grain or molasses); (b) adequate levels of calcium and phosphorus; (c) required level of trace mineral elements; (d) a nitrogen-sulfur ratio which is not wider than 15:1; (e) unidentified urea-protein synthesis factors (dehydrated alfalfa meal in dry urea mixes, and distillers' solubles in liquid supplements); (f) iodized salt, to improve the palatability and mask the taste of urea; (g) fortified with proper levels of synthetic vitamin A, to furnish a minimum of 20,000 IU of vitamin A daily for growing and finishing cattle; (h) fortified with vitamin D if cattle are confined, and (i) fortified with vitamin E if natural feedstuffs are low.

10. Do not feed urea or supplements containing urea to newly arrived or shipped-in cattle for a period of 21 to 28 days.

11. Do not feed urea to cattle that have been starved or off feed for 36 hours until they have had a chance to fill the rumen with feed.

12. For growing cattle, feed a maximum of 0.15 pound (68 grams) of urea daily.

13. For finishing steers or heifers on grain and roughage, do not feed more than 0.22 pound (100 grams) of urea per head daily.

14. Formulate complete cattle rations so that no more than 33% of the crude protein or nitrogen is derived from urea. Protein supplements may contain 85 to 90% of the protein from urea; but when blended with natural feedstuffs like grain and roughage, the total contribution of protein from urea is usually less than 33%.

15. Do not feed urea over and above the protein requirement; add only enough properly balanced urea supplement to meet the protein needs.

16. Urea should be either thoroughly mixed in a properly balanced supplement or incorporated in a complete ration.

17. If supplements containing urea are self-fed, the intake should be controlled by using a lick wheel for liquid supplements or incorporating high levels of salt and dry supplements.

(Also see Chapter 8, section entitled, "Nonprotein Nitrogen Sources.")

PROTEIN POINTERS

The following points should be taken into consideration to assure proper protein utilization by finishing beef cattle:

1. The protein content of the major feed ingredient(s) should be known so that the ration formulation can be precise.

2. Consideration should be given to the protein content and digestibility of the grain, because the grain is the most economical source of protein and supplies the largest percent of the ration protein.

3. Higher levels of protein are needed in all-concentrate rations.

4. When possible, use a roughage high in protein so as to lessen the amount of supplemental protein necessary.

5. Urea can be successfully used at a level up to one percent of the total finishing ration to replace natural protein, provided the ration formulation permits optimum utilization of the urea of the rumen microorganisms.

6. When excessive protein levels are fed, the overage is wasted and performance may actually be reduced with high performing animals.

7. Absolute protein requirements depend upon the age of the animal, and probably the energy of the ration.

8. When per head consumption of corn reaches a level of 18 pounds per day, no supplemental protein is required, although other additives such as minerals and vitamins are necessary. Thus, protein supplements are not needed in the late finishing stages of yearling cattle.

9. The production of single-cell protein (SCP) from solid waste as a protein supplement for feedlot cattle will likely increase in importance.

Minerals

Minerals play an important role in the nutrition of feedlot cattle. They help regulate the normal functions of the metabolic processes in the animal.

The amount of each mineral in the ration is important; both deficiencies and excesses are to be avoided. Therefore, an analysis is important, particularly where new feeds and feeds from new areas are involved. The following minerals have been established as dietary essentials for cattle:

● *Major minerals*—Calcium, phosphorus, magnesium, sodium, potassium, chlorine, and sulfur.

● *Trace minerals*—Cobalt, copper, iodine, iron, manganese, and zinc. Also, there is good evidence that selenium will move into the "essential" list.[3]

Where a complete mixed feed (roughage and concentrate combined) is fed to finishing cattle, it should contain 0.25 to 0.5 percent salt. Also, in the larger feedlots, the other needed minerals are usually incorporated in the ration as a special mineral supplement or in the protein supplement. For recommended kinds and allowances of minerals, see Chapter 8, Tables 8-7 and 8-8. Special attention needs to be given to trace minerals in areas where there is a deficiency of one or more of them, when poor quality roughage is fed, or when high- or all-concentrate rations are fed.

Even when minerals are added to the ration of finishing cattle, the author favors self-feeding them in addition. For this purpose, use a two-compartment mineral box, with (1) salt (iodized salt in iodine-deficient areas[4]) in one side, and (2) dicalcium phosphate, defluorinated phosphate, or a mixture of ⅓ salt (salt added for purpose of palatability) and ⅔ steamed bone meal, or a good commercial mineral mixture, in the other side. With this arrangement, if cattle need added minerals, they will consume them; if they don't need them, they'll pass them up. In particular, incoming feedlot cattle frequently crave minerals, due to deficiencies in their previous feeding. Such cattle, however, should be given only limited quantities of minerals until the danger of overeating has passed.

(Also see Chapter 8, section on "Minerals.")

Vitamins

Vitamins A, E, and in some cases vitamin D, should be added to feedlot rations.

The rumen organisms synthesize adequate B vitamins and vitamin K, and nothing is gained by adding these to feedlot rations of healthy cattle. Likewise, no benefit has been reported from supplemental vitamin C. Vitamin D is produced in the skin of animals in direct sunlight, but during cloudy, winter weather, or when cattle are confined, it should be added to the ration.

Feeders should watch for the following symptoms of vitamin A deficiency in feedlot cattle: rough hair coat, watery eyes, loose and watery droppings, edema (stovepipe legs), and night blindness.

Chapter 8, Table 8-11, gives the vitamin requirements of beef cattle,

---

[3]At this writing, the only approved method of selenium supplementation for cattle is through injection of the element by a veterinarian.

[4]Unless cane molasses is included in the ration, a trace-mineralized salt may well be used. Cane molasses appears to be a good source of most of the needed trace minerals, and many supply sufficient trace minerals in certain areas and with certain feeds.

whereas Table 30-5 gives the recommended levels of vitamin A for feedlot cattle, with overage for safety.

A common guideline on the level of vitamin A for feedlot cattle is to use 3,000 IU per pounds body weight, or 1,000 IU for each pound of total feed. One million IU of vitamin A will cost about 3 cents as a feed additive, or about 7 cents as an injectable.

TABLE 30-5

RECOMMENDED LEVELS OF VITAMIN A FOR FEEDLOT CATTLE

	Vitamin A/ Head/Day	Vitamin A/Ton of Supplement When It Is Fed at Level of—	
		1 Lb/Head/Da.	2 Lb/Head/Da.
	(IU)	*(IU)*	*(IU)*
Cattle on growing ration in winter	10,000	20,000,000	10,000,000
Cattle on full-fed finishing ration in winter	20,000	40,000,000	20,000,000
Cattle full-fed grain on pasture	20,000	40,000,000	20,000,000
Cattle fed in drylot in summer	30,000	60,000,000	30,000,000

When vitamin D is needed and added to the ration, it is recommended that 4,000 to 6,000 IU of it be given per head per day. This is approximately $1/6$ to $1/7$ the recommended level of vitamin A.

Where grains are heat processed for feedlot cattle, some research shows that it may be advisable to provide supplemental vitamin E. The National Research Council indicates that the requirement for vitamin E is about 7 to 27 IU per pound of ration dry matter for growing and finishing cattle; hence, feeding this level may be advisable where grain is subjected to heat processing.

(Also see Chapter 8, section on "Vitamins.")

Implants and Growth Stimulants[5]

Feed additives first made headlines in 1952 when Iowa State University researchers announced the results of cattle feeding trials indicating a major breakthrough in lowering feed usage and increasing weight gains by feeding the compound diethylstilbestrol (DES).

---

[5]This section and Table 30-6 were authoritatively reviewed by the following: Dr. T. W. Perry, Department of Animal Sciences, Purdue University, Lafayette, Ind.; Dr. Wise Burroughs, Department of Animal Science, Iowa State University, Ames, Iowa; Dr. Wilton W. Heinemann, Animal Scientist, Irrigated Agriculture Research and Extension Center, Washington State University, Prosser, Wash.; Dr. A. T. Ralston, Department of Animal Science, Oregon State University, Corvallis, Ore.; Dr. Dean E. Hodge, Director, Beef Cattle Research Division, Ralston Purina Company, St. Louis, Mo.; Dr. Jack E. Martin, Sterling Nutritional Service, Inc., Sterling, Colo.; Dr. T. M. Means, International Animal Research Coordinator, Eli Lilly and Company, Greenfield, Ind.; Dr. Calvin Drake, Aid, Inc., Wichita, Kan.; and Dr. Aaron L. Andrews, General Manager, Hess and Clark, Ashland, Ohio.

For the next 20 years, cattle feeders and consumers greatly benefited from the increased rate and efficiency of gains accruing from the use of this product. Feeding and implanting DES increased growth rate 10 to 15 percent; improved feed efficiency 10 percent; put feedlot cattle on the market 20 to 30 days sooner; decreased feedlot costs by 3 to 3.5 cents per pound; and lowered the retail price of beef by 5 to 6 cents per pound. But the Food and Drug Administration banned the use of both oral (January 1, 1973) and implanted (April 25, 1973) DES, for the reason that stilbestrol is a carcinogen (in large enough quantities it can produce cancer in men and animals); and the Delaney Amendment clearly states that foods can contain no residues (zero level) of carcinogens. Actually, the small amount of DES detected in the liver and kidney of beef cattle was not a human health hazard; DES never occurred in sufficient amounts to cause cancer in man or animals. Less than 2 parts per billion were detected in liver. At this level (2 ppb), a person would have to eat 26,666 pounds of liver to be equivalent to 24 mg of DES administered to women as a "morning after pill." Nevertheless, the law calls for the impossible—a zero level. Then, on January 24, 1974, the United States Court of Appeals invalidated the ban on DES and reinstated its use as an implant or feed additive. But the ruling of the court did nothing to change the Delaney Amendment. In the meantime, newer and more sophisticated assay techniques evolved which made it possible to detect the presence of DES residues in meat at lower levels than formerly. Thus, in one way or another, it appears that FDA will ban the use of DES for cattle and sheep.

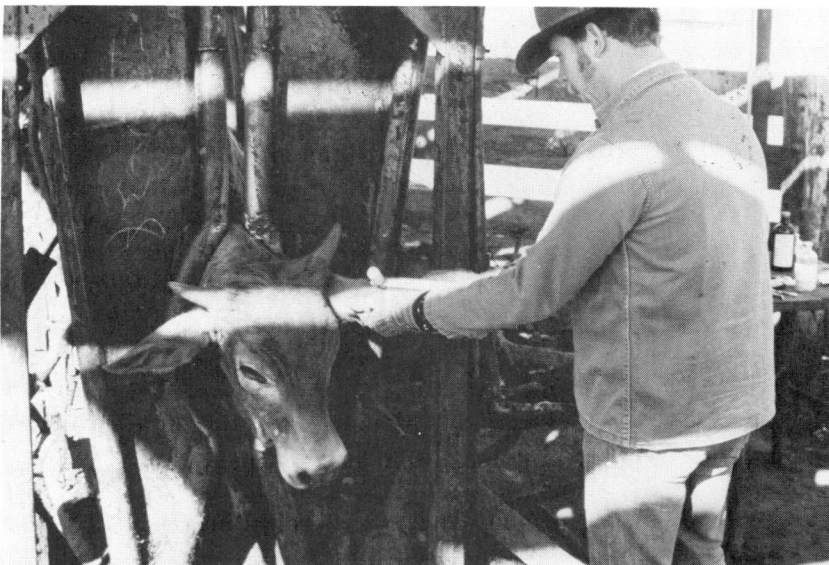

Fig. 30-5. Implanting a growth stimulant. (Courtesy, Benedict Feeding Co., Casa Grande, Ariz.)

Table 30-6 summarizes the growth stimulants that are presently available and can be used. All of these products have been shown to improve gain and feed efficiency of feedlot cattle significantly.

TABLE 30-6

IMPLANTS AND GROWTH STIMULANTS FOR FINISHING CATTLE

Class of Cattle	Additive	Method of Administering	Dosage	Cost	Increase in Daily Rate of Gain	Increase in Feed Efficiency	Effect on Carcass Quality	Other Comments	Withdrawal Period Prior to Slaughter
Finishing Steers	1. Antibiotic	Oral	10 mg/100 lb body wt. daily; or 70 to 75 mg/head daily.	0.3¢/day	6%	4%	Improves carcass quality slightly; more fat deposition and marbling.	Antibiotics will also reduce the disease level. More effective on high roughage rations than on high concentrate rations.	48 hours
	2. Ralgro (Zeranol)	Implant	36 mg resorcylic acid lactone	85¢/dose	10%	5-10%	No effect.	Nonestrogenic.	65 days.
	3. Rumensin	Oral	50-360 mg/head/day	1½¢/day	Gain not affected. Saves feed.	10%	No effect	Not a hormone. It results in more propionic acid and less butyric and acetic acids; hence, more energy.	No withdrawal required.
	4. Stilbestrol (diethylstilbestrol)	a. Oral b. Implant	10 mg/head/day 12-36 mg implant	0.3¢/day 30¢/dose (2 implants)	10-15% 10-15%	10% 10%	Stilbestrol produces more lean and less fat; hence, stilbestrol-treated cattle must be fed to heavier weights; otherwise they will grade lower.	Simultaneous use of stilbestrol in feed and as implants is not approved. Implants are available in 12 or 15 mg pellets.	14 days Cattle must not be slaughtered for at least 120 days following implantation.
	5. Synovex S (for steers)	Implant	200 mg progesterone 20 mg estradiol benzoate	80¢/dose	10-15%	5-10%	No effect.		60 days
Finishing Heifers	1. Antibiotic	Oral	10 mg/100 lb body wt. daily; or 70 to 75 mg/head daily.	0.3¢/day	6%	4%	Improves carcass quality slightly; more fat deposition and marbling.	Antibiotics will also reduce the disease level. More effective on high roughage than on high concentrate rations.	48 hours
	2. MGA	Oral	0.25 to 0.50 mg daily melengestrol acetate	1¢/day	11%	8%	MGA will lower the incidence of estrus in heifers and increase rate and efficiency of gain. It is not effective with pregnant heifers.	MGA is effective for heifers, but not for steers.	48 hours

(Continued)

Table 30-6 (Continued)

Class of Cattle	Additive	Method of Administering	Dosage	Cost	Increase in Daily Rate of Gain	Increase in Feed Efficiency	Effect on Carcass Quality	Other Comments	Withdrawal Period Prior to Slaughter
	3. Ralgro (Zeranol)	Implant	36 mg resorcyclic acid lactone	85¢/dose	10%	5-10%	No effect.	Nonestrogenic.	65 days
	4. Rumensin	Oral	50-360 mg/head/day	1½¢/day	Gain not affected. Saves feed.	10%	No effect.	Not a hormone. It results in more propionic acid and less butyric and acetic acids; hence, more energy.	No withdrawal required.
	5. Synovex H (for heifers)	Implant	200 mg testosterone propionate 20 mg estradiol benzoate	80¢/dose	10%	5-10%	No effect.	Recommended for use in heifers during last 60 to 150 days of the finishing period.	60 days
Suckling Calves	1. Antibiotic	Oral (in creep feed)	15 to 20 mg/100 lb body wt. daily	0.1¢/day	6%	4%		Antibiotics will also reduce the disease level.	48 hours
	2. Ralgro (Zeranol)	Implant	36 mg resorcyclic acid lactone	85¢/dose	10%	5-10%	No effect.	Nonestrogenic.	65 days
	3. Stilbestrol	Implant	15 mg	15¢/dose	12 lb heavier weaning wt.			Implant creep-fed calves when 60-90 days of age.	Cattle must not be slaughtered for at least 120 days following implantation

In considering the additives listed in Table 30-6, it should be noted that there is no evidence to indicate that the use of these products can or will alleviate the need for vigilant sanitation, improved nutrition, and superior management. Also, the benefits of each one must be weighed against its cost.

Other Methods of Improving Rate and Efficiency of Gain

Several other methods, in addition to additives, can be used to increase the rate and efficiency of gain of feedlot cattle. Among them are the following:

1. Feed young bulls (uncastrated males) instead of steers. The male hormones secreted by the testicles are excellent growth stimulants and will improve gain and feed efficiency by 10 to 15 percent. Alternatives to bulls that merit consideration are short scrotum bulls (induced cryptorchidiam), and Russian castrates. With these methods, testosterone is produced, yet, in comparison with bulls, the animals are easier to handle and may be carried to advanced ages without being labeled "bull beef."

2. Take advantage of the genetic improvement of beef cattle by crossbreeding and the introduction of genes from the exotic breeds. This offers one of the most permanent ways of increasing the weaning weight of calves and improving performance in feedlot cattle. The selection of fast gaining and efficient cattle within the straightbreds, and in crossbreeding, may improve the efficiency of performance of feedlot cattle by 10 percent or more.

3. Reduce the cost of producing beef by improving the quality of cattle rations through grain processing and nutritionally balanced protein supplements.

4. Eliminate internal and external parasites, and protect cattle against the common diseases. This will save millions of dollars for both cattle feeders and consumers.

5. Keep abreast of new developments, including the discovery of new growth stimulants.

Feed Preparation

Prior to 1960, very little attention was given to feed processing for commercial cattle production, other than grinding or crushing grain and chopping forage. But in recent years great progress has been made and many new techniques have been developed.

The preparation of roughages has received less attention than the preparation of grains. However, with increasing world food shortages, roughages will become more important. Under these circumstances, their processing will assume greater importance. Hence, roughage preparation will also be discussed herein.

GRAIN PREPARATION

Modern day fattening rations usually contain from 75 to 95 percent concentrate. Moreover, grains supply up to 90 percent of the usable energy of the ration. Thus, any improvement in the efficiency of utilization of grain will be reflected in improved performance and feed requirement of fattening cattle.

The success of grain processing must result from physical and/or chemical changes. Physical changes include moisture level, heat, pressure, and particle size. Chemical factors may include structural changes in the starch, protein, and fat of grains resulting in changes in digestibility and metabolic end products. In some cases, so-called physiochemical changes occur in that both physical and chemical alterations are simultaneously apparent. Rate of ingesta passage and site of digestion within the G.I. tract are both likely end results of physiochemical changes in processed grains.

The primary reasons for processing grains for feedlot cattle are:
• To increase digestibility.
• To increase palatability.
• To increase surface area for greater microbial activity.
• To give rumen microorganisms and digestive enzymes easier access to the starches and readily utilizable nutrients.
• To affect the rate of passage of feed through the digestive tract, or to affect rumen mobility by increasing the bulk through certain processing methods.
• To increase feed efficiency through a combination of the above factors.

Among the factors to consider when deciding on the grain processing method are the size of kernel, percentage of moisture, and percentage of concentrate in the ration.

When any of the dry processing methods are used, it is important that the kernel be broken, but that there be coarseness and relative freedom from fines.

• *Dry rolling*—For small feedlot operations, dry rolling of corn, barley, and wheat is preferable to grinding. However, it possesses some of the same weaknesses common to grinding—especially the presence of "fines."

• *Fatty acid-treated grain*—A recent development in preserving high-moisture grain is the use of certain so-called "short chain" fatty acids; namely, acetic (acetic acid is the predominate acid in vinegar) and propionic acids. Both of these organic acids have nutritive value and are products of metabolism in the rumen. Fatty acids preserve high-moisture grain without the necessity of limited oxygen storage (airtight silos); hence, acid-treated grain can be stored in open wooden bins without spoilage. The fatty acid acts as a mold inhibitor or fungicide for treated high-moisture grain. Limited studies to date indicate that acid-treated grain has approximately the same feeding value as high-moisture grain. It appears, therefore, that the fatty acid treatment of grain offers real promise as a practical way in which to preserve high-moisture grains for cattle feeding, while at the same time retaining the improved feeding value that has been shown for high-moisture grain.

• *Flaking*—Flaking is the modification of steam rolling in which the grain is subjected to steam for a longer period of time. Milo is generally flaked as follows: The grain is subjected to 20 pounds of steam pressure for 20 to 25 minutes, at approximately 205° F; then, at 18 to 20 percent moisture content, it is run through large rollers operated at ⅓ to ½ capacity and rolled to thin flakes. The end product has a distinct and pleasant aroma, resembling cooked cereal.

Fig. 30-6. Flaked sorghum (milo). (Courtesy, Benedict Feeding Co., Casa Grande, Ariz.)

The flaking process varies according to the grain. For example, corn is usually steamed for approximately 20 minutes at a temperature of 200° F with a moisture content of about 18 percent. The grain that responds the most to steam processing and flaking is milo. In comparison with dry rolling or grinding, cattle fed steam processed and flaked milo will gain from 0.25 to 0.5 lb more per head per day and require 5 to 10 percent less feed. In terms of improvement due to flaking, corn follows milo. Steam processing and flaking of barley and wheat appear to improve gain but not utilization of the grain. This is probably due to improved palatability and intake of the flaked product as compared to dry rolled or ground product.

● *Gelatinization*—Gelatinization is accomplished by expanding grain. The process usually involves grinding the grain, followed by heating with steam in order to soften it, then forcing the material through a steel tube by an auger. The softened material is then extruded through cone-shaped holes which are smaller where the feed enters and gradually enlarge where the feed is expelled. This results in a sudden release of pressure and the escaping of steam, which expands the grain. Feeding trials to date have given conflicting results from feeding gelatinized corn to fattening cattle, both from the standpoint of feed consumption and rate of gain. Thus, at this time, gelatinization cannot be recommended as a method of processing grain for finishing cattle.

● *Grinding*—Grinding may be the simplest and the least expensive method for preparing grains for feedlot cattle, but it is usually an undesirable method for large operations. Ground grains are usually dusty and are not readily consumed by cattle. This is especially true when rations contain as much as 80 percent grain. If rations contain less than 50 percent grain, the roughage portion tends to prevent the finely ground grain from settling out. Silage is also helpful in this regard. Except for small fattening operations or certain growing operations, the use of ground grain is not recommended. If grain must be ground, it should be ground coarsely.

• *High-moisture (early harvested) grain*—This refers to grain which is harvested at a moisture level of 25 to 30 percent. After the grain is harvested, it may be stored in either an upright or horizontal structure in the same manner as reconstituted grain. Normally it is stored whole in upright structures, then ground or rolled prior to feeding. In horizontal structures, it is usually ground prior to storage in order to obtain satisfactory compaction and exclude air. As would be expected, high-moisture grain has approximately the same feeding value as reconstituted grain. The chief advantage of early harvesting or high-moisture grain over harvesting at the conventional time and reconstituting is that it lessens bird damage and lodging.

Fig. 30-7. A front-end loader scoops up a huge load of high moisture corn (ground) in one of the four concrete-lined trench storage pits at the Farr Feeders, Inc., feedlot northeast of Greeley, Colo. The high moisture corn and corn silage are stored in the pits in large quantities (60,000 tons and 100,000 tons, respectively) to provide a year's continuous supply of feed. The two feeds make up the primary ingredients of the mixed ration, which, by computerized control in the feed mill, maintains a balanced nutritional diet for each animal. The storage pits are covered by polyethelene plastic and weighted down by tens of thousands of tires. This protects the feed from the weather while providing excellent fermentation conditions. (Courtesy, Farr Feeders, Inc., Greeley, Colo.)

Studies with grain sorghum indicate approximately an 8-20 percent improvement in feed conversion of early-harvested grain over dry ground grain, due to (1) much finer grinding, and (2) an increase in both protein and starch digestibility.

●*Jet-sploded grain*—This is a new process, the principle of which is based on a super-hot dry heat system. Hot air, up to 600° F, is pumped through jets in the heat exchanger, quickly heating the grain to the desired temperature. Pressure built up within the kernel will cause the grain to puff out upon reaching atmospheric pressure. The exploded product is then dry rolled. All grains can be processed with this system. Limited feedlot studies to date have shown that jet-sploded milo compares favorably with steam processed and flaked milo. Jet-sploding requires high pressure steam and a complicated piece of equipment. However, high capacity can be built into this system without difficulty, and quality control is automatic—essentially 100 percent of the grain explodes. This method of grain processing appears promising as a way in which to improve the feeding value of grain; thus, it should be thoroughly researched.

● *Micronizing*—Micronizing is a coined word used to describe a dry heat treatment of sorghum grain since microwaves emitted from infrared burners are used in the processing. In micronizing, grain sorghum is heated to 300° F by gas-fired infrared generators. Micronized grain is not popped. It is reduced to about 7 percent moisture, then rolled to produce a uniform, stable, dry, free-flowing product. Water is usually added just prior to feeding to adjust to a 10 percent moisture content. Micronized grain sorghum compares favorably to steam flaked grain sorghum, from the standpoint of rate and efficiency of gain. However, cost of processing favors the micronizing technique over steam flaking because of a lower initial cost in equipment. The main limitation to this method of processing is that current micronizing units have a relatively low capacity. For this reason, they appear to hold the most promise for small feeding operations.

●*Pelleting concentrates, and high-concentrate rations*—Sometimes high-concentrate rations (those with a relatively high proportion of concentrate to roughage) are pelleted. However, the practice is not recommended for high-grain finishing rations, because it reduces feed intake and gains slightly, and makes for more digestive disturbances. These disadvantages, plus the added cost of pelleting, are usually insufficient to counterbalance the 5 to 10 percent advantage in feed efficiency from pelleting.

The pelleting of concentrates in cattle feedlot rations is generally limited to the protein supplement. If the protein supplement contains urea, trace minerals, vitamins, stilbestrol, and/or antibiotics, it is usually pelleted.

●*Popping*—Popping sorghum for feedlot cattle was first done by a research worker, Dr. George F. Ellis, Jr., while he was on the staff of Texas Tech University, whose first popping equipment consisted of a skillet on the kitchen stove. Now, special popping equipment for feedlots is manufactured commercially.

Popping of grain sorghum involves dry heat, at a temperature of 300° to 310° F. The percentage of grain which is actually popped varies from about

13 to 45 percent, depending primarily on moisture content, temperature, and rate of flow through the machine.

As would be expected, popping greatly reduces the density of milo; normal grain weighs 48.8 pounds per cubic foot, in comparison with a weight of 5.8 pounds per cubic foot for completely popped grain. Cattle on popped grain consume much less of the product than cattle fed sorghum processed by other methods; consequently, they gain less weight per day. However, cattle fed popped grain sorghum require about 17 percent less feed per pound of gain than when cracked milo is fed.

• *Reconstituted grain*—Reconstituted grain is mature grain that is harvested at the normal moisture level (10 to 14% moisture), following which water is added to bring the moisture level to 25 to 30% and the wet product is stored in a suitable structure for 15 to 21 days prior to feeding. When stored in upright silos, the grain is stored whole, then rolled or ground at the time of removal from storage. When stored in horizontal silos, the grain must be ground when placed in storage in order to get satisfactory compaction to minimize inclusion of air. It would appear that an upright storage unit is the most satisfactory for reconstituting grain. Properly reconstituted milo and steam processed flaked milo will give similar results with fattening cattle. Corn is also greatly improved by reconstituting, but there appears to be less advantage from reconstituting barley or wheat. It is noteworthy that, unlike most other methods of processing, no gelatinization of the starch occurs in reconstituted grain, yet the utilization of the starch is similar to that of other processing methods. Also, protein utilization of reconstituted grain is higher than that of other processing methods.

• *Roasting*—In roasting, corn is heated to about 300° F. The roasted grain has a pleasant, "nutty" aroma and a puffed, carmelized appearance. Very few kernels are actually popped. However, there is some expansion during the roasting process; raw corn weighs 45 pounds per cubic foot, whereas the roasted corn weighs only 39 pounds per cubic foot. Also, the moisture content of the grain is decreased to 5 to 9%. Purdue University reports that for fattening cattle, roasting improves feed efficiency by 10% and increases weight gains by 14% over ground corn.

• *Whole shelled corn*—Cattle on dry, whole shelled corn gain an average of 5% faster and require 7% less feed per pound of gain than cattle on ground or rolled corn *when high-concentrate rations are fed*. However, processing appears to have some value for dry shelled corn in rations with 20% or more roughage content or when corn is very dry—less than 12% moisture.

Eliminating processing costs is the main advantage from feeding whole corn.

ROUGHAGE PREPARATION

In recent years, researchers and cattle feeders have been much interested in improved processing of grains. But little study has been made of roughage preparation, except from the standpoint of mechanizing and ease in mixing. With the increased competition for grain for human consumption around the world, it is expected that roughage preparation for feedlot cattle will assume greater importance.

The three common methods of roughage preparation are chopping, cubing, and pelleting. Each of these methods will be discussed briefly. However, a few generalizations are pertinent to all of these methods of preparation; namely—

1. In preparing forages, (a) avoid processing those with high moisture, which may heat and produce spontaneous combustion, and (b) avoid processing those in which there are foreign objects (wire and other hardware) which cattle may not be able to select out, and which may ignite a fire when being processed.

2. Processing forages does make for added cost, running from $2 to $10 a ton, depending on the method of processing. Therefore, each cattle feeder should apply his own cost figures, then determine which processing method would be most profitable to him. The important thing is that all costs be accounted for. For example, in computing the cost of baled hay, with which most processing methods are compared, such added "hidden" costs as losses in handling, shrinkage and wastage, grinding costs and losses, insurance, interest, and storage must be considered. Also, the age and grade of the cattle, other available feeds and prices, and starter vs finishing rations must be considered.

● *Chopping*—In comparison with forage fed in long form, chopped hay (1) is easier to handle and mechanize, (2) can be stored in smaller area at less cost, (3) is fed with less waste, and (4) may increase gain slightly. Lower quality and coarser roughages (like fodder or stover) usually benefit more from chopping than high quality and less coarse roughages.

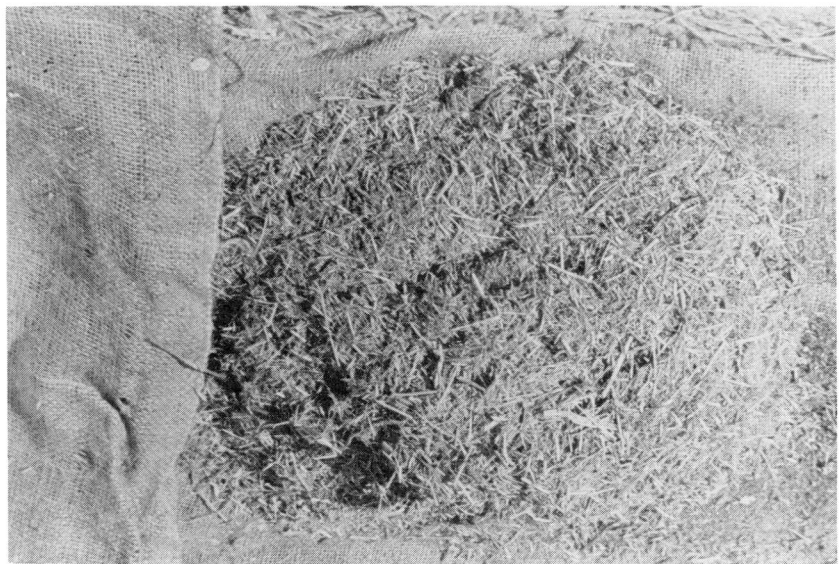

Fig. 30-8. Coarsely chopped alfalfa hay. (Courtesy, Benedict Feeding Co., Casa Grande, Ariz.)

The vast majority of hay fed to cattle in commerical feedlots is chopped. Chopping facilitates both mechanization and mixing. For cattle, roughages should be coarsely chopped—not less than 2 inches in length. Also, dustiness can be lessened by the use of fat, molasses, and moisture.

● *Long hay or fodder*—In general, the practice of using long hay and fodder is limited to small cattle feedlots, and to starting incoming cattle on feed. It makes for considerable waste and requires much labor.

● *Pelleting*—This involves compressing forage, preceded by grinding. The two biggest deterrents to pelleting are (1) fine grinding, and (2) cost. From the standpoint of the animal, pelleted forage should be chopped coarsely in order to allow for optimum cellulose digestion in the rumen and to alleviate the incidence of bloat. As a rule of thumb, one would be on the safe side if the forage were not chopped more finely than silage. Also, there is a cost factor; processors charge up to $10 per ton for an all-roughage pellet.

On the average, cattle on high-roughage (above 80% roughage) or all-roughage rations will eat about ⅓ more pellets than long or chopped hay, make about ½ to ¾ lb faster daily gains, and require 200 to 250 lb less feed per 100 lb of gain. Also, it is recognized that low quality roughages are improved most by pelleting.

● *Cubing*—This refers to the practice of compressing long or coarsely cut hay in cubes about 1 ¼ inch square and 2 inches long, with a bulk density of 30 to 32 lb per cubic foot. Cubes offer most of the advantages of pelleted forages, with few of the disadvantages. It alleviates fine grinding, and it facilitates automation in both haymaking and feeding. Cubing costs about $5 per ton more than baling.

Both cubing and pelleting will (1) simplify haymaking, (2) lessen transportation costs and storage space, (3) reduce labor, (4) make automatic hay feeding feasible, (5) decrease nutrient losses, and (6) eliminate dust.

With cubing or pelleting, the spread between high and low quality roughage is narrowed; that is, the poorer the quality of the roughage, the greater advantage from cubing or pelleting. This is so because such preparation assures complete consumption of the roughage. Also, cubing or pelleting, especially the latter, usually speeds up the passage of roughage through the digestive system.

Mixed Rations vs Feeding Roughage and Concentrate Separately

Most experiments and experiences have not shown any difference between mixed rations and the feeding of roughage and concentrates separately insofar as rate and efficiency of gain are concerned. However, a mixed ration has the following advantages:

1. It makes for greater efficiency in feeding and lessens the sorting at the feed bunk.

2. Where the roughage is relatively unpalatable, a mixed ration forces consumption.

3. Where it is desired to limit concentrate consumption, mixing with the roughage is desirable.

Fig. 30-9. A mixed ration, with forage and concentrates mixed together, containing 69.5 Mcal (or therms) per cwt. (Courtesy, Benedict Feeding Co., Casa Grande, Ariz.)

4. After cattle have become adjusted to the feedlot, a mixed ration makes it easier to get them on full feed.

Thus, each feeder must make his own decision on the matter of mixed vs feeding roughage and concentrate separately, with relative costs and other factors considered. Most large feedlots use completely mixed rations.

Fig. 30-10. Finishing cattle at feed. Feed accounts for approximately 80% of the cost of finishing cattle, exclusive of the purchase price of the feeders. (Courtesy, Benedict Feeding Co., Casa Grande, Ariz.)

QUESTIONS FOR STUDY AND DISCUSSION

1. Sketch the parts of a cow's stomach and describe the processes that occur in energy digestion and metabolism.

2. Explain why roughages compare more favorably with grain for maintenance than for production.

3. Formulate a balanced ration for 660-pound steers. Then, using net energy values, predict the number of pounds of the ration that must be consumed in order to produce a daily gain of 2.4 lb.

4. Under what circumstances, and in what quantities, would you use (a) cane molasses, or (b) fat in the ration of finishing cattle?

5. Under what circumstances would you feed an all-concentrate ration to finishing cattle?

6. Discuss one by-product feed that is being used by a cattle feeder of your acquaintance (or by a feeder with whom you will get acquainted). Among other things, determine the following relative to it: (a) price; (b) chemical composition; (c) quantity fed; and (d) replacement value.

7. List the conditions that are essential for proper urea utilization by feedlot cattle.

8. What minerals would you provide feedlot cattle, in what quantities would you provide them, and how would you provide them?

9. What vitamins would you provide feedlot cattle, in what quantities would you provide them, and how would you provide them?

10. What additive, if any, would you use for each (a) finishing steers, (b) finishing heifers, and (c) suckling calves, based on net returns?

11. Discuss methods, other than additives, for improving rate and efficiency of gains.

12. Discuss the advantages and disadvantages of each of the methods of feed preparation of (a) concentrates, and (b) roughages; then, indicate your preference, with justification for same.

SELECTED REFERENCES

Title of Publication	Author(s)	Publisher
Beef Cattle Science Handbook	Ed. by M. E. Ensminger	Agriservices Foundation, Clovis, Calif., pub. annually since 1964
Cattle Feeders Hand Book	R. M. Bonelli	Computer Publishing Co., Phoenix, Ariz., 1968
Feedlot, The	Ed. by I. A. Dyer, C. C. O'Mary	Lea & Febiger, Philadelphia, Penn., 1972
How to Make Money Feeding Cattle	L. H. Simerl B. Russell	United States Publishing Company, Indianapolis, Ind., 1958
Stockman's Handbook, The, Fourth Edition	M. E. Ensminger	The Interstate Printers & Publishers, Inc., Danville, Ill., 1970
Third Dimension of Cattle Feeding, The	J. M. Hutchison	General Management Services, Inc., Phoenix, Ariz., 1970

MANAGEMENT OF FEEDLOT CATTLE

Contents Page

Although it is not possible to arrive at any overall, certain formula for success in operating a cattle feedlot, those operators who have made money have paid close attention to the details of management.

There are many facets of cattle management. Some are applicable to both cow-calf and cattle feedlot operations. These are covered in Chapter 10 of this book; hence, they will not be repeated. Other management practices are unique to cattle feedlots.

Fig. 31-1. Management gives point and purpose to everything else. Jim Benedict, able and successful manager, and recipient of the coveted "Cattle Feeder-of-the-Year Award," shown (left) with a lot of finished cattle and with the feed mill in the background. (Courtesy, Benedict Feeding Co., Casa Grande, Ariz.)

BACKGROUNDING

Backgrounding is the preparation of cattle from weaning until placing on finishing rations. It involves maximum roughage consumption and moderate gains.

The growing of calves from weaning until placing on finishing rations is not new. Only the term "backgrounding" is new. Likewise, some new "wrinkles" have been added to the method of conducting it.

Growing of calves to the yearling stage for placement in feedlots was, and still is, known as growing stockers. Farmers have, for many years, fed high roughage rations to calves prior to marketing. Some ranch operators have, historically, retained calves for a second grazing season. Also, wintering cattle on small grain pasture is a well-established practice in the South and Lower Plains.

The term "backgrounding" came in with the development of large commercial cattle feedlots—outfits that usually had limited amounts of available roughage and other cheap feeds, and that had need for, on a year-round basis, growthy, but unfinished, cattle of a certain weight, usually within the range of 600 to 750 pounds. Today, there is renewed interest in backgrounded cattle, due to high grain prices and the need to produce more beef from roughage. Also, it isn't particularly efficient for large feedlots to tie up capital for feeding cattle where limited gains are involved.

Kinds of Backgrounding

Backgrounding of stockers and feeders can be divided into two systems: (1) backgrounding on pasture, in which calves or light yearlings are wintered and grazed, or grazed only, and sold as feeders in the late summer or fall; and (2) backgrounding in the drylot, in which the cattle are fed harvested roughage and grain and then transferred to another lot for finishing.

Who Does the Backgrounding?

Backgrounding is done by three different types of operators:

1. Cow-calf operators (farmers and ranchers) who either have a surplus of roughage or decide that it may be more profitable to market their calf crop at a later period.

2. Commercial finishing lots who do backgrounding as a means of assuring a supply of cattle to go into the finishing lot at the required times.

3. Specialized backgrounding lots. Such lots have evolved in recent years; and it appears that more operations of this kind will be needed in the future. Economics usually restrict these highly specialized operations to areas or large farming operations with high roughage feed producing capacities, yet close enough to large feeder cattle production points. Investment costs in backgrounding feedyards can be considerably lower than for finishing yards. Lot space requirements are lower on a per unit basis, because cattle are never carried beyond yearling weights. Also, milling facilities and grain storage requirements are much less than for finishing operations.

Kinds of Cattle to Background

Generally speaking, the English beef breeds are best suited for backgrounding purposes. This is because they should be grown to approximately 600 to 750 pounds before placing on finishing, or high energy, rations. Holsteins and some of the larger, growthier exotics are not well suited to backgrounding, unless heavy finishing weights are planned. They need to be placed on high energy rations at weaning time; otherwise, they will not finish out at desirable weights of 1,050 to 1,100 pounds—instead, they will be too heavy at market time.

Rate of Gain of Backgrounded Cattle

Properly backgrounded cattle should gain from 0.75 to 1.50 pounds per head per day. Cattle finishers object to cattle that have made higher gains, because it lessens, or eliminates, compensatory growth. That is, when put on high energy rations, animals that have been backgrounded so as to make minimal daily gains usually gain better than similar cattle that have been fed more liberally during the backgrounding period. For the latter reason, when contracting for backgrounding calves, feedlots commonly specify the kind of ration and the range in gains.

Profit Potential from Backgrounding

Profit is the goal in any type of feeding operation, whether it be backgrounding or finishing. The profit potential in a background system is dependent upon (1) purchase cost of feeders, (2) total cost of gain, (3) amount of gain, and (4) selling price of the backgrounded feeders. The most profitable backgrounding operations are generally in areas where feed costs are low, where overhead costs are minimal, and where the health program is superior.

Advantages of Backgrounding

The *advantages* of the backgrounding system include the following: (1) owners have the cattle when they make the most efficient gains; (2) it is well adapted to the use of roughage and by-product feeds; (3) it provides a way in which to make use of seasonal surplus labor or of buildings that are present on some farms or ranches; and (4) volume is more flexible than with cow-calf operations—that is, numbers can be easily changed to fit feed, labor, or economic outlook.

Disadvantages of Backgrounding

The *disadvantages* of backgrounding include the following: (1) high buying and selling skills are required, because there is no established market as with slaughter cattle; (2) buying, selling, and shrink costs must be absorbed by a limited amount of gain; (3) cost of gain must be kept down to offset negative margins that usually prevail; (4) high risk results from seasonal and yearly price fluctuations in feeder cattle; and (5) backgrounding operations cannot be located too far from finishing feedlots, because of high transportation costs.

It should be noted that backgrounding operations will have increasing competition for their finished products—calves ready to go into the finishing lot—from cow-calf men who wean off heavier calves. The latter is being accomplished by crossbreeding, along with heavier milking cows.

PEOPLE MANAGEMENT

Big feedlots must rely on hired labor, all or in part. Good help is hard to come by; it's scarce, in strong demand, and difficult to keep. Moreover, the labor situation is going to become more difficult in the years ahead. It matters little if the feedlot facilities, the feed, and the cattle are the best if the labor fails to feed them properly or to treat the sick. It is important, therefore, in the operation of the feedlot that laborers be recruited with care, that there be an organization chart and job description, and that there be an incentive basis for the help. These points are covered in Chapter 13 of this book.

HOUSEKEEPING AND REPAIR

A well-kept and attractive feedlot makes for better employee attitude, which in turn makes for better cattle care. Fences, gates, equipment, and roads should be kept in constant repair. Unused materials should be stored or repaired. The entire premises should be tidied up at all times. Pride in the physical plant will be reflected in pride in the work.

RECORDS

Complete and well-kept records are a must in the operation of a cattle feedyard, even though they require a lot of time and expense. Deficient records and deficient managers generally go hand in hand.

Record Forms

There is no limit to the number of different kinds of record forms that can be, and are, kept in a given feedlot. Also, there is little similarity in record forms between lots, due to differences between people, primarily managers and bookkeepers. The important things are that (1) record forms be so designed as to facilitate record keeping, with as much ease, efficiency, and accuracy as possible; and (2) records be kept.

Figs. 31-2 and 31-3 show two basic record forms; Fig. 31-2 is a Daily Record, whereas Fig. 31-3 is a Monthly, Cumulative, and Final Feed Summary. Many variations of these can be made.

Among other necessary records are the following:

1. Feed costs, with this record kept by individual pens.
2. Grain inventory.
3. Roughage inventory.
4. Feed projections ahead.
5. Cattle receiving and movement records.
6. Sick pen and movement records.
7. Sick pen costs.
8. Mortality slips and proof of death.
9. Maintenance and repair costs.
10. Routine office bookkeeping.
11. Customer billing for feed.
12. Closeout records.

Make 28-Day Test Weights

Twenty-eight-day test weights will not adversely affect the performance of feedlot cattle, provided the cattle are handled properly. Check weights should include a representative cross section of the cattle in the yard, including age, weight, type, background, and sex. Where it is not convenient, or it is not desired, to weigh an entire lot of cattle, "markers"—cattle of certain odd colors, animals with tail switches clipped, etc.—may be weighed. Also, it is important that weighing consistently be done at the same time of day, and that the lots be weighed in the same order, due to the effect of rumen fill.

MONTHLY, CUMULATIVE, AND FINAL FEED SUMMARY

Feedlot: _____ Period: _____

Pen No. _____ No. Head _____ Date Started _____ Date Closed _____

RATION:

Ration No.	Total Pounds	Price/Ton	Total Cost
	(lb)	$	$
#1			
#2			
#3			
#4			
#5			
#6			

FEED ANALYSIS:

Total feed fed _____ _____ lb

Total cost of feed to date _____ $ _____

Feed days (no. head × days) _____ _____ no.

Net weight out _____ _____ lb

Net weight in _____ _____ lb

Net gain _____ _____ lb

Av. weight out _____ _____ lb

Av. weight in _____ _____ lb

Feed per head per day _____ _____ lb

Cost per head per day _____ _____ ¢

Gain per head per day _____ _____ lb

Cost per lb gain _____ _____ ¢

Feed conversion (lb feed/lb gain) _____ _____ lb

OTHER COSTS:

Milling charges _____ _____

Mineral charges _____ _____

Medication _____ _____

Management _____ _____

Labor _____ _____

Physical plant (other than milling) _____ _____

Fig. 31-2. Form for Daily Record.

DAILY RECORD

Feedlot: _____ Pen No. _____ Date Started _____ Month _____ Day _____ Year _____

Day of Month	No.	Origin	Head In — Pur. Price	Head In — Total Pur. Wt.	Head In — Total Wt. at Lot	Head In — Av. Wt. at Lot	Death Losses (cause)	No.	Head Out — Total Wt.	Head Out — Av. Wt.	Daily Feed — Total Lb	Daily Feed — Lb/Head/Day	To	Sold — Price/Cwt	Carcass Grade	Comments
1																
2																
3																
4																
5																
6																
7																
8																
9																
10																
11																
12																
13																
14																
15																
16																
17																
18																
19																
20																
21																
22																
23																
24																
25																
26																
27																
28																
29																
30																
31																

Fig. 31-3. Form for Monthly, Cumulative, and Final Feed Summary.

MILL MANAGEMENT

Mill management includes many things; among them, the following: commodity scheduling and purchase; inventory control; maintenance, repair, and new construction; mill cleanliness; commodity and ration quality control; and milling to meet daily feed needs. Quality control refers to being able to deliver to the cattle on a consistent basis the same quality or intended ration composition.

Fig. 31-4. Cattle feed processing facilities at C & B Livestock Co., Hermiston, Ore. (Courtesy, Ron Baker, President, C & B Livestock Co., Hermiston, Ore.)

• *Commodity quality control*—All feed ingredients should be bought by specification and grade. Then, each of them should be analyzed in order (1) to determine if the ingredient received meets these specifications, and (2) to know the composition of the ingredients used in ration formulation.

A chemical analysis is no better than the sampling. Thus, state feed control officials should be contacted for publications detailing recommended methods of sampling commodities and the equipment to use. All commodity samples should be properly labeled, including the yard name, name of the sampler, date sample was taken, commodity name, vendor of the commodity, invoice number, and car or truck number.

• *Ration quality control*—Rations should be analyzed according to an established schedule. Most of the very large cattle feedlots sample rations daily, then prepare a weekly composite of each ration, which is submitted to the laboratory for analysis. Additionally, a complete analysis (proximate, calcium, phosphorus, and nitrate) is made of each ration monthly. Always label each ration with care. This should include the identity of the sample or ration number, feedyard name, name of sampler, and date.

FEED BUNK MANAGEMENT

Feed bunk management is a combination of management factors involved with obtaining maximum performance, minimum digestive disorders, and keeping cattle on feed. Feed bunk management and quality control are directly involved with obtaining maximum and economical performance from cattle. It should be every feeder's goal to obtain maximum feed intake of a consistently high quality ration, since both rate and efficiency of gain are directly related to nutrient intake.

Schedule for Getting Cattle on Feed

When new cattle arrive at the feedlot, the objective is to get them on full feed as rapidly as possible, without throwing them off feed. This is not easily accomplished because many factors influence the difficulties experienced in starting new cattle on feed, among them: (1) the length of time that the cattle have been without feed; (2) the kind of feed to which the cattle were accustomed prior to shipment; (3) the age of the cattle—young cattle adapt to a change in feed more easily than old cattle; (4) whether or not the cattle have been fed and watered out of troughs before; (5) the weather conditions; and (6) existing nutritional deficiencies.

Fig. 31-5. Calves that have been on feed 45 days, following shipment from South Texas to an Arizona feedlot. (Courtesy, Benedict Feeding Co., Casa Grande, Ariz.)

TRADITIONAL PROCEDURES OF GETTING OLDER CATTLE (NOT CALVES) ON FEED

When first brought into the feedlot, cattle that are not accustomed to grain may be started on feed by either of the following procedures:

1. Self-fed long grass hay (and/or corn or sorghum silage), and hand-fed concentrate according to the following schedule (with the cattle automatically lessening their self-fed hay consumption as the grain is increased):

First day—Feed 4 lb of concentrate/head/day, consisting of 2 lb of grain and 2 lb of protein supplement.

Daily increase—Step up the grain by one lb/head/day until cattle are receiving one lb/cwt body weight.

Increase every third day—After a level of one lb daily/cwt body weight is reached, make increases every third day as follows:

Calves—¼ lb

Yearlings—½ lb

2-year-olds—1 lb

2. Hand-fed a mixed ration of chopped grass hay (and/or corn or sorghum silage) and concentrate, with the proportion of roughage decreased and the grain increased according to the following schedule:

Day	Kind of Feed	Percent of Roughage
1	Grass hay and/or nonlegume silage	100
2-4	Grass hay plus starter	60-90
5-14	Starter ration	40-60
15-21	Transition ration	15-40
22-to market	Finisher ration	5-15

Although one of the above procedures may serve as a useful guide, it is recognized that no set of instructions can replace the cattle intuition and good judgment of an experienced feeder.

After cattle are on full feed, they may either be self-fed or hand-fed. Most large feedlots feed twice daily, barely letting the cattle clean up the previous feed before the next feeding.

SCHEDULE AND RATION FOR GETTING CALVES ON FEED

Most cattle feeders follow the procedure and type of ration given above for getting cattle on feed; they start them on a high roughage ration, then work them over to a high concentrate ration as they progress through the feeding program. However, based on University of California studies, it appears that for calves (not older cattle) a starting ration consisting of 28 % hay (roughage) and 72% concentrate is best. This is similar to the procedure outlined in point 2 above for older cattle, except that the calves are immediately started on a lower roughage (28%)—higher concentrate (72%) ration.

Among the problems encountered in new cattle are feed and water refusal, lactic acidosis, bloat, and diarrhea. Refusal of feed and water is generally due to the fact that the animals are not used to conventional troughs and/or the feed is so different.

Lactic acidosis generally results from feeding hungry cattle excessive levels of rapidly fermentable feeds. The condition is characterized by an accumulation of lactic acid in the rumen and a lowering of the pH in the blood and urine. The problem can be minimized by starting cattle on a high roughage ration and shifting them gradually to a high concentrate ration.

Bloat occasionally occurs in new cattle, although it is more frequent during the later stages of feeding. Bloat and diarrhea in new cattle can generally be prevented by feeding generous quantities of such roughages as straw, grass hay, cottonseed hulls, or corncobs.

Frequency of Feeding

Experiments and experiences show that increased frequency of feeding from 1 to 3 times per day improves performance above increased costs. For this reason, most commercial feedlot cattle are fed 3 times daily. Some experiments indicate that even more frequent feeding—more than 3 times daily—will produce slightly more rapid gains and result in greater feed efficiency. However, the improved performance is not enough to warrant the increased costs.

Amount to Feed; Full vs Limited Feeding

Feed intake is one of the key factors affecting feedlot performance. Perhaps no other factor has such overriding importance in determining rate and efficiency of gain, and, ultimately, the profit derived from feeding cattle. Of course, the reason for emphasis on high feed intake is that once a sufficient amount of the ration is consumed to meet the maintenance needs of a finishing animal, the remainder is converted to gain with remarkable efficiency. Thus, as shown in Fig. 31-6, by adding 4 lb to the daily feed intake of a 600-lb steer, rate of gain may be increased by $1^1/_{10}$ lb per day. Conversely, poor feed intake results in too high of a percentage of the total nutrients being expended for maintenance.

Thus, finishing cattle should receive a maximum ration over and above the maintenance requirements. In general, they will consume daily an amount (on an air-dry basis) equal to 2.5 to 3.0 percent of their liveweight. Feed intake will vary according to the condition of the cattle, the palatability of the feeds, the energy of the ration (in general, animals eat to meet their energy needs), the weather conditions, and the management practices. For example, older and more fleshy cattle consume less feed per hundred weight than do younger animals carrying less condition; thus, mature, overfinished steers will consume feeds in amounts equal to about 1.5 percent of their liveweight, whereas thin steers under 2 years of age will consume fully twice as much feed per unit liveweight.

Overfeeding is also undesirable, being wasteful of feeds and creating a health hazard. When overfeeding exists, there is usually considerable leftover feed and wastage, and there is a high incidence of bloat, founder, scours, and even death. Animals that suffer from mild digestive disturbances are commonly referred to as "off feed."

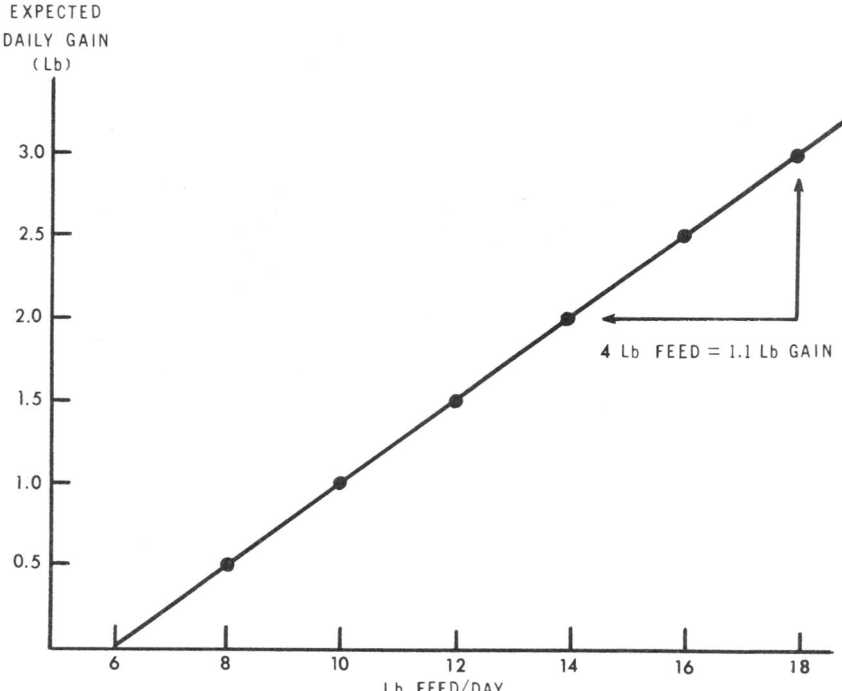

Fig. 31-6. Relationship of daily feed intake to rate of gain; 600-lb steers fed 85% concentrate ration. (Source: Pope, L. S., Assoc. Dean of Agriculture, Texas A & M University, College Station, Tex., Factors Affecting Feed Intake, *Great Plains Beef Feeding Handbook*)

Limited feeding means just what the name indicates—not giving the animals all they want. Limited feeding generally decreases the rate of gain, adversely affects feed conversion, and increases cost of gains. Under most conditions, cattle should be full-fed throughout the finishing period.

FACTORS AFFECTING FEED INTAKE

A number of factors play major roles in governing feed intake; among them, the following:

• *Age of cattle*—Calves consume a larger proportion of feed in proportion to their body weight than do older cattle. Table 31-1, which gives the average expected dry matter intakes per head daily for cattle, points up this situation.

• *Propionic acid in the blood curbs appetite*—All mammals, including man, have an "appetite center" at the base of the brain in the lateral region of the hypothalmus. Certain nerve cells actually regulate energy intake, causing the sensation of hunger, or preventing the animal from consuming too much. But there is a great species difference as to how much feed is enough—or too much. A growing boy "wolfs down" 8 to 9% of his body weight daily. He is outdone by a baby chick, which eats 10% of its body weight. A hog eats 5 to 6% of its body weight. But a fattening steer consumes

TABLE 31-1

AVERAGE EXPECTED DRY MATTER INTAKES
PER HEAD DAILY FOR CATTLE

Body Weight		Expected Dry Matter Intake
300	....	8.5
400	....	10.5
500	....	12.5
600	....	14.5
700	....	16.5
800	....	18.0
900	....	20.0
1,000	....	22.0
1,100	....	23.0
1,200	....	24.0

[1]From Great Plains *Beef Cattle Feeding Handbook*, GPE 1100.3.

only 2 to 3% of its body weight. Why the difference? It appears that propionic acid is the triggering mechanism which tells a steer when to stop eating. It acts on the nervous center of ruminants. Since increased grain levels tend to step up propionic acid, it is conjectured that grain rations, which increase propionic acid levels of the blood, are self-limiting when it comes to feed intake. For this reason, cattle eat fewer pounds per head daily of a high concentrate ration than of a high roughage ration, but both groups tend to take in about the same level of energy.

● *Rumen "fill"*—The "fill" in the rumen places a ceiling on feed intake. Since low quality roughages pass through the rumen at a slower rate than high quality roughages, they can limit the amount of total feed that the animal can consume in a 24-hour period. Of course, this is of little consequence with fattening-type rations, which normally contain limited amounts of roughages and roughages of good quality. The amount of concentrate in the ration has a marked effect on the total pounds of feed consumed daily. Thus, when on all-concentrate rations, the total energy intake of cattle over a 24-hour period is not much greater than with a bulky, low concentrate ration. Apparently, this is due to the regulation of total calorie intake by the ruminant.

● *Physical makeup of the ration*—Coarsely processed grains are more palatable to cattle than finely processed grains. Thus, they will eat more of such feed.

● *Energy density, or bulk, of the ration*—The weight per unit volume, or the bulk, of the ration affects total feed intake. That is, bulky rations which make for "fill" reduce dry matter consumption. Most feeders think of energy density in terms of roughage:concentrate ratio. When density is very low, as in a high roughage ration, animals simply cannot hold enough to make gains. As energy density increases, gain increases.

● *Heritability*—Experiments and experiences confirm that some cattle are better "eaters" than others—that they will consume more feed. This may be due either to a difference in rumen capacity and/or a difference in "threshold" for circulating metabolites in the bloodstream which affect the appetite center of the hypothalmus.

● *Heat stress*—Cattle feeders have long known that finishing cattle consume less during hot weather. This was confirmed by the California station in trials at the Imperial Valley Station. The lowering of feed consumption due to heat stress may be lessened by lowering the roughage in the diet, cooling the drinking water, providing shades, and adding higher levels of vitamin A.

● *Protein or phosphorus deficiency*—A protein deficiency can markedly reduce feed intake by depressing the rumen bacterial count and the rate of breakdown of feeds. A phosphorus deficiency can cause a reduction in feed intake, and even a depraved appetite.

● *Other factors affecting feed intake*—The feed intake of cattle is also affected by: (1) moisture level—very high moisture feeds reduce total dry matter intake; (2) dustiness—dusty feeds lower total feed intake; (3) lack of water—depriving cattle of water will markedly reduce feed intake; (4) frequency of feeding—more frequent feeding results in higher feed consumption; and (5) freshness of feed—cattle will consume more clean, fresh feed than stale feed.

Feed Regularly

Feedlot cattle should be fed at regular times each day, by the clock. This means that in the larger lots cattle in each alley should be fed in the same order each day. Be prompt—remember that cattle are creatures of habit.

Keep Bunks Clean

All feed bunks should be cleaned thoroughly at least once a week during dry weather, and as needed during stormy weather. No feed should ever be allowed to spoil in the feed bunk. Manure should be cleaned from feed bunks daily.

Weather Affects Eating and Drinking Habits

During hot weather, feedlot cattle "peak" their eating during early morning and again during the evening hours—when it is cool. With heat, night drinking increases. In cool weather, they eat more during the midday than when it's hot. The feeder should sense these changes in cattle eating habits and program their feeding accordingly.

Cattle eat more following a bad storm or a hot spell. Thus, at such times the bunks may be "slick" for two to three hours and the cattle may line up waiting to be fed. When this happens, the ration should be increased. By going to a higher roughage ration at these times, the problems from acidosis and laminitis can be minimized.

WATER

Water is the cheapest feed! Thus, cattle should have access to plenty of clean, fresh water at all times. They will consume 7 to 12 gallons per head per day. In cold climates, waterers should be equipped with heaters. Where the water supply is not limited by cost or volume, continuous-flow waterers are excellent. In order to keep the pathogen and algae content at a minimum, water tanks should be cleaned at least once a week in the winter and twice a week in the summer. In sick pens and pens of new cattle, the water tanks should be cleaned daily.

PROGRESSIVE CHANGES IN FEEDLOT CATTLE IN (1) RATE OF GAIN, (2) FEED CONSUMPTION, (3) FEED EFFICIENCY, AND (4) COST OF GAIN

It is important that cattle feeders be cognizant of the progressive changes that normally occur in feedlot cattle, from start to finish of the feeding period. Of course, it is recognized that many factors influence the degree of these changes.

The Arizona Station recorded the changes at 28-day intervals in 41 lots of feedlot cattle, with all lots taken from start to finish. Their findings are given in Table 31-2.

The Michigan Station used a slightly different approach to obtain changes at 28-day intervals. They fed 7 different lots of cattle, with each lot carrying a different length feeding period in increments of 28 days, ranging from 115 to 283 days on feed; and each group of cattle was closed out and slaughtered at the end of its feeding period. Their findings are given in Table 31-3.

Based on these two studies, the following deductions may be made relative to the progressive changes in feedlot cattle, from start to finish: (1) rate of gain decreases; (2) daily feed consumed per 100 lb of body weight decreases; (3) feed per 100 lb of gain increases; and (4) feed costs per 100 lb gain increases.

CULL OUT; TOP OUT

Obvious poor doers should be taken out early; and marketed at Standard grade. Where individual weighing can be made, consideration should be given to the practicality of individually tagging (with duplicate tags, one in each ear) and weighing incoming calves; weighing them again at the end of the grower-ration period and prior to going on finishing rations; then culling out the bottom 10 percent.

Also, cattle should be sold when they make their grade; thereby avoiding loss in efficiency, excess finish, and too heavy weights. Usually, it is unwise to challenge a sagging market by holding and feeding for a higher market. There is no need to put feed and labor into heavy cattle at a cash discount when younger cattle will use these resources more efficiently.

TABLE 31-2

AVERAGE DAILY GAIN, FEED PER POUND OF GAIN, AND FEED COSTS OF CATTLE AT 28-DAY INTERVALS[1]

28-Day Feeding Period	No. of Days on Feed	Pay Weight of Cattle on Feed			Average Daily Gain	Average Daily Feed	Lb of Air-Dry Feed per Lb of Gain	Cost per Lb of Gain Put on During the 28-Day Feeding Period				Accumulative Average Total Cost per Lb of Gain
		Beginning	Ending	Average				Feed Costs	Nonfeed Costs Exclusive of Interest on Cattle	Interest on Cattle	Total Cost of Gain	
	(days)	(lb)	(lb)	(lb)	(lb)	(lb)	(lb)	(¢)	(¢)	(¢)	(¢)	(¢)
1st	1-28	597	697	647	3.56	22.75	6.39	13.93	3.19	0.80	17.92	17.92
2nd	29-56	697	783	740	3.06	24.24	7.92	17.58	3.96	0.93	22.47	20.02
3rd	57-84	783	858	821	2.69	23.43	8.71	19.99	4.36	1.06	25.41	21.58
4th	85-112	858	923	890	2.30	23.83	10.36	23.78	5.18	1.24	30.20	23.29
5th	113-139	923	979	951	2.07	24.43	11.80	27.00	5.90	1.38	34.28	24.95

[1]Data collected by Harvey A. Meier and W. H. Hale, Arizona Agricultural Experiment Station, unpublished. Data embraces 9 experiments, 41 lots of yearling steers, and 640 animals. All cattle were started on relatively high roughage rations, with the roughage decreased as the period progressed.

TABLE 31-3

EFFECT OF LENGTH OF FEEDING PERIOD ON (1) RATE OF GAIN, (2)
FEED CONSUMPTION, (3) FEED EFFICIENCY, AND (4) COST OF GAIN[1]

	DAYS ON FEED						
	115	143	171	199	227	255	283
No. steers	8	8	8	8	8	8	8
Av. initial weight, ... lb	655	657	685	685	672	679	673
Av. final weight, ... lb	928	990	1037	1084	1133	1203	1224
Av. total gain, ... lb	273	333	352	399	461	524	551
Av. daily gain, ... lb	2.73	2.33	2.06	2.01	2.03	2.05	1.95
Daily feed/100 lb body wt., lb	2.60	2.44	2.28	2.19	2.20	2.11	2.05
Total feed/100 lb gain, ... lb	869	861	952	962	975	968	996
Feed cost/100 lb gain, ... $	15.80	15.64	17.30	17.54	17.81	17.69	18.21
Concentrate: roughage ratio	77:23	77:23	77:23	77:23	77:23	77:23	77:23

[1]Merkel, R. A., H. E. Henderson, and H. W. Newland, Effect of Length of Feeding Period on Rate, Composition and Cost of Gain, *Michigan Beef Cattle Day Report 1966*, Michigan State University, East Lansing, Mich., p. 10.

Fig. 31-7. Choice grade cattle topped out and on the way to market. Cattle should be sold when they make their grade. (Courtesy, Benedict Feeding Co., Casa Grande, Ariz.)

OVERFINISHING

Excessive finishing is undesirable, both from the standpoint of the producer and the consumer. Experienced cattle feeders are fully aware of the fact that to carry finishing cattle to an unnecessarily high finish is usually prohibitive from a profit standpoint. This is true because the gains in weight then consist chiefly of fat but little water. In addition, a very fat animal eats less heartily, with the result that a small proportion of the nutrients, over and above the maintenance requirement, is available for making body tissue.

Table 31-4 shows that the heavier the cattle, the more expensive the gains. Also, these figures point up (1) the importance of topping out finished cattle, rather than waiting until the entire lot is ready; and (2) the reason why it is generally wise to sell cattle when they are ready to go, rather than to hold for a higher market.

TABLE 31-4

APPROXIMATE COST OF ADDING 100 POUNDS OF BEEF
ON STEERS AT VARIOUS WEIGHTS

Steer Weight	If Cost of No. 2 Corn per Bushel Is				
	$2.50	$2.75	$3.00	$3.25	$3.50
400 lb	22.78	25.16	27.55	29.52	31.77
500 lb	23.89	26.40	28.77	33.83	33.38
600 lb	25.97	28.74	31.36	36.64	36.21
700 lb	29.03	32.18	35.32	41.07	40.52
800 lb	33.89	37.54	40.91	47.86	47.25
900 lb	40.69	45.10	49.09	52.68	56.81
1,000 lb	50.42	55.69	60.82	65.27	70.27
1,100 lb	64.03	70.81	77.32	82.88	89.38
1,200 lb	83.33	92.26	103.36	108.06	116.44

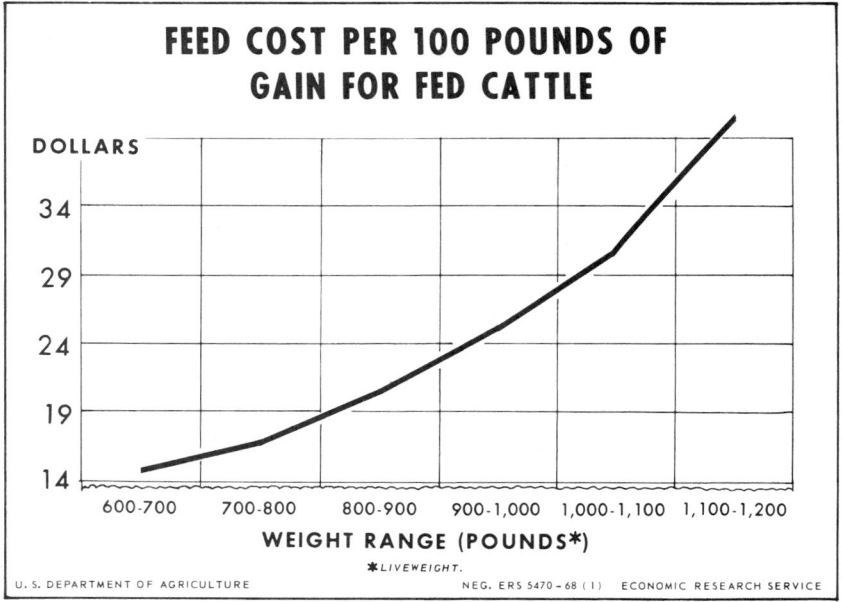

Fig. 31-8. This graph illustrates changes in feed conversion efficiency for cattle from normal feeder weights to slaughter weights. Note that feed costs per 100 lb gain more than double from 600-700 lb to 1,000-1,100 lb, and that the conversion efficiency ratio changes even more sharply when cattle pass 1,100 lb.

ABORTING HEIFERS

Frequently feedlot operators have need to abort heifers. When such occasions arise, they want to know how to do it.

Caesareans on feedlot heifers must be performed by a veterinarian; hence, they are expensive. Moreover, such surgery lowers the rate and efficiency of gain very considerably.

Several drugs can be used. Stilbestrol is the drug of choice, provided it is used in the early stages of pregnancy; but there is some uncertainty as to how long stilbestrol will be available. (See Chapter 30, section entitled, "Implants and Growth Stimulants.") Estriadol propionate, in large doses, can be used; but for best results it should be administered before the fourth month of pregnancy. The cortical steriods—such as azium, vetalog, flucort, and others—can be used in late pregnancy.

The best nondrug method is to pregnancy check each heifer within 2 to 3 weeks after arrival in the feedlot. Those under 4 months pregnant can be aborted by manually removing the corpus luteum on the ovary. Earlier pregnancies, 30 to 50 days, can be terminated by manually crushing the small fetus while pregnancy checking. Heifers over 4 months pregnant should be identified and sent to pasture for calving.

BULLERS OR RIDERS

There are two types of bullers or riders: (1) those that persistently ride other cattle; and (2) those that other cattle pick out and ride. The first are often bulls or stags. A buller that is being ridden by a number of animals can receive serious injuries, and may even be killed.

The buller problem became more prevalent with the advent of stilbestrol implants. Also, the problem is accentuated by the following:

1. Improper stilbestrol implantation. The pellet should be midway between the base and the tip of the ear. If it is near the base, absorption takes place rapidly and riding may result. Also, care should be taken to avoid crushing the implant, as this will also result in more rapid absorption and likely riding.

2. The consumption of feeds high in estrogens, like green chopped alfalfa.

Bullers or riders should be handled as follows:

1. Remove the implant(s).

2. Give 500 mg of Repositol Progesterone.

3. Put the buller in a heifer pen or a "buller pen." This works very well unless there are so many bullers that the bullers start riding the bullers.

4. After an elapse of about 72 hours from the time of removal of the implant(s) and the administration of progesterone, return the buller to his original pen. This return should be effected during the morning feeding period, so that others may not notice him. Then, if riding reoccurs, remove the buller before injury is inflicted.

MUD PROBLEM

University of California studies show that mud can reduce cattle gains by as much as 25 to 35 percent. Thus, it is important that the problem be minimized, especially in high rainfall areas. Good drainage is the first essential. This should be assured at the time the feedlot is located and constructed. Mounds, preferably perpendicular to the feed bunk, will provide cattle a dry place on which to lie down. Concrete aprons along the bunk will provide them with solid footing on which to stand and feed. Also, lessening of cattle density during the winter months—fewer animals per lot—is an effective method of controlling the mud problem. Thus, many feedlots plan to feed fewer cattle during the muddy season.

Bedding

Open feedlots are usually mounded, but not bedded. Sometimes sawdust or shavings are placed on top of the mounds.

Shelters and confined houses are usually bedded, unless a liquid manure system is used. Whether or not bedding is used, and the choice of kind of bedding, depend primarily on the local cost and availability of bedding, and labor available for cleaning. For a full discussion of bedding cattle, see Chapter 10 of this book, under the heading entitled "Bedding Cattle."

BIRD CONTROL

Birds are gluttons and filthy; hence, they should be controlled. In a 5-month study of a 12,000-head cattle feedlot in California, University of California researchers found that the birds ate between 200 to 500 pounds of feed each day, adding up to a total of 32,500 to 74,000 pounds for the 5-month winter season. The bird population in the feedlot ranged from 10,000 to 20,000, of all species, with the most abundant being the house sparrow, which was estimated at from 6,000 to 8,000. Other feed-consuming bird species identified were starlings, brewer blackbirds, red-wing blackbirds, and cow-birds.

Nationwide, starlings constitute the major feedlot bird problem. Some large commercial feedlots estimate their starling population at 100,000 per lot. Iowa feeders figure that starlings add $3 to $4 to the cost of each steer marketed. Some western feedlot operators compute the cost for overwintering each 1,000 starlings at $100; others estimate that starling nuisance and feed costs add 2¢ to the cost of each pound of gain.

In addition to feed consumption, birds contaminate much feed and spread diseases—to both animals and humans. The starling has been incriminated in the spread of coccidiosis among animals, transmissible gastric enteritis (TGE) in swine, and histoplasmosis in humans.

Recordings of distressed bird calls, carbide connons, and harassment or killing with guns achieve only partial control. Many chemicals and baits have been tested, and a few have been found to be effective. However, some states do not allow the use of chemicals in bird control. Therefore, before using any chemical, the cattle feeder should check with the appropriate Federal, state, and local departments of health. Also, chemicals should always be handled with care; they should neither come in contact with the skin nor be inhaled. Gloves and a respirator should be worn when mixing or handling them.

FLY CONTROL

The housefly is the most common type of fly found around cattle feedlots. It is a scavenger and does not feed on animals, but it does cause irritation and annoyance. Stable flies, which are blood feeders, may also be present in certain areas and certain feedlots.

Effective housefly control requires proper animal waste management and good feedlot sanitation. The basic objective in fly control is to eliminate possible sources of fly development. This can be accomplished by the following steps: (1) provide proper drainage and avoid wet spots; (2) remove manure immediately after a pen is vacated; and (3) remove manure and spilled feed at important fly breeding areas such as fence lines, feed bunks, hospital pens, horse pens, truck washing stations, and receiving and shipping areas. Chemical control should be used in conjunction with the proper waste management techniques, and not as a sole means of control. Residual and space sprays aid in reduction of adult flies; and larvicides may be applied to areas of intense larval development such as manure stockpiles, hospital, and horse pens.

Also, see Chapter 12, section entitled "Parasites of Beef Cattle."

HOGS FOLLOWING CATTLE

There was a time when hogs following feedlot cattle was commonplace. But the practice declined with the advent of large, specialized commercial feedlots of 1,000 head, or larger, capacity. Today, it is estimated that only 15 percent of the farmer-feeders of the Corn Belt have hogs following cattle,[1] and the practice is almost nonexistent in the large commercial cattle feedlots of the nation. Because farmer-feeders account for relatively small numbers of cattle, in comparison with large commercial feeders, it follows that very few cattle are followed by hogs. The primary reasons given by cattle feeders for a decline in the practice are: (1) feeds are being processed in a more sophisticated manner than formerly, with the result that few grains pass through cattle whole; (2) hogs tend to get cattle up and to get into troughs; (3) and increase in fenceline feeders, which won't keep hogs in; and (4) few hog-tight commercial feedlots.

Nevertheless, cattle feeders who have a convenient source of feeder pigs, who are not "allergic" to keeping hogs, and whose cattle lots are fenced hog-tight, can add to their net income by having hogs follow cattle.

[1]Based on a consensus of Chairmen of Animal Science Departments of Land Grant Colleges in the Corn Belt, obtained by the author in personal communications.

Fig. 31-9. Hogs following cattle. One pig should follow every one to three steers, the ratio of pigs varying with the kind and preparation of the feed and the age of the cattle. Sometimes the only profit obtained is in the gains made by pigs following cattle. (Courtesy, American Feed Manufacturers)

The following hog: cattle ratio is recommended, using 75- to 150-pound pigs:

	If Whole Shelled Corn Is Fed	If Ground or Rolled Grain Is Fed
	(Pig:Steer Ratio)	(Pig:Steer Ratio)
Calves	1:3	1:5
Yearlings	1:2	1:4
Two-year-olds	1:1½	1:3

For every 50 bushels of whole corn fed to yearling cattle, approximately 50 pounds of pork will be produced. Allowing 30¢ for hogs, and subtracting $5 per pig for protein and other costs, that's $10 per pig.

Pigs sometimes inflict injury on heifers (injuring the vulva when they are lying down); therefore, their use is generally limited to steers.

Sows may be used, but because of their size they may create problems from getting into the feed and water facilities.

CONDUCTING APPLIED FEEDLOT TESTS

When carefully conducted, and properly interpreted and used, feedlot trials can be a valuable adjunct in the operation of a large feedlot. Among their virtues, the feedlot operator can study area and feed differences.

Among their limitations, usually less accuracy and fewer controls than most university-conducted experiments. For the latter reason, most of them should be looked upon as applied tests or demonstrations *per se,* rather than carefully controlled, basic experiments; terminology which doesn't detract from their value, but which does place them in proper perspective.

The number of pens which a feedlot should devote to test work will vary according to the size of the operation and the number of treatments planned at one time.

There should always be a minimum of 2 lots for controls, plus 2 lots for each treatment evaluated. Generally, the 2 control lots should be fed the standard feedlot ration, and 2 lots should be given each treatment evaluated.

The local county extension agent should be invited to participate in the test; usually he'll welcome the opportunity.

The following procedure is recommended in conducting feedlot tests:

1. *Cattle*—The animals should be of uniform breeding, background, age and weight, and of the same sex. Use cattle owned by the operator, rather than custom-fed animals.

2. *Number per lot*—Ten head if individually weighed; 20 to 40 head, or more, if group weighed.

3. *Randomization*—Gate or chute cut; one per treatment, or not more than 5 at a time.

4. *Identity*—Preferably (a) apply a different brand to each lot, and (b) individually identify each animal with duplicate numbers—one in each ear. For the latter, use plastic ear tags, the numbers on which can be easily read at a distance.

5. *Variables*—Have only one variable in each pair of treatment lots. Let us suppose, for example, that in a given feedlot steers are now being implanted with Synovex S as standard procedure. However, the owner desires to determine if it would be practical to (a) switch to Ralgro implants, or (b) use a combination of both Ralgro implants and antibiotics. The design would be as follows:

	Lot	Treatment
Controls	1	Control (Standard ration with Synovex S implant)
	2	Control (Standard ration with Synovex S implant)
Treatment 1, Ralgro implants	3	Standard ration; Ralgro implant
	4	Standard ration; Ralgro implant
Treatment 2, Ralgro + antibiotics	5	Standard ration; Ralgro implant plus antibiotics
	6	Standard ration; Ralgro implant plus antibiotics

6. *Adjustment period*—After sorting cattle into test lots, allow a minimum adjustment period of 7 days; during which the cattle should be individually tagged and handled as necessary, and gradually accustomed to their new rations. In case of sickness, a longer adjustment period may be necessary—sometimes as much as 2 to 4 weeks.

7. *Weighing conditions*—Keep off feed and water overnight, then weigh the next morning. Weigh (preferably using a self-recording beam, so as to alleviate the human error) pens in the same order and at the same time each morning when (a) initiating the experiment, (b) at 28-day intervals, and (c) at the close of the test.

Also, weigh and record the amount of feed given to each lot of cattle; using a modified paired-feeding technique, in which the paired lots are limited in feed consumption to the lot consuming the least. Sometimes it is best to limit all lots (both controls and treatments) to the level of the lot consuming the least, although this will vary according to the treatment being evaluated.

8. *Carcass data*—Sell, or have custom slaughtered, with the stipulation that the slaughter plant provide individual (according to individual ear tags) (a) carcass weight and yield, and (b) Federal grade. If slaughter data cannot be obtained on all cattle, get it for as many as possible and of the same number from each lot.

9. *Summarize results*—At the end of the trial, summarize the results, using as criteria (a) rate of gain, (b) feed efficiency, and (c) carcass results.

10. *Determine the application*—If both lots of a given treatment are considerably better than the controls, decide (a) whether to repeat the test, or (b) adopt and use the new treatment throughout the feedlot. If the latter becomes the new standard, continue with it until a new and superior treatment evolves, based on new trials.

QUESTIONS FOR STUDY AND DISCUSSION

1. Outline a method for handling incoming feedlot cattle for the first three to four weeks in the feedlot.
2. What is the difference between (a) preconditioning, (b) backgrounding and (c) handling newly arrived feedlot cattle?
3. Detail the kind of feedlot records that should be kept.
4. How, and how frequently, should feedlot cattle be weighed?
5. Outline, step by step, a program for getting (a) yearling cattle and (b) calves on full feed.
6. Discuss the factors affecting feed intake of cattle.
7. How may weather affect the eating and drinking habits of cattle?
8. Outline the progressive changes that normally occur in feedlot cattle, from start to finish of the feeding period, including (a) rate of gain, (b) feed consumption, (c) feed efficiency, and (d) cost of gain.
9. How can you tell when cattle are ready to go to market?
10. Why is overfinishing undesirable?
11. Should feedlot heifers be aborted? If so, at what stage of pregnancy should it be done, and how should it be done?
12. Outline, step by step, a program for handling bullers or riders.
13. How may a feedlot mud problem be alleviated?
14. How would you control (a) birds and (b) flies in a cattle feedlot?
15. Would you recommend that hogs follow feedlot cattle? Justify your answer.
16. How would you design and conduct a cattle feedlot test?

SELECTED REFERENCES

Title of Publication	Author(s)	Publisher
Beef Cattle Science Handbook	Ed. by M. E. Ensminger	Agriservices Foundation, Clovis, Calif., pub. annually since 1964
Cattle Feeders Hand Book	R. M. Bonelli	Computer Publishing Co., Phoenix, Ariz., 1968
Feedlot, The	Ed. by I. A. Dyer, C. C. O'Mary	Lea & Febiger, Philadelphia, Penn., 1972
How to Make Money Feeding Cattle	L. H. Simerl B. Russell	United States Publishing Company, Indianapolis, Ind., 1959
Stockman's Handbook, The, Fourth Edition	M. E. Ensminger	The Interstate Printers & Publishers, Inc., Danville, Ill., 1970
Third Dimension of Cattle Feeding, The		General Management Services, Inc., Phoenix, Ariz., 1970

CHAPTER 32

FEEDLOT POLLUTION CONTROL

In recent years, there has been a worldwide awakening to the problem of pollution of the environment (air, water, and soil) and its effect on human health and on other forms of life. Much of this concern stemmed from the amount of manure produced by the sudden increase of animals in confinement. Certainly, there have been abuses of the environment (and it hasn't been limited to agriculture). There is no argument that such neglect should be rectified in a sound, orderly manner, but it should be done with a minimum disruption of the economy and lowering of the standard of living.

Fig. 32-1. A happy balance must be found! Perhaps it's preferable to have a little smoke in the air than no fire in the furnace, and to have a bit of runoff from feedlots than no meat on the table.

In altogether too many cases extreme environmentalists advocate policy changes and legislation that may in the end be detrimental to agriculture, to our food production potential, and to society in general. Frequently, these new messiahs have only used the data that support their theories about ecological doom. One of their favorite comparisons deals with the relative magnitude of the effect on the environment caused by animal manure, industrial waste, and municipal waste. Then, they add the "scare" to their story by citing the number of blue babies and suffocated fish caused by runoff from manure. In particular, they have incriminated cattle feedlots as major culprits. But many of their facts and figures have been in error. In order to set the records straight, and to assist cattlemen and others in controlling pollution to the maximum, this chapter is presented.

Invoking an old law (the Refuse Act of 1899, which gave the Corps of Engineers control over runoff or seepage into any stream which flows into navigable waters), the Federal Environmental Protection Agency (EPA) launched the program to control water pollution by requiring that all cattle feedlots which had 1,000 head or more the previous year must apply for a permit by July 1, 1971. The states followed suit; although differing in their regulations, all of them increased legal pressures for clean water and air. Then followed the Federal Water Pollution Control Act Amendments, enacted by Congress in 1972, charging the EPA with developing a broad national program to eliminate water pollution.

POLLUTION LAWS AND REGULATIONS

Both open lot and confinement cattle systems come under pollution regulations. Open lots present drainage and runoff problems. Confinement systems must be coordinated to disposal area in order that pollution is not created when storage pits are emptied.

Registration of facilities and a permit to operate are the primary requirements that the Federal Government, and the various states, are, or will be, using to insure that livestock wastes are properly handled. Since state regulations vary somewhat, it is recommended that the cattle feeder check into the regulations of the state in which he is operating or plans to operate. The Federal guidelines, which are rather broad, are as follows:

1. *Who must apply*—The basic provisions of the Federal regulations are (In some states two permits are needed—one state, the other Federal.):

a. Feedlots with 1,000 or more animal units (1,000 cattle, 700 mature dairy cattle, 2,500 swine weighing over 55 lb, 12,000 sheep or lambs, 55,000 turkeys, 180,000 laying hens, or 290,000 broilers) must obtain a permit.

b. Feedlots with fewer than 1,000 animal units, but more than 300 animal units (more than 300 cattle, 200 mature dairy cattle, 750 swine weighing over 55 lb, or 3,000 sheep), must obtain a permit if the facility discharges pollutants either in (1) a man-made conveyance constructed for the purpose, or (2) waters that pass through the confined area.

c. Feedlots with fewer than 300 animal units are not subject to the permit requirements except where so designated on a case-by-case basis.

Livestock confinement facilities include open feedlots, confined operations, stockyards, livestock auction barns, and buying stations. Also, regulations apply to any combination of species in the same feedlot.

Cropland runoff and manure spreading are not specifically regulated under the Federal Act, although they may be in the future.

2. *How to apply*—Forms may be secured from the offices of EPA and state environmental agencies, the County Agent, or the SCS district offices. Fill out a *Short Form B* and send it, along with $10 filing fee, to the EPA regional office. Then, either the Federal EPA or the state agency will make an on-the-site inspection. They will draft a proposed permit, put it on public notice, and give the applicant and the public 30 days to comment on it. Then, if there are no protests, the Federal discharge permit will be issued.

3. *Cost-sharing help*—In 1973-74, the Rural Environmental Assistance Program (REAP) provided cost-share payments on certain livestock waste storage and diversion facilities, with a limitation of 50% of the cost and a maximum of $2,500. The stockman should check on the availability of such funds in the future.

WHAT ARE THE CONCERNS?

Broadly speaking, there have been the following major concerns voiced by the environmentalists pertaining to animal pollution:

● *Nitrates and blue babies*—Nitrates are a compound form of nitrogen, usually found in the soil. Plants utilize nitrates in the growth process to secure nitrogen. However, when the amount of nitrogen contained within nitrates exceeds 10 parts per million (ppm) in underground water, there is a chance that the excess nitrates may cause a disorder in human babies, commonly known as "blue babies." The major concern, therefore, is that the subsurface water from a cattle feedlot, or other similar animal facility, might make for a nitrate buildup and cause blue babies. However, research evidence indicates that there is little nitrate buildup under active feedlots or runoff holding ponds. In fact, most studies indicate that there is actually a decrease in nitrates immediately beneath the surface of feedlots. There is a logical explanation of this phenomenon. The manure from the animals falls upon the feedlot surface. It is stirred by the animals' hooves. As the manure continues to build up, the bottom layer becomes compacted while the surface layer remains loose. Because there is still energy in the manure, microorganisms continue their metabolic process. The microorganisms in the surface layer can use the oxygen that is mixed in the manure. Nitrates are one of the waste products of the metabolism of the organisms in the surface layer. (See Fig. 32-2.)

But the microorganisms in the compacted lower layer do not have oxygen available to complete their metabolism. They are desperate for oxygen molecules. Thus, they strip the oxygen from the nitrate compounds. Nitrogen gas, which constitutes 75.8 percent of the air, is created.

FEEDLOT SURFACE

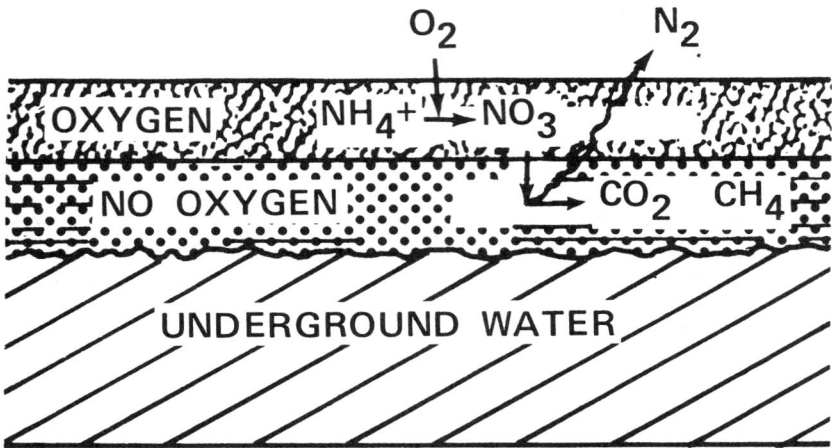

Fig. 32-2. Bacteria on feedlot surface utilize oxygen (O_2) and combine it with ammonium ions (NH_4+) to produce nitrates (NO_3). However, nitrates do not go into groundwater because the bacteria in the lower layer of manure do not have free oxygen. They must strip oxygen molecules from the nitrates. The result is free nitrogen (N_2). (Source: Paine, Myron D., Confined Animals and Public Environment, *Great Plains Beef Cattle Feeding Handbook*, GPE-7000, April, 1973)

An exception to the above phenomenon does exist in feedlots that are lightly stocked, or that are abandoned for a portion of the year without cleaning. Under these circumstances, natural processes break up the compacted lower layer of manure and allow nitrates to move downward toward groundwater. However, an adequately stocked, active feedlot results in conditions that reduce nitrate movement to groundwater; hence, it does not materially affect nitrates in ground water, and it does not cause blue babies.

• *Oxygen demand of runoff sludge suffocates fish*—In the 1960s, before feedlot control runoff measures were instituted, the runoff from the feedlots in the Great Plains area carried organic material that had a high oxygen demand. This runoff traveled as a sludge down the streams. The oxygen within the sludge was consumed. As the sludge passed along, fish were suffocated. The average pollutional loading on a stream was relatively minor, but the effect of a few hours without oxygen was spectacular. Cattlemen reacted swiftly. They constructed runoff control facilities in large commercial feedlots.

Fig. 32-3 shows fish kills in Kansas which were attributed to feedlots. Note that the largest number of fish killed occurred in the years 1964-67. Kansas regulations went into effect on January 1, 1968. The reduction of fish killed since 1968 is partially credited to swift compliance with the regulations.

• *Odors and dust*—Cattle feedlots located near centers of populations are having an increasing number of complaints lodged against them because of odors and dust. Lawsuits, based on the nuisance law, are being filed against them.

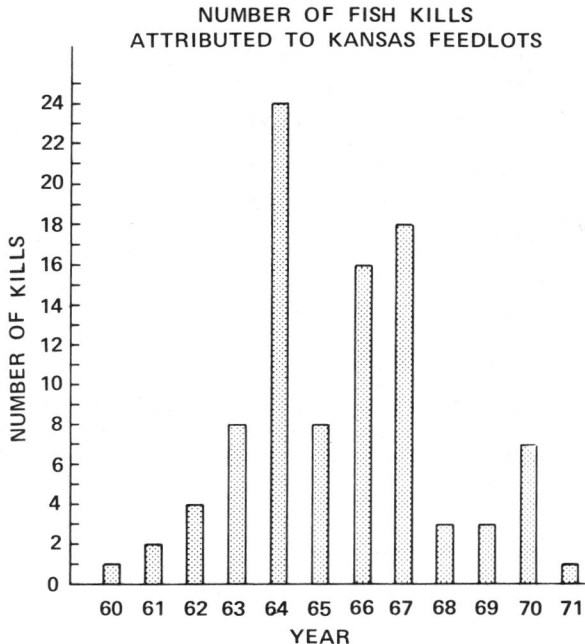

Fig. 32-3. Since the mid-1960s, fish kills caused by feedlots in Kansas have declined to previous low levels. (Source: Paine, Myron D., Confined Animals and Public Environment, *Great Plains Beef Cattle Feeding Handbook*, GPE-7000, April, 1973)

MANURE

A general discussion of the subject of "Cattle Manure" appears in Chapter 10 of this book; hence, the reader is referred thereto. But feedlot manure as such will be covered in the sections that follow.

As shown in Table 10-4 of Chapter 10, approximately 1,587,694,173 tons of manure are produced annually in the United States. It should be noted, however, that this is all manure from all classes of animals, and without regard to confinement, moisture, or the ration. Much of this 1.59 billion tons of manure is distributed by the animals themselves onto pasture and grazing land, with the result that it neither has to be hauled to the field nor is it a pollution problem.

Manure Produced vs Manure Handled

Facts pertinent to calculating the amount of the 1.59 billion tons of manure produced that must be hauled to the field follow:

1. Only 25% of U.S. animals are in confinement.

2. Most of the animals kept in confinement are fed for meat production. These animals utilize a high concentrate ration with a digestion between 75 to 90%. They eat a smaller amount of ration to secure their growth energy. The dry solid material produced by animals on a high concentrate ration is

about one-half (50%) of the dry solids coming from a dairy animal. Hence, a reduction factor for high energy ration needs to be made.

3. The estimated 1.59 billion pounds of manure produced annually is on a fresh manure basis, containing approximately 85% water. Following defecation, manure loses moisture by evaporation. Most manure that is hauled to the field has approximately 30 to 50% water. Thus, the manure that must be hauled may well be reduced by a factor from moisture evaporation of 50%.

Based on the above points, the calculations in Table 32-1 show the total U.S. manure production vs manure to be hauled:

TABLE 32-1

TOTAL U.S. MANURE PRODUCTION VS MANURE TO BE HAULED

Total Manure Produced	=	1,588 million tons
Reduction factor for confinement only		×0.25
Manure produced in confinement	=	397 million tons
Reduction factor for high energy ration		×0.50
Manure from high energy ration	=	198.5 million tons
Reduction factor for moisture evaporation		×0.50
Total weight of manure hauled	=	99.25 million tons

The 99.25 million tons of manure that must be hauled includes all classes of livestock, of which feedlot cattle are only a part.

Most feedlots estimate that one animal will produce about a ton of manure. Thus, the quantity is drastically reduced from the original quantity emitted by the animal (approximately 3.75 tons of fresh manure). The causes for this reduction are: (1) runoff, (2) evaporation, (3) leaching and percolation, and (4) decomposition. On the basis of one ton of manure produced per steer, at the time feedlots are cleaned, the 26.8 million feedlot cattle finished in 1972 produced that many tons of manure—26.8 million tons.

Solid Waste

As shown in the Table 32-1 calculations, approximately 100 million tons (99.25 million tons rounded off) total weight of manure must be hauled to the fields, or otherwise disposed of. As shown in Fig. 32-4, this compares with 110 million tons of industry solid waste and 250 million tons of residential solid waste.

Thus, the manure that must be hauled to the field, or otherwise disposed of, is only 21.7 percent of the total solid waste produced by municipalities, industries, and animal systems.

SOLID WASTE COMPARISONS BY WEIGHT

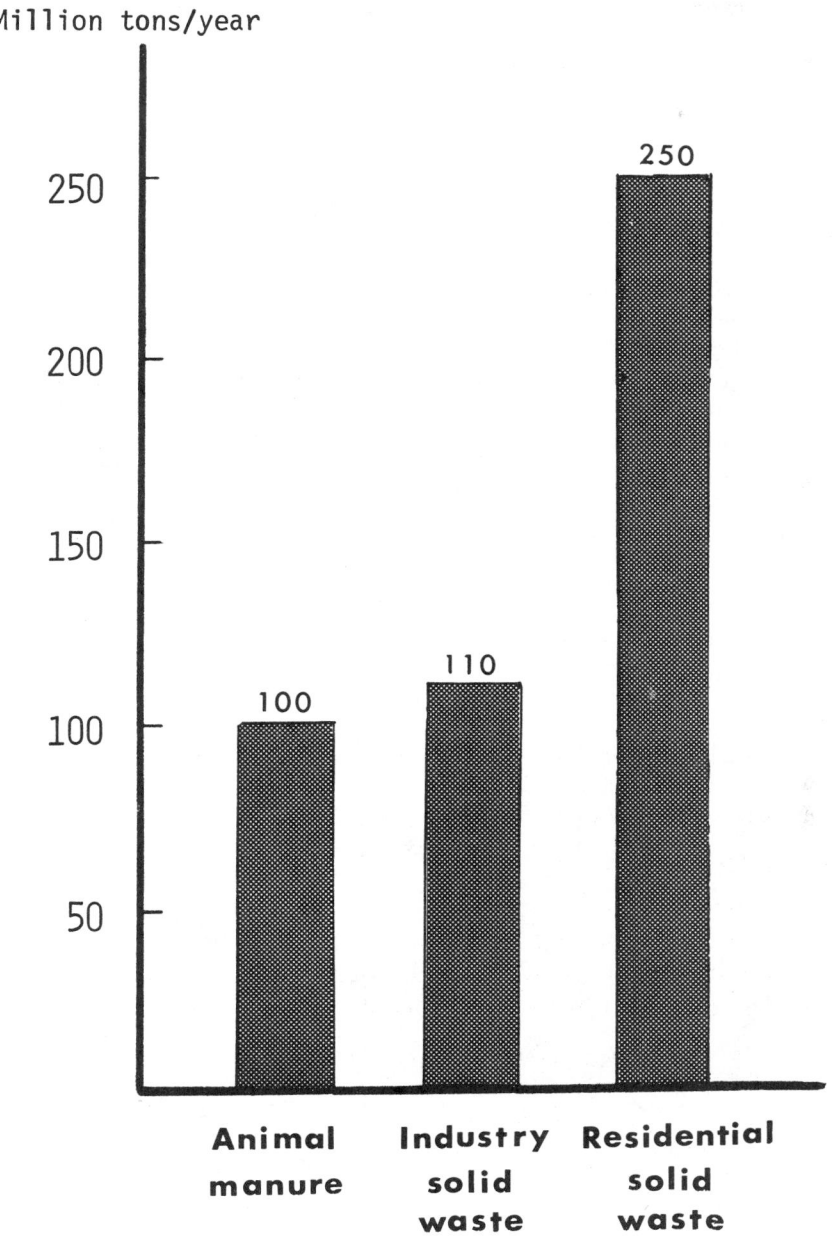

Fig. 32-4. The manure weight hauled from confined animals is only about ¼ of the solid waste weight hauled from cities and industries. (Adapted by the author from *Great Plains Beef Cattle Feeding Handbook*, GPE-7000, Confined Animals and Public Environment, by Myron D. Paine)

Manure Disposal Methods

The following methods are being used to dispose of manure:

1. Return to the soil, with or without composting.
2. Settling, flocculation, and dehydration, or other means of concentration with ultimate return to the soil.
3. Recycling, perhaps selectively, with or without processing as animal feed.
4. Incineration.
5. Generate energy.
6. Laissez-faire.

All of these methods have shortcomings. In many cases, the cost of waste disposal will exceed the value to the user. These costs may be reflected in price of product or in taxes.

It is noteworthy that the position of Food and Drug Administration (FDA) relative to the feeding of animal waste (manure) is set forth by requiring submission of the following three basic categories of information: (1) establishing nutritive value (or efficacy); (2) determining safety to animals; and (3) determining that food from animals consuming such product is safe for man. Research on the feeding value of manure is not discouraged by FDA, but the commercial marketing of manure is approved only within the framework given above.

HOW FEEDLOTS HANDLE THEIR MANURE

Most feedlot operators dispose of manure in two ways: (1) the solids are spread on land; and (2) detention ponds are used as a runoff control device.

Fig. 32-5. Farr Feeders, Inc., Greeley, Colorado, utilizes high-speed self-elevating scrapers, like the two pictured above, for pen and manure maintenance. This equipment is used to haul manure to a stockpile for later delivery to local farms. (Courtesy, Richard Farr)

• *The solids are applied to the land*—Most feedlot operators keep the manure moved out as a matter of animal health. After they move out a lot of cattle, they clean the pen. However, this soon results in a huge pile of manure, which must be disposed of. Most feedlot operators have solved this problem by developing close working relationships with local farmers, who take the manure for use as a fertilizer. Some of them are even paying as much as $2.00 per ton for it. In areas where farmers are not convinced of the fertilizer value, private negotiations to move the manure have been worked out in many ways.

With the heavy concentration of cattle in one location, and with many lots ranging from 20,000 up to 100,000 head, the question is being asked: "How high rates of manure can be applied to the land without depressing crop yields, making for nitrate problems in feed, or contributing excess nitrate to groundwater or surface streams?" Based on earlier studies in the Midwest, before the rise of commercial fertilizers, it would appear that, on most soils and in most areas, one can apply up to 20 tons of manure per acre, year after year, with benefit. Heavier applications can be made, but probably should not be repeated every year. Excessive manure applications can (1) increase the potential for polluting surface of groundwater; (2) result in nitrogen concentrations in forage that pose a threat to animal health; and (3) cause salts such as sodium chloride to accumulate in concentrations that are toxic to plants and detrimental to soil structure. Of course, salt accumulation can be controlled by lessening the salt content of the ration. Without doubt the maximum rate at which manure can be applied to the land will vary widely according to soil type, rainfall, and temperature.

State regulations differ in limiting the rate of manure application. Missouri limits it to 30 tons per acre on pasture, and 40 tons per acre on cropland. Indiana limits manure application according to the amount of nitrogen applied, with the maximum limit set at 225 pounds per acre per year. Nebraska requires only one-half acre of land for liquid manure disposal per acre of feedlot, which appears to be the least acreage for manure disposal required by any state.

Fig. 32-6. Loading solid feedlot manure by conveyor onto spreader trucks. (Courtesy, Benedict Feeding Co., Casa Grande, Ariz.)

It is estimated that a 1,000-pound beef animal will excrete 120 pounds of nitrogen per year.[1] Table 32-2 shows the average nitrogen losses expected on a year-round basis; whereas Table 32-3 shows the number of 1,000-pound beef animals necessary to provide 100 pounds of nitrogen annually.[2]

TABLE 32-2

ESTIMATED NITROGEN LOSS DURING
STORAGE, TREATMENT, AND HANDLING
FOR VARIOUS WASTE MANAGEMENT SYSTEMS

System[1]	1	2	3	4	5	6
Total N loss, %	84	66	78	61	34	57

[1]System 1—Oxidation ditch, anaerobic lagoon, irrigation or liquid spreading.
System 2—Deep pit storage, liquid spreading.
System 3—Anaerobic lagoon, irrigation or liquid spreading.
System 4—Aerobic lagoon, irrigation or liquid spreading.
System 5—Bedded confinement, solid spreading.
System 6—Open lot (with or without shelter), solid spreading, runoff collected and irrigated or spread.

TABLE 32-3

NUMBER OF ANIMAL UNITS[1] (AU) PER ACRE OF
DISPOSAL AREA TO PROVIDE 100 POUNDS OF NITROGEN ANNUALLY

System[2]	1	2	3	4	5	6
Beef-AU	5.2	2.5	3.8	2.1	1.3	1.9

[1]Animal unit = 1,000-pound beef animal.
[2]Same as Table 32-2, footnote 1.

Thus, in Indiana, which allows for a maximum application of 225 pounds of nitrogen per acre, the manure produced from 4.3 head of 1,000-pound steers (1.9 × 2.25=4.3) could safely be applied to one acre of land. Since a feedlot steer averages about 750 pounds throughout the feeding period (500 pounds in weight; 1,000 pounds out-weight) only ¾ of this amount of manure will be produced per animal.

Hence, one acre of Indiana land will take care of the production of 4.8 steers on full feed for a year. With a turnover of 2.25 steers per year, that means that the manure from 11 feedlot steers could be spread on 1 acre. An Indiana farmer feeding out 1,000 head of steers each year could spread all the manure on 90 acres of farmland.

As noted above, under the section entitled "Manure Produced vs Manure Handled," a total of 26.8 million tons of feedlot manure were produced in 1972, or one ton per animal. Assuming 2¼ turnover of cattle per feedlot per year, and the production of one ton of manure per animal, it would take

---

[1]*Area Needed for Land Disposal of Beef and Swine Wastes,* Iowa State University, Ames, Iowa, Pm-552 (Rev.), Aug., 1973.
[2]Ibid.

six animals a whole year to produce enough manure to apply at the rate of 13.5 tons per acre. Thus, as shown in Fig. 32-7, it takes the production of about 18 acres of crops to produce enough feed for 6 feedlot cattle for a year. Yet, those 6 animals produce only sufficient manure for 1 acre, if manure is applied at the rate of 13.5 tons per acre.

THE CYCLE: CROPS—→FEED—→MANURE (FERTILIZER)—→CROPS

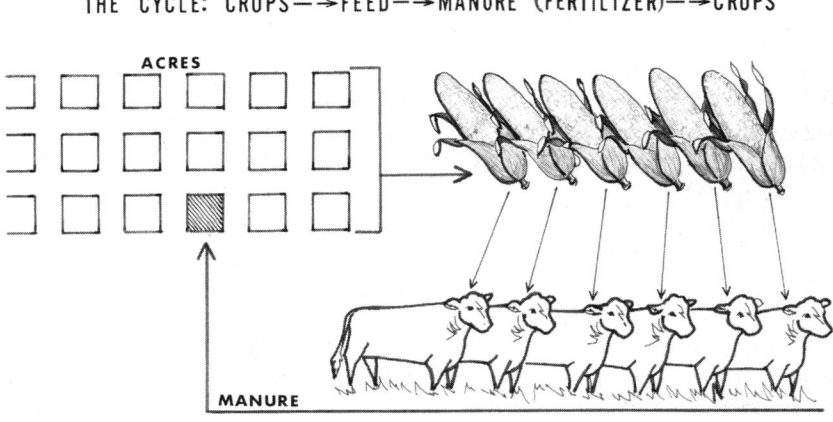

Fig. 32-7. Eighteen acres of crops produce feed for 6 beef cattle on full feed for a year. The manure from these cattle will provide enough nutrients to refertilize only 1 acre, at the rate of 13.5 tons per acre.

Another noteworthy statistic is this: Since it takes 6 steers to produce enough manure for 1 acre of land, the nation's 26.8 million steers (1972 number) would produce enough manure for 4.5 million acres, an area slightly smaller than the state of New Jersey.

• *Detention ponds catch runoff*—In addition to disposing of the solids as indicated above, many feedlot operators are now using a detention pond as a runoff control device. Of the pollution control systems presently in operation, the detention pond, or some variation of this method, is probably the most common. As the name implies, this system detains the runoff until it can be disposed of. The detention pond is usually constructed from earth. In some cases, it will require soil sealer to prevent leaching. In a detention pond, the solids will settle out and usually very little decomposition or bacterial digestion will occur. Thus, solids will eventually accumulate and will have to be disposed of.

Disposal can generally be made in one of two ways: (1) the material can be pumped into a "honey wagon" and spread on the land; or (2) it can be pumped through an irrigation system. In either case, feeders who are using this method try to dispose of the suspended solids as well as the liquid through the disposal system. Most of the systems are agitated in some way before pumping to get as many solids into suspension as possible.

Sometimes a "settling basin" is used in combination with the detention pond. This is a small, shallow (2 to 4 feet deep) detention pond that will allow the solids to settle out before reaching the larger detention pond. The basin should be constructed of concrete or partly of concrete so that the solids can easily be cleaned out with a tractor and loader.

DUST CONTROL

In dry areas, feedlot dust can be a problem. But dust can be minimized by (1) control of cattle density—the more cattle per pen, the more "in pen" water (urine) application; (2) water application; and (3) manure removal. Increasing cattle density to 70 to 80 square feet per head is effective in controlling dust; hence, this method should be used where shade area does not limit density. Also, water may be applied within lots by means of fence-line sprinklers. Dust from roadways, working alleys, and ground areas surrounding a feedlot may be controlled by water (water wagons and sprinklers), road oils, coarse gravel, or special chemicals designed for this purpose. Generally speaking, if the surface moisture of feedlots is maintained in excess of 20 percent, dust problems will be minimized. In addition to laying the dust, manure should be removed regularly, allowing less than one inch of a loose manure pad to remain.

ODOR CONTROL

Concentration of cattle into large feedlots tends to accentuate the odor problem. Also, odor may become more intense on calm, humid, warm days just after a rain.

Odor is a subjective and emotional issue. Like most human traits, reaction to odors varies a great deal among individuals. Solving an odor problem can be complicated by the fact that it smells strong to some people and weak to others, and that it is not considered unpleasant by all people. Just the discussion of odor may encourage overreaction by an urban public.

Most cattlemen are concerned about odor control because they realize that odor does leave the confines of the feedlot, and, at present, there appears to be no economical method of positive control.

Lack of oxygen causes feedlot odor. Cattle manure contains energy for metabolism. Microorganisms in the manure accomplish this metabolism. The metabolic process converts complex carbohydrates, proteins, and fats to simpler compounds. When oxygen is present, the end products of metabolism are heat, CO_2 and H_2O. This process is called aerobic metabolism.

The oxygen transfer rate into manure must exceed the demand in order to prevent odor. When oxygen demand exceeds the transfer rate, anaerobic metabolism produces odor compounds.

Another odor produced in feedlots is ammonia (NH_3). Ammonia escapes as gas from urine. It is lighter than air and has an irritating smell. A well-maintained feedlot surface will reduce odor transfer. Management practices that encourage aerobic conditions on the feedlot surface are desirable.

Types of Odor Control Agents

There are five main types of odor control agents:

1. *Masking agents*—These are mixtures of aromatic oils which cover the odor of manure with a stronger, more "pleasant" odor.

2. *Counteractants*—There are certain pairs of odors, which appear to cancel each other when mixed in the correct relative concentrations. When the two are sniffed together, the intensity of each odor is diminished. Certain mixtures of aromatic oils (counteractants) are used to neutralize the odor of manure.

3. *Deodorants*—A deodorant is a mixture of chemicals designed to "kill" the odor of manure. The deodorant action may also kill the bacteria which produce the odor.

4. *Digestive deodorants*—Digestive deodorants consist of a combination of digestive enzymes, aerobic, and anaerobic bacteria. The logic appears to be that the enzymes or bacteria will create a digestive process that eliminates the odor.

5. *Sagebrush*—Researchers at Colorado State University have discovered that feedlot odors are reduced or eliminated when small quantities of sagebrush are included in the ration of cattle. The volatile oils in sage brush reduce the digestive tract's bacterial population that produces the disagreeable odors.

Effectiveness of Odor Control

In terms of effectiveness, four types (exclusive of sagebrush) of odor control agents have been ranked[3] as follows: Masking agents are the most effective as covering agents, but, remember, effective masking agents do not reduce the odor—they may even create another odor problem. Counteractants are generally the second most effective class of products because they may reduce odor level. They may even be better than certain types of masking agents. Deodorants are not considered to be as effective in manure odor control as are the masking agents or counteractants. Digestive deodorizers are found to be the least effective for control of manure odors.

---

[3]Ranked by Myron D. Paine, in Chemical Control of Manure Odor, *Great Plains Beef Cattle Feeding Handbook*, GPE-7801.1, April, 1973.

QUESTIONS FOR STUDY AND DISCUSSION

1. How far should a nation go in pollution control laws and regulations, bearing in mind (a) that pollution control measures cost money—hence, they will increase product prices to consumers; (b) that the ultimate in pollution control will lower food production potential in some cases; and (c) that many of the good things of life which contribute to our high standard of living, such as electricity, make for pollution?

2. How do the animal unit relationships between species used by the Environmental Protection Agency (EPA) compare with those given in the Appendix of this book?

3. Who among stockmen must register and secure a permit to operate? Give the step-by-step procedure involved in securing approval.

4. Debate either the affirmative or negative of the following statements:
   a. The runoff of cattle feedlots, and other similar facilities, causes blue babies.
   b. The runoff of cattle feedlots, and other similar facilities, suffocates fish.

5. In arriving at the amount of manure that must be handled, the author assumed the following:
   a. That only 25% of U.S. animals are in confinement.
   b. That the dry solids of animals on high energy rations kept in confinement are equivalent to only 50% of the dry solids coming from a dairy animal.
   c. That 50% of the moisture of manure is lost before hauling.
   Challenge each of the above assumptions.

6. Fig. 32-4 clearly shows that both residential solid waste and industry solid waste exceed the total tonnage of manure that must be hauled to the field or otherwise disposed of. Then, why are the environmentalists pointing such an accusing finger at stockmen?

7. Discuss each of the six manure disposal methods listed in this chapter. Which two methods do you feel offer the most hope? Justify your choices.

8. The author's computations show that it takes the production of about 18 acres of crops to produce enough feed for 6 feedlot cattle for a year. Yet, these 6 animals produce only sufficient manure for 1 acre, *if* manure is applied at the rate of 13.5 tons per acre. Challenge these computations.

9. What tonnage of manure per acre can be applied to the land on an annual basis?

10. What's a detention pond? How does it work?

11. How would you recommend that feedlot dust be controlled?

12. How would you recommend that feedlot odors be controlled?

SELECTED REFERENCES

Title of Publication	Author(s)	Publisher
Agriculture and the Environment	Alberta Institute of Agrologists	Alberta Institute of Agrologists, Edmonton, Alberta, Canada, 1971
Agriculture and the Quality of our Environment	Ed. by N. C. Brady	American Association for the Advancement of Science, Washington, D.C., 1967
Closing Circle, The, Nature, Man, and Technology	B. Commoner	Alfred A. Knopf, Inc., New York, N.Y., 1971
Distribution of Nitrates and Other Water Pollutants Under Fields and Corrals in the Middle South Platte Valley of Colorado, ARS 41-134	B. A. Stewart F. G. Viets, Jr. G. L. Hutchinson W. D. Kemper F. E. Clark M. L. Fairbourn F. Strauch	U.S. Department of Agriculture, Washington, D.C., 1967

(Continued)

Title of Publication	Author(s)	Publisher
Environment, The, A National Mission for the Seventies	Editors of *Fortune*	Harper & Row, New York, N.Y., 1969
Environmental Science, Challenge for the Seventies	National Science Board	National Science Board, National Science Foundation, Washington, D.C., 1971
False Prophets of Pollution	R. M. Carleton	Trend Publication, Inc., Tampa, Florida, 1973
Guide to Environmental Research on Animals, A	National Research Council	National Academy of Sciences, Washington, D.C., 1971
Health Hazards of the Human Environment	World Health Organization	World Health Organization, Geneva, Switzerland, 1972
NFC Directory of Environmental Information Sources, Second Edition	Ed. by C. E. Thibeau	The National Foundation for Environmental Control, Inc., Boston, Mass., 1972
Our Living Land	Ed. by E. P. Essertier, V. S. Hart, E. R. Nicolai, R. Ritzenberg	U.S. Department of the Interior, Washington, D.C., 1971
Population, Resources, Environment, Issues in Human Ecology, Second Edition	P. R. Ehrlich A. H. Ehrlich	W. H. Freeman and Company, San Francisco, Calif., 1972
Principals of Animal Environment	M. L. Esmay	The Avi Publishing Company, Inc., Westport, Conn., 1969
Proceedings of the International Livestock Environment Symposium	American Society of Agricultural Engineers	American Society of Agricultural Engineers, St. Joseph, Mich., 1974
Water Encyclopedia, The	Ed. by D. K. Todd	Water Information Center, Inc., Port Washington, N.Y., 1970
Water Policies for the Future, The	National Water Commission	Water Information Center, Inc., Port Washington, N.Y., 1973

CHAPTER 33

CATTLE FEEDLOT HEALTH

Contents   **Page**

Loss from disease is greater in cattle feedlot operations than in any other type of cattle enterprise. The movement of cattle, stress conditions, methods of purchase, feeding of concentrated feeds, population density, sometimes unsanitary conditions, and the bigness and complexity of the operation all contribute to disease incidence; and disease incidence is directly proportional to population density.

Dr. Don Gill, Animal Scientist of Oklahoma State University, is authority for the statement that 1.5 million head of cattle died in feedlots during 1970, at an estimated loss of over $530 million.[1] Dr. Gill further estimated that losses from sickness run two to five times greater than actual death losses. He calculated that the combined losses from death and sickness add $20 to $25 per head onto the cost of every feedlot-finished animal. This situation can be greatly improved by (1) preconditioning, (2) moving cattle directly from the producer's farm or ranch to the feedlot (fewer than 20% of the cattle do so now), (3) reducing the time between ranch and feedlot, (4) lessening the amount of stress and exposure to infection during marketing and transportation periods, (5) providing the man receiving the cattle with more adequate medical and nutritional history of the cattle, (6) handling of incoming feedlot cattle properly, and (7) diagnosing and treating sick cattle early.

HANDLING NEWLY ARRIVED CATTLE

The most critical period for feeder cattle is the first 21 to 28 days in the feedlot. The following recommendations pertaining to incoming cattle will minimize death losses and maximize performance:

---

[1]Gill, Don, Oklahoma State University, Stillwater, Okla.; report given as Chairman of the National Preconditioning Committee, at the 1972 annual meeting of Livestock Conservation, Inc., as reported in *Beef*, June, 1972.

Fig. 33-1. Newly arrived, heavily stressed calves, trucked from South Texas to an Arizona feedlot. (Courtesy, Benedict Feeding Co., Casa Grande, Ariz.)

● *Provide clean, dry, comfortable quarters*—Whether it be an open lot or a building, incoming cattle should be provided with clean, dry, comfortable quarters. A dry and comfortable bed for resting is very essential because cattle are tired and have a low resistance to respiratory diseases.

● *Process upon arrival*—The relative merits of processing calves (1) at point of origin, (2) upon arrival at destination, or (3) two to three weeks after arrival are often debated. A common view is that stressed calves will not develop antibodies efficiently until recovered from stress; thus, vaccination should be delayed until the stress has been overcome.

In a well-designed experiment, involving 358 calves, the University of California provided the answer to this question. The calves originated in Texas, and were in transit 32 to 38 hours with no rest stops. In all loads, ⅓ of the calves were processed at origin, ⅓ upon arrival at destination, and ⅓ were delayed 2 to 3 weeks after arrival. Processing consisted of branding, castration, ear tagging, use of a pour-on grubicide, vaccination (IBR, P13, blackleg, and malignant edema), intramuscular injection of vitamins A, D, and E and one gram of oxytetracycline.

Recognizing that the effect of time and place of processing on the entire feeding period from purchase to slaughter is the important thing, rather than the effect on the first month, the California workers very wisely carried part of these cattle from arrival through slaughter. A total of 120 of these calves, with an average purchase weight of 210 pounds, were shipped from Houston, Texas, and were in transit 38 hours to El Centro, California. The results of the entire 344-day feeding period, from purchase to slaughter, are summarized in Table 33-1.

The California study showed the following relative to time of processing:

TABLE 33-1

PERFORMANCE FOR ENTIRE PERIOD[1]

Items Compared	Time of Processing		
	Origin	Arrival	Delayed
Number of calves	40	39	37
Purchase weight per head, lb	209	213	209
Net slaughter weight, lb	970	1,019	989
Average days fed	344	341	344
Daily weight gain, lb	2.21	2.36	2.27
Total feed consumed per head, lb			
72% receiving ration	166	182	166
55% growing ration	1,012	1,036	1,053
90% finishing ration	3,336	3,314	3,404
Totals	4,514	4,532	4,623
Average daily feed intake, lb	13.12	13.29	13.44
Feed per pound gain, lb	5.93	5.63	5.93
Carcass data:			
Yield, %	62.5	62.6	63.1
Quality grade score[2]	11.9	12.4	12.2
Cutability grade[3]	2.6	3.0	2.5
Receiving phase:			
Number of days	27	27	27
Gain from purchase weight, lb	37	47	36
Cost per pound of gain, ¢[4]	27.49	22.89	28.72
Postreceiving period:			
Number of days	317	314	317
Postreceiving gain, lb	724	759	744
Cost per pound of gain, ¢	31.51	30.03	31.41
Entire period, purchase to slaughter:			
Number of days	344	341	344
Total gain, lb	761	806	780
Cost per pound gain, ¢	31.32	29.62	31.28
Amount of cost per pound gain due to receiving period, ¢	1.34	1.33	1.33
Amount of cost per pound gain due to postreceiving period, ¢	29.98	28.29	29.95

[1]Lofgreen, G. P., University of California, El Centro, California, paper in *Beef Cattle Science Handbook*, Vol. 12; edited by M. E. Ensminger and published by Agriservices Foundation, Clovis, Calif., 1975.
[2]Quality grade score: Choice = 13; Low Choice = 12; High Good = 11.
[3]Cutability scored from 1 to 5, with 1 being the highest yield.
[4]Receiving costs include feed, processing, and medication. Feed costs used were $101, $85, and $111 per ton for the receiving, growing and finishing rations, respectively.

1. For the entire feeding period—from purchase to slaughter—calves processed upon arrival gained at a slightly faster rate and had a higher feed efficiency than calves which were processed immediately prior to shipment, or calves which were delayed-processed 13 days after arrival.

2. At the end of the 27-day receiving period (the first 27 days after arrival), the calves processed on arrival had a weight advantage of 10 lb over those processed at origin and 11 lb over those delayed-processed. At slaugh-

ter, these advantages had been increased to 45 and 26 lb, respectively. It appears, therefore, that if calves attain a weight gain advantage during the first month after arrival in the yard, they will retain that advantage throughout the entire feeding period.

3. Based on (a) rate of gain, (b) disease resistance, and (c) cost per pound of gain for feed, processing, and medication, these studies indicate that processing at arrival is best, and that processing at point of origin is preferable to delayed processing.

● *Provide clean fresh water*—give the cattle easy access to clean, fresh water because they are usually dehydrated and thirsty upon arrival and will drink water before they eat feed. Open water tanks are preferable to automatic water bowls because most farm and ranch cattle are accustomed to drinking from tanks or ponds.

● *Provide a palatable ration*—Feeding a palatable ration—one that cattle will start eating soon after they are unloaded in the feedlot—will reduce the incidence of shipping fever and make the cattle recover their weight loss more rapidly.

1. *Roughage*—The best roughage for newly arrived feedlot cattle is *long grass* hay, because it is very similar in composition and taste to the grass to which most feedlot cattle have been accustomed. Thus, cattle will usually eat long grass hay more quickly than any other roughage. In areas where grass hays are not available, or are too expensive to feed, any other non-legume roughage can be fed, such as corn silage, sorghum silage, cottonseed hulls, corncobs, or grass-legume hay that contains more grass than legumes. Above all, do not feed high quality alfalfa hay because it is too laxative and it will cause scouring which will trigger shipping fever. The same may be said relative to alfalfa haylage or alfalfa silage.

Fig. 33-2. Incoming cattle may be started on about 4 lb of concentrate/head/day, consisting of 2 lb of grain and 2 lb of protein supplement. (Courtesy, Ralston Purina Company, St. Louis, Mo.)

Corn silage of approximately 65 percent moisture content is an excellent feed for new cattle. If cattle do not eat the corn silage too well at the outset, the feeder should sprinkle a little grass hay on the top of it to encourage them to start eating.

2. *Concentrate*—Incoming cattle may be fed approximately 4 pounds of concentrate per head daily, with a breakdown between protein supplement and grains as follows:

a. Two pounds of a natural protein supplement, such as soybean oil meal, cottonseed meal, or a good commercial supplement, preferably with a little cane molasses added from the standpoint of palatability. The protein supplement should be fortified so as to provide 50,000 IU of vitamin A daily. For heavily stressed cattle, the protein supplement should also contain a high level of antibiotic, or a combination of antibiotic and a bactericidal agent such as sulfamethazine. The following level of antibiotic-sulfamethazine is recommended:

Feed 350 mg of Aureomycin plus 350 mg of sulfamethazine per head daily to newly arrived cattle for a period of 28 days. With the antibiotic-sulfamethazine treatment, shipping fever is practically alleviated.

Do not feed urea for the first 28 days after the cattle arrive. Starvation destroys the ability of the rumen to utilize urea or other nonprotein nitrogen and makes cattle more sensitive to urea toxicity. Therefore, it is not wise to put extra stress on cattle by using urea during this adjustment period.

b. Two pounds of cereal grain per head daily, with the grain processed in the usual manner. The grain level can be raised at the rate of 1.0 pound per head daily if it seems desirable.

It has been, and still is, common practice to start cattle on a high roughage ration, then gradually work them over to a high concentrate ration as they progress through the feeding program. However, based on University of California studies, it appears that for calves, a starting ration consisting of 72 percent concentrate and 28 percent hay (roughage) is best.

● *Satisfy mineral hunger*—Incoming cattle are usually hungry for minerals, especially if they have been on dry range forage. Thus, they should have access either to a mineral mixture consisting of two parts of dicalcium phosphate and one part of salt, or to a good commercial mineral.

● *Observe, isolate, and treat sick animals*—Newly arrived cattle should be observed at least twice daily. Sick animals should be removed and treated. Treating sick animals promptly, rather than waiting until tomorrow, may mean the difference between life and death. Animals that show clinical signs of shipping fever—sunken eyes, runny nose, drooling at the mouth, labored breathing, and/or weaving (unsteady gait)—should be isolated in a separate "sick pen" or "hospital."

Rest, fresh water, good feed, proper medication, and TLC (tender loving care) are the cardinal essentials for preventing shipping fever and death losses.

CATTLE FEEDLOT DISEASE AND PARASITE CONTROL PROGRAM

A written-down cattle feedlot health program should be developed in cooperation with the feedlot veterinarian. The following outline will serve as a useful guide for this purpose.

1. Process (dehorn, castrate, and vaccinate) incoming cattle soon after they are unloaded from the truck, rather than wait. At that time, also inject with vitamin A and implant with any approved growth stimulant intended to use.

Newly arrived cattle are usually vaccinated against bovine virus diarrhea, red nose, shipping fever, and lepto; and given high injections of vitamin A if they are stressed (250,000 to 1,000,000 IU; depending on size of cattle and the degree of stressing).

2. For the first 28 days after arrival, fortify the natural protein supplement (do not feed incoming cattle urea) with the following per head per day: 350 mg of Aureomycin plus 350 mg of sulfamethazine (Aureo S-700, which is a combination of an antibiotic and the bactericidal agent sulfamethazine) *plus* 50,000 IU of vitamin A.

3. Adminster plenty of TLC (tender loving care) to the sick.

4. Treat sick animals three times in the first 24 hours. It will reduce repeats and chronic illness.

5. Keep hospital chutes and pens clean and well bedded.

6. Disinfect chutes, tanks and syringes, and balling guns to prevent spread of disease.

7. Seek professional advice if there is no response to medication within 48 hours.

8. Autopsy all dead animals for an accurate diagnosis. If one specific problem is causing a lot of deaths, seek prevention.

9. Don't change rations too fast.

10. Have all members of the health team—detection, treatment, and convalescence—performing at top level.

11. Control parasites:

a. *Internal parasites*—Worm calves, if necessary, within 2 weeks after arrival in the feedlot, with one of the following drugs, used according to manufacturer's directions: phenothiazine, thiabendazole, levamisole, coumaphos, or haloxon.

Treat only (1) if animals appear to be heavily parasitized or are from areas where previous experience has shown that they are heavily parasitized, or (2) if 300 or more eggs per gram (epg) of dry feces are found.

b. *External parasites*—Treat for external parasites (commonly lice, grubs, flies, and ticks), if necessary, using a recommended insecticide and following the manufacturer's instructions on the label or container.

c. *Control of flies around the feedlots and feedmill*—Prevent flies by starting the following program early in the season: sanitation, prevention of breeding areas, and use of residual sprays and mists before the fly population builds up. Before using any chemical in a control program directed toward either the adult fly or the larvae, read the label carefully to determine the usage and restrictions of the material.

CARE OF SICK CATTLE

Proper care of sick cattle necessitates two things: (1) suitable hospital facilities; and (2) prompt and correct diagnosis and treatment. Hospital facilities are covered in Chapter 28 of this book; hence, the discussion at this point will be limited to diagnosis and treatment.

Diagnosis and treatment are very important in the health program of any cattle feedlot. The pen checkers and hospital technician cannot completely operate this phase of the program alone. The yard foreman must supervise it, although he has too many responsibilities to be completely responsible for the program. The consulting veterinarian must establish general policies and treat unusual and difficult cases, although he cannot examine every animal as it enters the hospital, simply because the health program costs would be too high. Thus, diagnosis and treatment is really a team approach. Each person involved should contribute to the program according to his responsibilities and abilities, thereby assuring the most effective and economical health program.

Diagnosis is the art or act of recognizing disease from its symptoms. Thus, correct diagnosis assumes that those responsible for the cattle feedlot health program recognize what is "normal." Then, any deviation from the normal should be reason for concern and further study. Practical experience in both the normal and abnormal is essential in arriving at a correct diagnosis.

Treatment should be left in the hands of the veterinarian. He should give instructions for the proper use of drugs and biologicals—the administration, dosage levels, indications, and contraindications of a given product, and signs and lesions of specific diseases. It is his responsibility to keep management current on the latest findings regarding research, new products, and disease conditions.

NUTRITIONAL DISEASES

Several metabolic disorders, or diseases, in feedlot cattle are attributable wholly or in part to the feeding regimen. Among the more prevalent ones are: acidosis, bloat, liver abscesses, and urinary calculi (water belly; urolithiasis).

Acidosis, or Lactic Acid Acidosis

Acidosis usually develops early during the fattening process when the ration is changed too rapidly from roughage to concentrate.

● *Cause*—Acidosis is caused by an increase in lactic acid-producing bacteria (both the D- and L-forms) and the rapid production of lactic acid. It commonly occurs when there is a sudden shift from a high roughage to a high concentrate ration. However, cattle maintained on high energy rations are constantly in a marginal state of acidosis due to the formation of lactic acid in the rumen flora. Thus, ingredient changes, poor mixing of grain in the ration, or faulty feeding can promote acute acidosis.

• *Symptoms and signs*—Marginal acidosis is characterized by poor performance and inconsistent feed ingestion. If ingredient changes or erratic feeding persist, acute acidosis may result, creating laminitis—and eventually "ski shoe" cattle. In severe cases, the rumen becomes immobilized, followed by increased pulse and respiration rate, variable rectal temperature, sunken eyes, loss of dermal elasticity, staggering, coma, and death.

• *Prevention and treatment*—Prevention consists in avoiding erratic feeding and abrupt ration changes. Different treatments have been used with varying degrees of success; among them: (1) removal of rumen contents and replacement by contents of an animal on a normal ration; (2) feeding a high level of penicillin (12 to 20 million units) to suppress lactic acid-producing bacteria; (3) drenching (or intravenous injection) with a solution of sodium bicarbonate to restore the acid base balance; (4) daily intramuscular administration of antihistamines and cortical steroids for each of several days to help prevent intoxication and laminitis; or (5) backing the cattle down on both amount and kind of feed (lessening the ration, and returning to the mix that was being used before trouble was encountered).

Bloat

Bloat in feedlot cattle can be costly, both in treatment and losses. A Kansas survey showed that 0.6% of the feedlot cattle bloated mildly to moderately, 0.2% bloated severely, and 0.1% died of bloat.[2]

• *Cause*—Bloat is usually considered to be a nutritional disease, because certain feeds ferment more readily and cause more bloat than others. The high concentrate rations of feedlot cattle increase the number of slime-producing bacteria in the rumen. Slime traps fermentation gas and produces bloat.

Feedlot bloat may occur during any month of the year. However, it is more common during hot, humid weather.

• *Symptoms and signs*—Bloat is characterized by a greatly distended paunch, noticeable on the left side in front of the hip bone.

• *Prevention and treatment*—When trouble strikes, alleviate all likely causes as quickly as possible. A high incidence of bloat frequently occurs in feedlot rations containing high levels of both barley and alfalfa; hence, replace some of the barley with dried beet pulp, oats, or ground corncobs, and replace part of the alfalfa with grass hay or straw. Also, less bloat is experienced on long or coarsely chopped hay than on finely ground hay; and less on well-flaked grain than on finely ground grain. In addition to rectifying these situations, allow access to a little coarse, long hay or straw.

---

[2]Meyer, R. M., *27th Kansas Formula Feed Conference Proceedings P. H1*. Kansas State University, Manhattan, Kan., 1972.

When bloat is encountered, increase the proportion of roughage in the ration if feasible. Also, consider adding poloxalene (Bloat Guard®) to the ration, according to the manufacturer's directions.

Time permitting, bloated animals should be drenched with one to two ounces of poloxalene (Therabloat®). Then, 10 minutes after treatment, the gas should be relieved with a stomach tube.

Puncturing of the paunch should be a last resort.

Liver Abscesses

Abscesses, as indicated by the name, are single or multiple abscesses on the liver, observed at slaughter. Usually the abscess consists of a central mass of necrotic liver surrounded by pus and a wall of connective tissue. At slaughter, most livers affected with abscesses are condemned for human food.

In some lots of cattle on high concentrate rations, as high as 75 percent of the livers have been condemned. On the average, however, it will probably run between 5 and 10 percent. Since the liver of a 1,000-pound steer weighs approximately 11 pounds, its condemnation represents a considerable monetary loss. The loss from reduced feed efficiency and gains may be even greater.

• *Cause*—The direct cause of most bovine liver abscesses is *Spherophorus necrophorus*, the same bacteria which causes foot rot. This organism, which is ever present in ruminal contents, penetrates the covering epithelium through points of injury, discontinuity, and necrosis. Factors back of rumenitis include: (1) rapid rate of change from a diet of roughage to one high in concentrate, (2) fattening with a diet containing more than 25 percent concentrate, (3) foreign body penetration of the wall, and (4) miscellaneous agents.

• *Symptoms and signs*—Liver abscesses generally go undetected until cattle are slaughtered. However, reduced feed intake and gains near the end of the feeding period may be indicative.

• *Prevention and treatment*—Liver abscesses can be greatly reduced, but not entirely eliminated, by continuous feeding during the fattening process of the commonly used antibiotics (Aureomycin, Terramycin, Bacitracin, or Tylan). Also, the incidence of rumenitis, and consequently liver abscesses, can be reduced by avoiding the contamination of feed with penetrating objects, such as nails, pieces of wire and glass, and by the gradual change from a ration of high roughage to a ration of high concentrate.

Urinary Calculi (Water Belly; Urolithiasis)

The term "urinary calculi" refers to mineral deposits in the urinary tract, which may block the flow of urine in the urethra, particularly in castrated male cattle. Prolonged blockage generally results in rupture of the urinary bladder or urethra, releasing urine into the surrounding tissue. This produces the condition commonly referred to as "water belly."

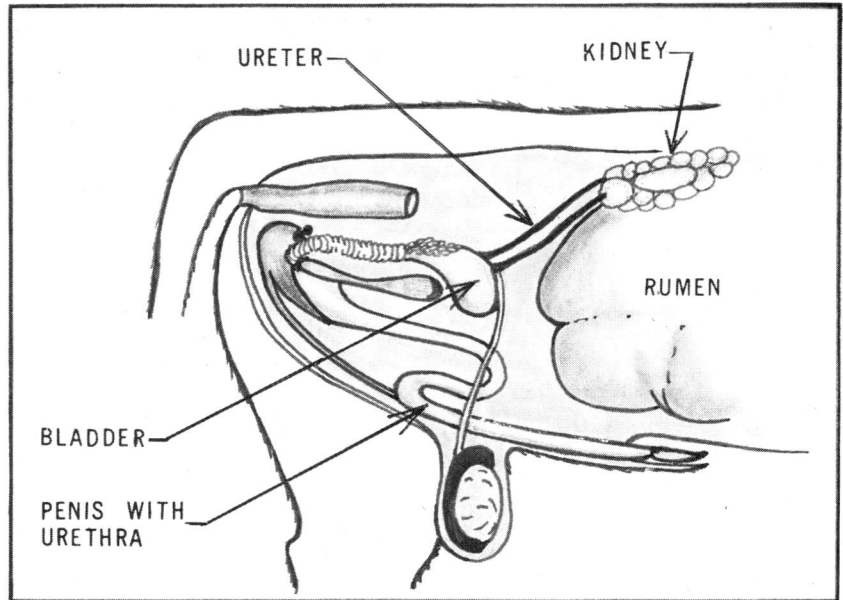

Fig. 33-3. Urinary tract of a bovine male. When the urethral tube becomes blocked, urine production continues causing extension and finally rupture of the bladder.

The mineral deposits can be of variable sizes, shapes, and composition. In cattle, the phosphatic type predominates under feedlot conditions, and the siliceous type occurs most frequently in range cattle.

● *Cause*—A high phosphorus-low calcium ration favors calculi formation; hence, the Ca:P ratio should be about 2:1. A high silica content in rations may also be a contributing factor.

Some grains and forages appear to have a tendency to increase the incidence of urinary calculi. These include high silica native grasses, wheat straw, sugar beet leaves or pulp, sorghums, and cottonseed meal. High dosages of diethylstilbestrol or a deficiency of vitamin A may be contributing factors.

● *Symptoms and signs*—At first, animals afflicted with urinary calculi appear restless and strain frequently in an unsuccessful attempt to urinate. They may repeatedly stamp their feet and kick their abdomen. If urinary blockage is not complete, urine may dribble slowly from the sheath. If the stone(s) fails to pass, there will likely be complete blockage of urine flow, followed by rupture of bladder and urethra and release of urine into the body cavity or surrounding tissues. Death follows.

● *Prevention and treatment*—Most materials and practices offering some degree of protection against phosphatic urinary calculi appear to include at least one of the following:

1. *A lowering of the urinary phosphorus level*—A calcium to phosphorus ratio of approximately 2:1 is recommended.

2. *Acidification of the urine*—This may be achieved by the feeding of acid forming salts. Ammonium chloride fed daily at the rate of 28.4-42.5 g (1.0 to 1.5 ounces) to fattening cattle is effective.

3. *An increase in urine volume*—An increase in urine volume is dependent upon an increase in water consumption. This can be achieved by including salt in the diet at a level of 3 to 4 percent.

The only treatment that has been of demonstrable value is the use of urinary tract relaxants that aid in keeping the urethra open and allow passage of mineral deposits. Surgery represents the most effective treatment, with the stone(s) removed at the point of blockage. In steers, the urethra may be bisected and brought to the outside of the body to bypass the constricted portion of the tract. After a short period of time to eliminate any tissue residue of urine, such animals are marketable.

INFECTIOUS DISEASES AND PARASITES

In the above discussion, the most common noninfectious diseases of feedlot cattle have been covered. Additionally, feedlot cattle are subject to (1) infectious diseases—those caused by specific entities such as bacteria or viruses, most of which are contagious; and (2) parasitic diseases—those caused by parasites. The infectious and parasitic diseases affecting feedlot cattle are the same as those that affect breeding cattle. Hence, they have all been covered together in this book (see Chapter 12).

QUESTIONS FOR STUDY AND DISCUSSION

1. Dr. Don Gill of Oklahoma State University calculated that the combined losses from death and sickness add $20 to $25 per head onto the cost of every feedlot finished animal. Outline how this cost could be lowered.

2. Detail a program for handling newly arrived cattle that will minimize death losses and maximize performance.

3. Traditionally, cattle are started on a high roughage ration, then gradually worked over to a high concentrate ration as they progress through the feeding program. However, experiments conducted by the California Station showed that for calves a starting ration consisting of 72 percent concentrate and 28 percent hay (roughage) is best. Which method do you favor—and why?

4. Outline a cattle feedlot disease and parasite control program.

5. Outline a fly control program for use around a cattle feedlot and feed mill.

6. Discuss the place of, and the relationship between, each (a) the feedlot personnel, and (b) the veterinarian, in the diagnosis and treatment of feedlot diseases.

7. Discuss the (a) cause, (b) symptoms, and (c) prevention and treatment of each of the following disorders, or diseases, in feedlot cattle:

   Acidosis
   Bloat
   Liver abscesses
   Urinary calculi

SELECTED REFERENCES

Title of Publication	Author(s)	Publisher
Diseases of Cattle, Second Edition	Ed. by W. J. Gibbons	American Veterinary Publications, Inc., Santa Barbara, Calif., 1963
Diseases of Feedlot Cattle, Second Edition	R. Jensen D. R. Mackey	Lea & Febiger, Philadelphia, Penn., 1971
Feedlot, The	Ed. by I. A. Dyer, C. C. O'Mary	Lea & Febiger, Philadelphia, Penn., 1972

PASTURE FINISHING (FATTENING)

Contents   **Page**

Pasture finishing of cattle has changed. Fewer and fewer steers are finished on grass alone, primarily because consumers no longer want heavy, grass-finished steers weighing 1,200 pounds or more. Young cattle grow but do not reach market finish under usual pasture conditions. Thus, it is impossible to finish cattle at early ages and light weights without either supplemental feeding on pasture and/or lot finishing at the end of the grazing season.

Fig. 34-1. Yearling Hereford steers from Texas, finishing on bluestem grass in the famous grazing area in the Kansas Flint Hills. (Courtesy, Kansas State University, Manhattan, Kan.)

Generally speaking, no cheaper method of harvesting forage crops has been devised than is afforded by harvesting directly by grazing animals. Moreover, even most seeded pastures last several years; thus, seeding costs may be distributed over the entire period. Naturally, the cash income to be derived from pastures will vary from year to year and from place to place depending upon such factors as market price levels, class of animals, soil, season, and the use of adapted varieties.

ADVANTAGES AND DISADVANTAGES OF GRAIN FEEDING ON PASTURE

The *advantages* of grain feeding cattle on pasture, compared to strictly feedlot finishing, are:

1. Pasture gains are cheaper because: (a) less grain is required per 100 pounds gain; (b) grass is a cheaper roughage than hay or silage; and (c) less expensive protein supplement is required. Generally speaking, comparisons of self-feeding on pasture vs drylot indicate that pasture saves about 100 lb of dry feed per 100 lb of gain. Thus, if we assume that it requires 500 lb of gain to finish a steer, then each steer would require 500 lb less feed on pasture than in drylot. If feed costs 5¢ per pound, that's a saving of $25 per steer on feed cost.

2. Less labor is required because the cattle gather their own roughage and the labor required for feeding roughage is eliminated. In brief, grass-finished cattle do their own harvesting. Furthermore, it may be possible to get satisfactory results with but one grain feeding each day in finishing on pasture, or the animals may be self-fed, with the caretaker merely filling the feeder at intervals.

3. Handling of manure is eliminated, the maximum fertility value of the manure is conserved, and there is no pollution problem. When pastures are utilized by livestock, approximately 80 percent of the plant nutrients of the crop is returned to the soil.

4. Pasture finishing eliminates any requirement for buildings.

5. Finishing cattle on pasture is especially adapted to the small feeder and to areas where some of the land should be kept in permanent pasture.

The *disadvantages* of finishing cattle on pasture, compared to strictly feedlot finishing, are:

1. Most feeder cattle are marketed in the fall rather than in the spring. Therefore, feeder steers purchased in the spring and intended for pasture finishing are usually scarce and high in price.

2. Though less labor is required, less labor is available. The cropping season is a rush season.

3. During the midsummer, the combination of heat and flies may cause much discomfort to the animals and reduce the gains made.

4. Pastures may become dry and parched, reducing the gains made during dry seasons.

5. The manure is usually dropped on permanent pastures year after year, which may result in the neglect of the other fields.

6. In many pastures, availability of shade and water does not present a problem. However, some areas are less fortunate in this regard.

After both the advantages and disadvantages of pasture finishing are considered, the availability of cheap, rough pasture land and the price of concentrates will usually be the determining factors in deciding upon the system to follow.

SYSTEMS OF PASTURE FINISHING

When cattle are finished on pasture, any one of the following systems may be employed:

1. Finishing on pastures alone—no concentrates being fed.
2. Limited grain allowance during the entire pasture period.
3. Full feeding during the entire pasture period.
4. Full or limited grain feeding on pasture following the period of peak pasture growth.
5. Short feeding (60 to 120 days) in the feedlot at the end of the pasture period.

The system of pasture finishing that will be decided upon will depend upon the age of the cattle, the quality of the pasture, the price of concentrates, the rapidity of gains desired, and the market conditions.

BASIC CONSIDERATIONS IN UTILIZING PASTURES FOR FINISHING CATTLE

The following points are basic in utilizing pastures for finishing cattle:

● *Moderate winter feeding makes for most effective pasture utilization*—The more liberally beef cattle are fed during the winter, the less will be their effective utilization of pasture the following summer—the less the compensatory gains. Generally speaking, for maximum utilization of pasture, stocker calves should be fed for winter gains not in excess of 1.25 pounds per head daily, and yearlings not in excess of 0.9 pounds.

● *Early pastures are "washy" but high in protein*—Cattle should not be turned to pasture too early. The first growth is extremely "washy," possessing little energy. However, the crude protein content of the forages is high during the early stages of growth and rapidly decreases as the forages mature. This would indicate the importance of pasturing rather heavily during the period of maximum growth in the spring and early summer.

● *Sudden changes are to be avoided*—Changes from drylot to pastures or from less succulent to more succulent pastures should be made with care; for grass is a laxative, and the cattle may shrink severely. Also, bloat may occur.

● *Time of starting grain feeding on pastures is determined by condition of cattle and quality of pastures*—Cattle that have been fed grain rather liberally through the winter and are in good condition should usually be fed grain from the beginning of the grazing period. On the other hand, if they have been roughed through the winter, it may be just as well to feed the grain only during the last 80 to 120 days of the grazing season, after the season of peak pasture growth. The latter recommendation is made because it is sometimes difficult to get animals to consume grain when an abundance of palatable forage is available. At peak pasture growth, the animals should be started on feed and brought to full feed as rapidly as possible.

● *Grain supplements on pastures usually make for larger daily gains and earlier marketing*—Young cattle (calves and yearlings) on summer pasture usually do not grow at their maximum potential due to energy and protein deficiencies in the feed at various times of the season. Thus, the addition of a grain supplement for cattle on pasture makes for larger daily gains and earlier marketing—either directly off grass or with a shorter drylot finishing period. The owner thus avoids late fall competition and lower prices of strictly grass cattle. Also, because cattle that are grain fed on pasture can be marketed over a wider period of time, there is greater flexibility in the operations. However, cattle that are supplemented on summer pasture often sell for less to go into feedlots because feedlot operators fear that they may not gain as rapidly as cattle that are not supplemented on pasture. To answer this question, the Nevada workers studied the subsequent feedlot performance of yearling steers receiving a feed supplement while on summer pasture vs nonsupplemented cattle.[1] The pasture-supplemented cattle were fed pellets on the ground 3 times daily, at an average daily rate of 2.18 lb per head per day, or slightly less than one-half percent of their body weight. The supplemented cattle outgained the nonsupplemented cattle by 0.38 lb daily, or a total of 41 lb; and each pound of supplement produced 5.7 lb of gain. At the end of the pasture season, both groups of steers were placed in the feedlot. Subsequent feedlot gains of the two groups were essentially the same—407 and 406 lb for the supplemented and controls, respectively; and an average daily gain of 3.3 lb by each group. These results indicate that yearling steers may be profitably supplemented on summer pasture at the above rates without adversely affecting their subsequent feedlot performance.

● *Whole corn preferred to rolled corn*—When self-feeding steers on pasture, whole corn is preferred to rolled corn for the following reasons: (1) slightly less feed is required per 100 lb gain; (2) it alleviates processing cost; and (3) it results in less incidence of founder and rumen parakeratosis because whole corn supplies some "roughness factor" in the ration to stimulate the rumen.

● *Protein supplement not needed on good pasture*—As long as pasture is green and growing, no supplemental protein is required. During drought periods and late fall when the grass matures, extra protein is needed. At such times, it is good business to add 1 pound of protein supplement to each 8 to 12 pounds of grain. Usually this will increase the rate and efficiency of gain.

● *Carrying capacity of pastures will vary*—The carrying capacity of pastures will vary with the amount of grain supplement, the quality of pasture, and the age and condition of the cattle. Because of these factors, the acreage per steer will vary all the way from 1 to 10.

● *Age is a factor*—Young cattle (yearlings) tend to grow as well as to fatten. Thus, older cattle (two years or older) will reach a high degree of finish on pastures alone. As good as the pastures are, it must be remembered that grass is still a roughage.

● *Minerals for cattle on pasture*—Salt is especially necessary when grass is being utilized. Finishing steers consume from ¾ to 1½ ounces of salt per

---

[1]Probert, Curt, Supplementing Yearling Steers on Summer Pasture and Its Subsequent Effect on Feedlot Performance, *Western Livestock Journal*, Aug. 1974, p. 228.

Fig. 34-2. Yearling steers grazing irrigated orchardgrass pasture which is at the proper stage of growth for maximum animal gains. Orchardgrass is one of the top yielding grasses for the northern humid states and the western irrigated areas. (Courtesy, Dr. Wilton W. Heinemann, Irrigated Agriculture Research and Extension Center, Washington State University, Prosser, Wash.)

Fig. 34-3. Two-year-old Hereford steers on pasture, showing good finish. (Courtesy, Oklahoma State University, Stillwater, Okla.)

head daily. Also, cattle on pasture should have free access to a mineral mixture composed of 2 parts of dicalcium phosphate and 1 part of iodized salt.

● *Species of grasses or legumes will vary*—The best species of grasses or legumes or grass-legumes mixtures to be seeded will vary according to the area, especially according to the soil and climatic conditions. Pasture yields vary greatly from area to area and season to season.

Temporary or supplemental pastures, such as Sudan grass or millet, are used for a short period and are usually more productive and palatable than permanent pastures. They are seeded for the purpose of providing supplemental grazing during the season when the regular permanent or rotation pastures are relatively unproductive.

Fig. 34-4. Yearling Hereford steers on Sudan grass pasture near Weeping Water, Nebraska. (Courtesy, Soil Conservation Service, USDA)

● *Grass vs grass-legume mixtures should be considered*—In general, where adapted legumes can be successfully grown—either alone or with grass mixtures—the results are superior to yields obtained from pure stands of the grasses. At Pullman, Washington, in a study of pure species of smooth brome and crested wheatgrass vs grass-alfalfa mixtures, it was found that: (1) the grasses produced an average of 87 pounds of beef per acre, whereas the grass-alfalfa mixtures averaged 223 pounds of beef per acre; (2) when based on forage yields at monthly intervals, the same mixtures produced 3 times as much oven-dry forage per acre as the pastures seeded to grasses alone; (3) the grass-legume mixtures provided a slightly longer grazing season; (4) the grass-legume mixtures provided a higher carrying capacity in terms of animals per acre; (5) the erosion-resisting characters of the soil were improved by the fibrous grass roots, both while the crop was growing and after the seeding had been plowed under; and (6) the addition of grasses to legumes tended to keep out cheatgrass and other undesirable plants.[2] The two latter points are based merely on careful observation, whereas the rest of the points were proved experimentally.

● *Grain-feeding will lengthen the grazing season*—At the Washington Agricultural Experiment Station, grain feeding cattle on pasture lengthened the grazing season by an average of 57 days.

● *Self-feeding vs hand-feeding on pasture*—Self-feeding grain on pasture has generally proved superior to hand-feeding, as the animals consume more feed, make more rapid gains, and return more profit.

● *Economy of grain feeding on pasture*—Whether or not it will be profitable to feed grain on pasture will depend primarily upon the price of grain, the premium paid for cattle of higher finish and grade, the season in which it is desired to market, and the area and quality of pasture.

The following practices will be helpful in reducing the bloat hazard:

1. Give a full feed of hay or other dry roughage before the animals are turned to legume pastures, to prevent the animals from filling too rapidly on the green material.

2. After the animals are once turned to pasture, they should be left there continuously. If they must be removed overnight or for longer periods, they should be filled with dry roughage before they are returned to pasture.

3. Mixtures that contain approximately half grasses and half legumes should be used.

Fig. 34-5. Yearling steers on alfalfa-orchardgrass, which they have grazed for about 120 days. Note stage of forage growth and condition of these 900-lb steers. (Courtesy, Dr. Wilton W. Heinemann, Irrigated Agriculture Research and Extension Center, Washington State University, Prosser, Wash.)

---

[2]Ensminger, M. E., *et al. Wash. Agr. Exp. Sta. Bull. No. 444.*

4. Water and salt should be conveniently accessible at all times.

5. The animals should not be allowed to become empty when they congregate in a drylot for shade or insect protection and then be allowed to gorge themselves suddenly on the green forage.

6. Many practical cattlemen feel that the bloat hazard is reduced by mowing alternate strips through the pasture, thus allowing the animals to consume the dry forage along with the pasture. Others keep in the pasture a rack well filled with dry hay or straw.

7. Because of the many serrations on the leaves, Sudan hay appears especially effective in preventing bloat when fed to cattle on legume pastures.

8. Consider the use of poloxalene, a nonionic surfactant, developed through research at Kansas State University, for the control of legume bloat in cattle. One such product is manufactured by Smith Kline & French Laboratories and marketed under the trade name Bloat Guard. Always use such products according to the manufacturer's directions. Bloat Guard is a granular product containing 53 percent active ingredient (poloxalene). It is cleared by the Food and Drug Administration for use as a top dressing on feed; it is fed at the rate of ⅔ ounce daily for animals under 1,000 pounds.

QUESTIONS FOR STUDY AND DISCUSSION

1. Why are so few cattle finished on pasture alone?

2. As grain becomes scarcer and higher in price, is it likely that more cattle will be grass finished, perhaps by supplemental grain feeding?

3. The finish and marbling requirements of the top Federal grades of beef were recently lowered. Does this favor more pasture finishing?

4. Using current feed prices, compute the value of grass on a per steer basis if it effects a saving of 100 pounds of dry feed per 100 pounds of gain. What additional advantages accrue from grain feeding cattle on pasture, compared to drylot finishing?

5. List and discuss the disadvantages of finishing cattle on pasture.

6. List and discuss each of the common systems of pasture finishing. What factors should determine the choice of the system?

7. Discuss each of the following basic points as they apply to pasture finishing:
   a. Moderate winter gains.
   b. Early, "washy" pasture.
   c. When to grain feed on pasture.
   d. Effect of supplementing young cattle on pasture on subsequent feedlot performance.
   e. Self-feeding whole corn vs rolled corn on pasture.
   f. The use of a protein supplement.
   g. Age of cattle.
   h. Grass vs grass-legume pastures.
   i. Self-feeding vs hand-feeding.
   j. Bloat control.

SELECTED REFERENCES

Title of Publication	Author(s)	Publisher
Beef Cattle, Sixth Edition	A. L. Neumann R. R. Snapp	John Wiley & Sons, Inc., New York, N.Y., 1969

(Continued)

Title of Publication	Author(s)	Publisher
Beef Cattle Production	K. A. Wagnon R. Albaugh G. H. Hart	The Macmillan Company, New York, N.Y., 1960
Beef Production in the South	S. H. Fowler	The Interstate Printers & Publishers, Inc., Danville, Ill., 1969
Beef Cattle Science Handbook	Ed. by M. E. Ensminger	Agriservices Foundation, Clovis, Calif., pub. annually since 1964.
Stockman's Handbook, The, Fourth Edition	M. E. Ensminger	The Interstate Printers & Publishers, Inc., Danville, Ill., 1970

FUTURE OF CATTLE FINISHING (FATTENING)

The challenge of the cattle feeding industry is to produce 69 percent more fed cattle by 1985 than in 1972. This is based on the following 1985 projections: 246.3 million people (vs 209.2 million in 1972); per capita beef consumption of 140 lb (vs 116.0 lb in 1972)[1]; carcass weights remaining the same (622 lb); and no increase in beef imports. This means that 45.2 million fed cattle will be needed in 1985 (vs 26.8 million fed cattle in 1972), an increase of 18.4 million or 69 percent. This challenge will be met. The only questions are how will it be met, and what feedlots will be there first with the most of the best?

Fig. 35-1. Cattle feeding of the future will be more sophisticated. This shows Farr Feeders' new $4 million feedlot, near Greeley, Colorado, capable of producing for market more than 80,000 head of cattle a year. The feedlot includes 160 pens, each engineered for drainage, encircled with steel fencing, and with a capacity of 250 head. There are 5½ miles of concrete bunks, with 10-foot concrete platforms on which the cattle stand while they feed. The computerized feed mill, with mixing capacity of 2,100 pounds per minute, is in the background. (Courtesy, Farr Feeders, Inc., Greeley, Colo.)

---

[1]Per capita beef consumption herein is computed on carcass weight basis, as is common in the meat trade and by the government.

The author's crystal ball shows that the cattle feeding industry in the future will be characterized by the following:

● *Lower profits per animal and greater difficulty in securing desirable financing*—Feedlot profits, on a per-head basis, will likely decrease in the years ahead, due to (1) overexpansion of feedlots, (2) shortage of and higher priced feeder cattle, and (3) higher priced feeds as animals compete with humans for grains. Hand in hand with this, capital will be more difficult to get.

● *Lots operating to capacity*—In the era of rapid expansion in the 1950s and 1960s, many lots were not operated to capacity, all or part of the year. Empty lots don't make money! Feedlots must stay 70 to 80 percent full to turn a profit. Hence, with smaller net profits per head in the future, feedlots will be forced to operate near capacity.

● *More integrated feedlots both horizontally and vertically*—There will be more horizontal integration (one feedlot merging with another feedlot) motivated by (1) the desire of lots to get bigger, through acquisition, (2) the better financed lots buying the weaker financed ones, especially during periods of financial stress, and (3) spreading the area risks (of such things as droughts and feed shortages) by locating in different areas, both within the United States (for example, one lot might be in the Corn Belt and another in Texas) and abroad (such as having one feedlot in the United States and another in Australia).

Also, greater efficiency will be achieved through increased vertical integration (the control or ownership of other levels of the functional system), with more feedlots owned by (1) large cow-calf operators and (2) packers. Increasingly, outside interests, with large amounts of outside capital, will engage in the cow-calf operations. Many of them will develop elaborate integrated systems which will permit them to retain ownership of the cattle through the feedlot—and even to the carcass hanging on the rail, and the retail meat market.

● *Lessened tax shelter advantages*—It seems reasonable to expect that the Internal Revenue Service will lessen the tax shelter advantages (primarily through limited partnerships and prepayment of feed) enjoyed by some "outside investors"—doctors, lawyers, merchants, chiefs.

● *Feedlot managers with superior business acumen*—The business aspects of the cattle feeding will become more important. More capital will be required, and there will be more competition for money available for lending; credit will be more important; production costs and labor problems will mount; inflation will be increasingly difficult to curb; computers will be used more extensively; beef futures will be more widely used; feeder calves will be in short supply; more grains will go for human consumption abroad, with the result that cattle feeders will increasingly find grains scarce and high in price. All these, and more, will call for top cattle feedlot managers with both superior cattle know-how and business acumen.

● *Increased computerized management*—Feedlot managers will make increasing use of computers in decision making, beyond their use in balancing rations and inventory of cattle and feed. Among other things, they will be used to determine when to buy and sell.

● *Greater emphasis on public relations*—The cattle feeding industry will develop a more sophisticated and costly public relations program on a continuing basis, rather than "put out fires." The need for such a PR program is driven home by such things as (1) picketers protesting the high cost of beef, and (2) fewer and fewer legislators with farm background.

● *Pollution control*—Environmentalists and neighbors will force pollution control. Feedlots will lessen harassments from these sources by better control of odors and dust and improved handling of manure.

● *Recycling manure*—Recycling of feedlot manure will be perfected on a practical basis. Also, manure will be fed to fish, as is now being done in China.

● *Fewer lots under 1,000 head capacity; more lots over 32,000 head capacity*—In recent years, there has been a decrease in small lots and an increase in big lots. In 1972, feedlots with capacity of less than 1,000 head decreased 7 percent, while lots with 32,000 head or more increased 30 percent. This trend to bigness will continue.

● *Increased mechanization*—Cattle feeding will become more mechanized, with the degree and expenditure for mechanization determined by the cost of labor replaced.

● *More sheltered or confined feeding*—In the past, open feedlots tended to locate in areas with low rainfall and mild winters, to lessen mud and energy requirements. As sheltered, confined feeding becomes more practical (because, in comparison with open lots, of lower (1) land costs and (2) labor costs, due to better adaptation to mechanization), weather will be eliminated as a factor in determining location.

● *More pasture finishing*—As grain becomes more scarce and higher in price, grass finishing will increase; particularly in the Southeast and on smaller farms.

● *Growth areas ahead*—The growth areas in cattle feeding and slaughtering to 1985 will be (1) the Southern Plains—Oklahoma, Texas, and New Mexico; and (2) the Central Plains—Colorado, western Nebraska, and western Kansas. However, impending water shortages may well affect the irrigated feed production of the Southern Plains. The Desert Southwest, the Western Corn Belt, and the Lake States will progressively become less important as cattle feeding areas as it becomes necessary to replace substantial investments in feeding and slaughtering facilities.

● *Further spreading over the country so as to be located near available feeds and cattle*—Certain highly concentrated feedlot areas appear to have approached the saturation point so far as proximity to available feed and cattle are concerned. When this point is reached, some other area will be favored for new lots.

● *Not being limited to sparsely human populated areas for environmental reasons*—In the past, feedlots have commonly located in sparsely populated areas in order to alleviate the harassment of environmentalists, most of whom are more interested in pollution control than cattle feeding. As we learn to control feedlot odors and handle manure, pollution will be lessened as a factor in determining location.

● *More feeders produced by cows kept for beef*—During the 1950s and 1960s, a very considerable proportion of increased feeder cattle numbers came from young animals of dairy breeding and animals that were formerly marketed as veal. These animals will continue to be fed, but little expansion from these sources can be expected. Thus, the increased feeders of the future must come from increased cows kept for beef, along with increased twinning.

● *More crossbreds*—Not only will there be more crossbreds, but many of them will carry breeding of the newer breeds, recently imported. As a result, feedlots of the future will have a great array of colors and breeding.

● *More efficient cattle being fed*—Feedlot cattle of the future will be bred for predictable rapid and efficient gains because of (1) being descended from selected, production-tested stock, and (2) their hybrid vigor, since more of them will be crossbreds.

● *Stockers changed*—Stockers will be changed in two ways, with some going each route: (1) Some will be carried on roughage to older ages and heavier weights, followed by shorter feedlot periods; and (2) others out of good milking cows and with heavy weaning weights will result in either eliminating or shortening the stocker stage.

● *Most feeders being preconditioned*—With heavy, but young, calves being weaned and sent directly to feedlots, rather than spending time as stockers, preconditioning will be more important than ever. Thus, there will be more preconditioning in the future.

● *Feed less total grain in feedlots*—This will be achieved in two ways: (1) by shortening the feeding period through putting heavier calves in the feedlot; and (2) by lessening the degree of finish.

● *More nonprotein nitrogen (NPN)*—Worldwide protein shortages will result in increased quantities of the oilseed proteins (especially soybean meal and cottonseed meal) being diverted for human consumption. As a result, greater quantities of nonprotein nitrogen will be fed to feedlot cattle. Also, more slow-release NPN products will be developed.

● *New additives*—New and effective noncarcinogenic growth stimulants will be developed and used.

● *More bulls being fed*—With increasing acceptance of "bullock" beef—which will come, just as it has all over Europe—more bulls will be fed, to obtain faster gains, greater feed efficiency, and more red meat.

● *More beef produced from grass and milk*—With rising feed costs due to animals competing with humans for the source of supply—around the world, a larger proportion of the product will be produced from grass and milk. This calls for better milking cows and heavier weaning weights.

● *More by-product feeds utilized in feedlots*—As human competition for grain increases around the world, more and more by-product feeds will be fed to finishing cattle. Of course, by-product feeds are being fed to cattle now. But many are not used. Some idea of the potential can be gained when it is realized that from one to three pounds of by-product feed are produced per pound of food produced.

● *Feedlot cattle producing beef to more exacting specifications*—This will include weight, lean-to-bone ratio, amount of "bark" or outside fat, and palatability and tenderness. To meet this need, the cattle feeder will be called upon to satisfy a narrow range in market demand; in turn, the cow-calf man will be asked to produce feeders of the right kind. This will be accomplished largely through (1) production testing and selection of the purebreds back of the crossbreds, since carcass quality is little affected by heterosis, and (2) complementary, or matching, of the purebred breeds that are used to produce crossbred feeders.

● *More beef being fabricated, boxed, and branded in packing plants*—In the future, more and more beef will be fabricated (cut and packaged in portion-ready cuts before shipping to retailers) in packing plants and warehouses, instead of the backrooms of supermarkets. This will result in (1) a saving in transportation (100 lb, or more, of fat and bone from each carcass unwanted by the supermarket will be left at the packing plant), (2) improved sanitation (beef packed in vacuum bags, rather than exposed in carcass or wholesale cuts), and (3) eliminating the present 250,000 inefficient fabricating units—the backrooms of the nation's supermarkets.

● *Increased use of frozen beef*—The prejudice against frozen beef began disappearing with the coming in of freezer lockers. In the future more beef will be frozen, thereby providing for a more orderly flow of the product.

● *More beef tenderizers being used*—The use of beef tenderizers prior to slaughter will become more widespread and effective.

● *Inroads made by simulated meats (synthetic meats, or meat analogs)*—The simulated meats, or meat analogs, especially those made from soybeans, will likely become more competitive with beef in the future, as their price becomes relatively more favorable and their taste and texture are improved. To meet this competition, the cattle industry of the future will place increasing emphasis on palatability.

● *Improved market efficiency*—Currently, reasonably adequate feeder cattle and slaughter cattle market information is available. However, each feedlot is at liberty to buy and sell when it chooses, with the result that there are inevitable market gluts and scarcities, with prices varying accordingly. There is need for market controls which will allow the feeder to retain most of his fierce independence, yet avoid wide swings in numbers and prices. Perhaps something in the pattern of the agricultural commodity boards of California is the answer.

● *Beef futures trading specifications that are more exacting*—As beef over the counter conforms to more exacting specifications, it would seem logical that such specifications should be reflected in beef futures trading. Thus, in addition to weight and grade, as now called for in beef futures trading, such things as guarantees of lean yield may be added.

● *Beef imports causing constant apprehension among beef producers*—While the United States consumes about 26.4 percent of the world's beef and veal, it produces 24.2 percent. The difference is made up through importation. Despite some apprehension on the part of cattlemen, beef importations will continue, especially from Australia. But worldwide beef shortages and limited grain-feeding potential in other countries preclude any real foreign threat to United States cattle feeders, although we shall continue to import considerable quantities of lean, frozen beef.

● *Finished beef exported to Japan and Europe*—Finished beef, produced in United States feedlots, and fabricated, packaged, and frozen in United States packing plants, will be transported via refrigerated jet freight and marketed in Japan and Europe, at higher prices than can be secured at home.

● *Less emphasis on carcass quality*—In a "sellers market" for beef, there is the likely hazard that carcass quality will be relegated to a position of minor importance. Should this happen to an appreciable degree, per capita beef consumption could eventually suffer. Cattlemen should be ever mindful of what happened to pork, which was replaced by beef as the preferred red meat in 1953, primarily because the quality of pork failed to respond to consumer preference for a leaner product.

QUESTIONS FOR STUDY AND DISCUSSION

1. What is your opinion of the following 1985 U.S. projections: 246.3 million people; per capita beef consumption of 140 pounds; and 45.2 million fed cattle?

2. Which type of integration will occur most frequently in cattle in the future—horizontal or vertical?

3. Would removal of the tax shelter advantages of cattle feeding be good or bad? Justify your answer.

4. Outline a public relations program for the U.S. cattle industry designed to educate consumers and alleviate unjustified beef picket protesting.

5. How will feedlot pollution be controlled in the future?

6. How will cattle feedlot manure be used in the future?

7. Will increased confinement feeding make for a geographical shift in cattle feeding, with more cattle fed in the northern states?

8. As grain becomes more scarce and higher in price, will there be more pasture finishing, perhaps with grain supplementation?

9. How will the stocker stage be changed in the future?

10. Will a higher proportion of feeders be produced by beef cows as such in the future, with a smaller proportion from dairy cows?

11. Will feedlots of the future have more crossbreds?

12. Will more bulls be fed in the future? What problems go hand in hand with feeding bulls?

13. Will more feeders be preconditioned in the future?

14. What suitable by-products of which you have knowledge are not now being used for cattle feed?

15. What forces favor more beef being fabricated, boxed, and branded in packing plants in the future?

16. Is there any basis for the prejudice against frozen beef? Does it differ from frozen beef in a locker?

17. What can, or should, the beef industry do to meet the likely increasing competition from synthetic meat analogs in the future?

18. What are our chances of becoming a beef exporting nation in the future?

19. Is there hazard of neglecting carcass quality during an era of worldwide beef shortages?

SELECTED REFERENCES

Title of Publication	Author(s)	Publisher
Beef Cattle Science Handbook	Ed. by M. E. Ensminger	Agriservices Foundation, Clovis, Calif., pub. annually since 1964
Feedlot, The	Ed. by I. A. Dyer, C. C. O'Mary	Lea & Febiger, Philadelphia. Penn., 1972

APPENDIX

SECTION I—COMPOSITION OF FEEDS[1]

Table I-3 (beginning on page 1482) gives the composition of feeds commonly used in beef cattle rations, for which the following information is pertinent to understanding:

1. *NRC nomenclature*—In Table I-3, names of the feeds are based on a scheme proposed by L. E. Harris.[2] The names, called NRC names, are designed to give a qualitative description of each product, where such information is available or pertinent. A complete NRC name consists of as many as eight components, written in linear form, with components separated by commas. The components are as follows:

a. Origin (or parent material)
b. Species, variety, or kind
c. Part eaten
d. Process(es) and treatment(s) to which product has been subjected
e. Stage of maturity
f. Cutting or crop
g. Grade or quality designations
h. Classification

Feeds of the same origin (and of the same species, variety, or kind, if one of these is stated) are grouped into eight classes, each of which is designated by a number in parentheses. The numbers and the classes they designate are as follows:

(1) Dry forages or dry roughages
(2) Pasture, range plants, and feeds fed green
(3) Silages
(4) Energy feeds
(5) Protein supplements
(6) Minerals
(7) Vitamins
(8) Additives

---

[1]This entire section, including Tables I-1 to I-6, was taken from *Nutrient Requirements of Beef Cattle*, Fifth Revised Edition, National Academy of Sciences, 1976, with slight adaptations by the author.

[2]Harris, L. E., "Symposium on Feeds and Meats Terminology, III, A System for Naming and Describing Feeds, Energy Terminology, and the Use of Such Information in Calculating Diets," *Journal of Animal Science*, Vol. 22, p. 535, 1963.

Feeds that in the dry state contain on the average more than 18% of crude fiber are classified as forages and roughages. Products that contain 20% or more of protein are classified as protein supplements. Products with less than 20% of protein are classified as energy feeds. (These guidelines are approximate, and there is some overlapping.)

Abbreviations have been devised for many of the terms used in the NRC system (Table I-1). Stage-of-maturity terms are given in Table I-2.

The following listings show how three feeds are described:

Components of Name	Feed No. 1	Feed No. 2	Feed No. 3
Origin (or parent material)	Corn	Soybean	Wheat
Species, variety, or kind	—	—	—
Part eaten	aerial pt	seed wo hulls	flour by-prod
Process(es) and treatment(s) to which product has been subjected	ensiled	solv-extd grnd	f-sift
Stage of maturity	—	—	—
Cutting or crop	—	—	—
Grade or quality designations	—	mx 3 fbr	mx 4 fbr
Classification	(3) (silages)	(5) (protein supplements)	(4) (energy feeds)

Thus, the NRC names of the three feeds are written as follows:

No. 1: Corn, aerial pt, ensiled, (3)
No. 2: Soybean, seed wo hulls, solv-extd grnd, mx 3 fbr, (5)
No. 3: Wheat, flour by-prod, f-sift, mx 4 fbr, (4)

The analytical data are expressed in the metric system (with the exception of the bushel weights of the cereal grains) and are shown on an as-fed as well as a dry basis. See Table I-4 for weight-unit conversion factors and Table I-5 for weight equivalents.

2. *Locating names in the table*—To locate in Table I-3 the NRC name of a feed, one must know its origin (the name of the parent material), and usually the variety and kind. The first word of each NRC name is the name of the parent material. For feeds of vegetable origin, the origin term is the name of the plant (e.g., alfalfa, barley, oats), not the word "plant."

If a reader is uncertain about the origin term that introduces an NRC name in Table I-3, he may find the term by referring to the common name of the feed in which he is interested. Common names appear in the table in their alphabetical place.

Names having the same origin term are arranged in an order that depends on whether the names include references to species, variety, or kind. Names lacking such references are arranged under the origin term as follows:

First: numerically, by classes.
Second (within a class): Alphabetically, by parts eaten.

Names that include references to species, variety, or kind are arranged under the origin term as follows:

First: alphabetically, by species, variety, or kind.
Second (within species, variety, or kind): numerically, by classes.
Third (within a class): alphabetically, by parts eaten.

Many feeds have names that were given to them by the Association of American Feed Control Officials (AAFCO), the Canada Feed Act (CFA), or the Canada Grain Act (CGA). In addition, some feeds have regional or local names. The reader will find these names in their alphabetical place, where they are cross-referenced to the NRC names: he will also find them under the NRC names.

A six-digit reference number is listed after the NRC name and other names. The number may be used as the "numerical name" of a feed when performing linear programming with electronic computers.

The common name of the parent material is followed by the scientific name (example: Alfalfa. *Medicago sativa*).

3. *Energy value of feeds*—It has not been possible to obtain apparent digestible energy (DE) and metabolizable energy (ME) values for all feedstuffs. In some of these cases, values have been estimated by using the following formulas:

ME (Mcal/kg) for cattle = DE (Mcal) $\times$ 0.82
TDN = ME $\div$ 3.6155 kcal

Net energy values for gain (NE_{gain}) have in some cases been estimated by using formulas developed by Lofgreen and Garrett.[3] The formulas are as follows:

Log F = 2.2577 − 0.2213 ME
NE_m = 77/F
NE_{gain} = 2.54 − 0.0314 F

The terms used in the above formulas are as follows:

ME is the metabolizable energy in Mcal/kg of dry matter (DM).
F is the grams of dry matter per unit of $W^{0.75}$ required to maintain energy equilibrium.
NE_m is the net energy for maintenance in Mcal/kg of DM.
NE_{gain} is the net energy for gain in weight in Mcal/kg of DM.

Pertinent energy terms are explained in Fig. I-1 and in Chapter 30 under the heading, "Expressing Energy Values of Feeds."

4. *Carotene conversion*—International standards for vitamin A activity as related to vitamin A and beta-carotene are as follows:

1 IU of vitamin A = 1 USP unit
= vitamin A activity of 0.300 μg of crystalline vitamin A alcohol, which corresponds to 0.344 μg of vitamin A acetate or 0.550 μg of vitamin A palmitate

Beta-carotene is the standard for provitamin A.

1 IU of vitamin A = 0.6 μg of beta-carotene
1 mg of beta-carotene = 1,667 IU of vitamin A

---

[3]Lofgreen, G. P., and W. N. Garrett, "A System for Expressing Net Energy Requirements and Feed Values for Growing and Finishing Beef Cattle," *Journal of Animal Science*, Vol. 27, pp. 793-806, 1968.

HEAT PRODUCTION — UTILIZATION OF ENERGY

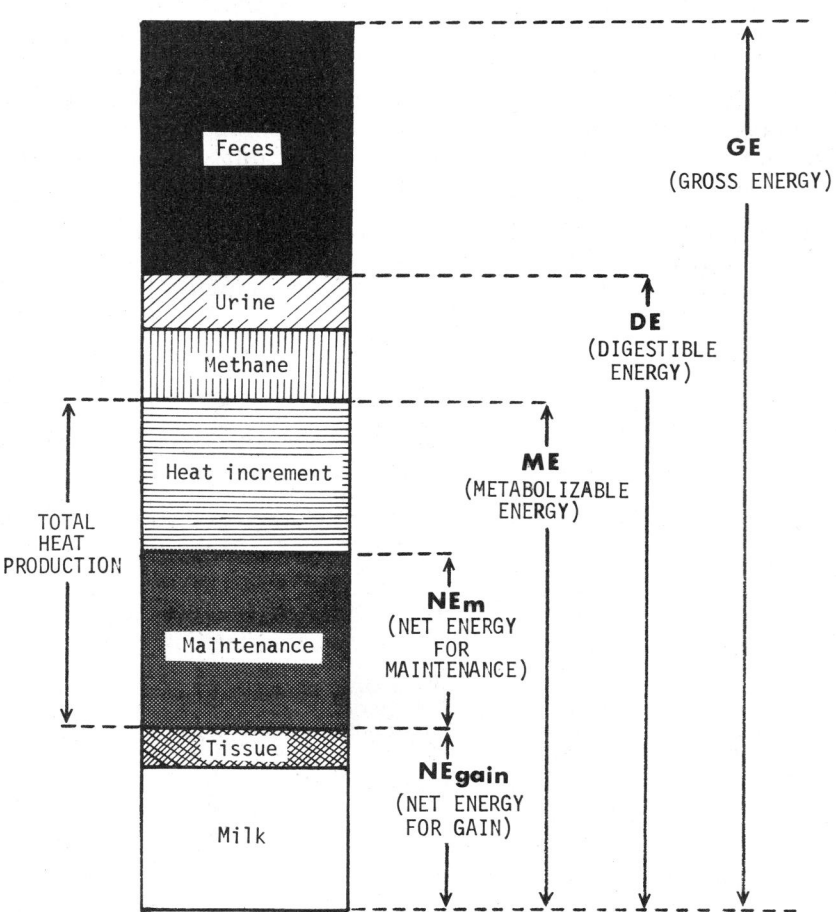

Fig. I-1. The total heat production and energy utilization of a lactating cow, adapted by the author from data of W. P. Flatt.

International standards for vitamin A are based on the utilization of vitamin A and β-carotene by the rat. Since beef cattle do not convert carotene to vitamin A in the same ratio as rats, it is suggested that the values in Table I-3 (when used in connection with Tables 8-11, 8-12, 8-18, and 8-19) be converted as follows:

Converting mg of β-carotene to IU of vitamin A	1 mg = 400 IU
IU of vitamin A activity (%)[4]	24.0
Factor for converting carotene to vitamin A	4.17

5. *Formulating rations*—This publication gives the nutrient requirements of beef cattle and feed composition on a dry matter basis.

Rations can be converted to an as-fed basis by taking the following steps:

a. $\dfrac{\text{Percent of ingredient in the dry ration}}{\text{Percent dry matter of ingredient}} \times 100$

   = Parts of ingredient in the ration as fed.

b. Parts of ingredient in the ration as fed are totaled.

c. $\dfrac{\text{Parts of ingredient}}{\text{Total parts}} \times 100$ = percent of the ingredient in the ration as fed

An example of a simple ration formulated from corn silage, ground ear corn, and a supplement is given in Table I-6. The ration was formulated to contain 1.15 Mcal of NE_{gain} per kg of dry matter.

The calculations for the last two columns in Table I-6 were as follows:

$\dfrac{55}{40} \times 100 = 137.50$

$\dfrac{40}{87} \times 100 = 45.98$

$\dfrac{5}{90} \times 100 = \underline{5.56}$

Total   189.04

$\dfrac{137.50}{189.04} \times 100 = 72.74$ percent of corn silage as fed

$\dfrac{45.98}{189.04} \times 100 = 24.32$ percent of ground ear corn as fed

$\dfrac{5.56}{189.04} \times 100 = 2.94$ percent of supplement as fed

The adequacy of the ration formulated by the procedure illustrated in Table I-6 can be evaluated as follows for a 300-kg finishing steer with a predicted gain of 1.1 kg per day:

Daily feed consumption	kg of dry matter (Table 8-18)	7.60
Daily NE_m required	Mcal (Table 8-20)	5.55
Feed needed for maintenance	kg of dry matter (5.55 ÷ 1.83)	3.03

---

[4]Calculated from carotene.

Feed left for gain	kg of dry matter (7.60−3.03)	4.57
NE$_{gain}$ available for deposition in gain	Mcal (4.57 × 1.15)	5.26
Expected gain	kg/day	1.20

These procedures can be followed in calculating a ration expressed in amounts per day as well as in calculating a ration on a percentage basis.

TABLE I-1

Abbreviations for Terms Used in Table I-3

AAFCO	Association of American Feed Control Officials	lb	pound(s)
		mech	mechanical
Can	Canadian	mech-extd	mechanically extracted, expeller-extracted, hydraulic-extracted, or old process
CE	Canadian Eastern		
CGA	Canada Grain Act	μg	microgram
CFA	Canada Feeds Act	mg	milligram
cp	chemically pure	mm	millimeter
CW	Canadian Western	mn	minimum
dehy	dehydrated	mx	maximum
extd	extracted	NRC	National Research Council
extn	extraction	ppm	parts per million
extn unspec	extraction unspecified	s-c	suncured
g	gram(s)	solv-extd	solvent-extracted
gr	grade	spp	species
grnd	ground	US	United States
ICU	International Chick Unit	USP	United States Pharmacopeia
IU	International Units	w	with
kcal	kilocalories	wo	without
kg	kilogram(s)	wt	weight

TABLE I-2

Stage-of-Maturity Terms Used in Table I-3

Preferred Maturity Term	Definition	Comparable Term
Germinated	Resumption of growth by the embryo in a seed after a period of dormancy.	Sprouted.
Early leaf	Stage at which the plant reaches ⅓ of its growth before blooming.	Fresh new growth, very immature.
Immature	Period between ⅓ and ⅔ of its growth before blooming (this may include fall aftermath).	Prebud stage, young before boot, before heading out.
Prebloom	Stage including the last third of growth before blooming.	Bud, bud stage, budding plants, in bud, preflowering, before bloom, heading to in bloom, boot, heads just showing.
Early bloom	Period between initiation of bloom up to stage at which ¹/₁₀ of the plants are in bloom.	Up to ¹/₁₀ bloom, initial bloom, heading out, in head.
Mid-bloom	Period during which ¹/₁₀ to ⅔ of the plants are in bloom.	Bloom, flowering plants, flowering, half bloom, in bloom.
Full bloom	When ⅔ or more of the plants are in bloom.	¾ to full bloom.
Late bloom	When blossoms begin to dry and fall and seeds begin to form.	Seed developing, 15 days after silking, before milk, early pod.
Milk stage	Seeds well formed, but soft and immature.	Post bloom to early seed, pod stage, early seed, in tassel, fruiting.
Dough stage	Stage at which the seeds are soft and immature.	Seeds dough, seed well developed, nearly mature.
Mature	Stage at which the plant would normally be harvested for seed.	Fruiting plants, fruiting, in seed, well matured, dough to glazing, kernels ripe.
Overripe	Stage after the plant is mature, seeds are ripe and initial weathering has taken place (applies mostly to range plants).	Late seed, ripe, very mature, well matured.
Dormant	Plants cured on the stem, seeds have been cast, and weathering has taken place (applies mostly to range plants).	Seeds cast, mature and weathered.

TABLE I-3

COMPOSITION OF FEEDS COMMONLY USED IN BEEF CATTLE RATIONS[1]

Name	Basis	DM (%)	ME (Mcal/kg)	NEm (Mcal/kg)	NEg (Mcal/kg)	NEmilk	TDN (%)	Protein (%)	Dig. pro. (%)	C.F. (%)	Ca (%)	Co (mg/kg)	Cu (mg/kg)	Fe (%)	Mg (%)	Mn (mg/kg)	P (%)	K (%)	S (%)	Z	Caro. (mg/kg)	Vit. E	Vit. D (IU/g)
ALFALFA, (Medicago sativa)																							
Aerial part, dehy grnd, mn 15% protein	as fed	93.1	2.05	1.22	.64	1.39	57	15.2	11.8	26.4	1.23	.177	10.4	.031	.29	29.0	.22	2.33	—	20.0	101.9	—	—
	dry	100.0	2.20	1.31	.69	1.49	61	16.3	12.7	28.5	1.32	.190	11.2	.033	.31	31.1	.24	2.50	—	21.5	109.5	—	—
Aerial part, dehy grnd, mn 17% protein	as fed	93.0	2.05	1.22	.64	1.41	57	17.8	14.0	24.3	1.33	.363	9.8	.046	.29	29.0	.24	2.49	—	16.0	161.2	—	440.0
	dry	100.0	2.20	1.31	.69	1.52	62	19.2	15.0	26.1	1.43	.390	10.6	.049	.31	31.2	.26	2.68	—	17.2	173.3	—	473.0
Hay, s-c, immature	as fed	89.1	1.84	1.21	.51	1.19	51	19.2	14.0	23.4	1.89	—	—	.018	.23	34.4	.27	2.01	.56	—	446.6	—	—
	dry	100.0	2.06	1.36	.57	1.34	57	21.5	15.0	26.3	2.12	—	—	.020	.26	38.6	.30	2.26	.63	—	501.2	—	—
Hay, s-c, prebloom	as fed	84.5	1.93	1.15	.64	1.33	53	16.4	9.9	24.1	1.06	—	—	.021	.21	29.0	.30	1.99	.53	—	—	—	—
	dry	100.0	2.28	1.36	.76	1.57	63	19.4	11.7	28.5	1.25	—	—	.025	.25	34.3	.36	2.36	.63	—	—	—	—
Hay, s-c, early bloom	as fed	90.0	2.11	1.22	.44	1.21	51	17.7	11.4	26.8	1.12	.080	12.1	.018	.27	28.4	.21	1.87	.27	—	161.6	23.4	1,792.0
	dry	100.0	2.35	1.35	.49	1.34	57	19.7	12.7	29.8	1.25	.090	13.4	.020	.30	31.5	.24	2.08	.30	—	179.5	26.0	1,991.0
Hay, s-c, mid-bloom	as fed	89.2	1.87	1.11	.53	1.23	52	15.6	11.1	27.3	1.33	—	13.7	.012	.27	21.1	.21	1.62	.27	—	23.3	—	—
	dry	100.0	2.10	1.24	.59	1.38	58	17.5	12.5	30.6	1.49	—	15.4	.013	.30	23.7	.24	1.82	.30	—	26.1	—	—
Hay, s-c, full bloom	as fed	87.7	1.81	1.07	.48	1.17	50	13.9	10.0	28.2	1.12	.109	11.8	.018	.31	29.6	.18	.48	—	—	21.5	—	—
	dry	100.0	2.06	1.22	.55	1.34	57	15.9	11.4	32.2	1.28	.124	13.4	.020	.35	33.7	.20	.55	—	—	24.5	—	—
Hay, s-c, mature	as fed	91.2	1.82	1.07	.43	—	50	12.4	8.7	34.2	—	—	—	—	—	—	—	—	—	—	—	—	—
	dry	100.0	2.00	1.17	.47	—	55	13.6	9.5	37.5	—	—	—	—	—	—	—	—	—	—	—	—	—
Aerial part, fresh	as fed	27.2	.60	.36	.16	—	17	5.2	4.1	7.4	.47	.020	2.7	.008	.07	13.7	.08	.55	.11	4.8	54.1	41.4	.05
	dry	100.0	2.21	1.32	.71	—	61	19.3	15.0	27.4	1.72	.090	9.9	.030	.27	50.5	.31	2.03	.39	17.6	198.9	152.1	.2
Aerial part, ensiled	as fed	30.4	.61	.36	.16	.40	16	5.4	3.6	9.2	.49	.050	2.9	.009	.10	15.3	.12	.73	.11	—	27.3	—	87.8
	dry	100.0	2.02	1.20	.53	1.30	56	17.8	11.9	30.4	1.61	.150	9.7	.030	.34	50.3	.38	2.40	.36	—	89.7	—	288.8
Aerial part, ensiled, early bloom, mn 50% dry matter	as fed	55.0	1.03	.60	.19	—	29	9.8	5.9	17.8	.89	—	—	—	—	—	.21	—	—	—	—	—	—
	dry	100.0	1.88	1.10	.35	—	52	17.9	10.7	32.4	1.61	—	—	—	—	—	.38	—	—	—	—	—	—
Aerial part, ensiled, early bloom, mn 30% mx 50% dry matter	as fed	38.5	.74	.44	.16	—	20	6.8	4.1	11.1	.58	—	—	—	—	—	.12	—	—	—	—	—	—
	dry	100.0	1.92	1.13	.41	—	53	17.6	10.6	28.6	1.52	—	—	—	—	—	.30	—	—	—	—	—	—
Aerial part, ensiled, early bloom, mx 30% dry matter	as fed	28.3	.54	.32	.11	—	15	5.2	3.2	8.2	.40	—	—	—	.10	—	.09	.67	—	—	18.7	—	.2
	dry	100.0	1.92	1.12	.39	—	53	18.6	11.2	28.9	1.40	—	—	—	.36	—	.32	2.36	—	—	51.6	—	.6
Aerial part, wilted ensiled	as fed	36.2	.77	.47	.25	.50	21	6.4	4.3	11.2	.50	—	3.4	.011	.12	18.5	.12	.85	—	—	—	—	—
	dry	100.0	2.12	1.31	.69	1.38	58	17.8	11.9	30.9	1.40	—	9.3	.030	.33	52.0	.32	2.36	—	—	—	—	—
Aerial part w molasses added, ensiled	as fed	32.2	.70	.41	.20	.53	19	5.6	3.9	9.3	.56	—	4.1	.010	.11	13.7	.10	.82	—	—	31.3	—	—
	dry	100.0	2.18	1.27	.63	1.41	60	17.5	12.2	28.8	1.74	—	12.6	.030	.34	42.6	.31	2.56	—	—	97.2	—	—
ALFALFA-BROME, SMOOTH (Medicago sativa-Bromus inermis)																							
Aerial part, fresh	as fed	21.6	.49	.29	.16	.34	14	4.2	3.2	5.5	.33	—	—	—	.08	—	.08	.84	—	—	—	—	—
	dry	100.0	2.28	1.35	.75	1.57	63	19.6	14.6	25.3	1.52	—	—	—	.35	—	.37	3.87	—	—	—	—	—
Aerial part, fresh, early bloom	as fed	21.6	.48	.28	.15	—	13	4.2	3.1	5.5	.33	—	—	—	.08	—	.08	.84	—	—	—	—	—
	dry	100.0	2.24	1.32	.71	—	62	19.6	14.4	25.3	1.52	—	—	—	.35	—	.37	3.87	—	—	—	—	—
Aerial part, ensiled, mn 50% dry matter	as fed	55.0	1.07	.63	.24	—	30	4.6	2.6	18.5	—	—	—	—	—	—	—	—	—	—	—	—	—
	dry	100.0	1.95	1.15	.43	—	54	8.4	4.8	33.7	—	—	—	—	—	—	—	—	—	—	—	—	—
Aerial part, ensiled, mn 30% mx 50% dry matter	as fed	46.5	.89	.52	.19	—	25	7.2	4.2	15.3	—	—	—	—	—	—	—	—	—	—	—	—	—
	dry	100.0	1.92	1.13	.40	—	53	15.5	9.0	33.0	—	—	—	—	—	—	—	—	—	—	—	—	—
Aerial part, ensiled, mx 30% dry matter	as fed	25.0	.50	.30	.12	—	14	3.8	2.2	7.7	—	—	—	—	—	—	—	—	—	—	—	—	—
	dry	100.0	1.99	1.18	.49	—	55	15.2	8.8	30.8	—	—	—	—	—	—	—	—	—	—	—	—	—

Feed	Basis																				
ALFALFA-ORCHARDGRASS, (Medicago sativa-Dactylis glomerata)																					
Aerial part, ensiled, mn 50% dry matter	as fed	61.0	1.19	.70	.27	—	33	9.9	5.7	18.6	—	—	—	—	—	—	—	—	—	—	—
	dry	100.0	1.95	1.15	.44	—	54	16.2	9.4	30.5	—	—	—	—	—	—	—	—	—	—	—
Aerial part, ensiled, mn 30% mx 50% dry matter	as fed	40.0	.78	.46	.18	—	22	6.9	4.0	12.6	—	—	—	—	—	—	—	—	—	—	—
	dry	100.0	1.95	1.15	.44	—	54	17.2	10.0	31.6	—	—	—	—	—	—	—	—	—	—	—
Aerial part, ensiled mx 30% dry matter	as fed	28.0	.55	.32	.13	—	15	4.8	2.8	8.8	—	—	—	—	—	—	—	—	—	—	—
	dry	100.0	1.95	1.16	.45	—	54	17.1	9.9	31.4	—	—	—	—	—	—	—	—	—	—	—
ALKALI SACATON, (Sporobolus airoides)																					
Aerial part, fresh, dormant	as fed	86.0	1.42	.85	.08	—	—	2.9	—	—	.58	—	—	—	—	.07	—	—	—	.3	—
	dry	100.0	1.65	.99	.09	—	—	3.4	—	—	.67	—	—	—	—	.08	—	—	—	.4	—
ANIMAL																					
Carcass residue, dry rendered dehy grnd, mn 9% indig material mx 4.4% phosphorus	as fed	93.5	2.22	1.62	1.07	1.91	71	53.4	48.6	2.3	7.94	.128	9.7	.044	.27	9.5	4.03	.55	.50	—	—
	dry	100.0	2.38	1.73	1.14	2.04	76	57.1	52.0	2.5	8.49	.137	10.4	.047	.29	10.2	4.31	.59	.53	—	—
Meat meal																					
Meat scrap																					
Carcass residue w bone, dry rendered dehy grnd, mn 9% indig material mn 4% phosphorus	as fed	94.0	2.44	1.51	.97	1.79	60.9	49.7	45.2	1.8	11.47	.183	1.5	.050	1.13	12.3	5.72	1.46	—	—	—
	dry	100.0	2.60	1.61	1.03	1.90	64.8	52.9	48.1	1.9	12.20	.195	1.6	.053	1.20	13.1	6.08	1.56	—	—	—
Meat and bone meal																					
Bone, cooked dehy grnd, mn 10% phosphorus	as fed	94.5	—	—	—	—	—	17.7	—	—	25.82	—	16.3	.084	.64	30.4	12.35	—	—	424.7	—
	dry	100.0	—	—	—	—	—	18.8	—	—	27.32	—	17.2	.088	.67	32.0	13.07	—	—	447.1	—
Feeding bone meal																					
Bone, steamed dehy grnd	as fed	95.0	—	—	—	—	15	12.1	8.2	2.0	28.98	.095	—	.005	—	—	13.60	.14	—	—	—
	dry	100.0	—	—	—	—	16	12.7	8.6	2.1	30.51	.100	—	—	—	—	14.31	.16	—	—	—
Bone meal steamed																					
Bone charcoal, retort-charred grnd	as fed	90.0	—	—	—	—	—	8.5	—	—	27.10	—	—	—	.53	—	12.73	—	—	—	—
	dry	100.0	—	—	—	—	—	9.4	—	10.1	30.11	—	—	—	.59	—	14.14	—	—	—	—
ANIMAL-POULTRY																					
Fat, heat rendered, mn 90% fatty acids mx 2.5% unsaponifiable matter mx 1% insoluble matter	as fed	99.5	4.64	4.55	2.61	—	—	—	—	—	—	—	—	—	—	—	—	—	—	—	—
	dry	100.0	4.66	4.57	2.62	—	—	—	—	—	—	—	—	—	—	—	—	—	—	—	—
Animal fat																					
BARLEY, (Hordeum vulgare)																					
Hay, s-c	as fed	87.3	1.76	1.07	.50	1.17	50	7.8	3.9	23.0	.25	—	—	—	—	34.2	.22	1.30	—	—	—
	dry	100.0	2.02	1.23	.57	1.34	57	8.9	4.4	26.4	.29	—	.17	—	—	39.2	.25	1.49	—	—	—
Straw	as fed	88.2	1.30	.89	.12	.66	36	3.6	.4	37.4	.30	—	.17	.026	—	15.2	.08	2.01	—	—	—
	dry	100.0	1.48	1.01	.14	.75	41	4.1	.5	42.4	.34	—	.19	.030	—	17.2	.09	2.28	—	—	—
Grain	as fed	89.0	2.67	1.90	1.25	2.05	72	11.6	8.5	5.3	.07	.089	5.8	.005	.12	7.9	.40	.49	—	15.3	9.8
	dry	100.0	3.00	2.13	1.40	2.30	81	13.0	9.6	6.0	.08	.100	6.5	.006	.14	8.9	.45	.55	—	17.2	11.0
Grain, Pacific coast	as fed	89.0	2.50	1.72	1.15	2.02	73	8.7	4.4	6.5	.06	—	—	—	—	—	.45	—	—	—	—
	dry	100.0	2.81	1.93	1.29	2.27	82	9.8	5.0	7.3	.07	—	—	—	—	—	.45	—	—	—	—
Grain, screenings	as fed	89.0	2.57	1.65	1.10	1.96	71	12.0	9.0	8.0	—	—	—	—	—	—	.40	—	—	—	—
	dry	100.0	2.89	1.85	1.23	2.20	80	13.5	10.1	9.0	—	—	—	—	—	—	.45	—	—	—	—
BEAN, (Phaseolus spp.)																					
Navy, seeds	as fed	90.0	2.70	1.76	1.18	2.07	75	22.9	20.2	4.2	.15	—	—	—	—	—	.57	1.70	—	—	—
	dry	100.0	3.00	1.96	1.31	2.30	83	25.4	22.4	4.7	.17	—	—	—	—	—	.63	1.89	—	—	—
BEEF - see Cattle																					

(Continued)

TABLE I-3 (Continued)

Name	Basis	DM (%)	ME (Mcal/kg)	NEm (Mcal/kg)	NEg (Mcal/kg)	NEmilk (Mcal/kg)	TDN (%)	Protein (%)	Dig. pro. (%)	C.F. (%)	Ca (%)	Co (mg/kg)	Cu (mg/kg)	Fe (%)	Mg (%)	Mn (mg/kg)	P (%)	K (%)	S (%)	Z (%)	Caro. (mg/kg)	Vit. E (mg/kg)	Vit. D (IU/g)
BEET, MANGELS, (Beta spp.) Roots	as fed	10.6	.30	.19	.13	.22	8	1.4	1.0	.8	.02	—	—	.002	.02	—	.02	.24	.02	—	—	—	—
	dry	100.0	2.86	1.80	1.20	2.12	78	13.2	9.8	7.5	.22	—	—	.022	.22	—	.22	2.28	.22	—	—	—	—
BEET, SUGAR, (Beta saccharifera) Aerial part w crowns, ensiled	as fed	20.7	.40	.24	.09	—	11	2.6	2.1	2.6	.48	—	—	—	—	—	.04	—	—	—	—	—	—
	dry	100.0	1.95	1.14	.42	—	54	12.7	10.0	12.6	2.32	—	—	—	—	—	.20	—	—	—	—	—	—
Molasses, mn 48% invert sugar min 79.5 degrees brix / Beet molasses Molasses	as fed	77.0	2.19	1.57	1.05	1.94	61	6.5	3.8	—	.12	.385	17.6	.008	.23	4.6	.03	4.77	.47	—	—	—	—
	dry	100.0	2.84	2.04	1.36	2.52	79	8.5	4.9	—	.16	.500	22.9	.010	.30	6.0	.03	6.20	.61	—	—	—	—
Pulp, dehy / Dried beet pulp	as fed	96.0	2.37	1.46	.94	1.73	66	9.1	3.9	18.3	.68	.102	12.5	.030	.27	35.0	.09	.19	—	.7	—	—	551.6
	dry	100.0	2.60	1.60	1.03	1.90	72	10.0	4.3	20.1	.75	.112	13.7	.033	.30	38.5	.10	.21	—	.8	—	—	606.2
Pulp, wet	as fed	10.0	.25	.15	.10	.18	7	.9	.4	2.0	.09	—	—	—	—	—	.01	.02	—	—	—	—	—
	dry	100.0	2.46	1.52	.95	1.75	68	9.0	4.0	20.0	.90	—	—	—	—	—	.10	.20	—	—	—	—	—
Pulp w molasses, dehy	as fed	92.0	2.47	1.87	1.23	1.81	68	9.1	6.0	13.0	.56	—	—	—	—	—	.07	1.64	—	—	—	—	—
	dry	100.0	2.68	2.03	1.34	1.97	74	9.9	6.5	14.1	.61	—	—	—	.14	—	.08	1.78	—	—	—	—	—
BERMUDAGRSSS, (Cynodon dactylon) Hay, s-c	as fed	91.1	1.42	.97	.23	—	39	8.1	4.4	27.0	.42	—	—	.026	.16	—	.18	1.34	—	—	117.2	—	—
	dry	100.0	1.56	1.06	.43	—	43	8.9	4.8	29.6	.46	—	—	.029	.17	—	.20	1.47	—	—	128.7	—	—
BERMUDAGRASS, COASTAL, (Cynodon dactylon) Hay, s-c	as fed	91.5	1.46	.99	.24	.80	40	8.7	4.7	29.7	.31	—	—	—	.15	—	.14	—	—	—	74.4	—	—
	dry	100.0	1.59	1.08	.26	.86	44	9.5	5.1	32.5	.34	—	—	—	.16	—	.16	—	—	—	81.7	—	—
BIRDSFOOT Trefoil—see Trefoil, birdsfoot																							
BLOOD MEAL—see Animal, blood, dehy grnd																							
BLUEGRASS, CANADA, (Poa compressa) Hay, s-c, immature	as fed	96.7	2.48	1.53	.97	1.80	69	16.7	11.7	24.9	.38	—	—	—	.16	76.6	.38	1.97	—	—	—	—	—
	dry	100.0	2.57	1.58	1.00	1.86	71	17.3	12.1	25.8	.39	—	—	—	.16	79.2	.39	2.04	—	—	—	—	—
Hay, s-c	as fed	93.4	2.20	1.32	.77	1.53	61	10.8	6.5	27.0	.28	—	—	—	.31	86.5	.27	1.48	—	—	—	—	—
	dry	100.0	2.35	1.41	.82	1.64	65	11.6	7.0	28.9	.30	—	—	—	.33	92.6	.29	1.59	—	—	—	—	—
Aerial part, fresh, immature	as fed	25.9	.67	.41	.26	.48	18	4.8	3.6	6.6	—	—	—	—	—	—	—	—	—	—	—	—	—
	dry	100.0	2.57	1.58	1.00	1.86	71	18.7	13.8	25.5	—	—	—	—	—	—	—	—	—	—	—	—	—
Aerial part, fresh	as fed	30.6	.77	.47	.30	.56	21	5.2	3.8	8.1	—	—	—	—	—	—	—	—	—	—	—	—	—
	dry	100.0	2.53	1.54	.97	1.83	70	17.0	12.3	26.4	—	—	—	—	—	—	—	—	—	—	—	—	—
BLUEGRASS, KENTUCKY, (Poa pratensis) Aerial part, fresh, immature	as fed	30.5	.79	.48	.31	—	22	5.3	3.8	7.6	.17	—	4.4	.009	.06	24.5	.14	.70	—	—	116.8	—	—
	dry	100.0	2.60	1.59	1.02	—	72	17.3	12.6	25.1	.56	—	14.1	.030	.18	80.3	.47	2.28	—	—	383.0	—	—
Aerial part, fresh, early bloom	as fed	35.7	.89	.54	.34	.64	25	5.9	4.3	9.8	.16	—	—	—	.04	—	.14	.72	—	—	—	55.7	—
	dry	100.0	2.49	1.51	.94	1.78	69	16.6	12.0	27.4	.46	—	—	—	.11	—	.39	2.01	—	—	—	156.1	—
BLUESTEM, (Andropogon spp.) Aerial part, fresh, immature	as fed	31.6	.73	.44	.25	—	20	3.5	2.3	9.1	.20	—	11.6	.022	—	26.3	.05	.43	—	—	69.3	—	—
	dry	100.0	2.31	1.38	.79	—	64	11.0	7.2	28.9	.63	—	36.8	.070	—	83.3	.17	1.35	—	—	219.2	—	—
Aerial part, fresh, mature	as fed	71.3	1.60	.95	.52	—	44	3.2	1.2	24.2	.28	—	11.5	.043	.04	26.2	.08	.36	—	—	—	—	—
	dry	100.0	2.24	1.33	.73	—	62	4.5	1.7	34.0	.40	—	16.1	.060	.06	36.8	.11	.51	—	—	—	—	—

BONE MEAL—see Animal
BONE BLACK—see Animal, bone charcoal
BONE CHAR—see Animal, bone charcoal
BRAN—see Wheat
BREWERS' DRIED GRAINS—see Grains
BREWERS' DRIED YEAST—see Yeast, brewers
CANE MOLASSES — See Sugarcane

Feed	Basis	(DM)													
BROME, (*Bromus spp.*) Hay, s-c	as fed	89.7	1.43	.97	.23	.77	39	10.6	4.5	28.7	—	—	—	—	—
	dry	100.0	1.59	1.08	.26	.86	44	11.8	5.0	32.0	.19	—	—	—	—
Aerial part, fresh, immature	as fed	32.5	.80	.49	.31	.57	22	6.6	4.9	7.8	.19	.12	.06	.015	10.9
	dry	100.0	2.46	1.52	.95	1.75	68	20.3	15.1	23.9	.59	.37	.18	.020	11.9
Aerial part, fresh, mature	as fed	56.1	1.32	.79	.46	—	36	3.6	2.3	18.5	.17	.15	—	—	—
	dry	100.0	2.35	1.41	.82	—	65	6.4	3.3	33.0	.30	.26	—	—	—
BUFFALOGRASS, (*Buchloe dactyloides*) Aerial part, fresh	as fed	47.7	1.01	.60	.30	—	28	4.5	2.9	13.2	.27	.08	.07	—	—
	dry	100.0	2.12	1.26	.63	—	59	9.5	6.0	27.7	.57	.16	.14	—	—
CACTUS, PRICKLYPEAR, (*Opuntia spp.*) Aerial part, fresh	as fed	17.1	.36	.22	.11	—	10	.8	.5	—	2.27	.01	.28	.015	—
	dry	100.0	2.13	1.28	.65	—	59	5.0	2.8	13.3	6.29	.08	1.65	.090	—
CALCIUM PHOSPHATE Dibasic, commerical / Dicalcium phosphate	as fed	96.0	—	—	—	—	—	—	—	—	22.20	17.90	—	—	—
	dry	100.0	—	—	—	—	—	—	—	—	23.13	18.65	—	—	—
CANARYGRASS, REED, (*Phalaris arundinacea*) Hay, s-c	as fed	91.3	1.84	1.09	.48	—	51	8.0	4.2	31.3	.31	.23	.24	.018	10.9
	dry	100.0	2.02	1.19	.52	—	56	8.8	4.6	34.3	.34	.25	.26	.020	11.9
Aerial part, fresh	as fed	25.8	.62	.37	.08	—	—	17.0	2.3	6.9	.10	.08	.08	—	—
	dry	100.0	2.39	1.45	.87	—	66	13.2	9.1	26.8	.40	.30	—	—	—
CARROT, (*Daucus spp.*) Roots, fresh	as fed	11.9	.35	.16	—	—	10	1.1	.6	1.1	—	.04	.02	.002	.0
	dry	100.0	2.96	1.37	.24	—	82	10.1	5.0	9.2	.42	.34	.17	.017	10.9
CATTLE, (*Bos spp.*) Milk, dehy, feed gr mx 8% moisture mn 26% fat / Dried whole milk, feed grade	as fed	93.7	4.40	4.30	1.88	.48	16	25.2	—	3.1	—	.68	—	—	—
	dry	100.0	4.70	4.59	2.01	4.01	130	26.9	—	25.8	—	.72	—	.017	.9
Milk, fresh	as fed	12.0	.56	.55	.24	.48	9	3.1	2.6	2.7	.12	.10	—	.002	.1
	dry	100.0	4.70	4.59	2.01	4.01	93	25.8	27.4	28.5	1.26	1.03	.11	.017	11.5
Milk, skimmed centrifugal	as fed	9.6	.32	.14	—	.26	—	2.7	—	33.5	1.26	1.03	—	.017	9
	dry	100.0	3.36	1.50	2.32	2.67	—	28.5	.2	35.6	1.34	1.10	.12	.005	12.2
Milk skimmed dehy, mx 8% moisture / Dried skimmed milk, feed grade	as fed	94.0	2.92	1.29	—	—	—	33.5	.2	—	1.26	1.67	.11	—	11.5
	dry	100.0	3.11	1.37	—	—	—	35.6	.2	—	1.34	1.78	.12	—	12.2
CITRUS, (*Citrus spp.*) Pulp, ensiled	as fed	19.5	.58	.42	.28	.49	17	1.4	.4	3.1	.40	.03	.03	.003	—
	dry	100.0	2.99	2.15	1.42	2.49	88	7.1	1.8	15.9	2.04	.15	.16	.016	5.7
Pulp wo tmes, shredded dehy	as fed	90.0	2.50	1.77	1.88	1.88	69	6.4	2.2	13.0	1.96	.12	.16	.016	6.3
	dry	100.0	2.78	1.97	1.32	2.09	77	7.1	2.5	14.4	2.18	.13	.18	.018	—

(Continued)

TABLE I-3 (Continued)

Name	Basis	DM (%)	ME (Mcal/kg)	NEm (Mcal/kg)	NEg (Mcal/kg)	NEmilk (Mcal/kg)	TDN (%)	Protein (%)	Dig. pro. (%)	C.F. (%)	Ca (%)	Co (mg/kg)	Cu (mg/kg)	Fe (%)	Mg (%)	Mn (mg/kg)	P (%)	K (%)	S (%)	Z (mg/kg)	Caro. (mg/kg)	Vit. E	Vit. D (IU/g)
Syrup, mn 45% invert sugar mn 71 degrees brix	as fed	65.0	1.81	1.28	.86	1.36	50	7.1	3.6	—	—	—	—	—	—	—	.16	.09	—	—	—	—	—
	dry	100.0	2.78	1.97	1.32	2.09	77	10.9	5.6	—	—	—	—	—	—	—	.25	.14	—	—	—	—	—
Citrus molasses	as fed	87.9	1.91	1.13	.58	1.28	53	12.9	8.2	—	1.31	—	72.8	.033	.14	26.0	.23	1.49	.18	88.8	164.4	—	—
	dry	100.0	2.17	1.29	.66	1.46	60	14.7	9.3	—	2.01	—	112.0	.050	.22	40.0	.26	1.70	.21	136.7	187.0	—	—
CLOVER, ALSIKE, (Trifolium hybridum) Hay, s-c	as fed	87.4	1.94	1.13	.58	1.28	53	14.8	10.3	26.4	1.15	—	5.3	.040	.28	102.8	.16	1.35	.24	15.5	—	—	—
	dry	100.0	2.22	1.29	.66	1.46	61	16.9	11.8	30.1	1.31	—	6.0	.045	.32	117.0	.18	1.54	.28	17.0	—	—	—
CLOVER, CRIMSON (Trifolium incarnatum) Hay, s-c	as fed	91.2	2.02	1.20	.63	1.36	56	18.8	11.8	24.2	1.24	—	—	.061	.24	149.7	.29	2.21	.20	—	—	—	—
	dry	100.0	2.21	1.31	.66	1.49	61	20.7	13.0	27.7	1.42	—	—	.070	.27	171.3	.32	2.42	.22	—	—	—	—
CLOVER, LADINO, (Trifolium repens) Hay, s-c	as fed	87.7	1.87	1.10	.54	1.24	52	13.1	7.8	22.0	1.56	.137	8.0	.055	.46	120.8	.19	1.54	.15	—	147.0	—	1,677.3
	dry	100.0	2.13	1.26	.62	1.41	59	14.9	8.9	24.1	1.71	.150	8.8	.060	.50	132.5	.22	1.76	.17	—	161.2	—	1,912.5
CLOVER, RED, (Trifolium pratense) Aerial part, fresh, early bloom	as fed	19.6	.50	.31	.19	.36	14	4.1	3.0	3.7	1.41	.132	9.8	.009	.40	57.6	.07	.49	—	15.1	32.3	—	—
	dry	100.0	2.53	1.56	.99	1.83	70	21.1	15.3	19.0	1.61	.150	11.2	.010	.45	65.7	.38	2.49	—	17.2	36.8	—	—
Aerial part, fresh, full bloom	as fed	27.7	.64	.38	.22	.44	18	4.1	2.7	8.2	—	—	—	—	.14	—	.08	.54	—	—	—	—	—
	dry	100.0	2.32	1.39	.80	1.60	64	14.9	9.7	29.6	—	—	—	—	.51	—	.27	1.96	—	—	—	—	—
Aerial part, fresh, cut	as fed	27.1	.63	.38	.22	.43	17	4.7	3.0	6.8	—	—	—	—	—	—	.12	.70	—	—	—	—	—
	dry	100.0	2.31	1.39	.80	1.60	64	17.3	11.2	25.1	—	—	—	—	—	—	.46	2.59	—	—	—	—	—
COCONUT, (Cocos nucifera) Meats, mech-extd grnd	as fed	93.0	2.74	1.75	1.16	2.07	75	20.7	16.8	11.3	.21	—	14.1	.132	.30	65.6	.62	1.54	.34	—	—	—	—
	dry	100.0	2.95	1.88	1.25	2.23	81	22.3	18.1	12.2	.23	—	15.2	.142	.33	70.6	.67	1.65	.37	—	—	—	—
Coconut meal, mechanical extracted; Copra meal, mechanical extracted; Meats, solv-extd grnd	as fed	92.0	2.51	1.53	.99	1.81	68	21.5	15.3	14.2	.18	—	—	—	—	—	.62	—	—	—	—	—	—
	dry	100.0	2.73	1.66	1.08	1.97	74	23.4	16.6	15.4	.19	—	—	—	—	—	.67	—	—	—	—	—	—
CORN, (Zea mays) Aerial part, s-c	as fed	82.4	1.94	1.19	.50	1.35	54	7.1	3.3	22.0	.25	.118	4.0	.088	.15	56.2	.15	.82	.12	—	3.6	—	—
	dry	100.0	2.36	1.44	.61	1.64	65	8.6	4.0	26.7	.30	.130	4.8	.010	.18	68.2	.18	.99	.14	—	4.4	—	—
Aerial part wo ears w husks, s-c mature	as fed	87.2	1.86	1.06	.48	—	51	5.1	1.9	32.4	.43	—	—	—	—	—	.08	—	—	—	—	—	—
	dry	100.0	2.13	1.21	.55	—	59	5.9	2.2	37.1	.49	—	—	—	—	—	.09	—	—	—	—	—	—
Cobs, grnd	as fed	90.4	1.54	.96	.23	—	42	2.5	—	32.4	.11	—	6.6	.021	.06	5.6	.04	.76	.42	—	.6	—	—
	dry	100.0	1.70	1.06	.25	—	47	2.8	—	35.8	.12	—	7.3	.023	.07	6.2	.04	.84	.47	—	.7	—	—
Aerial part, ensiled, mature, well-eared mn 50% dry matter	as fed	55.0	1.41	.86	.55	—	39	4.3	2.5	12.6	.15	—	—	—	—	—	.10	—	—	—	—	—	—
	dry	100.0	2.57	1.57	1.00	—	71	7.8	4.5	23.0	.27	—	—	—	—	—	.19	—	—	—	—	—	—
Aerial part, ensiled mature, well-eared mx 50% mn 30% dry matter	as fed	40.0	1.01	.62	.40	—	28	3.2	1.9	9.8	.11	—	—	—	—	—	.08	.42	—	—	—	—	—
	dry	100.0	2.53	1.56	.99	—	70	8.1	4.7	24.4	.27	—	—	—	—	—	.20	1.05	—	—	—	—	—

CLOVER, SWEET—see Sweetclover

Ground corncob

Feed	Basis	1	2	3	4	5	6	7	8	9	10	11	12	13	14	15	16	17	18	19	20	21	22
Aerial part, ensiled, mature, well-eared mx 30% dry matter	as fed	27.9	.71	.44	.28	—	20	2.3	1.4	7.3	.08	—	—	—	.05	—	.06	.26	—	—	—	—	33.2
	dry	100.0	2.53	1.56	.99	—	70	8.4	4.9	26.3	.28	.296	6.7	.007	.18	11.3	.21	.95	—	—	—	—	119.0
Aerial part wo ears wo husks, ensiled	as fed	27.2	.57	.34	.16	—	16	2.0	.8	8.7	.10	—	—	—	.08	—	.05	.39	—	—	—	—	—
	dry	100.0	2.10	1.24	.59	—	58	7.2	2.9	32.1	.38	.340	7.7	.008	.31	13.0	.19	1.43	—	—	—	—	—
Corn stover silage	as fed	43.4	1.17	.69	.45	.82	32	4.2	2.2	4.2	.03	—	—	—	.13	—	.12	—	—	—	—	—	—
	dry	100.0	2.70	1.60	1.03	1.90	75	9.8	5.1	9.7	.06	—	—	—	.15	—	.27	—	—	—	—	—	—
Ears w husks, ensiled	as fed	87.0	2.68	1.94	1.21	2.23	74	8.1	4.0	7.2	.04	—	—	—	—	—	.23	.44	—	—	—	—	—
	dry	100.0	3.08	2.23	1.39	2.56	85	9.3	4.6	8.3	.04	—	—	—	—	—	.26	.50	—	—	—	—	—
Ears, grnd																							
Corn and cob meal																								
Ear corn chop																								
Ground ear corn																								
CORN GRAIN—see Corn dent yellow																								
Grits by-product, mn 5% fat	as fed	90.6	3.12	2.22	1.40	—	86	10.7	7.2	5.0	.05	.060	14.6	.006	.24	14.6	.52	.67	.03	—	9.2	—	—
	dry	100.0	3.44	2.45	1.55	—	95	11.8	7.9	5.5	.06	.066	16.1	.007	.26	16.1	.58	.74	.03	—	10.1	—	—
Hominy feed	as fed	92.0	2.81	1.83	1.22	2.15	77	27.3	21.2	12.3	.09	.081	44.3	.022	.06	22.1	.40	.17	.42	—	—	—	—
	dry	100.0	3.05	1.99	1.33	2.34	84	29.7	23.1	13.4	.10	.088	48.2	.024	.07	24.0	.43	.18	.46	—	—	—	—
Distillers' grains, dehy	as fed	92.0	2.89	1.99	1.31	2.29	80	26.9	21.5	9.0	.19	.110	61.4	.031	.25	28.3	.78	.64	.29	—	—	—	—
	dry	100.0	3.14	2.16	1.42	2.49	87	29.2	23.4	9.8	.21	.120	66.8	.034	.27	30.8	.85	.70	.32	—	—	—	—
Corn distillers' dried grains																								
Distillers' grains w solubles, dehy, mn 75% original solids	as fed	93.0	2.89	2.00	1.32	2.32	80	29.3	21.0	3.5	.34	.196	83.1	.055	.64	73.8	1.38	1.75	.37	—	3.7	—	—
	dry	100.0	3.11	2.15	1.42	2.49	86	31.5	22.6	3.8	.37	.211	89.4	.059	.69	79.4	1.48	1.88	.40	—	4.0	—	—
Corn distillers' dried grains with solubles																								
Distillers' solubles, dehy	as fed	91.0	2.73	1.81	1.21	2.13	76	39.0	35.7	4.6	.15	.071	28.3	.040	.05	10.6	.46	.57	.22	—	.7	—	—
	dry	100.0	3.00	1.99	1.33	2.34	83	42.9	39.2	5.1	.16	.078	31.1	.044	.05	11.6	.51	.63	.24	—	.8	—	—
Corn distillers' dried solubles																								
Gluten, wet milled dehy	as fed	90.0	2.66	1.74	1.16	2.04	74	25.7	22.1	7.3	.44	.088	47.6	.046	.29	23.8	.77	.03	—	—	—	—	507.0
	dry	100.0	2.96	1.93	1.29	2.27	82	28.6	24.6	8.1	.49	.098	52.9	.051	.32	26.4	.86	.03	—	—	—	—	551.1
Corn glutenmeal																								
Gluten w bran, wet milled dehy																								
Corn gluten feed																								
CORN, DENT YELLOW (Zea mays indentata)																								
Grain, gr 2 US mn wt 54 lb per bushel	as fed	89.0	2.93	2.03	1.32	2.31	81	8.9	6.7	2.0	.02	—	—	—	—	—	.31	—	—	—	1.8	22.0	—
	dry	100.0	3.29	2.28	1.48	2.60	91	10.0	7.5	2.2	.02	—	—	—	—	—	.35	—	—	—	2.0	24.7	—
Grain, gr 3 US mn wt 52 lb per bushel	as fed	86.0	2.83	1.96	1.27	2.24	78	8.7	6.5	2.0	.02	—	—	.002	—	5.5	.25	—	—	—	—	—	—
	dry	100.0	3.29	2.28	1.48	2.60	91	10.1	7.6	2.3	.02	—	—	.002	—	6.4	.29	—	—	—	—	—	—
CORN, SWEET, (Zea mays saccharata)																								
Cannery residue, fresh	as fed	77.0	1.95	1.20	.76	1.07	54	6.8	3.8	17.3	—	—	—	—	—	—	.69	—	—	—	10.4	—	—
	dry	100.0	2.53	1.56	.99	1.39	70	8.9	5.0	22.5	—	—	—	—	—	—	.90	—	—	—	13.5	—	—
Corn, sweet, cannery refuse																								
Cannery residue, ensiled	as fed	29.4	.76	.48	.31	—	21	2.6	1.4	7.9	—	—	—	—	—	—	—	—	—	—	—	—	—
	dry	100.0	2.60	1.62	1.04	—	72	8.8	4.9	26.8	—	—	—	—	—	—	—	—	—	—	—	—	—
CORN, WHITE, (Zea mays)																								
Grits by-product, mn 5% fat	as fed	89.9	—	—	—	—	11	—	4.7	.05	.05	—	—	—	—	—	.99	—	—	—	—	—	—
	dry	100.0	—	—	—	—	12	—	5.2	.06	.06	—	—	—	—	—	1.10	—	—	—	—	—	—
White hominy feed																								
CORN-SOYBEAN, (Zea mays, Glycine max)																								
Aerial part, ensiled	as fed	26.1	.68	.42	.27	—	19	2.5	1.6	6.9	.22	—	—	—	—	—	.09	.29	—	—	—	—	—
	dry	100.0	2.60	1.60	1.03	—	72	19.7	6.0	26.3	.83	—	—	—	—	—	.36	1.11	—	—	—	—	—

(Continued)

TABLE I-3 (Continued)

Name	Basis	DM	ME	NEm	NEg	NEmilk	TDN	Protein	Dig. pro.	C.F.	Ca	Co	Cu	Fe	Mg	Mn	P	K	S	Z	Caro.	Vit. E	Vit. D
		(%)	(Mcal/kg)				(%)	(%)	(%)	(%)	(%)	(mg/kg)	(mg/kg)	(%)	(%)	(mg/kg)	(%)	(%)	(%)		(mg/kg)	(mg/kg)	(IU/g)
COTTON, (Gossypium spp.)																							
Bolls, s-c	as fed	91.0	1.56	.96	.18	.71	43	10.3	6.1	28.9	.61	—	—	—	.26	—	.10	2.32	—	—	—	—	—
	dry	100.0	1.71	1.06	.20	.78	47	11.3	6.7	31.8	.67	—	—	—	.28	—	.11	2.55	—	—	—	—	—
Seed hulls	as fed	90.3	1.69	.93	.17	.68	37	4.0	.2	40.6	.13	.020	14.2	.010	.13	106.3	.06	.76	—	19.9	—	—	—
Cottonseed hulls	dry	100.0	1.87	1.03	.19	.75	41	4.4	.2	45.0	.14	.020	15.7	.010	.14	117.7	.07	.84	—	22.0	—	—	—
Seeds, grnd	as fed	92.7	3.05	1.86	1.11	2.41	84	23.1	14.6	16.9	.14	—	—	—	—	—	.68	1.11	—	—	—	—	—
Cottonseed, whole, grnd	dry	100.0	3.29	2.01	1.20	2.60	91	24.9	15.7	18.2	.15	—	—	—	—	—	.73	1.20	—	—	—	—	—
Seeds, mech-extd grnd	as fed	92.4	1.87	1.43	.91	1.20	52	28.0	19.6	21.4	—	—	—	—	—	—	—	1.26	—	—	—	—	—
Whole pressed cottonseed, mechanical extracted	dry	100.0	2.02	1.55	.98	1.30	56	30.3	21.2	23.2	—	—	—	—	—	—	—	1.36	—	—	—	—	—
Seeds w some hulls mech-extd grnd, mn 36% protein mx 17% fiber mn 2% fat	as fed	93.5	2.47	1.55	1.01	2.47	68	36.6	31.3	15.7	.23	—	—	—	.50	—	.94	1.43	.26	—	—	—	—
Cottonseed meal, 36% protein	dry	100.0	2.64	1.66	1.08	2.64	73	39.1	33.5	16.8	.25	—	—	—	.53	—	1.00	1.53	.28	—	—	—	—
Seeds w some hulls, mech-extd grnd, min 41% protein mx 14% fiber mn 2% fat	as fed	94.0	2.40	1.70	1.13	1.99	73	42.0	34.0	11.1	.19	1.900	18.4	.009	.50	23.3	1.11	1.27	.40	79.0	—	—	—
Cottonseed meal, 41% protein	dry	100.0	2.56	1.81	1.20	2.12	78	44.7	36.2	11.8	.20	2.000	19.6	.010	.53	24.8	1.18	1.35	.43	84.1	—	—	—
Seeds w some hulls, solv-extd grnd, min 41% protein mx 14% fiber mn 0.5% fat	as fed	91.5	2.51	1.55	1.02	1.84	69	42.1	34.1	11.4	.16	1.900	18.2	.009	.48	21.9	1.06	1.26	.21	60.4	—	—	—
Cottonseed meal solvent extracted, 41% protein	dry	100.0	2.74	1.69	1.11	2.01	75	46.0	37.3	12.5	.18	2.100	19.9	.010	.52	23.9	1.16	1.38	.23	66.0	—	—	—
Seeds wo hulls, pre-press solv-extd grnd mn 50% protein	as fed	92.5	2.51	1.57	1.03	1.86	69	50.0	40.4	8.5	.16	.093	17.9	.011	.46	22.8	1.01	1.26	—	73.3	—	—	—
Cottonseed meal, pre-press solvent extracted, 50% protein	dry	100.0	2.71	1.70	1.11	2.01	75	54.0	43.7	9.2	.17	.100	19.4	.012	.50	24.6	1.09	1.36	—	79.2	—	—	—
COWPEA, (Vigna spp.)																							
Hay, s-c	as fed	90.5	1.93	1.23	.69	1.42	53	16.0	11.8	24.3	1.38	.060	—	.080	.37	439.0	.34	2.16	.32	—	—	—	—
	dry	100.0	2.13	1.36	.76	1.57	59	17.7	13.0	26.9	1.52	.070	—	.090	.41	485.1	.37	2.39	.35	—	—	—	—
DEFLUORINATED PHOSPHATE— see Phosphate defluorinated																							
DESERT MOLLY, (Kochia vestita)																							
Browse, fresh, dormant	as fed	80.0	1.52	.90	.30	—	—	7.2	—	—	1.90	—	—	—	—	—	.10	—	—	—	14.5	—	—
	dry	100.0	1.90	1.12	.38	—	—	9.0	—	—	2.37	—	—	—	—	—	.12	—	—	—	18.1	—	—
DICALCIUM PHOSPHATE—see Calcium phosphate, dibasic, comm																							
DISTILLERS' GRAINS—Corn, see Sorghum, grain variety, see Rye																							
DROPSEED, SAND, (Sporobolus cryptandrus)																							
Aerial part, fresh, dormant	as fed	86.0	1.78	1.05	.48	—	—	4.3	—	—	.49	—	—	—	—	—	.05	—	—	—	.3	—	—
	dry	100.0	2.07	1.22	.56	—	—	5.0	—	—	.57	—	—	—	—	—	.06	—	—	—	.4	—	—
Sand dropseed																							

Feed																						
EAR CORN CHOP—see Corn, ears, grnd																						
FAT—see Animal-poultry																						
FEEDING BONE MEAL—see Bone, cooked dehy grnd																						
FEEDING OATMEAL—see Oats, cereal by-product, mx 4% fiber																						
FESCUE, MEADOW, (*Festuca elatior*) Hay, s-c	as fed	88.5	1.95	1.18	.64	1.34	54	8.4	4.6	28.8	.44	—	—	.44	21.7	.20	1.41	—	—	63.2	—	—
	dry	100.0	2.20	1.33	.72	1.52	61	9.5	5.2	32.5	.50	—	—	.50	24.5	.22	1.60	—	—	71.4	—	—
Fescue hay, tall																						
FISH, MENHADEN, (*Brevoortia tyrannus*) Whole or cuttings, cooked mech-extd dehy grnd	as fed	92.0	2.47	1.53	.99	—	—	61.3	—	1.0	5.49	—	8.4	.056	25.7	2.81	—	—	—	—	—	—
	dry	100.0	2.68	1.66	1.08	—	—	66.6	—	1.1	5.97	—	9.1	.061	27.9	3.05	—	—	—	—	—	—
FLAX, (*Linum usitatissimum*) Seed, screenings	as fed	91.6	2.12	1.26	.72	1.47	59	15.9	8.9	13.3	.37	—	—	—	—	.43	—	—	—	—	—	—
	dry	100.0	2.31	1.38	.79	1.60	64	17.4	9.7	17.4	.40	—	—	—	—	.47	—	—	—	—	—	—
Seeds, mech-extd grnd, mx 0.5% acid insoluble ash	as fed	91.0	2.67	1.73	1.16	—	74	35.3	31.0	9.0	.44	.455	26.4	.017	39.4	.89	1.24	—	—	—	—	—
	dry	100.0	2.93	1.90	1.27	—	81	38.8	34.1	9.9	.48	.500	29.0	.019	43.3	.98	1.36	—	—	—	.2	—
Linseed meal, mech-extd																						
Linseed meal																						
Seeds, solv-extd grnd, mx .5% acid insoluble ash	as fed	91.0	2.50	1.57	1.04	1.87	69	35.1	30.9	9.0	.40	.180	25.7	.033	37.6	.83	1.38	—	—	—	—	—
	dry	100.0	2.75	1.73	1.14	2.05	76	38.6	34.0	9.9	.44	.200	28.2	.036	41.3	.91	1.52	—	—	—	—	—
Linseed meal, solvent extracted																						
Solvent extracted linseed meal																						
Seed screenings, extn unspecified grnd	as fed	91.3	1.94	1.15	.58	1.29	54	24.1	13.6	11.4	.44	—	—	—	—	.63	—	—	—	—	—	—
	dry	100.0	2.13	1.26	.63	1.41	59	26.4	14.9	12.5	.48	—	—	—	—	.69	—	—	—	—	—	—
Flax seed screenings oil feed																						
Seed screenings, mech-extd grnd	as fed	91.0	1.88	1.11	.51	1.79	52	15.8	10.9	12.0	.37	—	—	—	—	.43	—	—	—	—	—	—
	dry	100.0	2.06	1.22	.56	1.97	57	17.4	12.0	17.4	.41	—	—	—	—	.47	—	—	—	—	—	—
Flaxseed screenings meal, mech-extd																						
GALLETA, (*Hilaria jamesii*) Aerial part, fresh, dormant	as fed	86.0	1.13	.71	.00	—	—	4.7	—	—	.90	—	—	—	—	.06	—	—	.3	—	—	—
	dry	100.0	1.31	.83	.00	—	—	5.5	—	—	1.05	—	—	—	—	.07	—	—	.4	—	—	—
GLUTEN FEED—see Corn, gluten w bran																						
GRAIN SORGHUM—see Sorghum, grain variety																						
GRAINS																						
GRAINS SCREENINGS—see also Barley, grain screenings, Wheat, grain screenings																						
Screenings, refuse mx 100% small weed seeds chaff hulls dust scourings noxious seeds	as fed	90.3	2.04	.98	.29	1.01	56	12.5	7.8	21.9	.20	—	—	—	—	.20	—	—	—	—	—	—
	dry	100.0	2.26	1.09	.32	1.12	62	13.8	8.7	24.3	.22	—	—	—	—	.22	—	—	—	—	—	—

(Continued)

TABLE I-3 (Continued)

Name	Basis	DM (%)	ME (Mcal/kg)	NEm (Mcal/kg)	NEg (Mcal/kg)	NEmilk (Mcal/kg)	TDN (%)	Protein (%)	Dig. pro. (%)	C.F. (%)	Ca (%)	Co (mg/kg)	Cu (mg/kg)	Fe (%)	Mg (%)	Mn (mg/kg)	P (%)	K (%)	S (%)	Z	Caro. (mg/kg)	Vit. E	Vit. D (IU/g)
Refuse screenings — Screenings, uncleaned, mn 12% grain mx 3% wild oats mx 17% buckwheat and large seeds mx 68% small weed seeds chaff hulls dust scourings noxious seeds	as fed	91.5	2.15	1.29	.75	1.50	59	14.3	10.3	16.7	.37	—	—	—	—	—	.41	—	—	—	—	—	—
	dry	100.0	2.35	1.41	.82	1.64	65	15.6	11.3	18.3	.40	—	—	—	—	—	.45	—	—	—	—	—	—
Uncleaned screenings — Brewers' grains, dehy, mn 3% dried spent hops	as fed	92.0	2.20	1.31	.76	1.54	61	26.0	19.3	14.8	.28	.062	21.2	.025	.14	37.4	.49	.09	.30	—	—	—	—
	dry	100.0	2.40	1.42	.83	1.67	66	28.3	21.0	16.1	.30	.067	23.1	.027	.15	40.7	.53	.19	.33	—	—	—	—
Brewers' dried grains — GRAMA, (*Bouteloua* spp.) Aerial part, fresh, immature	as fed	41.0	.95	.57	.32	—	26	5.4	3.7	11.2	.22	—	2.2	—	—	15.5	.08	—	—	—	—	—	—
	dry	100.0	2.31	1.39	.79	—	64	13.1	9.0	27.2	.53	—	5.5	—	—	37.9	.19	—	—	—	—	—	—
Aerial part, fresh, mature	as fed	63.4	1.33	.79	.37	—	37	4.1	2.2	20.7	.22	.110	8.1	—	.08	30.0	.08	.22	—	—	19.3	—	—
	dry	100.0	2.10	1.24	.59	—	58	6.5	3.4	32.7	.34	.180	12.8	—	.13	47.4	.12	.35	—	—	30.4	—	—
GRASS-LEGUME Aerial part, ensiled	as fed	29.3	.59	.35	.15	.38	16	3.4	1.8	9.2	.23	—	—	—	—	—	.08	—	—	—	—	—	—
	dry	100.0	2.02	1.19	.50	1.30	56	11.8	6.0	31.4	.78	—	—	—	—	—	.28	—	—	—	—	—	—
Aerial part w molasses added, ensiled	as fed	30.0	.64	.37	.17	.40	17	3.4	1.7	9.3	.28	—	—	—	—	—	.08	—	—	—	—	—	—
	dry	100.0	2.15	1.22	.56	1.34	57	11.2	5.7	31.1	.92	—	—	—	—	—	.26	—	—	—	—	—	—
GROUNDNUT—see Peanut HOMINY FEED—see Corn, grits by-product JOHNSONGRASS—see Sorghum, Johnson grass																							
HOPS, (*Humulus* spp.) Spent, dehy	as fed	93.1	1.08	—	—	—	29.7	23.0	3.9	22.6	—	—	—	—	—	—	—	—	—	—	—	—	—
	dry	100.0	1.16	—	—	—	31.9	24.8	4.5	24.3	—	—	—	—	—	—	—	—	—	—	—	—	—
Dried spent hops KAFIR—see Sorghum kafir KENTUCKY BLUEGRASS—see Bluegrass, Kentucky																							
LESPEDEZA, (*Lespedeza* spp.) Hay, s-c, pre-bloom	as fed	92.1	2.10	1.25	.70	—	58	16.4	11.4	21.8	1.05	—	—	—	—	—	.24	—	—	—	—	—	—
	dry	100.0	2.28	1.36	.76	—	63	17.8	12.4	23.7	1.14	—	—	—	—	—	.26	—	—	—	—	—	—
Hay, s-c, early bloom	as fed	93.4	1.96	1.16	.55	—	54	14.5	9.7	27.6	1.15	—	—	.040	.26	191.6	.23	.93	—	—	—	—	—
	dry	100.0	2.10	1.24	.59	—	58	15.5	10.4	29.6	1.23	—	—	.040	.28	205.1	.25	1.00	—	—	—	—	—
Hay, s-c, mid-bloom	as fed	93.0	1.92	1.14	.51	1.25	53	14.6	9.8	28.6	1.11	—	—	.030	.25	—	.24	.98	—	—	—	—	—
	dry	100.0	2.06	1.22	.55	1.34	57	15.7	10.5	30.7	1.19	—	—	.030	.27	141.2	.26	1.05	—	—	—	—	—
Hay, s-c, full bloom	as fed	93.2	1.84	1.09	.44	—	51	12.5	7.9	28.9	.97	—	—	.030	.22	—	.21	.96	—	—	—	—	—
	dry	100.0	1.98	1.17	.47	—	55	13.4	8.5	31.0	1.04	—	—	.030	.24	151.5	.23	1.03	—	—	—	—	—
Aerial part, fresh, early bloom	as fed	25.0	.60	.37	.23	.43	17	4.1	3.0	8.0	.34	—	—	.006	.07	—	.05	.28	—	—	—	—	—
	dry	100.0	2.42	1.47	.90	1.71	67	16.4	11.8	32.0	1.35	—	—	.025	.27	30.2	.21	1.12	—	—	—	—	—
Aerial part, fresh, mature	as fed	35.5	.75	.58	.38	.68	26	4.5	3.1	15.9	.36	—	—	.010	.06	30.2	.11	.27	—	—	—	—	—
	dry	100.0	2.11	1.63	1.06	1.93	73	12.8	8.8	44.9	1.02	—	—	.020	.16	85.1	.31	.77	—	—	—	—	—

LIMESTONE
   Grnd, mn 33% calcium
   Limestone, grnd
LINSEED MEAL—see Flax
MANGEL—see Beet, mangels
MEADOW HAY—see Native plants, Intermountain
MEAT AND BONE MEAL—see Animal, carcass residue
MEAT MEAL—see Animal, carcass residue
MEAT SCRAP—see Animal, carcass residue
MILK—see Cattle, milk
MILO—see Sorghum, milo
MOLASSES—see Beet, sugar, molasses; see Citrus, syrup; see Sugarcane, molasses
MONOSODIUM PHOSPHATE—see Sodium phosphate, monobasic
NAPIERGRASS, (*Pennisetum purpureum*)
   Aerial part, fresh, pre-bloom
   Aerial part, fresh, late bloom
NATIVE PLANTS, INTERMOUNTAIN
   Hay, s-c (Meadow hay)
NATIVE PLANTS, MIDWEST
   Hay, s-c, immature / Prairie hay, immature
   Hay, s-c, mid-bloom / Prairie hay, mid-bloom
   Hay, s-c, full bloom / Prairie hay, full bloom
   Hay, s-c, late bloom / Prairie hay, late bloom
   Hay, s-c, milk stage / Prairie hay, milk stage
   Hay, s-c, mature / Prairie hay, mature
   Prairie hay, overripe

Feed	State	DM	C1	C2	C3	C4	C5	C6	C7	C8	C9	C10	C11	C12	C13	C14	C15
Limestone, grnd	as fed	100.0	—	—	—	—	—	—	—	—	—	—	—	—	—	—	—
	dry	100.0	—	—	—	—	—	—	—	—	—	—	—	—	—	—	—
Napiergrass, aerial part, fresh, pre-bloom	as fed	14.9	.33	.20	.11	.23	8	1.0	.5	5.1	.09	.06	—	—	—	—	—
	dry	100.0	2.20	1.36	.76	1.57	55	6.4	3.3	34.5	.60	.41	—	—	—	—	—
Napiergrass, aerial part, fresh, late bloom	as fed	23.0	.43	.25	.08	.26	12	1.8	.4	9.0	.08	.07	—	—	—	—	—
	dry	100.0	1.88	1.10	.35	1.15	52	7.8	1.9	39.0	.35	.30	—	—	—	—	—
Native plants, Intermountain, Hay, s-c	as fed	92.9	1.54	.93	.10	—	43	8.4	2.7	28.7	.53	.16	—	—	—	—	—
	dry	100.0	1.66	1.00	.11	—	46	9.1	2.9	30.9	.57	.17	—	—	—	—	—
Hay, s-c, immature / Prairie hay, immature	as fed	89.5	1.65	.98	.28	—	46	9.7	2.7	28.1	.44	.17	—	.010	.22	.97	—
	dry	100.0	1.84	1.09	.31	—	51	10.8	3.0	31.4	.49	.19	—	.010	.24	1.08	—
Hay, s-c, mid-bloom / Prairie hay, mid-bloom	as fed	91.0	1.65	.97	.26	—	46	7.4	3.7	29.2	.31	.19	—	—	—	—	18.3
	dry	100.0	1.81	1.07	.28	—	50	8.1	4.1	32.1	.34	.21	—	—	—	—	20.1
Hay, s-c, full bloom / Prairie hay, full bloom	as fed	83.3	1.53	.91	.26	—	42	6.3	1.7	27.5	—	—	—	—	—	—	7.2
	dry	100.0	1.84	1.09	.31	—	51	7.6	2.1	33.0	—	—	—	—	—	—	7.9
Hay, s-c, late bloom / Prairie hay, late bloom	as fed	91.3	1.62	.96	.22	—	45	6.0	2.0	29.7	.33	.12	—	—	—	—	—
	dry	100.0	1.77	1.05	.24	—	49	6.6	2.2	32.5	.36	.13	—	—	—	—	—
Hay, s-c, milk stage / Prairie hay, milk stage	as fed	91.9	1.60	.95	.18	—	44	4.4	.9	30.9	.36	.12	—	.010	.22	.63	—
	dry	100.0	1.74	1.03	.19	—	48	4.8	1.0	33.6	.39	.13	—	.010	.24	.68	—
Hay, s-c, mature / Prairie hay, mature	as fed	92.3	1.80	1.06	.41	—	50	4.2	1.2	31.2	.35	.08	.120	—	—	—	9.6
	dry	100.0	1.95	1.15	.44	—	54	4.6	1.3	33.8	.38	.09	.130	—	—	—	10.4
Prairie hay, overripe	as fed	91.5	1.78	1.05	.40	—	49	3.7	1.0	31.5	—	—	—	—	—	—	—
	dry	100.0	1.95	1.15	.44	—	54	4.0	1.1	34.4	—	—	—	—	—	—	—

(Continued)

TABLE I-3 (Continued)

Name	Basis	DM	ME	NEm	NEg	NEmilk	TDN	Protein	Dig. pro.	C.F.	Ca	Co	Cu	Fe	Mg	Mn	P	K	S	Z	Caro.	Vit. E	Vit. D
		(%)	(Mcal/kg)				(%)	(%)	(%)	(%)	(%)	(mg/kg)	(mg/kg)	(%)	(%)	(mg/kg)	(%)	(%)	(%)	(mg/kg)	(mg/kg)	(mg/kg)	(IU/g)
NEEDLEANDTHREAD, (Stipa comata)																							
Aerial part, fresh, dormant	as fed	86.0	1.42	.85	.08	—	—	3.4	—	—	.76	—	—	—	—	—	.06	—	—	—	.3	—	—
	dry	100.0	1.65	.99	.09	—	—	4.0	—	—	.88	—	—	—	—	—	.07	—	—	—	.4	—	—
OATS, (Avena sativa)																							
Hay, s-c	as fed	88.2	1.97	1.16	.62	1.31	53	7.5	3.6	27.2	.21	.060	3.9	.040	.16	80.8	.19	.83	—	—	89.1	—	—
	dry	100.0	2.17	1.31	.70	1.49	60	8.5	4.1	30.8	.24	.070	4.4	.050	.18	91.6	.22	.94	—	—	101.0	—	—
Straw	as fed	90.1	1.71	1.00	.32	1.04	47	3.9	1.3	36.9	.24	—	9.9	.020	.18	29.8	.09	2.01	.20	—	—	—	—
	dry	100.0	1.90	1.11	.35	1.15	52	4.3	1.4	41.0	.27	—	11.0	.020	.20	33.1	.10	2.23	.22	—	—	—	—
Aerial part, ensiled	as fed	31.7	.68	.40	.20	.45	19	3.1	1.4	10.0	.12	—	1.7	.000	.01	12.8	.10	1.08	—	—	37.9	—	—
	dry	100.0	2.13	1.27	.64	1.41	59	9.7	4.5	31.6	.37	—	5.5	.010	.02	40.4	.30	3.41	—	—	119.5	—	—
Cereal by-product, mx 4% fiber	as fed	91.0	3.46	2.11	1.36	2.40	90	15.9	11.0	2.9	.07	—	5.3	.027	.16	43.0	.46	.50	.26	—	—	—	—
	dry	100.0	3.58	2.32	1.50	2.64	99	17.5	12.1	3.2	.08	—	5.8	.030	.18	47.3	.51	.55	.29	—	—	—	—
Feeding oat meal																							
Oat middlings	as fed	89.0	2.40	1.54	1.02	1.82	66	12.2	8.7	10.6	.09	.057	8.3	.007	.16	43.2	.33	.43	.20	—	—	32.6	—
	dry	100.0	2.70	1.73	1.14	2.04	74	13.2	9.8	11.9	.10	.064	9.3	.008	.18	48.6	.37	.48	.23	—	—	36.6	—
Grain	as fed	91.2	3.12	1.60	1.06	1.91	70	9.0	4.6	11.0	.08	—	—	—	—	—	.30	—	—	—	—	—	—
	dry	100.0	3.42	1.76	1.16	2.09	77	9.9	5.1	12.1	.09	—	—	—	—	—	.33	—	—	—	—	—	—
Grain, Pacific coast	as fed	—	—	—	—	—	70	9.9	5.1	—	—	—	—	—	—	—	—	—	—	—	—	—	—
	dry	—	—	—	—	—	77	10.9	5.6	—	—	—	—	—	—	—	—	—	—	—	—	—	—
Groats	as fed	91.0	3.06	2.15	1.38	—	85	16.7	11.7	3.0	.07	—	6.4	—	.09	28.6	.43	.34	—	—	—	—	—
	dry	100.0	3.36	2.36	1.52	—	93	18.4	12.9	3.3	.08	—	7.0	—	.10	31.4	.47	.37	—	—	—	—	—
Oat groats																							
Hulled oats																							
OATS, WHITE, (Avena sativa)																							
Grain, gr 1 heavy US mn wt 36 lb per bushel mx 2% foreign material	as fed	90.0	2.49	1.56	1.03	1.84	69	12.2	9.1	10.3	—	—	—	—	—	—	—	—	—	—	—	—	—
	dry	100.0	2.77	1.73	1.14	2.04	77	13.5	10.1	11.4	—	—	—	—	—	—	—	—	—	—	—	—	—
Grain, gr 2 US mn wt 32 lb per bushel mx 3% foreign material	as fed	90.0	2.48	1.56	1.03	1.84	68	11.2	8.5	10.8	—	—	—	—	—	—	—	—	—	—	—	—	—
	dry	100.0	2.75	1.73	1.14	2.04	76	12.5	9.4	12.0	—	—	—	—	—	—	—	—	—	—	—	—	—
Grain, gr 3 US mn wt 30 lb per bushel mx 4% foreign material	as fed	91.0	2.47	1.57	1.04	1.86	68	12.6	9.5	11.9	—	—	—	—	—	—	—	—	—	—	—	—	—
	dry	100.0	2.71	1.73	1.14	2.04	75	13.8	10.4	13.1	—	—	—	—	—	—	—	—	—	—	—	—	—
Grain, gr 4 US mn wt 27 lb per bushel mn 5% foreign material	as fed	92.0	2.52	1.59	1.05	—	70	12.7	9.6	11.6	—	—	—	—	—	—	—	—	—	—	—	—	—
	dry	100.0	2.74	1.73	1.14	—	76	13.8	10.4	12.6	—	—	—	—	—	—	—	—	—	—	—	—	—
ORCHARDGRASS, (Dactylis glomerata)																							
Hay, s-c	as fed	88.3	1.82	1.08	.49	—	50	8.6	5.1	30.0	.40	.020	12.1	.010	.28	220.4	.33	1.85	.23	16.0	29.6	—	—
	dry	100.0	2.06	1.22	.55	—	57	9.7	5.8	34.0	.45	.020	13.7	.010	.32	249.6	.37	2.10	.26	18.1	33.5	—	—
Aerial part, fresh, immature	as fed	23.8	.56	.34	.20	.39	15	4.4	3.2	5.6	.13	—	—	.000	.08	29.4	.12	.63	.05	—	80.3	—	—
	dry	100.0	2.36	1.41	.82	1.64	65	18.4	13.5	23.4	.54	—	—	.020	.33	123.6	.50	2.64	.21	—	337.3	—	—
Aerial part, fresh, early bloom	as fed	27.5	.67	.44	.28	.46	19	3.2	2.1	9.0	.07	—	9.1	.020	—	28.6	.11	—	—	—	—	29.9	—
	dry	100.0	2.45	1.60	1.03	1.67	68	11.5	7.7	32.8	.25	—	33.1	.080	—	104.1	.39	—	—	—	—	108.9	—
OYSTERS, (Crassostrea spp., Ostrea spp.)																							
Shells, fine grnd, mn 33% calcium	as fed	100.0	—	—	—	—	—	1.0	—	—	38.05	—	—	.290	.30	133.3	.07	.10	—	—	—	—	—
	dry	100.0	—	—	—	—	—	1.0	—	—	38.05	—	—	.290	.30	133.3	.07	.10	—	—	—	—	—
Oyster shell flour																							

PEA, *(Pisum spp.)*																							
Aerial part wo seeds, ensiled	as fed	24.5	.52	.29	.13	.32	14	3.2	2.0	7.6	.32	—	—	—	—	—	.06	—	—	—	46.3	—	—
Pea vine silage	dry	100.0	2.12	1.20	.53	1.30	59	13.0	8.1	31.0	1.31	—	—	—	—	—	.24	—	—	—	188.9	—	—
PEANUT, *(Arachis hypogaea)*																							
Kernels, mech-extd grnd, mx 7% fiber	as fed	92.0	2.92	1.80	1.20	2.12	81	46.6	42.0	7.0	.17	—	—	—	.33	25.6	.57	1.15	.29	—	.2	—	—
Peanut meal	dry	100.0	3.17	1.96	1.31	2.30	88	50.6	45.6	7.6	.18	—	—	—	.36	27.8	.62	1.25	.32	—	.2	—	—
Kernels, solv-extd grnd, 45% protein	as fed	92.0	2.56	1.62	1.07	1.92	71	47.6	42.9	13.2	.20	—	—	—	.04	29.0	.65	—	—	—	—	—	—
Solvent extracted peanut meal Groundnut meal, solvent extracted	dry	100.0	2.78	1.76	1.16	2.09	77	51.8	46.6	14.3	.22	—	—	—	.04	31.5	.71	—	—	—	—	—	—
PHOSPHATE ROCK																							
Defluorinated grnd, mx 1 part fluorine per 100 parts phosphorus	as fed	99.8	—	—	—	—	—	—	—	—	33.00	—	—	.920	—	—	18.00	.09	—	—	—	—	—
Phosphate, defluorinated Defluorinated phsophate	dry	100.0	—	—	—	—	—	—	—	—	33.07	—	—	.922	—	—	18.04	.09	—	—	—	—	—
POTATO, *(Solanum tuberosum)*																							
Tubers, ensiled	as fed	25.1	.77	.46	.31	.54	21	1.8	0.7	0.8	.01	—	—	—	—	—	.06	—	—	—	—	—	—
	dry	100.0	3.07	1.83	1.22	2.15	85	7.1	2.7	3.0	.04	—	—	—	—	—	.23	—	—	—	—	—	—
Tubers, fresh	as fed	24.6	.70	.48	.32	—	19	2.2	1.6	.5	—	—	—	—	—	—	—	—	—	—	—	—	—
	dry	100.0	2.86	1.95	1.30	—	79	9.0	6.4	2.1	—	—	—	—	—	—	—	—	—	—	—	—	—
PRAIRIE HAY—see Native plants, Midwest																							
RAPE, *(Brassica spp.)*																							
Seeds, mech-extd grnd	as fed	93.6	2.60	1.57	1.02	1.84	69	32.0	27.5	12.4	.94	—	—	—	—	61.9	1.31	—	—	—	—	—	—
Rapeseed meal, expeller extracted	dry	100.0	2.78	1.67	1.09	1.97	74	34.2	29.4	13.2	1.00	—	—	—	—	66.1	1.40	—	—	—	—	—	—
Seeds, solv-extd grnd	as fed	90.3	2.25	1.36	.85	1.61	62	39.4	33.9	13.8	.40	—	—	—	—	—	.90	—	—	—	—	—	—
Rapeseed meal, solvent extracted	dry	100.0	2.49	1.51	.94	1.78	69	43.6	37.5	15.3	.44	—	—	—	—	—	1.00	—	—	—	—	—	—
REDTOP, *(Agrostis alba)*																							
Aerial part, fresh, full bloom	as fed	26.3	.59	.35	.19	—	16	2.1	1.3	6.6	—	—	—	—	—	—	—	—	—	—	40.1	—	—
	dry	100.0	2.24	1.33	.73	—	62	8.1	4.8	25.1	—	—	—	—	—	—	—	—	—	—	152.6	—	—
REFUSE SCREENINGS—see Grains, screenings																							
RICE, *(Oryza sativa)*																							
Bran w germ, dry milled, mx 13% fiber calcium carbonate declared above 3% mn	as fed	91.0	2.14	1.30	.77	1.52	59	13.0	8.5	11.3	.07	—	13.0	.019	.95	417.9	1.59	1.75	.18	30.1	—	—	—
Rice bran	dry	100.0	2.35	1.43	.85	1.67	65	14.3	9.3	12.4	.08	—	14.3	.021	1.04	459.2	1.75	1.92	.20	33.1	—	—	—
Groats, polished	as fed	89.0	2.84	1.78	1.18	2.08	78	7.2	3.0	.4	.02	—	2.9	.002	.02	10.9	.10	.09	.08	1.8	—	—	—
Rice, white, polished	dry	100.0	3.19	2.00	1.33	2.34	88	8.1	3.4	.4	.02	—	3.3	.002	.02	12.3	.11	.10	.09	2.0	—	—	—
RUSSIANTHISTLE, TUMBLING, *(Salsola kali tenuifolia)*																							
Hay, s-c	as fed	88.4	1.56	.93	.19	—	43	10.5	6.6	25.0	1.63	—	—	—	1.05	—	.26	—	—	—	—	—	—
	dry	100.0	1.77	1.05	.21	—	49	11.9	7.5	28.3	1.84	—	—	—	1.19	—	.29	—	—	—	—	—	—
Aerial part, fresh, dormant	as fed	80.0	1.42	.84	.18	—	40	11.8	7.8	—	2.64	—	—	—	—	—	.13	—	—	—	7.2	—	—
	dry	100.0	1.78	1.05	.22	—	50	14.7	9.7	—	3.30	—	—	—	—	—	.16	—	—	—	9.0	—	—

(Continued)

TABLE 1-3 (Continued)

Name	Basis	DM (%)	ME (Mcal/kg)	NEm (Mcal/kg)	NEg (Mcal/kg)	NEmilk (Mcal/kg)	TDN (%)	Protein (%)	Dig. pro. (%)	C.F. (%)	Ca (%)	Co (mg/kg)	Cu (mg/kg)	Fe (%)	Mg (%)	Mn (mg/kg)	P (%)	K (%)	S (%)	Z	Caro. (mg/kg)	Vit. E (mg/kg)	Vit. D (IU/g)
RYE, (*Secale cereale*)																							
Straw	as fed	88.9	1.00	.67	—	—	28	2.7	—	42.3	.25	—	—	—	.07	5.9	.09	.86	.10	—	—	—	—
	dry	100.0	1.12	.75	—	—	31	3.0	—	47.6	.28	—	—	—	.08	6.6	.10	.97	.11	—	—	—	—
Flour by-product, coarse sift, mx 8.5% fiber (Rye middlings)	as fed	90.0	2.83	1.90	1.26	—	78	15.5	13.0	4.6	.06	—	3.6	—	—	44.2	.63	.63	—	—	—	—	—
	dry	100.0	3.15	2.11	1.40	—	87	17.2	14.4	5.1	.07	—	4.0	—	—	49.1	.70	.70	—	—	—	—	—
Flour by-product mill run, mx 9.5% fiber (Rye mill run)	as fed	90.0	2.41	1.50	.98	—	67	15.5	13.4	5.0	.07	—	—	—	.23	—	.59	.83	—	—	—	—	—
	dry	100.0	2.68	1.67	1.09	—	74	19.1	14.9	5.6	.08	—	—	—	.26	—	.66	.92	—	—	—	—	—
Grain	as fed	89.0	2.73	1.82	1.21	—	76	11.9	9.4	2.0	.06	—	7.8	.008	.12	66.9	.34	.46	—	—	—	—	—
	dry	100.0	3.07	2.04	1.36	—	85	13.4	10.6	2.2	.07	—	8.8	.009	.13	75.2	.38	.52	—	—	—	—	—
Distillers' grains, dehy (Rye distillers' dried grains)	as fed	93.0	1.54	.92	.10	—	43	22.4	9.7	14.0	.13	—	—	—	.17	18.5	.41	.11	—	30.5	—	—	—
	dry	100.0	1.66	.99	.11	—	46	24.1	10.4	15.1	.14	—	—	—	.18	19.9	.44	.12	—	34.3	—	—	—
RYEGRASS, ITALIAN, (*Lolium multiflorum*)																							
Aerial part, fresh	as fed	24.3	.51	.32	.18	.37	14	4.0	3.8	5.5	.16	—	—	—	.08	—	.10	.49	—	—	—	—	—
	dry	100.0	2.11	1.33	.72	1.52	58	15.5	11.0	22.5	.65	—	—	—	.35	—	.41	2.00	—	—	—	—	—
SAFFLOWER, (*Carthamus tinctorius*)																							
Seeds	as fed	93.1	3.00	2.05	1.34	1.42	83	16.3	13.0	26.6	—	—	9.7	.050	.33	17.9	.60	—	—	—	—	—	—
	dry	100.0	3.22	2.20	1.44	1.52	89	17.5	14.0	28.6	—	—	10.7	.050	.36	19.7	.66	—	—	—	—	—	—
Seeds, mech-extd grnd (Safflower seed mech-extd; Safflower meal expeller extd)	as fed	91.0	1.85	1.11	.51	1.22	51	21.4	14.5	31.8	.24	—	—	—	—	—	—	.72	.06	40.0	—	—	—
	dry	100.0	2.03	1.22	.56	1.34	56	23.5	15.9	35.0	.26	—	—	—	—	—	—	.79	.06	44.0	—	—	—
Seeds, solv-extd grnd (Safflower seed, solv-extd; Safflower meal, solv-extd)	as fed	91.8	1.83	1.07	.44	—	50	21.4	17.2	32.3	.34	—	—	—	—	—	.84	—	—	—	—	—	—
	dry	100.0	1.99	1.17	.48	—	55	23.3	18.7	35.2	.37	—	—	—	—	—	.92	—	—	—	—	—	—
Seeds, wo hulls, solv-extd grmd (Safflower meal without hulls, solv-extd)	as fed	90.5	2.49	1.41	.90	—	69	44.4	37.4	8.5	.24	—	—	—	—	—	1.66	—	—	—	—	—	—
	dry	100.0	2.75	1.56	.99	—	76	49.1	41.3	9.4	.26	—	—	—	—	—	1.83	—	—	—	—	—	—
SAGE, BLACK (*Salvis mellitera*)																							
Browse, fresh, dormant	as fed	76.0	—	—	—	—	—	6.5	—	—	.46	—	—	—	—	—	.12	—	—	—	13.4	—	—
	dry	100.0	—	—	—	—	—	8.5	—	—	.60	—	—	—	—	—	.16	—	—	—	17.6	—	—
SAGEBRUSH, BIG, (*Artemisia tridentata*)																							
Browse, fresh dormant	as fed	76.0	—	—	—	—	—	7.1	—	—	.51	—	—	—	—	—	.14	—	—	—	12.2	—	—
	dry	100.0	—	—	—	—	—	9.4	—	—	.67	—	—	—	—	—	.18	—	—	—	16.1	—	—
SAGEBRUSH, BUD, (*Artemisia spinescens*)																							
Browse, fresh, early leaf	as fed	25.0	.50	.30	.12	—	—	4.3	—	—	.24	—	—	—	—	—	.08	—	—	—	6.0	—	—
	dry	100.0	2.01	1.18	.49	—	—	17.3	—	—	.97	—	—	—	—	—	.33	—	—	—	23.8	—	—
Browse, fresh, immature	as fed	28.0	.54	.32	.11	—	15	4.0	2.8	6.6	.30	—	—	—	.14	—	.08	—	—	—	—	—	—
	dry	100.0	1.92	1.13	.40	—	53	14.4	10.1	23.4	1.06	—	—	—	.49	—	.29	—	—	—	—	—	—

Feed																											
SAGEBRUSH, FRINGED, (*Artemisia frigida*) Browse, fresh, mid-bloom	as fed	43.2	.97	.58	.31	—	27	3.7	2.2	11.4	—	—	—	—	—	—	—	—	—	—	—	—	—	—	—	—	—
	dry	100.0	2.24	1.33	.72	—	62	8.5	5.1	26.5	—	—	—	—	—	—	—	—	—	—	—	—	—	—	—	—	—
Browse, fresh, mature	as fed	50.0	.96	.56	.20	—	26	3.0	1.5	15.8	—	—	—	—	—	—	—	—	—	—	—	—	—	—	—	—	—
	dry	100.0	1.92	1.13	.40	—	53	6.0	3.0	31.7	—	—	—	—	—	—	—	—	—	—	—	—	—	—	—	—	—
SALTBUSH, NUTTAL, (*Atriplex nuttallii*) Browse, fresh, dormant	as fed	75.0	.99	.62	—	—	—	5.4	—	—	1.66	—	—	—	.16	—	14.2										
	dry	100.0	1.32	.83	—	—	—	7.2	—	—	2.21	—	—	—	.21	—	19.0										
SALTBUSH, SHADSCALE, (*Atriplex confertifolia*) Browse, fresh, dormant	as fed	80.0	.70	.53	—	—	—	6.2	—	—	2.02	—	—	—	.07	—	15.7										
	dry	100.0	.88	.66	—	—	—	7.7	—	—	2.53	—	—	—	.09	—	19.6										
SALTGRASS, (*Distichlis spp.*) Hay	as fed	88.4	2.11	1.28	.77	—	58	7.8	4.1	27.6	—	—	—	—	—	—	—										
	dry	100.0	2.39	1.45	.87	—	66	8.8	4.6	31.2	—	—	—	—	—	—	—										
Aerial part, fresh, overripe	as fed	74.4	1.64	.98	.52	—	45	3.1	1.1	25.3	.17	—	.22	—	—	—	—										
	dry	100.0	2.21	1.31	.70	—	61	4.2	1.5	34.0	.23	—	.30	—	—	—	—										
Aerial part, fresh	as fed	74.4	1.70	1.01	.56	—	47	4.8	2.5	22.5	.16	.020	.21	115.2	—	.18	—										
	dry	100.0	2.28	1.36	.76	—	63	6.5	3.4	30.3	.21	.020	.28	154.8	—	.24	—										
SALTGRASS, DESERT, (*Distichlis stricta*) Aerial part, fresh	as fed	75.0	1.73	1.04	.59	—	48	4.4	2.2	22.3	.12	—	—	—	.07	—	—										
	dry	100.0	2.31	1.38	.79	—	64	5.9	2.9	29.7	.16	—	—	—	.09	—	—										
SAND DROPSEED—see Dropseed, sand																											
SCREENINGS—see Barley, grain screenings; see Grains, screenings; see Wheat, grain screenings																											
SEDGE, (*Carex spp.*) Hay, s-c	as fed	90.2	1.63	.96	.25	1.87	45	9.3	4.7	29.9	—	—	—	—	—	—	—										
	dry	100.0	1.81	1.07	.28	2.01	50	10.3	5.2	30.9	—	—	—	—	—	—	—										
SESAME, (*Sesamum indicum*) Seeds, mech-extd grnd	as fed	93.0	2.65	1.57	1.03	—	73	44.6	35.7	5.4	2.01	—	—	48.2	1.34	—	—										
	dry	100.0	2.85	1.69	1.11	—	79	48.0	38.4	5.8	2.16	—	—	51.8	1.44	—	—										
Sesame meal, expeller extd																											
SHORTS—see Wheat, flour by-product, coarse sift, mx 7% fiber																											
SKIMMED MILK—see Cattle, milk, skim																											
SODIUM PHOSPHATE Monobasic, NaH2PO4·H2O, technical	as fed	96.7	—	—											21.72												
	dry	100.0	—	—											22.46												
Monosodium phosphate																											
SODIUM TRIPOLYPHOSPHATE Commercial	as fed	96.0	—	—											24.94												
	dry	100.0	—	—											25.98												
Sodium tripolyphosphate																											
SORGHUM GRAIN—see Sorghum, grain variety; Sorghum kafir; Sorghum, milo																											

(Continued)

TABLE I-3 (Continued)

Name	Basis	DM	ME	NEm	NEg	NEmilk	TDN	Protein	Dig. pro.	C.F.	Ca	Co	Cu	Fe	Mg	Mn	P	K	S	Z	Caro.	Vit. E	Vit. D
		(%)	(Mcal/kg)	(Mcal/kg)	(Mcal/kg)	(Mcal/kg)	(%)	(%)	(%)	(%)	(%)	(mg/kg)	(mg/kg)	(%)	(%)	(mg/kg)	(%)	(%)	(%)		(mg/kg)		(IU/g)
SORGHUM, GRAIN VARIETY, (Sorghum vulgare)																							
Aerial part, s-c	as fed	90.3	1.87	1.12	.52	1.25	51	6.2	2.6	26.1	.56						.17						
	dry	100.0	2.07	1.24	.58	1.38	57	6.9	2.9	28.9	.62						.19						
Grain sorghum fodder, suncured																							
Aerial part wo heads	as fed	85.1	1.75	1.03	.47		48	4.5	1.5	27.7	.34						.09						
	dry	100.0	2.06	1.21	.55		57	5.3	1.8	32.6	.40						.11						
Grain sorghum stover, suncured																							
Aerial part, ensiled	as fed	29.4	.61	.36	.17		17	2.1	.6	7.7	.07						.05						
	dry	100.0	2.06	1.22	.57		57	7.3	2.0	26.3	.25						.18						
Grain sorghum fodder silage																							
Grain	as fed	89.0	2.67	1.74	1.17		74	11.1	6.3	2.0	.04	.123	9.6		.17	14.5	.31	.34		13.7			
	dry	100.0	3.00	1.96	1.31		83	12.5	7.1	2.2	.05	.138	10.8		.19	16.3	.35	.38		15.4			
Grain, mn 6% mx 9% protein	as fed	88.0	2.58	1.66	1.11		71	7.0	4.0	1.9													
	dry	100.0	2.93	1.89	1.26		81	7.9	4.5	2.1													
Grain, mn 9% mx 12% protein	as fed	88.0	2.54	1.64	1.09		70	10.3	5.9	2.1													
	dry	100.0	2.89	1.86	1.24		80	11.7	6.7	2.4													
Grain, mn 12% mx 15% pretein	as fed	88.0	2.48	1.58	1.05		69	11.4	6.5	1.8													
	dry	100.0	2.82	1.79	1.19		78	13.0	7.4	2.0													
Distillers' grains, dehy	as fed	94.0	2.78	1.81	1.21		77	31.2	24.9	12.0	.14						.59						
	dry	100.0	2.96	1.93	1.29		82	33.2	26.5	12.8	.15						.63						
Grain sorghum distillers' dried grains																							
SORGHUM, JOHNSONGRASS, (Sorghum halepense)																							
Hay, s-c	as fed	90.7	1.95	1.08	.46	1.18	54	6.9	3.3	30.0	.81			.054	.32		.27	1.22			38.7		
	dry	100.0	2.15	1.19	.51	1.30	59	7.6	3.6	33.1	.89			.060	.35		.30	1.35			42.7		
SORGHUM, KAFIR, (Sorghum vulgare)																							
Grain	as fed	90.0	2.34	1.44	.93		65	11.8	6.8	2.0	.04		6.3	.009		15.8	.33						
	dry	100.0	2.60	1.60	1.03		72	13.1	7.6	2.2	.04		7.0	.010		17.6	.37						
SORGHUM, MILO, (Sorghum vulgare)																							
Grain	as fed	89.0	2.57	1.65	1.10		71	11.0	6.3	2.2	.04	.089	14.1		.20	12.9	.29	.35					
	dry	100.0	2.89	1.85	1.23		80	12.4	7.1	2.2	.04	.100	15.8		.22	14.5	.33	.39					
Heads, chopped	as fed	90.0	2.48	1.56	1.03		68	10.0	5.7	7.0													
	dry	100.0	2.75	1.73	1.14		76	11.1	6.3	7.8													
SORGHUM, SORGO, (Sorghum vulgare saccharatum)																							
Aerial part, ensiled	as fed	26.0	.54	.32	.16	.36	15	1.6	.5	6.0	.09		8.1	.005	.07	17.5	.05	.29			6.7		
	dry	100.0	2.06	1.25	.61	1.38	57	6.8	1.8	23.3	.33		31.3	.020	.27	67.3	.20	1.12			25.9		
Sorghum, sorgo, fodder, silage																							
SORGHUM, SUDANGRASS, (Sorghum vulgare sudanense)																							
Hay, s-c	as fed	88.9	1.90	1.12	.56	1.25	52	8.6	3.7	27.3	.36	.110	32.7	.017	.31	81.3	.27	1.87	.05		4.7		
	dry	100.0	2.14	1.26	.63	1.41	59	9.7	4.2	30.7	.40	.124	36.8	.019	.35	91.5	.30	2.10	.06		5.3		
Aerial part, fresh, immature	as fed	17.6	.44	.28	.18	.32	12	3.0	2.1	5.4													
	dry	100.0	2.53	1.56	1.00	1.83	70	16.8	12.2	30.9													
Aerial part, fresh, mid-bloom	as fed	22.7	.52	.31	.18	.36	14	2.0	1.2	8.2													
	dry	100.0	2.28	1.37	.77	1.57	63	8.7	5.3	36.1													
Aerial part, ensiled	as fed	23.3	.48	.30	.15	.33	13	2.6	1.5	8.5	.12		8.5	.003	.11	23.0	.04	.72			6.0		
	dry	100.0	2.05	1.27	.64	1.41	57	11.3	6.5	36.6	.53		36.6	.014	.49	98.8	.19	3.07			26.0		

SORGO—see Sorghum, sorgo																							
SOYBEAN, *(Glycine max)*																							
Hay, s-c	as fed	89.2	1.66	.99	.32	1.03	46	13.1	8.1	33.4	1.15	.080	8.0	.027	.70	82.6	.20	.86	.23	21.4	31.8	23.8	633.1
	dry	100.0	1.86	1.11	.36	1.15	52	14.7	9.1	37.4	1.29	.090	9.0	.030	.79	92.6	.23	.97	.26	24.0	35.7	26.7	709.8
Hulls	as fed	91.3	2.11	.90	.06	.81	58	11.3	6.9	33.0	.54	—	—	—	—	12.7	.16	—	—	—	—	—	—
	dry	100.0	2.32	.98	.07	.89	64	12.4	7.6	36.1	.59	—	—	—	—	13.9	.17	—	—	—	—	—	—
Soybean hulls																							
Soybean flakes																							
Straw	as fed	87.6	1.34	.74	—	.56	37	4.6	1.2	38.8	1.39	—	—	—	.81	44.8	.05	.49	—	—	—	—	—
	dry	100.0	1.53	.85	—	.64	42	5.2	1.4	44.3	1.59	—	—	—	.92	51.1	.06	.56	—	—	—	—	—
Aerial part, ensiled	as fed	28.0	.55	.32	.13	.34	15	4.9	3.1	8.0	.39	—	2.6	.011	.11	31.8	.13	.26	—	—	21.7	—	—
	dry	100.0	1.96	1.16	.45	1.23	54	17.5	11.0	28.6	1.39	—	9.3	.040	.38	113.5	.46	.93	—	—	77.6	—	—
Seeds	as fed	90.0	2.97	2.17	1.38	2.44	83	37.5	33.8	5.2	.25	—	15.7	.008	.28	29.5	.57	1.59	.22	—	—	32.9	—
	dry	100.0	3.30	2.41	1.53	2.71	92	41.7	37.5	5.8	.28	—	17.4	.009	.31	32.8	.63	1.77	.24	—	—	36.6	—
Seeds, mech-extd grnd mx 7% fiber	as fed	90.0	2.76	1.85	1.23	2.14	76	42.0	35.7	6.1	.28	.155	18.0	.016	.25	32.3	.58	1.74	.33	—	.2	—	—
	dry	100.0	3.07	2.06	1.37	2.38	85	46.7	39.7	6.8	.31	.172	20.0	.018	.28	35.9	.65	1.93	.37	—	.2	—	—
Soybean meal, mech-extd																							
Seeds, solv-extd grnd, mx 7% fiber	as fed	89.0	2.65	1.72	1.15	1.98	73	46.6	42.0	5.2	.29	.093	14.2	.013	.27	27.3	.65	1.91	.43	—	.2	—	—
	dry	100.0	2.98	1.93	1.29	2.23	82	52.4	47.2	5.9	.33	.104	15.9	.015	.30	30.7	.73	2.15	.48	—	.2	—	—
Soybean meal, solvent extracted																							
SOYBRAN FLAKES—see Soybean hulls																							
SPENT BONE BLACK—see Animal, bone charcoal																							
SQUIRRELTAIL, *(Sitanion spp.)*																							
Aerial part, fresh, dormant	as fed	86.0	1.38	.83	.04	—	—	3.9	—	—	.58	—	—	—	—	—	.06	—	—	—	.9	—	—
	dry	100.0	1.61	.97	.05	—	—	4.5	—	—	.67	—	—	—	—	—	.07	—	—	—	1.1	—	—
SUDANGRASS—see sorghum, sudangrass																							
SUGARCANE, *(Saccharum officinarum)*																							
Molasses, dehy	as fed	96.0	2.70	1.71	1.13	—	65	10.3	—	5.0	—	—	—	—	—	—	—	—	—	—	—	—	—
	dry	100.0	2.81	1.78	1.18	—	68	10.7	—	5.2	—	—	—	—	—	—	—	—	—	—	—	—	—
Cane molasses, dried, Molasses, cane, dried																							
Molasses, mn 48% invert sugar mn 79.5 degrees brix	as fed	75.0	2.60	1.70	1.11	1.95	72	4.4	2.6	—	.79	.912	59.6	.019	.35	42.9	.08	3.02	.34	—	—	—	—
	dry	100.0	3.47	2.27	1.48	2.60	96	5.9	3.4	—	1.05	1.216	79.4	.025	.47	57.2	.11	4.02	.46	—	—	—	—
Cane molasses																							
Molasses, cane																							
SUNFLOWER, *(Helianthus spp.)*																							
Seeds wo hulls, mech-extd grnd	as fed	93.0	2.34	1.43	.90	1.70	65	41.0	36.5	13.3	.43	—	—	—	—	22.9	1.04	1.08	—	—	—	—	—
	dry	100.0	2.52	1.54	.97	1.83	70	44.1	39.3	14.3	.46	—	—	—	—	24.6	1.12	1.16	—	—	—	—	—
Sunflower meal without hulls mech-extd																							
Seeds wo hulls, solv-extd grnd	as fed	93.0	2.20	1.31	.77	1.52	60	46.8	41.7	10.8	—	—	—	—	—	—	—	—	—	—	—	—	—
	dry	100.0	2.36	1.41	.83	1.64	65	50.3	44.8	11.6	—	—	—	—	—	—	—	—	—	—	—	—	—
Sunflower meal without hulls, solvent extracted																							
SWEETCLOVER, *(Melilotus spp.)*																							
Hay, s-c	as fed	87.2	1.85	1.06	.48	1.17	51	14.3	9.8	26.2	1.26	—	8.7	.011	.22	102.8	.24	1.60	.39	—	86.2	—	—
	dry	100.0	2.12	1.22	.55	1.34	59	16.4	11.2	30.0	1.44	—	10.0	.013	.25	117.9	.27	1.84	.45	—	98.8	—	—
TIMOTHY *(Phleum pratense)*																							
Hay, s-c, pre-bloom	as fed	88.6	2.01	1.19	.65	1.35	56	10.1	5.5	28.9	.58	—	—	—	—	—	.30	—	—	—	—	—	—
	dry	100.0	2.27	1.34	.73	1.52	63	11.4	6.2	32.6	.66	—	—	—	—	—	.34	—	—	—	—	—	—

(Continued)

TABLE I-3 (Continued)

Name	Basis	DM (%)	ME (Mcal/kg)	NEm (Mcal/kg)	NEg (Mcal/kg)	NEmilk (Mcal/kg)	TDN (%)	Protein (%)	Dig. pro (%)	C.F. (%)	Ca (%)	Co (mg/kg)	Cu (mg/kg)	Fe (%)	Mg (%)	Mn (mg/kg)	P (%)	K (%)	S (%)	Z (%)	Caro (mg/kg)	Vit. E (mg/kg)	Vit. D (IU/g)
Hay, s-c, early bloom	as fed	87.7	1.89	1.10	.54	1.24	53	7.6	4.4	30.1	.46	—	—	—	—	—	.22	.81	—	—	—	11.4	1,386.2
	dry	100.0	2.16	1.26	.62	1.41	60	8.7	5.0	34.3	.53	—	—	—	—	—	.25	.92	—	—	—	13.0	1,580.6
Hay, s-c, mid-bloom	as fed	88.4	1.93	1.16	.61	1.32	53	9.3	5.0	30.7	.36	—	—	—	—	—	.17	—	—	—	47.2	—	—
	dry	100.0	2.18	1.31	.69	1.49	60	10.5	5.7	34.7	.41	—	—	—	—	—	.19	—	—	—	53.4	—	—
Hay, s-c, late bloom	as fed	88.0	1.87	1.09	.52	1.21	52	6.8	3.2	28.5	.33	—	—	—	.14	—	.16	—	—	—	8.5	—	—
	dry	100.0	2.12	1.24	.59	1.38	59	7.7	3.7	32.4	.38	—	—	—	.16	—	.18	—	—	—	9.7	—	—
Aerial part, fresh, pre-bloom	as fed	28.3	.62	.42	.26	.48	19	2.8	1.8	9.1	.08	—	—	—	.04	—	.08	.68	—	—	—	—	—
	dry	100.0	2.20	1.47	.90	1.71	67	9.9	6.3	32.1	.28	—	—	—	.15	—	.28	2.40	—	—	—	—	—
Aerial part, fresh, mid-bloom	as fed	28.1	.67	.40	.24	.47	18	2.6	1.5	9.4	.07	—	3.1	.004	.04	54.1	.07	.48	.04	—	—	—	—
	dry	100.0	2.39	1.43	.85	1.67	66	9.1	5.4	33.5	.25	—	11.2	.016	.15	192.5	.25	1.71	.13	—	—	—	—
Aerial part, ensiled	as fed	37.5	.81	.47	.23	.53	22	4.0	2.2	13.2	.22	—	2.1	.004	.06	33.8	.11	.63	—	—	29.6	—	—
	dry	100.0	2.15	1.26	.62	1.41	60	10.6	5.9	35.2	.58	—	5.5	.010	.15	90.2	.29	1.69	—	—	78.9	—	—
TORULA DRIED YEAST—see Yeast, torulopsis																							
TREFOIL, BIRDSFOOT, (Lotus corniculatus)																							
Hay, s-c	as fed	91.2	1.91	1.20	.63	1.36	53	14.6	9.8	27.0	1.60	—	—	—	—	—	.20	—	—	—	—	—	—
	dry	100.0	2.09	1.31	.69	1.49	58	16.0	10.8	29.6	1.75	—	—	—	—	—	.22	—	—	—	—	—	—
Aerial part, fresh	as fed	20.0	.54	.34	.22	—	15	5.6	4.6	2.6	.44	—	—	—	—	—	.05	.46	—	—	—	—	—
	dry	100.0	2.71	1.68	1.10	—	75	28.0	23.0	13.0	2.20	—	—	—	—	—	.25	2.30	—	—	—	—	—
TURNIP (Brassica rapa)																							
Roots, fresh	as fed	9.3	.29	.18	.12	.22	8	1.1	.6	1.0	.05	—	1.98	.001	.02	4.0	.02	.28	.04	—	—	—	—
	dry	100.0	3.10	1.97	1.31	2.34	86	11.4	6.4	11.1	.56	—	21.3	.011	.22	42.7	.28	2.99	.43	—	—	—	—
VETCH, (Vicia spp.)																							
Hay, s-c	as fed	88.2	1.95	1.17	.64	1.34	55	17.6	11.6	25.1	1.20	.309	8.7	.044	.24	53.7	.30	1.87	.13	—	—	—	—
	dry	100.0	2.21	1.33	.73	1.52	62	20.0	13.2	28.5	1.36	.350	9.9	.050	.27	60.9	.34	2.12	.15	—	—	—	—
WHEAT, (Triticum spp.)																							
Hay, s-c	as fed	85.9	2.05	1.24	.74	1.44	57	6.4	2.9	23.9	.15	—	—	—	.11	—	—	—	—	—	95.9	—	—
	dry	100.0	2.39	1.44	.86	1.67	66	7.5	3.4	27.8	.17	—	—	—	.12	—	—	—	—	—	111.6	—	—
Straw	as fed	90.1	1.57	.93	.17	—	43	3.2	.4	37.4	—	.036	3.0	.018	—	36.4	.07	1.00	.17	—	2.0	—	—
	dry	100.0	1.74	1.03	.19	—	48	3.6	.4	41.5	—	.040	3.3	.020	—	40.4	.08	1.11	.19	—	2.2	—	—
Aerial part, fresh, immature	as fed	21.5	.57	.35	.23	.42	16	6.1	4.8	3.7	.09	—	—	—	.04	—	.09	.75	.22	—	111.8	—	—
	dry	100.0	2.64	1.64	1.07	1.93	73	28.6	22.2	17.4	.42	—	—	—	.21	—	.40	3.50	.25	—	520.2	—	—
Bran, dry milled	as fed	89.0	2.28	1.36	.85	1.63	63	15.8	12.3	8.6	.14	.039	12.3	.017	.55	115.7	1.18	1.24	—	—	—	2.8	—
	dry	100.0	2.56	1.53	.96	1.83	71	17.8	13.8	9.7	.16	.044	13.8	.019	.62	130.0	1.32	1.39	—	—	—	3.1	—
Wheat bran																								
Bran																								
Flour by-product, coarse sift, mx 7% fiber	as fed	90.0	2.80	1.87	1.24	2.17	77	18.4	13.2	5.0	.11	.090	9.3	.010	.26	104.5	.76	.85	—	—	—	28.5	—
	dry	100.0	3.11	2.08	1.38	2.41	86	20.4	14.7	5.6	.12	.100	10.3	.011	.29	116.1	.84	.94	—	—	—	31.7	—
Wheat shorts, mx 7% fiber																								
Shorts, mx 8% fiber																								
Flour by-product, mx 9.5% fiber	as fed	90.0	2.70	1.76	1.18	2.07	75	17.2	12.2	8.0	.14	.090	22.0	.009	.37	118.4	.91	.97	.22	—	—	—	—
	dry	100.0	3.00	1.96	1.31	2.30	83	19.1	13.6	8.9	.16	.100	24.4	.010	.41	131.5	1.01	1.08	.25	—	—	—	—
Wheat middlings																								
Wheat standard middlings																								

Flour by-product mill run, mx 9.5% fiber	as fed	90.0	2.40	1.70	1.13	2.01	67	15.6	10.7	8.1	.10	.225	18.7	.010	.50	102.6	1.06	1.28	—	—	—	—	—
	dry	100.0	2.67	1.89	1.26	2.23	74	17.3	11.9	9.0	.11	.250	20.8	.011	.56	114.0	1.18	1.42	—	—	—	—	—
Wheat mill run																							
Grain	as fed	89.0	2.83	1.91	1.26	2.22	78	13.2	10.3	3.0	.23	.080	7.2	.005	.16	48.6	.29	.52	—	13.7	—	30.4	—
	dry	100.0	3.18	2.15	1.42	2.49	88	14.8	11.6	3.4	.26	.090	8.1	.006	.18	554.6	.33	.58	—	15.4	—	34.2	—
Grain, Pacific coast	as fed	89.2	2.84	1.92	1.27	—	78	9.9	7.7	2.7	.12	—	—	—	—	—	.30	—	—	—	—	—	—
	dry	100.0	3.18	2.15	1.42	—	88	11.1	8.6	3.0	.14	—	—	—	—	—	.34	—	—	—	—	—	—
Grain, thresher-run, wt mn 55 mx 60 lb per bushel mx 5% foreign material	as fed	88.0	2.77	1.86	1.23	—	76	15.8	12.3	2.6	—	—	—	—	—	—	—	—	—	—	—	—	—
	dry	100.0	3.15	2.11	1.40	—	87	18.0	14.0	2.9	—	—	—	—	—	—	—	—	—	—	—	—	—
WHEAT GRAIN—see also Wheat, hard Red spring; see Wheat, soft red winter																							
Grain screenings	as fed	89.0	2.08	1.57	1.04	1.86	58	14.5	9.8	7.3	.26	—	—	—	—	28.5	.37	—	—	—	—	—	—
	dry	100.0	2.34	1.76	1.17	2.09	65	16.3	11.0	8.2	.29	—	—	—	—	32.1	.42	—	—	—	—	—	—
WHEAT, *(Triticum spp.)*																							
Germ, grnd, mn 25% protein mn 7% fat	as fed	90.0	3.01	2.20	1.40	2.48	83.2	24.6	23.1	3.3	.06	—	9.4	.006	.26	127.1	.97	—	—	—	—	142.8	—
	dry	100.0	3.34	2.44	1.55	2.75	92.4	27.3	25.7	3.7	.07	—	10.4	.007	.29	141.2	1.08	—	—	—	—	158.7	—
Wheat germ meal																							
WHEAT, HARD RED SPRING, *(Triticum aestivum)*																							
Grain	as fed	89.2	2.82	1.92	1.27	2.22	78	13.5	8.8	2.4	.04	.124	8.2	.003	.11	50.7	.39	.36	—	38.4	—	14.4	—
	dry	100.0	3.16	2.15	1.42	2.49	88	15.1	9.9	2.7	.04	.140	9.2	.004	.13	56.9	.44	.40	—	43.0	—	16.2	—
WHEAT, HARD RED WINTER, *(Triticum aestivum)*																							
Grain	as fed	88.5	2.81	1.91	1.26	2.20	78	12.1	7.6	2.3	.04	.098	5.8	.003	.10	33.2	.38	.42	.15	47.2	—	14.3	—
	dry	100.0	3.18	2.16	1.42	2.49	88	13.7	8.6	2.6	.05	.111	6.6	.004	.12	37.5	.43	.48	.17	53.4	—	16.1	—
WHEAT, SOFT RED WINTER, *(Triticum aestivum)*																							
Grain	as fed	88.1	2.72	1.90	1.25	2.19	78	11.5	7.0	2.3	.05	.102	7.0	.003	.10	33.3	.37	.40	.11	42.0	—	15.6	—
	dry	100.0	3.09	2.16	1.42	2.49	88	13.0	8.0	2.7	.06	.116	8.0	.004	.11	37.8	.42	.45	.12	47.7	—	17.7	—
WHEATGRASS, *(Agropyron spp.)*																							
Aerial part, fresh, mature	as fed	60.5	1.36	.80	.43	—	38	5.4	3.4	17.5	.22	—	—	—	.00	—	.09	—	—	—	41.3	—	—
	dry	100.0	2.25	1.32	.71	—	62	9.0	5.5	29.0	.36	—	—	—	.09	—	.15	—	—	—	68.3	—	—
WHEATGRASS, CRESTED, *(Agropyron cristatum)*																							
Hay, s-c	as fed	92.3	1.78	1.14	.54	—	49	7.3	3.5	32.9	.24	.220	—	—	—	—	.14	—	—	—	20.6	—	—
	dry	100.0	1.93	1.24	.59	—	53	7.9	3.8	35.7	.26	.238	—	—	—	—	.15	—	—	—	22.3	—	—
Aerial part, fresh, immature	as fed	28.0	.77	.46	.30	—	21	7.3	5.7	5.8	.13	—	—	—	—	—	.09	—	—	—	121.2	—	—
	dry	100.0	2.76	1.65	1.08	—	76	26.3	20.2	20.7	.48	—	—	—	—	—	.32	—	—	—	433.6	—	—
Aerial part, fresh, early bloom	as fed	36.9	.75	.51	.29	—	21	2.7	1.5	12.2	—	—	—	—	—	—	—	—	—	—	—	—	—
	dry	100.0	2.04	1.38	.79	—	56	7.3	4.1	33.1	—	—	—	—	—	—	—	—	—	—	—	—	—
Aerial part, fresh, full bloom	as fed	50.0	1.12	.66	.36	—	—	—	—	—	—	—	—	—	—	—	—	—	—	—	76.7	—	—
	dry	100.0	2.24	1.33	.72	—	—	—	—	—	—	—	—	—	—	—	—	—	—	—	153.4	—	—
Aerial part, fresh, mature	as fed	60.0	1.18	.82	.46	—	32	3.3	1.5	23.2	.16	—	—	—	—	—	.09	—	—	—	45.2	—	—
	dry	100.0	1.96	1.36	.76	—	54	5.5	2.6	38.7	.27	—	—	—	—	—	.15	—	—	—	75.4	—	—
Aerial part, fresh, overripe	as fed	80.0	1.97	1.19	.74	—	—	2.5	.4	32.2	.22	.192	6.7	—	—	42.3	.06	—	—	—	.2	—	—
	dry	100.0	2.46	1.49	.92	—	—	3.1	.5	40.3	.27	.240	8.4	—	—	52.9	.07	—	—	—	.2	—	—
WHITE HOMINY FEED—see Corn, white grits by-product																							
WHOLE PLANT CORN SILAGE—see Corn, aerial part, ensiled																							

(Continued)

TABLE I-3 (Continued)

Name	Basis	DM	ME	NEm	NEg	NEmilk	TDN	Protein	Dig. pro.	C.F.	Ca	Co	Cu	Fe	Mg	Mn	P	K	S	Z	Caro.	Vit. E	Vit. D
		(%)		(Mcal/kg)			(%)	(%)	(%)	(%)	(%)	(mg/kg)	(mg/kg)	(%)	(%)	(mg/kg)	(%)	(%)	(%)	(%)	(mg/kg)		(IU/g)
WINTERFAT (Eurotia spp.) Aerial part, fresh, dormant	as fed	76.0	.90	.59	—	—	—	8.4	—	—	1.63	—	—	—	—	—	.09	—	—	—	12.8	—	—
	dry	100.0	1.19	.78	—	—	—	11.0	—	—	2.14	—	—	—	—	—	.12	—	—	—	16.8	—	—
YEAST, (Saccharomyces cerevisiae) Brewers' Saccharomyces, dehy grnd, mn 40% protein	as fed	93.7	2.67	1.66	1.10	1.99	73	44.7	41.1	2.7	.13	.184	33.1	.013	.23	5.7	1.46	1.72	.38	38.9	—	—	—
	dry	100.0	2.85	1.77	1.17	2.12	79	47.8	44.0	2.9	.14	.196	35.3	.014	.25	6.1	1.56	1.84	.41	41.5	—	—	—
Brewers' dried yeast YEAST, TORULOPSIS (Torulopsis utilis) Dehy, mn 40% protein	as fed	92.6	2.68	1.72	1.15	2.04	74	47.0	42.8	2.5	.57	—	13.4	.009	.13	12.8	1.68	1.88	—	98.7	—	—	—
	dry	100.0	2.89	1.86	1.24	2.20	80	51.0	46.4	2.7	.62	—	14.5	.010	.14	13.8	1.81	2.03	—	106.6	—	—	—
Torula dried yeast YELLOWBRUSH (Chrysothamnus stenophyllus) Browse, fresh, dormant	as fed	70.0	1.18	.70	.08	—	—	4.6	—	—	1.33	—	—	—	—	—	.07	—	—	—	3.2	—	—
	dry	100.0	1.68	1.00	.11	—	—	6.6	—	—	1.90	—	—	—	—	—	.10	—	—	—	4.6	—	—

[1]Adapted by Dr. Wilton W. Heinemann (Animal Scientist, Irrigation Experiment Station, Washington State University, Prosser, Wash.) from *Nutrient Requirements of Beef Cattle*, 5th Rev. Edition, National Academy of Sciences, Washington, D.C., 1976, Table 11.

TABLE I-4

Weight-Unit Conversion Factors

Units Given	Units Wanted	For Conversion Multiply by	Units Given	Units Wanted	For Conversion Multiply by
lb	g	453.6	µg/kg	µg/lb	0.4536
lb	kg	0.4536	Mcal	kcal	1,000.
oz	g	28.35	kcal/kg	kcal/lb	0.4536
kg	lb	2.2046	kcal/lb	kcal/kg	2.2046
kg	mg	1,000,000.	ppm	µg/g	1.
kg	g	1,000.	ppm	mg/kg	1.
g	mg	1,000.	ppm	mg/lb	0.4536
g	µg	1,000,000.	mg/kg	%	0.0001
mg	µg	1,000.	ppm	%	0.0001
mg/g	mg/lb	453.6	mg/g	%	0.1
mg/kg	mg/lb	0.4536	g/kg	%	0.1

TABLE I-5

Weight Equivalents

1 lb = 453.6 g = .4536 kg = 16 oz
1 oz = 28.35 g
1 kg = 1,000 g = 2.2046 lb
1 g = 1,000 mg
1 mg = 1,000 μg = .001 g
1 μg = .001 mg = .000001 g
1 μg per g or 1 mg per kg is the same as ppm

TABLE I-6

Example of Formulated Ration

Ingredient	Percentage of Dry Matter in Feeds as Fed	Amount in Ration (Dry-Matter Basis)			Amount in Ration (As-Fed Basis)	
		In Ration (%)	NE_m (Mcal/kg)	NE_{gain} (Mcal/kg)	Parts	In Ration (%)
Corn, aerial part, ensiled	40	55	0.86	0.54	137.50	72.8
Corn, ears, ground	87	40	0.84	0.56	45.98	24.3
Supplement	90	5	0.08	0.05	5.56	2.9
Total		100	1.78	1.15	189.04	100.0

SECTION II—ANIMAL UNITS

An animal unit is a common animal denominator, based on feed consumption. It is assumed that one mature cow represents an animal unit. Then, the comparative (to a mature cow) feed consumption of other age groups or classes of animals determines the proportion of an animal unit which they represent. For example, it is generally estimated that the ration of one mature cow will feed 5 mature ewes, or that 5 mature ewes equal 1.0 animal unit.

The original concept of an animal unit included a weight stipulation—an animal unit referred to a 1,000-lb cow, with or without a calf at side. Unfortunately, in recent years, the 1,000-lb qualification has been dropped. Certainly, there is a wide difference in the daily feed requirements of a 900-lb range cow and of a 1,500-lb exotic cow. Both will consume dry matter on a daily basis at a level equivalent to about 2 percent of their body weight.

Hence, a 1,500-lb cow will consume 50 percent more feed than a 1,000-lb cow.

Also, the period of time to be grazed has an effect on the total carrying capacity. For example, if an animal is carried for one month only, it will take one-twelfth of the total feed required to carry the same animal one year. For

TABLE II-1

ANIMAL UNITS

Type of Livestock	Animal Units
Cattle:	
Cow, with or without unweaned calf at side, or heifer 2 yrs. old or older	1.0
Bull, 2 yrs. old or older	1.3
Young cattle, 1 to 2 years	0.8
Weaned calves to yearlings	0.6
Horses:	
Horse, mature	1.3
Horse, yearling	1.0
Weanling colt or filly	0.75
Sheep:	
5 mature ewes, with or without unweaned lambs at side	1.0
5 rams, 2 yrs. old or over	1.3
5 yearlings	0.8
5 weaned lambs to yearlings	0.6
Swine:	
Sow	0.4
Boar	0.5
Pigs to 200 pounds	0.2
Chickens:	
75 layers or breeders	1.0
325 replacement pullets to 6 mo. of age	1.0
650 8-week-old broilers	1.0
Turkeys:	
35 breeders	1.0
40 turkeys raised to maturity	1.0
75 turkeys to 6 mo. of age	1.0

this reason, the term "animal unit months" is becoming increasingly important. So in addition to the weight factor, the time factor has a distinct bearing on the ultimate carrying capacity of a tract of land.

Table II-1 gives the animal units of different classes and ages of livestock.

SECTION III—WEIGHTS AND MEASURES[5]

From time to time, stockmen and those who counsel with stockmen have need to refer to such weights and measures as follow:

METRIC AND U.S. CUSTOMARY

Length

Unit	Is Equal to	
Metric System		(U.S.)
1 millimicron (mμ)	.000000001 meter	.000000039 in.
1 micron (μ)	.000001 meter	.000039 in.
1 millimeter (mm)	.001 meter	.0394 in.
1 centimeter (cm)	.01 meter	.3937 in.
1 decimeter (dm)	.1 meter	3.937 in.
1 meter (m)	1 meter	39.37 in.; 3.281 ft; 1.094 yd
1 hectometer (hm)	100 meters	328 ft, 1 in.
1 kilometer (km)	1,000 meters	3,280 ft, 10 in.; 0.621 mile
U.S. System		(metric)
1 inch (in.)		2.54 centimeters
1 hand[1]	4 in.	
1 foot (ft)	12 in.	30.48 cm; .305 meter
1 yard (yd)	3 feet	.914 meter
1 fathom[2]	6.08 feet	1.829 meters
1 rod (rd), pole, or perch	16½ ft; 5½ yd	5.029 meters
1 furlong	220 yd; 40 rods	201.168 meters
1 mile	5,280 ft; 1,760 yd; 320 rods; 8 furlongs	1.609 kilometers
1 knot or nautical mile	6,080 ft; 1.15 land miles	
1 league (land)	3 miles (land)	
1 league (nautical)	3 miles (nautical)	

[1]Used in measuring height of horses.
[2]Used in measuring depth at sea.

CONVERSIONS

To Change	To	Multiply by
inches	centimeters	2.54
feet	meters	.305
meters	inches	39.37
miles	kilometers	1.609
kilometers	miles	.621

---

[5]For additional conversion factors, or for greater accuracy, see *Misc. Publ. 233*, the National Bureau of Standards.

Surface or Area

Unit	Is Equal to	
Metric System		(U.S.)
1 sq. millimeter (mm²)	.000001 m²	.00155 sq. in.
1 sq. centimeter (cm²)	.001 m²	.155 sq. in.
1 sq. decimeter (dm²)	.01 m²	15.50 sq. in.
1 sq. meter (m²)	1 centare (ca)	1,550 sq. in.; 10.76 sq. ft; 1.196 sq. yd
1 are (a)	100 m²	119.6 sq. yd
1 hectare (ha)	10,000 m²	2.47 acres
1 sq. kilometer (km²)	1,000,000 m²	247.1 acres; .386 sq. mi
U.S. System		(metric)
1 sq. inch (sq. in.)	1 inch × 1 inch	6.452 cm²
1 sq. foot (sq. ft)	144 sq. in.	.093 m²
1 sq. yard (sq. yd)	1,296 sq. in.; 9 sq. ft	.836 m²
1 sq. rod (sq. rd)	272.25 sq. ft; 30.25 sq. yd	25.29 m²
1 rood	40 sq. rods	10.117 ares
1 acre (A)	43,560 sq. ft; 4,840 sq. yd; 160 sq. rd; 4 roods	4,046.87 m² or 0.4047 hectare
1 sq. mile (sq. mi)	640 acres	259 hectares or 2.59 sq. km
1 township	36 sections; 6 miles square	

CONVERSIONS

To Change	To	Multiply by
square inches	square centimeters	6.452
square centimeters	square inches	.155
square yards	square meters	.836
square meters	square yards	1.196

Volume

Unit	Is Equal to	(Liquid) (U.S.) / (Ounces)	(Cu. In.)	(metric)	(Dry)
Liquid and Dry:					
Metric System					
1 milliliter (ml)	.001 l	.271 dram (fl)			.061 cu. in.
1 centiliter (cl)	.01 l	.338 oz (fl)			.610 cu. in.
1 deciliter (dl)	.1 l	3.38 oz (fl)			
1 liter (l)	1,000 cc	33.81 fl oz or 2.1134 pt or 1.057 qt (fl) or 0.2642 gal			.908 qt
1 hectoliter (hl)	100 l	26.418 gal			2.838 bu
1 kiloliter (kl)	1,000 l	264.18 gal			1,308 cu. yd
U.S. System					
Liquid:					
1 teaspoon (t)	60 drops	1/6			
1 dessert spoon	2 t				
1 tablespoon (T)	3 t	1/2			
1 gill (gi.)	1/2 c.	4	7.22	118.29 ml	
1 cup (c.)	16 T	8	14.44	236.58 ml	
1 pint (pt)	2 c.	16	28.88	.47 l	
1 quart (qt)	2 pt	32	57.75	.95 l	
1 gallon (gal)	4 qt	8.34 lb	231	3.79 l	
1 barrel	31½ gal				
1 hogshead	2 barrels				
Dry:					
1 pint (pt)	½ qt		33.6	.55 l	
1 quart (qt)	2 pt		67.20	1.10 l	
1 peck (pk)	8 qt		537.61	8.81 l	
1 bushel (bu)	4 pk		2,150.42	35.24 l	

Unit	Is Equal to		
Solid:		(Cu. In.)	(metric)
Metric System			
1 cu. millimeter (mm³)	.001 cc		
1 cu. centimeter (cc)	1,000 mm	.061	
1 cu. decimeter (dm³)	1,000 cc	61.023	
1 cu. meter (m³)	1,000 dm³	35.315 cu. ft or 1.308 cu. yd	
U.S. System			
1 cubic inch (cu. in.)			16.387 cc
1 board foot (bd ft)	144 cu. in.		2,359.8 cc
1 cubic foot (cu. ft)	1,728 cu. in.		.028 mm³
1 cubic yard (cu. yd)	27 cu. ft		.765 mm³
1 cord	128 cu. ft		3.625 mm³

CONVERSIONS

To Change	To	Multiply by
ounces (fluid)	cubic centimeters	29.57
cu. centimeters	ounces (fluid)	.034
quarts	liters	.946
liters	quarts	1.057
cu. inches	cu. centimeters	16.387
cu. centimeters	cu. inches	.061
cu. yards	cu. meters	.765
cu. meters	cu. yards	1.308

Weight

Unit	Is Equal to	
Metric System		(U.S.)
1 microgram (mcg)	.001 mg	
1 milligram (mg)	.001 gram	.015 grain
1 centigram (cg)	.01 gram	.154 grain
1 decigram (dg)	.1 gram	1.543 grains
1 gram (g)	1,000 mg	.035 oz
1 dekagram (dkg)	10 grams	5.643 drams
1 hectogram (hg)	100 grams	3.527 oz
1 kilogram (kg)	1,000 grams	35.274 oz or 2.205 lb
1 ton	1,000 kg	2,204.6 lb or 1.102 tons (short) or 0.984 ton (long)
U.S. System		(metric)
1 grain	.037 dram	64.8 mg; .065 g
1 dram	.063 oz	1.772 g
1 ounce (oz)	16 drams	28.35 g
1 pound (lb)	16 oz	453.6 g or 0.4536 kg
1 hundredweight (cwt)	100 lb	
1 ton (short)	2,000 lb	907.18 kg or 0.907 (metric) ton
1 ton (long)	2,200 lb	1,016.05 kg or 1.016 (metric) ton
1 part per million (ppm)	1 microgram/gram 1 mg/l 1 mg/kg	.454 mg/lb .907 g/ton .0001 % .013 oz/gal
1 percent (%) (1 part in 100 parts)	10,000 ppm 10 g/l	1.28 oz/gal 8 lb/100 gal

CONVERSIONS

To Change	To	Multiply by
grains	milligrams	64.799
ounces (dry)	grams	28.35
pounds	grams	454
pounds (dry)	kilograms	.454
kilograms	pounds	2.2046
mg/lb	ppm	2.2
ppm	grams/ton	.908
grams/ton	ppm	1.1
mg/lb	grams/ton	2
grams/ton	mgm/lb	.5
grams/lb	grams/ton	2,000
grams/ton	grams/lb	.0005
grams/ton	lb/ton	.0022
lb/ton	grams/ton	453.6
grams/ton	%	.00011
%	grams/ton	9,072

Weights and Measures Per Unit

Unit	Is Equal to
Volume per unit area:	
1 liter/hectare	0.107 gal/acre
1 gal/acre	9.354 liters/hectare
Weight per unit area:	
1 kilogram/cm²	14.22 lb/sq. inch
1 kilogram/hectare	0.892 lb/acre
1 lb/sq. in.	0.0703 kilogram/cm²
1 lb/acre	1.121 kilograms/hectare
Area per unit weight:	
1 cm²/kg	0.0703 sq. in./lb
1 sq. in./lb	14.22 cm²/kg

Temperature

One Centigrade (C) degree is 1/100 the difference between the temperature of melting ice and that of water boiling at standard atmospheric pressure. One Centigrade degree equals 1.8°F.

One Fahrenheit (F) degree is 1/180 of the difference between the temperature of melting ice and that of water boiling at standard atmospheric pressure. One Fahrenheit degree equals 0.556°C.

To Change	To	Do This
Degrees Centigrade ...	Degrees Fahrenheit ...	Multiply by 9/5 and add 32
Degrees Fahrenheit ...	Degrees Centigrade ...	Subtract 32, then multiply by 5/9

Weights and Measures of Common Feeds

In calculating rations and mixing concentrates, it is usually necessary to use weights rather than measures. However, in practical feeding operations it is often more convenient for the farmer or rancher to measure the concentrates. The following tabulation will serve as a guide in feeding by measure:

TABLE III-1

WEIGHTS AND MEASURES OF COMMON FEEDS

Feed	Approximate weight	
	Lb per Quart	Lb per Bushel
Alfalfa meal	0.6	19
Barley	1.5	48
Beet pulp (dried)	0.6	19
Brewers' grain (dried)	0.6	19
Buckwheat	1.6	50
Buckwheat bran	1.0	29
Corn, husked ear	—	70
Corn, cracked	1.6	50
Corn, shelled	1.8	56
Corn meal	1.6	50
Corn-and-cob meal	1.4	45
Cottonseed meal	1.5	48
Cowpeas	1.9	60
Distillers' grain (dried)	0.6	19
Fish meal	1.0	35
Gluten feed	1.3	42
Linseed meal (old process)	1.1	35
Linseed meal (new process)	0.9	29
Meat scrap	1.3	42
Molasses feed	0.8	26
Oats	1.0	32
Oats, ground	0.7	22
Oat middlings	1.5	48
Peanut meal	1.0	32
Rice bran	0.8	26
Rye	1.7	56
Sorghum	1.8	56
Soybeans	1.8	60
Tankage	1.6	51
Velvetbeans, shelled	1.8	60
Wheat	1.9	60
Wheat bran	0.5	16
Wheat middlings, standard	0.8	26
Wheat screenings	1.0	32

SECTION IV—ESTIMATING BEEF CATTLE WEIGHTS FROM HEART GIRTH MEASUREMENTS

Cattle feeders who finish large numbers of animals have scales in their feedyards for use in determining in-weights of feeder cattle, out-weights of finished cattle, and interim weight gains of cattle while they're on feed. Likewise, both purebred breeders and large commercial cow-calf operators usually have scales. However, those with only a few head of cattle—such as 4-H Club and FFA members, and part-time farmers—may not have scales. Under such circumstances, a simple but reasonably accurate method of estimating body weight is very useful. Fortunately, cattle weights may be determined with reasonable accuracy by taking two body measurements (length and circumference), then applying a certain formula. Here is how it works:

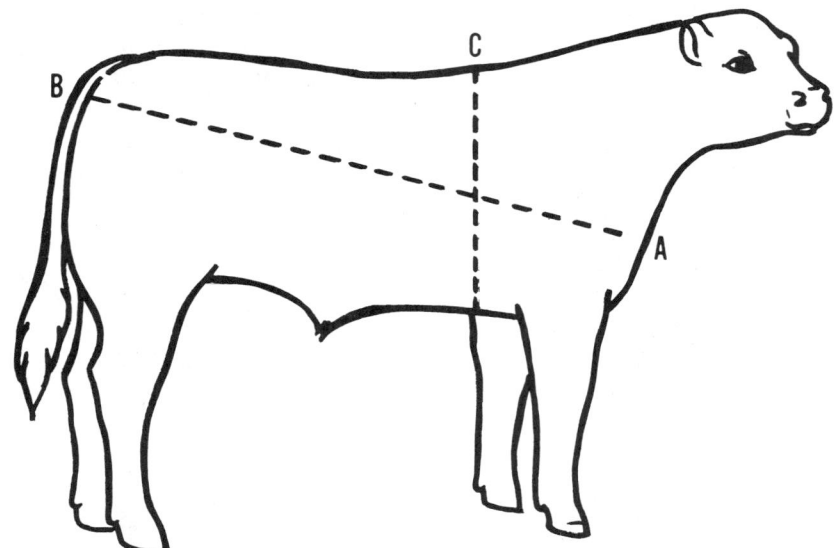

Fig. IV-1. How and where to measure beef cattle.

1. *Step 1*—Measure the length of body, from the point of shoulder to the point of rump (pin bone), in inches (distance A-B of Fig. IV-1).
2. *Step 2*—Measure the circumference (heart girth), from point slightly behind shoulder blade, thence down over fore-ribs and under body behind elbow (distance C of Fig. IV-1).
3. *Step 3*—Take the values obtained in Steps 1 and 2 and apply the following formula to caculate body weight:
   Heart girth × heart girth × body length ÷ 300 = weight in pounds.
4. *Example*—Assume that the heart girth measures 76 inches and the body length 66 inches. How much does the animal weigh?
   76 × 76 = 5,776
   5,776 × 66 = 381,216
   381,216 ÷ 300 = 1,270 pounds

SECTION V—GESTATION TABLE

The cattleman who has information relative to breeding dates can easily estimate parturition dates from Table V-1.

TABLE V-1

GESTATION TABLE FOR COWS

Date Bred	Date due, 283 days		Date Bred	Date due, 283 days
Jan. 1	Oct. 11		July 5	Apr. 14
Jan. 6	Oct. 16		July 10	Apr. 19
Jan. 11	Oct. 21		July 15	Apr. 24
Jan. 16	Oct. 26		July 20	Apr. 29
Jan. 21	Oct. 31		July 25	May 4
Jan. 26	Nov. 5		July 30	May 9
Jan. 31	Nov. 10		Aug. 4	May 14
Feb. 5	Nov. 15		Aug. 9	May 19
Feb. 10	Nov. 20		Aug. 14	May 24
Feb. 15	Nov. 25		Aug. 19	May 29
Feb. 20	Nov. 30		Aug. 24	June 3
Feb. 25	Dec. 5		Aug. 29	June 8
Mar. 2	Dec. 10		Sept. 3	June 13
Mar. 7	Dec. 15		Sept. 8	June 18
Mar. 12	Dec. 20		Sept. 13	June 23
Mar. 17	Dec. 25		Sept. 18	June 28
Mar. 22	Dec. 30		Sept. 23	July 3
Mar. 27	Jan. 4		Sept. 28	July 8
Apr. 1	Jan. 9		Oct. 3	July 13
Apr. 6	Jan. 14		Oct. 8	July 18
Apr. 11	Jan. 19		Oct. 13	July 23
Apr. 16	Jan. 24		Oct. 18	July 28
Apr. 21	Jan. 29		Oct. 23	Aug. 2
Apr. 26	Feb. 3		Oct. 28	Aug. 7
May 1	Feb. 8		Nov. 2	Aug. 12
May 6	Feb. 13		Nov. 7	Aug. 17
May 11	Feb. 18		Nov. 12	Aug. 22
May 16	Feb. 23		Nov. 17	Aug. 27
May 21	Feb. 28		Nov. 22	Sept. 1
May 26	Mar. 5		Nov. 27	Sept. 6
May 31	Mar. 10		Dec. 2	Sept. 11
June 5	Mar. 15		Dec. 7	Sept. 16
June 10	Mar. 20		Dec. 12	Sept. 21
June 15	Mar. 25		Dec. 17	Sept. 26
June 20	Mar. 30		Dec. 22	Oct. 1
June 25	Apr. 4		Dec. 27	Oct. 6
June 30	Apr. 9			

SECTION VI—ALL-TIME TOP SALES

Cattlemen and students frequently like to refer to the great sales in the history of the many breeds. Presented herewith is a summary of some of the record cattle sales, both for individual animals and consignment or herd sales.

TABLE VI-1

ALL-TIME TOP INDIVIDUAL SALES[1]

Breed	Year of Sale	Identity of Animal	Sex	Price	Private Treaty or Auction	Seller	Purchaser
				(Dollars)			
Angus[2]	1946	Erianna B 6th	Female	25,000	Auction	J. G. Tolan, Pleasant Plains, Ill.	Vaughn Bros., Albion, Mich.
	1948	Prince Sunbeam 249	Bull	60,000	Private Treaty	Sunbeam Farms, Miami, Okla.	Ellerslie Farm, Charlottesville, Va.
	1950	Prince Eric of Sunbeam	Bull	100,000	Private Treaty	L. L. O'Bryan Lakewood Farms, Mukwonago, Wisc.	Shadow Isle Farm, Red Bank, N.J.
	1951	Shadow Isle Prince 1259470	Bull	57,000	Auction	Shadow Isle Farm, Red Bank, N.J.	R. L. Smith Farms, Chillicothe, Mo.
		Eulimamere T-1245955	Female	25,500	Auction	J. G. Tolan, Pleasant Plains, Ill.	Kinloch Farm, Supply, Va.
	1952	Prince Sunbeam 249th	Bull	100,000 (½ int.)	Private Treaty	Dr. Armand Hammer, Shadow Isle Farm, Red Bank, N.J.	T. F. Murchison and Wm. A. Ljungdahl, San Antonio, Tex.
		Homeplace Gumer 5" 1340482	Female	38,000	Auction	Penney and James, Hamilton, Mo.	Taylor Bros., Essex, Mo.
	1953	Prince Esquire 2nd of Sunbeam	Bull	60,000	Auction	Sunbeam Farms, Miami, Okla.	F. W. Defoe, Mt. Pleasant Plantation, Andrews, S.C.
		Mahrapo Black Jestress	Cow & calf	32,500	Auction	Shadow Isle Farm, Red Bank, N.J.	R. L. Smith Farms, Chillicothe, Mo.
	1954	Prince 105th of TT.	Bull	115,000 (½ int.)	Private Treaty	Simon Angus Farms, Madison, Kan.	Byars Royal Oaks Farm, Tyler, and Four Wynnes Angus Farm, Kaufman, Tex.
		Barbarosa Essar 13th	Female	38,500	Auction	M. & L. Ranch, Burnet, Tex.	Hideaway Farm, Chester, N.J., and H. & L. Farms, Marlboro, N.J.
	1955	Homeplace Eileenmere 375th	Bull	87,000	Auction	Penney & James, Hamilton, Mo.	Thomas Staley, K. C., Mo., and C. A. Neilson, Hemet, Calif.
		Evermere 9 of RLS	Female	21,000	Auction	R. L. Smith, Chillicothe, Mo.	Kemp Ranch, Midlothian, Tex.
	1956	Prince 105 HR	Bull	55,000	Auction	Kemp Ranch, Midlothian, Tex.	Haystack Angus Ranch, Longmont, Colo.
		Evermere 6 of RLS	Female	2,000	Auction	Gallagher's Farm, Gainesville, Va.	Marion Harper, Jr., New York, N.Y.
	1957	Black Knight 98 of AV	Bull	37,000 (⅔ int.)	Auction	Kinloch Farm, Supply, Va.	Angus Valley Farms, Tulsa, Okla.
		Mahrapo Black Jestress	Female	16,100	Auction	R. L. Smith, Chillicothe, Mo.	Sugar Loaf Farm, Staunton, Va.
	1958	Ankony Bombardier	Bull	100,000 (½ int.)	Private Treaty	Ankony Farm, Rhinebeck, N.Y.	Haystack Ranch, Longmont, Colo.
		Elation's Eriskay	Female	9,000	Auction	Model Farms, Mundelein, Ill.	Treasure Acres, Blairsburg, Iowa
	1959	Elevate of Eastfield	Bull	75,000	Auction	T. H. Brewis Eastfield of Lempitlow, Kelso, Scotland	Lester Leachman, Claverack, N.Y.
		Ankonian 32233	Bull	35,000 (⅔ int.)	Auction	Ankony Farm, Rhinebeck, N.Y.	Hijos de Jose Firpo, Buenos Aires, Argentina
		Barbara M of Bolehill 4	Female	7,600	Auction	Brays Island Plantation, Yemassee, S.C.	Sinkola Plantation, Thomasville, Ga.
	1960	Ankonian Picador	Bull	45,000	Auction	Ankony Farm, Rhinebeck, N.Y.	Sr. Julio Perkins, Argentina
		Blackcap of Red Top 16th	Female	13,500	Auction	Treasure Acres, Blairsburg, Iowa	Sugar Loaf Farm, Stanton, Va.

(Continued)

TABLE VI-1 (Continued)

Breed	Year of Sale	Identity of Animal	Sex	Price	Private Treaty or Auction	Seller	Purchaser
				(Dollars)			
Angus (cont.)	1961	Ankonian Peerless	Bull	115,000 (½ int.)	Auction	Ankony Farm, Rhinebeck, N.Y.	Aberan Herd of Edgehill, Shadwell, Va.
		Swan Point Erianna 2	Female	7,500	Auction	Blue Sky Farm, Kearney, Mo.	Green Valley Farms, Willis, Tex.
	1962	Elevate of Eastfield	Bull	165,000 (½ int.)	Private Treaty	Aberan Herd of Edgehill, Va.	Jack and Ed Dick, Fishkill Plains, Wappingers Falls, N.Y.
		Ankony Barbara 38	Female	16,000	Auction	Ankony Farm, Rhinebeck, N.Y.	Lee Dan Farms, Titusville, N.J.
	1963	Ankonian President	Bull	140,000 (⅓ int.)	Auction	Ankony Farm, Rhinebeck, N.Y.	Black Watch Farms, Wappingers Falls, N.Y.
		Moles Hill Blue Lady 1243	Female	20,000	Auction	Moles Hill Farm, Sharm, Conn.	Allandale Farm, Kinpport, Tenn.
	1964	Escort of Manorhill	Bull	187,500 (⅓ int.)	Auction	Black Watch Farms, Wappingers Falls, N.Y.	Dixie Stock Farm, Lancaster, Ky.
		Black Watch Sammer 334	Female	25,000	Auction	Black Watch Farms, Wappingers Falls, N.Y.	Clan Ross Farms, Haywood, Va.
	1965	Penitent of Haymount	Bull	170,000 (½ int.)	Auction	Black Watch Farms, Wappingers Falls, N.Y.	M-Q Angus Ranch, Hastings, Neb.
		Genista of Mangatoro	Female	26,000	Auction	Garrett Wynne, Longmont, Colo.	Ankony Farm Rhinebeck, N.Y.
	1966	Ankonian Jingo	Bull	203,000 (⅓ int.)	Auction	Lamington Lodge Farm, Whitehouse Station, N.J. & Ankony Hyland Angus, Inc., Highmore, S.D.	White Hall Farm Charlottesville, N.C.
		Pure Pride 4 of Gaidrew	Female	40,000	Auction	Ankony Hyland, Angus, Inc., Rhinebeck, N.Y.	Angus Glen Farm, Unionville, Ont. Canada
	1967	Ankonian Prodigy	Bull	35,000	Auction	McDougal & Downs, Dissolution, Murray, Ky.	Teater Bros., Lancaster, Ky.
		Compton's Bardo	Female	20,000	Auction	Black Watch Farms, Hawk Springs, Wyo.	Nonpareil Angus, Ranch, Blackfoot, Ida.
	1968	Moles Hill Maximilian 1205	Bull	200,000 (⅓ int.)	Auction	Haystack Disp., Longmont, Colo.	Kliner Bell & Co., Beverly Hills, Calif.
		Meadow Lane Queen Harrison 3	Female	20,000	Auction	Fourche River, Ranch, Perryville, Ark.	Mahogany Farms, Williamston, Mich.
	1969	Black Watch Nobleman	Bull	210,000 (⅓ int.)	Auction	Embassy Farms, Walls, Miss.	El Capitan, Goleta, Calif.
		White Hall Jingo	Female	11,000	Auction	White Hall Farms, Charlottesville, Va.	Locmere Farms, Longview, Tex.
	1973	Ankony Camilla Anne JHL	Female	30,000	Auction	Sugar Loaf Farms, Staunton, Va.	Kilmaurs Farm, Warrenton, Va.
	1976	5-month-old son of Brad's Marshall Pride 1980	Bull	32,500 (½ int.)	Auction	Doran L. Bollman, Colfax, Iowa	H-Way Angus Farm, Elmwood, Ill.
		Angus-Chianina cross	Steer	20,375	Auction	Gregg Hilbert, Algona, Iowa	F. C. Knoebel, Denver, Colo.
Barzona	1973	BKR 2-162	Female	650	Private Treaty	Pioneer Beef Co.	
		BKR 2-97	Bull	1,000	Private Treaty	Ralph Rainwater	
Beefalo	1974	Joe's Pride	Bull	2,500,000	Private Treaty	Bud C. Basolo, Tracy, Calif.	Alga Holding Ltd., Canada

(Continued)

TABLE VI-1 (Continued)

Breed	Year of Sale	Identity of Animal	Sex	Price	Private Treaty or Auction	Seller	Purchaser
				(Dollars)			
Beefmaster	1973	Lot #75 Weber #57 Born 1968	Female	5,100	Auction	F. W. Weber, female sale, Charlotte, Tex.	W. B. Stokes, Encinale, Tex.
		Lot #4 LJ #139 Born 12/2/71	Bull	9,000	Auction	Lev A. Jasik & Son, Pleasonton, Tex.	Paul Swann, Dalton, Ga.
Brahma	1950	Emperor Jr., 4th	Bull	10,000	Private Treaty	H. O. Partin & Sons, Kissimmee, Fla.	J. P. James, Bartow, Fla.
	1956	King	Bull	10,000	Private Treaty	Lamar Beauchamp, Winter Haven, Fla.	E. E. O'Reilly, New Smyrna Beach, Fla.
	1962	5 Jumbo Manso	Bull	2,525	Auction	Pecan Acres, Simonton, Tex.	Ralph Johnston, Houston, Tex.
		ADS Duson's Lady Alma	Female	1,200	Auction	A. Duda & Sons, Cocoa, Fla.	Dr. Van de Merwe, South Africa
	1963	Imperator of LSU 52*	Bull	2,000	Auction	Louisiana State University, Baton Rouge, La.	Camp Stock Farm, Port Allen, La.
		Jumbo Eli Manso 678/1*	Bull	2,000	Auction	Pecan Acres, Simonton, Tex.	J. E. Tomeu, Venezuela
		Joan 11th of LSU	Female	620	Auction	Louisiana State University, Baton Rouge, La.	Bob Doland, Grand Chenier, La.
	1964	JDH Rex Le Manso	Bull	12,000	Private Treaty	J. D. Hudgins, Inc., Hungerford, Tex.	Code Brahman Stud, Hoopstad, South Africa
		LM Miss Rex Manso 346	Female	2,100	Auction	Lock Moore & Co., Lake Charles, La.	
	1965		Bull	1,300	Auction	Louisiana State University, Baton Rouge, La.	Harold Morris, Port Allen, La.
			Female	800	Auction	Louisiana State University, Baton Rouge, La.	Berry Farms, Keithville, La.
	1966	Imperator of LSU 62	Bull	6,900	Auction	Louisiana State University, Baton Rouge, La.	7-J Stock Farms, Madisonville, Tex.
	1967	Sugarland's Rexcrata 1	Bull	5,650	Auction	U.S. Sugar Corp., Clewiston, Fla.	I. W. Whitaker, Donie, Tex.
		Ester of Sugarland 24	Female	1,500	Auction	U.S. Sugar Corp., Clewiston, Fla.	Joe Pacheco, Costa Rica
	1968	Cherokee Arauto 1/285	Bull	9,000	Auction	E. O. Doggett, Tyler, Tex.	H. C. Koontz, Placedo, Tex.
		Miss Cherokee Arauto 1/113	Female	2,200	Auction	E. O. Doggett, Tyler, Tex.	Hodge Owen, Athens, Tex.
	1969	JDH Prem R Manso 6	Bull	10,000	Auction	J. D. Hudgins, Hungerford, Tex.	E L Ranches, San Antonio, Tex.
		JDH Lady Rex Kris Manso	Female	3,400	Auction	J. D. Hudgins, Hungerford, Tex.	Roy Seaberg, Dayton, Tex.
	1970	IW's Sugar Rexcrata 201	Bull	6,200	Auction	I. W. Whitaker & Sons, Donie, Tex.	W. M. Johnson, Buffalo, Tex.
		Miss Cherokee Princess	Female	5,000	Auction	McKellar Ranch, Mt. Pleasant, Tex.	Cherokee Ranch, Cherokee, Tex.
	1971	IW's Rexcrata 262	Bull	6,650	Auction	3-W Ranches, Donie, Tex.	Paul Mutschler, Smiley, Tex.
		Miss Angelina Red 077	Female	5,000	Auction	Winston 8 Ranch, Lufkin, Tex.	West Ward, Vanderbilt, Tex.
	1972	Sugarland's Loxacrata 54	Bull	25,000	Private Treaty	U.S. Sugar Corp., Clewiston, Fla.	Seaberg Ranch, Dayton, Tex.
		Miss Cerokee Arauto	Female	8,000	Auction	Cherokee Ranch, Cherokee, Tex.	Billy Powell, Troup, Tex.

(Continued)

TABLE VI-1 (Continued)

Breed	Year of Sale	Identity of Animal	Sex	Price	Private Treaty or Auction	Seller	Purchaser
				(Dollars)			
Brahma (cont.)	1973	RQ's Rex Crata 67	Bull	36,000	Private Treaty	Seaberg Ranch, Dayton, Tex.	Josbon Brahman Estates, South Africa
		Miss Rex Leonco 159 259058	Female	6,500	Private Treaty	Wendell M. Johnson, Buffalo, Tex.	Agropecuria el Regidor, Buenos Aires, Argentina
	1974	Mr. 3X Mucho Grande	Bull	50,000 (¼ int.)	Private Treaty	3X Ranch, Placedo, Tex.	El Regidor Ranch, Basie, Argentina
Brangus	1952	Clear Creek Duke 28th	Bull	10,000	Auction	Clear Creek Ranch, 2909 First Nat'l. Bldg., Oklahoma City, Okla.	F. B. Daniel, Orange, Va.
	1966	Sir Ned 3	Bull	12,500	Auction	Paul Davis, Oklahoma City, Okla.	K. S. Adams, Houston, Tex.
	1967	Rocky Joe	Bull	10,000	Private Treaty	Ray Moody, Rock Springs, Tex.	Dr. Herman Gardner, Willow Springs Ranch, Burton, Tex.
	1969	Texas Chief	Bull	15,000	Auction	Paul McConnell, Fulshear, Tex.	Brink's Brangus, Kerrville, Tex.
	1971	Beacon	Bull	26,500 (⅓ int.)	Auction	Brink's Brangus, Kerrville, Tex.	W. B. (Tiny) Harris, Columbus, Tex.
		Rocky Joe's Any	Female	5,100	Auction	Willow Springs Ranch, Burton, Tex.	Latimer Murfee, Cypress, Tex.
	1972	Beacon	Bull	62,500 (¼ int.)	Auction	Brink's Brangus, Kerrville, Tex.	A. Bering and K. Hubert, Houston, Tex.
		Sugar Queen	Female	13,000	Auction	Brink's Brangus, Kerrville, Tex.	Latimer Murfee, Cypress, Tex.
	1973	Van B's Chief	Bull	130,000 (¼ int.)	Auction	Brink's Brangus, Kerrville, Tex.	M. J. Stavola, Anthony, Fla.
		Miss BB Supreme	Female w/bull calf	25,000	Auction	Brink's Brangus, Kerrville, Tex.	Adolph Stieler, Comfort, Tex.
	1975	Titan	Bull	50,000 (¼ int.)	Auction	Brink's Brangus, Kerrville, Tex.	Kreuger Brangus, Crockett & Harris Farms, Columbus, Tex.
		Yucca's Lass 31-2	Female	8,600	Auction	Brink's Brangus, Kerrville, Tex.	Clayton Williams, Fort Stockton, Tex.
Charolais	1956		Female	5,400	Auction	A. M. Askew, Houston, Tex.	4T Ranches, Weslaco, Tex.
	1965	Fuyard	Bull	50,000 (½ int.)	Private Treaty	Am. Charolais Cattle Co., Homewood, Ill.	Keating Ranch, Molt, Mont.
	1966	Apollon 03163A05	Bull	84,000 (½ int.)	Private Treaty	Birchfield Farms, Ltd., British Columbia, Canada	Western Charolais Assn.
		Chata	Female	13,500	Auction	Ralph Hutchins, Raymondville, Tex.	Emerson E. Evans, Triple EEE Ranch, Gallipolis, Ohio
	1967	Architecte	Bull	45,500	Auction	J. A. Lawton, Sulphur, La.	Alexander Best, Toronto, Canada
		Miss La Blanca	Female	18,000	Auction	Ralph Hutchins, Raymondville, Tex.	Springfield Natchez Associates, Natchez, Miss.
	1968	Facco Uranium A 03	Bull	37,000 (½ int.)	Auction	Ralph Hutchins, Raymondville, Tex.	Lynch Cattle Co., Harlingen, Tex.; and Sam Countiss, Corpus Christi, Tex.
		Alezane	Female	77,500	Auction	F. W. (Bert) Wheeler, Camp Cooley Ranch, Easterley, Tex.	Nelson Bunker Hunt Ranches, Terrell, Tex.

(Continued)

TABLE VI-1 (Continued)

Breed	Year of Sale	Identity of Animal	Sex	Price	Private Treaty or Auction	Seller	Purchaser
				(Dollars)			
Charolais (cont.)	1969	Ankonian Pure Power	Bull	43,000	Auction	Ankony Charolais, Unionville, Ontario, Canada	W. E. Van Winkle, Gainesville, Fla.
		Charlotte	Female	100,000	Auction	Rocky Nelson's Riverview Ranch, Ozark, Mo.	Mr. and Mrs. Tom McLeod, Mar-K Charolais Ranch, High Springs, Fla.
	1970	FWT Director 741	Bull	9,500 (¼ int.)	Auction	Joe Hutchison, Shawnee Mission, Kan; and Schearbrook Land & Livestock, Inc., Ozark, Mo.	Aaron Thal, Atlanta, Ga.
		G. B. & S. Aronde A 03	Female	37,000	Auction	Gordon Banks & Sons, Sumner, Tex.	J & L Charolais, Clarksville, Tex.
	1971	ALS Fripon	Bull	8,200	Auction	Sanderson Charolais, Prince Albert, Saskatchewan, Canada	John Ragsdale, Booth, Tex.
		Brigette B69	Female	32,000	Auction	Charrouse Ranch, Okotoks, Alberta, Canada	Harvey's James Canyon Ranch, Genoa, Nev.
	1972	Wat-Cha Admiral Bingo 3C	Bull	18,000 (¾ int.)	Auction	Charles F. Watson, Wat-Cha Farms Ltd., Mt. Forest, Ontario, Canada	Alpine Charolais, Sylacauga, Ala.
		Cavalcade C 01	Female	30,000	Auction	Nelson Bunker Hunt Ranches, Terrell, Tex.	J. N. Roof & Family, Cleburne, Tex.
	1973	JCJ Polled Vanguard 197	Bull	32,900 (¼ int.)	Auction	JCJ Charolais Ranch, Big Island, Va.	Nash Farms, Louisburg, N.C.
		Blondine 6/F11	Female	47,000	Auction	Nelson Bunker Hunt Ranches, Terrell, Tex.	George Frisbie, Gypsum, Kan.
		RCC Royal Princess 37	Female	47,000	Auction	Royal Charolais Co., Uniontown, Penn.	Larry Gumberg, University Park, Penn.
	1975	Crestomere Alger	Bull	20,000 (¼ int.)	Auction	NB Hunt Ranches, Roanoke, Tex.	TKOC Charolais Corp.
		Magic Valley Urgande	Female	47,000	Auction	Rocking B Ranch, Athens, Tex.	Canada Coal Co., Morehead, Ky.
Chianina	1972	Miss Chianina U.S.A.	Female	20,000	Auction	Italian White Cattle Breeders, and Tannehill Cattle Co., Monterey, Calif.	Mr. and Mrs. James Wheeler, Twin Wheels Chianina Farm, Sedalia, Ky.
	1973	½-blood bull	Bull	7,700	Auction	Blue Sky Farm, Kearney, Mont.	Open Diamond Ranch, Bozeman, Mont.
		½-blood female	Female	27,000	Auction	Tannehill Cattle Co., Monterey, Calif.	Twin Wheels Farm, Sedalia, Ky.
	1974	Full-blood Italian mature bull	Bull	175,000	Auction	IWCB of North America, Edmonton, Alberta, Canada	Fillippo Assn., Dayton, Ohio
		Open ¾ Chianina female	Female	18,500	Auction	Black Champ Farm, Waxahachie, Tex.	Livestock Investors, Sunflower, Miss.
		Open ¾ Chianina female	Female	18,500	Auction	Twin Wheels Farm, Sedalia, Ky.	Dexter Farms & ICBS, Crescent City, Fla.
	1975	CL Caesar	Bull	17,800	Auction	CL Ranches Ltd., Calgary, Alberta, Canada	Chianina Investors, Palo Pinto, Calif.

(Continued)

TABLE VI-1 (Continued)

Breed	Year of Sale	Identity of Animal	Sex	Price	Private Treaty or Auction	Seller	Purchaser
				(Dollars)			
Chianina (cont.)		Miss Diaceto the First	Female	71,500	Auction	Kurly Hebb, Kaufman, Tex.	Little Chief Land & Cattle Co., Stillwater, Okla.
Devon	1973	Rahaim Mr. Champ (17097)	Bull	2,500	Auction	Shiar Rahaim Laurel, Miss.	Thomas R. Howell, M.D., Hope Farms, Laurel, Miss.
	1974	Devonacres Calgary 2085, Reg. #18322	Bull	3,000	Auction	Devonacres, Eagle Point, Ore.	Max Woodward, Eagle Point, Ore.
		Devonacres Miss Ideal 2040, Reg. #P31351	Female	2,000	Auction	Devonacres, Eagle Point, Ore.	New Market Land & Cattle Co., New Market, Md.
Dexter	1959	Atlantic Flash 2nd	Female	1,500	Private Treaty	E. C. Lard, Syosset, Long Island, N.Y.	Palmer H. Langdon, Westwood, N.Y.
Galloway	1965	DUR Contract Commander	Bull	5,400	Auction	Abbott Galloways, Athol, Ida.	Flying M Ranch, Merced, Calif.
	1972	White Lightning	Bull	4,100	Auction	Holt Ranch, Eureka, Kan.	Maricopa Acres, Harlington, New Brunswick, Canada
		200 701	Female	1,000	Auction	Holt Ranch, Eureka, Kan.	Roy Lemons, Los Lunas, N.M.
		200 717	Female	1,000	Auction	Holt Ranch, Eureka, Kan.	Roy Lemons, Los Lunas, N.M.
	1974	Netherbank Command	Bull	6,600	Auction	Patchgrove Galloways, Livelong, Sasketchewan, Canada	Ebony Farms, Sanger, Tex.; and Flying S Galloways, Woodward, Okla.
		T-Bones Squaw	Female	2,350	Auction	T-Bone Galloways, Carpenter, Wyo.	Harry Krug, Russell, Kan.
Gelbvieh	1973	Miss G & J 0260	Female	20,000	Auction	G & J, Inc., Altura, Minn.	Golden Glow Farms, St. Charles, Ill.
	1974	Houston	Bull	43,000	Auction	H & F Cattle Co., Edmonton; and Diamond V Ranch, High River, Alberta, Canada	Syndicate, Altura, Minn.
		½ blood-bred	Female	14,000	Auction	Gene Meisch, Altura, Minn.	Golden Glow Gelbvieh, St. Charles, Minn.
Hereford	1946	Real Silver Domino 44 3317191	Bull	52,000	Auction	Jack Turner, Fort Worth, Tex.	Hills and Dales, La Grange, Ga.
	1947	WHR Helmsman 89th 4635085	Bull	61,000	Auction	Wyoming Hereford Ranch, Cheyenne, Wyo.	Hi-Wan Ranch, Evergreen, Colo.
		T. T. Zato Heiress	Female	35,000	Auction	Thornton Hereford Ranch, Gunnison, Colo.	Gerald Montgomery, Madera, Calif.
	1949	Baca Duke 2nd	Bull	65,000	Auction	Albert Noe Farms, Pulaski, Tenn.	A. H. Karpe, Greenfield Hereford Ranch, Bakersfield, Calif.
	1950	Hillcrest Larry 4th 6,000,000	Bull	70,500	Auction	Hillcrest Farm, Chester, W. Va.	Chino Farms, Church Hill, Md.
	1951	M W Larry Domino 107	Bull	80,000 (½ int.)	Auction	Honey Creek Ranch, Grove, Okla.	J. S. Bridwell, Wichita Falls, Texas
		Baca Princess 63rd	Female	15,000	Auction	Freeman and Graves, Pulaski, Tenn.	E. S. Culver, Richland Farms, Pulaski, Tenn.
	1952	HC Larry Domino 12th	Bull	105,000 (½ int.)	Private Treaty	C. A. Smith, Chester, W. Va.	E. C. McCormick, Jr., Medina, Ohio

(Continued)

TABLE VI-1 (Continued)

Breed	Year of Sale	Identity of Animal	Sex	Price	Private Treaty or Auction	Seller	Purchaser
				(Dollars)			
Hereford (cont.)	1953	MW Larry Domino 83rd	Bull	33,333 (⅓ int.)	Private Treaty	Milky Way Hereford Ranch, Phoenix, Ariz.	Kirk's Valley Stream Hereford Farm, Valley Center, Calif.
	1954	Hillcrest Larry 25th	Bull	100,000	Auction	Hillcrest Farms, Chester, W. Va.	Mr. and Mrs. R. Fair, Fair Oaks Ranch, Boerne, Tex.
	1957	TR Royal Zato 27th	Bull	60,000 (¼ int.)	Private Treaty	Turner Ranch, Sulfur, Okla.	L & J Crusoe Ranch, Cheboygan, Mich.
	1958	TR Ameroyal Zato	Bull	50,000 (½ int.)	Private Treaty	Turner Ranch, Sulfur, Okla.	Burk Healey, Flying L Ranch, Davis, Okla.
	1963		Bull	52,000 (½ int.)	Private Treaty	Bridwell Hereford Ranch, Windthorst, Tex.	Hi-Point Farms, Brighton, Mich.
	1964	Real Silver Crest	Bull	45,000 (½ int.)	Auction	Jack L. Turner & Sons, Ft. Worth, Tex.	Big T Ranch, Longview, Tex.
		NP Miss Domino 227th	Female	4,050	Auction	Northern Pump Farm, McHenry, Ill.	Werner Herefords, Howell, Mich.
	1965	OHF Mill Iron 1st	Bull	25,000 (¼ int.)	Private Treaty	Honey Creek Ranch, Grove, Okla.	N Bar Herefords, Shuqualak, Miss.
		Silver Fashionette 73rd	Female	10,000	Auction	Bridwell Hereford Ranch, Windthorst, Tex.	Oxley Hereford Ranch, Pawhuska, Okla.
	1966	Winrock A 34th	Bull	37,000 (½ int.)	Auction	Winrock Farms, Sulphur, Okla.	Rancho Luna, Athens, Tex.
		Winrockette A 42nd	Female	5,500	Auction	Winrock Farms, Sulphur, Okla.	26 Bar Ranch, Stanfield, Ariz.
	1967	Silver Design 88th	Bull	52,500	Auction	Bridwell Hereford Ranch, Windthorst, Tex.	Oxley Hereford Ranch, Pawhuska, Okla.
		Lady Mischief 96th	Female	5,600	Auction	26 Bar Ranch, Stanfield, Ariz.	Dr. M. L. Barnes, Beaver Dam, Ky.
	1968	Winrock Master 21	Bull	102,000	Auction	Winrock Farms, Sulphur, Okla.	Lacy Herefords, Kansas City, Mo.
		Blanch Mischief XL 643rd	Female	5,100	Auction	XL Ranch, Longmont, Colo.	Burke Roche, Oak Park, Ill.
	1969	Upstream Golden R 164th	Bull	25,000 (¼ int.)	Private Treaty	LS Herefords, Ft. Davis, Tex.	Rocking Chair Ranch, Menard, Tex.
		26 Lady Mischief C 151st	Female	22,500	Auction	26 Bar Ranch, Stanfield, Ariz.	Emerald Acres, Ann Arbor, Mich.
	1970	RC Dan Mischief 45	Bull	20,000 (¼ int.)	Private Treaty	M & M Herefords, Midland, S.D.	J. C. Bieber & Sons, Onida, S.D.
		HHR Miss Vern 15 (& B.C.)	Female	6,000	Auction	McLean Ranch, Springfield, Mo.	Dr. Fenders Herefords
		26 Lady Misch C 237	Female	6,000	Auction	26 Bar Ranch, Stanfield, Ariz.	Emerald Acres, Ann Arbor, Mich.
		26 Queen C 280	Female	6,000	Auction	26 Bar Ranch, Stanfield, Ariz.	Emerald Acres, Ann Arbor, Mich.
	1971	26 Lad E 29	Bull	50,000 (¼ int.)	Auction	26 Bar Ranch, Stanfield, Ariz.	Rockbrook Farms, Montalba, Tex.
		26 Miss Royal Dhu D 233	Female	12,000	Auction	26 Bar Ranch, Stanfield, Ariz.	Emerald Acres, Ann Arbor, Mich.
	1972	26 Lad F 35	Bull	85,000 (¾ int.)	Auction	26 Bar Ranch, Stanfield, Ariz.	Rockbrook Farms, Montalba, Tex.
		26 Queen E 277	Female	30,000	Auction	26 Bar Ranch, Stanfield, Ariz.	Emerald Acres, Ann Arbor, Mich.
	1973	Evan Innovator	Bull	12,500 (¼ int.)	Auction	Bayers Hereford Ranch, Twin Bridges, Mont.	Keltner Herefords, Terry, Mont.

(Continued)

TABLE VI-1 (Continued)

Breed	Year of Sale	Identity of Animal	Sex	Price	Private Treaty or Auction	Seller	Purchaser
				(Dollars)			
Hereford (cont.)		26 Queen J 282	Female	10,000	Auction	26 Bar Ranch, Stanfield, Ariz.	Pick & Shovel Herefords, San Luis Obispo, Calif.
	1974	RC Dan Mischief 19	Bull	25,000 (¼ int.)	Private Treaty	Pine Creek Ranch, Salmon, Ida.	Bayers Hereford Ranch, Twin Bridges, Mont.
		Samie Tiberius 1640	Female	7,500	Auction	Jack V. Williams, San Angelo, Tex.	Rex Henderson, San Angelo, Tex.
	1975	BB Selkirk Lad 3808	Bull	41,000 (½ int.)	Auction	BB Cattle Co., Connell, Wash.	Degregory Export-Import Ltd., Buenos Aires, Argentina
	1976	Grand Slam	Bull	100,000 (½ int.)	Auction	Weise & Sons, Manning, Iowa; and Glenkirk Farms, Maysville, Mo.	Granite Hills Hereford Ranch, Llano, Tex.
		26 Lad E 29 Daughter	Female	6,600	Auction	26 Bar Ranch, Stanfield, Ariz.	T-Cross Ranches, Colorado Springs, Colo.
Limousin	1971	75%	Female	10,500	Auction	Baldco, North Platte, Neb.	H. E. Jones, Jacksonville, Ill.
	1972	Dandin C	Bull	176,000	Auction	CDA Research Station, Brandon, Manitoba, Canada	International Beef Breeders, Denver, Colo.
		Fabida	Female	30,000	Auction	Valiant Ranches, Okotoks, Alberta, Canada	LV Ranches, Erskine, Alberta, Canada
	1973	Rousseau	Bull	51,000	Auction	LV Ranches, Erskine, Alberta, Canada	Circle N Ranch, Grove, Okla.
		Harbinger	Female	50,000	Auction	Limousin Breeders Ltd., Oklahoma City, Okla.	Budrick Farms, Mannsville, Okla.
	1974	Email CIM 55	Bull	75,000	Auction	Agri-Cadre, Houston, Tex.	Curtiss Breeding Service, Cary, Ill.
		Hermine	Female	53,000	Auction	International Breeders Service, Inc., U.S.A.	Southeast Cattle Breeders, Ft. Lauderdale, Fla.
Maine-Anjou	1972	¾ bull calf	Bull	6,000	Auction	Tower Hill Cattle, Tower Hill, Ill.	G & F Cattle, Brownwood, Tex.
			Female	21,000	Auction	Johnson Brothers, Trail City, S.D.	Maverick 5 Outfit, Sheridan, Wyo.
	1973	Fullblood	Bull	102,000	Auction	Harold Biensch, Neilburg, Saskatchewan, Canada	Goliath Cattle Co., Elbert, Colo.; New Breeds Industries, Manhattan, Kan.; and Dr. D. Kerby, Brownwood, Tex.
		¾ blood	Female	43,500	Auction	Peters Cattle Co., Elk Creek, Neb.	Heaven II Farms, Carlisle, Ky.
	1974	PTR Epic 83 (E)	Bull	102,000	Auction	Harold Biensch, Neilburg, Saskatchewan, Canada	Goliath Cattle Co., Elbert, Colo.; New Breeds Industries, Manhattan, Kan.; Western Breeders Service, Balzac, Alberta, Canada; and Dr. D. R. Kerby, Brownwood, Tex.
		Cross Lanes Georgia	Female	100,000	Auction	Roger Chapman, Cross Lanes Herd, Reading, England	Cattle Genetics Enterprises, Brookline, Mo.

(Continued)

TABLE VI-1 (Continued)

Breed	Year of Sale	Identity of Animal	Sex	Price	Private Treaty or Auction	Seller	Purchaser
				(Dollars)			
March-igiana	1973	Miss Marchigiana #7	Female	23,600	Auction	W. E. Goodrick, Lucky Clover Farm, Peculiar, Mo.	Dr. A. Rajamannan, I.C.B.S., St. Paul, Minn.
	1974	Sally T #5H	Female	5,600	Auction	Ky-Tel Ranch, Weatherford, Tex.	Mr. and Mrs. Jack B. Tilson, Ruston, La.
Milking Shorthorn	1917	Lady of the Glen	Female	5,500	Auction	The Otis Herd, Willoughby, Ohio	A. T. Cole, Wheaton, Ill.
	1950	Revelex Daisy's Premier	Bull	9,000	Auction	R. H. Vigus, Herts, England	Last Chance Ranch, Lake Placid, N.Y.
		Lilydale Dagney Pride	Female	5,600	Auction	Lilydale Farm, Springfield, Mo.	Mystery Farm, Hope, R.I.
	1973	Gold Mine Pathfinder	Bull	1,475	Auction	Irving F. Meyer & Son, McGregor, Iowa	Thomas Bros., Huron, S.D.
		Greenvista S. V. P. P. Peggy	Female	3,400	Auction	Donald Brooks, Siloam Springs, Ark.	Jim Young, Rte. 1, Newville, Penn.
Murray Grey	1973	Bimbadeen Westward Ho	Bull	49,000	Auction	Murray Grey, USA Box 397 Troup, Tex.	Western Murray Greys, Inc. Thermopolis, Wyo.
	1974	Pine Leigh Indian Chief	Bull	26,800	Auction	Frank Harper, Palm Springs, Calif.	Iowa MG Stud, Union, Iowa
		Allonby	Female	23,000	Auction	Frank Harper, Palm Springs, Calif.	Great Plains Beef Enterprise, Billings, Mont.
Norwegian Red	1975	W. F. Viking Erik B 177	Bull	50,000	Auction	Worley Farms, Atwood, Kan.	Two Twenty Cattle Corp., Houston, Tex.
Pinzgauer	1974	Miss Alpine	Female	10,800	Auction	Alpine Cattle, Alamosa, Colo.	R & B Cattle Enterprises, Alsask, Saskatchewan, Canada
Polled Hereford	1947	Alf Choice Domino 6	Bull	35,000	Auction	J. M. Lewis & Sons, Larned, Kan.	A. G. Rolfe, Poolesville, Md.
	1949	M. H. Supreme Beauty	Female	10,700	Auction	Mousel and Hausler, Holbrook, Neb.	Mike Lewis Livestock Co., Henderson, Colo.
	1950	S. B. Letston Lady 4th	Female	11,000	Auction	A. G. Rolfe, Poolesville, Md.	C. C. Potter, Pottstown, Penn.
	1951	EER Victor Anxiety	Bull	42,000	Auction	Double E. Ranch, Senatobia, Miss.	R. C. Malone, Meridian, Miss.
		Helen Dandy Domino 10th	Female	13,000	Auction	Wilford Scott, Chadron, Neb.	Calvin Fowler, Franklin, Tenn.
	1952	Real Plato Domino Jr.	Bull	43,500	Auction	Fritz Kerbs & Sons, Otis, Kan.	Allen Engler & Sons, Topeka, Kan.
		EER Victoria Tone 25th	Female	14,000	Auction	E. E. Moore, Senatobia, Miss.	J. C. Lanham, Eastover, S.C.
	1956	Gold Co-Pilot	Bull	50,000 (½ int.)	Private Treaty	O. E. Kuhlmann, North Platte, Neb.	Hull-Dobbs Ranch, Walls, Miss.
	1957	Mischief Lady 16th	Female	17,000	Auction	Santa Fe River Ranch, Alachua, Fla.	Todiway Acres, Pike Road, Ala.
	1958	Mixer Return	Bull	40,000 (⅓ int.)	Private Treaty	C. E. Knowlton, Bellefontaine, Ohio, and O'Bryan Ranch, Hiattville, Kan.	Harold Huber, Schneider, Ind.
		Blanche Larry 11th	Female	18,100	Auction	Circle M Ranch, Senatobia, Miss.	Circle L Ranch, Greenwood, Ark.
	1960	CMR Rollotrend 5th	Bull	80,000 (¼ int.)	Auction	Circle M Ranch, Senatobia, Miss.	S. M. Stoody, Chino, Calif.

(Continued)

TABLE VI-1 (Continued)

Breed	Year of Sale	Identity of Animal	Sex	Price	Private Treaty or Auction	Seller	Purchaser
				(Dollars)			
Polled Hereford (cont.)	1961	EJH Royal Domino 12th	Bull	40,000 (½ int.)	Private Treaty	E. J. Haberer, New Lebanon, Ohio	Myrtlewood Farms, Lexington, Ky.
		CMR Supermaid	Female	16,000	Auction	Circle M Ranch, Senatobia, Miss.	Santa Fe River Ranch, Alachua, Fla.
	1963	CMR Superfactor	Bull	50,000 (¼ int.)	Auction	Circle M Ranch, Senatobia, Miss.	C. E. Knowlton, Bellefontaine, Ohio
		CMR Astermaid 64	Female	20,000	Auction	B. C. Barnes, Carrollton, Ga.	Otis Jester, Asheville, Ala.
	1964	CMR Superol 109th	Bull	45,000	Auction	Circle M Ranch, Senatobia, Miss.	Col. E. Brooke Lee, Silver Spring Farm, Damascus, Md.
		CLR Super Miss 6	Female	10,000	Auction	Circle L Ranch, Greenwood, Ark.	Scott Hereford Farm, Hickory Flat, Miss.
	1965	KA Rollo Mischief JR	Bull	42,500	Auction	Ken-Al Ranch, Columbus, Miss.	Mont-Vue Farms, Niota, Tenn.
		CMR Miss Larolamp 4	Female	13,000	Auction	Circle M Ranch, Senatobia, Miss.	Tucannon River Ranch, Pomeroy, Wash.
	1966	CMR Advance Lamp	Bull	38,500 (¼ int.)	Auction	Circle M Ranch, Senatobia, Miss.	M. H. Rich & Sons, Chatham, Miss.
		SS Miss Mischief 25B	Female	15,000	Auction	Tucannon River Ranch, Pomeroy, Wash.	Ed Somerville, Brandon, Manitoba, Canada
	1967	CMR Super Domino 164	Bull	26,000 (¼ int.)	Auction	Circle M Ranch, Senatobia, Miss.	Bedford Farms, Shelbyville, Tenn.
		CMR Blanchetrend 90	Female	10,100	Auction	Circle M Ranch, Senatobia, Miss.	M. H. Rich & Son, Chatham, Miss.
	1968	BB Modest Lamp 49	Bull	27,000 (¼ int.)	Auction	Beau Brehm L Ranch, Belle Rive, Ill.	El Rancho Grande, Cross City, Fla.
		CMR Blanchetrend 83	Female	24,600	Auction	Circle M Ranch, Senatobia, Miss.	Charles Auger, California, Mo.
	1969	Predominant 25U	Bull	60,000 (¼ int.)	Auction	American Herdsman Institute, Lee's Summit, Mo.	Paul Deer Farm, Springfield, Ohio
		CMR Blanchetrend 109	Female	20,000	Auction	Circle M Ranch, Senatobia, Miss.	M. H. Rich & Son, Chatham, Miss.
	1970	CMR Masterpiece 49	Bull	23,300 (¼ int.)	Auction	Circle M Ranch, Senatobia, Miss.	Farris N. Jobe, Romance, Ark.
		CMR Miss Supernew 36	Female	25,700	Auction	Circle M Ranch, Senatobia, Miss.	M. H. Rich & Son, Chatham, Miss.
	1971	AAB Superol	Bull	28,500 (¼ int.)	Auction	Anthony A. Buford, Caledonia, Mo.	Larry McShane, Conway, Mo.
		CPH Miss Woodrow 41-75	Female	6,500	Auction	Bedford Farms, Shelbyville, Tenn.	Neal Farms, Osgood, Ind.
	1972	Predominant 25U	Bull	50,000 (¼ int.)	Auction	American Herdsman Institute, Lee's Summit, Mo.	Chisholm Trail Ranch, Rhome, Tex.
		MVF Choice Miss J141	Female	4,300	Auction	American Herdsman Institute, Lee's Summit, Mo.	Mr. and Mrs. W. B. Williams, Colmar, Ill.
	1973	Big Sky Guy	Bull	100,000 (¼ int.)	Private Treaty	Little Beaver Creek Ranches, Missoula, Mont.	Dutton Brookfield, Kansas City, Mo.
		FR Miss Changer 791	Female	7,500	Auction	Clove Creek Ranch, Poughquag, N.Y.	Maxwell Evasts, Mt. Kisco, N.Y.
	1974	Big Sky Guy	Bull	100,000 (¼ int.)	Private Treaty	Little Beaver Creek Ranches, Missoula, Mont.	Heins and Sons, Split Butte Ranch, Rupert, Ida.

(Continued)

TABLE VI-1 (Continued)

Breed	Year of Sale	Identity of Animal	Sex	Price	Private Treaty or Auction	Seller	Purchaser
				(Dollars)			
Polled Shorthorn —see Shorthorn							
Red Angus	1973	White's Schimpp 501	Bull	15,000	Private Treaty	Garnett W. White & Sons, Spokane, Wash.	Ndhlovn Red Angus Stud, Gingindhlovu, Zululand, So. Africa
		Cow w/calf	Female	1,800	Auction	Beckton Stock Farm, Sheridan, Wyo.	Kirk Ranch, Frannie, Wyo.
	1974	CPR Royal 265	Bull	12,500	Auction	Roger L. Roper, Marietta, Okla.	Jack Dixon, Lewisville, Ark.
	1975	Leachman Dynamo 605	Bull	12,200 (½ int.)	Auction	Leachman Cattle Co., Billings, Mont.	Dan Nelson, and Dale Faylor, Haigler, Neb.
		BKT Laralelang 68 daughter	Female	2,600	Auction	Leachman Cattle Co., Billings, Mont.	Enid State School, Enid, Okla.
Red Brangus	1972	PFR Oscar's Improver 100/2	Bull	5,000	Auction	Paleface Ranch, Spicewood, Tex.	Pete Skeen, Troup, Tex.
		Miss Eig Eag 19/8	Female	1,400	Auction	Jud Alexander, Brenham, Tex.	Golden Rainbow Ranch, Lakeland, Fla.
	1973	PFR Oscar's Improver 100/3	Bull	14,000	Auction	Paleface Ranch, Spicewood, Tex.	Lee Gregory, Houston, Tex.
	1974	Miss PFR 102/0	Female	5,450	Auction	Paleface Ranch, Spicewood, Tex.	Rafter Two Ranch, Austin, Tex.
Red Poll	1942	Ruler's Defender 58808	Bull	1,550	Auction	A. B. Adams, Clifton, Ill.	Mark R. Musgrave, Springfield, Ohio
	1947	Sunshine's Starlight 98420	Female	1,200	Auction	A. E. Young, Meadville, Mo.	Ray L. Olson, Lincoln, Neb.
	1963	HNR Highlander 5th 89018	Bull	800 (½ int.)	Auction	Highlandview Ranch, Harrison, Ark.	Stee-John Farms, Holdenville, Okla.
	1965	Pinpur Prince 90764	Bull	1,450	Auction	Purdue Univ., Lafayette, Ind.	Armour & Co., BCI, Chicago, Ill.
	1969		Bull	1,200	Auction	Herman W. Ebers Estate, Seward, Neb.	Frank Duerst, Lyle, Minn.; and Harley P. Olson & Son, Oneida, Ill.
			Female	500	Auction	John and Frank Duerst, Lyle, Minn.	Mr. and Mrs. John E. Hunsicker, Jr., Shreveport, La.
	1970		Bull	1,350	Auction	H. W. Ebers Estate, Seward, Neb.	S. O. and R. S. Bourne, Noble, Ill.
			Female	975	Auction	Pinney Purdue Agr. Ctr., Wanatah, Ill.	R. A. Parish, Adona, Ark.
	1971		Bull	2,100	Auction	H. W. Ebers Estate, Seward, Neb.	Mr. and Mrs. John E. Hunsicker, Jr., Shreveport, La.
			Female	1,000	Auction	Pinney Purdue Agr. Ctr., Wanatah, Ill.	Dr. E. V. Spillman, Talking Rock, Ga.
	1972	Pinpur Broadcaster	Bull	4,200	Auction	Pinney Purdue Agr. Ctr., Wanatah, Ill.	J. R. Kitchen, Jr., Nashville, Tenn.
			Female	2,300	Auction	Pinney Purdue Agr. Ctr., Wanatah, Ill.	Dr. E. V. Spillman, Talking Rock, Ga.
	1974	Parkwood Charmer Kate III	Female	3,150	Auction	Michael E. Parks, Crawfordsville, Ind.	Mr. and Mrs. John E. Hunsicker, Jr., Shreveport, La.

(Continued)

TABLE VI-1 (Continued)

Breed	Year of Sale	Identity of Animal	Sex	Price	Private Treaty or Auction	Seller	Purchaser
				(Dollars)			
Salers	1975	Jet	Bull	100,000	Private Treaty	R. Moore and B. Parker, Huntington, Ore.	Rugged Red Cattle Co.
Santa Gertrudis	1950	Lot No. 32 (prior to regis. by new assn.)	Bull	10,000	Auction	King Ranch, Kingsville, Tex.	Edgar Brown Orange, Tex.
	1951	Lot No. 65 (prior to regis. by new assn.)	Bull	27,200	Auction	King Ranch, Kingsville, Tex.	Garvin Tankersley, Washington, D.C.
	1952	Lot No. 7	Bull	40,000	Auction	King Ranch, Kingsville, Tex.	Briggs Syndicate, San Antonio, Tex.
	1953		Bull	31,500	Auction	King Ranch, Kingsville, Tex.	Winthrop Rockefeller, Morrilton, Ark.
	1955	Dulcy	Bull	28,000	Auction	King Ranch, Kingsville, Tex.	Winrock Farms, Morrilton, Ark.
	1958	Gata	Female	6,000	Auction	R. W. Briggs, San Antonio, Tex.	R. J. Kleberg, Kingsville, Tex.
	1959		Bull	27,500	Auction	King Ranch, Kingsville, Tex.	Winrock Farms, Morrilton, Ark.
	1960		Bull	28,000	Auction	Wm. C. duPont, Wilmington, Del.	King Ranch, Kingsville, Tex.
	1961		Bull	59,000	Auction	Winthrop Rockefeller, Morrilton, Ark.	J. T. Flagg, Florence, Ala.
	1961		Female	15,000	Auction	Winthrop Rockefeller, Morrilton, Ark.	Col. C. Michael Paul, New York, N.Y.
	1962	WR Bigamo Jr.	Bull	17,500	Auction	Winrock Farms, Morrilton, Ark.	7-11 Ranch, Boerne, Tex.
		Novia	Female	10,000	Private Treaty	Vesper Ranch, Cotulla, Tex.	D. H. Boyd, Dallas, Tex.
	1963	WR 399/1	Bull	23,000	Auction	Winrock Farms, Morrilton, Ark.	Nine Bar Ranch, Cypress, Tex.
		WR 617/1	Female	8,350	Auction	Winrock Farms, Morrilton, Ark.	R. D. Keene, Orlando, Fla.
	1964	WR Bondigo 305	Bull	25,000 (¼ int.)	Private Treaty	Winrock Farms, Morrilton, Ark.	Forked Lightning Ranch, Pecos, N.M.
		WR 472	Female	7,000	Auction	Winrock Farms, Morrilton, Ark.	Los Jaboncillos Ranch, Premont, Tex.
	1965	9-2190	Bull	58,000	Auction	Nine Bar Ranch, Cypress, Tex.	King Ranch, Kingsville, Tex.
		9-cow	Female	5,000	Auction	Nine Bar Ranch, Cypress, Tex.	Hualalai Ranch, Kona, Hawaii
	1966	Masterpiece 511	Bull	62,500 (¼ int.)	Private Treaty	Mirasol Ranch, Uvalde, Tex.	Diamond P Ranch, Plano, Tex.
		7-11 cow	Female	6,700	Auction	7-11 Ranch, Boerne, Tex.	Nine Bar Ranch, Cypress, Tex.
	1967	Kings Ransom 1/466	Bull	30,000	Auction	Nine Bar Ranch, Cypress, Tex.	King Ranch, Kingsville, Tex.
		WR cow	Female	4,500	Auction	Winrock Farms, Morrilton, Ark.	Nine Bar Ranch, Cypress, Tex.
	1968	El Capitan 2-5055	Bull	31,000	Auction	Nine Bar Ranch, Cypress, Tex.	Winrock Farms & Roberto Berger, Guatamala City, Guatamala
		Queen Beth 3-6260	Female	6,200	Auction	Nine Bar Ranch, Cypress, Tex.	Hualalai Ranch, Kona, Hawaii
	1969	Jugador 101/7	Bull	32,000	Auction	Vesper Ranch, Cotulla, Tex.	Fair View Ranch, Elkton, Md.; and Silverbrook Farm, Middlebrook, Va.
		WR cow	Female	5,100	Auction	Winrock Farms, Morrilton, Ark.	K. D. Eatmon, Pompano Beach, Fla.
	1970	WR Cargado 6119	Bull	46,500	Auction	Winrock Farms, Morrilton, Ark.	King Ranch, Kingsville, Tex.

(Continued)

TABLE VI-1 (Continued)

Breed	Year of Sale	Identity of Animal	Sex	Price	Private Treaty or Auction	Seller	Purchaser
				(Dollars)			
Santa Gertrudis (cont.)		Mirasol 6-257	Female	19,000	Auction	Mirasol Ranch, Uvalde, Tex.	Hualalai Ranch, Kona, Hawaii
	1971	Fairview 025	Bull	15,800	Auction	Fairview Farms, Pleasanton, Tex.	Nine Bar Ranch, Cypress, Tex.
		WR cow	Female	5,100	Auction	Winrock Farms, Morrilton, Ark.	Nine Bar Ranch, Cypress, Tex.
	1972	El Capitan 3-0270	Bull	30,000	Auction	Nine Bar Ranch, Cypress, Tex.	Four G Ranch, Hempstead, Tex.
		Briggs cow	Female	5,300	Auction	Briggs Ranch, Catarina, Tex.	Three B Ranch, Lakeland, Fla.
	1973	No. 1-13	Bull	60,000	Auction	Fairview Farms, Pleasanton, Tex.	Marcham, Connally, Johnson & Cromeens, Tex.
		3-R-4-2359	Female	10,000	Auction	Nine Bar Ranch, Cypress, Tex.	Three B Ranch, Lakeland, Fla.
	1974	La Capitan 3-QE3-2351	Bull	105,000	Auction	Nine Bar Ranch, Cypress, Tex.	G. Anthony, Burton, S.C.
			Female	10,000	Auction	Chapparrosa Ranch, La Pryer, Tex.	Longfellow Ranch, Sanderson, Tex.
Scotch Highland	1964	LC Loretta's King	Bull	3,000 (½ int.)	Private Treaty	Keith L. Crew, Interior, S.D.	Armour & Co., Chicago, Ill.
Shorthorn	1873	8th Duchess of Geneva	Female	40,600	Auction	New York Mills, Utica, N.Y.	R. P. Davies, England
	1946	Pittodrie Upright	Bull	61,335	Auction (Perth, Scotland)	L. S. Pitcaple, Aberdeenshire, Scotland	Sni-A-Bar Farms, Grain Valley, Mo.
	1948		Bull	6,300	Auction	Johnson Disposal Sale, Rockwell City, Iowa	
			Female	3,600	Auction	International Polled Shorthorn Sale, Chicago, Ill.	
	1951	Leveldale Good News 2637691	Bull	16,500	Auction	L. E. Mathers & Son, Mason City, Ill.	John Alexander & Son, Aurora, Ill.
	1952	Calrossie Highland	Bull	29,000	Auction (Perth, Scotland)	John and D. P. Mac-Gillivray, Calrossie Ross-shire, scotland	A. R. Cross, Alberta, Canada
		Edellyn Rosewood 65th	Female	10,000	Auction	T. E. Wilson, Wilson, Ill.	Henry Knight Lexington, Ky.
	1953	Cruggelton Rainey	Bull	26,000	Auction	Scofield Ranch, Austin, Tex.	U. R. Bronk, Houston, Tex.
		Circle M Musil 3rd	Female	14,000	Auction	Scofield Ranch, Austin, Tex.	Roberta Dickson, Austin, Tex.
	1954	Max Coronet 5th	Bull	20,000	Private Treaty	L. W. Thieman, Concordia, Mo.	Avenel Farms, Bethesda, Md.
	1956	Bapton Constructor	Bull	45,000	Private Treaty	Cecil Moore, Aberdeenshire, Scotland	Louada Manor Farms, Peterborough, Canada
	1959	Louada Bonaparte	Bull	30,000	Auction	Louada Farm, Peterborough, Ont.	Sittyton Herd, Argentina
Shorthorn (and Polled Shorthorn)	1962	N.W. Rosewood Man o'War	Bull	6,250	Auction	Northwood Farms, Cary, Ill.	Schreiber Farms, St. Joseph, Mo.
		Glastullich Augusta Lyn	Female	5,500	Auction	Northwood Farms, Cary, Ill.	Tampel Farms, Wadsworth, Ill.
	1963	Ashbourne Scepter	Bull	10,000 (½ int.)	Private Treaty	W. L. Lyons Brown, LaGrange, Ky.	Lone Star Ranch, Houston, Tex.
		Acadia Queen of Beauty 8th	Female	4,400	Auction	Acadia Farms, Northfield, Ohio	Melbourne Farms, Big Rock, Ill.

(Continued)

TABLE VI-1 (Continued)

Breed	Year of Sale	Identity of Animal	Sex	Price	Private Treaty or Auction	Seller	Purchaser
				(Dollars)			
Shorthorn (and Polled Shorthorn) (cont.)	1964	Glamis Benefactor	Bull	20,000 (⅓ int.)	Private Treaty	Dave and Patricia Lorenz, K C Shorthorns, Kansas City, Mo.	B. Sterling Ambrose, Royalty Farms, Owensboro, Ky.
		Ashbourne Princess Mary	Female	3,800	Auction	Bilmar Farms, Grand Blanc, Mich.	B. Sterling Ambrose, Royalty Farms, Owensboro, Ky.
	1965	Melbourne Consort	Bull	7,700	Auction	Acadia Farms, Northfield, Ohio	Bilmar Farms, Grand Blanc, Mich.
		Melbourne Missie Eliza 2nd	Female	3,700	Auction	Acadia Farms, Northfield, Ohio	Fernwood Farm, Barrington, Ill.
	1966	Colomeadow Pride of 66	Bull	6,500 (⅔ int.)	Auction	National Western, Denver, Colo.	M. A. Rambo, Temple, Tex.
		Royalty Blood Royal	Female	5,000	Auction	Royalty Farms, Owensboro, Ky.	Darrell and Ted Bandy, Blue Mound, Ill.
	1967	Acadia Troubadour 70th	Bull	10,000 (½ int.)	Auction	Acadia Farms, Northfield, Ohio	Boll Weevil Plantation, Waynesboro, Ga.
		Boll Weevil Golden Lace 2nd	Female	2,600	Auction	Boll Weevil Plantation, Waynesboro, Ga.	Acadia Farms, Northfield, Ohio
	1968	Brentwood Band Master	Bull	4,600 (⅔ int.)	Auction	National Western, Denver, Colo.	Mrs. Robert G. Haley, Ventura, Calif.
		Bilmar Atlas Diadem	Bull	4,600 (⅔ int.)	Auction	Bilmar Farms, Grand Blanc, Mich.	Russell Held, Hinton, Iowa
		Sangamon Victoria 9th ×	Female	3,900	Auction	National Western, Denver, Colo.	Orville Leonardson, Camden, N.Y.
	1969	Kinnaber Leader 22nd ×	Bull	7,200 (⅔ int.)	Auction	KC Shorthorn Sale, Kansas City, Mo.	Carl Jernberg, Atlanta, Ga.
		Queen of Margaret 4th of KC	Female	3,350	Auction	KC Shorthorn Sale, Kansas City, Mo.	Carl Jernberg, Atlanta, Ga.
	1970	Sutherland Citation ×	Bull	13,200 (⅔ int.)	Auction	Polled Shorthorn Congress, Kansas City, Mo.	Windholme Farm, Orange, Va.
		Sutherland Countess 3rd ×	Female	3,000	Auction	Sutherland Sale	Sundance Farm, Anchorage, Ky.
	1971	Butte Lee Leader 82nd ×	Bull	3,500 (⅔ int.)	Auction	National Western, Denver, Colo.	Mar J Ranch, Montrose, Colo.
		Sweet Home Red Rose	Female	5,000	Auction	National Western, Denver, Colo.	Lewis Bristin, Yachats, Ore.
	1972	Mr. Thieman	Bull	26,000 (⅓ int.)	Private Treaty	Nold's Weston Shorthorns, Onaka, S.D.; and Kinaber Stock Farm, Souris, Manitoba, Canada	SMIP Ranch of Woodside, Calif.
		Thomas Victoria 101 B ×	Female	3,200	Auction	Bob-Lee Shorthorns, Fredericktown, Ohio	Windholme Farm, Orange, Va.
	1973	Weston Big Ben ×	Bull	5,500 (½ int.)	Auction	Bob-Lee Farms, Fredericktown, Ohio	T. Howard Snethen, DeWittville, N.Y.
		Bitter Root Beauty 2nd	Female	3,000	Auction	Coby Smith, Hamilton, Mont.	LaFayette Corp, Filer, Ida.; and Charles Measor, Tower City, N.D.

(Continued)

TABLE VI-1 (Continued)

Breed	Year of Sale	Identity of Animal	Sex	Price	Private Treaty or Auction	Seller	Purchaser
				(Dollars)			
Shorthorn (and Polled Shorthorn) (cont.)	1975	Mill Brook Ransom G9	Bull	10,000	Auction	Bob-Lee Shorthorns, Fredericktown, Ohio	Sutherland Farms, Prospect, Ky.
		Kinnaber Caroline	Female	3,000	Auction	Nold's Weston Shorthorns, Onaka, S.D.	Sangamon Farms, DeWittville, N.Y.
Simmental	1970	Amor	Bull	154,000	Private Treaty	Switzerland	Syndicate, Codding-Nova Ranch, Foraker, Okla.
	1971	Tattoo SSSS97B	Bull	1,750	Auction	F. R. Smith & Sons, Mountain View, Alberta, Canada	Codding Cattle Research, Foraker, Okla.
		ASA 201	Female	8,750	Auction	CM Cattle Co., Kirksville, Mo.	United Simmental Ltd., Cardston, Alberta, Canada
	1972	ASA 16698	Bull	40,500	Auction	Kingfield H. B. King, Ontario, Canada	Snowball Cattle, North Platte, Neb.
		ASA 16693	Female	12,500	Auction	Bar Eleven Simmental Ranch, Calgary, Alberta, Canada	Rio Vista Farms, San Antonio, Tex.
	1973	ASA 46136	Bull	52,000	Auction	C. Owen Rogers, Frenchman Butte, Saskatchewan, Canada	Agro Farms, Oklahoma City, Okla.
		ASA 46135	Female	20,000	Auction	OK Ranch, Ottawa, Kan.	Carl Houston, Jr., Lexington, Ky.
	1974	Kingfield Deschamps ASA 98260	Bull	100,000	Auction	Kingfield H.B. King, Ontario, Canada	American Breeders Service, DeForest, Wisc.
		Demby Miss Parisien 1503 ASA 98263	Female	62,000	Auction	Dembiczak Bros., Lake Alma, Saskatchewan, Canada	Emmons, Peterson, Phillips, Butcher, & Hannah, Lewistown, Mont.
	1975	SC Salvator 30D ASA 126624	Bull	52,000	Auction	Angus Glen Farm, Unionville, Ontario, Canada	Conner, Chapman-wood, Houston, Tex.
		ASA 163970	Female	54,000	Auction	Simmental Breeders Ltd., Cardston, Alberta, Canada	Rio Vista Farms, San Antonio, Tex.
Texas Longhorn	1965		Female	525	Auction	Wichita Wildlife Refuge, Cache, Okla.	YO Ranch, Mountain Home, Tex.
	1967		Bull	775	Auction	Wichita Wildlife Refuge, Cache, Okla.	YO Ranch, Mountain Home, Tex.
	1968		Bull	285	Auction	Wichita Wildlife Refuge, Cache, Okla.	Unknown.
			Female	315	Auction	Wichita Wildlife Refuge, Cache, Okla.	Unknown.
	1970		Bull	620	Auction	Ft. Niobrara	Unknown.
			Female	240	Auction	Ft. Niobrara	Unknown.
	1971		Bull	415	Auction	Wichita Wildlife Refuge, Cache, Okla.	W. B. Scott, Goliad, Tex.
			Female	325	Auction	Wichita Wildlife Refuge, Cache, Okla.	Arthur Brigh, LeGrand, Calif.

(Continued)

TABLE VI-1 (Continued)

Breed	Year of Sale	Identity of Animal	Sex	Price	Private Treaty or Auction	Seller	Purchaser
				(Dollars)			
Texas Longhorn (cont.)	1972		Bull	450	Auction	Wichita Wildlife Refuge, Cache, Okla.	Unknown.
			Female	455	Auction	Wichita Wildlife Refuge, Cache, Okla.	Unknown.
	1973	TLBAA417	Bull	1,900	Auction	Wichita Wildlife Refuge, Cache, Okla.	Joe Evans, Killeen, Tex.
		TLBAA1083	Female	950	Auction	Wichita Wildlife Refuge, Cache, Okla.	YO Ranch, Mountain Home, Tex.
	1974		Bull	4,600	Auction	YO Ranch, Mountain Home, Tex.	Alan Sparger, D'Hanis, Tex.
			Female	2,500	Auction		YO Ranch, Mountain Home, Tex.

[1]Where a part interest is specified, the price listed represents the sale price for the part interest only.
[2]The Angus sales herein reported were either obtained from or verified by the *Aberdeen-Angus Journal*.

TABLE VI-2

ALL-TIME TOP CONSIGNMENT OR HERD SALES

Breed	Year of Sale	Number of Animals	Average Price	Seller
			(Dollars)	
Angus	1946	46	5,614	Sunbeam Farms, Miami, Okla.
	1951	50	10,246	J. G. Tolan, Pleasant Plains, Ill.
	1952	62	6,464	Angus Valley Farms, Tulsa, Okla.
	1953	62	5,821	Sunbeam Farms, Miami, Okla.
	1956	56	5,825	Ankony Farm, Rhinebeck, N.Y.
	1961	37½	7,548	International Bull Sale, Chicago, Ill.
	1962	33	10,902	International Livestock Exposition, Chicago, Ill.
	1963	11⅚	11,885	International Livestock Exposition, Chicago, Ill.
	1964	75⅔	11,034	Black Watch Farms, Wappinger Falls, N.Y.
	1965	86⅓	10,325	Black Watch Farms, Wappinger Falls, N.Y.
	1966	58	9,156	Thieman Farms, Bixby, Okla.
	1967	45	4,020	Meadow Lane Farm, North Salem, N.Y.
	1968	44	5,685	Meadow Lane Farm, North Salem, N.Y.
	1969	65⅓	3,947	Colossal Event at Sugar Loaf, Staunton, Va.
	1970	58	6,874	Colossal Event, Staunton, Va.
	1971	64	4,186	Colossal Event, Staunton, Va.
	1972	60	3,186	Colossal Event, Staunton, Va.
	1973	9	8,030	Lettunich & Sons, Payette, Ida.
	1974	135	4,442	Sir William Farms, Hillsdale, N.Y.
Barzona	1973	287	577	Doug Bard, Williams, Ariz.
Beef Friesian	1974	58	3,083	National Western Beef Friesian Sale, Denver, Colo.

(Continued)

TABLE VI-2 (Continued)

Breed	Year of Sale	Number of Animals	Average Price	Seller
			(Dollars)	
Beefmaster	1973	108	1,926	Weber Beefmaster Sale, Mr. F. W. Weber, Charlotte, Tex.
	1974	160	1,185	Southeastern Satellite Sale, Jackson, Miss.
Braford	1973	3,500	371	Adams Ranch, Fort Pierce, Fla.
Brahman	1962	20	875	Texas Brahman Assn., 1st Halter Sale, Houston, Tex.
	1963	32	645	Louisiana Brahman Assn. 17th Annual Sale, Baton Rouge, La.
	1964	14	1,600	National Brahman Sale, Dallas, Tex.
	1965	23	662	Louisiana Brahman Assn. 19th Annual Sale, Baton Rouge, La.
	1966	24	1,291	Louisiana Brahman Assn. 20th Annual Sale, Baton Rouge, La.
	1967	21	1,615	National Brahman Sale, Houston, Tex.
	1968	21	1,211	Texas Brahman Assn. Sale, Houston, Tex.
	1969	16	1,823	Texas Brahman Assn. Sale, Houston, Tex.
	1970	21	1,725	Texas Brahman Assn. Sale, Houston, Tex.
	1971	19	2,171	National Brahman Sale, Baton Rouge, La.
	1972	20	2,779	National Brahman Sale, Dallas, Tex.
	1973	143	2,382	McKellar Ranch Production, Mt. Pleasant, Tex.
	1974	23	7,165	National Brahman Sale, Houston, Tex.
Brangus	1966	37	1,402	International Brangus Sale (IBBA)
	1967	29	1,398	International Brangus Sale
	1968	18	1,279	International Brangus Sale
	1969	28	2,156	International Brangus Sale
	1970	42	2,990	L. D. Brinkman, Kerrville, Tex.
	1971	47⅓	3,696	L. D. Brinkman, Kerrville, Tex.
	1972	36¾	8,822	Brink's Brangus, Kerrville, Tex.
	1973	50⁵/₁₂ lots	13,570	Brink's Brangus, Kerrville, Tex.
	1975	101¼	3,017	Brink's Brangus, Kerrville, Tex.
Brown Swiss	1961-62	35	822	14th Annual Imperial Sale
	1962-63	45	652	15th Annual Imperial Sale
	1963-64	146	2,614	Welcome in Farm Dispersal
	1964-65	114	702	C/B Farm Dispersal
	1965-66	32	980	The Imperial Sale
	1966-67	53	1,352	The Springtime Show Window Sale
	1967-68	57	716	The Springtime Show Window Sale
	1968-69	54	1,002	Top Acres Sale
	1969-70	101	708	Alphil Farms Dispersal
	1970-71	99	1,086	White Cloud Dispersal
	1971-72	30	1,141	Pennsylvania All-American Sale
	1973	40	2,061	Pacific Northwest Imperial Sale

(Continued)

TABLE VI-2 (Continued)

Breed	Year of Sale	Number of Animals	Average Price	Seller
			(Dollars)	
Charolais	1953	30	1,430	C. M. Frost, Houston, Tex.
	1956	30	2,960	A. M. Askew, Houston, Tex.
	1966	45½	4,407	Sale of Sales, Houston, Tex.
	1968	41¼	8,789	Sale of Sales, Houston, Tex.
	1969	66	13,027	Sale of Sales, Houston, Tex.
	1970	63¼	9,011	Sale of Sales, Houston, Tex.
	1971	33	9,182	Sale of Sales, Houston, Tex.
	1972	35	11,429	Sale of Sales, Houston, Tex.
	1973	35	14,717	Sale of Sales, Houston, Tex.
	1975	106¼	9,412	Rocking B Ranch, Athens, Tex.
Chianina	1972	148	3,570	Schearbrook Land & Livestock, Inc., Springfield, Mo.
	1973	55	6,322	National Chianina Sale, Kansas City, Mo.
	1974	53	4,750	National Chianina Sale, Kansas City, Mo.
Devon	1973	122	858	Western Devon Cattle Assn. Sale, Eagle Point, Ore.
	1974	58	847	3rd National Devon Sale, Jackson, Miss.
Galloway	1972	48	911	Mile Hi Sale, Denver, Colo.
	1973	10 bulls	1,297	Rocky Mountain Galloway Sale, Torrington, Wyo.
	1974	5 bred heifers	2,040	T Bone Galloways, Carpenter, Wyo.
Gelbvieh	1972	30 cows w/calves	1,500	Jack Schmitt, Geyser, Mont.
	1973	18	1,850	Mitch Dobson, Newkirk, Okla.
	1974	79	3,073	Second National Gelbvieh Sale, Fort Worth, Tex.
Hereford	1919	140	3,635	W. T. McCray, Kentland, Ind.
	1947	72	5,934	Wyoming Hereford Ranch, Cheyenne, Wyo.
	1952	76	5,306	Wyoming Hereford Ranch, Cheyenne, Wyo.
	1954	50	6,120	Turner Ranch, Sulfur, Okla.
	1964	48	2,040	Morlunda Herefords, Lewisburg, W. Va.
	1965	99½	3,073	Bridwell Hereford Ranch, Windthorst, Tex.
	1966	28½	6,546	Winrock Farms, Sulphur, Okla.
	1967	113	2,559	National Western Hereford Sale, Denver, Colo.
	1968	20	15,550	Twenty Grand Sale, Kansas City, Mo.
	1969	20	6,513	Twenty Grand Sale, Kansas City, Mo.
	1970	51	4,828	26 Bar Ranch, Stanfield, Ariz.
	1971	50¾	5,909	26 Bar Ranch, Stanfield, Ariz.
	1972	50¾	7,528	26 Bar Ranch, Stanfield, Ariz.
	1973	15¼	7,272	Twenty Grand Sale, Kansas City, Mo.
	1974	59	8,248	Cooper Holden, Great Falls, Mont.
Limousin	1972	78	2,839	NALF National Sale, Denver, Colo.
	1972	68	2,188	Limousin Cattle Co. of Oklahoma, Pauls Valley, Okla.
	1973	74	5,954	Limousin Cattle Co. of Oklahoma, Pauls Valley, Okla.
	1974	69	6,795	NALF National Sale, Denver, Colo.

(Continued)

TABLE VI-2 (Continued)

Breed	Year of Sale	Number of Animals	Average Price	Seller
			(Dollars)	
Maine-Anjou	1972	5 ½ bloods	4,300	Morris Palmer Ranch
	1973	81	4,794	Heaven II Farms
	1974	59	8,839	Astro Classic Maine-Anjou Sale, Houston, Tex.
	1975	73	7,515	Maine-Anjou National Sale, Denver, Colo.
Marchigiana (½)	1973	1 half	23,600	W. E. Goodrick, Peculiar, Mo.
	1973	57 halves	2,174	
Milking Shorthorn	1917	39	1,110	The Otis Herd, Willoughby, Ohio
	1950	42	1,056	National Milking Shorthorn Congress Sale, Springfield, Mo.
	1965	35	432	Lilydale Farm, Springfield, Mo.
	1966	39	517	Elmer Von Tungelin, Verden, Okla.
	1967	35	479	Sam Yoder, Shoemakersville, Penn.
	1968	31	499	Ernest Fox & Son, Keokuk, Iowa
	1969	28	464	Ernest Fox & Son, Keokuk, Iowa
	1970	21	633	Victor E. Odum & Sons, Wabash, Ind.
	1971	31	599	Clampitt & Lyon, Toledo, Iowa
	1972	30	837	Don McElhaney, Republic, Mo.
	1973	26	804	DeWitt C. and Donald E. Brooks, Siloam Springs, Ark.
	1974	54	1,212	Rye Bros., Avalon, Wisc.
Murray Grey	1974	72	627	Keefer Ranches, Ryegate, Mont.
Pinzgauer	1974	61	2,995	American Pinzgauer Assn. Sale
Polled Hereford	1951	56	4,421	Circle M Ranch, Senatobia, Miss.
	1952	50	5,061	Circle M Ranch, Senatobia, Miss.
	1953	50	7,230	Circle M Ranch, Senatobia, Miss.
	1954	51	4,548	Circle M Ranch, Senatobia, Miss.
	1956	50	4,034	Circle M Ranch, Senatobia, Miss.
	1960	50¼	4,709	Circle M Ranch, Senatobia, Miss.
	1961	50	4,934	Circle M Ranch, Senatobia, Miss.
	1962	51	6,416	Circle M Ranch, Senatobia, Miss.
	1963	50¼	7,965	Circle M Ranch, Senatobia, Miss.
	1964	50	6,515	Circle M Ranch, Senatobia, Miss.
	1965	51	6,433	Circle M Ranch, Senatobia, Miss.
	1966	50¼	6,082	Circle M Ranch, Senatobia, Miss.
	1967	50½	5,325	Circle M Ranch, Senatobia, Miss.
	1968	50	4,527	Circle M Ranch, Senatobia, Miss.
	1969	46¾	4,598	American Herdsman Institute, Kansas City, Mo.
	1970	50	4,294	Circle M Ranch, Senatobia, Miss.
	1971	50¼	3,818	Circle M Ranch, Senatobia, Miss.
	1972	51½	3,879	Circle M Ranch, Senatobia, Miss.
	1973	22¾	6,585	National Polled Hereford Sale, Fort Worth, Tex.
	1974	19	11,212	National Polled Hereford Sale, Denver, Colo.
Polled Shorthorn —see Shorthorn				

(Continued)

TABLE VI-2 (Continued)

Breed	Year of Sale	Number of Animals	Average Price	Seller
			(Dollars)	
Red Brangus	1972	100 bulls	1,200	Paleface Ranch, Spicewood, Tex.
	1973	100 bulls	1,650	Paleface Ranch, Spicewood, Tex.
	1974	100 bulls	950	Paleface Ranch, Spicewood, Tex.
		40 females	1,950	
Red Poll	1951	62	496	National Red Poll Consignment Sale
	1962	85	286	National Red Poll Consignment Sale
	1963	64	289	National Red Poll Consignment Sale
	1964	53	255	National Red Poll Consignment Sale
	1965	49	277	National Red Poll Consignment Sale
	1966	38	262	National Red Poll Consignment Sale
	1967	44	348	National Red Poll Consignment Sale
	1968	57	409	National Red Poll Consignment Sale
	1969	44	365	National Red Poll Consignment Sale
	1970	35	379	National Red Poll Consignment Sale
	1971	94	477	National Red Poll Consignment Sale
	1972	88	832	National Red Poll Consignment Sale
	1974	42	882	Indiana Red Poll Cattle Breeders' Assn. Sale
Santa Gertrudis	1951	25	8,506	King Ranch, Kingsville, Tex.
	1952	21	8,583	King Ranch, Kingsville, Tex.
	1953	24 bulls	8,440	King Ranch, Kingsville, Tex.
	1957	25 bulls	6,311	King Ranch, Kingsville, Tex.
	1958	24 bulls	6,033	King Ranch, Kingsville, Tex.
	1958	39	4,370	Winrock Farms, Morrilton, Ark.
	1959	25	8,690	King Ranch, Kingsville, Tex.
	1960	24	6,604	King Ranch, Kingsville, Tex.
	1961	40	5,727	Winrock Farms, Morrilton, Ark.
	1962	24 bulls	3,421	King Ranch, Kingsville, Tex.
	1963	25 bulls	3,747	King Ranch, Kingsville, Tex.
	1964	34	3,493	Nine Bar Ranch, Cypress, Tex.
	1965	30	7,662	Nine Bar Ranch, Cypress, Tex.
	1966	24½	7,661	Nine Bar Ranch, Cypress, Tex.
	1967	45	3,740	Nine Bar Ranch, Cypress, Tex.
	1968	41	4,102	Nine Bar Ranch, Cypress, Tex.
	1969	41	4,413	Winrock Farms, Morrilton, Ark.
	1970	19 bulls	11,467	King Ranch, Kingsville, Tex.
	1971	15 bulls	3,173	King Ranch, Kingsville, Tex.
	1972	14 bulls	10,428	King Ranch, Kingsville, Tex.
	1973	20 bulls	11,605	King Ranch, Kingsville, Tex.
Shorthorn (and Polled Shorthorn)	1873	109	3,504	New York Mills, Utica, N.Y.
	1919	39	5,140	Wm. Duthie of Collyne, Tarves, Aberdeenshire, Scotland
	1951	42	1,295	C. B. Teegardin & Sons, Ashville, Ohio
	1952	50	3,030	L. E. Mathers & Sons, Mason City, Ill.

(Continued)

TABLE VI-2 (Continued)

Breed	Year of Sale	Number of Animals	Average Price	Seller
			(Dollars)	
Shorthorn (and Polled Shorthorn) (cont.)	1953	34	3,715	Edellyn Farms, Wilson, Ill.
	1961	62	2,482	Northwood Farms, Cary, Ill.
	1962	62	1,506	U.S. Northwood Farms, Cary, Ill.
	1963	48⅓	1,577	U.S. Acadia Farms, Northfield, Ohio
	1964	47⅔	1,265	Bilmar Farms, Grand Blanc, Mich.
	1965	55	1,103	Acadia Farms, Northfield, Ohio
	1966	81⅓	1,151	Melbourne Sale, Big Rock, Ill.
	1967	58⅔	1,275	KC Production Sale, Kansas City, Mo.
	1968	43⅔	957	National Western, Denver, Colo.
	1969	55⅔	1,111	KC Shorthorn Sale, Kansas City, Mo.
	1970	45⅓	1,083	U.S. Polled Shorthorn Congress, Kansas City, Mo.
	1971	51	897	National Western, Denver, Colo.
	1972	81⅔	1,017	Bob-Lee Production Sale, Fredericktown, Ohio
	1973	91	1,255	Bob-Lee Farms and the R. Lee Johnson Family, Fredericktown, Ohio
	1975	51	1,167	National Western, Denver, Colo.
Simmental	1971	112	1,053	National Sale, Denver, Colo.
	1972	126	4,099	National Sale, Fort Worth, Tex.
	1973	97	6,932	National Sale, Denver, Colo.
	1974	86	14,222	National Sale, Louisville, Ky.
	1975	122	5,150	National Sale, Houston, Tex.
Texas Longhorn	1967	67	297	TLBAA, San Antonio, Tex.
	1968	82	177	TLBAA, San Antonio, Tex.
	1970	105	195	Wichita Wildlife Refuge, Cache, Okla.
	1971	52	255	Ft. Niobrara, Neb.
	1972	96	361	Wichita Wildlife Refuge, Cache, Okla.
	1973	95	567	Wichita Wildlife Refuge, Cache, Okla.
	1974	106	770	Consignment Sale—South Texas Longhorn Sale

SECTION VII—STATE COLLEGES OF AGRICULTURE

The stockman can obtain a list of available bulletins and circulars and other information regarding livestock, by writing to his state agricultural college. One should also write to the Superintendent of Documents, Washington, D.C., for lists of available bulletins and circulars. A list of the state agricultural colleges follows. Land-grant institutions are designated by an asterisk (*).

State	Address
Alabama	* School of Agriculture, Auburn University, Auburn, Alabama 36830 Tuskegee Institute, Tuskegee, Alabama 36088
Alaska	* Division of Agriculture, University of Alaska, Palmer, Alaska 99645
Arizona	College of Agriculture, University of Arizona, Tucson, Arizona 85721 Arizona State University, Tempe, Arizona 85281
Arkansas	* College of Agriculture, University of Arkansas, Fayetteville, Arkansas 72203

(Continued)

State	Address
California	* College of Agriculture and Environmental Sciences, University of California, Davis, California 95616
	California Polytechnic State University, San Luis Obispo, California 93401
	California Polytechnic State University, Kellogg-Voorhis, Pomona—San Dimas, California 91766
	Chico State College, Chico, California 95927
	School of Agricultural Sciences, California State University-Fresno, Fresno, California 93726
Colorado	* College of Agriculture, Colorado State University, Fort Collins, Colorado 80521
Connecticut	* College of Agriculture, University of Connecticut, Storrs, Connecticut 06268
Delaware	* School of Agriculture, University of Delaware, Newark, Delaware 19711
Florida	* College of Agriculture, University of Florida, Gainesville, Florida 32601
	Florida A&M Univeristy, Tallahassee, Florida 32307
Georgia	* College of Agriculture, University of Georgia, Athens, Georgia 30601
Hawaii	* Department of Agriculture, University of Hawaii, Honolulu, Hawaii 96822
Idaho	* College of Agriculture, University of Idaho, Moscow, Idaho 83843
Illinois	* The College of Agriculture, University of Illinois, Urbana, Illinois 61801
	The School of Agriculture, Southern Illinois University, Carbondale, Illinois 62903
	Illinois State University, Normal, Illinois 61761
	Western Illinois University, Macomb, Illinois 61455
Indiana	* School of Agriculture, Purdue University, Lafayette, Indiana 47907
Iowa	* The Division of Agriculture, Iowa State University, Ames, Iowa 50010
Kansas	* College of Agriculture, Kansas State University, Manhattan, Kansas 66502
Kentucky	* College of Agriculture, University of Kentucky, Lexington, Kentucky 40506
	Berea College, Berea, Kentucky 40403
	Morehead State University, Morehead, Kentucky 40351
	Murray State University, Murray, Kentucky 42072
	Western Kentucky State College, Bowling Green, Kentucky 42102
Louisiana	* The Louisiana State University, and A&M College, University Station, Baton Rouge, Louisiana 70803
	Frances T. Nicholls State College, Thibodaux, Louisiana 70301
	Grambling College, Grambling, Louisiana 71245
	Louisiana Polytechnic Institute, Ruston, Louisiana 71270
	McNeese State College, Lake Charles, Louisiana 70602
	Northeast Louisiana State College, Monroe, Louisiana 71201
	Northwestern State College of Louisiana, Natchitoches, Louisiana 71457
	Southeastern Louisiana State College, Hammond, Louisiana 70402
	Southern University and A&M College, Baton Rouge, Louisiana 70813
	The University of Southwestern Louisiana, Lafayette, Louisiana 70506
Maine	* College of Agriculture, University of Maine, Orono, Maine 04473
Maryland	* College of Agriculture, University of Maryland, College Park, Maryland 20740
Massachusetts	* School of Agriculture, University of Massachusetts, Amherst, Massachusetts 01002
Michigan	* College of Agriculture, Michigan State University, East Lansing, Michigan 48823
	Andrews University, Berrien Springs, Michigan 49104
Minnesota	* Department of Agriculture, University of Minnesota, University Farm, St. Paul, Minnesota 55101
Mississippi	* School of Agriculture, Mississippi State University, State College, Mississippi 39762
Missouri	* Division of Agricultural Sciences, University of Missouri, Columbia, Missouri 65201

(Continued)

State	Address
Montana	* Department of Agriculture, Montana State University, Bozeman, Montana 59715
Nebraska	* College of Agriculture, University of Nebraska, Lincoln, Nebraska 68503
Nevada	* College of Agriculture, University of Nevada, Reno, Nevada 89507
New Hampshire	* College of Agriculture, University of New Hampshire, Durham, New Hampshire 03824
New Jersey	* State College of Agriculture, Rutgers University, New Brunswick, New Jersey 08903
New Mexico	* College of Agriculture, New Mexico State University, University Park, New Mexico 88070
New York	* New York State College of Agriculture, Cornell University, Ithaca, New York 14850
North Carolina	* School of Agriculture, North Carolina State University, Raleigh, North Carolina 27607
	Agricultural and Technical College of North Carolina, Greensboro, North Carolina 27411
	Pembroke State College, Pembroke, North Carolina 28372
North Dakota	* College of Agriculture, North Dakota State University, Fargo, North Dakota 58102
Ohio	* The College of Agriculture, Ohio State University, Columbus, Ohio 43210
Oklahoma	* School of Agriculture, Oklahoma State University, Stillwater, Oklahoma 74074
	Panhandle A&M College, Goodwell, Oklahoma 73939
Oregon	* School of Agriculture, Oregon State University, Corvallis, Oregon 97330
Pennsylvania	* School of Agriculture, Pennsylvania State University, State College, University Park Station, Pennsylvania 16802
	Delaware Valley College of Science and Agriculture, Doylestown, Pennsylvania 18901
Puerto Rico	* College of Agriculture, University of Puerto Rico, Rio Piedras, Puerto Rico 00926
Rhode Island	* School of Agriculture, University of Rhode Island, Kingston, Rhode Island 02881
South Carolina	* College of Agricultural Sciences, Clemson University, Clemson, South Carolina 29631
South Dakota	* College of Agriculture, South Dakota State University, Brookings, South Dakota 57006
Tennessee	* College of Agriculture, University of Tennessee, Knoxville, Tennessee 37916
	Middle Tennessee State University, Murfreesboro, Tennessee 37130
	School of Agriculture, Tennessee State University, Nashville, Tennessee 37203
	Tennessee Technological Institute, Cooksville, Tennessee 38501
Texas	* Texas A&M University, College Station, Texas 77843
	Abilene Christian College, Abilene, Texas 79601
	Sam Houston State University, Huntsville, Texas 77340
	School of Agriculture, Prairie View A&M College, Prairie View, Texas 77445
	Sul Ross State College, Alpine, Texas 79830
	Texas A&I University, Kingsville, Texas 78363
Utah	* The College of Agriculture, Utah State University, Logan, Utah 84321
Vermont	* College of Agriculture, University of Vermont, Burlington, Vermont 05401
Virginia	* School of Agriculture, Virginia Polytechnic Institute, Blacksburg, Virginia 24060
	Virginia State College, Petersburg, Virginia 23806
Washington	* The Institute of Agricultural Sciences, Washington State University, Pullman, Washington 99163
West Virginia	* College of Agriculture, West Virginia University, Morgantown, West Virginia 26505
Wisconsin	* College of Agriculture, University of Wisconsin, Madison, Wisconsin 54306
	Wisconsin State University, River Falls, Wisconsin 54022

(Continued)

State	Address
Wyoming	* College of Agriculture, University of Wyoming, Laramie, Wyoming 82070

In Canada

Alberta	University of Alberta, Edmonton
British Columbia	University of British Columbia, Vancouver
Manitoba	University of Manitoba, Winnipeg
New Brunswick	University of New Brunswick, Fredericton
Ontario	University of Guelph, Guelph
Saskatchewan	University of Saskatchewan, Saskatoon
Quebec	Faculty d'Agriculture, University of Laval, Quebec City MacDonald College, Montreal

INDEX